Quo Vadis Computational Intelligence?

Studies in Fuzziness and Soft Computing

Editor-in-chief
Prof. Janusz Kacprzyk
Systems Research Institute
Polish Academy of Sciences
ul. Newelska 6
01-447 Warsaw, Poland
E-mail: kacprzyk@ibspan.waw.pl
http://www.springer.de/cgi-bin/search_book.pl?series=2941

Vol. 3. A. Geyer-Schulz
Fuzzy Rule-Based Expert Systems and Genetic Machine Learning, 2nd ed. 1996
ISBN 3-7908-0964-0

Vol. 4. T. Onisawa and J. Kacprzyk (Eds.)
Reliability and Safety Analyses under Fuzziness, 1995
ISBN 3-7908-0837-7

Vol. 5. P. Bosc and J. Kacprzyk (Eds.)
Fuzziness in Database Management Systems, 1995
ISBN 3-7908-0858-X

Vol. 6. E. S. Lee and Q. Zhu
Fuzzy and Evidence Reasoning, 1995
ISBN 3-7908-0880-6

Vol. 7. B. A. Juliano and W. Bandler
Tracing Chains-of-Thought, 1996
ISBN 3-7908-0922-5

Vol. 8. F. Herrera and J. L. Verdegay (Eds.)
Genetic Algorithms and Soft Computing, 1996
ISBN 3-7908-0956-X

Vol. 9. M. Sato et al.
Fuzzy Clustering Models and Applications, 1997
ISBN 3-7908-1026-6

Vol. 10. L. C. Jain (Ed.)
Soft Computing Techniques in Knowledge-based Intelligent Engineering Systems, 1997
ISBN 3-7908-1035-5

Vol. 11. W. Mielczarski (Ed.)
Fuzzy Logic Techniques in Power Systems, 1998
ISBN 3-7908-1044-4

Vol. 12. B. Bouchon-Meunier (Ed.)
Aggregation and Fusion of Imperfect Information, 1998
ISBN 3-7908-1048-7

Vol. 13. E. Orłowska (Ed.)
Incomplete Information: Rough Set Analysis, 1998
ISBN 3-7908-1049-5

Vol. 14. E. Hisdal
Logical Structures for Representation of Knowledge and Uncertainty, 1998
ISBN 3-7908-1056-8

Vol. 15. G. J. Klir and M. J. Wierman
Uncertainty-Based Information, 2nd ed., 1999
ISBN 3-7908-1242-0

Vol. 16. D. Driankov and R. Palm (Eds.)
Advances in Fuzzy Control, 1998
ISBN 3-7908-1090-8

Vol. 17. L. Reznik, V. Dimitrov and J. Kacprzyk (Eds.)
Fuzzy Systems Design, 1998
ISBN 3-7908-1118-1

Vol. 18. L. Polkowski and A. Skowron (Eds.)
Rough Sets in Knowledge Discovery 1, 1998
ISBN 3-7908-1119-X

Vol. 19. L. Polkowski and A. Skowron (Eds.)
Rough Sets in Knowledge Discovery 2, 1998
ISBN 3-7908-1120-3

Vol. 20. J. N. Mordeson and P. S. Nair
Fuzzy Mathematics, 1998
ISBN 3-7908-1121-1

Vol. 21. L. C. Jain and T. Fukuda (Eds.)
Soft Computing for Intelligent Robotic Systems, 1998, ISBN 3-7908-1147-5

Vol. 22. J. Cardoso and H. Camargo (Eds.)
Fuzziness in Petri Nets, 1999
ISBN 3-7908-1158-0

Vol. 23. P. S. Szczepaniak (Ed.)
Computational Intelligence and Applications, 1999
ISBN 3-7908-1161-0

Vol. 24. E. Orłowska (Ed.)
Logic at Work, 1999
ISBN 3-7908-1164-5

continued on page 498

Peter Sinčák · Ján Vaščák (Editors)

Quo Vadis Computational Intelligence?

New Trends and Approaches in Computational Intelligence

With a Foreword from Jim Bezdeck

With 132 Figures
and 48 Tables

Physica-Verlag
A Springer-Verlag Company

Ass. Prof. Dr. Peter Sinčák
Dr. Ján Vaščák
Technical University of Košice
Faculty of Electrical Engineering and Informatics
Letna 9/B
040 00 Košice
Slovakia
E-mail: sincak@tuke.sk
vascak@tuke.sk

ISSN 1434-9922
ISBN 3-7908-1324-9 Physica-Verlag Heidelberg New York

Cataloging-in-Publication Data applied for
Die Deutsche Bibliothek – CIP-Einheitsaufnahme
Sinčák, Peter; Vaščák, Ján: Quo Vadis Computational Intelligence? New Trends and Approaches in Computational Intelligence: with 48 tables / Peter Sinčák, Ján Vaščák. – Heidelberg; New York: Physica-Verl., 2000
(Studies in fuzziness and soft computing; Vol. 54)
ISBN 3-7908-1324-9

Physica-Verlag Heidelberg New York
a member of BertelsmannSpringer Science+Business Media GmbH

Printed in Germany

Hardcover Design: Erich Kirchner, Heidelberg

SPIN 10774732 88/2202-5 4 3 2 1 0 – Printed on acid-free paper

Quo Vadis Computational Intelligence? Perseus and the Gorgon: a Cautionary Tale

I guess Peter Sincak asked me to write a foreword for this volume because I am responsible for the popularization (but not invention) of the term "computational intelligence", abbreviated nowadays simply as CI. I have declined previous requests to write book forewords because they all seem to say the roughly the same thing, viz., "here you will find many watershed papers of great importance to this vital field, so please buy this book", or something in a similar vein. But my experience with books of collected papers is that most of the papers are pretty average, there are a handful of good ones, perhaps one or two excellent ones, and, of course, a few clunkers. I don't want to write stuff like this (but I just did, right?), editors don't want forewords that say this, and publishers probably would omit them anyway.

Since I haven't seen the papers in this book (only the titles and authors), I cannot assert that my general experience along these lines is corroborated by this collection of articles, but certainly the 40 or so volumes I have in my office would bear out my opinion about books in this genre. Whether the present volume will follow what I believe to be the general trend for books of similar ilk remains to be seen; you can decide for yourself after you read the papers. (The one I wrote for this volume is certainly interesting to me, for example; my self assessment of it, however, is that it is no better than average.) But the title of this particular volume is so intriguing that I decided to take a crack at the foreword anyway. So, here I am, writing a foreword for "Quo Vadis CI?" What I have to say has been simmering on a back burner in my mind for about 15 years. I will use my platform to speak less about the contents of the volume, and more about the underlying implication of the question in the title - where is computational intelligence headed, anyway?. The results will be unlike most forewords you have ever read (does anyone ever read them? I don't!). Will I regret it later? Probably. Will this stop me from writing it? Certainly not. Will it end my run of book forewords at exactly one - this one? Probably. I just hope Peter lets it go through as is. Here goes.

Quo Vadis? OK, I am in trouble right out of the blocks. True, I took two years of Latin in high school, but that was 45 years ago. I think "Quo Vadis" means "which way"?, but it might mean "where are we going"?, or it might mean something slightly different from either of these. In any case, it's a good question, but it's a question with an imprecise answer at best - not because none of us can see into the future (which we can't, so this is a perfectly fine reason for the imprecision), but because the object of the question - computational intelligence - is a pretty elusive notion in the first place.

The fact that you need only mention CI, and not the words it stands for anymore (just as everyone knows that AI means "artificial intelligence"), to describe or identify some area of activity you are interested in, serves notice of how wildly popular the term has become since I first used it just 8 years ago. But if I tell you I work in CI, can you guess what I really study or do for a living? I don't think so. The term now extends far beyond my original idea. To see this, here is the abstract from my 1992 paper [1]:

> "This paper concerns the relationship between neural-like computational networks, numerical pattern recognition and intelligence. Extensive research that proposes the use of neural models for a wide variety of applications has been conducted in the past few years. Sometimes the justification for investigating the potential of neural nets (NNs) is obvious. On the other hand, current enthusiasm for this approach has also led to the use of neural models when the apparent rationale for their use has been justified by what is best described as "feeding frenzy". In this latter instance there is at times a concomitant lack of concern about many "side issues" connected with algorithms (e.g., complexity, convergence, stability, robustness and performance validation) that need attention before any computational model becomes part of an operational system. These issues are examined with a view towards guessing how best to integrate and exploit the promise of the neural approach with other efforts aimed at advancing the art and science of pattern recognition and its applications in fielded systems in the next decade. A further purpose of the present paper is to characterize the notions of computational, artificial and biological intelligence; my hope is that a careful discussion of the relationship between systems that exhibit each of these properties will serve to guide rational expectations and development of models that exhibit or mimic "human behavior"."

Assume that we have reasonable quantitative definitions of computational adaptivity and computational fault tolerance to go with current notions of speed and error rate optimality. If these four properties are hallmarks of biologically intelligent systems (certainly there are many others), then I suggested that these should also be used to qualify computational intelligence. Thus, I made this definition of CI in 1992 [1]:

> "A system is computationally intelligent when it : deals only with numerical (low-level) data, has a pattern recognition component, does not use knowledge in the AI sense; and additionally, when it (begins to) exhibit (i) computational adaptivity; (ii) computational fault tolerance; (iii) speed approaching human-like turnaround, and (iv) error rates that approximate human performance."

In my view, then and now, an artificially intelligent system is a CI system whose added value comes from incorporating knowledge (tidbits) obtained from humans in a non-numerical way. If you like my definition, can you test your system for computational intelligence? Of course not. But if you describe properties of your system with terms such as these so that I can see what they mean, measure them, compare them, and correlate them with my understanding of their more commonly held usage in natural language, you will have done a real service for science, and I will better know what you have accomplished.

My purpose in [1, 4, and even 9, just two years ago] was directed towards elimination of the use of seductive semantics in scientific writing. I don't think it is useful (in fact, I think it is unintentionally misleading) to read that your algorithm "learns" or is "adaptive" unless you tell me in a technical way that is specific to your model what you mean by these words. Everyone seems to have such models - whose learns best? which are more adaptive? Writing that uses words such as these would be greatly improved if the meaning intended by authors was clearly specified, and that is really what I wanted to focus on - crisp technical writing. Thus, my original paper on CI was really directed towards the semantics of scientific description rather than the characterization of a scientific field. But then things got crazy.

You may notice the lack of mention of either fuzzy systems (FS) or evolutionary computation (EC) in my abstract and definition, and neither of these disciplines played an essential role in my first paper on this subject, but these two fields quickly became appended to neural networks (NNs) to form the triumvirate of disciplines currently residing under the umbrella term CI. Why? Well, Bob Marks wrote an editorial in 1993 that gave his perception the difference between CI and AI [2]. Marks gave his own interpretation of CI, and added some interesting (and in my view, misleading) data about the disciplines that might be in CI and AI. Bob's editorial was partly occasioned by the choice (again and alas, mine) of the name World Congress on Computational Intelligence (WCCI); and it was partly occasioned by internal IEEE politics, which need not concern us here.

The 1994 IEEE WCCI meeting combined the three annual international IEEE conferences on neural networks (ICNN, now the IJCNN again), fuzzy systems (FUZZ-IEEE) and evolutionary computation (ICEC) that are sponsored by the IEEE Neural Networks Council (NNC). And so, the triumvirate came to life. In keeping with the classics theme of this volume (Quo Vadis), let me call this triumvirate the CI Gorgon. In the Greek myth, the Gorgon represented three evil sisters, each of whom had a head composed of many snakes. Perseus, the son of Zeus and Dana, eventually beheaded Medusa, the most dangerous of the three sisters, by looking in a mirror and slaying her "over the shoulder", so to speak, so he would not be turned to stone. So, in the myth, the Gorgon represented evil, while Perseus was the champion of goodness. In this little update of that classic allegory, however, the roles will be reversed: the Gorgon (CI) is good, while Perseus, it will turn out, is evil. Can you guess who - or what - Perseus turns out to be? No? Is there a scientific Perseus that will appear to slay the three sisters? Or one of them? Or two of them? Or will the Gorgon triumph, carrying science along as a prize? I will try to answer this question, but let me preface my answer with the disclaimer that it is simply an opinion, not grounded in facts. Facts, when and if they exist - are only of secondary importance to possible answers to this question .

After the 1994 WCCI, the Gorgon grew, like an alien life form in a science fiction movie (or an evil beast in a Greek myth) - fast. Since 1994, there has been an amazing proliferation of usage of the term CI, as well as authors who offer their own explanation of it [3 -9]. There have been many other international CI conferences besides the 1998 and [future] 2002 IEEE WCCIs; there are dozens and dozens and dozens of organizational CI units in academia and industry such as "Institutes of CI", or "CI laboratory", or whatever; there are a bunch of CI journals, several textbooks, various proceedings and many monographs with CI in their titles; and there are now several universities in that award graduate degrees in the subject! (e.g., Royal Holloway, a college of the London University, awards the MSCI degree). This doesn't sound so bad, does it? For example, it got me a trip to a wonderful country, whose nearest neighbor (the Czech Republic) was my grandfather's homeland, and that's pretty nice.

The papers in this volume are a pretty fair snapshot of the relationship between the three sisters of the CI Gorgon: evolutionary computation has less representation than either neural networks or fuzzy systems. This seems natural, since EC is much less encompassing than neural networks, which in turn is an (arguably) less ambitious field in the sense of the applications areas it spans than fuzzy systems. Many authors write about the use of various EC strategies (e.g., genetic algorithms) as tools for tuning or optimizing NNs and FSs, but EC is not usually an end in itself - rather, it is a means to some end, usually in NNs or FSs. Many papers now use NNs or FSs as tools for "optimization of each other". For example, NNs are sometimes used to tune fuzzy rule bases, and conversely, fuzzification of standard NNs has now been discussed in many, many ways. So my first observation is that the three sisters are not equally dangerous. One (EC) really plays a supporting role for the other two, which in turn, can play supporting roles for each other, or can attack on their own. Nowadays, all three attack together wherever the term CI appears.

First, put aside your belief that CI is a well-defined term. None of the sisters are well-defined (can you really tell me what a fuzzy system is? - of course not - it is Medusa, the sister with the most dangerous snakes for a head), so the Gorgon itself will never be a widely agreed upon compendium of disciplines. CI is simply a convenient way to give a loose description of a bunch of stuff, some of which has scientific merit, but more importantly, most of which has market cachet (right now anyway). To address the question - Quo Vadis CI? - we have to come to grips with what CI really is. Here is the reason I offered in [9] for the exponential growth of the use of the term CI over the six years 1992-1998:

> "[But] I suspect that there are two main reasons. First, the technical community is somewhat disenchanted with (perceptions, anyway) of the basis of AI research. [I will argue here that] AI tackles really hard problems, and that goals may have been set unrealistically high in the early days of AI. And second, scientists and engineers have a certain hunger - maybe even a justifiable need - for new terms that will spark interest and help sell papers, grant proposals, research and development programs and even [perhaps this is the main point] products. These are the defining characteristics of the so-called buzzword, of which CI is currently a prime example. After all, computational neural networks in their best known form have been around since 1943 [10], evolutionary computation since 1954 [11], and fuzzy sets since 1965 [12]. Funding entities and journal editors get tired of the same old terms. Is my attitude about this a little too cynical ? Probably. But I think it's pretty accurate. "

My distinction between computational and artificial models based on the injection of knowledge tidbits from humans arose from a desire to specifically point out that feed-forward neural networks do not "learn" in the sense I think humans do (the authors of [8] think otherwise, and I think they are wrong, because I don't think biological intelligence in the whole can be replicated by non-biological systems). But [1, 4] have inspired many authors to jump into the "what is intelligence?" fray - a jump that for me - a low level nuts and bolts kind of guy - is fraught with peril. I don't think there is a safe landing place - at least not one we will ever agree to.

So, is computational intelligence more than just a buzzword? Others seem to think so, but I still don't - never will - know. The utility of a good buzzword is hard to overstate. But isn't there more to it than that? Well, sure. I think the real point of using the term "computational" is that it places the emphasis on models and methods that try to solve realistic problems, as opposed to "teaching machines how to think". For me, models and methods such as FSs, NNs and EC are enabling technologies for AI. Certainly many AI researchers have come to think of the three sisters this way.

Artificial intelligence sets some pretty lofty goals. The realization that most of these goals are just not attainable with our current understanding of the way the biological neural network functions to enable intelligence or intelligent behavior has left a terminology vacuum for those who want to back away from such grand objectives. The word computational is much less provocative than the word artificial, and really connotes a "feet on the ground" approach to problem solving. I think this is the real appeal of the term, and I think this is a good way to use it.

So, what's next for CI? Well, it's easier to ask - what's next for each of the sisters?, because the sisters are a little easier to comprehend than the entire

Gorgon. For EC, I suspect we will see a continuing evolution (pun intended) of useful ways to: optimize functions, improve programming methods, and to solve certain limited classes of problems. Interest in EC seems pretty stable right now, and being the most recently noticed of the sisters, EC probably has the longest life expectancy as a "marketable" new technology. I simply don't know enough about EC to say much more than this about its future.

NNs are pretty well established as the best way to approximate complex functions, and have a pretty good track record in real, fielded systems at this point, so there is no reason to make a dire prediction about their future: it is secure. But papers I read about "new" NNs aren't very exciting anymore, and most of the new scholarly work in this field will probably be very incremental in a scientific sense. I think it is easy to document a general decline in attendance at NN meetings, a decline in interest in NN journals, and a low-level sort of "yeah - so what else is new'? attitude among young scientists in most areas of NNs. The real action here will be to carry fairly well understood architectures and algorithms into the field (translation: make a few bucks on the technology). NNs are still doing well in Academia, but not nearly as well as they were a decade ago. This is one of the signals a mature technology emits: don't study me anymore, just use me (e.g., we all take complex numbers for granted; advances in this area are still made, but are relatively unnoticed by most of the world).

And fuzzy systems? This seems to me to be the most secure of the three sisters at this point (all the more remarkable, considering the age of this youngest but biggest sister - 35 years). The cornerstone application areas for fuzzy models - pattern recognition and control - enjoy much success as enabling technologies for many patented systems (you know about the fuzzy washing machines, fuzzy vacuum cleaners, fuzzy heat pumps, unmanned helicopters based on fuzzy controllers, bullet trains, post office systems, fuzzy financial software, blah, blah, blah). Moreover, these two main areas have now matured sufficiently that they are used in crossover applications. Examples: Takagi-Sugeno and Mamdani-Assilian fuzzy systems are used to solve pattern recognition problems with fuzzy rule-based classifiers; and fuzzy clustering is used in the design of many fuzzy rule-based systems [13].

So, this largest sister is well-developed and easy to spot. But interest in FSs seems to waning faster than the attention currently being paid to the other two sisters; conferences are disappearing (EUFIT), attendance at our "premier" meeting is falling annually (FUZZ-IEEE, 2000, 230), journal subscriptions are declining (the IEEE Transactions on Fuzzy Systems has dropped by about 5000 since its inaugural issue in 1993). In my experience, the quality of papers in this area is becoming woeful; and their direction has become one dimensional - control. According to one friend of mine who is an associate editor of a leading fuzzy journal, about 75the submissions sent to him for reviews are bad junk, and most of the remainder is good junk - but almost all of it is basically junk. (Don't ask me who it is - none of us want this editor to handle our papers, do we?) My own experience over six years as founding editor of the IEEE Transactions on Fuzzy Systems agrees with this observation - it is getting harder and harder to find good papers (much less great ones), good reviewers, and good associate editors. This sister is not dead, nor is she dying, but she has lost much of the glamour that attracted casual observers to this field when she was "new and exciting", and I might add - controversial. What's the problem? Aging movie stars simply lose their appeal; so it is with buzzwords.

Do I sound like Cassandra? Should you read my comments as "beware the Ides of Y2K"? Well, all three of the sisters will still provide interesting and important problems that academicians will be asked to solve. However, the three sisters, and many of their siblings in other areas of science (mathematics, physics, computer science, you name it), are under attack by a very important, very subtle, and very destructive force which has been at work in the past ten years or so, and which is growing stronger every day. This force - alas, this Perseus - may undermine the basic foundation which growth in science has relied upon since the time of the Greeks, and before the Greeks. And what is this dreaded enemy? Perseus is technology itself - personified in the main by our beloved computers. And just how, you are asking, do computers hurt CI? After all, they are the raison d'tre for m definition of CI! Isn't this a little like saying the parents are killing their own children? Well, perhaps it is - but is it so?

In my view, the educational system that humans have relied upon for thousands of years - viz., communication of information from the older generations to the younger ones via verbal, and then written, and finally printed, form is an endangered species that may soon be extinct. Here I refer to education in its broadest sense; fostering questions, nurturing ideas, developing creativity - in short, refining and improving our collective "intelligence". This trend is less noticeable in Europe and the Far East than it is in the United States - less noticeable, but it's there, and it's growing - fast - just like the Gorgon did.

In the USA, most children no longer learn to enjoy talking or reading or writing, and many never learn to do any of these competently, much less very well. They rarely learn spelling, grammar, or basic arithmetic. Many of their teachers, from Kindergarten to and through graduate school don't encourage them to think, to listen, to question, to analyze. These are skills just like those needed to excel in, say, soccer; they must be honed by constant, repetitive practice. Weaned on television, video games, internet shopping, chat rooms, and other forms of instant gratification provided by modern computers, many students arrive at American Universities with scant ability to read, much less understand, books; with no experience at deriving information from lectures; with no belief that practicing to think (homework) is useful; and most importantly, with little ability to communicate with each other. Most of them don't care that what I say is correct, because most of them will get degrees anyway, and they know it! This will never happen at your University, right?

"Distance learning" is a reality. In Pensacola, a smallish, relatively typical American city of about 300,000 persons, students can obtain a Bachelor's degree long distance in almost any subject from about a dozen universities that aren't even in our state. Why? It's profitable for the universities, and easier for the students. Indeed, master's degrees are now routinely available this way, and some schools even grant PhDs to students who have had very minimal contact with real people, real books, in short, with real life (OK, universities and their ivory towers are hardly good models of "real life", but what I mean here by real life is life itself).

Students in my classes tell me they want web pages, not books; they want video instruction, not lectures; they want email, not conversation. And they want good grades and good jobs, not a good education, without really working at it much. And you know what? That is exactly what most of them get. Young professors in my department - people I see in the hallway every day - prefer to send me email, rather than simply talk to me about whatever is on their minds. Here is Perseus, staring directly at the three sisters of CI, saying to them: " if your acolytes cannot learn elementary facts, and hence, will not learn advanced ideas, how will they maintain you? you will never achieve immortality".

I know many of the older snakes on all three heads of the sisters, and these older snakes are scared. Sure, this is just like always, you say? Sounds like the usual "It was better when I was a boy"? No surprise, "Old dogs can't learn new tricks"? Think so? I read that the amount of information on the planet doubles every twenty months now. Really? Useful information? Impossible! How do we find the progressively less dense useful trees in the growing forest of junk? Data mining? I doubt it. Does the number of professors double this often? Are the curricula of our educational systems doubling every twenty months, so that this information will feed and nourish the next generation of scientists and engineers as the older heads shed their skins?

This is the challenge I think the three sisters and all of their siblings really face. The computer is changing the way education is perceived, delivered and used. Consequently, it may also changing the ability of young scientists and engineers to learn to think, to be creative, to find new and exciting ways to solve new and exciting problems. The future of the three sisters, and more generally, all of science, is in the hands of the products of this new education system. Will they be able to carry science forward? Or are the professors and papers that this new system is beginning to produce currently caught in a hopeless maelstrom? Will science itself now begin (or, has it already begun?) a descent, recycled over and over, destined, like the unlucky eldest brother in Poe's classic story [14], to disappear into the vortex created by "electronic communication"? I don't know, and I won't be around long enough for you to have the last laugh either. Humans are wonderfully and remarkably adaptable creatures, and have always been, in the main, able to turn technological advances to their benefit. But I don't think humans have ever faced the type of challenge looming on the near horizon created by a fully networked planet. It could be heaven - or it could be hell. This is the topic I hope younger workers in the Gorgon will attack: how to cope with the information explosion and concomitant changes it is causing in the way we deliver education to younger generations.

Well, that's it. Peter asked me to take a shot at answering "Quo Vadis CI?", so that's what I've done. I have spent my life being educated, and being an educator. I love my profession, and I enjoy knowing that I have helped a few ardent hikers along the path of Science. It's been fun to reread Poe, the Greek myth about the Gargon, and think a little about where CI in particular, and Science in general, is headed. I'm neither a cynic, nor a pessimist, and I know that nothing is as constant as change. Maybe one of the papers in this volume provides a first step towards correcting the growing problems in education our computers have created. Maybe. I hope so. Read the papers and decide for yourself. Then you can rewrite this modern allegory so that it has a better ending!

Jim Bezdek
Summer, Y2K Pensacola, FL, USA

References

1. Bezdek, J. (1992). *On the relationship between neural networks, pattern recognition and intelligence,* Int. Jo. Approx. Reasoning, 6(2), 85-107.
2. Marks, R. (1993). *Intelligence : Computational versus Artificial,* IEEE Trans. Neural Networks, 4(5), 737-739.
3. Zurada, J., Marks, R. and Robinson, C. (1994). *Introduction to Computational Intelligence: Imitating Life,* ed. J. Zurada, R. Marks and C. Robinson, IEEE Press, Piscataway, NJ, v-xi.
4. Bezdek, J. C. (1994). *What is Computational Intelligence? in Computational Intelligence: Imitating Life,* ed. J. Zurada, R. Marks and C. Robinson, IEEE Press, Piscataway,1-12.
5. Fogel, D. (1995). *Review of Computational Intelligence Imitating Life,* ed. J. Zurada, B. Marks and C. Robinson, IEEE Press, Piscataway, NJ, IEEE Trans. Neural Networks, 6(6), 1562 - 1565.
6. Palaniswami, M., Attikiouzel, Y., Marks, R.J., Fogel, D. and Fukuda, T. (1995). *Introduction to Computational Intelligence : A Dynamic System Perspective,* ed. Palaniswami, M., Attikiouzel, Y., Marks, R.J., Fogel, D. and Fukuda, T., IEEE Press, Piscataway, NJ, 1-5.
7. Eberhart, R. (1995). *Computational intelligence : a snapshot, in Computational Intelligence : A Dynamic System Perspective,* ed. Palaniswami, M., Attikiouzel, Y., Marks, R.J., Fogel, D. and Fukuda, T., IEEE Press, Piscataway, NJ, 9-15.
8. Eberhart, R., Dobbins, R. W. and Simpson, P.K. (1996). *Computational intelligence PC tools,* in press, Academic Press Professional (APP), NY.

9. Bezdek, J. C. (1998). *Computational Intelligence Defined - by Everyone!,* in Computational Intelligence: Soft Computing and Fuzzy-Neuro Integration with Applications, eds., O. Kaynak, L. A. Zadeh, B. Turksen and I. J. Rudas, NATO ASI series F, v 162, Springer-Verlag, Berlin, 10-37.
10. Mccullogh W. and Pitts, W. (1943). *A logical calculus of the ideas immanent in nervous activity,* Bull. Math. Biophysics, 7, 115-133.
11. Barricelli, N. A. (1954). *Esempi Numerici di Processi di Evoluzione,* Methodos, 6(21-22), 45-68.
12. Zadeh, L. A. (1965). *Fuzzy Sets, Information and Control,* 8, 338-352.
13. J. C. Bezdek, J. Keller, R. Krishnapuram and N. R. Pal, *Fuzzy Models and Algorithms for Pattern Recognition and Image Processing,* Kluwer, Norwell, MA, 1999.
14. Poe, E. A. (1809-1849). *A Descent into the Maelstrom,* in The Collected Tales and Poems of Edgar Allan Poe, Random House, NY, 1992.

Preface

There are many different views about computational intelligence in the scientific community and its relationship with the notion of artificial intelligence. The notion of Computational Intelligence, given by Prof. James Bezdek, also has several different interpretations. The most important of all opinions is that computational intelligence is creating a framework through cooperation of different techniques, mainly neural networks, fuzzy systems, evolutionary computations and some other approaches based on computational techniques. This kind of approach is being extended and progressing to the notion of intelligent computing. These methods represent a bottom-up approach in the simulation of a cognitive conceptual abyss and have extensive biological inspiration. Contributions in this book are presented by a number of leading scientists in the field of neural networks, fuzzy systems and evolutionary computations. The papers indicate the direction of development of techniques integrated through the notion of Computational Intelligence. The aim of this book is to keep searching for the answer to an important question: "Quo Vadis Computational Intelligence?". I hope that it can be found in many papers of this multi-author book.

The book consists of four parts devoted to neural networks, fuzzy systems, miscellaneous CI techniques and hybrid systems. I am very proud to have a wonderful foreword "Quo Vadis Computational Inteligence ? Perseus and the Gorgon: a Cautionary Tale" written by Prof. James Bezdek for this book. Every part contains some leading scientists' contributions to Computational Intelligence.

In the part devoted to the neural networks Gail Carpenter presents a new generation of ART-family neural networks with distributed coding. K. Fukushima reports about new trends in using neural networks in vision. A very interesting contribution is made by V. Kurkova concerning the mathematical background of neural networks. B. Igelnik presents a new adaptive architecture for multivariate data analysis. E. Lotfy presents an interesting application of neural network in civil engineering.

In the part devoted to fuzzy systems, L. Zadeh wrote a principal paper about the new trends in Computing by words. This paper is followed by a contribution from J. Kacprzyk and S. Zadrozny, which is an extension of Zadeh's ideas in this field. E. Pap presents his views about the readiness of mathematics for computational intelligence. J. Bezdek wrote a very interesting paper about clustering techniques of incomplete data. K. Hirota deals with fuzzy expert system application in the trade domain. R. Mesiar reports about new trends in fuzzy calculus. V. Duch presents some interesting work in data understanding.

In the part devoted to miscellaneous CI techniques V. Kvasnička discusses a problem of evolutionary model of symbiosis. J. Kubalik presents new genetic operators with special capabilities and P. Ošmera wrote a paper about genetic algorithms with special features. There is a very interesting paper by M. Gams about the integration of intelligence and information in society.

In part four, which is devoted to hybrid systems, N. Kasabov wrote about the progress in his ECOS system to build an evolving connectionist system. H. Takagi presents an interesting work and discusses the future of CI from the point of view of 2 basic approaches: analytic and synthetic computational intelligence. C. Moraga presents systems of knowledge/rule extraction from neural network topology. H. Adeli wrote a paper about using wavelets in feature extraction procedures. Papers written by Sinčák, Smolikova & Zurada, Merenyi, Thackrah and Gedeon are about the development of various CI approaches and its testing on remote sensing data. L. Koczy presents an interesting technique for fuzzy controllers based on the implementation of the RBF approximation scheme. L. Rutkowski wrote an interesting paper about the flexible structure of neuro-fuzzy systems and S. C. Tan presents a rule extraction approach from adaptive neural network for fault diagnosis.

Let me express a strong belief that this book is a useful contribution to the worldwide effort for pushing leading edge technology towards usefulness in society. I am strongly convinced that the creator of mankind, **our Lord,** is supporting our mission and supervising our paces not to misuse technology against human-kind. Computational Intelligence should be about helping people and society to solve tasks more easily and effectively in every domain of our life.

I would like to express my gratitude to all authors and reviewers preparing this multi-author book. I also thank Prof. Janusz Kacprzyk and the people from Physica-Verlag for encouragement during the editing of this book. Finally, I would like to thank all my colleagues and my PhD and MSc. students for their help during this multi-author book preparation.

Dr. Peter Sinčák

Technical University of Košice, Slovakia

August, 2000

Contents

Part 4 : Hybrid Systems

Part 1

Neural Networks

ART Neural Networks: Distributed Coding and ARTMAP Applications

Gail A. Carpenter

Boston University, Department of Cognitive and Neural Systems, 677 Beacon Street, Boston, Massachusetts 02215 USA, gail@cns.bu.edu

Abstract. *ART (Adaptive Resonance Theory) neural networks for fast, stable learning and prediction have been applied in a variety of areas. Applications include airplane design and manufacturing, automatic target recognition, financial forecasting, machine tool monitoring, digital circuit design, chemical analysis, and robot vision. Supervised ART architectures, called ARTMAP systems, feature internal control mechanisms that create stable recognition categories of optimal size by maximizing code compression while minimizing predictive error in an on-line setting. Special-purpose requirements of various application domains have led to a number of ARTMAP variants, including fuzzy ARTMAP, ART-EMAP, Gaussian ARTMAP, and distributed ARTMAP. ARTMAP has been used for a variety of applications, including computer-assisted medical diagnosis. Medical databases present many of the challenges found in general information management settings where speed, efficiency, ease of use, and accuracy are at a premium. A direct goal of improved computer-assisted medicine is to help deliver quality emergency care in situations that may be less than ideal. Working with these problems has stimulated a number of ART architecture developments, including ARTMAP-IC [1]. This paper describes a recent collaborative effort, using a new cardiac care database for system development, has brought together medical statisticians and clinicians at the New England Medical Center with researchers developing expert systems and neural networks, in order to create a hybrid method for medical diagnosis. The paper also considers new neural network architectures, including distributed ART (dART), a real-time model of parallel distributed pattern learning that permits fast as well as slow adaptation, without catastrophic forgetting. Local synaptic computations in the dART model quantitatively match the paradoxical phenomenon of Markram-Tsodyks [2] redistribution of synaptic efficacy, as a consequence of global system hypotheses.*

Keywords. *Adaptive Resonance Theory, ART, ARTMAP, neural networks, medical prediction, redistribution of synaptic efficacy.*

1. ART and ARTMAP Neural Networks

Adaptive resonance theory originated from an analysis of human cognitive information processing and stable coding in a complex input environment [3,4]. An evolving series of ART neural network models have added new principles to the early theory and have realized these principles as quantitative systems that can be applied to problems of category learning, recognition, and prediction. Each ART network forms stable recognition categories in response to arbitrary input sequences with either fast or slow learning regimes. The first ART model, ART 1 [5], was an unsupervised learning system to categorize binary input patterns. ART 2 [6] and fuzzy ART [7] extend the ART 1 domain to categorize analog as well as binary input patterns.

Supervised ART architectures, called ARTMAP systems, self-organize arbitrary mappings from input vectors, representing features such as spectral values and terrain variables, to output vectors, representing predictions such as vegetation classes in a remote sensing application. Internal ARTMAP control mechanisms create stable recognition categories of optimal size by maximizing code compression while minimizing predictive error in an on-line setting. Binary ART 1 computations are the foundation of the first ARTMAP network [8], which therefore learns binary maps. When fuzzy ART replaces ART 1 in an ARTMAP system, the resulting fuzzy ARTMAP architecture [9] rapidly learns stable mappings between analog or binary input and output vectors.

2. Match-based Learning, Error-Based Learning, and Fast Learning

The central feature of all ART systems is a pattern matching process that compares the current input with a learned expectation produced by an active code, or hypothesis. ART matching leads either to a resonant state, which focuses attention and triggers learning, or to a self–regulating parallel memory search, which eventually leads to a resonant state, unless the network's memory capacity is exceeded. If the search ends at an established code, the memory representation may stay the same or may be refined to incorporate information from attended portions of the current input. If the search ends at a new code, the code's memory representation begins by learning the current input itself. This *match–based learning* process is the foundation of ART code stability. Match–based learning allows memories to change only when input from the external world is close enough to internal expectations, or when something completely new occurs. This feature makes ART and ARTMAP well suited to problems that require on–line learning of large and evolving databases.

Match–based learning is contrasted with *error–based learning*, which responds to a mismatch by sending the difference between a target output and an actual output toward zero, rather than by initiating a search for a better match. Error–based learning is naturally suited to problems such as adaptive control and the learning of sensory–motor maps, which require ongoing adaptation to present statistics. Neural networks that employ error–based learning include back propagation [10] and other multilayer perceptrons (MLPs).

Many ART applications use fast learning, whereby adaptive weights fully converge to equilibrium values in response to each input pattern. Fast learning enables a system to adapt quickly to inputs that occur only rarely but that may require immediate accurate recall. Remembering many details of an exciting movie is a typical example of fast learning. When the difference between actual output and target output defines "error," present inputs would drive out past learning, since fast learning zeroes the error on each input trial. Therefore fast learning destabilizes the memories of error–based models like back propagation. This feature restricts the domain of most MLPs to off–line applications with a slow learning rate.

3. Distributed Coding

In ART and ARTMAP networks, winner–take–all competitive activation supports stable coding, but this limiting case of competition may cause category proliferation when noisy inputs are trained with fast learning. In contrast, MLPs feature distributed McCulloch–Pitts activation, which promotes noise tolerance and code compression, but which causes catastrophic forgetting with fast learning. A recently introduced family of networks called *distributed ART* models combine the best of these two worlds: distributed activation enhances noise tolerance and code compression while new system dynamics retain the stable fast learning capabilities of winner–take–all (WTA) ART systems. With WTA coding, the unsupervised distributed ART model (dART) [11,12] reduces to fuzzy ART and the supervised distributed ARTMAP model (dARTMAP) [13] reduces to fuzzy ARTMAP. With distributed coding, these networks automatically apportion learned changes according to the degree of activation of each node, which permits fast as well as slow learning without catastrophic forgetting. A parallel distributed match–reset–search process also helps stabilize memory. The result of adaptation resembles long–term potentiation (LTP) for single-pulse or low-frequency test inputs but can resemble long-term depression (LTD) for higher frequencies. This dynamic is traced to dual computational properties of frequency-dependent and frequency-independent components of the coding signal. During learning, the frequency-independent component increases nonspecifically, for all inputs, while the frequency-dependent component becomes more selective, maximally favoring the current input.

The disappearance of LTP enhancement for high–frequency test inputs has been observed by Markram and Tsodyks [2] in the neocortex. Distributed ART features redistribution of synaptic efficacy at the local synaptic level as a consequence of necessary system hypotheses at the global pattern processing level. This "top-down" approach to understanding the Markram–Tsodyks data suggests, by example, how this apparently paradoxical phenomenon may actually be precisely the element needed to solve a critical pattern coding problem at a higher processing level. Both ART and dART models employ competitive learning schemes for code selection, and both are designed to stabilize learning. However, because ART networks use a classical steepest-descent paradigm called instar learning [14], these systems require winner-take-all coding to maintain stability with fast learning. A new learning law called the *distributed instar (dInstar)* allows dART code representations to be distributed across any number of network nodes. Both ART and dART also employ a preprocessing step called *complement coding* [7], which presents to the learning system both the original input vector and its complement. This device is analogous to on-cell/off-cell coding found in the early visual system. Complement coding solves a category proliferation problem pointed out by Moore [15]. It also suggests a computational solution to the tendency of redistribution of synaptic efficacy to enhance only low-frequency inputs: if an input component is consistently large with respect to a given code, then the network can embody this fact in the complementary component, which can also be enhanced since it will be consistently small.

The dynamic behavior of an individual dART synapse is seen in the context of its role in stabilizing distributed pattern learning, rather than as a primary hypothesis. Redistribution of synaptic efficacy here reflects a tradeoff between frequency-dependent and frequency-independent synaptic signal components which support a tradeoff between pattern selectivity and a nonspecific gain increase at the network level. Models that implement distributed coding via gain adaptation alone tend to suffer catastrophic forgetting and require slow or limited learning. In dART, each increase in frequency-independent synaptic efficacy is balanced by a corresponding decrease in frequency-dependent efficacy. The net result is redistribution, rather than nonspecific enhancement, of synaptic efficacy. The system uses this mechanism to enhance network response to a given pattern while suppressing the response to mismatched patterns. At the same time, the dART network learning law protects prior codes against catastrophic forgetting. It does so by formally replacing the traditional multiplicative weight with a *dynamic weight*, equal to the rectified difference between target node activation and an *adaptive threshold*, which embodies the long-term memory of the system [16]. The dynamic weight permits adaptation only at the most active coding nodes, which are limited in number due to competition at the target field. In addition, thresholds, which are initially zero, become increasingly resistant to change as they become larger. Note that, although thresholds following a minimal dInstar learning law can only increase monotonically, complement coding allocates two thresholds for each component of the original input, which allows the network to encode a full range of input features.

4. Rules, Applications, and Biological Substrates

ART principles have also been used to explain challenging behavioral and brain data in the areas of visual perception, visual object recognition, auditory source identification, variable–rate speech and word recognition, and adaptive sensory–motor control (e.g., [17,18]). One area of recent progress concerns how the neocortex is organized into layers. This new work suggests how "laminar computing" leads to intelligent behavior by modeling how bottom–up, top–down, and horizontal interactions are organized within the cortical layers. These interactions have thus far been studied within the visual cortex. Here, a model has been developed to show how visual cortex (1) stably develops circuits that match environmental constraints, and continues to refine this structure through adult learning; (2) binds or groups distributed information into coherent object representations; and (3) pays attention to important events (e.g., [19]). The mechanisms that govern (1) in the infant are proposed to lead to properties (2) and (3) in the adult. These results are clarifying how ART design principles are embedded within the neocortical circuits that subserve other types of intelligent behaviors, and open the way towards designing general–purpose vision systems that can autonomously learn optimal operating parameters in response to specialized image domains.

ART and dART systems are part of a rapidly growing family of attentive self–organizing systems that have evolved from the biological theory of cognitive information processing. ART modules have found their way into such diverse applications as industrial design and manufacturing, the control of mobile robots, face recognition, remote sensing land cover classification, target recognition, medical diagnosis, electrocardiogram analysis, signature verification, tool failure monitoring, chemical analysis, circuit design, protein/DNA analysis, 3–D visual object recognition, musical analysis, and seismic, sonar, and radar recognition (e.g., [20–22]). A recent book focuses on the implementation of ART systems as VLSI microchips [23].

Applications exploit the ability of an ART system to rapidly learn to classify large databases in a stable fashion, to calibrate confidence in a classification, and to focus attention upon those featural groupings that the system deems to be important based upon experience. The learned expertise of an ARTMAP system also translates to IF–THEN "rules." Within each recognition code, the expectation, or prototype represents a rule that predicts a given outcome. With WTA coding, these prototype vectors provide a transparent set of rules that characterize the decision–making process. ARTMAP neural networks have now provided new methodologies for medical database analysis. A case study of this method, applied to a cardiac database developed at the New England Medical Center (NEMC) [24] is introduced in the following section.

5. The New England Medical Center (NEMC) Modeling Project

A group of physicians and statisticians from the Division of Clinical Care Research at the New England Medical Center (NEMC) have created a database of Emergency Department patients who were considered for admission to the coronary care unit. A primary goal of the NEMC project is to develop methods to support a physician's decision making process. The project specifically aims to understand the utility and limitations of established and new modeling procedures and to promote their appropriate use in medical research, health care policy, and care assessment. These goals are accomplished through systematic investigation and rigorous evaluation of the relative predictive performance of the analyzed modeling methods. The project is being carried out as a collaboration between the physicians and statisticians who designed the NEMC cardiac database and researchers from Boston University and MIT.

5.1. The NEMC Database

The NEMC cardiac database consists of the records of 3,068 study subjects examined at the Emergency Departments (ED) of six participating New England hospitals. The database includes clinical features available to ED physicians, such as clinical presentation, history, physical findings, electrocardiogram, socio-demographic characteristics, and coronary-disease risk factors. Of the 3,068 subjects in the database, 15.7% were diagnosed with cardiac problems requiring hospitalization. These positive outcome cases fall into three categories: arrhythmia, hemodynamic condition, and ischemia. These positive categories are not mutually exclusive; for example, arrhythmia and ischemia are usually accompanied by a hemodynamic condition. The identity of subcategories among the positive outcome variable was not used during model development. That is, the dichotomous output in the NEMC database codes only whether a patient required hospitalization, without specifying particular medical conditions.

The NEMC database includes records for each patient that represent 32 clinical variables, 199 raw ECG variables, and 78 derived ECG variables. Clinical variables quantify features such as medical complaints at arrival to the hospital, age, gender, body-mass index, history of past disease, and medication. The 199 raw ECG variables for the NEMC database were handpicked from a large pool of describing an ECG cycle. Chosen variables describe the amplitude, duration, slope, and area for segments of interest from each of the 12 leads (e.g., Q wave amplitude and duration, QRS area and duration, ST slope). Several variables describing general aspects of the ECG cycle were also included (e.g., mean ventricular rate, mean QRS duration, mean QT interval).

The 78 derived ECG variables in the NEMC database were created to separate clinically important aspects from irrelevant features of the raw signal. These variables are thought to be less sensitive to random fluctuations than the raw signal. Derived ECG variables consist of four groups:

□ Five (5) derived ECG variables qualitatively describe the presence or absence of the following abnormalities in the cardiac cycle: Q waves, ST elevation, ST depression, T wave elevation, and T wave inversion. The derivation was based on recordings from all 12 leads.

□ Fifty five (55) derived lead-by-lead ECG variables describe whether the five abnormalities enumerated above appear in individual leads. If an abnormality is detected in a raw lead, the same type of abnormality should be registered in one of its contiguous leads. Otherwise, the abnormality is attributed to noise and the variable is set to zero.

□ Fifteen (15) summary regional ECG variables describe whether the locations of each of the five abnormalities, in the anterior, inferior, or lateral regions. For each region, at least two raw leads should have registered the abnormalities. Otherwise, the abnormality was ignored and the variable was set to zero.

□ Three (3) summary dichotomous ECG variables describe the presence or absence of right bundle branch block, left bundle branch block, and left ventricular hypertrophy. According to the NEMC researchers, these variables have been used in previous regression models to override the effect of ST elevation, although it is not clear what their direct predictive value might be.

5.2. ARTMAP in the NEMC Project

ARTMAP-IC [1], an extension of fuzzy ARTMAP [9] and ART-EMAP [25], was initially designed to solve a computational problems commonly encountered in medical modeling, including how to encode inconsistent cases, where identical patient records are associated with different outcomes. When the ARTMAP-IC system was introduced, its performance was evaluated on four benchmark medical databases. One of these databases was the Cleveland heart disease database from the UCI repository [26]. This database contained the records of 303 cardiology patients, 45.9% of which were diagnosed with heart disease. Each record had 13 attributes, including age, gender, heart rate, angina, ST depression, and ST slope. These initial simulation results demonstrated ARTMAP-IC's potential value for cardiac diagnosis.

Exploratory studies of the NEMC database indicated that ARTMAP-IC was not well suited for this problem. In particular, the low prevalence of positive outcomes (15.7%) rendered the system's instance counting feature counterproductive. The final successful ARTMAP algorithm for the NEMC project did, on the other hand, incorporate variations of the basic network that had been developed for other applications. This experience is typical: a given application usually benefits from certain model features but not others. A new model extension was also designed to improve the probability estimation capabilities of the ARTMAP system. Finally, although ARTMAP can handle an unlimited number of inputs, the significance of individual variables is clinically important. With 309 input variables in the NEMC database, variable selection and dimensionality reduction were important for purposes of interpretation. The project therefore included a novel method for estimation of the impact of individual input variables within the framework of ARTMAP networks. With this system, ARTMAP's generalization capabilities were seen to compare favorably to those of logistic regression and decision trees.

The logistic regression approach, on the other hand, appears to offer a simple model with reasonable discrimination and calibration capabilities. This claim of model simplicity is somewhat misleading because some of the derived variables, used as inputs to the regression, were laboriously handcrafted and have a great deal of complexity embedded in them. Still, one may argue that explicit variable derivation rules have certain advantages over complex self-organizing systems, such as ARTMAP or decision trees, because one can create rules encoding physicians' knowledge and diagnostic techniques. A counterargument is that self-organizing systems can discover unnoted patterns in the data and thus offer new diagnostic insights. A crucial test for broad acceptance of data-driven modeling approaches such as ARTMAP in the medical community would be the possibility of unraveling the information encoded in their complex structure. While an ARTMAP network is much easier to analyze and interpret than a standard backpropagation network, the high input dimensionality and the large number of category nodes pose ongoing challenges to structure visualization and rule extraction.

Acknowledgements: This research was supported in part by the Office of Naval Research (ONR N00014-95–1–0409 and ONR N00014–95–1–0657).

References

[1] Carpenter GA, Markuzon N. ARTMAP-IC and medical diagnosis: Instance counting and inconsistent cases. *Neural Networks* 1998; 11:323-336

[2] Markram H, Tsodyks M. Redistribution of synaptic efficacy between neocortical pyramidal neurons. *Nature* 1996; 382:807–810

[3] Grossberg S. Adaptive pattern classification and universal recoding, I: Parallel development and coding of neural feature detectors & II: Feedback, expectation, olfaction, and illusions. *Biological Cybernetics* 1976; 23:121-134 & 187-202

[4] Grossberg S. How does a brain build a cognitive code? *Psychological Review* 1980; 87:1-51

[5] Carpenter GA, Grossberg S. A massively parallel architecture for a self–organizing neural pattern recognition machine. *Computer Vision, Graphics, and Image Processing* 1987a; 37:54-115

[6] Carpenter GA, Grossberg S. ART 2: Self-organization of stable category recognition codes for analog input patterns. *Applied Optics* 1987b; 26:4919-4930

[7] Carpenter GA, Grossberg S, Rosen DB. Fuzzy ART: Fast stable learning and categorization of analog patterns by an adaptive resonance system. *Neural Networks* 1991; 4:759-771

[8] Carpenter GA, Grossberg S, Reynolds JH. ARTMAP: Supervised real-time learning and classification of nonstationary data by a self-organizing neural network. *Neural Networks* 1991; 4:565-588

[9] Carpenter GA, Grossberg S, Markuzon N, et al. Fuzzy ARTMAP: A neural network architecture for incremental supervised learning of analog multidimensional maps. *IEEE Transactions on Neural Networks* 1992; 3:698-713

[10] Werbos P. Beyond regression: New tools for prediction and analysis in the behavioral sciences. PhD thesis, Harvard University, Cambridge, Massachusetts, 1974

[11] Carpenter, GA. Distributed activation, search, and learning by ART and ARTMAP neural networks. In: *Proceedings of the International Conference on Neural Networks (ICNN'96): Plenary, Panel and Special Sessions*. IEEE Press, Piscataway, New Jersey, 1996, pp. 244-249

[12] Carpenter GA. Distributed learning, recognition, and prediction by ART and ARTMAP neural networks. *Neural Networks* 1997; 10:1473-1494

[13] Carpenter, GA, Milenova, BL, Noeske, BW. Distributed ARTMAP: a neural network for fast distributed supervised learning. *Neural Networks* 1998; 11:793-813

[14] Grossberg, S. Neural expectation.: Cerebellar and retinal analogs of cells fired by learnable or unlearned pattern classes. *Kybernetik*, 1972; 10:49-57

[15] Moore, B. (1989). ART 1 and pattern clustering. In: Touretzky D, Hinton G, Sejnowski T (eds) *Proceedings of the 1988 Connectionist Models Summer*

School. Morgan Kaufmann Publishers, San Mateo, California, 1989, pp. 174-185

[16] Carpenter, GA. A distributed outstar network for spatial pattern learning. *Neural Networks* 1994; 7:159-168

[17] Desimone R. Neural circuits for visual attention in the primate brain. In: Carpenter GA, Grossberg S (eds) *Neural Networks for Vision and Image Processing.* MIT Press, Cambridge, Massachusetts, 1992, pp 343-364

[18] Grossberg S. How does the cerebral cortex work? Learning, attention, and grouping by the laminar circuits of visual cortex. *Spatial Vision* 1999a; 12:163-186

[19] Grossberg S. The link between brain learning, attention, and consciousness. *Consciousness and Cognition* 1999b; 8:1-44

[20] Caudell TP, Smith SDG, Escobedo R, et al. NIRS: Large scale ART–1 neural architectures for engineering design retrieval. *Neural Networks* 1994; 7:1339-1350

[21] Christodoulou CG, Huang J, Georgiopoulos M, et al. Design of gratings and frequency selective surfaces using fuzzy ARTMAP neural networks. *Journal of Electromagnetic Waves and Applications* 1995; 9:17-36

[22] Soliz P, Donohoe, GW. Adaptive resonance theory neural network for fundus image segmentation. In: *Proceedings of the World Congress on Neural Networks (WCNN'96)*, Erlbaum, Hillsdale, New Jersey, 1996, pp 1180-1183

[23] Serrano–Gotarredona T, Linares–Barranco B, Andreou, AG. *Adaptive Resonance Theory Microchips: Circuit Cesign Techniques.* Kluwer Academic Publishers, Boston, 1998

[24] Carpenter GA, Kopco N, Milenova BL, et al. A neural network method for supervised learning and medical database analysis. *Statistics in Medicine*, to appear

[25] Carpenter GA, Ross WD. ART-EMAP: A neural network architecture for object recognition by evidence accumulation. *IEEE Transactions on Neural Networks* 1995; 6:805-818

[26] Murphy PM, Aha DW. UCI repository of machine learning databases. University of California, Department of Information and Computer Science, Irvine, California. http://www.ics.uci.edu/~mlearn/MLRepository.html, 1992

Neural Network Models for Vision

Kunihiko Fukushima

The University of Electro-Communications, Chofu, Tokyo 182-8585, Japan

Abstract: To capture and process visual information flexibly and efficiently from changing external world, the function of active and adaptive information processing is indispensable. Visual information processing in the brain can be interpreted as a process of eliminating irrelevant information from a flood of signals received by the retina. Self-organization of the neural network is another important function for flexible information processing. This paper shows two neural network models for these mechanisms: "recognition of partially occluded patterns" and "self-organization of shift-invariant receptive fields".

1 Introduction

When looking at an object, we do not passively accept the entire information available within our visual field, but actively gather only necessary information. We change the focus of our attention to the places that attract our interest. We capture information from there and process it selectively. Visual information processing in the brain can be interpreted as a process of eliminating irrelevant information from a flood of signals received by the retina.

Self-organization of the neural network is another important function for flexible visual information processing. The brain modifies its own structure to adapt to the environment.

This paper discusses the importance of active and adaptive processing in vision, showing two neural network models for these mechanisms: "recognition of partially occluded patterns" and "self-organization of shift-invariant receptive fields.

2 Recognition of Occluded Patterns

We often can read or recognize a letter or word contaminated by stains of ink, which partly occlude the letter. If the stains are completely erased and the occluded areas of the letter are changed to white, however, we usually have difficulty in recognizing the letter, which now have some missing parts. For example, the patterns in Fig. 1(a), in which the occluding objects are invisible, are almost illegible, but the patterns in (b), in which the occluding objects are visible, are much easier to read. Even after having watched patterns in (b) and having known what the occluding objects are, we still have difficulty in perceiving the original shapes of the patterns in (a) if the occluding objects are invisible.

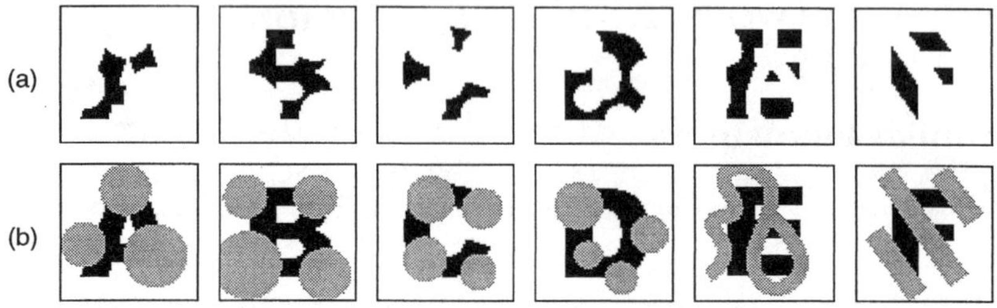

Fig. 1. Patterns partially occluded by (a) invisible and (b) visible masking objects [1]. These patterns are used to test the proposed neural network model.

This section proposes a hypothesis explaining why a pattern is easier to be recognized when the occluding objects are visible. A neural network model is constructed based on the proposed hypothesis and is simulated on a computer [1].

A pattern usually contains a variety of visual features, such as edges, corners, and so on. The visual system extracts, at an early stage, these local features from the input pattern and then recognizes it. When a pattern is partially occluded, a number of new features, which did not exist in the original pattern, are generated near the contour of the occluding objects. When the occluding objects are invisible, the visual system will have difficulty in distinguishing which features are relevant to the original pattern and which are newly generate by the occlusion. These irrelevant features will hinder the visual system from recognizing the occluded pattern correctly.

When the occluding objects are visible, however, the visual system can easily discriminate relevant from irrelevant features. The features extracted near the contours of the occluding objects are apt to be irrelevant to the occluded pattern. If the responses of feature extractors in charge of the area covered by the occluding objects are suppressed, the signals related to the disturbing irrelevant features will be blocked and will not reach higher stages of the visual system. Since the visual system usually has some tolerance to partial absence of local features of the pattern, it can recognize the occluded pattern correctly if the disturbing signals from the irrelevant features disappear.

We have constructed a neural network model based on this hypothesis. The model is an extended version of the *neocognitron* [2]. The neocognitron is a neural network model of the visual system, and has a hierarchical multilayered architecture. It acquires an ability to robustly recognize visual patterns through learning. It consists of layers of S-cells, which resemble simple cells in the primary visual cortex, and layers of C-cells, which resemble complex cells. In the whole network, layers of S-cells and C-cells are arranged alternately in a hierarchical manner. S-cells are feature-extracting cells, and input connections to S-cells are variable and are modified through learning. C-cells, whose input connections are fixed and unmodifiable, exhibit an approximate invariance to the position of the stimuli presented within their receptive fields. The C-cells in the highest stage work as recognition cells indicating the result of pattern recognition.

Figure 2 shows the architecture of the proposed neural network model, and an example of the response of the cells in the network is displayed in Fig. 3. The alphabetical characters shown in Fig. 4 have been used to train the network in the computer simulation. After finishing the learning, partially occluded alphabetical characters shown in Fig. 1 are presented to the network. Tests are made for two types of input images: in which the occluding objects are (a) invisible, and (b) visible.

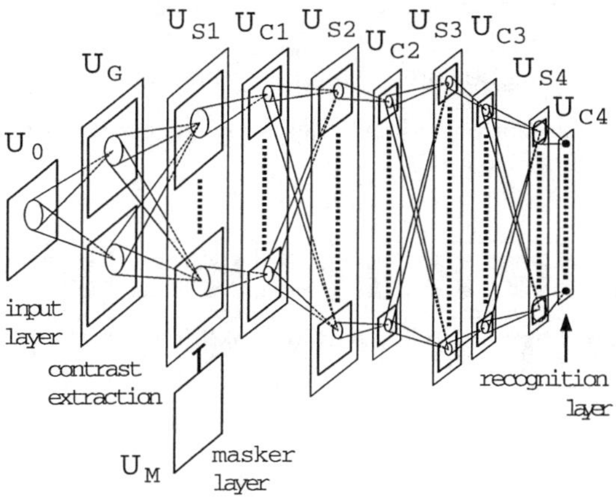

Fig. 2. The architecture of the neural network model that can recognize partially occluded patterns [1].

The stimulus pattern is presented to the input layer (photoreceptor layer) U_0. A layer of contrast-extracting cells (U_G), which correspond to retinal ganglion cells or LGN cells, follows layer U_0. Layer U_G has two cell-planes: a cell-plane consisting of on-center cells, which extract positive contrast of the input image, and a cell-plane consisting of off-center cells, which extract negative contrast. The output of layer U_G is sent to the S-cell layer of the first stage (U_{S1}). The S-cells of layer U_{S1}, which have been trained with supervised learning [2], extract edge components of various orientations from input image.

The network consists of four stages of S- and C-cell layers. The output of layer U_{Sl} (S-cell layer of the lth stage) is fed to layer U_{Cl}, and a blurred version of the response of layer U_{Sl} appears in layer U_{Cl}. The density of the cells in each cell-plane is reduced between layers U_{Sl} and U_{Cl}. In Fig. 3, the responses of S-cell layers are not displayed, because their responses are too thin to be observed in a small display. The response of an S-cell layer, however,

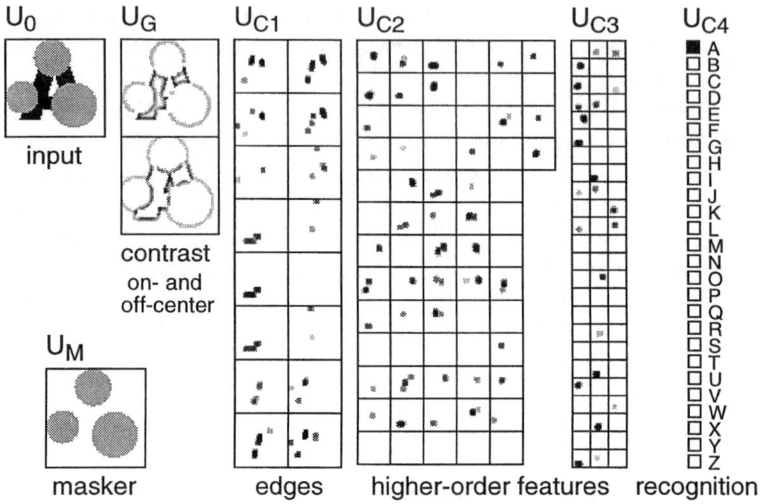

Fig. 3. Response of the network when the occluding objects are visible [1]. The occluded input pattern is correctly recognized as A.

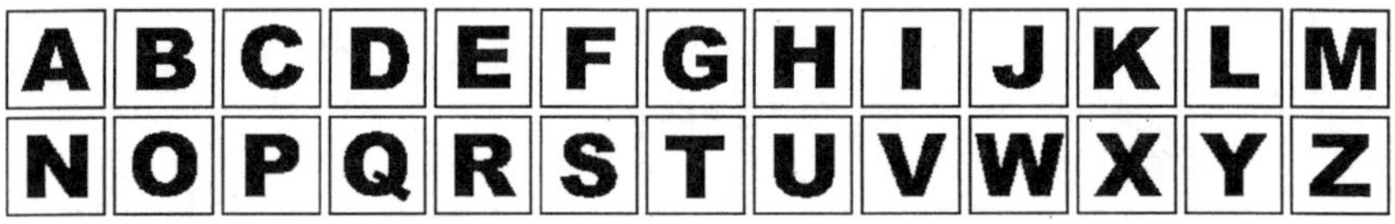

Fig. 4. The learning patterns used to train the network [1].

can easily be estimated from the blurred response of the corresponding C-cell layer. The S-cells of the intermediate stages (U_{S2} and U_{S3}) have been self-organized by unsupervised competitive learning in a similar way as for the conventional neocognitron.

The rightmost layer U_{C4}, which is the highest stage of the network, is the recognition layer, whose response shows the final result of recognition by the network. This layer has been trained to recognize all learning patterns correctly through supervised competitive learning [3].

The model has a layer (U_M) consisting of only one cell-plane. The layer, which is called the *masker layer* in this paper, detects and responds only to occluding objects. Its response is displayed in the lower left of Fig. 3. If the occluding objects are visible, their shape is detected and appears in this layer, in the same shape and at the same location as in the input layer U_0. (The mechanism of segmenting occluding objects from the input image is not the main issue here. In the computer simulation, occluding objects are segmented based on the difference in brightness between occluding objects and occluded patterns.)

Figure 3 shows an example of the response when occluding objects are visible. Since occluding objects are visible, the occluding objects are detected and appear in layer U_M. There are slightly diverging, and topographically ordered, inhibitory connections from masker layer U_M to all cell-planes of layer U_{S1}. The response to features irrelevant to the occluded pattern are suppressed in layer U_{S1} by the inhibitory signals from layer U_M. This can be seen from the response of layer U_{C1}, which is a blurred version of the response of layer U_{S1}. Although the responses of higher layers are slightly sparser than the responses to the unoccluded entire pattern, responses irrelevant to the target pattern scarcely appear. Hence, recognition cell corresponding to 'A' correctly responds in the recognition layer U_{C4}.

In contrast to the case shown in Fig. 3, if the occluding objects are invisible as shown in Fig. 1(a), no response appears in the masker layer U_M, and no inhibitory signals come from the layer. This usually results in an erroneous recognition of the target pattern.

Table 1 summarizes how the model recognizes the images of Fig. 1(a) and (b), in both cases when the masker layer U_M is working and not working.

Table 1. Result of recognition by the network, when the masker layer is working (proposed model), and not working (neocognitron). This table shows how the images in Figure 1 are recognized.

model	occluding objects	recognized as					
either model	none	A	B	C	D	E	F
	invisible (Fig. 1(a))	I	E	Z	E	L	G
with masker layer (proposed model)	visible (Fig. 1(b))	A	B	C	D	E	F
without masker layer (neocognitron)		Q	B	G	Q	E	X

As can be seen from the table, the network with a masker layer recognizes all patterns correctly, if the occluding objects are visible. If the occluding objects are invisible, however, all patterns are recognized erroneously because the masker layer cannot detects occluding objects and cannot send inhibitory signals to layer U_{S1}. The patterns with invisible occluding objects are hardly recognized correctly even by human beings. Our network thus responds to occluded patterns like human beings.

If the masker layer is removed from the network, the network becomes almost like the conventional neocognitron, and fails to recognize most of the occluded patterns even when the occluding objects are visible.

In the neural network model proposed in this section, the mechanism of segmenting occluding objects from the input image is abbreviated. It is a future problem how to detect occluding objects, especially when the occluding objects have the same brightness as the occluded pattern. The interaction between bottom-up and top-down signals like the one in the *selective attention model* [4], for example, will play an important role for this purpose.

3 Self-Organization of Shift-Invariant Receptive Fields

This section discusses a learning rule by which cells with shift-invariant receptive fields are self-organized [5]. With this learning rule, cells similar to simple and complex cells in the primary visual cortex are generated in a network. Training patterns that are presented during the learning are long straight lines of various orientations that sweep across the input layer.

To demonstrate the learning rule, we simulate a three-layered network that consists of an input layer, a layer of S-cells (or simple cells), and a layer of C-cells (or complex cells). The network, which is illustrated in Fig. 5, resembles the first stage of the *neocognitron*. In contrast to the neocognitron, however, input connections to C-cells, as well as S-cells, can be created through learning. The network neither has the architecture of cell-planes nor requires the condition of shared connections to realize shift-invariant receptive fields of C-cells.

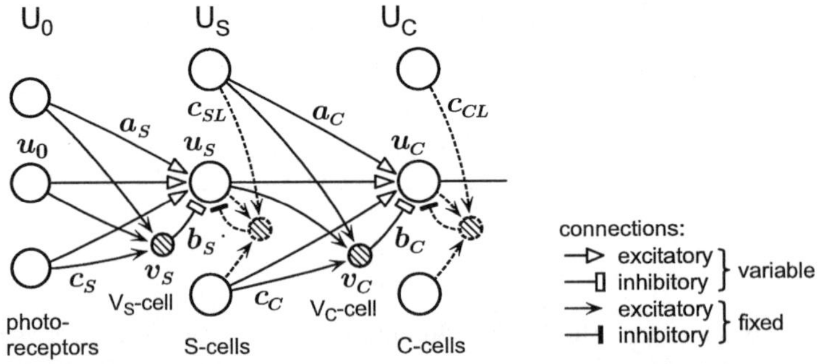

Fig. 5. Self-organization of shift-invariant receptive fields. Connections between cells in the network [5].

The learning rule of the S-cells is similar to the unsupervised learning for the conventional neocognitron. Each cell (S-cell) competes with other cells in its vicinity, and the competition depends on the instantaneous activities of the cells. Only winners of the competition have their input connections increased. The increment of each connection is proportional to the presynaptic activity.

In other words, LTP (long term potentiation) is induced in the input synapses of the winner S-cells.

If a straight line is presented during the learning phase, winners in the competition are generally distributed along the line. When the line is shifted to another location, another set of winners is chosen. Thus, after finishing a sweep of the line, S-cells whose receptive fields have the same preferred orientation as the line are generated and become distributed over the layer U_S. S-cells of other preferred orientations are generated by sweeps of lines of other orientations.

C-cells also compete with each other for the self-organization. In contrast to S-cells, however, the competition is based on the *traces* of their activities, not on their instantaneous activities. A trace is a kind of temporal average (or moving average) [6]. Once the winners are determined by the competition, the excitatory connections to the winners increase (LTP). The increment of each connection is proportional to the instantaneous presynaptic activity. At the same time, the losers have their excitatory input connections decreased (LTD). The decrement of each connection is proportional to both the instantaneous presynaptic activity and the instantaneous postsynaptic activity.

We will now discuss how the self-organization of the C-cell layer (U_C) progresses under this learning rule. If a line stimulus sweeps across the input layer, S-cells whose preferred orientation matches the orientation of the line become active. The timings of becoming active, however, differ among S-cells. For the creation of shift-invariant receptive fields of C-cells, it is desired that a single C-cell obtain strong excitatory connections from all of these S-cells in its connectable area.

Since winners are determined based on traces of outputs of C-cells, the same C-cell has a large probability to continue to be a winner throughout the period when the line is sweeping across its receptive field. Namely, if a C-cell once becomes a winner, the same C-cell will be apt to keep winning for a while after that because the trace of an output lasts for some time after the extinction of the output. Hence a larger number of S-cells will become connected to the C-cell. The larger the number of connected S-cells becomes, the larger the chance of winning becomes. Thus, the C-cell will finally acquire connections from all relevant S-cells.

The increase of inhibitory connections from V_C-cell to C-cell is important for preventing a single C-cell from coming to respond to lines of all orientations. The inhibitory connection to each C-cell is increased in such a way that the total amount of the excitatory inputs never exceeds the inhibitory input to the C-cell. In other words, the activity of a cell is always regulated so as not to exceed a certain value.

Figure 6 shows responses of the network that has finished self-organization. The responses of photoreceptors of U_0, S-cells of U_S and C-cells of U_C are displayed in the figure by the size of the small dots (filled squares).

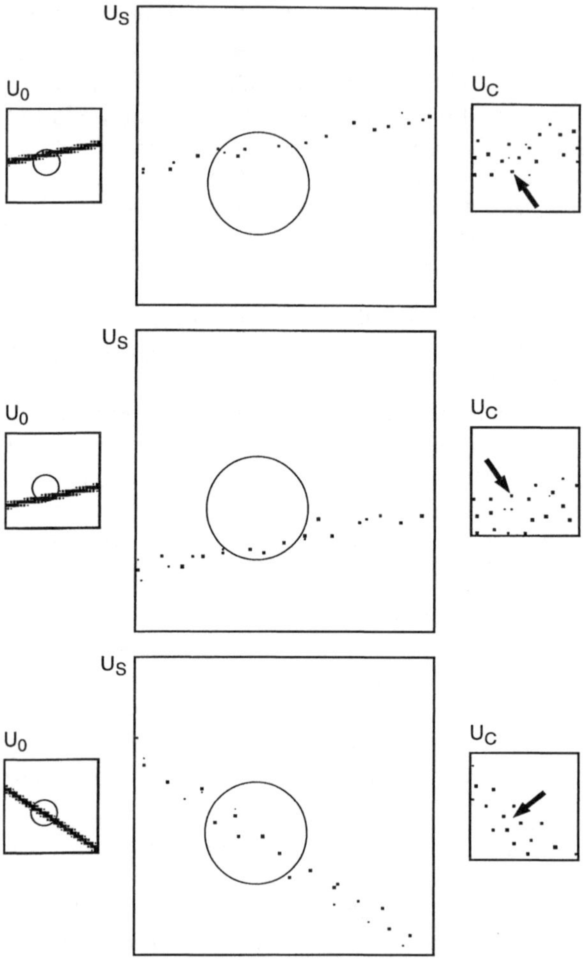

(a) The stimulus line is optimally oriented to the C-cell marked with an arrow.

(b) A stimulus line of the same orientation as in (a) is presented at a different location.

(c) The stimulus line is rotated to another orientation.

Fig. 6. Responses of the network that has finished self-organization [5].

If a line is presented to a location on the input layer as shown in Fig. 6(a), a number of S-cells respond to the line, and consequently C-cells that receive excitatory connections from these S-cells become active. Let us watch, for example, an active C-cell marked with an arrow in layer U_C. The circles in U_S and U_0 show the connectable area and the effective receptive field of the C-cell, respectively. When comparing the sizes of these areas, note that layers U_S and U_C, which have a larger number of cells, are displayed in this figure on a larger scale than layer U_0.

If the line shifts to a new location as shown in Fig. 6(b), other S-cells become active because S-cells are sensitive to stimulus location. In layer U_C, however, several C-cells that were active in Fig. 6(a) are still active for this shifted stimulus. For example, the C-cell marked with an arrow continues responding. This shows that the C-cell exhibits a shift-invariance within the receptive field. When the line is rotated to another orientation as shown in Fig. 6(c), this C-cell is of course silent even if the line is presented in its receptive field.

Figure 7 shows typical orientation tuning curves for S- and C-cells. A stimulus line is rotated at various orientations, and is swept across the input layer. The abscissa of a curve is the line orientation, and the ordinate is the total response of the cell during a sweep of the line in each orientation. Incidentally, the curve for the C-cell in this figure is taken from the C-cell marked with the arrow in Fig. 6.

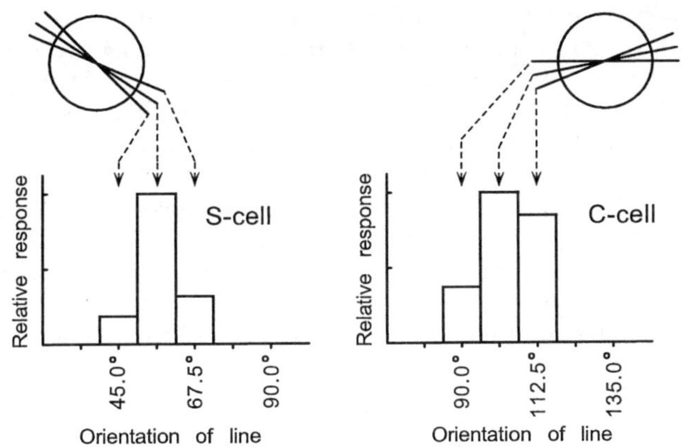

Fig. 7. Typical orientation-tuning curves for S- and C-cells [5].

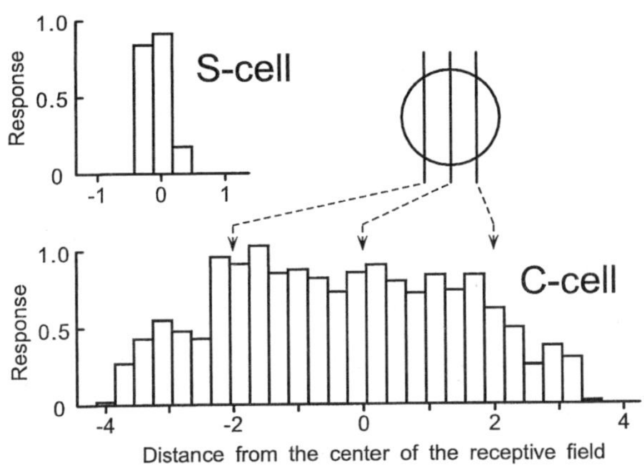

Fig. 8. Typical responses of S- and C-cells to the shift of an optimally oriented line [5]. The responses are taken from the same cells as in Fig. 7.

Figure 8 shows the response of the same cells to the shift of an optimally oriented line. It can be seen that the C-cell responds to the line within a wide range of locations and shows shift-invariance within its receptive field.

The computer simulation has thus demonstrated that cells with shift-invariant receptive fields are successfully generated if lines stimuli of various orientations sweep across the input layer during the learning.

References

1. Fukushima, K. (2000): Recognition of occluded patterns: a neural network model. (in Japanese). Technical Report of IEICE, No. **NC99**-170
2. Fukushima, K. (1988): Neocognitron: a hierarchical neural network capable of visual pattern recognition. Neural Networks **1**, 119–130
3. Fukushima, K., Nagahara, K., Shouno, H. (1997): Training neocognitron to recognize handwritten digits in the real world. In: Mirenkov, N., Gu, Q.-P., Peng, S, Sedukhin, S. (Eds.): pAs'97 (The Second Aizu International Symposium on Parallel Algorithms/Architectures Synthesis, Aizu-Wakamatsu, Japan), IEEE Computer Soc Press, Los Alamos, 292–298
4. Fukushima, K. (1987): Neural network model for selective attention in visual pattern recognition and associative recall. Applied Optics **26**, 4985–4992
5. Fukushima, K. (1999): Self-organization of shift-invariant receptive fields. Neural Networks **12**, 791–801
6. Földiák, P. (1991): Learning invariance from transformation sequences. Neural Computation **3**, 194–200

Rates of Approximation by Neural Networks

Věra Kůrková

Institute of Computer Science, Academy of Sciences of the Czech Republic
Pod vodárenskou věží 2, P.0. Box 5, 182 07 Prague 8, Czechia
e-mail vera@cs.cas.cz

Abstract: A theoretical framework for investigation of complexity of feedforward networks measured by the number of hidden units is presented in the context of nonlinear approximation theory. Upper bounds on rates of approximation by neural networks are derived in terms of certain norms depending on types of computational units.

1 Introduction

Neurocomputing devices are now based on mathematical properties and engineering experience rather than biological analogy. The simple computational components, from which such machines (and their models) are assembled, may still be called "artificial neurons" even when they utilize designs entirely different from the original threshold perceptron. Artificial neural systems are characterized not only by the nature of computational units, but by their interconnection into networks and the capability for problem solving based on adaptation.

Experience has shown that simple architectures with relatively few computational units can achieve surprisingly good performance on many moderately sized problems. However, as the size of the problem is scaled upward, solutions may fail to hold because of physical and electronic constraints. A major theoretical issue is *complexity*, which must be considered with respect to feasibility of a particular network implementation. While engineering and physics may constrain their achievability, some designs can be mathematically proven to have desirable computational capabilities. Of particular interest are results showing that the number of computational units does not necessarily need to increase with the dimensionality of the problem. Such results supplement experience with design criteria based on conceptual models.

Mathematical formalization of neurocomputing models has shown that many types of *feedforward networks* (including all standard types that are popular in applications as well as many others that may not have been considered by experimentalists) are "universal", in the sense that when they are sufficiently elaborate, then their parameters can be adjusted to recognize or classify patterns of any reasonable type (for a survey of results on neural networks as universal approximators see e.g., Pinkus [21] and Kůrková [13]).

Each universal computational model together with a type of its implementation generate a hierarchy of complexity on computational tasks. Theoretical computer science has been investigating such a complexity hierarchy with respect to Turing machines, while the rich variety of universal neurocomputing models generates many other measures of complexity. Currently, feedforward networks are mostly simulated on classical computers. For such simulations, the limiting factor is the *number of hidden units*. Complexity measured by the number of computational units can be studied in the context of approximation theory in terms of *rates of approximation*. Generally, rates of approximation characterize the trade-off between accuracy of approximation and complexity of approximating functions. When such functions are from a parameterized family, then their complexity depends on the length of a properly defined parameter vector (corresponding, for example, to the degree of a polynomial or a rational function, the number of knots in a spline or the number of hidden units in a neural network).

In this paper, we present a theoretical framework for investigation of some aspects of complexity of feedforward networks in the context of nonlinear approximation theory in terms of approximation by variable-basis functions, i.e., approximation by linear combinations of n-tuples of elements of a given set of functions. This approximation scheme includes free-nodes splines, free-poles rational functions and feedforward multilayer neural networks with a single linear output unit. In the case of one-hidden-layer networks, a variable basis corresponds to computational units in the hidden layer.

We survey recent results on rates of variable-basis approximation: upper bounds on such rates in terms of certain norms, methods of estimation of such norms and characterization of sets of multivariable functions with dimension-independent rates of approximation.

The paper is organized as follows. Section 2 contains basic concepts and notations concerning feedforward neural networks and approximation in normed linear spaces. Section 3 is devoted to dimension-independent rates of approximation studied in the context of variable-basis approximation. In 3.1, upper bounds on rates of variable-basis approximation are derived, and in 3.2, these bounds are formulated in terms of a norm tailored to a given set of basis functions. Tightness is these bounds in investigated in 3.3. In 3.4, there are discussed methods of derivation of upper and lower estimates of the norms, in terms of which bounds on rates of variable-basis approximation are stated. In Section 5, tools developed in the previous sections are applied to perceptron networks. Section 4 contains a brief conclusion.

2 Preliminaries

2.1 Feedforward neural networks

Feedforward neural networks compute parametrized sets of functions depending on the type of computational units as well as the type of their interconnections. *Computational units* compute functions depending on two vector variables: an *input vector* and a *parameter vector*. Generally, such units compute functions of the form $\phi : \mathcal{R}^p \times \mathcal{R}^d \to \mathcal{R}$, where ϕ corresponds to the *type of the unit*, p and d to the dimension of a *parameter* and an *input space*, resp., where $\mathcal{R}$ denotes the set of real numbers. Since in practical applications, all inputs are bounded, it is sufficient to study networks with inputs in some compact (closed and bounded) subset of $\mathcal{R}$; usually, results are formulated for inputs in the interval $[0, 1]$.

We call one-hidden-layer networks with hidden units computing a function ϕ and a single linear output unit *ϕ-networks*. Thus ϕ-networks compute functions of d variables of the form $\sum_{i=1}^{n} w_i \phi(\mathbf{a}_i, .)$, where all $w_i \in \mathcal{R}$, all $\mathbf{a}_i \in \mathcal{R}^p$ and $n \in \mathcal{N}_+$ ($\mathcal{N}_+$ denotes the set of all positive integers and $\mathcal{N}$ the set of natural numbers).

Standard types of hidden units are perceptrons. A *perceptron* with an activation function $\psi : \mathcal{R} \to \mathcal{R}$ computes functions of the form $\phi((\mathbf{v}, b), \mathbf{x}) = \psi(\mathbf{v} \cdot \mathbf{x} + b) : \mathcal{R}^{d+1} \times \mathcal{R}^d \to \mathcal{R}$, where $\mathbf{v} \in \mathcal{R}^d$ is an *input weight* vector and $b \in \mathcal{R}$ is a *bias*. By $P_d(\psi) = P_d(\psi, J) = \{\psi(\mathbf{v} \cdot \mathbf{x} + b) : J^d \to \mathcal{R}; \mathbf{v} \in \mathcal{R}^d, b \in \mathcal{R}\}$ we denote the set of functions on J^d computable by ψ-perceptron networks.

The most common activation functions are *sigmoidals*, i.e. functions $\sigma : \mathcal{R} \to [0, 1]$ such that $\lim_{t \to -\infty} \sigma(t) = 0$ and $\lim_{t \to \infty} \sigma(t) = 1$. There are used both continuous sigmoidals (like the logistic sigmoid $\frac{1}{1+e^{-t}}$ or the hyperbolic tangent) and the discontinuous Heaviside function ϑ defined by $\vartheta(t) = 0$ for $t < 0$ and $\vartheta(t) = 1$ for $t \geq 0$.

2.2 Approximation by variable-basis functions

Approximation by feedforward neural networks can be studied in a more general context of *approximation by variable-basis functions*. In this subsection, we introduce basic concepts and tools for investigation of such an approximation scheme.

By a *linear space* we mean a linear space over real numbers. Recall that a *Hilbert space* is a normed linear space with the norm induced by an inner product. For $J \subset \mathcal{R}$ and $p \in [1, \infty]$, we denote by $(\mathcal{L}_p(J^d), \|.\|_p)$ the linear space of all $\mathcal{L}_p$-functions on J^d with the $\mathcal{L}_p$-norm defined by $\|f\|_p = \sqrt[p]{\int_{J^d} f^p(x)dx}$. By $(\mathcal{C}(J^d), \|.\|_\mathcal{C})$ is denoted the space of all continuous functions on J^d with the supremum norm.

Let $(X, \|.\|)$ be a normed linear space, then $B_r(\|.\|)$ denotes the ball of radius r with respect to the norm $\|.\|$, i.e., $B_r(\|.\|) = \{f \in X; \|f\| \leq r\}$.

If G is a subset of $(X, \|.\|)$ and $c \in \mathcal{R}$, then we denote $cG = \{cg; g \in G\}$ and $G(c) = \{wg; g \in G, w \in \mathcal{R} \,\&\, |w| \leq c\}$. If $G = G(1)$, G is called *balanced*.

By $G^\perp$ is denoted the *orthogonal complement* of G.

The *closure* of G is denoted by $cl\, G$ and defined as $cl\, G = \{f \in X; (\forall \varepsilon > 0)(\exists g \in G)(\|f - g\| < \varepsilon)\}$. G is *dense* in $(X, \|.\|)$ if $cl\, G = X$.

The *Chebyshev radius* of a subset G of a linear space X with respect to a norm $\|.\|$ on X is defined as $\sup_{g \in G} \|g\|$.

The *linear span* of G, denoted by *span* G, is the set of all linear combinations of its elements, i.e., $span\, G = \{\sum_{i=1}^{n} w_i g_i; w_i \in \mathcal{R}, g_i \in G, n \in \mathcal{N}_+\}$; $span_n G$ denotes the set of all linear combinations of at most n elements of G, i.e., $span_n G = \{\sum_{i=1}^{n} w_i g_i; w_i \in \mathcal{R}, g_i \in G\}$.

The *convex hull* of G, denoted by *conv* G, is the set of all convex combinations of its elements, i.e., $conv\, G = \{\sum_{i=1}^{n} a_i g_i; a_i \in [0,1], \sum_{i=1}^{n} a_i = 1, g_i \in G, n \in \mathcal{N}_+\}$; $conv_n\, G$ denotes the set of all convex combinations of n elements of G, i.e., $conv_n\, G = \{\sum_{i=1}^{n} a_i g_i;\ a_i \in [0,1], \sum_{i=1}^{n} a_i = 1, g_i \in G\}$.

If M is a subset of a normed linear space $(X, \|.\|)$ and $f \in X$ then $\|f - M\| = \inf_{g \in M} \|f - g\|$ denotes the *distance* of f from M.

Suitability of an approximating set M for approximation of functions from a set Y can be investigated using estimates of the *worst-case error*, mathematically formalized by the concept of *deviation of* Y *from* M, defined as $\delta(Y, M) = \delta(Y, M, (X, \|.\|)) = \sup_{f \in Y} \|f - M\| = \sup_{f \in Y} \inf_{g \in M} \|f - g\|$.

Linear approximation theory investigates approximation by finite-dimensional linear subspaces that are often generated by first n elements of a linearly independent subset G of $(X, \|.\|)$ with a *fixed ordering*. For example, when G is the set of powers $\{x^k; k \in \mathcal{N}\}$, then the linear space generated by its first n elements is the set of all polynomials of order at most $n - 1$.

To describe a theoretical lower bound on linear approximation, Kolmogorov introduced a concept of n-width (which became later called Kolmogorov width) defined as

$$d_n(Y) = \inf_{X_n} \sup_{f \in Y} \|f - X_n\|,$$

where the infimum is taken over all n-dimensional subspaces X_n of X.

Nonlinear approximation by variable-basis functions investigates approximation by linear combinations of all n-tuples of elements of some set G (usually linearly independent). In contrast to linear approximation, where approximating functions are from a single n-dimensional linear subspace, in the case of variable-basis approximation, approximating functions are members of $span_n G$, which is the *union of* n*-dimensional subspaces* generated by all n-tuples of elements of a set G.

The set of input/output functions of ϕ-networks with n hidden units is $span_n G_\phi$, where $G_\phi = \{\phi(\mathbf{a}, \cdot); \mathbf{a} \in \mathcal{R}^p\}$. Also multilayer networks with a single linear output unit and n units in the last hidden layer belong to this approximation scheme; however, they compute functions from $span_n G$ with G depending on the number of units in the previous hidden layers.

We shall denote the deviation from $span_n G$ by $\delta_{G,n}$, i.e.,

$$\delta_{G,n}(Y) = \delta(Y, span_n G).$$

The basic properties of $\delta_{G,n}$ can be derived easily from its definition (see Kůrková and Sanguineti [16]).

Proposition 1. *Let $(X, \|.\|)$ be a normed linear space, G and Y be its subsets. Then for all positive integers n,*
(i) for any $c \in \mathcal{R}$ $\delta_{G,n}(cY) = |c|\delta_{G,n}(Y)$;
(ii) $\delta_{G,n}(cl\, Y) = \delta_{G,n}(Y)$;
(iii) $\delta_{G,n}(conv\, Y) = \delta_{G,n}(Y)$;
(iv) if $Y' \subseteq Y$ then $\delta_{G,n}(Y') \leq \delta_{G,n}(Y)$.

Rates of approximation characterize the trade-off between accuracy of approximation and complexity of approximating functions. When approximating functions are from a parametrized family, then their complexity can be measured by the length of a parameter vector (corresponding, for example, to the degree of a polynomial or a rational function, the number of knots in a spline, or the number of hidden units in a neural network). Such a parametrized family can be represented as a sequence of sets of functions with parameter vectors of increasing length. In traditional approximation schemes (like polynomials or exponentional sums), these sets are *finite-dimensional subspaces* of increasing dimensionality, while in the case of variable-basis approximation, they form *unions of finite-dimensional subspaces* that are generated by an increasing number of elements of a given set of basis functions.

Let $\{M_n; n \in \mathcal{N}_+\}$ be a sequence of nested subsets of a normed linear space $(X, \|.\|)$, then *rate of approximation* of $f \in X$ by $\{M_n; n \in \mathcal{N}_+\}$ is measured by the speed of decrease of the distance $\|f - M_n\| = \inf_{h \in M} \|f - h\|$. When $\bigcup_{n \in \mathcal{N}_+} M_n$ is dense in $(X, \|.\|)$, then for each $f \in X$, the sequence $\{\|f - M_n\|; n \in \mathcal{N}_+\}$ converges to zero. But for practical applications, this convergence has to be sufficiently fast to guarantee a desired accuracy of approximation for n small enough, so that all functions from M_n are implementable.

In the case of functions of d variables, it often happens that the deviation is of order $\mathcal{O}\left(\frac{1}{\sqrt[d]{n}}\right)$. It means that to achieve an accuracy within ε, approximating functions of complexity of order $\left(\frac{1}{\varepsilon}\right)^d$ are required. Such an exponentional dependence of complexity on the number of variables is called the *curse of dimensionality*. For example, in $(\mathcal{L}_p([0,1]^d), \|.\|_p)$, the Kolmogorov n-width of balls in certain norms exhibits the curse of dimensionality (see e.g., Pinkus [20]). When complexity of approximating functions does not depend on the number of variables d, then the approximation is called *dimension-independent*.

3 Dimension-independent rates of approximation

3.1 Maurey-Jones-Barron's theorem

In applications, functions of hundreds of variables have been approximated quite well by networks with only a moderate number of hidden units (see e.g., Sejnowski and Rosenberg's [24] NETtalk). However, estimates of rates of approximation by neural networks derived from constructive proofs of the universal approximation property have not been able to explain such successes – the constructions used in such arguments lead to networks with complexity growing exponentially with the number of input units.

Some understanding why many high-dimensional tasks can be performed quite efficiently by neural networks with a moderate number of hidden units has been found by Jones [7], who has constructed incremental approximants with "dimension-independent" rates of convergence. The same estimate of rates of approximation has been earlier proved by Maurey by a probabilistic argument (it has been quoted by Pisier [22], see also Barron [2]). Barron [2] has improved Jones' [7] upper bound and applied it to neural networks. The following theorem presents this upper bound in a slightly reformulated way; its proof is a simplification of Barron's argument.

Theorem 1. *Let $(X, \|.\|)$ be a Hilbert space, b a positive real number, G a subset of X such that for every $g \in G$ $\|g\| \leq b$, and let $f \in cl\, conv\, G$. Then for every positive integer n,*

$$\|f - conv_n G\| \leq \sqrt{\frac{b^2 - \|f\|^2}{n}}.$$

Proof. Since the distance from $conv_n G$ is continuous on X (see e.g., [23]), it is sufficient to verify the statement for $f \in conv\, G$. For a sequence $\{g_i; i \in \mathcal{N}_+\}$ of elements of G consider barycenters $f_n = \frac{1}{n}\sum_{i=1}^n g_i$. Then setting $e_n = \|f - f_n\|$ we get $e_{n+1}^2 = \|\frac{n}{n+1}(f - f_n) + \frac{1}{n+1}(f - g_{n+1})\|^2 = \frac{n^2}{(n+1)^2}e_n^2 + \frac{2n}{(n+1)^2}(f - f_n)\cdot(f - g_{n+1}) + \frac{1}{(n+1)^2}\|f - g_{n+1}\|^2$.

Let $f = \sum_{j=1}^m a_j g_j$ be a representation of f as a convex combination of elements of G. Then for every n we can choose $g_{n+1} \in G$ such that the last two terms in the above expression of e_{n+1}^2 are bounded by $\frac{b^2 - \|f\|^2}{(n+1)^2}$. Indeed, $\sum_{j=1}^m a_j \left(\frac{2n}{(n+1)^2}(f - f_n)\cdot(f - g_j) + \frac{1}{(n+1)^2}\|f - g_j\|^2\right) = \frac{2n}{(n+1)^2}(f - f_n)\cdot(f - \sum_{j=1}^m a_j g_j) + \frac{1}{(n+1)^2}\left(\|f\|^2 - 2f\cdot(\sum_{j=1}^m a_j g_j) + \sum_{j=1}^m a_j\|g_j\|^2\right) = \frac{1}{(n+1)^2}(\sum_{j=1}^m a_j g_j - \|f\|^2) \leq \frac{1}{(n+1)^2}(b^2 - \|f\|^2)$. Analogously, one can show that there exists $g_1 \in G$ such that $e_1^2 = \|f - g_1\|^2 \leq b^2 - \|f\|^2$.

Thus setting $c = b^2 - \|f\|^2$, we get $e_1^2 \leq c$ and $e_{n+1}^2 \leq \frac{n^2}{(n+1)^2}e_n^2 + \frac{c}{(n+1)^2}$. One can check easily by induction that this recursive formula implies that $e_n^2 \leq \frac{c}{n}$. Indeed, $e_{n+1}^2 \leq \frac{n^2}{(n+1)^2}\frac{c}{n} + \frac{c}{(n+1)^2} = \frac{c}{n+1}$.

The same upper bound as in Theorem 1 has been derived by Maurey using a probabilistic argument (see Pisier [22], also Barron [2]), in which a representation of a function f as an element of the set $conv\, G$, $f = \sum_{j=1}^m a_j g_i$, defines a finite probability distribution P on the set G by $P(g = g_j) = a_j$. Then for $f_n = \sum_{i=1}^n \frac{1}{n}h_i$, a random variable corresponding to the barycenter of an n-tuple of elements of G chosen randomly with respect to the probability P, the mean value $\|f - f_n\|^2$ is bounded from above by $\frac{b^2-\|f\|^2}{n}$. Hence there must exist at least one n-tuple $h_1, \ldots, h_n \in G$, for which $\|f - \sum_{i=1}^{n} \frac{1}{n} h_i\| \leq \sqrt{\frac{b^2-\|f\|^2}{n}}$.

Note that both these arguments prove that rates bounded from above by $\frac{b^2-\|f\|^2}{n}$ can be even achieved when barycenters of n-tuples of elements of G (with repetitions) are used as approximators. More precisely, they show that $\|f - bar_n G\| \leq \sqrt{\frac{b^2-\|f\|^2}{n}}$, where $bar_n G = \{f \in X; f = \sum_{i=1}^{k} \frac{j_i}{n} g_i, k \in \mathcal{N}_+, k \leq n, \sum_{i=1}^{n} j_i = n, \forall i = 1, \ldots n \; g_i \in G\}$.

Moreover, as a byproduct, the above constructive proof of Theorem 1 gives bounds on rates of convergence of incremental learning algorithms (i.e., algorithms that in each step add a new hidden unit, see e.g., Kůrková [12]).

Note that similar estimates of the distance from $conv_n G$ as in Theorem 1 are not restricted to Hilbert spaces – Darken et al. [3] extended Maurey-Jones-Barron's upper bound to $\mathcal{L}_p$-spaces for $p \in (1, \infty)$ with a slightly worse rate of approximation (of order $\mathcal{O}(n^{-\frac{1}{q}})$, where $q = \max(p, \frac{p}{p-1})$); there also exist extensions for the supremum norm (see e.g., Barron [1], Girosi [5], Gurvits and Koiran [6], Makovoz [18], Kůrková, Savický and Hlaváčková [15]).

3.2 Variation with respect to a set of functions

Since $conv_n G \subseteq span_n G$, the upper bound from Theorem 1 also applies to rates of approximation by $span_n G$. When G is not closed up to multiplication by scalars, $conv\, G$ is a proper subset of $span\, G$ and hence also $cl\, conv\, G$ is a proper subset of $cl\, span\, G$. Thus density of $span\, G$ in $(X, \|.\|)$ does not guarantee that Theorem 1 can be applied to all elements of X. Since $conv_n G(c) \subset span_n G(c) = span_n G$ for any $c \in \mathcal{R}$, replacing the set G by $G(c) = \{wg; w \in \mathcal{R}, |w| \leq c, g \in G\}$, we can apply Theorem 1 to all elements of $\cup_{c \in \mathcal{R}_+} cl\, conv\, G(c)$. This approach can be mathematically formulated in terms of a norm tailored to a set G (in particular, to sets G_ϕ corresponding to various computational units ϕ).

Let $(X, \|.\|)$ be a normed linear space and G be its subset, then *G-variation* (variation with respect to G) is defined as the Minkowski functional of the set $cl\, convG(1)$, i.e.,

$$\|f\|_G = \inf\{c \in \mathcal{R}_+; f \in cl\, conv\, G(c)\}.$$

G-variation has been introduced by Kůrková [11], as an extension of Barron's [1], concept of variation with respect to half-spaces (for functions of one variable, variation with respect to half-spaces coincides, up to a constant, with the notion of total variation studied in the integration theory).

G-variation is a norm on the subspace $\{f \in X; \|f\|_G < \infty\} \subseteq X$. The closure in its definition depends on the topology induced on X by the norm

$\|.\|$; however, when X is finite-dimensional, G-variation does not depend on the choice of a norm on X, since all norms on a finite-dimensional space are topologically equivalent. The following basic properties of G-variation can be derived easily from its definition.

Proposition 2. *Let $(X, \|.\|)$ be a normed linear space, and let G and F be its subsets. Then*
(i) for all $f \in X$, $\|f\| \leq \|f\|_G \sup_{g\in G} \|g\|$;
(ii) if $f \in span\, G$, then $\|f\|_G = \min\{\sum_{i=1}^{m} |w_i|; f = \sum_{i=1}^{m} w_i g_i, m \in \mathcal{N}_+, g_i \in G, w_i \in \mathcal{R}\}$;
(iii) $\|.\|_G \leq c\|.\|_F$ if and only if for all $h \in F$, $\|h\|_G \leq c$.

The following theorem is a reformulation of Theorem 1 in terms of G-variation (see Kůrková [11]).

Theorem 2. *Let $(X, \|.\|)$ be a Hilbert space and G be its subset. Then for every $f \in X$ and for every positive integer n,*

$$\|f - span_n G\| \leq \sqrt{\frac{(s_G\|f\|_G)^2 - \|f\|^2}{n}},$$

where s_G is the Chebyshev radius of G with respect to $\|.\|$.

Since $0 \in span_n G$, for all $f \in X$, $\|f - span_n G\| \leq \|f\|$. Thus Theorem 2 gives a nontrivial upper bound only when $\|f\|^2 \geq \frac{(s_G\|f\|_G)^2 - \|f\|^2}{n}$, or equivalently $\frac{\|f\|}{s_G\|f\|_G} \geq \frac{1}{\sqrt{n+1}}$ (for example, when $s_G = 1$ and $\|f\|_G = 1$, this means that $\|f\| \geq \frac{1}{\sqrt{n+1}}$).

Theorem 2 implies upper bounds on the deviation from $span_n G$, $\delta_{G,n}$, of the unit ball in G-variation and of its subsets defined by constraints on the value of the norm $\|.\|$.

Corollary 1. *Let $(X, \|.\|)$ be a Hilbert space, G be its subset and let $0 \leq r \leq s_G$, where s_G is the Chebyshev radius of G. Then for every positive integer n,*

$$\delta_{G,n}(B_1(\|.\|_G)) \leq \frac{s_G}{\sqrt{n}};$$

$$\delta_{G,n}(\{f \in B_1(\|.\|_G); \|f\| \geq r\}) \leq \sqrt{\frac{s_G^2 - r^2}{n}}.$$

Thus in $(\mathcal{L}_2([0,1]^d, \|.\|_2)$, all functions from the unit ball in G_ϕ-variation can be approximated within $\frac{s_{G_\phi}}{\sqrt{n}}$ by ϕ-networks with n hidden units *independently on the number of variables d*. However, with increasing number of variables, the condition of being in the unit ball in G_ϕ-variation becomes more and more constraining.

3.3 Approximation by orthonormal variable basis

Besides of being a generalization of the concept of total variation, G-variation is also an extension of the notion of l_1-norm. If G is an orthogonal basis of a separable Hilbert space $(X, \|.\|)$, then the *l_1-norm with respect to G* is defined as $\|f\|_{1,G} = \sum_{g \in G} |f \cdot g|$. It follows from Proposition 2 that for G a finite orthonormal basis of X, $\|f\|_G = \sum_{g \in G} |f \cdot g|$. The following proposition describes the relationship between G-variation and l_1-norm with respect to G (see Kůrková and Sanguineti [16]).

Proposition 3. *Let $(X, \|.\|)$ be a separable Hilbert space, and let G be its orthogonal basis, then $\|.\|_G \leq \|.\|_{1,G}$. When, in addition, $\|.\|$ and $\|.\|_G$ are equivalent, then $\|.\|_G = \|.\|_{1,G}$.*

Thus, when G is an orthogonal basis of X, then the unit ball in G-variation contains the unit ball in the l_1-norm with respect to G. For example, the unit ball in variation with respect to the Fourier basis contains the unit ball in the l_1-norm with respect to this basis.

When G is orthonormal, an upper bound on the deviation of the unit ball in G-variation from $span_n G$ stated in Corollary 1 can be improved to a tight bound. The following theorem is based on an argument by Mhaskar and Micchelli [19] as improved in Kůrková, Savický and Hlaváčková [15] and Kůrková and Sanguineti [17].

Theorem 3. *Let $(X, \|.\|)$ be an infinite-dimensional separable Hilbert space and G be its orthonormal basis. Then for every positive real number b and every positive integer n,*

$$\delta_{G,n}(B_b(\|.\|_G)) = \delta_{G,n}(B_b(\|.\|_{1,G})) = \frac{b}{2\sqrt{n}}.$$

Proof. By Proposition 3, $\delta_{G,n}(B_b(\|.\|_G)) = \delta_{G,n}(B_b(\|.\|_{1,G}))$ and by Proposition 1, $b\,\delta_{G,n}(B_1(\|.\|_G)) = \delta_{G,n}(B_b(\|.\|_G))$. So it is sufficient to verify that $\delta_{G,n}(B_1(\|.\|_{1,G})) = \frac{1}{2\sqrt{n}}$.

To derive the upper bound, for $f \in B_1(\|.\|_{1,G})$ reorder G in such a way that $f = \sum_{i=1}^{\infty} w_i g_i$, where for all $i \in \mathcal{N}_+$, $|w_i| \geq |w_{i+1}|$ and $g_i \in G$. Set $f_n = \sum_{i=1}^{n} w_i g_i$. Then $\delta_{G,n}(B_1(\|.\|_{1,G}))^2 \leq \|f - span_n G\|^2$. Then $\|f - span_n G\|^2 = \|f - f_n\|^2 = \sum_{i=n+1}^{\infty} w_i^2 \leq |w_{n+1}| \sum_{i=n+1}^{\infty} |w_i|$. Since $\sum_{i=n+1}^{\infty} |w_i| \leq 1 - n|w_{n+1}|$ we have $\|f - span_n G\|^2 \leq |w_{n+1}|(1 - n|w_{n+1}|)$. Setting $t = |w_{n+1}|$, we get $\delta_{G,n}(B_1(\|.\|_{1,G}))^2 \leq t(1 - nt)$ for all $t \in [0,1]$. The right side of this inequality achieves its maximum equal to $\frac{1}{4n}$ for $t = \frac{1}{2n}$. Thus $\delta_{G,n}(B_1(\|.\|_{1,G})) \leq \frac{1}{2\sqrt{n}}$.

To verify the lower bound, let $f_n = \frac{1}{2n} \sum_{i=1}^{2n} g_i$. Then $\|f_n\|_{1,G} = 1$ and $\delta_{G,n}(B_1(\|.\|_{1,G})) \geq \|f_n - span_n G\|$. By orthonormality of G, $\|f_n - span_n G\| \geq \|\frac{1}{2n} \sum_{i=1}^{n} g_i\| = \frac{1}{2\sqrt{n}}$.

For G a countable infinite orthonormal basis, Theorem 3 improves the upper bound stated in Corollary 1 to an exact value of the deviation of the unit ball in G-variation from $span_nG$. In contrast to Theorem 2, which expresses an upper bound in terms of both $\|f\|_G$, and $\|f\|$, Theorem 3 does not take $\|f\|$ into account. A tight bound on the distance from $span_nG$ in terms of both norms, $\|.\|$ and $\|.\|_G$, have been derived by Kůrková, Savický and Hlaváčková [15] for G an orthonormal subset of a finite-dimensional Hilbert space, and extended by Kůrková and Sanguineti [17] to infinite-dimensional spaces. These results show that for a general set G, Maurey-Jones-Barron's bound cannot be improved substantially: in the orthogonal case, it can at most be improved by a factor depending on the ratio of the two above-mentioned norms, but the term $\frac{1}{\sqrt{n}}$ remains essentially unchanged; it only is replaced by $\frac{1}{2\sqrt{n-1}}$.

3.4 Upper and lower bounds on variation

Properties of balls in variation corresponding to standard hidden units are not yet well understood. In some cases, such balls can be described in terms of images of balls in $\mathcal{L}_1$-norm under certain integral transformations.

A *quadrature* is a rule that allows computation of integrals using only their values at prescribed points as well as weighting coefficients. Quadratures of suitable integral representations can be used to estimate G_ϕ-variations. More precisely, G_ϕ-variation can be estimated from above when a function to be approximated can be represented by an integral formula corresponding metaphorically to a ϕ-network with a continuum of hidden units. For proof of the following theorem as well as its extension to $\mathcal{L}_p$-norms see Kůrková, Kainen and Kreinovich [14].

Theorem 4. *Let d,p be positive integers, $J \subseteq \mathcal{R}^d$ and $f \in (\mathcal{C}(J).\|.\|_{\mathcal{C}})$ be any function which can be represented as $f(\mathbf{x}) = \int_Y w(\mathbf{y})\phi(\mathbf{x},\mathbf{y})d\mathbf{y}$, where $Y \subseteq \mathcal{R}^p$, $w \in \mathcal{C}(Y)$ is compactly supported, and let $G_\phi = \{\phi(.,\mathbf{y}) : J \to \mathcal{R}; \mathbf{y} \in Y\}$. Then $\|f\|_{G_\phi} \leq \int_Y |w(\mathbf{y})|d\mathbf{y}$.*

Note that the integral equation assumed as the hypothesis of Theorem 4 is a *linear integral equation of the first kind with the kernel ϕ*. So G_ϕ-variation of $T_\phi(w)$, where the integral operator T_ϕ is defined by $T_\phi(w) = \int_Y w(\mathbf{y})\phi(\mathbf{x},\mathbf{y})d\mathbf{y}$, is bounded from above by the $\mathcal{L}_1$-norm of w. Various authors have derived upper bounds on rates of approximation by neural networks using suitable integral transforms: Barron [2] and Girosi and Anzellotti [4] have used weighted Fourier transforms, while Kůrková, Kainen and Kreinovich [14] have applied a version of Radon transform.

Lower bounds on G-variation can be obtained from the following geometrical characterization of G-variation (for its proof see Kůrková, Savický and Hlaváčková [15]).

Theorem 5. *Let $(X, \|.\|)$ be a Hilbert space and G be its nonempty subset. Then for every $f \in X$,*
$\|f\|_G = \sup_{h \in S} \frac{|f \cdot h|}{\sup_{g \in G} |g \cdot h|}$, *where* $S = \{h \in X - G^{\perp}; \|h\| = 1\}$.

In particular, when f is not orthogonal to G, then $\|f\|_G \geq \frac{\|f\|}{\sup_{g \in G} |f \cdot g|}$. Thus functions that are "almost orthogonal" to G have large G-variations. Theorem 5 have been used by Kůrková, Savický and Hlaváčková [15] to describe functions with variations growing exponentially with the number of hidden units. For such functions, the upper bounds on rates of approximation following from Maurey-Jones-Barron's theorem depend on d exponentially. However, these results only give exponentially growing *upper bounds* on rates of approximation; finding whether these large upper bounds are tight seems to be a difficult task related to some open problems in the theory of complexity of Boolean circuits.

In the classical (linear) approximation theory, there is a great advantage of continuity of approximation operators that allows one to derive lower bounds on the worst-case errors by methods of algebraic topology (see e.g., Pinkus [20]). Such arguments cannot be transferred to variable-basis approximation, since geometrical properties (nonconvexity) of approximating sets do not allow continuity of approximation operators (see Kainen, Kůrková and Vogt [8], [9], [10]).

4 Rates of approximation by perceptron networks

To apply to perceptron networks the tools developed in the previous sections, we shall consider approximation by $span_n P_d(\psi)$.

Notice that the set $P_d(\vartheta, J)$ of functions computable by Heaviside perceptrons is equal to the *set of characteristic functions of half-spaces* of $\mathcal{R}^d$ restricted to J^d. Indeed, $\vartheta(\mathbf{v} \cdot . + b)$ restricted to J^d is equal to the characteristic function of $H^{+}_{\mathbf{v},b} = \{\mathbf{x} \in J^d; \mathbf{v} \cdot \mathbf{x} + b \geq 0\}$. We shall write H_d, instead of $P_d(\vartheta, [0,1])$, call H_d-variation shortly *variation with respect to half-spaces* and denote it by $\|.\|_{H_d}$.

In $\mathcal{L}_p$-spaces with $p \in [1, \infty)$, for all sigmoidal functions σ, $\|.\|_{H_d} = \|.\|_{P_d(\sigma)}$, i.e., variation with respect to half-spaces is equal to variation with respect to perceptrons with any sigmoidal activation (see Kůrková, Kainen and Kreinovich [14]). Thus estimates of rates of approximation derived from Maurey-Jones-Barrron's bound do not distinguish between various sigmoidals. Corollary 1 implies that all functions within a ball of a fixed radius in variation with respect to half-spaces can be approximated by perceptron networks with any sigmoidal activation function with rates of order $\mathcal{O}(\frac{1}{\sqrt{n}})$ independently on the number of variables.

Description of some subsets of such balls can be derived using suitable integral representations. By an approach reminiscent of Radon transform but based directly on distributional techniques, Kůrková, Kainen and Kreinovich [14] have shown that if d is odd and f is a compactly supported function on $\mathcal{R}^d$ whith continuous d-th order partial derivatives, then f satisfies the following integral equation with the Heaviside function as a kernel:

$$f(\mathbf{x}) = a_d \int_{S^{d-1}\times\mathcal{R}} w_f(\mathbf{e}, b)\vartheta(\mathbf{e}\cdot\mathbf{x}+b)d\mathbf{e}db,$$

where $a_d = (-1)^{k-1}(1/2)(2\pi)^{1-d}$ for $d = 2k+1$, $w_f = \int_{H_{\mathbf{e},b}} D_e^{(d)} f(\mathbf{y})d\mathbf{y}$, $D_e^{(d)} f$ is the directional derivative of f in the direction $\mathbf{e}$ iterated d times, $d\mathbf{e}$ is the $(d-1)$-dimensional volume element on the sphere $S^{d-1} \subset \mathcal{R}^d$, and $d\mathbf{y}$ is likewise on the hyperplane $H_{\mathbf{e},b} = \{\mathbf{y} \in \mathcal{R}^d; \mathbf{e}\cdot\mathbf{y}+b = 0\}$. Thus by Theorem 4, variation with respect to half-spaces of any sufficiently smooth function f can be estimated from above by the $\mathcal{L}_1$-norm of the weighting function w_f.

Sometimes, it is easier to estimate variation with respect to a given set of interest using variation with respect to a suitable orthonormal set, since such a variation is related to the $\mathcal{L}_1$-norm (see Propositions 2 and 3. In particular, variation with respect to half-spaces can be estimated in terms of a suitable spectral norm. Barron [2] considered sets Γ_c defined as $\Gamma_c = \{f \in \mathcal{L}_2([0,1]^d), \|.\|_2); c_f \leq c\}$, where $c_f = \int_{\mathcal{R}^d} \|\omega\|_2 \,|\tilde{f}(\omega)|\, d\omega$, $\tilde{f}$ is the Fourier transform of f, and $\|\omega\|_2 = \sqrt{\omega\cdot\omega}$ denotes the l_2-norm of the frequency ω. Applying a theorem similar to Theorem 4 to a weighted Fourier transform, he has proved that for all sigmoidal activations $\delta_{P_d(\sigma),n}(\Gamma_c) \leq \frac{c}{\sqrt{n}}$.

For discussion of strength and weakness of the method of estimation of rates of approximation by sigmoidal perceptron networks by means of variation with respect to an orthonormal basis see Kůrková, Savický and Hlaváčková [15]. They have given examples of classes of functions having variations with respect to half-spaces growing with the number of variables only polynomially, while their spectral norms are growing exponentially. Moreover, they have demonstrated limitations of "dimension-independent" approximation by neural networks by describing a class of functions, that have variations with respect to half-spaces growing exponentially with the number of variables. It is an open question, whether such functions can be approximated within a small error by one-hidden-layer Heaviside perceptron networks with the number of hidden units depending on the number of variables only polynomially.

5 Conclusion

Many classification or pattern recognition tasks can be (using a suitable coding of inputs and outputs) formulated as mappings between subsets of multidimensional vector spaces. The goal of a computer scientist modelling such tasks is to find a simple device, that is capable on the basis of a change of parameters to classify or recognize a great variety of patterns. When parameters can be adjusted using some learning algorithm, such a device can perform pattern recognition or classification without an explicit formulation of rules or feature extraction.

Mathematical formalization of neurocomputing models have shown that many types of feedforward networks satisfy these requirements. Provided such networks contain sufficiently many computational units, their parameters can be adjusted to obtain as an input/output function a good approximation of almost any mapping between subsets of multidimensional spaces. In neural network terminology, such classes of networks are called *universal approximators*. But universality has a price: it cannot be proven within reasonable bounds on complexity. Each kind of network implementation determines a different measure of complexity. Currently, feedforward networks are mostly simulated on classical computers. For such simulations, the limiting factor is the *number of hidden units*. In some cases, this number can be considerably reduced by an appropriate choice of coding of outputs allowing quite large tolerance in accuracy due to robustness of error-correction (best-guess).

When network complexity reduction due to output coding is not sufficient, another type of computational units and structure of their connections has to be found that can solve a given task within feasible limits of network complexity. Some guidelines for choice of a type of network can be derived from mathematically grounded theory of complexity of neural networks. Its recent achievements include estimates of rates of approximation by various types of feedforward networks, description of "dimension-independent" approximators and comparisons of complexity requirements of approximation by neural networks with such requirements for other approximation methods.

Acknowledgement

This work was partially supported by GA ČR grant number 201/99/0092 and by GA ČR grant number 201/00/1482.

References

1. Barron, A. R. (1992). Neural net approximation. In *Proceedings of the 7th Yale Workshop on Adaptive and Learning Systems* (pp. 69–72).
2. Barron, A. R. (1993). Universal approximation bounds for superposition of a sigmoidal function. *IEEE Transactions on Information Theory*, **39**, 930–945.
3. Darken, C., Donahue, M., Gurvits, L. & Sontag, E. (1993). Rate of approximation results motivated by robust neural network learning. In *Proceedings of the 6th Annual ACM Conference on Computational Learning Theory* (pp. 303–309). New York: ACM.

4. Girosi, F., & Anzellotti, G. (1993). Rates of convergence for radial basis function and neural networks. In *Artificial Neural Networks for Speech and Vision* (pp. 97–113). London: Chapman & Hall.
5. Girosi, F. (1995). Approximation error bounds that use VC-bounds. In *Proceedings of ICANN'95* (pp. 295–302). Paris: EC2 & Cie.
6. Gurvits, L. & Koiran, P. (1997). Approximation and learning of convex superpositions. *Journal of Computer and System Sciences*, **55**, 161–170.
7. Jones, L. K. (1992). A simple lemma on greedy approximation in Hilbert space and convergence rates for projection pursuit regression and neural network training. *Annals of Statistics*, **20**, 608–613.
8. Kainen, P. C., Kůrková, V. & Vogt, A. (1999). Approximation by neural networks is not continuous. *Neurocomputing*, **29**, 47–56.
9. Kainen, P. C., Kůrková, V. & Vogt, A. (2000). Continuity of approximation by neural networks in $\mathcal{L}_p$-spaces. *Annals of Operational Research* (to appear).
10. Kainen, P. C., Kůrková, V. & Vogt, A. (2000). Geometry and topology of continuous best and near best approximations. *Journal of Approximation Theory* (to appear).
11. Kůrková, V. (1997). Dimension-independent rates of approximation by neural networks. In *Computer-Intensive Methods in Control and Signal Processing: Curse of Dimensionality* (Eds. Warwick, K., Kárný, M.) (pp. 261–270). Boston: Birkhauser.
12. Kůrková, V. (1998). Incremental approximation by neural networks. In *Complexity: Neural Network Approach.* (Eds. Warwick, K., Kárný, M., Kůrková, V.) (pp. 177–188). London: Springer.
13. Kůrková, V. (2000). Neural networks as nonlinear approximators. In *Proceedings of NC 2000* (to appear).
14. Kůrková, V., Kainen, P. C. & Kreinovich, V. (1997). Estimates of the number of hidden units and variation with respect to half-spaces. *Neural Networks*, **10**, 1061–1068.
15. Kůrková, V., Savický , P. & Hlaváčková, K. (1998). Representations and rates of approximation of real-valued Boolean functions by neural networks. *Neural Networks*, **11**, 651–659.
16. Kůrková, V. & Sanguineti, M. (2000). Tools for comparing neural network and linear approximation. Submitted.
17. Kůrková, V. & Sanguineti, M. (2000). Bounds on rates of variable-basis and neural-network approximation. Submitted.
18. Makovoz, Y. (1998). Uniform approximation by neural networks. *Journal of Approximation Theory*, **95**, 215–228.
19. Mhaskar, H. N. & Micchelli, C. A. (1994). Dimension-independent bounds on the degree of approximation by neural networks. *IBM Journal od Research and Development*, **38**, 277–283.
20. Pinkus, A. (1986). *n-Width in Approximation Theory.* Berlin: Springer.
21. Pinkus, A. (1998). Approximation theory of the MLP model in neural networks. *Acta Numerica*, **8**, 277–283.
22. Pisier, G. (1981). Remarques sur un resultat non publié de B. Maurey. In *Seminaire d'Analyse Fonctionelle* I., n.12, Ecole Polytechnique.
23. Singer, I. (1970). *Best Approximation in Normed Linear Spaces by Elements of Linear Subspaces*, Berlin: Springer.
24. Sejnowski, T. J. & Rosenberg, C. (1987). Parallel networks that learn to pronounce English text. *Complex Systems*, **1**, 145–168.

Neural Classifier Using Hyperspheres

Marcel Jirina jr[1]. and Marcel Jirina[2]

[1] Czech Technical University Prague, Faculty of Electrical Engineering, Technicka 2, 166 27 Prague 6 - Dejvice, Czech Republic

[2] Institute of Computer Science, Academy of Sciences of the Czech Republic, Pod vodarenskou vezi 2, 182 07 Prague 8 - Liben, Czech Republic

Abstract: This paper discussed a new approach to classification of high-dimensional data using special neural network classifier Growing hyperspheres neural network (GHS net) based on a data representation by means of hyperspheres. Architecture of the network and the process of learning and recalling is described. Main features of the new network are discussed and a short comparison with other neural classifiers is made. Results of a task classifying data from a gamma telescope are briefly mentioned.

1 Introduction

There is a wide range of different classifiers. The main two criterions for classification are: criterion of minimal error and criterion of minimal distance. A representative of classifiers of minimal error is the well-known Bayes's classifier, see e.g. [2], [3]. The second group of neural classifiers using a criterion of minimal distance is widely represented, see e.g. [1], [2], [3], [4].

An interesting approach to data classification is Spport Vector Machine [6]. The main idea of the technique is to separate classes with a surface that maximizes the margin between them. Hyperplanes are used for the separation. For linearly non-separable problems a mapping of input patterns to higher dimensional spaces is performed and then a linear classification is done.

In difference to other classifiers which uses complex hypersurfaces for separation of classes the GHS neural network classifier uses hyperspheres. This principle gives rather simple, understandable and quantifiable insight to classification.

The research has been carried out in part under the support of the Czech Ministry of Education grants No. VS 96047 and in part under the support of the Czech Ministry of Industry and Commerce in the project No. RP-4210/69/97 "Cooperation of the Czech Republic with CERN".

2 Growing Hyperspheres Neural Network (GHS)

The main idea of the suggested neural network classifier is based on an extraction of learning patterns of the same class within a hypersphere. The center of the hypersphere and its radius determine an area, which contains learning patterns of the same class. A disjunction of these hyperspheres demarcates an area of one class that contains all patterns of this class and separates the learning patterns from patterns of different classes. An advantage of such an approach is that there is a simple representation of hyperspheres. Each node of the network contains n central coordinates of a hypersphere and its radius. A structure of the GHS network is a simple linked list of nodes.

For a classification in k classes, k groups of hyperspheres can be used. However for k classes only k-1 groups of hyperspheres are sufficient. The last, say k-th, class is just the one that does not belong to any other, i.e. previous k-1, classes.

Once we find the hyperspheres in a learning process we can simply classify new unknown patterns in individual classes. We run through all stored nodes and find a node whose hypersphere contains this pattern. An appropriate class assigned to this hypersphere is just a class to which the pattern belongs.

2.1 Learning Algorithm of GHS

The main idea of the learning process consists in moving of hyperspheres in areas of higher concentration of patterns in a defined direction and coincidental increasing of their diameters. A hypersphere can increase its diameter only if a number of internal patterns grows or keeps constant.

The learning algorithm works this way. Let n be a space dimension, k be a number of classes, $N_1,\ldots,N_k$ be numbers of patterns in individual classes and $N=\sum_{i=1}^{k} N_i$ be total number of patters in a learning set. The learning process will go over all k classes and for each class the same learning subprocedure will run. The learning will finish after processing all classes. A properly chosen order of processed classes could significantly reduce a total number nodes used for learning set representation and thus the network size.

A subprocedure works this way. First we randomly chose a pattern from a set containing patterns that belong to the first class. This pattern will represent a center $\vec{w}=(w_1,\ldots,w_n)$ of a hypersphere. A primary radius r^* of the hypersphere is found as a distance between the hypersphere center $\vec{w}$ and the nearest pattern from any different class $\vec{h}=(h_1,\ldots,h_n)$. Such selected radius is not optimal because we want to trace a boundary between some classes approximately in a half of a free space occurring between the classes. This fact can be easily corrected. Using the just constructed hypersphere we find a pattern $\vec{g}=(g_1,\ldots,g_n)$ within the hypersphere that is nearest to the pattern $\vec{h}$. The final radius r is than recalculated according to formula

$$r = \sqrt{\sum_{i=1}^{n} \left(w_i - \frac{h_i + g_i}{2} \right)^2} .$$

A number of patterns within the hypersphere with center $\vec{w}$ and radius r is counted and all these parameters are included in a new node.

It is probable that the hypersphere is not placed properly. Therefore a shift of the hypersphere to a new better position that would maximize the number of internal patterns is necessary. The shift is derived from the hypersphere radius *r*. The new center $\vec{w}^* = \left(w_1^*, \ldots, w_n^*\right)$ of the hypersphere is $w_i^* = w_i + d \cdot r \cdot s_i$, $i = 1, \ldots, n$, where d is a shift constant and $\vec{s} = \left(s_1, \ldots, s_n\right)$ is a unit directional vector whose coordinates are calculated by formula

$$s_i = \frac{w_i - h_i}{\sqrt{\sum_{j=1}^{n} \left(w_j - h_j\right)^2}} .$$

In this moment we have a new center $\vec{w}^*$ of the hypersphere, but we need recalculate its radius. The calculation of the new radius runs the same way as mentioned above for the initial hypersphere. For this new hypersphere a number of internal patterns is computed.

Now it is necessary to decide if the new shifted hypersphere is better than the original one. The quality of a hypersphere is measured by number of internal patterns. If new hypersphere has a greater number of patterns, we continue with its expansion. If number of patterns within the new hypersphere is less or equal than in the original hypersphere then we do the following steps. In this place it is necessary to emphasize that we should not simply stop the hypersphere shift.

In Fig. 1 an expansion of a two-dimensional hypersphere (circle) in the direction *a* after several shifts is shown. The last expansion of a of a hypersphere with center $\vec{w}^*$ interferes a different class and thus the maximal permitted hypersphere is that with center $\vec{w}$. It is obvious that such a hypersphere is not placed optimally and should expand in the direction *b* to include more patterns.

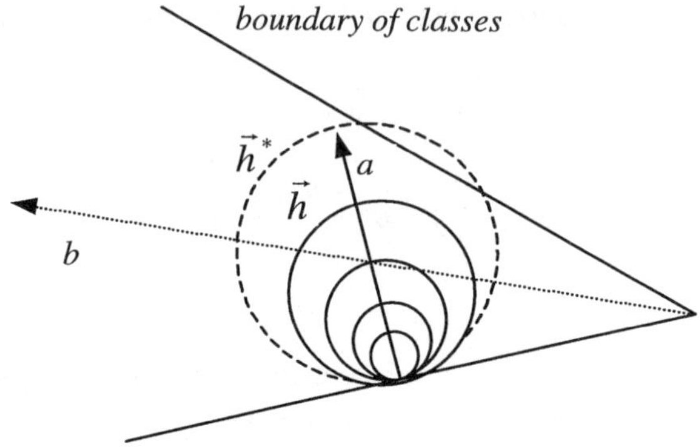

Fig. 1. A two-dimensional hypersphere expansion in a corner

The idea of further expansion consists in allowing a hypersphere expansion even if new shifted hyperspheres are temporarily worse (in the sense of a number of internal patterns) during these several learning steps. This allows a hypersphere to bounce from a boundary with a different class and return back to a space of the current class. For example in Fig. 1 the hypersphere will bounce zigzag from the boundaries and thus move in the direction *b*. A number of temporary steps allowed is a control parameter of GHS. Generally we can say that the higher number of allowed temporary steps the better fitting of a hypersphere in a space is done.

3 Conclusion

A use of a hypersphere as a separation surface is very natural because usually real data with common features belonging to one class are often concentrated near to themselves around some center and form a nearly hypersphere shape. A class separation hypersurface for more complicated data is a disjunction of individual hyperspheres. A further important advantage is a low memory demand. A center coordinates and radius determine each hypersphere.

The GHS network was tested for classification of particle physics data into two classes. The data was obtained from CERN [5] and describe physical phenomena observed in gamma telescope. Experimental results have shown that the GHS network is better than the best-known neural classifiers, [1], [2], [3], [4]. An only disadvantage is a possible overlearning of the GHS network and thus lower ability of generalization.

References

1. Lippmann R. P. (1987). An Introduction to Computing with Neural Nets. IEEE ASSP Magazine
2. Beale R., Jackson T. (1990). Neural Computing: An Introduction. Adam Hilger, Bristol, Philadelphia, New York
3. Rojas R. (1996). Neural Networks: A Systematic Introduction. Springer, Berlin Heidelberg
4. Kohonen T. (1995). Self-Organizing Maps. Springer, Berlin Heidelberg
5. Bock R. K. (1999). Personal communications. CERN
6. Osuna E. E., Freund R., Girosi F. (1997). Support Vector Machines: Training and Applications. MIT AI Memo 1602
7. Carpenter G. A, Grossberg S., Reynolds J. H. (1991). ARTMAP: Supervised Real-Time Learning and Classification of Nonstationary Data by a Self-Organizing Neural Network. Neural Networks, 4, pp. 565-588

Building Predictive Models on Complex Symbolic Sequences via a First-Order Recurrent BCM Network with Lateral Inhibition

Peter Tiňo, Michal Stančík, and Ľubica Beňušková

Department of Computer Science and Engineering, Slovak Technical University, Ilkovičova 3, 81219 Bratislava 1, Slovakia

Abstract: We use a recurrent version of the Bienenstock, Cooper and Munro network (RBCMN) with lateral inhibition [2] to map histories of symbols into activations in the recurrent layer. After training the networks, we construct finite-context predictive models on top of recurrent activations by grouping close activation patterns via a vector quantization. Predictive models extracted from RBCMNs are compared with those extracted from "classical" recurrent networks (RNN) trained in a supervised regime to perform the next-symbol prediction. As a test bed we use two complex symbolic sequences with rather deep memory structures. Surprisingly, the BCM-based model (RBCMNs are trained in an unsupervised mode) has a comparable performance to its RNN-based counterpart. This can be explained by the familiar information latching problem in recurrent networks when longer time spans are to be latched [3, 4]. We argue that BCM-based models correspond to variable memory length Markov models.

1 Introduction

In our recent study of hybrid symbolic-neural time series modeling [1], we trained the second-order RNNs on complex symbolic sequences and then extracted finite state neural-based predictive models by using only recurrent part of the trained networks and discarding the non-recurrent RNN layers. Loosely speaking, the recurrent part of RNN is used to code histories of seen symbols into vectors in the RNN state space (the activation space of recurrent units in the RNN). Using this transformation, simple prediction models can be built by partitioning the RNN state space into a finite set of compartments (representing abstract prediction states) and then approximating the state-conditional next-symbol probabilities by a piece-wise constant functions (constant over the partition compartments). In practical terms, this means driving the trained RNN with fixed weights once more with the training sequence, saving the activation vectors in the recurrent part of the network and then quantizing the recurrent neurons' activation stream (e.g. using K-means). Then, we identify abstract prediction states with the resulting quantization centers. Driving the RNN through the training sequence once again, at each time step, we identify the quantization center (prediction state) to which the current recurrent activation belongs and collect the state-conditional frequencies of the next symbols in turn. We found that such finite state neural-based prediction machines often outperform the mother RNN they were extracted from. It is important to note that the RNNs were trained in a *supervised* setting to organize its state space for the task of next-symbol prediction.

In this paper, we investigate an *unsupervised* alternative to the state space organization. In particular, we use a recurrent version of the BCM with lateral inhibition [2] to implement the transformation of histories of seen symbols into activations in the recurrent layer. The study was initiated with an intuition that, due to the familiar information latching problem [4], the corresponding transformations in "classical" RNN may not be optimal for sequences exhibiting a deep memory structure. Non-recurrent BCM networks with the so-called feed-forward lateral inhibition are known to perform projection pursuit of the input space into the space of BCM activations [5, 6]. The (extended) input to recurrent BCM network (RBCMN) consists of the code of the current input symbol and the current network state, which is the activation vector of BCM recurrent neurons from the previous time step (Fig. 1a). Recurrent BCM networks then perform a *time-conditional projection pursuit*, i.e. organize the state space so that the extended input vectors (current input + current state) project into the next states unveiling the most prominent departures from the Gaussian distribution in the extended input. After training with real time recurrent learning (RTRL), we treat RBCMN in the same way as recurrent parts of RNN in our previous work. We extract finite state prediction machines through vector quantization of the recurrent layer activations.

We test the RNN- and BCM-based predictive models on two complex symbolic sequences, namely, on the sequence of quantized activity changes of a chaotic laser, and on the binary Feigenbaum sequence known for its deep memory structure [3]. Both investigated types of RNNs are of the 1st order, i.e. they do not contain the 2nd order neurons (Fig. 1). The models are compared using normalized negative log-likelihood on unseen continuations of the sequences.

2 First-order BCM RNN with lateral inhibition

The output of the ith neuron in the 1st-order RBCMN comprised of n neurons with lateral inhibition of the constant strength μ between each pair of neurons (Fig. 1a) is [2]:

$$c_i(t+1) = \sigma\Big\{\Big[w_{ij}(t) - \mu\sum_{\beta\neq i} w_{\beta j}(t)\Big] d_j(t) + \Big[m_{ik}(t) - \mu\sum_{\beta\neq i} m_{\beta k}(t)\Big] c_k(t)\Big\} \quad (1)$$

where σ denotes the sigmoid activation function, i.e. $\sigma(x) = 1/(1+exp(-\lambda x))$ with the slope λ. The input symbol at time t is coded by the binary input vector $\mathbf{d}(\mathrm{t})=\{d_j(t)\}$. The lateral inhibition is of the so-called feed-forward type [5]. Synaptic weights w_{ij} and m_{ik} are updated at each time step t, i.e. after each presentation of a consecutive symbol in the temporal sequence. The weight changes are derived by means of the gradient descent minimization of the total loss function $L(t) = \sum L_i(t)$ [5], where

$$L_i(t) = -\left\{\frac{1}{3}c_i^3(t) - \frac{1}{4}E[c_i^2(t)]\ c_i^2(t)\right\} \tag{2}$$

where $E[c_i^2(t)] = \theta_M^i(t)$ is the moving synaptic modification threshold of the ith neuron calculated as [7]

$$\theta_M^i(t) = \frac{1}{\tau}\int_{-\infty}^{t} c_i^2(t')e^{-\frac{t-t'}{\tau}}dt' \tag{3}$$

where τ is the averaging period. Relation between $\theta_M^i(t)$ and the average square of the ith neuron activity guarantees three important features of the model: (i) $\theta_M^i(t)$ is always ≥ 0, (ii) sliding of $\theta_M^i(t)$ guarantees the upper boundedness of synaptic weights without placing artificial constraints on them, and (iii) if all the synaptic weights approached zero, $\theta_M^i(t)$ must also eventually approach zero, thereby allowing for the possible recovery (potentiation) of synaptic weights. Thus

$$\frac{dw_{ij}(t+1)}{dt} = -\eta\frac{\partial L(t+1)}{\partial w_{ij}(t)} = \eta\Big[\sum_{\alpha}\phi_\alpha(t+1)\frac{\partial c_\alpha(t+1)}{\partial w_{ij}(t)}\Big] \tag{4}$$

where η is the learning speed. $\phi_\alpha(t) = c_\alpha(t)[c_\alpha(t) - \theta_M^\alpha(t)]$ is the synaptic modification function for the αth neuron. When the neuronal activity $0 < c_\alpha(t) < \theta_M^\alpha(t)$, all its active synapses weaken. On the other hand, when $c_\alpha(t) > \theta_M^\alpha(t)$, all its active synapses potentiate. As for the partial derivatives of the neuronal activity according to the weights, we get [2]

$$\frac{\partial c_\alpha(t+1)}{\partial w_{ij}(t)} = \sigma'(t+1)\Big\{[\delta_{i\alpha}^{Kron.} - (1-\delta_{i\alpha}^{Kron.})\mu]\ d_j(t) + \sum_{\gamma}\Big[m_{\alpha\gamma}(t) - \mu\sum_{\beta\neq\alpha}m_{\beta\gamma}(t)\Big]\ \frac{\partial c_\gamma(t)}{\partial w_{ij}(t-1)}\Big\} \tag{5}$$

Analogical equations to (4) and (5) can be written with respect to m_{ik} simply by replacing w_{ij} with m_{ik}.

3 "Classical" first-order RNN

The output of the mth neuron in the 1st-order RNN (Fig. 1b) with synaptic weight matrices between layers **v**, **q**, **g**, **w** and **m**, is:

$$o_m(t) = \sigma\Big(\sum_{k} v_{mi}(t)h_i(t) + \vartheta_m^o\Big) \tag{6}$$

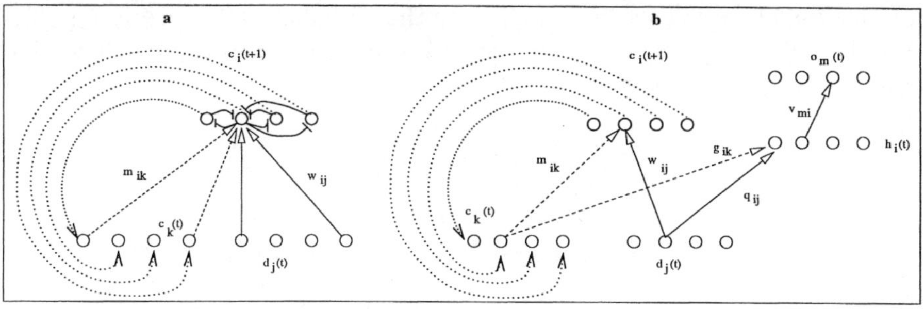

Fig. 1. (a) A 1st-order recurrent BCM neural network with lateral inhibition. (b) A "classical" 1st-order RNN for processing time series.

where ϑ_m^o is the threshold of the m-th neuron in the output layer.

The activations $h_i(t)$ of hidden non-recurrent neurons are determined by

$$h_i(t) = \sigma\Big(\sum_j q_{ij}(t)d_j(t) + \sum_k g_{ik}(t)c_k(t) + \vartheta_i^h\Big) \tag{7}$$

Here, the threshold of the i-th neuron is denoted by ϑ_i^h. The input symbol at time t is coded by the binary input vector $\mathbf{d}(t) = \{d_j(t)\}$.

The activations of recurrent neurons (network state) at the next time step, $c_i(t+1)$, are computed as follows:

$$c_i(t+1) = \sigma\Big(\sum_j w_{ij}(t)d_j(t) + \sum_k m_{ik}(t)c_k(t) + \vartheta_i^c\Big) \tag{8}$$

The threshold term is denoted by ϑ_i^c.

At each time step t, the RNN is trained to predict the next symbol in the sequence. Training is performed via the RTRL [8] minimization of the error function:

$$E(t) = \frac{1}{2}\sum_m (r_m(t) - o_m(t))^2 \tag{9}$$

where the desired (required) output vector $\mathbf{r}(t)=\{r_m(t)\}$ is the binary code of the next symbol in the training sequence.

4 Building predictive models

First we trained both RBCMN and RNN on a long sequence of quantized activity changes of a real laser in a chaotic regime. The chaotic laser produces periods of oscillations with increasing amplitude, followed by sudden, difficult to predict, activity collapses. Thus, relatively predictable subsequences are followed by highly unpredictable events requiring a deeper memory. The laser sequence of 10^4 differences Δ_t between successive activations was quantized into a symbolic sequence $S = \{s_t\}$ over the alphabet $\mathcal{A} = \{a, b, c, d\}$, with $A = 4$ symbols. The symbols correspond to low and high positive/negative activity changes such that

$$s_t \equiv \begin{cases} a \ (normal\ up), & \text{if } 0 \leq \Delta_t < \rho_2 \\ b \ (high\ up), & \text{if } \rho_2 \leq \Delta_t \\ c \ (normal\ down), & \text{if } \rho_1 \leq \Delta_t < 0 \\ d \ (deep\ down), & \text{if } \Delta_t < \rho_1 \end{cases} \tag{10}$$

where the parameters ρ_1 and ρ_2 correspond to 10% and 90% sample quantiles, respectively. The number of positive Δ_ts is approximately equal to the number of negative Δ_ts. The first 8000 symbols and the remaining 2000 symbols formed the training and test sequences, S_{train} and S_{test}, respectively.

Second, we applied both networks to the Feigenbaum binary ($A = 2$) sequence over the alphabet $\mathcal{A} = \{a, b\}$ [9]. It can be described by the following iterative composition rule:

$$\omega_0 = b, \quad \omega_1 = ba, \quad \omega_{n+1} = \omega_n \omega_{n-1} \omega_{n-1}, \quad n \geq 2 \tag{11}$$

Very specialized prediction contexts are needed to model this sequence. We created a large portion of the Feigenbaum sequence and trained both a RNN and a RBCM on the first 260 000 symbols. The following 200 000 symbols formed S_{test}.

Now we describe construction of predictive models from the trained networks. After training, all synaptic weights are fixed and stored. Next, sliding through the training sequence, for each symbol in S_{train}, using these weights, we record the recurrent activations $\mathbf{c}(\mathrm{t+1}) = \{c_i(t+1)\}$. This creates an activation sequence S_{train}^{act} for each network. Then we perform a vector quantization on S_{train}^{act} via the K-means clustering for N quantization centers. In our predictive models, the quantization centers are identified with predictive states. To calculate the state-conditional probabilities, we associate with each center A counters, one for each symbol from the input alphabet $\mathcal{A}$. Sliding through the recurrent activations' sequence, S_{train}^{act}, the closest center C is found for the current activation vector. Next, we identify the next symbol s_{next} in S_{train}. For the center C, the counter associated with s_{next} is raised by one. After seeing the whole training sequence S_{train}, for each quantization center (prediction state) C, we normalize the counters to calculate the conditional next-symbol probabilities $P_{\mathcal{M}}(\cdot|C)$. Each quantization center now represents a predictive state.

On the test sequence $S_{test} = s_1 s_2 ... s_t ... s_m$, the next-symbol probabilities are determined as follows: for each $t = 1, 2, ..., m-1$, given the network state $\mathbf{c}(t) = \{c_i(t)\}$, the symbol $s_t \in S_{test}$ drives the network to a new recurrent activation vector $\mathbf{c}(\mathrm{t+1}) = \{c_i(t+1)\}$. We find the quantization center C_{t+1} closest to $\mathbf{c}(\mathrm{t+1})$. The next-symbol probabilities are $P_{\mathcal{M}}(s_{t+1}|C_{t+1})$.

For each model $\mathcal{M}$, we evaluated its performance by means of the normalized negative log-likelihood NNL on the test sequence S_{test}:

$$NNL_{\mathcal{M}}(S_{test}) = \frac{-\sum_{t=1}^{m-1} log_A P_{\mathcal{M}}(s_{t+1}|C_{t+1})}{m} \tag{12}$$

where the base of the logarithm is the number of symbols A in the alphabet $\mathcal{A}$. The $NNL_{\mathcal{M}}(S_{test})$ measures the total amount of "statistical surprise" induced by the model [10].

The metric in vector quantization and nearest-center detection is Euclidean distance.

5 Results

The code vectors for the laser symbols were $\mathbf{d}_a$ = (1,0,0,0), $\mathbf{d}_b$ = (0,1,0,0), etc., and for the Feigenbaum binary symbols, they were $\mathbf{d}_a$ = (1,0), $\mathbf{d}_b$ = (0,1). NNLs were calculated for different numbers of predictive centers, i.e. $N = 1, 2, 5, 10, 15, 20, 30, 60, 100, 150, 200, 300$.

The values of parameters of the 1st-order RBCMN were set to: number of (recurrent) neurons $n = 4$, $\tau = 100$ iterations, $\eta = 0.01$, $\mu = 0.2$, $\lambda = 0.1$ for the unipolar '0–1' sigmoid. The results were not much different for τ = 10–10^3 iterations and μ = 0.1–0.3. However, λ must have been small. The learning was stopped after the weights have not been changing noticeably for quite a while ($\sim 10^3$ iterations). This usually happened after 5 epochs of training.

The values of parameters of the 1st-order RNN were set to: $n = 4$ neurons for all the layers, $\eta = 0.15$, momentum $= 0.08$, and $\lambda = 1$ for the unipolar '0–1' sigmoid. The initial (reset) activations of recurrent neurons were randomly generated from a uniform distribution over [-0.5, 0.5]. Thirty epochs were used to train the RNN on the laser sequence and five epochs for the Feigenbaum sequence.

For both networks, the initial weights were uniformly generated from the interval of [-1, 1].

In Fig. 2, we present NNLs for the laser and Feigenbaum test sequences. All presented values are the averages over 10 vector quantization runs in our predictive model construction (see section 4). Also shown are the corresponding standard deviations.

6 Discussion

The main theme of this paper is related to representation of symbolic sequences by points in a vector space (see [11]). Association map between the recurrent and output layers in RNN is usually smooth and so by training RNN to perform the next-symbol prediction, the network is forced to map

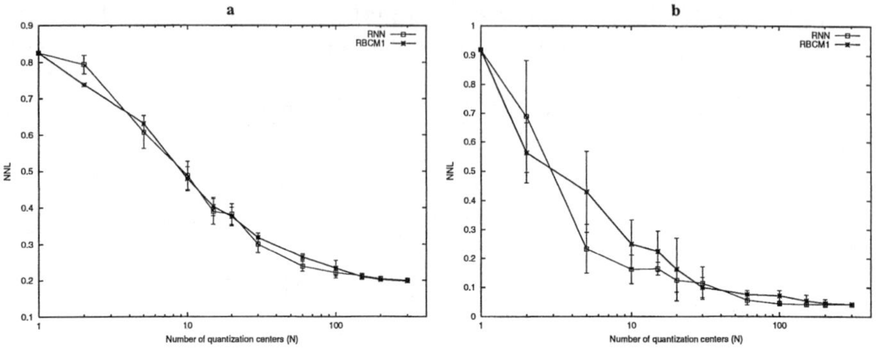

Fig. 2. Normalized negative log-likelihoods (NNL) for the (a) laser test sequence, (b) Feigenbaum test sequence. Results of the predictive model built upon the 1st-order BCM RNN with lateral inhibition are indicated by RBCM1. RNN denotes results of the classical 1nd-order RNN.

sequences of observed symbols close to each other if they lead to similar next-symbol distributions. This intuition can be used to extract simple finite state predictive models via detection of activation clusters in the RNN recurrent layer – "close" activation vectors correspond to subsequences yielding similar prediction probabilities. In other words, the activation clusters correspond to "abstract prediction states" found by RNN. Actually, the extracted models outperform their mother RNN (see [1]).

We were interested whether an *unsupervised* construction of a mapping from symbolic sequences into points in a vector space can lead to comparable predictive models. In this case, there is no explicit driving force in the form of the target next symbols. We chose to work with a recurrent form of a biologically motivated 1st-order BCM network that performs a time-conditional projection pursuit. The input symbols are coded using the one-of-n encoding scheme, so there is no a-priori metric structure imposed on the input alphabet. The recurrent BCM network (RBCMN) tries to project the current state $\mathbf{c}(t)$ (recurrent activations from the previous time step) that codes the symbols it has observed up to time step $t-1$, extended with the code $\mathbf{d}_{s_t}$ of the current input symbol s_t, so that the projections $\mathbf{c}(t+1)$ (states at time $t+1$) are "non-trivially" spread out in the recurrent activation space. Here, non-triviality means non-gaussianity.

Our results are quite surprising. On two chaotic sequences with relatively deep memory structures, the supervised and unsupervised approaches yield similar predictive model performances. When more than 20 states is used in the predictive models, performances of the RBCMN-based and the RNN-based models are almost the same. We got similar results also with other forms of spatial representations of temporal structures in symbolic sequences (see [3]). One of the reasons may be that even though the "classical" RNN are potentially very powerful in building state space representations suited to the specific task (e.g. next-symbol prediction), in practice it is difficult to train them to span large time intervals due to the information latching problem [4].

We will loosely explain the RBCMN coding scheme. Consider two identical RBCMNs, both initially in the same state $\mathbf{c}_1(1)=\mathbf{c}_2(1) = \mathbf{c}(1)$. The networks were already trained on a long sequence coming from an ergodic source $\mathcal{S}$. The first RBCMN will be fed by a symbolic stream $U = u_1u_2u_3...$, the second receives its input from a stream $V = v_1v_2v_3....$ The streams U, V come from the source $\mathcal{S}$.

Assume the length of the longest common prefix of U and V is $m-1$, i.e. $u_1u_2...u_{m-1}=v_1, v_2...v_{m-1}$ and $u_m \neq v_m$. Then, for the first $m-1$ time steps, the state sequence in both networks will be the same (see eq. (1)). Upon seeing the m-th symbol, states of the two networks diverge, $\mathbf{c}_1(m+1) \neq \mathbf{c}_2(m+1)$, and by the projection pursuit principle implemented by the networks, will remain divergent while the corresponding symbols in the stream continuations $u_{m+1}u_{m+2}...$ and $v_{m+1}v_{m+2}...$ differ. This follows from the fact that the network state $\mathbf{c}(t)$ is contained in the open hypercube $(0,1)^n$, and input symbols are encoded by one-of-m codes $(0,...,0,1,0,...,0)$. By forming the extended input (current symbol + current state), the current state $\mathbf{c}(t)$ is injected into the subspace of the space of extended inputs, that corresponds to the current input symbol. Different input symbols initiate projections of $\mathbf{c}(t)$ into orthogonal subspaces, enhancing non-gaussianity in the extended input space. Such projections, when transformed into $\mathbf{c}(t+1)$ by projection pursuit, land far from each other.

Consider two scenarios:

1. After $m+\ell$ time steps, the streams U, V again coincide in a block of length L: $u_{m+\ell+1}...u_{m+\ell+L}=v_{m+\ell+1}...v_{m+\ell+L}$.
2. The corresponding symbols in $u_{m+\ell+1}...u_{m+\ell+L}$ and $v_{m+\ell+1}...v_{m+\ell+L}$ are different.

By the projection pursuit principle, after $m+\ell+L$ steps, the recurrent activations $\mathbf{c}_1(m+\ell+L+1)$ and $\mathbf{c}_2(m+\ell+L+1)$ in the two networks will lie closer in the first scenario than in the second one. Also, it is reasonable to expect that longer common blocks in U and V will lead to closer corresponding network states.

By vector quantizing the state activations of the RBCMN driven with the training sequence $S = s_1s_2...$, we group together network states lying close to each other. Now, assume that the state vectors $\mathbf{c}(t_1+1)$ and $\mathbf{c}(t_2+1)$ appearing in the recurrent layer at times t_1 and t_2, respectively, are "close". The states $\mathbf{c}(t_1+1)$ and $\mathbf{c}(t_2+1)$ code the histories $s_1s_2...s_{t_1}$ and $s_1s_2...s_{t_2}$ of input symbols up to times t_1 and t_2, respectively. As outlined above, since the two states are close, the sequences $s_1s_2...s_{t_1}$ and $s_1s_2...s_{t_2}$ share the same suffix. Moreover, the distance between the states tends to reflect the length of the common suffix. Smaller distances correspond to longer common suffixes (i.e. recent symbols). Hence by vector quantizing the network state activations, we construct the predictive states on the basis of common suffixes. In other words, for next-symbol prediction at time t, only a block of recently seen symbols (up to time t) is considered. This corresponds to the basic assumption behind the Markov models. In our case, however, the Markov order is not fixed, but the memory depth associated with the quantization centers can be of variable length. Dense clusters, quantized by many quantization centers, correspond to deep-memory contexts. A sparsely inhabited region represented by a single center corresponds to a shallow memory.

In this sense, the predictive models extracted from RBCMNs by vector-quantizing their state space correspond to the class of variable memory length Markov models (VLMMs) [10].

Acknowledgments
Supported by the VEGA grants 2/6018/99 and 1/7611/20.

References

1. Tiňo P. and Köteleš M. (1999) Extracting finite state representations from recurrent neural networks trained on chaotic symbolic sequences. *IEEE Transactions on Neural Networks* **10**, 284–302.
2. Bachman C.M., Musman S.A., Luong D. and Shultz A. (1994) Unsupervised BCM projection pursuit algorithms for classification of simulated radar presentations. *Neural Networks* **7**, 709–728.
3. Tiňo P. and Dorffner G. (1999) Building predictive models from spatial representations of symbolic sequences. (Will appear in *Advances in Neural Information Processing Systems 12*, S. A. Solla, T. K. Leen, K-R. Müller (eds.), MIT Press, Cambridge, 2000.)
4. Bengio Y, Simard P. and Frasconi P. (1994) Learning long-term dependencies with gradient is difficult. *IEEE Transactions on Neural Networks* **5**, 157–166.
5. Intrator N. and Cooper L.N. (1992) Objective function formulation of the BCM theory of visual cortical plasticity: statistical connections, stability conditions. *Neural Networks* **5**, 3–17.
6. Dotan Y. and Intrator N. (1998) Multimodality exploration by an unsupervised projection pursuit neural network. *IEEE Transactions on Neural Networks* **9**, 464–472.
7. Beňušková L., Diamond M.E. and Ebner F.F. (1994) Dynamic synaptic modification threshold: computational model of experience-dependent plasticity in adult rat barrel cortex. *Proc. Natl. Acad. Sci. USA* **91**, 4791–4795.
8. Williams R.J. and Zipser D. (1989) A learning algorithm for continually running fully recurrent neural networks. *Neural Computation* **1**, 270–280.
9. Freund J., Ebeling W. and Rateitschak K. (1996) Self-similar sequences and universal scaling of dynamical entropies. *Physical Review E* **5**, 5561–5566.
10. Ron D., Singer E. and Tishby N. (1996) The power of amnesia: learning probabilistic automata with variable memory length. *Machine Learning* **25**, 117–149.
11. Tiňo P. (1999) Spatial representation of symbolic sequences through iterative function systems. *IEEE Transactions on Systems, Man, and Cybernetics Part A: Systems and Humans* **29**, pp. 386-392.

Network of Integrate-and-Fire Neurons Using Rank Order Coding : How to Implement Hebbian Learning

Laurent Perrinet[1], Arnaud Delorme[2], Manuel Samuelides[1], and Simon Thorpe[2]

[1] ONERA-DTIM, 2 Av. E. Belin, BP 4025, F-31055 Toulouse, France, http://www.cert.fr/fr/dtim/m2sn-vision.html
[2] CERCO/CNRS, 133 Route de Narbonne, F-31062 Toulouse, France, http://www.cerco.ups-tlse.fr/~modele/

Abstract: To fit biological constraints on the speed of processing in the visual system [1], we propose a code based on the relative date of discharge across a group of neurons instead of a classical rate code. First, we set up the model and explore the computational benefits and limits of the rank coding / decoding scheme we constructed. Second, we propose a hebbian-like learning rule, that performs good fitting of the network to the statistics of the inputs. Particularly, we observe the emergence to weight mappings analogous to orientation-sensitive receptive fields from the V1 cortex area for natural images.

Keywords : Rate coding, spikes, integrate-and-fire neuron, asynchrony, hebbian learning, statistics of natural images.

1 The need for a new coding scheme

1.1 Visual processing : neuro-psychological data

Neuropsychological experiments have recently shown the incompatibility of the classical rate code introduced in 1926 by Adrian and Zotterman[2]. Actually, the speed of processing in the visual system can be as short as 150 ms for a complex task involving at least 10 processing layers [1] leaving less than 15 ms for the neuron to process and transmit the information.
The information conveyed by the neurons is often supposed to be carried by the mean firing rates observed on spike trains generated with a Poisson process. However, the neuron has a limited bandwith and we have to propose a new code to account for the speed of processing of biological neural networks.

1.2 The solution of asynchrony : introducing spiking neurons

A solution is to use the asynchronous characteristics of the neural code, as in the analog-to-delay conversion characteristic of retinal ganglion cells by using integrate-and-fire like models of neurons: generally, the stronger the

input intensity, the quicker the neuron would fire.
The unit of information is then the generated action potential or *spike*. It's shape and duration is constant throughout all the neurons and therefore, they are relevant by their relative timing.
This model is nowadays very extensively studied ([3], [4]) on computationnal and biological sides. It outlines that the information carried by the spikes is not only coded by the rate of the spike trains, but also shows properties of this coding, like the benefits of the sparseness [5].

1.3 Analog to Rank coding

At this point, we propose that :

- each sensory input is emitting at most one spike, the most relevant inputs emitting first,
- somatic integration is not dependent on time but on the number of spikes received by the efferent neuron,
- when the somatic integration reaches a threshold, the neuron propagates a spike to the efferent neurons
- each neuron emits only one spike

This model is easy understandable for a network of layers of neurons, which we'll study in this paper (see the following example in Fig. 1), but may be used for reccurent networks.

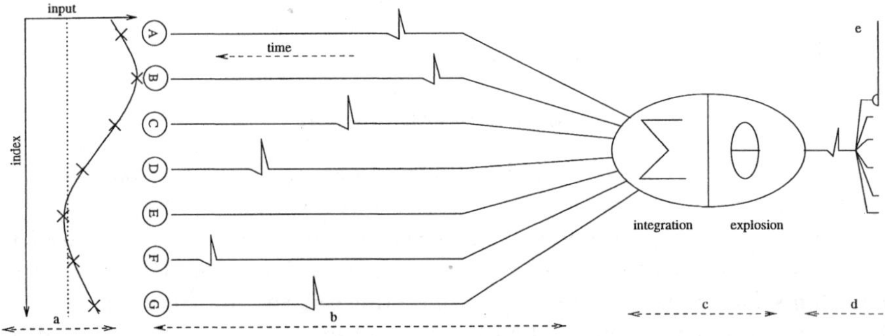

Fig. 1. Analog to rank coding. A discrete analog input is converted in **(a)** emititing at most one spike per index in the dendrites **(b)**. On this example, the information is simply $B > A > C > G > D > F$. The spikes are propagated and integrated in the soma **(c)**. A spike is emitted when the somatic potentiation reaches the threshold and is then transmitted via the axon **(d)** to the efferent neurons **(e)** to do the same processing **(b) to (e)**.

By those assumptions, the information carried by the neural code is the order of spike arrivals rather than the exact latency [6]. But this coding is not only an analog to rank conversion : the propagation is dynamic between the different layers of neurons, parallel in the layer and interruptible for each neuron. This model was implemented using an artificial neural network called SpikeNET [7] which proved to be as effective as classical models but with quicker processing speed.

2 Propagation scheme and neural network topology

2.1 Asynchronous feed-forward propagation

Considering propagation, the spike wave is emitted in a feed-forward fashion to the connected neurons of the efferent layer, and therefore we consider emitting fields in comparison with the classical view of receptive fields. In this view, receptive and emitting fields have homologous geometry. In our case we admit that neurons have same receptive fields topology accross the image, and therefore this translation invariance leads to a central symmetric relation between receptive and emittive fields.
Also, on the computational side, this type of coding is highly parallel : the spikes are propagated until aim is achieved, time is over or all spikes were fired, in opposition to the classical sequential propagation from the first to the last layer. Each fired spike may be considered like a computer thread emitting to his efferent neurons.
Lastly, it is dynamically interruptible (i.e. once a neuron spiked, the calculation can be stopped on this neuron) and it generates sparse representation of the signal. The consequence is that the process is therefore well suited for demanding tasks like visual processing.

2.2 Rule for Rank coding

Mathematically, for a given neuron, the nth spike arriving at the soma is integrated to the potential V by adding the synaptic weight $w_{ordo(n)}$ of the last spiking dendrite of index $ordo(n)$ modulated by a decreasing function $Mod(.)$ of the rank order of the spike : Beginning with $V(0) = 0$ and $Mod(0) = 1$, we have for $n > 1$,

$$V(n) = V(n-1) + w_{ordo(n)}.Mod(n) \tag{1}$$

Therefore, we index time by the spike arrivals and not continuously. The neuron emits itself a spike on his axon when it reaches the threshold θ.
For the example of Fig. 1, we'll have $V(1) = 0 + w_B.Mod(0)$, $V(2) = V(1) + w_A.Mod(1)$ and so on, until the potential V reaches θ. The network propagates then this spike to the next layer and if so to the following layers of the network.

2.3 Computational benefits

Numerous theoretical results have been found. Firstly, we showed that rank order coding is invariant with any increasing monotonous transformation of the pixel intensities (the order is conserved), which is analoguous to the relative invariance towards contrast of natural vision.
To analyze integration in the soma, we used rank linear statistics, linking the potential value with the Spearman correlation coefficient [8]: the neurons measure dynamically a non-linear correlation between the input and the weight pattern. We then proved that under certain conditions, the distribution of potential for random inputs (i.e. permutations of the ranks) is gaussian with known moments (permutationnal central limit theorem). This enables us to determine with precise selectivity the threshold θ.
Another important aspect of this coding is the equalization of the signal coded by rank order : the distribution of the ordered signal is uniform. This corresponds to an information theoretic optimization of the information transmission.

3 Hebbian learning in an asynchronous neural network

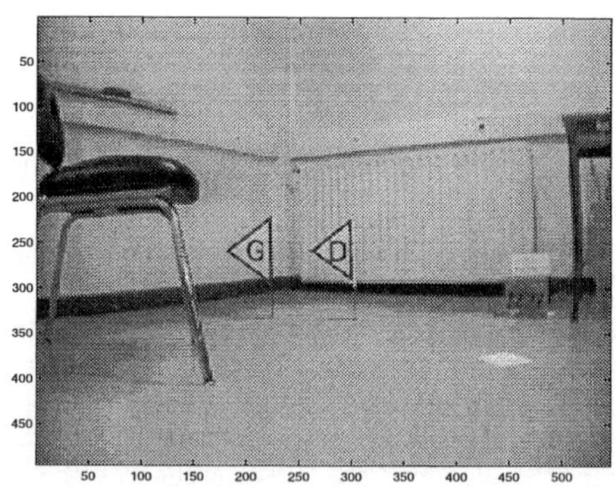

Fig. 2. Sample image used in the experience taken from the camera of a robotic vehicle

3.1 Principle

Due to its simplicity and biological plausibility, we have chosen to implement hebbian-like learning: the synaptic weights are reinforced towards the value of modulation at the value of the rank order. For this model of learning, we proved that under certain assumptions, the learning algorithm was converging in the quadratic function space [9]. We also used this type of rule to update

the modulation function. We then proved that for natural images, $Mod(.)$ converges towards the cumulative gaussian distribution.
For the first layers of the network, the learning process is unsupervised: "the winner takes all" and therefore, only the first spiking neuron of the layer is updated. Moreover, we used lateral inhibition and an adaptive rule for the threshold so that it becomes more selective during the learning process of each neuron (and therefore with its "maturity"), allowing the different neurons to compete during this process.

3.2 Results

We implemented those algorithms on MATLAB for modeling purposes and then on the SpikeNET technology allowing the computation with a bigger size of the learning database. With used different types of images (textures, benchmark natural images, ...) and particularly images from a robot vehicle (see Fig. 2). The topology of the network we used is linked to the topology in the primate's visual system. On the first layer, the image is transformed by ganglion-like convolution (derivate of gaussian) giving a contrast image. Then, this image is emitting spikes in two parallel network layers, one for high contrast (corresponding to ON-center / OFF-surround) and one for high negative contrast.
The third layer, corresponding to LGN or V1 shows the emergence of V1 like type of direction selective receptor fields on the first layer of our network for those images (see Fig. 3). Those results are relatively independant of the type of image but varry with the parameters of our coding.

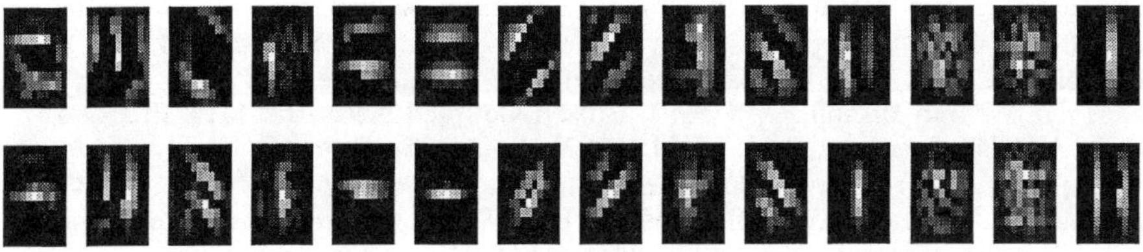

Fig. 3. Emergence of V1-like receptive fields (RFs). The upper (resp. lower) row corresponds to the positive (resp. negative) contrasts. RFs are corresponding column by column. This 'snapshot' of the weight map is taken during the learning process, showing the emergence of direction sensitive receptors, that some RFs are redundant and finally that some remain random

3.3 Discussion

Actually, biological learning mechanisms are similar to the one we used in the present model. They have been shown to depend on very small differ-

ence in latencies between the input and the output cell discharge date [10]. Based on this observation, we implemented an adapted rule introducing the time of emission of the spike. Particularly, we studied the convergence of the algorithm with synthesized images (gratings), textures (from the Brodatz database) and natural images and showed similar results as with the first hebbian rule.
Analysis of the results showed, as we expected, that as we learn on the spike really emitted, the learned patterns have a gradually more sparse representation and more selective response. Using information theory, we then studied the respective performances of the algorithms, putting into evidence the similar properties of transmission of the information. Actually, those receptors are optimal to generate a sparse and non redundant representation of the image allowing compression of the spatial correlation of different levels. For instance, at the first level, neighboring pixels are more probable to be at the same value in comparison with distant pixels, on the second level, neighboring edges are more probable to point at same directions. Classically, those filters are decorrelating kernels to allow maximization of the information transmission [11]. In our model of asynchronous propagation of spikes with hebbian learning, this information theoretic optimization is still sound and is combined with the advantages of the fast processing allowed by the rank order scheme. Lastly, the latter scheme is biologically more plausible and computationally more effective.

References

1. Thorpe, S.J., Fize, D., and Marlot, C. (1996), "Speed of processing in the human visual system", *Nature*, 381, 520-52.
2. Adrian, E. and Zotterman, Y. (1926). The impulses produced by sensory nerve-endings, part . The response of a single end-organ. *J. Physiology*, 61, 151-171.
3. Rieke, F. et al. (1998). Spikes, exploring the neural code. MIT Press.
4. Maas, W. and Bishop, C., (1999). Pulsed Coupled Networks. MIT Press.
5. B. A. Olshausen and D. J. Field (1997). Sparse Coding with an Overcomplete Basis Set: A strategy employed in V1. *Vision Research*, Vol. 37, pp. 3311-3325.
6. Gautrais, J., and Thorpe, S.J. (1998). Rate Coding vs Temporal Order Coding: a theorical approach.*Biosystems*, 48(1-3), 57-65.
7. Delorme, A., Gautrais, J., Van Rullen, R. and Thorpe, S.J.(1999). SpikeNET : a simulator for modeling large networks of integrate and fire neurons. *Neurocomputing, 26-27*, 989-996
8. Spearman, (1904).The proof and measurement of association between two things, *Am. J. of Psychology*, 15, 72-101.
9. M. Samuelides, S. Thorpe and E. Veneau, Implementing Hebbian Learning in a Rank-Based Neural Network, *Lecture Notes in Computer Science*, 1997, vol. 1327, p. 145
10. Markram, H. et al. (1997). Regulation of Synaptic Efficacy by Concidence of Postsynaptic APs and EPSPs. *Science*, 275:213-215.
11. Attick, J. (1992). Could Information Theory Provide an Ecological Theory of Sensory Processing? *Network: Comp. in Neural Systems*, vol.3, no. 2, 213-252.

Pulse-Coupled Neural Networks for Image Smoothing and Segmentation

Alexei N. Skourikhine

Safeguards Systems Group, MS E541, Los Alamos National Laboratory

Los Alamos, NM 87545, USA, E-mail: alexei@lanl.gov

Abstract: Pulse-coupled neural networks (PCNNs) based on Eckhorn's model of the cat visual cortex find many applications in image processing, including smoothing, segmentation, and edge extraction. These applications are supported by synchronization of neuron activity, where each neuron corresponds to each pixel in the input image. The synchronization relies on similarities in the gray scale and the linking between neurons. The paper reports on an evaluation based on histogram analysis of PCNN applications to smoothing and segmentation of gray-scale images and also considers the effect of feedback from global inhibitory neuron on the segmentation.

Keywords: pulse-coupled neural networks, image processing, segmentation, image smoothing

1 Image Processing Using Pulse-Coupled Neural Networks

A PCNN is a biologically inspired algorithm for image processing [1,2]. It is to a very large extent based on the Eckhorn model of the cat visual cortex [3,4]. The typical neuron of the PCNN is shown in Fig. 1. The equations for a single iteration of the PCNN are

$$
\begin{aligned}
&F_{ij}[n]=e^{-\alpha_F}F_{ij}[n-1]+S_{ij}+V_F\sum_{kl}m_{ijkl}Y_{kl}[n-1]+I(t)\\
&L_{ij}[n]=e^{-\alpha_L}L_{ij}[n-1]+V_L\sum_{kl}w_{ijkl}Y_{kl}[n-1]\\
&U_{ij}[n]=F_{ij}[n](1+\beta L_{ij}[n])\\
&Y_{ij}[n]=\begin{cases}1, & \text{if } U_{ij}[n]>\Theta_{ij}[n]\\ 0, & \text{otherwise}\end{cases}\\
&\Theta_{ij}[n]=e^{-\alpha_\Theta}\Theta_{ij}[n-1]+V_\Theta Y_{ij}[n]
\end{aligned}
\tag{1}
$$

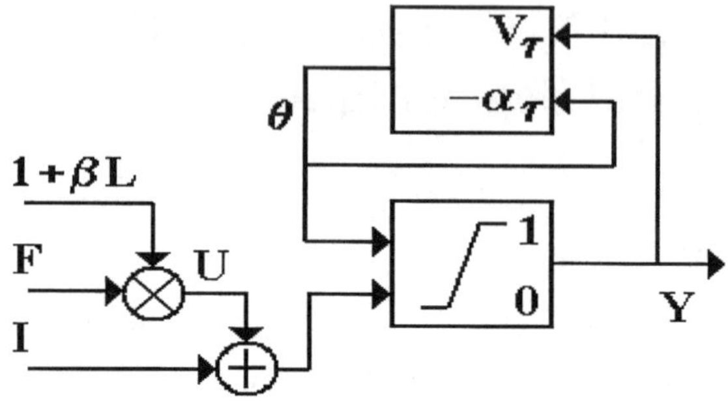

Fig. 1. The basic PCNN neuron.

where S is the input signal, F is the feed, L is the link, U is the internal activity, Y is the pulse output, and Θ is the dynamic threshold. The weight matrices M and W are local interconnections and β is the linking constant. I is the inhibition term that is determined by the total activity of the network. The output values of all neurons are summed up, negated, and fed back to each neuron of the neural network.

The basic simplified structure of the pulse-coupled neural network processor for a 2-D input image is shown in Fig. 2. An input gray-scale image is composed of $M \times N$ pixels. This image can be represented as an array of $M \times N$ normalized intensity values. Then the array is fed in at the $M \times N$ inputs of PCNN. If initially all neurons are set to 0, the input results in activation of all of the neurons at a first iteration. The threshold of each neuron, Θ, significantly increases when the neuron fires, then the threshold value decays. When the threshold falls below the respective neuron's potential (U) the neuron fires again, which raises the threshold. The process continues creating binary pulses for each neuron. While this process goes on, neurons encourage their neighbors to fire simultaneously in a way that is supported through interconnections. The firing neurons begin to communicate with their nearest neighbors, which in turn communicate with their neighbors. The result is an autowave that expands from active regions. Thus, if a group of neurons is close to firing, then one neuron can trigger the group. Due to linking between neurons, the pulsing activity of invoked neurons leads to the synchronization between groups of neurons corresponding to subregions of the image that have similar properties and produces a temporal series of binary images. This synchronization supports image segmentation.

It is often better to reduce noise in images by filtering before segmentation. PCNNs can also be used for image smoothing, which is achieved through iterated modification of the original intencity levels of an image. Adjustment of a pixel's intensity depends on the history of how in time a pixel's neighbors fire. If a given pixel does not fire simultaneously with the majority of its neighbors, then its intensity is examined. If a pixel fires after the majority of its neighborhood, its intensity is increased, otherwise it is decreased by a small value. This procedure results in both a reduction of the noise contained in the image and the number of pixel intensities represented in the image.

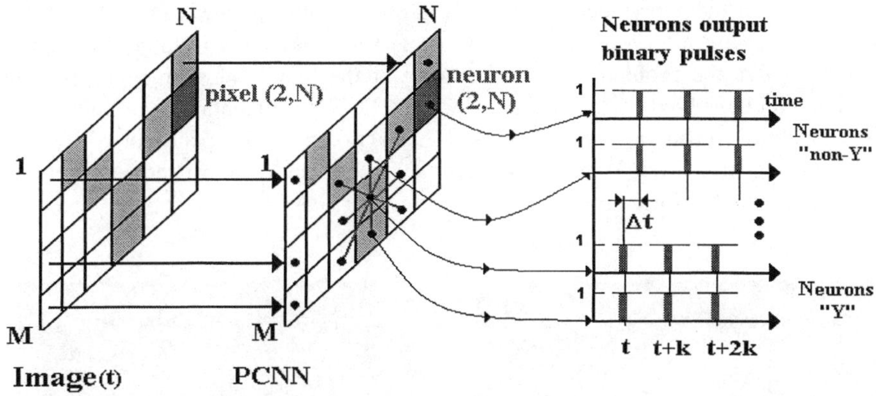

Fig. 2. Image processing using a pulse-coupled neural network.

2 Results of Simulation Experiments

By utilizing a PCNN, the gray-scale image (Fig. 3 (a), whose histogram is shown in Fig. 4) is processed by the PCNN algorithm to produce a series of binary images containing the segmentation results. Histograms of the binary images mapped to the original image are shown in Fig. 5 and Fig. 6. Underlying ordering of the PCNN segmentation is obvious from these figures: the brighter the image's region the sooner it fires and is segmented. These histograms also show the difference between the results produced by PCNN without inhibition and PCNN with inhibition. The results of image smoothing are shown in Fig. 3 (b) and Fig. 7.

Our evaluation was performed on a 600 MHz Pentium with 768 Mbytes of RAM and Windows NT. It generally took 0.32 s to produce a single binary image. One run through the whole image required six iterations to have all the image pixels fired in the shown experiment. Thus the minimum total time to process a gray-scale image of 512×512 pixels size is about 2 s.

3 Conclusions

PCNN is useful for image smoothing, segmentation, and edge extraction. However, the algorithm's speed is crucial, if real-time image processing is desired. Hardware implementation is most appropriate for real-time applications.

The PCNN complexity is the proper setting of the various parameters of the network, such as linking parameters, thresholds, and interconnection matrices. Adjustment of these parameters, depending on the input image, can make PCNN image processing more efficient. At the same time, having adjustable parameters does not solve the problem of interpretation of the results of segmentation. This problem can be solved by the teamwork of the PCNN segmentation module, high-level features grouping modules and recognizer, and specifically by the development of a system with the interactive hierarchy of different layers of processing visual information.

(a) (b)

Fig. 3. PCNN smoothing. **(a)** Original image of the size 512×512 pixels. **(b)** Original image after smoothing.

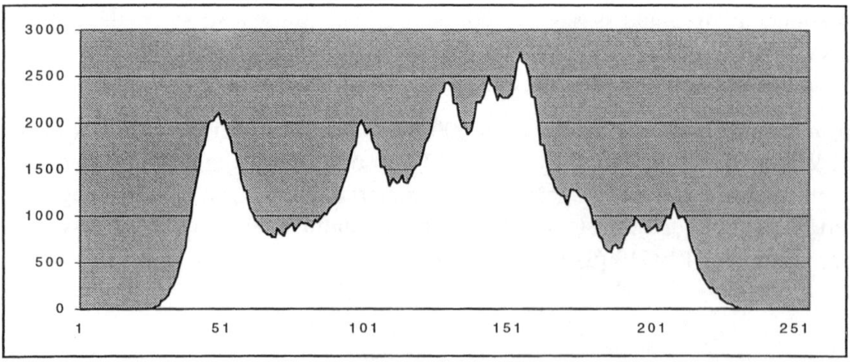

Fig. 4. Histogram of the image shown on Fig. 3.

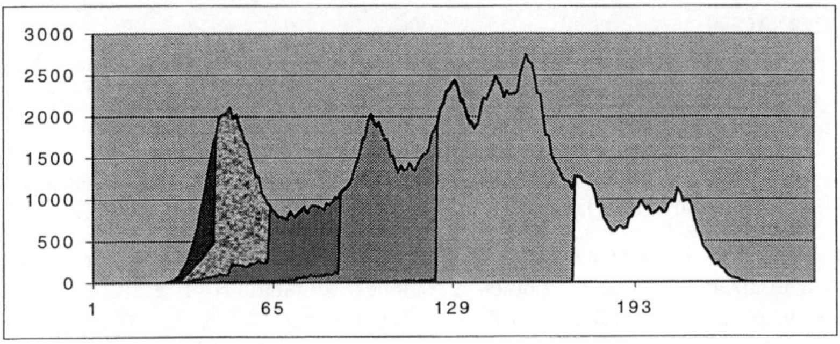

Fig. 5. Six histogram regions (from brightest to darkest) correspond to six binary images produced by PCNN without inhibition.

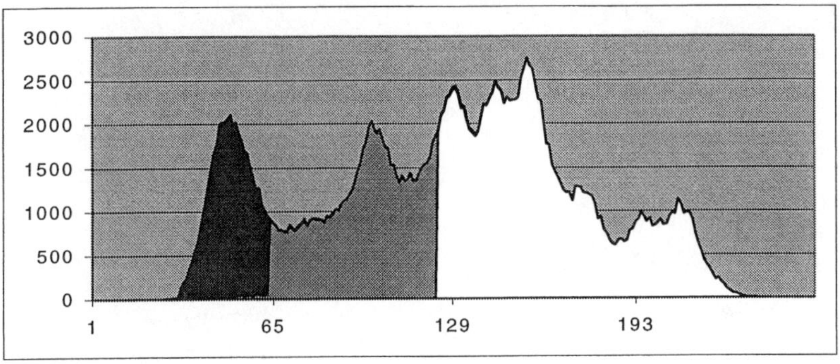

Fig. 6. Three histogram regions (from brightest to darkest) correspond to three binary images produced by PCNN with inhibition.

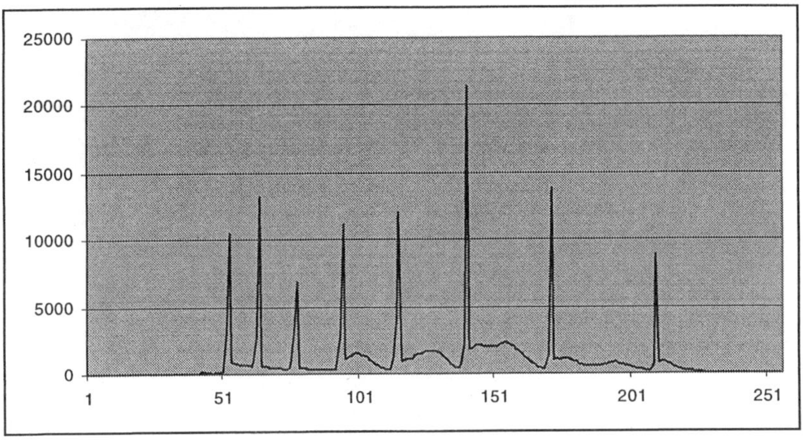

Fig. 7. Histogram of the smoothed image shown in Fig. 3 (b).

References

1. J. L. Johnson, M. L. Padgett. (1999) PCNN models and applications. IEEE Trans. on Neural Networks, **10**, 3, 480–498.

2. T. Lindblad, J. Kinser. (1998) Image Processing using Pulse-Coupled Neural Networks. Springer-Verlag.

3. R. Eckhorn, H. J. Reitboeck, M. Arndt, P. Dicke. (1990) Feature linking via synchronization among distributed assemblies: Simulations of results from cat visual cortex. Neural Comp., **2**, 293–307.

4. M. Stoecker, H. J. Reitboeck, R. Eckhorn. (1996) A neural network for scene segmentation by temporal coding. Neurocomp., **11**, 123–134.

Some New Adaptive Architectures for Learning, Generalization, and Visualization of Multivariate Data

Boris Igelnik

Case Western Reserve University, Cleveland OH 44106, USA

Abstract. We present a general adaptive architecture for learning, and generalization of multivariate data, which comprises a finite number of adaptive learning computational schemes combined by a common algorithm for learning and generalization. Some new schemes different from traditional neural networks or RBF networks are described. These are the network with complex weights and the network for learning Lennard-Jones potentials. Side by side with a manual choice of appropriate adaptive learning computational scheme an example of automatic and continuous mode is presented. We describe as well a novel perspective method for visualization of multivariate data in lower dimensions. Selective examples of applications of these techniques and illustrative mathematical examples are given.

Keywords: adaptive learning, generalization, visualization, complex weights, stochastic optimization.

1 Introduction

There is a number of adaptive computational schemes (let us call them nets) for approximating multivariate mappings with use in regression and classification tasks. The incomplete list of them includes neural networks and radial basis functions expansions (RBF) [1], [2], projection pursuit nets[3], hinging hyperplanes [4], probabilistic nets [5], random nets [6], high-order nets [7], wavelets [8]. The mathematical model for some of these nets has the following form

$$\tilde{f}(x)=\sum_{n=1}^{N} a_n g\left[\sum_{i=1}^{d} w_{ni}\psi(x_i, c_{ni})\right] \tag{1}$$

where $x=(x_1,\dots x_d)\in D\subset \mathbf{R}^d$, D is a closed bounded set in $\mathbf{R}^d$, without loss of generality taken as the standard unit cube $D=\{x=(x_1,\dots x_d)\mid 0\le x_1\le 1,\dots,0\le x_d\le 1\}$.

The computational (and analytical) model $\tilde{f}$ approximates an unknown function *f*, defined and continuous on *D*. The parameter $a = (a_1,...a_N)$, *the external parameter*, and the parameters $w_n = (w_{n1},...w_{nd})$ and $c_n = (c_{n1},...c_{nd})$, *the internal parameters*, are adjustable on the data, as well as the number of nodes *N*. The univariate function *g* is called the *external* or *activation function*. The univariate functions $\psi_i, i = 1,...d$ are called the *internal functions*. In the current state of this field they are of the same shape for all functions *f* from the class of functions, defined and continuous on *D*, in approximations such as nonlinear perceptron, RBF nets or wavelets.

The multitude of computational models reflects the fact: none of them can be uniformly better than all other models. For example, use of homogeneous basis functions inevitably leads to inefficiency for some applications. It should be noticed as well, that the Kolmogorov's superposition theorem, which gives the most theoretically efficient representation of a multivariate continuous function through superpositions and sums of univariate functions [9], requires external function dependent on data. Although this theorem suggests that data-independent internal functions exist (for a given dimension of data) their implementation is a task of extremely high complexity. Therefore, the nets with adaptive shape external and internal functions might address the demand for parsimonious and accurate adaptive approximation architectures. The idea of adaptive shape univariate external and internal representations closely interlaces with the idea of the Functional-Link net [10] that appropriately chosen basis functions can be combined linearly for efficient learning and generalization.

Having these facts in mind we have suggested and successfully applied the ensemble approach (EA) for learning and generalization [11], [12] for some tasks. The EA uses a mathematical model which is more general than (1):

$$\tilde{f}(x) = \sum_{n=1}^{N} a_n g\left[\varphi(x, w_n, c_n)\right] \tag{2}$$

where *g* is a univariate external function, φ is a multivariate internal representation, a, w_n, c_n are adjustable parameters. One of the features of the EA is that it has a finite but expandable set of external functions (currently containing logistic function, hyperbolic tangent, Gaussian, second derivative of Gaussian, thin plate function and cube) and a set of internal representations (currently nonlinear perceptron, RBF, and product of univariate neurons). This feature gives an opportunity to adjust not only the parameters but the type of basis functions for a particular application. The architecture of the EA is described in section 2.

Recently we have suggested and tested, both on mathematical examples and applications, two nets with new internal representations: the net with complex weights (CWN) and the net for learning Lennard-Jones potentials (LJ). These nets were included in the general architecture of the EA. The presentation of the CWN and LJ are given in sections 3 and 4 respectively.

Currently the type of basis function is discretely and manually adjusted, but we will describe in section 5 our first experiments with automatic and continuous adjustment mode.

In section 6 selective application and mathematical examples of learning and generalization are given.

In section 7 we present a new method for visualization of multivariate data in a lower dimension. In a number of examples with simulated data of different geometric structure this method has shown uniformly better performance than five other known methods of visualization. This method seems to be helpful in the preprocessing stage of different learning procedures.

Future work is addressed in section 8.

2 The Ensemble Approach

EA basic ideas. EA is a method for training and generalization the nets described by equation (2). Unlike the gradient methods of optimization for adjusting parameters of the model the task of optimization is divided into two stages in EA: the recursive linear regression (RLR) [13] and the adaptive stochastic optimization (ASO) [12]. The ensemble of nets with randomly chosen internal parameters is generated. For each net from the ensemble the values of the external parameters are optimized by RLR. The optimization of the internal parameters is made through the stochastic search over the ensemble. Thus, the two stages of optimization in EA, both global, are RLR and the stochastic search.

However, the simple stochastic search [14] is a slow procedure. That is why we replaced them by the adaptive stochastic optimization (ASO). In ASO the ensemble of internal parameters, generated by adaptive random generator (ARG), is divided into a number of portions. The distribution of each univariate component of an internal parameter in each portion is learned using current information about nets' performance in the previous portions of the ensemble. Starting with the uniform distribution of the internal parameters in the first portion, we correct the distribution in subsequent portions on criteria of the minimal training error. This is shown in Fig. 1.

Different modes of EA. EA can operate in sequential or non-sequential, local or non-local modes. In sequential mode the training is performed one node at a time. It starts with a simplest net $f_0(x)=a_0$ and calculates the optimal value of a_0. Suppose the optimal net (the best net in the ensemble) with n - 1 nodes has been built

$$f_{n-1}(x)=a_0+\sum_{i=1}^{n-1}a_i g\left[\varphi(x,w_i,c_i)\right]. \quad (3)$$

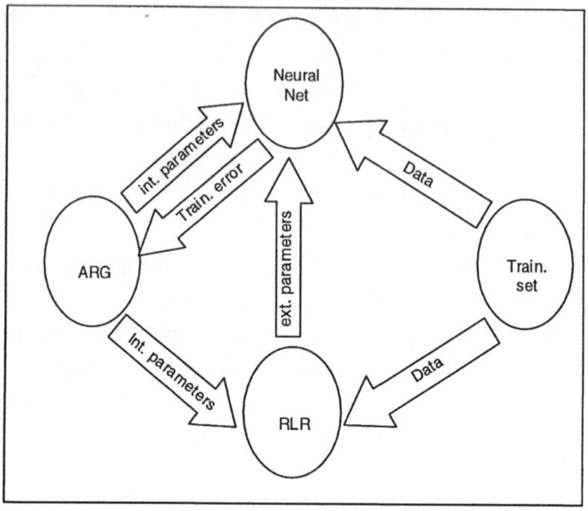

Fig. 1. Schematic representation of the EA

In building the optimal net with *n* nodes

$$f_n(x) = a_0 + \sum_{i=1}^{n} a_i g\left[\varphi(x, w_i, c_i)\right] \tag{4}$$

the internal parameters $w_i, c_i, i = 1,...n-1$ retain their values from the previous step and only the internal parameters w_n, c_n are optimized. Thus, in this mode the ensemble contains the internal parameters of only one node. However, the external parameters $a_0, a_1,...a_{n-1}$ in (4) are different from those in (3) because they are recalculated by RLR. The use of RLR is especially efficient in this mode. The essential decrease of the size of searching space makes the sequential procedure faster than non-sequential one. The same reason makes the sequential procedure theoretically less accurate. The justification for using sequential mode lies in the following result (proved for neural nets with sigmoidal activation functions only) [15].

The upper bound of the training error can be achieved by an iterative sequence of approximations of the form

$$f_n(x) = \alpha_n f_{n-1}(x) + a_n g\left[\varphi(x, w_n, c_n)\right] \tag{5}$$

Thus, even with more restrictions on the search space a near-optimal accuracy can be achieved. We, however, have made a practical correction to this theoretical result because of its asymptotic nature.

The non-sequential mode assumes learning the internal parameters of all nodes simultaneously. Theoretically speaking it has an advantage over the sequential mode in accuracy. Practically, however, this advantage can be implemented only for nets with a small number of nodes. In particular we have recently used the non-sequential mode for learning Lennard-Jones potentials in a multi-atom system [16]. This problem can be solved using relatively small number of nodes.

The non-local procedure is the standard one when one net, trained on the whole training set, is used for prediction for all patterns in the testing set. The local net builds separate net for each testing pattern by training only on a subset of training set, consisting of K nearest neighbors to the testing pattern. In the problems where time of testing is not crucial the local mode may give more accurate results in prediction than non-local one. In particular we have used local net in the formers-non-formers problem with advantage. We, however, recommend use of local net with caution, because it destroys continuity of the mapping on the whole input space. Local net is not appropriate, as well, if one intends to make use of the net as an analytical model.

Different external and internal functions. The basic types of internal representation are: traditional nonlinear perceptron (P), RBF (R), and product of univariate neurons (U), as shown in Fig. 2, with expandable list of external functions. The user of EA has an opportunity to manually choose the type of architecture.

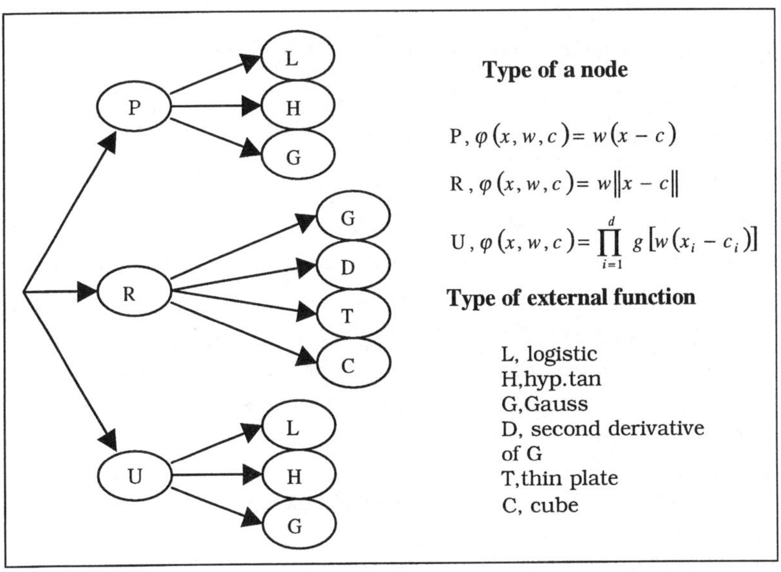

Fig. 2. Basic nets in EA

Recently we added two new types of architecture: a net with complex coefficients (CWN) and a net for learning Lennard-Jones potentials (LJ). These architectures are shown in Fig 3. CWN uses same set of external functions as RBF, LJ uses a special type of a node described in section 4.

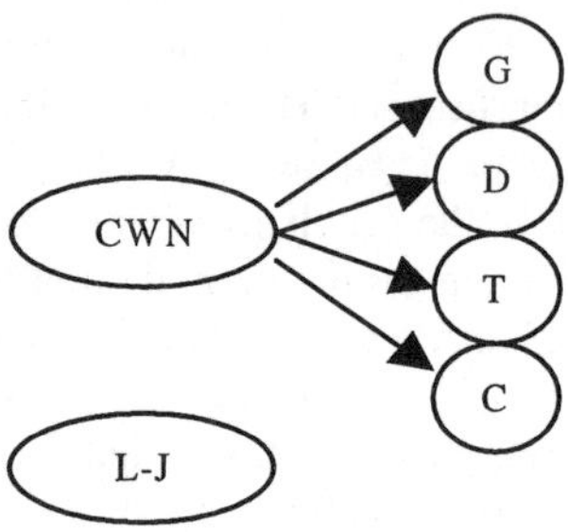

Type of a node

$$\text{CWN},\ \varphi(x,w,c,\theta)=w\sqrt{\sum_{j=1}^{d} e^{i\theta_j}(x_j-c_j)\sum_{j=1}^{d} e^{-i\theta_j}(x_j-c_j)}$$

$$\text{L-J},\ \varphi(r,c,a)=\sum_{i=1}^{M} a_i r^{-c_i}$$

Fig. 3. Additional nets in EA

3 The Net with Complex Weights (CWN)

The CWN computational model [17] is of the following form

$$\tilde{f}(x)=a_0+\sum_{n=1}^{N} a_n g[<w_n,x-c_n>\cdot<\overline{w}_n,x-c_n>], \tag{6}$$

where the parameters a_n are real numbers, c_n are real vectors, while the parameters w_n are complex vectors. In (6) $\overline{w}_n$ stands for complex conjugate of w_n and $<w_n,x-c_n>$ for inner product of two vectors w_n and $x-c_n$. Unlike the neural, or RBF net, or Kolmogorov's net, the internal representation in a basis function is not a weighted sum of univariate functions but constitutes a quadratic function of d variables with *cross-product* terms. These high-order terms are introduced in a parsimonious way. Instead of $d(d+1)/2+1$ parameters for general quadratic function only $2d+1$ parameters are used.

The use of CWN is justified by its universal approximation capability and efficiency in solving important classification tasks.

The neural networks use with complex parameters are also described in [18]. Unlike [17], the authors of [18] make use of complex analytic and non-analytic activation functions and the architecture of nonlinear multilayer perceptron and their method of training is different from the EA. However, they obtained similar results. Namely, the universal approximation capability of nets with complex parameters, savings in computation time, and improved efficiency, were demonstrated in different applications compared with nets with real parameters.

The initial incentive for considering CWN was its possible implementation with quantum devices. However, currently the quantum device and quantum integrated circuit technologies are not sufficiently developed to enable implementation of the CWN architecture in hardware.

4 The Net for Learning Lennard-Jones Potentials

The LJ net has been especially built for a solution of the following problem. The energy of interaction E between two atoms with the distance r between them can be described through the Lennard-Jones potentials as

$$E = a_1 r^{-c_1} + a_2 r^{-c_2} .$$

For this simple system the values of the parameters a_1, a_2, c_1, c_2 are known. For those values of the parameters the system of two atoms has one stable state and, therefore, the energy as a function of distance has one minimum. We considered the system with many atoms of two types, α and β, with one and more than one stable state. For simplicity of notations consider here only the system with one stable state. A net can represent the energy of this system as

$$\tilde{f}(r,c,) = a_1 \sum_{i=1}^{M_{\alpha\beta}} r_i^{-c_1} + a_2 \sum_{i=1}^{M_{\alpha\beta}} r_i^{-c_2} + a_3 \sum_{i=M_{\alpha\beta}+1}^{M_{\alpha\beta}+M_\alpha} r_i^{-c_3} + a_4 \sum_{i=M_{\alpha\beta}+1}^{M_{\alpha\beta}+M_\alpha} r_i^{-c_4} + a_5 \sum_{i=M_{\alpha\beta}+M_\alpha+1}^{M_{\alpha\beta}+M_\alpha+M_\beta} r_i^{-c_5} + a_6 \sum_{i=M_{\alpha\beta}+M_\alpha+1}^{M_{\alpha\beta}+M_\alpha+M_\beta} r_i^{-c_6} \tag{7}$$

where $M_{\alpha\beta}, M_\alpha, M_\beta$ are the number of pairs of type $\alpha\beta, \alpha, \beta$ respectively, r_i is the distance between two atoms from the i-th pair. The training data is a set of vectors $(r_1, r_2, ...r_M, E)$ and the task is to evaluate the parameters $a_i, c_i, i = 1,...6$. The specific of this task is that, although the number of inputs can be large, the number of nodes is limited to 6. This circumstance allowed using non-sequential mode with the advantage in accuracy. In general case, when the number of stable states is not known, the number of nodes is not limited. Therefore, the non-sequential mode may lose that advantage.

5 The Automatic and Continuous Adjustment of the Shape of Internal Representation

A first step was made in the direction of automatic and continuous mode of choosing the internal representation in the net architecture. Our experiments have shown that combining subsets of nodes of very different nature in one net not only does not improve performance of the resulting net but can drastically decrease it. To illustrate the direction we are moving now consider a simple example with the nonlinear perceptron and RBF structure of a net node. Both can be written in a unified form

$$nodeOutput = g\left[\sum_{i=1}^{d} w_i\varphi\left(x_i - c_i\right)\right],$$
$$\varphi(x) = \begin{cases} x, & \text{for NLP} \\ x^2, & \text{for RBF} \end{cases} \tag{8}$$

Considering two functions φ_1 and φ_2, where $\varphi_1(x) = x,\ \varphi_2(x) = x^2$, as two points in the functional space $L_\infty[0,1]$ we can define a subset of points (functions) in this space which contains φ_1 and φ_2 and a set of intermediate points described by one continuously changing parameter An arbitrary point in this subset has the equation

$$\varphi(x,\Delta) = 4\Delta x^2 + (1-4\Delta)x, \tag{9}$$

where

$$0 \le \Delta \le 0.25. \tag{10}$$

The 4 plots corresponding equation (9) are shown in Fig. 4 with values of $\Delta = 0, 0.0833, 0.1666,$ and 0.25 respectively (from top to buttom).

Thus, introducing only one additional adjustable parameter Δ we were able to adaptively adjust the shape of internal representation and obtain a net with nodes of variable shape. Applying this net to formers-nonformers problem, described in the next section, we were able to decrease the error from 0.5% to 0.3%, a substantial achievement for this task. It is possible in this manner to introduce more variety in the shape of internal representation by the use of the Lagrange interpolating polynomials of higher degree. If, for example, we choose the Lagrange polynomial of 3 degree then by continuous change of 2 parameters a continuum of polynomial type shapes including strait line, parabola, and cubic polynomial curve will be obtained.

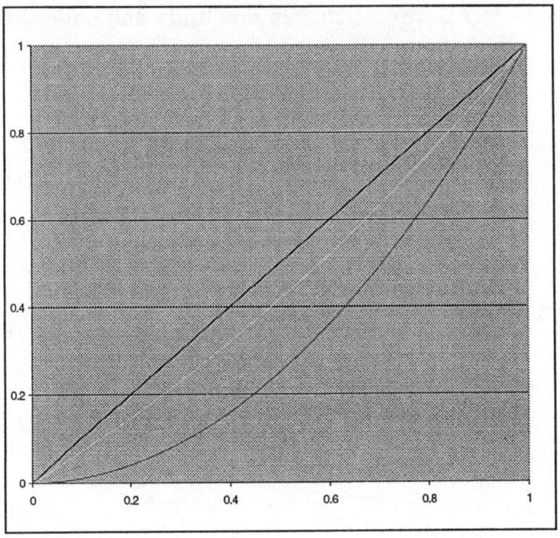

Fig. 4. Four curves from the continuum of curves between straight line and parabola

Thus, introducing only one additional adjustable parameter Δ we were able to adaptively adjust the shape of internal representation and obtain a net with nodes of variable shape. Applying this net to formers-nonformers problem, described in the next section, we were able to decrease the error from 0.5% to 0.3%, a substantial achievement for this task. It is possible in this manner to introduce more variety in the shape of internal representation by the use of the Lagrange interpolating polynomials of higher degree. If, for example, we choose the Lagrange polynomial of 3 degree then by continuous change of 2 parameters a continuum of polynomial type shapes including strait line, parabola, and cubic polynomial curve will be obtained.

6 Selective Application and Mathematical Examples

A number of applications of techniques described above as well as mathematical examples were reported in [12] and [17]. In this paper we selectively describe 2 examples demonstrating capabilities of the CWN architecture.

Example 1. Two helixes. We consider a difficult task to discern the data placed on two helixes close to each other. P = 10000 points are randomly placed on two helixes. The coordinates of these points are calculated by the equations

$$x = R\cos t,\ y = R\sin t,\ z = Vt$$
$$x = R\cos t,\ y = R\sin t,\ z = V(t+\pi),$$

where the parameter t, $0 \le t \le 2\pi N$ is chosen randomly and uniformly. This is shown in Fig. 5

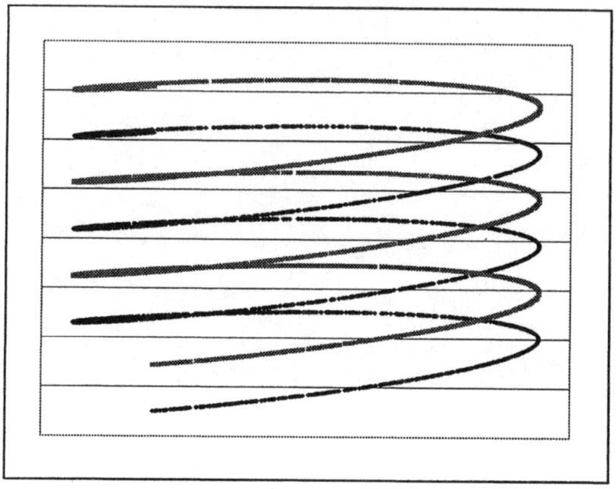

Fig. 5. Data on 2 helixes

We compare performance of two neural nets, CWN and RBF. The outputs of the nets take values 0 or 1 depending on the helix where the point, with coordinates equals to the input, is placed. The number of loops N equals to 12 or 15, $V = 1$, and $R = 8$ or 12, 2500 patterns of the data were used for testing. The algorithm for training is local sequential. The parameter *NB* is the maximum number of nearest neighbors. For each testing pattern a net with recursively increasing number of nodes is trained on the set of *NB* nearest neighbors belonging to the training set (7500 patterns). The training stops if the training error becomes less than *THRESHOLD* (which was 0.05) or the number of nodes becomes larger than *NB*. The complexity of task, as a rule, will increase with *NB* increasing in the range of parameters we used. The typical results of experiments are shown in Fig.6

The comparison between CWN (thick curves) and RBF are made in accuracy in testing and time for training and testing with varying NB. All graphs have demonstrated that CWN is superior than RBF, both in accuracy and time, when the number of nearest neighbors becomes larger than some value of NB. That means that CWN is superior for more difficult versions of this task. Then the task is easier and can be solved with small and approximately same average number of nodes the advantage of RBF over CWN in computational time for one node plays the major role. More easier tasks can be solved sometimes more accurately and efficiently by RBF.

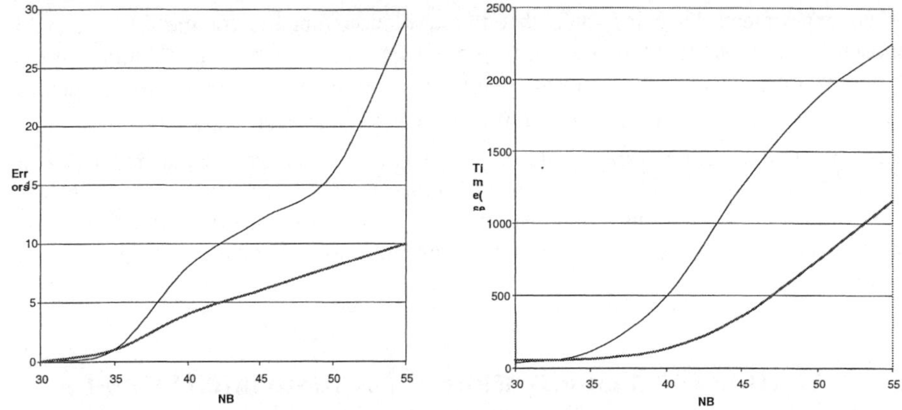

Fig. 6. Error and time versus NB. Case $R = 12$, $N = 12$

Example 2. *The formers - non-formers problem.* A body of data constitutes 6358 patterns of ternary systems (systems of 3 chemical elements) with 15 features of the elements in the system, 5 for each element. These are Zunger radius, valence, melting temperature, Mendeleev number, and electrical negativity. For each system it is known whether it can or can not form a compound. This information is available through long and expensive experimentation and lengthy calculations. The task is to build a neural net, which can accurately predict possible formation of a compound for a new system, not available in the database. It was found empirically [28], [29] that the Mendeleev number is superior to other features in this task. The comparison between CWN net and the RBF net in accuracy was made using only Mendeleev numbers as inputs for two nets, both using the ensemble approach for learning and generalization. Different modes of training and testing were tried, non-local sequential (NLS), non-local non-sequential (NLNS), and local sequential (LS) with 40 nearest neighbors. The results of testing on the subset of data consisting of 1589 patterns (4769 patterns were used as a training set) are shown in Table 1. CWN demonstrates an obvious superiority over RBF net.

Table 1. The results of testing

Node type	Mode type	# of misclassifications	Accuracy in %
CWN	NLS	48	97.0
RBF	NLS	60	96.3
CWN	NLNS	60	96.3
RBF	NLNS	72	95.5
CWN	LS	15	99.1
RBF	LS	40	97.5

Our experiments have indicated that the activation function for the CWN can be chosen from the same set of functions as for RBF net. In particular the minimum of generalization error is achieved with the "thin plate" activation function $f(t)=t^2\log(t)$. Finding out that CWN + LS is the best choice we continued experimentation adding other features to Mendeleev number. Better results were obtained only using all 5 properties (15 inputs). 99.5% accuracy in testing was achieved for the same training and testing sets. As was mentioned in section 5 a further improvement of performance was made through use of adaptive shape internal representation.

7 A Method for Visualization of Multivariate Data in a Lower Dimension

To support described above methods of learning and generalization in preprocessing we developed a new method of visualization of multivariate data in a lower dimension (primarily in 2D). This method is related to the group of methods constructing a mapping $y=f(x,W), x\in\mathbf{R}^D, y\in\mathbf{R}^d$ from a high-dimensional space $\mathbf{R}^D$ to a lower-dimensional space $\mathbf{R}^d$. The mapping is parameterized by the vector-parameter *W*, is taken in the form of a net, and is adjusted on the set of data according to explicit topological criteria. The examples of these methods are the Variance Constrain (VC) and the Ratio Distance Constrain (RC) methods [19], [20]. In another group of methods the criteria for training the mapping has no explicit topological meaning. The examples of these are well-known the Karhunen-Loeve transformation (KL), the Self-Organizing Map (SOM), and the Auto-Associative net (AA). In the first group of methods a topological criteria takes a form of preserving in $\mathbf{R}^d$ the distances between any two points in $\mathbf{R}^D$. However this task is infeasible since the number of adjustable parameters for building the mapping does not exceed *dP*, while the number of distances to preserve equals to $P(P-1)/2$ (*P* is the number of data points), essentially greater number. Therefore, one should concentrate on preserving only the most important distances. These are the nearest neighbors for any given data point. However introducing the nearest neighbors in the objective function for optimization of the parameters of the mapping makes this function non-differentiable, excluding use of efficient gradient methods of optimization. Taking into account all these difficulties we suggested another way of introducing the order of distances into the objective function. Let write down the objective function for a pair (x_p, x_q) in the following form

$$e_{pq}(W)=\left(\left\|y(x_p,W)-y(x_q,W)\right\|-\left\|x_p-x_q\right\|\right)^2 * \\ \left(1-G\left[\mu_p\left\|y(x_p,W)-y(x_q,W)\right\|\right]\right). \tag{11}$$

The second factor in the right hand side of the equation (11) is called a filter, G is a monotonically increasing function with values between 0 and 1. In particular we have chosen $G(x)=1/(1+e^{-x})$. The overall objective function is defined as

$$e(W)=\frac{2}{P(P-1)}\sum_{p=1}^{P}\sum_{q,q\neq p}e_{pq}(W). \tag{12}$$

Thus, the function $1-G(x)$ is decreasing with x increasing, and $1-G(x)$ is approaching zero when x is approaching infinity. It plays a role of a filter that cut out the less important square distance differences (first factor in the right hand side of formula (11)) in the objective function, making the task of optimization more feasible. The factor μ_p in the same formula changes the slope of the curve allowing a regulating the degree of suppressing these less important distances. Note that the filter works automatically without sorting out the distances in the output space.

The algorithm of training is basically the gradient descent with the initial approximation estimated by the adaptive stochastic optimization step from the EA. The details of the training algorithm will be described elsewhere.

This method of visualization was favorably compared with five other methods the KL, SOM, AA, VC, and RC mentioned above on a number of examples, including mathematical examples and real data of dimensions from 3 to 12. This data was visualizes in 2D. In the following two illustrative examples we describe and illustrate the original data in 3D and then give 6 pictures of visualization by the KL, SOM, AA, VC, and RC and by our method which we call Distance Conservation with Filtering (DCF). For first five methods we used the package developed by Zhou Meng, former PhD student of Y.-H. Pao, for DCF we used EXCEL.

Example 1. The data constitutes 1000 points on two helixes and was described in the example1, section 6. Visualization of the data by 4 methods is given in Fig. 7

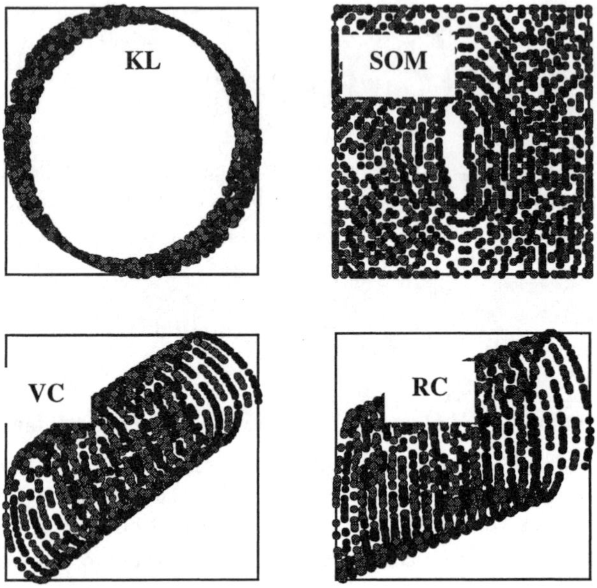

Fig. 7. Visualization of data by KL, SOM, VC, and RC

Visualization of data by AA and DCF is given in Fig. 8

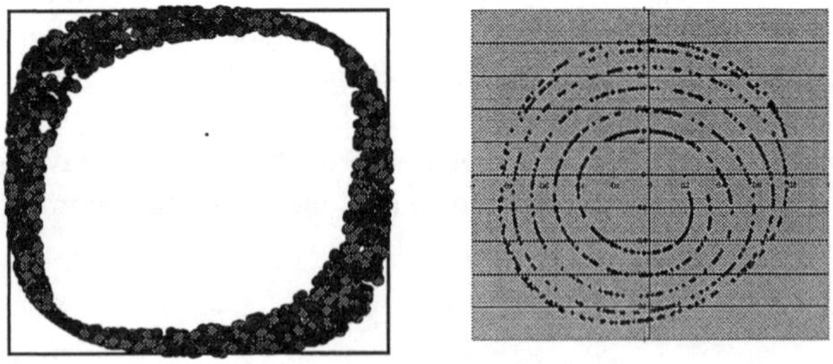

Fig. 8. Visualization of data by AA (left) and DCF (right)

Example 2. Data (1000 points) on the surface of hyperbolic paraboloid $z = x^2 - y^2$.
Visualization of data by all 6 methods is shown in Fig. 9

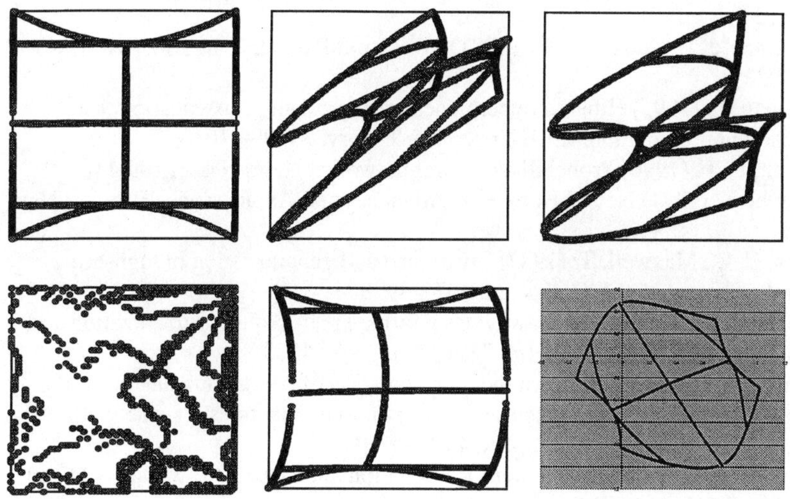

Fig. 9. Visualization of data by 6 methods KL, VC, RC, SOM, AA, and DCF (from left to right and from top to bottom).

Obviously DCF performed better than 5 other methods in both examples.

8 Conclusion and Future Work

The new architectures of adaptive computational schemes, described in this paper, have a solid motivation and have demonstrated a visible advantage on some other known schemes in performance and efficiency in a number of applications. Our future work will be concentrated both on applications of these architectures and on theoretical development of a new, completely adaptive architecture.

References

1. Haykin S. (1994) Neural Networks. A Comprehensive Foundation. Macmillan, New York
2. Bishop C. M. (1995) Neural Networks for Pattern Recognition. Clarendon, Oxford
3. Friedman, J. H., Stueltzle, W. (1981) Projection Pursuit. Jour. Am. Stat. Assos. **76**, 817-823
4. Breiman, L. (1993) Hinging hyperplanes for regression, classification, and function approximation. IEEE Trans. Inf. Theory. **39**, 999-1013
5. Specht, D. F. (1990) Probabilistic neural networks. Neur. Net. **3**, 109-118
6. Gelenbe E. (1991) Neural Networks: Advances and Applications, Elsevier, New York
7. Giles, C. L., Maxwell, T. (1987) Invariance and generalization in high-order neural networks. Appl. Optics. **26**, 4972-4978
8. Daubechies, I. (1990) The wavelet transform, Time-frequency localization and signal analysis. IEEE Trans. Inf. Theory. **36**,. 961-1005
9. Kolmogorov, A. N. (1957) On the representation of continuous functions of many variables by superposition of continuous functions of one variable and addition. Dokl. Akad. Nauk SSSR. **114**, 953-956
10. Pao Y-H. (1989) Adaptive Pattern Recognition and Neural Networks. Addison-Wesley, Reading, MA
11. Igelnik B., Pao, Y.-H. (1995) Stochastic choice of basis functions and adaptive function approximation. IEEE Trans. Neur. Net. **6**, 1320-1329
12. Igelnik, B., Pao, Y.-H. et al. (1999) The ensemble approach to neural network learning and generalization. IEEE Trans. Neur. Net. **10**, 19-30
13. Albert A. (1972) Regression and the Moore-Penrose Pseudoinverse. Academic Press, New York
14. Kalos M., Witlock P. A. (1986) Monte Carlo Methods. Wiley, New York
15. Barron, A., R. (1993) Universal approximation bounds for superpositions of a sigmoidal function. IEEE Trans. Inf. Theory. **39**, 930-945
16. Wert C. A., Thomson R. M. (1970) Physics of Solids, McGraw-Hill, New York
17. Igelnik, B., Tabib-Azar, M. et al. (1999) A net with complex weights. Submitted to IEEE Trans. Neur. Net.
18. Arena P., Fortuna L. et al. (1998) Neural Networks in Multidimensional Domains. Springer, New York
19. Pao, Y.-H., Shen C. Y. (1997) Visualization of Pattern Data through Learning of Nonlinear Variance-Constrained Dimension-Reduction Mapping. Pattern Recognition, **30**, 1705-1717
20. Pao, Y.-H., Meng, Z. (1998) Visualization and the Understanding of Multidimensional Data. Engineering Applications of Artificial Intelligence, **11**, 659-668

Generalized Radial-Basis Function Neural Networks for Data Mining of Symbolic Data

S.Papadimitriou[1,2], A.Bezerianos[1], L.Vladutu[1], S.Mavroudi[1], G. Pavlides[2]

[1] Dept. of Medical Physics, School of Medicine, Univ of Patras, 26500 Patras, Greece, *email:* bezer@patreas.upatras.gr

[2] Dept. of Comp. Eng. and Informatics, Univ of Patras,26500 Patras, Greece, *email:* stergios@heart.med.upatras.gr

Abstract:The application of neural networks to domains involving prediction and classification of symbolic data requires a reconsideration and a careful definition of the concept of distance between patterns. This distance in addition to provide information about the proximity of patterns should also obey to some mathematical criteria in order to be applicable. Traditional distances are inadequate to access the differences between the symbolic patterns. This work proposes the utilization of a statistically extracted distance measure at the context of the Generalized Radial Basis Function networks. The regularization potential of the Generalized Radial Basis Function networks can be realized with this type of distance. The obtained performances are significantly better than the traditional nearest neighbour schemes in many data mining problems.

Keywords: Neural Network Learning, Data Mining, Symbolic Data Classification, Radial Basis Functions

1 Introduction

The emergence of neural network technology [1,2,3,4] offers valuable insight to confront complicated data mining problems. However, the application of neural networks although it was proved very successful in problem domains, as signal processing and pattern recognition has not yielded adequate performances in the domain of data mining of symbolic data. Patterns arising both from commercial databases and from many engineering databases (as those that describe biosequences) involve data defined over a space that lacks the fundamental properties of distance metric spaces.

This work proposes a distance metric for expressing the distance between values of features in symbolic domains. This metric owns some geometric properties that make it effective at the context of the regularization formulation of the Generalized Radial Basis Function (GRBF) networks. Regularization techniques impose the learning of a smooth functional from the network [2,3,4,6]. Therefore, it is justifiable to expect from the network to be able of learning the underlying smooth dependence of the outcomes on the attributes, despite the presence of noise that induces the perturbation. The potential of this distance metric to regularize the solution of the GRBF networks is the theoretical justification of the improved performance related to the simple nearest neighbor schemes.

The paper proceeds as follows: Section 1 reviews the Generalized Radial Basis Function Networks (GRBFs). Section 2 presents the proposed distance metric that has been proved quite effective. Section 3 discusses how the statistical distance is fitted at the context of GRBF networks. Section 4 presents some results from the application of the proposed GRBF design in practical problems. Finally, the conclusions are presented along with directions for future work.

2 Generalized Radial Basis Function Networks

The Generalized Radial Basis Function (GRBF) networks explore the Tikhonov's regularization theory for obtaining generalization performance. They try to obtain a tradeoff between a term that measures the fitness of the solution to the training set and one that evaluates the smoothness of the solution. Denoting $\mathbf{x}_i$, d_i, $\mathbf{F}(\mathbf{x}_i)$ the input vectors, the desired responses and the corresponding realizations of the network respectively this tradeoff can be formulated with a cost function as [1, 2] :

$$C(F) = C_s(F) + \lambda C_r(F) \tag{1}$$

The $C_s(F)$ is the standard error term that accounts for the fitting to the training set in the least squares error sense, λ is a positive real number called the *regularization parameter* and $C_r(F)$ is the regularized term that favors the smoothness of the solution. Techniques for designing GRBF networks for optimizing the performance of applications of diverse fields are described in [9,10,11]. However, these designs are based on the usual continuous numeric distance metric and cannot directly access the requirements of the symbolic data.

3 The Statistical Distance Metric (SDM)

The key problem for applications involving symbolic features is the definition of the distance metric. In domains where features are numeric, it is straghtforward to compute the distance between two points in the pattern space in terms of a geometric distance (e.g. Euclidean, Manhatan). Indeed, the traditional RBF learning algorithms have been formulated and operate effectively in numeric domains with such distances. However, when the features are symbolic (as is usually in bioinformatics and in commercial data mining applications) the utilization of the traditional types of distances yields inadequate performance. Two common approaches for handling symbolic information is the *overlap* method and the *orthogonal* representation [5, 7]. However, these distance metrics oversimplify the pattern space, ignoring all the information presented with the training set and therefore they yield (as expected) poor performance.

In order to be able to obtain an effective formulation of the distances between patterns with symbolic feature values we have adapted the distance measure proposed in [7]. This statistical distance measure takes into account the overall similarity of classification of all instances for each possible value of each feature. This method extracts with a statistical approach from the training set, a matrix that defines the distances between all possible values of a given feature. Therefore, a separate matrix for each feature is obtained. The distance measure for a specific feature is defined according to the following equation:

$$d(V_A, V_B) = \sum_{i=1}^{N} \left| \frac{C_{A_i}}{C_A} - \frac{C_{B_i}}{C_B} \right|^k \tag{2}$$

In the equation above, V_A and V_B denote two possible values for the feature, e.g. for the DNA promoter data they will be two nucleotides. The distance between the values is the sum over all the N classes. For example, for the DNA promoter example (discussed below) there are two classes, either the sequence is a promoter (i.e. a sequence that initiates a process called transcription) or not. The number of patterns for which the value V_A (V_B) is classified to class i, is denoted by C_{A_i} (C_{B_i}). Also, the total number of patterns of class *A* (*B*) is denoted by C_A (C_B), and k is a constant usually set to 1. These counts are computed over all patterns of the training set. It becomes easily evident that the more correlated are the classifications of patterns pertaining to two values for a feature the smallest is their statistical distance computed with equation. Therefore for feature values belonging to training set patterns with similar classifications a small statistical distance will be computed. Equation accounts for the overall similarity of classification of all training instances for each possible class. The distance $D(X,Y)$ between two patterns X,Y is obtained by a weighted sum of distances between the values of the individual features of these patterns (obtained from equation).

4 Generalized RBFs with the Statistical Distance Metric (SDM)

The SDM was used as a distance between patterns in the context of GRBF networks. One prerequisite for the application of this distance type is to have enough training data for the accurate construction of the SDM space. In contrast to example based nearest neighbor learning schemes, the GRBF learns a smooth functional that weights the contribution of each example subject to the requirements imposed by the regularizing term for the smoothness of the solution. This fact is the theoretical explanation for the superior performance of GRBF networks related to the Instance Based Learning (IBL) schemes. At the performed simulation experiments we have validated that the GRBF can improve the results over simple nearest neighbor schemes that are also based on a similar formulation of the distance metric. Specifically, we have tested also the PEBLS system [12], that is an exemplar based learning system based on a similar type of statistical distance. The performance obtained with the PEBLS system depends on the specification of many parameters (e.g. number of neighbors considered, the voting scheme, the heuristics used to adjust the weight of each example). On the contrary, the GRBFs can be trained with a formally defined procedure that involves a few heuristics for the selection of the network parameters that do not affect in general significantly the network performance. Also, the GRBFs use a formal basis to obtain generalization performance based on the Tikhonov's regularization theory. The improvement of GRBFs over the PEBLS system has been verified in most of the performed experiments.

A parameter of particular importance is the region of influence of the Gaussian Radial Basis Function kernels that is determined by their spread parameter. This problem becomes more complicated within the domain of statistical distances and the heuristic suggestions of [1] to compute the spread σ as $\sigma = d_{\max} / \sqrt{2m}$ where $d_{\max}$ is the maximum distance and m the number of RBF centers has not been proved effective in practice. In order to obtain an effective setting for the spread parameter, a sensible approach is to obtain at a first step an estimate of the average distance d_{av}, of patterns within the space defined with the SDM. Then the region of influence of the RBF kernels is designed by requiring that at a particular distance *Spread* from the RBF center expressed in units of d_{av}, the attenuation of influence is decreased by a. Mathematically, this requirement is formulated as: $\exp(-DF \cdot d_{av} \cdot Spread) = a$ and therefore the required parameter DF is derived as $DF = -\log(a)/(Spread \cdot d_{av})$. Values of these parameters that realize good results are for example *Spread=5* and *a=0.01,* meaning that at a distance from an RBF center 5 times larger than the average distance between patterns, the influence of the RBF function attenuates with a factor of 0.01.

Statistical distances like the proposed one have been used in the past at the context of nearest neighbor classifiers as the PEBLS system [12]. These systems can cope well with symbolic attributes, but for the numeric ones they perform at a first stage discretization of the numeric ranges. Thereafter, they apply the distance extraction algorithm on the discretized numeric attributes, that with the discretization behave like the symbolic ones. On the other hand the GRBF networks have the inherent ability to handle numeric attributes effectively. Consequently, by using a heterogeneous type of distance that for symbolic features is the Statistical Distance Metric defined with (2) and for the numeric is a traditional distance metric (e.g. Euclidean, Manhatan) the GRBFs can cope well with any feature type. This ability is an advantage of GRBFs over machine learning systems capable to handle efficiently only symbols of symbolic type as the ID3 and the AQ [13].

5 Applications

We have applied the GRBF based solutions to a variety of data mining problems both from the engineering domain and from the commercial databases domain. The first application concerns the prediction of promoter sequences [5]. This task involves predicting whether or not a given subsequence of a DNA sequence is a promoter, i.e. a sequence of genes that initiates a process called transcription. The data set contains 106 examples, 53 of which were positive examples (promoters) and the rest negative ones. A training pattern consists of a sequence of 57 nucleotides (features) from the alphabet *a, c, g* and *t* with the respective classification (promoter or not promoter). Since the available number of patterns were small the classification performance was tested with the leave-one-out methodology, i.e. repeatedly trials have been performed by training on 105 examples and testing on the remaining one. The computed performance was 2/106 (i.e. an average of 2 errors over 106 trials) versus 4/106 for a competitive experiment that used the KBANN neural network model [8].

Another application concerns a different domain: the data mining of commercial databases. A large training set is extracted from a database kept by a mobile telecommunications company. This set contains 62000 records concerning some attributes of its customers (e.g. job, area, sex, and pay method) and the classification in terms of their quality as customers. This classification is in six classes, ranging from 0 to 5 in terms of increasing customer quality. The objective for the learning system was to perform well at predicting the class of a new customer from its attributes and therefore to guide the company strategy. For this problem the presented statistical distance with GRBF learning has obtained the best classification performance, around 60% success at the prediction of the customer class. In contrast a nearest neighbourhood classification scheme based on the same distance obtains only around 30% and a Self-Organizing Map (SOM) operating with the traditional distance types with a numerical coding of feature values (it is not easy to adapt the statistical distance at the context of SOM) obtains a performance around 42%.

6 Conclusions

Neural network algorithms for learning are very effective in domains in which all features have numeric values. At these domains, the examples are treated as points and distance metrics obeys to standard definitions. However the usual domain of data mining applications is the symbolic domain. The utilization of the traditional distance metrics for data mining with neural networks usually results in incompetent results. The paper has adapted a SDM for application in the context of the Generalized Radial Basis Function (GRBF) neural networks. This distance metric extends the area of effectiveness of GRBF neural networks to the symbolic world. The results indicate that the generalisation potential of neural networks can be utilised for patterns with symbolic features when the learning and evaluation algorithms are designed with the statistically extracted distance metric.

References

[1] Simon Haykin, *Neural Networks,* Prentice Hall International Inc., Second Edition, 1999.

[2] T. Poggio and F. Girosi. Regularization algorithms for learning that are equivalent to multilayer perceptrons, *Science 247,* pp:978-982, 1990.

[3] T. Poggio, and F. Girosi, "Networks for approximation and learning", *Proceedings of the IEEE,* vol. 78, pp. 1481-1497, 1990.

[4] Vapnik., V. N., *Statistical Learning Theory,* New York, Wiley, 1998.

[5] Pierre Baldi, Soren Brunak, *Bioinformatics,* MIT Press, 1998.

[6] Federico Girosi, "An Equivalence Between Sparse Approximation and Support Vector Machines", *Neural Computation,* 10:6, pp. 1455-1480, 1998.

[7] C. Stanfill, D. Waltz, "Toward memory-based reasoning", *Communications of the ACM,* 29:12 , pp.1213-1228, 1986.

[8] G. Towell, J. Shavlik, M. Noordewier, "Refinement of approximate domain theories by knowledge-based neural networks", *Proceedings Eight National Conference on Artificial Intelligence,* pp. 861-866, Menlo Park, CA:AAAI Press, 1990.

[9] S. Papadimitriou, A. Bezerianos, A. Bountis, "Radial Basis Function Networks as Chaotic Generators for Secure Communication Systems", *International Journal On Bifurcation and Chaos,* Vol. 9, No. 1, pp. 221-232, 1999.

[10] S. Papadimitriou, A. Bezerianos, "Nonlinear analysis of the performance and reliability of wavelet singularity detection based denoising for Doppler ultrasound fetal heart rate signals", *International Journal On Medical Informatics.* 53, pp. 43-60, 1999.

[11] A. Bezerianos, S. Papadimitriou, D. Alexopoulos, "Radial Basis Function Neural Networks for the Characterization of Heart Rate Variability Dynamics", *Artificial Intelligence in Medicine 15,* pp. 215-234, 1999.

[12] S. Cost S Salzberg, "A wighted nearest neighor algorithm for learning with symbolic features", *Machine Learning, 10* (1), pp. 57-78,1993

[13] Miroslav Kubat, Ivan Bratko, Ryszard S. Michalski, "A Review of Machine Learning Methods", In *Machine Learning and Data Mining: Methods and Applications",* Edited by R. S. Michalski, I. Bratko, M. Kubat, Jon Wiley & Sons, pp. 3-69,1998.

Data Generation Using Sensitivity Analysis

A. P. Engelbrecht

Department of Computer Science, University of Pretoria, South Africa

Abstract: This paper presents a new approach to the generation of training data for supervised neural networks. Sensitivity analysis is used to find the most informative regions in input space. Knowledge of these informative regions can be used to remove redundant training patterns and to generate additional training patterns withing the informative regions. Preliminary experimental results show that this apporach to data generation holds much promise.

1 Introduction

One of the major problems when agents need to learn from data is to obtain sufficient representative data for the training set. For many real-world problems it is difficult to obtain enough observations to train agents. The other side of the coin is also true. May real-world applications do have an abundance of training data, of which substantial parts are redundant. Such redundant data increases the time complexity of the training step, and cause worse performance compared to training on the most informative patterns alone [1,2,4]. This paper presents an approach to locate the most influential parts of input space as obtained from the given data. Knowledge of these influential spaces can help in two ways:

- The removal of redundant training patterns such that the agent focusses only on the most informative training patterns [1,2,4].
- The generation of additional data within these influential regions, thereby presenting the network with more information to more accurately learn the given task.

This approach to data reduction and data generation makes use of a sensitivity analysis algorithm to locate decision boundaries formed over input space [3]. The decision boundary algorithm is based on the fact that boundaries are located at the point in input space where a small perturbation to an input value causes a change in output value.

For the purpose of this paper we consider neural network agents with classification tasks only. Only feedforward neural networks with one hidden layer is considered. Extension to more than one hidden layer is straigthforward. Section 2 gives an explanation of the approach to detect decision boundaries. Section 3 explains the approach to locate the most informative regions of input space and how that knowledge can be used to remove redundant training patterns. Preliminary experimental results are given in section 4.

2 Locating Decision Boundaries

Under the assumptions that

Assumption 1: A NN implements a differentiable mapping $\mathcal{F}_{NN}(D_T; W) : \mathrm{R}^I \to \mathrm{R}^K$, where D_T is the training set, W represents the weights, I is the dimension of input space, and K is the dimension of output space.

Assumption 2: The activation functions f_{o_k} in the output layer are monotonically increasing, bounded, and produces a binary discrete output $o_k^{(p)} = f_{o_k}(net_{o_k}^{(p)})$ for each output unit o_k, representing a specific class. $(net_{o_k}^{(p)} = \sum_{j=1}^{J+1} w_{kj} y_j^{(p)}$, where y_{J+1} is the hidden layer bias unit)

the following definitions of decision boundaries are given [3,4]:

Definition 1. Axis-parallel Decision Boundary: If there exists an input parameter value $z_i^{(p)}$ and a small perturbation Δz_i of $z_i^{(p)}$ such that, for any output unit o_k, $f_{o_k}(z_1^{(p)}, \cdots, z_i^{(p)}, \cdots, z_I^{(p)}) \neq f_{o_k}(z_1^{(p)}, \cdots, z_i^{(p)} + \Delta z_i, \cdots, z_I^{p)})$, then a decision boundary is located in the range $[z_i^{(p)}, z_i^{(p)} + \Delta z_i]$ of input parameter z_i, where p denotes a single pattern.

Definition 2. Non-axis-parallel Decision Boundary: If there exist two input parameter values $z_{i_1}^{(p)}$ and $z_{i_2}^{(p)}$ with $z_{i_1}^{(p)} < z_{i_2}^{(p)}$ such that for $z_i^{(p)} \in [z_{i_1}^{(p)}, z_{i_2}^{(p)}]$, a small perturbation Δz_i of $z_i^{(p)}$, and any output unit o_k, $f_{o_k}(z_1^{(p)}, \cdots, z_i^{(p)}, \cdots, z_I^{(p)}) \neq f_{o_k}(z_1^{(p)}, \cdots, z_i^{(p)} + \Delta z_i, \cdots, z_I^{(p)})$, then a decision boundary spans over the range of values $[z_{i_1}^{(p)}, z_{i_2}^{(p)}]$ of input parameter z_i, where p denotes a single pattern.

For the above, if continuous activation functions such as the sigmoid function are used in the hidden and output layers, the class to which a pattern belongs is determined from a rule such as

$$o_k^{(p)} = \begin{cases} 1 \text{ if } f_{o_k}(net_{o_k}^{(p)}) \geq 0.7 \\ 0 \text{ if } f_{o_k}(net_{o_k}^{(p)}) \leq 0.3 \end{cases}$$

which ensures a binary output. To ensure that output values are produced close to 1 or 0, a sigmoid function with large steepness λ is usually used, for example $\lambda \geq 5$ in

$$f_{o_k}(net_{o_k}) = \frac{1}{1 + e^{-\lambda net_{o_k}}}$$

From definition 1, a decision boundary is located at the point in input space where a small perturbation to a value of an input unit causes the value of an output unit to change from one class to another. Similarly, definition 2 defines a range of input parameter values over which a decision boundary is formed.

The approach is justified from a first order Taylor expansion of f_{o_k} around $z_i^{(p)}$. For any differentiable function f_{o_k}, the characteristics of that function under small perturbations Δz_i of its input parameter z_i are expressed by the series

$$f_{o_k}(z_1^{(p)},\cdots,z_i^{(p)}+\Delta z_i,\cdots,z_I^{(p)}) = f_{o_k}(z_1^{(p)},\cdots,z_i^{(p)},\cdots,z_I^{(p)}) + \Delta z_i \frac{\partial f_{o_k}}{\partial z_i^{(p)}} + \frac{1}{2}\Delta z_i^2 \frac{\partial^2 f_{o_k}}{\partial z_i^{(p)2}} + \cdots \quad (1)$$

Under the assumption of small Δz_i,

$$\lim_{\Delta z_i \to 0} (\frac{1}{2}\Delta z_i^2 \frac{\partial^2 f_{o_k}}{\partial z_i^{(p)2}} + \cdots) = 0 \quad (2)$$

and equation (1) can be reduced to

$$f_{o_k}(z_1^{(p)},\cdots,z_i^{(p)}+\Delta z_i,\cdots,z_I^{(p)}) \approx f_{o_k}(z_1^{(p)},\cdots,z_i^{(p)},\cdots,z_I^{(p)}) + \Delta z_i \frac{\partial f_{o_k}}{\partial z_i^{(p)}} \quad (3)$$

The second term in equation (3) determines whether the value of an output unit changes. That is, the higher the value of $\frac{\partial f_{o_k}}{\partial z_i^{(p)}}$, the greater the chance that $f_{o_k}(z_1^{(p)},\cdots,z_i^{(p)},\cdots,z_I^{(p)}) \neq f_{o_k}(z_1,\cdots,z_i^{(p)}+\Delta z_i,\cdots,z_I^{(p)})$. Therefore, patterns with high $\frac{\partial f_{o_k}}{\partial z_i^{(p)}}$ values lie closest to decision boundaries [1,3,4].

Figure 1 illustrates the approach for the *iris* classification problem. This figure plots the sensitivity values for each pattern with respect to the class *Iris Versicolor*. Peaks in the figure illustrates the location of the decision boundaries. These peaks coincides with the boundaries as obtained from rule extraction algorithms [5].

3 Data Pruning and Generation

The first-order sensitivity analysis in section 2 assigns a "measure of closeness" to boundaries. Patterns that lie close to boundaries form the most informative regions of input space, since the decision boundaries are defined through these patterns. This section illustrates an approach to exclude training patterns that are not within these informative regions and to generate new training patterns from these regions.

The most informative patterns are selected after training on all available information (redundant patterns can also be pruned during training as explained in [1,4]), using

$$\Psi(\boldsymbol{\Phi}_\infty) = (1-\beta)\overline{\boldsymbol{\Phi}}_\infty \quad (4)$$

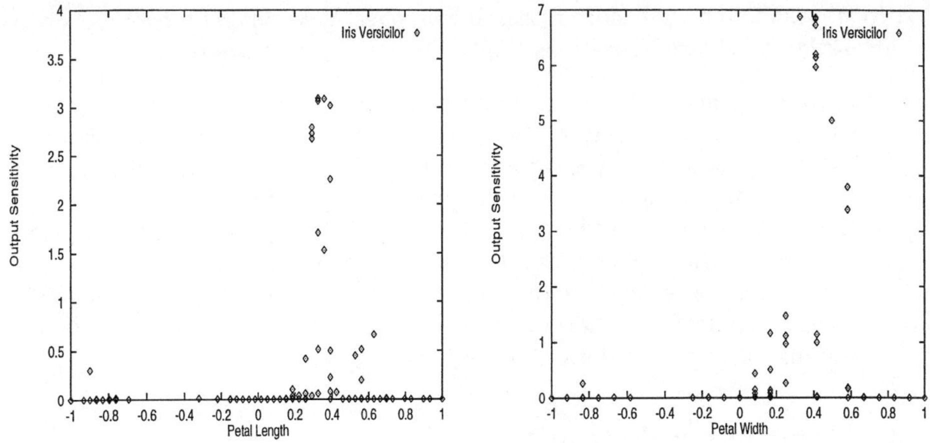

(a) Boundary position over *petal length* (b) Boundary position over *petal width*

Fig. 1. Boundaries formed for *Iris Versicolor*

where β controls the range around decision boundaries from which patterns are selected. The larger the value of β, the more patterns will be selected. $\overline{\Phi}_\infty$ is the average pattern informativeness,

$$\overline{\Phi}_\infty = \frac{\sum_{p=1}^{P_C} \Phi_\infty^{(p)}}{P_C} \tag{5}$$

Patterns with informativeness a factor less than the average informativeness over all patterns are therefore selected.

In the above, $\Phi_\infty^{(p)}$ is the informativeness of pattern p, and $\boldsymbol{\Phi}_\infty$ is the vector of pattern informativeness values. The informativeness of a pattern is computed using first-order sensitivity analysis, i.e.

$$\Phi_\infty^{(p)} = \max_{k=1,\cdots,K}\{|S_{o,k}^{(p)}|\} \tag{6}$$

with

$$S_{o,k}^{(p)} = \frac{\partial o_k}{\partial z_i^{(p)}} \tag{7}$$

Additional training patterns can be generated within the space of informative patterns. The generation of additional data may improve performance, since more information is used to refine the position of decision boundaries. The approach to data generation followed in this paper is to duplicate the informative patterns and to add Gaussian noise with a zero mean and small variance to these patterns. The reader should note that this is a preliminary investigation, and that more research will investigate different approaches to generating new data. Questions that still need to be answered are

- Is it really necessary to generate new data? Experimental results presented in the next section illustrates that training on just the boundary patterns produces very good results. However, in cases where the original data is limited, the duplication approach will be helpful.
- How many times should data be duplicated? The answer to this question depends on the problem, and the size of the original data set.
- What should the variance of the noise be? Here the answer also depends on the problem. Too large variances may obscure the objective of focussing on just the informative regions.

4 Experimental Results

This section illustrates the approach to data removal and generation on the *iris, wine, glass* and *diabetes* problems from the UCI repository. Training used gradient descent and 4-3-3, 13-3-3, 9-15-6 and 8-15-1 architectures for each problem respectively. The number of hidden units was obtained using the pruning algorithm in [6]. Table 1 summarizes the results obtained. A concervative value of $\beta = 0.8$ has been used. For the original data set, D_T and D_G respectively denotes the training and test sets. For the last two columns, D_T denotes the set of boundary or generated patterns, D_G denotes the original training set, and D_V denotes the original test set. The percentages express the number of patterns correctly classified. The results show that training on just the boundary patterns presents similar or improved performance compared to training on all the data. For the generated training data we have comparable results. Most notable are the results for the *wine* and *diabetes* problems for which improved performance have been obtained. Significant improvements resulted for the *diabetes* problem with a noise variance of 0.1.

5 Conclusions

This paper presented an approach to remove uninformative patterns from training sets, and to generate additional training data in the regions of decision boundaries. Preliminary results showed that the approach holds promise. More research is needed to investigate the influence of the β parameter, the factor of duplication of decision boundary patterns, and the variance of the noise injected to the duplicated boundary patterns. Further research will also investigate the use of non-stochastic learning approaches, such as scaled conjugate gradient, where the network is not influenced by oscillations due to weight updates after each pattern presentation. In stochastic learning using gradient descent these oscillations are even more evident, since training considers patterns around decision boundaries for which a large error results per

Problem	Original Data D_T	D_G	Boundary Patterns D_T	D_G	D_V	Duplicated Patterns D_T	D_G	D_V
iris			24 out of 106 patterns			variance = 0.05		
epoch 100	96.2%	95.6%	62.5%	79.2%	75.6%	79.2%	92.5%	86.7%
epoch 200	96.2%	95.6%	95.8%	89.6%	95.6%	87.5%	92.5%	86.7%
wine			76 out of 125 patterns			variance = 0.05		
epoch 100	95.2%	96.2%	88.2%	89.6%	96.2%	97.4%	96.8%	98.1%
epoch 200	98.4%	96.2%	100%	95.2%	98.1%	100%	99.2%	100%
glass			102 out of 150 patterns			variance = 0.05		
epoch 1000	64.7%	71.9%	49.0%	56.0%	68.8%	64.2%	64.7%	67.2%
epoch 2000	67.3%	78.1%	63.7%	70.0%	75.0%	72.5%	73.3%	76.6%
diabetes			419 out of 490 patterns			variance = 0.05		
epoch 500	58.2%	62.4%	50.4%	53.4%	57.9%	55.7%	61.7%	63.9%
epoch 1000	62.0%	65.2%	58.9%	60.7%	66.5%	66.3%	70.5%	74.0%
						variance = 0.1		
epoch 500						78.4%	77.6%	82.2%
epoch 1000						89.3%	87.8%	91.3%

Table 1. Data Removal and Generation Results

pattern. Therefore large weight adjustments compared to patterns away from boundaries. Further work will also investigate online data generation to allow the generation of new training patterns during training.

References

1. AP Engelbrecht, I Cloete, *Selective Learning using Sensitivity Analysis*, IEEE World Congress on Computational Intelligence, International Joint Conference on Neural Networks, Anchorage, Alaska, 1998, pp 1150-1155.
2. AP Engelbrecht, I Cloete, *Incremental Learning using Sensitivity Analysis*, IEEE International Joint Conference on Neural Networks, Washington DC, USA, 1999, paper 380.
3. AP Engelbrecht, *Sensitivity Analysis for Decision Boundaries*, Neural Processing Letters, Vol 10, 1999, pp 1-14.
4. AP Engelbrecht, *Sensitivity Analysis of Multilayer Neural Networks*, PhD dissertation, Department of Computer Science, University of Stellenbosch, South Africa, 1999.
5. AP Engelbrecht, HL Viktor, *Rule Improvement through Decision Boundary Detection using Sensitivity Analysis*, the International Work Conference on Neural Networks (IWANN'99), 2-4 June, Alicante: Spain, Lecture Notes in Computer Science (Volume 1607), Springer-Verlag, Berlin: Germany, pp 78-84.
6. AP Engelbrecht, L Fletcher, I Cloete, *Variance Analysis of Sensitivity Information for Pruning Feedforward Neural Networks*, IEEE International Joint Conference on Neural Networks, Washington DC, USA, 1999, paper 379.

Applying Neural Networks in Case-Based Reasoning Adaptation for Cost Assessment of Steel Buildings

Essam Amin M. Lotfy[1] and Ahmed Sameh Mohamed[2]

[1] Products Resaerch & Development, Zamil Steel Building Co. Egypt, 30 St. 270 New Maddi, Cairo, eamin@ritsec1.com.eg

[2] Departement of Computer Science, The American University in Cairo, 113 Kasr El Aini St., Cairo 11511 Egypt, sameh@aucegypt.edu

Abstract: In is paper, a new technique in complex adaptation is investigated. By presenting the closest retrieved cases to a neural network, it learns about the domain of the problem being solved. The new problem is then fed to the trained neural network and the output becomes the solution to that problem. The methodology is applied to a problem in the steel construction and the sought output is the cost estimation of pre-engineered steel buildings.Several experiments are conducted to prove that these steps are successful. System verification is done and shows that both the system and the methodology are successful to develop a complete adaptation mechanism in CBR.

Keywords: Case-Based Reasoning, Case Adaptation, Complex Case Adaptation, Neural Networks

1 Introduction

Case adaptation, a central component in case based reasoning, is often considered the most difficult part of a case-based reasoning system. The difficulties arise from the fact that adaptation often does not converge, especially if it is not done in a systematic way. This problem, sometimes termed the assimilation problem, is especially pronounced in the case-based design problem-solving domain where a large set of constraints and features are processed. Furthermore, in the steel building design domain, multiple cases must be considered in conjunction, in order to solve a new problem, resulting in the difficulty of how to efficiently combine the cases into a global solution for the new problem (Purvis & Pu, 1995).

Steel contractors are very competitive in the market to provide low cost pre-engineered[1] buildings in fast construction time. They may be faced with the problem of deciding the bidding price of a building or several buildings owned by one customer. They cannot make full design and bill of material of the building before they win the bid. So they make cost estimation of the building based on their own experience in previous buildings. If the building cost is underestimated, a construction enterprise may suffer a tremendous loss and eventually fail. If the information about the previously executed projects could be captured, stored and reused as a part of the estimation process, many estimation problems could be solved more easily and accurately. This information does not contain general knowledge. It is instead created to be valid for a specific design situation. This kind of information is termed cases. The rational way to use old design cases when solving a new design problem is the case-based reasoning and is the aim of this paper. This paper has dual objectives. The first is investigating the feasibility of integrating neural networks with case-based reasoning to come up with strong knowledge-intensive model of complex adaptation. The second is building a CBR system that assists steel contractors in developing fast and accurate cost estimates for new projects.

The next two sections deal with Pre-Engineered steel buildings, CBR, and Neural Networks. After a section with the technical description of system development, the followed sections are mainly implementation issues treating all the CBR issues. Then the section on result evaluation discusses the obtained results. Finally the evaluation of the system and the conclusion sections judge whether the system has been successful in marrying CBR and Neural Networks with the Pre-Engineered building cost estimation.

2 The Pre-Engineered Steel Buildings

Although, they are pre-engineered, Pre-Engineered buildings need a lengthy design process to accurately calculate all of their component material. The ultimate requirement of this lengthy design process is to produce ***safe*** and ***economic*** building. To produce a safe design is so easy. One should apply the design code equations to get safe dimensions of a building. However, economic design of a building encounters a very important aspect called shape optimization. Figure 1 illustrates four alternate shapes of one Pre-Engineered steel building.

---

[1] Pre-engineered buildings are parameterized buildings with known architectural shapes. Customers need to fill these parameters to get an appropriate building that fits their needs.

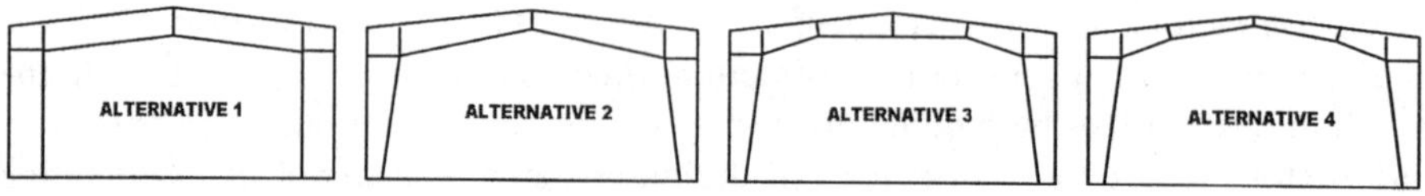

Figure 1: Four Alternative Shapes of One P.E.B.

Because the stress variation in the main structural element, frames can take several shapes. A good design is the one that produces a safe frame with as minimum weight as possible. To do it, one engineer may perform several trials with full calculations, and that is why the process of designing a building might take weeks or months even with the use of computer-aided design software (CAD).

Most of the Pre-Engineered steel buildings have some similarities in their parameters, and design engineers have calculated them in many cases over and over again. If the information about previously executed projects could be captured, stored and reused as a natural part of the estimation process, the cost assessment will be done easily and accurately. Much of the information about old projects is available in the large contracting companies whether in hardcopy or in electronic databases. For example, Zamil Steel Building Co. has a database of previously executed projects, which has over 8000 records. So it will be appropriate if this data is collected and converted into a case-base to allow automated cost estimation systems.

3 Domain Analysis & Case Representation

Each building in our model is considered a case. Cases can be represented in a variety of forms using the full range of data representation. An object-oriented model is selected to represent the cases.

A major advantage of object-oriented model is the use of one-to-one mapping between the model and the real world objects. So, for a Pre-Engineered steel building, the key class is the building itself as shown in the class diagram. Other classes can be divided into three categories.

Building sub-systems category includes roof monitors, fascias, roof extensions, expansion joints, and canopies. Building accessories category includes doors, windows, ventilators, and louvers. Additional systems category includes cranes and mezzanines. One more class, which does not belong to the previous categories, is the loads acting on the building.

A major relationship in the Pre-Engineered steel building domain is the "Has" relationship. The domain has one aggregate class that owns other classes. The owner class is the client class and the owned class is the supplier class.

Defining attributes that describe different classes is considered simple task if the classes are similar to real world classes. For example the class building has width, length, height, and roof slope. These real parameters will be attributes in the class building. All other classes will be cured in the same manner.

4 Data Acquisition

Much of the effort in building a case-based system goes into case collection. If it is impossible or especially difficult to collect cases, then case based reasoning will be too difficult, or impossible, to apply.

Cases need to be collected according to the needs of the system, that is, based on its goals. Cases in the case library should provide as much coverage as possible about achieving reasoning goals. The first pass at building a case library is often unlikely to provide full coverage. So refinement of this pass is required to achieve the required coverage (Kolodner, 1993).

Fortunately, the available data from Zamil Steel projects is adequate to build good case library of the Pre-Engineered steel building. The actual case-base is refined and a case-base consists of 480 cases of real projects is built.

4.1 Assigning Importance Values to Case Features

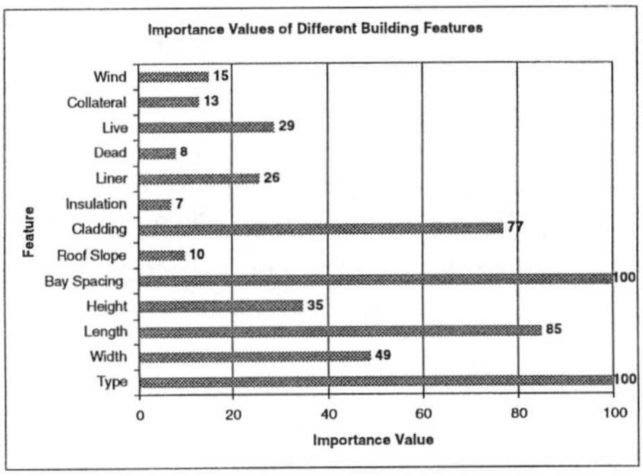

Figure 2: Importance Values of Building Features

The weight assigned to each feature of the case tells how much attention to pay to matches and mismatches in the field when computing the distance measure of a case. One way to assign importance values is to have a human expert assign them as the case library is being built. The expert might have some feeling about which dimension and combinations of dimensions make good predictors (Kolodner, 1993). Another way to assign importance values is to do a statistical evaluation of a known corpus of cases to determine which dimension predicts different outcomes and/or solutions best. Those that are good predictors are then assigned higher importance for matching (Kolodner, 1993). Both methods are applied to assign importance values to different features in the case-base.

Figure 2 illustrates the assigned values of the different building features. The figure reflects the experience and the statistical study of the different parameters. The figure shows that 'Bay-Spacing' and 'Type' are the most important features. Following these features are the 'Length' and 'Cladding'. 'Dead' and 'Insulation' have the least importance over all values.

5 Case Indexing & Retereival

The intended CBR system is about to suggest an accurate cost estimation for a new steel building, given all the building attributes. The assumption underlying case-based reasoning is that, if the building attributes of two different buildings are similar, their cost should be similar. In fact, this is seldom to be done in the steel building domain because of the wide variations of the Pre-Engineered building attributes. Instead, several closest match cases should be retrieved and adapted. The system's accuracy, precision, speed, and efficiency thus depend on its ability to recall a set of cases that are significantly similar to the new problem case.

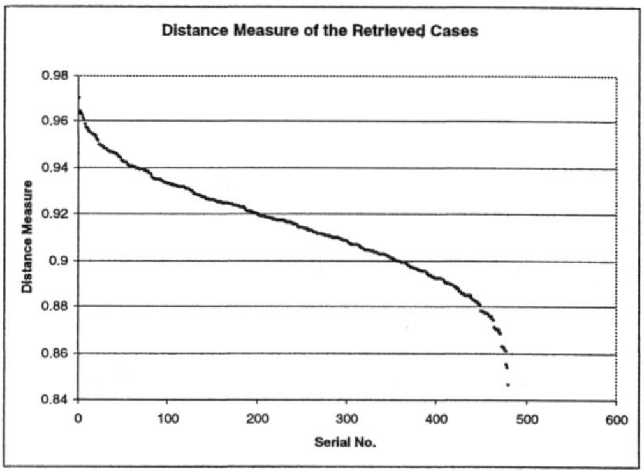

Figure 3: Distance Measure of the Retrieved Cases

Figure 3 shows the results of the distance measure between the 480 cases and one query case. The numbers in the X-axis is just a serial number of the cases and the Y-axis represents the distance measure. Cases are sorted according to the distance measure.

6 Case Adaptation

As illustrated above the Pre-Engineered building system has four main class categories: the main building with loads, the building sub-systems, the mezzanine systems, and the crane systems. The experiments done in this section are divided into two main parts. Part one focuses only on the building system with loads. The second part examines the decomposition of cases and applies it on the building system and the mezzanine system.

In the second part, the problem is completely decomposed into subproblems. Then each of the subproblems is interpreted as a query case, and for each one closest match cases can be retrieved. The retrieval process gives two different sets of cases each one corresponds to one subproblem. Each set is encoded and fed into a neural network to get its solution. Finally, the two partial solutions are composed to make the final solution. Figure 4 illustrates the decomposition of cases in complex case adaptation.

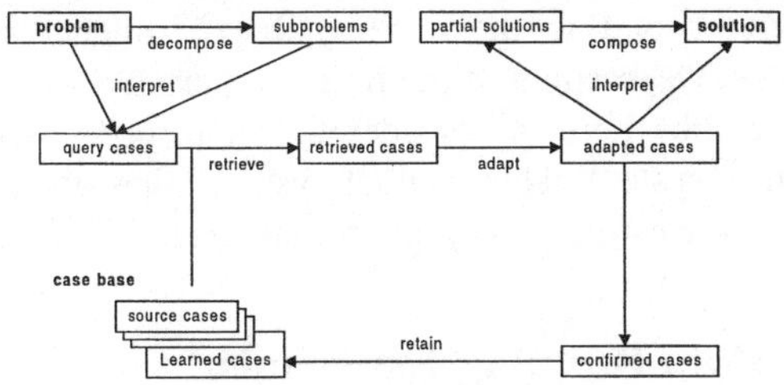

Figure 4: Decomposition of Cases in Complex Adaptation

Neural computing applications are usually designed and optimized using an iterative approach (try- adjust- try again). Therefore, the initial design task is a starting point for an iterative process of design refinements. Its result is not a single design but a set of design options together with a specifications for the experiments intended to resolve the competing design issues.

After the iterative process, the used neural network has one input layer with 11 units, two hidden layers, the first has 5 units, the second has 4 units, and one output unit. All activation functions are sigmoid, and the network is fully interconnected. The learning rate (η) ranges from 0.15 to 0.25, and the momentum (α) equals to 0.5.

6.1 Training the Neural Network

On-line training is used to deal with changing conditions. Since the retrieved cases change with time, on-line training or periodic retaining of specialized architectures is necessary. Two important points are taken into account during the installation of the on-line training.

The first point is the stability of the training algorithm. No sudden changes in the weights of the neural network have been discovered. In each experiment, the network is initialized and it converges at the end of the experiment with almost the same root mean square error. Therefore, there is no significant risk from the on-line training. This is because the closeness of the input patterns. These patterns are all close to one case that is the query case. In ordinary networks, data may include diverse of input patterns that may cause sudden change in weights during the training.

The second point is that the on-line training may be very expensive, costly, and too slow to respond to the users. All experiments are done on ordinary personal computer. The response time for all experiments ranges from 15 to 60 minutes. The response time here is considered fast response with respect to the manual cost estimation system. This manual system takes about two to three working days to complete the estimation of an ordinary Zamil Steel project.

6.2 Main Building Experiments

The following four experiments illustrate the neural network training which is conducted for eleven input parameters described above and 30 cases for each query. Because building type and roof slope are constant for all selected buildings, they are execluded from the threteen features showd in figure 2. Table 1 illustrates the network parameters and training results.

Table 1: Main Building Experiments Training Results

	(A)	(B)	(C)	(D)
Input units	11	11	11	11
Hidden Layers	2 (5,4)	2 (5,4)	2 (5,4)	2 (5,4)
Output Unit	1	1	1	1
Learning rate (η)	0.25-0.15	0.25-0.15	0.25-0.15	0.25-0.15
Momentum (α)	0.5	0.5	0.5	0.5
Activation Functions	Sigmoid	Sigmoid	Sigmoid	Sigmoid
Epoch Length	30	30	30	30
Iterations	125120	253305	248480	240000
Root Mean Square Error (RMS)	0.00256	0.0032	0.0034	0.00247
Target RMS:	0.001	0.001	0.001	0.001

6.2.1 Experiment (E) Complex Case Adaptation

In this experiment, The mezzanine system is separated from its building to generate two other sub-problems: the building system and the mezzanine system. Two techniques are applied incremental-CBR (Cunningham, et al., 1998) in the case retrieval and decomposition for the case adaptation. The incremental CBR mechanism can initiate case retrieval with a structural case description that is used to retrieve a matching subset of the case-base. This is done here as the mezzanine system can query the case-base and retrieve a subset of the mezzanine cases, and so the building system does. As the decomposition technique, described above, is applied, the two sets of cases are then fed into two neural networks. Table 2 illustrates the network parameters and training results for the decomposed system into main building and a mezzanine system.

Table 2: Experiment (E) Training Results

	Main Building	Mezzanine
Input units	11	6
Hidden Layers	2 (5,4)	2 (3,2)
Output Unit	1	1
Learning rate (η)	0.25-0.18	0.25-0.15
Momentum (α)	0.5	0.5
Activation Functions	Sigmoid	Sigmoid
Epoch Length	30	30
Iterations	425820	47105
Root Mean Square Error (RMS)	0.003	0.002
Target RMS	0.001	0.001

6.3 Testing the Neural Network

For the testing task, post-processing of the neural network output is required. Post-processing is done the opposite way the pre-processing is done. The neural output is decoded by multiplying the output with the cost of the reference case. This gives back the cost of the solution. Table 3 illustrates the testing results of the five experiments.

Table 3: Neural Network Results

Experiments		A	B	C	D	E
	Building Cost	893293	1363900	2402380	744648	1178023
CBR	Retrieved Cost	756933	1421148	2475792	646401	967860
Neural Network	RMS	0.017854	0.029208	0.025104	0.010738	B=0.0243 M=0.0179
	Estimated Cost	844258	1313315	2333379	715387	B=831806 M=272264 T=1104070

B stands for Building, M stands for Mezzanine, T stands for Total

The table shows that for the four experiments A, B, C, and D the estimated cost enhances the value of the closest match cost. The deviation of the estimated cost from the actual cost is acceptable in the range of the hundred thousands Saudi Reyals (the currency used in Zamil Steel database). This is also applicable for experiment E in the two sub-systems.

Note that the cost is measured here in Saudi Reyals. One may say that if the projects in the available case-base are captured two years ago, these figures should give incorrect prices. This is correct if the stored cost is the actual cost, but the stored cost here is the price book cost. Large contracting companies often calculate the cost of their projects by using the price book system. Price book is a book that contains prices for every item used in the project as well as an equation to calculate it after certain number of years. So, the prices included in this book are considered constant prices. However, companies never sell their buildings using the prices obtained form the price book directly because they are too high in order to ensure steady state. Yet, they have to give their customers a discount over these prices according to the market status. The results in this paper give the price book cost of the building. Sales engineers have to apply an appropriate discount to the output price.

7 System Verification & Results Evaluation

7.1 Check Retrieval Accuracy

A case should exactly match itself. This means that if the case-base is queried with one of its cases, it should gives the same case with distance measure equals 100%. This is shown in Figure 5. The only case hits the value 1.0 is the query case. If a case does not retrieve itself or does not match exactly, there is something wrong in the retrieval algorithm (Watson, 1996).

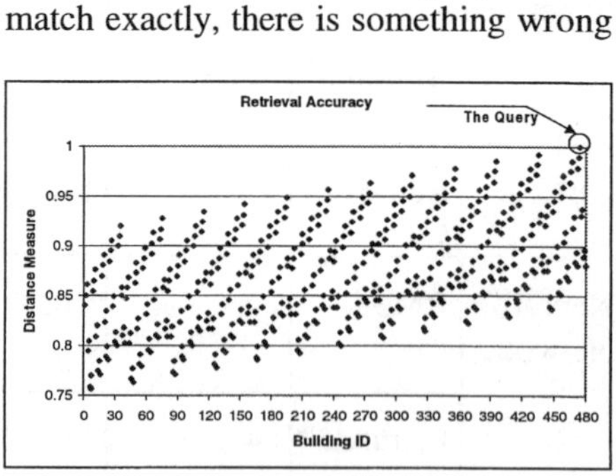

Figure 5: Logical Validity of the CBR

7.2 Check Retrieval Consistency

If exactly the same search has been performed twice, the same source cases should be retrieved with the same accuracy. If this does not happen, then there is something wrong with the retrieval algorithm (Watson, 1996).

7.2.1 Check for Case Duplication

A case should exactly match itself, but should not be identical to other cases. Duplicate cases will not harm decision making but may slow the system down or cause other administrative problems (Watson, 1996). This is clearly shown in Figure 5. There is one just case that matches itself.

7.2.2 Check Case Coverage

To check the case coverage, the standard deviation of the numerical features should be as low as possible. The mean of the feature value should also be near to middle of the feature range (Watson, 1996). Table 4 illustrates the standard deviation (SD), mean, median of the some different numerical features of the building. The SD of the 'Width' and 'Length' seem to be high with respect to that of the ' Height'. The reason is the 'Width' feature ranges from 15.00 to 80.00 meters and the 'Length' ranges from 30.00 to 120.00 meters while the 'Height' ranges from 6.00 to 12.00 meters. This is also applicable for 'Bay Spacing'. The table shows that the median and the mean of the features are so close.

Table 4: Case Coverage Check

Test	Width	Height	Length	Bay Spacing
(SD)	17.33854	1.43415	20.87982	1.061767
Median	44.00	8.70	60	7.00
Mean	44.9375	8.35	64.6875	7.5

7.3 Global System Verification Tests

The steps described above have all concentrated on individual aspects of a CBR system and its performance, retrieval accuracy, consistency, and case coverage that undoubtedly affects the system's performance. However, it is also vitally important to verify the overall performance of the system. The steps of this test are as follows (Watson, 1996):

- Did the system retrieve a useful set of cases?
- Was the retrieval time acceptable?
- Was the adaptation successful?

Table 5 has the answer of the above questions.

Table 5: Global System Verification Check

Experiments		A	B	C	D	E
Building	Cost	893293	1363900	2402380	744648	1178023
Retrieval	Retrieved Cost	756933	1421148	2475792	646401	967860

Adaptation	Estimated Cost	844258	1313315	2333379	715387	1104070
%	Retrieved/ Original	84.7	104	103	86	82
	Adapted/ Original	94.5	96	97	96	94

From the above table the answer to the first question is yes. It is clear that retrieved closest match case is close to the source case. In some experiments, it is very close. The answer of the second question is not more than 3 seconds. All the retrieval experiments take about 3 seconds to get all cases sorted according their distance measure. The answer to the third question is also yes. The adaptation results are clearly very good. The estimated building cost is close to the original building with a range of 4%. In the manual estimation process, there is over estimation up to 10% of the building, this percent will cover the 4% difference of the estimated cost.

In conclusion, after all the experiments, the system is successful in getting a correct estimation of the five buildings. This proves the concept of applying neural networks in case-based reasoning adaptation.

8 Conclusion

A new technique in case-based reasoning adaptation is successfully developed. The technique applies neural computing in the most difficult part of the CBR process, which is case adaptation. The application of the new technique is done within a real world problem that faces a lot of contracting companies, which is the cost estimation of Pre-Engineered steel buildings.

Several experiments are conducted to prove that these steps are successful. Both un-decomposed and decomposed CBR experiments were conducted. The system was successful in getting correct estimation to five different queries (the last one was decomposed CBR testing). The system was verified by conducting several tests in order to measure retrieval accuracy, retrieval consistency, case duplication, case coverage, and the overall system performance. System verification shows that both the system and the methodology are successful to develop a complete complex adaptation mechanism in CBR.

References

1. Aamodt, A., & Plaza, E. (1994). Case-Based Reasoning: Foundational Issues, Methodological Variation and System Approaches. *AICOM*, Vol.7, No.1.

2. Abd Rabou, E. A. (1999). *Applying Neural Networks in Case-Based Reasoning Adaptation for Cost Assessment of Steel Buildings.* Ms.C. Thesis submitted to the American University in Cairo, School of Science and Engineering.

3. Achofiled, C. & Shepperd, M. (1995*). Software Support for Cost Estimation by Analogy.* ESCOM 6, Rolduc. Available: http://dec.bournemouth.ac.uk/ dec_ind/decind22/web/ESCOM95.html

4. Aha, D. W., & Ram, A. (1995). Adaptation of Knowledge for Reuse. *Papers from the 1995 Fall Symposium.* California: The AAAI Press.

5. Bonzano, A. (1998). *ISAC: a CBR system for decision support in air traffic control.* Ph.D. Thesis submitted to the University of Dublin, Trinity College.

6. Booch, G. (1994). *Object-oriented analysis and design.* Redwood City, CA: The Benjamin/ Cummings Publishing Company, Inc.

7. Carnahan, B., & Smith, A. E. (1997). Pittnet [Computer Software, Free Copy]. Department of Industrial Engineering, University of Pittsburgh.

8. Carnahan, B., & Smith, A. E. (1997). *User's Guide to Pittnet Neural Network Educational Software.* Department of Industrial Engineering, University of Pittsburgh.

9. Chester, D. (1990). Why two hidden layers are better than one. *Proceedings of the International Joint Conference on Neural Networks.* Washington, D.C., Neural Network Society.

10. Cunningham, P., & Symth, B., & Bonzano, A. (1998). An Incremental Retrieval Mechanism for Case-Based Electronic Fault Diagnosis. *Knowledge-Based Systems Journal, Jan.*

11. Daniel, R., & James, H., & Garret, J. (1991). Neural computing for intelligent structural systems. *Proceedings of the International Workshop on Intelligent Systems.*

12. Danon (1995). *WinNN, 0.97* [Computer Software, Registered Copy].

13. DTI NeuroComputing Web, (1993). [On-line], Available: http://www.clients.globalweb.co.uk/nctt/guide/

14. Durkin, J. (1994). *Expert systems design and development.* New York: Macmillan Publishing Company.

15. Flemming, U. & Coyne, R. & Snyder, J. (1993). SEED: A Software Environment to Support the Early Phases in Building Design. *In ARECDAO93 Proc. IV Int. Conf. On Computer Aided Design in Architecture and Civil Engineering.* Barcelona, Spain.

16. Gebhardt, F., & Vob, A., & Grather, W., & Schmidt-Belz, B. (1997). *Reasoning with Complex Cases.* Boston: Kluwer Academic Publishers.

17. Hammond, K. (1986). CHEF: A model of case-based planning. *In Proceedings of AAAI-86.* Cambridge, MA: AAAI Press/MIT Press.

18. Hegazy, T. & Fazio, P. & Moselhi, O. (1994). Developing practical neural network applications using back-propagation. *Microcomputers in Civil Engineering.* Vol. 9.

19. Hegazy, T. & Moselhi, O. (1994). Analogy-Based Solution to Markup Estimation Problem. *Journal of Computing in Civil Engineering.* Vol. 8, No. 1.

20. Hinrichs, T. R. (1992). *Problem solving in open worlds: A case study in design.* Northvale, NJ: Erlbaum.

21. Kolodner, J. L. (1993). *Case-Based Reasoning,* California: Morgan Kaufmann Publishers.

22. Koton, P. (1988). Integrating case-based and causal reasoning. *In Proceeding of the Tenth annual Conference of the Cognitive Science Society.* Northvale, NJ:Erlbaum.

23. Kriegsman, M. & Barletta, I. R. (1993). Building a Case-Based Help Desk Application. *IEEE Expert,* 8(6).

24. Leake, D. B. (1995). Combining rules and cases to learn case adaptation. *Proceeding of the Seventeenth Annual Conference of the Cognitive Science Society.*

25. Leake, D. B. (1996). *Case-based reasoning: experiences, lessons, and future directions.* Menlo Park: AAAI Press/ MIT Press.

26. Li, H. (1995). Neural Network Based Cost Estimation Model. System Design. *Journal of Australian Institute of Building Papers.*

27. Li, H. (1995). Neural Network Models for Intelligent Support of Mark-Up Estimation. *International Journal of Engineering Construction Management.*

28. Maher, M. & Gomez de Silva Garza, A. (1996). Design case adaptation using genetic algorithms. *Computing in Civil Engineering.* American Society of Civil Engineering.

29. Maher, M. & Gomez de Silva Garza, A. (1996). Developing case-based reasoning for structural design. *IEEE Expert,* 11(3).

30. Maher, M. & Gomez de Silva Garza, A. (1996). The adaptation of structural system designs using genetic algorithms. *Proceedings of the International Conference on Information Technology in Civil and Structural Engineering Design—Taking Stock and Future Dir.*

31. Manago, M. & Bergmann, R. & et. al. (1994). *CASUEL: a common case representation language.* ESPRIT project 6322,Task 1.1, Deliverable D1. Available FTP: ftp://ftpagr.informatik.uni-kl.de/pub/CASUEL/Documentation/Casuel2.01.ps.gz

32. Moselhi, O. & Hegazy, T. (1993). DBID: Analogy-Based DSS for Bidding in Construction. *Journal of Construction Engineering and Management,* Vol. 119, No. 3.

33. Pao, Y. H. (1989). *Adaptive pattern recognition and neural networks.* Reading, Ma: Addison Wesley.

34. Polyak, B.T. (1964). *Some methods of speeding up the convergence of iteration methods.* Z. Vycisl. Mat. i Mat. Fiz., 4.

35. Purvis, L.& Pu, P. (1995). Adaptation using constraint satisfaction techniques. In Veloso, M. & Asmodt, A. (eds) *Topics in Case Based Reasoning, Proceedings of the First International Conference on Case Based Reasoning, LNAI Series.* Springer Verlag.

36. Reategui, E. & Campbell, J. A. & Borghetti, S. (1995). Using a neural network to learn general knowledge in a case-based system. *Case-Based Reasoning Research and Development.*

37. Rich, E. & Knight, K. (1991). *Artificial Intelligence 2nd Edition.* New York: McGraw-Hill, Inc.

38. Riesbeck, C. K., & Schank, R. C. (1989). *Inside case-based reasoning.* Hillsdale, New Jersey: Lawrence Erlbaum Associates Publishers.

39. Rose 98 (1998). *Rational Rose 98.* [Computer Software, Evaluation Copy].

40. Rumelhart, D.E., Hinton, G.E., and Williams, R.J. (1986). Learning internal representations by error propagation. Rumelhart, D.E. and McClelland, J. L., eds. (1986). *Parallel Distributed Processing: Explorations in the Microstructure of Cognition.* Vol. 1, 318-362, Cambridge, MA: The MIT Press.

41. TECHINNO (1995). *CBR-Works* [Computer Software, Evaluation Copy].

42. Tsatsoulis, C. & Change, Q. & Wei, H. (1997). Integrating case-based reasoning and decision theory. *IEEE Expert*, 11(4).

43. Watson, I. (1994). The Case for Case-Based Reasoning, *Proceedings of EPSRC/DRAL*, November 1994.

44. Watson, I. (1997). *Applying Case-Based Reasoning: Techniques for Enterprise Systems.* California: Morgan Kaufmann Publishers.

45. White, I. (1994). *Using the Booch Method: A Rational Approach.* California: The Benjamin/ Cummings Publishing Company, Inc.

46. Zamil Steel (1997). Product Brochure: Pre-Engineered Buildings. Zamil Steel Building Co. LTD.

47. Zamil Steel (1998). Technical Manual: Pre-Engineered Buildings. Zamil Steel Building Co. LTD.

A Projection Approach for Arabic Handwritten Characters Recognition

Khalid Saeed

Computer Engineering Department, Faculty of Computer Science,
Bialystok University of Technology, Bialystok, Poland.
e-mail: aidabt@ii.pb.bialystok.pl

Abstract: This article presents a trial to apply the idea of projection on feature vectors extracting for the sake of Arabic letters description and classification. The method was tested on letters written by different hands. The results showed it is possible to successfully apply the method on words and hence texts recognition after segmentation. The efficiency of the algorithm reached 99 % in some cases especially those of isolated letters written by one person. This method was used as one of three agents interacting to increase the efficiency of recognition [1].

Keywords: Arabic Handwritten, Recognition, Projection approach.

1 Introduction

Although there exist several algorithms, methods and very interesting approaches for Arabic or Farsi alphabet recognition [2,3,4], we have not met a perfect one (or at least no author had declared so) with zero restrictions and 100% efficiency, even for ideal conditions assumed by the authors themselves. In this paper the author tries to add some new elements which may help find a general and simple method for recognizing both machine printed and handwritten scripts.

The presented algorithm is based on some old, unfortunately unpublished, ideas and thoughts of the author [College Project, 1974] in which image points were projected on two axes (Fig. 1a) or four (Fig. 1b). Afterwards, in his paper [5] Burr suggested a more interesting method of projection in which the letter is assumed to be projected in such a way to fill a 13-segment shape (Fig. 1c). However, our experiments showed that the optimal number of projection segments is 12, situated in the the way shown in Fig.1d. The projected pixels of the letter will then construct the required feature vectors for classification and description.

Manuscript is supported by Bialystok University of Technology - Rector's Project.

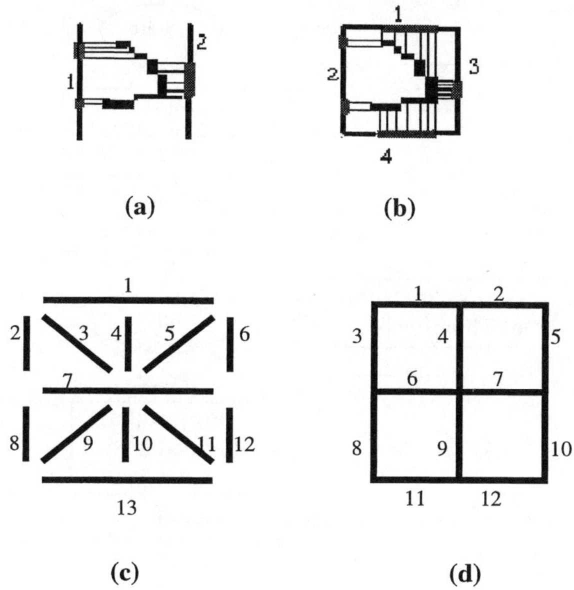

The idea of Projection: (a) and (b) the Arabic letter د - D (Dal) projected in two different elementary ways, (c) the projection axes suggested by Burr, (d) the axes for the projection used in this paper.

2 Theoretical Investigation

For the purpose of recognition, segmentation should be done first. This separates the letters of the word. Next, the letter is thinned to one-pixel width skeleton [6] since human never writes with the same thickness. Now, the script is ready to input with to the following algorithm.

2.1 Algorithm

The algorithm, on the basis of which the computer program is worked out (Fig. 3 shows its Flow Chart) and hence the recognition is achieved, depends mainly on the projection method according to Fig. 1d. First the processed letter is projected without dots, even if they exist. Then an initial decision is taken, classifying the scripts with dots to one of the following groups: (ب, ن, ت, ث), (ج, ح, خ), (ر, ز), (د, ذ), (س, ش), (ص, ض), (ع, غ), (ط, ظ), (ف, ق). The final classification takes place when fixing the number of dots (one, two or three) and their place (over or under the letter). This decreases the number of classifying groups of projected scripts, and hence increases the efficiency of the algorithm by eliminating the similar shapes belonging to different classifying groups.

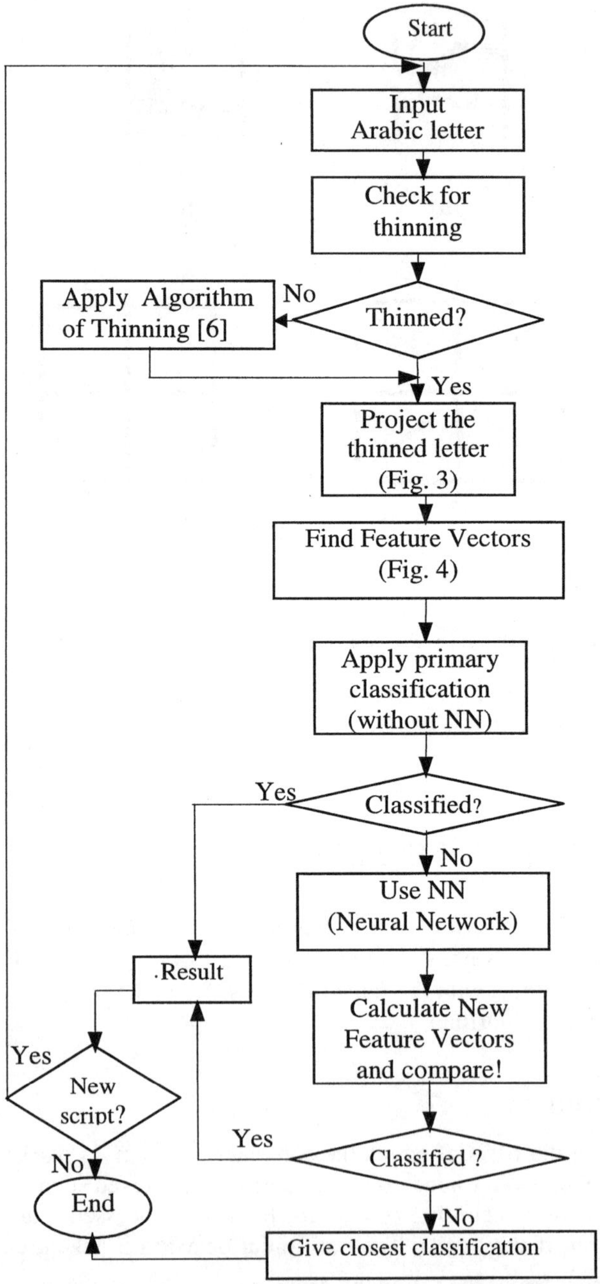

Fig. 2 Computer Flow Chart of the given Algorithm.

2.2 How does it work?

- Each point of the script is projected on the nearest two segments of Fig. 1d, one horizontal and the second vertical.
- The 12 segments are loaded by the projected on them points.
- This segment loading forms a vector whose elements are the lengths of those loaded segments.
- The dimension of this vector depends on the number of chosen segments.
- The procedure of projecting the script ص is shown in Fig. 3.

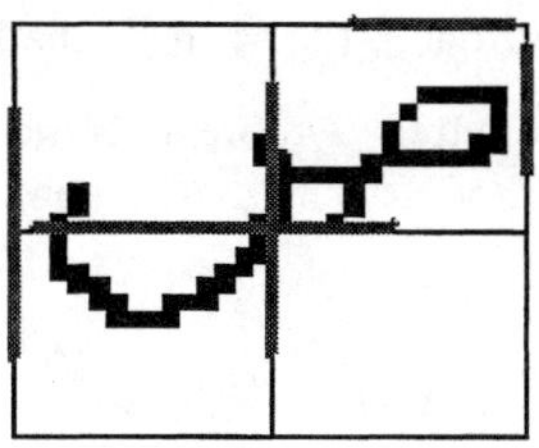

Fig. 3 The letter ص - Ssad projected according to the algorithm.

As a result of projection we obtain the feature vector which consists of 12 units, some of which can have zero value.

2.3 Neural Network

An artificial *Neural Network* was used [7,8] to increase the efficiency of the criterion. The feature vectors are used to train the neural network, which in turn gives the last decision of classification and recognition in a fast and efficient way. The applied multi-layer NN is trained by supervised-learning method of back-propagation [9], following the RPROP (Resilient backPROPagation) algorithm presented by Riedmiller and Braun [10]. The change in weights vector is given by:

$$\Delta W_{ij}(k) = -\eta_{ij}^{(k)} signum(\frac{\partial E(W(k))}{\partial W_{ij}})$$

where $E(W)$ – objective function,

$$\eta_{ij}^{(k)} = \begin{cases} \min(a\eta_{ij}^{(k-1)}, \eta_{\max}) & for \quad S_{ij}(k)S_{ij}(k-1) > 0 \\ \min(b\eta_{ij}^{(k-1)}, \eta_{\min}) & for \quad S_{ij}(k)S_{ij}(k-1) < 0 \\ \eta_{ij}^{(k-1)} & otherwise \end{cases}$$

with $a = 1.2$, $b = 0.5$, $\eta_{min} = 10^{-6}$, $\eta_{max} = 50$ and

$$S_{ij}(k) = \frac{\partial E(W(k))}{\partial W_{ij}}$$

3 Example

Although Neural Networks are essential in classifying some letters (Fig. 2), the following example gave the same results with and without NN, assuring the feasibility of the Projection Approach. Fig. 4 tests the letters (ص, ض) belonging to the same classifying group. The letter ض represents the same group of the letter ص, differing in a dot above it.

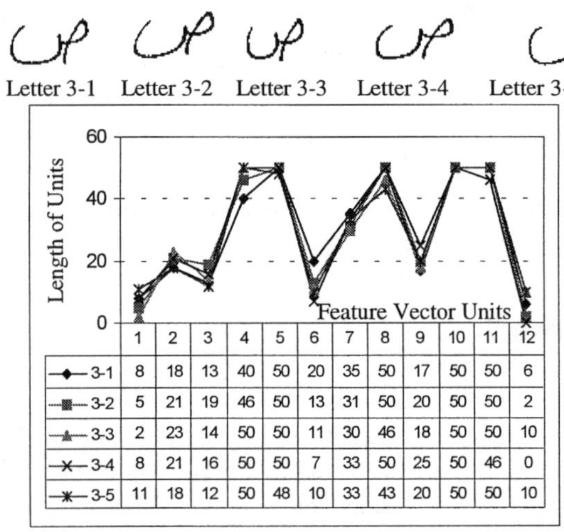

	1	2	3	4	5	6	7	8	9	10	11	12
3-1	8	18	13	40	50	20	35	50	17	50	50	6
3-2	5	21	19	46	50	13	31	50	20	50	50	2
3-3	2	23	14	50	50	11	30	46	18	50	50	10
3-4	8	21	16	50	50	7	33	50	25	50	46	0
3-5	11	18	12	50	48	10	33	43	20	50	50	10

Fig. 4 Feature vectors for five differently handwritten letters Ssad -ص.

4 Conclusions

After many experiments the method proved to be of high efficiency. Together with the thinning algorithms [6] it presents a practical, efficient and unaffected by distortions system for recognition of Arabic letters - both printed and handwritten. Observing feature vectors one can conclude that differences between vectors corresponding to fonts of letters belonging to the same group are minimal and hence the scripts can be recognised very easily, whereas those of different groups are different. These are the main useful results and achievements in this paper. Moreover, on the basis of the model applied here and some new approaches in the theory of scripts recognition, an interesting method of modelling [11] has been worked out to classify and precisely recognise Arabic letters.

5 References

[1] K.Saeed, "Three-Agent System for Cursive-Scripts Recognition," Proc. CVPRIP'2000 Computer Vision, Pattern Recognition and Image Proc., The 5th Joint Conf. on Information Sciences JCIS'2000, Feb.27th – March 3rd 2000, N. Jersey, USA.

[2] M. Ghuwar, "Modelling and Recognition of Arabic Scripts," Ph.D. Thesis, Institute of Computer Science, Polish Academy of Sciences, Warsaw, February 1997.

[3] R. Niedzielski, "A Criterion for Recognition of Machine written Cursive Scripts," M.Sc. Thesis, Faculty of Computer Science, Technical University of Bialystok, Bialystok, July 1999.

[4] M. Dehgham, K Faez, "Handwritten Farsi Character Recognition Using Evolutionary Fuzzy Clustering," Proc. of Eusipco-98, 9th European Signal Processing Conference, pp. 423-426, Rhode's, Greece, Sep. 8-11, 1998.

[5] D. Burr, "Experiments on Neural Net Recognition of Spoken and Written Text, IEEE Trans. On Acoustic, Speech and Signal Proc., 36, no. 7, 1162-1168, 1988.

[6] K. Saeed, R. Niedzielski, "Experiments on Thinning of Cursive-Style Alphabets," Inter. Conf. on Information Technologies ITESB'99, June 24-25, Mińsk 1999.

[7] K. Saeed, M. Nalewajko, "Recognition of Cursive Scripts by Projection - A Neural Approach," Proceedings of 6th Conf. of PTSK on Simulationin Research and Development (in Polish), pp. 65-67, Aug. 25-27, Bialystok-Bialowieza 1999.

[8] M. J. Nalewajko, "Universal system for recognition of scripts using artificial neural networks," M.Sc. Thesis, Faculty of Computer Science, Technical University of Bialystok, Bialystok, July 1999.

[9] S. Osowski, "Neural Networks - Algorithmic Approach," (in Polish), WNT, Warszawa 1996.

[10] M. Riedmiller, H. Braun, "RPROP – A Fast Adaptive Learning Algorithm," Technical Report, Karlsruhe University, 1992.

[11] K. Saeed, A. Dardzińska, "Ibn Muqla Arabic Letters Model for a New Classification Approach." Submitted to FQAS'2000 – 4th Int. Conerence on Flexible Query Answering Systems, Warsaw 2000.

Neural Network Modelling of the Potential Redistribution of Plant Species in Response to Climate Change

R.G. Pearson[1], T.P. Dawson[1], P.M. Berry[1], P.A. Harrison[1] and R. Lampinen[2]

[1] Environmental Change Institute, University of Oxford, UK

[2] Finnish Museum of Natural History, Botanical Museum, P.O. Box 7, FIN-00014 Helsinki University, Finland

Abstract:This paper explores the feasibility of using artificial neural networks for the modelling of plant species distributions and their response to climate change. A neural network has been successfully trained to recognise bioclimatic envelopes based upon climatic and physical data, and current European species distributions. The trained network has proved to be a powerful tool for predicting potential future changes to species distributions in Great Britain. It is proposed that the methodology developed is potentially applicable to many different species of flora.

Keywords: neural networks, biogeography, climate change.

1 Introduction

Artificial neural networks (ANNs) are computer systems that have the ability to 'learn' output data from sets of input patterns. Such systems have been widely employed in the physical and chemical sciences, but have only recently been applied in the environmental field (for example Lek and Guegan, 1999; Manel et al, 1999; Zealand et al, 1999; Mastrorillo et al, 1997). This study explores the feasibility of using neural network modelling as an alternative approach for the modelling of plant species distributions and their response to climate change. The response of individual plant species to climate change has been modelled before, notably Huntley et al (1995) who used climate response surfaces to define bioclimatic envelopes based on three environmental variables, but the use of artificial neural networks for this purpose is a new line of research. The ANN approach is essentially correlative and has the advantage over existing methodologies of using a wider range of input variables, having no requirement for detailed physiological data of individual species and being robust to 'noise' (Dawson et al, 1998). The non-statistical nature of the ANN methodology also enables the accommodation of individual species that occur in more than one climate space.

Thus, careful selection of relevant input variables has enabled a neural network to be trained to recognise bioclimatic envelopes based upon climatic and physical data, and the considerable data available on the presence/absence of species over Great Britain and Europe. The trained network has proved to be a powerful tool for predicting changing species distributions, and also an explanatory tool for highlighting important driving forces. The model has been given the acronym SPECIES (Spatial Estimator of Climate Impacts on the Envelope of Species) and is currently being developed and utilised at the Environmental Change Institute, University of Oxford. The model has been developed as part of a UK government-funded project studying the integrated impacts of climate change on agriculture, water resources and biodiversity. However, the particular results presented here for *Fagus sylvatica* are not part of that project. The modelling methodology discussed is to be applied to a number of different species.

2 Methodology

The current European distribution of *Fagus sylvatica* (beech) has been obtained as presence/absence data and mapped on a 0.5 ^{o} grid. This current distribution is used as the target output during network training. However, to avoid presenting the ANN with simply presence or absence data, the distribution was interpolated to a 10 km grid (Kriging geostatistical gridding method; Oliver and Webster, 1990) to give a distribution based on probability of occurrence. This smoothing assisted network training by ensuring that 'presences' (a probability of occurrence of 1) did not neighbour 'absences' (probability of occurrence 0) when environmental conditions at each location would be similar.

Environmental inputs which are thought to have direct physiological roles in limiting the ability of plants to survive and grow have been calculated from climate and soils data at both European (0.5 ^{o} grid) and GB (5 km national grid) scales. Five variables, based on the fundamental requirements of temperature and moisture, are used as network inputs: absolute minimum temperature, maximum temperature, growing degree days above 5°C, soil moisture deficit and soil moisture surplus.

Model training is carried out at the European scale so as to include the full climate range of the species. The trained network is then downscaled in its application to Great Britain. This ensures that, when applied to future climate scenarios, the network is not used to extrapolate outside its training dataset. Thus, the interpolated European species distribution and environmental inputs were normalised and the full dataset (consisting of 4688 training patterns) split into three randomly selected and equally sized datasets (for training, validating and testing). The training set has then been back-propagated through a feed-forward network with 5 input nodes, 11 nodes in a single hidden layer, and one output node. Back-propagation employed the quadratic (mean squared error) function:

$$E(t) = {}^1\!/_2\, \Sigma(d_j(t) - y_j(t))^2$$

where E(t) is the global error function at discrete time t, $y_j(t)$ is the predicted network output at t, and $d_j(t)$ is the desired network output at t, and connection weights were adjusted using the generalised delta-rule, which reads as follows:

$$\Delta w_{ij} = \eta \delta_j o_i$$

$$\delta_j = f^1_j(net_j)(t_j - o_j) \text{ if unit j is an output unit}$$

$$\text{or } \delta_j = f^1_j(net_j)\, \Sigma_k \delta_k w_{jk} \text{ if unit j is a hidden unit}$$

where:

η = learning parameter (a constant - typically in the range 0.1 to 1.0)

δ_j = error (difference between the real output and the teaching input) of unit j

t_j = teaching input of unit j

o_i = output of the preceding unit i

i = index of a predecessor to the current unit j with link w_{ij} from i to j

j = index of the current unit

k = index of a successor to the current unit j with link w_{ij} from j to k.

Initial connection weights were selected randomly in the range 2.0 to -2.0, and the learning parameter (η) was set at 0.2. The validation set was used to test the network every 100 cycles. This aided identification of when the network began to overgeneralise and learning could be stopped at the minimum of the validation set error (Fig 1).

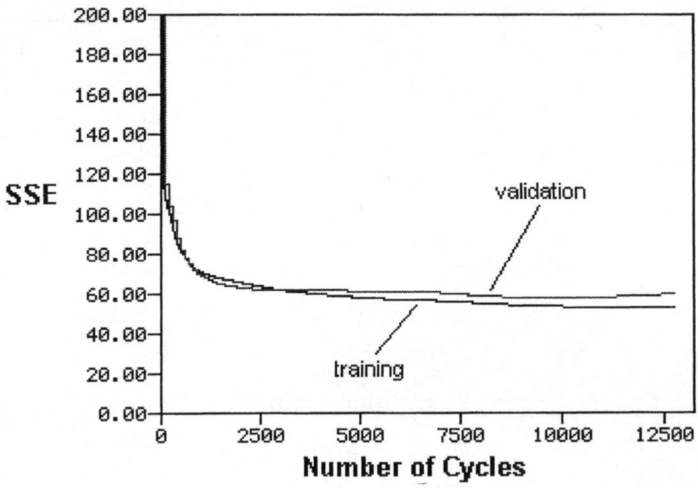

Fig. 1.

Close study of the validation set error revealed that a minimum summed squared error (SSE) was reached after 10,000 iterations (at which point mean squared error = 0.037). A new network was thus trained for this number of cycles.

3 Model Testing

The trained network was then tested on the remaining 'unseen' third of the full dataset. The predicted outputs show good association with the target outputs, yielding a Pearson correlation coefficient of 0.89. However, a more comprehensive test of the model is provided by its accuracy in simulating the observed species distribution. The full European dataset was thus propagated through the trained network and the outputs mapped. Due to the distribution smoothing, it was necessary at this stage to determine a threshold probability of occurrence that best represents the actual distribution. This was achieved through calculation of the Kappa statistic (Monserud, 1990; Light, 1971) for threshold probabilities increasing in increments of 0.1. A maximum correlation of 0.814 between the simulated and actual distributions is given when a threshold probability of occurrence of 0.6 and above is used to represent 'presence' (Figure 2). The model is thus found to compare favourably with more traditional biogeographical models.

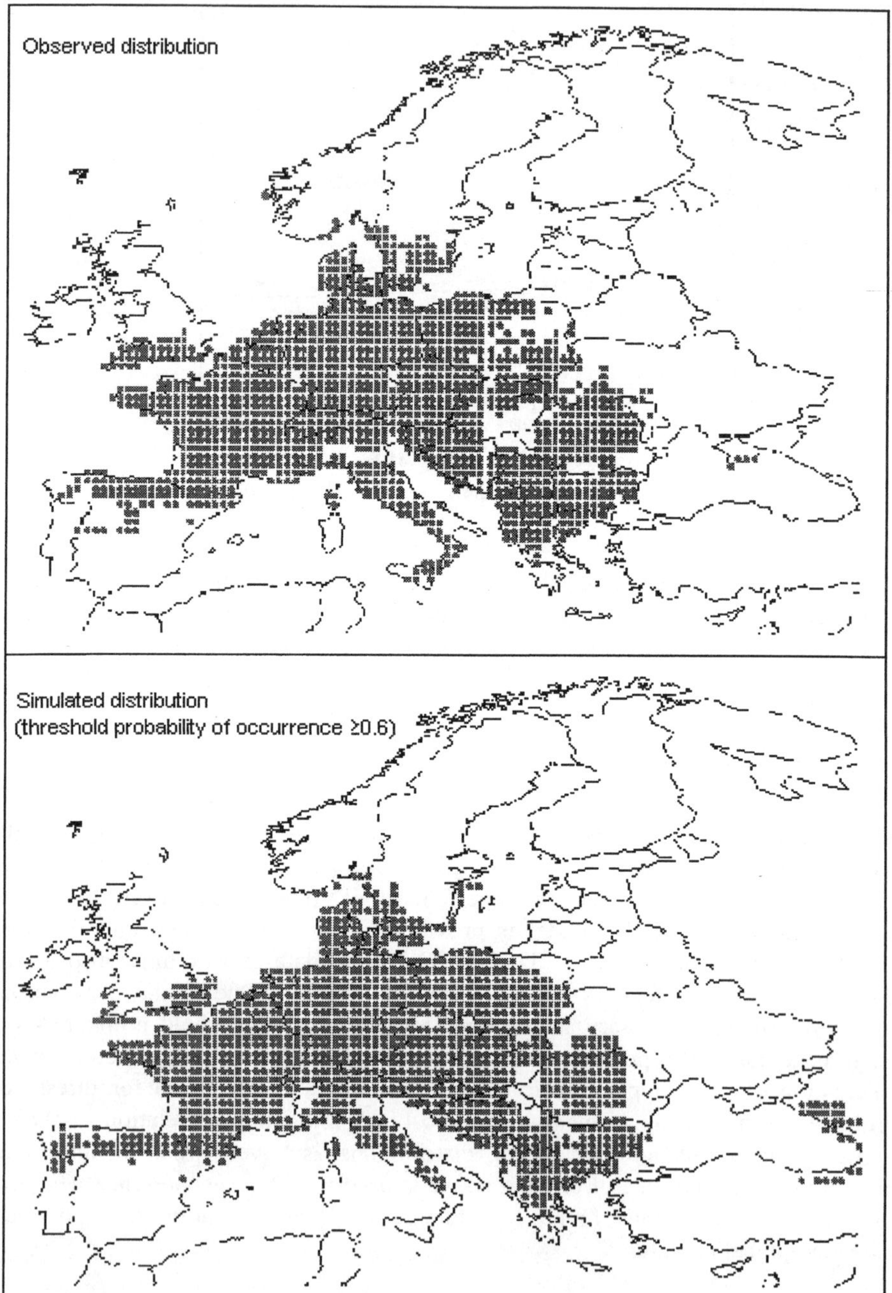
Observed distribution
Simulated distribution
(threshold probability of occurrence ≥0.6)

Fig. 2.

4 Future Predictions

In order to use the network to predict future distributions under climate change, the environmental inputs for the UK were recalculated using future scenario climate data based on the modelling experiments of the UK Hadley Centre (HadCM2 model; Johns et al, 1997). The scenarios span a range of climate sensitivities and emissions scenarios (based on IS92a; IPCC, 1992) and are for the 2020s and 2050s (high and low). The future scenario datasets (again to the 5 km national grid) have then been forward propagated through the trained network to give predicted future distributions of *Fagus sylvatica* (a threshold probability of occurrence of ≥0.6 is applied as determined earlier). The predicted distributions under the 2020 and 2050 high climate change scenarios are shown in Figures 3b and 3c. Figure 3a shows the current UK distribution as simulated by propagating the baseline UK inputs through the network.

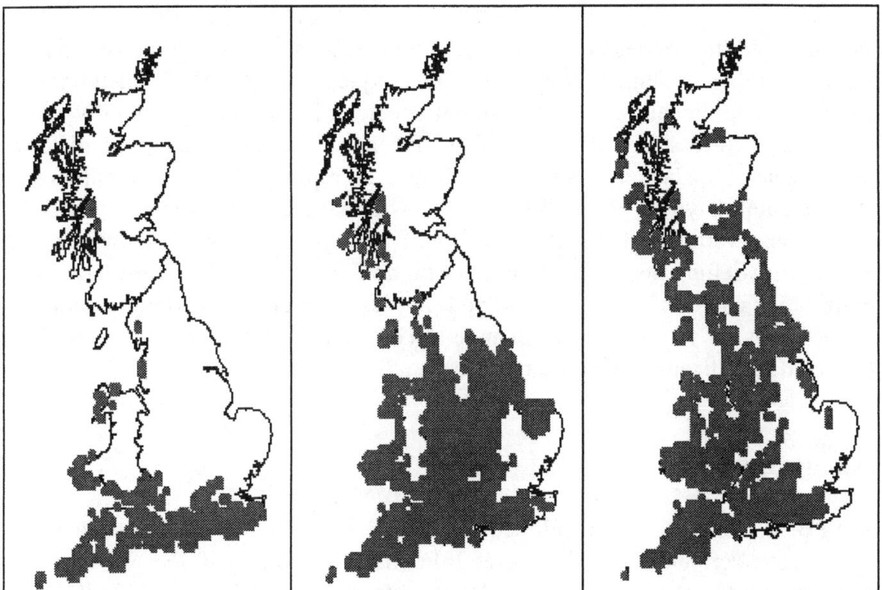

a, Simulated current distribution

b, Predicted distribution under 2020 high scenario

c, Predicted distribution under 2050 high scenario

Fig. 3.

5 Discussion and Conclusion

Comparison of Figures 3a, b and c reveals that the potential range of Fagus sylvatica in Great Britain will expand significantly under climate warming. Similar results showing northward expansion are expected for other species, though for some species already favouring northern climates, northward migration may lead to a contracting distribution in Great Britain. It must be emphasised that the future distributions presented are *potential* distributions and take no account of factors that may limit the species dispersal. Limiting factors include migration rates, barriers to migration (including urban areas and agricultural belts), competition from other species and suitable habitats for growth. The current model does not account for such factors, but the use of a land-use mask could assist in this respect.

It is concluded that ANNs are a feasible, and possibly superior, alternative to more traditional biogeographical modelling techniques. The ANN methodology developed is applicable to many different species of flora, and potentially fauna. Of particular interest are species which give a good representation of different climate spaces, endangered species and species representative of a number of different habitat types. Modelling results will have important implications for species conservation and the preservation of biological diversity. However, this black-box modelling approach should not be applied without due consideration of the limitations and uncertainties involved, not least the great uncertainties inherent in predictions of future climate change. The model outputs must be carefully interpreted as indicators of future trends, not as concrete predictions.

Acknowledgement

The SPECIES model was developed as part of a jointly funded project between Ministry of Agriculture, Fisheries and Food, Department of the Environment Transport and the Regions and UK Water Industry Research *"CC0337: Regional climate change impact & response studies in East Anglia and the North West (REGIS)"*. The description of the model is presented here with the permission of the sponsors. The model run for Fagus sylvatica was not part of the REGIS project.

References

Dawson, T.P., Curran, P.J. and Plummer, S.E. (1998) The biochemical decomposition of slash pine needles from reflectance spectra using neural networks. Int. Journal of Remote Sensing, 19 (7), pp. 1433-1438.

Huntley, B., Berry, P.M., Cramer, W. and McDonald, A.P. (1995) Modelling present and potential future ranges of some European higher plants using climate response surfaces. Journal of Biogeography, 22, pp.967-1001.

Intergovernmental Panel on Climate Change (1992) Climate Change 1992: the supplementary report to the IPCC scientific assessment. Cambridge University Press.

Johns, T.C., Carnell, R.E., Crossley, J.F., Gregory, J.M., Mitchell, J.F.B., Senior, C.A., Tett, S.F.B. and Wood, R.A. (1997) The second Hadley Centre coupled ocean-atmosphere GCM: model description, spinup and validation. Climate Dynamics, 13, pp.103-134.

Lek, S. and Guegan, J.F. (1999) Artificial neural networks as a tool in ecological modelling, an introduction. Ecological Modelling, 120, pp.65-73.

Light, R.J. (1971) Measures of response agreement for qualitative data: some generalizations and alternatives. Psychological Bulletin, 76 (5), pp. 365-377.

Manel, S., Dias, J-M. and Ormerod, S.J. (1999) Comparing discriminant analysis, neural networks and logistic regression for predicting species distributions: a case study with a Himalayan river bird. Ecological Modelling, 120, pp.337-347.

Masrtorillo, S., Lek, S., Dauba, F. and Belaud, A. (1997) The use of artificial nerual networks to predict the presence of small-bodied fish in a river. Freshwater Biology, 38, pp. 237-246.

Monserud, R.A. (1990) Methods for comparing global vegetation maps. Working Paper WP-90-40, International Institute for Applied Systems Analysis (IIASA), Laxenburg, Austria.

Oliver, M.A and Webster, R. "Kriging: a method of interpolation for geographical information systems", Int. J. Geographical Information Systems, 1990, 4 (3), pp. 313-332

Zealand, C.M., Burn, D.H. and Simonovic, S.P. (1999) Short term streamflow forecasting using artificial neural networks. Journal of Hydrology, 214, pp. 32-48.

Part 2

Fuzzy Systems

Fuzzy Logic = Computing with Words*

Lotfi A. Zadeh

Computer Science Division and Electronics Research Laboratory
Dept. of Electrical and Electronics Engineering and Computer Science
University of California
Berkeley, CA 94720-1776, USA
E-mail: zadeh@cs.berkeley.edu

To Professor Michio Sugeno, who has contributed so much and in so many ways to the development of fuzzy logic and its applications

Abstract. As its name suggests, computing with words, CW, is a methodology in which words are used in place of numbers for computing and reasoning. The point of this note is that fuzzy logic plays a pivotal role in CW and vice-versa. Thus, as an approximation, fuzzy logic may be equated to CW.

There are two major imperatives for computing with words. First, computing with words is a necessity when the available information is too imprecise to justify the use of numbers. And second, when there is a tolerance for imprecision which can be exploited to achieve tractability, robustness, low solution cost and better rapport with reality. Exploitation of the tolerance for imprecision is an issue of central importance in CW.

In CW, a words is viewed as a label of a granule, that is, a fuzzy set of points drawn together by similarity, with the fuzzy set playing the role of a fuzzy constraint on a variable. The premises are assumed to be expressed as propositions in a natural language. For purposes of computation, the propositions are expressed as canonical forms which serve to place in evidence the fuzzy constraints that are implicit in the premises. Then, the rules of inference in fuzzy logic are employed to propagate the constraints from premises to conclusions.

At this juncture, the techniques of computing with words underlie -- in one way or another -- almost all applications of fuzzy logic. In coming years, computing with words is likely to evolve into a basic methodology in its own right with wide-ranging ramifications on both basic and applied levels.

---

* Research supported in part by NASA Grant NCC 2-275 and the BISC (Berkeley Initiative in Soft Computing) Program. Reprinted from IEEE Transactions on Fuzzy Systems, **4**, pp. 103-111, 1996 by permission of IEEE.

1 Introduction

Fuzzy logic has come of age. Its foundations have become firmer, its applications have grown in number and variety; and its influence within the basic sciences -- and especially in mathematical and physical sciences -- has become more visible and more substantive.

And yet, there are two questions that are still frequently raised: (a) What is fuzzy logic?; and (b) What can be done with fuzzy logic that cannot be done equally well with other methodologies, e.g., predicate logic, probability theory, neural network theory, Bayesian networks and classical control.

The title of this note is intended to suggest a succinct answer: The main contribution of fuzzy logic is a methodology for computing with words (CW). No other methodology serves this purpose.

What follows is an elaboration on this suggestion. A fuller exposition of the methodology of CW will appear in a forthcoming paper.

Needless to say, there is more to fuzzy logic than a methodology for CW. Thus, strictly speaking, the equality in the title of this note should be an inclusion; using the equality serves to accentuate the importance of computing with words as a branch of fuzzy logic.

2 What is Computing with Words?

In its traditional sense, computing involves for the most part manipulation of numbers and symbols. By contrast, humans employ mostly words in computing and reasoning, arriving at conclusions expressed as words from premises expressed in a natural language or having the form of mental perceptions. As used by humans, words have fuzzy denotations. The same applies to the role played by words in computing with words (CW).

The concept of CW is rooted in several papers starting with my 1973 paper "A New Approach to the Analysis of Complex Systems and Decision Processes," in which the concepts of a linguistic variable and granulation were introduced. The concepts of a fuzzy constraint and fuzzy constraint propagation were introduced in "The Calculus of Fuzzy Restrictions," (1974), and developed more fully in "A Theory of Approximate Reasoning," (1979) and "Outline of a Computational Approach to Meaning and Knowledge Representation Based on a Concept of a Generalized Assignment Statement," (1986). Application of fuzzy logic to meaning representation and its role in test-score semantics are discussed in "PRUF - A Meaning Representation Language for Natural Languages," (1978), and "Test-Score Semantics for Natural Languages and Meaning-Representation via PRUF" (1981).

Although the foundations of computing with words were laid some time ago, its evolution into a distinct methodology in its own right reflects many advances in our understanding of fuzzy logic and soft computing -- advances which took place within the past few years. (See References and Related Papers.) A key aspect of CW is that it involves a fusion of natural languages and computation with fuzzy variables. It is this fusion that is likely to result in an evolution of CW into a basic methodology in its own right, with wide ranging ramifications and applications.

We begin our exposition of CW with a few definitions. It should be understood that the definitions are dispositional, that is, admit of exceptions.

The point of departure in CW is the concept of a granule. In essence, a granule is a fuzzy set of points having the form of a clump of elements drawn together by similarity. A word, *w*, is a label of a granule, *g*, and conversely *g* is the denotation of *w*. A word may be atomic, as in *young*, or composite, as in *not very young*. Unless stated to the contrary, a word will be assumed to be composite. The denotation of a word may be a higher order predicate, as in a Montague grammar (Partee 1976).

In CW, a granule, *g*, which is the denotation of a word, *w*, is viewed as a fuzzy constraint on a variable. A pivotal role in CW is played by fuzzy constraint propagation from premises to conclusions. It should be noted that, as a basic technique, constraint propagation plays important roles in many methodologies, especially in mathematical programming, constraint programming and logic programming. (See References and Related Papers.)

As a simple illustration, consider the proposition *Mary is young*. In this case, *young* is the label of a granule *young*. (Note that for simplicity the same symbol is used both for a word and its denotation.) The fuzzy set *young* plays the role of a fuzzy constraint on the age of Mary.

As a further example consider the propositions

p_1 = *Carol lives near Mary*

and

p_2 = *Mary lives near Pat.*

In this case, the words *lives near* in p_1 and p_2 play the role of fuzzy constraints on the distances between the residences of Carol and Mary, and Mary and Pat, respectively. If the query is: How far is Carol from Pat?, an answer yielded by fuzzy constraint propagation might be expressed as p_3, where

p_3 = *Carol lives not far from Pat*

More about fuzzy constraint propagation will be said at a later point.

A basic assumption in CW is that information is conveyed by constraining the values of variables. Furthermore, information is assumed to consist of a collection of propositions expressed in natural or synthetic language.

A basic generic problem in CW is the following:

> We are given a collection of propositions expressed in a natural language which constitute the *initial data set*, or IDS for short.
>
> From the initial data set we wish to infer an answer to a query expressed in a natural language. The answer, also expressed in a natural language, is

referred to as the *terminal data set*, or TDS for short. The problem is to derive TDS from IDS.

A few problems will serve to illustrate these concepts. At this juncture, the problems will be formulated but not solved:

1. Assume that a function, f, $f: U \rightarrow V$, $X \in U$, $Y \in V$, is described in words by the fuzzy if-then rules:

   f: if X is small then Y is small
   if X is medium then Y is large
   if X is large then Y is small

What this implies is that f is approximated to by the fuzzy graph f^* (Fig. 1), where

$$f^* = small \times small + medium \times large + large \times small$$

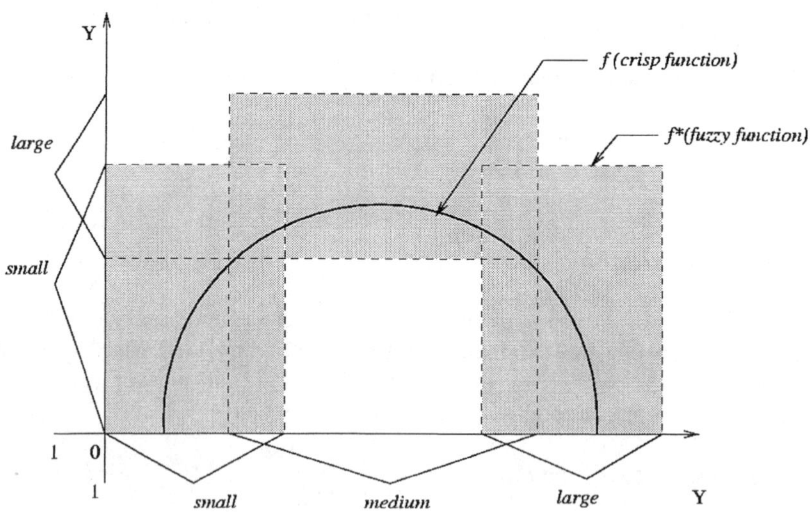

Figure 1. f^* is a fuzzy graph which approximates a function f

In f^*, + and × denote, respectively, the disjunction and cartesian product. An expression of the form $A \times B$, where A and B are words, will be referred to as a *cartesian granule*. In this sense, a fuzzy graph may be viewed as a disjunction of cartesian granules. In essence, a fuzzy graph serves as an approximation to a function or a relation (Zadeh 1974,1994).

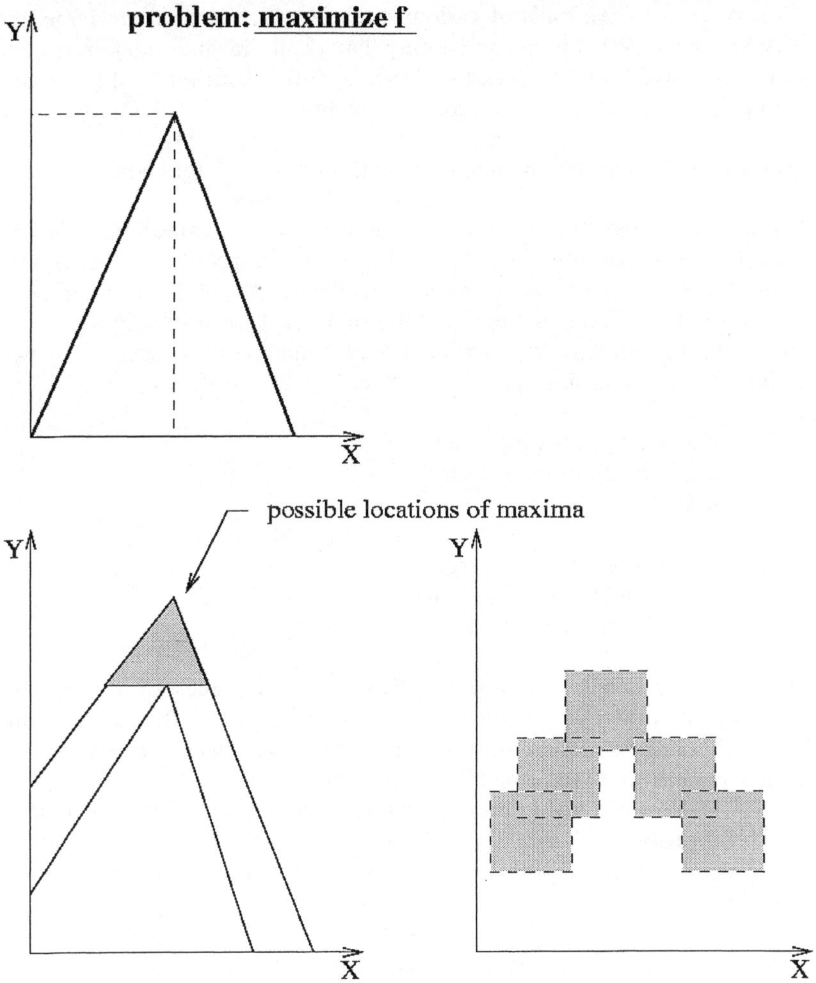

Figure 2. Maximization of a crisp function; an interval-valued function; and a fuzzy graph

In the example under consideration, the IDS consists of the fuzzy rule-set *f*. The query is: What is the maximum value of f^* (cf. Fig. 2)?

More broadly, the problem is: How can one compute an attribute of a function, *f*, e.g., its maximum value or its area or its roots if *f* is described in words as a collection of fuzzy if-then rules?

2. A box contains ten balls of various sizes of which several are large and a few are small. What is the probability that a ball drawn at random is neither large nor small? In this case, the IDS is a verbal description of the contents of the box: The TDS is the desired probability.

3. A less simple example of computing with words is the following.

Let X and Y be independent random variables taking values in a finite set $V=\{v_1,\ldots,v_n\}$ with probabilities $p_1,\ldots, p_n$ and $q_1,\ldots, q_n$, respectively. For simplicity of notation, the same symbols will be used to denote X and Y and their generic values, with p and q denoting the probabilities of X and Y, respectively.

Assume that the probability distributions of X and Y are described in words through the fuzzy if-then rules:

P: if *X* is small then p is small
if *X* is medium then p is large
if *X* is large then p is small

and

Q: if *Y* is small then q is large
if *Y* is medium then q is small
if *Y* is large then q is large

where the granules small, medium and large are the values of the linguistic variables X and Y in their respective universes of discourse. In the example under consideration, those rule-sets constitute the IDS. Note that small in P need not have the same meaning as small in Q, and likewise for medium and large.

The query is: How can we describe in words the joint probability distribution of X and Y? This probability distribution is the TDS.

For convenience, the probability distributions of X and Y may be represented as fuzzy graphs:

P: *small* × *small* + *medium* × *large* + *large* × *small*
Q: *small* × *large* + *medium* × *large* + *large* × *large*

with the understanding that the underlying numerical probabilities must add up to unity.

Since X and Y are independent random variables, their joint probability distribution (P,Q) is the product of P and Q. In words, the product may be expressed as (Zadeh 1994):

(P,Q): *small* × *small* × (*small* * *large*) + *small* × *medium* × (*small* * *small*) +
+ *small* × *large* × (*small* * *large*) + ... + *large* × *large* ×(*small* * *large*)

where * is the arithmetic product in fuzzy arithmetic (Kaufmann and Gupta 1986, Mares 1994). In this example, what we have done, in effect, amounts to a derivation of a linguistic characterization of the joint probability distribution of X

and Y starting with linguistic characterizations of the probability distribution of X and the probability distribution of Y.

A few comments are in order. In linguistic characterizations of variables and their dependencies, words serve as the values of variables and play the role of fuzzy constraints. In this perspective, the use of words may be viewed as a form of granulation, which in turn may be regarded as a form of fuzzy quantization.

Granulation plays a key role in human cognition. For humans, it serves as a way of achieving data compression. This is one of the pivotal advantages accruing through the use of words in human, machine and man-machine communication.

In the final analysis, the rationale for computing with words rests on two major imperatives. First, computing with words is a necessity when the available information is too imprecise to justify the use of numbers. And second, when there is a tolerance for imprecision which can be exploited to achieve tractability, robustness, low solution cost and better rapport with reality.

The conceptual structure of computing with words is schematized in Fig. 3a and Fig. 3b. Basically, CW may be viewed as a confluence of two related streams: fuzzy logic and test-score semantics, with the latter based on fuzzy logic. The point of contact is the collection of canonical forms of the premises, which are assumed to be propositions expressed in a natural language (NL). The function of canonical forms is to explicitate the implicit fuzzy constraints which are resident in the premises. With canonical forms as the point of departure, fuzzy constraint propagation leads to conclusions in the form of induced fuzzy constraints. Finally, the induced constraints are translated into a natural language through the use of linguistic approximation (Zadeh 1973, Mamdani and Gaines 1981).

In computing with words, there are two core issues that arise. First is the issue of representation of fuzzy constraints. More specifically, the question is: How can the fuzzy constraints which are implicit in propositions expressed in a natural language be made explicit. And second is the issue of fuzzy constraint propagation, that is, the question of how can fuzzy constraints in premises be propagated to conclusions.

These are the issues which are addressed in the following.

3 Representation of fuzzy constraints and canonical forms

Our approach to the representation of fuzzy constraints is based on test-score semantics (Zadeh 1978, 1981). In outline, in this semantics, a proposition, p, in a natural language is viewed as a network of fuzzy (elastic) constraints. Upon aggregation, the constraints which are embodied in p result in an overall fuzzy constraint which can be represented as an expression of the form

X is R

where R is a constraining fuzzy relation and X is the constrained variable. The expression in question is called a *canonical form*. The function of a canonical form is to place in evidence the fuzzy constraint which is implicit in p.

This is represented schematically as

$\mathrm{p} \rightarrow X$ is R

in which the arrow $\rightarrow$ denotes explicitation. The variable X may be vector-valued and/or conditioned.

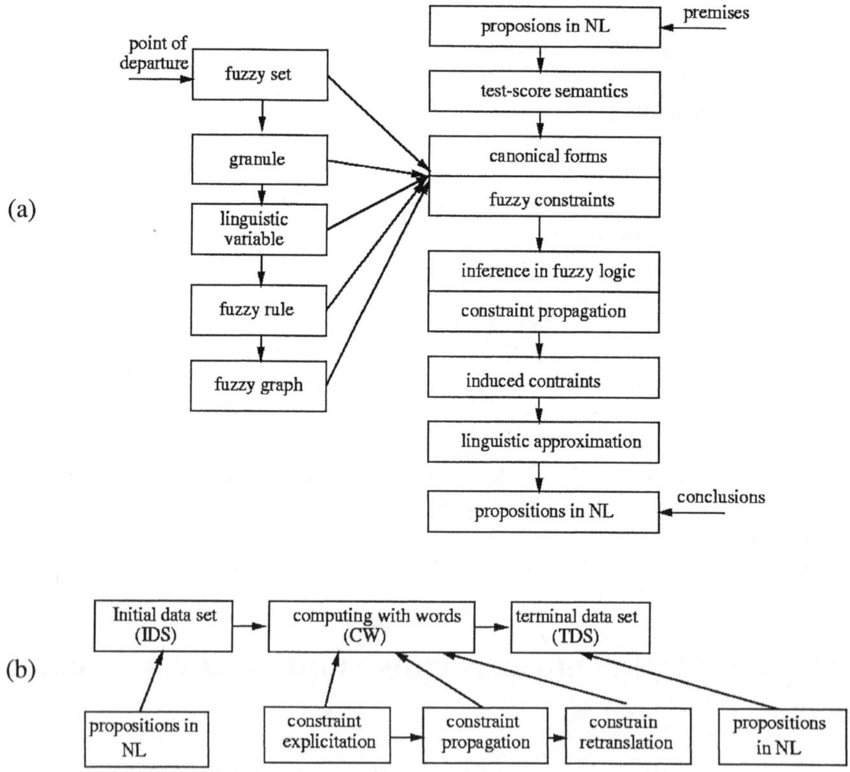

Figure 3. Conceptual structure of CW. Figure 3b is a more detailed version of Figure 3a.

In this perspective, the meaning of p is defined by two procedures. The first procedure acts on a so-called explanatory database, ED, and returns the constrained variable, X. The second procedure acts on ED and returns the constraining relation, R.

An explanatory database is a collection of relations in terms of which the meaning of p is defined. The relations are empty, that is, they consist of relation names, relations attributes and attribute domains, with no entries in the relations. When there are entries in ED, ED is said to be *instantiated* and is denoted EDI. EDI may be viewed as a description of a possible world in possible world semantics (Cresswell 1973), while ED defines a collection of possible worlds, with each possible world in the collection corresponding to a particular instantiation of ED.

As a simple illustration, consider the proposition

p = *Mary is not young*

Assume that the explanatory database is chosen to be

$$ED = POPULATION[Name; Age] + YOUNG[Age; \mu]$$

in which POPULATION is a relation with arguments Name and Age; YOUNG is a relation with arguments Age and μ; and + is the disjunction. In this case, the constrained variable is the age of Mary, which in terms of ED may be expressed as

$$X = \text{Age(Mary)} =_{\text{Age}} \text{POPULATION[Name=Mary]}$$

This expression specifies the procedure which acts on ED and returns X. More specifically, in this procedure, Name is instantiated to Mary and the resulting relation is projected on Age, yielding the age of Mary.

The constraining relation, R, is given by

$$R = ({}^{2}\text{YOUNG})$$

which implies that the intensifier *very* is interpreted as a squaring operation, and the negation *not* as the operation of complementation.

Equivalently, R may be expressed as

$$R = \text{YOUNG [Age; } 1\text{-}\mu^2 \text{]}$$

As a further example, consider the proposition

p = *Carol lives in a small city near San Francisco*

and assume that the explanatory database is

$$ED = \text{POPULATION[Name;Residence]} + \text{SMALL[City; } \mu] + \text{NEAR[City1; City2; } \mu]$$

In this case

$$X = \text{Residence(Carol)} =_{\text{Residence}} \text{POPULATION[Name=Carol]}$$

and

$R =$ SMALL[City; μ] $\cap_{\text{City1}}$ NEAR[City2=San Francisco]

In R, the first constituent is the fuzzy set of small cities; the second constituent is the fuzzy set of cities which are near San Francisco; and $\cap$ denotes the intersection of these sets.

So far we have confined our attention to constraints of the form

X is R

In fact, constraints can have a variety of forms. In particular, a constraint -- expressed as a canonical form -- can be conditional, that is, of the form

if X is R then Y is S

which may also be written as

Y is S if X is R

The constraints in question will be referred to as *basic.*

For purposes of meaning representation, the richness of natural languages necessitates a wide variety of constraints in relation to which the basic constraints form an important though special class. The so-called generalized constraints (Zadeh 1986) contain the basic constraints as a special case and are defined as follows.

A generalized constraint is represented as

X isr R

where isr, pronounced "ezar," is a variable copula which defines the way in which R constrains X. More specifically, the role of R in relation to X is defined by the value of the discrete variable r. The values of r and their interpretations are defined below:

e:	equal	(abbreviated to =)
d:	disjunctive (possibilistic)	(abbreviated to blank)
c:	conjunctive	
p:	probabilistic	
λ:	probability value	
u:	usuality	
rs:	random set	
rsf:	random fuzzy set	
fg:	fuzzy graph	
ps :	rough set (Pawlak Set)	
..:	...	

As an illustration, when $r=e$, the constraint is an equality constraint and is abbreviated to =. When r takes the value d, the constraint is *disjunctive* (possibilistic) and is abbreviated to is, leading to the expression

X is R

in which R is a fuzzy relation which constrains X by playing the role of the possibility distribution of X. More specifically, if X takes values in a universe of discourse, $U=\{u\}$, then $Poss\{X=u\} = \mu_R(u)$, where μ_R is the membership function of R, and Π_X is the possibility distribution of X, that is, the fuzzy set of its possible values. In schematic form:

$$X \text{ is } R \quad \rightarrow \Pi_X = R$$
$$\rightarrow Poss\{X=u\} = \mu_R(u)$$

Similarly, when R takes the value c, the constraint is *conjunctive.* In the case,

X isc R

means that if the grade of membership of u in R is μ, then $X = u$ has truth value μ. For example, a canonical form of the proposition

p= *John is proficient in English, French and German*

may be expressed as

Proficiency(John) isc (1/English + 0.7/French + 0.6/German)

in which 1.0, 0.7 and 0.6 represent, respectively, the truth values of the propositions *John is proficient in English, John is proficient in French* and *John is proficient in German.*

When $r = p$, the constraint is *probabilistic.* In this case,

X isp r

means that R is the probability distribution of X. For example

X isp $N(m,\sigma^2)$

means that X is normally distributed with mean m and variance σ^2. Similarly,

X isp (0.2\a + 0.5\b + 0.3\c)

means that X is a random variable which takes the values, a, b and c with respective probabilities 0.2, 0.5 and 0.3.

The constraint

X isu R

is an abbreviation for

usually(X is R)

which in turn means that

Prob{X is R} is *usually*

In this expression X is R is a fuzzy event and *usually* is its fuzzy probability, that is, the possibility distribution of its crisp probability.

The constraint

X isrs P

is a random set constraint. This constraint is a combination of probabilistic and possibilistic constraints. More specifically, in a schematic form, it is expressed as

$$\frac{\begin{array}{c} X \text{ isp } P \\ (X,Y) \text{ is } Q \end{array}}{Y \text{ isrs } R}$$

where Q is a joint possibilitistic constraint on X and Y, and R is a random set. It is of interest to note that the Dempster-Shafer theory of evidence is, in essence, a theory of random set constraints.

In computing with words, the starting point is a collection of propositions which play the role of premises. In most cases, the canonical forms of these propositions are constraints of the basic, disjunctive type. In a more general setting, the constraints are of the generalized type, implying that the explicitation of a proposition, p, may be represented as

$p \rightarrow X$ isr R

where X isr R is the canonical form of p.

As in the case of basic constraints, the canonical form of a proposition may be derived through the use of *test-score semantics*. In this context, the depth of p is, roughly, a measure of the effort that is needed to explicitate p, that is, to translate p into its canonical form. In this sense, the proposition X isr R is a surface constraint (depth=zero), with the depth of explicitation increasing in the downward direction (Fig. 4). Thus a proposition such as *Mary is young* is shallow, whereas a proposition *it is not very likely that there will be a substantial increase in the price of oil in the near future*, is not.

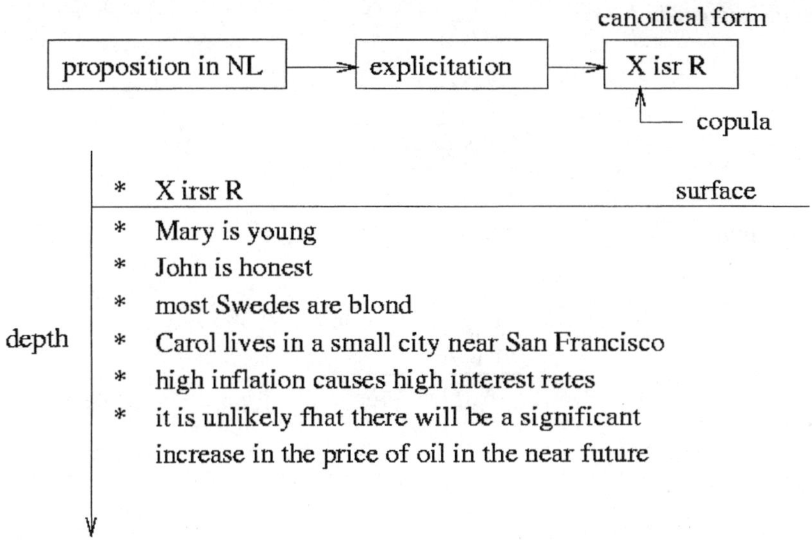

Figure 4. Depth of expliciation

Once the propositions in the initial data set are expressed in their canonical forms, the groundwork is laid for fuzzy constraint propagation. This is a basic part of CW which is discussed in the following section.

4 Fuzzy constraint propagation and the rules of inference in fuzzy logic

The rules governing fuzzy constraint propagation are, in effect, the rules of inference in fuzzy logic. In addition to these rules, it is helpful to have rules governing fuzzy constraint modification, The latter rules will be discussed at a later point in this section.

In a summarized form, the rules governing fuzzy constraint propagation are the following [A and B are fuzzy relations. Disjunction and conjunction are defined, respectively, as max and min, with the understanding that, more generally, they could be defined via t-norms and s-norms (Klir and Yuan 1995).]:

- *conjunctive rule 1*

$$\frac{\begin{array}{l} X \text{ is } A \\ X \text{ is } B \end{array}}{X \text{ is } A \cap B}$$

- *conjunctive rule 2* $(X \in U,\ Y \in B,\ A \subset U,\ B \subset V)$

$$\frac{\begin{array}{l} X \text{ is } A \\ Y \text{ is } B \end{array}}{(X, Y) \text{ is } A \times B}$$

- *disjunctive rule 1*

$$\frac{\begin{array}{l} X \text{ is } A \\ X \text{ is } B \end{array}}{X \text{ is } A \cup B}$$

- *disjunctive rule 2* $(A \subset U,\ B \subset V)$

$$\frac{\begin{array}{l} X \text{ is } A \\ Y \text{ is } B \end{array}}{(X, Y) \text{ is } A \times V \cup U \times B}$$

where $A \times V$ and $U \times B$ are cylindrical extensions of A and B, respectively.

- *conjunctive rule for isc*

$$\frac{\begin{array}{l} X \text{ isc } A \\ X \text{ isc } B \end{array}}{X \text{ isc } A \cup B}$$

- *disjunctive rule for isc*

$$\frac{\begin{array}{l} \quad X \text{ isc } A \\ \text{or} \\ \quad X \text{ isc } B \end{array}}{X \text{ isc } A \cap B}$$

- *projective rule*

$$\frac{(X,Y) \text{ is } A}{Y \text{ is } \mathrm{proj}_v A}$$

where $\mathrm{proj}_v = \sup_u A$,

- *surjective rule*

$$\frac{X \text{ is } A}{(X,Y) \text{ is } A \times V}$$

Derived rules

- *compositional rule*

$$\frac{\begin{array}{l} X \text{ is } A \\ (X,Y) \text{ is } B \end{array}}{Y \text{ is } A \circ B}$$

where $A \circ B$ denotes the composition of A and B.

- *extension principle* (mapping rule)

$$\frac{X \text{ is } A}{f(X) \text{ is } f(A)}$$

where f: $U \rightarrow V$, and $f(A)$ is defined by

$$\mu_{f(A)} = sup_{u \mid v = f(u)}\, \mu_A(u)$$

- *inverse mapping rule*

$$\frac{f(X) \text{ is } A}{X \text{ is } f^{-1}(A)}$$

where $\mu_{f\text{--}1(A)}(u) = \mu_A(f(u))$,

- *generalized modus ponens*

$$\frac{\begin{array}{l} X \text{ is } A \\ \text{If } X \text{ is } B \text{ then } Y \text{ is } C \end{array}}{Y \text{ is } A \circ ((\neg B) \oplus C)}$$

where the bounded sum $\neg B \oplus C$ represents Łukasiewicz's definition of implication,

- *generalized extension principle*

$$\frac{f(X) \text{ is } A}{q(X) \text{ is } q(f^{-1}(A))}$$

where $\mu_q(v) = sup_{u\,|\,v=f(u)}\, \mu_A((q(u))$.

The generalized extension principle plays a pivotal role in fuzzy constraint propagation.

- *syllogistic rule* (Zadeh 1981)

$$\frac{\begin{array}{l} Q_1\ A\text{'s are } B\text{'s} \\ Q_2\ (A \text{ and } B)\text{'s are } C\text{'s} \end{array}}{(Q_1 \otimes Q_2)\ A\text{'s are } (B \text{ and } C)\text{'s}}$$

where Q_1 and Q_2 are fuzzy quantifiers; A, B, and C are fuzzy relations; and $Q_1 \otimes Q_2$ is the product of Q_1 and Q_2 in fuzzy arithmetic,

- *constraint modification rules* (Zadeh 1972,1979)

X is $mA \rightarrow X$ is $f(A)$

where m is a modifier such as *not*, *very*, *more or less*, etc. and $f(A)$ defines the way in which m modifies A. Specifically,

if $m = not$ then $f(A) = A'$ (complement)
if $m = very$ then $f(A) = {}^2A$ (left square)

where $\mu^2{}_A(u)=[\mu_A(u)]^2$. This rule is a convention and should not be construed as a realistic approximation to the way in which the modifier *very* functions in a natural language.

- *probability qualification rule* (Zadeh 1979)

(X is A) is $\Lambda \rightarrow P$ is Λ

where X is a random variable taking values in U with probability density $p(u)$; Λ is a linguistic probability expressed in words like *likely*, *not very likely*, etc.; and P is the probability of the fuzzy event X is A, expressed as

$$P = \int_U \mu_A(u)p(u)du$$

The primary purpose of this summary is to underscore the coincidence of the principal rules governing fuzzy constraint propagation with the principal values of inference in fuzzy logic. Of necessity, the summary is not complete and there are many specialized rules which are not included. Furthermore, most of the rules in the summary apply to constraints which are of the basic, disjunctive type. Further development of the rules governing fuzzy constraint propagation will require an extension of the rules of inference to generalized constraints.

As was alluded to in the summary, the principal rule governing constraint propagation is the generalized extension principle which in a schematic form may be represented as

$$\frac{f(X_1,\ldots,X_n) \text{ is } A}{q(X_1,\ldots,X_n) \text{ is } q(f^{-1}(A))}$$

In this expression, $X_1,\ldots,X_n$ are database variables; the term above the line represents the constraint induced by the IDS; and the term below the line is the TDS expressed as a constraint on the query $q(X_1,\ldots,X_n)$. In the latter constraint, by $f^{-1}(A)$ we denote the preimage of the fuzzy relation A under the mapping f: $U \rightarrow V$,, where A is a fuzzy subset of V and U is the domain of $f(X_1,\ldots,X_n)$.

Expressed in terms of the membership functions of A and $q(f^{-1}(A))$, the generalized extension principle reduces the derivation of the TDS to the solution of the constrained maximization problem

$$\mu_{q(X_1,\ldots,X_n)}(v) = \sup_{(u_1,\ldots,u_n)}(\mu_A(f(u_1,\ldots,u_n)))$$

in which $u_1,\ldots,u_n$ are constrained by

$$v = q(u_1,\ldots,u_n)$$

The generalized extension principle is simpler than it appears. An illustration of its use is provided by the following example.

The IDS is:

most Swedes are tall

The query is:

What is the average height of Swedes?

The explanatory database consists of a population of N Swedes, $Name_1$, ..., $Name_N$. The database variables are h_1, ..., h_N, where h_i is the height of $Name_i$, and the grade of membership of $Name_i$ in *tall* is $\mu_{tall}(h_i)$, $i = 1, \ldots, n$.

The proportion of Swedes who are tall is given by

$$\frac{1}{N}\sum_{i=1}^{N}\mu_{tall}(h_i)$$

from which it follows that the constraint on the database variables induced by the IDS is

$$\frac{1}{N}\sum_{i=1}^{N}\mu_{tall}(h_i) \text{ is } most$$

In terms of the database variables, h_1, ..., h_N, the average height of Swedes is given by

$$h_{ave} = \frac{1}{N}\sum_{i=1}^{N} h_i$$

Since the IDS is a fuzzy proposition, h_{ave} is a fuzzy set whose determination reduces to the constrained maximization problem

$$\mu_{h_{ave}}(v) = \sup_{h_1,\ldots,h_N} (\mu_{most}(\frac{1}{N}\sum_{i=1}^{N}\mu_{tall}(h_i)))$$

subject to the constraint

$$v = \frac{1}{N}\sum_{i=1}^{N} h_i$$

It is possible that approximate solutions to problems of this type might be obtainable through the use of genetic-algorithm-based methods.

A key point which is brought out by this example and the preceding discussion is that explicitation and constraint propagation play pivotal roles in CW. What is important to recognize is that there is a great deal of computing with numbers in CW which takes place behind a curtain, unseen by the user. Thus, what matters is that in CW the IDS is allowed to be expressed in a natural language. No other methodology offers this facility. As an illustration of this point, consider the following problem.

A box contains ten balls of various sizes of which several are large and a few are small. What is the probability that a ball drawn at random is neither large nor small?

To be able to answer this question it is necessary to be able to define the meanings of *large*, *small*, *several large balls* and *neither large nor small.* This is a problem in semantics which falls outside of probability theory, neurocomputing and other methodologies.

There are two observations which are in order. First, in using fuzzy constraint propagation rules in computing with words, application of the extension principle reduces, in general, to the solution of a nonlinear program. What we need - and do not have at present - are approximate methods of solving such programs which are capable of exploiting the tolerance for imprecision. Without such methods, the cost of solutions may be excessive in relation to the imprecision which is intrinsic in the use of words. In this connection, an intriguing possibility is to use genetic-algorithms-based methods to arrive at approximate solutions to constrained maximization problems.

Second, given a collection of premises expressed in a natural language, we can in principle express them in their canonical forms and thereby explicitate the implicit fuzzy constraints. For this purpose, we have to employ test-score semantics. However, in test-score semantics we do not have at present effective algorithms for the derivation of canonical forms without human intervention. This is a problem that remains to be addressed.

5 Concluding remarks

The main purpose of this note is to draw attention to the centrality of the role of fuzzy logic in computing with words and vice-versa.

In coming years, computing with words is likely to emerge as a major field in its own right. In a reversal of long-standing attitudes, the use of words in place of numbers is destined to gain respectability. This is certain to happen because it is becoming abundantly clear that in dealing with real-world problems there is much to be gained by exploiting the tolerance for imprecision. In the final analysis, it is the exploitation of the tolerance for imprecision that is the prime motivation for CW. The role model for CW is the human mind.

References and Related Publications

Berenji, H.R (1994) "Fuzzy Reinforcement Learning and Dynamic Programming," in *Fuzzy Logic in Artificial Intelligence, Proc. IJCAI '93 Workshop*, ed. A.L. Ralescu, Berlin: Springer-Verlag, pp. 1-9.

Black, M. (1963), "Reasoning with Loose Concepts," *Dialog 2*, pp. 1-12.

Bosch, P. (1978), "Vagueness, Ambiguity and All the Rest," in *Sprachstruktur, Individuum und Gesselschaft*, eds. M. Van de Velde and W. Vandeweghe, Tubingen: Niemeyer.

Bowen, J., Lai, R., and Bahler D. (1992), "Fuzzy Semantics and Fuzzy Constraint Networks", *Proc. of the 1st IEEE Conf. on Fuzzy Systems*, San Francisco, pp. 1009-1016.

Bowen, J., Lai, R., and Bahler D. (1992), "Lexical Imprecision in Fuzzy Constraint Networks", *Proc. of the National Conf. on Artificial Intelligence*, pp. 616-620.

Cresswell, M.J. (1973), *Logic and Languages*, London: Methuen.

Dubois, D., Fargier, H., and Prade H. (1994), "Propagation and Satisfaction of Flexible Constraints", in *Fuzzy Sets, Neural Networks, and Soft Computing*, eds. R.R. Yager, L.A. Zadeh, New York: Von Nostrand Reinhold, pp. 166-187.

Dubois, D., Fargier, H., and Prade H. "Possibility Theory in Constraint Satisfaction Problems: Handling Priority, Preference and Uncertainty", to appear in *Applied Intelligence Journal*.

Dubois, D., Fargier, H., and Prade H. (1993), "The Calculus of Fuzzy Restrictions as a Basis for Flexible Constraint Satisfaction," *Proc. of the 2nd IEEE Int. Conf. on Fuzzy Systems*, San Francisco, pp. 1131-1136.

Freuder, E.C. and Snow P. (1990), "Improved Relaxation and Search Methods for Approximate Constraint Satisfaction with a Maximin Criterion", *Proc. of the 8th Biennial Conf. on the Canadian Society for Computational Studies of Intelligence*, Ontario, pp. 227-230.

Goguen, J.A. (1969), "The Logic of Inexact Concepts," *Synthese 19*, pp. 325-373.

Hobbs, J.R. (1978), "Making Computation Sense of Montague's Intensional Logic," *Artificial Intelligence 9*, pp. 287-306.

Katai, O., Matsubara, S., Masuichi, H., Ida, M., et. al. (1992), "Synergetic Computation for Constraint Satisfaction Problems Involving Continuous and Fuzzy Variables by Using Occam," in *Transputer/Occam, Proc. of the 4th Transputer/Occam Int. Conf.*, eds. S. Noguchi and H. Umeo, Amsterdam: IOS Press, pp. 146-160.

Kaufmann, A. and Gupta M.M. (1985), *Introduction to Fuzzy Arithmetic: Theory and Applications*, New York: Von Nostrand.

Klir, G. and Yuan, B. (1995), *Fuzzy Sets and Fuzzy Logic*, New Jersey: Prentice Hall.

Lano, K. (1991), "A Constraint-Based Fuzzy Inference System," in *EPIA 91, 5th Portuguese Conf. on Artificial Intelligence*, eds. P. Barahona, L.M. Pereira, and A. Porto, Berlin: Springer-Verlag, pp. 45-59.

Lodwick, W.A. (1990), "Analysis of Structure in Fuzzy Linear Programs," *Fuzzy Sets and Systems, 38(1)*, pp. 15-26.

Mamdani, E.H. and Gaines, B.R. (1981), Eds., *Fuzzy Reasoning and its Applications*, London.

Mares, M. (1994), *Computation Over Fuzzy Quantities*, Boca Raton: CRC Press.

Novak, V (1991), "Fuzzy Logic, Fuzzy Sets, and Natural Languages," *Int. J. of General Systems 20(1)*, pp. 83-97.

Novak, V., Ramik, M., Cerny, M., and Nekola, J. (1992), Eds., *Fuzzy Approach to Reasoning and Decision-Making*, Boston: Kluwer.

Oshan, M.S., Saad, O.M., and Hassan, A.G. (1995), "On the Solution of Fuzzy Multiobjective Integer Linear Programming Problems with a Parametric Study," *Advances in Modelling & Analysis A, 24(2)*, pp. 49-64.

Partee, B. (1976), *Montague Grammar*, New York: Academic Press.

Qi G. and Friedrich, G. (1992), "Extending Constraint Satisfaction Problem Solving in Structural Design," in *Industrial and Engineering Applications of Artificial Intelligence and Expert Systems, 5th Int. Conf., IEA/AIE-92*, eds. F. Belli and F.J. Radermacher, Berlin: Springer-Verlag, pp. 341-350.

Rosenfeld, A., Hummel, R.A., and Zucker, S.W. (1976), "Scene Labeling by Relaxation Operations," *IEEE Trans. on Systems, Man and Cybernetics, 6*, pp. 420-433.

Sakawa, M., Sawada, K., and Inuiguchi, M. (1995), "A Fuzzy Satisficing Method for Large-Scale Linear Programming Problems with Block Angular Structure," *European Journal of Operational Research, 81(2)*, pp. 399-409.

Tong S.C., (1994), "Interval Number and Fuzzy Number Linear Programming," *Advances in Modelling & Analysis A, 20(2)*, pp. 51-56.

Yager R.R. (1989), "Some extensions of constraint propagation of label sets," *Int. J. of Approximate Reasoning, 3*, pp. 417-435.

Zadeh, L.A. (1972), "A fuzzy-set-theoretic interpretation of linguistic hedges," *J. of Cybernetics* 2, pp. 4-34.

Zadeh, L.A. (1973), "Outline of a New Approach to the Analysis of Complex System and Decision Processes," *IEEE Trans. on Systems, Man, and Cybernetics*, SMC-3, pp.28-44.

Zadeh, L.A. (1974), "On the Analysis of Large Scale Systems," Systems Approaches and Environment Problems, ed. H. Gottinger, Gottingen: Vandenhoeck and Ruprecht, pp. 23-37.

Zadeh, L.A. (1975), "Calculus of Fuzzy Restrictions," in *Fuzzy Sets and Their Applications to Cognitive and Decision Processes*, eds. L.A. Zadeh, K.S. Fu, M. Shimura, New York: Academic Press, pp. 1-39.

Zadeh, L.A. (1978), "PRUF - a Meaning Representation Language for Natural Languages," *Int. J. Man-Machines Studies*, 10, pp. 395-460.

Zadeh, L.A. (1979), "Fuzzy Sets and Information Granularity," in *Advances in Fuzzy Set Theory and Applications*, eds. M. Gupta, R.Ragade and R. Yager, Amsterdam: North-Holland, pp. 3-18.

Zadeh, L.A. (1979), "A Theory of Approximate Reasoning," *Machine Intelligence 9*, eds. J. Hayes, D. Michie, and L.I. Mikulich, New York: Halstead Press, pp. 149-194.

Zadeh, L.A. (1981), "Test-Score Semantics for Natural Languages and Meaning Representation via PRUF," *Empirical Semantics*, ed. B. Rieger, W. Germany: Brockmeyer, pp. 281-349. Also *Technical Report Memorandum 246*, AI Center, SRI International, Menlo Park, CA.

Zadeh, L.A. (1986), "Outline of a Computational Approach to Meaning and Knowledge Representation Based on a Concept of a Generalized Assignment Statement," Proc. of the Int. Seminar on Artificial Intelligence and Man-Machine Systems, eds. M. Thoma and A. Wyner, Heidelberg: Springer-Verlag, pp. 198-211.

Zadeh, L.A. (1994), "Fuzzy Logic, Neural Networks and Soft Computing," *Communications of the ACM*, 37(3), pp. 77-84.

Computing with Words: Towards a New Generation of Linguistic Querying and Summarization of Databases

Janusz Kacprzyk and Sławomir Zadrożny

Systems Research Institute, Polish Academy of Sciences
ul. Newelska 6, 01-447 Warsaw, Poland
Email: {kacprzyk,zadrozny}@ibspan.waw.pl

Abstract. In this paper we present the use of a new paradigm of computing with words (cf. Zadeh's paper in this volume) in linguistic querying and data mining. The former is meant as an extension of conventional database querying for making possible a direct use of linguistic (fuzzy) terms, relations, etc. in queries as, e.g., implemented in the authors' FQUERY for Access package [cf. Kacprzyk and Zadrożny (1995-2000)]. The latter is meant as the derivation of linguistic summaries of databases in the sense of Yager (1982, 1991) [cf. Kacprzyk and Yager (1999), Kacprzyk, Yager and Zadrożny (2000)]. Moreover, we will mention the use of some algorithms for the derivation of association rules [cf. Agrawal and Srikant (1994)] to find database summaries of more advanced forms.

1. Introduction

Zadeh's [cf. Zadeh's paper in this volume, and Zadeh and Kacprzyk (1999)] new paradigm of computing with words (and its closely related manipulation with perceptions) has opened new vistas in various fields [cf. the papers in Zadeh and Kacprzyk (1999) and Wang (2000)], notably in those related to computational intelligence.

In this paper, first, we present the use of this paradigm in database querying using imprecise (linguistic, and fuzzy) specifications [cf. Bosc and Kacprzyk (1995)], notably in a perspective of add-ons to conventional databases implemented in the authors' FQUERY for Access package [cf. Kacprzyk and Zadrożny (1995 – 2000)]. Second, we present the use of this paradigm in the derivation of linguistic summaries of databases in the sense of Yager (1982, 1991), and its further extensions along the lines of Kacprzyk and Yager (2000) and Kacprzyk, Yager and Zadrożny (2000). We will illustrate the discussion on a real-world example of the derivation of linguistic database summaries of sales data at a computer retailer. Third, we briefly mention how the concept of and algorithms for the derivation of so-called association rules [cf. Agrawal and Srikant (1994)] may be applied to derive linguistic database summaries of more advanced forms better suitable for a human user.

2. Computing with words in database querying

2.1 Types of linguistic terms in flexible queries

Basically, the construction of a database query consists in spelling out conditions that should be met by the data sought. Very often, the meaning of these conditions is deeply rooted in natural language, i.e., their original formulation is available in the form of natural language utterances. It is then, often with difficulty, translated into mathematical formulas requested by the traditional query languages. For example, looking for a suitable house in a real estate agency database one may prefer a *cheap* one. In order to pose a query, the concept of "cheap" has to be expressed by an interval of prices. The bounds of such an interval will usually be rather difficult to assess. Thus, a tool to somehow define the notion of "being cheap" may essentially ease the construction of a query. The same definition may be then used, in other queries referring to this concept, also in the context of other words, as, e.g., *very*. The words of this kind, interpreted as so-called *modifiers*, modify the meaning of the original concept in a way that may be assumed context-independent and expressed by a strict mathematical formula.

Usually, a query comprises more than just one condition. In such a case, the user may require various combinations of conditions to be met, including:

- all conditions required (corresponds to the use of logical AND connective),
- any one condition met (corresponds to the use of logical OR connective),

In the first case, meeting each condition separately does contribute to the overall satisfaction of our query only in case all other conditions are also met. In the second case, meeting any, just one condition, is enough to satisfy the query. Sometimes a more flexible scheme of *aggregation*, positioned somewhere in between the two above extreme cases, may be useful. That is, we would like to consider each condition as determining a *partial* satisfaction of the query. Then, only meeting a predefined proportion of, e.g., *most*, of the conditions provides for a *complete* satisfaction of the query. Moreover, we can differentiate among the conditions as to their contribution to the overall satisfaction of the query. Some of them may be more *important* than others. Such flexible aggregation operators may be modeled using fuzzy (linguistic) quantifiers. In the framework of this model the importance degrees of the entities being aggregated may also be taken into account.

The linguistic terms and their possible parameters which may be used in queries may be classified in the following way:

1) terms representing inherent imprecision of some queries' conditions, including:
   a) numeric fuzzy values,

   b) modifiers,
   c) fuzzy relations,
   d) fuzzy sets of scalar values
2) terms corresponding to the non-standard aggregation operators, including:
   a) fuzzy (linguistic) quantifiers (or possibly OWA operators),
   b) importance coefficients

The query languages of classical DBMSs do not provide any means for representing such linguistic terms. Thus, in accordance with the *computing with words* paradigm, we will try to devise some extension of a classical query language, the SQL. In this section we will present a possible extension of its syntax. Obviously, this is not enough to extend the language alone – one must also provide for a proper interpretation of particular new linguistic constructs accompanied with some scheme for their representation, elicitation and manipulation.

We will focus our attention on the well-known SELECT...FROM...WHERE command of the SQL. More precisely, we will deal only with its WHERE clause. Starting with its somehow simplified version, e.g., excluding subqueries, we propose the following additions to the usual syntax of this clause providing for the direct use of linguistic terms:

<WHERE-clause> ::= WHERE <condition>
<condition> ::= <linguistic quantifier> <sequence of subconditions>
<sequence of subconditions> ::= <subcondition> |
        <subcondition> OR <sequence of subconditions>
<subcondition> ::= <importance coefficient> <linguistic quantifier>
        <sequence of atomic conditions>
<sequence of atomic conditions> ::= <atomic condition> |
        <atomic condition> AND <sequence of atomic conditions>
<atomic condition> ::= <attribute> = <modifier> <fuzzy value> |
        <attribute> <fuzzy relation> <attribute> |
        <attribute> <fuzzy relation> number> |
        <single-valued-attribute> IN <fuzzy-set constant> |
        <multi-valued-attribute> <compatibility operator> <fuzzy-set constant>
<attribute> ::= <numeric field>
<linguistic quantifier> ::= <OWA-tag> <quantifier name>
<OWA-tag> ::= OWA |
<modifier> ::= VERY | MORE OR LESS | RATHER | NOT |
<compatibility operator> ::= possible matching | necessary matching |
        Jackard compatibility

We will discuss particular categories of linguistic terms on examples referring to a hypothetical database of a real estate agency. Particular houses are characterized by: price, land area, location (region), number of bedrooms and bathrooms and other life quality indicators as, e.g., an overall assessment of the environment, transportation infrastructure or shopping facilities.

Atomic condition. The basic building block of a query condition is the *atomic condition*. Basically, it contains a name of an attribute and a constraint imposed on the value of this attribute. Such a constraint may be a traditional, crisp one as, e.g., in "*Price* <= *200,000* ". It may also employ one of linguistic terms as, e.g.:

1. *Price* = ***low*** (numeric fuzzy value)
2. *Land area* = ***very large*** (numeric fuzzy value + modifier)
3. *Price* ***is not much greater than*** *250,000* (fuzzy relation)
4. *Location belongs to* ***favorite regions*** (fuzzy set constant)
5. ***Life quality indicators*** *are* ***compatible,*** *with* ***high quality of life pattern*** (multi-valued-attribute + fuzzy set constant)

Numeric fuzzy values are to be used in connection with numeric fields as, e.g., with the field *Price*. Meaning of such a linguistic term is intuitively obvious, although rather subjective. Thus, it should be possible for each user to define his or her meaning of the linguistic term *low*. On the other hand, it would be advantageous to make it possible to use an already once defined term, like *low*, in various fields. Numeric fuzzy values may be accompanied by *modifiers* as, e.g., *very*, that directly correspond to the similar structures of the natural language.

Fuzzy relations make it possible to soften rigidness of crisp relations. In the third example given above, the atomic condition employs the *much greater than* fuzzy relation accompanied with the negation operator treated as a modifier. Thus, such a condition will accept the price of, e.g., 255,000, which seems to be much more practical than treating 250,000 as a sharp limit.

The examples discussed so far employed linguistic terms to be used along with the numeric data. The fourth example introduces a *fuzzy set constant* which is similar to numeric fuzzy values but meant to be used with scalar data. In this example, the *favorite regions* constant represents the user's preferences as to the location of the house sought. The concept of favorite regions will quite often turn out to be fuzzy, i.e., some regions will be perceived by the user as the best location, some will be completely rejected, and the rest will be acceptable to a degree. Obviously, such a concept is highly subjective.

Finally, the fifth example presents the concept of a *multi-valued attribute*. This concept emerged from the application of our fuzzy querying interface against a database of questionnaires on RTD institutions (see, e.g., Kacprzyk and Zadrożny, 1997c, Zadrożny and Kacprzyk, 1996). It can be best explained on the example given there. In brief, assume that each RTD institution is described by a set of obvious attributes, as, e.g., name, address etc. Usually, such an institution deals with a number of scientific disciplines. Some of them belong to the mainstream of activities of the institution, other may be more or less important. Thus, such an attribute, let us call it *subjects of activity*, may take on a set of values instead of just one as it is required in the classic relational approach. Moreover, quite often it may be advantageous to treat such a set as a fuzzy set, i.e., the institute's mainstream scientific disciplines will be assigned the membership degree 1 and the rest of them an appropriate degree from (0,1). Then, the user looking for an RTD institution in our database, may specify in his or her query another, possibly fuzzy, set of disciplines. Such a query should also employ one of the supported *compatibility operators*, making it possible to calculate the matching of (fuzzy) sets provided in the query and each of the relevant records from the database. Various compatibility operators may satisfy different requirements of the user. Namely, keeping up with our example, the user may be looking for an institution dealing:

- with at least one subject of activity specified in a query,
- with all activities specified in a query (and maybe some more),
- only with activities specified in the query, not necessarily all,
- with most activities specified in the query and not many other.

Obviously, appropriate queries may be constructed without calling for multi-valued attributes and compatibility operators concepts, but such a construction would be rather cumbersome.

We should stress here the difference between this concept of a multi-valued attribute and similar concepts used in the fuzzy databases. In the similarity based as well as in the possibility based models of the fuzzy database, a field may take on a set as its value. However, in both models, a set is used due to the imprecise knowledge of the real, *single* value of the field. Our multi-valued attribute, on the other hand, takes a set as its value by definition. As was mentioned earlier, we introduced the concept of a multi-valued attribute for practical reasons. This makes the use of our fuzzy querying interface more suitable in some situations. Although it seems to go beyond the paradigm of fuzzy querying against a crisp database, in fact it does not. As it is explained in what follows, multi-valued attribute is somehow a *virtual* concept and does not require any changes in the underlying, traditional DBMS. This virtual nature of our multi-valued attribute resembles the concept of an ambiguous attribute (cf. Nomura et al., 1992).

A traditional atomic condition, i.e., one without any linguistic terms, may be satisfied by a given record from the database (or, more precisely, by a given field from a row/record of a table in question) or not. Thus, binary logic applies. In the opposite case, an atomic condition may be satisfied to a *degree*. This is the essence of the linguistic approach. For example, the price may be definitely *low* or definitely *not low*, but also intermediate cases do exist.

<u>Sequence of atomic conditions and a subcondition.</u> The *sequence of atomic conditions* is just a conjunction of atomic conditions. Due to the fact that particular atomic conditions may be satisfied to a degree, we need to employ some generalization of the classical AND logical connective. For the fuzzy logic based approach, the natural choice is the *min* operator.

In the context considered here, the classical logical connective AND may be perceived as an aggregation operator. Namely, it aggregates the truth values of atomic conditions. This may however be sometimes counter-intuitive. Let us consider two cases. First, where all but one of n atomic conditions are met, and the second where just one atomic condition is satisfied. In both cases, the result of aggregation will be *False* (or 0). The *min* operator, making it possible to aggregate non-binary logical values, still preserves this behavior of the AND connective. In order to achieve a flexibility of the aggregation postulated at the beginning of this section, a linguistic quantifier may be applied for selected sequence of atomic conditions. Then, it replaces the AND connective as the aggregation operator. Linguistic quantifiers may be defined and interpreted in two ways, as it is explained in what follows. Here, we should only add that a *default* linguistic quantifier in the OWA operator form is automatically attached to each sequence of atomic conditions used in a query without explicitly specified quantifier. Originally, this default quantifier acts exactly like the AND connective, but may be easily manipulated by the user so as to reach a suitable aggregation behavior.

Each sequence of atomic conditions may be additionally assigned an importance coefficient. That way, the user may vary the degree to which given sequence contributes to the overall satisfaction degree of the whole query.

Finally, the sequence of atomic conditions, possibly accompanied by a linguistic quantifier and an importance coefficient, is called a *subcondition.*

Sequence of subconditions and a condition. The *sequence of subconditions* is the disjunction of subconditions. This structuring of various elements of the condition adheres to the scheme assumed in Microsoft Access. In fact, using this package, the user composes the query's condition in the disjunctive normal form.

As in case of the AND connective, the OR connective is replaced by the corresponding *max* operator of fuzzy logic. Again, sometimes it may be advantageous to apply another, more flexible scheme of aggregation. In such a case, a linguistic quantifier may be employed. The way a selected quantifier works may be guided by importance coefficients, possibly present in particular subconditions.

2.2. Representation of linguistic terms

Basically, all linguistic terms are represented as fuzzy sets. Obviously, the universe used to define particular fuzzy set depends on the type of their corresponding linguistic term. Namely, in order to represent terms used along with the *numeric* fields (attributes) we use trapezoidal fuzzy numbers. It seems obvious, that these fuzzy numbers are defined over the domains of corresponding fields. Namely, if the *price* of the house ranges from, say, 10,000 up to 100,000 USD, then the fuzzy value *low*, to be used in the atomic condition "price is low", should be defined as a fuzzy number over the interval [10,000, 100,000]. However, then if we strictly stick to this scheme, we have to define *low* separately for each field which we are going to use together with this linguistic term. In order to avoid this apparent inconvenience, we propose to use "context-free" definitions of numeric linguistic terms. Namely, the corresponding fuzzy numbers are defined over the universal interval, e.g., [-10, 10]. Additionally, for each relevant field, its range of variability, the [LowerLimit, UpperLimit] interval, has to be provided before a given linguistic term may be used along with this field in an atomic condition. Then, during the calculation of a matching degree the definitions of relevant fuzzy numbers are translated into the domains of corresponding fields. In some cases, defining fuzzy numbers over the universal interval may be cumbersome. Thus, we provide the possibility to use, both, context-free as well as context-dependent definitions of fuzzy numbers.

The same approach is adopted for dealing with *fuzzy relations*. Basically, a fuzzy relation is a fuzzy set defined in the universe of the Cartesian product of two domains. In the context considered here, these are the domains of one or two *numeric* attributes involved. Obviously, it would be rather tedious to define a fuzzy relation directly on this Cartesian product. Thus, in our approach as the universe for the definition of a fuzzy relation we use the set of possible values of the difference of values of the two attributes involved. More precisely, the same universal interval [-10,10] is used for the definition which is then translated into the real interval of possible differences during the matching degree calculation. The definitions again take the form of trapezoidal fuzzy numbers.

The idea of a *universal* domain may also be applied for the definition of a linguistic quantifier. In fact, in this case it is then even more essential. Basically, linguistic quantifiers are defined in terms of Zadeh's (1983) approach. Again, the quantifier is represented as a fuzzy number – this time not trapezoidal but similar, defined by a piecewise linear membership function. The linguistic quantifier may be treated as an aggregation operator. Briefly speaking, in order to aggregate, we have to carry out the normalization using the number of propositions being aggregated, *N*. In case of our querying language syntax, the propositions correspond to the *atomic conditions* or *subconditions*. Thus, the number *N* will usually vary essentially between different queries. This, again, requires separate definitions of linguistic quantifiers for all queries posed. The solution is similar to that applied for numeric fuzzy terms but this time [0, 10] plays the role of the universal interval. We assume that the membership function of the fuzzy linguistic quantifier is piecewise linear (cf. Zadeh, 1983).

In our approach, a linguistic quantifier has to be defined in the sense of Zadeh (1983). Then, for the matching degree calculation two options exist. First, the quantifier's original form may be employed directly. Second, an original definition may be automatically translated in order to obtain a corresponding Yager's OWA operator (cf. Yager, 1988, Yager and Kacprzyk, 1997). In the latter case, an OWA operator may be further fine tuned by the user so as to achieve a required aggregation functionality.

The linguistic terms envisaged for the use along with scalar attributes, i.e., *fuzzy set constants*, have to be defined directly in the appropriate universe of discourse corresponding to the domain of a given scalar attribute. More precisely, the set of given attribute values is taken as the universe of discourse. Thus, these constants are, by definition, context dependent. Their discrete membership functions are defined as arrays of numbers. Fuzzy set constants may be defined for any field of character (text) type. Since such a constant is dedicated for a selected field, no additional data about this field is required. This is in contrast to the case of context-independent numeric fuzzy values, where appropriate lower and upper limits have to be provided. Fuzzy set constants may also be defined for - and used together with - *multi-valued attributes*. Obviously, these attributes have to be separately defined, according to the explanation given earlier. The definition consists in indicating the list of fields which may then be treated jointly as a virtual attribute.

Basically, information regarding all linguistic terms is stored in some additional tables in the host relational database. Thus, it may be treated as part of *metadata* usually maintained by the relational database management system.

2.3. Computation of a matching degree with words (linguistic terms)

The answer to a classical query is a set of records matching the query, i.e., meeting conditions specified in the query. Classically, the notion of matching is binary: a record satisfies conditions or not. In case of flexible queries containing ambiguous and imprecise linguistic terms it is natural to consider a *degree of matching*. Here we briefly present algorithms for calculation of matching degree.

Real life corporate databases may contain hundreds of thousands of records and more. Queries have to be processed in a highly optimized way by, e.g., exploiting all existing and relevant indexes. However, for our purposes we will adopt more simplified a view of this procedure. Namely, we assume that query processing consists in scanning all relevant records, calculating for each of them the matching degree. Thus, while processing a query, a DBMS does the following:

1) fetches a record from the database,
2) evaluates all (or only a part of them, depending on the query structure and adopted optimization scheme) atomic conditions of the query using attributes' values present in the record,
3) aggregates partial matching degrees obtained in the previous step, thus producing the overall matching degree sought,
4) if this matching degree is high enough, includes the record in the answer to the query,
5) moves to the next record,
6) if there are no more records, stop,
7) otherwise, proceeds to step 1.

This procedure does not require any comments, except for steps 2 and 3. In what follows we explain the calculations carried out in these steps. We omit obvious operations on crisp data and focus on the computing with words, i.e., linguistic terms, possibly appearing in the query.

We assume the following notation:

AT_1, AT_2 – attributes,
$v(AT)$ – value of attribute AT in given record,
$[LL_{AT}, UL_{AT}]$ – lower and upper limit for the values of attribute AT (as defined by the user),
τ_{AT} – mapping from the interval $[LL_{AT}, UL_{AT}]$ onto the *universal interval* [-10,10]
τ_{AT1AT2} – mapping from the interval $[LL_{AT1}\text{-}UL_{AT2}, UL_{AT1}\text{-}LL_{AT2}]$ onto the *universal interval* [-10,10],
FV, μ_{FV} – numeric fuzzy value and its membership function; μ_{FV}: $[-10, 10] \rightarrow [0,1]$,
MOD, η – modifier and its associated function; η: $[0,1] \rightarrow [0,1]$,

FR, μ_{FR} – fuzzy relation and its membership function; μ_{FR}: [-10, 10] → [0,1],
FS, μ_{FS} – fuzzy set constant and its membership function defined on a custom domain,
MVAT – multi-valued attribute,
μ_{MVAT} – value of multi-valued attribute, i.e., membership function of corresponding fuzzy set
COMPOP – compatibility operator (label)
$\psi_{COMPOP}(\cdot,\cdot)$ – function providing semantics for given compatibility operator (see comments under the table)
Q, μ_Q – fuzzy (linguistic) quantifier and its membership function; μ_Q: [0, 10] → [0,1]
τ_N – mapping from the interval [0,N] onto the *universal interval* [0,10]; N is the number of atomic conditions in a subcondition or the number of subconditions in the whole query
md_i – partial matching degree (for *i*-th atomic condition or subcondition)
κ_i – importance coefficient of the *i*-th subcondition

In the following table, we collected the formulas used to obtain a matching degree for atomic conditions containing particular linguistic terms:

Linguistic term type	Atomic condition form	Formula for the calculation Of matching degree
Numeric fuzzy value	AT = FV	$\mu_{FV}(\tau_{AT}(v(AT))$
Numeric fuzzy value with modifier	AT = MOD FV	$\eta(\mu_{FV}(\tau_{AT}(v(AT)))$
Fuzzy relation	AT_1 FR AT_2	$\mu_{FR}(\tau_{AT_1AT_2}(v(AT_1)-v(AT_2)))$
Fuzzy relation	AT FR number	$\mu_{FR}(\tau_{ATAT}(v(AT)-number))$
Fuzzy set constant	AT in FS	$\mu_{FS}(v(AT))$
Fuzzy set constant with multi-valued attribute	MVAT COMPOP FS	$\psi_{COMPOP}(\mu_{MVAT},\mu_{FS})$

The mappings τ are defined also outside the intervals previously mentioned. Namely, for arguments less than the lower limit of the interval they take on –10 as their value. Similarly, for arguments higher than the upper limit they take on 10. These additional assumptions are needed because it is possible that, e.g., v(AT) for a certain record in a database lies outside of that interval. In other words, the interval corresponding to the range of variability of a particular attribute, as perceived and specified by the user, need not be identical with the actual range of variability of this attribute in the database.

Compatibility operators measure the compatibility of the value of a multi-valued attribute and a fuzzy set constant. The former may be a crisp set or a fuzzy set represented by a characteristic or membership function, respectively. The characteristic function may be treated as a special case of the membership function, thus, in what follows, we assume the latter is given. Various compatibility operators are applicable depending on the intended meaning of a query. In a table below, we present formulas for computing ψ_{COMPOP} for four operators, whose use we advocate.

Compatibility operator	Formula for $\psi_{\text{COMPOP}}(\mu_{\text{MVAT}}, \mu_{\text{FS}})$
Degree of possibility of matching	$\Pi(\text{FS} \mid \text{MVAT}) = \sup_{u \in U} \min(\mu_{\text{FS}}(u), \mu_{\text{MVAT}}(u))$
Degree of necessity of matching	$\text{N}(\text{FS} \mid \text{MVAT}) = \inf_{u \in U} \max(1 - \mu_{\text{FS}}(u), \mu_{\text{MVAT}}(u))$
Degree of necessity of matching 2	$\text{N}(\text{MVAT} \mid \text{FS}) = \inf_{u \in U} \max(1 - \mu_{\text{MVAT}}(u), \mu_{\text{FS}}(u))$
Generalized Jaccard coefficient	$\lvert FS \cap D \rvert / \lvert FS \cup D \rvert = \sum_{u \in U} (\mu_{\text{FS}}(u) \wedge \mu_{\text{MVAT}}(u)) / \sum_{u \in U} (\mu_{\text{FS}}(u) \vee \mu_{\text{MVAT}}(u))$

The compatibility operator based on the degree of necessity of matching is not symmetric as to the arguments. Thus, its second version, given in the third entry of the table, has a different semantic which may be appropriate for some queries.

Finally, the partial matching degrees calculated for atomic conditions (subconditions) are aggregated in the following way:

$$md = \mu_Q(\tau_N(\frac{1}{N}\sum_{i=1}^{N} md_i)) \tag{1}$$

where *md* stands for subcondition (overall) matching degree.

For the overall matching degree calculation, the above formula may be extended to take into account the importance coefficients of particular subconditions (cf. Zadeh, 1983):

$$md = \mu_Q(\tau_N(\frac{\sum_i (md_i \wedge \kappa_i)}{\sum_i \kappa_i})) \tag{2}$$

In some cases other means of grasping the meaning of linguistic quantification (aggregation) are more suitable. For example, the OWA operators due to Yager (1988) [cf. Yager and Kacprzyk's (1997) volume] provide more flexibility in defining the range of aggregation operators.

3. Linguistic summaries using fuzzy logic with linguistic quantifiers

Here we will briefly present the basic Yager's (1982) approach to the linguistic summarization of sets of data. This will provide a point of departure for our further analysis of more complicated and realistic summaries.

In Yager's (1982) approach, we have:

- V is a quality (attribute) of interest, e.g. salary in a database of workers,
- $Y = \{y_1, \ldots, y_n\}$ is a set of objects (records) that manifest quality V, e.g. the set of workers; hence $V(y_i)$ are values of quality V for object y_i,
- $D = \{V(y_1), \ldots, V(y_n)\}$ is a set of data (the "database" on question)

A *summary* of data set consists of:

- a summarizer S (e.g. young),
- a quantity in agreement Q (e.g. most),
- truth (validity) T - e.g. 0.7,

as, e.g., "T(*most* of employees are *young*)=0.7".

Basically, given a set of data D, we can hypothetize any appropriate summarizer S and any quantity in agreement Q, and the assumed measure of truth (validity) will indicate the truth (validity) of the statement that Q data items satisfy the statement (summarizer) S.

First, we should comment on the form of the basic elements of the summary, i.e. the summarizer, quantity in agreement, and how to calculate the degree of truth (validity).

3.1. On the form of the summarizer

Since the only fully natural and human consistent means of communication for the humans is natural language, then we assume that the summarizer S is a linguistic expression semantically represented by a fuzzy set. For instance, in our example a summarizer like "young" would be represented as a fuzzy set in the universe of discourse as, say, {1, 2, ..., 90}, i.e. containing possible values of the human age, and "young" could be given as, say, a fuzzy set with a nonincreasing membership function in that universe such that, in a simple case of a piecewise linear membership function, the age up to 35 years is for sure "young", i.e. the grade of membership is equal to 1, the age over 50 years is for sure "not young", i.e. the grade of membership is equal to 0, and for the ages between 35 and 50 years the grades of membership are between 1 and 0, the higher the age the lower its corresponding grade of membership. Clearly, the meaning of the summarizer, i.e. its corresponding fuzzy set is in practice subjective, and may be either predefined or elicited from the user when needed.

Such a simple one-attribute-related summarizer exemplified by „young" does well serve the purpose of introducing the concept a a linguistic summary, hence it was assumed by Yager (1982). However, it is of a lesser practical relevance. It can be extended, for some confluence of attribute values as, e.g, "*young* and *well paid*", and then to more complicated combinations.

Clearly, when we try to linguistically summarize data, the most interesting are non-trivial, *human-consistent* summarizers (concepts) as, e.g.:

- *productive* workers,
- *stimulating* work environment,
- *difficult* orders, etc.

involving complicated *combinations of attributes*, e.g.: a hierarchy (not all attributes are of the same importance), the attribute values are ANDed and/or ORed, *k* out of *n*, *most*, etc. of them should be accounted for, etc.

The generation and processing of such non-trivial summarizers needs some specific tools and techniques to be discussed later.

3.2. On the form of the quantity in agreement

The quantity in agreement, *Q*, is a proposed indication of the extent to which the data satisfy the summary. Once again, a precise indication is not human consistent, and a linguistic term represented by a fuzzy set is employed.

Basically, two types of such a linguistic quantity in agreement can be used:

- absolute as, e.g., „about 5", „more or less 100", „several", and
- relative as, e.g., „a few", „more or less a half", „most", almost all"etc.

Notice that the above linguistic expressions are the so-called fuzzy linguistic quantifiers (cf. Zadeh, 1983, 1985) that can be handled by fuzzy logic.

As for the fuzzy summarizer, also in case of a fuzzy quantity in agreement, its form is subjective, and can be either predefined or elicited from the user when needed.

3.3. Calculation of the truth (validity) of a linguistic summary

Basically, the calculation of the truth (validity) of the basic type of a linguistic summary considered in this section is equivalent to the calculation of the truth value (from the unit interval) of a linguistically quantified statement (e.g., „*most* of the employees are *young*"). This may be done by two most relevant techniques using either Zadeh's (1983) calculus of linguistically quantified statements [cf. Zadeh and Kacprzyk (1992)] or Yager's (1988) OWA operators [cf. Yager and Kacprzyk (1997)]; for a survey, see Liu and Kerre (1998).

A linguistically quantified proposition, exemplified by "most experts are convinced", is written as "Qy 's are F" where Q is a linguistic quantifier (e.g., most), $Y=\{y\}$ is a set of objects (e.g., experts), and F is a property (e.g., convinced). Importance B may be added yielding "QBy 's are F", e.g., "most (*Q*) of the important (*B*) experts (*y*'s) are convinced (*F*)". The problem is to find truth(Qy 's are F) or truth(QBy 's are F), respectively, knowing truth(y is F), $\forall y \in Y$ which is done here using Zadeh's (1983) fuzzy-logic-based calculus of linguistically quantified propositions.

Property F and importance B are fuzzy sets in Y, and a (proportional, nondecreasing) linguistic quantifier Q is assumed to be a fuzzy set in [0,1] as, e.g.

$$\mu_Q(x) = \begin{cases} 1 & \text{for } x \geq 0.8 \\ 2x - 0.6 & \text{for } 0.3 < x < 0.8 \\ 0 & \text{for } x \leq 0.3 \end{cases} \tag{3}$$

Then, due to Zadeh (1983)

$$\text{truth}(Qy\text{'s are } F) = \mu_Q[\tfrac{1}{n}\textstyle\sum_{i=1}^{n}\mu_F(y_i)] \tag{4}$$

$$\text{truth}(QBy\text{'s are } F) = \mu_Q[\textstyle\sum_{i=1}^{n}(\mu_B(y_i) \wedge \mu_F(y_i)) / \sum_{i=1}^{n}\mu_B(y_i)] \tag{5}$$

An OWA operator [Yager (1988); Yager and Kacprzyk (1997)] of dimension p is a mapping $F:[0,1]^p \to [0,1]$ if associated with F is a weighting vector $W = [w_1,\ldots,w_p]^T$, $w_i \in [0,1], w_1 + \cdots + w_p = 1$, and

$$F(x_1,\ldots,x_p) = w_1 b_1 + \cdots w_p b_p = W^T B \tag{6}$$

where b_i is the i-th largest element among $x_1,\ldots,x_p$, $B = [b_1,\ldots,b_p]$.

The OWA weights may be found from the membership function of Q due to [cf. Yager (1988)]:

$$w_i = \begin{cases} \mu_Q(i) - \mu_Q(i-1) & \text{for i} = 1,\ldots,p \\ \mu_Q(0) & \text{for i} = 0 \end{cases} \tag{7}$$

The OWA operators can model a wide array of aggregation operators (including linguistic quantifiers), from $w_1 = \cdots = w_{p-1} = 0$ and $w_p = 1$ which corresponds to "all", to $w_1 = 1$ and $w_2 = \cdots = w_p = 0$ which corresponds to " at least one", through all intermediate situations.

An important issue is related with the OWA operators for importance qualified data. Suppose that we have $A = [a_1,\ldots,a_p]$, and a vector of importances $V = [v_1,\ldots,v_p]$ such that $v_i \in [0,1]$ is the importance of $a_i, i = 1,\ldots,p$, $v_1 + \cdots v_p = 1$. The case of an *ordered weighted averaging operator with importance qualification*, denoted OWA_Q is not trivial. In Yager's (1988) approach to be used here, which seems to be highly plausible (though is sometimes criticized), some redefinition of the OWA's weights w_i's into $\overline{w}_i$'s is performed, and (6) becomes

$$F_I(x_1, \ldots, x_p) = \overline{w}_1 b_1 + \cdots \overline{w}_p b_p = \overline{W}^T B \tag{8}$$

where

$$w_j = \mu_Q \left(\frac{\sum_{k=1}^{j} u_k}{\sum_{k=1}^{p} u_k} \right) - \mu_Q \left(\frac{\sum_{k=1}^{j-1} u_k}{\sum_{k=1}^{p} u_k} \right) \tag{9}$$

where u_k is the importance of b_k, i.e. the k-largest element of A (i.e. the corresponding v_k).

The basic validity criterion, i.e. the truth of a linguistically quantified statement given by (4) and (5), is certainly the most important in the general framework assumed. However, it does not grasp all aspects of a linguistic summary. Some attempts at devising other quality (validity) criteria will be briefly surveyed following Kacprzyk and Yager (2000).

First, Yager (1982, 1991) proposed a measure of informativeness that may be summarized as follows. Suppose that we have a data set, whose elements are from a measurement space X. One can say that the data set itself is its own most informative description, and any other summary implies a loss of information. So, a natural question is whether a particular summary is informative, and to what extent.

It turns out that the degree of truth used so far is not a good measure of informativeness (cf. Yager, 1982, 1991). Let the summary be characterized by the triple (S, Q, T), and let a related summary be characterized by the triple (S^c, Q^c, T), such that S^c is the negation of S, i.e. $\mu_S{}^c(.) = 1 - \mu_S(.)$, and $\mu_Q{}^c(.) = 1 - \mu_Q(.)$.

Then Yager (1982, 1991) proposes the following measure of informativeness of a summary

$$I = [T \cdot SP(Q) \cdot SP(S)] \vee [(1-T) \cdot Sp(Q^c) Sp(S^c)] \tag{10}$$

where $SP(Q)$ is the specificity of Q given as

$$SP(Q) = \int_0^1 \frac{1}{\text{card} Q_\alpha} d_\alpha \tag{11}$$

where Q_α is the α-cut of Q and card(.) is the cardinality of the respective set; and similarly for Q^c, S, S^c.

This measure of informativeness results from a very plausible reasoning which can be found, e.g., in Yager (1982, 1991).

Unfortunately, though the above measure of informativeness is plausible and a considerable step forward, it is by no means a definite solution. First, let us briefly mention George and Srikanth's (1996) proposal. Suppose that a linguistic summary of interest involves more than 1 attribute (e.g., „age", „salary" and „seniority" in the case of employees). Basically, for the same set of data, two summaries are generated:

- a *constraint descriptor* which is the most specific description (summary) that fits the largest number of tuples in the relation (database) involving the attributes in question,
- a *constituent descriptor* which is the decsription (summary) that fits the largest subset of tuples with the condition that each tuple attribute value takes on at least a threshold value of membership.

George and Srikanth (1996) use these two summaries to derive a fitness function (goodness of a summary) that is later used for deriving a solution (a best summary) via a genetic algorithm they employ. This fitness function represent a compromise between the most specific summary (corresponding to the constraint descriptor) and the most general summary (corresponding to the constituent descriptor).

Then, Kacprzyk (1999a, b), and Kacprzyk and Strykowski (1999a, b) propose some additional measures that have been further developed by Kacprzyk and Yager (2000).

For convenience of the reader, let us briefly repeat some basic notation. We have a data set (database) D that concerns some objects (e.g. employees) $Y = \{y_1, ..., y_n\}$ described by some attribute V (e.g. age) taking on values in a set $X = \{x_1, x_2, ...\}$ exemplified by {20, 21, ..., 100} or {very young, young, ..., old, very old}. Let $d_i = V(y_i)$ denote the value of attribute V for object y_i. Therefore, the data set to be summarized is given as a table

$$D = [d_1, ..., d_n] = [V(y_1), V(y_2), ..., V(y_n)] \tag{12}$$

In a more realistic case the data set is described by more than one attribute, and let $V = \{V_1, V_2, ..., V_m\}$ be a set of such attributes taking values in X_i, i=1, ..., m; $V_j(y_i)$ denotes the value of attribute V_j for object y_i, and attribute V_j takes on its values from a set X_j.

The data set to be summarized is therefore:

$$D = \{[V_1(y_1), V_2(y_1), ..., V_m(y_1)], [V_1(y_2), V_2(y_2), ..., V_m(y_2)], ..., [V_1(y_n), V_2(y_n), ..., V_m(y_n)]\} \tag{13}$$

In this case of multiple attributes the description (summarizer) S is a family of fuzzy sets $S = \{S_1, S_2, ... S_m\}$ where S_i is a fuzzy set in X_i, $i = 1, ..., m$. Then, $\mu_S(y_i)$ may be defined as:

$$\mu_S(y_i) = \min_{j \in \{1,2,...,n\}}[\mu_{S_j}(V_j(y_i))] \tag{14}$$

So, having S, we can calculate the truth value T of a summary for any quantity in agreement. However, to find a best summary, we should calculate T for each possible summarizer, and for each record in the database in question. This computational prohibitive for virtually all non-trivial databases and number of attributes.

A natural line of reasoning would be to either limit the number of attributes of interest or to limit the class of possible summaries by setting a more specific description by predefining a „narrower" description (e.g. very young , young and well paid, etc. employees). This will limit the search space.

We will deal here with the second option. The user can limit the scope of a linguistic summary to, say, those for which the attribute „age" takes on the value „young (employees)" only, i.e. to fix the summarizer related to that attribute. That is, this will correspond to the searching of the database using the query w_g equated with the fuzzy set in X_g corresponding to „young" related to attribute V_g (i.e. age), i.e. characterized by $\mu_{w_g}(.)$. In such a case, $\mu_S(y_i)$ given by (14) becomes

$$\mu_S(y_i) = \min_{j\in\{1,2,\dots,m\}}[\mu_{Sj}(V_j(y_i)) \wedge \mu_{w_g}(V_g(y_i)], \qquad i=1, \dots, n \tag{15}$$

where " $\wedge$ " is the minimum (or, more generally, a t-norm), and then

$$r = \frac{\sum_{i=1}^{n} \mu_S(y_i)}{\sum_{i=1}^{n} \mu_{w_g}(V_g(y_i))} \tag{16}$$

and $T=\mu_Q(r)$.

Now, we will present the 5 new quality measures of quality of linguistic database summaries introduced in Kacprzyk (1999a, b), Kacprzyk and Strykowski (1999a, b), and Kacprzyk and Yager (2000):

- a truth value [which basically corresponds to the degree of truth of a lingustically quantified proposition representing the summary given by, say, (4) or (5)],
- a degree of imprecision (fuzziness),
- a degree of covering,
- a degree of appropriateness,
- a length of a summary,

and these degrees will now be formally defined.

For notational simplicity later on, let us rewrite (15) and (16) as:

$$\mu_S(d_i) = \min_{j\in\{1,2,\dots,m\}}[(\mu_{Sj}(V_j(y_i))], \qquad i=1, \dots, n \tag{17}$$

and

$$r = \frac{\sum_{i=1}^{n} [\mu_S(y_i) \wedge \mu_{w_g}(V_g(y_i))]}{\sum_{i=1}^{n} \mu_{w_g}(V_g(y_i))} \tag{18}$$

The **degree of truth**, T_1, is the basic validity criterion introduced in the source Yager's (1982, 1991) works and commonly employed. It is clearly equal to

$$T_1 = \mu_Q(r) \tag{19}$$

and (17) results clearly from Zadeh's (1983) calculus of linguistically quantified propositions.

The **degree of imprecision (fuzziness, specificity)** is an obvious and important validity criterion. Basically, a very imprecise (fuzzy) linguistic summary (e.g. on almost all winter days the temperature is rather cold) has a very high degree of truth yet it is not useful.

Suppose that description (summarizer) S is given as a family of fuzzy sets $S=\{s_1, s_2, \ldots s_m\}$. For a fuzzy set s_j, $j=1, \ldots, m$, we can define its degree of fuzziness as, e.g.:

$$\text{in}(s_j) = \frac{\text{card}\{x \in X_j : \mu_{s_j}(x) > 0\}}{\text{card}\, X_j} \tag{20}$$

where card denotes the cardinality of the corresponding (nonfuzzy) set. That is, the „flatter" the fuzzy set s_j the higher the value of in(s_j). The degree of imprecision (fuzziness), T_2, of the summary – or, in fact, of S – is then defined as:

$$T_2 = 1 - \sqrt[m]{\prod_{j=1,\ldots,m} \text{in}(s_j)} \tag{21}$$

Notice that the degree of imprecision T_2 depends on the form of the summary only and not on the database, that is its calculation does not require the searching of the database (all its records) which is very important.

The **degree of covering**, T_3, is defined as

$$T_3 = \frac{\sum_{i=1}^{n} t_i}{\sum_{i=1}^{n} h_i} \tag{22}$$

where:

$$t_i = \begin{cases} 1 & \text{if } \mu_S(y_i) > 0 \quad \text{and} \quad \mu_{w_g}(V_g(y_i)) > 0 \\ 0 & \text{otherwise} \end{cases}$$

$$h_i = \begin{cases} 1 & \text{if } \mu_{w_g}(V_g(y_i)) > 0 \\ 0 & \text{otherwise} \end{cases}$$

The degree of covering says how many objects in the data set (database) corresponding to the query w_g are „covered" by the particular summary, i.e. to the particular description S. Its interpretation is simple as, e.g., if it is equal to 0.15, then this means that 15% of the objects are consistent with the summary in question. The value of this degree depends clearly on the contents of the database.

The **degree of appropriateness** is the most relevant degree of validity. To present its idea, suppose that the summary containing the description (fuzzy sets) $S = (S_1, S_2, \ldots, S_m)$ is partitioned into m partial summaries each of which encompasses the particular attributes $S_1, S_2, \ldots, S_m$, such that each partial summary corresponds to one fuzzy value only, then if we denote:

$$S_j(y_i) = \mu_{Sj}(V_j(y_i)) \tag{23}$$

then

$$r_j = \frac{\sum_{i=1}^{n} h_i}{n}, \quad j=1, \ldots, n \tag{24}$$

where:

$$h_i = \begin{cases} 1 & \text{if } S_j(y_i) > 0 \\ 0 & \text{otherwise} \end{cases}$$

and the degree of appropriateness, T_4, is defined as:

$$T_4 = \text{abs}\left(\prod_{j=1,\ldots,m} r_j - T_3\right) \tag{25}$$

The degree of appropriateness means that if we have a database concerning the employees. If 50% of them are less than 25 year old and 50% are highly qualified, then we may expect that 25% of the employees would be less than 25 year old and highly qualified; this would correspond to a typical, fully expected situation. However, if the degree of appropriateness is, say, 0.39 (i.e. 39% are less than 25 years old and highly qualified), then the summary found reflects an interesting, not fully expected relation in our data. This degree describes therefore how characteristic for the particular database the summary found is. T_4 is very important because, for instance, a trivial summary like „100 % of sale is of any articles" has full validity (truth) if we use the traditional degree of truth but its degree of appropriateness is equal 0 which is correct.

The **length** of a summary is relevant because a long summary is not easily comprehensible by the human user. This length, T_5, may be defined in various way, and the below form has proven to be useful:

$$T_5 = 2\,(0.5^{\text{card}\,S}) \tag{26}$$

where card S is the number of elements in S.

Now, the (total) degree of validity, T, of a particulat linguistic summary is defined as the weighted average of the above 5 degrees of validity, i.e.:

$$T = T(T_1, T_2, T_3, T_4, T_5; w_1, w_2, w_3, w_4, w_5) = \sum_{i=1,2,\ldots,5} w_i T_i \tag{27}$$

And the problem is to find an optimal summary, $S^* \in \{S\}$, such that

$$S^* = \arg\max_S \sum_{i=1,2,\ldots,5} w_i T_i \tag{28}$$

where: $w_1,\ldots,w_5$ are weights assigned to the particular degrees of validity, with values from the unit interval, the higher the more important such that $\sum_{i=1,2,\ldots,5} w_i = 1$.

The definition of weights, $w_1,\ldots,w_5$, is a problem in itself, and will not be dealt with in more detail. The weights can be predefined or elicited from the user. In the case study presented later on the weights are determined by using the well-known Saaty's (1980) AHP (analytical hierarchy process) approach that works well in the problem considered.

3.4. An example of implementation

The proposed data summarization procedure was implemented on a sales database of a computer retailer in Southern Poland [cf. Kacprzyk (1999a, b), Kacprzyk and Strykowski (1999a, b)].

The basic structure of the database is as shown in Table 1.

Table 1. Structure of the database

Attribute name	Attribute type	Description
Date	Date	Date of sale
Time	Time	Time of sale transaction
Name	Text	Name of the product
Amount (number)	Numeric	Number of products sold in the transaction
Price	Numeric	Unit price
Commission	Numeric	Commission (in %) on sale
Value	Numeric	Value = amount (number) x price; of the product
Discount	Numeric	Discount (in %) for transaction
Group	Text	Product group to which the product belongs
Transaction value	Numeric	Value of the whole transaction
Total sale to customer	Numeric	Total value of sales to the customer in fiscal year
Purchasing frequency	Numeric	Number of purchases by customer in fiscal year
Town	Text	Town where the customer lives

First, after some initialization, we provide the parameters concerning mainly: definition of attributes and the subject, definition of how the results should be presented, definition of parameters of the method (i.e. a genetic algorithm or, seldom, full search). Then, we initialize the search, and obtain the results shown in the next tables in which the consecutive columns contain: the summary, values of the 4 indicators, i.e. the degrees of appropriateness, covering, truth, and fuzziness (the length is not accounted for in our simple case), and finally the weighted average (with weights chosen after some learning and fine tuning phase).

Let us analyze some exemplary results. Suppose that we are interested in relations between the groups of products and times of sale. Then, we obtain the results as in Table 2.

In Table 3 some of the obtained linguistic summaries expressing relations between the attributes: size of customer, regularity of customer (purchasing frequency), date of sale, time of sale, commission, group of product and day of sale. This is an example of the most sophisticated form of linguistic summaries allowed in the system. These sets of most valid summaries (normally, not just one summary) will give much inside into relations between the attributes chosen.

Table 2. Linguistic summaries expressing relations between the groups of products and times of sale

Summary	Degree of appropriateness	Degree of covering	Degree of validity
	Degree of imprecision	Weighted average	
About 1/3 of sales of computers is by the end of year	0,0999 0,2010	0,3009 0,1274	0,2801
About ½ of sales in autumn is of accessories	0,0642 0,4095	0,4737 0,1143	0,4790
About 1/3 of sales of network elements is in the beginning of year	0,0733 0,2124	0,2857 0,0982	0,1957
Very few sales of network elements is by the end of year	0,0833 0,2010	0,1176 0,0980	0,0929
Very few sales of software is in the beginning of year	0,0768 0,2124	0,1355 0,0929	0,0958
About ½ of sales in the beginning of year is of accessories	0,0348 0,4095	0,4443 0,0860	0,4343
About 1/3 of sales in the summer is of accessories	0,0464 0,2745	0,3209 0,0853	0,3092
About 1/3 of sales of peripherals is in the spring period	0,0507 0,2525	0,3032 0,0809	0,2140
About 1/3 of sales of software is by the end of year	0,0446 0,2010	0,2455 0,0768	0,2258
About 1/3 of sales of network elements is in the spring period	0,0458 0,2525	0,2983 0,0763	0,2081
About 1/3 of sales in the summer period is of components	0,0336 0,2745	0,3081 0,0745	0,3081

Table 3. Linguistic summaries expressing relations between the attributes: size of customer, regularity of customer (purchasing frequency), date of sale, time of sale, commission, group of product and day of sale

Summary	Degree of appropriateness Degree of imprecision	Degree of covering Weighted average	Degree of validity
Much sales on Saturday is about noon with a low commission	0,3843 0,2748	0,6591 0,3863	0,3951
Much sales on Saturday is about noon for bigger customers	0,3425 0,4075	0,7500 0,3648	0,4430
Much sales on Saturday is about noon	0,3133 0,4708	0,7841 0,3564	0,4654
Much sales on Saturday is about noon for regular customers	0,3391 0,3540	0,6932 0,3558	0,4153
A few sales for regular customers is with a low commission	0,3882 0,5837	0,1954 0,3451	0,1578
A few sales for small customers is with a low commission	0,3574 0,5837	0,2263 0,3263	0,1915
A few sales for one-time customers is with a low commission	0,3497 0,5837	0,2339 0,3195	0,1726
Much sales for small customers is for nonregular customers	0,6250 0,1458	0,7709 0,5986	0,5105

4 Linguistic summaries from the perspective of a fuzzy logic based querying interface

In the previous section we studied the summarization independently, and here we will restate it in the fuzzy querying context. We start with the reinterpretation of (4) and (5). Thus, (4) formally expresses a statement:

"*Most records match query S*" (29)

where S replaces F in (4) since we refer here directly to the concept of a summarizer.

We assume a standard meaning of the query as a set of conditions on the values of fields from the database's tables, connected with AND and OR. We allow for fuzzy terms in a query which implies a degree of matching from [0,1] rather than a yes/no matching. So, a query S defines a fuzzy subset (fuzzy property) on the set of the records, and the membership of them is determined by their matching degree with the query.

Similarly, (5) may be interpreted as expressing a statement of the following type:

"Most records meeting conditions F match query S" (30)

Thus, (30) says something only about a subset of records taken into account by (29). That is, in database terminology, *F* corresponds to a *filter* and (30) claims that *most* records passing through *F* match query *S*. Moreover, since the filter may be fuzzy, a record may pass through it to a degree from [0,1].

We seek, for a given database, propositions of the type (5), interpreted as (30) that are highly true, and they consist of three elements: a fuzzy filter *F* (optional), a query *S*, and a linguistic quantifier *Q*. There are two limit cases, where we:

- do not assume anything about the form of any of these elements,
- assume concrete forms of the fuzzy filter and query, and look only for a linguistic quantifier *Q*.

In the first case data summarization will be extremely time-consuming, but may produce interesting results. In the second case the user has to guess a good candidate formula for summarization but the evaluation is fairly simple being equivalent to the answering of a (fuzzy) query. Thus, the second case refers to the summarization known as *ad hoc queries*, extended with an automatic determination of a linguistic quantifier.

In between these two extreme cases, there are different types of summaries, with various assumptions on what is given and what is sought. In case of a linguistic quantifier the situation is simple: it may be given or sought. In case of a fuzzy filter *F* and a fuzzy query *S*, more possibilities exist as both *F* and *S* consist of simple conditions, each stating what *value* a *field* should take on, and connected using logical connectives. Here we assume that the table(s) of interest for summarization are fixed.

We will use the following notation to describe what is given or what is sought with respect to the fuzzy filter *F* and query *S* (below *A* stands for either *F* or *S*):

- A - all is given (or sought), i.e., attributes, values and the structure,
- A^{fc} - attributes and structure are given, but values are left out,
- A^{v} - denotes sought left out values referred to in the above notation, and
- A^{f} - only a set of attributes is given, the other elements are sought.

Using the above notation we may propose the classification of the summaries shown in Table 4.

Thus, we distinguish 5 main types of data summarization. Type 1 may be easily produced by a simple extension of fuzzy querying as proposed and implemented in our FQUERY for Access package. Basically, the user has to construct a query - a candidate summary. Then, it has to be determined what is the fraction of the rows matching this query and what linguistic quantifier best denotes this fraction. The primary target of this type of summarization is certainly to propose such a query that a large proportion, e.g., *most*, of the rows satisfies it. On the other hand, it may be interesting to learn that only *few* rows satisfy some meaningful query. A Type 2 summary is a straight extension of Type 1 summaries by adding a fuzzy filter. As soon as a fuzzy querying engine deals with fuzzy filters, the computational complexity of this type of summaries is the same as for Type 1. For more on these types of summaries, see for instance Anwar, Beck and Navathe (1992), or Kacprzyk and Zadrożny (1998)

Table 4. Classification of linguistic summaries

Type	Given	Sought	Remarks
1	S	Q - sought	Simple summaries through ad-hoc queries
2	$S\ F$	Q - sought	Conditional summaries through ad-hoc queries
3	$Q\ S^{fc}$	S^{v}	Simple value oriented summaries
4	$Q\ S^{fc}\ F$	S^{v}	Conditional value oriented summaries
5	nothing	$S\ F\ Q$	General fuzzy rules

The summaries of Type 3 require much more effort. A primary goal of this type of summary is to determine typical (exceptional) values of an attribute. In such a special case, query S consists of only one simple condition built of the attribute whose typical (exceptional) value is sought, the '=' relational operator and a placeholder for the value sought. For example, using the following summary in a context of personal data:

Q = "*most*" and S = "age=?" ("?" denotes placeholder mentioned above)

we look for a typical value of the age of the employees.

Then, we try to find such a (possibly fuzzy) value that the query matches to a high degree Q of the rows. Depending on the category of the Q used as, e.g., *most* versus *few*, typical or exceptional values are sought, respectively. Some more considerations are required as in some cases all values may turn out to be exceptional and none to be typical. This type of summaries may be used with more complicated, regular queries but it may quickly become computationally infeasible (due to the combinatorial explosion) and the interpretation of results becomes vague. A Type 4 summary may produce typical (exceptional) values for some, possibly fuzzy, subset of rows. From the computational point of view, the same remarks apply as for Type 1 versus Type 2 summaries.

A Type 5 summary represents the most general form considered here. In its full version this type of summaries is to produce fuzzy rules describing dependencies between specific values of particular attributes. Here the use of the filter is essential, in contrast to the previous types where it was optional. The very meaning of a fuzzy rule obtained is that if a row meets a filter's condition, then it meets also the query's conditions - this corresponds to a classical IF-THEN rule. For a general form of such a rule it is difficult to devise an effective and efficient generation algorithm. Full search may be acceptable only in case of restrictively limited sets of rule building blocks, i.e. attributes and their possible values. Here, some genetic algorithm based approaches may be employed [cf. George and Srikanth (1996)] to alleviate the computational complexity, additional assumptions may also be made. For example, some sets of relevant (interesting, promising, etc.) attributes for the query (S^f) and the filter (F^f) may be selected in advance. Some constraints may also be put on the structure of the query S and filter F (in terms of the number of logical connectives allowed). Another, important special case of Type 5 summaries refers to the situation where the query (S) is fixed and only the filter (F) and quantifier (Q) are sought, i.e. we look for causes of given data features. For example, we may set in a query that profitability of a venture is *high* and look for the characterization of ventures (rows) securing such a high profitability.

5 Association rules as a special case of database summaries

The concept of data mining presented in the previous sections features an attractive uniform theoretical background referring to the calculus of linguistically quantified propositions. This paves the way for a consistent use of linguistic terms in database summaries. Unfortunately, in the original Yager's version as well as in the previous section there are rather limited clues as to the devising of efficient algorithms generating summaries of large data sets. Simplest types of summaries are directly computationally tractable. Another general approach calls for the use of genetic algorithms [cf. George and Srikanth (1996)]. Thus, in this section looking for efficient algorithms we investigate the similarity of our Type 5 summaries to what is known in the literature as *association rules* [cf. Agrawal and Srikant (1994)].

Originally, the association rules were defined for binary valued attributes in the following form:

$$A_1 \wedge A_2 \wedge \ldots \wedge A_n \rightarrow A_{n+1} \tag{31}$$

Thus, such an association rule states, that if in a database row all the attributes from the set $\{A_1, A_2, .., A_n\}$ take on value 1, then also the attribute A_{n+1} is expected to take on value 1 (remember that all attributes are assumed to take on their values from the set $\{0,1\}$).

A row in a database (table) is said to *support* a set of attributes $\{A_i\}_{i \in I}$ if all attributes from the set take on in this row value 1. There are two measures of the quality of an association rule used:

- the *support* of a rule (31) is the fraction of the number of rows supporting the set of attributes $\{A_i\}$, $i \in \{1,..,n+1\}$ in a database (table), and

- the *confidence* of a rule in a database (table) is the fraction of the number of rows supporting the set of attributes $\{A_i\}$, $i \in \{1,..,n+1\}$ among all rows supporting the set of attributes $\{A_i\}$, $i \in \{1,..,n\}$.

Thus, while the support determines a statistical significance of a rule, the confidence measures its strength in the database. Usually, we are interested in rules having values of the support measure above some minimal threshold and a high value of the confidence measure. This form of association rules was motivated by their early applications for a so-called customer's basket analysis.

A number of efficient algorithms for finding association rules exist, see, e.g. Agrawal and Srikant (1994), Mannila et al. (1994). The original concept of the association rule essentially evolved over time but still the same algorithms are applicable. The extensions to the initial form of the association rule include:

- the right-hand side, like the left-hand side, may contain a conjunction of the attributes instead of just one attribute,
- many-valued scalar values and their hierarchies may be used in rules [cf. Srikant and Agrawal (1995)],
- numerical, real-valued attributes may be used in rules [cf. Srikant and Agrawal (1996)]
- some constraints may be imposed on combinations of attributes used in rules [cf. Srikant et al. (1997)]]

The concept of the association rule fits into our general format of summaries of Type 5 as follows. The left-hand and right-hand sides correspond to the filter F and the query S, respectively. In our fuzzy querying framework of Section 2, a query may be composed of a set of simple conditions combined using the classical AND and OR logical connectives as well as linguistic quantifiers. In comparison, the structure of the filter and the query in case of association rules is rather limited but this secures the existence of efficient algorithms for rules generation.

The truth-value of the summary, defined by (4), corresponds to the confidence measure of a rule. Both definitions become identical when we assume the linguistic quantifier in (4) to be a so-called uniform quantifier.

A fuzzy querying interface may offer practical ways to circumvent some problems the "classical" search for the association rules faces [cf. Agrawal and Srikant (1994), Srikant and Agrawal (1996), Srikant et al. (1997)]. For example, when dealing with a numerical attribute, one has to partition its domain into a number of intervals. Then, treating each interval as an additional binary attribute, one can employ known algorithms for the generation of association rules. The question remains how to partition the domain. Srikant and Agrawal (1996) advocate the partition into equal intervals but admit that in some cases this may fail. Miller and Yang (1997) claim that cluster analysis should be used to determine a proper partition. Moreover, they suggest to group attributes provided a meaningful distance in corresponding multi-dimensional spaces may be defined. Then, a condition of a rule may require that the values of a group of attributes lie in a cluster.

The use of numerical fuzzy values itself alleviates the problem of a proper partition of an attributes domain. The domain may be covered by a number of overlapping fuzzy values making the partitioning more flexible. Lee and Lee-Kwang (1997) discuss the use of classical algorithms to produce association rules referring to fuzzy values. A fuzzy querying interface makes it even simpler as we have a dictionary of fuzzy values at hand, ready to be used in the rules. Moreover, these fuzzy values are familiar for the user as they were defined and tested by him or her in some fuzzy queries. Thus, the meaning of rules produced using such fuzzy values should be much more clear to the user. Context independence of numerical fuzzy values, mentioned in Section 2, is here of special importance as all these values may be used along with all attributes.

Our implementation of association rules is based on the Agrawal and Srikant's AprioriTid algorithm [cf. Agrawal and Srikant (1994)]. Basically, the original algorithm (for binary data) consists in finding *large itemsets*. An itemset, is a set of items (attributes) which may be interpreted as a logical formula comprising a conjunction of the member items. A record in the database is said to *support* an itemset if the corresponding conjunction "is true" for this record, i.e. all member items (attributes) take on value 1 in this record. An itemset containing *k* items is called a *k*-itemset. The algorithm starts with the evaluation of 1-itemsets. These itemsets which are not supported by sufficient number of records are deleted. Then, the 2-itemsets are evaluated. Thanks to the specific scheme [cf. Agrawal and Srikant (1994)] not all combinations of pairs of items are taken into account. Generally, the evaluation of the *k*-itemsets is done in two steps. First, promising *k*-itemsets are generated from large (*k*-1)-itemsets. Then, all generated *k*-itemsets are evaluated against all records in the database. Finally, all *k*-itemsets (for each *k*) are a basis for the generation of rules. More precisely, for each large itemset, *LIT*, all possible proper subsets of items, *PREM*, are considered as a premise of a hypothetical rule – the rest of items constitute the consequence of such a rule. Notice that both the premise and the consequence of the rule are large itemsets. The rule is accepted if its *confidence*, calculated as the fraction of records supporting the whole large itemset *LIT* among the records supporting the premise *PREM*, is higher than a pre-specified *confidence threshold*. Observe that due to the definition of large itemsets, the former number of records is always less or equal to the latter one; hence the confidence of a rule belongs to [0,1].

In our implementation we deal with real valued attributes. Thus, following previously mentioned approaches, for each such an attribute we introduce a number of artificial attributes (items) which may be treated as binary attributes. For example, the attribute PRICE may be replaced with the following artificial attributes: PRICEIsLow, PRICEIsMedium and PRICEIsHigh. The meaning of these attributes is obvious: a record supports, for example, PRICEIsLow if the value of the original attribute PRICE in this record falls into an interval identified as 'Low'. Instead of the intervals of values, employed in non-fuzzy approaches, we use fuzzy values (linguistic terms), see also Lee and Lee-Kwang (1997). The use of fuzzy values implies that a record supports a given artificial attribute to a degree from [0,1]. It is possible to use such a degree directly or to "defuzzify" it employing a threshold.

We will sketch the pilot implementation illustrating it on an example.

Step 1. Selection of the attributes and fuzzy values

In all but trivial cases, the use of all attributes and fuzzy values for rules construction is not recommended. Thus, the user has to choose some of them. In the current implementation, the user simply builds a query referring somehow to all attributes and fuzzy values, which should be taken into account.

Then, the user initiates the data summarization and the system automatically performs the rest of steps.

Step 2. Construction of the artificial attributes (items)

For each pair - attribute and fuzzy value - the system creates an artificial attribute (item). Each generated item constitutes a separate 1-itemset.

Step 3. Calculation of the support for 1-itemsets

This step requires a full search of the database. For each record, the membership degrees, *MD*, of the values of relevant attributes in fuzzy sets corresponding to items generated in Step 2 are calculated. Here, the fuzzy querying module is employed to calculate the membership degree, *MD*, sought. Only items for which the corresponding membership degree is greater than the *matching threshold* are considered as supported by given record. Thus, as a result of the database search we obtain for each item its *support*, i.e. the total number of records supporting it. The *fuzzy support* is also computed. While for the sum corresponding to the support each record contributes 0 or 1, for the fuzzy support each record contributes the value of membership degree *MD*, unless it is below the matching threshold - then the record contributes 0.

Step 4 Pruning of the set of 1-itemsets

We drop all itemsets with the support lower than a *support threshold*. The reasons for that are twofold. First of all, only rules that may be checked against large enough subsets of records are interesting. Second, such a pruning makes it possible to limit the complexity of further calculations. Additionally, we also drop itemsets with the support higher than another threshold, *item drop threshold*. The reason is that items present in almost all records will contribute nothing interesting to the produced rules. As the result we obtain the large 1-itemsets.

Step 5 Construction of the new transaction table

The calculation of the support for 1-itemsets is rather expensive and grows for *k*-itemsets, *k*>1. Thus, following the idea of Apriori Tid algorithm, we create special *transaction table* containing for each record all 1-itemsets supported by this record to the degree higher than matching threshold. It saves the computation time, but at the expense of needed extra storage space.

SET k = 2

Step 6 Generate *k*-itemsets

Basically, they are generated from large (*k*-1)-itemsets as in original Apriori Tid algorithm. Briefly, pairs of large (*k*-1) itemsets of the form (we follow here the notation of (31)) $A_1 \wedge A_2 \wedge ... \wedge A_{k-1}$ and $B_1 \wedge B_2 \wedge ... \wedge B_{k-1}$, where $A_i = B_i$ for $i=1,..,k-2$ are sought. Then, the new *k*-itemset of the form $A_1 \wedge A_2 \wedge ... \wedge A_{k-1} \wedge B_{k-1}$ is generated. In the original algorithm, the rules generated this way are additionally tested and possibly eliminated before the evaluation done in Step 7. On the other hand, due to the approach adopted here, we add another *k*-itemset generation limitation, namely the items A_{k-1} and B_{k-1} have to correspond to different original attributes (remember that each item is based on the pair (attribute,fuzzy value) - see Step 2).

Step 7 Calculate support for all *k*-itemsets

The procedure is as in the Step 3, but this time the transaction table generated for (*k*-1)-itemsets is employed.

Step 8 Pruning of the set of k-itemsets (as in Step 4)

As the result we obtain the large *k*-itemsets.

IF the set of *k*-itemsets is empty, then GOTO Step 10

Step 9 Construction of the new transaction table

As in Step 5 but this time the transaction table is generated for *k*-itemsets

SET k = k + 1; GOTO Step 6

Step 10 Generate rules from large *l*-itemsets, $l=1,..,k-1$

This completes a brief presentation of our implementation of the association rules generation algorithm. Its distinguishing features are: the use of linguistic (fuzzy) terms and the integration with the querying interface.

References

Agrawal R. and Srikant R. (1994) Fast algorithms for mining association rules, in: Proceedings of the 20th International Conference on Very Large Databases, Santiago, Chile, 1994.

Anwar T.M., Beck H.W. and Navathe S.B. (1992) Knowledge mining by imprecise querying: A classification based system, in: Proceedings of the International Conference on Data Engineering, Tampa, USA, 622-630.

Bosc P. and Kacprzyk J., Eds. (1995) Fuzziness in Database Management Systems. Physica-Verlag, Heidleberg,

Bosc P., L. Lietard and O. Pivert (1995) Quantified statements and database fuzzy querying. In P. Bosc and J. Kacprzyk (Eds.): Fuzziness in Database Management Systems. Physica-Verlag, Heidelberg, pp. 275-308.

Bosc P. and Pivert O. (1992) Fuzzy querying in conventional databases. In L.A. Zadeh and J. Kacprzyk (Eds.): Fuzzy Logic for the Management of Uncertainty. Wiley, New York, pp. 645-671.

George R. and Srikanth R. (1996) Data summarization using genetic algorithms and fuzzy logic. In F. Herrera and J.L. Verdegay (Eds.): Genetic Algorithms and Soft Computing. Physica-Verlag, Heidelberg, pp. 599-611.

Kacprzyk J. (1999a) An Interactive Fuzzy Logic Approach to Linguistic Data Summaries, Proceedings of NAFIPS'99 - 18th International Conference of the North American Fuzzy Information Processing Society - NAFIPS, IEEE, pp. 595-599.

Kacprzyk J. (1999b) A New Paradigm Shift from Computation on Numbers to Computation on Words on an Example of Linguistic Database Summarization, In: N. Kasabov (ed.): Emerging Knowledge Engineering and Connectionist-Based Information Systems – Proceedings of ICONIP/ANZIIS/ANNES'99, University of Otago, pp. 179-179.

Kacprzyk J. and Strykowski P. (1999a) Linguistic data summaries for intelligent decision support, In: R. Felix (ed.): Fuzzy Decision Analysis and Recognition Technology for Management, Planning and Optimization – Proceedings of EFDAN'99, pp. 3-12.

Kacprzyk J. and Strykowski P. (1999b) Linguistic Summaries of Sales Data at a Computer Retailer: A Case Study. Proceedings of IFSA'99 (Taipei, Taiwan R.O.C), vol. 1 pp. 29-33.

Kacprzyk J. and Yager R.R. (2000) Linguistic Summaries of Data Using Fuzzy Logic. International Journal of General Systems (in press)

Kacprzyk J., Yager R.R. and S. Zadrożny (2000) A fuzzy logic based approach to linguistic summaries of databases. International Journal of Applied Mathematics and Computer Science (in press).

Kacprzyk J. and Zadrożny S. (1995) FQUERY for Access: fuzzy querying for a Windows-based DBMS, in P. Bosc and J. Kacprzyk (eds.): Fuzziness in Database Management Systems. Physica-Verlag: Heidelberg, pp. 415-433.

Kacprzyk J. and Zadrożny S. (1997a) Fuzzy queries in Microsoft Access v. 2, in D. Dubois, H. Prade and R.R. Yager (Eds.): Fuzzy Information Engineering - A Guided Tour of Applications, Wiley, New York, 1997, pp. 223 - 232.

Kacprzyk J. and Zadrożny S. (1997b) Implementation of OWA operators in fuzzy querying for Microsoft Access. In: R.R. Yager and J. Kacprzyk (Eds.) The Ordered Weighted Averaging Operators: Theory and Applications, Kluwer, Boston, 1997, pp. 293 - 306.

Kacprzyk and Zadrożny S. (1997c) Flexible querying using fuzzy logic: An implementation for Microsoft Access, in T. Andreasen, H. Christiansen and H.L. Larsen (eds.): Flexible Query Answering Systems, Kluwer, Boston, 1997, pp. 247-275.

Kacprzyk J. and Zadrożny S. (1998a) Data Mining via Linguistic Summaries of Data: An Interactive Approach, in T. Yamakawa and G. Matsumoto (eds.): Methodologies for the Conception, Design and Application of Soft Computing (Proceedings of IIZUKA'98, Iizuka, Japan), pp. 668-671.

Kacprzyk J. and Zadrożny S. (1998b) On sumarization of large datasets via a fuzzy-logic-based querying add-on to Microsoft Access", in: Intelligent Information Systems VII (Malbork, Poland), IPI PAN, Warsaw, 1998, 249-258.

Kacprzyk J. and Zadrożny S. (1999) On Interactive Linguistic Summarization of Databases via a Fuzzy-Logic-Based Querying Add-On to Microsoft Access. W: Bernd Reusch (Ed.)(Red.): Computational Intelligence: Theory and Applications, Springer-Verlag, Heidelberg, pp. 462-472.

Kacprzyk J. and Zadrożny S. (2000a) On combining intelligent querying and data mining using fuzzy logic concepts. In: G. Bordogna and G. Pasi (ed.): Recent Research Issues on the Management of Fuzziness in Databases, Physica - Verlag, Heidelberg and New York (in press).

Kacprzyk J. and Zadrożny S. (2000b): Using Fuzzy Querying over the Internet to Browse Through Information Resources. In: P.P. Wang(Red.): Computing with Words, Wiley, New York (in press).

Kacprzyk J., Zadrożny S. and Ziółkowski A. (1989) FQUERY III+: a 'human consistent` database querying system based on fuzzy logic with linguistic quantifiers. Information Systems 6, 443 - 453.

Kacprzyk J. and Ziółkowski A. (1986) Database queries with fuzzy linguistic quantifiers. IEEE Transactions on Systems, Man and Cybernetics SMC - 16, 474 - 479.

Kacprzyk J., Zadrożny S. and Ziółkowski A. (1989) FQUERY III+: a 'human consistent` database querying system based on fuzzy logic with linguistic quantifiers. Information Systems 6, 443 - 453.

Larsen H.L. and Yager R.R. (1993) The use of fuzzy relational thesauri for classificatory problem solving in information retrieval and expert systems. IEEE Trans. On Syst., Man and Cybern. SMC-23, 31 - 41.

Lee J.-H. and Lee-Kwang H. (1997), An extension of association rules using fuzzy sets, in: Proceedings of the Seventh IFSA World Congress, 1997, Prague, Czech Republic. Vol. 1, 399-402.

Liu Y. and Kerre E.E. (1998) An overview of fuzzy quantifiers. (I) Interpretations. Fuzzy Sets and Systems 95, 1 – 21.

Mannila H., Toivonen H. and Verkamo A.I. (1994) Efficient algorithms for discovering association rules, in: U.M. Fayyad and R. Uthurusamy (eds.) Proceedings of the AAAI Workshop on Knowledge Discovery in Databases, Seattle, USA, 1994, 181-192.

Miller R.J. and Yang Y. (1997) Association rules over interval data,. in: Proceedings of the ACM SIGMOD International Conference on the Management of Data, Tucson, USA, 1997, 452-461.

Miyamoto S. (1990) Fuzzy Sets in Information Retrieval and Cluster Analysis. Kluwer Academic Publishers. Dordrecht, Boston, London.

Nomura T., Odaka T., Ohki N., Yokoyama T. And Matsuhita Y. (1992) Generating ambiguous attributes for fuzzy queries. Proceedings of IEEE International Conference on Fuzzy Systems (FUZZ-IEEE '92), 1992, pp. 753-760.

Petry F.E. (1996) Fuzzy Databases: Principles and Applications. Kluwer, Boston.

Pons O., Vila M.A. and Kacprzyk J. (Eds.): Knowledge Management in Fuzzy Databases, Physica-Verlag, Heidelberg, New York.

Rasmussen D. and Yager R.R. (1997a) A fuzzy SQL summary language for data discovery, In: Dubois, D., Prade, H. and Yager, R.R. (eds.): Fuzzy Information Engineering: A Guided Tour of Applications, Wiley, New York, pp. 253-264, 1997.

Rasmussen D. and Yager R.R. (1997b) SummarySQL-A fuzzy tool for data mining, Intelligent Data Analysis - An International Journal 1 (Electonic Publication), URL-http//:www-east.elsevier.com/ida/browse/96-6/ida96-6.htm.

Rasmussen D. and Yager R.R. (1997c) Fuzzy query language for hypothesis evaluation, in Andreasen T., H. Christiansen and H. L. Larsen (eds.) *Flexible Query Answering Systems*. Kluwer, Boston/Dordrecht/London, 23-43.

Rasmussen D. and Yager R.R. (1999) Finding fuzzy and gradual functional dependencies with summarySQL, Fuzzy Sets and Systems 106, 131-142.

Saaty T.L. (1980) The Analytic Hierarchy Process: Planning, Priority Setting, Resource Allocation.. McGraw-Hill, New York.

Srikant R. and Agrawal R. (1995) Mining generalized association rules, in: Proceedings of the 21st International Conference on Very Large Databases", Zurich, Switzerland, 1995.

Srikant R. and Agrawal R. (1996) Mining quantitative association rules in large relational tables, in: Proceedings of the ACM-SIGMOD 1996 Conference on Management of Data, Montreal, Canada, 1996.

Srikant R., Vu Q. and Agrawal R. (1997) Mining association rules with item constraints, in: Proceedings of the 3rd International Conference on Knowledge Discovery in Databases and Data Mining, Newport Beach, USA, 1997.

Wang P.P. (2000) Computing with Words. Wiley, New York.

Yager R.R. (1982) A new approach to the summarization of data. Information Sciences, 28, 69 – 86.

Yager R.R. (1988) On ordered weighted avaraging operators in multicriteria decision making. IEEE Trans. on Systems, Man and Cybern. SMC-18, 183-190.

Yager R.R. (1991) On linguistic summaries of data, In: Piatetsky-Shapiro, G. and Frawley, B. (eds.): Knowledge Discovery in Databases, MIT Press, Cambridge, MA, pp. 347-363.

Yager, R.R. (1994) Linguistic summaries as a tool for database discovery, Proceedings of Workshop on Flexible Query-Answering Systems, Roskilde University, Denmark, pp. 17-22.

Yager, R.R. (1995) Fuzzy summaries in database mining, Proceedings of the 11th Conference on Artificial Intelligence for Applications (Los Angeles, USA), pp. 265-269.

Yager, R.R. (1996) Database discovery using fuzzy sets, International Journal of Intelligent Systems 11, 691-712.

Yager R.R. and Kacprzyk J., Eds. (1997) The Ordered Weighted Averaging Operators: Theory and Applications. Kluwer, Boston.

Yager R.R. and Kacprzyk J. (1999) Linguistic Data Summaries: A Perspective. Proceedings of IFSA'99 Congress (Taipei), vol. 1, 44-48

Yager, R. R. and Rubinson T. C. (1981) Linguistic summaries of data bases, Proceedings of IEEE Conf. on Decision and Control (San Diego, USA), pp. 1094-1097.

Yazici A. and George R. (1999) Fuzzy Database Modeling. Physica-Verlag, Heidelberg and New York.

Zadeh L.A. (1983) A computational approach to fuzzy quantifiers in natural languages. Computers and Maths with Appls. 9, 149-184.

Zadeh L.A. (1985) Syllogistic reasoning in fuzzy logic and its application to usuality and reasoning with dispositions. IEEE Transaction on Systems, Man and Cybernetics, SMC-15, 754--763.

Zadeh L.A. (1987) A computational theory of dispositions, *International Journal of Intelligent Systems* 2, 1987, 39-64.

Zadeh L.A. and Kacprzyk J., Eds. (1992) Fuzzy Logic for the Management of Uncertainty, Wiley, New York.

Zadeh L.A. and Kacprzyk J., Eds. (1999) Computing with Words in Information/Intelligent Systems 1. Foundations, 2 Applications. Physica-Verlag, Heidelberg and New York.

Zadrożny S. and Kacprzyk J. (1996) Multi-valued fields and values in fuzzy querying via FQUERY for Access, Proceedings of FUZZ–IEEE'96 - Fifth International Conference on Fuzzy Systems (New Orleans, USA), vol. 2, pp. 1351 - 1357.

Zadrożny S. and Kacprzyk J. (1999) On database summarization using a fuzzy querying interface, Proceedings of IFSA'99 World Congress (Taipei, Taiwan R.O.C.), vol. 1, pp. 39-43.

Is the Today's Mathematics Ready for the Computational Intelligence?

Endre Pap

Institute of Mathematics, University of Novi Sad, 21000 Novi Sad, Yugoslavia,
e-mail: pap@unsim.ns.ac.yu , pape@eunet.yu

Abstract: There is presented unified mathematical tool (we call it pseudo-analysis) for treating problems with uncertainties, nonlinearity and optimization. A procedure for finding a global maximum based on the idea of Lebesgue integral is presented. There are given applications in general decision making (utility theory, engineering design).It is characterized the families of operations involved in generalized mixtures and integration, due to a previous result on the characterization of the pairs of continuous t-norm and t-conorm such that the former is conditionally distributive over the latter. What is obtained is a family of mixtures that combine probabilistic and possibilistic mixtures via a threshold.

1 Introduction

Which kind of mathematics is needed for Computational Intelligence (CI)? Today we already have some machinery as triangular norms, conorms, aggregation operators, pseudo-analysis, fuzzy measures. This paper is a continuation of the paper [30]. We shall speak about applications in general decision making (utility theory, engineering design). There is also a feedback of CI in Mathematics (nonlinear equations, global optimization). We shall use an unified mathematical tool (we call it pseudo-analysis) for treating problems with uncertainties, nonlinearity and optimization (Baccelli et al. [3], Klement et al. [14], Maslov et al. [20], Pap [29]-[33], Wang, Klir [43]).

We present first in section 2 some general real operations which forms a base for pseudo-analysis presented in section 3. We give in section 4 a procedure for finding a global maximum based on the idea of Lebesgue integral, α-level, approximation and statistics. We present in section 5 some elements of decision making under uncertainty in engineering design. In order to generalize decision theory to non probabilistic uncertainty, one approach is to generalize mixture sets, given in section 6. It is based on a restricted family of t-conorm/ t-norm pairs which are very special ordinal sums. Any practically useful theory of pseudo-additive measures must use such special pairs of operations in order to extend the additivity property, and the notion of probabilistic independence.

2 Triangular norms and related operations

We shall start with the operations on the real interval $[0, 1]$.

A *triangular norm* T (t-norm briefly) is a function $T : [0,1]^2 \to [0,1]$ such that it is commutative, associative,monotone, and 1 is the unit element.

Example 1. The following are the most important t-norms

$$T_{\mathbf{M}}(x,y) = \min(x,y) \ , \ T_{\mathbf{P}}(x,y) = xy \ ,$$

$$T_{\mathbf{L}}(x,y) = \max(0, x+y-1) \ , \ T_{\mathbf{D}}(x,y) = \begin{cases} \min(x,y) & \text{if } \max(x,y) = 1 \\ 0 & \text{otherwise.} \end{cases}$$

The corresponding dual operation is given by

A *triangular conorm* S (t-conorm briefly) is a function $S : [0,1]^2 \to [0,1]$ such that it is commutative, associative,monotone, and 0 is the unit element.

Example 2. The following are the most important t-conorms :

$$S_{\mathbf{M}}(x,y) = \max(x,y) \ , \ S_{\mathbf{P}}(x,y) = x + y - xy \ ,$$

$$S_{\mathbf{L}}(x,y) = \min(1, x+y) \ , \ S_{\mathbf{D}}(x,y) = \begin{cases} \max(x,y) & \text{if } \min(x,y) = 0 \\ 1 & \text{otherwise.} \end{cases}$$

Many other important t-norms and t–conorms can be found in Klement et al. [14], Schweizer et al. [40]. The basic connectives in fuzzy logic and operations by fuzzy sets are based on triangular norms and triangular conorms ([9], [14]).

We extend now the considered interval for the previous operations. Let [a,b] be a closed (in some cases semiclosed) subinterval of $[-\infty, +\infty]$. We consider here a total order $\leq$ on $[a,b]$ (although it can be taken in the general case a partial order). The operation $\oplus$ (*pseudo-addition*) is a function $\oplus : [a,b] \times [a,b] \to [a,b]$ which is commutative, nondecreasing, associative and has a zero element, denoted by 0.

Let $[a,b]_+ = \{x : x \in [a,b], \ \ x \geq 0\}$. The operation $\odot$ (*pseudo-multiplication*) is a function $\odot : [a,b] \times [a,b] \to [a,b]$ which is commutative, positively nondecreasing, i.e. $x \leq y$ implies $x \odot z \leq y \odot z$, $z \in [a,b]_+$, associative and for which there exist a unit element $1 \in [a,b]$, i.e., for each $x \in [a,b]$ we have $1 \odot x = x$.

We suppose, further, $0 \odot x = 0$ and that $\odot$ is a distributive pseudo-multiplication with respect to $\oplus$, i.e.

$$x \odot (y \oplus z) = (x \odot y) \oplus (x \odot z).$$

The structure $([a,b], \oplus, \odot)$ is called a *semiring.*

In this paper we will consider only special semirings with the following continuous operations:

Case I) (i) $x \oplus y = \min(x,y), \ \ x \odot y = x + y$, on the interval $]-\infty, +\infty]$. We have $0 = +\infty$ and $1 = 0$.

(ii) $x \oplus y = \max(x,y), \ \ x \odot y = x + y$, on the interval $[-\infty, +\infty[$. We have $0 = -\infty$ and $1 = 0$.

Case II) Semirings with pseudo-operations defined by monotone and continuous generator g (Aczél [1], Golan [10], Mesiar et al. [25], Pap [29],[14]).

In this case we will consider only strict pseudo - addition, i.e., such that the function $\oplus$ is continuous and strictly increasing on $]a,b[\times]a,b[$.

By representation theorem for each strict pseudo - addition $\oplus$ there exists a monotone function g (generator for $\oplus$) , $g:[a,b]\to[-\infty,\infty]$ (or with values in $[0,\infty]$) such $g(0)=0$ and $u\oplus v=g^{-1}(g(u)+g(v))$. Using a generator g of strict pseudo-addition $\oplus$, we can define pseudo-multiplication $\odot: u\odot v = g^{-1}(g(u)g(v))$.

Case III (i) Let $\oplus=\max$ and $\odot=\min$ on the interval $[-\infty,+\infty]$.

(ii) Let $\oplus=\min$ and $\odot=\max$ on the interval $[-\infty,+\infty]$.

Relaxation of the distributivity property gives further important applications. A t-norm T is *conditionally distributive* over a t-conorm S if for all $x,y,z\in[0,1]$ we have

$$(CD)\quad T(x,S(y,z))=S(T(x,y),T(x,z))\quad \text{whenever}\quad S(y,z)<1.$$

Complete characterization of the family of continuous pairs (S,T) which satisfy the condition (CD) is obtained in [14]:

A continuous t-norm T is conditionally distributive over a continuous t-conorm S if and only if there exist $a\in[0,1]$, a strict t-norm T' and a nilpotent t-conorm S^* such that the additive generator s^* of S^* satisfying $s^*(1)=1$ is also a multiplicative generator of T' such that $T=(<0,a,T_1>,<a,1,T'>)$ where T_1 is an arbitrary continuous t-norm and $S=(<a,1,S^*>)$. We denote by $(<S_M,S^*>,<T_1,T'>)_a$ the pair of continuous t-conorm S and t-norm T.

Example 3. The extreme case $a=0$ for the pair $(<S_M,S_L>,<T_1,T_P>)_0$, where T_1 is arbitrary, gives us the pair S_L and T_P. The other extreme case $a=1$ for the pair $(<S_M,S_L>,<T_1,T_P>)_1$, where T_1 is an arbitrary continuous t-norm, gives us the pair S_M and T_1. Let $0<a<1$. Then we have the pair $(<S_M,S_L>,<T_1,T_P>)_a$

The aggregation operators are closely related to previously considered operations. In many systems (specially intelligent) the aggregation of incoming data plays the main role. The aggregation operators form a fundamental part of multi-criteria decision making, engineering design, expert systems, pattern recognition, neural networks, fuzzy controllers, genetic algorithms, etc., see Grabisch et al. [11], Klement et al. [13,14], Klir et al. [16].

Fuzzy neural networks give more information with respect to the classical neural networks, which are more or less Black Boxes (see Nauck et. al. [26]). Since the t–norm T_M can be obtained as a limit of a family of continuous Archimedean t–norms (see Klement et al. [14]) , taking enough big value for this parameter we obtain satisfactory approximation. In the Archimedean case (t–norm or t–conorm) we have an additive generator h. For example, if the activation function of the neuron u_j is given by by a differentiable function h which is induced by a t–norm or t–conorm , then for activation a_{u_i} by neuron u_i we have

$$\frac{\partial net_{u_j}^{(p)}}{\partial a_{u_i}^{(p)}}=\frac{h'(a_{u_i}^{(p)})}{h'(h^{(-1)}(\sum_k h(a_{u_k}^{(p)})))},$$

where the sum goes through all neurons u_k which have connection with the neuron u_j (see [26]).

3 Pseudo-Analysis

We want to stress here some of the advantages of the pseudo–analysis. We cover with one theory and so with unified methods problems (usually non-linear) from many different fields (system theory, optimization, control theory, differential equations, difference equations, etc.). Important fact is that this approach gives also solutions in the form which are not achieved by other theories (for example, Bellman difference equation, Hamilton–Jacobi equation with non–smooth Hamiltonians, see [29,34]). In some cases it enables for mostly nonlinear equations to obtain exact solutions in the similar form as for the linear equations. Some obtained principles as for example the pseudo-linear superposition principle allows us to transfer the methods of linear equations to the nonlinear equations.

Pseudo-analysis uses many mathematical tools from different fields as functional equations, variational calculus, measure theory, functional analysis, optimization theory, semiring theory, etc. and still is in the developing form. There are many investigations from the theoretical point of view as well as in many new applications (Baccelli et al. [3], Grabisch et al. [11], Klement et al. [14], Maslov et al. [20], Pap [29], Wang et al. [43]).

Now we restrict ourselves on special non–additive set functions on so called decomposable measures.This measures are based on pseudo–addition and serve as an important tool in pseudo–analysis.

Let X be a non-empty set. Let Σ be a $\sigma-$algebra of subsets of X. A set function $m : \Sigma \to [a, b]_+$ (or semiclosed interval) is a $\oplus-$decomposable measure if there hold $m(\emptyset) = 0$ (if $\oplus$ is not idempotent) ;
$m(A \cup B) = m(A) \oplus m(B)$ for $A, B \in \Sigma$ such that $A \cap B = \emptyset$. In the case when $\oplus$ is idempotent, it is possible that m is not defined on an empty set. A $\oplus-$decomposable measure m is $\sigma - \oplus-$decomposable if

$$m(\bigcup_{i=1}^{\infty} A_i) = \bigoplus_{i=1}^{\infty} m(A_i)$$

hold for any sequence $\{A_i\}$ of pairwise disjoint sets from Σ.

Let m be a $\sigma - \oplus-$decomposable measure. We define the characteristic function with values in a semiring

$$\chi_A(x) = \begin{cases} 0, & x \notin A \\ 1, & x \in A \end{cases} .$$

A mapping $e : X \to [a, b]$ is an elementary (measurable) function if it has the following representation $e = \bigoplus_{i=1}^{\infty} a_i \odot \chi_{A_i}$ for $a_i \in [a, b]$ and $A_i \in \Sigma$ disjoint if $\oplus$ is not idempotent.

The pseudo-integral of a bounded measurable function $f : X \to [a, b]$, for which, if $\oplus$ is not idempotent for each $\varepsilon > 0$ there exists a monotone ε−net in $f(X)$, is defined by

$$\int_X^{\oplus} f \odot dm = \lim_{n\to\infty} \int_X^{\oplus} \varphi_n(x) \odot dm,$$

where (φ_n) is the sequence of elementary functions constructed in Pap [29].

We have introduced in the paper Pap et al. [34] (see also Pap [33], Pap et al. [36], Maslov et al. [20], Bacceli et al. [3]) the notion of the pseudo-convolution of functions. The*pseudo-convolution* of two functions f and h with respect to a $\oplus$-decomposable measure m is given in the following way

$$(f \star h)_m(x) = \int_{[0,x]}^{\oplus} f(x-t) \odot dm_h,$$

where $m_h = m$ in the case of sup-decomposable measure $m(A) = \sup_{x\in A} h(x)$ (and in the case of inf-decomposable measure $m(A) = \inf_{x\in A} h(x)$), and $dm_h = h \odot dm$ in the case of $\oplus$-decomposable measure m, where $\oplus$ has an additive generator g and $g \circ m$ is the Lebesgue measure (g-calculus).

The pseudo-character of the group $(G, +)$, $G \subset$ R is a continuous map $\xi : G \to P$, of the group $(G, +)$ into semiring $(P, \oplus, \odot)$, with the property

$$\xi(x + y) = \xi(x) \odot \xi(y), \quad x, y \in G.$$

Pseudo-Laplace transform $\mathcal{L}^{\oplus}(f)$ of a function f is defined by

$$(\mathcal{L}^{\oplus} f)(z) = \int_{G\cap[0,\infty)}^{\oplus} (\xi(x, z) \odot f(x))dm,$$

where ξ is the pseudo-character . For cases I) and III) (section 2) we consider also pseudo-Laplace transform taking in the pseudo-integral the whole G instead of $G \cap [0, \infty)$.

We give some examples, where the numerator correspond to the section 2: I) (i) $(\mathcal{L}^{\oplus} f)(z) = \inf_{x\geq 0}(-xz + f(x))$, II) g-Laplace transform

$$(\mathcal{L}^{\oplus} f)(z) = g^{-1}(\int_0^{\infty} e^{-xz} g(f(x))dx).$$

We have proved in [34] the pseudo-exchange formula

$$\mathcal{L}^{\oplus}(f_1 \star f_2) = \mathcal{L}^{\oplus}(f_1) \odot \mathcal{L}^{\oplus}(f_2).$$

The obtained results are applied in some optimization problems. There are some pseudo-integral representations of nonlinear operators which occurs in dynamical programming (Baccelli et al. [3], Golan [10], Maslov et al. [20], Pap [29]).

The non-additive integrals (Sugeno, Choquet and their generalizations, see [29]) can be applied in the *decision (subjective, multi–criteria) theory* as an aggregation operator, see [11]. These integrals have the advantage with respect to other aggregation operators that the non–additivity of the considered measures takes into account the interaction between criteria.

4 Global maximum

We briefly describe a procedure for finding the approximated global maximum, more details in [35], applied on a membership function. As it is well-known the knowledge of the height of a membership function can be used in the defuzzyfication procedure (see [9]).

Let A be a fuzzy subset of a crisp set R^n. If μ_A is the corresponding membership function to the fuzzy subset A, then for $\alpha \in [0,1]$ *the α-level set* (α-cut) A_α is defined by $A_\alpha = \{x \mid \mu_A(x) \geq \alpha\}$. Then we have $A_0 = \mathrm{R}^n$ and $\mu_A = \mu_B$ if and only if $A_\alpha = B_\alpha$ for every $\alpha \in [0,1]$. We can reconstruct μ_A by $\{A_\alpha\}_{\alpha\in[0,1]}$. Namely, if 1_{A_α} is the characteristic function of the crisp set A_α then we have $\mu_A(x) = \sup_{\alpha\in[0,1]}\{\min(\alpha, 1_{A_\alpha}(x))\}$.

We define a function $H_{\mu_A} : [0,1] \to [0,\infty]$ by $H_{\mu_A}(\alpha) = m(A_\alpha)$, where $m(A_\alpha)$ is the Lebesgue measure of the α-level set A_α of the fuzzy subset A given by the membership function μ_A.

Our method of finding the global maximum is based on the properties of a special function H_{μ_A} which we correspond to the membership function μ_A. By the presented method for finding the height of the membership function μ_A it is enough to find the zero of the function H_{μ_A}. Unfortunately the analytical form of the function H_{μ_A}, which can serve as a direct tool for finding the height of the membership function μ_A, is available only for very limited cases, usually for simpler μ_A and lower dimension n. For this reason, we have to proceed with approximation and extrapolation of the function H_{μ_A}. To approximate the integral which represents the function H_{μ_A} we will use statistical methods. We give here an algorithm for the whole procedure.

The algorithm for finding the height of the membership function (see [35]):

1. step. Randomize the coordinates. a) We choose randomly (with the same probabilities) the coordinates $x_1, x_2, ..., x_n$ of $x = (x_1, x_2, ..., x_n)$.

b) We find the value $\mu_A(x_1, x_2, ..., x_n)$.

c) We repeat k-times the procedures a) and b), denoting by $x^1, x^2, ..., x^k$ the corresponding values of x , and we obtain the set of values $\{\mu_A(x^1), \mu_A(x^2), ..., \mu_A(x^k)\}$ which we denote by S_k.

2. step. The function $H^{st}(\alpha)$. a) We find $\max S_k$ and $\min S_k$.

b) We divide the closed interval $[\frac{\max S_k + \min S_k}{2}, \max S_k]$ on l equal subintervals, i.e.,

$$\frac{\max S_k + \min S_k}{2} = \alpha_0 < \alpha_1 < ... < \alpha_l = \max S_k$$

and

$$\alpha_i = \frac{\max S_k + \min S_k}{2} + i\Delta\alpha$$

for $i = 1, 2, ..., l$, where $\Delta\alpha = \frac{\max S_k - \min S_k}{2l}$.
c) We find the numbers $r_1, r_2, ..., r_l$, where r_i means that r_i times occurs the situation $\mu_A(x) \geq \alpha_i$, where $\mu_A(x) \in S_k$.
d) We find $H^{st}(\alpha_i) = \frac{r_i}{k}$ for $i = 1, 2, ..., l$.

3. step. The maximum $\alpha_{\max}$ of μ_A. a) We find the numbers $E_1, E_2, E_3, E_4, E_{0,1}, E_{1,1}, E_{2,1}$ given by

$$E_1 = E_1(\alpha) = \frac{1}{l}\sum_{i=1}^{l} \alpha_i, \quad E_2 = E_2(\alpha) = \frac{1}{l}\sum_{i=1}^{l} \alpha_i^2,$$

$$E_3 = E_3(\alpha) = \frac{1}{l}\sum_{i=1}^{l} \alpha_i^3, \quad E_4 = E_4(\alpha) = \frac{1}{l}\sum_{i=}^{l} \alpha_i^4,$$

$$E_{0,1} = E_{0,1}(\alpha) = \frac{1}{l}\sum_{i=1}^{l} H^{st}(\alpha_i), \quad E_{1,1} = E_{1,1}(\alpha) = \frac{1}{l}\sum_{i=1}^{l} \alpha_i H^{st}(\alpha_i),$$

$$E_{2,1} = E_{2,1}(\alpha) = \frac{1}{l}\sum_{i=1}^{l} \alpha_i^2 H^{st}(\alpha_i).$$

b) We solve the system of linear equations

$$E_4 a_2 + E_3 a_1 + E_2 a_0 = E_{2,1}, \; E_3 a_2 + E_2 a_1 + E_1 a_0 = E_{1,1}$$

$$E_2 a_2 + E_1 a_1 + E_0 a_0 = E_{0,1},$$

where we take $E_0 = 1$.
c) We find the roots α_1^0 and α_2^0 of the equation $a_2\alpha^2 + a_1\alpha + a_0 = 0$.
d) If the roots are complex numbers we start again with the 1. step. For real roots we find $\min(\alpha_1^0, \alpha_2^0)$, which is the possible height (global maximum) $\alpha_{\max}$ of the membership function μ_A.

4. step. The coordinates $x_1^{\max}, x_2^{\max}, ..., x_n^{\max}$ of the global maximum (the height).
a) We find for each α_i $i = 1, 2, ..., l$ the coordinates x_s^j , $s = 1, 2, ..., n$; $j = 1, 2, ..., r_i$ for which the inequalities $\mu_A(x_1^j, x_2^j, ..., x_n^j) \geq \alpha_i$, $j = 1, 2, ..., r_i$, hold.
b) We find for each α_i, $i = 1, 2, ..., l$ the values $x_s^\star(\alpha_i)$ $s = 1, 2, ..., n$; $i = 1, 2, ..., l$, by

$$x_s^\star(\alpha_i) = \frac{\sum_{j=1}^{r_i} x_s^j}{r_i}.$$

c) We find $X_{0,1}, X_{1,1}, X_{2,1}$ by

$$X_{0,1} = \frac{1}{l}\sum_{i=1}^{l} x_s^\star(\alpha_i),\ X_{1,1} = \frac{1}{l}\sum_{i=1}^{l} \alpha_i x_s^\star(\alpha_i),\ X_{0,1} = \frac{1}{l}\sum_{i=1}^{l} \alpha_i^2 x_s^\star(\alpha_i).$$

d) For each s, $s = 1, 2, ..., n$, we solve the system of linear equations

$$E_4 b_2^s + E_3 b_1^s + E_2 b_0^s = X_{2,1},\ E_3 b_2^S + E_2 b_1^s + E_1 b_0^s = X_{1,1}$$

$$E_2 b_2^s + E_1 b_1^s + E_0 b_0^s = X_{0,1},$$

where E_0, E_1, E_2, E_3 are from *3. step* a).
e) We find the coordinates of the maximum $(x_1^{\max}, x_2^{\max}, ..., x_n^{\max})$ by

$$x_s^{\max}(\alpha_{\max}) = b_2^s \alpha_{\max}^2 + b_1^s \alpha_{\max} + b_0^s$$

for $s = 1, 2, ..., n$.

5. step. Comparison. a) We find the global maximum $\mu_A(x_1^{\max}, x_2^{\max}, ..., x_n^{\max})$ and denote it by $\mu_A^{\max}$.
b) We compare $\mu_A^{\max}$ and $\alpha_{\max}$ (obtained in *3. step*). If $\alpha_{\max} \geq \mu_A^{\max}$ we repeat the whole algorithm.

We have presented a universal method for finding the global maximum of a function . The advantages of the method is applicability on multi-dimensional cases and even on non-continuous cases, when classical gradient methods are not applicable. The presented procedure (with some improvements) can be applied also in other optimization problems.

5 Engineering design

The advances of new computers as teraflop machines, wireless communication, next-generation Internet, will enable concurrent engineering teams to obtain the main design criteria early in engineering design process. This will enable a rapid accommodation of new engineering design paradigms and improvement in the design process.

Engineering design is recently recognized as a decision-making process at the core. Engineering design is conducted with incomplete and imperfect information, yet most traditional design approaches treat the design problem are deterministic. The proposed research is to develop tools for decision making under risk and uncertainty and apply the tools to engineering design, see [37]. Decision making under uncertainty and risk goes back to Bernoulli, 1738. It has three main elements : identification of the options, determination of expectations on each option and the expression of values. The main decision rule is : the preferred decision is the option whose expectation has the highest value. Classical decision theory (Luce and Raiffa [18]) separates expectations and values - a common mistake is to make them equal. The decision making involve options, expectations and values. An engineering decision cannot be done in the absence of human values, whereas problems in the science are solved in the absence of options and values. The purpose of values in decision making is to rank order alternatives. This ranking is managed by a preference relation, which is connected by the usual real order relation through the utility function . The person who has a transitive preference relation usually is called rational, in the opposite case he is irrational. The famous Arrow's Impossibility Theorem (Arrow [2]) states that a group consisting only of rational individuals need not exhibit transitive preferences.

John von Neumann and Oscar Morgenstern [27] determined a value measure to rank order alternatives under uncertainties and risk. Their utility is based on a lottery, where there are more desired outcomes and less desired outcomes. They proposed that if the probability of the more desired outcome occurring tends unity, the utility of lottery should approach the utility of the more desired outcome, and conversely. Hence the utility of the lottery should lie between the utilities of the less and more desired outcomes. J. von Neumann and O. Morgenstern developed from this notion and six axioms [27] an utility theory. The main result is that the utility of a lottery is the sum of the utilities of all possible outcomes of the lottery weighted by their probabilities of occurrence. Based on von Neumann and Morgenstern axioms Hazelrigg [12] has introduced an axiomatic approach to engineering design.

A decision is made under the risk if the only available knowledge related the outcome states is the probability distribution. This can be used in the optimization of the utility function. If the knowledge about the probabilities of the outcome is unknown, then the decision have to be made under uncertainty. In engineering design one of the most critical problems is the preliminary design decision when the design is imprecise and most costly . In such situations, fuzzy decision making can be used to handle this vagueness. This fuzziness can be modeled in different ways: fuzzy sets (membership function) (Otto, Antonsson [28] (The Method of Imprecision), Zimmermann, Sebastian [42]) , fuzzy measures (Choquet and Sugeno integrals) (Grabisch et al. [11], Pap [29], Wang, Klir [43]) . There are other design methodologies as optimization tools (linear, nonlinear, integer programming , multi-objective optimization - e.g., with weighted sum technique), probability methods. Aggregation operators play important role in many different approaches to decision making (Otto, Antonsson [28], Grabisch et al. [11], Klement et al. [14]).

Engineering decision making need more general mathematical models, which involve also non-additive measures. Previously used additive probability measures could not model some situations as e.g. the Ellsberg Paradox.

6 Hybrid utility function

A basic notion in probability theory is independence. The main issue in probabilistic independence is the existence of special events $A_1, \ldots, A_n$ such that

$P(A_1 \cap \ldots \cap A_n) = \prod_{i=1}^n P(A_i)$. Such events are called independent events. In order to preserve the computational advantages of independence, any operation $*$ for which it could be established that $P(A_1 \cap \ldots \cap A_n) = *_{i=1}^n P(A_i)$, would do. However the Boolean structure of sets of events and the additivity of the probability measure, impose considerable constraint on the choice of operation $*$.

In the paper [7] is studied the possible operations $*$ when changing P for a pseudo-additive (decomposable) measure based on a t-conorm S. A first remark is that it is natural to require that $*$ be a continuous triangular norm. If $A = X$ is a sure event, then A and X are independent, and it follows that $m(A \cap X) = m(A) * m(X) = m(A) * 1 = m(A)$. Commutativity and associativity of $*$ reflect the corresponding properties for conjunctions. It is also very natural that $*$ be non-decreasing in each place and continuous. We try to find which triangular norms can be used for extending the notion of independence for pseudo-additive measures in the sense of a prescribed triangular conorm. Since the term independence has a precise meaning in probability theory, we shall speak of separability in the framework of S-measures. Two events A and B are said to be *$*$-separable* if $m(A \cap B) = m(A) * m(B)$ for a triangular norm $*$. It turns out by [7] that the only reasonable pseudo-additive measures admitting of an independence-like concept, are based on conditionally distributive pairs (S, T) of conorms and t-conorms, namely:

(a) probability measures (and $* =$ product);
(b) possibility measures (and $*$ is any t-norm);
(c) suitably normalized hybrid set-functions m such that there is $a \in]0, 1[$ which gives for A and B disjoint

$$m(A \cup B) = \begin{cases} m(A) + m(B) - a & \text{if } m(A) > a, m(B) > a, \\ \max(m(A), m(B)) & \text{otherwise,} \end{cases}$$

and for separability:

$$m(A \cap B) = \begin{cases} a + \frac{(m(A)-a)(m(B)-a)}{1-a} & \text{if } m(A) > a, m(B) > a, \\ a \cdot T_1(\frac{m(A)}{a}, \frac{m(B)}{a}) & \text{if } m(A) \leq a, m(B) \leq a, \\ \min(m(A), m(B)) & \text{otherwise,} \end{cases}$$

Any probability distribution on a finite set X can be represented as a sequence of binary lotteries. A binary lottery is 4-uple (A, α, x, y) where $A \subset X$ and $\alpha \in [0, 1]$ such that $P(A) = \alpha$, and it represents the random event that yields x if A occurs and y otherwise. Let P be a probability on X such that $p_i = P(x_i), x_i \in X$. Assume $X = \{x_1, x_2, x_3\}$ then p can be described by the following. The binary tree is obtained as follows: First partition X into $\{x_1\}$ and $\{x_2, x_3\}$ with probabilities p_1 and $p_2 + p_3$, respectively, then partition $X \setminus \{x_1\}$ into $\{x_2\}$ and $\{x_3\}$ with probabilities $p_2/(p_2+p_3)$ and $p_3/(p_2+p_3)$, respectively. The two trees are equivalent, provided that the probability of x_i is calculated by performing the product of weights on the path from the root of the tree until the leaf x.

More generally, suppose m is a S-measure on $X = \{x_1, x_2, x_3\}$ and $m_i = m(\{x_i\})$. Suppose we want to decompose the ternary tree into the binary tree so that they are equivalent. Then the reduction of lottery property enforces the following equations

$$S(v_1, v_2) = 1, \quad T(\mu, v_1) = m_2, \quad T(\mu, v_2) = m_3,$$

where T is the triangular norm that expresses separability for S-measures. The first condition expresses normalization (with no truncating effect for t-conorm S allowed). If these equations have unique solutions, then by iterating this construction, any distribution of a S-measure can be decomposed into a sequence of binary lotteries. This property is basic in probability theory since it explains why probability trees can be used as a primitive notion for developing the notion of probability after Shafer [39]. Turning the S-measure into a sequence of binary trees leads to the necessity of solving the following system of equations

$$\alpha_1 = T(\mu, v_1),\ \alpha_2 = T(\mu, v_2),\ S(v_1, v_2) = 1 \tag{1}$$

for given α_1 and α_2. Assuming that $T_1 = \min$ we have solved (1) completely in [7] and exhibited the analytical forms of (μ, v_1, v_2).

We define the set $\Phi_{S,a}$ of ordered pairs (α, β) in the following way

$$\Phi_{S,a} = \{(\alpha, \beta) \mid (\alpha, \beta) \in]a, 1[, \alpha + \beta = 1 + a\}$$

$$\cup\{(\alpha, \beta) \mid \min(\alpha, \beta) \leq a, \max(\alpha, \beta) = 1\}.$$

A *hybrid mixture set* is a quadruple $(\mathcal{G}, M, T, S)$ where $\mathcal{G}$ is a set, (S, T) is a pair of continuous t-conorm and t-norm, respectively, which satisfy the condition (CD) and $M : \mathcal{G}^2 \times \Phi_{S,a} \to \mathcal{G}$ is a function (hybrid mixture operation) given by

$$M(x, y; \alpha, \beta) = S(T(\alpha, x), T(\beta, y)).$$

It is enough to restrict to the case $(< S_M, S_L >, < T_1, T_P >)_a$. Then it is easy to verify that M satisfies the axioms **M1-M5** on $\Phi_{S,a}$, where

M1. $M(x, y; 1, 0) = x$;
M2. $M(x, y; \alpha, \beta) = M(y, x; \beta, \alpha)$;
M3. $M(M(x, y; \alpha, \beta), y; \gamma, \delta) = M(x, y; T(\alpha, \gamma), S(T(\beta, \gamma), T(\delta, 1)))$.
M4. $M(x, x; \alpha, \beta) = x$.
M5. $M(M(x, y; \alpha, \beta), M(x, y; \gamma, \delta); \lambda, \mu)$

$$= M(x, y; S(T(\alpha, \lambda), T(\gamma, \mu)), S(T(\beta, \lambda), T(\delta, \mu)))$$

holds for all $x, y \in \mathcal{G}$ and all $(\alpha, \beta), (\gamma, \delta), (\lambda, \mu) \in \Phi_{S,a}$.

This kind of mixtures exhausts the possible solutions to **M1-M5**.

Let us show the main appeal of **M1-M5**. Let (S, T) be a pair of continuous t-conorm and t-norm, respectively, of the form $(< S_M, S_L >, < T_1, T_P >)_a$. Let

u_1, u_2 be two utilities taking values in the unit interval $[0,1]$ and let μ_1, μ_2 be two degrees of plausibility from $\Phi_{S,a}$. Then we define *the optimistic hybrid utility function* by means of the hybrid mixture as

$$U(u_1, u_2; \mu_1, \mu_2) = S(T(u_1, \mu_1), T(u_2, \mu_2)).$$

In the paper [7] it is examined in details this utility function. Although the above description of optimistic hybrid utility is rather complex, it can be easily explained, including the name optimistic, see [7]. Putting together the results of this paper, the utility of a n-ary lottery can be computed by decomposing the S- measure into a sequence of binary trees and applying the above computation scheme for hybrid utility recursively from the bottom to the top of the binary tree expansion.

More details and proofs of theorems stated in this paper can be found in [7]. It remains a pending problem to find a corresponding axiomatization for hybrid utility as was done for classical utility theory in [27] and possibility utility theory in [6].

The pair of t-conorm S and t-norm T which satisfies (CD) and the corresponding S-measure also give a basis for an integration theory, so called (S,T)-integral, see [14,15].

Cox's well-known theorem, see [4,38], which justifies the use of probability for treating uncertainty, was discussed in many papers. Some relaxed axioms have been proposed, enabling non-additive functions to be admissible solutions. It is clear that the family of pseudo-additive measures exhibited in this paper is worth studying in Cox relaxed framework, since a natural form of conditioning can be defined on the basis of the triangular norm in the pair (S,T) satisfying (CD), leading to an almost regular independence notion exhibited here.

7 Conclusions

We have illustrated only on few examples the existing mathematical machinery applied in CI, but also the feed-back on mathematics itself. Finally, combining the methods of pure mathematics and CI, benefit both parts!

References

1. J. Aczél, *Lectures on Functional Equations and their Applications,* Academic Press, New York, 1969.
2. K.J. Arrow, *Social Choice and Individual Values*, 2. ed., Yale University Press, New Haven, CT, 1963.
3. F. Baccelli, G. Cohen, G.J. Olsder and J.-P. Quadrat, *Synchronization and Linearity: an Algebra for Discrete Event Systems,* John-Wiley and Sons, New York, 1992.
4. R. Cox, Probability, frequency, and reasonable expectation, American Journal of Physics 14 (1) (1946) , 1-13

5. B. De Baets, A. Marková-Stupňanová, Analytical expressions for the addition of fuzzy intervals, Fuzzy Sets and Systems 91 (1997) 203-214.
6. D. Dubois , L. Godo, H. Prade, A. Zapico, Making decision in a qualitative setting: from decision under uncertainty to case-based decision.(Eds. A.G. Cohn, L. Schubert, S.C. Shapiro) Proceedings of the Sixth International Conference "Principles of Knowledge Representation and Reasoning" (KR ' 98), Morgan Kaufman Publishers, Inc, San Francisco,1998, 594-605
7. D. Dubois , E. Pap, H. Prade, Hybrid probabilistic-possibilistic mixtures and utility functions, (Eds. B.de Baets, J. Fodor, P. Perny) Preferences and Decisions under Incomplete Knowledge , Springer-Verlag (to appear).
8. D. Dubois, H. Prade, *Possibility theory: qualitative and quantitative aspects,* The Handbook of Defeasible Reasoning and Uncertainty Management Systems Vol I (Ed. Ph. Smets), Kluwer,
9. D. Dubois, H. Prade, *Fuzzy sets and systems: Theory and applications*, Academic Press, New York, 1980.
10. J. S. Golan, *The Theory of Semirings with Applications in Mathematics and Theoretical Computer Science,* Pitman Monographs and Surveys in Pure and Appl. Math.. 54, Longman, New York, 1992.
11. M. Grabisch, H. T. Nguyen, E. A. Walker, Fundamentals of Uncertainity Calculi with Application to Fuzzy Inference, Kluwer Academic Publishers, Dordrecht/Boston/London, 1995.
12. G. A. Hazelrigg, An Axiomatic framework for Engineering Design, Technical repor, 1997.
13. E. P. Klement, R. Mesiar and E. Pap, On the Relationship of Associative Compensatory Operators to Triangular Norms and Conorms, Inter. J. Uncertainty, Fuzziness and Knowledge Based Systems 4 (1996), 129-144.
14. E. P. Klement, R. Mesiar, E. Pap, *Triangular Norms,* Kluwer Academic Publishers, Dordrecht, 2000.
15. E. P. Klement, R. Mesiar, E. Pap, (S, U)-integral. Proc. EUSFLAT - 99, Palma de Mallorca, pages 371-374, 1999.
16. G.J. Klir, B. Yuan, *Fuzzy Sets and Fuzzy Logic, Theory and Applications*, Prentice Hall, New Jersey, 1995.
17. A. Kolesárová, Similarity preserving t-norm-based additions of fuzzy numbers, Fuzzy Sets and Systems 91 (1997) 215-230.
18. R.D. Luce, H. Raiffa, *Games and Decisions*, John Wiley & Sons, Inc, New York, 1957.
19. A. Marková-Stupňanová, A note to the idempotent functions with respect to pseudo-convolution, Fuzzy Sets and Systems 102 (1999), 417-421.
20. V.P. Maslov, S.N. Samborskij (eds.), *Idempotent Analysis,* Advances in Soviet Mathematics 13, Amer. Math. Soc., Providence, Rhode Island, 1992.
21. R. Mesiar, Shape preserving additions of fuzzy intervals, Fuzzy Sets and Systems 86 (1997) 73-78.
22. R. Mesiar, Triangular-norm-based addition of fuzzy intervals, Fuzzy Sets and Systems 91 (1997) 231-238.
23. R. Mesiar, E. Pap, Different interpretations of triangular norms and related operations, Fuzzy Sets and Systems 96 (1998) 183-189.
24. R. Mesiar, E. Pap, Idempotent integral as limit of g-integrals, Fuzzy Sets and Systems 102 (1999), 385-392.
25. R. Mesiar, J. Rybarik, PAN–operations structure, Fuzzy Sets and Systems 74 (1995) 365-369.

26. D. Nauck, F. Klawonn, R. Kruse, *Neuronale Netze und Fuzzy– Systeme*. Braunschweig/Wiesbaden, Vieweg, 1994.
27. von Neumann, J., Morgenstern, O., *Theory of Games and Economic Behavior*, Princeton Univ. Press, Princeton, NJ, 1944.
28. K. Otto and E. Antonsson, Trade-Off Strategies in Engineering Design, *Research in Engineering Design* 3 (2) (1991), 87-104.
29. E. Pap, *Null-Additive Set Functions,* Kluwer Academic Publishers, Dordrecht-Boston-London, 1995.
30. E. Pap, Pseudo-analysis as a mathematical base for soft computing, Soft Computing 1 (1997), 61-68.
31. E. Pap, Decomposable measures and nonlinear equations, Fuzzy Sets and Systems 92 (1997), 205-222.
32. E. Pap, Applications of Decomposable Measures, chapter in Handbook "Mathematics of Fuzzy Sets-Logic, Topology and Measure Theory" Volume 2 (Eds. U. Höhle, S.E. Rodabaugh),Kluwer Academic Publishers, Dordrecht- Boston-London, 1999, 675-700.
33. E. Pap, Pseudo-convolution and Its Applications, (Eds. M. Grabisch, T. Murofushi, M. Sugeno) Fuzzy Measures and Integrals, Theory and Applications, Physica-Verlag (Springer-Verlag Company), 2000, 171-204.
34. E. Pap, N. Ralević, Pseudo-Laplace Transform, Nonlinear Analysis 33 (1998) 553-550.
35. E. Pap, D. Surla, Lebesgue measure of α-cuts approach for finding the height of the membership function, Fuzzy Sets and Systems 111 (2000), 341-350.
36. E. Pap, I. Štajner, Generalized pseudo-convolution in the theory of probabilistic metric spaces, information, fuzzy numbers, optimization, system theory, Fuzzy Sets and Systems 102 (1999), 393-415.
37. E. Pap, Aggregation operators in the engineering design, Edited volume "Aggregation operators", Springer-Verlag (to appear).
38. Paris, J.B., *The Uncertain Reasoner's Companion, A Mathematical Perspective*, Cambridge, U.K.: Cambridge University Press, 1994.
39. Shafer, G., *The Art of Causal Conjecture*, The MIT Press, Cambridge, Massachusetts, 1996.
40. B. Schweizer, A. Sklar, *Probabilistic Metric Spaces,* North-Holland, Amsterdam, 1983.
41. L. A. Zadeh, The concept of a linguistic variable and its applicatin to approximate reasoning - Part I, Inform. Sci. 8 (1975) 199-249.
42. H.–J. Zimmermann and H.–J. Sebastian, Optimization and Fuzziness in Problems of Design and Configuration, Proc. of the Second IEEE International Conference on Fuzzy Systems, Pergamon Press,1993, 1237-1240.
43. Z. Wang , G. J. Klir, *Fuzzy Measure Theory,* Plenum Press, New York, 1992.

Some New Ways to Cluster in Incomplete Data

James C. Bezdek[1, A] and Richard J. Hathaway[2, B]

[1]Department of Computer Science
University of West Florida, Pensacola, FL, USA
[2]Mathematics and Computer Science Department
Georgia Southern University, Statesboro, GA, USA

Abstract: The problem of clustering in a real s-dimensional data set $X = \{x_1,\ldots,x_n\} \subset R^s$ *is considered. Usually, each observation (or datum) consists of numerical values for all s features (such as height, length, etc.), but sometimes data sets contain vectors that are missing one or more feature values. For example,* x_k *might have the form* $x_k = (254.3, ?, 333.2, 47.44, ?)^T$*, where the second and fifth feature values are missing. Clustering algorithms such as fuzzy c-means are not directly applicable to incomplete data. Four strategies that extend fuzzy c-means to incomplete data sets are given, three of which involve modified versions of the fuzzy c-means algorithm. One of the four methods relies on a new technique we call tri-level alternating optimization. Some theoretical material is provided for this new method, numerical convergence properties of the new algorithms are discussed, and all approaches are tested using artificially generated incomplete data sets.*

Keywords: Clustering, Missing data, Incomplete data, Fuzzy c-means, Tri-level Optimization.

1 Introduction

We discuss clustering a set of numerical object data $X = \{x_1,\ldots, x_n\} \subset R^s$ into c clusters, $1 < c < n$. The jth component of feature vector x_k is denoted by x_{jk}. In some situations feature vectors in X can have missing components [1, 2]. Any datum with some (but not all) missing feature values is referred to as an *incomplete datum*, e.g., $x_5 = (1.23, 3.14, ?, 4.66, ?)^T$, where x_{35} and x_{55} are missing. A data set with at least one incomplete datum is referred to as an *incomplete data set*; otherwise, it is called complete.

[1,A]Research supported by ONR Grant 00014-96-1-0642; corresponding author.
[2,B]Research supported in part by the Faculty Research Committee of Georgia Southern University.

Theoretical approaches to pattern recognition with incomplete data based on probabilistic assumptions data back to at least 1962 [3]. An important empirical study was done by Dixon [1] is summarized in [2]. It may be that estimates of the missing values are desired [4, 5], or that predicted values of new data are sought [6], but often the goal is to simply use the available data to perform some analysis [7, 8]. Two of our approaches produce estimates of missing values as a by-product of the calculations. Two other useful sources of information for statistical approaches to analysis of incomplete data are [9, 10].

We begin with a discussion of *visualization* of incomplete data sets. When $X \subset R^2$, it is possible to represent both complete and incomplete data sets with scatter diagrams. To do this, we represent an incomplete datum with missing second component such as $x_k = (x_{1k},?)^T$ as a vertical line with horizontal component x_{1k}; and when the first component is missing, as for $x_j = (?, x_{2j})^T$, we represent x_j by a horizontal line with vertical component x_{2j}. The incomplete null datum (?,?) cannot be represented in such a plot. In higher dimensions, an incomplete datum could correspond to a line, plane, etc., depending on the number of missing feature values. Figure (1a) corresponds to a complete data set. Figure (1b) depicts an incomplete version of the data, where 25% of the (20) feature values are missing. In this example, three values of the first feature and two of the second feature are missing. Notice the increased difficulty in visually identifying the presence (and location) of the two clusters when a substantial percentage of feature values are missing.

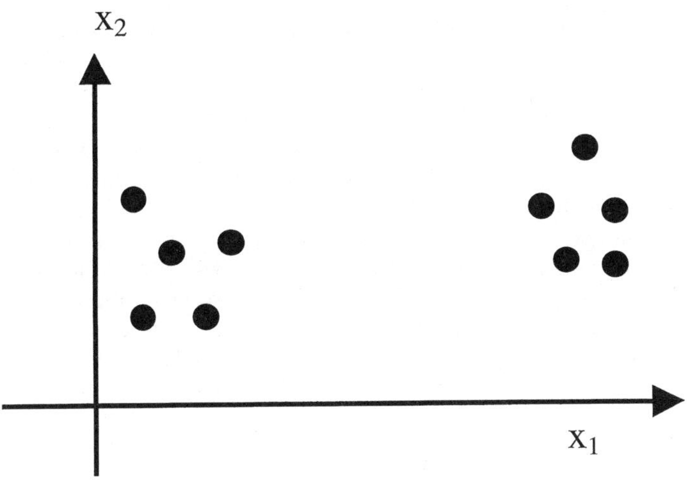

Figure 1. (a) Complete data

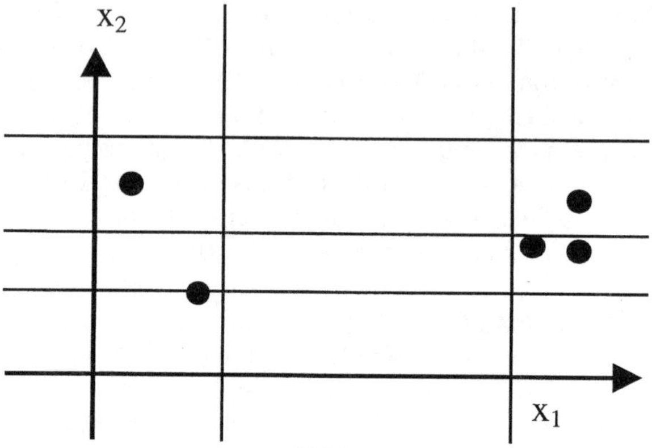

(1b)

Figure 1 (b) Incomplete data with 25 % of feature values missing.

An interesting extreme case of the visual difficulty demonstrated in Figure 1 involves what we "virtual clusters". When *every* vector in X has just one feature value, we call X *maximally incomplete*. A maximally incomplete data set $X \subset R^2$ has exactly 50% of its feature values missing. For this case a plot such as the right side of Figure 1 will contain *no* points; instead, it will consist entirely of a lattice of n horizontal/vertical lines as shown in Figure 2. More generally, a maximally incomplete data set $X \subset R^s$ is missing exactly $\frac{100(s-1)}{s}$ % of its feature values. It can be impossible to find good estimates for the cluster centers if X is maximally incomplete.

Each of the plots in Figure 2 depicts a maximally incomplete two-dimensional data set, which, if complete, would have two *visual clusters*. These views were obtained by starting with complete data, whose general positions are indicated by the circles; and then for each datum, designating one of its two feature values as "missing". Is it possible to recover reasonable estimates of the original (complete) data *cluster centers* from a maximally incomplete set? The answer is yes for the data underlying the left view in Figure 2, but is no for the data that produce the right view, where two additional "virtual clusters" are indicated. Seven different cluster structures, involving from 2 to 4 clusters, could produce the incomplete data scatter diagram in view (2b).

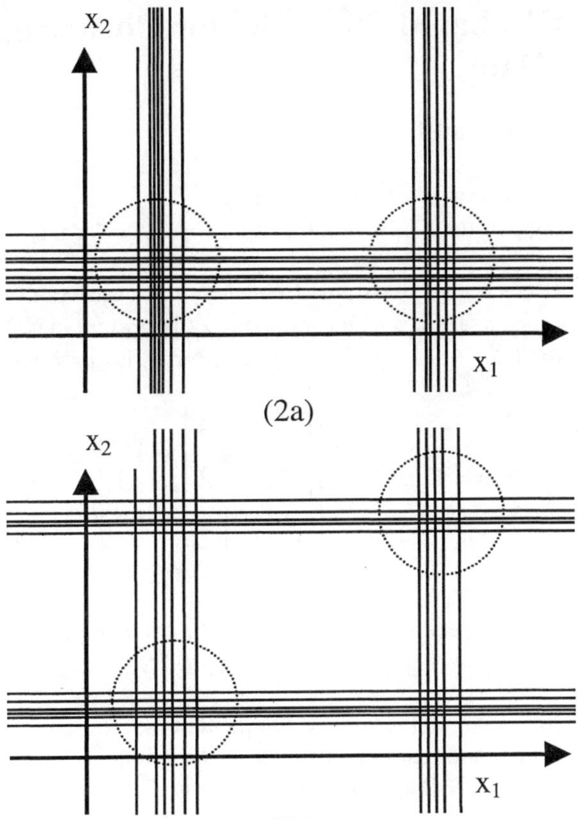

(2b)

Figure 2. Two maximally incomplete data sets.

There is only one possible choice for c in view (2a), viz., c = 2, whereas c could be 2, 3 or 4 in view (2b). Borrowing a term commonly used in estimation of the joint parameters of a mixture density, we say that a maximally incomplete data set is *c-identifiable* when (the parameter) c and c corresponding cluster centers are uniquely recoverable. Figure 2 makes the intuitive idea of c-identifiability clear; but attempts by us to characterize this notion in mathematical terms have (so far) been unsuccessful.

This paper introduces four new strategies for clustering incomplete data sets. Three of these consist of new adaptations of the *fuzzy c-means* (FCM) algorithm [11], and all four provide estimates of the locations of cluster centers and fuzzy partitions of the data. Section 2 fixes our notation, reviews FCM, gives the new algorithms, and discusses their theoretical convergence properties. In Section 3 the four approaches are compared using several artificially generated two-dimensional data sets. We examine the accuracy of the computed cluster centers and the correctness of the clusters produced by each of the methods. Section 4 summarizes our results and conclusions.

2 Four FCM-based Strategies for Clustering Incomplete Data

The notation that will be used:

x_k=k^{th} s-dimensional data vector, for $1 \le k \le n$; (1a)
x_{jk}=j^{th} feature value of the kth data vector, for $1 \le j \le s$, $1 \le k \le n$; (1b)
$X=\{x_1,\ldots,x_n\}$; (1c)
$X_W=\{x_k \in X \mid x_k$ is a complete datum$\}$ (the *whole-data* subset of X) ; (1d)
$X_P = \{ x_{jk}$ for $1 \le j \le s$, $1 \le k \le n \mid$ the value for x_{jk} is *present* in X $\}$; and (1e)
$X_M = \{ x_{jk} = ?$ for $1 \le j \le s$, $1 \le k \le n \mid$ the value for x_{jk} is *missing* from X$\}$. (1f)

For example, let s = 3, n = 4 and $X = \left\{ \begin{bmatrix} 2 \\ ? \\ ? \end{bmatrix}, \begin{bmatrix} 3 \\ 2 \\ 4 \end{bmatrix}, \begin{bmatrix} 2 \\ 1 \\ ? \end{bmatrix}, \begin{bmatrix} 1 \\ 1 \\ 2 \end{bmatrix} \right\}$. Then $X_W = \left\{ \begin{bmatrix} 3 \\ 2 \\ 4 \end{bmatrix}, \begin{bmatrix} 1 \\ 1 \\ 2 \end{bmatrix} \right\}$, $X_M = \{x_{12}, x_{13}, x_{33}\}$ and $X_P = \{x_{11} = 2, x_{21} = 3, x_{22} = 2, x_{23} = 4, x_{31} = 2, x_{32} = 1, x_{41} = 1, x_{42} = 1, x_{43} = 2\}$. Notice that the available feature values in X_P and X are the same, and that only a subset of these feature values is available in X_W. In this sense, X_P and X contain more "information" than X_W.

The FCM model has been widely used, adapted and generalized [12]. The fuzzy c-means clustering algorithm is used in a pure or modified form for all four approaches given in this section. FCM simultaneously produces a cxn matrix representation U of the fuzzy cluster substructure of X along with prototypes for each of the c clusters via an iterative scheme for solving:

$$\min_{(U,v)} \left\{ J_m(U,v) = \sum_{i=1}^{c} \sum_{k=1}^{n} U_{ik}^{m} \| x_k - v_i \|_A^2 \right\}, \text{ where} \tag{2}$$

$m > 1$ is the fuzzification parameter; $\| x \|_A^2 = x^T A x$ is a vector A-norm,

$$U \in M_{fcn} = \left\{ U \in R^{c \times n} : U_{ik} \in [0,1], \forall i,k; \sum_{i=1}^{c} U_{ik} = 1 \ \forall k; \sum_{k=1}^{n} U_{ik} > 0 \ \forall i \right\}, \text{ and}$$

(3)

$v = \{v_1, \ldots, v_c\} \subset R^s.$

(4)

When $A = I_{s \times s}$, $\|x\|_A = \|x\|_2$, the Euclidean norm. The elements of the membership matrix U represent the (relative) degrees of membership of each datum in each cluster; i.e., U_{ik} = the degree to which x_k belongs to the ith cluster. The set M_{fcn} in (3) is referred to as the set of *nondegenerate* (meaning each cluster is non-empty) *fuzzy partition matrices*. The vector v_i is the prototype or center of the ith cluster. The FCM algorithm for solving (2) alternates optimizations of J_m over the U and v variables. This method of *alternating optimization* (AO) has been well studied [13, 14] and typically provides iterate sequences that converge q-linearly to local (and sometimes global) minimizers. The details of one version of FCM are now given. Other versions arise if we initialize with $U^{(0)}$ rather than $v^{(0)}$; or if we check $\left\|U^{(r+1)} - U^{(r)}\right\|$ rather than $\left\|v^{(r+1)} - v^{(r)}\right\|$ for termination.

FCM-1 Fix $m > 1$, $1 < c < n$ and $\varepsilon > 0$. Pick $v^{(0)} \subset R^s$, an $\| \ \|_A$ norm for J_m, and a termination norm $\| \ \|$ for FCM. Then at step r, r = 0, 1, 2,...:

FCM-2 Calculate $U^{(r+1)} = \underset{U \in M_{fcn}}{\arg\min} \left\{ J_m(U, v^{(r)}) \right\}$ for $1 \le i \le c$ and $1 \le k \le n$:

$$U_{ik}^{(r+1)} = \left(D_{ik}^{1/(1-m)} \right) \Big/ \left(\sum_{j=1}^{c} D_{jk}^{1/(1-m)} \right) \quad \text{where}$$

$$D_{ik} = \left\| x_k - v_i^{(r)} \right\|_A^2 . \qquad (5)$$

FCM-3 Calculate $v^{(r+1)} = \underset{v}{\arg\min} \left\{ J_m(U^{(r+1)}, v) \right\}$ for $1 \le i \le c$ and $1 \le j \le s$:

$$v_{ij}^{(r+1)} = \left(\sum_{k=1}^{n} \left(U_{ik}^{(r+1)} \right)^m x_{kj} \right) \Big/ \left(\sum_{k=1}^{n} \left(U_{ik}^{(r+1)} \right)^m \right).$$

(6)

FCM-4 Compare $v^{(r+1)}$ to $v^{(r)}$ using $\left\| v^{(r+1)} - v^{(r)} \right\| < \varepsilon$. If true, then stop. Otherwise, set r = r+1 and return to FCM-2.

- FCM uses every value in X, so it cannot be directly applied to incomplete data.

2.1 Whole data strategy (WDS). If the proportion of incomplete data is small, delete the incomplete data vectors and apply FCM to X_W. The application of FCM to X_W is the *whole data strategy* (WDS); the algorithm is WDSFCM. Set a fairly stringent limit on the number of vectors that are cast out of X because they have one or more missing values. Our belief is that the WDS should be used only if $n_p/n \geq 0.75$, where $n_p = |X_P|$ and $n = |X|$. The WDS may provide good estimates of the prototypes $v = \{v_1,\ldots,v_c\}$, but it won't provide membership information for vectors in $(X - X_W)$. Memberships of missing vectors can be estimated, however, using a nearest-prototype classification scheme based on partial distances from each incomplete datum to each of the computed cluster centers. WDSFCM enjoys all the standard numerical convergence properties of FCM, since it is exactly FCM

2.2 Partial distance strategy (PDS). The top strategy recommended by Dixon [1] when the WDS cannot be justified is to calculate partial (squared Euclidean) distances using all available (i.e., non-missing) feature values, and then scale them by the reciprocal of the proportion of components used. We call this the *partial distance strategy* (PDS). Here is an example of the D_{ik} calculation in (5) using this method for a pair of 5-vectors:

$$D_{ik} = \left\| x_k - v_i^{(r)} \right\|_2^2 = \left\| (1 \quad ? \quad ? \quad 4 \quad ?)^T - (5 \quad 6 \quad 7 \quad 8 \quad 9)^T \right\|_2^2 = \frac{5}{(5-3)} \left((1-5)^2 + (4-8)^2 \right) \quad (7)$$

The general formula for the partial distance calculation of D_{ik} is given by:

$$D_{ik} = \frac{s}{I_k} \sum_{j=1}^{s} (x_{kj} - v_{ij})^2 I_{kj} \quad \text{, where} \quad (8a)$$

$$I_{kj} = \begin{cases} 0 & \text{if } x_{kj} \in X_M \\ 1 & \text{if } x_{kj} \in X_P \end{cases} \quad \text{for } 1 \leq j \leq s \text{ and } 1 \leq k \leq n, \text{ and } I_k = \sum_{j=1}^{s} I_{kj}. \quad (8b)$$

The *partial distance strategy fuzzy c-means algorithm* (PDSFCM), is obtained by making two modifications to the FCM algorithm given earlier for the special case of $\| x \|_A = \| x \|_2$. They are: (1) calculate D_{ik} in (5) for incomplete data according to (8); and (2) replace the calculation of $v^{(r+1)}$ in (6) with

$$v_{ij}^{(r+1)} = \left(\sum_{k=1}^{n} \left(U_{ik}^{(r+1)} \right)^m I_{kj} \, x_{kj} \right) \Bigg/ \left(\sum_{k=1}^{n} \left(U_{ik}^{(r+1)} \right)^m I_{kj} \right) \qquad (9)$$

This approach enjoys the standard convergence properties of FCM because it is an instance of alternating optimization [13]. The global convergence theory of Zangwill [15] guarantees us that any limit point of a PDSFCM iteration sequence will either be a local minimizer, or at worst a saddle point, of the PDSFCM functional. The local result states that if we initialize the iteration sequence sufficiently near a local minimizer satisfying the second order sufficiency conditions for a minimizer, then the iteration sequence will converge q-linearly to that minimizer. (This is the standard FCM theory.)

2.3 Optimal completion strategy (OCS). The third approach to FCM clustering of incomplete data is based on an *optimal completion strategy* (OCS). In this approach we view the missing components (X_M) as additional unknowns in the FCM model. Our strategy is to estimate the missing values in a way that leads to the smallest possible value of J_m given the available data X_P. The OCS modification of FCM, referred to here as OCSFCM, is obtained by adding an optimization step over the X_M "variables" during each iteration. Here is the *Optimal Completion Strategy Fuzzy c-Means Algorithm:*

OCSFCM-1 Fix $m > 1$, $1 < c < n$ and $\varepsilon > 0$. Pick $v^{(0)} \in R^{s \times c}$, an $\| \, \|_A$ norm for J_m, and a termination norm $\| \, \|$ for FCM. Pick initial values for cluster means $v^{(0)} \subset R^s$ and missing data $X_M^{(0)}$. Then at step r, r = 0, 1, 2, ... :

OCSFCM-2 For $1 \le i \le c$ and $1 \le k \le n$,

calculate $U^{(r+1)} = \underset{U \in M_{fcn}}{\arg\min} \left\{ J_m(U, v^{(r)}, X_M^{(r)}) \right\}$ using

$$U_{ik}^{(r+1)} = \left(D_{ik}^{1/(1-m)} \right) \Bigg/ \left(\sum_{j=1}^{c} D_{jk}^{1/(1-m)} \right), \text{ where}$$

$$D_{ik} = \left\| x_k - v_i^{(r)} \right\|_A^2 . \qquad (11)$$

OCSFCM-3 For $1 \le i \le c$ and $1 \le k \le n$, calculate $v^{(r+1)} = \arg\min_{v} \left\{ J_m(U^{(r+1)}, v, X_M^{(r)}) \right\}$ using

$$v_{ij}^{(r+1)} = \left(\sum_{k=1}^{n} \left(U_{ik}^{(r+1)}\right)^m x_{kj} \right) \Bigg/ \left(\sum_{k=1}^{n} \left(U_{ik}^{(r+1)}\right)^m \right) \quad (12)$$

OCSFCM-4 Compare $v^{(r+1)}$ to $v^{(r)}$ using $\left\| v^{(r+1)} - v^{(r)} \right\| < \varepsilon$. If true, then stop. Otherwise, go to OCSFCM-5.

OCSFCM-5 Calculate $X_M^{(r+1)} = \arg\min_{X_M} \left\{ J_m(U^{(r+1)}, v^{(r+1)}, X_M) \right\}$ for all $x_{kj} \in X_M$ using:

$$x_{kj}^{(r+1)} = \left(\sum_{i=1}^{c} \left(U_{ik}^{(r+1)}\right)^m v_{ij}^{(r+1)} \right) \Bigg/ \left(\sum_{i=1}^{c} \left(U_{ik}^{(r+1)}\right)^m \right). \quad (13)$$

Now set $r = r + 1$ and return to step OCSFCM-2.

Zeroing the gradient of $J(x_k) = \sum_{i=1}^{c} U_{ik}^m \left\| x_k - v_i \right\|_A^2$ with respect to the missing values of x_k leads to (13). Steps OCSFCM-2, 3, and 5 show that OCSFCM is a *tri-level alternating optimization* (TLAO) of $J_m(U,v,X_M)$. Convergence properties of TLAO algorithms are fully developed in [16]. Briefly, any limit point of an OCSFCM iteration sequence will be a minimizer of $J_m(U,v,X_M)$, or at worst, will be a special type of saddle point that is optimal in each of the three variables while holding the other two fixed. The iteration sequence will always converge q-linearly to any minimizer of $J_m(U,v,X_M)$ satisfying the second order sufficiency conditions if initialized sufficiently near it. Since this is a new idea in non-linear programming, we state the main results here, and refer readers to [16] for complete proofs.

Math Insert : A thumbnail sketch of TLAO theory. Suppose $f: R^s \to R$ is twice differentiable, partition $w \in R^s$ as $w = (x,y,z)$, with $x \in R^p$, $y \in R^q$, and $w \in R^r$, $p + q + r = s$, so $f: R^p \times R^q \times R^r \to R$. The TLAO procedure is one way to estimate solutions of the problem

$$\min_{w \in R^s} f(w) \quad (14)$$

Here is the basic iteration scheme that seeks extrema of f with tri-level alternating optimization:

TLAO-1. Pick an initial iterate $w^{(0)} = (x^{(0)}, y^{(0)}, z^{(0)})$ and stopping criterion. For example, choose a vector norm $\|\cdot\|$ and termination threshold ε, and stop when $\|w^{(k+1)} - w^{(k)}\| \le \varepsilon$ or when k > T, where T is a maximum iteration limit. Set k = 0.

TLAO-2. Compute $$x^{(k+1)} = \arg\min_{x \in R^p} f(x, y^{(k)}, z^{(k)}) \quad (15)$$

TLAO-3. Compute $$y^{(k+1)} = \arg\min_{y \in R^q} f(x^{(k+1)}, y, z^{(k)}) \quad (16)$$

TLAO-4. Compute $$z^{(k+1)} = \arg\min_{z \in R^r} f(x^{(k+1)}, y^{(k+1)}, z) \quad (17)$$

TLAO-5. If $\|w^{(k+1)} - w^{(k)}\| \le \varepsilon$ or k > T, then quit; otherwise, set k = k+1 and go to TLAO-2.

Corresponding to (15-17)
define X: $R^q \times R^r \to R^p$, Y: $R^p \times R^r \to R^q$, and Z: $R^p \times R^q \to R^r$, as :

$$X(y,z) = \arg\min_{x \in R^p} f(x, y, z) \quad (18)$$

$$Y(x,z) = \arg\min_{y \in R^q} f(x, y, z) \quad (19)$$

$$Z(x,y) = \arg\min_{z \in R^r} f(x, y, z) \quad (20)$$

Assume without loss that the local minimizer of interest is $(0,0,0) \in R^p \times R^q \times R^r$. Under reasonable assumptions, X, Y, and Z are continuously differentiable near $(0,0) \in R^q \times R^r, R^p \times R^r, R^p \times R^q$:

- **Lemma 2.1** Let f: $R^p \times R^q \times R^r \to R$, $f''(w)$ be the s×s Hessian of f, and suppose

(i) f is C^2 in a neighborhood of (0,0,0)

(ii) $f''(0,0,0)$ is positive definite

; and

(iii) (0,0,0) is a local minimizer of f

Then in some neighborhood of $(0,0) \in R^q \times R^r$, the minimizing function X(y,z) in (5) exists and is continuously differentiable. Similar results hold for Y(x,z) and Z(x,y).

- **Proof.** [16]

Define the mapping S: $R^p \times R^q \times R^r \rightarrow R^p \times R^q \times R^r$ corresponding to one iteration through steps TLAO-2, 3 and 4. Lemma 2.1 implies that S is continuously differentiable in a neighborhood of (0,0,0) with S(0,0,0) = (0,0,0). Let $\rho(S'(0,0,0))$ denote the *spectral radius* of $S'(0,0,0)$. The fundamental property needed to establish convergence of a TLAO sequence is that $\rho(S'(0,0,0)) < 1$:

- **Lemma 2.2** Let f: $R^p \times R^q \times R^r \rightarrow R$ satisfy the conditions of Lemma 2.1 and let S: $R^p \times R^q \times R^r \rightarrow R^p \times R^q \times R^r$ be defined by (9). Then $\rho(S'(0,0,0)) < 1$.
- **Proof** [16]

The main result for local convergence is essentially an adaptation of Ostrowski's theorem, 8.1.7 in [17].

- **Theorem 2.1** Let w* be a local minimizer of f: $R^s \rightarrow R$ for which $f''(w^*)$ is positive definite, and let f be C^2 on a neighborhood of w*. Then there is a neighborhood N(w*) of w* such that for any $w^{(0)} \in N(w^*)$, the corresponding TLAO iteration sequence $\{w^{(k)}\}$ defined using S as $w^{(k+1)} = S(w^{(k)})$ converges q-linearly to w*.

- **Proof** [16]

This completes our sketch of the emerging theory of tri-level optimization, which provides a sound theoretical basis for our OCSFCM algorithm.

2.4 Nearest prototype strategy (NPS). The last approach uses the *nearest prototype strategy* (NPS) and can be described as a simple modification of OCSFCM. We call this approach NPSFCM and define it as OCSFCM with Step OCSFCM-5 replaced by:

NPSFCM-5 Calculate $X_M^{(r+1)}$ for all $x_{kj} \in X_M$ using partial distances D_{ik} in (8) by:

$$x_{kj}^{(r+1)} = v_{ij}^{(r+1)} \text{ where } D_{ik} = \min\{D_{1k}, D_{2k}, \ldots, D_{ck}\} \tag{20}$$

Now set r = r + 1 and return to step OCSFCM-2.

In the rare case that an incomplete datum x_k is equidistant to two or more prototypes, a tie-breaking rule is needed to get $x_{kj}^{(r+1)}$ in (20). While NPSFCM terminated in all numerical tests, we have not established theoretically that this procedure will converge.

3 Numerical examples

In this section we compare the WDS, PDS, OCS and NPS versions of FCM using four artificially generated incomplete data sets. First a complete data set X is chosen. Then X is modified to obtain an incomplete data set by randomly selecting a specified percentage of its components to be designated as missing. The complete data sets are shown in Figures 3- 6.

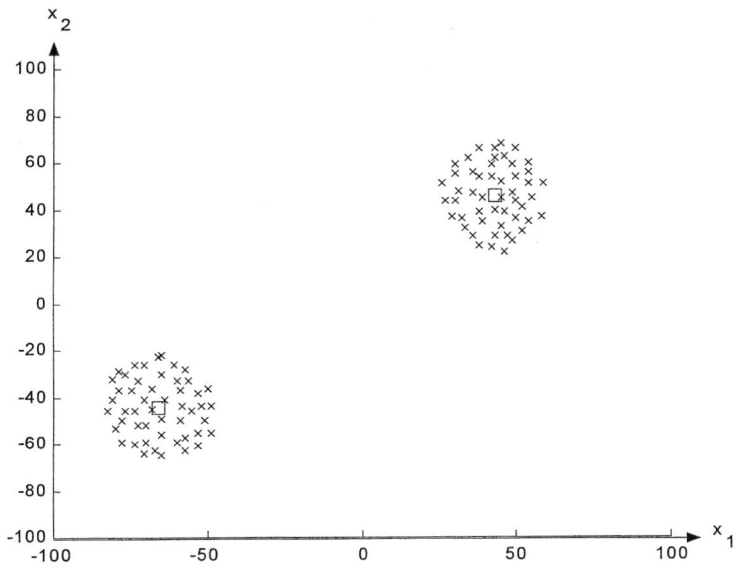

Figure 3. DS: Two Diagonally-Positioned Small Clusters. Full data v* represented by □.

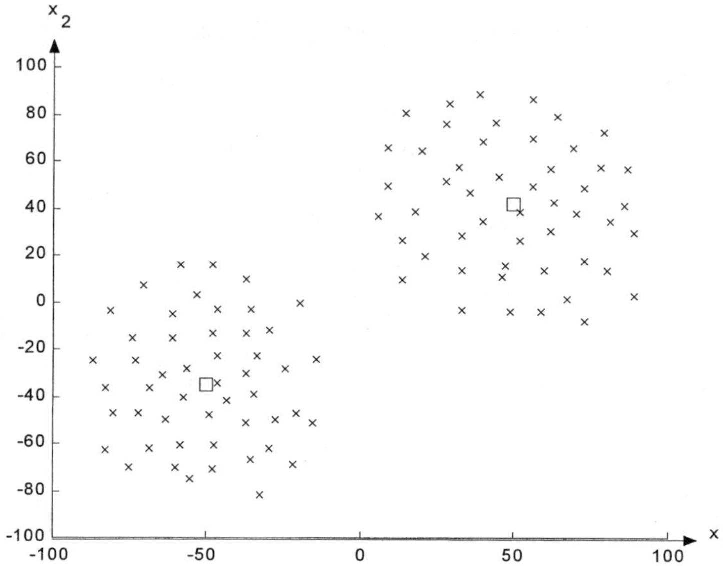

Figure 4. DL: Two Diagonally-Positioned Large Clusters. Full data v* represented by □.

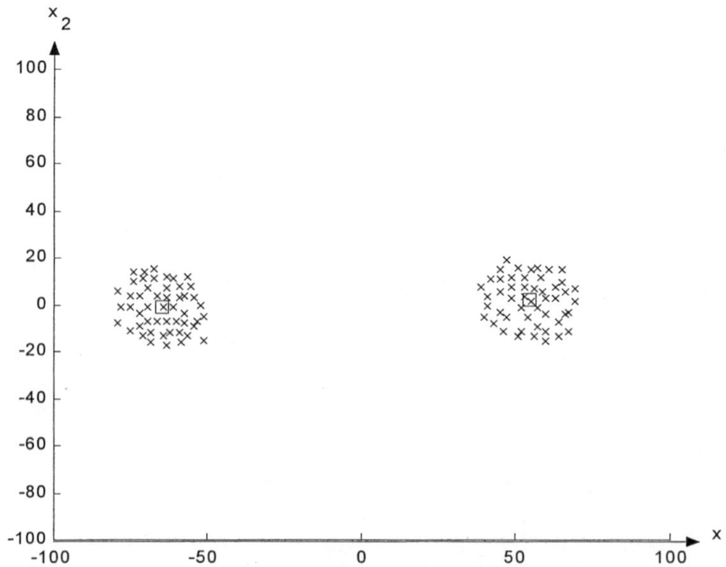

Figure 5. HS: Two Horizontally-Positioned Small Clusters. Full data v* represented by □.

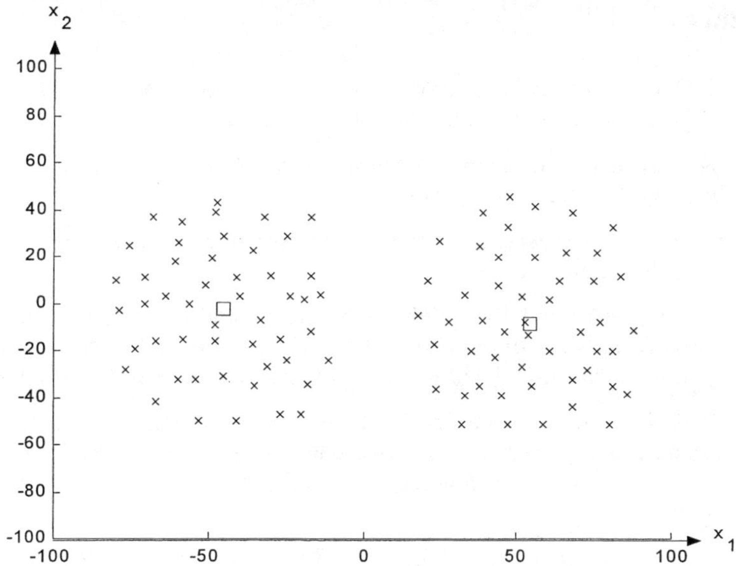

Figure 6. HL: Two Horizontally-Positioned Large Clusters. Full data v* represented by □.

We chose four cases involving small and large clusters having horizontal and diagonal cluster placement. Since our data is in R^2, maximally incomplete sets occur when the percentage of missing values is 50%. The random selection of missing feature values is constrained so that: (C1) each original feature vector x_k retains at least one component, and (C2): each feature has at least one value present in the incomplete data set The number of data in each of the four (complete) artificial data sets is n = 100, evenly divided ($n_1 = n_2 = 50$) between the two clusters. To compare resubstitution errors, we crisply labeled points in each data set as class 1 or 2, using the obvious visual clusters.

The full-data, globally optimal FCM prototypes v* are represented by □ in Figures 3-6. The initialization $v^{(0)}$ used in all cases is the set v*. The additional initialization $X_M^{(0)}$ needed for the OCS and NPS versions is obtained using v* for v in NPSFCM-5, equation (20). The stopping criterion $\left\| v^{(r+1)} - v^{(r)} \right\| < \varepsilon$ was:

$$\max\left(\left| v_{ji}^{(r+1)} - v_{ji}^{(r)} \right|, \text{ for } 1 \le j \le s \text{ and } 1 \le i \le c \right) \le 0.00001 \qquad (21)$$

There was significant variation in the results from trial to trial, so Tables 1-4 list averages obtained over 50 trials for each row in a table. The same sample is used in each trial for each of the four approaches so that the results are directly

References

[1] J.K. Dixon, Pattern recognition with partly missing data, *IEEE Transactions on Systems, Man and Cybernetics*, 9, 617-621, 1979.

[2] A.K. Jain and R.C. Dubes, *Algorithms for Clustering Data*. Englewood Cliffs, NJ: Prentice-Hall, 1988.

[3] G.S. Sebestyen, *Decision-making Processes in Pattern Recognition*. New York: McMillan, 74, 1962.

[4] N.G. Zagoruiko and V.N. Yolkina, Inference and data tables with missing values, *Handbook of Statistics*, P.R. Krishnaiah and L.N. Kanal, eds., Amsterdam: North-Holland Publishing Company, 2, 493-500, 1982.

[5] R.C.T. Lee, J.R. Slagle and C.T. Mong, Application of clustering to estimate missing data and improve data integrity, *Proceedings of the 2nd International Software Engineering Conference*, San Francisco, CA, 539-544, 1976.

[6] J.C. Bezdek and K. Solomon, Simulation of implicit numerical characteristics using small samples, *Proceedings of ICASRC*, ed. G.E. Lasker, Pergamon, NY, VI, 2773-2784, 1981.

[7] J. Kittler, Classification and incomplete pattern vectors using modified discriminant functions, *IEEE Transactions on Computers*, C-27, 367-375, 1978.

[8] C.Z. Janikow, Fuzzy decision trees: issues and methods, *IEEE Transactions on Systems, Man and Cybernetics*, B28, 1-14, 1998.

[9] J.L. Schafer, *Analysis of Incomplete Multivariate Data*. London, UK: Chapman & Hall, 1997.

[10] R.J.A. Little and D.B. Rubin, *Statistical Analysis with Missing Data*. New York: John Wiley & Sons, 1987.

[11] J.C. Bezdek, *Pattern Recognition with Fuzzy Objective Function Algorithms*. New York: Plenum, 1981.

[12] J.C. Bezdek, J.M. Keller, R. Krishnapuram and N.R. Pal, *Fuzzy Models and Algorithms for Pattern Recognition and Image Processing*. Norwell, MA: Kluwer, 1999.

[13] J.C. Bezdek, R.J. Hathaway, R.E. Howard, C.A. Wilson and M.P. Windham, Local convergence analysis of a grouped variable version of coordinate descent, *Journal of Optimization Theory and Applications*, 54, 471-477, 1987.

[14] J.C. Bezdek, R.J. Hathaway, M.J. Sabin and W.T. Tucker, Convergence theory for fuzzy c-means: counterexamples and repairs, *IEEE Transactions on Systems, Man and Cybernetics*, SMC-17, 873-877, 1987.

[15] W. Zangwill, *Nonlinear Programming: A Unified Approach.* Englewood Cliffs, NJ: Prentice-Hall, 1969.

[16] R.J. Hathaway and J.C. Bezdek, Local convergence of tri-level alternating optimization, in review, *Journal of Optimization Theory and Applications*, 1999.

[17] J. M. Ortega, *Numerical Analysis: A Second Course*, Academic Press, New York, 1972.

Table 1. Averaged Results of 50 Trials using Incomplete DS Data Sets

% Missing	Mean Number of Iterations to Termination				Mean Number of Misclassifications (Training Error)				Mean Prototype Error			
	WDS	PDS	OCS	NPS	WDS	PDS	OCS	NPS	WDS	PDS	OCS	NPS
0	1	1	1	1	0	0	0	0	0.00	0.00	0.00	0.00
10	4.84	4.94	7.36	7.36	0	0	0	0	1.42	0.97	0.96	0.96
20	4.98	5.04	9.88	9.86	0	0	0	0	2.55	1.54	1.53	1.53
30	4.98	5.40	12.88	12.90	0	0	0	0	3.78	1.96	1.92	1.92
40	5.02	5.78	17.34	17.32	0	0	0	0	6.05	2.59	2.53	2.53
50	N/A	6.44	22.92	22.88	N/A	0	0	0	N/A	3.12	3.02	3.02

Table 2. Averaged Results of 50 Trials using Incomplete DL Data Sets

% Missing	Mean Number of Iterations to Termination				Mean Number of Misclassifications (Training Error)				Mean Prototype Error			
	WDS	PDS	OCS	NPS	WDS	PDS	OCS	NPS	WDS	PDS	OCS	NPS
0	1	1	1	1	0	0	0	0	0.00	0.00	0.00	0.00
10	8.82	8.98	10.80	10.72	0.92	0.90	0.90	0.90	3.85	2.53	2.59	2.59
20	9.06	10.36	13.68	13.62	2.06	2.06	2.06	2.06	5.71	4.02	3.82	3.92
30	9.74	12.28	17.90	17.80	3.04	3.00	3.00	2.94	8.54	5.60	5.05	5.27
40	10.22	15.58	21.58	21.28	3.98	3.82	3.88	3.80	14.53	6.52	5.54	5.82
50	N/A	22.58	27.74	27.96	N/A	5.10	5.28	5.06	N/A	9.96	6.79	7.92

Table 3. Averaged Results of 50 Trials using Incomplete HS Data Sets

% Missing	Mean Number of Iterations to Termination				Mean Number of Misclassifications (Training Error)				Mean Prototype Error			
	WDS	PDS	OCS	NPS	WDS	PDS	OCS	NPS	WDS	PDS	OCS	NPS
0	1	1	1	1	0	0	0	0	0.00	0.00	0.00	0.00
10	4.66	7.18	7.42	7.42	3.98	4.00	4.02	4.00	1.23	1.04	1.26	1.27
20	4.90	11.32	10.16	10.14	8.68	8.64	8.62	8.64	2.13	1.90	2.53	2.54
30	4.84	14.58	13.16	13.22	13.46	13.38	13.32	13.30	2.82	3.31	4.37	4.40
40	4.96	15.32	17.56	17.68	18.78	17.00	16.68	16.76	4.77	6.12	6.65	6.68
50	N/A	16.56	23.04	23.08	N/A	21.04	20.96	21.02	N/A	9.70	9.14	9.16

Table 4. Averaged Results of 50 Trials using Incomplete HL Data Sets

% Missing	WDS	PDS	OCS	NPS	WDS	PDS	OCS	NPS	WDS	PDS	OCS	NPS
0	1	1	1	1	0	0	0	0	0.00	0.00	0.00	0.00
10	10.46	15.72	13.14	13.06	4.50	4.46	4.34	4.34	3.67	3.30	4.58	4.59
20	11.20	25.96	16.76	16.70	9.46	9.42	8.58	8.58	6.08	7.19	9.55	9.61
30	11.10	25.58	19.92	19.96	14.08	13.20	13.08	12.88	7.87	12.75	15.59	15.75
40	13.10	18.02	23.98	24.08	20.08	18.10	17.92	17.82	14.78	21.21	21.82	22.05
50	N/A	21.32	26.70	26.98	N/A	21.18	21.26	20.96	N/A	30.40	28.37	28.68

comparable. The smallest entries in each row (i.e., the "best" values in this study) are highlighted by shading.

The 2nd through 5th columns of Tables 1-4 give the average number of iterations required termination each of the four schemes. For 0% missing data, all approaches reduce to regular FCM, and therefore terminate after 1 iteration. The middle columns in the four tables give the average number of misclassifications (resubstitution or training error) obtained when the terminal prototypes are used to relabel each of the input feature vectors according to its nearest prototype. The last four columns in the tables give the average Frobenius norm distance between terminal prototypes $v^{(t)}$ and full prototypes v^*, $\left\|v^{(t)} - v^*\right\|_F$ =

$$\left(\sum_{j=1}^{s}\sum_{i=1}^{c}\left|v_{ji}^{(t)} - v_{ji}^{*}\right|^2\right)^{1/2}$$

Although the mean prototype error in Tables 1 and 2 for diagonally placed clusters increases with the percentage of missing values, this criterion yields little to distinguish any of the four methods - they are essentially equal (row by row). Thus, when the data are *clearly* well-structured, all four methods are produce fairly similar results. The best choice in this case might be to pick the WDS or PDS, which both terminate more rapidly than the OCS and NPS. For diagonally placed clusters, the OCS version produces the most accurate terminal prototypes $v^{(t)}$, seen by comparisons among the last four columns of Tables 1 and 2. The simple WDS approach consistently requires the fewest number of iterations to converge, and OCS and NPS are the most expensive in this respect. For DL, the lowest misclassification errors are consistently obtained by NPS. The problem of virtual clusters (cf. Figure 2) never occurred for diagonal placement of clusters because we use good initialization (v^*).

For the two experiments with horizontally-placed clusters that are summarized in Tables 3 and 4, WDS continues to terminate most quickly. NPS and OCS are the overall best performers in terms of misclassification errors. PDS and WDS do best in terms of terminal prototype accuracy, as measured by the last four columns of Tables 3 and 4.

Throughout the experiments the approach with the smallest mean prototype error does not necessarily produce the smallest number of misclassifications. Another point worth mentioning is that the training error rates for the horizontal clusters (Tables 3 and 4) are *much* higher with all four methods than the corresponding error rates for the two sets of diagonally placed clusters. For example, the *maximum* error rate from Tables 1 and 2 is 5.28%, committed by OCS on data set DL with *half* of its values missing! On the other hand, this error rate (5.28%) is exceeded by all four *methods* for both HS and HL when only 20% of their values are missing. And the *maximum* error rate for the horizontally aligned clusters in

Tables 3 and 4 is 21.26% (OCS, 50% row of Table 4); this is more than 4 times the maximum for Tables 1 and 2! So, while Figure 2 might suggest that "diagonal" clusters are "harder" than "horizontal" clusters, the numerical evidence is to the contrary.

4 Conclusions

We considered four different approaches for fuzzy c-means clustering in incomplete data sets: the whole data strategy (WDS); the partial distance strategy (PDSFCM); the optimal completion strategy (OCSFCM); and the nearest prototype strategy (NPSFCM). Convergence properties of PDSFCM and OCSFCM exactly parallel those for standard FCM. We outlined a new theory of tri-level alternating optimization that supports convergence analysis for the OCS method. In practice, all four techniques terminated in all cases, but theoretical convergence results for NPSFCM are missing. An important implication of our numerical experiments is that although the simple approach of deleting incomplete data (WDSFCM) works fine for small percentages of missing data (less than, say 15-20%), the other approaches are generally superior if a larger proportion of data is incomplete. Without exception, OCS produced the most accurate terminal prototypes in the case of maximally incomplete data sets.

Several interesting questions and possible areas of generalization occur to us. There are incomplete forms of data other than the missing feature values case discussed here. For example, it may be that the exact value x_{kj} is unknown but is known to satisfy $x_{kj} \geq 10$. This type of data could arise in an aborted experiment that attempted to record the time required for some event to occur for object k. In this case, PDS and NPS adaptations are not obvious, but a reasonable OCS approach is to simply add the constraint $x_{kj} \geq 10$ to the optimization over X_M. (These constraints are easily implemented and add very little computational overhead to what is already required in the missing feature value case.) Because of examples such as this, we view the generalization potential of OCS to be greater than that of the other strategies.

Our current work is aimed at understanding the relationship between OCS and the expectation-maximization (EM) algorithm for computing statistical maximum-likelihood estimates in the case of incomplete data. Both EM and the obvious extensions of OCS are general approaches that essentially replace some missing part of the data with estimates. The optimal completion strategy as applied to FCM can be interpreted in two ways. Here we have emphasized that it estimates the missing data X_M in a way that leads to the smallest possible value for J_m. Another interpretation, based on examination of Step OCSFCM-5, is that we are replacing unknown data with an estimate corresponding to a fuzzy mean. (A somewhat similar estimation technique was employed in [6].) This second interpretation is provocatively similar to the EM approach of replacing unknown data with expected values.

Fuzzy Legal Expert System (FLES) for CISG

Kaoru Hirota[1], Mingqiang Xu[2], and Yasufumi Takama[1]

[1] Tokyo Institute of Technology, Yokohama 226-8502, Japan
[2] Docomo Engineering, Tokyo, Japan

Abstract: A fuzzy legal expert system(FLES) is constructed, where the target law is the United Nations Convention on Contracts for the International Sale of Goods (CISG). The FLES is composed of fuzzy legal case-based reasoning(FLCBR) module and fuzzy legal argument(FLA) module. The former provides a primary study on vague legal concepts. The latter is the extension of the former, and can make an argument between plaintiff and defendant. It aims to classify vague legal concepts in the CISG and support the education for the beginners of international law.

1 Architecture of FLES

Statutory rules are different from the logical ones because statutory rules consist of vague legal concepts. Study of statutory rules is difficult for the beginner of law students. An intelligent fuzzy legal expert system (FLES) is constructed to support the law students in studying the vague concepts in the CISG. It explains what the meaning of vague legal concept in a query case is.

Study of legal vague concept is considered to be a two-stage process including the understanding of query case and the related statutory rules, as well as the basic skill of legal argument with cases. Similar precedents are very useful for the understanding of the vague concept. That is why case-based reasoning is used in legal reasoning. Argument is an important characteristic of human intelligence. Especially in law, which is an adversarial domain, for the development of the conventional case-based reasoning, an argument made from the viewpoints of plaintiff and defendant is very useful for the study of vague legal concepts. Therefore, the proposed FLES is composed of FLCBR and FLA modules. In the first stage, the fuzzy legal case-based reasoning(FLCBR) is used. After this stage, users can know the vague concepts and the related case and statutory rules to some extent, in the second stage, the fuzzy legal argument(FLA) is used to give a deeper interpretation. FLA is a high-level legal reasoning because it involves the legal argument. FLCBR is regarded to be the basis of FLA. But, if user has enough knowledge of vague concepts and query case, the first stage of FLES - FLCBR - is not necessarily performed. So, FLCBR and FLA are designed to be performed independently.

The architecture of FLES is shown in Figure 1.

• Case Base: contains the precedents selected from the CLOUT (Case Law on UNCITRTL (United Nations Commission on International Trade Law) Texts). They are represented by the textual case and case representation including the hierarchical fuzzy frame and the fuzzy factor hierarchy.

• User Interface: consists of integrated tools for case display, case analysis and output of explanation.

• Computational Model: includes the computational models in the two modules.

In the case base of FLES, the case representation is a hybrid knowledge representation that can be appropriate for the two kinds of legal reasoning.

The user interface of FLES offers a set of menu-driven input and output windows, displaying the case representation of the precedent in case base, the case analysis for query case, as well as the inference process.

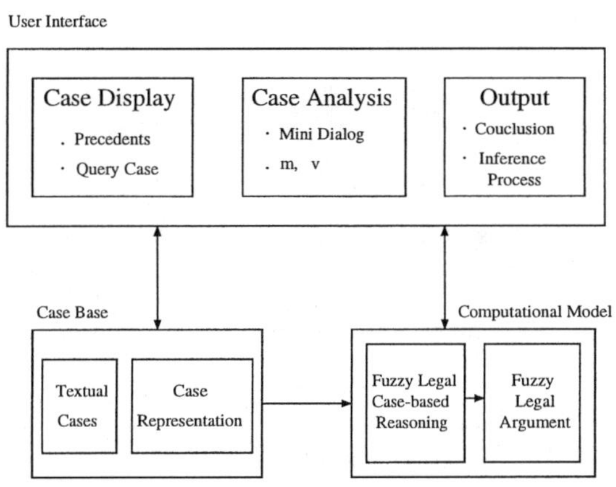

Fig. 1. Architecture of FLES

• Case Display consists of precedents in case base and query case. The case base is composed of a textual summary, hierarchical fuzzy frame and fuzzy factor hierarchy for each precedent. The textual cases of this system consist of eight precedents. All precedents selected here are relevant for the formation of a contract. All of these originate from CLOUT.

• Case Analysis is conducted by a Mini Dialog. A query case is given by making the judgment in terms of the membership and vagueness concepts that are represented by the linguistic expression introduced in the last chapter.

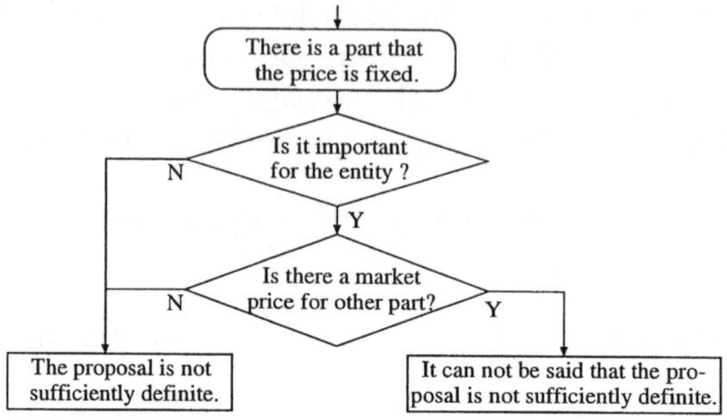

Fig. 4. The Case Rule for the Verification of Price of Entity

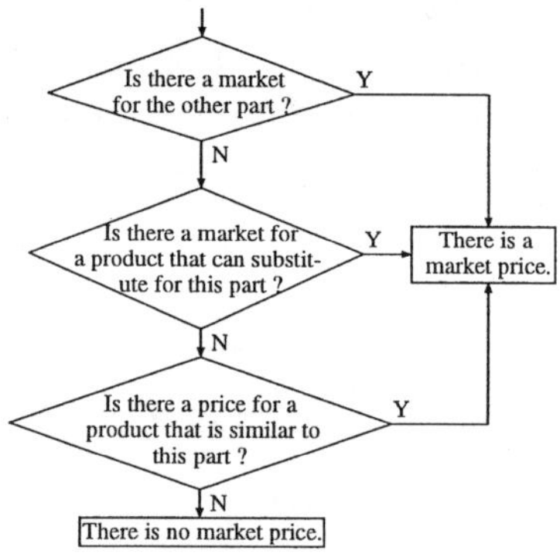

Fig. 5. The Case Rule for the Verification of Market Price

The case rule, being a hierarchical structure, usually consists of more than one stage.

Fuzzy Frames of Jet Engine Case

The issue "The proposal is sufficiently definite", involved in the article 14 of the CISG, can be used to determine whether the proposal is sufficiently definite, which is relevant to goods, quantity and price. The issue, features and case rules of *Jet Engine Case* are described by fuzzy frames shown in Fig.6.

5 Retrieval

For retrieving the most similar precedent from a case base, a method of two-stage retrieval is proposed. In the first stage, called Issue Index, a set of precedents, in the case base, which seems to be relevant to the query case, is hunt. In the second stage, that is based on the similarity measure between features, the most promising case is further retrieved from the collection of the

Issue

The proposal is sufficiently definite	No

Features

	m	v
The goods are indicated	1.0	0.0
The quantity is fixed	1.0	0.0
The eneity price is fixed	0.0	0.0
A part price is fixed	1.0	0.0

Case Rules

	m	v
It is an important part	1.0	0.0
The other part has the market price	0.25	0.0
It has the product that can be substitute	0.25	0.5
It has the product that is similar	0.25	1.0

Fig. 6. Fuzzy Frame Representation about Jet Engine Case

retrieved cases. The two stages are based on the hierarchical representation of the cases.

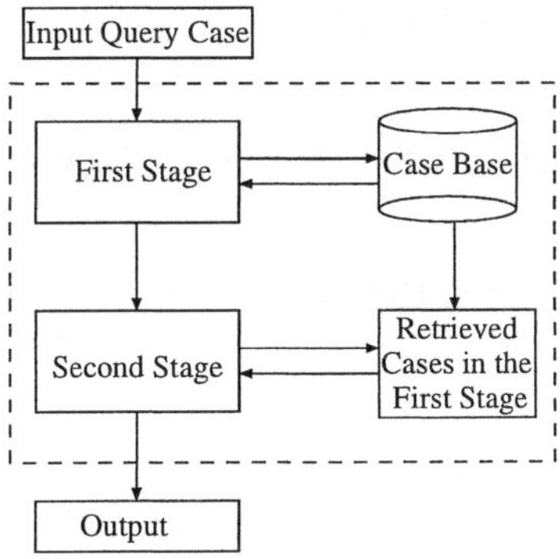

Fig. 7. Two-stage Retrieval

Issue Index

In FLBCR, a case is structured as concept-centered and each concept corresponds to an issue. An issue may include several other issues. Each issue is related to the corresponding features and case rules. So, the retrieval and inference are organized by issues. According to the issue, and the conclusion of the issue that the user is interested in, the related cases can be retrieved. The issues index can improve the utilization of the index in case retrieval, especially since it can smoothly narrow down cases to a single issue, to meet the users' needs from different viewpoints.

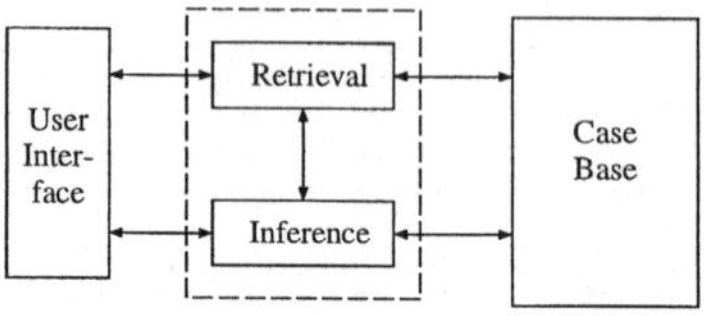

Fig. 3. The Structure of the Fuzzy Legal CBR Module

Each precedent is constructed by using the following three components: issues, features and case rules. For a query case, it is necessary to retrieve, from the case base, the precedents that are similar to it. The retrieval is made in terms of issues and features. Features are properties of a case that are not concerned with the details, but correspond to the abstract properties of a case. Usually features give a quick but coarse description of a case. The case rules of the retrieved precedents are used to infer the results that can be obtained from the retrieved cases, namely, whether the conclusion of a query case is the same as the retrieved precedent's or not.

The case base and user interface of FLCBR are parts of the case base and user interface of FLES, respectively.

4 Representation of Jet Engine Case

The *Jet Engine Case*, introduced in the last chapter, will be represented in detail in this section.

Issue

The *Jet Engine Case* is focused at the effectiveness of the offer. In order to judge it, a vague legal concept in the CISG 14, i.e. "A proposal is sufficiently definite" must be inferred. So, this vague legal concept is a point of argument and is consequently regarded as an issue of the case. According to the issue, the *Jet Engine Case* can be simplified as follows:

Event: proposal
Description of event:
The goods are jet engine systems.
The quantity of engine systems can be calculated by the quantity of planes that will be purchased.
Concerning the price:
There is no description about the prices of Boeing jet engine systems.
The price of Boeing jet engine is fixed.
The jet engine system includes a support package, services and so on.

Features

On the basis of Article 14 of the CISG, the Hungarian Supreme Court declared that the proposal was not sufficiently definite, because the price of jet engine system was not fixed and jet engine system had no market prices[1].

According to the sentence, it is clear that the goods have been "indicated". It is also clear that the "quantity" has been "fixed". The price of entity, that consists of several parts, is not fixed and it needs expert's interpretation. We can extract the features from the *Jet Engine Case* as follows:

Issue: The proposal is sufficiently definite

Judgment: No

Features: The goods are indicated
The quantity is fixed
The entity price is not fixed,
The part price is fixed

The fuzziness in the judgment on the existence of the above features can be expressed through the membership m and the vagueness v.

Case Rules

The features are not sufficient to describe the case deeply, The verification of the entity price is further interpreted as follows: If the entity price doesn't exist, it is necessary to verify whether there is a part for which the price is fixed. If there is no part for which the price is fixed, we can know that "The proposal is not sufficiently definite." If there is a part for which the price is fixed, the importance of this part for the entity needs to be verified. If this part is not important, it is considered that "The proposal is not sufficiently definite." If this part is important for the entity then, for the other part, namely the part for which the price is not fixed, it needs to be determined whether there is a market price or not.

The process of the verification of the market price is as follows: If there is a market for this part, it is considered that there is a market price for this part. If there is no market for this part, it needs to be verified whether there is a market for a product that can substitute this part or not. If there is a market, it is considered that there is a market price for this part. If there is no market then, what needs to be verified is whether there is a price for a product that is similar to this one. If there is a price, it is considered that there is a market price for this part. If there is no price, it is considered that there isn't a market price for this part.

The process of interpretation of the price is shown in Fig.4 and Fig. 5. These are the case rules for the verification of the price of entity in *Jet Engine Case.*

• Output consists of conclusion of the inference process and the final conclusion. It is of benefit to the students studying the legal reasoning by FLES.

The CISG is an international law, in order to apply our system to international exchange, the user interface of this system is developed in several languages.

The heart of the architecture of FLES is the legal reasoning approach in the computational model. FLES uses FLCBR in the model to retrieve similar precedents from the case base, and judge the adaptation of the retrieved case to the query case. Then, a legal argument is generated by FLA to deeply interpret the classification of vague legal concept from the viewpoints of plaintiff and defendant. The characteristics of the two kinds of legal reasoning used in this system will be further discussed in the next section.

2 Characteristics of the Two Reasonings

Two kinds of the legal reasoning approaches in FLCBR and FLA mentioned before are employed in the proposed system FLES. The characteristics of these two reasonings are summarized in Table 1.

Table 1. Characteristics of the Two Reasonings in FLCBR and FLA

	FLCBR	FLA
Knowledge Representation	Fuzzy Frame	Fuzzy Factor Hierarchy
Similarity	Distance-Based	Structural
Inference	Case Rules	Argument Moves
Function	Study of Rules, Cases	Skill of Argument
Viewpoints	Goal-Driven	Plaintiff, Defendant
Structure	Simple	Complex

FLCBR is a typical case-based reasoning, appropriate for retrieving a similar case form a large-scale case base, and knowing if the query case has a conclusion of the retrieved precedent. It is suitable for the case base from which the features and case rules can be obtained. It is useful for users to study the knowledge of the query case, the related precedents and the vague concepts.

Legal vague concepts have a clear kernel, and an uncertain border region, so there is room for argument. The fuzzy legal CBR is further extended by considering the legal argument. The training of law education involves learning to make arguments for and against the application of statutory rules, and for and against the previous case with respect to the query case. Argument has long been the main topic of AI and Law researches. Two types of arguments usually exist. One is rule-based, while the other is case-based. Our study is focused on the case-based argument.

FLA is a legal argument module based on the proposed structural similarity measure that is combination of similarity measure theory and fuzzy theory. The viewpoints of plaintiff and defendant are both considered, thus the conclusion is a set of the similarity and dissimilarity measures. It is applicable to let students know the legal knowledge from the sides of the plaintiff and defendant, and learn the skill of legal argument.

Generally, the fuzzy legal argument needs a fuzzy factor hierarchy that describes the issues in a adversarial way. Accordingly, a structural similarity is measured in both symbolic and numeric ways. So, the knowledge representation and the similarity measure are more complex in FLA than those of the FLCBR, usually are time-consuming.

3 Fuzzy Legal Case–based Reasoning

The schema of CBR in law is shown in Fig.2.

On an issue I, the event a on a precedent is described by $A(a)$, it leads to the conclusion $B(a)$. An event a' of a query case is described as $A(a')$. If $A(a')$ is similar to $A(a)$ on I, the conclusion $B(a')$ of the event a' in query case is also similar to the conclusion of event a in case A, namely, it is the $B(a)$.

Issue: I
Event of precedent : a
Event of query case : a'
Precedent : A(a) ---> Conclusion B(a)
a' is similar to a : A (a') ≒ A (a)

The conclusion on I : B (a') ≒ B(a)

Fig. 2. Schema of CBR in Law

In dealing with the vague legal concepts, CBR also involves the fuzziness, including the description of $A(a)$ and $A(a')$, as well as the similarity measure that $A(a')$ is similar to $A(a)$.

For capturing the fuzziness of the legal inference made by CBR, fuzzy sets are employed in the presented FLCBR module that is composed of four parts, i.e. case base, retrieval, inference and interface modules (Fig.3). Fuzzy theory is applied for the case representation in the case base, retrieval and inference. In case representation, fuzziness is described by fuzzy sets. In retrieval and inference, similarity measures are made by considering fuzziness.

In the first stage, the cases can be indexed by the article number and the names of issues, as well as the conclusion of cases. This stage uses a simple crisp matching.

Similarity Measure between Features

In the second stage, the similarity between issues is measured by the similarity between the features of the cases, that are represented by fuzzy frames. The similarity is measured by means of similarity measure between fuzzy sets introduced in the last section.

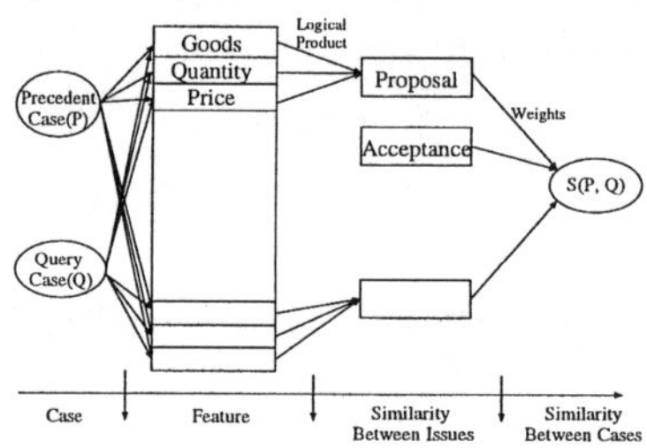

Fig. 8. Similarity Measure in the Retrieval

Moreover, sometimes, not only a single issue is to be compared between two cases. The algorithm applied for more than one relevant issue should be considered. The result of comparison between two cases is different if users address different aims. So, the weight is introduced in the case retrieval, and the similarity between cases is considered to be the weighted summation of the similarity between issues (Fig.8).

6 Inference Based on Case Rule

Inference based case rule can also be divided into two stages. In the first stage, according to the case rule, the inference is made by the correspondence of the facts of a query case and elements of the case rule, and is expressed by YES or NO. The judgment of YES or NO is based on the fuzzy membership function. In the second stage, the inference by means of the similarity between the query case and the precedent is done.

Conclusions of these two stages are compared with the conclusion of the precedent. If they are identical with the conclusion of the precedent, the query case has the same conclusion as the precedent. If they are not identical, the query case can't arrive at the same conclusion as the precedent.

The process of inference is shown in Figure 9.

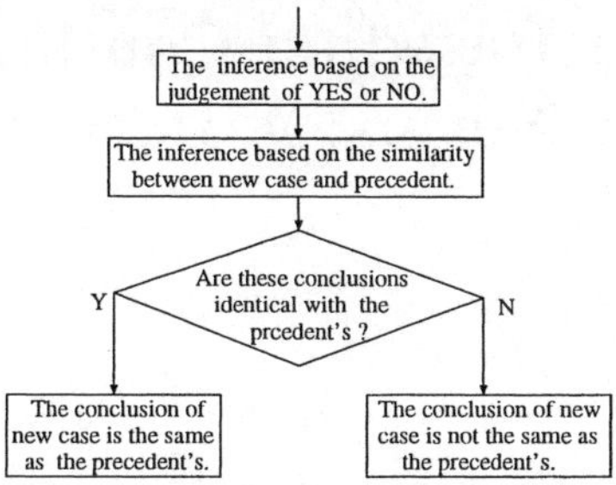

Fig. 9. The Process of Inference Based on Case Rule

Inference Based on a Yes/No Judgment

The judgment on the correspondence of facts of the query case and elements of the rules is represented by fuzzy membership function. According to the case rules of Figures 4 and 5, a Yes/No judgment is necessary for the inference by case rules. Here, the judgment is determined by the center of gravity of a fuzzy membership function.

Let Q_i be a fuzzy set that describes the judgment of element of case rule for the query case shown in Figure 10.

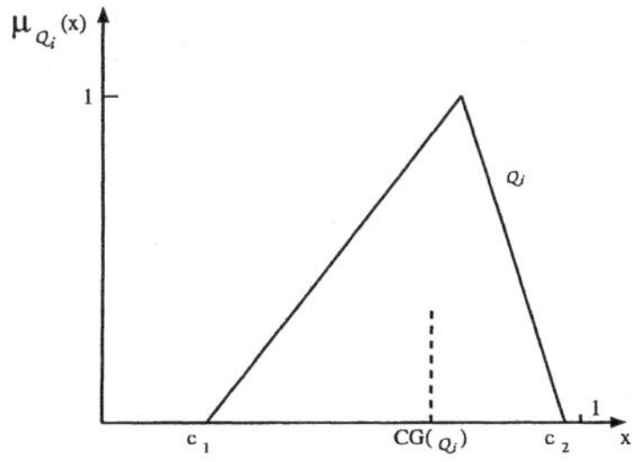

Fig. 10. The Center of Gravity of Fuzzy Set

Membership function of Q_i is μ_{Q_i}. The center of gravity of Q_i can be calculated by

$$CG(Q_i) = \frac{\int_{c1}^{c2} x\mu_{Q_i}(x)dx}{\int_{c1}^{c2} \mu_{Q_i}(x)dx}. \tag{1}$$

The center of gravity of Q_i belongs to [0, 1]. When the value of the center of gravity is near 1, it can be considered that the judgment is YES. When the value of the center of gravity is close to 0, it can be considered that the judgment is NO. The value 0.7 is selected as the threshold. When the value of the center of gravity is greater than 0.7, the judgment is YES. When the value of center of gravity is less than or equal to 0.7, the judgment is NO.

Fuzzy Calculus Based on Continuous Archimedean t-Norms: A Basis for Intelligent Computing

Radko Mesiar[1,2]

[1] Department of Mathematics, Faculty of Civil Engineering, Slovak University of Technology Bratislava, Radlinského 11, 813 68 Bratislava, Slovakia
[2] ÚTIA AV ČR, P.O.Box 18, 182 08 Prague, Czech Republik, E-mail: mesiar @leo.svf.stuba.sk

Abstract: Intelligent computing with quantified data is based on fuzzy arithmetics. Mostly, fuzzy arithmetics based on min-t-norm is applied. To avoid the linear increase of uncertainty during calculations, alternative fuzzy arithmetics based on continuous Archimedean t-norms such as product or Łukasiewicz t-norm has been developed. Conditions under which T-based summation possesses the minimal growth of uncertainty (and hence coincides with drastic product summation) are shown. An explicit output formula for summation of fuzzy numbers with related shapes is given, generalizing several particular known results. A special attention is paid to the summation preserving the linearity of input data allowing to control the output uncertainty.

1 Introduction

Many intelligent techniques working with fuzzy quantities are based on the relevant fuzzy calculus developed from the Zadeh idea of the extension principle. Fuzzy arithmetic based on the limit t-norms is well established. Fuzzy calculus based on other basic t-norms (product T_P, Łukasiewicz t-norm T_L) was discussed in eighties, e.g., by Badard [1,2], Dubois and Prade [4,5]. A deeper investigation in the general framework of continuous Archimedean t-norms-based fuzzy calculus has started in nighties by Fullér and Keresztfalvi [7], and continued by several authors, e.g., Hong and Hwang [10–12], Mesiar [27–29], Kolesárová [19–21], Marková [24,25] and others.

In this contribution we give an overview of the most important results of T-based fuzzy calculus, where T is supposed to be a continuous Archimedean triangular norm. For its direct applications in computation of compositional rule of inference see, e.g., [8,9,22]. Several other applications can be found in linear programming area, possibilistic reasoning and other domains, see [6,18].

2 Fuzzy numbers and basic shapes

The standard axiomatization of fuzzy numbers (i.e., special fuzzy subsets of the real line) applied in several domains requires their normality, convexity, and (upper semi-) continuity. For the simplicity, we will deal with continuous membership functions only, and we identify fuzzy quantities and their membership functions.

Definition 1. A mapping $\varphi : [0, \infty[\to [0, 1]$ which is continuous, non–increasing and $\varphi(x) = 1$ if and only if $x = 0$, will be called a basic shape. The set of all basic shapes will be denoted by $\mathcal{S}$.

Proposition 1. *Each fuzzy number A can be represented as $A = (a, b, \varphi_1, \varphi_2)$, where $a, b \in \overline{\mathrm{R}}$ are crisp numbers and $\varphi_1, \varphi_2 \in \mathcal{S}$ are basic shapes, namely*

$$A(x) = \begin{cases} \varphi_1(a - x) & \text{if } x \le a \\ \varphi_2(x - b) & \text{if } x \ge b \\ 1 & \text{otherwise,} \end{cases}$$

It is evident that $Ker\, A = [a, b] \cap \mathrm{R}$. Further, it is straightforward that for any $\varphi \in \mathcal{S}$ and $\alpha \in]0, \infty[$, the function $\varphi^{(\alpha)} : [0, \infty[\to [0, 1]$, defined by

$$\varphi^{(\alpha)}(x) = \varphi\left(\frac{x}{\alpha}\right), \tag{1}$$

is also a basic shape.
For a given t-norm T, T-sum $A \oplus_T B$ of two fuzzy numbers A and B is defined as

$$A \oplus_T B(x) = \sup_{y+z=x} T\left(A(y), B(z)\right) .$$

However, if $A = (a, b, \varphi_1, \varphi_2)$, $B = (c, d, \varphi_3, \varphi_4)$, then by known results from fuzzy arithmetics [6,29] we have the following claim.

Proposition 2. *Let $A = (a, b, \varphi_1, \varphi_2)$, $B = (c, d, \varphi_3, \varphi_4)$. Then*

$$A \oplus_T B = (a + c, b + d, \varphi_1 \oplus_T \varphi_3, \varphi_2 \oplus_T \varphi_4). \tag{2}$$

Consequently, the T-sum of fuzzy numbers can be described by the corresponding sums of (extended) real numbers and T-sums of basic shapes.

3 T-sums of basic shapes based on continuous Archimedean t-norms

Recall that a t-norm $T : [0, 1]^2 \to [0, 1]$ is an associative, commutative, non–decreasing mapping with neutral element 1. Four basic t-norms are minimum and they are shown in the "Selected Precedent" windows. For example, the case 4 (*Jet Engine case*) is the most similar to the query case (cf. Figure 11).

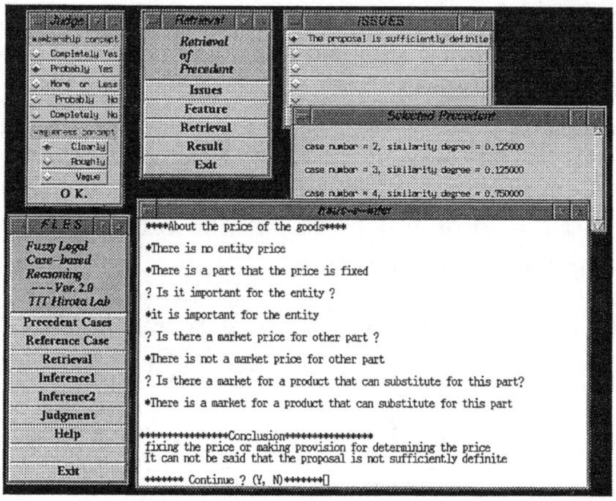

Fig. 12. The Inference by Case Rules

Then the case rules of *Jet Engine Case* is judged by users.

In terms of the fact of query case, users can select the fuzzy linguistic variables to answer the case rules (cf. Figure 12). The conclusion on the basis of the judgment is derived and displayed in "inference window".

The result is different with the inputs selected by the user. A user can see how small changes in the problem could lead to different results. It is helpful for users to know that results are changed by the different inputs. It also helps users to understand the meaning of the statutory rules of the CISG and the meaning of precedents and query case.

FLCBR is constructed on the basis of the comments of the researchers on the legal domain and the law students [8].

The viewpoints of plaintiff and defendant are not emphasized in FLCBR. Legal argument made by plaintiff and defendant is also an important reasoning method in law. A fuzzy legal argument considering the viewpoints of both plaintiff and defendant is the subject for the future studies.

Acknowledgement

This work is supported in part by the Japanese Ministry of Education, Science, and Culture: "Development of Legal CAI System using Internet Technology (1999 ~ 2001), " and is mainly from a part of Xu's PhD dissertation "Similarity–based Fuzzy Legal Reasoning and its Applications to the CISG."

References

1. MingQiang Xu, Kaoru Hirota, Hajime Yoshino (1998) A Fuzzy Theoretical Approach to Representation and Inference of Cases in CISG, Artificial Intelligence and Law, To Appear

2. MingQiang Xu, Kaoru Hirota, Hajime Yoshino (1998) Learning a Vague Concept and Making Argument from Examples by Fuzzy Factors Hierarchy in the Legal Reasoning System, 5th International Conference and Soft Computing and Information/Intelligence Systems, Iizuka'98, Japan

3. Ronald R. Yager (1997) Case-based reasoning, Fuzzy Systems Modeling and Solution Composition, Case-based Reasoning Reserch and Development(Eds. David B. Leake), Springer, 633–642

4. Koyiqi Yano (1997) Distance Space and Topology Structure, Koritsu Press, 141–142 (Japanese)

5. Hajime Yoshino (1997) On the Logical Foundations of Compound Predicate Formula Legal Knowledge Representation, Artificial Intelligence and Law, Vol.5, No.1–2, 1–20

6. Hajime Yoshino, MingQiang Xu, Kaoru Hirota (1997) Representation and Inference of Cases with Fuzziness in CISG, 4th International Workshop on a Legal Expert System for the CISG, Australia, 5–9

7. Hajime Yoshino, MingQiang Xu, Kaoru Hirota (1997) A Fuzzy Judgment Approach to inference of case in CISG, 6th International Conference on Artificial Intelligence and Law, Australia, 60–64

8. Hajime Yoshino et al. (1995–1998) Proc.of the Symposium on the Development of Legal Expert System (Japanese)

After all the correspondences of the elements of case rules and facts of query cases are judged, the conclusion whether the legal requirement occurs or not is inferred according to the case rule.

Inference Based on Similarity Measure

The case rule about an issue is represented by several frames.

If the degree of similarity is greater than the threshold given in advance, the conclusion of frame FR_q of query case is the same as that of frame FR_p of the precedent. For example, in a precedent, if there is a conclusion that "The proposal is sufficiently definite", the conclusion of the query case is also "The proposal is sufficiently definite." If the degree of similarity is less than the given threshold, the conclusion of frame FR_q can't arrive at the same conclusion as that of FR_p. This does not mean that the query case has the conclusion opposite to the precedent's. It is necessary to infer the conclusion using the other precedents for this query case.

7 Experiment

The query case *Cultivator* [8], that is devised for the legal reasoning of the CISG, is cited for the experiment based on our fuzzy legal case-based reasoning.

The query case is as follows:

1) On April 1, company C in New York dispatched a letter containing an offer to the business branch of a Japanese company D in Hamburg, the content of which is that C sells a set of cultivator(the price of the tractor itself is $50,000 to D. The tractor should be equipped with a rake, which is product of company E. The farming machinery is delivered by a U.S. cargo ship).

2) The letter reached D on April 8.

3) On April 9, D telephoned C to tell "I accept your offer, but you should transport the machinery by a Japanese container"

There are several issues as to whether the contract is concluded. Whether the proposal from C is effective or not is one of them. On the condition that "The proposal is sufficiently definite," the proposal is effective.

According to the issue, this case can be simply described as follows:

Event: proposal
Description of event:
The goods are cultivator.
The quantity of cultivator is one.
Concerning the price:

The price of the tractor is fixed.
The price of a set of cultivator is not fixed.
The cultivator contains a rake.

There are eight precedents in the case base for the experiment:

case1: Experiment tube affair
case2: Screw affair
case3: Leather affair
case4: Jet engine affair
case5: Car affair
case6: Shoes affair
case7: Tyre affair
case8: Electronic parts affair

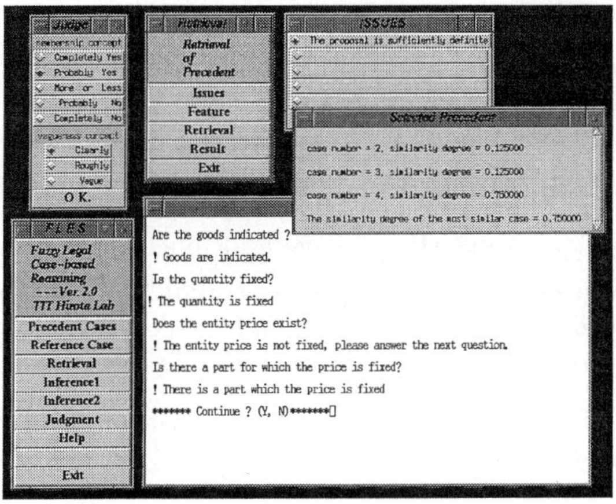

Fig. 11. The Retrieval of Precedents

Articles of precedents can be read from precedents menu. All cases are also presented by a hierarchical frame structure that is composed of issues, features and case rules. The slot values of the frame are described with fuzzy membership value.

In the query case *Cultivator Case*, the controversy between the plaintiff and defendant i.e. whether a proposal is sufficiently definite, can be considered as an issue. The retrieval procedure is divided into two stages. In the first stage, pursuant to issue index, the cases dealing with CISG 14 are picked from the precedents. The precedents in the case base are narrowed down to a part of the precedents. As a result of this example, case2, case3 and case4 are searched for in the first stage. In the second stage, similarities between the query case and the retrieved precedents in the first stage are measured

T_M, product T_P, Łukasiewicz t-norm T_L and drastic product T_D, which are defined by:

$$\begin{aligned} T_M(x,y) &= \min(x,y) \\ T_P(x,y) &= xy \\ T_L(x,y) &= \max(0, x+y-1) \\ T_D(x,y) &= \begin{cases} x & \text{if } y=1 \\ y & \text{if } x=1 \\ 0 & \text{else.} \end{cases} \end{aligned}$$

Each continuous t-norm T can be represented as an ordinal sum with Archimedean summands [17,30] and thus because of [3], it is enough to discuss min-based fuzzy calculus which is well known and discussed by many authors, see e.g. [6,23,29], and the fuzzy calculus based on continuous Archimedean t-norms. Note that each continuous Archimedean t-norm T is isomorphic either to the algebraic product T_P or to the Łukasiewicz t-norm T_L, and that it is generated by an additive generator.

Denote by f an additive generator of the considered continuous Archimedean t–norm T, i.e., a continuous, strictly decreasing function $f : [0,1] \to [0,\infty]$ with $f(0) = 1$, such that

$$T(x,y) = f^{-1}\left(\min\left(f(0), f(x) + f(y)\right)\right), \quad (x,y) \in [0,1]^2.$$

Note that f is unique up to the positive multiplicative constant.

Recall that the product t-norm T_P is generated, e.g., by $f_P(x) = -\log x$, and the Łukasiewicz t-norm T_L by the (normed) generator $f_L(x) = 1 - x$. Now, we show when the T-sums reduce to the T_D-sums.

Theorem 1. *Let T be a continuous Archimedean t–norm with an additive generator f and let φ, η be basic shapes. Let $f \circ \varphi$ and $f \circ \eta$ be concave functions. Then*

$$\varphi \oplus_T \eta = \max(\varphi, \eta) = \varphi \oplus_{T_D} \eta.$$

Proof. The proof follows from the fact that any concave function defined on a closed interval has its minimum on the boundary.

Put, e.g., $\varphi_L(x) = \max(0, 1-x)$, then $\varphi_L^{(\alpha)}(x) = \max(0, 1 - \frac{x}{\alpha})$ and for $T = T_L$, by the above theorem we have

$$\varphi_L^{(\alpha)} \oplus_T \varphi_L^{(\beta)} = \varphi_L^{(\max(\alpha,\beta))},$$

i.e., the shape of incoming fuzzy numbers is preserved. Note that the same result holds true if we put $T = T_D$ and consequently, the result is valid for any T between T_D and T_L, i.e., whenever $T \le T_L$.

Similarly, for $\varphi(x) = \exp(-x)$ and $T = T_P$, or for weaker t-norms, we obtain

$$\varphi^{(\alpha)} \oplus_T \varphi^{(\beta)} = \varphi^{(\alpha+\beta)}(x) = \exp\left(\frac{-x}{\max(\alpha,\beta)}\right).$$

Another example: take the product t-norm T_P and two shapes $\varphi(x) = e^{-x}$ and $\eta(x) = e^{-\sqrt{x}}$. Then

$$\varphi \oplus_{T_P} \eta(x) = \max(\varphi(x), \eta(x)) = \begin{cases} e^{-x} & \text{if } x \in [0,1] \\ e^{-\sqrt{x}} & \text{if } x > 1. \end{cases}$$

Note that the concavity of composites $f \circ \varphi$ and $f \circ \eta$ in Theorem 1 can be relaxed to the subadditivity of these composites, i.e., it is enough to require

$$f \circ \varphi(u + v) \leq f \circ \varphi(u) + f \circ \varphi(v)$$

for all $u, v \in [0, \infty[$ and similarly for $f \circ \eta$, see [12,26].

The next result is rather general and covers all till now known relevant results.

Theorem 2. *Let T be a continuous Archimedean t–norm with an additive generator f. Let $K : [0, \infty[\to [0, \infty[$ be a strictly increasing convex function with $K(0) = 0$. Then for each $\alpha \in]0, \infty[$, the function $\varphi_{(\alpha)} : [0, \infty[\to [0, 1]$ defined by*

$$\varphi_{(\alpha)}(x) = f^{-1}\left(\min\left(f(0), \alpha K\left(\frac{x}{\alpha}\right)\right)\right),$$

is a basic shape.
Moreover, for any $\alpha, \beta \in]0, \infty[$ we have

$$\varphi_{(\alpha)} \oplus_T \varphi_{(\beta)} = \varphi_{(\alpha+\beta)}.$$

Proof. The proof is essentially based on the properties of convex functions defined on closed intervals. For more details we recommend [29].

A simple example of an appropriate function K fitting Theorem 2 is the function $K(x) = x^p$ with $p > 1$. Note that then for $\alpha \in]0, \infty[$,

$$\alpha K\left(\frac{x}{\alpha}\right) = \alpha \left(\frac{x}{\alpha}\right)^p = \frac{x^p}{\alpha^{p-1}}\ .$$

Consequently, basic shape $\varphi_c(x) = f^{-1}\left(\min(f(0), \frac{x^p}{c})\right)$ with $c \in]0, \infty[$ can be identified with $\varphi = \varphi_{(\alpha)}$, where $c = \alpha^{p-1}$. However, then $\varphi_c \oplus \varphi_d = \varphi_e$, where $c = \alpha^{p-1}$, $d = \beta^{p-1}$, and $e = (\alpha + \beta)^{p-1}$, i.e.,

$$e = \left(c^{1/(p-1)} + d^{1/(p-1)}\right)^{p-1}.$$

Another example: Let $\Phi = (\varphi, \varphi)$, where $\varphi(x) = \exp(-x^2)$. Consider the product t–norm T_P. Then clearly,

$$f \circ \varphi^{(\alpha)}(x) = -\log \exp\left(-\frac{x^2}{\alpha^2}\right) = \frac{x^2}{\alpha^2}\ .$$

Connecting Fuzzy and Statistical Analysis with Qualitative Maps

Vincent Vigneron

CEMIF, Lab. des Systèmes Complexes
40 rue du Pelvoux
91020 Evry Courcouronnes cedex, France.
email : vvigne@iup.univ-evry.fr

Abstract: Despite many applications of fuzzy sets nowadays (Artificial Intelligence, automatic control, pattern recognition, etc), statistical analysis is a domain where they are rarely used: probably because the output of statistical methods have to be carefully analysed by a human expert. The aim of this paper is to show how it is possible to make crisp data compatible with fuzzy data. The way to make them compatible is to use a unique scale model – i.e. *category model like Multiple Correspondence Analysis (MCA). MCA is a descriptive multidimensional method which can investigate several empirical situations being described by a categorial variable set. It is a linear based model. But it has the advantage to highlight non-linear relationship.*

When MCA is not appropriate due to the intrinsic structure of the data, one can try to use neural models. One can use for example Kohonen maps to solve some of the interesting problems which are encountered both in data analysis and in fuzzy inference systems: how to realize a rapid and robust classification based on the qualitative variables, how to vizualize the classes, their differences and their homogeneity, how to put in evidence the most important explanatory variables.

1 Introduction

In the design of fuzzy systems, the most important thing is to decide the parameters of membership functions (MFs). There are several problems in determining the number of MFs –this point is related to system identification– the centers and width of MFs, and the slope of MFs. Most of the time, manual trial processes are required to evaluate whether the parameters are suitable or not (see Sugeno [9]).

If the parameters of MFs are arbitrarily fixed, the tuning process always takes time to optimize the system performance. Another problem is that the tuning procedures (such as least-squares) are easily trapped in local minima. Remark that an effective partition of the input space enhances the speed in both *learning* and *application* phases [3]. Several different methods of fuzzy clustering have been proposed to determine class membership for each sample by determining the class prototype with the smallest distance (or highest similarity). Bezdek suggested that interesting algorithms could result from the allocation of fuzzy membership to the input vector thus affording fuzzy decisions based on fuzzy labels [1]. Fuzzy $k-$nearest neighbors proposed by Keller [6] produces a fuzzy classification rule by assigning class memberhip to a sample vector rather that assigning the vector to a particular class. Kohonen algorithm could be adapted to find the centers of membership functions [7]. However, it cannot deal with fuzzy variables and it cannot guarantee that the membership functions can partition the overall universe adequately and may effect worse results.

In this paper, we propose to transform a set of *fuzzified data* in a set of *categorical data* (termed *crisp* data or *qualitative* data). This form of data is easier to handle with classical statistical tools such as Correspondence analysis (CA) or Multiple Correspondence analysis (MCA). Nevertheless, we refine the solution using Kohonen algorithm and show it is capable to find optimal membership functions. The remaining part of the article is organized as follows: Section 2 propose a tricky transformation of a fuzzy set. Sections 3 and 4 describe most used methodology to map qualitative data on a Kohonen grid. In section 5, both strategies are compared in a simulation study.

2 Comparing Crisp/Fuzzy data

Let us denote by μ the membership function given in Fig.1a and by $\hat{\mu}$ the corresponding trivial membership function given in Fig.1b. Fig.1a and Fig.1b describe the same situation but more *information* is present in Fig.1a than in Fig.1b in the sense of information. The problem is to find a injunctive transformation $\mathcal{D}$ which transform the space described by μ in the space partitionned by $\hat{\mu}$. Formally, we look for $\mathcal{D}$ such that $\hat{\mu} = \mathcal{D}(\mu)$. $\mathcal{D}$ "should" transform any fuzzy data set in $\{0,1\}$ set in minimizing the information exchange between the fuzzy and the crisp representations.

Assume the statistical model which generates the vector observation $\boldsymbol{x}$ is

$$\boldsymbol{x} = A\boldsymbol{z} \tag{1}$$

where $\boldsymbol{z} \in \{0,1\}^N$. The aim of such a model is to recover the unobserved $\boldsymbol{z}$ from the observations $\boldsymbol{x}$ when there is no knowledge of the matrix $A \in \mathrm{R}^{N\times N}$.

We only use strictly deterministic transformation for Eq.1 in this article. The fundamental assumption made on the vector $\boldsymbol{z}$ is that it has individual components in the set $\{0,1\}$: in other words, individual components are mutually independent. Cardoso [2] proves that the exploitation of sample cumulants of distributions of stochastic processes leads to a linear orthogonal transformation which is optimal in the least square sense. The first step consist in forming the data covariance matrix $\Gamma = \mathrm{E}\{\boldsymbol{x}\boldsymbol{x}^T\}$ and then decomposing this using a standard decomposition technique such as eigenvalue decomposition so that $M = C\Lambda C^{-1}$, where Λ is he diagonal matrix whose entries are the eigenvalues and C is the matrix whose columns are the respective eigenvectors.The orthonormalisation consists in projecting the received data onto the eigenvectors such that $\boldsymbol{y} = C\boldsymbol{x}$. Hence $\mathrm{E}\{\boldsymbol{y}\boldsymbol{y}^T\} = I$.

References

1. Badard,R. (1982): The law of large numbers for fuzzy processes and the estimation problem. Information Sciences 28, 161-178
2. Badard, R. (1984): Fixed point theorems for fuzzy numbers. Fuzzy Sets and Systems 13, 291-302
3. De Baets,B., Marková–Stupňanová,A. (1997): Analytical expressions for addition of fuzzy intervals. Fuzzy Sets and Systems 91, 203–213
4. Dubois,D., Prade,H. (1981): Additions of interactive fuzzy numbers. IEEE Trans. Automat. Control 26, 926–936
5. Dubois,D., Prade,H. (1988): Fuzzy numbers: An overview. In: Bezdek,J.C.: Analysis of Fuzzy Information. CRC Press, Boca Raton, 3–39
6. Dubois,D., Kerre,E.E., Mesiar,R., Prade,H. (2000): Fuzzy Interval Analysis. In: Dubois,D., Prade,H. (Eds.): Fundamentals of Fuzzy Sets 1. Kluwer, Dordrecht, 483-581
7. Fullér,R., Keresztfalvi,T. (1992): t-Norm based addition of fuzzy intervals. Fuzzy Sets and Systems 51, 155-159
8. Fullér,R., Zimmermann,H–J. (1992): On computation of the compositional rule of inference under triangular norms. Fuzzy Sets and Systems 51, 267–275
9. Hong,D.H., Hwang,S.Y. (1994): On the compositional rule of inference under triangular norms. Fuzzy Sets and Systems 66, 25–38
10. Hong.D.H., Hwang S.Y. (1994): On the convergence of T-sum of L-R fuzzy numbers. Fuzzy Sets and Systems 63, 175-180
11. Hong,D.H. (1995): A note on t-norm-based addition of fuzzy intervals. Fuzzy sets and Systems 75, 263-278
12. Hong,D.H. : Some results on the additions of fuzzy intervals. Fuzzy Sets and Systems, accepted for publication
13. Keresztfalvi,T., Kovács,M. (1992): g,p–fuzzification of arithmetic operations. Tatra Mountains Math. Publ. 1, 65-71
14. Klement,E.P., Mesiar,R. (1997): Triangular norms. Tatra Mountains Math. Publ. 13, 169–194
15. Klement,E.P., Mesiar,R., Pap,E. (1997): A characterization of the ordering of continuous t–norms. Fuzzy Sets and Systems 86, 189–195
16. Klement,E.P., Mesiar,R., Pap,E. (1999): Quasi- and pseudo–inverses of monotone functions, and construction of t–norms. Fuzzy Sets and Systems 104, 3–13
17. Klement,E.P., Mesiar,R., Pap,E. (2000): Triangular Norms. Kluwer, (to appear)
18. Klir,J.G, Bo Yuan (1995): Fuzzy Sets and Fuzzy Logic: Theory and Applications. Prentice Hall.
19. Kolesárová,A. (1995): Triangular norm-based addition of linear fuzzy numbers. Tatra Mountains Math. Publ. 6, 75–81
20. Kolesárová,A. (1997): Similarity preserving t–norm-based additions of fuzzy numbers. Fuzzy Sets and Systems 91, 215–229
21. Kolesárová,A. (1998): Triangular norm-based addition preserving linearity of T-sums of linear fuzzy intervals. Mathware and Soft Computing 1, 91–98
22. Kolesárová,A., Kerre,E.E. (2000): Compositional rule of inference based on triangular norms. In: DaRuan, Kerre,E.E (Eds.): Fuzzy If-Then Rules in Computational Intelligence: Theory and Applications. Kluwer Academic Press, to appear
23. Mareš,M. (1994): Computation over fuzzy quantities. CRC Press, Boca Raton

24. Marková,A. (1995): Additions of $L - R$ fuzzy numbers. Busefal 63, 25–29
25. Marková,A. (1997): T-sum of $L - R$ fuzzy numbers. Fuzzy Sets and Systems 85, 379–384
26. Marková–Stupňanová,A. (1997): Idempotents of T–addition of fuzzy numbers. Tatra Mountains Math. Publ. 12, 65-72
27. Mesiar,R. (1996): A note to the T-sum of $L - R$ fuzzy numbers. Fuzzy Sets and Systems 79, 259–261
28. Mesiar,R. (1997): Shape preserving additions of fuzzy intervals. Fuzzy Sets and Systems 86, 73–78
29. Mesiar,R. (1997): Triangular norm-based addition of fuzzy intervals. Fuzzy Sets and Systems 91, 231–237
30. Schweizer B., Sklar,A. (1963): Associative functions and abstract semigroups. Publ. Math. Debrecen 10, 69–81
31. Zadeh,L.A. (1975): The concept of a linguistic variable and its applications to approximate reasoning, Part 1, 2 and 3. Information Science 8, 199–249, 8, 301–357, 9, 43–80

Put $K(x) = x^2$. The function K satisfies all conditions given in Theorem 4. Then

$$\varphi_{(\alpha^2)}(x) = f^{-1}\left(\alpha^2\left(\frac{x}{\alpha^2}\right)^2\right) = \exp\left(-\frac{x^2}{\alpha^2}\right) = \varphi^{(\alpha)}.$$

Therefore

$$\varphi^{(\alpha)} \oplus_{T_P} \varphi^{(\beta)} = \varphi_{(\alpha^2)} \oplus_{T_P} \varphi_{(\beta^2)} = \varphi_{(\alpha^2+\beta^2)} = \varphi^{(\sqrt{\alpha^2+\beta^2})}.$$

Recall that fuzzy quantities $\Phi(\mu, \mu, \sigma^2, \sigma^2) = G(\mu, \sigma^2)$ are called Gaussian fuzzy numbers. They are given by the formula

$$G(\mu, \sigma^2)(x) = \exp\left(-\frac{(x-\mu)^2}{\sigma^2}\right),$$

and from the above formula we immediately obtain

$$G(\mu_1, \sigma_1^2) \oplus_{T_P} G(\mu_2, \sigma_2^2) = G(\mu_1 + \mu_2, \sigma_1^2 + \sigma_2^2).$$

4 Linearity preserving calculus of basic shapes

The piecewise linear basic shape $\varphi_L^{(\alpha)}(x) = \max(0, 1 - \frac{x}{\alpha})$, already discussed in the previous section, is a background of so called linear fuzzy numbers (triangular or trapezoidal) which are very popular in many application areas. Linearity preserving T-sums were investigated for instance by Keresztfalvi and Kovács [13], Kolesárová [19,21] and Mesiar [28]. Because of the importance of the topic, we recall these results and give some new ones in the language of basic shapes.

Theorem 3.
(i) $\varphi_L^{(\alpha)} \oplus_T \varphi_L^{(\beta)} = \varphi_L^{(\max(\alpha,\beta))}$ *if and only if* $T \leq T_L$.
(ii) $\varphi_L^{(\alpha)} \oplus_T \varphi_L^{(\beta)} = \varphi_L^{(\alpha+\beta)}$ *if and only if* $T = T_M$.

The constants α, β can be understood as spreads describing the uncertainty of linear fuzzy numbers $\varphi_L^{(\alpha)}, \varphi_L^{(\beta)}$. Then the T_L-based sum gives the result with the least possible output uncertainty (equal to maximal input uncertainty), while T_M-based sum leads to the linear increase of uncertainties. Next result offers us a tool for controlling the growth of output uncertainty in between the above mentioned limits.

Now, let us consider the system $\{T_s^Y\}_{s\in]0,\infty[}$ of Yager's t–norms which are defined by

$$T_s^Y(x,y) = \max\left(0, 1 - ((1-x)^s + (1-y)^s)^{1/s}\right), \quad (x,y) \in [0,1]^2$$

and whose normed additive generators are given by $f_s(x) = (1-x)^s$, $x \in [0,1]$.

Theorem 4. *Let $T = T_s^Y, s > 1$ be the Yager t–norm. Then*

$$\varphi_L^{(\alpha)} \oplus_T \varphi_L^{(\beta)} = \varphi_L^{(\gamma)},$$

where

$$\gamma = (\alpha^r + \beta^r)^{1/r}, \quad \textit{and} \quad r = \frac{s}{s-1}.$$

Proof. We put $K(x) = x^s$, $x \in [0, \infty[$ and apply Theorem 2 for $T = T_s^Y$, $s > 1$, i.e., for the additive generator $f_s(x) = (1-x)^s$. Then

$$\varphi_{(\alpha)}(x) = \max\left(0, 1 - \frac{x}{\alpha^{(s-1)/s}}\right),$$

i.e., $\varphi_L^{(\alpha)} = \varphi_{(\alpha^r)}$ and thus

$$\varphi_L^{(\alpha)} \oplus_T \varphi_L^{(\beta)} = \varphi_{(\alpha^r)} \oplus_T \varphi_{(\beta^r)} = \varphi_{(\alpha^r+\beta^r)} = \varphi_{(\gamma^r)} = \varphi^{(\gamma)}.$$

So, e.g., if $s = 2$, then $r = 2$ and $\gamma = \sqrt{\alpha^2 + \beta^2}$. It is evident that if $s \to \infty$, then $r \to 1$ and $\gamma \to \alpha + \beta$. Note that for $s \to \infty$ we also have $T_s^Y \to T_M$. Similarly, if $s \to 1^+$, then $r \to \infty$ and $\gamma \to \max(\alpha, \beta)$. However, in that case $T_s^Y \to T_L$.
Note that the above result holds in the class of continuous Archimedean t-norms with convex generators, or we can say in the class of Archimedean copulas, only for the Yager t-norms. Whether there are some t-norms out of this class for which the previous theorem is valid, remains an open problem. In the affirmative case a fractal–like structure of relevant t-norms is expected. Similar results hold also in the case of shapes $\varphi(x) = \max(0, 1-x^\epsilon)$, $\epsilon \in]0, \infty[$. Put, e.g., $\epsilon = 2$, i.e., $\varphi(x) = \max(0, 1-x^2)$. Then for any $T \le T_{0.5}^Y$ we have

$$\varphi^{(\alpha)} \oplus_T \varphi^{(\beta)} = \varphi^{(\max(\alpha,\beta))}$$

and for $T = T_s^Y$ with $s > 0.5$ we obtain

$$\varphi^{(\alpha)} \oplus_T \varphi^{(\beta)} = \varphi^{(\gamma)},$$

where

$$\gamma = (\alpha^r + \beta^r)^{1/r} \quad \text{and} \quad r = \frac{2s}{2s-1}.$$

5 Conclusions

The fuzzy arithmetics of basic shapes based on continuous Archimedean t-norms has been discussed. Several applications in approximate reasoning can be expected.

Acknowledgment

This work was supported by the grants VEGA 2/6087/99, 1/7146/20 and GAČR 402/99/0032.

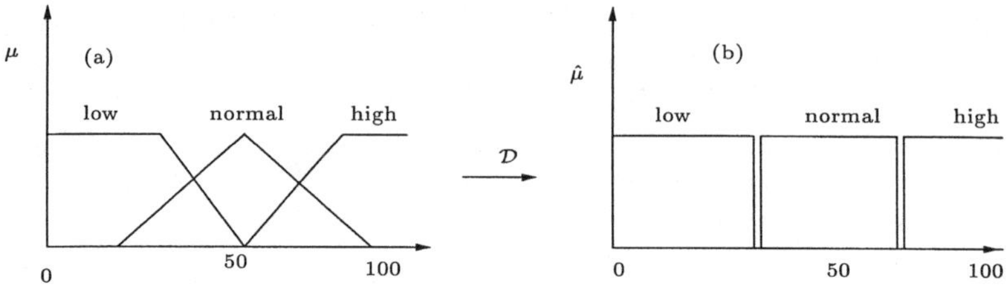

Fig. 1. Comparison between crisp and fuzzy representation of class membership.

3 Algebraic properties of MCA

One of the principal encountered difficulties relates to the classification of the individuals[1] when the size of the data is very large. Let us give some notations. We consider the situation in which $n-$ *sample* of individuals that answer to Q questions (variables). Each question has m_q possible answers (or modalities). The individuals answer each question q $(1 \leq q \leq Q)$ by choosing only one modality among the m_q modalities. If we assume that $Q = 3$ and $m_1 = 3$, $m_2 = 2$ and $m_3 = 4$, then an answer of an individual could be $(0,1,0|0,1|0,0,0,1)$ in the case of *crisp data*, where 1 corresponds to the chosen modality for each question, or $(.2,.77,.03|.4,.6|.19,.33,.17,.31)$ for *fuzzy data*. Let us denote by M the total number of all the modalities: $M = \sum_{q=1}^{Q} m_q$. To simplify, we can enumerate all the modalities from 1 to M and denote by Z_i, $(1 \leq i \leq M)$ the column vector constructed by the n answers to the i-th modality. The k-th element of the vector Z_i is 1 or 0, according to the choice of the individual k (it is 1 if and only if the individual k has chosen the modality i). Let us denote by $K_{(n \times M)}$ the matrix with n rows and M columns which corresponds to the complete disjonctive table :

$$K = \{k_{ij}\} \text{ where } k_{ij} = \begin{cases} 1 & \text{if individual } i \text{ chooses modality } j, \\ 0 & \text{otherwise.} \end{cases}$$

The marginals of the rows of K are constant equal to the number Q of questions $k_{i\cdot} = \sum_{j=1}^{M} k_{ij} = Q$. This matrix K gives the complete data and is called *complete disjunctive table* (see Tab.1). It is essential if we want to remember who answered what. But if we only have to study the *relations between the Q variables* (or questions), we can sum up the data in a crosstabulations table, called *Burt matrix*, defined by $B = K^T K$, where K^T is the transposed matrix of K (see Tab.??).

The matrix B is a $(M \times M)$ symetric matrix and is composed of $Q \times Q$ blocks, such that the $(q \times r)$ block B_{qr} $(for\ \ 1 \leq q, r \leq Q)$ contains the n answers

[1] The word "individual" is a generic term that represent any statistical qualitative object like household, dogs, cars ...

$K_{(12\times3)}=$

m_1	m_2	m_3
0 1 0	0 1	0 0 0 1
0 1 0	1 0	0 0 1 0
0 0 1	1 0	0 1 0 0
1 0 0	0 1	0 0 0 1
1 0 0	0 1	0 0 1 0
0 1 0	0 1	0 0 1 0
0 0 1	1 0	1 0 0 0
1 0 0	1 0	1 0 0 0
0 1 0	1 0	0 1 0 0
0 1 0	0 1	0 0 1 0
0 0 1	0 1	0 1 0 0
1 0 0	1 0	0 0 0 1

$B_{(9\times9)}=$

4 0 1	2 2	1 0 1 2
0 5 0	2 3	0 1 3 1
1 0 3	2 1	1 2 0 0
2 2 2	6 0	1 2 1 1
2 3 1	0 6	0 1 3 2
1 0 1	2 0	2 0 1 0
0 1 2	2 1	0 3 0 0
1 3 0	1 3	1 0 4 0
2 1 0	1 2	0 0 0 3

Table 1. Example of a disjunctive table $K_{(12\times3)}$ and the corresponding Burt table $B_{(9\times9)}$ obtained from $K_{(12\times3)}$.

to the question r. The block B_{qq} is a diagonal matrix, whose diagonal entries are the numbers of individuals who have respectively chosen the modalities $1,...,m_q$, for the question q. The Burt table $B_{(M\times M)}$ has to be seen as a generalized contingency table, when more than 2 kinds of variables are to be studied simultaneously (see for example [8]). In this case we loose a part of the information about the individuals answers but we keep the information regarding the relations between the modalities of the qualitative variables. Let us denote by b_{ij} the entries of the matrix B, whatever are the questions which contain the modalities i or j. The total sum of all the entries of B is $b = \sum_{i,j} b_{ij} = Q^2n$. One defines successively

- the table F of the relative frequencies, with entry $f_{ij} = \frac{k_{ij}}{b}$,
- the margins with entry $f_{i.} = \sum_j f_{ij}$ or $f_{.j} = \sum_i f_{ij}$,
- the table P of the profiles which sum to 1, with entry $P_{ij} = \frac{f_{ij}}{f_{i.}}$.

The matrix B is a $M \times M$ symetric matrix and is composed of $Q \times Q$ blocks, such that the $(q \times r)$ block B_{qr} $(\forall\ 1 \leq q, r \leq Q)$ contains the N answers to the question r. The block B_{qq} is a diagonal matrix, whose diagonal entries are the numbers of individuals who have respectively chosen the modalities $1, \ldots, m_q$, for the question q. The matrix D is composed of Q^2 blocks with the same diagonal elements as B. The Burt table $B_{(M\times M)}$ has to be seen as a generalized contingency table, when more than 2 kinds of variables are to be studied simultaneously (see for example [8]). In this case we loose a part of the information about the individuals answers but we keep the information regarding the relations between the modalities of the qualitative variables. Each row of the matrix B characterizes *a modality of a question* (or variable). Let us denote by b_{ij} the entries of the matrix B, whatever are the questions which contain the modalities i or j. The total sum of all the entries of B is $b = \sum_{i,j} b_{ij} = Q^2N$. The next step in Multiple Correspondance Analysis

(MCA) is to assign weights to the dimensions. One defines successively

- the table F of the relative frequencies, with entry $f_{ij} = \frac{b_{ij}}{b}$,
- the margins with entry $f_{i.} = \sum_j f_{ij}$ or $f_{.j} = \sum_i f_{ij}$,
- the table P of the profiles which sum to 1, with entry $P_{ij} = \frac{f_{ij}}{f_{i.}}$.

In MCA, a weighted euclidean distance is used, where the dimension weights are equal to $(\frac{\sum_j f_{ij}}{\sum_{ij} f_{ij}})^{-1/2}$, that is the inverse of the square root of the column means. The weighted euclidean distance between row profiles k and ℓ is given by

$$d_{k\ell} = \left(\sum_j (\frac{f_{ij}}{\sqrt{f_{.j} f_{k.}}} - \frac{f_{\ell j}}{\sqrt{f_{.j} f_{\ell.}}})^2\right)^{\frac{1}{2}}. \quad (2)$$

So it is equivalent to compute euclidean distances between row profiles of a matrix C whose entry is $c_{k\ell} = \frac{f_{k\ell}}{\sqrt{f_{.\ell} f_{k.}}}$.

4 Crisp/Fuzzy maps

It is possible at this stage to use a Kohonen algorithm on "decomposable" sets to get a non-linear representation of the 3-way contingency table, as it has been already proposed by [5]. The main tool is a Kohonen network, generally a two-dimensional grid (see for example Fig.??) with $n \times n$ units, called *code-vectors*, which has the same dimension as the input space. Let $\boldsymbol{c}$ be an input pattern (row-vector from matrix C) selected at random from the input space. Let the weight vector j be denoted by

$$\boldsymbol{w}_j = \{w_{j1}, \ldots, w_{jm}\}^T,\ j = 1, \ldots, n^2.$$

To find the best match of the input vector $\boldsymbol{c}$ with the weight w_j, compare the inner product $\boldsymbol{w}_j^T \boldsymbol{c}$ for $j = 1, \ldots, n^2$ and select the largest. The algorithm is summarized in Tab.2 where $\Lambda(i(\boldsymbol{c}), t)$ is the neighborhood function, which depends of the number of iterations and the winning unit [7] or [4].

To represent the individuals on the same Kohonen map, we proceed the following way: we use the rows of the matrix K. For example, the individual $\boldsymbol{c}$ corresponding to the row-vector $\boldsymbol{k}_j$ will be affected to the unit $i(\boldsymbol{k})$ such that $i(\boldsymbol{k}) = \arg\min_s \|\boldsymbol{\omega}_s^* - \boldsymbol{k}_j\|$ where ω_s^* is the final value of the weight-vector ω after the trainig step.

5 Example section

This exemple is taken from the SAS examples (version 6.0). The data consist in the characterization of 80 individuals by the four following variables: Age **(Old, Young)** $m_1 = 2$,

1. Choose random values to initialize the weight vectors $\boldsymbol{w}_j^{(0)}$
2. Draw a sample $\boldsymbol{c} = (\frac{f_{1j}}{\sqrt{f_{.j}f_{1.}}}, \dots, \frac{f_{Mj}}{\sqrt{f_{.j}f_{M.}}})$ according to a probability distribution π,
3. Find the best matching unit $i(\boldsymbol{c})$ at time step n among all the lattice by using minimum-distance euclidean criterion:
$i(\boldsymbol{c}) = \arg\min_j \|\boldsymbol{w}_j - \boldsymbol{c}\|, \ \ j = 1, \dots, n^2$
4. Adjust the weights of all units by using the update formula

$$\boldsymbol{\omega}_j^{(t+1)} = \begin{cases} \boldsymbol{\omega}_j^{(t)} - \epsilon(t)(\boldsymbol{\omega}^{(t)} - \boldsymbol{c}), & \forall i \in \Lambda(i(\boldsymbol{c}), t) \\ \boldsymbol{\omega}_j^{(t)}, & \forall i \notin \Lambda(i(\boldsymbol{c}), t) \end{cases}$$

5. Continue on step 2 until no noticable change in the feature map is observed

Table 2. Kohonen algorithm.

Sex (Male, Female) $m_2 = 2$,
Height (*Tall, Short*) $m_3 = 2$,
Hair (White, Brown, Blond) $m_4 = 3$.

(a) Crisp data

Blond		**Young**	*Tall*
Female			Brown
			Male
White	**Old**	*Short*	

(b) Fuzzy data

Blond		*Tall*	**Young**
Female			Brown
	Short		Male
White	**Old**		

Table 3. Kohonen maps on modalities given crisp and fuzzified characteristics.

We train a 4 × 4 Kohonen grid with 500 iterations. The resulting grids using fuzzified data (Table 3.a) or crisp data (Table 3.b) give us the good partionning.

6 Provisional conclusion

In this paper, we answer two important questions concerning the *optimal number of clusters* and the *the fuzzy identification of membership functions* respectively : the optimal number of clusters can be found by hierarchical clustering, the problems of structure identification can be solved using a Kohonen algorithm applied on crisp (qualitative) data or on fuzzy data. These provides an effective and robust approach for identifying fuzzy models from real data.

References

1. Bezdek J.C. *Pattern recognition with fuzzy objective function algorithms*, Plenum Press, New-York, 1987.

2. Cardoso, J.-F.: "Source separation using Higher-order cumulants" *Proc. IEEE ICASSP*, (1997) pp.2109-2112.
3. Chen M.S. and Wang S.W.: "Fuzzy clustering analysis for optimizing fuzzy membership functions", *Fuzzy sets and systems* 103 (1999) 239-254.
4. Cottrell M., Fort and J.C, Pagès G.,: Two or three things that we know about the Kohonen algorithm, In *Proc of ESANN 94*, M.Verleysen ED., D Facto, Bruxelles
5. Ibbou S. and Cottrell M., "Multiple Correspondence analysis of a crosstabulations matrix using the Kohonen algorithm", In *Proc. of ESANN 99*, M. Verleysen Eds., Springer.
6. Keller J.M., Gray M.R. and Gigens J.A.: "A fuzzy $k-$nearest neighbor algorithm", *IEEE Trans. System Man Cybernet.* **SMC-15**, 580-585, 1995.
7. Kohonen T., *Self-organisation and Associative Memory*, Springer, 1989.
8. Lebart L., Morineau A., Piron M. *Statistique exploratoire multidimensionnelle*, Dunod, Paris, 1995.
9. Sugeno M. and Kang G.T., "structure identification of fuzzy model", *Fuzzy sets and systems* 28 (1988) 15-33.

Robust Estimations in Fuzzy Linear Regression Models

Štefan Varga

Department of Mathematics, Faculty of Chemical Technology, Slovak Technical University, Radlinského 9, 812 34 Bratislava, Slovakia

Abstract : Some robust estimations of the unknown parameters in the classical linear regression model are generalized onto the linear regression model with fuzzy parameters and onto the linear regression model with fuzzy variables.

Keywords : Fuzzy numbers, Fuzzy statistics and data analysis, Fuzzy regression, Robust estimations, Observed values, Fitted values

1 Introduction

In the classical linear regression model

$$y = a_1f_1(x) + a_2f_2(x) + \dots + a_mf_m(x) \tag{1}$$

where $f_i(x)$ are known functions of the input variable x, y is an output variable and $\mathbf{a} = (a_1, a_2, \dots, a_m)^T$ is the vector of unknown parameters are used both classical estimations and robust estimations. These estimations are obtained on the basis of observed data $(x_1, y_1), (x_2, y_2), \dots, (x_n, y_n)$ for the pair of the variables (x, y) by means minimizing suitable functions of the residuals

$$r_i(\mathbf{a}) = y_i - a_1f_1(x_i) - a_2f_2(x_i) - \dots - a_mf_m(x_i) \tag{2}$$

$(i = 1, 2, \dots, n)$. Classical least squares estimations give us good results in the model (1) when there are not outliers among the observed data. In the opposite case it is better to use some robust estimations (see for example [2], [7], [8]) which subdue the influence of the outliers on the estimated parameters. The most known estimations of the unknown parameters $\mathbf{a} = (a_1, a_2, \dots, a_m)^T$ in the model (1) are defined bellow.

A. The classical least squares (**LS**) estimation (not robust) of the vector of unknown parameters $\mathbf{a} = (a_1, a_2, \dots, a_m)^T$ in the model (1) is

$$\hat{\mathbf{a}}^{LS} = \arg\min_{\mathbf{a}\in\Theta} \sum_{i=1}^{n} r_i^2(\mathbf{a}) \tag{3}$$

B. **M** – estimation (robust, see for example [2]) of the vector of unknown parameters $\mathbf{a} = (a_1, a_2, \dots, a_m)^T$ in the model (1) is

$$\hat{\mathbf{a}}^{M} = \arg\min_{\mathbf{a}\in\Theta} \sum_{i=1}^{n} \rho(r_i(\mathbf{a})) \tag{4}$$

where $\rho(x)$ is a function which increase to the infinity more slowly than the function x^2. Example of this function is $\rho(x) = \text{abs}(x)$.

C. Least median of squares (**LMS**) estimation (robust, see for example [7]) of the vector of unknown parameters $\mathbf{a} = (a_1, a_2, \dots, a_m)^T$ in the model (1) is

$$\hat{\mathbf{a}}^{LMS} = \arg\min_{\mathbf{a}\in\Theta} \text{med}\left(r_i^2(\mathbf{a})\right) \tag{5}$$

where med $(r_i^2(\mathbf{a}))$ is the median of the squares of the residuals.

D. Least trimmed squares (**LTS**) estimation (robust, see for example [2]) of the vector of unknown parameters $\mathbf{a} = (a_1, a_2, \dots, a_m)^T$ in the model (1) is

$$\hat{\mathbf{a}}^{LTS} = \arg\min_{\mathbf{a}\in\Theta} \sum_{i=1}^{q} r_{(i)}^2(\mathbf{a}) \tag{6}$$

where $r^2_{(1)}(\theta) \leq r^2_{(2)}(\theta) \leq \dots \leq r^2_{(n)}(\theta)$ are the ordered residuals and q is a whole number from the interval [n/2, n]. The LTS - estimations are obtained by minimizing the sum of the squares of q least residuals (choice of q see for example [2]).

The least squares estimations (3) are generalized and obtained in the linear regression model with fuzzy parameters [9] and in the linear regression model with fuzzy data [10]. In this paper the robust estimations (4), (5), (6) are generalized onto the both fuzzy linear regression models. These models are fuzzy because there are used fuzzy numbers and operations with them.

2 Fuzzy Numbers and Operations with Them

In the both fuzzy regression models are used only symmetric triangular fuzzy numbers

$$A = \langle a, s \rangle$$

where a is the center and s is the spread of the fuzzy number A defined by the membership function

$$A(x)=\begin{cases}1-\dfrac{|x-a|}{s} & a-s\le x\le a+s\\ 0 & \text{otherwise}\end{cases}$$

For addition fuzzy numbers and multiplication fuzzy numbers with real numbers we used arithmetic based on the Yager t-norm [3]. If $A_1 = < a_1 , s_1 >$, $A_2 = < a_2 , s_2 >$ are two symmetric triangular fuzzy numbers and k a real number then

$$A_1 + A_2 = \left\langle a_1 + a_2 \,,\, \sqrt[w]{s_1^{\,w} + s_2^{\,w}} \right\rangle$$

(7)

$$k\,A_1 = \left\langle k\,a_1 \,,\, \sqrt[w]{|k|}\; s_1 \right\rangle$$

(w is a parameter ; $w \ge 1$). Observed and estimated data of the output variable Y in the both fuzzy regression models are fuzzy numbers. For estimations of unknown parameters we need a suitable distance of them. This distance of two fuzzy numbers can be the Diamond distance $d_D(A_1, A_2)$ [1] defined by the formula as follows:

(8)

$$d_D^2\,(A_1\,,A_2)=\frac{1}{3}\Big[\big(a_1-s_1-(a_2-s_2)\big)^2+(a_1-a_2)^2+\big(a_1+s_1-(a_2+s_2)\big)^2\Big]=$$
$$=(a_1-a_2)^2+\frac{2}{3}(s_1-s_2)^2$$

3 Robust Estimations in the Linear Regression Model with Fuzzy Parameters

Assume the linear regression model

$$Y = A_1f_1(x) + A_2f_2(x) + \ldots + A_mf_m(x) \qquad (9)$$

where x is a real - valued (crisp) input variable, $f_1(x), \ldots , f_m(x)$ are known real functions, Y is a fuzzy output variable and the parameters $A_1, A_2, \ldots , A_m$ are unknown symmetric triangular fuzzy numbers

$$A_i = < a_i , s_i > \qquad (10)$$

with the center a_i and the spread s_i ($i = 1, 2, \ldots , m$). Further assume that in the real point x_i is the observed value $< y_i , z_i >$ - symmetric triangular fuzzy number with the center y_i and the spread z_i ($i = 1, 2, \dots, m$) and the observed data we can write in the form

x	x_1	x_2	...	x_n
Y	$< y_1 , z_1 >$	$< y_2 , z_2 >$	...	$< y_n , z_n >$

The estimated value $Y(x_i)$ is a symmetric triangular fuzzy number and we get it from the right side of the fuzzy number (9) using the operations (7)

$$Y(x_i) = \left\langle \mathbf{f}_i^T \mathbf{a} , \sqrt[w]{\left|\mathbf{f}_i^T\right| \mathbf{s}_w} \right\rangle \tag{11}$$

where $\mathbf{f}_i^T = (f_1(x_i), f_2(x_i), \dots , f_m(x_i))$, $|\mathbf{f}_i^T| = (|f_1(x_i)|, |f_2(x_i)|, \dots , |f_m(x_i)|)$, $\mathbf{a} = (a_1, a_2, \dots , a_m)^T$, $\mathbf{s_w} = (s_1^w, s_2^w, \dots , s_m^w)^T$, $\mathbf{s} = (s_1 , s_2 , \dots , s_m)^T$ (T denotes the operation of transposition). The residuals r_i $(\mathbf{a}, \mathbf{s})$ ($i = 1, 2, \dots , n$) dependent on the vectors $\mathbf{a}, \mathbf{s}$ are defined by the Diamond distance (8) between the estimated value $Y(x_i)$ and the observed value $< y_i , z_i >$

$$r_i^2(\mathbf{a}, \mathbf{s}) = d^2\left(Y(x_i), < y_i , z_i >\right) = \left(\mathbf{f}_i^T \mathbf{a} - y_i\right)^2 + \frac{2}{3}\left(\sqrt[w]{\left|\mathbf{f}_i^T\right| \mathbf{s}_w} - z_i\right)^2 \tag{12}$$

and $r_{(i)}(\mathbf{a}, \mathbf{s})$ are ordered residuals for which

$$r_{(1)}(\mathbf{a}, \mathbf{s}) \le r_{(2)}(\mathbf{a}, \mathbf{s}) \le \dots \le r_{(n)}(\mathbf{a}, \mathbf{s}) \tag{13}$$

Outcome 1. Let us consider the linear regression model (9) with unknown fuzzy parameters $A_i = < a_i , s_i >$ ($i = 1, 2, \dots, m$).

a) **M** estimations of the vectors $\mathbf{a} = (a_1, a_2, \dots , a_m)^T$ and $\mathbf{s} = (s_1, s_2, \dots , s_m)^T$ we get by minimizing

$$\sum_{i=1}^{n} \rho(r_i^2(\mathbf{a}, \mathbf{s})) \tag{14}$$

where for example $\rho(x) = x^2/2$ ($|x| \le A$) ; $\rho(x) = A|x| - A^2/2$ ($|x| \le A$) [2].

b) Least median of squares (**LMS**) estimations of the vectors $\mathbf{a} = (a_1, a_2, \dots , a_m)^T$ and

$\mathbf{s} = (s_1, s_2, \dots , s_m)^T$ we get by minimizing

$$\operatorname*{med}_i \left(r_i^2(\mathbf{a}, \mathbf{s})\right) \tag{15}$$

c) Least trimmed squares (**LTS**) estimations of the vectors $\mathbf{a} = (a_1, a_2, \dots, a_m)^T$ and $\mathbf{s} = (s_1, s_2, \dots, s_m)^T$ we get by minimizing (choice of $q \in [n/2, n]$ see [2])

$$\sum_{i=1}^{q} r_{(i)}^2(\mathbf{a}, \mathbf{s}) \tag{16}$$

Theorem 1. When the observed values $< y_i , z_i >$ of the output variable Y in the fuzzy regression model (9) are crisp ($z_i = 0$, $i = 1, 2, \dots, n$) then the robust estimations (14), (15), (16) of the vectors **a**, **s** are equal to the robust estimations (4), (5), (6) in the classical regression model (1).

Proof. The both parts of the residuals (12) are positive and if $z_i = 0$ we will get the minimum some function of residuals for $(| \mathbf{f}_i^T | s_w)^{1/w} = 0$. It is obvious that the estimations of all spreads s_i must be equal zero and the residuals (12) will be the same as the residuals (2) in the classical regression model (1).

4 Robust Estimations in the Linear Regression Model with Fuzzy Variables

Assume the linear regression model

$$Y = a_1 X_1 + a_2 X_2 + \dots + a_m X_m \tag{17}$$

where X_i ($i = 1, 2, \dots, m$) and Y are fuzzy variables which values are symmetric triangular fuzzy numbers and $a_1, a_2, \dots, a_m$ are unknown crisp parameters. The observed values of the input variable X_i are assuming in the form

$$X_{ij} = < x_{ij} , s_{ij} > \; ; \; j = 1, 2, \dots, n \tag{18}$$

where x_{ij} is the center and s_{ij} is the spread of the symmetric triangular fuzzy number X_{ij}. Like the observed values of the output variable Y is in the form

$$Y_j = < y_j , s_j > \; ; \; j = 1, 2, \dots, n \tag{19}$$

where y_j is the center and s_j the spread of the symmetric triangular fuzzy number Y_j. It is not hard to show (7) that the estimated (fitted) value

$$Y^{(j)} = a_1 X_{1j} + a_2 X_{2j} + \dots + a_m X_{mj}$$

$j = 1, 2, \dots, n$, is the symmetric triangular fuzzy number

$$Y^{(j)} = < \mathbf{x}_j^T \mathbf{a}, (\mathbf{s}_j^T |\mathbf{a}|)^{1/w} > \tag{20}$$

with the center $\mathbf{x}_j^T \mathbf{a}$ and the spread $(\mathbf{s}_j^T |\mathbf{a}|)^{1/w}$ where $\mathbf{x}_j^T = (x_{1j}, x_{2j}, \dots, x_{mj})$, $\mathbf{s}_j^T = (s_{1j}^w, s_{2j}^w, \dots, s_{mj}^w)$, $\mathbf{a} = (a_1, a_2, \dots, a_m)^T$, $|\mathbf{a}| = (|a_1|, |a_2|, \dots, |a_m|)^T$.

Using the Diamond distance (8) between the observed value Y_j (19) and the estimated value $Y^{(j)}$ (20); $j = 1,\ldots, n$ we will get the crisp residuals

$$r_j^2(\mathbf{a}) = d^2\left(Y^{(j)}, Y_j\right) = \left(\mathbf{x}_j^T\mathbf{a} - y_j\right)^2 + \frac{2}{3}\left(\sqrt[w]{\mathbf{s}_j^T|\mathbf{a}|} - s_j\right)^2 \tag{21}$$

and the robust estimations of the vector of unknown parameters $\mathbf{a} = (a_1, \ldots, a_m)^T$ are in the outcome 2.

Outcome 2. Let us consider the linear regression model (17) with fuzzy variables and with the vector of unknown crisp parameters $\mathbf{a} = (a_1, a_2, \ldots, a_m)^T$.

a) **M** estimation of the vector $\mathbf{a} = (a_1, a_2, \ldots, a_m)^T$ is

$$\hat{\mathbf{a}}^M = \arg\min_{\mathbf{a}} \sum_{j=1}^{n} \rho\left(r_j(\mathbf{a})\right) \tag{22}$$

b) Least median of squares (**LMS**) estimation of the vector $\mathbf{a} = (a_1, a_2, \ldots, a_m)^T$ is

$$\hat{\mathbf{a}}^{LMS} = \arg\min_{\mathbf{a}} \operatorname{med}_{j}\left(r_j^2(\mathbf{a})\right) \tag{23}$$

c) Least trimmed squares (**LTS**) estimation of the vector $\mathbf{a} = (a_1, a_2, \ldots, a_m)^T$ is

$$\hat{\mathbf{a}}^{LTS} = \arg\min_{\mathbf{a}} \sum_{j=1}^{k} r_{(j)}^2(\mathbf{a}) \tag{24}$$

Example 1. Let triangular symmetric fuzzy numbers

$$< 1, 0.5 >, \quad < 1, 0.5 >, \quad < 3, 1 >$$
$$< 2, 0.5 >, \quad < 1, 0.5 >, \quad < 4, 1 >$$
$$< 1, 0.5 >, \quad < 2, 0.5 >, \quad < 5, 1 >$$
$$< 2, 0.5 >, \quad < 2, 0.5 >, \quad < 0, 1 >$$

are the observed data for the triple of the variables (X_1, X_2, Y) same as in Example 1 in [10] only $y_4 \neq <6, 1>$ but $y_4 = <0, 1>$ (y_4 is an outlier).

a) Find the least squares (LS) estimations of the crisp parameters a_1, a_2 of the dependence $Y = a_1 X_1 + a_2 X_2$.

b) Find the robust least trimmed squares (LTS) estimations of the crisp parameters a_1, a_2 of the dependence $Y = a_1 X_1 + a_2 X_2$ $(w = 1)$.

Solution. a) In this example $n = 4$, $\mathbf{x}_j^T = (x_{1j}, x_{2j})$, $\mathbf{s}_j^T = (s_{1j}, s_{2j})$; i = 1, 2, 3, 4, $\mathbf{a} = (a_1, a_2)^T$, $|\mathbf{a}| = (|a_1|, |a_2|)^T$ and therefore we will minimize (7) in [10].

The LS estimations are

$$\mathbf{a_1 = 0.377}\ , \quad \mathbf{a_2 = 1.377}\ .$$

b) In this example $k = 3$, $\mathbf{x}_j^T = (x_{1j}, x_{2j})$, $\mathbf{s}_j^T = (s_{1j}, s_{2j})$; i = 1, 2, 3, $\mathbf{a} = (a_1, a_2)^T$, $|\mathbf{a}| = (|a_1|, |a_2|)^T$ and therefore we will minimize (24) in the form

$$\sum_{j=1}^{3} r_{(j)}^2(\mathbf{a}) =$$

$$(a_1+a_2-3)^2 + 2/3\ (0.5a_1+0.5a_2 - 1)^2 + \ldots + (2a_1+2a_2-5)^2 + 2/3\ (0.5a_1+0.5a_2 - 1)^2$$

$$= \boxed{13/2\ a_1^2 + 13/2\ a_2^2 + 11\ a_1a_2 - 34\ a_1 - 36\ a_2 + 52}$$

The minimum was obtained for a_1, a_2 which are the solution of the linear system

$$13\ a_1 + 11\ a_2 = 34$$

$$11\ a_1 + 13\ a_2 = 36$$

The LTS estimations of the crisp parameters a_1, a_2 of the dependence $Y = a_1 X_1 + a_2 X_2$ are

$$\mathbf{a_1 = 0.9583}\ , \quad \mathbf{a_2 = 1.9583}\ .$$

The observed values and the fitted values are in the table 1.

Table 1

X_{1j}	X_{2j}	Observed value Y_j	Fitted value (crisp) $Y^{(j)}$	Fitted value (robust $Y^{(j)}$
(0.5 ; 1 ; 1.5)	(0.5 ; 1 ; 1.5)	(2 ; 3 ; 4)	(0.88 ; 1.75 ; 2.63)	(1.46 ; 2.92 ; 4.38)
(1.5 ; 2 ; 2.5)	(0.5 ; 1 ; 1.5)	(3 ; 4 ; 5)	(1.25 ; 2.13 ; 3.01)	(2.42 ; 3.88 ; 5.33)
(0.5 ; 1 ; 1.5)	(1.5 ; 2 ; 2.5)	(4 ; 5 ; 6)	(2.25 ; 3.13 ; 4.01)	(3.42 ; 4.88 ; 6.33)
(1.5 ; 2 ; 2.5)	(1.5 ; 2 ; 2.5)	(5 ; 6 ; 7)	(2.63 ; 3.51 ; 4.39)	(4.38 ; 5.83 ; 7.29)

References

[1] A. Bárdossy, L. Duckstein, Fuzzy Rule – Based Modeling with Applications to Geophysical, Biological and Eng. Systems, CRC Press, Boca Raton, 1995.

[2] P. J. Huber, Robust Statistics, Wiley, New York, 1981.

[3] G. J. Klir, B. Yuan, Fuzzy Sets and Fuzzy Logic - Theory and Applications, Prentice – Hall PTR, Upper Saddle River, NJ, USA, 1995.

[4] A. Kolesárová, Triangular norm-based addition preserving linearity of T - sums of linear fuzzy intervals, Mathware & Soft Comp., Vol. 5, No.1 (1998) 91- 98.

[5] M. Mareš, Computation Over Fuzzy Quantities, CRC Press, Boca Raton, 1994.

[6] R. Mesiar, Shape preserving additions of fuzzy intervals, Fuzzy Sets and Systems 86 (1997), 73 – 78.

[7] P. J. Rousseeuw, Least median of squares regression, JASA 79 (1984), 871-880

[8] Š. Varga, Robust estimations in statistics, Mathematical statistics and numerical mathematics on universities, Kálnica (1998), 86 - 114.

[9] Š. Varga, M. Šabo, Linear regression with fuzzy parameters, Fuzzy Sets and Systems (In print, 2000).

[10] Š. Varga, M. Šabo, Linear regression with fuzzy variables. Fuzzy Sets and Systems (In print, 2000).

On Construction of Distance Measure, Similarity Measure and Evolutionary Operators

Anikó Szakál, Katalin Lőrincz

Budapest Polytechnic

1081 Budapest, Népszínház u. 8, Hungary

Abstract: In this paper new methods for constructing distance measure, similarity measure and generalized operators are outlined. To construct distance and similariry measures the concepts of elementary distance and elementery similarity functions are introduced. The general form of these functions and the way to use them for measure construction are discussed. By means of the elementary distance function a new method for constructing evolutionary operators is given.

Keywords: Fuzzy entropy, distance measure, similarity measure, generalized operators

1 Introduction

Original fuzzy set theory was formulated in terms of Zadeh's standard operations of intersection, union and complement. Since 1965 for each of these operations several classes of operators, satisfying appropriate axioms, have been introduced. By accepting some basic conditions, a broad class of set of operations for union and intersection is formed by t-operators [1].

In many applications where for example compensation behavior is required these conventional operators do not work well. Recently in order to get rid of this and other disadvantages several generalizations have been introduced.

Rudas [10] has introduced the concept of evolutionary operators or with other words the maximum and minimum distance operators with respect to e, where e is an arbitrary element of the unit interval [0,1]. The special cases $e = 0$ and $e = 1$ lead to the max and min operators. The new operators are evolutionary types in the sense that if e is increasing starting from zero till $e = 1$ the min operator is developing into the max operator, while on the other side the max is transformed into the min operator.

In this paper new methods for constructing distance and similarity measures are intruduced by using the concepts of elementary similarity and distance functions. It is shown that these functions can be used to construct the most and less similar evolutionary operators.

The paper is constructed as follows. First the definition of fuzzy entropy and the concept of the elementary entropy function are introduced, by recalling the basic theorems of Rudas [9] regarding the construction of elementary entropy functions and fuzzy entropies. Using this approach next the elementary distance and similarity functions are introduced and the related theorems are prooved. Finally the construction of the new operators are outlined.

Throughout this paper the following notations will be used; X is the universal set, X^F is the class of all fuzzy subsets of X, $\Re^+$ is the set of non negative real numbers, $\overline{A}$ is the fuzzy complement of $A \in X^F$ and $|A|$ is the cardinality of A.

2 Entropy and elementary entropy function

Rudas et.al. have introduced the concept of elementary entropy function and its properties [3], [4], [5], [7]. For our further discussion these basic results are recalled in this paragraph.

Definition. Let X be a universal set and A is a fuzzy subset of X defined as

$$A = \{(x_i, \mu_A(x_i)) | x_i \in X, \mu_A(x_i) \in [0,1]\}. \tag{1}$$

The *fuzzy entropy* is a function

$$e : X^F \to \Re^+ \tag{2}$$

which satisfies the following axioms:

AE 1. $e(A) = 0$ if A is a crisp set.

AE 2. If $A \prec B$ then $e(A) \le e(B)$; where $A \prec B$ means that A *is sharper than* B.

AE 3. $e(A)$ assumes the maximum value if and only if A is *maximally fuzzy*.

AE 4. $e(A) = e(\overline{A}),\ \forall A \in X^F$.

That is, the function e assigns a value $e(A)$ to each fuzzy subset A of the universe X that characterises the degree of fuzziness of A.

Let us specify **AE 2.** and **AE 3.** as follows

AES 2. A *is sharper than* B in the following sense:

$\mu_A(x) \le \mu_B(x)$ for $\mu_B(x) \le \frac{1}{2}$ and

$\mu_A(x) \ge \mu_B(x)$ for $\mu_B(x) \ge \frac{1}{2}$, for all $x \in X$.

AES 3. A is defined maximally fuzzy when

$\mu_A(x) = \frac{1}{2}\ \forall x \in X$.

Definition. Let A be a fuzzy subset of X and define the following function

$$\varphi_A : x \mapsto \begin{cases} \mu_A(x) \text{ if } \mu_A(x) \le \frac{1}{2} \\ 1 - \mu_A(x) \text{if } \mu_A(x) > \frac{1}{2} \end{cases} \quad (3)$$

Definition. Denote Φ_A the fuzzy set generated by φ_A

$$\Phi_A = \{(x, \varphi_A(x)) | x \in X\}. \quad (4)$$

Theorem. (Rudas, Kaynak [3]. [4], [5].). The $g(|\Phi_A|)$ is an entropy, where $g : \Re \to \Re$ is a monotonically increasing real function and $g(0) = 0$.

Definition. Let A be a fuzzy subset of X. f is said to be an *elementary fuzzy entropy function* if the cardinality of the fuzzy set

$$\Phi_A = \{(x, f(x)) | x \in X, f(x) \in [0,1]\} \quad (5)$$

is an entropy of A.

It is obvious that φ_A is an elementary entropy function.

Theorem. (Rudas, Kaynak [5]). The function $f(\varphi_A)$ is also an elementary entropy function if f has the following properties:

1. $f : [0, \frac{1}{2}] \to [0,1]$,

2. $f(0) = 0$,

3. f is strickly monotonically increasing in $[0, \frac{1}{2}]$.

Theorem. (Rudas, Kaynak [5]). The function $h(\mu_A)$ is also an elementary entropy function if h has the following properties:

1. $h : [0,1] \to [0,1]$,

2. $h(0)=0,\ h(1)=0$,

3. h is strickly monotonically increasing in $[0,\frac{1}{2}]$, and monotonically decreasing in $[\frac{1}{2},0]$,

4. $h\left(\frac{1}{2}\right)$ is a unique maximum of h.

3 Distance measure and elementary distance function

Based on the concept of elementary entropy function the elementary distance function in introduced that can be used to construct distance measure.

Definition. The *distance measure* is a function

$$\mathrm{D}: X^F \times X^F \rightarrow \Re^+$$

which satisfies the following axioms:

AD 1. $D(A,A)=0,\ \forall A \in X^F$.

AD 2. $D(A,B)=D(B,A),\ \forall A,B \in X^F$

AD 3. $D(C,\overline{C})=\max\limits_{A,B\in X^F} D(B,A),\ \forall C \in X^P$

AD 4.

If $A \subset B \subset C$, then $D(A,B) \le D(A,C)$, and $D(B,C) \le D(A,C),\ \forall A,B,C \in X^F$.

Definition. Let A and B are fuzzy subset of X and define the following function

$$d_{A,B}: \mathrm{x}\ \mathrm{a}\ \begin{cases} |\varphi_A(x)-\varphi_B(x)| \text{ if } \mu_A(x),\mu_B(x) \le \frac{1}{2} \text{ or } \mu_A(x),\mu_B(x) > \frac{1}{2} \\ 1\text{-}\varphi_A(x)\text{-}\varphi_B(x) \text{otherwise} \end{cases}. \qquad (6)$$

Definition. Denote $\Lambda_{A,B}$ the fuzzy set generated by $d_{A,B}$, i.e.

$$\Lambda_{A,B}=\{(x,d_{A,B}(x))|x \in X, d_{A,B}(x) \in [0,1]\}.$$

Theorem. The $g(|\Lambda_{A,B}|)$ is a distance measure, where $g:\Re \rightarrow \Re$ is a monotonically increasing real function and $g(0)=0$.

Proof. We have to prove that $g(|\Lambda_{A,B}|)$ satisfies the axioms of the distance measure.

Based on the definition of the elementary distance function it is easy to verify that $d_{A,A}(\mathrm{x})=0$ and $d_{A,B}(x)=d_{B,A}(x)$, $g(\Lambda_{A,B}(x))=g(\Lambda_{B,A}(x))$.

If A is a crisp set then $\varphi_A(x)=0, \forall \mathrm{x}\in \mathrm{X}$, so $d_{A,\bar{A}}(\mathrm{x})=1$ which provides the maximum value of $g(|\Lambda_{A,B}|)$.

If $A\subset B\subset C$ then $\mu_A(x)\le \mu_B(x)\le \mu_C(x)$.

Suppose that $\mu_C(x)\le \frac{1}{2}$. In this case $\varphi_A(x)\le \varphi_B(x)\le \varphi_C(x)$ and $|\varphi_A(x)-\varphi_B(x)|\le|\varphi_A(x)-\varphi_C(x)|$ and $|\varphi_B(x)-\varphi_C(x)|\le|\varphi_A(x)-\varphi_C(x)|$.

If $\mu_A(x)>\frac{1}{2}$ then $\varphi_A(x)\ge\varphi_B(x)\ge\varphi_C(x)$ and the statement holds.

If $\mu_A(x)\le\mu_B(x)\le\frac{1}{2}, \mu_C(x)>\frac{1}{2}$ then $d_{A,B}(x)=|\mu_B(x)-\mu_A(x)|$ and
$d_{A,C}(x)=1-\varphi_C(c)-\varphi_A(x)=\mu_C(x)-\mu_A(x)\ge d_{A,B}(x)$,
$d_{B,C}(x)=1-\varphi_B(c)-\varphi_C(x)=\mu_C(x)-\mu_B(x)\le d_{A,C}(x)$

If $\mu_A(x)\le\frac{1}{2}$, $\mu_B(x), \mu_C(x)>\frac{1}{2}$ then $d_{A,B}(x)=1-\varphi_B(x)-\varphi_A(x)=\mu_B(x)-\mu_A(x)$, $d_{A,C}(x)=\mu_C(x)-\mu_A(x)\ge d_{A,B}(x)$, $d_{B,C}(x)=\mu_C(x)-\mu_B(x)\le d_{A,C}(x)$. ■

Definition. Let A and B be fuzzy subset of X. f is said to be an *elementary fuzzy distance function* if the cardinality of the fuzzy set

$$\Lambda_{A,B}=\{(x,f(x))|x\in X, f(x)\in[0,1]\}$$

is a distance measure of A.

It is obvious that $d_{A,B}$ is an elementary distance function.

Theorem. The function $f(d_{A,B})$ is also an elementary distance function if f has the following properties:

1. $f:[0,1]\to[0,1]$,
2. $f(0)=0,\ f(1)=1$
3. f is strickly monotonically increasing in $[0,1]$.

Proof. We have to prove that the cardinality of the fuzzy set $\Lambda_{A,B}=\{(x,f(d_{A,B}(x)))|x\in X\}$ satisfies the axioms of the distance measure.

Since $d_{A,A}(\mathrm{x})=0$ and $d_{A,B}(x)=d_{B,A}(x)$ it is obvious that the axioms **AD 1** and **AD 2** are satisfied.

If A is a crisp set then $\varphi_A(x)=0, \forall x \in X$, so $f\left(d_{A,\bar{A}}(x)\right)=1$ which provides the maximum value of $\left|\Lambda_{A,B}\right|$.

Because of the monotonity of f it is easy to see that the axiom **AD 4** is satisfied. ■

4 Similarity measure and elementary similarity function

Definition. The *similarity measure* is a function

$$S : X^F \times X^F \rightarrow \Re^+$$

which satisfies the following axioms:

AS 1. $S(C,\overline{C})=0,\ \forall C \in X^P$

AS 2. $S(A,B)=S(B,A),\ \forall A,B \in X^F$

AS 3. $S(A,A)=\max\limits_{B,C\in X^F} S(B,C),\ \forall A \in X^F$

AS 4. If $A \subset B \subset C$, then $S(A,B) \geq S(A,C)$ and , $S(B,C) \geq S(A,C) \forall A,B,C \in X^F$.

Definition. Let A and B are fuzzy subset of X and define the following function

$$s_{A,B} : x \mapsto 1 - d_{A,B}(x). \tag{7}$$

Definition. Denote $\Gamma_{A,B}$ the fuzzy set generated by $s_{A,B}$, i.e.

$$\Gamma_{A,B} = \left\{\left(x, s_{A,B}(x)\right) \middle| x \in X, s_{A,B}(x) \in [0,1]\right\}.$$

Theorem. The $g\left(\left|\Gamma_{A,B}\right|\right)$ is a similarity measure, where $g : \Re \rightarrow \Re$ is a monotonically decreasing real function and $g(0)=0$.

Proof. We have to prove that $g\left(\left|\Gamma_{A,B}\right|\right)$ satisfies the axioms of the similarity measure.

1. $s_{A,\bar{A}}(x)=1-d_{A,\bar{A}}(x)=0, \forall A \in X^F$.

2. $d_{A,B}(x)=d_{B,A}(x)$ so $s_{A,B}(x)=s_{B,A}(x)$.

3. $d_{A,A}(x)=0$ so $s_{A,A}(x)=1$ which is the maximum value of the function s.

4. If $A \subset B \subset C$, then $s_{A,B} = 1 - d_{A,B} \geq s_{A,C} = 1 - \mathrm{d}_{\mathrm{A,C}}$ and

$s_{B,C} = 1 - d_{B,C} \geq s_{A,C} = 1 - \mathrm{d}_{\mathrm{A,C}}$, $\forall A, B, C \in X^F$. ■

Definition. Let A and B be fuzzy subset of X. f is said to be an *elementary fuzzy similarity function* if the cardinality of the fuzzy set

$$\Lambda_{A,B} = \{(x, f(x)) | x \in X, f(x) \in [0,1]\}$$

is a similarity measure of A.

It is obvious that $s_{A,B}$ is an elementary similarity function.

Theorem. The function $f(s_{A,B})$ is also an elementary similarity function if f has the following properties:

1. $f : [0,1] \to [0,1]$,
2. $f(0) = 0, \quad f(1) = 1$
3. f is strickly monotonically increasing in $[0,1]$.

Proof. We have to prove that the cardinality of the fuzzy set $\Lambda_{A,B} = \{(x, f(s_{A,B}(x))) | x \in X\}$ satisfies the axioms of the similarity measure.

1. $f(s_{A,\bar{A}}(x)) = f(0) = 0, \forall A \in X^F$.
2. $f(d_{A,B}(x)) = f(d_{B,A}(x))$ so $f(s_{A,B}(x)) = f(s_{B,A}(x))$.
3. $d_{A,A}(\mathrm{x}) = 0$ so $f(s_{A,A}(\mathrm{x})) = 1$ which is the maximum value of the function f.
4. Because of the monotonity of f **AS 4** is holds. ■

5 Most and less similar operators

Let e be an arbitrary element of the closed unit interval [0,1], A and B two arbitrary fuzzy sets of the universe X, and denote by $d_{A,e}$ and $d_{B,e}$ the elementary distance functions.

Definition The membership function of the *less similar minimum operator of A and B with respect to* $e \in [0,1]$ is defined as

$$\mu_e^{LS\min}(\mu_A(x), \mu_B(x)) = \begin{cases} \mu_A(x), & \text{if } d_{A,e}(x) > d_{B,e}(x) \\ \mu_B(x) & \text{if } d_{A,e}(x) < d_{B,e}(x). \\ \min(\mu_A(x), \mu_B(x)), & \text{if } d_{A,e}(x) = d_{B,e}(x) \end{cases} \tag{8}$$

Definition The membership function of the *less similar maximum operator with respect to* $e \in [0,1]$ is defined as

$$\mu_e^{LS\max}(\mu_A(x), \mu_B(x)) = \begin{cases} \mu_A(x), & \text{if } d_{A,e}(x) > d_{B,e}(x) \\ \mu_B(x) & \text{if } d_{A,e}(x) < d_{B,e}(x) \\ \max(\mu_A(x), \mu_B(x)), & \text{if } d_{A,e}(x) = d_{B,e}(x) \end{cases}. \tag{9}$$

Definition The membership function of the *most similar minimum operator with respect to* $e \in [0,1]$ is defined as

$$\mu_e^{MS\min}(\mu_A(x), \mu_B(x)) = \begin{cases} \mu_A(x), & \text{if } d_{A,e}(x) < d_{B,e}(x) \\ \mu_B(x) & \text{if } d_{A,e}(x) > d_{B,e}(x) \\ \min(\mu_A(x), \mu_B(x)), & \text{if } d_{A,e}(x) = d_{B,e}(x) \end{cases}. \tag{10}$$

Definition The membership function of the *most similar maximum operator with respect to* $e \in [0,1]$ is defined as

$$\mu_e^{MS\max}(\mu_A(x), \mu_B(x)) = \begin{cases} \mu_A(x), & \text{if } d_{A,e}(x) < d_{B,e}(x) \\ \mu_B(x) & \text{if } d_{A,e}(x) > d_{B,e}(x) \\ \max(\mu_A(x), \mu_B(x)), & \text{if } d_{A,e}(x) = d_{B,e}(x) \end{cases}. \tag{11}$$

6 Conclusions

The class of evolutionary operators is a new way in generalization of t-operators. In this paper a new method is given to construct these operators. For this purpose the concept of elementary distance and similarity functions are introduced. As a second result of the paper it is shown how these functions can be used to construct distance and similarity measures.

References

[1] Klir, G. J. and Folger T. A.: *Fuzzy sets, Uncertainty, and Information*, Prentice-Hall International Editions, USA, 1988.

[2] Kosko, B.: *Neural Networks and Fuzzy Systems*. Prentice-Hall International, USA, 1992.

[3] Imre Rudas, O. Kaynak: New Types of Generalized Operations. *Computational Intelligence: Soft Computing and Fuzzy-Neuro Integration with Applications. Springer NATO ASI Series. Series F: Computer and Systems Sciences, Vol. 192. 1998. (O. Kaynak, L. A. Zadeh, B. Türksen, I. J. Rudas editors),* pp. 128-156.

[4] Imre Rudas, M. O. Kaynak: Entropy-Based Operations on Fuzzy Sets. *IEEE Transactions on Fuzzy Systems,* vol. 6, no. 1, February 1998. pp. 33-40.

[5] Imre Rudas, M. O. Kaynak: Minimum and maximum fuzziness generalized operators *Fuzzy Sets and Systems* 98 (1998) 83-94.

[6] Okyay Kaynak, Imre Rudas: Soft computing methodologies and their fusion in mechatronic products. *Computing & Control Engineering Journal,* April 1995, pp. 68-72.

[7] Imre Rudas: Generalization of Fuzzy t-operators. *Proceedings of the Bánki Donát Politechnic Jubilee International Conference*, Budapest, 1999, pp. 51-55.

[8] Imre Rudas, Márta Takács: Generalized Mamdani-Type Fuzzy Controllers. *Proceedings of Eurofuse-Sic '99*, Budapest, 1999, pp. 162-165

[9] Imre Rudas: Towards the generalization of t-operators: a distance based approach. Proceedings of the International Conference on Intelligent Information Systems 1999, (IIS'99). Varazdin, September 24-26, 1999.

[10] Imre Rudas: Evolutionary Operators. *Proceedings of INES' 99, IEEE International Conference on Intelligent Engineering Systems*, Poprad, High Tatras, Stara Lesna, Slovakia, November 1-3, 1999, pp. 331-335.

Computational intelligence methods and data understanding

Włodzisław Duch[1] and Yoichi Hayashi[2]

[1] Department of Computer Methods, Nicholas Copernicus University, Grudziadzka 5, 87-100 Toruń, Poland.
WWW: http://www.phys.uni.torun.pl/kmk

[2] Department of Computer Science, Meiji University, Kawasaki, Japan.

Abstract: Experts in machine learning and fuzzy system frequently identify understanding the data with the use of logical rules. Reasons for inadequacy of crisp and fuzzy rule-based explanations are presented. An approach based on analysis of probabilities of classification $p(C_i|\mathbf{X};\rho)$ as a function of the size of the neighborhood ρ of the given case $\mathbf{X}$ is presented. Probabilities are evaluated using Monte Carlo sampling or – for some models – using analytical formulas. Coupled with topographically correct visualization of the data in this neighborhood this approach, applicable to any classifiers, gives in many cases a better evaluation of the new data than rule-based systems. Two real life examples of such interpretation are presented.

1 Introduction

Classification and prediction are the two most common applications of computational intelligence (CI) methods, i.e. methods designed for solving problems that are effectively non-algorithmic. Although sometimes classification by a black box is sufficient – if it is reliable – domain experts use software to help them to understand (or explain) the data. The goal of data mining, or Knowledge Discovery in Databases (KDD), is more ambitious than understanding of new data. Knowledge discovery requires analysis of relations in the whole database, trying to create a symbolic domain theory, while understanding new data may be done locally, for example by pointing that it is typical for a known category. Most methods used for of classification are unable to do that. They rarely provide confidence bounds, so the user is not able to evaluate how reliable is the answer given by the system. Rule-based systems are not much better in this respect if continuos values of attributes are used. Near the rule decision borders reliability of the system should decrease and alternative diagnoses should appear in a smooth way. Statistical methods are oriented towards global characterization of numerical aspects of the data for a whole class of objects, providing means and variances instead of symbolic, understandable descriptions.

Explanation and reliability assessment are in some applications equally important as the accuracy of classification and prediction. Rule-based explanation has been a strong point of the machine learning (ML) inductive systems [1,2]. Although ML algorithms work well with simple data building models and learning from structured data is still rather difficult. Inductive Logic Programming (IPL) is considered to be the best approach in such cases, but for large databases it may be too slow [3]. Molecular biology and genomics are two very important fields in which theory building, rather than classification models, seem to be necessary. Molecules are not described in a fixed-dimensional feature spaces. By analogy to the database terms one may call typical data analysis problems defined in fixed feature spaces "flat", while those problems that require construction of a new set of attributes, relating the description to known domain theories, as "relational".

Leaving relational problems aside for the flat data problems crisp logical rules provide adequate explanation only if all attributes have a few discrete values. Continuous values require discretization, or defining some "receptive fields" that may be used in rule conditions. Many books have been written about the advantages of fuzzy rules, soft versions of neural and other CI algorithms have been published (cf. [4] or [5]), but it still remains to be seen whether these algorithms will have a lasting impact on computational intelligence field. So far no extensive comparison of fuzzy classification methods with other methods on well-known data (such as the Stalog datasets [6]) have been made, and little has been done in applications to difficult, real-world problems such as handwritten character or speech recognition.

An important advantage of the fuzzy and other rule-based approaches to data analysis is their ability to provide explanation for the action of the system in form of understandable rules. The goal of the present paper is to analyze whether such explanations are really satisfactory and to propose a universal approach, applicable in connection with any classifier, that may provide an explanation. In the next section advantages and disadvantages of fuzzy explanations are presented. An alternative approach allowing for explanation of the data, based on calculation of probabilities as a function of the input uncertainty and visualization of the neighborhood of the case analyzed is presented in the third section. Two real-life examples are presented in the fourth section and a short discussion concludes this paper.

2 Fuzzy explanations

Acceptable explanations depend on the particular domain and the type of data. Fuzzy set theory [7] is a method to deal with imprecise information. Since our language is discrete while our senses provide continuous information some form of categorization or discretization must be done to allow for communication. Categorization can be modeled at a high level as clusterization of percepts. An explanation of the sensory data is usually given by an assignment of a percept to some category and the ability to quote the reasons for such assignment. Reasoning refers to a background knowledge, which in the case of our brains is very complex and highly specialized. For example, the ability to recognize faces depends on the activity of neocortex in the inferotemporal gyrus region of the right lobe, containing group of neurons that specialize in high-level object recognition. Explanations referring to features analyzed by lower levels of visual system functions, such as particular shapes, colors, movements, are not only unnatural, but ultimately impossible. All important work enabling the final classification is done at the pre-processing level, defining complex transformations that allow us to recognize a particular pattern as a face. Thus explanation should rely on sophisticated transformation of input features and rarely may be reduced to rules with conditions using large receptive fields defined by some membership functions applied to raw measurements. These receptive fields are usually called linguistic terms or linguistic variables [8].

Although crisp logical rules are the easiest to interpret there are several problems with explanations based on such rules [9]:

- Only one class is predicted as the winner.
- Some cases may be rejected as unclassified without justification.
- Crisp rules are not stable against small perturbations of input values.
- Non-gradient optimization methods should be used because the cost function based on the number of errors is discontinuous.

Fuzzy systems overcome these problems using continuous membership functions, but do they provide an adequate explanation of the data? Fuzzy membership functions allow for a natural description of many features, providing large receptive fields and thus simplifying the classification, so in principle they could provide good explanations. In practice fuzzy rules are rarely successful in explaining the data. For example, in the two recent books on neurofuzzy systems [4,5] many fuzzy rules for different problems have been derived, but would a domain expert find any of these rules useful? Problem of assessing the reliability of classification and of understanding the data are not addressed in these and other books on the subject. It is highly doubtful that fuzzy or even neurofuzzy systems would be able to compete with other classifiers on the ground of their high accuracy only. Reasons for this apparent failure of fuzzy adaptive systems to provide explanation of data are related to the problems of:

- setting up and understanding the linguistic variables;
- complexity of rules the system generate;
- large number of rules generated by some systems;
- problems of finding the simplest set of rules due to overparametrization of adaptive fuzzy systems;
- existence of equivalent sets of rules (or sets of similar accuracy).

Quoting a rule counts as an explanation only if the number of rules is relatively small and their accuracy is sufficiently high. Fuzzy and neurofuzzy systems apply membership functions to the input features, filtering information passed to the classifier by creating large receptive fields. The multilayer perceptron (MLP) networks [10] use their first hidden layer to define combinations of inputs that – treated as new input features – should transform the classification problem to a separable one. The output neurons provide separating hyperplanes in the transformed space. RBF networks [10] perform initial clusterization of the data using all features provided. None of these approaches is really directed at providing explanations at a higher level, with a transformed set of features. For that purpose some restrictions on meaningful transformations of input features should be set first, combinations of these features determined by an MLP-type layer, linguistic variables extracted by a constructive network with transfer function capable of local response (such as the bicentral functions [11]), and finally the outputs should be combined by a rule layer. The C-MLP2LN network [12,13] has an aggregation layer (A-layer), linguistic variables layer (L-layer), a rule layer (R-layer) and an output layer. This network was used previously in data mining tasks [13] but it is also well suited as a general architecture facilitating explanation of data.

Fuzzy rules involve a number of parameters determining position and shape of the membership functions. To avoid overparameterization systems based on fuzzy logic frequently use a fixed set of membership functions, with predetermined shapes. Defining linguistic variables in such context-independent way amounts in effect to a regular partitioning of the whole input space into convex regions. This approach suffers from the curse of dimensionality [10], since with k linguistic variables in d dimensions the number of possible input combinations is k^d. Fuzzy rules simply pick up those areas in the input space that contain vectors from a single class. Without the possibility to adapt membership functions to individual clusters in a single rule fuzzy rules do not allow for optimal description of these clusters. Much better results may be obtained with context-dependent linguistic variables created by some neurofuzzy systems [8,13].

Overparameterization of fuzzy systems and lack of regularization methods [10] commonly used in neural networks is responsible for high complexity of generated rules and difficulties of finding the simplest set of rules. Let us take as an example the Wisconsin breast cancer data [14] (available from the UCI machine learning repository [15]). This dataset describes properties of cancerogenic cells using 9 attributes with integer values in the range 1-10 (for example, feature f_2 is "clump thickness" and f_8 is "bland chromatin"). Fuzzy clustering methods or neurofuzzy systems do not produce membership functions with reasonable linguistic interpretation. The NEFCLASS neurofuzzy system [4], one of the best systems of its kind, combined with fuzzy clustering, produces after simplification 4 rules of the form:

$$\begin{aligned}\text{IF}\ \ & f_1 = l \wedge f_2 = l \wedge f_3 = l \wedge f_4 = l \wedge f_4 = l \wedge f_5 = s \wedge \\ & f_6 = l \wedge f_7 = l \wedge f_8 = l \wedge f_9 = s \quad \text{THEN } \textit{malignant}\end{aligned}$$

Most fuzzy systems will produce much more complex sets of rules, but even the rule shown above given as an explanation of observed data, with specific membership functions for each attribute, may be too difficult for medical doctors to understand and use. The simplest rules for this dataset found for the malignant class using the C-MLP2LN are:

$$f_2 \geq 7 \vee f_7 \geq 6 \qquad (95.6\%)$$

These rules cover 215 malignant cases and 10 benign cases, achieving overall accuracy (including the ELSE condition) of 94.9%. They could not be found using decision trees (C4.5 [16] and SSV [17]) and other systems, but they are the easiest to understand to the medical experts.

Understanding the data may require the simplest description possible. This is evident in another medical example of the Ljubliana cancer data [15], containing 286 cases, of which 201 are no-recurrence-events (70.3%) and 85 are recurrence-events (29.7%). Each case is described by 9 attributes, with 2 to 13 different values each. A single logical rule for the recurrence-events has been found by C-MLP2LN system:

$$\text{involved nodes}{=}\neg[0,2] \wedge \text{ Degree-malignant} = 3$$

With ELSE condition for the second class this rule gives over 77% accuracy in crossvalidation tests. The rule is easy to interpret: recurrence is expected if the number of involved nodes is greater than two and the cells are highly malignant. More complex and more accurate sets of rules may be found [13] for these datasets. Unfortunately the more complex the rules are the less likely it is that they represent interesting knowledge. This is true especially for smaller datasets, where accidental correlations easily occur and rules cover small number of cases. An adequate understanding of the data may require generation of several equivalent sets of rules.

In general the more complex the CI system is, the more difficult finding the simplest solution will be, since all adaptive methods require some form of minimization or iterative optimization procedures that are less reliable for larger number of parameters. Neurofuzzy systems frequently introduce additional fuzzification and defuzzification layers, increasing the number of parameters significantly. These layers are sometimes designed by hand so their parameters are fixed, but in effect suboptimal linguistic variables are used, the system is far from being automatic and the hidden layer performing classification has anyway more parameters, because each continuos input is replaced by several quasi-discrete inputs.

Suppose that in a 2-dimensional problem two classes are separated with a decision border (using the data rescaled to [0,1] interval) $x_1 + x_2 = 1$, i.e. the first class is in the lower corner of the unit square and the second outside. A single neuron will handle this problem with 2 weights and 1 bias. It would be difficult, if not impossible, to find a membership functions for the N variables – for example by evolving trapezoidal or other membership functions using genetic algorithms – that could separate approximate correctly a plane

$x_1 + x_2 + \ldots x_n = 1$. Thus although in principle neural networks and fuzzy systems are equivalent [18] in practice there will be problems for which an approximation using any fuzzy system will converge very slowly. The same is true for the MLP networks which for some data distributions may suffer from slow convergence. The importance of a proper choice of the transfer functions in neural networks has been realized only quite recently [11].

Undoubtedly there are many industrial applications of fuzzy logic, although they seem to be restricted to control problems [19]. Since global control models are impossible to construct, the space of control parameters is divided into subregions where appropriate actions are fired by rules; in essence fuzzy control is based on local learning in multiple regions and the rules are simply identifying these regions. The best way to solve such problems would be to construct neural systems with weights $W(X)$ parametrized by the input vector X, i.e. to construct neural networks $NN(X; W(X))$ which represent a family of mappings that change in different regions of the input space. For example, in a larger modular network $W(X) = WG(X; P_k, \sigma_{x0})$ may be defined for a number of reference points P_k in the input space. Networks appropriate for this region of space may then perform specific actions. The architecture should be similar to the constructive RBF type of networks (cf. [10,20,21]) with appropriate action conditions performed by the output nodes.

3 Black box systems, assessment of reliability and data explanation

Neural systems usually do not provide explanation or even an assessment of reliability of their decisions. The same is true for other CI methods, such as clusterization, nearest neighbors or discriminant analysis (including Support Vector Machines) methods. Explanation or understanding of new data does not require full data mining. It may be sufficient to relate the new case to known cases, estimate probabilities for different classes, show how these probabilities depend on possible inaccuracies of measurements and visualize the data (this is called "an explanatory data analysis" in the statistical literature, cf. [22]). It is worthwhile to start searching for the simplest description of the data in form of crisp logical rules; it that fails more complex fuzzy rules should be produced and if that fails – i.e. if rules are not sufficiently accurate or they become too complex – more complex, black box classification models should be tried. In all cases given a new vector **X** for evaluation classification probabilities should be calculated and a diagnostic map should be provided, showing this case in relation to others.

Finding the simplest description of data involves the following steps:

- Create good features by combining the existing ones; use methods of feature selection, principal components (PCA), non-linear PCA [23], independent components (ICA), or strong regularization enforcing skeletonization of the input-hidden layer connections in neural network.

- Select linguistic variables [8] as subsets of discrete values or as interval $[X_k, X'_k]$.
- Extract logical rules from the data using neural, machine learning or statistical techniques.
- Optimize linguistic variables ($[X_k, X'_k]$ intervals) using the extracted rules and exploring the reliability/rejection rate tradeoff.
- Repeat the procedure until a stable set of rules is found.

These steps were described in details in [8,9,12,13]. The simplicity/accuracy tradeoff is explored first: rough description of the data is made using appropriate regularization parameters, with a hierarchy of more complex models following, until optimal complexity (from the generalization point of view) of the rule base is found. After selecting satisfactory level of data description another tradeoff is explored, between the reliability of predictions and the rejection rate of the classifier (number of vectors in the "don't know" class). Let $\mathcal{P}(C_i, C_j)$ be the confusion matrix for a rule-based classifier M (the model M containing intervals $[X_k, X'_k]$ and subsets defining linguistic variables). The number of wrong predictions $\min_M \left[\sum_{i\neq j} \mathcal{P}(C_i, C_j)\right]$ should be minimized simultaneously with maximization of the predictive power $\max_M [\text{Tr } \mathcal{P}(C_i, C_j)]$ of the classifier over all adaptive parameters of the model M. The following cost function $E(M)$ is minimized without constraints:

$$E(M) = \gamma \sum_{i\neq j} \mathcal{P}(C_i, C_j) - \text{Tr } \mathcal{P}(C_i, C_j) \geq -n \tag{1}$$

where the parameter γ determines a tradeoff between reliability and rejection rate. Sets of rules of lower reliability make larger number of errors but have low rejection rate. These two tradeoffs: simplicity/accuracy and reliability/rejection rate cannot be determined automatically since they depend on the goals of the user.

Crisp rule based system does not allow to evaluate the probabilities of classification and, since $\mathcal{P}(C_i, C_j)$ depends in a discontinuous way on the parameters of linguistic variables non-gradient minimization methods are required. Real input values $\mathbf{X}$ are obtained by measurements that are carried with finite precision. The brain uses not only large receptive fields for categorization, but also small receptive fields to extract feature values. Instead of a crisp number X a Gaussian distribution $G_X = G(Y; X, S_X)$ centered around X with dispersion S_X should be used. Probabilities $p(C_i|\mathbf{X}; M)$ may be computed for any classification model M by performing a Monte Carlo sampling from the joint Gaussian distribution for all continuous features $G_{\mathbf{X}} = G(\mathbf{Y}; \mathbf{X}, \mathbf{S}_X)$. Dispersions $\mathbf{S}_X = (s(X_1), s(X_2) \ldots s(X_N))$ define the volume of the input space around $\mathbf{X}$ that has an influence on computed probabilities. One way to "explore the neighborhood" of $\mathbf{X}$ and see the probabilities of alternative classes is to increase the fuzziness $\mathbf{S}_X$ defining $s(X_i) = (X_{i,max} - X_{i,min})\rho$, where the parameter ρ defines a percentage of fuzziness relatively to the range of X_i values.

With increasing ρ values the probabilities $p(C_i|\mathbf{X};\rho,M)$ change. Even if a crisp rule-based classifier is used non-zero probabilities of classes alternative to the winning class will gradually appear. The way in which these probabilities change shows how reliable is the classification and what are the alternatives worth remembering. If the probability $p(C_i|\mathbf{X};\rho,M)$ changes rapidly around some value ρ_0 the case $\mathbf{X}$ is near classification border and an analysis of $p(C_i|\mathbf{X};\rho_0,s_i,M)$ as a function of each $s_i = s(X_i), i = 1\ldots N$ is needed to see which features have strong influence on classification. Displaying such probabilites allows for a detailed evaluation of the new data also in cases where analysis of rules is too complicated. A more detailed analysis of these probabilities based on *confidence intervals* and *probabilistic confidence intervals* has recently been presented by Jankowski [24,9]. Confidence intervals are calculated individually for a given input vector while logical rules are extracted for the whole *training set*. Confidence intervals measure maximal deviation from the given feature value X_i (assuming that other features of the vector $\mathbf{X}$ are fixed) that do not change the most probable classification of the vector $\mathbf{X}$. If this vector lies near the class border the confidence intervals are narrow, while for vectors that are typical for their class confidence intervals should be wide. These intervals facilitate precise interpretation and allow to analyze the stability of sets of rules.

For some classification models probabilities $p(C_i|\mathbf{X};\rho,M)$ may be calculated analytically. For the crisp rule classifiers [25] a rule $R_{[a,b]}(X)$, which is true if $X \in [a,b]$ and false otherwise, is fulfilled by a Gaussian number G_X with probability:

$$p(R_{[a,b]}(G_X) = T) \approx \sigma(\beta(X-a)) - \sigma(\beta(X-b)) \tag{2}$$

where the logistic function $\sigma(\beta X) = 1/(1+\exp(-\beta X))$ has $\beta = 2.4/\sqrt{2}s(X)$ slope. For large uncertainty $s(X)$ this probability is significantly different from zero well outside the interval $[a,b]$. Thus crisp logical rules for data with Gaussian distribution of errors are equivalent to fuzzy rules with "soft trapezoid" membership functions defined by the difference of the two sigmoids, used with crisp input value. The slope of these membership functions, determined by the parameter β, is inversely proportional to the uncertainty of the inputs. In the C-MLP2LN neural model [13] such membership functions are computed by the network "linguistic units" $L(X;a,b) = \sigma(\beta(X-a)) - \sigma(\beta(X-b))$. Relating the slope β to the input uncertainty allows to calculate probabilities in agreement with the Monte Carlo sampling. Another way of calculating probabilities, based on the softmax neural outputs $p(C_j|\mathbf{X};M) = O_j(\mathbf{X})/\sum_i O_i(\mathbf{X})$ has been presented in [9].

After uncertainty of inputs has been taken into account probabilities $p(C_i|\mathbf{X};M)$ depend in a continuous way on intervals defining linguistic variables. The error function:

$$E(M, \mathbf{S}) = \frac{1}{2} \sum_{\mathbf{X}} \sum_{i} (p(C_i|\mathbf{X}; M) - \delta(C(\mathbf{X}), C_i))^2 \tag{3}$$

depends also on uncertainties of inputs $\mathbf{S}$. Several variants of such models may be considered, with Gaussian or conical (triangular-shaped) assumptions for input distributions, or neural models with bicentral transfer functions in the first hidden layer. Confusion matrix computed using probabilities instead of the number of yes/no errors allows for optimization of Eq. (1) using gradient-based methods. This minimization may be performed directly or may be presented as a neural network problem with a special network architecture. Uncertainties s_i of the values of features may be treated as additional adaptive parameters for optimization. To avoid too many new adaptive parameters optimization of all, or perhaps of a few groups of s_i uncertainties, is replaced by common ρ factors defining the percentage of assumed uncertainty for each group.

This approach leads to the following important improvements for any rule-based system:

- Crisp logical rules provide basic description of the data, giving maximal comprehensibility.
- Instead of 0/1 decisions probabilities of classes $p(C_i|\mathbf{X}; M)$ are obtained.
- Inexpensive gradient method are used allowing for optimization of very large sets of rules.
- Uncertainties of inputs s_i provide additional adaptive parameters.
- Rules with wider classification margins are obtained, overcoming the brittleness problem of some rule-based systems.

Wide classification margins are desirable to optimize the placement of decision borders, improving generalization of the system. If the vector $\mathbf{X}$ of an unknown class is quite typical to one of the classes C_k increasing uncertainties s_i of X_i inputs to a reasonable value (several times the real uncertainty, estimated for a given data) should not decrease the $p(C_k|\mathbf{X}; M)$ probability significantly. If this is not the case $\mathbf{X}$ may be close to the class border and analysis of $p(C_i|\mathbf{X}; \rho, s_i, M)$ as a function of each s_i is needed. These probabilities allow to evaluate the influence of different features on classification. If simple rules are available such explanation may be satisfactory. Otherwise to gain understanding of the whole data a similarity-based approach to classification and explanation is worth trying. Prototype vectors R_i are constructed using a clusterization, dendrogram or a decision tree algorithm and a similarity measure $D(\mathbf{X}, \mathbf{R})$ is introduced. Positions of the prototype vectors R_i, parameters of the similarity measures $D(\cdot)$ and other adaptive parameters of the system are then optimized using a general framework for similarity-based methods [26]. This approach includes radial basis function

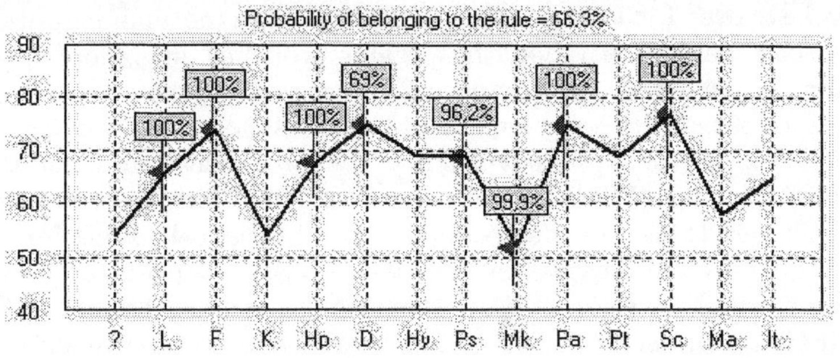

Fig. 1. An example of a rule with Gaussian uncertainties of conditions displayed.

networks, clusterization procedures, vector quantization methods and generalized nearest neighbor methods as special examples. An explanation in this case is given by pointing out to the similarity of the new case **X** to one or more of the prototype cases R_i.

The final step for data explanation requires visualization of the data in the neighborhood of **X**. Although there are many interesting methods of vizualization (cf. [22]) its seems that the most appropriate here are the methods based on multi-dimensional scaling (MDS, [27]). In essence these methods try to present a map of N-dimensional objects in two (or three) dimensions, trying to preserve all distance relations $D(\mathbf{X}_i, \mathbf{X}_j)$ in the original space. This is achieved by minimization of some form of topographical measure of distortion, for example the Stress $S(\mathbf{x})$ introduced by Kruskal:

$$S(\mathbf{x}) = \sum_{ij}^{N_t} w_{ij} \cdot \left(\hat{D}(\mathbf{X}_i, \mathbf{X}_j) - d(x_i, x_j)\right)^2 \tag{4}$$

where $d(x_i, x_j)$ are distances in the low-dimensional target space and w_{ij} are user-defined weights. Similar formula is used [28] to map a single point relatively to an existing map. An interactive program for such vizualization has been written [28], allowing for creation of maps (cf. Fig. 4) by zooming on the neighborhood of **X**. The zoomed MDS diagrams may be correlated with estimations of probability $p(C_i|\mathbf{X};\rho,M)$ and with more detailed $p(C_i|\mathbf{X};\rho,s_i,M)$ graphs presenting confidence intervals, giving better evaluation of new data.

4 Real-life examples

Two examples are shown here: psychometric data [29] and the Telugu vovel data [5]. In the first case rule-based approach to data understanding has been successful, in the second a prototype-based approach followed by visualization has been necessary.

The psychometric data [29] has 14 continuous attributes obtained by combining answers to questionnaires. They measure tendencies towards hypochondria, depression, hysteria, psychopathy, paranoia, schizophrenia, etc. The datasets have over 1000 cases belonging to 27 classes (normal, neurotic, drug addicts, schizophrenic, psychopaths, organic problems, malingerers, persons with criminal tendencies etc.) determined by expert psychologists. Crisp logical rules for these data were generated using C4.5 classification tree [16], a very good classification system which may generate logical rules, and the Feature Space Mapping (FSM) neural network [20,21], since these two systems were the easiest to use on such complex data. Statistical estimation of generalization by 10-fold crossvalidation gave for crisp unoptimized FSM rules 82-85% correct answers, for C4.5 decision tree the accuracy was in the 79-84% range. If Gaussian uncertainty is included rule conditions become fuzzy (see Fig. 1) and optimization of probabilities improves the FSM crossvalidation results up to 90-92%. The IncNet model, an ontogenic neural network [30], obtained 93-95% accuracy in crossvalidation tests. Although it is not a rule-based system calculation of probabilities and visualization described below allows to evaluate the new data quite well also with such black-box systems.

Fig. 2 shows the probabilities of a well-behaved case belonging to the "organic problem class" as a function of ρ. The two plots in Fig. 3 show the probability as a function of two feature values for the case in which a sharp change in probabilities with increasing ρ has been observed, essentially mixing psychopathy, paranoia and schizophrenia classes. Probabilities for the most probable class at the measured $\mathbf{X}$ value and for the second most probable class are displayed. Seeing the mixture of schizophrenia and paranoia in Fig. 3 an expert may inferred that it is the case of schizoidal paranoia, which was not present as one of the classes in the data base.

The Telugu vovel data [5] give an example for which no rule-based understanding of the data is possible. Each vovel is described by intensities of 3 formants, therefore one may view the data in 3 dimensions and notice a strong overlap among the 6 classes – in Fig. 4 MDS maps of this data are displayed. The results of some tests are collected in Table 1. Unfortunately the

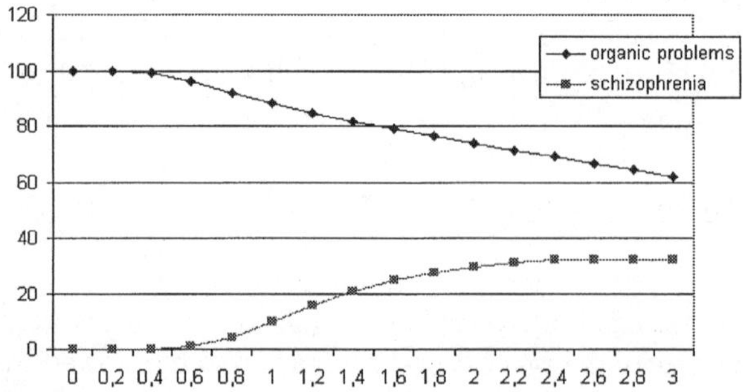

Fig. 2. Probabilities of different classes as a function of assumed uncertainties ρ shown in percents of the input feature range; this case is uniquely classified as "organic problems".

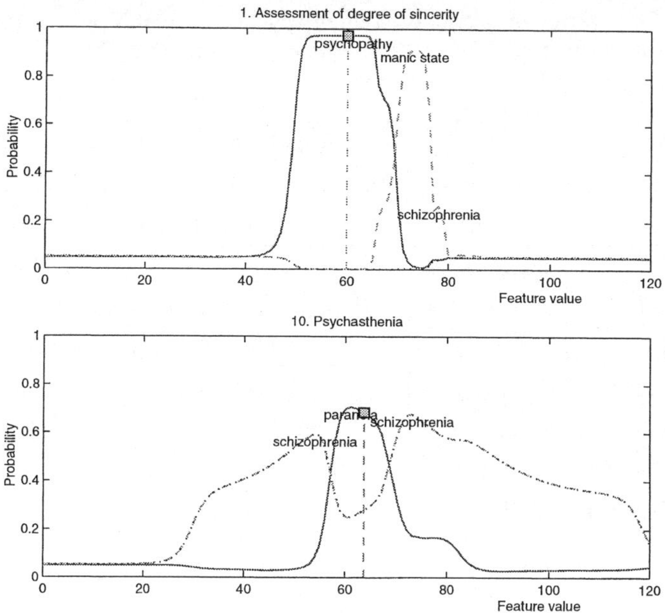

Fig. 3. Psychopathy is quite probable here, but small perturbations of feature 1 change it into a maniac state. In the lower picture paranoia diagnosis lies within the schizophrenia class.

variance has not been given in ref. [5] and since only two-fold crossvalidation was done for some models it may be quite large (even well-behaved models show on this dataset differences exceeding 5% due to the data partitioning). Using fuzzy rules or other sophisticated logical description in this case does not explain anything. A possible explanation of what makes one speech sound different than the other could either invoke a rule – but the number of rules produced by different systems is rather large and they are not too accurate, so no good rules exist here – or evaluate similarity to a prototype case.

FSM [20,21] neurofuzzy system with rectangular functions produces a large (75 or more) number of neurons, giving the 10-fold stratified crossvalidation (10xCV) accuracy of about 80%. The simplest FSM prototype-based solutions with Gaussian weighting of the Euclidean distance has 6 prototypes (one per class) and in the 10xCV test it gives 72.4% accuracy (74% on the whole dataset). Although this accuracy is not the highest 6 prototypes seem to be sufficient to explain main features of this data. Adding more neurons to the FSM network (the algorithm is constructive and the number of neurons is regulated by the desired accuracy) is equivalent to using more than one prototype per class.

Table 1. Comparison of results on Telugu vovel data. 10xCV means 10-fold stratified crossvalidation test, Manhattan is the distance function.

System	Accuracy	Remarks
kNN	88.1± 0.4%	k=3, Manhattan, 10xCV, this paper
FSM	87.4 ± 0.5%	Gauss, about 65 neurons, 10xCV, this paper
SSV	86.0%	22 rules, 10xCV, this paper
kNN	86.1± 0.6%	k=3, Euclidean, 2xCV, this paper
FSM	85.2 ± 1.2%	Gauss, over 40 neurons, 2xCV, this paper
MLP	84.6 %	2xCV, 10 neuronw, Ref. [5]
Fuzzy MLP	84.2 %	2xCV, 10 neuronw, Ref. [5]
SSV	83.3± 0.9 %	2xCV, beam search, this paper
Bayes Classifier	79.2 %	2xCV, Ref. [5]
Fuzzy Kohonen	73.5 %	2xCV, Ref. [5]

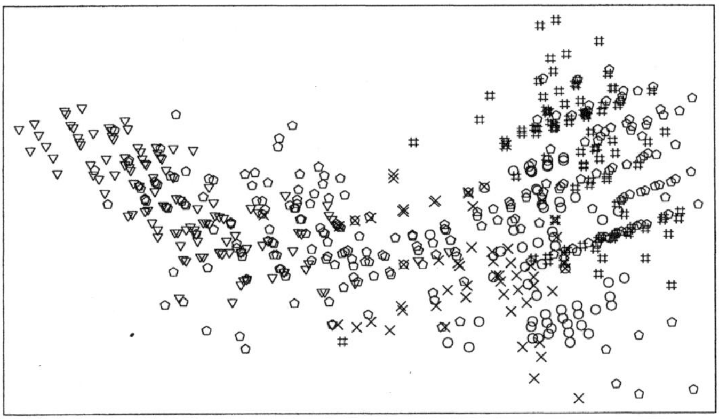

Fig. 4. The MDS map of the Telugu vovel data – classes strongly overlap.

5 Discussion

Acceptable explanations depend on the particular domain and the type of data analyzed. Machine Learning community has focused on artificial cases with a few symbolic attributes, for which simple logical rules are sufficient. Rule-based classifiers are useful only if rules are reliable, accurate, stable and sufficiently simple to be understood. In real data mining problems many continuous-valued attributes may be present and large sets of rules may be needed. Most classifiers are unstable [31] and lead to rules that are significantly different if the training set is slightly changed. Such rules contain little useful information and in fact may be rather misleading. Even if stable and robust rules are found [13] the user should be warned about potential misclassifications, other probable classification possibilities and influence of each feature on the classification probability.

Interpretation of new cases requires something more than a set of rules and yes/no decisions. Through correlation of rules, investigation of class probabilities in the neighborhood of a new case, confidence intervals for interesting features and interactive visualization the new data may be understood much better in relation to the whole database of known cases. For some classifiers probabilities $p(C_k|\mathbf{X};\rho,M)$ may be estimated analytically, for other classifiers Monte Carlo estimations are necessary. Since one new case is evaluated at a time the cost of such simulations is not so important. In practical applications users are interested in relevant features and may rarely be satisfied with answers to questions "why" based on quotation of logical rules belonging to a complex set. Similarity to prototypes, or case-based interpretation, is an alternative to logical rule-based systems. Expert systems in medical and other fields should help to understand and evaluate new cases, not just classify it.

Acknowledgement: The kNN program was written by Karol Grudziński, FSM by Rafał Adamczak, SSV by Krzysztof Grabczewski, Fig. 3 are from Norbert Jankowski, and Fig. 3 was made with program written by Antoine Naud, all of the Dept. of Computer Methods, Nicholas Copernicus University.

References

1. Michalski, R.S. (1983) A Theory and Methodology of Inductive Learning. Artificial Intelligence **20**, 111–161
2. Michalski, R.S., Bratko, I. and Kubat, M. (eds.), Machine Learning and Data Mining: Methods and Applications, London, John Wiley and Sons, 1997.
3. Piatetsky-Shapiro, G. (2000) Knowledge Discovery in Databases: 10 years after. SIGKDD Explorations **1**: 59–61
4. Nauck, D., Klawonn, F., Kruse, R. (1997) Foundations on Neuro-Fuzzy Systems. J. Wiley, New York
5. Pal, S.K. and Mitra S. (1999) *Neuro-Fuzzy Pattern Recognition.* J. Wiley, New York
6. Michie, D., Spiegelhalter, D. J. Taylor, C. C. (eds.) (1994) Machine Learning, Neural and Statistical Classification. Ellis Horwood, New York.
7. Kosko B. (1992) Neural Networks and Fuzzy Systems. Prentice Hall
8. Duch, W., Adamczak, R., Grabczewski, K. (1999) Neural optimization of linguistic variables and membership functions. Int. Conference on Neural Information Processing (ICONIP'99), Perth, Australia, Nov. 1999, Vol. II, pp. 616–621
9. Duch, W., Jankowski, N., Adamczak, R., Grabczewski, K. (2000) Optimization and Interpretation of Rule-Based Classifiers. Intelligent Information Systems IX, Springer Verlag (in print)
10. Haykin, S. (1994) Neural networks: a comprehensive foundations. MacMillian.
11. Duch, W., Jankowski, N. (1999) New neural transfer functions. Neural Computing Surveys **2**, 639–658

12. Duch, W., Adamczak, R., Grabczewski, K. (1998) Extraction of logical rules from backpropagation networks.
13. Duch, W., Adamczak, R., Grabczewski, K. (in print) Methodology of extraction, optimization and application of crisp and fuzzy logical rules. IEEE Transactions on Neural Networks. Neural Processing Letters **7**, 1-9
14. Bennett K.P., Mangasarian, O.L. (1992) Robust linear programming discrimination of two linearly inseparable sets. Optimization Methods and Software **1**, 23–34
15. Mertz, C.J., Murphy, P.M. UCI repository of machine learning databases, http://www.ics.uci.edu/pub/machine-learning-data-bases.
16. Quinlan J.R. (1993) C4.5: Programs for machine learning. San Mateo, Morgan Kaufman
17. Grąbczewski, K., Duch, W. (1999) A general purpose separability criterion for classification systems. 4th Conf. on Neural Networks and Their Applications, Zakopane, Poland, May 1999, pp. 203–208
18. Buckley, J.J., Hayashi, Y., Czogala, E. (1993) On the Equivalence of Neural Nets and Fuzzy Expert Systems. Fuzzy Sets and Systems **53**, 129–134.
19. Yen, J., Langari, R., Zadeh, L.A. (eds) Industrial Applications of Fuzzy Logic and Intelligent Systems. IEEE Press, New York, 1995.
20. Duch, W., Diercksen, G.H.F. (1995) *Feature Space Mapping as a universal adaptive system*, Computer Physics Communication **87**, 341–371
21. Duch, W., Adamczak, R., Jankowski, N. (1997) New developments in the Feature Space Mapping model. 3rd Conf. on Neural Networks, Kule, Poland, Oct. 1997, pp. 65-70
22. Kay, J.W., Titterington, D.M., (1999) Statistics and neural networks. Oxford University Press, U.K.
23. Hastie T., Stuetzle, W. (1989) Principal curves. J. Am. Stat. Assoc. **84**, 502–516
24. Jankowski, N., (1999) Ontogenic neural networks and their applications to classification of medical data. PhD thesis (in Polish), Department of Computer Methods, Nicholas Copernicus University, Toruń, Poland
25. Duch, W., Adamczak, R., Grabczewski, K. (1999) Methodology of extraction, optimization and application of logical rules. Intelligent Information Systems VIII, Ustroń, Poland, 14-18.06.1999, pp. 22-31
26. Duch, W. (1998) A framework for similarity-based classification methods, Intelligent Information Systems VII, Malbork, Poland, June 1998, pp. 288–291
27. Kruskal, J. B (1964) Non metric multidimensional scaling : a numerical method. Psychometrika **29**, 115–129.
28. Naud, A., Duch, W., (2000) Interactive data exploration using MDS mapping. 5th Conf. on Neural Networks and Soft Computing, Zakopane, Poland, June 2000 (in print)
29. Duch W., Adamczak R., Grabczewski K. (1999) Neural methods for analysis of psychometric data. Proc. of Enginnering Applications of Neural Networks (Duch W, ed.), Warsaw, Poland, Sept. 1999, pp. 45-50
30. Jankowski, N., Kadirkamanathan, V. (1997) Statistical control of RBF-like networks for classification. 7th Int. Conf. on Artificial Neural Networks, Lausanne, Switzerland 1997, pp 385-390, Springer Verlag.
31. Breiman L. (1998) Bias-Variance, regularization, instability and stabilization. In: C. Bishop (ed.) Neural Networks and Machine Learning. Springer Verlag

On the Model-Theoretic Semantics of Fuzzy Set Logic for Dealing with Inconsistency

Tru Hoang Cao

Artificial Intelligence Group, Department of Engineering Mathematics, University of Bristol, United Kingdom BS8 1TR, Tru.Cao@bristol.ac.uk

Abstract: Inconsistency is likely to exist in a knowledge base built up from various sources of information. In such a case, inconsistency is often local and thus should not cause the entire knowledge base to collapse. For a fuzzy logic system to be able to deal with local inconsistency, an appropriate model-theoretic semantics is required. This paper analyses and shows that the model-theoretic semantics approach of possibilistic logic, which has recently become popular to be applied to the development of fuzzy set logic whose formulas can have fuzzy set constants, cannot deal with local inconsistency. We then introduce the annotated logic-based approach as a solution, and discuss on the difference and relation between the two approaches.

1 Introduction

Inconsistency can exist in knowledge bases. It is likely when a knowledge base is built up from various sources of information and when uncertainty and fuzziness are present. For example, in acquiring knowledge from experts, it is very common for the experts to disagree on several points. Dubois, Lang and Prade ([7]) also argued that it is not fully satisfactory to have a logic that deals with uncertainty but does not allow inconsistency. Further, inconsistency when occuring is often local rather than global on a whole knowledge base and thus it should not cause the entire system to collapse.

However, it appears that most of the works so far on fuzzy logic systems do not address the issue or do not assume the presence of inconsistency. Here, our attention is focused on the most expressive type of fuzzy logic systems whose formulas can have fuzzy set constants, which we call *fuzzy set logic*. Examples of fuzzy set logic systems are those of [13], [12], [2], [10], [8], [14], [6], and [1].

In another aspect, it is desirable that a fuzzy logic system have a formal and well-founded semantics. In particular, it should have a model-theoretic semantics so that the soundness and the completeness of inference rules can be justified. Among the current formal approaches, the one based on possibilistic logic ([7]) has become popular.

However, as first analysed in [5] and presented in Section 2 below, the possibilistic logic-based (PL-based) model-theoretic semantics approach cannot deal with local inconsistency. Then, in Section 3, we introduce the annotated logic-based (AL-based) approach developed in [4] as a solution, and discuss on the difference and relation between the two approaches. Finally, Section 4 presents concluding remarks of the paper.

2 The Possibilistic Logic-Based Approach

Firstly, we recall that in possibilistic logic each classical logic formula is associated with either a possibility degree or a necessity degree, which is a value in [0, 1]. An interpretation I is defined by a possibility distribution π_I on the set Ω of all possible worlds, which are classical logic interpretations. The denotation functions Π and N then respectively define the possibility degree and the necessity degree of a classical logic formula p with respect to I as follows:

$$\Pi(p) = sup_{\omega\in\Omega}\{\pi_I(\omega) \mid \omega \models p\}$$

$$N(p) = 1 - \Pi(\neg p)$$

where $\neg$ is the negation connective in classical logic and $\omega \models p$ means that ω satisfies p with respect to classical logic semantics.

For formulas with fuzzy set constants, Esteva, Garcia-Calvés and Godo ([9]) extended the model-theoretic semantics of possibilistic logic by using many-valued logic interpretations, instead of classical logic ones, as possible worlds, to which a possibility distribution is assigned. That is, in the authors' fuzzy truth-valued logic, an interpretation I is also defined by a possibility distribution π_I on the set Ω of all possible worlds. However, while in possibilistic logic a proposition p is to be or not to be satisfied by a possible world $\omega \in \Omega$, in fuzzy truth-valued logic p has a truth degree $T_p(\omega) \in [0, 1]$. Then, the denotation of p with respect to I is defined to be the fuzzy truth value τ, which is a fuzzy set on [0, 1], such that $\forall u \in [0, 1]$: $\mu_\tau(u) = sup_{\omega\in\Omega}\{\pi_I(\omega) \mid u = T_p(\omega)\}$.

Godo and Vila ([10]) adapted the model-theoretic semantics of fuzzy truth-valued logic by defining the denotation of a fuzzy proposition p (i.e., one that involves fuzzy set constants) with respect to an interpretation I as the relative necessity degree of T_p given π_I defined by:

$$N(T_p \mid \pi_I) = inf_{\omega\in\Omega}\{min\{1, 1 - \pi_I(\omega) + T_p(\omega)\}\} \qquad \text{(Def. 1)}$$

satisfying the intuitive property that $N(T_p \mid \pi_I) = 1$ if and only if $\pi_I \subseteq_f T_p$ (i.e., π_I is a fuzzy subset of T_p). Meanwhile, Alsinet, Godo and Sandri ([1]) used another definition of relative necessity degrees as follows:

$$N(T_p \mid \pi_I) = inf_{\omega\in\Omega}\{max\{1 - \pi_I(\omega), T_p(\omega)\}\} \qquad \text{(Def. 2)}$$

A necessity-valued formula $(p\ \alpha)$ where $\alpha \in [0, 1]$, which expresses "p is certain at least to degree α", is said to be satisfied by I if and only if $N(T_p \mid \pi_I) \geq \alpha$.

The main and common point of the above model-theoretic semantics extending possibilistic logic's one is that an interpretation is defined by a single possibility distribution on the set of all possible worlds. That approach cannot actually deal with local inconsistency, because when such a possibility distribution is subnormal, it imposes a global inconsistency on all propositions involved in possible worlds. As such, it rules out interpretations in that only some propositions are inconsistent, whence a set of formulas with inconsistent ones may have arbitrary logical consequences.

For example, let us consider a set S of two necessity-valued formulas (age(John, A) 1) and (height(Mary, B) 1), where A and B are fuzzy sets on domains $U = [0, 120]$ and $V = [0, 2.4]$, respectively. A possible world can be $\omega =$

{age(John, 30), height(Mary, 1.7),...}, whence the truth degrees of age(John, A) and height(Mary, B) with respect to ω are $T_{\text{age(John, } A)}(\omega) = \mu_A(30)$ and $T_{\text{height(Mary, } B)}(\omega) = \mu_B(1.7)$, where μ_A and μ_B are the membership functions of A and B, respectively.

Now suppose that A is subnormal and B is normal, that is, S has an inconsistency about John's age only. Let I be a model of S. Then I satisfies (age(John, A) 1), i.e., $N(T_{\text{age(John, } A)} \mid \pi_I) = 1$. If (Def. 1) is applied, one has $\pi_I \subseteq_f T_{\text{age(John, } A)}$. Since A is subnormal, $T_{\text{age(John, } A)}$ and thus π_I are also subnormal, where $sup_{\omega\in\Omega}\{\pi_I(\omega)\} \leq sup_{\omega\in\Omega}\{T_{\text{age(John, } A)}(\omega)\} = sup_{u\in U}\{\mu_A(u)\} < 1$. Therefore, I satisfies any (height(Mary, B^*) 1) such that $\forall v \in V$: $\mu_{B^*}(v) \geq sup_{u\in U}\{\mu_A(u)\}$, because $N(T_{\text{height(Mary, } B^*)} \mid \pi_I) = 1$. Since I is an arbitrary model of S, that means any such (height(Mary, B^*) 1) is a logical consequence of S, regardless of the given fact (height(Mary, B) 1). Similarly, if (Def. 2) is applied, then $\pi_I(\omega) = 0$ for every $\omega \in \Omega$, whence $N(T_p \mid \pi_I) = 1$ for any T_p and thus any formula is a logical consequence of S!

This can be regarded as a non-classical logic version of the classical logic problem that no inconsistency is allowed in a model, whence a set of formulas with inconsistent ones has no model and thus everything becomes its logical consequence. In fact, to overcome this problem of classical logic was the initial motivation of the work on annotated logic (e.g. [11]), whereby a set of formulas with inconsistent ones still has models and does not arbitrarily entail everything.

Dubois, Lang and Prade ([7]) presented a remedy for this problem of possibilistic logic as follows. Given a set of necessity-valued formulas of the form $(p\ \alpha)$, the inconsistency degree of the whole set is evaluated. Then only derived formulas $(r\ \beta)$ with β being strictly greater than that inconsistency degree are allowed as *non-trivial* logical consequences of the set. For example, the inconsistency degree of the set $\{(p\ \alpha), (\neg p\ \alpha), (r\ \beta)\}$ is α, and if $\beta \leq \alpha$ then the deduction of $(r\ \beta)$ is not allowed.

This is not reasonable when $(r\ \beta)$ is actually from the set and has nothing to do with p and the local inconsistency of p. Moreover, when there is just a proposition with inconsistency degree 1 in a set of formulas, the problem will be the same as that of classical logic. That is, one has to either accept everything as a logical consequence of the set or allow nothing deduced from it. The system thus collapses, either way. In fact, the authors' solution is based on the property that any arbitrary logical consequence $(r\ \beta)$ of a set of formulas, i.e., one regardless of the formulas in the set, is always with β being less than or equal to the inconsistency degree α of the set. However, as exemplified above, the inverse does not hold and, thus, the condition $\beta > \alpha$ is sufficient but not necessary for a derived formula $(r\ \beta)$ being a non-arbitrary logical consequence of the set.

We argue that, neither such the defined inconsistency degree of a set of formulas nor any other measure can be used to decide whether a logical consequence of the set is arbitrary or not. The problem is radical in the model-theoretic semantics approach of possibilistic logic, which rules out local inconsistency in its definition of interpretations. Moreover, the authors' remedy has a high computation cost. As proved therein, computing the inconsistency degree of a set of propositional necessity-valued formulas requires a logarithmic number of calls to the satisfiability problem, which is NP-complete for propositional logic and undecidable for first-order predicate logic ([3]).

3 The Annotated Logic-Based Approach

In Cao ([4, 5]), we extended classical annotated logic programs to fuzzy ones whose clauses have the following form:

Obj: $H \leftarrow Obj_1$: $B_1 \wedge Obj_2$: $B_2 \wedge \ldots \wedge Obj_n$: B_n [α]

where Obj, Obj_1, …, Obj_n called *objects* are first-order predicate logic atoms or terms, H, B_1, …, B_n are fuzzy sets possibly on different domains, and $\alpha \in [0, 1]$ represents the certainty lower bound of the clause. Below is an example annotated fuzzy logic program, where *very true*, *true*, *not expensive*, and *fairly cheap* are linguistic labels of fuzzy sets:

buy(John, x): *very true* $\leftarrow$ like(John, x): *true* $\wedge$ price(x): *not expensive*
like(John, #36): *very true*
price(#36): *fairly cheap*.

Assuming x and #36 to stand for cars, the rule expresses "If it is *true* that John likes a car and the car's price is *not expensive*, then it is *very true* that he buys it". The first fact expresses "It is *very true* that John likes car #36". The second fact expresses "The price of car #36 is *fairly cheap*".

Extending the model-theoretic semantics of classical annotated logic programs, an interpretation of a fuzzy annotated logic program is defined as a mapping from the given set of all ground objects to the given set of all fuzzy set constants. A ground object can be mapped to a subnormal fuzzy set to represent a local inconsistency about that object only. As such, a program containing locally inconsistent clauses still has models that do not satisfy arbitrary clauses, and thus the program does not arbitrarily entail everything. The SLD-resolution style proof procedure developed therein is sound and complete with respect to the defined declarative semantics.

We now show the difference and relation between interpretations of the PL-based approach and the AL-based approach. For simplicity, let us consider a language with a finite set of n variables $x_1, x_2, \ldots, x_n$ that take values in domains U_1, $U_2, \ldots, U_n$, respectively. In practice, each variable x_i can stand for age(John) or height(Mary) in the example in Section 2, or buy(John, #36) or price(#36) in the example above, for instance.

Then, in the PL-based approach, a possible world $\omega \in \Omega$ maps each x_i to a value $u_i \in U_i$, and thus can be represented by $(u_1, u_2, \ldots, u_n) \in U_1 \times U_2 \times \ldots \times U_n$. As such, a possibility distribution on Ω, defining an interpretation, can be considered as a joint possibility distribution $\pi_{(x_1, x_2, \ldots, x_n)}$ of $(x_1, x_2, \ldots, x_n)$ on $U_1 \times U_2 \times \ldots \times U_n$. Meanwhile, in the AL-based approach, an interpretation maps each x_i to a fuzzy set on U_i or, in other words, it is defined by n marginal possibility distributions π_{x_1}, π_{x_2}, …, π_{x_n} of $x_1, x_2, \ldots, x_n$ on $U_1, U_2, \ldots, U_n$, respectively. For example, Fig.1 below illustrates a PL-based interpretation and an AL-based interpretation in the case $n = 2$, $x_1 =$ age(John), and $x_2 =$ height(Mary).

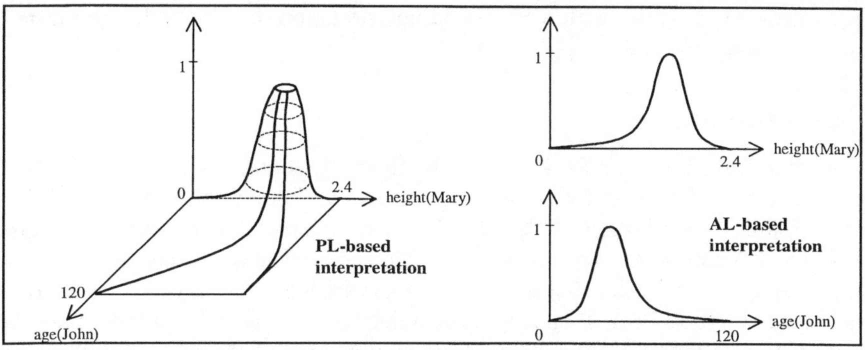

Fig. 1. The difference between PL-based and AL-based interpretations

Therefore, a PL-based interpretation can be obtained from an AL-based interpretation by applying the conjunction principle ([15]) as follows:

$\forall u_1 \in U_1 \forall u_2 \in U_2 \ldots \forall u_n \in U_n$: $\pi_{(x_1,x_2,\ldots,x_n)}(u_1, u_2, \ldots, u_n) = min_{i=1,n}\{\pi_{x_i}(u_i)\}$.

Meanwhile, an AL-based interpretation can be obtained from a PL-based interpretation by applying the projection principle as follows:

$\forall u_i \in U_i$: $\pi_{x_i}(u_i) = sup_{(u_1,\ldots,u_{i-1},u_{i+1},\ldots,u_n)\in U_1\times\ldots\times U_{i-1}\times U_{i+1}\times\ldots\times U_n}\{\pi_{(x_1,x_2,\ldots,x_n)}(u_1, u_2, \ldots, u_n)\}$

for every i from 1 to n.

Furthermore, the denotation of a fuzzy proposition p with respect to a PL-based interpretation I, as defined by (Def. 1) or (Def. 2) in Section 2, is actually equal to the relative necessity degree of the joint possibility distribution induced by p of the variables in p given the one obtained by the projection of π_I defining I on those variables. For a proof, without loss of generality, suppose that the variables in p are $x_1, x_2, \ldots, x_k$ $(1 \le k \le n)$ and the joint possibility distribution induced by p is π_p on $U_1\times U_2\times \ldots \times U_k$. Using (Def. 1), one has:

$$
\begin{aligned}
& N(T_p \mid \pi_I) \\
= \; & inf_{\omega\in\Omega}\{min\{1, 1 - \pi_I(\omega) + T_p(\omega)\}\} \\
= \; & inf_{(u_1,u_2,\ldots,u_n)\in U_1\times U_2\times\ldots\times U_n}\{min\{1, 1 - \pi_{I(x_1,x_2,\ldots,x_n)}(u_1, u_2, \ldots, u_n) + \pi_p(u_1, u_2, \ldots, u_k)\}\} \\
= \; & inf_{(u_1,u_2,\ldots,u_k)\in U_1\times U_2\times\ldots\times U_k}\, inf_{(u_{k+1},\ldots,u_n)\in U_{k+1}\times\ldots\times U_n}\{min\{1, 1 - \pi_{I(x_1,x_2,\ldots,x_n)}(u_1, u_2, \ldots, u_n) \\
& \qquad + \pi_p(u_1, u_2, \ldots, u_k)\}\} \\
= \; & inf_{(u_1,u_2,\ldots,u_k)\in U_1\times U_2\times\ldots\times U_k}\{min\{1, 1 - sup_{(u_{k+1},\ldots,u_n)\in U_{k+1}\times\ldots\times U_n}\{\pi_{I(x_1,x_2,\ldots,x_n)}(u_1, u_2, \ldots, u_n)\} \\
& \qquad + \pi_p(u_1, u_2, \ldots, u_k)\}\} \\
= \; & inf_{(u_1,u_2,\ldots,u_k)\in U_1\times U_2\times\ldots\times U_k}\{min\{1, 1 - \pi_{I(x_1,x_2,\ldots,x_k)}(u_1, u_2, \ldots, u_k) + \pi_p(u_1, u_2, \ldots, u_k)\}\} \\
= \; & N(\pi_p \mid \pi_{I(x_1,x_2,\ldots,x_k)}).
\end{aligned}
$$

The proof with (Def. 2) is similar.

The significance of this result is that, in previous works of the PL-based approach, AL-based interpretations can be used instead of PL-based ones, overcoming the discussed local inconsistency problem. Whereby, for evaluating fuzzy proposition denotations, such a $\pi_{I(x_1,x_2,\ldots,x_k)}$ in the proof above is obtained by the conjunction of the marginal possibility distributions $\pi_{x_1}, \pi_{x_2}, \ldots, \pi_{x_k}$ of an AL-based interpretation, which is normal if all $\pi_{x_1}, \pi_{x_2}, \ldots, \pi_{x_k}$ are normal, regardless of

$\pi_{x_{k+1}}, \ldots, \pi_{x_n}$. Meanwhile, with a PL-based interpretation, $\pi_{I(x_1,x_2,\ldots,x_k)}$ is subnormal if π_I is subnormal.

4 Conclusion

We have analysed that the PL-based model-theoretic semantics approach for fuzzy set logic cannot deal with local inconsistency, as an interpretation is defined by a joint possibility distribution of all variables in a universe of discourse. Meanwhile, the AL-based approach overcomes the problem by defining an interpretation as a set of marginal possibility distributions of the variables. We have also shown the relation between PL-based and AL-based interpretations, and proved that AL-based interpretations can be applied to previous works of the PL-based approach, in particular for evaluating the denotation of a fuzzy proposition with respect to an interpretation.

Our work using the AL-based approach has, however, considered only Horn-like clauses. In order to obtain a full-fledged fuzzy set logic that can deal with inconsistency, the semantics of general formulas possibly with negation and disjunction connectives has to be studied and defined. Further, for greater expressiveness, the logic should be extended with variable certainty weights on formulas as in [1]. These are among the topics that we are currently investigating.

References

1. Alsinet, T., Godo, L. and Sandri, S. (1999), On the semantics and automated deduction for PLFC, a logic of possibilistic uncertainty and fuzziness. In Proceedings of the 15th Conference on Uncertainty in Artificial Intelligence.
2. Baldwin, J.F., Martin, T.P. and Pilsworth, B.W. (1995), *Fril - Fuzzy and Evidential Reasoning in Artificial Intelligence*. Research Studies Press.
3. Börger, E. 1989. *Computability, Complexity, Logic*. North-Holland Publishing Company.
4. Cao, T.H. (2000), Annotated fuzzy logic programs. *International Journal for Fuzzy Sets and Systems*, 113 (2), 277-298.
5. Cao, T.H. (1999), Foundations of order-sorted fuzzy set logic programming in predicate logic and conceptual graphs. PhD Thesis, University of Queensland.
6. Cao, T.H. and Creasy, P.N. (1998), Fuzzy order-sorted logic programming in conceptual graphs with a sound and complete proof procedure. In Mugnier, M.L. and Chein, M. (eds), *Conceptual Structures: Theory, Tools and Applications*, LNAI 1453, Springer-Verlag, pp. 270-284.
7. Dubois, D., Lang, J. and Prade, H. (1994), Possibilistic logic. In Gabbay, Dov M. et al. (eds), *Handbook of Logic in Artificial Intelligence and Logic Programming*, Vol. 3, Oxford University Press, pp. 439-514.
8. Dubois, D., Prade, H. and Sandri, S. (1998), Possibilistic logic with fuzzy constants and fuzzily restricted quantifiers. In Martin, T.P. and Arcelli Fontana, F. (eds), *Logic Programming and Soft Computing*, Research Studies Press, pp. 69-90.

9. Esteva, F., Garcia-Calvés, P. and Godo, L. (1994), Relating and extending semantical approaches to possibilistic reasoning. *International Journal of Approximate Reasoning*, 10, 311-344.
10. Godo, L. and Vila, L. (1995), Possibilistic temporal reasoning based on fuzzy temporal constraints. In Proceedings of the 14th International Joint Conference on Artificial Intelligence, pp. 1916-1922.
11. Kifer, M. and Lozinskii, E.L. 1992. A logic for reasoning with inconsistency. *Journal of Automated Reasoning*, 9, 179-215.
12. Li, D. and Liu, D. (1990), *A Fuzzy Prolog Database System*. Research Studies Press.
13. Umano, M. 1987. Fuzzy set Prolog. In Preprints of the 2nd International Fuzzy Systems Association Congress, pp. 750-753.
14. Virtanen, H.E. (1998), Linguistic logic programming. In Martin, T.P. and Arcelli Fontana, F. (eds), *Logic Programming and Soft Computing*, Research Studies Press, pp. 91-110.
15. Zadeh, L.A. 1979. A theory of approximate reasoning. *Machine Intelligence*, 9, 149-194.

Fuzzy Data Mining from Multidimensional Databases

Anne Laurent, Bernadette Bouchon-Meunier, Anne Doucet, Stéphane Gançarski, and Christophe Marsala

LIP6, mailbox 169, Université Pierre et Marie Curie - Paris 6, 4 place Jussieu 75252 Paris Cedex 05, France – E-mail: Anne.Laurent@lip6.fr

Abstract: Most of the existing learning systems work on data that are stored in poorly structured files. This approach prevents them from dealing with data from real world, which is often heterogeneous and massive and which requires database management tools. In this article, we propose an original solution to data mining which integrates a fuzzy learning tool that constructs fuzzy decision trees with a multidimensional database management system.

1 Introduction

Fuzzy decision tree based methods provide good tools to discover knowledge from data. They are equivalent to a set of *if-then* rules and are *declarative* since the classification they propose may be explained. Moreover the use of fuzzy set theory allows the treatment of numerical values in a more natural way. But most existing solutions to construct decision trees use files and it is wellknown that this approach is reasonable only if the amount of data used for knowledge discovery is rather small (e.g. fits in core memory). Often, these methods are not appropriate for data mining, which aims at discovering non trivial knowledge from very large real-world data stored in data warehouses which require specific tools such as database management systems (DBMS).

When data are stored in a DBMS, there is a need to retrieve efficiently data relevant to data mining processes based on inductive learning methods. This is the reason why the integration of inductive learning algorithms is a crucial point in order to build efficient data mining applications. On-Line Analytical Processing (OLAP) provides means to implement solutions for extracting relevant knowledge (e.g. aggregated data values) from databases.

However, it has been shown that the relational data model does not fit OLAP approach and a new model has emerged: multidimensional databases [5]. Multidimensional databases and OLAP provide means to execute very complex queries, working on a large amount of data that are processed on an aggregated level, and not as the individual level of records. This kind of databases offer performance advantages, especially for multidimensional aggregate computations, which are highly interesting if we consider the automatic inductive learning such as, for instance, decision tree construction from data.

In this paper, we propose a new approach to perform OLAP-based mining using multidimensional DBMS and fuzzy decision trees. In Section 2 and in Section 3, we recall basic notions regarding multidimensional databases, OLAP, and fuzzy decision trees. In Section 4, we present our system of construction of fuzzy decision trees from a multidimensional database by means of a DBMS. In Section 5, we highlight the interest of the use of fuzzy decision trees in a data mining process. Implementation and obtained results on a large database are presented in Section 6, and we conclude with some future work.

2 Multidimensional databases and OLAP

The relational model of data provides efficient tools to store and deal with large amount of data at an individual level. The OLAP terminology has been proposed by Codd [5] for the kind of technologies providing means to collect, store, and deal with multidimensional data, with a view to deploying analysis processes. In the OLAP framework, data are stored in data hypercubes (simply called cubes). A cube is a set of data organized as a multidimensional array of values representing measures over several dimensions. Hierarchies may be defined on dimensions to organize data on more than one level of aggregation.

A cube B is designated by means of m dimensions, each dimension associated with a domain D_i of values, and by a set of elements $S_B(d_1, ..., d_m)$ (with $d_i \in D_i$, $i = 1, ..., m$) belonging to a set V of values. Algebraic operations have been defined on hyper cubes (e.g. [1]) in order to visualize and analyze them: roll up, drill down, slice, dice, rotate, switch, split nest, push and more conventional operations such as join, union, etc.

Consolidation is used in order to speed up queries. It consists in precomputing all or part of the cube with an aggregation function. Some of the most expensive computations are made before the query and their result is stored in the database in order to be used when needed (see [4], [6], [8]).

3 Fuzzy decision trees

Decision trees are well-known tools to represent knowledge and several inductive learning methods exist to construct a decision tree from a training set of data [9]. The use of fuzzy set theory enhances the understandability of decision trees when considering numerical attributes. Moreover, it enables to take into account imprecise values. Decision trees can be generalized into fuzzy decision trees when considering fuzzy values as labels of edges of the tree. Thus, classical methods have been adapted to handle fuzzy values either during their construction or when classifying new cases [2].

Let $\mathcal{A} = \{A_1, ..., A_N\}$ be a set of attributes and let $\mathcal{C} = \{c_1, ..., c_K\}$ be a set of classes. $\mathcal{A}$ and $\mathcal{C}$ enable the construction of *examples*: each example

e_i is composed by a *description* (a N-tuple of attribute-value pairs (A_j, v_{jl})) associated with a particular class c_k from $\mathcal{C}$.

Given a training set $\mathcal{E} = \{e_1, ..., e_n\}$ of examples, a (fuzzy) decision tree is constructed from the root to the leaves, by successive partitioning of the training set into subsets. The construction process can be split in 3 fundamentals steps [3]. An attribute is selected thanks to a *measure of discrimination* H (*step 1*) that orders the attributes according to their accuracy with regards to the class [7]. The partitioning is done by means of a splitting strategy P (*step 2*). A stopping criterion T enables us to stop splitting a set and to construct a leaf in the tree (*step 3*). To build fuzzy decision trees, the commonly used H measure is the star-entropy measure defined as:

$$H^*(\mathcal{C}|A_j) = -\sum_{l=1}^{L} P^*(v_{jl}) \sum_{k=1}^{K} P^*(c_k|v_{jl}) \log(P^*(c_k|v_{jl}))$$

where $v_{j1}, ..., v_{jL}$ are values from $\mathcal{E}$ for attribute A_j. This measure is obtained from the Shannon measure of entropy by introducing Zadeh's probability measure P^* of fuzzy events.

The algorithm to construct fuzzy decision trees is implemented in the Salammbô system. The output of the system is a fuzzy decision tree that can be considered as a set of discovered classification rules [3].

4 Fuzzy decision tree construction from a multidimensional database

The multidimensional database management system may either send only basic information on data, or compute complex fuzzy operations and aggregations, or any of the intermediate solutions.

In the first case, the integration of other inductive learning applications will be easier. They indeed all require at lowest level simple aggregate computation such as *count* (*e.g.* frequentist computations). So the exchanged basic statistics we propose will still be needed. All the other more complex computations will remain outside of the multidimensional database management system which will thus be generic.

On the other hand, the second solution enhances the aggregates' computations (*e.g.* computation of complex functions as, for instance, entropy measure of a set) since multidimensional database management systems are designed for this kind of operations.

In our system, an interface is implemented that consists in exchanging, for each node of the tree, statistics on the data associated with the current node. These data may either be singleton values or intervals, which enable us to construct fuzzy decision trees. These statistics are computed using OLAP queries to extract a sub-cube from the database and fetching statistics on it with aggregation functions. As detailed prevously, they either concern basic *count* on the cells in the extracted data cube or may require more complex calculus, such as entropy computation. These statistics are then used by Salammbô to choose the best attribute to go ahead in the construction of the tree. Then *step 1* of the decision tree construction process (see part 3) is replaced by the following process (see Fig. 1):

step 1.1 Salammbô sends a query specifying the current node to develop in the tree. This query is composed by a set of questions on values of attributes that compose the path from the root to this node.

step 1.2 The interface program queries the data cube and returns for each attribute value, for each class value, the corresponding statistics computed from instances of the database that fulfill the Salammb' query.

step 1.3 Salammbô receives the statistics and chooses the best attributes. Depending on the solution, exchanged statistics are either frequentist count of instances, or discrimination measure of each attribute.

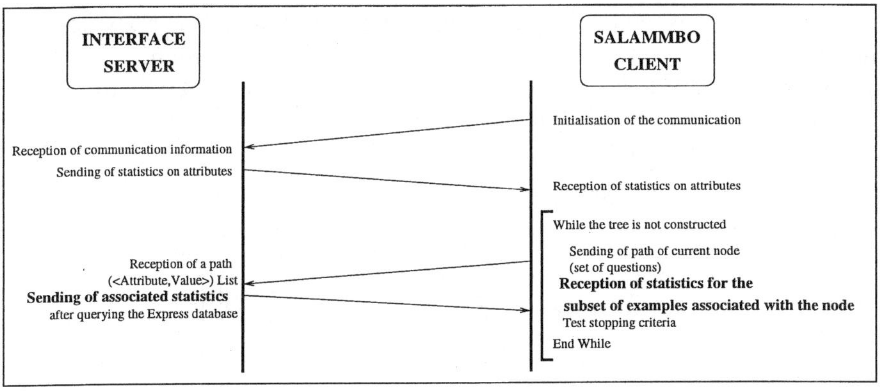

Fig. 1. Processus of Fuzzy Decision Tree Construction

5 Interest of fuzzy knowledge

As previously mentioned, a fuzzy decision tree is equivalent to a fuzzy rule base. In real-world applications, to describe numerical data, or to handle imprecise data, fuzzy knowledge has been shown more understandable than classical knowledge. The fuzzy decision tree constructed from a training set composed by numerical data is rather smaller, in size, than classical decision tree that can be constructed from the same training set. So, the interest of fuzzy decision trees lies in their greater understandability than classical decision trees. Moreover, experiments have been conducted that highlight the greater generalisation power of such trees than classical ones.

We place ourselves in the process of knowledge discovery to help users or experts in understanding data and to highlight high-level relations between data. In this kind of process, the interest of the induced knowledge lies in its understandability. The induced rules are given to experts of the domain and are used as synthetic knowledge of the whole database. This approach of using induced knowledge is similar to the approach based on the computation of statistics on the database to gain understandability of the data. On the other hand, such discovered knowledge can be used later in the database management system itself. For instance, fuzzy decision trees can be used as a set of integrity constraints. Or, it can be associated with the database as external knowledge that enables particular queries to give more information on data. Moreover, in OLAP sessions, fuzzy rules can be given to users.

6 Implementation and results

This system has been developed through a client/server architecture, including a multidimensional database management system (Oracle Express) on SunOS, an Oracle relational database server on a PC computer under Microsoft Windows NT, a C++ interface on Windows NT, and Salammbô running on a second SunOS station (Fig. 2).

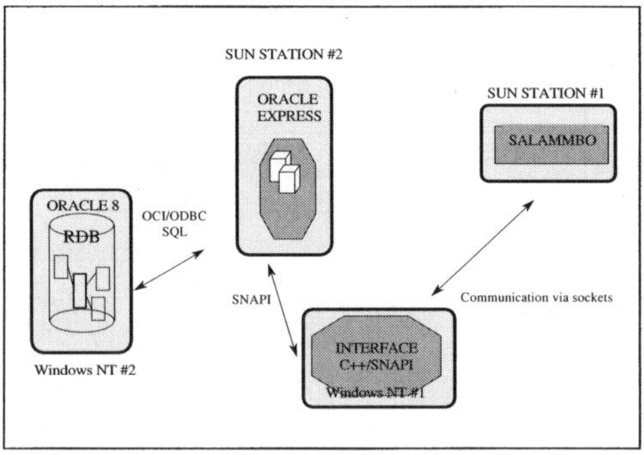

Fig. 2. Proposed architecture

This proposed system was evaluated on data from a data warehouse of the French Educational Department, containing individual results in high school certificate through two years (about one million records). Instances of discovered rules are:

R1: If the high school is public, and the certificate category is economics, then the proportion of candidates that succeed with mention (rather good, good, or very good) is small.

R2: If the high school is private, and the academic year is 1996, and the specialty of the student is foreign Language 2, and diploma is "STI", then the proportion of candidates that succeed with mention (rather good, good, or very good) is big.

7 Conclusion and future work

In this paper, a new approach is presented to build fuzzy decision trees from a data warehouse using the OLAP technology. This approach leads to the definition of a client/server architecture that associates the Salammbô software, a fuzzy decision tree construction system, with *Oracle Express*, a multidimensional DBMS.

Results are promising and encourage us to initiate further research, among which the tests with other databases, further comparisons between the different solutions we have presented here to exchange statistics, the integration of other learning methods, the implementation of a more friendly user interface and the handling of fuzzy data (imprecise and/or uncertain) in the hypercubes.

Acknowledgments

This work is supported by the LIP6 (Computer Science Department of the University Paris 6) and has been made possible thanks to the database provided by the French Educational Department.

References

1. Agrawal, Gupta A., and Sarawagi S., Modeling Multidimensional Databases in Proc. of the 13th Int. Conf. on Data Engineering, Birmingham, U.K., 1997.
2. Bouchon-Meunier B., Marsala C., Ramdani M., "Learning from Imperfect Data", in Fuzzy Information Engineering: a Guided Tour of Applications, D. Dubois and H. Prade and R. R. Yager eds, John Wileys and Sons pub., chap. 8, pp. 139-148, 1997.
3. Bouchon-Meunier B., Marsala C. "Learning Fuzzy Decision Rules", in Fuzzy Sets in Approximate Reasoning and Information Systems, J. Bezdek, D. Dubois and H. Prade eds, Kluwer Academic Pub., Handbooks on Fuzzy Sets Series, chap. 4, pp. 279-304, 1999.
4. M.S. Chen, J. Han, and P.S. Yu, "Data Mining: An Overview from a Database Perspective", IEEE Trans. on Knowledge and Data Engineering, pp 866-883, 1996.
5. Codd E.F., Codd S.B. and Salley C.T., "Providing OLAP (On-Line Analytical Processing) to User-Analysts: An IT Mandate", 1993.
6. V. Harinarayan, A. Rajaraman, J. Ullman, "Implementing data cubes efficiently", in Proc. of the 1996 ACM SIGMOD Int. Conf. on Management of Data, pp 205-216, 1996.
7. Marsala C., Bouchon-Meunier B., and Ramer A., Hierarchical Model for Discrimination Measures, in Proceedings of the IFSA'99 World Congress, Tapei, Taiwan, pp. 339–343, August 1999.
8. "An introduction to OLAP, Multidimensional Terminology and Technology", Pilot White Paper, http://www.pilotsw.com/olap/olap.htm#dsgl, 1999.
9. Quinlan J.R., "Induction of Decision Trees", in Machine Learning 1:1, pp. 86–106, 1986.

Evaluation of the Accuracy of Merged Land Cover Classifications Made Using a Fuzzy Integral

Graham Thackrah[1], **Mike Barnsley**[2], **and Paul Pan**[3]

[1] Environmental Modelling and Earth Observation Group,
University of Wales Swansea, Singleton Park, Swansea, SA2 8PP.
email: ggthack@swansea.ac.uk

[2] Postal address as above,
email: m.barnsley@swansea.ac.uk

[3] Department of Maritime Studies and International Transport,
University of Wales Cardiff, CF10 3YP.
email:pan@cardiff.ac.uk

Abstract: This paper describes results of the analysis of merging land cover classifications using the Sugeno fuzzy integral [2]. The analysis is being carried out to determine how data sets that have a similar information content may be combined, using the Sugeno fuzzy integral, to improve the overall classification accuracy. It is assumed that data sets with a similar information content will not add significantly to the accuracy of a land cover classification produced from their combination. To test this, the results of the combination of neural network based classifications of two data sets with similar information contents are compared to that of adding a third data set with a different information content. It is shown that the classification produced from the combination of data sets with similar information contents is at least as good as that produced by either on their own. The addition of a data set with a different information content did not improve the classification results in this case, a number of explanations are suggested.

Keywords: Fuzzy integral, classification integration.

1 Introduction

1.1 Study Site

Kenfig National Nature Reserve (Kenfig NNR) is a sand dune system in south Wales. The site has several levels of protective status under UK and European nature conservation legislation. Under these legal directives there is a requirement for regular monitoring of the criteria for notification of protective status.

Kenfig NNR is approximately 600 hectares in size, a medium sized site that would be relatively time consuming to survey by humans. Research is being undertaken to investigate the potential for airborne remote sensing in monitoring and mapping the site in terms of land cover.

There are both spectral and topographical data sets available for the site. As there is a strong relationship between topography and land cover type at sand dune sites in general and this site in particular [7], it is hoped that by classifying each data set and using combinations of the results, a more accurate classification of the site will be produced than from a classification of a single (topographic or spectral) data set. Lone classification of spectral and topographic data sets produce multiple competing hypotheses of land cover type, it is hoped that by fusing these a more accurate classification will result.

1.2 Combining Classifications

Classifications of image data may be achieved in many ways, a popular method in current times is the use of an artificial neural network (ann). In the case of processing image data for classification, the input vector of an ann is usually the vector of values of bands for a pixel in the image. The output vector is some expression of a feature (in the broadest sense) of interest that we wish to identify, in this case the feature is a land cover class.

Methods for combining classifications range from the simplest, such as voting methods, simple statistical methods such as averaging and composite data set methods, to more complex methods. The largest flaw with most of the simple methods is that they treat each data source equally. The more complex methods may take into account information on the accuracy of each classification in identifying each class when fusing the classification outputs.

In the present case, where multiple data sets exist, and training data for the site are available, it is possible to assess the accuracy of each classification in identifying each class of interest. Using this information we can construct a method to combine data in a manner preferential to the classification that is the more accurate for a particular class.

1.3 Application of the Sugeno Fuzzy Integral

The Sugeno fuzzy integral may be used for data fusion. It is possible to use this fuzzy integral for fusing the results of a number of classifications of the same data source [3] or for fusing the results of classifications of multiple data sources as in the current case and [4].

The application of the fuzzy integral in this case relies on the ability to obtain the accuracy of each classification in correctly identifying each class of interest. Once these figures are obtained they are used to construct a g_λ fuzzy measure. A g_λ fuzzy measure is calculated for each class as follows,

$$\lambda + 1 = \prod_{i=1}^{n} (1 + \lambda g_i) \tag{1}$$

where n is equal to the number of classifications (sources in this case) and g is a measure of the accuracy of the ith source in identifying the class of interest. Lambda is then defined as the unique root of the above equation greater than $^{-}1$.

The Sugeno fuzzy integral is then calculated for each pixel and performs a fusion of the classifications as follows,

$$e = \max\left[\min\left(h\left(x_i\right), g\left(A_i\right)\right)\right] \quad (2)$$

where,

$$g\left(A_i\right) = g_i + g\left(A_{i-1}\right) + \lambda g_i g\left(A_{i-1}\right) \quad (3)$$

and $h\left(x_i\right)$ is a function such that $h : X \rightarrow \{0, 1\}$ (where X is the evaluation of a pixel by source i for a given class) and $h\left(x_1\right) \geq h\left(x_2\right) \geq h\left(x_n\right)$, if not, X is rearranged such that the condition holds.

The properties and derivation of the fuzzy integral and g_λ fuzzy measures are discussed primarily in [2].

1.4 The use of a fuzzy integral in fusing classifications from multiple sources

In the specific case of using a fuzzy integral for fusing the results of a classification of multiple sources of remotely sensed data, there are a number of interesting implications. The first of these is the need for each data source to be classified into the same classes. This may not always be appropriate as it is somewhat unlikely that the classification of a data set based on elevation data could be adequately classified into the same classes as a multispectral data set.

Another implication is discussed in [1] where a discussion is made of the interactions between data sets when combined. In [1] a trivial example is given of a measure of the performance of several students when examined in three subject areas. By not taking into account the fact that students taking subjects like mathematics and physics are likely to produce similar results in both, one may think a student who performs well in both is disproportionally as clever as a student who performs well in both mathematics and english.

Disregarding the fact that student cleverness is likely to be a highly subjective subject, the application to the current case is obvious. Where two similar spectral data sets are combined with a third data set of topographic variables, by regarding the combination of any two data sets of equal importance the same as a combination of any other two data sets, one is not likely to give the correct weight to the contribution of those combinations.

In order to investigate the effect of this, the current paper compares the results of combining two spectral data sets using the Sugeno fuzzy integral with that of combining two spectral data sets with a third topographic data set. The method used here of constructing the fuzzy measure does not necessarily allow for the modelling of the interaction between data sets outlined above, if the results of combining data sets are poor using this method then it is suggested that an alternative method of constructing a fuzzy measure be applied where data set interaction is likely to be significant.

2 Data and Methods

2.1 Data Sets

Data sets available for the study site are multi–spectral image data acquired by two different sensors (Compact Airborne Spectrographic Imager–CASI and the Airborne Thematic Mapper–ATM), and digital elevation data acquired by a LIDAR instrument. The two spectral instruments cover similar, but not identical, areas of the electromagnetic spectrum (see table 1) and one would thus not expect to improve on the accuracy of the classification of either alone by aggregating the results of both.

ATM	436	493.5	561.5	614.5	660.5	726	839	962	1692.5	2244	1075	NA	NA	NA
CASI	489	510	555	597	625	662	672	683	691	702	711	750	856	878

Table 1. CASI and ATM central wavelengths of bands (micrometers)

The elevation data retrieved by the LIDAR instrument were converted to aspect and slope values (both in degrees) and a composite three band image of these three values used in the classification. Due to the poor ability of the topographic data set to be adequately classified into the same classes as the spectral data however, a composite data set including NDVI (normalised difference vegetation index, a function of the reflectance in the red band and the near infra red band) as a fourth band was constructed. This was necessary in order for the classification results to produce suitable solutions to equation 1.

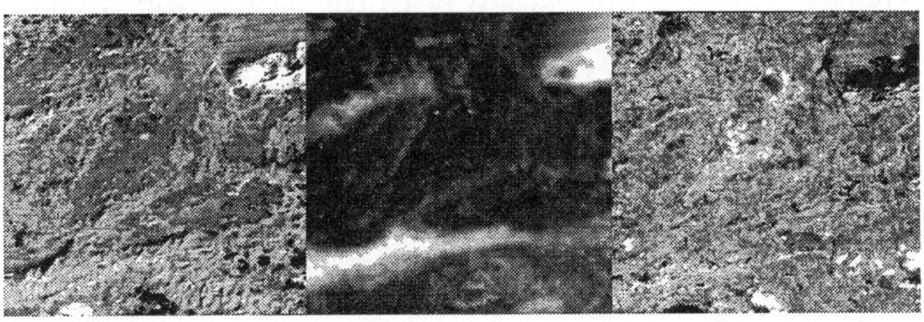

Fig. 1. Left: ATM image (band 5), Middle: elevation data, Right: NDVI.

2.2 Classification using ANNs

Artificial neural networks are used in this study as a classification method that produces a membership value for each cover type for each pixel of interest. Other methods could be used, maximum likelihood classifiers produce a similar output. In an earlier study anns were used to simulate data of different information contents through differential learning regimes [5], in this study they are used simply as a method of classifying multi–dimensional data and producing 'membership' values for each pixel for each class. The use of the output of a neural network as a membership value relies on it's interpretation as an *a posteriori* Bayesian probability [3] although some question the validity of this [6]. The accuracy of each classification was assessed using ground truth data available for the site, co–registered to the three data sets.

2.3 Classification Fusion

The results of the classifications of each data source were used in the construction of various fuzzy measures. The classification results of the two spectral data sets were used to construct one fuzzy measure. The classification results of the ATM data and topographic/ndvi composite data set were used to construct another fuzzy measure and finally all three data sets together were used to construct the final fuzzy measure. Fusions of these three combinations were performed using the fuzzy integral as in [2].

3 Results and Discussion

Results from simulated data sets suggest that a fuzzy integral allows aggregation of classification results producing a classification that is at least as good as the single best lone classification [5]. Results presented here in table 2 show that carrying out fusions with real data supports this finding at least for the case of similar data sets. The fusion of two different data sets produces a worse fused classification than the best of the constituent classifications. Furthermore, the addition of all three classifications produces a slightly better result though it is still worse than both the case of the two spectral data sets and the single best classification.

Co–registration of the data sets is critical for the correct functioning of the fuzzy integral as the assessment of accuracy of the classifications is dependant on the same ground truth data being applicable to each classification. If the ground truth data is in error then false g_λ fuzzy measures would be constructed and the fuzzy integral would perform an incorrect fusion of data. This is likely to lead to a reduction in the accuracy of the final classification. Secondly, the method of constructing the fuzzy measure could be flawed if there is expected to be a high degree of similarity between two or more data sets. The interaction between data sets in this case could produce unexpected results when they are combined. It is unsure if either of these would produce the unexpected lowering of accuracy as seen in the case of the three data source fusion.

A	C	D	A C	A D	A D C	
67.51	57.95	38.88	67.49	61.60	63.30	overall accuracy (percent)
0.5378	0.3777	0.1845	0.4785	0.4349	0.5337	kappa coefficient

Table 2. Classification accuracies, A=ATM, C=CASI, D=DEM

4 Conclusions

Modelling the interaction between sources in the manner suggested in [1] may improve the results in this case though this is not conclusively proved here. The lower classification accuracy produced with the combination of a topographical data set and two spectral data sets is disappointing and needs further investigation. The effect of registration of the separate data sets is also worth further note as it is expected to influence the accuracy measures of each classification and hence the g_λ fuzzy measure.

Acknowledgements

Dr. Clive Hurford, Rob Davies of Bridgend Borough Council, the National Assembly of Wales and the Environment agency for supplying the LIDAR data and NERC for supplying the spectral data and for supporting this research through a CASE studentship (Ref. GT22/10/EO/97).

References

1. Grabisch, M. (1996) The application of fuzzy integrals in multicriteria decision making. European Journal of Operational Research. **89**, 445–456
2. Sugeno, M. (1977) Fuzzy Measures and Fuzzy Integrals: A Survey, in Fuzzy Automata and Decision Processes (*North Holland*). 89–102.
3. Cho, S. B., Kim, J. H. (1995) Combining multiple neural networks by fuzzy integral for robust classification. IEEE Transactions on Systems, Man and Cybernetics. **25**, 380–384.
4. Tahani, H., Keller, J. M. (1990) Information fusion in computer vision using the fuzzy integral. IEEE Transactions on Systems, Man and Cybernetics. **20**, 733–741.
5. Thackrah, G. N., Barnsley, M, J., and Pan, P. (1999) Merging land cover classifications from artificial neural networks using the Sugeno fuzzy integral. Proceedings of the Remote Sensing Society (RSS99), Cardiff, 71–78
6. Barr, S. (1998) *pers comm.*
7. Davies, R. (2000) *pers comm.* Warden at Kenfig NNR.

Part 3

Miscellaneous CI Techniques

An Evolutionary Model of Symbiosis

Vladimír Kvasnička

Department of Mathematics, Slovak Technical University
812 37 Bratislava, Slovakia
email: kvasnic@cvt.stuba.sk

Abstract. The purpose of the present communication is to discuss an evolutionary model of symbiosis based on an assumption that simpler prokaryotic organisms are spontaneously grouped together into stable more complex eukaryotic organisms. Each prokaryotic organism is represented by a binary chromosome that corresponds to a solution of a given local optimization problem. Eukaryotic organisms are represented by binary chromosomes that are created by a concatenation of binary chromosomes of its prokaryotic constituents. Fitness of prokaryotic chromosomes is determined by a quality of solution represented by the chromosomes. If fitness of these chromosomes is greater than a threshold value, then they may take part in a concatenation process, which produces larger eukaryotic binary chromosomes.. Emerged eukaryotic chromosomes may take part in a new concatenation processes such that larger and larger eukaryotic chromosomes are formed.

1 Introduction

In biology [1-3], a symbiosis means the living together of different organisms. A special case of the symbiosis is the mutualism, where symbionts (organisms in symbiotic relationship) are mutually beneficial. Symbiosis can lead to the symbiogenesis, i.e. the creation of new species via the genetic integration of symbionts (eukaryotic cells have a symbiogenic origin). In modern evolutionary biology, the symbiosis is regarded as an integral part of evolution and a fundamental source of innovation. Recently, this interesting biological concept was studied by R. A. Watson and J. B. Pollack in their paper [4] entitled *How Symbiosis Can Guide Evolution*, where they used a slightly modified approach originally used by Hinton and Nowlan [5] in their seminal paper on explanation of mechanism of Baldwin effect. The Hinton and Nowlan concept of "lifetime plasticity" of cognitive device of organisms is replaced by a Watson and Pollack concept of "lifetime interaction" between organisms. Their model is composed of two periods: *First period*, symbiotic groups find the solution of a problem more quickly than the solution can be found by a single organism. *Second period*, after a group has found the solution, the evolution of individual organisms is continued towards their ability to be independent of other members of a group.

The main shortcoming of Watson and Pollack theory of symbiosis is its biological implausibility; in particular, an assumption that after second period organisms are again fully independent. This is evidently implausible for evolutionary biology, where it is postulated that symbiosis belongs to main sources of increase of complexity in biotic systems [1].

The purpose of the present communication is to discuss an alternative evolutionary model of symbiosis endowed with a feature that simpler organisms are grouped together into stable more complex organisms. The grouping of organism is represented in our evolutionary model by an assumption that simpler (prokaryotic) chromosomes can be concatenated into larger (eukaryotic) chromosomes. A crucial position in the present approach play constraints under which two chromosomes (either prokaryotic or eukaryotic) are grouped together. The constraints are specified in such a way, that both constituents should have fitness closely related to a threshold value and their conjoined size (a number of prokaryotic constituents) is smaller than a prescribed value.

2 Theory

Let us assume that we have p independent "local" optimization subproblems

$$f_i : \{0,1\}^n \rightarrow [0,\infty) \quad (i = 1,2,\ldots,p) \tag{1a}$$

$$\boldsymbol{x}_{opt}^{(i)} = \arg \min_{\boldsymbol{x}\in\{0,1\}^n} f_i(\boldsymbol{x}) \ , \ f_i\left(\boldsymbol{x}_{opt}^{(i)}\right) = 0 \tag{1b}$$

where each "local" function f_i is defined over binary strings of the fixed length n, their nonnegative functional values are scaled so that all functions have a zero minimum. A composition of the above n "subfunctions" gives a "global" function

$$f = \sum_{i=1}^{p} f_i : \{0,1\}^{n\times p} \rightarrow [0,\infty) \tag{2a}$$

$$f(\boldsymbol{x}) = \sum_{i=1}^{p} f_i\left(\boldsymbol{x}^{(i)}\right) \tag{2b}$$

where the variable $\boldsymbol{x}$ is a binary vector of the length $p\times n$ determined as a "direct sum" $\boldsymbol{x} = \boldsymbol{x}^{(1)} \oplus \boldsymbol{x}^{(2)} \oplus \ldots \oplus \boldsymbol{x}^{(p)}$. A "global" minimization problem of the function f is

$$\boldsymbol{x}_{opt} = arg \ \min_{\boldsymbol{x}\in\{0,1\}^{p\times n}} f(\boldsymbol{x}) \tag{2c}$$

A solution of this minimization problem of f is simply determined by

$$f(\boldsymbol{x}_{opt}) = \sum_{i=1}^{p} f_i\left(\boldsymbol{x}_{opt}^{(i)}\right) = 0 \tag{3a}$$

$$\boldsymbol{x}_{opt} = \boldsymbol{x}_{opt}^{(1)} \oplus \boldsymbol{x}_{opt}^{(2)} \oplus \ldots \oplus \boldsymbol{x}_{opt}^{(p)} \tag{3b}$$

There is possible to use two different strategies how to perform the minimization problem (2c), see Fig. 1. The *first* approach to the minimization (2c) is straightforward, we look for its global minimum in a space of all binary vectors of the length $p\times n$. The *second* approach utilizes the structure (2b) of f. Since the function f is a composition of p parts that are fully independent, its minimization problem (2c) may be realized by p independent "local" optimization subproblems (1b). A computational demand of this second approach is substantially smaller than a computational demand of (2c). An optimization problem (2c) performed in $p\times n$-dimensional space of binary vectors is substituted in the second approach by p

smaller optimization subproblems performed only in n-dimensional space of binary vectors.

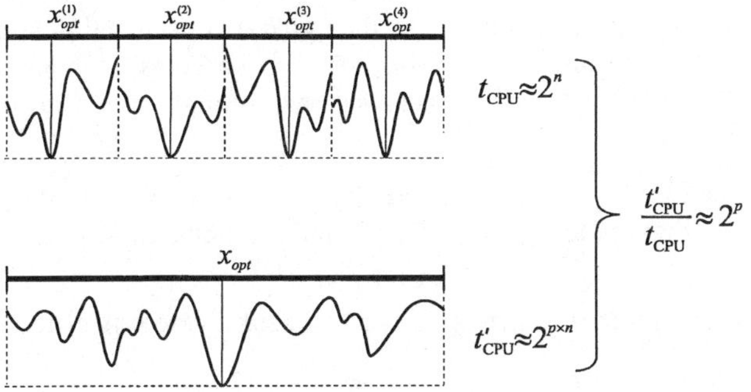

Figure 1. The upper diagram represents p independent optimization subproblems that are solved over chromosomes specified by binary vectors of the length n. Since all optimization subproblems are solved simultaneously (evolutionary parallelism), a CPU time necessary for a simultaneous solution of all subproblems is proportional to 2^n. The lower diagram corresponds to a "global" optimization problem solved over chromosomes that are specified by binary vectors of the length $p \times n$, its CPU time is proportional to $2^{p \times n}$. An increase of complexity is specified by a ratio of CPU times that are assigned to the "global" optimization and p simultaneous "local" optimization subproblems, this ratio is proportional to 2^p.

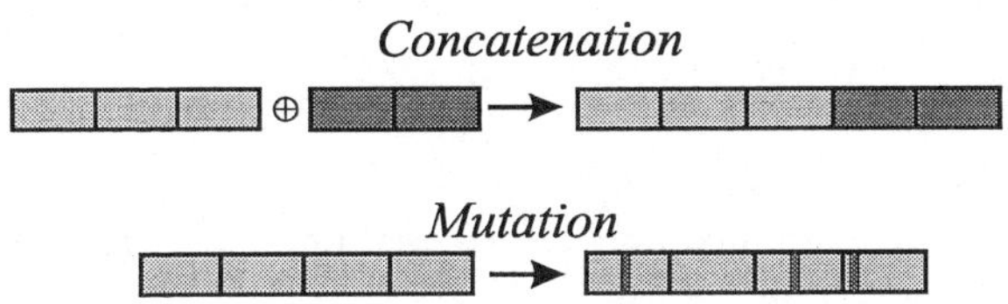

Figure 2. Diagrammatic illustration of two different operations that transform chromosomes (prokaryotic and/or eukaryotic) onto another chromosomes. The first operator corresponds to a *concatenation*, where two chromosomes (that are either both prokaryotic or eukaryotic or one of them is prokaryote while another one is eukaryote) are grouped together in an eukaryote chromosome. The second operator called the *mutation*, corresponds to stochastic changes of single binary entries (these changes, flippings, are specified by a probability P_{mut}).

The above discussion about an efficiency of solution of the "global" minimization problem (2c) offers a basic idea of our theory of symbiosis. Let us postulate that the basic evolutionary goal is to solve the "global" minimization problem (2c) in the whole space of $p \times n$-dimensional binary vectors. The used strategy, called the "symbiosis", is such that in the first period "local" optimization subproblems (1b) are solved by elementary "prokaryotic organisms" (represented by binary chromosomes of the length n). In the second period the obtained "local" solutions are integrated into more complex "eukaryotic organisms" that correspond to two or more optimization subproblems . A final goal of evolution will be an "eukaryote" represented by a chromosome that represents a solution of the optimization problem (2c). In other words, "prokaryotes" are used for an evolutionary solution of "local" optimization subproblems (1b), their integration into more complex "eukaryotes" corresponds to solutions of the "global" optimization subproblems. A final "eukaryote" , composed of p different "prokaryotes" , is represented by a chromosome that may be assigned to a (suboptimal) solution of the whole optimization problem (2c).

Let a chromosome $\boldsymbol{x}$ be specified by a binary vector composed of r subchromosomes that are assigned to "local" subproblems $i_1, i_2,..., i_r$.

$$\boldsymbol{x} = \boldsymbol{x}^{(i_1)} \oplus \boldsymbol{x}^{(i_2)} \oplus ... \oplus \boldsymbol{x}^{(i_r)} \tag{4}$$

If $r \geq 2$, then the chromosome corresponds to an "eukaryote" specified by indices i_1, $i_2,..., i_r$.

We introduce the following two stochastic operations that transform chromosomes onto some other chromosomes (see Fig. 2):

(1) ***Concatenation***, two chromosomes $\boldsymbol{x}^{(i)}$ and $\boldsymbol{x}^{(j)}$ (that are either both prokaryotic, or both eukaryotic, or one of them is prokaryotic while another one is eukaryotic) are concatenated into an eukaryotic chromosome

$$z = O_{concat}\left(\boldsymbol{x}^{(i)}, \boldsymbol{x}^{(j)}\right) = \boldsymbol{x}^{(i)} \oplus \boldsymbol{x}^{(j)} \tag{5}$$

(2) ***Mutation***, for a chromosome $\boldsymbol{x} = \boldsymbol{x}^{(i_1)} \oplus ... \oplus \boldsymbol{x}^{(i_r)}$ a few randomly selected binary entries are flipped

$$\tilde{\boldsymbol{x}} = O_{mut}\left(\boldsymbol{x}\right) = \tilde{\boldsymbol{x}}^{(i_1)} \oplus ... \oplus \tilde{\boldsymbol{x}}^{(i_r)} \tag{6}$$

$$\tilde{\boldsymbol{x}}^{(i_k)} = O_{mut}\left(\boldsymbol{x}^{(i_k)}\right) \tag{7}$$

where O_{mut} is a *stochastic* mutation operator that randomly changes binary entries of $\boldsymbol{x}$ (specified by a probability P_{mut}).

Fitness of a chromosome $\boldsymbol{x} = \boldsymbol{x}^{(i_1)} \oplus ... \oplus \boldsymbol{x}^{(i_r)}$ is determined by

$$F\left(\boldsymbol{x}\right) = \min_k \left(\frac{1}{\delta + f_{i_k}\left(\boldsymbol{x}^{(i_k)}\right)} \right) - \left(p + r - 2d\right) \times \omega \tag{8}$$

where δ is a small positive number. A fitness of an eukaryotic chromosome is specified by the smallest fitness of its constituents (loosely speaking, a chain is as strong as its weakest link), and d is a number of *different* subproblems that are included in $\boldsymbol{x}$. The second negative term on the right hand side corresponds to a *penalization*, where ω is a small positive number. This terms decreases the fitness by a value that depends on the number (p-d) of subproblems that are not included in $\boldsymbol{x}$ and a number (r-d) of repetition of some subproblems in $\boldsymbol{x}$. A maximal fitness of chromosomes composed of d different subchromosomes is determined by $(1/\delta)$-$(p\text{-}d) \times \omega$, for $p=d$ (i.e. chromosome is composed of all subproblems and none repetition there exists) we get (1/δ). This means that by a proper choice of δ we may control the maximal value of fitness.

A concatenation operator (5) is applied stochastically in the course of evolutionary process onto a pair of chromosomes that are quasirandomly selected from a population, where the probabilities of selections are proportional to fitness of selected chromosomes. Let us have two chromosomes $\boldsymbol{x}$ and $\boldsymbol{y}$, their fitness is denoted by $F(\boldsymbol{x})$ and $F(\boldsymbol{y})$, respectively. A concatenation operator is applied to these chromosomes with a probability that is specified as follows:

(1) Both chromosomes are prokaryotic

$$prob_{p-p}(\boldsymbol{x},\boldsymbol{y}) = e^{-\beta_{prokar}\left|F(\boldsymbol{x})+F(\boldsymbol{y})-2\vartheta_{prokar}\right|} \tag{9a}$$

(2) Both chromosomes are eukaryotic

$$prob_{e-e}(\boldsymbol{x},\boldsymbol{y}) = e^{-\beta_{eukar}\left|F(\boldsymbol{x})+F(\boldsymbol{y})-2\vartheta_{eukar}\right|} \tag{9b}$$

(3) Chromosome $\boldsymbol{x}$ is eukaryotic and chromosome $\boldsymbol{y}$ is prokaryotic

$$prob_{e-e}(\boldsymbol{x},\boldsymbol{y}) = e^{-\beta_{eukar}\left|F(\boldsymbol{x})-\vartheta_{eukar}\right|} \times e^{-\beta_{prokar}\left|F(\boldsymbol{y})-\vartheta_{prokar}\right|} \tag{9c}$$

where ϑ_{eukar} and ϑ_{prokar} are threshold fitness values for eukaryotic and prokaryotic chromosomes, respectively. If values of fitness of both chromosomes are closely related to the threshold ϑ, then chromosomes $\boldsymbol{y}$ and $\boldsymbol{x}$ are concatenated; in the opposite case, the concatenation is applied with a probability specified by (9). A positive parameter β (called the slope) determines a sharpness of nonlinearity of the probability calculation. If this parameter tends to great values ($\beta\rightarrow\infty$), then the probability is equal either to zero (for fitness values that are different from the thresholds) or to one (for the fitness values that are equal to thresholds). According to this rule, we may say, that the concatenation process (a formal analog of grouping of different organisms together) is applied only when both chromosomes have fitness from the very thin interval of values somewhere about the prescribed threshold value ϑ.

An evolutionary approach to a simulation of emergence of eukaryotic chromosomes is performed by a simple evolutionary algorithm outlined in Fig. 3. Two quasiindependent populations are used, where the first one is composed of prokaryotic chromosomes while the second one is composed of eukaryotic chromosomes. Prokaryotic chromosomes are evolved by applying a very simple version of evolutionary algorithm, where the reproduction process is composed of a quasirandom proportional selection and selected chromosomes are simply copied to a new population, with the copy process including mutations (specified by a probability $P_{mut,prokaryot}$). An evolution of the second population composed of eukaryotic chromosomes is of much more complex nature than that one for the first population. In particular, at the beginning of evolutionary process the population is sequentially filled by simple eukaryotic chromosomes that are formed from pairs of prokaryotic chromosomes of good quality (their fitness is closely related to a threshold ϑ_{prokar}). These simple eukaryotic chromosomes are then mutually combined by concatenation operations onto larger and larger eukaryotic chromosomes. Of course, their size has an upper bound specified by an integer p (a number of optimization subproblems). Independently of these processes, in the population of eukaryotes runs standard evolution formally similar to that one specified above for the population of prokaryotes.

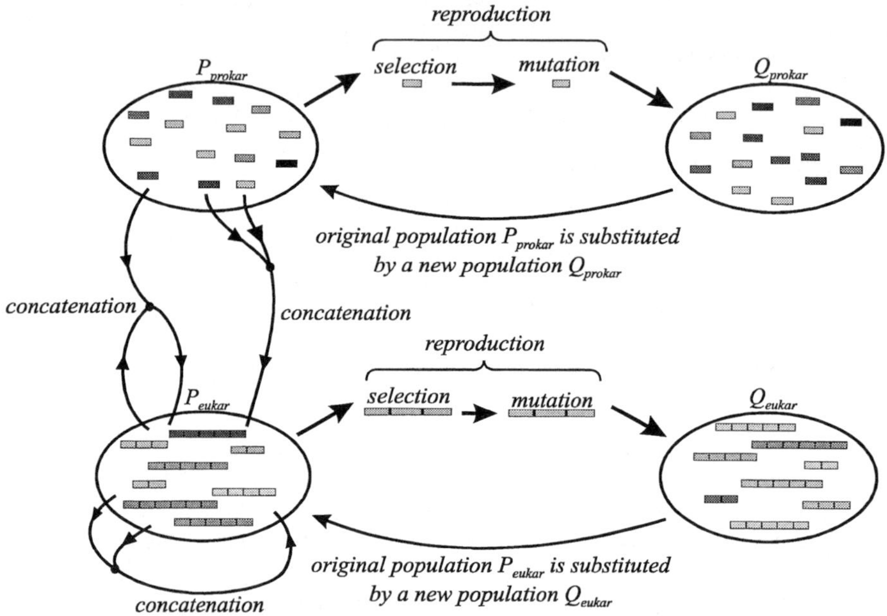

Figure 3. A diagrammatic visualization of a simple model of simultaneous evolution of two populations that are composed of prokaryotic and eukaryotic chromosomes, respectively. An evolutionary process of these two populations is not fully independent, they interact in a one-way mode, one or two randomly selected prokaryotic chromosomes are participating in concatenation processes and the resulting product is shifted to the eukaryotic population. We may say that the eukaryotic population is coevolved with the prokaryotic population. If a new population Q composed of offspring of reproduction process has the same number of individuals as an original population P, then the population P is updated by the new population Q, $P \leftarrow Q$. We have to emphasize that a formation of eukaryotic chromosomes by a concatenation process is applied only for a stage when the number of chromosomes in eukaryotic population is smaller than a prescribed maximal value. On the other hand, if the eukaryotic population has already achieved a prescribed maximal number of its elements, then an inflow of new eukaryotic chromosomes that are created by concatenation process is stopped and the eukaryotic population is evolved fully independently of the prokaryotic population.

Simulation calculations were performed for parameters that are listed in Table 1. Most important parameters that considerably affect the results are slope parameters β_{prokar} , β_{eukar} and thresholds ϑ_{prokar}, ϑ_{eukar} from the probabilities (9) that specify performances of concatenation operators. In particular the parameters β_{prokar} and ϑ_{prokar} specify very strictly applications of concatenations to prokaryotic chromosomes. For instance, if the threshold ϑ_{eukar}=(1/δ) and β_{prokar} has a great positive value (say β_{prokar}=10), then only those prokaryotes are concatenated that correspond to exact (or almost exact) solutions of optimization problems (1b). It means that a fitness of produced eukaryote will be specified by $(1/\delta)+\omega(p\text{-}2)$. Its further evolution does not require any application of mutation (its prokaryotic constituents are exact). Such eukaryote may be potentially used as an input constituent for concatenations that produce bigger eukaryotic chromosomes composed at least of four prokaryotic constituents. On the other hand, if slope parameter β_{prokar} is equal to a small positive number, then the selection of prokaryotes to concatenations is not such rigorous as in the previous case, now also prokaryotes with fitness substantially smaller than $(1/\delta)$ have a chance to be concatenated into an eukaryote.

Table 1. List of basic parameters and their numerical values

No.	Parameter	Meaning
1	n=10	The length of binary string of each subproblem, see eq. (1)
2	p=10	The number of subproblems, see eq. (2)
3	N_{prokar}=2000	Size of population of prokaryotes, kept fixed through the whole evolution
4	$0 \leq N_{eukar} \leq 2000$	Size of population of eukaryotes, monotonously increased during the whole evolution
5	$P_{mut,prokar}$=0.01	Probability of bit mutations in prokaryotic chromosomes
6	$P_{mut,eukar}$=0.001	Probability of mutations in eukaryotic chromosomes
7	β_{prokar}=1.0	Slope parameter for prokaryotes that specifies probabilities (9a) and (9c)
8	β_{eukar}=0.01	Slope parameter for eukaryotes that specifies probabilities (9b) and (9c)
9	ϑ_{prokar}=99.99	Threshold value for prokaryotes that specifies probabilities (9a) and (9c)
10	ϑ_{eukar}=85.00	Threshold value for eukaryotes that specifies probabilities (9b) and (9c)
11	δ=0.01	A coefficient from the fitness definition (8), its inverse value $1/\delta$ determines a maximal value of the fitness 100
12	ω=3.0	Penalization parameter
13	N=10 and K=3	Parameters of Kauffman's function (see Appendix A)

Functions f_i (for i=1,2,...,p) from (1a) are determined in our simulation calculations as Kaufman's ragged functions KN [6] (see Appendix). These binary functions are highly multimodal and deceptive, and therefore they represent nontrivial optimization tasks for our simple realizations of optimization problems (1b) and (2c).

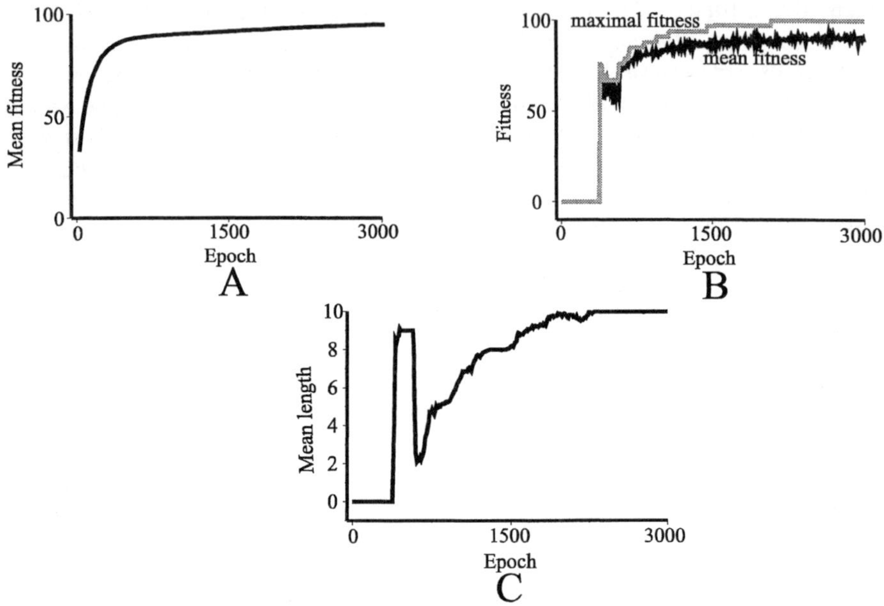

Figure 4. The first diagram A corresponds to a plot of the mean value of fitness of prokaryotic chromosomes for the evolution of population P_{prokar} composed of 3000 epochs. A mean fitness of prokaryotic chromosomes monotonously increases to a maximal value, when almost all prokaryotic chromosomes correspond to correct solutions of corresponding optimization subproblems. Diagram B displays the maximal and mean fitness of eukaryotic chromosomes. After 2000 epochs the population reaches an evolutionary state where exist chromosomes that have fitness equal to its maximal possible value 100.00. The plot of the maximal fitness manifests a stair increasing mode with relatively long period between two subsequent stair steps. This observation may be simply interpreted as an example of neutral evolution [7]. Diagram C shows a plot of the mean value of length of eukaryotic chromosomes, after 1000 epochs almost all eukaryotic chromosomes are composed of all ten prokaryotic subchromosomes. At the first period of evolution there exists a very fast increase of the mean length of eukaryotic chromosomes, then after this period, the mean length is rapidly falling down to a value slightly greater than two. This peak of the mean-length plot corresponds to a fact that at the beginning of evolution there was a fast process of creation of imperfect eukaryotes that contained some prokaryotic subchromosomes more than once. These eukaryotes then become extinct because of their relatively small fitness.

Numerical results are summarized in Fig. 4. An evolution of a population composed of prokaryotic chromosomes is displayed in diagram A. We see that for our simple version of evolutionary algorithm an optimization of all 10 types of subproblems does not represent any serious problem. Starting from 1000 epoch there exist prokaryotic chromosomes that are good solutions of optimization subproblems (1b) and therefore they have a great chance to take part in a concatenation process of two prokaryotic chromosomes into a one eukaryotic chromosome. This first stage of evolution of a population composed of eukaryotic chromosomes is characterized by a "giant resonance" of the mean length. This means that in the beginning of evolution there are created by concatenation eukaryotic chromosomes with maximal or slightly smaller length where some prokaryotic part appears more than once. These eukaryotic chromosomes are penalized (see eq. (8)) and therefore they become extinct in forthcoming epochs of evolution of eukaryotic population. Proper eukaryotic chromosomes (without repeating prokaryotic constituents) are evolved with greater and greater length, starting from 2000 epoch almost all eukaryotes in population have the length 10 and the their mean fitness is closely related to the maximal fitness value 100. Fig. 5 shows two different plots of mean fitness of eukaryotic chromosomes that were performed for different values of slope parameter β_{prokar}. We see that if eukaryotic chromosomes are created from prokaryotic chromosomes of smaller fitness (plot A), then forthcoming evolution of eukaryotic chromosomes is unable to ensure a considerable increase of their fitness (cf. plot B). This observation is in full agreement with our initial assumption that an optimization in a space of eukaryotic population is much more difficult problem than an optimization in a space of prokaryotic chromosomes (see Fig. 1).

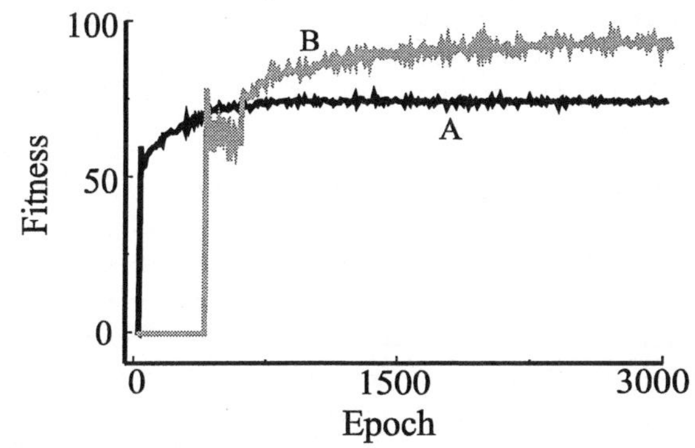

Figure 5. Two different plots for mean eukaryotic chromosomes. Plot A corresponds to an evolution of prokaryotic population with the slope parameter β_{prokar}=0.1, that is a selection of prokaryotic chromosomes (that take part in concatenation process creating initial eukaryotic chromosomes) is much less "selective" than in our original calculations (see Fig. 4, diagram B, where β_{prokar}=1.0). It means that initial eukaryotic chromosomes are evaluated by substantially smaller fitness than those ones appearing in population evolved with a greater slope parameter β_{prokar}=1.0 (see plot B). Furthermore, the eukaryotic chromosomes are of much more complex nature than the prokaryotic chromosomes and therefore there is very realistic to expect that evolution of eukaryotic population is much harder problem than an evolution of prokaryotic population.

3 Summary and discussion

The purpose of this communication is a presentation of a simple evolutionary model of symbiosis. In order to set-up the model we discuss a complex optimization problem that can be decomposed onto independent optimization subproblems. There is possible to expect that an optimization of initial complex problem can be advantageously performed in such a way that simpler optimization subproblems are solved and then these solution are linked together (concatenated) in one complex solution. This simple heuristic approach to a solution of complex optimization problem is much more efficient than a straightforward "brute-force" direct optimization of the whole complex problem. Therefore, we have divided an evolution of emergence of eukaryotic chromosomes into two stages. In the first stage of evolution single prokaryotic chromosomes are evolved such that there is expected an emergence of a population composed of prokaryotic chromosomes assigned to high-quality solutions of optimization subproblems. In the second stage, pairs of prokaryotic chromosomes (that were evolved in the first stage of evolution) are quasirandomly selected (a probability of selection is proportional to their fitness) to a concatenation process, where simplest eukaryotic chromosomes (composed of two prokaryotic constituents) are created. These eukaryotes are further evolved in such a way that there is allowed a prolongation process, which consists in a quasirandom selection either two eukaryotic chromosomes or one prokaryotic and one eukaryotic chromosome, and these chromosomes are concatenated into a lengthier chromosome. Moreover, this concatenation process is accompanied by standard Darwinian's evolution of eukaryotic chromosomes, i.e. there exists a reproduction process based on proportional selection of chromosomes and their simple reproduction with random mutations.

Two independent calculations were performed. The first one is based on very rigorous selection of prokaryotic chromosomes to the concatenation process and the second one applied a substantially weaker selection of prokaryotic chromosomes (i.e. for a concatenation processs there are selected chromosomes with smaller fitness than in the first case). We have observed that an evolution of eukaryotic population cannot substantially improve eukaryotic chromosomes, since, as was already mentioned, an optimization problem of lengthy eukaryotic chromosomes is much more combinatorially complex than an evolutionary optimization of single prokaryotic chromosomes. This very important observation implies that constituents of eukaryotic organisms (that have an origin in prokaryotes) could not be substantially evolutionary improved. What can be expected is only the so-called *bricolage* [8], which repairs some unimportant accidents caused by a genetic drift.

Acknowledgement

This work was supported by the grants # 1/4209/97 and # 1/5229/98 of the Scientific Grant Agency of Slovak Republic. As usual I have benefited from discussions with Jiří Pospíchal, who also criticized the manuscript and called attention to errors in the manuscript.

Appendix. Kauffman's *NK* objective functions [6]

Let $\boldsymbol{x}$ be a binary vector composed of N entries

$$\alpha = (\alpha_1\alpha_2...\alpha_N) \in \{0,1\}^N \tag{A1}$$

Each entry x_i has K randomly generated neighborhoods (positions, not binary entries), these neighbor positions form a 2-dimensional table $\boldsymbol{T}=(T_{ij})$, where a row index is $i\in[1,n]$ and a column index is $j\in[1,K+1]$. Row entries of the table $\boldsymbol{T}$ satisfy the following two conditions:

(1) Entries of each row form an increasing sequence

$$T_{i1} < T_{i2} < ... < T_{iK} < T_{iK+1} \tag{A2}$$

(2) Each row contains just one entry equal to its row index i

$$\exists j \in [1, K+1] : T_{ij} = i \tag{A3}$$

A binary vector $\beta = (\beta_1\beta_2...\beta_{K+1}) \in \{0,1\}^{K+1}$ of the length $K+1$ may be interpreted as an nonnegative integer $int(\beta)$ ranged by $0 \leq int(\beta) \leq 2^{K+1}-1$. Let us introduce a vector χ composed of 2^K random numbers with uniform distribution from the interval [0,1]

$$\chi = \left(\chi(0), \chi(1), ..., \chi\left(2^{K+1}-2\right), \chi\left(2^{K+1}-1\right)\right) \tag{A4}$$

Vector χ can be interpreted as a lookup table where each vector $\beta = (\beta_1\beta_2...\beta_{K+1})$ is evaluated by a random number taken from the interval [0,1]

$int(\beta)$	β_1	β_2	..	β_{K+1}	χ
0	0	0	..	0	$\chi(0)$
1	0	0	..	1	$\chi(1)$
...	..	..	..	..	..
2^{K+1}-1	1	1	..	1	$\chi(2^{K+1}-1)$

Kauffman's *NK* function maps binary vectors $\boldsymbol{x}$ of the length N onto positive real number from the interval [0,1], this mapping is determined with respect to the table $\boldsymbol{T}$ and the vector χ as follows

$$F(\boldsymbol{x}) = \frac{1}{N}\sum_{i=1}^{N} \chi\left(int\left(x_{T_{i1}} x_{T_{i2}} ... x_{T_{iK}} x_{T_{ik+1}}\right)\right) \tag{A5}$$

Finally, an objective function $f(\boldsymbol{x})$ defined over binary vectors of the length N with zero global minimum (see conditions (1a) and (1b)) may be simply defined as follows

$$f(\boldsymbol{x}) = \left|F(\boldsymbol{x}) - F_{min}\right| \tag{A6a}$$

$$F_{min} = \arg \min_{\boldsymbol{x}\in\{0,1\}^N} F(\boldsymbol{x}) \tag{A6b}$$

It has a zero global minimum as was required by (1b). An illustrative plot of Kaufman's function for K=3, N=10 is outlined in Fig. A1.

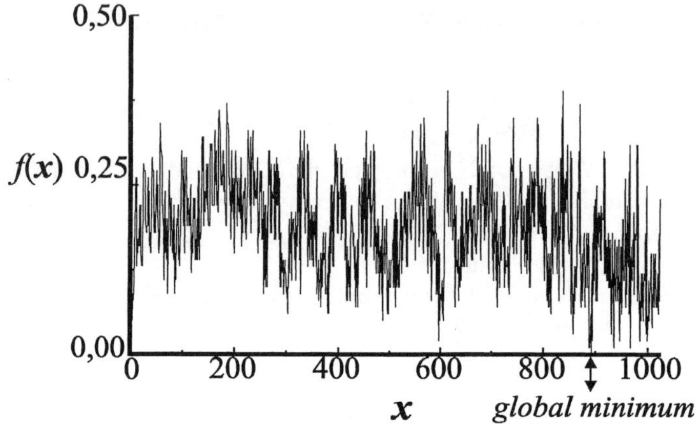

Figure A1. An illustrative plot of Kauffman's function for K=3 and N=10. We see that this objective function corresponds to an extremely multimodal function with zero global minimum f_{opt}=0.0 for $\boldsymbol{x}_{opt}$=882=*int*(1101111100).

References

1. Margulis, L.: *Symbiosis in Cell Evolution*. Second edition. W.H. Freeman, New York, 1992.
2. Frank, S. A.: The origin of synergistic symbiosis. *Journal of Theoretical Biology*, **176** (1995), 403-410.
3. Frank, S. A.: Models of symbiosis. *American Naturalist*, **150** (1997), S80-S99.
4. Watson, R. A., Pollack, J. B.: How Symbiosis Can Guide Evolution. *Fifth European Conference on Artificial Life*. Eds. Dario Floreano, Jean-Daniel Nicoud, Francesco Mondada. Springer-Verlag, Berlin, 1999, pp. 29-38.
5. Hinton, G. E. and Nowlan, S. J.: How Learning Can Guide Evolution. *Complex Systems*, **1** (1987), 495-502.
6. Kauffman, S. A.: *The Origin of Order. Self-Organization and Selection in Evolution*. Oxford University Press, New York, 1993, pp. 40-47.
7. Kimura, M.: *The Neutral Theory of Molecular Evolution*. Cambridge University Press, Cambridge, UK, 1983.
8. Jacob, F.: *Le Jeu des possibless, Essai sur la diversité du vivant*. Libraire Arthéme Fayard, Paris, 1987.

New Genetic Operator with Enhanced Capabilities

Jiří Kubalík, Jiří Lažanský

The Gerstner Laboratory, Czech Technical University, Technická 2, 166 27 Prague 6, Czech Republic

Abstract: The paper describes an enhancement of the traditional 2-point crossover operator used for binary representation in genetic algorithms. This operator preserves a schema common to both parent chromosomes. The enhancement of its functionality is in a modified treatment of this common schema. The offspring produced by the modified operator is partially randomised so that it contains both the common schema and its binary complement. It helps to prevent the genetic algorithms from getting stuck in a local optimum and enhance the exploration of the search space beyond the limits imposed by the classical operator's functionality. The partially randomised 2-point crossover operator has been tested on a number of different problems and compared to the original 2-point operator.

Keywords: Genetic algorithms, schema, building block, deception, search space

1 Genetic Algorithms, Schemata and Deception

Genetic algorithms (GAs) are probabilistic search and optimisation techniques, which operate on a population of chromosomes, representing potential solutions of the given problem. In standard GA (SGA), binary strings represent the chromosomes. Each chromosome is assigned a fitness value expressing its quality reflecting the given objective function. In the main loop of SGA, chromosomes are reproduced and recombined to generate a new population of chromosomes, i.e. new sample points from hopefully more promising parts of the search space. This is repeatedly performed until some given termination-condition is fulfilled. The best chromosome encountered so far is then considered as the found solution.

The basic analysis of SGA behaviour is based on a notion of a schema as a template, which matches certain class of chromosomes. As the population evolves some good schemata are represented by increasing number of chromosomes whilst bad schemata disappear. 0s and 1s of those good schemata constitute building blocks, which represent important components of the final solution. The optimum solution emerges when these building blocks are mixed together in an optimal way in some chromosome. It can be seen that GAs simultaneously carry out exploitation of the promising regions found so far and exploration of other areas for potentially better solution.

SGA very often converge to a false solution that may happen for several reasons. One of them is an insufficient used population size causing lack of genetic material to evolve the optimal chromosome. Problems with convergence can also be caused by improperly designed genetic operators, which can not maintain an optimal balance between exploitation and exploration of the SGA. SGA also fail to find a global optimum due to so-called *deception* included in the solved problem. This phenomenon concerns the fitness function and its improper mapping on the binary string. The deception appears when low-order schemata, which cover some non-optimal solution have higher fitness than their complementary schemata covering the global optimum. It results in a situation that the good-looking building blocks mislead the SGA and cause its convergence to a false solution. For more about deception see (Goldberg 1989), (Whitley, 1991).

2 Partially Randomised Cross-over Operator (PRX)

This section introduces an enhancement of the functionality of classical crossover operators working on the binary representation. The original 1-point, 2-point and uniform operators are designed so that both offspring inherit the all bits, which are the same in both parents' chromosomes. Those bits represent a common schema of the parents. Remaining bits are filled in according to a rule specific for each operator. As the population converges, the common schema occupies a growing portion of the parent chromosomes and the crossover can produce little new. The promising building blocks are mixed together but their inner configuration cannot be reset any more. Though any of those operators can be used, an enhancement of the 2-point crossover is described in this paper.

The enhancement of the crossover lies in a modified treatment of the common schema of the parents' chromosomes. The PRX operator does not strictly preserve the common schema of the parents. In one of the two generated offspring, a portion of the common schema is inverted. Thus, the population is not saturated with superior building blocks but also with their randomly chosen binary complements. The diversity of the population is permanently maintained, which helps to preserve the SGA from getting stuck in a local optimum and enhance the exploration of the search space beyond the limits imposed by the pure evolution.

PRX operator works in two steps. First, the offspring chromosomes are created by a standard 2-point crossover. In the second step, one of the offspring is partially changed so that in a randomly chosen section of length l_{PRX} the bits belonging to the common schema, are inverted. Each of the N_{CS} common-schema bits in the given section is changed with a probability $P_{PRX} = l_{PRX}/(2 \cdot N_{CS})$. Neither it means that a constant number of bits nor bits of the same part of the representation are changed. Instead, randomly chosen subset of bits in randomly chosen region of the chromosome is affected by each application of the randomisation. It can be considered as a permanent re-initialisation of the population genotype at the level of schemata of predefined length l_{PRX}.

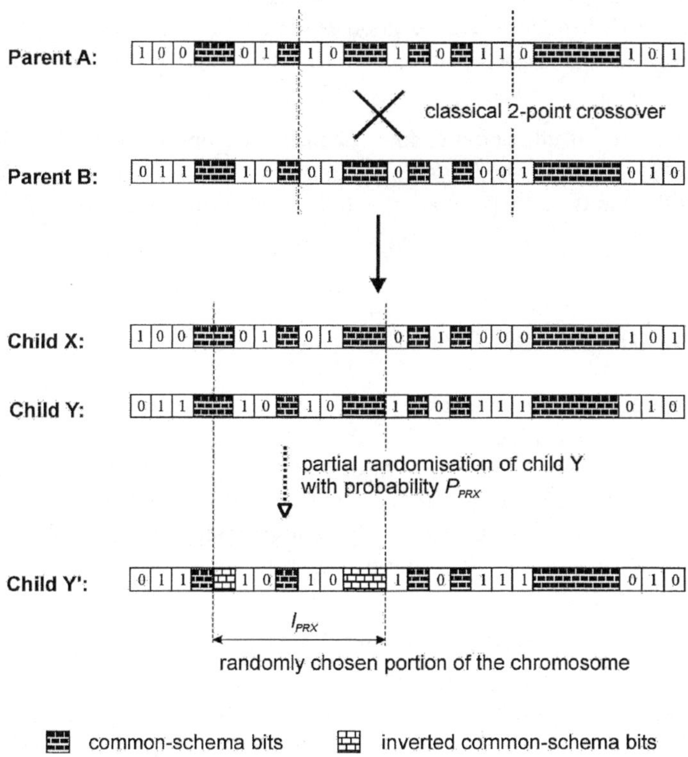

Fig.1. Partially randomised 2-point crossover

It is clear that the PRX operator will perform best on such problems, where the related bits, constituting the fundamental building blocks, will be situated close each other within the chromosome. In the other case, the positive effect of the randomisation ceases. Note that the same assumption holds for the 2-point crossover used as the basic operator. Despite this condition, the set of possible applications still covers a large range of suitable problems. An example is the problems where the structure or interpretation of the chromosome is not unknown. Some of the experiments carried out on such problems are treated further.

The randomisation will not only reintroduce a useful formation of bits into the chromosome but it will cause damages, as well. The radical changes imposed on the generated chromosomes could represent a severe obstacle for GA to effectively converge to the desired solution. For this reason a replacement strategy equipped with an auxiliary rule to keep a balance between the initialisation power of the randomisation and the level of undesirable disruptions is used as well. It serves as an arbiter responsible for the proper composition of a new population. The rule determines whether the randomised individual will be inserted into the new population or not as follows:

> If any of the offspring is at least as good as the better parent than both offspring will be used in the new population, otherwise the better parent and better child will be used.

Note that an effect of utilisation of this enhanced strategy is twofold (1) it ensures that the quality of the created population is preserved and (2) at least one of the offspring will be in the new population and thus the evolution is still in motion.

3 Experiments and Results

First, let's present experiments carried out on a test-bed composed of 50 copies of order-4 fully deceptive function *DF*3, which is taken from (Whitley, 1991). The function has a deceptive attractor 0000 of the fitness 10, which is surrounded by four strings of quite high fitness (1000-28, 0100-27, 0010-26, 0001-25). The global optimum string 1111 has value 30. The global optimum of the whole problem is 1500. The value of l_{PRX} was 10 so the partial randomisation was operating over 3 adjacent building blocks.

The results in Table 1 show the dramatic difference between the original and partially randomised 2-point crossovers' performance. When the original 2-point crossover was used the total optimal solution was not found in any out of the 50 experiments. The number of the optimal building blocks in the solutions was about 20 in average. On the contrary the PRX operator was even 100 % successful with the improved replacement strategy.

Table 1. Table of results obtained on the problem composed of 50 copies of *DF*3. The results are in the format *fbest* / *#success*, where *fbest* is the average best-of-run value and *#success* is the number of completely solved cases out of the 50 runs.

2-point		PRX 2-point	
Replacement simple	Replacement with rule	Replacement simple	Replacement with rule
1440 / 0	1435 / 0	1499.8 / 45	1500 / 50

Another series of experiments have been carried out on the test-beds which use De Jong's (De Jong, 1975) nonlinear function over 2 variables *F*2(*x*,*y*) as the basic building blocks. In the case of the problem *simple* F2 the chromosome consists of 10 pairs of *x*, *y* parameters each of them is contributing with its *F*2(*x*,*y*) value to the fitness of the chromosome. Problem *modified* F2 is a bit harder, it consists of 10 triples x_1-$y_{1,2}$-x_2 where each contributes to the fitness with the value $F2(x_1,y_{1,2})$ +

$F2(x_2,y_{1,2})$. It means that the parameter $y_{1,2}$ is nonlinearly bound with the parameter x_1 as well as with the corresponding parameter x_2. So once the value of $y_{1,2}$ is changed both parameters x_1 and x_2 should be optimised. In both cases *simple* F2 and *modified* F2, a minimal fitness is searched for and the optimal value is 0.0. Again the value of l_{PRX} was 10 so only a part of the considered building blocks can be affected by the randomisation.

Besides the fact that the PRX crossover is again better than the original 2-point crossover one can see how the utilisation of the enhanced replacement strategy improved the behaviour of the PRX crossover.

Table 2. Table of results obtained (1) on *simple* F2 problem made of 10 copies of F2 function and (2) on the *modified* F2 problem consisting of 10 pairs of F2 functions bound via *y* parameter. The results are in the format *fbest/stdev* derived from 100 runs.

	2-point		PRX 2-point	
Problem	Replacement simple	Replacement with rule	Replacement simple	Replacement with rule
Simple F2	3.47 / 1.51	3.61 / 1.67	1.55 / 0.50	0.81 / 0.26
Modified F2	11.2 / 4.6	10.9 / 4.9	3.1 / 1.0	2.3 / 0.7

The last experiments presented here were carried out on a simple job shop scheduling problem (JSSP). It assumes *N* jobs to be processed on *M* machines (in our case *N*=*M*=6). The schedules are encoded in the binary strings according to the representation introduced in (Nakano, 1991). A chromosome consists of $N\cdot(N\text{-}1)\cdot M/2$ bits, each carrying priority information of certain job-pair on a given machine. Despite all the bits are strongly bounded each other there are groups of *M* adjacent bits, which describe the priority situation on particular machine. The partial randomisation has been focused directly on those blocks so that it is applied on portions of the chromosome related to 2 randomly selected machines.

Table 3. Results obtained on the simple JSSP.

2-point	PRX 2-point
0, 1, 14, 85	6, 1, 19, 74

There are the statistics of found schedules of the total length 55, 57, 58, and 59 out of 100 performed experiments in the Table 3. The schedules of length 59 represent strong local optima. So the interesting schedules are those of length 55, 57, and 58. It is important that the PRX 2-point crossover was able to find the optimal schedule of the length 55 in 6 cases out of 100 experiments and the number of found schedules with length 58 was higher as well.

4 Conclusions

The paper describes an enhancement of the functionality of the classical 2-point crossover operator working on a binary representation. The enhancement lies in a modified treatment of the common schema of the parent-pair's chromosomes. The offspring produced by the modified operators is partially randomised so that it contains both the common schema and its binary complement. The paper presents some experiments which show how the partially randomised 2-point operator outperforms the original 2-point operator on such problems where the spatially compact building blocks can be identified.

Acknowledgements

This research was supported by the State Department of Education within the frame of the project MSM 212300014.

References

1. De Jong K. (1975): An Analysis of the Behavior of a Class of Genetic Adaptive Systems. PhD Dissertation. Dept. of Computer and Communication Sciences. Univ. of Michigan. Ann Arbor.

2. Goldberg D. E. (1989): Genetic Algorithms in Search, Optimisation, and Machine Learning. Addison-Wesley.

3. Kubalík, J., Lažanský, J. (2000): Partially randomised Crossover Operators, In: Advances in Soft Computing, 93-98, R. John & R. Birkenhead (Eds.), Heidelberg: Physica-Verlag, ISBN 3-7908-1257-9.

4. Nakano R., Yamada T. (1991): Conventional Genetic Algorithms for Job Shop Scheduling Problems. In: Proceedings of the Fourth International Conference on Genetic Algorithms, 474-479. Morgan Kaufmann. California.

5. Whitley D. (1991): Fundamentals Principles of Deception in Genetic Search. Foundations of Genetic Algorithms, 221-241. G. Rawlins, ed., Morgan Kaufmann.

Genetic Methods for Point Labeling on Maps

Attila Álmos and Annamária R. Várkonyi-Kóczy

Department of Measurement and Information Systems,
Budapest University of Technology and Economics,
Műegyetem rkp. 9, Budapest, Hungary, H-1521
Email: aalmos@mit.bme.hu

Abstract: During the last years cartography has got into the focus of information technology and computer science. The hardness of the problem stems in the fact that some of the important tasks to be solved in this field proved to be computationally hard problems. One of them is the map labeling, i.e., the arranging of the textual information belonging to the map features.

Traditionally the label placement is solved based on human intelligence and knowledge. The known computer algorithms can work only on simplified versions of the task. The results are much weaker compared with the human-made labelings. In this paper such genetic algorithm based methods are proposed, which can take this NP-hard optimization problem in its complexity. The presented techniques provide more freedom for the label positioning, thus, resulting in higher quality results. The fitness function to evaluate the labelings can consider all the relevant aspects taking into cartographic experts' account.

1 Introduction

The map labeling problem is a special type of the resource allocation problems. The labels on the map compete with each other for their area. They should be assigned to their map features satisfying given rules. The most important of them is to avoid the overlaps between labels or between a label and the objects on the map.

As there are three main groups of the map features, three types of labels should be placed: point-, line- and area-type labels. In the literature of the research field the point labeling problem dominates. It can be defined in easier way than the other groups of label placing, which contain more complicated requirements. Although, hand the point labeling techniques can be applied in the latter cases, as well.

In the examined problem point-elements of the map should be labeled. The main goal is to maximize the number of non-overlapping labels. This type of task is often called fixed size labeling. In this paper the label sizes are predefined individually according to the assigned text and text style. The labels are placed in an axis parallel rectangle given by width and height.

Each point (site) has a priority determining how to choose among them when there are unavoidable conflicts. Beside overlaps there exist other aspects which should be considered, for example, association between sites and labels.

2 Existing Approaches

The basic task in the considered problem is the placing of labels in the neighborhood of their sites. The most frequently applied simplification used in the solutions is that the allowed positions of the labels are reduced. The resulting algorithms handle only discrete, limited label locations, e.g., they allow only the 4 regular positions, which are preferred in cartography (see Fig. 1). These simpler forms of the problem have some disadvantages. Their main weakness is that they reduce the degree of labeling freedom. So the provided results may have weaker quality, because of the frequent overlaps, especially in high density regions. The map labeling problem, even in these simple cases proved to be NP-hard problems [1]. The methods must apply heuristics or other non-classical techniques. Some methods are based on combinatorial approaches, e.g., [2] presents a graph based algorithm applying rules to find results, while the method in [3] uses heuristics to trace the task back to a 2-SAT problem. Other approaches use stochastic optimization techniques, e.g., simulated annealing (SA) [4] or genetic algorithms (GA) [5].

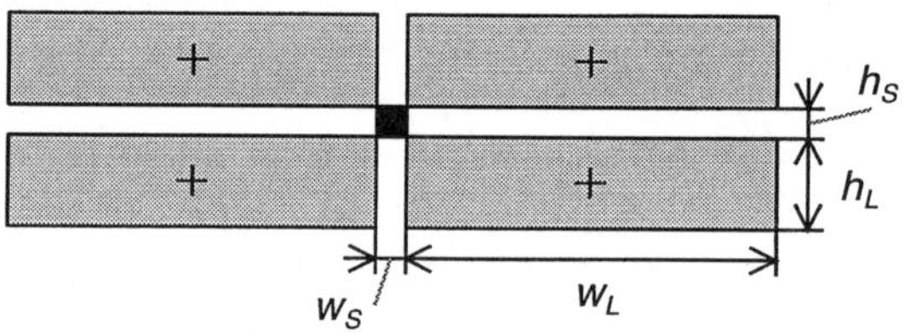

Fig. 1. Basic locations around the site

An advanced group of methods allows greater freedom for label placing. Their position can be put continuously in the neighborhood of the site, so these approaches can find better labelings, because they can reduce the overlaps more efficiently. In [6] an energy minimization method is presented. It attempts to eliminate conflicts applying an interesting idea using repulsive forces between labels. The disadvantage of the algorithm is that it can be easily stuck in a local minimum especially in the mentioned difficult regions. This is a common problem of the above approaches. The stochastic optimization techniques, among others GA, can offer a good answer this latter weakness.

3 The Improved Genetic Algorithm for Point Labeling

Nowadays there already exist some genetic based methods to solve the point labeling problem (see e.g. [7]), but they have some weaknesses. In this paper authors propose an improved GA based algorithm for point labeling to eliminate these disadvantages as presented in the followings.

One of the GA's key elements is the representation, i.e., the coding of label placement in chromosomes. In the proposed method continuous label positioning is suggested because of its advantages described above. As the most general technique, it allows placing the labels in any directions relatively to their sites, and the distances between sites and labels can change in a tolerance interval. The representation of the label positions is solved by two parameters: $\alpha_i, \beta_i \in [0, 1]$. β is a "distance" parameter determining a rectangle for the center of label (see Fig. 2). The nearest one is at value 0, while the outermost rectangle is at 1. This distance is limited by x_{md} and y_{md} giving maximum distance values in the directions of the axes. α can be viewed as some kind of "angle" parameter fixing the position of the center on the above rectangle. The 0 position belongs to the lower left corner and as going around the border the α value changes from 0 to 1.

During the mapping of α and β preference functions are applied. These well-adjusted functions give greater weights for the preferred positions (see Fig. 1) and preferred distances. Further benefits of this technique are that the mappings are continuous.

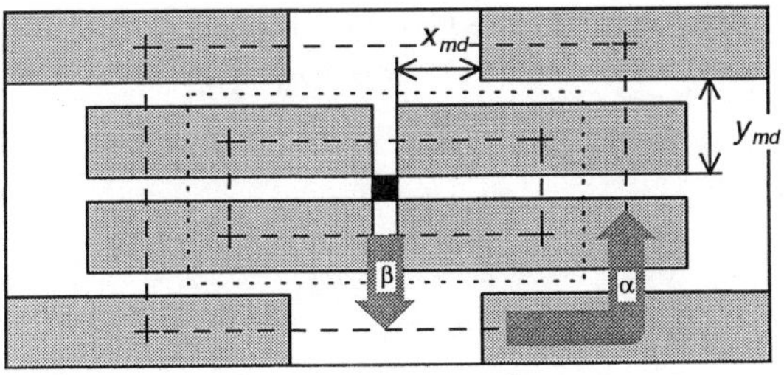

Fig. 2. Label positioning applied in the new GA based algorithm. *Dotted rectangle* belongs to a given β value, while *dashed rectangles* represent values 0 and 1

A further relevant component of the GA is the fitness function (objective function). In the algorithm the fitness function is calculated as a punishment function, and its minimization is the goal. The function considers the following relevant aspects first discussed in [8]:

- Conflicts (overlaps) between two labels, between a label and a site or the map border. The number of overlaps and the overlapped areas are summarized separately and calculated with a weight function.
- Association between the label and the site. Each label should be the closest to its own site. The distances between labels and sites are calculated, and punishment value is derived from bad distances.
- Relative position to the site. The deviations from the preferred positions are considered by a distance value.
- Overall quality. A statistical analysis examines first of all the label densities locally. The density values, which reach a level, are punished.

The set punishment values harmonize with the priorities of the aspects. The most important, i.e., having the highest priority and thus the highest punishment, is the first on the list, while the least important is the last one.

The genetic algorithm is initialized randomly in the search space. For the crossover different techniques were tested. Beside the uniform operation, which swaps parameters between the parents exploiting their benefits, another method was also tested to attempt to average the parameters of parents.

For the mutation the simple random operation as at initialization and an improved version was also used, which considers the conflict area related to each label.

4 Experiments and Conclusions

The presented algorithm was tested on several maps containing site sets. The results of these runs show that the method can be applied for wide range of the point labeling problem. Particularly outstanding performance was presented in high density regions of maps. Fig. 3 shows a part of a map with non-overlapping label placement of the GA based method. Similar density of sites causes label ejection in case of other approaches.

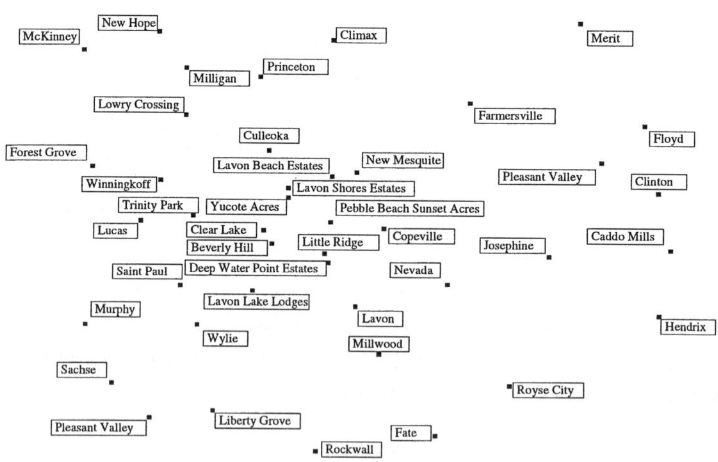

Fig. 3. Map labeling result of GA method

Comparing with other methods this algorithm can provide at least such a good solution as the presented approach in the literature. The computational complexity of the described algorithm can be higher than the other methods, but this disadvantage of the algorithm is not a relevant aspect in the map labeling problem field. However, the presented advanced mutation shows better performance, than the standard random mutation, causing faster convergence, thus reducing the computational requirement (see Fig. 4 and Fig. 5).

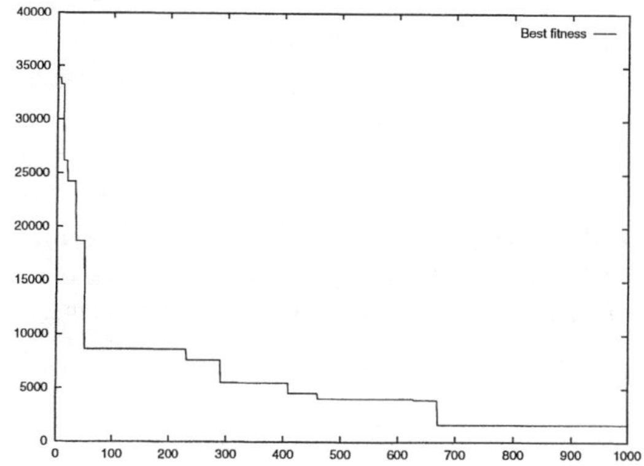

Fig. 4. Best fitness values by generations. Simple random mutation

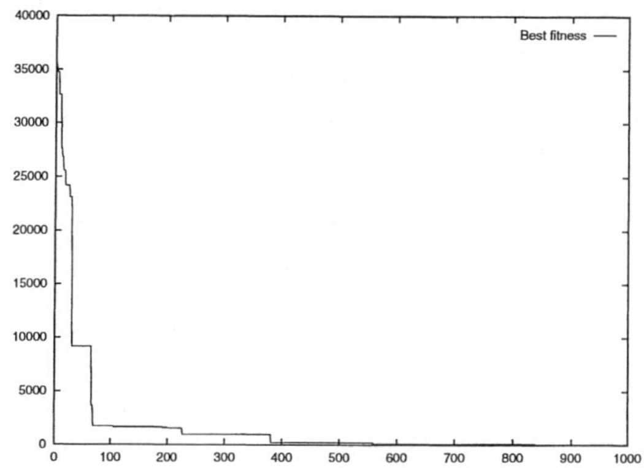

Fig. 5. Best fitness values by generations. Advanced mutation

The advantage of the GA based method shows itself in those mentioned difficult circumstances, where the sites are placed very closed to each other. Other methods in these cases cannot give good results, i.e., their placements contain labels with overlaps, so labels must be eliminated. The better performance is followed by the higher degree of label placement freedom. The method can provide good results even in such cases when the site density is near the theoretical limit.

A further advantage of the algorithm is that it can consider several aspects of the label placement. On the other hand when the method works on dense areas, the secondary aspects keep in the shade against the conflict avoidance. It causes esthetically weaker results.

In the future a method to eliminate the above weakness will be proposed. The algorithm should be able to distinguish between the rare and dense regions. Further improvement will also be planned to make the method more efficient.

Acknowledgements

The software for this work used the GAlib genetic algorithm package, written by Matthew Wall at the Massachusetts Institute of Technology.

Test data was used from freely accessible source database of GNIS (Geographic Names Information System, http://mapping.usgs.gov/www/gnis).

References

1. Marks, J., Shieber, S. (1992) The Computational Complexity of Cartography Label Placement. Technical Report TR-25-92, Harvard University, Cambridge, Massachusetts
2. Wagner, F., Wolff, A. (1998) A Combinatorial Framework for Map Labeling. In Sue H. Whitesides, editor, Proceedings of the Symposium on Graph Drawing '98, vol. 1547 of Lecture Notes in Computer Science, pages 316-331. Springer-Verlag
3. Wagner, F., Wolff, A. (1995) An Efficient and Effective Approximation Algorithm for the Map Labeling Problem. In Proceedings of the 3rd Annual European Symposium on Algorithms (ESA 1995), pages 420-433
4. Christensen, J., Marks, J., Shieber, S. (1995) An Empirical Study of Algorithms for Point-feature Label Placement. ACM Transactions on Graphics, 14(3):203-232
5. van Dijk, S., Thierens, D., de Berg, M. (1998) Robust Genetic Algorithms for High Quality Map Labeling. Technical Report UU-CS-1998-41, Department of Computer Science, Utrecht University
6. Hirsch, S. A. (1982) An Algorithm for Automatic Name Placement around Point Data. The American Cartographer, 9(1):5-17
7. Preuß, M. (1998) Solving Map Labeling Problems by Means of Evolution Strategies. Master's thesis, Fachbereich Informatik, Universität Dortmund
8. van Dijk, S., van Kreveld, M., Strijk, T., Wolff, A. (1999) Towards an Evaluation of Quality for Label Placement Methods. In Proceedings of the 19th International Cartographic Conference, pages 905-913, Ottawa.

Genetic Algorithms with Diploid Chromosomes and Sexual Reproduction

P. Ošmera, J. Roupec, R. Matoušek

Institute of Automation and Computer Science, Brno University of Technology, Technická 2, 616 69 Brno, Czech Republic
{osmera, roupec, matousek}@uai.fme.vutbr.cz

Abstract: A diploid genetic algorithm with sexual reproduction is a genetic algorithm (GA) using double candidates for each gene within a singleindividual, and it uses a form of dominance mechanisms to decide which gene is active in the phenotype. We designed a new version of a diploid genetic algorithm and investigated it along with a simple test problem that is known to be difficult for haploid GAs. Diploid chromosomes increase the efficiency and robustness of GAs, and they can better track optimal parameters in achanging environment. The adaptive significance of sexuality in GAs ispresented. This type of GAs can be used for solving very difficult economicproblems that need high adaptability to dynamic models.

Keywords: genetic algorithms, diploid chromosomes, sexual reproduction, adaptability, changing environment

1 Introduction

Scientific discussion of evolution date back than 200 years [1]. Darwin suggested that slight variation among individuals significantly affects the change that a given individual will survive and reproduce. He called this differential reproductive success of varying individuals *natural selection.*

Gregor Mendel (1822-1884) accurately observed patterns of inheritance and proposed a mechanism to account for some of the patterns. Genes determine individual traits. Various kinds of offspring appear in proportion that can be predicted from Mendel's laws. We often use the term Mendelian genetics to refer to the most basic patterns of inheritance in sexually reproducing organisms with more then one chromosome. Mendel gave his classic paper *Experiments in Plant Hybrids* for *Natural Science Society* in Brno in 1865.

In diploid organisms, chromosomes come in pairs. One of each chromosome pair derives from each parent. In many animals, including us, sex is determined either by a single sex chromosome or by a pair of them.

Sex is determined by differences in the chromosomes; but such determination operates in a bewildering variety of ways [7]. Most mammals evolved sociality via the familial route. Some present-day mammals represent the first stage, in which solitary females or male-female pairs care for their young. Social organization in mammals is dependent on the group size, as follows:

Group size:	**social unit:**
1 or 2	pair
2-12	male with small harem
2-100	territory-defending male with harem
up to thousands	herd in which males defend females
up to thousands	herd with male dominance hierarchy

2 Models with Diploid Chromosomes

Many GAs are implemented on a population consisting of haploid individuals. However, in nature, many living organisms have more than one chromosome and there are mechanisms used to determine dominant genes. Diploid GAs or polyploid (multiploid) GAs seems to confer a number of advantages on an optimized system, mainly by enhancing population diversity, and retains previously discovered useful solutions. In early research [5], [6], we introduced a diploid chromosome pair mapped to a particular phenotype using a XOR dominance map, which is coded in the chromosome itself.

Ryan [4] uses so called shades (diploid chromozomes without dominance – additive diploid chromozomes) which can outperform all other methods in tracking the environment. Avoiding dominance means that both phenotypes have an equally likely change of expression. This also permits parents of the same phenotype to produce a child of a different phenotype.

An evolution of chromosomes on a fitness landscape varying through time can be implemented as time-dependent objective functions used in GAs. It may be demonstrated that standard GAs with haploid chromosomes are unable to correctly locate optimal solutions for time-dependent objective functions. This shortcoming will be simply surmounted when haploid chromosomes are substituted with diploid chromosomes [2,3]. Usually, diploid versions of GAs are sufficiently effective and robust to locate correct solutions (that are closely related or even equal to global ones) for fitness landscapes that are varying through the time of evolution. Naturally, the quality of adaptation depends on the speed of landscape changes through the time.

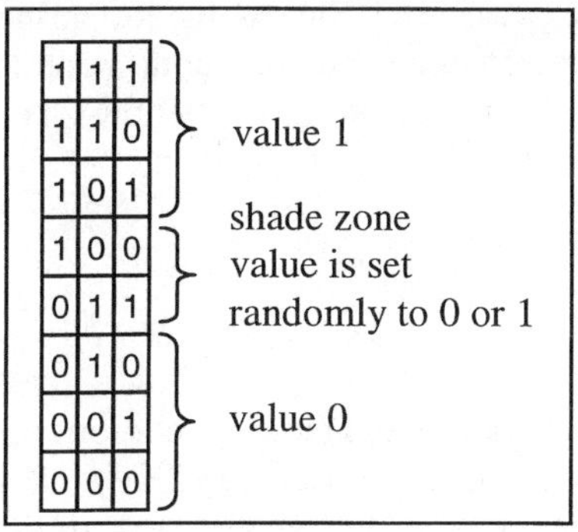

Fig. 1

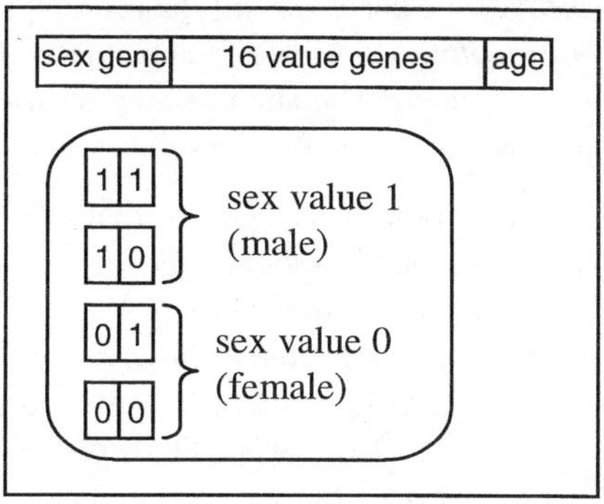

Fig. 2

3 Dynamic Environment

We use a simple dynamic objective function to demonstrate the efficiency and robustness of the present modification of the GA. We use a simple dynamic objective function:

$$g_1(x,t)=1-e^{-200(x-c(t))^2}, \quad \textbf{(1)}$$

determined for $\mathbf{x}\in[2000,2090]$ and $t\in\{0,1,...,199\}$. The function $c(t)$ is determined by $c(t)=10\lceil t/20\rceil$, where $\lceil\xi\rceil$ is an integer part of ξ, e.g. $\lceil 2.6\rceil = 2$. Function $c(t)$ corresponds to a step function:

$$c(t)=2000+k(t \bmod 20). \quad \textbf{(2)}$$

Values of **x** are of 16 bits length. Mapping from real-valued parameters to the binary bit strings in GA were performed using a Grey coding.

Usually every bit of a real value is represented by one bit of the gene. Genes can also contain redundantly coded information, a possible approach using so called shades is described in [4]. Shades can be considered to be one of possible representations of multiploid genes. The mapping between gene bit strings and real values using shades is shown on Fig. 1 (gene length is three bits, these bits represent value of one bit length). In the "shade zone" the value is set randomly (only once for every gene, this value is constant during the whole lifetime of the gene and can be changed only if a mutation occures).

The structure of the chromozome used in described GA is shown on Fig. 2. Sex gene is of diploid type, the mapping of gene values to sex is also shown on Fig. 2. In the reproduction, the sex genes also participate on crossover. The sex genes are not mutated. Value genes use the shade technology shown on Fig. 1. Uniform crossover is used for all genes including sex gene (each bit of the offspring chromozome is selected separately from corresponding bits of both parents chromozomes). The age is a counter used for the *death* operation (see bellow). The age is not mutated nor crossed over.

Three modifications of GA were used for testing:

A. *GA not using shades and sex*

B. *GA using shades and sex*

B.1 *Territory defending male strategy*

The best member of "male" part of the population is chosen for crossover with 80% probability, otherwise is chosen randomly. The second parent is chosen randomly from the second ("female") part of the population.

B.2 *Pairs*

Parents for crossover are chosen sequentially according to their fitness (both males and females), every member of the population is used only once in one iteration.

GAs we use have the following scheme:

1. *Generation of the initial population*

   In the beginning the whole population is generated randomly, the members are sorted by the fitness (descendently). In modifications which use a sex two parts of the population are generated separately – male and female.

2. *Mutation*

   The mutation is applied to each gene with the same probability, all GAs described here use $p_{mut} = 0.05$. The mutation of the gene means the inversion of one randomly selected bit in the gene.

3. *Death*

   Classical GA uses two main operation – crossover and mutation (the other operation should be migration – see [8]). In GAs described in this paper we use the third operation – *death.* Every member of the population has additional information – age. The age is represented by a simple counter which is incremented in every GA iteration. If age of any member reaches the preset lifetime limit, it "dies" and is immediatelly replaced by a new randomly generated member.

4. *Sorting by fitness*

5. *Crossover*, go to step 2

   In crossover we do not replace all members of the population. Only the number of individuals corresponding to the quarter of the population is generated by the crossover. Created individuals are sorted into the corresponding places in the population according to their fitness in such a way that the size of the population remains the same (sex of the individuals is respected). Newly created offsprings of low fitness do not have to be involved in the population.

The function (1) was used as the objective function for all three modifications of GA. Two values of **k** parameter (function 2) were chosen: 10 and 20, that means, that two sets of computations were made. The sum of mean squares $s=\sum_{i=0}^{199}\left(x_i^{best}-c(t)\right)^2$ in 200 iterations (every iteration corresponds to one time tic in (2)) was computed, computation was repeated 100 times for every algorithm modification and for values of the lifetime limit of the chromozome from 1 to 20. Fig. 3 to 8 show how the sum of mean squares depends on different values of age limit (three curves in graphs on Fig. 3 to 8 represent the worst, the medium and the best result obtained). On Figures 3 to 8 the different sensitivity of three versions of GA for the lifetime limit is shown. Results for selected values of the lifetime limit are shown in Tab. 1, the results for unlimited lifetime is in the Table 1 too.

4 Conclusions

As one can see, the GAs using diploid chromozomes and sexual reproduction are less sensitive for the lifetime limit and the haploid "asexual" GA is highly sensitive for the lifetime limit. The comparison of results obtained by the A version of GA for unlimited timelife with results of the same version of GA for limited timelife shows that simple and fast GA can approach results comparable with complex GAs using diploid chromozomes and sexual reproduction only when the death operator is used. The death operator seems to be a powerfull way to increase the speed and adaptability of the GA.

Acknowledgments

This work has been supported by MŠMT grant No: CZ J22/98 260000013

Table 1 – Medium values of sum of mean square

lifetime limit	GA version and value of k					
	A, k=10	A, k=20	B1, k=10	B1, k=20	B2, k=10	B2, k=20
2	169	540	9318	19376	9075	17891
5	103	286	341	625	254	379
7	84	220	183	278	149	306
10	99	305	107	197	115	192
15	413	3093	261	246	111	228
25	871	6058	172	234	120	261
unlimited	4826	5829	1072	1773	1392	1087

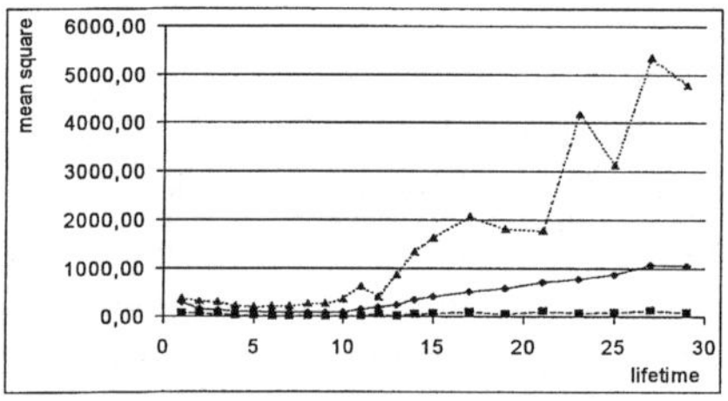

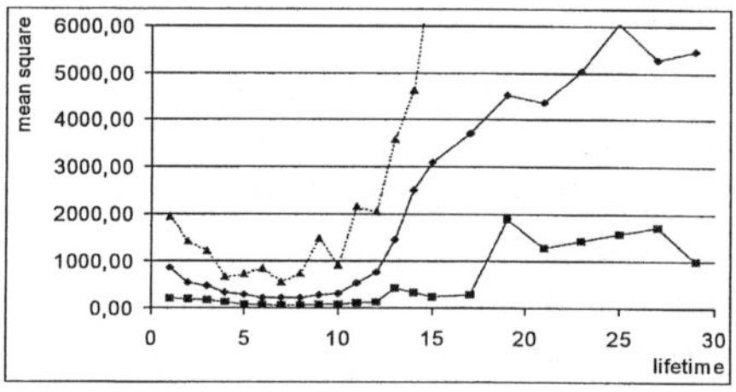

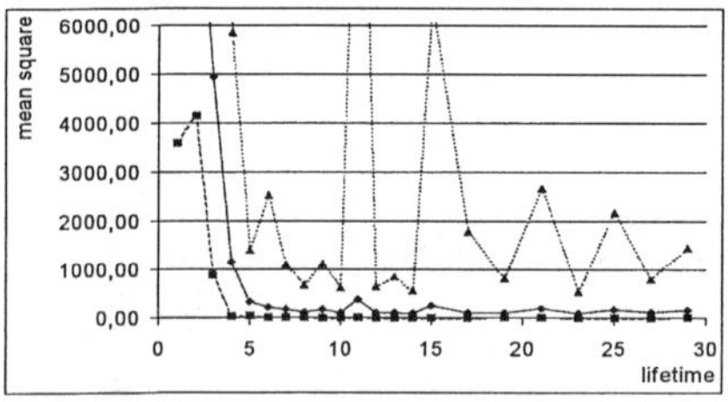

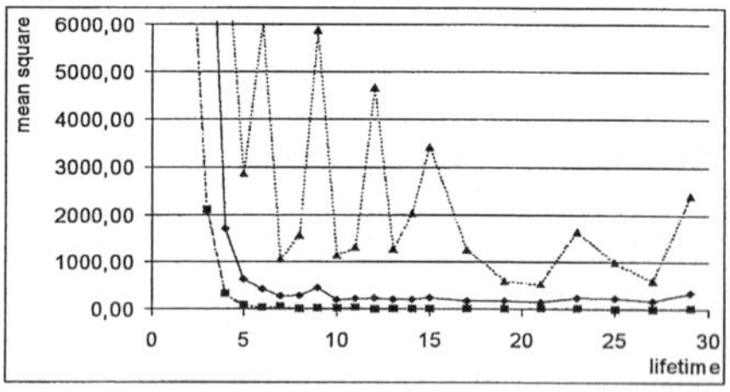

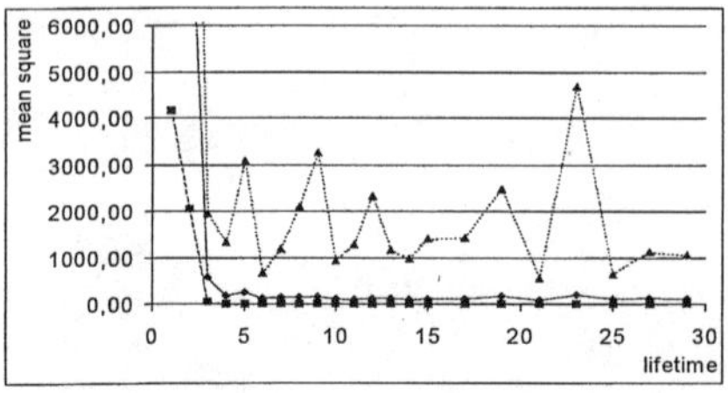

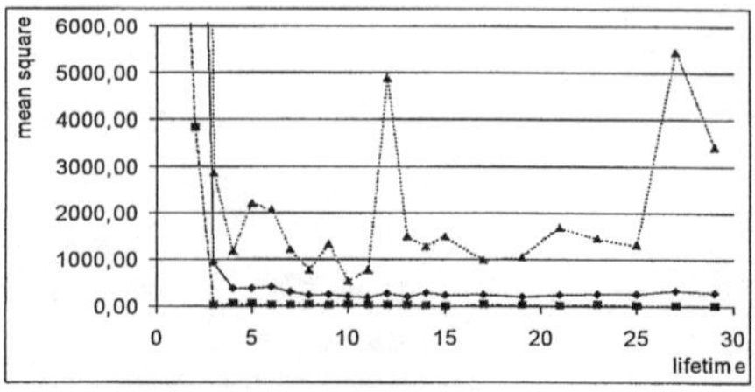

References

[1] Purves, W. K. - Orians, G. H. - Heller, H. C.: *Life, the science of biology*, Sinauer Associates, Inc., 1992.

[2] Goldberg, D. E.: *Nonstationary Function Optimization with Dominance and Diploidy*, Proceedings of ICGA2, 1987.

[3] Goldberg, D. E.: *Genetic Algorithms in Search, Optimization, and Machine Learning*, Addison-Wesley, 1989.

[4] Ryan, C.: *Shades. Polygenic Inheritance Scheme*, MENDEL'97, Brno, Czech Republic, p. 140 – 147, June 1997.

[5] Ošmera, P.: *An Application of Genetic Algorithms with Diploid Chromosomes*, Proceedings of MENDEL'98, p. 86 – 89, Brno, Czech Republic, 1998.

[6] Ošmera, P. - Kvasnička, V. - Pospíchal, J.: *Genetic Algorithms with Diploid Chromosomes*, Proceedings of MENDEL'97, p. 111 – 116, 1997.

[7] Ošmera, P. - Roupec, J.: *Sexual Genetic Algorithms with Death Operator.* Proceedings of MENDEL'2000, Brno, Czech Republic, 2000.

[8] Ošmera, P. – Šimoník, I. – Roupec, J.: *Multilevel distributed genetic algorithms.* Proceedings of the International Conference IEE/IEEE on Genetic Algorithms. Sheffield, 1995.

Artificial Life Simulation of Living Algae Cells

Marek Polak[1], Jana Kadukova[2]

[1]Technical University in Kosice, Department of Cybernetics and Artificial Intelligence, Letna 9, 04223 Kosice, Slovak Republic, polak@alife.tuke.sk

[2]Technical University in Kosice, Institute of Metallurgy and Materials, Letna 9, 04223 Kosice, Slovak Republic, kadukova@hfnov.tuke.sk

Abstract - Living algae cells and its resistance mechanisms against toxic metals are used for sorption of arsenic or other toxic metals from contaminated waters. The process of detoxification takes a long time and it is not very variable. The important part is also the mutation level caused by the impact of toxic metals on living cells. With artificial life simulation we are testing and running a simulation that could at least partially substitute the real experiments with real cells and probably allow experiments that are not possible due to time consumption or money claim. The Swarm simulation system, we use for implementation and simulation of this biological model, is very suitable for implementing rules for each agent and world, to produce a multi agent simulation with artificial life base - emergent behaviour. In last version Swarm supports also a Java programming language so, it makes it platform independent simulating system. Primary we are focused on alga Chlorella Kessleri due to the currently running biological experiments with this specie. Disadvantage is that for tune up our simulation model we need more output data that is needed for biological result of experiment so it cost more time and extra work. Comparative results are not yet available due to the development of simulation, which is still in testing phase. It is possible to use this model with minor modifications for other kinds of species.

Keywords: Artificial Life, Swarm, Algae, Chlorella Kessleri, Sorption Mechanisms

1. Introduction

1.1 Biological rules

Ability of organisms including microorganisms to survive depends, besides other factors also, on their adaptability to deal with toxic properties of various materials present in their environment. Microorganisms have adapted to higher toxicity in environment by evolving mechanisms to maintain low intracellular concentrations of toxic pollutant [1]. This adaptation to resist toxic substances became the component part of their genome, coded by nuclear chromosomes. According to the resistance mechanisms location, it is possible to divide them into three groups:

- Mechanisms localised on the cell wall
- Mechanisms localised on the cell wall and cytoplasm membrane
- Intracellular localised mechanisms

Concerning the sorption ability of alga biomass, we decided to study biosorption by green algae - *Chlorella kessleri. Chlorella kessleri (Ch. kessleri)* belongs to the family *Chlorellaceae*, subfamily *Chlorelloideae* and genus *Chlorella.* The small differences between the strain *Chlorella kessleri* and *Chlorella vulgaris* are not significant in our experiments (e.g. chloroplast shape). The cells are spherical, relatively small with a smooth cell wall, without mucilage. One or more mantel-shaped chloroplast are present with a distinct basal pyrenoid. Usually 4 autospores are formed and released by opening in mother cell wall [2] and the mother cell dies (it depends on the strain of the cell).

1.2. ALife model

Artificial life (ALife) is a rapidly growing field of scientific research linking computer science and biology. It tries to understand how lifelike processes can be embodied in computer programs. Constructing biological phenomena out of artificial components rather than breaking natural life forms down into their components does this. This is a typical bottom-up approach, which is fundamentally different from top-down approach typically used in traditional artificial intelligence (AI) and therefore ALife is sometimes called “the new AI”. The ALife is connected with some important terms: self-replication, evolution (by mutation and selection), emergency (phenomena at a certain level arise from complex interactions at lower level), parallelism, information (uninterpreted - genotype and interpreted – phenotype). The true evolution must be open-ended, i.e. not governed by a fixed fitness function.

Our ALife model is designed to match the biological model as close as possible at the certain level as published in [3]. Basic ideas for our simulations were published in [4,5]. Individual agent with defined metabolism, reproduction mechanism, chloroplast and cell wall represents each cell of real *Ch. kessleri.* Each agent has also its environmental (space) and system resources. Cell cycle is the same as with the living cells except that our simulation doesn't include any DNA, yet. The primary source of energy is external light and dissolved nutrition in the contaminated water. Each agent (cell) has a set of rules for each action. Internal metabolism is provided by an amount of internal energy of the cell. Each action (reproduction, sorption and higher level of toxic metal in environment) causes the decrement of internal energy or its increment (light, "eating"). The lost of energy leads to the dead of the cell and frees its environmental and system resources. It is very simple but useful mechanism. The quality of simulation – e.g. it approach to reality is mainly dependent, on the quality of relationship between each entity of simulation. In our model we are yet mainly focused on implementation of sorption mechanism at cell wall, especially cell surface sorption and partially intracellular sorption mechanisms [6]. Each agent provides this mechanism, it depends on agent's internal energy, concentration of arsenic in certain place where agent is located, and on its "cell surface". The amount of arsenic that cells could resist this way is limited like in real living cells and it varies with the strain of the cells.

2. Implementation

All implementation is based on ALife object based simulation system Swarm and its libraries [7]. Simulating environment is defined as multi-layer lattice with different integer values in each cell representing food, kind of heavy metal, presence of Ch. kessleri. Each cell could occupy only place where no other cell is. During reproduction (usually 4 autospores, its random 2, 4 or 8) its daughter cells will occupy the closest positions to mother's cell. Reproduction is possible only if there is a free place near mother cell. During this process, the energy of mother cell is divided between offspring and mother cell will die - free its resources (memory, space, and CPU time). Each cell is visualised as a small circle, foods is shown as square points of grey scale and concentration of heavy metal is represented by the contrast of the red colour. Food and heavy metal representation are implemented in two-dimensional discrete lattice with diffusion provided by Swarm libraries. Due to immobility of algae cells the slow flowing of environment is needed. In real experiments, in vitro shaking with substrate provides this or in bigger scale by basins with algae, cultures and slow flowing contaminated water. Important part of the simulation are rules for environment which concerns the temperature, intensity of light, concentration of arsenic, diffusion of arsenic, size of the world, number of cells, concentration of oxygen, nitrogen, carbon, phosphorus (e.g. food). On Fig. 1 is shown the entire simulation environment running under Linux operating system.

3. Experiments

Experiments are now in their first stage. We are tuning-up our artificial life model and its rules regarding the real experiments result. We suggest to use a genetic algorithm in the process of searching the proper input values. These values are not provided in real experiments (e.g. internal constants of our model, increment/decrement of internal energy etc.) but it is time consuming task due to hundreds (probably up to thousands) running simulations needed. *In vitro* experiments require special attention in case of obtaining more complex experimental data with focus on usage for debug our artificial cells. It concerns also about another reachable data arsenic concentration, light intensity, and concentration of oxygen, nitrogen, carbon and phosphorus. Experiments *in vitro* takes much longer time than *in silico* so the final application will not be available until results from all experiments will be compared and computer model will be debugged and tuned.

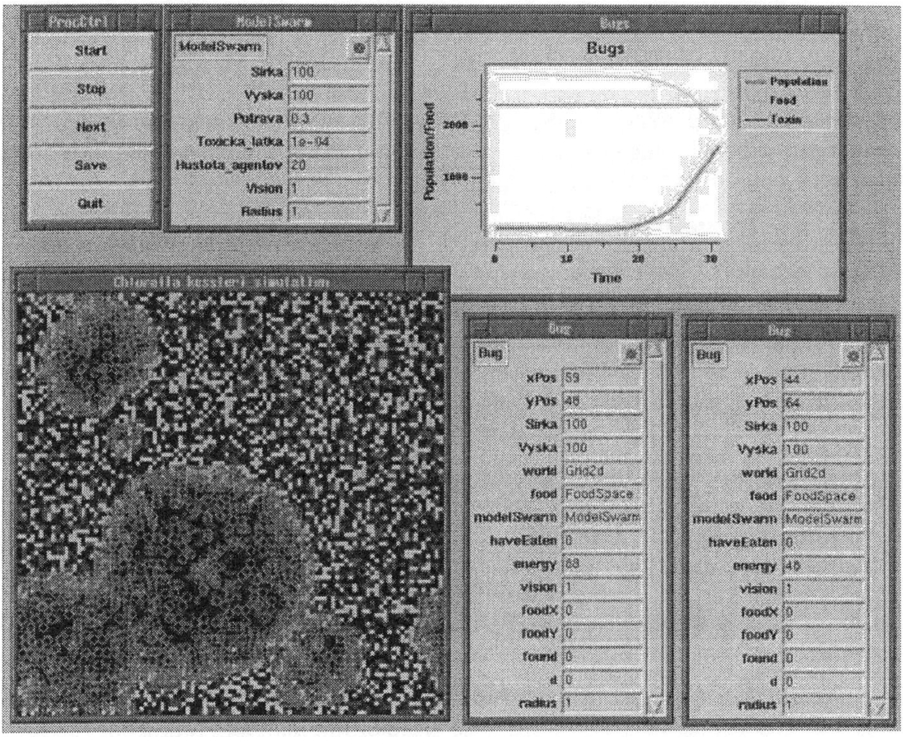

Fig. 1. Simulation of living algae cells in Swarm multi-agent system.

4 Conclusions

This paper covers approach to the practical use of artificial life principles and multi-agent simulators for modelling such a complex population as the population of living cells could be. Our aim is to build such a simulation that could at least partially substitute *in vitro* and *in vivo* experiments with saving, money, time and work. We assume that with slight modifications the final simulation could fit also another cell oriented experiments in different research fields.

Acknowledgement

This work is supported by grant VEGA (1/5032/98) for Integration of intelligent technologies.

References

[1] Wood, J. M., Wang, H-K.: Microbial Resistance to Heavy Metals, In: Environmental Inorganic Chemistry (proceedings), VCH Publishers, Inc. Deerfield Beach, Florida, p. 487-512, 1985.

[2] Hindak, F.: Studies on the chlorococcal algae (Chlorophyceae). III,VEDA, Bratislava, p.177-181, 1984.

[3] Csonto, J., Polak, M.: Artificial Life Simulation of Alga Chlorella Kessleri, INES Proceedings 1999, Poprad – High Tatras, Slovakia

[4] Kreft, J.U., Booth, G., Wimpenny, J. W. T.: BacSim, a simulator for individual-based modeling of bacterial colony growth, Microbiology 144, p.3275-3287 http://www.socgenmicrobiol.org.uk/MIC/144/3275/1443275H.HTM

[5] Zvirinsky, P.: MAG: Metabolising Agents simulation. Project, Report,Department of Computer and Information Science, Lade, NTNU Trondheim, Norway, 1999, http://www.ifi.ntnu.no/~zvirinsk/projects.htm

[6] C. Langton, N. Minar, R. Burkhart, M. Askenazi, *The Swarm simulation system*, http://www.santafe.edu/projects/swarm, 1995, Santa Fe Institute, Santa Fe

[7] Csonto, J., Polak, M., Kadukova, J.: Artificial life simulation of living alga cells and its sorption mechanisms. Special Issue of International Journal of Medical Informatics, Elsevier, accepted.

Information and Intelligence Fusion

Matjaz Gams

Jozef Stefan Institute, Jamova 39, 1000 Ljubljana, Slovenia
Phone: +386 61 1773 900, Fax: +386 61 219 385
matjaz.gams@ijs.si, http://ai.ijs.si/mezi/matjaz.html

Abstract: We analyse twelve information society laws and their influence on our society. Countries successfully transforming their society and economy proliferate, others lag behind. Overall, human civilisation quickly changes due to the introduction of information activities in all aspect of our lives. It is claimed that information society initiates additional promotion of intelligent systems and even the emergence of primitive network intelligence demonstrated through intelligent assistants on the Internet. These programs are essential for humans because without intelligent assistants we can not cope with information overload. Amongst other motivations for the introduction of intelligent systems we find the most important the technological growth of the Internet.

Keywords: information society, information overload, intelligent systems, and intelligent agents.

1 Introduction

Human society has moved from primitive hunting-agrarian society into currently most advanced information society. We are facing unseen well-fare, but at the same time our civilisation behaves irrationally - naively burning natural resources, overpopulating the Earth and neglecting plans to inhabit other planets.

In a way similar, the Internet is a quagmire and gold mine at the same time [1]. While Internet decentralisation enables seemingly equal rights when using sending or receiving data, the Internet gets overloaded with low-quality information especially with the forthcoming commercialisation. As Internet becomes a massive media like TV, the quality of information is more and more at the level of below-average users.

But certain phenomena in the information age deserve intensive study. In this presentation, the emphasis will be on the information society laws and their influence on human civilisation. The main hypothesis is that information overload in the information society causes an essential need for intelligent systems, and intelligent agents [2, 3] in particular. Without these and further advanced intelligent information assistants, we will face barriers of our performance. It is similar to the process in previous civilisation progresses - human army greatly advanced its effectiveness when acquiring first horses and later tanks. Workers incredibly increased their effectiveness when powered machined took over most of their simple physical work. In the post-industrial age, our performance additionally increased due to introduction of computers. But now we face one step further - the blind, deaf and stupid computer [4], however fast, is not competitive any more. Now we need personalised intelligent assistants to enable us cope with enormous amounts of information. As with mechanical machines, simple repetitive tasks are the most perspective for quick introduction of artificial assistants.

2 Information Society

Human society can be grouped into a couple of generations (see Table 1) regarding the major activity [5].

Table 1: Human generations

Generation no.	Generation name	Main object
I.	*agrarian*	*food*
II.	*industrial*	*production*
III.	*post-industrial*	*services*
IV.	*information*	*information*

Table 1 indicates the social change. Our human society is moving from post-industrial society and economy into information society and information-technology dominated economy. Changes essentially influence the way we work and live. By 2002, it is predicted that over 80 million Europeans will have access to the Internet, and that 5 % of EU gross domestic product will be affected by the use of digital systems. Great trends are expected in e-commerce [6]. In 1998, nearly $8 billion sales were generated in USA by around 9 million American households. In comparison, only $1.2 billion were accounted for online shopping in Europe. Europe was and is significantly lagging behind USA. But in 1999, Europe progressed much faster. Forecasts predict that 500.000 e-commerce-related jobs will be created within the next few years. Each year, Web sales to consumers are expected to grow by 70%. This growth is expected to be exponential for a couple of forthcoming years.

Table 2 represents the major way in which humans use computers. The first generation was based on primitive machine-oriented interfaces coupled with hardware. Improved programming was first in the form of assembly languages, then based on higher-level programming languages such as Fortran, Pascal or C. Current modern programming languages such as Java or JavaScript have many similarities with these languages, especially with C. However, programming demands high intellectual skills. With massive use of computers, specialised tools emerged for common tasks. Word-processing systems, spreadsheets, systems for drawing etc. are simple to use. This generation is dominated by the introduction of PCs and the emergence of Microsoft. Today we still use several tools on PC's and on the Internet, even programming languages, but the dominant orientation is not knowledge about tools or programming. The essence in the information age is about handling information. The essence is in *infosphere* - the world where one lives on information. Knowledge how to find, use and create information is the major factor of success.

Table 2: Human-computer generations

Generation no.	Generation name	Main object
I.	*machine-level*	*hardware*
II.	*simple programming*	*assembler*
III.	*programming*	*advanced programming*
IV.	*tools*	*databases, spreadsheets*
V.	*information society*	*information services, Internet*

3 Information Society Laws

The technological basis of the information age is the Internet with its constant exponential growth [7]. The pace of progress is so quick and unpredictable in details that we can not determine future in detail. What we can do is to identify major information society laws and discuss potential consequences [5, 8, 9, 10].

3.1 Moore's Law

Moore's Law (http://www.whatis.com/mooresla.htm) is the best known law in computer science and engineering. It describes a constant exponential trend in chip properties, i.e. density of components per surface unit. The chip density/capacity doubles in a time span from 1.5 to 2 years depending on the type of particular performance of a chip (see Figure 1). The formula is:

$$\text{Performance(new)} = \text{Performance(old)} \; * \; 1.5^{\text{time}}$$

where "time" is the number of years.

Figure 1: Moore's law for chips shows exponential growth.

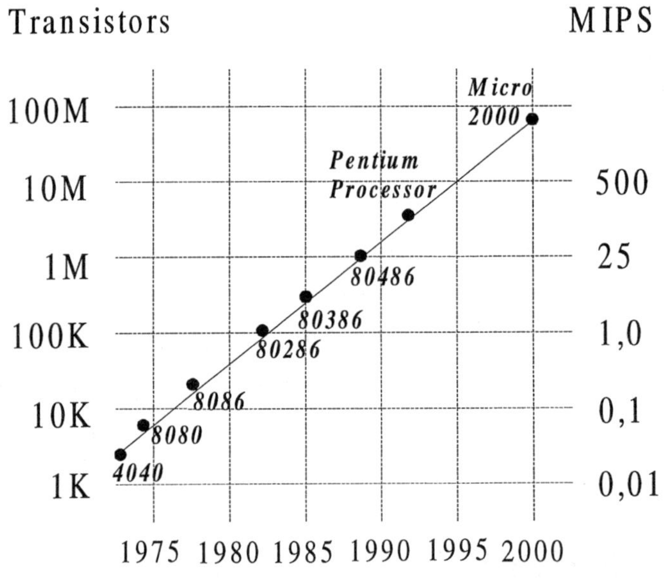

The basic property of the law - the constant exponential growth - remains unchanged over several decades [11, 12]. It will continue for several decades before hitting several unbreakable laws - e.g. the speed of light. Note that exponential growth is commonly presented in logarithmic scales in order to represent it with better visual effect.

Consequences of the law: This is the basic law that determines the growth of several key computer functions. Chips with higher density enable faster performance and better memory capacities. Other components such as hard-disk capacity continue to grow exponentially as well. As a consequence, computers become faster and with better storage capabilities at the same price. New, smaller and lighter versions of computers emerge - palmtops, handheld organisers etc. Advanced microelectronic products find new functions in relation to networks, TVs, other components.

These days, PCs with several 10 GB are something "normal" yet typing one char per second would take us 1000 years. Primary memory with several 100M storage enable storing and sorting large lists without use of slow hard disk thus changing the way algorithms perform. We have enough memory capabilities and enough computing power that PCs on average most of their time wait idle. As a consequence, we are less and less concerned with internal functions of the computer because we don't have to. But we need computers to become more user-friendly and more intelligent assistants.

3.2 Metcalfe's Law

Metcalfe's Law (http://www.cuug.ab.ca/~branderr/csce/ metcalfe.html) says that the value of a network is proportional to the square of the number of nodes, connected by the network:

$$\text{Value} = K * \text{nodes}^2$$

In other words, the bigger the net, the square bigger the value. "K" is a constant.

Consequences: The Internet (or WWW) is the network of networks. Since the value of a network increases with growing number of nodes per square, the Internet is enormously more valuable than lots of small local network even with the same number of users. The globalisation trend is obvious.

3.3 Sidgemore's Law

Sidgemore's Law determines the growth of traffic over nets. The law says that the traffic doubles every three months:

$$\text{Traffic(new)} = \text{Traffic(old)} * 2^{(4 \text{ time})}$$

Consequences: The law says that the information flow through networks also increases exponentially (remember Moore's law). This is already one of the essential reasons for information overflow.

3.4 Andreesen's Law

Andreesen's Law says that the cost of bandwidth is dropping exponentially and inversely proportional to Sidgemore's law:

$$Cost(new) = Cost(old) * 1/2^{(4\ time)}$$

Consequences: Several technologies in computers and electronics enable exponential growth, be it chips, computers, hard disks or bandwidth. It is rather obvious that without exponential growth in capabilities information flow could not be exponential as claimed in the previous Sidgemore's law.

3.5 Lewis/Flemig's Law

Lewis/Flemig's Law describes the network type of capitalism [8]. It denotes nearly "friction-free economy" in the sense that there is small marginal cost and a huge shelf space. The exponential growth indicates that huge profits will award a genuine new market idea. But in addition to quick rise, an exponential decline is expected when new, more advanced systems appear on the market.

The equation describing the law is:

$$MarketShare(time) = 1/(1+ K * B * time)$$

where "K" is a constant. The "B" parameter denotes the learning parameter.

Consequences: Economy in information society puzzles even economists. On the surface, nothing changes, only the type of activities is slightly different. Instead of industrial activities, information services prevail. People and other material goods remain more or less unchanged in numbers. But the speed of business increases as competition becomes global. An interesting idea on the Internet might very quickly attract large amount of users and cash. But very soon another idea in the global Internet cybernetic world becomes attractive and overtakes the throne.

3.6 Put on the Internet all your information and information activities

This law means that it is cheaper to put information and information activities on the Web sooner than later [7]. Not only it is indeed cheaper and more cost-effective than when done in a standard way, it is also the only way to go along with competition.

Consequences: We are doomed to jump into information society, to use the Internet, like it or not. Information economy thrives on information and their value is most precious on the Internet.

3.7 The cyber-world doubles money in circulation

The introduction of the cyber-world in effect tends to double activities and capital. It means that economic growth can be expected even though no such major trend in material production is observed [5, 10].

Humans used to live in Descartes' dualistic two worlds: the physical and the mental. After the introduction of computers, another world emerges - the electronic or cybernetic world. After putting the first programs and text files on the computer, traces of our mental and physical world started to appear in electronic form. Since then, the cyber-copy of our world matures with exponential growth.

The growth is based on computer capability growth as well as on several previously mentioned laws, e.g. about information economy. It also guarantees further progress of the developed world despite saturation in other human activities, which are related to classical material world. Interesting to notice, countries with fast progress, USA or Finland in particular, correspond with overwhelming introduction of information economy. Downfalls on the other hand, e.g. in Japan, correspond to slow introduction of information society.

Many economists write about overblown markets, about e-companies not producing profits. That may be so, but these economists seem to apply their old criteria and parameters to evaluate a new trend - information economy in information society. Old approaches don't show any progress and no support for additional money in circulation. But the reason for increases is in the cyber-world, not in classical improved production or other economic indicators.

Another important trend is that our information systems on computers are becoming more and more a cyber-copy of ourselves. This copying is purely mechanical, in form of text files or pictures. The active intelligent component is missing, however, the context, background and demand for intelligent assistant or maybe even an electronic clone-copy of a human is becoming more and more evident.

3.8 Advanced information education and knowledge is needed

Due to essential changes in the way we do business, there is a big gap between existing and needed knowledge in practically all profiles [13].

The gap is especially big in regards to top executives, politicians and other leaders without information education. These professions are largely oriented towards dealing with other people, e.g. politicians tend to get support of a common voter, therefore, their primary orientation is not towards progress and information society. All over the world, there is a big gap compared to the desired knowledge for successful managing and leadership, but the biggest gap is in not in the most developed countries. Several leaders of most advanced countries in the world like USA or some EU countries (Clinton, Blaire) have intensively propagated information society in their election programmes, but an average political leader has not even mentioned these issues.

In general, it is advised that all top decision-makers once per couple of years attend specialised courses and/or hire a special information-educated deputy.

Information society is often met with scepticism especially at not well-educated middle-aged humans working in the same professions for decades. For them, the introduction of information activities often means unemployment or at least a pressure to start learning again. Most of the unemployed people are those without diploma while the need for workers with information education constantly increases. This is the reason while countries like USA or Germany apply special politics for workers with this type of education. According to estimates of American IT association (http://www.acm.org/technews/current/homepage.html), 10 million Americans are employed in the IT (information technologies) field. In 2000, 1.6 million new high-tech jobs are expected to be created.

In general, more educated people also possess more knowledge about information society. According to US statistics, an average person with university diploma earns about twice as much as those without high-school diploma. An average unemployed person is low-educated.

3.9 Information society is a world's society of partners

In any society there are hierarchies, owners and "users". In information society, all humans with Internet access are partners.

All material world, e.g. land, is partitioned between owners and non-owners. Only air and offshore sea is international without owners. Similarly, Internet has no owners. Anybody with Internet access can receive and submit information. This is not something capitalists or governments want. It is a result of academic research, a by-product of intellectual, technological power and human free will.

Among essential institutions of democratic societies are civil institutions [14]. Information society is by definition a civil society although governmental institutions typically implement it. An example would be Clinton's advocating of information highway or several governmental information society projects in Europe; e.g. Bangemann's reports (http://europa.eu.int/comm/dg03/speechba.htm). In less developed countries, lacking richness of civil society structures developed in decades of Western democracy, the introduction of Internet is a major inhibitor of faster progress.

In this respect, science and technology have again given something precious to humanity. In public and among politicians, these facts are too easily forgotten, especially in Eastern and Central European countries in transition where economy troubles caused substantial downgrading of governmental funding.

3.10 The Internet is the most democratic and free media in the world

Information society promotes democracy and enables new advanced forms of democracy [15]. In any society there are some hierarchical structures responsible for performing the decision process. It is often said that democratic countries are those that vote president democratically. This seems quite far from truth since once elected, president or political party may not fulfil wishes of fellows countrymen. There are some notable exceptions such as Switzerland with its regular referendums, but current average political solution in the world is still very far from the general idea of ideal democracy. With electronic voting, referendums could become an essential mechanism in every truly democratic country.

The Internet is also much more democratic compared to other mass media due to simple fact that its technical abilities are more advanced. This was legally established with the American Supreme Court decisions about pornography and free speech on the Internet. In the simplest way it can be observed as a fact that pornography (inside "reasonable" limits) is allowed on the Internet and not on public TV. It also means that anarchy and even criminal organisations can exploit this freedom, but the freedom of speech is accompanied by the fact that in such a case sooner or later we are going to hear things we don't like.

It is hard to imagine dictatorship and information blackout in the Internet era. The Internet just breaks the information barriers apart. Therefore, the Internet is the most democratic media humans ever had. While countries differ in their social and economic order, the Internet enhances democracy and civil society regardless of their previous level.

3.11 Globalisation introduces and destroys monopolies

Microsoft is a company that overthrew IBM by adopting to the new generation of tools (See Table 2) and becoming a world monopolist in PC software [5, 16]. Currently, Microsoft covers over 90% of all operating systems on computers sold worldwide. Will it enhance its monopoly to the next computer-human information-society generation as presented in Table 2? Looking to rules 3.9 and 3.10 one would say that the globalisation effect in democracy would likely cause a conflict between the monopolist and the democratic Internet society. Indeed, that has happened with recent Microsoft trial. Mighty Microsoft has already paid the price for its old-thinking leadership. Monopoly is acceptable at the level of word-processing, but the open-minded Internet society is not tolerant to monopolies of any kind. Microsoft's Bill Gates has become the richest man in the world by a superb technological vision and aggressive monopolistic behaviour. Understanding information society relations seems to be a problem even for the most advanced technical geeks of the past tools generation. Applying old approaches in the new society never worked too well.

3.12 Side effect of information society is information overload

In infosphere we have to cope with more and more information from one month to another in order just to stay competitive. As a consequence, the information overload causes disappearance of free time; it causes the brain overload and decrease in classical human social life.

According to Table 3 we are approaching terminal velocity [5]. E.g., the generation duration is shortening to irrational short periods. An end to this phenomenon must be in sight!

But the relevant approaching limit is not the light-speed limit saturating the Moore's law. It is a human limited speed of accepting new information. In the information overload we simply can not receive the amount of information we ought to due to its exponential growth (see rules 3.1, 3.3).

Just one decade ago projects lasted for several years. Now, according to this and previous laws, e.g., 3.5, we have to finish projects in months if not in weeks. If we don't, somebody else will. This is the conflict we are facing. We can not work 48 hours per day to finish our work in time. We can not conceive and give birth in 9 weeks. The conflict between human and information clock is paramount.

Table 3: Information overload causes the first major block in information society.

SOCIETY	DURATION	COMMUNICATION SPEED
agrarian	3000-5000 years	3-5 km/h human

industrial	300-500	30-50 horse, car
post-industrial	30-50	300-500 aeroplane
information	3-5	3000-5000 network

Note that data in Table 3 are slightly modified to show the major trend.

4 Summary: Information-Age Intelligence

Information society is our future and hope for the future. There have been many technological innovations that spurred human progress in giant leaps. For example, we speak of the "iron age" historic period. These days we speak of the "information age" or of the "information-society age". Not only new technology changes the way we live in the technical sense - the changes are essential also in the way society functions [15]. At the same time, the world of computer systems we use is rapidly changing due to the massive introduction of information activities. The trend is towards more user-friendly intelligent systems [17].

An important improvement in information society is the emergence of network intelligence that represents a natural step in the computational evolution heading towards more helpful, adaptive and creative programs. Artificial intelligence [18] has long been known to oscillate from overoptimistic to overpessimistic periods. The current curve seems to be on raise again. The major public break-through has been the Kasparov vs. Deep Blue chess duel with millions watchers on the Internet [19]. While this was a big media event that substantially changed the way humans treated artificial intelligence, the practical change behind public awareness is based on major numbers of intelligent systems [20] and agents [21] massively used throughout the Internet.

Look at the laws of the information age! They strongly indicate that we are approaching the terminal velocity because we can not cope anymore with the information overflow. We urgently need intelligent systems to help us manage the enormous amount of information. We can create intelligent systems and agents as intelligent assistants to help us. Maybe soon, these artificial autonomous and social entities will inhabit the Internet as Bradshaw proposes [22].

More needed, though, is active help. We can not work for 48 hours a day to fulfil our demands. If we only could clone ourselves a couple of times and make cloned assistants work for us. But then again, that is similar to what some of intelligent systems do already. Seven days a week they work 24 hours with thousands of users at the same time. They might still be dumb, but their ability is growing. Although far from approaching levels of true intelligence, these systems are going to the more and more helpful by becoming more and more technologically advanced and intelligent. Practically all rules of the information age lead us to the same conclusion.

References

[1] Etzioni, O. (1996): The WWW: Quagmire or Gold Mine?. Communications of the ACM 39, 11, 65-68

[2] Bradshaw, J. M. (Ed.) (1997): Software Agents. AAAI Press, MIT Press

[3] O'Leary, D. E. (1997): A Lack of Knowledge at the Top. IEEE Expert, November, 2

[4] Maes, P. (1997): Humanising the global computer. IEEE Internet Computing, July/August, 10-30

[5] Lewis, T. G. (1999): Microsoft Rising ... and other tales of Sillicon Valley, IEEE Computer Society

[6] Maes, P., Gutmann, R. H., Moukas, A. G. (1999): Agents that Buy and Sell: Transforming Commerce as we Know It. Communications of the ACM 42, 3, March, ACM Press, 81-91

[7] Petrie, C. J., Rutkovski, A. M., Zacks, M. etc. (1998): Dimensioning the Internet. IEEE Internet Computing, April, 8-9

[8] Lewis, T. (1997): Anderseen's laws vs. the techno treadmill. IEEE Internet Computing, September, 93-96

[9] Metcalfe, B. (1997): What's Wrong with the Internet?. IEEE Internet Computing, 6-8

[10] Gams, M. (1998): Information Society Laws. In: Gams, M. (Ed.): Information Society 98, Državna založba Slovenije (in Slovene), 1- 4

[11] Moore, G. E. (1975): Progress in digital integrated electronics. Technical Digest of 1975 International Electronic Devices Meeting 11

[12] Hamilton, S. (1999): Taking Moore's Law into the Next Century. IEEE Computer, Januar, 43-48

[13] O'Leary, D. E. (1997): A Lack of Knowledge at the Top. IEEE Expert, November, 2

[14] Borenstein, N. S. (1998): Whose Net is it Anyway?. Communications of the ACM, April, 19-21

[15] Negroponte, N. (1998): A Wired Worldview. EU RTD info, March, 28-30

[16] Hamilton, S. (1998): Inside Microsoft Research. Computer, January, 51-58

[17] Goonatilake, S., Treleaven, P. (Eds.) (1995): Intelligent Systems for Finance and Business. Wiley

[18] Davis, R. (1998): What Are Intelligence? And Why?. AI Magazine, Spring, 91-110

[19] Krol, M. (1999): Have We Witnessed a Real-life Turing Test?. Computer, March, 27-30

[20] Buchanan, B., Uthurusamy, S. (Eds.) (1999): Innovative Applications of Artificial Intelligence. AI Magazine, Spring

[21] Joshi, A., Singh, M. P., Ma, M. (Eds.) (1999): Multiagent Systems on the Net and Agents in E-commerce. Communications of the ACM 42, 3, March

[22] Bradshaw, J. M. (1999): Steps towards the permanent colonization of cyberscape. In: Liu, J., Zhong, N. (Eds.) Intelligent Agents Technology, World Scientific, 8

On Post-Modularity and Emergence from a Grammar-Theoretic Point of View[1]

Jozef Kelemen

Institute of Computer Science, Silesian University 746 01 Opava, Czech Republic, and Department of Applied Informatics, University of Economics 852 35 Bratislava, Slovakia, email: kelemen@{fpf.slu.cz, euba.sk}

Abstract: A possibility of capturing some aspects of a new paradigm of systems design – called post-modularity by Stein in 1997 – in the theoretical framework of grammar systems developed in the theory of formal grammars and languages, is sketched and related to (decentralized) rule-based systems studied in artificial intelligence and knowledge engineering research. It is argued that the phenomenon appearing in language generation by grammar systems belongs, according the emergence test invented by Ronald, Sipper and Capcarrére in 1999, to the category of emergent phenomena.

Keywords: artificial intelligence, rule-based systems, formal languages, grammar systems, colonies, modularity (of systems), post-modularity, emergence

1. Introduction

The history of modern computing can be divided into three main periods:

(1) The *period of sequential computing* on one-processor machines.

(2) The *period of parallel computing* on multiprocessor machines.

(3) The *period of decentralized computing* using more or less freely cooperating or communicating processors generating solutions of given problems as an emergent effect of their individual functioning and mutual interactions.

Computing systems of the first period has been built from components which were defined through their careful functional (input/output) specifications with respect the functional specification of the desired system. The modularity resulting from such an approach lies in the core of the traditional functional modularity of systems. However, the experiences form different fields dealing with some computing systems of the third period as well as with some *prima face* non-computational systems of this period – like behavior-based (collective) robotics (Connell, 1990), (Arkin, 1997, 1998), (Brooks, 1999), artificial life (Langton,

[1] Research supported by the Grant Agency of the Czech Republic, Grant No. 201/99/1086.

1995), multi-agent systems (Ferber, 1999), artificial economies (Epstein, Axtell, 1996), etc. – result in considering another possibility how to set up systems from components. In these fields usually the larger or the more complex systems evolve incrementally, usually by adding new components to the existing ones (and by deleting some others). In many cases this is the only possible way how to change the systems architecture at all. The communication among these components is usually very opportunistic and guided by desired global behavior of the resulting systems. Consequently, the modularity of such systems is principally different from those of the previous ones.

Discussions on the *emergence* (e.g. Holland, 1998) and esp. on the *emergent computation* (e.g. Forrest, 1991) collect and confront the ideas appearing in the third period of development of our view of computing. Despite of the fact that – as S. Forrest pointed out (Forrest, 1991) – the concept of emergent computation cannot contribute magical computational properties, she advocates a way of thinking about the design of computational systems that could potentially lead to radically different architectures which are more robust and efficient than current designs.

Changes in the basic view of computing have influenced the development of different theories of computational systems in order to understand the possibilities and limitations provided by them. In (Stein, 1997), the term *post-modularity* is used in order to characterize the architecture of such kind of systems. To understand systems modularized in this new way is a highly actual challenge for theoretical computer science. Traditional theories cannot be used for plausible description of systems modularized in such a new way nor for making any rigorous predictions of their behavior. Rather than imagine and develop completely new conceptual approaches to understand the new modularity theoretically, we propose to reuse at least some of the existing ones. There are several reasons to proceed in such a way: the existence of working experiences, the well prepared and skillful researchers, and the necessary relations of the new subject to the traditional approaches are some among the most important ones.

The aim of this contribution is to illustrate the possibility of how to deal with post-modular systems in the formal framework of the theory of grammar systems, a field which satisfy the previously listed requirements. We will concentrate to the computing systems of the third period of development of computing. A possibility of capturing some aspects of a new paradigm of systems design – the so called *post-modularity* – in the theoretical framework of *grammar systems* – a sub-field of the theory of formal grammars and languages – will be sketched and related to (decentralized) rule-based systems studied in artificial intelligence and knowledge engineering research. It will be argued that the phenomenon appearing in *language generation by grammar systems* belongs, according the emergence test invented in (Ronald et al., 1999), to the category of *emergent phenomena.*

2. Rule-based systems

In order to focus into some technical detail, in this contribution we will deal with rule-based systems. A *rule-based systems* will be considered (in its simplest form usually discussed in the literature on knowledge systems) as a systems consisting of :

(1) a finite *data-base* of some items,

(2) a finite set of *rules* replacing or rewriting the corresponding items by/to another ones, and

(3) an *interpreter* which makes the rewritings effectively possible and decides about what rule will be used when.

The structure of the data-base we will consider as a string of items. We will consider some items appearing in the data-base as some target, terminal items which cannot be rewritten by any production rule. The structure of the rules will be supposed in the form left-hand side $\rightarrow$ right-hand side Rules change strings (the data-base) by replacing items appearing in their left-hand sides by items (or strings of items) appearing in their right-hand sides. The interpreter will select rules non-deterministically among those which are just applicable to the currant state or the data-base. If there is no such rule, the rewriting process ends. The basic function of such simple version of rule-based systems consists in transforming some initial state(s) of the data-base interpreting rules into some required terminal form(s) of it.

Now we give a very simple example: Our items will be the symbols *S, a* and *b*, the initial state of the data-base will containing the symbol *S*, and the finite set of rules $\{S \rightarrow aSb,\ S \rightarrow ab\}$ will be at hand for rewritings. We can ask, because of some practical reason, for the possibility of generating the set of all strings in the common form $a^n b^n$ (for arbitrary positive *n*) over the set $\{a, b\}$ of terminal symbols by this rule-based system. How to design such a system?

3. Traditional modularity of rule-based systems

To generate the required set of strings let us imagine a module presented by the rule $S \rightarrow aSb$, which changes the contents of the data-base, and makes also the activity of the second module possible, which is presented by the rules $S \rightarrow ab$. The first module/rule will work sequentially during the rewriting process in order to generate a string of structure of the form *aaa....aSbbb....b*. The *S* will be, at the end of the derivation process, rewritten into *ab* by the second rule. In such a way, for arbitrary given positive *i* the interpreter is capable to generate the string of symbols *a* and *b* of the above required form. Note that each rule plays in the above sketched process its own precisely defined functional role.

This style of thinking and the resulting system decomposition are examples of systems decomposition into functionally specified modules. For instance, all traditional formal grammars widely studied in the literature on theoretical computer science are in certain sense examples of functionally modularized generative systems for generating (infinite) sets of symbol strings – formal languages. But while functional decomposition is certainly a legitimate architecture principle, it is not the only one applied in the present time computer science.

4. Rule-based systems and formal grammars

The examples of rule-based systems given in the previous two sections are for those familiar with the basics of the theory of formal grammars and languages similar to formal grammars.

If, for the situation described in the example in the sections 2 and 3, we consider the finite set $\{S, a, b\}$ as a total alphabet, the finite set $\{a, b\}$ as a terminal alphabet (or an alphabet of terminal symbols), the finite set of rewriting rules $\{ S \rightarrow aSb, S \rightarrow ab\}$, and the symbol S as the starting non-terminal symbol, these items define completely a structure called *formal grammar* in the theory of formal grammars and languages. This formal grammar, say G, generates – in the sense specified formally e.g. in (Salomaa, 1973) – the corresponding *formal language* over the terminal alphabet $\{a, b\}$, namely the infinite *language*

$$L(G) = \{ a^n b^n \mid n \geq 1\}.$$

The process of how the generation of the so called *terminal words* (elements) of $L(G)$ proceeds using the rewriting rules of G consists in the following three main steps of the string generation:

1. The string of symbols aSb (or ab, and all the porocess ends with that) over the total alphabet of G (a so called *sentential form*) is generated from the starting non-terminal S of G.

2. Then the strings of terminal symbols a and b are recursively generated using the rule $S \rightarrow aSb$.

3. In the third basic step the generative process consists in finalizing the generation of a finite string of terminal symbols of G using the rewriting rules $S \rightarrow ab$.

Note, that the ordering of the above mentioned basic steps cannot be changed. In each basic step some conditions are prepared for starting the next one, and what some basic step will be executed, is strictly predefined and cannot be changed without substantial change of the set defined by the grammar G. In this sense, we can consider traditional formal grammars as generative devices which reflects the paradigm of traditional modularization as sketched in the Section 3.

5. Post-modular rule-based systems

L. Stein (1997) provides several reasons to suppose that many systems are neither analyzable nor synthesable from the previously sketched positions. This follows mainly from the interactive nature of some non-computational systems (like behavior-based robots) but increasingly also from many computational systems (like software agents or computer networks). The common characteristics of such systems is their massive mutual interaction and/or their massive interaction with dynamically changing environments in which there are (physically or informationally) situated.

Generally, an *interactive system* is – according Stein – one in which information is continually being sensed and behavior is continually being generated. Interactive systems exploit properties of their environment to generate future behaviors. They are always operating concurrently with their environment. While traditional computer procedures – while executing – are blissfully ignorant of their environment, writes Stein, an interactive system, in contrast, is always responding to its environment. It means also, that interactive systems are opportunistic in the sense that they depend on simpler systems without necessarily preserving modular-functional encapsulation with all of the redundancies, interdependence, etc. Fortunately, this does not mean that we must give up all our conviction in principled analysis or synthesis. Stein (1997) provides three principles – *imagination*, *representational alignment through shared grounding*, and *incremental adaptation* – for analyzing and synthesis of so called *post-modular systems*. However, the basic principle derived from experiences e.g. in behavior-based robotics, software agent technology or computer networking and open information system design, is that the post-modular systems are set up opportunistically from parts working in the massive interaction with their environment(s) where *interactive* systems exploit their environments to generate future behaviors, and that the behavior of the whole post-modular system emerges from this interaction.

6. On post-modularity of grammar systems and colonies

In the traditional theory of formal grammars and languages the languages are considered as possibly infinite objects generated by finite number of rules of corresponding formal grammars. *Grammar systems* as presented in (Csuhaj-Varjú et al., 1994) or in (Dassow et al., 1997) represent another possibility. Grammar systems define languages through finite sets of formal grammars working in shared strings of symbols (sentential forms). Many variants of grammar systems are studied having different motivations mainly in distribute, cooperative, or multi-agent symbol processing systems.

Perhaps the simplest variant of grammar systems is a *colony*. Originally, colonies have been invented in (Kelemen, Kelemenová, 1992) as formal generative models of purely reactive agents studied in robotics (Connell, 1990), (Brooks, 1999). Colonies as well-formalized language generating devices have been developed during the nineties in several directions in lot of papers, e.g. (Dassow et al., 1993), (Kelemenová, Csuhaj-Varjú, 1994), (Paun, 1995), (Baník, 1996), (Sosík, Štýbnar, 1997), (Martín-Vide, Paun, 1998, 1999), (Kelemenová, 1999), (Sosík, 1999), (Gašo, 2000), (Kelemen et al., 2000a) etc. In (Kelemen 1997a, 1997b) colonies are connected with the field of reactive robotics, while in (Kelemen, Paun, 2000) with rule-based systems. This contribution may be considered as a continuation of looking for certain further interplays in this direction.

Formally, a *colony* C is a triple (R, V, T) set up from:

- a finite set R of *components* $R_i \in R$ of C,
- a finite set V of symbols called the *total alphabet* of C,
- and a finite *terminal alphabet* $T \subseteq V$ of the colony C.

Components $R_i \in R$ $(1 \leq i \leq n)$ of a colony C are *regular grammars generating finite languages* and operating on a common shared string of symbols – the *environment* of the colony – without any explicitly predefined strategy of cooperation of the components. A terminal symbol of one component can occur as a non-terminal symbol of another one.

An *environment* of a colony is formed by strings of symbols from a finite alphabet V (the environment is, in fact, the structure identical with the so called *sentential form* known in the classical theory of formal grammars and languages). Let us suppose that strings are modified only by sequential activities of components of a colony[2]. Because of the lack of any predefined strategy of cooperation of components, each component can participate in the rewriting of the current string, whenever its start-symbol is present in the actual string. Conflicts are solved non-deterministically, as it is usual in the classical theory of formal grammars.

The *language* L_C generated by a colony C is formed as a subset of all strings over the terminal alphabet T, which are generated by the components from R of the colony C from a given starting string. A terminal symbol of one component can occur as a non-terminal symbol of another one, so the possible cooperation of components of the colony allows to generate by colonies substantially more than finite languages.

As an example of a colony C, let us consider:

- A two elements set R of components R_1 working over the total alphabet $\{A, B, X, a\}$, the terminal alphabet $\{a, B\}$ with rules $\{A \rightarrow aX, X \rightarrow B\}$, and R_2

---

[2] In (Dassow et al., 1993) the parallel case – the so called *parallel colonies* with two different kind of parallelism – is invented and discussed.

working over the total alphabet $\{A, B, b, Y\}$, using the set of rules $\{B \rightarrow b, B \rightarrow AY, Y \rightarrow b\}$, over the terminal alphabet $\{A, b\}$.

- The total alphabet $V = \{A, B, a, b, X, Y\}$.
- The terminal alphabet $T = \{a, b\}$.

It is easy to see that – starting with the symbol A – the colony C generates the language

$$L(C) = \{ a^n b^n \mid n \geq 1\}.$$

It is interesting to realize that $L(C)$ is an *infinite* language, while both of the two components R_1 and R_2 of the colony C generate two *finite* languages, namely the languages $L(R_1) = \{aB\}$ and $L(R_2) = \{b, Ab\}$, only.

Grammar systems and colonies are formal devices which express an important features of post-modular systems postulated by L. Stein (1997) – namely the *representational alignment through shared grounding.*

Realize first, that components of a grammar system or of a colony are really situated in their environments, of course in very simple way. However, if „situatedness" is understood as exploration of the environment for generating components own behavior, in the case of grammar systems or colonies components we are confronted exactly with that.

Second, making possible for the components of a grammar system or of a colony to interact with a shared environment – with the shared sentential form, in this case – a we can obtain (lot of proved theorems – e. g. in (Csuhaj-Varjú et al., 1994) – show this possibility in the theory of grammar systems and colonies) completely different behaviors (languages) as the behaviors (languages) of the components. In the case of colonies this fact is very spectacular, because from the interaction of components – which generate finite languages only – with a shared environment we receive infinite behaviors. Moreover, having in the mind the fact that an arbitrary finite language is a regular language in the *Chomsky hierarchy* of languages, colonies define on the base of such finite languages the whole family of context-free languages – a considerably larger family of languages in this hierarchy containing all the family of regular languages. Putting some additional – simple and very natural – conditions on the capabilities of components and without changing their individual generative capacities, the generative capacity of colonies can be enlarged also to some of the context-sensitive languages, cf. (Kelemen, Kelemenová, 1992), (Baník, 1996), (Kelemen et al., 2000a).

So, colonies enable us to prove formally the fact that, as Stein (1997) states, we can capitalize on the observation that interactive systems situated in a shared environment often rely on some of the environment regularities, putting aspects of two systems into alignment, to bootstrap a shared „understanding" between the two systems[3].

More spectacular – but technically more complicated – are the so called *eco-grammar systems* which express some life-like properties of decentralized symbol systems (Csuhaj-Varjú et al., 1997). Eco-grammar systems represent a modification of grammar systems in several aspects. One among the important feature of eco-grammar systems is that the environment in which the components work has it own dynamics which is independent on the components activities. Components are formal grammars influencing the environment (represented by a string of symbols). In some variant of eco-grammar systems the direct interactions between components are alowed, too. Moreover, some internal changes (development) of components is provided in dependence on the environment states and dynamics.

7. On emergence in colonies

Observing the behavior of a post-modular system (similarly as that of a grammar system) we mantion that the complex behavior of the whole system comes from the simpler behaviors of its massively interacting components. As in Section 3 we have already stated, „...the behavior of the whole *post-modular system* emerges from this interaction". Now, we will argue in short for the suitable use of the word *emergence* in the context sketched in this contribution. We will do that by applying a test which indicates whether some phenomena appearing in grammar systems (considered to be a model of post-modular systems) emerges or not.

E. M. A. Ronald et al. (1999) have formulated a *test of emergence* trying to offer an operant definition of emergence for artificial life experiments. The requirements putted onto a systems in which the phenomenon of emergence appears are the following:

Design. The designer designs the systems by describing *local* interactions between components in a language L_1.

Observations. The observer describes global behaviors of the running system using a language L_2.

Surprise. The language of design L_1 and the language of observation L_2 are distinct, and the causal link between the elementary interactions programmed in L_1 and the observations observed in L_2 is non-obvious.

Imagine the situation described in the example in Section 3 and analyze the *design* process in more details, first. Imagine a designer (or more designers) who „program(s)" (defines) the particular simple modules, say R_1 and R_2, for generating some required finite sets of strings. The designer is satisfied because the

[3] More technical results on the post-modular nature of colonies can be found in (Kelemen et al., 2000b).

constructed modules work fits the given requirements. Using both of these generative devices in isolation, they generate the simple sum of the two behaviors – the *finite* language *{aB, Ab, b}*. No surprise during the observation of such kind of system created from the (just described) *isolated* modules.

Now, imagine that the designer(s) put the modules into a shared environment consisting of a string *AB* . What they will *observe*? The global behavior of the whole systems will be now an *infinite* language (instead of the *finite* one from the previous case), the language $\{ a^n b^n \mid n \geq 1\}$. However, the design of the two components remains unchanged! So, while the language of the *designer* has been concerned with *finite* behaviors, the language of the *observer* of the systems created in a rather simple way form the designed components will be concerned with *infinite* behaviors (languages).

What about the *surprise*? The impression of a surprise can become from realizing the just demonstrated finite/infinite gap between the behavior of the above mentioned two systems set up from the same components but working in different environments (starting with different starting strings). But theoretically, some more surprising facts are proved concerning colonies. As we have mentioned already, colonies – devices set up from regular grammars generating finite languages – define the whole family of context-free languages, and if they work with some simple and very „natural" additional restrictions, they generative capacity can be enlarged also to some of the context-sensitive languages. Moreover, in the case of the finite languages some problems seems to be highly actual, for instance the complexity of defining such languages, and some of them are completely trivial, like problems concerning decidability, computability, etc. Some questions become to be meaningful only in the case of infinite languages. In other words, there are two (not completely distinct, of course) languages for talking (and thinking) about finite languages and devices generating them, and the infinite ones.

If we understood the Ronald's and his colleagues (1999) *test of emergence* well, we could state that colonies satisfy it, and that they offer a quite simple and theoretically well grounded example of how the test can be applied. In such a case, moreover, we can demonstrate the appearance of the phenomenon of the emergence in theoretically (mathematically) well described formal systems, too, not only in more or less experimental situations referred in (Ronald et al., 1999).

References:

1. Arkin, R. C. (Ed.): *Robot Colonies*. Kluwer, Boston, Mass., 1997
2. Arkin, R. C.: *Behavior-Based Robotics*. MIT Press, Cambridge, Mass., 1998
3. Baník, I.: Colonies with positions. *Comp. and Artif. Intel.* **15** (1996) 141-154
4. Brooks, R. A.: *Cambrian Intelligence*. MIT Press, Cambridge, Mass., 1999

5. Connell, J. H.: *Minimalist Mobile Robotics – A Colony Architecture for a Mobile Robot*. Academic Press, New York, 1990

6. Csuhaj-Varjú, E., Dassow, J., Kelemen, J., Paun, Gh.: *Grammar Systems*. Gordon and Breach, London, 1994

7. Csuhaj-Varjú, E., Kelemen, J., Kelemenová, A., Paun, Gh.: Eco-grammar systems – a grammatical framework for lifelike interactions. *Artificial Life* **3** (1997) 1-28

8. Dassow, J., Kelemen, J., Paun, Gh.: On parallelism in colonies. *Cybernetics and Systems* **24** (1993) 37-49

9. Dassow, J., Paun, Gh., Rozenberg, G.: Grammar systems. In: *Handbook of Formal Languages, vol. 2.* (G. Rozenberg, A. Salomaa, eds.) Springer-Verlag, Berlin, 1997, pp. 155-214

10. Epstein, J. M., Axtell, R.: *Growing Artificial Societies – Social Science from the Bottom Up*. The MIT Press, Cambridge, Mass., 1996

11. Ferber, J.: *Multi-Agent Systems – An Introduction to Distributed Artificial Intelligence*. Addison-Wesley, Harlow, 1999

12. Forrest, S.: Emergent computation – self-organizing, collective, and cooperative phenomena in natural and artificial computing networks. In: *Emergent Computation* (S. Forrest, Ed.). The MIT Press, Cambridge, Mass., 1991, pp. 1-11

13. Gašo, J.: Unreliable colonies – the sequential case. *Journal of Automata, Languages and Combinatorics* **5** (2000) 31-44

14. Holland, J. H.: *Emergence – from Chaos to Order*. Addison-Wesley, Reading, Mass., 1998

15. Kelemen, J.: Colonies as models of reactive systems. In: *New Trends in Formal Languages*. (Gh. Paun, A. Salomaa, eds.) Springer-Verlag, Berlin, 1997a, pp. 220-235

16. Kelemen, J.: Colonies – grammars of reactive systems. In: *Artificial Intelligence and Information-Control Systems of Robots '97*. (I. Plander, Ed.) World Scientific, Singapore, 1997b, pp. 27-40

17. Kelemen, J., Kelemenová, A.: A grammar-theoretic treatment of multiagent systems. *Cybernetics and Systems* **23** (1992) 621-633

18. Kelemen, J., Kelemenová, A., Martín-Vide, C., Mitrana, V.: Colonies with limited activation of components. *Theoretical Computer Science* 2000a (accept.)

19. Kelemen, J., Kelemenová, A., Mitrana, V.: Neo-modularity and colonies. In: *Sequences, Grammars, Languages I – Where Do Mathematics, Computer Science, Linguistics and Biology Meet*. Kluwer, Dordrecht, 2000b (accept.)

20. Kelemen, J., Paun, Gh.: Robustness af decentralized knowledge systems – a grammar-theoretic view. *Journal of Exp. & Theor. Artificial Intelligence* **12** (2000) in press

21. Kelemenová, A.: Timing in colonies. In: *Grammatical Models of Multi-Agent Systems*. (Gh. Paun, A. Salomaa, eds.) Gordon and Breach, London, 1999, pp. 136-143

22. Kelemenová, A., Csuhaj-Varjú, E.: Languages of colonies. *Theoretical Computer Science* **134** (1994) 119-130

23. Langton, Ch. (Ed.): *Artificial Life – An Overview.* MIT Press, Cambridge, Mass., 1995

24. Martín-Vide, C., Paun, Gh.: PM-colonies. *Comp. and Artif. Intel.* **17** (1998) 553-582

25. Martín-Vide, C., Paun, Gh.: New topics in colonies theory. *Grammars* **1** (1999) 209-223

26. Ronald, E. M. A., Sipper, M., Capcarrére, M. S.: Design, observation, surprise! A test of emergence. *Artificial Life* **5** (1999) 225-239

27. Salomaa, A.: *Formal Languages.* Academic Press, New York, 1973

28. Sosík, P.: Parallel accepting colonies and neural networks. In: *Grammatical Models of Multi-Agent Systems.* (Gh. Paun and A. Salomaa, eds.) Gordon and Breach, London, 1999, pp. 144-156

29. Sosík, P., Štýbnar, L.: Grammatical inference of colonies. In: *New Trends in Formal Languages.* (Gh. Paun, A. Salomaa, eds.) Springer-Verlag, Berlin, 1997, pp. 236-246

30. Stein, L. A.: Post-modular systems – architectural principles for cognitive robotics. *Cybernetics and Systems* **28** (1997) 471-487

Part 4

Hybrid Systems

Review of Integration Strategies in Neural Hybrid Systems

Jiří Šíma*

Institute of Computer Science, Academy of Sciences of the Czech Republic,
Pod vodárenskou věží 2, 182 07 Prague 8, Czech Republic, *sima@cs.cas.cz*

Abstract: In the present paper both the explicit and implicit neural approaches to representing the knowledge base are compared from various points of view. Then several integration strategies to eliminate the drawbacks of both technologies that have been identified among the existing prototypical neural hybrid systems are reviewed. Thus a basic taxonomy of neural hybrid systems is proposed.

1 Comparison of Symbolic and Neural Approaches

The representation of large amounts of knowledge in computers ensuring their integrity, consistency, and effective exploitation is one of the main issues of artificial intelligence. The knowledge representation is usually specified by a data format in which the information is stored in the knowledge base. The format should handle a natural inaccurate and incomplete character of human knowledge. Clearly, the representation also determines of how the knowledge is processed by an inference engine.

In a classical approach the representation is based on a symbolic expression of human knowledge, skills, and experience, which is called the *explicit knowledge representation* because the respective data format ensures that the information is comprehensible to specialists at least. Typical examples of explicit knowledge representation, e.g. in expert systems [2,8,24] include IF–THEN rules, inference networks, frames, automata, tables with probability-related information, etc. whose processing is based on fuzzy logic, Bayesian theory, for example.

Nevertheless, the human knowledge can be often hardly expressed in a computer format. Typically, a specialist is unable to formulate her/his knowledge and experience using rules because, in fact, human experts do not usually apply formal logic to each situation, but they rather associate the new case with some old pattern to derive a solution. Moreover, some problem domains naturally include numerical data instead of rules, e.g. visual images, signals, etc. This data requires additional processing to be really understood and thus they represent the *implicit knowledge*.

Therefore an alternative approach suggests to exploit (artificial) *neural networks* [27] for the implicit knowledge representation [5,10,16,30]. This represents the main focus of the present paper. The neural knowledge base is

* Research supported by grants GA ČR No. 201/00/1489, GA AS CR B2030007.

automatically learned from example inferences. Thus the information is implicitly distributed into numerical weight parameters within a neural network and cannot easily be read without additional processing. The inference engine simply evaluates the function value of the neural network for a given input and produces the corresponding output while the network is reminiscent of a 'black box'.

The explicit and implicit neural approaches to knowledge representation can be compared from several points of view:

Knowledge Acquisition and Editing: In the explicit knowledge representation typically a knowledge engineer with the aid of an expert from the problem domain formulates particular rules, estimates their confidences and debugs the resulting knowledge base (e.g. appropriate order of rules and numerical confidence refinement) in order to achieve a reasonable performance of the inference engine. This is a very time-consuming process (even years) and further editing can cause side-effects. In neural networks an expert is generally not necessary and the knowledge is implicitly contained in a training set which consists of representative sample inferences sufficiently covering the problem domain. The main advantage of the neural approach is that building a neural knowledge base is automated by learning from the training set, which may take only a few weeks or months. Moreover, additional learning to debug the inference is possible in this case.

Task Size: In the explicit representation the knowledge base can include plenty of rules for solving very large and complex problems. On the contrary, in neural networks the learning task for larger problems (thousands of neurons and patterns) can be intractable [31]. Therefore the size of the neural knowledge base is limited and single neural networks are suitable only for partial subproblems with restricted domains.

Partial Matching: The inference engine is sometimes able to exploit partially satisfied rules to infer reasonable conclusions by using e.g. fuzzy logic when no condition of all rules in the knowledge base is entirely satisfied and the conditions match the facts only partially. In neural systems this problem arises when the presented input has not occurred in the training set. Nevertheless, the neural networks are especially suitable for partial matching because of their good generalization capability.

Incomplete Information: In the explicit knowledge representation incomplete or even inaccurate input information is handled by using confidence calculations or fuzzy logic. The system builds gradually the internal model of the case being currently solved on the basis of a possibly incomplete input and provides partial hypotheses and conclusions. In neural networks, all inputs (including unknown and irrelevant ones) have to be specified to perform the neural inference. This makes operating with incomplete information difficult or even impossible [6]. Sometimes, the input or output neuron state may be interpreted as a confidence in the corresponding facts.

Explanation Capabilities: The explicit knowledge representation provides perfect explanation capabilities. The inference process is transparent and comprehensible due to the explicitness of its logical steps. On the other hand, the implicit neural representation makes such an explanation difficult or even impossible.

It appears from the preceding comparison that the advantages of explicit and implicit knowledge representations are complementary. In the next section the integration strategies of these two approaches in neural hybrid systems will be surveyed.

2 Neural Hybrid Systems

In many applications different types of knowledge occur for which particular representations are needed to be integrated in one system at various levels (e.g. loosely or tightly coupled). In this section, the existing hybrid systems that make use of the advantages of both the symbolic and neural knowledge representations are inspected in order to extract general integration principles [3,5,16]. The underlying ideas are summarized into a list of several integration strategies which are completed by relevant references. This review also provides a basic taxonomy of neural hybrid systems:

Divide and Conquer: A large problem is divided into pieces which are solved separately by using the most appropriate methods. A subtask, in which the rules are known and justification of the inference is required, is suitable for a symbolic representation. The other part of the problem where only data is available, can be processed by neural network technology. Sometimes, a neural solution does not even need any explanation because the solution itself gives reasons for the underlying conclusions, e.g. the optimization process is justified by its result.

Embedded Neural Network: Neural networks are embedded into a (fuzzy) rule-based system to implement some rules or their parts whereas the remaining rules represent explicit knowledge. Such hybrid systems are controlled by a procedural inference engine which preserves explanation capabilities. For example, a neural network is employed for a rule matching process which is especially suitable for pattern matching where logical expressions cannot be applied (e.g. visual or signal inputs); or the application of a rule can be executed by a neural network. This can even include neural learning and adaptation to the current situation. In addition, even some parts of the conventional inference engine can be realized by neural networks. For example, the selection of the most effective rule of the applicable ones can be implemented by a neural network architecture that responds with a single-best-response category (e.g. Kohonen networks). Examples of the embedded approach include hybrid system shells COLE [17], (Fuzzy)COPE [14,15], FEL–EXPERT [7], or a hybrid robotic control system [11].

Neural Implementation of Explicit Knowledge: The explicit knowledge is transformed into a neural network to make use of advantages like parallelism, competition, recurrent architecture, gradient minimization, partial matching, additional learning, etc. The neurons represent microconcepts, propositions, etc. and the connections among them express their causal relationships, logical dependencies, etc. Thus the explicit knowledge representation is preserved although the neural network is utilized. The inference is implemented by neural computational dynamics which can be interpreted in logical terms due to the knowledge explicitness. Hence, justification of results is guaranteed in such systems. Examples of this approach include the RUBICON system [29], fuzzy cognitive maps [19] and others [18,25].

Incorporating Rules into Neural Networks: The explicit knowledge is incorporated into a neural network after its training in order to improve the generalization capability while preserving the implicit neural knowledge. This strategy is suitable when an example data is available and also several isolated explicit rules are known whose instances cannot be all included into the training set to ensure a perfect consistency of the neural network with respect to these rules. Such rules can even be essential for neural system applicability because breaking them is inadmissible or dangerous in practice. A so-called rule network can be implemented which is integrated with a usual neural net so that the rules have a higher priority [22].

Rule Extraction from Neural Networks: Neural networks are used for explicit knowledge extraction to replace human specialists for knowledge acquisition. A neural network is trained from data to solve the problem acceptable enough and then it is analyzed to extract, e.g., a set of rules from the trained network, possibly with the aid of the training set. After being refined or extended by a human expert, the resulting rules form an explicit knowledge base, or can be exploited for neural system explanation or knowledge debugging. Several explicit knowledge extraction techniques have been developed for neural networks [9,12,26,28].

Fuzzy-Rule Completion in Neural Networks: The neural implementation of explicit knowledge is combined with rule-extraction methods to complete fuzzy rules and to automate the computation and debugging of their confidences by neural learning. The fuzzy rules that represent a first-cut problem solution are wired into a neural network so that the introduced non-standard neurons compute, e.g., the underlying fuzzy logic operations and the connections among them are labeled with confidences, etc. which are initialized according to these rules. The confidence parameters are restricted (or even fixed) within specific intervals to preserve their fuzzy-quantity interpretation. The neural architecture can also be completed (or even entirely constructed) by 'empty' fuzzy rules with random parameters. Then the neural network is trained from example data to complete and refine the original explicit knowledge. For this purpose, the learning algorithm is tailored to this specific neural network function. The rules can also generate additional training patterns in order to strengthen the initial knowledge. Finally, the adapted (or even new) rules are straightforwardly extracted from the trained neural network. Sometimes the rules are not required and the explicit knowledge is used only for neural network initialization. This combined approach is widely applied [1,4,13,20,21,23,32,33].

Neural Expert Systems: A neural network architecture is enriched by heuristics analyzing the compact neural representation in order to handle incomplete information and to provide simple explanation of inferences which are the features required of any expert system [10,30].

Several integration strategies to represent knowledge in neural hybrid systems have been identified, reviewed and illustrated by examples of existing prototypical systems. This review is useful not only for basic classification of neural hybrid systems but it can also help to select the most appropriate method of knowledge representation for particular domains.

References

1. Berenji, H. R., Khedkar, P. (1992) Learning and tuning fuzzy logic controllers through reinforcements. *IEEE Transactions on Neural Networks* **3**, 724–740
2. Buchanan, B., Shortliffe, E. (1984) *Rule-Based Expert Systems.* Addison–Wesley, Reading, MA
3. Caudill, M. (1991) Expert networks. *Byte* **16**, 108–116
4. Chen, S. (1987) Automated reasoning on neural networks: A probabilistic approach. In *Proceedings of the IEEE International Conference on Neural Networks (San Diego)* **II**, 373–378
5. Cloete, I., Zurada, J. M. (Eds.) (2000) *Knowledge-Based Neurocomputing.* The MIT Press, Cambridge, MA
6. Drucker, H. (1990) Implementation of minimum error expert system. In *Proceedings of the International Joint Conference on Neural Networks IJCNN'90 (San Diego)* **I**, 291–296
7. Družecký, P. (1992) Expert system for medical diagnostics. Master thesis, Czech Technical University, Prague
8. Duda, R. O., Reboh, R. (1983) AI and decision making: The PROSPECTOR experience. In Reitman, W. (Ed.) *Artificial Intelligence Applications for Business.* Ablex Publishing Co., Norwood, NJ
9. Enbutsu, I., Baba, K., Hara, N. (1991) Fuzzy rule extraction from a multilayered neural network. In *Proceedings of the International Joint Conference on Neural Networks IJCNN'91 (Seattle)* **II**, 461–465
10. Gallant, S. I. (1993) *Neural Network Learning and Expert Systems.* The MIT Press, Cambridge, MA
11. Handelman, D. A., Lane, S. H., Gelfand, J. J. (1989) Integration of knowledge-based system and neural network techniques for autonomous learning machines. In *Proceedings of the International Joint Conference on Neural Networks IJCNN'89 (Washington)* **I**, 683–688
12. Ishikawa, M. (1996) Structural learning with forgetting. *Neural Networks* **9**, 509–521

13. Jang, J. S. (1992) Self-learning fuzzy controllers based on temporal back propagation. *IEEE Transactions on Neural Networks* **3**, 714–723
14. Kasabov, N. K. (1993) Hybrid connectionist production systems: An approach to realising fuzzy expert systems. *Journal of Systems Engineering* **1**, 15–21
15. Kasabov, N. K. (1995) Hybrid connectionist fuzzy production systems: Toward building comprehensive AI. *Intell. Automation and Soft Computing* **1**, 351–360
16. Kasabov, N. K. (1996) *Foundations of Neural Networks, Fuzzy Systems, and Knowledge Engineering.* The MIT Press, Cambridge, MA
17. Kasabov, N. K., Petkov, S. H. (1992) Neural networks and logic programming—A hybrid model and its applicability to building expert systems. In Neumann, B. (Ed.) *Proceedings of the* 10th *European Conference on Artificial Intelligence ECAI'92 (Vienna)*, John Wiley & Sons, 287–288
18. Kasabov, N. K., Shishkov, S. I. (1993) A connectionist production system with partial match and its use for approximate reasoning. *Connection Science* **5**, 275–305
19. Kosko, B. (1987) Adaptive inference in fuzzy knowledge networks. In *Proceedings of the IEEE International Conference on Neural Networks (San Diego)* **II**, 261–268
20. Lacher, R. C., Hruska, S. I., Kuncicky, D. C. (1992) Back-propagation learning in expert networks. *IEEE Transactions on Neural Networks* **3**, 62–72
21. Lin, C. T., Lee, C. S. G. (1991) Neural-network-based fuzzy logic control and decision system. *IEEE Transactions on Computers* **40**, 1320–1336
22. Mazný, M. (1995) Integrating rule-based and neural approaches to expert system design. Master thesis, Charles University, Prague
23. McMillan, C., Mozer, M. C., Smolensky, M. (1991) Learning explicit rules in a neural network. In *Proceedings of the International Joint Conference on Neural Networks IJCNN'91 (Seattle)* **II**, 83–88
24. Miller, R. A., Pople, H. E., Myers, J. D. (1982) INTERNIST-I, an experimental computer-based diagnostic consultant for general internal medicine. *New England Journal of Medicine* **307**, 468–476
25. Narazaki, H., Ralescu, A. L. (1992) A connectionist approach for rule-based inference using improved relaxation method. *IEEE Transactions on Neural Networks* **3**, 741–751
26. Omlin, C. W., Giles, C. L. (1996) Extraction of rules from discrete-time recurrent neural networks. *Neural Networks* **9**, 41–52
27. Rojas, R. R. (1996) *Neural Networks: A Systematic Introduction.* Springer, Berlin
28. Saito, K., Nakano, R. (1988) Medical diagnostic expert system based on PDP model. In *Proceedings of the IEEE International Conference on Neural Networks (San Diego)* **I**, 255–262
29. Samad, T. (1988) Towards connectionist rule-based systems. In *Proceedings of the IEEE International Conference on Neural Networks (San Diego)* **II**, 525–532
30. Šíma J. (1995) Neural expert systems. *Neural Networks* **8**, 261–271
31. Šíma J. (1996) Back-propagation is not efficient. *Neural Networks* **9**, 1017–1023
32. Yager, R. R. (1994) Modelling and formulating fuzzy knowledge bases using neural networks. *Neural Networks* **7**, 1273–1283
33. Yang, Q., Bhargava, V. K. (1990) Building expert systems by a modified perceptron network with rule-transfer algorithms. In *Proceedings of the International Joint Conference on Neural Networks IJCNN'90 (San Diego)* **II**, 77–82

Evolving and Evolutionary Connectionist Systems for On-line Learning and Knowledge Engineering

Nikola Kasabov

Department of Information Science, University of Otago, P.O Box 56, Dunedin, New Zealand, Phone: +64 3 479 8319, fax: +64 3 479 8311, nkasabov@otago.ac.nz

Abstract: This paper makes a revision of the major principles, applications and publications on the evolving connecionist systems (ECOS) paradigm. ECOS can learn and improve continuously over time. They can adapt quickly, possibly in a real time, to new incoming data form a changing environment; they can learn in a "lifelong" learning mode; they are "able to explain" what they have learned in a form of rules. The papercompares ECOS with other AI models, evolutionary models included, and outlines directions for further research.

Keywords: evolving connectionist systems; neural networks; adaptive systems.

1 On-Line Learning and Knowledge Engineering

The complexity and dynamics of real-world problems, such as adaptive speech recognition and language acquisition, adaptive intelligent prediction and control systems, intelligent agent-based systems; adaptive agents on the Web; mobile robots; visual monitoring systems and multi-modal information processing, and many more require sophisticated methods and tools for building on-line, adaptive, knowledge-based systems (IS). Such systems should be able to: (1) *learn fast* from a large amount of data (using fast training); (2) *adapt incrementally* in an on-line mode; (3) dynamically create new modules – have open structure; (4) *memorise information* that can be used later; (5) *interact continuously* with the environment in a "life-long" learning mode; (6) deal with *knowledge* (e.g. rules), as well as with data; (7) adequately represent in their structure *space and time.*

Several methods and systems have been developed so far that meet some of the criteria above and that have influenced the development of the evolving connectionist systems – ECOS - paradigm and the evolving fuzzy neural networks (EFuNN) [1-38], for example: *adaptive learning* [32,33,19,27]; *incremental learning* [29,35,28]; *local learning* [20,31]; *lifelong learning* [24]; *on-line learning* [22,23,24,26,32-35]; *constructivist structural learning* [35] that is supported by biological facts [29]; *selectivist structural learning*; hybrid constructivist/selectivist structural learning; *knowledge-based learning neural networks (KBNN)* [36].

The ECOS and the EFuNN models discussed here belong to all the above groups. The models are called evolving because of the nature of the structural growth and structural adaptation of the whole evolving connectionist system (ECOS). In terms of on-line neuron allocation, the EFuNN model is similar to the Resource Allocating Network (RAN) suggested by Platt [32] and improved in other related models [33]. The RAN model allocates a new neuron for a new input example (x, y) if the input vector x is distant in the input space from the already allocated radial basis neurons, and also the output error evaluation y-y', where y' is the produced by the system output for the input vector x, is above another threshold. Otherwise, neurons will be adapted to the example (x,y) through a gradient descent algorithm. In terms of adaptive optimisation of many individual linear units EFuNN is close to the Receptive Field Weight regression (RFWR) model by Schaal and Atkeson [35]. EFuNNs have also similarities with the Fritzke's Growing Cell Structures and Growing Neural Gas models [23], and with other dynamic radial basis function networks (RBFN) [31,37,19] in terms of separating the unsupervised learning, which is performed first, from the supervised learning, applied next, in a two-tyre structure. The EFuNN learning algorithm though differs from the above in many aspects, mainly in the employment of simpler and faster learning modes, more flexibility when evolving internal structures and representations, knowledge-base orientation. A comparative analysis between EFuNNs and other similar models on benchmark problems shows that while EFuNNs are comparable with the other methods in terms of accuracy of the obtained results, they are much faster, more controllable, and evolve more compact internal representations. EFuNNs suggest a new neuro-fuzzy systemic approach that employs more sophisticated supervised/unsupervised, knowledge-based learning methods. EFuNNs allow for rule insertion and rule extraction at any time of the operation of the system. The functionality of EFuNNs can be fully utilised when EFuNNs are used as elements of the ECOS framework of an adaptive, intelligent, knowledge-based system.

The on-line, knowledge-based type of learning in the ECOS and EFuNNs allow for building dynamic knowledge-based systems where the knowledge-base is built as the system operates. This approach is called here on-line knowledge engineering. It is also the main feature of the evolving neuro-fuzzy systems ENFUS that are fuzzy systems consisting of the Takagi Sugeno type fuzzy rules and are dynamically created in an on-line mode [2].

2 Evolving Connectionist Systems (ECOS) – Major Principles

The ECOS paradigm was published first in [4,8] and then elaborated in other publications [5,7,9-11]. It allows an intelligent system to be automatically created from incoming data. An ECOS consists of nodes (units) that perform pre-defined functions, and connections between them. The system has a minimal initial structure that includes preliminary input and output nodes and few preliminary connections. Data is allowed to flow into the system so that if an input data vector is associated with a desired output vector, the system stores this association into a new node and creates new connections. Nodes and connections are created automatically to reflect the data distribution. The system's structure is dynamically changing after each data item is introduced.

The number of the input and output variables in ECOS can vary during the learning process thus allowing for more (or less) input and output variables to be introduced at any stage of the learning process. Input and output variables can have 'missing values' at any time of the learning process. If there is no output vector associated with an input vector, the system produces its own output vector (its own solution). If the desired output vector became known afterwards, the system will adjust its structure to produce this output, or one close to it, next time the same input vector is presented. The system continuously and adaptively learns from data to associate inputs to outputs and to cluster the data trough allocating nodes to represent exemplars of data.

The learning process in ECOS is achieved through interaction with the environment which supplies the data flow and reacts to the output produced by the system.

In addition the system provides the knowledge it has learned in the form of IF-THEN rules [1]. The ECOS framework implies that a system evolves through its operation in an interactive way. The more data are presented to the system, the more the system evolves. The learning process is on-line, life-long.

Nodes and connections in an ECOS system can be created, modified, merged, and pruned, in a self-organising manner, similar to how the human brain learns through creating and wiring neuronal structures. The system's structure grows or shrinks depending on the incoming data distribution and pre-set parameters. Through the process of evolving from data the system learns the rules of its own behaviour. The rules that constitute the system's knowledge can be reported and/or extracted at any time of the system operation.

Here the main principles of ECOS are summarised:

(1) fast learning from a large amount of data, e.g. through one-pass training;
(2) adaptation in an on-line mode, where new data is incrementally accommodated;
(3) 'open' structure, where new features (relevant to the task) can be introduced at any stage of the system's operation, e.g., the system creates "on the fly" new inputs, new outputs, new modules and connections;
(4) memorisation of data exemplars for a further refinement, or for information retrieval;
(5) learn and improve through active interaction with other IS and with the environment in a multi-modular, hierarchical fashion;
(6) adequately represent space and time in their different scales; have parameters that represent short-term and long-term memory, age, forgetting, etc.;
(7) deal with knowledge in its different forms (e.g., rules; probabilities); analyse itself in terms of behaviour, global error and success; "explain" what the system has learned and what it "knows" about the problem it is trained to solve; make decisions for a further improvement.

Biological motivations for ECOS are presented in [3].

3 Evolving Fuzzy Neural Networks (EFuNNs) and Evolving Neuro-fuzzy Systems (ENFUS)

One realisation of ECOS is the evolving fuzzy neural network EFuNN. EFuNNs were first introduced in [8]. EFuNNs are models for evolving supervised learning from data that have five-layer structure where nodes and connections are created/connected as data examples are presented. An optional short-term memory layer can be used through a feedback connection from the rule (or 'case') node layer.

The input layer of an EFuNN represents input variables. The second layer of nodes (fuzzy input neurons, or fuzzy inputs) represents fuzzy quantisation of each input variable space. For example, two fuzzy input neurons can be used to represent "small" and "large" fuzzy values. Different membership functions (MF) can be attached to these neurons (triangular, Gaussian, etc.). The number and the type of MF can be dynamically modified in an EFuNN. New neurons can evolve in this layer if, for a given input vector, the corresponding variable value does not belong to any of the existing MF to a degree greater than a membership threshold. A new fuzzy input neuron, or an input neuron, can be created during the adaptation phase of an EFuNN. The task of the fuzzy input nodes is to transfer the input values into membership degrees to which they belong to the MF. The third layer contains rule (case) nodes that evolve through supervised or unsupervised learning. Examples of the latter are given in [33,34]. The rule nodes represent prototypes (exemplars, clusters) of input-output data associations, graphically represented as an association of hyper-spheres from the fuzzy input and fuzzy output spaces.

Each rule node r in an EFuNN is defined by two vectors of connection weights – W1(r) and W2(r), the latter being adjusted through supervised learning based on the output error, and the former being adjusted through unsupervised learning based on similarity measure within a local area of the problem space. The fourth layer of neurons represents fuzzy quantization for the output variables, similar to the input fuzzy neurons representation. The fifth layer represents the real values for the output variables.

The EFuNN evolving algorithm includes also: aggregation of rule nodes (i.e. merging rule nodes); pruning of rule nodes and connections, and other operations.

In the aggregation procedure two thresholds are used T1 and T2, and every two rule nodes for which the distance in the input space is less than T1 and the distance in the output space is less than T2get aggregated. A normalised distance as a measure for the similarity between two vectors (two rule nodes represented by their connection weight vectors W1 and W2) is used in such a manner that providing the two vectors have been normalised and all their values are inside the interval [0,1], the distance is also inside this interval.

EFuNNs can operate in several modes depending on the number of the rule nodes for which activation is propagated to the output layer:

- One of n (1-of-n);
- Many of n (m-of-n)

Another development of the ECOS paradigm is ENFUS [2] - a system that uses fuzzy rules of Takagi-Sugeno type, rather than the EFuNN Zadeh-Mamdany rules.

The unsupervised version of EFuNN does not use output nodes. Instead, it evolves rule nodes based on similarities between the already evolved nodes and new data. One version of such systems is the evolving SOM (ESOM) used in [13].

Simulators of different types of EFuNNs, ESOMs and ENFUS are available from the WWW htp://divcom.otago.ac.nz/infosci/kel/CBIIS.html. EFuNNs are incorporated as models of software environments for knowledge engineering, such as the Repository of the Intelligent Connectionist Base Information Systems (RICBIS) and FuzzyCOPE/4 [12] also available from this site.

4 ECOS versus/and Evolutionary Systems

ECOS and evolutionary computation (EC) systems are built on similar, but different principles. EC is concerned with population-based search and optimisation of individual systems through generations of populations. EC is applied to the optimisation of different structures and processes, one of them being the connectionist structures and connectionist learning processes. EC, and the genetic algorithms (GA) in particular, include in principal a stage of development of the individual systems, so that a system develops, evolves through interaction with the environment that is also based on the genetic material embodied in the system. This process of development has been in many cases ignored or neglected as insignificant from the point of view of the long process of generating hundreds and hundreds of generations each of them containing hundreds and thousands of individuals.

ECOS deal with the process of interactive, on-line adaptive learning of a single system that evolves from the incoming data. The system can either have its parameters (genes) pre-defined, or self-optimised during the learning process starting from some initial values. There are several ways in which EC and ECOS can be inter-linked. For example, it is possible to use EC to optimise the parameters of an ECOS at a certain time of their operation, or to use the methods of ECOS for the development of the individual systems (individuals) as part of the global EC process.

5 Applications of ECOS

ECOS and EFuNNs have been used in several projects of the Knowledge Engineering Laboratory at the University of Otago as learning tools in the following application oriented projects:

- adaptive and robust speech recognition;
- modelling speech and language acquisition [18];
- robot speech control;

- adaptive video mode detection [16];
- multimodal auditory and visual information processing;
- image classification [15];
- medical time series analysis [17];
- adaptive robot navigation [6];
- adaptive agent systems [11];
- adaptive process control [2];
- learning sequences of DNA data;
- building adaptive expert systems;
- decision making and prediction [13,14];

6 Future Research

The following are directions for future research on ECOS and other adaptive on-line systems for learning and knowledge, manipulation:

(1) Elaboration on the mathematical theory of ECOS (universality; convergence, etc)
(2) Developing different types and modifications of ECOS (e.g. spatial-temporal, etc)
(3) Building hybrid models that include ECOS and HMM
(4) Using ECOS for a multi-modal information processing
(5) Building robust to noise, adaptive speech recognisers.
(6) Building on-line learning systems on the WWW.

ECOS combine both known AI features and new features that make them promising techniques for the future development of intelligent systems and information sciences.

Acknowledgements.

The development of ECOS and their applications is supported through grant UOO808 funded by the Foundation for Research, Science and Technology, New Zealand.

References

ECOS references:

1. Kasabov, N and B. Woodford, Rule insertion and rule extraction from evolving fuzzy neural networks: algorithms and applications for building adaptive, intelligent expert systems, 1999 IEEE International Fuzzy Systems Conference Proceedings, Seoul, August 1999, v.III, 1406-1409

2. Kasabov, N. and Q.Song. ENFUS: Evolving neuro-fuzzy systems, *IEEE Transactions on Fuzzy Systems,* (1999) submitted
3. Kasabov, N. Brain-like functions in evolving connectionist systems for on-line, knowledge-based learning, in: T. Kitamura (ed) Modeling brain functions for intelligent systems, World Scientific (2000)
4. Kasabov, N. ECOS - A framework for evolving connectionist systems and the 'eco' training method, in: S.Usui and T.Omori (eds) *Proceedings of ICONIP'98 - The Fifth International Conference on Neural Information Processing,* Kitakyushu, Japan, 21-23 October 1998, IOS Press, vol.3, 1232-1235
5. Kasabov, N. Evolving connectionist and fuzzy connectionist systems – theory and applications for adaptive, on-line intelligent systems, In: *Neuro-Fuzzy Techniques for Intelligent Information Systems,* N. Kasabov and R.Kozma, eds. Heidelberg, Physica Verlag (1999) 111-146
6. Kasabov, N. Evolving connectionist and fuzzy connectionist systems for on-line adaptive decision making and control, in*: Advances in Soft Computing - Engineering Design and Manufacturing,* R. Roy, T. Furuhashi and P.K. Chawdhry (Eds.) Springer-Verlag, London Limited, 1999, 638 pages
7. Kasabov, N. Evolving fuzzy neural networks for on-line, knowledge based learning. *IEEE Transactions on Man, Systems and Cybernetics,* (submitted 2000), published also as TR99-02, Department of Information Science, University of Otago
8. Kasabov, N. Evolving fuzzy neural networks - algorithms, applications and biological motivation, in: T.Yamakawa and G.Matsumoto (eds) Methodologies for the Conception, Design and Application of Soft Computing, World Scientific, 1998, 271-274
9. Kasabov, N. Evolving fuzzy neural networks: Theory and Applications for on-line adaptive prediction, decision making and control, *Australian Journal of Intelligent Information Processing Systems,* 5 (3): 154-160 (1998
10. Kasabov, N. The ECOS framework and the 'eco' training method for evolving connectionist systems. *Journal of Advanced Computational Intelligence* (1998) vol.2, No.6, 195-202
11. Kasabov, N. Theory and applications of evolving connectionist agents and systems, *Proceedings of the 1998 international conference on Neural Networks and Brain (NN&B),* Beijing, October 27-30 (1998), Publishing House of Electronics Industry, China, 668-671
12. Kasabov, N**.,** D. Tuck, and M. Watts, Implementing Knowledge and Data Fusion in a Versatile Software Environment for Adaptive Learning and Decision-Making, in: Proc. of the Int. Conference on Data Fusion, San Hose, July 1999 (1999)
13. Kasabov, N., D.Deng, L.Erzegovezi, M.Fedrizzi, A. Beber, D.Deng, Hybrid Intelligent Decision Support Systems and Applications for Risk Analysis and Prediction, International conference on intelligent systems for investment decision making, Bond University, Gold Cost, December (1999)
14. Kasabov, N., L.Erzegovezi, M.Fedrizzi, A. Beber, D.Deng, Hybrid Intelligent Decision Support Systems and Applications for Risk Analysis and Prediction of Evolving Economic Clusters in Europe, in: "Future directions for intelligent

systems and information sciences", N.Kasabov (ed) Physica Verlag (Springer Verlag), 2000

15. Kasabov, N., S.Israel, B.Woodford, Methodology and evolving connectionist architecture for image pattern recognition, in: S.Pal, Ghosh and Kundu (eds) *Soft computing and image processing,* Heidelberg, Physica-Verlag (Springer Verlag) (2000)
16. Koprinska, I. and N. Kasabov An Application of Evolving Fuzzy Neural Network for Compressed Video Parsing, In: Emerging Knowledge Engineering and Conectionist-based Systems (Proceedings of the ICONIP/AMZIIS/ANNES'99 Workshop "Future directions for intelligent systems and information sciences, Dunedin, 22-23 Nov.1999), N.Kasabov and K.Ko (eds), 96-102
17. Swope, J.A., Kasabov, N., Williams, M., Neuro-fuzzy modelling of heart rate signals and applications to diagnostics, in: Szepanjuk ed, Fuzzy Systems in Medicine, Physica Verlag (2000) 519-542
18. Taylor, J. and N.Kasabov Modelling the Emergence of Speech and Language through Evolving Connectionist Systems, in: "Future directions for intelligent systems and information sciences", N.Kasabov (ed) Physica Verlag (Springer Verlag), 2000

Related references:

19. Blanzieri, E., P.Katenkamp, "Learning radial basis function networks on-line", in: Proc. of Intern. Conf. On Machine Learning, 37-45 (1996)
20. Bottu and Vapnik, "Local learning computation", Neural Computation, 4, 888-900 (1992)
21. Edelman, G., Neuronal Darwinism: The theory of neuronal group selection, Basic Books (1992).
22. Freeman, J., D. Saad, "On-line learning in radial basis function networks", Neural Computation vol. 9, No.7 (1997).
23. Fritzke, B. "A growing neural gas network learns topologies", Advances in Neural Information Processing Systems, vol.7 (1995)
24. Gaussier, T., and S. Zrehen, "A topological neural map for on-line learning: Emergence of obstacle avoidance in a mobile robot", In: From Animals to Animats No.3, 282-290, (1994).
25. Hech-Nielsen, R. "Counter-propagation networks", IEEE First int. conference on neural networks, San Diego, vol.2, pp.19-31 (1987)
26. Heskes, T.M., B. Kappen, "On-line learning processes in artificial neural networks", in: Math. foundations of neural networks, Elsevier, Amsterdam, 199-233, (1993).
27. Kohonen, T., Self-Organizing Maps, second edition, Springer Verlag, 1997
28. Mandziuk, J., Shastri, L. "Incremental class learning approach and its application to hand-written digit recognition, Proc. of the fifth int. conf. on neuro-information processing, Kitakyushu, Japan, Oct. 21-23, 1998
29. McClelland, J., B.L. McNaughton, and R.C. Reilly "Why there are Complementary Learning Systems in the Hippocampus and Neo-cortex: Insights from the Successes and Failures of Connectionist Models of Learning and Memory", CMU TR PDP.CNS.94.1, March, (1994).
30. Mitchell, M.T., "Machine Learning", MacGraw-Hill (1997)

31. Moody, J., Darken, C., Fast learning in networks of locally-tuned processing units, Neural Computation, 1(2), 281-294 (1989)
32. Platt, J., "A resource allocating network for function interpolation, Neural Computation, 3, 213-225 (1991)
33. Rosipal,R., M.Koska, I.Farkas, "Prediction of chaotic time-series with a resource-allocating RBF network, Neural Processing Letters, 10:26 (1997)
34. Saad, D. (ed) On-line learning in neural networks, Cambridge University Press, 1999
35. Schaal, S. and C. Atkeson, "Constructive incremental learning from only local information, Neural Computation, 10, 2047-2084 (1998)
36. Towel, G., J. Shavlik, and M. Noordewier, "Refinement of approximate domain theories by knowledge-based neural networks", Proc. of the 8th National Conf. on Artificial Intelligence AAAI'90, Morgan Kaufmann, 861-866 (1990).
37. Topchy, A., O.Lebedko, V.Miagkikh and N.Kasabov, "Adaptive training of radial basis function networks based on co-operative evolution and evolutionary programming", in: Progress in connectionist-based information systems, N.Kasabov et al (eds), Springer, 253-258 (1998)

Computational Intelligence with Human Capability

Hideyuki Takagi

Kyushu Institute of Design, Fukuoka, 815-8540 Japan
takagi@kyushu-id.ac.jp, http://www.kyushu-id.ac.jp/~takagi

Abstract: We discuss a new approach to computational intelligence research. First, we overview the cooperative technologies of neural networks, fuzzy systems, and evolutionary computations (EC) as the trend in computational intelligence research during the 1990s. Second, we state that this research is based on the analytical approach that analyzes and models human and biological functionality as computational intelligence tools. Third, we discuss a synthetic approach that combines the different strengths of the technologies in computational intelligence and human capabilities as an emerging trend. Finally, we overview the research on an interactive EC (IEC) that combines the human evaluation and EC optimization as an example of the synthetic approach. We also introduce Visualized IEC which combines an EC search with the human ability to evaluate and also perform a global search.

1 Introduction

Many technologies have been proposed in computational intelligence research. Some of representative technologies are neural networks (NNs), fuzzy systems (FSs), and evolutionary computation (EC); some are categorized as artificial intelligence; several new proposals model immune systems, ant colony systems, particle swarms, cultural algorithms, and other systems; new research areas such as artificial life, chaos, and complex systems, have gained prominence in computational intelligence. The typical technologies of NN, FS, and EC were independently researched until the 1980s. Since 1988, as research activities have increased, interest in fusing NN, FS, and EC technologies has also been rapidly increasing. These cooperative technologies were used in commercial products and industrial systems during the 1990s. One computational intelligence research direction is discussed in section 2.

Computational intelligence aims to simulate human intelligence or life in computers. It is therefore natural to study living things, modeling biological mechanisms that work well and refining these models as technologies in computational intelligence.

However, this is not the only approach for computational intelligence research. The conventional approach focuses on modeling biological functions and researching the outputs of the model as computational intelligence technologies. An approach to adopt human functions are superior to modeling only one part of a total system. The approach to deal with computational intelligence systems as a cooperative system with humans was unusual in comparison to the number of computational intelligence models developed by the conventional approach. We respectively refer to these two different research approaches as the synthetic approach and the analytical approach in section 3.

In this paper, we stress that computational intelligence research should be conducted and expanded while maintaining a balancing between these two approaches. Interactive EC is introduced in section 4 as an example of human-centered or human-cooperative research in computational intelligence.

2 Research and Development of NN, FS, and EC

One historical topic in computational intelligence during the past decade was the cooperative technologies of NN, FS, and EC. Several models were proposed and applied to consumer products and industrial systems. In this section, we examine the research and development [25] as samples of *computational intelligence without humans* and compare it with the main subject of this paper discussed in sections 3 and 4.

FS design using NN In 1988, NN-driven Fuzzy Reasoning [23] triggered the neuro-fuzzy boom. The goal was to determine the shape of fuzzy partitioned input sub-spaces that correspond to rule areas using an NN. The NN outputs corresponded to the rule strengths for each fuzzy rule. This model is used to control a rolling mill developed by Hitachi [17]. The uneven surface of a metal reel is recognized by an NN, the control value for each fuzzy rule is weighted by the outputs of the NN that corresponds to the antecedent parts of fuzzy rules, and all rule outputs are aggregated to calculate the final control value for 20 rolls of the rolling mill [29].

The second method, introduced in 1989, allows NNs to tune the parameters of FSs, especially the shape parameters of membership functions [5]. While NN is used only in the development phase, FSs are implemented in the production phase. Consumer products developed using this method have been on the market since 1991; some products include washing machines, vacuum cleaners, rice cookers, clothes driers, dishwashers, electric thermo-flasks, inductive heating cookers, toaster ovens, kerosene fan heaters, refrigerators, electric fans, range-hoods, photocopiers, and so on [25]

FS design using GA Samsung refrigerators, introduced to the market in 1994, use two FSs developed by GA [9]. The first FS estimates the temperature distribution inside the refrigerator, which the second FS uses as an input to determine a spout position of chilled air. Samsung also applied this method to their washing machines and introduced the products to the market in 1995 [9]. Variable speed washing machine motors are controlled by an FS, and the membership functions of the FS were designed by GA.

LG Electronics dishwashers, rice cookers, and microwave ovens adopt neuro-fuzzy models designed by GA [22]. Other LG Electronics products that adopt FSs designed by GA include refrigerators, washing machines, and vacuum cleaners. They also applied this method to acquire fuzzy control rules for a plant process that experiences long delays.

Structured NNs based on a fuzzy rule structure NARA is a structured NN constructed according to the structure of a fuzzy rule base that describes a priori knowledge for a given task [24]. The NARA consists of an NN that corresponds to all antecedent parts of a fuzzy rule base and sub NNs that correspond to each consequent part. The complexity of the given task decreases when a priori knowledge of the task is embedded into the NN structure.

The NARA is used in FAX ordering systems for electric shops. When retailers order electric goods, they usually send a hand-written order form. The FAX machine of a wholesaler sends FAX images of the order forms to the NARA, and the NARA sends the results of hand-written character recognition to their distributors [29].

Combinations of NN and FS In addition to the aforementioned NN developing FS combination, three other combinations of NN and FS are found in commercial products [25].

The first combination is an independent use of NN and FS in a same system where there is no cooperation between NN and FS. Matsushita Electric's air conditioners use an FS to control a compressor and an NN to estimate a PMV value as an index of comfort, but there is no direct cooperation between them.

The second combination is the use of an NN correcting the outputs of an FS or vice versa. This combination model is used to increase the performance of washing machines manufactured by Hitachi, Sanyo, and Toshiba as well as microwave ovens also manufactured by Toshiba.

The third is a cascading combination of an NN and an FS. An electric fan manufactured by Sanyo estimates the position of a remote controller and changes the direction of its rotating neck. An FS first estimates the distance between the fan body and the remote controller, and an NN estimates the direction using the FS's output of estimated distance and the outputs from other sensors. This model is also used in a pulp factory to recover chemicals. An NN inputs image and categorizes the edge shape of a pile in a tank, and an FS determines the parameters of a PID controller to maximize the chemical recovery.

NN learning and configuration by GA LG Electronics developed air conditioners that embeds NN and GA to realize a user trainable function.

RCE type of NN in the air conditioner inputs sensor information and determines a control value. When a user indicates to change its cooling characteristics, a GA modifies the number of neurons and weights of the NN.

GA fitness function consisting NN Hydroponics systems adopt a GA to determine the best time pattern for draining or supplying the system with water and an NN to determine a maximum photo-synthetic rate of a live plant [15]. In this example, the NN plays the role of a fitness function for the GA. A similar combination is found in an alcohol fermentation system[8]. Mizutani et al.'s computer color recipe uses GA with a fitness function consisting of two NNs and a rule base [14].

GA controlled by FS Dynamic parametric GA whose parameters are dynamically optimized by FS was proposed [12]. Although there is no specific method to determine the values of GA parameters such as population size, crossover rate, and mutation rate, we have a qualitative knowledge. An FS that describes the knowledge monitors the fitness values and any changes and modifies the GA parameters.

3 Human-Centered Computational Intelligence

3.1 Analytical approach vs. synthetic approach

The research approach of computational intelligence was to construct functionally and biologically inspired models of learning, reasoning, memory, associative memory, and evolution. As the research progressed, models such as NN and EC were developed; knowledge representation, knowledge acquisition, and fuzziness came to be understood; these research results have been applied to several tasks. This analytical approach will not lose its importance in future computational intelligence research.

However, this is not the only approach in computational intelligence research. Computational intelligence that can perform several application tasks is strongly needed for engineering technologies. The outputs of the computational intelligence research based on the analytical approach have been applied to pattern recognition, control, scheduling, optimization and other tasks and yielded good results. Performance was the primary acceptance criterion for application tasks rather than a basis in the analytical approach.

In contrast to the analytical approach, consider the synthetic approach. A conventional approach, especially an AI approach, was to functionally model human capabilities and to realize them in computers. However, human capabilities often outperforms their models. Especially since it is difficult to model human preference, feeling, emotion, sentiment, passion and so on that operate on a higher level than the cognitive layer.

Accounting for the difficulty of modeling such human psychological functions, there are cooperative computational intelligence systems in which a human is embedded outperform systems that only use human functional models. This research attitude is called the synthetic approach in this paper.

We emphasize two different approaches and suggest that they should work closely together for future researches in computational intelligence. The direction of future researches on the analytical approach will be to further study humans or biology. The direction of researches on the synthetic approach will be to study human-computer interaction, psychology, and subjective aspects such as preference and emotion.

The synthetic approach is a personal optimization that considers personal preference and a sense of values. Classical psychology deals with three layers of sensation, perception, and cognition. Distinctive feature of the psychological research does not deal with the value of external information. However, personalization is required in some engineering applications, and we must deal with the *KANSEI* layer that exists at a higher level than the cognitive layer and deals with subjective preference and value. It is difficult for our current research to model the *KANSEI*. The conventional analytical approach copes with such tasks by using the learning function of NN. However, a synthetic system with a user must outperform a system with only approximate user models. We show such technical examples in next section.

3.2 An example of human-centered technology

One of computational intelligence technologies that are based on a synthetic approach by embedding human into a system is interactive EC. The interactive EC is a technology that combines the human ability to evaluate and an EC ability to optimize (see Fig. 1.) It is a system optimization technology based on human preferences and values.

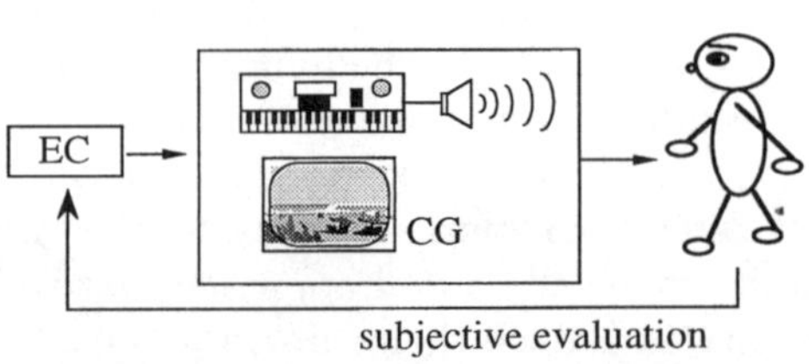

Fig. 1. Interactive EC. EC optimizes the parameters of application tasks according to the subjective evaluation of a human.

The human ability to evaluate is superior to that of computer. A person is better than models of his or her preference and sense of values for external information. Conversely, EC that algorithmically searches in n-D space is superior to a human search that tends to be random in an invisible space. Interactive EC is based on the view that the combination of both abilities and is superior to either one.

Currently, there are two courses of interactive EC research. One is the application-oriented research in which the interactive EC is applied to several tasks and evaluated its performance in each application field. The other is an interface research that reduces the fatigue of interactive EC users.

The application fields of the interactive EC are art, engineering, education and entertainment. Applications in the artistic field include creating CG arts such as drawing lines of insects, plants, and animal, CG rendering, CG animated motion, montage pictures, 3-D CG lighting design support system, industrial design for suspension bridges concrete arch dams, cars, and one-piece dresses, melody generation for jazz session, music rhythm generation, and so on. Applications in the engineering field include speech processing and speech prosody control, database retrieval, data mining, virtual reality, control, image enhancement, and others. Applications in education or entertainment include writing support system, toy robot control, painting software, game software, and others.

The purpose of the interactive EC interface research is to reduce human fatigue due to the cooperative work between a human with a tireless computer. Interface research include discrete fitness values input method [26,27,19], prediction of human evaluation characteristics [28,3,18,20], increasing an EC population size behind a fewer number of individuals displayed to a user [16], accelerating GA convergence, especially in early generations for the interactive EC [6], and active human intervention to EC search [4,34,37].

See details overviews of these applications and interface and rich references whose number is close to 100 in the references of [30–32].

4 Interactive EC

4.1 Applications of the interactive EC

In this section, some applications in our lab are introduced as example applications of the interactive EC, while section 3.2 reviewed the status quo of interactive EC research.

3-D CG lighting support system 3-D CG is a simulation of photographs and its expression can be dramatically changed by lighting just as in photography. Although it is difficult for CG designers without experience in lighting techniques to design a CG expression with lighting, it is not difficult for them to evaluate whether a given CG's expression is suitable for a design concept. Next, we applied the interactive EC to the lighting design support.

Manual lighting design and the interactive EC-based design are compared and evaluated through subjective tests [1,2]. The task is to realize a cheerful, gloomy, heroic, or villainous expression on the same CG woman by changing the brightness, type, and positions of three light sources in 3-D space. Fig. 2 shows one lighting design example of a same subject.

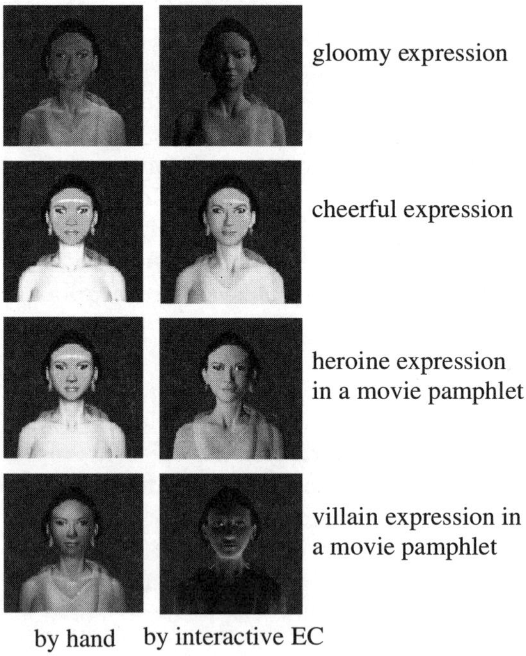

Fig. 2. Examples of 3-D CG lighting expression obtained by hand and interactive EC for given lighting motives. Original images are in color.

The subjective tests have shown that the interactive EC-based lighting support system is especially useful for beginners but is relatively limited for skilled CG designers.

Hearing aid fitting The performance of hearing aids has been increased recently because of digital signal processing, but the process of fitting a hearing aid has not changed much; rather, fittings have become more difficult due to the increased number of parameters of digital hearing aids.

The fundamental problem of conventional fitting methods is that a fitting engineer must adjust the parameters of a hearing aid for its user, the only person who can correctly determine whether it is fitted properly. Unfortunately, users cannot optimize their own hearing aids unless they are a hearing aid engineer. The only information that the conventional fitting methods can use is the user's measured hearing characteristics and the user's feedback in natural language.

IEC Fitting is the hearing aid fitting method based on the interactive EC [21,35]. A hearing impaired person evaluates his or her hearing with their hearing aid, and an EC optimizes the parameters of the hearing aid; the fitting is directly conducted by how the user hears and what the user prefers (see Fig. 3.) This feature is essentially different from conventional methods. Other advantages of the IEC Fitting is that any acoustic sound is usable for fitting, while the conventional methods require to use pure tones or band noise that is unusual in daily life.

The experimental evaluation with hearing impaired persons has shown that the IEC Fitting was superior to the conventional loudness compensation method for both user preference and VCV syllable articulation tests [35].

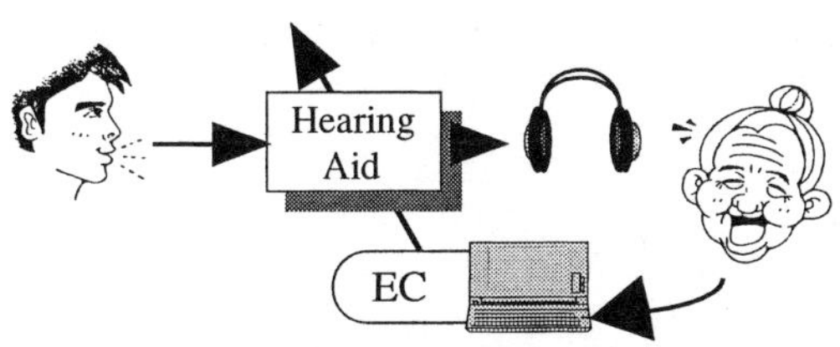

Fig. 3. IEC Fitting method that a hearing impaired person evaluates and EC optimizes the hearing aid parameters.

VR (Virtual reality) An Interactive EC can be used to analyze feeling sensations. Suppose a VR arm wrestling robot and human oppose each other [38]. When the control rules of the robot is suitable, the human can feel as if the opposing robot arm were human, which is the VR of force sensation.

To analyze the factor that causes the VR sense, we first obtained control rules to beat an opponent; 20,000 rules were obtained by a classifier system that reinforces and refines the rules when the rules push an human opponent. Second, the 20,000 rules were compiled into 8 fuzzy rules that are easier to be analyzed [7]. We used GA to obtain the fuzzy rules from the 20,000 crisp rules.

Finally, we modify the fuzzy control rules for winning to those for VR using an interactive EC (see Fig. 4.) Only a human player can evaluate whether the control rules of the arm wrestling robot provides him or her the feel VR of force sensory. The interactive EC is suitable to modify the fuzzy control rules to increase the human evaluation and satisfaction. It is expected that the factor of VR can be analyzed by comparing the difference between the initial fuzzy rules for wining and those finally obtained by the interactive EC.

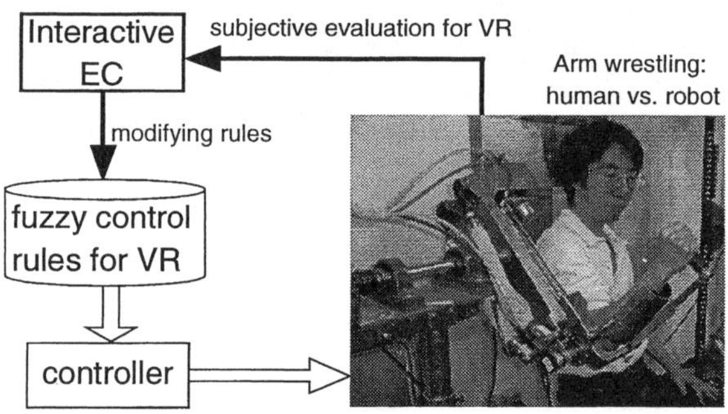

Fig. 4. An arm wrestling robot and an interactive EC system to obtain fuzzy control rules for virtual reality.

Media database retrieval Suppose one needed to find images or music for certain purpose from a huge database or the internet. Retrieving images or music for a certain purpose does not mean retrieving a specific image or music, but retrieving images or music that is suitable or preferable. In such cases, it is difficult to use keywords for retrieving and the retriever may not know what images or music is in the database or websites.

Interactive EC can be used for this type of database retrieval [33]. EC searches in a physical feature space and the human retriever evaluates how the impression of searched media matches his or her mental impression. Since EC searches in a feature space and a human searches in a psychological space, we need a mapping system between two spaces. We adopt an NN and GA for the mapping and construct an interactive retrieval system (see Fig. 5.) We can also design a media converter by adopting a common psychological space for the impression of image, music, and other media [36]; when we play music for a computer, the computer displays image whose impression is similar to the music, and vice versa.

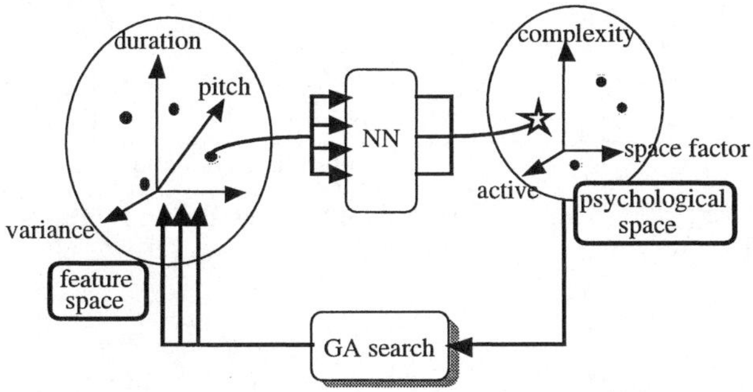

Fig. 5. Media database retrieval based on impression conducted by mapping by NN and reverse mapping by GA.

4.2 Visualized IEC

The role of a human user in the interactive EC is only to evaluate the fitness for EC search; the roles of the user and the EC are closely related each other but completely separated. In a conventional 2-D visualized-based searching [13,10,11], the human undertakes both evaluation and searching. While algorithmic EC search in n-D space is much better than human search in general, humans have an excellent capability to grasp the whole distribution of individuals in a visible space from macro level, which is better than EC because the EC search is based on mating using the limited number of parent individuals. Here, there is space where humans can cooperate with the EC searching rather than sharing the extreme searching and evaluation roles.

Visualized IEC is an interactive EC that allows a human to directly join the EC search (see Fig. 6) [37]. To use the human search capability, the Visualized IEC provides the user a visualized landscape of the n-D searching space by mapping the past individuals from the n-D space to a 2-D space. Although all geometric information on the n-D space cannot be kept on the mapped 2-D space, the visualized distribution of the individuals helps our observation and estimation of the area where the global optimum is located.

We compared the convergence performances of the Visualized EC and normal EC instead of the Visualized IEC vs. interactive EC, on determining the global optimum of Schaffer's second function:

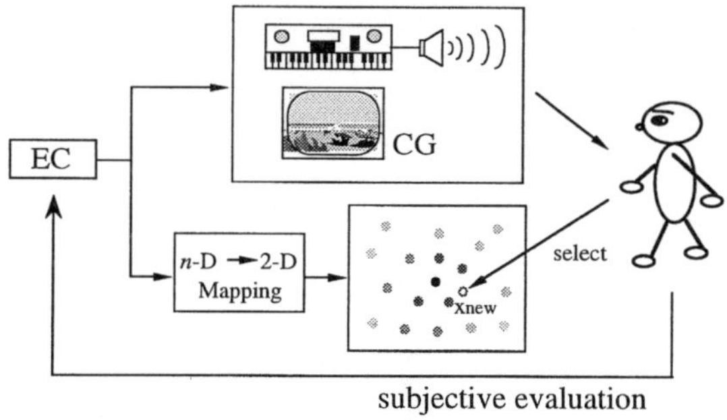

Fig. 6. Visualized IEC. The user not only evaluates EC individuals but also joins the EC search by pointing the possible global optimum.

$\sqrt[4]{\sum_{i=1}^{n} x_i^2}\left\{\sin^2\left(50\sqrt[10]{\sum_{i=1}^{n} x_i^2}\right)+1\right\}$, where $n = 3$ and 5 in our experiment. As a human user does not give subjective evaluation to EC in this experiment and we used GA, this comparison should be called GA vs. Visualized GA if an exact expression is required. We used self-organizing maps among several mapping methods [10].

Fig. 7 shows the average convergence characteristics of five subjects. One of the three best searching points that a subject selected in each generation replaced the worst individuals that EC searched and became a new member of individuals. This experiment showed that the convergence of the Visualized GA with a population size of 20 is equivalent to that of a normal GA with a population size between 100 and 1,000; in other words, the Visualized IEC can be expected to converge five times faster or more than normal IEC [37].

5 Conclusion

We discussed the future direction of computational intelligence research. We categorized computational research into two types, the analytical approach and the synthetic approach, and emphasized the importance to balance the two approaches. Currently, most of computational intelligence research has been biased toward the analytical approach. To achieve a balance, computational intelligence research needs to introduce the human factors, such as human-computer interaction dynamics or psychology. Although we only showed interactive EC and Visualized IEC as examples of the synthetic approach, there are several aspects and trials. We hope that this paper triggers

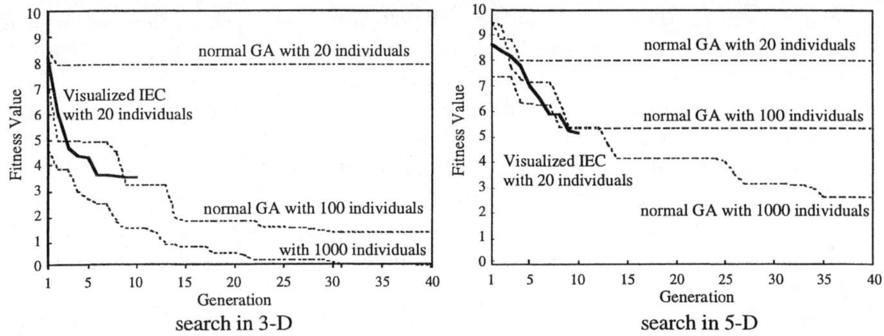

Fig. 7. Average convergence characteristics of the Visualized GA with 20 population size and normal GA with population sizes of 20, 100, and 1,000. Benchmark function is the Schaffer's second function in (a) 3-D and (b) 5-D spaces.

the increase in the number of researches on human factors and leads the progress of computational intelligence research.

References

1. Aoki, K., Takagi, H.: "3-D CG lighting with an interactive GA," 1st Int'l Conf. on Conventional and Knowledge-based Intelligent Electronic Systems (KES'97), Adelaide, Australia (May, 1997) 296–301.
2. Aoki, K., Takagi, H.: "Interactive GA-based design support system for lighting design in 3-D computer graphics," Trans. of IEICE, **J81-DII (7)** (1998) 1601–1608 (*in Japanese*).
3. Biles, J.A., Anderson, P.G., Loggi, L.W.: "Neural network fitness functions for a musical IGA," Int'l ICSC Symposia on Intelligent Industrial Automation and Soft Computing (IIA'96/SOCO'96) (March, 1996) B39–44.
4. Caldwell, C., Johnston, V.S.: "Tracking a criminal suspect through "face-space" with a genetic algorithm," 4th Int'l Conf. on Genetic Algorithm (ICGA'91), San Diego, CA, US: Morgan Kaufmann Publisher, CA, USA (1991) 416–421.
5. Ichihashi, H., Watanabe, T.: "Learning control by fuzzy models using a simplified fuzzy reasoning," J. of Japan Society for Fuzzy Theory and Systems, **2(3)** (1990) 429–437 (*in Japanese*).
6. Ingu, T., Takagi, H.: "Accelerating a GA convergence by fitting a single-peak function," IEEE Int'l Conf. on Fuzzy Systems (FUZZ-IEEE'99), Seoul, Korea (Aug., 1999) 1415–1420.

7. Kamohara, S., Takagi, H., Takeda, T.: "Control rule acquisition for an arm wrestling robot," IEEE Int'l Conf. on System, Man, Cybernetics (SMC'97), **5**, Orlando, Florida, USA (Oct., 1997) 4227–4231.
8. Karim, M., Rivera, S.L.: "Use of recurrent neural networks for bioprocess identification in on-line optimization by micro-genetic algorithms," American Control Conf., Chicago, IL, USA (June, 1992) 1931–1932.
9. Kim, J.W., Ahn, L.S., Yi, Y.S.: "Industrial applications of intelligent control at Samsung Electric Co., – in the Home Appliance Devision –," in (eds.) W. Chiang and J. Lee *Fuzzy Logic for the Applications to Complex Systems,* World Scientific Publishing (1995) 478–482.
10. König, A.: "A survey of multivariate data projection, visualization and interactive analysis," 5th Int'l Conf. on Soft Computing (IIZUKA'98), Iizuka, Fukuoka, Japan: World Scientific, Singapore (1998) 55–59.
11. König, A., Blutner, F. E., Eberhardt, M., Wenzel, R.: "An acoustic data base navigator for the interactive analysis of psycho-acoustic sound archives," 5th Int'l Conf. on Soft Computing (IIZUKA'98), Iizuka, Fukuoka, Japan: World Scientific, Singapore (1998) 60–63.
12. Lee, A.M., Takagi, H.: "Dynamic Control of Genetic Algorithms Using Fuzzy Logic Techniques," 5th Int'l Conf. on Genetic Algorithms (ICGA'93), Urbana Champaign, IL USA (July, 1993) 76–83.
13. Marks, J., Andalman, B., Beardsley, P.A., et al.: "Design Galleries: A General Approach to Setting Parameters for Computer Graphics and Animation," ACM 24th Int'l Conf. on Computer Graphics and Interactive Techniques (SIGGRAPH'97), Los Angeles, CA, USA (Aug. 1997) 389–400.
14. Mizutani, E., Takagi, H., Auslander, M.: "A Cooperative System based on Soft Computing Methods to Realize Higher Precision of Computer Color Recipe Prediction," Proc. of SPIE - The Int'l Society for Optical Engineering, *2492* (April, 1995) 303–314.
15. Morimoto, T., Takeuchi, T., Hashimoto, Y.: "Growth optimization of plant by means of the hybrid system of genetic algorithm and neural network," Int'l J. Conf. on Neural Networks (IJCNN'93), Nagoya, Japan, *3* (Oct., 1993) 2979–2982.
16. Nagao, M., Yamamoto, M., Suzuki, K. Ohuchi, A.: "Evaluation of the image retrieval system using interactive genetic algorithm," J. of Japanese Society for Artificial Intelligence, **13(5)** (1998) 720–727 (*in Japanese*).
17. Nakajima, M., Okada, T., Hattori, S., Morooka, Y.: "Application of pattern recognition and control technique to shape control of the rolling mill," Hitachi Review, *75(2)* (1993) 9–12 (*in Japanese*).
18. Ohsaki, M. Takagi, H., Ingu, T.: "Methods to reduce the human burden of interactive evolutionary computation," Asia Fuzzy System Symposium (AFSS'98), Masan, Korea (June, 1998) 495–500.
19. Ohsaki, M., Takagi, H., Ohya, K.: "An input method using discrete fitness values for interactive GA," J. of Intelligent and Fuzzy Systems, **6** (1998) 131–145.
20. Ohsaki, M., Takagi, H.: "Improvement of presenting interface by predicting the evaluation order to reduce the burden of human interactive EC operators," IEEE Int'l Conf. on System, Man, Cybernetics (SMC'98), San Diego, CA, USA (Oct., 1998) 1284–1289.

21. Ohsaki, M., Takagi, H.: "Application of interactive evolutionary computation to optimal tuning of digital hearing aids," 5th Int'l Conf. on Soft Computing (IIZUKA'98), Iizuka, Fukuoka, Japan: World Scientific, Singapore (Oct., 1998) 849–852.
22. Shin, M.S., Lee, K.L., Lim, T.L., Wang, B.H.: "Applications of evolutionary computations at LG Electrics," in (eds.) W. Chiang and J. Lee *Fuzzy Logic for the Applications to Complex Systems,* World Scientific Publishing (1995) 483–488.
23. Takagi, H. and Hayashi, I., "NN-driven Fuzzy Reasoning," Int'l J. of Approximate Reasoning (Special Issues of IIZUKA'88), *5(3)* (1991) 191–212.
24. Takagi, H., Suzuki, N., Kouda, T., Kojima, Y.: "Neural networks designed on Approximate Reasoning Architecture," IEEE Trans. on Neural Networks, *3(5)* (1992) 752–760.
25. Takagi, H.: "Applications of neural networks and fuzzy logic to consumer products," in (eds.) J. Yen, et al., *Industrial Applications of Fuzzy Control and Intelligent Systems*, Ch.5, IEEE Press, Piscataway, NJ, USA (1995) 93–106.
26. Takagi, H., Ohya, K.: "Discrete fitness values for improving the human interface in an interactive GA", IEEE 3rd Int'l Conf. on Evolutionary Computation (ICEC'96), Nagoya, Aichi, Japan (May, 1996) 109–112.
27. Takagi, H., Ohya, K., Ohsaki, M.: "Improvement of input interface for interactive GA and its evaluation," Int'l Conf. on Soft Computing (IIZUKA'96), Iizuka, Fukuoka, Japan: World Scientific, Singapore (Sept./Oct.1996) 490–493.
28. Takagi, H.: "Interactive GA for system optimization: problems and solution," 4th European Congress on Intelligent Techniques and Soft Computing (EUFIT'96), Aachen, Germany (Sept., 1996) 1440–1444.
29. Takagi, H., Lee, M. A.: "Integration of fuzzy systems and neural networks, and fuzzy systems and genetic algorithms," Ch.2, in (eds.) Jain, L.C., Jain, R.K. *Hybrid Intelligent Engineering Systems* World Scientific (Singapore) (May, 1997)
30. Takagi, H.: "Interactive evolutionary computation: system optimization based on human subjective evaluation," IEEE Int'l Conf. on Intelligent Engineering Systems (INES'98), Vienna, Austria, (Sept., 1998) 1–6.
31. Takagi, H.: "Interactive evolutionary computation – cooperation of computational intelligence and human *KANSEI* —," 5th Int'l Conf. on Soft Computing (IIZUKA'98), Iizuka, Japan: World Scientific, Singapore (Oct., 1998) 41–50.
32. Takagi, H., Unemi, T., Terano, T.: "Perspective on interactive evolutionary computing," J. of Japan Society for Artificial Intelligence, **13(5)** (1998) 692–703 (*in Japanese*).
33. Takagi, H., Cho, S.-B., Noda, T.: "Evaluation of an IGA-based image retrieval system using wavelet coefficients," IEEE Int'l Conf. on Fuzzy Systems (FUZZ-IEEE'99), Seoul, Korea (Aug., 1999) 1775–1780.
34. Takagi, H., Kishi, K.: "On-line knowledge embedding for interactive EC-based montage system," 3rd Int'l Conf. on Knowledge-Based Intelligent Information Engineering Systems (KES'99), Adelaide, Australia (Aug./Sept., 1999) 280–283.
35. Takagi, H., Ohsaki, M.: "IEC-based hearing aids fitting," IEEE Int'l Conf. on System, Man, and Cybernetics (SMC'99), Tokyo, Japan, vol.III (Oct., 1999) 657–662.

36. Takagi, H., Noda, T., Cho, S.-B.: "Psychological space to hold impression among media in common for media database retrieval system," IEEE Int'l Conf. on System, Man, and Cybernetics (SMC'99), Tokyo, Japan, **VI** (Oct., 1999) 263–268.
37. Takagi, H.: "Active user intervention in an EC search," Int'l Conf. on Information Sciences (JCIS2000), Atlantic City, NJ, USA, (Feb./Mar., 2000) 995–998.
38. Takeda, T., Kamohara, S.: "Construction of VR system to the arm wrestling with virtual person," Trans. of IEICE, **J79-A (2)** (1996) 489–497 (*in Japanese*).

Neuro-Fuzzy Modeling of Compensating Systems

Claudio Moraga

Department of Computer Science, Computer Engineering and Computing Education, University of Dortmund, 44221 Dortmund, Germany

Abstract:. There are fuzzy inference systems for which the connection of premises by means of t-norms is not adequate. Compensating aggregation operators are required. The paper shows how to extract this kind of fuzzy if-then rules from numerical data of behaviour with the help of neural networks. Furthermore the method allows to extract fuzzy if-then rules from any neural network using activation functions from a given class (which includes the logistic function).

1 Introduction

Neuro-fuzzy modeling is concerned with the extraction of models from numerical data representing the behaviour of a system. The models in this case are rule-based and use the formalism of fuzzy logic, *i.e.* they consists of sets of fuzzy "if - then" rules with possibly several premises. The learning capability of feedforward neural networks supports the model extraction if the architecture of the network, once properly trained, may be translated into rules without loss of information. This idea has been thoroughly studied by several authors starting with the beginning of the nineties (see *e.g.* [1], [2], [3], [4], [5], [6], [7], [8]) and continues to be an important research area [9]. This kind of modeling has a two-fold purpose: First, to provide a model that allows to control and possibly to forecast the behaviour of the underlying unknown system. Second, to provide a model to better understand the system. The former, sets demands on accuracy to meet the requirements of a given application. The latter, demands that the rules be simple to interpret (see *e.g.* [10]). Emphasis in obtaining a high numerical accuracy may deteriorate the interpretability of the rules (see *e.g.* [11], [12]). A pioneering result of the early years is, without doubts, the one represented by the system ANFIS [Jan 93], which is presently available in MatLab®. ANFIS belongs to the class of rules extracting systems using a *decompositional* strategy [13], where rules are extracted at the level of individual nodes within the neural network and then aggregating these rules to form global behaviour descriptions. From the point of view of fuzzy logic, such a system corresponds to the so-called FITA scheme ("*first inference, then aggregation*") [14], [15]. In ANFIS and all other systems using a similar approach, the main steps of a fuzzy inference *i.e.*, fuzzyfication of the input physical variables and computation of the degree of satisfaction of the available linguistic terms, followed by the conjunction of the premises (if there are more than one) as well as the actual fuzzy inference, concluding (if needed) with defuzzyfication, are realized in sequentially ordered layers of a neural network with an architecture such that the weights to be adjusted in the network (usually by means of a gradient descent algorithm) have a meaning as parameters of the rules to be extracted. ANFIS and other similar systems assume that like in most fuzzy inference systems, the premises will be conjuncted by means of t-norms [16], [17]. Furthermore, since in order to apply the gradient descent algorithm every single transfer function on input-output paths must be differentiable, the t-norm *product* has been always used. There is a large class of problems where aggregation of the premises with t-norms is however not appropriate. Such systems will be studied in this paper and are called here *compensating systems*. In the next section an evaluation of ANFIS will be done and suggestions for its improvement will be discussed. In section 3 compensating systems will be illustrated and a neural network-based method to extract adequate fuzzy rules without loss of information will be explained in details. Furthermore it will be shown that the proposed method may be applied to already existing trained neural networks and may also be used to obtain rule bases with chained rules.

2 Decompositional extraction of rules

Systems for decompositional extraction of rules like ANFIS use a feedforward neural network where the elementary processors of the first hidden layer have a bell-shaped activation function (mostly a Gauss function) and simulate a linguistic term of the linguistic variable [18] associated to the corresponding physical variable giving the input values. The effect of training the neural network upon this elementary processor is the tuning of the position and width of the bell, thus shaping the linguistic term realized by that elementary processor. For a given input, the value of the output signal of the processor corresponds to the degree of satisfaction of that particular input to the linguistic term. The second hidden layer comprises only analoge multipliers to implement the conjunction of the degree of satisfaction of coupled premises. These processors have no parameters to be tuned during the training of the neural network. Their realization is highly probably based on the 4-quadrants method used in the "good old times" of analogue computers: $x \cdot y = \frac{1}{4}[(x+y)^2 - (x-y)^2]$. The structure of the following layers depends on the kind of inference being pursued. For assistant expert systems it might be convenient to obtain fuzzy conclusions, meanwhile for automatic control purposes, crisp outputs will be needed and the methods of Takagi and Sugeno [19] or Tsukamoto [20] will be favoured (see *e.g.* [21]). The scheme of additive fuzzy systems [22] should presently also be considered. Very good *numerical* results have been obtained with ANFIS-type systems.

In spite of the attractive features mentioned above, these systems have also some critical aspects that must be discussed. ANFIS requires from the user an *a priori* decision on the number of linguistic terms to be used by every linguistic variable. If the user does not have the advice of an expert, he will have to work based on trial-and-error. Furthermore it may be the case that the neural network may be trained to obtain a very small mean square error -(model with a very good *numerical* accuracy)- but the resulting "linguistic terms" exhibit a large degree of overlapping thus impairing a clear interpretation. To obtain both numerical accuracy and good interpretability of the rules some kind of "constrained gradient descent" algorithm should be developed. Such an algorithm is however still missing. A way of avoiding the problem would be to preprocess the training data with a clustering algorithm [23], [24], [25] and based on this result, decide the number of linguistic terms per variable [26]. Clustering results may be used to directly identify premises, as done in the NARA-system [6], [27], [28]; however leading to rules of a high level of abstraction -(" *if the premises* p_1, p_2, ..., p_k *satisfy the conditions to belong to cluster* c_j *then...*")- which are usually difficult to understand. A way to follow the idea of clustering but to avoid the extraction of rules with a high level of abstraction is to use an evolutionary design of the neural network layers constituting the front end of ANFIS [29], [30], [31]. The second aspect to be critically analyzed is the fact that ANFIS uses t-norms as conjunction of the premises and, by doing this, induces a rigid grid-like partition of the input space. The blocks of the fuzzy grid are sorounded by "valleys" resulting from the crossing of neighbour linguistic terms. This is typical of a fuzzy partition; however if the rigid grid splits a region of the input space (which should constitute only one block) the valley will contribute an error to the corresponding rule. This phenomenon has been called "hypothesis of rigid granularity" [32]. This may be alleviated if an evolutionary design of the front end is done, including the evolutionary choice of the activation functions to support multiresolution. See in figure 1 a problem that has 5 blocks. The typical good ANFIS solution would have to consider 3 linguistic terms for each variable, leading to a partition with 9 blocks (and consequently, 9 rules), as shown at the left hand side. Each input is connected to three processors each, that realize the corresponding (three) linguistic terms. The coloured processors are the multipliers, from where it is clear, which bolcks are realized in two parts. An evolutionary design of the front end (using a library of activation functions) could implicitly support multiresolution if the fitness function has as goal to minimize the number of product-processors (which is equivalent to minimizing the number of rules). This is shown in figure 1b. The coloured processors in the first hidden layer show that five liguistic term per variable will be learnt, where the colours show wich (pair of) linguistic terms identify each block. The white processors (of the first hidden layer) learn to use a bell-shaped activation function, meanwhile all others use simple sigmoide activation functions. From the point of view of a processing model, both networks have the same number of nodes, which might be interpreted as a sign of similar complexity. Within a hardware frame however, it is simple to see that five 4-quadrant multipliers are much less complex than nine and an output linear combination of five hidden signals is less complex than the linear combination of nine. Both networks represent an additive fuzzy system [22]. The latter is however more accurate than the former since it does not produce extra "valleys" spliting the blocks.

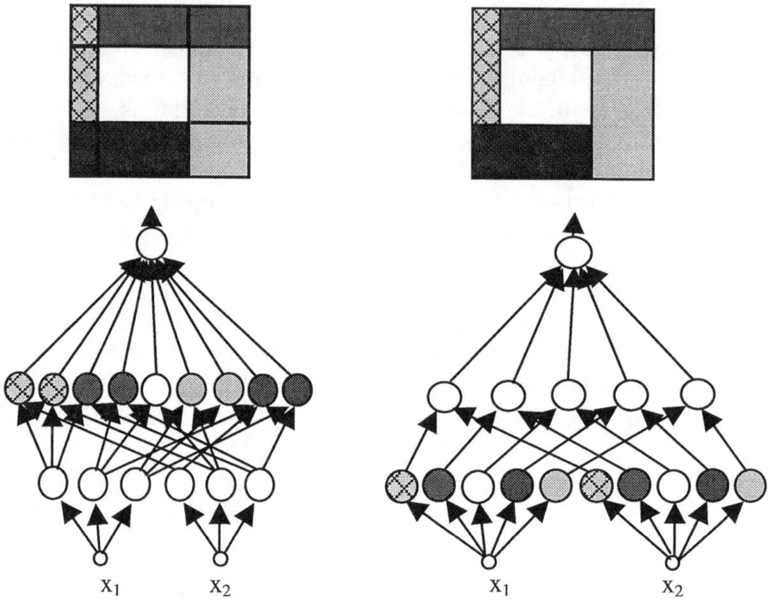

Fig. 1: (a) Classical ANFIS solution of the 5 blocks problem. (b) Evolutionary front-end for ANFIS to solve the 5 blocks problem

3 Compensating systems

Consider the case of a manager of a football (soccer) team, looking for a good forward player. He proceeds following the "standard" rule of thumb:

*"**if** the player runs (with the ball) 100 m in 10 sec **and** he is a dribbling king **and** where he sees an unprotected spot of the net he shoots the ball **and** the transfer fee is reasonable **then** go get him!"*

Furthermore, assume the manager has been reading about fuzzy logic and wants to give it a try. He gives to the linguistic variables representing the premises degrees of membership between 0.5 and 1 when they are increasingly getting close to satisfy the assertion and between 0.5 and 0 as they get increasingly far from the acceptable "reasonable minimum" (which is graded with 0.5). The manager has also read that t-norms *product* and *minimum* are "everyone's first choice". Not feeling confortable with the intricacies of implications, the manager chooses the simplest of the Takagi-Sugeno systems: the joint "strength" of the premises is taken as the strength of the conclusion. He has been lucky to discover three young very talented players. His study of the situation to take a decision is summarized in table 1. After watching the results he was quite dissappointed and confused: his business experience and his common sense told him that he would never take a decision as exemplified in the table! Using the t-norm product, he would have to reject all three candidates (all three satisfied the premises with a strength well below 0.5). Using the t-norm minimum it would be the same to choose the best or the weakest goal-shooter, the most or the least expensive player. Actually one could choose blind! The moral of the story is, that the manager had to do with a typical compensating system, where t-norms fail to model a human decision-making, because they can process neither compromises nor implicit priorities.

Table 1: decision table based on t-norms

	Player 1	Player 2	Player 3	Player 1	Player 2	Player 3
Speed	0.8	0.7	0.9	0.8	0.7	0.9
Dribbling	0.8	0.5	0.8	0.8	0.5	0.8
Shooting	0.9	0.9	0.5	0.9	0.9	0.5
Transfer fee	0.5	0.9	0.7	0.5	0.9	0.7
Result	0.288	0.283	0.252	0.5	0.5	0.5
t-norm	product			minimum		

Compensating systems are such systems where the goal of satisfying several premises at the same time may be better expressed by means of *aggregation* operators [33], [34], [35] rather than with t-norms.

Is it possible to do neuro-fuzzy modeling in the case of compensating systems? Is it possible to design and train a neural network, such that its architecture parameters may characterize the components of a fuzzy rule base using some aggregation operator to combine the degrees of satisfaction of the premises? The first question has an unrestircted positive answer (see discussion below). The second, has a positive answer, however presently restricted to generalizations of the *symmetric sum* [33], introduced in [36] and [37].

Definition 1 [37]:

Let f: $\mathsf{R} \to (0,1)$ be continuous, monotone increasing and invertible. Furthermore $\lim_{x \to \infty} f(x) = 1$; $\lim_{x \to -\infty} f(x) = 0$ and $\forall x \in \mathsf{R}$, $f(-x) = 1 - f(x)$, (from where $f(0) = ½$). Then f is called "S-function".

It is simple to see that among others, the logistic function $f(x) = [1+ \exp(-x)]^{-1}$ as well as the asymmetric Elliot function [38] $f(x) = ½[1 + x/(1 + |x|)]$, the generalized logistic function $f(x) = [1 + k^{-x}]^{-1}$, $k \in \mathsf{R}$, $k > 1$ [36] and the trigonometric functions $f(x) = \frac{1}{2} + (1/\pi)\arctan(\varpi x)$ [34], $f(x) = 1 - (1/\pi)\cot^{-1}(\varpi x)$ [37] and $f(x) = \frac{1}{2}[1 + \tanh(\varpi x)]$, with $\varpi \in \mathbb{R}_+$, are S-functions.

S-functions have the following properties [37]:

i) S-functions are allover differentiable, which makes them adequate as activation functions in neural networks tuned with a gradient descent algorithm.

ii) S-functions may be used to represent soft half-trapezoidal fuzzy sets.

iii) In the notation of [39], S-functions are *squashing* functions. If they are used as activation functions in the hidden layer of a feedforward neural network with linear output, these networks fulfill the conditions for universal approximation [39], [40].

iv) Since additionally S-functions are invertible, neural networks using S-functions allover as activation functions are also universal approximators [41].

Definition 2:

Let f be an S-function. For all x_1 and x_2 in $\mathbb{R}$ let $f(x_1) = v_{x1}$ and $f(x_2) = v_{x2}$. The aggregation operator generated by f is given by:

$$f(x_1) \otimes f(x_2) =_{def} f\left(f^{-1}(v_{x1}) + f^{-1}(v_{x2}) \right) = f(x_1 + x_2)$$

It has been shown [34] that this operators maps the real line into the open interval (0,1), it is associative, commutative, strict monotone in $(0,1) \times (0,1)$ and continuous (except at the points (0,1) and (1,0)). Furthermore it has been shown [37] that $((0,1), \otimes)$ is an Abelian group where the neutral element is ½ and the inverse of an element x of (0,1) is 1-x. It becomes apparent that inversion in the group corresponds to a strong negation [42] (actually, to the most used negation) in the context of fuzzy logic.

Lemma 1:

For all x, y in (0,1) if $x, y > \frac{1}{2}$ then $x \otimes y > \max(x, y)$ and if $x, y < \frac{1}{2}$ then $x \otimes y < \min(x,y)$.

Proof:

See figure 2 and definition 2. Similarly for the case "<"

Corollary 1.1:

For all x, y in (0,1) if $x > \frac{1}{2} > y$, then $x > (x \otimes y) > y$ and the operator has a *compensating* character.

Corollary 1.2:

For all x, y in [½, 1) the ⊗-operator behaves like a t-conorm and for all x, y in (0, ½], the ⊗-operator behaves like a t-norm.

Corollary 1.3:

Let f and g be S-functions. If for all x in (0,1) $h(x) = f(x) \otimes g(x)$ or $h(x) = \gamma f(x) + (1-\gamma)g(x)$, where $\gamma \in [0,1]$, then h is also an S-function.

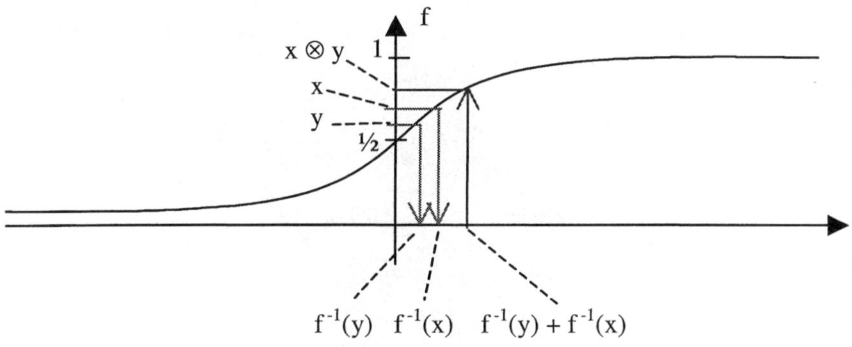

Fig. 2: Graphical representation of the situation x, y > ½ of the Lemma

It becomes apparent that the closer to ½ is the satisfaction degree of a premise, the weaker will be its influence on the aggregation of the premises with the ⊗-operator. In other words, the weaker the statement of a premise, the less it will affect the final conclusion.

Figure 3 illustrates the S-function $f(x+y) = [1 + k^{-(x+y)}]^{-1}$ with k=5 and k=1.5 as well as the ⊗-operator induced by f (which is independent of the value of k).

Applying the ⊗-operator to the players decision problem the results would be as shown in table 2. The ⊗-selection of the best-shooter, even though not the fastest forward and an average dribbler, but however the less expensive player, "makes sense".

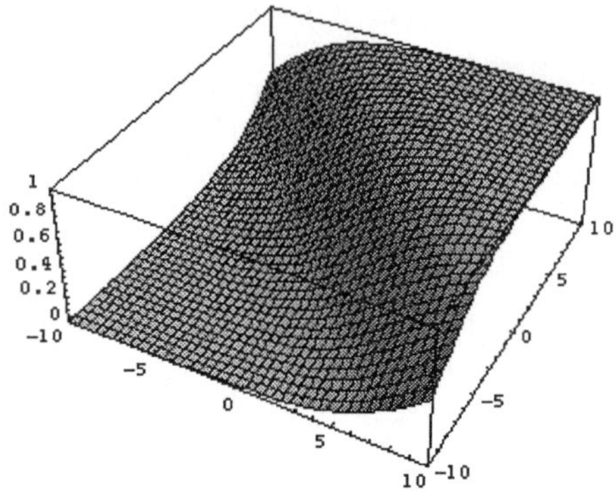

(a)

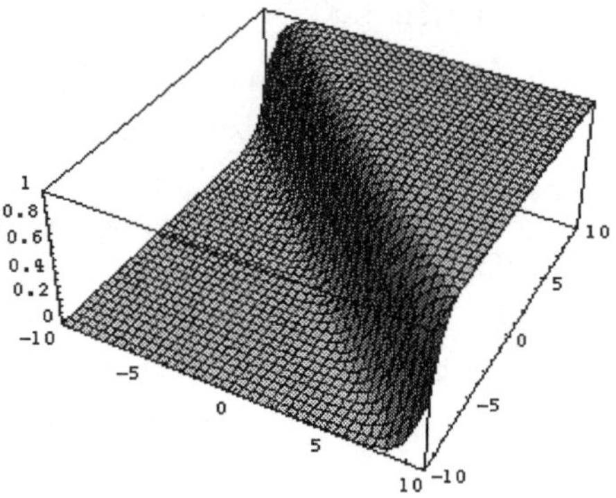

(b)

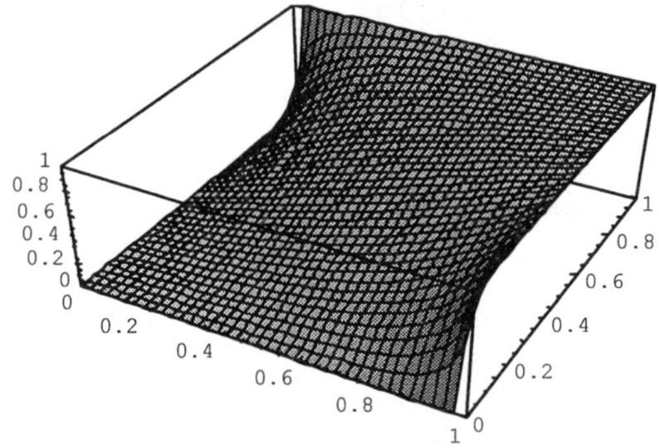

(c)

Fig. 3: (a) $f(x+y) = [1 + (5)^{-(x+y)}]^{-1}$; (b) $f(x+y) = [1 + (1.5)^{-(x+y)}]^{-1}$; (c) ⊗-operator induced by f.

Table 2. Player selection with ⊗			
	Player 1	Player 2	Player 3
Speed	0.8	0.7	0.9
Dribbling	0.8	0.5	0.8
Shooting	0.9	0.9	0.5
Transfer fee	0.5	0.9	0.7
Result	0.9931	**0.9947**	0.9882
Connective	⊗-operator		

Consider a feedforward neural network with one hidden layer such that all nodes have an S-funtion as activation function and a linear output node. (Since S-activation functions satisfy the requirements for universal approximation, it will always be possible to build a neural network to model a system defined on a compact subset of the reals.) It will be shown below how the ⊗-operator allows rule extraction from a neural network. Moreover, it allows rule-extraction from *already existing* trained feedforward neural networks, independently of the fact that the system may be compensating or not!

Let f be the S-activation function of a hidden node of a feedforward neural network with n real-valued inputs. Its output is given by:

$$h = f(w_1 x_1 + w_2 x_2 + \dots + w_n x_n) , \tag{1}$$

where $w_1, \dots, w_n$ are real-valued weights. (The case of a bias is included as a weight associated to an extra input "variable" with the constant value 1). According to definition 2 , the output of the node may be expressed as:

$$h = f(w_1 x_1) \otimes f(w_2 x_2) \otimes \dots \otimes f(w_n x_n) \tag{2}$$

The weights of the variables may be associated as a new parameter of the S-activation functions, modifying their slope. Accordingly, equation (2) turns into:

$$h = f^{(w1)}(x_1) \otimes f^{(w2)}(x_2) \otimes \dots \otimes f^{(wn)}(x_n) \tag{3}$$

where for i = 1, ..., n

$$f^{(wi)}(x_i) =_{def} f(w_i x_i)$$

Since $f^{(wi)}(x_i)$ may be given the interpretation of a right (left) open fuzzy set on x_i (depending on whether w_i is positive (negative)), then equation (3) represents the generalized symmetric sum of the degree of satisfaction of the corresponding fuzzy sets. For the case of the classical logistic function, this method was disclosed in [43] and for the generalized logistic function, in [36].

Equation (3) may be given the following interpretation:

*" **if** x_1 is bounded by T_1 **and** ... **and** x_n is bounded by T_n **then** h is given by the (generalized) symmetric sum of their degrees of satisfaction"*

where "bounded" denotes "*at least*" if the corresponding weight w is positive or "*at most*" if w is negative, and T_j is the linguistic label associated to the fuzzy set represented by $f^{(wj)}$. Since these fuzzy sets are only asymptotically normal, the value of the argument with membership 0.9 will be used as reference to apply the conditions "at least" or "at most", respectively [36].

Figure 4 shows a neural network with one hidden layer with S-activation functions and a linear output node. The rule extraction at the first node is illustrated at the right hand side. It becomes apparent that the network represents a *compensating* fuzzy additive system, *i.e.* the output rule is a linear combination of the rules generated at the hidden nodes. Furthermore, it is easy to see that such a rule extraction may be done upon already existing trained feedforward neural networks, since no special architecture requirements are posed. The more complicated the topology of a feedforward neural network though, the heavier the burden on the interpretability of the possibly chained extracted rules.

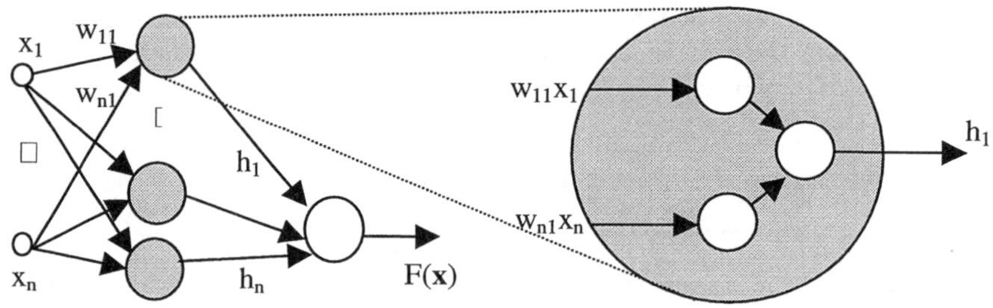

Fig. 4: (Left) The neural network view. (Right) The fuzzy logic interpretation

Example

The *"Diabetes Problem"* [44], [45] is well known in the machine learning and neural networks communities. Women of the Pima-indians have to be classified as diabetes positive or not, on the basis of 8 parameters like age, number of children, weight, sugar concentration in blood, blood pressure, glucose tolerance test and others. It is a classification problem with 576 learning vectors of diagnostic data (together with their correct classification) and 192 test vectors. It appears to be a problem with *missing information,* since the error rate of the best 10 systems reported in [44] ranges from 22.3% to 25.8%. Among these, there is a feedforward neural network achieving a classification error of 24.8% on the test set.

Assuming that the goal were to obtain a set of *classification rules* and an ANFIS-like system were used, then if *only two linguistic terms* for each of the eight variables (parameters) were considered, this would lead to $2^8 = 256$ rules and if *three linguistic terms* per variable were considered, $3^8 = 6561$ rules. A simple feedforward neural network with a single hidden layer with five nodes using the logistic activation function and a linear output node has been trained to do the classification reaching a classification error of 24,67% (average over 10 runs). Using the rule-extraction method discussed above, it is possible to extract *five* rules to do the classification. The output of the system being a linear combination of the rules, that is characteristic of a fuzzy additive system.

Let the output of the i-th hidden processor be:

$$h_i = f\left(\sum_{j=1}^{8} w_{ji} p_j + b_i \right),$$

where $1 \leq i \leq 5$, f denotes the logistic function, p_j, $1 \leq j \leq 8$ are the input parameters, w_{ji} denotes the weight associated to the j-th parameter entering the i-th hidden node and b_i represents a possible bias at the i-th node.

For all i and j as above let $f^{(w_{ji})}(p_j) =_{def} f(w_{ji}p_j)$. Moreover let T_{ji} be the linguistic term corresponding to the fuzzy set defined by $f^{(w_{ji})}$, then the i-th ("hidden") rule has the following structure:

$$R_i: \textbf{\textit{if}}\ p_1 \textit{ is bounded by } T_{1i} \textbf{\textit{ and}} \dots \textbf{\textit{and}}\ p_8 \textit{ is bounded by } T_{8i}$$

$$\textbf{\textit{then}}\ h_i = \left(\bigotimes_{j=1}^{8} f^{(w_{ji})}(p_j) \right) \otimes f(b_i)$$

It may be noticed that the complexity of interpretation of such a rule is comparable to that of a Takagi-Sugeno system only that instead of a linear combination of the inputs, a *compensating* combination of the inputs constitute the conclusion.

4 Conclusions

It has been shown that there are systems that are not appropriately modeled with the help of t-norms, but with compensating aggregation operators. Furthermore a one to one relation between neural networks using S-activation functions and rules combinig premises with a generalized symmetric summation has been proven. This allows not only a convenient way of neuro-fuzzy modeling compensating systems, but also to extract (compensating) rules from existing neural networks. If the hidden layer has been minimized either by means of a pruning strategy or an evolutionary algorithm, this will lead to a correspondingly reduced number of reasonably understandable rules.

References

1 Glorennec P.Y., Barret C., Brunet M., (1992): Application of Neuro–Fuzzy Networks to identification and control of nonlinear dynamic systems. *Proc. Int. Conference on Information Processing and Management of Uncertainty in Knowledge–based Systems (IPMU)*, 507–510, Palma de Mallorca

2 Horikawa S.I., Furuhashi T., Uchikawa Y., (1992): A new type of Fuzzy Neural Network for Linguistic Fuzzy Modeling. *Proc. 2nd. Int. Conference on Fuzzy Logic and Neural Networks.* 1053–1056, Iizuka, Japan

3 Jang J.S.R., (1993) : ANFIS: Adaptive Network based Fuzzy Inference System. *IEEE Trans. on Systems, Man and Cybernetics* **23**, (3), 665–685

4 Jang J.S.R., Sun C.T., (1995): Neuro–fuzzy Modeling and Control. *Proceedings IEEE* **83**, (3), 378–406

5 Keller J.M., Tager R.R., Tahani H., (1992): Neural Network implementation of fuzzy logic. *Fuzzy Sets and Systems* **45**, (1), 1–12

6 Takagi H., Hayashi I., (1991): NN-driven fuzzy reasoning. *Int. Journal of Approximate Reasoning* **5**; (3), 191–212

7 Yi H.J., Oh K.W., (1992): Neural Network based Fuzzy Production Rule Generation and its application to an Approximate Reasoning Approach. *Proceedings 2nd. Int. Conference on Fuzzy Logic and Neural Networks.* 333–336, Iizuka, Japan

8 Han J., Moraga C., (1996): Parametric Feedforward Network based Adaptive Fuzzy Modeling. *Proc. Int. Symp. Intelligent Industrial Automation and Soft Computing*, B–159–165, Reading UK, ICSC Academic Press, Canada

9 Tickle A.B., Maire F., Bologna G., Diederich J., (1999): Extracting the knowledge embedded within trained artificial neural networks: defining the agenda. *Proceedings International Symposium on Soft Computing (SOCO'99)*, 732-738, Genova, Italy. ICSC Academic Press, Canada

10 Craven M., (1996): Extracting comprehensible models from trained neural networks. PhD Thesis, University of Wisconsin, Madison, Wisconsin, USA

11 Vergara V., Moraga C., (1997): Optimization of fuzzy models by global numeric optimization. In: *Fuzzy Model Identification* (H. Hellendorn, D. Driankov, Eds.), 251-278. Springer, Berlin

12 Rotshtein A., (1999): Design and tuning of fuzzy rule-based systems for medical diagnosis. In: *Fuzzy and Neuro-Fuzzy Systems in Medicine* (H.-N. Teodorescu, A. Kandel, L.C. Jain, Eds.), 243-289. CRC Press, Boca Raton, FLA.

13 Andrews R., Diederich J., Tickle A.B., (1995): A survey and critique of techniques for extracting rules from trained artificial neural networks. *Knowledge-based Systems* **8** (6), 373-389

14 Temme K.-H., Thiele H., (1995): On the Correctness of the Principles FATI and FITA and their Equivalence. *Proceedings VI International Fuzzy Systems Association World Congress*, São Paulo, Brazil

15 Temme K.-H., Thiele H., (1996): On the chaining of if-then rule bases interpreted by the principle FITA. *Proceedings Fourth European Congress on Intelligent Techniques and Soft Computing EUFIT*, 939-945, Aachen, Germany

16 Menger K., (1942): Statistical Metrics. *Proceedings National Academy of Sciences USA* **28**, 535–537

17 Schweizer B., Sklar A., (1983): *Generalized Metric Spaces.* North Holland, Amsterdam

18 Zadeh L.A., (1975): The concept of a Linguistic Variable and ist application to approximate reasoning. *Information Sciences* **1**, (8), 199–249; **2**, (8), 301–357; **3**, (9), 43–80

19 Takagi H., Sugeno M., (1985): Fuzzy identification of systems and its application to modeling and control. *IEEE Trans. on Systems, Man and Cybernetics* **15**, (1), 116-132

20 Tsukamoto Y., (1979): An approach to fuzzy reasoning method. In: *Advances in Fuzzy Set Theory and Applications* (R.R. Yager, M.M. Gupta, R.K. Ragade, Eds.) 137-149. North Holland, Amsterdam

21 Brahim K., (1993): Neuro-Fuzzy Inferenz-Systeme. In: *Fuzzy Logic. Theorie und Praxis* (B. Reusch, Ed.), 176-186, Springer, Heidelberg

22 Kosko B., (1994): Fuzzy systems as universal approximators. *IEEE Trans. Computers* **43**, (11), 1324–1333

23 Tsao E.C., Bezdek J.C., Pal N.R., (1994): Fuzzy Kohonen Clustering Network.

Pattern Recognition **27**, (5), 757–764

24 Bezdek J.C., (1981): *Pattern Recognition with fuzzy objective function algorithms.* Plenum Press, New York

25 Duran B.S., Odell P.L., (1974): Cluster Analysis (A Survey). *Lecture Notes in Economics and Mathematical Systems* **100**, Springer, Berlin

26 Jang J.S.R., (1997): ANFIS architecture. In: *Neuro-fuzzy and Soft Computing* (J.S. Jang, C.-T. Sun, E. Mizutani, Eds.), Prentice Hall, New Jersey

27 Takagi H., Susuki N., Koda T., Kojima Y., (1992): Neural Networks designed on approximate reasoning architecture and their applications. *IEEE Trans. Neural Networks* **3**, (5), 752–760

28 Moraga C., Han J., (1996): Optimized neural nets for neuro–adaptive fuzzy modeling. *Proc. Symp. Qualitative Systems Modeling, Qualitative Fault Diagnosis, and Fuzzy Logic and Control,* 50–55. (S-JEP 07759–94 Tempus), T.U. Budapest

29 Heider R., Moraga C., (1996): Evolutionary synthesis of neural networks by means of a graph-grammar. *Proceedings International Panel Conference on Soft and Intelligent Computing,* 119–126. Budapest, (http://www.dr-heider.de/html/synthesis.html), ISBN 963 420 510 0,

30 Heider R., (1997): *Evolutionaere Synthese neuronaler Netze unter Verwendung einer Graphgrammatik.* Shaker Verlag, Aachen,

31 Mayer H.A., Fürlinger K., Strapetz M., (1999): Extraction of compact rule sets from evolutionary designed artificial neural networks. *Proc. 1999 Int`l Conference on Systems, Man, and Cybernetics.* I-420-424, Tokyo, Japan. IEEE-CS-Press

32 Strehl K., Moraga C., Temme K.-H., Stankovic R.S., (2000): Fuzzy Decision Diagrams for the representation analysis and optimization of rule bases. *Proc. 30th Int´l Symposium on Multiple-valued Logic.* Portland, Oregon. IEEE-CS-Press

33 Silvert W., (1979): Symmetric summation: A class of operations on fuzzy sets. *IEEE Trans. on Systems, Man and Cybernetics* **9**, 659-667

34 Dombi J., (1982): Basic concepts for a theory of evaluation: The aggregative operator. *European Jr. Operation Research* **10**, 282-293

35 Dubois D., Prade H., (1985): A review of fuzzy set aggregation connectives. *Information Sciences* **36**, 85-121

36 Temme K.-H., Heider R., Moraga C., (1999): Generalized neural networks for fuzzy modeling. *Proc., Int'l Conference of the European Society of Fuzzy Logic and Technology, EUSFLAT'99.* Palma de Mallorca, Spain

37 Moraga C., Temme K.-H., (2000): S-neural networks are fuzzy models. *Proceedings 8th Int'l Conference on Information Processing and Management of Uncertainty in Knowledge Based Systems (IPMU)*, Madrid, Spain

38 Georgiou G.M., (1992): Parallel distributed processing in the complex domain. Dissertation, Dept. C. Sc., Tulane University, New Orleans, Lousiana

39 Hornik K., Stinchcombe M., White H., (1989): Multilayer Feedforward Networks are universal approximators. *Neural Networks* **2**, 359–366

40 Funahashi K.I., (1989): On the approximate realization of continuous mappings by neural networks. *Neural Networks* **2**, 183–192

41 Moraga C. (1997): Properties of Parametric Feedforward Networks. *Proceedings XXIII Conferencia Latinoamericana de Informática*, 861–870, Valparaíso, Chile

42 Alsina C., Trillas E., Valverde L., (1983): On some logical connectives for fuzzy sets theory. *Jr. of Mathematical Analysis and Applications* **93**, 15-26

43 Benítez J.M., Castro J.L., Requena I., (1997): Are neural networks black boxes? *IEEE Trans. on Neural Networks* **8**, 1156-1163

44 Michie D., Spiegelhalter D.J., Taylor C.C., (1994): *Machine Learning, neural and statistical classification.* Ellis Horwood Ltd.

45 Prechelt L., (1994): PROBEN-1 - A set of benchmarks and benchmarking rules for neural network training algorithms. Technical Report 21/24, Faculty of Computer Science, University of Karlsruhe, Germany. (ftp://ftp.ira.uka.de/pub/papers/techreports /1994/1994-21.ps.gz)

Wavelets to Enhance Computational Intelligence

Hojjat Adeli and Abhish Samant

Department of Civil and Environmental Engineering and Geodetic Science, The Ohio State University, 470 Hitchcock Hall 2070 Neil Avenue, Columbus, Ohio 43210, U.S.A.

Abstract: In this Keynote Lecture it is shown how wavelets can be used to enhance computational intelligence. This is explained through a complicated pattern recognition problem, the automatic detection of traffic incidents in freeways. The dimensionality of the training input data is high and the embedded incident characteristics are not easily detectable. A computational model is presented for automatic traffic incident detection using discrete wavelet transform, linear discriminant analysis, and neural networks. Wavelet transform and linear discriminant analysis are used for feature extraction, de-noising, and effective preprocessing of data before an adaptive neural network model is used to make the traffic incident detection. For incidents with duration of more than five minutes, the incident detection model yields a detection rate of nearly 100% and false alarm rate of about 1% for two- or three-lane freeways.

Keywords: fuzzy logic, incident detection, intelligent transportation systems, neural networks, wavelets

1 Introduction

In recent years computational intelligence approaches including neurocomputing, evolutionary computing, and fuzzy logic have been employed successfully to solve complex problems. Adeli and his associates have advocated a multi-paradigm problem solving approach to solve increasingly complicated real-life problems accurately and efficiently (Adeli and Hung, 1995; Adeli Park, 1998; Adeli and Kumar, 1999).

The performance and efficiency of neural networks algorithms depend on two main factors: the size of the network and the size of the training set. The computational complexity increases exponentially with an increase in the size of the network. Furthermore, with an increase in the size of the network the size of the training set has to be increased exponentially in order to achieve the same level of accuracy. This "*double exponential complexity*" is a major and fundamental challenge for application of neural network to complicated real life pattern recognition problems in terms of both accuracy and efficiency (CPU time) (Wu and Adeli, 2000).

To reduce the dimension of the input space without any significant loss of information we propose a wavelet-based feature extraction approach with a number of advantages including a substantial reduction in (Wu and Adeli, 2000)

a) the computational complexity (that is the network becomes easier to train),
b) the size of the network,
c) the required number of training examples (that means more accurate generalization), and
d) the influence of the noise in the training patterns.

2 Wavelet Transform

The wavelet transform is found to be an effective tool in signal and image processing due to its attractive properties such as time-frequency localization (obtaining a signal at particular time or frequency), multi-rate filtering (differentiating the signals having various frequencies), scale-space analysis (extracting features at various locations in space at different scales), and multi-resolution analysis (Burrus et al., 1998; Daubechies, 1992; Farge et al., 1993; Jameson et al., 1996; Mallat, 1998; Vetterli and Kovacevic, 1995; Wickerhauser, 1994). Using these properties one can extract the desired features from an input signal characterized by certain local properties in time and space.

3 Wavelet-Based Feature Extraction for Intelligent Freeway Systems

Reliable automatic detection of traffic incidents is required in intelligent freeway systems. Travel time delays occur due to lane blockages and the corresponding reduction in the capacity of the freeway following the incident. Since there are lots of traffic fluctuations in the traffic flow for various reasons the incident and incident-free decision regions cannot be divided easily. This is the main cause for the poor performance of the existing traffic incident detection algorithms.

Until early 90s, the two-station comparative algorithms such as California algorithm (Payne and Tignore, 1978) were widely used, where the differences between traffic flow parameters (e.g. traffic volume and occupancy) at upstream and downstream stations are used for the detection of the operating problems in the traffic flow. Persaud et al. (1990) proposed a single station algorithm known as McMaster algorithm, where congestion is detected using traffic volume, occupancy, and vehicle speed (if available) values at a single station.

Though these computational models were easy to implement, they could not achieve the desired level of accuracy. The traffic incident detection problem can be viewed as recognizing incident patterns from the incident-free patterns. To take into account the uncertainty and imprecision inherent in the incident detection, researchers have recently explored the use of new computing approaches such as fuzzy logic (Zadeh, 1978) and neural networks (Adeli and Hung, 1995; Adeli and Park, 1998) to improve the incident detection rate with simultaneous reduction in false alarms (Cheu and Ritchie; Stephanedes and Liu, 1993; Hsiao et al., 1994). Neural networks are known as a powerful method for pattern recognition and classification. The price to pay for their adaptive learning capability is often the need for large computational resources when the problem is complicated requiring a large network and a large number of training instances.

In this paper we present the outline of a computational model for automatic traffic incident detection using discrete wavelet, linear discriminant analysis, and neural networks. Details of the computational models and extensive parametric studies will be presented in two forthcoming articles (Adeli and Samant, 2000; Samant and Adeli, 2000). To eliminate the false alarms an effective incident detection algorithm must be able to extract features from the traffic patterns, which are related to the incident.

Actual traffic data obtained from the sensors on the freeways are not well suited as a direct input to a neural network model to be used to detect incidents. The dimensionality of the training input data is generally high as various traffic parameters (e.g. traffic volume and occupancy) at different locations (e.g. upstream and downstream of the various incident locations) and at many instances of time are required to be inputted, and the embedded incident characteristics may not be easily detectable. Also, the training of a neural network incident detection algorithm requires input patterns containing sufficient incident data. Thus, effective pre-processing of the sensory data is essential before they can be used in a neural network model.

In this work, we perform feature extraction in two steps. In the first step, the data is filtered and the high frequency signals representing noise, which may not be related to an incident, are removed using wavelet transform. In the second step, the features are enhanced using linear disciminant analysis. The feature extraction algorithm also helps reduce the dimensionality of the input space to a neural network model without any significant loss of related traffic information.

We view the traffic flow as a signal, with traffic incidents as well as other traffic patterns such as traffic bottleneck or compression wave having different time-space properties. Most of the previous incident detection algorithms performed unsatisfactorily because they can not distinguish the traffic incident patterns from other similar traffic patterns such as recurrent congestion, and specially the compression wave, consistently. We use the wavelet transform to extract the specific features distinguishing such traffic patterns as it can extract features from different time scales having different resolutions quite effectively. For the traffic incident detection problem, we consider various traffic data (e.g. traffic volumes and occupancies at various locations) recorded at a fixed time interval (e.g. 20-30 seconds). Each of these data series can be represented by a series.

To extract the traffic incident pattern from the traffic data we perform multi-resolution analysis of the wavelet transforms of traffic patterns. Multi-resolution analysis involves dividing the original *signal* (e.g. traffic volume or occupancy) into *signals* having different frequencies and time localizations and analyzing the signal in different scales (Ruskai et al., 1992). Digital filtering of the input signal is carried out by convoluting the signal with another set of numbers known as the filter coefficients or impulse responses (Mallat, 1998) and the downsampling process involving decimation of some of the input data, such as the alternate data points.

4 Linear Discriminant Analysis

We use a linear discriminant analysis to reduce the dimensionality of the problem as well as to improve the generalization capability of the pattern classifier while at the same time reducing its computational processing requirements. The linear discriminant analysis achieves feature extraction through linear mapping of the input space to the output space. The most popular and commonly used linear discriminant classifiers are Fisher Linear Discriminant Classifier (FLD) and Nearest Mean Classifier (NMC) or Euclidean Distance Classifier. The construction procedure for both classifiers is almost the same with minor differences in the end (Fukunaga, 1990; McLachlan, 1992; Ruskai et al., 1992).

In the incident detection problem, the incident and incident-free data may not have identical distribution around their class means. Consequently, the covariance part has to be considered for optimal linear classification. The standard FLD takes into account the covariance part, but linear classification using FLD involves inversion of within-class covariance matrix, which is often an ill-conditioned matrix. This problem is overcome by adding some constant value to the diagonal elements of the covariance matrix as follows (Skurichina and Duin, 1998).

5 Adaptive Conjugate Gradient Neural Network Learning Model

In this work we use the adaptive conjugate gradient neural network learning algorithm of Adeli and Hung (Adeli and Hing, 1994) which combines the conjugate gradient method originally proposed by Fletcher and Reeves (1964) and modified by Powell (1986) with an inexact line search with three criteria for finding the optimal search direction. The conjugate gradient method is based on the steepest descent method where weight changes are made along the direction resulting in the maximum decrease in the system error. Determination of the step length of a gradient-based optimization algorithm has a significant impact on its efficiency. A very accurate or “exact” line search requires many function evaluations thus making the algorithm prohibitively and unnecessarily expensive.

An appropriate inexact line search algorithm can determine the step length, within a small percentage of that found based on an exact search. Adeli and Hung use the backtracking inexact line search algorithm of Dennis and Schnable (1983), the step length selection terminating criterion of Armijo (1966) to ensure the step length is not too large, the terminating criterion of Goldstein (1967) to ensure the step length is not too small, and the direction convergence criterion of Nocedal (1990) to ensure that the descent direction is always generated.

6 Results and Conclusion

In this work we use the traffic simulation package TSIS/CORSIM developed by ITT Systems and Sciences Corporation (http://www.fhwa-tsis.com) to simulate the freeway traffic needed to train the incident detection algorithm. This simulation package allows you to simulate the road conditions for a given grade, curvature, or maximum allowable speed as well as the traffic conditions such as traffic flow, incident location, percentage lane blockage, and duration of the incident.

The traffic incident detection results and parametric studies using the new computational model are presented in a forthcoming article (Samant and Adeli, 2000). The simulated data used for testing include 45 incidents with traffic volume varied from 300 to 2000 vehicles/hour per lane. The application of linear discriminant analysis was investigated in two different ways, as a linear classifier and as a feature enhancer. As a linear classifier, it was applied to all the data series simultaneously using a single data point from each data series, without using neural networks. As a feature enhancer, it was applied to each data series separately and the resulting traffic parameter values are used as input to the neural network model. The results show that linear discriminant analysis by itself is a poor classifier for the problem at hand.

Using the traffic volume and occupancy as input data, the best incident detection results were obtained employing the adaptive conjugate neural network learning algorithm after the filtering and preprocessing of data by discrete wavelet transform and linear discriminant analysis. The traffic data are first filtered using discrete wavelet transform and multi-resolution analysis, and the high-resolution components are discarded. The low and medium resolution components are found to be sufficient for representing the traffic flow.

Our feature extraction model is a two-step algorithm consisting of discrete wavelet transform and linear discriminant analysis. In order to investigate their relative contribution in feature extraction, we also used discrete wavelet transform as the sole feature extractor. A comparison of the data indicate that most of the feature extraction capability is due to discrete wavelet transform. Linear disciminant analysis has a smaller contribution toward improving the incident detection. One can say it has a fine tuning effect for reducing the false alarm rates.

In summary, we have developed a robust incident detection computational model and algorithm through adroit integration of three different computational approaches: signal processing and wavelet transform, statistical linear discriminant analysis, and artificial neural networks. For incidents with duration of more than five minutes, the algorithm yields a detection rate of nearly 100% and false alarm rate of about 1% for two- or three-lane freeways. For incidents with duration of less than 5 minutes, the incident detection rate for two- or three-lane freeways is about 98% with a false alarm rate of about 1%.

Acknowledgment

This paper is based on the research sponsored by Ohio Department of Transportation, Federal Highway Administration, and The Ohio State University Center for Intelligent Transportation Research. The authors are also grateful to Mr. George Saylor of Ohio Department of Transportation (DOT), Mr. Leonard Palek of Minnesota DOT, and Phil Carter of Arizona DOT for their help in obtaining actual traffic data and Dr. Henry Lieu of Federal Highway Administration for providing access to the traffic simulation package (TSIS/CORSIM) to test the computational models developed in this research.

References

1. Adeli, H. and Hung, S.L. (1994), "An adaptive conjugate gradient algorithm for efficient training of neural networks", *Applied Mathematics and Computation*, Vol. 62, pp. 81-102.
2. Adeli, H. and Hung, S.L. (1995), *Machine Learning - Neural Networks, Genetic Algorithms, and Fuzzy System*, John Wiley, New York.
3. Adeli, H. and Kumar, S. (1999), *Distributed Computer-Aided Engineering*, CRC Press, Boca Raton, Florida.
4. Adeli, H. and Park, H.S. (1998), *Neurocomputing in Design Automation*, CRC Press, Boca Raton, Florida.
5. Adeli, H. and Samant, A.M. (2000), "An adaptive conjugate gradient neural network-wavelet model for traffic incident detection", *Computer-Aided Civil and Infrastructure Engineering*, Vol. 15, to be published.
6. Armijo, L. (1966), "Minimization of functions having Lipshitz-continuous first partial derivatives", *Pacific Journal of Mathematics*, Vol. 16, No. 1, pp. 1-3.
7. Burrus, C.S., Gopinath, R.A. and, Guo, H. (1998), *Introduction to Wavelets and Wavelet Transforms: A Primer,* Prentice Hall, Upper Saddle River, New Jersey.
8. Cheu, R.L. and Ritchie, S.G. (1995), "Automated detection of lane-blocking freeway incidents using artificial neural networks", *Transportation Research* 3C(6), pp. 371-388.
9. Daubechies, I. (1992), *Ten Lectures on Wavelets*, SIAM, Philadelphia, Pennsylvania.
10. Dennis, J.E., Jr., and Schnable, R.B. (1983), *Numerical Methods for Unconstrained Optimization and Nonlinear Equations*, Prentice Hall, Englewood Cliffs, New Jersey.
11. Farge, M., Hunt, J.C.R., and Vassilicos, J.C., Ed. (1993), *Wavelets, Fractals, and Fourier Transform*, The Institute of Mathematics and its Applications Conference series, Clarendon Press, Oxford.
12. Fletcher, R. and Reeves, R.M. (1964), "Function minimization by conjugate gradients", *The Computer Journal*, Vol. 7, No. 2, pp. 149-160.
13. Fukunaga, K. (1990), *Introduction to Statistical Pattern Recognition*, 2nd Ed., Academic Press, New York.
14. Goldstein, A.A. (1967), *Constructive Real Analysis*, Harper & Row, New York.

15. Hsiao, C.H., Lin, C.T., and Cassidy, M. (1994), "Application of fuzzy logic and neural networks to automatically detect freeway traffic incidents", *Journal of Transportation Engineering*, Vol. 120, No. 5, pp. 753-771.

16. Jameson, L.M., Hussaini, M.Y., Erlebacher, M., Eds. (1996), *Wavelets: Theory and Applications*, ICASE/LaRC Series in Computational Science and Engineering, Oxford University Press, New York.

17. Kollias, S. and Anastassiou, D. (1989), "An adaptive least-squares algorithm for the efficient training of artificial neural networks", *IEEE Transactions on Circuits and Systems*, Vol. 36, No. 8, pp. 1092-1101.

18. Mallat, S.G. (1998), *Wavelet Tour of Signal Processing*, Academic Press, London.

19. McLachlan, G.J. (1992), *Discriminant Analysis and Statistical Pattern Recognition*, John Wiley, New York.

20. Nocedal, J. (1990), "The performance of several algorithms for a large scale unconstrained optimization", in Coleman T.F., and Li, Y., Eds., *Large-Scale Numerical Optimization*, Society for Industrial and Applied Mathematics, Philadelphia, pp. 138-151.

21. Payne, H.J., and Tignor, S.C. (1978), "Freeway incident-detection algorithms based on decision trees with states", *Transportation Research Records* 682, TRB, National Research Council, Washington, DC, pp.30-37.

22. Persaud, B.N., Hall, F.L., and Hall, L.M. (1990), "Congestion identification aspects of the McMaster incident detection algorithm", *Transportation Research Records* 1287, TRB, National Research Council, Washington, DC, pp.167-175.

23. Powell, M.J.D. (1986), "Convergence properties of algorithms for nonlinear optimization", *SIAM Review*, Vol. 28, No. 4, pp. 487-500.

24. Ruskai, M.B., Beylkin, G., Coifman, R., Daubechies, I., Mallat, S., Meyer, Y., and Raphael L. (1992), *Wavelets and their Applications*, Jones and Bartlett Publishers, Boston.

25. Samant, A. and Adeli, H. (2000), "Feature extraction for traffic incident detection using wavelet transform and linear discriminant analysis", *Computer-Aided Civil and Infrastructure Engineering*, Vol. 15, to be published.

26. Skurichina, M. and Duin, R. (1998), "Bagging for linear classifiers", *Pattern Recognition,* Vol. 31, No. 7, pp. 909-930.

27. Stephanedes, Y.J. and Liu, X. (1995), "Artificial neural networks for freeway incident detection", *Transportation Research Records* 1494, TRB, National Research Council, Washington, DC, pp. 91-97.

28. Vetterli, M. and Kovacevic, J. (1995), *Wavelets and Subband Coding*, Prentice Hall, Inc., Englewood Cliffs, New Jersey.

29. Wickerhauser, M.V. (1994), *Adapted Wavelet Analysis from Theory to Software,* A K Peters, Ltd., Wellesley, Massachusetts.

30. Wu, M. and Adeli, H. (2000), "Wavelet-Neural Network Model for Automatic Traffic Incident Detection", to be published.

31. Zadeh, L. (1978), "Fuzzy set as a basis for a theory of possibility", *Fuzzy Sets and Systems*, Vol. 1, No. 1, NY, pp.3-28.

Multispectral Image Processing Using Computational Intelligence[1]

P. Sinčák[1], N. Kopčo[1,2], M. Hrickova [1],S. Kaleta[1], H. Veregin[3]

[1]Computational Intelligence Group, Laboratory of AI, Department of Cybernetics and AI, Faculty of Electrical Engineering and Informatics, TU Košice, Slovakia
E-mail: cig@neuron-ai.tuke.sk, http://neuron-ai.tuke.sk/cig,

[2]Boston University, Center for Neural Sciences, Boston , USA ,
E-mail: kopco@bu.edu ,

[3]Department of Geography, University of Minnesota, Minneapolis, MN, USA
E-mail: veregin@x.geog.umn.edu

Abstract: The paper deals with the application computational intelligence tools for multispectral image classification. Pattern Recognition scheme is a global approach where a classification part is playing an important role to achieve the highest classification accuracy. Multispcetral images are data mainly used in remote sensing and this kind of classification is very difficult to make an accuracy assessment of classification results. There is a feedback problem to adjust the parts of pattern recognition scheme. Precise classification accuracy assessment is almost not possible to obtain, because it is a very laborious procedure. In the paper simple neural networks for multispectral image classification, ARTMAP neural networks as more sophisticated tools for classification, neurogenetic systems like ECOS for image classification and also modular and multi-classifier approach to achieve the highest classification accuracy on multispectral images are presented. There is a strong link to progress in computer technology which gives much better conditions for modeling more sophisticated classifiers for multispectral images.

Keywords: multi-spectral satellite images, pattern recognition principles, classifier design, classification accuracy assessment, contihency tables, Backpropagation neural networks, fuzzy BP neural networks, Art and Artmap neural networks, Neurogenetic systems as classifiers, Modular neural networks, multi-classifier methods for image classification.

---

[1] This project is supported by Slovak National Science Agency under project # 9433 "Computational Intelligence for decision procedures" , 1999-2001

1 Introduction

Remote sensing is an important part of modern technology and its application possibilities include environmental monitoring, something that is very important for the future of humankind. Image processing can be the most important part of this system and determines the quality accuracy of information retrieval from the images. Generally we have to take into consideration the basic pattern recognition principles where classification plays an important role. These principles are presented in the following figure:

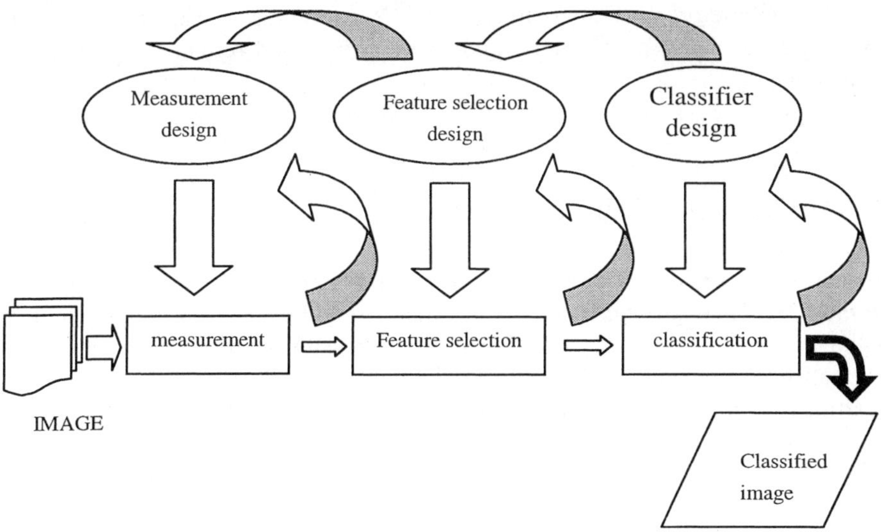

Figure 1: Basic principles of pattern recognition

As it is clear from the above figure, the determination of feature space is playing a key role in overall pattern recognition principles. The feature selection and classification are based on this information and the level of ability to approximate the nonlinear discrimination function determinates the success of classification procedure. A very important aspect of this procedure is accuracy assessment of classification results. The contingency analysis is a proper way of accuracy assessment and it reflects the multiclass results of classification procedure.

The important aspects of the pattern recognition principle are the adaptability of the overall procedure and the equal importance of all steps in the procedure including feature space determination and feature selection procedure. This pattern recognition principle reveals such questions as follows:

- Is it more useful to "invest" more time to the best feature selection procedure?

- Or is it more important to "invest" more time to develop the most sophisticated classification procedure for any type of features?

These two questions are playing the key-role in the future of research in the classification domain. Certainly these two problems are closely connected and usually feature selection techniques assume a specific type of classifier. We can find a number of techniques dealing with the first approach and a number of approaches dealing with the second problem. Basically the first approach is based on the assumption that a simple and fast classification procedure is used while the second problem is related to the assumption of a simple and almost raw data or data with minor preprocessing procedures. Certainly the combination of both approaches could be fruitful and could bring some good results for real world application.

2 Motivation of the Project

This part of the project reflects the second question and it is searching for the best and most highly sophisticated classifiers which would be able to handle the highly nonlinear classification problems. The most successful approaches are based on pure statistical methods. Traditional approaches to this task are based on statistical classification and pattern analysis methods, the most important being the Maximum Likelihood classifier [2]; and on various rule-based algorithms [3]. The statistical methods have several useful properties, the most important one being their optimal behavior if several assumptions are fulfilled. The disadvantages of these methods include their poor performance if the assumptions are not fulfilled, as well as their large complexity (in terms of training time and memory requirements).

Among these approaches, artificial neural networks play a significant role [4], and among them the Error Back-propagation algorithm [5] is the predominant method. There are also some modifications of these approaches e.g. fuzzy BP approach and some others. There are also several reports of application of the *Adaptive Resonance Theory (ART)* neural networks in this domain [6, 7, 8], most of them applying fuzzy ARTMAP. Using Neurogenetic systems is a very interesting and challenging domain. Important works are presented in [14] where they attempt to utilize and combine the main advantages of both neural networks and evolutionary approach to achieve the best classification results. Further the modular systems are playing an important role. Important work has been done by Jordan using the excellent idea to divide problems into sub-problems by considering sub-modules and solving them parallel and with higher accuracy. The modular approach seems to be very interesting also from the aspect of building hardware implementations of the proposed systems. Also the idea of the multi-classifier approach seems to be interesting. The same task is given to more types of classifiers and then these classification results are combined and conflated into integration of classification results. The accuracy of this conflation could be higher because the best results of all classification approaches are taken into consideration. They based on the maximum likelihood approach, which is based on the assumption of gaussian aposterior probability measures.

3 Neural Network Classifiers

Neural Networks are considered as universal aproximators of the unknown joint function. Classification can be considered as a mapping procedure from the measured or given feature space in the space of discrete classes. There is no assumption of any probability distribution of data and we call such systems "Model Free Systems". The training procedure is able to determine the parameters of the neural networks (usually synaptic weights) to approximate the mapping functions.

3.1 Simple Neural Network Classifiers

One of the simplest neural networks is a multilayer perceptron with simple Backpropagation adaptation rules. The approximation power of this neural network is very good but the problems with topology design and convergence can cause some problems, especially with data having high nonlinear relations between classes. Some extension or better application of these methods was done with using fuzzy class identification using simple BP percentron. The point is in changing the fuzzy class definition and training set for simple feed forward BP neural networks. The problem of multi-cluster classes is not taken into consideration. The training set has been changed where the first part of ordered couples in the training set were at the point in the feature space and the second part of that couple were a value of the membership function of input to the fuzzy class(set) A. So the training set for fuzzy class A looks like :

$$A \in \{[x_1, \mu_A(x_1)], \ldots, [x_n, \mu_A(x_n)]\} \tag{1}$$

According to this approach the input of the neural network is associated with output which is as follows:

$$\{\mu_A(x_n), \mu_B(x_n), \mu_C(x_n), \ldots \mu_N(x_n)\} \tag{2}$$

So each input is trained with its relations to each class of interests using membership function's values to each fuzzy class. This input output mapping is trained by conventional BP approach. Re-scaling the values of membership functions does some enhancement of differences between fuzzy classes. More info about this approach can be found in Lin.

3.2 ARTMAP Neural Network Classifiers

ARTMAP neural networks belong to the class of neural networks called Adaptive Resonance Theory (ART), a theory of cognitive information processing in the human brain [9]. Based on this theory, a whole family of neural network algorithms was developed. These neural networks were shown to give a very good performance in applications involving clustering, classification, and pattern recognition. When compared to statistical and other neural-network-based clustering/classification algorithms, these networks usually obtain very good classification accuracy, while securing proven stability and a high level of compression in the system.

An overview of the process of development of ARTMAP neural networks can be found in [8]. From the point of view of this study, the currently available ARTMAP classification systems can be divided into two groups. First, systems based on (or systems that are a modification of) fuzzy ARTMAP algorithm (e.g., ARTMAP-IC, ART-EMAP, etc., see [10]). All these systems share the property that they prefer data clusters distributed into hyper-rectangles in feature space. In these systems the basic properties of the original ARTMAP design (stability, proven convergence, fast on-line learning) are preserved, but they also have well-known disadvantages, e.g., noise sensitivity and tendency to category proliferation. The other group is based on the Gaussian ARTMAP neural network [11]. In this group of networks, preferably identifying Gaussian-shaped clusters, the stability and fast on-line learning properties of the fuzzy ARTMAP networks are traded for emphasis on the ability of the system to generalize and for its decreased sensitivity to noise in the input data.

Structurally, every ARTMAP network (fuzzy ARTMAP or Gaussian ARTMAP) can be divided into two parts. The first part, represented by an ART module, dynamically generates units, each identifying a single data cluster in feature space. This part can be used autonomously for cluster analysis of a given data set. The second part serves to identify each of the clusters found in the data with one of the classes defined in the data set.

A detailed description of fuzzy ARTMAP (FA), first of the algorithms analyzed in this study, can be found in many previously published studies. For a description directly related to the processing of data from remote sensing, the reader is referred to publications [6, 7, and 8]. From the point of view of this study, the most important property of this system is that the subsystem identifying clusters in feature space, preferably identifies the clusters in which patterns that are distributed as hyper-rectangles.

The second algorithm, Gaussian ARTMAP (GA), is described in detail, e.g., in [7 and 11]. Its main feature is that it preferably identifies clusters with Gaussian distribution, in which the co-variance (off-diagonal) coefficients in the co-variance matrix describing the cluster are fixed to zero. This restriction was imposed on the Gaussian ARTMAP system for computational purposes, the reason being that with this kind of representation each cluster-identifying node is described by $2*M+1$ parameters, where M is the dimensionality of feature space. This memory requirement is only slightly worse than memory requirements of fuzzy ARTMAP networks ($2*M$). But it is much lower than that of the Extended Gaussian ARTMAP. On the following figure are the basic principles of ARTMAP neural network.

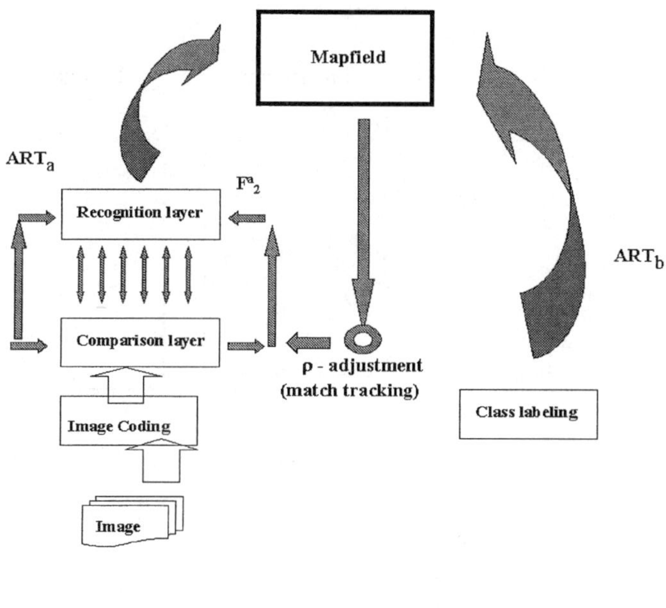

Figure 2: Basic topology of ARTMAP NN

Most of ARTMAP neural networks use so called complementary coding which is improtant for discrimination curves determination in the feature space. While Fuzzy ARTMAP makes rectancles, Gaussian ARTMAP makes ellipsoids as discrimination curves in the feature space. Illustration of this fact can be found in figures XY. In { } MF- ARTMAP approach is presented and offers a fuzzy cluster indentification in the feature space and so for each input, the values of membership functions of the input to the each fuzzy class is calculated. The fuzzy class is a union of fuzzy clusters belonging to a considered class defined by a training set e.g.

$$CL = \{\bigcup_{i=1}^{n} A_i\} \tag{3}$$

Generally we can consider the fuzzy class as a set of fuzzy sets A_i representing the variety of the numerical representation of the class. Relation between μ_{CL} (x) and μ_A(x) must be as follows:

$$\mu_{CL}(X) = \max_{i\ =\ 1,\ n}(\mu_{Ai}(X)) \tag{4}$$

where A_i is a fuzzy cluster which belongs to class CL and n is number of fuzzy clusters creating class CL. The MF-ARTMAP is intended to be such a tool to calculate values of membership functions of X to each class of interest in feature space.

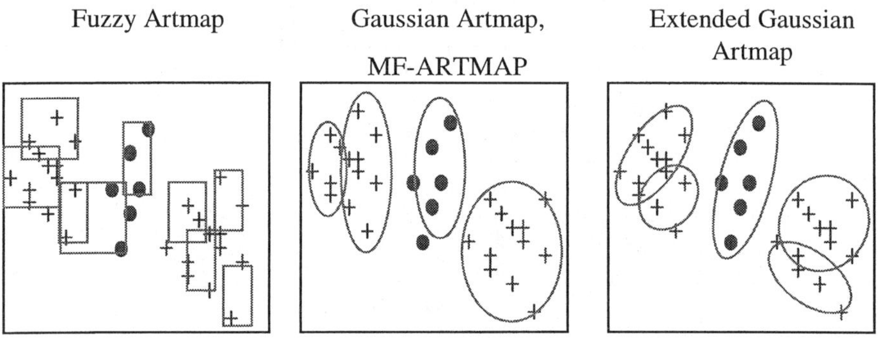

Figure 3: Representation of discrimination curves in the feature space

Results of mentioned methods on benchmark data, as well as on the real world data, are presented in a later section of this paper.

4 Neurogenetic Systems as Classifiers

Evolutionary computation represents a wide area of methods where genetic algorithms play an important role. The advantages of these approaches are rather in optimization problems. Neurogentics systems are based on utilization of advantages of both neural networks with their learning abilities and genetic algorithms with their optimization abilities. The result of this cooperation is synthesis in hybrid system with the ability to find the best topology of neural network using genetic algorithm to determine it.

4.2 Evolving Fuzzy Neural Network

We used the Evolving fuzzy neural network (EfuNN) system developed by {Kasabov} and is part of his ECOS strategy to create an intelligent system with many important features. Basically, EFuNN has a five-layer structure. The input layer represents input variables. The second layer of nodes represents fuzzy quantization of each input variable space. The third contains rule (case) nodes that evolve through supervised/unsupervised learning. The rule nodes represent exemplars of input-output data associations, graphically represented as an association of hyper-spheres from the fuzzy input and fuzzy output spaces. Each rule node r is defined by two vectors of connection weight $W_{1r}(t)$ a $W_{2r}(t)$. The latter is being adjusted through supervised learning based on the output error, and the former being adjusted through unsupervised learning based on similarity measure within a local area of the problem space. It is expressed through the following adaptation rules:

$$W_{1r}(t+1) = W_{1r}(t) + \gamma_1 DIST \tag{5}$$

where DIST = $(W_{1r}(t\text{-}1) - W_{1r}(t))$ and also

$$W_{2r}(t+1) = W_{2r}(t) + \gamma_2 Err \tag{6}$$

where Err = $(W_{2r}(t\text{-}1) – W_{2r}(t))$.

The fourth layer of neurons represents fuzzy quantization for the output variables. The fifth layer represents the real values for the output variables. The structure of ECOS can be found in the following figure 4. It can be seen that ECOS is highly modular approach and it is based on strategy developed by Kasabov. Basically out approach used more evolutionary computation for determination the right neural network topology. Generally ECOS is a very progressive approach with biological motivation in Brain and modular structure of the intelligent system.

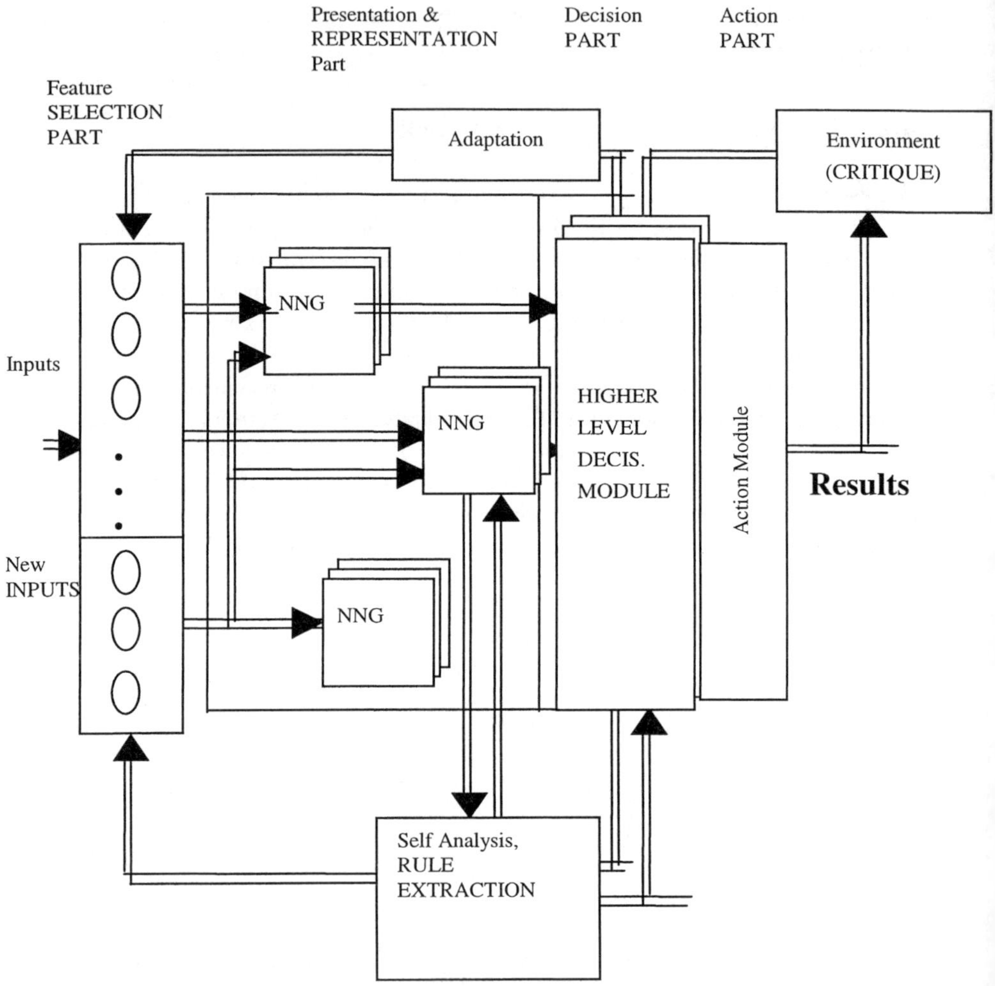

Figure 4: Basic topology of ECOS

4.3 Using Evolutionary Algorithm to Find a Neural Network Topology

The aim of the evolutionary algorithm (EA) described in this paper is to find neural network (NN), or more NNs, with the best classification results. The vector (chromosome) of the eleven numbers was chosen to represent the NN, as an individual in EA. Example of this NN representation with a brief description of each artificial gen is as follows:

$$Gen = \{mf, sthr, errthr, lr1, lr2, prune, old, ep, thra1, thra2, nag\} \tag{7}$$

Where mf - number of the fuzzy membership function (we used values between 2-15); sthr - sensitivity threshold (0,4-0,9); errthr - error threshold(0,01-0,1); lr1 – the first learning rate (0-1.0); lr2 – the second learning rate (0-1.0); prune - pruning parameter (if<0, no prunning); old - pruning parameter for nodes; ep - number of training cycles(1-3); thra1 – the first aggregation threshold (0-1.0); thra2 – the second aggregation threshold (0-1.0); nag - mumber of epochs, after which an aggregation will be applied (aprx. 80). In the population, its own chromosome represents each NN. The train and test sets are for each NN the same. Chromosome's values are input values (each NN in the population has its own input values) for learning algorithm. Each NN of the population is trained and tested. Fitness of the NN is given as its total accuracy on the testing set. There is a brief description of EA used here:

1. Create initial random population, it used only 5 NNs to fill the population, because of the experiment duration.
2. Evaluate fitness of each member (each NN) trained, tested and its total accuracy is its fitness function.
3. Is the stopping criterion satisfied? if YES - stop EA, NO - continue. The used stopping criterion was 100% fitness
4. Selection using the selection method to determine the parents. It was usually a used tournament with 2 members.
5. Create offspring, that means perform crossover and mutation to create the new members (new NNs)of the population. Crossover was used with 0.4 probability, and mutation with 0.8 probability.
6. Evaluate the fitness of each offspring and each new member trained, tested and evaluated.
7. Replacement means replace the old population with the new one. Usually, elitism was used (it makes sure, that the fittest chromosome of a population is passed on the next generation unchanged)
8. Increase the number of generations and go to step no. 3

Usually we can get more than 1 neural network with optimal results on testing sets and with various topologies. It seems to be convenient to choose the minimum topology neural network to reduce computational demands which are a bit high for these procedures. High computational processing pays off with higher classification accuracy on test data subset. Results of mentioned methods on benchmark data and on the real world data are presented in later section of this paper.

5 Modular Neural Network as Classifiers

The notion of modular neural network hes been known for many years. It is a very promising idea of solving complex problems by its distribution into more sub-problems that are easy to solve. Basically the "divide and conquer" principles are usually used in modular neural networks. There are some difficult questions about the separability problems and discrimination hyper-planes determination. The key problem is in making an answer to the following problem:

Is it easier to separate one particular class from the feature space or to identify more classes among each other? So in fact the question is about difficulties of dichotomous classification comparing the multi-class approach. The first impression could be that dichotomous classification is always easier than the multi-class approach, but it is very difficult to conclude in general. For investigating these ideas, a parallel ARTMAP approach was designed and tested. The basic philosophy of this approach is illustrated in the following figure 5.

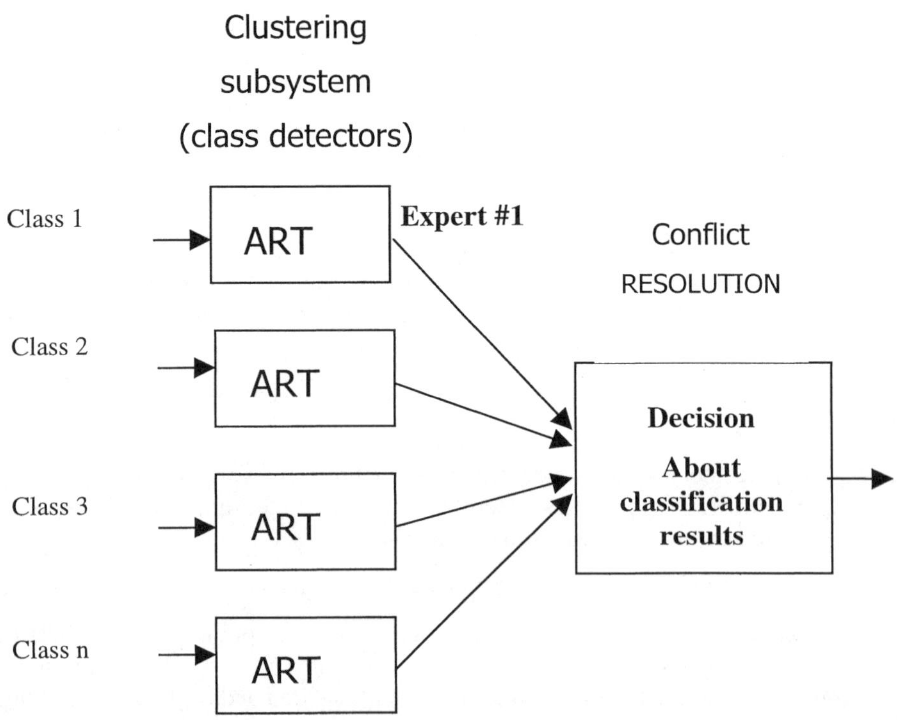

Figure 5: The basic principle of Parallel ARTMAP.

Each Expert is making dichotomous classification with the output related to input as follows: "Yes" or "No" in sense input belongs to associated class or input does not belong to the associated class. On the second level the conflicts are processed in case that more than one expert will identify input as belonging to its class. The conflict solution can be various and we used the values of activation on the expert neural networks. Each sub-net is trained on patterns from a single class using cross-validation: - 9/10 - estimation set, - 1/10 - validation set. Training is repeated for different values of ρ. The result of identifying the proper vigilance parameter ρ is illustrated in the following figure 6.

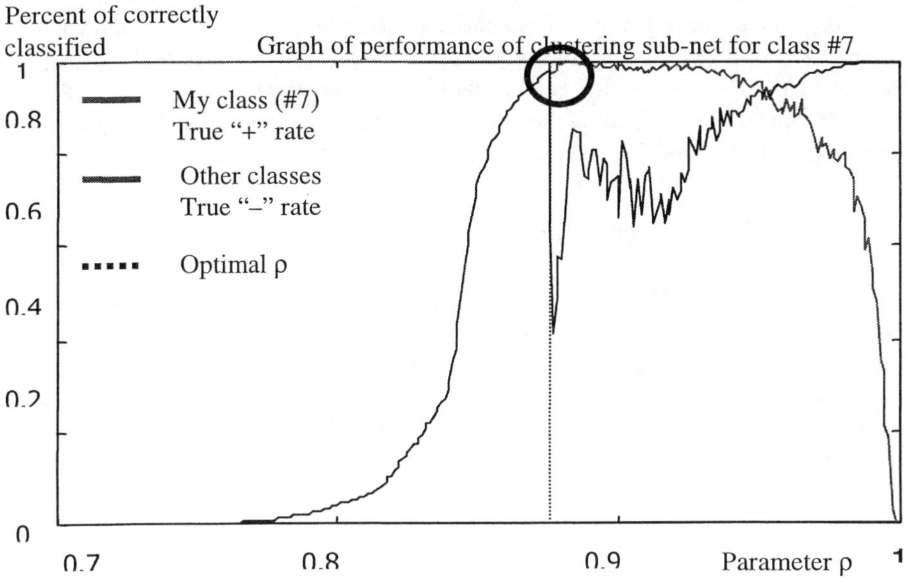

Figure 6: Identification of vigilence parameter

A similar procedure is done for every sub-net with aim to identify the proper vigilance parameter for each sub-net. From the experiments the following rule has been pointed out :

"Choose ρ which corresponds to the the point just before the first dip in the true positive graph" (look on the graph)

This rule assures that each detector will correctly identify almost all of the patterns belonging to it, while minimizing the false alarm rate.

Classification results using this policy were not easy to interpret. Concerning the large databases this approach seems to be interesting. It has explicit incremental learning so in the case of adding a new class a new expert is involved in the overall procedure. In spite of current unsuccessful results we assume this approach to be interesting and research should be focused to parallel classification approaches. Some partial results were achieved with the MF-ARTMAP classification approach. This method utilizes the values of membership functions to solve conflict resolution.

6 Experimental Results

The experiments using the benchmark, as well as real-world data, were done during this project. The aim was a comparative analysis of known CI systems with those modified or developed during the research, namely MF-Artmap, GA-ECOS, Paralel ARTMAP. Basicly the same real-world data were used and so the comparison can be done assuming the same training and testing data, which means the same amount of knowledge was used for classification procedure.

6.1 Experiments on Benchmark Data

There were 2 benchmark data used for testing classification results for comparative purposes. Circle in the rectangle and double spriral problem were used for dichotomial classification testing. These two benchmark data are most used for estimating the classifier level of sophistication. If results of these benchmark data are good there is a good assumption that the classifier will be successful in main applications.

Circle in square:

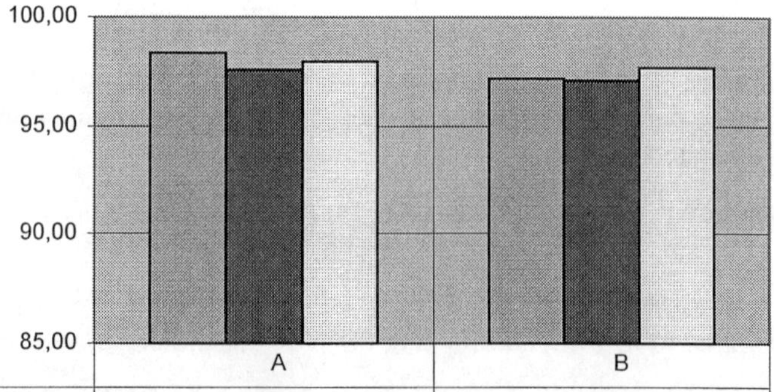

	A	B
Results of Fyzzy ARTMAP	98,34	97,20
Results of MF ARTMAP	97,62	97,08
Results of GA-ECOS	97,94	97,66

Two spirals:

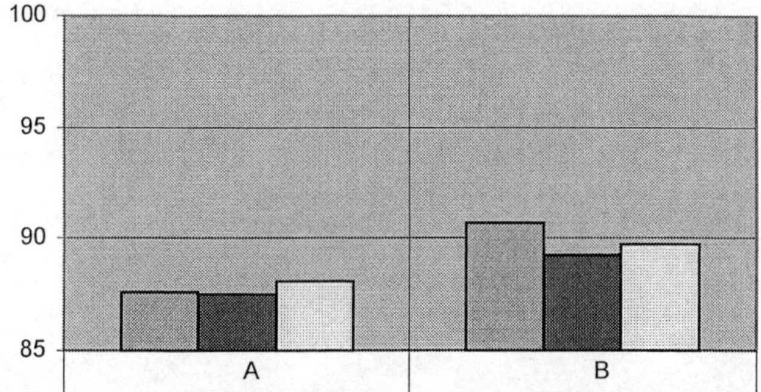

	A	B
Results of Fyzzy ARTMAP	87,59	90,74
Results of MF ARTMAP	87,54	89,24
Results of GA-ECOS	88,05	89,75

Figure 7: Graph of classification accuracy

6.2 Experiments on Real-World Data

Experiments were done on benchmark and real-world data. Basically the behaviors of the methods were observed on multispectral image data with the aim to obtain the best classification accuracy on the test data subset. The Košice data consists of a training set of 3164 points in the feature space and of a test set of 3167 points of the feature space. A point in the feature space has 7 real-valued coordinates of the feature space normalized into the interval (0,1) and 7 binary output values. The class of a fact is determined by that output which has a value of one; the other six output values are zero. The data represents 7 attributes of the color spectrum sensed from Landsat satellite. The representation set was determined by the geographer and was supported by ground verification procedure. The main goal was landuse identification using the most precise classification procedure for achieving accurate results. The image was taken over the eastern Slovakia region particularly from the City of Kosice region. There were seven classes of interest picked up for classification procedure as it can be seen on the figure 8. Results of GA ECOS and MF/ARTMAP can be found in figures 9 and 10 respectivelly. Results on Fuzzy Artmap were very similar what can be observed also in the figure 11 where results are presented in the graphs. Results on Paralel ARTMAP were significantly worse but we hope that this approach has logical background and sense for classification in the feature space.

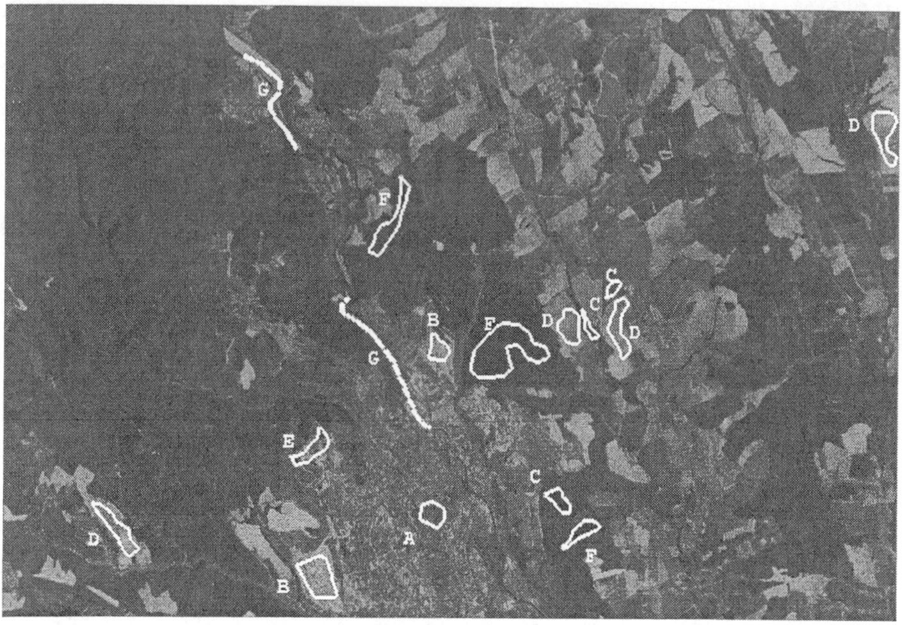

Fig. 8 Original image. Highlighted areas were classified by an expert (A – urban area, B – barren fields, C – bushes, D – agricultural fields, E – meadows, F – forests, G – water)

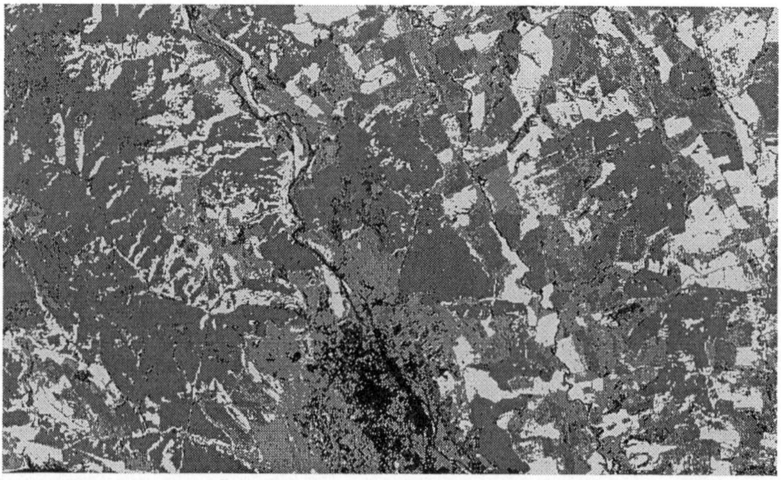

Fig. 9 Classified images by GA-ECOS

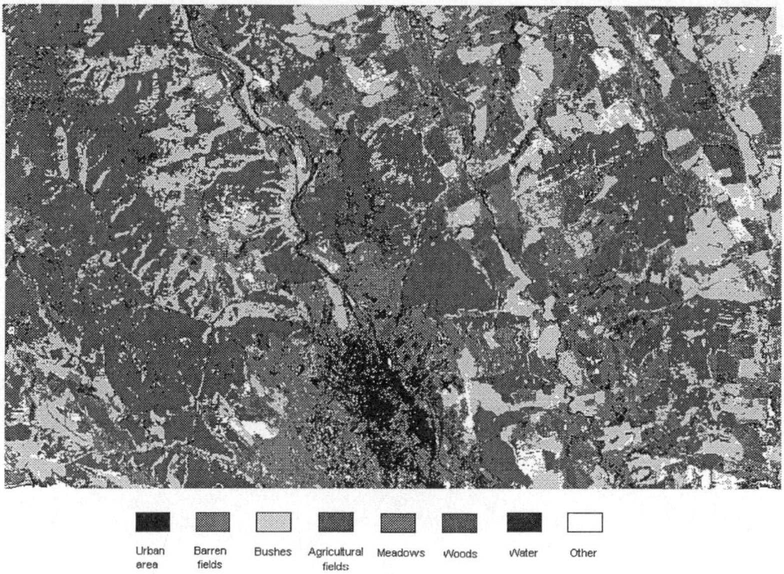

Fig. 10 Classified images by MF ARTMAP

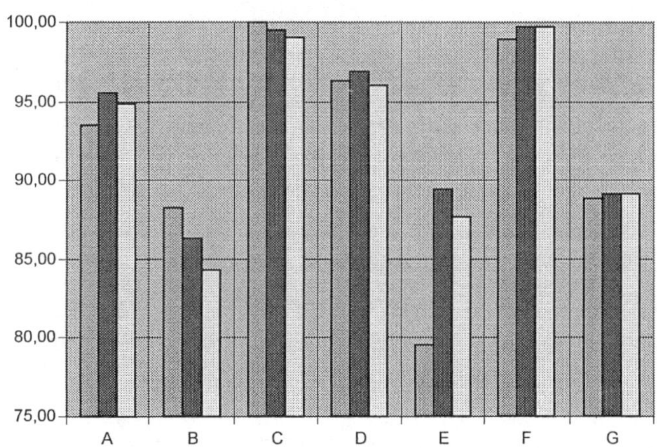

	A	B	C	D	E	F	G
Results of Fyzzy ARTMAP	93,51	88,30	100,00	96,33	79,55	98,97	88,84
Results of GA-ECOS	95,51	86,36	99,51	96,90	89,47	99,74	89,19
Results of MF ARTMAP:	94,87	84,30	99,03	96,01	87,72	99,74	89,19

Figure 11 : Classification Results on real-world data

7 Conclusion

The paper deals with selected methods from computational intelligence domain for multi-spectral image processing. There are several CI approaches with potential benefits over methods from different domains. The MF-Artmap developed on the basis of Artmap principles using identification of fuzzy clusters in the feature space was tested on benchmark and real-world data. This developed approach has comparable results to fuzzy Artmap but gives more information about fuzzy relation to any point in the feature space to fuzzy classes consisting of fuzzy clusters. Some study was made also for modular neural networks using parallel artmap approach for classification purposes. In spite of the fact that results were not very successful it is a worthy direction of research for the future.

References

[1] P.M. Atkinson and A.R.L. Tatnall, "Neural networks in remote sensing," *Int. J. Remote Sensing,* vol. 18, 4, 1997, 711-725.

[2] J. Richards, Remote sensing digital image analysis: An introduction. Springer Verlag: Berlin, 1993.

[3] B.G. Lees and K. Ritman, "Decision-tree and rule-induction approach to integration of remotely sensed and GIS data in mapping vegetation in disturbed or hilly environments," *Environmental Management*, vol. 15, 1991, pp. 823-831.

[4] C.H. Chen, "Trends on information processing for remote sensing," in *Proceedings of the International Geoscience and Remote Sensing Symposium (IGARSS)*, vol. 3, Aug. 3-8 1997, pp. 1190-1192.

[5] P. Werbos, *Beyond Regression: New Tools for Prediction and Analysis in the Behavioral Sciences*, Ph.D. thesis, Harvard University, 1974.

[6] G.A. Carpenter, M.N. Gjaja, S. Gopal, and C.E. Woodcock, "ART neural networks for remote sensing" Vegetation classification from Landsat TM and terrain data," *IEEE Transactions on Geoscience and Remote Sensing*, vol. 35, no. 2, 1997, pp. 308-325.

[7] N. Kopčo, P. Sinčák, and H. Veregin, "Extended Methods for Classification of Remotely Sensed Images Based on ARTMAP Neural Networks," in *Computational Intelligence – Theory and Applications (Lecture Notes in Computer Science 1625), Proceedings of International Conference "The 6-th Fuzzy days,"* Dortmund, Germany, May 1999, pp. 206-219.

[8] H. Veregin, P. Sinčák, N. Kopčo, "Conflation techniques in multispectral image processing," *Geocarto Int*, March. 2000, pp 11-19.

[9] S. Grossberg, "Adaptive pattern classification and universal recoding, I: Feedback, expectation, olfaction, and illusions," *Biological Cybernetics,* vol. 23, 1976, pp. 187-202.

[10] G.A. Carpenter, B.L. Milenova, and B.W. Noeske, "Distributed ARTMAP: a neural network for fast distributed supervised learning," *Neural Networks,* vol. 11, no. 5, Jul. 1998, pp. 793-813.

[11] J.R. Williamson, "Gaussian ARTMAP: A neural network for fast incremental learning of noisy multidimensional maps," *Neural Networks,* vol. 9, 1996, pp. 881-897.

[12] R.K. Cunningham, Learning and recognizing patterns of visual motion, color, and form, *Unpublished Ph.D. thesis, Boston University, Boston*, MA: 1998.

[13] R. Duda and P. Hart, Pattern Classification and Scene Analysis, *Wiley, New York: 1973.*

[14] N. Kasabov, R. Kozma: Neuro-Fuzzy Techniques for Intelligent Information Systems, *Springer Verlag,* 1999, ISBN 3-79081187-4

[15] M. Anthony, P L. Barlett: Neural Network Learning : Theoretical Foundations, *Cambridge University Press*, 1999, ISBN 0521 57353

Classification of Multispectral Data Using Fuzzy Neural Networks

Renata Smolikova, Mostafa G.-H. Mostafa, Jacek M. Zurada*, and Aly A. Farag

Electrical and Computer Engineering Department, University of Louisville, Louisville KY 40292, USA.
Emails:
(mostafa,farag)@cvip.uofl.edu, (r0smol01,jmzura02)@athena.louisville.edu

Abstract: In this paper we propose a fuzzy neural network (FNN) methodology for the supervised classification of Landsat multispectral images. Instead of using direct radiance values, we generate feature vectors from the data that take into account the effect of the neighborhood on the pixel classificaton. The FNN architecture deals with fuzzy inputs and crisp weights to account for the large variation in the sample data. Results are compared with the classification results of a conventional "crisp" feedforward neural network.

Keywords: image classification, multispectral data, Landsat images, fuzzy neural network, feedforward neural network, backpropagation learning

1 Introduction

Recent developments in hyperspectral and multispectral sensing modalities have enabled the acquisition of high dimensional data. Classification of such data is an intensive research area in the remote sensing community [1]. Typical hyperspectral data include variability in the radiance signature of a specific material due to many environmental factors. In addition, data collected from sensors are subject to noise from many sources. This contributes to the variation in pixel intensities for the same material at the same spectral band. This uncertainty in the data is the main source of classification errors.

A number of techniques have been considered for the analysis of multispectral data [9]. Neural networks can be seen as an intermediate approach between statistical and structural methods, although they resemble the former more than the latter. It has been shown that neural networks may yield better classification results than conventional multispectral classification methods (e.g. maximum likelihood classification) [2]. In the learning process of neural networks, imprecise information is often encountered. Fuzzy logic [1]

This work has been supported in part by the DOD Grant N00014-97-1-1076.

*Sponsored also in part by the Systems Research Institute, Polish Academy of Science, Newelska 6, 01-447 Warsaw, Poland.

provides useful concepts and tools to deal with such uncertainty inherent in remote sensing spectral data. Therefore, fuzzy set theory has been incorporated into neural network architectures to improve their ability to handle imprecise data.

Numerous fuzzy neural network models have been proposed in literature. Fuzzy neural networks have proven to be a good tool for pattern classification [3], [5], [7]. We experimented with a FNN based on a backpropagation (BP) learning algorithm with momentum for classification of fuzzy inputs represented by fuzzy numbers [7]. We compared classification results of the FNN with those from a conventional feedforward neural network trained using BP learning algorithm with momentum. It is demonstrated that the FNN may classify patterns on the borders of two classes more accurately.

2 Fuzzy neural network architecture

A multilayer feedforward neural network with fuzzy input vectors represented by fuzzy numbers, crisp weights, and a unipolar sigmoidal transfer function was used for the classification of images of interest. The training set consisted of the fuzzy input vector and non-fuzzy target output for indicating correct classes. A fuzzy neural network with this architecture ([7]) maps a fuzzy input vector to a fuzzy output vector. A cost function is defined by the fuzzy input vector and corresponding non-fuzzy target output, and a learning algorithm is derived from this cost function.

Multispectral data show large variation in the radiance values for the same object. Also, these variations are not constant for all bands. This data structure suggests a triangular membership function. Hence, data representation and FNN training can be achieved as follows.

An input to a fuzzy neural network is a fuzzy vector $A_k = (A_{k1}, \ldots, A_{kn})$, $k = 1, \ldots, s$ where A_{ki}, $i = 1, \ldots, n$ is a symmetric triangular fuzzy number with center a and spread b with a membership function

$$\mu_A(x) = \max\{1 - \frac{|a-x|}{b}, 0\}. \tag{1}$$

Considering the α cut of fuzzy number A defined by

$$A_\alpha = \{x : \mu_A(x) \geq \alpha\} \, \text{for} \, 0 < \alpha \leq 1, \tag{2}$$

α cuts are simply intervals on the x axis and thus, for each α we have $A_\alpha = [a_1^{(\alpha)}, a_2^{(\alpha)}] = [a - (1-\alpha)b, a + (1-\alpha)b]$.

Parameterizing the shape of a fuzzy number by α is a convenient way to compute with fuzzy numbers because it transforms fuzzy arithmetic operations into interval operations.

The inputs to the FNN are intervals $A_{ki} = [a_{1_{ki}}^{(\alpha)}, a_{2_{ki}}^{(\alpha)}]$. The fuzzy input–output relation of each unit of the neural network is fully explained in [7].

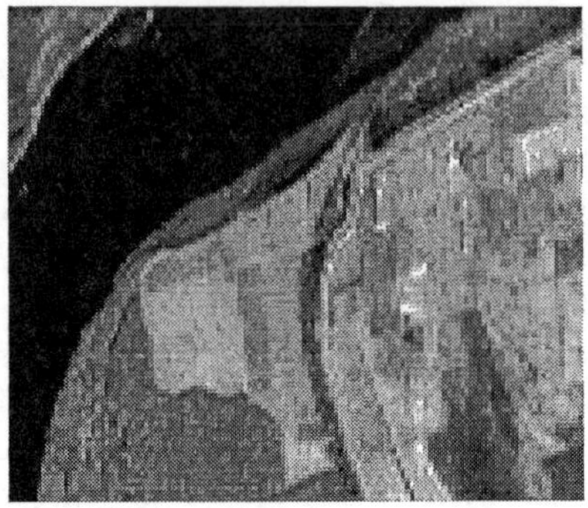

Fig. 1. *Original image (band 7).*

The training of the neural network involves minimizing the cost function. The weights are changed according to momentum learning rules. Using the cost function, the neural network is trained for a fixed value of α. The biases are adjusted in the same manner as the weights ([7]).

3 Results and discussion

3.1 Classifying Landsat data with standard neural networks

In this part of the study, the aim was to classify multispectral Landsat data using feedforward neural networks (NN). The dataset of 7 spectral bands was selected from a section 128×128 pixels of a Landsat TM scene of the United States and obtained from NASA's observatorium web page. The scene represents a section of the Mississippi river and an inland region. The original image for spectral band 7 is shown in Fig. 1.

The aim of the classification was to distinguish three classes: water, built-up land and coastal land. Several samples for every class from 6 bands were taken for training and testing of the classification accuracy of the neural network. In this specific case, band 6 did not seem to provide clear information. A sample is defined as a block of $N \times N$ pixels. In this study, the window size was 6×6 pixels. Two features - mean and standard deviation - were extracted for every sample.

The samples were analyzed using a three–layer feedforward network with a BP learning algorithm with momentum. The learning constant and momentum constant were chosen as $\eta = 0.5$ and $\beta = 0.9$, respectively. The neural network consisted of 12 inputs (2 for each band), 12 hidden units and 3 outputs. There were 36 training and 30 testing patterns. SNNS (Stuttgart Neural Network Simulator) was used to implement the standard neural network.

After learning, all training patterns were classified correctly. The confusion matrix of the classification, computed prior to testing, is shown in Table

1. The average percentage correctly classified was 96.30%. Considering these results, the feedforward NN performed very well with a very small number of missclassified patterns.

	Class 1	Class 2	Class 3
Class 1	88.89	0.00	11.11
Class 2	0.00	100.00	0.00
Class 3	0.00	0.00	100.0

Table 1. Percentage of correct classification of testing patterns using the standard NN.

3.2 Classifying Landsat data with fuzzy neural networks

To examine the behaviour of the FNN, we took several samples from every class and extracted two features - mean (a in Eqn. (1)) and standard deviation (b in Eqn. (1)). Using these features, we created the inputs to the FNN. We trained the neural network with 36 patterns. The BP learning algorithm with momentum with the learning constant $\eta = 0.5$ and the momentum constant $\beta = 0.9$ was used. The FNN had 6 input intervals, 12 hidden units and 3 crisp targets.

After training with $\alpha = 0.3$, the average correct classification percentage was 97.22%. Training with smaller α cuts resulted in a much larger missclassification. However, there was no missclassification with α cuts equal to 0.4 and higher. The results are as shown in Table 2.

	Class 1	Class 2	Class 3
Class 1	100.00	0.00	0.00
Class 2	0.00	91.67	8.33
Class 3	0.00	0.0	100.00

(a) $\alpha = 0.3$

	Class 1	Class 2	Class 3
Class 1	100.00	0.00	0.00
Class 2	0.00	100.00	0.00
Class 3	0.00	0.0	100.00

(b) $\alpha = 0.4$

Table 2. Percentage of correct classification of training patterns using the FNN with different α cuts.

After testing the FNN with 30 patterns and $\alpha = 0.3$, the average of 90.24% of the testing patterns was classified correctly. As with the training patterns, there was no missclassification with α cuts equal to 0.4 and higher. The results are shown in Table 3.

The differences in results demonstrate that the FNN is very sensitive to training parameters. Employing several α values in the training process may improve results.

	Class 1	Class 2	Class 3
Class 1	88.89	0.00	11.11
Class 2	0.00	81.82	18.18
Class 3	0.00	0.00	100.00

(a) $\alpha = 0.3$

	Class 1	Class 2	Class 3
Class 1	100.00	0.00	100.00
Class 2	0.00	100.00	00.00
Class 3	0.00	0.00	100.00

(b) $\alpha = 0.4$

Table 3. Percentage of correct classification of testing patterns using the FNN with different α cuts.

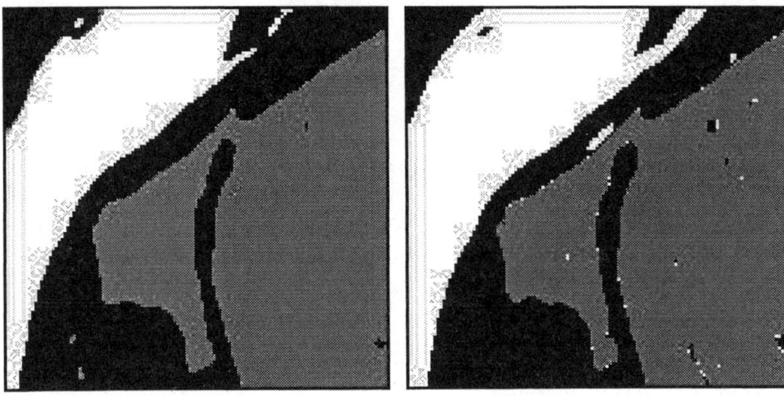

(a) Classification using NN (b) Classification using FNN

Fig. 2. *Results of classification with NN and FNN. Water is represented as white, built-up land as gray and coast as black.*

The classified outputs of both network architectures are shown in Figs. 2 a and b.

4 Conclusions

In our study we concentrated on classification of Landsat data using standard and fuzzy neural networks. As seen in Fig. 2, both architectures performed well in classifying the images. Many α cut values were tried, but $\alpha = 0.4$ produced the lowest errors during training and testing. It is felt that α cut values closer to 1 reduce some of the fuzziness of the NN. In performing our experiments, it was also seen that the successful performance of FNNs is very sensitive to different parameters and the number of hidden neurons. Initial experiments with 5 such neurons resulted in very poor classification.

The standard NN appeared to classify the built-up land better than the FNN. However, the FNN seemed to better classify smaller, more imprecise features. These features are visible in the original image (Fig. 1), and include the small island in the upper left quadrant. Furthermore, the tiny features present in the FNN image (not in the NN image) may represent objects of interest to those interpreting the images.

Because of the good performance of both architectures, it is recommended that both networks be used in classification tasks. An appropriate data fusion approach could incorporate the results of both networks. Furthermore, alternative FNN learning algorithms and architectures, such as fuzzy ARTMAP ([3]) may improve performance.

Acknowledgments

The first author would like to thank Mr. Mark Wachowiak for his help and useful comments.

References

1. Bezdek J. C., Pal S. K. (1992) Fuzzy Model for Pattern Recognition. IEEE, Piscataway, NJ.
2. Bischof H., Schneider W., Pinz A.-J. (1992) Multispectral Classification of Landsat-Images Using Neural Networks. IEEE Transactions on Geoscience and Remote Sensing **30**, 482–489.
3. Chen S., Chen C., Cherng S., Fang C., Chang K. (1997) Neural-Fuzzy Classification for Segmentation of Remotely Sensed Images. IEEE Transactions on Signal Processing **45**(11), 2639–2654.
4. Fullér R. (1999) Introduction to Neuro-Fuzzy System. Heidelberg:Physica–Verlag.
5. Lee H., Wang W. (1994) A neural network architecture for classification of fuzzy inputs. Fuzzy Sets and Systems **63**, 159–173.
6. Haykin S. (1994) Neural Networks: A Comprehensive Foundation. New York:Macmillan.
7. Ishibuchi H., Fujioka R., Tanaka H. (1993) Neural Networks that Learn from Fuzzy If–Then Rules. IEEE Transactions on Fuzzy Systems **1**(2), 85–97.
8. Ishibuchi H., Kwon K., Tanaka H. (1995) A learning algorithm of fuzzy neural networks with triangular fuzzy weights. Fuzzy Sets and Systems **71**, 277–293.
9. Jimenez L., Landgrebe D. (1998) Supervised classification in high dimensional space: Geometrical, statistical, and asymptotical properties of multivariate data. IEEE Trans. on System Man and Cybernetics, 28C(1):39.
10. Lee H. M., Wang W. T. (1994) A neural network architecture for classification of fuzzy inputs. Fuzzy Sets and Systems **63**, 159–173.
11. Tsapatsoulis N., Schnorrenberg F., Pattichis C., Kollias S. (1997) An Image Analysis System for Automated Detection of Breast Cancer Nuclei Proc. of the Int. Conference on Image Processing, Washington DC.
12. Zurada J. M. (1992) Introduction to artificial neural systems. Boston:PWS

“Precision Mining” of High-Dimensional Patterns with Self-Organizing Maps: Interpretation of Hyperspectral Images

Erzsébet Merényi

University of Arizona, Lunar and Planetary Laboratory, Tucson, AZ 85721, USA

Abstract: Utilization of remote sensing multi- and hyperspectral imagery has rapidly been increasing in numerous areas of economic and scientific significance. Hyperspectral sensors, in particular, provide the detailed spectral signatures that are known to uniquely characterize and identify minerals, soils, rocks, plants, aquatic species, man-made and other surface materials. This opens up tremendous possibilities for resource exploration and management, environmental monitoring, natural hazard prediction, and more. However, conventional analysis methods often prove inadequate for the exploitation of the wealth of information in hyperspectral images, because of the high dimensionality of the intricate spectra. Artificial Neural Networks hold the promise to revolutionize this area by meeting mathematical and other challenges that traditional techniques fail at. In this paper, we discuss the powers of Self-Organization for the detailed interpretation of hyperspectral images, using the full spectral dimensionality.

1 Remote Spectral Images as High-Dimensional Data

Airborne and satellite-borne spectral imaging has become one of the most advanced tools for collecting vital information about the surface covers of Earth and other planets. The utilization of these data includes areas such as mineral exploration, land use, forestry, ecosystem management; assessment of natural hazards, water resources, environmental contamination, biomass and productivity; and many other activities of economic significance, as well as prime scientific pursuits such as looking for possible sources of past or present life on other planets. The number of applications has dramatically increased in the past ten years with the advent of imaging spectrometers, that greatly surpass traditional multi-spectral sensors (*e.g.*, Landsat Thematic Mapper). Imaging spectrometers can resolve the known, unique, discriminating spectral features of minerals, soils, rocks, and vegetation. While a multi-spectral sensor samples a given wavelength window (typically the 0.4 – 2.5 μm range in the case of Visible and Near-Infrared surface reflectance imaging) with several broad bandpasses, leaving large gaps between the bands, imaging spectrometers sample a spectral window contiguously with very narrow badpasses (Fig. 1). Hyperspectral technology is in great demand because direct identification of surface compounds is possible without prior field work, for materials with known spectral signatures.

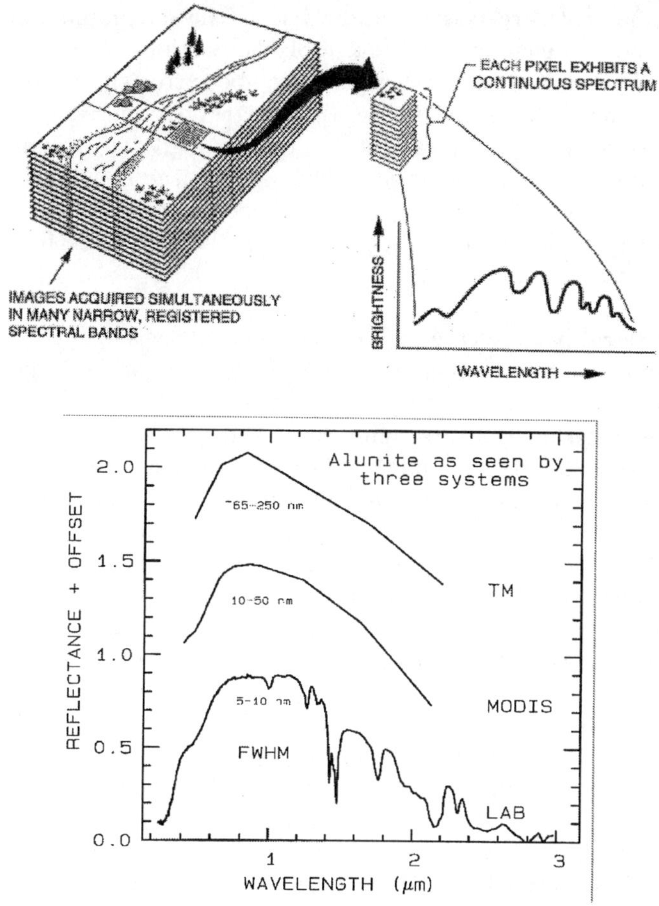

Fig. 1. Left: The concept of hyperspectral imaging. Figure from [1]. Right: The spectral signature of the mineral alunite as seen through the 6 broad bands of Landsat TM, as seen by the moderate spectral resolution sensor MODIS (20 bands in this region), and as measured in laboratory. Hyperspectral sensors such as AVIRIS of NASA/JPL [2] produce spectral details comparable to laboratory measurements.

Formally, the vector $S^{x,y} = (S_1^{x,y}, ..., S_{NB}^{x,y})$, where $S_k^{x,y}$ is the data value in the kth image band ($k = 1, ..., NB$) at pixel location (x, y), is called a spectrum. It is a characteristic pattern (Fig. 1) that provides a clue to the surface material(s) within pixel (x, y). NB denotes the number of image bands. $x = 1, ..., Xmax$ and $y = 1, ..., Ymax$, where $Xmax$ and $Ymax$ are the spatial size of the image in the x and y direction, respectively. The feature space spanned by Visible-Near Infrared reflectance spectra is $[0 - noise, U + noise]^{NB} \subset \Re^{NB}$ where $U > 0$ represents an upper limit of the measured scaled reflectivity and *noise* is the maximum value of noise across all spectral channels and image pixels. Sections of this space can be very densely populated while other parts may be extremely sparse, depending on the materials in the scene and on the spectral bandpasses of the sensor.

Great spectral detail comes at a cost of very high data volume and complexity. It poses new mathematical challenges in the clustering and classification of images with high spectral dimensionality. The specific problems associated with remote sensing spectral image analyses arise from any combination of the following:

- The spectral patterns are high dimensional (dozens $\leq NB \leq$ hundreds);
- The number of data points (image pixels) can be as large as several millions;
- Surface materials that are significantly different for an application may be distinguished by very subtle differences in their spectral patterns;
- Given the richness of data, the goal is to separate many cover classes, or to separate materials with subtle spectral differences;
- The pixels can be mixed: Several different materials may contribute to the spectral signature detected from each pixel;
- Very little training data may be available for some classes; and classes may be represented very unevenly.

Noise is far less problematic than the intricacy of the actual spectral patterns, because of the high Signal-to-Noise Ratios (500 – 1,500) that present-day hyperspectral imagers provide. For this discussion, we will omit noise issues, and additional effects such as atmospheric distortions, illumination geometry and albedo variations in the scene, because these can be addressed through well-established procedures prior to clustering or classification.

ANNs have been gaining recognition as powerful answers to the above challenges. Many sucessful applications were published, that analyze lower dimensional spectral data (such as Landsat TM). See, for example, [3], and references therein. Few have attempted ANN (or any) classification of hyperspectral data into a large number of classes even though that is where the shortcomings of classical methods become a severe limiting factor in accomplishing scientific or technical objectives. For hyperspectral images dimensionality reduction is frequently accepted in order to accomodate the data to the capability of traditional methods. This, however, often results in undesirable loss of information, preventing "precision mining": discovery of new, interesting classes, or discrimination of classes with subtle spectral differences. Thus the analysis negates the purpose of the sophisticated sensors.

The rest of the paper is structured as follows: In Section 2, the test data set, used throughout this paper, is introduced. Section 3 discusses hyperspectral dimensionality. Section 4 and 5 present results of clustering and classification of the test data, using the full hyperspectral dimensionality; and in Setion 6, correctness and relaiabilty issues are discussed.

2 Test Data Set: The Lunar Crater Volcanic Field AVIRIS Image

In this paper a VIS-NIR (0.4 - 2.5 μm), 224-band, 30 m/pixel AVIRIS image of the Lunar Crater Volcanic Field (LCVF), Nevada, USA, is used. The LCVF is one of NASA's remote sensing test sites, where comprehensive field studies were conducted [4]. In addition, much accumulated field knowledge and independent geologic research by individuals provide excellent and detailed basis for evaluation of the results presented here [5].

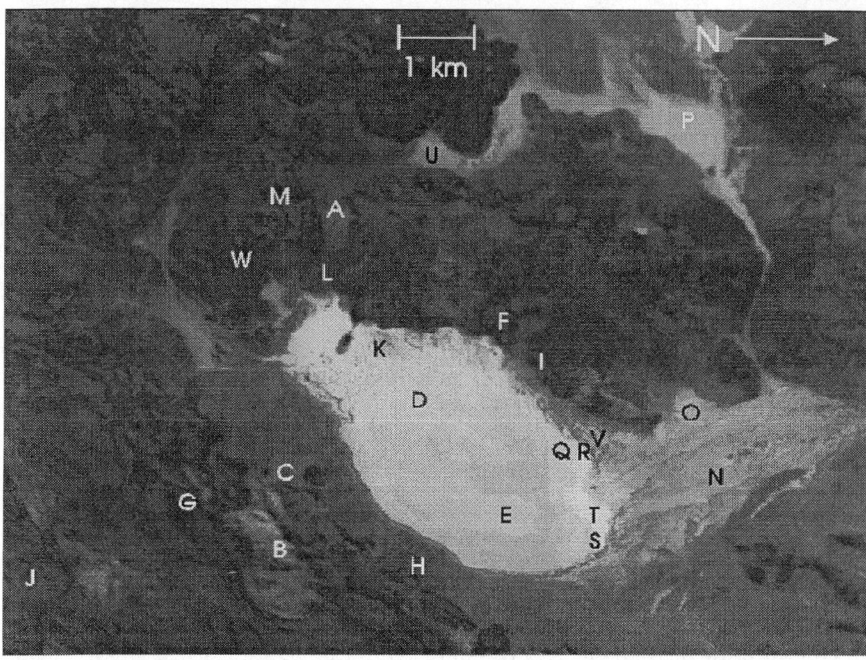

Fig. 2. The Lunar Crater Volcanic Field. Color composite from an AVIRIS, 1994 image. The original image comprises 224 image bands over the 0.4 - 2.5 μm range, 512 x 614 pixels, altogether 140 Mbytes of data. Labels indicate different cover types described in the text. The ground resolution is 30 m/pixel.

Fig. 2 shows a natural color composite of the LCVF with locations representative of 23 cover classes marked by their respective class labels. This 512 x 614 pixel AVIRIS image of a 10 x 12 square km area contains, among other materials, volcanic cinder cones (class A, reddest peaks) and weathered derivatives thereof such as ferric oxide rich soils (L, M, W), basalt flows of various ages (F, G, I), a dry lake divided into two halves of sandy (D) and clayey composition (E); a small rhyolitic outcrop (B); and some vegetation at the lower left corner (J), and along washes (C). Alluvial material (H), dry (N,O,P,U) and wet (Q,R,S,T) playa outwash with sediments of various clay content as well as other sediments (V) in depressions of the mountain slopes, and basalt cobble stones strewn around the playa (K) form a challenging series of spectral signatures for pattern recognition (see in [6]). A long, NW-SE trending scarp, straddled by the label G, borders the vegetated area. On this color composite, containing information from only three image bands, many of the cover type variations do not manifest. They will become visible in the cluster and class maps below. The image comprises 140 Mb of data. After atmospheric correction and removal of excessively noisy bands (saturated water bands and overlapping detector channels), 194 image bands remained. These 194-dimensional spectra are the input patterns in the following analyses.

3 The Dimensionality of Hyperspectral Patterns

The spectral dimensionality of hyperspectral images is not well understood. Many believe that hyperspectral images are highly redundant because of band correlations. Others maintain an opposite view, which also manifests in the vigorous development of hyperspectral sensors and commercialization of hyperspectral data services in the last several years. Few investigations exist yet into the intrinsic dimensionality (ID) of hyperspectral images. Linear methods such as PCA or determination of mixture model endmembers using either of the two major approaches [7] [8] usually yield 3–8 "effective components". Bruske [9] finds the spectral ID of the LCVF AVIRIS image (Fig. 2) to be between 4 and 7, using a non-linear Neural Net based approach, Optimally Topology Preserving Maps. These surprisingly low numbers, that increase with improved sensor performance [10], result from using statistical thresholds for the determination of what is "relevant", regardless of application dependent criteria.

The number of relevant components increases dramatically when specific goals are considered such as what cover classes should be separated or what known properties of the surface can be postulated. Pendock [11], using an associative neural network, extracted 20 linear mixing endmembers from a 50-band (2.0-2.5 μm) segment of an AVIRIS image (Fig. 2), requiring only that a rather general surface texture criterium be met. Benediktsson et al. [12] performed feature extraction on an AVIRIS geologic scene, which resulted in 35 bands. They used an ANN (the same network that performed the classification itself) for Decision Boundary Feature Extraction (DBFE). The DBFE is claimed to preserve all features that are necessary to achieve the same accuracy as in the original data space, by a given classifier for predetermined classes. However, no comparison of classification accuracy was made using the full spectral dimension, to support the DBFE claim. In this particular study a relatively low number of classes, 9, were of interest, and the question posed was to find the number of features to describe those classes. Separation of a higher number of classes may require more features.

It is not clear how feature extraction should be done in order to preserve relevant information in hyperspectral images. Later in this paper it is demonstrated that extraction of about 30 features from our LCVF image by any of several methods leads to a loss of a number of the originally determined 23 cover classes. Wavelet compression studies on an earlier image of the the same AVIRIS scene [13] conclude that various schemes and compression rates affect different spectral classes differently, and none was found overall better than another, within 25% – 50% compressions (retaining 75% – 50% of the wavelet coefficients). In a study on simulated, 201-band spectral data, [14] show slight accuracy increase across classifications on 20-band, 40-band, and 60-band subsets. However, they base the study on only two vegetation classes, the feature extraction is a progressive hierarchical subsampling of the spectral bands, and there is no comparison with using the full, 201-band case. Comparative studies using full spectral resolution and many classes are lacking, in general, because few methods can cope with such high-dimensional data technically, and the ones that are capable (such as Minimum Distance, Parallel Piped) often perform too poorly to merit consideration.

Undesirable loss of relevant information can result using any of these feature extraction approaches. In any case, finding an optimal feature extraction requires great preprocessing efforts just to taylor the data to available tools. An alternative is to develop capabilities to handle the full spectral information. Analysis of unreduced data is important for exploration and novelty detection (such as in the case of hard-earned data in planetary exploration); as well as to allow for the distinction of significantly greater number of cover types, according to the potential provided by modern imaging spectrometers.

For example, remote monitoring of the Colorado River ecosystem in Grand Canyon, Arizona, USA, requires discrimination and identification (spatial mapping) of about seventy plant and aquatic species, in order to provide a meaningful information service for 26 stakeholders in businesses of dam operations, electric power generation, recreation, tourism, cultural affairs of native American tribes, and scientific research. Some of the species exhibit very slight but systematic spectral differences that are captured in hyperspectral data. These differences would most probably be lost in a PCA or a wavelet transformation, and possibly in more sophisticated algorithms. Preliminary SOM-based classification, utilizing the full spectral information, was successful in separating several "doubtful" cover types [15]. This example represents a modern emerging paradigm that is based on the expectation of continuous availability of (satellite-based) hyperspectral imagery and that the information content of such data, if extracted properly, can satisfy a wide range of needs simultaneously.

4 Effects of Dimensionality Reduction and Inadequate Tools

A systematic study was conducted on the LCVF image (Fig. 2), to simultaneously assess loss of information due to reduction of spectral dimensionality, and to compare performances of several traditional and an SOM-based hybrid ANN classifier. 23 geologically relevant classes, as indicated in Fig. 2, were used. These represent a great variety in terms of spatial extent, the similarity of spectral signatures [6], and the number of available training samples. The full study is described in [16]. Here the most dramatic finding is presented.

Fig. 6, top panel, shows the best classification, produced by an SOM-hybrid ANN using all 194 spectral bands. This ANN first learns in an unsupervised mode, during which the input data are clustered in the hidden SOM layer. After the SOM converges, the output layer is allowed to learn class labels. The preformed clusters in the SOM greatly aid in accurate and sensitive classification, by helping prevent the learning of inconsistent class labels. Detailed description of this classifier is given in several previous scientific studies, which produced improved interpretation of high-dimensional spectral data compared to earlier analyses [17] [18] [19]. Training samples for the supervised classifications in our study were selected based on knowledge of and interest in geologic units. The SOM hidden layer was not evaluated and used for identification of spectral types, prior to training sample determination. So, Fig. 6 reflects the geologist's view of the desirable segmentation.

In order to apply Maximum Likelihood and other covariance based classifiers, the number of spectral channels needed to be reduced to 30, since the maximum number of training spectra that could be identified for *all* classes was 31. Dimensionality reduction was performed in several ways, including PCA, equidistant subsampling, and band selection by a domain expert. Band selection by domain expert proved most favorable. Fig. 6, bottom panel, shows the Maximum Likelihood classification on the LCVF data, reduced to 30 bands. A number of classes (notably the ones with subtle spectral differences, such as N, Q, R, S, T, V, W) were entirely lost. Class K (basalt cobbles) disappeared from most of the edge of the playa, and only traces of B (rhyolitic outcrop) remained. Class G and F were greatly overestimated. Although the ANN classifier produced better results (not shown here) on the same 30-band reduced data set than the Maximum Likelihood, a marked drop in accuracy occurred compared to classification on the full data set. This emphasizes that accurate mapping of "interesting", spatially small geologic units is possible from full hyperspectral information and with appropriate tools.

It should be noted that a Backpropagation (BP) network is very difficult to train for the combination of such high number of inputs and output classes. This author has not seen hyperspectral images classified by a BP network, at full spectral resolution. In contrast, the SOM-hybrid ANN above scales up gracefully and performs the task with relative ease. Inclusion of unsupervised learning phases was recently shown to increase the classification accuracy of a BP network too [20].

5 Discovery in Hyperspectral Images with SOMs

Let's examine now how the SOM preforms in terms of detection of clusters in high-dimensional data. Fig. 3 displays a 40 x 40 SOM that was generated by a modification of the original SOM [21], the conscience algorithm of DeSieno [22]. The input data space was the entire 194-band LCVF image. Groups of neurons, altogether 32, that were found to be sensitized to groups of similar spectra in the 194-dimensional LCVF input data, are indicated by various colors. The boundaries of these clusters were determined by a somewhat modified version of the Ultsch and Simeon method [23]. Areas where no data points (spectra) were mapped are the grey corners with uniformly high fences, and are relatively small. The black background in the SOM lattice shows areas that have not been evaluated for cluster detection. The spatial locations of the image pixels mapped onto the groups of neurons in Fig. 3, are shown in the same colors in Fig. 4. Color coding for clusters that correspond to classes or subclasses of those in Fig. 6, top, is the same as in Fig. 6, to show similarities. Colors for additional groups were added.

The first observation is the striking match between the supervised ANN class map in Fig. 6, top panel, and this clustering: the SOM detected all classes that were known to us as meaningful geological units. The "discovery" of classes B (rhyolitic outcrop, white), F (young basalt flows, dark grey and black, some shown in the black ovals), G (a different basalt, exposed along the scarp, dark blue, one segment outlined in the white rectangle), K (basalt cobbles, light blue, one segment shown in the black rectangle), and other spatially small classes such as the series of playa deposits (N, O, P, Q, R, S, T) is significant. This is the capability we need for sifting through high-volume, high-information-content data to be alerted to interesting, novel, or hard-to-find units. The second observation is that the SOM detected more, spatially coherent, clusters than the number of classes that we trained for in Fig. 6. The SOM's view of the data is more refined and more precise than that of the geologist's. For example, class A (red in Fig. 6) is split here into a red (peak of cinder cones) and a dark orange (flanks of cinder cones) cluster, that make geologic sense. The maroon cluster to the right of the red and dark orange clusters at the bottom of the SOM fills in some areas that remained unclassified (bg) in the ANN class map, in Fig. 6. An example is the arcuate feature at the base of the cinder cone in the white oval, that apparently contains a material different enough to merit a separate spectral cluster. This material fills other areas too, consistently at the foot of cinder cones (another example is seen in the large black oval). Observation of further refinements are left to the reader.

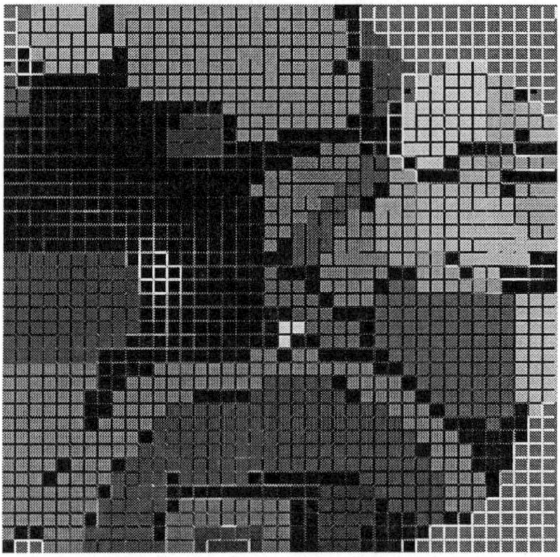

Fig. 3. Clusters identified in a 40 x 40 SOM. The SOM was trained on the entire 194-band LCVF image, using the DeSieno [22] algorithm.

6 Correctness, Reliability, and Precision of SOM Analyses

The correctness of the SOM mapping is a serious concern. For small data sets such as in [17], and for a low number of clusters, topology violation can be detected by semi-manual examination of the map, and the situation remedied by using a larger lattice or often by just retraining the SOM starting from a different random initial state. For large data sets, the effort needed to detect the ill conditions are great, and repeated runs are expensive. Hyperspectral dimension is of particular concern because very few SOM studies exist on such high-dimensional data, to draw experience from [24]. A measure that expresses the level of topology violation in mapping an n-dimensional data space onto lower dimension has been constructed by Villmann, Bauer et al. [25] [26]. This work is based on findings by Martinetz and Schulten [27],

Fig. 4. The clusters from Fig. 3 remapped to the original spatial image, to show where the different spectral types originated from. The relatively large, light grey areas correspond to the black, unevaluated parts of the SOM in Fig. 3. Ovals and rectangles highlight examples discussed in the text.

and carry further in that it provides a formula that can be computed from the weights and neighborhood relations, and immediately alert for entanglement of clusters. The work of Villman et al. also provides relief from topology violation by automatically growing the SOM lattice [28] [29]. Applications to 6-band Landsat images worked well [30]. Further work is underway to evaluate the powers of this method for hyperspectral imagery [31].

A mapping that matches the probability density (*pdf*) of the input data is preferable, because it represents regions of the input space with resolutions appropriate for their relative densities. For this, the map magnification factor $m(x)$, which is the number of neurons representing a subsection dx of the input space, is required to be proportional to the *pdf*. The original SOM algorithm does not meet the above requirement and tends to overrepresent regions of low density, and underrepresent regions of high density [32]. This could prevent sensitive separation of spectral clusters with subtle signature differences, such as the sandy and clayey part of the playa (classes D and E), or the splitting of class A into the red and dark orange clusters (Fig. 6, bottom panel and Fig. 4). The *conscience algorithm* of DeSieno [22] adjusts the winning frequencies of the SOM neurons so as to ensure proper spreading (compaction) of information that is dense (sparse) in the input feature space. Update of the weight vectors is given by $w_i^{t+1} = w_i^t + \alpha^t(S - w_i^t)$, for $i \in N_c$ and $w_i^{t+1} = w_i^t$ otherwise, where c and N_c denote, respectively, the index and the the index set of the neighbourhood of the winning neuron PE_c. $c = arg\min_i(\|S - w_i\| - B_i)$ where the bias term $B_i = \gamma(1/M - F_i^t)$ adjusts the distance between the incoming pattern S and w_i based on the historic winning frequency F_i of PE_i. M is the number of PEs in the Kohonen lattice. F_i is updated along with the weights, according to $F_i^{t+1} = F_i^t + \beta(\delta_{N_c} - F_i^t)$ where δ_{N_c} is 1 for $i \in N_c$, 0 otherwise. α, β and γ are user controlled parameters decreasing in time. In his new book, van Hulle [33] points out that adding a conscience algorithm to the SOM does not equate to equiprobabilistic mapping, in general. However, for *very high dimensions*, a minimum distortion quantizer (such as the conscience algorithm) approaches an equiprobable quantizer [33] (page 93). Fig. 3, 4, and 5 seem to support this statement, although precise evaluation is difficult because of the long time needed to reach a mature state of the SOM for large hyperspectral images.

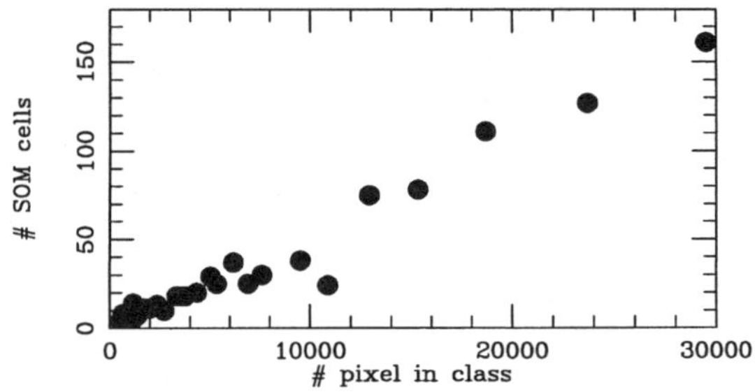

Fig. 5. Relationship between the number of image pixels in the clusters in Fig. 4, and the number of SOM neurons representing them in Fig. 3.

Visualization of the SOM's knowledge is also crucial for "precision mining". Detection of clusters — as a post-processing step after SOM convergence — hinges on how well the SOM's internal view of the data space is communicated through knowledge extraction and representation. This has been targeted by several works [23] [34] [35] [36] [37]. However, ours is the first application to hyperspectral images. Closest in spirit to the browsing capabilities of [37] and the WEBSOM as used in [24], our applications require entirely different engineering solutions because of the specifics of spectral-spatial image data. In our facilities, developed at the Lunar and Planetary Laboratory, University of Arizona, interactive cluster determination tools include "knobs" for on-the-fly adjustment of fence scaling and thresholding in the Ultsch-Simeon representation; changing the underlying map from "plain" (as in Fig. 3) to "density map" (where the number of data points mapped on a grid cell is indicated by grey scale or monochrome color intensities); switching between separate or combined viewing of map and fences; chang-

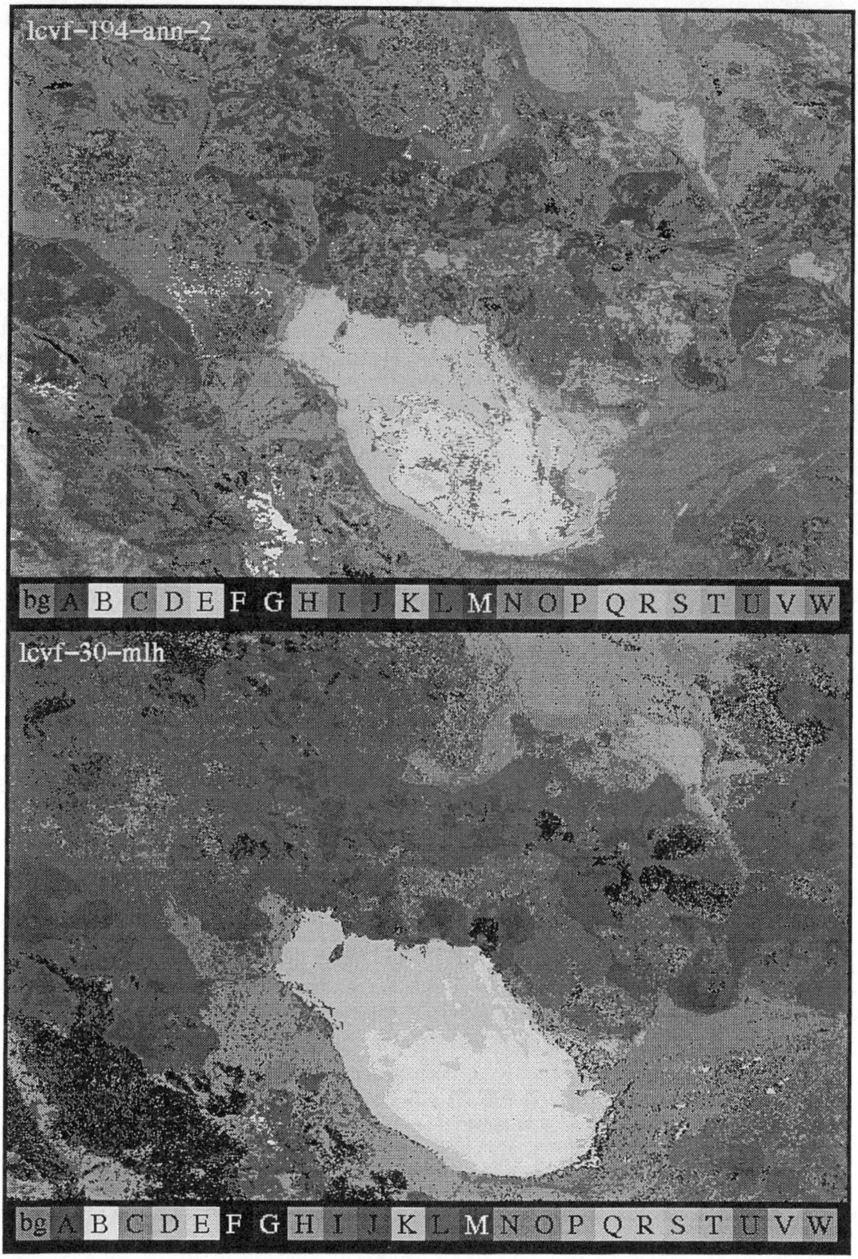

Fig. 6. Top: SOM-hybrid ANN classification of the LCVF scene, using 194 image bands. Bottom: Maximum Likelihood classification of the LCVF scene. 30, strategically selected bands were used due to the limited number of training samples. Considerable loss of class distinction occurred compared to the ANN classification. 'bg' stands for background (unclassified pixels).

ing colors, and size of grid cells and fences, for clearest possible viewing, etc. Different representations such as [35] are also planned. The ultimate goal is automation of cluster extraction. Alternative approaches such as [38] are being evaluated.

Acknowledgements

Software development for this work has been supported by the Applied Information Systems Research Program of the Office of Space Science, NASA, NAG54001. NeuralWare and Khoros packages were utilized. I am grateful to Dr. Thomas Villmann for valuable discussions on various aspects of SOMs. Many thanks are due to Dr. William H. Farrand for providing field and geologic knowledge in evaluating results for the LCVF; and to many interdisciplinary collaborators for their participation in the large-scale, real-life applications that provide the test bed for this research.

References

1. Campbell, J. (1996) Introduction to Remote Sensing. The Guilford Press
2. Green, R. O. Ed. (1996) Summaries of the 6th Annual JPL Airborne Geoscience Workshop, **1** AVIRIS Workshop, Pasadena, CA, March 4–6, 1996
3. Carpenter, G. A., Gjaja, M. N., et al. (1997) ART Neural Networks for Remote Sensing: Vegetation Classification from Landsat TM and Terrain Data, IEEE. Trans. Geosci. and Remote Sens. **35**(2), 308–325
4. Arvidson, R. E., Dale-Bannister, M. et al. (1991) Archiving and distribution of Geologic Remote Sensing Field Experiment Data. EOS, Transactions of the American Geophysical Union, **72**(17), 176
5. Farrand, W. H. (1991) VIS/NIR Reflectance Spectroscopy of Tuff Rings and Tuff Cones. Ph.D. Thesis, University of Arizona
6. Merényi, E., (1998) Self-Organizing ANNs for Planetary Surface Composition Research, Proc. 6th European Symposium on Artificial Neural Networks, ESANN'98, Bruges, Belgium, April 22–24, 1998, 197–202
7. Adams, J. B., Smith, M. O., and Gillespie, A. R. (1993) Imaging spectroscopy: Interpretation based on spectral mixture analysis, In Remote Geochemical Analysis: Elemental and Mineralogical Composition. C.M. Peters and P.A.J. Englert, Eds., Cambridge University Press, New York, 145–166
8. Research Systems Inc. (1997) ENVI v.3 User's Guide, RSI, Boulder, 614 pp.
9. Bruske, J., and Merényi, E. (1999) Estimating the Intrinsic Dimensionality of Hyperspectral Images. Proc. 7th European Symposium on Artificial Neural Networks, ESANN'98, Bruges, Belgium, April 21–23, 1999, 105–110
10. Green, R. O. and Boardman, J. (2000) Exploration of the relationship between Information content and Signal-to-Noise ratio and spatial resolution. Proc. 9th AVIRIS Earth Science and Applications Workshop, February 23–25, Pasadena, CA (in print).
11. Pendock, N. (1999) A Simple Associative Neural Network for Producing Spatially Homogeneous Spectral Abundance Interpretations of Hyperspectral Imagery. Proc. 7th European Symposium on Artificial Neural Networks, ESANN'98, Bruges, Belgium, April 21–23, 1999, 99–104

12. Benediktsson, J. A., Sveinsson, J. R., et al. (1994) Classification of Very-High-Dimensional Data with Geological Applications, Proc. MAC Europe 91, Lenggreis, Germany, 4–6 October, 1994, 13–18
13. Moon, T., and Merényi, E. (1995) Classification of hyperspectral images using wavelet transforms and neural networks. Proc. of the Annual SPIE Conference, July 9–14, San Diego, CA **2569**
14. Benediktsson, J. A., Swain, P. H., et al. (1990) Classification of Very High Dimensional Data Using Neural Networks, IGARSS'90 10th Annual International Geoscience and Remote Sensing Symp. **2**, 1269
15. Merényi, E., Farrand, W. H. et al. (2000) Studying the Potential for Monitoring Colorado River Ecosystem Resources Below Glen Canyon Dam Using Low-Altitude AVIRIS Data. Proc. 9th AVIRIS Earth Science and Applications Workshop, February 23–25, Pasadena, CA (in print).
16. Merényi, E., Farrand, W. H. et al. (2000) Efficient Geologic Mapping from Hyperspectral Images with Artificial Neural Networks Classification: a Comparison to Conventional Tools. In preparation for submission to IEEE TGARS
17. Howell, E. S., Merényi, E., Lebofsky, L. A. (1994) Classification of Asteroid Spectra Using a Neural Network Jour. Geophys. Res. **99**, 10,847–10,865
18. Merényi, E., Howell, E. S., et al. (1997) Prediction of Water In Asteroids from Spectral Data Shortward of 3 Microns. ICARUS **129**, 421–439
19. Merényi, E., Singer, R. B., Miller, J. S. (1996) Mapping of Spectral Variations On the Surface of Mars From High Spectral Resolution Telescopic Images, ICARUS **124**, 280–295
20. Fardanesh, M. T., and Ersoy, O. K. (1998) Classification Accuracy Improvement of Neural Network Classifiers by Using Unlabeled Data. IEEE Trans. Geosci. and Remote Sens.**36**(3), 1020–1025
21. Kohonen, T. (1997) Self-Organizing Maps. Springer Series in Information Sciences, **30**, Springer, Berlin, Heidelberg, New York, 1995, 1997
22. DeSieno, D. (1988) Adding a Conscience to Competitive Learning. Proc. ICNN, New York, July 1988 **I**, 117–124
23. Ultsch, A. and Simeon, H. P. (1990) Kohonen's Self Organizing Feature Map for Exploratory Data Analysis. Proc. INNC-90-PARIS **I**, 305–308
24. Kohonen, T. (1997) Exploration of Very Large Databases by Self-Organizing Maps. Proc. IEEE ICNN'97 **I**, PL1–6
25. Bauer, H.U., Herrmann, M., Villmann, Th. (1997) Topology Preservation in Neural Maps of Real World Data. Proc. 5th European Symposium on Artificial Neural Networks, ESANN'97, Bruges, Belgium, April 22–24, 1997, 205–210
26. Villmann, Th., Herrmann, R.Der, and Martinetz, Th. (1997) Topology Preservation in Self-Organizing Feature Maps: Exact Definition and Measurement. IEEE Trans. on Neural Networks **8**(2), 256–266
27. Martinetz, Th., and Schulten, K. (1994) Topology Representing Networks. Neural Networks **7**(3), 507–522
28. Bauer, H.U. and Villmann, Th. (1997) Growing a Hypercubical Output Space in a Self-Organizing Feature Map. IEEE Trans. on Neural Networks **8**(2), 218–226
29. Villmann, Th. and Bauer, H.-U. (1998) Applications of the growing self-organizing map. Neurocomputing **21**, 91–100
30. Villmann, T. (1999) Benefits and Limits of the Self-Organizing Map and its Variants in the Area of Satellite Remote Sensoring Processing. Proc. 7th Euro-

pean Symposium on Artificial Neural Networks, ESANN'98, Bruges, Belgium, April 21–23, 1999, 111–116

31. Villmann, T. and Merényi, E. (2000) Extensions and Modifications of SOM and its Application in Satellite Remote Sensing Processing. Accepted to Proc. 2nd Int'l Computer Science Conventions Symposium on Neural Computation, NC'2000, May 23–26, 2000, Berlin, Germany
Conference on Geologic Remote Sensing, San Antonio, TX, 9–12 May,
32. Haykin, S. (1995) Neural Networks. A Comprehensive Foundation. McMillan
33. van Hulle, M. M. (2000) Faithful Representations and Topographic Maps: From Distortion- to Information-Based Self-Organization. Wiley & Sons
34. Mao, J. and Jain, A. K. (1995) Artificial Neural Networks for Feature Extraction and Multivariate Data Projection. IEEE Trans. on Neural Networks **6**(2), 296–317
35. Cottrell, M. and de Bodt, E. (1996) A Kohonen Map Representation to Avoid Misleading Interpretations. Proc. European Symposium on Artificial Neural Networks, Bruges, Belgium, 22–24 April, 1996
36. Kraaijveld, M. A., Mao, J. and Jain, A. K. (1995) A Nonlinear Projection Method Based on Kohonen's Topology Preserving Maps. IEEE Trans. on Neural Networks **6**(3), 548–559
37. Kaski, S., Nikkila, J. and Kohonen, T. (1998) Methods for Interpreting a Self-Organizing Map in Data Analysis. Proc. 6th European Symposium on Artificial Neural Networks, ESANN'98, Bruges, Belgium, April 22–24, 1998, 185–190
38. Fyfe, C. and Baddeley, R. (1995) Non-linear data structure extraction using simple hebbian networks. Biol. Cybern. **72**, 533–541

The Symbolic Side of a Neuro-Fuzzy System

Arthur Ramer[1] and Maria do Carmo Nicoletti[1,2]

[1]CSE - University of New South Wales, Sydney 2052, Australia
[2]On leave from Universidade Federal de São Carlos – SP, Brazil

Abstract. *One of the advantages of a particular model of a neuro-fuzzy (nf) system is that this classifier can be interpreted in the form of linguistic rules, which are considered easy to understand. This paper empirically investigates a nf system concerning its symbolic representation. It presents a comparative evaluation of a nf system and a traditional symbolic system in a few public knowledge domains.*

1 Introduction

The main concern of this paper is the symbolic aspect of a nf system known as NEFCLASS [4,5,6] which is a tool for learning fuzzy rules and fuzzy sets by supervised learning. NEFCLASS is based on a generic 3-layer perceptron model and can be considered a fuzzy rule-based system represented by a network architecture and trained as such. In order to accomplish this goal the results obtained with NEFCLASS in a few public knowledge domains will be compared with the results obtained using a conventional learning system known as CN2 [2].

2 The NEFCLASS and CN2 Systems

NEFCLASS *(NEuron Fuzzy CLASSification)* is part of a collection of software made available by the University of Magdeburg as tools for fuzzy processing and can be downloaded from the Web site in Germany http://fuzzy.cs.uni-magdeburg.de/. NEFCLASS implements a supervised learning procedure where the net is trained with a set of pre-classified training examples (training set), each of them belonging to one of a number of distinct crisp classes. The training examples are expected to be described as vectors of crisp numerical values. The first layer of units of the 3-layer fuzzy-perceptron represents the pattern features – in this layer there should be as many units as the number of features that describe the training instances. The hidden layer is composed of rule units that represent the fuzzy rules. The number of output units in the third layer is given by the number of different classes in the training set. Note that a NEFCLASS system is a specialization of a 3-layer fuzzy perceptron, where: 1) the weights associated with connections between rule units and output units can only be 0 or 1; 2) a rule unit should only be connected to one output unit and 3) there are no two rule units connected to input units that have the same connection weights. After the training, the resulting 3-layer perceptron can be transformed into a set of fuzzy rules. When operating NEFCLASS the user has first to define the initial fuzzy sets associated with each attribute that describe the training set and also provide the system with a number k, to limit the maximum number of rule nodes the perceptron will have. The results of the experiments with NEFCLASS were obtained using the defaults of the system, except for the number of epochs (we used 1000). We also used the option "same for all units" with the value 3, when defining the number of fuzzy sets associated with each input unit.

The CN2 algorithm is described in [2] and the system that implements it, also called CN2, is available for download in many sites (see, for example, http://www.cs.utexas.edu/users/pclark). This system is a rule-induction program for crisp data which takes a set of examples (vectors of attribute-values and an associated class) and generates a set of rules that generalise them. As with NEFCLASS, it allows the user to evaluate the accuracy of a set of rules in terms of a set of pre-classified examples (testing set). The CN2 system implements ideas from both machine learning families of algorithms, the AQ [3] and the TDIDT [8]. The rules induced by this system have the following syntax: *if <complex> then <class>* where <complex> is defined as a conjunction of selectors, a selector being a basic test of attribute. In this context, a selector or a complex covers an example if the expression they represent is true for that example. The last rule CN2 induces is a 'default rule' which predicts the most commonly occurring class in the training data. When using the induced rules for classifying new examples, this system tries each of the existing rules, until it finds one whose conditions are satisfied by the example being classified. The resulting class prediction of this rule is then assigned as the class of that example. If no induced rules are satisfied, the final default rule assigns the most common class to the new example. CN2 has various control parameters that can be tuned by the user. In the experiments, we used their default values.

3 Experiments

We have used NEFCLASS and CN2 in a few knowledge domains. The data related to the performance of NEFCLASS in two of these domains (Pima & Bupa) have been discussed in [7]. The choice of these domains was basically determined by two characteristics they have: the type of attributes, since we were interested only in numerical attributes and the absence of missing values for attributes. Table 1 lists the knowledge domains used in the experiments. In the Bupa Liver Disorders the instances are classified into two classes: patients with or without liver disorders. The Wine recognition data describes the results of chemical analysis of wines derived from three different vineyards. The Habermam's Survival Data contains data on the survival of patients who had undergone surgery for breast cancer. The Pima Indians Diabetes contains data classified into two classes, depending upon whether the patient shows (or not) signs of diabetes according to the World Health Organization. The Glass Identification Database contains instances that describe six different types of glass. Since these domains are well-known datasets and part of the UCI Repository [1], we present in Table 1 only the information that is relevant to the work described in this paper. More details about these datasets can be found in the UCI site.

For each of these datasets the following procedure was repeated 10 times. The original training dataset was randomly divided into two sets: a new training set – TrS (containing 75% of the original) and a testing set – TeS (containing the remaining 25%). So, for each knowledge domain we generated 10 pairs of datasets TrS-TeS. The TrS was then used for inducing knowledge and subsequently, the associated testing set TeS was used for evaluating the performance of the induced knowledge when classifying new examples. Each of the 10 pairs TrS–TeS was used by both NEFCLASS and CN2 for training and testing respectively. Table 1 also shows the number of training instances and testing instances used in each domain.

Table 1. Number of instances, attributes and classes per domain

Domain	|Original|	|TrS|	|TeS|	|attributes|	|classes|
Bupa (Bu)	345	260	85	6	2
Wine (Wi)	178	134	44	13	3
Haberman (Ha)	306	230	76	3	2
Pima (Pi)	768	580	188	8	2
Glass (Gl)	214	161	53	9	6

In the following tables, *NR:* number of rules created; *mTrS:* number of misclassified instances in TrS, *mTeS:* number of misclassified instances in TeS, *o_ac*: overall accuracy of the rules (concerning both sets TrS and TeS). With respect to performance, Table 2 and 3 show the average values of the 10 trials we obtained with NEFCLASS and CN2 respectively, in the five datasets. As figures have been rounded some results may have a little disparity.

Table 2. NEFCLASS

domain	NR	mTrS	o_ac (%)	mTeS	o_ac (%)	k
Bu	46.6	102.3	60.66	38.4	54.83	100
Wi	131.2	10.9	91.87	4.3	90.23	150
Ha	18	62.1	73.0	19.7	74.08	30
Pi	30	189.4	67.34	63.4	66.28	30
Gl	30	71.5	55.60	26.2	50.57	30

Table 3. CN2

domain	NR	mTrS	o_ac (%)	mTeS	o_ac (%)
Bu	28.6	29.5	88.65	30	64.7
Wi	7.6	0.7	99.51	3.3	92.52
Ha	46	19.9	91.35	22.2	70.8
Pi	42.9	56.1	90.32	49.8	73.51
Gl	20.6	23	85.77	18.2	65.67

It can be seen in Table 2 that for the last two datasets the system reached the value specified for k when constructing the hidden layer of units. In a previous experiment using Pi [7], we realized that by allowing a greater value for k the number of rules increased considerably but their performance improved only slightly.

The if-then rules created by NEFCLASS contain, in their condition part, all the attributes that describe the training instances. In the case of the Wi domain, for example, each of the 131.2 (on average) rules has 13 tests of attributes. In some situations, if the set of rules and fuzzy sets associated to attributes are not carefully analysed and refined by the user, the 'interpretability' of the induced rules becomes highly questionable. On the other hand, the selection of the attributes that are relevant to a classification problem is part of the CN2 algorithm and, for this reason, only attributes that effectively contribute to characterizing a class are part of the rule that predicts it. Table 4 lists, for each domain, the average number of rules per tests of attributes induced by CN2. It can be seen that on average, for the 5 domains, the number of tests of attributes is relatively low compared to NEFCLASS, making the interpretability of the corresponding set of rules an easy task.

Table 4. CN2 – Average number of rules per number of tests of attributes (t)

Domain	6 t	5 t	4 t	3 t	2 t	1 t
Bu	0.5	2.4	10.1	9.3	4.0	2.3
Wi	0	0.1	0.6	1.5	4	1.4
Ha	0.3	2.5	14.9	21.7	6.1	0.5
Pi	1	4.9	14.1	14.6	8.1	0.2
Gl	0.2	2.2	4.2	5.9	6.2	1.9

In [6] the authors describe the use of a fuzzy clustering method as a previous step to using NEFCLASS. The identified clusters were interpreted as fuzzy rules (by means of projections to each of the dimensions) and were used for initializing NEFCLASS with the goal of refining this initial set of rules. By using the clustering method it was possible to identify one attribute that did not contribute to the classification and, consequently, to reduce the dimension of the search space and improve interpretability.

Using the information provided by CN2 we were able to use NEFCLASS in a more efficient way in order to obtain a more concise set of rules described by a smaller set of attributes and consequently, more "interpretable". This can be seen in the Wine domain. Table 5 shows the "relevance" (number of tests involving the attribute/number of rules that describe a class) of an attribute for characterising a class. This information was used to eliminate four attributes (v_5, v_6, v_8 and v_9) from the training set and induce the rules from the reduced training set. Table 6 shows the improved results in this domain.

Table 5. CN2 – Relevance of attributes for describing a class (Wi)

attrib.	class_1	class_2	class_3	attrib.	class_1	class_2	class_3
v_1	0.481	0.345	0.15	*v_8*	0	0	0
v_2	0.148	0.241	0	*v_9*	0.037	0.034	0.05
v_3	0.185	0.276	0	*v_10*	0	0.345	0.65
v_4	0.259	0.034	0	*v_11*	0.037	0.034	0.35
v_5	0.074	0.069	0	*v_12*	0.185	0.276	0.05
v_6	0.037	0	0	*v_13*	0.630	0.483	0.05
v_7	0.148	0.276	0.6				

Table 6. Performance of NEFCLASS in the Wine domain without v_5, v_6, v_8 and v_9

NR	mTrS	o_ac (%)	mTeS	o_ac (%)
116.7	5.3	96.04	3.5	92.04

4 Conclusions

We would like to point out that, in spite of aiming for the same goal, the NEFCLASS and CN2 systems serve different purposes, implement different strategies and deal with different types of data. However, the results obtained with CN2 can be used to improve the results obtained with NEFCLASS in relation to interpretability and performance. CN2's results can also be used to reduce the number of attributes that describe a particular class but this has not been investigated in this paper. As can be seen in Table 5, attributes v_2, v_3 and v_4 are irrelevant for expressing class_3 and consequently these attributes can be eliminated from the rules generated by NEFCLASS, that describe this class. In the other domains, the results given by CN2 show that all the attributes are relevant for inducing the classification rules.

Acknowledgments: Maria do Carmo Nicoletti wishes to thank Brenda Padgett for her comments and to FAPESP for her postdoctoral scholarship (Proc. 98/05547-6).

References

1. Blake, C. L., Merz, C. J. (1999) UCI Repository of machine learning databases Irvine, CA: University of California, Dept. of Information and Computer Science. [http://www.ics.uci.edu/~mlearn/MLRepository.html]

2. Clark, P., Niblett, T. (1989) The CN2 induction algorithm. Machine Learning 3, 261-283

3. Michalski, R.S. (1983) A theory and methodology of inductive learning. R. S. Michalski, J. G. Carbonnel & T. M. Mitchell (eds.) Machine Learning: an artificial intelligence approach. Vol. 1, Tioga Publishing Co., Palo Alto, CA, pp 83-134

4. Nauck, D., Kruse, R. (1995) NEFCLASS – a neuro-fuzzy approach for the classification of data. In: Proc. ACM Symposium on Applied Computing, Nashville, Feb. 26-28, ACM Press, 461-465

5. Nauck, D., Kruse, R. (1997) A neuro-fuzzy method to learn fuzzy classification rules from data. Fuzzy Sets and Systems 89, pp 277-288

6. Nauck, D., Kruse, R. (1997) What are neuro-fuzzy classifies? In: Proc. Seventh International Fuzzy Systems Association World Congress IFSA'97, Vol. IV, Academia Prague, pp 228-233

7. Nicoletti, M. C., Ramer, A. (1999) Experiments in inducing fuzzy knowledge using a neuro-fuzzy system. In: Proc. ICONIP'99 , N. Zealand, pp 244-249

8. Quinlan, J. R., Compton, P. J., Horn, K. A. & Lazarus, L. (1986) Inductive knowledge acquisition: a case study. In: Proc. of the 2nd Australian Conference on Applications of ES, Sydney, pp 183-203

Indicators of Input Contributions: Comparing Magnitude and Sensitivity Measures

Tamás D. Gedeon

School of Information Technology, Murdoch University
Perth 6150 Australia E-mail: t.gedeon@murdoch.edu.au

Abstract: The problem of data encoding and feature selection for training back-propagation neural networks is well known. The basic principles are to avoid encrypting the underlying structure of the data, and to avoid using irrelevant inputs. This is not easy in the real world, where we often receive data which has been processed by at least one previous user. The data may contain too many instances of some class, and too few instances of other classes. Real data sets often include many irrelevant or redundant fields. This paper compares the use of the weight matrix of the trained neural network itself to determine which inputs are significant, with the use of significant measures. A number of techniques are introduced, including some of our own, and some from the literature. We present our experience and results on some satellite data augmented by a terrain model. The task was to predict the forest supra-type based on the available information. A brute force technique eliminating randomly selected inputs was used to validate our approach.

Keywords: Neural Networks, Feature Selection, Input Contributions, Sensitivity Analysis.

1 Introduction

The initial network topology was 16-10-5, being sixteen inputs, ten hidden neurons, and five output neurons. The raw data for this study comes from an area in the Nullica State Forest on the south coast of New South Wales, Australia. The available information is from a rectangular grid of 179,831 pixels 30 m by 30 m, and is a vector of 16 values [1]. Each pixel has a value for altitude, aspect, slope, geology, topographic position, rainfall, temperature (from a terrain model derived from soil maps, aerial photography and so on), and Landsat TM bands 1 to 7. The outputs are the forest supra-type, being *scrub*, *dry sclerophyll*, *wet-dry sclerophyll*, *wet sclerophyll*, and *rainforest*. For the purpose of training 190 detailed sample plots have been surveyed [4]. This data gives us classifications for 190 of the pixels in the field area, and have used 150 for training the neural network and retained 40 for testing. The network was trained using error-backpropagation [5]. All connections are from units in one level to units in the next level, with no lateral, backward or multi-layer connections. Each unit is connected to each unit in the preceding layer by a simple weighted link. The network is trained using a training set of input patterns with desired outputs, using the back-propagation of error measures. The network is tested using a validation set of patterns which are never seen by the network during training and thus can provide a good measure of the generalization capabilities of the network. Thus all results quoted in this paper are for the test set. We have used the basic sigmoid logistic activation function, $y = (1+e^{-x})^{-1}$ though this is not essential to the substance of our results.

2 Analysis Techniques

Garson [2] proposed the following measure for the proportional contribution of an input to a particular output:

$$G_{ik} = \frac{\sum_{j=1}^{nh} \frac{w_{ij}}{\sum_{p=1}^{ni} w_{pj}} w_{jk}}{\sum_{q=1}^{ni} \left(\sum_{j=1}^{nh} \frac{w_{qj}}{\sum_{p=1}^{ni} w_{pj}} w_{qj} \right)}$$

A disadvantage of this approach is that during the summation process, positive and negative weights can cancel their contribution, which leads to inconsistent results.

Wong, Gedeon and Taggart [6] used the following measure for the contribution of an input to a neuron in the hidden layer:

$$P_{ij} = \frac{|w_{ij}|}{\sum_{p=1}^{ni} |w_{pj}|}$$

Milne [3] commented that the sign of the contribution is lost, and proposed the following measure:

$$M_{ik} = \frac{\sum_{j=1}^{nh} \frac{w_{ij}}{\sum_{p=1}^{ni} |w_{pj}|} w_{jk}}{\sum_{q=1}^{ni} \left(\sum_{j=1}^{nh} \left| \frac{w_{qj}}{\sum_{p=1}^{ni} |w_{pj}|} w_{qj} \right| \right)}$$

The measure introduced here is an extension of our technique [6]. We can define a measure P_{jk} for the contribution of a hidden neuron to an output neuron similar to the measure P_{ij} used above:

$$P_{jk} = \frac{|w_{jk}|}{\sum_{r=1}^{nh} |w_{rk}|}$$

The contribution of an input neuron to an output neuron is then:

$$Q_{ik} = \sum_{r=1}^{nh} (P_{ir} \times P_{rk})$$

The benefit of this approach is that the magnitude of the contribution is disentangled from the sign of the contribution.

The magnitude of contributions is significant in indicating whether an input is important, while the sign of contribution is largely irrelevant in the decision to remove or retain an input, and is recoverable in any case from the raw data by simple statistical methods.

Each of the above techniques could be extended to networks with larger numbers of hidden layers than the topology used in this experiment.

3 Brute Force Analysis

The brute force approach is to eliminate inputs and to compare the results with the predictions. Eliminating only 1 input produced inconsistent results, hence 2 inputs were eliminated. With 16 inputs, there are 120 ways to choose 2 inputs to remove. Four networks with the same topology (14-10-5) were trained for each of the 120 possibilities.

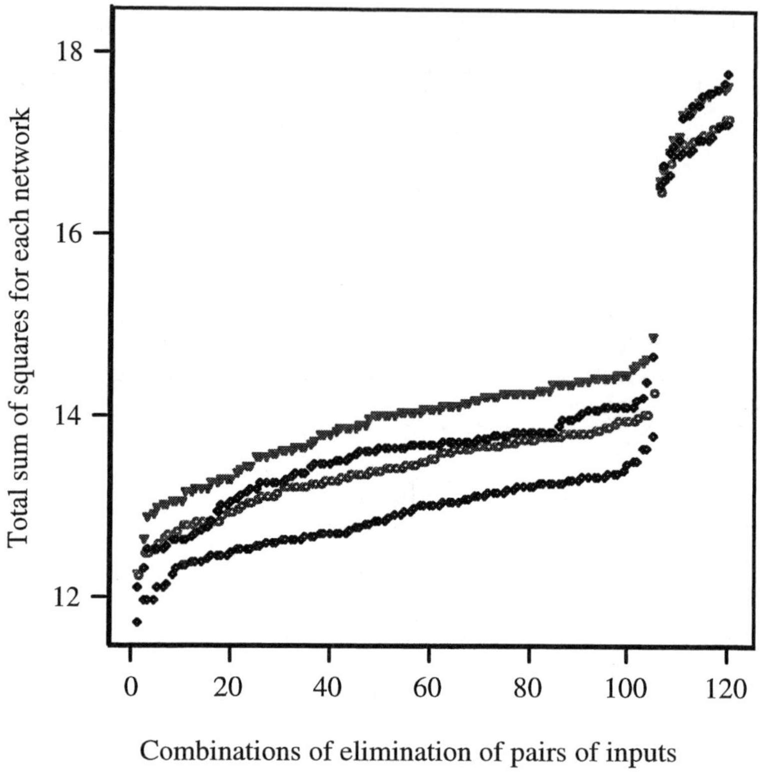

The above graph shows the results on the best total sum of squares (tss) value on the test set for each of the 480 networks run. The initial weights for each of the 4 runs was generated for the full 16 input network, and the appropriate weights excluded when 2 inputs were eliminated. Thus, each run had largely the same initial (random) starting weights. This has some small effect on all of the descendant networks, as shown by the consistent (minor) overall differences between the curves The total sum of squares (tss) values are sorted into increasing order, as there is no meaningful one dimensional scale on which to represent the removal of pairs of inputs. The discontinuity in the values plotted demonstrates the point at which there was some significant degradation of the neural network prediction, which correlated well with the tss values. The right hand part of the graph and the leftmost part of the longer curve were used to determine the most and least significant inputs, respectively, by calculating the average rank for the tss values for each combination of inputs.

4 Sensitivity Analysis

Sensitivity analysis is a simple method of finding the effect an input has on the output of the network. The relationship of an input neuron i and an output neuron k is found by determining the impact that a small change in i has on k. If drastic change occurs i is considered to be one of the key factors in producing the current activation value of k.

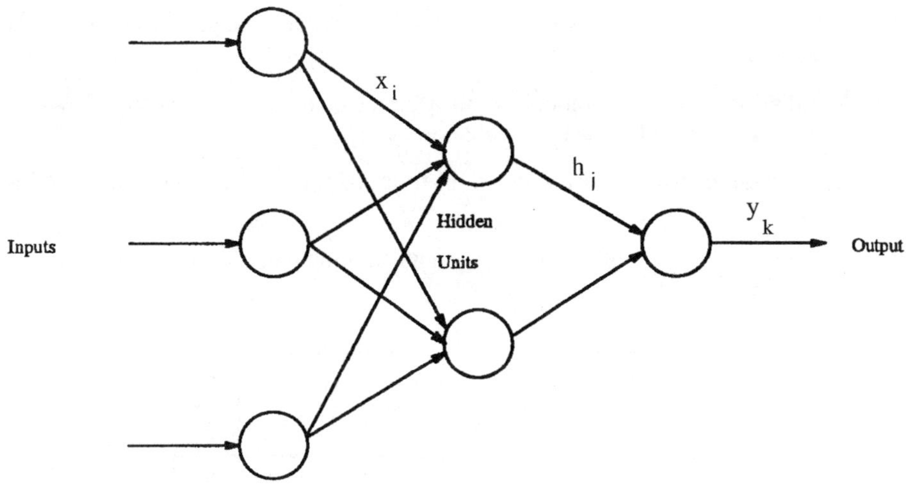

Given the above (simplified) network structure, we find the rate of change of an output neuron y_k with respect to an input neuron x_i by calculating the derivative dy_k/dx_i using the chain rule of differentiation.

$$\frac{dy_k}{dx_i} = \frac{dy_k}{dU_{k2}} \cdot \frac{dU_{k2}}{dh_j} \cdot \frac{dh_j}{dU_{j1}} \cdot \frac{dU_{j1}}{dx_i}$$

$$= f'(U_{k2}).f'(U_{j1}).\sum_j w_{jk}.w_{ij}$$

This method has also been used with the assumption that the product f'(Uk2).f'(Uj1) is constant for all k and j [7]. The influence of x_i on y_k could thus statically be determined from the weight matrix of the trained network. Unfortunately in all the domains we have tried, this assumption does not hold [8, 9] and thus we must continue with the computationally more expensive approach of calculation the effect of perturbations.

We experimented with a range of increasingly computationally expensive methods to determine the magnitude of change in value of outputs to particular inputs. These were:

1) Perturbation of single inputs on a single pattern, all elements 0.5.

2) Perturbation of single inputs on a single pattern, all elements the actual average value for that input as calculated from the training patterns.

3) Perturbation of single inputs on all training patterns, deltas accumulated for each input.

4) Perturbation of pairs of inputs on all training patterns, deltas accumulated for each input involved in each perturbation event.

5) Perturbation of triplets of inputs on all training patterns, deltas accumulated as above.

6) Perturbation of quadruplets of inputs on all training patterns, deltas accumulated as above.

Perturb	0.5	Ave	Δ1	Δ2	Δ3	Δ4
Most significant	8	8	8	8	8	8
	15	5	7	7	7	7
	12	12	13	13	13	13
	6	6	15	15	15	15
	2	15	12	12	12	12
	...	...	...	...	...	
Least significant	7	10	10	3	16	9
	10	16	3	11	3	16
	16	1	11	16	5	11
	1	7	1	10	10	3
	11	11	16	2	2	14

5 Comparison of Results

The contributions of inputs to each of the 5 outputs were averaged to determine the significance of inputs to the entire task the network is solving.

For clarity, the comparisons are made between the brute force technique and the calculated measures on the top and bottom thirds of ordering.

model	B	Q	G	M	Δ2
Most significant	5	11	2	11	8
	1	10	4	15	7
	10	2	6	7	13
	14	12	7	13	15
	11	14	11	12	12
	...	...	...	...	...
Least significant	7	13	9	8	3
	8	5	13	10	11
	13	4	14	2	16
	9	8	1	4	10
	6	7	15	5	2

From the table it is clear that the model Q is 3/5ths in accord with the brute force method B for both the most and least significant inputs, while both of the other methods me only 1/5ths in accord on either end of the significance scale.

The sensitivity ranking found on perturbing pairs of inputs is also shown. This choice is to match the brute force elimination of pairs of inputs. Then is very little accord between the brute force results and for sensitivity!

The following diagrams show the changes in the 5 most significant inputs on over training.

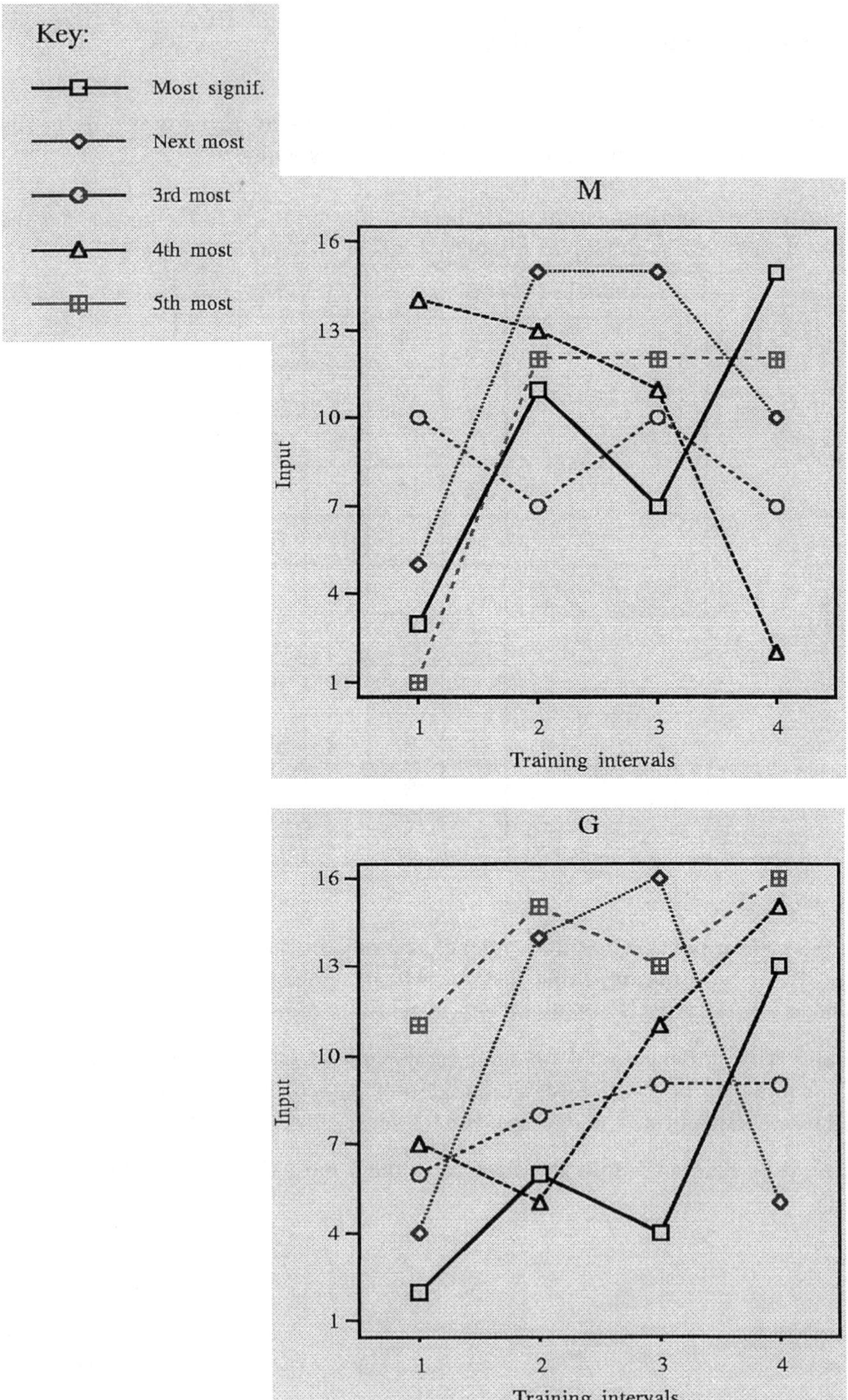
Key:
Most signif.
Next most
3rd most
4th most
5th most
M
Input
Training intervals
G
Input
Training intervals

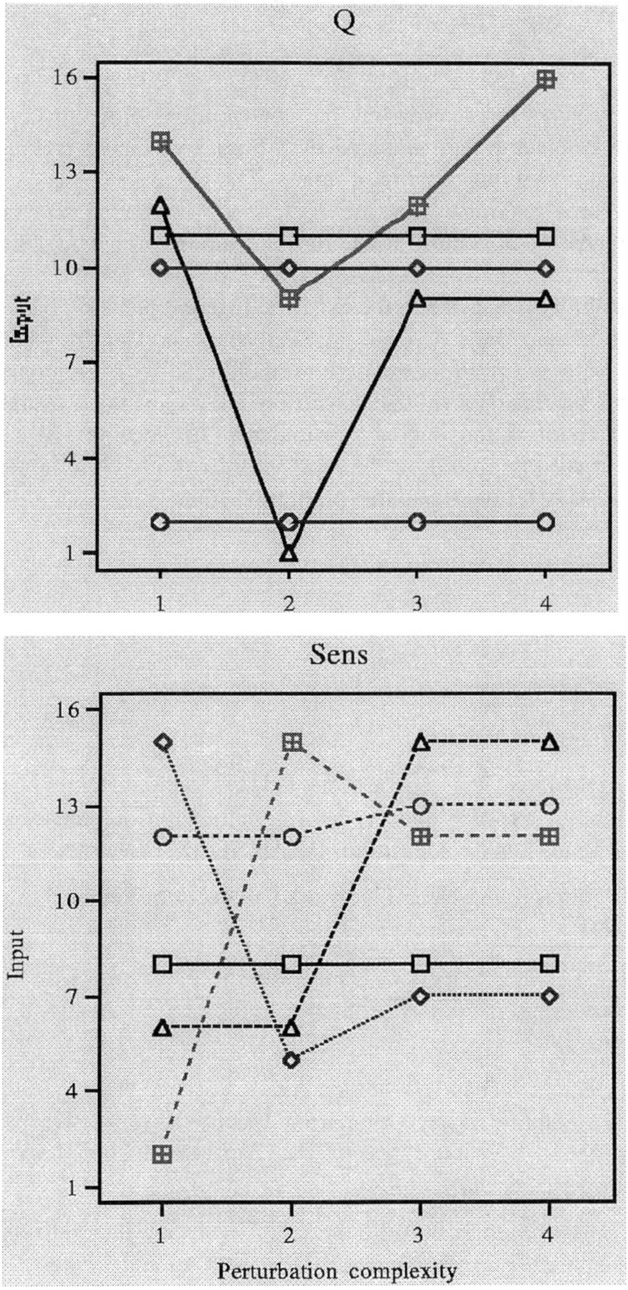
Q
Input
16
13
10
7
4
1
1
2
3
4
Sens
Input
16
13
10
7
4
1
1
2
3
4
Perturbation complexity

Conclusion

A number of measures for determining the contributions different inputs make to the outputs w e compared, and validated by a brute force input ranking technique eliminating all combinations of pairs of inputs. The Q measure was 60% in accord with the brute force ranking, while the comparison measures were only 20% in accord. The Q measure also more stable during training, which suggests that there is closer coupling to the network behaviour over time than with the other measures. Note that this stability is not due to the use of a different network, but is a result of the choice of formula solely. A sensitivity analysis as performed, which produced results which were not well correlated to the brute force approach. We can conclude at least for this data set that sensitivity analysis is not a useful technique. An unexpected result is the degree of similarity between the M measure and sensitivity, being 80% in accord on the most important, and 40% in accord on the least important. We will investigate this problem further.

References

REFERENCES [l] Bustos, RA and Gedeon, TD "Decrypting Neural Network Data: A GIS Case Study," Proceedings of the International Conference on Artificial Neural Networks and Generic Algorithms (ICANNGA), 4 pages, Alés, 1995.

[2] Garson, GD "Interpreting Neural Network Connection Weights," Al Expert, pp. 47-S1, April, 1991.

[3] Milne, LK "Feature Selection Using Neural Networks with Contribution Measures," Proceedings of the Australian Conference on Artificial Intelligence (AI'95), Canberra, 1995.

[4] Milne, LK, Gedeon, TD and Skidmore, AK "Classifying Dry Sclerophyll Forest from Augmented Satellite Data: Comparing Neural Network, Decision Tree & Maximum Likelihood," Proceedings of the Australian Conference on Neural Networks, pp. 160.163, Sydney, 1995.

[5] Rumelhart, DE, Hinton, GE, Williams, RJ, "Learning internal representations by error propagation," in Rumelhart, DE, McClelland, Parallel distributed processing, vol. 1, MIT Press, 1986.

[6] Wong, PM, Gedeon, TD and Taggart, U "An Improved Technique in Porosity Prediction: A Neural Network Approach," IEEE Transactions on Geoscience and Remote Sensing, vol. 33, n. 4, pp. 971-980, 1995.

[7] Hora, N, Enbutsu, I and Baba,K "Fuzzy role extraction from a multilayer neural net," Proc. IEEE, vol. 2, pp. 461-465, 1991.

[8] Gedeon, TD and Turner, H "Explaining student grades predicted by a oral network," Proceedings International Joint Conference on Neural Networks, pp. 609-612, Nagoya, 1993.

[9] Gedeon, TD "Indicators of Hidden Neuron Functionality: the Weight Matrix versus Neuron Behaviour," Australian Journal of Intelligent Information Processing Systems, vol. 3, no 2, pp. 1-9, 1996.

On Fuzzy Controllers Implementing RBF Approximation Scheme*

László T. Kóczy[1,2], Domonkos Tikk[1,2], and Tamás D. Gedeon[2]

[1] Dept. of Telecom. & Telematics
Budapest University of Technology and Economics
H-1117 Budapest, Pázmány sétány 1/d, Hungary
[2] School of IT, Murdoch University
South Street, Murdoch, 6148 W.A., Australia

Abstract: Both general fuzzy systems and most of the neural networks are universal approximators in the sense that they are capable of approximating any continuous function with arbitrary accuracy with respect to, e.g., the supremum norm. It means that these techniques share approximation capabilities. However, the way they captures the underlying transfer function is different. Fuzzy systems operating with if-then rules have the advantage of easy linguistic interpretability, while neural networks can adapt learning methods to improve their performance according to a training data set. We point out in this paper that several fuzzy controllers implement one of the typical neural networks (having radial basis activation functions), and hence, their combination may alloy the the advantageous properties of the two techniques.

1 Introduction

One of the most important and intriguing goals of Artificial Intelligence is to construct methods, models and algorithms to cope with very complex systems, often lacking known analytical models. Such approaches have attempted to follow examples of natural intelligence and biological organisms. The classical AI approach was based on symbolic logic and its tautologies, such as *Modus Ponens, Hypothetical Syllogism*, etc. The former is the simple logical scheme that works when an implication and its antecedent part are known to be true; then, its consequent part is concluded to also be true:

$$\frac{\begin{array}{l}A \to B\\ A\end{array}}{B}$$

A great disadvantage of this approach is that knowledge stored in the form of implications does not relate directly to the structure of the state

* This research was supported by the Australian Research Council, the Hungarian Ministry of Education (MKM) FKFP 0235/1997 and FKFP 0422/1997 and the National Science Research Fund (OTKA) 019671 and 030655.

space of the problem in question. Most physical phenomena can be modeled in a state space where each dimension is a subset of $\mathbb{R}^n$, usually a hyper interval $X = X_1 \times X_2 \times \cdots \times X_n = \prod_{i=1}^{n}[L_i, U_i]$ ($\prod$ denoting Cartesian product). Consequently, there is an ordering $\prec_i \subset X_i \times X_i$ $(i = 1, \ldots, n)$ in every dimension of X, and a partial ordering $\prec$ in X itself, so that $\underline{x}_1 \prec \underline{x}_2 \Leftrightarrow x_{i1} \prec_i x_{i2}$ for all $i = 1, \ldots, n$. (Throughout this work, as denotation convention, we use henceforth letter n for the dimension of the input space, $\underline{x}$, and $\underline{\underline{A}}$ for vector x and matrix A, respectively.) Further, there is a norm $||\cdot||$ in X, usually the Euclidean norm. When applying a symbolic logical approach to a system in the state space X, it is necessary to quantify this space, so that every "quantum" or interval is denoted by a symbol (see Figure 1). The symbols A_{i_k} represent the whole, kth, interval in X_i, and there is no way to differentiate among values within the same quantum or interval. This fact results in a certain coarseness of the model that can be refined by applying smaller intervals and more symbols. It is a more essential problem, however, that the relative position of these symbols to each other cannot be expressed at this level, e.g. the fact that $A_{i_k} \prec_i A_{i_{k+1}} \prec_i A_{i_{k+2}}$ (i.e. in the sense that the typical values or centres are ordered according to $\prec_i$). The same applies to relations like $||A_{i_k} - A_{i_{k+1}}|| < ||A_{i_k} - A_{i_{k+2}}||$, as the norm and the distance between two logical symbols are senseless without the additional information referring to the metrics of the space ("$<$" meaning here the usual ordering on real numbers).

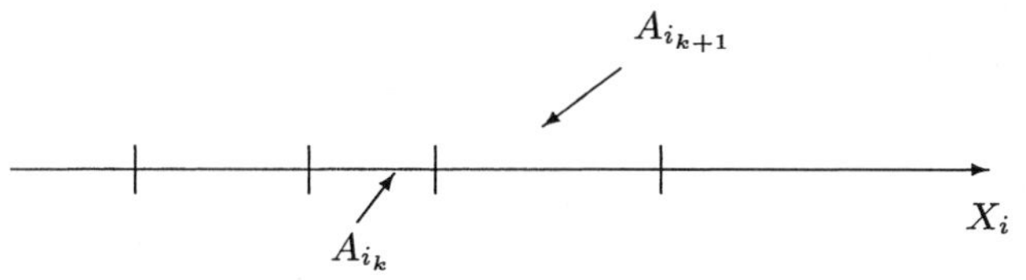

Fig. 1. Quantification and symbol correspondence

The knowledge base consisting of symbolic rules has the form

$$R = \{R_j | j = 1, \ldots, r\}, \quad R_j = A_j \rightarrow B_j$$

(r is the number of the rules), where $A_j = \prod_{i=1}^{n} A_{ji}$, and each A_{ji} is a logic symbol in X_i (the antecedent of the jth rule in the ith dimension). A fact or observation in such a system has the form $A^* = \prod_{i=1}^{n} A_i^*$, and it must match exactly with the antecedent in one of the R_j $(j = 1, \ldots, r)$ in order to make it possible to apply e.g. the *Modus Ponens* in order to obtain a conclusion (necessarily equal to one of the rule consequents).

It was an essential step towards establishing a means for describing the internal structure of the state space, when the use of fuzzy symbols was proposed [1]. By these, based on the information present in the membership functions attached to each symbol, the intervals close to each other were merging into single "fuzzy intervals" or "fuzzy numbers" By the possibility of these membership functions' partial overlapping, the partial semantic overlapping of the meaning of "close by" symbols could be formally expressed (see Figure 2).

A similar feature is inherent for artificial neural networks, where sub-symbolic information is contained in the activation functions and weights. It is obvious that sub-symbolic models have a smaller number of symbols, and consequently, a smaller number of rules in the base, although in this case each symbol has a much more complicated description in the form of an additional membership function. Let us assume that there are T_i symbols used in dimension X_i. Then the number of rules in the base is $|R| = \prod_{i=1}^{n} T_i$. Let $T_i \leq T$ for all $i = 1, \ldots, n$, then the number of rules in the base is at most $|R| \leq T^n$. Let c be the rate of reduction of the number of symbols when fuzzy membership functions are introduced, and let e be the rate of extension of the amount of information belonging to a single symbol. Then, instead of approximately pieces of information, only $\left(\frac{T}{c}\right)^n e$ will be needed. However, $\left(\frac{T}{c}\right)^n e < T^n$ holds, whenever $c > e^{\frac{1}{n}}$, i.e. when n is large enough ($n > \frac{\log e}{\log c}$). Obviously, the introduction of fuzzy symbols leads to the reduction of computational complexity when the number of state space dimensions is large enough.

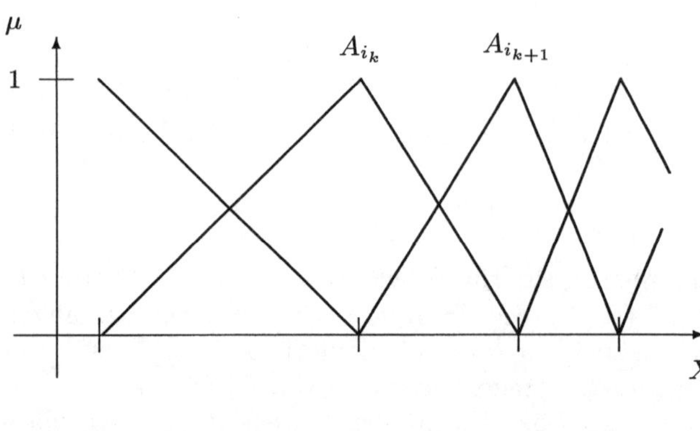

Fig. 2. Symbols with fuzzy membership functions

The reason for this effective reduction is that the use of fuzzy membership functions utilizes the knowledge about the structure of the space and introduces a kind of interpolation.

Calculations about the explicit interpolation functions with triangular and trapezoidal membership functions in various fuzzy reasoning and control systems can be found in [2].

A common feature of such fuzzy rule bases is that the antecedents form a collection of fuzzy sets, where their kernels (either single points or intervals) represent typical values of $\underline{x}$, and between these kernels the membership functions cover the space, so that for every $\underline{x} \in X$ there is at least one such antecedent, which is true over $\underline{x}$ at least to a fixed degree $\alpha > 0$. (Usually $\alpha \geq 0.5$.) If this latter condition is satisfied, it is true that there is at least one rule for every x, which is at least as true as it is false. (If $\alpha > 0.5$ is satisfied then there is at least one rule which is more true than false.) If

$$\forall x_i \in X_i,\ i = 1, \ldots, n\ \exists j\ (j \in \{1, \ldots, r\}) : A_{ji}(x_i) \geq \alpha > 0$$

then $\{A_{ji} | j = 1, \ldots, r\}$ form an α-cover of X_i. If also

$$\forall \underline{x} \in X,\ \exists j\ (j \in \{1, \ldots, r\}) : A_j(\underline{x}) \geq \alpha > 0$$

is fulfilled, then $\{A_j | j = 1, \ldots, r\}$ is an α-cover of X. In the original Mamdani-algorithm [3], e.g. the antecedents formed a cover for approximately $\alpha = 0.75$. In most current applications however $\alpha = 0.5$ is satisfactory.

If $R = \{R_j | j = 1, \ldots, r\}$ is a *full set of rules*, then for arbitrarily chosen n rules rules $R_{j_1}, R_{j_2}, \ldots, R_{j_n} \in R$, ($A_{j_\ell,i}$ being the ith projection of A_{j_ℓ}, the antecedent of rule R_{j_ℓ}, to the sub-space X_i $(i, \ell = 1, \ldots, n)$)

$$\exists m \in \{1, \ldots, r\} : A_m = A_{j_1,1} \times A_{j_2,1} \times \cdots \times A_{j_n,n}.$$

holds. This means that the rule antecedents form a "grid" in X as illustrated in Figure 3. Full rule bases contain $|R| = \prod_{i=1}^{n} r_i$ rules, where r_i stands for the number of different antecedent projections appearing in dimension X_i. It is reasonable to use here k subscripts, where the ith one indicates that the antecedent of the given rule generates the ith projection in X_i if all projections are ordered along each component axis. In most practical applications the rule set is full, unless some rules in the original set are omitted in order to decrease the size of the rule base.

In most cases the antecedents also form a *Ruspini-partition* of X, meaning that

$$\forall \underline{x} \in X : \sum_{j=1}^{r} A_j(\underline{x}) = 1$$

Full rule bases with the property of Ruspini-partition are usually 0.5-covers.

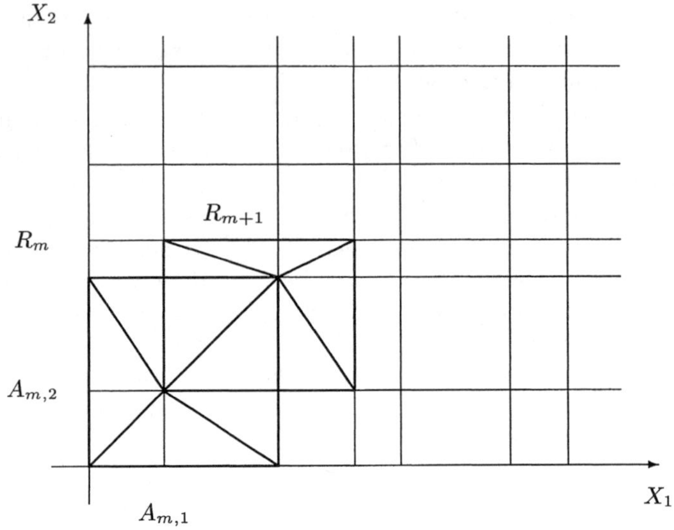

Fig. 3. Rules represented by their supports in a full rule base in two-dimensional space

2 Fuzzy rule based controllers

It is obvious that although the introduction of fuzzy membership functions for describing the sub-symbolic information behind the symbols leads to the reduction of the overall computational complexity, whenever k is large enough the reduction is not radical [8]. There are further possibilities to decrease the effective value of T, by applying rule bases where the antecedents do not form an α-cover of the input state space.

The original idea of reasoning and control within fuzzy rule bases was proposed by Zadeh [4], and was called the Compositional Rule of Inference (CRI) and had the disadvantage of running directly in the k-dimensional input variable space, while being able however to describe multi-dimensional membership function distributions of arbitrary shape. Its modified version, the Mamdani-algorithm [3] applied the projections of the antecedents and thus returned to the form of rules $R_j = A_j \to B_j$, where $A_j = \prod_{i=1}^{n} A_{ji}$, meaning that only such A_j could be used that were the cylindrical closures of some membership functions A_{ji}, each being of the type $A_{ji} : X_i \to [0,1]$. This method offers much better computational speed.

The idea of the Mamdani-algorithm is to use the heights of the intersections of the observation with the antecedents and weighing the consequents with these, e.g. by simply "cutting them" at the given height:

$$B^* = \max_j \{\min \{\text{height}(A^*(\underline{x}), A_j(\underline{x})), B_j(y)\}\}, \tag{1}$$

where $\text{height}(S(\underline{x})) = \max_{\underline{x}} S(\underline{x})$. In the Larsen-method, the modification is that the consequents are multiplied with these heights:

$$B^* = \max_j \{\text{height}(A^*(\underline{x}), A_j(\underline{x}))B_j(y)\}.$$

In both cases, y^* is obtained by defuzzification, e.g. in the most popular techniques, calculating the centre of gravity of the modified consequents. This latter means that instead of taking the maximum, in effect the sum of the membership functions will be considered:

$$B^* = \sum_{j=1}^{r} \text{height}(A^*(\underline{x}), A_j(\underline{x}))B_j(y),$$

which is in case of a crisp singleton observation identical with

$$B^*_{\text{eff}} = \sum_{j=1}^{r} A_j(\underline{x}^*)B_j(y).$$

This latter is the most popular formula used for fuzzy control applications.

3 Interpolation in sparse fuzzy rule bases

If for given α the antecedents form no α-cover, the rule base is α-sparse. If they do not form a cover for any $\alpha > 0$, the rule base is sparse. The CRI, the Mamdani method and its variants, the Takagi–Sugeno method [5], which applies linear functions of $\underline{x}$ in the consequent part, and all other fuzzy control approaches based on some form of degree of matching between observation and antecedents cannot be used for sparse rule systems.

Instead, a more general notion of the degree of similarity was introduced, based on the family of *α-distances between two convex and normal fuzzy sets* (CNF sets), which is defined as

$$\begin{aligned}\widetilde{d}(A_1, A_2) &= \{d_{\alpha,C}(A_1, A_2),\ \alpha \in (0,1], C \in \{L, U\}\} \\ &= \{\|A_{1\alpha L} - A_{2\alpha L}\|, \|A_{1\alpha U} - A_{2\alpha U}\|, \alpha \in (0,1]\}\end{aligned} \tag{2}$$

subscripts L and U denoting the minimum and maximum of the respective α-cuts. For simplicity, the distance belonging to fixed and L or U will be denoted by $d_{\alpha,C}(A_1, A_2)$. The conditions for the existence of the fuzzy distance set is that both fuzzy sets are CNF, and that they are comparable in the sense of $\prec$.

Using this notion of distance (2), the fuzzy similarity set can be defined as

$$\begin{aligned}\widetilde{s}(A_1, A_2) &= \{s_{\alpha,C}(A_1, A_2),\ \alpha \in (0,1], C \in \{L, U\}\} \\ &= \left\{\frac{1}{d_{\alpha,C}(A_1, A_2)}, \alpha \in (0,1]\right\}\end{aligned} \tag{3}$$

the elements of the similarity degree set being the reciprocals of the elements of the distance set.

The basic idea of the rule interpolation is formulated in the *Fundamental Equation of Rule Interpolation* (FERI):

$$D(A^*, A_1) : D(A^*, A_2) = D(B^*, B_1) : D(B^*, B_2). \tag{4}$$

In this equation A^* and B^* denote the observation and the corresponding conclusion, while $R_1 = A_1 \to B_1$, $R_2 = A_2 \to B_2$ are the rules to be interpolated, such that $A_1 \prec A^* \prec A_2$ and $B_1 \prec B_2$. If D denotes the Euclidean distance between two symbols, the solution for B^* results in simple linear interpolation. If $D = \widetilde{d}$ (the fuzzy distance family), linear interpolation between corresponding α-cuts is performed.

A more general form of (4) gives

$$B^*_\alpha = \sum_{j=1}^{r} s_\alpha(A^*, A_j) B^*_{j\alpha}, \quad \alpha \in (0,1]$$

where s_α is some α-cut related similarity degree, e.g., the fuzzy similarity (3) obtained from the reciprocal distances of the α-cuts. This similarity can be considered as an extended "degree of matching", and these similarity degrees replace the degrees $A_j(\underline{x}^*)$ or height$(A_j \wedge A^*)$ ($\wedge$ denoting the appropriate t-norms) which are used in the Mamdani, Larsen, (Takagi–) Sugeno and other classical fuzzy reasoning algorithms. These methods can be substituted by fuzzy interpolation (KH interpolation [6]), which also functions in sparse rule bases. The Mamdani type algorithms are thus replaced by

$$B^*_\alpha = \frac{\sum_{j=1}^{r} s_\alpha(A^*, A_j) B_{j\alpha C}}{\sum_{j=1}^{r} s_\alpha(A^*, A_j)} = \frac{\sum_{j=1}^{r} \frac{1}{d_{\alpha,C}(A^*,A_j)} B_{j\alpha C}}{\sum_{j=1}^{r} \frac{1}{d_{\alpha,C}(A^*,A_j)}} \tag{5}$$

where the normalized degree of similarity for fixed α and C is the reciprocal distance of the observation from the corresponding antecedent, divided by the sum of all these distances. It must be noted, however, that B^* reconstructed directly from the above α-cuts does not always exist, as various abnormalities in the shape of the conclusion might necessitate some transformations, which eventually result in obtaining sub-normal conclusions (cf. [6]). It should be noted that in practical applications, it is enough to do calculations for $\alpha \in \mathcal{B}$, the breakpoint set of the membership functions, which is four points altogether [7] in the case of the most widely applied trapezoidal functions $\mathcal{B} = \{0, 1; L, U\}$.

In all these approaches, the approximated mapping is "fuzzy set to fuzzy set":

$$\widetilde{\Psi} : \widetilde{P}(X) \to \widetilde{P}(Y)$$

where $\widetilde{P}(Z)$ denotes the fuzzy power set of Z, i.e. all fuzzy sets of the universe of discourse Z, so that a fuzzy observation is always mapped into a fuzzy conclusion by the respective rule base and inference engine: $\widetilde{\Psi}(A^*) = B^*$. In the special case where the observation is a crisp singleton, we have $\widetilde{\Psi}(\underline{x}^*) = B^*$. In control applications the conclusion set should always be defuzzified, so that $y^* = \text{defuzz}\,(B^*)$. With both crisp singleton inputs and outputs, we get $y^* = \text{defuzz}\left(\widetilde{\Psi}(\underline{x}^*)\right) = \Psi(\underline{x}^*)$. (Some of the most important Ψ-s are explicitly given in [2].) In this sense, fuzzy rule based controllers can be considered as multivariate real function approximators.

The respective interpolative extension for the equation for the Takagi–Sugeno algorithm is

$$y^* = \frac{\sum_{j=1}^{r} s_\alpha(A^*, A_j)(b_{j_1}\underline{x}^* + b_{j_0})}{\sum_{j=1}^{r} s_\alpha(A^*, A_j)}$$

where $\forall j \in \{1, r\} : b_{j_1} = 0$ in the case of the Sugeno-controller. Here, the similarity degree is an overall one that can be derived from the reciprocal of an overall distance of the two fuzzy sets in question, e.g. by defining it as

$$s(A^*, A_j) = \frac{1}{\int\limits_{\alpha_L \in (0,1] \wedge \alpha_u \in (0,1]} d_{\alpha,C}(A^*, A_j)\mathrm{d}\alpha}. \tag{6}$$

In the case of trapezoidal membership functions (6) can be simplified to

$$s(A^*, A_j) = \frac{4}{d_{0,L}(A^*, A_j) + d_{1,L}(A^*, A_j) + d_{0,U}(A^*, A_j) + d_{1,U}(A^*, A_j)}.$$

because both the left and right pairs of flanks of the antecedents define trapezoidal areas. This latter interpolation has the advantage of obtaining directly defuzzified consequents, which fact eliminates the problem of the abnormal conclusion shape. This interpolation method will be investigated in more detail in the future.

The investigation of the KH interpolation has shown that using a more radical similarity degree improves the convergence properties tremendously. If s is defined as

$$s_{\alpha,C}(A^*, A_j) = \frac{\frac{1}{d^n_{\alpha,C}(A^*,A_j)}}{\sum_{j=1}^{r} \frac{1}{d^n_{\alpha,C}(A^*,A_j)}}, \tag{7}$$

it will be guaranteed that when the consequent membership functions are known ε with accuracy, the conclusion function will be also calculable with $\delta(\varepsilon)$ accuracy [9]. Due to this result, the input-output function of the general KH interpolation obtained by means of the radical similarity defined in (7),

$$B^*_{\alpha,C} = \frac{\sum_{j=1}^{r} \frac{1}{d^n_{\alpha,C}(A^*,A_j)} B_{j\alpha C}}{\sum_{j=1}^{r} \frac{1}{d^n_{\alpha,C}(A^*,A_j)}} \tag{8}$$

possesses the universal approximation property for every fixed α, i.e., every continuous function defined on a compact set can be approximated arbitrarily accurately with respect to the norm L_p ($p \in \{[1, \infty)\} \cup \infty$) with functions of this form. By this, the mathematical stability of the method can be formulated simply as follows: If the antecedents or the observation of a rule base change only slightly, it does not cause a significant change in the conclusion.

General KH interpolation has strong connections with the so-called Balzs–Shepard interpolator operator well known in approximation theory. This operator is also capable of approximating real valued continuous functions arbitrarily well, where the convergence of the approximation depends on the modulus of continuity (a sort of smoothness factor) of the approximated real function. It has been shown that for every fixed value of α, the function generated by the general KH interpolation is a Balzs–Shepard operator. On the other hand, for practically important piecewise linear membership functions it can be proved that it is sufficient to calculate the conclusion for the elements of the breakpoint set of the membership functions, and in the regions between the breakpoint levels the conclusion can be approximated well by linear functions. Using these results, general KH interpolation can be considered as a generalized function approximator being able to approximate fuzzy-to-fuzzy mappings on the Cartesian product of compact universes of discourse.

There are further methods for rule interpolation that we do not discuss here in detail (cf. [10–12]).

4 Radial basis functions in neural networks

An alternative approach where sub-symbolic information is added is the artificial neural network model. Three-layer feedforward neural networks using *radial basis functions*(RBF) are similarly function approximators as are rule based fuzzy systems. Park and Sandberg proved that any continuous function f could be arbitrarily well approximated on a compact subset $X \subset R^n$ in L_∞ norm [13]:

$$f(\underline{x}) = \sum_{i=1}^{r} \lambda_i \varphi(\|\underline{x} - \underline{\xi}_i\|) \tag{9}$$

where $\varphi : R^+ \to R$, $\underline{\xi}_i$ are called the centres of the functions and are real numbers. If there are r known points of the function to be approximated, $\{y_i = F(\underline{x}_i), i = 1, \ldots, r\}$, by substituting them into (1.9), a set of linear equations for is obtained [14]:

$$\begin{pmatrix} y_1 \\ \vdots \\ y_r \end{pmatrix} = \begin{pmatrix} \varphi_{11} & \cdots & \varphi_{1r} \\ \vdots & \ddots & \vdots \\ \varphi_{r1} & \cdots & \varphi_{rr} \end{pmatrix} \begin{pmatrix} \lambda_1 \\ \vdots \\ \lambda_r \end{pmatrix} \tag{10}$$

where

$$\varphi_{ij} = \varphi(\|\underline{x}_i - \underline{\xi}_j\|)$$

By denoting $\underline{\underline{\Phi}} = \begin{pmatrix} \varphi_{11} & \dots & \varphi_{1r} \\ \vdots & \ddots & \vdots \\ \varphi_{r1} & \dots & \varphi_{rr} \end{pmatrix}$, $\underline{\lambda} = \begin{pmatrix} \lambda_1 \\ \vdots \\ \lambda_r \end{pmatrix}$, and $\underline{y} = \begin{pmatrix} y_1 \\ \vdots \\ y_r \end{pmatrix}$, (10) is simply written as

$$\underline{y} = \underline{\underline{\Phi}}\underline{\lambda}$$

From here, the solutions for λ_i can be obtained by

$$\underline{\lambda} = \underline{\underline{\Phi}}^{-1}\underline{y}$$

if $\underline{\underline{\Phi}}$ is regular. For a rather large set of various functions it has been proved that $\underline{\underline{\Phi}}$ is always regular (see e.g. [15]), if $\underline{\xi}_i$ $(i = 1, \dots, r)$ are distinct vectors. For example, x, $x^2 \log x$, x^3, $(x^2 + c)^\alpha$, $(x^2 + c)^{-1/2}$, and e^{-cx^2} $(c > 0,\ 0 < \alpha < 1)$. These functions belong to two different groups: *localized* functions which satisfy $\varphi(\underline{z}) \to 0$, when $z \to 0$, and so $\underline{\underline{\Phi}}$ is positive definite, while *non-local* functions $\varphi(z)$ become unbounded, when $z \to 0$, and so $\underline{\underline{\Phi}}$ is not positive definite, having $r - 1$ negative eigenvalues and only one positive one [16].

The RBF approach can be extended to multiple output mappings, and more general approximation, where the number of known points rather larger than the number of functions φ_i, using the Moore–Penrose pseudo inverse (cf. [14,17]):

$$\underline{\underline{\Phi}}^+ = (\underline{\underline{\Phi}}^T\underline{\underline{\Phi}})^{-1}\underline{\underline{\Phi}}^T,$$

where $\underline{\underline{\Phi}}^+\underline{\underline{\Phi}} = I_r$, the $r \times r$ identity matrix.

Three-layer feedforward neural networks using radial basis functions and with fully interconnected layers are similarly suitable function approximators as the fuzzy systems discussed in section 2. The number of units in the input layer should be n, the number of input variables. Each connection between the input layer and the hidden layer should be assigned a value ξ_{ij}, where $i \in \{1, \dots, r\}$ indicates the subscript of the corresponding unit centre, and $j \in \{1, \dots, n\}$ refers to the respective component. (In the basic case $r = n$.) The ith hidden unit computes the norm $\|\underline{x}^* - \underline{\xi}_i\|$, and its mapping by φ. The output layer consists of a single unit in the one output variable case, and this unit is connected to the ith hidden unit by a weight λ_i, and is linear, so it generates the value $f(\underline{x}^*) = \sum_{i=1}^{r} \varphi(\|\underline{x}^* - \underline{\xi}_i\|)$ (see Figure 4). Multiple unit output layers can similarly implement multiple output variable cases.

5 The connection of fuzzy algorithms and RBF

Comparing the formulae of various fuzzy interpolation techniques and RBF based feedforward networks, some similarities can be detected. Fuzzy reasoning and control systems are approximators on two different levels. They

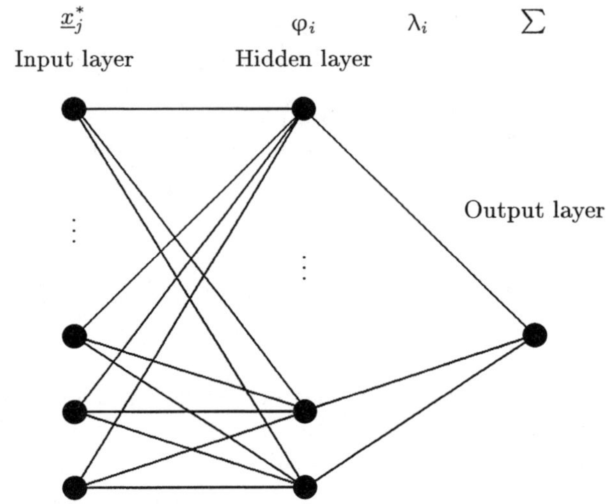

Fig. 4. Three-layer feedforward neural network with RBF activation functions in the hidden layer

approximate primarily fuzzy-to-fuzzy mappings in the following sense:

$$S : \widetilde{P}(X) \to \widetilde{P}(Y),$$

S is the generalization of the notion of a set-to-set function. As every fuzzy set is described by a membership function, for every given $\alpha \in (0,1]$ and fixed C („lower or upper end"), S realizes a function

$$S_{\alpha,C} : X \to Y,$$

and fuzzy interpolation means the solution of an ordinary interpolation problem for every $S_{\alpha,C}$. The r rules define r vectors $(A_{i\alpha C}, B_{i\alpha C})$, so that the interpolation problem is equivalent to fitting a function to these r control points. Fuzzy interpolation means finding a family of functions $y = F(\underline{x}; \alpha, C)$, $\alpha \in (0,1]$, $C \in \{L, U\}$interpolating these families of points. For a fixed α-cut and C, the problem to solve is identical with that of ordinary interpolation.

On the other hand, if a fuzzy system is considered with crisp observations and defuzzified outputs, the whole system is equivalent to a single approximator. Because of the different defuzzification techniques are rather complicated to be expressed in terms of the approximation error involved, in this context we refer to the explicit formulae of the most frequently used fuzzy controllers [2], but will not discuss this aspect here in detail.

Getting back to the first type of problem, it can be recognized that Mamdani control has a similarity to RBF approximation, but is less specific concerning the type of functions used.

Obviously, the original Mamdani method is rather different from the RBF approach, as in expression (1) not addition but a max operator combines the functions height($A^*(\underline{x}), A_j(\underline{x})$, and also the "weighting" is done by min rather than multiplication. So, the formula (1) can be written like:

$$F(\underline{x}; \alpha, C) = \max_{j=1}^{r}\{\min\{\lambda_j(\alpha, C), A_j(\underline{x})\}\}, \tag{11}$$

where $\lambda_j(\alpha, C) = B_{j\alpha C}(y)$ if B^* is calculated point by point. There is a striking analogy with the formulae of the RBF-s, especially as in the theory of possibility measures, the max operator replaces the addition and min might play a similar role as multiplication in probability theory. So, the above equation can be interpreted as a "***possibilistic linear combination***" of the consequent function points. An additional essential difference is however the lack of radial behavior.

According to many results, Mamdani type approximators are universal [18–20], at least in the sense that their full versions including defuzzification can arbitrarily well approximate an arbitrary continuous function over a compact domain. It is well worth investigating in the future whether this statement remains true for arbitrary continuous fuzzy-to-fuzzy mappings as well.

If Larsen interpolation is applied with centre of gravity defuzzification, the equivalent conclusion can be written point by point as

$$F(\underline{x}; \alpha, C) = \sum_{j=1}^{r} \lambda_j(\alpha, C) A_j(\underline{x}), \tag{12}$$

where $\lambda_j(\alpha, C)$ is the same as in (11) for the Mamdani formula. Here, the only difference to the RBF case is the non-radial nature of the general antecedent function.

Proposition 1. *If in the Larsen algorithm all antecedent functions are symmetric around the centre of the kernel: $A_i(\underline{\xi}_i - \delta) = A_i(\underline{\xi}_i + \delta)$, and further, if $A_i(\underline{x})$ $(i = 1, \ldots, r)$ differ from each other only in a constant factor, this special fuzzy controller implements an RBF approximator for every fixed α.*

It has been shown in [2], that the transfer function of symmetric Sugeno controllers is identical with the transfer function of Larsen algorithm using the centre of gravity method. Exploiting this functional equivalence, we can have the following

Corollary 1. *Sugeno controllers with symmetric shaped antecedent sets around the centre of the kernel: $A_i(\underline{\xi}_i - \delta) = A_i(\underline{\xi}_i + \delta)$, and further, if $A_i(\underline{x})$ $(i = 1, \ldots, r)$ differ from each other only in a constant factor, this special fuzzy controller implements an RBF approximator for every fixed α, as well.*

Considering the results of [20], it would be interesting to investigate whether in general, any convex and normal fuzzy (CNF) membership function could generate a suitable RBF with positive definite matrix $\underline{\underline{\Phi}}$, where the only necessary condition for the radial basis function is that $\varphi(0) = 1$, $\varphi(z) \to 0$, when $z \to 0$ and is monotone decreasing.

Considering now the formulae of the linear KH interpolation,

$$F(\underline{x};\alpha,C) = \sum_{j=1}^{2} \lambda_j(\alpha,C)\varphi(\underline{x}),$$

where $\varphi(x) = \dfrac{1}{d_{\alpha,C}(A^*(\underline{x}),A_j(\underline{x}))}$ and $\lambda_j(\alpha,C) = \dfrac{B_{j\alpha C}}{\sum_{j=1}^{2} \frac{1}{d_{\alpha,C}(A^*(\underline{x}),A_j(\underline{x}))}}$, is obviously an RBF approach for every fixed level and end point.

Similar constructions apply for the general KH interpolation, where

$$F(\underline{x};\alpha,C) = \sum_{j=1}^{r} \lambda_j(\alpha,C)\varphi(\underline{x}),$$

$\varphi(x) = \dfrac{1}{d_{\alpha,C}(A^*(\underline{x}),A_j(\underline{x}))}$ and $\lambda_j(\alpha,C) = \dfrac{B_{j\alpha C}}{\sum_{j=1}^{r} \frac{1}{d_{\alpha,C}(A^*(\underline{x}),A_j(\underline{x}))}}$, for every fixed level and end point.

Analogously, for the stabilized KH interpolator also similar constructions apply, where we take the sth power of the distance function in the previous expressions, where $s \geq n$ and n stands for the number of variables.

Hence we have the following result:

Proposition 2. *The general KH interpolation implements an RBF approximator for every fixed α. Here, due to the construction of $\varphi(\underline{x})$ and $\lambda_j(\alpha,C)$, the vectors $\underline{\xi}_i$ $(i = 1,\ldots,r)$ need not to be distinct to assure the regularity of the coefficient matrix $\underline{\underline{\Phi}}$.*

A very similar statement can be said about the stabilized version of the KH interpolation, where however the distance function is raised to the power s ($s \geq n$ and n is the number of variables). Consequently we have:

Proposition 3. *The stabilized general KH interpolation also implements an RBF approximator for every fixed α. Here, due to the construction of and $\varphi(\underline{x})$ and $\lambda_j(\alpha,C)$, the vectors $\underline{\xi}_i$ $(i = 1,\ldots,r)$ need not to be distinct to assure the regularity of the coefficient matrix $\underline{\underline{\Phi}}$.*

6 Conclusion

In this paper new examples have been given for the approximate functional equivalence of fuzzy and neural models. We pointed out that under certain conditions Larsen type controllers and Sugeno type controllers implement RBF approximation scheme.

We have also shown that the (stabilized) general fuzzy KH interpolation and the RBF approximation scheme (usually implemented by RBF networks); both being considered as universal approximators, can be connected straightforwardly. The general fuzzy KH interpolators can be considered as a special RBF approximation scheme for fixed α-cuts $\alpha \in (0,1]$, and $C \in \{L, U\}$.

Acknowledgement

The authors want to thank to János Levendovszky for his valuable suggestions on the connection of KH interpolation and RBF functions.

References

1. L. A. Zadeh. (1965) Fuzzy sets, Information and Control **8**(3), 338–353.
2. L. T. Kóczy and M. Sugeno. (1996) Explicit functions of fuzzy control systems. International Journal of Uncertainty, Fuzziness and Knowledge-Based Systems **4**, 515–535.
3. E. H. Mamdani and S. Assilian. (1975) An experiment in linguistic synthesis with a fuzzy logic controller. Int. J. of Man Machine Studies **7**, 1–13.
4. L. A. Zadeh. (1973) Outline of a new approach to the analysis of complex systems and decision processes. IEEE Trans. on SMC **3**, 28–44.
5. M. Sugeno and T. Takagi. (1985) Fuzzy identification of systems and its applications to modeling and control. IEEE Trans. on SMC **15**, 116–132.
6. L. T. Kóczy and K. Hirota. (1993) Approximate reasoning by linear rule interpolation and general approximation. Internat. J. Approx. Reasoning **9**, 197–225.
7. Y. Shi and M. Mizumoto. (1995) Some considerations on Kóczy's interpolative reasoning method. Proc. of the FUZZ-IEEE/IFES'95, Yokohama, 2117–2122.
8. L. T. Kóczy. (1995) Algorithmic aspects of fuzzy control. Internat. J. of Approx. Reasoning **12**, 159–217.
9. D. Tikk, I. Joó, L. T. Kóczy, P. Várlaki and B. Moser. Stability of interpolative fuzzy KH -controllers. To appear in Fuzzy Sets and Systems.
10. T. D. Gedeon and L. T. Kóczy. (1996) Conservation of fuzziness in rule interpolation. Proc. of the Symp. on New Trends in Control of Large Scale Systems, Herlany, vol. 1, 13–19.
11. P. Baranyi and T. D. Gedeon. (1996) Rule interpolation by spatial geometric representation. Proc. of the IPMU'96, Granada, 483–488.
12. D. Tikk and P. Baranyi. Comprehensive analysis of a new fuzzy rule interpolation method. To appear in IEEE Trans. on Fuzzy Systems.
13. J. Park and I. W. Sandberg. (1991) Universal approximation using radial-basis-function networks. Neural Computation **3**, 246–257.
14. K. Hlavačková and R. Neruda. (w.y.) Radial basis function networks. Inst. of Information and Computer Science, Czech Academy of Sciences, Prague, report.

15. A. C. Faul and M. J. D. Powell. (1999) Krylov subspace methods for radial basis function interpolation. DAMPT 1999/NA11, University of Cambridge, 25p.
16. C. A. Micchelli. (1986) Interpolation of scattered data: Distance matrices and conditionally positive definite functions. Constructive Approximation **2**, 11–22.
17. D. S. Broomhead and D. Lowe. (1988) Multivariable functional interpolation and adaptive networks. Complex Systems **2**, 321–355.
18. L. X. Wang. (1991) Fuzzy systems are universal approximators. Proc. of the IEEE Int. Conf. on Fuzzy Systems, San Diego, 1163–1169.
19. B. Kosko. (1991) Fuzzy systems as universal approximators. Proc. of the IEEE Int. Conf. on Fuzzy Systems, San Diego, 1153–1162.
20. J. Castro. (1995) Fuzzy logic controllers are universal approximators. IEEE Trans. on SMC **25**, 629–635.

Flexible Structures of Neuro – Fuzzy Systems

Leszek Rutkowski, Krzysztof Cpałka

Technical University of Czestochowa, Al. Armii Krajowej 36, Czestochowa, Poland

Abstract: This paper presents an approach to design flexible neuro - fuzzy systems. The flexibility is introduced by softening the OWA aggregating operators. Simulation results illustrate the effect of softening parameters on truck baker - upper control problem.

1 Introduction

In the last decade several approaches have been developed to incorporate the concept of fuzzy logic into various schemas of neural networks. An interested reader is referred to the excellent monographs and textbooks [1 - 5, 7] and the recent survey [6]. Neuro – fuzzy systems combine learning abilities of neural networks and natural language description of fuzzy systems. The structure of such systems depends on chosen fuzzy implication (Mamdani method or logical method), rule aggregation methods and defuzzification techniques. In the paper we adopt the concept of Yager and Filev [8] to develop flexible structures of neuro – fuzzy systems. To our best knowledge such concept has not been yet presented in the context of the neuro – fuzzy systems. The flexible structure is characterized by "softing" the fuzzy operators:

1. the logical AND connective that is used to connect the antecedents in the individual rules,
2. fuzzy implication operators,
3. the aggregation procedures.

Most structures of neuro - fuzzy systems are based on an application of the T - norm operator in the form

$$T\{a,b\} = \min\{a,b\} \tag{1a}$$

or

$$T\{a,b\} = ab \tag{1b}$$

and the S - norm operator expressed by

$$S\{a,b\}=\max\{a,b\} \quad (2)$$

We generalize symbols (1a), (1b) and (2) for more than two arguments

$$T\{a_1,a_2,\ldots,a_n\}=\overset{n}{\underset{i=1}{T}}\{a_i\}=a_1\overset{T}{*}a_2\overset{T}{*}\ldots\overset{T}{*}a_n \quad (3)$$

Analogously

$$S\{a_1,a_2,\ldots,a_n\}=\overset{n}{\underset{i=1}{S}}\{a_i\}=a_1\overset{S}{*}a_2\overset{S}{*}\ldots\overset{S}{*}a_n \quad (4)$$

The soft versions of operators (3) and (4) are given by (see [8])

$$\tilde{T}\{a_1,a_2,\ldots,a_n\}=\frac{1}{n}(1-\alpha)\sum_{i=1}^{n}a_i+\alpha\overset{n}{\underset{i=1}{T}}\{a_i\} \quad (5)$$

and

$$\tilde{S}\{a_1,a_2,\ldots,a_n\}=\frac{1}{n}(1-\beta)\sum_{i=1}^{n}a_i+\beta\overset{n}{\underset{i=1}{S}}\{a_i\} \quad (6)$$

where $0\le\alpha\le1$ and $0\le\beta\le1$.

In the paper we derive flexible neuro - fuzzy structures and compare them with standard neuro – fuzzy systems. The performance of the neuro - fuzzy systems will be illustrated and compared on the inverted pendulum control problem.

2 Flexible Fuzzy Inference Systems

In this paper, we consider multi - input, single - output fuzzy systems mapping $\mathbf{X}\to\mathbf{Y}$, where $\mathbf{X}\subset\mathbf{R}^n$ and $\mathbf{Y}\subset\mathbf{R}$, see Fig. 1.

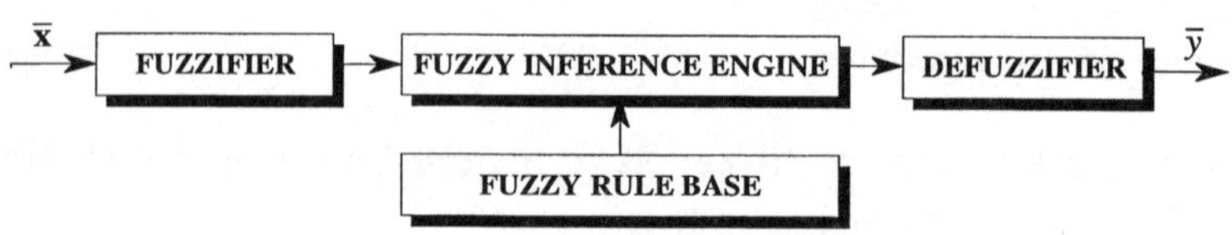

Fig. 1. Fuzzy - logic system.

The fuzzifier performs a mapping from the observed crisp input space $\mathbf{X}\subset\mathbf{R}^n$ to the fuzzy sets defined in $\mathbf{X}$. The most commonly used fuzzifier is the singleton fuzzifier which maps $\bar{\mathbf{x}}=[\bar{x}_1,...,\bar{x}_n]\in\mathbf{X}$ into a fuzzy set $A'\subset\mathbf{X}$ characterized by the membership function

$$\mu_{A'}(\mathbf{x})=\begin{cases}1 & \text{if} \quad \mathbf{x}=\bar{\mathbf{x}}\\ 0 & \text{if} \quad \mathbf{x}\neq\bar{\mathbf{x}}\end{cases} \tag{7}$$

The fuzzy rule base consists of a collection of N fuzzy IF - THEN rules in the form

$$R^{(k)}: \text{IF } x_1 \text{ is } A_1^k \text{ AND } x_2 \text{ is } A_2^k \text{ AND}\dots\text{AND } x_n \text{ is } A_n^k \text{ THEN } y \text{ is } B^k \tag{8}$$

or

$$R^{(k)}: \text{IF } \mathbf{x} \text{ is } A^k \text{ THEN } y \text{ is } B^k \tag{9}$$

where $\mathbf{x}=[x_1,...,x_n]\in\mathbf{X}$, $y\in\mathbf{Y}$ and $A^k=A_1^k\times A_2^k\times...\times A_n^k$, B^k, $k=1,...,N$ are fuzzy sets characterized by membership functions $\mu_{A^k}(\mathbf{x})$ and $\mu_{B^k}(y)$, respectively.

The fuzzy inference determines a mapping from the fuzzy sets in the input space $\mathbf{X}$ to the fuzzy sets in the output space $\mathbf{Y}$. Each on N rules (9) determines a fuzzy set $\bar{B}^k\subset\mathbf{Y}$ given by the compositional rule of inference

$$\bar{B}^k=A'\circ\left(A^k\to B^k\right) \tag{10}$$

Fuzzy sets $\bar{B}^k$, according to the formula (10) are characterized by membership functions expressed by the *sup - star* composition

$$\mu_{\bar{B}^k}(y)=\sup_{\mathbf{x}\in\mathbf{X}}\left\{\mu_{A'}(\mathbf{x})*\mu_{A_1^k\times..\times A_n^k\to B^k}(\mathbf{x},y)\right\} \tag{11}$$

where $*$ can be any operator in the class of T - norms. It is easily seen that for a crisp input $\bar{\mathbf{x}}\in\mathbf{X}$, i.e. a singleton fuzzifier (7), formula (11) becomes

$$\mu_{\bar{B}^k}(y)=\mu_{A_1^k\times..\times A_n^k\to B^k}(\bar{\mathbf{x}},y) \tag{12}$$

The aggregation operator, applied in order to obtain the fuzzy set B' based on fuzzy sets $\bar{B}^k$, is the T - norm or S - norm operator, depending on the type of fuzzy implication.

Consider the boolean (or Kleene - Dienes) fuzzy implication operator given by

$$\mu_{A^k\to B^k}(\mathbf{x},y)=\max\left\{1-\mu_{A^k}(\mathbf{x}),\mu_{B^k}(y)\right\} \tag{13}$$

In view of (6) the soft version of (13) takes the form

$$\mu_{A^k\to B^k}(\mathbf{x},y)=\frac{1}{2}(1-\beta)\left(1-\mu_{A^k}(\mathbf{x})+\mu_{B^k}(y)\right)+\beta\max\left\{1-\mu_{A^k}(\mathbf{x}),\mu_{B^k}(y)\right\} \tag{14}$$

where $0\leq\beta\leq1$. In this case the aggregation is performed by making use of operator (5), i.e.

$$\mu_{B'}(y)=\mathop{\tilde{T}}_{k=1}^{N}\left\{\mu_{\overline{B}^k}(y)\right\} \tag{15}$$

where $\mu_{\overline{B}^k}(y)$ is given by formula (12).

The defuzzifier performs a mapping from fuzzy sets $\overline{B}^k$ or fuzzy set B' to a crisp point $\overline{y}$ in $\mathbf{Y}\subset\mathbf{R}$. The COA (center of area) method is defined by following formula

$$\overline{y}=\frac{\int_{\mathbf{Y}} y\mu_{B'}(y)dy}{\int_{\mathbf{Y}}\mu_{B'}(y)dy} \tag{16}$$

3 Flexible Neuro - Fuzzy Systems

The structure of our system is determined by the following lemma.

Lemma. Consider the fuzzy logic system with discrete version of defuzzifier (16), fuzzy implication operator (14) and output membership functions satisfying

$$\mu_{B^r}\left(\overline{y}^r\right)=1 \tag{17}$$

If the rules are aggregated by

$$\mu_{B'}(y)=\frac{1}{N}(1-\alpha_A)\sum_{k=1}^{N}\mu_{\overline{B}^k}(y)+\alpha_A\min_{k=1,\dots,N}\left\{\mu_{\overline{B}^k}(y)\right\} \tag{18}$$

and the firing strength of rules is soften by

$$\mu_{A_1\times A_2\times..\times A_n}(x_1,x_2,\dots,x_n)=\frac{1}{n}(1-\alpha_F)\sum_{i=1}^{n}\mu_{A_i}(x_i)+\alpha_F\min_{i=1..n}\left\{\mu_{A_i}(x_i)\right\} \tag{19}$$

where $0\le\alpha_A\le1$ and $0\le\alpha_F\le1$, then

$$\overline{y}=\frac{\sum_{r=1}^{N}\overline{y}^r\cdot\mu_{B'}\left(\overline{y}^r\right)}{\sum_{r=1}^{N}\mu_{B'}\left(\overline{y}^r\right)} \tag{20}$$

where $\overline{y}^r$ denotes centers of the membership functions $\mu_{B^r}(y)$, i.e. for $r=1,\dots,N$,

$$\mu_{B^r}\left(\overline{y}^r\right)=\max_{y\in\mathbf{Y}}\left\{\mu_{B^r}(y)\right\} \tag{21}$$

and

$$
\begin{aligned}
\mu_{B'}\left(\bar{y}^r\right) = & \frac{1}{N}\left(1-\alpha_A\right)\left(\begin{array}{l}
\frac{1}{2}(1-\beta)\left(2-\frac{1}{n}\left(1-\alpha_F\right)\sum_{i=1}^{n}\mu_{A_i^r}\left(\bar{x}_i\right)-\alpha_F \min_{i=1,\ldots,n}\left\{\mu_{A_i^r}\left(\bar{x}_i\right)\right\}\right)+\beta+ \\
+\sum_{\substack{k=1\\k\neq r}}^{N}\left(\begin{array}{l}
\frac{1}{2}(1-\beta)\left(1-\frac{1}{n}\left(1-\alpha_F\right)\sum_{i=1}^{n}\mu_{A_i^k}\left(\bar{x}_i\right)-\alpha_F \min_{i=1,\ldots,n}\left\{\mu_{A_i^k}\left(\bar{x}_i\right)\right\}+\mu_{B^k}\left(\bar{y}^r\right)\right)+ \\
+\beta \max\left\{1-\frac{1}{n}\left(1-\alpha_F\right)\sum_{i=1}^{n}\mu_{A_i^k}\left(\bar{x}_i\right)-\alpha_F \min_{i=1,\ldots,n}\left\{\mu_{A_i^k}\left(\bar{x}_i\right)\right\},\mu_{B^k}\left(\bar{y}^r\right)\right\}
\end{array}\right)
\end{array}\right)+ \\
& +\alpha_A \min\left\{\begin{array}{l}
\frac{1}{2}(1-\beta)\left(2-\frac{1}{n}\left(1-\alpha_F\right)\sum_{i=1}^{n}\mu_{A_i^r}\left(\bar{x}_i\right)-\alpha_F \min_{i=1,\ldots,n}\left\{\mu_{A_i^r}\left(\bar{x}_i\right)\right\}\right)+\beta, \\
\min_{\substack{k=1,\ldots,N\\k\neq r}}\left\{\begin{array}{l}
\frac{1}{2}(1-\beta)\left(1-\frac{1}{n}\left(1-\alpha_F\right)\sum_{i=1}^{n}\mu_{A_i^k}\left(\bar{x}_i\right)-\alpha_F \min_{i=1,\ldots,n}\left\{\mu_{A_i^k}\left(\bar{x}_i\right)\right\}+\mu_{B^k}\left(\bar{y}^r\right)\right)+ \\
+\beta \max\left\{1-\frac{1}{n}\left(1-\alpha_F\right)\sum_{i=1}^{n}\mu_{A_i^k}\left(\bar{x}_i\right)-\alpha_F \min_{i=1,\ldots,n}\left\{\mu_{A_i^k}\left(\bar{x}_i\right)\right\},\mu_{B^k}\left(\bar{y}^r\right)\right\}
\end{array}\right\}
\end{array}\right\}
\end{aligned} \tag{22}
$$

Corollary. If $\alpha_F = \beta = \alpha_A = 1$, then

$$
\bar{y} = \frac{\sum_{r=1}^{N}\bar{y}^r \cdot \left(\min_{\substack{k=1,\ldots,N\\k\neq r}}\left\{\max\left\{1-\min_{i=1,\ldots,n}\left\{\mu_{A_i^k}\left(\bar{x}_i\right)\right\},\mu_{B^k}\left(\bar{y}^r\right)\right\}\right\}\right)}{\sum_{r=1}^{N}\left(\min_{\substack{k=1,\ldots,N\\k\neq r}}\left\{\max\left\{1-\min_{i=1,\ldots,n}\left\{\mu_{A_i^k}\left(\bar{x}_i\right)\right\},\mu_{B^k}\left(\bar{y}^r\right)\right\}\right\}\right)} \tag{23}
$$

4 Simulation results

The purpose of the simulation is to illustrate the effect of softening parameters α_F, β and α_A on the truck trajectories in the truck backer - upper control problem [7]. The truck trajectories, for parameters given in table 1, are shown in Fig. 2.

It is easily seen that parameters $\alpha_F = 0.85$, $\beta = 0.80$ and $\alpha_A = 0.70$ assure the best performance of the system with zero mistakes in 16 trials. Observe that trajectories in Fig. 2a correspond to a standard neuro - fuzzy system given by formula (23) and are characterized by the worse performance.

Table1. Flexible parameters in system (20)

Number	α_F	β	α_A	**Mistakes/Trials**
a)	1.00	1.00	1.00	5/16
b)	0.85	1.00	1.00	4/16
c)	1.00	0.80	1.00	4/16
d)	1.00	1.00	0.70	1/16
e)	**0.85**	**0.80**	**0.70**	**0/16**

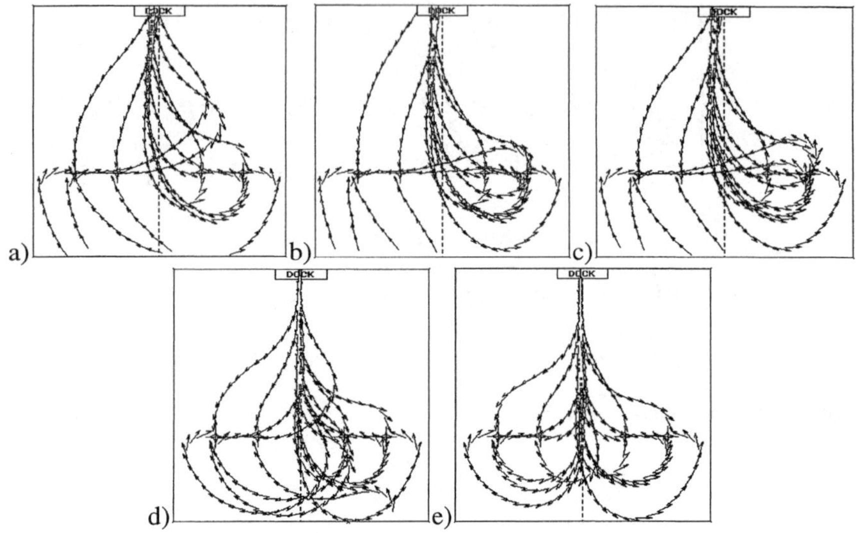

Fig. 2. Truck trajectories.

References

1. Jang J. S. R., Sun C. T., and Mizutani E. (1997) Neuro – Fuzzy and Soft Computing, Prentice Hall, Englewood Cliffs.

2. Kasabov N. (1996) Foundations of Neural Networks, Fuzzy Systems and Knowledge Engineering, The MIT Press, Cambridge.

3. Lin C. T., Lee C. S. G. (1996) Neural Fuzzy Systems – A Neuro – Fuzzy Synergism to intelligent Systems, Prentice Hall, New Jersey.

4. Nauck D., Klawonn F. and Kruse R. (1997) Foundations of Neuro – Fuzzy Systems, John Wiley & Sons, England.

5. Rutkowska D., Piliński M., Rutkowski L. (1997) Neural Networks, Genetic Algorithms and Fuzzy Systems, PWN, Warszawa (in Polish).

6. Sushmita M., Hayashi Y. (2000) Neuro – Fuzzy Rule Generation: Survey in Soft Computing Framework, IEEE Transactions on Neural Networks, to be published.

7. Wang L. X. (1994) Adaptive Fuzzy Systems and Control, Prentice Hall, Englewood Cliffs.

8. Yager R. R., Filev D. P. (1994) Essentials of Fuzzy Modeling and Control, John Wiley & Sons, New York.

A Review of Soft Computing and Gynaecological Cancer

A. A. Odusanya, M. O. Odetayo, R. N. G. Naguib, and D. Petrovic

Biomedical Computing Research Group (BIOCORE)
School of Mathematical and Information Sciences
Coventry University, Priory Street
Coventry, CV1 5FB, UK

Abstract: Soft computing has been utilized in a number of domains varying from stock market prediction, data mining, control, and medical tasks such as diagnosis, prognosis, and monitoring. This is a review of soft somputing in the cancer subdomain of gynaecological cancer. An overview of the major categories of soft computing techniques is presented, and instances of these in gynaecological cancer research are discussed. An insight into the plausible makeup of gynaecological data, for current and future systems is presented. The review concludes by highlighting ways of employing untapped areas of soft computing in gynaecological cancer.

1 Introduction

1.1 Gynaecological Cancer

Cancer refers to the abnormal growth of cells in any part of the body, and is malignant. Various types of cancers exist, including [14]: lung, skin, breast, colon, prostrate, bladder, stomach, rectum, ovary, cervix, and uterus.

Gynaecological cancer covers the following areas of cancer occurrences in the reproductive system of women, and includes the following regions: cervix, vulva, vagina, uterus (endometrium), ovary, and fallopian tube.

1.2 Soft computing (sub-symbolic artificial intelligence)

Three key areas exist in the area of soft computing: Artificial Neural Networks, Fuzzy Logic systems, and Evolutionary Algorithms, mainly Genetic Algorithms.

Artificial Neural Networks (ANNs) ANNs are based on the neural structure of the brain, consisting of an interconnected mesh of a combination of synapses, and soma, computationally regarded as weights and nodes. Usually what is learnt is not known, but learning can be guaranteed in most cases.

The most common type of ANN is the feedforward neural network, categorised into supervised and unsupervised ANNs. Supervised ANNs rely on a metric of the difference between the value of the desired target and that proferred by the system. Unsupervised ANNs search for clusters and relationships amongst the information presented. Supervised ANN algorithms are typically the backpropagation and the radial basis function. Examples of unsupervised ANN algorithms are: self-organising maps, adaptive resonance theory and hebbian networks [29], [9], and [15].

Fuzzy Logic (FL) systems FL systems are approximate systems utilised for modelling real life systems with focus on the imprecise nature (as opposed to precise in many computational mathematical models,) of real-life information. FL systems are premised on the notion of fuzzy sets [33].

A FL system has a series of rules comprising of an antecedent and a consequent, combined as if-then semantics.

An antecedent is a conjunction of input variables, each as an expressed and evaluatable degree of a fuzzy set (also known as a membership function). A consequent is a single output variable as an expressed and evaluatable degree of some fuzzy set.

A FL system comprises of four stages in its operation (with various methods establishing a specific system): fuzzification, inference ([5], [15], [29], and [32]), composition/aggregation ([5], often combined with the defuzzification process [32]), and defuzzification ([32], [15], [5], [29], and [23]).

Evolutionary Algorithms (EAs) EAs are nondeterministic algorithms that use key aspects of the evolution process as the basis for searching a solution space. Evolutionary algorithms have been described [2], and are classified into three main classes: evolutionary strategies, evolutionary programming, and genetic algorithms.

For each evolutionary algorithm a fitness function is associated that determines the degree of feasibility of each individual in a given population. Individuals are recombined and mutated, usually on a generational basis, until the best fit individual representing the solution to the problem is found.

2 Gynaecological cancer and soft computing research

The incidence of gynaecological cancer in the UK is not as high as other forms of cancer in the UK. The incidence of cancer of the ovary is given as 0.03%, while the incidence of cancer of the cervix as 0.02%, and cancer of the uterus as 0.02% [14], and the three (cancer of the ovary, cervix and uterus) form the most frequent occurrences of gynaecological cancer in the UK. And as such other cancer types have been the focus of a lot of research involving soft computing techniques. Major areas of focus include prostrate cancer and breast cancer [1], [11], [16], [24], [25], and [26].

Soft computing research into gynaecological cancer has only been embarked on recently, and would continue to attract attention, as more successes are reported.

It is worthy of note that there has been no reported instance of the successful use of EAs or FL systems in this area.

2.1 ANN methods

The main focus of soft computing researchers in gynaecological cancer research, has been ANNs: the main learning algorithm being the backpropagation algorithm for feedforward networks. For example, backpropagation ANNs have been successfully deployed in PAPNET, a cytological screening system for cancer of the cervix reported by [4], [7], [20], and [27]. Backpropagation ANNs have been successfully used for classification in automated screening for cancer of the cervix [21]. Backpropagation ANNs have also been used successfully for cancer of the ovary, as reported by Biagiotti et al [3].

Cancer of the cervix has traditionally been investigated by means of cervical smears, which are images, and these images are digitized and classified. Mehdi et al [21] reported that they prefered using two-thirds the number of input nodes as the hidden nodes, and this worked well for classification of cervical smears. The one reported instance of research into cancer of the ovary has used a set of indicators or markers, and has reported better performance of the ANN used in comparison with multiple logistic regression analysis, three layers were used: the input layer, one hidden layer, and an output layer. The hidden layer had three neurons. The backprogation learning algorithm seems to be the learning algorithm of choice.

2.2 Gynaecological cancer data makeup

The information to be analysed in gynaecological cancer could either be images (smears) on one hand or indicators, variables, parameters or markers on the other. Image analysis on smears requires identifying abnormal cells. A number of variables are useful for analysis of gynaecological oncology. Medical literature do report favourite indicators useful for analysis, but there tend to be conflicts amongst the preferences.

Davidson et al [6] favour the combined use of Ber-EP4, B72.3, and BG8, while CA125 and CEA are not favoured.

Kagie et al [12] discount p53 in squamous cell carcinoma of the vulva as a prognostic parameter.

The European Group on Tumour Markers [8], has pointed out key markers in order of importance for three cancer categories

- cancer of the ovary - CA125 (but, with limited sensitivity "for early stage disease"), CA15.3, OVX1, CA72.4, TPA, and TPS.
- cancer of the cervix - Squamous Cell Carcinoma Antigen (SCCA) (also, with limited sensitivity for Stage I disease, but over 90% sensitivity for Stage IV disease), CEA, hCG beta core fragment (hCGßcf), TPA, and CYFRA 21-1.

- cancer of the endometrium (uterus) - CA125 (the same sensitivity as in cancer of the ovary, and of the cervix).

The choice of the above markers is supported by others: Meier et al [22], and Lundqvist et al [18] (adding the significance of TPA and CEA in ovarian carcinoma, and SCC in endometrial carcinoma, and the following statement is noted: "the incidence of death was highest among those who had TPA elevated (91%) followed by neopterin (86%) and CRP (76%)."), and Sarandakou et al [28], (also adding the significance of TPS in ovarian carcinoma and cervical adenocarcinoma, and CA125 in cervical adenocarcinoma).

Biagiotti et al [3] have a set of variables (some of which are called Doppler parameters) out of which, some are selected via multiple logistic regression (MLR): Age, Mean diameter, Multilocularity, Papillary projections, Random echogenicity, Flow detected, Peak systolic velocity, and Resistance index.

3 Conclusion

While soft computing has been widely employed in prostrate cancer, and breast cancer, there has been reduced use in areas of gynaecological cancer, most especially the soft computing fields of EAs and FL systems.

EAs are good optimizers, and robust search engines. FL systems lend themselves well to prognosis and diagnosis: Wang and Mendel [30] have proved that a fuzzy logic system can act as a universal approximator, i.e., it can map from input space to output space to any level of accuracy. We are of the view that a combination of an EA and an FL would work well with the EA optimizing the FL to perform the required decision task. It would be necessary to develop rules for the FL. Rules in FL systems are traditionally linguistically sourced. Other methods include via EAs [17], ANNs [10], and alternative deterministic routes such as that reported in [30], and [31]. Kaymak et al [13] mentions that fuzzy rules/models can be generated via supervised techniques such as gradient learning, or unsupervised techniques such as fuzzy clustering. A neurofuzzy system could be utilized in place of an FL system. Gradient learning, is adopted by many neurofuzzy implementations and is akin to the traditional backpropagation learning algorithm [19]. An evolutionary algorithm can be designed to generate rules for prognostic/diagnostic purposes.

The authors would experiment in future, with EAs and FL systems in gynaecological cancer.

References

1. Andrés, C., Peña, -R., and Sipper, M. (1999) A fuzzy-genetic approach to breast cancer diagnosis. Artificial Intelligence in Medicine **17**(2), 131–155

2. Bäck, T. (1996) Evolutionary algorithms in theory and practice: evolutionary strategies evolutionary programming genetic algorithms. Oxford University Press
3. Biagiotti, R., Desli, C., Vanzi, E., and Gacci, G. (1999) Predicting ovarian malignancy: application of artificial neural networks to transvaginal and color doppler flow US. Radiology **210**, 399–403
4. Boon, M. E., Kok, L. P., Nygaard-Nielsen, M., Holm, K., and Holund, B. (1994) Neural network processing of cervical smears can lead to a decrease in diagnostic variability and an increase in screening efficacy: a study of 63 false-negative smears. Modern Pathology **7**(9), 957–961
5. Cox, E. The fuzzy systems handbook. (1994) A practitioner's guide to building, using, and maintaining fuzzy systems. Academic Press, Inc.
6. Davidson, B., Risberg, B., Kristensen, G., Kvalheim, G., Emilsen, E., Bjåmer, A., and Berner, A. (1999) Detection of cancer cells in effusions from patients diagnosed with gynaecological malignancies. Evaluation of five epithelial markers. Vichows Arch **435**, 43–49
7. Doornewaard, H., Van de Seijp, H., Woudt, J. M. C., Van der Graaf, Y., and Van den Tweel, J. G. (1997) Negative cervical smears before CIN 3/carcinoma. Reevaluation with the PAPNET testing system. Acta Cytologica **41**, 74–78
8. European Group on Tumour Markers. (1999) Tumour markers in gynaecological cancers - EGTM recommendations. Anticancer Research **19**, 2685–2820
9. Freeman, J. A., and Skapura, D. M. (1991) Neural networks algorithms, applications, and programming techniques, 1st ed. Addison-Wesley Publishing Company, Inc.
10. Ichimura, T., Tazaki, E., and Yoshida, K. (1995) Extraction of fuzzy rules using neural networks with structure level adaptation – verification to the diagnosis of hepatobiliary disorders. Int. J. Biomed. Comput. **40**(2), 139–46
11. Jain, R., Mazumdar, J., and Moran, W. (1998) Application of fuzzy–classifier system to coronary artery disease and breast cancer. Australasian Physical & Engineering Sciences in Medicine **21**(3), 141–7
12. Kagie, M. J., Kenter, G.G., Tollenaar, R. A. E. M., Hermans, J., Trimbos, J. B., and Fleuren, G. J. (1997) p53 overexpression, a frequent observation in squamous cell carcinoma of the vulva and in various synchronous vulvar epithelia, has no value as a prognostic parameter. International Journal of Gynecological Pathology **16**, 124–130
13. Kaymak, U., Babuska, R., Setnes, M., Verbruggen, H. B., and van Nauta Lemke, H. R. (1997) Methods of simplification of fuzzy models. IN: Ruan, D. ed. Intelligent Hybrid Systems, Fuzzy Logic, Neural Networks, Genetic Algorithms, Kluwer Academic Publishers. pp. 91–108
14. Kitchener, H. C. (1999) Gynaecological cancer. Postgrad. Med. J. **75**, 332–338
15. Kosko, B. (1992) Neural networks and fuzzy systems. A dynamical systems approach to machine intelligence. 1st ed. Prentice-Hall International Editions
16. Kovalerchuk, B., Triantaphyllou, E., Ruiz, J. F., and Clayton, J. (1997) Fuzzy logic in computer-aided breast cancer diagnosis: analysis of lobulation. Artificial Intelligence in Medicine **11**, 75–85
17. Lee, M, -R. (1998) Generating fuzzy rules by genetic method and its application. International Journal on Artificial Intelligence Tools **7**(4), 399–413
18. Lundqvist, E. Å., Nordström, L., Sjövall, K., and Eneroth, P. (1989) Evaluation of seven different markers for the establishment of tumour marker panels in gynecologic malignancies. Eur. J. Gyneac. Oncol. **X**(6), 395–405

19. Maguire, L. P., McGinnity, T. M., and McDavid, L. J. A fuzzy neural network for approximate fuzzy reasoning. IN: Ruan, D. ed. Intelligent Hybrid Systems, Fuzzy Logic, Neural Networks, Genetic Algorithms, Kluwer Academic Publishers. pp. 35–58
20. Mango, L. J. (1994) Computer-assisted cervical cancer screening using neural networks. Cancer Letters, **77**, 155–162
21. Mehdi, B., Stacey, D., and Harauz, G. (1994) A hierarchical neural network assembly for classification of cervical cells in automated screening. Analytical Cellular Pathology, **7**, 171–180
22. Meier, W., Eiermann, P., Stieber, P., Schneider, A., Fateh-Moghadam, A., and Hepp, H. (1989) Experiences with SCC antigen, a new tumour marker for cervical carcinoma. Eur. J. Cancer Clin. Oncol. **25**(11), 1555–1559
23. Mizumoto, M., and Shi, Y. (1997) A new approach of neurofuzzy learning algorithm. IN: Ruan, D. ed. Intelligent Hybrid Systems, Fuzzy Logic, Neural Networks, Genetic Algorithms, Kluwer Academic Publishers. pp. 109–129
24. Naguib, R. N. G., Adams, A. E., Horne, C. H. W., Angus, B., Smith, A. F., Sherbet, G. V., and Lennard, T. W. J. (1997) Prediction of nodal metastasis and prognosis in breast cancer: a neural model. Anticancer Research, **17**, 2735–2742
25. Naguib, R. N. G., Hamdy, F. C. (1998) A general regression neural network analysis of prognostic markers in prostrate cancer. Neurocomputing, **19**, 145–150
26. Naguib, R. N. G., Robinson, M. C., Neal, D. E., and Hamdy, F. C. (1998) Neural network analysis of combined conventional and experimental prognostic markers in prostrate cancer: a pilot study. British Journal of Cancer, **78**(2), 246–250
27. Ouwerkerk-Noordam, E., Boon, M. E., and Beck, S. (1994) Computer-assisted primary screening of cervical smears using the PAPNET method: comparison with conventional screening and evaluation of the role of the cytologist. Cytopathology, **5**, 211–218
28. Sarandakou, A., Phocas, I., Botsis, D., Sikiotis, K. Rizos, D., Kalambokis, D., Trakakis, E., and Chryssikopoulos, A. (1998) Tumour-associated antigens CEA, CA125, SCC and TPS in gynaecological cancer Eur. J. Gynaecol. Oncol. **XIX**(1), 73–7
29. Takagi, H. Introduction to fuzzy systems, neural networks, and genetic algorithms. IN: Ruan, D. ed. Intelligent hybrid systems. Fuzzy logic, neural networks, and genetic algorithms. 1st ed. Kluwer Academic Publishers. pp. 3–33
30. Wang, L. -X., and Mendel, J. M. (1991) Generating fuzzy rules from numerical data, with applications. University of Southern California, Signal and Image Processing Institute, USC-SIPI REPORT #169
31. Wang, L. -X., and Mendel, J. M. (1992) Generating fuzzy rules by learning from examples. IEEE Transactions on Systems, Man, and Cybernetics, **22**(6), 1414–1427
32. Yuan, Y., and Suarga, S. (1995) On the integration of neural networks and fuzzy logic systems. IN: IEEE International Conference on Systems, Man and Cybernetics. pp. 452–7
33. Zadeh, L. A. (1965) Fuzzy sets. Information and Control, **8**, 338–353

Rule Extraction from an Adaptive Neural Network for Fault Diagnosis

Shing Chiang Tan, Chee Peng Lim*

School of Industrial Technology
Universiti Sains Malaysia
11800 Penang, Malaysia

Abstract: Advances in neural computing have opened up the way for artificial neural networks to be employed as a practical tool for solving fault detection and diagnosis problems in complex processes. In this paper, applicability of the Fuzzy ARTMAP neural network as an intelligent learning system to fault detection and diagnosis in a power generation plant is examined. The process under scrutiny is the Circulating Water (CW) system in the power plant station, with specific attention to the detection of blockage in the system. Serious blockage in the CW system will result in a critical failure that will affect the efficiency of heat transfer in the condensers, and consequently will affect the amount of power generated. Several experiments have been conducted to investigate the setting of a group of parameters towards classification results and rule extraction from the Fuzzy ARTMAP network in this case study.

Keywords: Fuzzy ARTMAP, Circulating Water system, fault detection, diagnosis problems, classification

1 Introduction

Artificial neural networks have gained their fame as intelligent systems owing to their intrinsic parallelism, adaptability, and ability to handle noisy data. They can learn complex associations and relationship directly from data. They can also handle fusion of multiple sources of data and information. Following these attractive features, neural-network-based systems have been employed as intelligent fault diagnosis tools to identify and distinguish between faulty and normal operating conditions in complex processes and systems [1]. Nonetheless, most of the neural-network-based fault diagnosis systems suffer from the so-called 'black-box' phenomenon, i.e. it is difficult to extract domain knowledge encoded in the trained network to explain its predictions [2]. Users, who wish to be convinced with the reasoning of the predicted outcome, are often reluctant to use the system if the network is unable to provide explanatory facilities to justify how a prediction is reached. Hence, we investigate the applicability of the Fuzzy ARTMAP [3] (FAM) neural network, which has been enhanced with rule extraction capabilities [4], to fault diagnosis problems in a power generation plant.

---

* Corresponding author: Dr. CP Lim, School of Industrial Technology, Universiti Sains Malaysia, 11800, Penang, Malaysia.
Tel: +604-6577888 Ext 2200, Fax: +604-6573678, Email: cplim@usm.my

The rules extraction technique enables derivation of informative rules that link antecedents to their consequences, and that provide explanations to the network predictions.

In this paper, a study on the use of FAM for fault diagnosis using data collected from a power generation plant is described. The system being monitored in the power station is the Circulating Water (CW) system [5]. The function of the CW system is to supply a sufficient and continuous amount of cooling water to the main turbine condenser to condense steam from the turbine exhaust and other steam flows into the condensers. In this study we monitor the operating conditions of the condensers as the cooling water flowing through the system. The source of the cooling water is supplied from the sea. There may be mud, and fine solid materials such as seaweed, shells, and sand, which have passed through the filters in the system, and consequently, they may enter and block the condenser tubes. Hence, the prime attention of this study is focused on blockage detection, i.e., either significant blockage or insignificant/no blockage, in the condenser tubes.

2 Fuzzy ARTMAP

FAM is an incremental learning network that encompasses a pair of Fuzzy ART [6] modules, ART_a and ART_b, in which they are interconnected via a map field, as shown in Fig. 1. Each ART module comprises two layers of nodes: F_1 is the input layer; F_2 is a dynamic layer that encodes prototypes of input patterns, and that allows the creation of new nodes when necessary.

Fig. 1 The architecture of FAM.

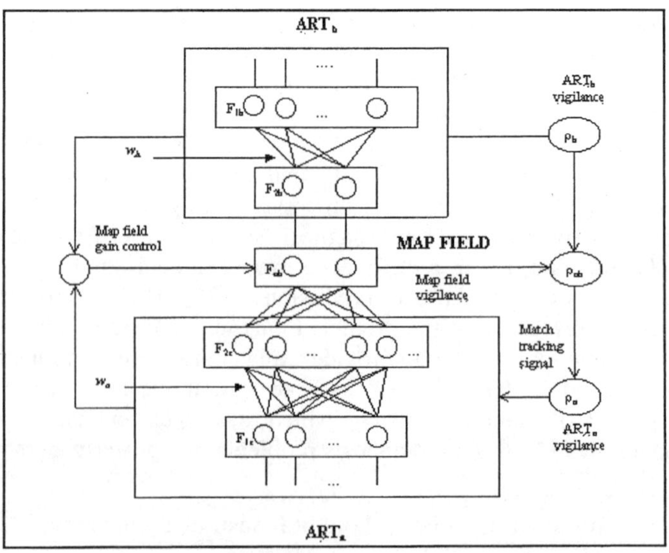

On presentation of a new input-target pattern pair, ART_a and ART_b receive the input and target patterns, respectively. A pattern matching cycle, which incorporates a two-stage hypothesis selection and test process [7], occurs in each ART network. For the input pattern in ART_a, a feedforward pass is initiated to select a prototype that is most similar to the input pattern as the winning node in F_{2a}. The degree of similarity is measured by a choice function [3], as follows.

$$\frac{|A \wedge w_{a-j}|}{\alpha_a + |w_{a-j}|} \tag{1}$$

where A is the input pattern in complement-coded format [3]; α_a is the choice parameter of ART_a; w_{a-j} is the weight vector of the j-th F_{2a} node.

A feedback pass then ensues. The winning prototype (represented by J) has to go through a similarity check, called the vigilance test [3], against a threshold, as follows.

$$\frac{|A \wedge w_{a-J}|}{A} \geq \rho_a \tag{2}$$

where w_{a-J} is the winning J-th node in F_{2a}; ρ_a is ART_a vigilance parameter. If the vigilance test is not satisfied, a new cycle of feedforward selection and feedback test is carried out until the selected winning prototype is able to pass the vigilance test. If no such node exists, a new node will be created in F_{2a} to code the input pattern. The same pattern matching cycle occurs simultaneously in ART_b for the target pattern.

After resonance has occurred in ART_a and ART_b, the winning node in F_{2a} sends a prediction to ART_b via the map field. A map field vigilance test is used to confirm the prediction. If the map field vigilance test fails, it indicates that the winning prototype of ART_a has made a wrong prediction of the target class in ART_b. Consequently, match-tracking [3] is initiated to trigger a search in ART_a. Otherwise, learning takes places by updating the weights of the ART_a winning prototype, according to the following equation [3]:

$$w_{a-J}^{(new)} = \beta_a \left(A \wedge w_{a-J}^{(old)}\right) + \left(1 - \beta_a\right) w_{a-J}^{(old)} \tag{3}$$

where β_a is the ART_a learning rate. Note that the above is a concise account on FAM. Interested readers are referred to [3] for further detail of FAM.

In the context of rule extraction, an additional algorithm for rule pruning and weight quantization is needed. The purpose of rule pruning is to remove F_{2a} category prototypes that have low confidence factors [4]. An F_{2a} category prototype will be eliminated when its confidence factor is lower than a user-defined pruning threshold. The confidence factor of each F_{2a} category prototype is expressed in terms of its usage frequency in a training set and its predictive accuracy in a predicting set, as shown below [4]:

$$CF_j = \gamma U_j + (1-\gamma)A_j \qquad (4)$$

where U_j is the usage of node *j*; A_j is the accuracy of node *j*; $\gamma \in [0,1]$ is a weighting factor.

Weight quantization is then used to convert real-valued weights of F_{2a} category nodes into feature values. With the inclusion of weight quantization techniques to the FAM network, it is said to be able to provide semantic rules and their interpretation for users. According to [4], there are two common techniques of quantization, i.e., quantization by truncation and quantization by round-off. In this paper, we have employed the former technique of weight quantization, as denoted below [4]:

$$V_q = (q-1)/Q \qquad (5)$$

where $q = 1...Q$; Q denotes quantization steps. A detailed explanation on rule extraction from FAM can be found in [4].

3 Experiments

This section highlights the application of FAM to fault diagnosis and condition monitoring of the CW system in a power station. The program of FAM with rule extraction capabilities is obtainable from [8]. The data samples were collected on 80MWatts of the target power generation. There were a total of 2439 samples in the database in which 1313 cases indicated significant cooling water blockage in the CW system. Each data sample comprised 12 measurements of temperature and pressure at various inlet and outlet points of the condenser, as shown in Table 1. The data samples were equally divided into training, prediction, and test sets. Several experiments were conducted to evaluate the performance of FAM in predicting the occurrence of blockage in the CW system.

Table 1. Abbreviations of the parameters relating to the CW system.

LPT A	Low Pressure Cylinder Exhaust Temperature A
LPT B	Low Pressure Cylinder Exhaust Temperature B
GEN	Generator
CWIT A	Condenser Circulating Water Inlet Temperature A
CWIT B	Condenser Circulating Water Inlet Temperature B
CWOT A	Condenser Circulating Water Outlet Temperature A
CWOT B	Condenser Circulating Water Outlet Temperature B
CWIP A	Condenser Circulating Water Inlet Pressure A
CWOP A	Condenser Circulating Water Outlet Pressure A
CWIP B	Condenser Circulating Water Inlet Pressure B
CWOP B	Condenser Circulating Water Outlet Pressure B
VAC	Condenser vacuum

The experiments were repeated five times in fast learning mode using multiple-epoch training [3], by varying ART_a baseline vigilance, $\bar{\rho}_a$ from 0.0 to

0.6. The pruning threshold τ, and quantization level Q were, respectively, set to 0.0 and 5. The results in terms of accuracy rates for the three data sets were averaged across five independent runs, and are shown in Table 2.

From Table 2, we can see that $\overline{\rho}_a$ between 0.4 and 0.5 gave the best results. Further tests were conducted to fine-tune $\overline{\rho}_a$ using a smaller division between 0.4 and 0.5. Table 3 depicts the results. The best accuracy rates were produced by $\overline{\rho}_a = 0.49$. This $\overline{\rho}_a$ value was used in subsequent experiments.

Next, the pruning threshold τ was varied from 0.1 to 0.9, to prune the trained FAM network in order to facilitate rule extraction. The best results of the pruned network fell within the range of 0.6 to 0.8. The network was further fine-tuned so as to derive rules with higher accuracy rates. Tables 4 and 5 summarize all the results. By setting $\tau = 0.65$, accuracy rates of 93.41/92.92/ 93.41 percents were obtained for the training/prediction/test sets, respectively.

Table 2. The accuracy rates of blockage detection for the training/ prediction/test sets, with varying ART_a baseline vigilance values.

$\overline{\rho}_a$	Average of Accuracy (%)		
	Train	Predict	Test
0	88.04	86.35	87.60
0.1	88.04	86.35	87.60
0.2	88.04	86.35	87.60
0.3	88.04	86.35	87.60
0.4	88.34	87.08	88.09
0.5	88.31	86.27	88.09
0.6	88.09	86.40	87.68

Table 3. The accuracy rates of blockage detection for the training/ prediction/test sets, with smaller range of ART_a baseline vigilance.

$\overline{\rho}_a$	Average of Accuracy (%)		
	Train	Predict	Test
0.41	88.09	86.40	87.68
0.43	88.09	86.40	87.68
0.45	88.09	86.40	87.68
0.47	90.55	88.93	90.11
0.49	91.02	89.72	90.73
0.51	90.36	88.93	89.74
0.53	90.04	88.61	89.42

Table 4. The accuracy rates of blockage detection for tbe training /prediction/test sets with varying pruning threshold values.

τ	Average of Accuracy(%)		
	Train	Predict	Test
0.1	91.02	89.72	90.73
0.2	91.02	89.72	90.73
0.3	91.02	89.72	90.73
0.4	90.63	89.59	90.43
0.5	90.60	89.62	90.43
0.6	90.60	89.62	90.43
0.7	93.38	92.89	93.38
0.8	81.55	80.32	80.15
0.9	72.82	70.92	70.68

Table 5. The accuracy rates of blockage detection for the training/ prediction/test sets with smaller range of the pruning threshold.

τ	Average of Accuracy(%)		
	Train	Predict	Test
0.61	90.50	89.45	90.26
0.63	90.26	89.25	89.25
0.65	93.41	92.92	93.41
0.67	93.38	92.89	93.38
0.69	93.38	92.89	93.38
0.71	93.38	92.89	93.38
0.73	93.38	92.89	93.38
0.75	87.68	86.59	87.60
0.77	84.60	83.42	83.86
0.79	81.55	80.32	80.15

Table 6. Rules extraction from the FAM network for blockage detection.

Rule 1		Rule 2	
IF		IF	
	LPT A = very low to high		LPT A = very low to medium
And	LPT B = very low to high	And	LPT B = very low to medium
And	GEN = low to very high	And	GEN = very low to high
And	CWIT A= medium to very high	And	CWIT A= no conclusive finding
And	CWIT B= very low to high	And	CWIT B= low to high
And	CWOT A= very low to medium	And	CWOT A= very low to high
And	CWOT B= very low to medium	And	CWOT B= low to high
And	CWIP A= very low to medium	And	CWIP A= very low to high
And	CWOP A= very low to high	And	CWOP A= low to high
And	CWIP B= very low to medium	And	CWIP B= low to medium
And	CWOP B= very low	And	CWOP B= medium to very high
And	VAC = very low to high	And	VAC = very low to medium
THEN		THEN	
Significant blockage		No significant blockage	

A quantization step of 5 was employed to describe the antecedents in terms of *very high, high, medium, low,* and *very low*, that were subsequently linked to a consequence, in *if-then* form. From the experiment with $\overline{\rho}_a = 0.49$, $Q = 5$, and $\tau = 0.65$, a total of 6 rules was extracted from FAM based on the highest accuracy rate from 5 runs. Table 6 shows two of the derived rules, with confidence factors more than 0.9. The key feature of blockage detection in the CW system is CWOP B, and the detection is supported by other parameters. If the value of CWOP B is *very low*, it implies insufficient flow of the fresh circulating water in the condensers, as a result of significant blockage in the CW system. Otherwise, the system is in healthy operating condition. We can also find that if inlet and outlet pressures for both condenser A and B are *very low*, then the system is most probably in significant blockage condition. Besides, LPT A and LPT B may be raised to *high*, and CWOT A and CWOT may be *very low*, under this inefficient condition. All the extracted rules are conformed with the experts' opinions in maintaining the CW system.

4 Summary

A simple and clear methodology has been employed to investigate the applicability of the FAM network to fault diagnosis in the CW System in a power generation station. From the experiments, the FAM network has shown its potentials in predicting blockage conditions in the condensers, and in extracting rules with satisfactory results. Nevertheless, we discover that there are overlaps among the antecedents with different consequences in rules. This might be owing to the deficiency of the rule extraction algorithm to find and to identify conflict-free rules. Our future work will concentrate on the incorporation of a shrinking/growing algorithm into the FAM network to improve its rule extraction capabilities.

Acknowledgments

The authors are grateful to Prai Power Station, TNB Generation Sdn. Bhd for the provision of the database used in this research work. The authors also acknowledge the research grants provided by Universiti Sains Malaysia as well as the Ministry of Science, Technology, and the Environment Malaysia (No. 06-02-05-8002) that have in part resulted in this article.

References

[1] Polycarpou M. M., and Helmicki A. J.(1995): Automated Fault Detection and Accomodation: A Learning System Approach. IEEE Trans on System, Man, and Cybernetics, **25**, No. 11, 1447-.1458.

[2] Downs, J., Harrison, R. F., and Cross, S. S., (1995): Evaluating a neural network decision-support tool for the diagnosis of breast cancer. Proceedings of the European Society for Artificial Intelligence in Medicine Fifth European Conference on Artificial Intelligence in Medicine, AIME-95, Pavia, Italy. 239-250.

[3] Carpenter, G.A., Grossberg, S., et al (1992).: Fuzzy ARTMAP: A Neural Network Architecture for Incremental Supervised Learning of Analogue Multidimensional Maps. IEEE Trans on Neural Networks, **3**, 698-712.

[4] Carpenter, G.A., and Tan, A.H. (1995): Rule Extraction: From Neural Architecture to Symbolic Representation. Connection Science, **7**, 3-27.

[5] System Description and Operating Procedures (1999), Prai Power Station Stage 3, Vol. 14.

[6] Carpenter G.A., Grossberg S., Rosen D.B. (1991): Fuzzy ART: Fast Stable Learning and Categorization of Analogue Patterns by An Adaptive Resonance System. Neural Networks, **4**, 759-771.

[7] Carpenter G.A., Grossberg S. (1987): A Massively Parallel Architecture for A Self-organizing Neural Pattern Recognition Machine. Computer Vision, Graphics and Image Processing, **37**, 54-115.

[8] http://www.cs.cmu.edu/Group/AI/areas/fuzzy/systems/artmap/0.html

Druck: Strauss Offsetdruck, Mörlenbach
Verarbeitung: Schäffer, Grünstadt

Studies in Fuzziness and Soft Computing

Vol. 25. J. Buckley and Th. Feuring
Fuzzy and Neural: Interactions and Applications, 1999
ISBN 3-7908-1170-X

Vol. 26. A. Yazici and R. George
Fuzzy Database Modeling, 1999
ISBN 3-7908-1171-8

Vol. 27. M. Zaus
Crisp and Soft Computing with Hypercubical Calculus, 1999
ISBN 3-7908-1172-6

Vol. 28. R. A. Ribeiro, H.-J. Zimmermann, R. R. Yager and J. Kacprzyk (Eds.)
Soft Computing in Financial Engineering, 1999
ISBN 3-7908-1173-4

Vol. 29. H. Tanaka and P. Guo
Possibilistic Data Analysis for Operations Research, 1999
ISBN 3-7908-1183-1

Vol. 30. N. Kasabov and R. Kozma (Eds.)
Neuro-Fuzzy Techniques for Intelligent Information Systems, 1999
ISBN 3-7908-1187-4

Vol. 31. B. Kostek
Soft Computing in Acoustics, 1999
ISBN 3-7908-1190-4

Vol. 32. K. Hirota and T. Fukuda
Soft Computing in Mechatronics, 1999
ISBN 3-7908-1212-9

Vol. 33. L. A. Zadeh and J. Kacprzyk (Eds.)
Computing with Words in Information/Intelligent Systems 1, 1999
ISBN 3-7908-1217-X

Vol. 34. L. A. Zadeh and J. Kacprzyk (Eds.)
Computing with Words in Information/Intelligent Systems 2, 1999
ISBN 3-7908-1218-8

Vol. 35. K. T. Atanassov
Intuitionistic Fuzzy Sets, 1999
ISBN 3-7908-1228-5

Vol. 36. L. C. Jain (Ed.)
Innovative Teaching and Learning, 2000
ISBN 3-7908-1246-3

Vol. 37. R. Słowiński and M. Hapke (Eds.)
Scheduling Under Fuzziness, 2000
ISBN 3-7908-1249-8

Vol. 38. D. Ruan (Ed.)
Fuzzy Systems and Soft Computing in Nuclear Engineering, 2000
ISBN 3-7908-1251-X

Vol. 39. O. Pons, M. A. Vila and J. Kacprzyk (Eds.)
Knowledge Management in Fuzzy Databases, 2000
ISBN 3-7908-1255-2

Vol. 40. M. Grabisch, T. Murofushi and M. Sugeno (Eds.)
Fuzzy Measures and Integrals, 2000
ISBN 3-7908-1255-2

Vol. 41. P. Szczepaniak, P. Lisboa and J. Kacprzyk (Eds.)
Fuzzy Systems in Medicine, 2000
ISBN 3-7908-1263-4

Vol. 42. S. Pal, G. Ashish and M. Kundu (Eds.)
Soft Computing for Image Processing, 2000
ISBN 3-7908-1217-X

Vol. 43. L. C. Jain, B. Lazzerini and U. Halici (Eds.)
Innovations in ART Neural Networks, 2000
ISBN 3-7908-1270-6

Vol. 44. J. Aracil and F. Gordillo (Eds.)
Stability Issues in Fuzzy Control, 2000
ISBN 3-7908-1277-3

Vol. 45. N. Kasabov (Ed.)
Future Directions for Intelligent Information Systems on Information Sciences, 2000
ISBN 3-7908-1276-5

Vol. 46. J. N. Mordeson and P. S. Nair
Fuzzy Graphs and Fuzzy Hypergraphs, 2000
ISBN 3-7908-1286-2

Vol. 47. Ernest Czogała† and Jacek Łęski
Fuzzy and Neuro-Fuzzy Intelligent Systems, 2000
ISBN 3-7908-1289-7

Vol. 48. Masatoshi Sakawa
Large Scale Interactive Fuzzy Multiobjective Programming, 2000
ISBN 3-7908-1293-5

Vol. 49. Ludmila I. Kuncheva
Fuzzy Classifier Design, 2000
ISBN 3-7908-1298-6

Vol. 50. Fabio Crestani and Gabriella Pasi
Soft Computing in Information Retrieval, 2000
ISBN 3-7908-1299-4

Vol. 51. János Fodor, Bernard De Baets and Patrice Perny (Eds.)
Preferences and Decisions under Incomplete Knowledge, 2000
ISBN 3-7908-1303-6

Vol. 52. Etienne E. Kerre and Mike Nachtegael (Eds.)
Fuzzy Techniques in Image Processing, 2000
ISBN 3-7908-1304-4

CHAPTER X. RIEMANN SURFACE

CHAPTER XI. FUCHSIAN GROUP

CHAPTER VIII. CLUSTER SET OF A MEROMORPHIC FUNCTION

CHAPTER IX. CONFORMAL MAPPING

PART I. Simply connected domains

PART II. Multiply connected domains

CHAPTER VI. AHLFORS' THEORY OF COVERING SURFACES

CHAPTER VII. BOREL'S DIRECTION

PART I. Meromorphic functions for $|z|<\infty$

PART II. Meromorphic functions in $|z|<1$

Blessings From Grant

By

Carole J. Gariepy

With cartoons and Stories By

Gerry Gariepy

Branden Books, Boston

Library of Congress Cataloging-in-Publication Data

Names: Gariepy, Carole J., author. | Gariepy, Gerry, author.
Title: Blessings from Grant / by Carole J. Gariepy ; with cartoons and stories by Gerry Gariepy.
Description: Boston : Branden Books, 2017.
Identifiers: LCCN 2017005997 (print) | LCCN 2017010682 (ebook) | ISBN 9780828326575 (E-Book) | ISBN 9780828326568 (pbk. : alk. paper) | ISBN 9780828326575 (ebook)
Subjects: LCSH: Gariepy, Carole J.,--Health. | Brain-damaged children--Biography. | Brain damage--Patients--Biography. | Brain-damaged children--Family relationships. | Parent and child.
Classification: LCC RJ496.B7 (ebook) | LCC RJ496.B7 G373 2017 (print) | DDC 617.4/81044083--dc23
LC record available at https://lccn.loc.gov/2017005997

ISBN 9780828326568 Paperback
ISBN 9780828326575 E-Book

Branden Books PO
Box 812094
Wellesley NA 02482

Orders:
www.brandenbooks.com
branden@brandenbooks.com
Amazon.com

Contents

Acknowledgments

Linda Langevin and Joan Johnson are two dear friends who were a tremendous help with the book. Just like many who willingly participated in Grant's life to bring his development as far as possible, these friends were at my side to put the finishing touches on this book.

Linda edited the cartoon images from the books Gerry made for Grant that showed years of loving use. When she finished, the pages looked like new. The work took many hours and made it possible to use them in the book. She also enhanced photos and helped with the layout design. When you're good with a computer, you can really make a difference! I told her that watching her work was like watching a magician.

Joan designed the cover, the third book cover she designed for me. She's an excellent artist. I love the way she put Grant "on top of the world" with a long strand of beads. She depicted how he feels when he's making a necklace for someone -- "on top of the world."

It's wonderful to have good friends at your side who are competent, interested, and a pleasure to work with. Thank you both. I couldn't have done it without you.

Last, but certainly not least, I need to acknowledge my husband Gerry. Writing this book was his idea and when he suggested it, I was immediately excited about the project. I really should say -- completing a story I started about 40 years ago, and it's a good thing I had begun writing it then because the words and notes from so long ago preserved details that would have been forgotten. Besides inspiring the project, Gerry read my daily writings and made helpful suggestions. Incorporating some of the delightful story books he made for Grant about his adventures is a fun addition. Thank you, Gerry, for initiating this project and your big part in it. This book took us on an inspiring past-to-the-present journey. The words flowed so easily, it almost seemed like the book wrote itself.

This book is gratefully dedicated
to the
People of Barre, Massachusetts

Introduction

There's a good reason, both present and past, why the people in the town of Barre, Massachusetts deserve to have this book dedicated to them.

In the mid 1800's the townspeople welcomed the building of a school for the "feeble-minded" to their town where Dr. Brown could carry out new methods that showed the country what a difference patience and kindness could make in the lives of people that had been previously labeled as "idiots." It was a time when people with disabilities were despised, cruelly treated, housed in dark crowded buildings, even chained up and put on display for people to laugh at. Dr. Brown constructed a fine complex of buildings near the center of town, large houses surrounded by lovely landscaped grounds and buildings for recreational activities. The residents received good nutritious food and medical attention. He believed that under favorable conditions they could learn self care, social graces and have job responsibilities -- gardens to tend to, a barn with animals to care for, a shop to do woodworking. He was successful and his methods were a model that was slowly adopted in many places throughout the country. Barre needs to feel proud for welcoming this school at a time when most places would have protested loudly against it.

When I moved to Barre in 1961, I met Jackie White. He was one of the best-known people in town, called by some "the mayor," by others "the town crier." He was a friendly young man with mental disabilities who made his daily rounds stopping briefly at all the businesses around the common. He entered places with a cheery hello and gathered news that he passed on to those further down his route. Everyone welcomed him and shared the news of the day. And he was a Boy Scout forever, a loyal Scout. Boy Scouting years end at 18, but this young man was just as much a part of the Barre troop when he was 30 as when he was 12. No one questioned him, he was completely accepted and enjoyed the meetings and camping trips. The boys liked him and learned

valuable life lessons about treating everyone with kindness and patience.

So, when Grant came along, the town was ready to give another person their love and support. The attitude of the Barre people made a difference in his life and ours, just as it did in the lives of many before us who faced similar circumstances.

8

1. Will Grant be normal?

You have to have faith that things will work out,
maybe not how you planned, but just how it's meant to be.
Kristin Harmel

Fifty-three years ago when Grant was born with brain damage, I never would have expected that someday I'd be saying he has been a blessing in our lives. Yes, truly a blessing. He brought people and experiences to us we never would have had. Because of him, we learned to look at life differently, we learned patience, acceptance, understanding, and we became less judgmental. The challenges Grant presented made us stronger spiritually and appreciative of the most important things in life, the things we take for granted -- a normal child, good health, family, friends, a caring community, good medical services, and a God who gives us strength to carry on. I feel a pull to write about raising Grant, how he changed our lives and benefitted us in so many ways. This writing has given Gerry 85 and me 79 an opportunity to reminisce about the journey.

I remember the anticipation of our first child, the hopes the plans, the expectations. We had been married two years, Gerry was happy with his new job at Norton Company in Worcester, and piece-by-piece we were furnishing our home in the community of Barre, Massachusetts, a small New England town of 5000 where Gerry grew up and where I was teaching fifth grade. We spent our weekends during the warm months camping throughout sections of New England and enjoyed longer trips during vacation weeks. In the winter, we frequently headed for the mountains to ski. We joyously accepted and took for granted the good things life had to offer and looked forward to having children to participate in it. It was 1964, I was twenty-six, Gerry thirty-one. We were ready to start a family.

I stopped teaching at the end of my seventh month of pregnancy but did daily tutoring to help fill the long days of waiting. Bob Allen, our jesting neighbor, predicted our baby would

be a boy born on March 12th or 16th, either on his or his wife Elinor's birthday, both dates overdue for me. Well, he was right! Grant was born on March 12th, on Elinor's birthday. Little did we realize then what an important part she and her family were to play in her birthday baby's life.

I left for the hospital on March 10th with labor pains about ten minutes apart. Gerry spent the night in a chair in the waiting room expecting an announcement momentarily from the doctor, but instead he was startled to be awakened by me telling him that the pains had ceased and he might as well go to work and I'd call when they started up again. Drowsily, off he went.

The doctor had appointments during the day and simply told me to walk and labor should resume soon. I spent the entire day more like an expectant father than a mother as I paced the hospital corridors without a sign of pain.

Both Gerry and the doctor returned around suppertime and the doctor started inducing me with an intramuscular injection saying he didn't know what was keeping that baby. The nurse, an old friend, went off duty at eleven that night saying she was sorry she wouldn't be there for my delivery but that it wouldn't be much longer.The contractions were now only a few minutes apart and Gerry and I happily timed them until I was taken into the labor room. (Husbands rarely participated during birth in those days, 1964.) Gerry dozed once more in the chair only to be awakened several hours later by me once again announcing that the pains had stopped. The poor exhausted fellow dragged off to work.

The doctor then decided to induce me intravenously and about eight hours later, with my old nurse friend back on duty, Grant finally was born. I have read since then that protracted labor is often a cause of brain damage, but at that time we thought nothing could possibly be wrong with our handsome, healthy-looking son who slept and nursed contentedly. He was pink, plump, and a picture of Gerry.

We just naturally assumed his normalcy, even a few weeks later when I noticed a different feel when I held my friend's new baby who seemed stronger; Grant was more limp but I liked his cuddliness. However, soon it was apparent that he was not perfect.

His eyes turned inward. Initially, I paid no attention as many books emphasize that a child's eyes won't focus properly for several months, but my mother was alarmed and convinced me that I should have him checked. She pointed out that most babies' eyes focus correctly much of the time, but Grant's were always crossed.

Grant a few months old

When he was six months old, we took him to an ophthalmologist who diagnosed strabismus and far-sightedness AND prescribed glasses for him. Imagine a six-month-old with glasses! Of course, it was the topic of conversation everywhere we went, and the beginning of many trips to the optician's office. The glasses were his favorite toy. He threw them, he pulled and twisted them, he rolled over on them, and he used them as a teething ring. One set of frames only lasted two days, not to mention the many replacements of scratched plastic lenses. We soon adjusted to this situation, and people started calling him the "little professor." That title was pretty impressive and helped us overcome any worry we felt about that early problem. It's wonderful what some kind supportive words can do.

At ten months of age Grant had the first of a two-step operation on his eye muscles. The operation was successful and his eyes were somewhat straighter. Later the eye problem was insignificant in comparison to those yet to develop.

My mother then gave alarm to new fears about Grant. His back and neck were not strong. He was not sitting alone. In Dr. Spock's "Bible" that every mother relied on in the 1960's, it said that many babies, especially fat, placid ones, don't sit alone until as late as one year. Grant seemed to fit into that category ... we thought, we hoped. Mom had raised five kids, but we were in denial of her message that more-and-more appeared to be true.

My parents, Jim and Gwen Lombard

Nevertheless, we surrounded him with pillows and tried practicing sitting him up every day. About the age of one he could sit alone but slumped forward and lost his balance easily. At this point we knew he was below average and yet he was progressing and alert. He was responsive to people and objects. He recognized familiar people and cried when strangers held him. He beamed with delight when we brought out his favorite toy pelican. Our family doctor dismissed his slowness with, *Just wait till he gets going. That's when your troubles will begin.* Well, we waited and tried teaching him to crawl. Every night Gerry pushed his legs and I pulled his arms. At fourteen months he began to locomote a little on his stomach. He'd struggle toward a cookie or toy by pulling with his little arms. His legs kicked but were of little help in the effort.

It was at this time that we were blessed with a second son, Cortland. What a relief it was to see those perfectly straight eyes that never strayed! He felt sturdy when we held him and soon supported his head. I don't think anyone can appreciate a well child as much as parents of a handicapped one can.

At nineteen months our physician became concerned because Grant was only able to crawl slowly on his stomach and his sitting position was still slumped and unsteady. He referred us to a competent pediatrician in Worcester for a complete check-up. We were anxiously looking forward to this testing.

After a hasty observation, the pediatrician dealt the tactless and dismal blow that Grant was so hopelessly brain damaged that he'd never be able to walk or have much mental development. I was completely shocked and heartbroken. I tried desperately to explain some of Grant's achievements: *He understands things, Doctor, I'm sure. He knows all his toys by name. He knows about the toilet; he doesn't tell me to take him, but wetness bothers him so I take him routinely and he performs. He tries to say some words.* I wanted to continue but the doctor's disbelieving eyes stopped me. His only remark was, *Children are my field. I've seen thousands and I know the signs when I see them. As far as saying some words goes, even a parrot can mimic.*

I was defeated. I was no longer composed enough to discuss anything. I felt empty and hopeless. He did add that if we wished another doctor's viewpoint he'd make an appointment for me with a pediatric neurologist in Boston. I nodded that I would like the appointment. He left and the secretary called Boston and gave me an appointment card for one week from that day, quite different from today when we have to wait a month or more to see a specialist.

I gained a new perspective on doctors that day. I realized the importance of treating the whole person, the huge effect on how the presentation of a diagnosis can affect a person's mental state and ability to cope. Caring and gentleness combined with honesty can make a world of difference. We would hear some discouraging reports in the future but, thank goodness, none were dealt to us in that manner.

I held Grant close as I carried him to the car after the awful experience. His face wore a happy smile as he rested his head on my shoulder. In his total innocence, he could neither realize nor alter the effect this terrible condition would have on his life. Poor Grant. What can we do for him? How will this affect Cort's life? Why has this happened to us? My mind was filled with torment. That drive from Worcester to Barre was the longest I've ever had.

At last I arrived home and was relieved to find Gerry there. Quickly and probably as tactlessly as the pediatrician did, I poured out the foreboding report to him. Gerry, who's always calm and cool, couldn't control his emotions as he clutched Grant to him.

Somehow Gerry's reaction strengthened me. I firmly said that I didn't believe this diagnosis and again went through Grant's accomplishments to convince both of us he was all right. That evening we tried to discuss Grant's problem objectively. We vacillated from believing the doctor totally incorrect to thinking we were unrealistic about his disability and unwilling to accept the truth. We decided not to tell anyone, not even family, until we had things verified. I knew I'd cry if I talked about it. Whatever the problem was, I had to deal with it myself before I could talk to others. We even tried to evade the issue ourselves during the seven anxious days ahead. However, whenever he responded to

something we considered intelligent, we were quick to say, *Did you see that? A mentally retarded child wouldn't do that. The pediatrician may know signs but he doesn't know Grant.*

Finally, the day came for our Boston appointment. I packed Grant's books and some toys and met Gerry at work. We had a late afternoon appointment which was ideal so Grant could nap in the car on the way and arrive rested and ready for his best performance at the doctor's office. However, to our disappointment, he didn't nap and we worried that he would be sleepy and irritable and not show all he could do. But, to our relief, he performed at his best, and greatly to our emotional benefit, the doctor who did the testing was a caring person.

The very warm personable man tested for almost two hours -- tapping joints with his rubber hammer, moving each limb, having Grant try various coordination skills, talking to him for response and giving him a complete physical examination. Grant was good and cooperated as much as he could. He pointed to pictures in books and tried to stack blocks and fit two objects together. The diagnosis we received that day, though vague, bolstered our spirits tremendously. The doctor said Grant's problem was something like cerebral palsy in that his muscle tone wasn't quite right but that he would walk by age two. The doctor added that Grant probably always would be a bit clumsy in athletics and handwriting BUT his mental level seemed average. We were elated! *What did athletics or handwriting matter? There's a tremendous range in such abilities even with the normal child,* we told ourselves. This doctor concluded by showing us exercises to turn Grant's hands and feet outward because he tended to turn them in. These exercises were to be performed several times daily until he walked. He also suggested that we see a local orthopedist every three months to have his hands and feet checked, and then return to him in a year. We thanked him and nearly danced to the car. In our minds, our normal boy was given back to us -- at least for awhile.

For the next three months we religiously did Grant's exercises, and watched for strides toward walking ... which didn't appear. Grant could move more quickly on his stomach but still sat unsteadily. We tried not to feel alarmed and kept hoping for a

miraculous change in two months when he would be two. Inwardly it felt like I was on an emotional roller coaster. If he couldn't sit well, how could he walk? Could such a big change occur in two short months?

During this period, my father-in-law gave us a book with an amazing new concept, How to Teach Your Baby to Read by Glenn Doman. Being a teacher, I was excited about the idea, something that might be good for him. A mother with a hurting child will grasp at any straw to help her child, and after reading the book, I thought it would be a fun exercise for both of us. Grandad's main reason for suggesting this for Grant was to provide him with needed mental exercise as he was unable to explore and fulfill the natural curiosities that a child his age had. With both a motor and eye problem, his life and environment were very limited. Grandad said, *His mind must be stimulated!*

Gerry's parents, Armand and Elisabeth Gariepy
with Grant and Cort

Gerry's dad was a well-known motivation speaker, full of ideas to excite people to succeed, and that reading portion of the day proved to be very enjoyable for both Grant and me. It was thrilling to see his excitement when he saw me reach for the word cards and his response in touching the parts of his body and objects that were indicated on the cards. He learned about thirty words during a three-month period. I was thankful for this project. The accomplishments he made with it gave us something to hold onto because the magic walking age of two had gone by and there were still no strides toward achieving it.

Grant identifying the word "cheek"

It was time for our first appointment with the orthopedist in Worcester. Every time we had an appointment with a new professional, our thoughts and emotions vacillated between hope

and fear. Would the prognosis be bright or dismal? Would there be an answer why Grant was not making strides toward walking? Also, we were beginning to wonder why he wasn't making verbal progress, still just uttering a few sounds that were meaningful only to us. Hopefully we would get some answers.

Grant had another complete exam. The doctor recommended that I bring him to the Cerebral Palsy Clinic for therapy, but avoided telling me whether or not he thought that cerebral palsy was the problem. He only said, *I've seen lots worse cases and this therapy will help him.*

By the time Grant started attending the Clinic, he had progressed from crawling on his stomach to awkwardly creeping on his hands and knees. His therapy consisted of spending most of the morning in an electric walking contraption. He was placed in a harness which went around on a track. It was hoped that the motion would induce him to use his legs to walk, but Grant just let them drag as he rode around and around the room. The therapists were lovely ladies and dedicated to these children, and I'm sure the social hour with snacks that followed the therapy benefitted him, an opportunity he'd never had before, but he was never stimulated to use his legs.

The parents were encouraged to leave during these sessions, which gave me some one-on-one time with Cort, our one-year-old, who hadn't had much individual time. We went to the zoo, shopped, and visited friends who lived nearby. So often our healthy children get neglected when we have a special needs child whose demands are so great. We have to make a conscious effort to keep a good balance and I think I failed at that, especially at the beginning. Later, I accepted that Grant would never be normal, no matter how hard I tried, and my other children's needs must be met, and I worked harder to meet that goal.

After attending the Clinic for six months Grant stopped using the few words he attempted to use and had completely lost interest in reading. The head therapist suggested speech therapy at a rehabilitation center, a weekly session we promptly added to our busy schedule. The speech therapist was good and Grant liked him and enjoyed the visits, but never made any meaningful sounds.

We discontinued it after several months and were advised to take him for a psychological evaluation, a factor the therapist thought might account for his regression. Of course, the suggestion of visiting a psychologist was upsetting because that made us think we were somehow to blame. That's even harder to bear than thinking it was a birth problem, especially when we thought we were doing our best. But, if that were the case, we thought we'd better find out and change whatever had to be changed.

We attended a prescribed guidance clinic for ten weeks, Gerry, Grant and I. Grant was nearing his third birthday, still having regular therapy at the Clinic, and in all ways doing more poorly than the previous year. He crept less, sat no better, and seemed to be living in his own little world. He no longer enjoyed his word cards and had little interest in books.

After observing and testing Grant during the ten week period, the evaluating team presented us with their conclusions and suggestions saying, *Grant has emotional and psychological problems stemming from the rapid development of his brother and the realization that he cannot compete. Grant has crawled back into his shell. Also, Grant senses pressure from his mother who spends too much time trying to help him do and learn things.* What a blow! His environment was causing this terrible thing to happen.

My eyes welled with tears. I could stop giving Grant so much attention but what could I do about Cort? I certainly couldn't stop his development. The psychologists prescribed that Grant stop the cerebral palsy therapy and be placed in a class for retarded children in their school. They added, *It will be a daily relief for you, Mother.* Mothers sometimes need relief, but this wasn't what I was looking for. I didn't feel we really had gotten to the bottom of the problem. I needed an answer. It felt like we were "sitting on a fence," the most uncomfortable place to be. What should we do? We thanked the psychologists for their time and told them we would consider their recommendation and call before summer if we wanted him registered.

Everyone looks at the world differently. What's right for one may not be the right answer for another. I remember when we first learned about Grant's physical and mental limitations, a friend

thought we should place him in an institution. Another even brought us literature on possible placements. Those people cared and gave advice, wanting to make life easier for us. However, we couldn't do that, but that doesn't mean it wouldn't be the right choice for someone else. We can't judge. We don't know what it's like to walk in another person's shoes. Everyone is different and everyone's situation is different. We have to respect those differences. I say,

Express compassion instead of judgment.

When we got home that day, a letter had arrived in the mail that brought us new hope ... just when we needed a new straw to grasp for.

2. Searching for an answer

Love always protects, always trusts,
always hopes, always perseveres.
1Corinthians 13:7

The letter was from Aunt Antonia "Tony" Grant, Gerry's dear aunt in Utica, New York for whom Grant was named. She had recently read an article in Reader's Digest about the work of Doctors Doman and Delacato at the Institute for Rehabilitation of Brain-Injured, Inc., a clinic near Philadelphia that had developed a program which was accomplishing wonders with brain-damaged children. At the end of the letter, Aunt Tony added that she and Uncle Doug would like us to investigate this treatment and if we felt it could help Grant, they wanted to finance it as their special project. They had no children and were pleased that we named our son to honor them, and being very caring people, they shared our worry about his development.

We read the article. Wow! The children cited sounded like Grant. We wrote to the center for more information and requested an appointment with Dr. Spitz, the doctor affiliated with the clinic who did the medical evaluations to decide whether or not the Institute's program would be beneficial. Dr. Spitz, a neurosurgeon, was already renowned in the medical world for the invention of the VJ shunt that has given a normal life to children with hydrocephalus (water on the brain). It was an exciting prospect to possibly become connected with such an accomplished doctor.

While waiting for a response, we planned how Gerry could use one of his vacation weeks for this appointment, and my mother agreed to take care of Cort. A spring trip to historic Philadelphia would be ideal. We would combine the appointment with a much needed pleasure trip. About a week later the envelope with Dr. Spitz's return address arrived. We anxiously opened it and couldn't believe what the appointment card revealed. It read "October" -- six months to wait!

We decided not to enroll Grant in the class for the retarded; the program had not impressed us anyway and seemed more like a day-care situation. We optimistically waited and prayed that Dr. Spitz would help and enlighten us. A fall trip to the area would be just lovely.

Finally the day came. Because of the long drive, we left the day before the appointment. From Barre, Massachusetts to Media, Pennsylvania, the suburb of Philadelphia where the clinic was located, it would take at least eight hours, especially with Gerry at the helm who consistently drives at least ten miles under the speed limit. This habit is great on scenic back roads but -- the New Jersey Turnpike! Oh, well, that's my calm, relaxed husband. The trip took a little over ten hours.

We had reservations in a motel near the clinic. It was comfortable and we ate the supper I packed as it would have been difficult to take Grant into a restaurant. We retired, full of anticipation for the next day.

What a scene the huge waiting room presented! We'd seen problem cases during our search, but nothing like this. There was one child in a wheelchair with the most enormous head you could imagine, so heavy that she couldn't support it, and her eyes, like many we would see, were vague and lost to another world. There was a hyperkinetic youngster who was in perpetual motion, his movements awkward, and his opened drooling mouth uttered guttural sounds. Grant was at that time three and one-half years old, small for his age, still unsteady in his sitting, drooling, and clumsy with his hand dexterity. He couldn't talk, had limited vision even with his glasses, and his legs and feet were tight together and quite undeveloped. There were many patients there, each with individual difficulties but most were similar to Grant in that they were being sheltered and soothed in their parents' laps, some of them limp, some stiff-jointed, and most of them unable to relate to the outside world, but their needs and desires were understood by the loving ones who held them. One case that grieved me deeply was the young man on a stretcher whose wife fed him strained foods and liquids through a straw. He'd been in an accident and was speechless and paralyzed, but his eyes were alert. They were

too young to have been married long. In fact, their problem which resulted from the accident had to be very difficult to face, both physically and emotionally. Having a normal person and losing him or her to a dysfunctional physical or mental state has to be much harder to bear. We never knew Grant as a normal child so we were spared that adjustment.

When Grant's name was called, Gerry carried him into the office and I had my defensive speech on Grant's capabilities carefully rehearsed in my mind, praying not to be turned away with another pronouncement of his hopelessness. A huge man with a kindly face stood to shake our hands. I was amazed to see how quick, graceful and agile this enormous man's movements were. I judged him to be in his forties. Later we were to learn that Dr. Spitz had five children, one with spina bifida, and the youngest a hydrocephalic child whom he adopted when the child's adoptive parents discovered they'd been given a defective youngster and rejected her. We came to know him as a genius with deep compassion, a blend equipping him perfectly for his work.

He proceeded to do a thorough physical of Grant and I began to deliver my well-memorized speech. I feared the worst when he stopped me shortly, then went on quietly to say, *Mrs. Gariepy, please do not be on the defensive about Grant. There is no one who knows a child better than the parents and I value their opinions highly. You realize he has a problem and that is what we'll check for. Other questions I'll need answered will be asked and your responses believed. We will be honest with one another.* Wow! I liked this man! We quietly observed the examination, the usual tapping of the joints with the rubber hammer and moving the limbs, hands, and feet, and when he finished he spoke seriously of what he suspected: *I can't be sure of Grant's problem until I perform further tests. I suspect from reading his records about his present condition that he has brain damage that is getting progressively worse. We will have to hospitalize him to carry out the necessary tests in order to make a diagnosis. But I assure you, his problem is not emotional.* With that we were escorted to a secretary who made hospital arrangements for Grant to be admitted in two weeks.

Although we were relieved that his condition wasn't worsening because of being emotionally disturbed, the thought of progressive brain damage sounded scary. Nevertheless, an amazing calm came over us while we spent the rest of the week touring old Philadelphia, it was a peaceful relieved feeling we hadn't felt for two years. At last, we found a doctor that we believed would get to the bottom of things.

Both Gerry and I love history, and Grant was good and easy to manage in his stroller as we enjoyed the walking tour of the historic city and relived the early days of the founding fathers of our country. After returning home, we busily prepared for the rapidly approaching date of Grant's hospitalization. As Gerry couldn't take more time from work, Wendy, my sister, accompanied Grant and me to the hospital in Philadelphia, a long drive that grew shorter over the next few years. It's amazing how a new route seems to take forever and a familiar one shortens when you make it regularly.

I fought back tears as we left him in a crib with the only familiarities being his favorite toys. He didn't cry, but looked helpless and pathetic as he lay in bed and his questioning eyes searched the strange room. I wished I could tell him why and what was happening. It was the beginning of learning about tough love, the many things we have to do that may cause temporary hurt in order to achieve the better life that may lie ahead.

Grant's tests included a spinal tap to remove a sample of the fluid that surrounds the brain and then several days later, a brain biopsy to remove some brain tissue. Both would be examined by a pathologist. The results would enable Dr. Spitz to make a diagnosis. After a ten day stay at the hospital, Wendy and I brought home a sorry-looking Grant, his body was thinner and his head shaved and bandaged to cover the two holes that were drilled to do the biopsy procedure. It was good to have him home. He was weaker for awhile, but good natured and loving as he always was. I worked at fattening him up again as we waited for an appointment to return after the test results were in.

During this time of waiting, I had another source of emotional help. I talked with a minister who was also a psychologist and counselor. I told him my problems, my fears for Grant's future, and my concern for the effects on Cort's life and our own. We needed somehow to put things in place. Grant was really running our lives. I was seeing the situation and wanting to do the best I could. At the same time, I was feeling sorry for myself. I was sure a minister, a person of compassion, would have some sympathetic words for me, but he replied with words that totally surprised and shocked me. *I'm happy for you!* he declared. *Anyone can be a good Christian under ideal circumstances. You have a cross to bear. Now let's see how you'll handle it. Will you crumble with self-pity or rise with the proper attitude to meet the problem and provide a good Christian home for your family?* At first, I thought his answer was very cruel, but after I thought about it a lot, I decided firmly that self-pity never ever gets anyone anywhere. In fact, I came to realize that when we pity ourselves, we become pitiful; also if we pitied Grant, he would become pitiful. I determinedly decided -- from this day forward, I would meet the problem squarely, deal with it the best I could, and not allow it to completely fill my thoughts. No matter what, our home would have a normal happy atmosphere without self-pity. The minister's words, though more severe than ones I'd ever feel comfortable using, turned out to be just the tonic I needed at the time.

Self-pity is the most destructive of all narcotics.
Sebastian Horsley

The call came shortly, and we returned to Dr. Spitz. Gerry joined me for this trip as we felt we might need each other to shoulder the news that would be forthcoming.

Dr. Spitz's grim face told us immediately that the news was not going to be good. *Grant has a brain lesion that has destroyed about one-third of his brain,* he said. *A lesion is like an open sore and causes him to have sub-clinical seizures that are killing more brain cells all the time.* He added quickly, though, *I am experimenting with a treatment for this problem using a drug*

called Depo-Medrol. The procedure involves removing a small amount of spinal fluid and replacing it with this drug. It will be done every two weeks for awhile and then once a month, gradually increasing the dosage from 10 milligrams to 100 milligrams, then decreasing it again. The entire program will take about eighteen months. It appears that some lesions are being arrested by this treatment though it is too early to be sure. We will do an EEG (a brain wave test) each time he comes to see if the seizures are stopping. If he seems to be responding, I'll refer him to the Institute as I feel he could benefit greatly.

What alternative did we have? Without trying the experimental drug, Grant's brain cells would all be killed by the seizures. We had already seen regression taking place. There was no alternative. We had to begin the treatment, our only hope for the future. Grant's series of spinal injections began that day. As we were leaving, Dr. Spitz comically added, *Don't worry about the one-third brain damage; some folks function ok with just half a brain, especially politicians.*

Bless Aunt Tony and Uncle Doug; the bills were soon to loom into the thousands of dollars, such experimental medicines were not covered by insurance. We were so thankful to be able to enter extensive treatment programs and not be burdened with financial worries. When we read about fundraisers for worthy causes, we respond knowing someone else is being blessed.

I have a wonderful mother, who today at 100 is still alive and well. She willingly cared for Cort when we went for appointments. Often she invited his young cousin Mathew to visit also and they became good buddies. Gerry's mother whom I, along with the children, called Grammy, passed away in 1981. She was eager to help and volunteered to drive on most of the trips for the spinal shots. She was a willing, cheerful and tireless companion and the frequency of the trips posed no problem to her. And, fortunately, unlike her son, Grammy traveled at the speed limit. We always drove down the day before, had the shot the next day and returned home.

It makes me smile to remember the attention we drew walking from the clinic to the car. After receiving the shot, I was told to hold Grant upside down for about two hours so that by gravity most of the drug would get to the brain as that drug was readily absorbed in the spinal column. Many staring faces watched me as I walked with Grant's feet over my shoulder and his head dangling at my waist. Those people probably had a unique conversation piece at suppertime telling about the weird woman they saw carrying her child upside down.

Another problem. The drug made Grant violently sick to his stomach. Poor Grammy's nice car. Being the staunch supporter she was, she never complained about the odor, but my stomach felt queasy. After the first episode, we brought plastic bags and then resolved the problem entirely with fasting. We didn't give Grant anything to eat until after the effects of the shot had worn off. It was a long day of hunger when sometimes Dr. Spitz was delayed and our wait might be several hours.

But, we never complained about a wait. We knew this man's schedule was very demanding and that the emergency he was trying to meet could have been Grant's. With every visit, we grew more hopeful as each EEG report seemed more encouraging. Then, after a few months we received the good word. *I believe the treatment may work. We'll schedule you to begin the Doman-Delacato program at the Institute,* said the good doctor. The words we had been hoping for! I felt a new destiny for Grant and for us.

3. A hopeful therapy

Rejoice in your hope, patient in tribulation,
constant in prayer.
Romans 12:12

We loved winter. We were married in winter, and the January day we set out for the Institute was much like our wedding day -- a bright sunny day, cold and crisp with fresh white snow glistening on the landscape. We were invigorated, excited about the beginning of a new important venture. Grant would be four in March.

We deposited a cheerful Cort at Mom's again, this time for him to enjoy a three-day vacation with her and Grampa Jim. His vacations were becoming more and more frequent. He was two and one-half; his baby years had disappeared fast, too fast.

We reflect on how our outlook changes with the years. I think back on that drive to Pennsylvania and how we watched the snow dwindle to slush as we went south and we remarked on how fortunate we were to live in a place cold enough to be able to enjoy winter sports and scenery. Today we seek out warm places during the winter. Change comes to our bodies, our thinking and our activities; change is something we can't stop. Life has its seasons too, a time for different things.

The only thing constant is change.
Heraclitus

That drive and stay at the motel was our first day of the new beginning, the second was our introduction to the Institute's program. Twelve other families were with us for the initial presentation. Dr. Doman, an orthopedist, showed a film and delivered a long lecture about the course on which all of us were about to embark. We learned a lot about the brain. We learned that there are about ten billion brain cells and that dead cells cannot be regenerated but that live cells are versatile and can be trained to do

many jobs. For example, if the part of the brain that normally controls balance is damaged, through proper and constant programming, another part of the brain will take over that activity.

Dr. Doman explained how he originally taught brace-making. He and Dr. Delacato had wondered about the possibility of training the brain, the source of the deformed and undeveloped outer extremities, rather than training the limbs. In other words, treat the cause instead of the effect. They began to observe babies and their movements and found that normal children's basic movements began and progressed according to a definite pattern. They began to experiment and developed their program by putting brain-damaged children through those patterns. We were very encouraged with their statistics that showed about 86% of the people programmed showed significant improvement and 33% became normal.

We were fascinated by the film of a boy who'd suffered extensive brain damage from an auto injury. He was what most people might refer to as a "vegetable." The film showed the patterning, the crawling, and all the eye and body stimulations bringing normalcy back to the boy. In my mind's eye, I could picture Grant running, talking, seeing! It was unnecessary in my case when they kept emphasizing the importance of carrying out the program completely, that every possible movement must be purposeful every day, all day. I was prepared to work industriously. I would do everything I could to produce the results we'd seen in the film.

A question and answer period followed, and with all the pages of notes I had from the day and evening's sessions, one would have thought I'd taken a semester's course in brain damage therapy. Fortunately, I have those notes or I wouldn't be able to write with this detail today. It was after <u>midnight</u> when we retrieved Grant from the babysitter they'd provided. Filled with great expectations, we wearily headed for our motel room.

After a short sleep and breakfast, we headed back to the Institute for training to carry out a program designed for Grant's individual needs. There were several evaluators, Mr. Linton was ours. He was a perceptive, thorough and friendly man whom we

liked immediately. He checked Grant's creeping, his vision, his hearing (which was excellent), his hand dexterity, his speech (or lack of it), his legs and their reflexes -- every inch from head to toe. He charted the results on a big graph which had dark lines representing where a normal, slow, average and superior child would rank. Grant's grade was far below that of even a slow child his age. Each time we returned, Mr. Linton would fill in this chart and we could compare the rate of development under the program with his past rate of development.

The evaluation completed, we were ready to begin. I was prepared to work like a beaver but I don't believe any beaver had to put in the days that we obviously were going to have ahead. Not only would we have to work hard, but Mr. Linton said we'd have to solicit the aid of a coordinator, sixty-three regular patterners, and a sizeable list of people for substitutions. We were very independent people. Mom and Grammy had been helping, but we hadn't involved others. Besides some close friends, we hadn't even shared the extent of Grant's disabilities with people. We didn't want pity for us or for Grant. We hoped we could quietly work things out. We would soon come to realize that when we hide problems, people hide from us. When we are open and talk about our problems, people come forward with support. It would not be easy for us to ask, but we knew we would have to depend on many to fulfill the needs of the program.

The coordinator, Mr. Linton explained, *has the biggest job. That person would be responsible to schedule the patterners, receive all calls, and make substitutions when necessary. It must be a person who's home a lot and is a good organizer. It's a big responsibility and the successful running of the program rests largely on her shoulders.* Then he continued about the patterners: *Any able-bodied person can pattern, from teenagers to the elderly. They will work in groups of three, pattern for ten minutes, wait thirty minutes and then pattern again for another ten minutes. It involves about one hour a week for each volunteer. You will need three teams, nine people a day, seven days a week. It is most important that these people are reliable.*

With that task in mind we were placed around a high, rather narrow padded table with a line drawn down the middle. Grant was centered on this line with Gerry placed at his head, Mr. Linton on one side and me on the other. Then Mr. Linton proceeded to teach us to pattern, Gerry turning his head as Mr. Linton and I moved the arms and legs. At first, we felt awkward but we soon had Grant moving rhythmically in the first basic baby movement. When Mr. Linton felt satisfied that each of us could perform all three positions perfectly, he proceeded to instruct us in the rest of the program.

A huge variety of highly stimulating activities for all his senses were included in the program. Grant was to have two salt baths a day. To do this, I was instructed to make a thick paste of salt mixed with a little water which was to be rubbed in a firm circular motion over his hands, fingers, arms, shoulders, back, legs, feet and buttocks. Then he had to be rinsed thoroughly with cool, then warm, then cool water. After briskly rubbing him dry, he was to have an alcohol rub. This treatment looked torturous but it did highly stimulate his whole body and the pink color he glowed afterward gave him a healthy look indeed.

To develop auditory skills we were told to talk about twenty times as much as usual to him, which was to become inevitable with all the new people about to enter into his life. Also, we were told to use a variety of tones -- loud, soft, loving, sad, happy, etc. We were to read ten stories a day, five with simple plots and five written in rhyme, such as Dr. Seuss books. Much of the reading would be carried out by the patterners, and the books these people loaned made his selection quite magnificent. Cort, too, was to benefit from that activity as he was always on the other side of the reader when story time came. I'm sure that's why they are book-lovers to this day.

I would have to gag Grant twenty-four times a day to stimulate muscles used in speech, to be done with a tongue depressor in six sessions with four gags each session. I always carried out this activity over the bathroom sink and never directly following a meal. I got to be quite expert at getting the depressor out in time but neither Grant nor I ever relished this aspect of the program.

To develop tongue muscles we were told to dry the inside of his mouth with a towel and then fill his mouth FULL of peanut butter. What a job it was for him to work away to consume that thick, gooey mass! That had to be done three times a day. Surprisingly, Grant didn't mind the chore and went about the task willingly. He still loves peanut butter, but I wish I'd taken pictures of some of the patterners who watched him struggle with the mouthful; their sympathetic facial contortions matched his.

Despite all the gagging, peanut butter, and speech release activity, his speech didn't blossom. However, never getting discouraged, Mr. Linton had other tricks to try. He instructed me to hold Grant comfortably, then to yawn in a calm, relaxed and tired fashion, a yawn that should inspire him to yawn also. When I'd gotten him to yawn after me, I was to yawn and add a word. (Yawn -- "Hi," or Yawn -- "Daddy"). After many unsuccessful days of yawning, I told him to open his mouth and yawn. He tried to duplicate that, seemingly to humor his foolish tired mother but the idea of trying to say a word after it never got through. Finding a way to give birth to his speech was certainly the most difficult and frustrating effort, and to this day it's often difficult for him to get out his words.

Grant had very poor hand dexterity. He was clumsy grasping with his whole hand. He had no thumb-finger coordination so picking up small objects was practically impossible. He had mastered eating with a spoon by holding it tightly in his fist and forcing it flat down into his bowl, eating whatever food spilled into it. We fed him the remains he couldn't get. Mr. Linton started him on manual competence activities, first picking up large objects from a flat, smooth surface, which we were directed to have him practice equally with both hands, then as he progressed to have the objects get smaller until he was able to pick up small, flat items such as paper clips or pennies using his thumb and finger. He was to practice these skills for one hour each day, divided into six ten-minute periods. His progress in this area was absolutely fantastic. The patterners added so much creativity in their conversation and

enthusiasm to those routine tasks, motivation we were sure contributed to his gains.

Right from the beginning Grant loved the activities involving the sense of touch. I enjoyed this game too, his delight was contagious. We were instructed to begin by showing him two or three large objects (a comb, a spoon, a small car), and saying their names while putting them into a cloth bag. Then I was to ask him to feel in the bag and find the object asked for. He rapidly became quite adept at doing it and we progressed to a greater number and smaller objects (buttons, screws, rings, paper clips). With a gleam of confidence on his face, he'd reach into the bag, feel carefully for a few serious moments, and then with pride produce his selection. He spent five, five-minute periods on these tactile activities daily. I would never have believed on the day of his initial evaluation when he had difficulty even handling a large object that in a few short months he would have progressed so far. The patterners rejoiced with me. What a support system those folks were!

The Doman-Delacato reading program I had worked on with Grant years before was assigned. I was to print six new words, each on separate pieces of cardboard, every week. He was to learn a new word each day from Monday through Saturday and be tested on Sunday. Each day he was to have three review sessions of the words taught earlier that week. I looked forward to trying this, remembering how he'd enjoyed it before the time of his regression. Because he couldn't talk and tell me what a word said, we used objects. For example, I would show Grant the word "car" and say it while showing him a little car. Then I'd place the car in a basket with other small objects and at intervals pop the word in front of him and ask him to get me what it said. He responded well and at the end of each week there would be six objects in the basket that he could match quite well with the word cards. We often brought out review words and objects and were really amazed at how well he'd remember them.

Our little professor was still wearing glasses and having regular check-ups with the ophthalmologist. He was also still too abusive in the way he used them and repairs were needed regularly. Gerry, who most appropriately nicknamed Grant's glasses "The Flexible

Flyers," became quite expert at maintaining them after we started programming when it was impossible to find time for the trek to the optician's office. At first, Gerry's tools were rather unorthodox, just wire and tape, but they did hold the glasses together. Then he bought a repair kit and learned to do quite a professional fix. We're happy to report that Grant is careful with his glasses today and they last perfectly from one annual check-up to the next.

Mr. Linton said that Grant was to watch a flashing red light in a dark room that was timed to be on five seconds, then off five seconds. That exercise was to be done for five minutes, five times a day. More eye exercises would follow. Grant's vision improved and his eyes became straight. He still needs glasses but he never needed the second corrective surgery the doctor said would be necessary.

Crawl and creep! Creep and crawl! It seemed that this would be how we would be spending most of our days for some time to come. Mr. Linton prescribed TWO HOURS of crawling each day through a box which would have to be constructed to be sixteen feet long, two and one-half feet wide, only fourteen inches high and the bottom surface lined with rubber. The top would be closed in by lacing a rope through eye screws on the sides of the box, thereby enabling me to keep Grant moving properly but preventing him from being able to lift himself up from his stomach. Crawling flat on one's stomach is a very tedious way of locomotion but on the rubber which created so much friction, it was an unbelievable chore! I fully appreciated Grant's struggle when I watched the tremendous effort it took for normal children to get through that box.

Another two hours were prescribed for creeping on his hands and knees. This was to be done on a carpeted surface. One good outcome I could foresee from all that activity might be the purchase of the new carpet I'd been wanting. Gerry, a born conservative, never believed in buying anything that wasn't absolutely necessary, and the old rug that came with the house was not in poor condition, but I wanted to replace it because it didn't match our furniture. With all the creeping and a house full of

people most of the time, I was happily predicting the fate of the old carpet.

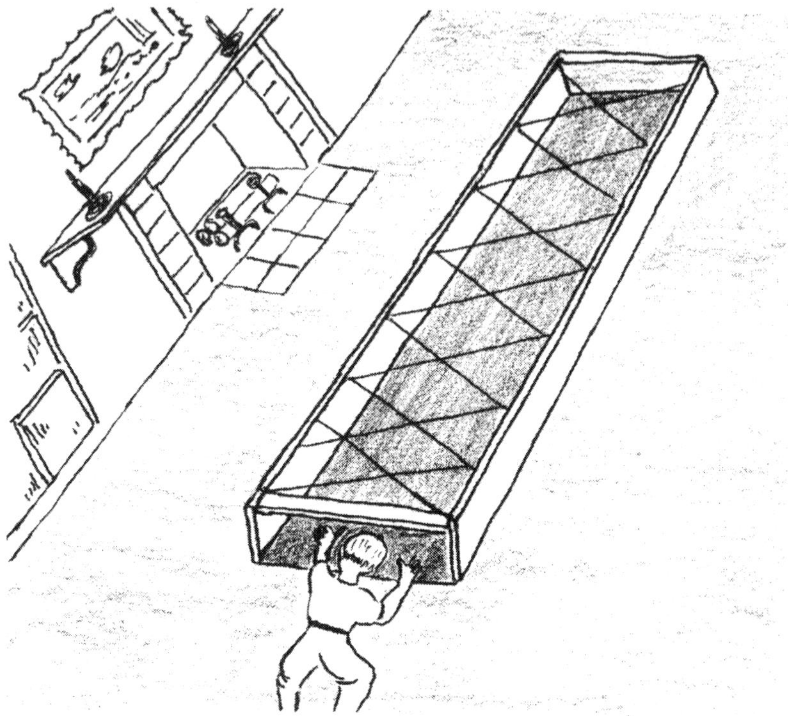

Gerry's drawing of the crawl box

Mr. Linton said that Grant should wear a light shirt and shorts, no shoes or socks. Well, Grant, who had been very inactive all his life, had soft delicate skin on his knees and the initial wear-and-tear from all the creeping on a carpet caused painful soreness. I had to watch his knees carefully and when there were signs of redness, we'd put on long pants for awhile so his knees wouldn't crack and bleed. How they would sting during the salt baths! But I did think those baths contributed to his fast healing. After several months that soft skin was converted to tough elephant hide.

Mr. Linton ended our evaluating session with a handshake, his encouraging best wishes, and a helpful packet of literature describing and illustrating all he had presented to us so we'd have references and instructions of all we needed when we got home. We thanked him, crossed over to Dr. Spitz's office to get the next spinal injection, and somewhat overwhelmed, headed for Barre.

A coordinator! Sixty-three people! Substitutes! A patterning table! A crawl box! A timed flashing light! Wow! It would take a lot to carry out that program.

The one thing we hadn't estimated during the ride home was the people of Barre. We couldn't possibly have imagined the incredible response we'd get from the article I put in the local Barre Gazette telling about the program and our needs for assistance. Volunteers responded immediately, not only from Barre but also from surrounding communities, some people we knew and many new faces, all who would soon be an important part of our lives. Unbelievably, all our needs were met and we were ready to begin the program in less than ten days. It felt like a miracle had happened. Experiencing that response was truly a blessing from Grant. His needs brought out so much goodness in people.

4. Need for many volunteers

We had the three things Joseph Addison said were essential
to attain happiness --
something to do
something to love
something to hope for

We had known Bob and Elinor Allen for a long time. They attended our wedding, we became neighbors when we bought their house and they moved across the street to the new one they built, Gerry and Bob were both members of the Kiwanis Club, Grant was born on Elinor's birthday creating another bond with them, and their daughter Lucy was our first babysitter. Then, when we needed a coordinator for Grant's program, we both agreed that Elinor would be a very capable and probably a very willing candidate for the job. When I approached her about taking on this complex task, not only was she willing, but eager, and she efficiently coordinated the program from beginning to end for almost two years, and also put in a great deal of time substituting when other patterners couldn't make it.

Elinor had been fully aware of Grant's problems and our years of searching for medical for help. She was interested in him and they related well to each other. Grant has a postcard collection, which now has become quite vast, and he spends a great deal of time reviewing his mail, a hobby that benefits us too, keeping us up to date on everyone's travels. But, Elinor's cards have special appeal because she always writes about something disastrous or dramatic that occurred on her trips, and her accounts of these incidents so delightfully written get read over and over. Could the reason for their good communication be because they were born under the same stars??? For many years his favorite books were about the disasters Curious George and Amelia Bedelia got themselves into. And, to this day, he likes to hear about disasters. Several friends make a point to have some dramatic episode to tell him ... a plumbing problem they had, some mischief caused by a

pet, car trouble. Recently our friend Kathleen's house repair turned into a much bigger project than was anticipated. When the carpenter regrettably told her that it would even entail having to take down the wall on one side of the room, he was surprised when she suddenly started to laugh. He was sure she hadn't heard correctly, and then probably thought she was a little crazy when she said, *What a great disaster story I have for Grant!* Very likely, Elinor's stories initiated his love of disasters.

Lucy, Grant's first babysitter

To initiate our campaign for people to pattern we submitted the article in our local newspaper, and Elinor made a plea before various groups, including Girl Scouts and Rainbow Girls. People enlisted from all walks of life, from teenagers to senior citizens, relatives, friends and strangers, even priests and brothers studying to be priests from the local Novitiate.

Herbert "Grampa" Rice, actually Gerry's distant cousin, who was an excellent woodworker, decided that outfitting us with a patterning table would be his contribution. It was sturdy and beautifully made and promptly replaced our dining room table which along with its chairs went to the spare room upstairs. We knew we weren't going to be doing much fancy dining for awhile anyway.

To cover the table we needed a foam pad topped with a smooth durable fabric. We decided to check uptown at Mr. Rucidlo's upholstery shop for some vinyl material. I had white in mind. That was the color I'd seen at the Institute and it had a nice sanitary look. However, the upholsterer didn't have white in stock, but said he could get some quickly. As he was making out the order, my shock at the cost of quality vinyl per yard must have shown. A sudden gleam of remembrance came into his eyes as he said, *I did a seat for a truck driver some time ago and have a large piece left. No one seems to want that color. It's a terrific quality, and if you can use it, it's yours -- free.* He went into a back room and came out with the tough, sizeable piece. I immediately understood why it never sold but knew it would serve our purpose well. I'll bet no other child was patterned on deep purple! We knew that somewhere there was a trucker we could thank for that donation which held up perfectly throughout the program.

The flashing light requirement was also handled by Grampa Rice. He volunteered his unsuspecting brother, who lived out-of-town and was unaware of the project. An electrical engineer, he readily accepted the challenge and a precision instrument was soon produced. As a dark room was necessary for watching the light, we decided the best place for the device would be our bathroom which had no outside windows. We placed a small chair in the bathroom, a few feet from the light, and thought how good these five-minute

breaks would be for me as Grant watched the flasher. However, the light was all too fascinating and Grant was not satisfied with sitting at a distance to watch it. He would immediately slip out of the chair to explore. We couldn't place him on the toilet seat because inevitably, with his unsteadiness, he'd surely crash onto the hard tile floor. So, for awhile I sat holding him and together we watched the light flash. So much for my break! As the program progressed, we came up with more efficient ideas and one of the best plans was to substitute the small chair for a potty chair with a snap-on tray to keep Grant imprisoned for the five-minute watch. That became toilet time too. Grant still didn't communicate when he had to go to the bathroom, but I was timing-trained and he cooperated. He was much like a little dog who performs at each tree, no matter how frequently he comes upon one. Grant always responded when put on his chair, so the flasher achieved two activities in one. Later, I thought, *Why not gag him, dry his mouth and have him work on his peanut butter while he was in the bathroom, pottying and developing his eyes?*

It had always been difficult separating Grant's tight-together legs to fit around the deflector while putting him on the potty chair. We were amazed at how fast the tightness lessened and how much more flexible he became after he started patterning. This change made it much easier to dress him. In some areas, improvements could be noticed almost weekly, others were much slower.

Charlie Allen, our next door neighbor who owned a local factory, a man who got right to the point of whatever was on his mind, stopped by and said, *I hear you need a special box for Grant's therapy. Give me the instructions and we'll have it made down at the shop.* He allowed no fanfare of thanks, just took the instructions as if the act was simply a part of his civic duty, and a few days later the much appreciated contraption was delivered by Harold Dorsey, the company carpenter. Luckily the living room was long enough for the sixteen-foot object. To complete the box, we ordered the rubber needed to line the bottom of it from a Philadelphia company Mr. Linton had told us about.

We moved the couch and what chairs could still fit in the room back against the walls. The crawl box became the focal point. The remainder of our living room furniture joined the dining room set upstairs to complete a rather unique decor. Anyway, the spare room was no longer empty.

Shortly, the rubber arrived and Gerry nailed it in place. Then we enclosed the crawl box by lacing clothesline rope through the eye screws along the top. Gerry added a little stain to the plain boards to give it a more finished look and it was quite a sight to behold. I'll never forget when some of the brothers from the Novitiate stopped by for a short visit with a guest priest from Canada. The priest could only speak French and as I only spoke English, our conversation was limited to facial pleasantries. The brothers brought him to see the child so many of them were interested in helping, but during this particular brief visit Grant did not use the crawl box. The next time the brothers came they couldn't wait to tell me, *When we left, Father said the program we were working on was very interesting and worthwhile, but that he'd never seen such a piece of furniture as that huge modern coffee table you have in the middle of the room.*

We had always led a close quiet family life. Suddenly, our whole life style changed. We became a community project with our lives exposed to all. Privacy was a thing of the past. We were grateful for such whole-hearted support, but we wondered how we were going to like this new way of living -- seven days a week.

5. Patterning program

If you light a lamp for someone else,
it will also brighten your path.
Buddha

Life did change, a change that touched my heart. It was truly inspirational to watch those people of varied ages and backgrounds united to work with great dedication for a life-changing purpose. They not only capably carried out the necessary patterning but they encouraged us and added fun and creativeness to the program. They never made me feel they were inconvenienced or burdened by all the time and energy their commitments took, rather they seemed to enjoy their contribution and thrilled with us when Grant progressed. He had a whole team pulling for him. How could I have carried out all those routine tasks alone? Also -- those were the days when I heard all the local news ... and maybe a bit of gossip too.

The following was Grant's first daily program, similar to schedules we were to have for many months:

7:50 - gag, flashing light
8:00 - crawl, creep
8:15 - speech, new word
8:20 - crawl creep
8:30 - eat, dress
8:45 - manual competence
9:00 - new word, word review
9:10 - tactile
9:15 - story
9:20 - crawl, creep
9:30 - pattern
9:40 - crawl, creep
9:55 - manual competence
10:05 - story
10:15 - pattern

10:25 - crawl, creep
10:45 - gag, peanut butter
10:55 - new word, speech
11:00 - crawl, creep
11:15 - tactile
11:20 - crawl, creep
11:40 - flashing light, gag (I made lunch)
11:45 - crawl, creep
12:00 - new word, word review
12:05 - eat
12:15 - story while eating
1:15 - new word, word review
1:20 - tactile
1:30 - pattern
1:40 - crawl, creep
1:55 - manual competence
2:05 - story
2:15 - pattern
2:25 - crawl, creep
2:40 - speech
2:45 - story
2:55 - crawl, creep
3:10 - speech
3:15 - gag, peanut butter
3:20 - new word
3:25 - crawl, creep
3:45 - tactile
3:50 - gag, flashing light
4:00 - pattern
4:10 - crawl, creep
4:25 - manual competence
4:35 - story
4:45 - pattern
4:55 - crawl, creep
5:15 - new word
5:20 - gag, story
5:30 - crawl, creep

5:45 - manual competence
6:00 - speech, new word
6:05 - gag, crawl, creep
6:20 - supper
6:40 - flashing light
6:45 - tactile
6:50 - crawl, creep
7:05 - story
7:10 - speech
7:15 - new word, review
7:20 - crawl, creep
7:35 - gag, peanut butter, flashing light
7:45 - bed

As Gerry and I review this schedule today, we wonder how we did it.

God gave us burdens, also shoulders.
Yiddish Proverb

I'm reminded of what my mother said recently after spending an afternoon with the great grandchildren, *It's a good thing people have children when they're young.* We were young and energetic then, and met the demands with no complaints. Hope carried us forward. The program was a full day for sure. Gerry, always helpful and cooperative, cheerfully pitched in when he arrived home at 5:15 P.M. so I could prepare supper and, after eating he completed the evening's activities with Grant while I did household chores.

Gerry is joining me in this book project, just as he joined me years ago in the patterning program. He reads and edits every evening after I finish writing, so we're reliving the program days together. Later in the book, I am including some of the cartoons and stories he wrote for Grant that I think readers will enjoy.

The first week of programming was spent mostly in explaining the program and training the patterners how to move Grant's limbs and head properly and in rhythm. Moving his rigid, unyielding arms and legs steadily for ten minutes was not an easy job, especially at first when the exercise, which soon would become automatic, required concentration and the use of some dormant muscles. We always put the person who considered herself the "muscleman" of the group on the left side because that side was stiffer, more deformed and difficult to move. His head was clasped in front of his ears and turned back and forth, a motion to stimulate balance. It sounds simple but his head became quite heavy during the ten minutes of lifting and turning. Many reported how their arms and backs ached afterward. But, no one quit.

I've always cherished a poem written by Grant's cousin Beth when she was ten years old. It reflects her view of his life. She was his youngest patterner.

My Cousin

I have a little cousin, and his name is Grant.
He crawls and does all kinds of things and never says I can't."
Seven days a week! His schedule is so tight.
He works so hard, every day, you'd think he'd use up all his might!
He crawls and creeps, he reads and tries to talk,
And by the end of this year he might start to walk.
He likes all kinds of animals, cats and dogs, birds and fish,
And to have one as a pet someday, just might be his wish.
He has to watch a flashing light to get muscles in his eyes,
And all the other work he does he really, really tries.
Every word I said here is very, very true,
So I hope you will believe me, everyone else seems to.
Now I've told you all about him, I hope you understood,
'Cause Grant's a very nice boy and very, very good.

Cousin Beth holding Grant, Steve with Cort and cousins in back, Craig, Derek, and Scott

Grant relaxed and enjoyed the patterning, and the patterners chatted to him and amongst themselves as they did it. Once everyone was trained, I had ten minutes of freedom! And, it was amazing how much I could accomplish during that break -- washing the dishes, hanging out a load of wash, making beds, or playing with Cort.

It was interesting to note the predicaments some people found themselves in when they were unable to use their hands for the ten-minute patterning period. One dignified lady had a problem that dismayed her greatly. Pearl's nose dripped whenever she was nervous, and her fear that it would occur when she was patterning and helpless to dry it invariably caused it to drip. Later, when the team was well acquainted, it became a good joke, but it was no joking matter in the beginning. Mary Alice's glasses always slipped and by the end of each session they were usually balancing precariously on the tip of her nose. At times, she knew they'd never make it and I was called upon to push them back into position. She did try to remember to remove them beforehand, but they were such a permanent part of her that her intentions were often forgotten.

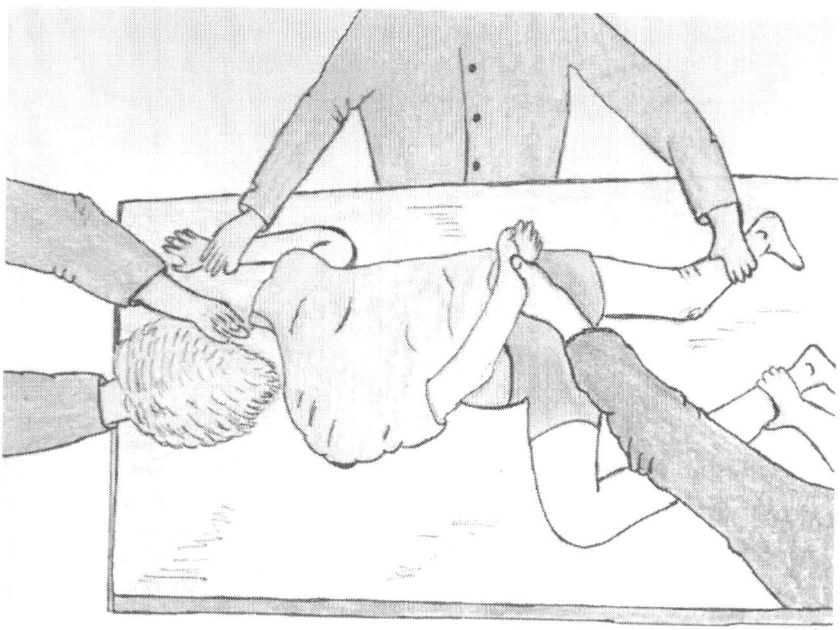

Gerry's drawing of Grant being patterned

Be sure to notice when you see a baby creeping that the right hand moves forward with the left leg, then the left hand with the right leg. These movements come naturally to a normal child, but didn't come naturally to Grant. Three patterners worked on Grant at a time. One person moved the head back and forth as the person on each side moved his hands and legs. They did this exercise rhythmically to teach his brain the correct patterns of movement. Grant was nearing his fourth birthday when we started this rigorous program.

Many people got itchy and, of course, the more one thinks about an itch that can't be scratched, the itchier one gets. When the timing bell went off, it was not uncommon to see people sigh with relief and scratch themselves as if they had a case of poison ivy.

A common complaint was what the activity did to hairdos. By the time the ten-minute patterning timer went off, many eyes were covered. The hair gradually worked its way forward while the bent bodies worked.

As soon as the timer rang, Grant always sat up to have a short visit with the people he grew to love, but when Ettie patterned, Grant did not jump up. From the first, she always gave him a backrub after patterning and soon he connected her with that reward and would lie restfully on the table awaiting his treat. It became obvious that he knew each person and expected certain responses from them. That ability pleased us and I'm sure the patterners must have felt their individual value in his life.

Olive was always bruised with cuts or scrapes she'd gotten at her job. Grant was fascinated with her bandages and listened sympathetically to the tales of woe as she explained the injuries. I'll never forget the horror I felt the day she came in with her leg swathed in bandages. I couldn't imagine how she could even walk and asked, *What happened?* She whispered, *It's nothing. You know how Grant likes to look at my band aids. Well, I had no cuts this week and I didn't want to disappoint him so I did this up.* Grant's eyes opened in wonderment at the sight, knowing a huge disaster story must be forthcoming.

Many outfits in Jessie's wardrobe had belts that tied with ends hanging down. When Grant was fascinated and pulled on them, I commented, *They're like tails, aren't they Grant?* He noticed them every week but little was made of it until a few weeks later when he was searching and crying while pulling on Jessie's dress. She was puzzled and upset, but I understood immediately. Jessie had not worn a dress with ties. *I think he misses your tails,* I said. From that day on whether the outfit warranted it or not, Jessie came attired in tails. At holiday time her tails became very special because they were decorated with witches and pumpkins or turkeys and Pilgrims, tinsel and holly, or valentine hearts -- whatever the season warranted. She'd conceal them sometimes and great would be his joy when he found them.

Adults were very helpful with manual competence activities to develop hand dexterity and reading stories to him, but mine was the most tedious part of the program -- the four hours of creeping and crawling. He crept up on the rug beside the crawl box and then crawled down inside it. Usually I motivated him with tidbits of food at the end of each crawl. I crept beside him many hours each day, enticing him toward that bite of applesauce or morsel of graham cracker. Of course, the reward had to be small or he'd soon be filled up and his desire gone. He ate all his desserts in this fashion.

What a delight it was when the energetic teenagers arrived in the afternoon! Grant really traveled for them. They had creeping races; they made back-bend bridges for him to creep under and tunnels with their arms for him to go through; they brought toy cars and other objects for him to hold at the end of a crawl through the box. They rough-housed and sometimes brought friends, having fun with each other while taking their mission seriously. It was a good healthy fun time for Grant and he worked hard for them.

Cort had quite naturally slipped into the role of big brother. No emphasis had been made of ages and, apparently realizing his greater maturity, he treated Grant as a younger brother. He was interested and helpful with the activities that we described as "getting Grant's muscles." We hadn't explained to three-year-old

Cort about Grant's brain-damage. To him, this routine was just a way of life. It wasn't until our patterner Carol Staiti returned with her newborn baby after missing a few weeks of patterning that we realized that Cort needed an explanation. After he quietly viewed the beautiful new child, he thoughtfully said to me, *Mom, after Grant learns to walk, let's give the crawl box to Mrs. Staiti so she can help her baby get muscles.* We realized he thought everyone had to go through that ritual to learn to walk and talk. (Parenthood must have looked like quite an ordeal to him.)

Although Cort's way of life was obviously different, he was by no means neglected. We always allowed him to stay up a little later than Grant to give him individual time and attention, and either Gerry or I made a point to do something special with him on weekends when we took turns working with Grant. Wonderfully, the patterners did a great deal to enrich and fill his life. Betty took him home to play with her son Steve after patterning on Thursdays, and Aunt Shirley took him on Fridays to play with Cousin Derek. Many others invited him to join with their children for special outings or for an afternoon's play. When the brothers from the Novitiate came, one or two regularly spent time between patterning sessions to build blocks or play games with him. Many women read stories and teenagers took him for walks. Like Grant, he enjoyed the constant flow of people, which may account for his present outgoing sociability.

One day, while I was in the dining room substituting for a patterner who was unable to come at the last minute, I looked out the window to keep an eye on Cort playing under a crabapple tree, and as I watched, he climbed up the tree. The achievement moved me to tears. How blessed to be a normal child, to be able to so naturally perform these wonderful skills. Needless to say, the ladies patterning with me were alarmed at the sudden outburst of crying. I'm sure they felt I'd emotionally collapsed from the strain of Grant's program. We all kept on patterning, but I felt their desire to comfort me and their hopelessness to do so as we quietly continued the exercise until I composed myself and explained. They were relieved. They understood. Such a deep feeling of

gratitude for a child's natural ability would not have occurred to me years earlier when I took so many good things for granted.

We have learned not to take the good things in life for granted. At that time, I felt thankful for Cort's ability to climb a tree. Recently we were reminded of another natural ability when Gerry's brother Dick called to tell us the rehab exercises have helped his wife Shirley's balance to improve. What joy was in his voice! She has Parkinson's disease, an illness she and Dick have been dealing with valiantly for several years. Balance is something very important that most of us take for granted. Through Grant's struggles, Shirley's struggles, and that of other friends we have come to be thankful for the everyday things that don't cost money but do make all the difference in the world in our lives. Just as our minister Stephanie once said, *The most important things in life are not the things you can buy.* Let's not take them for granted. They are blessings.

Just as happiness increases when we share it with others, so does anxiety decrease when shared with friends. Our lives were eased when Grant's problems were known and understood by all. People want to help, are kind, and accept the cares of others when the bearer is open and straight-forward. When we try to hide or cover up problems, people shy away. They know something is wrong but because we don't talk about it, they hesitate to reach out to us for fear of hurting our feelings. When we try to conceal, we only hurt ourselves. How good it was for us to be bonded with so many in our work with Grant!

Thank goodness, the world has many wonderful physicians. But occasionally, there are some whose advice causes stress and suffering. We experienced it when the doctor told us Grant's problems were psychological, so when a distraught mother who heard about Grant called saying she had a problem with her five-year-old daughter, I was glad to provide a caring ear. She wanted to discuss what we planned to do about Grant's education. Hesitantly, she said her daughter had been diagnosed with ... Down syndrome. Her doctor told her to *Bluff it. Don't tell anyone. She's not too bad.* For five years, the poor woman carried the

burden of covering up for her child's slowness to protect her from society's discovering her unspeakable condition.

I invited her to come for a visit. She brought her daughter, a very gentle, well behaved child. It was obvious she had been lovingly raised. I realized this Mom had to face the situation if she were to move forward. I talked openly about Grant's problems and asked questions about her daughter, using the term Down syndrome with acceptance and respect to let her know that it was not something to be ashamed of. Gerry and I were fortunate that the Institute's program forced us to share Grant's needs. It gave us a freeing feeling. This woman's doctor told her to hide it giving her an imprisoning feeling. Awful. When a mother is made to feel shame for her child, how can that child feel happy? We became friends and good supports for each other. Her daughter attended a special school, later worked in a sheltered workshop, matured well, and was happy. And, her mother was happy too. We all have problems of one sort or another. The following quote helped me a lot, and I think everyone can benefit from thinking about it:

It's not what happens to you, but how you react to it that matters.
Epictetus

A very important revelation came to me during a patterning session when I heard some young mothers discussing their own children, and one of them said, *When I get home, I'm going to hug my children. I've never really appreciated them before or thought about how fortunate I am that they were born okey.* Grant had brought a special awakening into their lives. It made me think that maybe the world needs a Grant now and then to help us better appreciate His more perfect works. A blessing from Grant.

6. Outreach from afar

Where there is a human being,
there is an opportunity for kindness.
Lucius Annaeus

We learned that the goodness of people was immeasurable. A stranger's voice on the telephone said, *I'm Anita Towle and I live in Newtown Square, Pennsylvania, which is only about a ten-minute drive from the Institute. I learned about Grant from relatives in Barre, and to thank God for our four healthy children, my family would like you to share meals with us and stay at our home when you come for your medical visits.*

I couldn't believe my ears or imagine imposing on someone, especially someone I didn't know. I told her how difficult it would be with Grant, his feeding, needing a crib for sleeping, how frequently we came, and that I always brought a companion to help with him during the ride. Anita was not dissuaded. She was convinced that her family could ease our trips. It was something she wanted to do. Grant had appointments scheduled with Mr. Linton and Dr. Spitz a few days following that conversation, and Anita drove to the Institute to meet us after we finished.

Her personality matched her voice on the telephone -- cheerful, energetic, caring. I liked her immediately. We had a delightful surprise in store when we followed her back to her home in Newtown Square which, like Media is a large bustling suburb of Philadelphia. When we turned off the busy highway into her driveway, a hillside with the most beautiful pastoral scene you could imagine was spread before us -- pastures with peacefully grazing Guernsey cows, several barns, outbuildings, and old stucco houses nestled snugly along a meandering lane. Even though we were to go there many times in the coming years, we always paused at the first turn to absorb the quiet vista that never lost its enchanting effect ... and the family beyond was just as welcoming as the setting.

Anita introduced us to the family -- her husband Win, two teenage daughters, a teenage son, and a newly-married daughter who was visiting while her husband was away on business. We learned very quickly that this family had full busy lives -- Anita was a teacher, Win managed the fantastic farm which was owned by a DuPont family, the teens had school activities, and all of them were involved in church functions. Nevertheless, they were never too busy to reach out to meet a need. Happiness and goodwill pervaded their home. Some words that always stayed with me were said in a sermon by Rev. Illingworth, the minister who married us 55 years ago. He said:

If you ever feel blue, go out and do something good for someone.
It will not only bring happiness to the person,
it will also make you feel good.

I don't think the Towle family had a chance to feel blue. It is said that you get what you give, and they gave a lot.

From the first, Grant was lovingly accepted. They got their old crib down from the attic and set it up permanently in the guest room until Grant was able to sleep in a regular bed. We had supper and breakfast with them. No fuss was made, we felt like part of the family, including helping them with dishes or whatever chore the family might be doing.

Their family life inspired us and trips became like ones to visit dear friends or relatives rather than the cold stays with picnic lunches in a motel room that we'd previously had. Win took us to the barn where we caressed the baby calves and enjoyed watching the cows being milked. Gerry was fascinated as Win explained the intricacies of breeding a prize herd.

For the next two years, the Towle's home was our motel. Their goodness indelibly touched our lives, and the friendship that developed lasted until Anita's death two years ago, Win had passed several years prior. Their daughter Pat lives in New Hampshire, and when they went there, they also included a stop to see us. No trip here was complete for Anita without seeing Grant, and their love for each other never dimmed.

During the first year of programming, we had a surprise from another stranger. We received an envelope in the mail with a check and small slip of paper stating, *This is for Grant. I am praying for him.* The address on the corner of the check showed that he was from Wisconsin, a big mystery! I wrote him a thank you letter and described Grant's problems and progress. We never heard from him again but the mystery was solved later when I told the Towles about the check and note. Their daughter Joyce, who was in college, gave a talk about Grant in her speech class, and a young man from Wisconsin, who was handicapped from a motorcycle accident, heard it. He understood Grant's situation and decided he wanted to help in what way he could -- prayer and money, a powerful combination. After class, he asked Joyce for Grant's address. We were inspired and grateful for that compassionate man, also thankful to know he had a strong faith to sustain him on his own difficult path. We used the check to pay for the next evaluation fee.

A word from the heart goes straight to the heart.
Abbe Huvelin

Every month we journeyed to Media for the spinal shot and EEG, and every third month the trip also included an evaluation appointment. Gerry always accompanied me on the evaluation trips but a friend, relative, and sometimes a patterner joined me for the regular monthly visits. The EEG's continued to show more healing of the lesion and the line on the graph which had previously risen slowly began to climb more rapidly. Everyone involved in the program anxiously awaited the news when we went for appointments and shared our joy when we returned with a good report.

Between appointments with the doctor and the Institute, we and other families sometimes spent long hours in the waiting room. As we waited we talked with folks from all over the country. We were all in comparable circumstances and conversation came easy among us. We learned that many had been through similarly

unsatisfying searches with their children and were there as a final hope. It was during these sharing sessions that we learned the important value of support systems. Our shared stories helped one another to adjust and cope. Sometimes when we face a difficult situation, we feel alone. It's a comfort to share with people who are facing similar challenges. We were not alone.

With each evaluation many activities were changed, reduced or eliminated and new exercises added. The mask was added during the early period and used for many months. Doctors Doman and Delacato estimated that about fifty percent of the oxygen we breathed was utilized by the brain. Mr. Linton said Grant's breathing was shallow and that the masking exercise would stimulate deeper breathing. The mask was a plastic bag-like device that went from his nose to under his chin, and a rubber band went around his head to hold it tightly over his nose and mouth. When he breathed and exhaled, the small bag quickly filled with carbon dioxide, and he had to inhale deeply to draw in oxygen through the small opening in a tube at the bottom. Grant had to use this every thirty minutes for one minute, eighteen times a day. He seemed to be gasping for air by the time the one minute was up.

Grant's crawling developed well but his left side was much slower to develop than his right. As a result, he was given his most disliked and difficult task. For one hour a day he had to crawl through the box with a splint on his right leg. Pushing over that resistant rubber with his much undeveloped left limb while dragging the right leg wrapped in a heavy wooden splint was a terrible chore! Every inch forward took a great deal of persuasion on my part, but I was determined to have him develop to his greatest potential and we persevered, an ordeal that wasn't easy for either of us, for him physically, for me emotionally. It surely was an exercise in tough love. And, I'm glad to say that the exercise did help. The left side became more flexible and less deformed. We learned that it took a great deal of firmness along with love and patience to help a handicapped person.

Parallel bars and an overhead ladder were added to the program, and a few more pieces of furniture were delegated to the spare room upstairs. Even a foreign visitor with the wildest imagination

could not have mistaken those additions to our living room as modern furniture. The room had the definite appearance of a gymnasium or health spa.

The parallel bars were low to aid him in walking. When most children first get on their feet, they start walking sideways while moving around the playpen or holding onto furniture. The same was true for Grant. He wanted to walk sideways holding onto only one bar and it was really difficult for him to move straight forward holding onto two. We walked behind him, bending over to hold his hands on the bars, and as we slowly moved his hands ahead on the bars, bit by bit, he followed with his feet. He looked so adorable to us standing upright, wearing the little sneakers that were prescribed for this exercise. After a slow start, he practically got so he could race up and down as he held onto the bars. After only being on the program for seven months, he began to take some steps independently, only five or six, but an unbelievable accomplishment. We could hardly wait to tell Mr. Linton.

However, the next meeting with Mr. Linton was as astonishing as Grant's walking accomplishment. Bursting as we were to tell the good news, we entered the office with the usual handshake and were dumbfounded when Mr. Linton greeted us with, *When did Grant start walking?* Did he know this was forthcoming? Could he tell by Grant's appearance? Could he read it on our faces? We never knew, but our surprising news was no surprise to that perceptive man.

Grant's overall progress was fantastic in so many ways. He could use his thumb and fingers quite adeptly. He could take apart and put together nuts and bolts. He could draw circles. Walking on his knees and elephant walking on his hands and toes were added to further develop balance. His vision improved and eye exercises continued. He had been on the program for nine months, exercises with peanut butter and gagging were continued ... but still no talking.

The program was reduced from seven days a week to six, with the requirement that the day off MUST be used for an outing. Except for our visits to the Institute, we had done nothing but programming every day for nine months. What a great prescription

it was, not just to have a day off but a day off with the requirement that we have an outing! Enthusiastically, we planned each week how the special day would be spent. One patterner even compiled a long list of places and events that would make good family excursions. We would pack a big lunch and leave behind household chores to joyously absorb natural places, country fairs, and historic points of interest, the weather often determining the final choice of activity. The outings were great for Cort, but we doubted that Grant grasped much from them. However, we felt the change of routine and participation with the family was good for him. Also, it was good to acclimate him into a more normal lifestyle of not constantly being the center of attention. But, the truth of the matter is that the day off probably benefitted me the most.

Selfishness. Being selfish is not a quality we would usually consider desirable, but sometimes it can be a virtue. A fulltime caregiver is always nurturing, always giving, always thinking about someone else. It's a God-given role, one that is very rewarding, a time that allows us to grow very close to a person, to see the inner qualities a person possesses, to grow in love. When my brother-in-law, Dick, who cares for his wife, told me *,I love Shirley more now than I ever did,* I was reminded of when I empathized with our former neighbor Charlie Allen years ago about the demands of his bedridden wife Gladys, he said, *I wouldn't give up these years for anything.* Fortunately, both of these dedicated husbands had help so they could leave the house for short breaks. Caregivers need to think of themselves. They need nurturing in order to keep nurturing others. Without caring for ourselves, we wouldn't be able to do our jobs well or cheerfully. Asking for or accepting help is not something to be ashamed of, and when other people come in, it gives our loved ones an opportunity to visit with new faces and expand their limited lives.

Gerry was wonderful about taking over Grant's exercises after he got home from work, and I can remember even requesting to mow the lawn so I could get out of the house for awhile. Caregivers should never feel guilty about needing relief. We must

be fed in order to be cheerful feeders. We need a bit of selfishness. The one-day off each week was a tonic for me.

On Christmas Day at Grammy's house, all the family noted the great strides Grant had made. For the first time, he could open packages and explore through the house. One time, when he was missing for a short time, we found him in the kitchen beside Grammy's breakfast cupboard. There he happily sat surrounded by a huge mound of cereal, opening and dumping box after box. I was not pleased to see him in the midst of such a mess and the cause of so much waste, but the rest of the family laughed and was delighted to see how far his manual competence activities had brought him. Uncle Dick halted clean-up operations until he had properly photographed the achievement. This photo of Grant is now one of my favorites though the scene at the moment was not.

Because his pituitary gland had been inactive, Grant was undersize and growth changes were extremely slow. After one year of programming, the tests showed that his pituitary gland had begun to function normally. His best little green outfit that he'd worn for two years suddenly was too small. I was truly amazed at how the stimulation of the program changed his growth pattern.

At that time, our schedule was reduced to a regular five-day work week with weekends off, one day reserved for an outing and one day to be spent at home. We had been enjoying our outings for three months, but had not had a day at home with Grant for one year without programming him. What a difference! That previously helpless little boy had become able to open drawers, unscrew bottle caps, take out pots and pans, etc., etc., etc. He made up for all the mischief he'd never gotten into and we couldn't take our eyes off him for a minute. He was four years old and didn't know how to play or what to do with himself. It was strange for him to be at home with just the family. He missed the patterners and all the attention they gave him. The day at home was really a traumatic change. We realized that great behavioral adjustments lay ahead, but our biggest concern was that Grant was still not talking although we had industriously worked on every exercise Mr. Linton suggested.

I remember how cute it was when Cousin Beth decided that perhaps she could teach him to say "ouch" while he was having his salt bath. She believed the occasion warranted that word and she'd "ouch" dramatically as I scrubbed away. Unfortunately, her efforts were in vain.

Mr. Linton resorted to an unusual experiment in hopes of initiating speech. We were ordered to have no talking in the house from 9 A.M. to 5 P.M. every day for two weeks. It was hoped that the silent treatment would induce Grant to talk. Mr. Linton said he'd had good results from other patients by using that method.

Our first thought was Cort, who was a perpetually chattering three-year-old. How could we quiet him? My parents solved that dilemma with a two-week vacation for him at their house. The experiment began. I put a big sign on the door, "NO TALKING FROM 9 A.M. to 5 P.M." The first group of patterners to arrive looked at the sign in bewilderment and when I rushed out to explain, they asked, *Have you stopped patterning?*

How strange it was to see groups of women gathered together during that two-week period without exchanging a word! Many men would have claimed it an impossibility, but I witnessed its achievement. And, poor Grant was totally confused and upset by the absence of fun, coaxing, and loving words.

The funniest reaction came from the milkman. In those days, milkmen made door-to-door deliveries, and he always brought my milk inside and put it in the refrigerator. He didn't know about Grant's program and probably wondered about the steady stream of people coming and going, and then after I put up the "NO TALKING" sign, he stopped bringing in the milk. I decided to watch for the truck and tell him it was still all right to bring in the milk. When I went to the door and whispered, *It's ok to bring in the milk,* he gave me the most bewildered look and asked, *What kind of religion are you anyway?* When I whispered back that we were trying to teach our child to talk, he shook his head and left, more puzzled than ever, but he did resume bringing in the milk.

The strain of the two silent weeks only brought forth tears from our usually happy Grant. No words. The experiment went on his speech record, along with all the other efforts, as a total failure.

At the next evaluation, Mr. Linton had another suggestion. He told us to try having Grant say "Hi" to the patterners while they remained in the kitchen, and tell them not to enter the living room until they heard his greeting. Now that was motivation ... and it did work! They'd call *Hi* over and over but wouldn't come in as I explained to Grant that he had to greet them so they'd know he was happy to see them and then would come in. Soon he started saying a soft *Huh* and gradually it got louder as they'd encourage him by saying they didn't hear it yet. From that day onward, speech came slowly, but steadily, increasing from a few words to short sentences. His sentence structure, clarity and vocabulary improved through the years, but speech remains a struggle to this day. Sometimes the words come out quite smoothly, but often it takes time and effort to get out what he wants to say. Nevertheless, we are grateful for this milestone that almost seemed insurmountable. Life without communication has to be like an imprisonment. I often think of people who had a stroke and were left unable to talk. You can see the intelligence in their eyes and desire to communicate. What awful frustration they have to feel to be unable to express their thoughts. We've learned from Grant that a person with hesitant speech needs a relaxed environment with a very patient and encouraging listener who will allow plenty of time for the person to find his words.

It was spring, Grant had just turned five. He was three and one-half when he started the program at the Institute and amazing changes had happened in that one-and-one-half-year period! The EEG readings showed that Grant's lesion was healed. No longer would he need to see Dr. Spitz for spinal shots. No longer would his brain damage worsen. No longer would he need patterning. His program was reduced to five hours a day for the next five months, exercises for us to continue at home to maintain the skills he developed and to further develop his balance and talking. Our big assignment was to enroll him in a nursery school in the fall.

All good news, although we wondered what school would be willing to "tackle" Grant whose care was as demanding as that of an eighteen-month-old baby. When we arrived home, I called the patterners to thank them and tell them the good report that their

service and goals had been successfully completed. Everyone shared our joy in the success of the great effort, many had loyally been with Grant right from the beginning. Those dedicated people had brought a lot to our lives, not just their able bodies to pattern, but their spirit, their hope, their caring. I felt like crying when some said their feelings were mixed -- happy about the outcome but sad to see their participation come to an end. We knew that many eyes and ears would be following Grant's progress for years to come, the benefits of a small town.

The five hours of programming alone with Grant during the following five months were long lonely days, and our search for a nursery school was discouraging. No one felt they could give him the time he would demand and most felt he'd upset their whole program.

I understood that he'd be difficult to handle in a group situation. He took a great deal of my time at home. He didn't entertain or work well by himself. Other than Cort, who catered to his needs, he hadn't interacted with other children. His social relationships had been with the patterners. His whole life had been very individualized and sheltered. He knew nothing about living in or with the outside world. How could he fit into a school program?

Then I met Sister Irene.

7. Starting school

Never be so busy as not to think of others.
Mother Teresa

We are not Catholic. We both were reared in Congregational churches and when one of Grant's former patterners suggested that we investigate the new Montessori School in neighboring Petersham, I didn't feel optimistic about Grant's chances of being accepted. The school was run by the Sisters of Assumption, and though the school was interdenominational, I wondered if the nuns would accept a challenging child like him, especially from a non-Catholic family. I had never met any nuns but had heard stories of their rigid discipline from some friends who attended Catholic schools. I imagined them to be very formidable women.

Not feeling sure if I would be comfortable having Grant attend a school with such teachers, I decided to request an appointment to observe a class in session. Wow, was I wrong about judging nuns! Another lesson learned -- not to pre-judge someone or something I know nothing about. How could women who devote their entire lives to teaching children and serving God be anything but good people? The nuns who disciplined my friends probably administered punishments that were needed.

I was thoroughly impressed with the warm, happy, stimulating atmosphere and amazed at the fantastic accomplishments of the youngsters -- the reading at such a young age, the spelling, the mathematical concepts. What a feat to teach such a class! Sister Irene, a small, agile, middle-aged woman, was in constant motion, moving about the room cheerfully and patiently helping, motivating and checking on all the individual and small group activities taking place. The children were so absorbed in their work that they became unaware of my presence. My thoughts went to Cort and how he'd thrive in that environment, but could they possibly consider accepting Grant with his limited speech, unsteady walking, need of assistance in toileting, and unpredictable behavior? I hardly dared to hope.

During the interview that followed the observation I explained our need to place our brain-damaged son Grant in a school situation and also my desire to register Cort as well. Sister Irene was very interested and set a date to visit our home to observe me working with Grant on his program.

As always when we had a new opportunity, I prayed that Grant would give his best performance. Fortunately, he went through his routine cooperatively -- crawling, creeping, masking and manual competence. I explained his past experiences and the program that brought him to his present level of development. Her desire and eagerness to help were obvious. She said, *This is what we're here for -- to help wherever we can.* Not only did she want him to come to her school, but she added that perhaps she could get some financial assistance for him. I could hardly believe my ears. I wouldn't have been surprised if she had said we would have to pay a greater fee if they accepted him. It turned out that the school did not give scholarships for handicapped children but her effort in pursuing one moved us greatly and displayed her whole-hearted desire to help. Great dedication is definitely essential when dealing with a child with multiple problems. A bright child will achieve almost in spite of the teacher but when a retarded child achieves something, the instructor can truly feel rewarded for a job well done.

His acceptance, especially at such a fine school, gave us tremendous relief. During that summer, Grant and I had a good opportunity to get better acquainted with Sister Irene when she and another Montessori teacher, Sister Jeanne, accompanied us on the trip to Pennsylvania when I took him for his next appointment. The sisters had trained in Philadelphia and were anxious to visit their Montessori instructor, so I provided them with transportation and they helped with Grant in the car. He responded delightedly to these women who read books to him and tried very hard to converse and understand his unclear speech. They stayed with their instructor and Grant and I stayed with the Towles. I picked them up the next day after our appointment for a pleasant ride home. I was filled with thankful thoughts that our two sons would be in the

hands of these lovely people, and I'm sure Sister Irene was gaining insight into her forthcoming mission with Grant.

Grant did not attend during the opening weeks of school. Sister Irene wanted that time to get the class adjusted to the school routine and also to prepare them for the special boy who would be joining them. And, how well she did prepare them! Grant was received like a little prince. He was lovingly accepted right from the start and became everyone's project. The children wanted to wait on him, hang up his jacket and hold his hand. In their eagerness to help, they allowed him to do very little independently, and it was a chore for Sister Irene to gradually wean him from all his helpers to do tasks alone. Importantly, he was never rejected. For children to learn compassion at such a young age is a wonderful lesson and Sister Irene taught it well. Also, a great deal of credit goes to the parents as there was never a complaint from them about the extra time and effort Grant required.

Grant and Cort with Sister Irene

Years later, a mature teenage girl approached us while we were dining at one of the children's favorite restaurants. She introduced herself and said she was one of Grant's classmates in Montessori School. We would not have recognized her, but she and her family recognized Grant, whose distinct features remained quite unchanged. How surprised I was when she told us that the small photo of him that we pasted on a paper heart for a Valentine's Day card was still on her bureau! I realized that the kind loving place for him in the children's hearts made a lasting impression.

Besides the children, several others helped Sister Irene meet Grant's needs. Mr. Duteau, the caretaker at the convent, carried Grant down the stairs. Grant had mastered the stairs at home on his hands and knees, but the people at school were cautious because the steps were hard and uncarpeted. Some of the retired sisters worked with him in the classroom to give him the extra time he needed to master certain skills. His favorite times were with the nurses on visits to the infirmary. They stimulated his speech and increased his vocabulary by talking with him about their work and the equipment they used. To this day, he talks about those nurses, Sister Judith and Sister Jacqueline.

Some of the sisters, though, in their loving sympathy would spoil Grant if Sister Irene wasn't watchful. One time she had to leave the classroom to get something and another sister stayed with him. When she returned, she found the other sister rocking him and saying, *Poor little boy, he was so thirsty. He asked for water and drank a whole glass.* That was a no-no. Because of the brain surgery, the program required him to be on a restricted liquid diet, and whenever Grant found himself with unsuspecting people, he'd try to inveigle a drink from them. He loved to go to the sink, and Sister Irene would spot him sucking on the sponge. The water bubbler was also a place that required guarding.

The Montessori method was designed to appeal to all the senses. I remember way back when I was in college that professors said, *The more senses you can appeal to when teaching a subject, the faster it will be learned and the longer remembered.* I saw that emphasis in the Doman-Delacato program and again with Montessori. The Montessori methods that get such fantastic results

from normal children, were perfect too for a child with special needs who so desperately needed extra stimulation.

With both boys happily settled in school, I decided to apply for the remedial reading teaching position in the local school system. It was part-time and fit perfectly with my free-morning schedule. Not only did the job pay for the boys' schooling, but being back in a classroom was fulfilling for me. The days were full with my job and Grant's five-hour afternoon program, but with the boys in school and weekends free, our routine became more normal.

Grant made striking progress in coordination and speech, but his concentration on any given task developed very slowly. Dr. Spitz prescribed the drug Ritalin to increase his attention span, a new drug that held great hopes for children with attention deficit. Grant took the medication for a few months to give it a fair trial, but it caused behavioral extremes, from becoming overly stimulated, causing him to sleep little, then growing physically lethargic. He often stayed awake until one or two o'clock in the morning, and then was drooped in a chair or across a table the next day. The drug dosage was reduced and eventually stopped. Sister Irene had other ideas for increasing his concentration and one activity did help -- stringing beads. Grant loved stringing beads ... and still loves it.

During his year at Montessori School, his progress leveled off. His evaluations at the Institute showed only slight improvement, improvements he likely could have made on his own with the lesion healed. We were still continuing the five-hour afternoon program but the slow advancements were discouraging. Gerry, my realistic husband, urged me to stop programming, a thought that scared me. I knew the Institute held our only hope for Grant's normalcy and severing that tie was difficult to face. We continued the exercises during several months of indecision and talked with other families when we went for evaluation appointments. Some families had been programming for years with children who had not nearly attained Grant's level. These families told me that their children too had made great strides at first but in many cases there still was a long way to go. Though some did progress to a normal range, it was obvious that Grant and many others would never

attain that level. Even though changes were slow the past two years, it was really tough for me to "face the facts," as Gerry put it, that Grant had benefitted all he could from the program.

Gerry also discussed the importance of not neglecting our normal child or sacrificing all our time and energy for the child who could never achieve a normal life. In our experiences with other parents of retarded children, both of us had often recognized this blind dedication. We didn't want to be blind to the situation, but it was very hard to stop. Parents become so emotionally involved with handicapped children who need them so much. Ann Landers said:

Some people believe holding-on and hanging-in-there are signs of great strength. However, there are times when it takes much more strength to know when to let go and then do it.

Gerry thought of the words in Kenny Rogers' song, "The Gambler:"

You gotta know when to hold 'em,
Know when to fold 'em.

Finally, after three years of regular visits, including almost two years of patterning, we made the difficult decision to stop the program. Though Grant was not normal, we were grateful for a much improved son. He had made great achievements, two really amazing ones. He could walk, though his balance was not good. He had a limp but did quite well while in motion, but when he stopped, he had to hold on to a something or he would often fall, something he became was very adept at, as he could fall without hurting himself. We always said he would have been a great football player as he wouldn't get hurt when tackled. The other great achievement was that he could talk His speech wasn't clear but it was still improving. His healed brain lesion gave him the opportunity to continue to grow and develop, and we hoped he would improve enough physically and mentally so he could fit into

society in some capacity. Our outlook for his future was lowered but still hopeful.

That summer with Grant six years old and Cort five, we planned our first vacation trip, a camping vacation on Cape Cod. Lucy, Grant's former babysitter, who was home from college, joined us to help with the boys. We all delighted in the National Seashore scenery and soaked up the sun, built sand castles, and frolicked in the vigorous Atlantic. It was there that the episode we later called "The Hairy Man Story" occurred:

Grant, who is always drawn to something unusual, ventured off while the rest of us were lying on our blankets. We couldn't have taken our eyes off him for more than a few minutes but he could creep at record-breaking speeds when he had the notion. We were terrified when we found him missing until a quick survey revealed him only a few blankets away gently massaging the chest of the hairiest man I'd ever seen. The man was just lying there with his eyes closed, and Grant was completely intrigued with his thick coating of black curly hair. We couldn't tell whether the man was annoyed or amused by his young masseur. We watched and debated about who should reclaim him. Lucy was elected. We decided that no man could be angry with a striking young blond in a bikini ... and he wasn't. We wished we could have seen the man's face when Grant first arrived and started rubbing his chest. Was he awakened from a nap? He must have been startled, and then perhaps looked at Grant and decided the absorbed masseur was not threatening. Maybe he settled back and was enjoying the rubdown.

Anyway, that was Grant's first unique adventure out in the real world, there were many more to come.

Grant and Cort returned to Montessori School in the fall, but I didn't return to teaching because I was pregnant. For the first time in six years, I had no therapy schedule, clinical testings, or evaluations to attend. It was a marvelously free feeling. My only apprehension was the thought, *Would this baby be all right?* We hoped and prayed, and the sisters at the convent prayed too.

At this time, we learned about Sister Alice, a speech therapist at Mercy Centre in Worcester who was well-known for her outstanding work with retarded children. Grant was making good

strides in speech but his words were understood only by those who were attuned to his special language. Not only were his words unclear, but he stammered and struggled to release the message trapped in his mind. We felt that perhaps he might benefit from this service, so I enrolled him for a half-hour lesson once a week. He made progress under Sister Alice and began a friendship with another nun who would become an important part of his life.

On February 12 our third son, Barton, arrived. We thanked God for our healthy child and were pleased that he would be sharing a birth date with an important President, Abraham Lincoln. When the doctor came in to see me after I was settled in my room, he commented on how radiant I was and said that it wasn't usual for a mother to be so happy after delivering a third child of the same sex. He didn't realize how unimportant the sex was in comparison to hearing that all was well with the baby. It's amazing how our experiences affect our outlook on life.

What a joy it was to see Bart move so naturally the way we had patterned Grant! Later, he crawled and crept with the movements we had spent so many hours training Grant to do. Watching him made us appreciate how well the Institute had studied motor development. It was thrilling to watch our baby develop and touching to see how he was loved by his brothers.

We will forever be indebted to Sister Irene for so willingly and lovingly accepting and helping Grant at his most difficult stage of schooling. When I thanked her for all she'd done, the reply she made was so typical of that humble woman, *I considered it a privilege to have worked with such a special child as Grant. Working with him was an enriching experience for which I am very grateful.* We maintained a wonderful friendship with her until she became ill a few years ago and was moved to another facility. We visited at the convent and she joined us every year to celebrate Grant's birthday.

Today, children with special needs are integrated into the regular classroom. That was not the situation in 1971. He would have to go to a special school with facilities to meet his needs. The school hunt was on once more.

Rev. Illingworth holding Bart on the day of his baptism, Grant working hard to hold his balance for the picture

8. Teaching the whole child

Life presents many choices,
the choices we make determines our future.
Catherine Pulsifer

Choosing the right school is always a serious matter, especially for someone who has some kind of special needs. We were not considering sending Grant to a boarding school because we felt he'd benefit most from growing up in a family situation. He needed a school with a program emphasizing the development of the whole child. He was very lacking socially. He preferred to interact with adults, like the folks who catered to him during the patterning program. And, at Montessori School, Sister Irene said that though the children had been wonderful to him, he chiefly sought out adult company, wanting to hear the patient, lengthy explanations they gave him. She said that some children felt upset because Grant did not seem to love them as much as they loved him. He needed to learn to work and play with his peers.

I visited several schools in the area that had classrooms for special needs kids, but none seemed quite right for him. Then I wondered about Mercy Centre where he was still doing speech therapy with Sister Alice. The methods she used were very impressive -- enthusiastic motivation, warmth, and discipline. Teachers make the school, and it would take very special ones to meet his many needs. Her speech classes were part of the curriculum in a school for mentally retarded children ... but the school did <u>not</u> have a program for children with physical handicaps, and Grant was affected both physically and mentally. Nevertheless, I asked for an appointment to tour the facility.

I was given an appointment with Sister Eileen, the teacher who taught the six and seven-year-olds, the class Grant would enter if he went there. She struck me as a serious, rather formidable person, kind of like the nuns my friends had described about their Catholic teachers. But, as we proceeded from class to class, I observed an atmosphere of caring and stimulation blended with

good control. The children responded happily to the stern woman who was guiding me, and I could detect a loving interest under her firm facade. I was impressed. It was easy to understand why Sister Eileen was assigned the responsibility of piloting these young children into their first formal classroom routine. The job was very demanding, and without love and dedication, she never could have achieved it. An outstanding teacher is truly a blessing, and fortunately for Grant he would have many.

Often I felt an atmosphere of pathos when I was with a group of disabled or retarded children, but that wasn't what I felt at Mercy Centre. The atmosphere was vibrant with happiness. I could sense that these children had found a haven in which to grow and develop proudly. Pride, yes. Good self-esteem is the best gift we can give a child.

I am reminded of a friend whose grandson lost an arm when he was a small child, a terrible disability for sure. But, is he ashamed or embarrassed about his appearance? No. I'm sure his family was heartbroken about the accident, but he was loved, raised without pity, went to a business college, and is as self-assured, outgoing, and successful as anyone without a disability. He even builds furniture in his spare time. The battle is half won if you think the way my father-in-law Armand J. Gariepy emphasized in his motivational talks:

If you think you can, you can.

A parent can sense when something is just right for her child, and I knew that Grant's best chances for obtaining the development he needed was there, at Mercy Centre. But, again, that great apprehension arose in me. Would they consider taking him? He was different from their other students. He was physically handicapped. His comprehension seemed equal to the others, but they were not affected physically. We had to try.

We filled out the application for admission and anxiously awaited their response. We were soon notified that a home interview with the family was the next step.

Sister Eileen called and a weekend appointment when we would all be at home was scheduled. I was still a bit in awe of her and didn't know what to expect. (We later learned she was often referred to as "The Sergeant.") I prayed she would approve of us. The visit actually was very easy. She watched Cort and Grant play, took a peek at Bart who slept peacefully throughout the call, and asked such questions as: *Do you think Grant will be able to manage the stairs?* and *Though the classes are comprised of children from various religious backgrounds, religion is a part of the curriculum and Mass is held each morning. Being non-Catholic, what are your feelings about this?* We told her that we had no fears of Grant using the stairs, that he used the stairs independently at home. He didn't always mount them in the smoothest manner but he held on tightly to the railing and was careful. We told her we tried hard not to be overly protective and endeavored to have him do as much on his own as possible. We had no hesitation regarding religion either. Although some Catholic rituals differed from our Protestant services, we could only foresee good things coming from the emphasis on Christian teachings.

Sister Eileen added that the school was interested in working with children whose parents would join cooperatively with their efforts. Just as with normal children, when the parent groups join with the teachers to achieve certain goals, the combined efforts build a better school. We welcomed the concept of being able to work closely with Grant's teachers.

I had told Gerry about my visit to Mercy Centre, and after Sister Eileen's cordial visitation, he joined with me in my enthusiasm about it being the right place for Grant. We felt optimistic about the interview and rejoiced when the letter of acceptance arrived a few weeks later! We relaxed and enjoyed the rest of the summer with the knowledge and promise of the excellent educational opportunity that lie ahead for Grant in the fall.

Unless a woman is the mother of twins, she doesn't usually have two children starting their first full day of school at the same time, but that's how it was for me. Cort began first grade on the same day that Grant started at Mercy Centre. Cort walked to a nearby

intersection to wait for his bus and Grant was picked up at the door by a van that transported children with special needs or teens taking vocational courses at city schools. That first day seemed never-ending for me, both happy and sad, delighted with their readiness for this new venture and lonely from the emptiness that pervaded the house. When they attended Montessori, mornings were short, interrupted with transportation, teaching, and the needs of a small baby. Now Bart was a happy, easy, seven-month-old and I had no teaching or transporting to do. I found myself wondering how they were doing, especially Grant. I kept checking the clock and as the time for dismissal drew near, I watched the street for Cort to appear. How great it was when he came into view, skipping down the hill with papers waving in his hand and then listening to the exciting report of his day!

Shortly afterward, Grant's van arrived and he emerged proudly displaying a star on the back of his hand. Sister Eileen had told us how important these stars were and instructed us to respond heartily when one was earned and to be disappointed with its absence. She did add that all children would receive stars on the first days of school to encourage them. I responded happily and told Grant how proud I was. He was obviously happy and pleased with himself. No school day is ever as exciting as that first day is.

After the initial letdown of the boys leaving the nest, I began to enjoy and found satisfaction in a new lifestyle. I felt free. Bart was a good- natured flexible baby and easy to take places. I did things with other young mothers, I did more cooking, worked in the garden, learned to preserve vegetables and make jellies and pickles. If you're a person who likes to be busy, you quickly find new ways to fill your days, and it was wonderful to be involved in things outside the home -- the ladies' group at church, and later teaching Sunday School and being a Cub Scout den mother.

For Sister Eileen, Grant was not an easy pupil. She strived hard to have him function with the group rather than on an individual basis. At Montessori, Grant was the only special child and he received a great deal of attention, just as he had at home. At Mercy Centre, he was only one of many special children and did not respond well in group activities. He tried many ways to get Sister

Eileen's attention and probably the most successful was when he learned to vomit at will. That was difficult to ignore. She tried a variety of methods to control his behavior and anyone with less determination would have given up. Isolating him from her and the rest of the class was the discipline that was most effective. On one visit to the classroom, a very conspicuous object caught my eye -- a two-sided screen that enclosed a corner of the room, kind of like a little prison cell. I shuddered and hesitantly asked, *Is that for Grant?* She responded, *Yes, it's the handiwork of the custodian and it's a very successful contrivance.* I'm reminded of Winston Churchill's words when I think about Sister Eileen's work with children:

Never, never, never give up.

Grant hated solitary confinement. His vomiting stopped.

Sister Eileen was very positive in her approach, always complimenting and rewarding good improvement. When Grant started behaving better and completing his work within the group with less individual encouragement, she decided to select him for a most special treat. He and a few other children were chosen to make an honored visit to see the Bishop at his home. Social graces were begun in the first classes at Mercy Centre and stressed throughout the entire school curriculum. The teachers were proud of their accomplishments and took real pride in showing off their children. A visit to see the Bishop was the top reward. Sister spent a lot of time in preparation. She gave them lessons in etiquette on how to meet and talk with the Bishop. They practiced shaking hands, kissing his ring, and saying "Pleased to meet you."

We knew about the upcoming visit and told Grant how proud we were that he was going to have such a special treat. It was obvious that he felt very important about it, and Sister Eileen hoped that by emphasizing his grown-up behavior that it would encourage still more improvement and eliminate some of his silly attention-getting performances.

Finally, the special day came. The children were dressed up, primed, and excited about the great honor. They arrived at the house and together walked up to the door. The kind-faced Bishop opened the door to greet them, holding a cigar in his hand, and Grant, whose speech was not usually easy to understand, stated clearly, *Cigars stink!* Poor Sister Eileen saw all her carefully rehearsed words evaporate, but the good man of God saved the occasion with his reply, *You're right,* and he quickly extinguished his smoke. The introductions didn't go as planned, but it was a day well-remembered by the sisters and the Bishop. Grant did have special outings in the years ahead, but none back to see him.

The annual Christmas play was one of the Mercy Centre highlights. Sister Alice wrote the play using a Christmas theme adapted to the abilities of the children. Every child took part. It was the topic of conversation with the children for about a month before the presentation and they seemed to feel as important as great Hollywood stars.

Grant talked about being a shepherd. I couldn't imagine him on a stage performing in a play. We received a message instructing us to have him at the school about an hour early to allow time for make-up and getting into his costume. We left him at a designated area and proceeded to find seats and read the program which one of the students handed us.

A quick survey of the leaflet showed a nativity scene at the end. This, we decided, was probably Grant's shepherd scene. We wondered how he'd manage walking in a long shepherd robe with his poor balance. But, we relaxed and began to marvel at the accomplishment that unfolded before us. True, the children's speech was not always clear, but the theme of the simple plot was easy to understand and the songs and dances were amazingly well choreographed.

At last the finale came and the curtain opened to display the nativity scene with the serious faces of Joseph and Mary admiring the infant doll she held as the angels, Wise Men and shepherds looked on adoringly. Seated on a low stool was the little shepherd so familiar to us. I was relieved to see him sitting as the choir began singing the verses of "Silent Night" for I knew he could

never have held his balance for that length of time standing. Then, to my dismay, the little shepherd soon stopped looking adoringly at the small infant and began gazing about the stage and hall. The glint in his eyes spelled trouble, and I silently prayed that he'd contain himself throughout the sacred scene, but everyone soon saw how rapidly Grant could creep even when attired in shepherd robes. I closed my eyes fearing the worst as he raced across the stage laden with microphone wires. Fortunately, the always alert Sister Alice suddenly appeared from behind the curtain and speedily scooped him into her arms, planted a tender kiss on his cheek, and the two quickly exited. Once again, after the brief interruption, our attention focused on the historic event that has so wonderfully influenced and shaped the lives of people in a large part of the world.

Hardly an eye was dry as the play concluded, and when the Bishop mounted the stairs of the stage to congratulate the happy children who surrounded him, the words of Jesus came to my mind,

Let the little ones come unto me.

Often and deservingly, the Mercy Centre play was televised, and every year there usually was some unrehearsed act, but thankfully Grant did not star in that kind of role again.

Very often the giver of a service benefits more than the person who receives it. We are all compensated with a warm satisfying feeling when we know we've brought help or a ray of sunshine into a needy person's life. We knew a woman named Gussie, short for Augusta, and she would regularly be seen walking down the sidewalks in Barre with a basket under her arm containing freshly-made custards to bring to a shut-in or someone not feeling well. Her neighbor who only used the whites of eggs provided her with the yolks to make custards. Gussie didn't have a lot of money but her giving filled her life with happiness. All of us know people like her who are ready with a helping hand, and these people are happy people.

When Grant and Cort were entering their second year at school and Bart was one and one-half years old, I was asked to home-tutor an eighth grade boy with hemophilia. Feeling a reluctance to accept the responsibility, I didn't give an immediate response. I loved my freedom, and Bart, my best reason for refusing the job, was still very young. However, when the principal added, *The family will be very happy to have you bring your youngster with you when you come,* my excuse for rejecting the job was gone. Also, my conscience reminded me about how generous patterners had been with their time to meet our needs and how selfish I'd be if I didn't answer this plea for help. Little did I realize when I first consented to do the tutoring how much I personally would benefit from the experience.

I quickly identified with Chris and his family's struggle with his hemophilia. Though he was a normal boy in every other way, the concern for his safety and worry about his future had to weigh heavily on his parents' minds. His internal bleeding could be triggered simply by a quick movement, a jolt, or even such regular activities as walking.

When I met Chris, he was only able to attend school about fifty percent of the time. Often the swelling in a joint would be as large as a softball. Naturally, it was immensely painful and the joint, often the knee, would be paralyzed, crippling him until the blood was absorbed.

His mother would call me when he was having a bleed and unable to go to school. Often he went to the hospital for a transfusion, and then would recover at home for several days. During that healing period, Bart and I would go to school to pick up the assignments and drive to Chris's house to teach him the lessons. The sessions were two hours, and while we worked, Bart busily amused himself at a special large drawer filled with toys that Chris's mother prepared for him.

Some days when Chris was unable to go to school, but well enough to move on crutches, I invited him to come to my house for the day while his parents, who were teachers, were at school. In that way, I was free to do my chores while he took a test or did written work, and he liked the change of location.

It was during that winter that our house was finally getting a much-needed redecorating job. It had been nine years since we'd moved in and some rooms hadn't had new paint or wallpaper for many years before that. It was a special time we'd saved and planned for. Most importantly, the well-worn living room rug that didn't match our stuff was finally being replaced! When you decorate your home, it really feels like it's yours. We hired a friend who did beautiful work to do the job. This man, like me, found Chris an inspiration. Chris never complained about the pain and suffering he had to endure. It always amazed me to hear him talk about his interests and ambitions as though he had no obstacle to deter him, and he was very enthused about all the stages of the painter's preparation and work. For Chris to have a break to visit with the painter and watch the redecorating process was great motivation for him to have a good work session.

His family never sheltered or "babied" him, which I believe attributed to his cheerfulness and toughness. I realized a child with special needs must be tough and tried hard to follow their example. They are bound to face harder knocks in life than normal children, and their parents have to help prepare them or they will surely crumble and constantly need a protective environment.

In the spring, Chris's mother and I enrolled both Grant and Chris in the Easter Seal swimming program at a nearby YMCA, and they both benefitted from the experience. The staffing was excellent. The instructor had volunteer aides to work with each individual. Sometimes there were two aides for each youngster. The instructor was perceptive about each person's potential and directed the volunteers to help each to achieve his best. It was obvious that the young girl who worked with Grant found her vocational calling doing that volunteer service as she later went to college and studied physical therapy. Grant learned to swim under water. He never did master swimming above water, but it was amazing how long he could hold his breath. His strokes never became coordinated, he kind of swerved from side-to-side through the underwater as a snake might move, but it was smooth to watch and he looked so relaxed. He never looked relaxed walking, always was an effort to maintain balance. Swimming was his sport and he

loved it. We called his method "the Grant stroke." The Easter Seal program got him started and provided an enjoyable weekly outing. For Chris, the non-restrictive exercise that swimming allowed was excellent for a child with hemophilia.

The Home Infusion Program became developed the following year so Chris no longer needed my help. He carried two small bottles, one contained a powder with coagulant that had been extracted from a blood donor's contribution, and the other held sterilized water. By mixing the two and placing them in a syringe, he could give himself an injection. A fantastic breakthrough! No more painful bleeds and deformities for people with that dreadful condition. We probably all know someone whose life has been changed by medical research, Chris with the coagulant for his blood and Grant with the Depo-Medrol that healed his brain lesion. Blessings.

I felt happiness and gratification when I helped Chris and his family, the kind of joy Gussie must have felt making custards for people.

You get what you give.

Also, the association with Chris and his family inspired me then and in the years to come. As Grant entered adolescence, his behavior became a great trial for me. He had uncontrollable silliness that made it hard for him to talk, or walk, or eat without choking. He was especially out of control when he was excited, so sad when someone he loved came to visit and it was hard for him to interact in an acceptable way. Often, I reminded myself of Chris's family's patience and struggle and prayed that we too could have such understanding, faith, and courage to meet our challenge.

9. Counseling support

All of us, at certain moments in our lives,
need to take advice and receive help from other people.
Alexis Carrel

It's fortunate that often when we feel at the end of our rope, just the right person appears. Grant was eleven, I was at the end of my rope when Jim Howard came into out lives. Before I introduce Jim, let me explain why I was feeling discouraged.

Grant was still attending Mercy Centre. His academic progress was satisfactory, but dealing with his silliness was very frustrating both at home and at school. Laughter, laughter, laughter. It was seldom over something funny, it could be over anything, even something sad. I found myself wishing I could hear him fuss or cry. We pretty much gave up having outside activities because if we went someplace, we had to spend most of the time trying to keep Grant under control, and if we had a babysitter, so we could go out, Grant would not settle down for her, and his behavior would be worse the next day because he was tired.

Our friends and families were sympathetic and understanding as Grant mastered new ways to get attention. He became very attracted to jewelry, and still loves jewelry; stringing beads has become his lifetime passion. Back then, if he were left alone with a visitor for a short time, it was not unusual for me to return and find the person stripped of her jewelry and Grant decorated with necklaces, bracelets, rings and watches. People wanted to please him, but I feared broken watches or even a broken finger as he would pull quite vigorously on their rings. Often friends came overly adorned with jewelry so he'd have something spectacular to enjoy. Once, Aunt Sue brought him a small box containing ten play rings. His delight was fantastic, and he insisted on wearing all of them to school.

One "ring" incident I enjoy reminiscing about created a reunion with my Cousin Bev. Actually, Bev is not my real cousin, but the cousin of my cousin. We hadn't seen each other since we'd played

together when we were pre-schoolers. After she was married, she and her family camped on weekends at the local campground and recreation area where I took my boys swimming. She had learned from our mutual aunt that we went there and hoped one day that we'd meet but doubted the possibility in such a large place. Then, one time when my eyes were off Grant for an instant, he spied a woman in the wading pool with three rings on one hand. The temptation was too great and he thrashed over to her at top speed and pulled at the rings. Poor unsuspecting Bev wondered who's terrible child it was, and then thought, *Could this be Carole's boy that Aunt Pearl told me about?* I rushed over to rescue the poor stranger from Grant's grasp and was surprised to hear her ask, *Are you Carole Gariepy?*

It's an ill wind that blows no good.
John Heywood

Bev and I spent many pleasant afternoons visiting together while watching our children swim, and a real bonus for me from the reconnection was that she introduced me to a hobby I've enjoyed ever since ... quilting.

The jewelry magnetism waned thanks to the efforts of Mr. Hurley, an instructor at Mercy Centre. He had a "man-to-man" talk with Grant and promised him a coffee outing once a week if he would wear only one, masculine style ring at a time and stop pulling on people's rings. The motivation worked. He looked forward to the weekly reward and resisted the behavior that had previously brought him a lot of attention.

Asking about car parts was another attention-getting behavior. When he saw someone come into the driveway, he flew to the driver to ask routine questions: *Where are the emergency lights?* Kindhearted people would operate lights, blow the horn, move the seats back and forth, clean the windshield, and answer a myriad of car questions until we came to relieve them. Grant wanted interaction but didn't know how to communicate in a normal fashion, a problem that is improved but still exists.

If I took my eyes off Grant for a minute when we were away from home, he would "take off." One day he gave us quite a scare when he disappeared at the local pool. I checked the two pools. Cort checked the men's room. I checked the women's room. We both checked the snack bar. He was nowhere to be seen, yet he had been laying on the towel just moments before. Finally, I spied him wrapped up like a little mummy on a woman's blanket being soothed and rubbed. When I reached him, she said, *He told me he was so-o-o cold.* Grant continued that tactic till he developed a better one.

When we were camping with friends for a weekend in Rhode Island, we decided to play a game of miniature golf. We got the balls and clubs and were about to start when we noticed Grant was missing. It didn't take long for us to locate him as we heard him crying loudly to a nearby woman, *I can't find my mother!* To his disappointment, we recovered him quickly, only to discover that he was starting a new and successful method for getting attention whenever he could manage an escape from us. Strangers hugged and displayed sympathy for the poor lost handicapped child, and Grant soaked up the attention. It was hard to keep my eye on him every second when I had two others that needed watching too. Fortunately, unlike Grant, they were quite good about staying where they were supposed to be. Today we laugh about some of those situations, but at the time, it was not fun.

We love vacations. I always tell people that vacations, especially family vacations, are an investment in lasting memories. But vacations with Grant were a trial. When he was younger, we left him with my parents or friends on a few occasions, but the magnitude of the behavioral problem seemed greater when he was bigger, and we didn't feel we could ask someone to bear the burden that he had become, and no respite services were available at that time.

We wondered about using one of the state institutions for the mentally retarded for respite care. Would that be the answer? There were three such institutions in our area. We decided to investigate the possibility. Much to our dismay, they refused, saying that we could place Grant there on a permanent basis, but

not for just a week. To think that they would admit him permanently as a state charge but not let us pay board to have him taken care of for one week! Unbelievable. If we needed respite service, we felt others must also. We contacted our local state representative who understood the ridiculousness of the situation and thought a change to the outdated rules should be made. He soon notified us that the legislature agreed to his proposal, and Grant spent our next vacation at the state hospital in Monson.

The experience was not the best. The buildings were old, crowded, and drab, and the activities were limited, but his needs were met. Since then, the state has developed excellent respite facilities with programs that include appealing activities as well as good care, places where families can feel good about leaving their kids when they need a break or to take a vacation. Grant went to the respite house in nearby Fitchburg once or twice a year and loved it.

The state institutions have been pretty much closed down, and the clients are integrated into the community. They are supported in group homes, private homes and apartments. Many who were institutionalized as youngsters are now able to function satisfactorily in the community or in sheltered workshops.

One very good benefit did come from Grant's stay at the state institution. At the intake interview, we discussed his behavioral problems with Mr. O'Shea, the man in charge of the respite arrangements, and the following week when we returned to get Grant, he told us that he'd given the behavior problems some thought and told us about Jim Howard who might be able to help.

Jim was a student in the psychology graduate program at the University of Massachusetts and needed a case study for a semester to complete his degree requirements. We met with Jim and the head of the department to discuss our concerns, and they optimistically talked about ways they may be able to effect some improvements. Jim was in his early twenties, fair complexioned, average height, slim, long dark wavy hair, and was clad in jeans as were most college students. His manner was soft-spoken and pleasant. We were doubtful that great changes would result from his service but were willing to give anything a try. Jim made

arrangements to begin the following week and would come three or four days a week, usually in the afternoon after Grant arrived home from school, and added that sometimes he would go to Mercy Centre during the day.

When he arrived at the house, he instructed me to pretend he was not there so he could watch Grant "as a fly on the wall." Grant approached him with questions as soon as he entered, but because Jim's answers were brief and with no conversation between Jim and me, Grant soon regarded him as an uninteresting person, forgot about his presence, and behaved naturally. We went about our normal routine and Jim wrote in his black notebook. It amazed me how much he found noteworthy about our everyday life, but he wrote with a flurry as he observed Grant and after every little exchange between him and his brothers, or him and me, or him and visitors.

After several days of note taking, Jim said he would like to observe Grant shopping with me at the local market. He followed behind me to the store in his car as he didn't want Grant to see him there and slipped in quietly after we started down the first aisle. Grant's balance was secure as he pushed the shopping cart with me guiding it in back of him. It amused me to see Jim busily taking notes as he peered around the ends of the rows, or from behind a display case to spy on our shopping. However, as we progressed, I failed to notice him anymore and thought how cleverly he hid because I couldn't detect his place of concealment. I paid for our groceries and we drove home. I was surprised that Jim didn't drive in soon after us. The story he told when he did arrive was another great adventure to add to Grant's growing repertoire.

It seems that the clerks had some questioning thoughts about the long-haired young man who moved stealthily from aisle to aisle with no obvious intent to buy anything, he just peered around and wrote in a black notebook. Definitely a suspicious character. He was escorted to the manager's office where he was questioned by the manager and the policeman who had been called. I'm sure the two of them weren't expecting to hear the unusual account that Jim was about to tell to explain his dubious movements. Luckily, he had the black notebook to prove his alibi about scouting a

young brain-damaged boy to help correct his behavioral problems. He was released. Grant still likes to talk about that story, and I'll bet it's one the manager and policeman told their families about at suppertime that night.

Grant enjoyed the following months. Jim took him to the library, stores, and restaurants, always focusing on ways to achieve acceptable behavior, and as he learned methods that induced more serious and sensible responses from Grant, he taught them to us. He told me that my hovering over Grant to keep him from bothering people was not good. I wrote down what he said so I could share it with people who visited us, and am glad I kept his words of wisdom: *Instead of constantly restraining Grant, train your friends and relatives. People who think they're being kind to Grant when they allow him to pull on rings or answer foolish questions were actually encouraging poor behavior. And, as long as outsiders permit it, Grant will continue. You are uneasy and no one really enjoys the actions, even though, for brief periods, people are tolerant. Instruct people to expect and demand normal behavior. People want to help. They are condoning poor behavior to be kind and helpful. They will be demanding if you make them aware that it will help Grant. Behavior won't improve all at once, but he will learn that certain kinds of behavior get favorable response and will adjust accordingly.* I thought a lot about Jim's closing comments:

To a great degree, we get from life or from a person what we expect.
Expect normal behavior.

Jim's analysis made a great deal of sense. Grant had never behaved normally and we had grown not to expect it. We began telling people about our new goal and how we needed their help in attaining it. Some felt cruel and found it difficult to be firm or to refuse his request to wear their jewelry. We'd hear answers like, *I don't mind. It's all right. He doesn't bother us.* They felt sorry for him and had to be convinced that the approach would be good for him. Others were expert in handling him immediately and I

wondered why they hadn't been more demanding of him sooner ... probably they feared hurting our feelings.

Grant's behavior still has room for improvement, but his educators, the workshop staff, and Gerry and I have all strived to meet the goals set by Jim. Progress has been slow, but steady. I'm reminded of one of my father-in-law's greatest quotes:

Success by the yard is hard
But a cinch by the inch.

Grant's behavioral strides were surely by the inch. He still has difficulty maintaining control when he is excited. At those times, he will get silly and have trouble getting out his words, but under routine situations at home or at outside events and restaurants, he does quite well and is a pleasure to be with.

Another important lesson we learned from Jim's counseling was in developing Grant's independence. He said, *Don't be home tomorrow when Grant arrives from school. Come ten or fifteen minutes later.* I protested saying it was too risky. We had never left Grant alone. He was unable to use the telephone to call or get help if he needed it, and I knew he'd feel frightened. Jim retorted, *He is almost twelve years old, old enough to be alone for a short time. You will have a problem on your hands when he's eighteen if you don't start building independence while he's young. As far as risk goes, yes, there is a risk. But, you take an even greater risk when you let Cort go to a friend's house on his bicycle.* Jim was right again. That night I instructed Grant on what to do if I weren't home when he got there. I told him to stay in the den with his toys and books. When the next afternoon came, Bart and I waited at the top of the hill for ten long minutes, then rushed home to see how Grant fared. He was a little sad but was in the den with his books and toys. That was another first. We had been very protective of him, but it was time to gradually let go and allow him the opportunity to become as independent and responsible as possible. I thought of what Charles Lindbergh said:

What kind of man would live where there is no daring?
I don't believe in taking foolish chances,
But nothing can be accomplished without taking any chance at all.

Staying alone for a short period was a big step ahead for Grant, and as he grew older, he became comfortable being alone for longer and longer periods, also he had trustworthy behavior in the house when he was alone. Like a computer, he always did what he was programmed to do.

I felt quite desperate when I met Jim. He saw us as only an outsider could. The lessons he taught were invaluable. I learned:

The more normally a retarded person is treated by everyone,
The more normal his behavior will be.

I also humbly learned that everyone can benefit from some good counseling. We don't always see the whole picture clearly when we're too close to it.

10. Orthopedic problems

Don't dwell on what went wrong.
Instead focus on what to do next.
Denis Waitley

Grant's behavior slowly improved, but his walking worsened. His heel cords were getting tighter and tighter, to the point where he could not put his feet down flat. He was walking on his toes and often his left foot turned inward when he stepped. The nurse at Mercy Centre informed us about an excellent new orthopedic service where doctors from Massachusetts General Hospital were conducting clinics in area hospitals. I called and we were given an appointment at the clinic in Gardner which was the one nearest to us.

We were impressed with the competent and thorough manner of Dr. Wyman's examination. He told us that spinal deformities had developed because of Grant's lop-sided posture, something commonly seen in people with cerebral palsy. His weakness was on his left side and that side of his body was more undeveloped. His left shoulder sloped lower and his whole body lowered when he stepped with his left foot, causing him to limp. The doctor added that surgery could not remedy the spinal deformity, but braces could help his walking by forcing his feet into the proper position, making it easier to balance with his entire foot on the ground rather than having to balance on his toes. He prescribed steel knee-length braces that would be attached to high leather shoes.

Dr. Wyman also informed us that Grant was eligible for the services of a home physical therapist. He said that it would be impossible to correct his multiple problems but asked if there were some aspect that we felt could make a big difference in Grant's life if it could be improved. I immediately thought of his inability to get up from the floor to a standing position. When he lost his balance, which he did frequently, especially on uneven terrain, he had to creep to a wall, tree, or building to pull himself upright

again. There was a large field next to our house where the boys often played and if Grant fell there, the creeping distance could be fifty feet to reach an object he could use to get back on his feet. That laborious creeping trek usually meant permanently grass-stained pants. If he fell in the driveway, usually the creeping resulted in holes in the knees of his pants and rubber worn off his sneakers, which constantly had to be replaced because of holes in the toes. It was really awful to look out the window and see him creeping when the conditions were muddy in the spring or wet following a rainstorm. Although Grant never cried or complained, yes, he was tough, but it had to be degrading for a twelve-year-old boy to have to creep in order to get back to a standing position. His brothers helped him if they were outside, an easy task for Cort, and it really warmed my heart to see little Bart rush to him and rigidly brace himself so Grant could use him as a support for pulling himself up.

Dr. Wyman referred us to Barbara Bush (not the future First Lady), a physical therapist who would begin one-hour monthly visits to our home with the task of teaching Grant to stand from the floor unassisted. We had faithfully tried to teach him that skill for a long time but were unsuccessful so didn't feel too optimistic about what seemed quite an impossible endeavor, but hoped it could be possible as it would make a huge difference in his life. Barbara worked with him for the next two years. First, she pushed and pulled his limbs to understand the extent of his disability and assess the best approach to help him learn to stand. Right from the start, she was encouraging about the possibility of reaching the goal. She taught me simple exercises to do to develop greater flexibility and balance and she visited Mercy Centre to explain her program and ask for their cooperation in not allowing Grant to hold on to people when standing still. She said his balance was adequate to support himself and that holding or leaning on people was a habit he'd developed from past needs. Grant felt insecure standing without aid, but through constant reminders at home and school, he improved a lot, though sometimes, especially when excited or if the stance was for too long, he'd move in circles to maintain his stability. After <u>two years</u> of monthly visits from

Barbara and regular visits to the orthopedic clinic, the objective was fulfilled. What perseverance she had! By getting upright on his knees, then lifting one knee up with his foot underneath on the floor, he placed his two hands on that knee, and pushing on the knee, he stood up! The whole family was so excited. But try as he would, it was another month before he was able to repeat the performance. With great hope and encouragement from us all, he kept practicing and before too long, standing up from the floor or grass became a regular act. I thought of Barbara and Grant when I read Charles Spurgeon's quote:

By perseverance the snail reached the ark.

Success is no accident. When a slow learner or retarded child masters something, we know the teaching has been good.

Now, back to the leg braces. Grant wore them throughout the two years of therapy with Barbara and for a year afterward while he continued regular check-ups at the orthopedic clinic. They did keep his feet flat so his walking improved, but they did not accomplish all the changes the doctors hoped for.

Adjusting to braces was a painful experience for a boy with tight heel cords. We put them on for only short periods at first, gradually increasing the time until he could wear them all day. Besides forcing his feet into a new and uncomfortable position, the dreadful shoes were heavy and unyielding, a big change from the light flexible sneakers Grant had worn from the time he took his first steps. He did not like those braces and, mischievous boy that he was, he found good hiding places for his "leg irons," the name Gerry coined for them. Many mornings we frantically searched for the "irons" as Grant giggled through the breakfast hour, and his satisfaction was great if we couldn't locate them by the time the bus came and he could wear his beloved sneakers to school that day. Later, I'd locate them in the clothes dryer, the washer, the clothes hamper, the toy box, the waste basket, or some other unlikely place. The time he put them in my scrubbing pail under the laundry sink, I didn't find them for three days.

It was at this period in Grant's life that Gerry started making little books for him, books that would both amuse and help him feel good about himself and overcome obstacles along the way. "Grant and His Leg Irons" was the first of what has become an extensive series of stories depicting the episodes in Grant's life. They are very much enjoyed and have served worthwhile purposes. Anyone who knows Grant is familiar with them as he shows them to everyone. I am going to incorporate a few of them as I complete this story. In the books, Gerry's words tell the story and the illustrations have humorous exaggerations. I will type the written parts under the pictures he drew. The following is the story of the leg irons:

Braces were prescribed for Grant.

Dad goes to pick up the new braces at the shop. Mr. Michael says, "Grant's leg irons are all ready. Be sure to put them on tight for good support."

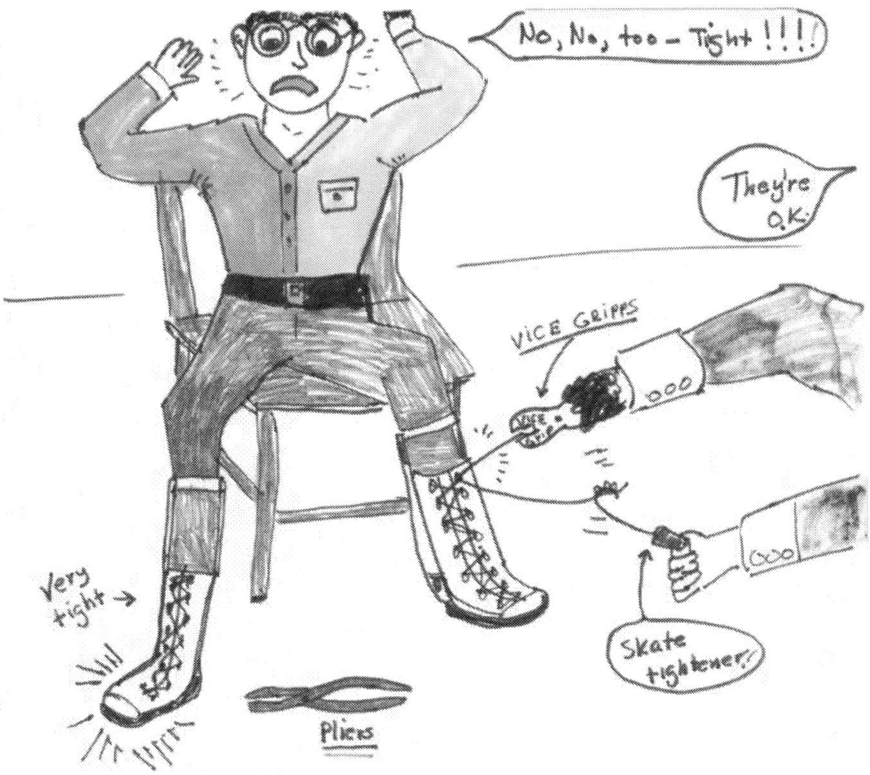

First morning wearing the new irons. After breakfast Dad gets the irons out and puts them on. To be sure they are tight enough he uses tools -- vice grips, pliers, and even skate tighteners!!

The next day - Grant hides the leg irons in the toy box -- Dad can't find them the next morning ... so -- Grant goes to school in his sneakers!

At school Grant was going downstairs to the shop. Then - - big crash all the way down. Grant landed at the bottom of the steps right on his bottom. Mr. Marshall sees that Grant does not have his irons on and is very upset ...

Now, when old Grant goes out he can walk anywhere, even up curbs where there are no railings. All the sisters and teachers and students see how well Grant is doing now that he wears his irons all the time.

****

The braces were not a permanent fix. When they were taken off, his feet quickly went back to the deformed position with his toes stiffly pointed down. At age fifteen, the doctors decided to do surgery on both ankles to lengthen the heel cords. The surgery was done at Massachusetts General Hospital, and there is no more delightful way to tell that story than the illustrations and explanations in Gerry's book.

Ol' Grant likes to walk and he does pretty well getting here and there. But he does one funny thing -- he walks on his toes and this gets him tired quickly and is a problem with good balance -- in fact ...

Grant has a real problem stopping and standing still. So when he wants to talk to someone, he needs to circle around until he can get down flat.

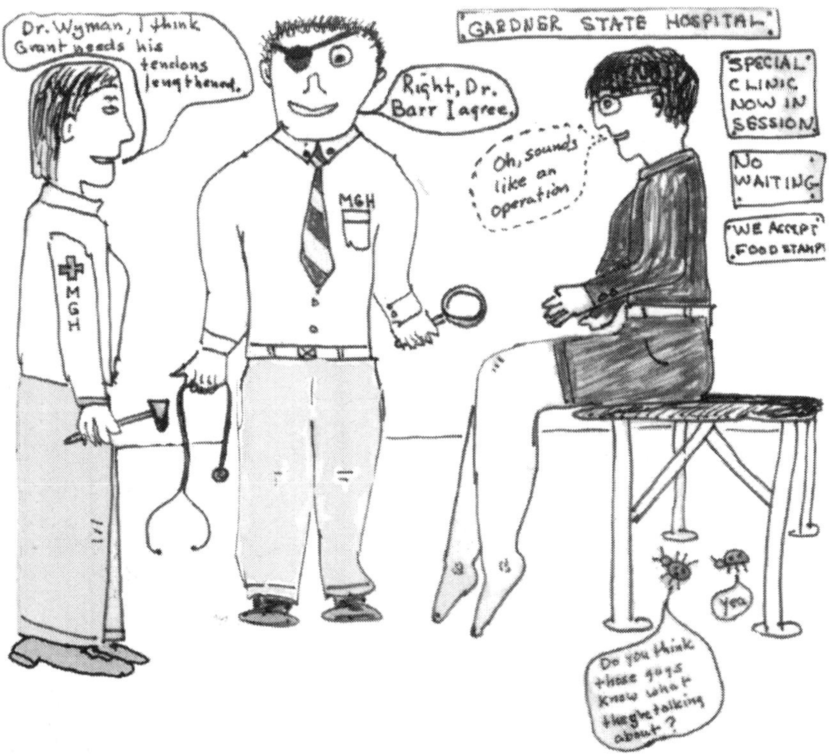

At his next examination by the MGH team over in Gardner, Dr. Barr decides to do an operation to make Grant's ankle tendons longer. They hope this will allow him to walk flat and have better balance. Grant wonders at first but then thinks it's a good idea.

Arrangements were made to go to Mass. General for the operation. They sent out a book telling all about the hospital. Grant studied it every day until he knew everything about MGH that was worth knowing. Grant's attitude for the visit was super!

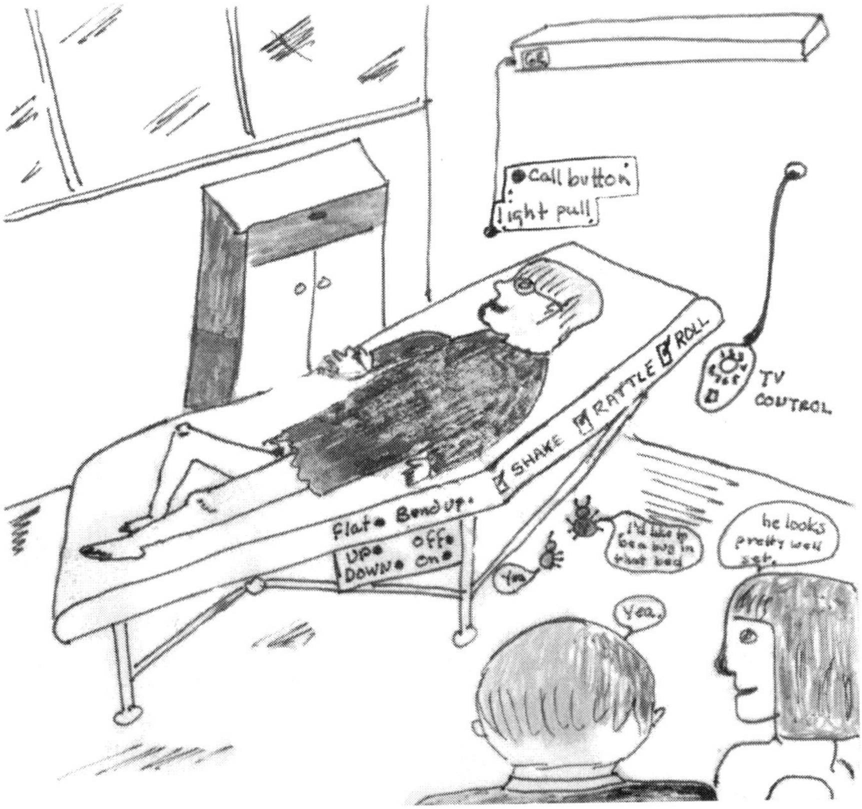

One fine September morning we left for Boston. After a nice drive we found MGH and went to the admissions department. Grant was checked in and the nurses showed Grant all about his room and bed. He learned to operate all the controls and switches.

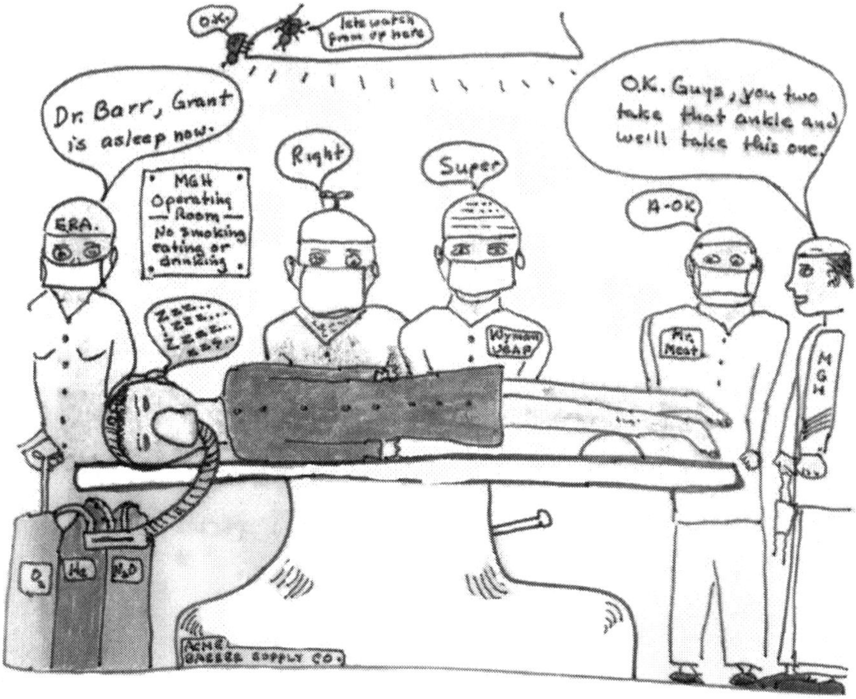

The next morning Grant was taken to the operating room. The anesthetist gave him gas and he went to sleep. Then the surgeons came in to do the operation. There were four of them, two for each ankle so he would be done quickly.

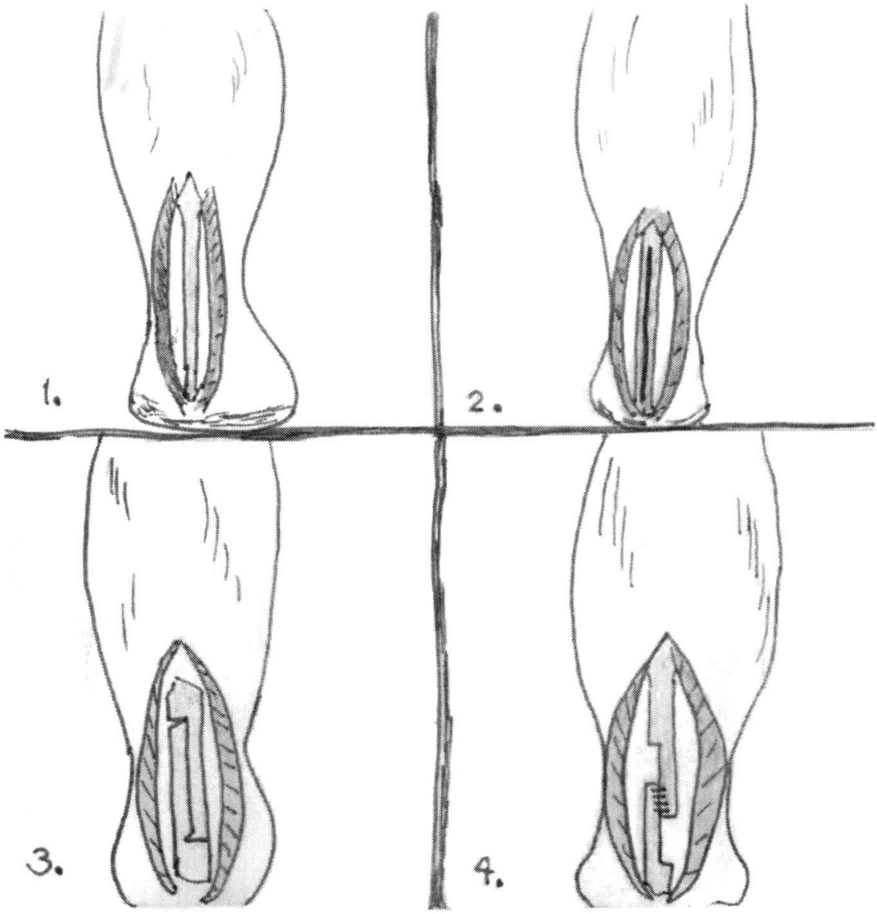

Here's how the operation went.
First --open up the Achilles tendon in Grant's ankle.
Then -- make a longitudinal incision in the tendon.
Third --Make two lateral incisions from each side, top and bottom.
Finally -- Lengthen the cord and suture together.

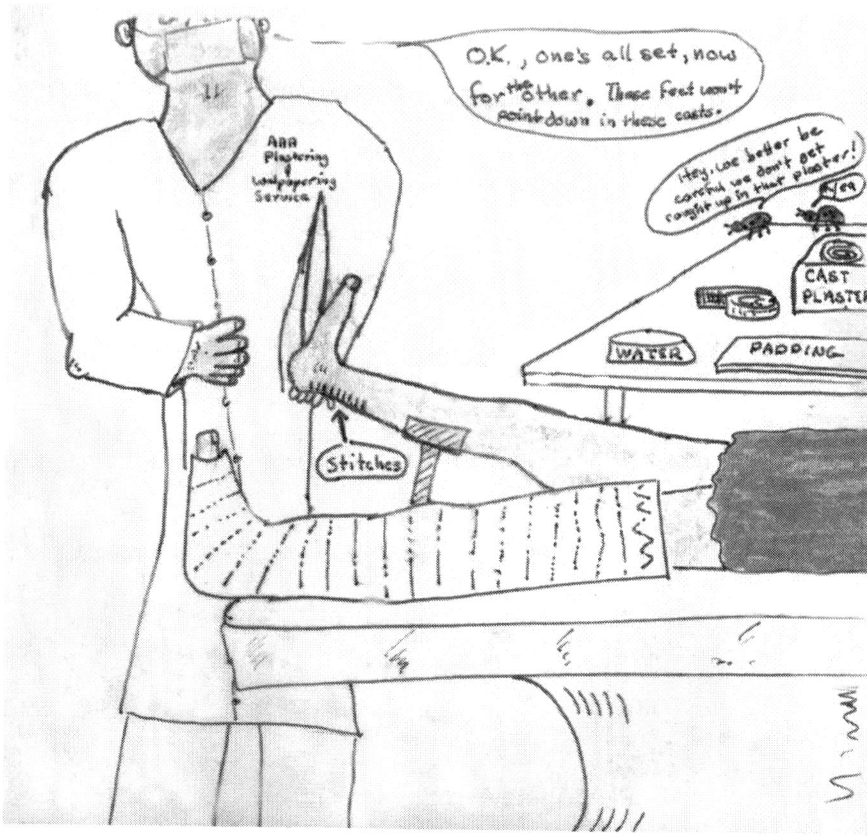

After the operation was all over, they put his legs in two huge plaster casts. This was to be sure the feet healed in a straight manner. Then ol' *Grant* was taken back to his room by his nurses. Now for recuperation.

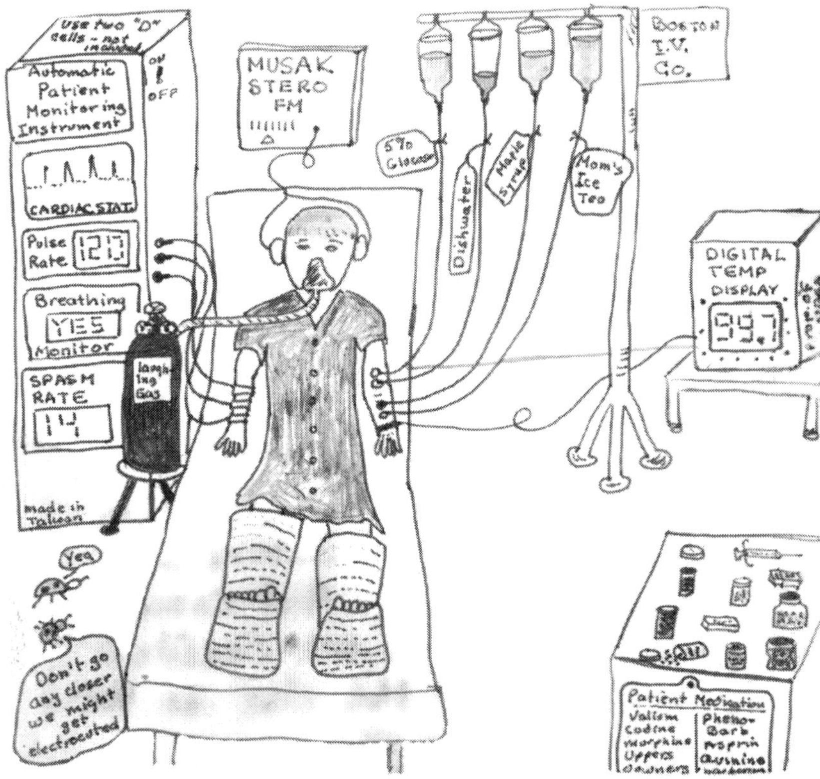

Grant was taken back to his room -- 406 in Burnham Building where all the orthopedic kids stay. He was really wiped out after his surgery. So zonked out he didn't want to eat or talk or anything. He was sad! But the nurses took really good care of him, hitched him up to I.V. and wired him up to lots of other stuff. Then Beth came to see Grant and he was very happy to see her. He felt a little better.

Well, pretty soon Grant was doing much better. They took off the wires, hoses, I.V. etc. and he began to recuperate. Then he had lots of visitors. After a few more days it was time for Grant to be released. He was only there one week. So, on another fine day, Mom, Dad, Bart, and Anthony came in to pick him up.

We decided to set him up on a chaise lounge in the den so he'd be close to everything. We padded it with pads and puffs and even a sheepskin fleece that Audrey Stevens brought to make it soft. Then Grant slept in the living room on the rollaway bed so he wouldn't have to go upstairs. Then all sorts of people came to see him. They brought cards and presents. Grant's class from Mercy Centre even came.

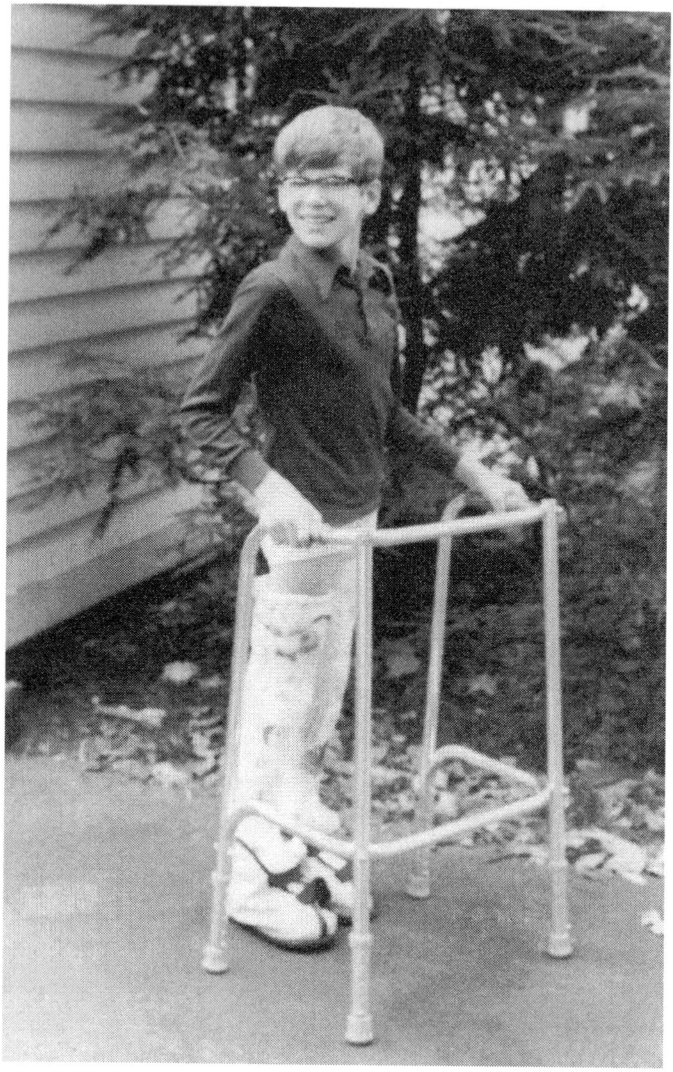

Grant in casts after operation

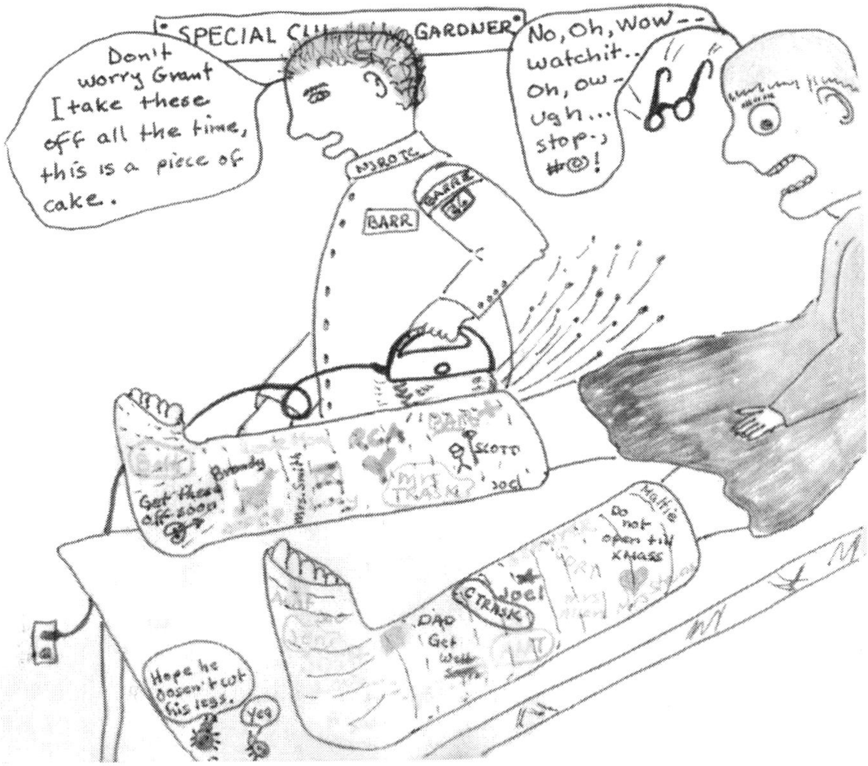

After six weeks recuperation in his casts, Grant's ankles were healed up. The casts had kept his feet flat during this time. Grant had done a lot of walking with his walker to keep his strength up. Now it was time for the casts to come off. By now they were decorated with names, pictures, and notes put on in many colors by Grant's visitors. Well, Mom took him back to see Dr. Barr in Gardner and the doctor picked up an electric saw and proceeded to remove the casts. Wow -- Grant was shook when he heard and watched that saw -- he thought he might lose his legs. But no, the casts came off easily. But now Grant's legs were very stiff.

Yup, when those casts came off, his legs could not straighten out and he couldn't stand at all. The doctor said, "We'll need some therapy (rhymes with Gariepy) for this guy." Well, Grant went to Holden to learn the exercises for loosening up those muscles and also to have a whirlpool treatment. Those hot swirling baths really relaxed his legs and he could straighten them out afterward. Instead of going to Holden every day, Mom arranged with Mr. Trask to use the whirlpool at the high school and that saved a lot of time. Well, Grant progressed pretty fast and got so he could stand up and walk with his walker.

So, Grant was all through with the wheelchair by now and it could be given back. He really could get around with the four legged walker and he worked on balance. The next time Grant saw Dr. Barr in Gardner he said, "Time now to go to long arm crutches." So Mom took Grant to Worcester to see Smiley Michael and got these fancy walking sticks that support you at the wrist as well as by your hands. Grant had to learn another set up tricks to handle these crutches and before long he was going anywhere he wished. He even went down the driveway, across the street, and down the lane to the Bob Allen's.

And then one day a few days before Christmas we were in the den working with Grant by holding onto his arms and walking with him and when we were through, I said, "Grant, try to walk over here without your crutches." Well, he did, he walked across the den all alone! From then on it was all uphill, he got better and better. He got some new braces for his ankles and started to go places. Mrs. Haynes invited him for tea. He had high tea with Mrs. Allan, and even went to a New Year's Party at Aunt Sue's and Grammy's. This was a good Christmas present for us all. The operation was a success!!

Family photo, Grant age 16

Following the operation, Grant's life fell into a more normal routine. People weren't responding to his foolish questions and tolerating his silly behavior. He still doesn't have perfect control, but it helped so much to get people outside the home involved. His family and the school and the community were Grant's world, and there's real power when everyone in one's world works for a common goal. Wouldn't it be wonderful if the goal of all nations was to work for peace!

There is power in unity and there is power in numbers.
Martin Luther King

11. Steps toward independence

Knowledge brings understanding and acceptance.
Erica Spindler

Grant loves his birthday, and at age 16, because of his more "grown-up" behavior, a very special outing was added to his celebration ... lunch with his speech teacher Sister Alice. For weeks preceding his birthday, they discussed and decided where they would dine. And when the big day finally arrived, Grant dressed in his best clothes to prepare for the very personal observance that made him feel so important and grown up.

People who go the extra mile make such a difference in the world.

Of course, we still had our family party at home. We have the tradition of the birthday person picking his or her favorite dessert for the birthday cake, and Grant always chose Baked Alaska. And, to make the day extra special, we invited Elinor Allen (who shared the same birthday), Lucy, Judy Potter Fayre, Sister Irene from Montessori School, and two patterners who stayed close in his life -- Mrs. Hale and Mrs. Tuttle.

Though we made Grant's birthday special, the extra effort and enthusiasm was to stimulate ourselves as well as him. His birthdays also had a discouraging aspect – we were reminded of his age and how far behind he was for his years. Throughout the year we accepted Grant as he was and thought little of his chronological age … until birthday time. Then, we had to purposefully encourage ourselves to be thankful for whatever development he made and be especially thankful that throughout it all, he's been a happy person.

Elinor Allen "Mrs. Bob" and Grant with their baked Alaska

The Massachusetts Chapter 766 law instituted in 1974 changed the lives and outlook for special needs children and their families. It guaranteed equal educational services for everyone. Previously, parents with a child like Grant, who couldn't fit into a public school system, were burdened with the choice of a high-cost private school, cloistering the child away at home, or placing him in an institution. After passing 766, the cost of the child's educational needs were met, whether at a public or private facility. The state's cost under this law was less than the previous expense of lifetime care in an institution, and it has given many individuals the opportunity to develop into useful people in society. And when special needs people became a visible part of society, their problems became better understood and better accepted. No longer were they people who should be hidden away.

We are all God's children.

The school year was great with full days for Grant, but the summer days were long. He had no friends except his brothers and

few activities other than the few hours a day when I took the boys swimming at the local pool. So, it was a blessing when the state's Chapter 766 law included funding for summer programs for children with special needs. Grant was able to attend summer school at Mercy Centre. Finances for the six-week program also included transportation. The opportunity to be with his peers, to participate in swimming, crafts, games, and field trips afforded him the first of many excellent summer experiences. Ever since then, the highlight of his year has been his summer weeks at camp, most recently at Grotonwood, a Christian camp for special adults.

Grant began the pre-vocational program at Mercy Centre, a program that stressed self-care, emotional and social development, and preparation for a job in a sheltered workshop. All aspects were geared toward readiness for life and work. The instructors were called bosses. Subjects included motor development skills – handwriting and shoe-tying, not easy tasks for someone with poor hand coordination. He did enjoy handwriting and began writing letters to people, even to me. His writing has always looked like it was done by a pre-school child, but he did master the ability of writing and loves doing it. I don't think a day passes today that he doesn't write to someone or about something.

Shoe-tying was a different story. The "bosses" had an amazing method of teaching the skill, a slow process, and Grant learned it but never got the laces tied tightly, also he didn't really like the job of tying his shoes but he tried.

He liked audio and reading classes and did well in both. Audio classes stressed the important ability to listen and follow instructions, and the reading curriculum included phonetic and sight vocabulary, similar to that of any school. We read with Grant from the time he was little and had him read to us, and he has advanced to a second grade level, a really good achievement. Reading to a child is so important. It develops good learning and listening skills, stimulates an interest in reading, and offers a time of close bonding between parent and child.

Math classes were money related. Adding 1 + 1 was not meaningful to those kids but $1 +$1 was. Grant is weak in math, but he does know money denominations and has some concept of

the value. A lot of trust goes to store cashiers when special kids go shopping. They could easily be cheated, but if such a thing happened, the person I'd pity most is the one who would treat them unfairly. Jesus said,

What you do unto the least of mine, you do unto me.

Great stress was placed on – "When on the job, stick to it!" Grant developed excellent work ethics in that pre-workshop training program. He took his job seriously and never wanted to miss a day or be late. And at the end of the week, he was proud to bring home his pay envelope with a check for $1.

In keeping with the routine of regular work places, Mercy Centre had a coffee break for their employees. (They actually served cocoa.) The children, then teenagers, filed by the "coffee" dispenser to fill their cups, then sat at a table to drink and socialize in a proper manner. Grant could not carry his cup back to a table unassisted as his awkward movements made him spill it, so one of the bosses carried it for him. Later he was trained to ask a peer to carry it for him, as he would have to do in a real-life situation. We all have to learn to advocate for ourselves and ask for help if we need it, and most people are very happy to lend a helping hand. The coffee break lasted twenty minutes. Of course, everyone liked the break, which also served as a time-awareness period. It is important for them, and for us all, to learn the importance of being on time. If they didn't get to the coffee area and get their drink promptly, they missed out on the treat. An evaluation session followed the break when they had a discussion of the child's proper or improper manners, someone's friendliness toward another who might be alone, or about their conversational tones that improved or needed improvement. Coffee break was another growing and learning experience.

Gerry wrote "Things I Can Do Now" that compiled the achievements Grant began to work on at age 16. The stories make him feel very proud, and the humorous exaggerations in the drawings provide good conversation pieces with his readers about

what he absolutely would not do. It's probably his most beloved book, and consequently shows signs of wear.

This chapter tells about Grant's job. After he finished his academic work at Mercy Centre, Grant needed to get a job. He applied for work in the workshop at Mercy Centre and they took him on trial for six weeks. He had to prove himself - that he punched in on time, got back from coffee and lunch break on time, kept right to work, did what the boss said, and didn't quit early. Well, he

passed and got hired - FULL TIME JOB! We had a big celebration at home for this special event. Now Grant goes to work every day and even cashes his pay check at the Gulf Station up the street.

Nobody went to the Gulf Station to cash a check, but that was an easy safe walking destination for Grant, and the proprietor there was happy to be the first to accommodate Grant's out-of-the-house ventures. Later, Grant mastered the walk to the bank, and again they cashed his one-dollar check, even though it was a requirement that only people with accounts were able to cash checks. We were fortunate to live in a small town. He had become a town project and everyone seemed to be cheering for his development. The town was good to him and I somewhat think he was good for the town; everyone was working together to help a young boy to have a life.

Grant learned to go uptown without any grown-ups. Bart walked with him at first, then walked behind him to be sure Grant did it safely. Grant is very careful to stay out of the road all the way until he gets to the Gulf Station. There he watches both ways and if all is clear he walks across to Cumberland's to do his shopping. He likes to buy candy with the money he earns. He does the shopping all by himself. When he's done, he goes back very carefully the same way.

We cooperated with the school in giving him responsibilities at home. A person who contributes to the functioning of the home feels an important part of the family. It's satisfying to be needed. Emptying the dishwasher was his first job. I operated the dishwasher each day after lunch, so it was ready for him to unload when he arrived home from school. Holding onto the counter with one hand to steady himself, he carefully removed the dishes and put away the ones that went on low shelves and stacked those that went up high. Later he learned to fill the dishwasher after supper. We highly praised him for his work and let him know what a difference it made for me.

Grant learned to empty the dishwasher. He unloads the racks and puts the dishes, pans, pots, glasses, and silverware in neat piles on the counter or carefully on the shelves or in their spots in the kitchen drawers. He inspects all the dishes to be sure they have been washed well and if not he puts them back for another run.

Cort and Bart earned an allowance for doing their chores so it was time to put Grant on the payroll too. We paid him twenty-five cents every Friday. One week when he was sick and Friday came, I explained that there would be no pay that week because he didn't do any work. After he felt better, I found him emptying the dishwasher, and when I said I was glad he felt better, he asked, "Am I back on the payroll now?"

At first, we felt limited in the chores we could give him because of his physical problems, but he surprised us in all he could manage. Next I taught him how to take clothes off the clothesline. He held the clothes with one hand, removed the clothespin with the other, then took a big circular route to the basket with the article. (The circular route was necessary because he couldn't move to the side or backwards without losing his balance.) It was a long slow job, but he stuck to it and successfully could do all the items except sheets. Later, he learned to hang out the wash. I wish I had taken a photo to show how my clothesline looked because it was a sight to behold, but the clothes dried and he had one more job accomplishment.

Mom would bring out the clothes basket and put it near the line. Then Grant would sort out the clothes and carefully hang them up. Grant learned to be very neat and keep the freshly washed clothes clean.

Routine is very important for Grant to function well, as it is for lots of children, especially those with any special challenge. We had him pick out his clothes before going to bed, and he'd carefully slide the belt through the loops so the pants were ready for him to put on in the morning, not an easy job so we soon bought him pants with an elastic waist that he could manage more easily, both for dressing and bathroom purposes. Also, we changed his shirts to pullovers that didn't need buttoning. This page in the book tells the beginning of that learning experience and Grant worked very hard to succeed.

In this saga we describe Grant's ability to take care of himself. He has learned to dress himself in all respects. This includes things such as picking out shirt and pants that match, socks that match, being able to button up things correctly, putting belts in loops and buckling up, snapping or buttoning pants, putting on coats and mittens. Grant could get himself ready alone. A big step to independence.

One aspect of his self-care took a lot of practice – shaving. And, Gerry, being of the old school didn't believe in teaching him with an electric razor, it had to be with a regular safety razor. He bought Grant a razor and an aerosol can of shaving lotion and they patiently set to work on shaving lessons. (Gerry, of course, still used the William's shaving soap in a mug and his badger hair brush, so Grant did have a step-up from that.) It was a long session every night, they never would have had time in the morning. Grant was careful and especially liked using the lotion. He would really lather up, kind of looked like Santa Claus before he'd begin applying the razor, and being right-handed, the right side of his face got a good close shave, including shaving high over his ears. His grooming was always out of balance, long sideburns and some whiskers on the left side and a high hairline and smooth shave on the right, but he did it himself and was proud of it, so we let it go and he did gradually improve. As writer Richard Carlson would say,

Don't sweat the small stuff...

This chapter logs about how Grant learned to shave. When it was time for Grant to shave we discussed the situation and decided that he should do this himself and NOT have someone else shave him. Grant learned to wet his face down with hot water and lather up with the shaving brush. This softens up the whiskers. Now after the whiskers are soft and wet, Grant uses a safety razor and shaves all over twice to be sure he's got everything. He always shaves the sideburns even and downwards. He uses care not to cut himself. When he's all done, he cleans up his equipment and puts it away. Then he washes up his face and gets all the shaving lather off. Last of all he checks the job and be sure everything is perfect.

This chapter tells the story of how Grant has learned to eat with good manners. Many things are important while eating, such as - keep your left hand in your lap, break your bread before eating it. Also, use your utensils, not your hands and be careful not to spill your milk. And he also knows that you must keep your elbows off the table so they won't get in the food. And keep your hands clean so your glasses will stay clean. Grant has gotten so good with manners that he has been invited to join Mercy Centre's Supper Club where they get to go out to eat.

I never ceased to be amazed at the details everyone has to learn in order to fit into society acceptably, things we just absorb from everyday life, things we just take for granted. But for Grant, all these things had to be taught and I never ceased to be amazed how Mercy Centre addressed them all.

12. Community involvement

The game of life is not so much in holding a good hand as playing a poor hand well.

I like that quotation by H.L. Leslie. We all know people who fit into that category. Many people have physical or mental afflictions. Some people make the affliction their lives, and others have a full life in spite of the affliction. Grant was dealt a poor hand, but thankfully we were seeing him develop the attitudes and skills to play his hand well.

He had many falls, but he learned how to fall without getting hurt. He'd trip or lose his balance when walking on uneven ground or in slippery conditions. But, like a tough football player, he'd get himself up and keep going. And, only once do we remember hearing him complain about his limitations and that was the day we stopped at the waterslide park on our way home from the beach, something Cort and Bart and their friends wanted to do. When we saw the hundreds of steps they had to take to get to the top of the slide, we realized it would be too difficult and unsafe for Grant, but there had been many activities his brothers did where he could not participate – street hockey, baseball, ice skating, skiing, etc. He never said anything about not playing those games so we didn't think much about him not going down the waterslide with them. Gerry, Grant and I watched from the bottom and cheered as they splashed into the pool in front of us, and they excitedly commented about the fun of it as they gathered their mats and raced up the steps for another run. After a few times down, Grant turned to us and said, *Handicap stinks*. One thing Grant does like is swimming and he felt the loss of not being able to join them. There was nothing we could do but agree and say that there are times when all of us wish we could do something we can't do, but our hearts truly hurt for him. Nevertheless, only one complaint in 53 years is a good attitude record.

His next big step was going to the post office, which was located only a short distance from the bank where he cashed his check. We knew it would be an inconvenience for the post office workers as Grant would not be able to work the combination to open our box, and he would have to go to the counter and ask them to get the mail for him. I humbly approached Marilyn Skerry, the Post Mistress, about Grant learning to do that errand and her response was more than enthusiastic. Like everyone else in town, she was delighted to be a part of his development.

Ask and you shall receive.

Jesus must have talked a lot about the necessity of people making their needs known because the previous Biblical quote is found in the books of Matthew, Luke, and John.

We do have to advocate for those with special needs. We have found that whenever it's possible, people want to help, but they are not mind readers. We have to let them know what help is needed. However, we also have to be prepared to accept refusals without feeling hurt if someone can't comply with our requests. People know the limitations of what they can do and we have to appreciate that.

I took Grant to the post office to introduce him to Mrs. Skerry and together we explained the routine. I appreciated how she talked to him, not talking down to him, but treating him with dignity and respect. We talked about the big responsibility he was about to have and the confidence we had in his ability to carry it out. He was serious and seemed to mature as the job was discussed. He had been walking uptown to cash his check and buy candy, but this was his first real independent duty.

Our friend Elisabeth Hyder made him a sturdy mail bag with a draw string so the mail would not fall out. Bart quietly observed him on the first trip and reported that he did just fine. Then Grant carried out the job every day for over ten years, close to a one mile hike round-trip, a real aerobic workout for someone with an unsteady jerky gait.

Over time, the post office trip took longer and longer. The post office people became his favorite friends and he also struck up conversations with townspeople who stopped by to pick up their mail. Everyone knew Grant. Sometimes when I was introduced to a new person in town, immediately upon hearing my last name, they'd ask if I were Grant's mom. Everyone didn't know us, but everyone knew Grant. I asked Mrs. Skerry if his lingering stays were a problem, and she told me not to worry that they would take care of it and tell him it was time to take his mail home if he was staying too long. She said he was obedient when they reminded him.

Grant still wasn't good at making proper conversation. He would ask or tell the same stories over and over to everyone he met at the post office. Luckily, all the people he asked to come to his house and read the book his father wrote didn't take his invite seriously, but one man did. On that particular day, the man he asked to come to his house did come. I was surprised to see him drive in and wondered what he was there for. He came to the door and said Grant had invited him to come for tea and read his book. Wow! That was very nice of him to take Grant seriously and take the time to stop for a visit. I made tea and they sipped it together as he read to Grant. Then he thanked me for the tea and thanked Grant for the invite and said he enjoyed his book. It was really lovely, but made me realize I'd be running a tea house if everyone took his invites as seriously as that man did.

We showed Grant what advertising flyers were and told him he could throw them in the wastebasket at the post office rather than try to stuff them all in his mail bag. He did that carefully, but throwing things away gave him one idea that was not good. I was working as a tour guide at the Worcester Art Museum and sometimes wasn't home when Grant arrived from Mercy Centre. It was no problem as he could just get his mail bag and head off on his errand, but he didn't like me not being there. So, he got the idea that if he threw away all the mail that came from the museum that I wouldn't go there any more. Well, one day when I went to the museum for a meeting, I arrived to find that the meeting was

changed to the next day. They said a notice had been mailed about the changed date. Also, I learned that there were other mailings I should have received. I wondered about it and knew there was only one answer. Grant. So I talked with Mrs. Skerry who talked with Grant about mail fraud. She told him: *I understand you are tampering with the United States mail. It is illegal and ... a jail sentence! But I won't call the police this time, but if it happens again* ... From then on, the art museum mail came with regularity.

Mrs. Willey lived up the street from us. She was not well and asked the post office people if they knew anyone who might be able to deliver her mail. Well, they immediately thought of Grant! They talked to her about him and, of course, she had seen him walk by on his daily mission. She agreed to give him a try and called me about it. We felt so good that someone would have that trust in him. Elisabeth made him another bag so he could keep the mailings separate and Grant had a paying job, $1.50 a week.

Then another post office job came his way, putting on the lights in the mailbox room. Several times people had gone to the counter and asked to have the lights turned on so they could see well enough to do their combinations. Mrs. Skerry thought – we need the lights turned on just about the time Grant arrives. We'll ask him to do that job. No one else probably had such a set schedule and would have loved the job more. He did it dutifully and was well rewarded. The post office workers had a special night out to eat together every year at Christmastime, and they decided that Grant should join them as his lighting job had made him a post office worker. Mr. Stevens, the rural delivery man, picked him up every year for that treat. As I think of all the wonderful Barre people, how could we not dedicate this book to them.

This chapter reviews Grant's mission with the post office. After Grant was walking very well, we decided he could do the daily mail run. He got a special homemade mail bag from Mrs. Hyder. Then he would go to the post office and Mrs. Skerry would give him the mail. Later,

when the new "key" boxes came in Grant learned how to open old 444 all by himself and now he keeps the key in his bag. He is very careful to put all the mail in his bag (except the bills which he throws away) and to close the box tightly. He also can handle the post office duties such as paying the box rent, buying stamps, and even mailing letters to Aita, our exchange student in Switzerland. Grant enjoys this daily hike uptown as he sees lots of his friends especially the people at 01005 such as Mrs. Skerry, Mrs. Dwelly, Mr. Stevens, Christy, Mrs. Cohen, and Ann Marie. Also, Grant was recommended for a special new job by the post office staff. This was

to pick up the mail for Mrs. Willey who lives on James Street. He got another mail bag from Mrs. Hyder so as to keep the mail separate and keep the key safe. A big responsibility.

Mercy Centre continued to add activities that helped with social development and preparedness for a job in a workshop. Telling time, it was important to know the times of day that mattered in their lives. Grant never learned to tell time in detail but he learned 8:00 was time for work, 12:00 was lunch time, 2:00 was time to go home, and 4:00 was time for "Sesame Street," a show he watched for years and enjoyed the humor, and that program was followed by "Mr. Rogers" whose reassuring words meant a lot to him, *I like you just the way you are.* Now, with digital watches, telling time is easy for him and everyone. No gift was more special to Grant than his first digital watch. Wearing that watch became a part of his grown-up self image. He's never without it.

The social studies class at Mercy Centre taught them how to relate to and get along with people. The lessons included scenarios that applied to interactions at school, work, home, and in the community. A desire to be a pleasing person was emphasized. Grant was nearing 20, and at that age people with special needs are well aware of their limitations. Mercy Centre acknowledged and discussed each individual's shortcomings, stressing that everyone has weaknesses but also strengths. The bosses analyzed worthy strengths and talents with each child. Developing a good self-image was a major objective at the school, as it needs to be a major objective for every parent and teacher. Every child needs discipline, but "putting a child down" is never the way to do it. The world could learn important "building a child up" lessons from Mercy Centre. They helped the children aspire to be the best they could be and feel good about themselves. They followed Abraham Lincoln's advice:

Whatever you are, be a good one.

Fridays were special at Mercy Centre. In the morning the children swam at the Y.M.C.A., the only sport Grant could do and he loved it. In the afternoon they had dancing, not an easy activity for him or a pleasant one for his partner who might get stomped on or pulled down if he lost his balance. However, his dedicated instructor found a sideways position for Grant to hold his partner so he could dance without fear of crushing his partner's foot or tripping over it. One day I was curious about the dancing I'd heard so much talk about and told him I felt like dancing. The mannerly procedure that followed amazed me. He politely asked, *May I have this dance?* When I answered, *Yes,* he took my arm, led me to the center of the den, and earnestly danced, then escorted me back to the chair. (I wondered if Cort conducted himself as well when he went to dances, and Bart who would soon be entering junior high could certainly take some lessons from Grant.) Refreshments were served during the dancing social, always discussing proper manners and taking conservative amounts of food. No aspect of life was left to chance. All the little details would determine a person's acceptability in society.

Cort and Bart both led full active lives. They were always kind to Grant and never displayed any dislike or resentment, but I wondered if they felt any embarrassment or regret having a brother who was different, so I decided to ask them, thinking it may be something we needed to discuss. I was so touched with their responses and am glad I wrote them down in the notebook I kept. Bart said, *I have a lot of fun with Grant.* He was the one who shadowed Grant to be sure he was safe on new ventures. And, Cort's response was, *Life would not be as good without Grant.* I wondered how he could say that and asked, *Why?* He thought a moment and said, *Well, Grant's always here, and he's always glad to see me.* I realized how true that statement was. Grant was always home (except for school when they were gone too) and always excited to see them come home. We all feel good to receive a warm welcome wherever we are. In later years, it was obvious by Cort's and Bart's deeds how their lives were marked by having a special brother.

13. Sheltered workshop

Strength and growth come only through continuous effort ...
Napoleon Hill

Grant was twenty-one and his Mercy Centre education ended. He'd been connected there since he was six, first with speech lessons, then in the school, and lastly in the pre-workshop program. It was his second home. He was secure and happy there, he felt loved. Now it was time for him to move ahead into his adult life. How would he adjust?

He transferred to GAAMHA (Gardner Athol Area Mental Health Association), a sheltered workshop in Gardner, Massachusetts and is still there today. We made a big deal about him being ready to have a real job and earn a good salary. We talked about how well he'd succeed, that he would work hard and do just what the bosses told him to do. We were concerned about how that change would be for him but we needn't have worried. He was well prepared for a sheltered workshop and was comfortable and happy there right away. His routine was similar to what it had been – van pick-up in the morning, day job at the workshop, post office after he got home, and regular evening schedule. The only change in the home routine was that Cort had finished high school and was in the Navy, stationed in California. Grant knew about the change Cort made, and we related that to the change he was making, we had two grown-up young men.

The jobs at the workshop are actual jobs. One of the staff members visits companies throughout the area and contracts with them to have their simple routine tasks done in the workshop. It is a help to the companies and provides employment for those in the workshop. We've seen Grant attach pads on top of a marked spot on the inside of hockey helmets, also place plastic spoons in individual slots, and when all the slots were filled, it was the right number to be packaged. He had jobs sorting silverware, doing nut and bolt assembly, and gluing boxes to be used for packaging computers. The staff is very creative in providing the tools so the

clients can do the jobs and closely supervises them to be sure the tasks are done correctly.

Salaries are based on piece work. For example: They timed how many 25-count groups of spoons a normal person could have ready for packaging in an hour, and based on a minimum pay scale of maybe $7.00 an hour, when the client achieved that amount, he would earn $7.00. Grant was a very slow worker, but accurate. Typically, he earned about $3.00 a week. He had a real job and was proud of it.

We all enjoyed going to a mall. Bart would bring a friend, and their favorite attraction was the Arcade. Games on handheld computers hadn't been invented yet but Pac Man and some other games were on the Arcade machines, and they loved an opportunity to play them. Grant followed along with us to do shopping and we all met at Friendly's for supper. A nice outing for everyone.

Then, one evening after we returned home from the mall, Gerry and I contemplated about Grant going to Friendly's on his own. It would be another big step toward independence for him and an experience in the outside world where the people he interacted with didn't know him. We knew it would be something he would love doing so we talked with him about it and began role playing. Grant would sit at the kitchen table pretending he was at Friendly's and Gerry, pretending to be the waiter, would go over to greet him, hand him a menu, and ask what he would like to drink. Then, when Gerry returned with the make-believe drink (an empty glass), he asked if he were ready to order. Grant always ordered the same thing in those days when we went to Friendly's – clam chowder, grilled cheese sandwich, and a small coke. Gerry pretended to write down the order, left for awhile, then returned with empty dishes of the make-believe meal. We emphasized about saying please and thank you, also eating with his best manners. In those days, that order cost under three dollars, so if he brought three dollars with him, he would have enough to pay for the meal and leave a tip, which we told him was very necessary for the waiter. The next step was to practice him going to the cashier to pay. We rehearsed many, many times. Grant enjoyed the role playing and

learned the steps well. It was like having rehearsals for a theatrical production, and him finally being ready for the performance.

Grant spent little on candy for two weeks, saved three dollars from his pay check, and we were off to the mall. When we arrived at Friendly's, we pointed out the bench outside the restaurant where he should wait for us when he finished, and casually said goodbye to him and told him we'd be shopping while he ate. Did we go off shopping? Of course not. We just hovered in a place outside the restaurant where we could watch and wonder how he'd make out. We kind of felt like we were sending a kindergartener off to a restaurant by himself. Would the waiter be able to understand him? Would he take him seriously? Would Grant behave well? Would he leave a tip? Would he pay? That first experience was a memorable one and Gerry made a book about it, "The Super Sundae Story."

When the family went out to eat Grant always wanted to go to Friendly's. So when it came time for Grant to go to a restaurant all by himself he chose that one. The big event took place at the Hampshire Mall. Mom and Dad went over all the things that Grant should do when he was on his own.

Now Grant walked over to Friendly's and went in, found an empty stool at the counter and sat down. After awhile the waiter came over and took his order. Grant knew just what he wanted and spoke slowly and clearly so the waiter could write it all down on the order pad. Then Grant sat back and relaxed while his meal was being prepared.

During this wait the man sitting next to Grant finished up his dinner and ordered dessert. Well, Grant was amazed when he saw how huge it was and he couldn't keep his eyes off that treat. But then his meal came and Grant used his best manners to eat his chowder, sandwich and his drink.

Now when he was done, the waiter came back and while collecting the dirty dishes he asked Grant if he would like anything else. Well, Grant was still thinking about that luscious bowl of dessert and decided to order one.

He couldn't believe his eyes when it came - a Super Sundae. It was in a huge glass bowl and it had five scoops of different ice cream on top of bananas. And over all that was melted chocolate and whipped cream plus nuts and cherries to top it off. Well, Grant took his time and enjoyed every bite.

All had gone well up to now but when the waiter brought the bill and Grant looked at it and then at how much money he had … he knew he was in trouble. Because the Super Sundae was super he didn't have enough money. Grant was worried, what was he going to do now?

Well, lucky for Grant, Mom and Dad were nearby all the time, sort of watching to see how he would do at a restaurant all alone for the first time. When they saw his problem, they came to the rescue and gave him some more money to pay the bill and leave a tip. Afterwards they discussed how we always want to have enough money to pay for what we want before we go in a store.

After that experience, he never exceeded his financial limit again, and we did go shopping while he ate, knowing that he would behave correctly at the restaurant and be sitting on the mall bench waiting for us after he finished. To this day, Friendly's is his favorite restaurant. However, he did have another bad experience

at the mall that started while Grant was innocently taking a walk before he went to Friendly's for supper… another growing experience for him. We all know –

A lot of learning comes from our mistakes and misunderstandings.

Gerry told about it in "The Jason Story:"

This is another adventure in the life of Grant Gariepy. It took place one Saturday afternoon when the family decided to go on an outing to the Hampshire Mall in Amherst, Massachusetts to do shopping, looking, eating, and other things. We drove there - Mom and I in front and Grant, Bart, and his friend Anthony (better known as "Ant") in the back seat.

After we parked the car and went into the mall we discussed what we all were going to do. One of the nice things about the mall is that there are many different things to do so everyone can be on his own and enjoy himself. Grant loves this mall for he can take a walk, have his supper and sit near the fountain and watch things while he waits for us.

Well, we all separated to do the things we wished. Mom and I went to the shoe store. Mom tried on some shoes. Then we heard a noise outside in the main hall of the mall. We recognized it at once ... it was *Grant*! But it was a terrible noise, awful, horrible. Squeaking and roaring and squealing and everything. What was happening? We couldn't imagine so we rushed outside to see.

Now imagine this. Grant was running down the broad aisle with his arms flailing, his face red and making all sorts of wild sounds. He was weaving through the crowd at top speed and then we noticed a plainclothesman following him with a walkie-talkie radio. This spy just kept following Grant - not stopping him but continuing to talk into his radio. We rushed out of the shoe store and went after the two of them.

We caught up to Grant and stopped him and asked, "What's wrong?" Then the security man came up and we asked him what was going on. He said that a boy named "JASON" was lost and that Grant matched his description and he tried to ask Grant about it but Grant <u>ran away.</u> He said, *"Is this Jason?"* We said, *"No, this is Grant."* The spy left and we reviewed this episode with old Grant.

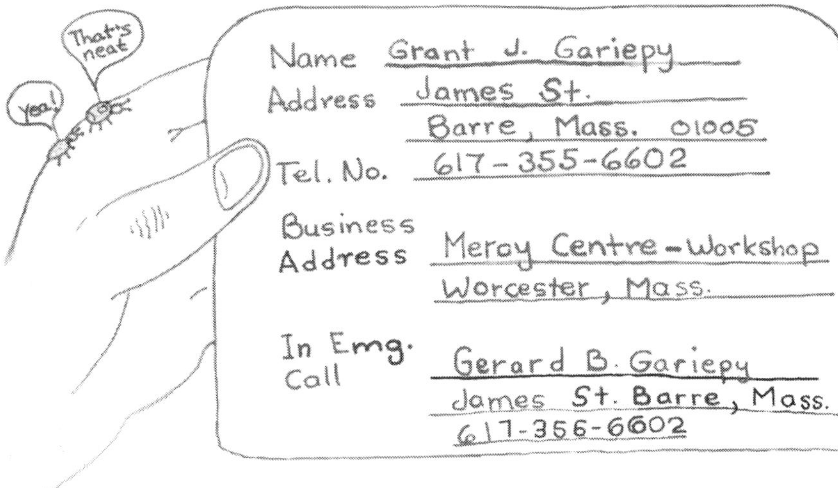

We talked about how police and security men are our friends and it's good to listen to their questions and then answer them. And to never, never run away because then they might think something is wrong. We talked about how to identify yourself with an ID card. Later we made up a complete ID card for Grant to keep in his wallet.

Grant has learned a lot of important lessons from his books, and everyone has fun talking with him about Gerry's exaggerated drawings. Gerry would have Grant sit and pose as he drew, just as an acclaimed artist would have his sitters do. It was a fun venture for both of them, and they talked about the accomplishment or incident being depicted as the book was being created.

Grant's life fell into a regular routine -- no need to find new medical help, school, or workshop. A social worker monitored his workshop, his situation at home, and overall progress. Grant's life was full and he was content, the trip to the post office being the highlight of his day. He was the only child still living at home. Cort completed his four-year tour of duty in the Navy and both he

and Bart were enrolled in college, Cort at San Diego State and Bart at Worcester Polytechnic Institute.

Grant's trip to the post office expanded. He became acquainted with many folks along his route, especially Mrs. Tuttle and Mrs. Keenan. We didn't know until several years later how often those ladies invited him in for a visit and gave him a glass of soda. Grant could always detect a soft heart, and he took advantage of it. When those ladies were inside in the winter, he'd ring the doorbell and tell them he was cold so he could be invited in for a visit and snack. We were quite shocked when we learned about those visits and called to tell the ladies not to invite him in, but they protested and said they liked to have him come. Even more surprising was a call from a new family who moved into a house up the street from us. She said, *I want my children to meet Grant so they can learn to accept and be comfortable with people with special needs. Could he come for supper some evening?* I could hardly believe my ears. Some people had steered their children away from Grant, sheltering them from someone who was different. This new family wanted to have their children see the problems some people have to face. Their parents' modeling of acceptance and compassion were life lessons for them. Grant went for supper there several times before they moved away, and I'm sure the kindnesses those children learned have stayed with them.

We all live in a different world and have to face differing circumstances and challenges. Life is not perfect for anyone. We gain a deeper understanding and acceptance of people when we open ourselves to their lives and realize what it's like to "walk in their shoes." My psychology professor Dr. Bozicas said:

Everyone is where they are at for very good reasons.

I remember an instance when Cort came home from school all upset. He said the boy in his class who had come home with him to play the previous day had mimicked Grant in school that day – mimicking the way Grant walked and stammering with unclear speech. Cort was very hurt and embarrassed. My only explanation for him was that we can tell a lot about people by the way they

respond to Grant. I talked with him about his friends who had shown kindness toward Grant and how the way they treated him shows the feelings in their hearts. Years later, when we went to Cort's wedding in San Diego, we met the young woman who was his Big Sister in college. When he joined the Christian fraternity on campus, each new pledge picked a Big Brother and Big Sister who were upper class members of the fraternity or the sister sorority, people who would be there for support when they first started college. His sorority choice was Rita. He had talked about her, said she was a sweet helpful person, and asked me to help him pick out a Christmas present for her. But, it wasn't until we went to his wedding a few years later that we saw that she was confined to a wheelchair. Cort had never mentioned her handicap. But, did her handicap keep her from dancing? No. Cort pushed and twirled the wheelchair and nothing made my heart happier than to see the fun she was having and the outreach Cort had. I realized his connection with her was another blessing from Grant.

Bart too was blessed by Grant. When he moved to Oregon to teach, he connected with a group home to do volunteer work teaching a swimming class to special needs boys. He had served as waterfront director at Boy Scout camp during the summer when he was in college so helping those guys was in an area where he'd had a lot of experience. Bart was always a close buddy to Grant, and found joy in relating to him and his school or work mates. Whenever he was home, the visit always included an outing to Friendly's with Grant, so working with the swimming program was a natural outreach for him.

We were proud that both boys felt a desire to see, serve, and support others in the ways they did.

14. Value of having a hobby

Who pleasure gives, shall joy receive.
Benjamin Franklin

Making necklaces is the activity that most distinguishes Grant, and no recipient of his beads could be happier to receive the gift than he is to give it.

He started stringing beads at Montessori School when he was six, an activity to help increase his attention span. He strung large wooden beads onto a heavy shoelace-type string. Not only did the activity increase his attention span, but it was also good for his hand-eye coordination. Best of all, he loved doing it and found a lifetime hobby.

Being unable to play the games and sports normal children did, Grant needed an activity to fill his hours at home, so when we learned about his interest in beading, we bought some large ones for him to string. He enjoyed doing it and was able to do it faster and faster so we got him a packet of smaller colorful wooden ones, about the size of marbles, to further challenge him, and were surprised at how quickly he mastered stringing them. Then, we had an idea that has fashioned his life. We bought some clasps so he could make a necklace for his dear patterning friend Mrs. Tuttle who maintained a friendship with him. He was delighted with the idea of making a present for her, his first necklace and his first handmade gift for someone. We called and told her Grant had a surprise and asked her to stop by when it was convenient. Anyone in Barre who knew Mrs. Tuttle would know that she would come the very next day, a very caring woman, and she was thrilled with the necklace and put it on right away. I was surprised several years later when she pulled the wooden necklace out from under her shirt and told me that she wore it everyday and, amazingly, the unfinished wooden beads had taken on a lovely luster from the oils on her skin. I was moved to see how much she treasured his work.

Grant with Mrs. Tuttle

And speaking of Mrs. Tuttle, she brought back an old-time holiday to Barre. She made May baskets for her family, her friends, shut-ins, and businesses in town. Grant was one of the folks on her giving list, a list that numbered about 80. She saved cookie boxes through the year and cut them down to be the base for the crêpe paper designs she would use to decorate them. She even made her own flour paste for attaching the paper. The generous-size baskets were filled with peanuts, store candies, her homemade fudge and penuche, and topped off with a May Day poem. It was a beautiful, meaningful, and delicious piece of artwork straight from the heart. Grant loved that gift and her May Day visit. May Day was as important to him as Christmas. When the basket arrived, Grant listened patiently as I read the poem, then he explored the sweet contents. But one year, when the basket

arrived, his behavior really upset us. When Mrs. Tuttle gave him the basket, he started pulling out all the candy and peanuts. I was angry and embarrassed by his lack of appreciation and the way he so carelessly tore apart the carefully packaged candy. I said quite severely, *Stop that, Grant. That's not nice.* He turned to me almost in tears and said, *I can't find the poem.* No one is more a creature of habit than Grant. Mrs. Tuttle had not made up poems that year. Her other recipients didn't complain but for Grant, that poem was part of the ritual. Mrs. Tuttle made sure it was never missing again. Until she died last year at age 99, she still made a few May baskets, and Grant was one of the lucky ones to get one.

May baskets saved through the years.

Mrs. Tuttle's Poem

The hanging of May Baskets by children is an old New England tradition. It is a token of love and friendship, an announcement of spring and good cheer. Starting with May Day, children place gaily filled May Baskets on the doorstop of a friend's home, shout "May Basket" and run away. May Baskets are not only given to children but also are delivered to loved ones, to invalids and shut-ins.

Anyway, Mrs. Tuttle's excitement about his necklace stimulated Grant to want to make beads for others … and production started. Mrs. Bob was next, then Lucy, then Nana, then Aunt Sue, Aunt Jo, Aunt Wendy, Beth, and on and on. Anyone who was kind to him received a necklace. Then, we bought him some smaller-sized beads, and again he strung them successfully

and got faster at doing it. We were amazed and impressed with his ability. We set up a muffin pan with beads and placed them on a stool at the end of the couch and every evening after supper and for many hours on weekends, he contentedly worked on his necklaces, and we began to realize that he had a real passion for that hobby. So healthy to have a hobby. Everyone needs something to feel passionate about and we were thankful Grant had found his specialty.

When it was obvious that this hobby appeared to be quite lasting, we decided to buy him some even smaller beads that were made of glass or ceramic so he could make some good quality necklaces. When I went shopping, I discovered how expensive beads were and realized that his hobby was going to be a costly one. A friend said she's seen lots of necklaces in thrift shops, and shopping there has been the perfect solution for obtaining a bead supply, also at flea markets. Gerry made a page in "Things I Can Do Now" that shows how important the beading has been.

In this chapter we're going to tell about Grant's hobby, which is making necklaces. Grant learned how to string beads when he was working on improving his hand-eye coordination. Then he got some marble-sized beads and had fun stringing and restringing them. Then Mom bought some smaller wooden beads at Spags discount store and Grant made his first necklace and gave it to his friend Mrs. Tuttle. Well, he really enjoyed doing this and used his own money to buy more beads and clasps. Now Grant has made necklaces for many people and can use very tiny beads. Mom ties the end of his string to a clasp so the beads won't fall off. He never leaves his needles around and is careful with his muffin pan where he keeps his beads so it doesn't get spilled.

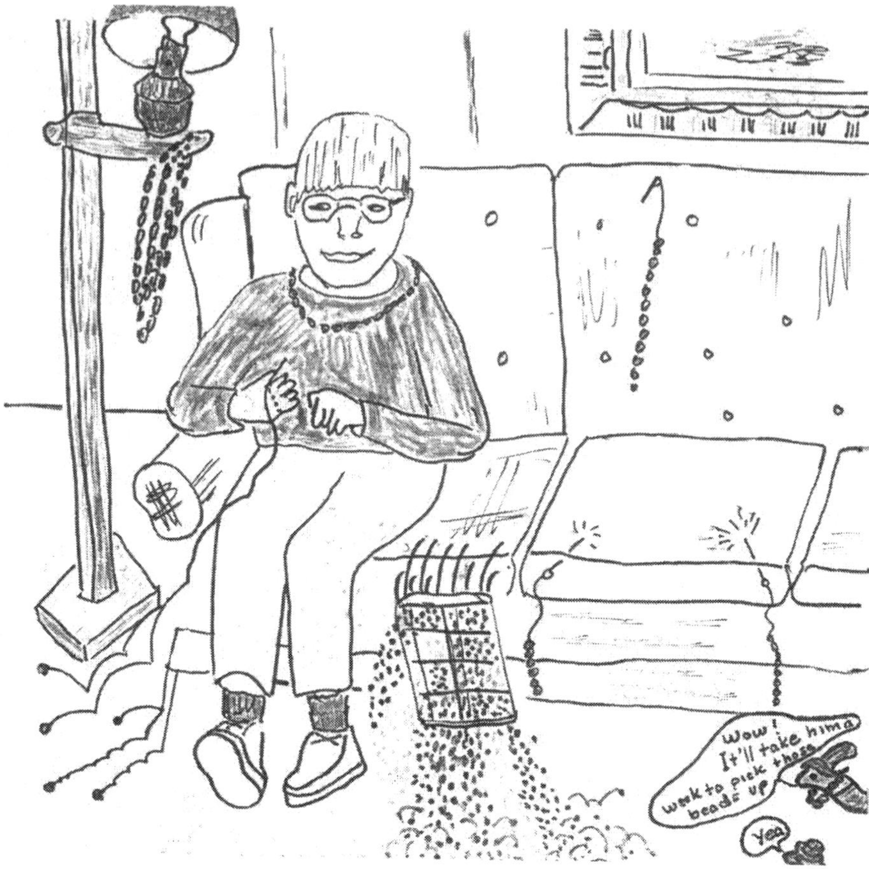

There were a few spills, but very few, and when they happened Gerry and I would be down on our hands and knees retrieving the beads that were as precious to him as real gem stones. Considering his clumsiness, it's amazing there weren't more accidents. We knew he did his best. There are few of us who function at 100 percent, I know I could always do more, learn more, help more, but I think Grant functions at his very best.

About twenty-five years ago the state offered three grants to special needs adults who could produce a craft that was good enough to be sold, and they had agreements with some shops that would carry the person's product and sell it without taking a profit

for themselves. Gerry and I were not aware of the grant opportunity, but the GAAMHA workshop had received notices and entered some of Grant's necklaces. (Of course, Grant had made some for his bosses so they had samples to show.) Well, he was one of the winners! We were notified about it and told he had $200 to spend on supplies and two craft shops would carry his necklaces and sell them for $6 each. Fantastic! We were so proud and talked with him about being in business. It was hard to believe what stringing those big beads at Montessori School would lead to.

Grant and I went shopping at a craft shop that had lots of beading supplies. $200 to spend. We bought beading thread, needles, clasps, and lots of beads, carefully choosing colors that would look pretty together. It was obvious that he was going to need more than the muffin pan to hold his beads so we bought some plastic boxes with individual compartments. Throughout the shopping outing I talked with him about the grant and the special job he had making beautiful necklaces to sell.

When we got home, we put the beads in the new boxes and got ourselves ready for production. My job has always been to thread the needle, attach the clasp to the end, and then together we'd decide on the design, often following a person's color request – for example, two black beads, then three pearls or maybe three dark blues with a silver in between each bead, then two light blues with a silver in between them. Grant completes the necklace following the design and then I tie the finished end to the clasp. That has been our routine since the beginning and continues today. With the new business venture in mind, we got right to work and I set up six necklaces for him to string. His love for the hobby has never dimmed, but his bead business was a short-lived sad story.

A disappointing story on one hand, yet a tender moving story on the other. Family and friends heard about Grant winning the grant. Imagine – a grant for Grant! Everyone knew he was about to earn money doing what he liked doing most. It's a wonderful thing when a person loves his or her job. I recently read about an elderly man who worked ten-to-twelve hours a day throughout his working years, and he said he loved his job so much that he felt he never worked a day in his life. Oh, if everyone could find the job

that suited them so well! You'd think it would be like that for Grant with his beading. But it wasn't.

Grant with beads

When he was all set to go, Grant sat down and started to string the beads – slowly with no zeal, and Grant was always enthusiastic and eager to begin a new necklace. We couldn't understand what was wrong. We reminded him many times about the honor to win the award and the six dollars he would earn for each necklace, but he was not motivated. After he finally finished one, with our constant urging, a friend stopped by and said she needed a blue necklace for a special occasion. Well, Grant was eager to begin that one right away and his enthusiasm returned. Then, we were back to the business ones and back to pushing him to do it, when it occurred to us the reason why Grant didn't enjoy his new job. We realized he did not know who to think about when he was making a necklace to sell at the craft shop. Always previously, when he was making a necklace for someone, that person was constantly on his mind. It was the person he talked about and wanted to hear stories about. Making a necklace to sell had no meaning for him. The money motivation had no value to him. The value in making a necklace was in the happiness it gave him to make it for someone he knew or heard about. So, we gave up the business venture. We felt guilty about all the beading supplies he bought with the grant money but had to justify it in our minds knowing that he used it for good purposes even if he didn't get established in business.

The following incident is another example of how Grant is like a computer. We have to work at programming him to learn certain behaviors, skills, or routines and then when he learns them, he follows them exactly. After the failure of his business enterprise, we had to reprogram his thinking about pay. He had always given away the necklaces he made, but soon after that period he made one for Mrs. Cooley and asked her to pay $6 when he gave her the necklace. She gladly reached into her purse and handed him a ten-dollar bill. I protested but she said it was beautiful and worth it. Grant wanted the money and she wouldn't take it back. We realized how much we had stressed the opportunity of earning money for his necklaces, and the six dollars had gotten programmed into his head. I lectured him about how nice Mrs. Cooley was, that she was one of his patterners, a good church friend, and that she had given him a nice toy. He stayed stuck on

the six dollars, so I did something drastic, something one sometimes has to do to make an impression. I took away his beads for a month … and he hadn't gone a day without working on them for a long time. It was a terrible penalty, and he begged Dad to take away his punishment. Gerry thought it was a bit severe, but being an old school teacher, I stuck to my sentence. I talked to him about all the people who were good to him and how he should be kind to others. Grant was reprogrammed. He never charged again. Also, he gave the money back to Mrs. Cooley, who was reluctant to accept it until I explained the situation.

His necklaces have brought him on a good journey. It warms my heart to see him stringing them. He has to work hard to put them on, bead by bead, and after sliding each one down the string, he holds it up to admire it. He loves watching it grow just as he has loving thoughts about the person who will be receiving it.

Grant stringing beads with his niece Brianna

In recent years, the workshop had him make necklaces for women in a center for abused women, and on a few occasions they brought him to the center to deliver them. All of us are touched by acts of caring, but those caring acts are especially significant for people who are having a tough time, and when the outreach of kindness comes from someone who is also having his own struggles, it feels especially meaningful. Grant's outreach gives him happiness and makes him feel needed.

Anything you're good at contributes to happiness.
Bertrand Russell

Our minister Pastor Stephanie Flynn gave Grant a new very important position; she appointed him a "missionary." When she knows a woman that she thinks would benefit from an act of kindness, she writes a note to Grant telling him about the woman's struggle, maybe grieving over the loss of a loved one or recovering from an illness, and asks him if he could make a necklace to help cheer her up. Often she adds a color the woman likes or the color of her eyes to give him an idea about the color to select. Immediately when he gets the letter about the need, the request from the minister becomes his priority. He quickly tells me that a new one has to be set up, and he puts aside whatever necklace he's been working on. We all need to feel a sense of worth and doing that missionary work fulfills that need for Grant. With his handicap, it could easily have happened that he would not possess the good feeling of the self worth that he has, but being able to do something important for others gives him a purpose of value and we are grateful to Stephanie for satisfying that need.

It's not how much we give
But how much love we put into giving.
Mother Teresa

Grant handing Pastor Stephanie
the missionary necklace he made.
She is wearing a necklace made by him.

Two years ago, Pastor Stephanie had a special Sunday service recognizing people who reach out to others, and she acknowledged several who reach out in selfless ways – Barbara who makes exquisite quilts to give away, sometimes even for people she doesn't know other than the story of the difficulties they are facing, and to Marilyn who works with puppies that will be trained to be service dogs. A man using a service dog attended the service to explain the value the dog has been in his life. And, Stephanie also included Grant and told about his outreach. When we think of the sort of news that makes headlines, all the violence and scandal, I think how much better it would be for the world if reporters sought out stories about the good acts people do, and there are so many good people. We are influenced by what we read, and how much better it would be for readers to read about the kinds of stories our pastor emphasizes.

Speaking of good caring people, I think of our Massachusetts State Senator Anne Gobi who has so many responsibilities from the large district of people and towns that she serves. But, when she learned about Grant's gift of necklaces, she sent him a magnificent certificate of recognition from the Senate. It's a pretty special political servant who takes the time to reach out to the lowly. She has been thanked … with a necklace, of course.

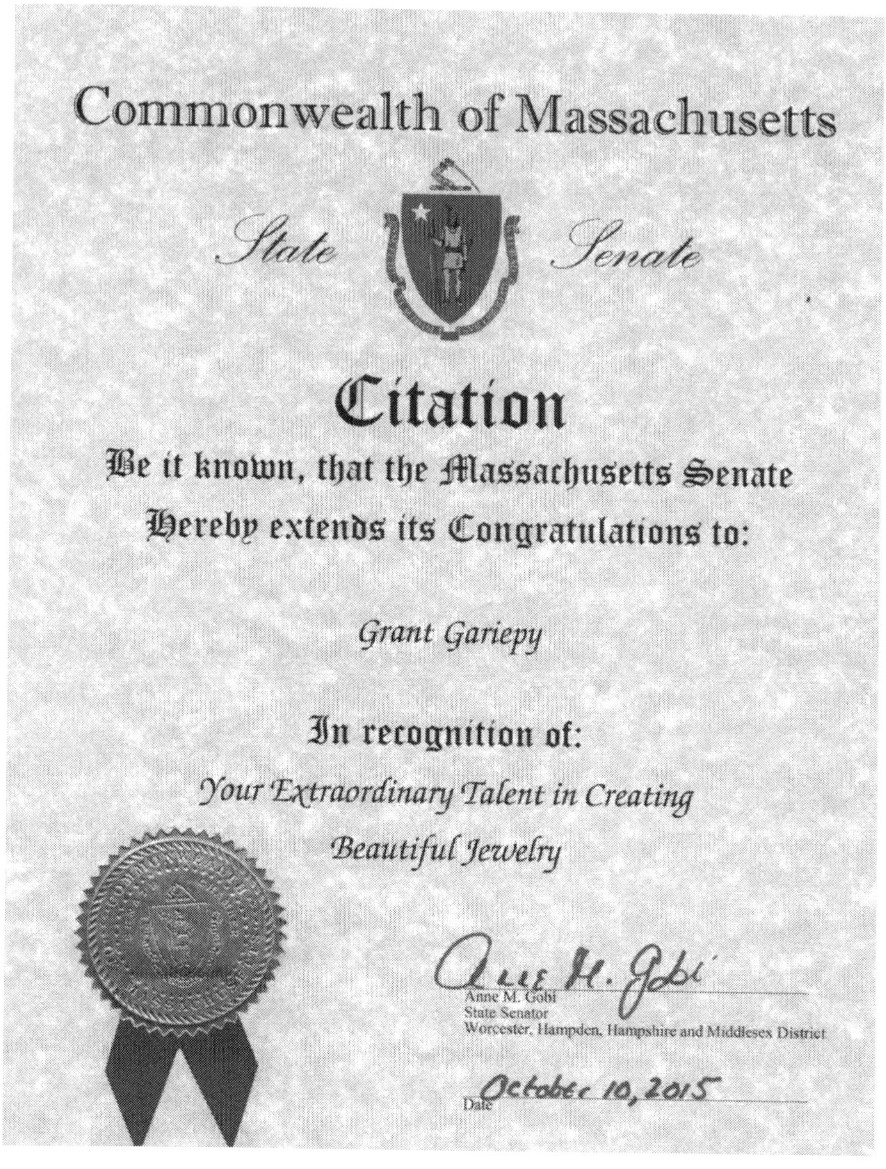

Certificate of Commendation from Massachusetts Senate

15. Move to group home

Thanks to Barre people Grant can move ahead

Those were the front page headlines written by *Barre Gazette* Editor Marilyn Haynes in the weekly newspaper on March 2, 1995. And the article began:

Grant Gariepy is 30, on his own now, living in a group home in Winchendon. Because he was so much a part of Barre, it was with some reluctance the townspeople said good-bye when he moved away last fall.

No more will Grant be there at the post office when people go in to pick up their mail. His friendly smile and greeting, "Hi. Why don't you come for a visit some time?" is conspicuously absent now as people pass in the post office lobby.

No more will people near the center of town watch Grant struggle up James Street and on across the Common with his mail bag. He won't be there to inspire us with his enthusiasm, patience and determination. But he has left Barre with a gift – the ability to appreciate what we have, and not take it for granted.

Grant was born with a birth defect that crippled both his mind and his body. When he speaks, it's sometimes difficult to understand him. His gait is uneven and he appears to have trouble keeping his balance. His knowledge of worldly things is limited.

But as much as one notices these limitations, one notices more what Grant has to give. He has a keen memory and once introduced to people, he never forgets their names. He never complains. His attitude is always upbeat and friendly.

Corry Trask, Grant's neighbor for many years said, "He has a special kind of gift. He doesn't pass judgment on anyone. He loves everyone with an unconditional love, and he behaves toward people as Christ himself would want us all to behave toward one another."

We thought about the following Bible passage as plans were being made for Grant to move ahead with his life.

There is a time for everything,
and a season for every activity under the heavens.
Ecclesiastes 3

Yes, there is a time for every season, a time to raise your child, and a time to let him go. That's how it is with our normal children. We raise them, give them our best guidance and advice, and then let them go off into the world to make their own way – college, military, job, marriage, parenting. But, it's not easy to let go of a child with special needs. It kind of feels like you're sending a kindergartener out into the world, unprepared both mentally and physically. But what is the alternative? We spent a lot of time talking with Grant's social worker about the pros and cons.

The social worker talked with us about parents whose whole life centered around taking care of their handicapped child, how they sacrificed everything to meet his needs, and how difficult it was for the adult child to adjust when the parents were no longer able to care for him. In comparison, he presented a scenario of the adult special needs child going out into the world while his family is still there to give support, just as we are when our normal children move on. It's not saying goodbye to them, it's watching them move ahead while we're still there for them to come home to visit, to call, to offer support, to love. We had a choice. What was best for Grant? What was best for us?

Cort had finished a tour in the Navy, graduated from college, and was married and living in California, and our youngest son Bart had been gone for six years, moving on when he began college, so Grant, age 30, was the only child living at home. Gerry was still working, but retirement was not far off, a time when we hoped to do some traveling. How would we feel about Grant then? We didn't want the joy we'd always felt about him turning to feelings of him being a burden. Also, we understood how grown-up he could feel if he moved on to his own home while having our support and regular presence in his life to help him adjust. He saw

how his brothers moved ahead with their lives and were still close to us and an important part of our lives. With our social worker's guidance, we began to look into living facilities and there are many excellent ones today that meet a wide range of needs. Fortunately, the old institutional living is a thing of the past.

Social worker, Barbara Perry had the job of setting up group homes for clients who would be leaving their home for the first time. First of all, clients are grouped together according to their level of functioning and compatibility so they will be happy and safe living together, a lengthy sorting-out process. Then when Grant and two other men were selected to be together, Barb spent a lot of time slowly testing out the relationships and developing bonding – afternoon outings, day trips, a weekend together, then time at the house to select their own bedrooms, curtains, rugs, etc. Everything to get them enthused and ready for the move. Grant became more excited after each occasion together. He was about to be a grown-up!

We talked with him about the furniture he would bring, his bedroom set, a chair, a bureau, and very importantly his books, especially the ones Dad made. Other people showed their enthusiasm by contributing to his furnishings – Mary gave him a desk for his room, Max carved a wall figure for him to hang, Bertyne gave him a lamp, and Alice and Sarah gave him an address book that he still uses 23 years later so he can send Christmas cards to his old friends. Barbara and the guys held an open house shortly after they moved in and a large number of people from Barre attended, even some of the post office workers. Barre people supported him when he lived in town and helped him move ahead. The transition went smoothly and fortunately was in the same district as his workshop so he didn't have to make a change there.

The location was a big plus. The Winchendon Post Office was just down the street from the house! Learning about that sure made Grant happy, and Barb took him to the post office right away to get his own box. Of course, he was used to always finding mail in our box at home, but his box? I could imagine it being empty most of the time. We started sending him notes, and some friends did too,

but everyday? Well, kindness must be a trait of small town postal workers because when the Winchendon staff saw how important the postal stop was for him, they made sure there was something in his box every day -- a flyer, some ad, something. And, Barb told Grant how important this mail was so they could shop for bargains.

Roommates Chris, Grant and Ed

Grant learned how to use the telephone. He had answered the phone at home but he didn't make calls himself. Barb thought it was important that he learn how to call us. A good idea to help with his adjustment. He learned it quite quickly, and he did call ... every day. We looked forward to the calls. It was comforting to hear his voice and that he sounded happy. He told us about his day at his new home and the workshop and we kept him abreast of news and happenings back home.

Friday and Saturday were double days of activity for him, Friday after work was the post office and the bank for cashing his check, and Saturday morning was the post office and the donut shop. The donut shop was a new activity and he loved it. He and Ed went together and Grant always got a Boston Cream Pie donut. The owner told them to come in the late morning when they weren't too busy so the waitresses would have extra time to spend with them. It's a good thing walking was strenuous for him and

used up all the extra calories he was enjoying. Sunday was special too, they walked to church.

We picked Grant up at the workshop and brought him home one afternoon a week to visit and have supper with us, also there was time to set up necklaces for him to work on at his new home.

Setting up necklaces

Sometimes we went to his house on weekends so we could join him on a jaunt to the donut shop or to church. It was pleasing to see how well he was received. The donut shop people, even the owner, greeted us and told us how happy they were to have Grant and Ed come to their shop. And, the people at church were just as solicitous. The couple who sat in front of them said that the men always sat in the same pew and had a nice visit with them before and after the service. (It's hard for me to think of Grant as a man, to me he's always my boy.) Grant always could seek out a caring

heart, and if a person wasn't responsive to him, he went on to someone else. The couple in front of them was kind so they always sat behind them. The minister said some kind words and showed warm respect as he shook their hands at the end of the service. Those visits to the donut shop and church gave us the opportunity to express our appreciation for their outreach and acceptance. More blessings.

John and Mary at Grant's 50th birthday party

Grant always joined the family for holidays, and his favorite was Thanksgiving when we all go to John and Mary Allen's home. John is a special man in Grant's life. Usually Grant gravitates to women who are more patient and understanding of him ... kind of a maternal thing. The men we know are good to Grant but often don't know how to really relate to him. John does, and not by talking down to him, but talking up in a man-to-man fashion. I always feel a little sorry for Mary who cooks such special meals

and is so kind to him, but if John is around, he has Grant's focus. And, that is a typical trait Grant has. He focuses on one person at a time, he doesn't interact well with a group. Usually there's one person who really stands out in his mind and that person gets his full attention. We see that also when Pastor Stephanie or Cousin Beth are nearby. I wonder what he would do if John, Pastor Stephanie, and Beth were all together in a room with him.

Grant was happy and well adjusted in his new living situation, but he was slowly starting to fail physically. Years ago, doctors told us he would probably be confined to a wheelchair by the time he was 30. Well, he fooled them, he was nearing 40 and still walking, but his deformities were defeating his will to maintain the level of mobility he had achieved. He was having more falls. Luckily, he didn't break any bones but his skin often had black and blue areas that revealed another tumble. Did he complain? No. He kept plugging along, but "the handwriting was on the wall." He had to have help. The orthopedic doctor prescribed a walker.

How did Grant do with a walker? Terrible. He lifted it when he took a step and the results were -- Grant toppled on top of the walker, more injurious than when he walked without it. Grant was used to lifting his arms high when he walked, actually kind of flailing them to keep balance, keeping his hands down on the walker was unnatural for him. Change was not easy. The doctor set him up with a physical therapist to train him in the usage. He improved but not consistently. It was Gerry who had the solution. He can fix most anything, instead of "jury-rigged," Cort says everything around our house is "Gerry-rigged." Well, he Gerry-rigged the walker. He got a heavy steel cylinder-shaped weight that was used in the framework of old windows so they could be raised. He taped the weight onto the front of Grant's walker so he was unable to lift it. After several weeks, he adjusted to walking without lifting his hands and the mission was accomplished. Gerry removed the weight and Grant safely and proudly walked with his new device, and we named the walker his Cadillac.

Grant with walker

But, walking wasn't the only problem he faced. The organization managing the group home collapsed and it was closed down. Another transition was ahead for Grant. He was placed in a private home with a very nice family. The state offers an attractive salary for the care of a special needs adult, and many placements have worked out very satisfactorily. However, even though Grant is a very good person, his care is demanding, not easy for a homemaker who also has to meet the demands of others in her family, and after several months he was transferred to another home, then another. All were good homes and gave him good care. We understood. Seven days a week, twenty-four hours a day was not easy. After he was transferred the second time, I had asked the social worker if he could go to a group home where there are many staff people who come in to work a shift then back home, a job that is not a 24-hour one, much less demanding. We were told that

there were no openings in our area. When he was about to be transferred for the fifth time, I requested a group home again and once again told there was nothing available.

We had to do something. Maybe the answer could lie with a congressman. I'd heard of the wonderful changes that these folks have brought about in people's lives. Well, Gerry didn't like that idea. He does not like to ask for special favors, but when you have a handicapped child who's in a bad situation, you have to step up and be his advocate. He certainly couldn't advocate for himself. I called Senator Steve Brewer and Grant's life soon changed. I don't know how it was accomplished, but within two weeks, Grant was placed in a group home managed by HMEA, Horace Mann Educational Associates, and he's been there ever since. He's very happy and we're very happy. It's true when they say ...

A mother is about as happy as her least happy child.

The group home is in Gardner, and Grant lives there with three other men. There is a house manager and many staff people who work various shifts. Some are college students working part time as they pursue careers in social work or other related fields, and the experience they get in the group home helps prepare them for their future work. All the faces add variety to Grant's life, and his love spreads out to them all. Often he will call to tell us that Jen will take him shopping this weekend, or Rebecca worked yesterday, or Stephanie is on duty tomorrow. The current manager, Teresa, has been on the staff for many years and maintains the pleasant caring spirit and care that has prevailed. Ashley and Mike are central managers who keep many HMEA houses in check, and Murray the social worker keeps good communication between the house and workshop. Grant is blessed. We are blessed. Thank you, HMEA. Thank you, Senator Brewer.

16. Changes, hurdles, blessings

My philosophy of life is
to be the very best person that I am able to be in all respects --
be a good friend and a good citizen,
make the most of all opportunities,
and face and overcome obstacles in the best way I can.

I wrote that statement in one of my papers 50 years ago when I was studying at Worcester State Teachers College in 1958. Little did I realize then the obstacles that would lie ahead.

Every one thinks his sack is heaviest;
None knows the weight of another's burden.
George Herbert

And, the longer I live, the more I realize that we all have hurdles to face. Difficult challenges affect families who have to deal with chronic illnesses, addictions or accidents that happen in their children's lives or the lives of loved ones. We feel a loved one's need. We feel a loved one's pain. We are parents and want to fix the problems. We have to learn in what ways we can be helpful, in what ways our actions enable the person to stay stuck in his or her problems, and we have to know when it's time to let go and move on with our lives and not neglect other family members, knowing there is nothing more we can do. Whether the challenges we face are physical, mental, alcohol, drugs, marriage, child, parent, grievance, or other issues, we all have trials to deal with.

God grant me the serenity to accept the things I cannot change,
the courage to change the things I can,
and the wisdom to know the difference.
Reinhold Niebuhr

We had a new hurdle to overcome. Grief. Bart died in an accident at age 31, and anyone who loses a child will understand the heartbreak. A book could be written, and many have been, about facing the death of a child, but this is a book about Grant so I will just write a couple of meaningful quotes that were sent to us at the time. I hope they will touch someone else with a deep grief.

Don't cry because he died
Smile because he lived.

And the following poem:

When tomorrow comes without me,
Don't think we're far apart,
For every time you think of me
I'm right there in your heart.

We don't forget a loved one, nor should we. Time helps with healing but not forgetting. Grant greatly felt the loss of his brother. Whenever Bart came home, there was always a special outing just for the two of them.

Grant still talks about him, and some of Bart's friends keep his memory alive by keeping in contact with us. Jed, his friend from childhood and his former college roommate Mike both bring their families to visit every summer, college friend Rachel used his name for a middle name for one of her sons, Harold, a friend throughout his life, buys a seat for Bart when he attends an annual concert they used to attend together. He recently sent a photo of "the empty seat." We treasure these pieces that remain. Help your friends with losses by sharing good memories, letting them know you remember too.

Grant with Bart

Gerry's favorite photo of "My Three Sons"

Grant's physical problems continued to worsen, and it has been heart-wrenching to see his skills deteriorate. His balance got worse and his legs weakened. Even though he used the walker, he once again was having falls. Also, he started having toileting accidents.

His hands became less steady and he couldn't hold the small beads to guide them onto the string. He couldn't manage his eating utensils as well and began having digestive problems. By the time he was 50, he was confined to a wheelchair. He held out 20 years longer than was predicted. Grant has grit.

The saddest part for us was that we can't bring him home to visit anymore. We, as much as Grant, looked forward to his weekly visit, but we no longer could physically manage him. Nevertheless, we continued our vow not to show him pity. Being handicapped is OK but being pitiful is not. We determined to be positive, encouraging, and proud of whatever he was able to achieve, because most of all, we wanted him to be happy. The family's attitude toward a person's disability very strongly affects his attitude toward himself. I can't tell you how many times through the years I motivated myself with Abraham Lincoln's words of wisdom:

We will be just about as happy as we make up our minds to be.

I said to Grant, *I used to wonder what I would cook for you when you came home. Now we get to go to Friendly's every week, a treat for us all.* I talk with the group home manager about what day is convenient for them to transport Grant to the restaurant each week, and they cheerfully drive him there in the handicapped accessible van and come for him when we finish eating. They want their clients to be happy and have as full lives as possible, and the Friendly's people all know Grant and greet him like a prince. He feels special, and we have a good outing.

"*When life gives you lemons, make lemonade.*"
Elbert Hubbard

He's mastered operating the electric wheelchair, something we were very impressed about, and he continues to make necklaces with larger beads even though it sometimes takes him 10 minutes to put on a bead. He spends a lot of time every day on his necklaces, and his "missionary" work of making necklaces for

Pastor Stephanie to give to people "to cheer them up" is still his top priority. So good for a needy person to be able to help someone else.

You just can't beat the person who never gives up.
Babe Ruth

Grant is no longer able to keep up with the demands needed to work in the workshop division of GAAMHA and was transferred to the division called STREAM, which means Success Through Relationships, Employment, And Meaningful experiences. I recently spent a morning observing the STREAM program, a morning that filled me with thankfulness and smiles. I say thankfulness because I'm so grateful Grant is able to attend such an outstanding day program, and I include "smiles" because one of the staff members said that she could be having a "down" day before she got to work, but as soon as she's with these special people, all with developmental disabilities of some kind, they bring smiles to her life.

The STREAM is a new program, only two years old. Director Tina explained that it's individually tailored to give each person the most enriching life opportunities possible. I observed this aspect in the first session of the day as the clients met with their support staff person to make out their schedule for the following week. The program is well organized with diverse choices that include both in-house and community-related activities. The direct client support person, Don, worked with Grant and three other clients to organize their week. He read the options for three morning sessions and two afternoon ones, and each client made their selections, Don sometimes gently guiding or encouraging them to include an activity that would be a beneficial or fun choice. The well-rounded program emphasizes stimulating options that help with social development and community involvement. Grant always immediately knew what he wanted -- the computer lab, money skills, word games, writing, out to eat, and best of all a visit to the Red Apple Farm.

I noted how serious and interested the clients were as they prepared for the upcoming week, also their patience and respect for each other. It was heart warming to see the clients treated with dignity, kindness, and respect as they made their choices. They felt secure and their opinions valued. I learned from Grant's case manager that this day program is a model being observed and incorporated by other day services, many of which have a very limited in-house program. STREAM certainly is one worth emulating.

Case Manager Cynthia and Client Support Person Don

Grant's first session was Computer Lab, a choice he would pick every day if he could and he always signs in to one of the many beading sites. He selects the site he wants and when it comes up he's "lost to the rest of the world," so engrossed he is in watching professional bead makers perform his favorite hobby.

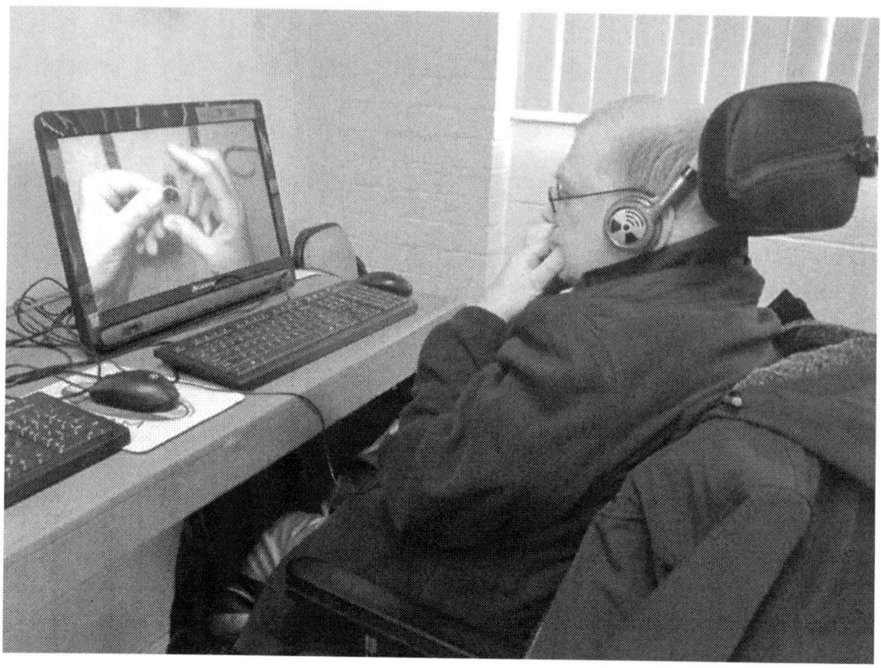

Grant watching beading on the computer

We went from there to a craft session to make a Christmas decoration. As I observed the innocence of these clients, I thought about how raising a child with developmental disabilities spared us from some of the worries one has with raising normal kids.

-- no teenage trials and tribulations.
-- no worry about him being out late at night
--no worry about him hanging around with the wrong crowd
-- no worry about him driving too fast
-- no worry about drugs or alcohol
-- no worry about effects of social media

Grant is kind of like having a young child forever, one who loves and trusts us.

I also thought about how the interactions of the clients and staff would be good examples for the world. They are accepting of everyone -- non-judgmental and courteous; they are positive and expect goodness; they are not intimidated nor intimidating; they are happy; they are at peace. They would not look for the sensationalism or scandal that make headlines in the newspapers. Their world would look for the multitude of good acts that don't make the headlines.

As I was leaving the STREAM, one of the clients came up to me with an outstretched hand and shook mine saying, *It was nice to meet you.* Like his staff worker, I was filled with smiles.

Another good thing in Grant's life is that he has not lost contact with his old friends. We usually invite a friend to join us at Friendly's, and he tells us who he'd like to have come. Grant loves that sociability, and we do too, and very often he will bring one of the books Dad wrote so they can read about his adventures. A new social benefit he has added in recent years is a telephone connection. He has regulars who call him every month. Some have set days, like Beth who always calls on the first of the month, others call when it's convenient. Those calls are very important for him, and he calls us to report the news. *Beth is busy getting ready for a concert ... Lisa's daughter is getting married in August ... Wendy said Eunice's dog swallowed her telephone ... Susan said Larry broke his arm ... Judy is going on a cruise ... Lucy said Mrs. Bob doesn't feel well ... Mrs. Pollard had a terrible leak in the bathroom and had to call the plumber.*

The day when Mrs. Pollard was telling Grant about the leak, her grandson Tim came in and while he sat on the couch waiting for her to finish her telephone conversation, he listened to her exclaiming, *The toilet overflowed ... the water ran all over the bathroom and out onto the rug in the bedroom ... the rug had to be taken up to dry ... I may have to buy a new rug ... my house is a mess with all the bedroom furniture moved out ... the cat is upset ... it happened in the middle of the night ... I'm so tired ...* The grandson couldn't believe how she was dramatically going on and

on and on, and when she hung up he asked, *WHO were you talking to?* She explained how Grant likes disaster stories ... and she had told him a good one. Of course, Grant quickly called us. He always calls to tell us the news from his callers. I say, *Thank goodness for you, Grant, for keeping us up to date on what's going on with our friends.*

Correspondence is important in his life. Grant's earlier joy was going to the post office. But rural delivery is the service at the group home, so going to the mailbox is the first thing he does when he gets off the van in the afternoon after work. Mrs. Cohen, one of the retired Barre post office workers, who moved to Florida sends him mail very regularly. All the drawers in his desk are filled with the postcards and notes he's received through the years and he can recall who sent most of them just by looking at the picture on the front. He sends Christmas cards to his friends, a writing effort that takes him about a month. He starts writing them in late October so they will be ready to mail out on Thanksgiving. Usually his cards are the first ones his friends receive, and he's rewarded by receiving many in return.

Grant really cares about the people he knows and wants to keep up on what is happening in their lives. If he hears someone is sick, he calls us to ask how they are doing He calls regularly to keep track of our doings, and when he knows we are going to be visiting with someone, he will call the next day and say, *Tell me all about _______.* And if he calls and gets our answering machine, when we call him back, he always asks, *Where were you?*

He regularly writes notes to me, some he signs from himself and some he signs from Dad. The ones from him are always sweet and tender, the ones signed from Dad are about cancelling a trip or telling that he didn't like a trip we had taken. Grant does not like it when we go on vacation, and he knows I'm the one who plans trips, so he writes notes from Dad to me either saying he wants to cancel an upcoming trip or telling me he didn't like the one we went on. I always pretend to be upset with Dad about his negative attitude: *What! You didn't like the trip.* And, of course, Gerry goes along with the note pretending he did write it. Grant has great

creativity and good humor and he loves to watch our dramatic response to the notes. Here is a letter signed by Grant to me:

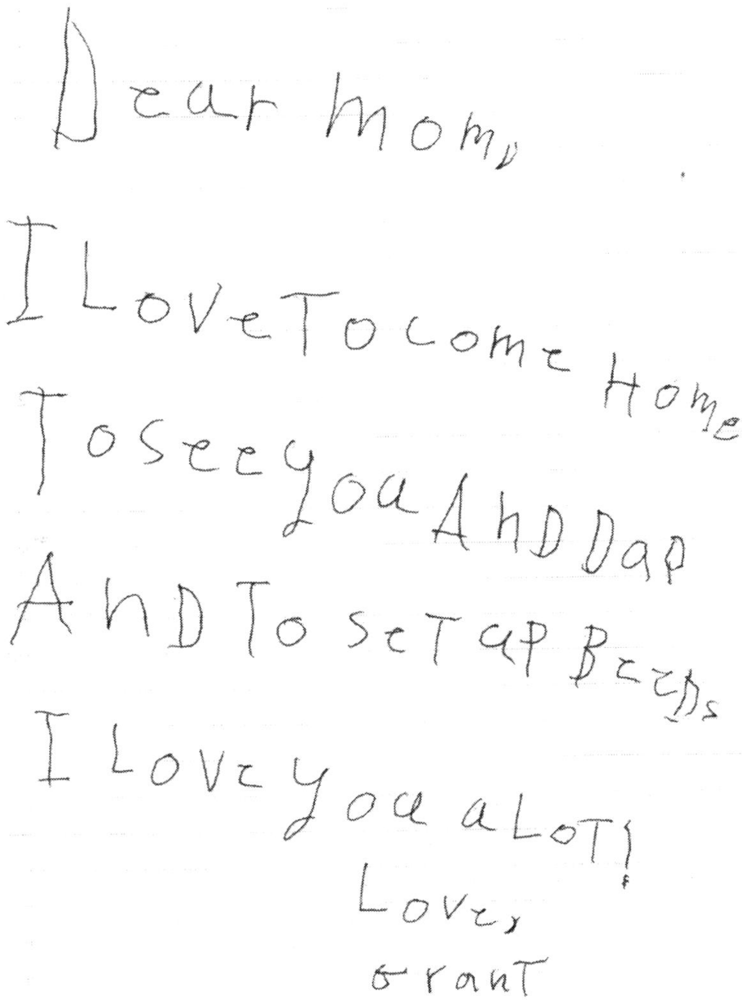

Dear mom,

I Love To come Home To see you And Dad And To set up Beds

I Love you a Lot!

Love,

Grant

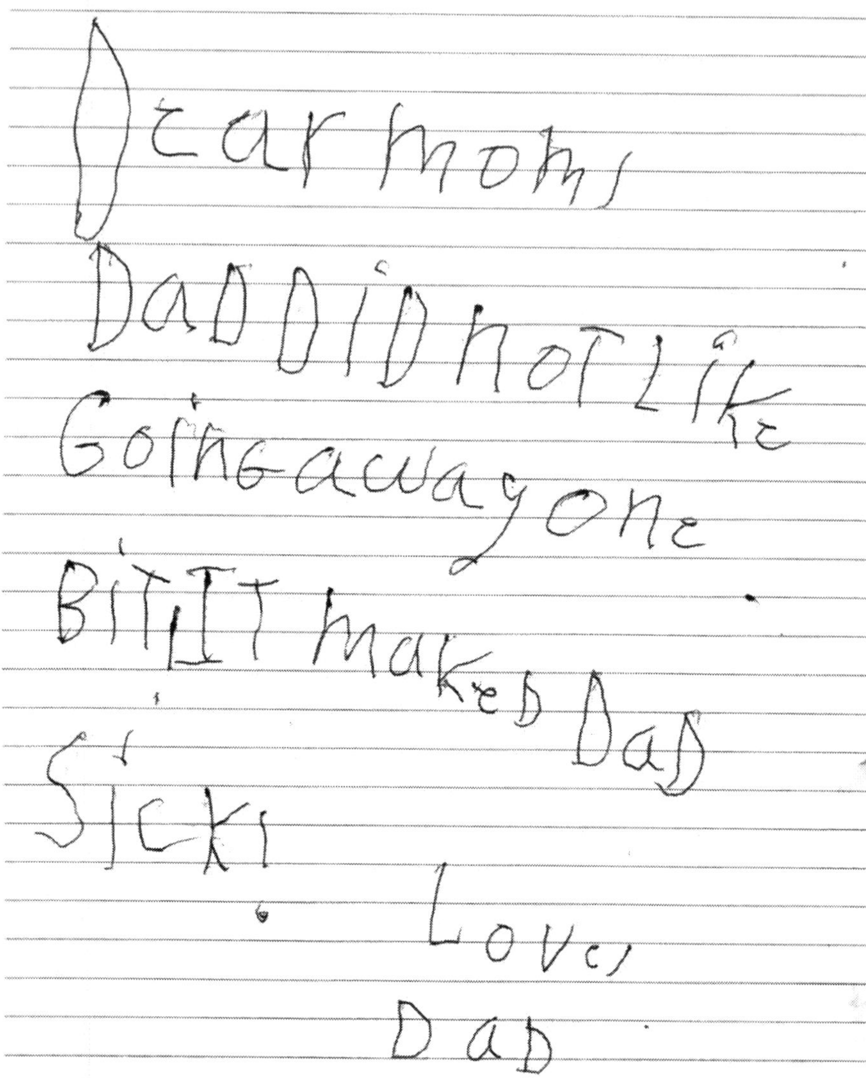
Dear mom,
Dad did not like
Going away one
Bit It maked Dad
Sick.
Love,
Dad

Letter signed "from Dad" to me

Drawing by Grant

His house manager has set up an email account for him so we can send notes and pictures when we do go away, and he tells her what news to write to us. The computer keeps us close even when we travel.

We don't know what the next days will bring, but we know Grant will give each day his best. No one can ask for more than that. He's not only a blessing in our lives, he's also an inspiration -- the way he always focuses on the positive. No matter what, he always looks at life with his glass half full, never half empty.

We appreciate all the people who played a part in making his life a happy one, all the positive, encouraging, and kind acts. And our appreciation goes to all of you who have reached out to someone in your community who needed a caring soul to make a difference.

God is great. God is good. People are great. People are good.

83135852R00107

Made in the USA
Columbia, SC
08 December 2017

CHAPTER I

DIRICHLET PROBLEM

1. Subharmonic- and superharmonic-functions

Let D be a domain on a closed Riemann surface F, spread over the z-sphere and $\phi(z)$ be a continuous function, defined in D. If for any $z_0 \in D$, there corresponds $\rho=\rho(z_0)$, such that a disc $|z-z_0|\leqq\rho$ is contained in D and for any $0\leqq r<\rho$,[1)]

$$\phi(z_0)\leqq\frac{1}{2\pi}\int_0^{2\pi}\phi(z_0+re^{i\theta})d\theta, \quad 0\leqq r<\rho , \tag{1}$$

then $\phi(z)$ is called *a subharmonic function in D.*

If

$$\phi(z_0)\geqq\frac{1}{2\pi}\int_0^{2\pi}\phi(z_0+re^{i\theta})d\theta, \quad 0\leqq r<\rho , \tag{2}$$

then $\phi(z)$ is called *a superharmonic function in D.*

A harmonic function is subharmonic and superharmonic. We shall study subharmonic functions under a more general definition in Chapter II, but for the purpose of this chapter, we adopt the above definition.

Let $f(z)$ be regular in a finite domain D on the z-plane, then $|f(z)|^p$ $(p>0)$ and $|z|^a|f(z)|^p$ $(a>0,\ p>0)$ are subharmonic in D. If $z=0$ does not belong to D, then $|z|^a\cdot|f(z)|^p$ $(p>0)$ is subharmonic in D, even if $a<0$.

Proof. If $f(z_0)=0$, then (1) is satisfied. If $f(z_0)\neq 0$, then $f(z)\neq 0$ in $|z-z_0|\leqq\rho$, if $\rho>0$ is sufficiently small, then $(f(z))^p$ is regular in $|z-z_0|<\rho$, hence

$$(f(z_0))^p=\frac{1}{2\pi}\int_0^{2\pi}(f(z_0+re^{i\theta}))^p d\theta, \quad 0\leqq r<\rho ,$$

so that

$$|f(z_0)|^p\leqq\frac{1}{2\pi}\int_0^{2\pi}|f(z_0+re^{i\theta})|^p d\theta, \quad 0\leqq r<\rho .$$

The other part can be proved similarly.

Theorem I. 1. (i) *If $\phi(z)$ is subharmonic, then $-\phi(z)$ is superharmonic and if $\phi(z)$ is superharmonic, then $-\phi(z)$ is subharmonic.*

(ii) *If $\phi_i(z)$ $(i=1,2,\cdots,n)$ is sub(super)harmonic, then $\sum_{i=1}^{n} c_i\phi_i(z)$ $(c_i>0)$ is sub(super)harmonic.*

---

1) If $z_0=\infty$, or a branch point of F, then we consider on the plane of the local parameter at z_0. (p. 469).

(iii) *If* $\phi_i(z)$ $(i=1, 2, \cdots, n)$ *is subharmonic, then*

$$\phi(z)=\mathrm{Max}(\phi_1(z), \cdots, \phi_n(z)) \text{ is subharmonic.}$$

If $\psi_i(z)$ $(i=1, 2, \cdots, n)$ *is superharmonic, then*

$$\psi(z)=\mathrm{Min}(\psi_1(z), \cdots, \psi_n(z)) \text{ is superharmonic.}$$

This can be proved easily from the definition.

Theorem I. 2. *Let* D *be a compact domain and* Γ *be its boundary.*

(i) *If* $\phi(z)$ *is subharmonic in* D *and continuous in* $\overline{D}$ *and* $\phi(z)\leqq M$ *on* Γ, *then* $\phi(z)\leqq M$ *in* D.

(ii) *If* $\psi(z)$ *is superharmonic and continuous in* $\overline{D}$ *and* $\psi(z)\geqq m$ *on* Γ, *then* $\psi(z)\geqq m$ *in* D.

(iii) *Let* $u(z)$ *be harmonic in* D *and continuous in* $\overline{D}$.

If $\phi(z)\leqq u(z)$ *on* Γ, *then* $\phi(z)\leqq u(z)$ *in* D.

If $\psi(z)\geqq u(z)$ *on* Γ, *then* $\psi(z)\geqq u(z)$ *in* D.

Hence a subharmonic function attains its maximum on Γ and a superharmonic function attains its minimum on Γ.

Proof. (i) We put $M=\underset{z\in\overline{D}}{\mathrm{Max}}\,\phi(z)$ and suppose that at an inner point z_0, $\phi(z_0)=M$. We assume that $\phi(z)\not\equiv$ const. Let E be the set of z, such that $\phi(z)=M$, then E is a closed set and since $\phi(z)\not\equiv$ const., E has a boundary point $z_1\in D$. Let $|z-z_1|\leqq r$ be contained in D, then

$$M=\phi(z_1)\leqq \frac{1}{2\pi}\int_0^{2\pi}\phi(z_1+re^{i\theta})d\theta \ .$$

Since on $|z-z_1|=r$, there is an arc, which lies outside of E, on which $\phi(z)<M$, we have $\frac{1}{2\pi}\int_0^{2\pi}\phi(z_1+re^{i\theta})d\theta<M$, which is absurd. Hence $\phi(z)$ attains its maximum on Γ, so that if $\phi(z)\leqq M$ on Γ, then $\phi(z)\leqq M$ in D.

(iii) Since $\phi(z)-u(z)$ is subharmonic and $\leqq 0$ on Γ, we have $\phi(z)-u(z)\leqq 0$, or $\phi(z)\leqq u(z)$ in D. ——

Since if $f(z)$ is regular, $|f(z)|^p (p>0)$ is subharmonic, we have

Theorem I. 3. *If* $f_i(z)$ $(i=1, 2, \cdots, n)$ *is regular in* D *and continuous in* $\overline{D}$, *then* $\sum_{i=1}^{n}|f_i(z)|^{p_i}$ $(p_i>0)$ *attains its maximum on the boundary of* D.

Theorem I. 4. *Let* $\phi(z)$ *be subharmonic in* D *and* $|z-z_0|\leqq r$ *be contained in* D, *then*

$$\phi(z_0)\leqq \frac{1}{2\pi}\int_0^{2\pi}\phi(z_0+re^{i\theta})d\theta \ .$$

If $\psi(z)$ *is superharmonic in* D, *then*

$$\psi(z_0)\geqq \frac{1}{2\pi}\int_0^{2\pi}\psi(z_0+re^{i\theta})d\theta \ .$$

Proof. Let $u(z)$ be harmonic in $|z-z_0|<r$, such that $u=\phi$ on $|z-z_0|=r$, then by Theorem I. 3,

$$\phi(z_0)\leqq u(z_0)=\frac{1}{2\pi}\int_0^{2\pi}u(z_0+re^{i\theta})d\theta=\frac{1}{2\pi}\int_0^{2\pi}\phi(z_0+re^{i\theta})d\theta\,. \text{ ——}$$

2. Upper- and lower-functions

1. Let D be a domain, contained in a closed Riemann surface F, spread over the z-sphere and Γ be its boundary and $v(z)$ be a continuous function in D. Let Δ be a disc, contained in D and C be its bounding circle. We define $v_\Delta(z)$ in D as follows.

$$v_\Delta(z)=v(z) \quad \text{in } D-\Delta, \qquad v_\Delta(z)=w(z) \quad \text{in } \Delta, \tag{1}$$

where $w(z)$ is harmonic in Δ, such that $w=v$ on C.

Then for two continuous functions $v(z)$ and $u(z)$ in D,

(i) $\alpha v_\Delta(z)+\beta u_\Delta(z)=(\alpha v(z)+\beta u(z))_\Delta$, where α, β are constants.

(ii) If $v(z)\geqq u(z)$ in D, then $v_\Delta(z)\geqq u(z)$ in D.

Lemma 1. *If $v(z)$ is superharmonic in D, then $v(z)\geqq v_\Delta(z)$.*
If $v(z)$ is subharmonic in D, then $v(z)\leqq v_\Delta(z)$.

Proof. To prove the first part, it suffices to prove that $v(z)\geqq v_\Delta(z)$ in Δ.

Since $v=v_\Delta$ on C and v_Δ is harmonic in Δ, we have by Theorem I. 2, $v(z)\geqq v_\Delta(z)$ in Δ.

Lemma 2. *If $v(z)$ is super(sub)harmonic in D, then $v_\Delta(z)$ is super-(sub)harmonic in D.*

Proof. Since $v_\Delta(z)$ is harmonic in Δ and $v_\Delta=v$ in $D-\Delta$, it suffices to prove that v_Δ is superharmonic on C. If $z_0\in C$ and $|z-z_0|\leqq r\in D$, then by Lemma 1,

$$v_\Delta(z_0)=v(z)\geqq\frac{1}{2\pi}\int_0^{2\pi}v(z_0+re^{i\theta})d\theta\geqq\frac{1}{2\pi}\int_0^{2\pi}v_\Delta(z_0+re^{i\theta})d\theta\,. \text{ ——}$$

2. Let $f(\zeta)$ be a given bounded function on the boundary Γ of D. We put for $\zeta_0\in\Gamma$,

$$\underline{f}(\zeta_0)=\varliminf_{\zeta\to\zeta_0} f(\zeta)\quad,\quad \bar{f}(\zeta_0)=\varlimsup_{\zeta\to\zeta_0} f(\zeta)\,, \tag{1}$$

where $\zeta\to\zeta_0$ on Γ.

If $m\leqq f(\zeta)\leqq M$ on Γ, then

$$m\leqq\underline{f}(\zeta)\leqq\bar{f}(\zeta)\leqq M \quad \text{on } \Gamma\,. \tag{2}$$

If $f(\zeta)$ is continuous at ζ_0, then $\underline{f}(\zeta_0)=\bar{f}(\zeta_0)=f(\zeta_0)$.

Now we define *the upper function* $\psi(z)$ and *the lower function* $\phi(z)$ as follows.

(i) $\psi(z)$ is continuous and superharmonic in D.

(ii) $\underline{\lim}_{z\to\zeta}\psi(z)\geqq\bar{f}(\zeta)$, $\zeta\in\Gamma$, $z\in D$.

Similarly

(i) $\phi(z)$ is continuous and subharmonic in D.

(ii) $\overline{\lim}_{z\to\zeta}\phi(z)\leqq\underline{f}(\zeta)$, $\zeta\in\Gamma$, $z\in D$.

Since $\psi(z)\equiv M$ is an upper function and $\varphi(z)\equiv m$ is a lower function, the sets of upper functions and lower functions are not empty.

Theorem I. 5. *upper function* $\geqq$ *lower funcion.*

Proof. Let $\psi(z)$ be an upper function and $\phi(z)$ be a lower function and put $\omega(z)=\psi(z)-\phi(z)$, then $\omega(z)$ is superharmonic and at $\zeta\in\Gamma$, $\underline{\lim}_{z\to\zeta}\omega(z)\geqq\underline{\lim}_{z\to\zeta}\psi(z)-\overline{\lim}_{z\to\zeta}\phi(z)\geqq\bar{f}(\zeta)-\underline{f}(\zeta)\geqq0$ $(z\in D)$, so that $\omega(z)\geqq0$ in D, or $\psi(z)\geqq\phi(z)$ in D.

Theorem I. 6. *If* $\psi_i(z)$ $(i=1,2,\cdots,n)$ *are upper functions, then*

$$\psi(z)=\mathrm{Min}\,(\psi_1(z),\cdots,\psi_n(z))$$

is an upper function.

If $\phi_i(z)$ $(i=1,2,\cdots,n)$ *are lower functions, then*

$$\phi(z)=\mathrm{Max}\,(\phi_1(z),\cdots,\phi_n(z))$$

is a lower function.

Proof. $\psi(z)$ is continuous and superharmonic in D. Since $\underline{\lim}_{z\to\zeta}\psi_i(z)\geqq\bar{f}(\zeta)$, $\underline{\lim}_{z\to\zeta}\psi(z)\geqq\bar{f}(\zeta)$ $(\zeta\in\Gamma)$, so that $\psi(z)$ is an upper function. Similarly $\phi(z)$ is a lower function.

Theorem I. 7. *If* $\psi(z)$ *is an upper function, then* $\psi_\Delta(z)$ *is an upper function. If* $\phi(z)$ *is a lower function, then* $\phi_\Delta(z)$ *is a lower function.*

3. Dirichlet problem.

1. The following proof is due to Perron [1], simplified by Radó and F. Riesz [1]. Let $f(\zeta)$ be a bounded function, defined on the boundary Γ of a domain D, contained in a closed Riemann surface, such that $m\leqq f(\zeta)\leqq M$ on Γ and $\mathfrak{F}$ be the family of all upper functions $\psi(z)$ and put

$$u(z)=\inf_{\psi\in\mathfrak{F}}\psi(z)\ , \tag{1}$$

then since $\underline{\lim}_{z\to\zeta}\psi(z)\geqq\bar{f}(\zeta)\geqq m$ on Γ, we have $\psi(z)\geqq m$ in D, so that $u(z)\geqq m$ in D and since $\psi(z)\equiv M$ is an upper function, we have

$$m\leqq u(z)\leqq M \quad \text{in } D\ . \tag{2}$$

We shall prove

Theorem I. 8. $u(z)$ *is harmonic in* D.

We call $u(z)=u_f(z)$ *the solution of Dirichlet problem for* D *with the boundary value* f.

Similarly if $\mathfrak{F}$ be the family of all lower functions $\phi(z)$ and put

$$u(z)=\sup_{\phi\in\mathfrak{F}} \phi(z) , \tag{1*}$$

then $u(z)$ is harmonic in D.

2. First we shall prove two lemmas.

Lemma 1. *Let z_i $(i=1, 2, \cdots)$ be given points in D, then there exist upper functions $\psi^{(n)}(z)$, which satisfy the following conditions:*

(i) $\psi^{(1)}(z)\geqq\psi^{(2)}(z)\geqq\cdots\geqq\psi^{(n)}(z)\geqq\cdots$ *in* D,

(ii) $\lim\limits_{n\to\infty}\psi^{(n)}(z_i)=u(z_i)$ $\quad(i=1, 2, \cdots)$.

Proof. By the definition of $u(z)$, there exist upper functions $\psi_i^{(n)}(z)$, such that

$$\lim_{n\to\infty}\psi_i^{(n)}(z_i)=u(z_i) \quad (i=1, 2, \cdots) . \tag{1}$$

Let

$$\psi^{(n)}(z)=\text{Min } \psi_i^{(k)}(z) \quad (i, k=1, 2, \cdots, n) , \tag{2}$$

then $\psi^{(n)}(z)$ is an upper function, so that by the definition

$$\psi^{(n)}(z)\geqq u(z) , \tag{3}$$

$$\psi^{(1)}(z)\geqq\psi^{(2)}(z)\geqq \cdots \geqq\psi^{(n)}(z)\geqq\cdots . \tag{4}$$

Since $\psi_i^{(n)}(z_i)\geqq\psi^{(n)}(z_i)\geqq u(z_i)$, we have $\lim\limits_{n\to\infty}\psi^{(n)}(z_i)=u(z_i)$ $(i=1, 2,\cdots)$.

Lemma 2. *Let Δ be a disc, contained in D and z_i $(i=1, 2, \cdots)$ be given points in Δ, then there exists a harmonic function $U(z)$ in Δ, such that $U(z_i)=u(z_i)$ $(i=1, 2, \cdots)$.*

Proof. Let $\psi^{(n)}(z)$ be the upper functions of Lemma 1, then $\psi_\Delta^{(n)}(z)$ is an upper function, and

$$\psi_\Delta^{(1)}(z)\geqq\psi_\Delta^{(2)}(z)\geqq \cdots \geqq\psi_\Delta^{(n)}(z)\geqq\cdots, \quad \psi^{(n)}(z)\geqq\psi_\Delta^{(n)}(z)\geqq u(z) ,$$

so that

$$\lim_{n\to\infty}\psi_\Delta^{(n)}(z_i)=u(z_i) \quad (i=1, 2, \cdots) .$$

By Harnack's theorem,

$$\lim_{n\to\infty}\psi_\Delta^{(n)}(z)=U(z)$$

is harmonic in Δ and $U(z_i)=u(z_i)$ $(i=1, 2, \cdots)$. ——

3. Now we shall prove Theorem I. 8.

Let Δ be a disc, contained in D and we shall prove that $u(z)$ is harmonic in Δ. Let z_i $(i=1, 2, \cdots)$ be all rational points in Δ. Then by Lemma 2, there exists a harmonic function $U(z)$ in Δ, such that $U z_i)=u(z_i)$ $(i=1, 2, \cdots)$. Let ζ be any point of Δ, then similarly by Lemma 2, there exists a harmonic function $V(z)$ in Δ, such that $V(z_i)=u(z_i)$ $(i = 1, 2, \cdots)$ and $V(\zeta) = u(\zeta)$, so that $U(z_i) = V(z_i)$ $(i = 1, 2, \cdots)$.

Since U and V are harmonic in Δ, $U\equiv V$, so that $U(\zeta)=u(\zeta)$, hence $u(z)\equiv U(z)$, so that $u(z)$ is harmonic in Δ. ——

4. Regular- and irregular-points

1. Let $\zeta_0\in\Gamma$ and D_ρ be the part of D, contained in $|z-\zeta_0|<\rho$. If there exists a function $\Omega(z)$, which satisfies the following conditions (i), (ii), (iii), then $\Omega(z)$ is called *a barrier at* ζ_0.

(i) $\Omega(z)>0$ and is continuous and superharmonic in D_{ρ_0}.

(ii) $\lim_{z\to\zeta_0}\Omega(z)=0, \quad z\in D.$

(iii) On the part of $|z-\zeta_0|=\rho(\leqq\rho_0)$, contained in D,

$$\Omega(z)\geqq a=a(\rho)>0 .$$

Let $\zeta_0\in\Gamma$ and $u(z)$ be the solution of Dirichlet problem for D with the boundary value f and put

$$\underline{u}(\zeta_0)=\varliminf_{z\to\zeta_0} u(z), \quad \bar{u}(\zeta_0)=\varlimsup_{z\to\zeta_0} u(z), \quad z\in D . \tag{1}$$

If $\underline{u}(\zeta_0)=\bar{u}(\zeta_0)$, then $\lim_{z\to\zeta_0} u(z)$ exists, when $z\to\zeta_0$ from the inside of D.

Theorem I. 9. *If a barrier $\Omega(z)$ exists at $\zeta_0\in\Gamma$, then*

$$\underline{f}(\zeta_0)\leqq\underline{u}(\zeta_0)\leqq\bar{u}(\zeta_0)\leqq\bar{f}(\zeta_0) .$$

Hence if $f(\zeta)$ is continuous at ζ_0, then $\lim_{z\to\zeta_0} u(z)=f(\zeta_0)$.

Proof. Let Γ_ρ be the part of Γ, contained in $|z-\zeta_0|\leqq\rho$. Since $\underline{f}(\zeta)$ is lower semi-continuous and $\bar{f}(\zeta)$ is upper semi-continuous, we can determine $\rho=\rho(\varepsilon)$ for any $\varepsilon>0$, such that

$$\bar{f}(\zeta_0)-\varepsilon\leqq\bar{f}(\zeta)\leqq\underline{f}(\zeta)\leqq\underline{f}(\zeta_0)+\varepsilon, \quad \zeta\in\Gamma_\rho . \tag{1}$$

We define $w(z)$ in D by the condition:

$$w(z)=\begin{cases}\text{Min}\,(C\Omega(z),\ 1) & \text{in } D_\rho ,\\ 1 & \text{in } D-D_\rho ,\end{cases} \tag{2}$$

where $C>0$ is a constant, such that $Ca(\rho)\geqq1$, then $w(z)$ is continuous and $0<w(z)\leqq1$ in D, $w=1$ on $|z-\zeta_0|=\rho$, $\lim_{z\to\zeta_0} w(z)=0$ and we can prove easily that $w(z)$ is superharmonic in D.

Now

$$\varphi(z)=\underline{f}(\zeta_0)-\varepsilon-Kw(z), \quad K>0 , \tag{3}$$

is subharmonic in D. Since $w(z)>0$, we have by (1),

$$\varlimsup_{z\to\zeta}\varphi(z)\leqq\underline{f}(\zeta_0)-\varepsilon\leqq\underline{f}(\zeta), \quad \zeta\in\Gamma_\rho . \tag{4}$$

Since $w(z)=1$ outside of D_ρ, if we choose $K>0$ sufficiently large, then by (3),

$$\overline{\lim_{z\to\zeta}}\, \varphi(z) \leqq \underline{f}(\zeta), \qquad \zeta\in\Gamma-\Gamma_\rho\,. \tag{5}$$

Hence by (4), (5), $\overline{\lim_{z\to\zeta}}\, \varphi(z)\leqq \underline{f}(\zeta)$ on Γ, so that $\varphi(z)$ is a lower function.

Similarly

$$\psi(z)=\bar{f}(\zeta_0)+\varepsilon+Kw(z), \qquad K>0\,, \tag{6}$$

is an upper function, if $K>0$ is sufficiently large.

Since $\lim_{z\to\zeta_0} w(z)=0$, we have by (3), $\lim_{z\to\zeta_0}\varphi(z)=\underline{f}(\zeta_0)-\varepsilon$ $(z\in D)$, so that if we choose $\rho_1<\rho$ sufficiently small, then

$$\varphi(z)>\underline{f}(\zeta_0)-2\,\varepsilon, \qquad z\in D_{\rho_1}\,.$$

Since upper function $\geqq$ lower function, we have by the definition of $u(z)$ ((1), p. 4), $u(z)\geqq\varphi(z)$, so that

$$u(z)>\underline{f}(\zeta_0)-2\,\varepsilon, \qquad z\in D_{\rho_1}\,,$$

hence $\underline{u}(\zeta_0)\geqq\underline{f}(\zeta_0)-2\,\varepsilon$. Since $\varepsilon>0$ is arbitrary, we have

$$\underline{u}(\zeta_0)\geqq\underline{f}(\zeta_0)\,. \tag{7}$$

Since $\lim_{z\to\zeta_0}\psi(z)=\bar{f}(\zeta_0)+\varepsilon$, $z\in D$ and $u(z)\leqq\psi(z)$, we have $\bar{u}(\zeta_0)\leqq\bar{f}(\zeta_0)+\varepsilon$. Since $\varepsilon>0$ is arbitrary, we have

$$\bar{u}(\zeta_0)\leqq\bar{f}(\zeta_0)\,. \tag{8}$$

Hence by (7), (8),

$$\underline{f}(\zeta_0)\leqq\underline{u}(\zeta_0)\leqq\bar{u}(\zeta_0)\leqq\bar{f}(\zeta_0)\,. \text{ ---}$$

2. Let $f(\zeta)$ be a bounded function on Γ, which is continuous at $\zeta_0\in\Gamma$ and $u(z)$ be the solution of Dirichlet problem for D with the boundary value f. If for all such $f(\zeta)$, $\lim_{z\to\zeta_0}u(z)=f(\zeta_0)$, then ζ_0 is called *a regular point* and *an irregular point*, if otherwise. Hence we have

Theorem I. 10. *Let $\zeta_0\in\Gamma$. If a barrier exists at ζ_0, then ζ_0 is a regular point.*

We shall prove the converse in Theorem III. 34.

Theorem I. 11. *Let $\zeta_0\in\Gamma$. If there exists a continuum $\in\Gamma$, which contains ζ_0, then there exists a barrier at ζ_0, so that ζ_0 is a regular point.*

Proof. Let $0<\rho<1$, $z-\zeta_0=x+iy$, $z\in D_\rho$ and put $\log(z-\zeta_0)=p(x,y)+iq(x,y)$, then $p(x,y)<0$ in D_ρ. By the hypothesis, we can choose $\rho>0$ so small that $\log(z-\zeta_0)$ is one-valued in D_ρ. Then

$$\Omega(z)=\Re\left(\frac{-1}{\log(z-\zeta_0)}\right)=\frac{-p}{p^2+q^2}>0$$

satisfies the condition of the barrier, so that there exists a barrier at ζ_0, hence ζ_0 is a regular point.

As a special case, we have

Theorem I. 12. *Let D be a domain on a closed Riemann surface, which is bounded by a finite number of Jordan curves. Let $f(\zeta)$ be a given continuous function on the boundary Γ of D. Then there exists a harmonic function in D, which takes the given value f on Γ.*

3. The condition (iii) in the definition of barrier is not necessary, as the following theorem shows.

Theorem I. 13. *Let D be a bounded domain on the z-plane and Γ be its boundary and $\zeta_0 \in \Gamma$. If there exists a positive continuous superharmonic function $w(z)>0$ in D_{ρ_0}, such that $\lim_{z\to\zeta_0} w(z)=0$ $(z\in D)$, then there exists a barrier at ζ_0, so that ζ_0 is a regular point.*

Bouligand [1].

Proof. (Brelot [1]). We put $K=\operatorname{Max}_{\zeta\in\Gamma}|\zeta-\zeta_0|$. Let $\gamma_\rho: |z-\zeta_0|=\rho$ $(0<\rho<\rho_0)$ be a circle about ζ_0 and γ_ρ^* be its part, contained in D. We choose a closed set e on γ_ρ^*, such that

$$|\gamma_\rho^*-e|\leqq \frac{2\pi\rho^2}{K}, \tag{1}$$

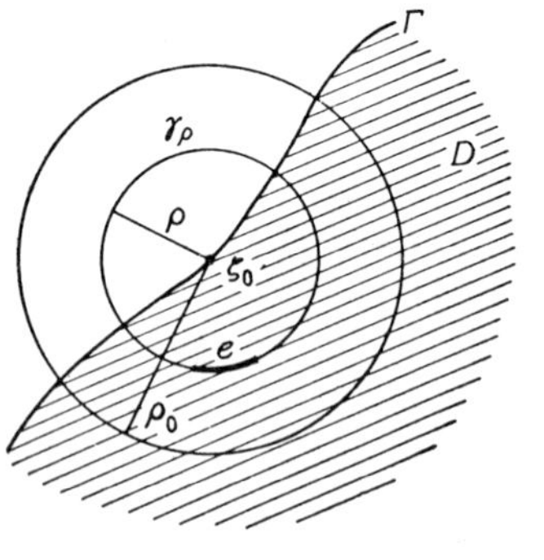

where | | denotes the linear measure.

We put

$$\operatorname{Min}_{z\in e} w(z)=m>0. \tag{2}$$

Let $v(z)$ be harmonic in γ_ρ, such that

$$\left.\begin{array}{l} v=K \text{ on } \gamma_\rho^*-e, \\ v=0 \text{ on } \gamma_\rho-(\gamma_\rho^*-e). \end{array}\right\} \tag{3}$$

Then

$$0<v(\zeta_0)\leqq K\frac{|\gamma_\rho^*-e|}{2\pi\rho}\leqq\rho. \tag{4}$$

Let $f(\zeta)=|\zeta-\zeta_0|$ $(\zeta\in\Gamma)$, then $0\leqq f\leqq K$ on Γ. Let $\varphi(z)$ be a lower function for f, then since $\overline{\lim}_{z\to\zeta}\varphi(z)\leqq f(\zeta)\leqq K$ on Γ, $\varphi(z)\leqq K$ in D and

$$\overline{\lim_{z\to\zeta}}\,\varphi(z)\leqq\rho, \quad \zeta\in\Gamma_\rho. \tag{5}$$

Let $\mathfrak{F}$ be the family of all lower functions $\varphi(z)$ for f and put

$$u(z)=\sup_{\varphi\in\mathfrak{F}}\varphi(z), \tag{6}$$

then $u(z)$ is harmonic in D and $0\leqq u\leqq K$ in D.

Since $|z-\zeta_0|$ is subharmonic, $|z-\zeta_0|$ is a lower function, so that

$$|z-\zeta_0|\leqq u(z)\leqq K \text{ in } D. \tag{7}$$

Now in D_ρ,

$$\omega(z)=\varphi(z)-\rho-\frac{Kw(z)}{m}-v(z) \tag{8}$$

is subharmonic and since $w>0$, $v>0$, we have by (5),

$$\overline{\lim_{z\to\zeta}}\,\omega(z)\leqq 0, \qquad \zeta\in\Gamma_\rho . \tag{9}$$

If $z\in\gamma_\rho^*-e$, then $v=K$, $\varphi\leqq K$, so that $\omega\leqq 0$.

If $z\in e$, then $\varphi(z)\leqq K$, $\dfrac{Kw}{m}\geqq K$, so that $\omega\leqq 0$.

Hence $\overline{\lim}\,\omega(z)\leqq 0$ on the boundary of D_ρ, so that $\omega(z)\leqq 0$ in D_ρ, or

$$\varphi(z)\leqq\rho+\frac{Kw(z)}{m}+v(z) \quad \text{in } D_\rho ,$$

hence by (6),

$$u(z)\leqq\rho+\frac{Kw(z)}{m}+v(z) \quad \text{in } D_\rho . \tag{10}$$

Since $\lim_{z\to\zeta_0} w(z)=0$ and $v(\zeta_0)\leqq\rho$, we have $\overline{\lim_{z\to\zeta_0}}\,u(z)\leqq 2\rho$.

Since $\rho>0$ is arbitrary, $\lim_{z\to\zeta_0} u(z)=0$, hence by (7), $u(z)$ is a barrier at ζ_0.

5. Dirichlet principle

1. Let $u(z)$ be a real valued continuous function in a domain D, contained in a closed Riemann surface, spread over the z-sphere, with piecewise continuous partial derivatives of the first order. Let Γ be the boundary of D. The Dirichlet integral $D[u]$ of u is defined by

$$D[u]=D_D[u]=\iint_D\left(\left(\frac{\partial u}{\partial x}\right)^2+\left(\frac{\partial u}{\partial y}\right)^2\right)dxdy, \quad z=x+iy . \tag{1}$$

If $v(z)$ has the same property, then we define $D[u, v]$ by

$$D[u, v]=D[v, u]=\iint_D\left(\frac{\partial u}{\partial x}\frac{\partial v}{\partial x}+\frac{\partial u}{\partial y}\frac{\partial v}{\partial y}\right)dxdy . \tag{2}$$

Hence $D[u]=D[u, u]$ and

$$D[u+v]=D[u]+2D[u, v]+D[v] . \tag{3}$$

By Schwarz's inequality, we have

$$|D[u, v]|\leqq\sqrt{D[u]D[v]}, \tag{4}$$

so that

$$\sqrt{|D[u+v]|}\leqq\sqrt{D[u]}+\sqrt{D[v]}. \tag{5}$$

If u has continuous derivatives of the second order, we have by the Green's formula,

$$D[u, v]+\iint_D v\Delta u \; dxdy=-\int_\Gamma v\frac{\partial u}{\partial\nu}ds , \tag{6}$$

where ν is the inner normal of Γ. Hence if u is harmonic in D, then

$$D[u]=-\int_\Gamma u\frac{\partial u}{\partial\nu}\,ds,\qquad D[u,v]=-\int_\Gamma v\frac{\partial u}{\partial\nu}\,ds\,. \tag{7}$$

Lemma 1. *Let $w(z)$ be continuous in a circular ring Δ: $0<\rho\leqq|z|\leqq1$ and have piece-wise continuous partial derivatives of the first order in $\rho<|z|<1$ and its Dirichlet integral $D[w]=D_\Delta[w]<\infty$. Let $u(z)$ be harmonic in $\rho<|z|<1$ and continuous in $\rho\leqq|z|\leqq1$, such that $u=w$ on $|z|=\rho$ and $|z|=1$. Then*

$$D[u]\leqq D[w]\,.$$

If w is not harmonic in Δ, then $D[u]<D[w]$.

Proof. Let in $\rho<|z|<1$,

$$u(z)=u(re^{i\theta})=$$
$$A\log r+a_0+\sum_{k=1}^{\infty}(a_kr^k+a_{-k}r^{-k})\cos k\theta+\sum_{k=1}^{\infty}(b_kr^k+b_{-k}r^{-k})\sin k\theta^{1)}, \tag{1}$$

where

$$\left.\begin{aligned}
a_0&=\frac{1}{2\pi}\int_0^{2\pi}u(e^{i\theta})d\theta=\frac{1}{2\pi}\int_0^{2\pi}w(e^{i\theta})d\theta\,,\\
a_k+a_{-k}&=\frac{1}{\pi}\int_0^{2\pi}u(e^{i\theta})\cos k\theta\,d\theta=\frac{1}{\pi}\int_0^{2\pi}w(e^{i\theta})\cos k\theta\,d\theta\,,\\
b_k+b_{-k}&=\frac{1}{\pi}\int_0^{2\pi}u(e^{i\theta})\sin k\theta\,d\theta=\frac{1}{\pi}\int_0^{2\pi}w(e^{i\theta})\sin k\theta\,d\theta\,.
\end{aligned}\right\} \tag{2}$$

$$\left.\begin{aligned}
A\log\rho+a_0&=\frac{1}{2\pi}\int_0^{2\pi}u(\rho e^{i\theta})d\theta=\frac{1}{2\pi}\int_0^{2\pi}w(\rho e^{i\theta})d\theta\,,\\
a_k\rho^k+a_{-k}\rho^{-k}&=\frac{1}{\pi}\int_0^{2\pi}u(\rho e^{i\theta})\cos k\theta\,d\theta=\frac{1}{\pi}\int_0^{2\pi}w(\rho e^{i\theta})\cos k\theta\,d\theta\,,\\
b_k\rho^k+b_{-k}\rho^{-k}&=\frac{1}{\pi}\int_0^{2\pi}u(\rho e^{i\theta})\sin k\theta\,d\theta=\frac{1}{\pi}\int_0^{2\pi}w(\rho e^{i\theta})\sin k\theta\,d\theta\,.
\end{aligned}\right\} \tag{2*}$$

We put

$$u_n(z)=A\log r+a_0+\sum_{k=1}^{n}(a_kr^k+a_{-k}r^{-k})\cos k\theta+\sum_{k=1}^{n}(b_kr^k+b_{-k}r^{-k})\sin k\theta\,, \tag{3}$$

then $\Delta u_n=0$ and

$$\frac{\partial u_n}{\partial r}=\frac{A}{r}+\sum_{k=1}^{n}k(a_k\,r^{k-1}-a_{-k}\,r^{-k-1})\cos k\theta+\sum_{k=1}^{n}k(b_k\,r^{k-1}-b_{-k}\,r^{-k})\sin k\theta\,,$$

so that

$$\frac{1}{\pi}\int_0^{2\pi}u_n(e^{i\theta})\left(\frac{\partial u_n}{\partial r}\right)_{r=1}d\theta=2\,a_0\,A+\sum_{k=1}^{n}k(a_k-a_{-k}^2)+\sum_{k=1}^{n}k(b_k^2-b_{-k}^2)\,,$$

---

1) Let $v(z)$ be the conjugate harmonic function of $u(z)$, then $f(z)=u(z)+iv(z)-A\log z$, $\left(A=\frac{1}{2\pi}\int_{|z|=r}dv\ (\rho<r<1)\right)$ is one-valued and regular in $\rho<|z|<1$, so that $f(z)=\sum_{-\infty}^{\infty}a_nz^n$, hence $u(z)=A\log r+\Re\left(\sum_{-\infty}^{\infty}a_nz^n\right)$.

and by (2),

$$\frac{1}{\pi}\int_0^{2\pi} w(e^{i\theta})\left(\frac{\partial u_n}{\partial r}\right)_{r=1} d\theta = \frac{A}{\pi}\int_0^{2\pi} w(e^{i\theta})\,d\theta + \sum_{k=1}^{n} k(a_k - a_{-k})\frac{1}{\pi}\int_0^{2\pi} w(e^{i\theta})\cos k\theta\,d\theta +$$

$$\sum_{k=1}^{n} k\,(b_k - b_{-k})\frac{1}{\pi}\int_0^{2\pi} w(e^{i\theta})\sin k\theta\,d\theta = 2a_0A + \sum_{k=1}^{n} k\,(a_k^2 - a_{-k}^2) + \sum_{k=1}^{n} k(b_k^2 - b_{-k}^2)\,,$$

hence

$$\int_0^{2\pi}\Big(w(e^{i\theta}) - u_n(e^{i\theta})\Big)\left(\frac{\partial u_n}{\partial r}\right)_{r=1} d\theta = 0\,.$$

Similarly

$$\int_0^{2\pi}\Big((w(\rho e^{i\theta}) - u_n(\rho e^{i\theta})\Big)\left(\frac{\partial u_n}{\partial r}\right)_{r=\rho} d\theta = 0\,.$$

Hence by the Green's formula, we have $D_\Delta[w-u_n, u_n]=0$, so that

$$D_\Delta[w] = D_\Delta[u_n] + D_\Delta[w-u_n] \geqq D_\Delta[u_n] \geqq D_{\Delta'}[u_n]\,,$$

where Δ' is a ring domain, contained entirely in Δ. Hence if we make $n\to\infty$ and then $\Delta'\to\Delta$, we have

$$D[u] \leqq D[w]\,. \tag{4}$$

We shall prove that if $D[w]=D[u]$, then $w\equiv u$.

Since $v(z)=w(z)+\lambda(u(z)-w(z))$ satisfies the same condition as $w(z)$,

$$D[u] \leqq D[v] = D[w] + 2\lambda D[w, u-w] + \lambda^2 D[u-w]\,.$$

Hence if $D[w]=D[u]$, then $2\lambda D[w, u-w]+\lambda^2 D[u-w]\geqq 0$.

Since λ is arbitrary, we have $D[w, u-w]=0$, so that $D[u]=D[w]+D[u-w]$. Hence $D[u-w]=0$, or $u-w\equiv\text{const}$. Since it vanishes on $|z|=\rho$ and $|z|=1$, we have $u-w\equiv 0$, or $w\equiv u$. Hence if w is not harmonic in Δ, then $D[u]<D[w]$. ——

Similarly we can prove

Lemma 2. *Let $w(z)$ be continuous in $|z|\leqq 1$ and have piece-wise continuous partial derivatives of the first order in $|z|<1$ and its Dirichlet integral $D[w]<\infty$. Let $u(z)$ be harmonic in $|z|<1$ and continuous in $|z|\leqq 1$, such that $u=w$ on $|z|=1$. Then*

$$D[u] \leqq D[w].$$

If w is not harmonic in $|z|<1$, then $D[u]<D[w]$.

Lemma 3. *Let $u_n(z)$ $(n=1, 2, \cdots)$ be harmonic in a domain D, contained in a closed Riemann surface, $D[u_n]\leqq K$ $(n=1, 2, \cdots)$ and at a point $z_0\in D$, $u_n(z_0)=0$, then $\{u_n(z)\}$ forms a normal family.*

Proof. Let $\Delta: |z-z_0|\leqq\rho$ be a disc, contained in D and $v_n(z)$ be the conjugate harmonic function of u_n in Δ, which we normalize, such that $v_n(z_0)=0$ and let $f_n(z)=u_n+iv_n$, then $f_n(z)$ is regular in Δ and $f_n(z_0)=0$.

Since $|f_n'(z)|^2$ is subharmonic, we have for any point z in $|z-z_0|\leqq\rho_1<\rho$,

$$|f_n'(z)|^2 \leqq \frac{1}{\pi(\rho-\rho_1)^2}\int_0^{\rho-\rho_1}\int_0^{2\pi}|f_n'(z+re^{i\theta})|^2 r dr d\theta \leqq \frac{D[u_n]}{\pi(\rho-\rho_1)^2} \leqq \frac{K}{\pi(\rho-\rho_1)^2},$$

$$|f_n'(z)| \leqq \sqrt{\frac{K}{\pi}}\cdot\frac{1}{\rho-\rho_1} \quad \text{in } |z-z_0|\leqq\rho_1, \tag{1}$$

so that

$$|u_n(z)| \leqq |f_n(z)| \leqq \int_0^z |f_n'(z_0+re^{i\theta})|dr \leqq \sqrt{\frac{K}{\pi}}\cdot\frac{\rho_1}{\rho-\rho_1} \quad \text{in } |z-z_0|\leqq\rho_1, \tag{2}$$

and for any two points z_1, z_2 in $|z-z_0|\leqq\rho_1$,

$$|u_n(z_1)-u_n(z_n)| \leqq |f_n(z_1)-f_n(z_n)| \leqq \int_{z_1}^{z_2}|f_n'(z)|\,|dz| \leqq \sqrt{\frac{K}{\pi}}\cdot\frac{|z_1-z_2|}{\rho-\rho_1}. \tag{3}$$

Hence $\{u_n(z)\}$ is uniformaly bounded and equi-continuous in $|z-z_0|\leqq\rho_1$, and from this, we see that $\{u_n(z)\}$ is uniformly bounded and equi-continuous in any compact domain $D_1\subset D$, so that by Arzelà's theorem, $\{u_n(z)\}$ forms a normal family.

2. Now we shall prove the following.

Theorem I. 14. (Dirichlet principle). *Let D be a domain, contained in a closed Riemann surface F, spread over the z-sphere, whose boundary Γ consists of a finite number of disjoint Jordan curves: $\Gamma=\sum_{i=1}^{k}\Gamma_i$. Suppose that there exists a function $w_0(z)$ on D, which is continuous in its closure $\bar{D}$ and has piece-wise continuous partial derivatives of the first order and its Dirichlet integral $D[w_0]$ is finite. Let $\mathfrak{F}$ be the family of $w(z)$, which has the same property as $w_0(z)$ and $w=w_0$ on Γ and put*

$$d=\inf_{w\in\mathfrak{F}} D[w].$$

Then there exists a harmonic function $u(z)\in\mathfrak{F}$, such that

$$D[u]=d. \qquad \text{Courant [2].}$$

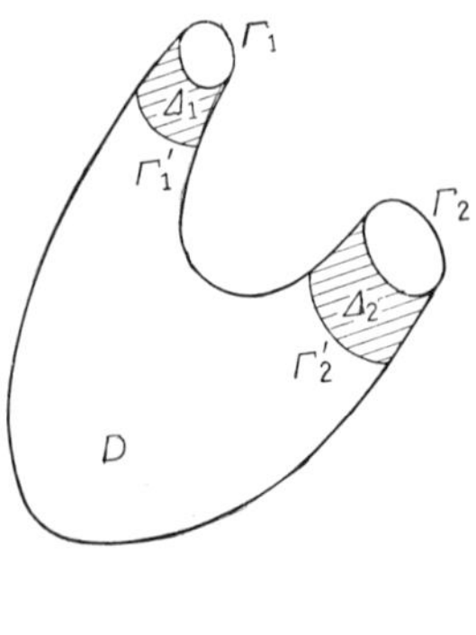

Proof. (Tsuji [52]). By the definition of d, there exist $w_n\in\mathfrak{F}$ $(n=1, 2, \cdots)$, such that $D[w_n]\to d$. Let Γ_i' be an analytic Jordan curve in D, which lies in a small neighbourhood of Γ_i, such that Γ_i', Γ_i bound a doubly connected domain Δ_i, where Δ_i, Δ_j $(i\neq j)$ are disjoint.

Let $u_n^{(i)}(z)$ be a harmonic function in Δ_i, such that $u_n^{(i)}=w_n$ on Γ_i' and Γ_i. We map Δ_i conformally on a circular ring V_i: $0<\rho_i<|\zeta|<1$ on the ζ-plane, such that Γ_i', Γ_i

correspond to $|\zeta|=\rho_i$, $|\zeta|=1$ respectively[1] and consider $u_n^{(i)}(z)$ and $w_n(z)$ as functions of ζ in V_i and apply Lemma 1, then since the Dirichlet integral is invariant by conformal mapping, we have $D_{\Delta_i}[u_n^{(i)}]\leqq D_{\Delta_i}[w_n]$. Hence if we define $\tilde{w}_n(z)$ in D by $\tilde{w}_n=u_n^{(i)}$ in Δ_i $(i=1, 2, \cdots, k)$, $\tilde{w}_n=w_n$ in $D-\sum_{i=1}^{k}\Delta_i$, then $D[\tilde{w}_n]\leqq D[w_n]$ and $\tilde{w}_n\in\mathfrak{F}$, so that $D[\tilde{w}_n]\to d$. Hence we assume that

$$w_n(z) \text{ is harmonic in } \Delta_i \ (i=1 2, \cdots, k)\,. \tag{1}$$

If we put $w_n(z)=v_n(\zeta)$ in V_i: $\rho_i<|\zeta|<1$, then since $v_n(\zeta)-v_1(\zeta)$ is harmonic in V_i and vanishes on $|\zeta|=1$, it is harmonic in V_i^*: $\rho_i<|\zeta|<\dfrac{1}{\rho_i}$ and

$$D_{V_i^*}[v_n-v_1]=2\,D_{V_i}[v_n-v_1]=2\,D_{\Delta_i}[w_n-w_1]\leqq 2(\sqrt{D_{\Delta_i}[w_n]}+\sqrt{D_{\Delta_i}[w_1]})^2\,,$$

so that $D_{V_i^*}[v_n-v_1]$ is bounded for $n=1, 2, \cdots$ and since $v_n-v_1=0$ on $|\zeta|=1$, by Lemma 3, we can select a partial sequence from n, which we denote by n again, such that v_n-v_1 and hence v_n converges uniformaly in V_i^*, hence returning to Δ_i, we have

$$\lim_n w_n(z)=w(z) \text{ uniformly in } \Delta_i \ (i=1, 2, \cdots, n)\,. \tag{2}$$

$w(z)$ is harmonic in $\sum_{i=1}^{k}\Delta_i$ and is continuous on Γ and $w=w_1=w_0$ on Γ.

Let γ_i be an analytic Jordan curve in Δ_i, which separates Γ_i', Γ_i in Δ_i and let Ω_0 be the connected sub-domain of D, which is bounded by $\sum_{i=1}^{k}\gamma_i$ and contains $\sum_{i=1}^{k}\Gamma_i'$, then by (2),

$$D_\Omega[w]=\lim_n D_\Omega[w_n], \quad \text{where } \Omega=D-\Omega_0\,. \tag{3}$$

Let $\varphi_n(z)$ be harmonic in Ω_0, such that $\varphi_n=w_n$ on $\sum_{i=1}^{k}\gamma_i$. Since by the hypothesis (1), $w_n(z)$ is harmonic on γ_i and $\varphi_n-w_n=0$ on γ_i, we see that φ_n-w_n and hence φ_n is harmonic in a neighbourhood U_i of γ_i, so that $\dfrac{\partial\varphi_n}{\partial\nu}$ is continuous on γ_i, where ν is the inner normal with respect to Ω_0, hence by the Green's formula,

$$D_{\Omega_0}[w_n-\varphi_n, \varphi_n]=-\sum_{i=1}^{k}\int_{\gamma_i}(w_n-\varphi_n)\frac{\partial\varphi_n}{\partial\nu}ds=0\,,$$

so that

$$D_{\Omega_0}[w_n]=D_{\Omega_0}[\varphi_n]+D_{\Omega_0}[w_n-\varphi_n]\,,$$

hence

$$D_{\Omega_0}[\varphi_n]\leqq D_{\Omega_0}[w_n]\,. \tag{4}$$

---

1) This can be proved easily as follows. Let $u(z)$ be harmonic in Δ_i, such that $u=0$ on Γ_i, $u=-1$ on Γ_i' and $v(z)$ be its conjugate harmonic function. Let $\int_{\Gamma_i'}\dfrac{\partial u}{\partial\nu}ds=\alpha>0$, where ν is the inner normal of Γ_i' with respect to Δ_i and put $f(z)=\dfrac{2\pi}{\alpha}(u+iv)$, then $\zeta=e^{f(z)}$ is the required mapping function.

Since by (2), w_n converges uniformly in U_i and by (4), $D_{U_i}[\varphi_n - w_n]$ is bounded for $n=1, 2, \cdots$ and $\varphi_n - w_n = 0$ on γ_i, $\varphi_n - w_n$ and hence φ_n converges uniformly in U_i, so that if we put

$$\varphi(z) = \lim_n \varphi_n(z) ,$$

then $\varphi(z)$ is harmonic in $\Omega_0 + \sum_{i=1}^k U_i$ and $D_{\Omega_0}[\varphi] = \lim_n D_{\Omega_0}[\varphi_n]$, so that by (4),

$$D_{\Omega_0}[\varphi] \leqq \varliminf_n D_{\Omega_0}[w_n] . \tag{5}$$

Hence if we define $u(z)$ in D, by $u=\varphi$ in Ω_0, $u=w$ in $D-\Omega_0$, then by (3), (5),

$$D[u] \leqq \lim_n D[w_n] = d .$$

Since $u \in \mathfrak{F}$, we have

$$D[u] = d . \tag{6}$$

u is harmonic in D, except on $\sum_{i=1}^k \gamma_i$. We shall prove that u is harmonic on γ_i. Suppose that u is not harmonic at a point z_0 on γ_i. Let Δ be a circular disc about z_0, contained in D and $v(z)$ be a harmonic function in Δ, such that $v=u$ on the bounding circle of Δ. Then by Lemma 2, $D_\Delta[v] < D_\Delta[u]$, so that if we put $\tilde{u} = v$ in Δ, $\tilde{u} = u$ in $D - \Delta$, then $\tilde{u} \in \mathfrak{F}$ and $D[\tilde{u}] < D[u] = d$, which is absurd. Hence u is harmonic on $\sum_{i=1}^k \gamma_i$, hence in D. ——

Remark. If $w \in \mathfrak{F}$ and $D[w] = d$, then by Lemma 2, w is harmonic in D, so that $w \equiv u$, hence such w is unique.

6. Green's function

1. Let D be a domain, contained in a closed Riemann surface F, spread over the z-sphere, whose boundary Γ consists of a finite number of disjoint Jordan curves: $\Gamma = \sum_{i=1}^k \Gamma_i$. Let $z=a$ belong to D. Then *the Green's function* $g(z, a)$ *of* D, with a as its pole, is defined by the following condition.

$g(z, a)$ is harmonic in D, except at a, where if we put

$$g(z, a) = \log \frac{1}{|z-a|} + \omega(z) ,$$

then $\omega(z)$ is harmonic at a and $g(z, a) = 0$ at $z \in \Gamma$.

We shall prove

Theorem I. 15. *The Green's function* $g(z, a)$ *exists.*

Proof. We assume that $a=0$. First suppose that $\{\Gamma_i\}$ are analytic Jordan curves. Let $1 > r_0 > r_1 > \cdots > r_n \to 0$ and C_n be a circle $|z| = r_n$ and we denote the inside of C_n by (C_n). We assume that $(C_0) \subset D$.

Let $u_n(z)$ be harmonic in $D-(C_n)$, such that

$$u_n=0 \text{ on } \Gamma, \quad u_n=\log\frac{1}{r_n} \text{ on } C_n\,. \tag{1}$$

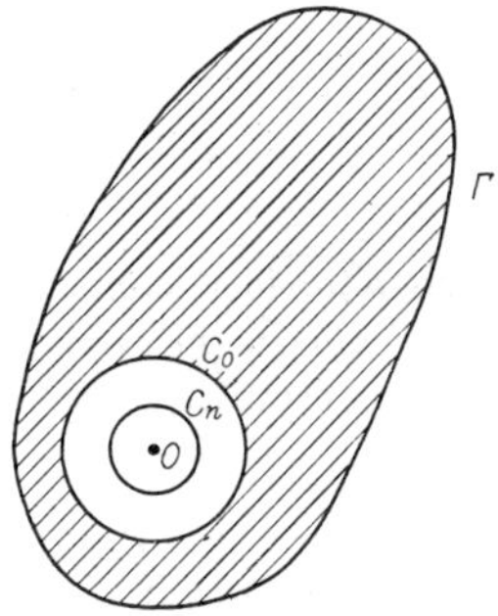

We apply the Green's formula to u_n, u_0 for the domain $D-(C_0)$, then since $u_0=0$, $u_n=0$ on Γ,

$$\int_{C_0}\left(u_n\frac{\partial u_0}{\partial r}-u_0\frac{\partial u_n}{\partial r}\right)ds=0, \qquad r=|z|\,,$$

so that

$$\int_{C_0}u_n\left(-\frac{\partial u_0}{\partial r}\right)ds=\log\frac{1}{r_0}\int_{C_0}\left(-\frac{\partial u_n}{\partial r}\right)ds\,. \tag{2}$$

Similarly by the Green's formula, applied to u_n, $\log 1/r$ for the circular ring (C_n, C_0), bounded by C_n, C_0, we have

$$\int_{C_0}\left(u_n\frac{\partial \log 1/r}{\partial r}-\log\frac{1}{r_0}\frac{\partial u_n}{\partial r}\right)ds=\int_{C_n}\left(u_n\frac{\partial \log 1/r}{\partial r}-\log\frac{1}{r_n}\frac{\partial u_n}{\partial r}\right)ds\,,$$

hence

$$-\frac{1}{r_0}\int_{C_0}u_n\,ds-\log\frac{1}{r_0}\int_{C_0}\frac{\partial u_n}{dr}ds=-2\pi\log\frac{1}{r_n}-\log\frac{1}{r_n}\int_{C_n}\frac{\partial u_n}{\partial r}ds$$

$$=-2\pi\log\frac{1}{r_n}-\log\frac{1}{r_n}\int_{C_0}\frac{\partial u_n}{\partial r}ds\,,$$

so that

$$\left(\log\frac{1}{r_n}-\log\frac{1}{r_0}\right)\int_{C_0}\left(-\frac{\partial u_n}{\partial r}\right)ds=2\pi\log\frac{1}{r_n}-\frac{1}{r_0}\int_{C_0}u_n\,ds<2\pi\log\frac{1}{r_n}\,,$$

$$\int_{C_0}\left(-\frac{\partial u_n}{\partial r}\right)ds\leqq\frac{2\pi\log 1/r_n}{\log r_0/r_n}\leqq\frac{2\pi\log 1/r_1}{\log r_0/r_1} \quad (n\geqq 1)\,.$$

Hence by (2),

$$\int_{C_0}u_n\left(-\frac{\partial u_0}{\partial r}\right)ds\leqq 2\pi A, \qquad A=\frac{\log 1/r_0\log 1/r_1}{\log r_0/r_1} \quad (n\geqq 1)\,. \tag{3}$$

Since $-\dfrac{\partial u_0}{\partial r}\geqq a>0$[1] on C_0, if we put

1) By a suitable linear transformation, it suffices to prove the following. *Let $u(z)=u(x+iy)$ be harmonic in $|z|<\rho$, such that $u>0$ for $y>0$, $u=0$ for $y=0$, then* $\left(\dfrac{\partial u}{\partial y}\right)_{z=0}>0$. *Proof.* Let $u(re^{i\theta})=\sum_{n=1}^{\infty}a_nr^n\sin n\theta$, then $a_1=\left(\dfrac{\partial u}{\partial y}\right)_{z=0}\geqq 0$. If $a_1=0$, let $a_1=a_2=\cdot=a_{k-1}$, $a_k\neq 0$ $(k\geqq 2)$, then since $\sin k\theta>0$ for $0<\theta<\dfrac{\pi}{k}$ and $\sin k\theta<0$ for $\dfrac{\pi}{k}<\theta<\dfrac{2\pi}{k}$, if r_0 is sufficiently small, there exists θ_0 $(0<\theta_0<\pi)$, such that $u(r_0e^{i\theta_0})<0$, which is absurd, hence $a_1>0$.

$$m_n = \operatorname*{Min}_{z \in C_0} u_n(z) , \tag{4}$$

then by (3),

$$m_n \leqq B = \frac{A}{a r_0} \quad (n \geqq 1) . \tag{5}$$

Let $\Delta_\rho : |z| \leqq \rho \, (< r_0)$ be a disc and $D_\rho^* = D - \Delta_\rho$, then since $u_n > 0$, we see by (5) and Harnack's theorem, that

$$0 < u_n(z) \leqq K(\rho) \text{ in } D_\rho^*, \; n \geqq n_0 , \tag{6}$$

where $K(\rho)$ is a constant, which depends on ρ only.

We choose $\rho_1 > \rho_2 > \cdots > \rho_\nu \to 0$ for ρ, then by (6), we can find a partial sequence from n, which we denote by n again, such that

$$\lim_n u_n(z) = u(z) \tag{7}$$

uniformly in the wider sense in $D - \{0\}$.
$u(z)$ is harmonic in D, except at $z=0$. We shall prove that $u=0$ on Γ.

Let by (6),

$$0 < u_n(z) \leqq K = \text{const.} \quad \text{on } C_0 \; (n = 0, 1, 2, \cdots) , \tag{8}$$

then $K \geqq \log 1/r_0$. By the maximum principle, we have

$$0 < u_n(z) \leqq K u_0(z) / \log \frac{1}{r_0} \quad \text{in } D - (C_0) ,$$

so that

$$0 \leqq u(z) \leqq K u_0(z) / \log \frac{1}{r_0} \quad \text{in } D - (C_0) . \tag{9}$$

Since $u_0 = 0$ on Γ, we have $u=0$ on Γ.

Next we shall prove that $u(z) - \log \frac{1}{r}$ is harmonic at $z=0$.

Since $u_n(z) - \log \frac{1}{r} = 0$ on C_n, we have in (C_n, C_0),

$$\left| u_n(z) - \log \frac{1}{r} \right| \leqq \operatorname*{Max}_{z \in C_0} \left| u_n(z) - \log \frac{1}{r} \right| \leqq K - \log \frac{1}{r_0} \leqq K ,$$

so that

$$\left| u(z) - \log \frac{1}{r} \right| \leqq K \quad \text{in } 0 < |z| \leqq r_0 .$$

Hence $u(z) - \log \frac{1}{r}$ is harmonic at $z=0$, so that $u(z) = g(z, 0)$ is the Green's function of D, with $z=0$ as its pole.

We assumed that $\{\Gamma_i\}$ are analytic Jordan curves. In the general case, let D^* be a domain, contained in F, which contains D and whose boundary Γ^* consists of a finite number of analytic Jordan curves and let $g^*(z, 0)$ be its Green's function. We exhaust D by a sequence of domains D_n, such that $\bar{D}_n \subset D_{n+1}$ and the boundary Γ_n of D_n consists of a finite number of analytic Jordan curves and let $g_n(z, 0)$ be the Green's function of D_n, then by the maximum principle, $g_n(z, 0)<g_{n+1}(z, 0)<g^*(z, 0)$, so that

$$\lim_n g_n(z, 0)=g(z, 0) , \ z\in D \tag{10}$$

exists.

We see easily that $g(z, 0)-\log\frac{1}{r}$ is harmonic at $z=0$. Next we shall prove that $g(z, 0)=0$ on Γ. Let $\zeta_0\in\Gamma$ and $D(\rho)$, $D_n(\rho)$ be the part of D, D_n, which lies in $|z-\zeta_0|<\rho$ respectively. Since Γ_i is a Jordan curve, there exists a barrier $\Omega(z)$ of $D(\rho)$ at ζ_0, such that $\Omega(z)>0$ in $D(\rho)$ and is superharmonic and $\lim_{z\to\zeta_0}\Omega(z)=0$, $\Omega(z)\geqq a=a(\rho)$ on $|z-\zeta_0|=\rho$ $(\leqq\rho_0)$. Let $g_n(z, 0)\leqq K$ in $D_n(\rho)$ $(n=1, 2, \cdots)$, then since $g_n=0$ on Γ_n and $\Omega(z)>0$ is superharmonic, by the maximum principle,

$$\Omega(z)\geqq\frac{a(\rho)}{K}g_n(z, 0) \quad \text{in } D_n(\rho) ,$$

so that

$$\Omega(z)\geqq\frac{a(\rho)}{K}g(z, 0) \quad \text{in } D(\rho) ,$$

hence $\lim_{z\to\zeta_0} g(z, 0)=0$. Hence $g(z, 0)$ is the Green's function of D, with $z=0$ as its pole.

Remark. We assumed that the boundary of D consists of a finite number of Jordan curves. If D is any domain, contained in a closed Riemann surface and Γ be its boundary and if a barrier exists at a point $\zeta_0\in\Gamma$, then by the above proof, we see that $\lim_{z\to\zeta_0} g(z, 0)=0$.

Theorem I. 16. $g(a, b)=g(b, a)$, $\quad a\in D$, $b\in D$.

Proof. Let Γ_δ be the niveau curve: $g(z, a)=\delta>0$, then Γ_δ is the boundary of a domain D_δ, in which $g(z, a)>\delta$. Γ_δ consists of a finite number of analytic Jordan curves and $\Gamma_\delta\to\Gamma$ as $\delta\to 0$. By the Green's formula,

$$\int_{\Gamma_\delta}\frac{\partial g(z, a)}{\partial\nu}\,ds=2\pi , \tag{1}$$

where ν is the inner normal of Γ_δ with respect to D_δ.

We choose δ so small that b lies in D_δ and $0<g(z, b)<\varepsilon$ on Γ_δ, where $\varepsilon\to 0$ with $\delta\to 0$. We apply the Green's formula: $\int\left(u\frac{\partial v}{\partial\nu}-v\frac{\partial u}{\partial\nu}\right)ds=0$ to $u=g(z, a)$, $v=g(z, b)$ for the domain, bounded by Γ_δ and two small circles about a and b and then make the radii of these circles tend to zero, then we have

$$g(a, b)-g(b, a)=\frac{1}{2\pi}\int_{\Gamma_\delta}\left(g(z, a)\frac{\partial g(z, b)}{\partial\nu}-g(z, b)\frac{\partial g(z, a)}{\partial\nu}\right)ds. \tag{2}$$

Now

$$\int_{\Gamma_\delta} g(z,a)\frac{\partial(z,b)}{\partial\nu}ds=\delta\int_{\Gamma_\delta}\frac{\partial g(z,b)}{\partial\nu}ds=2\pi\delta\ ,$$

$$0<\int_{\Gamma_\delta} g(z,b)\frac{\partial g(z,a)}{\partial\nu}ds<\varepsilon\int_{\Gamma_\delta}\frac{\partial g(z,a)}{\partial\nu}ds=2\pi\varepsilon\ ,$$

so that if we make $\delta\to 0$, $\varepsilon\to 0$, we have

$$g(a,b)=g(b,a). \text{ —} \tag{3}$$

As an application, we shall prove

Theorem I. 17. *Let D be a simply connected domain, contained in a closed Riemann surface, spread over the z-sphere, whose boundary Γ is a continuum. Then D can be mapped conformally on $|w|<1$.*

Proof. Let $g(z,a)$ be the Green's function of D and $h(z,a)$ be its conjugate harmonic function, then we see easily that by $w=e^{-(g+ih)}$, D is mapped conformally on $|w|<1$.

2. Now we shall define *the Green's function of an open Rieman surface.* Let F be an open Riemann surface, spread over the z-sphere. We exhaust F by a sequence of compact Riemann surfaces F_n $(n=0,1,2,\cdots)$, where $\bar{F}_n\subset F_{n+1}$ and the boundary Γ_n of F_n consists of a finite number of analytic Jordan curves. Let $a\in F_0$ and $g_n(z,a)$ be the Green's function of F_n, with a as its pole, then by the maximum principle, $g_n(z,a)<g_{n+1}(z,a)$ on F_n, so that by Harnack's theorem,

$$\text{(i)}\ \lim_n g_n(z,a)\equiv+\infty,\quad\text{or}\quad\text{(ii)}\ \lim_n g_n(z,a)=g(z,a) \tag{1}$$

uniformly in the wider sense on F.

In case (i), we say that there exists no Green's function on F and in case (ii), we call $g(z,a)$ *the Green's function of F.* $g(z,a)$ is harmonic on F, except at a, where if we put

$$g(z,a)=\log\frac{1}{|z-a|}+\omega(z)\ , \tag{2}$$

then $\omega(z)$ is harmonic at a.

We call $\omega(a)$ *the Robin's constant of F at a.*

Whether the case (i), or (ii) occurs is independent of the choice of a, which is proved as follows. Let $a\neq b$ be two points of F and we choose $z_0(\neq a,\ \neq b)$ on F. Then by Harnack's theorem, applied to $g_n(z,z_0)>0$ $(n\geqq n_0)$, we have

$$A\,g_n(a,z_0)\leqq g_n(b,z_0)\leqq B\,g_n(a,z_0)\ ,$$

or

$$A\,g_n(z_0,a)\leqq g_n(z_0,b)\leqq B\,g_n(z_0,a)\ ,$$

where $A>0$, $B>0$ are constants, independent of n, so that

$$A \lim_n g_n(z_0, a) \leqq \lim_n g_n(z_0, b) \leqq B \lim_n g_n(z_0, a) .$$

Hence $\lim_n g(z_0, b)=\infty$, or $<\infty$, according as $\lim_n g_n(z_0, a)=\infty$, or $<\infty$.— F is called *of null boundary,* or *of positive boundary,* according as there exists no Green's function on F, or does exist a Green's function on F and we denote

$$F \in O_G, \quad F \in P_G \tag{3}$$

respectively.

From the remark on Theorem I. 15, we have

Theorem I. 18. *Let F be a Riemann surface of positive boundary and $g(z, a)$ be its Green's function. Let Γ be the compact boundary of F and $\zeta_0 \in \Gamma$. If there exists a barrier at ζ_0, then*

$$\lim_{z \to \zeta_0} g(z, a)=0, \qquad z \in F .$$

3. If F is a closed Riemann surface, then there exists no function $u(z, a)$ on F, which is harmonic, except at a, where if we put $u(z, a) =\log \frac{1}{|z-a|}+\omega(z)$, then $\omega(z)$ is harmonic at a.

For, suppose that there exists such $u(z, a)$. Let $\Delta: |z-a|<\rho$ be a disc and $C: |z-a|=\rho$ be its bounding circle. Then $\int_C \frac{\partial u}{\partial \nu} ds=2\pi$, where ν is the inner normal of C with respect to Δ. But since u is harmonic on $F-\Delta$, $\int_C \frac{\partial u}{\partial \nu} ds=0$, which is absurd. Hence there exists no such $u(z, a)$ on F.

Let F be a closed Riemann surface and we take off a point z_0 from F and put $F'=F-\{z_0\}$, then F' is an open Riemann surface. Then there exists no Green's function on F'. For, if there exists a Green's function $g(z, a)$ $(a \neq z_0)$ of F', then since g is bounded in a neighbourhood of z_0, g is harmonic at z_0, so that g satisfies the above condition, which is absurd. Hence there exists no Green's function on F'. By means of this, we shall prove

Theorem I. 19. *Let F be a closed Riemann surface of genus $p=0$, spread over the z-sphere, then F can be mapped conformally on the whole w-sphere $|w| \leqq \infty$.*

Proof. Let $z_0 (\neq 0)$ be a point of F and $\Delta_n: |z-z_0| \leqq r_n$ $(r_1>r_2>\cdots>r_n \to 0)$ be a disc about z_0 and $F_n=F-\Delta_n$. Let $g_n(z, 0)$ be the Green's function of F_n and $h_n(z, 0)$ be its conjugate harmonic function, then by

$$w=\varphi_n(z)=e^{-(g_n+ih_n)}, \qquad \varphi_n(0)=0,$$

F_n is mapped conformally on $|w|<1$, hence by

$$w=f_n(z)=\varphi_n(z)/\varphi_n'(0), \qquad f_n(0)=0,\ f_n'(0)=1\,,$$

F_n is mapped conformally on $|w|<R_n$, where $R_n=1/|\varphi_n'(0)|=e^{\omega_n(0)}$, $\omega_n(0)$ being the Robin's constant of F_n at $z=0$.

Since $g_n(z,0)<g_{n+1}(z,0)$, we have $\omega_n(0)\leqq\omega_{n+1}(0)$, so that $R_n\leqq R_{n+1}$, hence

$$\lim_n R_n=R, \qquad 0<R\leqq\infty\,,$$

exists. Since by Koebe's distortion theorem, $\{f_n(z)\}$ forms a normal family, we may assume that

$$w=\lim_n f_n(z)=f(z), \qquad f(0)=0,\ f'(0)=1\,,$$

exists uniformly in the wider sense on $F'=F-\{z_0\}$. Then by $w=f(z)$, F' is mapped conformally on $|w|<R$. If $R<\infty$, then there exists a Green's function on F', which is absurd. Hence $R=\infty$, so that $w=f(z)$ maps F' conformally on $|w|<\infty$. We see easily that $w=f(z)$ maps F conformally on $|w|\leqq\infty$.

4. Let D be a domain, contained in a closed Riemann surface, spread over the z-sphere and Γ be its boundary. We suppose that Γ consists of a finite number of analytic Jordan curves, or analytic Jordan arcs: $\Gamma=\sum_{i=1}^{k}\Gamma_i$, where if Γ_i, Γ_{i+1} are analytic Jordan arcs and meet at ζ_i (vertex), then they make an angle π at ζ_i. Let $g(z,z_0)$ be the Green's function of D. Then we shall prove

Theorem I. 20. *Let ζ be any point of Γ and ν be the inner normal of Γ at ζ, then* $\dfrac{\partial g(\zeta,z_0)}{\partial\nu}$ $(z_0\in D)$ *is continuous and* >0 *on Γ and*

$$\int_\Gamma \frac{\partial g(\zeta,z_0)}{\partial\nu}|d\zeta|=2\pi\,.$$

Let $\zeta_0\in\Gamma$ and α be a sub-arc of Γ, which contains ζ_0 as an inner point, then

$$\lim_{z_0\to\zeta_0}\int_\alpha \frac{\partial g(\zeta,z_0)}{\partial\nu}|d\zeta|=2\pi, \qquad \lim_{z_0\to\zeta_0}\int_{\Gamma-\alpha} \frac{\partial g(\zeta,z_0)}{\partial\nu}|d\zeta|=0\,.$$

Proof. We assume that $\zeta_0=0$ and Γ touch the real axis at $z=0$. Let D_ρ be the part of D, contained in $|z|<\rho$. We choose $\rho>0$ so small that z_0 lies outside of D_ρ and D_ρ is simply connected and we assume that D_ρ lies in the upper half-plane. We map D_ρ conformally on the half-disc: $|w|<1$, $y>0$ on the $w=x+iy$-plane, such that $z=0$ corresponds to $w=0$ and the part Γ' of Γ, contained in D_ρ becomes a diameter: $-1<x<1$, $y=0$ of $|w|=1$ and let $z=z(w)$ be the mapping function and put $g(z,z_0)=G(w)=G(x,y)$, then $G(w)$ is harmonic in a half-disc $|w|<1$, $y>0$ and $G=0$ on the real axis, so that it can be continued harmonically into $y<0$ across the real axis, so that G is harmonic in $|w|<1$ and by the foot note (p. 15).

$$\left(\frac{\partial G(x, y)}{\partial y}\right)_{y=0}=G_y(x, 0)>0, \quad -1<x<1. \tag{1}$$

If $\zeta\in\Gamma'$ corresponds to $w=x$, then

$$\frac{\partial g(\zeta, z_0)}{\partial\nu}\left|\frac{dz}{dw}\right|_{w=x}=G_y(x, 0)>0. \tag{2}$$

Since by Theorem IX. 8, even if $\zeta_0=0$ is a vertex of Γ, $\left|\frac{dz}{dw}\right|_{w=x}$ is continuous and >0 on $-1<x<1$, $\frac{\partial g(\zeta, z_0)}{\partial\nu}$ is continuous and >0 on Γ'. Hence $\frac{\partial g(\zeta, z_0)}{\partial\nu}$ is continuous and >0 on Γ.

Let $\Gamma_\delta: g(z, z_0)=\delta>0$ be the niveau curve of g, then

$$\int_{\Gamma_\delta}\frac{\partial g(\zeta, z_0)}{\partial\nu}|d\zeta|=2\pi.$$

Since $\frac{\partial g(\zeta, z_0)}{\partial\nu}$ is continuous on Γ, if we make $\delta\to 0$, then we have

$$\int_{\Gamma}\frac{\partial g(\zeta, z_0)}{\partial\nu}|d\zeta|=2\pi. \tag{3}$$

Next we shall prove the second part. Let z_0 lie in D_ρ, and w_0 be its image in $|w|<1$, $y>0$.

If we put

$$g(z, z_0)=G(w, w_0)=\log\left|\frac{w-\overline{w}_0}{w-w_0}\right|+H(w, w_0), \tag{4}$$

then $H(w, w_0)$ is harmonic in $|w|<1$. Since $\lim_{w_0\to 0} G(w, w_0)=0$ and $\lim_{w_0\to 0}\log\left|\frac{w-\overline{w}_0}{w-w_0}\right|=0$ uniformly on $|w|=r<1$, $\lim_{w_0\to 0} H(w, w_0)=0$ uniformly on $|w|=r$, hence uniformly in $|w|\leqq r$, so that $\lim_{w_0\to 0} H_y(x, 0; w_0)=0$ uniformly in $|x|\leqq r_1<r$, hence

$$\lim_{w_0\to 0}\int_{-r_1}^{r_1}H_y(x, 0, w_0)dx=0. \tag{5}$$

Now

$$\left(\frac{\partial}{\partial y}\log\left|\frac{w-\overline{w}_0}{w-w_0}\right|\right)_{y=0}=\frac{2y_0}{(x-x_0)^2+y_0^2} \quad (w_0=x_0+iy_0).$$

We may assume that α corresponds to a segment $-r_1\leqq x\leqq r_1$, $y=0$, then

$$\int_\alpha\frac{\partial g(\zeta, z_0)}{\partial\nu}|d\zeta|=\int_{-r_1}^{r_1}\left(\frac{\partial}{\partial y}\log\left|\frac{w-\overline{w}_0}{w-w_0}\right|\right)_{y=0}dx+\int_{-r_1}^{r_1}H_y(x, 0, w_0)dx,$$

so that by (5)

$$\lim_{z_0\to 0}\int_\alpha \frac{\partial g(\zeta, z_0)}{\partial \nu}|d\zeta| = \lim_{w_0\to 0}\int_{-r_1}^{r_1} \frac{2y_0}{(x-x_0)^2+y_0^2}dx = 2\pi . \tag{6}$$

Hence from

$$\int_\alpha \frac{\partial g(\zeta, z_0)}{\partial \nu}|d\zeta| + \int_{\Gamma-\alpha} \frac{\partial g(\zeta, z_0)}{\partial \nu}|d\zeta| = \int_\Gamma \frac{\partial g(\zeta, z_0)}{\partial \nu}|d\zeta| = 2\pi ,$$

we have

$$\lim_{z_0\to 0}\int_{\Gamma-\alpha} \frac{\partial g(\zeta, z_0)}{\partial \nu}|d\zeta| = 0 . \tag{7}$$

Remark. We assumed that Γ_i, Γ_{i+1} make an inner angle π at ζ_i. If Γ_i, Γ_{i+1} make an inner angle $\alpha_i\pi (0<\alpha_i\leqq 2)$ at ζ_i, then by $w=(z-\zeta_i)^{1/\alpha_i}$, we map a neighbourhood U_i of ζ_i conformally on a neighbourhood V_i of $w=0$, then the images of Γ_i, Γ_{i+1} make an inner angle π at $w=0$. If we identify U_i with V_i, then since $\frac{\partial g}{\partial \nu}ds$ is invariant by conformal mapping, the general case is reduced to the case $\alpha_i=1$, so that the same result holds for such a domain.

From the theorem, we have easily

Theorem I. 21. *Under the same condition as Theorem* I. 20, *let* $f(\zeta)$ *be a bounded measurable function, defined on* Γ *and put*

$$u(z) = \frac{1}{2\pi}\int_\Gamma f(\zeta)\frac{\partial g(\zeta, z)}{\partial \nu}|d\zeta| ,$$

then $u(z)$ *is a bounded harmonic function in* D *and if* $f(\zeta)$ *is continuous at* $\zeta_0\in\Gamma$, *then*

$$\lim_{z\to\zeta_0} u(z) = f(\zeta_0) , \qquad z\in D .$$

7. Selberg's theorem on Green's function

Theorem I. 22. *Let* D *be a domain on the* z*-plane and* $g(z, z_0)=g_D(z, z_0)$ *be its Green's function with* z_0 *as its pole. Let* S *be the intersection of* D *with a circle* $|z|=r$ *and* $r\theta(r)$ *be its linear measure, then*

$$\int_S g(re^{i\theta}, z_0)d\theta \leqq \pi^2 \tan\frac{\theta(r)}{4} . \qquad \text{Selberg [2].}$$

Proof. First we consider a special case, where D is the complement of a closed set S^* on $|z|=r$, which consists of a finite number of closed arcs on $|z|=r$, so that S consists of a finite number of open arcs on $|z|=r$. Let for $|z|<r$,

$$\omega(z) = \omega(te^{i\varphi}) = \frac{1}{2\pi}\int_S \frac{r^2-t^2}{r^2-2rt\cos(\psi-\varphi)+t^2}d\psi , \tag{1}$$

then $\omega(z)$ is harmonic and $0<\omega<1$ in $|z|<r$, $\omega=1$ on S, $\omega=0$ at inner

points of S^*. We see easily that

$$\omega(z) \leqq \frac{1}{2\pi}\int_{-\frac{\theta(r)}{2}}^{\frac{\theta(r)}{2}} \frac{r^2-t^2}{r^2-2rt\cos\chi+t^2}\,d\chi = \frac{1}{\pi}\int_0^{\frac{\theta(r)}{2}} \frac{r-t^2}{r^2-2rt\cos\chi+t^2}\,d\chi\,.$$

Since

$$1 = \frac{1}{\pi}\int_0^{\pi} \frac{r^2-t^2}{r^2-2rt\cos\chi+t^2}\,d\chi\,,$$

$$\frac{1-\omega(z)}{r-t} \geqq \frac{1}{\pi}\int_{\frac{\theta(r)}{2}}^{\pi} \frac{r+t}{r^2-2rt\cos\chi+t^2}\,d\chi\,,$$

so that on S,

$$\frac{\partial\omega}{\partial r} \geqq \frac{1}{\pi r}\int_{\frac{\theta(r)}{2}}^{\pi} \frac{d\chi}{1-\cos\chi} = \frac{1}{\pi r}\cot\frac{\theta(r)}{4}\,. \tag{2}$$

We assume that $|z_0|<r$, then by the Green's formula,

$$\int_{|z|=r}\left(g\frac{\partial\omega}{\partial r}-\omega\frac{\partial g}{\partial r}\right)ds = 2\pi\omega(z_0)\,,$$

so that, since g and ω vanish on S^*, and $\omega=1$ on S

$$\int_S g\frac{\partial\omega}{\partial r}\,ds = \int_S \frac{\partial g}{\partial r}\,ds + 2\pi\omega(z_0)\,. \tag{3}$$

Since $\omega=1$ on S, $\omega(z)$ can be continued harmonically across S into $|z|>r$ by putting $\omega(z^*)-1=-(\omega(z)-1)$, or $\omega(z)+\omega(z^*)=2$, where z^* is the inverse point of z with respect to the circle $|z|=r$.

Now S^* consists of two sides S^*_+, S^*_-, where S^*_+ belongs to the boundary of $|z|>r$ and S^*_- to that of $|z|<r$. Since $\omega=0$ on S^*_-, we have $\omega=2$ on S^*_+, so that $\frac{\omega(z)}{2}=0$ on S^*_-, $\frac{\omega(z)}{2}=1$ on S^*_+, *i, e,* $\frac{\omega(z)}{2}$ is the harmonic measure of S^*_+ with respect to D[1], hence

$$\frac{\omega(z_0)}{2} = \frac{1}{2\pi}\int_{S^*_+}\frac{\partial g}{\partial r}\,ds\,.$$

Since g is harmonic in $|z|>r$,

$$\int_{S^*_+}\frac{\partial g}{\partial r}\,ds + \int_S \frac{\partial g}{\partial r}\,ds = 0\,,$$

so that

$$\frac{\omega(z_0)}{2} = -\frac{1}{2\pi}\int_S \frac{\partial g}{\partial r}\,ds\,,$$

or

$$\int_S \frac{\partial g}{\partial r}\,ds = -\pi\omega(z_0)\,,$$

1) For the definition of harmonic measure, c.f. p. 111.

hence by (3),

$$\int_S g \frac{\partial \omega}{\partial r} ds = \pi\omega(z_0) \leqq \pi , \tag{4}$$

so that by (2),

$$\int_S g\, d\theta \leqq \pi^2 \tan \frac{\theta(r)}{4} . \tag{5}$$

We assumed that $|z_0|<r$. If $|z_0|>r$, then we make a transformation $\zeta=r^2/z$ and apply (5), then we see that (5) holds for any $z_0 \in D$. We assumed that S consists of a finite number of open arcs on $|z|=r$. We see easily that (5) holds for any open set S on $|z|=r$.

Now we can prove the general case as follows. Let D be any domain and S^* be the complement of S with respect to $|z|=r$ and D^* be the complement of S^* with respect to the z-plane, then $D \subset D^*$, so that $g_D(z, z_0) \leqq g_{D^*}(z, z_0)$, hence by (5),

$$\int_S g_D(re^{i\theta}, z_0) d\theta \leqq \int_S g_{D^*}(re^{i\theta}, z_0)\, d\theta \leqq \pi^2 \tan \frac{\theta(r)}{4} . \tag{6}$$

Theorem I. 23. *Let D be a domain on the $z=x+iy$-plane, which does not contain $z=\infty$ and $g(z, z_0)$ be its Green's function. Let S be the intersection of D with the real axis and L be its linear measure, then*

$$\int_S g(x, z_0)\, dx \leqq \frac{\pi}{2} L ,$$

where the equality holds, only when $z_0=\alpha$ is real and D coincides with the complement of two half-lines: $\alpha + \frac{L}{2} \leqq x \leqq \infty$ and $-\infty \leqq x \leqq \alpha - \frac{L}{2}$.

Selberg [3].

We remark that if in Theorem I. 22, we take the circle $|z-ir|=r$ instead of the circle $|z|=r$ and make $r \to \infty$, then we obtain a worse result:

$$\int_S g(x, z_0)\, dx \leqq \frac{\pi^2}{4} L .$$

Proof. First we consider a special case, where D is the complement of a closed set S^* on the real axis, such that its complement S consists of a finite number of open intervals on the real axis: $(x_1, x_2), (x_3, x_4), \cdots, (x_{2n-1}, x_{2n})$ $(x_1<x_2<x_3<x_4< \cdots <x_{2n-1}<x_n)$.

Let $f(z)$ be an analytic function in D, whose real part is g, such that

$$\Re f(z) = g(z, z_0) , \tag{1}$$

then $f(z)$ is not one-valued, but by the principle of reflexion, it can be continued analytically across the slits belonging to S^*. Hence the one-valued branch of $f(z)$ in $\Im z>0$ is regular at $z=\infty$ and by a suitable normalization, we assume that $f=0$ at $z=\infty$. Since $f(z)$ is purely imaginary on the part $|x|>\text{Max}(|x_1|, |x_{2n}|)$ of the real axis,

$$f(z)=\frac{iA}{z}+\cdots \text{ at } z=\infty, \quad A=\text{real}. \tag{2}$$

Hence

$$g(z, z_0)=\frac{Ay}{x^2+y^2}+O\left(\frac{1}{|z|^2}\right) \quad \text{at } z=\infty,\ y>0\,, \tag{3}$$

so that
$$A>0\,.$$

Suppose that $\Im z_0\leqq 0$, then by (2) and Cauchy's fundamental theorem, we have

$$\int_{-\infty}^{\infty} g(x+iy_1, z_0)\,dx=\int_{-\infty}^{\infty} g(x+iy_2, z_0)\,dx \quad (0<y_1<y_2). \tag{4}$$

Since

$$\lim_{y_1\to 0}\int_{-\infty}^{\infty} g(x+iy_1, z_0)\,dx=\int_S g(x, z_0)\,dx\,,$$

$$\lim_{y_2\to\infty}\int_{-\infty}^{\infty} g(x+iy_2, z_0)\,dx=\lim_{y_2\to\infty}\int_{-\infty}^{\infty}\frac{Ay_2}{x^2+y_2^2}\,dx=\pi A\,,$$

we have

$$\int_S g(x;\,D)\,dx=\pi A\,. \tag{5}$$

Next we shall evaluate $A>0$. First suppose that $z_0=\alpha$ is real and $x_{2m-1}<\alpha<x_{2m}$. Since as easily seen, $f(z)$ is an Abelian integral of the third kind of the Riemann surface of two sheets, whose branch points are $\{x_\nu\}$ and since $f'(z)=-\dfrac{iA}{z^2}+\cdots$ at $z=\infty$, we have

$$f'(z)=\frac{-iA\,P(z)}{(z-\alpha)\sqrt{U(z)}}\,, \tag{6}$$

where

$$U(z)=\prod_{\nu=1}^{2n}(z-x_\nu), \qquad P(z)=z^{n-1}+c_1 z^{n-2}+\cdots+c_{n-1}\,. \tag{7}$$

We shall prove that $P(z)$ is a real polynomial. Let $f(z)=g(x, y)+ih(x, y)$, then if $x>0$ is sufficiently large, $\left(\dfrac{\partial g(x, y)}{\partial y}\right)_{y=0}>0$, so that $\left(\dfrac{\partial h(x, y)}{\partial x}\right)_{y=0}<0$. Since $\left(\dfrac{\partial g(x, y)}{\partial x}\right)_{y=0}=0$, we have $-\dfrac{1}{i}f'(x)=-\left(\dfrac{\partial h(x, y)}{\partial x}\right)_{y=0}>0$, so that $P(x)>0$, if $x>0$ is sufficiently large, hence $P(z)$ is a real polynomial.

Starting from $z=\infty$, we integrate (6) on the x-axis, avoiding x_ν by a small semi-circle about x_ν, which lies in $y>0$, then we see easily that

$dh=0$ on S, so that $f'(x)=g'(x,0)$ on S, hence the zero points $\{\xi_\nu\}$ of $P(z)$ coincide with the zero points of $\frac{dg}{dx}$, so that by Rolle's theorem, $x_{2\nu-1}<\xi_\nu<x_{2\nu}$ $(\nu\neq m)$, hence

$$P(z)=(z-\xi_1)\cdots(z-\xi_{m-1})\,(z-\xi_{m+1})\cdots(z-\xi_n)\,. \tag{8}$$

Since $f'(z)$ has a pole of the first order at $z=\alpha$ with the residue -1, we have

$$A=\left|\frac{\sqrt{U(\alpha)}}{P(\alpha)}\right|=\Big[(\alpha-x_{2m-1})\,(x_{2m}-\alpha)\,\Pi_1\,\Pi_2\Big]^{\frac{1}{2}}, \tag{9}$$

where

$$\Pi_1=\prod_{\nu=1}^{m-1}\frac{(\alpha-x_{2\nu-1})(\alpha-x_{2\nu})}{(\alpha-\xi_\nu)^2},\quad \Pi_2=\prod_{\nu=m+1}^{n}\frac{(x_{2\nu-1}-\alpha)(x_{2\nu}-\alpha)}{(\xi_\nu-\alpha)^2}.$$

In Π_1,

$$\frac{(\alpha-x_{2\nu-1})(\alpha-x_{2\nu})}{(\alpha-\xi_\nu)^2}\leqq\frac{\alpha-x_{2\nu-1}}{\alpha-x_{2\nu}},$$

in Π_2,

$$\frac{(x_{2\nu-1}-\alpha)(x_{2\nu}-\alpha)}{(\xi_\nu-\alpha)^2}\leqq\frac{x_{2\nu}-\alpha}{x_{2\nu-1}-\alpha},$$

so that

$$\Pi_1\leqq\prod_{\nu=1}^{m-1}\frac{\alpha-x_{2\nu-1}}{\alpha-x_{2\nu}}=\prod_{\nu=1}^{m-1}\left(1+\frac{x_{2\nu}-x_{2\nu-1}}{\alpha-x_{2\nu}}\right),\quad \Pi_2\leqq\prod_{\nu=m+1}^{n}\left(1+\frac{x_{2\nu}-x_{2\nu-1}}{x_{2\nu-1}-\alpha}\right).$$

Since if $a_1>0$, $a_2>0$, $0<a_2\leqq A_1-A_2$, $A_1>A_2$,

$$\left(1+\frac{a_1}{A_1}\right)\left(1+\frac{a_2}{A_2}\right)\leqq1+\frac{a_1+a_2}{A_2},$$

we have

$$\left(1+\frac{x_2-x_1}{\alpha-x_2}\right)\left(1+\frac{x_4-x_3}{\alpha-x_4}\right)\leqq1+\frac{(x_2-x_1)+(x_4-x_3)}{\alpha-x_4},$$

so that

$$\Pi_1\leqq1+\frac{1}{\alpha-x_{2m-2}}\sum_{\nu=1}^{m-1}(x_{2\nu}-x_{2\nu-1})\leqq1+\frac{1}{\alpha-x_{2m-1}}\sum_{\nu=1}^{m-1}(x_{2\nu}-x_{2\nu-1}),$$

$$(\alpha-x_{2m-1})\,\Pi_1\leqq\sum_{\nu=1}^{m-1}(x_{2\nu}-x_{2\nu-1})+(\alpha-x_{2m-1})\,. \tag{10}$$

Hence if we denote the linear measure of the part of S, contained in $(-\infty,\alpha)$ by $l(\alpha)$, then $(\alpha-x_{2m-1})\,\Pi_1\leqq l(\alpha)$.

Similarly $(x_{2m}-\alpha)\Pi_2\leqq L-l(\alpha)$, so that by (9),

$$A\leqq\sqrt{l(\alpha)\,(L-l(\alpha))}\leqq L/2\,,$$

hence by (5),

$$\int_S g(x,\alpha)\,dx\leqq\frac{\pi}{2}L\,. \tag{11}$$

If $\Im z_0 \leqq 0$, we put

$$u=u(z_0)=\int_S g(x,\, z_0)\, dx\,,$$

then u is harmonic in $\Im z_0<0$ and $u=0$ on S^*, so that by the maximum principle, $u(z_0)\leqq \operatorname*{Max}_{\alpha\in S} u(\alpha)\leqq \frac{\pi}{2}L$, hence if $\Im z_0\leqq 0$,

$$\int_S g(x,\, z_0)\, dx \leqq \frac{\pi}{2} L\,. \tag{12}$$

From the symmetry, (12) holds if $\Im z_0 \geqq 0$. We have assumed that S consists of a finite number of open intervals on the real axis, but by the limiting process, (12) holds for any open set S on the real axis. Hence the theorem is proved, if D is the special domain, defined in the begining. The general case can be proved similarly as Theorem I. 22 and from the proof, we see that the equality holds only for the special domain of the theorem.

8. Neumann's problem

Let D be a domain, contained in a closed Riemann surface F, spread over the z-sphere, whose boundary Γ consists of a finite number of analytic Jordan curves or analytic Jordan arcs: $\Gamma=\sum_{i=1}^{k}\Gamma_i$, such that if two Jordan arcs Γ_i, Γ_{i+1} meet at ζ_i (vertex), then they make an inner angle $\alpha_i\pi(0<\alpha_i\leqq 2)$ at ζ_i. Let $f(\zeta)$ be a given bounded function on Γ, which is continuous, except at the vertices ζ_i, where $f(\zeta)$ may be discontinuous, such that

$$|f(\zeta)|\leqq M \text{ on } \Gamma \tag{1}$$

and satisfies the following condition:

$$\int_\Gamma f(\zeta)|d\zeta|=0\,. \tag{2}$$

Then there exists a solution of *the Neumann's problem* as follows.

Theorem I. 24. *There exists a harmonic function $u(z)$ in D, which is continuous in $\bar{D}$, such that*

(i) $$|u(z)|\leqq k_1 M \quad \text{in } \bar{D},$$

where $k_1=k_1(D)$ is a constant, which depends on D only.

(ii) $$\frac{\partial u}{\partial \nu}\to f(\zeta), \quad \zeta\in\Gamma,\ \zeta\neq\zeta_i\ (i=1, 2, \cdots, k),$$

when $z\to\zeta$ along the inner normal ν of Γ at ζ, where $\frac{\partial u}{\partial \nu}$ is the derivative of u in the direction ν.

(iii) *Let $D[u]$ be the Dirichlet integral of u in D, then $D[u]$ can be expressed as*

$$D[u]=-\int_\Gamma u\frac{\partial u}{\partial\nu}|d\zeta|=-\int_\Gamma u(\zeta)f(\zeta)|d\zeta|,$$

so that

$$D[u]\leqq k_2M^2,$$

where $k_2=k_2(D)$ is a constant, which depends on D only.

The existence of such $u(z)$ is proved usually by means of integral equations. We shall give a direct proof as follows.

Proof. (Tsuji [59]). We may assume that $M=1$, so that

$$|f(\zeta)|\leqq 1 \text{ on } \Gamma, \qquad \int_\Gamma f(\zeta)|d\zeta|=0. \tag{1}$$

First we assume that $\alpha_i=1$ at every vertex ζ_i and $z=0$ belongs to D and is not a branch point of F. Let $g(z, 0)$ be the Green's function of D, with $z=0$ as its pole, then by Theorem I. 20, $\dfrac{\partial g(\zeta, 0)}{\partial\nu}$ is continuous and >0 on Γ, where ν is the inner normal of Γ at ζ. Hence if we put

$$\varphi(\zeta)=f(\zeta)/\frac{\partial g(\zeta, 0)}{\partial\nu}, \tag{2}$$

then $\varphi(\zeta)$ is bounded on Γ and

$$\int_\Gamma \varphi(\zeta)\frac{\partial g(\zeta, 0)}{\partial\nu}|d\zeta|=\int_\Gamma f(\zeta)|d\zeta|=0. \tag{3}$$

We put

$$v(z)=\frac{1}{2\pi}\int_\Gamma \varphi(\zeta)\frac{\partial g(\zeta, z)}{\partial\nu}|d\zeta|, \qquad z\in D, \tag{4}$$

then $v(z)$ is harmonic in D, $v(0)=0$ and

$$\lim_{z\to\zeta} v(z)=\varphi(\zeta), \qquad \zeta\in\Gamma. \tag{5}$$

Let $h=h(z, 0)$ be the conjugate harmonic function of $g=g(z, 0)$ and put

$$u(z)=\int_0^z v(t)dg(t, 0), \tag{6}$$

where we integrate on the niveau curve $dh=0$ of h, which connects $z=0$ to z. Since $v(0)=0$, the integral is finite. First we must prove that $u(z)$ is one-valued on D, such that if there exist two curves C_1, C_2, which connect $z=0$ to z, on which $dh=0$, then

$$\int_{C_1} v(t)dg(t, 0)=\int_{C_2} v(t)dg(t, 0).$$

It suffices to prove that if C is a closed curve, on which $dh=0$, then

$$\int_C v(t)dg(t, 0)=0\,. \tag{7}$$

To prove this, we put

$$w(z)=\int_C g(t, z)\,dg(t, 0) \tag{8}$$

and we shall prove that $w(z)\equiv 0$.

First we assume that C does not pass through $z=0$, then $w(z)$ is continuous on C, so that $w(z)$ is continuous in D. We shall prove that $w(z)$ is superharmonic on D. To prove it, it suffices to prove that $w(z)$ is superharmonic at every point of C. Let z_0 be a point of C, which is different from double points of the curve $dh=0$. Let $U(z_0)$ be its neighbourhood and C_1 be the part of C, which lies in $U(z_0)$ and $C_2=C-C_1$. Then

$$w(z)=\int_{C_1} g(t, z)\,dg(t, 0)+\int_{C_2} g(t, z)\,dg(t, 0)=w_1(z)+w_2(z)\,. \tag{9}$$

By $\zeta=x+iy=g(z, 0)+ih(z,0)-(g(z_0, 0)+ih(z_0, 0))$, we map $U(z_0)$ conformally on a neighbourhood V of $\zeta=0$, such that C_1 becomes a segment: $-\delta\leqq x\leqq\delta$, $y=0$ on the real axis and for $z\in U(z_0)$, we put $w_1(z)=W_1(\zeta)$, $g(t, z)=G(x, \zeta)$, then

$$w_1(z)=W_1(\zeta)=\int_{-\delta}^{\delta} G(x, \zeta)dx\,.$$

Since $G(x, \zeta)=\log\dfrac{1}{|x-\zeta|}+$ (harmonic function), we see that $W_1(\zeta)$ is superharmonic in V, so that $w_1(z)$ is superharmonic in $U(z_0)$. Since $w_2(z)$ is harmonic in $U(z_0)$, $w(z)$ is superharmonic in $U(z_0)$. Next suppose that z_0 coincides with a double point of the curve $dh=0$ and let $U(z_0)$ be its neighbourhood, which contains only one double point z_0 of the curve $dh=0$. Then $w(z)$ is superharmonic in $U(z_0)$, except at z_0, but since $w(z)$ is continuous, we see easily that $w(z)$ is superharmonic at z_0. Hence $w(z)$ is superharmonic on C, so that in D. Since from (8), $w=0$ on Γ, by the maximum principle, we have $w\geqq 0$ in D, but $w(0)=0$, so that $w(z)\equiv 0$. Since

$$w(z)=\int_C g(t, z)\,dg(t, 0)=\int_C g(z, t)\,dg(t, 0)\equiv 0\,,$$

we have

$$\int_C \frac{\partial g(\zeta, t)}{\partial\nu}\,dg(t, 0)=0, \qquad \zeta\in\Gamma\,,$$

so that

$$\int_C v(t)dg(t,0)=\frac{1}{2\pi}\int_\Gamma \varphi(\zeta)\,|d\zeta|\int_C \frac{\partial g(\zeta,t)}{\partial \nu}dg(t,0)=0\,.$$

Hence (7) is proved, if C does not pass through $z=0$.

Next suppose that C passes through $z=0$. Let Γ_λ be the niveau curve: $g(z,0)=\text{const.}=\lambda>0$, then if λ is sufficiently large, Γ_λ is a Jordan curve, surrounding $z=0$, so that C meets Γ_λ at two points z_1, z_2. Let C_λ be the part of C, which lies inside of Γ_λ and put

$$w_\lambda(z)=\int_{C-C_\lambda} g(t,z)\,dg(t,0)\,,\qquad z\in D\,, \tag{10}$$

then since

$$w_\lambda(0)=\int_{C-C_\lambda} g(t,0)\,dg(t,0)=\frac{1}{2}(g^2(z_2,0)-g^2(z_1,0))=\frac{1}{2}(\lambda^2-\lambda^2)=0\,,$$

we can prove similarly as above that $w_\lambda(z)\equiv 0$ and

$$\int_{C-C_\lambda} v(t)\,dg(t,0)=0\,.$$

Hence if we make $\lambda\to\infty$, we have

$$\int_C v(t)\,dg(t,0)=0\,.$$

Hence (7) is proved, so that $u(z)$ is one-valued in D. We see easily that $u(z)$ is continuous in $\bar{D}$, so that $|u(z)|\leq k_1$ in $\bar{D}$, where k_1 is a constant, which depends on D only.

Next we shall prove that $u(z)$ is harmonic in D. Let z be a point of D, which is different from double points $\{a_i\}$ of the curve $dh=0$, then since $\left[\frac{\partial v}{\partial g}\right]_{z=0}=0$, we have from (6),

$$\frac{\partial^2 u}{\partial g^2}=\frac{\partial v}{\partial g}\,,\qquad \frac{\partial u^2}{\partial h^2}=\int_0^z \frac{\partial^2 v}{\partial h^2}dg=-\int_0^z\frac{\partial^2 v}{\partial g^2}dg=-\frac{\partial v}{\partial g}\,,$$

so that $\Delta u=\frac{\partial^2 u}{\partial g^2}+\frac{\partial^2 u}{\partial h^2}=0$, hence $u(z)$ is harmonic at z.

Since $u(z)$ is bounded in a neighbourhood of a_i, $u(z)$ is harmonic at a_i, hence $u(z)$ is harmonic in D. We see from (2), (5), that if $z\to\zeta\in\Gamma$ along the inner normal ν of Γ at ζ, then

$$\frac{\partial u(z)}{\partial\nu}=v(z)\frac{\partial g(z,0)}{\partial\nu}\to\varphi(\zeta)\frac{\partial g(\zeta,0)}{\partial\nu}=f(\zeta)\,. \tag{11}$$

We can prove easily that

$$D[u]=-\int_\Gamma u\frac{\partial u}{\partial\nu}\,|d\zeta|=-\int_\Gamma u(\zeta)f(\zeta)|d\zeta|\,, \tag{12}$$

so that $D[u] \leqq k_2$, where k_2 is a constant, which depends on D only.

Hence the theorem is proved, if $\alpha_i=1$ at every vertex ζ_i. If two Jordan arcs Γ_i, Γ_{i+1} meet at ζ_i at an inner angle $\alpha_i\pi(\alpha_i \neq 1)$, then by $x=(z-\zeta_i)^{1/\alpha_i}$, we map a neighbourhood $U(\zeta_i)$ of ζ_i conformally on a neighbourhood V_i of $x=0$, such that the images γ_i, γ_{i+1} of Γ_i, Γ_{i+1} meet at an angle π at $x=0$. We define $f^*(x)$ on $\gamma_i+\gamma_{i+1}$ by

$$f^*(x)|dx|=f(\zeta)|d\zeta| .$$

If we identify V_i with $U(\zeta_i)$ and take $f^*(x)$ instead of $f(\zeta)$ at every vertex ζ_i, where $\alpha_i \neq 1$, then the general case is reduced to the special case, where $\alpha_i=1$ at every vertex ζ_i. Then the modified $f(\zeta)$ is not bounded on Γ in general, but is integrable and $\int_\Gamma f(\zeta)|d\zeta|=0$, so that we need no change in the proof. Hence the theorem is proved in the general case.

9. Mixed boundary value problem

Theorem I. 25. *Let D be a domain, contained in a closed Riemann surface, spread over the z-sphere, whose boundary Γ consists of a finite number of disjoint analytic Jordan curves: $\Gamma=\sum_{i=1}^{m}\Gamma_i+\sum_{j=1}^{n}C_j$. Let there be given continuous functions $f_i(\zeta)$ on Γ_i and $g_j(\zeta)$ on C_j. Then there exists a harmonic function $u(z)$ in D, such that*

$$u=f_i \text{ on } \Gamma_i, \qquad \frac{\partial u}{\partial \nu}=g_j \text{ on } C_j \qquad (i=1, 2, \cdots, m, j=1, 2, \cdots, n),$$

where ν is the inner normal of C_j.

Proof. To prove the theorem, first we shall define *the Schottky prolongation of D with respect to $\sum_{j=1}^{n} C_j$* as follows.

We consider D as consist of two faces, the upper and the lower face and let $\dot{D}$ be the upper face and $\underset{\cdot}{D}$ be the lower one. If we consider Γ_i as belong to $\dot{D}$, or to $\underset{\cdot}{D}$, then we denote it by $\dot{\Gamma}_i, \underset{\cdot}{\Gamma}_i$ respectively. We join $\dot{D}$ and $\underset{\cdot}{D}$ along $\sum_{j=1}^{n} C_j$, then we obtain a surface $\tilde{D}=\dot{D}+\underset{\cdot}{D}$, whose boundary consists of $\sum_{i=1}^{m}\dot{\Gamma}_i+\sum_{i=1}^{m}\underset{\cdot}{\Gamma}_i$. By defining suitably the parameter neighbourhood of a point $\in C_j$, $\tilde{D}$ can be considered as a Riemann surface. We call $\tilde{D}$ the Schottky prolongation of D with respect to $\sum_{j=1}^{n} C_j$.

Now we shall prove the theorem. First we consider two special cases: (i) $u=f_i$ on Γ_i $(i=1, 2, \cdots, m)$, $\frac{\partial u}{\partial \nu}=0$ on C_j $(j=1, 2, \cdots, n)$.

(ii) $u=0$ on $\Gamma_i (i=1, 2, \cdots, m)$, $\frac{\partial u}{\partial \nu}=g_j$ on C_j $(j=1, 2, \cdots, n)$.

First consider the case (i). Let $\tilde{D}$ be the Schottky pronlongation of D with respect to $\sum_{j=1}^{n} C_j$. We solve the Dirichlet problem for $\tilde{D}$ with the boundary value f_i on $\dot{\Gamma}_i$ and $\underset{\bullet}{\Gamma}_i$ $(i=1, 2, \cdots, m)$ and let $u(z)$ be its solution. Let $\dot{z} \in \dot{D}$, $\underset{\bullet}{z} \in \underset{\bullet}{D}$ be two points, whose projection is z, then $u(\dot{z})=u(\underset{\bullet}{z})$. To see this, we put $v(z)=\frac{1}{2}(u(\dot{z})+u(\underset{\bullet}{z}))$, then $v(z)$ is also a solution of the Dirichlet problem, hence by the uniqueness of the solution, we have $v(z)\equiv u(z)$. Since $v(\dot{z})=v(\underset{\bullet}{z})$, we have $u(\dot{z})=u(\underset{\bullet}{z})$, so that $\frac{\partial u}{\partial \nu}=0$ on C_j $(j=1, 2, \cdots, n)$, hence $u(z)$ is the solution of the case (i).

Next consider the case (ii). Let $\tilde{D}$ be the Schottky prolongation of D with respect to $\sum_{i=1}^{m} \Gamma_i$. We solve the Neumann's problem for $\tilde{D}$, with the boundary value: $\frac{\partial u}{d\nu}=g_j$ on $\dot{C}_j$, $\frac{\partial u}{\partial \nu}=-g_j$ on $\underset{\bullet}{C}_j$ $(j=1,2, \cdots, n)$, then the condition of the solvability of the Neumann's problem is satisfied. Let $u(z)$ be its solution, then $v(z)=u(z)+u(\underset{\bullet}{z})$ is the solution of the Neumann's problem for $\tilde{D}$ with the boundary value $\frac{\partial v}{\partial \nu}=0$, so that $v\equiv$const.. If we normalize $u(z)$, such that $u(\zeta_0)=0$ at a point $\zeta_0 \in \Gamma_1$, then $v(\zeta_0)=0$, so that $v\equiv 0$. Since $\dot{z}=\underset{\bullet}{z}$ on Γ_i, we have $u=0$ on Γ_i and $\frac{\partial u}{\partial \nu}=g_j$ on C_j, so that $u(z)$ is the solution of the case (ii).

Next consider the general case. Let $u_1(z)$ be the solution of (i) and $u_2(z)$ be that of (ii), then $u(z)=u_1(z)+u_2(z)$ is the solution of the general case.

CHAPTER II

SUBHARMONIC FUNCTIONS

1. Carathéodory's outer measure

Let R_n be an n-dimentional Euclidean space and e be any set in R_n. If a set function $\mu^*(e) \geqq 0$ satisfies the following conditions (i), (ii), (iii), (iv), then μ^* is called *a Carathéodory's outer measure.*

We denote the empty set by 0.

(i) $\mu^*(0)=0$.

(ii) If $e_1 \subset e_2$, then $\mu^*(e_1) \leqq \mu^*(e_2)$.

(iii) If $e=\sum_{n=1}^{\infty} e_n$, then $\mu^*(e) \leqq \sum_{n=1}^{\infty} \mu^*(e_n)$.

(iv) If the distance of e_1, e_2 is positive, then $\mu^*(e_1+e_2)=\mu^*(e_1)+\mu^*(e_2)$.

Then we can prove that $\mu^*(e)$ is a totally additive set function for Borel sets (Saks [1]), hence

Theorem II. 1. *Let $\{e_n\}$ be disjoint Borel sets and $e=\sum_{n=1}^{\infty} e_n$, then*

$$\mu^*(e)=\sum_{n=1}^{\infty} \mu^*(e_n) .$$

Theorem II. 2. *Let $\mu(e) \geqq 0$ be a totally additive set function, defined for Borel sets. Let e be any bounded set in R_n. We cover e by at most a countable number of open spheres Δ_i $(i=1, 2, \cdots)$ and put*

$$\mu^*(e)=\inf \sum_i \mu(\Delta_i) , \quad e \subset \sum_i \Delta_i ,$$

then $\mu^*(e)$ is a Carathéodory's outer measure and if e is a Boerl set, then

$$\mu^*(e)=\mu(e) .$$

Proof. From $e \subset \sum_i \Delta_i$, we have $\mu(e) \leqq \sum_i \mu(\Delta_i)$, so that by the definition,

$$\mu(e) \leqq \mu^*(e) . \tag{1}$$

Let Δ be an open sphere, then if we take $\Delta_i=\Delta$, then by the definition, $\mu^*(\Delta) \leqq \mu(\Delta)$, and by (1), $\mu(\Delta) \leqq \mu^*(\Delta)$, hence

$$\mu^*(\Delta)=\mu(\Delta) . \tag{2}$$

Let e be a bounded Borel set aud Δ be an open sphere, which contains e, then $\Delta - e$ is a Borel set so that $\mu(\Delta)=\mu(e)+\mu(\Delta-e)$ and by Theorem II. 1, $\mu^*(\Delta)=\mu^*(e)+\mu^*(\Delta-e)$, hence by (2),

$$\mu(e)+\mu(\Delta-e)=\mu^*(e)+\mu^*(\Delta-e) . \tag{3}$$

Since by (1), $\mu(e) \leqq \mu^*(e)$, $\mu(\Delta-e) \leqq \mu^*(\Delta-e)$, we have $\mu^*(e)=\mu(e)$. —— From the theorem, we have

Theorem II. 3. *Let $\mu_1(e) \geqq 0$, $\mu_2(e) \geqq 0$ be two totally additive set functions, defined for Borel sets. If $\mu_1(\Delta)=\mu_2(\Delta)$ for any open sphere Δ, then $\mu_1(e)=\mu_2(e)$ for any Borel set e.*

2. Selection theorem

Let $\mu(e) \geqq 0$ be a totally additive set function, defined for Borel sets e. Let E be a Borel set and $\bar{E}$ be its closure and $\mathring{E}$ be its open kernel, which consists of inner points of E, then $\Gamma(E)=\bar{E}-\mathring{E}$ is the boundary of E. If

$$\mu(\Gamma(E))=\mu(\bar{E}-\mathring{E})=\mu(\bar{E})-\mu(\mathring{E})=0\,, \tag{1}$$

then E is called *regular with respect to* μ.

Hence if E is regular with respect to μ, then since $\mu(\mathring{E}) \leqq \mu(E) \leqq \mu(\bar{E})$, we have

$$\mu(E)=\mu(\bar{E})=\mu(\mathring{E}). \tag{2}$$

Let $\mu_n(e) \geqq 0$ $(n=1, 2, \cdots)$ be a sequence of totally additive set functions, defined for Borel sets e. If there exists a totally additive set function $\mu(e) \geqq 0$, defined for Borel sets e, such that for any Borel set E, which is regular with respect to μ,

$$\lim_n \mu_n(E)=\mu(E)\,, \tag{3}$$

then we say that μ_n *converges to* μ.

Theorem II. 4. (Selection theorem). *Let $\mu_n(e) \geqq 0$ $(n=1, 2, \cdots)$ be a sequence of totally additive set functions, defined for Borel sets e, such that for any Borel set e,*

$$0 \leqq \mu_n(e) \leqq K \quad (n=1, 2, \cdots),$$

then we can select a partial sequence μ_{n_k} from μ_n, which converges to a totally additive set function $\mu \geqq 0$. Frostman [1].

Proof. For the sake of brevity, we assume that the sets lie in a square Δ on the xy-plane. We divide the sides of Δ into 2^n equal parts, then we obtain 4^n equal squares. Hence for $n=1, 2, \cdots$, we have a countable number of squares and the totality of sets, which are sums of a finite number of such squares, is countable, which we denote by $I_1, I_2, \cdots$. Let $\bar{I}_j$, $\mathring{I}_j$ be the closure and the open kernel of I_j respectively.

By the hypothesis,

$$0 \leqq \mu_n(\bar{I}_j) \leqq K \quad (n=1, 2, \cdots)\,.$$

Hence by the diagonal method, we can find μ_{n_k}, such that $\lim \mu_{n_k}(\bar{I}_j)$ and $\lim_k \mu_{n_k}(\mathring{I}_j)$ $(j=1, 2, \cdots)$ exist. By changing the notation, we assume that

$$\lim_n \mu_n(\bar{I}_j), \quad \lim_n \mu_n(\mathring{I}_j) \ (j=1, 2, \cdots) \ \text{exist.} \tag{2}$$

Let F be a closed set and O be an open set, then we define $\mu(F)$, and $\mu(O)$ by

$$\mu(F)=\inf_{F\subset\mathring{I}_j}(\lim_n \mu_n(\mathring{I}_j)), \quad \mu(O)=\sup_{\bar{I}_j\subset O}(\lim_n \mu_n(\bar{I}_j)). \tag{3}$$

If $F\subset O$, then there exists I_j, such that $F\subset\mathring{I}_j\subset\bar{I}_j\subset O$, so that by then definition, $\mu(F)\leqq\lim_n \mu_n(\mathring{I}_j)\leqq\lim_n \mu_n(\bar{I}_j)\leqq\mu(O)$, hence

$$\mu(F)\leqq\mu(O), \quad F\subset O\,. \tag{4}$$

Since $\bar{I}_j\subset O$, we have $\mu_n(\bar{I}_j)\leqq\mu_n(O)$, hence $\lim_n \mu_n(\bar{I}_j)\leqq\varliminf_n \mu_n(O)$, so that by (3),

$$\mu(O)\leqq \varliminf_n \mu_n(O)\,. \tag{5}$$

Since $F\subset\mathring{I}_j$, we have $\mu_n(F)\leqq\mu_n(\mathring{I}_j)$, hence $\varlimsup_n \mu_n(F)\leqq\lim_n \mu_n(\mathring{I}_j)$, so that by (3),

$$\varlimsup_n \mu_n(F)\leqq \mu(F)\,. \tag{6}$$

Hence from (4), (6), (3),

$$\mu(O)\geqq \sup_{F\subset O}\mu(F)\geqq \sup_{F\subset O}(\varlimsup_n \mu_n(F))\geqq \sup_{\bar{I}_j\subset O}(\lim_n \mu_n(\bar{I}_j))=\mu(O)\,,$$

so that

$$\mu(O)=\sup_{F\subset O}\mu(F)\,. \tag{7}$$

From (4), (5), (3),

$$\mu(F)\leqq \inf_{F\subset O}\mu(O)\leqq \inf_{F\subset O}(\varliminf_n \mu_n(O))\leqq \inf_{F\subset\mathring{I}_j}(\lim_n \mu_n(\mathring{I}_j))=\mu(F)\,,$$

$$\mu(F)=\inf_{O\supset F}\mu(O)\,. \tag{8}$$

Next for any bounded set e, we define $\mu(e)$ by

$$\mu(e)=\inf_{e\subset O}\mu(O)\,. \tag{9}$$

By (7), (8), we can prove easily that $\mu(e)$ is a Carathéodory's outer measure, hence $\mu(e)$ is a totally additive set function, defined for Borel sets e. If e is a Borel set, which is regular with respect to μ, then by (5), (6),

$$\mu(e)=\mu(\mathring{e})\leqq \varliminf_n \mu_n(\mathring{e})\leqq\varliminf_n \mu_n(e)\leqq \varlimsup_n \mu_n(e)\leqq \varlimsup_n \mu_n(\bar{e})\leqq\mu(\bar{e})=\mu(e)\,,$$

hence

$$\lim_n \mu_n(e)=\mu(e). \text{ ——}$$

3. Semi-continuous function

Let E be a bounded closed set in R_n and $-\infty\leqq f(P)<\infty$ be a function, defined on E. If at $P_0\in E$,

$$\overline{\lim_{P\to P_0}} f(P) \leqq f(P_0)\,, \quad P\in E\,, \tag{1}$$

then $f(P)$ is called *upper semi-continuous at* P_0. If $f(P)$ is upper semi-continuous at every point of E, then $f(P)$ is called upper semi-continuous on E.

Similarly, if $-\infty < f(P) \leqq \infty$ on E and at $P_0 \in E$,

$$\underline{\lim_{P\to P_0}} f(P) \geqq f(P_0), \quad P\in E, \tag{2}$$

then $f(P)$ is called *lower semi-continuous at* P_0 and if $f(P)$ is lower semi-continuous at every point of E, then $f(P)$ is called lower semi-continuous on E.

If $f(P)$ is upper semi-continuous, then $-f(P)$ is lower semi-continuous.

Theorem II. 5. (*R. Baire*). *Let E be a bounded closed set in R_n and $f(P)$ be upper semi-continuous on E, such that $f(P)\leqq M$ on E, then there exist continuous functions $\{\varphi_k(P)\}$ in R_n, such that*

(i) $M \geqq \varphi_1(P) \geqq \varphi_2(P) \geqq \cdots \geqq \varphi_k(P) \geqq \cdots$ *in* R_n.

(ii) $\lim_k \varphi_k(P) = f(P)$ *on* E.

If $f(P)$ is lower semi-continuous on E, such that $f(P)\geqq m$ on E, then there exist continuous functions $\{\psi_k(P)\}$ in R_n, such that

(i) $m \leqq \psi_1(P) \leqq \psi_2(P) \leqq \cdots \leqq \psi_k(P) \leqq \cdots$ *in* R_n.

(ii) $\lim_k \psi_k(P) = f(P)$ *on* E.

Proof. (Hausdorff [1]). We shall prove the second part. Let $f(P)$ be lower semi-continuous on E, such that $f(P)\geqq m$ on E.

We define $\psi_k(Q)$ in R_n by

$$\psi_k(Q) = \inf_{P\in E} (f(P) + k\,\overline{PQ}) \quad (k=1, 2, \cdots)\,. \tag{1}$$

Since $f(P)\geqq m$ on E,

$$m \leqq \psi_k(Q) \leqq \psi_{k+1}(Q) \quad \text{in } R_n\,. \tag{2}$$

We shall prove that $\psi_k(Q)$ is continuous in R_n. Let Q_1, Q_2 be two points of R_n and let $\overline{Q_1Q_2}=\delta$. Then by (1), there exists $P_0\in E$, such that

$$\psi_k(Q_1) > f(P_0) + k\overline{P_0Q_1} - \delta\,,$$

and by the definition,

$$\psi_k(Q_2) \leqq f(P_0) + k\,\overline{P_0Q_2}\,.$$

Since $\overline{P_0Q_2} \leqq \overline{P_0Q_1} + \overline{Q_1Q_2} = \overline{P_0Q_1} + \delta$,

$$\psi_k(Q_2) \leqq f(P_0) + k\,\overline{P_0Q_1} + k\,\delta < \psi_k(Q_1) + (k+1)\delta\,.$$

Similarly $\psi_k(Q_1) < \psi_k(Q_2) + (k+1)\delta$, so that

$$|\psi_k(Q_1)-\psi_k(Q_2)|<(k+1)\ \overline{Q_1Q_2}\ , \tag{3}$$

hence $\varphi_k(Q)$ is continuous in R_n.

If $P_0 \in E$, then by (1),

$$\psi_k(P_0)=\inf_{P\in E}(f(P)+k\ \overline{PP_0}) \leqq f(P_0)+k\ \overline{P_0P_0}=f(P_0)\ ,$$

hence by (2),

$$m \leqq \lim_k \psi_k(P_0) \leqq f(P_0) \quad \text{on } E\ . \tag{4}$$

For any $\varepsilon>0$, we can determine $\rho=\rho(\varepsilon)$, such that

$$f(P)>f(P_0)-\varepsilon, \quad \text{if } P\in E, \quad \overline{PP_0}<\rho\ . \tag{5}$$

If $P\in E$, $\overline{PP_0}\geqq\rho$, then

$$f(P)+k\ \overline{PP_0}\geqq m+k\rho\ , \tag{6}$$

so that by (5), (6), if k is sufficiently large, for any $P\in E$,

$$f(P)+k\ \overline{PP_0}>f(P_0)-\varepsilon\ ,$$

hence

$$\psi_k(P_0)=\inf_{P\in E}(f(P)+k\overline{PP_0}) \geqq f(P_0)-\varepsilon\ ,$$

$$\lim_k \psi_k(P_0) \geqq f(P_0)-\varepsilon\ .$$

Since $\varepsilon>0$ is arbitrary, we have

$$\lim_k \psi_k(P_0) \geqq f(P_0) \quad \text{on } E\ , \tag{7}$$

hence by (4),

$$\lim_k \psi_k(P_0)=f(P_0) \quad \text{on } E\ .$$

4. Subharmonic function

Let $u(z)$ be defined in a domain D on the z-plane. If $u(z)$ satisfies the following conditions (i), (ii), (iii), then $u(z)$ is called *a subharmonic function in* D[1].

(i) $-\infty\leqq u<\infty$, $u\not\equiv-\infty$.

(ii) $u(z)$ is upper semi-continuous.

(iii) Let D_1 be any Jordan domain, contained in D, whose boundary Γ consists of a finite number of disjoint Jordan curves. Let $H(z)$ be harmonic in D_1 and continuous in $\bar{D}_1$, such that $H\geqq u$ on Γ, then

$$H(z)\geqq u(z) \text{ in } D_1.$$

In the following, $u(z)$ denotes a subharmonic function in D.

1) Similarly, if $u(z)$ satisfies the following condition, then $u(z)$ is called *a superharmonic function in D*.

(i) $-\infty<u\leqq\infty$, $u\not\equiv\infty$.

(ii) $u(z)$ is lower semi-continuous.

(iii) If $H\leqq u$ on Γ, then $H(z)\leqq u(z)$ in D_1.

Hence if u is subharmonic, then $-u$ is superharmonic.

Let E be a bounded closed set in D, then

$$u(z) \leqq K < \infty \text{ on } E. \tag{1}$$

For, if otherwise, there exists $z_n \in E$, $(n=1,2,\cdots)$, such that $u(z_n) \to \infty$. We may assume that $\{z_n\}$ converges to a point $z_0 \in E$, then by (ii), $u(z_0)=\infty$, which contradicts (i). ——

By (ii) and Theorem II. 5, there exist continuous functions $\varphi_n(z)$ $(n=1, 2, \cdots)$, such that

$$\varphi_1(z) \geqq \varphi_2(z) \geqq \cdots \geqq \varphi_n(z) \to u(z). \tag{2}$$

Let $D_1 \subset D$ be a Jordan domain, with the boundary Γ and $H_n(z)$ be a harmonic function in D_1, such that $H_n = \varphi_n$ on Γ, then

$$H_1(z) \geqq H_2(z) \geqq \cdots \geqq H_n(z) \geqq u(z) \quad \text{in } D_1.$$

Since $u \not\equiv -\infty$, by Harnack's theorem,

$$\lim_n H_n(z) = \overline{H}(z) \geqq u(z) \quad \text{in } D_1, \tag{3}$$

where $\overline{H}(z)$ is harmonic in D_1. We call $\overline{H}(z)$ *the best harmonic majorant of* $u(z)$ *in* D_1.

Theorem II. 6. *If a disc* $|z-z_0| \leqq r$ *is contained in* D *and* $u(z_0) > -\infty$, *then* $u(z)$ *is integrable on the circle* $|z-z_0|=r$ *and in the disc* $|z-z_0| \leqq r$ *and if we put*

$$L(u; z_0, r) = \frac{1}{2\pi}\int_0^{2\pi} u(z_0 + re^{i\theta})d\theta,$$

$$A(u; z_0, r) = \frac{1}{\pi r^2}\iint\limits_{|z-z_0|\leqq r} u(z) r dr d\theta, \qquad z = z_0 + re^{i\theta},$$

then

$$-\infty < u(z_0) \leqq L(u; z_0, r) < \infty, \qquad -\infty < u(z_0) \leqq A(u; z_0, r) < \infty.$$

Proof. Let $\overline{H}(z)$ be the best harmonic majorant of u in $|z-z_0| \leqq r$ such that if we define $H_k(z)$ for $|z-z_0| \leqq r$ as before, then

$$\lim_k H_k(z) = \overline{H}(z) \geqq u(z) \quad \text{in } |z-z_0| \leqq r.$$

Since $u(z) \leqq H_k(z)$,

$$u(z_0) \leqq H_k(z_0) = L(H_k; z_0, r). \tag{1}$$

Since $H_1 \geqq H_2 \geqq \cdots \geqq H_k \to u$ on $|z-z_0|=r$, $H_1 - H_k \geqq 0$ increases with k and $H_1 - H_k \to H_1 - u \geqq 0$ on $|z-z_0|=r$, so that by Lebesgue's theorem,

$$\lim_k \int_0^{2\pi} (H_1(z) - H_k(z))d\theta = \int_0^{2\pi} (H_1(z) - u(z))d\theta, \qquad z = z_0 + re^{i\theta}. \tag{2}$$

Since

$$\frac{1}{2\pi}\int_0^{2\pi} (H_1(z) - H_k(z))d\theta = H_1(z_0) - H_k(z_0) \leqq H_1(z_0) - u(z_0),$$

$$\frac{1}{2\pi}\int_0^{2\pi} (H_1(z) - u(z))d\theta \leqq H_1(z_0) - u(z_0) < \infty,$$

so that $H_1(z)-u(z)$ and hence $u(z)$ is integrable on $|z-z_0|=r$ and hence by (2),

$$\lim_k L(H_k;\, z_0,\, r)=L(u;\, z_0\, r)\,, \tag{3}$$

so that by (1),

$$-\infty<u(z_0)\leqq L(u\,;\, z_0,\, r)\,. \tag{4}$$

Similarly from $u(z_0)\leqq H_k(z_0)=A(H_k;\, z_0,\, r)$, we can prove that $u(z)$ is integrable in $|z-z_0|\leqq r$. Since $u(z_0)\leqq L(u\,;\, z_0,\, \rho)$ $(0\leqq\rho\leqq r)$,

$$-\infty<u(z_0)=\frac{2}{r^2}u(z_0)\int_0^r \rho d\rho\leqq\frac{2}{r^2}\int_0^r L(u\,;\, z_0,\, \rho)\,\rho d\rho=A(u\,;\, z_0,\, r)\,. \tag{5}$$

Theorem II. 7. *$u(z)$ is integrable on any bounded closed domain, contained in D, so that $u>-\infty$ almost everywhere in D.* (c. f. Theorem II. 26.)

Proof. Since $u\not\equiv-\infty$, there exists $z_0\in D$, such that $u(z_0)>-\infty$. Let $\varDelta$ be a disc, contained in D, with z_0 as its conter, then $u(z)$ is integrable in $\varDelta$, so that $u>-\infty$ almost everywhere in $\varDelta$. From this, we can prove that u is integrable in any closed domain, contained in D.

Theorem II. 8. *If a circle $|z-z_0|=r$ is contained in D, then $u(z)$ is integrable on $|z-z_0|=r$, even if the disc $|z-z_0|\leqq r$ contains points, which do not belong to D.*

Proof. Choose $0<r_1<r<r_2$, such that $r_1\leqq|z-z_0|\leqq r_2$ is contained in D. We define $H_k(z)$ for $r_1\leqq|z-z_0|\leqq r_2$, then $H_k(z)\to\overline{H}(z)\not\equiv-\infty$ $(k\to\infty)$. Since $H_k(z)$ is harmonic in $r_1<|z-z_0|<r_2$, by the Green's formula,

$$\int_0^{2\pi}\frac{\partial H_k(z_0+\rho e^{i\theta})}{\partial\rho}\,\rho d\theta=\text{const.}\ (r_1<\rho<r_2)\,,$$

so that

$$L(H_k;\, z_0,\, \rho)=a_k\log\rho+b_k\ \ (r_1\leqq\rho\leqq r_2)\,. \tag{1}$$

Similarly

$$L(\overline{H};\, z_0,\, \rho)=a\log\rho+b\qquad(r_1\leqq\rho\leqq r_2)\,. \tag{2}$$

Since $H_k\to\overline{H}$ by decreasing, we have by Lebesgue's theorem, $L(H_k;\, z_0,\, \rho)\to L(\overline{H}\,;\, z_0,\, \rho)$, so that $\lim_k a_k=a$, $\lim_k b_k=b$, hence

$$L(u;\, z_0,\, r)\ =\lim_k L(H_k\,;\, z_0,\, r)=L(\overline{H};\, z_0,\, r)=a\log r+b,$$

so that u is integrable on $|z-z_0|=r$.

Theorem II. 9. *If $0<r_1\leqq|z-z_0|\leqq r_2$ is contained in D, then $L(u\,;\, z_0,\, r)$ is a continuous convex function of $\log r$ $(r_1\leqq r\leqq r_2)$.*

Proof. We define $H_k(z)$ for $r_1\leqq|z-z_0|\leqq r_2$, then as before,

$$\lim_k L(H_k;\, z_0,\, r)=a\log r+b \quad (r_1\leqq r\leqq r_2)\,, \tag{1}$$

$$\left.\begin{aligned} L(u\,;\, z_0,\, r_1)&=\lim_k L(H_k;\, z_0,\, r_1)=a\log r_1+b,\\ L(u\,;\, z_0,\, r_2)&=\lim_k L(H_k;\, z_0,\, r_2)=a\log r_2+b.\end{aligned}\right\} \tag{2}$$

Since $u\leqq H_k$, $L(u\,;\, z_0,\, r)\leqq L(H_k;\, z_0,\, r)$, so that

$$L(u\,;\, z_0,\quad)\leqq \lim_k L(H_k\,;\, z_0,\, r)=a\log r+b \quad (r_1\leqq r\leqq r_2)\,. \tag{3}$$

Hence from (2), (3), $L(u\,;\, z_0,\, r)$ is a convex function of $\log r$ and since $L(u\,;\, z,\, r)$ is bounded upwards, it is continuous.

Theorem II. 10. *If* $|z-z_0|\leqq r_0$ *is contained in* D, *then* $L(u\,;\, z_0,\, r)$ *is an increasing function of* r $(0<r\leqq r_0)$ *and*

$$\lim_{r\to 0} L(u;\, z_0,\, r)=u(z_0)\,.$$

Proof. Let $r_1<r_2\leqq r_0$ and we define $H_k(z)$ for $|z-z_0|\leqq r_2$, then

$$L(H_k;\, z_0,\, r_2)=H_k(z_0)=L(H_k;\, z_0,\, r_1)\geqq L(u;\, z_0,\, r_1)\,,$$

so that for $k\to\infty$, we have

$$L(u;\, z_0,\, r_2)\geqq L(u;\, z_0,\, r_1) \qquad (r_2>r_1)\,, \tag{1}$$

hence $L(u;\, z_0,\, r)$ is an increasing function of r.

From $u(z_0)\leqq L(u;\, z_0,\, r)$, we have $u(z_0)\leqq \lim_{r\to 0} L(u;\, z_0,\, r)$ and from the upper semi-continuity of u, $\lim_{r\to 0} L(u;\, z_0,\, r)\leqq u(z_0)$, so that

$$\lim_{r\to 0} L(u;\, z_0,\, r)=u(z_0)\,. \tag{2}$$

Theorem II. 11. *If* $|z-z_0|\leqq r_0$ *is contained in* D, *then* $A(u\,;\, z_0,\, r)$ *is an increasing convex function of* $\log r$ $(0<r\leqq r_0)$ *and*

$$u(z_0)\leqq A(u\,;\, z_0,\, r)\leqq L(u\,;\, z_0,\, r)\,,$$

$$\lim_{r\to 0} A(u\,;\, z_0,\, r)=u(z_0)\,.$$

Proof. For $0<r\leqq r_0$,

$$A(u\,;\, z_0,\, r)=\frac{2}{r^2}\int_0^r L(u;\, z_0,\, \rho)\rho d\rho\,. \tag{1}$$

Since $L(u;\, z_0,\, r)$ is a continuous function of r and

$$\frac{2}{r^2}\int_{\frac{(k-1)r}{n}}^{\frac{kr}{n}} \rho d\rho=\frac{2k}{n^2}-\frac{1}{n^2}\,,$$

we have by (1),

$$A(u;\, z_0,\, r)=\lim_{n\to\infty}\sum_{k=1}^{n} L\left(u;\, z_0,\, \frac{kr}{n}\right)\frac{2k}{n^2}. \tag{2}$$

Since $L\left(u\,;\, z_0,\, \dfrac{kr}{n}\right)\dfrac{2k}{n^2}$ is an increasing convex function of $\log r$.

$A(u; z_0, r)$ is an increasing convex function of $\log r$.

Since $L(u; z_0, r)$ is an increasing function of r, we have from (1),

$$u(z_0) \leqq A(u; z_0, r) \leqq L(u; z_0, r),$$

so that by Theorem II. 10,

$$\lim_{r \to 0} A(u; z_0, r) = u(z_0). \tag{3}$$

Remark. From $\lim_{r \to 0} A(u; z_0, r) = u(z_0)$ and the upper semi-continuity of u, we have

$$u(z_0) = \overline{\lim_{z \to z}} u(z). \tag{4}$$

Theorem II. 12. *If $u(z)$ is subharmonic in D and attains a local maximum at a point $z_0 \in D$, then $u \equiv$ const. in a neighbourhood of z_0.*

Proof. Let $|z-z_0| \leqq r_0$ be contained in D and suppose that $u(z) \leqq u(z_0)$ in $|z-z_0| \leqq r_0$, then $A(u; z_0, r_0) \leqq u(z_0)$, but $u(z_0) \leqq A(u; z_0, r_0)$, so that $u(z_0) = A(u; z_0, r_0)$. If there exists z_1 in $|z-z_0| < r_0$, such that $u(z_1) < u(z_0)$, then by the upper semi-continuity of u, $u(z) < u(z_0) - \varepsilon$ $(\varepsilon > 0)$ in a neighbourhood U of z_1. Since U has a positive measure, $A(u; z_0, r_0) < u(z_0)$, which is absurd. Hence $u(z) = u(z_0)$ in $|z-z_0| < r_0$.

Theorem II. 13. *Let $u(z)$ be defined in D and satisfy the following conditions* (i), (ii), (iii):

(i) $-\infty \leqq u < \infty$, $u \not\equiv -\infty$.

(ii) *u is upper semi-continuous in D.*

(iii) *For any $z_0 \in D$, there corresponds $r_0 = r(z_0)$, such that $|z-z_0| \leqq r_0$ is contained in D and for any $0 \leqq r \leqq r_0$,*

$$u(z_0) \leqq L(u; z_0, r) \quad (\text{or } u(z_0) \leqq A(u; z_0, r)).$$

Then $u(z)$ is subharmonic in D.

Proof. Let $D_1 \subset D$ be a Jordan domain, such that $\overline{D}_1 \subset D$, whose boundary Γ consists of a finite number of disjoint Jordan curves. Let $H(z)$ be harmonic in D_1, such that $H \geqq u$ on Γ, then we have to prove that $H \geqq u$ in D_1. Since $v(z) = u(z) - H(z)$ is upper semi-continuous, it attains its maximum M in $\overline{D}_1$, so that let E be the set of z, such that $v(z) = M$, then E is a closed set. Suppose that $M > 0$, then since $v \leqq 0$ on Γ, E contains a boundary point $z_0 \in D_1$, where $v(z_0) = M$. Let $|z-z_0| \leqq r_0$ be contained in D_1, then by the condition (iii), for any $0 \leqq r \leqq r_0$,

$$\begin{aligned} M = v(z_0) = u(z_0) - H(z_0) &\leqq L(u; z_0, r) - L(H, z_0, r) \\ &= L(u-H, z_0, r) = L(v; z_0, r). \end{aligned}$$

Since z_0 is a boundary point of E, for a suitable $0 < r \leqq r_0$, there is an arc of $|z-z_0| = r$, which lies outside of E, on which $v < M - \varepsilon$ $(\varepsilon > 0)$, so that $M=L(v; z_0, r)<M$, which is absurd. Hence $M\leqq 0$, or $H(z)\geqq u(z)$ in D_1.

Similarly, we can prove that if $u(z_0)\leqq A(u; z_0, r)$, then u is subharmonic in D.

Theorem II. 14. *If $u_n(z)\,(n=1, 2, \cdots)$ are subharmonic in D and $\lim\limits_n u_n(z)=u(z)$ uniformly in the wider sense in D, then $u(z)$ is subharmonic in D.*

Proof. $u(z)$ is upper semi-continuous and from $u_n(z_0)\leqq L(u_n; z_0, r)$, we have $u(z_0)\leqq L(u; z_0, r)$.

Theorem II. 15. *If $u_i(z)\,(i=1, 2, \cdots, n)$ are subharmonic in D, then*

$$u(z)=\mathrm{Max}\,(u_1(z), \cdots, u_n(z))$$

is subharmonic in D.

Proof. Since $u_i(z)\leqq K$ $(i=1, 2, \cdots, n)$ on a bounded closed set in D, $u<\infty$ and $u(z)$ is upper semi-continuous and for some i,

$$u(z_0)=u_i(z_0)\leqq L(u_i; z_0, r)\leqq L(u; z_0, r)\,.$$

Theorem II. 16. *If $u_n(z)\,(n=1, 2, \cdots)$ are subharmonic in D and $u_1(z)\geqq u_2(z)\geqq\cdots\geqq u_n(z)\to u(z)$, then $u(z)$ is subharmonic in D.*

Proof. We can prove easily that $u(z)$ is upper semi-continuous and if $u(z_0)>-\infty$, then $-\infty<u(z_0)\leqq u_n(z_0)\leqq L(u_n; z_0, r)$, so that

$$-\infty<u(z_0)\leqq L(u; z_0, r)\,.$$

Theorem II. 17. *If $u(z)$ has continuous partial derivatives of the second order and $\Delta u=\dfrac{6^2 u}{\partial x^2}+\dfrac{\partial^2 u}{\partial y^2}>0$ in D, then $u(z)$ is subharmonic in D.*

Similarly, if $\Delta u<0$ in D, then $u(z)$ is superharmonic in D.

Proof. Let $D_1\subset D$ be a Jordan domain and Γ be its boundary. Let $H(z)$ be harmonic in D_1, such that $H\geqq u$ on Γ, then we have to prove that $H\geqq u$ in D_1. Let $v(z)=u(z)-H(z)$, then $v\leqq 0$ on Γ and $\Delta v=\Delta u-\Delta H=\Delta u>0$. Suppose that v has a positive maximum at an inner point $z_0\in D_1$, then $\Delta v\leqq 0$ at z_0, which contradicts $\Delta v>0$. Hence $v\leqq 0$, or $H(z)\geqq u(z)$ in D_1.

Theorem II. 18. *Let $u(z)$ have continuous partial derivatives of the second order in D. Then $u(z)$ is subharmonic in D, when and only when $\Delta u\geqq 0$ in D.*

Proof. If $u(z)$ is subharmonic in D, then we shall prove that $\Delta u\geqq 0$. Suppose that $\Delta u<0$ at $z_0\in D$, then $\Delta u<0$ in a neighbourhood U of z_0, so that u is superharmonic in U, hence u is harmonic in U, so that

$\Delta u=0$, which is absurd. Hence $\Delta u \geqq 0$ in D.

Next suppose that $\Delta u \geqq 0$ in D and put $u^*=u+\varepsilon(x^2+y^2)$ $(\varepsilon>0)$, then $\Delta u^*>0$, so that u^* is subharmonic, hence $u=\lim_{\varepsilon\to 0} u^*$ is subharmonic in D.

5. Approximation theorem

Theorem II. 19. *Let $u(z)$ be subharmonic in D and D_1 be a subdomain of D, such that $\overline{D}_1\subset D$, then there exists $\rho>0$, such that for any $z_0\in D_1$, a disc: $|z-z_0|\leqq 2\rho$ is contained in D. If we put for $z\in D_1$,*

$$A_\rho(u;z)=\frac{1}{\pi\rho^2}\iint_{|\zeta-z|\leq\rho} u(\zeta)rdrd\theta, \qquad \zeta=z+re^{i\theta},$$

then $A(u;z)$ is a continuous subharmonic function of z in D_1.

Proof. Since $u(z)$ is integrable on any bounded measurable set E in D_1, the integral of u on E tends to zero, with $mE\to 0$, so that $\lim_{z\to z_0} A_\rho(u;z)=A_\rho(u;z_0)$ $(z_0\in D_1)$, hence $A_\rho(u;z)$ is a continuous function of z in D_1.

Next we shall prove that $A_\rho(u;z)$ is subharmonic in D_1. Let $z_0\in D_1$ and we assume that $z_0=0$ and put $u(z)=u(x,y)$ $(z=x+iy)$, then if $|z|<\rho$,

$$u(x,y)\leqq\frac{1}{\pi\rho^2}\iint_{\xi^2+\eta^2\leq\rho^2} u(x+\xi,y+\eta)d\xi\,d\eta\,,$$

so that by taking the integral mean on $|z|\leqq\rho$, we have

$$A_\rho(u,0)=\frac{1}{\pi\rho^2}\iint_{x^2+\eta^2\leq\rho^2} u(x,y)dxdy\leqq\frac{1}{\pi\rho^2}\iint_{x^2+y^2\leq\rho^2} dxdy\frac{1}{\pi\rho^2}\iint_{\xi^2+\eta^2\leq\rho^2} u(x+\xi,y+\eta)d\xi d\eta.$$

Since $u\leqq$const. in D_1, we may interchange the order of integration, so that

$$A_\rho(u;0)\leqq\frac{1}{\pi\rho^2}\iint_{\xi^2+\eta^2\leq\rho^2} d\xi\,d\eta\frac{1}{\pi\rho^2}\iint_{x^2+y^2\leq\rho^2} u(\xi+x,\eta+y)dx\,dy\,.$$

Since $A_\rho(u;\xi,\eta)=\frac{1}{\pi\rho^2}\iint_{x^2+y^2\leq\rho^2} u(\xi+x,\eta+y)dxdy$, we have

$$A_\rho(u,0)\leqq\frac{1}{\pi\rho^2}\iint_{\xi^2+\eta^2\leq\rho^2} A_\rho(u;\xi,\eta)d\xi d\eta\,.$$

Hence by Theorem II. 13, $A_\rho(u,z)$ is subharmonic in D_1. ——

Suppose that for any $z_0\in D_1$, a disc $|z-z_0|\leqq n\rho$ is contained in D and put

$$A_\rho^{(1)}(u;z)=A_\rho(u;z),\ A_\rho^{(2)}(u;z)=A_\rho(A_\rho^{(1)}(u;z);z),\ \cdots,$$
$$A_\rho^{(n)}(u;z)=A_\rho(A_\rho^{(n-1)}(u;z);z)\,, \tag{1}$$

then $A_\rho^{(n)}(u; z)$ is a continuous subharmonic function in D_1 and

$$u(z) \leqq A_\rho^{(1)}(u; z) \leqq A_\rho^{(2)}(u; z) \leqq \cdots \leqq A_\rho^{(n)}(u; z). \tag{2}$$

Since by Theorem II. 12, $u(z) \leqq A_\rho(u; z) \leqq A_{\rho_1}(u; z)$ $(\rho<\rho_1)$, we have

Theorem II. 20. $u(z) \leqq A_\rho^{(n)}(u; z) \leqq A_{\rho_1}^{(n)}(u; z)$ $(\rho<\rho_1)$,

$$\lim_{\rho\to 0} A_\rho^{(n)}(u; z) = u(z) .$$

Theorem II. 21. **(Approximation theorem).** *Let $u(z)$ be subharmonic in D and put for $z \in D_1$, $u_\rho^{(3)}(z) = A_\rho^{(3)}(u; z)$, then $u_\rho^{(3)}(z)$ is subharmonic in D_1 and has continuous partial derivatives of the second order and*

$$u(z) \leqq u_\rho^{(3)}(z), \qquad \lim_{\rho\to 0} u_\rho^{(3)}(z) = u(z) \quad \textit{by decreasing.}$$

Proof. Let $f(z)$ be a continuous function in D and put

$$A_\rho(f; z) = \frac{1}{\pi\rho^2} \int_0^\rho \int_0^{2\pi} f(z+re^{i\theta})\, r dr d\theta, \qquad z=x+iy , \tag{1}$$

then

$$\frac{\partial A_\rho(f; z)}{\partial x} = \frac{1}{\pi\rho} \int_0^{2\pi} f(z+\rho e^{i\theta}) \cos\theta \, d\theta, \qquad \frac{\partial A_\rho(f; z)}{\partial y} = \frac{1}{\pi\rho} \int_0^{2\pi} f(z+\rho e^{i\theta}) \sin\theta d\theta.$$

Hence if $f(z)$ is continuous, then $A_\rho(f; z)$ has continuous partial derivatives of the first order.

If $f(z)$ has continuous partial derivatives of the first order f_x, f_y, then by (1),

$$\frac{\partial A_\rho(f; z)}{\partial x} = A_\rho(f_x; z), \qquad \frac{\partial A_\rho(f; z)}{\partial y} = A_\rho(f_y; z) , \tag{2}$$

and since f_x, f_y are continuous, $\dfrac{\partial A_\rho(f_x; z)}{\partial x}$, $\dfrac{\partial A_\rho(f_y; z)}{\partial y}$ has continuous partial derivatives of the first order, so that $A_\rho(f; z)$ has continuous partial derivatives of the second order.

If we apply this result to $f(z) = A_\rho^{(1)}(u; z)$, then since $A_\rho^{(1)}(u; z)$ is continuous, $A_\rho^{(2)}(u; z)$ has continuous partial derivatives of the first order, so that $u_\rho^{(3)}(z) = A_\rho^{(3)}(u; z)$ has continuous partial derivatives of the second order.

Theorem II. 22. *Let $u(z)$ be subharmonic in D and $D_1 \subset D$ be a subdomain of D, such that for any $z_0 \in D_1$, a disc: $|z-z_0| \leqq 3\rho_0$ is contained in D. Let E be a bounded closed set in D_1, then for any $0<\rho \leqq \rho_0$,*

$$0 \leqq I(\rho) = \iint_E \Delta u_\rho^{(3)} \, dx dy \leqq M = \text{const..}$$

Proof. By Borel's covering theorem, we may assume that E is a disc: $|z-z_0| \leqq r_0$. Since $u_\rho^{(3)}$ is subharmonic, $\Delta u_\rho^{(3)} \geqq 0$, so that $I(\rho) \geqq 0$.

To prove that $I(\rho)\leqq M$, it suffices to prove that $\overline{\lim_{\rho\to 0}}\, I(\rho)<\infty$. Since $L(u_\rho^{(3)}; z_0, r)$ is a convex function of $\log r$, we have by the Green's formula,

$$I(\rho)=\iint_{|z-z_0|\leqq r_0} \Delta u_\rho^{(3)}\, dxdy=\int_0^{2\pi} \frac{\partial u_\rho^{(3)}}{\partial r_0} r_0\, d\theta=r_0 \frac{\partial}{\partial r_0}\int_0^{2\pi} u_\rho^{(3)}\, d\theta$$

$$=2\pi \frac{dL(u_\rho^{(3)}; z_0, r_0)}{d\log r_0}\leqq 2\pi\ \frac{L(u_\rho^{(3)}; z_0, 2r_0)-L(u_\rho^{(3)}; z_0, r_0)}{\log 2r_0-\log r_0},$$

so that

$$\overline{\lim_{\rho\to 0}}\, I(\rho)\leqq 2\pi \frac{L(u; z_0, 2r_0)-L(u; z_0, r_0)}{\log 2}<\infty\, .$$

6. Logarithmic potential

Let E be a bounded closed set on the z-plane and $d\mu(a)\geqq 0$ be a positive mass distribution[1] of finite total mass on E, then

$$u(z)=\int_E \log \frac{1}{|z-a|}\, d\mu(a) \tag{1}$$

is called *a logarithmic potential, due to the mass distribution* μ.

Theorem II. 23. *$u(z)$ is lower semi-continuous and superharmonic for* $|z|<\infty$.

Proof. $u(z)$ is harmonic outside of E, so that $u\not\equiv\infty$. Let E be contained in $|z|\leqq R$, then for any $a\in E$, $|z-a|\leqq|z|+R$, so that

$$\int_E \log \frac{1}{|z-a|}\, d\mu(a) \geqq \mu(E) \log \frac{1}{|z|+R}>-\infty\, ,$$

hence

$$-\infty<u\leqq\infty\, .$$

For a fixed $\rho>0$, we define $|z-a|_\rho$ by

$$|z-a|\ =|z-a|, \text{ if } |z-a|\geqq\rho\, , \quad |z-a|_\rho=\rho, \text{ if } |z-a|\leqq\rho\, , \tag{1}$$

then $\log\dfrac{1}{|z-a|_\rho}$ is a continuous function of two variables z and a and by Theorem II. 13, we see easily that $\log\dfrac{1}{|z-a|_\rho}$ is a superharmonic function of z. We put

$$u_\rho(z)=\int_E \log \frac{1}{|z-a|_\rho}\, d\mu(a)\, , \tag{2}$$

then $u\ (z)$ is a continuous function of z and since $|z-a|_{\rho_1}\leqq|z-a|_\rho$ $(\rho_1<\rho)$,

$$u_\rho(z)\leqq u_{\rho_1}(z), \text{ if } 0<\rho_1<\rho\, . \tag{3}$$

---

1) In the following, a positive mass distribution denotes a totally additive set function $\geqq 0$, defined for Borel sets.

We decompose E into a finite number of disjoint sub-sets E_i $(i=1, 2, \cdots, n)$ of diameter $<\delta$ and take $a_i \in E_i$, then

$$u_\rho(z)=\lim_{\delta\to 0}\sum_{i=1}^{n}\log\frac{1}{|z-a_i|_\rho}\mu(E_i) . \tag{4}$$

$\log\dfrac{1}{|z-a_i|_\rho}\mu(E_i)$ and hence $\sum_{i=1}^{n}\log\dfrac{1}{|z-a_i|_\rho}\mu(E_i)$ is superharmonic, so that that $u_\rho(z)$ is superharmonic and by (3), $u_\rho(z)\to u(z)$ $(\rho\to 0)$ by increasing, so that $u(z)$ is lower semi-continuous and, by Theorem II. 16, is superharmonic.

7. Some lemmas

Lemma 1. *Let $f(x)$, $f_n(x)$ $(n=1, 2, \cdots)$ be convex functions of x, defined in $[a, b]$, such that $\lim_n f_n(x)=f(x)$.*

Let $x_0 (a<x_0<b)$ be such a point, that $f'(x_0)$, $f_n'(x_0)$ $(n=1, 2, \cdots)$ exist[1], *then $\lim_n f_n'(x_0)=f'(x_0)$.*

Proof. We may assume that $\lim_n f_n'(x_0)=\alpha$ exists, since otherwise, we take a suitable sub-sequence from n. If $\alpha>f'(x_0)$, then for α_1, such that $\alpha>\alpha_1>f'(x_0)$, $f_n'(x_0)>\alpha_1>f'(x_0)$ $(n\geqq n_0)$, so that since $f_n(x)$ is convex, $f_n(x)-f_n(x_0)\geqq\alpha_1(x-x_0)$ $(x_0<x<b)$, so that $f(x)-f(x_0)\geqq\alpha_1(x-x_0)$, hence $f'(x_0)\geqq\alpha_1$, which is absurd. If $\alpha<f'(x_0)$, then for $\alpha<\alpha_1<f'(x_0)$, $f_n'(x_0)<\alpha_1<f'(x_0)$ $(n\geqq n_0)$, so that $f_n(x_0)-f_n(x)\leqq\alpha_1(x_0-x)$ $(a<x<x_0)$, hence $f(x_0)-f(x)\leqq\alpha_1(x_0-x)$, so that $f'(x_0)\leqq\alpha_1$, which is absurd. Hence $\alpha=f'(x_0)$.

Lemma 2. *Let $f(x)$ be a convex function, defined in $[a, b]$ and x_0, $x_\nu (\nu=1, 2, \cdots)$ be points of (a, b), such that $x_\nu\to x_0$ and $f'(x_0)$, $f'(x_\nu)$ $(\nu=1, 2, \cdots)$ exist, then $\lim_\nu f'(x_\nu)=f'(x_0)$.*

Proof. We put $f_\nu(x)=f(x+x_\nu-x_0)$ and apply Lemma 1. —— Let $u(z)$ be subharmonic in D and z be any point of D, such that a disc $\varDelta_\rho(z)$ about z of radius ρ is contained in D and put

$$L(u; z, r)=\frac{1}{2\pi}\int_0^{2\pi}u(z+re^{i\theta})\,d\theta, \quad 0<r\leqq\rho .$$

If $\dfrac{rdL(u; z, r)}{dr}$ exists, we call $\varDelta_r(z)$ *a regular disc.* Since $L(u; z, r)$ is an increasing convex function of $\log r$, $r\dfrac{dL(u; z, r)}{dr}\geqq 0$ exists, except at most a countable number of values of r.

Lemma 3. *There exists a positive mass distribution $\mu\geqq 0$ in D, such that for any regular disc $\varDelta_r(z)$,*

1) Being convex functions, $f'(x)$, $f_n'(x)$ exist, except at most a countable number of points.

$$\mu(\Delta_r(z)) = r\frac{dL(u;\, z,\, r)}{dr} \geqq 0\,.$$

Proof. For a sufficiently small $\rho>0$, we define $u_\rho^{(3)}(z)$ as Theorem II. 21 and put

$$L(u_\rho^{(3)};\, z,\, r) = \frac{1}{2\pi}\int_0^{2\pi} u_\rho^{(3)}(z+re^{i\theta})d\theta\,. \tag{1}$$

Then $L(u_\rho^{(3)};\, z,\, r)$ is an increasing convex function of $\log r$ and $L(u_\rho^{(3)};\, z,\, r) \to L(u;\, z,\, r)$ $(\rho\to 0)$ by decreasing, so that by Lemma 1, for a regular disc $\Delta_r(z)$,

$$\lim_{\rho\to 0} r\frac{dL(u_\rho^{(3)};\, z,\, r)}{dr} = r\frac{dL(u;\, z,\, r)}{dr}\,. \tag{2}$$

Let Ω be any sub-domain of D, such that $\bar{\Omega}\subset D$ and whose boundary Γ consists of a finite number of disjoint analytic Jordan curves. We choose $\rho_0>0$, such that for any $z\in\Omega$, $\Delta_{\rho_0}(z)$ is contained in D.

Let $0<\rho\leqq\rho_0$, then since $\Delta u_\rho^{(3)}\geqq 0$, we define a positive mass $\mu_\rho\geqq 0$ in Ω by

$$0\leqq\mu_\rho(\Omega) = \frac{1}{2\pi}\iint_\Omega \Delta u_\rho^{(3)}\, dxdy = \frac{1}{2\pi}\int_\Gamma \frac{\partial u_\rho^{(3)}}{\partial\nu} ds\,, \tag{3}$$

where ν is the outer normal of Γ. Then by Theorem II. 22,

$$0\leqq\mu_\rho(\Omega)\leqq M=M(\Omega), \quad 0\leqq\rho\leqq\rho_0\,, \tag{4}$$

where $M(\Omega)$ is a constant, which depends on Ω only.

$\mu_\rho(\Omega)$ is a finitely additive function of a domain Ω. Let e be any bounded set in D. We cover e by at most a countable number of regular discs $\{\Delta_{r_\nu}(z_\nu)\}$ and put

$$\mu_\rho^*(e) = \inf \sum \mu_\rho(\Delta_{r_\nu}(z_\nu))\,, \tag{5}$$

then μ_ρ^* is a Carathéodory's outer measure, so that it is a totally additive set function for Borel sets $e\subset D$. As the proof of Theorem II. 2, we have $\mu_\rho^*(\Delta_r(z)) = \mu_\rho(\Delta_r(z))$ for a regular disc $\Delta_r(z)$, contained in D. For a Borel set e, we write $\mu_\rho(e)$ instead of $\mu_\rho^*(e)$. In virtue of (4), by Theorem II. 4, we can find $\rho_\nu\to 0$, such that μ_{ρ_ν} converges to a totally additive set function $\mu\geqq 0$, defined for Borel sets.

We shall prove that if $\Delta_r(z)$ is a regular disc, then its boundary Γ does not contain a positive μ-mass. Now by (2), if $\Delta_r(z)$ is a regular disc,

$$\mu_\rho(\Delta_r(z)) = \frac{1}{2\pi}\int_\Gamma \frac{\partial u_\rho^{(3)}}{\partial\nu} ds = r\frac{dL(u_\rho^{(3)};\, z,\, r)}{dr} \to r\frac{dL(u;\, z,\, r)}{dr} \quad (\rho\to 0). \tag{6}$$

By Lemma 2, for any $\varepsilon>0$, we can choose regular values $r_1<r_2$ $(r_1<r<r_2)$, such that

$$0 \leqq r_2 \frac{dL(u;\, z,\, r_2)}{dr_2} - r_1 \frac{dL(u;\, z,\, r_1)}{dr_1} < \varepsilon\,.$$

Hence by (6),

$$0 \leqq \mu_\rho(\varDelta_{r_2}(z)) - \mu_\rho(\varDelta_{r_1}(z)) \leqq \varepsilon, \quad \text{if } 0 < \rho \leqq \rho(\varepsilon)\,,$$

so that

$$0 \leqq \mu(\varDelta_{r_2}(z)) - \mu(\varDelta_{r_1}(z)) \leqq \varepsilon\,.$$

Since $\varepsilon > 0$ is arbitrary, the boundary Γ of $\varDelta_r(z)$ does not contain a positive μ-mass, so that by Theorem II. 4, $\mu_{\rho_\nu}(\varDelta_r(z)) \to \mu(\varDelta_r(z))$ $(\rho_\nu \to 0)$, hence by (6), for a regular disc $\varDelta_r(z)$,

$$\mu(\varDelta_r(z)) = r \frac{dL(u;\, z,\, r)}{dr} \geqq 0\,. \tag{7}$$

8. F. Riesz' decomposition theorem

1. We shall prove the following

Theorem II. 24. (**F. Riesz' decomposition theorem**). *Let $u(z)$ be subharmonic in a domain D, then there exists a unique positive mass distribution $\mu \geqq 0$, defined for Borel sets in D, such that for any sub-domain Ω of D, such that $\overline{\Omega} \subset D$,*

$$u(z) = v(z) - \int_\Omega \log \frac{1}{|z-a|}\, d\mu(a), \qquad z \in \Omega\,,$$

where $v(z)$ is harmonic in Ω. F. Riesz [2].

Proof. (Tsuji [28]). Let μ be the positive mass distribution in D, defined by Lemma 3 and we define $v(z)$, $w(z)$ for $z \in \Omega$ by

$$u(z) = v(z) - \int_\Omega \log \frac{1}{|z-a|}\, d\mu(a), \quad w(z) = \int_\Omega \log \frac{1}{|z-a|}\, d\mu(a)\,. \tag{1}$$

$$u(z) = v(z) - w(z), \qquad z \in \Omega\,.$$

We shall prove that $v(z)$ is harmonic in Ω. We choose $\rho_0 > 0$ so small that for any $z \in \Omega$, a disc $\varDelta_{\rho_0}(z)$ is contained in D and let

$$L(w;\, z,\, r) = \frac{1}{2\pi} \int_0^{2\pi} w(z + re^{i\theta})\, d\theta, \qquad 0 < r \leqq \rho_0, \qquad z \in \Omega\,. \tag{2}$$

Suppose that $z=0$ belongs to Ω, so that

$$L(w;\, 0,\, r) = \frac{1}{2\pi} \int_0^{2\pi} w(re^{i\theta})\, d\theta, \qquad 0 < r \leqq \rho_0\,. \tag{3}$$

Let Ω be contained in $|z| < R$ and put

$$\omega(r) = \int_{|a|<r} d\mu(a) \qquad (0 < r < R)\,. \tag{4}$$

Since

$$\frac{1}{2\pi}\int_0^{2\pi} \log|re^{i\theta}-a|\,d\theta = \mathrm{Max}\,(\log r,\ \log|a|)\,,$$

we have

$$\begin{aligned} L(w;\,0,\,r) &= -\frac{1}{2\pi}\int_\Omega d\mu(a)\int_0^{2\pi}\log|re^{i\theta}-a|d\theta \\ &= -\log r\,\omega(+0)-\int_{+0}^{R}\mathrm{Max}(\log r,\ \log t)d\omega(t) \\ &= -\log r\,\omega(+0)-\log r\int_{+0}^{r}d\omega(t)-\int_r^R \log t\,d\omega(t) \\ &= -\omega(R)\log R+\int_r^R\frac{\omega(t)dt}{t}\,, \end{aligned}$$

or

$$L(w;\,0,\,r) = -\mu(\Omega)\log R+\int_r^R\frac{\omega(t)dt}{t}\,. \tag{5}$$

If $\varDelta_r(0)$ is a regular disc, then its boundary $|z|=r$ does not contain a positive μ-mass, so that $\omega(t)$ is continuous at $t=r$, hence we have from (5)

$$-r\frac{dL(w,\,0,\,r)}{dr}=\omega(r)=\mu(\varDelta_r(0))=r\frac{dL(u;\,0,\,r)}{dr}\ (0<r<\rho_0)\,. \tag{6}$$

Hence (6) holds, except at most a countable number of values of r.

Being a convex function of $\log r$, $-L(w;\,0,\,r)$, $L(u;\,0,\,r)$ are absolutely continuous functions of r in any closed interval, contained in $(0,\,\rho_0)$ and by (6), their derivatives coincide almost everywhere, so that $L(u;\,0,\,r)+L(w;\,0,\,r)\equiv$ const., hence

$$L(v;\,0,\,r)=L(u;\,0,\,r)+L(w;\,0,\,r)\ \equiv\text{const.}\ (0<r\leqq\rho_0).$$

Similarly for any $z\in\Omega$,

$$L(v;\,z,\,r)=L(u;\,z,\,r)+L(w;\,z,\,r)\equiv\text{const.}\quad(0<r\leqq\rho_0)\,, \tag{7}$$

where const. depends on z. Since

$$A_\rho(u;\,z)=\frac{2}{\rho^2}\int_0^\rho L(u;\,z,\,r)rdr,\qquad A_\rho(w;\,z)=\frac{2}{\rho^2}\int_0^\rho L(w;\,z,\,r)\,rdr\,,$$

we have

$$A_\rho(v,\,z)=A_\rho(u;\,z)+A_\rho(w;\,z)\equiv\text{const.}\quad(0<\rho\leqq\rho_0)\,.$$

Since u and $-w$ are subharmonic, $\lim_{\rho\to 0} A\ (u;\,z)=u(z)$, $\lim_{\rho\to 0} A_\rho(w;\,z)=w(z)$, so that

$$A_\rho(v;\,z)=\lim_{\rho\to 0} A_\rho(v;\,z)=u(z)+w(z)=v(z)\ (0<\rho\leqq\rho_0)\,. \tag{8}$$

Since $A_\rho(v;\,z)$ is a continuous function of z, $v(z)$ is a continuous function of z and $A_\rho(v;\,z)=v(z)$, so that $v(z)$ is harmonic in Ω.

Next we shall prove that such μ is unique. Suppose that there exist μ_1, μ_2, such that

$$u(z)=v_1(z)-\int_\Omega \log\frac{1}{|z-a|}\,d\mu_1(a)=v_2(z)-\int_\Omega \log\frac{1}{|z-a|}\,d\mu_2(a), \qquad z\in\Omega,$$

where $v_1(z)$, $v_2(z)$ are harmonic in Ω. If we put $v=v_1-v_2$, $\mu=\mu_1-\mu_2$, then

$$v(z)=-\int_\Omega \log\frac{1}{|z-a|}\,d\mu(a) \tag{9}$$

is harmonic in Ω.

Suppose that $z=0$ belongs to Ω and if we put

$$\omega(r)=\int_{|a|<r} d\mu(a)=\int_{|a|<r} d\mu_1(a)-\int_{|a|<r} d\mu_2(a)=\omega_1(r)-\omega_2(r)\,,$$

then by (9), (5),

$$v(0)=\frac{1}{2\pi}\int_0^{2\pi} v(re^{i\theta})d\theta=L(v;\,0,\,r)=\mu(\Omega)\log R-\int_r^R \frac{\omega(t)}{t}\,dt\,. \tag{10}$$

Being a difference of two increasing functions, $\omega(r)$ is continuous, except at most a countable number of values of r. If $\omega(r)$ is continuous at r, then by differentiating the both sides of (10), we have $\omega(r)=0$, or $\omega_1(r)=\omega_2(r)$, hence $\mu_1(\Delta_r(0))=\mu_2(\Delta_r(0))$.

Since $\omega_i(r+0)=\omega_i(r)$ $(i=1, 2)$, we have $\mu_1(\Delta_r(0))=\mu_2(\Delta_r(0))$ for any $r(0<r<\rho)$, hence by Theorem II. 3, $\mu_1\equiv\mu_2$. ——

From the proof, we have

Theorem II. 25. *Let $\mu=\mu_1-\mu_2$ be a difference of two positive mass distributions $\mu_1\geqq 0$, $\mu_2\geqq 0$, defined in a domain D, such that $\mu_1(D)<\infty$, $\mu_2(D)<\infty$ and put*

$$u(z)=\int_D \log\frac{1}{|z-a|}\,d\mu(a)\,.$$

*Let Δ be a disc, contained in D. If $u(z)$ is harmonic in Δ, or $u\equiv$*const. *almost everywhere in Δ, then $\mu(\Delta)=0$.*

As an application of the decomposition theorem, we shall prove

Theorem II. 26. *Let $u(z)$ be subharmonic in D and C be an analytic curve in D, then $u(z)$ is integrable on C.*

Proof. Let C be contained in a compact sub-domain Ω of D and let

$$u(z)=v(z)-\int_\Omega \log\frac{1}{|z-a|}\,d\mu(a), \qquad z\in\Omega\,.$$

Then

$$\int_C |u(z)||dz|\leqq\int_C |v(z)|\,|dz|+\int_\Omega d\mu(a)\int_C\left|\log\frac{1}{|z-a|}\right||dz|<\infty\,. \text{ ——}$$

2. We assumed that D is a plane region, but if D is a domain,

contained in a closed Riemann surface, the proof is the same and we have

Theorem II. 27. *Let D be a domain, contained in a closed Riemann surface and u(z) be subharmonic in D. Then there exists a unique positive mass distribution $\mu\geqq 0$ in D, such that for any compact sub-domain Ω of D,*

$$u(z)=v(z)-\int_{\Omega} g(z, a)\,d\mu(a), \qquad z\in\Omega\,,$$

where v(z) is harmonic in Ω and g(z, a) is the Green's function of Ω, with a as its pole.

Theorem II. 28. *If the boundary Γ of Ω consists of a finite number of disjoint analytic Jordan curves, then*

$$u(z)=\frac{1}{2\pi}\int_{\Gamma} u(\zeta)\frac{\partial g(\zeta, z)}{\partial\nu}\,|d\zeta|-\int_{\Omega} g(z, a)\,d\mu(a), \qquad z\in\Omega\,,$$

where ν is the inner normal of Γ at ζ.

Proof. We choose a compact sub-domain Ω_0 of D, such that $\bar{\Omega}\subset\Omega_0$, then in Ω_0,

$$u(z)=v_0(z)-\int_{\Omega_0} g_0(z, a)\,d\mu(a), \qquad z\in\Omega_0\,, \tag{1}$$

where $v_0(z)$ is harmonic in Ω_0 and $g_0(z, a)$ is the Green's function of Ω_0.

If $z\in\Omega$, then

$$\begin{aligned}\frac{1}{2\pi}\int_{\Gamma} u(\zeta)\frac{\partial g(\zeta, z)}{\partial\nu}\,|d\zeta|&=\frac{1}{2\pi}\int_{\Gamma} v_0(\zeta)\frac{\partial g(\zeta, z)}{\partial\nu}\,|d\zeta|\\ &\quad-\frac{1}{2\pi}\int_{\Omega_0} d\mu(a)\int_{\Gamma} g_0(\zeta, a)\frac{\partial g(\zeta, z)}{\partial\nu}\,|d\zeta|\\ &=v_0(z)-\frac{1}{2\pi}\int_{\Omega_0} d\mu(a)\int_{\Gamma} g_0(\zeta, a)\frac{\partial g(\zeta, z)}{\partial\nu}\,|d\zeta|\\ &=u(z)+\int_{\Omega_0} g_0(z, a)\,d\mu(a)-\frac{1}{2\pi}\int_{\Omega_0} d\mu(a)\int_{\Gamma} g_0(\zeta, a)\frac{\partial g(\zeta, z)}{\partial\nu}\,|d\zeta|\,.\end{aligned} \tag{2}$$

If $a\in\Omega_0-\bar{\Omega}$, then since $g_0(z, a)$ is harmonic in $\bar{\Omega}$,

$$\frac{1}{2\pi}\int_{\Gamma} g_0(\zeta, a)\frac{\partial g(\zeta, z)}{\partial\nu}\,|d\zeta|=g_0(z, a)\,. \tag{3}$$

If $a\in\Gamma$, we choose $a'\in\Omega_0-\bar{\Omega}$ in a small neighbourhood of a, then (3) hold for a', so that if we make $a'\to a$, then (3) hold for a. Hence

$$\frac{1}{2\pi}\int_{\Gamma} g_0(\zeta, a)\frac{\partial g(\zeta, z)}{\partial\nu}\,|d\zeta|=g_0(z, a)\,, \qquad a\in\Omega_0-\Omega\,. \tag{4}$$

If $a\in\Omega$, let γ_a: $|\zeta-a|=r$ be a circle about a and γ_z: $|\zeta-z|=r$ be that about $z(\neq a)$, such that γ_a, γ_z are contained in Ω and disjoint each other.

By the Green's formula ,

$$\int_\Gamma\left(g_0(\zeta,a)\frac{\partial g(\zeta,z)}{\partial\nu}-g(\zeta,z)\frac{\partial g_0(\zeta,a)}{\partial\nu}\right)|d\zeta|+\int_{\gamma_a}\left(g_0(\zeta,a)\frac{\partial g(\zeta,z)}{\partial\nu}\right.$$

$$\left.-g(\zeta,z)\frac{\partial g_0(\zeta,a)}{\partial\nu}\right)|d\zeta|+\int_{\gamma_z}\left((g_0(\zeta,a)\frac{\partial g(\zeta,z)}{\partial\nu}-g(\zeta,z)\frac{\partial g_0(\zeta,a)}{\partial\nu}\right)|d\zeta|=0,$$

where $g(\zeta,z)=0$ in the first integral.

If we make $r\to 0$, then we have easily

$$\frac{1}{2\pi}\int_\Gamma g_0(\zeta,a)\frac{\partial g(\zeta,z)}{\partial\nu}|d\zeta|=g_0(z,a)-g(z,a),\quad a\in\Omega. \tag{5}$$

Hence by (4), (5),

$$\int_{\Omega_0}g_0(z,a)\,d\mu(a)-\frac{1}{2\pi}\int_{\Omega_0}d\mu(a)\int_\Gamma g_0(\zeta,a)\frac{\partial g(\zeta,z)}{\partial\nu}|d\zeta|=\int_{\Omega_0}g_0(z,a)d\mu(a)$$

$$-\int_{\Omega_0-\Omega}g_0(z,a)\,d\mu(a)-\int_\Omega(g_0(z,a)-g(z,a))\,d\mu(a)=\int_\Omega g(z,a)\,d\mu(a),$$

so that by (2),

$$u(z)=\frac{1}{2\pi}\int_\Gamma u(\zeta)\frac{\partial g(\zeta,z)}{\partial\nu}|d\zeta|-\int_\Omega g(z,a)\,d\mu(a).\ \text{—}$$

CHAPTER III

POTENTIAL THEORY

1. Maximum principle for logarithmic potential

Let E be a bounded closed set on the z-plane and $\mu \geqq 0$ be a positive mass distribution on E of finite total mass and consider the logarithmic potential:

$$u(z) = \int_E \log \frac{1}{|z-a|} \, d\mu(a) .$$

Then $u(z)$ is lower semi-continuous and superharmonic for $|z| < \infty$.

Theorem III. 1. (**Maximum principle**). *If $u(z) \leqq K$ on E, then $u(z) \leqq K$ for $|z| < \infty$.* Maria [1].

Proof. Let D be the complement of E. Since $u(z)$ is harmonic in D, except at $z=\infty$, where $u=-\infty$, it suffices to prove that

$$\overline{\lim_{z \to a_0}} u(z) \leqq K, \qquad z \in D , \tag{1}$$

where a_0 is any boundary point of D.

Let $U_\rho : |z-a_0| < \rho$ and E_ρ be the part of E, contained in $\overline{U}_\rho$.

Since $u \leqq K$ on E, a single point does not contain a positive mass, so that for any $\varepsilon > 0$, we can choose $\rho > 0$ so small that $\mu(E_\rho) < \varepsilon$. Let D_ρ be the part of D, contained in U_ρ and $z \in D_\rho$. We choose $z_1 \in E_\rho$, such that $|z-z_1| \leqq |z-a|$ for any $a \in E$, then

$$|z_1 - a| \leqq |z-z_1| + |z-a| \leqq 2|z-a| \text{ for any } a \in E_\rho,$$

so that

$$\int_{E_\rho} \log \frac{1}{|z-a|} \, d\mu(a) \leqq \mu(E_\rho) \log 2 + \int_{E_\rho} \log \frac{1}{|z_1-a|} \, d\mu(a) \leqq$$

$$\varepsilon \log 2 + \int_{E_\rho} \log \frac{1}{|z_1-a|} \, d\mu(a)$$

$$= \varepsilon \log 2 + \int_E \log \frac{1}{|z_1-a|} \, d\mu(a) - \int_{E-E_\rho} \log \frac{1}{|z_1-a|} \, d\mu(a) \leqq$$

$$\varepsilon \log 2 + K - \int_{E-E_\rho} \log \frac{1}{|z_1-a|} \, d\mu(a) ,$$

$$\int_{E_\rho} \log \frac{1}{|z-a|} \, d\mu(a) + \int_{E-E_\rho} \log \frac{1}{|z_1-a|} \, d\mu(a) \leqq \varepsilon \log 2 + K . \tag{2}$$

Since $\lim_{z \to a_0} |z_1 - z| = 0$, if z lies sufficiently near to a_0,

$$\int_{E-E_\rho} \log \frac{1}{|z_1-a|}\, d\mu(a) > \int_{E-E_\rho} \log \frac{1}{|z-a|}\, d\mu(a) - \varepsilon ,$$

hence by (2),

$$u(z) = \int_{E_\rho} \log \frac{1}{|z-a|}\, d\mu(a) + \int_{E-E_\rho} \log \frac{1}{|z-a|}\, d\mu(a) < \varepsilon \log 2 + K + \varepsilon .$$

Since $\varepsilon > 0$ is arbitrary, $\overline{\lim}_{z\to a_0} u(z) \leqq K$. ——

Theorem III. 2. *If $u(z)$ is continuous at $a_0 \in E$, considered as a function on E, then $u(z)$ is continuous at a_0, considered as a function in a full neighbourhood of a_0.*

Proof. If we take $K = u(z_1)$ in the above proof, then if $|z-a_0| < \rho_0 < \rho$, $u(z) < \varepsilon \log 2 + u(z_1) + \varepsilon$. Since $u(z)$ is continuous on E at a_0, if ρ_0 is small, $u(z_1) < u(a_0) + \varepsilon$, so that $u(z) < \varepsilon \log 2 + 2\varepsilon + u(a_0)$. Since $\varepsilon > 0$ is arbitrary, $\overline{\lim}_{z\to a_0} u(z) \leqq u(a_0)$. Since $u(z)$ is lower semi-continuous, $u(a_0) \leqq \underline{\lim}_{z\to a_0} u(z)$, so that

$$\lim_{z\to a_0} u(z) = u(a_0) .$$

2. Capacity. Conductor potential

Let E be a bounded closed set on the z-plane and $\mu \geqq 0$ be a positive mass distribution on E of total mass 1. We consider the energy integral:

$$I(\mu) = \iint_E \log \frac{1}{|a-b|}\, d\mu(a) d\mu(b), \qquad \mu(E) = 1 \tag{1}$$

and put

$$V = \inf_\mu I(\mu) \tag{2}$$

We take $K > 1$ so large that $|a-b| \leqq K$ for any $a \in E$, $b \in E$, then $K/|a-b| \geqq 1$, so that

$$I(\mu) = \iint_E \log \frac{K}{|a-b|}\, d\mu(a) d\mu(b) - \log K \geqq -\log K ,$$

hence

$$-\infty < V \leqq \infty .$$

We assume that $V < \infty$. Then by (2), there exists μ_n, such that $I(\mu_n) \to V$, then since $\mu_n(E) = 1$, by Theorem II. 4, we may assume that $\mu_n \to \mu$ and put

$$I(\mu) = \iint_E \log \frac{1}{|a-b|}\, d\mu(a)\, d\mu(b), \qquad \mu(E) = 1 , \tag{3}$$

then

$$V \leqq I(\mu). \tag{4}$$

Let for $\rho > 0$,

$$|a-b|_\rho = |a-b|, \text{ if } |a-b| \geqq \rho, \qquad |a-b|_\rho = \rho, \text{ if } |a-b| \leqq \rho ,$$

then since $\log \frac{1}{|a-b|_\rho}$ is a continuous function of two variables a, b, we have

$$\iint_E \log \frac{1}{|a-b|_\rho} d\, u(a)\, d\mu(b) = \lim_n \iint_E \log \frac{1}{|a-b|_\rho} d\mu_n(a)\, d\mu_n(b)$$

$$\leqq \lim_n \iint_E \log \frac{1}{|a-b|} d\mu_n(a)\, d\mu_n(b) = V .$$

Hence if we make $\rho \to 0$, then we have

$$I(\mu) = \iint_E \log \frac{1}{|a-b|} d\mu(a)\, d\mu(b) \leqq V ,$$

so that by (4)

$$V = I(\mu) = \iint_E \log \frac{1}{|a-b|} d\mu(a)\, d\mu(b), \qquad \mu(E) = 1 . \tag{5}$$

We call μ *the equilibrium distribution of* E and

$$u(z) = \int_E \log \frac{1}{|z-a|} d\mu(a), \qquad \mu(E) = 1 , \tag{6}$$

the conductor potential of E.

The uniqueness of μ will be proved in Theorem III. 31.

The carrier E^* *of* μ is defined as the set of $a \in E$, such that any small neighbourhood U of a contains a positive μ-mass. E^* is a closed sub-set of E. From (6), we have

$$u(z) = \int_{E^*} \log \frac{1}{|z-a|} d\mu(a), \qquad \mu(E^*) = 1 \tag{7}$$

and

$$\int_E u(z)\, d\mu(z) = \int_{E^*} u(z)\, d\mu(z) = V . \tag{8}$$

We define *the capacity* $\gamma(E)$ of E by

$$\gamma(E) = e^{-V}, \qquad V = \log 1/\gamma(E) , \tag{9}$$

so that if $V < \infty$, then $\gamma(E) > 0$ and if $V = \infty$, then $\gamma(E) = 0$.

Let $E_1 \subset E_2$ be two bounded closed sets, then from the definition, we have $V(E_1) \geqq V(E_2)$, so that $\gamma(E_1) \leqq \gamma(E_2)$.

The capacity $\gamma(E)$ of any set E is defined by

$$\gamma(E) = \sup_{F \subset E} \gamma(F) , \tag{10}$$

where F is a bounded closed sub-set of E[1).]

1) The capacity thus defined is called also the inner capacity, since the inner measure of a set is defined similarly. The outer capacity $\gamma^*(E)$ of E is defined by $\gamma^*(E) = \inf_{E \subset O} \gamma(O)$, where O is an open set, which contains E. In this book, capacity means always the inner capacity.

In this book, almost everywhere means, except a set of Lebesgue measure zero and "almost everywhere" means, except a set of capacity zero.

From the definition, we have

Theorem III. 3. *If* $E_1 \subset E_2$, *then* $\gamma(E_1) \leqq \gamma(E_2)$.

Theorem III. 4. *Let* T: $z' = az + b$ *be a linear transformation and* $E_1 = T(E)$, *then* $\gamma(E_1) = |a|\ \gamma(E)$.

Theorem III. 5. *If* E *is a continuum, then* $\gamma(E) > 0$, *so that a set of capacity zero does not contain a continuum.*

Proof. We may assume that E lies on the upper half of the $z = x + iy$-plane and the projection of E on the x-axis is a segment L. Through any point α of L, we draw a parallel to the y-axis and let a be the point of intersection of this parallel with E, such that $\Im a > 0$ is the smallest. We associate this a to α and we define $d\mu(a)$ on E by $d\mu(a) = d\alpha/|L|$, so that $\int_E d\mu(a) = 1$. If α, β correspond to a, b of E, then $|\alpha - \beta| \leqq |a - b|$, so that

$$\log 1/\gamma(E) = V(E) \leqq \iint_E \log \frac{1}{|a-b|}\, d\mu(a)\, d\mu(b) \leqq \frac{1}{L^2} \iint_L \log \frac{1}{|\alpha-\beta|}\, d\alpha\, d\beta < \infty,$$

hence
$$\gamma(E) > 0\,.$$

Theorem III. 6. *Let* E *be a bounded closed set of positive capacity and* μ *be its equilibrium distribution and* E^* *be its carrier, then*

$$\gamma(E^*) = \gamma(E)\,.$$

Proof. Let $\gamma(E) = e^{-V}$, $\gamma(E^*) = e^{-V^*}$, then

$$V = I(\mu) = \iint_{E^*} \log \frac{1}{|a-b|}\, d\mu(a)\, d\mu(b) \geqq V^* \quad \text{by definition,}$$

hence $\gamma(E^*) \geqq \gamma(E)$, but since $E^* \subset E$, $\gamma(E^*) \leqq \gamma(E)$, so that $\gamma(E^*) = \gamma(E)$.

Theorem III. 7. *Let* E *be a bounded closed set of positive capacity and* $\nu \geqq 0$ *be a positive mass distribution on* E *of total mass 1, such that*

$$I(\nu) = \iint_E \log \frac{1}{|a-b|}\, d\nu(a)\, d\nu(b) < \infty\,.$$

Then a Borel sub-set e *of* E, *which is of capacity zero, does not contain a positive* ν-*mass, i, e,* $\nu(e) = 0$.

Proof. We may assume that E is contained in $|z| \leqq 1/2$, then

$\log \dfrac{1}{|a-b|} \geqq 0$, $a \in E$, $b \in E$, so that

$$\iint_e \log \frac{1}{|a-b|}\, d\nu(a)\, d\nu(b) \leqq \iint_E \log \frac{1}{|a-b|}\, d\nu(a)\, d\nu(b) < \infty .$$

Suppose that $\nu(e)>0$, then for a suitable closed sub-set F of e, $\nu(F)>0$, hence

$$\iint_F \log \frac{1}{|a-b|}\, d\nu(a)\, d\nu(b) \leqq \iint_e \log \frac{1}{|a-b|}\, d\nu(a)\, d\nu(b) < \infty ,$$

so that $\gamma(F)>0$, hence $\gamma(e) \geqq \gamma(F)>0$, which is absurd. Hence $\nu(e)=0$.

Theorem III. 8. *If E_n $(n=1, 2, \cdots)$ is a Borel set of capacity zero, then $E=\sum_{n=1}^{\infty} E_n$ is of capacity zero.*

Since a set, which consists of only one point is of capacity zero, a at most countable set is of capacity zero.

Proof. If $\gamma(E)>0$, then there exists a bounded closed sub-set F of E, such that $\gamma(F)>0$, so that if μ is the equilibrium distribution of F,

$$\iint_F \log \frac{1}{|a-b|}\, d\mu(a)\, d\mu(b) < \infty, \qquad \mu(F)=1 .$$

Since $1=\mu(F) \leqq \sum_{n=1}^{\infty} \mu(F.\, E_n)$, $\mu(F.E_n)>0$ for a certain n. We may assume that F is contained in $|z| \leqq 1/2$, then

$$\iint_{F.\, E_n} \log \frac{1}{|a-b|}\, d\mu(a)\, d\mu(b) \leqq \iint_F \log \frac{1}{|a-b|}\, d\mu(a)\, d\mu(b) < \infty .$$

Since $\gamma(F.\, E_n)=0$ and $\mu(F.\, E_n)>0$, this contradicts the above theorem, hence

$$\gamma(E)=0 .$$

Theorem III. 9. *Let E be a bounded closed set and O_n $(n=1, 2, \cdots)$ be an open set, which contains E and is bounded by a finite number of disjoint Jordan curves, such that $O_n \supset O_{n+1}$, $\lim_n O_n=E$, then*

$$\lim_n \gamma(\bar{O}_n)=\gamma(E) .$$

Proof. We put $E_n=\bar{O}_n$ and we may assume that E_n lies in $|z| \leqq 1/2$. Since $E \subset E_{n+1} \subset E_n$, $\gamma(E) \leqq \gamma(E_{n+1}) \leqq \gamma(E_n)$, so that

$$\gamma(E) \leqq \lim_n \gamma(E_n) . \tag{1}$$

First suppose that $\gamma(E)>0$ and μ be the equilibrium distribution of E and μ_n be that of E_n, then

$$\log 1/\gamma(E_n)=V_n=\iint_{E_n} \log \frac{1}{|a-b|}\, d\mu_n(a)\, d\mu_n(b), \qquad \mu_n(F_n)=1 .$$

By Theorem II. 4, we may assume that $\mu_n \to \nu$, $\nu(E)=1$. Since $\log \frac{1}{|a-b|} \geqq 0$, for $a \in E$, $b \in E$, we have by Fatou's lemma and the definition of V,

$$\log 1/\gamma(E) = V \leqq \iint_E \log \frac{1}{|a-b|} d\nu(a)\, d\nu(b)$$

$$\leqq \lim_n \iint_{E_n} \log \frac{1}{|a-b|} d\mu_n(a)\, d\mu_n(b) = \log 1/\lim_n \gamma(E_n)\,,$$

so that

$$\gamma(E) \geqq \lim_n \gamma(E_n)\,, \tag{2}$$

hence by (1), (2), $\lim_n \gamma(E_n) = \gamma(E)\,.$

If $\gamma(E)=0$, then we can prove easily that $\lim_n \gamma(E_n)=0$.

Theorem III. 10. *Let E be a bounded closed set and $m(E)$ be its measure, then*

$$\gamma(E) \geqq \sqrt{m(E)/\pi e}\,.$$

Hence if $\gamma(E)=0$, then $m(E)=0$.

Proof. We may assume that $m(E)>0$. We define R by $\pi R^2 = m(E)$ and let $K: |z| \leqq R$ be a disc, then $m(K)=m(E)$.

$$\iint_K \log \frac{1}{r}\, r\, dr\, d\theta = \int_0^R \int_0^{2\pi} \log \frac{1}{r}\, r\, dr\, d\theta = \pi R^2 \left(\log \frac{1}{R} + \frac{1}{2} \right)$$

$$= m(E) \log \sqrt{\pi e/m(E)}. \tag{1}$$

Now

$$\iint_K \log \frac{1}{r}\, r\, dr\, d\theta = \iint_{K.E} \log \frac{1}{r}\, r\, dr\, d\theta + \iint_{K-K.E} \log \frac{1}{r}\, r\, dr\, d\theta$$

$$\geqq \iint_{K.E} \log \frac{1}{r}\, r\, dr\, d\theta + \log \frac{1}{R} \Big(m(K) - m(K.\,E) \Big). \tag{2}$$

$$\iint_E \log \frac{1}{r}\, r\, dr\, d\theta = \iint_{E.K} \log \frac{1}{r}\, r\, dr\, d\theta + \iint_{E-E.K} \log \frac{1}{r}\, r\, dr\, d\theta$$

$$\leqq \iint_{E.K} \log \frac{1}{r}\, r\, dr\, d\theta + \log \frac{1}{R} (m(E) - m(E.\,K))\,. \tag{3}$$

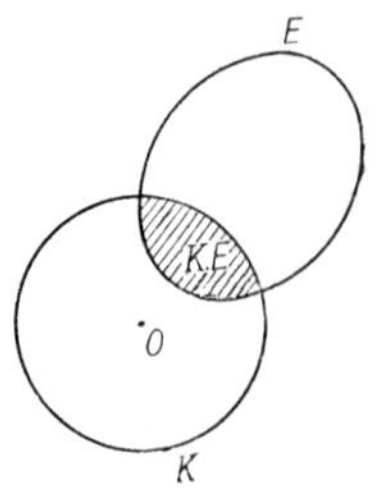

Since $m(E)=m(K)$, we have by (2), (3),

$$\iint_E \log \frac{1}{r}\, r\, dr\, d\theta \leqq \iint_K \log \frac{1}{r}\, r\, dr\, d\theta = m(E) \log \sqrt{\pi e/m(E)}\,. \tag{4}$$

Hence if $d\sigma(a)$ is the surface element, then by (4),

$$\iint \log \frac{1}{|a-b|} d\sigma(a)\, d\sigma(b) \leqq (m(E))^2 \log\sqrt{\pi e/m(E)}\,.$$

Hence if we put $d\mu(a)=d\sigma(a)/m(E)$, then $\mu(E)=1$, so that by the definition of V,

$$\log 1/\gamma(E)=V \leqq \iint_E \log \frac{1}{|a-b|} d\mu(a)\, d\mu(b) \leqq \log\sqrt{\pi e/m(E)},$$

or

$$\gamma(E) \geqq \sqrt{m(E)/\pi e}\,.$$

Theorem III. 11. *Let C be a rectifiable Jordan arc and E be a closed set on C and $L(E)$ be its linear measure, measured by the arc length of C. If $L(E)>0$, then $\gamma(E)>0$, hence if $\gamma(E)=0$, then $L(E)=0$.*

Proof. Let a_0 be one of two end points of C and a be any point of C and $s(a)$ be the arc length of $\widehat{a_0 a}$. Then for almost all a on C, $\lim_{b\to a}\left|\frac{s(b)-s(a)}{b-a}\right|=1$ (Saks [1]), so that if $L(E)>0$, then by Egoroff's theorem, there exists a closed sub-set E_1 of E, such that $L_1=L(E_1)>0$ and $\lim_{b\to a}\left|\frac{s(b)-s(a)}{b-a}\right|=1$ uniformly for $a\in E_1$, hence there exists a constant $K>1$, such that

$$\left|\frac{s(b)-s(a)}{b-a}\right| \leqq K \text{ for any } a\in E_1,\ b\in E_1\,. \tag{1}$$

Let $a\in E_1$ be a fixed point and b be a variable point on E_1 and put $s=s(b)-s(a)$, then similarly as Theorem III. 10, we can prove that

$$\int_{E_1} \log \frac{1}{|s|} ds \leqq \int_{-\frac{L_1}{2}}^{\frac{L_1}{2}} \log \frac{1}{|s|} ds = L_1 \log \frac{2}{L_1} + L_1\,,$$

or

$$\int_{E_1} \log \frac{1}{|s(b)-s(a)|} ds(b) \leqq L_1 \log \frac{2}{L_1} + L_1\,,$$

hence

$$\iint_{E_1} \log \frac{1}{|s(b)-s(a)|} ds(a)\, ds(b) \leqq L_1^2 \log \frac{2}{L_1} + L_1^2\,. \tag{2}$$

If we put $d\mu(a)=ds(a)/L_1$, then $\mu(E_1)=1$ and

$$\iint_{E_1} \log \frac{1}{|s(b)-s(a)|} d\mu(a)\, d\mu(b) \leqq \log \frac{2}{L_1} + 1\,, \tag{3}$$

hence by (1),

$$\log 1/\gamma(E_1)=V_1 \leqq \iint_{E_1} \log \frac{1}{|a-b|} d\mu(a)\, d\mu(b) \leqq \log K+\log \frac{2}{L_1}+1=\log \frac{2Ke}{L_1},$$

so that

$$\gamma(E) \geqq \gamma(E_1) \geqq \frac{L_1}{2Ke}>0 .$$

3. Fundamental theorem

Theorem III. 12. (Fundamental theorem). *Let E be a bounded closed set of positive capacity and $V=1/\gamma(E)$ and $u(z)$ be its conductor potential. Then*

(i) $u(z) \leqq V$ *on the whole z-plane,*

(ii) $u(z)=V$ *"almost everywhere" on E,*

such that $u(z)=V$ on E, except an F_σ-set of capacity zero. Frostman [1].

Proof. First we shall prove that

$$u(z) \geqq V \quad \text{"almost everywhere" on } E .$$

Let A be the set of $z \in E$, such that $u(z)<V$ and suppose that $\gamma(A)>0$. Let A_n be the set of $z \in E$, such that $u(z) \leqq V-\frac{1}{n}$ $(n=1, 2, \cdots)$, then

$$A_1 \subset A_2 \subset \cdots \subset A_n \rightarrow A .$$

Since $u(z)$ is lower semi-continuous, A_n is a closed set.

Since $A=\sum_n A_n$ and $\gamma(A)>0$, by Theorem III. 8, $\gamma(A_n)>0$ for a certain n, hence for a suitable $\varepsilon>0$, there exists a closed sub-set E_1 of E, such that

$$u(z)<V-2\varepsilon \text{ on } E_1, \quad \gamma(E_1)>0 . \tag{1}$$

Let μ be the equilibrium distribution of E and E^* be its carrier, then since $\int_{E^*} u\, d\mu=V$, there exists a point $a_0 \in E^*$, such that $u(a_0)>V-\varepsilon$, hence by (1), a_0 lies outside of E_1 and by the lower semi-continuity of u,

$$u(z)>V-\varepsilon \quad \text{in } U(a_0) , \tag{2}$$

where $U(a_0)$ is a neighbourhood of a_0, which we choose so small that $U(a_0)$ and E_1 have a positive distance. Since $a_0 \in E^*$,

$$\mu(U(a_0))=m>0 . \tag{3}$$

Since $\gamma(E_1)>0$, there exists a positive mass distribution $\sigma \geqq 0$ on E_1, such that

$$I(\sigma)=\iint_{E_1} \log \frac{1}{|a-b|} d\sigma(a)\, d\sigma(b)<\infty, \quad \sigma(E_1)=m>0 . \tag{4}$$

Now we define a mass distribution σ_1 on E by

$$\sigma_1=\sigma \text{ on } E_1, \quad \sigma_1=-\mu \text{ in } U(a_0), \quad \sigma_1=0 \text{ elsewhere} , \tag{5}$$

then

$$\sigma_1(E_1)=m, \quad \sigma(U(a_0))=-m, \quad \sigma_1(E)=0\,,$$

$$I(\sigma_1)=\iint_E \log\frac{1}{|a-b|}\,d\sigma_1(a)\,d\sigma_1(b)\neq\infty\,, \tag{6}$$

and for $0<\eta<1$,

$$\mu+\eta\sigma_1\geqq 0, \qquad \int_E d(\mu+\eta\sigma_1)=1\,.$$

By (1), (2), (6), we have if $\eta>0$ is small,

$$\delta I=I(\mu+\eta\sigma_1)-I(\mu)=2\eta\int_E u d\sigma_1+\eta^2 I(\sigma_1)=2\eta\left(\int_{E_1} u d\sigma_1+\int_{U(a_0)} u d\sigma_1\right)+\eta^2 I(\sigma_1)$$

$$<2\eta[m(V-2\varepsilon)-m(V-\varepsilon)]+\eta^2 I(\sigma_1)=-\eta(2m\,\varepsilon-\eta\, I(\sigma_1))<0\,,$$

which contradicts the definition of $I(\mu)$, hence

$$u(z)\geqq V \quad \text{"almost everywhere" on } E. \tag{7}$$

Next we shall prove that

$$u(z)\leqq V \quad \text{on } E^*\,. \tag{8}$$

Suppose that $u(a_0)>V$ at a point $a_0\in E^*$, then by the lower semi-continuity of u,

$$u(z)>V+\varepsilon \quad \text{in } U(a_0), \quad \varepsilon>0\,. \tag{9}$$

Since $a_0\in E^*$, $\mu(U(a_0))>0$ and since a set of capacity zero does not contain a positive mass (Theorem III. 7), we have by (7),

$$V=\int_{E^*U(a_0)} u d\mu+\int_{E^*-E^*U(a_0)} u d\mu>(V+\varepsilon)\,\mu(U(a_0))+V[1-\mu(U(a_0)]=V+\varepsilon\mu(U(a_0))>V\,,$$

which is absurd, so that $u(z)\leqq V$ on E^*, hence by the maximum principle, $u(z)\leqq V$ on the whole z-plane, so that by (7), $u(z)=V$ "almost everywhere" on E. Since $u(z)$ is lower semi-continuous, the set of $z\in E$, such that $u(z)<V$ is an F_σ-set. ——

By $u(z)\leqq V$ and the lower semi-continuity of u, we have

Theorem III. 13. *If $u(a_0)=V$, then $u(z)$ is continuous at a_0.*

Theorem III. 14. *If a_0 is an inner point of E, then $u(a_0)=V$, hence*

$$\mu(E^0)=0,$$

where E^0 is the open kernel of E.

Proof. Let $U(a_0)$ be a neighbourhood of a_0, which is contained in E^0, then since u is superharmonic,

$$\frac{1}{|U(a_0)|}\iint_{U(a_0)} u\,dxdy \leqq u(a_0)\leqq V\,. \tag{1}$$

Since $u(z)=V$ "almost everywhere", hence almost everywhere in $U(a_0)$ (Theorem III. 10), the left hand side of (1) is V, so that $u(a_0)=V$, hence $u(z)=V$ at every point of E^0, so that by Theorem II. 25, $\mu(E^0)=0$.

Theorem III. 15. *Let E be a bounded closed set of positive capacity and $\nu\geqq 0$ be a positive mass distribution on E of total mass 1 and*

$$v(z)=\int_E \log\frac{1}{|z-a|}d\nu(a), \qquad \nu(E)=1 .$$

Then

$$\inf_{z\in E} v(z)\leqq V\leqq \sup_{z\in E} v(z), \qquad V=\log 1/\gamma(E) .$$

Proof. Let

$$u(z)=\int_E \log\frac{1}{|z-a|}d\mu(a), \qquad \mu(E)=1 ,$$

be the conductor potential of E. We may assume that E lies in $|z|\leqq 1/2$, then $\log\dfrac{1}{|a-b|}\geqq 0$, $a\in E$, $b\in E$, so that by Tonelli's theorem,

$$\int_E u d\nu=\int_E v d\mu .$$

Since $u\leqq V$, $\int_E u d\nu\leqq V$, so that $\int_E v\,d\mu\leqq V$, hence

$$\inf_{z\in E} v(z)\leqq V . \tag{1}$$

To prove that $\sup_{z\in E} v(z)\geqq V$, we may assume that $\sup_{z\in E} v(z)<+\infty$. Then $I(\nu)<\infty$, hence by Theorem III. 7, a set of capacity zero does not contain a positive ν-mass and since $u=V$ "almost everywhere" on E, $\int_E u d\nu=V$, hence $\int_E v d\mu=V$. From this we have

$$\sup_{z\in E} v(z)\geqq V . \tag{2}$$

Theorem III. 16. *Let $v(z)$ be the same as Theorem* III. 15, *then the set E_1 of $z\in E$, such that $v(z)=+\infty$ is of capacity zero.*

Proof. Suppose that $\gamma(E_1)>0$ and we may assume that E_1 is a closed set. Let $u_1(z)$ be the conductor potential of E_1 and μ_1 be its equilibrium distribution, then as before,

$$\int_E u_1 d\nu=\int_{E_1} v\,d\mu_1=+\infty .$$

Since $u_1\leqq V_1=\log 1/\gamma(E_1)$, we have $\int_E u_1 d\nu<\infty$, which

Hence the theorem.

Theorem III. 17. *Let* E_n $(n=1, 2, \cdots)$ *be a Borel set, which is contained in* $|z|\leqq 1/2$ *and* $E=\sum_{n=1}^{\infty} E_n$, *then*

$$\frac{1}{\log 1/\gamma(E)}\leqq\sum_{n=1}^{\infty}\frac{1}{\log 1/\gamma(E_n)}\,.$$

Proof. We may assume that E and E_n $(n=1, 2, \cdots)$ are closed sets and $\gamma(E)>0$, $\gamma(E_n)>0$. Let

$$u(z)=\int_E \log\frac{1}{|z-a|}\,d\mu(a), \qquad \mu(E)=1\,,$$

be the conductor potential of E, then since by the hypothesis,

$$\log\frac{1}{|z-a|}\geqq 0,\ a\in E,\ z\in E,$$

we have

$$\log 1/\gamma(E)=V(E)\geqq u(z)\geqq u_n(z)=\int_{E_n}\log\frac{1}{|z-a|}\,d\mu(a),\ z\in E_n\,,$$

so that

$$V(E)\geqq \sup_{z\in E_n} u_n(z)\geqq V(E_n)\,\mu(E_n)\,, \qquad \text{(Theorem III. 15)},$$

hence

$$\sum_{n=1}^{\infty}\frac{V(E)}{V(E_n)}\geqq\sum_{n=1}^{\infty}\mu(E_n)\geqq\mu(E)=1\,,$$

or

$$\frac{1}{V(E)}\leqq\sum_{n=1}^{\infty}\frac{1}{V(E_n)}\,,$$

$$\frac{1}{\log 1/\gamma(E)}\leqq\sum_{n=1}^{\infty}\frac{1}{\log 1/\gamma(E_n)}\,.$$

Theorem III. 18. *Let E and e be Borel sets. If* $\gamma(e)=0$, *then*

$$\gamma(E+e)=\gamma(E), \qquad \gamma(E-e)=\gamma(E)\,.$$

Proof. We may assume that $E+e$ lies in $|z|\leqq 1/2$, then if we put $E_1=E+e$, $\dfrac{1}{\log 1/\gamma(E_1)}\leqq\dfrac{1}{\log 1/\gamma(E)}+\dfrac{1}{\log 1/\gamma(e)}=\dfrac{1}{\log 1/\gamma(E)}$, $\gamma(E_1)\leqq\gamma(E)$. Since $E\subset E_1$, $\gamma(E)\leqq\gamma(E_1)$, so that $\gamma(E_1)=\gamma(E)$. Similarly $\gamma(E-e)=\gamma(E)$.

4. Hausdorff's measure

Let $h(t)>0$ be an increasing function of t $(0<t\leqq 1)$, $h(0)=0$. Let e be a bounded set on the z-plane. We cover e by at most a countable number of squares of sides $d_i<\rho$, whose sides are parallel to the coordinates axes and put

$$H(\rho)=\inf\sum_i h(d_i)\,, \tag{1}$$

then $H(\rho)$ increases, if ρ decreases, so that

$$\lim_{\rho\to 0} H(\rho)=h^*(e) \quad (0\leqq h^*(e)\leqq\infty) \tag{2}$$

exists. $h^*(e)$ is called *the Hausdorff's measure of e*. Since as easily seen, $h^*(e)$ is a Carathéodory's outer measure, $h^*(e)$ is a totally additive set function for Borel sets.

If $h(t)=t^\alpha$ $(\alpha>0)$, then $h^*(e)$ is called *the α-dimensional measure* and if $h(t)=\dfrac{1}{\log 1/t}$, $h^*(e)$ is called *the logarithmic measure*, which we write $\lambda(e)$ instead of $h^*(e)$.

Lemma. *Let E be a closed set, contained in a unit square Q_0 of side 1 on the z-plane with its center at $z=0$ and $h^*(E)>0$. Then there exists a positive mass distribution $\mu\geqq 0$ on E of finite total mass, such that for any point $a\in E$,*

$$\mu(\Delta(r, a))\leqq 36\, h(r) \quad (0<r\leqq 1),$$

where $\Delta(r, a)$ is a circular disc of radius r about a.

Proof. First we remark the following. We cover E by a finite number of squares of sides d_i, without the restriction $d_i<\rho$, then since $h^*(E)>0$, we see easily that

$$\sum_i h(d_i)\geqq\eta>0,$$

where η is a constant, independent of the choice of covering squares. We divide Q_0 into 4^n equal squares $\{\overline{Q}_n^{(j)}\}$ of sides $\delta_n=1/2^n$ $(n=0, 1, 2, \cdots)$. Let

$$\overline{Q}_n^{(j)}:\ \frac{k}{2^n}\leqq x\leqq\frac{k+1}{2^n},\quad \frac{l}{2^n}\leqq y\leqq\frac{l+1}{2^n},\qquad z=x+iy, \tag{1}$$

where k, l are integers, then we define a half-open square $Q_n^{(j)}$ by

$$Q_n^{(j)}:\ \frac{k}{2^n}\leqq x<\frac{k+1}{2^n},\quad \frac{l}{2^n}\leqq y<\frac{l+1}{2^n}, \tag{2}$$

then $Q_n^{(j)}$, $Q_n^{(j')}$ $(j\neq j')$ are disjoint.

We consider only such $Q_n^{(j)}$, which contain points of E and let $\mathfrak{M}_n$ be the set of such $Q_n^{(j)}$ and let $\mathfrak{M}_n=\{Q_n^{(j)}\}$.

Now we define a positive mass distribution $\mu_n\geqq 0$ in Q_0 as follows. The density of μ_n at every point of a square $Q_n^{(j)}\in\mathfrak{M}_n$ is constant, such that

$$\mu_n(Q_n^{(j)})=h(\delta_n). \tag{3}$$

If for a certain $Q_{n-1}^{(j)}\in\mathfrak{M}_{n-1}$, $\mu_n(Q_{n-1}^{(j)})>h(\delta_{n-1})$, then we reduce μ_n in $Q_{n-1}^{(j)}$ by dividing the density by a suitable positive constant>1, such that

$$\mu_n(Q_{n-1}^{(j)})=h(\delta_{n-1}). \tag{4}$$

If $\mu_n(Q_{n-1}^{(j)}) \leqq h(\delta_{n-1})$, then we do not perform the reduction. Then

$$\left.\begin{aligned} \mu_n(Q_n^{(j)}) &\leqq h(\delta_n) \quad \text{for any } Q_n^{(j)} \in \mathfrak{M}_n, \\ \mu_n(Q_{n-1}^{(j)}) &\leqq h(\delta_{n-1}) \quad \text{for any } Q_{n-1}^{(j)} \in \mathfrak{M}_{n-1}\,. \end{aligned}\right\} \tag{5}$$

Next if for a certain $Q_{n-2}^{(j)} \in \mathfrak{M}_{n-2}$, $\mu_n(Q_{n-2}^{(j)}) > h(\delta_{n-2})$, then we reduce μ_n in $Q_{n-2}^{(j)}$ again, such that $\mu_n(Q_{n-2}^{(j)}) = h(\delta_{n-2})$ and if $\mu_n(Q_{n-2}^{(j)}) \leqq h(\delta_{n-2})$, we do not perform the reduction. We continue the reduction till we attain Q_0, then

$$\mu_n(Q_0) \leqq h(1)\,, \tag{6}$$

$$\mu_n(Q_i^{(j)}) \leqq h(\delta_i) \quad \text{for any } Q_i^{(j)} \in \mathfrak{M}_i \ (i=0, 1, 2, \cdots, n)\,. \tag{7}$$

From the construction of μ_n, we see easily that any point $a \in E$ is contained in a certain $Q_i^{(j)} \in \mathfrak{M}_i$, such that

$$\mu_n(Q_i^{(j)}) = h(\delta_i)\,. \tag{8}$$

If there are several such $Q_i^{(j)}$, then we choose the largest one among them and we denote it by $Q_i^{(j)}$, then $\{Q_i^{(j)}\}$ are disjoint and

$$h(1) \geqq \mu_n(Q_0) = \sum_i \mu_n(Q_i^{(j)}) = \sum_i h(\delta_i) \geqq \eta > 0 \quad (n=0, 1, 2, \cdots)\,, \tag{9}$$

so that by Theorem II. 4, we can select a partial sequence μ_{n_k} from μ_n, which converges to a positive mass distribution $\mu \geqq 0$ on E, such that

$$h(1) \geqq \mu(E) \geqq \eta > 0\,. \tag{10}$$

Let a be any point of E and $\Delta(r, a)$ be a circular disc about a. We choose n, such that $\frac{1}{2^n} \leqq r < \frac{1}{2^{n-1}}$. If $\frac{1}{2^n} = \delta_n \leqq r < r' < \delta_{n-1} = \frac{1}{2^{n-1}}$, then $\Delta(r', a)$ has common points with at most 9 squares of $\mathfrak{M}_{n-1}$, so that $\Delta(r', a)$ has common points with at most 36 squares of $\mathfrak{M}_n$, so that by (5),

$$\mu_n(\Delta(r', a)) \leqq 36\, h(\delta_n) \leqq 36\, h(r)\,,$$

hence

$$\mu(\Delta(r, a)) \leqq \lim_k \mu_{n_k}(\Delta(r', a)) \leqq 36\, h(r)\,.$$

Theorem III. 19. *Suppose that* $\int_0^1 \frac{h(t)}{t}\, dt < \infty$. *If E is a bounded closed set, such that $h^*(E) > 0$, then*

$$\gamma(E) > 0\,.$$

Hence if $\gamma(E)=0$, *then* $h^*(E)=0$. Frostman [1].

Proof. We may assume that E is contained in a unit square Q_0 with its center at $z=0$ and let $\mu \geqq 0$ be the positive mass distribution on E, defined by the lemma and put

$$v(z) = \int_E \log \frac{1}{|z-a|}\, d\mu(a)\,, \quad \mu(E) > 0\,. \tag{1}$$

Let z be any point of E and $\Delta(r, z)$ be a disc of radius r about z and put $\Omega(r)=\mu(\Delta(r, z))$, then by the lemma, $\Omega(r)\leqq 36\, h(r)$, so that

$$v(z)=\int_0^1 \log\frac{1}{r}\,d\Omega(r)=\left[\log\frac{1}{r}\,\Omega(r)\right]_0^1+\int_0^1\frac{\Omega(r)}{r}dr$$
$$\leqq\left[\log\frac{1}{r}\,\Omega(r)\right]_0^1+36\int_0^1\frac{h(r)}{r}\,dr\,.$$

Since

$\int_0^1\frac{h(t)}{t}dt<\infty$, $\lim_{r\to 0} h(r)\log 1/r=0$, so that $\lim_{r\to 0}\log\frac{1}{r}\,\Omega(r)=0$, hence

$$v(z)\leqq 36\int_0^1\frac{h(r)}{r}\,dr\,. \qquad (2)$$

Hence $v(z)$ is bounded on E, so that

$$I(\mu)=\iint_E\log\frac{1}{|a-b|}\,d\mu(a)\,d\mu(b)=\int_E v(b)d\mu(b)<\infty\,,$$

hence $\gamma(E)>0$. ——

If $\int_0^1\frac{h(t)dt}{t}=\infty$, then from $h^*(E)>0$, it does not follow that $\gamma(E)>0$ in general, as the following theorem shows.

Theorem III. 20. *Let E be a bounded closed set and $\lambda(E)$ be its logarithmic measure. If $\lambda(E)<\infty$, then $\gamma(E)=0$.* Erdös and Gillis [1].

Proof. (Tsuji [58]). First we remark the following. Let a be any point on the z-plane and C_r: $|z-a|=r$ be a circle about a and $\mathfrak{M}_a$ be the set of r, such that C_r intersects E, then since $\lambda(E)<\infty$, $\lambda(\mathfrak{M}_a)<\infty$, so that $\mathfrak{M}_a$ does not contain an interval[1)], hence for any r, there exists r' in its any small neighbourhood, such that C_r' does not intersect E. Hence for any $\varepsilon>0$, we can cover E by a finite number of open circular discs K_i $(i=1, 2, \cdots, n)$ of diameter $d_i<\rho$, such that the bounding circle C_i of K_i does not contain a point of E and

$$\lambda(E)-\varepsilon<\lambda_n=\sum_{i=1}^n\frac{1}{\log 1/d_i}<\lambda(E)+\varepsilon\,, \qquad (1)$$

where, without loss of generality, we define $\lambda(E)$ by means of discs of diameter $d_i<\rho$, instead of squares of sides $d_i<\rho$.

If we put $E_n=\sum_{i=1}^n\overline{K}_i$, then E_n is a closed set and $E\subset E_n$.

1) Let J be a closed interval, then we shall prove that $\lambda(J)=\infty$. We cover J by intervals $\{I_i\}$, then for any large $K>0$, if $|I_i|<\rho$ and ρ is sufficiently small, $\frac{1}{\log 1/|I_i|}\geqq K|I_i|$, so that $\sum_i\frac{1}{\log 1/|I_i|}\geqq K\sum_i|I_i|\geqq K|J|$. From this, we have $\lambda(J)=\infty$

We denote $A-A.B$ by $A-B$ and put $\Delta_1=\overline{K}_1$, $\Delta_2=(\overline{K}_1+\overline{K}_2)-\overline{K}_1$, $\Delta_3=(\overline{K}_1+\overline{K}_2+\overline{K}_3)-(\overline{K}_1+\overline{K}_2)$ etc.. We may assume that $\Delta_i\neq 0$, since if $\Delta_i=0$, we except K_i from our consideration. Then E_n can be decomposed into n disjoint sets $\Delta_i\subset\overline{K}_i$:

$$E_n=\sum_{i=1}^{n}\Delta_i\,, \tag{2}$$

where Δ_i consists of a finite number of connected domains, each of which is bounded by circular arcs, belonging to $\sum_{i=1}^{n}C_i$. Hence the boundary of Δ_i does not contain a point of E. Since $\Delta_i\subset\overline{K}_i$, if we denote the diameter of Δ_i by $\delta(\Delta_i)$, then $\delta(\Delta_i)\leqq d_i$. We put

$$v_n(z)=\sum_{i=1}^{n}\frac{1}{|\Delta_i|\log 1/d_i}\int_{\Delta_i}\log\frac{1}{|z-a|}\,d\sigma(a), \tag{3}$$

where $|\Delta_i|$ is the area of Δ_i and $d\sigma(a)$ is the area element.

Hence if we put

$$d\nu_n(a)=\frac{1}{|\Delta_i|\log 1/d_i}\,d\sigma(a) \quad \text{in } \Delta_i\,, \tag{4}$$

then

$$v_n(z)=\int_{E_n}\log\frac{1}{|z-a|}\,d\nu_n(a)\,, \tag{5}$$

where

$$\int_{E_n}d\nu_n(a)=\sum_{i=1}^{n}\frac{1}{\log 1/d_i}=\lambda_n\,. \tag{6}$$

We may assume that E is contained in $|z|<1/2$, then if $\rho>0$ is sufficiently small, E_n is contained in $|z|<1/2$, so that

$$\log\frac{1}{|z-a|}>0 \quad \text{for any } a\in E_n,\ z\in E_n.$$

If $z\in\Delta_i$, then $\log\dfrac{1}{|z-a|}\geqq\log\dfrac{1}{d_i}$ for any $a\in\Delta_i$, so that by (3)

$$v_n(z)\geqq\frac{1}{|\Delta_i|\log 1/d_i}\int_{\Delta_i}\log\frac{1}{|z-a|}\,d\sigma(a)\geqq 1\,, \qquad z\in\Delta_i\,,$$

hence

$$v_n(z)\geqq 1 \quad \text{on } E_n, \quad \text{a fortiori, on } E. \tag{7}$$

Suppose that $\gamma(E)>0$ and let

$$u(z)=\int_E\log\frac{1}{|z-a|}\,d\mu(a), \qquad \mu(E)=1\,, \tag{8}$$

be the conductor potential of E, such that

$$u(z)\leqq V=\log 1/\gamma(E) \quad \text{for any } z. \tag{9}$$

From $\int_{E_n} u\,d\nu_n = \int_E v_n\,d\mu$, we have by (9), (6), (7),

$$\log\frac{1}{\gamma(E)}\cdot\lambda_n \geqq 1, \quad 0<\gamma(E)\leqq e^{-\frac{1}{\lambda_n}}.$$

Since $\varepsilon>0$ is arbitrary, we have

$$0<\gamma(E)\leqq e^{-\frac{1}{\lambda}}, \quad \text{so that } \lambda(E)>0\,. \tag{10}$$

Hence if $\lambda(E)=0$, then $\gamma(E)=0$, which is proved by Lindeberg [1].

To prove the general case, we suppose that

$$0\leqq\lambda(E)<\infty, \quad \gamma(E)>0 \tag{11}$$

and we shall deduce a contradiction.

We cover E by n open circular discs K_i $(i=1, 2, \cdots, n)$ of diameter $d_i<\rho$, such that the bounding circle C_i of K_i does not contain a point of E and let

$$\lambda_n=\sum_{i=1}^{n}\frac{1}{\log 1/d_i}<A, \quad A=\lambda(E)+1\,, \tag{12}$$

and we define $E_n=\sum_{i=1}^{n}\varDelta_i$ as before.

Next we cover E by m open circular discs K_{ij} of diameter $d_{ij}<\rho'<\rho$, such that the bounding circle C_{ij} of K_{ij} does not contain a point of E and

$$\lambda_m=\sum_{i,j}\frac{1}{\log 1/d_{ij}}<A\,. \tag{13}$$

Since the boundary of $\varDelta_i$ does not contain a point of E, if we take ρ' sufficiently small, then we may assume that $K_{ij}\subset\varDelta_i$.

We put $E_m=\sum_{i,j}\overline{K}_{ij}$ and we decompose E_m into m disjoint sets $\varDelta_{ij}\in\overline{K}_{ij}$ as before:

$$E_m=\sum_{i,j}\varDelta_{ij}\,, \quad \varDelta_{ij}\subset\varDelta_i\,, \tag{14}$$

where the boundary of $\varDelta_{ij}$ does not contain a point of E. We put

$$\left.\begin{aligned} &v_m(z)=\sum_{i,j}\frac{1}{|\varDelta_{ij}|\log 1/d_{ij}}\int_{\varDelta_{ij}}\log\frac{1}{|z-a|}d\sigma(a)=\int_{E_n}\log\frac{1}{|z-a|}d\nu_m(a)\,,\\ &\text{where}\\ &d\nu_m(a)=\frac{1}{|\varDelta_{ij}|\log 1/d_{ij}}d\sigma(a) \ \text{ in } \varDelta_{ij}, \quad \int_{E_m}d\nu_m(a)=\sum_{i,j}\frac{1}{\log 1/d_{ij}}=\lambda_m\,. \end{aligned}\right\} \tag{15}$$

If $z\in\varDelta_{ij_0}$, then since $\log\left|\frac{1}{z-a}\right|\geqq\log\frac{1}{d_{ij_0}}$ for any $a\in\varDelta_{ij_0}$,

$$\frac{1}{|\varDelta_{ij_0}|\log 1/d_{ij_0}}\int_{\varDelta_{ij_0}}\log\frac{1}{|z-a|}d\sigma(a)\geqq 1\,, \tag{16}$$

and since $z\in\Delta_i$, $\log\dfrac{1}{|z-a|}\geqq\log\dfrac{1}{d_i}$ for any $a\in\Delta_{ij}$ $(\subset\Delta_i)$, so that

$$\frac{1}{|\Delta_{ij}|\log 1/d_{ij}}\int_{\Delta_{ij}}\log\frac{1}{|z-a|}\,d\sigma(a)\geqq\frac{\log 1/d_i}{\log 1/d_{ij}}\ (j\neq j_0)\,, \tag{17}$$

hence by (16), (17), if $z\in\Delta_{ij_0}$,

$$v_m(z)\geqq 1+\log\frac{1}{d_i}\sum_{j(j\neq j_0)}\frac{1}{\log 1/d_{ij}}=1+\log\frac{1}{d_i}\sum_j\frac{1}{\log 1/d_{ij}}-\frac{\log 1/d_i}{\log 1/d_{ij_0}}\,.$$

If we choose ρ' sufficiently small, such that

$$\frac{\log 1/d_i}{\log 1/d_{ij}}<\delta<1 \quad \text{for any } i, j,$$

then if $z\in\Delta_{ij_0}$,

$$v_m(z)\geqq 1-\delta+\log\frac{1}{d_i}\sum_j\frac{1}{\log 1/d_{ij}}\,. \tag{18}$$

Hence from $\int_{E_m}u\,d\nu_m=\int_E v_m\,d\mu=\sum_i\int_{\Delta_i}v_m\,d\mu$ and (9), (15), (18), we have

$$\log\frac{1}{\gamma(E)}\cdot\lambda_m\geqq 1-\delta+\sum_i\mu(\Delta_i)\log\frac{1}{d_i}\sum_j\frac{1}{\log 1/d_{ij}}\,. \tag{19}$$

Now $\{\Delta_i\}$ is decomposed into two groups $\{\Delta_i'\}$, $\{\Delta_i''\}$, where

$$\mu(\Delta_i')\log\frac{1}{d_i}\geqq\frac{1}{2\lambda_n}\,,\qquad \mu(\Delta_i'')\log\frac{1}{d_i}<\frac{1}{2\lambda_n}\,,$$

then from (19),

$$\log\frac{1}{\gamma(E)}\cdot\lambda_m\geqq 1-\delta+\frac{1}{2\lambda_n}\sum_i{}'\sum_j\frac{1}{\log 1/d_{ij}}\,, \tag{20}$$

where $\sum_i{}'$ denotes that we sum up for all $\{\Delta_i'\}$. Since

$$\sum_i\mu(\Delta_i'')<\frac{1}{2\lambda_n}\sum_i\frac{1}{\log 1/d_i}=\frac{1}{2}\,,$$

and $\sum_i\mu(\Delta_i')+\sum_i\mu(\Delta_i'')=1$, we have

$$\sum_i\mu(\Delta_i')>\frac{1}{2}\,. \tag{21}$$

Let

$$w_i(z)=\int_{\Delta_i'}\log\frac{1}{|z-a|}\,d\nu_m(a)\,,\qquad u_i(z)=\int_{\Delta_i'}\log\frac{1}{|z-a|}\,d\mu(a)\,, \tag{22}$$

then since $\quad w_i(z)\geqq\int_{\Delta_{ij}}\log\dfrac{1}{|z-a|}\,d\nu_m(a)\geqq 1\,,\quad$ if $z\in\Delta_{ij}\subset\Delta_i'$,

$$\int_{\Delta_i'}u_i\,d\nu_m=\int_{\Delta_i'}w_i\,d\mu\geqq\mu(\Delta_i')\,,$$

and

$$\int_{\Delta'_i} u_i d\nu_m \leqq \int_{\Delta'_i} u d\nu_m \leqq V \int_{\Delta'_i} d\nu_m = V \sum_j \frac{1}{\log 1/d_{ij}}, \qquad V = \log 1/\gamma(E),$$

so that by (21),

$$\sum_i{}' \sum_j \frac{1}{\log 1/d_{ij}} \geqq \frac{1}{V} \sum_i \mu(\Delta_i') > \frac{1}{2V}.$$

Since $\lambda_n < A$, $\lambda_m < A$, we have from (20), by putting $\kappa = 1/4AV$,

$$\log \frac{1}{\gamma(E)} \cdot \lambda_m \geqq 1-\delta + \frac{1}{4\lambda_n V} > 1-\delta + \frac{1}{4AV} = 1-\delta+\kappa > (1-\delta)(1+\kappa),$$

$$\log 1/\gamma(E) > (1-\delta)(1+\kappa)/A. \tag{23}$$

Next we cover E by p open circular discs K_{ijk} of diameter $d_{ijk} < \rho'' < \rho'$, such that the bounding circle C_{ijk} of K_{ijk} does not contain a point of E and

$$\lambda_p = \sum_{i,j,k} \frac{1}{\log 1/d_{ijk}} < A.$$

By taking ρ'' sufficiently small, we may assume that $K_{ijk} \subset \Delta_{ij}$ and put $E_p = \sum_{i,j,k} \overline{K}_{ijk}$ and as before, we decompose E_p into p disjoint sets $\Delta_{ijk} \subset \overline{K}_{ijk}$:

$$E_p = \sum_{i,j,k} \Delta_{ijk}, \quad \Delta_{ijk} \subset \Delta_{ij},$$

and put

$$v_p(z) = \sum_{i,j,k} \frac{1}{|\Delta_{ijk}| \log 1/d_{ijk}} \int_{\Delta_{ijk}} \log \frac{1}{|z-a|} d\sigma(a) = \int_{E_p} \log \frac{1}{|z-a|} d\nu_p(a), \tag{24}$$

where

$$d\nu_p(a) = \frac{1}{|\Delta_{ijk}| \log 1/d_{ijk}} d\sigma(a) \quad \text{in } \Delta_{ijk}.$$

If $z \in \Delta_{ij_0k_0}$, then $z \in \Delta_{ij_0} \subset \Delta_i$, so that we have as before,

$$v_p(z) \geqq 1-\delta + \log \frac{1}{d_{ij_0}} \sum_k \frac{1}{\log 1/d_{ij_0k}} + \log \frac{1}{d_i} \sum_{j,k(j \neq j_0)} \frac{1}{\log 1/d_{ijk}}$$

$$= 1-\delta + \log \frac{1}{d_{ij_0}} \sum_k \frac{1}{\log 1/d_{ij_0k}} + \log \frac{1}{d_i} \sum_{j,k} \frac{1}{\log 1/d_{ijk}} - \log \frac{1}{d_i} \sum_k \frac{1}{\log 1/d_{ij_0k}}$$

$$= 1-\delta + \log \frac{1}{d_{ij_0}} \left(1 - \frac{\log 1/d_i}{\log 1/d_{ij_0}}\right) \sum_k \frac{1}{\log 1/d_{ij_0k}} + \log \frac{1}{d_i} \sum_{j,k} \frac{1}{\log 1/d_{ijk}}$$

$$\geqq 1-\delta + (1-\delta) \log \frac{1}{d_{ij_0}} \sum_k \frac{1}{\log 1/d_{ij_0k}} + \log \frac{1}{d_i} \sum_{j,k} \frac{1}{\log 1/d_{ijk}}. \tag{25}$$

From this, we have as before,

$$\log \frac{1}{\gamma(E)} \cdot \lambda_p \geqq 1-\delta + (1-\delta) \sum_{i,j} \mu(\Delta_{ij}) \log \frac{1}{d_{ij}} \sum_k \frac{1}{\log 1/d_{ijk}}$$

$$+\sum_i \mu(\Delta_i)\log\frac{1}{d_i}\sum_{j,k}\frac{1}{\log 1/d_{ijk}} \geqq 1-\delta+(1-\delta)\kappa+\kappa>(1-\delta)(1+2\kappa),$$

$$\log 1/\gamma(E)>(1-\delta)(1+2\kappa)/A\,. \tag{26}$$

Similarly after N steps, we have

$$\log 1/\gamma(E)>(1-\delta)(1+N\kappa)/A \quad (N=1,2,\cdots), \tag{27}$$

which is absurd, since $\gamma(E)>0$. Hence if $\lambda(E)<\infty$, then $\gamma(E)=0$.

5. Transfinite diameter. Tchebycheff's constant

1. Let E be a bounded closed set on the z-plane, which contains infinitely many points. We take n points z_i $(i=1, 2, \cdots, n)$ on E and put

$$V(z_1, \cdots, z_n)=\prod_{i<k}^{1,\cdot,n}|z_i-z_k|, \qquad V_n=\operatorname*{Max}_{z_i\in E} V(z_1, \cdots, z_n)\,, \tag{1}$$

$$d_n=\sqrt[\binom{n}{2}]{V_n}\,. \tag{2}$$

Theorem III. 21. $d_{n+1}\leqq d_n$, *so that* $\lim_n d_n=\tau(E)$ *exists.* Fekete [1].

$\tau(E)$ is called *the transfinite diameter of E.*

Proof. Let $V(z_1, \cdots z_{n+1})$ attain its maximum V_{n+1} at ξ_i $(i=1, 2,\cdots,n+1)$ on E, then

$$V_{n+1}=\prod_{i<k}^{1,\cdot,n+1}|\xi_i-\xi_k|=|(\xi_1-\xi_2)(\xi_1-\xi_3)\cdots(\xi_1-\xi_{n+1})|\,V(\xi_2,\ldots,\xi_{n+1})$$
$$\leqq|(\xi_1-\xi_2)(\xi_1-\xi_3)\cdots(\xi_1-\xi_{n+1})|\,V_n\,.$$

Similarly

$$V_{n+1}\leqq|(\xi_2-\xi_1)\cdots(\xi_2-\xi_{n+1})|\,V_n, \;\cdots,\; V_{n+1}\leqq|(\xi_{n+1}-\xi_1)\cdots(\xi_{n+1}-\xi_n)|\,V_n\,.$$

If we multiply these ineqalities, then we have

$$V_{n+1}^{n+1}\leqq V_n^{n+1}\,V_{n+1}^2, \qquad V_{n+1}^{n-1}\leqq V_n^{n+1}, \text{ or } d_{n+1}\leqq d_n. \text{——}$$

2. Next we consider polynomials of the form $P(z)=z^n+c_1z^{n-1}+\cdots+c_n$ and put

$$\mu(P)=\operatorname*{Max}_{z\in E}|P(z)|, \qquad m_n=\inf_P \mu(P)\,, \tag{1}$$

then there exists $T_n(z)=z^n+a_1z^{n-1}+\cdots+a_n$, such that

$$\mu(T_n)=\operatorname*{Max}_{z\in E}|T_n(z)|=m_n\,, \tag{2}$$

which can be proved as follows. Let $P_\nu(z)=z^n+c_1^{(\nu)}z^{n-1}+\cdots+c_n^{(\nu)}$, such that $\mu(P_\nu)\to m_n$, then $|P_\nu(z)|\leqq M$ $(\nu=1, 2, \cdots)$ on E. Let $\xi_1, \cdots, \xi_n$ be n points on E, then

$$\xi_i^n+c_1^{(\nu)}\xi_i^{n-1}+\cdots+c_n^{(\nu)}=P_\nu(\xi_i)\ (i=1, 2, \cdots, n)\,.$$

Since $|P_\nu(\xi_i)|\leqq M$, we see that $c_j^{(\nu)}$ $(j=1, 2, \cdots, n)$ are bounded for $\nu=1, 2, \cdots$. Hence we can find a partial sequence from $\{c_j^{(\nu)}\}$, which we denote by $\{c_j^{(\nu)}\}$ again, such that $\lim_\nu c_j^{(\nu)}=a_j$ and put $T_n(z)=z^n+a_1 z^{n-1}+\cdots+a_n$, then $\mu(T_n)=m_n$.

$T_n(z)$ is called *the Tchebycheff's polynomial of E of order n.*

Theorem III. 22. *Let* $|T_n(z_j)|=m_n$ *at* $z_j (j=1, 2, \cdots, \nu)$ *on* E, *then*

$$\nu\geqq n+1.$$

Proof. Let $T_n(z_j)=m_n e^{i\alpha_j}$ $(j=1, 2, \cdots, \nu)$ and suppose that $\nu\leqq n$. Then there exists a polynomial $Q(z)=q_1 z^{n-1}+\cdots+q_n$, such that $Q(z_j)=T_n(z_j)$ $(j=1, 2, \cdots, \nu)$. Let $U(z_j)$: $|z-z_j|<\rho$. If ρ is sufficiently small, then for any $z\in E.U(z_j)$, $w=T_n(z)$ lies in a sector S_j: $|w|\leqq m_n$, $|\arg w-\alpha_j|<\pi/6$. We see easily that if $w_1\in S_j$, $w_2\in S_j$, then $|w_1-w_2|<\operatorname{Max}(|w_1|, |w_2|)$. We choose $0<\omega<1$ so small that for any $z\in E.U(z_j)$, $\omega Q(z)$ lies in S_j and $\omega|Q(z)|\leqq|T_n(z)|$, so that

$$|T_n(z)-\omega Q(z)|<\operatorname{Max}(|T_n(z)|, \omega|Q(z)|)=|T_n(z)|\leqq m_n \text{ in } E.U(z_j) \quad (1)$$
$$(j=1, 2, \cdots, \nu).$$

For any z on E, which lies outside of $\sum_{j=1}^{\nu} U(z_j)$, $|T_n(z)|\leqq m_n'<m_n$ and $|Q(z)|\leqq K=$const., so that if ω is sufficiently small,

$$|T_n(z)-\omega Q(z)|\leqq m_n'+\omega K<m_n, \quad (2)$$

hence by (1), (2), $|T_n(z)-\omega Q(z)|<m_n$ on E, which contradicts the definition of m_n, hence $\nu\geqq n+1$. ——

Theorem III. 23. *$T_n(z)$ is unique.* Tonelli [1].

Proof. Suppose that there are two such polynominals $T_n'(z)$, $T_n''(z)$, then $T_n(z)=\frac{1}{2}(T_n'(z)+T_n''(z))$ is also a Tchebycheff's polynominal of order n. $|T_n(z_j)|=m_n$ at least at $n+1$ points $z_j (j=1, 2\cdot\cdot, n+1)$ on E and since $|T_n'(z_j)|\leqq m_n$, $|T_n''(z_j)|\leqq m_n$, we have

$$m_n=|T_n(z_j)|\leqq\frac{1}{2}(|T_n'(z_j)|+|T_n''(z_j)|)\leqq m_n,$$

so that $|T_n''(z_j)|=|T_n'(z_j)|=m_n$, hence $|T_n'(z_j)+T_n''(z_j)|=2|T_n(z_j)|=2m_n=|T_n'(z_j)|+|T_n''(z_j)|$, so that $T_n'(z_j)=T_n''(z_j)$ $(j=1, 2, \cdots, n+1)$, hence

$$T_n'(z)\equiv T_n''(z). \text{ ——}$$

From the definition of $T_n(z)$, we can prove easily

Theorem III. 24. *All zero points of $T_n(z)$ lie in the smallest convex set, which contains E.*

Theorem III. 25. *Let* $m_n=\operatorname{Max}_{z\in E}|T_n(z)|$, $r_n=\sqrt[n]{m_n}$, *then*

$$\lim_n r_n=\rho(E) \text{ exists.} \qquad \text{Fekete [1].}$$

$\rho(E)$ is called *the Tchebycheff's constant of E.*

Lemma. *If* $a_{m+n} \leqq a_m + a_n$ $(m, n=1, 2, \cdots)$, *then* $\lim_n a_n/n$ *exists.*

Proof. Let $\inf_n a_n/n = \alpha$ and we may assume that $\alpha > -\infty$. Then for any $\varepsilon > 0$, there exists m, such that $a_m/m < \alpha + \varepsilon$. Now any natural number n can be written in the form $n = qm + r$ $(0 \leqq r \leqq m-1)$, so that if we put $a_0 = 0$, then $a_n = a_{qm+r} \leqq a_{qm} + a_r \leqq q a_m + a_r$, hence

$$\alpha \leqq \frac{a_n}{n} \leqq \frac{q a_m + a_r}{qm+r} = \frac{a_m}{m} \cdot \frac{qm}{qm+r} + \frac{a_r}{n} < (\alpha+\varepsilon)\frac{qm}{qm+r} + \frac{a_r}{n},$$

$$\overline{\lim_n}\, a_n/n \leqq \alpha + \varepsilon ,$$

Since $\varepsilon > 0$ is arbitrary, $\overline{\lim_n}\, a_n/n \leqq \alpha$, hence $\lim_n a_n/n = \alpha$. ——

Proof of the theorem. Let $T_n(z) = \prod_{i=1}^{n} (z - a_i)$, then by the definition of $T_n(z)$,

$$\log \frac{1}{r_n} = \operatorname*{Min}_{z \in E} \frac{1}{n} \sum_{i=1}^{n} \log \frac{1}{|z-a_i|} = \operatorname*{Max}_{|z_i|<\infty} \operatorname*{Min}_{z \in E} \frac{1}{n} \sum_{i=1}^{n} \log \frac{1}{|z-z_i|} .$$

If $T_m(z) = \prod_{j=1}^{m} (z - b_j)$, then $T_m(z) T_n(z) = z^{m+n} + \cdots$, so that

$$(m+n) \log \frac{1}{r_{m+n}} = \operatorname*{Min}_{z \in E} \log \frac{1}{|T_{m+n}(z)|} \geqq \operatorname*{Min}_{z \in E} \log \frac{1}{|T_m(z) T_n(z)|}$$

$$\geqq \operatorname*{Min}_{z \in E} \sum_{j=1}^{m} \frac{1}{|z-b_j|} + \operatorname*{Min}_{z \in E} \sum_{i=1}^{n} \log \frac{1}{|z-a_i|} = m \log \frac{1}{r_m} + n \log \frac{1}{r_n},$$

$(m+n) \log r_{m+n} \leqq m \log r_m + n \log r_n$, hence if we take $a_n = n \log r_n$ in the lemma, we see that $\lim_n r_n$ exists.

3. We shall prove

Theorem III. 26. $\tau(E) = \rho(E) = \gamma(E)$.

$\tau(E) = \rho(E)$ was proved by Fekete [1] and $\tau(E) = \gamma(E)$ by Szegö [2].

Proof. By the definition,

$$\log \frac{1}{d_n} = \operatorname*{Min}_{z_i \in E} \frac{1}{\binom{n}{2}} \sum_{i<k}^{1,\cdot,n} \log \frac{1}{|z_i - z_k|}, \tag{1}$$

$$\log \frac{1}{r_n} = \operatorname*{Min}_{z \in E} \frac{1}{n} \sum_{i=1}^{n} \log \frac{1}{|z-a_i|} = \frac{1}{n} \operatorname*{Max}_{|z_i|<\infty} \operatorname*{Min}_{z \in E} \sum_{i=1}^{n} \log \frac{1}{|z-z_i|}, \tag{2}$$

where $$T_n(z) = \prod_{i=1}^{n} (z - a_i) .$$

If suffices to prove that

(i) $\gamma(E) \leqq \rho(E)$, (ii) $\rho(E) \leqq \tau(E)$, (iii) $\tau(E) \leqq \gamma(E)$.

(i) $\gamma(E) \leqq \rho(E)$.

Let D_n be the lemniscate domain: $v(z)=\frac{1}{n}\sum_{i=1}^{n}\log\frac{1}{|z-a_i|}\geqq\log\frac{1}{r_n}$, then by (2), $E\subset D_n$. Now $v(z)$ can be written in the form:

$$v(z)=\int_{D_n}\log\frac{1}{|z-a|}d\omega(a),\qquad \omega(D_n)=1\,,$$

where $d\omega(a_i)=1/n$ $(i=1, 2, \cdots, n)$ and $d\omega(a)=0$ elsewhere. Hence by Theorem III. 15,

$$\log\frac{1}{r_n}=\inf_{z\in D_n} v(z)\leqq V(D_n)=\log 1/\gamma(D_n)\,,$$

so that $\gamma(D_n)\leqq r_n$. Since $E\subset D_n$, $\gamma(E)\leqq\gamma(D_n)\leqq r_n$, hence $\gamma(E)\leqq\rho(E)$.

(ii) $\rho(E)\leqq\tau(E)$.

We take $n+1$ points z_i $(i=1, 2, \cdots, n+1)$ on E, such that

$$\binom{n+1}{2}\log\frac{1}{d_{n+1}}=\sum_{i<j}^{1,\cdot,n+1}\log\frac{1}{|z_i-z_j|}=\frac{1}{2}\sum_{i,j(i\neq j)}^{1,\cdot,n+1}\log\frac{1}{|z_i-z_j|}$$

$$=\frac{1}{2}\sum_{k=1}^{n+1}\sum_{i(i\neq k)}^{1,\cdot,n+1}\log\frac{1}{|z_k-z_i|}\,. \tag{3}$$

If we fix $z_1, \cdots, z_{k-1}, z_{k+1}, \cdots, z_{n+1}$, then by (1),

$$\sum_{i(i\neq k)}^{1,\cdot,n+1}\log\frac{1}{|z_k-z_i|}=\operatorname{Min}_{z\in E}\sum_{i(i\neq k)}^{1,\cdot,n+1}\log\frac{1}{|z-z_i|},\quad\text{so that by (2),}$$

$$\sum_{i(i\neq k)}^{1,\cdot,n+1}\log\frac{1}{|z_k-z_i|}\leqq n\log\frac{1}{r_n},\quad\text{hence by (3),}$$

$$\binom{n+1}{2}\log\frac{1}{d_{n+1}}\leqq\frac{n(n+1)}{2}\log\frac{1}{r_n},\quad r_n\leqq d_{n+1},\ \text{so that } \rho(E)\leqq\tau(E).$$

(iii) $\tau(E)\leqq\gamma(E)$.

By Theorem III. 9, for any $\delta>0$, there exists an open set D, which contains E, such that $\gamma(\bar{D})\leqq\gamma(E)+\delta$. We take n points z_i $(i=1, 2, \cdots, n)$ on E and let Δ_i: $|z-z_i|\leqq\varepsilon$ $\left(\varepsilon=\frac{1}{\sqrt{\pi n}}\right)$ be a disc about z_i, then $\sigma(\Delta_i)=1/n$, where $\sigma(\Delta_i)$ denotes the area of Δ_i. We take n so large that Δ_i $(i=1, 2, \cdots, n)$ are contained in D. We define the density $\rho_i(a)$ $(i=1, 2, \cdots, n)$ in D, such that $\rho_i(a)=1$ for any point $a\in\Delta_i$ and $\rho_i(a)=0$ elsewhere and put $\rho(a)=\sum_{i=1}^{n}\rho_i(a)$, then

$$\int_D\rho(a)d\sigma(a)=1,$$

where $d\sigma(a)$ is the surface element.

Hence

$$\log\frac{1}{\gamma(E)+\delta}\leqq\log\frac{1}{\gamma(\bar{D})}=V(\bar{D})\leqq\iint_{\bar{D}}\log\frac{1}{|a-b|}\rho(a)\rho(b)d\sigma(a)d\sigma(b)$$

$$=\sum_{j=1}^{n}\int_{\Delta_j}d\sigma(a)\sum_{i=1}^{n}\int_{\Delta_i}\log\frac{1}{|a-b|}d\sigma(b)\,. \tag{4}$$

Since for a fixed a, $\log\frac{1}{|a-b|}$ is a superharmonic function of b,

$$\int_{\Delta_i}\log\frac{1}{|a-b|}d\sigma(b)\leqq\sigma(\Delta_i)\log\frac{1}{|a-z_i|}=\frac{1}{n}\log\frac{1}{|a-z_i|}\,,$$

hence

$$\log\frac{1}{\gamma(E)+\delta}\leqq\frac{1}{n}\sum_{i,j}^{1,\cdot,n}\int_{\Delta_j}\log\frac{1}{|a-z_i|}d\sigma(a)\,. \tag{5}$$

Since $\log\frac{1}{|a-z_i|}$ is a superharmonic function of a, we have

$$\int_{\Delta_j}\log\frac{1}{|a-z_i|}d\sigma(a)\leqq\sigma(\Delta_j)\log\frac{1}{|z_j-z_i|}=\frac{1}{n}\log\frac{1}{|z_j-z_i|},\ \text{if } j\neq i,$$

and if $j=i$,

$$\int_{\Delta_i}\log\frac{1}{|a-z_i|}d\sigma(a)=\pi\varepsilon^2\log\frac{1}{\varepsilon}+\frac{\pi\varepsilon^2}{2}=O\left(\frac{\log n}{n}\right),$$

so that by (5)

$$\log\frac{1}{\gamma(E)+\delta}\leqq\frac{1}{n}\left(\frac{2}{n}\sum_{i<j}^{1,\cdot,n}\log\frac{1}{|z_i-z_j|}+O(\log n)\right)$$

$$=\frac{2}{n^2}\sum_{i<j}^{1,\cdot,n}\log\frac{1}{|z_i-z_j|}+O\left(\frac{\log n}{n}\right).$$

If we choose z_i, such that

$$\sum_{i<j}^{1,\cdot,n}\log\frac{1}{|z_i-z_j|}=\binom{n}{2}\log\frac{1}{d_n}\,,$$

then

$$\log\frac{1}{\gamma(E)+\delta}\leqq\frac{n-1}{n}\log\frac{1}{d_n}+O\left(\frac{\log n}{n}\right).$$

Hence if we make $n\to\infty$ and then $\delta\to 0$, then we have $\tau(E)\leqq\gamma(E)$.

6. Evans' theorem

Theorem III. 27. *Let E be a bounded closed set of capacity zero on* *the z-plane. Then there exists a positive mass distribution $\mu \geqq 0$ on E of total mass 1, such that*

$$u(z)=\int_E \log \frac{1}{|z-a|} d\mu(a), \quad \mu(E)=1 ,$$

tends to $+\infty$, when z tends to any point of E. Evans [1].[1)]

We call $u(z)$ *the Evans' function with respect to E.*

Proof. We take n points z_i $(i=1, 2, \cdots, n)$ on E and put

$$V(z_1, \cdots, z_n)=\prod_{i<k}^{1,\cdot,n} |z_i-z_k|, \quad V_n=\operatorname*{Max}_{z_i \in E} V(z_1, \cdots, z_n), \quad d_n=\sqrt[\binom{n}{2}]{V_n}, \tag{1}$$

then

$$d_n \to \tau(E)=\gamma(E)=0 \quad \text{as } n\to\infty . \tag{2}$$

We take n points a_ν $(\nu=1, 2, \cdots, n)$ on E and put

$$P(z)=\prod_{\nu=1}^{n} (z-a_\nu) \ (a_\nu \in E), \quad \mu^*(P)=\operatorname*{Max}_{z\in E} |P(z)|, \tag{3}$$

$$m_n{}^*=\inf_P \mu^*(P) .$$

We remark that in the case of Tchebycheff's polynomial, a_ν may be any points on the z-plane.

We can prove easily that there exists $T_n^*(z)=\prod_{\nu=1}^{n}(z-a_\nu^0)$ $(a_\nu^0 \in E)$, such that

$$m_n^*=\mu^*(T_n^*)=\operatorname*{Max}_{z\in E} |T_n{}^*(z)| . \tag{4}$$

Let V_{n+1} be attained at $n+1$ points a_i $(i=1, 2, \cdots, n+1)$ on E, then $V_{n+1}=V(a_1, \cdots, a_{n+1})=|(a_1-a_2)(a_1-a_3)\cdots (a_1-a_{n+1})| V(a_2, \cdots, a_{n+1})$.

If $P(z)=(z-a_2)(z-a_3)\cdots (z-a_{n+1})$, then by the definition of V_{n+1} and $m_n{}^*$,

$$|(a_1-a_2)(a_1-a_3)\cdots (a_1-a_{n+1})|=\operatorname*{Max}_{z\in E} |P(z)| \geqq m_n{}^* . \tag{5}$$

The similar relation holds, if we take a_i instead of a_1, hence multiplying $n+1$ similar inequalities, we have

$$V_{n+1}^2 \geqq m_n^{*n+1}, \quad d_{n+1} \geqq \sqrt[n]{m_n{}^*} ,$$

so that by (2),

$$\lim_{n\to\infty} \sqrt[n]{m_n{}^*}=0 . \tag{6}$$

Now we consider

1) Evans' theorem is enunciated in the form of Newtonian potential, but his proof can be modified to the logarithmic potential. The following proof is due to Noshiro [3].

$$v_n(z)=\frac{1}{n}\log\frac{1}{|T_n^*(z)|}=\frac{1}{n}\sum_{\nu=1}^{n}\log\frac{1}{|z-a_\nu^0|}=\int_E\log\frac{1}{|z-a|}d\mu_n(a)\,,\quad (7)$$
$$\mu_n(E)=1\,,$$

where $d\mu_n(a)=\frac{1}{n}$ at $a=a_\nu^0$ ($\nu=1, 2, \cdots, n$) and $d\mu_n(a)=0$ elsewhere.

Then by (4),

$$v_n(z)\geqq M_n^*=\frac{1}{n}\log\frac{1}{m_n^*}\quad \text{on } E\,. \qquad (8)$$

Since by (6), $\lim_n M_n^*=\infty$, we can find n_ν ($\nu=1, 2, \cdots$), such that $M_{n_\nu}^*\geqq 2^\nu$, hence if we put $u_\nu(z)=\frac{1}{2^\nu}v_{n_\nu}(z)$, then $u_\nu(z)\geqq 1$ on E and by (7)

$$u(z)=\sum_{\nu=1}^{\infty}u_\nu(z) \qquad (9)$$

is a potential function of positive mass distribution of total mass 1 on E and since $u_\nu(z)\geqq 1$ on E, $u(z)=+\infty$ on E. Since $u(z)$ is lower semicontinuous, $\lim u(z)=+\infty$, when z tends to any point of E.

7. Maximum principle for harmonic function

1. We shall prove the following maximum principle for harmonic function.

Theorem III. 28. (Maximum principle). *Let D be a domain on a closed Riemann surface F and Γ be its boundary and e be a set of capacity zero on Γ. Let $u(z)$ be a harmonic, or a continuous subharmonic function in D, which is bounded upwards in D. If $\overline{\lim_{z\to\zeta}}\,u(z)\leqq M$, when z tends to any point $\zeta\in\Gamma-e$ from the inside of D, then $u(z)\leqq M$ in D.*

Similarly, if $u(z)$ is a harmonic, or a continuous superharmonic function in D, which is bounded downwards in D, and $\underline{\lim_{z\to\zeta}}\,u(z)\geqq m$ ($\zeta\in\Gamma-e$), then $u(z)\geqq m$ in D.

Corollary. *Let $u(z)$ be a bounded harmonic function in D and $\lim_{z\to\zeta}u(z)=0$, when z tends to any point $\zeta\in\Gamma-e$, then $u\equiv 0$.*

Proof. We shall prove the first part. Since the theorem is evident, if $e=0$, we assume that $e\neq 0$. Suppose that $\sup_{z\in D}u(z)=M'>M$, then there exist $z_n\in D$ ($n=1, 2, \cdots$), such that $u(z_n)=M'-\eta_n>M$, $\eta_n\to 0$. We may assume that $z_n\to z_0$, then $z_0\in e$ and we may assume that z_0 is different from a branch point of F[1] and let U: $|z-z_0|\leqq\rho<1/2$ be contained in a schlicht sheet of F. Let D_n be the sub-domain of D, in which $u>M_n=M'-2\eta_n$ and contains z_n, then $u=M_n$ on the part Γ_n^* of the boundary

1) If otherwise, we consider on the plane of the local parameter at z_0.

Γ_n of D_n, which lies in D. D_n cluster to z_0, if $n\to\infty$, hence we assume that D_n lies in U. Let e_n be the intersection of Γ_n and Γ, then e_n is a closed sub-set of e, so that is of capacity zero and is contained in U.

Let $v(z)$ be the Evans' function with respect to e_n and put

$$w(z)=u(z)-\delta v(z), \quad \delta>0 .$$

Since D_n lies in U, $v(z)>0$ in D_n, so that $w\leqq u=M_n$ on Γ_n^*. Since $\lim v(z)=+\infty$, when z tends to any point of e_n, $w\leqq M_n$ on Γ_n, so that by the maximum principle, $w\leqq M_n$ in D_n. Since $\delta>0$ is arbitrary, $u\leqq M_n$ in D_n, which is absurd, hence the theorem.

Theorem III. 29. *Let E be a bounded closed set of capacity zero on the z-plane and U be its neighbourhood. Let $u(z)$ be a bounded harmonic (regular) function in $U-E$, then $u(z)$ is harmonic (regular) on E.*

Proof. We may assume that U is bounded by a Jordan curve C and $u(z)$ is harmonic and bounded in $U-E$. Let $v(z)$ be the solution of the Dirichlet problem for U, with the boundary value u on C, then $w(z)=u(z)-v(z)$ is a bounded harmonic function in $U-E$ and $w=0$ on C, so that by Theorem III. 28. Cor., $w\equiv 0$, or $u\equiv v$ in U, hence $u(z)$ is harmonic on E.

Theorem III. 30. *Let E be a bounded closed set of capacity zero on the z-plane and U be its neighbourhood. Let $u(z)$ be a subharmonic function in $U-E$, which is bounded upwards. If we define $u(a)$ at any point $a\in E$ by $u(a)=\overline{\lim\limits_{z\to a}}\, u(z)$ $(z\in U-E)$, then $u(z)$ becomes a subharmonic function in U and such a subharmonic prolongation is unique.*

Brelot [2].

Proof. By the definition of $u(a)$ at $a\in E$, $u(z)$ becomes an upper semi-continuous function in U. Let D be a Jordan domain in U, whose boundary Γ consists of a finite number of Jordan curves. Let $H(z)$ be harmonic in D and continuous in $\bar{D}$, such that $u\leqq H$ on Γ, then it suffices to prove that at any point $z\in D$,

$$u(z)\leqq H(z) . \tag{1}$$

We may assume that $\bar{D}$ contains a point of E and let E_1 be the sub-set of E, contained in $\bar{D}$, then E_1 is a bounded closed set of capacity zero, so that by Evans' theorem, there exists a positive mass distribution $d\mu$ on E of total mass 1, such that

$$v(z)=\int_{E_1}\log\frac{1}{|z-a|}\,d\mu(a)$$

tends to $+\infty$, if z tends to any point of E_1. By a suitable linear transformation, we may assume that U is contained in $|z|<1/2$, then $v>0$ in U.

Now the niveau curve Γ_r: $v(z)=\text{const.}=r>0$ consists of a finite number of Jordan curves, which cluster to E_1, if $r\to\infty$. Let D_r be the domain, which is bounded by $\Gamma+\Gamma_r$ and let $H_r(z)$ be a harmonic function in D_r, which is continuous in $\overline{D}_r$, such that $H_r=H$ on Γ, and $u\leqq H_r$ on Γ_r, then since u is subharmonic in $D-E$, at any point $z\in D_r$,

$$u(z)\leqq H_r(z)\,. \tag{2}$$

Let $u(z)<K$ in $U-E$, then we may assume that $H_r<K$ on Γ_r. For any $\varepsilon>0$, we take $r>0$ so large that $K\leqq H(z)+\varepsilon v(z)$ on Γ_r. Since $v>0$, we have $H_r(z)<H(z)+\varepsilon v(z)$ on $\Gamma+\Gamma_r$, so that $u(z)\leqq H_r(z)<H(z)+\varepsilon v(z)$ in D_r. Since $\varepsilon>0$ is arbitrary, we have

$$u(z)\leqq H(z) \quad \text{in } D-E_1\,. \tag{3}$$

Since at any point $a\in E_1$, which belongs to D,

$$u(a)=\overline{\lim_{z\to a}}\,u(z)\leqq\overline{\lim_{z\to a}}\,H(z)=H(a)\,,$$

we have

$$u(z)\leqq H(z) \quad \text{in } D\,. \tag{4}$$

Hence $u(z)$ is subharmonic in U.

Let $u_1(z)$ be another subharmonic prolongation of u, such that $u_1=u$ in $U-E$, then by Remark (p. 41), $u_1(a)=\overline{\lim}_{z\to a}\,u_1(z)=\overline{\lim}_{z\to a}\,u(z)=u(a)$ $(a\in E)$, so that $u_1\equiv u$, hence such a prolongation is unique. ——

2. Let E be a bounded closed set on the z-plane, then its complement consists of at most a countable number of connected domains $\{D_\nu\}+D_\infty$, where D_∞ contains $z=\infty$. Let Γ^* be the boundary of D_∞, which we call *the outer boundary of E*. The complement of Γ^* consists of at most a countable number of connected domains $\{\Delta_\nu\}+D_\infty$.

Theorem III. 31. *Let E be a bounded closed set of positive capacity on the z-plane and $u(z)$ be its conductor potential and μ be its equilibrium distribution and E^* be its carrier. Then*

(i) *the μ-mass lies on the outer boundary Γ^* of E and $\gamma(\Gamma^*-E^*)=0$, so that*

$$\gamma(E)=\gamma(E^*)=\gamma(\Gamma^*)\,.$$

(ii) *$u(z)=V$ in Δ_ν, where $V=\log 1/\gamma(E)$.*

Proof. $u(z)$ is a bounded harmonic function in D_ν and $u=V$ "almost everywhere" on its boundary, hence $u=V$ in D_ν (Theorem III. 28. Cor.). Since $u=V$ at inner points of E (Theorem III. 14), $u=V$ "almost everywhere" in Δ_ν, hence almost everywhere in Δ_ν, so that $\mu(\Delta_\nu)=0$ (Theorem II. 26), hence $u(z)$ is harmonic and $u=V$ in Δ_ν and the μ-mass lies on Γ^*, so that $E^*\subset\Gamma^*$.

Next we shall prove that $\gamma(\Gamma^*-E^*)=0$. We put $E_1=\Gamma^*-E^*$ and suppose that $\gamma(E_1)>0$, then E_1 contains a closed sub-set F of positive capacity. Since F and E^* have a positive distance, $u<V$ on F, but since $\gamma(F)>0$, F contains a point z_0, such that $u(z_0)=V$, which is absurd. Hence $\gamma(\Gamma^*-E^*)=0$, so that by Theorem III. 18,

$$\gamma(E)=\gamma(E^*)=\gamma(\Gamma^*) .$$

Theorem III. 32. *Let E be a bounded closed set of positive capacity on the z-plane, then its equilibrium distribution μ is unique.*

Proof. Suppose that there are two such μ_1, μ_2 and put

$$u_i(z)=\int_E \log\frac{1}{|z-a|}\,d\mu_i(a), \qquad \mu_i(E)=1 \quad (i=1, 2) .$$

$$u(z)=u_1(z)-u_2(z)=\int_E \log\frac{1}{|z-a|}\,d\mu(a), \qquad \mu=\mu_1-\mu_2 .$$

Then $u(z)$ is a bounded harmonic function in D_∞ and since $u_1=u_2=V$ "almost everywhere" on the boundary Γ^* of D_∞, $u\equiv 0$ in D_∞. Since $u_1=u_2=V$ in Δ_ν, $u=0$ in Δ_ν, so that $u=0$ "almost everywhere", hence almost everywhere on the z-plane, so that $\mu\equiv 0$ (Theorem II. 26), hence

$$\mu_1\equiv\mu_2 .$$

8. Kellogg's theorem

Theorem III. 33. *Let E be a bounded closed set of positive capacity on the z-plane and Γ^* be its outer boundary, then Γ^* contains infinitely many regular points of the Dirichlet problem for D_∞ and the set $\wedge$ of irregular points on Γ^* is of capacity zero.*

Kellogg [2].

We shall prove in Theorem III. 38, that $\wedge$ is an F_σ-set.

Proof. Since $\gamma(\Gamma^*)=\gamma(E)>0$, Γ^* contains infinitely many points a, such that $u(a)=V$. We shall prove that a is a regular point of the Dirichlet problem for D_∞. Since $u(a)=V$, $u(z)$ is continuous at a, so that $\lim_{z\to a}(V-u(z))=0$. Since $V-u(z)>0$ is harmonic in $U(a)\cdot D_\infty$, where $U(a)$ is a neighbourhood of a, by Theorem I. 13, there exists a barrier at a, so that a is a regular point of the Dirichlet problem for D_∞.

Since $u(a)=V$ "almost everywhere" on Γ^*, the set of irregular points on Γ^* is of capacity zero.

Theorem III. 34. *Let D be a domain on a closed Riemann surface F and Γ be its boundary and $\gamma(\Gamma)>0$. Let $\zeta_0 \in \Gamma$, then ζ_0 is a regular point of the Dirichlet problem for D, when and only when there exists a barrier at ζ_0,*

Hence *the regularity is a local property.*

Proof. By Theorem I, 10, we have only to prove that if ζ_0 is a regular point, then there exists a barrier at ζ_0. Let Γ_ρ be the part of Γ, which lies in $|z-\zeta_0|\leqq\rho$. We define $f(\zeta)$ on Γ, such that $f(\zeta)=\frac{1}{\rho}|\zeta-\zeta_0|$ on Γ_ρ and $f(\zeta)=1$ on $\Gamma-\Gamma_\rho$. Then $f(\zeta)$ is continuous on Γ. Since $\gamma(\Gamma)>0$, Γ contains infinitely many regular points, so that we take $\rho>0$ so small that $\Gamma-\Gamma_\rho$ contains a regular point. We solve the Dirichlet problem for D with the boundary value $f(\zeta)$ and let $u(z)$ be its solution. Then since $u=1$ at regular points on $\Gamma-\Gamma_\rho$, $u\not\equiv 0$, so that $u>0$ in D. Since ζ_0 is a regular point, $\lim\limits_{z\to\zeta_0} u(z)=0$, hence if we take $w(z)=u(z)$ in Theorem I. 13, then we see that there exists a barrier at ζ_0.

Theorem III. 35. *Let D be a domain on a closed Riemann surface F and Γ be its boundary. Then the Green's function $g(z, z_0)$ of D exists, when and only when $\gamma(\Gamma)>0$.*

Proof. We may assume that z_0 is different from a branch point of F. Suppose that a Green's function $g(z, z_0)$ of D exists, and $\gamma(\Gamma)=0$, then since $g(z, z_0)$ is bounded in a neighbourhood of Γ, $g(z, z_0)$ is harmonic on Γ, hence $g(z, z_0)$ is harmonic on F, except at z_0, where if we put $g(z, z_0)-\log\frac{1}{|z-z_0|}=\omega(z)$, then $\omega(z)$ is harmonic at z_0, but as proved in p. 19, there exists no such $g(z, z_0)$, hence if the Green's function $g(z, z_0)$ of D exists, then $\gamma(\Gamma)>0$. Next we shall prove that if $\gamma(\Gamma)>0$, then the Green's function $g(z, z_0)$ of D exists.

Let Δ: $|z-z_0|\leqq\rho$ be a schlicht disc, contained in D and C: $|z-z_0|=\rho$ be its bounding circle. We solve the Dirichlet problem for $D-\Delta$, with the boundary value 0 on C and 1 on Γ and let $u(z)$ be its solution, then since $\gamma(\Gamma)>0$, Γ contains a regular point ζ_0, at which $u(\zeta_0)=1$, hence $u\not\equiv 0$, so that $0<u<1$ in $D-\Delta$.

Let $D_0\subset D_1\subset\cdots\subset D_n\to D$ be an exhaustion of D, where $D_0=\Delta$ and the boundary Γ_n of D_n consists of a finite number of analytic Jordan curves. Let $g_n(z, z_0)$ be the Green's function of D_n, then

$$\int_C \left(u \frac{\partial g_n}{\partial \nu} - g_n \frac{\partial u}{\partial \nu} \right) ds + \int_{\Gamma_n} \left(u \frac{\partial g_n}{\partial \nu} - g_n \frac{\partial u}{\partial \nu} \right) ds = 0 ,$$

where ν is the inner normal with respect to $D_n - \Delta$.

Since $g_n = 0$ on Γ_n, and $u = 0$ on C, we have

$$\int_C g_n \frac{\partial u}{\partial \nu} ds = \int_{\Gamma_n} u \frac{\partial g_n}{\partial \nu} ds \leqq \int_{\Gamma_n} \frac{\partial g_n}{\partial \nu} ds = 2\pi ,$$

so that

$$\int_C g \frac{\partial u}{\partial \nu} ds \leqq 2\pi ,$$

where $g = g(z, z_0) = \lim_n g_n(z, z_0)$, hence $g \neq \infty$, so that the Green's function $g(z, z_0)$ of D exists.

Theorem III. 36. *Let D be a domain on a closed Riemann surface and Γ be its boundary and $\gamma(\Gamma) > 0$ and $g(z, z_0)$ be the Green's function of D. Let $\zeta_0 \in \Gamma$, then ζ_0 is a regular point of the Dirichlet problem for D, when and only when*

$$\lim_{z \to \zeta_0} g(z, z_0) = 0, \quad z \in D .$$

Proof. If ζ_0 is a regular point, then a barrier exists at ζ_0, so that by Theorem I. 18, $\lim_{z \to \zeta_0} g(z, z_0) = 0$. If $\lim_{z \to \zeta_0} g(z, z_0) = 0$, then since $g(z, z_0) > 0$ in D and is harmonic in a neighbourhood of ζ_0, by Theorem I. 13, ζ_0 is a regular point.

Theorem III. 37. *Let E be a bounded closed set of positive capacity on the z-plane and $g(z, \infty)$ be the Green's function of D_∞ with $z = \infty$ as its pole, then*

$$g(z, \infty) = V - u(z),$$

where $V = \log 1/\gamma(E)$ and $u(z)$ is the conductor potential of E.

Proof. Let $D_0 \subset D_1 \subset \cdots \subset D_n \to D_\infty$ be an exhaustion of D_∞, where D_n contains $z = \infty$ and its boundary Γ_n consists of a finite number of analytic Jordan curves and $g_n(z, \infty)$ be the Green's function of D_n, then $\lim_n g_n(z, \infty) = g(z, \infty)$ is that of D_∞.

Now $V - u(z) - g_n(z, \infty)$ is harmonic in D_n and $V - u(z) > 0$ in D_∞, $g_n = 0$ on Γ_n, so that by the maximum principle, $V - u(z) \geqq g_n(z, \infty)$ in D_n, hence

$$V - u(z) \geqq g(z, \infty) \quad \text{in } D_\infty .$$

Let a be a point on the boundary Γ^* of D_∞, such that $u(a) = V$, then $u(z)$ is continuous at a, so that $\lim_{z \to a} (V - u(z)) = 0$, hence $\lim_{z \to a} g(z, \infty) = 0$, hence $V - u(z) - g(z, \infty)$ is a bounded harmonic function in D_∞, which vanishes "almost everywhere" on Γ^*, so that $V-u(z)-g(z,\infty)\equiv 0$, or

$$g(z,\infty)=V-u(z). \text{ ——}$$

Let

$$u(z)=\int_E \log\frac{1}{|z-a|}d\mu(a), \qquad \mu(E)=1\,,$$

then

$$g(z,\infty)=V-u(z)=\log|z|+V-\int_E \log\left|\frac{z}{z-a}\right| d\mu(a)\,,$$

hence

$$\omega(\infty)=\lim_{z\to\infty}(g(z,\infty)-\log|z|)=V\,,$$

where $\omega(\infty)$ is the Robin's constant of D_∞ at $z=\infty$. Hence

Corollary. *Let E be a bounded closed set of positive capacity on the z-plane, then*

$$\omega(\infty)=V=\log 1/\gamma(E)\,,$$

where $\omega(\infty)$ is the Robin's constant of D_∞ at $z=\infty$.

Theorem III. 38. *Let E be a bounded closed set of positive capacity on the z-plane and $u(z)$ be its conductor potential. Let a_0 be a point on the outer boundary Γ^* of E, then a_0 is a regular point of the Dirichlet problem for D_∞, when and only when*

$$u(a_0)=V=\log 1/\gamma(E)\,.$$

Hence the set of irregular points on Γ^* is an F_σ-set of capacity zero.

Proof. If $u(a_0)=V$, then by Theorem III. 36, 37, a_0 is a regular point, hence it remains to prove that if a_0 is a regular point, then $u(a_0)=V$. Suppose that a_0 is a regular point, then

$$\lim_{z\to\infty} g(z,\infty)=\lim_{z\to\infty}(V-u(z))=0, \qquad z\in D_\infty\,. \tag{1}$$

Let U be a neighbourhood of a_0, then since u is superharmonic,

$$\frac{1}{|U|}\iint_U u(z)\,dxdy\leqq u(a_0)\,. \tag{2}$$

Since $u=V$ "almost everywhere" in the complement of D_∞, we see from (1), (2), that $V\leqq u(a_0)$, but $u(a_0)\leqq V$, hence $u(a_0)=V$.

Hence the set of irregular points on Γ^* is a set of a, such that $u(a)<V$, so that it is an F_δ-set of capacity zero, since u is a lower semi-continuous function.

9. Mapping radius

1. We shall prove

Theorem III. 39. *Let E be a bounded continuum on the z-plane. If we map D_∞ conformally on $|w|>R$ on the w-plane by $w=w(z)$, such that*

$$w=w(z)=z+a_0+\frac{a_1}{z}+\cdots \text{ at } z=\infty,$$

then

$$R=\gamma(E).$$

R is called *the mapping radius of E.*

Proof. Let $u(z)$ be the conductor potential of E and put

$$v(z)=u(z)-\log\frac{1}{|w(z)|},$$

then $v(z)$ is a bounded harmonic function in D_∞ and $v=V-\log 1/R$ on Γ^*, so that $v(z)\equiv\text{const.}=V-\log 1/R$, but since $v(\infty)=0$, const.$=0$, so that $V-\log 1/R=0$, or $R=\gamma(E)$.

Corollary 1. *Let $\Delta: |z|\leqq R$ be a disc and $C: |z|=R$ be a circle, then*

$$\gamma(\Delta)=\gamma(C)=R.$$

Corollary 2. *Let α be an arc on a circle $|z|=R$, whose central angle is θ, then*

$$\gamma(\alpha)=R\sin\theta/4.$$

Corollary 3. *Let J be a rectilinear segment of length L, then*

$$\gamma(J)=L/4.$$

Proof. We may assume that J lies on the real axis and J: $-\frac{L}{2}\leqq x\leqq\frac{L}{2}$, $y=0$. Then by $z=\frac{L}{4}(v+1/v)$, the outside of J is mapped on $|v|>1$, so that if we put $w=Lv/4$, then by $z=w+L^2/16w$, the outside of J is mapped on $|w|>L/4$ and $w=z+\cdots$ at $z=\infty$, hence $\gamma(J)=L/4$.

Corollary 4. *Let E be a bounded closed set on the x-axis and L be its linear measure, then* $\gamma(E)\geqq L/4$.

Proof. Let E be contained in $[a, b]$ and $L(x)$ be the linear measure of the part E_x of E, which is contained in $[a, x]$, then by $t=L(x)$, E is mapped on an interval J: $0\leqq t\leqq L$ on the t-axis. We take n points $0=t_1<t_2<\cdots<t_n=L$ on J and let $t_i=L(x_i)$, $x_i\in E$, then $|t_i-t_k|\leqq|x_i-x_k|$,

so that by comparing the transfinite diameters, we have $L/4=\gamma(J)\leqq\gamma(E)$.

Corollary 5. *Let E be a continuum on the z-plane and $\delta(E)$ be its diameter, then $\gamma(E)\geqq\delta(E)/4$.*

Proof. By taking the coordinates axes suitably, we may assume that E is projected on a segment: $a\leqq x\leqq b$, $b-a=\delta(E)$ of length $\delta(E)$ on the x-axis.

We take n points $a=x_1<x_2<\cdots<x_n=b$ and through x_i, we draw a parallel to the y-axis, then since E is a continuum, it meets E at a point z_i, then $|z_i-z_k|\geqq|x_i-x_k|$, so that by comparing the transfinite diameters, we have by Cor. 3, $\gamma(E)\geqq\delta(E)/4$.

Similarly we can prove

Corollary 6. *Let E be a bounded closed set on the $z=x+iy$-plane. Let $E^{(x)}$ be the set of r on the positive x-axis, such that the circle $|z|=r$ intersects E and L be its linear measure, then*

$$\gamma(E)\geqq\gamma(E^{(x)})\geqq L/4\ ,$$

where the equality holds, only when E is a segment of length L on the line: $\arg z=\text{const.}$

Hence if $\gamma(E)=0$, then $L=0$.

2. As an application of Cor. 6, we shall prove the following extension of Koebe's 1/4-disc theorem.

Theorem III. 40. *Let $w=f(z)=z+a_2z^2+\cdots$ be regular in $|z|<1$ and D_f be the set of values $w=f(z)\,(|z|<1)$ and E be the set of $r(\geqq0)$, such that the whole circle $|w|=r$ is contained in D_f, then*

$$mE\geqq1/4\ ,$$

where the equality holds, only when

$$w=f_\alpha(z)=\frac{z}{(1+ze^{i\alpha})^2}\ ,$$

which maps $|z|<1$ conformally on the w-plane, excepted a half-line $L_{-\alpha}$: $w=\rho e^{-i\alpha}$ $(1/4\leqq\rho\leqq\infty)$. Hayman [1].

Hence if $f(z)\neq f_\alpha(z)$, then there exists $r\geqq1/4$, such that the whole circle $|w|=r$ is contained in D_f.

First we shall prove a lemma.

Lemma. *Let $w=f(z)=\dfrac{1}{z}+a_0+a_1z+\cdots$ be meromorphic in $|z|<1$*

and E_f *be the complement of* D_f, *then* E_f *is a bounded closed set and*

$$\gamma(E_f)\leqq 1 .$$

Proof. Let $g(w, \infty)$ be the Green's function of D_f and $v(z)=g(f(z), \infty)-\log\frac{1}{|z|}$, then since $f(z)$ may have a pole in $0<|z|<1$, $v(z)$ is superharmonic in $|z|<1$ and $\underline{\lim}\, v(z)\geqq 0$, when z tends to $|z|=1$, so that $v(z)\geqq 0$ in $|z|<1$, hence $v(0)\geqq 0$.

Since $v(0)$ is the Robin's constant of D_f at $w=\infty$, which is equal to $\log 1/\gamma(E_f)$ (Theorem III. 37, Cor.), we have

$$\gamma(E_f)\leqq 1 .$$

Proof of the theorem.

If we apply the lemma to

$$v(z)=\frac{1}{w}=\frac{1}{f(z)}=\frac{1}{z}+\cdots, \tag{1}$$

then $\gamma(E_v)\leqq 1$. Let M' be the set of $r\,(\geqq 0)$, such that the circle $|v|=r$ intersects E_v, then by Theorem III. 39. Cor. 6, $1\geqq\gamma(E_v)\geqq\frac{1}{4}\,mM'$, $mM'\leqq 4$. Let M be the complement of M' on the positive r-axis, then we see easily that

$$mE=\int_M \frac{dr}{r^2}.$$

Since $mM'\leqq 4$,

$$mE=\int_M \frac{dr}{r^2}\geqq\int_4^\infty \frac{dr}{r^2}=\frac{1}{4}. \tag{2}$$

If $mE=1/4$, then $mM'=4$, $\gamma(E_v)=1$ and M' is a segment $[0, 4]$. From $\gamma(E_v)=mM'/4$, we see that E_v is a segment: $v=\rho e^{i\alpha}\,(0\leqq\rho\leqq 4)$, so that E_f is a half-line $L_{-\alpha}$: $w=\rho e^{-i\alpha}\,(1/4\leqq\rho\leqq\infty)$. We map D_f conformally on $|\zeta|<1$ by $w=\frac{\zeta}{(1+\zeta e^{i\alpha})^2}$, then $\zeta=z+\cdots$, so that by Schwarz's lemma, $\zeta\equiv z$, hence

$$w=\frac{z}{(1+ze^{i\alpha})^2}.$$

10. Balayage

1. Let D be a bounded domain on the z-plane and Γ be its boundary. We exhaust D by a sequence of domains $D_1\subset D_2\subset\cdots\subset D_n\to D$, where the boundary Γ_n of D_n consists of a finite number of analytic Jordan curves. Let $g_n(z, z_0)$ $(z_0\in D_n)$ be the Green's function of D_n with z_0 as its pole, then

$$g_n(z, z_0)=\log\frac{1}{|z-z_0|}-\frac{1}{2\pi}\int_{\Gamma_n}\log\frac{1}{|z_0-a|}\frac{\partial g_n(a, z)}{\partial\nu_a}\,|da|, \tag{1}$$

where ν_a is the inner normal of Γ_n at a.

Since $g_n(z, z_0)=g_n(z_0, z)$,

$$g_n(z, z_0)=\log\frac{1}{|z-z_0|}-\frac{1}{2\pi}\int_{\Gamma_n}\log\frac{1}{|z-a|}\frac{\partial g_n(a, z_0)}{\partial\nu_a}|da|. \quad (2)$$

Hence if we put

$$d\mu_{z_0}^n(a)=\frac{1}{2\pi}\frac{\partial g_n(a, z_0)}{\partial\nu_a}|da|\geqq 0\,, \quad (3)$$

then

$$g_n(z, z_0)=\log\frac{1}{|z-z_0|}-\int_{\Gamma_n}\log\frac{1}{|z-a|}d\mu_{z_0}^n(a), \qquad z\in D_n\,, \quad (4)$$

$$\int_{\Gamma_n}d\mu_{z_0}^n(a)=1\,.$$

By Theorem II. 4, we can select n_ν, such that $\mu_{z_0}^{n_\nu}\to\mu_{z_0}$ $(\nu\to\infty)$, hence

Theorem III. 41. *Let $g(z, z_0)$ be the Green's function of D, then*

$$g(z, z_0)=\log\frac{1}{|z-z_0|}-\int_{\Gamma}\log\frac{1}{|z-a|}d\mu_{z_0}(a), \quad z_0\in D,\ z\in D\,,$$

$$\int_{\Gamma}d\mu_{z_0}(a)=1\,.$$

μ_{z_0} is called *the mass of balayage of a unit mass, located at z_0.*

Now in $\mu_{z_0}^{n_\nu}\to\mu_{z_0}$ $(\nu\to\infty)$, $\{n_\nu\}$ depends on z_0 in general. We shall prove that we can find $\{n_\nu\}$, such that $\mu_{z_0}^{n_\nu}\to\mu_{z_0}$ for any $z_0\in D$.

Let z_k $(k=1, 2, \cdots)$ be the set of all rational points in D, then by the diagonal method, we can find $\{n_\nu\}$, such that $\mu_{z_k}^{n_\nu}\to\mu_{z_k}$ $(\nu\to\infty)$ for any $k=1, 2, \cdots$. Let $a_0\in\Gamma$ and U: $|z-a_0|<\rho$ be its neighbourhood and $\Gamma(\rho)$, $\Gamma_n(\rho)$ be the part of Γ, Γ_n, which lie in U respectively.

If we put

$$v_n(z)=\int_{\Gamma_n(\rho)}d\mu_z^n(a)=\frac{1}{2\pi}\int_{\Gamma_n(\rho)}\frac{\partial g_n(a, z)}{\partial\nu_a}|da|\,, \quad (1)$$

then $v_n(z)$ $(n=1, 2, \cdots)$ is a harmonic function in D_n and is uniformly bounded in any compact domain $\subset D$, so that they form a normal family. Since $\lim_\nu v_{n_\nu}(z_k)$ $(k=1, 2, \cdots)$ exists, we have for any $z_0\in D$,

$$v(z_0)=\lim_\nu v_{n_\nu}(z_0)=\lim_\nu\int_{\Gamma_{n_\nu}(\rho)}d\mu_{z_0}^{n_\nu}(a)=\int_{\Gamma(\rho)}d\mu_{z_0}(a) \quad (2)$$

exists and uniformly in the wider sense in D, so that $v(z_0)$ is a harmonic function of $z_0\in D$. Hence $\mu_{z_0}^{n_\nu}\to\mu_{z_0}$ for any $z_0\in D$. We say that *$d\mu_{z_0}(a)$ is a harmonic function of z_0*, which means that $\int_{\Gamma(\rho)}d\mu_{z_0}(a)$ is a harmonic function of z_0 for any $\rho>0$.

2. Since $g(z, z_0)=g(z_0, z)$,

$$g(z, z_0)=\log\frac{1}{|z-z_0|}-\int_\Gamma \log\frac{1}{|z_0-a|}\,d\mu_z(a), \quad z_0\in D,\ z\in D. \tag{1}$$

Let Ω be the complement of $\bar{D}$, then for a fixed $z\in\Omega$, $\log\dfrac{1}{|z-z_0|}$ is a harmonic function of $z_0\in D$, so that

$$\log\frac{1}{|z-z_0|}=\frac{1}{2\pi}\int_{\Gamma_n}\log\frac{1}{|z-a|}\,\frac{\partial g_n(a, z_0)}{\partial\nu_a}\,|da|\,,$$

hence

$$\log\frac{1}{|z-z_0|}=\int_\Gamma \log\frac{1}{|z-a|}\,d\mu_{z_0}(a), \quad \text{if} \quad z_0\in D,\ z\in\Omega\,. \tag{2}$$

Hence if we define $g(z, z_0)=0$ for $z_0\in D$, $z\in\Omega$, then $g(z, z_0)$ can be expressed by the same expression as follows:

$$g(z, z_0)=\log\frac{1}{|z-z_0|}-\int_\Gamma \log\frac{1}{|z-a|}\,d\mu_{z_0}(a), \qquad z_0\in D\,, \tag{3}$$
$$z\in D, \text{ or } z\in\Omega\,.$$

Theorem III. 42. *μ_{z_0} is unique.*

Proof. Suppose that there are two μ'_{z_0}, μ''_{z_0}, such that

$$g(z, z_0)=\log\frac{1}{|z-z_0|}-\int_\Gamma \log\frac{1}{|z-a|}\,d\mu'_{z_0}(a)$$
$$=\log\frac{1}{|z-z_0|}-\int_\Gamma \log\frac{1}{|z-a|}\,d\mu''_{z_0}(a)\,, \qquad z\in D, \text{ or } z\in\Omega\,,$$

then if we put $\mu_{z_0}=\mu'_{z_0}-\mu''_{z_0}$, we have for any $z\in D$, $z\in\Omega$,

$$\int_\Gamma \log\frac{1}{|z-a|}\,d\mu_{z_0}(a)=0, \quad z_0\in D\,. \tag{1}$$

Since $g(z, z_0)=0$ "almost everywhere" on Γ, (1) holds almost everywhere on the z-plane, so that by Theorem II. 26, $\mu_{z_0}\equiv 0$, or $\mu'_{z_0}\equiv\mu''_{z_0}$.——

3. Let $a_0\in\Gamma$ and U: $|z-a_0|<\rho$ be its neighbourhood. If $\lim\limits_{z\to a_0}\int_U d\mu_z(a)=1$ $(z\in D)$ for any small U, then we write $d\mu_z\to 1$, as $z\to a_0$.

Theorem III. 43. *Let $a_0\in\Gamma$, then a_0 is a regular point of the Dirichlet problem for D, when and only when*

$$d\mu_z\to 1, \quad \textit{as } z\to a_0\,.$$

Proof. Since

$$g(z, z_0)=\log\frac{1}{|z-z_0|}-\int_\Gamma \log\frac{1}{|z_0-a|}\,d\mu_z(a)\,, \quad z\in D\,.$$

if $d\mu_z\to 1$ as $z\to a_0$, then $\lim\limits_{z\to a_0} g(z, z_0)=0$, so that by Theorem III. 36, a_0 is a regular point.

If a_0 is a regular point, then $\lim\limits_k g(z_k, z_0)=0$ $(z_k\to a_0)$. We may assume that $\mu_{z_k}\to\mu^*(k\to\infty)$, hence from

$$g(z_k, z_0)=\log\frac{1}{|z_k-z_0|}-\int_\Gamma \log\frac{1}{|z_0-a|}\,d\mu_{z_k}(a)\,,$$

we have

$$\log\frac{1}{|a_0-z_0|}=\int_\Gamma \log\frac{1}{|z_0-a|}\,d\mu^*(a)\,.$$

Let $b\in\Gamma$ $(b\neq a_0)$ and V be its neighbourhood, which does not contain a_0, then $\log\frac{1}{|a_0-z_0|}$ is a harmonic function of $z_0\in V$, so that $\mu^*(V)=0$, hence $d\mu_z\to 1$ as $z\to a_0$. ——

By Theorem III. 42, we can express the solution of the Dirichlet problem by means of μ_z as follows.

Theorem III. 44. *Let D be a bounded domain on the z-plane and Γ be its boundary and $f(a)$ be a given continuous function on Γ and put*

$$u(z)=\int_\Gamma f(a)d\mu_z(a)\,,$$

then $u(z)$ is a bounded harmonic function in D and if $a_0\in\Gamma$ is a regular point, then $\lim_{z\to a_0}u(z)=f(a_0)$. Hence $\lim_{z\to a}u(z)=f(a)$ "almost everywhere" on Γ.

4. By the theorem, we can prove the following Wiener's theorem simply as follows. Suppose that $f(a)$ is continuous on Γ, then by Lebesgue's theorem, there exists a continuous function $F(z)$ on the whole z-plane, such that $F(a)=f(a)$ on Γ. We exhaust D by a sequence of domains $D_1\subset D_2\subset\cdots\subset D_n\to D$, where the boundary Γ_n of D_n consists of a finite number of analytic Jordan curves. Let $u_n(z)$ be the solution of the Dirichlet problem for D_n, with the boundary value F on Γ_n, then

$$u_n(z)=\frac{1}{2\pi}\int_{\Gamma_n} F(a)\frac{\partial g_n(a, z)}{\partial\nu_a}|da|,\qquad z\in D_n\,,$$

so that

$$\lim_n u_n(z)=u(z)=\int_\Gamma f(a)\,d\mu_z(a),\qquad z\in D\,.$$

Since the right hand side is independent of the choice of $F(z)$ and $\{D_n\}$, we have

Theorem III. 45. *$\lim_n u_n(z)=u(z)$ exists, where $u(z)$ is independent of the choice of $F(z)$ and $\{D_n\}$ and can be expressed by*

$$u(z)=\int_\Gamma f(a)d\mu_z(a).$$

Wiener [2].

11. Elliptic capacity

1. Let K be the Riemann sphere of diameter 1, which touches the

z-plane at $z=0$ and $[a, b]=\dfrac{|a-b|}{\sqrt{(1+|a|^2)(1+|b|^2)}}$ be the spherical distance of two points a, b. Let E be a closed set on K and $\mu\geqq 0$ be a positive mass distribution on E of total mass 1 and put

$$\mathrm{I}(\mu)=\iint_E \log\frac{1}{[a, b]}\,d\mu(a)\,d\mu(b), \quad \mu(E)=1\,, \tag{1}$$

$$V=\inf_{\mu} I(\mu)\,. \tag{2}$$

If $V<\infty$, then as in §2, we can prove that there exists μ, such that

$$V=I(\mu)=\iint_E \log\frac{1}{[a, b]}\,d\mu(a)\,d\mu(b), \quad \mu(E)=1\,. \tag{3}$$

We call μ *the elliptic equlibrium distribution of E* and

$$u(z)=\int_E \log\frac{1}{[z, a]}\,d\mu(a), \quad \mu(E)=1\,, \tag{4}$$

the elliptic conductor potential of E[1].

$u(z)$ is lower semi-continuous and since

$$\varDelta \log\frac{\sqrt{1+|z|^2}}{|z-a|}=\frac{2}{(1+|z|^2)^2}>0\,,$$

$u(z)$ is subharmonic outside the mass, where $\varDelta$ denotes the Lapacian. Hence the maximum principle holds for $u(z)$.

We define *the elliptic capacity $\gamma_+(E)$ of E* by

$$\gamma_+(E)=e^{-V}, \quad V=\log 1/\gamma_+(E)\,, \tag{5}$$

and the elliptic capacity $\gamma_+(E)$ of any set E by

$$\gamma_+(E)=\sup_{F\subset E}\gamma_+(F)\,,$$

where F is a closed sub-set of E.

We can prove easily that if $E_1\subset E_2$, then $\gamma_+(E_1)\leqq\gamma_+(E_2)$ and $\gamma_+(E)=0$ and $\gamma(E)=0$ are equivalent. Since $u(z)$ is subharmonic outside of E, it attains its maximum on E, so that similarly as Theorem III. 12, we can prove

Theorem III. 46. *Let E be a closed set of positive capacity on K and $u(z)$ be its elliptic conductor potential, then $u(z)\leqq V$ on the whole sphere K and $u=V$ on E, except an F_σ-set of capacity zero.*

Theorem III. 47. *$u(a)=V$ at an inner point a of E and the density of μ at a is $1/\pi$.*

---

1) The conductor potential and the capacity, defined in §2, may be called the parabolic conductor potential and the parabolic capacity.

Hence the open kernel E° of E contains a positive μ-mass. (c.f. Theorem III. 14).

Proof. By a rotation of K, we may assume that $z=0$ is an inner point of E and put

$$\omega(r)=\int_{|a|<r} d\mu(a)\,,$$

then

$$u(re^{i\theta})=\int_E \log\frac{\sqrt{(1+r^2)(1+|a|^2)}}{|re^{i\theta}-a|}\,d\mu(a)$$

$$=\log\sqrt{1+r^2}+\int_E \log\frac{\sqrt{1+|a|^2}}{|re^{i\theta}-a|}\,d\mu(a)\,, \tag{1}$$

$$u(0)=\int_0^\infty \log\frac{\sqrt{1+t^2}}{t}\,d\omega(t)\leqq V<\infty\,, \tag{2}$$

so that

$$\lim_{t\to0}\omega(t)\log\frac{1}{t}=0.$$

Since

$$\frac{1}{2\pi}\int_0^{2\pi}\log|re^{i\theta}-a|\,d\theta=\mathrm{Max}\,(\log r,\ \log|a|),$$

we have

$$\frac{1}{2\pi}\int_0^{2\pi}u(re^{i\theta})\,d\theta=\log\sqrt{1+r^2}+\int_0^\infty(\log\sqrt{1+t^2}-\mathrm{Max}\,(\log r,\ \log t))\ d\omega(t)$$

$$=\log\sqrt{1+r^2}+\int_0^r\log\frac{\sqrt{1+t^2}}{r}\,d\omega(t)+\int_r^\infty\log\frac{\sqrt{1+t^2}}{t}\,d\omega(t)$$

$$=\log\sqrt{1+r^2}+\int_0^r\log\frac{\sqrt{1+t^2}}{r}\,d\omega(t)+\int_0^\infty\log\frac{\sqrt{1+t^2}}{t}\,d\omega(t)-\int_0^r\log\frac{\sqrt{1+t^2}}{t}\,d\omega(t)$$

$$=\log\sqrt{1+r^2}+u(0)+\int_0^r\log\frac{t}{r}\,d\omega(t)=\log\sqrt{1+r^2}+u(0)-\int_0^r\frac{\omega(t)}{t}\,dt. \tag{3}$$

Hence

$$\lim_{r\to0}\frac{1}{2\pi}\int_0^{2\pi}u(re^{i\theta})\,d\theta=u(0)\,,$$

so that

$$\lim_{r\to0}\frac{1}{\pi r^2}\int_0^r\int_0^{2\pi}u(re^{i\theta})\,rdrd\theta=u(0)\,. \tag{4}$$

Let a disc $|z|<\rho$ be contained in E, then since $u=V$ "almost everywhere" in $|z|<\rho$, we have from (4), $u(0)=V$, so that $u=V$ at inner points of E. Hence the left hand side of (3) is equal to V, if $0\leqq r\leqq\rho$, so that

$$\int_0^r\frac{\omega(t)dt}{t}=\log\sqrt{1+r^2}\,.$$

From this, we have $\omega(r)=\dfrac{r^2}{1+r^2}(0\leqq r<\rho)$. Since $\dfrac{\pi r^2}{1+r^2}$ is the area of the spherical disc on K, which corresponds to $|z|\leqq r$, the density of μ at $z=0$ is $1/\pi$. ——

Hence if $E=\bar{D}$ is a closed domain on K, which is bounded by a Jordan curve Γ, so that $\bar{D}=D+\Gamma$, then

$$\mu(D)=|D|/\pi, \quad \mu(\Gamma)=1-|D|/\pi,$$

where $|D|$ denotes the spherical area of D.

Theorem III. 48. *μ is unique.*

Proof. Suppose that there are two μ_1, μ_2 and put

$$u_i(z)=\int_E \log \frac{1}{[z, a]}\, d\mu_i(a), \quad \mu_i(E)=1 \quad (i=1, 2),$$

then if we put $\mu=\mu_1-\mu_2$,

$$u(z)=u_1(z)-u_2(z)=\int_E \log \frac{1}{|z-a|}\, d\mu(a)+\text{const.}.$$

By a suitable rotation of K, we may assume that E is projected on a bounded closed set on the z-plane. Then $u(z)$ is harmonic outside of E. Since $u=V$ at inner points of E and "almost everywhere" on the boundary of the complement of E, $u(z)=0$ almost everywhere on the whole z-plane, so that by Theorem II. 26, $\mu\equiv 0$, or $\mu_1\equiv\mu_2$.

2. Let E be a closed set on K, which contains infinitely many points. We take n points $z_i (i=1, 2, \cdots, n)$ on E and put

$$V(z_1, \cdots, z_n)=\prod_{i<k}^{1,\cdot,n} [z_i, z_k], \quad V_n=\underset{z_i\in E}{\mathrm{Max}}\, V(z_1, \cdots, z_n), \tag{1}$$

$$d_n=\sqrt[\binom{n}{2}]{V_n}. \tag{2}$$

Then similarly as Theorem III. 21, we can prove

Theorem III. 49. $d_{n+1}\leqq d_n$, *so that* $\lim\limits_n d_n=\tau_+(E)$ *exists.*

We call $\tau_+(E)$ *the elliptic transfinite diameter of* E.

Let $P(z)=\prod\limits_{i=1}^{n} [z, z_i]\ (z_i\in K)$ and

$$\mu(P)=\underset{z\in E}{\mathrm{Max}}\, P(z), \quad m_n=\inf_P \mu(P), \tag{1}$$

then there exists $T_n(z)=\prod\limits_{i=1}^{n} [z, a_i]\ (a_i\in K)$, such that

$$m_n=\mu(T_n)=\underset{z\in E}{\mathrm{Max}}\, T_n(z). \tag{2}$$

We call $T_n(z)$ *the elliptic Tchebycheff's polynomial of* E *of order* n. Similarly as Theorem III. 25, we can prove

Theorem III. 50. *If we put* $\sqrt[n]{m_n}=r_n$, *then* $\lim\limits_n r_n=\rho_+(E)$ *exists.*

We call $\rho_+(E)$ *the elliptic Tchebycheff's constant of* E.

Theorem III. 51. $\tau_+(E)=\rho_+(E)=\gamma_+(E)$. Tsuji [16].

Proof. By the definition,

$$\log\frac{1}{d_n}=\underset{z_i\in E}{\mathrm{Min}}\frac{1}{\binom{n}{2}}\sum_{i<k}^{1,\dots,n}\log\frac{1}{[z_i,\,z_k]},\tag{1}$$

$$\log\frac{1}{r_n}=\underset{z\in E}{\mathrm{Min}}\frac{1}{n}\sum_{i=1}^{n}\log\frac{1}{[z,\,a_i]}=\underset{z_i\in K}{\mathrm{Max}}\ \underset{z\in E}{\mathrm{Min}}\frac{1}{n}\sum_{i=1}^{n}\log\frac{1}{[z,\,z_i]},$$

where
$$T_n(z)=\prod_{i=1}^{n}[z,\,a_i]\,.\tag{2}$$

It suffices to prove that

(i) $\gamma_+(E)\leqq\rho_+(E)$, (ii) $\rho_+(E)\leqq\tau_+(E)$, (iii) $\tau_+(E)\leqq\gamma_+(E)$.

(i) $\gamma_+(E)\leqq\rho_+(E)$.

Let D_n be the lemniscate domain: $\frac{1}{n}\sum_{i=1}^{n}\log\frac{1}{[z,\,a_i]}\geqq\log\frac{1}{r_n}$, then we can prove similarly as Theorem III. 26, $\gamma_+(E)\leqq\gamma_+(D_n)\leqq r_n$, so that
$$\gamma_+(E)\leqq\rho_+(E)\,.$$

(ii) $\rho_+(E)\leqq\tau_+(E)$ can be proved similarly as Theorem III. 26.

(iii) $\tau_+(E)\leqq\gamma_+(E)$.

By a suitable rotation of K, we may assume that E is projected on a bounded closed set on the z-plane. With the same notation as the proof of Theorem III. 26, if $d\sigma(a)$ be the surface element on the z-plane, then

$$\log\frac{1}{\gamma_+(E)+\delta}\leqq\log\frac{1}{\gamma_+(\bar{D})}=V(\bar{D})\leqq\iint_{\bar{D}}\log\frac{1}{[a,\,b]}\rho(a)\,\rho(b)\,d\sigma(a)\,d\sigma(b)$$
$$=\sum_{j=1}^{n}\int_{\Delta_j}d\sigma(a)\sum_{i=1}^{n}\int_{\Delta_i}\log\frac{1}{[a,\,b]}\,d\sigma(b)\,.\tag{3}$$

Now
$$\int_{\Delta_i}\log\frac{1}{[a,\,b]}\,d\sigma(b)=\int_{\Delta_i}\log\frac{\sqrt{1+|a|^2}}{|a-b|}\,d\sigma(b)+\int_{\Delta_i}\log\sqrt{1+|b|^2}\,d\sigma(b)\,.\tag{4}$$

Since for a fixed a, $\log\frac{\sqrt{1+|a|^2}}{|a-b|}$ is a superharmonic function of b,

$$\int_{\Delta_i}\log\frac{\sqrt{1+|a|^2}}{|a-b|}\,d\sigma(b)\leqq\sigma(\Delta_i)\log\frac{\sqrt{1+|a|^2}}{|a-z_i|}=\frac{1}{n}\log\frac{\sqrt{1+|a|^2}}{|a-z_i|}\,.\tag{5}$$

If we put $z_i=r_ie^{\sqrt{-1}\cdot\theta_i}$, $b=z_i+re^{\sqrt{-1}\cdot\theta}$, then

$$\int_{\Delta_i}\log\sqrt{1+|b|^2}\,d\sigma(b)=\frac{1}{2}\int_0^{\varepsilon}\int_0^{2\pi}\log\left(1+r_i^2+r^2+2r\,r_i\cos(\theta-\theta_i)\right)rdrd\theta$$
$$=\frac{1}{n}\log\sqrt{1+r_i^2}+\frac{1}{2}\int_0^{\varepsilon}\int_0^{2\pi}\log\left(1+\frac{r^2+2r\,r_i\cos(\theta-\theta_i)}{1+r_i^2}\right)rdrd\theta$$

$$=\frac{1}{n}\log\sqrt{1+r_i^2}+\int_0^\varepsilon\int_0^{2\pi}\frac{r\,r_i\cos(\theta-\theta_i)}{1+r_i^2}\,rdrd\theta+\int_0^\varepsilon\int_0^{2\pi}O(r^2)\,rdrd\theta$$

$$=\frac{1}{n}\log\sqrt{1+r_i^2}+\int_0^\varepsilon\int_0^{2\pi}O(r^2)\,rdrd\theta=\frac{1}{n}\log\sqrt{1+r_i^2}+O(\varepsilon^4)$$

$$=\frac{1}{n}\log\sqrt{1+r_i^2}+O\left(\frac{1}{n^2}\right), \tag{6}$$

since
$$\int_\theta^{2\pi}\cos(\theta-\theta_i)d\theta=0\,.$$

Hence by (4), (5), (6),

$$\int_{\Delta_i}\log\frac{1}{[a,b]}\,d\sigma(b)\leqq\frac{1}{n}\log\frac{\sqrt{(1+|a|^2)(1+|z_i|^2)}}{|a-z_i|}+O\left(\frac{1}{n^2}\right)$$

$$=\frac{1}{n}\log\frac{1}{[a,z_i]}+O\left(\frac{1}{n^2}\right), \tag{7}$$

hence by (3), since $\sigma(\Delta_i)=1/n$,

$$\log\frac{1}{\gamma_+(E)+\delta}\leqq\frac{1}{n}\sum_{i,j}^{1,\cdot,n}\int_{\Delta_j}\log\frac{1}{[a,z_i]}\,d\sigma(a)+O\left(\frac{1}{n}\right). \tag{8}$$

If $j\neq i$, then by (7),

$$\int_{\Delta_j}\log\frac{1}{[a,z_i]}\,d\sigma(a)\leqq\frac{1}{n}\log\frac{1}{[z_i,z_j]}+O\left(\frac{1}{n^2}\right). \tag{9}$$

If $j=i$, then similarly as the proof of Theorem III. 26, we have

$$\int_{\Delta_i}\log\frac{1}{[a,z_i]}\,d\sigma(a)=O\left(\frac{\log n}{n}\right), \tag{10}$$

hence by (8), (9), (10),

$$\log\frac{1}{\gamma_+(E)+\delta}\leqq\frac{1}{n}\left(\frac{2}{n}\sum_{i<j}^{1,\cdot,n}\log\frac{1}{[z_i,z_j]}+O(\log n)\right)+O\left(\frac{1}{n}\right)$$

$$=\frac{2}{n^2}\sum_{i<j}^{1,\cdot,n}\log\frac{1}{[z_i,z_j]}+O\left(\frac{\log n}{n}\right). \tag{11}$$

From this, we have $\tau_+(E)\leqq\gamma_+(E)$, similarly as Theorem III. 26.

12. Hyperbolic capacity

Let a, b be two points in $|z|<1$, then we define the hyperbolic distance $[a, b]$ of a, b by $[a,b]=\left|\frac{a-b}{1-\bar{a}b}\right|$, hence $[a,b]<1$ and $[a,b]=[b,a]$. Let S be a linear transformation, which makes $|z|<1$ invariant, then

$$[S(a),\ S(b)]=[a,\ b].$$

Let E be a closed set, contained in $|z|<1$ and $\mu\geqq 0$ be a positive mass distribution on E of total mass 1 and put

$$I(\mu)=\iint_E \log \frac{1}{[a, b]}\, d\mu(a)\, d\mu(b), \qquad \mu(E)=1\,, \tag{1}$$

$$V=\inf_{\mu} I(\mu)\,, \tag{2}$$

then $0<V\leqq\infty$. If $V<\infty$, then there exists μ, such that

$$V=I(\mu)=\iint_E \log \frac{1}{[a, b]}\, d\mu(a)\, d\mu(b), \qquad \mu(E)=1\,. \tag{3}$$

We call μ *the hyperbolic equilibrium distribution of* E and

$$u(z)=\int_E \log \frac{1}{[z, a]}\, d\mu(a)=\int_E \log\left|\frac{1-\bar{a}z}{z-a}\right| d\mu(a), \qquad \mu(E)=1\,, \tag{4}$$

the hyperbolic conductor potential of E.

We define *the hyperbolic capacity* $\gamma_-(E)$ *of* E by

$$\gamma_-(E)=e^{-V}, \qquad V=\log 1/\gamma_-(E)\,, \tag{5}$$

and the hyperbolic capacity $\gamma_-(E)$ of any set E in $|z|<1$ by

$$\gamma_-(E)=\sup_{F\subset E} \gamma_-(F)\,,$$

where F is a closed sub-set of E.

If $E_1\subset E_2$, then $\gamma_-(E_1)\leqq\gamma_-(E_2)$ and $\gamma_-(E)=0$ and $\gamma(E)=0$ are equivalent. Similarly we Theorem III. 12, we can prove

Theorem III. 52. *Let E be a closed set of positive capacity in $|z|<1$ and $u(z)$ be its hyperbolic conductor potential, then $u(z)\leqq V$ in $|z|<1$ and $u=V$ on E, except an F_σ-set of capacity zero and $u=V$ at an inner point of E. The hyperbolic equilibrium distribution μ of E is unique.*

Let E be a closed set, contained in $|z|<1$. Then similarly as §1, we can define by means of the hyperbolic metric $[a, b]$, *the hyperbolic transfinite diameter* $\tau_-(E)$ and *the hyperbolic Tchebycheff's constant* $\rho_-(E)$ *of* E. Then

Theorem III. 53. $\tau_-(E)=\rho_-(E)=\gamma_-(E)$. Tsuji [16].

Proof. We can prove similarly as Theorem III. 26, (i) $\gamma_-(E)\leqq\rho_-(E)$ and (ii) $\rho_-(E)\leqq\tau_-(E)$. Hence it remains to prove that (iii) $\tau_-(E)\leqq\gamma_-(E)$. With the same notation as before, we have

$$\log\frac{1}{\gamma_-(E)+\delta} \leqq \sum_{j=1}^{n}\int_{\Delta_j} d\sigma(a)\sum_{i=1}^{u}\int_{\Delta_i} \log \frac{1}{[a, b]}\, d\sigma(b)\,. \tag{1}$$

Since $\log \dfrac{1}{[a, b]}=\log\left|\dfrac{1-\bar{a}b}{a-b}\right|$ is a superharmonic function of b,

$$\int_{\Delta_i} \log \frac{1}{[a, b]} d\sigma(b) \leqq \sigma(\varDelta_i) \log \frac{1}{[a, z_i]} = \frac{1}{n} \log \frac{1}{[a, z_i]},$$

so that

$$\log \frac{1}{\gamma_-(E)+\delta} \leqq \frac{1}{n} \sum_{i,j}^{1,\cdot,n} \int_{\Delta_j} \log \frac{1}{[a, z_i]} d\sigma(a) . \tag{2}$$

If $j \neq i$, then

$$\int_{\Delta_j} \log \frac{1}{[a, z_i]} d\sigma(a) \leqq \frac{1}{n} \log \frac{1}{[z_i, z_j]}. \tag{3}$$

If $j = i$, then

$$\int_{\Delta_i} \log \frac{1}{[a, z_i]} d\sigma(a) = \int_{\Delta_i} \log \frac{1}{|a-z_i|} d\sigma(a) + \int_{\Delta_i} \log |1-a\bar{z}_i| \, d\sigma(a)$$

$$= O\left(\frac{\log n}{n}\right) + \sigma(\varDelta_i) \log |1-z_i\bar{z}_i| = O\left(\frac{\log n}{n}\right). \tag{4}$$

From this, we have $\tau_-(E) \leqq \gamma_-(E)$ as before.

13. Modulus of a ring domain

1. Let $\varDelta$ be a ring domain on the ζ-plane, which is bounded by two continua E_1, E_2. Then we can map $\varDelta$ conformally on a circular ring: $0<r<|w|<1$ on the w-plane, such that E_1 is mapped on $|w|=1$ and E_2 on $|w|=r$. Then

$$M = M(\varDelta) = \log \frac{1}{r} \tag{1}$$

is called *the modulus of a ring domain* $\varDelta$.

Evidently M is a conformal invariant and we obtain the same value r, if we interchange E_1 and E_2. If we map the outside of E_1 conformally on $|z|<1$, then E_2 is mapped on a continuum E, contained in $|z|<1$. Then E and $|z|=1$ bound a ring domain D. By the conformal invariance of M, $$M(\varDelta) = M(D) .$$

Theorem III. 54. *Let E be a continuum, contained in $|z|<1$, then E and $|z|=1$ bound a ring domain D. If we map D conformally on a circular ring: $0<r<|w|<1$ on the w-plane, such that $|z|=1$ is mapped on $|w|=1$ and E on $|w|=r$, then*

$$r = \gamma_-(E), \qquad M(D) = \log 1/\gamma_-(E).$$

Proof. Let

$$u(z) = \int_E \log \left| \frac{1-\bar{a}z}{z-a} \right| d\mu(a), \qquad \mu(E)=1$$

be the hyperbolic conductor potential of E, then since every point of E is a regular point of the Dirichlet problem, $u = V = \log 1/\gamma_-(E)$ on E. Let $v(z)$ be the conjugate harmonic function of $u(z)$, then if C be a

Jordan curve, surrounding E and contained in D,

$$-\int_C dv = -\int_C \frac{dv}{ds}\,ds = -\int_C \frac{\partial u}{\partial \nu}\,ds = -\int_E d\mu(a)\int_C \frac{\partial}{\partial \nu}\log\left|\frac{1-\bar{a}z}{z-a}\right||dz|$$

$$=\int_E d\mu(a)\int_C d\arg(z-a) = 2\pi\int_E d\mu(a) = 2\pi\,,$$

where ν is the outer normal of C, hence by $w=e^{-(u+iv)}$, D is mapped conformally on $r<|w|<1$, where $r=e^{-V}=\gamma_-(E)$ and $|z|=1$ is mapped on $|w|=1$ and E on $|w|=r$. ——

2. Let Δ be a ring domain on the z-plane, which is bounded by two continua E_1, E_2 and let Δ_1 be the simply connected domain, which is bounded by E_1 and contains E_2. If we map Δ_1 conformally on $|\zeta|<1$, such that E_1 is mapped on $|\zeta|=1$, then E_2 is mapped on a continuum E, contained in $|\zeta|<1$, which bounds with $|\zeta|=1$ a ring domain D, then

$$M(\Delta)=M(D)=\log 1/\gamma_-(E)\,. \tag{1}$$

Let $g(z, a)$ be the Green's function of Δ_1. If z, a correspond to ζ, α in $|\zeta|<1$, then $g(z, a)=\log\left|\dfrac{1-\bar{\alpha}\zeta}{\zeta-\alpha}\right|$, so that from the definition of $V=\log 1/\gamma_-(E)$ and (1), we have

Theorem III. 55. *Let Δ be a ring domain on the z-plane, which is bounded by two continua E_1, E_2 and Δ_1 be the simply connected domain, which is bounded by E_1 and contains E_2. Let $g(z, a)$ be the Green's function of Δ_1. We distribute a positive mass μ on E_2 of total mass 1 and put*

$$I(\mu)=\iint_{a,b\in E_2} g(a, b)\,d\mu(a)\,d\mu(b), \qquad \mu(E_2)=1\,,$$

$$V=\inf_{\mu} I(\mu)\,,$$

then $$M(\Delta)=V\,,$$ *so that*

$$M(\Delta)\leqq I(\mu)\,.$$

There exists μ_0, such that $I(\mu_0)=V$.

3. As an application, we shall prove the following Grötzsch's theorem. Let $1<r<R\leqq\infty$ and $\Omega(r, R)$ be the ring domain on the $z=x+iy$-plane, which is bounded by a unit circle $|z|=1$ and a segment $L(r, R)$: $r\leqq x\leqq R$, $y=0$.

Theorem III. 56. *Let Δ be a ring domain on the z-plane, which lies outside of $|z|=1$ and is bounded by $|z|=1$ and a continuum E, which intersects two circles $|z|=r$ and $|z|=R$ $(1<r<R\leqq\infty)$. Then*

$$M(\varDelta)\leqq M(\Omega), \qquad \Omega=\Omega(r,R),$$

where the equality holds, only when $\varDelta$ is obtained from Ω by a rotation about $z=0$. Grötzsch [1].

Proof. (Tsuji [45]). $\log\left|\dfrac{1-\bar{a}z}{z-a}\right|$ ($|a|>1$, $|z|>1$) is the Green's function of $|z|>1$, so that by the above theorem, there exists a positive mass distribution $\mu^*\geqq 0$ on the segment $L(r, R)$ of total mass 1, such that

$$M(\Omega)=\iint\limits_{L(r,R)}\log\left|\frac{1-\alpha\beta}{\alpha-\beta}\right|d\mu^*(\alpha)\,d\mu^*(\beta).$$

Let a be the point of E, which lies on the circle $|z|=\alpha$ and $\arg a\geqq 0$ is the smallest. We associate this a to $\alpha=|a|$ and put $d\mu(a)=d\mu^*(\alpha)$, then $d\mu(a)$ is a positive mass distribution on E of total mass 1, so that

$$M(\varDelta)\leqq\iint\limits_{E}\log\left|\frac{1-\bar{a}b}{a-b}\right|d\mu(a)\,d\mu(b)\leqq\iint\limits_{E}\log\left|\frac{1-|a||b|}{|a|-|b|}\right|d\mu(a)\,d\mu(b)^{1)}$$

$$=\iint\limits_{L(r,R)}\log\left|\frac{1-\alpha\beta}{\alpha-\beta}\right|d\mu^*(\alpha)\,d\mu^*(\beta)=M(\Omega).$$

We see easily that the equality holds, only when $\varDelta$ is obtained from Ω by a rotation about $z=0$.

4. We shall prove some theorems on $M(\varDelta)$.

Theorem III. 57. *Let $\varDelta$ be a ring domain on the $z=\tau e^{i\theta}$-plane, which does not contain $z=0$ and $z=\infty$ and separates $z=0$ and $z=\infty$. Let*

$$\lambda(\varDelta)=\iint\limits_{\Delta}\frac{d\tau d\theta}{\tau}$$

be its logarithmic surface measure. Then

$$M(\varDelta)\leqq\frac{1}{2\pi}\lambda(\varDelta),$$

where the equality holds, only when $\varDelta$ is a circular ring of the form:

$$R_1<|z|<R_2.$$

Proof. We map $\varDelta$ conformally on $r<|w|<1$ on the $w=\rho e^{i\varphi}$-plane by $z=z(w)$, where $M(\varDelta)=\log\dfrac{1}{r}$, then considering the image of $|w|=\text{const.}=\rho$ ($r<\rho<1$) on the $\log z$-plane, we see easily that

$$2\pi\leqq\int_0^{2\pi}\left|\frac{d\log z}{dw}\right|\rho d\varphi, \qquad w=\rho e^{i\varphi}, \tag{1}$$

---

1) $\left|\dfrac{1-\bar{a}b}{a-b}\right|^2-1=\dfrac{(|a|^2-1)(|b|^2-1)}{|a-b|^2}\leqq\dfrac{(|a|^2-1)(|b|^2-1)}{(|a|-|b|)^2}=\left(\dfrac{(1-|a||b|}{|a|-|b|}\right)^2-1$, so that

$$\left|\frac{1-\bar{a}b}{a-b}\right|\leqq\left|\frac{1-|a|\,|b|}{|a|-|b|}\right|.$$

so that

$$4\pi^2 \leqq \int_0^{2\pi} d\varphi \int_0^{2\pi} \left| \frac{d\log z}{dw} \right|^2 \rho^2 d\varphi = 2\pi \int_0^{2\pi} \left| \frac{d\log z}{dw} \right|^2 \rho^2 d\varphi , \tag{2}$$

$$2\pi \frac{1}{\rho} \leqq \int_0^{2\pi} \left| \frac{d\log z}{dw} \right|^2 \rho d\varphi .$$

Hence

$$2\pi M(\varDelta) = 2\pi \log \frac{1}{r} = 2\pi \int_r^1 \frac{d\rho}{\rho} \leqq \int_r^1 \int_0^{2\pi} \left| \frac{d\log z}{dw} \right|^2 \rho d\rho\, d\varphi = \int_r^1 \int_0^{2\pi} \frac{|z'(w)|^2}{|z|^2} \rho d\rho d\varphi$$

$$= \iint_\varDelta \frac{1}{\tau^2} \tau\, d\tau\, d\theta = \iint_\varDelta \frac{d\tau d\theta}{\tau} = \lambda(\varDelta) . \tag{3}$$

If the equality holds in (3), then the equality holds in (1) and (2). If the equality holds in (1), then $2\pi = \int |d\log z|$ on the image of $|w| = \text{const.} = \rho$, so that the image of $|w| = \rho$ is a circle $|z| = \text{const.}$ If the equality holds in (2), then $\left| \frac{d\log z}{dw} \right| \rho = \left| \frac{d\log z}{d\log w} \right|$ is a function of ρ only, so that $\frac{d\log z}{d\log w} = \text{const.}$, $z = w^k$ ($k = \text{const.}$). We see easily that $k = \pm 1$, hence $z = \text{const.}\, w$, or $z = \text{const.}/w$, hence $\varDelta$ is a circular ring of the form: $R_1 < |z| < R_2$.

Theorem III. 58. *Let $\varDelta$ be a ring domain on the z-plane, which is bounded by two continua E_1, E_2 and $\varDelta_1$, $\varDelta_2$ be two disjoint ring domains $\in \varDelta$, such that $\varDelta_1$ is bounded by E_1 and a continuum E'_1 and $\varDelta_2$ is bounded by E_2 and a continuum E_2'. Then*

$$M(\varDelta_1) + M(\varDelta_2) \leqq M(\varDelta),$$

where the equality holds, only when $E_1' = E_2' (= E)$ and if we map $\varDelta$ conformally on $r < |w| < 1$, then E is mapped on a circle $|w| = \rho$ $(r < \rho < 1)$.

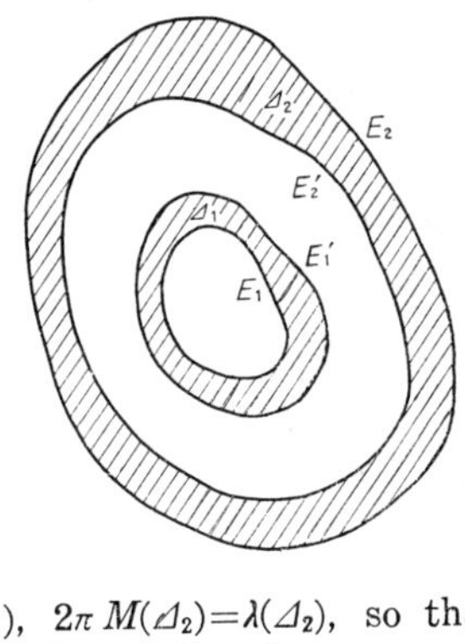

Proof. We may assume that $\varDelta$ is a circular ring: $r < |z| < 1$, where $M(\varDelta) = \log \frac{1}{r}$. Then $\lambda(\varDelta) = 2\pi \log \frac{1}{r} = 2\pi M(\varDelta)$.

Since $2\pi M(\varDelta_1) \leqq \lambda(\varDelta_1)$, $2\pi M(\varDelta_2) \leqq \lambda(\varDelta_2)$, we have $2\pi(M(\varDelta_1) + M(\varDelta_2)) \leqq \lambda(\varDelta_1) + \lambda(\varDelta_2) \leqq \lambda(\varDelta) = 2\pi M(\varDelta)$, so that $M(\varDelta_1) + M(\varDelta_2) \leqq M(\varDelta)$.

If the equality holds, then $2\pi M(\varDelta_1) = \lambda(\varDelta_1)$, $2\pi M(\varDelta_2) = \lambda(\varDelta_2)$, so that $\varDelta_1$, $\varDelta_2$ are circular rings, concentric with $\varDelta$ and $\varDelta_1 + \varDelta_2 = \varDelta$. ——

From the theorem, we have

Theorem III. 59. *Let $\varDelta_1$, $\varDelta_2$ be two ring domains on the z-plane,* *such that* $\Delta_1 \subset \Delta_2$ *and* Δ_1 *separates the two bounding continua of* Δ_2, *then* $M(\Delta_1) \leqq M(\Delta_2)$, *where the equality holds, only when* $\Delta_1 = \Delta_2$.

14. Capacity potential

Let E be a bounded closed set of positive capacity on the z-plane and

$$u(z) = \int_E \log \frac{1}{|z-a|} d\mu(a), \qquad \mu(E) = 1\,, \tag{1}$$

be its conductor potential. We assume that $\gamma(E) < 1$, $V = \log 1/\gamma(E) > 0$.

We put $d\nu = d\mu/V$ and

$$v(z) = \int_E \log \frac{1}{|z-a|} d\nu(a), \qquad \nu(E) = 1/V\,. \tag{2}$$

$v(z)$ is called *the capacity potential of* E.

Since $u \leqq V$ and $V > 0$,

$$v(z) \leqq 1 \text{ for any } z \quad \text{and} \quad v = 1 \text{ "almost everywhere" on } E. \tag{3}$$

By Theorem III. 38, we have

Theorem III. 60. *Let* a_0 *be a point on the outer boundary* Γ^* *of* E, *then* a_0 *is a regular point of the Dirichlet problem for* D_∞, *when and only when*

$$v(a_0) = 1.$$

Lemma. *Let* E *be a bounded closed set of positive capacity on the* z*-plane and* $a_0 \in \Gamma^*$. *Let* E'_ρ *be the part of* E, *which is contained in* $|z - a_0| < \rho$ *and* $E_\rho = E - E'_\rho$. *Let* $u(z)$, $u_\rho(z)$ *be the conductor potential and* $v(z)$, $v_\rho(z)$ *the capacity potential of* E *and* E_ρ *respectively, then*

$$\lim_{\rho \to 0} u_\rho(a_0) = u(a_0), \qquad \lim_{\rho \to 0} v_\rho(a_0) = v(a_0)\,.$$

Proof. Let

$$u(z) = \int_E \log \frac{1}{|z-a|} d\mu(a) = \int_{E'_\rho} \log \frac{1}{|z-a|} d\mu(a) + \int_{E_\rho} \log \frac{1}{|z-a|} d\mu(a)\,,$$
$$\mu(E) = 1\,. \tag{1}$$

$$u_\rho(z) = \int_{E_\rho} \log \frac{1}{|z-a|} d\mu_\rho(a), \qquad \mu_\rho(E_\rho) = 1\,. \tag{2}$$

We put

$$\tilde{u}_\rho(z) = \frac{1}{\mu(E_\rho)} \int_E \log \frac{1}{|z-a|} d\mu(a)\,, \tag{3}$$

then since $\lim_{\rho \to 0} \int_{E_\rho} \log \dfrac{1}{|a_0 - a|} d\mu(a) = 0$ and $\lim_{\rho \to 0} \mu(E_\rho) = 1$,

$$\tilde{u}_\rho(a_0)=\frac{1}{\mu(E_\rho)}\int_E \log\frac{1}{|a_0-a|}d\mu(a)-\frac{1}{\mu(E_\rho)}\int_{E'_\rho}\log\frac{1}{|a_0-a|}d\mu(a)$$
$$\to\int_E \log\frac{1}{|a_0-a|}d\mu(a)=u(a_0) \quad (\rho\to 0)\,. \tag{4}$$

Now $u(z)-\tilde{u}_\rho(z)$ is a bounded harmonic function outside of E_ρ and $u_\rho=V_\rho$ "almost everywhere" on E_ρ, where $V_\rho=\log 1/\gamma(E_\rho)$ and $|\tilde{u}(z)-V|<\varepsilon$ "almost everywhere" on E_ρ, where $V=\log 1/\gamma(E)$ and $\varepsilon\to 0$ with $\rho\to 0$, so that $|u_\rho(z)-\tilde{u}_\rho(z)|<|V-V_\rho|+\varepsilon$ "almost everywhere" on E_ρ, hence by the maximum principle, $|u_\rho(a_0)-\tilde{u}_\rho(a_0)|<|V-V_\rho|+\varepsilon$, so that

$$\lim_{\rho\to 0}(u_\rho(a_0)-\tilde{u}(a_0))=0\,. \tag{5}$$

Hence by (4), (5), $\lim\limits_{\rho\to 0} u_\rho(a_0)=u(a_0)$, so that $\lim\limits_{\rho\to 0} v_\rho(a_0)=v(a_0)$.

Theorem III. 61. *Let E be a bounded closed set of positive capacity on the z-plane and a_0 be a point on the outer boundary Γ^* of E. Let $0<\lambda<1$ and E_n be the part of E, contained in $\lambda^{n+1}\leqq|z-a_0|\leqq\lambda^n$ $(n=0, 1, 2, \cdots)$ and $v_n(z)$ be the capacity potential of E_n, then a_0 is a regular point of the Dirichlet problem for D_∞, when and only when*

$$\sum_n v_n(a_0)=\infty\,.$$

de la Vallée-Poussin [1].

Proof. We may assume that $a_0=0$ and let

$$v_n(z)=\int_{E_n}\log\frac{1}{|z-a|}d\nu_n(a), \qquad \nu_n(E_n)=1/V_n\,, \tag{1}$$

where $V_n=\log 1/\gamma(E_n)$.

(i) First suppose that

$$\sum_n v_n(0)<\infty\,, \tag{2}$$

and we shall prove that $z=0$ is an irregular point. We take $n_0>1$ so large that $\lambda^{n_0}<1/2$ and $\sum\limits_{n=n_0}^{\infty} v_n(0)<1/2$ and we may assume that

$$E=\sum_{n=n_0}^{\infty}E_n+\{0\},$$

then E is contained in $|z|<1/2$.

Let

$$v(z)=\int_E \log\frac{1}{|z-a|}d\nu(a), \qquad \nu(E)=1/V \tag{3}$$

be the capacity potential of E and put

$$v^*(z)=\sum_{n=n_0}^{\infty}v_n^*(z), \qquad v_n^*(z)=\int_{E_n}\log\frac{1}{|z-a|}d\nu(a)\,. \tag{4}$$

Since E_n, E_{n+1} are not disjoint in general, we have easily

$$\frac{1}{2}v^*(z)\leqq v(z)\leqq v^*(z)\,. \tag{5}$$

Since $v_n=1$ "almost everywhere" on E_n and $0\leqq v_n^*\leqq 1$ on E_n,

$$\nu(E_n)=\int_{E_n} v_n d\nu=\int_{E_n} v_n^* \, d\nu_n \leqq \nu_n(E_n) .$$

Since from (1), $v_n(0)\geqq n \log 1/\lambda.\ \nu_n(E_n)$ and from (4), $v_n^*(0)\leqq(n+1) \log 1/\lambda.\ \nu(E_n)$, we have

$$v_n^*(0)\leqq \frac{n+1}{n} v_n(0)\leqq 2v_n(0) \qquad (n\geqq n_0) , \tag{6}$$

hence

$$v(0)\leqq v^*(0)=\sum_{n=n_0}^{\infty} v_n^*(0)\leqq 2\sum_{n=n_0}^{\infty} v_n(0)<1 ,$$

so that by Theorem III. 60, $z=0$ is an irregular point.

(ii) Next suppose that

$$\sum_n v_n(0)=\infty , \tag{7}$$

and we shall prove that $z=0$ is a regular point. First we shall prove that λ can be replaced by $\kappa=\lambda^k$, where k is any positive integer. Let $\kappa=\lambda^k$ and $\widetilde{E}_s$ be the part of E, contained in $\kappa^{s+1}\leqq|z|\leqq\kappa^s$ $(s=1, 2, \cdots)$ and let

$$\widetilde{v}_s(z)=\int_{\widetilde{E}_s} \log \frac{1}{|z-a|} d\widetilde{\nu}_s(a) \tag{8}$$

be the capacity potential of $\widetilde{E}_s$.

If $ks\leqq n\leqq ks+k-1$, then $E_n\subset\widetilde{E}_s$, so that $\widetilde{v}_s=1$ "almost everywhere" on E_n and $v_n\leqq 1$ on $\widetilde{E}_s$, hence

$$\nu_n(E_n)=\int_{E_n} \widetilde{v}_s d\nu_n=\int_{\widetilde{E}_s} v_n \, d\widetilde{\nu}_s\leqq\widetilde{\nu}_s(\widetilde{E}_s) .$$

From this, we have as (6),

$$v_n(0)\leqq \text{const. } \widetilde{v}_s(0), \qquad \text{if } ks\leqq n\leqq ks+k-1 ,$$

so that

$$\sum_s \widetilde{v}_s(0)=\infty .$$

Hence by taking k sufficiently large, we may assume that λ is any small number. We shall use this remark in the sequel. From (7), we may assume, without loss of generality, that

$$\sum_{n=1}^{\infty} v_{2n}(0)=\infty , \tag{9}$$

and by taking λ small, $E=\sum_{n=2}^{\infty} E_n+\{0\}$ is contained in $|z|<1/2$ and put

$$E'=\sum_{n=1}^{\infty} E_{2n}+\{0\} .$$

Let

$$v'(z)=\int_{E'} \log \frac{1}{|z-a|} \, d\nu'(a) \tag{10}$$

be the capacity potential of E', then since E_{2n}, E_{2n+2} are disjoint.

$$v'(z)=\sum_{n=1}^{\infty} v'_{2n}(z), \quad v'_{2n}(z)=\int_{E_{2n}} \log \frac{1}{|z-a|}\, d\nu'(a) . \tag{11}$$

We put $E^{(n)}=\sum_{i=2}^{2n} E_i$, $E'^{(n)}=\sum_{i=1}^{n} E_{2i}$ and let $u^{(n)}(z)$, $u'^{(n)}(z)$ be the conductor potential of $E^{(n)}$, $E'^{(n)}$ respectively, then $V^{(n)}-u^{(n)}(z)\leqq V'^{(n)}-u'^{(n)}(z)$ "almost everywhere" on $E^{(n)}$, so that by the maximum principle,

$$V^{(n)}-u^{(n)}(0)\leqq V'^{(n)}-u'^{(n)}(0)\ , \quad V^{(n)}=V(E^{(n)}), \qquad V'^{(n)}=V(E'^{(n)}) .$$

Hence by the lemma, if we make $n\to\infty$, we have

$$V-u(0)\leqq V'-u'(0), \quad V(1-v(0))\leqq V'(1-v'(0)), \quad V=V(E), \quad V'=V(E'), \tag{12}$$

where $v(z)$, $v'(z)$ is the capacity potential of E, E' respectively.

Suppose that $z=0$ is an irregular point, then $v(0)<\alpha<1$, so that

$$V(1-\alpha)\leqq V'(1-v'(0)) .$$

We may assume that λ is so small that $\log\frac{1}{1-\lambda}<V(1-\alpha)$, then

$$\beta=\frac{1}{V'}\log\frac{1}{1-\lambda}+v'(0)<1 . \tag{13}$$

Let $a_{2i}\in E_{2i}$, $a_{2j}\in E_{2j}$ $(i\neq j)$.

If $i<j$, then $\left|\frac{a_{2j}}{a_{2i}}\right|\leqq\lambda$, so that $\left|\frac{a_{2i}-a_{2j}}{a_{2i}}\right|\geqq 1-\lambda$.

If $i>j$, then $\left|\frac{a_{2j}}{a_{2i}}\right|\geqq\frac{1}{\lambda}$, so that $\left|\frac{a_{2i}-a_{2j}}{a_{2i}}\right|\geqq\frac{1}{\lambda}-1\geqq 1-\lambda$.

Hence in any case,

$$\frac{1}{1-\lambda}\frac{1}{|a_{2i}|}\geqq\frac{1}{|a_{2i}-a_{2j}|}, \quad \log\frac{1}{1-\lambda}+\log\frac{1}{|a_{2i}|}\geqq\log\frac{1}{|a_{2i}-a_{2j}|} \qquad (i\neq j),$$

so that if ν' is defined by (10), then

$$\log\frac{1}{1-\lambda}\int_{E_{2i}} d\nu'(a)+\int_{E_{2i}}\log\frac{1}{|a|}\,d\nu'(a)\geqq\int_{E_{2i}}\log\frac{1}{|a-a_{2j}|}\,d\nu'(a) \quad (i\neq j). \tag{14}$$

Since $\int_{E'} d\nu'(a)=1/V'$, summing up for $i=1, 2, \cdots$, we have by (11)

$$\beta=\frac{1}{V'}\log\frac{1}{1-\lambda}+v'(0)\geqq v'(a_{2j})-v'_{2j}(a_{2j}) . \tag{15}$$

Since $v'(a_{2j})=1$ "almost everywhere" on E_{2j},

$$v'_{2j}(a_{2j})\geqq 1-\beta>0 \qquad \text{"almost everywhere" on } E_{2j} .$$

Since, if $v_n(z)$ is defined by (1), $v_{2j}(z)=1$ "almost everywhere" on E_{2j}, we have

$$\nu'(E_{2j})=\int_{E_{2j}} v_{2j}\, d\nu'=\int_{E_{2j}} v'_{2j}\, d\nu_{2j}\geqq(1-\beta)\,\nu_{2j}(E_{2j}) .$$

From this, we have as (6)

$$v'_{2j}(0) \geqq \text{const. } v_{2j}(0) ,$$

so that by (9)

$$1 \geqq v'(0) = \sum_{j=1}^{\infty} v'_{2j}(0) \geqq \text{const.} \sum_{j=1}^{\infty} v_{2j}(0) = \infty ,$$

which is absurd. Hence $z=0$ is a regular point.

15. Criteria of regularity

Theorem III. 62. *Let E be a bounded closed set of positive capacity on the z-plane and a_0 be a point on the outer boundary Γ^* of E. Let $0<\lambda<1$ and E_n be the part of E, contained in $\lambda^{n+1} \leqq |z-a_0| \leqq \lambda^n$, then a_0 is a regular point of the Dirichlet problem for D_∞, when and only when*

$$\sum_n \frac{n}{\log 1/\gamma_n} = \infty , \quad \gamma_n = \gamma(E_n) .$$ Wiener [1].

Proof. Let

$$v_n(z) = \int_{E_n} \log \frac{1}{|z-a|} \, d\nu_n(a), \qquad \nu_n(E_n) = 1/V_n, \qquad V_n = \log 1/\gamma_n ,$$

be the capacity potential of E_n, then

$$\frac{n \log 1/\lambda}{\log 1/\gamma_n} \leqq v_n(a_0) \leqq (n+1) \frac{\log 1/\lambda}{\log 1/\gamma_n} ,$$

hence the theorem follows from Theorem III. 61. ——

Since $x \log \dfrac{1}{x}$ is an increasing function of x in $(0, 1/e)$ and $\gamma_n \leqq \lambda^n$ (Theorem III. 39. Cor. 1), we have

$$\gamma_n \log 1/\gamma_n \leqq \lambda^n \log 1/\lambda^n, \qquad \frac{\gamma_n}{\lambda^n} \leqq \frac{n}{\log 1/\gamma_n} \log 1/\lambda \qquad (n \geqq n_0) ,$$

hence we have

Corollary 1. *If $\sum_n \dfrac{\gamma_n}{\lambda^n} = \infty$, then a_0 is a regular point.*

Corollary 2. *Let E_r be the part of E, contained in $|z-a_0| \leqq r$ and $\gamma(r) = \gamma(E_r)$. If $\overline{\lim_{r\to 0}} \dfrac{\log 1/r}{\log 1/\gamma(r)} > 0$, especially, if $\overline{\lim_{r\to 0}} \dfrac{\gamma(r)}{r} > 0$, then a_0 is a regular point.*

$\overline{\lim_{r\to 0}} \dfrac{\gamma(r)}{r}$ is called *the upper capacity density of E at a_0*.

Proof. Let

$$v_r(z) = \int_{E_r} \log \frac{1}{|z-a|} \, d\nu(a), \qquad \nu(E_r) = \frac{1}{\log 1/\gamma(r)} , \tag{1}$$

be the capacity potential of E_r.

If we take $r=\lambda^N\,(0<\lambda<1)$, where N is a positive integer and put

$$v_r^*(a_0)=\sum_{n=N}^{\infty} v_n{}^*(a_0),\qquad v_n^*(a_0)=\int_{E_n}\log\frac{1}{|a-a_0|}\,d\nu(a),\qquad (2)$$

$$E_n:\ \lambda^{n+1}\leqq|z-a_0|\leqq\lambda^n,$$

then

$$\frac{1}{2}v_r{}^*(a_0)\leqq v_r(a_0)\leqq v_r{}^*(a_0).$$

Suppose that a_0 is an irregular point, then $\sum_n v_n(a_0)<\infty$, where $v_n(z)$ is the capacity potential of E_n. By (6) of the proof Theorem III. 61, $v_n^*(a_0)\leqq\text{const.}v_n(a_0)$, so that $v_r(a_0)\leqq v_r^*(a_0)=\sum_{n=N}^{\infty}v_n^*(a_0)\leqq\text{const.}\sum_{n=N}^{\infty}v_n(a_0)<\infty$, hence $\lim_{r\to0}v_r(a_0)=0$. Since by (1),

$$v_r(a_0)\geqq\frac{\log 1/r}{\log 1/\gamma(r)},$$

we have $\lim_{r\to0}\dfrac{\log 1/r}{\log 1/\gamma(r)}=0$, hence if $\overline{\lim_{r\to0}}\dfrac{\log 1/r}{\log 1/\gamma(r)}>0$, then a_0 is a regular point.

Corollary 3. *Let ω_n be the surface measure of E_n: $\lambda^{n+1}\leqq|z-a_0|\leqq\lambda^n$. If $\sum_n\dfrac{\omega_n}{\lambda^{2n}}=\infty$, then a_0 is a regular point.*

Proof. By Theorem III. 10, $\sqrt{\omega_n}\leqq\sqrt{\pi e}\gamma_n$ and since $\omega_n\leqq\pi\lambda^{2n}$, $\dfrac{\omega_n}{\lambda^{2n}}=\dfrac{\sqrt{\omega_n}}{\lambda^n}\cdot\dfrac{\sqrt{\omega_n}}{\lambda^n}\leqq\sqrt{\pi}\dfrac{\sqrt{\pi e}\gamma_n}{\lambda^n}=\pi\sqrt{e}\cdot\dfrac{\gamma_n}{\lambda^n}$, hence $\sum_n\dfrac{\gamma_n}{\lambda^n}=\infty$, so that by Cor.1, a_0 is a regular point. ——

From Corollary 3, we have easily

Corollary 4. *Let $\omega(r)$ be the surface measure of the part E_r of E, contained in $|z-a_0|\leqq r$. If $\int_0^1\dfrac{\omega(r)}{r^3}\,dr=\infty$, then a_0 is a regular point.*

Corollary 5. *Let E be a bounded closed set of positive capacity on the z-plane and $z=0$ lies on the outer boundary Γ^* of E. Let $E^{(x)}$ be the set on the positive x-axis, defined in Theorem* III. 39. Cor. 6 *and $E^{(x)}(r)$ be its part, contained in $0\leqq x\leqq r$ and $\mu(r)$ be its linear measure. If $\int_0^1\dfrac{d\mu(r)}{r}=\infty$, then $z=0$ is a regular point.* Beurling [1].

Proof. Let $L_n=\int_{\lambda^{n+1}}^{\lambda^n}d\mu(r)\ (0<\lambda<1)$, then by Theorem III. 39. Cor. 6, $L_n\leqq4\gamma_n$. Since

$$\int_{\lambda^{n+1}}^{\lambda^n}\frac{d\mu(r)}{r}\leqq\frac{L_n}{\lambda^{n+1}}\leqq\frac{4}{\lambda}\cdot\frac{\gamma_n}{\lambda^n},$$

we have $\int_0^1 \frac{d\mu(r)}{r} \leqq \frac{4}{\lambda} \sum_{n=0}^{\infty} \frac{\gamma_n}{\lambda^n}$, so that $\sum_n \frac{\gamma_n}{\lambda^n} = \infty$, hence by Cor. 1, $z=0$ is a regular point.

16. General Cantor set

1. General linear Cantor set.

Let I be a closed interval on the x-axis. We take k ($\geqq 2$) disjoint closed intervals I_{i_1} ($i_1=1, 2, \cdots, k$) in I and k disjoint closed intervals $I_{i_1 i_2}$ ($i_2=1, 2, \cdots, k$) in I_{i_1} and proceed similarly, then after n steps, we obtain k^n closed intervals $I_{i_1\cdots i_n}$ ($i_1, \cdots, i_n=1, 2, \cdots, k$), such that $I_{i_1\cdots i_n} \subset I_{i_1\cdots i_{n-1}}$ ($i_n=1, 2, \cdots, k$).

We put

$$E=\prod_{n=1}^{\infty}\left(\sum_{i_1\cdots i_n}^{1,\cdot,k} I_{i_1\cdots i_n}\right), \tag{1}$$

which is a perfect non-dense set.

We denote the length of an interval J by $|J|$ and the shortest distance of two intervals J_1, J_2 by $\rho(J_1, J_2)$.

We assume that there exist $0<a<1$, $0<b<1$, such that for $n=1, 2, \cdots$,

(i) $|I_{i_1\cdots i_{n-1} i_n}| \geqq a\,|I_{i_1\cdots i_{n-1}}|$ $\quad (i_n=1, 2, \cdots, k)$,

(ii) $\rho(I_{i_1\cdots i_{n-1}\mu}, I_{i_1\cdots i_{n-1}\nu}) \geqq b\,|I_{i_1\cdots i_{n-1}}|$ $\quad (\mu, \nu=1, 2, \cdots, k,\ \mu\neq\nu)$.

Then we call E *a general linear Cantor set.*

Theorem III. 63. *Let E be a general linear Cantor set, then*

$$mE=0, \qquad \gamma(E)\geqq a^{\frac{1}{k-1}} b|I|>0.$$

At every point of E, the upper capacity density is positive, so that every point of E is a regular point of the Dirichlet problem. Tsuji [40].

Proof. By (ii), $I_{i_1\cdots i_{n-1}} - \sum_{i_n=1}^{k} I_{i_1\cdots i_n}$ contains an interval J, such that $|J| \geqq b\,|I_{i_1\cdots i_{n-1}}|$, so that

$$|I_{i_1\cdots i_{n-1}}| \geqq |J| + \sum_{i_n=1}^{k} |I_{i_1\cdots i_n}| \geqq b\,|I_{i_1\cdots i_{n-1}}| + \sum_{i_n=1}^{k} |I_{i_1\cdots i_n}|,$$

hence $\sum_{i_n=1}^{k} |I_{i_1\cdots i_n}| \leqq (1-b)\,|I_{i_1\cdots i_{n-1}}|$. From this, we have

$$mE \leqq \sum_{i_1\cdots i_n}^{1\cdots k} |I_{i_1\cdots i_n}| \leqq (1-b)^n |I| \to 0, \qquad n\to\infty,$$

hence

$$mE=0.$$

By (i)

$$|I_{i_1\cdots i_n}| \geqq a^n |I| \qquad (i_1, \cdots, i_n=1, 2, \cdots, k), \tag{1}$$

hence by (ii)

$$\rho(I_{i_1\cdots i_{n-1}\mu}, I_{i_1\cdots i_{n-1}\nu}) \geqq a^{n-1} b\,|I| \qquad (\mu\neq\nu). \tag{2}$$

Let M be a bounded closed set on the x-axis, then by Theorem III. 26, there exist x_i ($i=1, 2, \cdots, n$) on M, such that

$$d_n(M)=\sqrt[\binom{n}{2}]{\prod_{i<k}^{1\cdots n}|x_i-x_k|}\to\gamma(M)\,,\quad n\to\infty\,.$$

We put $E_n=\sum_{i_1\cdots i_n}^{1\cdots k} I_{i_1\cdots i_n}$ and take N points $x^{\nu}_{i_1\cdots i_n}$ $(\nu=1, 2, \cdots, N)$ on each $I_{i_1\cdots i_n}$, such that

$$\sqrt[\binom{N}{2}]{\prod_{\mu<\nu}^{1\cdots N}|x^{\mu}_{i_1\cdots i_n}-x^{\nu}_{i_1\cdots i_n}|}\to\gamma(I_{i_1\cdots i_n}),\qquad N\to\infty\,. \tag{3}$$

Since there are k^nN points $x^{\nu}_{i_1\cdots i_n}$ on E_n, we have by the definition of d_n (Theorem III. 21),

$$\left((d_{k^nN}(E_n)\right)^{\binom{k^nN}{2}}\geqq\prod_{i,j}^{1\cdots k}\prod_{\mu,\nu}^{1\cdots N}\left|x^{\mu}_{i_1\cdots i_n}-x^{\nu}_{j_1\cdots j_n}\right|=\Pi,\qquad\text{say.} \tag{4}$$

Now Π consists of $(n+1)$ factors $\Pi=\Pi_n\Pi_{n-1}\cdots\Pi_1\Pi_0$, where Π_n is formed with pairs of points, which lie in the same $I_{i_1\cdots i_n}$ and Π_{n-1} is formed with pairs of points, which lie in the same $I_{i_1\cdots i_{n-1}}$ and belong to $I_{i_1\cdots i_{n-1}j}$ and $I_{i_1\cdots i_{n-1}j'}$ $(j\neq j')$ respectively and Π_{n-2} is formed with pairs of points, which lie in the same $I_{i_1\cdots i_{n-2}}$ and belong to $I_{i_n\cdots i_{n-2}j}$ and $I_{i_1\cdots i_{n-2}j'}$ $(j\neq j')$ respectively and finally Π_0 is formed with pairs of points, which lie in I and belong to I_j and $I_{j'}$ $(j\neq j')$ respectively. By (3),

$$\lim_{N\to\infty}\sqrt[\binom{N}{2}]{\Pi_n}=\prod_{i_1\cdots i_n}^{1,\cdot,k}\gamma(I_{i_1\cdots i_n})\,. \tag{5}$$

Since the capacity of an interval J is $|J|/4$ (Theorem III. 39. Cor. 3), we have by (1), $\gamma(I_{i_1\cdots i_n})\geqq a^n|I|/4$, so that

$$\lim_{N\to\infty}\sqrt[\binom{N}{2}]{\Pi_n}\geqq\left(\frac{a^n|I|}{4}\right)^{k^n}. \tag{6}$$

By (2), $|x^{\mu}_{i_1\cdots i_n}-x^{\nu}_{j_1\cdots j_n}|\geqq a^{n-1}b\,|I|$ in Π_{n-1} and the number of such pairs is $\binom{k}{2}N^2k^{n-1}=\frac{(k-1)}{2}N^2k^n$, so that

$$\Pi_{n-1}\geqq(a^{n-1}b\,|I|)^{\frac{(k-1)}{2}N^2k^n}\,. \tag{7}$$

Since $|x^{\mu}_{i_2\cdots i_n}-x^{\nu}_{j_1\cdots j_n}|\geqq a^{n-2}b\,|I|$ in Π_{n-2} and the number of such pairs is $\binom{k}{2}(k\,N)^2k^{n-2}=\frac{(k-1)}{2}N^2k^{n+1}$, we have

$$\Pi_{n-2}\geqq(a^{n-2}b\,|I|)^{\frac{(k-1)}{2}N^2k^{n+1}}, \tag{8}$$

and finally

$$\Pi_0\geqq(b\,|I|)^{\frac{(k-1)}{2}N^2k^{2n-1}}\,. \tag{9}$$

Since $S_n=(n-1)+(n-2)k+\cdots+(n-n)k^{n-1}=\frac{1}{k-1}\left(\frac{k^n-1}{k-1}-n\right)$, we have

$$\Pi_{n-1}\Pi_{n-2}\cdots\Pi_0\geqq a^{\frac{N^2}{2}k^n\left(\frac{k^n-1}{k-1}-n\right)}(b\,|I|)^{\frac{N^2}{2}k^n(k^n-1)}. \tag{10}$$

Since $\binom{k^n N}{2} \sim \frac{N^2 k^{2n}}{2}$ $(N \to \infty)$, if we make $N \to \infty$, then we have by (4), (6), (10),

$$(\gamma(E_n))^{k^{2n}} \geqq \left(\frac{a^n|I|}{4}\right)^{k^n} a^{k^n\left(\frac{k^n-1}{k-1}-n\right)} (b\,|I|)^{k^n(k^n-1)} . \tag{11}$$

Since $\gamma(E_n) \to \gamma(E)$ $(n \to \infty)$, if we make $n \to \infty$ in (11), then we have

$$\gamma(E) \geqq a^{\frac{1}{k-1}} b\,|I| > 0 . \tag{12}$$

Let $x_0 \in E$, then there exist $\{I_{i_1 \cdots i_n}\}$, such that $x_0 \in I_{i_1}$, $x_0 \in I_{i_1 i_2}$, $\cdots$ $x_0 \in I_{i_1 \cdots i_n}, \ldots$. Let $E_{i_1 \cdots i_n}$ be the part of E, which is contained in $I_{i_1 \cdots i_n}$, then by (12),

$$\gamma(E_{i_1 \cdots i_n}) \geqq a^{\frac{1}{k-1}} b\,|I_{i_1 \cdots i_n}| ,$$

hence the upper capacity density of E at x_0 is positive, so that by Theorem III. 62. Cor. 2, x_0 is a regular point of the Dirichlet problem.

Remark. The ordinary Cantor set is defined as follows. Let I: $0 \leqq x \leqq 1$ and I_1: $0 \leqq x \leqq x_1$, I_2: $x_2 \leqq x \leqq 1$ $(x_1 < x_2)$, such that

$$|I_1| = |I_2| = \frac{1}{2p}, \qquad x_2 - x_1 = 1 - \frac{1}{p} (p > 1) .$$

We perform the similar operations on I_1 and I_2 and proceed similarly, then after n steps, we obtain 2^n intervals $I_{i_1 \cdots i_n}$ $(i_1 \cdots i_n = 1, 2)$. Then $E = \prod_{n=1}^{\infty} (\sum_{i_1 \cdots i_n}^{1,2} I_{i_1 \cdots i_n})$ is the ordinary Cantor set. In this case, $k=2$, $a = \frac{1}{2p}$, $b = 1 - \frac{1}{p}$, $|I| = 1$, so that $a^{\frac{1}{k-1}} b\,|I| = \frac{1}{2p}\left(1 - \frac{1}{p}\right)$, hence

$$\gamma(E) \geqq \frac{1}{2p}\left(1 - \frac{1}{p}\right) > 0 .$$ R. Nevanlinna [6].

2. General planar Cantor set.

In the following, $\delta(M)$ denotes the diameter of a set M and $\rho(M_1, M_2)$ the shortest distance of two sets M_1, M_2. Let Δ be a simply connected Jordan domain on the z-plane. We take k $(\geqq 2)$ disjoint simply connected Jordan domains Δ_{i_1} $(i_1 = 1, 2, \cdots k)$ in Δ and proceed similarly as in the general linear Cantor set, then after n steps, we obtain k^n disjoint simply connected Jordan domains $\Delta_{i_1 \cdots i_n}$ $(i_1, \cdots, i_n = 1, 2, \cdots, k)$. We put

$$E = \prod_{n=1}^{\infty} (\sum_{i_1 \cdots i_n}^{1, \cdots, k} \Delta_{i_1 \cdots i_n}) . \tag{1}$$

We assume that there exist $0 < a < 1$, $0 < b < 1$, such that for $n = 1, 2, \cdots$,

(i) $\delta(\Delta_{i_1 \cdots i_n}) \geqq a\,\delta(\Delta_{i_1 \cdots i_{n-1}})$ $(i_n = 1, 2, \cdots, k)$,

(ii) $\rho(\Delta_{i_1 \cdots i_{n-1}\mu}, \Delta_{i_1 \cdots i_{n-1}\nu}) \geqq b\,\delta(\Delta_{i_1 \cdots i_{n-1}})$ $(\mu, \nu = 1, 2, \cdots, k, \mu \neq \nu)$. (2)

Then we call E *a general planar Cantor set.*

By means of Theorem III. 39. Cor. 5, we can prove similarly as Theorem III. 63 the following

Theorem III. 64. *Let E be a general planar Cantor set, then*

$$\gamma(E) \geqq a^{\frac{1}{k-1}} b\, \delta(\Delta) > 0 .$$

At every point of E, the upper capacity density is positive, so that every point of E is a regular point of the Dirichlet problem.

3. Singular set of a linear group of Schottky type.

Let Δ be an infinite domain on the z-plane, which contains $z=\infty$ and G be a linear group of Schottky type, whose elements are linear functions $f_n(z)=\dfrac{a_n z+b_n}{c_n z+d_n}$ $(n=0,1,2,\cdots)$, which transform Δ on itself. Let D_0 be its fundamental domain. We assume that D_0 contains $z=\infty$ and is bounded by $p\,(2\leqq p<\infty)$ pairs of analytic Jordan curves C_i, C_i' $(i=1,2,\cdots,p)$, where C_i, C_i' are equivalent by G. Let $\{D_\nu\}$ be the equivalents of D_0 by G, then D_ν cluster to a perfect non-dense set E.

We call E *the singular set of G.*

Theorem III. 65. *Let E be the singular set of a linear group of Schottky type, then*

$$mE=0, \quad \gamma(E)>0.^{1)}$$ Myrberg [3].

At every point of E, the upper capacity density is positive, so that every point of E is a regular point of the Dirichlet problem.

First we shall prove a lemma. Tsuji [40].

Lemma. *Let C be an analytic Jordan curve on the z-plane and D be a doubly connected domain, which contains C and is bounded by two Jordan curves C_1, C_2, such that C separates C_1 and C_2, and $w=f(z)$ be regular and schlicht in $\overline{D}$. Let Γ be the image of C on the w-plane, $l(\Gamma)$ be its length and $\delta(\Gamma)$ be its diameter, then*

$$l(\Gamma)\leqq \kappa\delta(\Gamma),$$

where $\kappa>0$ is a constant, which depends on C and D only and is independent of $f(z)$.

1) From the proof, we see that $\gamma(E)>0$, if $f_n(z)$ are meromorphic and schlicht in Δ, such that $f_n(\Delta)=\Delta$ and D_0 contains boundary curves other than C_i, C_i' $(i=1,2)$. Hence *if G is a group of Schottky type, whose elements $f_n(z)$ are meromorphic and schlicht in Δ, such that $f_n(\Delta)=\Delta$ and the boundary of its fundamental domain D_0 contains at least two pairs of equivalent Jordan curves, then $\gamma(E)>0$, where E is the boundary of Δ.*

Proof. Let z_0 be a point on C, then by Koebe's distortion theorem,

$$A|f'(z_0)| \leqq |f'(z)| \leqq B|f'(z_0)|, \qquad z \in C, \tag{1}$$

where $A>0$, $B>0$ are constants, which depend on C and D only, so that

$$l(\Gamma)=\int_C |f'(z)|\,|dz| \leqq \text{const.}\,|f'(z_0)|. \tag{2}$$

Let C' be an analytic Jordan curve in D, which bounds with C a doubly connected domain $\varDelta$ and let Γ' be its image on the w-plane. We choose C', such that Γ' lies inside of Γ. We take $w \in \Gamma$, $w' \in \Gamma$, such that $|w-w'|$ is equal to the shortest distance of Γ, Γ', then the segment $\overline{ww'}$ lies in a doubly connected domain, which is bounded by Γ, Γ', so that its image γ on the z-plane lies in $\varDelta$. Then by Koebe's distortion theorem,

$$\delta(\Gamma) \geqq |w-w'| = \int_\gamma |f'(z)|\,|dz| \geqq \text{const.}\,|f'(z_0)|. \tag{3}$$

Hence by (2), (3), we have $l(\Gamma) \leqq \text{const.}\ \delta(\Gamma)$.

Proof of the theorem.

Let z_0 be an inner point of D_0 and K_0 be a circular disc about z_0, which is contained in D_0 and $\{K_n\}$ be its equivalents by G, then since $\{K_n\}$ do not overlap, the sum of areas of K_n is finite, so that by Koebe's 1/4-disc-theorem, we have

$$\sum_{n=1}^{\infty} |f'_n(z_0)|^2 < \infty.$$

Let $C_n^{(i)}$ be the image of C_i $(i=1, 2, \cdots, p)$ by $w=f_n(z)$ $(n=1, 2, \cdots)$ and $L_n^{(i)}$ be its length, then by Koebe's distortion theorem, $L_n^{(i)} \leqq \text{const.}\,|f'_n(z_0)|$, so that $\sum_n [L_n^{(i)}]^2 < \infty$. If we denote the inside of $C_n^{(i)}$ by $\varDelta_n^{(i)}$ and its area by $A_n^{(i)}$, then by the isoperimetric inequality, $4\pi A_n^{(i)} \leqq [L_n^{(i)}]^2$, hence $\sum_{i=1}^{p} \sum_{n=0}^{\infty} A_n^{(i)} < \infty$. From this, we see that $mE=0$. Now we consider the totality of equivalents of C_i, C_i' $(i=1, 2, \cdots, p)$, which are contained inside of $C=C_1$, say. Then they can be written in the form $C_{i_1 \cdots i_n}$ $(i_1, \cdots, i_n=1, 2, \cdots, q$ $(q=2p-1)$, $n=1, 2, \cdots)$, such that if we denote the inside of $C_{i_1 \cdots i_n}$ by $\varDelta_{i_1 \cdots i_n}$, then $\varDelta_{i_1 \cdots i_n} \subset \varDelta_{i_1 \cdots i_{n-1}}$ $(i=1, 2, \cdots, q)$. Let z_0 be an inner point of D_0, then by Koebe's distortion theorem, for any z on C_i, C_i' $(i=1, 2, \cdots, p)$,

$$A|f'_n(z_0)| \leqq |f'_n(z)| \leqq B|f'_n(z_0)|,$$

where $A>0$, $B>0$ are constants. Hence if $C_{i_1 \cdots i_n}$ is the image of one of C_i, C_i' by $w=f_\nu(z)$ and $L_{i_1 \cdots i_n}$ be its length, then

$$\text{const.}\,|f'_\nu(z_0)| \leqq L_{i_1 \cdots i_n} \leqq \text{const.}\,|f'_\nu(z_0)|.$$

Hence by the lemma,

$$\delta(\Delta_{i_1\cdots i_n}) \geqq \text{const. } L_{i_1\cdots i_n} \geqq \text{const. } |f'_\nu(z_0)| \geqq \text{const. } L_{i_1\cdots i_{n-1}}$$
$$\geqq \text{const. } \delta(\Delta_{i_1\cdots i_{n-1}}) \quad (i_n=1, 2, \cdots, q). \qquad (1)$$

Next we shall prove that

$$\rho(\Delta_{i_1\cdots i_{n-1}\mu}, \Delta_{i_1\cdots i_{n-1}\nu}) \geqq \text{const. } \delta(\Delta_{i_1\cdots i_{n-1}}) \quad (\mu, \nu=1, 2, \cdots, q, \ \mu\neq\nu). \qquad (2)$$

Let d be the shortest distance of $C_{i_1\cdots i_{n-1}\mu}$, $C_{i_1\cdots i_{n-1}\nu}$ and let $z_\mu \in C_{i_1\cdots i_{n-1}\mu}$, $z_\nu \in C_{i_1\cdots i_{n-1}\nu}$, such that $|z_\mu - z_\nu| = d$. If the open segment $\overline{z_\mu z_\nu}$ does not intersect equivalents of $\{C_i\}$, $\{C_i'\}$, then $\overline{z_\mu z_\nu}$ is contained in an equivalent D_m of D_0. Let γ be the equivalent of the segment $\overline{z_\mu z_\nu}$ in D_0 by $w=f_m(z)$, then by (1),

$$\rho(\Delta_{i_1\cdots i_{n-1}\mu}, \Delta_{i_1\cdots i_{n-1}\nu}) = d = |z_\mu - z_\nu| = \int_\gamma |f'_m(z)||dz| \geqq \text{const. } |f'_m(z_0)|$$
$$\geqq \text{const. } \delta(\Delta_{i_1\cdots i_{n-1}}). \qquad (3)$$

We see easily that (3) holds, if the segment $\overline{z_\mu z_\nu}$ intersects the equivalents of $\{C_i\}$, $\{C_i'\}$ other than $C_{i_1\cdots i_{n-1}\mu}$, $C_{i_1\cdots i_{n-1}\nu}$. Hence by (1), (2), E is a general planar Cantor set, so that the theorem follows from Theorem III. 64. Similarly we can prove

Theorem III. 66. *Let C_i ($i=1, 2, \cdots, p$, $3 \leqq p < \infty$) be p disjoint circles on the z-plane, which lie outside each other. We invert C_i into C_j ($i \neq j$) and we perform the inversions indefinitely, then we obtain infinitely many circles, clustering to a perfect non-dense set E.*

Then $mE=0$, $\gamma(E)>0$ and at every point of E, the upper capacity density is positive, so that every point of E is a regular point of the Dirichlet problem.

17. Harmonic measure

1. Let D be a domain on the z-plane and Γ be its boundary and α be a closed set on Γ. We solve the Dirichlet problem for D with the boundary value 1 on α and 0 on $\Gamma-\alpha$ and let $\omega=\omega(z, \alpha; D)$ be its solution, then except a set of capacity zero, $\omega=1$ on α and $\omega=0$ on $\Gamma-\alpha$. We call ω *the harmonic measure of α with respect to D.*

Evidently

$$\omega(z, \alpha; D) = 1 - \omega(z, \Gamma\text{-}\alpha; D). \qquad (1)$$

If $D_1 \subset D$ is a sub-domain of D, whose boundary Γ_1 contains α, then by the maximum principle, we have easily

$$\omega(z, \alpha; D_1) \leqq \omega(z, \alpha; D). \qquad (2)$$

In several problems, it is important to obtain the majoration of the harmonic measure. We shall prove a theorem on the majoration of the harmonic measure as follows.

Let D be an infinite domain on the z-plane, which contains $z=0$ and $z=\infty$ belongs to its boundary Γ. The part of D, which lies in $|z|<r$ consists of at most a countable number of connected domains. Let D_r be such one, which contains $z=0$ and let θ_r be the part of the circle $|z|=r$, which lies in D and belongs to the boundary of D_r and $r\theta(r)$ be its linear measure. Let Γ_r be the part of Γ, which belongs to the boundary of D_r, so that $\Gamma_r+\theta_r$ is the whole boundary of D_r.

We define $\theta^*(r)$ as follows. If the circle $|z|=r$ intersects Γ, then we put $\theta^*(r)=\theta(r)$ and if the circle $|z|=r$ does not intersect Γ, so that it is contained entirely in D, then we put $\theta^*(r)=\infty$. Hence if a disc $|z|<r_0$ belongs to D, then

$$\theta^*(r)=\infty \quad \text{for } 0\leqq r<r_0\,.$$

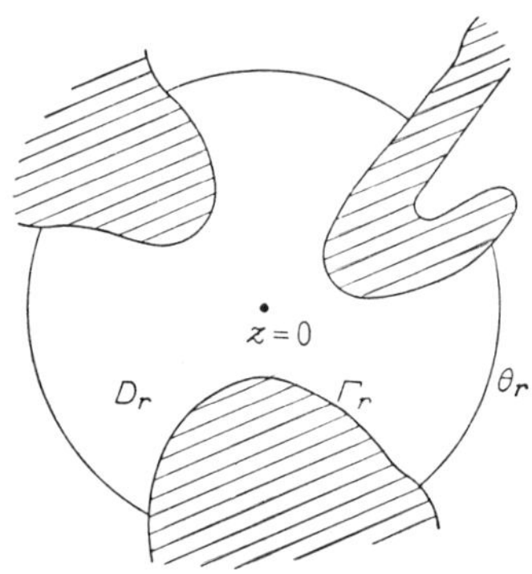

Theorem III. 67. *Let $u_r(z)$ be the harmonic measure of θ_r with respect to D_r, then*

$$u_r(0)\leqq \frac{3}{\sqrt{1-\kappa}}e^{-\pi\int_0^{\kappa r}\frac{dr}{r\theta^*(r)}} \quad (0<\kappa<1)\,.$$

Tsuji [22].

For the proof, we use the following

Wirtinger's inequality. *Let $f(x)$, $f'(x)$ be continuous in $[a, b]$ and $f(a)=f(b)=0$, then*

$$\int_a^b f'^2(x)dx\geqq \frac{\pi^2}{(b-a)^2}\int_a^b f^2(x)dx\,.$$

Proof. We may assume that $a=0$, $b=\pi$, so that we have to prove that

$$\int_0^\pi f'^2(x)\,dx \geqq \int_0^\pi f^2(x)\,dx\,.$$

We define $f(x)$ in $-\pi\leqq x\leqq 0$ by putting $f(x)=-f(-x)$, then $f(x)$ is a continuous odd function in $[-\pi, \pi]$ and $f(-\pi)=f(0)=f(\pi)=0$.

Hence $f(x)$ can be expanded in the form:

$$f(x)=\sum_{n=1}^\infty a_n\sin nx, \qquad \text{so that} \qquad f'(x)=\sum_{n=1}^\infty na_n\cos nx\,.$$

Hence by Parseval's theorem,

$$\frac{2}{\pi}\int_0^\pi f'^2(x)dx=\frac{1}{\pi}\int_{-\pi}^\pi f'^2(x)dx=\sum_{n=1}^\infty n^2a_n^2\geqq \sum_{n=1}^\infty a_n^2=\frac{1}{\pi}\int_{-\pi}^\pi f^2(x)dx$$

$$=\frac{2}{\pi}\int_0^{\pi} f^2(x)dx .$$

Proof of the theorem. (i) First we assume that Γ consists of at most a countable number of analytic curves, which do not cluster at a finite distance. Then θ_r consists of a finite number of circular arcs θ_r^i $(i=1, 2,\cdots, n_r)$ and let $r\theta_i(r)$ be its linear measure, then $\theta(r)=\sum_{i=1}^{n_r}\theta_i(r)$. Let $0\leqq\rho<r$, then from the definition of $\theta(\rho)$, $\theta(\rho)$ is continuous, except at a finite number of points $0<\rho_1<\cdots<\rho_n<r$, where $\theta(\rho_i-0)=\theta(\rho_i)<\theta(\rho_i+0)$.

We put $u(z)=u_r(z)$ and with Carleman we consider

$$m(\rho)=\frac{1}{2\pi}\int_{\theta_\rho} u^2(\rho e^{i\theta})d\theta \quad (0<\rho<r) . \tag{1}$$

If the circle $|z|=\rho$ intersects Γ, then $u=0$ at the end points of θ_ρ, so that we may differentiate (1) by ρ $(\neq\rho_i)$ under the integral sign,

$$\frac{\partial m(\rho)}{\partial \log\rho}=\frac{1}{\pi}\int_{\theta_\rho} u\frac{\partial u}{\partial \log\rho}d\theta , \tag{2}$$

$$\frac{\partial^2 m(\rho)}{\partial\log\rho^2}=\frac{1}{\pi}\int_{\theta_\rho}\left(\left(\frac{\partial u}{\partial\log\rho}\right)^2+u\frac{\partial^2 u}{\partial\log\rho^2}\right)d\theta=\frac{1}{\pi}\int_{\theta_\rho}\left(\left(\frac{\partial u}{\partial\log\rho}\right)^2-u\frac{\partial^2 u}{\partial\theta^2}\right)d\theta$$

$$=\frac{1}{\pi}\int_{\theta_\rho}\left(\left(\frac{\partial u}{\partial\log\rho}\right)^2+\left(\frac{\partial u}{d\theta}\right)^2\right)d\theta>0 . \tag{3}$$

Hence $m(\rho)$ is a convex function of $\log\rho$ in (ρ_i, ρ_{i+1}).

At $\rho=\rho_i$, $m(\rho)$ is discontinuous, such that $m(\rho_i-0)=m(\rho_i)<m(\rho_i+0)$. Let ν be the outer normal of Γ with respect to D, then since $u=0$ on Γ_ρ, if we apply the Green's formula for D_ρ, then we have

$$\rho m'(\rho)=\frac{\partial m(\rho)}{\partial\log\rho}=\frac{1}{\pi}\int_{\theta_\rho} u\frac{\partial u}{\partial\rho}\rho d\theta=\frac{1}{\pi}\int_{\Gamma_\rho+\theta_\rho} u\frac{\partial u}{\partial\nu}ds$$

$$=\frac{1}{\pi}\iint_{D_\rho}\left(\left(\frac{\partial u}{\partial x}\right)^2+\left(\frac{\partial u}{\partial y}\right)^2\right)dxdy, \quad z=x+iy .$$

Hence if we put

$$S(\rho)=\frac{1}{\pi}\iint_{D_\rho}\left(\left(\frac{\partial u}{\partial x}\right)^2+\left(\frac{\partial u}{\partial y}\right)^2\right)dxdy , \tag{4}$$

then

$$m'(\rho)=\frac{S(\rho)}{\rho}>0, \quad \rho\neq\rho_i . \tag{5}$$

Hence $m(\rho)$ is an increasing function of ρ in (ρ_i, ρ_{i+1}) and since $m(\rho_i-0)=m(\rho_i)<m(\rho_i+0)$, $m(\rho)$ is an increasing function of ρ $(0\leqq\rho<r)$, so that

$$m(r)-m(0)\geqq\int_0^r m'(\rho)\,d\rho=\int_0^r\frac{S(\rho)}{\rho}\,d\rho\,. \tag{6}$$

From (2), we have

$$\left(\frac{\partial m(\rho)}{\partial\log\rho}\right)^2\leqq\frac{1}{\pi^2}\int_{\Theta_\rho}u^2\,d\theta\int_{\Theta_\rho}\left(\frac{\partial u}{\partial\log\rho}\right)^2 d\theta=\frac{2m(\rho)}{\pi}\int_{\Theta_\rho}\left(\frac{\partial u}{\partial\log\rho}\right)^2 d\theta\,,$$

so that

$$\frac{1}{\pi}\int_{\Theta_\rho}\left(\frac{\partial u}{\partial\log\rho}\right)^2 d\theta\geqq\frac{1}{2m(\rho)}\left(\frac{\partial m(\rho)}{\partial\log\rho}\right)^2. \tag{7}$$

Suppose that the circle $|z|=\rho$ intersects Γ, then θ_ρ consists of a finite number of circular arcs θ_ρ, such that $\theta_\rho=\sum_i\theta_\rho$ and let $\rho\theta_i(\rho)$ be the arc length of θ_ρ. Since $u=0$ at the end points of θ_ρ, we have by Wirtinger's inequality,

$$\int_{\Theta_\rho^i}\left(\frac{\partial u}{\partial\theta}\right)^2 d\theta\geqq\frac{\pi^2}{[\theta_i(\rho)]^2}\int_{\Theta_\rho^i}u^2 d\theta\geqq\frac{\pi^2}{[\theta^*(\rho)]^2}\int_{\Theta}u^2\,d\theta\,,$$

hence summing up for i,

$$\frac{1}{\pi}\int_{\Theta_\rho}\left(\frac{\partial u}{\partial\theta}\right)^2 d\theta\geqq\frac{\pi}{[\theta^*(\rho)]^2}\int_{\Theta_\rho}u^2\,d\theta=\frac{2\pi^2}{[\theta^*(\rho)]^2}\,m(\rho)\,. \tag{8}$$

Hence from (3), (7), (8), we have

$$\frac{\partial^2 m(\rho)}{\partial\log\rho^2}\geqq\frac{1}{2m(\rho)}\left(\frac{\partial m(\rho)}{\partial\log\rho}\right)^2+\frac{1}{2}\left(\frac{2\pi}{\theta^*(\rho)}\right)^2 m(\rho)\,. \tag{9}$$

If the circle $|z|=\rho$ does not intersect Γ and is contained entirely in D, then $\theta^*(\rho)=\infty$ and from (3), (7), we have

$$\frac{\partial^2 m(\rho)}{\partial\log\rho^2}\geqq\frac{1}{2m(\rho)}\left(\frac{\partial m(\rho)}{\partial\log\rho}\right)^2,$$

hence (9) holds in general. Hence if we put

$$t=\log\rho,\qquad \mu(t)=m(\rho),\qquad \Phi(t)=\frac{2\pi}{\theta^*(\rho)},\qquad t_i=\log\rho_i\,, \tag{10}$$

then (9) becomes

$$\mu''(t)\geqq\frac{\mu'(t)^2}{2\mu(t)}+\frac{1}{2}\Phi^2(t)\,\mu(t)\,, \tag{11}$$

so that if we put $\lambda(t)=\log\mu(t)$, then $\lambda'(t)^2+2\lambda''(t)\geqq\Phi^2(t)$, so that

$$\left(\frac{\mu''(t)}{\mu'(t)}\right)^2=\left(\lambda'(t)+\frac{\lambda''(t)}{\lambda'(t)}\right)^2\geqq\lambda'(t)^2+2\lambda''(t)\geqq\Phi^2(t)\,.$$

Since in $(t_i,\ t_{i+1})$, $\mu''(t)>0$ by (11) and $\mu'(t)>0$ by (5), we have

$$\frac{\mu''(t)}{u'(t)}>0\quad \text{in } (t_i,\ t_{i+1})\,,$$

so that

$$\frac{\mu''(t)}{\mu'(t)} \geqq \Phi(t) \quad \text{in } (t_i,\ t_{i+1}). \tag{12}$$

Since by (5),

$$\mu'(t) = \frac{\partial m(\rho)}{\partial \log \rho} = S(\rho) \quad (\rho \neq \rho_i),$$

we have $\mu'(t_i - 0) < \mu'(t_i + 0)$, so that if $t < \tau$,

$$\log \mu'(\tau) - \log \mu'(t) \geqq \int_t^\tau \frac{\mu''(t)}{\mu'(t)}\, dt \geqq \int_t^\tau \Phi(t) dt,$$

$$\mu'(\tau) \geqq \mu'(t)\, e^{\int_t^\tau \Phi(t) dt} \quad (t < \tau). \tag{13}$$

Next for $\rho \leqq \sigma \leqq r$, we put

$$t = \log \rho,\ \tau = \log \sigma,\ t_0 = \log r.$$

If $\rho \leqq \kappa r\ (0 < \kappa < 1)$, then from (13),

$$1 \geqq \mu(t_0) \geqq \mu(t_0) - \mu(t) \geqq \int_t^{t_0} \mu'(\tau) d\tau \geqq \mu'(t) \int_t^{t_0} e^{\int_t^\tau \Phi(t) dt}\, d\tau$$

$$= \mu'(t) \int_\rho^r e^{2\pi \int_\rho^\sigma \frac{d\rho}{\rho\theta^*(\rho)}} \frac{d\sigma}{\sigma} \geqq \mu'(t) \int_{\kappa r}^r e^{2\pi \int_\rho^\sigma \frac{d\rho}{\rho\theta^*(\rho)}} \frac{d\sigma}{\sigma} \geqq (1-\kappa) \mu'(t)\, e^{2\pi \int_\rho^{\kappa r} \frac{d\rho}{\rho\theta^*(\rho)}},$$

$$\mu'(t) \leqq \frac{1}{1-\kappa}\, e^{-2\pi \int_\rho^{\kappa r} \frac{d\rho}{\rho\theta^*(\rho)}} \quad (0 < \kappa < 1). \tag{14}$$

From (11), $\mu''(t) \geqq \frac{1}{2} \Phi^2(t) \mu(t)$ and since $\Phi^2(t) = \frac{4\pi^2}{\theta^*(\rho)^2} \geqq \frac{2\pi}{\theta^*(\rho)}$, we have

$$\mu''(t) \geqq \frac{\pi m(\rho)}{\theta^*(\rho)},$$

and since $m(\rho)$ is an increasing function of ρ,

$$\mu'(\tau) \geqq \mu'(\tau) - \mu'(t) \geqq \int_t^\tau \mu''(t) dt \geqq \pi \int_\rho^\sigma \frac{m(\rho)}{\rho\theta^*(\rho)}\, d\rho \geqq \pi m(\rho) \int_\rho^\sigma \frac{d\rho}{\rho\theta^*(\rho)}.$$

Hence if $\sigma \leqq \kappa r$, we have from (14),

$$\frac{1}{1-\kappa}\, e^{-2\pi \int_\sigma^{\kappa r} \frac{d\rho}{\rho\theta^*(\rho)}} \geqq \mu'(\tau) \geqq \pi m(\rho) \int_\rho^\sigma \frac{d\rho}{\rho\theta^*(\rho)},$$

or

$$\frac{1}{1-\kappa}\, e^{2\pi \int_\rho^\sigma \frac{d\rho}{\rho\theta^*(\rho)} - 2\pi \int_\rho^{\kappa r} \frac{d\rho}{\rho\theta^*(\rho)}} \geqq \frac{m(\rho)}{2} \cdot 2\pi \int_\rho^\sigma \frac{d\rho}{\rho\theta^*(\rho)}. \tag{15}$$

If $2\pi \int_\rho^{\kappa r} \frac{d\rho}{\rho\theta^*(\rho)} > 1$, then we choose σ $(\rho < \sigma < \kappa r)$, such that $2\pi \int_\rho^\sigma \frac{d\rho}{\rho\theta^*(\rho)} = 1$, then

$$m(\rho) \leqq \frac{2e}{1-\kappa} e^{-2\pi \int_{\rho}^{\kappa r} \frac{d\rho}{\rho \theta^*(\rho)}}. \tag{16}$$

If $2\pi \int_{\rho}^{\kappa r} \frac{d\rho}{\rho\theta^*(\rho)} \leqq 1$, then since $m(\rho) \leqq 1$ and $\frac{2}{1-\kappa} > 1$, (16) holds, hence (16) holds in general, so that

$$u_r(0) = \sqrt{m(0)} \leqq \frac{3}{\sqrt{1-\kappa}} e^{-\pi \int_0^{\kappa r} \frac{d\rho}{\rho\theta^*(\rho)}} \quad (0<\kappa<1). \tag{17}$$

(ii) Next we consider the general case. We exhaust D by a sequence of domains $D^1 \subset D^2 \subset \cdots \subset D^n \to D$, where the boundary Γ^n of D^n satisfies the condition of (i). Let $u_r^n(z)$ be the harmonic measure of θ_r^n with respect to D_r^n and $\theta_n{}^*(\rho)$ be defined as before, then since $\frac{1}{\theta_n{}^*(\rho)} \geqq \frac{1}{\theta^*(\rho)}$, we have by (17),

$$u_r^n(0) \leqq \frac{3}{\sqrt{1-\kappa}} e^{-\pi \int_0^{\kappa r} \frac{d\rho}{\rho\theta_n^*(\rho)}} \leqq \frac{3}{\sqrt{1-\kappa}} e^{-\pi \int_0^{\kappa r} \frac{d\rho}{\rho\theta^*(\rho)}}. \tag{18}$$

By the maximum principle, $u_r^n(z) \leqq u_r^{n+1}(z) \leqq u_r(z)$ in D_r^n, so that $\lim_n u_r^n(z) = v_r(z)$ exists and $v_r(z) \leqq u_r(z)$ in D_r. By Harnack's theorem, $v_r(z)$ is harmonic in D_r. Since $u_r=0$ "almost everywhere" on Γ_r, we see that $v_r=0$ "almost everywhere" on Γ_r and since $u_r=v_r=1$ on θ_r, $u_r=v_r$ "almost everywhere" on the whole boundary of D_r, so that $v_r \equiv u_r$, hence if we make $n \to \infty$ in (18), then we have

$$u_r(0) \leqq \frac{3}{\sqrt{1-\kappa}} e^{-\pi \int_0^{\kappa r} \frac{d\rho}{\rho\theta^*(\rho)}} \quad (0<\kappa<1). \tag{19}$$

Corollary. *If $z \in D$ and $|z| < \frac{\kappa r}{2}$, then*

$$u_r(z) \leqq \frac{9}{\sqrt{1-\kappa}} e^{-\pi \int_{2|z|}^{\kappa r} \frac{d\rho}{\rho\theta^*(\rho)}} \quad (0<\kappa<1).$$

Proof. Let $\varDelta$ be a disc about $z=0$ of radius $2|z|$ and we join $\varDelta$ to D and put $\tilde{D} = D + \varDelta$ and $\tilde{u}_r(z)$ be the harmonic measure of θ_r with respect to $\tilde{D}_r$. We define $\tilde{\theta}^*(\rho)$ for $\tilde{D}$, then $\tilde{\theta}^*(\rho) = \infty$ for $0 \leqq \rho < 2|z|$ and $\tilde{\theta}^*(\rho) = \theta^*(\rho)$ for $2|z| \leqq \rho < \kappa r$, so that

$$\tilde{u}_r(0) \leqq \frac{3}{\sqrt{1-\kappa}} e^{-\pi \int_0^{\kappa r} \frac{d\rho}{\rho\tilde{\theta}^*(\rho)}} = \frac{3}{\sqrt{1-\kappa}} e^{-\pi \int_{2|z|}^{\kappa r} \frac{d\rho}{\rho\theta^*(\rho)}}.$$

If we put $2|z| = r_0$, then

$$\tilde{u}_r(z) = \frac{1}{2\pi} \int_0^{2\pi} \tilde{u}_r(r_0 e^{i\theta}) \frac{r_0{}^2 - |z|^2}{|z - r_0 e^{i\theta}|^2} d\theta \leqq \frac{r_0{}^2 - (r_0/2)^2}{(r_0 - r_0/2)^2} \frac{1}{2\pi} \int_0^{2\pi} \tilde{u}_r(r_0 e^{i\theta}) d\theta = 3\,\tilde{u}_r(0).$$

Since by the maximum principle, $u_r(z)\leqq\tilde{u}_r(z)$, we have

$$u_r(z)\leqq\tilde{u}_r(z)\leqq 3\,\tilde{u}_r(0)\leqq\frac{9}{\sqrt{1-\kappa}}\,e^{-\pi\int_{2|z|}^{\kappa r}\frac{d\rho}{\rho\theta^*(\rho)}}.$$

Remark. In the definition of D_r, we assumed that $z=0$ belongs to D. If $z=0$ does not belong to D, let D_r be the connected sub-domain of D, which lies in $|z|<r$ and contains a fixed point z_0 of D, then from the above proof, we see that the same relation holds for D.

2. As an application of Theorem III. 67. Cor., we shall prove the following theorem, by means of which we can prove easily Carleman-Ahlfors' theorem on the number of asymptotic values of an integral function of finite order.

Theorem III. 68. *Let D be an infinite domain on the z-plane and Γ be its boundary. Let $f(z)$ be regular in D and $|f(z)|\leqq 1$ on Γ and $M(r;D)=\underset{z\in D,\ |z|=r}{\mathrm{Max}}|f(z)|$. If there exists $z_0\in D$, such that $|f(z_0)|>1$, then*

$$\log\log M(r;D)\geqq\pi\int_1^{\kappa r}\frac{dr}{r\theta^*(r)}-\text{const.}\quad(0<\kappa<1)\,.$$

Hence if $\varlimsup_{r\to\infty}\left(\pi\int_1^{\kappa r}\frac{dr}{r\theta^*(r)}-\log\log M(r;D)\right)=\infty$, *then*

$$|f(z)|\leqq 1\quad \textit{in } D\,.$$

Proof. Since $\log^+|f(z)|$ is subharmonic and vanishes on Γ, we have by the maximum principle and Theorem III. 67. Cor., if $\kappa r>r_0=2|z_0|$,

$$0<\log^+|f(z_0)|\leqq\log M(r;D)\,u_r(z_0)\leqq\text{const.}\log M(r;D)\,e^{-\pi\int_{r_0}^{\kappa r}\frac{dr}{r\theta^*(r)}},$$

so that

$$\log\log M(r;D)\geqq\pi\int_1^{\kappa r}\frac{dr}{r\theta^*(r)}-\text{const.}\quad ——$$

As a special case, we have the following Phragmen-Lindelöf's theorem.

Theorem III. 69. *Let $f(z)$ be regular in an angular domain D: $|\arg z|<\frac{\alpha}{2}$ and $|f(z)|\leqq 1$ on its boundary Γ and $M(r;D)=\underset{|\theta|\leqq\alpha/2}{\mathrm{Max}}|f(re^{i\theta})|$. If for any $\delta>0$,*

$$M(r;D)<e^{\delta r^\rho}\quad\left(\rho=\frac{\pi}{\alpha}\right)\quad \textit{for } r\geqq r_0(\delta)\,,$$

then

$$|f(z)|\leqq 1\quad \textit{in } D\,.$$

Proof. In this case, $\theta^*(r)=\alpha$, so that

$$\pi\int_1^{\kappa r}\frac{dr}{r\theta^*(r)}=\frac{\pi}{\alpha}\log r+\text{const.}\,,$$

and

$$\log\log M(r;\ D) \leqq \log\delta + \frac{\pi}{\alpha}\log r \quad (r \geqq r_0(\delta))\,.$$

Since $\delta>0$ is arbitrary,

$$\lim_{r\to\infty}\left(\pi\int_1^{\kappa r}\frac{dr}{r\theta^*(r)} - \log\log M(r;D)\right) = \infty\,,$$

so that

$$|f(z)| \leqq 1 \quad \text{in } D\,.$$

18. Dirichlet problem with unbounded boundary values

1. Let D be an infinite domain on the z-plane, which contains $z=0$ and $z=\infty$ lie on its boundary Γ. Let $\varphi(\zeta)$ be a given continuous function on Γ. In the usual Dirichlet problem, $\varphi(\zeta)$ is assumed to be bounded on Γ. If $\varphi(\zeta)$ is unbounded, then there exists, in general, no harmonic function in D, which takes the given value $\varphi(\zeta)$ on Γ.

We consider a special case, where $\varphi(\zeta)=|\zeta|^k$ $(k>0)$ and shall prove

Theorem III. 70. *If*

$$\underline{\lambda}(D) = \varliminf_{r\to\infty}\frac{2\pi}{\log r}\int_1^r\frac{dr}{r\theta^*(r)} > 2k\,,$$

then there exists a harmonic function $u(z)$ in D, which takes the value $r^k(|z|=r)$ on Γ and

$$|z|^k \leqq u(z) \quad \text{in } D.$$

If further

$$\underline{\lambda}(D) \geqq \underline{\lambda}^*(D) = \varliminf_{r/a\to\infty}\frac{2\pi}{\log r/a}\int_a^r\frac{dr}{r\theta^*(r)} > 2k\,,$$

then

$$|z|^k \leqq u(z) \leqq \text{const.}|z|^k \quad \text{in } D.$$

Proof. If we take $\kappa=1/2$ in Theorem III. 67. Cor., then we have

$$u_r(z) \leqq \text{const. } e^{-\pi\int_{2|z|}^{r/2}\frac{dr}{r\theta^*(r)}}, \qquad |z|<r/4\,. \tag{1}$$

If $\underline{\lambda}(D)>2k$, then

$$\pi\int_1^{r/2}\frac{dr}{r\theta^*(r)} > k_1\log r \ \ (k_1>k) \qquad r\geqq r_0,$$

so that

$$u_r(z) \leqq \frac{\text{const.}}{r}, \qquad r\geqq r_0\,, \tag{2}$$

where const. is independent of r, but depends on z.

Let Γ_r be the part of Γ, which is contained in $|z|\leqq r$ and $v_r(z)$ be a bounded harmonic function in D, such that

$$v_r=0 \text{ on } \Gamma_r, \quad v_r=1 \text{ on } \Gamma-\Gamma_r , \tag{3}$$

then by the maximum principle, $v_r(z)\leqq u_r(z)$ in D_r, so that by (2), $v_r(z)\leqq \dfrac{\text{const.}}{r^{k_1}}$, hence the integral

$$u(z)=k\int_0^\infty v_r(z)r^{k-1}dr \tag{4}$$

converges and represents a harmonic function in D.

Let $a_0(|a_0|=r_0)$ be a regular point of Γ, then if z tends to a_0 from the inside of D, $\lim_{z\to a_0} v_r(z)=v_r(a_0)$, where $v_r(a_0)=1$, if $r<r_0$ and $v_r(a_0)=0$, if $r_0<r$. Since $v_r(z)\leqq\text{const.}/r^{k_1}$ $(k_1>k)$, we can prove easily that

$$\lim_{z\to a_0} u(z)=k\int_0^\infty v_r(a_0)r^{k-1}dr=k\int_0^{r_0} r^{k-1}dr=r_0{}^k . \tag{5}$$

Hence $u(z)=r^k$ on Γ at regular points.

Let $V_R(z)$ be a harmonic function in D_R, such that $V_R(z)=|z|^k$ on the whole boundary of D_R, then since $|z|^k$ is subharmonic, by the maximum principle,

$$|z|^k\leqq V_R(z)\leqq R^k u_R(z)+u(z) \quad \text{in } D_R$$

Since by (2), $\lim_{R\to\infty} R^k u_R(z)=0$, we have

$$|z|^k\leqq u(z) \quad \text{in } D . \tag{6}$$

Hence the first part is proved.

Next suppose that $\underline{\lambda}^*(D)>2k$, then if $r/|z|\geqq\rho_0$,

$$\pi\int_{2|z|}^{r/2} \frac{dr}{r\theta^*(r)} >k_1 \log r/|z| \quad (k_1>k) , \tag{7}$$

so that by (1),

$$u_r(z)\leqq\text{const.}\,(|z|/r)^{k_1} , \text{ if } r\geqq\rho_0|z| . \tag{8}$$

Since $0\leqq v_r(z)\leqq1$ and $v_r(z)\leqq u_r(z)$, we have

$$u(z)\leqq k\int_0^{\rho_0|z|} r^{k-1}dr+\int_{\rho_0|z|}^\infty v_r(z)r^{k-1}dr \leqq (\rho_0|z|)^k+\text{const.}\int_{\rho_0|z|}^\infty (|z|/r)^{k_1}r^{k-1}dr$$

$$\leqq(\rho_0|z|)^k+\text{const.}\,|z|^k\int_{\rho_0}^\infty \frac{dt}{t^{1+k_1-k}} \leqq\text{const.}|z|^k \quad (r=|z|t) ,$$

hence

$$u(z)\leqq\text{const.}|z|^k \quad \text{in } D . \tag{9}$$

2. As an application of the theorem, we shall prove

Theorem III. 71. *Let D be an infinite domain on the z-plane and $z=\infty$ lie on its boundary Γ. Let $f(z)$ be regular in D and*

$$M(r;\,D)=\operatorname*{Max}_{z\in D,\ |z|=r}|f(z)|.$$

Suppose that

$$\overline{\lim_{r\to\infty}}\frac{\log\log M(r;\,D)}{\log r}=\rho \qquad (0<\rho<\infty)\,.$$

(i) *If* $\log|f(z)|\leqq|z|^{\rho-\varepsilon}\,(0<\varepsilon<\rho)$ *on* Γ, *or* (ii) *If* $\overline{\lim\limits_{r\to\infty}}\dfrac{\log M(r;\,D)}{r^\rho}=\infty$ *and* $\log|f(z)|\leqq\text{const.}\,|z|^\rho$ *on* Γ, *then*

$$\underline{\lambda}^*(D)=\lim_{r/a\to\infty}\frac{2\pi}{\log r/a}\int_a^r\frac{dr}{r\theta^*(r)}\leqq 2\rho\,.$$

Proof. First we consider the case (i). We put $\rho_1=\rho-\varepsilon$. Suppose that $\underline{\lambda}^*(D)>2\rho$, then $\underline{\lambda}^*(D)>2\rho_1$. By Theorem III. 70, there exists a harmonic function $u(z)$ in D, such that $u=|z|^{\rho_1}$ on Γ and

$$|z|^{\rho_1}\leqq u(z)\leqq\text{const.}\,|z|^{\rho_1} \quad \text{in } D\,. \tag{1}$$

By the hypothesis, there exists a point z_0 in D, such that $\log|f(z_0)|>|z_0|^{\rho-\frac{\varepsilon}{2}}$, where $|z_0|$ is arbitrarily large, so that

$$\log|f(z_0)|-u(z_0)>0\,. \tag{2}$$

Let D_r be the part of D, contained in $|z|<r$ and $u_r(z)$ be the harmonic measure of θ_r, with respect to D_r, then since $\log|f(z)|-u(z)\leqq 0$ on Γ and $\log|f(z)|-u(z)$ is subharmonic, by the maximum principle, $\log|f(z)|-u(z)\leqq\log M(r;\,D)\,u_r(z)$ in D_r, so that by Theorem III. 67. Cor.,

$$0<\log|f(z_0)|-u(z_0)\leqq\log M(r;\,D)u_r(z_0)\leqq\text{const.}\,\log M(r;\,D)\,e^{-\pi\int_{2|z_0|}^{r/2}\frac{dr}{r\theta^*(r)}},$$

hence

$$\pi\int_{2|z_0|}^{r/2}\frac{dr}{r\theta^*(r)}\leqq\log\log M(r;\,D)+\text{const.}\,,$$

so that

$$\underline{\lambda}^*(D)\leqq\underline{\lambda}(D)\leqq\bar{\lambda}(D)=\overline{\lim_{r\to\infty}}\frac{2\pi}{\log r}\int_1^r\frac{dr}{r\theta^*(r)}\leqq 2\rho\,,$$

which contradicts the hypothesis, hence $\underline{\lambda}^*(D)\leqq 2\rho$.

Next suppose that $\overline{\lim\limits_{r\to\infty}}\dfrac{\log M(r;D)}{r^\rho}=\infty$ and $\log|f(z)|\leqq k|z|^\rho$ on Γ, where k is a constant. If $\underline{\lambda}^*(D)>2\rho$, then there exists a harmonic function $u(z)$ in D, such that $u=k|z|^\rho$ on Γ and $u(z)\leqq\text{const.}\,|z|^\rho$ in D. Then there exists $z_0\in D$, such that $\log|f(z_0)|-u(z_0)>0$. From this, the proof is the same as above and we have $\underline{\lambda}^*(D)\leqq 2\rho$.

19. Wiman's theorem

1. Let $f(z)$ be an integral function of order ρ $(0<\rho<1/2)$ and

$$M(r)=\underset{|z|=r}{\mathrm{Max}}|f(z)|, \qquad m(r)=\underset{|z|=r}{\mathrm{Min}}|f(z)|,$$

then Wiman [1] proved that there exist $r_1<r_2<\cdots<r_\nu\to\infty$, such that $m(r_\nu)\to\infty$. To prove a more precise theorem, first we shall introduce three kinds of density as follows.

Let E be a measurable set on the positive x-axis and $\varphi(x)$ be its characteristic function and put

$$\underline{\delta}(E)=\varliminf_{r\to\infty}\frac{1}{r}\int_0^r\varphi(r)dr\,, \qquad \overline{\delta}(E)=\varlimsup_{r\to\infty}\frac{1}{r}\int_0^r\varphi(r)dr\,, \tag{1}$$

$$\underline{\lambda}(E)=\varliminf_{r\to\infty}\frac{1}{\log r}\int_1^r\frac{\varphi(r)}{r}dr\,, \qquad \overline{\lambda}(E)=\varlimsup_{r\to\infty}\frac{1}{\log r}\int_1^r\frac{\varphi(r)}{r}dr\,, \tag{2}$$

$$\underline{\lambda}^*(E)=\varliminf_{r/a\to\infty}\frac{1}{\log r/a}\int_a^r\frac{\varphi(r)}{r}dr\,, \quad \overline{\lambda}^*(E)=\varlimsup_{r/a\to\infty}\frac{1}{\log r/a}\int_a^r\frac{\varphi(r)}{r}dr\,. \tag{3}$$

We call (1) the lower (upper) density of E, (2) the lower (upper) logarithmic density of E, (3) the lower (upper) strong logarithmic density of E. Evidently

$$0\leqq\underline{\delta}(E)\leqq\overline{\delta}(E)\leqq1,\ 0\leqq\underline{\lambda}(E)\leqq\overline{\lambda}(E)\leqq1,\ 0\leqq\underline{\lambda}^*(E)\leqq\overline{\lambda}^*(E)\leqq1\,, \tag{4}$$

$$\underline{\delta}(E)+\overline{\delta}(C(E))=1,\ \underline{\lambda}(E)+\overline{\lambda}(C(E))=1,\ \underline{\lambda}^*(E)+\overline{\lambda}^*(C(E))=1\,, \tag{5}$$

where $C(E)$ denotes the complement of E on the positive x-axis.

Lemma. $0\leqq\underline{\delta}(E)\leqq\underline{\lambda}^*(E)\leqq\underline{\lambda}(E)\leqq\overline{\lambda}(E)\leqq\overline{\lambda}^*(E)\leqq\overline{\delta}(E)\leqq1$.

Proof. Since $\overline{\lambda}(E)\leqq\overline{\lambda}^*(E)$, we shall prove that $\overline{\lambda}^*(E)\leqq\overline{\delta}(E)$. Let $\overline{\delta}(E)=\alpha$, then for any $\varepsilon>0$,

$$\mu(r)=\int_0^r\varphi(r)dr\leqq r(\alpha+\varepsilon), \quad \text{if } r\geqq r_0=r_0(\varepsilon)>1\,.$$

If $1\leqq a\leqq r_0<r$, then since $\mu(r)\leqq r$,

$$\int_a^r\frac{\varphi(r)}{r}dr\leqq\int_1^{r_0}\frac{dr}{r}+\int_{r_0}^r\frac{d\mu(r)}{r}\leqq r_0+\left[\frac{\mu(r)}{r}\right]_{r_0}^r+\int_{r_0}^r\frac{\mu(r)dr}{r^2}$$

$$\leqq r_0+1+(\alpha+\varepsilon)\int_{r_0}^r\frac{dr}{r}\leqq r_0+1+(\alpha+\varepsilon)\log\frac{r}{a} \qquad (1\leqq a\leqq r_0<r)\,. \tag{1}$$

If $r_0\leqq a<r$, then

$$\int_a^r\frac{\varphi(r)dr}{r}=\int_a^r\frac{d\mu(r)}{r}=\left[\frac{\mu(r)}{r}\right]_a^r+\int_a^r\frac{\mu(r)dr}{r^2}\leqq1+(\alpha+\varepsilon)\log\frac{r}{a}\ (r_0\leqq a<r). \tag{2}$$

From (1), (2), we have $\overline{\lambda}^*(E)\leqq\alpha=\overline{\delta}(E)$, so that by (5), $\underline{\delta}(E)\leqq\underline{\lambda}^*(E)$.

2. Now we shall prove the following **Precise form of Wiman's theorem.**

Theorem III. 72. *Let $f(z)$ be an integral function of order ρ*

$(0<\rho<1/2)$. *Then*

(i) $\bar{\lambda}^*[E(\log m(r)>r^{\rho-\varepsilon})]$[1] $\geqq 1-2\rho>0 \qquad (0<\varepsilon<\rho)$.

(ii) *If* $\overline{\lim_{r\to\infty}}\dfrac{\log M(r)}{r^\rho}=\infty$, *then for any* $k>0$,

$$\bar{\lambda}^*[E(\log m(r)>kr^\rho)]\geqq 1-2\rho>0 .$$ Tsuji [18].

Besicovitch [1] and Pennyquick [1] proved that $\bar{\delta}[E(\log m(r)>r^{\rho-\varepsilon})]\geqq 1-2\rho$, which follows from the theorem by the lemma.

Proof. We may assume that $f(0)=1$, then since $\rho<1$,

$$f(z)=\prod_{n=1}^{\infty}\left(1-\frac{z}{a_n}\right). \qquad (1)$$

Since

$$\prod_{n=1}^{\infty}\left|1-\frac{r}{|a_n|}\right|\leqq m(r)\leqq M(r)\leqq\prod_{n=1}^{\infty}\left(1+\frac{r}{|a_n|}\right),$$

we may assume that $a_n>0$ $(n=1, 2, \cdots)$, so that

$$m(r)=\prod_{n=1}^{\infty}\left|1-\frac{r}{a_n}\right|=|f(r)|, \qquad M(r)=\prod_{n=1}^{\infty}\left(1+\frac{r}{a_n}\right)=f(-r) . \qquad (2)$$

We put $\rho_1=\rho-\varepsilon$ and $E_1=E(\log m(r)\leqq r^{\rho_1})$. Then since $E(\log m(r)>r^{\rho_1})$ and E_1 are complementary sets, it suffices to prove that $\underline{\lambda}^*(E_1)\leqq 2\rho$.

Now E_1 is a closed set on the positive real axis and $\log|f(r)|\leqq r^{\rho_1}$ on E_1, so that if we denote the outside of E_1 by D, then by Theorem III. 71,

$$\underline{\lambda}^*(D)=\underline{\lim}_{r/a\to\infty}\frac{2\pi}{\log r/a}\int_a^r\frac{dr}{r\theta^*(r)}\leqq 2\rho .$$

If $\varphi(r)$ denotes the characteristic function of E_1, then $\dfrac{2\pi}{r\theta^*(r)}=\dfrac{\varphi(r)}{r}$, so that

$$\underline{\lambda}^*(E_1)=\underline{\lim}_{r/a\to\infty}\frac{1}{\log r/a}\int_a^r\frac{\varphi(r)dr}{r}\leqq 2\rho ,$$

hence
$$\bar{\lambda}^*[E(\log m(r)>r^{\rho-\varepsilon})]\geqq 1-2\rho.$$

Next suppose that $\overline{\lim_{r\to\infty}}\dfrac{\log M(r)}{r^\rho}=\infty$. Let $E_1=E(\log m(r)\leqq kr^\rho)$, then E_1 is a closed set on the positive real axis and $\log|f(r)|\leqq kr^\rho$ on E_1, so that by Theorem III. 71, we have similarly as above that

$$\underline{\lambda}^*(E_1)\leqq 2\rho, \quad \text{or} \quad \bar{\lambda}^*[E(\log m(r)>kr^\rho)]\geqq 1-2\rho .$$

3. Next we shall prove the following analogous theorem concerning $M(r)$.

Theorem III. 73. *Let $f(z)$ be an integral function of finite order $\rho>0$, then for any $\varepsilon>0$,*

1) $E[\log m(r)>r^{\rho-\varepsilon}]$ denotes the set of r, such that $\log m(r)>r^{\rho-\varepsilon}$.

$$\overline{\lambda}^*[E(\log M(r)>r^{\rho-\varepsilon})]=1\,.$$

If $\overline{\lim\limits_{r\to\infty}}\dfrac{\log M(r)}{r^{\rho}}=\infty$, *then for any* $k>0$,

$$\overline{\lambda}^*[E(\log M(r)>kr^{\rho})]=1\,. \qquad \text{Tsuji [18].}$$

Proof. Let $\rho_1=\rho-\varepsilon$ and $E_1=E(\log M(r)\leqq r^{\rho_1})$, then E_1 is a closed set on the positive real axis, which consists of disjoint closed intervals $[r_\nu, r_\nu']$ ($\nu=1, 2, \cdots$), ($r_1<r_1'<r_2<r_2'<\cdots$), so that

$$\log|f(z)|\leqq r^{\rho_1} \text{ on } |z|=r \quad (r_\nu\leqq r\leqq r_\nu')\,. \tag{1}$$

Let Δ_ν: $r_\nu\leqq|z|\leqq r_\nu'$ be a circular ring and we take off from Δ_ν^0 a part: $r_\nu\leqq|z|\leqq r_\nu'$, $|\arg z|\leqq\delta$ ($\delta>0$) and let Δ_ν be the remaining domain and put $\Delta=\sum\limits_{\nu=1}^{\infty}\Delta_\nu$. Let D be the complement of Δ, then D is an infinite domain and $\log|f(z)|\leqq|z|^{\rho_1}$ on its boundary Γ, so that by Theorem III. 71,

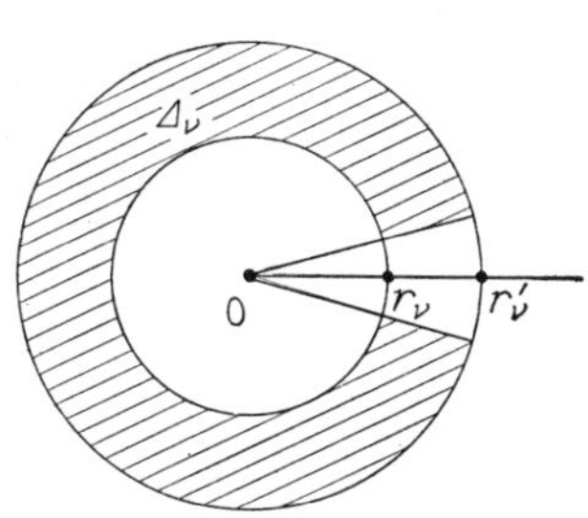

$$\underline{\lim}_{r/a\to\infty}\frac{2\pi}{\log r/a}\int_a^r\frac{dr}{r\theta^*(r)}\leqq 2\rho\,.$$

Since $\dfrac{2\delta}{r\theta^*(r)}=\dfrac{\varphi(r)}{r}$, where $\varphi(r)$ is the characteristic function of E_1,

$$\underline{\lambda}^*(E_1)=\underline{\lim}_{r/a\to\infty}\frac{1}{\log r/a}\int_a^r\frac{\varphi(r)dr}{r}\leqq\frac{2\delta\rho}{\pi}\,.$$

Since $\delta>0$ is arbitrary, we have $\underline{\lambda}^*(E_1)=0$, so that $\overline{\lambda}^*[E(\log M(r)>r^{\rho-\varepsilon})]=1$. Similarly we can prove the second part.

Remark. In Theorem III. 72 and 73, $\overline{\lambda}^*$ can not be replaced by $\overline{\lambda}$, since for any $\rho\,(0<\rho<1/2)$, there exists an integral function of order ρ, such that

$$\overline{\lambda}[E(\log m(r)>r^{\rho-\varepsilon})]<1-2\rho\,, \qquad \text{if } 0<\varepsilon<\rho(1-2\rho)\,,$$

and for any $\rho>0$, there exists an integral function of order ρ, such that

$$\overline{\lambda}[E(\log M(r)>r^{\rho-\varepsilon})]<1, \qquad \text{if } 0<\varepsilon<\mathrm{Min}\left(\frac{\rho^2}{1+\rho},\ \frac{\rho}{2}\right). \quad \text{Tsuji [18]}$$

20. Capacity of a set in the space of regular functions

Let D be a domain on the z-plane, which is bounded by a Jordan curve and Ω be the space of uniformly bounded regular functions $\{g(z)\}$ in D, such that

$$|g(z)|<M<\infty \tag{1}$$

for any $z\in D$ and $g\in\Omega$. If $g\in\Omega$, $h\in\Omega$, $0<\lambda<1$, then $\lambda g+(1-\lambda)h\in\Omega$, so that Ω is a convex space. Let $D'\subset D$ be a closed Jordan domain, contained in D, then we define the metric $\rho(g, h)$ in Ω by

$$\rho(g, h)=\operatorname*{Max}_{z\in D'}|g(z)-h(z)| \, . \tag{2}$$

We see easily that $\rho(g, h)$ satisfies the triangle inequality. $\rho(g, h)$ depends on the choice of D', but it is of no importance in the following consideration. By this metric, we define an open set and a cluster point of a set and a closed set and hence a Borel set in Ω.

Since by (1), $\{g(z)\}$ is a normal family, there holds in Ω an analogue of Weierstrass-Bolzano's theorem for point sets.

We map D conformally on $|t|<1$ and let $g=\sum_{n=0}^{\infty}a_n t^n(|t|<1)$, then by taking a_n as complex rational numbers, we see that Ω has a countable bases, so that there holds in Ω analogues of covering theorems of Lindelöf and Heine-Borel for point sets.

Let $\mathfrak{M}$ be a closed set in Ω and $d\sigma(g)\geqq 0$ be a positive mass distribution on $\mathfrak{M}$ of total mass 1, defined for Borel sub-sets of $\mathfrak{M}$ and for a fixed $z\in D'$, we consider a potential:

$$\chi_z(\zeta)=\int_{\mathfrak{M}}\log\frac{1}{|\zeta-g(z)|}\,d\sigma(g), \qquad \sigma(\mathfrak{M})=1, \qquad z\in D' \, . \tag{3}$$

Let M_z be the set of points $\{g(z)\}$ $(g\in\mathfrak{M})$, then M_z is a bounded closed set. Let $a\in M_z$ and $\mathfrak{M}_a$ be the set of $g\in\mathfrak{M}$, such that $g(z)=a$ and we define a positive mass distribution $d\nu(a)\geqq 0$ on M_z by $d\nu(a)=\sum_{g\in\mathfrak{M}_a}d\sigma(g)$, then $\nu(M_z)=\int_{M_z}d\nu(a)=\int_{\mathfrak{M}}d\sigma(g)=1$ and

$$\chi_z(\zeta)=\int_{M_z}\log\frac{1}{|\zeta-a|}\,d\nu(a), \qquad \nu(M_z)=1 \, , \tag{4}$$

so that $\chi_z(\zeta)$ is a logarithmic potential, due to a positive mass distribution on M_z, hence if we put

$$\sup_{\zeta\in M_z,\, z\in D'}\chi_z(\zeta)=J(\sigma) \, , \tag{5}$$

then by the maximum principle for logarithmic potential, we have

$$\int_{\mathfrak{M}}\log\frac{1}{|\zeta-g(z)|}\,d\sigma(g)\leqq J(\sigma) \quad \text{for any } \zeta \text{ and } z\in D' \, . \tag{6}$$

Let

$$\inf_{\sigma} J(\sigma)=V(\mathfrak{M}) \, , \tag{7}$$

and we define the capacity $\Gamma(\mathfrak{M})$ of $\mathfrak{M}$ by

$$\Gamma(\mathfrak{M})=e^{-V(\mathfrak{M})}, \qquad V(\mathfrak{M})=\log 1/\Gamma(\mathfrak{M}). \tag{8}$$

We denote the capacity of a point set M by $\gamma(M)$ as before.

If $\Gamma(\mathfrak{M})>0$, $V(\mathfrak{M})<\infty$, then there exists σ_n $(n=1, 2, \cdots)$, such that $J(\sigma_n)\to V(\mathfrak{M})$, hence we can select a partial sequence from σ_n, which converges to σ, so that by (6) and Fatou's lemma, we have for any ζ and $z\in D'$,

$$\chi_z(\zeta)=\int_{\mathfrak{M}}\log\frac{1}{|\zeta-g(z)|}\,d\sigma(g)\leqq V(\mathfrak{M})<\infty, \qquad \sigma(\mathfrak{M})=1. \tag{9}$$

We call σ *the equilibrium distribution of* $\mathfrak{M}$.

If $\mathfrak{M}$ is not a closed set, then we define its capacity $\Gamma(\mathfrak{M})$ by

$$\Gamma(\mathfrak{M})=\sup_{\mathfrak{M}_1\subset\mathfrak{M}}\Gamma(\mathfrak{M}_1)\,, \tag{10}$$

where $\mathfrak{M}_1$ is a closed sub-set of $\mathfrak{M}$.

From the definition, if $\mathfrak{M}_1\subset\mathfrak{M}_2$, then $\Gamma(\mathfrak{M}_1)\leqq\Gamma(\mathfrak{M}_2)$.

Theorem III. 74. *Let* $\mathfrak{A}_n\,(n=1, 2, \cdots)$ *be Borel sets of capacity zero, then* $\mathfrak{A}=\sum_{n=1}^{\infty}\mathfrak{A}_n$ *is of capacity zero.*

Proof. Suppose that $\Gamma(\mathfrak{A})>0$, then $\mathfrak{A}$ contains a closed sub-set $\mathfrak{M}$ of positive capacity. Let σ be its equilibrium distribution, such that for any ζ and $z\in D'$,

$$\int_{\mathfrak{M}}\log\frac{1}{|\zeta-g(z)|}\,d\sigma(g)\leqq V(\mathfrak{M})<\infty, \qquad \sigma(\mathfrak{M})=1\,.$$

Since $1=\sigma(\mathfrak{M})\leqq\sum_{n=1}^{\infty}\sigma(\mathfrak{M}.\,\mathfrak{A}_n)$, there exists a certain n_0, such that $\sigma(\mathfrak{M}.\,\mathfrak{A}_{n_0})>0$. We may assume that $\zeta\in M_z$, then $|\zeta-g(z)|<2M$, hence

$$\int_{\mathfrak{M}\cdot\mathfrak{A}_{n_0}}\log\frac{1}{|\zeta-g(z)|}\,d\sigma(g)\leqq\int_{\mathfrak{M}\cdot\mathfrak{A}_{n_0}}\log\frac{2M}{|\zeta-g(z)|}\,d\sigma(g)+\log\frac{1}{2M}$$

$$\leqq\int_{\mathfrak{M}}\log\frac{2M}{|\zeta-g(z)|}\,d\sigma(g)+\log\frac{1}{2M}=\int_{\mathfrak{M}}\log\frac{1}{|\zeta-g(z)|}\,d\sigma(g)\leqq V(\mathfrak{M})\,,$$

so that $\Gamma(\mathfrak{M}.\,\mathfrak{A}_{n_0})>0$, which is absurd, since being a sub-set of $\mathfrak{A}_{n_0}$, $\Gamma(\mathfrak{M}.\,\mathfrak{A}_{n_0})=0$, hence $\Gamma(\mathfrak{A})=0$.

Theorem III. 75. *Let* $\mathfrak{M}$ *be a closed set of positive capacity in* Ω *and for a fixed* $z\in D'$, M_z *be the bounded closed set of points* $\{g(z)\}(g\in\mathfrak{M})$, *then*

$$\gamma(M_z)\geqq\Gamma(\mathfrak{M})>0, \quad z\in D'\,.$$

Hence if $\gamma(M_{z_0})=0$ for one $z_0\in D'$, then $\Gamma(\mathfrak{M})=0$.

Let $\mathfrak{M}=\{g\}$ and $\mathfrak{M}_1$ be the set of $g_1(z)=(z-z_0)g(z)$, where z_0 is a fixed point of D', then $g_1(z_0)=0$ for any $g_1 \in \mathfrak{M}_1$, so that $\Gamma(\mathfrak{M}_1)=0$. Now $g_1=(z-z_0)g$ is an affine transformation in Ω, so that a set of positive capacity in Ω becomes a set of zero capacity by an affine transformation, but this is not the case for point sets. In spite of this inconvinience, we adopt the above definition of capacity.

Proof. Let σ be the equilibrium distribution of $\mathfrak{M}$, such that for any ζ and $z \in D'$,

$$\chi_z(\zeta)=\int_{\mathfrak{M}} \log \frac{1}{|\zeta-g(z)|}\, d\sigma(g) \leqq V(\mathfrak{M})<\infty, \qquad \sigma(\mathfrak{M})=1 . \tag{1}$$

Since $\chi_z(\zeta)$ is a logarithmic potential, due to a positive mass distribution on M_z, we see from (1) that $\gamma(M_z)>0$.

Let $d\mu(\zeta)$ be the equilibrium distribution of M_z of total mass 1, then

$$u(a)=\int_{M_z} \log \frac{1}{|a-\zeta|}\, d\mu(\zeta)=V(M_z)=\log 1/\gamma(M_z) \quad \text{on } M_z , \tag{2}$$

except a set M_0 of a, which is of capacity zero.

Let $\mathfrak{M}_0$ be the set of g, such that $g(z) \in M_0$, then $\sigma(\mathfrak{M}_0)=0$, since if otherwise, we can prove that $\gamma(M_0)>0$ from (1), which is absurd. Hence by (1) and (2),

$$V(M_z)=\int_{\mathfrak{M}} d\sigma(g) \int_{M_z} \log \frac{1}{|\zeta-g(z)|}\, d\mu(\zeta) \leqq V(\mathfrak{M}) ,$$

so that

$$\gamma(M_z) \geqq \Gamma(\mathfrak{M})>0 .$$

Theorem III. 76. *Let $g_\alpha(z)=g(z, \alpha) \not\equiv$ const. be a regular function of two complex variables z and α for $z \in D$ and $\alpha \in \Delta$, where Δ is a Jordan domain and suppose that for any fixed $z \in D$, $\dfrac{\partial g(z, \alpha)}{\partial \alpha} \not\equiv 0$ as a function of α. Let M be a closed set in Δ and $\mathfrak{M}$ be the set of $\{g_\alpha\}(\alpha \in M)$, then $\Gamma(\mathfrak{M})>0$, when and only when $\gamma(M)>0$.*

Proof. Since M is a closed set, $\mathfrak{M}$ is a closed set in Ω.

(i) First suppose that $\gamma(M)>0$ and let μ be its equilibrium distribution of total mass 1, such that for any ζ,

$$u(\zeta)=\int_M \log \frac{1}{|\zeta-\alpha|}\, d\mu(\alpha) \leqq V(M)<\infty, \qquad \mu(M)=1 . \tag{1}$$

For a fixed $z \in D'$, we put

$$\chi_z(\zeta)=\int_M \log \frac{1}{|\zeta-g_\alpha(z)|}\, d\mu(\alpha), \qquad z \in D' . \tag{2}$$

Let M_z be the bounded closed set of points $\{g_\alpha(z)\}$ $(\alpha \in M)$ and put

$$\sup_{\zeta \in M_z} \chi_z(\zeta) = K_z \,, \tag{3}$$

then since $\chi_z(\zeta)$ is a logarithmic potential, due to a positive mass distribution on M_z, we have by the maximum principle for logarithmic potential,

$$\chi_z(\zeta) \leqq K_z \quad \text{for any } \zeta \,. \tag{4}$$

If $\zeta \in M_z$, then $\zeta = g_\beta(z)$ for a certain $\beta \in M$, hence

$$K_z = \sup_{\beta \in M} \int_M \log \frac{1}{|g_\beta(z) - g_\alpha(z)|} \, d\mu(\alpha) \,. \tag{5}$$

Since by the hypothesis, for any fixed $z \in D'$, $\dfrac{\partial g(z, \alpha)}{\partial \alpha} \not\equiv 0$, we see that for any $\beta \in M$, there exists a finite number of $\alpha \in M$, $\alpha_1, \cdots, \alpha_n$, $(\alpha_n = \beta)$ say, such that

$$g(z, \beta) = g(z, \alpha_i) \ (i = 1, 2, \cdots, n), \ z \in D' \,, \tag{6}$$

where if α_1 is a k-ple root of (6), then we put $\alpha_1 = \alpha_2 \cdots = \cdots = \alpha_k$. We see easily that $n \leqq n_0$, where n_0 is independent of $z \in D'$ and $\beta \in M$. We draw a disc Δ_i of radius ρ $(0 < \rho < 1)$ about α_i and put

$$M^* = M - M \sum_{i=1}^{n} \Delta_i \,.$$

If $\alpha \in M^*$, then

$$|g_\beta(z) - g_\alpha(z)| \geqq A > 0 \,, \tag{7}$$

where $A > 0$ is a constant, independent of $z \in D'$ and $\beta \in M$, so that

$$\int_{M^*} \log \frac{1}{|g_\beta(z) - g_\alpha(z)|} \, d\mu(\alpha) \leqq \log 1/A \,. \tag{8}$$

If $\alpha \in \sum_{i=1}^{n} \Delta_i$, let $\alpha \in \Delta_1, \cdots, \alpha \in \Delta_k$ $(k \leqq n)$, then by (6),

$$|g_\beta(z) - g_\alpha(z)| \geqq B \prod_{i=1}^{k} |\alpha - \alpha_i| \geqq B_1 \prod_{i=1}^{n} |\alpha - \alpha_i| \,, \tag{9}$$

where $B > 0$ and $B_1 > 0$ are constants, independent of $z \in D'$ and $\beta \in M$, so that

$$\begin{aligned} \int_{M \cdot \sum_{i=1}^{n} \Delta_i} \log \frac{1}{|g_\beta(z) - g_\alpha(z)|} \, d\mu(\alpha) &\leqq \sum_{i=1}^{n} \int_{M \cdot \Delta_i} \log \frac{1}{|\alpha_i - \alpha|} \, d\mu(\alpha) + \log 1/B_1 \\ &\leqq \sum_{i=1}^{n} \left(\int_M \log \frac{1}{|\alpha_i - \alpha|} \, d\mu(\alpha) + \log 1/\rho \right) + \log 1/B_1 \\ &\leqq n_0 (V(M) + \log 1/\rho) + \log 1/B_1 \,. \end{aligned} \tag{10}$$

Hence by (8), (10), we have

$$\int_M \log \frac{1}{|g_\beta(z) - g_\alpha(z)|} \, d\mu(\alpha) \leqq K \,,$$

where $K>0$ is a constant, independent of $z\in D'$ and $\beta\in M$, hence $K_z\leqq K$, so that by (3), for any ζ and $z\in D'$,

$$\int_M \log\frac{1}{|\zeta-g_\alpha(z)|}\,d\mu(\alpha)\leqq K\,.$$

Hence if we put $d\sigma(g_\alpha)=d\mu(\alpha)$, then for any ζ and $z\in D'$,

$$\int_{\mathfrak{M}} \log\frac{1}{|\zeta-g_\alpha(z)|}\,d\sigma(g_\alpha)\leqq K\,, \tag{11}$$

so that

$$\Gamma(\mathfrak{M})>0\,.$$

(ii) Next suppose that $\gamma(M)=0$, then since for a fixed $z_0\in D'$, $|g(z_0,\alpha_1)-g(z_0,\alpha_2)|\leqq K_0|\alpha_1-\alpha_2|$ (K_0=const.), we see, by means of the transfinite diameter, that $\gamma(M_{z_0})=0$, hence by Theorem III, 75, $\Gamma(\mathfrak{M})=0$.

CHAPTER IV

POISSON INTEGRAL

1. Poisson integral

Let $f(\theta)$ be integrable in $[-\pi, \pi]$ and

$$u(z)=u(re^{i\theta})=\frac{1}{2\pi}\int_{-\pi}^{\pi} f(\varphi)\frac{1-r^2}{1-2r\cos(\varphi-\theta)+r^2}d\varphi \quad (0\leqq r<1),$$

be the Poisson integral with respect to $f(\theta)$. $u(z)$ is harmonic in $|z|<1$.

Theorem IV. 1. *For almost all $e^{i\theta}$ on $|z|=1$, $\lim_{z\to e^{i\theta}} u(z)=f(\theta)$ uniformly, when $z\to e^{i\theta}$ from the inside of any fixed Stolz domain, whose vertex is at $e^{i\theta}$.* Fatou [1].

Proof. By Lebesgue's theorem, for almost all θ,

$$\lim_{t\to 0}\frac{1}{t}\int_0^t |f(\theta+t)-f(\theta)|dt=0. \tag{1}$$

Suppose that (1) holds at $\theta=0$, then

$$\lim_{t\to 0}\frac{1}{t}\int_0^t |f(t)-f(0)|dt=0.$$

Hence if we put

$$F(t)=|f(t)-f(0)|, \qquad \Phi(t)=\int_0^t F(t)dt, \tag{2}$$

then

$$|\Phi(t)|\leqq\varepsilon(t)|t|, \qquad |\Phi(t)|\leqq \text{const}\,|t| \quad (|t|\leqq\pi), \tag{3}$$

where $\lim_{t\to 0}\varepsilon(t)=0$.

Let Δ be a Stolz domain, whose vertex is at $z=1$. If $z=re^{i\theta}\in\Delta$, then $|\theta|\leqq\text{const.}(1-r)$, so that if

$$1-r=\delta\sqrt{\varepsilon(\delta)}, \qquad \text{then} \qquad |\theta|\leqq\text{const.}\,\delta\sqrt{\varepsilon(\delta)} \quad (\delta>0). \tag{4}$$

Since

$$\frac{1}{2\pi}\int_{-\pi}^{\pi}\frac{1-r^2}{1-2r\cos(\varphi-\theta)+r^2}d\varphi=1,$$

we have

$$|u(z)-f(0)|\leqq\frac{1}{2\pi}\int_{-\pi}^{\pi} F(\varphi)\frac{1-r^2}{1-2r\cos(\varphi-\theta)+r^2}d\varphi$$
$$=\frac{1}{2\pi}\int_{-\delta}^{\delta}+\frac{1}{2\pi}\int_{\delta}^{\pi}+\frac{1}{2\pi}\int_{-\pi}^{-\delta}=I+II+III. \tag{5}$$

Since

$$1-2r\cos\theta+r^2=(1-r)^2+4r\sin^2\frac{\theta}{2}\geqq(1-r)^2+a^2\theta^2\geqq(1-r)^2\,, \qquad (6)$$

$(a=\text{const.})$

we have by (3), (4), (6),

$$I\leqq\frac{1-r^2}{2\pi(1-r)^2}\int_{-\delta}^{\delta}F(\varphi)d\varphi\leqq\frac{(1+r)\delta\varepsilon(\delta)}{\pi(1-r)}\leqq\frac{2}{\pi}\sqrt{\varepsilon(\delta)}\to0 \quad \text{as } \delta\to0\,. \qquad (7)$$

In II, if δ is small, then by (4), $|\theta|\leqq\delta/2$, so that $|\theta|\leqq\varphi/2$, $\varphi-\theta\geqq\varphi-|\theta|\geqq\varphi/2$, hence by (3), (6),

$$II\leqq\frac{1-r^2}{2\pi}\int_{\delta}^{\pi}\frac{F(\varphi)d\varphi}{a^2(\varphi-\theta)^2}\leqq\frac{2(1-r^2)}{\pi a^2}\int_{\delta}^{\pi}\frac{F(\varphi)d\varphi}{\varphi^2}$$

$$\leqq\frac{4(1-r)}{\pi a^2}\left(\left[\frac{\Phi(\varphi)}{\varphi^2}\right]_{\delta}^{\pi}+2\int_{\delta}^{\pi}\frac{\Phi(\varphi)}{\varphi^3}d\varphi\right)\leqq\text{const.}\,(1-r)+\text{const.}\,(1-r)\int_{\delta}^{\pi}\frac{d\varphi}{\varphi^2}$$

$$\leqq\text{const.}\,(1-r)+\text{const.}\,\frac{1-r}{\delta}\leqq\text{const.}\,\delta\sqrt{\varepsilon(\delta)}+\text{const.}\,\sqrt{\varepsilon(\delta)}\to0 \quad \text{as } \delta\to0\,.$$

Similarly $III\to0$ as $\delta\to0$, hence $u(z)\to f(0)$ uniformly, when $z\to1$ in Δ.

Theorem IV. 2. *If $f(\theta)$ is continuous at θ_0, then $\lim_{z\to e^{i\theta_0}}u(z)=f(\theta_0)$, when $z\to e^{i\theta_0}$ from the inside of $|z|<1$. If $f(\theta)$ is continuous on $\alpha\leqq\theta\leqq\beta$, then $\lim_{z\to e^{i\theta}}u(z)=f(\theta)$ uniformly on $\alpha\leqq\theta\leqq\beta$.* Schwarz [1].

Proof. We may assume that $\theta_0=0$, so that if $|\varphi|\leqq\delta=\delta(\varepsilon)$, then

$$|f(e^{i\varphi})-f(1)|<\varepsilon\,. \qquad (1)$$

We have as before

$$|u(z)-f(1)|\leqq\frac{1}{2\pi}\int_{-\pi}^{\pi}|f(e^{i\varphi})-f(1)|\frac{1-r^2}{1-2r\cos(\varphi-\theta)+r^2}d\varphi$$

$$=\frac{1}{2\pi}\int_{|\varphi|\leqq\delta}+\frac{1}{2\pi}\int_{\delta\leqq|\varphi|\leqq\pi}=I+II\,, \qquad z=re^{i\theta}\,. \qquad (2)$$

By (1),

$$I=\frac{\varepsilon}{2\pi}\int_{-\delta}^{\delta}\frac{1-r^2}{1-2r\cos(\varphi-\theta)+r^2}d\varphi\leqq\frac{\varepsilon}{2\pi}\int_{-\pi}^{\pi}\frac{1-r^2}{1-2r\cos(\varphi-\theta)+r^2}d\varphi=\varepsilon\,. \qquad (3)$$

In II, if $|\theta|\leqq\delta/2$, then $|\varphi-\theta|\geqq|\varphi|-|\theta|\geqq\delta/2$ and $1-2r\cos(\varphi-\theta)+r^2\geqq\sin^2(\delta/2)$, so that if $1-r\leqq\eta(\varepsilon)$,

$$II\leqq\frac{1-r^2}{2\pi\sin^2\frac{\delta}{2}}\int_{-\pi}^{\pi}(|f(e^{i\varphi})|+|f(1)|)d\varphi\leqq\text{const.}\,(1-r)<\varepsilon\,. \qquad (4)$$

Since $\varepsilon>0$ is arbitrary, $u(z)\to f(1)$, when $z\to 1$ from the inside of $|z|<1$. The second part can be proved similarly by means of the uniform continuity of $f(\theta)$ on $[\alpha, \beta]$.

Theorem IV. 3. *Suppose that $f(\theta)$ is discontinuous at θ_0, such that $f(\theta_0+0)$, $f(\theta_0-0)$ exist. Let $l_\psi(e^{i\theta_0})$ be a segment through $e^{i\theta_0}$, making an angle ψ $(0<\psi<\pi)$ with the positive tangent of $|z|=1$ at $e^{i\theta_0}$.*

If $z\to e^{i\theta_0}$ along $l_\psi(e^{i\theta_0})$, then

$$u(z)\to f(\theta_0+0)+\frac{\psi}{\pi}(f(\theta_0-0)-f(\theta_0+0)) .$$

Hence the limiting value is a linear function of ψ, which takes the value $f(\theta_0+0)$ at $\psi=0$ and $f(\theta_0-0)$ at $\psi=\pi$.
First we shall prove a lemma.

Lemma. *Let*

$$v(z)=v(re^{i\theta})=\frac{1}{2\pi}\int_0^{2\pi}\varphi\frac{1-r^2}{1-2r\cos(\varphi-\theta)+r^2}d\varphi .$$

If z lies on $l_\psi(1)$, then

$$v(z) = 2\psi .$$

Proof. $v(z)=\dfrac{a_0}{2}+\sum_{n=1}^{\infty}r^n(a_n\cos n\theta+b_n\sin n\theta)$, $\quad z=re^{i\theta}$,

where, since $\dfrac{1-r^2}{1-2r\cos(\varphi-\theta)+r^2}=\Re\left(\dfrac{e^{i\varphi}+re^{i\theta}}{e^{i\varphi}-re^{i\theta}}\right)=1+2\sum_{n=1}^{\infty}r^n\cos n(\varphi-\theta)$,

$$a_n=\frac{1}{\pi}\int_0^{2\pi}\varphi\cos n\varphi\, d\varphi , \qquad b_n=\frac{1}{\pi}\int_0^{2\pi}\varphi\sin n\varphi\, d\varphi ,$$

so that $a_0=2\pi$, $a_n=0$ $(n\geqq 1)$, $b_n=-2/n$, hence if $z\in l_\psi(1)$,

$$v(z)=\pi-2\sum_{n=1}^{\infty}\frac{r^n\sin n\theta}{n}=\pi+2\Im(\log(1-z))=2\psi .$$

Proof of the theorem.

We assume that $\theta_0=0$. If we put $F(\varphi)=f(\varphi)-\dfrac{\varphi}{2\pi}(f(2\pi-0)-f(+0))$, then $F(+0)=F(2\pi-0)=f(+0)$, so that if we define $F(0)$ by $f(+0)$, then $F(\varphi)$ is continuous at $\varphi=0$, hence if we put

$$w(z)=\frac{1}{2\pi}\int_0^{2\pi}F(\varphi)\frac{1-r^2}{1-2r\cos(\varphi-\theta)+r^2}d\varphi ,$$

then $w(z)\to F(0)=f(+0)$, when $z\to 1$ from the inside of $|z|<1$. Since $u(z)=w(z)+\dfrac{f(2\pi-0)-f(+0)}{2\pi}v(z)$, we have by the lemma, if $z\to 1$ on $l_\psi(1)$,

$$u(z)\to f(+0)+\frac{\psi}{\pi}(f(2\pi-0)-f(+0)) .$$

Theorem IV. 4. *If $f(\theta)\in L^p$ $(p\geqq 1)$, then*

$$\lim_{r\to 1}\int_0^{2\pi}|u(re^{i\theta})-f(\theta)|^p d\theta=0\,.$$

Proof. By Hölder's inequality,

$$|u(re^{i\theta})-f(\theta)|^p\leqq\frac{1}{2\pi}\int_{-\pi}^{\pi}|f(\varphi)-f(\theta)|^p\frac{1-r^2}{1-2r\cos(\varphi-\theta)+r^2}d\varphi$$
$$=\frac{1}{2\pi}\int_{-\pi}^{\pi}|f(\theta+t)-f(\theta)|^p\frac{1-r^2}{1-2r\cos t+r^2}dt\,,$$

hence if we put

$$F(t)=\int_0^{2\pi}|f(\theta+t)-f(\theta)|^p d\theta\,, \tag{1}$$

then

$$\int_0^{2\pi}|u(re^{i\theta})-f(\theta)|^p d\theta\leqq\frac{1}{2\pi}\int_{-\pi}^{\pi}F(t)\frac{1-r^2}{1-2r\cos t+r^2}dt$$
$$=\frac{1}{2\pi}\int_{|t|\leqq\delta}+\frac{1}{2\pi}\int_{\delta\leqq|t|\leqq\pi}=I+II\,. \tag{2}$$

Since $\lim_{t\to 0}F(t)=0$, $|F(t)|<\varepsilon$, if $|t|\leqq\delta=\delta(\varepsilon)$, so that

$$I\leqq\frac{\varepsilon}{2\pi}\int_{-\delta}^{\delta}\frac{1-r^2}{1-2r\cos t+r^2}dt\leqq\frac{\varepsilon}{2\pi}\int_{-\pi}^{\pi}\frac{1-r^2}{1-2r\cos t+r^2}dt=\varepsilon\,. \tag{3}$$

Since in II, $1-2r\cos t+r^2\geqq 4r\sin^2 t/2\geqq 4r\sin^2\delta/2$,

$$II\leqq\text{const.}\,(1-r)<\varepsilon\,,\quad\text{if } 1-r \text{ is small.} \tag{4}$$

Since $\varepsilon>0$ is arbitrary, we have from (2), (3), (4),

$$\lim_{r\to 1}\int_0^{2\pi}|u(re^{i\theta})-f(\theta)|^p d\theta=0\,.$$

Theorem IV. 5. *Let*

$$u(z)=u(re^{i\theta})=\frac{1}{2\pi}\int_{-\pi}^{\pi}f(\varphi)\frac{1-r^2}{1-2r\cos(\varphi-\theta)+r^2}d\varphi\,.$$

If $f'(\theta_0)\neq\infty$ exists, then

$$\frac{\partial u}{\partial\theta}\to f'(\theta_0)\qquad\textit{uniformly,}$$

when $z\to e^{i\theta_0}$ from the inside of any fixed Stolz domain Δ, whose vertex is at $e^{i\theta_0}$.

If $f(\theta)$ is monotone increasing in a neighbourhood of θ_0 and $f'(\theta_0)=+\infty$, then $\partial u/\partial\theta\to+\infty$ uniformly, when $z\to e^{i\theta_0}$ from the inside of Δ.

If $f'(\theta)$ is continuous on $\alpha\leqq\theta\leqq\beta$ and $\alpha<\alpha_1<\beta_1<\beta$, then $\partial u/\partial\theta\to f'(\theta)$ uniformly on $[\alpha_1,\beta_1]$, when $z\to e^{i\theta}$ from the inside of $|z|<1$. Fatou [1].

First we shall prove two lemmas.

Lemma 1. *Let*

$$v(z)=v(re^{i\theta})=\frac{1}{2\pi}\int_{-\pi}^{\pi}\varphi\frac{1-r^2}{1-2r\cos(\varphi-\theta)+r^2}d\varphi\,,$$

then $\partial v/\partial\theta \to 1$, *when* $z \to 1$.

Proof. If we put

$$v(re^{i\theta})=\frac{a_0}{2}+\sum_{n=1}^{\infty} r^n(a_n \cos n\theta+b_n \sin n\theta)\,,$$

then $a_n=0$, $b_n=2(-1)^{n-1}/n$, so that

$$v(re^{i\theta})=2\left(r\sin\theta-r^2\frac{\sin 2\theta}{2}+\cdots\right),$$

hence if $z\to 1$,

$$\frac{\partial v}{\partial\theta}=2(r\cos\theta-r^2\cos 2\theta+\cdots)=2\Re\left(\frac{z}{1+z}\right)\to 1\,.$$

Lemma 2. $\dfrac{2r}{1+r}=-\dfrac{1}{2\pi}\displaystyle\int_{-\pi}^{\pi} t\frac{d}{dt}\left(\frac{1-r^2}{1-2r\cos t+r^2}\right)dt\,.$

Proof of the theorem.

We may assume that $\theta_0=0$, $f(0)=0$. First we assume that $f'(0)=0$. Let Δ be a Stolz domain, whose vertex is at $z=1$ and $z=re^{i\theta}\in\Delta$, then

$$|\theta|\leqq K(1-r),\quad K=\text{const.}\,. \tag{1}$$

$$\begin{aligned}\frac{\partial u}{\partial\theta}&=\frac{1}{2\pi}\int_{-\pi}^{\pi} f(\varphi)\frac{\partial}{\partial\theta}\left(\frac{1-r^2}{1-2r\cos(\varphi-\theta)+r^2}\right)d\varphi\\&=-\frac{1}{2\pi}\int_{-\pi}^{\pi} f(\varphi)\frac{\partial}{\partial\varphi}\left(\frac{1-r^2}{1-2r\cos(\varphi-\theta)+r^2}\right)d\varphi\\&=-\frac{1}{2\pi}\int_{-\pi}^{\pi} f(\theta+t)\frac{\partial}{\partial t}\left(\frac{1-r^2}{1-2r\cos t+r^2}\right)dt\\&=\frac{1}{2\pi}\int_{-\pi}^{\pi} f(\theta+t)\frac{(1-r^2)\,2r\sin t}{(1-2r\cos t+r^2)^2}dt\\&=\frac{1}{2\pi}\int_{|t|\leqq\delta}+\frac{1}{2\pi}\int_{\delta\leqq|t|\leqq\pi}=I+II\,.\end{aligned} \tag{2}$$

$$II\leqq\text{const.}\,(1-r)\to 0\,,\qquad \text{when}\quad z\to 1\,. \tag{3}$$

In I, we put $f(\theta+t)=(\theta+t)\eta$, then $|\eta|<\varepsilon$, if $|\theta+t|$ is small.

$$\begin{aligned}I=\frac{1}{2\pi}\int_{-\delta}^{\delta}\theta\eta\frac{(1-r^2)\,2r\sin t}{(1-2r\cos t+r^2)}dt+\frac{1}{2\pi}\int_{-\delta}^{\delta}t\eta\,\frac{(1-r^2)\,2r\sin t}{(1-2r\cos t+r^2)}dt\\=I_1+I_2\,.\end{aligned} \tag{4}$$

$$\begin{aligned}|I_1|&\leqq\frac{1}{2\pi}\int_{-\delta}^{\delta}\frac{|\theta\eta|(1-r^2)\,2r|\sin t|}{(1-2r\cos t+r^2)^2}dt\leqq\frac{|\theta|\varepsilon}{\pi}\int_0^{\delta}\frac{(1-r^2)\,2r\sin t}{(1-2r\cos t+r^2)^2}dt\\&\leqq\frac{|\theta|\varepsilon}{\pi}\int_0^{\pi}\frac{(1-r^2)\,2r\sin t}{(1-2r\cos t+r^2)^2}dt=-\frac{|\theta|\varepsilon}{\pi}\int_0^{\pi}\frac{d}{dt}\left(\frac{1-r^2}{1-2r\cos t+r^2}\right)dt\\&\leqq\frac{\text{const.}\,|\theta|\varepsilon}{1-r}\leqq\text{const.}\,K\varepsilon\to 0\,,\quad \text{if } z\to 1 \text{ in } \Delta\,.\end{aligned} \tag{5}$$

$$|I_2| \leqq \frac{1}{2\pi}\int_{-\delta}^{\delta} |t\eta| \frac{(1-r^2)2r|\sin t|}{(1-2r\cos t+r^2)^2} dt \leqq -\frac{\varepsilon}{2\pi}\int_{-\pi}^{\pi} t\frac{d}{dt}\left(\frac{1-r^2}{1-2r\cos t+r^2}\right)dt$$

$$\leqq \frac{2r\varepsilon}{1+r} \text{ (Lemma 2)} \to 0, \quad \text{if } z\to 1 \text{ in } \varDelta. \tag{6}$$

Hence by (2), (3), (5), (6),

$$\frac{\partial u}{\partial \theta} \to 0 \quad \text{uniformly, if } z\to 1 \text{ from the inside of } \varDelta. \tag{7}$$

If $f'(0) \neq 0$, we put $F(\varphi) = f(\varphi) - f'(0)\varphi$, then $F(0)=0$, $F'(0)=0$, so that if $v(z)$ is the harmonic function of Lemma 1, then

$$u(z) - f'(0)v(z) = \frac{1}{2\pi}\int_{-\pi}^{\pi} F(\varphi)\frac{1-r^2}{1-2r\cos(\varphi-\theta)+r^2}d\varphi .$$

Hence $\frac{\partial u}{\partial \theta} - f'(0)\frac{\partial v}{\partial \theta} \to 0$, so that by Lemma 1,

$$\frac{\partial u}{\partial \theta} \to f'(0) \text{ uniformly, if } z\to 1 \text{ from the inside of } \varDelta. \tag{8}$$

Hence the case $f'(0) \neq \infty$ is proved. Next suppose that $f(\theta)$ is monotone increasing in a neighbourhood of $\theta=0$ and $f'(0)=+\infty$ and we shall prove that $\partial u/\partial\theta \to +\infty$, if $z\to 1$ from the inside of $\varDelta$.

$$\frac{\partial u}{\partial \theta} = \frac{1}{2\pi}\int_0^{\pi} (f(\theta+t)-f(\theta-t))\frac{(1-r^2)\,2r\sin t}{(1-2r\cos t+r^2)^2}dt$$

$$= O(1-r) + \frac{1}{2\pi}\int_0^{\delta} (f(\theta+t)-f(\theta-t))\frac{(1-r^2)2r\sin t}{(1-2r\cos t+r^2)^2}dt . \tag{9}$$

$$(0<\delta<1)$$

Since $|\theta| \leqq K(1-r)$, we have if $t \geqq 2K(1-r)$, then $\theta+t \geqq 0$, $\theta-t \leqq 0$ and $\theta+t \geqq t-|\theta| \geqq t/2$, so that

$$f(\theta+t)-f(\theta-t) \geqq f(\theta+t)-f(0) \geqq Nt , \quad (2K(1-r) \leqq t \leqq \delta) ,$$

where $N\to\infty$, with $\delta\to 0$.

Since $f(\theta+t)-f(\theta-t) \geqq 0$ $(0\leqq t\leqq \delta)$ and $1-2r\cos t+r^2 \leqq (1-r)^2+a^2t^2$ ($a=$const.), we have by (9)

$$\frac{\partial u}{\partial \theta} \geqq O(1-r) + \text{const.}\, N\int_{2K(1-r)}^{\delta} \frac{(1-r)t^2}{((1-r)^2+a^2t^2)^2}dt$$

$$= O(1-r) + \text{const.}\, N\int_{2K}^{\frac{\delta}{1-r}} \frac{\tau^2}{(1+a^2\tau^2)^2}d\tau \to \text{const.}\, N\int_{2K}^{\infty} \frac{\tau^2}{(1+a^2\tau^2)^2}d\tau \;(r\to 1). \tag{10}$$

Since N is arbitrary,

$$\frac{\partial u}{\partial \theta} \to +\infty \text{ uniformly, if } z\to 1 \text{ from the inside of } \varDelta. \tag{11}$$

Next suppose that $f'(\theta)$ is continuous on $\alpha \leqq \theta \leqq \beta$. Let J be the arc

$\alpha \leqq \theta \leqq \beta$ on $|z|=1$ and J' be its complement and J_1 be the arc $\alpha_1 \leqq \theta \leqq \beta_1$ $(\alpha<\alpha_1<\beta_1<\beta)$ on $|z|=1$, then if $\theta \in J_1$,

$$\frac{\partial u}{\partial \theta}=\frac{1}{2\pi}\int_{-\pi}^{\pi} f(\varphi)\frac{(1-r^2)2r\sin(\varphi-\theta)}{(1-2r\cos(\varphi-\theta)+r^2)^2}d\varphi=\frac{1}{2\pi}\int_J+\frac{1}{2\pi}\int_{J'}=I+II\,, \tag{12}$$

where

$$\lim_{z\to e^{i\theta}} II=0 \text{ uniformly on } J_1\,. \tag{13}$$

$$\begin{aligned} I&=-\frac{1}{2\pi}\int_J f(\varphi)\frac{\partial}{\partial\varphi}\left(\frac{1-r^2}{1-2r\cos(\varphi-\theta)+r^2}\right)d\varphi \\ &=\frac{1}{2\pi}\left[-f(\varphi)\frac{1-r^2}{1-2r\cos(\varphi-\theta)+r^2}\right]_\alpha^\beta+\frac{1}{2\pi}\int_J f'(\varphi)\frac{1-r^2}{1-2r\cos(\varphi-\theta)+r^2}d\varphi\,. \end{aligned}$$

The first term tends to zero, if $z\to e^{i\theta}\in J_1$ and by Schwarz's theorem, the second term tends to $f'(\theta)$ uniformly on J_1, if $z\to e^{i\theta}$ from the inside of $|z|<1$, so that

$$\frac{\partial u}{\partial\theta}\to f'(\theta) \quad \text{uniformly on } J_1\,, \tag{14}$$

when $z\to e^{i\theta}$ from the inside of $|z|<1$.

Remark. In the case $f'(\theta_0)=+\infty$, we assumed that $f(\theta)$ is monotone increasing in a neighbourhood of θ_0. If we do not assume the monotony of $f(\theta)$, then from the proof, we see easily that $\partial u/\partial\theta\to+\infty$, if $z\to e^{i\theta_0}$ along the radius of $|z|=1$ through $e^{i\theta_0}$.

2. Fatou's theorem

Theorem IV. 6. *Let $u(z)$ be a bounded harmonic function in $|z|<1$, then for almost all $e^{i\theta}$ on $|z|=1$, $\lim_{z\to e^{i\theta}} u(z)=u(e^{i\theta})$ exists uniformly, when $z\to e^{i\theta}$ from the inside of any fixed Stolz domain, whose vertex is at $e^{i\theta}$.*

Fatou [1].

Proof. Let $0<\rho<1$, then for $|z|<\rho<1$,

$$u(z)=\frac{1}{2\pi}\int_{-\pi}^{\pi} u(\rho e^{i\varphi})\frac{\rho^2-r^2}{\rho^2-2\rho r\cos(\varphi-\theta)+r^2}d\varphi\,. \tag{1}$$

$$z=re^{i\theta}\,,\quad 0\leqq r<\rho\,.$$

We may assume that $0\leqq u(z)\leqq M$ in $|z|<1$.

For a Borel set e on $|z|=1$, we define an additive set function $\mu_\rho(e)$ by

$$\mu_\rho(e)=\int_e u(\rho e^{i\theta})d\theta\,, \tag{2}$$

then $0\leqq\mu_\rho(e)\leqq Mme\leqq M\cdot 2\pi$, so that by Theorem II. 4., we can find $\rho_\nu\to1$, such that $\mu_{\rho_\nu}\to\mu$, where μ is an additive set function and since $0\leqq\mu_\rho(e)\leqq Mme$, we have $0\leqq\mu(e)\leqq Mme$, so that μ is an absolutely continuous additive set function. Since (1) can be written in the form:

$$u(z)=\frac{1}{2\pi}\int_{|z|=1}\frac{\rho_\nu^2-r^2}{\rho_\nu^2-2\rho_\nu r\cos(\varphi-\theta)+r^2}d\mu_{\rho_\nu}(\varphi)\,,\quad |z|<\rho_\nu\,,$$

we have

$$u(z)=\frac{1}{2\pi}\int_{|z|=1}\frac{1-r^2}{1-2r\cos(\varphi-\theta)+r^2}d\mu(\varphi)\,,\quad |z|<1\,. \tag{3}$$

Since μ is an absolutely continuous additive set function, $d\mu(\varphi)$ can be written in the form $d\mu(\varphi)=f(\varphi)d\varphi$, where $f(\varphi)$ is a bounded measurable function, such that $0\leqq f(\varphi)\leqq M$, hence

$$u(z)=\frac{1}{2\pi}\int_{-\pi}^{\pi}f(\varphi)\frac{1-r^2}{1-2r\cos(\varphi-\theta)+r^2}d\varphi\,, \tag{4}$$

so that the theorem follows from Theorem IV. 1. ——

From the theorem, we have

Theorem IV. 7. *Let $f(z)$ be a bounded regular function in $|z|<1$, then for almost all $e^{i\theta}$ on $|z|=1$, $\lim_{z\to e^{i\theta}} f(z)=f(e^{i\theta})$ exists uniformly, when $z\to e^{i\theta}$ from the inside of any fixed Stolz domain, whose vertex is at $e^{i\theta}$.*

Fatou [1].

Theorem IV. 8. *Let $f(z)$ be a bounded regular function in $|z|<1$.*

(i) *If $f(z)\not\equiv 0$ and $f(z)=a_pz^p+\cdots$, $a_p\neq 0$, then*

$$\log|a_p|\leqq\frac{1}{2\pi}\int_0^{2\pi}\log|f(e^{i\theta})|d\theta\,.$$

Especially, if $f(0)\neq 0$, then

$$\log|f(0)|\leqq\frac{1}{2\pi}\int_0^{2\pi}\log|f(e^{i\theta})|d\theta\,,$$

hence $|\log|f(e^{i\theta})||$ is integrable on $|z|=1$. Szegö [1].

(ii) *If $|f(e^{i\theta})|\leqq m$ on a set E on $|z|=1$ and $|f(e^{i\theta})|\leqq M$ on its complement, then*

$$|a_p|\leqq m^{\frac{mE}{2\pi}}M^{1-\frac{mE}{2\pi}}\,.$$

Proof. We may assume that $|f(z)|<1$ in $|z|<1$. First we assume that $f(0)\neq 0$. Since $\log 1/|f(z)|$ is superharmonic in $|z|<1$,

$$\log\frac{1}{|f(0)|}\geqq\frac{1}{2\pi}\int_0^{2\pi}\log\frac{1}{|f(re^{i\theta})|}d\theta\quad(0<r<1)\,.$$

Since $\log 1/|f(re^{i\theta})|>0$ and $\to\log 1/|f(e^{i\theta})|$ as $r\to 1$, we have by Fatou's lemma,

$$\log\frac{1}{|f(0)|}\geqq\frac{1}{2\pi}\int_0^{2\pi}\log\frac{1}{|f(e^{i\theta})|}d\theta\,,$$

$$\log|f(0)| \leqq \frac{1}{2\pi}\int_0^{2\pi} \log|f(e^{i\theta})|d\theta . \tag{1}$$

If $f(0)=0$ and $f(z)=a_p z^p+\cdots$, $a_p \neq 0$, $p \geqq 1$, then we consider $\frac{f(z)}{z^p}$ instead of $f(z)$. ——

From the theorem, we have

Theorem IV. 9. *Let $f(z)$ be a bounded regular function in $|z|<1$. If $f(e^{i\theta})=0$ on a set E of positive measure on $|z|=1$, then $f(z)\equiv 0$.*

F. and M. Riesz [1].

3. Remark on Fatou's theorem

In Fatou's theorem, the non-tangential approach to $|z|=1$ is essential, since the following theorem holds.

Theorem IV. 10. *Let C be any Jordan arc in $|z|<1$, which touches $|z|=1$ at $z=1$ internally and C_θ be the curve, obtained from C by rotating C about $z=0$ through an angle θ, which touches $|z|=1$ at $e^{i\theta}$. Then there exists a bounded harmonic function $u(z)$ and a bounded regular function $f(z)$ in $|z|<1$, such that for almost all $e^{i\theta}$, if z tends to $e^{i\theta}$ along C_θ, then $\lim u(z)$ and $\lim f(z)$ does not exist.*

Littlewood [2], Zygmund [1].

First we shall prove a lemma.

Lemma. *Let $0<\eta<1$,*

$$u(r)=\frac{1}{2\pi}\int_{-\eta}^{\eta} \frac{1-r^2}{1-2r\cos\theta+r^2}d\theta , \quad 0\leqq r<1 .$$

Then

$$u(r)>\frac{1}{\pi}, \quad \textit{if} \quad 1-r\leqq\eta .$$

Proof. Let $\delta=1-r$, then if $\delta\leqq\eta$,

$$u(r)\geqq\frac{1}{2\pi}\int_{-\delta}^{\delta}\frac{1-r^2}{(1-r)^2+4r\sin^2\theta/2}d\theta>\frac{2\delta}{2\pi}\cdot\frac{\delta(1+r)}{\delta^2+r\delta^2}=\frac{1}{\pi} .$$

Proof of the theorem.

First suppose that the radius of $|z|=1$: $\arg z=\theta$ $(0\leqq\theta\leqq\theta_0)$ meets C at one point and the equation of C in polar coordinates be

$$C: \ 1-r=\chi(\theta) \quad (0\leqq\theta\leqq\theta_0) , \tag{1}$$

where $\chi(\theta)>0$ is a non-decreasing function of θ.

Since C touches $|z|=1$ at $z=1$,

$$\eta(\theta)=\chi(\theta)/\theta\to 0 \qquad \text{as} \quad \theta\to+0 . \tag{2}$$

We subdivide $(0, 2\pi)$ into n equal parts $I_n^{(1)}, \cdots, I_n^{(n)}$ of length $2\pi/n$ and let $J_n^{(i)}$ $(i=1, 2, \cdots, n)$ be an open interval of length $2\chi\left(\frac{2\pi}{n}\right)$, concentric with $I_n^{(i)}$. Then

$$B_n = J_n^{(1)} + \cdots + J_n^{(n)} \tag{3}$$

is an open set and its measure is $2n\chi\left(\frac{2\pi}{n}\right)=4\pi\eta\left(\frac{2\pi}{n}\right)\to 0$ as $n\to\infty$, hence for any $\varepsilon>0$, we can select $n_1<n_2<\cdots<n_k<\cdots\to\infty$, such that

$$B = B_{n_1} + B_{n_2} + \cdots + B_{n_k} + \cdots \tag{4}$$

is of measure $<\varepsilon$. Let A be the complement of B on $|z|=1$, then A is a perfect non-dense set on $|z|=1$, whose measure is $\geqq 2\pi-\varepsilon$.

Let $\theta_0 \in A$ be any point of A and $(\theta_0+\alpha_k, \theta_0+\beta_k)$ be the interval $J_{n_k}^{(i_k)}$ among $\{J_{n_k}^{(i)}\}_{i=1,2\cdots}$, which lies nearest to θ_0 on the right. Its length is $2\chi\left(\frac{2\pi}{n_k}\right)$ and the distance d_k of its mid-point ξ_k to θ_0 is $\leqq 2\pi/n_k$, which is the distance of mid-points of two consecutive intervals of $\{J_{n_k}^{(i)}\}$.

Since $\chi(\theta)$ is non-decreasing,

$$\chi(d_k) \leqq \chi\left(\frac{2\pi}{n_k}\right). \tag{5}$$

Let $h(\theta)$ be the characteristic function of B and $u(z)$ be the Poisson integral, formed with $h(\theta)$, then $u(z)$ exceeds the Poisson integral $u_k(z)$, formed with the characteristic function of $J_{n_k}^{(i_k)}$. Let the radius of $|z|=1$ through ξ_k meet C_{θ_0} at a point ζ_k, then $1-|\zeta_k|=\chi(d_k)\leqq\chi\left(\frac{2\pi}{n_k}\right)$, so that by the lemma, $u_k(\zeta_k)>1/\pi$, hence

$$u(\zeta_k) > u_k(\zeta_k) > \frac{1}{\pi}. \tag{6}$$

Since A is a perfect non-dense set, there are infinitely many such $J_{n_k}^{(i_k)}$, converging to θ_0, so that if z tends to $e^{i\theta_0}$ along C_{θ_0}, then

$$\overline{\lim}\, u(z) > \frac{1}{\pi}. \tag{7}$$

Next we shall prove that for almost all $\theta \in A$,

$$\underline{\lim}\, u(z) = 0, \quad \text{when } z\to e^{i\theta} \text{ along } C_\theta. \tag{8}$$

Since $h(\theta)=0$, if $\theta\in A$, by Fatou's theorem, for almost all $\theta\in A$, $\lim\limits_{r\to 0} u(re^{i\theta})=0$. Hence by Egoroff's theorem, for any $\delta>0$, there exists a perfect subset A' of A of measure $>mA-\delta$, such that $\lim\limits_{r\to 1} u(re^{i\theta})=0$ uniformly on A'. Let $\theta_0\in A'$, then, except at most a countable set of θ_0, there exist $\{\theta_\nu\}\in A'$, such that $\theta_\nu\to\theta_0$ on the right, so that the radius of $|z|=1$ through $e^{i\theta_\nu}$ intersects C_{θ_0}, hence $\underline{\lim}\, u(z)=0$, when $z\to e^{i\theta_0}$ along C_{θ_0}. Since $\delta>0$ is arbitrary, this holds for almost all $\theta_0\in A$. Hence by (7), for almost all $\theta_0\in A$, if $z\to e^{i\theta_0}$ along C_{θ_0}, then

$$\overline{\lim}\, u(z) - \underline{\lim}\, u(z) > \frac{1}{\pi}. \tag{9}$$

We take $\varepsilon_1>\varepsilon_2>\cdots>\varepsilon_\nu\to 0$ for ε in (4) and let A_ν be the corresponding A of measure $\geqq 2\pi-\varepsilon_\nu$ and $h_\nu(\theta)$, $u_\nu(z)$ be the corresponding $h(\theta)$, $u(z)$ and put for a positive integer $q>1$,

$$H(\theta)=\sum_{\nu=1}^{\infty}\frac{h_\nu(\theta)}{q^\nu}\,, \qquad U(z)=\sum_{\nu=1}^{\infty}\frac{u_\nu(z)}{q^\nu}\,, \tag{10}$$

then since $0\leqq h_\nu(\theta)\leqq 1$, $0\leqq u_\nu(z)\leqq 1$, the series converge uniformly.

Let E_ν be the set of points $\theta\in A_\nu$, such that the oscillation of $u_\nu(z)$ along C_θ is $\leqq\frac{1}{\pi}$, then by (9), E_ν is a null set, so that $E=\sum_{\nu=1}^{\infty}E_\nu$ is a null set, hence $A^*=\sum_{\nu=1}^{\infty}A_\nu-E$ is of measure 2π. Let $\theta_0\in A^*$ and n be the least ν, such that $\theta_0\in A_\nu$, then θ_0 belongs to the complement of $A_1+\cdots+A_{n-1}$. This complement is an open set and each of $h_1(\theta),\cdots,$ $h_{n-1}(\theta)$ vanishes on it, hence $u_1(z),\cdots,u_{n-1}(z)$ tends to zero, if $z\to e^{i\theta_0}$ along C_{θ_0}.

By (9), the oscillation of $u_n(z)/q^n$ along C_{θ_0} exceeds $1/\pi q^n$ and the oscillation of the remainder $\sum_{\nu=n+1}^{\infty}u_\nu(z)/q^\nu$ does not exceed $1/q^{n+1}+1/q^{n+2}+\cdots$ $=1/(q-1)q^n$, since each of $u_\nu(z)$ is contained between 0 and 1. Hence the oscillation of $U(z)$ along C_{θ_0} exceeds $1/\pi q^n-1/(q-1)q^n>0$, if $q\geqq 5$, so that $U(z)$ does not tend to any limiting value, if $z\to e^{i\theta_0}$ along C_{θ_0}. This proves the first part of the theorem concerning harmonic functions, if C is of the form (1). In the general case, the radius of $|z|=1$: $\arg z=\theta$ may intersect C at more than one point $z_1(\theta)$, $z_2(\theta)$, $\cdots$. Let

$$\rho(\theta)=\operatorname*{Min}_{\nu}|z_\nu(\theta)|\,, \qquad \chi^*(\theta)=\sup_{0<\varphi\leqq\theta}(1-\rho(\varphi))\,, \tag{11}$$

then $\chi^*(\theta)$ is non-decreasing and we can prove easily that $\chi^*(\theta)/\theta\to 0$ as $\theta\to 0$. In the definition of B, we replace $\chi(\theta)$ by $\chi^*(\theta)$ and the rest of the proof is unchanged. Hence the first part of the theorem is proved. The second part of the theorem concerning bounded regular functions follows from the first part by putting $f(z)=e^{u+iv}$, where v is the conjugate harmonic function of u.

4. Poisson integral of two variables

Lemma. *Let $f(x, y)$ be a bounded measurable function of x, y, defined in a square Δ: $0\leqq x\leqq 1$, $0\leqq y\leqq 1$ and Q: $|x-x_0|\leqq\delta$, $|y-y_0|\leqq\delta'$ be a rectangle, contained in Δ. Then for almost all $(x_0, y_0)\in\Delta$,*

$$\lim_{\delta\to 0,\,\delta'\to 0}\frac{1}{|Q|}\iint_Q|f(x, y)-f(x_0, y_0)|\,dx\,dy=0\,,$$

where $|Q|=4\delta\delta'$ denotes the area of Q.

Proof. By the strong density theorem of Saks [1], we can prove easily that for almost all $(x_0, y_0) \in \Delta$,

$$\lim_{\delta\to 0,\delta'\to 0} \frac{1}{|Q|}\iint_Q f(x, y)dx\,dy = f(x_0, y_0)\,. \quad \text{Saks [1].} \quad (1)$$

Let $\{r\}$ be the set of all rational numbers r and N_r be the null set in Δ, such that

$$\lim_{\delta\to 0,\delta'\to 0} \frac{1}{|Q|}\iint_Q |f(x, y)-r|dx\,dy = |f(x_0, y_0)-r| \qquad (2)$$

does not hold, so that $N=\sum_r N_r$ is a null set in Δ.

Let α be any real number and r a rational number, then since

$$|f(x, y)-r|-|r-\alpha| \leqq |f(x, y)-\alpha| \leqq |f(x, y)-r|+|r-\alpha|\,,$$

we have if $(x_0, y_0) \in \Delta - N$,

$$|f(x_0, y_0)-r|-|r-\alpha| \leqq \overline{\lim_{\delta\to 0\ \delta'\to 0}} \frac{1}{|Q|}\iint_Q |f(x, y)-\alpha|dx\,dy \leqq |f(x_0, y_0)-r|+|r-\alpha|,$$

so that if we make $r\to\alpha$, then we have

$$\lim_{\delta\to 0,\delta'\to 0} \frac{1}{|Q|}\iint_Q |f(x, y)-\alpha|dx\,dy = |f(x_0, y_0)-\alpha|\,, \quad (x_0, y_0) \in \Delta - N\,. \quad (3)$$

Hence if we put $\alpha = f(x_0, y_0)$, then we have the lemma. ——

Theorem IV. 11. *Let $f(\theta, \varphi)$ be a bounded measurable function on a torus Θ: $0\leqq\theta\leqq 2\pi$, $0\leqq\varphi\leqq 2\pi$ and*

$$u(z, w) = \frac{1}{4\pi^2}\iint_\Theta f(\theta', \varphi')\frac{(1-r^2)(1-\rho^2)}{(1-2r\cos(\theta'-\theta)+r^2)(1-2\rho\cos(\varphi'-\varphi)+\rho^2)}d\theta' d\varphi'\,,$$

$$z = re^{i\theta}\,, \quad w = \rho e^{i\varphi}\,, \quad 0\leqq r<1\,, \quad 0\leqq\rho<1\,.$$

Then for almost all $(\theta, \varphi) \in \Theta$, $u(z, w) \to f(\theta, \varphi)$ uniformly, when $z\to e^{i\theta}$, $w\to e^{i\varphi}$ from the inside of any fixed Stolz domain, whose vertex is at $e^{i\theta}$, $e^{i\varphi}$ respectively. Tsuji [14].

Proof. By the lemma, for almost all $(\theta_0, \varphi_0) \in \Theta$,

$$\lim_{\delta\to 0,\delta'\to 0} \frac{1}{\delta\delta'}\int_{-\delta}^{\delta}\int_{-\delta'}^{\delta'} |f(\theta_0+\theta, \varphi_0+\varphi)-f(\theta_0, \varphi_0)|d\theta\,d\varphi = 0\,. \qquad (1)$$

Let (1) hold at $\theta_0=0$, $\varphi_0=0$ and we shall prove that $u(z, w) \to f(0, 0)$ uniformly, when $z\to 1$, $w\to 1$ from the inside of any Stolz domain Δ, whose vertex is at $z=1$, $w=1$ respectively. We put

$$F(\theta, \varphi) = |f(\theta, \varphi)-f(0, 0)|\,, \qquad \int_{-\delta}^{\delta}\int_{-\delta'}^{\delta'} F(\theta, \varphi)d\theta\,d\varphi = \delta\delta'\varepsilon(\delta, \delta')\,, \quad (2)$$

then $\varepsilon(\delta, \delta')\to 0$, if $\delta\to 0$, $\delta'\to 0$.

If $z=re^{i\theta}$, $w=\rho e^{i\varphi}$ lie in Δ, then $|\theta|\leqq\text{const.}(1-r)$, $|\varphi|\leqq\text{const.}(1-\rho)$, so that if

$$1-r=\delta\sqrt[3]{\varepsilon(\delta, \delta')}\,,\qquad 1-\rho=\delta'\sqrt[3]{\varepsilon(\delta, \delta')}\,, \tag{3}$$

then

$$|\theta|\leqq\text{const.}\,\delta\sqrt[3]{\varepsilon(\delta, \delta')}\,,\qquad |\varphi|\leqq\text{const.}\,\delta'\sqrt[3]{\varepsilon(\delta, \delta')}\,. \tag{4}$$

Since

$$1=\frac{1}{4\pi^2}\iint_{\Theta}\frac{(1-r^2)(1-\rho^2)}{(1-2r\cos(\theta'-\theta)+r^2)(1-2\rho\cos(\varphi'-\varphi)+\rho^2)}d\theta' d\varphi'\,,$$

we have

$$\begin{aligned}&|u(z, w)-f(0, 0)|\leqq\\ &\frac{1}{4\pi^2}\iint_{\Theta}F(\theta', \varphi')\frac{(1-r^2)(1-\rho^2)}{(1-2r\cos(\theta'-\theta)+r^2)(1-2\rho\cos(\varphi'-\varphi)+\rho^2)}d\theta' d\varphi'\\ &\qquad =I+II+III+IV\,,\end{aligned} \tag{5}$$

where

$$\left.\begin{aligned}I&=\frac{1}{4\pi^2}\iint_{|\theta'|\leqq\delta,\ |\varphi'|\leqq\delta'}\,, & II&=\frac{1}{4\pi^2}\iint_{|\theta'|\leqq\delta,\ \delta'\leqq|\varphi'|\leqq\pi}\,,\\ III&=\frac{1}{4\pi^2}\iint_{\delta\leqq|\theta|\leqq\pi,\ |\varphi'|\leqq\delta'}\,, & IV&=\frac{1}{4\pi^2}\iint_{\delta\leqq|\theta'|\leqq\pi,\ \delta'\leqq|\varphi'|\leqq\pi}\,.\end{aligned}\right\} \tag{6}$$

Since $1-2r\cos\theta+r^2\geqq(1-r)^2+a^2\theta^2\geqq(1-r)^2$ ($a=\text{const.}$) and $1-r^2\leqq 2(1-r)$,

$$\begin{aligned}I&\leqq\frac{(1-r)(1-\rho)}{\pi^2}\int_{-\delta}^{\delta}\int_{-\delta'}^{\delta'}\frac{F(\theta', \varphi')d\theta' d\varphi'}{(1-r)^2(1-\rho)^2}\\ &=\frac{1}{\pi^2(1-r)(1-\rho)}\int_{-\delta}^{\delta}\int_{-\delta'}^{\delta'}F(\theta', \varphi')d\theta' d\varphi'=\frac{\delta\delta'\varepsilon(\delta, \delta')}{\pi^2(1-r)(1-\rho)}\\ &=\frac{\sqrt[3]{\varepsilon(\delta, \delta')}}{\pi^2}\to 0\,,\quad\text{if}\quad \delta\to 0\,,\quad \delta'\to 0\,.\end{aligned} \tag{7}$$

By (4), we choose δ, δ' so small that $|\theta|\leqq\delta/2$, $|\varphi|\leqq\delta'/2$, then in II, $|\varphi'-\varphi|\geqq|\varphi'|-|\varphi|\geqq|\varphi'|/2$, hence if $0\leqq F(\theta, \varphi)\leqq M$ on Θ,

$$\begin{aligned}II&\leqq\frac{M(1-r)(1-\rho)}{\pi^2}\int_{|\theta'|\leqq\delta}\frac{d\theta'}{(1-r)^2+a^2(\theta'-\theta)^2}\int_{\delta'\leqq|\varphi'|\leqq\pi}\frac{d\varphi'}{a^2(\varphi'-\varphi)^2}\\ &\leqq\frac{2M(1-r)(1-\rho)}{\pi^2}\int_{-\infty}^{\infty}\frac{dt}{(1-r)^2+a^2t^2}\int_{\delta'}^{\infty}\frac{d\varphi'}{a^2(\varphi'/2)^2}\\ &=\frac{8M(1-\rho)}{\pi a^3\delta'}=\frac{8M\sqrt[3]{\varepsilon(\delta, \delta')}}{\pi a^3}\to 0\,,\quad\text{if}\ \delta\to 0\,,\ \delta\to 0\,.\end{aligned} \tag{8}$$

Similarly

$$III \to 0\,, \quad \text{if} \quad \delta \to 0\,, \quad \delta' \to 0\,. \tag{9}$$

$$\begin{aligned} IV &\leqq \frac{M(1-r)(1-\rho)}{\pi^2} \int\limits_{\delta \leqq |\theta'| \leqq \pi} \frac{d\theta'}{a^2(\theta'-\theta)^2} \int\limits_{\delta' \leqq |\varphi'| \leqq \pi} \frac{d\varphi'}{a^2(\varphi'-\varphi)^2} \\ &\leqq \frac{M(1-r)(1-\rho)}{\pi^2} \int\limits_{\delta \leqq |\theta'| \leqq \pi} \frac{d\theta'}{a^2(\theta'/2)^2} \int\limits_{\delta' \leqq |\varphi'| \leqq \pi} \frac{d\varphi'}{a^2(\varphi'/2)^2} \\ &\leqq \frac{64M(1-r)(1-\rho)}{\pi^2 a^4} \int_\delta^\infty \frac{d\theta'}{\theta'^2} \int_{\delta'}^\infty \frac{d\varphi'}{\varphi'^2} = \frac{64M(1-r)(1-\rho)}{\pi^2 a^4 \delta\delta'} \\ &= \frac{64M[\varepsilon(\delta, \delta')]^{\frac{2}{3}}}{\pi^2 a^4} \to 0\,, \quad \text{if} \quad \delta \to 0\,, \quad \delta' \to 0\,. \end{aligned} \tag{10}$$

Hence $u(z, w) \to f(0, 0)$ uniformly, when $z \to 1$, $w \to 1$ from the inside of Δ.

Theorem IV. 12. *Let $u(z, w)$ be a harmonic function of each variable z and w in $|z|<1$, $|w|<1$ and bounded, such that $|u(z, w)| \leqq M$ for $|z|<1$, $|w|<1$. Then for almost all $(\theta, \varphi) \in \Theta$,*

$$\lim_{z \to e^{i\theta}, w \to e^{i\varphi}} u(z, w) = u(e^{i\theta}, e^{i\varphi}) \text{ exists uniformly,}$$

when $z \to e^{i\theta}$, $w \to e^{i\varphi}$ from the inside of any fixed Stolz domain, whose vertex is at $e^{i\theta}$, $e^{i\varphi}$ respectively and $u(z, w)$ can be expressed by

$$u(z, w) = \frac{1}{4\pi^2} \iint\limits_{\Theta} u(e^{i\theta'}, e^{i\varphi'}) \frac{(1-r^2)(1-\rho^2)}{(1-2r\cos(\theta'-\theta)+r^2)(1-2\rho\cos(\varphi'-\varphi)+\rho^2)}\, d\theta' d\varphi'\,,$$

$$z = re^{i\theta}\,, \quad w = \rho e^{i\varphi}\,.$$

Proof. For a fixed $w = \rho e^{i\varphi}$ $(0<\rho<1)$, $u(z, \rho e^{i\varphi})$ is a harmonic function of z in $|z|<1$, so that for $|z|<R<1$,

$$u(z, w) = \frac{1}{2\pi} \int_{-\pi}^{\pi} u(Re^{i\theta'}, \rho e^{i\varphi}) \frac{R^2-r^2}{R^2-2Rr\cos(\theta'-\theta)+r^2} d\theta'\,, \quad z = re^{i\theta}\,, \quad w = \rho e^{i\varphi}\,.$$

Since $u(Re^{i\theta'}, w)$ is a harmonic function of w in $|w|<1$, we have for $\rho<R<1$,

$$u(Re^{i\theta'}, \rho e^{i\varphi}) = \frac{1}{2\pi} \int_{-\pi}^{\pi} u(Re^{i\theta'}, Re^{i\varphi'}) \frac{R^2-\rho^2}{R^2-2R\rho\cos(\varphi'-\varphi)+\rho^2} d\varphi'\,,$$

so that

$$u(z, w) = \frac{1}{4\pi^2} \iint\limits_{\Theta} u(Re^{i\theta'}, Re^{i\varphi'}) \frac{(R^2-r^2)(R^2-\rho^2)}{(R^2-2Rr\cos(\theta'-\theta)+r^2)(R^2-2R\rho\cos(\varphi'-\varphi)+\rho^2)} d\theta' d\varphi'\,,$$

$$|z|<R\,, \quad |w|<R\,. \tag{1}$$

We may assume that $0 \leqq u(z, w) \leqq M$ for $|z|<1$, $|w|<1$. We define an additive set function $\mu_R(e)$ $(0<R<1)$, defined for a Borel set e on Θ by

$$\mu_R(e)=\iint_e u(Re^{i\theta}, Re^{i\varphi})d\theta\, d\varphi\ ,$$

then $0\leqq\mu_R(e)\leqq Mme\leqq 4\pi^2 M$, so that by Theorem II. 4., we can select R_n, such that $R_1<R_2<\cdots<R_n\to 1$, $\mu_{R_n}\to\mu$, where μ is an absolutely continuous additive set function, such that $0\leqq\mu(e)\leqq Mme$. If we take $R=R_n$ in (1), then since $u(R_n e^{i\theta'}, R_n{}^{i\varphi'})d\theta' d\varphi'=d\mu_{Rn}(\theta', \varphi')$, if we make $n\to\infty$, then we have

$$u(z, w)=\frac{1}{4\pi^2}\iint_\Theta \frac{(1-r^2)(1-\rho^2)}{(1-2r\cos(\theta'-\theta)+r^2)(1-2\rho\cos(\varphi'-\varphi)+\rho^2)}d\mu(\theta', \varphi')\ . \quad (2)$$

$$|z|<1\ ,\quad |w|<1\ .$$

Since μ is absolutely continuous, $d\mu(\theta, \varphi)$ can be written in the form $d\mu(\theta, \varphi)=f(\theta, \varphi)d\theta d\varphi$, where $f(\theta, \varphi)$ is a bounded measurable function on Θ, hence

$$u(z, w)=\frac{1}{4\pi^2}\iint_\Theta f(\theta', \varphi')\frac{(1-r^2)(1-\rho^2)}{(1-2r\cos(\theta'-\theta)+r^2)(1-2\rho\cos(\varphi'-\varphi)+\rho^2)}d\theta' d\varphi'\ , \quad (3)$$

so that by Theorem IV. 11., for almost all $(\theta, \varphi)\in\Theta$, $u(z, w)\to f(\theta, \varphi)$ uniformly, when $z\to e^{i\theta}$, $w\to e^{i\varphi}$ from the inside of any Stolz domain, whose vertex is at $e^{i\theta}$, $e^{i\varphi}$ respectively.——

We can prove easily the following theorem.

Theorem IV. 13. *Let $f(z, w)$ be a bounded regular function of two complex variables z, w in $|z|<1$, $|w|<1$. Then for almost all $(\theta, \varphi)\in\Theta$,*

$$\lim_{z\to e^{i\theta},\, w\to e^{i\varphi}} f(z, w)=f(e^{i\theta}, e^{i\varphi}) \text{ uniformly,}$$

when $z\to e^{i\theta}$, $w\to e^{i\varphi}$ from the inside of any fixed Stolz domain, whose vertex is at $e^{i\theta}$, $e^{i\varphi}$ respectively.

(i) *If $f(z, w)\not\equiv 0$, then $|\log|f(e^{i\theta}, e^{i\varphi})||$ is integrable on Θ, such that if $f(0, 0)\neq 0$,*

$$\log|f(0, 0)|\leqq\frac{1}{4\pi^2}\iint_\Theta \log|f(e^{i\theta}, e^{i\varphi})d\theta\, d\varphi\ .$$

Hence if $f(e^{i\theta}, e^{i\varphi})=0$ on a set E of positive measure on Θ, then $f(z, w)\equiv 0$.

(ii) *If $|f(e^{i\theta}, e^{i\varphi})|\leqq m$ on a set E on Θ and $|f(e^{i\theta}, e^{i\varphi})|\leqq M$ on its complement, then*

$$|f(0, 0)|\leqq m^{\frac{mE}{4\pi^2}} M^{1-\frac{mE}{4\pi^2}}\ .$$

5. Positive harmonic function in a unit circle

Theorem IV. 14. *Let $u(z)$ be a positive harmonic function in $|z|<1$, then $u(z)$ can be expressed in the form:*

$$u(z)=u(re^{i\theta})=\frac{1}{2\pi}\int_{-\pi+0}^{\pi+0}\frac{1-r^2}{1-2r\cos(\varphi-\theta)+r^2}d\chi(\varphi)\,, \qquad z=re^{i\theta}\,,$$

where $\chi(\theta)$ is a non-decreasing function of θ in $(-\infty, \infty)$ and if we normalize it, such that $\chi(0)=0$, $\chi(\theta-0)=\chi(\theta)$, then it is unique.
If $0\leqq\chi'(\theta_0)\leqq\infty$, then

$$\lim_{z\to e^{i\theta_0}} u(z)=\chi'(\theta_0) \quad \textit{uniformly}\,,$$

when $z\to e^{i\theta_0}$ from the inside of any fixed Stolz domain, whose vertex is at $e^{i\theta_0}$.

Since $\chi'(\theta)<\infty$ for almost all θ,

$\lim_{z\to e^{i\theta}} u(z)=u(e^{i\theta})<\infty$ *exists almost everywhere on* $|z|=1$.

Proof. Let $|z|<\rho<1$, then

$$u(z)=\frac{1}{2\pi}\int_{-\pi}^{\pi}u(\rho e^{i\varphi})\frac{\rho^2-r^2}{\rho^2-2\rho r\cos(\varphi-\theta)+r^2}d\varphi\,, \tag{1}$$

$$z=re^{i\theta}\,, \quad 0\leqq r<\rho\,.$$

We define an additive set function $\mu_\rho(e)$, defined for a Borel set e on $|z|=1$ by

$$\mu_\rho(e)=\int_e u(\rho e^{i\theta})d\theta\geqq 0\,,$$

then

$$\mu_\rho(e)\leqq\int_0^{2\pi}u(\rho e^{i\theta})d\theta=2\pi u(0)\,.$$

Hence by Theorem II. 4., we can find $\rho_\nu\to 1$, such that $\mu_{\rho_\nu}\to\mu$, where μ is an additive set function, hence by (1), we have as the proof of theorem IV. 6,

$$u(z)=\frac{1}{2\pi}\int_{|z|=1}\frac{1-r^2}{1-2r\cos(\varphi-\theta)+r^2}d\mu(\varphi)\,, \qquad |z|<1\,. \tag{2}$$

If we put

$$\chi(\theta)=\int_{-\pi+0}^{\theta+0}d\mu(\varphi)\,, \tag{3}$$

then $\chi(\theta)$ is a non-decreasing function of θ in $(-\infty, \infty)$ and

$$u(z)=\frac{1}{2\pi}\int_{-\pi+0}^{\pi+0}\frac{1-r^2}{1-2r\cos(\varphi-\theta)+r^2}d\chi(\varphi)\,. \tag{4}$$

Hence

$$u(z)=\frac{1}{2\pi}\left[\frac{(1-r^2)\chi(\varphi)}{1-2r\cos(\varphi-\theta)+r^2}\right]_{-\pi+0}^{\pi+0} -\frac{1}{2\pi}\int_{-\pi}^{\pi}\chi(\varphi)\frac{\partial}{\partial\varphi}\left(\frac{1-r^2}{1-2r\cos(\varphi-\theta)+r^2}\right)d\varphi\,.$$

$$=\frac{1}{2\pi}\left[\frac{(1-r^2)\chi(\varphi)}{1-2r\cos(\varphi-\theta)+r^2}\right]_{-\pi+0}^{\pi+0}+\frac{1}{2\pi}\int_{-\pi}^{\pi}\chi(\varphi)\frac{\partial}{\partial\theta}\left(\frac{1-r^2}{1-2r\cos(\varphi-\theta)+r^2}\right)d\varphi. \tag{5}$$

Since $\chi(\theta)$ is a non-decreasing function of θ, $\chi'(\theta)<\infty$ exists almost everywhere. If $0\leqq\chi'(\theta_0)\leqq\infty$, then by Theorem IV. 5,

$$\lim_{z\to e^{i\theta_0}} u(z)=\chi'(\theta_0) \quad \text{uniformly}\,, \tag{6}$$

if $z\to e^{i\theta_0}$ from the inside of any fixed Stolz domain, whose vertex is at $e^{i\theta_0}$.

Next we shall prove the uniqueness of χ. Since $\chi(\theta)$ is non-decreasing, $\chi(\theta)$ is continuous, except at most a countable number of values of θ. Let $\chi(\theta)$ be continuous at θ_1, θ_2 $(\theta_1<\theta_2)$, then $d\chi(\theta_1)=0$, $d\chi(\theta_2)=0$ and

$$\int_{\theta_1}^{\theta_2} u(re^{i\theta})d\theta=\int_{-\pi+0}^{\pi+0} d\chi(\varphi)\frac{1}{2\pi}\int_{\theta_1}^{\theta_2}\frac{1-r^2}{1-2r\cos(\varphi-\theta)+r^2}d\theta\,. \tag{7}$$

Since

$$\lim_{r\to 1}\frac{1}{2\pi}\int_{\theta_1}^{\theta_2}\frac{1-r^2}{1-2r\cos(\varphi-\theta)+r^2}d\theta=\begin{cases}1, & \text{if } \theta_1<\varphi<\theta_2\\ 0, & \text{if } \varphi \text{ lies outside of } [\theta_1,\theta_2]\end{cases}$$

and $d\chi(\theta_1)=0$, $d\chi(\theta_2)=0$, we have

$$\lim_{r\to 1}\int_{\theta_1}^{\theta_2} u(re^{i\theta})d\theta=\int_{\theta_1}^{\theta_2} d\chi(\varphi)=\chi(\theta_2)-\chi(\theta_1)\,. \tag{8}$$

Hence $\chi(\theta_2)-\chi(\theta_1)$ is unique, if $\chi(\theta)$ if continuous at θ_1, θ_2. From this, we can prove easily that if we normalize $\chi(\theta)$, such that $\chi(0)=0$, $\chi(\theta-0)=\chi(\theta)$, then $\chi(\theta)$ is unique. ——

Now $\chi(\theta)$ can be decomposed into the form: $\chi(\theta)=\psi(\theta)+\nu(\theta)$, where $\psi(\theta)$ is an absolutely continuous increasing function and $\nu(\theta)$ is a singular increasing function of θ. If $\nu\not\equiv 0$, then there exists θ_0, such that $\nu'(\theta_0)=+\infty$. Hence

Corollary 1. *If $\chi(\theta)$ is not absolutely continuous, then there exists θ_0, such that $\lim_{z\to e^{i\theta_0}} u(z)=+\infty$ uniformly, if $z\to e^{i\theta_0}$ from the inside of any fixed Stolz domain, whose vertex is at $e^{i\theta_0}$.*

Corollary 2. $$\lim_{r\to 1}\int_0^{2\pi}|u(re^{i\theta})-u(e^{i\theta})|d\theta=\int_{-\pi+0}^{\pi+0} d\nu(\theta)\,,$$ *hence*

$$\lim_{r\to 1}\int_0^{2\pi}|u(re^{i\theta})-u(e^{i\theta})|d\theta=0\,,$$

when and only when $\chi(\theta)$ is absolutely continuous, so that $u(z)$ can be expressed by a Poisson integral.

Proof. Since $\psi(\theta)$ is absolutely continuous, $d\psi(\varphi)=f(\varphi)d\varphi$, where $f\in L$, so that

$$u(z)=v(z)+w(z)\,,$$

where

$$v(z)=\frac{1}{2\pi}\int_{-\pi}^{\pi}f(\varphi)\frac{1-r^2}{1-2r\cos(\varphi-\theta)+r^2}d\varphi\,,$$

$$w(z)=\frac{1}{2\pi}\int_{-\pi+0}^{\pi+0}\frac{1-r^2}{1-2r\cos(\varphi-\theta)+r^2}d\nu(\varphi)\,.$$

$v(e^{i\theta})=f(\theta)$ almost everywhere and since $\nu'(\theta)=0$ almost everywhere, $w(e^{i\theta})=0$ almost everywhere, so that $u(e^{i\theta})=f(\theta)$ almost everywhere. Hence for almost all θ,

$$w(re^{i\theta})-|v(re^{i\theta})-f(\theta)|\leqq|u(re^{i\theta})-u(e^{i\theta})|\leqq w(re^{i\theta})+|v(re^{i\theta})-f(\theta)|\,,$$

so that

$$2\pi w(0)-\int_0^{2\pi}|v(re^{i\theta})-f(\theta)|d\theta\leqq\int_0^{2\pi}|u(re^{i\theta})-u(e^{i\theta})|d\theta$$

$$\leqq 2\pi w(0)+\int_0^{2\pi}|v(re^{i\theta})-f(\theta)|d\theta\,.$$

Since by Theorem IV. 4, $\lim_{r\to1}\int_0^{2\pi}|v(re^{i\theta})-f(\theta)|d\theta=0$, we have

$$\lim_{r\to1}\int_0^{2\pi}|u(re^{i\theta})-u(e^{i\theta})|d\theta=2\pi w(0)=\int_{-\pi+0}^{\pi+0}d\nu(\theta)\,.$$

Theorem IV. 15. *Let $f(z)=u(z)+iv(z)$ be regular in $|z|<1$ and $u>0$ in $|z|<1$. Then for almost all $e^{i\theta}$,*

$$\lim_{z\to e^{i\theta}}f(z)=f(e^{i\theta})\ (\neq0,\ \neq\infty)$$

exists uniformly, when $z\to e^{i\theta}$ from the inside of any fixed Stolz domain, whose vertex is at $e^{i\theta}$ and $f(z)$ can be expressed in the form:

$$f(z)=\frac{1}{2\pi}\int_{-\pi+0}^{\pi+0}\frac{e^{i\theta}+z}{e^{i\theta}-z}d\chi(\theta)+iv(0)\,,$$

where $\chi(\theta)$ is a non-decreasing function of θ in $(-\infty,\infty)$ and if we normalize it, such that $\chi(0)=0$, $\chi(\theta-0)=\chi(\theta)$, then it is unique.

Proof. Since $u>0$ in $|z|<1$,

$$u(z)=\frac{1}{2\pi}\int_{-\pi+0}^{\pi+0}\frac{1-r^2}{1-2r\cos(\varphi-\theta)+r^2}d\chi(\varphi)=\frac{1}{2\pi}\int_{-\pi+0}^{\pi+0}\Re\left(\frac{e^{i\varphi}+z}{e^{i\varphi}-z}\right)d\chi(\varphi)\,,$$

$$z=re^{i\theta}\,,$$

so that

$$f(z)=\frac{1}{2\pi}\int_{-\pi+0}^{\pi+0}\frac{e^{i\varphi}+z}{e^{i\varphi}-z}d\chi(\varphi)+iv(0)\,. \qquad (1)$$

Since $u>0$ in $|z|<1$, $F(z)=\dfrac{f(z)-1}{f(z)+1}$ is regular and $|F(z)|<1$ in $|z|<1$, so that by Fatou's theorem, for almost all $e^{i\theta}$ on $|z|=1$, $\lim_{z\to e^{i\theta}}F(z)=F(e^{i\theta})$ exists uniformly, when $z\to e^{i\theta}$ from the inside of any fixed Stolz domain, whose vertex is at $e^{i\theta}$. If we apply Theorem IV. 9 to $F(z)-1$ and

$F(z)+1$, then we see that $F(e^{i\theta})\neq 1, -1$ for almost all $e^{i\theta}$.

Since $f(z)=\dfrac{1+F(z)}{1-F(z)}$, we see that $\lim_{z\to e^{i\theta}} f(z)=f(e^{i\theta})=\dfrac{1+F(e^{i\theta})}{1-F(e^{i\theta})}\neq 0, \infty$, almost everywhere.

Theorem IV. 16. *Let $u(z)$ be harmonic in $|z|<1$ and*

$$L(r)=\int_0^{2\pi} |u(re^{i\theta})|d\theta=O(1), \qquad 0\leqq r<1 .$$

Then $u(z)$ can be expressed as a difference of two non-negative harmonic functions in $|z|<1$, so that for almost all $e^{i\theta}$, $\lim_{z\to e^{i\theta}} u(z)=u(e^{i\theta})$ exists uniformly, when $z\to e^{i\theta}$ from the inside of any fixed Stolz domain, whose vertex is at $e^{i\theta}$.

Conversely, if $u(z)$ is a difference of two non-negative harmonic functions in $|z|<1$, then $L(r)=O(1)$, $0\leqq r<1$.

Proof. We put

$$u^+ = \mathrm{Max}\,(u, 0) ,$$

then

$$u=u^+-(-u)^+ , \quad |u|=u^++(-u)^+ , \quad u^+\leqq|u| , \quad (-u)^+\leqq|u| ,$$

so that

$$\int_0^{2\pi} u^+(re^{i\theta})d\theta=O(1) , \qquad \int_0^{2\pi} (-u(re^{i\theta}))^+d\theta=O(1) . \tag{1}$$

For any Borel set e on $|z|=1$, we define additive set functions $\mu_\rho^{(1)}, \mu_\rho^{(2)}$ by

$$\mu_\rho^{(1)}(e)=\int_e u^+(\rho e^{i\theta})d\theta\geqq 0 , \quad \mu_\rho^{(2)}(e)=\int_e (-u(\rho e^{i\theta}))^+d\theta\geqq 0 , \quad 0<\rho<1 . \tag{2}$$

Then by (1) and Theorem II. 4, we can find $\rho_\nu\to 1$, such that $\mu_{\rho_\nu}^{(1)}\to\mu^{(1)}\geqq 0$, $\mu_\nu^{(2)}\to\mu^{(2)}\geqq 0$, hence as the proof of Theorem IV. 6., we have

$$u(z)=u^{(1)}(z)-u^{(2)}(z)=\frac{1}{2\pi}\int_{-\pi+0}^{\pi+0}\frac{1-r^2}{1-2r\cos(\varphi-\theta)+r^2}d\mu(\varphi) , \tag{3}$$

where

$$u^{(1)}(z)=\frac{1}{2\pi}\int_{-\pi+0}^{\pi+0}\frac{1-r^2}{1-2r\cos(\varphi-\theta)+r^2}d\mu^{(1)}(\varphi) ,$$

$$u^{(2)}(z)=\frac{1}{2\pi}\int_{-\pi+0}^{\pi+0}\frac{1-r^2}{1-2r\cos(\varphi-\theta)+r^2}d\mu^{(2)}(\varphi) , \quad \mu=\mu^{(1)}-\mu^{(2)} .$$

$u^{(1)}(z)$ and $u^{(2)}(z)$ are non-negative harmonic functions in $|z|<1$.

Conversely, if u is a difference of two non-negative harmonic functions $u=u^{(1)}-u^{(2)}$, $u^{(1)}\geqq 0$, $u^{(2)}\geqq 0$ in $|z|<1$, then $|u|\leqq u^{(1)}+u^{(2)}$, so that

$$\int_0^{2\pi} |u(re^{i\theta})|d\theta\leqq\int_0^{2\pi} u^{(1)}(re^{i\theta})d\theta+\int_0^{2\pi} u^{(2)}(re^{i\theta})d\theta=2\pi(u^{(1)}(0)+u^{(2)}(0)) .$$

Theorem IV. 17. *Let $u(z)$ be harmonic in $|z|<1$ and*

$$\int_0^{2\pi} |u(re^{i\theta})|^p d\theta = O(1)\,. \qquad 0 \leqq r < 1\,, \quad p>1\,,$$

then

$$u(z)=\frac{1}{2\pi}\int_{-\pi}^{\pi} f(\varphi)\frac{1-r^2}{1-2r\cos(\varphi-\theta)+r^2}d\varphi\,, \qquad z=re^{i\theta}\,,$$

where $f \in L^p$, so that for almost all $e^{i\theta}$, $\lim_{z\to e^{i\theta}} u(z)=u(e^{i\theta})$ exists uniformly, when $z \to e^{i\theta}$ from the inside of any fixed Stolz domain, whose vertex is at $e^{i\theta}$ and

$$\lim_{r\to 1}\int_0^{2\pi} |u(re^{i\theta})-u(e^{i\theta})|^p d\theta = 0\,.$$

Proof. Since $p>1$, $\int_0^{2\pi} |u(re^{i\theta})|d\theta = O(1)$, so that

$$u(z)=\frac{1}{2\pi}\int_{-\pi+0}^{\pi+0} \frac{1-r^2}{1-2r\cos(\varphi-\theta)+r^2}d\mu(\varphi)\,, \tag{1}$$

where $\mu = \lim_{\nu} \mu_{\rho_\nu}$,

$$\mu_{\rho_\nu}(e) = \int_e u(\rho_\nu e^{i\theta})d\theta\,. \tag{2}$$

Let $\varphi(e)$ be the characteristic function of e, then since $\varphi^q=\varphi$, we have for $i=1, 2$,

$$|\mu_{\rho_\nu}(e)| \leqq \int_0^{2\pi} \varphi(e^{i\theta})|u(\rho_\nu e^{i\theta})|d\theta \leqq \left(\int_0^{2\pi} \varphi(e^{i\theta})d\theta\right)^{1/q}\left(\int_0^{2\pi} |u(\rho_\nu e^{i\theta})|^p d\theta\right)^{1/p}$$

$$\leqq \text{const.}\ \sqrt[q]{me} \quad \left(\frac{1}{p}+\frac{1}{q}=1\right),$$

so that $|\mu(e)| \leqq \text{const.}\ \sqrt[q]{me}$, hence μ is an absolutely continuous additive set function, so that $d\mu(\varphi)$ can be written in the form $d\mu(\varphi)=f(\varphi)d\varphi$, where $f \in L$, so that

$$u(z)=\frac{1}{2\pi}\int_{-\pi}^{\pi} f(\varphi)\frac{1-r^2}{1-2r\cos(\varphi-\theta)+r^2}d\varphi\,,$$

hence for almost all $e^{i\theta}$, $\lim_{z\to e^{i\theta}} u(z)=f(\theta)$ uniformly, when $z\to e^{i\theta}$ from the inside of any fixed Stolz domain, whose vertex is at $e^{i\theta}$.

By Fatou's lemma,

$$\int_0^{2\pi} |f(\theta)|^p d\theta = \int_0^{2\pi} |u(e^{i\theta})|^p d\theta \leqq \underline{\lim}_{r\to 1}\int_0^{2\pi} |u(re^{i\theta})|^p d\theta < \infty\,,$$

hence $f \in L^p$, so that by Theorem IV. 4.,

$$\lim_{r\to 1}\int_0^{2\pi} |u(re^{i\theta})-u(e^{i\theta})|^p d\theta = 0\,.$$

6. Positive harmonic function in a half-plane

Theorem IV. 18. *Let $u(z)=u(x+iy)$ be a positive harmonic function in a half-plane $x>0$, then*

$$u(z)=\frac{x}{\pi}\int_{-\infty}^{\infty}\frac{d\chi(s)}{|z-s|^2}+cx \qquad c\geqq 0\,,$$

where $\chi(s)$ is a non-decreasing function of s and if we normalize it, such that $\chi(0)=0$, $\chi(s-0)=\chi(s)$, then it is unique and $c\geqq 0$ is a constant, which is unique.

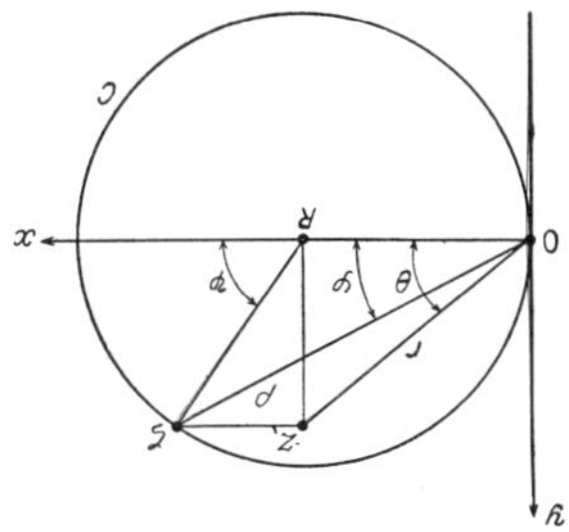

Proof. (Tsuji [2]). First we assume that $u(z)$ is harmonic at $z=0$. We express u by a Poisson integral on a circle C: $|z-R|=R$. Since $\rho=2R\cos\varphi$, $\psi=2\varphi$, $r\cos\theta=x$ in the figure,

$$u(z)=u(re^{i\theta})=\frac{1}{2\pi}\int_{-\pi}^{\pi}u(\zeta)\frac{R^2-|z-R|^2}{|z-\zeta|^2}\,d\psi=$$

$$\frac{1}{\pi}\int_{-\frac{\pi}{2}}^{\frac{\pi}{2}}u(\zeta)\frac{2Rr\cos\theta-r^2}{4R^2\cos^2\varphi-4Rr\cos\varphi\cos(\varphi-\theta)+r^2}\,d\varphi=$$

$$\frac{x}{\pi}\left(1-\frac{r^2}{2Rx}\right)\int_{-\frac{\pi}{2}}^{\frac{\pi}{2}}u(\zeta)\frac{2Rd\varphi}{4R^2\cos^2\varphi-4Rr\cos\varphi\cos(\varphi-\theta)+r^2}\,. \tag{1}$$

We take $R_1<R_2<\cdots<R_n\to\infty$ for R and put for a fixed L,

$$\int_{-\frac{\pi}{2}}^{\frac{\pi}{2}}u(\zeta)\frac{2R_nd\varphi}{4R_n^2\cos^2\varphi-4R_nr\cos\varphi\cos(\varphi-\theta)+r^2}=\int_{-\frac{\pi}{2}}^{-\frac{\pi}{2}+\frac{L}{2R_n}}+\int_{-\frac{\pi}{2}+\frac{L}{2R_n}}^{\frac{\pi}{2}-\frac{L}{2R_n}}+\int_{\frac{\pi}{2}-\frac{L}{2R_n}}^{\frac{\pi}{2}}$$

$$=I_n+II_n+III_n\,. \tag{2}$$

If we denote the arc length on the circle C_n: $|z-R_n|=R_n$ from $z=0$ to $\zeta=\rho e^{i\varphi}$ by s, then s varies from 0 to L in III_n and $2R_nd\varphi=-ds$.

We put $u(\zeta)=u_n(s)$ in III_n, then by (1), (2) and $u>0$, we have

$$III_n=\int_0^L\frac{u_n(s)ds}{|z-\zeta|^2}\leqq\frac{\pi u(z)}{x(1-\varepsilon_n)}\,, \qquad \varepsilon_n=\frac{r^2}{2R_nx}\,.$$

Since $|z-\zeta|\leqq\text{const.}=a$ in III_n,

$$\int_0^L u_n(s)ds\leqq\frac{a^2\pi u(z)}{x(1-\varepsilon_n)} \qquad (n=1,2,\cdots)\,.$$

Hence for $n=1,2,\cdots$,

$$0\leqq\chi_n(s)=\int_0^s u_n(s)ds\leqq\text{const.} \qquad (0\leqq s\leqq L)\,,$$

and $\chi_n(s)$ is a non-decreasing function of s $(0\leqq s\leqq L)$.

Hence by Helly's selection theorem, we can find a partial sequence from $\chi_n(s)$, which we denote by $\chi_n(s)$ again, such that $\lim_n \chi_n(s)=\chi(s)\geqq 0$, $\chi(0)=0$ $(0\leqq s\leqq L)$ exists. Then $\chi(s)$ is a non-decreasing function of s $(0\leqq s\leqq L)$. Similarly from I_n, we can define a non-decreasing function $\chi(s)\leqq 0$ for $-L\leqq s\leqq 0$, so that $\chi(s)$ is a non-decreasing function for $-L\leqq s\leqq L$ and $\chi(0)=0$. Hence

$$\lim_n (I_n+III_n)=\int_{-L}^{L}\frac{d\chi(s)}{|z-is|^2},$$

so that from (1), (2),

$$u(z)=\frac{x}{\pi}\left(\int_{-L}^{L}\frac{d\chi(s)}{|z-is|^2}+\lim_n II_n\right). \tag{3}$$

Now in II_n, $\cos\varphi\geqq\cos\left(\frac{\pi}{2}-\frac{L}{2R_n}\right)=\sin\frac{L}{2R_n}\geqq\frac{L}{\pi R_n}$, so that $2R_n\cos\varphi\geqq\frac{2L}{\pi}$, hence for a large L,

$$II_n=(1+\varepsilon(L))\int_{-\frac{\pi}{2}+\frac{L}{2R_n}}^{\frac{\pi}{2}-\frac{L}{2R_n}}\frac{u(\zeta)}{2R_n\cos^2\varphi}\,d\varphi, \tag{4}$$

where $\lim_{L\to\infty}\varepsilon(L)=0$.

We take $0<L_1<L_2<\cdots<L_k\to\infty$ for L and taking again a converging partial sequence from the corresponding $\chi(s)$, we get from (3), (4),

$$u(z)=\frac{x}{\pi}\left(\int_{-\infty}^{\infty}\frac{d\chi(s)}{|z-is|^2}+\lim_k\lim_n\int_{-\frac{\pi}{2}+\frac{L_k}{2R_n}}^{\frac{\pi}{2}-\frac{L_k}{2R_n}}u(\zeta)\frac{d\varphi}{2R_n\cos^2\varphi}\right), \tag{5}$$

where the double limit is evidently a constant, independent of z, which we denote by πc, then

$$u(z)=\frac{x}{\pi}\int_{-\infty}^{\infty}\frac{d\chi(s)}{|z-is|^2}+cx, \qquad c\geqq 0. \tag{6}$$

We assumed that u is harmonic at $z=0$. If u is not harmonic at $z=0$, then we express it by a Poisson integral on the circle $|z-(R_n+\delta_n)|=R_n$ $(\delta_n>0)$ and make $\delta_n\to 0$ with $R_n\to\infty$, then we arrive at the same result.

Since $u(1)<\infty$, we have

$$\int_{|s|\geq 1}\frac{d\chi(s)}{s^2}<\infty. \tag{7}$$

From this, we can prove easily that

$$\lim_{x\to\infty}\frac{u(x)}{x}=c. \tag{8}$$

Hence c is unique. Next we shall prove the uniqueness of χ.

Let $\chi(s)$ be continuous at a, b $(a<b)$, then $d\chi(a)=0$, $d\chi(b)=0$.

$$\begin{aligned}\int_a^b u(x+it)dt&=\int_{-N}^{N} d\chi(s)\frac{1}{\pi}\int_a^b \frac{x}{x^2+(t-s)^2}dt\\&+\int_{|s|\geqq N} d\chi(s)\frac{1}{\pi}\int_a^b \frac{x}{x^2+(t-s)^2}dt+cx(b-a)\\&=I+II+cx(b-a)\,.\end{aligned} \tag{9}$$

By (7), $\lim_{N\to\infty} II=0$ and

$$\lim_{x\to 0}\frac{1}{\pi}\int_a^b \frac{xdt}{x^2+(t-s)^2}=\begin{cases}1, & \text{if } a<s<b,\\ 0, & \text{if } s \text{ lies outside of } [a,b]\,.\end{cases}$$

Since $d\chi(a)=0$, $d\chi(b)=0$, we have $\lim_{x\to 0} I=\chi(b)-\chi(a)$, so that

$$\lim_{x\to 0}\int_a^b u(x+it)dt=\chi(b)-\chi(a)\,. \tag{10}$$

Since $\chi(s)$ is non-decreasing, $\chi(s)$ is continuous, except at most a countable number of values of s, hence by (10), we see easily that $\chi(s)$ is unique, if we normalize it, such that $\chi(0)=0$, $\chi(s-0)=\chi(s)$.——

If $u=0$ on the imaginary axis, then by (10), $\chi(s)\equiv$const., so that $u(z)=cx$. Hence

Corollary. *Let $u(z)=u(x+iy)$ be a positive harmonic function in $x>0$ and $u=0$ on the imaginary axis, then $u(z)=cx$, $c\geqq 0$.*

Remark 1. If we put

$$m(r)=\int_{-\frac{\pi}{2}}^{\frac{\pi}{2}} u(re^{i\theta})\cos\theta\, d\theta\,, \tag{1}$$

then we can prove that

$$\frac{m(r)}{r}=\int_r^\infty \frac{\psi(s)ds}{s}+\frac{\pi}{2}c\,,\qquad \psi(s)=\chi(s)-\chi(-s)\,. \tag{2}$$

Hence $\frac{m(r)}{r}$ *is a non-increasing function of r and a concave function of $1/r^2$ and* $\lim_{r\to\infty}\frac{m(r)}{r}=\frac{\pi}{2}c$.

If $u=0$ on the imaginary axis, then $u=cx$, so that $\frac{m(r)}{r}\equiv\frac{\pi}{2}c$.

Remark 2. By the same method, we can prove an analogous theorem for a positive harmonic function $u(x_1,\cdots,x_n)$ of n variables in a half-space $x_1>0$. Dinghas [4].

Theorem IV. 19. *Let $f(z)=f(x+iy)=u+iv$ be regular in $x>0$ and*

$u>0$, then

$$\lim_{z\to\infty}\frac{f(z)}{z}=c\,,\qquad \lim_{z\to\infty}f'(z)=c\,,\qquad 0\leqq c<\infty\,,$$

uniformly, when $z\to\infty$ from the inside of any fixed angular domain $\Delta:|\arg z|\leqq\varphi_0<\pi/2$.

Proof. Since $u>0$,

$$u(z)=\frac{1}{\pi}\int_{-\infty}^{\infty}\frac{x}{|z-is|^2}d\chi(s)+cx\,,\qquad 0\leqq c<\infty\,.$$

Since $\Re\left(\frac{1}{z-is}\right)=\frac{x}{|z-is|^2}$,

$$\begin{aligned}f(z)&=\frac{1}{\pi}\int_{-\infty}^{\infty}\left(\frac{1}{z-is}-\frac{1}{1-is}\right)d\chi(s)+cz+\text{const.}\\&=\frac{1-z}{\pi}\int_{-\infty}^{\infty}\frac{d\chi(s)}{(z-is)(1-is)}+cz+\text{const.}\,,\end{aligned}\tag{1}$$

where $\frac{-1}{1-is}$ is added to secure the convergence of the integral.

For any $\varepsilon>0$, we choose $N>0$ and then $|z|$ $(z\in\Delta)$ so large that

$$\int_{|s|\geq N}\frac{d\chi(s)}{s^2}<\varepsilon\,,\qquad \int_{-N}^{N}\frac{d\chi(s)}{|(z-is)(1-is)|}<\varepsilon\,,\tag{2}$$

then since for any $z\in\Delta$, $|z-is|\geqq\cos\varphi_0\,|s|$, we have

$$\int_{-\infty}^{\infty}\frac{d\chi(s)}{|(z-is)(1-is)|}\leqq\frac{1}{\cos\varphi_0}\int_{|s|\geq N}\frac{d\chi(s)}{s^2}+\int_{-N}^{N}\frac{d\chi(s)}{|(z-is)(1-is)|}<\varepsilon/\cos\varphi_0+\varepsilon\,.$$

Since $\varepsilon>0$ is arbitrary, we have from (1),

$$\lim_{z\to\infty}\frac{f(z)}{z}=c\quad\text{uniformly, when } z\to\infty \text{ in } \Delta\,.\tag{3}$$

Similarly from

$$f'(z)=-\int_{-\infty}^{\infty}\frac{d\chi(s)}{(z-is)^2}+c\,,$$

we can prove that

$$\lim_{z\ \infty}f'(z)=c\quad\text{uniformly, when } z\to\infty \text{ in } \Delta.\tag{4}$$

Theorem IV. 20. *Let $u(z)=u(x+iy)$ be a positive harmonic function in a half-plane $x>0$. Let C be a Jordan arc, which ends at $z=0$ and is contained in angular domain: $|\arg z|\leqq\varphi_0<\pi/2$. If u is bounded on C, then u is bounded in a sector Δ: $|\arg z|\leqq\varphi<\pi/2$, $|z|\leqq1$ for any $\varphi<\pi/2$.*

If $u\to0$ (or $+\infty$), when $z\to0$ along C, then $u\to0$ (or $+\infty$) uniformly, when $z\to0$ from the inside of Δ. Tsuji [51]

Proof. Let $0<u\leqq M$ on C and $z_0=x_0+iy_0\in C$. Since $u>0$, by means of Poisson integral, we have easily

$$u(z) \leqq \frac{1+1/2}{1-1/2} u(z_0) \leqq 3M \quad \text{in } |z-z_0| \leqq x_0/2\,, \tag{1}$$

so that

$$u(z_1) \leqq 3M\,, \qquad z_1 = x_0 + iy_1\,, \quad y_1 = y_0 \pm x_0/2\,.$$

Hence

$$u(z) \leqq 3^2 M \quad \text{in } |z-z_1| \leqq x_0/2\,, \tag{2}$$

so that

$$u(z_2) \leqq 3^2 M\,, \qquad z_2 = x_0 + iy_2\,, \quad y_2 = y_1 \pm x_0/2\,.$$

Repeating the same argument, we see that

$$u(z) \leqq 3^n M \tag{3}$$

on the segment: $z = x_0 + iy \left(|y-y_0| \leqq \dfrac{nx_0}{2} \right)$.

Since $|y_0| \leqq \tan\varphi_0 \cdot x_0$, the segment contains a real point $z = x_0$, if n is large, so that u is bounded on the positive real axis in a neighbourhood of $z=0$. If we repeat the same argument, starting from the point on the positive real axis, then we see that u is bounded in a sector: $|\arg z| \leqq \varphi < \pi/2$, $|z| \leqq 1$, for any $\varphi < \pi/2$.

The second part can be proved similarly.

Theorem IV. 21. *Let $u(z) = u(x+iy)$ be a positive harmonic function in a half-plane $x>0$. If $\lim_{r\to 0} u(re^{i\alpha}) = \lim_{r\to 0} u(re^{i\beta}) = \omega$ $(|\alpha| < \pi/2,\ |\beta| < \pi/2)$ exist, then $\lim_{r\to 0} u(re^{i\theta}) = \omega$ uniformly for $|\theta| \leqq \varphi_0 < \pi/2$ for any $\varphi_0 < \pi/2$.*

Loomis [1].

Proof. By the above theorem, u is bounded in a sector: $|\arg z| \leqq \varphi_0 < \pi/2$, $|z| \leqq 1$, where φ_0 is chosen, such that $|\alpha| < \varphi_0$, $|\beta| < \varphi_0$. Since u is bounded in a sector: $\alpha \leqq \arg z \leqq \beta$, $|z| \leqq 1$ and $\lim_{r\to 0} u(re^{i\alpha}) = \lim_{r\to 0} u(re^{i\beta}) = \omega$, we see that $\lim_{r\to 0} u(re^{i\theta}) = \omega$ uniformly for $\alpha \leqq \theta \leqq \beta$. Let D: $|\arg z| \leqq \varphi_0$, $1/2 \leqq |z| \leqq 1$ and consider a family of harmonic functions $u_\tau(z) = u(\tau z)$ $(0<\tau<1)$ in D, then since $u_\tau(z)$ is uniformly bounded, it is a normal family in D. Since $u_\tau(z) \to \omega$ $(\tau \to 0)$ in its partial domain: $\alpha \leqq \arg z \leqq \beta$, $\dfrac{1}{2} \leqq |z| \leqq 1$, $u_\tau(z) \to \omega$ $(\tau \to 0)$ in D, or $\lim_{r\to 0} u(re^{i\theta}) = \omega$ uniformly for $|\theta| \leqq \varphi_0 < \pi/2$.

7. Carathéodory-Toeplitz's theorem

1. Let $f(z) = \dfrac{a_0}{2} + \sum_{n=1}^{\infty} a_n z^n = u(z) + iv(z)$ $(a_0 = \text{real})$ be regular in $|z|<1$ and $u>0$, then by Theorem IV. 15,

$$f(z) = \frac{1}{2\pi} \int_{-\pi+0}^{\pi+0} \frac{e^{i\theta}+z}{e^{i\theta}-z}\, d\chi(\theta)\,, \tag{1}$$

where $\chi(\theta)$ is a non-decreasing function of θ. Hence

$$a_n=\frac{1}{\pi}\int_{-\pi+0}^{\pi+0} e^{-in\theta}d\chi(\theta) \qquad (n=0, 1, 2, \cdots). \tag{2}$$

Hence if we put $a_{-\nu}=\bar{a}_\nu$, then

$$H_n(x)=H(x_0, x_1, \cdots, x_n)=\sum_{\mu,\nu=0}^{n} a_{\mu-\nu}x_\nu x_\mu$$
$$=\frac{1}{\pi}\int_{-\pi+0}^{\pi+0} |x_0+x_1e^{i\theta}+\cdots+x_ne^{in\theta}|^2 d\chi(\theta)\geqq 0, \tag{3}$$

so that $H_n(x)$ is a non-negative Hermitian form for $n=0, 1, 2, \cdots$.

We remark that the coefficient matrix of $H_n(x)$ is

$$H_n=\begin{pmatrix} a_0 & a_1 & \cdots & a_n \\ \bar{a}_1 & a_0 & \cdots & a_{n-1} \\ \cdots & \cdots & \cdots & \cdots \\ \bar{a}_n & \bar{a}_{n-1} & \cdots & a_0 \end{pmatrix}. \tag{4}$$

Theorem IV. 22. *Let* $f(z)=\frac{a_0}{2}+\sum_{n=1}^{\infty}a_nz^n=u(z)+iv(z)$ (a_0=real).

(i) *If* $f(z)$ *is regular and* $\Re(f(z))\geqq 0$ *in* $|z|<1$, *then* $H_n(x)\geqq 0$ *for* $n=0, 1, 2, \cdots$.

(ii) *If* $H_n(x)\geqq 0$ *for* $n=0, 1, 2, \cdots$, *then* $f(z)$ *is regular and* $\Re(f(z))\geqq 0$ *in* $|z|<1$.

If $H_0(x), \cdots, H_{k-1}(x)$ *are positive definite and* $H_k(x)$ *is positive semi-definite, then* $f(z)$ *is of the special form:*

$$f(z)=\sum_{\nu=1}^{k}\frac{r_\nu}{2}\cdot\frac{1+\varepsilon_\nu z}{1-\varepsilon_\nu z}, \qquad r_\nu>0, \ |\varepsilon_\nu|=1, \ \varepsilon_i\neq\varepsilon_j \ (i\neq j),$$

so that $\Re(f(z))=0$ *on* $|z|=1$. Carathéodory [1], Toeplitz [1].

We remark that if $H_{k-1}(x)$ is positive definite, then $H_0(x), \cdots, H_{k-2}(x)$ are positive definite.

2. For the proof, we use the following Fejér's theorem.

Lemma. *Let* $\tau(\theta)=\lambda_0+\sum_{\nu=1}^{n}(\lambda_\nu\cos\nu\theta+\mu_\nu\sin\nu\theta)\geqq 0$ *in* $[0, 2\pi]$, *then* $\tau(\theta)$ *can be expressed in the form* $\tau(\theta)=|\gamma_0+\gamma_1e^{i\theta}+\cdots+\gamma_ne^{in\theta}|^2$. Fejér [1].

Proof. Since $\cos\theta=\frac{z+z^{-1}}{2}$, $\sin\theta=\frac{z-z^{-1}}{2i}$ $(z=e^{i\theta})$,

$$z^n\tau(\theta)=G(z)=u_0+u_1z+\cdots+u_nz^n+\cdots+u_{2n}z^{2n}, \tag{1}$$

where $u_\nu=\bar{u}_{2n-\nu}$ $(\nu=0, 1, 2, \cdots, n-1)$, so that zero points of $G(z)$, which do not lie on $|z|=1$ appear in pairs z_ν, $1/\bar{z}_\nu$ with the same multiplicity, hence if $\{\zeta_\mu\}$ are zero points on $|z|=1$, then

$$G(z)=\text{const.}\ z^r\prod_{\mu=1}^{k}(z-\zeta_\mu)\prod_{\nu=1}^{s}(z-z_\nu)(z-1/\bar{z}_\nu) \quad (r+k+2s=2n). \tag{2}$$

The zero point $\zeta_\mu=e^{i\theta_\mu}$ corresponds to a zero point θ_μ of $\tau(\theta)$.

By differentiating $\tau(\theta)=e^{-in\theta}G(e^{i\theta})$ by θ, we see that the multiplicity of ζ_μ is equal to that of θ_μ for $\tau(\theta)$.

Since $\tau(\theta)\geqq 0$, the multiplicity of θ_μ is even, so that $\prod_{\mu=1}^{k}(z-\zeta_\mu)=\prod_{\mu=1}^{k/2}(z-\zeta_\mu)^2$, hence

$$\tau(\theta)=|\tau(\theta)|=|G(e^{i\theta})|=\text{const.}\prod_{\mu=1}^{k/2}|e^{i\theta}-\zeta_\mu|^2\prod_{\nu=1}^{s}|e^{i\theta}-z_\nu|^2 .$$

From this we see that

$$\tau(\theta)=|\gamma_0+\gamma_1e^{i\theta}+\cdots+\gamma_ne^{in\theta}|^2.$$

3. *Proof of the theorem.*

We have only to prove (ii). If $H_n(x)\geqq 0$ for $n=0,1,2,\cdots$, then $H_n(x_0,0,\cdots,0,x_n)\geqq 0$, so that its determinant $\begin{vmatrix} a_0 & a_n \\ \bar{a}_n & a_0 \end{vmatrix}=a_0^2-|a_n|^2\geqq 0$, hence $|a_n|\leqq|a_0|$, so that $f(z)$ is regular in $|z|<1$.

Next we shall prove that $u\geqq 0$ in $|z|<1$.

If we put $a_n=\alpha_n+i\beta_n$, then $u(re^{i\theta})=\dfrac{a_0}{2}+\sum_{n=1}^{\infty}r^n(\alpha_n\cos n\theta-\beta_n\sin n\theta)$, so that

$$a_\nu\rho^{|\nu|}=\frac{1}{\pi}\int_0^{2\pi}u(\rho e^{i\theta})e^{-i\nu\theta}d\theta \qquad (0<\rho<1)\quad(\nu=0,\pm1,\pm2,\cdots), \qquad (1)$$

hence

$$H_n^{(\rho)}(x)=\sum_0^n a_{\mu-\nu}\rho^{|\mu-\nu|}x_\nu\bar{x}_\mu=\frac{1}{\pi}\int_0^{2\pi}u(\rho e^{i\theta})|x_0+x_1e^{i\theta}+\cdots+x_ne^{in\theta}|^2d\theta . \qquad (2)$$

Now

$$F(z)=\frac{1+\rho z}{2(1-\rho z)}=\frac{1}{2}+\sum_{n=1}^{\infty}\rho^nz^n$$

is regular and $\Re(F(z))>0$ in $|z|<1$, so that

$$F(z)=\frac{1}{2\pi}\int_{-\pi+0}^{\pi+0}\frac{e^{i\theta}+z}{e^{i\theta}-z}d\chi(\theta)\ , \quad \text{hence} \quad \rho^{|\nu|}=\frac{1}{\pi}\int_{-\pi+0}^{\pi+0}e^{-i\nu\theta}d\chi(\theta)\ , \qquad (3)$$

where $\chi(\theta)$ is a non-decreasing function of θ, hence by (2) and the hypothesis,

$$H_n^{(\rho)}(x)=\frac{1}{\pi}\int_{-\pi+0}^{\pi+0}\left(\sum_0^n a_{\mu-\nu}x_\nu e^{i\nu\theta}\cdot\bar{x}_\mu e^{-i\mu\theta}\right)d\chi(\theta)=\frac{1}{\pi}\int_{-\pi+0}^{\pi+0}H(\xi_0,\cdots,\xi_n)d\chi\geqq 0\ ,$$
$$(\xi_\nu=x_\nu e^{i\nu\theta})$$

so that by (2),

$$H_n^{(\rho)}(x)=\frac{1}{\pi}\int_0^{2\pi}u(\rho e^{i\theta})|x_0+x_1e^{i\theta}+\cdots+x_ne^{in\theta}|^2d\theta\geqq 0 \quad (0<\rho<1)\ . \qquad (4)$$

Now for any θ_0, $|u(\rho e^{i\theta})-u(\rho e^{i\theta_0})|<\varepsilon$, if $|\theta-\theta_0|<\delta=\delta(\varepsilon)$.

We define a positive continuous function $g(\theta)$ with period 2π as follows.

(i) $g(\theta)=\text{const.}=M>0$ in $|\theta-\theta_0|\leqq\delta \pmod{2\pi}$,

(ii) $g(\theta)=\text{const.}=2\varepsilon$, outside of $|\theta-\theta_0|\leqq\delta' \pmod{2\pi}$ $(\delta'>\delta)$,

(iii) $g(\theta)$ is linear in $[\theta_0-\delta', \theta_0-\delta]$ and $[\theta_0+\delta, \theta_0+\delta']$, such that

$$g(\theta_0-\delta')=2\varepsilon\ ,\quad g(\theta_0-\delta)=M\ ,\quad g(\theta_0+\delta)=M\ ,\quad g(\theta_0+\delta')=2\varepsilon\ .$$

We choose $M>0$ so large and $\delta'-\delta$ so small that

$$\int_{\theta_0-\delta}^{\theta_0+\delta} g(\theta)d\theta=\pi\ ,\qquad \int_J g(\theta)d\theta<4\pi\varepsilon\ ,$$

where J is the complement of $|\theta-\theta_0|\leqq\delta$ with respect to $[0, 2\pi]$.

Now we approximate $g(\theta)$ by a trigonometrical polynomical $\tau(\theta)$ of order n, such that $|g(\theta)-\tau(\theta)|<\varepsilon$ in $[0, 2\pi]$, then $\tau(\theta)>g(\theta)-\varepsilon>0$ in $[0, 2\pi]$, so that by the lemma, $\tau(\theta)=|x_0+x_0e^{i\theta}+\cdots+x_ne^{in\theta}|^2$ with suitable x_ν. Hence by (4),

$$0\leqq H_n^{(\rho)}(x)=\frac{1}{\pi}\int_0^{2\pi} u(\rho e^{i\theta})\tau(\theta)\,d\theta=\frac{1}{\pi}\int_0^{2\pi} u(\rho e^{i\theta})g(\theta)d\theta+O(\varepsilon)$$

$$=\frac{1}{\pi}\int_{\theta_0-\delta}^{\theta_0+\delta} u(\rho e^{i\theta})g(\theta)d\theta+O(\varepsilon)=u(\rho e^{i\theta_0})+O(\varepsilon)\,.$$

Since $\varepsilon>0$ is arbitrary, we have $u(\rho e^{i\theta_0})\geqq 0$, so that

$$\Re(f(z))=u(z)\geqq 0 \qquad \text{in}\quad |z|<1\,. \tag{5}$$

Hence the first part is proved. Since $u\geqq 0$ in $|z|<1$,

$$f(z)=\frac{1}{2\pi}\int_{-\pi+0}^{\pi+0}\frac{e^{i\theta}+z}{e^{i\theta}-z}d\chi^*(\theta)\ , \tag{6}$$

where $\chi^*(\theta)$ is a non-decreasing function of θ.

Suppose that $H_0(x), \cdots, H_{k-1}(x)$ are positive definite and $H_k(x)$ is positive semi-definite, then there exist $(x_0', \cdots, x_k')\neq(0, \cdots, 0)$, such that $H_k(x_0', \cdots, x_k')=0$. Hence by (6)

$$H_k(x')=\frac{1}{\pi}\int_{-\pi+0}^{\pi+0}|x_0'+x'e^{i\theta}+\cdots+x_k'e^{ik\theta}|^2 d\chi^*(\theta)=0\,. \tag{7}$$

If $\chi^*(\theta)$ is increasing at θ_0 and $x_0'+x_1'e^{i\theta_0}+\cdots+x_k'e^{ik\theta_0}\neq 0$, then $H_k(x')>0$, which is absurd. Hence if $\chi^*(\theta)$ is increasing at θ_0, then $x_0'+x_1'e^{i\theta_0}+\cdots+x_k'e^{ik\theta_0}=0$. Since $x_0'+x_1'z+\cdots+x_k'z^k=0$ has at most k roots on $|z|=1$, $\chi^*(\theta)$ is increasing at θ_ν $(\nu=1, 2, \cdots, j\leqq k)$ and is constant outside of $\{\theta_\nu\}$, so that

$$H_{k-1}(x)=\sum_{\nu=1}^{j} r_\nu|x_0+x_1e^{i\theta_\nu}+\cdots+x_{k-1}e^{i(k-1)\theta_\nu}|^2\ ,\quad r_\nu=d\chi^*(\theta_\nu)/\pi>0\,. \tag{8}$$

If $j\leqq k-1$, then the system of linear equations in $x_0, \cdots, x_{k-1}$:

$$x_0+x_1e^{i\theta_\nu}+\cdots+x_{k-1}e^{i(k-1)\theta_\nu}=0 \qquad (\nu=1, 2, \cdots, j)$$

has a solution $(x_0'', \cdots, x_{k-1}'')\neq(0, \cdots, 0)$, so that $H_{k-1}(x'')=0$, which contradicts the hypothesis. Hence $j=k$, so that by (6), $f(z)$ is of the special form:

$$f(z)=\sum_{\nu=1}^{k}\frac{r_\nu}{2}\cdot\frac{1+\varepsilon_\nu z}{1-\varepsilon_\nu z}\,, \qquad r_\nu>0\,,\quad \varepsilon_\nu=e^{-i\theta_\nu}\,,\quad \varepsilon_i\neq\varepsilon_j\ (i\neq j)\,, \tag{9}$$

hence for $n=0, 1, 2, \cdots,$

$$a_n=r_1\varepsilon_1^n+\cdots+r_k\varepsilon_k^n\,, \tag{10}$$

$$H_n(x)=\sum_{\nu=1}^{k}r_\nu|x_0+x_1\varepsilon_\nu^{-1}+\cdots+x_n\varepsilon_\nu^{-n}|^2\,. \tag{11}$$

Conversely, if $f(z)$ is of the form (9), then by (11), $H_0(x), \cdots, H_{k-1}(x)$ are positive definite and $H_k(x)$ is positive semi-definite.——

4. Now

$$\delta_n=\delta(a_0, \cdots, a_n)=\begin{vmatrix} a_0 & a_1 & \cdots & a_n \\ \bar{a}_1 & a_0 & \cdots & a_{n-1} \\ \cdots & \cdots & \cdots & \cdots \\ \bar{a}_n & \bar{a}_{n-1} & \cdots & a_0 \end{vmatrix} \tag{1}$$

is the determinant of $H_n(x)$. Hence if $H_n(x)$ is positive definite, then

$$\delta_0>0\,,\quad \delta_1>0\,,\quad \cdots,\quad \delta_n>0\,, \tag{2}$$

and conversely, if (2) holds, then $H_n(x)$ is positive definite. Hence

Theorem IV. 23. *Let* $f(z)=\frac{a_0}{2}+\sum_{n=1}^{\infty}a_nz^n$ *(a_0=real). Then $f(z)$ is regular and $\Re(f(z))\geqq 0$ in $|z|<1$, when and only when*

(i) $\delta_n>0$ $(n=0, 1, 2, \cdots)$, or (ii) $\delta_0>0, \cdots, \delta_{k-1}>0, \delta_k=\delta_{k+1}=\cdots=0$.

In case (ii), $f(z)$ is of the special form of Theorem IV. 22.

Theorem IV. 24. *Let $a_0, a_1, \cdots, a_n$ (a_0=real) be $n+1$ given complex numbers, such that $H_n(x)=\sum_0^n a_{\mu-\nu}x_\nu x_\mu$ is non-negative.*

Then there exists a regular function $f(z)$ in $|z|<1$, such that $\Re(f(z))\geqq 0$ in $|z|<1$ and

$$f(z)=\frac{a_0}{2}+a_1z+\cdots+a_nz^n\,, \qquad \bmod z^{n+1}\,.$$

If $H_n(x)$ is positive semi-definite, then such $f(z)$ is unique and is of the special form of Theorem IV. 22., where $k\leqq n$. Carathéodory [1].

Proof. We consider $(a)=(a_0, \cdots, a_n)$ as a point in an $(2n+1)$-dimensional space and with Carathéodory, we consider a domain $\mathfrak{R}_n$:

$$\mathfrak{R}_n:\ \delta(a_0)>0\,,\quad \delta(a_0, a_1)>0\,,\quad \cdots,\quad \delta(a_0, \cdots, a_n)>0\,. \tag{1}$$

If $(a)\in\mathfrak{R}_n$, then $H_n(x)$ is positive definite. From this, we see easily that $\mathfrak{R}_n$ is a convex domain. Its boundary consists of points:

$$\left.\begin{array}{l}\delta(a_0)>0\ ,\quad \delta(a_0, a_1)>0\ ,\quad \cdots,\quad \delta(a_0, a_1, \cdots, a_{k-1})>0\ ,\\ \delta(a_0, a_1, \cdots, a_k)=\delta(a_0, a_1, \cdots, a_{k+1})=\cdots=0\ ,\end{array}\right\} \tag{2}$$

for some $k \leqq n$, since the boundary point corresponds to a positive semi-definite form $H_n(x)$. Suppose that $H_n(x)$ is non-negative, then (a) lies in $\mathfrak{R}_n$, or on its boundary, so that there exists in its ε-neighbourhood an inner point (a') of $\mathfrak{R}_n$, such that

$$\delta_0'=\delta(a_0')>0\ ,\quad \delta_1'=\delta(a_0', a_1')>0\ ,\quad \cdots,\quad \delta_n'=\delta(a_0', \cdots, a_n')>0\ .$$

Let

$$F(z)=\delta(a_0', a_1', \cdots, a_n', z)=Az\bar{z}+Bz+\bar{B}\bar{z}+C\ , \tag{3}$$

where A, C are real. Then $A=-\delta_{n-1}'\neq 0$. Since by Jacobi's theorem[1], $\begin{vmatrix}\delta_n' & B\\ \bar{B} & \delta_n'\end{vmatrix}=\delta'_{n-1}C=-AC$, $|B|^2-AC=\delta_n'^2>0$, $F(z)=0$ is a real circle. Hence there exists a_{n+1}', such that $F(a_{n+1}')=0$. Then $(a_0', a_1', \cdots, a_{n+1}')$ belongs to the boundary of $\mathfrak{R}_{n+1}$, so that $\begin{vmatrix}a_0' & a_{n+1}'\\ \bar{a}_{n+1}' & a_0'\end{vmatrix}\geqq 0$, $|a_{n+1}'|\leqq|a_0'|$. Hence if we make $a_0'\to a_0, \cdots, a_n'\to a_n$, then the corresponding a_{n+1}' are bounded, so that we can select a convergent sequence from $\{a_{n+1}'\}$, such that $a_{n+1}'\to a_{n+1}$, then $(a_0, a_1, \cdots, a_{n+1})$ belongs to the boundary of $\mathfrak{R}_{n+1}$, so that $H_{n+1}(x)$ is positive semi-definite. Similarly we can find a_{n+2}, a_{n+3}, $\cdots$, such that $H_{n+2}(x)$, $H_{n+3}(x)$, $\cdots$ are positive semi-definite, so that $|a_\nu|\leqq|a_0|$ $(\nu=1, 2, \cdots)$, hence

$$f(z)=\frac{a_0}{2}+\sum_{n=1}^{\infty}a_n z^n=\frac{a_0}{2}+a_1 z+\cdots+a_n z^n\ ,\qquad \mathrm{mod}\ z^{n+1}, \tag{4}$$

is regular in $|z|<1$ and by Theorem IV. 22, $\Re(f(z))\geqq 0$ in $|z|<1$.

Next we shall prove that if $H_n(x)$ is positive semi-definite, then such $f(z)$ is unique. Suppose that $H_0(x), H_1(x), \cdots, H_{k-1}(x)$ $(k\leqq n)$ are positive definite and $H_k(x)$ is positive semi-definite, then $f(z)$ is of the special form of Theorem IV. 22, so that $a_\nu=r_1\varepsilon_1^\nu+\cdots+r_k\varepsilon_k^\nu$ $(\nu=0, 1, \cdots)$. Let

$$F_k(x)=\begin{vmatrix}a_0 & a_1 & \cdots & a_{k-1} & a_k\\ a_1 & a_0 & \cdots & a_{k-2} & a_{k-1}\\ \cdots & \cdots & \cdots & \cdots & \cdots\\ \bar{a}_{k-1} & \bar{a}_{k-2} & \cdots & a_0 & a_1\\ 1 & x & \cdots & x^{k-1} & x^k\end{vmatrix}=\delta(a_0, a_1, \cdots, a_{k-1})x^k+\cdots\ , \tag{5}$$

---

1) Let $D=\begin{vmatrix}a_{11} & \cdots & a_{1n}\\ \cdots & \cdots & \cdots\\ a_{n1} & \cdots & a_{nn}\end{vmatrix}$ and A_{ik} be the cofactor of a_{ik}, then

$$\begin{vmatrix}A_{11} & A_{12}\\ A_{21} & A_{22}\end{vmatrix}=D\begin{vmatrix}a_{33} & \cdots & a_{3n}\\ \cdots & \cdots & \cdots\\ a_{n3} & \cdots & a_{nn}\end{vmatrix}.$$

then

$$F_k(x)=\begin{vmatrix} r_1 & r_2 & \cdots & r_k & 0 \\ r_1\varepsilon_1^{-1} & r_2\varepsilon_2^{-1} & \cdots & r_k\varepsilon_k^{-1} & 0 \\ \cdots & \cdots & \cdots & \cdots & \cdots \\ r_1\varepsilon_1^{-(k-1)} & r_2\varepsilon_2^{-(k-1)} & \cdots & r_k\varepsilon_k^{-(k-1)} & 0 \\ 0 & 0 & \cdots & 0 & 1 \end{vmatrix}\cdot\begin{vmatrix} 1 & \varepsilon_1 & \cdots & \varepsilon_1^k \\ 1 & \varepsilon_2 & \cdots & \varepsilon_2^k \\ \cdots & \cdots & \cdots & \cdots \\ 1 & \varepsilon_k & \cdots & \varepsilon_k^k \\ 1 & x & \cdots & x^k \end{vmatrix}.$$

Hence $\varepsilon_1, \cdots, \varepsilon_k$ are roots of $F_k(x)=0$, so that $\varepsilon_1, \cdots, \varepsilon_k$ are unique and since $r_1, \cdots, r_k$ are solutions of a system of linear equations:

$$a_\nu=r_1\varepsilon_1^\nu+\cdots+r_k\varepsilon_k^\nu \qquad (\nu=0, 1, 2, \cdots, k-1)\,,$$

$r_1, \cdots, r_k$ are unique, so that $f(z)$ is unique.

8. I. Schur's theorem

1. Let $f(z)=\sum_{n=0}^{\infty} a_n z^n$ and put

$$A_n=\begin{pmatrix} a_0 & a_1 & \cdots & a_n \\ 0 & a_0 & \cdots & a_{n-1} \\ \cdots & \cdots & \cdots & \cdots \\ 0 & 0 & \cdots & a_0 \end{pmatrix}, \qquad \bar{A}'_n=\begin{pmatrix} \bar{a}_0 & 0 & 0 & \cdots & 0 \\ \bar{a}_1 & \bar{a}_0 & 0 & \cdots & 0 \\ \cdots & \cdots & \cdots & \cdots & \cdots \\ \bar{a}_n & \bar{a}_{n-1} & \bar{a}_{n-2} & \cdots & \bar{a}_0 \end{pmatrix}, \tag{1}$$

and E_n be the unit matrix of order $n+1$ and put

$$\gamma_n=\gamma(a_0, a_1, \cdots, a_n)=\begin{vmatrix} E_n & A_n \\ \bar{A}'_n & E_n \end{vmatrix}, \tag{2}$$

which is a determinant of order $2(n+1)$.

Theorem IV. 25. *Let $f(z)=\sum_{n=0}^{\infty} a_n z^n$. Then $f(z)$ is regular and $|f(z)|\leqq 1$ in $|z|<1$, when and only when*

(i) $\gamma_n>0\,(n=0, 1, 2, \cdots)$, *or* (ii) $\gamma_0>0, \gamma_1>0, \cdots, \gamma_{k-1}>0,\ \gamma_k=\gamma_{k+1}=\cdots=0$.

In case (ii), *$f(z)$ is of the special form:*

$$f(z)=\varepsilon\prod_{\nu=1}^{k}\frac{z-\omega_\nu}{1-\bar{\omega}_\nu z}\,, \qquad |\varepsilon|=1\,, \quad |\omega_\nu|<1\,,$$

so that $\qquad |f(z)|=1$ *on* $|z|=1$. I. Schur [1].

2. First we shall prove a lemma. (I. Schur [1]).

Lemma. *Let P, Q, R, S be four matrices of order n, such that $PR=RP$ and let $|M|$ be the determinant of the following matrix M of order $2n$:*

$$M=\begin{pmatrix} P & Q \\ R & S \end{pmatrix}.$$

Then $|M|$ is equal to the determinant of the matrix $PS-RQ$ of order n.

Proof. If $|P|\neq 0$, then denoting a unit matrix of order n by E and a null matrix of order n by 0, we have

$$\begin{pmatrix} P^{-1} & 0 \\ -RP^{-1} & E \end{pmatrix}\begin{pmatrix} P & Q \\ R & S \end{pmatrix}=\begin{pmatrix} E & P^{-1}Q \\ 0 & S-RP^{-1}Q \end{pmatrix},$$

so that $|P^{-1}|\cdot|M|=|S-RP^{-1}Q|$, hence

$$|M|=|P|\cdot|S-RP^{-1}Q|=|PS-PRP^{-1}Q|=|PS-RQ|. \qquad (1)$$

If $|P|=0$, consider $f(\varepsilon)=|P+\varepsilon E|$, then $f(\varepsilon)=\varepsilon^n+\cdots\not\equiv 0$ and since $f(0)=|P|=0$, $f(\varepsilon)=|P+\varepsilon E|\neq 0$, if $\varepsilon>0$ is small. If we put

$$M_1=\begin{pmatrix} P+\varepsilon E & Q \\ R & S \end{pmatrix}, \qquad (2)$$

then $(P+\varepsilon E)R=R(P+\varepsilon E)$ and $|P+\varepsilon E|\neq 0$, so that by (1), $|M_1|=|(P+\varepsilon E)S-RQ|$, hence if we make $\varepsilon\to 0$, then we have

$$|M|=|PS-RQ|.$$

3. *Proof of the theorem.*

We put $F(z)=\dfrac{1+f(z)}{1-f(z)}=\dfrac{c_0}{2}+\sum_{n=1}^{\infty} c_n z^n$,

$$1+f(z)=\sum_{n=0}^{\infty} p_n z^n, \qquad 1-f(z)=\sum_{n=0}^{\infty} q_n z^n, \qquad (1)$$

then since $f(z)=\dfrac{F(z)-1}{F(z)+1}$, $f(z)$ is regular and $|f(z)|\leqq 1$ in $|z|<1$, when and only when $F(z)$ is regular and $\Re(F(z))\geqq 0$ in $|z|<1$.

Similarly as A_n (p. 159 (1)), we define C_n, P_n, Q_n for $F(z)$, $1+f(z)$, $1-f(z)$ respectively, then

$$P_n=C_nQ_n, \quad C_n=P_nQ_n^{-1}, \quad \bar{C}_n'=(\bar{Q}_n^{-1})'\bar{P}_n'=\bar{Q}_n'^{-1}\bar{P}_n',$$

hence if we put $H_n=C_n+\bar{C}_n'$ then

$$H_n=\begin{pmatrix} \Re(c_0) & c_1 & \cdots & c_n \\ \bar{c}_1 & \Re(c_0) & \cdots & c_{n-1} \\ \cdots & \cdots & \cdots & \cdots \\ c_n & \bar{c}_{n-1} & \cdots & \Re(c_0) \end{pmatrix},$$

so that

$$|H_n|=\delta(\Re(c_0), c_1, \cdots, c_n) \qquad (2)$$

is the determinant, defined in Theorem IV. 23.

Since $H_n=P_nQ_n^{-1}+\bar{Q}_n'^{-1}\bar{P}_n'$,

$$\bar{Q}_n'H_nQ_n=\bar{Q}_n'P_n+\bar{P}_n'Q_n, \qquad (3)$$

and from (1), $P_n=E_n+A_n$, $Q_n=E_n-A_n$, hence by (3),

$$\bar{Q}_n'H_nQ_n=(E_n-\bar{A}_n')(E_n+A_n)+(E_n+\bar{A}_n')(E_n-A_n)=2(E_nE_n-\bar{A}_n'A_n).$$

The right hand side is a matrix of order $(n+1)$ and $|\bar{Q}_n'||Q_n|=|1-a_0|^{2(n+1)}$, so that by the lemma, we have

$$|1-a_0|^{2(n+1)}|H_n|=2^{n+1}|E_nE_n-\bar{A}_n'A_n|=2^{n+1}\begin{vmatrix} E_n & A_n \\ \bar{A}_n' & E_n \end{vmatrix}=2^{n+1}\gamma(a_0, a_1, \cdots, a_n),$$

hence by (2),

$$\delta(\Re(c_0), c_1, \cdots, c_n) = \frac{2^{n+1}}{|1-a_0|^{2(n+1)}} \gamma(a_0, a_1, \cdots, a_n) . \tag{4}$$

Hence the first part (i) of the theorem follows from Theorem IV. 23. In case (ii), by Theorem IV. 23,

$$F(z) = \sum_{\nu=1}^{k} \frac{r_\nu}{2} \cdot \frac{1+\varepsilon_\nu z}{1-\varepsilon_\nu z} + ic , \qquad r_\nu > 0 , \quad |\varepsilon_\nu| = 1 , \quad 1 \leqq k \leqq n ,$$

where c is a real constant, so that $f(z) = \dfrac{F(z)-1}{F(z)+1}$ is a rational function of order k.

Since $\Re(F(z))=0$ on $|z|=1$, we see that $|f(z)|=1$ on $|z|=1$. Then the zero point of $f(z)$ corresponds to $F(z)=1$ and the pole of $f(z)$ corresponds to $F(z)=-1$ and since -1 and 1 lie symmetric to the imaginary axis, if z_ν be a zero point of $f(z)$, then $1/\bar{z}_\nu$ is a pole of $f(z)$. Since $f(z)$ is regular in $|z|<1$, all zero points of $f(z)$ lie in $|z|<1$. Let $\omega_1, \cdots, \omega_k$ be zero points of $f(z)$ in $|z|<1$, then $1/\bar{\omega}_1, \cdots, 1/\bar{\omega}_k$ are poles of $f(z)$, so that $f(z)$ is of the form:

$$f(z) = \varepsilon \prod_{\nu=1}^{k} \frac{z-\omega_\nu}{1-\bar{\omega}_\nu z} , \qquad |\varepsilon|=1 , \quad |\omega_\nu|<1 . \tag{5}$$

9. Conjugate function

Let $f(\theta)$ be a real function, which is integrable in $[0, 2\pi]$ and

$$F(z) = u(z) + iv(z) = \frac{1}{2\pi} \int_0^{2\pi} f(\varphi) \frac{e^{i\varphi}+z}{e^{i\varphi}-z} d\varphi , \qquad v(0)=0 , \tag{1}$$

$$z = re^{i\theta} , \quad 0 \leqq r < 1 ,$$

then

$$u(z) = \frac{1}{2\pi} \int_0^{2\pi} f(\varphi) \frac{1-r^2}{1-2r\cos(\varphi-\theta)+r^2} d\varphi = \frac{a_0}{2} + \sum_{n=1}^{\infty} (a_n \cos n\theta + b_n \sin n\theta) r^n ,$$

$$v(z) = -\frac{1}{\pi} \int_0^{2\pi} f(\varphi) \frac{r \sin(\varphi-\theta)}{1-2r\cos(\varphi-\theta)+r^2} d\varphi = \sum_{n=1}^{\infty} (-b_n \cos n\theta + a_n \sin n\theta) r^n .$$

Theorem IV. 26. *For almost all* $e^{i\theta}$, $\lim_{z \to e^{i\theta}} v(z) = v(e^{i\theta})$ $(\neq \infty)$ *exists uniformly, when* $z \to e^{i\theta}$ *from the inside of any fixed Stolz domain, whose vertex is at* $e^{i\theta}$.

$v(e^{i\theta}) = \bar{f}(\theta)$ is called *the conjugate function of* $f(\theta)$.

Proof. Since $f(\theta)$ can be expressed as a difference of two non-negative integrable functions, $u(z)$ can be expressed as a difference of two non-negative harmonic functions: $u(z) = u_1(z) - u_2(z)$, $u_1 \geqq 0$, $u_2 \geqq 0$.

Let $v_j(z)$ $(j=1,2)$ be the conjugate harmonic function of $u_j(z)$, then $v(z)=v_1(z)-v_2(z)$. Hence if we put $F_j(z)=u_j(z)+iv_j(z)$ $(j=1,2)$, then $F_j(z)$ is regular and $\Re(F(z))\geqq 0$ in $|z|<1$, so that by Theorem IV. 15, for almost all $e^{i\theta}$, $\lim_{z\to e^{i\theta}} F(z)=F_j(e^{i\theta})$ $(\neq\infty)$ exists uniformly, when $z\to e^{i\theta}$ from the inside of any fixed Stolz domain, whose vertex is at $e^{i\theta}$. Hence for almost all $e^{i\theta}$, $\lim_{z\to e^{i\theta}} v(z)=\lim_{z\to e^{i\theta}} v_1(z)-\lim_{z\to e^{i\theta}} v_2(z)$ exists uniformly, when $z\to e^{i\theta}$ from the inside of any fixed Stolz domain, whose vertex is at $e^{i\theta}$.

Theorem IV. 27. *If*

$$\lim_{t\to 0}\frac{1}{t}\int_0^t (f(\theta+t)-f(\theta-t))dt=0\,,$$

then

$$\lim_{r\to 1}\left(v(re^{i\theta})+\frac{1}{2\pi}\int_\varepsilon^\pi (f(\theta+t)-f(\theta-t))\cot\frac{t}{2}dt\right)=0\,,$$

$$\varepsilon=\sin^{-1}(1-r)\,.$$

Hence for almost all θ,

$$\bar{f}(\theta)=v(e^{i\theta})=-\frac{1}{2\pi}\int_0^\pi (f(\theta+t)-f(\theta-t))\cot\frac{t}{2}dt\,.$$

Plessner [1].

Proof. We put

$$\varphi(t)=f(\theta+t)-f(\theta-t)\,,\qquad \Phi(t)=\int_0^t\varphi(t)dt\,,\tag{1}$$

then by the hypothesis,

$$\lim_{t\to 0}\frac{\Phi(t)}{t}=0\,,\tag{2}$$

which holds for almost all θ in $[0, 2\pi]$ by Lebesgue's theorem.

$$v(re^{i\theta})=-\frac{1}{\pi}\int_{-\pi}^{\pi}\frac{r\sin t}{1-2r\cos t+r^2}f(\theta+t)dt=-\frac{1}{\pi}\int_0^\pi\frac{r\sin t}{1-2r\cos t+r^2}\varphi(t)dt$$

$$=-\frac{1}{\pi}\int_0^\varepsilon\quad-\frac{1}{\pi}\int_\varepsilon^\pi.$$

Since

$$\frac{r\sin t}{1-2r\cos t+r^2}=\frac{1}{2}\cot\frac{t}{2}-\frac{(1-r)^2\cot\dfrac{t}{2}}{2(1-2r\cos t+r^2)}\,,$$

if we put $\varepsilon=\sin^{-1}(1-r)$, then

$$v(re^{i\theta})+\frac{1}{2\pi}\int_\varepsilon^\pi\varphi(t)\cot\frac{t}{2}dt=$$

$$\frac{-1}{\pi}\int_0^\varepsilon\frac{r\sin t}{1-2r\cos t+r^2}\varphi(t)dt+\frac{(1-r)^2}{2\pi}\int_\varepsilon^\pi\frac{\cot\dfrac{t}{2}}{1-2r\cos t+r^2}\varphi(t)dt=-I+II.\tag{3}$$

Since $\frac{r\sin t}{1-2r\cos t+r^2}$ is an increasing function of t in $(0, \varepsilon)$[1], we have by the second mean value theorem[2],

$$I=\frac{r\sin\varepsilon}{\pi(1-2r\cos\varepsilon+r^2)}\int_\xi^\varepsilon\varphi(t)dt=\frac{r\sin\varepsilon}{\pi(1-2r\cos\varepsilon+r^2)}(\Phi(\varepsilon)-\Phi(\xi)),\quad 0<\xi<\varepsilon.$$

Since $1-2r\cos t+r^2=(1-r)^2+4r\sin^2 t/2\geqq 4r\sin^2 t/2$,

$$|I|\leqq\frac{\sin\varepsilon(|\Phi(\varepsilon)|+|\Phi(\xi)|)}{4\pi\sin^2\varepsilon/2}\leqq\frac{1}{2\pi}\left(\frac{|\Phi(\varepsilon)|}{\sin\varepsilon/2}+\frac{|\Phi(\xi)|}{\sin\xi/2}\right),$$

so that by (2)

$$\lim_{r\to1} I=0. \tag{4}$$

$$II=\frac{(1-r)^2}{2\pi}\left[\frac{\cot\frac{t}{2}\Phi(t)}{1-2r\cos t+r^2}\right]_\varepsilon^\pi-\frac{(1-r)^2}{2\pi}\int_\varepsilon^\pi\frac{d}{dt}\left(\frac{\cot\frac{t}{2}}{1-2r\cos t+r^2}\right)\Phi(t)dt$$

$$=-\frac{(1-r)^2\cot\frac{\varepsilon}{2}\Phi(\varepsilon)}{2\pi(1-2r\cos\varepsilon+r^2)}+$$

$$\frac{(1-r)^2}{4\pi}\int_\varepsilon^\pi\frac{1}{(1-2r\cos t+r^2)\sin^2 t/2}\left(1+\frac{2r\sin^2 t}{1-2r\cos t+r^2}\right)\Phi(t)dt$$

$$=II_1+II_2. \tag{5}$$

Since $1-2r\cos t+r^2\geqq(1-r)^2$, we have by (2)

$$|II_1|\leqq\frac{\cot\frac{\varepsilon}{2}|\Phi(\varepsilon)|}{2\pi}\to0,\qquad \varepsilon\to0. \tag{6}$$

$$1+\frac{2r\sin^2 t}{1-2r\cos t+r^2}\leqq1+\frac{\sin^2 t}{2\sin^2 t/2}=1+2\cos^2 t/2\leqq 3,$$

$(1-r)^2\leqq(1-r)(1-r^2)$, $\quad 1-r=\sin\varepsilon$, $\quad 1-r\leqq 2\sin\frac{\varepsilon}{2}$, so that

$$|II_2|\leqq\frac{3(1-r)}{4\pi\sin\varepsilon/2}\int_\varepsilon^\pi\frac{1-r^2}{1-2r\cos t+r^2}\left|\frac{\Phi(t)}{\sin t/2}\right|dt$$

$$\leqq 3\cdot\frac{1}{2\pi}\int_{-\pi}^\pi\frac{1-r^2}{1-2r\cos t+r^2}\left|\frac{\Phi(t)}{\sin t/2}\right|dt. \tag{7}$$

Since (7) is the Poisson integral of $F(t)=3\left|\frac{\Phi(t)}{\sin t/2}\right|$ and $F(0)=0$, we have by Schwarz's theorem,

---

1) $\frac{d}{dt}\left(\frac{\sin t}{1-2r\cos t+r^2}\right)=\frac{(1+r^2)\cos t-2r}{(1-2r\cos t+r^2)^2}$. If $0\leqq t\leqq\varepsilon$, then $\sin t\leqq 1-r\leqq\frac{1-r^2}{1+r^2}$, so that $\cos t\geqq\frac{2r}{1+r^2}$, hence $\frac{d}{dt}\left(\frac{\sin t}{1-2r\cos t+r^2}\right)\geqq0$.

2) If $\varphi(x)$ is an increasing function of x in $[a, b]$, then

$$\int_a^b\varphi(x)f(x)dx=\varphi(a)\int_0^\xi f(x)dx+\varphi(b)\int_\xi^b f(x)dx\quad(a<\xi<b).$$

$$\lim_{r\to 1} II_2 = 0 . \tag{8}$$

Hence by (3), (4), (5), (6), (8), the theorem is proved.——

From Theorem IV. 26 and 27, we have

Corollary. *Let $f(\theta)$ be integrable in $[0, 2\pi]$, then for almost all θ,*

$$\int_0^\pi \frac{f(\theta+t)-f(\theta-t)}{t} dt \ (\neq\infty) \text{ exists.}$$

Theorem IV. 28. *Let $f(z)$ be integrable on $|z|=1$ and put*

$$F(z)=\frac{1}{2\pi i}\int_{|\zeta|=1}\frac{f(\zeta)}{\zeta-z}\,d\zeta\,, \quad |z|<1 .$$

Then for almost all $e^{i\theta}$, $\lim_{z\to e^{i\theta}} F(z)=F(e^{i\theta})$ $(\neq\infty)$ exists uniformly, when $z\to e^{i\theta}$ from the inside of any fixed Stolz domain, whose vertex is at $e^{i\theta}$.

$F(z)$ is called *the Cauchy's integral with respect to $f(z)$.*

Proof. Let $f(e^{i\varphi})=f_1(\varphi)+if_2(\varphi)$, where f_1 and f_2 are real functions and $F(z)=u(z)+iv(z)$, $z=re^{i\theta}$, then

$$u(z)=\frac{1}{4\pi}\int_0^{2\pi} f_1(\varphi)\frac{1-r^2}{1-2r\cos(\varphi-\theta)+r^2}d\varphi$$

$$+\frac{1}{2\pi}\int_\theta^{2\pi} f_2(\varphi)\frac{r\sin(\varphi-\theta)}{1-2r\cos(\varphi-\theta)+r^2}d\varphi+\frac{1}{4\pi}\int_0^{2\pi} f_1(\varphi)d\varphi\,, \tag{1}$$

$$v(z)=-\frac{1}{2\pi}\int_0^{2\pi} f_1(\varphi)\frac{r\sin(\varphi-\theta)}{1-2r\cos(\varphi-\theta)+r^2}d\varphi$$

$$+\frac{1}{4\pi}\int_0^{2\pi} f_2(\varphi)\frac{1-r^2}{1-2r\cos(\varphi-\theta)+r^2}d\varphi+\frac{1}{4\pi}\int_0^{2\pi} f_2(\varphi)d\varphi\,. \tag{2}$$

Hence for almost all $e^{i\theta}$, $\lim_{z\to e^{i\theta}} F(z)=F(e^{i\theta})=u(e^{i\theta})+iv(e^{i\theta})$ exists uniformly, when $z\to e^{i\theta}$ from the inside of any fixed Stolz domain, whose vertex is at $e^{i\theta}$, where

$$u(e^{i\theta})=\frac{1}{4\pi}\int_0^\pi (f_2(\theta+t)-f_2(\theta-t))\cot\frac{t}{2}dt+\frac{1}{4\pi}\int_0^{2\pi} f_1(\varphi)d\varphi+\frac{f_1(\theta)}{2}\,, \tag{1*}$$

$$v(e^{i\theta})=-\frac{1}{4\pi}\int_0^\pi (f_1(\theta+t)-f_1(\theta-t))\cot\frac{t}{2}dt+\frac{1}{4\pi}\int_0^{2\pi} f_2(\varphi)d\varphi+\frac{f_2(\theta)}{2}\,. \tag{2*}$$

Hence

$$F(e^{i\theta})=\frac{1}{4\pi i}\int_0^\pi (f(e^{i(\theta+t)})-f(e^{i(\theta-t)}))\cot\frac{t}{2}dt+\frac{1}{4\pi}\int_0^{2\pi} f(e^{i\varphi})d\varphi+\frac{f(e^{i\theta})}{2}\,. \tag{3}$$

Theorem IV. 29. *If in Theorem* IV. 28., *$F(z)\equiv 0$ for $|z|>1$, i.e., if*

$$\int_{|\zeta|=1} f(\zeta)\,\zeta^n d\zeta=0 \quad (n=0, 1, 2, \cdots),$$

then $F(e^{i\theta})=f(e^{i\theta})$ *almost everywhere on* $|z|=1$.

Proof. Let $z=re^{i\theta}$ $(0\leqq r<1)$ and $z^*=1/\bar{z}$, then by the hypothesis,

$$\int_{|\zeta|=1}\frac{f(\zeta)}{\zeta-z^*}\,d\zeta=0\,,$$

so that

$$F(z)=\frac{1}{2\pi i}\int_{|\zeta|=1}f(\zeta)\Big(\frac{1}{\zeta-z}-\frac{1}{\zeta-z^*}\Big)\,d\zeta=\frac{1}{2\pi}\int_0^{2\pi}f(e^{i\varphi})\frac{1-r^2}{1-2r\cos(\varphi-\theta)+r^2}\,d\varphi,$$

hence $F(e^{i\theta})=f(e^{i\theta})$ almost everywhere on $|z|=1$.

Theorem IV. 30. *If* $f(\theta)\in\mathrm{Lip}\,\alpha$ $(0<\alpha<1)$[1], *then* $\bar{f}(\theta)\in\mathrm{Lip}\,\alpha$.

Privaloff [1].

Proof. Since $f\in\mathrm{Lip}\,\alpha$, we have for any θ,

$$\bar{f}(\theta)=-\frac{1}{2\pi}\int_0^{\pi}(f(\theta+t)-f(\theta-t))\cot\frac{t}{2}dt\,,$$

so that

$$\bar{f}(\theta)=-\frac{1}{2\pi}\int_{-\pi}^{\pi}(f(\theta+t)-f(\theta))\cot\frac{t}{2}dt\,,$$

$$\bar{f}(\theta+h)=-\frac{1}{2\pi}\int_{-\pi}^{\pi}(f(\theta+t)-f(\theta+h))\cot\frac{t-h}{2}dt\,.$$

Now

$$\bar{f}(\theta)=-\frac{1}{2\pi}\Big(\int_{-\pi}^{-2h}+\int_{2h}^{\pi}\Big)(f(\theta+t)-f(\theta))\cot\frac{t}{2}dt+R_1\,, \qquad (1)$$

$$\bar{f}(\theta+h)=-\frac{1}{2\pi}\Big(\int_{-\pi}^{-2h}+\int_{2h}^{\pi}\Big)(f(\theta+t)-f(\theta+h))\cot\frac{t-h}{2}dt+R_2\,, \qquad (2)$$

where

$$R_1=-\frac{1}{2\pi}\int_{-2h}^{2h}(f(\theta+t)-f(\theta))\cot\frac{t}{2}dt=O\Big(\int_{-2h}^{2h}|t|^{\alpha-1}dt\Big)=O(h^\alpha)\,,$$

$$R_2=-\frac{1}{2\pi}\int_{-2h}^{2h}(f(\theta+t)-f(\theta+h))\cot\frac{t-h}{2}dt=O\Big(\int_{-2h}^{2h}|t-h|^{\alpha-1}dt\Big)=O(h^\alpha)\,.$$

Hence if we write $f(\theta+t)-f(\theta+h)=f(\theta+t)-f(\theta)-(f(\theta+h)-f(\theta))$ in (2),

$$\begin{aligned}\bar{f}(\theta+h)-\bar{f}(\theta)&=-\frac{1}{2\pi}\Big(\int_{-\pi}^{-2h}+\int_{2h}^{\pi}\Big)(f(\theta+t)-f(\theta))\Big(\cot\frac{t-h}{2}-\cot\frac{t}{2}\Big)dt\\&\quad+\frac{1}{2\pi}(f(\theta+h)-f(\theta))\int_{2h}^{\pi}\Big(\cot\frac{t-h}{2}-\cot\frac{t+h}{2}\Big)dt+O(h^\alpha)\\&=I+II+O(h^\alpha)\,. \qquad (3)\end{aligned}$$

Since for $2h\leqq|t|\leqq\pi$,

1) i. e. $|f(\theta+h)-f(\theta)|\leqq$ const. $|h|^\alpha$ for any θ and h.

$$\cot\frac{t-h}{2}-\cot\frac{t}{2}=\sin\frac{h}{2}\Big/\sin\frac{t-h}{2}\sin\frac{t}{2}=O\left(\frac{h}{|t|^2}\right),$$

$$I=O\left(h\int_{2h}^{\pi}t^{\alpha-2}\,dt\right)=O(h^\alpha). \tag{4}$$

Since

$$\int_{2h}^{\pi}\left(\cot\frac{t-h}{2}-\cot\frac{t+h}{2}\right)dt=\left[2\log\frac{\sin(t-h)/2}{\sin(t+h)/2}\right]_{2h}^{\pi}=O(1),$$

$$II=O(f(\theta+h)-f(\theta))=O(h^\alpha). \tag{5}$$

Hence by (3), (4), (5),

$$\bar{f}(\theta+h)-\bar{f}(\theta)=O(h^\alpha).$$

Theorem IV. 31. *Let $F(z)=u(z)+iv(z)$ $(v(0)=0)$ be regular in $|z|<1$, then*

$$\int_0^{2\pi}|v(re^{i\theta})|^p d\theta\leqq M_p\int_0^{2\pi}|u(re^{i\theta})|^p d\theta,\quad 0\leqq r<1,\quad p>1,$$

where M_p is a constant, which depends on p only. M. Riesz [1].

For the proof, we use the following lemmas.

Lemma 1. *Let $g(\theta)$ be continuous in $[0,2\pi]$ and $g(0)=g(2\pi)$ and $1/p+1/q=1$, $p>1$, $q>1$. If for any $f(\theta)$, which is continuous in $[0,2\pi]$ and $f(0)=f(2\pi)$,*

$$\left|\int_0^{2\pi}f(\theta)g(\theta)d\theta\right|\leqq A\left(\int_0^{2\pi}|f(\theta)|^p d\theta\right)^{1/p},\qquad A=\text{const.}, \tag{$*$}$$

then

$$\int_0^{2\pi}|g(\theta)|^q d\theta\leqq A^q.$$

Proof. By taking $f\operatorname{sign}g$ instead of f, we may assume that $g\geqq 0$ in $[0,2\pi]$. If we take $f=g^{q/p}$, then by the hypothesis,

$$\int_0^{2\pi}g^q(\theta)d\theta=\int_0^{2\pi}g^{1+(q/p)}(\theta)d\theta=\int_0^{2\pi}f(\theta)g(\theta)d\theta\leqq A\left(\int_0^{2\pi}f^p(\theta)d\theta\right)^{1/p}$$
$$=A\left(\int_0^{2\pi}g^q(\theta)d\theta\right)^{1/p},$$

so that

$$\int_0^{2\pi}g^q(\theta)d\theta\leqq A^q.$$

Remark. Since $f(\theta)$, which satisfies the condition of the lemma, can be approximated by trigonometrical polynomials, if $(*)$ holds for any trigonometrical polynomials $f(\theta)$, then $\int_0^{2\pi}|g(\theta)|^q d\theta\leqq A^q$. We use this remark in the sequel.

Lemma 2. *If for any* $F(z)=u(z)+iv(z)$ $(v(0)=0)$, *which is regular in* $|z|<1$,

$$\int_0^{2\pi}|v(re^{i\theta})|^p d\theta \leqq M_p^p \int_0^{2\pi}|u(re^{i\theta})|^p d\theta\,, \qquad p>1\,,$$

then

$$\int_0^{2\pi}|v(re^{i\theta})|^q d\theta \leqq M_p^q \int_0^{2\pi}|u(re^{i\theta})|^q d\theta\,, \qquad 1/p+1/q=1\,.$$

Proof. Let $u_1(\theta)=\dfrac{\alpha_0}{2}+\sum_{\nu=1}^{n}(\alpha_\nu\cos\nu\theta+\beta_\nu\sin\nu\theta)$,

$$v_1(\theta)=\sum_{\nu=1}^{n}(-\beta_\nu\cos\nu\theta+\alpha_\nu\sin\nu\theta)\,,$$

then by Parseval's theorem, we have

$$\int_0^{2\pi}u_1(\theta)v(re^{i\theta})d\theta=-\int_0^{2\pi}u(re^{i\theta})v_1(\theta)d\theta\,,$$

so that by the hypothesis,

$$\left|\int_0^{2\pi}u_1 v\,d\theta\right|=\left|\int_0^{2\pi}u v_1 d\theta\right|\leqq\left(\int_0^{2\pi}|u|^q d\theta\right)^{1/q}\left(\int_0^{2\pi}|v_1|^p d\theta\right)^{1/p}$$

$$\leqq M_p\left(\int_0^{2\pi}|u|^q d\theta\right)^{1/q}\left(\int_0^{2\pi}|u_1|^p d\theta\right)^{1/p},$$

which holds for any trigonometrical polynomials u_1, hence by the remark of Lemma 1, if we take $A=M_p\left(\int_0^{2\pi}|u|^q d\theta\right)^{1/q}$, then we have

$$\int_0^{2\pi}|v|^q d\theta\leqq M_p^q\int_0^{2\pi}|u|^q d\theta\,.$$

Proof of the theorem. (Stein [1]).

First we assume that $1<p\leqq 2$.

Now at z, where $u(z)\neq 0$, $F(z)\neq 0$,

$$\Delta|u|^p=p(p-1)|u|^{p-2}|F'|^2\,,$$
$$\Delta|F|^p=p^2|F|^{p-2}|F'|^2\,, \qquad (1)$$

where Δ denotes the Laplacian.

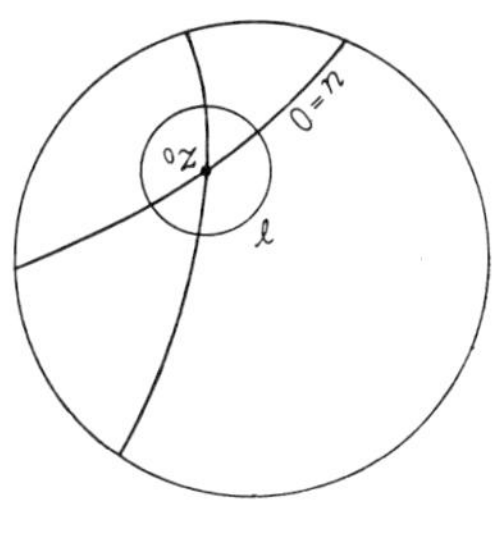

By the niveau curve $u=0$, the inside of $|z|<r$ is decomposed into a finite number of connected domains. Let z_0 be a multiple point of the curve $u=0$, then in a neighbourhood of z_0,

$$u(z)=\rho^k(a_k\cos k\theta+b_k\sin k\theta)+\cdots\,,$$
$$z-z_0=\rho e^{i\theta}\,, \qquad k\geqq 1\,,$$

so that on the circle $\gamma: |z-z_0|=\rho$,

$$\int_{\Gamma}\frac{\partial|u|^p}{\partial\rho}ds=p\int_0^{2\pi}|u|^{p-1}\frac{\partial|u|}{\partial\rho}\rho d\theta=O(\rho^{kp})\to 0\,,\qquad \rho\to 0\,. \tag{2}$$

If we apply the Green's formula: $\iint\Delta|u|^p dxdy=\int\frac{\partial|u|^p}{\partial\nu}ds$ $(z=x+iy)$ for the domain, which is obtained from $|z|<r$ by taking off the discs about multiple points of the curve $u=0$ and then make the radii of these discs tend to zero, we have by (2),

$$\int_{|z|=r}\frac{\partial|u|^p}{\partial r}rd\theta=\iint_{|z|<r}\Delta|u|^p dxdy=p(p-1)\iint_{|z|<r}|u|^{p-2}|F'|^2 dxdy\,. \tag{3}$$

Similarly we have

$$\int_{|z|=r}\frac{\partial|F|^p}{\partial r}rd\theta=\iint_{|z|<r}\Delta|F|^p dxdy=p^2\iint_{|z|<r}|F|^{p-2}|F'|^2 dxdy\,. \tag{4}$$

Since $|F|\geqq|u|$, $p-2\leqq 0$, we have $|F|^{p-2}\leqq|u|^{p-2}$, so that

$$\int_{|z|=r}\frac{\partial|F|^p}{\partial r}rd\theta\leqq\frac{p^2}{p(p-1)}\int_{|z|=r}\frac{\partial|u|^p}{\partial r}rd\theta\,,$$

$$\int_0^{2\pi}\frac{\partial|F|^p}{\partial r}d\theta\leqq\frac{p}{p-1}\int_0^{2\pi}\frac{\partial|u|^p}{\partial r}d\theta\,,$$

hence integrating by r, we have

$$\int_0^{2\pi}|F(re^{i\theta})|^p d\theta-2\pi|F(0)|^p\leqq\frac{p}{p-1}\int_0^{2\pi}|u(re^{i\theta})|^p d\theta-\frac{2\pi p|u(0)|^p}{p-1}\,.$$

Since $F(0)=u(0)$,

$$\int_0^{2\pi}|v(re^{i\theta})|^p d\theta\leqq\int_0^{2\pi}|F(re^{i\theta})|^p d\theta\leqq\frac{p}{p-1}\int_0^{2\pi}|u(re^{i\theta})|^p d\theta-2\pi|u(0)|^p\left(\frac{p}{p-1}-1\right)$$

$$\leqq\frac{p}{p-1}\int_0^{2\pi}|u(re^{i\theta})|^p d\theta\qquad(1<p\leqq 2)\,. \tag{5}$$

Hence the case $1<p\leqq 2$ is proved. If $p>2$, then we choose $q>1$, by the condition $1/p+1/q=1$, then $1<q<2$, so that

$$\int_0^{2\pi}|v(re^{i\theta})|^q d\theta\leqq\frac{q}{q-1}\int_0^{2\pi}|u(re^{i\theta})|^q d\theta=p\int_0^{2\pi}|u(re^{i\theta})|^q d\theta\,,$$

hence by Lemma 2,

$$\int_0^{2\pi}|v(re^{i\theta})|^p d\theta\leqq p^{p/q}\int_0^{2\pi}|u(re^{i\theta})|^p d\theta\quad(p>2)\,. \tag{6}$$

Theorem IV. 32. *If $f(\theta)\in L^p$ $(p>1)$, then $\bar{f}(\theta)\in L^p$.*

Proof. Let $F(z)=u(z)+iv(z)$ $(v(0)=0)$ and

$$u(z)=\frac{1}{2\pi}\int_0^{2\pi}f(\varphi)\frac{1-r^2}{1-2r\cos(\varphi-\theta)+r^2}d\varphi\,,\qquad z=re^{i\theta}\,,$$

then by Hölder's inequality,

$$\int_0^{2\pi} |u(re^{i\theta})|^p \, d\theta \leqq \int_0^{2\pi} |f(\theta)|^p d\theta \,,$$

so that

$$\int_0^{2\pi} |v(re^{i\theta})|^p d\theta \leqq M_p \int_0^{2\pi} |u(re^{i\theta})|^p d\theta \leqq M_p \int_0^{2\pi} |f(\theta)|^p d\theta \,,$$

hence by Fatou's lemma,

$$\int_0^{2\pi} |\bar{f}(\theta)|^p d\theta = \int_0^{2\pi} |v(e^{i\theta})|^p d\theta \leqq M_p \int_0^{2\pi} |f(\theta)|^p d\theta \,.$$

10. Littlewood's theorem on subharmonic functions in $|z|<1$

1. First we shall prove two lemmas.

Lemma 1. *Let* $\frac{1}{2} \leqq r < 1$, $a = \rho e^{i\varphi}$ $(0 \leqq \rho < 1)$, *then*

$$\log \left| \frac{1-\bar{a}r}{r-a} \right| \leqq \begin{cases} \dfrac{A(1-\rho)(1-r)}{\varphi^2} & \text{in general,} \\[2mm] \dfrac{A(1-\rho)(1-r)}{(1-r)^2+\varphi^2}\,, & \text{if } \dfrac{1+r}{2} \leqq \rho < 1\,, \\[2mm] \dfrac{A(1-\rho)}{1-r} \log \dfrac{1-r}{|\varphi|}\,, & \text{if } \rho \leqq \dfrac{1+r}{2}\,, \quad |\varphi| \leqq \dfrac{1-r}{2}\,, \end{cases}$$

where $A > 0$ *is a constant.*

Proof. In the proof, we denote a constant by the same letter A. Let

$$J = \log \left| \frac{1-\bar{a}r}{r-a} \right|, \tag{1}$$

then

$$J = \frac{1}{2} \log \left| \frac{1-\bar{a}r}{r-a} \right|^2 = \frac{1}{2} \log \left(1 + \frac{(1-\rho^2)(1-r^2)}{|r-a|^2} \right) \leqq \frac{(1-\rho^2)(1-r^2)}{2|r-a|^2}$$

$$\leqq \frac{2(1-\rho)(1-r)}{(r-\rho)^2+4\rho r \sin^2 \varphi/2} \leqq \frac{A(1-\rho)(1-r)}{(r-\rho)^2+\varphi^2} \leqq \frac{A(1-\rho)(1-r)}{\varphi^2}\,. \tag{2}$$

If $\frac{1+r}{2} \leqq \rho < 1$, then $\rho - r \geqq \frac{1-r}{2}$, so that

$$J \leqq \frac{A(1-\rho)(1-r)}{(1-r)^2+\varphi^2}\,. \tag{3}$$

If $\rho \leqq \frac{1+r}{2}$, $|\varphi| \leqq \frac{1-r}{2}$, then since $\frac{1-\rho}{1-r} \geqq \frac{1}{2}$, $\frac{1-r}{|\varphi|} \geqq 2$,

$$J \leqq \frac{1}{2} \log\left(1+\frac{A(1-\rho)(1-r)}{\varphi^2}\right)=\frac{1}{2} \log\left(1+\frac{A(1-\rho)}{1-r}\left(\frac{1-r}{\varphi}\right)^2\right)$$
$$\leqq \frac{A(1-\rho)}{1-r} \log \frac{1-r}{|\varphi|} . \tag{4}$$

Lemma 2. *Let* $\lambda(\theta) \geqq 0$ *be an increasing function of* θ *in* $0 \leqq \theta \leqq \pi$, $\lambda(0)=0$, *such that* $\overline{\lim}_{\theta\to 0} \frac{\lambda(\theta)}{\theta} \leqq m$, *then*

$$\overline{\lim_{\delta\to 0}}\, \delta \int_\delta^\pi \frac{d\lambda(\theta)}{\theta^2} \leqq 2m , \quad \overline{\lim_{\delta\to 0}}\, \delta \int_0^\delta \frac{d\lambda(\theta)}{\delta^2+\theta^2} \leqq \frac{\pi}{4} m , \quad \overline{\lim_{\delta\to 0}}\, \frac{1}{\delta} \int_0^\delta \log \frac{\delta}{\theta} d\lambda(\theta) \leqq m .$$

Proof.

$$\delta \int_\delta^\pi \frac{d\lambda(\theta)}{\theta^2}=\delta\left[\frac{\lambda(\theta)}{\theta^2}\right]_\delta^\pi+2\delta\int_\delta^\pi \frac{\lambda(\theta)}{\theta^3} d\theta \leqq O(\delta)+2\delta\int_\delta^\pi \frac{m\theta+o(\theta)}{\theta^3} d\theta=2m+o(1) .$$

$$\delta \int_0^\delta \frac{d\lambda(\theta)}{\delta^2+\theta^2}=\delta\left[\frac{\lambda(\theta)}{\delta^2+\theta^2}\right]_0^\delta+2\delta\int_0^\delta \frac{\theta\lambda(\theta)}{(\delta^2+\theta^2)^2} d\theta \leqq \frac{\lambda(\delta)}{2\delta}+2\delta\int_0^\delta \frac{m\theta^2+o(\theta^2)}{(\delta^2+\theta^2)^2} d\theta$$
$$\leqq \frac{m}{2}+o(1)+2\int_0^1 \frac{m+o(1)}{(1+t^2)^2} t^2 dt=\frac{\pi}{4} m+o(1) .$$

$$\frac{1}{\delta}\int_0^\delta \log \frac{\delta}{\theta} d\lambda(\theta)=\frac{1}{\delta}\left[\log \frac{\delta}{\theta} \lambda(\theta)\right]_0^\delta+\frac{1}{\delta}\int_0^\delta \frac{\lambda(\theta)}{\theta} d\theta \leqq \frac{1}{\delta}\int_0^\delta \frac{m\theta+o(\theta)}{\theta} d\theta$$
$$=m+o(1) .$$

2. Now we shall prove Littlewood's theorem.

Theorem IV. 33. *Let*

$$w(z)=\int_{|a|<1} \log\left|\frac{1-\bar{a}z}{z-a}\right| d\mu(a) ,$$

where μ *is a positive mass distribution in* $|z|<1$, *such that*

$$\int_{|a|<1} (1-|a|) d\mu(a) < \infty .$$

Then

$$\lim_{r\to 1} w(re^{i\theta})=0 \quad \textit{for almost all } e^{i\theta},$$

and

$$\lim_{r\to 1} \int_0^{2\pi} w(re^{i\theta}) d\theta=0 .$$ Littlewood [4].

Proof. We put $d\sigma(a)=(1-|a|)d\mu(a)$, then $\int_{|a|<1} d\sigma(a)<\infty$ and

$$w(z)=\int_{|a|<1} \log\left|\frac{1-\bar{a}z}{z-a}\right| \frac{d\sigma(a)}{1-|a|} . \tag{1}$$

We choose $0<\rho_1<\rho_2<\cdots<\rho_\nu<\cdots<1$, such that

$$\int_{\rho_\nu \leqq |a|<1} d\sigma(a) \leqq 1/4^\nu \qquad (\nu=1, 2, \cdots) . \tag{2}$$

Let $\lambda_\nu(\theta)$ be the σ-mass, contained in a domain: $-\pi<\arg z<\theta$, $\rho_\nu\leqq|z|<1$, then $\lambda_\nu(\theta)$ is an increasing function of θ, so that $\lambda_\nu'(\theta)$ exists almost everywhere and by (2),

$$\int_{-\pi}^{\pi}\lambda_\nu'(\theta)d\theta\leqq\lambda_\nu(\pi)\leqq 1/4^\nu,$$

so that if E_ν^* be the set of $e^{i\theta}$, such that $\lambda_\nu'(\theta)$ does not exist, or if it exists, $\lambda_\nu'(\theta)\geqq 1/2^\nu$, then $mE_\nu^*\leqq 1/2^\nu$, so that if we put $E^*=\lim\limits_{\nu}(E_\nu^*+E_{\nu+1}^*+\cdots)$, then $mE^*=0$, hence if we denote its complement $C(E^*)$ on $|z|=1$ by E, then $mE=2\pi$.

If $e^{i\theta}\in E$, then $e^{i\theta}\in C(E_\nu^*)$ $(\nu\geqq\nu_0)$, so that

$$\lambda_\nu'(\theta)\leqq 1/2^\nu \qquad (\nu\geqq\nu_0)\,. \tag{3}$$

We shall prove that if $e^{i\theta}\in E$, then

$$\lim_{r\to1} w(re^{i\theta})=0\,. \tag{4}$$

Suppose that $z=1$ belongs to E and we shall prove that $\lim\limits_{r\to1}w(r)=0$. Let $\rho_\nu<r<1$ and we define four domains $\varDelta_i$ $(i=1,2,3,4)$ as follows.

$\varDelta_1$: $|z|\leqq\rho_\nu$, $\qquad\varDelta_2$: $\rho_\nu\leqq|z|<1$, $\dfrac{1-r}{2}\leqq|\arg z|\leqq\pi$,

$\varDelta_3$: $\rho_\nu\leqq|z|\leqq\dfrac{1+r}{2}$, $|\arg z|\leqq\dfrac{1-r}{2}$, $\qquad\varDelta_4$: $\dfrac{1+r}{2}\leqq|z|<1$, $|\arg z|\leqq\dfrac{1-r}{2}$.

Then

$$w(r)\leqq\int_{\Delta_1}\log\left|\frac{1-\bar{a}r}{r-a}\right|\cdot\frac{d\sigma(a)}{1-|a|}+\int_{\Delta_2}+\int_{\Delta_3}+\int_{\Delta_4}$$
$$=I+II+III+IV\,, \tag{5}$$

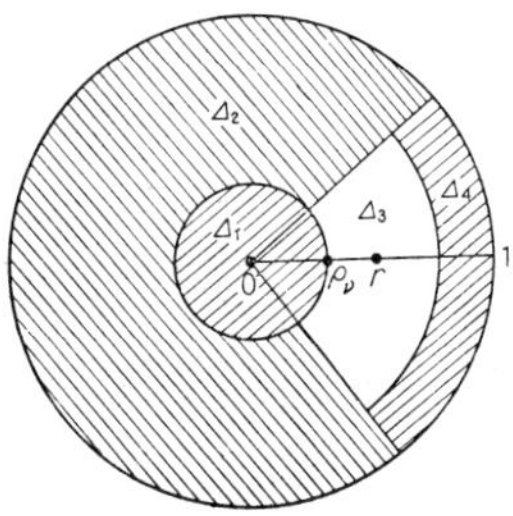

where

$$\lim_{r\to1} I=0\,.$$

We put $\chi_\nu(\theta)=\lambda_\nu(\theta)-\lambda_\nu(0)$, then since $\lambda_\nu'(0)$ exists, the positive real axis does not contain a positive σ-mass, so that $\chi_\nu(0)=0$, $\lim\limits_{\theta\to0}\chi_\nu(\theta)=0$ and

$$\chi_\nu'(0)=\lambda_\nu'(0)\leqq 1/2^\nu \qquad (\nu\geqq\nu_0)\,. \tag{6}$$

By Lemma 1,

$$II\leqq\text{const.}\,(1-r)\int_{\frac{1-r}{2}\leqq|\theta|\leqq\pi}\frac{d\chi_\nu(\theta)}{\theta^2},$$

so that by Lemma 2,

$$\overline{\lim_{r\to1}}\, II\leqq\text{const.}\,\chi_\nu'(0)\leqq\text{const.}\,1/2^\nu \qquad (\nu\geqq\nu_0)\,.$$

Similarly

$$\overline{\lim_{r\to 1}}\, III \leqq \text{const.}\ \overline{\lim_{r\to 1}}\, \frac{1}{1-r}\int_{-\frac{1-r}{2}}^{\frac{1-r}{2}} \log\frac{1-r}{|\theta|}\, d\chi_\nu(\theta) \leqq \text{const.}\ \chi_\nu'(0) \leqq \text{const.}\ 1/2^\nu\,,$$

$$\overline{\lim_{r\to 1}}\, IV \leqq \text{const.}\ \overline{\lim_{r\to 1}}\,(1-r)\int_{-\frac{1-r}{2}}^{\frac{1-r}{2}} \frac{d\chi_\nu(\theta)}{(1-r)^2+\theta^2} \leqq \text{const.}\ \chi_\nu'(0) \leqq \text{const.}\ 1/2^\nu\,.$$

Hence

$$\overline{\lim_{r\to 1}}\, w(r) \leqq \text{const.}\ 1/2^\nu \qquad (\nu \geqq \nu_0)\,.$$

Since ν is arbitrary, we have

$$\lim_{r\to 1} w(r) = 0. \tag{7}$$

If we put

$$\Omega(r) = \int_{|a|<r} d\mu(a)\,, \tag{8}$$

then

$$\int_{|a|<r} (1-|a|)\, d\mu(a) = \Omega(+0) + \int_{+0}^{r} (1-t)\, d\Omega(t) = (1-r)\Omega(r) + \int_0^r \Omega(t)\,dt \geqq \int_0^r \Omega(t)\,dt\,,$$

so that

$$\int_0^1 \Omega(t)\,dt < \infty\,, \text{ hence } \lim_{r\to 1}(1-r)\Omega(r) = 0\,. \tag{9}$$

By

$$\frac{1}{2\pi}\int_0^{2\pi} \log|re^{i\theta}-a|\,d\theta = \text{Max}\,(\log r,\ \log|a|)\,,$$

$$\int_0^{2\pi} \log|1-\bar{a}re^{i\theta}|\,d\theta = 0 \qquad (|a|<1,\ r<1)\,, \tag{10}$$

we have

$$\frac{1}{2\pi}\int_0^{2\pi} w(re^{i\theta})\,d\theta = -\int_{|a|<1} d\mu(a)\frac{1}{2\pi}\int_0^{2\pi} \log|re^{i\theta}-a|\,d\theta =$$

$$-\log r\ \Omega(+0) - \int_{+0}^{1} \text{Max}\,(\log r,\ \log t)\,d\Omega(t) = \int_r^1 \frac{\Omega(t)\,dt}{t}\,,$$

so that by (9),

$$\lim_{r\to 1}\int_0^{2\pi} w(re^{i\theta})\,d\theta = 0\,. \tag{11}$$

Theorem IV. 34. *Let $u(z)$ be subharmonic in $|z|<1$, such that*

$$\int_0^{2\pi} |u(re^{i\theta})|\,d\theta = O(1)\,, \qquad 0 \leqq r < 1\,,$$

then

$$u(z) = v(z) - w(z),$$

where $v(z)$ is harmonic in $|z|<1$, such that

$$\int_0^{2\pi} |v(re^{i\theta})|\,d\theta = O(1)\,, \qquad 0 \leqq r < 1\,,$$

and

$$w(z)=\int_{|a|<1}\log\left|\frac{1-\bar{a}z}{z-a}\right|d\mu(a)\,,$$

where μ is a positive mass distribution in $|z|<1$, such that

$$\int_{|a|<1}(1-|a|)d\mu(a)<\infty\,.$$

Hence for almost all $e^{i\theta}$,

$$\lim_{r\to1}u(re^{i\theta})=u(e^{i\theta})\ (\neq\infty)\ \textit{exists.}$$

$v(z)$ is the least harmonic majorant of u, such that if $v_\rho^*(z)$ be harmonic in $|z|<\rho<1$, such that $v_\rho^*=u$ on $|z|=\rho$, then

$$\lim_{\rho\to1}v_\rho^*(z)=v(z)\,.$$ Littlewood [3].

Proof. By F. Riesz' decomposition theorem, for $|z|<\rho<1$,

$$u(z)=v_\rho(z)-w_\rho(z)\,, \tag{1}$$

where $v_\rho(z)$ is harmonic in $|z|<\rho$ and

$$w_\rho(z)=\int_{|a|<\rho}\log\left|\frac{1-\bar{a}z}{z-a}\right|d\mu(a)\,, \tag{2}$$

where μ is a positive mass distribution in $|z|<1$. Let

$$\Omega(r)=\int_{|a|<r}d\mu(a)\,, \tag{3}$$

then by (10) (p. 172), we have for $r<\rho$,

$$\frac{1}{2\pi}\int_0^{2\pi}w_\rho(re^{i\theta})d\theta=-\int_{|a|<\rho}d\mu(a)\frac{1}{2\pi}\int_0^{2\pi}\log|re^{i\theta}-a|d\theta=-\log r\,\Omega(+0)$$
$$-\int_{+0}^{\rho}\mathrm{Max}(\log r,\log t)d\Omega(t)=-\log\rho\cdot\Omega(\rho)+\int_r^\rho\frac{\Omega(t)dt}{t}\qquad(r<\rho)\,. \tag{4}$$

Hence if we put

$$L(r,u)=\frac{1}{2\pi}\int_0^{2\pi}u(re^{i\theta})d\theta\,, \tag{5}$$

then

$$L(r,u)=v_\rho(0)+\log\rho\cdot\Omega(\rho)-\int_r^\rho\frac{\Omega(t)dt}{t}\,,$$

so that except at most a countable set of r, where $\Omega(r)$ is discontinuous,

$$rL'(r,u)=\Omega(r)\,. \tag{6}$$

By the hypothesis, $L(r,u)=O(1)$ and since $L(r,u)$ is an increasing function of r,

$$\int_{r_0}^{r}L'(r,u)dr\leqq L(r)-L(r_0)=O(1)\,,\qquad 0<r_0\leqq r<1\,,$$

so that

$$\int_0^1 \Omega(r)dr<\infty\,, \qquad \text{hence} \quad \lim_{r\to 1}(1-r)\Omega(r)=0\,. \tag{7}$$

Hence if we put $d\sigma(a)=(1-|a|)d\mu(a)$, then by (7), we can prove that

$$\int_{|a|<1} d\sigma(a)<\infty.$$

Since $\log\left|\dfrac{1-\bar{a}z}{z-a}\right|\leqq\dfrac{2(1-|a|)(1-|z|)}{|z-a|^2}$, we have

$$w_\rho(z)=\int_{|a|<\rho}\log\left|\frac{1-\bar{a}z}{z-a}\right|\frac{d\sigma(a)}{1-|a|}\leqq 2(1-|z|)\int_{|a|<\rho}\frac{d\sigma(a)}{|z-a|^2}\,, \tag{8}$$

so that we can make $\rho\to 1$ in (1) and we have

$$u(z)=v(z)-w(z)\,, \tag{9}$$

where $v(z)$ is harmonic in $|z|<1$ and

$$w(z)=\int_{|a|<1}\log\left|\frac{1-\bar{a}z}{z-a}\right|d\mu(a)\,. \tag{10}$$

Hence by Theorem IV. 33, $\lim_{r\to 1} w(re^{i\theta})=0$ for almost all $e^{i\theta}$, and

$$\lim_{r\to 1}\int_0^{2\pi} w(re^{i\theta})d\theta=0\,. \tag{11}$$

Let $v_\rho^*(z)$ be harmonic in $|z|<\rho<1$, such that $v_\rho^*=u$ on $|z|=\rho$ and put $\omega_\rho(z)=v(z)-v_\rho^*(z)$, then $\omega_\rho(z)$ is harmonic in $|z|<\rho$ and $\omega_\rho=w$ on $|z|=\rho$, so that by means of Poisson integral, we have for $r<\rho$,

$$|\omega_\rho(re^{i\theta})|\leqq\frac{1}{2\pi}\cdot\frac{\rho+r}{\rho-r}\int_0^{2\pi}|\omega_\rho(\rho e^{i\theta})|d\theta=\frac{1}{2\pi}\cdot\frac{\rho+r}{\rho-r}\int_0^{2\pi}w(\rho e^{i\theta})d\theta\to 0\,, \quad \rho\to 1\,,$$

hence $\lim_{\rho\to 1}\omega_\rho(z)=0$, so that

$$\lim_{\rho\to 1} v_\rho^*(z)=v(z)\,. \tag{12}$$

Since for $r<\rho$,

$$\int_0^{2\pi}|v_\rho^*(re^{i\theta})|d\theta\leqq\int_0^{2\pi}|v_\rho^*(\rho e^{i\theta})|d\theta=\int_0^{2\pi}|u(\rho e^{i\theta})|d\theta=O(1)\,,$$

$$\int_0^{2\pi}|v(re^{i\theta})|d\theta=O(1)\,, \quad 0\leqq r<1\,. \tag{13}$$

Hence by Theorem IV. 16, for almost all $e^{i\theta}$, $\lim_{z\to e^{i\theta}} v(z)=v(e^{i\theta})$ $(\neq\infty)$ exists uniformly, when $z\to e^{i\theta}$ from the inside of any Stolz domain, whose vertex is at $e^{i\theta}$. Hence for almost all $e^{i\theta}$,

$$\lim_{r\to 1} u(re^{i\theta})=\lim_{r\to 1} v(re^{i\theta})-\lim_{r\to 1} w(re^{i\theta})=v(e^{i\theta}).$$

3. In Theorem IV. 34, the radial approach can not be replaced by the approach from the inside of a Stolz domain, since there exists a

bounded subharmonic function $u(z)$ in $|z|<1$, such that for almost all $e^{i\theta}$, $\lim_{z\to e^{i\theta}} u(z)$ does not exist, when $z\to e^{i\theta}$ from the inside of any Stolz domain $\Delta(e^{i\theta})$, whose vertex is at $e^{i\theta}$. We shall construct such $u(z)$ as follows.

Let $r_\nu=1-1/2^\nu$ $(\nu=1,2,\cdots)$ and $z_\nu^j=r_\nu e^{\frac{2\pi j}{4^\nu}i}$ $(j=0,1,2,\cdots,4^\nu-1)$. We define a positive mass distribution $d\mu(a)$ in $|z|<1$, such that $d\mu(z_\nu^j)=1/16^\nu$ and $d\mu(a)=0$ elsewhere, then $\int_{|a|<1} d\mu(a)=\sum_{\nu=1}^{\infty} 1/4^\nu<\infty$.

Let

$$w(z)=-\int_{|a|<1}\log\left|\frac{1-\bar{a}z}{z-a}\right|d\mu(a),$$

then $w(z)$ is subharmonic in $|z|<1$ and $w(z_\nu^j)=-\infty$, so that if we put $u(z)=\operatorname{Max}(-1, w(z))$, then $u(z)$ is subharmonic and $-1\leqq u\leqq 0$ in $|z|<1$ and $u(z_\nu^j)=-1$. By Theorem IV. 34, $\lim_{r\to 1} w(re^{i\theta})=0$ for almost all $e^{i\theta}$, so that $\lim_{r\to 1} u(re^{i\theta})=0$ for almost all $e^{i\theta}$. Let $e^{i\theta}$ be such a point, then since there are infinitely many z_ν^j in $\Delta(e^{i\theta})$, such that $u(z_\nu^j)=-1$, $\lim_{z\to e^{i\theta}} u(z)$ does not exist, when $z\to e^{i\theta}$ from the inside of $\Delta(e^{i\theta})$. ——

There arises a question, in Theorem IV. 33, whether $\lim_{z\to e^{i\theta}} w(z)=0$, if $z\to e^{i\theta}$ along a line l_ψ through $e^{i\theta}$, making an angle ψ $\left(-\frac{\pi}{2}<\psi<\frac{\pi}{2}\right)$ with the inner normal of $|z|=1$ at $e^{i\theta}$, except a null set of ψ. This is proved by Huber (Compositio Math. 13), but for the sake of brevity, we assume that $\Omega(r)=O(1)$ and shall prove

Theorem IV. 35. *Let*

$$w(z)=\int_{|a|<1}\log\left|\frac{1-\bar{a}z}{z-a}\right|d\mu(a), \qquad \int_{|a|<1} d\mu(a)<\infty.$$

Let $l_\psi(e^{i\theta})$ be a line through $e^{i\theta}$, making an angle ψ $\left(-\frac{\pi}{2}<\psi<\frac{\pi}{2}\right)$ with the inner normal of $|z|=1$ at $e^{i\theta}$, then there exists a set E of measure 2π on $|z|=1$, such that if $e^{i\theta}\in E$, then for almost all ψ,

$$\lim_{z\to e^{i\theta}} w(z)=0, \qquad \textit{when } z\to e^{i\theta} \textit{ along } l_\psi(e^{i\theta}).$$

Proof. (Tsuji [54]). We put $d\sigma(a)=(1-|a|)d\mu(a)$, then since $\Omega(r)=\int_{|a|<r} d\mu(a)=O(1)$, $\int_{r\leqq|a|<1} d\sigma(a)=O(1-r)$, so that if $r_\nu=1-1/2^\nu$ $(\nu=1,2,\cdots)$, then $\int_{r_\nu\leqq|a|<1} d\sigma(a)\leqq \text{const.}\,1/2^\nu$. We decompose the circular ring: $r_\nu\leqq|z|<1$ into 2^ν equal domains ω_ν^j: $r_\nu\leqq|z|<1$, $\frac{2\pi}{2^\nu}j\leqq\arg z\leqq\frac{2\pi}{2^\nu}(j+1)$, $(j=0,1,2,\cdots,2^\nu-1)$.

Then $$\sum_{j=0}^{2^\nu-1}\sigma(\omega_\nu^j)\leqq \text{const.}\,1/2^\nu.$$

Let N be the number of ω_ν^j, such that $\sigma(\omega_\nu^j)\geqq 1/2^{\nu(1+\delta)}$ $(0<\delta<1)$, then $N\leqq$ const. $2^{\nu\delta}$. We join to such ω_ν^j two adjacent ω_ν^{j-1}, ω_ν^{j+1} and put $\omega_\nu^{*j}=\omega_\nu^{j-1}+\omega^j+\omega_\nu^{j+1}$. Let E_ν^* be the set of arcs on $|z|=1$, which belong to the boundary of $\{\omega_\nu^{*j}\}$, then $mE_\nu^*\leqq\frac{2\pi}{2^\nu}\cdot 3N\leqq$ const. $1/2^{\kappa\nu}$ $(\kappa=1-\delta>0)$.

Hence if we put $E^*=\lim_\nu(E_\nu^*+E_{\nu+1}^*+\cdots)$, then $mE^*=0$, so that if we denote its complement $C(E^*)$ on $|z|=1$ by E, then $mE=2\pi$. If $e^{i\theta}\in E$, then $e^{i\theta}\in C(E_\nu^*)$ $(\nu\geqq\nu_0)$, so that the domain

$$\omega_\nu(\theta):\ \ r_\nu<|z|<1\,,\ \ \theta-\frac{\pi}{2^\nu}\leqq\arg z\leqq\theta+\frac{\pi}{2^\nu}$$

is contained in $\omega_\nu^j+\omega_\nu^{j+1}$ for a certain j, such that $\sigma(\omega^j)\leqq 1/2^{\nu(1+\delta)}$, $\sigma(\omega_\nu^{j+1})\leqq 1/2^{\nu(1+\delta)}$, hence

$$\sigma(\omega_\nu(\theta))\leqq 2/2^{\nu(1+\delta)}\qquad(0<\delta<1)\,,\quad(\nu\geqq\nu_0)\,. \tag{1}$$

We assume that $z=1$ belongs to E and let l_ψ be a line through $z=1$, making an angle ψ $\left(-\frac{\pi}{2}<\psi<\frac{\pi}{2}\right)$ with the inner normal of $|z|=1$ at $z=1$ and we shall prove that $\lim_{z\to1}w(z)=0$, when $z\to1$ along l_ψ for almost all ψ.

Let Δ_ν be the common part of $|z|<1$ and $|z-1|<1/2^\nu$ $(\nu=1,2,\cdots)$, then by (1),

$$\sigma(\Delta_\nu)\leqq 2/2^{\nu(1+\delta)}\qquad(0<\delta<1)\,,\quad(\nu\geqq\nu_0)\,. \tag{2}$$

Let $0<\psi_0<\pi/2$ be fixed and $\lambda_\nu(\psi)$ $(-\psi_0<\psi<\psi_0)$ be the σ-mass, contained in a sector, which is bounded by a circle $|z-1|=1/2^\nu$ and two lines $l_{-\psi_0}$, l_ψ.

Since $\lambda_\nu(\psi)$ is an increasing function of ψ, $\lambda_\nu'(\psi)$ exists almost everywhere and

$$\int_{-\psi_0}^{\psi_0}\lambda_\nu'(\psi)d\psi\leqq\lambda_\nu(\psi_0)\leqq\sigma(\Delta_\nu)\leqq 2/2^{\nu(1+\delta)}\,. \tag{3}$$

Let e_ν^* be the set of ψ, such that $\lambda_\nu'(\psi)$ does not exist, or if it exists $\lambda_\nu'(\psi)\geqq 1/2^\nu$, then by (3), $me_\nu^*\leqq 2/2^{\nu\delta}$, so that if we put $e^*=\lim_\nu(e_\nu^*+e_{\nu+1}^*+\cdots)$, then $me^*=0$. Let e be its complement with respect to $(-\psi_0,\psi_0)$, then we shall prove that if $\psi_1\in e$, then

$$\lim_{z\to1}w(z)=0\,,\quad\text{when } z\to1 \text{ along } l_{\psi_1}. \tag{4}$$

Let $z\in l_{\psi_1}$ and $1/2^\nu\geqq|z-1|\geqq 1/2^{\nu+1}$. If we denote the complement of Δ_{ν_0} with respect to $|z|<1$ by $\Delta_{\nu_0}^*$ and $\Delta_{\nu,\nu'}=\Delta_\nu-\Delta_{\nu'}$ $(\nu<\nu')$, then

$$\begin{aligned}w(z)&=\int_{\Delta_{\nu_0}^*}\log\left|\frac{1-\bar a z}{z-a}\right|\frac{d\sigma(a)}{1-|a|}+\int_{\Delta_{\nu_0,\nu-1}}+\int_{\Delta_{\nu-1,\nu+2}}+\int_{\Delta_{\nu+2}}\\&=I+II+III+IV\,.\end{aligned} \tag{5}$$

First we shall evaluate III. Let $\Delta'_{\nu-1,\nu+2}$ be the part of $\Delta_{\nu-1,\nu+2}$, which lies between $l_{-\psi_0}$, l_{ψ_0} and $\Delta''_{\nu-1,\nu+2}$ be its complement with respect to $\Delta_{\nu-1,\nu-2}$, then

$$III=\int_{\Delta'_{\nu-1,\nu+2}}\log\left|\frac{1-\bar{a}z}{z-a}\right|\frac{d\sigma(a)}{1-|a|}+\int_{\Delta''_{\nu-1,\nu+2}}\log\left|\frac{1-\bar{a}z}{z-a}\right|\frac{d\sigma(a)}{1-|a|}$$
$$=III'+III''. \qquad (6)$$

Since in III', $1-|a|\geqq \text{const.}\,1/2^\nu$,

$$III'\leqq \text{const.}\,2^\nu\int_{\Delta'_{\nu-1,\nu+2}}\log\left|\frac{1-\bar{a}z}{z-a}\right|d\sigma(a).$$

Now $\log\left|\frac{1-\bar{a}z}{z-a}\right|=\frac{1}{2}\log\left(1+\frac{(1-|a|^2)(1-|z|^2)}{|z-a|^2}\right)$ and in III', $1-|a|\leqq\text{const.}/2^\nu$, $1-|z|\leqq \text{const.}/2^\nu$ and if a lies on l_ψ $(\psi\neq\psi_1)$, $|z-a|\geqq|z-1|\sin|\psi-\psi_1|\geqq$ const. $|\psi-\psi_1|/2^\nu$, so that

$$\log\left|\frac{1-\bar{a}z}{z-a}\right|\leqq\frac{1}{2}\log\left(1+\frac{\text{const.}}{(\psi-\psi_1)^2}\right)\leqq\text{const.}\log\frac{\pi}{|\psi-\psi_1|}.$$

Since if $\nu\geqq\nu_0$, $\lambda'_{\nu-1}(\psi_1)$ exists, l_{ψ_1} does not contain a positive σ-mass, so that

$$III'\leqq\text{const.}\,2^\nu\int_{\psi_1}^{\psi_0}\log\frac{\pi}{\psi-\psi_1}d\lambda_{\nu-1}(\psi)+\text{const.}\,2^\nu\int_{-\psi_0}^{\psi_1}\log\frac{\pi}{\psi_1-\psi}d\lambda_{\nu-1}(\psi). \qquad (7)$$

Since $\psi_1\in e$, $\psi_1\in C(e^*_\nu)$ $(\nu\geqq\nu_0)$, so that

$$\lambda'(\psi_1)\leqq 1/2^\nu \qquad (\nu\geqq\nu_0). \qquad (8)$$

Since by (3), $\lambda_\nu(\psi_0)\leqq 2/2^{\nu(1+\delta)}$,

$$\int_{\psi_1}^{\psi_0}\log\frac{\pi}{\psi-\psi_1}d\lambda_{\nu-1}(\psi)=\left[\log\frac{\pi}{\psi-\psi_1}(\lambda_{\nu-1}(\psi)-\lambda_{\nu-1}(\psi_1))\right]_{\psi_1}^{\psi_0}$$
$$+\int_{\psi_1}^{\psi_0}\frac{\lambda_{\nu-1}(\psi)-\lambda_{\nu-1}(\psi_1)}{\psi-\psi_1}d\psi\leqq\frac{\text{const.}}{2^{\nu(1+\delta)}}+\int_{\psi_1}^{\psi_1+1/2^\nu}\frac{\lambda_{\nu-1}(\psi)-\lambda_{\nu-1}(\psi_1)}{\psi-\psi_1}d\psi$$
$$+\int_{\psi_1+1/2^\nu}^{\psi_0}\frac{\lambda_{\nu-1}(\psi)-\lambda_{\nu-1}(\psi_1)}{\psi-\psi_1}d\psi.$$

If $\nu-1\geqq\nu_0$, then by (8),

$$\frac{\lambda_{\nu-1}(\psi)-\lambda_{\nu-1}(\psi_1)}{\psi-\psi_1}\leqq\frac{\lambda_{\nu_0}(\psi)-\lambda_{\nu_0}(\psi_1)}{\psi-\psi_1}\leqq\text{const.}\,\lambda'_{\nu_0}(\psi_1)\leqq\text{const.}/2^{\nu_0},$$

hence

$$\int_{\psi_1}^{\psi_1+1/2^\nu}\frac{\lambda_{\nu-1}(\psi)-\lambda_{\nu-1}(\psi_1)}{\psi-\psi_1}d\psi\leqq\text{const.}/2^{\nu+\nu_0}.$$

Since $\lambda_{\nu-1}(\psi)-\lambda_{\nu-1}(\psi_1)\leqq\lambda_{\nu-1}(\psi_0)\leqq\text{const.}/2^{\nu(1+\delta)}$ by (3), we have

$$\int_{\psi_1+1/2^\nu}^{\psi_0} \frac{\lambda_{\nu-1}(\psi)-\lambda_{\nu-1}(\psi_1)}{\psi-\psi_1} d\psi \leqq \frac{\text{const.}}{2^{\nu(1+\delta)}} \int_{1/2^\nu}^{\psi_0-\psi_1} \frac{dt}{t} \leqq \text{const.}\ \nu/2^{\nu(1+\delta)}.$$

Hence

$$\int_{\psi_1}^{\psi_0} \log \frac{\pi}{\psi-\psi_1}\, d\lambda_{\nu-1}(\psi) \leqq \text{const.}\ \nu/2^{\nu(1+\delta)} + \text{const.}/2^{\nu+\nu_0}.$$

Similarly we have the same evaluation for $\int_{-\psi_0}^{\psi_1} \log \dfrac{\pi}{\psi-\psi_1} d\lambda_{\nu-1}(\psi)$, so that by (7),

$$III' \leqq \text{const.}\ \nu/2^{\nu\delta} + \text{const.}\ 1/2^{\nu_0}. \tag{9}$$

In III'', $|z-a| \geqq |z-1| \sin|\psi_0-\psi_1| \geqq \text{const.}/2^\nu$, $1-|z| \leqq \text{const.}/2^\nu$, so that $\log\left|\dfrac{1-\bar a z}{z-a}\right| \leqq \dfrac{2(1-|a|)(1-|z|)}{|z-a|^2} \leqq \text{const.}\ 2^\nu(1-|a|)$, hence by (2),

$$III'' \leqq \text{const.}\ 2^\nu \int_{\Delta_{\nu-1}} d\sigma(a) \leqq \text{const.}/2^{\nu\delta}, \tag{10}$$

so that

$$III \leqq \text{const.}\ \nu/2^{\nu\delta} + \text{const.}/2^{\nu_0}. \tag{11}$$

In IV, $|z-a| \geqq \text{const.}\ 1/2^\nu$, so that similarly we have

$$IV \leqq \text{const.}\ 2^\nu \sigma(\Delta_{\nu+2}) \leqq \text{const.}\ 1/2^{\nu\delta}. \tag{12}$$

Now

$$II = \sum_{-\nu_0+1}^{\nu-1} \int_{\Delta_{j-1,j}} \log\left|\frac{1-\bar a z}{z-a}\right| \frac{d\sigma(a)}{1-|a|}.$$

If $a \in \Delta_{j-1,j}$, then $|z-a| \geqq 1/2^j - 1/2^\nu \geqq \text{const.}/2^j$, so that

$$\int_{\Delta_{j-1,j}} \log\left|\frac{1-\bar a z}{z-a}\right| \frac{d\sigma(a)}{1-|a|} \leqq 2\int_{\Delta_{j-1,j}} \frac{(1-|z|)d\sigma(a)}{|z-a|^2}$$

$$\leqq \text{const.}\ \frac{2^{2j}}{2^\nu} \sigma(\Delta_{j-1,j}) \leqq \text{const.}\ \frac{2^{2j}}{2^\nu} \cdot \frac{1}{2^{j(1+\delta)}} = \frac{\text{const.}}{2^\nu} \cdot 2^{j(1-\delta)},$$

hence

$$II \leqq \text{const.}/2^{\nu\delta}. \tag{13}$$

Hence by (5), (11), (12), (13),

$$w(z) \leqq I + \text{const.}\ \nu/2^{\nu\delta} + \text{const.}/2^{\nu_0} \quad (\nu \geqq \nu_0). \tag{14}$$

Since $\lim_{z\to 1} I = 0$ and ν_0 is arbitrary, we have

$$\lim_{z\to 1} w(z) = 0, \quad \text{if } z \to 1 \text{ along } l_{\psi_1}.$$

Theorem IV. 36. *Let*

$$u(z) = \int_{|a|<1} \log \frac{1}{|z-a|} d\mu(a), \quad \int_{|a|<1} d\mu(a) < \infty.$$

Then there exists a set E of measure 2π on $|z|=1$, such that if $e^{i\theta} \in E$, *then* $\lim_{z\to e^{i\theta}} u(z)=u(e^{i\theta})\neq\infty$ *exists, when* $z\to e^{i\theta}$ *along* $l_\psi(e^{i\theta})$ *for almost all* ψ $\left(-\frac{\pi}{2}<\psi<\frac{\pi}{2}\right)$, *where* $u(e^{i\theta})$ *is independent of* ψ.

Proof. If we put

$$w(z)=\int_{|a|<1}\log\left|\frac{1-\bar{a}z}{z-a}\right|d\mu(a)\,,\qquad v(z)=\int_{|a|<1}\log\left|\frac{1}{1-\bar{a}z}\right|d\mu(a)\,,\qquad (1)$$

then $u(z)=w(z)+v(z)$, where $v(z)$ is harmonic in $|z|<1$.

Let Δ: $|\arg z|<\pi/6$, $|z|<1$ be a sector and Δ^* be its complement with respect to $|z|<1$, then $v(z)=v_1(z)+v_2(z)$, where

$$v_1(z)=\int_{a\in\Delta^*}\log\left|\frac{1}{1-\bar{a}z}\right|d\mu(a)\,,\qquad v_2(z)=\int_{a\in\Delta}\log\left|\frac{1}{1-\bar{a}z}\right|d\mu(a)\,.\qquad (2)$$

$v_1(z)$ is harmonic on the arc C: $|\arg z|<\pi/6$ on $|z|=1$ and if $z\in\Delta$, $a\in\Delta$, then $|1-\bar{a}z|<1$, so that $v_2(z)$ is a positive harmonic function in Δ. Hence by mapping Δ conformally on the inside of a unit circle, we see that for almost all $e^{i\theta}$ on C, $\lim_{z\to e^{i\theta}} v_2(z)$ $(\neq\infty)$ exists, when $z\to e^{i\theta}$ non-tangentially to $|z|=1$, so that $\lim_{z\to e^{i\theta}} v(z)=v(e^{i\theta})$ $(\neq\infty)$ exists almost everywhere on C, when $z\to e^{i\theta}$ non-tangentially to $|z|=1$. If we take a sector $\Delta(\theta)$: $|\arg z-\theta|<\frac{\pi}{6}$, $|z|<1$ instead of Δ, then we see that $\lim_{z\to e^{i\theta}} v(z)=v(e^{i\theta})$ $(\neq\infty)$ exists almost everywhere on $|z|=1$, when $z\to e^{i\theta}$ non-tangentially to $|z|=1$. Hence the theorem follows from $u(z)=w(z)+v(z)$ and Theorem IV. 35. ——

Theorem IV. 37. *Let $u(z)$ be subharmonic in $|z|<1$, such that*

$$\int_0^{2\pi}|u(re^{i\theta})|^p d\theta=O(1)\,,\qquad 0\leqq r<1\,,\qquad p>1\,.$$

Then

$$\lim_{r\to1}\int_0^{2\pi}|u(re^{i\theta})-u(e^{i\theta})|^q\,d\theta=0\,,\qquad 1<q<p\,.$$

Littlewood [3].

Proof. Since $p>1$, $\int_0^{2\pi}|u(re^{i\theta})|d\theta=O(1)$, so that $\lim_{r\to1} u(re^{i\theta})=u(e^{i\theta})$ exists for almost all $e^{i\theta}$ and $u(z)=v(z)-w(z)$, where $v(z)$ is harmonic in $|z|<1$ and

$$w(z)=\int_{|a|<1}\log\left|\frac{1-\bar{a}z}{z-a}\right|d\mu(a)\,.\qquad (1)$$

Let $v_\rho^*(z)$ be harmonic in $|z|<\rho<1$, such that $v_\rho^*=u$ on $|z|=\rho$, then $\lim_{\rho\to1} v_\rho^*(z)=v(z)$ and for $r\leqq\rho$,

$$\int_0^{2\pi}|v_\rho^*(re^{i\theta})|^p\,d\theta\leqq\int_0^{2\pi}|v_\rho^*(\rho e^{i\theta})|^p\,d\theta=\int_0^{2\pi}|u(\rho e^{i\theta})|^p\,d\theta=O(1)\,,$$

so that

$$\int_0^{2\pi} |v(re^{i\theta})|^p \, d\theta = O(1) . \tag{2}$$

Hence by Theorem IV. 4,

$$\lim_{r\to 1} \int_0^{2\pi} |v(re^{i\theta}) - v(e^{i\theta})|^q \, d\theta = 0 \qquad (1<q<p) . \tag{3}$$

By (2) and $u=v-w$, we have

$$\int_0^{2\pi} |w(re^{i\theta})|^p \, d\theta = O(1) . \tag{4}$$

Let $\alpha = \dfrac{p-1}{p-q}$, $\beta = \dfrac{p-1}{q-1}$, then $\dfrac{1}{\alpha} + \dfrac{1}{\beta} = 1$ and $q = \dfrac{1}{\alpha} + \dfrac{p}{\beta}$, so that by (4),

$$\begin{aligned} \int_0^{2\pi} |w(re^{i\theta})|^q \, d\theta &= \int_0^{2\pi} |w(re^{i\theta})|^{1/\alpha} |w(re^{i\theta})|^{p/\beta} d\theta \\ &\leqq \left(\int_0^{2\pi} |w(re^{i\theta})| d\theta \right)^{1/\alpha} \left(\int_0^{2\pi} |w(re^{i\theta})|^p \, d\theta \right)^{1/\beta} \\ &\leqq \text{const.} \left(\int_0^{2\pi} |w(re^{i\theta})| d\theta \right)^{1/\alpha} . \end{aligned}$$

Since $\lim_{r\to 1} \int_0^{2\pi} |w(re^{i\theta})| d\theta = 0$ (Theorem IV. 33), we have

$$\lim_{r\to 1} \int_0^{2\pi} |w(re^{i\theta})|^q \, d\theta = 0 . \tag{5}$$

Since for almost all $e^{i\theta}$, $\lim_{r\to 1} u(re^{i\theta}) = u(e^{i\theta}) = v(e^{i\theta})$ and

$$|u(re^{i\theta}) - u(e^{i\theta})|^q \leqq 2^q (|v(re^{i\theta}) - v(e^{i\theta})|^q + |w(re^{i\theta})|^q) ,$$

we have by (3), (5),

$$\lim_{r\to 1} \int_0^{2\pi} |u(re^{i\theta}) - u(e^{i\theta})|^q \, d\theta = 0 .$$

11. Hardy-Littlewood's maximal theorem

1. First we shall prove some lemmas.

Lemma 1. *Let $f(x) \geqq 0$ be measurable in $[0, a]$, such that $\int_0^a f^p(x) dx < \infty$, $p>1$. Then*

$$\int_0^a \left(\frac{1}{x} \int_0^x f(t) dt \right)^p dx \leqq \left(\frac{p}{p-1} \right)^p \int_0^a f^p(x) dx .$$

Proof. Let $1/p + 1/q = 1$ $(p>1, q>1)$, then

$$F(x)=\int_0^x f(x)dx \leqq x^{1/q}\left(\int_0^x f^p(x)dx\right)^{1/p} \qquad (0<x<a)\,, \tag{1}$$

so that $F(x)^p x^{1-p} \leqq \int_0^x f^p(x)dx$, hence

$$\lim_{x\to 0} F^p(x)x^{1-p}=0\,. \tag{2}$$

$$\int_\xi^a \left(\frac{F(x)}{x}\right)^p dx = -\frac{1}{p-1}\int_\xi^a (F(x))^p \frac{dx^{1-p}}{dx}dx$$
$$\leqq \frac{F^p(\xi)\xi^{1-p}}{p-1}+\frac{p}{p-1}\int_\xi^a \left(\frac{F(x)}{x}\right)^{p-1} f(x)dx \qquad (0<\xi<a)\,,$$

so that by (2),

$$\int_0^a \left(\frac{F(x)}{x}\right)^p dx \leqq \frac{p}{p-1}\int_0^a \left(\frac{F(x)}{x}\right)^{p-1} f(x)dx$$
$$\leqq \frac{p}{p-1}\left(\int_0^a \left(\frac{F(x)}{x}\right)^p dx\right)^{1/q}\left(\int_0^a f^p(x)dx\right)^{1/p},$$

hence

$$\int_0^a \left(\frac{F(x)}{x}\right)^p dx \leqq \left(\frac{p}{p-1}\right)^p \int_0^a f^p(x)dx\,.$$

Lemma 2. *Let E be a bounded closed set on the x-axis. Suppose that at every point $x \in E$, there be given a set $\mathfrak{M}_x$ of open intervals $\{J(x)\}$, with x as its mid-point, such that $\mathfrak{M}=\sum_{x\in E}\mathfrak{M}_x$ is compact*[1]. *Then we can select a finite number of intervals from $\mathfrak{M}$, which cover E and overlap at most twice.*

Proof. Let $|J(x)|$ be the length of $J(x)$ and put $r(x)=\frac{1}{2}|J(x)|$. Let $r_0=\sup_{x\in E} r(x)$, then since $\mathfrak{M}$ is compact, there exists $x_0 \in E$ and an interval $J(x_0) \in \mathfrak{M}_{x_0}$, such that $\frac{1}{2}|J(x_0)|=r_0$. If $E_1=E-E\cdot J(x_0)\neq 0$, let $r_1=\sup_{x\in E_1} r(x)$ then there exists $x_1 \in E_1$ and an open interval $J(x_1) \in \mathfrak{M}_{x_1}$, such that $\frac{1}{2}|J(x_1)|=r_1$ and proceed similarly, then we obtain a decreasing sequence of bounded closed sets E_ν. We shall prove that $E_\nu=0$, if $\nu>n_0$, for a certain n_0. Suppose that $E_\nu\neq 0$ $(\nu=1,2,\cdots)$, then since E_ν is a bounded closed set, $A=\prod_{\nu=1}^{\infty} E_\nu \neq 0$.

Let $\xi_0 \in A$, then since $\xi_0 \in E_\nu$, $\frac{1}{2}|J(\xi_0)| \leqq r_\nu$, which is absurd, since as easily seen, $r_\nu \to 0$ $(\nu\to\infty)$. Hence $E_\nu=0$, if $\nu>n_0$ for a certain n_0. Hence $E\subset J(x_0)+J(x_1)+\cdots+J(x_{n_0})$. Next we shall prove that $\{J(x_\nu)\}$

1) i.e., if $x_\nu(\in E)\to x_0(\in E)$ and $J(x_\nu)(\in \mathfrak{M}_{x_\nu})\to$ an open interval J_0, then $J_0 \in \mathfrak{M}_x$.

overlap at most twice. Suppose that $\xi\in J(x_{n_1})$, $\xi\in J(x_{n_2})$ ($x_{n_1}<x_{n_2}$, $n_1<n_2$), then since $r_{n_1}\geqq r_{n_2}$, ξ belongs to the right half of $J(x_{n_1})$. If $x_{n_3}>x_{n_2}$ ($n_3>n_2$), then $|\xi-x_{n_3}|\geqq r_{n_2}\geqq r_{n_3}$, so that ξ does not belong to $J(x_{n_3})$, hence $\{J(x)\}$ overlap at most twice.

2. Now we shall prove the following

Theorem IV. 38. (Hardy-Littlewood's maximal theorem). *Let $f(x)\geqq 0$ be a measurable function on a unit circle $K: z=e^{ix}$ $(-\pi\leqq x\leqq\pi)$, such that*

$$\int_{-\pi}^{\pi} f^p(x)dx<\infty\,, \qquad p>1\,.$$

Let

$$\theta(x)=\sup_{0<r\leqq\pi}\frac{1}{2r}\int_{-r}^{r} f(x+t)dt\,,$$

then

$$\int_{-\pi}^{\pi}\theta^p(x)dx\leqq\left(\frac{2p}{p-1}\right)^p\int_{-\pi}^{\pi} f^p(x)dx\,.$$

Hardy-Littlewood [3].

Proof. (Tsuji [63]). Let $E(f>y)$ $(y\geqq 0)$ be the set of e^{ix} on K, such that $f(x)>y$ and $|E(f>y)|$ be its measure, then we shall prove that

$$\int_{-\pi}^{\pi} f(x)dx=\int_0^{\infty}|E(f>y)|dy\,. \tag{1}$$

We put $\lambda(y)=|E(f>y)|$. First suppose that $f(x)$ is bounded, such that $0\leqq f(x)\leqq M$, then $\lambda(M)=0$,

$$\int_{-\pi}^{\pi} f(x)dx=-\int_0^M y\,d\lambda(y)=\int_0^M\lambda(y)dy\,.$$

If $f(x)$ is unbounded, then since $\int_{E(f>y)} f(x)dx\geqq y\lambda(y)$, $\lim_{y\to\infty} y\lambda(y)=0$. Let $f_n(x)=\mathrm{Min}\,(f(x), n)$ $(n=1, 2, \cdots)$, then

$$\int_{-\pi}^{\pi} f_n(x)dx=-\int_0^n y\,d\lambda(y)=-n\lambda(n)+\int_0^n\lambda(y)dy\,,$$

so that

$$\int_{-\pi}^{\pi} f(x)dx=\lim_n\int_{-\pi}^{\pi} f_n(x)dx=\int_0^{\infty}\lambda(y)dy\,.$$

Now $x=\frac{1}{2}|E(f>y)|$ is a non-increasing function of y, such that $0\leqq x\leqq x_0$, where $x_0=\frac{1}{2}|E(f>0)|$.

Let $y=\mu(x)$ $(0\leqq x\leqq x_0)$ be its inverse function, such that $\mu(x+0)=\mu(x)$.

We define $f^*(x)$ in $[-\pi, \pi]$ as follows.

$$f^*(x)=\begin{cases}\mu(x)\,, & 0\leqq x\leqq x_0\,,\\ 0\,, & x_0<x\leqq \pi\,, \quad \text{if } x_0<\pi\,.\end{cases} \tag{2}$$
$$f^*(-x)=f^*(x)\,.$$

Then $f^*(x)$ is a non-increasing function of x in $[0, \pi]$. We see easily that $2x=|E(f>y)|=|E(f^*>y)|$, so that by (1)

$$\int_{-\pi}^{\pi} f(x)dx=\int_{-\pi}^{\pi} f^*(x)dx=2\int_0^{\pi} f^*(x)dx\,. \tag{3}$$

From the definition, we see that if $0\leqq f_1(x)\leqq f(x)$ on K, then $f_1^*(x)\leqq f^*(x)$.

Let E be any measurable set on K. We define $f_E(x)$, such that $f_E(x)=f(x)$ on E and $f_E(x)=0$ on its complement, then $f_E(x)\leqq f(x)$, so that $f_E^*(x)\leqq f^*(x)$, hence

$$\int_E f(x)dx=\int_{-\pi}^{\pi} f_E(x)dx=2\int_0^{|E|/2} f_E^*(x)dx\leqq 2\int_0^{|E|/2} f^*(x)dx\,,$$
$$\int_E f(x)dx\leqq 2\int_0^{|E|/2} f^*(x)dx\,. \tag{4}$$

Let

$$\Theta(x)=\frac{1}{2x}\int_{-x}^{x} f^*(t)dt=\frac{1}{x}\int_0^{x} f^*(t)dt\,, \qquad 0<x\leqq\pi\,. \tag{5}$$

Then $\Theta(x)$ is a non-increasing function of x. We shall prove that

$$|E(\theta>y)|\leqq 2|E(2\Theta>y)|\,, \qquad y\geqq 0\,.$$

Now by the definition of $\theta(x)$, $O=E(\theta>y)$ is an open set and at every point $x\in O$, there exists a set $\mathfrak{M}_x$ of open intervals $J(x)$ with x as its mid-point, such that

$$\int_{J(x)} f(t)dt\geqq y|J(x)|\,, \quad x\in O\,. \tag{6}$$

Let O^* be the complement of O on K, then at every point $\xi\in O^*$, $\theta(\xi)\leqq y$, hence by Lebesgue's theorem, for almost all $\xi\in O^*$,

$$f(\xi)=\lim_{r\to 0}\frac{1}{2r}\int_{-r}^{r} f(\xi+t)dt\leqq y\,,$$

so that

$$\int_{J(x)\cdot O^*} f(t)dt\leqq y|J(x)\cdot O^*|, \qquad x\in O\,.$$

Now by (6)

$$\int_{J(x)} f(t)dt=\int_{J(x)\cdot O} f(t)dt+\int_{J(x)\cdot O^*} f(t)dt\geqq y|J(x)|=y(|J(x)\cdot O|+|J(x)\cdot O^*|)\,,$$

so that

$$\int_{J(x)\cdot O} f(t)dt \geqq y|J(x)\cdot O| , \quad x \in O . \tag{7}$$

Let F be a closed sub-set of O, such that $|F|>|O|-\varepsilon$ $(\varepsilon>0)$, then $\mathfrak{M}=\sum_{x\in F}\mathfrak{M}_x$ is compact, so that by Lemma 2, we can select a finite number of intervals $\{J(x_\nu)\}$ from $\mathfrak{M}$, which cover F and overlap at most twice. If we apply (7) to $J(x_\nu)$ and add them, then we have

$$2\int_O f(t)dt \geqq \sum_\nu \int_{J(x_\nu)\cdot O} f(t)dt \geqq y\sum_\nu |J(x_\nu)\cdot O| \geqq y\sum_\nu |J(x_\nu)\cdot F| \geqq y|F| > y(|O|-\varepsilon) ,$$

hence if we make $\varepsilon\to 0$, we have

$$2\int_O f(t)dt \geqq y|O| ,$$

or

$$2\int_{E(\theta>y)} f(t)dt \geqq y|E(\theta>y)| . \tag{8}$$

Hence by (4),

$$\frac{1}{|E(\theta>y)|/2}\int_0^{|E(\theta<y)|/2} f^*(t)dt \geqq \frac{1}{|E(\theta>y)|}\int_{E(\theta>y)} f(t)dt \geqq \frac{y}{2} . \tag{9}$$

Since $\Theta(x)$ is a non-increasing function of x in $[0, \pi]$, $|E(\Theta \geqq y/2)|$ is the maximum value of x, such that

$$\Theta(x)=\frac{1}{x}\int_0^x f^*(t)dt \geqq \frac{y}{2} ,$$

hence by (9),

$$\frac{|E(\theta>y)|}{2} \leqq |E(\Theta \geqq y/2)| = |E(2\Theta \geqq y)| ,$$

so that

$$|E(\theta>y+\varepsilon)| \leqq 2|E(2\Theta \geqq y+\varepsilon)| , \qquad \varepsilon>0 .$$

If we make $\varepsilon\to 0$, then we have

$$|E(\theta>y)| \leqq 2|E(2\Theta>y)| , \qquad y \geqq 0 . \tag{10}$$

Hence if we put $\lambda(y)=|E(\theta>y)|$, $\Lambda(y)=|E(2\Theta>y)|$, then

$$\lambda(y) \leqq 2\Lambda(y) , \qquad y \geqq 0 . \tag{11}$$

Since $|E(\theta^p>y)|=|E(\theta>y^{1/p})|=\lambda(y^{1/p})$, we have by (1),

$$\int_{-\pi}^{\pi} \theta^p(x)dx=\int_0^\infty \lambda(y^{1/p})dy \leqq 2\int_0^\infty \Lambda(y^{1/p})dy , \tag{12}$$

and since $|E((2\Theta)^p>y)|=|E(2\Theta>y^{1/p})|=\Lambda(y^{1/p})$,

$$2^p\int_0^\pi \Theta^p(x)dx=\int_0^\infty \Lambda(y^{1/p})dy , \tag{13}$$

so that by (12), (13), (5), Lemma 1 and (3),

$$\int_{-\pi}^{\pi}\theta^p(x)dx \leqq 2\cdot 2^p\int_0^{\pi}\Theta^p(x)dx \leqq 2\left(\frac{2p}{p-1}\right)^p\int_0^{\pi} f^{*p}(x)dx$$
$$=\left(\frac{2p}{p-1}\right)^p\int_{-\pi}^{\pi} f^{*p}(x)dx=\left(\frac{2p}{p-1}\right)^p\int_{-\pi}^{\pi} f^p(x)dx\,.$$

3. As an application of the theorem, we shall prove

Theorem IV. 39. *Let $w(z)\geqq 0$ be a non-negative subharmonic function in $|z|<1$, which is continuous on $|z|=1$.*
Let $S(\theta,\alpha)$ $(0<\alpha<\pi/2)$ be a sector: $|\arg(1-ze^{-i\theta})|<\alpha$, $\frac{1}{2}\leqq|z|<1$ and

$$M(\theta,\alpha)=\sup_{z\in S(\theta,\alpha)} w(z)\,,$$

then

$$\int_{-\pi}^{\pi} M^p(\theta,\alpha)d\theta \leqq A(p,\alpha)\int_{-\pi}^{\pi} w^p(e^{i\theta})d\theta\,,\qquad p>1\,,$$

where $A(p,\alpha)$ is a constant, which depends on p and α only.

Hardy-Littlewood [3].

Proof. Let $u(z)$ be harmonic in $|z|<1$, such that $u=w$ on $|z|=1$, then $w\leqq u$ in $|z|<1$, so that we assume that $w(z)$ is harmonic in $|z|<1$. Then if we put $w(e^{i\theta})=f(\theta)$,

$$w(z)=w(re^{i\theta})=\frac{1}{2\pi}\int_{-\pi}^{\pi} f(\varphi)\frac{1-r^2}{1-2r\cos(\varphi-\theta)+r^2}d\varphi\,. \tag{1}$$

Since $1-2r\cos(\varphi-\theta)+r^2\geqq(1-r)^2+a^2(\varphi-\theta)^2$ $(a=\text{const.})$,

$$w(z)\leqq \text{const.}\int_{-\pi}^{\pi} f(\varphi)\frac{1-r}{(1-r)^2+(\varphi-\theta)^2}d\varphi\,. \tag{2}$$

Let $\theta(x)$ be defined as in Theorem IV. 38, then we shall prove that

$$M(0,\alpha)\leqq A\theta(0)\,, \tag{3}$$

where $A=A(\alpha)$ is a constant, which depends on α only.

If $z=re^{i\theta}\in S(0,\alpha)$, then $0\leqq|\theta|\leqq k(1-r)$, where $k=k(\alpha)$ is a constant, which depends on α only.

We may assume that $\theta\geqq 0$. Let $J=[\theta-2k(1-r),\ \theta+2k(1-r)]$ be an arc on $|z|=1$ and J^* be its complement, then

$$\int_{-\pi}^{\pi} f(\varphi)\frac{1-r}{(1-r)^2+(\varphi-\theta)^2}d\varphi=\int_J+\int_{J^*}=I+II\,. \tag{4}$$

Since $|\theta|\leqq k(1-r)$, J is contained in $[-3k(1-r),\ 3k(1-r)]$, so that

$$I\leqq\int_{-3k(1-r)}^{3k(1-r)} f(\varphi)\frac{1-r}{(1-r)^2+(\varphi-\theta)^2}d\varphi\leqq\frac{1}{1-r}\int_{-3k(1-r)}^{3k(1-r)} f(\varphi)d\varphi\leqq 6k\theta(0)\,. \tag{5}$$

In II, we put $\tilde{f}(\varphi)=f(\varphi)+f(-\varphi)$, $F(\varphi)=\int_0^{\varphi}\tilde{f}(\varphi)d\varphi$, then $F(\varphi)\leqq 2\varphi\cdot\theta(0)$.

Since $|\varphi|\geqq k(1-r)$ in II, we have $|\varphi-\theta|\geqq|\varphi|-k(1-r)$, so that

$$II\leqq\int_{k(1-r)}^{\pi}\tilde{f}(\varphi)\frac{1-r}{(1-r)^2+(\varphi-k(1-r))^2}d\varphi$$

$$\leqq\text{const.}\,F(\pi)+\int_{k(1-r)}^{\pi}\frac{F(\varphi)}{\varphi}\cdot\varphi\left|\frac{\partial}{\partial\varphi}\left(\frac{1-r}{(1-r)^2+(\varphi-k(1-r))^2}\right)\right|d\varphi$$

$$\leqq\text{const.}\,\theta(0)+4\theta(0)\int_{k(1-r)}^{\pi}\frac{\varphi(1-r)(\varphi-k(1-r))}{((1-r)^2+(\varphi-k(1-r))^2)^2}d\varphi$$

$$\leqq\text{const.}\,\theta(0)+4\theta(0)\int_{k}^{\infty}\frac{t(t-k)}{(1+(t-k)^2)^2}dt\leqq\text{const.}\,\theta(0)\,,\quad\varphi=(1-r)t\,,\quad(6)$$

where const. is a constant, which depends on α only.

Hence by (2), (4), (5), (6), we have

$$M(0,\alpha)\leqq A\theta(0)\,,$$

where $A=A(\alpha)$ is a constant, which depends on α only.

Similarly we have $M(x,\alpha)\leqq A\theta(x)$ for any point e^{ix} on $|z|=1$, hence

$$\int_{-\pi}^{\pi}M^p(x,\alpha)dx\leqq A^p\int_{-\pi}^{\pi}\theta^p(x)dx\leqq\left(\frac{2Ap}{p-1}\right)^p\int_{-\pi}^{\pi}f^p(x)dx$$

$$=\left(\frac{2Ap}{p-1}\right)^p\int_{-\pi}^{\pi}w^p(e^{ix})dx\,.\text{——}$$

If we apply the theorem on $w(z)=|f(z)|^{\lambda/2}$, $p=2$, then we have

Theorem IV. 40. *Let $f(z)$ be regular in $|z|<1$ and continuous on $|z|=1$ and*

$$M(\theta,\alpha)=\sup_{z\in S(\theta,\alpha)}|f(z)|\,,$$

then

$$\int_{-\pi}^{\pi}M^\lambda(\theta,\alpha)d\theta\leqq A(\alpha)\int_{-\pi}^{\pi}|f(e^{i\theta})|^\lambda d\theta\,,\qquad\lambda>0\,,$$

where $A(\alpha)$ is a constant, which depends on α only.

4. As an application of Theorem IV. 39, we shall prove

Theorem IV. 41. *Let $w(z)\geqq0$ be a non-negative subharmonic function in $|z|<1$, such that*

$$\iint_{|z|<1}|w(z)|^p dxdy<\infty\,,\qquad z=x+iy\,,\quad p>1\,.$$

Then for almost all θ,

$$w(z)=o\left(\frac{1}{\sqrt[p]{1-|z|}}\right)\qquad\textit{uniformly,}$$

as $z\to e^{i\theta}$ from the inside of any fixed Stolz domain, whose vertex is at $e^{i\theta}$.

Gehring [1].

Proof. Let $0<\alpha<\frac{\pi}{2}$ and $S^{(\alpha)}(1, \theta)$ be the domain, which is bounded by two tangents from $e^{i\theta}$ to the circle $|z|=\sin\alpha$ and the larger arc of this circle, which is bounded by two points of contact. It suffices to prove that for almost all θ,

$$\lim_{z\to e^{i\theta}}(1-|z|)w^p(z)=0 \quad \text{uniformly,} \tag{1}$$

as $z\to e^{i\theta}$ in $S^{(\alpha)}(1, \theta)$.

Let $S_r^{(\alpha)}(1, \theta)$ $(\sin\alpha<r<1)$ be the part of $S^{(\alpha)}(1, \theta)$, contained in $|z|<r$ and $S^{(\alpha)}(\rho, \theta)$ $(0<\rho<1)$ be the domain, which is obtained from $S^{(\alpha)}(1, \theta)$ by the affine transformation $z'=\rho z$. We choose ρ and β, such that $\rho=\sqrt{r}$, $\sin\beta=\sqrt{r}$, then since $r<\rho<1$, $\rho\sin\beta=r$,

$$S_r^{(\alpha)}(1, \theta)\subset S^{(\beta)}(\rho, \theta) .$$

Hence if we put

$$\left.\begin{aligned} U(r, \theta)&=\sup_{z\in S_r^{(\alpha)}(1,\theta)} w(z) , \\ W(\rho, \theta)&=\sup_{z\in S^{(\beta)}(\rho,\theta)} w(z) , \end{aligned}\right\} \tag{2}$$

then

$$U(r, \theta)\leqq W(\rho, \theta) ,$$

hence by Theorem IV. 39,

$$\int_0^{2\pi} U^p(r, \theta)d\theta\leqq\int_0^{2\pi} W^p(\rho, \theta)d\theta\leqq C_1\int_0^{2\pi} w^p(\rho e^{i\theta})d\theta , \qquad C_1=A(p, \alpha) ,$$

so that

$$\int_{\sin\alpha}^1\int_0^{2\pi} U^p(r, \theta)r\,dr\,d\theta\leqq C_2\int_0^1\int_0^{2\pi} w^p(\rho e^{i\theta})\rho\,d\rho\,d\theta<\infty , \qquad C_2=\text{const.}\,C_1 .$$

Hence for almost all θ,

$$\int_0^1 U^p(r, \theta)dr<\infty , \quad \text{so that} \quad \lim_{r\to 1}\int_r^1 U^p(r, \theta)dr=0 . \tag{3}$$

Since $U(r, \theta)$ is an increasing function of r,

$$(1-r)U^p(r, \theta)\leqq\int_r^1 U^p(r, \theta)dr ,$$

so that by (3),

$$\lim_{r\to 1}(1-r)U^p(r, \theta)=0 , \tag{4}$$

which implies (1).——

If we apply the theorem on $w(z)=|f(z)|^{\lambda/2}$, $p=2$, then we have

Theorem IV. 42. *Let $f(z)$ be regular in $|z|<1$, such that*

$$\iint_{|z|<1} |f(z)|^{\lambda}\,dx\,dy<\infty\,, \qquad \lambda>0\,.$$

Then for almost all θ,

$$f(z)=o\left(\frac{1}{\sqrt[\lambda]{1-|z|}}\right) \quad \textit{uniformly},$$

as $z\to e^{i\theta}$ from the inside of any fixed Stolz domain, whose vertex is at $e^{i\theta}$.

Especially, if

$$\iint_{|z|<1}|f'(z)|^2\,dx\,dy<\infty\,, \tag{1}$$

then for almost all θ,

$$f'(z)=o\left(\frac{1}{\sqrt{1-|z|}}\right). \tag{2}$$

Since any simply connected domain can be mapped conformally on a bounded domain by an elementary transformation, we have from (1), (2), the following theorem.

Theorem IV. 43. *Let $f(z)$ be regular and schlicht in $|z|<1$, then for almost all $e^{i\theta}$,*

$$f'(z)=o\left(\frac{1}{\sqrt{1-|z|}}\right) \quad \textit{uniformly},$$

as $z\to e^{i\theta}$ from the inside of any fixed Stolz domain, whose vertex is at $e^{i\theta}$. Seidel-Walsh [1].

12. Non-negative subharmonic function in a half-plane

1. First we shall prove some lemmas.

Lemma 1. *Let Δ_ρ: $|z|<\rho$, $\Re z>0$ be a half-disc and C_ρ: $|z|=\rho$, $\Re z\geqq 0$ be a semi-circle and*

$$G_\rho(z, a)=\log\left|\frac{z+\bar a}{z-a}\cdot\frac{\rho^2-z\bar a}{\rho^2+za}\right| \quad (a\in\Delta_\rho)\,,$$

be the Green's function of Δ_ρ. Then if $r<\rho$,

(i) $$\rho\frac{\partial G_\rho(\rho e^{i\varphi}, re^{i\theta})}{\partial\nu}=4(\lambda\cos\varphi\cos\theta+\lambda^2\sin 2\varphi\sin 2\theta+\lambda^3\cos 3\varphi\cos 3\theta+\cdots),$$
$$\lambda=r/\rho<1\,,$$

where ν is the inner normal of C_ρ at $z=\rho e^{i\varphi}$.

(ii) $\displaystyle\int_{-\frac{\pi}{2}}^{\frac{\pi}{2}} \frac{\partial G_\rho(\rho e^{i\varphi}, re^{i\theta})}{\partial\nu}\rho\cos\theta\, d\theta=\frac{2\pi r}{\rho}\cos\varphi$.

Proof. If $z=\rho e^{i\varphi}$, $a=re^{i\theta}$, then

$$\rho\frac{\partial G_\rho(z, a)}{\partial\nu}=-\Re\left(z\frac{d}{dz}\log\frac{z+\bar a}{z-a}\cdot\frac{\rho^2-z\bar a}{\rho^2+za}\right)=\Re\left(\frac{-z}{z+\bar a}+\frac{z}{z-a}+\frac{z\bar a}{\rho^2-z\bar a}+\frac{za}{\rho^2+za}\right)$$

$$=\Re\left(\frac{-z}{z+\bar a}+\frac{z}{z-a}+\frac{\rho^2}{\rho^2-z\bar a}-\frac{\rho^2}{\rho^2+za}\right)=2\Re\left(\frac{1}{1-\lambda e^{i(\theta-\varphi)}}-\frac{1}{1+\lambda e^{i(\theta+\varphi)}}\right)$$

$$=4(\lambda\cos\varphi\cos\theta+\lambda^2\sin 2\varphi\sin 2\theta+\lambda^3\cos 3\varphi\cos 3\theta+\cdots)\,. \tag{1}$$

Since

$$\int_{-\frac{\pi}{2}}^{\frac{\pi}{2}}\cos^2\theta\, d\theta=\pi/2\,,\qquad \int_{-\frac{\pi}{2}}^{\frac{\pi}{2}}\sin 2n\theta\cos\theta\, d\theta=0\,,$$

$$\int_{-\frac{\pi}{2}}^{\frac{\pi}{2}}\cos(2n+1)\theta\cos\theta\, d\theta=0\qquad (n=1, 2, \cdots)\,,$$

we have

$$\int_{-\frac{\pi}{2}}^{\frac{\pi}{2}}\frac{\partial G_\rho(\rho e^{i\varphi}, re^{i\theta})}{\partial\nu}\rho\cos\theta\, d\theta=\frac{2\pi r}{\rho}\cos\varphi\,. \tag{2}$$

Lemma 2. *Let $G(z, a)=\log\left|\dfrac{z+\bar a}{z-a}\right|$ ($z=re^{i\theta}$, $a=\tau e^{i\varphi}$) be the Green's function of the half-plane $x>0$ ($z=x+iy$), then*

(i) $G(z, a)\leqq G(e^{i\theta}, e^{i\varphi})$.

(ii) *If* $|\theta|\leqq\theta_0<\pi/2$, *then*

$$G(z, a)\leqq K(\theta_0)\frac{r\Re(a)}{|a|^2}(G(e^{i\theta}, e^{i\varphi})+1)\,,$$

where $K(\theta_0)=8/\cos^2\theta_0$.

Proof. Since (i) can be proved easily, we shall prove (ii). Let $\Re z=x$, $\Re a=\xi$, then $x\geqq r\cos\theta_0$.

$$G(z, a)=\frac{1}{2}\log\left(1+\frac{4x\xi}{|z-a|^2}\right)\leqq\frac{2x\xi}{|z-a|^2}\leqq\frac{2r\Re(a)}{|z-a|^2}\,. \tag{1}$$

If $|z-a|\geqq|a|\dfrac{\cos\theta_0}{2}$, then by (1),

$$G(z, a)\leqq\frac{8r\Re(a)}{\cos^2\theta_0|a|^2}\leqq\frac{8r\Re(a)}{\cos^2\theta_0|a|^2}(G(e^{i\theta}, e^{i\varphi})+1)\,. \tag{2}$$

If $|z-a|\leqq|a|\dfrac{\cos\theta_0}{2}$, then $\left|\dfrac{1}{a}-\dfrac{1}{z}\right|\leqq\dfrac{\cos\theta_0}{2|z|}$, so that

$$\frac{\xi}{|a|^2}\geqq\frac{x}{|z|^2}-\frac{\cos\theta_0}{2|z|}\geqq\frac{\cos\theta_0}{|z|}-\frac{\cos\theta_0}{2|z|}=\frac{\cos\theta_0}{2|z|},$$

so that $\dfrac{2r\Re(a)}{\cos\theta_0|a|^2}\geqq 1$, hence

$$G(z,a)\leqq G(e^{i\theta},e^{i\varphi})\leqq\frac{2r\Re(a)}{\cos\theta_0|a|^2}(G(e^{i\theta},e^{i\varphi})+1)\,. \tag{3}$$

Hence by (2), (3), we have

$$G(z,a)\leqq K(\theta_0)\frac{r\Re(a)}{|a|^2}(G(e^{i\theta},e^{i\varphi})+1)\,,\qquad K(\theta_0)=8/\cos^2\theta_0\,.$$

2. Now we shall prove

Theorem IV. 44. *Let $u(z)=u(x+iy)\not\equiv 0$ be a non-negative subharmonic function in a half-plane $x>0$, which vanishes continuously on the imaginary axis. Let*

$$m(r)=m(r,u)=\int_{-\frac{\pi}{2}}^{\frac{\pi}{2}}u(re^{i\theta})\cos\theta\,d\theta\,,\qquad 0<r<\infty\,,$$

then

(i) *$m(r)/r$ is a continuous non-decreasing function of r and is a convex function of $1/r^2$. Hence*

$$\lim_{r\to\infty}\frac{m(r)}{r}=c\,,\qquad 0<c\leqq\infty\,,$$

exists.

(ii) *If $0<c<\infty$, then*

A. $$u(z)=kx-\int_{\Re a>0}\log\left|\frac{z+\bar a}{z-a}\right|d\mu(a)\,,\qquad k=\frac{2c}{\pi}\,,$$

where μ is a positive mass distribution in $x>0$, such that

$$\int_{\Re a>0}\frac{\Re(a)d\mu(a)}{|a|^2}<\infty\,.$$

B. *Except a set of θ of capacity zero,*

$$\lim_{r\to\infty}\frac{u(re^{i\theta})}{r}=k\cos\theta\,.$$

That $m(r)/r$ is a non-decreasing function of r was proved by Ahlfors [4] and the proof was simplified by Dinghas [1] and (B) was proved by Ahlfors and Heins [1]. Compare this theorem with Remark 1 (p. 151).

As a special case, we have

Theorem IV. 45. *Let $f(z)=f(x+iy)$ be regular in a half-plane $x>0$ and continuous and $|f(z)|\leqq 1$ on the imaginary axis.*
Suppose that $\log^+|f(z)|\not\equiv 0$ *and let*

$$m(r)=\int_{-\frac{\pi}{2}}^{\frac{\pi}{2}} \log^+|f(re^{i\theta})|\cos\theta\, d\theta\,, \qquad 0<r<\infty\,,$$

then $m(r)/r$ is a continuous non-decreasing function of r and is a convex function of $1/r^2$.

Proof. (Tsuji [60]). If we put $u=0$ for $x<0$, then u becomes a subharmonic function for $|z|<\infty$. Let μ be the positive mass distribution, defined by the Riesz' decomposition theorem, then since $u=0$ for $x<0$, $\mu=0$ for $x<0$. Let Δ_ρ: $|z|<\rho$, $\Re z>0$ be a half-disc and C_ρ: $|z|=\rho$, $\Re z\geqq 0$ be a semi-circle and L_ρ: $z=iy$ $(-\rho\leqq y\leqq\rho)$ be a segment on the imaginary axis, then $\Gamma_\rho=C_\rho+L_\rho$ is the boundary of Δ_ρ and $G_\rho(z,a)=\log\left|\frac{z+\bar{a}}{z-a}\cdot\frac{\rho^2-z\bar{a}}{\rho^2+za}\right|$ is the Green's function of Δ_ρ. Since $u=0$ on L_ρ, we have by Theorem II. 29,

$$u(z)=v_\rho(z)-w_\rho(z) \quad \text{in } \Delta_\rho\,, \tag{1}$$

where

$$v_\rho(z)=\frac{1}{2\pi}\int_{-\frac{\pi}{2}}^{\frac{\pi}{2}} u(\rho e^{i\varphi})\frac{\partial G_\rho(\rho e^{i\varphi}, z)}{\partial\nu}\rho\, d\varphi\,, \tag{2}$$

$$w_\rho(z)=\int_{\Delta_\rho} G_\rho(z,a)d\mu(a)\,. \tag{3}$$

Let

$$m(r,v_\rho)=\int_{-\frac{\pi}{2}}^{\frac{\pi}{2}} v_\rho(re^{i\theta})\cos\theta\, d\theta\,, \qquad m(r,w_\rho)=\int_{-\frac{\pi}{2}}^{\frac{\pi}{2}} w_\rho(re^{i\theta})\cos\theta\, d\theta\,.$$

$$(0<r<\rho)$$

then

$$m(r,u)=m(r,v_\rho)-m(r,w_\rho)\,. \tag{4}$$

By (2) and Lemma 1,

$$\begin{aligned} m(r,v_\rho)&=\frac{1}{2\pi}\int_{-\frac{\pi}{2}}^{\frac{\pi}{2}} u(\rho e^{i\varphi})d\varphi\int_{-\frac{\pi}{2}}^{\frac{\pi}{2}}\frac{\partial G_\rho(\rho e^{i\varphi}, re^{i\theta})}{\partial\nu}\rho\cos\theta\, d\theta \\ &=\frac{r}{\rho}\int_{-\frac{\pi}{2}}^{\frac{\pi}{2}} u(\rho e^{i\varphi})\cos\varphi\, d\varphi=\frac{rm(\rho,u)}{\rho}\,, \end{aligned}$$

so that

$$\frac{m(r, v_\rho)}{r} = \frac{m(\rho, u)}{\rho} \qquad (0 < r \leqq \rho). \tag{5}$$

Now

$$w_\rho(z) = \int_{\Delta_\rho} \log \left| \frac{z + \bar{a}}{z - a} \right| d\mu(a) - \int_{\Delta_\rho} \log \left| \frac{\rho^2 + za}{\rho^2 - z\bar{a}} \right| d\mu(a) = w'_\rho(z) - w''_\rho(z). \tag{6}$$

Let $z = re^{i\theta}$, $a = \tau e^{i\varphi}$, then we can prove similarly as Lemma 1, that if $\tau \leqq r \leqq \rho$,

$$\int_{-\frac{\pi}{2}}^{\frac{\pi}{2}} \log \left| \frac{z + \bar{a}}{z - a} \right| \cos\theta \, d\theta = \frac{\pi\tau\cos\varphi}{r} = \pi \frac{\Re(a)}{r},$$

and if $r \leqq \tau \leqq \rho$,

$$\int_{-\frac{\pi}{2}}^{\frac{\pi}{2}} \log \left| \frac{z + \bar{a}}{z - a} \right| \cos\theta \, d\theta = \frac{\pi r\cos\varphi}{\tau} = \pi \frac{r\Re(a)}{|a|^2},$$

hence

$$\frac{m(r, w'_\rho)}{r} = \frac{\pi}{r^2} \int_{|a| < r} \Re(a) d\mu(a) + \pi \int_{r \leqq |a| < \rho} \frac{\Re(a) d\mu(a)}{|a|^2}. \tag{7}$$

Since

$$\int_{-\frac{\pi}{2}}^{\frac{\pi}{2}} \log \left| \frac{1 + (r\tau/\rho^2) e^{i(\theta+\varphi)}}{1 - (r\tau/\rho^2) e^{i(\theta-\varphi)}} \right| \cos\theta \, d\theta = \pi \frac{r\tau\cos\varphi}{\rho^2} = \pi \frac{r\Re(a)}{\rho^2},$$

we have

$$\frac{m(r, w''_\rho)}{r} = \frac{\pi}{\rho^2} \int_{|a| < \rho} \Re(a) d\mu(a), \tag{8}$$

so that by (6), (7), (8),

$$\frac{m(r, w_\rho)}{r} = \frac{\pi}{r^2} \int_{|a| < r} \Re(a) d\mu(a) + \pi \int_{r \leqq |a| < \rho} \frac{\Re(a) d\mu(a)}{|a|^2} - \frac{\pi}{\rho^2} \int_{|a| < \rho} \Re(a) d\mu(a).$$

Hence if we put

$$\Omega(r) = \int_{|a| < r} \Re(a) d\mu(a), \tag{9}$$

then by the partial integration, we have easily

$$\frac{m(r, w_\rho)}{r} = 2\pi \int_r^\rho \frac{\Omega(t)}{t^3} dt. \tag{10}$$

so that by (4), (5), (10),

$$\frac{m(r, u)}{r} = \frac{m(\rho, u)}{\rho} - 2\pi \int_r^\rho \frac{\Omega(t)}{t^3} dt \quad (0 < r \leqq \rho). \tag{11}$$

Hence $m(r, u)/r$ is a continuous non-decreasing function of r and since $\frac{d(m(r, u)/r)}{d(1/r^2)}=-\pi\Omega(r)$, $m(r, u)/r$ is a convex function of $1/r^2$.

From (11), we have

$$2\pi\int_r^\rho \frac{\Omega(t)}{t^3}dt\leqq\frac{m(\rho, u)}{\rho},$$

so that

$$2\pi\int_0^\rho \frac{\Omega(t)}{t^3}dt\leqq\frac{m(\rho, u)}{\rho}. \tag{12}$$

Since $m(r, u)/r$ is a non-decreasing function of r,

$$\lim_{r\to\infty} m(r, u)/r=c, \qquad 0<c\leqq\infty, \tag{13}$$

exists. If $0<c<\infty$, then by (11),

$$\frac{m(r, u)}{r}=c-2\pi\int_r^\infty \frac{\Omega(t)}{t^3}dt, \tag{14}$$

and by (12),

$$2\pi\int_0^\infty \frac{\Omega(t)}{t^3}dt\leqq c. \tag{15}$$

Hence $\lim_{r\to 0}\Omega(r)/r^2=0$, $\lim_{r\to\infty}\Omega(r)/r^2=0$ and from this, we have by the integration by parts

$$\int_{|a|<\infty}\frac{\Re(a)d\mu(a)}{|a|^2}=2\int_0^\infty \frac{\Omega(t)}{t^3}dt<\infty. \tag{16}$$

By Lemma 1, if $z=re^{i\theta}$, $\lambda=r/\rho<1$, then

$$\begin{aligned} v_\rho(z)&=\frac{1}{2\pi}\int_{-\frac{\pi}{2}}^{\frac{\pi}{2}} u(\rho e^{i\varphi})\frac{\partial G_\rho(\rho e^{i\varphi}, re^{i\theta})}{\partial\nu}\rho\, d\varphi \\ &=\frac{2}{\pi}\int_{-\frac{\pi}{2}}^{\frac{\pi}{2}} u(\rho e^{i\varphi})\left(\frac{r}{\rho}\cos\varphi\cos\theta+O\left(\frac{r}{\rho}\right)^2\right)d\varphi, \end{aligned}$$

so that

$$\lim_{\rho\to\infty} v_\rho(z)=\frac{2cr\cos\theta}{\pi}=kx, \qquad k=2c/\pi. \tag{17}$$

Since

$$\log\left|\frac{\rho^2+z\bar{a}}{\rho^2-za}\right|=\frac{1}{2}\log\left(1+\frac{4\rho^2x\xi}{|\rho^2-za|^2}\right)\leqq\frac{2\rho^2x\xi}{|\rho^2-za|^2}\leqq\frac{2|z|\xi}{(\rho-|z|)^2},$$

$$x=\Re z, \quad \xi=\Re a,$$

$$w_\rho''(z) \leqq \frac{2|z|}{(\rho-|z|)^2}\int_{|a|<\rho} \Re(a)d\mu(a) = \frac{2|z|\rho^2}{(\rho-|z|)^2}\frac{\Omega(\rho)}{\rho^2},$$

hence $\lim_{\rho\to\infty} w_\rho''(z)=0$, so that

$$u(z)=kx-w(z), \qquad k=2c/\pi, \tag{18}$$

where

$$w(z)=\int_{|a|<\infty} \log\left|\frac{z+\bar{a}}{z-a}\right| d\mu(a) = \int_{|a|<\infty} G(z, a)d\mu(a), \tag{19}$$

$G(z, a)$ being the Green's function of the half-plane $x>0$.

Next we shall prove that, except a set of θ of capacity zero,

$$\lim_{r\to\infty} \frac{w(re^{i\theta})}{r}=0. \tag{20}$$

By Lemma 2, if $|\theta|\leqq\theta_0<\pi/2$ and $\arg a=\varphi$, we have for any $r_0>0$,

$$\frac{w(re^{i\theta})}{r}=\frac{1}{r}\int_{|a|<r_0} G(re^{i\theta}, a)d\mu(a)+\frac{1}{r}\int_{r_0\leqq|a|<\infty} G(re^{i\theta}, a)d\mu(a)$$

$$\leqq\frac{1}{r}\int_{|a|<r_0} G(re^{i\theta}, a)d\mu(a)+K(\theta_0)\int_{r_0\leqq|a|<\infty} \frac{\Re(a)}{|a|^2}(G(e^{i\theta}, e^{i\varphi})+1)d\mu(a).$$

Hence if we put

$$\chi(\theta)=\overline{\lim_{r\to\infty}} \frac{w(re^{i\theta})}{r}, \tag{21}$$

then if $|\theta|\leqq\theta_0<\pi/2$,

$$\chi(\theta)\leqq K(\theta_0)\int_{r_0\leqq|a|<\infty} \frac{\Re(a)}{|a|^2}(G(e^{i\theta}, e^{i\varphi})+1)d\mu(a). \tag{22}$$

Suppose that $\chi(\theta)>0$ on a set E of positive capacity on $|z|=1$, then by taking a suitable closed sub-set, we may assume that E is a closed set, contained in an angular domain $|\arg z|\leqq\theta_0<\pi/2$.

Let ν be the equilibrium distribution of E and

$$U(a)=\int_E \log\left|\frac{a+e^{-i\theta}}{a-e^{i\theta}}\right| d\nu(\theta)=\int_E \log\left|\frac{e^{i\theta}+\bar{a}}{e^{i\theta}-a}\right| d\nu(\theta)=\int_E G(e^{i\theta}, a)d\nu(\theta), \quad \nu(E)=1,$$

be the conductor potential of E, such that $U(a)\leqq V<\infty$ for any a. Then

$$\int_E \chi(\theta)d\nu(\theta)\leqq K(\theta_0)(V+1)\int_{r_0\leqq|a|<\infty} \frac{\Re(a)d\mu(a)}{|a|^2}.$$

Since $\int_0^\infty \frac{\Re(a)d\mu(a)}{|a|^2}<\infty$, the right hand side tends to zero, if $r_0\to\infty$, hence $\int_E \chi(\theta)d\nu(\theta)=0$, which is absurd. Hence $\chi(\theta)=0$, or (20) holds, except a set of θ of capacity zero. Hence $\lim_{r\to\infty}\frac{u(re^{i\theta})}{r}=k\cos\theta$, except a set of θ of capacity zero.

CHAPTER V

NEVANLINNA'S THEORY OF MEROMORPHIC FUNCTIONS

1. First fundamental theorem

1. Let $[a, b]=\dfrac{|a-b|}{\sqrt{(1+|a|^2)(1+|b|^2)}}$ be the spherical distance of two points a, b.

For $a>0$, we put $\log^+ a=\operatorname{Max}(\log a, 0)$, then

$$\log a=\log^+ a-\log^+\frac{1}{a}, \qquad \log^+(a_1, \cdots, a_n)\leqq\sum_{\nu=1}^{n}\log^+ a_\nu,$$

$$\log^+(a_1+\cdots+a_n)\leqq\sum_{\nu=1}^{n}\log^+ a \ +\log n\ .$$

Let $w(z)$ be meromorphic for $|z|<R(\leqq\infty)$. If $R<\infty$, then by taking z/R instead of z, we may assume that $R=1$, so that we assume that

$$R=\infty, \text{ or } R=1\ .$$

Let $n(r, a)$ be the number of zero points of $w(z)-a$ in $|z|<r<R$, where if $a=\infty$, $n(r, a)$ denotes the number of poles of $w(z)$ in $|z|<r$.

We define $m(r, a)$ (Schmiegungsfunktion) and $N(r, a)$ (Anzahlfunktion), which play an important rôle in Nevanlinna's theory, as follows.

$$m(r, a)=\frac{1}{2\pi}\int_0^{2\pi}\log\frac{1}{[w(re^{i\theta}), a]}d\theta\ , \tag{1}$$

so that

$$m(r, a)=\frac{1}{2\pi}\int_0^{2\pi}\log\frac{\sqrt{(1+|w(re^{i\theta})|^2)(1+|a|^2)}}{|w(re^{i\theta})-a|}d\theta, \quad a\neq\infty,$$

$$=\frac{1}{2\pi}\int_0^{2\pi}\log\sqrt{1+|w(re^{i\theta})|^2}\,d\theta, \qquad a=\infty\ .$$

$$N(r, a)=\int_0^r\frac{n(r, a)-n(+0, a)}{r}dr+n(+0, a)\log r+C\ , \tag{2}$$

where C is a constant, which is determined by the condition:

$$\lim_{r\to 0}(m(r, a)+N(r, a))=0\ . \tag{3}$$

Since $[w, a]\leqq 1$, $\qquad m(r, a)\geqq 0.$

We remark that

$$m(r, a)=\frac{1}{2\pi}\int_0^{2\pi}\log^+\frac{1}{|w(re^{i\theta})-a|}d\theta+O(1), \quad a\neq\infty$$
$$=\frac{1}{2\pi}\int_0^{2\pi}\log^+|w(re^{i\theta})|d\theta+O(1), \qquad a=\infty .$$

Theorem V. 1. (First fundamental theorem). *Let $w(z)$ be meromorphic in $|z|<R$, then*

$$T(r)=m(r, a)+N(r, a) \quad (0\leqq r<R),$$ Nevanlinna [1].

where

$$T(r)=\int_0^r \frac{S(r)}{r}dr ,$$
$$S(r)=\frac{A(r)}{\pi}=\frac{1}{\pi}\iint_{|z|\leq r}\left(\frac{|w'(z)|}{1+|w(z)|^2}\right)^2 rdrd\theta, \qquad z=re^{i\theta},$$

$A(r)$ being the area of the image of $|z|\leqq r$ on the w-sphere.

Ahlfors [1], Shimizu [1].

Hence $T(r)$ is an increasing convex function of $\log r$.

$T(r)$ is called *the characteristic function of $w(z)$.*

Proof. (Ahlfors [5]). By the argument principle,

$$\frac{dm(r, a)}{dr}-\frac{dm(r, b)}{dr}=\frac{1}{2\pi}\int_0^{2\pi}\frac{d}{dr}\log\left|\frac{w-b}{w-a}\right|d\theta$$
$$=\frac{1}{2\pi r}\int_{|z|=r}d\arg\left(\frac{w-b}{w-a}\right)=\frac{n(r, b)-n(r, a)}{r}.$$

Since $\frac{dN(r, a)}{dr}=\frac{n(r, a)}{r}$, we have

$$\frac{dm(r, a)}{dr}+\frac{dN(r, a)}{dr}=\frac{dm(r, b)}{dr}+\frac{dN(r, b)}{dr}.$$

Hence by the condition (3), we have $m(r, a)+N(r, a)=m(r, b)+N(r, b)$, so that $T(r)=m(r, a)+N(r, a)$ depends on r only.

Let $d\omega(a)$ be the surface element on the w-sphere K, then

$$\iint_K d\omega(a)=\pi,$$

so that

$$T(r)=\frac{1}{\pi}\iint_K m(r, a)d\omega(a)+\frac{1}{\pi}\iint_R N(r, a)d\omega(a) .$$

Since $\iint_K\log\frac{1}{[w, a]}d\omega(a)$ is independent of w, we have

$$T(r)=\frac{1}{\pi}\iint_K N(r,a)\,d\omega(a)+\text{const.}=\frac{1}{\pi}\int_0^r\frac{dr}{r}\iint_K n(r,a)\,d\omega(a)+\text{const.}$$

Since

$$\frac{1}{\pi}\iint_K n(r,a)\,d\omega(a)=\frac{A(r)}{\pi}=S(r),\qquad T(r)=\int_0^r\frac{S(r)}{r}dr+\text{const.}.$$

Since by (3), $\lim_{r\to 0}T(r)=0$, we see that const. $=0$, so that

$$T(r)=\int_0^r\frac{S(r)}{r}dr\,.$$

2. If we denote the characteristic function of $w(z)$ by $T(w,r)$, then we have easily

Theorem V. 2.

(i) $T(1/w,r)=T(w,r)$,

(ii) $T(w_1+\cdots+w_n,r)\leqq T(w_1,r)+\cdots+T(w_n,r)+O(1)$,

(iii) $T(w_1\cdots w_n,r)\leqq T(w_1,r)+\cdots+T(w_n,r)+O(1)$,

(iv) $T\left(\dfrac{\alpha w+\beta}{\gamma w+\delta},r\right)=T(w,r)+O(1)\,.$

Proof. (i), (ii), (iii) follow from the definition of $T(r)$, (iv) can be proved easily for special linear functions $w+\alpha$, kw, $1/w$ and since any linear function can be decomposed into such special linear functions, (iv) holds in general.

Theorem V. 3. *Let $w(z)$ be meromorphic for $|z|<\infty$. If $w(z)$ is a rational function, then*

$$T(r)=O(\log r)\,.$$

If $z=\infty$ is an essential singularity of $w(z)$, then

$$\lim_{r\to\infty}\frac{T(r)}{\log r}=\infty\,.$$

Proof. If $w(z)$ is a rational function, then $A(r)=O(1)$, so that $T(r)=O(\log r)$. If $z=\infty$ is an essential singularity of $w(z)$ and is a cluster point of poles of $w(z)$, then $n(r,\infty)\to\infty$, as $r\to\infty$. Since $m(r,\infty)\geqq 0$,

$$T(r)\geqq N(r,\infty)=\int_0^r\frac{n(r,\infty)-n(+0,\infty)}{r}dr+n(+0,\infty)\log r+\text{const.}$$

$$\geqq\int_{\sqrt{r}}^r\frac{n(r,\infty)}{r}dr+O(1)\geqq n(\sqrt{r},\infty)\int_{\sqrt{r}}^r\frac{dr}{r}+O(1)=\frac{n(\sqrt{r},\infty)}{2}\log r+O(1),$$

hence

$$\lim_{r\to\infty}\frac{T(r)}{\log r}=\infty\,.$$

If $z=\infty$ is not a cluster point of poles of $w(z)$, then $w(z)$ is regular for $r_0\leqq|z|<\infty$. We choose z_1 arbitrary and U_1 be its neighbourhood and U_1^* be its image on the w-plane. Since by Weierstrass' theorem, the set of values, taken by $w=w(z)$ is everywhere dense on the w-plane, there exists z_2, which lies outside of U_1 and $w(z_2)\in U_1^*$. We choose a neighbourhood U_2 of z_2, so small that its image U_2^* lies in U_1^* Similarly we can find z_ν $(\nu=1, 2, \cdots)$ $(z_\nu\to\infty)$ and its neighbourhood U_ν and its image U_ν^*, such that $U_1^*\supset U_2^*\supset\cdots$. We choose U_ν so small that U_ν^* converges to a point a, then $w(z)-a$ has infinitely many zero points, so that $n(r, a)\to\infty$, as $r\to\infty$. From this, we can prove similarly as above that $\lim_{r\to\infty}\dfrac{T(r)}{\log r}=\infty$.

Theorem V. 4. *Let $w(z)$ be meromorphic for $|z|<\infty$ and*

$$L(r)=\int_0^{2\pi}\frac{|w'(re^{i\theta})|}{1+|w(re^{i\theta})|^2}r d\theta$$

be the length of the image of $|z|=r$ on the w-sphere and

$$\Lambda(r)=\int_1^r\frac{L(r)}{r}dr\,.$$

Let $r_0\,(>1)$ be such that $T(r_0)>1$. If $r\,(\geqq r_0)$ lies outside of a sequence of intervals $\{J_\nu\}$, such that $\sum_\nu\int_{J_\nu}\dfrac{dr}{r\log r}<\infty$, then

$$\Lambda(r)\leqq\sqrt{T(r)}\log T(r)\,.$$

Dinghas [2].

Proof. By Schwarz's inequality,

$$L(r)\leqq\left[2\pi r\int_0^{2\pi}\left(\frac{|w'(re^{i\theta})|}{1+|w(re^{i\theta})|^2}\right)^2 r d\theta\right]^{\frac{1}{2}}=\sqrt{2\pi^2 r\,S'(r)}\,,$$

so that

$$\Lambda(r)=\int_1^r\frac{L(r)}{r}dr\leqq\sqrt{2\pi^2}\int_1^r\sqrt{\frac{S'(r)}{r}}dr\,,$$

hence

$$\Lambda(r)^2\leqq 2\pi^2\int_1^r\frac{dr}{r}\int_1^r S'(r)dr\leqq 2\pi^2\log r\,S(r)=2\pi^2 r\log r\frac{dT(r)}{dr}\,. \tag{1}$$

The set of r, such that $\Lambda(r)>\sqrt{T(r)}\log T(r)$ consists of intervals $\{J_\nu\}$. If $r\geqq r_0$ lies outside of $\{J_\nu\}$, then

$$\Lambda(r)\leqq\sqrt{T(r)}\log T(r)\,. \tag{2}$$

If $r\in J_\nu$ and $r\geqq r_0$, then by (1),

$$T(r)\log^2 T(r)<\Lambda^2(r)\leqq 2\pi^2 r\log r\frac{dT(r)}{dr}\,,$$

so that

$$\sum_{\nu}\int_{J_\nu}\frac{dr}{r\log r}<2\pi^2\int_{T(r_0)}^{\infty}\frac{dT}{T\log^2 T}<\infty\,. \qquad (3)$$

2. On $m(r, a)$.

Theorem V. 5. *Let $w(z)$ be meromorphic for $|z|<R\leqq\infty$, such that $\lim_{r\to R} T(r)=\infty$. Then there exist $r_1<r_2<\cdots<r_n\to R$ and sets E_n: $E_n\supset E_{n+1}$, $\gamma_+(E_n)\leqq 1/T^2(r_n)$, such that if a lies outside of E_n, then*

$$m(r, a)\leqq 4\sqrt{T(r)}\log T(r), \qquad r_n<r\leqq R.$$

Hence except a set of a of capacity zero,

$$\lim_{r\to R} m(r, a)/T(r)=0.$$

Proof. Since $\lim_{r\to R} T(r)=\infty$, there exists $r_1<r_2<\cdots<r_\nu\to R$, such that

$$T(r_\nu)=T(r_{\nu-1})+2\sqrt{T(r_\nu)}\log T(r_\nu)\,. \qquad (1)$$

Since $x-2\sqrt{x}\log x$ is an increasing function of $x\geqq x_0>0$, $T(r)-2\sqrt{T(r)}\log T(r)$ is an increasing function of $r\geqq r_0>0$.

Let E be a closed set of positive capacity on the w-sphere K and $u(w)$ be its elliptic conductor potential:

$$u(w)=\int_E \log\frac{1}{[w, a]}\,d\mu(a), \qquad \mu(E)=1\,, \qquad (2)$$

such that $u(w)\leqq V=\log 1/\gamma_+(E)$ on K. Hence

$$\int_E m(r, a)\,d\mu(a)=\frac{1}{2\pi}\int_0^{2\pi}d\theta\int_E \log\frac{1}{[w(re^{i\theta}), a]}\,d\mu(a)\leqq V\,, \qquad (3)$$

so that by the first fundamental theorem,

$$T(r)\leqq V+\int_E N(r, a)\,d\mu(a)\,. \qquad (4)$$

Let e_ν be the set of a, such that

$$N(r_\nu, a)\leqq T(r_\nu)-2\sqrt{T(r_\nu)}\log T(r_\nu), \qquad a\in e_\nu\,, \qquad (5)$$

and $V_\nu=\log 1/\gamma_+(e_\nu)$. Suppose that $\gamma_+(e_\nu)>0$, then if we put $r=r_\nu$, $E=e_\nu$ in (4), then we have

$$2\sqrt{T(r_\nu)}\log T(r_\nu)\leqq V_\nu\,. \qquad (6)$$

Hence if we put $E_n=e_n+e_{n+1}+\cdots$ and $V_n^*=\log 1/\gamma_+(E_n)$, then similarly as Theorem III. 17, we have

$$\frac{1}{V_n^*}\leqq\sum_{\nu=n}^{\infty}\frac{1}{V_\nu}\leqq\frac{1}{2}\sum_{\nu=n}^{\infty}\frac{1}{\sqrt{T(r_\nu)}\log T(r_\nu)}\,. \qquad (7)$$

Now by (1),

$$\frac{1}{\sqrt{T(r_\nu)}\log T(r_\nu)}=\frac{T(r_\nu)-T(r_{\nu-1})}{\sqrt{T(r_\nu)}\log T(r_\nu)\cdot 2\sqrt{T(r_\nu)}\log T(r_\nu)}$$
$$=\frac{T(r_\nu)-T(r_{\nu-1})}{2T(r_\nu)\log^2 T(r_\nu)}\leqq\frac{1}{2}\int_{r_{\nu-1}}^{r_\nu}\frac{dT(r)}{T(r)\log^2 T(r)},$$

so that

$$\frac{1}{V_n^*}\leqq\frac{1}{2\sqrt{T(r_n)}\log T(r_n)}+\frac{1}{4}\int_{r_n}^{\infty}\frac{dT(r)}{T(r)\log^2 T(r)}$$
$$=\frac{1}{2\sqrt{T(r_n)}\log T(r_n)}+\frac{1}{4\log T(r_n)}\leqq\frac{1}{2\log T(r_n)}\quad(n\geqq n_0).$$

Hence

$$\gamma_+(E_n)\leqq 1/T^2(r_n)\qquad(n\geqq n_0). \tag{8}$$

If a lies outside of E_n, then

$$N(r_\nu, a)\geqq T(r_\nu)-2\sqrt{T(r_\nu)}\log T(r_\nu),\qquad \nu=n,n+1,\cdots. \tag{9}$$

If $r_\nu\leqq r\leqq r_{\nu+1}$, then by (1),

$$N(r, a)\geqq N(r_\nu, a)\geqq T(r_\nu)-2\sqrt{T(r_\nu)}\log T(r_\nu)$$
$$\geqq T(r)-2\sqrt{T(r)}\log T(r)+[T(r_\nu)-2\sqrt{T(r_\nu)}\log T(r_\nu)]$$
$$-[T(r_{\nu+1})-2\sqrt{T(r_{\nu+1})}\log T(r_{\nu+1})]=$$
$$=T(r)-2\sqrt{T(r)}\log T(r)-2\sqrt{T(r_\nu)}\log T(r_\nu)\geqq T(r)-4\sqrt{T(r)}\log T(r). \tag{10}$$

Hence by (9), (10), if a lies outside of E_n,

$$N(r, a)\geqq T(r)-4\sqrt{T(r)}\log T(r),\qquad r_n\leqq r<R,$$

or

$$m(r, a)\leqq 4\sqrt{T(r)}\log T(r),\qquad r_n\leqq r<R. \tag{11}$$

By the definition of E_n, $E_n\supset E_{n+1}$, so that if we put $\lim_n E_n=E$, then by (8), $\gamma(E)=0$ and if a lies outside of E, then $\lim_{r\to R} m(r, a)/T(r)=0$.

Theorem V. 6. *Let $w=w(z)$ be meromorphic in $|z|<R$ and F be the Riemann surface, generated by $w=w(z)$ on the w-sphere K.*

Let Δ: $|w-a|<\rho$ be a disc and F_ρ be the part of F, which lies above Δ, then F_ρ consists of at most a countable number of connected pieces. If F_ρ consists of only compact pieces and the number of sheets of each piece is $\leqq N<\infty$, then

$$m(r, a)=O(\log r). \qquad \text{Selberg [2].}$$

If F_ρ consists of only simply connected compact pieces and the number of sheets of each piece is $\leqq N<\infty$, then

$$m(r, a)=O(1). \qquad \text{Collingwood [1].}$$

We remark that if $w(z)$ is regular in $|z|<R$, then any compact piece of F_ρ is simply connected.

Proof. The image of F_ρ on the z-plane consists of only compact domains Δ_ν ($\nu=1, 2, \cdots$). Let the circle $|z|=r$ intersect Δ_ν ($\nu=1, 2, \cdots, n(r)$) in a set θ_ν of arcs and let $r\theta_\nu(r)$ be its linear measure, then

$$m(r, a)=\frac{1}{2\pi}\sum_{\nu=1}^{n(r)}\int_{\theta_\nu}\log\frac{1}{|w(re^{i\theta})-a|}d\theta+O(1)\,. \tag{1}$$

Let z_ν^i ($i=1, 2, \cdots, n_\nu \leqq N$) be the image of a in Δ_ν and $g_{\Delta_\nu}(z, z_\nu^i)$ be the Green's function of Δ_ν with z_ν^i as its pole, then

$$\log\frac{\rho}{|w(z)-a|}=\sum_{i=1}^{n_\nu} g_{\Delta_\nu}(z, z_\nu^i)\,, \qquad z\in\Delta_\nu\,. \tag{2}$$

By Theorem I. 22,

$$\int_{\theta_\nu} g_{\Delta_\nu}(re^{i\theta}, z_\nu^i)\,d\theta \leqq \pi^2\tan\frac{\theta_\nu(r)}{4}\,.$$

We may assume that $\theta_1(r)\geqq\theta_2(r)\geqq\cdots$, then $\theta_\nu(r)\leqq\pi$ ($\nu\geqq 2$), so that $\tan\frac{\theta_\nu(r)}{4}\leqq\frac{\theta_\nu(r)}{\pi}$ ($\nu\geqq 2$), hence

$$\sum_{\nu=2}^{n(r)}\sum_{i=1}^{n_\nu}\int_{\theta_\nu} g_{\Delta_\nu}(re^{i\theta}, z_\nu^i)\,d\theta \leqq N\pi\sum_{\nu=2}^{n(r)}\theta_\nu(r)\leqq 2N\pi^2\,. \tag{3}$$

$\theta_1(r)$ may be greater than π. For a large $r>0$, we may assume that Δ_1 does not contain $z=0$. Let r_0 be the minimum distance of the boundary of Δ_1 from $z=0$, then since $\log\left|\frac{r_0^2-z_0 z}{r_0(z-z_0)}\right|$ ($z_0\in\Delta_1$) is the Green's function of $|z|>r_0$, by the maximum principle, we have

$$g_{\Delta_1}(z, z_0)\leqq\log\left|\frac{r_0^2-\bar{z}_0 z}{r_0(z-z_0)}\right|, \qquad z_0\in\Delta_1\,,\ z\in\Delta_1\,. \tag{4}$$

Since if $r\geqq r_0$, $\int_0^{2\pi}\log\left|\frac{r_0^2-\bar{z}_0\,re^{i\theta}}{r_0(re^{i\theta}-z_0)}\right|d\theta=2\pi\,\mathrm{Min}\left(\log\frac{|z_0|}{r_0},\ \log\frac{r}{r_0}\right)$,[1)] we have

$$\int_{\theta_1} g_{\Delta_1}(re^{i\theta}, z_0)\,d\theta\leqq 2\pi\log\frac{r}{r_0}\,,$$

so that

$$\sum_{i=1}^{n_1} g_{\Delta_1}(re^{i\theta}, z_1^i)\,d\theta\leqq 2\pi N\log\frac{r}{r_0}\,. \tag{5}$$

Hence from (1), (2), (3), (5), we have $m(r, a)=O(\log r)$.

Next suppose that all Δ_ν are simply connected, then by $\zeta=\log z=\log r+i\theta$, we map Δ_ν conformally on a simply connected domain D_ν on the ζ-plane and apply Theorem I. 23, then we have $m(r, a)=O(1)$.

1) We use: $\int_0^{2\pi}\log|re^{i\theta}-a|d\theta=2\pi\,\mathrm{Max}\,(\log r,\ \log|a|)$.

3. Function of bounded type

Let $w(z)$ be meromorphic in $|z|<1$ and $T(r)=O(1)$ $(0\leqq r<1)$, then $w(z)$ is called *a function of bounded type*.

Theorem V. 7. *Let $w(z)$ be meromorphic in $|z|<1$ and $T(r)=O(1)$ and $\{z_\nu(a)\}$ be zero points of $w(z)-a$ in $|z|<1$, then $\sum_\nu(1-|z_\nu(a)|)<\infty$.*

Proof. $$n(r, a)(1-r)\leqq 2\int_r^{\frac{r+1}{2}}\frac{n(r, a)}{r}dr=2N\left(\frac{r+1}{2}, a\right)+O(1)$$
$$\leqq 2T\left(\frac{r+1}{2}\right)+O(1)=O(1)\,.$$
$$\sum_{|z_\nu(a)|<r}(1-|z_\nu(a)|)=\int_0^r(1-t)dn(t, a)\leqq(1-r)n(r, a)+\int_0^r n(t, a)\,dt$$
$$\leqq O(1)+\int_{\frac{1}{2}}^r\frac{n(t, a)dt}{t}=O(1)+N(r, a)\leqq O(1)+T(r)=O(1).$$

Theorem V. 8. *Let $w(z)$ be meromorphic in $|z|<1$ and $T(r)=O(1)$, then $w(z)$ can be expressed in the form: $w(z)=w_1(z)/w_2(z)$, where $w_i(z)$ $(i=1, 2)$ is regular and $|w_i(z)|<1$ in $|z|<1$.*

Conversely, if $w(z)$ is of this form, then $T(r)=O(1)$. Nevanlinna [6].

Proof. If $w(z)$ is of the form $w(z)=w_1(z)/w_2(z)$, then since $w_i(z)$ is regular in $|z|<1$, $N(w_i, r, \infty)=O(1)$ and since $m(w_i, r, \infty)=O(1)$, $T(w_i, r)=O(1)$, so that $T(r)=T(w, r)=T(w_1, r)+T(w_2, r)+O(1)=O(1)$.

Next suppose that $T(r)=O(1)$. We may assume that $w(0)\neq 0, \infty$, since, if otherwise, we divide $w(z)$ by a suitable z^k. Let $\{a_\nu\}$ be zero points and $\{b_\nu\}$ be poles of $w(z)$ in $|z|<1$, then since $\sum_\nu(1-|a_\nu|)<\infty$, $\sum_\nu(1-|b_\nu|)<\infty$ (Theorem. V. 7),

$$\pi_1(z)=\prod_\nu\frac{|a_\nu|}{a_\nu}\cdot\frac{a_\nu-z}{1-\bar{a}_\nu z}, \quad \pi_2(z)=\prod_\nu\frac{|b_\nu|}{b_\nu}\cdot\frac{b_\nu-z}{1-b_\nu z} \tag{1}$$

are convergent in $|z|<1$ and $\pi_i(z)$ $(i=1, 2)$ is regular and $|\pi_i(z)|<1$ in $|z|<1$.[1] If we put $F(z)=w(z)\pi_2(z)/\pi_1(z)$, then $F(z)$ is regular and $\neq 0$ in $|z|<1$ and

$$T(F, r)=T(w, r)+T(\pi_1, r)+T(\pi_2, r)+O(1)=O(1)\,,$$

so that $T(1/F, r)=T(F, r)=O(1)$. Hence

$$\int_0^{2\pi}\log^+|F(re^{i\theta})|d\theta=O(1), \quad \int_0^{2\pi}\log^+\left|\frac{1}{F(re^{i\theta})}\right|d\theta=O(1)\,. \tag{2}$$

Let for $0<\rho<1$

1) If $w(z)$ has no zero points in $|z|<1$, then we put $\pi_1(z)\equiv 1$ and similarly for $\pi_2(z)$.

$$u_\rho(z)=\frac{1}{2\pi}\int_0^{2\pi}\log^+|F(\rho e^{i\varphi})|\frac{\rho^2-r^2}{\rho^2-2\rho r\cos(\varphi-\theta)+r^2}d\varphi,\quad z=re^{i\theta},\quad (3)$$
$$(0\leqq r<\rho<1)$$

then $u_\rho(z)$ is harmonic in $|z|<\rho$ and $u_\rho=\log^+|F|$ on $|z|=\rho$. Since $\log^+|F(z)|$ is subharmonic, we have $\log^+|F(z)|\leqq u_\rho(z)$ in $|z|\leqq\rho$.
If $0<\rho_1<\rho_2<1$, then on $|z|=\rho_1$, $u_{\rho_1}(z)=\log^+|F(z)|\leqq u_{\rho_2}(z)$, so that $u_{\rho_1}(z)\leqq u_{\rho_2}(z)$ in $|z|\leqq\rho_1$, hence $u_\rho(z)$ is an increasing function of ρ and since $u_\rho(z)\geqq 0$ and

$$u_\rho(0)=\frac{1}{2\pi}\int_0^{2\pi}\log^+|F(\rho e^{i\varphi})|d\varphi=O(1)\,,\quad (4)$$

by Harnack's theorem, $\lim_{\rho\to 1} u_\rho(z)=u(z)\geqq 0$ $(|z|<1)$ exists and $u(z)$ is harmonic in $|z|<1$. Similarly if we put

$$u_\rho^*(z)=\frac{1}{2\pi}\int_0^{2\pi}\log^+\left|\frac{1}{F(\rho e^{i\varphi})}\right|\frac{\rho^2-r^2}{\rho^2-2\rho r\cos(\varphi-\theta)+r^2}d\varphi\,,\quad (5)$$

then $\lim_{\rho\to 1} u_\rho^*(z)=u^*(z)$ $(|z|<1)$ exists and $u^*(z)$ is harmonic in $|z|<1$.

Since $F(z)\neq 0$, $\log|F(z)|$ is harmonic in $|z|<1$, so that

$$\log|F(z)|=\frac{1}{2\pi}\int_0^{2\pi}\log|F(\rho e^{i\varphi})|\frac{\rho^2-r^2}{\rho^2-2\rho r\cos(\varphi-\theta)+r^2}d\varphi\,.$$

Since $\log|F|=\log^+|F|-\log^+\left|\frac{1}{F}\right|$, we have

$$\log|F(z)|=u_\rho(z)-u_\rho^*(z)\to u(z)-u^*(z),\quad \rho\to 1,$$

so that

$$\log|F(z)|=u(z)-u^*(z),\qquad u\geqq 0,\ u^*\geqq 0 \text{ in } |z|<1\,.\quad (6)$$

Let $v(z)$, $v^*(z)$ be the conjugate harmonic function of u, u^* respectively, then $\log F(z)=(u+iv)-(u^*+iv^*)$, hence if we put $g_1(z)=e^{-(u^*+iv^*)}$, $g_2(z)=e^{-(u+iv)}$, then $g_i(z)$ $(i=1,2)$ is regular and $|g_i(z)|<1$ in $|z|<1$ and $F(z)=g_1(z)/g_2(z)$, so that if we put $w_1(z)=\pi_1(z)\,g_1(z)$, $w_2(z)=\pi_2(z)\,g_2(z)$, then $w(z)=w_1(z)/w_2(z)$, where $w_i(z)$ $(i=1,2)$ is regular and $|w_i(z)|<1$ in $|z|<1$. ——

By Theorem IV. 8 and 9 and Theorem V. 8, we have

Theorem V. 9. *Let $w(z)$ be meromorphic in $|z|<1$ and $T(r)=O(1)$, then for almost all $e^{i\theta}$, $\lim_{z\to e^{i\theta}} w(z)=w(e^{i\theta})$ $(\neq 0,\ \neq\infty)$ exists uniformly, when $z\to e^{i\theta}$ from the inside of any fixed Stolz domain, whose vertex is at $e^{i\theta}$ and $|\log|w(e^{i\theta})||$ is integrable on $|z|=1$.*

4. Order of a meromorphic function

(i) Let $w(z)$ be meromorphic for $|z|<\infty$, then its *order* ρ is defined by

$$\overline{\lim_{r\to\infty}}\frac{\log T(r)}{\log r}=\rho\,.\quad (1)$$

We can prove easily that if $\rho<\infty$, then for any $\varepsilon>0$,

$$\int^{\infty}\frac{T(r)}{r^{\rho+1+\varepsilon}}dr<\infty, \qquad \int^{\infty}\frac{T(r)}{r^{\rho+1-\varepsilon}}dr=\infty. \tag{2}$$

$w(z)$ is said *of convergence*, or *of divergence type*, according as

$$\int^{\infty}\frac{T(r)}{r^{\rho+1}}dr<\infty, \quad \text{or} \quad \int^{\infty}\frac{T(r)}{r^{\rho+1}}dr=\infty. \tag{3}$$

By Theorem V. 2, *if* $w_1(z)$, $w_2(z)$ *are of order* $\leqq\rho$, *then* $w_1(z)+w_2(z)$, $w_1(z)w_2(z)$, $w_1(z)/w_2(z)$ *are of order* $\leqq\rho$.

Let $\{z_\nu(a)\}$ be zero points of $w(z)-a$, then we can prove easily that

$$\int^{\infty}\frac{N(r,a)}{r^{\lambda+1}}dr, \quad \int^{\infty}\frac{n(r,a)}{r^{\lambda+1}}dr, \quad \sum_{\nu}\frac{1}{|z_\nu(a)|^{\lambda}} \quad (\lambda>0) \tag{4}$$

are convergent or divergent at the same time.

Since $N(r,a)\leqq T(r)$, we have from (2), (4),

Theorem V. 10. *Let* $w(z)$ *be a meromorphic function of finite order* ρ *for* $|z|<\infty$, *then* $\sum_{\nu}\frac{1}{|z_\nu(a)|^{\rho+\varepsilon}}<\infty$ *for any* $\varepsilon>0$.

(ii) Let $w(z)$ be meromorphic in $|z|<1$. We define its *order* ρ by

$$\varlimsup_{r\to1}\log T(r)/\log\frac{1}{1-r}=\rho. \tag{1*}$$

If $\rho<\infty$, then for any $\varepsilon>0$,

$$\int^{1}T(r)(1-r)^{\rho-1+\varepsilon}dr<\infty, \quad \int^{1}T(r)(1-r)^{\rho-1-\varepsilon}dr=\infty. \tag{2*}$$

$w(z)$ is said *of convergence*, or *of divergence type*, according as

$$\int^{1}T(r)(1-r)^{\rho-1}dr<\infty, \quad \text{or} \quad \int^{1}T(r)(1-r)^{\rho-1}dr=\infty. \tag{3*}$$

If $w_1(z)$, $w_2(z)$ *are of order* $\leqq\rho$, *then* $w_1(z)+w_2(z)$, $w_1(z)w_2(z)$, $w_1(z)/w_2(z)$ *are of order* $\leqq\rho$.

We can prove easily that

$$\int^{1}N(r,a)(1-r)^{\lambda-1}dr, \quad \int^{1}n(r,a)(1-r)^{\lambda}dr, \quad \sum_{\nu}(1-|z_\nu(a)|)^{\lambda+1} \quad (\lambda>0) \tag{4*}$$

are convergent or divergent at the same time.

Hence from (2*), (4*), we have

Theorem V. 11. *Let* $w(z)$ *be a meromorphic function of finite order* ρ *in* $|z|<1$, *then* $\sum_{\nu}(1-|z_\nu(a)|)^{\rho+1+\varepsilon}<\infty$ *for any* $\varepsilon>0$.

Theorem V. 12. *Let* $w(z)$ *be an integral function and* $M(r)=\operatorname{Max}_{|z|=r}|w(z)|$, *then*

$$\overline{\lim_{r\to\infty}} \frac{\log\log M(r)}{\log r} = \overline{\lim_{r\to\infty}} \frac{\log T(r)}{\log r} = \rho .$$

Proof. Since $w(z)$ has no poles, $N(r, \infty)=O(1)$, so that

$$T(r)=m(r, \infty)+N(r, \infty)=\frac{1}{2\pi}\int_0^{2\pi}\log^+|w(re^{i\theta})|d\theta+O(1)$$

$$\leqq \log M(r)+O(1) . \tag{1}$$

Since $\log^+|w(z)|$ is subharmonic, if $r_1=2r$, then

$$\log^+|w(re^{i\theta})| \leqq \frac{1}{2\pi}\int_0^{2\pi}\log^+|w(r_1e^{i\varphi})|\frac{r_1^2-r^2}{r_1^2-2r_1r\cos(\varphi-\theta)+r^2}d\varphi$$

$$\leqq \frac{r_1+r}{r_1-r}\cdot\frac{1}{2\pi}\int_0^{2\pi}\log^+|w(r_1e^{i\theta})|d\theta \leqq 3\,T(r_1)+O(1)=3\,T(2r)+O(1) ,$$

hence

$$\log M(r)\leqq 3\,T(2r)+O(1) , \tag{2}$$

so that by (1), (2),

$$T(r)-O(1)\leqq \log M(r)\leqq 3\,T(2r)+O(1) . \tag{3}$$

From this, we have $\displaystyle\overline{\lim_{r\to\infty}} \frac{\log\log M(r)}{\log r} = \overline{\lim_{r\to\infty}} \frac{\log T(r)}{\log r}$. ——

Similarly we can prove

Theorem V. 13. *Let $w(z)$ be meromorphic in $|z|<1$ and put*

$$\overline{\lim_{r\to 1}}\log T(r)/\log\frac{1}{1-r}=\rho, \qquad \overline{\lim_{r\to 1}}\log\log M(r)/\log\frac{1}{1-r}=\rho^* ,$$

then

$$\rho\leqq\rho^*\leqq\rho+1 .$$

There exists $w(z)$, such that $\rho\neq\rho^*$. For example, if $w(z)=e^{\frac{1}{(1-z)^k}}$ $(k>1)$, then $\rho^*=k$, $\rho=k-1$.

5. Second fundamental theorem

Let $w(z)$ be meromorphic in $|z|<R\leqq\infty$ and ζ be a pole of order $\nu(\zeta)$, then $\nu(\zeta)-1$ is called the multiplicity of ζ. Let

$$\pi(r)=\sum_{|\zeta|<r}(\nu(\zeta)-1), \text{ added for all poles } \zeta \text{ in } |z|<r<R . \tag{1}$$

Since a pole of the k-th order of $w(z)$ is a pole of the $(k+1)$-th order of $w'(z)$ and $2k-(k+1)=k-1$, if we denote the number of poles of $w'(z)$ in $|z|<r$ by $n'(r, \infty)$, then

$$\pi(r)=2n(r, \infty)-n'(r, \infty) . \tag{2}$$

Let $\xi(a)$ $(a\neq\infty)$ be a zero point of $w(z)-a$ of order $\nu(\xi(a))$, then $\nu(\xi(a))-1$ is called the multiplicity of $\xi(a)$. Let

$$n_1(r, a) = \sum_{|\xi(a)|<r} (\nu(\xi(a))-1), \text{ added for all } \xi(a) \text{ in } |z|<r. \tag{3}$$

Since $w'(z)$ has a zero point of order $\nu(\xi(a))-1$ at $\xi(a)$, if we denote the number of zero points of $w'(z)$ in $|z|<r$ by $n'(r, 0)$, then

$$n'(r, 0) = \sum_a n_1(r, a), \text{ added for all } a \neq \infty . \tag{4}$$

If we put

$$n_1(r) = \pi(r) + n'(r, o) , \tag{5}$$

then by (2),

$$n_1(r) = n'(r, o) + 2\, n(r, \infty) - n'(r, \infty) . \tag{6}$$

$n_1(r)$ *is the sum of multiplicities of all multiple values of* $w(z)$ *in* $|z|<r$, so that $n_1(r) \geqq 0$ and $n_1(r)$ is an increasing function of r. We put

$$N_1(r) = \int_0^r \frac{n_1(r) - n_1(+0)}{r} dr + n_1(+0) \log r + C_1 , \tag{7}$$

where C_1 is a constant, which we shall determine later.

Theorem V. 14. (Second fundamental theorem).

A. *If* $w(z)$ *is meromorphic in* $|z|<\infty$, *then*

$$(q-2)T(r) \leqq \sum_{\nu=1}^{q} N(r, a_\nu) - N_1(r) + O(T(r)) + O(\log r), \quad q \geqq 3,$$

except in a set of intervals $\{J_\nu\}$, *such that* $\sum_\nu \int_{J_\nu} r^k dr < \infty \quad (k \geqq -1)$.

If $T(r) = O(r^\rho)$ $(\rho>0)$, *then for all* r,

$$(q-2)\ T(r) \leqq \sum_{\nu=1}^{q} N(r, a_\nu) - N_1(r) + O(\log r) .$$

B. *If* $w(z)$ *is meromorphic in* $|z|<1$, *then for any* $\varepsilon>0$,

$$(q-2)T(r) \leqq \sum_{\nu=1}^{q} N(r, a_\nu) - N_1(r) + O(\log T(r)) + (k+\varepsilon) \log \frac{1}{1-r},$$

except in a set of intervals $\{J_\nu\}$, *such that* $\sum_\nu \int_{J_\nu} \frac{dr}{(1-r)^k} < \infty \quad (k \geqq 1)$.

If $T(r) = O\left(\frac{1}{(1-r)^\rho}\right)$ $(\rho>0)$, *then for all* $r<1$,

$$(q-2)T(r) \leqq \sum_{\nu=1}^{q} N(r, a_\nu) - N_1(r) + O\left(\log \frac{1}{1-r}\right).$$

Nevanlinna [1].

Proof. (Ahlfors [5]). Let K be the w-sphere and $d\omega(a)$ be the surface element of K at a. Let $\rho(a) \geqq 0$ be a function, defined on K, which is continuous, except at a finite number of points, such that

$$\iint_K \rho(a) d\omega(a) = 1 . \tag{1}$$

From $T(r) = m(r, a) + N(r, a)$, we have

$$T(r)=\iint_K m(r,a)\,\rho(a)\,d\omega(a)+\iint_K N(r,a)\rho(a)d\omega(a)\,.$$

Hence if we put

$$m_\rho(r)=\iint_K m(r,a)\rho(a)d\omega(a),\quad N_\rho(r)=\iint_K N(r,a)\rho(a)d\omega(a)\,, \tag{2}$$

then

$$T(r)=m_\rho(r)+N_\rho(r)\,. \tag{3}$$

We define $n_\rho(r)$ by

$$\frac{dN_\rho(r)}{dr}=\frac{n_\rho(r)}{r},$$

then

$$n_\rho(r)=\iint_K n(r,a)\,\rho(a)d\omega(a)=\iint_{|z|\leq r}\left(\frac{|w'(z)|}{1+|w(z)|^2}\right)^2\rho(w(z))rdrd\theta\,,\quad z=re^{i\theta}. \tag{4}$$

If we put

$$\lambda(r)=\int_0^{2\pi}\left(\frac{|w'(re^{i\theta})|}{1+|w(re^{i\theta})|^2}\right)^2\rho(w(re^{i\theta}))\,d\theta\,, \tag{5}$$

then

$$n_\rho(r)=\int_0^r\lambda(r)rdr\,. \tag{6}$$

From

$$\frac{\lambda(r)}{2\pi}=\frac{1}{2\pi}\int_0^{2\pi}\left(\frac{|w'(re^{i\theta})|}{1+|w(re^{i\theta})|^2}\right)^2\rho(w(re^{i\theta}))\,d\theta\,,$$

we have[1)]

$$\log\frac{\lambda(r)}{2\pi}\geqq\frac{1}{2\pi}\int_0^{2\pi}\log\left[\left(\frac{|w'(re^{i\theta})|}{1+|w(re^{i\theta})|^2}\right)^2\rho(w(re^{i\theta}))\right]d\theta$$

$$=\frac{1}{\pi}\int_0^{2\pi}\log\frac{|w'(re^{i\theta})|}{1+|w(re^{i\theta})|^2}\,d\theta+\frac{1}{2\pi}\int_0^{2\pi}\log\rho(w(re^{i\theta}))\,d\theta.$$

Hence if we put

$$\mu(r)=\frac{1}{2\pi}\int_0^{2\pi}\log\frac{|w'(re^{i\theta})|}{1+|w(re^{i\theta})|^2}\,d\theta\,, \tag{7}$$

then

$$\log\frac{\lambda(r)}{2\pi}\geqq 2\mu(r)+\frac{1}{2\pi}\int_0^{2\pi}\log\rho\,(w(re^{i\theta}))\,d\theta\,,$$

---

1) We use the well known inequality:

$$\log\left(\frac{1}{a-b}\int_a^b f(x)dx\right)\geqq\frac{1}{a-b}\int_a^b\log f(x)\,dx.$$

or

$$\frac{1}{2\pi}\int_0^{2\pi}\log\rho(w(re^{i\theta})\,d\theta \leqq -2\mu(r)+\log\frac{\lambda(r)}{2\pi}. \tag{8}$$

Since

$$2m(r,\infty)=\frac{1}{2\pi}\int_0^{2\pi}\log(1+|w(re^{i\theta})|^2)\,d\theta\,,$$

we have

$$\mu(r)=\frac{1}{2\pi}\int_0^{2\pi}\log|w'(re^{i\theta})|d\theta-2\,m(r,\infty)\,,$$

$$\mu'(r)=\frac{1}{2\pi r}\int_{|z|=r} d\arg w'(re^{i\theta})-2\frac{dm(r,\infty)}{dr}$$

$$=\frac{n'(r,0)-n'(r,\infty)}{r}-2\frac{dm(r,\infty)}{dr}.$$

Hence

$$\mu'(r)=\frac{n'(r,0)}{r}-\frac{n'(r,\infty)}{r}-2\frac{dm(r,\infty)}{dr}.$$

From $n_1(r)=n'(r,0)-n'(r,\infty)+2n(r,\infty)$ and $T(r)=N(r,\infty)+m(r,\infty)$, we have

$$2T'(r)=\frac{2n(r,\infty)}{r}+\frac{2dm(r,\infty)}{dr}$$

$$=\frac{n_1(r)}{r}-\left(\frac{n'(r,0)}{r}-\frac{n'(r,\infty)}{r}-2\frac{dm(r,\infty)}{dr}\right)=\frac{n_1(r)}{r}-\mu'(r)\,,$$

so that

$$\mu'(r)=\frac{n_1(r)}{r}-2T'(r)\,. \tag{9}$$

Let

$$N_1(r)=\int^r\frac{n_1(r)dr}{r}+C_1\,, \tag{10}$$

where C_1 is a constant, which is determined by the condition:

$$\lim_{r\to 0}[\mu(r)-N_1(r)]=0\,, \tag{11}$$

then from (9), $\mu(r)=N_1(r)-2T(r)$, so that by (8),

$$\frac{1}{2\pi}\int_0^{2\pi}\log\rho(w(re^{i\theta}))d\theta\leqq 2(2T(r)-N_1(r))+\log\frac{\lambda(r)}{2\pi}. \tag{12}$$

Now we define $\rho(w)$ by

$$\log\rho(w)=2\sum_{\nu=1}^{q}\log\frac{1}{[w,a_\nu]}-\alpha\log\left(\sum_{\nu=1}^{q}\log\frac{1}{[w,a_\nu]}\right)-2C\quad(\alpha>1). \tag{13}$$

where C is a constant, which is determined by the condition, $\iint_K \rho(w)d\omega(w)=1$, which is possible, since $\alpha>1$, $\iint_K \rho(w)\,d\omega(w)<\infty$.

Since

$$\frac{1}{4\pi}\int_0^{2\pi}\log\rho(w(re^{i\theta}))\,d\theta+C=\sum_{\nu=1}^{q}m(r,a_\nu)-\frac{\alpha}{4\pi}\int_0^{2\pi}\log\left(\sum_{\nu=1}^{q}\log\frac{1}{[w(re^{i\theta}),a_\nu]}\right)d\theta,$$

we have by (12),

$$\sum_{\nu=1}^{q}m(r,a_\nu)-\frac{\alpha}{4\pi}\int_0^{2\pi}\log\left(\sum_{\nu=1}^{q}\log\frac{1}{[w(re^{i\theta}),a_\nu]}\right)d\theta$$
$$\leqq 2T(r)-N_1(r)+\frac{1}{2}\log\frac{\lambda(r)}{2\pi}+C\,. \tag{14}$$

If we put

$$m(r)=\sum_{\nu=1}^{q}m(r,a_\nu)\,, \tag{15}$$

then

$$\log m(r)=\log\left(\frac{1}{2\pi}\int_0^{2\pi}\sum_{\nu=1}^{q}\log\frac{1}{[w(re^{i\theta}),a_\nu]}d\theta\right)$$
$$\geqq\frac{1}{2\pi}\int_0^{2\pi}\log\left(\sum_{\nu=1}^{q}\log\frac{1}{[w(re^{i\theta}),a_\nu]}\right)d\theta\,,$$

so that by (14),

$$m(r)-\frac{\alpha}{2}\log m(r)\leqq 2T(r)-N_1(r)+\frac{1}{2}\log\frac{\lambda(r)}{2\pi}+C\,. \tag{16}$$

Since by the first fundamental theorem, $\log m(r)=O(\log T(r))$,

$$m(r)\leqq 2T(r)-N_1(r)+\frac{1}{2}\log\frac{\lambda(r)}{2\pi}+O(\log T(r)).$$

Since $m(r)=\sum_{\nu=1}^{q}m(r,a_\nu)=qT(r)-\sum_{\nu=1}^{q}N(r,a_\nu)$, we have

$$(q-2)T(r)\leqq\sum_{\nu=1}^{q}N(r,a_\nu)-N_1(r)+\frac{1}{2}\log\frac{\lambda(r)}{2\pi}+O(\log T(r))\,. \tag{17}$$

Next we shall evaluate $\lambda(r)$. Since by (6),

$$n_\rho(r)=\int_0^r\lambda(t)t\,dt,\qquad N_\rho'(r)=\frac{n_\rho(r)}{r}=\frac{1}{r}\int_0^r\lambda(t)t\,dt\,,$$

$$N_\rho(r)-N_\rho(r_0)=\int_{r_0}^r\frac{dr}{r}\int_0^r\lambda(t)t\,dt,\qquad r_0>0\,, \tag{18}$$

and since $T(r)=N_\rho(r)+m_\rho(r)$,

$$N_\rho(r)-N_\rho(r_0)=T(r)-T(r_0)-m_\rho(r)+m_\rho(r_0)<T(r)+m_\rho(r_0),$$

so that by (18),

$$\int_{r_0}^{r}\frac{dr}{r}\int_{0}^{r}\lambda(t)tdt<T(r)+m_\rho(r_0) .$$

Hence if we put

$$A(r)=\int_0^r\lambda(t)tdt,\qquad B(r)=\int_{r_0}^{r}\frac{A(r)dr}{r}, \tag{19}$$

then

$$B(r)<T(r)+m_\rho(r_0) . \tag{20}$$

(i) First suppose that $w(z)$ is meromorphic for $|z|<\infty$. The set of r, such that

$$\lambda(r)>r^{k-1}(A(r))^2,\quad k\geqq -1 , \tag{21}$$

consists of disjoint intervals $\{I_\nu\}$. Then in I_ν,

$$dA(r)=\lambda(r)rdr>r^k(A(r))^2dr,$$

so that

$$\sum_\nu\int_{I_\nu}r^k\,dr\leqq\sum_\nu\int_{I_\nu}\frac{dA(r)}{(A(r))^2}\leqq\int^\infty\frac{dA}{A^2}<\infty.$$

Next let in a set of disjoint intervals $\{I'_\nu\}$,

$$A(r)>r^{k+1}(B(r))^2, \tag{22}$$

then in I'_ν,

$$dB(r)=\frac{A(r)dr}{r}>r^k(B(r))^2dr ,$$

so that

$$\sum_\nu\int_{I'}r^k\,dr\leqq\sum_\nu\int_{I'}\frac{dB(r)}{(B(r))^2}<\int^\infty\frac{dB}{B^2}<\infty.$$

Hence if we put $\{J_\nu\}=\{I_\nu\}+\{I'_\nu\}$, then

$$\sum_\nu\int_{J_\nu}r^kdr<\infty,\quad k\geqq -1 , \tag{23}$$

and from (21), (22), if r lies outside of $\{J_\nu\}$, we have by (20),

$$\lambda(r)\leqq r^{k-1}(A(r))^2\leqq r^{3k+1}(B(r))^4\leqq r^{3k+1}(T(r)+m_\rho(r_0))^4,$$

so that

$$\log\lambda(r)=O(\log r)+O(\log T(r)) . \tag{24}$$

(ii) Next suppose that $w(z)$ is meromorphic in $|z|<1$. Let for any $\varepsilon>0$,

$$\lambda(r)>\frac{(A(r))^{1+\varepsilon}}{r(1-r)^k},\quad k\geqq 1 , \tag{25}$$

in a set of disjoint intervals $\{I_\nu\}$, then in I_ν,

$$dA(r)=\lambda(r)rdr>\frac{(A(r))^{1+\varepsilon}}{(1-r)^k}dr\,,$$

so that

$$\sum_\nu\int_{I_\nu}\frac{dr}{(1-r)^k}<\int^\infty\frac{dA}{A^{1+\varepsilon}}<\infty.$$

Next let

$$A(r)>\frac{r(B(r))^{1+\varepsilon}}{(1-r)^k} \tag{26}$$

in a set of disjoint intervals $\{I'_\nu\}$, then in I'_ν,

$$dB(r)=\frac{A(r)dr}{r}>\frac{(B(r))^{1+\varepsilon}}{(1-r)^k}dr\,,$$

so that

$$\sum_\nu\int_{I'_\nu}\frac{dr}{(1-r)^k}<\int^\infty\frac{dB}{B^{1+\varepsilon}}<\infty.$$

Hence if we put $\{J_\nu\}=\{I_\nu\}+\{I'_\nu\}$, then

$$\sum_\nu\int_{J_\nu}\frac{dr}{(1-r)^k}<\infty,\quad k\geqq 1\,. \tag{27}$$

If r lies outside of $\{J_\nu\}$, then

$$\lambda(r)\leqq\frac{(A(r))^{1+\varepsilon}}{r(1-r)^k}\leqq\frac{1}{r(1-r)^k}\left(\frac{r(B(r))^{1+\varepsilon}}{(1-r)^k}\right)^{1+\varepsilon}\leqq\frac{(B(r))^{1+\varepsilon'}}{(1-r)^{2k+\varepsilon'}}\,,$$

where $\varepsilon'\to 0$, with $\varepsilon\to 0$.
Hence if r lies outside of $\{J_\nu\}$, then we have from (20),

$$\log\lambda(r)\leqq O(\log T(r))+(2k+\varepsilon)\log\frac{1}{1-r}\,,\quad k\geqq 1,\ \varepsilon>0\,. \tag{28}$$

Hence by (17), (24), (28), the first part is proved.

Next we shall prove the second part.

(i) Suppose that $w(z)$ is meromorphic for $|z|<\infty$ and $T(r)=O(r^\rho)$, then

$$N(r,a)\leqq T(r)=O(r^\rho)\,,$$

so that

$$O(r^\rho)\geqq N(2r,a)-N(r,a)=\int_r^{2r}\frac{n(r,a)dr}{r}\geqq n(r,a)\int_r^{2r}\frac{dr}{r}=n(r,a)\log 2\,,$$

hence $n(r,a)=O(r^\rho)$. If we take $k=\rho-1$ in (23), then

$$\sum_\nu\int_{J_\nu}r^{\rho-1}dr<\infty\,.$$

Since $\log T(r)=O(\log r)$, if r lies outside of $\{J_\nu\}$, then

$$(q-2)T(r) \leqq \sum_{\nu=1}^{q} N(r, a_\nu) - N_1(r) + O(\log r) . \tag{29}$$

If r lies in J_ν, let r' be the right hand end point of J_ν, then

$$r'^{\rho} - r^{\rho} = \rho \int_r^{r'} r^{\rho-1} dr = O(1), \qquad \log r' = \log r + O(1) .$$

$$N(r', a) - N(r, a) = \int_r^{r'} \frac{n(r, a) dr}{r} = O\left(\int_{J_\nu} r^{\rho-1} dr\right) = O(1) ,$$

$$N(r', a) \leqq N(r, a) + O(1) .$$

Since $T(r)$, $N_1(r)$ are increasing functions of r and (29) holds for r',

$$(q-2)T(r) + N_1(r) \leqq (q-2)T(r') + N_1(r') \leqq \sum_{\nu=1}^{q} (r', a_\nu) + O(\log r')$$
$$\leqq \sum_{\nu=1}^{q} N(r, a_\nu) + O(\log r) .$$

Hence (29) holds for any r.

(ii) Next suppose that $w(z)$ is meromorphic in $|z| < 1$ and $T(r) = O(1/(1-r)^{\rho})$, then $N(r, a) \leqq T(r) = O(1/(1-r)^{\rho})$.

Let $r_1 = \dfrac{r+1}{2}$, then $N(r_1, a) = O(1/(1-r_1)^{\rho}) = O(1/(1-r)^{\rho})$,

so that

$$O\left(\frac{1}{(1-r)^{\rho}}\right) \geqq N(r_1, a) - N(r, a) = \int_r^{r_1} \frac{n(r, a) dr}{r} \geqq n(r, a) \int_r^{r_1} dr$$

$$\geqq n(r, a)\ (r_1 - r) = n(r, a) \frac{1-r}{2} ,$$

hence $n(r, a) = O\left(\dfrac{1}{(1-r)^{\rho+1}}\right)$. We take $k = \rho + 1$ in (27), then

$$\sum_\nu \int_{J_\nu} \frac{dr}{(1-r)^{\rho+1}} < \infty .$$

If r lies outside of $\{J_\nu\}$, then since $\log T(r) = O(\log 1/1-r))$,

$$(q-2)\ T(r) \leqq \sum_{\nu=1}^{q} N(r, a_\nu) - N_1(r) + O\left(\log \frac{1}{1-r}\right). \tag{30}$$

If r lies in J_ν, let r' be the right hand end point of J_ν, then

$$\frac{1}{(1-r')^{\rho}} - \frac{1}{(1-r)^{\rho}} \leqq \rho \int_{J_\nu} \frac{dr}{(1-r)^{\rho+1}} = O(1), \quad \log \frac{1}{1-r'} = \log \frac{1}{1-r} + O(1) ,$$

$$N(r', a) - N(r, a) = \int_r^{r'} \frac{n(r, a) dr}{r} = O\left(\int_{J_\nu} \frac{dr}{(1-r)^{\rho+1}}\right) = O(1) ,$$

$$N(r', a) \leqq N(r, a) + O(1) .$$

Since (30) holds for r', we have

$$(q-2)T(r)+N_1(r) \leqq (q-2)T(r')+N_1(r') \leqq \sum_{\nu=1}^{q} N(r', a_\nu)+O\left(\log \frac{1}{1-r'}\right)$$

$$\leqq \sum_{\nu=1}^{q} N(r, a_\nu)+O\left(\log \frac{1}{1-r}\right).$$

Hence (30) holds for any $r<1$.

6. On $m\left(\frac{w'}{w}, r, \infty\right)$.

Theorem V. 15. *Let $w(z)$ be meromorphic in $|z|<R$ and*

$$m\left(\frac{w'}{w}, r, \infty\right)=\frac{1}{2\pi}\int_0^{2\pi}\log^+\left|\frac{w'(re^{i\theta})}{w(re^{i\theta})}\right|d\theta, \quad 0 \leqq r<R.$$

(i) *If $R=\infty$, then*

$$m\left(\frac{w'}{w}, r, \infty\right)=O(\log r)+O(\log T(r)),$$

except in a set of intervals $\{J_\nu\}$, such that $\sum_\nu \int_{J_\nu} r^k dr<\infty$ $(k \geqq -1)$.

(ii) *If $R=1$, then*

$$m\left(\frac{w'}{w}, r, \infty\right)=O\left(\log \frac{1}{1-r}\right)+O(\log T(r)),$$

except in a set of intervals $\{J_\nu\}$, such that $\sum_\nu \int_{J_\nu} \frac{dr}{(1-r)^k}<\infty$ $(k \geqq 1)$.

Proof. In the proof of the second fundamental theorem, we take

$$\rho(a)=\frac{(1+r^2)^2}{2\pi^2 r^2(1+(\log r)^2)}, \quad a=re^{i\theta}, \tag{1}$$

then since $d\omega(a)=\frac{r\,dr\,d\theta}{(1+r^2)^2}$, we have

$$\iint_K \rho(a)d\omega(a)=\frac{1}{\pi}\int_0^\infty \frac{dr}{r(1+(\log r)^2)}=1.$$

As (5) of the proof of the second fundamental theorem, we put

$$\lambda(r)=\int_0^{2\pi}\left(\frac{|w'|}{1+|w|^2}\right)^2\rho(w)\,d\theta=\frac{1}{2\pi^2}\int_0^{2\pi}\frac{|w'|^2}{|w|^2(1+(\log|w|)^2)}d\theta,$$

$$w=w(re^{i\theta}) \tag{2}$$

then

$$\log^+(\pi\lambda(r))=\log^+\left(\frac{1}{2\pi}\int_0^{2\pi}\frac{|w'|^2}{|w|^2(1+(\log|w|)^2)}\,d\theta\right)$$

$$\geqq \frac{1}{2\pi}\int_0^{2\pi} \log^+ \frac{|w'|^2}{|w|^2(1+(\log|w|)^2)} d\theta - \log 2^{1)} ,$$

so that

$$\int_0^{2\pi} \log^+ \frac{|w'|^2}{|w|^2(1+(\log|w|)^2)} d\theta = O(\log \lambda(r)) + O(1) . \tag{3}$$

Now

$$m\left(\frac{w'}{w}, r, \infty\right) = \frac{1}{4\pi}\int_0^{2\pi} \log^+ \frac{|w'|^2}{|w|^2} d\theta$$

$$\leqq \frac{1}{4\pi}\int_0^{2\pi} \log^+ \frac{|w'|^2}{|w|^2(1+(\log|w|)^2)} d\theta + \frac{1}{2\pi}\int_0^{2\pi} \log \sqrt{1+(\log|w|)^2}\, d\theta , \tag{4}$$

$$\frac{1}{2\pi}\int_0^{2\pi} \log \sqrt{1+(\log|w|)^2}\, d\theta \leqq \log\left(\frac{1}{2\pi}\int_0^{2\pi} \sqrt{1+(\log|w|)^2}\, d\theta\right)$$

$$\leqq \log\left(\frac{1}{2\pi}\int_0^{2\pi} \log^+|w|\, d\theta + \frac{1}{2\pi}\int_2^{2\pi} \log^+ \left|\frac{1}{w}\right| d\theta + O(1)\right)$$

$$\leqq \log (T(w, r) + T(1/w, r) + O(1)) = \log (2T(w, r) + O(1)) = O(\log T(w, r)) . \tag{5}$$

Hence by (3), (4), (5), we have

$$m\left(\frac{w'}{w}, r, \infty\right) = O(\log \lambda(r)) + O(\log T(r)) . \tag{6}$$

If we put the evaluation of $\log \lambda(r)$ from the proof of the second fundamental theorem into (6), then we have the theorem.

7. Theorems of Picard type

From the second fundamental theorem and Theorem V. 3, we can prove easily the following Picard's theorem.

Theorem V. 16. *Let $w(z)$ be meromorphic for $r_0 \leqq |z| < \infty$ and have an essential singularity at $z=\infty$, then $w(z)$ takes any value infinitely*

1) Let $f(x) \geqq 0$, $f_1(x) = \text{Max}\,(f(x), 1)$, then

$$\log\left(\frac{1}{b-a}\int_a^b f_1(x)dx\right) \geqq \frac{1}{b-a}\int_a^b \log f_1(x)dx = \frac{1}{b-a}\int_a^b \log^+ f(x)dx.$$

Since $f_1(x) \leqq f(x)+1$,

$$\log^+\left(\frac{1}{b-a}\int_a^b f_1(x)dx\right) \leqq \log^+\left(\frac{1}{b-a}\int_a^b f(x)dx + 1\right)$$

$$\leqq \log^+\left(\frac{1}{b-a}\int_a^b f(x)dx\right) + \log^+ 1 + \log 2 = \log^+\left(\frac{1}{b-a}\int_a^b f(x)dx\right) + \log 2 ,$$

so that

$$\log^+\left(\frac{1}{b-a}\int_a^b f(x)dx\right) \geqq \frac{1}{b-a}\int_a^b \log^+ f(x)dx - \log 2.$$

often, with two possible exceptions.

More generally, let $g(z)$ be a meromorphic function for $r_0 \leqq |z| \leqq \infty$, then $w(z)-g(z)$ has infinitely many zero points, with two possible exceptions for g.

Proof. The second part can be proved as follows. Let $g_\nu(z)$ $(\nu=1, 2, 3)$ be meromorphic for $r_0 \leqq |z| \leqq \infty$ and suppose that $w(z)-g_\nu(z)$ $(\nu=1, 2, 3)$ has only a finite number of zero points and put

$$v(z)=\frac{w(z)-g_1(z)}{w(z)-g_2(z)}\cdot\frac{g_3(z)-g_2(z)}{g_3(z)-g_1(z)},$$

then $v(z)$ is meromorphic for $r_0 \leqq |z| < \infty$ and has an essential singularity at $z=\infty$ and takes three values $0, 1, \infty$ only a finite number of times, which contradicts the first part.

Theorem V. 17. *Let $w(z)$ be meromorphic in $|z|<1$ and take $p \geqq 3$ values a_ν $(\nu=1, 2, \cdots, p)$ only a finite number of times, then*

$$\overline{\lim_{r\to 1}}\, T(r)/\log\frac{1}{1-r} \leqq \frac{1}{p-2}.$$

Hence if

$$\overline{\lim_{r\to 1}}\, T(r)/\log\frac{1}{1-r}=\infty,$$

then $w(z)$ takes any value infinitely often, with two possible exceptions.

Ahlfors [5].

Proof. By the hypothesis, $N(r, a_\nu)=O(1)$. If we take $k=1$ in the second fundamental theorem, then

$$\sum_\nu \int_{J_\nu} \frac{dr}{1-r} < \infty.$$

If r lies outside of $\{J_\nu\}$, then for any $\varepsilon>0$,

$$(p-2)T(r) \leqq O(\log T(r))+(1+\varepsilon)\log\frac{1}{1-r}. \tag{1}$$

Since the theorem is evident, if $T(r)=O(1)$, we assume that $\lim_{r\to 1} T(r)=\infty$, then if $1-r$ is sufficiently small and $p-2-\varepsilon>0$ and if r lies outside of $\{J_\nu\}$,

$$(p-2-\varepsilon)T(r) \leqq (1+\varepsilon)\log\frac{1}{1-r}. \tag{2}$$

If $r \in J_\nu$ and r' be the right hand end point of J_ν, then

$$\log\frac{1}{1-r'}-\log\frac{1}{1-r} \leqq \int_{J_\nu}\frac{dr}{1-r}=O(1).$$

Since (2) holds at r',

$$(p-2-\varepsilon)T(r)\leqq(p-2-\varepsilon)T(r')\leqq(1+\varepsilon)\log\frac{1}{1-r'}\leqq(1+2\varepsilon)\log\frac{1}{1-r}. \quad (3)$$

Hence (3) holds for any r ($r_0\leqq r<1$) and since $\varepsilon>0$ is arbitrary, we have

$$\overline{\lim_{r\to1}}\, T(r)/\log\frac{1}{1-r}\leqq\frac{1}{p-2}.$$

8. Canonical product for a meromorphic function for $|z|<\infty$

Let $w(z)$ be a meromorphic function of finite order ρ for $|z|<\infty$ and $a_n\neq0$ ($n=1, 2, \cdots$) be its zero points, then by Theorem V. 10,

$$\sum_n \frac{1}{r_n^{\rho+\varepsilon}}<\infty \qquad \text{for any } \varepsilon>0, \qquad r_n=|a_n|\,. \quad (1)$$

Let $\mu\geqq0$ be *the convergence exponent of* r_n, such that

$$\sum_n \frac{1}{r_n^{\mu-\varepsilon}}=\infty, \qquad \sum_n \frac{1}{r_n^{\mu+\varepsilon}}<\infty \qquad \text{for any } \varepsilon>0, \quad (2)$$

then by (1),

$$\mu\leqq\rho\,. \quad (3)$$

Let $p\geqq0$ be an integer, such that

$$\sum_n \frac{1}{r_n^{p}}=\infty, \qquad \sum_n \frac{1}{r_n^{p+1}}<\infty\,. \quad (4)$$

If μ is not an integer, then $p=[\mu]$ and if μ is an integer, then

$p=\mu$, or $p=\mu-1$, according as $\sum_n \frac{1}{r_n^{\mu}}=\infty$, or $\sum_n \frac{1}{r_n^{\mu}}<\infty$. (5)

Hence in any case $p\leqq\mu$, so that by (3),

$$p\leqq\mu\leqq\rho\,. \quad (6)$$

We define *the canonical product* $P(z)$ *formed with* $\{a_n\}$ as follows.

If $p=0$, or $\sum_n \frac{1}{r_n}<\infty$, then

$$P(z)=\prod_{n=1}^{\infty}\left(1-\frac{z}{a_n}\right). \quad (7)$$

If $p\geqq1$, then

$$P(z)=\prod_{n=1}^{\infty}\left(1-\frac{z}{a_n}\right)e^{\frac{z}{a_n}+\frac{1}{2}\left(\frac{z}{a_n}\right)^2+\cdots+\frac{1}{p}\left(\frac{z}{a_n}\right)^p}. \quad (7^*)$$

By (4), $P(z)$ is an integral function and $P(a_n)=0$ ($n=1, 2, \cdots$).

Theorem V. 18. $\log^+|P(z)|\leqq 2^{p+2}r^{\mu+\varepsilon}\sum_{n=1}^{\infty}\frac{1}{r_n^{\mu+\varepsilon}}$, $(|z|=r,\quad |a_n|=r_n)$, *where* $\varepsilon=0$, if $p=\mu-1$ and $0<\varepsilon\leqq 1-(\mu-[\mu])$, *if* $p\neq\mu-1$.

Proof. We put

$$\Phi(z, a_n)=\log\left(1-\frac{z}{a_n}\right)+\frac{z}{a_n}+\frac{1}{2}\left(\frac{z}{a_n}\right)^2+\cdots+\frac{1}{p}\left(\frac{z}{a_n}\right)^p, \tag{1}$$

then

$$\log^+|P(z)|\leqq \sum_{\frac{r}{r_n}\geq\frac{1}{2}}\Re^+(\Phi(z, a_n))+\sum_{\frac{r}{r_n}<\frac{1}{2}}|\Phi(z, a_n)|=\sum\nolimits_1+\sum\nolimits_2 . \tag{2}$$

In $\sum_2$,

$$|\Phi(z, a_n)|\leqq \frac{1}{p+1}\left(\frac{r}{r_n}\right)^{p+1}+\frac{1}{p+2}\left(\frac{r}{r_n}\right)^{p+2}+\cdots$$

$$\leqq\left(\frac{r}{r_n}\right)^{p+1}\left(1+\frac{r}{r_n}+\left(\frac{r}{r_n}\right)^2+\cdots\right)\leqq\left(\frac{r}{r_n}\right)^{p+1}\left(1+\frac{1}{2}+\left(\frac{1}{2}\right)^2+\cdots\right)$$

$$=2\left(\frac{r}{r_n}\right)^{p+1}. \tag{3}$$

If $p=\mu-1$, then

$$|\Phi(z, a_n)|\leqq 2\left(\frac{r}{r_n}\right)^{\mu}. \tag{4}$$

If $p\neq\mu-1$ and $0<\varepsilon\leqq 1-(\mu-[\mu])$, then $0\leqq p+1-(\mu+\varepsilon)$, hence by (3),

$$|\Phi(z, a_n)|\leqq 2\frac{r^{\mu+\varepsilon}}{r_n^{p+1}}\cdot r^{p+1-(\mu+\varepsilon)}\leqq 2\frac{r^{\mu+\varepsilon}}{r_n^{p+1}}\cdot\left(\frac{r_n}{2}\right)^{p+1-(\mu+\varepsilon)}\leqq 2\left(\frac{r}{r_n}\right)^{\mu+\varepsilon}. \tag{5}$$

From (4), (5), we have

$$\Sigma_2|\Phi(z, a_n)|\leqq 2\,\Sigma_2\left(\frac{r}{r_n}\right)^{\mu+\varepsilon}, \tag{6}$$

where $\varepsilon=0$, if $p=\mu-1$ and $0<\varepsilon\leqq 1-(\mu-[\mu]$, if $p\neq\mu-1$.

In Σ_1,

$$\Re^+(\Phi(z, a_n))\leqq\log^+\left|1-\frac{z}{a_n}\right|+\frac{r}{r_n}+\frac{1}{2}\left(\frac{r}{r_n}\right)^2+\cdots+\frac{1}{p}\left(\frac{r}{r_n}\right)^p.$$

Since $\log^+\left|1-\frac{z}{a_n}\right|\leqq\log\left(1+\frac{r}{r_n}\right)\leqq\frac{r}{r_n}$,

$$\Re^+(\Phi(z, a_n))\leqq 2\left(\frac{r}{r_n}\right)^p\left(1+\frac{r_n}{r}+\cdots+\left(\frac{r_n}{r}\right)^{p-1}\right)\leqq 2\left(\frac{r}{r_n}\right)^p(1+2+\cdots+2^{p-1})$$

$$\leqq 2^{p+1}\left(\frac{r}{r_n}\right)^p. \tag{7}$$

If $p=\mu-1$, then

$$\Re^+(\Phi(z, a_n))\leqq 2^{p+1}\left(\frac{r}{r_n}\right)^{\mu}\cdot\frac{r_n}{r}\leqq 2^{p+2}\left(\frac{r}{r_n}\right)^{\mu}.$$

If $p \neq \mu-1$ and $0<\varepsilon \leqq 1-(\mu-[\mu])$, then $\mu+\varepsilon-p \geqq 0$, $\mu+\varepsilon+1 \leqq p+2$, so that by (7),

$$\Re^{+}(\Phi(z, a_n)) \leqq 2^{p+1}\left(\frac{r}{r_n}\right)^{\mu+\varepsilon} \cdot \left(\frac{r_n}{r}\right)^{\mu+\varepsilon-p} \leqq 2^{p+1}\left(\frac{r}{r_n}\right)^{\mu+\varepsilon} \cdot 2^{\mu+\varepsilon-p} \leqq 2^{p+2}\left(\frac{r}{r_n}\right)^{\mu+\varepsilon}.$$

Hence

$$\Sigma_1 \Re^{+}(\Phi(z, a_n)) \leqq 2^{p+2} \Sigma_1 \left(\frac{r}{r_n}\right)^{\mu+\varepsilon}, \tag{8}$$

where $\varepsilon=0$, if $p=\mu-1$ and $0<\varepsilon \leqq 1-(\mu-[\mu])$, if $p \neq \mu-1$.

Hence from (2), (6), (8), we have

$$\log^{+}|P(z)| \leqq 2^{p+2} r^{\mu+\varepsilon} \sum_{n=1}^{\infty} \frac{1}{r_n^{\mu+\varepsilon}}. \tag{9}$$

Theorem V. 19. *Let $\mu \geqq 0$ be the convergence exponent of r_n ($|a_n|=r_n$) and C_n: $|z-a_n|=\frac{1}{r_n^{\mu+\varepsilon}}$ be a circle about a_n. If z lies outside of $\{C_n\}$ and $|z|=r \geqq 2$, then*

$$\log^{+} \frac{1}{|P(z)|} \leqq \text{const. } r^{\mu+\varepsilon} \log r, \qquad |z|=r,$$

where $\varepsilon=0$, if $p=\mu-1$ and $0<\varepsilon \leqq 1-(\mu-[\mu])$, if $p \neq \mu-1$.

Proof. If we put

$$\Psi(z, a_n)=\log \frac{a_n}{z-a_n}-\left(\frac{z}{a_n}+\cdots+\frac{1}{p}\left(\frac{z}{a_n}\right)^p\right), \tag{1}$$

then

$$\log^{+} \frac{1}{|P(z)|} \leqq \sum_{\frac{r}{r_n} \geqq \frac{1}{2}} \Re^{+}(\Psi(z, a_n))+\sum_{\frac{r}{r_n}<\frac{1}{2}} |\Psi(z, a_n)|=\Sigma_1+\Sigma_2. \tag{2}$$

As the proof of Theorem V. 18, we have

$$\Sigma_2 |\Psi(z, a_n) \leqq 2 \Sigma_2 \left(\frac{r}{r_n}\right)^{\mu+\varepsilon}. \tag{3}$$

In Σ_1,

$$\Re^{+}(\Psi(z, a_n)) \leqq \log^{+}\left|\frac{a_n}{z-a_n}\right|+\frac{r}{r_n}+\frac{1}{2}\left(\frac{r}{r_n}\right)^2+\cdots+\frac{1}{p}\left(\frac{r}{r_n}\right)^p$$

$$\leqq \log^{+}\left|\frac{a_n}{z-a_n}\right|+2^{p+2}\left(\frac{r}{r_n}\right)^{\mu+\varepsilon}. \tag{4}$$

If z lies outside of C_n ($n=1, 2, \cdots$) and $r \geqq 2$, then

$$\log^{+}\left|\frac{a_n}{z-a_n}\right| \leqq \log^{+} r_n^{\mu+1+\varepsilon} \leqq \log (2r)^{\mu+1+\varepsilon}=(\mu+1+\varepsilon) \log 2r.$$

Since in Σ_1, $r/r_n \geqq 1/2$, we have

$$\mathfrak{R}^+(\Psi(z, a_n)) \leqq \text{const. } \log r \,.\, \left(\frac{r}{r_n}\right)^{\mu+\varepsilon},$$

so that

$$\Sigma_1 \mathfrak{R}^+(\Psi(z, a_n)) \leqq \text{const. } r^{\mu+\varepsilon} \log r \, \Sigma_1 \frac{1}{r_n^{\mu+\varepsilon}} \,. \tag{5}$$

Hence by (2), (3), (5), we have

$$\log^+ \frac{1}{|P(z)|} \leqq \text{const. } r^{\mu+\varepsilon} \log r \,.$$

Theorem V. 20. *Let ρ_0 be the order of $P(z)$, then ρ_0 is equal to the convergence exponent of $|a_n|$:*

$$\rho_0 = \mu \quad (\leqq \rho) \,.$$

Proof. Since $\mu \leqq \rho_0$((3) p. 216), we have only to prove that $\rho_0 \leqq \mu$. If we put $M(r) = \operatorname{Max}_{|z|=r} |P(z)|$, then by Theoıem V. 18, $\log M(r) \leqq$ const. $r^{\mu+\varepsilon}$, so that $\rho_0 = \overline{\lim}_{r\to\infty} \frac{\log \log M(r)}{\log r} \leqq \mu + \varepsilon$. Since $\varepsilon > 0$ is arbitrary, we have

$$\rho_0 \leqq \mu \,.$$

Theorem V. 21. *Let $w(z)$ be a meromorphic function of finite order ρ for $|z|<\infty$, then $w(z)$ can be expressed in the form: $w(z) = w_1(z)/w_2(z)$, where $w_i(z)$ $(i=1, 2)$ is an integral function of order $\leqq \rho$.*

Proof. Let $P(z)$ be the canonical product, formed with poles $a_n \neq 0$ of $w(z)$. If $w(z)$ is regular at $z=0$, then we put $w_2(z) = P(z)$. If $w(z)$ has a pole of the ν-th order at $z=0$, then we put $w_2(z) = z^\nu P(z)$. Then $w_2(z)$ is an integral function of order $\leqq \rho$. If we put $w_1(z) = w(z) w_2(z)$, then $w_1(z)$ is an integral function of order $\leqq \rho$, so that $w(z) = w_1(z)/w_2(z)$ is the desired decomposition.

Theorem V. 22. *Let $w(z)$ be an integral function of finite order ρ. Let $P(z)$ be the canonical product, formed with zero points $a_n \neq 0$ of $w(z)$, then*

$$w(z) = z^\nu P(z)\, e^{Q(z)} \,,$$

where $\nu \geqq 0$ is the order of zero of $w(z)$ at $z=0$ and $Q(z)$ is a polynomial of degree $\leqq \rho$. Hadamard [1].

Proof. If we put $F(z) = w(z)/z^\nu P(z)$, then $F(z)$ is an integral function and $\neq 0$, so that if we put $\log F(z) = Q(z)$, then $Q(z)$ is an integral function. We put $M(r) = \operatorname{Max}_{|z|=r} \mathfrak{R}(Q(z))$, then since $F(z)$ is of order $\leqq \rho$, we see easily that $M(r) = O(r^{\ +\varepsilon})$, hence by an extension of Liouville's theorem, $Q(z)$ is a polynomial of degree $\leqq \rho$.

Corollary. *Let $w(z)$ be an integral function of order<1. If $w(z)$ is not a ploynomial, then $w(z)$ has infinitely many zero points.*

9. Order of the derivative of a meromorphic function for $|z|<\infty$

Theorem V. 23. *Let $w(z)$ be a meromorphic function of order ρ $(\leqq\infty)$ for $|z|<\infty$ and ρ' be the order of $w'(z)$, then $\rho'=\rho$.* Whittaker [1].

Proof. (Tsuji [24], Komatu [3]). (i) First suppose that $w(z)$ is an integral function and put $M(r)=\underset{|z|=r}{\text{Max}}|w(z)|$, $M_1(r)=\underset{|z|=r}{\text{Max}}|w'(z)|$. Let $|w(re^{i\theta_0})|=M(r)$, then

$$M(r)=|w(re^{i\theta_0})|\leqq|w(0)|+\int_0^r|w'(te^{i\theta_0})|dt\leqq|w(0)|+rM_1(r)\,. \tag{1}$$

Let $|w'(z_1)|=M_1(r)$, $z_1=re^{i\theta_1}$, then

$$M_1(r)=|w'(z_1)|=\left|\frac{1}{2\pi i}\int_{|\zeta-z_1|=r}\frac{w(\zeta)d\zeta}{(\zeta-z_1)^2}\right|\leqq\frac{M(2r)}{r}, \tag{2}$$

hence by (1), (2),

$$\frac{M(r)-|w(0)|}{r}\leqq M_1(r)\leqq\frac{M(2r)}{r}, \tag{3}$$

so that

$$\rho=\overline{\lim_{r\to\infty}}\frac{\log\log M(r)}{\log r}=\overline{\lim_{r\to\infty}}\frac{\log\log M_1(r)}{\log r}=\rho'. \tag{4}$$

(ii) Next suppose that $w(z)$ has poles and $\rho<\infty$, then by Theorem V. 21, $w(z)=w_1(z)/w_2(z)$, where $w_i(z)$ $(i=1, 2)$ is an integral function of order$\leqq\rho$, hence from $w'(z)=\dfrac{w_1'(z)\,w_2(z)-w_2'(z)w_1(z)}{(w_2(z))^2}$ and (4), we have

$$\rho'\leqq\rho<\infty\,. \tag{5}$$

Next we shall prove that $\rho\leqq\rho'$. By the first fundamental theorem, $T(w, r)=m(w, r, \infty)+N(w, r, \infty)$, $T(w', r)=m(w', r, \infty)+N(w', r, \infty)$.

Since a pole of the k-th order of $w(z)$ is a pole of the $(k+1)$-th order of $w'(z)$,

$$N(w, r, \infty)\leqq N(w', r, \infty)+O(1)\leqq T(w', r)+O(1)=O(r^{\rho'+\varepsilon})\ (\varepsilon>0)\,. \tag{6}$$

Let $P(z)$ be the canonical product, formed with poles $a_n\neq0$ of $w'(z)$, then $w'(z)=Q(z)/z^\nu P(z)$ $(\nu\geqq0)$, where $Q(z)$ is an integral function of order$\leqq\rho'$.

Since the convergence exponent μ' of $|a_n|$ is $\leqq\rho'$, we see from Theorem V. 19, that if z lies outside of circles C_n: $|z-a_n|=1/|a_n|^{\mu'+\varepsilon}\,(\varepsilon>0)$

$(n=1, 2, \cdots)$, then $|P(z)|>e^{-r^{\rho'+\varepsilon}}$ $(r\geqq r_0(\varepsilon))$. Since $Q(z)$ is of order$\leqq\rho'$, if z lies outside of $\{C_n\}$, then

$$|w'(z)|<e^{r^{\rho'+\varepsilon}}, \quad r\geqq r_0(\varepsilon) . \tag{7}$$

Since the sum of radii of C_n is finite, there exists θ_0, such that a half-line L: $z=te^{i\theta_0}$ $(r_0\leqq t<\infty)$ lies outside of $\{C_n\}$.

Let J_n: $-1/|a_n|^{\mu'+\varepsilon}\leqq r-|a_n|\leqq 1/|a_n|^{\mu'+\varepsilon}$ be an interval on the positive real axis, then since the sum of lengths of J_n is finite, there exists $r>r_0$, which lies outside of J_n $(n=1, 2, \cdots)$, so that the circle $|z|=r$ lies outside of C_n $(n=1, 2, \cdots)$.

Let $\zeta_0=r_0e^{i\theta_0}$, $z_0=re^{i\theta_0}$ and $z=re^{i\theta}$ be any point on the circle $|z|=r$, then

$$|w(z)|=|w(re^{i\theta})|\leqq|w(\zeta_0)|+\int_{\zeta_0}^{z_0}|w'(z)|\,|dz|+\int_{z_0}^{z}|w'(z)|\,|dz| , \tag{8}$$

where we integrate from ζ_0 to z_0 on L and from z_0 to z on $|z|=r$. Since the path of integration lies outside of $\{C_n\}$, (7) holds on the path of integration, so that for any $z=re^{i\theta}$ on $|z|=r$, we have

$$|w(re^{i\theta})|<e^{r^{\rho'+\varepsilon}}, \quad r\geqq r_0(\varepsilon) .$$

Hence

$$m(w, r, \infty)=\frac{1}{2\pi}\int_0^{2\pi}\log^+|w(re^{i\theta})|d\theta+O(1)=O(r^{\rho'+\varepsilon}) , \tag{9}$$

so that by (6), if r lies outside of $\{J_n\}$, then

$$T(w, r)=O(r^{\rho'+\varepsilon}) . \tag{10}$$

Since the sum of lengths of J_n is finite, if r belongs to J_n and r is sufficiently large, then there exists r_1 $(r\leqq r_1\leqq 2r)$, which lies outside of $\{J_n\}$, so that

$$T(w, r)\leqq T(w, r_1)=O(r_1^{\rho'+\varepsilon})=O(r^{\rho'+\varepsilon}) .$$

Hence (10) holds for any $r\geqq r_0$, so that $\rho\leqq\rho'$, hence $\rho'=\rho$, if $\rho<\infty$. If $\rho'<\infty$, then from $\rho\leqq\rho'$, we have $\rho<\infty$, so that if $\rho=\infty$, then $\rho'=\infty$, hence $\rho'=\rho$ in general.

10. Canonical product for a meromorphic function in $|z|<1$

Let $w(z)$ be a meromorphic function of finite order ρ in $|z|<1$ and $a_n\neq 0$ be its zero points, then by Theorem V. 11,

$$\sum_n(1-|a_n|)^{\rho+1+\varepsilon}<\infty \quad \text{for any } \varepsilon>0 . \tag{1}$$

First we define *the convergence exponent* $\mu\geqq 0$ *of* $|a_n|$ as follows. If $\sum_n(1-|a_n|)<\infty$, then $\mu=0$.

If $\sum_n (1-|a_n|)=\infty$, then

$$\sum_n (1-|a_n|)^{\mu+1-\varepsilon}=\infty, \quad \sum_n (1-|a_n|)^{\mu+1+\varepsilon}<\infty \text{ for any } \varepsilon>0 . \tag{2}$$

Hence by (1),

$$0\leqq\mu\leqq\rho . \tag{3}$$

Let $p\geqq1$ be a positive integer, such that

$$\sum_n (1-|a_n|)^p=\infty, \quad \sum_n (1-|a_n|)^{p+1}<\infty . \tag{4}$$

If μ is not an integer, then $p=[\mu]+1$ and if μ is an integer, then $p=\mu$, or $p=\mu+1$, according as $\sum_n (1-|a_n|)^{\mu+1}<\infty$, or $\sum_n (1-|a_n|)^{\mu+1}=\infty$. (5) Hence in any case $p-1\leqq\mu$, so that by (3),

$$p-1\leqq\mu\leqq\rho . \tag{6}$$

Now we shall define *the canonical product $P(z)$ formed with $\{a_n\}$* as follows.

If $\sum_n (1-|a_n|)<\infty$, then

$$P(z)=\prod_{n=1}^{\infty}\left(1-\frac{1-|a_n|^2}{1-\bar{a}_n z}\right)=\prod_{n=1}^{\infty}\frac{\bar{a}_n(a_n-z)}{1-\bar{a}_n z}. \tag{7}$$

$P(z)$ is regular and $|P(z)|<1$ in $|z|<1$ and $P(a_n)=0$ $(n=1, 2, \cdots)$.

If $\sum_n (1-|a_n|)=\infty$, then

$$P(z)=\prod_{n=1}^{\infty}\left(1-\frac{1-|a_n|^2}{1-\bar{a}_n z}\right)e^{\frac{1-|a_n|^2}{1-\bar{a}_n z}+\frac{1}{2}\left(\frac{1-|a_n|^2}{1-\bar{a}_n z}\right)^2+\cdots+\frac{1}{p}\left(\frac{1-|a_n|^2}{1-\bar{a}_n z}\right)^p} . \tag{7*}$$

Then by (4), $P(z)$ is regular in $|z|<1$ and $P(a_n)=0$ $(n=1, 2, \cdots)$.

Theorem V. 24. *Suppose that $\sum_n (1-|a_n|)=\infty$, then*

$$\log^+|P(z)|\leqq 2^{p+1}\sum_{n=1}^{\infty}\left|\frac{1-|a_n|^2}{1-\bar{a}_n z}\right|^{\mu+1+\varepsilon},$$

where $\varepsilon=0$, if $p=\mu$ and $0<\varepsilon\leqq1-(\mu-[\mu])$, if $p\neq\mu$.

Proof. We put

$$\Phi(z, a_n)=\log\left(1-\frac{1-|a_n|^2}{1-\bar{a}_n z}\right)+\frac{1-|a_n|^2}{1-\bar{a}_n z}+\cdots+\frac{1}{p}\left(\frac{1-|a_n|^2}{1-\bar{a}_n z}\right)^p, \tag{1}$$

then

$$\log^+|P(z)|\leqq \sum_{\frac{1-|a_n|^2}{|1-\bar{a}_n z|}\geq\frac{1}{2}} \Re^+(\Phi(z, a_n))+\sum_{\frac{1-|a_n|^2}{|1-\bar{a}_n z|}<\frac{1}{2}} |\Phi(z, a_n)|=\Sigma_1+\Sigma_2 . \tag{2}$$

In $\sum_2$,

$$|\Phi(z, a_n)| \leqq \frac{1}{p+1}\left|\frac{1-|a_n|^2}{1-\bar{a}_n z}\right|^{p+1} + \frac{1}{p+2}\left|\frac{1-|a_n|^2}{1-\bar{a}_n z}\right|^{p+2} + \cdots$$

$$\leqq \left|\frac{1-|a_n|^2}{1-\bar{a}_n z}\right|^{p+1}\left(1+\left|\frac{1-|a_n|^2}{1-\bar{a}_n z}\right| + \left|\frac{1-|a_n|^2}{1-\bar{a}_n z}\right|^2 + \cdots\right)$$

$$\leqq \left|\frac{1-|a_n|^2}{1-\bar{a}_n z}\right|^{p+1}\left(1+\frac{1}{2}+\frac{1}{2^2}+\cdots\right) = 2\left|\frac{1-|a_n|^2}{1-\bar{a}_n z}\right|^{p+1}. \tag{3}$$

If $p=\mu$, then

$$|\Phi(z, a_n)| \leqq 2\left|\frac{1-|a_n|^2}{1-\bar{a}_n z}\right|^{\mu+1}. \tag{4}$$

If $p \neq \mu$, then $p-(\mu+\varepsilon) \geqq 0$, if $0<\varepsilon \leqq 1-(\mu-[\mu])$, so that

$$|\Phi(z, a_n)| \leqq 2\left|\frac{1-|a_n|^2}{1-\bar{a}_n z}\right|^{p+1} = 2\left|\frac{1-|a_n|^2}{1-\bar{a}_n z}\right|^{\mu+1+\varepsilon} \cdot \left|\frac{1-|a_n|^2}{1-\bar{a}_n z}\right|^{p-(\mu+\varepsilon)}$$

$$\leqq 2\left|\frac{1-|a_n|^2}{1-\bar{a}_n z}\right|^{\mu+1+\varepsilon}\left(\frac{1}{2}\right)^{p-(\mu+\varepsilon)} \leqq 2\left|\frac{1-|a_n|^2}{1-\bar{a}_n z}\right|^{\mu+1+\varepsilon}. \tag{5}$$

Hence by (4), (5),

$$\sum_2 |\Phi(z, a_n)| \leqq 2\sum_2 \left|\frac{1-|a_n|^2}{1-\bar{a}_n z}\right|^{\mu+1+\varepsilon}, \tag{6}$$

where $\varepsilon=0$, if $p=\mu$ and $0<\varepsilon\leqq 1-(\mu-[\mu])$, if $p\neq\mu$.

Next we shall evaluate $\sum_1$.

Since $\log\left|1-\frac{1-|a_n|^2}{1-\bar{a}_n z}\right| = \log\left|\frac{\bar{a}_n(a_n-z)}{1-\bar{a}_n z}\right| \leqq 0$, we have

$$\Re^+(\Phi(z, a_n)) \leqq \left|\frac{1-|a_n|^2}{1-\bar{a}_n z}\right| + \frac{1}{2}\left|\frac{1-|a_n|^2}{1-\bar{a}_n z}\right|^2 + \cdots + \frac{1}{p}\left|\frac{1-|a_n|^2}{1-\bar{a}_n z}\right|^p$$

$$\leqq \left|\frac{1-|a_n|^2}{1-\bar{a}_n z}\right|^p\left(1+\left|\frac{1-\bar{a}_n z}{1-|a_n|^2}\right| + \cdots + \left|\frac{1-\bar{a}_n z}{1-|a_n|^2}\right|^{p-1}\right)$$

$$\leqq \left|\frac{1-|a_n|^2}{1-\bar{a}_n z}\right|^p (1+2+\cdots+2^{p-1}) \leqq 2^p\left|\frac{1-|a_n|^2}{1-\bar{a}_n z}\right|^p. \tag{7}$$

If $p=\mu$, then since $\left|\frac{1-\bar{a}_n z}{1-|a_n|^2}\right| \leqq 2$ in Σ_1,

$$\Re^+(\Phi(z, a_n)) \leqq 2^p\left|\frac{1-|a_n|^2}{1-a_n z}\right|^p = 2^p\left|\frac{1-|a_n|^2}{1-\bar{a}_n z}\right|^{\mu+1} \cdot \left|\frac{1-\bar{a}_n z}{1-|a_n|^2}\right|$$

$$\leqq 2^{p+1}\left|\frac{1-|a_n|^2}{1-\bar{a}_n z}\right|^{\mu+1}. \tag{8}$$

If $p\neq\mu$, then $0<\mu+1+\varepsilon-p\leqq 1$, if $0<\varepsilon\leqq 1-(\mu-[\mu])$, so that

$$\Re^+(\Phi(z, a_n)) \leqq 2^p \left|\frac{1-|a_n|^2}{1-\bar{a}_n z}\right|^p = 2^p \left|\frac{1-|a_n|^2}{1-\bar{a}_n z}\right|^{\mu+1+\varepsilon} \cdot \left|\frac{1-\bar{a}_n z}{1-|a_n|^2}\right|^{\mu+1+\varepsilon-p}$$

$$\leqq 2^p \left|\frac{1-|a_n|^2}{1-\bar{a}_n z}\right|^{\mu+1+\varepsilon} \cdot 2^{\mu+1+\varepsilon-p} \leqq 2^{p+1} \left|\frac{1-|a_n|^2}{1-\bar{a}_n z}\right|^{\mu+1+\varepsilon}, \tag{9}$$

so that

$$\sum\nolimits_1 \Re^+(\Phi(z, a_n)) \leqq 2^{p+1} \sum\nolimits_1 \left|\frac{1-|a_n|^2}{1-\bar{a}_n z}\right|^{\mu+1+\varepsilon}, \tag{10}$$

where $\varepsilon=0$, if $p=\mu$ and $0<\varepsilon\leqq 1-(\mu-[\mu])$, if $p\neq\mu$.

Hence from (2), (6), (10), we have

$$\log^+|P(z)| \leqq 2^{p+1} \sum_{n=1}^{\infty} \left|\frac{1-|a_n|^2}{1-\bar{a}_n z}\right|^{\mu+1+\varepsilon}.$$

Theorem V. 25. *Let $\mu\geqq 0$ be the convergence exponent of $|a_n|$ and C_n: $|z-a_n|=(1-|a_n|^2)^{\mu+4}$ be a circle about a_n. If z lies outside of C_n $(n=1, 2, \cdots)$ and $1/2\leqq|z|<1$, then*

$$\log^+ \frac{1}{|P(z)|} \leqq \text{const.} \log \frac{1}{1-|z|} \sum_{n=1}^{\infty} \left|\frac{1-|a_n|^2}{1-\bar{a}_n z}\right|^{\mu+1+\varepsilon},$$

where if $\sum_n (1-|a_n|)<\infty$, then $\mu=0$, $\varepsilon=0$ and if $\sum_n (1-|a_n|)=\infty$, then $\varepsilon=0$, if $p=\mu$ and $0<\varepsilon\leqq 1-(\mu-[\mu])$, if $p\neq\mu$.

Proof. (i) First suppose that $\sum_n (1-|a_n|)=\infty$, then

$$P(z) = \prod_{n=1}^{\infty} \left(1-\frac{1-|a_n|^2}{1-\bar{a}_n z}\right) e^{\frac{1-|a_n|^2}{1-\bar{a}_n z} + \cdots + \frac{1}{p}\left(\frac{1-|a_n|^2}{1-\bar{a}_n z}\right)^p}. \tag{1}$$

Hence if we put

$$\Psi(z, a_n) = \log \frac{1-\bar{a}_n z}{\bar{a}_n(a_n-z)} - \left(\frac{1-|a_n|^2}{1-\bar{a}_n z} + \cdots + \frac{1}{p}\left(\frac{1-|a_n|^2}{1-\bar{a}_n z}\right)^p\right), \tag{2}$$

then

$$\log^+ \frac{1}{|P(z)|} \leqq \sum_{\left|\frac{1-|a_n|^2}{1-\bar{a}_n z}\right| \geq \frac{1}{2}} \Re^+(\Psi(z, a_n)) + \sum_{\left|\frac{1-|a_n|^2}{1-\bar{a}_n z}\right| < \frac{1}{2}} |\Psi(z, a_n)| = \Sigma_1 + \Sigma_2. \tag{3}$$

As the proof of Theorem V. 24, we have

$$\sum\nolimits_2 |\Psi(z, a_n)| \leqq 2 \sum\nolimits_2 \left|\frac{1-|a_n|^2}{1-\bar{a}_n z}\right|^{\mu+1+\varepsilon}. \tag{4}$$

In $\sum_1$,

$$\Re^+(\Psi(z, a_n)) \leqq \log \left|\frac{1-\bar{a}_n z}{\bar{a}_n(a_n-z)}\right| + 2^{p+1} \left|\frac{1-|a_n|^2}{1-\bar{a}_n z}\right|^{\mu+1+\varepsilon}. \tag{5}$$

If z lies outside of C_n: $|z-a_n|=(1-|a_n|^2)^{\mu+4}$ $(n=1, 2, \cdots)$, then in Σ_1,

$$\left|\frac{1-\bar{a}_n z}{\bar{a}_n(a_n-z)}\right| \leqq \frac{|1-\bar{a}_n z|}{|\bar{a}_n|(1-|a_n|^2)^{\mu+4}} \leqq \frac{|1-\bar{a}_n z|}{|\bar{a}_n|\cdot\left|\frac{1-\bar{a}_n z}{2}\right|^{\mu+4}}$$

$$\leqq \frac{\text{const.}}{|1-\bar{a}_n z|^{\mu+3}} \leqq \frac{\text{const.}}{(1-|z|)^{\mu+3}},$$

$$\log\left|\frac{1-\bar{a}_n z}{\bar{a}_n(a_n-z)}\right| \leqq \text{const.}+(\mu+3)\log\frac{1}{1-|z|}. \tag{6}$$

Since in $\sum_1$, $\left|\frac{1-|a_n|^2}{1-\bar{a}_n z}\right| \geqq \frac{1}{2}$, we have from (5), (6), if $1/2 \leqq |z| < 1$,

$$\textstyle\sum_1 \Re^+(\Psi(z, a_n)) \leqq \text{const. } \log\frac{1}{1-|z|}\sum_1\left|\frac{1-|a_n|^2}{1-\bar{a}_n z}\right|^{\mu+1+\varepsilon}. \tag{7}$$

Hence by (3), (4), (7), if z lies outside of $\{C_n\}$ and $1/2 \leqq |z| < 1$,

$$\log^+\frac{1}{|P(z)|} \leqq \text{const. } \log\frac{1}{1-|z|}\sum_{n=1}^{\infty}\left|\frac{1-|a_n|^2}{1-\bar{a}_n z}\right|^{\mu+1+\varepsilon}. \tag{8}$$

(ii) Next suppose that $\sum_n (1-|a_n|) < \infty$, then $\mu=0$ and

$$P(z) = \prod_{n=1}^{\infty}\frac{\bar{a}_n(a_n-z)}{1-\bar{a}_n z}, \tag{9}$$

so that

$$\log^+\frac{1}{|P(z)|} \leqq \sum_{\left|\frac{1-|a_n|^2}{1-\bar{a}_n z}\right| \geq \frac{1}{2}} \log\left|\frac{1-\bar{a}_n z}{\bar{a}_n(a_n-z)}\right| + \sum_{\left|\frac{1-|a_n|^2}{1-\bar{a}_n z}\right| < \frac{1}{2}} \log\left|\frac{1-\bar{a}_n z}{\bar{a}_n(a_n-z)}\right| = \textstyle\sum_1 + \sum_2. \tag{10}$$

In $\sum_2$,

$$\log\left|\frac{1-\bar{a} z_n}{\bar{a}_n(a_n-z)}\right| \leqq \left|\log\frac{1-\bar{a}_n z}{\bar{a}_n(a_n-z)}\right| \leqq \left|\frac{1-|a_n|^2}{1-\bar{a}_n z}\right| + \frac{1}{2}\left|\frac{1-|a_n|^2}{1-\bar{a}_n z}\right|^2 + \cdots$$

$$\leqq 2\left|\frac{1-|a_n|^2}{1-\bar{a}_n z}\right|,$$

so that

$$\textstyle\sum_2 \log\left|\frac{1-\bar{a}_n z}{\bar{a}_n(a_n-z)}\right| \leqq 2\sum_2\left|\frac{1-|a_n|^2}{1-\bar{a}_n z}\right|. \tag{11}$$

Next we evaluate Σ_1.

If z lies outside of $C_n: |z-a_n| = (1-|a_n|^2)^4$ and $1/2 \leqq |z| < 1$, then in $\sum_1$,

$$\log\left|\frac{1-\bar{a}_n z}{\bar{a}_n(a_n-z)}\right| \leqq \log\left|\frac{1-\bar{a}_n z}{\bar{a}_n(1-|a_n|^2)^4}\right| \leqq \log\frac{|1-\bar{a}_n z|}{|\bar{a}_n|\left(\frac{|1-\bar{a}_n z|}{2}\right)^4}$$

$$= \log\frac{16}{|\bar{a}_n||1-\bar{a}_n z|^3} \leqq \text{const. } \log\frac{1}{1-|z|}.$$

Since $\left|\frac{1-|a_n|^2}{1-\bar{a}_n z}\right| \geqq \frac{1}{2}$ in $\sum_1$, we have

$$\sum_1 \log\left|\frac{1-\bar{a}_n z}{\bar{a}_n(a_n-z)}\right| \leqq \text{const.} \log\frac{1}{1-|z|}\sum_1\left|\frac{1-|a_n|^2}{1-\bar{a}_n z}\right|. \tag{12}$$

Hence by (10), (11), (12), if z lies outside of $\{C_n\}$ and $1/2 \leqq |z| < 1$, then

$$\log^+\frac{1}{|P(z)|} \leqq \text{const.} \log\frac{1}{1-|z|}\sum_{n=1}^{\infty}\left|\frac{1-|a_n|^2}{1-\bar{a}_n z}\right|. \tag{13}$$

Lemma. *Let* $J=\int_{-\pi}^{\pi}\frac{d\theta}{|1-re^{i\theta}|^{\lambda+1}}$, $\lambda \geqq 0$, *then*

$$J=O\left(\frac{1}{(1-r)^{\lambda}}\right), \text{ if } \lambda>0, \quad J=O\left(\log\frac{1}{1-r}\right), \text{ if } \lambda=0.$$

Proof.

$$J=\int_{|\theta|\leqq 1-r}\frac{d\theta}{|1-re^{i\theta}|^{\lambda+1}}+\int_{1-r\leqq|\theta|\leqq\pi}\frac{d\theta}{|1-re^{i\theta}|^{\lambda+1}}=J_1+J_2. \tag{1}$$

(i) If $\lambda>0$,

$$J_1 \leqq \frac{1}{(1-r)^{\lambda+1}}\int_{|\theta|\leqq 1-r}d\theta=\frac{2}{(1-r)^{\lambda}}. \tag{2}$$

Since $|1-re^{i\theta}|^2=1-2r\cos\theta+r^2 \geqq (1-r)^2+a^2\theta^2$, $\quad a=\text{const.}$,

$$J_2=2\int_{1-r}^{\pi}\frac{d\theta}{|1-re^{i\theta}|^{\lambda+1}} \leqq 2\int_{1-r}^{\pi}\frac{d\theta}{((1-r)^2+a^2\theta^2)^{(\lambda+1)/2}} \leqq \frac{2}{(1-r)^{\lambda}}\int_1^{\infty}\frac{dt}{(1+a^2t^2)^{(\lambda+1)/2}}$$

$$=O\left(\frac{1}{(1-r)^{\lambda}}\right). \tag{3}$$

Hence

$$J=O\left(\frac{1}{(1-r)^{\lambda}}\right). \tag{4}$$

(ii) If $\lambda=0$, then $J_1=O(1)$ and

$$J_2=2\int_{1-r}^{\pi}\frac{d\theta}{|1-re^{i\theta}|} \leqq 2\int_{1-r}^{\pi}\frac{d\theta}{\sqrt{(1-r)^2+a^2\theta^2}} \leqq 2\int_0^{\frac{\pi}{1-r}}\frac{dt}{\sqrt{1+a^2t^2}}$$

$$=O\left(\log\frac{1}{1-r}\right).$$

Hence

$$J=O\left(\log\frac{1}{1-r}\right). \tag{5}$$

Theorem V. 26. *Suppose that* $\sum_n (1-|a_n|)=\infty$. *Let* ρ_0 *be the order of* $P(z)$, *then* ρ_0 *is equal to the convergence exponent of* $|a_n|$:

$$\rho_0=\mu \ (\leqq \rho) .$$

Proof. Since $\mu \leqq \rho_0$, we have only to prove that $\rho_0 \leqq \mu$. Let $T(r)$ be the characteristic function of $P(z)$, then since $P(z)$ is regular in $|z|<1$, we have by Theorem V. 24 and the lemma, since $\mu+\varepsilon>0$,

$$T(r)=\frac{1}{2\pi}\int_0^{2\pi} \log^+|P(re^{i\theta})|d\theta+O(1)\leqq \text{const.}\sum_{n=1}^{\infty}\int_0^{2\pi}\left|\frac{1-|a_n|^2}{1-\bar{a}_n re^{i\theta}}\right|^{\mu+1+\varepsilon} d\theta+O(1)$$

$$\leqq \frac{\text{const.}}{(1-r)^{\mu+\varepsilon}},$$

where $\varepsilon=0$, if $p=\mu$ and $\varepsilon>0$ is arbitrarily small, if $p\neq\mu$, so that $\rho_0\leqq\mu$, hence $$\rho_0=\mu. \text{ —— }$$
Similarly as Theorem V. 21, we can prove

Theorem V. 27. *Let* $w(z)$ *be a meromorphic function of finite order* ρ *in* $|z|<1$, *then* $w(z)$ *can be expressed in the form:* $w(z)=w_1(z)/w_2(z)$, *where* $w_i(z)$ $(i=1, 2)$ *is regular and of order* $\leqq\rho$ *in* $|z|<1$.

11. Order of the derivative of a meromorphic function in $|z|<1$

First we shall prove a lemma.

Lemma. *Let* $\{a_n\}$ *be points in* $|z|<1$ *and* $\mu\geqq 0$ *be its convergence exponent, then*

$$\sum_{r\leqq|a_n|<1}(1-|a_n|)^{\mu+4}=O((1-r)^2) .$$

Hence if C_n: $|z-a_n|=(1-|a_n|^2)^{\mu+4}$ *be a circle about* a_n, *and if* $1-r_0$ *is small, then for any* r $(r_0\leqq r<1)$, *there exists* r' $\left(r\leqq r'\leqq \frac{r+1}{2}\right)$, *such that the circle* $|z|=r'$ *lies outside of* C_n $(n=1, 2, \cdots)$.

Proof. Let $n(r)$ be the number of a_n, contained in $|z|<r<1$, then since $\sum_n (1-|a_n|)^{\mu+1+\varepsilon}<\infty$, we have $n(r)=O\left(\frac{1}{(1-r)^{\mu+1+\varepsilon}}\right)$, $0<\varepsilon<1$.

Let $0<r<\rho<1$, then

$$\sum_{r\leqq|a_n|\leqq\rho}(1-|a_n|)^{\mu+4}=\int_r^{\rho}(1-t)^{\mu+4}dn(t)\leqq(1-\rho)^{\mu+4}n(\rho)+(\mu+4)\int_r^{\rho}(1-t)^{\mu+3}n(t)dt .$$

Since $(1-\rho)^{\mu+4}n(\rho)\to 0$, as $\rho\to 1$, we have

$$\sum_{r\leq|a_n|<1}(1-|a_n|)^{\mu+4}\leqq(\mu+4)\int_r^1(1-t)^{\mu+3}\,n(t)\,dt=O\left(\int_r^1(1-t)^{2-\varepsilon}dt\right)$$
$$=O((1-r))^{3-\varepsilon}=O((1-r)^2)\,.$$

Theorem V. 28. *Let $w(z)$ be a meromorphic function of order $\rho(\leqq\infty)$ in $|z|<1$ and ρ' be the order of $w'(z)$, then $\rho'=\rho$.* Tsuji [53].

Proof. (i) First we suppose that $w(z)$ is regular in $|z|<1$ and $\rho<\infty$. Let $T(w, r)$ be the characteristic function of $w(z)$, then by Theorem V. 15,

$$m\left(\frac{w'}{w}, r, \infty\right)=\frac{1}{2\pi}\int_0^{2\pi}\log^+\left|\frac{w'(re^{i\theta})}{w(re^{i\theta})}\right|d\theta=O\left(\log\frac{1}{1-r}\right)+O(\log T(w, r))$$
$$=O\left(\log\frac{1}{1-r}\right), \tag{1}$$

except in a set of intervals $\{J_\nu\}$, such that $\sum_\nu\int_{J_\nu}\frac{dr}{(1-r)^{1+\rho+\varepsilon}}<\infty$.

Hence

$$T\left(\frac{w'}{w}, r\right)=m\left(\frac{w'}{w}, r, \infty\right)+N\left(\frac{w'}{w}, r, \infty\right)+O(1)$$
$$=m\left(\frac{w'}{w}, r, \infty\right)+N(w, r, 0)+O(1)\leqq m\left(\frac{w'}{w}, r, \infty\right)+T(w, r)+O(1)$$
$$=O\left(\frac{1}{(1-r)^{\ +\varepsilon}}\right), \quad \varepsilon>0\,, \tag{2}$$

except in $\{J_\nu\}$. Since $T\left(\frac{w'}{w}, r\right)$ is an increasing function of r, similarly as p. 212, we see that (2) holds without exception.

Since $\varepsilon>0$ is arbitrary, $\frac{w'(z)}{w(z)}$ is of order $\leqq\rho$, so that $w'(z)=w(z)\cdot\frac{w'(z)}{w(z)}$ is of order $\leqq\rho$, hence

$$\rho'\leqq\rho<\infty\,. \tag{3}$$

Next we shall prove that $\rho'\geqq\rho$. Since $w'(z)$ is of order $\rho'<\infty$,

$$\frac{1}{2\pi}\int_0^{2\pi}\log^+|w'(re^{i\theta})|d\theta=T(w', r)+O(1)=O\left(\frac{1}{(1-r)^{\rho'+\varepsilon}}\right), \quad \varepsilon>0\,. \tag{4}$$

Since $\log^+|w'|$ is subharmonic, by means of Poisson integral on $|z|=\frac{r+1}{2}$ and (4), we have

$$\mathfrak{M}(r)=\operatorname*{Max}_{|z|=r}\log^+|w'(z)|=O\left(\frac{1}{(1-r)^{\rho'+1+\varepsilon}}\right). \tag{5}$$

If we put

$$M_1(r, \theta)=\max_{0\leqq t\leqq r}|w'(te^{i\theta})| , \tag{6}$$

then

$$|w(re^{i\theta})|\leqq|w(0)|+\int_0^r|w'(te^{i\theta})|dt\leqq|w(0)|+M_1(r, \theta) ,$$

so that

$$T(w, r)=\frac{1}{2\pi}\int_0^{2\pi}\log^+|w(re^{i\theta})|d\theta+O(1)\leqq\frac{1}{2\pi}\int_0^{2\pi}\log^+M_1(r, \theta)d\theta+O(1)$$

$$\leqq\left(\frac{1}{2\pi}\int_0^{2\pi}(\log^+M_1(r, \theta))^{1+\varepsilon}d\theta\right)^{\frac{1}{1+\varepsilon}}+O(1) . \tag{7}$$

Since $\log^+M_1(r, \theta)=\max_{0\leqq t\leqq r}\log^+|w'(te^{i\theta})|$ and $\log^+|w'(z)|\geqq 0$ is a non-negative subharmonic function, we have by Theorem IV. 39 and (4), (5),

$$\int_0^{2\pi}(\log^+M_1(r, \theta))^{1+\varepsilon}d\theta\leqq\text{const.}\int_0^{2\pi}(\log^+|w'(re^{i\theta})|)^{1+\varepsilon}d\theta$$

$$\leqq\text{const.}\int_0^{2\pi}\log|w'(re^{i\theta})|(\mathfrak{M}(r))^{\varepsilon}d\theta\leqq\frac{\text{const.}}{(1-r)^{\varepsilon(\rho'+1+\varepsilon)}}\int_0^{2\pi}\log^+|w'(re^{i\theta})|d\theta$$

$$\leqq\frac{\text{const.}}{(1-r)^{\varepsilon(\rho'+1+\varepsilon)}}\cdot\frac{1}{(1-r)^{\rho'+\varepsilon}} . \tag{8}$$

Hence by taking $\varepsilon>0$ sufficiently small, we have from (7)

$$T(w, r)=O\left(\frac{1}{(1-r)^{\rho'+\delta}}\right) \qquad \text{for any } \delta>0 , \tag{9}$$

hence $\rho\leqq\rho'$, so that $\rho'=\rho$, if $\rho<\infty$. If $\rho'<\infty$, then from $\rho\leqq\rho'$, we have $\rho<\infty$, so that if $\rho=\infty$, then $\rho'=\infty$, hence $\rho'=\rho$ in general.

(ii) Next suppose that $w(z)$ has poles in $|z|<1$. First suppose that $\rho<\infty$. Then by Theorem V. 27, $w(z)=w_1(z)/w_2(z)$, where $w_i(z)$ $(i=1, 2)$ is regular and of order $\leqq\rho$ in $|z|<1$, hence $w_i'(z)$ is of order $\leqq\rho$.

Since

$$w'(z)=\frac{w_1'(z)w_2(z)-w_2'(z)w_1(z)}{(w_2(z))^2},$$

we have

$$\rho'\leqq\rho . \tag{1}$$

Next we shall prove that $\rho'\geqq\rho$. Let $a_n\neq 0$ be poles of $w'(z)$ and $P(z)$ be the canonical product, formed with $\{a_n\}$, then by Theorem V. 26, $P(z)$ is of order $\leqq\rho'$ and

$$w'(z)=Q(z)/z^{\nu}P(z), \qquad \nu\geqq 0 , \tag{2}$$

where $Q(z)$ is regular and of order $\leqq\rho'$ in $|z|<1$.

Let $\mu\geqq 0$ be the convergence exponent of $|a_n|$ and C_n: $|z-a_n|=(1-|a_n|^2)^{\mu+4}$ be a circle about a_n, then by the lemma, for any small $1-\tau$, there exist r_0 and $r\left(\tau\leqq r_0<r\leqq\frac{r_0+1}{2}<1\right)$, such that the circles $|z|=r_0$ and $|z|=r$ lie outside of C_n $(n=1, 2, \cdots)$. We take off the insides of $\{C_n\}$, which lie in $r_0<|z|<r$ from $r_0<|z|<r$ and let D be the remaining part. Then D consists of a finite number of connected domains. Since $\sum_n (1-|a_n|^2)^{\mu+4}<\infty$, there exists a connected one Δ among them, which contains the circles $|z|=r_0$ and $|z|=r$ on its boundary Γ. Let $L(r_0, r, \theta)$ be the segment: $z=te^{i\theta}$ $(r_0\leqq t\leqq r)$. We modify $L(r_0, r, \theta)$ into $\widetilde{L}(r_0, r, \theta)$ as follows.

If $L(r_0, r, \theta)$ does not intersect $\{C_n\}$, then we put $\widetilde{L}(r_0, r, \theta)=L(r_0, r, \theta)$. If $L(r_0, r, \theta)$ intersects $\{C_n\}$, then the part of $L(r_0, r, \theta)$, which lies in Δ, consists of an odd number of segments: $\overline{z_0z_1}, \overline{z_2z_3}, \cdots, \overline{z_{2n}z_{2n+1}}$, where $z_\nu=t_\nu e^{i\theta}$ $(r_0=t_0<t_1<\cdots<t_{2n+1}=r)$. There exists an arc α_1 of Γ, which connects z_1 to z_2 and there exists an arc α_2 of Γ, which connects z_3 to z_4 and similarly we define $\alpha_3, \cdots$, then we put

$$\widetilde{L}(r_0, r, \theta)=\overline{z_0z_1}+\alpha_1+\overline{z_2z_3}+\alpha_2+\overline{z\ z_5}+\cdots+\overline{z_{2n}z_{2n+1}}\,. \tag{3}$$

$\widetilde{L}(r_0, r, \theta)$ lies outside of $\{C_n\}$ and connects $z_0=r_0e^{i\theta}$ to $z=re^{i\theta}$, so that

$$|w(re^{i\theta})|\leqq|w(r_0e^{i\theta})|+\int_{\widetilde{L}(r_0,r,\theta)}|w'(z)||dz|=|w(r_0e^{i\theta})|+\int_{\widetilde{L}(r_0,r,\theta)}|Q(z)/z^\nu P(z)||dz|$$

$$\leqq|w(r_0e^{i\theta})|+\left[\int_{\widetilde{L}(r_0,r,\theta)}|Q(z)|^2|dz|\int_{\widetilde{L}(r_0,r,\theta)}1/|z^\nu P(z)|^2|dz|\right]^{\frac{1}{2}},$$

hence

$$\int_0^{2\pi}\log^+|w(re^{i\theta})|d\theta\leqq\int_0^{2\pi}\log^+\left(\int_{\widetilde{L}(r_0,r_0\theta)}|Q(z)|^2|dz|\right)d\theta$$

$$+\int_0^{2\pi}\log^+\left(\int_{\widetilde{L}(r_0,r,\theta)}1/|P(z)|^2\,|dz|\ \ d\theta+O(1)=\mathrm{I}+\mathrm{II}+O(1)\,. \tag{4}$$

We shall prove that if $1-r_0$ is sufficiently small, then $\widetilde{L}(r_0, r, \theta)$ is contained in a Stolz domain $\omega(e^{i\theta}, \pi/6)$: $|\arg(1-z\,e^{-i\theta})|<\pi/6$.

Let $z\in\widetilde{L}(r_0, r, \theta)$ and suppose that $z\in\alpha_\nu$. Since $(1-|a_n|^2)^{\mu+4}$ is the radius of C_n and by the lemma, $\sum_{r_0\leqq|a_n|<1}(1-|a_n|^2)^{\mu+4}=O((1-r_0)^2)$, we see that the diameter of α_ν is $O((1-r_0)^2)$, so that if $1-r_0$ is sufficiently small. z lies in $\omega(e^{i\theta}, \pi/6)$, hence $\widetilde{L}(r_0, r, \theta)\in\omega(e^{i\theta}, \pi/6)$. From this, we see easily that $\widetilde{L}(r_0, r, \theta)$ is contained in a Stolz domain $\omega(r'e^{i\theta}, \pi/3)$, $\left(r'=\frac{r+1}{2}\right)$, which is bounded by two lines through $r'e^{i\theta}$, making an angle $\pi/3$ with the radius of the circle $|z|=r'$ through $r'e^{i\theta}$.

Let $\omega_{r_0}(r'e^{i\theta}, \pi/3)$ be the part of $\omega(r'e^{i\theta}, \pi/3)$, which is contained in $r_0 \leqq |z| \leqq r'$ and put

$$M(r', \theta) = \operatorname*{Max}_{z\in\omega_{r_0}(r'e^{i\theta}, \pi/3)} |Q(z)|, \tag{5}$$

then by Theorem IV. 39,

$$\int_0^{2\pi} (\log^+ M(r', \theta))^{1+\varepsilon} d\theta \leqq \text{const.} \int_0^{2\pi} (\log^+ |Q(r'e^{i\theta})|)^{1+\varepsilon}\, d\theta, \quad \varepsilon>0 .$$

Hence

$$\mathrm{I} = \int_0^{2\pi} \log^+ \left(\int_{\tilde{L}(r_0, r, \theta)} |Q(z)|^2 |dz|\right) d\theta \leqq \text{const.} \int_0^{2\pi} \log^+ M(r', \theta)\, d\theta$$

$$\leqq \text{const.} \left(\int_0^{2\pi} (\log^+ M(r', \theta))^{1+\varepsilon} d\theta\right)^{\frac{1}{1+\varepsilon}} \leqq \text{const.} \int_0^{2\pi} (\log^+ |Q(r'e^{i\theta})|)^{1+\varepsilon} d\theta. \tag{6}$$

Since $Q(z)$ is of order $\leqq \rho'$, we have from (6), similarly as (8) (p. 229),

$$\mathrm{I} = O\left(\frac{1}{(1-r')^{\rho'+\delta}}\right) = O\left(\frac{1}{(1-r)^{\rho'+\delta}}\right) \quad \text{for any } \delta>0 . \tag{7}$$

Next we shall evaluate II.

Since $\tilde{L}(r_0, r, \theta) \in \omega(r'e^{i\theta}, \pi/3)$ $\left(r' = \dfrac{r+1}{2}\right)$, we see easily that for any $z \in \tilde{L}(r_0, r, \theta)$,

$$\frac{1}{|1-\bar{a}_n z|} \leqq \frac{\text{const.}}{|1-\bar{a}_n r' e^{i\theta}|}. \tag{8}$$

Since by Theorem V. 25, if z lies outside of $\{C_n\}$,

$$\log^+ \frac{1}{|P(z)|} \leqq \text{const.} \log \frac{1}{1-|z|} \sum_{n=1}^{\infty} \left|\frac{1-|a_n|^2}{1-\bar{a}_n z}\right|^{\mu+1+\varepsilon}, \tag{9}$$

and $\tilde{L}(r_0, r, \theta)$ lies outside of $\{C_n\}$, if we put

$$\mathfrak{M}(r, \theta) = \operatorname*{Max}_{z\in\tilde{L}(r_0, r, \theta)} \frac{1}{|P(z)|}, \tag{10}$$

then from (8), (9), we have

$$\log \mathfrak{M}(r, \theta) \leqq \text{const.} \log \frac{1}{1-r} \sum_{n=1}^{\infty} \left|\frac{1-|a_n|^2}{1-\bar{a}_n r' e^{i\theta}}\right|^{\mu+1+\varepsilon}, \tag{11}$$

so that by the lemma (p. 226)

$$\Pi = \int_0^{2\pi} \log^+ \left(\int_{\tilde{L}(r_0, r, \theta)} 1/|P(z)|^2 |dz|\right) d\theta \leqq \text{const.} \int_0^{2\pi} \log \mathfrak{M}(r, \theta)\, d\theta$$

$$\leqq \begin{cases} \text{const.} \log \dfrac{1}{1-r} \cdot \dfrac{1}{(1-r)^{\mu+\varepsilon}}, & \text{if } \mu+\varepsilon>0 , \\[2ex] \text{const.} \left(\log \dfrac{1}{1-r}\right)^2, & \text{if } \mu+\varepsilon=0 . \end{cases} \tag{12}$$

Since $\mu\leqq\rho'$, we have by (4), (7), (12),

$$m(w, r, \infty)=\frac{1}{2\pi}\int_0^{2\pi}\log^+|w(re^{i\theta})|d\theta=O\left(\frac{1}{(1-r)^{\rho'+\delta}}\right), \quad \delta>0 .$$

Hence if the circle $|z|=r$ lies outside of $\{C_n\}$, then

$$T(w, r)=m(w, r, \infty)+N(w, r, \infty)\leqq m(w, r, \infty)+N(w', r, \infty)$$

$$\leqq m(w, r, \infty)+T(w', r)=O\left(\frac{1}{(1-r)^{\rho'+\delta}}\right). \tag{13}$$

By the lemma, for any $r(r_0\leqq r<1)$, there exists $r'\left(r\leqq r'\leqq\frac{r+1}{2}\right)$, such that the circle $|z|=r'$ lies outside of $\{C_n\}$, so that for any r $(r_0\leqq r<1)$ and $\delta>0$,

$$T(w, r)\leqq T(w, r')=O\left(\frac{1}{(1-r')^{\rho'+\delta}}\right)=O\left(\frac{1}{(1-r)^{\rho'+\delta}}\right),$$

hence $\rho\leqq\rho'$, so that $\rho'=\rho$, if $\rho<\infty$. We can prove as in (i), that $\rho'=\infty$, if $\rho=\infty$, hence $\rho'=\rho$ in general.

12. Direct transcendental singularity

1. Let K: $|w|<R$ be a disc on the w-plane. For two points a, b of K, we define a metric (a, b) as follows. If $b=0$, then

$$(a, 0)=\frac{2R|a|}{R^2+|a|^2}, \quad 0\leqq(a, 0)<1 . \tag{1}$$

If $b\neq0$, let $U_a(w)=\dfrac{R^2(w-a)}{R^2-\bar{a}w}$, then

$$(a, b)=(U_a(b), 0)=\frac{2R|U_a(b)|}{R^2+|U_a(b)|^2}=\frac{2R|b-a|/|R^2-\bar{a}b|}{1+R^2|b-a|^2/|R^2-\bar{a}b|^2}. \tag{2}$$

Hence $(a, b)=(b, a)$ and we can prove easily that $(a, c)<(a, b)+(b, c)$ and

$$(U(a),\ U(b))=(a, b)\ , \tag{3}$$

where $U(w)$ is a linear transformation, which makes $|w|<R$ invariant. Since

$$\frac{1}{(w,\ a)}=\frac{R^2+|U_a(w)|^2}{2R|U_a(w)|}=1+\frac{(R-|U_a(w)|)^2}{2R|U_a(w)|}\geqq1\ ,$$

and $|U_a(w)|=R$ on $|w|=R$, $\log\dfrac{1}{(w, a)}$ $(\geqq0)$ and its normal derivative vanish on $|w|=R$.

If $w=w(z)$ is a regular function of z and $|w(z)|<R$, $w(z)\neq0$, then since

$$\Delta \log \frac{1}{(w(z), 0)} = \frac{4R^2|w'(z)|^2}{(R^2+|w(z)|^2)^2} \geqq 0 ,$$

$\log \frac{1}{(w(z), 0)}$ is a subharmonic function of z.

2. Let Δ be an infinite domain on the z-plane and Γ be its boundary. We assume that Γ consists of at most a countable number of analytic curves, which do not cluster at a finite distance and contains a curve, extending to infinity.

Let $w=w(z)$ be a one-valued regular function in Δ, such that

$$|w(z)|<R \text{ and } w\neq 0 \text{ in } \Delta , \qquad |w(z)|=R \text{ on } \Gamma. \tag{1}$$

If $r>r_0>0$, then the circle $|z|=r$ has common points with Δ. Let $\Delta_r(r>r_0)$ be the part of Δ, which lies in $|z|<r$, then Δ_r consists of a finite number of connected domains. Let θ_r be the part of the circle $|z|=r$, contained in Δ and $r\theta(r)$ be its length, then $\theta(r)$ is a continuous function of $r(>r_0)$. Let Γ_r be the part of Γ, which lies in $|z|\leqq r$, so that $\Gamma_r+\theta_r$ is the whole boundary of Δ_r.

We put

$$S(r ; \Delta)=\frac{1}{\pi}\iint_{\Delta_r}\left(\frac{|w'(z)|}{1+|w(z)|^2}\right)^2 rdrd\theta, \qquad z=re^{i\theta}, \tag{2}$$

$$T(r ; \Delta)=\int_{r_0}^{r}\frac{S(r;\Delta)}{r}\,dr , \tag{3}$$

and shall prove the following

Theorem V. 29. $\log T(r ; \Delta)\geqq \pi\int_{r_0}^{\kappa r}\frac{dr}{r\theta(r)}-\text{const.} \quad (0<\kappa<1)$.

Tsuji [34].

Proof. We put

$$u(z)=\log\frac{1}{(w(z), 0)}=\log\frac{R^2+|w(z)|^2}{2R|w(z)|}\geqq 0 , \tag{1}$$

then by the above remark,

$$u=0, \qquad \frac{\partial u}{\partial \nu}=0 \quad \text{on } \Gamma , \tag{2}$$

where ν is the outer normal of Γ with respect to Δ and

$$\Delta u=\frac{\partial^2 u}{\partial x^2}+\frac{\partial^2 u}{\partial y^2}=\frac{4R^2|w'|^2}{(R^2+|w|^2)^2}\geqq 0, \qquad \frac{\partial^2 u}{\partial \log r^2}+\frac{\partial^2 u}{\partial \theta^2}=r^2\Delta u\geqq 0 .$$

$$z=x+iy=re^{i\theta} . \tag{3}$$

We put

$$\mu(r)=\frac{1}{2\pi}\int_{\theta_r} u(re^{i\theta})\,d\theta\,, \tag{4}$$

then in virtue of (2), we have by the Green's formula,

$$r\mu'(r)=\frac{1}{2\pi}\int_{\theta_r}\frac{\partial u}{\partial r}\,rd\theta=\frac{1}{2\pi}\int_{\Gamma_r+\theta_r}\frac{\partial u}{\partial\nu}\,ds$$

$$=\frac{1}{2\pi}\iint_{\Delta_r}\Delta u\,dxdy=\frac{2R^2}{\pi}\iint_{\Delta_r}\frac{|w'|^2}{(R^2+|w|^2)^2}\,dxdy\leqq\text{const.}\iint_{\Delta_r}\frac{|w'|^2}{(1+|w|^2)^2}\,dxdy$$

Since $\mu'(r)>0$, $\mu(r)$ is an increasing function of r and

$$\mu(r)\leqq\text{const.}\int_{r_0}^{r}\frac{dr}{r}\iint_{\Delta_r}\frac{|w'|^2}{(1+|w|^2)^2}\,dxdy+\text{const.}$$

$$\leqq\text{const. } T(r;\Delta)+\text{const.}. \tag{5}$$

If we put

$$m(r)=\frac{1}{2\pi}\int_{\theta_r}[(u(re^{i\theta})]^2 d\theta\,, \tag{6}$$

then by (2),

$$\frac{dm(r)}{d\log r}=\frac{1}{\pi}\int_{\theta_r}u\frac{\partial u}{\partial\log r}\,d\theta=\frac{1}{\pi}\int_{\theta_r}u\frac{\partial u}{\partial r}\,rd\theta=\frac{1}{\pi}\int_{\Gamma_r+\theta_r}u\frac{\partial u}{\partial\nu}\,ds$$

$$=\frac{1}{\pi}\iint_{\Delta_r}\left(\left(\frac{\partial u}{\partial x}\right)^2+\left(\frac{\partial u}{\partial y}\right)^2\right)dxdy+\frac{1}{\pi}\iint_{\Delta_r}u\Delta u\,dxdy>0\,. \tag{7}$$

Hence $m(r)$ is an increasing function of r. By (3), we have

$$\frac{d^2m(r)}{d\log r^2}=\frac{1}{\pi}\int_{\theta_r}\left(\left(\frac{\partial u}{\partial\log r}\right)^2+u\frac{\partial^2 u}{\partial\log r^2}\right)d\theta$$

$$=\frac{1}{\pi}\int_{\theta_r}\left(\left(\frac{\partial u}{\partial\log r}\right)^2+r^2u\Delta u-u\frac{\partial^2 u}{\partial\theta^2}\right)d\theta\geqq\frac{1}{\pi}\int_{\theta_r}\left(\left(\frac{\partial u}{\partial\log r}\right)^2-u\frac{\partial^2 u}{\partial\theta^2}\right)d\theta$$

$$=\frac{1}{\pi}\int_{\theta_r}\left(\left(\frac{\partial u}{\partial\log r}\right)^2+\left(\frac{\partial u}{\partial\theta}\right)^2\right)d\theta>0\,. \tag{8}$$

From this, we have as Theorem III. 67,

$$\frac{d^2m(r)}{d\log r^2}\geqq\frac{1}{2\,m(r)}\left(\frac{d\,m(r)}{d\log r}\right)^2+\frac{1}{2}\left(\frac{2\pi}{\theta(r)}\right)^2 m(r)\geqq\frac{d\,m(r)}{d\log r}\cdot\frac{2\pi}{\theta(r)}\,. \tag{9}$$

Since $\dfrac{d\,m(r)}{d\log r}>0$, we have by integrating,

$$\log\frac{dm(r)}{d\log r}-\log\frac{dm(r_0)}{d\log r_0}\geqq 2\pi\int_{r_0}^{r}\frac{dt}{t\theta(t)},$$

$$rm'(r)\geqq \text{const. } e^{2\pi\int_{r_0}^{r}\frac{dt}{t\theta(t)}}. \tag{10}$$

Since $m(r)$ is an increasing function of r, we have for any $0<\alpha<1$,

$$m(r)\geqq m(r)-m(r_0)\geqq\int_{r_0}^{r}m'(r)dr\geqq\text{const.}\int_{r_0}^{r}e^{2\pi\int_{r_0}^{r}\frac{dt}{t\theta(t)}}\cdot\frac{dr}{r}$$

$$\geqq\text{const.}\int_{\alpha r}^{r}e^{2\pi\int_{r_0}^{r}\frac{dt}{t\theta(t)}}\cdot\frac{dr}{r}\geqq\text{const. } e^{2\pi\int_{r_0}^{\alpha r}\frac{dt}{t\theta(t)}}\int_{\alpha r}^{r}\frac{dr}{r}$$

$$\geqq\text{const. } e^{2\pi\int_{r_0}^{\alpha r}\frac{dt}{t\theta(t)}}. \tag{11}$$

Let $u(z)$ attain its maximum on θ_r at $z_0=re^{i\theta_0}$, then

$$(u(re^{i\theta_0}))^2\geqq\frac{1}{2\pi}\int_{\theta_r}(u(re^{i\theta}))^2\,d\theta=m(r)\geqq\text{const. } e^{2\pi\int_{r_0}^{\alpha r}\frac{dr}{r\theta(r)}},$$

so that

$$u(re^{i\theta_0})\geqq\text{const. } e^{\pi\int_{r_0}^{\alpha r}\frac{dr}{r\theta(r)}}. \tag{12}$$

Let for $r<\rho$,

$$v(z)=v(re^{i\theta})=\frac{1}{2\pi}\int_{\theta_\rho}u(\rho e^{i\varphi})\frac{\rho^2-r^2}{\rho^2-2\rho r\cos(\varphi-\theta)+r^2}d\varphi,$$

then $v=u$ on θ_ρ and $v>0$, $u=0$ on Γ_ρ and since v is harmonic and u is subharmonic, we have by the maximum principle, $u\leqq v$ in $\varDelta_\rho$. Hence if we put $\rho=\lambda r$ ($\lambda>1$), then by (5),

$$u(re^{i\theta_0})\leqq v(re^{i\theta_0})\leqq\frac{\lambda+1}{\lambda-1}\cdot\frac{1}{2\pi}\int_{\theta_\rho}u(\rho e^{i\theta})\,d\theta=\frac{\lambda+1}{\lambda-1}\mu(\lambda r)$$

$$\leqq\text{const. } T(\lambda r\,;\varDelta)+\text{const.},$$

so that by (12),

$$T(\lambda r\,;\varDelta)\geqq\text{const. } e^{\pi\int_{r_0}^{\alpha r}\frac{dr}{r\theta(r)}}-\text{const.},$$

hence if we write r instead if λr and put $\kappa=\dfrac{\alpha}{\lambda}<1$, then we have

$$\log T(r\,;\varDelta)\geqq\pi\int_{r_0}^{\kappa r}\frac{dr}{r\theta(r)}-\text{const.}. \tag{13}$$

3. Let $w=w(z)$ be a meromorphic function for $|z|<\infty$ and $z=z(w)$ be its inverse function and F be the Riemann surface, generated by $w=w(z)$ on the w-plane. Let (w_0) be an accessible boundary point of F, whose projection on the w-plane is w_0. Then there exists a curve C on the z-plane, extending to ∞, such that if z tends to ∞ on C, then $w(z)$

tends to w_0. Let F_ρ be a connected piece of F, which lies above a disc: $|w-w\,|<\rho$ and contains (w_0) on its boundary, then F_ρ is mapped on a connected domain $\varDelta$ on the z-plane, such that $|w(z)-w_0|<\rho$ in $\varDelta$ and $|w(z)-w_0|=\rho$ on its boundary. If $w(z)\neq w_0$ in $\varDelta$ for a certain $\rho>0$, then (w_0) is called *a direct transcendental singularity of* $z(w)$ and if otherwise, (w_0) is called *an indirect transcendental singularity of* $z(w)$.

Theorem V. 30. *If* $w(z)$ *is a meromorphic function of finite order* ρ *for* $|z|<\infty$ *and* n *be the number of direct transcendental singularities of* $z(w)$, *then*

$$n\leqq 2\rho,\ \text{if } 1\leqq\rho<\infty;\qquad n\leqq 1,\ \text{if } 0\leqq\rho<1.$$

Ahlfors [3].

Proof. Suppose that there are n direct transcendental singularities (a_i) $(i=1, 2, \cdots, n)$ of $z(w)$ and suppose that $n\geqq 2$. Let F_i be a connected piece of F, which lies above a disc: $|w-a_i|<\eta$ and contains (a_i) on its boundary and let $\varDelta_i$ be the image of F_i on the z-plane. We take $\eta>0$ so small that $\varDelta_i$ are disjoint each other and $w(z)\neq a_i$ in $\varDelta_i$. Since $n\geqq 2$, $\varDelta_i$ contains a boundary curve, extending to ∞, $|w(z)-a_i|<\eta$ in $\varDelta_i$ and $|w(z)-a_i|=\eta$ on its boundary.

We define $\theta_i(r)$, $T(r;\varDelta_i)$ for $\varDelta_i$, as in Theorem V. 29, then

$$\log T(r;\varDelta_i)\geqq\pi\int_{r_0}^{\kappa r}\frac{dr}{r\theta_i(r)}-\text{const.}\qquad(0<\kappa<1)\,. \tag{1}$$

Let $T(r)$ be the characteristic function of $w(z)$, then since $w(z)$ is of order ρ, we have for any $\varepsilon>0$,

$$\log T(r;\varDelta_i)\leqq\log T(r)\leqq(\rho+\varepsilon)\log r,\qquad r\geqq r_1(\varepsilon)>r_0\,, \tag{2}$$

so that from (1), (2),

$$n(\rho+\varepsilon)\log r\geqq\pi\int_{r_0}^{\kappa r}\frac{dr}{r}\sum_{i=1}^{n}\frac{1}{\theta_i(r)}-\text{const.}.$$

Since $n^2=\left(\frac{\sqrt{\theta_1}}{\sqrt{\theta_1}}+\cdot\cdot+\frac{\sqrt{\theta_n}}{\sqrt{\theta_n}}\right)^2\leqq(\theta_1+\cdot\cdot+\theta_n)\left(\frac{1}{\theta_1}+\cdot\cdot+\frac{1}{\theta_n}\right)\leqq 2\pi\left(\frac{1}{\theta_1}+\cdot\cdot+\frac{1}{\theta_n}\right)$,

we have

$$n(\rho+\varepsilon)\log r\geqq\frac{n^2}{2}\int_{r_0}^{\kappa r}\frac{dr}{r}-\text{const.}\geqq\frac{n^2}{2}\log r-\text{const.}\,.$$

Since $\varepsilon>0$ is arbitrary, we have $n\leqq 2\rho$.

Since we assumed that $n\geqq 2$, we have $\rho\geqq 1$, hence

$$n\leqq 2\rho,\ \text{if } \rho\geqq 1 \text{ and } n\leqq 1,\ \text{if } 0\leqq\rho<1.\ \text{——}$$

Remark. If $w(z)$ is of infinite order, then we have the following theorem, which is a special case of a more general Theorem X. 16.

Theorem V. 31. *Let $w(z)$ be a meromorphic function of infinite order for $|z|<\infty$, then the projection of direct transcendental singularities of its inverse function $z(w)$ on the w-plane is of capacity zero.*

13. Meromorphic function in a neighbourhood of a closed set of capacity zero

1. Let E be a bounded closed set of capacity zero on the z-plane, contained in a domain D, which is bounded by a Jordan curve C. Let $w=w(z)$ be one-valued and meromorphic in $D^*=\overline{D}-E$ and every point of E be an essential singularity of $w(z)$.

Since $\gamma(E)=0$, let

$$u(z)=\int_E \log\frac{1}{|z-a|}\,d\mu(a), \qquad \mu(E)=1\,, \tag{1}$$

be the Evans' function with respect to E, such that $u\to+\infty$, if z tends to any point of E. Let $\theta(z)$ be its conjugate harmonic function and put

$$\zeta=\zeta(z)=e^{u+i\theta}=re^{i\theta}, \qquad r=r(z), \qquad \theta=\theta(z)\,. \tag{2}$$

$r(z)$ and $\theta(z)$ play the same rôle as $|z|$ and $\arg z$ in the theory of meromorphic functions for $|z|<\infty$. In the following, we write r, θ instead of $r(z)$, $\theta(z)$.

Let C_r be the niveau curve: $r(z)=\text{const.}=r$, then C_r consists of a finite number of analytic Jordan curves, clustering to E, as $r\to\infty$. Let C_r be contained in D, if $r\geqq r_0$ and D_r be the domain, bounded by C and $C_r\,(r\geqq r_0)$. Then

$$\int_{C_r} d\theta=\int_{C_r}\frac{\partial u}{\partial\nu}\,ds=2\pi\int_E d\mu(a)=2\pi\,, \tag{3}$$

where we integrate on C_r in the positive sense with respect to D_r and ν the outer normal of C_r with respect to D_r and ds the arc element of C_r.

Since $\dfrac{\partial u}{\partial\nu}\geqq 0$ on C_r and $d\theta=\dfrac{\partial u}{\partial\nu}ds\geqq 0$, we have for any sub-arc C_r' of C_r,

$$\int_{C_r'} d\theta\leqq\int_{C_r} d\theta=2\pi\,. \tag{4}$$

We put

$$m(r,a)=\frac{1}{2\pi}\int_{C_r}\log\frac{1}{[w(z),a]}\,d\theta\,, \tag{5}$$

$$N(r,a)=\int_{r_0}^{r}\frac{n(r,a)}{r}dr\,, \tag{6}$$

$$T(r,a)=m(r,a)+N(r,a)\,, \tag{7}$$

where $n(r, a)$ denotes the number of zero points of $w(z)-a$, contained in D_r. Then we shall prove the following analogue of Nevanlinna's first fundamental theorem.

Theorem V. 32. $T(r, a)=T(r)+O(\log r)$,

$$T(r)=\int_{r_0}^{r}\frac{S(r)}{r}dr,\quad S(r)=A(r)/\pi\,,$$

$A(r)$ being the area of the image of D_r by $w=w(z)$ on the w-sphere.

$$\lim_{r\to\infty}\frac{T(r)}{\log r}=\infty\,.$$

We call $T(r)$ *the characteristic function of* $w(z)$, which is an increasing convex function of $\log r$.

The theorem was proved by Hällström [1] by means of the niveau curves of Selberg's function (Selberg [1]) and independently by Tsuji [11] by means of the niveau curves of Evans' function.

Proof. Considering $w(z)$ as a function of $\zeta=re^{i\theta}$, we have

$$\frac{dm(r,a)}{dr}-\frac{dm(r,b)}{dr}=\frac{1}{2\pi}\int_{C_r}\frac{\partial}{\partial r}\log\left|\frac{w-b}{w-a}\right|d\theta=\frac{1}{2\pi r}\int_{C_r}d\arg\left(\frac{w-b}{w-a}\right)$$

$$=\frac{1}{2\pi r}\int_{C+C_r}d\arg\left(\frac{w-b}{w-a}\right)-\frac{1}{2\pi r}\int_{C}d\arg\left(\frac{w-b}{w-a}\right)=\frac{n(r,b)-n(r,a)}{r}+O\left(\frac{1}{r}\right),$$

so that

$$T(r,a)=T(r,b)+O(\log r)\,. \tag{1}$$

Let $d\omega(a)$ be the surface element on the w-sphere K, then

$$T(r,a)=\frac{1}{\pi}\int_K T(r,b)\,d\omega(b)+O(\log r)=\int_{r_0}^{r}\frac{S(r)}{r}dr+O(\log r)$$

$$=T(r)+O(\log r)\,. \tag{2}$$

To prove that $\lim\limits_{r\to\infty} T(r)/\log r=\infty$, first we remark that the set of values, taken by $w=w(z)$ in D^* is everywhere dense on K. For if otherwise, there exists a point a_0, such that $1/(w(z)-a_0)$ is bounded in a neighbourhood of E, so that, since $\gamma(E)=0$, $1/(w(z)-a_0)$ is regular on E and hence $w(z)$ is meromorphic on E, which contradicts the hypothesis, hence the set of values, taken by $w=w(z)$ is everywhere dense on K. From this as Theorem V. 3, we can prove easily that there exists a point a_0, such that $n(r, a_0)\to\infty$, as $r\to\infty$, so that

$\lim_{r\to\infty} T(r, a_0)/\log r=\infty$, hence by (2), $\lim_{r\to\infty} T(r)/\log r=\infty$.

Theorem V. 33. *Under the same condition on $w(z)$ as Theorem V. 32, $w(z)$ takes any value infinitely often in D^*, except a set of capacity zero.* Nevanlinna [5], Kametani [1].

Proof. Let M be the set of values a, such that $n(r, a)=O(1)$ and suppose that $\gamma(M)>0$. Then taking a suitable sub-set of M, we may assume that M is a bounded closed set and for any $a\in M$, $n(r, a)\leqq n_0<\infty$ $(r_0\leqq r<\infty)$. Let

$$u(w)=\int_M \log\frac{1}{[w, a]}\,d\nu(a), \qquad \nu(M)=1\,, \tag{1}$$

be the elliptic conductor potential of M, such that $u(w)\leqq V<\infty$ on K. From

$$T(r)=m(r, a)+N(r, a)+O(\log r)\,,$$

and $u(w)\leqq V$, $n(r, a)\leqq n_0$ $(a\in M)$, we have

$$T(r)=\int_M m(r, a)\,d\nu(a)+\int_M N(r, a)d\nu(a)+O(\log r)=O(\log r)\,, \tag{2}$$

which contradicts Theorem V. 32, hence the theorem. ——

2. We shall prove a more general theorem as follows.[1)]

Theorem V. 34. *Let Ω be the space of uniformly bounded regular functions $\{g(z)\}$ in $D_0(\supset\bar{D})$, then under the same condition on $w(z)$ as Theorem V. 32, $w(z)-g(z)$ has infinitely many zero points in D^*, except a set of g, which is of capacity zero.* Tsuji [64].

Proof. We put

$$m(r, g)=\frac{1}{2\pi}\int_{C_r}\log\frac{1}{[w(z), g(z)]}\,d\theta\,, \tag{1}$$

$$N(r, g)=\int_{r_0}^{r}\frac{n(r, g)}{r}\,dr\,, \tag{2}$$

$$T(r, g)=m(r, g)+N(r, g)\,, \tag{3}$$

where $n(r, g)$ denotes the number of zero points of $w(z)-g(z)$ in D_r.

We shall prove that

$$T(r, g)=T(r)+O(\log r).$$

We have

$$\frac{d\,m(r, 0)}{dr}-\frac{d\,m(r, g)}{dr}=\frac{1}{2\pi}\int_{C_r}\frac{\partial}{\partial r}\log\left|\frac{w(z)-g(z)}{w(z)}\right|d\theta$$

1) c.f. Theorem V. 16.

$$-\frac{1}{2\pi}\frac{\partial}{\partial r}\int_{C_r}\log\sqrt{1+|g(z)|^2}\,d\theta .$$

Considering as a function of $\zeta=re^{i\theta}$, by the argument principle, we have as Theorem V. 32,

$$\frac{1}{2\pi}\int_{C_r}\frac{\partial}{\partial r}\log\left|\frac{w(z)-g(z)}{w(z)}\right|d\theta=\frac{1}{2\pi r}\int_{C_r}d\arg\left(\frac{w(z)-g(z)}{w(z)}\right)$$
$$=\frac{n(r, g)-n(r, 0)}{r}+O\left(\frac{1}{r}\right),$$

so that

$$\frac{d\,m(r, 0)}{dr}+\frac{n(r, 0)}{r}=\frac{d\,m(r, g)}{dr}+\frac{n(r, g)}{r}$$
$$-\frac{1}{2\pi}\frac{\partial}{\partial r}\int_{C_r}\log\sqrt{1+|g|^2}\,d\theta+O\left(\frac{1}{r}\right),$$

hence integrating we have, $T(r, 0)=T(r, g)+O(\log r)$, so that by Theorem V. 32,

$$T(r, g)=T(r)+O(\log r). \tag{4}$$

Let $\mathfrak{M}$ be a set of positive capacity in Ω and suppose that for any $g\in\mathfrak{M}$, $w(z)-g(z)$ has only a finite number of zero points in $D^*=D-E$, then we may assume that $\mathfrak{M}$ is a closed set and for any $g\in\mathfrak{M}$, $n(r, g)\leqq n_0<\infty$, where n_0 is independent of g and r so that $N(r, g)=O(\log r)$, hence by (4),

$$T(r)=m(r, g)+O(\log r), \qquad g\in\mathfrak{M} . \tag{5}$$

Let σ be the equilibrium distribution of $\mathfrak{M}$, such that for any ζ and $z\in D$,

$$\int_{\mathfrak{M}}\log\frac{1}{|\zeta-g(z)|}\,d\sigma(g)\leqq V(\mathfrak{M})<\infty, \qquad \sigma(\mathfrak{M})=1 . \tag{6}$$

Then by (5),

$$T(r)=\int_{\mathfrak{M}}T(r)d\sigma(g)=\int_{\mathfrak{M}}m(r, g)d\sigma(g)+O(\log r)$$
$$=\frac{1}{2\pi}\int_{C_r}d\theta\int_{\mathfrak{M}}\log\frac{\sqrt{1+|w(z)|^2}}{|w(z)-g(z)|}\,d\sigma(g)+O(\log r) . \tag{7}$$

Let $|g(z)|<M$ for any $z\in D$ and $g\in\mathfrak{M}$ and $C_r=C_r'+C_r''$, where $|w(z)|<2M$ on C_r' and $|w(z)|\geqq 2M$ on C_r'', then since

$$\log\frac{\sqrt{1+|w(z)|^2}}{|w(z)-g(z)|}=O(1) \qquad \text{on } C_r'',$$

we have by (6) and (7),

$$T(r)=\frac{1}{2\pi}\int_{C'_r} d\theta \int_{\mathfrak{M}} \log \frac{1}{|w(z)-g(z)|}\, d\sigma(g)+O(\log r)=O(\log r)\ ,$$

which is absurd, since $\lim_{r\to\infty} T(r)/\log r=\infty$, hence the theorem.

Theorem V. 35. *Let $g_\alpha(z)=g(z, \alpha)\not\equiv$ const. be a regular function of two complex variables z and α for $z\in D_0$ $(\supset\bar{D})$, $\alpha\in\Delta$, where Δ is a Jordan domain, then under the same condition on $w(z)$ as Theorem V. 32, $w(z)-g_\alpha(z)$ has infinitely many zero points in $D^*=D-E$, except a set of α, which is of capacity zero.*

Proof. If for any $z\in D$, $\dfrac{\partial g(z, \alpha)}{\partial\alpha}\not\equiv 0$ as a function of α, then the theorem is a special case of Theorem V. 34 in virtue of Theorem III. 75, hence we assume that for a certain $z\in D$, $\dfrac{\partial g(z, \alpha)}{\partial\alpha}\equiv 0$, then there exists a finite number of $z\in D$, $z_1, \cdots, z_n$, say, such that

$$g(z, \alpha)=\prod_{i=1}^{n}(z-z_i)\, g_1(z, \alpha)+h(z)\ , \tag{1}$$

where $g_1(z, \alpha)$ is a regular function of two complex variables z and α for $z\in D_0$, $\alpha\in\Delta$, such that for any $z\in D$, $\dfrac{\partial g_1(z, \alpha)}{\partial\alpha}\not\equiv 0$ and $h(z)$ is a regular function of z in D_0. If we define $w_1(z)$ by

$$w(z)=\prod_{i=1}^{n}(z-z_i)\, w_1(z)+h(z)\ , \tag{2}$$

then $w_1(z)$ is one-valued and meromorphic in D^* and every point of E is an essential singularity of $w_1(z)$, hence by Theorem V. 34 and Theorem III. 75, $w_1(z)-g_1(z, \alpha)$ and hence $w(z)-g(z,\alpha)=\prod_{i=1}^{n}(z-z_i)\,(w_1(z)-g_1(z,\alpha))$ has infinitely many zero points in D^*, except a set of α, which is of capacity zero.

3. We shall prove the following extension of Gross' theorem for meromorphic functions for $|z|<\infty$. (Gross [1])

Theorem V. 36. *Under the same condition on $w(z)$ as Theorem V. 32, let $z=z(w)$ be the inverse function of $w=w(z)$ and F be the Riemann surface, generated by $w=w(z)$ on the w-plane. Let $z(w)$ be regular at w_0, then we can continuate $z(w)$ analytically along the half-line: $w=w_0+te^{i\theta}$ $(0\leqq t<\infty)$ indefinitely, except a null set of θ-values.*

We say that F has *the Gross-property.* This theorem is a special case of a more general Theorem X. 14.

Proof. For a given θ, we continuate $z(w)$ analytically along a half-line: $w=w_0+te^{i\theta}$ $(0\leqq t<\infty)$, till we meet an algebraic, or a transcendental singularity of $z(w)$, thus we get the Mittag-Leffler's star-region H, with respect to w_0.

The edge of H is an algebraic, or a transcendental singularity of $z(w)$.

Let $\varDelta$ be the image of H on the z-plane. We denote the part of H, which lies in $|w-w_0|<R$ by $H^{(R)}$ and $\varDelta^{(R)}$ be its image on the z-plane. Let $\zeta=r(z)e^{i\theta(z)}$ and C_r: $r(z)=\text{const.}=r$ be defined as before and $\varDelta_r^{(R)}$ be the part of $\varDelta^{(R)}$, which lies outside of C_r and C_r^* be the part of C_r, which belongs to the boundary of $\varDelta_r^{(R)}$ and σ_r^* be its image in $H^{(R)}$ and $L(r)$ be its linear measure, then by putting $w(z)=F(\zeta)$, we have

$$L^2(r)=\left(\int_{C_r^*}|F'(\zeta)|rd\theta\right)^2\leqq 2\pi r\int_{C_r^*}|F'(\zeta)|^2 rd\theta=2\pi r\frac{dA(r)}{dr},$$

where $A(r)$ is the area of the image of $\varDelta_r^{(R)}$ in $H^{(R)}$. Hence

$$\int_{r_0}^{r}\frac{L^2(r)}{r}dr\leqq 2\pi A(R)\leqq 2\pi^2R^2 .$$

From this, we see that there exists $r_1<r_2<\cdots<r_n\to\infty$, such that $L(r_n)\to 0$. Hence the set of θ-values of edges of $H^{(R)}$, which are transcendental singularities of $z(w)$ is of meausre zero. Taking $R_1<R_2<\cdots<R_n\to\infty$ for R, we see that the set of θ-values of edges of H, which are transcendental singularities of $z(w)$ is of measure zero. Since there are at most a countable number of algebraic singularities of $z(w)$, the set of θ-values of edges of H is of measure zero.

CHAPTER VI

AHLFORS' THEORY OF COVERING SURFACES

1. Constants d_0, h_0, $d_0(\beta)$, $h_0(\beta)$

Let F_0 be a regular surface of finite genus p in an Euclidean space, which is bounded by a finite number of analytic Jordan curves $\{\Lambda_i\}$.

(i) **Constants d_0 and h_0.**

There exist constants $d_0>0$, $h_0>0$, which depend on F_0 only and satisfy the following condition. Let C be any analytic Jordan curve on F_0, or any analytic Jordan arc on F_0, whose both end points lie on the same Λ_i and let L be the length of C. If $d_0>0$ is sufficiently small and $L<d_0$, then in the first case, C bounds a simply connected domain $D\subset F_0$ and in the second case, C bounds with Λ_i a simply connected domain $D\subset F_0$. In each case, we call D the domain, bounded by C and we denote its area by $|D|$. Then

$$|D|<h_0L\,, \qquad \text{if} \quad L<d_0\,, \tag{1}$$

where $h_0>0$ is a constant, which depends on F_0 only.

(ii) **Constants $d_0(\beta)$ and $h_0(\beta)$.**

Let β be an analytic Jordan arc on F_0, which may lie on one of $\{\Lambda_i\}$. Then there exist constants $d_0(\beta)$, $h_0(\beta)$, which depend on β only and satisfy the following condition.

Let C be any analytic Jordan curve, or any analytic Jordan arc on F_0, which satisfies the condition (i) and L be its length and D be the domain, bounded by C. We denote the part of β, which lies in D by $D(\beta)$ and $|D(\beta)|$ be its length. Then

$$|D(\beta)|<h_0(\beta)L\,, \qquad \text{if} \quad L<d_0(\beta)<d_0\,, \tag{2}$$

where $d_0(\beta)$ and $h_0(\beta)$ are constants, which depend on β only.

2. Ahlfors' covering theorem

Let F be a finite covering surface of F_0, and consist of a finite number of sheets and be bounded by a finite number of analytic Jordan curves. The part of the boundary of F, which does not lie above the boundary of F_0, is called *the relative boundary of F*.

Let D be a domain on F_0, which is bounded by a finite number of analytic Jordan curves and $F(D)$ be the part of F, which lies above D. We denote the area of F_0, F, $F(D)$ by $|F_0|$, $|F|$, $|F(D)|$ respectively. We put

$$S=\frac{|F|}{|F_0|}, \qquad S(D)=\frac{|F(D)|}{|D|}. \tag{1}$$

Let β be an analytic Jordan arc on F_0, which may lie on the boundary of F_0 and $F(\beta)$ be the part of F, which lies above β and $|F(\beta)|$ be its length and put

$$S(\beta)=\frac{|F(\beta)|}{|\beta|}, \tag{2}$$

where $|\beta|$ denotes the length of β.

Theorem VI. 1. **(Ahlfors' covering theorem).** *For any finite covering surface F of F_0,*

$$\text{(I)} \quad |S-S(D)|<h\frac{L}{|D|}, \qquad \text{(II)} \quad |S-S(\beta)|<h(\beta)L,$$

where $h>0$ is a constant, which depends on F_0 only, $h(\beta)>0$ is a constant, which depends on β only and L is the length of the relative boundary of F.

Ahlfors [6].

Proof. First we shall prove (I). We decompose F into a finite number of sheets $F=\sum_{\nu=1}^{n} F_\nu$, where F_ν is the set of points of F_0, above which there are at least ν points of F. Let L_ν be the length of the relative boundary of F_ν, then

$$|F|=\sum_{\nu=1}^{n}|F_\nu|, \qquad L=\sum_{\nu=1}^{n}L_\nu. \tag{1}$$

Similarly

$$|F(D)|=\sum_{\nu=1}^{n}|F_\nu(D)|. \tag{2}$$

Since F is schlicht relatively to F_0, $0\leqq\frac{|F_\nu|}{|F_0|}\leqq 1$, $0\leqq\frac{|F_\nu(D)|}{|D|}\leqq 1$, so that if $L\geqq d_0$, then

$$\left|\frac{|F_\nu|}{|F_0|}-\frac{|F_\nu(D)|}{|D|}\right|<1\leqq\frac{|L_\nu|}{d_0}. \tag{3}$$

Next consider the case $L_\nu<d_0$. Now F_ν consists of a finite number of disjoint domains: $F_\nu=\sum_i F_\nu^{(i)}$ and the relative boundary of $F_\nu^{(i)}$ consists of a finite number of disjoint curves $\{C_{\nu,k}^{(i)}\}_{k=1,2,\cdots}$. Let $L_{\nu,k}^{(i)}$ be the length of $C_{\nu,k}^{(i)}$, then $L_\nu=\sum_i\sum_k L_{\nu,k}^{(i)}$.

If $L_\nu<d_0$, then $L_{\nu,k}^{(i)}<d_0$, so that by the definition of d_0, $C_{\nu,k}^{(i)}$ bounds a domain $D_{\nu,k}^{(i)}$, such that $|D_{\nu,k}^{(i)}|<h_0L_{\nu,k}^{(i)}$.

There occurs two cases: either (i) each $F_\nu^{(i)}$ $(i=1,2,\cdots)$ is small, such that $F^{(i)}\subset D_{\nu,k}^{(i)}$ for some k, or (ii) a certain $F_\nu^{(i_0)}$ is large, such that $F_\nu^{(i_0)}=F_0-\sum_k D_{\nu,k}^{(i_0)}$.

In case (i),

$$|F_\nu|=\sum_i |F_\nu^{(i)}|\leqq\sum_i\sum_k |D_{\nu,k}^{(i)}|<h_0\sum_i\sum_k L_{\nu,k}^{(i)}=h_0L_\nu\ ,$$

so that

$$\frac{|F_\nu|}{|F_0|}<\frac{h_0L_\nu}{|F_0|}\leqq\frac{h_0L_\nu}{|D|}\ ,\qquad \frac{|F_\nu(D)|}{|D|}\leqq\frac{|F_\nu|}{|D|}<\frac{h_0L_\nu}{|D|}\ , \tag{4}$$

hence

$$\left|\frac{|F_\nu|}{|F_0|}-\frac{|F\ (D)|}{|D|}\right|<\frac{h_0L_\nu}{|D|}\ . \tag{5}$$

In case (ii),

$$0\leqq|F_0|-|F\ |\leqq|F_0|-|F_\nu^{(i_0)}|\leqq\sum_k|D_{\nu,k}^{(i_0)}|<h_0L_\nu\ , \tag{6}$$

so that

$$\left|\frac{|F\ |}{|F_0|}-1\right|<\frac{h_0L}{|F_0|}\leqq\frac{h_0L_\nu}{|D|}\ . \tag{7}$$

Since by (6), $0\leqq|D|-|F_\nu(D)|\leqq|F_0|-|F_\nu|<h_0L_\nu$,

$$\left|\frac{|F_\nu(D)|}{|D|}-1\right|\leqq h\frac{L_\nu}{|D|}\ , \tag{8}$$

hence by (7), (8), we have the same relation (5).

From (3), (5), we have

$$\left|\frac{|F_\nu|}{|F_0|}-\frac{|F_\nu(D)|}{|D|}\right|<\mathrm{Max}\left(\frac{1}{d_0},\frac{h_0}{|D|}\right)L_\nu$$

$$=\mathrm{Max}\left(\frac{|D|}{d_0},\ h_0\right)\cdot\frac{L}{|D|}\leqq\mathrm{Max}\left(\frac{|F_0|}{d_0},\ h_0\right)\cdot\frac{L_\nu}{|D|}=h\frac{L_\nu}{|D|}\ , \tag{9}$$

where

$$h=\mathrm{Max}\left(\frac{|F_0|}{d_0},\ h_0\right).$$

Summing up (9) for $\nu=1,2,\cdots$, we have

$$\text{(I)}\qquad |S-S(D)|=\left|\frac{|F|}{|F_0|}-\frac{|F(D)|}{|D|}\right|<h\frac{L}{|D|}\ .$$

Next we shall prove (II). Let $F_\nu(\beta)$ be the part of F_ν, which lies above β, then $|F_\nu(\beta)|=\sum_i|F_\nu^{(i)}(\beta)|$.

Since $0\leqq\dfrac{|F_\nu|}{|F_0|}\leqq1$, $0\leqq\dfrac{|F_\nu(\beta)|}{|\beta|}\leqq1$, we have if $L_\nu\geqq d_0(\beta)$,

$$\left|\frac{|F_\nu|}{|F_0|}-\frac{|F_\nu(\beta)|}{|\beta|}\right|<1\leqq\frac{L_\nu}{d_0(\beta)}\ . \tag{10}$$

If $L_\nu < d_0(\beta)$, we consider two cases as before.

In case (i), since $L_{\nu,k}^{(i)} \leqq L_\nu < d_0(\beta)$, we have by the definition of $d_0(\beta)$,

$$|D_{\nu,k}^{(i)}(\beta)| < h_0(\beta) L_{\nu,k}^{(i)},$$

so that

$$|F_\nu(\beta)| = \sum_i |F_\nu^{(i)}(\beta)| \leqq \sum_i \sum_k |D_{\nu,k}^{(i)}(\beta)| < h_0(\beta) \sum_i \sum_k L_{\nu,k}^{(i)} = h_0(\beta) L_\nu ,$$

hence

$$\frac{|F_\nu(\beta)|}{|\beta|} < \frac{h_0(\beta)}{|\beta|} L_\nu . \tag{11}$$

Since by (4), $\dfrac{|F_\nu|}{|F_0|} < \dfrac{h_0 L_\nu}{|F_0|}$, we have

$$\left| \frac{|F_\nu|}{|F_0|} - \frac{|F_\nu(\beta)|}{|\beta|} \right| < \operatorname{Max}\left(\frac{h_0(\beta)}{|\beta|}, \frac{h_0}{|F_0|} \right) L_\nu , \qquad \text{if} \quad L_\nu < d_0(\beta) . \tag{12}$$

In case (ii),

$$0 \leqq |\beta| - |F_\nu(\beta)| \leqq |\beta| - |F_\nu^{(i_0)}(\beta)| \leqq \sum_k |D_{\nu,k}^{(i_0)}(\beta)| < h_0(\beta) \sum_k L_{\nu,k}^{(i_0)} \leqq h_0(\beta) L_\nu ,$$

hence $\left| \dfrac{|F_\nu(\beta)|}{|\beta|} - 1 \right| < \dfrac{h_0(\beta)}{|\beta|} L_\nu$ and $\left| \dfrac{|F_\nu|}{|F_0|} - 1 \right| < \dfrac{h_0 L_\nu}{|F_0|}$ by (7), so that we have the same relation (12). Hence from (10), (12), we have

$$\left| \frac{|F_\nu|}{|F_0|} - \frac{|F_\nu(\beta)|}{|\beta|} \right| < h(\beta) L_\nu , \tag{13}$$

where

$$h(\beta) = \operatorname{Max}\left(\frac{1}{d_0(\beta)}, \frac{h_0(\beta)}{|\beta|}, \frac{h_0}{|F_0|} \right) .$$

Summing up (13) for $\nu = 1, 2, \cdots$, we have

(II) $$|S - S(\beta)| < h(\beta) L .$$

3. Ahlfors' fundamental theorem

Let F_0 be a regular surface in the Euclidean space, which is bounded by a finite number of analytic Jordan curves, some of which may reduce to single points and F be a finite covering surface of F_0.

We decompose F into a finite number of triangles and let α_0, α_1, α_2 be the number of vertices, sides (which do not lie on the boundary of F) and triangles in the triangulation. Then $\rho = \rho(F) = -\alpha_0 + \alpha_1 - \alpha_2$ is called *the Euler's characteristic of* F. If F is of planar character[1] and is bounded by q disjoint Jordan curves, then $\rho(F) = q - 2$.

1) A surface is called of planar character, if it is decomposed into two parts by any ring cut of F.

If F is decomposed into two parts F_1, F_2 by a ring cut, then

$$\rho(F)=\rho(F_1)+\rho(F_2), \tag{1}$$

and if F is decomposed into $\{F_\nu\}$ by a finite number of ring cuts or cross cuts, then

$$\rho(F)=\sum_\nu \rho(F_\nu)+n(\sigma), \tag{2}$$

where $n(\sigma)$ denotes the number of cross cuts.

Now we shall prove the following **Ahlfors' fundamental theorem.**

Theorem VI. 2. *Let F be a finite covering surface of a basic surface F_0, then*

$$\rho^+ \geqq \rho_0 S - hL, \qquad \rho^+ = \mathrm{Max}\,(\rho, 0),$$

where $\rho=\rho(F)$, $\rho_0=\rho(F_0)$, $S=|F|/|F_0|$ and L is the length of the relative boundary of F and $h>0$ is a constant, which depends on F_0 only.

Ahlfors [6].

Proof. Let G be a sub-domain of F and let ∂G be the part of the boundary of G, which belongs to the relative boundary Γ of F.

(i) First suppose that F_0 is of planar character and is bounded by $q\,(\geqq 3)$ disjoint analytic Jordan curves $\Lambda_i\,(i=1, 2, \cdots, q)$, so that $\rho_0=q-2$. Without loss of generality, we assume that $|F_0|=1$, so that

$$S=\frac{|F|}{|F_0|}=|F|, \qquad |F_0|=1. \tag{1}$$

We connect Λ_i, Λ_{i+1} $(\Lambda_{q+1}=\Lambda_1)$ by an analytic Jordan arc $\beta_i\,(i=1,2,\cdots,q)$, where β_i, β_j $(i\neq j)$ are disjoint, then by $\{\beta_i\}$, F_0 is decomposed into two simply connected domains $F_0^{(1)}$, $F_0^{(2)}$. We choose $\{\beta_i\}$, such that $|F_0^{(1)}|=|F_0^{(2)}|=1/2$.

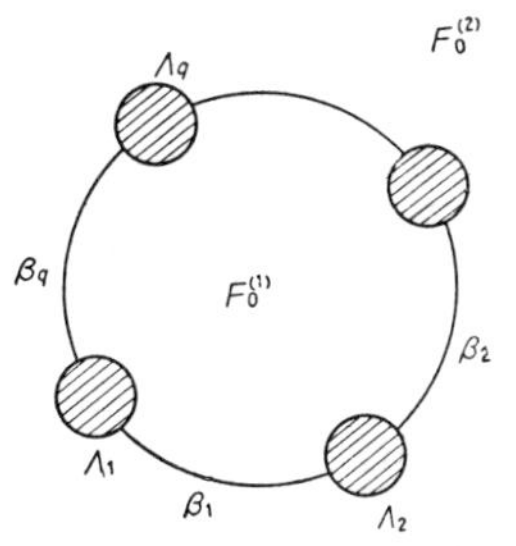

The part of F, which lies above one of $F_0^{(1)}$, $F_0^{(2)}$, $F_0^{(1)}$ say, consists of a finite number of connected domains $\{G\}$. Considering G as a covering surface of $F_0^{(1)}$, we put

$$S(G)=\frac{|G|}{|F_0^{(1)}|}=2|G|. \tag{2}$$

Similarly we define $S(G)$ for the part of F, which lies above $F_0^{(2)}$.

Let $G(\beta_i)$ be the set of curves on the boundary of G, which lie above β_i and put

$$S(G(\beta_i))=\frac{|G(\beta_i)|}{|\beta_i|} \quad (i=1, 2, \cdots, q). \tag{3}$$

Then, since G is a covering surface of $F_0^{(1)}$, or $F_0^{(2)}$, we have by Theorem VI. 1 and (2),

$$\begin{aligned} &|2|G|-S(G(\beta_i))|<h|\partial G|, \\ &|S(G(\beta_i))-S(G(\beta_j))|<h|\partial G| \qquad (i, j=1, 2, \cdots, q), \end{aligned} \tag{4}$$

where $h>0$ is a constant.

By cross cuts $\{\sigma\}$, which lie above $\{\beta_i\}$, F is decomposed into a finite number $N(G)$ of connected domains $\{G_\nu\}$, which lie above $F_0^{(1)}$ or $F_0^{(2)}$. Then

$$\rho=n(\sigma)+\sum_\nu \rho(G_\nu) ,$$

where $n(\sigma)$ is the number of cross cuts $\{\sigma\}$.

Since $\rho(G\ )\geqq -1$, we have

$$\rho\geqq n(\sigma)-N(G) . \tag{5}$$

Since each σ belongs to two adjacent $\{G_\nu\}$, (5) can be written in the form:

$$\rho\geqq\sum_{G_\nu}\left(\frac{n_\nu}{2}-1\right), \tag{6}$$

where n_ν is the number of σ, which belong to the boundary of G_ν.

Let $G_1^{(1)}, \cdots, G_{k_1}^{(1)}$ be such G_ν, for which $n_\nu=1$, then such G_ν abutts on $F-G_\nu$ along a single cross cut σ_ν, which we call *the cross cut corresponding to* G_ν *with respect to* F.

Now we take off all such $G_\nu^{(1)}$ $(\nu=1, 2, \cdots, k_1)$ from F and let $F^{(1)}$ be the remaining surface. Let G'_ν be defined for $F^{(1)}$ as G_ν for F, then $\sum_{G_\nu}\left(\frac{n_\nu}{2}-1\right)=\sum_{G'_\nu}\left(\frac{n'_\nu}{2}-1\right)$, so that by (6),

$$\rho\geqq\sum_{G'_\nu}\left(\frac{n'_\nu}{2}-1\right). \tag{7}$$

Let $G_1^{(2)}, \cdots, G_{k_2}^{(2)}$ be such G'_ν, for which $n'_\nu=1$. We take off all such G'_ν from $F^{(1)}$ and let $F^{(2)}$ be the remaining surface. We proceed similarly, as long as there are G_ν, for which $n_\nu=1$. Then after a finite number of steps, either (i) we obtain a surface, with no cross cuts above $\{\beta_i\}$, or (ii) we obtain a surface, such that $n_\nu\geqq 2$ for all G_ν.

Let $F^{(n)}$ be the surface, obtained after n steps, then by the definition,

$$F=F^{(1)}+\sum_{\nu=1}^{k_1}G_\nu^{(1)} , \quad F^{(1)}=F^{(2)}+\sum_{\nu=1}^{k_2}G_\nu^{(2)} , \quad \cdots , \quad F^{(n-1)}=F^{(n)}+\sum_{\nu=1}^{k_n}G_\nu^{(n)} ,$$

$$F=F^{(n)}+\sum_{s=1}^{n}\sum_{\nu}G_\nu^{(s)} , \tag{8}$$

where $G_\nu^{(s)}$ has only one corresponding cross cut, with respect to $F^{(s-1)}$. We shall prove that

$$|F|\leqq|F^{(n)}|+hL , \qquad L^{(n)}\leqq hL , \tag{9}$$

where $L^{(n)}$ is the length of the relative boundary of $F^{(n)}$ and $h>0$ is a constant, which depends on F_0 only.

Now $\sum_{s=1}^{n}\sum_{\nu} G_{\nu}^{(s)}$ consists of a finite number N of disjoint connected domains $\{G_{\mu}^{*}\}$:

$$\sum_{s=1}^{n}\sum_{\nu} G_{\nu}^{(s)}=\sum_{\mu=1}^{N} G_{\mu}^{*} . \tag{10}$$

Let $G_{\mu}^{*}=G_1+\cdots+G_m$, where $G_k(1\leqq k\leqq m)$ is some one of $\{G_{\nu}^{(s)}\}$.

Let σ_k be the cross cut corresponding to G_k and lie above β_j, say, and put

$$\lambda(\sigma_k)=\frac{|\sigma_k|}{|\beta_j|} , \tag{11}$$

then by (4),

$$\lambda(\sigma_k)\leqq S(G_k(\beta_j))\leqq S(G(\beta_i))+h|\partial G_k| \qquad (i=1, 2, \cdots, q) ,$$

so that

$$q\lambda(\sigma_k)\leqq\sum_{i=1}^{q} S(G_k(\beta_i))+qh|\partial G_k| . \tag{12}$$

Since each σ_k belongs to at most two G_ν $(1\leqq\nu\leqq m)$,

$$\sum_{k=1}^{m}\sum_{i=1}^{q} S(G_k(\beta_i))\leqq 2\sum_{k=1}^{m}\lambda(\sigma_k) ,$$

so that from (12) by writing h instead of $qh/q-2$, we have

$$\sum_{k=1}^{m}\lambda(\sigma_k)\leqq h\sum_{k=1}^{m}|\partial G_k|=h|\partial G_{\mu}^{*}| .$$

Since $\sum_{k=1}^{m}\lambda(\sigma_k)=\sum_{i=1}^{q} S(G_{\mu}^{*}(\beta_i))$, we have

$$\sum_{i=1}^{q} S(G_{\mu}^{*}(\beta_i))\leqq h|\partial G_{\mu}^{*}| .$$

Summing up for $\mu=1, 2, \cdots, N$, we have

$$\sum_{\mu=1}^{N}\sum_{i=1}^{q} S(G_{\mu}^{*}(\beta_i))\leqq h\sum_{\mu=1}^{N}|\partial G_{\mu}^{*}|\leqq hL . \tag{13}$$

Let

$$\kappa=\text{Max}\,(|\beta_1|, \cdots, |\beta_q|) ,$$

then

$$L^{(n)}\leqq L+\kappa\sum_{\mu=1}^{N}\sum_{i=1}^{q} S(G_{\mu}^{*}(\beta_i))\leqq L+\kappa hL=(1+\kappa h)L .$$

Hence writing h instead of $1+\kappa h$, we have

$$L^{(n)}\leqq hL . \tag{14}$$

If we apply (4) on G_k, we have $2|G_k|\leqq S(G_k(\beta_1))+h|\partial G_k|$, so that

$$2|G_\mu^*|=2\sum_{k=1}^m|G_k|\leqq\sum_{k=1}^m S(G_k(\beta_1))+h\sum_{k=1}^m|\partial G_k|\leqq 2S(G_\mu^*(\beta_1))+h|\partial G_\mu^*| , \quad (15)$$

hence by (13), (15),

$$|F|=|F^{(n)}|+\sum_{\mu=1}^N|G_\mu^*|\leqq|F^{(n)}|+\sum_{\mu=1}^N S(G_\mu^*(\beta_1))+h\sum_{\mu=1}^N|\partial G_\mu^*|$$
$$\leqq|F^{(n)}|+hL+hL=F^{(n)}+2hL ,$$

so that writing h instead of $2h$, we have

$$|F|\leqq|F^{(n)}|+hL . \quad (16)$$

By (14), (16), (9) is proved.

First suppose that after n steps, we obtain $F^{(n)}$, which has no cross cuts above $\{\beta_i\}$, then $F^{(n)}$ lies above one of $F_0^{(1)}$, $F_0^{(2)}$, $F_0^{(1)}$ say. Then $F^{(n)}$ does not cover $F_0^{(2)}$, so that if we apply Theorem VI. 1, by taking $F=F^{(n)}$, $D=F_0^{(2)}$, then since $S(D)=0$, we have $|F^{(n)}|\leqq hL^{(n)}$, so that by (9),

$$S=|F|\leqq|F^{(n)}|+hL\leqq hL^{(n)}+hL\leqq 2hL .$$

Hence by a suitable choice of h, $(q-2)S-hL\leqq 0$, so that

$$\rho^+\geqq(q-2)S-hL . \quad (17)$$

Hence in this case, the theorem is proved.

Next suppose that after n steps, we obtain $F^{(n)}$, such that $n_\nu\geqq 2$ for all G_ν. In virtue of (6), (7) and (9), by taking $F^{(n)}$ instead of F, we may assume that all $n_\nu\geqq 2$ in (6).

We remark the following. If we take $F^{(n)}$ as F, then the relative boundary Γ of F has arcs of positive length above $\{\beta_i\}$. We modify F slightly into F', such that the relative boundary of F' has no arcs of positive length above $\{\beta_i\}$. If the theorem holds for F', then making $F'\to F$, we see that the theorem holds for F. Hence we assume that the relative boundary Γ of F has no arcs of positive length above $\{\beta_i\}$ and $n_\nu\geqq 2$ for all G_ν in (6).

We consider two cases: either (i) $2\leqq n_\nu\leqq q-1$ for all G_ν, or (ii) there exists a certain G_ν, for which $n_\nu\geqq q$.

In case (i), in virtue on the hypothesis on Γ, each G_ν does not cover a certain β_i $(1\leqq i\leqq q)$, so that $S(G_\nu(\beta_i))=0$, hence by (4),

$$2|G_\nu|\leqq S(G_\nu(\beta_i))+h|\partial G_\nu|=h|\partial G_\nu| , \quad (18)$$

so that

$$S=|F|=\sum_\nu|G_\nu|\leqq h\sum_\nu|\partial G_\nu|=hL ,$$

hence we have (17) as before.

In case (ii), since $n_\nu/2\geqq 1$, we have by (6),

$$\rho \geqq \sum_{n_\nu \geqq q} \left(\frac{n_\nu}{2} - 1 \right) \geqq \frac{q-2}{2q} \sum_{n_\nu \geqq q} n_\nu \,. \tag{19}$$

Since by (4),

$$S(G_\nu(\beta_i)) \geqq 2|G_\nu| - h|\partial G_\nu| \qquad (i=1, 2, \cdots, q),$$

and Γ does not contain arcs of positive length above $\{\beta_i\}$,

$$n_\nu \geqq \sum_{i=1}^{q} S(G_\nu(\beta_i)) \geqq 2q|G_\nu| - qh|\partial G_\nu| \,,$$

so that

$$\rho \geqq \frac{q-2}{2q} \sum_{n_\nu \geqq q} n_\nu \geqq (q-2) \sum_{n_\nu \geqq q} |G_\nu| - hL \,. \tag{20}$$

For G_ν, for which $2 \leqq n_\nu \leqq q-1$, we have by (18), $|G_\nu| \leqq h|\partial G_\nu|$, so that

$$\sum_{2 \leqq n_\nu \leqq q-1} |G_\nu| \leqq hL \,,$$

hence

$$S = |F| = \sum_{n_\nu \geqq q} |G_\nu| + \sum_{2 \leqq n_\nu \leqq q-1} |G_\nu| \leqq \sum_{n_\nu \geqq q} |G_\nu| + hL \,,$$

$$\sum_{n_\nu \geqq q} |G_\nu| \geqq S - hL \,. \tag{21}$$

Hence by (20), (21), we have

$$\rho \geqq (q-2)S - hL \,. \tag{22}$$

Hence theorem is proved, if F_0 is of planar character.

(ii) Next suppose that F_0 is not of planar character and of genus $p \geqq 1$. By p disjoint ring cuts $C_1, \cdots, C_p$, we make F_0 into a surface F_0^*, which is of planar character, then

$$\rho_0 = \rho(F_0) = \rho(F_0^*) = \rho_0^* \,. \tag{23}$$

Let $\{\sigma\}$ be the set of curves on F, which lie above $\{C_i\}$, then σ is a cross cut, or a ring cut of F. By $\{\sigma\}$, F is decomposed into a finite number of connected surfaces $\{F_\nu\}$. Then

$$\rho = \rho(F) = \sum_\nu \rho(F_\nu) + n(\sigma) \,, \tag{24}$$

where $n(\sigma)$ is the number of cross cuts of F among $\{\sigma\}$.

Since if $\rho(F_\nu) < 0$, then $\rho(F_\nu) = -1$, we have

$$\rho = \sum_\nu \rho^+(F_\nu) - N_1 + n(\sigma) \,, \tag{25}$$

where N_1 is the number of simply connected F_ν. We shall prove that

$$\rho^+ \geqq \sum_\nu \rho^+(F_\nu) \,. \tag{26}$$

Since a connected surface is decomposed into at most two connected surfaces by a cross cut, we see easily that if F is multiply connected, i.e., if $\rho \geqq 0$, then $N_1 \leqq n(\sigma)$, so that $\rho^+ \geqq \sum_\nu \rho^+(F_\nu)$.

If F is simply connected, i.e., if $\rho = -1$, then each F_ν is simply connected, so that $\rho(F_\nu) = -1$, $\rho^+(F_\nu) = 0$, hence $\rho^+ \geqq \sum_\nu \rho^+(F_\nu)$. Hence (26) holds in general.

Since F_0^* is of planar character, $\rho_0^*=\rho_0$ and F_ν is a covering surface of F_0^*, we have

$$\rho^+(F_\nu)\geqq\rho_0 S_\nu - hL_\nu\ , \tag{27}$$

where L_ν is the length of the relative boundary of F_ν with respect to F_0^*, so that $\sum_\nu L_\nu\leqq L$, hence summing up (27) for $\nu=1,2,\cdots$, we have by (26),

$$\rho^+\geqq\rho_0 S-hL\ . \tag{28}$$

Hence the theorem is proved.

4. Simply connected finite covering surfaces of a sphere

Let K be a sphere of diameter 1 and F be a simply connected finite covering surface of K and put

$$S=\frac{|F|}{|K|}=\frac{|F|}{\pi}\ . \tag{1}$$

Let D_ν $(\nu=1,2,\cdots,q)$ be $q(\geqq 3)$ disjoint simply connected domains on K, each of which is bounded by an analytic Jordan curve. We take off $\{D_\nu\}$ from K and let $F_0=K-\sum_{\nu=1}^{q}D_\nu$ be the remaining domain. Then

$$\rho(F_0)=q-2.$$

Now $F(D_\nu)$ consists of a finite number of connected surfaces:

$$F(D_\nu)=\sum_k F^i_{\nu,k}+\sum_k F^z_{\nu,k}\ , \tag{2}$$

where $F^i_{\nu,k}$ has no relative boundary, with respect to D_ν, which is called *an island* (Insel) and $F^z_{\nu,k}$ has such one, which is called *a peninsula* (Zunge).

Theorem VI. 3. *Let F be a simply connected finite covering surface of a sphere K and n_ν be the number of simply connected islands in $F(D_\nu)$ $(\nu=1,2,\cdots,q)$, then*

$$\sum_{\nu=1}^{q}n_\nu\geqq(q-2)S-hL\ ,$$

where $h>0$ is a constant, which depends on $F_0=K-\sum_{\nu=1}^{q}D_\nu$ only and L is the length of the boundary of F.

Proof. We take off from F all peninsulas $\{F^z_{\nu,k}\}$ above $\{D_\nu\}$ and let F' be the remaining surface:

$$F=\sum_{\nu=1}^{q}\sum_k F^z_{\nu,k}+F'\ . \tag{1}$$

Next we take off from F' all islands $\{F^i_{\nu,k}\}$ above $\{D_\nu\}$ and let F'' be the remaining surface:

$$F'=\sum_{\nu=1}^{q}\sum_{k} F^{i}_{\nu,k}+F'' . \tag{2}$$

F'' consists of a finite number of connected surfaces;

$$F'' = \sum_{\mu} F''_{\mu} , \tag{3}$$

so that

$$F'=\sum_{\nu=1}^{q}\sum_{k} F^{i}_{\nu,k}+\sum_{\mu} F''_{\mu} . \tag{4}$$

F''_{μ} is a covering surface of F_0. Since F' is decomposed into $\{F^{i}_{\nu,k}\}$ and $\{F''_{\mu}\}$ by ring cuts, we have

$$\rho(F')=\sum_{\nu=1}^{q}\sum_{k} \rho(F^{i}_{\nu,k})+\sum_{\mu} \rho(F''_{\mu}) . \tag{5}$$

Since F is simply connected, F' consists of a finite number $N(F')$ of disjoint simply connected surfaces, so that

$$\rho(F')=-N(F') .$$

Let $N(F'')$ be the number of simply connected F''_{μ}, then

$$-N(F')=\rho(F')=\sum_{\nu=1}^{q}\sum_{k} \rho(F^{i}_{\nu,k})+\sum_{\mu} \rho^{+}(F''_{\mu})-N(F'') .$$

Since

$$\sum_{\nu=1}^{q}\sum_{k} \rho(F^{i}_{\nu,k})=\sum_{\nu=1}^{q}\sum_{k} \rho^{+}(F^{i}_{\nu,k})-\sum_{\nu=1}^{q} n_{\nu} ,$$

we have

$$\sum_{\nu=1}^{q} n_{\nu}=\sum_{\nu=1}^{q}\sum_{k} \rho^{+}(F^{i}_{\nu,k})+\sum_{\mu} \rho^{+}(F''_{\mu})+N(F')-N(F'') . \tag{6}$$

We see easily that $N(F')-N(F'')\geqq 0$, hence

$$\sum_{\nu=1}^{q} n_{\nu}\geqq\sum_{\mu} \rho^{+}(F''_{\mu}) . \tag{7}$$

Since F''_{μ} is a covering surface of F_0, by the fundamental theorem,

$$\rho^{+}(F''_{\mu})\geqq(q-2)S''_{\mu}-hL''_{\mu} ,$$

where $S''_{\mu}=|F''_{\mu}|/|F_0|$ and L''_{μ} is the length of the relative boundary of F''_{μ} with respect to F_0.

Now $\sum_{\mu} S''_{\mu}=S''=|F''|/|F_0|$, $|S''-S|<hL$ (Theorem VI. 1) and $\sum_{\mu} L''_{\mu}\leqq L$, so that

$$\sum_{\mu} \rho^{+}(F''_{\mu})\geqq(q-2)S-hL ,$$

hence by (7)

$$\sum_{\nu=1}^{q} n_{\nu}\geqq(q-2)S-hL . \tag{8}$$

Theorem VI. 4. *Under the same condition as Theorem* VI. 3, *let* n'_ν *be the number of simply connected islands in* $F(D_\nu)$, *which consist of only one sheet (schlicht disc), then*

$$\sum_{\nu=1}^{q} n'_\nu \geqq (q-4)S - hL .$$

If there are no simply connected islands in $F(D_q)$, *then*

$$\sum_{\nu=1}^{q-1} n'_\nu \geqq (q-3)S - hL .$$

Proof. Let n_ν be the number of simply connected islands in $F(D_\nu)$, and n''_ν be that of such ones, which consist of more than one sheet, then $n_\nu = n'_\nu + n''_\nu$ and $S(D_\nu) \geqq n'_\nu + 2n''_\nu$, so that

$$S + hL \geqq S(D_\nu) \geqq n'_\nu + 2n''_\nu = 2(n'_\nu + n''_\nu) - n'_\nu = 2n_\nu - n'_\nu ,$$
$$S + n'_\nu + hL \geqq 2n_\nu , \tag{1}$$

hence by Theorem VI. 3,

$$qS + \sum_{\nu=1}^{q} n'_\nu + qhL \geqq 2\sum_{\nu=1}^{q} n_\nu \geqq 2(q-2)S - 2hL ,$$

so that

$$\sum_{\nu=1}^{q} n'_\nu \geqq (q-4)S - hL , \tag{2}$$

with a suitable $h>0$.

If there are no simply connected inslands in $F(D_q)$, then $n_q=0$, so that from (1),

$$(q-1)S + \sum_{\nu=1}^{q-1} n'_\nu + (q-1)hL \geqq 2\sum_{\nu=1}^{q-1} n_\nu = 2\sum_{\nu=1}^{q} n_\nu \geqq 2(q-2)S - 2hL ,$$

so that

$$\sum_{\nu=1}^{q-1} n'_\nu \geqq (q-3)S - hL . \tag{3}$$

Theorem VI. 5. *Under the same condition as Theorem* VI. 3, *suppose that every simply connected island in* $F(D_\nu)$ *has at least* m_ν $(1 \leqq m_\nu \leqq \infty)$ *sheets, where* $m_\nu = \infty$ *means that there are no simply connected islands in* $F(D_\nu)$, *then*

$$\sum_{\nu=1}^{q} \left(1 - \frac{1}{m_\nu}\right) \leqq 2 + h\frac{L}{S} .$$

Proof. Let n_ν be the number of simply connected islands in $F(D_\nu)$, then if $m_\nu < \infty$, $S(D_\nu) \geqq m_\nu n_\nu$, so that by Theorem VI. 1,

$$n_\nu \leqq \frac{S(D\ )}{m_\nu} \leqq \frac{S}{m_\nu} + \frac{hL}{m_\nu} \leqq \frac{S}{m_\nu} + hL . \tag{1}$$

If $m_\nu=\infty$, then $n_\nu=0$, so that (1) holds, hence by Theorem VI. 3, we have

$$\sum_{\nu=1}^{q}\left(1-\frac{1}{m_\nu}\right)\leqq 2+h\frac{L}{S}\,. \tag{2}$$

Theorem VI. 6. *Let Δ: $|w|<R$ be a disc on the w-plane and D_ν ($\nu=1,2,\cdots,q$) be $q(\geqq 3)$ disjoint simply connected domains in Δ and F be a finite covering surface of Δ, which is of planar character, such that the relative boundary of F with respect to Δ contains at most one isolated closed curve, then with the same notation as Theorem VI. 3 and 4, where S is defined with respect to Δ, we have*

$$\sum_{\nu=1}^{q} n_\nu \geqq (q-2)S-hL\,,\qquad \sum_{\nu=1}^{q} n'_\nu \geqq (q-4)S-hL\,.$$

If F is simply connected, then

$$\sum_{\nu=1}^{q} n_\nu \geqq (q-1)S-hL\,,\qquad \sum_{\nu=1}^{q} n'_\nu \geqq (q-2)S-hL\,,$$

where $h>0$ is a constant, which depends on $F_0=\Delta-\sum_{\nu=1}^{q} D_\nu$ only and L is the length of the relative boundary of F with respect to Δ.

Proof. With the same notation as the proof of Theorem VI. 3, if F' consists of a finite number of connected surfaces $\{F'_j\}$, then $\rho(F')=\sum_j \rho^+(F'_j)-N(F')$, so that the equation (6) (p. 253) becomes

$$\begin{aligned}\sum_{\nu=1}^{q} n_\nu &= \sum_{\nu=1}^{q}\sum_{k}\rho^+(F^i_{\nu,k})-\sum_j \rho^+(F'_j)+\sum_\mu \rho^+(F''_\mu)+N(F')-N(F'')\\ &\geqq -\sum_j \rho^+(F'_j)+\sum_\mu \rho^+(F''_\mu)\,.\end{aligned} \tag{1}$$

Let the boundary of F'_j consist of N'_j disjoint closed curves, then since F and hence F'_j is of planar character, $\rho(F'_j)=N'_j-2$, so that $\sum_j \rho^+(F'_j)=\sum_j (N'_j-2)^+$. Let β: $|w|=R$ be the bounding circle of Δ and $S(\beta)=\dfrac{|F(\beta)|}{|\beta|}$. Since F is of planar character, F can be mapped topologically on a schlicht domain Ω. Hence if we identify F with Ω, then by the hypothesis, we see easily that $\sum_j (N'_j-2)^+\leqq S(\beta)\leqq S+hL$ (Theorem VI. 1), so that $\sum_j \rho^+(F'_j)\leqq S+hL$, hence

$$\sum_{\nu=1}^{q} n_\nu \geqq \sum_\mu \rho^+(F''_\mu)-S-hL\,. \tag{2}$$

Since $\rho(F_0)=(q+1)-2=q-1$, we have as the proof of Theorem VI. 3,

$$\sum_{\nu=1}^{q} n_\nu \geqq (q-2)S-hL\,. \tag{3}$$

If F is simply connected, then $\rho^+(F'_j)=0$, so that

$$\sum_{\nu=1}^{q} n_\nu \geqq \sum_\mu \rho^+(F''_\mu) \tag{4}$$

and from this, we have as Theorem VI. 3

$$\sum_{\nu=1}^{q} n_\nu \geqq (q-1)S - hL . \tag{5}$$

The part of the theorem, concerning n'_ν can be proved similarly as Theorem VI. 4.

5. Regular exhaustion in Ahlfors' sense

Let F be an open Riemann surface, spread over a sphere K. If F can be exhausted by a sequence of compact Riemann surfaces: $F_0 \subset F_1 \subset \cdots \subset F_n \to F$, where $\overline{F}_n \subset F_{n+1}$ and the boundary Γ_n of F_n consists of a finite number of analytic Jordan curves, such that

$$\lim_n \frac{L_n}{|F_n|} = 0 , \tag{1}$$

where L_n is the length of Γ_n and $|F_n|$ the spherical area of F_n, then we say that F is *regularly exhaustible in Ahlfors' sense.*[1]

Thorem VI. 7. *Let $w=w(z)$ be a trenscèndental meromorphic function for $|z|<\infty$, or a meromorphic function in $|z|<1$, such that $\overline{\lim_{r\to 1}}\, T(w, r)/\log \frac{1}{1-r} = \infty$, then the Riemann surface F, generated by $w=w(z)$ on the w-sphere, is regularly exhaustible in Ahlfors' sense.*

Proof. Let $w(z)$ be a transcendental meromorphic function for $|z|<\infty$ and put

$$A(r) = \iint_{|z|\leqq r} \left(\frac{|w'(z)|}{1+|w(z)|^2}\right)^2 r\,dr\,d\theta , \quad L(r) = \int_{|z|=r} \frac{|w'(z)|}{1+|w(z)|^2} r\,d\theta , \quad z=re^{i\theta} . \tag{1}$$

By Schwarz's inequality, we have

$$L^2(r) \leqq 2\pi r \frac{dA(r)}{dr} . \tag{2}$$

Since by Picard's theorem, $w(z)$ takes any value infinitely often, with two possible exceptions, $\lim_{r\to\infty} A(r) = \infty$.

Let $L(r) > (A(r))^{3/4}$ in a sequence of intervals $\{J_\nu\}$, then by (2),

$$\sum_\nu \int_{J_\nu} \frac{dr}{r} \leqq 2\pi \int^{\infty} \frac{dA(r)}{A(r)^{3/2}} < \infty ,$$

so that there exists $r_1 < r_2 < \cdots < r_n \to \infty$, such that $L(r_n) \leqq (A(r_n))^{3/4}$, so that $L(r_n)/A(r_n) \to 0$. Hence F is regularly exhaustible in Ahlfors' sense.

Next suppose that $w(z)$ is meromorphic in $|z|<1$ and F is not regularly exhaustible in Ahlfors' sense, such that for $r_0 \leqq r < 1$, $A(r) \leqq kL(r)$ (k=const.), then by (2),

1) If F is a covering surface of a basic domain F_0 on K, then L_n denotes the length of the relative boundary of F_n.

$$1-r<\int_r^1 \frac{dr}{r} \leqq 2\pi k^2 \int_r^1 \frac{dA(r)}{A(r)^2} \leqq 2\pi k^2 \frac{1}{A(r)},$$

so that $A(r)=O\left(\frac{1}{1-r}\right)$, $T(w,r)=O\left(\log\frac{1}{1-r}\right)$, hence if $\overline{\lim_{r\to 1}} T(w,r) \log\frac{1}{1-r} =\infty$, then F is regularly exhaustible in Ahlfors' sense, and by Theorem V. 17, $\lim_{r\to 1} A(r)=\infty$.

Theorm VI. 8. *Let F be a simply connected open Riemann surface, spread over the w-sphere K, such that $|F|=\infty$ and is regularly exhaustible in Ahlfors' sense, then*

(i) *Above one of three disjoint simply connected domains on K, there are infinitely many simply connected islands in F.*

Hence F covers any point of K infinitely often, with two possible exceptions.

(ii) *Above one of five disjoint simply connected domains on K, there are infinitely many schlicht discs in F.*

(iii) *If there exists a point w_0, which is not covered by F, then above one of three disjoint simply connected domains on K, which do not contain w_0, there are infinitely many schlicht discs in F.*

Ahlfors [6].

Proof. To prove (i), we take $q=3$ in Theorem VI. 3. To prove (ii), we take $q=5$ in Theorem VI. 4. To prove (iii), we take $q=4$ and a simply connected domain, which contains w_0 as D_4 in Theorem VI. 4. ——

Hence by Theorem VI. 7, we have

Corollary. *If $w(z)$ is a trascendental meromorphic function for $|z|<\infty$, or a meromorphic function in $|z|<1$, such that $\overline{\lim_{r\to 1}} T(w,r) \log\frac{1}{1-r}=\infty$, then the Riemann surface F, generated by $w=w(z)$ on the w-sphere, has the properties* (i), (ii).

If $w(z)$ is an transcendental integral function, then the Riemann surface F, generated by $w=w(z)$ on the w-sphere has the property (iii), *where $w_0=\infty$.*

We shall prove a more precise theorem as follows.

Theorem VI. 9. *Let $w(z)$ be a transcendental meromorphic function for $|z|<\infty$ and F be the Riemann surface, generated by $w=w(z)$ on the w-sphere K. Let Δ: $|w-w_0|<\rho$ be a disc and F_ρ be a connected piece of F, which lies above Δ, where we assume that F is non-compact.*

Then $|F_\rho|$ is infinite and F_ρ is regularly exhaustible in Ahlfors' sense and

(i) *Above of one of three disjoint simply connected domains in*

Δ, there are infinitely many simply connected islands in F_ρ. Hence F_ρ covers any point of Δ infinitely often, with two possible exceptions.

(ii) *Above one of five disjoint simply connected domains in Δ, there are infinitely many schlicht discs in F_ρ.* Kunugui [5].

If F_ρ is simply connected, then

(i) *Above one of two disjoint simply connected domains in Δ, there are infinitely many simply connected islands in F_ρ. Hence F_ρ covers any point of Δ infinitely often, with one possible exception.*

(ii) *Above one of three disjoint simply connected domains in Δ, there are infinitely many schlicht discs in F_ρ.* Noshiro [2].

Proof. Let D be the image of F_ρ on the z-plane, then since F_ρ is non-compact, D is an infinite domain. Let D^r be the part of D, which lies in $|z|<r$ and F_ρ^r be its image on F_ρ and $A(r)$ be the spherical area of F_ρ^r, then we shall prove that

$$|F_\rho|=\lim_{r\to\infty} A(r)=\infty\,. \tag{1}$$

First we assume that the boundary of D contains a curve, extending to ∞. Let θ_r be the part of the circle $|z|=r$, contained in D and $r\theta(r)$ be its linear measure. We put

$$S(r\,;\Delta)=\frac{A(r)}{\pi},\qquad A(r)=\iint\limits_{D^r}\left(\frac{|w'(z)|}{1+|w(z)|^2}\right)^2 r\,dr\,d\theta\,,\qquad z=re^{i\theta}\,, \tag{2}$$

$$T(r\,;\Delta)=\int_{r_0}^{r}\frac{S(r\,;\Delta)}{r}dr\,,\quad r_0>0\,.$$

Suppose that $A(r)=O(1)$, then $T(r\,;\Delta)=O(\log r)$ and there exists a point a_0 in Δ, which is covered finite times by F_ρ. By a linear transformation, which makes Δ invariant, we may assume that $a_0=0$, so that $w(z)\neq 0$ in the part of D, which lies outside of $|z|=r_0$, if r_0 is sufficiently large, hence Theorem V. 29 can be applied, so that since $\theta(r)\leqq 2\pi$,

$$\log T(r\,;\Delta)\geqq\pi\int_{r_0}^{\kappa r}\frac{dr}{r\theta(r)}-\text{const.}\geqq\frac{1}{2}\log r-\text{const.}\qquad(0<\kappa<1)\,,$$

$T(r:\Delta)\geqq\text{const.}\sqrt{r}$, which contradicts $T(r\,;\Delta)=O(\log r)$. Hence $\lim_{r\to\infty} A(r)=\infty$. Next suppose that the boundary of D contains no curve, extending to ∞, so that the boundary of D consists of a countable number of disjoint Jordan curves, clustering to ∞. Let

$$L(r)=\int_{\theta_r}\frac{|w'(z)|}{1+|w(z)|^2}r\,dr\,\theta\,, \tag{3}$$

then $L^2(r)\leqq 2\pi r\dfrac{dA(r)}{dr}$. Hence if $A(r)=O(1)$, then $\displaystyle\int^\infty\frac{L^2(r)}{r}dr<\infty$, so that there exists $r_1<r_2<\cdots<r_\nu\to\infty$, such that $L(r_\nu)\to 0$.

Let β be the bounding circle $|w-w_0|=\rho$ of Δ and $F_\rho^r(\beta)$ be the part of F_ρ^r, which lie above β and $|F_\rho^r(\beta)|$ be its length and $S(r,\beta)=\frac{|F_\rho^r(\beta)|}{|\beta|}$, then by Theorem VI. 1,

$$\left|\frac{A(r)}{|\Delta|}-S(r\,;\beta)\right|<hL(r)\,,$$

where $h>0$ is a constant. Since $A(r)=O(1)$ and $L(r_\nu)\to 0$, we have $S(r_\nu,\beta)=O(1)$, which is absurd, since β is covered infinitely often by F_ρ. Hence $\lim_{r\to\infty} A(r)=\infty$. By means of $L^2(r)\leqq 2\pi r\frac{dA(r)}{dr}$, we can prove similarly as Theorem VI. 7, that F_ρ is regularly exhaustible in Ahlfors' sense. This being established, the remaining part of the theorem can be proved by means of Theorem VI. 6.

6. Bloch's theorem and related theorems

Lemma. *Let $w=f(z)$ be regular and $|f(z)|<M$ in $|z|<1$, then*

$$|f(z)|\geqq\frac{M(|f(0)|-Mr)}{M-|f(0)|r}\,,\qquad |f'(z)|\leqq\frac{M^2-|f(z)|^2}{M(1-r^2)}\,.\qquad |z|=r<1\,.$$

Proof. First suppose that $M=1$, then by Schwarz's lemma, $\left|\frac{f(z)-f(0)}{1-\bar{f}(0)f(z)}\right|\leqq|z|=r$ and from this, we have

$$|f(z)|\geqq\frac{|f(0)|-r}{1-|f(0)|r}\,. \tag{1}$$

From $\left|\frac{f(z)-f(z_1)}{1-\bar{f}(z_1)f(z)}\right|\leqq\left|\frac{z-z_1}{1-\bar{z}_1 z}\right|$, or $\left|\frac{f(z)-f(z_1)}{z-z_1}\right|\leqq\left|\frac{1-\bar{f}(z_1)f(z)}{1-\bar{z}_1 z}\right|$,

we have by making $z_1\to z$,

$$|f'(z)|\leqq\frac{1-|f(z)|^2}{1-|z|^2}\,. \tag{2}$$

In the general case, we consider $f(z)/M$ instead of $f(z)$.

Theorem VI. 10. *Let $w=f(z)=z+a_2z^2+\cdots$ be regular and $|f(z)|<M$ in $|z|<1$. Then $f(z)$ is schlicht in $|z|<\rho_0=\frac{1}{M+\sqrt{M^2-1}}$ and the image of $|z|<\rho_0$ on the w-plane is a star-region with respect to $w=0$.*

Let $\mathfrak{F}$ be the family of $f(z)$, which satisfy the above condition, then ρ_0 is the greatest value of ρ, such that any function $\in\mathfrak{F}$ is schlicht in $|z|<\rho$. Dieudonné [1].

Proof. (Landau [1]). Suppose that $f(a_1)=f(a_2)=\alpha$, $|a_1|\leqq|a_2|=\rho<1$. Then $g(z)=\frac{M^2(\alpha-f(z))}{M^2-\bar{\alpha}f(z)}$ is regular and $|g(z)|<M$ in $|z|<1$ and $g(a_1)=g(a_2)=0$, hence $h(z)=\frac{g(z)(1-\bar{a}z)(1-\bar{a}_2z)}{(z-a_1)(z-a_2)}$ is regular and $|h(z)|\leqq M$ in $|z|<1$, so that $|h(0)|=\left|\frac{g(0)}{a_1a_2}\right|=\left|\frac{\alpha}{a_1a_2}\right|\leqq M$, hence

$$|\alpha|\leqq M|a_1a_2|\leqq M\rho^2 . \tag{1}$$

If we apply the lemma on $\varphi(z)=f(z)/z$ ($\varphi(0)=1$), then since $|\varphi(z)|\leqq M$, we have $M\geqq|\varphi(0)|=1$ and

$$|f(z)|\geqq\frac{Mr(1-Mr)}{M-r}, \tag{2}$$

so that $|\alpha|=|f(a_2)|\geqq\frac{M\rho(1-M\rho)}{M-\rho}$. Hence by (1),

$$\frac{M\rho(1-M\rho)}{M-\rho}\leqq M\rho^2 . \tag{3}$$

Since $M\geqq 1$, $0<\rho<1$, we have $M-\rho>0$, hence from (3), $\rho^2-2\rho M+1\leqq 0$, so that

$$\rho\geqq M-\sqrt{M^2-1}=\frac{1}{M+\sqrt{M^2-1}} . \tag{4}$$

Hence $f(z)$ is schlicht in $|z|<\rho_0=\frac{1}{M+\sqrt{M^2-1}}$.

Now $f_0(z)=\frac{Mz(1-Mz)}{M-z}=z+\cdots$ $(M>1)$ is regular and $|f(z)|<M$ in $|z|<1$ and $f'(\rho_0)=0$, so that $f(z)$ is not schlicht in $|z|<\rho_0+\varepsilon$ $(\varepsilon>0)$, hence ρ_0 is the greatest value of ρ, such that any function $\in\mathfrak{F}$ is schlicht in $|z|<\rho$.

Next we shall prove that the image of $|z|<\rho_0$ on the w-plane is a star-region with respect to $w=0$. For that, it suffices to prove that $\Re\left(z\frac{f'(z)}{f(z)}\right)>0$ in $|z|<\rho_0$, or it suffices to prove that

$$\left|z\frac{\varphi'(z)}{\varphi(z)}\right|<1 \quad \text{in } |z|<\rho_0 , \qquad \varphi(z)=f(z)/z . \tag{5}$$

Since $|\varphi(z)|\leqq M$, $\varphi(0)=1$, we have by the lemma, $|\varphi(z)|\geqq\frac{M(1-Mr)}{M-r}$, so that

$$|\varphi'(z)|\leqq\frac{M^2-|\varphi(z)|^2}{M(1-r^2)}\leqq\frac{M^2-M^2\left(\frac{1-Mr}{M-r}\right)^2}{M(1-r^2)}=\frac{M(M^2-1)}{(M-r)^2},$$

hence

$$\left|z\frac{\varphi'(z)}{\varphi(z)}\right|\leqq\frac{(M^2-1)r}{(M-r)(1-Mr)} .$$

The inequality $\frac{(M^2-1)r}{(M-r)(1-Mr)}<1$ is satisfied, if $r<\frac{1}{M+\sqrt{M^2-1}}=\rho_0$.

Hence if $|z|<\rho_0$, then $\Re\left(z\dfrac{f'(z)}{f(z)}\right)>0$.——

Since $R=\dfrac{1}{2M}<\dfrac{1}{M+\sqrt{M^2-1}}$, if we put $z=R\zeta$, $f(z)/R=F(\zeta)$, then $v=F(\zeta)=\zeta+\cdots$ is regular and schlicht in $|\zeta|<1$, so that by Koebe's theorem, the image of $|\zeta|<1$ on the v-plane contains a schlicht disc $|v|<\dfrac{1}{4}$, hence we have

Corollary. *Let $w=f(z)=z+\cdots$ be regular and $|f(z)|<M$ in $|z|<1$, then the image of $|z|<1$ on the w-plane contains a schlicht disc:*

$$|w|<\frac{1}{8M}.$$

Theorem VI. 11. *Let $w=f(z)$ be meromorphic in $|z|<1$ and $\dfrac{|f'(0)|}{1+|f(0)|^2}=1$ and F be the Riemann surface, generated by $w=f(z)$ on the w-sphere K. Then F contains a schlicht spherical disc of radius $d>0$, where d is a numerical constant.* Tsuji [5].

Proof. We may assume that $f(z)$ is meromorphic on $|z|=1$ and put

$$\max_{|z|\leq 1}\frac{|f'(z)|}{1+|f(z)|^2}(1-|z|^2)=M. \tag{1}$$

Since $\dfrac{|f'(0)|}{1+|f(0)|^2}=1$, we have

$$1\leqq M<\infty. \tag{2}$$

Let M be attained at z_0, such that $\dfrac{|f'(z_0)|}{1+|f(z_0)|^2}(1-|z_0|^2)=M$, then $|z_0|<1$. If we put

$$Z=\frac{z-z_0}{1-\bar{z}_0 z}, \qquad f(z)=F(Z), \tag{3}$$

then

$$\frac{|F'(Z)|}{1+|F(Z)|^2}(1-|Z|^2)=\frac{|f'(z)|}{1+|f(z)|^2}(1-|z|^2)\leqq M, \quad \frac{|F'(0)|}{1+|F(0)|^2}=M. \tag{4}$$

Let $\dfrac{F'(0)}{1+|F(0)|^2}=Me^{i\theta}$. We rotate the w-sphere K, such that $F(0)=f(z_0)$ becomes $w=0$, then

$$G(Z)=e^{-i\theta}\frac{F(Z)-F(0)}{1+\overline{F(0)}F(Z)}=MZ+\cdots \tag{5}$$

is meromorphic in $|Z|\leqq 1$. Since by (4),

$$\frac{|G'(Z)|}{1+|G(Z)|^2}=\frac{|F'(Z)|}{1+|F(Z)|^2}\leqq\frac{M}{1-|Z|^2},$$

we have by integrating along the radius of $|Z|=1$,

$$\int_0^Z \frac{|G'(Z)|}{1+|G(Z)|^2}|dZ| \leqq \frac{M}{2} \log \frac{1+|Z|}{1-|Z|} .$$

Since the left hand side is the length of the curve on K, which corresponds to the radius of $|Z|=1$, if $\frac{M}{2}\log\frac{1+|Z|}{1-|Z|}\leqq\frac{\pi}{4}$, or $|Z|\leqq\frac{e^{\pi/2M}-1}{e^{\pi/2M}+1}$, then $G(Z)$ lies on the lower half of K, so that

$$|G(Z)|\leqq 1 , \qquad \text{if} \quad |Z|\leqq\frac{e^{\pi/2M}-1}{e^{\pi/2M}+1} . \tag{6}$$

Since $M\geqq 1$,

$$\frac{e^{\pi/2M}-1}{e^{\pi/2M}+1}\geqq\frac{1}{e^{\pi/2}+1}\left(\frac{\pi}{2M}+\frac{1}{2}\left(\frac{\pi}{2M}\right)^2+\cdots\right)\geqq\frac{1}{e^{\pi/2}+1}\cdot\frac{\pi}{2M}=\frac{k}{M}, \tag{7}$$

$$k=\frac{\pi}{2(e^{\pi/2}+1)} .$$

Hence $G(Z)=MZ+\cdots$ is regular and $|G(Z)|<1$ for $|Z|<k/M$. Hence if we put $x=MZ/k$, $G(Z)/k=H(x)$, then $H(x)=x+\cdots$ is regular and $|H(x)|<1/k$ for $|x|<1$, so that by Theorem VI. 10. Cor., the image of $|x|<1$ on the H-plane contains a schlicht disc $|H|<k/8$, hence the image of $|x|<1$ on the G-plane contains a schlicht disc Δ: $|G|<k^2/8=\rho$.

Since Δ is projected on a spherical disc of radius $d=\frac{\rho}{1+\rho^2}$ on K, the Riemann surface F^* of $v=G(Z)$ contains a schlicht spherical disc of radius d. Since the Riemann surface F, generated by $w=f(z)$, is obtained from F^* by a rotation of K, F contains a schlicht spherical disc of radius d, where d is a numerical constant.——

From the theorem, we have the following Bloch's theorem.

Theorem VI. 12. *Let $w=f(z)=z+\cdots$ be regular in $|z|<1$, then the Riemann surface F, generated by $w=f(z)$ on the w-plane, contains a schlicht disc of radius $d>0$, where d is a numerical constant.*

Bloch [1].

7. A criterion on normal family

Theorem VI. 13. *Let $w=f(z)$ be meromorphic in $|z|<R$ and F be the Riemann surface, generated by $w=f(z)$ on the w-sphere K. Let D_ν $(\nu=1, 2, \cdots, q)$ $(q\geqq 3)$ be q disjoint simply connected domains on K and $1\leqq m_\nu\leqq\infty$ be a positive integer or ∞, such that*

$$\sum_{\nu=1}^{q}\left(1-\frac{1}{m_\nu}\right)>2 .$$

If every simply connected island of F, which lies above D_ν, has at *least m_ν sheets*[1), *then*

$$R\leqq\kappa\frac{1+|f(0)|^2}{|f'(0)|}, \qquad \frac{|f'(z)|}{1+|f(z)|^2}\leqq\frac{\kappa R}{R^2-|z|^2}, \qquad |z|<R,$$

where $\kappa>0$ is a constant, which depends on $D_1, \cdots, D_q$ only.

Proof. It can be proved easily that if $\sum_{\nu=1}^{q}\left(1-\frac{1}{m_\nu}\right)>2$, then

$$\sum_{\nu=1}^{q}\left(1-\frac{1}{m_\nu}\right)-2\geqq\frac{1}{42}, \tag{1}$$

where the minimum value 1/42 is attained, when $m_1=2$, $m_2=3$, $m_3=7$, $m_4=m_5=\cdots=m_q=1$.

First suppose that

$$\frac{|f'(0)|}{1+|f(0)|^2}=1, \tag{2}$$

and put

$$S(r)=\frac{1}{\pi}\iint_{|z|\leqq r}\left(\frac{|f'(z)|}{1+|f(z)|^2}\right)^2 r dr d\theta, \qquad L(r)=\int_0^{2\pi}\frac{|f'(z)|}{1+|f(z)|^2}r\, d\theta, \tag{3}$$

$$z=re^{i\theta}, \quad 0\leqq r<R,$$

then by Schwarz's inequality, $L^2(r)\leqq 2\pi^2 r\dfrac{dS(r)}{dr}$.

By Theorem VI. 5, there exists a constant $h>0$, which depends on $D_1, \cdots, D_q$ only, such that

$$\sum_{\nu=1}^{q}\left(1-\frac{1}{m_\nu}\right)\leqq 2+h\frac{L(r)}{S(r)},$$

so that

$$\delta=\frac{1}{42}\leqq\sum_{\nu=1}^{q}\left(1-\frac{1}{m_\nu}\right)-2\leqq h\frac{L(r)}{S(r)},$$

hence

$$\frac{\delta^2}{h^2}(S(r))^2\leqq(L(r))^2\leqq 2\pi^2 r\frac{dS(r)}{dr},$$

so that if $R>1$,

$$\log R=\int_1^R\frac{dr}{r}\leqq\frac{2\pi^2h^2}{\delta^2}\int_1^R\frac{dS(r)}{(S(r))^2}\leqq\frac{2\pi^2h^2}{\delta^2}\cdot\frac{1}{S(1)}. \tag{4}$$

By (2) and Theorem VI. 11, $S(1)\geqq h_0>0$, where $h_0>0$ is a numerical constant, hence

$$R\leqq\kappa=\exp(2\pi^2h^2/\delta^2h_0). \tag{5}$$

If $R\leqq 1$, then (5) holds evidently, hence the theorem is proved, if $\dfrac{|f'(0)|}{1+|f(0)|^2}=1$. In the general case, let $\dfrac{|f'(0)|}{1+|f(0)|^2}=\alpha$ and first suppose

1) $m=\infty$ means that there are no simply connected islands of F above D_ν.

that $\alpha>0$ and put $z=x/\alpha$, $f(z)=F(x)$. Then $F(x)$ is meromorphic in $|x|<R\alpha$ and $\dfrac{|F'(0)|}{1+|F(0)|^2}=1$, so that $R\alpha\leqq\kappa$, or

$$R\leqq\frac{\kappa}{\alpha}=\kappa\frac{1+|f(0)|^2}{|f'(0)|}, \qquad \frac{|f'(0)|}{1+|f(0)|^2}\leqq\frac{\kappa}{R}, \tag{6}$$

which holds evidently, if $\alpha=0$.

For any fixed z_0 ($|z_0|<R$), we put $\zeta=\dfrac{R^2(z-z_0)}{R^2-\bar{z}_0 z}$, $f(z)=F(\zeta)$, then $F(\zeta)$ satisfies the condition of the theorem, so that $\dfrac{|F'(0)|}{1+|F(0)|^2}\leqq\dfrac{\kappa}{R}$.

Since $|f'(z_0)|=\dfrac{R^2}{R^2-|z_0|^2}|F'(0)|$, we have $\dfrac{|f'(z_0)|}{1+|f(z_0)|^2}\leqq\dfrac{\kappa R}{R^2-|z_0|^2}$. —

Since $\dfrac{\kappa R}{R^2-|z|^2}$ does not depend on $f(z)$, we have

Theorem VI. 14. *Let $\mathfrak{F}$ be the family of $f(z)$, which satisfy the condition of Theorem* VI. 13, *then $\mathfrak{F}$ is normal.*

Especially, if $f(z)\neq a_i$ $(i=1, 2, 3)$ in $|z|<R$, then $\mathfrak{F}$ is normal.

Theorem VI. 15. *Let $w=f(z)$ be meromorphic in $|z|<R$ and the Riemann surface F, generated by $w=f(z)$ on the w-sphere K, is a covering surface of a closed Riemann surface Φ of genus $p\geqq 2$, then*

$$R\leqq\kappa\frac{1+|f(0)|^2}{|f'(0)|}, \qquad \frac{|f'(z)|}{1+|f(z)|^2}\leqq\frac{\kappa R}{R^2-|z|^2}, \qquad |z|<R,$$

where $\kappa>0$ is a constant, which depends on Φ only.

Hence the family $\mathfrak{F}$ of such $f(z)$ is normal. Ahlfors [6].

Proof. $\rho_0=2(p-1)$ is the Euler's characteristic of Φ. Let N be the number of sheets of Φ, then $|\Phi|=N\pi$. Let F_r be the Riemann surface, generated by $w=f(z)$ ($|z|\leqq r<R$) on K and $L(r)$ be the length of the image of $|z|=r$ and put

$$S(r)=\frac{|F_r|}{|\Phi|}=\frac{|F_r|}{N\pi},$$

and $\rho(r)$ be the Euler's characteristic of F_r, then by the fundamental theorem, $\rho^+(r)\geqq\rho_0 S(r)-hL(r)$, where $h>0$ is a constant, which depends on Φ only. Since F_r is simply connected, $\rho^+(r)=0$, so that $S(r)\leqq\dfrac{h}{\rho_0}L(r)$. From this, we can prove the theorem, similarly as Theorem VI. 13.

8. Uniformization of algebraic functions

Theorem VI. 16. *Let Φ be a closed Riemann surface of genus $p\geqq 2$, spread over the w-sphere K. Then there exists no function $w=f(z)$,* *which is one-valued and meromorphic for* $r_0 \leqq |z| < \infty$ *and has an essential singularity at* ∞, *such that the Riemann surface* F, *generated by* $w=f(z)$ *on the w-sphere is a covering surface of* Φ. Picard [1].

Proof. Suppose that there exists such $w=f(z)$ and let F_r be the image of $r_0 \leqq |z| \leqq r$ on K and $\pi S(r)$ be its spherical area and $L(r)$ be the length of the image of $|z|=r$. Since $z=\infty$ is an essential singularity of $f(z)$, by Theorem VI. 7, there exist $r_1 < r_2 < \cdots < r_n \to \infty$, such that $L(r_n)/S(r_n) \to 0$, but from the proof of Theorem VI. 15, $S(r)=O(L(r))$, which is absurd, hence the theorem. ——

From the theorem, we have

Theorem VI. 17. *The universal covering surface of a closed Riemann surface* Φ *of genus* $p \geqq 2$ *can not be mapped conformally on* $|z| < \infty$, *so that it can be mapped conformally on* $|z| < 1$.

Theorem VI. 18. *Let* Φ *be a closed Riemann surface of genus* $p \geqq 2$, *spread over the w-sphere* K. *Then there exists no function* $w=f(z)$, *which is one-valued and meromorphic in a neighbourhood* U *of a bounded closed set* E *of capacity zero, every point of which is an essential singularity of* $f(z)$, *such that the Riemann surface* F, *generated by* $w=f(z)$ *on* K, *is an unramified covering surface of* Φ.[1)]

Proof. Suppose that there exists such $w=f(z)$. Let $u(z)$ be the Evans' function with respect to E:

$$u(z)=\int_E \log \frac{1}{|z-a|} d\mu(a)\,, \qquad \mu(E)=1\,, \tag{1}$$

which tends to $+\infty$, if z tends to any point of E.

Let C_r be its niveau curve: $u(z)=\text{const.}=\log r$, which consists of a finite number of analytic Jordan curves. Let D_r be the domain, bounded by C_{r_0} and C_r $(r_0 < r)$, where we may assume that C_{r_0} coincides with the boundary of U.

Let $A(r)$ be the spherical area of the image of D_r by $w=f(z)$ and $S(r)=A(r)/N\pi$, where N is the number of sheets of Φ and $L(r)$ be the length of the image of C_r. Since by Theorem V. 33, $f(z)$ takes any value infinitely often, except a set of capacity zero, $\lim\limits_{r\to\infty} A(r)=\infty$.

Let $\theta(z)$ be the conjugate harmonic function of $u(z)$ and put $\zeta=e^{u+i\theta}=re^{i\theta}$, $f(z)=F(\zeta)$, then since

$$L(r)=\int_{C_r} \frac{|F'(\zeta)|}{1+|F(\zeta)|^2} r\, d\theta\,,$$

we can prove similarly as Theorem VI. 7, that there exist

1) i.e., F has no branch point relatively to Φ.

$r_1<r_2<\cdots<r_n\to\infty$, such that

$$\frac{L(r_n)}{S(r_n)}\to 0\,. \tag{2}$$

Hence F is regularly exhaustible in Ahlfors' sense.

Let C_r consists of $n(r)$ disjoint Jordan curves: $C_r=\sum_{i=1}^{n(r)} C_r^i$ and let $L^i(r)$ be the length of the image Λ_r^i of C_r^i. Since by the hypothesis, F is an unramified covering surface of Φ, and every point of E is an essential singularity of $f(z)$, we see that Λ_r^i is not homotop null, so that $L^i(r)\geqq a>0$, where $a>0$ is a constant, which depends on Φ only. Hence

$$L(r)\geqq an(r)\,. \tag{3}$$

Let F_r be the image of D_r, then F_r is a covering surface of Φ, so that by the fundamental theorem,

$$\rho^+(r)\geqq \rho_0 S(r)-h(L(r)+L(r_0))\,, \tag{4}$$

where $\rho(r)$ is the Euler's characteristic of F_r and $\rho_0=2(p-1)$ is that of Φ and $h>0$ is a constant, which depends Φ on only.

Since $\rho(r)=n(r)-1\leqq n(r)\leqq L(r)/a$, we have from (4), $S(r)=O(L(r))$, which contradicts (2), hence the theorem.

9. Schottky's theorem and related theorems

1. Let G be the modular group of linear transformations S: $z'=\dfrac{az+b}{cz+d}$, where a, b, c, d are integers, such that $ad-bc=1$. Then $z'=S(z)$ transforms the upper half of the z-plane into itself and as well known, its fundamental domain D_0 is as follows.

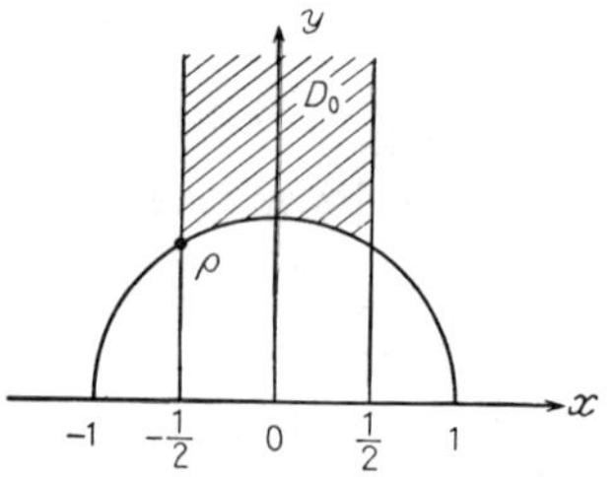

$$D_0:\begin{cases}|z|>1\,, & -\dfrac{1}{2}\leqq x<\dfrac{1}{2}\,,\\ |z|=1\,, & -\dfrac{1}{2}\leqq x\leqq 0\,,\quad z=x+iy\,.\end{cases}$$

Let D_0': $|z|>1$, $-\dfrac{1}{2}<x<0$ be the left half of D_0. We map D_0' conformally on the upper half of the w-plane by $w=J(z)$, such that

$$J(\rho)=0\,,\quad J(i)=1\,,\quad J(\infty)=\infty\,,\qquad \text{where}\quad \rho=\frac{-1+i\sqrt{3}}{2}\,.$$

Then $J(z)$ can be continued analytically on the upper half of the z-plane, such that $J(S(z))=J(z)$ for any $S\in G$. $J(z)$ has a zero point of the third order at $z=\rho$ and $J(z)-1$ has a zero point of the second order at $z=i$ and if we put $t=e^{2\pi iz}$, $J(z)=F(t)$, then $F(t)$ has a pole of the first order at $t=0$.

2. First we shall prove two lemmas.

Lemma 1. $|J(z)|<e^{c\left(y+\frac{1}{y}\right)}$, $\quad z=x+iy$, $y>0$, *where $c>0$ is a constant.*

Proof. (i) $y\geqq\sqrt{3}/2$.
We put $t=e^{2\pi iz}$, $J(z)=F(t)$, then $0<|t|=e^{-2\pi y}\leqq e^{-\pi\sqrt{3}}<1$.
Since $F(t)$ has a pole of the first order at $t=0$,

$$|J(z)|=|F(t)|\leqq\frac{c_1}{|t|}=c_1 e^{2\pi y}<e^{cy}<e^{c\left(y+\frac{1}{y}\right)}, \tag{1}$$

where $c_1>0$, $c>0$ are constants.

(ii) $0<y<\sqrt{3}/2$.

By a suitable $S\in G$, $z'=S(z)=\dfrac{az+b}{cz+d}\in D_0$. If we put $z=x+iy$, $z'=x'+iy'$, then

$$y'=\frac{y}{(cx+d)^2+c^2y^2}.$$

If $c=0$, then from $ad-bc=1$, we have $a=1$, $d=1$, so that $z'=z+b$, hence $y'=y<\sqrt{3}/2$, which is absurd. Hence $c\neq 0$. Since c is an integer, $|c|\geqq 1$, so that $y'\leqq\dfrac{y}{c^2y^2}\leqq\dfrac{1}{y}$. Since $y'\geqq\sqrt{3}/2$, we have by (i), $|J(z')|<e^{cy'}$, so that

$$|J(z)|=|J(z')|<e^{cy'}\leqq e^{\frac{c}{y}}<e^{c\left(y+\frac{1}{y}\right)}. \tag{2}$$

Lemma 2. *If $z=x+iy\in D_0$, then*

$$y\leqq B\log(|J(z)|+2),$$

where $B>0$ is a constant.

Proof. We put $t=e^{2\pi iz}$, $J(z)=F(t)$. Since $z\in D_0$, $y\geqq\sqrt{3}/2$, so that $0<|t|=e^{-2\pi y}\leqq e^{-\pi\sqrt{3}}$. Since $F(t)$ has a pole of the first order at $t=0$,

$$|J(z)|+2=|F(t)|+2>\frac{1}{B_1|t|}=\frac{e^{2\pi y}}{|B_1|}, \qquad B_1=\text{const.},$$

hence

$$2\pi y<\log B_1+\log(|J(z)|+2). \tag{1}$$

If we choose a constant $B>0$, so large that $\log B_1<(2\pi B-1)\log 2$, then

$$\log B_1+\log(|J(z)|+2)<(2\pi B-1)\log 2+\log(|J(z)|+2)$$
$$<(2\pi B-1)\log(|J(z)|+2)+\log(|J(z)|+2)=2\pi B\log(|J(z)|+2),$$

hence by (1),

$$y\leqq B\log(|J(z)|+2). \text{ ——} \tag{2}$$

3. Now we shall prove the following precise form of **Schottky's theorem.**

Theorem VI. 19. *Let* $f(z)=a_0+a_1z+\cdots$ *be regular and* $f(z)\neq 0, 1$ *in* $|z|<1$, *then*

$$|f(z)|\leqq e^{\frac{A\log(|a_0|+2)}{1-r}}, \qquad |z|=r<1,$$

where $A>0$ *is a constant.* Bohr-Landau [1].

Proof. The inverse function $z=\nu(w)$ of $w=J(z)$ is infinitely many valued. Let $\nu(w)$ be such a branch, that $\nu(a_0)=c_0\in D_0$ and put

$$F(z)=\nu(f(z))=c_0+c_1z+\cdots, \qquad f(z)=J(F(z)), \tag{1}$$

then since $f(z)\neq 0, 1$, $F(z)$ is regular and $\Im(F(z))>0$ in $|z|<1$, so that if we put $G(z)=\dfrac{F(z)-c_0}{F(z)-\bar{c}_0}$, then $G(z)$ is regular and $|G(z)|<1$ in $|z|<1$, $G(0)=0$, so that by Schwarz's lemma,

$$|G(z)|=\left|\frac{F(z)-c_0}{F(z)-\bar{c}_0}\right|\leqq r, \qquad |z|=r<1. \tag{2}$$

From (2), we see easily that

$$\Im(c_0)\frac{1-r}{1+r}\leqq\Im(F(z))\leqq\Im(c_0)\frac{1+r}{1-r}.$$

Since $\Im(c_0)\geqq\sqrt{3}/2$,

$$\frac{\sqrt{3}}{4}(1-r)\leqq\Im(F(z))\leqq\frac{2\Im(c_0)}{1-r}. \tag{3}$$

Since $c_0\in D_0$, by Lemma 2,

$$\Im(c_0)\leqq B\log(|J(c_0)|+2)=B\log(|a_0|+2),$$

so that

$$\frac{\sqrt{3}}{4}(1-r)\leqq\Im(F(z))\leqq\frac{2B\log(|a_0|+2)}{1-r}. \tag{4}$$

Hence by Lemma 1,

$$|f(z)|=|J(F(z))|\leqq e^{c(\Im(F(z))+1/\Im F(z))}\leqq e^{c\left(\frac{2B\log(|a_0|+2)}{1-r}+\frac{4}{\sqrt{3}(1-r)}\right)}$$
$$=e^{\frac{2c(\sqrt{3}B\log(|a_0|+2)+2)}{\sqrt{3}(1-r)}}. \tag{5}$$

If we choose $A>0$, so large that $2\leqq\left(\frac{\sqrt{3}}{2c}A-\sqrt{3}B\right)\log 2$, then

$$\sqrt{3}B\log(|a_0|+2)+2<\sqrt{3}B\log(|a_0|+2)+\left(\frac{\sqrt{3}}{2c}A-\sqrt{3}B\right)\log(|a_0|+2)$$
$$=\frac{\sqrt{3}}{2c}A(\log(|a_0|+2),$$

hence by (5),

$$|f(z)|\leqq e^{\frac{A\log(|a_0|+2)}{1-r}}.$$

Theorem VI. 20. *Let $w=w(z)$ be meromorphic in $|z|<1$ and $w(z)\neq a_\nu$ $(\nu=1,2,3)$, then*

$$T(w,r)\leqq A\log\frac{1}{1-r}, \qquad 0\leqq r<1,$$

where $A>0$ is a constant, which depends on a_1, a_2, a_3 only.

Proof. If we take in Theorem VI. 13, D_ν as a simply connected domain, which contains a_ν, then $m_\nu=\infty$, so that $\sum_{\nu=1}^{3}\left(1-\frac{1}{m_\nu}\right)=3>2$, hence

$$\frac{|w'(z)|}{1+|w(z)|^2}\leqq\frac{\kappa}{1-r^2}, \qquad |z|=r<1,$$

where $\kappa>0$ is a constant, which depends on a_1, a_2, a_3 only.

Hence

$$S(r)=\frac{1}{\pi}\iint_{|z|\leqq r}\left(\frac{|w'(z)|}{1+|1w(z)|^2}\right)^2 rdrd\theta\leqq\frac{\kappa^2r^2}{1-r^2},$$

so that

$$T(w,r)=\int_0^r\frac{S(r)}{r}dr\leqq\frac{\kappa^2}{2}\log\frac{1}{1-r^2}\leqq\frac{\kappa^2}{2}\log\frac{1}{1-r}.$$

4. We shall prove the following theorem, which plays an important rôle in Chapter V.

Theorem VI. 21. *Let $w=w(z)$ be meromorphic in $|z|<1$ and the number of zero points of $\prod_{\nu=1}^{3}(w(z)-a_\nu)$ in $|z|<1$ be $\leqq n$, where multiple zeros being counted only once, then*

$$S(r)\leqq n+\frac{A}{1-r}, \qquad 0\leqq r<1,$$

where

$$S(r)=\frac{1}{\pi}\iint_{|z|\leqq r}\left(\frac{|w'(z)|}{1+|w(z)|^2}\right)^2 rdrd\theta, \qquad z=re^{i\theta},$$

and $A>0$ is a constant, which depends on a_1, a_2, a_3 only.

Dufresnoy [1], Tsuji [20].

Proof. We take off three points a_1, a_2, a_3 from the w-sphere K and let F_0 be the remaining surface, then $\rho_0=\rho(F_0)=1$. We take off the zero points of $\prod_{\nu=1}^{3}(w(z)-a\ )$ from $|z|<1$ and let D be the remaining domain and D_r be its part, contained in $|z|<r<1$ and F_r be the image of D_r on K, then F_r is a covering surface of F_0, so that, since $\rho_0=1$, by the fundamental theorem,

$$\rho^+(r) \geqq S(r) - hL(r) , \tag{1}$$

where $\rho(r)$ is the Euler's characteristic of F_r, $L(r)$ the length of the image of $|z|=r$ on K and $h>0$ is a constant, which depends on a_1, a_2, a_3 only. Since by the hypothesis, $\rho^+(r) \leqq n$, we have

$$S(r) - n \leqq hL(r) . \tag{2}$$

If $S(r')-n>0$ for all r' $(r \leqq r' < 1)$, then from $(L(r))^2 \leqq 2\pi^2 r \frac{dS(r)}{dr}$, we have

$$(S(r')-n)^2 \leqq h^2(L(r')^2 \leqq 2\pi^2 h^2 r' \frac{dS(r')}{dr'} ,$$

so that

$$1-r < \int_r^1 \frac{dr'}{r'} \leqq 2\pi^2 h^2 \int_r^1 \frac{dS(r')}{(S(r')-n)^2} \leqq \frac{2\pi^2 h^2}{S(r)-n} ,$$

$$S(r) \leqq n + \frac{2\pi^2 h^2}{1-r} . \tag{3}$$

If $S(r')-n<0$ for some r' $(r \leqq r' < 1)$, then $S(r) \leqq S(r') \leqq n$, so that (3) holds in general.——

By means of Theorem VI. 3 and 4, we can prove similarly the following

Theorem VI. 22. *Let $w=w(z)$ be meromorphic in $|z|<1$ and F be the Riemann surface, generated by $w=w(z)$ on the w-sphere K.*

(i) *Let D_1, D_2, D_3 be three disjoint simply connected domains on K. If the total number of simply connected islands of F, above D_1, D_2, D_3 be $\leqq n$, then*

$$S(r) \leqq n + \frac{A}{1-r} , \quad 0 \leqq r < 1 ,$$

where A is a constant, which depends on D_1, D_2, D_3 only.

(ii) *Let $D_1, \cdots, D_5$ be five disjoint simply connected domains on K. If the total number of schlicht islands of F above $D_1, \cdots, D_5$ be $\leqq n'$, then*

$$S(r) \leqq n' + \frac{A}{1-r} , \quad 0 \leqq r < 1 ,$$

where A is a constant, which depends on $D_1, \cdots, D_5$ only.

CHAPTER VII

BOREL'S DIRECTION

1. Borel's theorem

1. Let $f(z)$ be a meromorphic function of finite order ρ for $|z|<\infty$, then by Theorem V. 10,

$$\sum_\nu \frac{1}{|z_\nu(a)|^{\rho+\varepsilon}}<\infty \quad \text{for any } \varepsilon>0,$$

where $\{z_\nu(a)\}$ are zero points of $f(z)-a$.

Theorem VII. 1. *Let $f(z)$ be a meromorphic function of finite order ρ for $|z|<\infty$, then*

$$\text{(i)} \qquad \sum_\nu \frac{1}{|z_\nu(a)|^{\rho-\varepsilon}}=\infty \quad \textit{for any } \varepsilon>0.$$

with two possible exceptions for a.

If $f(z)$ is of divergence type, then

$$\text{(ii)} \qquad \sum_\nu \frac{1}{|z_\nu(a)|^{\rho}}=\infty\,.$$

with two possible exceptions for a. Borel [1].

Proof. Suppose that $\sum_\nu \frac{1}{|z(a_i)|^{\rho-\varepsilon}}<\infty$ $(i=1, 2, 3)$, then $\int^\infty \frac{N(r,a_i)}{r^{\rho+1-\varepsilon}}dr<\infty$. By the second fundamental theorem, $T(r)\leqq\sum_{i=1}^{3} N(r, a_i)+O(\log r)$, so that $\int^\infty \frac{T(r)}{r^{\rho+1-\varepsilon}}dr<\infty$. From this, we have $\overline{\lim_{r\to\infty}}\frac{\log T(r)}{\log r}\leqq\rho-\varepsilon<\rho$, which is absurd. Hence (i) is proved. Similarly (ii) can be proved.——

2. Let $f(z)$ be a meromorphic function of finite order ρ in $|z|<1$, then by Theorem V. 11,

$$\sum(1-|z\,(a)|)^{\rho+1+\varepsilon}<\infty \quad \text{for any } \varepsilon>0.$$

Similarly as Theorem VII. 1, we can prove

Theorem VII. 2. *Let $f(z)$ be a meromorphic function of finite order ρ in $|z|<1$, then*

$$\text{(i)} \qquad \sum_\nu (1-|z_\nu(a)|^{\rho+1-\varepsilon})=\infty \quad \textit{for any } \varepsilon>0,$$

with two possible exceptions for a.

If $f(z)$ is of divergence type, then

(ii) $$\sum_{\nu}(1-|z_\nu(a)|)^{\rho+1}=\infty\,,$$

with two possible exceptions for a.

Part I. Meromorphic functions for $|z|<\infty$

1. Analogue of the second fundamental theorem for an angular domain

1. Let $w(z)$ be meromorphic in an angular domain Δ_0: $|\arg z|\leqq\alpha_0$ and Δ: $|\arg z|\leqq\alpha(<\alpha_0)$ be an angular domain, contained in Δ_0. Let $\Delta_0(r)$, $\Delta(r)$ be the part of Δ_0, Δ, which is contained in $|z|\leqq r$ respectively and put

$$S(r;\Delta)=S(w,r;\Delta)=\frac{1}{\pi}\iint_{\Delta(r)}\left(\frac{|w'(z)|}{1+|w(z)|^2}\right)^2 rdrd\theta\,,\qquad z=re^{i\theta}\,,\tag{1}$$

$$T(r;\Delta)=T(w,r;\Delta)=\int_0^r\frac{S(r;\Delta)}{r}dr\,.\tag{2}$$

Let $n(r,a;\Delta_0)$ be the number of zero points of $w(z)-a$, contained in $\Delta_0(r)$, multiple zeros being counted only once and put

$$N(r,a;\Delta_0)=N(w,r,a;\Delta_0)=\int_1^r\frac{n(r,a;\Delta_0)}{r}dr\,.\tag{3}$$

Then we shall prove the following analogue of the second fundamental theorem.

Theorem VII. 3. $$S(r;\Delta)\leqq 3\sum_{i=1}^{3}n(2r,a_i;\Delta_0)+O(\log r)\,,$$

$$T(r;\Delta)\leqq 3\sum_{i=1}^{3}N(2r,a_i;\Delta_0)+O((\log r)^2)\,.$$

Proof. Let $k=\sqrt{2}$, $r_\nu=k^\nu$ $(\nu=0,1,2,\cdots)$ and for $\nu\geqq 2$,

$$\left.\begin{aligned}\Delta_\nu:&\quad|\arg z|\leqq\alpha\,,\qquad r_{\nu-1}\leqq|z|\leqq r_\nu\,,\\ \Delta_\nu^0:&\quad|\arg z|\leqq\alpha_0\,,\qquad r_{\nu-2}\leqq|z|\leqq r_{\nu+1}\end{aligned}\right\}\tag{1}$$

and put

$$S_\nu=\frac{1}{\pi}\iint_{\Delta_\nu}\left(\frac{|w'(z)|}{1+|w(z)|^2}\right)^2 rdrd\theta\,.\tag{2}$$

Let n_ν^0 be the number of zero points of $\prod_{i=1}^{3}(w(z)-a_i)$ in Δ_ν^0, multiple zeros being counted only once. If we map Δ_ν^0 conformally on $|\zeta|<1$, such that the center of Δ_ν^0 becomes $\zeta=0$, then the image of Δ_ν is contained in $|\zeta|<\kappa<1$, where κ is a constant, which is independent of ν.

Hence if we apply Theorem VI. 21, then we have

$$S_\nu\leqq n_\nu^0+\frac{A}{1-\kappa}\,,$$

where $A>0$ is a constant, which depends on a_1, a_2, a_3 only. Hence

$$\sum_{\nu=2}^{n} S_\nu \leqq \sum_{\nu=2}^{n} n_\nu^0 + O(n) = \sum_{\nu=2}^{n} n_\nu^0 + O(\log r_n) . \tag{3}$$

$\sum_{\nu=2}^{n} S_\nu = S(r_n; \Delta) - S(r_1; \Delta)$ and since Δ_ν^0 overlap at most 3-times, $\sum_{\nu=2}^{n} n_\nu^0 \leqq 3 \sum_{i=1}^{3} n(r_{n+1}, a_i; \Delta_0)$, so that by (3),

$$S(r_n; \Delta) \leqq 3 \sum_{i=1}^{3} n(r_{n+1}, a_i; \Delta_0) + O(\log r_n) . \tag{4}$$

If $r_{n-1} \leqq r \leqq r_n$, then $S(r; \Delta) \leqq S(r_n; \Delta)$ and $r_{n+1} = k^2 r_{n-1} = 2r_{n-1} \leqq 2r$, so that

$$S(r; \Delta) \leqq 3 \sum_{i=1}^{3} n(2r, a_i; \Delta_0) + O(\log r)) , \tag{5}$$

hence

$$T(r; \Delta) \leqq 3 \sum_{i=1}^{3} N(2r, a_i; \Delta_0) + O((\log r)^2) . \text{——} \tag{6}$$

2. If we apply the theorem for angular domains Δ: $0 \leqq \arg z \leqq 2\pi$ and Δ_0: $-\varepsilon \leqq \arg z \leqq 2\pi + \varepsilon$ $(\varepsilon > 0)$, then we have

Theorem VII. 4. **(Modified second fundamental theorem).** *Let $w(z)$ be meromorphic for $|z| < \infty$, then*

$$T(r) \leqq 6 \sum_{i=1}^{3} N(2r; a_i) + O((\log r)^2) , \qquad 1 \leqq r < \infty ,$$

where $T(r)$, $N(r, a)$ are Nevanlinna's functions for $w(z)$.

We remark that in the second fundamental theorem, r lies outside of certain exceptional intervals $\{J_\nu\}$. As an application, we shall prove.

Theorem VII. 5. *Let $w(z)$ be meromorphic for $|z| < \infty$. If $\sum_\nu \frac{1}{|z_\nu(a_i)|^k} < \infty$, $k > 0$ $(i = 1, 2, 3)$, then $\int^\infty \frac{T(r)}{r^{k+1}} dr < \infty$, so that $w(z)$ is of order $\leqq k$.*

Proof. By the hypothesis, $\int^\infty \frac{N(r, a_i)}{r^{k+1}} dr < \infty$ $(i = 1, 2, 3)$, so that by Theorem VII. 4, $\int^\infty \frac{T(r)}{r^{k+1}} dr < \infty$, hence $w(z)$ is of order $\leqq k$.

2. Borel's direction

1. We shall prove the following **Valiron's theorem.**

Theorem VII. 6. *Let $f(z)$ be a meromorphic function of finite order $\rho > 0$ for $|z| < \infty$. Then there exists a direction J: $\arg z = \alpha$, which satisfies the following condition.*

Let Δ be any small angular domain, which contains J and $\{z_\nu(a; \Delta)\}$ be zero points of $w(z) - a$ in Δ, multiple zeros being counted only once, then

(i) $$\sum_{\nu} \frac{1}{|z_\nu(a;\Delta)|^{\rho-\varepsilon}}=\infty \quad \text{for any } \varepsilon>0,$$

with two possible exceptions for a.

If f(z) is of divergence type, then

(ii) $$\sum_{\nu} \frac{1}{|z_\nu(a;\Delta)|^{\rho}}=\infty,$$

with two possible exceptions for a. Valiron [1].

J is called *a Borel's direction of f(z).*

Proof. (Tsuji [20]). Now

$$\int^{\infty} \frac{T(r)}{r^{k+1}}dr=\infty, \qquad k>0, \tag{1}$$

where $k=\rho-\varepsilon$ ($\varepsilon>0$) in general and $k=\rho$, if $f(z)$ is of divergence type.

By dividing the unit circle into 2^n equal parts, we see that there exist angular domains Δ_n of magnitude $2\pi/2^n$, bounded by two lines through $z=0$, such that $\Delta_1\supset\Delta_2\supset\cdots\supset\Delta_n\supset\cdots$ and

$$\int^{\infty} \frac{T(r;\Delta_n)}{r^{k+1}}dr=\infty \qquad (n=1,2,\cdots). \tag{2}$$

Let Δ_n converge to a direction J: $\arg z=\alpha$, then for any small angular domain Δ: $|\arg z-\alpha|<\delta$, which contains J,

$$\int^{\infty} \frac{T(r;\Delta)}{r^{k+1}}dr=\infty. \tag{3}$$

Let Δ_0: $|\arg z-\alpha|<2\delta$ and apply Theorem VII. 3, then

$$\int^{\infty} \frac{T(r;\Delta)}{r^{k+1}}dr\leqq 3\sum_{i=1}^{3}\int^{\infty} \frac{N(2r,a_i;\Delta_0)}{r^{k+1}}dr+\text{const.}. \tag{4}$$

If $\sum_{\nu} \frac{1}{|z_\nu(a_i;\Delta_0)|^{k}}<\infty$ $(i=1,2,3)$, then $\int^{\infty} \frac{N(r,a_i;\Delta_0)}{r^{k+1}}dr<\infty$, so that $\int^{\infty} \frac{T(r;\Delta)}{r^{k+1}}dr<\infty$, which contradicts (3). Hence $\sum_{\nu} \frac{1}{|z_\nu(a;\Delta_0)|^{k}}=\infty$, with two possible exceptions for a.

If $f(z)$ is of divergence type, then $k=\rho$, so that J satisfies the condition of the theorem. If $f(z)$ is of convergence type, then we take $k=\rho-\varepsilon_n$ ($\varepsilon_1>\varepsilon_2>\cdots>\varepsilon\to 0$) and let J_n be the corresponding direction and J be one of the limiting positions of $\{J_n\}$, then J satisfies the condition of the theorem. ——

2. To prove a more precise theorem, first we shall prove a lemma.

Lemma. *Let E be a closed set, contained in* $|z|\leqq 1$ *and* $0<\rho<1$. *Then we can cover E by N discs* K_i $(i=1,2,\cdots,N)$ *of radius* ρ *with center* $z_i\in E$, *such that* $N\leqq 30/\rho^2$ *and let* $\widetilde{K}_i$ *be a disc of radius* 2ρ, *concentric with K_i, then $\{\tilde{K}_i\}$ overlap at most m_0-times, where m_0 is a numerical constant.*

Proof. We cover the z-plane by a net of regular triangles $\{\Delta\}$, whose vertex is $z_{mn}=m\rho+n\rho e^{\frac{\pi}{3}i}(m, n=0, \pm 1, \pm 2, \cdots)$. Then $|\Delta|=\frac{\sqrt{3}}{4}\rho^2$, where $|\Delta|$ is the area of Δ. Let $\Delta_1, \cdots, \Delta_N$ be such ones, which contain points of E, then since Δ_i is contained in $|z|\leqq 1+\rho$ and $0<\rho<1$,

$$N\leqq\frac{\pi(1+\rho)^2}{|\Delta|}=\frac{4\pi}{\sqrt{3}}\left(1+\frac{1}{\rho}\right)^2\leqq\frac{4\pi}{\sqrt{3}}\left(\frac{1}{\rho}+\frac{1}{\rho}\right)^2=\frac{16\pi}{\sqrt{3}\,\rho^2}<\frac{30}{\rho^2}.$$

We take a point $z_i\in E$ in Δ_i and draw a disc K_i of radius ρ about z_i, then K_i contains Δ_i, so that $\sum_{i=1}^{N}K_i$ covers E. Let $\tilde{K}_i$ be a disc of radius 2ρ, concentric with K_i, then we see that $\{\tilde{K}_i\}$ overlap at most m_0-times, where m_0 is a numerical constant.

Theorem VII. 7. *Let $f(z)$ be a meromorphic function of finite order $\rho>0$ for $|z|<\infty$. Let C: $z=z(t)$ $(0\leqq t\leqq\infty)$ be a simple curve, which connects $z=0$ to $z=\infty$, such that $z(0)=0$, $z(\infty)=\infty$ and for $\delta>0$, let $\Delta(\delta)$ be the domain, which is covered by discs: $|z-z(t)|<\delta|z(t)|$ $(0\leqq t\leqq\infty)$ and $\Delta_\theta(\delta)$ be the domain, obtained from $\Delta(\delta)$ by rotating an angle θ about $z=0$.*

Let $\{z_\nu(a;\Delta_\theta(\delta))\}$ be zero points of $f(z)-a$ in $\Delta_\theta(\delta)$, multiple zeros being counted only once. Then there exists θ_0, such that for any $\delta>0$,

$$\text{(i)} \qquad \sum_\nu\frac{1}{|z_\nu(a;\Delta_{\theta_0}(\delta))|^{\rho-\varepsilon}}=\infty, \qquad \varepsilon>0.$$

with two possible exceptions for a.

If $f(z)$ is of divergence type, then

$$\sum_\nu\frac{1}{|z_\nu(a;\Delta_{\theta_0}(\delta))|^{\rho}}=\infty,$$

with two possible exceptions for a.

Proof. As the proof of Theorem VII. 6, let $\int^\infty\frac{T(r)}{r^{k+1}}dr=\infty$, $k>0$, then we see easily that there exists θ_0, such that for any $\delta>0$,

$$\int^\infty\frac{T(r,\Delta_\theta(\delta))}{r^{k+1}}dr=\infty. \qquad (1)$$

We put $\Delta_{\theta_0}(\delta)=\Delta(\delta)$ and let $\Delta_\nu(\delta)$ be the part of $\Delta(\delta)$, contained in $2^{\nu-1}\leqq|z|\leqq 2^\nu$ and $\Delta_\nu^0(2\delta)$ be that of $\Delta(2\delta)$, contained in $2^{\nu-2}\leqq|z|\leqq 2^{\nu+1}$. By transforming $\Delta_\nu(\delta)$ into a closed set in $|\zeta|\leqq 1$ by $\zeta=z/2^\nu$ and applying the lemma, $\Delta_\nu(\delta)$ can be covered by $N\leqq 30/\delta_1^2$ discs $K_{\nu,i}$ $(i=1,2,\cdots,N)$ of radius $2^\nu\delta_1$ about $z_\nu^i\in\Delta_\nu(\delta)$ $(0<\delta_1<\delta)$ and the discs $\tilde{K}_{\nu,i}$ of radius $2^{\nu+1}\delta_1$, concentric with $K_{\nu,i}$, overlap at most m_0-times. Let

$$S_\nu=\frac{1}{\pi}\iint\limits_{\Delta_\nu(\delta)}\left(\frac{|f'(z)|}{1+|f(z)|^2}\right)^2 rdrd\theta, \qquad S_{\nu,i}=\frac{1}{\pi}\iint\limits_{K_{\nu,i}}\left(\frac{|f'(z)|}{1+|f(z)|^2}\right)^2 rdrd\theta, \quad (2)$$

and n_ν^0, $n_{\nu,i}^0$ be the number of zero points of $\prod_{i=1}^{3}(f(z)-a_i)$ in $\Delta_\nu^0(2\delta)$, $\tilde{K}_{\nu,i}$ respectively, multiple zeros being counted only once, then

$$\sum_{i=1}^{N} n_{\nu,i}^0 \leqq m_0 n_\nu^0, \quad \text{if } \delta_1>0 \text{ is sufficiently small.}$$

By Theorem VI. 21, $S_{\nu,i}\leqq n_{\nu,i}^0+\text{const.}$, so that

$$S_\nu \leqq \sum_{i=1}^{N} S_{\nu,i} \leqq \sum_{i=1}^{N} n_{\nu,i}^0+O(N)\leqq m_0 n_\nu^0+\text{const.} . \quad (3)$$

From this, we have as the proof of Theorem VII. 3,

$$S(r, \Delta(\delta))\leqq 3m_0\sum_{i=1}^{3} n(2r, a_i; \Delta(2\delta))+O(\log r) , \quad (4)$$

and we can prove the theorem similarly as Theorem VII. 6. ——

3. By means of Theorem VI. 22, we can prove the following theorem, similarly as Theorem VII. 6.

Theorem VII. 8. *Let $w=f(z)$ be a meromorphic function of finite order $\rho>0$ for $|z|<\infty$.*

Then there exists a direction J: $\arg z=\alpha$, which has the following property. Let Δ: $|\arg z-\alpha|<\delta$ be any angular domain, which contains J and $F_r(\Delta)$ be the Riemann surface, generated by $w=f(z)$ on the w-sphere K, when z varies in Δ_r: $|\arg z-\alpha|<\delta$, $0\leqq|z|\leqq r$. Let D be a simply connected domain on K and $n(r, D; \Delta)$ be the number of simply connected islands in $F_r(\Delta)$ above D and $n_1(r, D; \Delta)$ be that of schlicht discs in $F_r(\Delta)$ above D, then

(i) $$\int^\infty \frac{n(r, D; \Delta)}{r^{k+1}}dr=\infty$$

holds for at least one of any three disjoint simply connected domains on K and

(ii) $$\int^\infty \frac{n_1(r, D; \Delta)}{r^{k+1}}dr=\infty$$

holds for at least one of any five disjoint simply connected domains on K, where $k=\rho-\varepsilon$ $(\varepsilon>0)$ in general and $k=\rho$, if $f(z)$ is of divergence type.

J is called *an Ahlfors' direction of $f(z)$.*

We remark that

$$\int^{\infty}\frac{n(r, D)}{r^{\rho+1+\varepsilon}}dr<\infty\,, \qquad \varepsilon>0\,,$$

for any simply connected domain D on K, where n(r, D) is the number of (not necessaly simply connected) islands in F_r above D, F_r being the Riemann surface, generated by $w=f(z)$ on K, when z varies in $|z|\leqq r$.
Dinghas [3].

Proof. By Ahlfors' covering theorem,

$$n(r, D)\leqq S(r)+hL(r)\leqq S(r)+h(L^2(r)+1)\,. \tag{1}$$

Since by Schwarz's inequality,

$$\int_0^r\frac{L^2(r)dr}{r}\leqq 2\pi^2S(r)=2\pi^2rT'(r)\,.$$

we have by (1),

$$N(r, D)=\int^r\frac{n(r, D)dr}{r}\leqq T(r)+O(rT'(r))+O(\log r)\,,$$

so that

$$\int^r\frac{N(r, D)}{r^{\rho+1+\varepsilon}}dr\leqq\int^r\frac{T(r)}{r^{\rho+1+\varepsilon}}dr+O\Bigl(\int^r\frac{T'(r)}{r^{\rho+\varepsilon}}dr\Bigr)+O(1)=O\Bigl(\int^r\frac{T'(r)}{r^{\rho+\varepsilon}}dr\Bigr)+O(1)$$

$$=O\Bigl(\frac{T(r)}{r^{\rho+\varepsilon}}\Bigr)+O\Bigl(\int^r\frac{T(r)}{r^{\rho+1+\varepsilon}}dr\Bigr)+O(1)=O(1)\,,$$

hence by (4) (p. 204)

$$\int^{\infty}\frac{n(r, D)}{r^{\rho+1+\varepsilon}}dr<\infty. \tag{2}$$

3. A lemma

Lemma. *Let $w(z)$, $g_i(z)$ $(i=1, 2, 3, 4)$ be meromorphic functions for* $|z|<\infty$ *and*

$$f(z)=\frac{g_1(z)w(z)+g_2(z)}{g_3(z)w(z)+g_4(z)}\,.$$

Let $\varDelta\subset\varDelta_0$ be two angular domains:

$$\varDelta:\ |\arg z|\leqq\alpha\,, \qquad \varDelta_0:\ |\arg z|\leqq\alpha_0 \qquad (\alpha<\alpha_0)\,,$$

and $\varDelta(r)$, $\varDelta_0(r)$ be the part of $\varDelta$, $\varDelta_0$, which is contained in $1\leqq|z|\leqq r$ respectively and

$$S(f, r; \varDelta)=\frac{1}{\pi}\iint\limits_{\varDelta(r)}\Bigl(\frac{|f'(z)|}{1+|f(z)|^2}\Bigr)^2 rdrd\theta\,, \qquad z=re^{i\theta}\,.$$

Then

$$S(f, r; \varDelta)\leqq 27S(w, 64r; \varDelta_0)+O\Bigl(\int_1^{128r}\frac{T(g, r)}{r}dr\Bigr), \quad T(g, r)=\sum_{i=1}^4 T(g_i, r)\,,$$

where $T(g_i, r)$ is the characteristic function of $g_i(z)$.

Proof. First we consider two special cases:

(i) $f(z)=w(z)+g(z)$, (ii) $f(z)=g(z)w(z)$.

(i) $f(z)=w(z)+g(z)$.

Let $r_\nu=2^\nu$ $(\nu=0, 1, 2\cdots)$ and for $\nu\geqq 2$,

$$\left.\begin{aligned}\Delta_\nu:&\ |\arg z|\leqq\alpha, && r_{\nu-1}\leqq|z|\leqq r_\nu,\\ \Delta_\nu^0:&\ |\arg z|\leqq\alpha_0, && r_{\nu-2}\leqq|z|\leqq r_{\nu+1}.\end{aligned}\right\}\ (\alpha<\alpha_0) \tag{1}$$

If we map Δ_ν^0 conformally on $|\zeta|<1$, such that the center of Δ_ν^0 becomes $\zeta=0$, then the image of Δ_ν is contained in $|\zeta|\leqq\kappa<1$, where κ is a constant, which is independent of ν.

Let D_ν be the image of $|\zeta|\leqq\dfrac{1+\kappa}{2}$ in Δ_ν^0, then $\Delta_\nu\subset D_\nu\subset\Delta_\nu^0$. We put

$$M_\nu=\frac{1}{|D_\nu|}\iint_{D_\nu}\log\sqrt{1+|g|^2}\,rdrd\theta,\qquad z=re^{\theta}, \tag{2}$$

where $|D_\nu|$ is the area of D_ν. Then we see easily that

$$\iint_{|\zeta|\leqq\frac{1+\kappa}{2}}\log\sqrt{1+|g|^2}\,\rho\,d\rho\,d\varphi\leqq AM_\nu,\qquad A=\text{const.},\quad \zeta=\rho e^{i\varphi}. \tag{3}$$

Let $B>0$ and E be the set of ρ in $\left[\kappa, \dfrac{1+\kappa}{2}\right]$, such that

$$\int_{|\zeta|=\rho}\log\sqrt{1+|g|^2}\,d\varphi>BM_\nu,$$

then

$$BM_\nu\int_E d\rho\leqq\int_E d\rho\int_{|\zeta|=\rho}\log\sqrt{1+|g|^2}\,d\varphi\leqq\frac{1}{\kappa}\int_E\rho\,d\rho\int_{|\zeta|=\rho}\log\sqrt{1+|g|^2}\,d\varphi$$
$$\leqq\frac{1}{\kappa}\iint_{|\zeta|\leqq\frac{1+\kappa}{2}}\log\sqrt{1+|g|^2}\,\rho\,d\rho\,d\varphi\leqq\frac{AM_\nu}{\kappa},$$

so that

$$\int_E d\rho<\frac{A}{B\kappa},$$

hence if $B>0$ is sufficiently large, then there exist $\rho_1<\rho_2$, such that

$$\kappa\leqq\rho_1\leqq\kappa+\frac{1-\kappa}{6},\qquad \kappa+\frac{2(1-\kappa)}{6}\leqq\rho_2\leqq\frac{1+\kappa}{2}, \tag{4}$$

$$\int_{|\zeta|=\rho_1}\log\sqrt{1+|g|^2}\,d\varphi\leqq BM_\nu,\qquad \int_{|\zeta|=\rho_2}\log\sqrt{1+|g|^2}\,d\varphi\leqq BM_\nu. \tag{5}$$

Let

$$\left.\begin{aligned} S(f;\Delta_\nu)&=\frac{1}{\pi}\iint_{\Delta_\nu}\left(\frac{|f'(z)|}{1+|f(z)|^2}\right)^2 r\,dr\,d\theta\,, \\ S(F,\rho)&=\frac{1}{\pi}\iint_{|\zeta|\leqq\rho}\left(\frac{F'(\zeta)}{1+|F(\zeta)|^2}\right)^2 \rho\,d\rho\,d\varphi\,, \end{aligned}\right\}\quad F(\zeta)=f(z), \tag{6}$$

then

$$S(f;\Delta_\nu)\leqq S(F,\rho_1)\leqq S(F,\rho_2)\leqq S(f;\Delta_\nu^0)\,.$$

By Nevanlinna's first fundamental theorem,

$$\begin{aligned} \int_{\rho_1}^{\rho_2}\frac{S(F,\rho)}{\rho}d\rho&=\frac{1}{2\pi}\int_{|\zeta|=\rho_2}\log\sqrt{1+|w+g|^2}\,d\varphi-\frac{1}{2\pi}\int_{|\zeta|=\rho_1}\log\sqrt{1+|w+g|^2}\,d\varphi \\ &\quad+\int_{\rho_1}^{\rho_2}\frac{n(w+g,\rho,\infty)}{\rho}d\rho\,, \end{aligned}\tag{7}$$

$$\begin{aligned} \int_{\rho_1}^{\rho_2}\frac{S(w,\rho)}{\rho}d\rho&=\frac{1}{2\pi}\int_{|\zeta|=\rho_1}\log\sqrt{1+|w|^2}\,d\varphi-\frac{1}{2\pi}\int_{|\zeta|=\rho_1}\log\sqrt{1+|w|^2}\,d\varphi \\ &\quad+\int_{\rho_1}^{\rho_2}\frac{n(w,\rho,\infty)}{\rho}d\rho\,. \end{aligned}\tag{8}$$

Now by (5),

$$\begin{aligned} \frac{1}{2\pi}\int_{|\zeta|=\rho_2}\log\sqrt{1+|w+g|^2}\,d\varphi&\leqq\frac{1}{2\pi}\int_{|\zeta|=\rho_2}\log\sqrt{1+|w|^2}\,d\varphi \\ +\frac{1}{2\pi}\int_{|\zeta|=\rho_2}\log\sqrt{1+|g|^2}\,d\varphi+1&\leqq\frac{1}{2\pi}\int_{|\zeta|=\rho_2}\log\sqrt{1+|w|^2}\,d\varphi+O(M_\nu)+1\,, \end{aligned}$$

$$\begin{aligned} \frac{1}{2\pi}\int_{|\zeta|=\rho_1}\log\sqrt{1+|w|^2}\,d\varphi&=\frac{1}{2\pi}\int_{|\zeta|=\rho_1}\log\sqrt{1+|w+g-g|^2}\,d\varphi \\ &\leqq\frac{1}{2\pi}\int_{|\zeta|=\rho_1}\log\sqrt{1+|w+g|^2}\,d\varphi+\frac{1}{2\pi}\int_{|\zeta|=\rho_1}\log\sqrt{1+|g|^2}\,d\varphi+1 \\ &\leqq\frac{1}{2\pi}\int_{|\zeta|=\rho_1}\log\sqrt{1+|w+g|^2}\,d\varphi+O(M_\nu)+1\,, \end{aligned}$$

so that

$$\begin{aligned} &\frac{1}{2\pi}\int_{|\zeta|=\rho_2}\log\sqrt{1+|w+g|^2}\,d\varphi-\frac{1}{2\pi}\int_{|\zeta|=\rho_1}\log\sqrt{1+|w+g|^2}\,d\varphi \\ &\leqq\frac{1}{2\pi}\int_{|\zeta|=\rho_2}\log\sqrt{1+|w|^2}\,d\varphi-\frac{1}{2\pi}\int_{|\zeta|=\rho_1}\log\sqrt{1+|w|^2}\,d\varphi+O(M_\nu)+2\,, \end{aligned}$$

hence by (7), (8),

$$\begin{aligned} \int_{\rho_1}^{\rho_2}\frac{S(F,\rho)}{\rho}d\rho\leqq\int_{\rho_1}^{\rho_2}\frac{S(w,\rho)}{\rho}d\rho+\int_{\rho_1}^{\rho_2}\frac{n(w+g,\rho,\infty)}{\rho}d\rho-\int_{\rho_1}^{\rho_2}\frac{n(w,\rho,\infty)}{\rho} \\ +O(M_\nu)+2\,, \end{aligned}\tag{9}$$

$$\int_{\rho_1}^{\rho_2}\frac{n(w+g,\rho,\infty)}{\rho}d\rho\leqq\int_{\rho_1}^{\rho_2}\frac{n(w,\rho,\infty)}{\rho}d\rho+\int_{\rho_1}^{\rho_2}\frac{n(g,\rho,\infty)}{\rho}d\rho$$

$$\leqq \int_{\rho_1}^{\rho_2} \frac{n(w, \rho, \infty)}{\rho} d\rho + O(n(g, \rho_2, \infty)) , \tag{10}$$

where $n(g, \rho, \infty)$ is the number of poles of g in $|\zeta| \leqq \rho$.

Now the image of $|\zeta| \leqq \rho_2$ in Δ_ν^0 is contained in $|z| \leqq \tau_\nu < r_{\nu+1}$, such that $r_{\nu+1}/\tau_\nu \geqq \text{const.} > 1$, so that if we denote the number of poles of $g(z)$ in $|z| \leqq r$ by $n(g, r, \infty)$, then $n(g, \rho_2, \infty) \leqq n(g, \tau_\nu, \infty)$, so that by (9), (10),

$$\int_{\rho_1}^{\rho_2} \frac{S(F, \rho)}{\rho} d\rho \leqq \int_{\rho_1}^{\rho_2} \frac{S(w, \rho)}{\rho} d\rho + O(M_\nu) + O(n(g, \tau_\nu, \infty)) + 2 .$$

Since

$$S(F, \rho_1) \log \frac{\rho_2}{\rho_1} \leqq \int_{\rho_1}^{\rho_2} \frac{S(F, \rho)}{\rho} d\rho , \qquad \int_{\rho_1}^{\rho_2} \frac{S(w, \rho)}{\rho} d\rho \leqq S(w, \rho_2) \log \frac{\rho_2}{\rho_1} ,$$

we have

$$S(F, \rho_1) \leqq S(w, \rho_2) + O(M_\nu) + O(n(g, \tau_\nu, \infty)) + O(1) .$$

Since $S(f, \Delta_\nu) \leqq S(F, \rho_1)$, $S(w, \rho_2) \leqq S(w, \Delta_\nu^0)$, we have

$$S(f, \Delta_\nu) \leqq S(w, \Delta_\nu^0) + O(M_\nu) + O(n(g, \tau_\nu, \infty)) + O(1) . \tag{11}$$

Now

$$M_\nu = \frac{1}{|D_\nu|} \iint_{D_\nu} \log \sqrt{1+|g|^2}\, r dr d\theta \leqq \frac{1}{|D_\nu|} \iint_{\Delta_\nu^0} \log \sqrt{1+|g|^2}\, r dr d\theta$$

$$\leqq \frac{\text{const.}}{4^\nu} \int_{r_{\nu-2}}^{r_{\nu+1}} r\, dr \int_{|z|=r} \log \sqrt{1+|g|^2}\, d\theta \leqq \text{const.}\ T(g, r_{\nu+1}) , \tag{12}$$

and

$$T(g, r_{\nu+1}) + O(1) \geqq \int_{\tau_\nu}^{r_{\nu+1}} \frac{n(g, r, \infty)}{r} dr \geqq n(g, \tau_\nu, \infty) \log \frac{r_{\nu+1}}{\tau_\nu}$$

$$\geqq \text{const.}\ n(g, \tau_\nu, \infty) ,$$

so that

$$n(g, \tau_\nu, \infty) \leqq \text{const.}\ T(g, r_{\nu+1}) , \tag{13}$$

hence

$$\int_{r_{\nu+1}}^{r_{\nu+2}} \frac{T(g, r)}{r} dr \geqq \text{const.}\ T(g, r_{\nu+1}) \geqq \text{const.}\ n(g, \tau_\nu, \infty) , \tag{14}$$

so that from (11), (12), (14), we have

$$S(f, \Delta_\nu) \leqq S(w, \Delta_\nu^0) + O\left(\int_{r_{\nu+1}}^{r_{\nu+2}} \frac{T(g, r)}{r} dr\right), \qquad f = w + g . \tag{15}$$

(ii) $f(z) = g(z)\, w(z)$.

We choose ρ_1 in $\left[\kappa, \kappa + \frac{1-\kappa}{6}\right]$ and ρ_2 in $\left[\kappa + \frac{2(1-\kappa)}{6}, \frac{1+\kappa}{2}\right]$ as before, such that

$$\int_{|\zeta|=\rho_2} \log \sqrt{1+|g|^2}\, d\varphi = O(M_\nu) , \qquad \int_{|\zeta|=\rho_1} \log \sqrt{1+\frac{1}{|g|^2}}\, d\varphi = O(M_\nu^*) , \tag{16}$$

where

$$M_\nu^* = \frac{1}{|D_\nu|}\iint_{D_\nu} \log\sqrt{1+\frac{1}{|g|^2}}\, r dr d\theta \,. \tag{17}$$

Then

$$\frac{1}{2\pi}\int_{|\zeta|=\rho_2} \log\sqrt{1+|wg|^2}\, d\varphi \leqq \frac{1}{2\pi}\int_{|\zeta|=\rho_2} \log\sqrt{1+|w|^2}\, d\varphi$$
$$+\frac{1}{2\pi}\int_{|\zeta|=\rho_2} \log\sqrt{1+|g|^2}\, d\varphi \leqq \frac{1}{2\pi}\int_{|\zeta|=\rho_2} \log\sqrt{1+|w|^2}\, d\varphi + O(M_\nu)\,,$$
$$\frac{1}{2\pi}\int_{|\zeta|=\rho_1} \log\sqrt{1+|w|^2}\, d\varphi = \frac{1}{2\pi}\int_{|\zeta|=\rho_1} \log\sqrt{1+\left|\frac{wg}{g}\right|^2}\, d\varphi$$
$$\leqq \frac{1}{2\pi}\int_{|\zeta|=\rho_1} \log\sqrt{1+|wg|^2}\, d\varphi + \frac{1}{2\pi}\int_{|\zeta|=\rho_1} \log\sqrt{1+\frac{1}{|g|^2}}\, d\varphi$$
$$\leqq \frac{1}{2\pi}\int_{|\zeta|=\rho_1} \log\sqrt{1+|wg|^2}\, d\varphi + O(M_\nu^*)\,,$$

so that

$$\frac{1}{2\pi}\int_{|\zeta|=\rho_2} \log\sqrt{1+|wg|^2}\, d\varphi - \frac{1}{2\pi}\int_{|\zeta|=\rho_1} \log\sqrt{1+|wg|^2}\, d\varphi$$
$$\leqq \frac{1}{2\pi}\int_{|\zeta|=\rho_2} \log\sqrt{1+|w|^2}\, d\varphi - \frac{1}{2\pi}\int_{|\zeta|=\rho_1} \log\sqrt{1+|w|^2}\, d\varphi$$
$$+O(M_\nu)+O(M_\nu^*)\,. \tag{18}$$

Since as (12),

$$M_\nu^* \leqq \text{const. } T\left(\frac{1}{g}, r_{\nu+1}\right) = \text{const. } T(g, r_{\nu+1})\,,$$

we have (15), so that

$$S(f; \Delta_\nu) \leqq S(w; \Delta_\nu^0) + O\left(\int_{r_{\nu+1}}^{r_{\nu+2}} \frac{T(g, r)}{r} dr\right), \qquad f=w+g\,, \quad f=gw\,. \tag{19}$$

If we sum up (19) for $\nu=1, 2, \cdots, n$, then since Δ_ν^0 overlap at most 3-times,

$$S(f, r_n; \Delta) \leqq 3S(w, r_{n+1}; \Delta_0) + O\left(\int_1^{r_{n+2}} \frac{T(g, r)}{r} dr\right).$$

If $r_{n-1} \leqq r \leqq r_n$, then $S(f, r; \Delta) \leqq S(f, r_n; \Delta)$ and since $r_{n+1} \leqq 4r$, $r_{n+2} \leqq 8r$,

$$S(f, r; \Delta) \leqq 3S(w, 4r; \Delta_0) + O\left(\int_1^{8r} \frac{T(g, r)}{r} dr\right), \quad f=w+g\,, \quad f=gw\,. \tag{20}$$

(iii) Now we consider the general case: $f=\dfrac{g_1 w+g_2}{g_3 w+g_4}$. Then

$$f = h_1 + \frac{h_2}{w+h_3}, \quad \text{where } h_1 = \frac{g_1}{g_3}, \quad h_2 = \frac{g_2 g_3 - g_1 g_4}{g_3^2}, \quad h_3 = \frac{g_4}{g_3}, \tag{21}$$

so that

$$f=w_1+h_1\,,\quad w_1=h_2w_2\,,\quad w_2=1/w_3\,,\quad w_3=w+h_3\,. \tag{22}$$

Let

$$\left.\begin{array}{ll} \varDelta\ :\ |\arg z|\leqq\alpha\ , & \varDelta_1:\ |\arg z|\leqq\alpha_1\ , \\ \varDelta_2:\ |\arg z|\leqq\alpha_2\ , & \varDelta_0:\ |\arg z|\leqq\alpha_0\ , \end{array}\right\}\ (\alpha<\alpha_1<\alpha_2<\alpha_0) \tag{23}$$

then by (20),

$$S(f,r;\varDelta)\quad\leqq 3S(w_1,r_1;\varDelta_1)+O\left(\int_1^{2r_1}\frac{T(h_1,r)}{r}dr\right),\quad r_1=4r\,,$$

$$S(w_1,r_1;\varDelta_1)\leqq 3S(w_2,r_2;\varDelta_2)+O\left(\int_1^{2r_2}\frac{T(h_2,r)}{r}dr\right),\quad r_2=4r_1\,,$$

$$S(w_2,r_2;\varDelta_2)=S(w_3,r_2;\varDelta_2)\,,$$

$$S(w_3,r_2;\varDelta_2)\leqq 3S(w,r_3;\varDelta_0)+O\left(\int_1^{2r_3}\frac{T(h_3,r)}{r}dr\right),\quad r_3=4r_2=64r\,.$$

Since $T(h_i,r)=O(T(g,r))$ $(i=1,2,3)$, we have

$$S(f,r;\varDelta)\leqq 27S(w,64r;\varDelta_0)+O\left(\int_1^{128r}\frac{T(g,r)}{r}dr\right). \tag{24}$$

4. Biernacki-Rauch's theorem

Theorem VII. 9. *Let $f(z)$ be a meromorphic function of finite order $\rho>0$ for $|z|<\infty$, then there exists a direction J: $\arg z=\alpha$, which satisfies the following condition. Let $\varDelta$: $|\arg z-\alpha|<\delta$ be any small angular domain, which contains J and $g(z)$ be a meromorphic function for $|z|<\infty$ and $\{z_\nu(f=g;\varDelta)\}$ be zero points of $f(z)-g(z)$, contained in $\varDelta$, multiple zeros being counted only once.*

(i) *If $g(z)$ is of order $<\rho$, then*

$$\sum_\nu\frac{1}{|z_\nu(f=g;\varDelta)|^{\rho-\varepsilon}}=\infty\,,\qquad \varepsilon>0\,,$$

with two possible exceptions for g. Biernacki [1].

(ii) *If $f(z)$ is of divergence type and $\int^\infty\frac{T(g,r)}{r^{\rho+1}}dr<\infty$, then*

$$\sum_\nu\frac{1}{|z\ (f=g;\varDelta)|^{\rho}}=\infty\,,$$

with two possible exceptions for g. Rauch [1].

Proof. (Tsuji [20] (III)). Now $\int^\infty\frac{T(f,r)}{r^{k+1}}\,dr=\infty$, $k>0$, where $k=\rho-\varepsilon$ $(\varepsilon>0)$ in general and $k=\rho$, if $f(z)$ is of divergence type, then as the proof of Theorem VII. 4, there exists a direction J: $\arg z=\alpha$, such that for any small angular domain $\varDelta$, which contains J,

$$\int^{\infty}\frac{T(f,r;\Delta)}{r^{k+1}}dr=\infty. \tag{1}$$

We suppose that $\alpha=0$ and let

$$\Delta:\ |\arg z|<\delta\,,\quad \Delta_1:\ |\arg z|<2\delta\,,\quad \Delta_0:\ |\arg z|<3\delta\quad (\delta>0). \tag{2}$$

Let $g_i(z)$ $(i=1,2,3)$ be meromorphic functions for $|z|<\infty$, such that $\int^{\infty}\frac{T(g_i,r)}{r^{k+1}}dr<\infty$, then if we put $T(g,r)=\sum_{i=1}^{3}T(g_i,r)$, then

$$\int^{\infty}\frac{T(g,r)}{r^{k+1}}dr<\infty. \tag{3}$$

We shall prove that for one of g_i,

$$\sum_{\nu}\frac{1}{|z_\nu(f=g_i;\Delta_0)|^k}=\infty\,.$$

We put

$$w(z)=\frac{f(z)-g_1(z)}{f(z)-g_3(z)}\cdot\frac{g_2(z)-g_3(z)}{g_2(z)-g_1(z)}\,, \tag{4}$$

then

$$f(z)=\frac{h_1(z)w(z)+h_2(z)}{h_3(z)w(z)+h_4(z)}\,, \tag{5}$$

where

$$h_1=g_3(g_2-g_1)\,,\quad h_2=g_1(g_3-g_2)\,,\quad h_3=g_2-g_1\,,\quad h_4=g_3-g_2\,,$$

so that $T(h_i,r)=O(T(g,r))$ $(i=1,2,3,4)$. Hence by the lemma,

$$S(f,r;\Delta)\leqq 27S(w,64r;\Delta_1)+O(\Phi(r))\,, \tag{6}$$

$$\Phi(r)=\int_1^{128r}\frac{T(g,r)}{r}dr\,,$$

so that

$$T(f,r;\Delta)\leqq 27T(w,64r;\Delta_1)+O(\Psi(r))\,, \tag{7}$$

$$\Psi(r)=\int_1^r\frac{\Phi(r)}{r}dr\,.$$

We shall prove that

$$\int_1^r\frac{\Psi(r)}{r^{k+1}}dr=O(1)\,. \tag{8}$$

For

$$\int_1^r\frac{\Psi(r)}{r^{k+1}}dr=\left[\frac{-\Psi(r)}{kr^k}\right]_1^r+\frac{1}{k}\int_1^r\frac{\Phi(r)}{r^{k+1}}dr\leqq\frac{1}{k}\int_1^r\frac{\Phi(r)}{r^{k+1}}dr$$
$$=\frac{1}{k}\left[-\frac{\Phi(r)}{kr^k}\right]_1^r+\frac{1}{k^2}\int_1^r\frac{\Phi'(r)}{r^k}dr$$
$$\leqq\frac{1}{k^2}\int_1^r\frac{\Phi'(r)}{r^k}dr\leqq\text{const.}\int_1^{128r}\frac{T(g,r)}{r^{k+1}}dr=O(1)\,,$$

so that by (7), (1),

$$\int^{\infty}\frac{T(w,r;\Delta_1)}{r^{k+1}}dr=\infty\,. \tag{9}$$

Since by Theorem VII. 3,

$T(w,r;\Delta_1)\leqq 3[N(w,2r,0;\Delta_0)+N(w,2r,1;\Delta_0)+N(w,2r,\infty;\Delta_0)]+O((\log r)^2)$,

we see from (9), that one of

$$\int^{\infty}\frac{N(w,r,0;\Delta_0)}{r^{k+1}}dr\,,\quad \int^{\infty}\frac{N(w,r,1;\Delta_0)}{r^{k+1}}dr\,,\quad \int^{\infty}\frac{N(w,r,\infty;\Delta_0)}{r^{k+1}}dr$$

is ∞. Without loss of generality, we assume that

$$\int^{\infty}\frac{N(w,r,0;\Delta_0)}{r^{k+1}}dr=\infty. \tag{10}$$

Since

$$n(w,r,0;\Delta_0)\leqq n(r,f=g_1;\Delta_0)+n(r,g_2=g_3;\Delta_0)+\sum_{i=1}^{3}n(r,g_i=\infty;\Delta_0)\,,$$

we have

$$\begin{aligned}N(w,r,0;\Delta_0)&\leqq N(r,f=g_1;\Delta_0)+N(r,g_2=g_3;\Delta_0)+\sum_{i=1}^{3}N(r,g_i=\infty;\Delta_0)\\&\leqq N(r,f=g_1;\Delta_0)+O(T(g,r))\,,\end{aligned}$$

so that by (10) and the hypothesis on g,

$$\int^{\infty}\frac{N(r,f=g_1;\Delta_0)}{r^{k+1}}dr=\infty\,,\quad\text{or}\quad \sum_{\nu}\frac{1}{|z_\nu(f=g_1;\Delta_0)|^k}=\infty\,. \tag{11}$$

Hence for any small angular domain Δ_0, which contains J,

$$\sum_{\nu}\frac{1}{|z\,(f=g;\Delta_0)|^k}=\infty\,, \tag{12}$$

with two possible exceptions for g.

If $f(z)$ is of divergence type, then $k=\rho$, so that J satisfies the condition of the theorem. If $f(z)$ is of convergence type, we take $k=\rho-\varepsilon_n$ ($\varepsilon_1>\varepsilon_2>\cdots>\varepsilon_n\to 0$) and let J_n be the corresponding direction and J be one of the limiting position of $\{J_n\}$, then J satisfies the condition of the theorem.

5. Meromorphic function in an angular domain

1. Let $w(z)$ be meromorphic in an angular domain Δ_0: $|\arg z|\leqq\alpha_0$, where $|\Delta_0|=2\alpha_0$ is its magnitude. The order ρ of $w(z)$ in Δ_0 is defined by

$$\varlimsup_{r\to\infty}\frac{\log T(w,r;\Delta_0)}{\log r}=\rho\,. \tag{1}$$

If $\rho<\infty$, then

$$\int^{\infty}\frac{T(w,r;\Delta_0)}{r^{\rho+1+\varepsilon}}dr<\infty\,,\quad \int^{\infty}\frac{T(w,r;\Delta_0)}{r^{\rho+1-\varepsilon}}dr=\infty\,,\qquad \varepsilon>0\,. \tag{2}$$

Lemma. *Let $w(z)$ be meromorphic in an angular domain $\Delta_0: |\arg z| \leqq \alpha_0$ and $\Delta: |\arg z| \leqq \alpha\ (<\alpha_0)$ be an angular domain, contained in Δ_0, then for $1 \leqq r < \infty$,*

$$\int_1^r \frac{N(w, r, a; \Delta)}{r^{k+1}} dr \leqq \text{const.} \int_1^{\lambda r} \frac{T(w, r; \Delta_0)}{r^{k+1}} dr + \text{const.}, \quad k \geqq k_0 = \frac{\pi}{|\Delta_0|},$$

where
$$\lambda = 4/\cos\left(\frac{\pi}{2} \cdot \frac{\alpha}{\alpha_0}\right).$$

Proof. First we assume that $\alpha_0 = \pi/2$, so that Δ_0 is a half-plane. By $\zeta = \frac{z-1}{z+1}$, $z = \frac{1+\zeta}{1-\zeta}$, we map Δ_0 on $|\zeta| < 1$ and put $w(z) = v(\zeta)$,

$$S(v, \rho) = \frac{1}{\pi} \iint_{|\zeta| \leqq \rho} \left(\frac{|v'(\zeta)|}{1 + |v(\zeta)|^2} \right)^2 \rho\, d\rho\, d\varphi, \qquad 0 \leqq \rho < 1, \quad \zeta = \rho e^{i\varphi}. \tag{1}$$

The image of $|\zeta| = \rho$ is a circle $\left|\frac{z-1}{z+1}\right| = \rho$, which is contained in a sector $\Delta_0(r)$: $|\arg z| < \frac{\pi}{2}$, $|z| < r$, where $r = \frac{1+\rho}{1-\rho}$, $\rho = \frac{r-1}{r+1}$, so that $S(v, \rho) \leqq S(w, r; \Delta_0)$ and

$$\text{const.} \frac{dr}{r^2} \leqq \frac{d\rho}{\rho} \leqq \text{const.} \frac{dr}{r^2} \quad (1/2 \leqq \rho < 1), \tag{2}$$

hence

$$\int_{\frac{1}{2}}^{\rho} \frac{S(v, \rho)}{\rho} d\rho \leqq \text{const.} \int_1^r \frac{S(w, r; \Delta_0)}{r^2} dr. \tag{3}$$

Let $n(v, \rho, a)$ be the number of zero points of $v(\zeta) - a$ in $|\zeta| < \rho$, then by the first fundamental theorem,

$$\int_{\frac{1}{2}}^{\rho} \frac{n(v, \rho, a)}{\rho} d\rho \leqq T(v, \rho) + \text{const.} = \int_0^{\rho} \frac{S(v, \rho)}{\rho} d\rho + \text{const.}$$
$$\leqq \text{const.} \int_1^r \frac{S(w, r; \Delta_0)}{r^2} dr + \text{const.}. \tag{4}$$

If $1-\rho$ is sufficiently small, then we see easily that the image of $|\zeta| < \rho$ contains a domain: $|\arg z| \leqq \alpha\left(<\frac{\pi}{2}\right)$, $r_0 \leqq |z| \leqq \frac{r}{2} \cos\alpha$, with a suitable $r_0 > 0$, hence if Δ: $|\arg z| \leqq \alpha$, then

$$n(v, \rho, a) \geqq n\left(w, \frac{r}{2}\cos\alpha, a; \Delta\right) - \text{const.},$$

so that by (2),

$$\int_{\frac{1}{2}}^{\rho} \frac{n(v, \rho, a)}{\rho} d\rho \geqq \text{const.} \int_1^r \frac{n\left(w, \frac{r}{2}\cos\alpha, a; \Delta\right)}{r^2} dr - \text{const.},$$

hence by (4),

$$\int_1^r \frac{n\left(w,\ \frac{r}{2}\cos\alpha,\ a;\ \Delta\right)}{r^2}dr \leqq \text{const.}\int_1^r \frac{S(w,\ r;\ \Delta_0)}{r^2}dr+\text{const.},$$

or

$$\int_1^r \frac{n(w,\ r,\ a;\ \Delta)}{r^2}dr \leqq \text{const.}\int_1^{\mu r} \frac{S(w,\ r;\ \Delta_0)}{r^2}dr+\text{const.}. \qquad (5)$$

$$\mu=2/\cos\alpha\ .$$

If $\alpha_0 \neq \pi/2$, then by $x=z^{\frac{\pi}{2\alpha_0}}$, we map Δ_0 on a half-plane Δ_0^* : $|\arg x|<\pi/2$ on the x-plane, then Δ is mapped on an angular domain Δ^* : $|\arg x|<\beta\left(\beta=\frac{\pi}{2}\cdot\frac{\alpha}{\alpha_0}\right)$. We put $w(z)=u(x)$, $z=re^{i\theta}$, $x=te^{i\varphi}$, then by (5),

$$\int_1^t \frac{n(u,\ t,\ a;\ \Delta^*)}{t^2}dt \leqq \text{const.}\int_1^{\kappa t} \frac{S(u,\ t;\ \Delta_0^*)}{t^2}dt+\text{const.},\quad \kappa=2\Big/\cos\left(\frac{\pi}{2}\cdot\frac{\alpha}{\alpha_0}\right).$$

Since $t=r^{k_0}\left(k_0=\frac{\pi}{2\alpha_0}\right)$, we have

$$\int_1^r \frac{n(w,\ r,\ a;\ \Delta)}{r^{k_0+1}}dr \leqq \text{const.}\int_1^{\kappa r} \frac{S(w,\ r;\ \Delta_0)}{r^{k_0+1}}dr+\text{const.}.$$

Hence if we put

$$N_0(r)=\int_1^r \frac{n(w,\ r,\ a;\ \Delta)}{r^{k_0+1}}dr,\qquad T_0(r)=\int_1^r \frac{S(w,\ r;\ \Delta_0)}{r^{k_0+1}}dr, \qquad (6)$$

then

$$N_0(r) \leqq \text{const.}\ T_0(\kappa r)+\text{const.}. \qquad (7)$$

If $k=k_0+\delta$ $(\delta>0)$, then by (7)

$$\int_1^r \frac{n(w,\ r,\ a;\ \Delta)}{r^{k+1}}dr=\int_1^r \frac{dN_0(r)}{r^\delta}=\left[\frac{N_0(r)}{r^\delta}\right]_1^r+\delta\int_1^r \frac{N_0(r)}{r^{\delta+1}}dr$$

$$\leqq \text{const.}\ \frac{T_0(\kappa r)}{r^\delta}+\text{const.}\int_1^{\kappa r} \frac{T_0(r)}{r^{\delta+1}}dr+\text{const.}. \qquad (8)$$

$$\int_1^{\kappa r} \frac{S(w,\ r;\ \Delta_0)}{r^{k+1}}dr=\int_1^{\kappa r} \frac{dT_0(r)}{r^\delta}=\left[\frac{T_0(r)}{r^\delta}\right]_1^{\kappa r}+\delta\int_1^{\kappa r} \frac{T_0(r)}{r^{\delta+1}}dr$$

$$\geqq \frac{T_0(\kappa r)}{(\kappa r)^\delta}+\delta\int_1^{\kappa r} \frac{T_0(r)}{r^{\delta+1}}dr-\text{const.}.$$

Hence by (8), (7),

$$\int_1^r \frac{n(w,\ r,\ a;\ \Delta)}{r^{k+1}}dr \leqq \text{const.}\int_1^{\kappa r} \frac{S(w,\ r;\ \Delta_0)}{r^{k+1}}dr+\text{const.}.\quad (\kappa \geqq \kappa_0). \qquad (9)$$

Now by the definition,

$$N(w,\ r,\ a;\ \Delta)=\int_1^r \frac{n(w,\ r,\ a;\ \Delta)}{r}dr,\qquad T(w,\ r;\ \Delta_0)=\int_1^r \frac{S(w,\ r;\ \Delta_0)}{r}dr, \qquad (10)$$

so that by (9),

$$\int_1^r \frac{N(w, r, a; \Delta)}{r^{k+1}} dr = \left[\frac{-N(w, r, a; \Delta)}{kr^k} \right]_1^r + \frac{1}{k} \int_1^r \frac{n(w, r, a; \Delta)}{r^{k+1}} dr$$

$$\leqq \frac{1}{k} \int_1^r \frac{n(w, r, a; \Delta)}{r^{k+1}} dr + \text{const.} \leqq \text{const.} \int_1^{\kappa r} \frac{S(w, r; \Delta_0)}{r^{k+1}} dr + \text{const.}, \quad (11)$$

$$\int_1^{\kappa r} \frac{T(w, r; \Delta_0)}{r^{k+1}} dr = \left[\frac{-T(w, r; \Delta_0)}{kr^k} \right]_1^{\kappa r} + \frac{1}{k} \int_1^{\kappa r} \frac{S(w, r; \Delta_0)}{r^{k+1}} dr,$$

hence by (11),

$$\int_1^r \frac{N(w, r, a; \Delta)}{r^{k+1}} dr \leqq \text{const.} \frac{T(w, \kappa r; \Delta_0)}{r^k} + \text{const.} \int_1^{\kappa r} \frac{T(w, r; \Delta_0)}{r^{k+1}} dr. \quad (12)$$

Since

$$\frac{T(w, \kappa r; \Delta_0)}{r^k} \leqq \text{const.} \int_1^{2\kappa r} \frac{T(w, r; \Delta_0)}{r^{k+1}} dr,$$

if we put $\lambda = 2\kappa$, then we have from (12) and (7), if $k \geqq k_0$,

$$\int_1^r \frac{N(w, r, a; \Delta)}{r^{k+1}} dr \leqq \text{const.} \int_1^{\lambda r} \frac{T(w, r; \Delta_0)}{r^{k+1}} dr + \text{const.} \text{ ——} \quad (13)$$

2. From the lemma and 4 (p. 204), we have

Theorem VII. 10. *If $w(z)$ is a meromorphic function in an angular domain Δ_0: $|\arg z| \leqq \alpha_0$, such that*

$$\int^\infty \frac{T(w, r; \Delta_0)}{r^{k+1}} dr < \infty, \qquad k \geqq \pi/|\Delta_0|,$$

then for any a,

$$\sum_\nu \frac{1}{|z_\nu(a; \Delta)|^k} < \infty,$$

where $\{z_\nu(a, \Delta)\}$ are zero points of $w(z) - a$, contained in an angular domain Δ: $|\arg z| < \alpha$ $(< \alpha_0)$.

Theorem VII. 11. *Let $w(z)$ be meromorphic in an angular domain Δ_0: $|\arg z| \leqq \alpha_0$ and Δ: $|\arg z| \leqq \alpha$ $(< \alpha_0)$ be an angular domain, contained in Δ_0.*

If for a certain a_0,

$$\sum_\nu \frac{1}{|z_\nu(a_0; \Delta)|^\rho} = \infty, \qquad \rho > \pi/|\Delta_0|,$$

then Δ contains a Borel's direction of order ρ of divergence type[1].

Valiron [3].

1) A direction J: $\arg z = \alpha$ is called *a Borel's direction of order ρ of divergence type*, if for any small angular domain ω, which contains J, $\sum_\nu \frac{1}{|z_\nu(a; \omega)|^\rho} = \infty$, with two possible exceptions for a.

Proof. Let $\varDelta_1$: $|\arg z| \leqq \alpha_1$ $(\alpha<\alpha_1<\alpha_0)$ be an angular domain, such that $\rho>\pi/|\varDelta_1|$. Since by the hypothesis, $\int^{\infty} \frac{N(w, r, a_0; \varDelta)}{r^{\rho+1}} dr=\infty$, we have by the lemma,

$$\int^{\infty} \frac{T(w, r; \varDelta_1)}{r^{\rho+1}} dr=\infty .$$

From this, we can prove the theorem similarly as Theorem VII. 6.

Part II. Meromorphic functions in $|z|<1$

1. Analogue of the second fundamental theorem for an angular domain

1. Let $w(z)$ be meromorphic in $|z|<1$ and $\varDelta \subset \varDelta_0$ be two angular domains, each of which is bounded by two lines, through $z=1$, which do not touch $|z|=1$ and $\varDelta(r)$, $\varDelta_0(r)$ be the part of $\varDelta$, $\varDelta_0$, which lies in $r_0 \leqq |z| \leqq r<1$ respectively, where $r_0 \geqq 1/2$ is so chosen that the circle $|z|=r_0$ meets the both sides of $\varDelta$ and $\varDelta_0$. We put

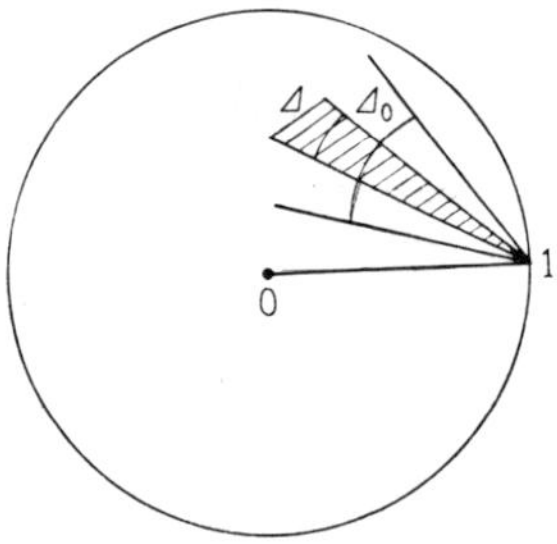

$$\left.\begin{aligned} S(r; \varDelta) &= \frac{1}{\pi} \iint_{\Delta(r)} \left(\frac{|w'(z)|}{1+|w(z)|^2}\right)^2 r dr d\theta , \\ T(r; \varDelta) &= \int_{r_0}^{r} \frac{S(r)}{r} dr , \qquad z=re^{i\theta}. \end{aligned}\right\} \quad (1)$$

Let $n(r, a; \varDelta_0)$ be the number of zero points of $w(z)-a$ in $\varDelta_0(r)$, multiple zeros being counted only once and

$$N(r, a; \varDelta_0)=\int_{r_0}^{r} \frac{n(r, a; \varDelta_0)}{r} dr . \qquad (2)$$

Then we shall prove

Theorem VII. 12. $S(r; \varDelta) \leqq 3 \sum_{i=1}^{3} n\left(\frac{r+3}{4}, a_i; \varDelta_0\right)+O\left(\log \frac{1}{1-r}\right)$,

$$T(r; \varDelta) \leqq 21 \sum_{i=1}^{3} N\left(\frac{r+3}{4}, a_i; \varDelta_0\right)+O(1) .$$

Proof. Let $r_\nu=1-\frac{1-r_0}{2^\nu}$ $(\nu=0, 1, 2, \cdots)$ and for $\nu \geqq 2$, $\varDelta$ be the part of $\varDelta$, which lies in $r_{\nu-1} \leqq |z| \leqq r_\nu$ and $\varDelta_\nu^0$ be that of $\varDelta_0$, which lies in $r_{\nu-2} \leqq |z| \leqq r_{\nu+1}$ and

$$S_\nu=\frac{1}{\pi} \iint_{\Delta_\nu} \left(\frac{|w'(z)|}{1+|w(z)|^2}\right)^2 r dr d\theta , \qquad z=re^{i\theta} . \qquad (1)$$

Let n_ν^0 be the number of zero points of $\prod_{i=1}^{3}(w(z)-a_i)$ in Δ_ν^0, multiple zeros being counted only once.

We map Δ^0 conformally on $|\zeta|<1$, such that the point $z_\nu\left(|z_\nu|=\dfrac{r_\nu+r_{\nu-1}}{2}\right)$ on the bisector of Δ becomes $\zeta=0$, then the image of Δ_ν is contained in $|\zeta|<\kappa<1$, where κ is a constant, independent of ν.

If we apply Theorem VI. 21, then $S_\nu \leqq n_\nu^0+K$ (K=const.), so that

$$\sum_{\nu=2}^{n} S_\nu \leqq \sum_{\nu=2}^{n} n_\nu^0+Kn \leqq \sum_{\nu=2}^{n} n_\nu^0+O\left(\log\frac{1}{1-r_n}\right). \tag{2}$$

Since $\sum_{\nu=2}^{n} S_\nu=S(r_n;\Delta)-S(r_1;\Delta)$ and Δ_ν^0 overlap at most 3-times, $\sum_{i=2}^{n} n_\nu^0 \leqq 3\sum_{i=1}^{3} n(r_{n+1}, a_i; \Delta_0)$, so that by (2),

$$S(r_n;\Delta)\leqq 3\sum_{i=1}^{3} n(r_{n+1}, a_i; \Delta_0)+O\left(\log\frac{1}{1-r_n}\right).$$

If $r_{n-1}\leqq r\leqq r_n$, then $S(r;\Delta)\leqq S(r_n;\Delta)$ and $r_{n+1}=\dfrac{r_{n-1}+3}{4}\leqq\dfrac{r+3}{4}$, hence

$$S(r;\Delta)\leqq 3\sum_{i=1}^{3} n\left(\frac{r+3}{4}, a_i; \Delta_0\right)+O\left(\log\frac{1}{1-r}\right), \tag{3}$$

so that

$$T(r;\Delta)\leqq 12\sum_{i=1}^{3}\int_{t_0}^{\frac{r+3}{4}}\frac{n(t, a_i; \Delta_0)}{4t-3}\,dt+O(1), \quad t_0=\frac{r_0+3}{4}\geqq\frac{1/2+3}{4}=\frac{7}{8}.$$

Since $4t-3\geqq\dfrac{4t}{7}$, if $t\geqq 7/8$, we have

$$T(r;\Delta)\leqq 21\sum_{i=1}^{3} N\left(\frac{r+3}{4}, a_i; \Delta_0\right)+O(1). \text{ —— } \tag{4}$$

2. Let $w(z)$ be meromorphic in $|z|<1$ and $\Delta\subset\Delta_0$ be two sectors, as shown in the figure, where $0<\rho_0<\rho<1$ and $\Delta(r)$, $\Delta_0(r)$ be the part of Δ, Δ_0, which lies in $r_0\leqq|z|\leqq r<1$ respectively, where $r_0=\dfrac{1+\rho}{2}$.

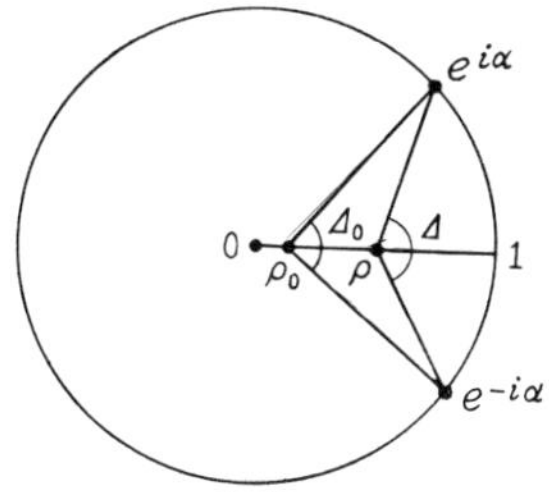

We define $S(r;\Delta)$, $T(r;\Delta)$, $n(r, a; \Delta_0)$, $N(r, a; \Delta_0)$ as before, then

Theorem VII. 13.

$$S(r;\Delta)\leqq 9\sum_{i=1}^{3} n\left(\frac{r+3}{4}, a_i; \Delta_0\right)+O\left(\frac{1}{1-r}\right),$$

$$T(r;\Delta)\leqq 63\sum_{i=1}^{3} N\left(\frac{r+3}{4}, a_i; \Delta_0\right)+O\left(\log\frac{1}{1-r}\right).$$

Proof. Let $r_\nu=1-\dfrac{1-r_0}{2^\nu}$ $(\nu=0, 1, 2, \cdots)$ and for $\nu\geqq 2$,

$$\left.\begin{aligned} \Delta_{\nu,s}:&\ \frac{(s-1)\delta}{2^\nu}\leqq\arg z\leqq\frac{s\delta}{2^\nu}\ , &\quad r_{\nu-1}\leqq|z|\leqq r_\nu\ ,\\ \Delta^0_{\nu,s}:&\ \frac{(s-2)\delta}{2^\nu}\leqq\arg z\leqq\frac{(s+1)\delta}{2^\nu}\ , &\quad r_{\nu-2}\leqq|z|\leqq r_{\nu+1}\ , \end{aligned}\right\} \quad (1)$$

$$(s=0, \pm 1, \pm 2, \cdots)$$

where we choose $\delta>0$ so small that, if $\Delta_{\nu,s}$ has common points with Δ, then $\Delta^0_{\nu,s}$ is contained in Δ_0 and the range of s is such that $\Delta_{\nu,s}$ has common points with Δ, so that the number of s is $O(2^\nu)$.

Let

$$S_{\nu,s}=\frac{1}{\pi}\iint_{\Delta_{\nu,s}}\left(\frac{|w'(z)|}{1+|w(z)|^2}\right)^2 rdrd\theta\ , \qquad z=re^{i\theta}, \tag{2}$$

and $n^0_{\nu,s}$ be the number of zero points of $\prod_{i=1}^{3}(w(z)-a_i)$ in $\Delta^0_{\nu,s}$, multiple zeros being counted only once. If we map $\Delta^0_{\nu,s}$ conformally on $|\zeta|<1$, such that the center of $\Delta^0_{\nu,s}$ becomes $\zeta=0$, then the image of $\Delta_{\nu,s}$ is contained in $|\zeta|\leqq\kappa<1$, where $\kappa>0$ is a constant, independent of ν and s. Hence by Theorem VI. 21, $S_{\nu,s}\leqq n^0_{\nu,s}+K$ (K=const.), so that

$$\sum_{\nu=2}^{n}\sum_{s}S_{\nu,s}\leqq\sum_{\nu=2}^{n}\sum_{s}n^0_{\nu,s}+O(2^n)=\sum_{\nu=2}^{n}\sum_{s}n^0_{\nu,s}+O\left(\frac{1}{1-r_n}\right).$$

$\sum_{\nu=2}^{n}\sum_{s}S_{\nu,s}=S(r_n;\Delta)-S(r_1;\Delta)$ and since $\Delta^0_{\nu,s}$ overlap at most 9-times,

$$\sum_{\nu=2}^{n}\sum_{s}n^0_{\nu}\leqq 9\sum_{i=1}^{3}n(r_{n+1}, a_i; \Delta_0)\ ,$$

so that

$$S(r_n;\Delta)\leqq 9\sum_{i=1}^{3}n(r_{n+1}, a_i; \Delta_0)+O\left(\frac{1}{1-r_n}\right). \tag{3}$$

If $r_{n-1}\leqq r\leqq r_n$, then $S(r;\Delta)\leqq S(r_n;\Delta)$ and $r_{n+1}\leqq\dfrac{r+3}{4}$, so that

$$S(r;\Delta)\leqq 9\sum_{i=1}^{3}n\left(\frac{r+3}{4}, a_i; \Delta_0\right)+O\left(\frac{1}{1-r}\right). \tag{4}$$

From this, we have as Theorem VII. 12,

$$T(r;\Delta)\leqq 63\sum_{i=1}^{3}N\left(\frac{r+3}{4}, a_i; \Delta_0\right)+O\left(\log\frac{1}{1-r}\right). \text{ ——} \tag{5}$$

3. Let $w(z)$ be meromorphic in $|z|<1$ and $\Delta\subset\Delta_0$ be two angular domains:

$$\Delta: |\arg z|<\alpha\ , \qquad \Delta_0: |\arg z|<\alpha_0 \qquad (\alpha<\alpha_0). \tag{1}$$

Let $\Delta(r)$, $\Delta_0(r)$ be the part of Δ, Δ_0, which lies in $|z|\leqq r<1$ respectively and

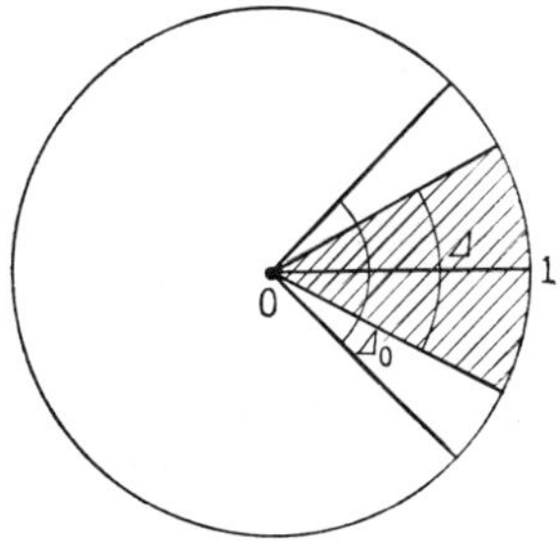

$$S(r;\varDelta)=\frac{1}{\pi}\iint\limits_{\Delta(r)}\left(\frac{|w'(z)|}{1+|w(z)|^2}\right)^2 r\,dr\,d\theta\,, \quad (2)$$

$$T(r;\varDelta)=\int_{r_0}^{r}\frac{S(r;\varDelta)}{r}dr \quad (r_0>0)\,. \quad (3)$$

Let $n(r, a; \varDelta_0)$ be the number of zero points of $w(z)-a$ in $\varDelta_0(r)$, multiple zeros being counted only once, and

$$N(r, a; \varDelta_0)=\int_{r_0}^{r}\frac{n(r, a; \varDelta_0)}{r}dr\,, \quad (4)$$

then

Theorem VII. 14. $S(r;\varDelta)\leqq 9\sum_{i=1}^{3} n\left(\frac{r+3}{4}, a_i; \varDelta_0\right)+O\left(\frac{1}{1-r}\right)$,

$$T(r;\varDelta)\leqq 63\sum_{i=1}^{3} N\left(\frac{r+3}{4}, a_i; \varDelta_0\right)+O\left(\log\frac{1}{1-r}\right).$$

Proof. Let $r_\nu=1-\frac{1-r_0}{2^\nu}$ $(\nu=0, 1, 2, \cdots)$ and for $\nu\geqq 2$,

$$\left.\begin{array}{lll}\varDelta_{\nu,s}: & \dfrac{(s-1)\delta}{2^\nu}\leqq \arg z\leqq \dfrac{s\delta}{2^\nu}\,, & r_{\nu-1}\leqq |z|\leqq r_\nu\,, \\ \varDelta^0_{\nu,s}: & \dfrac{(s-2)\delta}{2^\nu}\leqq \arg z\leqq \dfrac{(s+1)\delta}{2^\nu}\,, & r_{\nu-2}\leqq |z|\leqq r_{\nu+1} \\ & & (s=0, \pm 1, \pm 2, \cdots)\end{array}\right\}$$

and proceed similarly as the proof of Theorem VII. 13.——

If we apply the theorem for the angular domains $\varDelta$: $0\leqq \arg z\leqq 2\pi$ and $\varDelta_0$: $-\varepsilon\leqq|\arg z|\leqq 2\pi+\varepsilon$ $(\varepsilon>0)$, then we have

Theorem VII. 15. (Modified second fundamental theorem). *Let $w(z)$ be meromorphic in $|z|<1$, then*

$$T(r)\leqq 126\sum_{i=1}^{3} N\left(\frac{r+3}{4}, a_i\right)+O\left(\log\frac{1}{1-r}\right), \qquad 0\leqq r<1\,,$$

where $T(r)$, $N(r, a)$ are Nevanlinna's functions for $w(z)$.

From this and (4*) (p. 204), we can prove easily

Theorem VII. 16. *Let $w(z)$ be meromorphic in $|z|<1$. If $\sum_{\nu}(1-|z_\nu(a_i)|)^{k+1}<\infty$ $(k>0)$ $(i=1, 2, 3)$, then $\int_0^1 T(r)(1-r)^{k-1}dr<\infty$, so that $w(z)$ is of order $\leqq k$.*

2. A lemma

Lemma. *Let $w(z)$, $g_i(z)$ $(i=1, 2, 3, 4)$ be meromorphic in $|z|<1$ and*

$$f(z)=\frac{g_1(z)w(z)+g_2(z)}{g_3(z)w(z)+g_4(z)}\,.$$

Let $\varDelta\subset\varDelta_0$ be two angular domains of Theorem VII. 12, *then*

$$S(f, r; \varDelta)\leqq 27S\left(w, \frac{r+63}{64}; \varDelta_0\right)+O\left(\int_0^{\frac{r+127}{128}} \frac{T(g, r)}{(1-r)^2}dr\right), \quad T(g, r)=\sum_{i=1}^{4} T(g_i, r)\,,$$

where $T(g_i, r)$ *is the characteristic function of* $g_i(z)$.

The same relation holds, if $\varDelta\subset\varDelta_0$ *are sectors of Theorem* VII. 13, *where* 27 *is replaced by* 729.

Proof. Let $\varDelta\subset\varDelta_0$ be two angular domains of Theorem VII. 12. First we consider two special cases: (i) $f(z)=w(z)+g(z)$, (ii) $f(z)=g(z)w(z)$. First we consider the case (i) $f(z)=w(z)+g(z)$.

Let $r\ =1-\dfrac{1-r_0}{2}$ $(\nu=0, 1, 2, \cdots)$ and $\varDelta_\nu(\nu\geqq 2)$ be the part of $\varDelta$, which lies in $r\ _{-1}\leqq|z|\leqq r_\nu$ and $\varDelta_\nu^0$ be that of $\varDelta_0$, which lies in $r_{\nu-2}\leqq|z|\leqq r_{\nu+1}$. If we map $\varDelta_\nu^0$ conformally on $|\zeta|<1$, such that the point $z\left(|z_\nu|=\dfrac{r_{\nu-1}+r_\nu}{2}\right)$ on the bisector of $\varDelta_0$ becomes $\zeta=0$, then the image of $\varDelta_\nu$ is contained in $|\zeta|\leqq\kappa<1$, where $\kappa>0$ is a constant, independent of ν. Let D_ν be the image of $|\zeta|\leqq\dfrac{1+\kappa}{2}$ in $\varDelta^0$ and put

$$M_\nu=\frac{1}{|D\ |}\iint_{D_\nu}\log\sqrt{1+|g|^2}\,rdrd\theta \tag{1}$$

and proceed similarly as the proof of the lemma (p. 277).

Since the image of $|\zeta|\leqq\rho_2$ in $\varDelta_\nu^0$ is contained in $|z|\leqq\tau_\nu<r_{\nu+1}$, such that $r\ _{+1}-\tau\ \geqq$const.$/2^\nu$, we have as before (p. 280),

$$S(f; \varDelta_\nu)\leqq S(w; \varDelta_\nu^0)+O(M_\nu)+O(n(g, \tau_\nu, \infty))+O(1)\,. \tag{2}$$

Now

$$M\ =\frac{1}{|D_\nu|}\iint_{D_\nu}\log\sqrt{1+|g|^2}\ rdrd\theta\leqq\frac{1}{|D_\nu|}\iint_{\Delta_\nu^0}\log\sqrt{1+|g|^2}\,rdrd\theta$$

$$\leqq\text{const.}\,2^\nu\int_{r_{\nu-2}}^{r_{\nu+1}} r\,dr\int_{|z|=r}\log\sqrt{1+|g|^2}\,d\theta\leqq\text{const.}\,T(g, r_{\nu+1})\,, \tag{3}$$

$$T(g, r_{\nu+1})+O(1)\geqq\int_{\tau_\nu}^{r_{\nu+1}}\frac{n(g, r, \infty)}{r}\,dr\geqq n(g, \tau_\nu, \infty)(r_{\nu+1}-\tau_\nu)$$

$$\geqq\text{const.}\,\frac{n(g, \tau_\nu, \infty)}{2^\nu}\,,$$

so that

$$n(g, \tau_\nu, \infty)\leqq\text{const.}\,2\ \ T(g, r_{\nu+1})\leqq\text{const.}\int_{r_{\nu+1}}^{r_{\nu+2}}\frac{T(g, r)}{(1-r)^2}dr\,, \tag{4}$$

hence by (2), (3), (4),

$$S(f; \varDelta\ )\leqq S(w; \varDelta_\nu^0)+O\left(\int_{r_{\nu+1}}^{r_{\nu+2}}\frac{T(g, r)}{(1-r)^2}dr\right), \quad f=w+g\,. \tag{5}$$

Similarly as before, we can prove that (5) holds, if $f=gw$.

We sum up (5) for $\nu=0, 1, 2, \cdots, n$, then since $\varDelta_\nu^0$ overlap at most

3-times,

$$S(f, r_n; \Delta)\leqq 3S(w, r_{n+1}; \Delta_0)+O\left(\int_0^{r_{n+2}}\frac{T(g, r)}{(1-r)^2}dr\right). \qquad (6)$$

If $r_{n-1}\leqq r\leqq r_n$, then $S(f, r; \Delta)\leqq S(f, r_n; \Delta)$ and $r_{n+1}=\dfrac{r_{n-1}+3}{4}\leqq\dfrac{r+3}{4}$, $r_{n+2}=\dfrac{r_{n-1}+7}{8}\leqq\dfrac{r+7}{8}$, so that

$$S(f, r; \Delta)\leqq 3S\left(w, \frac{r+3}{4}; \Delta_0\right)+O\left(\int_0^{\frac{r+7}{8}}\frac{T(g, r)}{(1-r)^2}dr\right), \qquad f=w+g\ , \quad f=gw\ . \qquad (7)$$

Now we consider the general case $f=\dfrac{g_1w+g_2}{g_3w+g_4}$, then as before

$$f=w_1+h_1\ , \quad w_1=h_2w_2\ , \quad w_2=1/w_3\ , \quad w_3=w+h_3\ , \qquad (8)$$

where $T(h_i, r)=O(T(g, r))$ $(i=1, 2, 3)$.

Let $\Delta\subset\Delta_1\subset\Delta_2\subset\Delta_0$ be four angular domains, each of which is bounded by two lines through $z=1$, then by (7),

$$S(f, r; \Delta)\leqq 3S(w_1, r_1; \Delta_1)+ O\left(\int_0^{r_1'}\frac{T(h_1, r)}{(1-r)^2}dr\right), \quad r_1=\frac{r+3}{4}, \quad r_1'=\frac{r+7}{8},$$

$$S(w_1, r_1; \Delta_1)\leqq 3S(w_2,r_2 ; \Delta_2)+O\left(\int_0^{r_2'}\frac{T(h_2, r)}{(1-r)^2}dr\right), \quad r_2=\frac{r_1+3}{4}, \quad r_2'=\frac{r_1+7}{8},$$

$$S(w_2, r_2; \Delta_2)=S(w_3, r_2; \Delta_2)\ ,$$

$$S(w_3, r_2; \Delta_2)\leqq 3S(w, r_3; \Delta_0)+O\left(\int_0^{r_3'}\frac{T(h_3, r)}{(1-r)^2}dr\right), \quad r_3=\frac{r_2+3}{4}=\frac{r+63}{64},$$

$$r_3'=\frac{r_2+7}{8}=\frac{r+127}{128}.$$

Since $T(h_i, r)=O(T(g, r))$ $(i=1, 2, 3)$, we have

$$S(f, r; \Delta)\leqq 27S\left(w, \frac{r+63}{64}, \Delta_0\right)+O\left(\int_0^{\frac{r+127}{128}}\frac{T(g, r)}{(1-r)^2}dr\right). \qquad (9)$$

Hence the lemma is proved, if $\Delta\subset\Delta_0$ are two angular domains of Theorem VII. 12. If $\Delta\subset\Delta_0$ are two sectors of Theorem VII. 13, then we take $\Delta_{\nu,s}$, $\Delta_{\nu,s}^0$ instead of Δ_ν, Δ_ν^0, which are defined in the proof of Theorem VII. 13. Then (3) is replaced by $M_\nu\leqq \text{const.}\,2^\nu\, T(g, r_{\nu+1})$. Since $\Delta_{\nu,s}^0$ overlap at most 9-times, we have the same relation (7), where 3 is replaced by 9, so that in (9), 27 is replaced by $9^3=729$.

3. Analogue of Biernacki-Rauch's theorem

Theorem VII. 17. *Let $f(z)$ be a meromorphic function of finite order $\rho>0$ in $|z|<1$. Then there exists a points z_0 on $|z|=1$ and a line J through z_0, directed inward of $|z|<1$, which may coincide with the tangent of $|z|=1$ at z_0 and satisfies the following condition.*

Let ω be any small angular domain, which contains J and is bounded by two lines through z_0 and $g(z)$ be a meromorphic function in $|z|<1$ and $\{z_\nu(f=g, \omega)\}$ be zero points of $f(z)-g(z)$ in ω, multiple zeros being counted only once.

(i) *If $g(z)$ is of order $<\rho$, then*

$$\sum_\nu (1-|z_\nu(f=g;\omega)|)^{\rho+1-\varepsilon}=\infty\,, \qquad \varepsilon>0\,,$$

with two possible exceptions for g.

(ii) *If $f(z)$ is of divergence type and $\int_0^1 T(g,r)(1-r)^{\rho-1}dr<\infty$, then*

$$\sum_\nu (1-|z_\nu(f=g;\omega)|)^{\rho+1}=\infty\,,$$

with two possible exceptions for g. Tsuji [46].

Proof. Now

$$\int_0^1 T(f,r)(1-r)^{k-1}dr=\infty\,, \qquad k>0\,, \tag{1}$$

where $k=\rho-\varepsilon$ $(\varepsilon>0)$ in general and $k=\rho$, if $f(z)$ is of divergence type. By dividing the unit circle into 2^n equal parts, we see that there exists a direction L: $\arg z=\theta_0$, such that for any small angular domain Δ : $|\arg z-\theta_0|<\delta$, which contains L,

$$\int_0^1 T(f,r;\Delta)(1-r)^{k-1}dr=\infty\,. \tag{2}$$

We assume that $\theta_0=0$ and denote the sector: $|\arg z|<\delta$, $|z|<1$ by the same letter Δ. We draw two lines through $z=1$, making an angle $<\pi/2$ with the negative real axis and let Ω be the angular domain, bounded by these two lines and we denote the common part of Ω and Δ by the same letter Ω. Then Δ consists of three parts: $\Delta=\Omega+\Omega'+\Omega''$, as shown in the figure. Hence by (2), one of

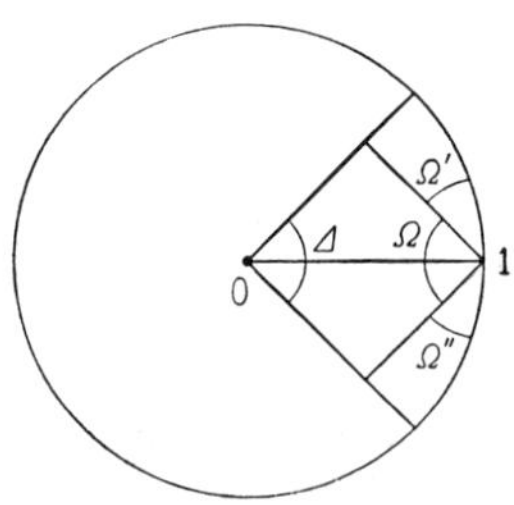

$$\int_0^1 T(f,r;\Omega)(1-r)^{k-1}dr\,,$$

$$\int_0^1 T(f,r;\Omega')(1-r)^{k-1}dr\,,$$

$$\int_0^1 T(f,r;\Omega'')(1-r)^{k-1}dr$$

is ∞. First suppose that

$$\int_0^1 T(f,r;\Omega)(1-r)^{k-1}dr=\infty\,. \tag{3}$$

By dividing the angular domain Ω into 2^n equal parts by lines through $z=1$, we see that there exists a line $J\in\Omega$ through $z=1$, such that for any small angular domain ω, which contains J and is bounded by two lines through $z=1$,

$$\int_0^1 T(f, r; \omega)(1-r)^{k-1} dr = \infty . \tag{4}$$

Let $\omega \subset \omega_1 \subset \omega_0$ be three angular domains, whose common vertex is at $z=1$. Let $g_i(z)$ $(i=1, 2, 3)$ be three meromorphic functions in $|z|<1$, such that

$$\int_0^1 T(g_i, r)(1-r)^{k-1} dr < \infty , \tag{5}$$

hence if we put $T(g, r)=\sum_{i=1}^{3} T(g_i, r)$, then

$$\int_0^1 T(g, r)(1-r)^{k-1} dr < \infty . \tag{6}$$

We shall prove that for one of g_i,

$$\sum_{\nu} (1-|z_\nu(f=g_i; \omega_0)|)^{k+1} = \infty .$$

We put

$$w(z)=\frac{f(z)-g_1(z)}{f(z)-g_3(z)} \cdot \frac{g_2(z)-g_3(z)}{g_2(z)-g_1(z)} , \qquad f(z)=\frac{h_1(z)w(z)+h_2(z)}{h_3(z)w(z)+h_4(z)} , \tag{7}$$

then $T(h_i, r)=O(T(g, r))$ $(i=1, 2, 3, 4)$, so that by the lemma (p. 291),

$$S(f, r; \omega) \leqq 27\, S\left(w, \frac{r+63}{64}, \omega_1\right)+O(\Phi(r)) , \tag{8}$$

$$\Phi(r)=\int_0^{\frac{r+127}{128}} \frac{T(g, r)}{(1-r)^2} dr .$$

Hence

$$T(f, r; \omega) \leqq \text{const. } T\left(w, \frac{r+63}{64}; \omega_1\right)+O(\Psi(r)) , \tag{9}$$

$$\Psi(r)=\int_0^r \Phi(r) dr .$$

We shall prove that

$$\int_0^r \Psi(r)(1-r)^{k-1} dr = O(1) . \tag{10}$$

For

$$\begin{aligned}\int_0^r \Psi(r)(1-r)^{k-1} dr &= \left[-\frac{(1-r)^k}{k}\Psi(r)\right]_0^r + \frac{1}{k}\int_0^r (1-r)^k \Phi(r) dr \\ &\leqq \frac{1}{k}\int_0^r (1-r)^k \Phi(r) dr \leqq \frac{1}{k(k+1)}\int_0^r (1-r)^{k+1} \Phi'(r) dr \\ &\leqq \text{const.} \int_0^1 T(g, r)(1-r)^{k-1} dr < \infty .\end{aligned}$$

Hence by (4), (9), (10), we have

$$\int_0^1 T(w, r; \omega_1)(1-r)^{k-1} dr = \infty . \tag{11}$$

If we apply Theorem VII. 12 for $\omega_1 \subset \omega_0$, $a_1=0$, $a_2=1$, $a_3=\infty$, then from (11), we see that one of

$$\int_0^1 N(w, r, 0; \omega_0)(1-r)^{k-1} dr\,, \qquad \int_0^1 N(w, r, 1; \omega_0)(1-r)^{k-1} dr\,,$$
$$\int_0^1 N(w, r, \infty; \omega_0)(1-r)^{k-1} dr$$

is ∞. Without loss of generality, we assume that

$$\int_0^1 N(w, r, 0; \omega_0)(1-r)^{k-1} dr = \infty\,. \tag{12}$$

Since by (7),

$$n(w, r, 0; \omega_0) \leqq n(r, f=g_1; \omega_0) + n(r, g_2=g_3; \omega_0) + \sum_{i=1}^{3} n(r, g_i=\infty; \omega_0)\,,$$

$$\begin{aligned} N(w, r, 0; \omega_0) &\leqq N(r, f=g_1; \omega_0) + N(r, g_2=g_3; \omega_0) + \sum_{i=1}^{3} N(r, g_i=\infty; \omega_0) \\ &\leqq N(r, f=g_1; \omega_0) + O(T(g, r))\,, \end{aligned} \tag{13}$$

we have from (12), (6) and (4*) (p. 204),

$$\int_0^1 N(r, f=g_1; \omega_0)(1-r)^{k-1} dr = \infty\,, \quad \text{or} \quad \sum_\nu (1-|z_\nu(f=g_1; \omega_0)|)^{k+1} = \infty\,.$$

Hence

$$\sum_\nu (1-|z\ (f=g; \omega_0)|)^{k+1} = \infty\,, \tag{14}$$

with two possible exceptions for g.

If $\int_0^1 T(f, r; \Omega)(1-r)^{k-1} dr < \infty$, then we may assume that

$$\int_0^1 T(f, r; \Omega')(1-r)^{k-1} dr = \infty\,. \tag{15}$$

Let ζ_0, a be defined as in the figure and ζ_1 be a point on $|z|=1$, which lies symmetric to $z=1$ with respect to the line $\overline{0\zeta_0}$.

Let a_0 be a point on the line $\overline{0\zeta_0}$, such that $a=\frac{1}{2}(a_0+\zeta_0)$.

Let Σ_0, Σ be the sectors, defined as in the figure. Then by (15),

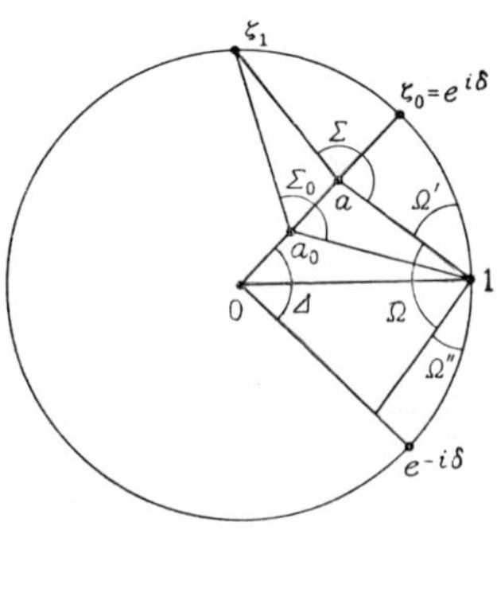

$$\int_0^1 T(f, r; \Sigma)(1-r)^{k-1} dr = \infty\,. \tag{16}$$

By means of (16) and Theorem VII. 13, we can prove similarly as before, that

$$\sum_\nu (1-|z_\nu(f=g; \Sigma_0)|)^{k+1} = \infty\,, \tag{17}$$

with two possible exceptions for g.

We denote the angular magnitude of Ω by $|\Omega|$, so that $0<|\Omega|<\pi$.

Let $\Omega_1 \subset \Omega_2 \subset \cdots \subset \Omega_n \subset \cdots$ be angular domains as Ω, such that $|\Omega_n|\to\pi$ and suppose that for $n=1, 2, \cdots$,

$$\int_0^1 T(f,r;\Omega_n)(1-r)^{k-1}dr<\infty\,,$$

then without loss of generality, we may assume that

$$\int_0^1 T(f,r;\Omega_n')(1-r)^{k-1}dr=\infty \qquad (n=1,2,\cdots)\,. \tag{18}$$

Let J be the positive tangent of $|z|=1$ at $z=1$, then from (17), we see that for any small angular domain ω, which contains J,

$$\sum_\nu (1-|z\,(f=g;\omega)|)^{k+1}=\infty\,, \tag{19}$$

with two possible exceptions for g.

Hence from (14), (19), in any case, there exists a line J through $z=1$, such that for any small angular domain ω, which contains J,

$$\sum_\nu (1-|z\,(f=g;\omega)|)^{k+1}=\infty\,, \tag{20}$$

with two possible exceptions for g.

If $f(z)$ is of divergence type, then $k=\rho$, so that J satisfies the condition of the theorem. If $f(z)$ is of convergence type, then we take $k_n=\rho-\varepsilon_n$, $\varepsilon_1>\varepsilon_2>\cdots>\varepsilon_n\to 0$, then we see that there exists a line L: $\arg z=\theta_0$, such that for any small angular domain Δ, which contains L,

$$\int_0^1 T(f,r;\Delta)(1-r)^{k_n-1}dr=\infty \qquad (n=1,2,\cdots)\,.$$

Let J_n be the corresponding direction through $z_0=e^{i\theta_0}$ and J be one of the limiting positions of $\{J_n\}$, then J satisfies the condition of the theorem.

4. Function of the class $\overline{\lim\limits_{r\to 1}}\, T(f,r)\Big/\log\dfrac{1}{1-r}=\infty$

Theorem VII. 18. *Let $f(z)$ be meromorphic in $|z|<1$, such that*

$$\overline{\lim_{r\to 1}}\, T(f,r)\Big/\log\frac{1}{1-r}=\infty\,.$$

Let $g(z)$ be meromorphic in $|z|<1$, such that $T(g,r)=O(1)$, then there exists a point z_0 on $|z|=1$ and a line J through z_0, directed inward of $|z|<1$, which may coincide with the tangent of $|z|=1$ at z_0 and satisfies the following condition. Let ω be any small angular domain, which contains J and is bounded by two lines through z_0, then $f(z)-g(z)$ has infinitely many zero points in ω, with two possible exceptions for g. More precisely,

$$\overline{\lim_{r\to 1}}\, N(r,f=g;\omega)\Big/\log\frac{1}{1-r}=\infty,$$

with two possible exceptions for g. Tsuji [46].

Proof. Similarly as the proof of Theorem VII. 17, there exists a line L: $\arg z=\theta_0$ ($z_0=e^{i\theta_0}$), such that for any small angular domain Δ: $|\arg z-\theta_0|<\delta$, which contains L,

$$\overline{\lim_{r\to 1}}\, T(f, r; \Delta)\Big/\log\frac{1}{1-r}=\infty\,. \tag{1}$$

We define Ω, Ω', Ω'' as before, then one of

$$\overline{\lim_{r\to 1}}\, T(f, r; \Omega)\Big/\log\frac{1}{1-r}\,, \qquad \overline{\lim_{r\to 1}}\, T(f, r; \Omega')\Big/\log\frac{1}{1-r}\,,$$
$$\overline{\lim_{r\to 1}}\, T(f, r; \Omega'')\Big/\log\frac{1}{1-r}$$

is ∞. First suppose that

$$\overline{\lim_{r\to 1}}\, T(f, r; \Omega)\Big/\log\frac{1}{1-r}=\infty\,. \tag{2}$$

Then as before, there exists a line $J\in\Omega$ through z_0, such that for any small angular domain ω, which contains J and is bounded by two lines through z_0,

$$\overline{\lim_{r\to 1}}\, T(f, r; \omega)\Big/\log\frac{1}{1-r}=\infty\,. \tag{3}$$

Let $\omega\subset\omega_1\subset\omega_0$ be three angular domains, with the common vertex z_0 and $g_i(z)$ ($i=1, 2, 3$) be three meromorphic functions in $|z|<1$, such that $T(g_i, r)=O(1)$, then $T(g, r)=\sum_{i=1}^{3} T(g_i, r)=O(1)$. We put

$$w(z)=\frac{f(z)-g_1(z)}{f(z)-g_3(z)}\cdot\frac{g_2(z)-g_3(z)}{g_2(z)-g_1(z)}\,, \qquad f(z)=\frac{h_1(z)w(z)+h_2(z)}{h_3(z)w(z)+h_4(z)}\,, \tag{4}$$

then $T(h_i, r)=O(T(g, r))=O(1)$ ($i=1, 2, 3, 4$).

By the lemma (p. 291),

$$S(f, r; \omega)\leqq 27\, S\left(w, \frac{r+63}{64}; \omega_1\right)+O(\Phi(r))\,, \tag{5}$$

$$\Phi(r)=\int_0^{\frac{r+127}{128}}\frac{T(g, r)}{(1-r)^2}dr=O\left(\frac{1}{1-r}\right),$$

so that

$$T(f, r; \omega)\leqq \text{const.}\, T\left(w, \frac{r+63}{64}; \omega_1\right)+O\left(\log\frac{1}{1-r}\right).$$

Hence by (3),

$$\overline{\lim_{r\to 1}}\, T(w, r; \omega_1)\Big/\log\frac{1}{1-r}=\infty\,. \tag{6}$$

Hence by Theorem VII. 12, for one of $N(w, r, 0; \omega_0)$, $N(w, r, 1; \omega_0)$. $N(w, r, \infty; \omega_0)$, $N(w, r, 0; \omega_0)$ say,

$$\varlimsup_{r\to 1} N(w, r, 0; \omega_0)\Big/\log\frac{1}{1-r}=\infty\,. \tag{7}$$

Similarly as the proof of Theorem VII. 17, we have

$$N(w, r, 0; \omega_0)\leqq N(r, f=g_1; \omega_0)+O(T(g, r))=N(r, f=g_1; \omega_0)+O(1)\,,$$

so that

$$\varlimsup_{r\to 1} N(r, f=g; \omega_0)\Big/\log\frac{1}{1-r}=\infty\,, \tag{8}$$

with two possible exceptions for g.

If $T(f, r; \Omega)=O\left(\log\frac{1}{1-r}\right)$ for any Ω, then we can prove as before, that the tangent J of $|z|=1$ at z_0 satisfies the condition of the theorem.

5. Function of the class $\lim_{r\to 1} T(f, r)=\infty$

Theorem VII. 19. *Let $f(z)$ be meromorphic in $|z|<1$, such that*

$$\lim_{r\to 1} T(f, r)=\infty\,,$$

then $f(z)$ takes any value infinitely often, except a set of capacity zero.

More precisely, there exists a point z_0 on $|z|=1$, such that $f(z)-a$ has infinitely many zero points in any small neighbourhood $U(z_0)$ of z_0, except a set of a, which is of capacity zero.

Proof. First we shall prove the first part. Suppose that $n(r, a)=O(1)$ for any value a of a set M of positive capacity on the w-sphere K, then we see easily that there exists a closed sub-set E of M, which is of positive capacity, such that for any $a\in E$, $n(r, a)\leqq n_0<\infty$ $(0\leqq r<1)$.

Let $u(w)$ be the elliptic conductor potential of E:

$$u(w)=\int_E \log\frac{1}{[w, a]}d\mu(a)\,, \qquad \mu(E)=1\,, \tag{1}$$

such that $u(w)\leqq V<\infty$ on K. Hence by the first fundamental theorem, we have

$$T(r)=\int_E m(r, a)\,d\mu(a)+\int_E N(r, a)d\mu(a)\leqq V+\int_E N(r, a)d\mu(a)=O(1)\,,$$

which is absurd, hence the first part is proved. Next we shall prove the second part.

By dividing the unit circle $|z|=1$ into 2^n equal parts, we see that there exists a line J: $\arg z=\theta_0$, such that for any small angular domain $\varDelta$: $|\arg z-\theta_0|<\delta$, which contains J,

$$\lim_{r\to 1} T(f, r; \varDelta)=\infty\,. \tag{2}$$

Without loss of generality, we assume that $\theta_0=0$ and let

$$\varDelta: |\arg z|<\delta\,, \qquad \varDelta_0: |\arg z|<2\delta\,, \qquad |z|<1\,, \tag{3}$$

be two sectors. We map $\varDelta_0$ conformally on $|\zeta|<1$, such that $z=0, 1/2, 1$ correspond to $\zeta=-1, 0, 1$ respectively and put $f(z)=F(\zeta)$, $\zeta=\rho e^{i\varphi}$.

Let r be the greatest value, such that a domain $|\arg z|\leqq\delta$, $0<r_0\leqq|z|\leqq r$ is cóntained in the image of $|\zeta|<\rho<1$, then

$$S(F, \rho)\geqq S(f, r; \varDelta)-O(1)\,.$$

We see easily that $r\to1$ with $\rho\to1$ and $\dfrac{d\rho}{\rho}\geqq\text{const.}\dfrac{dr}{r}$, so that

$$T(F, \rho)\geqq\text{const.}\ T(f, r; \varDelta)-O(1)\,. \tag{4}$$

Hence by (2), $\lim\limits_{\rho\to1} T(F, \rho)=\infty$, so that by the first part, $F(\zeta)$ takes any value infinitely often in $|\zeta|<1$, except a set of capacity zero. Since $\delta>0$ is arbitrary, returning to the z-plane, we see that $f(z)$ takes any value infinitely often in any small neighbourhood of $z=1$, except a set of capacity zero. Hence the second part is proved. ——

Similarly as Theorem V. 34 and 35, we can prove the following theorems.

Theorem VII. 20. *Let Ω be the space of uniformly bounded regular functions $\{g(z)\}$ in $|z|<R$ $(R>1)$, then under the same condition as Theorem* VII. 19, *there exists a point z_0 on $|z|=1$, such that $f(z)-g(z)$ has infinitely many zero points in any small neighbourhood $U(z_0)$ of z_0, a fortiori, in $|z|<1$, except a set of g, which is of capacity zero.*

Theorem VII. 21. *Let $g_\alpha(z)=g(z, \alpha)$ be a regular function of two complex variables z and α for $|z|<R$ $(R>1)$ and $\alpha\in\varDelta$, where $\varDelta$ is a Jordan domain, then under the same condition as Theorem* VII. 19, *there exists a point z_0 on $|z|=1$, such that $f(z)-g_\alpha(z)$ has infinitely many zero points in any small neighbourhood $U(z_0)$ of z_0, a fortiori, in $|z|<1$. except a set of α, which is of capacity zero.*

CHAPTER VIII

CLUSTER SET OF A MEROMORPHIC FUNCTION

1. Principle of majoration by harmonic measure

Let D be a bounded domain on the z-plane and Γ be its boundary, which consists of a finite number of Jordan curves and let α be a sub-arc of Γ. Let $f(z)$ be regular and bounded in D, such that

$$|f(z)| \leqq m \text{ on } \alpha, \quad |f(z)| \leqq M \text{ on } \Gamma-\alpha. \tag{1}$$

Let $\omega(z)=\omega(z, \alpha; D)$ be the harmonic measure of α with respect to D, such that ω is harmonic in D, $\omega=1$ on α, $\omega=0$ on $\Gamma-\alpha$.

Since $\log|f(z)|$ is subharmonic, by the maximum principle, we have

$$\log|f(z)| \leqq \omega(z, \alpha; D) \log m+(1-\omega(z, \alpha; D)) \log M, \tag{2}$$

or

$$|f(z)| \leqq m^{\omega(z, \alpha; D)} M^{1-\omega(z, \alpha; D)}. \tag{3}$$

We call (2) and (3) **the principle of majoration by harmonic measure**. We shall show some applications of this principle in the following.

Theorem VIII. 1. *Let D be a bounded simply connected domain on the z-plane, which is contained in a sector: $|\arg z|<\frac{\alpha\pi}{2}$, $|z|<R$. The boundary Γ of D consists of two parts: $\Gamma=L+\Lambda$, where L consists of segments on two lines OA: $\arg z=\frac{\alpha\pi}{2}$ and OB: $\arg z=-\frac{\alpha\pi}{2}$ and Λ consists of Jordan arcs, whose end points lie on OA, or on OB.*

Let $f(z)$ be regular and bounded in D, such that

$$|f(z)| \leqq M \text{ on } L, \quad |f(z)| \leqq m \text{ on } \Lambda \ (m \leqq M).$$

If $z=r>0$ belongs to D, then

$$|f(r)| \leqq M^{1-\left(\frac{r}{R}\right)^{\frac{1}{\alpha}}} m^{\left(\frac{r}{R}\right)^{\frac{1}{\alpha}}}.$$ Carleman [2].

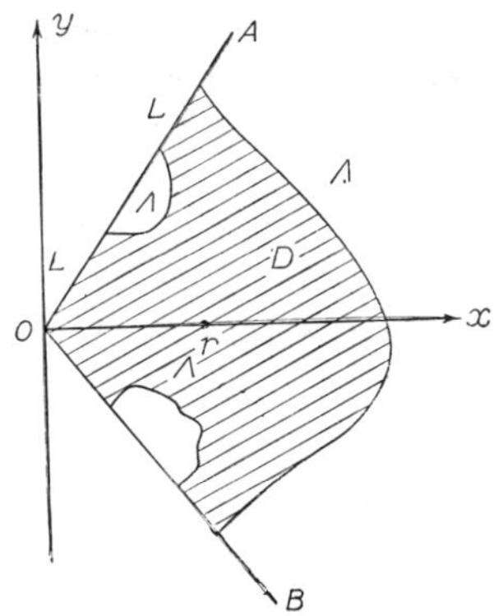

Proof. First suppose that $\alpha=1$, $R=1$, $M=1$, so that $0<m \leqq 1$. Then D is contained in a half-disc Δ: $|\arg z|<\pi/2$, $|z|<1$. Let C be the half-circle: $|z|=1$, $|\arg z|<\pi/2$, then

$$\omega(z, C; \Delta)=\frac{2\theta}{\pi}, \quad \theta=\arg\left(\frac{i-z}{i+z}\right). \tag{1}$$

Since $D \subset \Delta$, we have by the maximum principle,

$$\omega(z, C; \Delta) \leqq \omega(z, \Lambda; D), \quad z \in D,$$

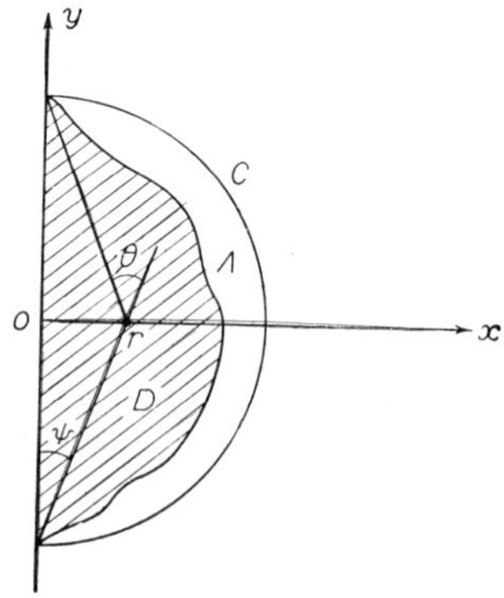

and since $\log M=0$, $\log m \leqq 0$, we have

$$\log |f(r)| \leqq \omega(r, \Lambda; D) \log m$$

$$\leqq \omega(r, C; \Delta) \log m = \frac{2\theta}{\pi} \log m$$

$$= \frac{4\psi}{\pi} \log m = \frac{4}{\pi} \tan^{-1} r . \log m . \quad (2)$$

Since $\frac{4}{\pi} \tan^{-1} r \geqq r$, if $0 \leqq r \leqq 1$[1], and $\log m \leqq 0$, we have

$$|f(r)| \leqq m^r . \quad (3)$$

In the general case, we put $\zeta = \left(\frac{z}{R}\right)^{\frac{1}{\alpha}}$, $F(\zeta) = f(z)$, $\tau = \left(\frac{r}{R}\right)^{\frac{1}{\alpha}}$ and apply (3) to $\frac{F(\zeta)}{M}$, then we have

$$|f(r)| = |F(\tau)| \leqq M . \left(\frac{m}{M}\right)^{\tau} = M^{1-\left(\frac{r}{R}\right)^{\frac{1}{\alpha}}} m^{\left(\frac{r}{R}\right)^{\frac{1}{\alpha}}} . \text{ ---}$$

Theorem VIII. 2. *Let C_n $(n=1, 2, \cdots)$ be Jordan arcs in $|z|<1$, which converges to $|z|=1$, such that two fixed radii OA, OB of $|z|=1$ intersect C_n $(n=1, 2, \cdots)$. Let $f(z)$ be regular and bounded in $|z|<1$ and $\max_{C_n} |f(z)| = m_n \to 0$ $(n \to \infty)$, then $f(z) \equiv 0$.*

Proof. Let D_n be the domain, bounded by OA, OB and C_n and apply Theorem VIII. 1 on D_n, then $f(z)=0$ at every point of the bisector OC of OA, OB, hence $f(z) \equiv 0$.

Theorem VIII. 3. *Let D be a simply connected domain on the z-plane, which lies in an angular domain: $o < \arg z < \alpha$ and its boundary Γ consists of a segment AB on the positive real axis and a Jordan arc Λ, which connects A, B, where A may coincide with the origin O. Let $f(z)$ be regular and bounded in D, such that*

$$|f(z)| \leqq M \text{ on } AB, \quad |f(z)| \leqq m \text{ on } \Lambda \ (m \leqq M).$$

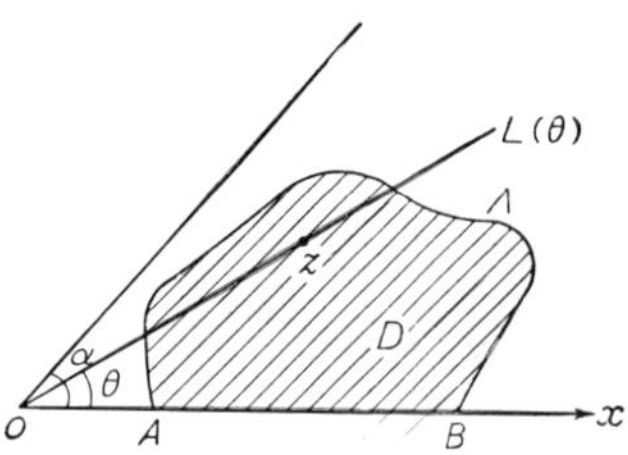

Let $L(\theta)$: $\arg z = \theta$ be a half-line, which penetrates into D, then at $z \in L(\theta)$,

$$|f(z)| \leqq M^{1-\frac{\theta}{\alpha}} m^{\frac{\theta}{\alpha}} .$$

Proof. Let $\omega(z)$ be the harmonic

---

1) If we put $\varphi(r) = \frac{4}{\pi} \tan^{-1} r$, then $\varphi(r)$ is a concave function of $r \geqq 0$ and $\varphi(0)=0$, $\varphi(1)=1$, so that $\varphi(r) \geqq r$ $(0 \leqq r \leqq 1)$.

measure of AB with respect to D, then

$$\log|f(z)|\leqq\omega(z)\log M+(1-\omega(z))\log m=\omega(z)(\log M-\log m)+\log m\,.$$

Since D is contained in $0<\arg z<\alpha$, we have $0\leqq\omega(z)\leqq1-\dfrac{\arg z}{\alpha}$ and $\log M-\log m\geqq0$, so that

$$\log|f(z)|\leqq\left(1-\frac{\arg z}{\alpha}\right)(\log M-\log m)+\log m$$
$$=\left(1-\frac{\arg z}{\alpha}\right)\log M+\frac{\arg z}{\alpha}\log m\,,$$

hence on $L(\theta)$,

$$|f(z)|\leqq M^{1-\frac{\theta}{\alpha}}\,m^{\frac{\theta}{\alpha}}\,.$$

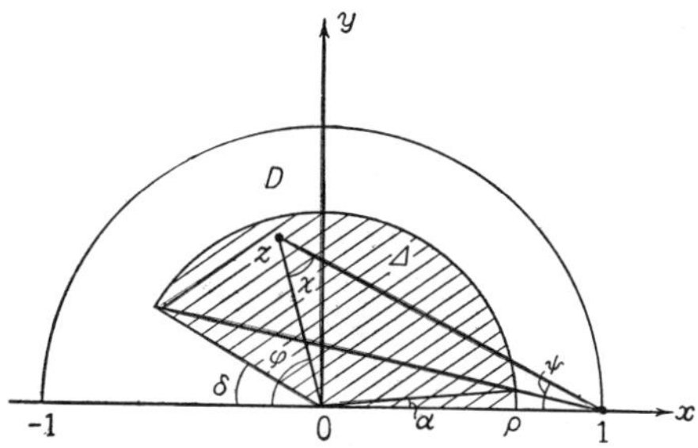

Lemma. *Let D be a half-disc: $|z|<1$, $0<\arg z<\pi$ and $\omega(z)$ be the harmonic measure of the segment $0\leqq x\leqq1$, $y=0$ with respect to D.*

Let $\varDelta: |z|<\rho<1$, $0<\arg z<\pi-\delta$ $(\delta>0)$ be a sector, then

$$\omega(z)\geqq\frac{\alpha}{\pi}=\frac{2}{\pi}\tan^{-1}\left(\frac{1-\rho}{1+\rho}\tan\frac{\delta}{2}\right)^{1)}\quad\text{in }\varDelta,$$

where α is defined in the figure.

Proof. We put $\mu=\operatorname*{Min}_{z\in\Delta}\omega(z)$ and we shall prove that μ is attained at $z=\rho e^{i(\pi-\delta)}$.

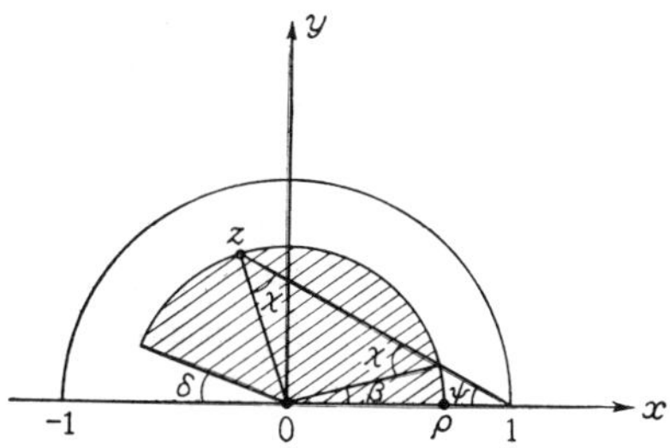

From the above figure, we have

$$\omega(z)=\frac{1}{\pi}(\varphi-2\psi)=\frac{1}{\pi}(\chi-\psi),\quad(1)$$

so that

$$\mu=\operatorname*{Min}_{0\leqq\theta\leqq\pi-\delta}\omega(\rho e^{i\theta}).\quad(2)$$

If z lies on $|z|=\rho$, then since $\chi-\psi=\beta$ from the figure, we see that $\omega(z)$ becomes minimum at $z=\rho e^{i(\pi-\delta)}$ and $\mu=\dfrac{\alpha}{\pi}$, so that

1) $y=\dfrac{\rho\sin\delta}{1+\rho\cos\delta}(1-x)$ is the line, through $z=1$ and $z=\rho e^{i(\pi-\delta)}$. Since $z=\rho e^{i\alpha}$ lies on this line, $\rho\sin\alpha=\dfrac{\rho\sin\delta}{1+\rho\cos\delta}(1-\rho\cos\alpha)$, so that $\rho\sin(\alpha+\delta)=\sin\delta-\sin\alpha$. From this, we have

$$\tan\frac{\alpha}{2}=\frac{1-\rho}{1+\rho}\tan\frac{\delta}{2}\,.$$

$$\omega(z) \geqq \frac{\alpha}{\pi} \quad \text{in} \quad \varDelta. \text{ ——}$$

By the lemma and the principle of majoration by harmonic measure, we have

Theorem VIII. 4. *Let D, $\varDelta$ be the same as the lemma and $f(z)$ be regular and bounded in D, such that*

$|f(z)| \leqq M$ *in* D, $|f(z)| \leqq m$ $(\leqq M)$ *on the segment*: $0 \leqq x \leqq 1$, $y=0$.

Then

$$|f(z)| \leqq M^{1-\frac{\alpha}{\pi}} m^{\frac{\alpha}{\pi}} \quad \text{in } \varDelta.$$

From this, we have

Theorem VIII. 5. *Let $f(z)$ be regular and bounded in a sector $\varDelta: 0 \leqq \arg z \leqq \theta_0$, $0<|z| \leqq R$. If $\lim_{z\to 0} f(z)=a$, when $z \to 0$ on the positive real axis, then $\lim_{z\to 0} f(re^{i\theta})=a$ uniformly for $0 \leqq \theta \leqq \theta_1$ $(<\theta_0)$.* Montel [1].

Theorem VIII. 6. *Let $f(z)$ be regular and bounded in a domain D, which is bounded by a Jordan curve C and $z=0$ belongs to C, then by $z=0$, the part of C, which lies in a neighbourhood of $z=0$ is decomposed into two parts C_1, C_2. If $\overline{\lim_{z\to 0}} |f(z)| \leqq m$, when $z \to 0$ along C_1 and C_2, then $\overline{\lim_{z\to 0}} |f(z)| \leqq m$, when $z \to 0$ from the inside of D.* Lindelöf [1].

Proof. We may assume that $|f(z)| \leqq 1$ in D and $0<m<1$. Let $C_1(r)$, $C_2(r)$, $D(r)$ be the part of C_1, C_2, D, which lies in $|z| \leqq r$.

For any $\varepsilon>0$, we take $r_0>0$ so small that $|f(z)|<m+\varepsilon$ on $C_1(r_0)$ and $C_2(r_0)$. Let $\omega(z)$ be the harmonic measure of $C_1(r_0)+C_2(r_0)$ with respect to $D(r_0)$, then $|f(z)| \leqq (m+\varepsilon)^{\omega(z)}$ in $D(r_0)$. Since $\lim_{z\to 0} \omega(z)=1$, when $z \to 0$ from the inside of D, we have $\overline{\lim_{z\to 0}} |f(z)| \leqq m+\varepsilon$ and since $\varepsilon>0$ is arbitrary, $\overline{\lim_{z\to 0}} |f(z)| \leqq m$, when $z \to 0$ from the inside of D. ——

If we apply the theorem to $f(z)-a$, then we have

Theorem VIII. 7. *Under the same condition as Theorem* VIII. 6, *if $\lim_{z\to 0} f(z)=a$, when $z \to 0$ on C_1 and C_2, then $\lim_{z\to 0} f(z)=a$ uniformly, when $z \to 0$ from the inside of D.* Lindelöf [1].

Theorem VIII. 8. *Under the same condition as Theorem* VIII. 6, *if $\lim_{z\to 0} f(z)=a$, when $z \to 0$ on C_1 and $\lim_{z\to 0} f(z)=b$, when $z \to 0$ on C_2, then $a=b$, so that $\lim_{z\to 0} f(z)=a$ uniformly, when $z \to 0$ from the inside D.*

Lindelöf [1].

Proof. $F(z)=(f(z)-a)(f(z)-b)$ is regular and bounded in D and $\lim_{z\to 0} F(z)=0$, when $z\to 0$ along C_1 and C_2, so that $\lim_{z\to 0} F(z)=0$ uniformly, when $z\to 0$ from the inside of D. Suppose that $a\neq b$, We consider a lemniscate Λ: $|(w-a)(w-b)|=\eta>0$ on the $w=f(z)$-plane, then if $\eta>0$ is small, Λ separates a and b. Let γ_r be the part of a circle $|z|=r$, contained in D and Γ_r be its image on the w-plane, then if r is small, Γ_r connects a point a_1, which lies in a small neighbourhood of a with a point b_1, which lies in a small neighbourhood of b, so that Γ_r intersects Λ at a point w_1, such that $|(w_1-a)(w_1-b)|=\eta$, which is absurd, since if $w_1=f(z_1)$, $F(z_1)=(w_1-a)(w_1-b)\to 0$, when $r\to 0$, so that $a=b$.

Theorem VIII. 9. *Let D: $0<\arg z<\pi$, $0<|z|<1$ be a half-disc and C be a Jordan arc in D, which connects $z=0$ to a point z_0 on $|z|=1$. Let $f(z)$ be regular and bounded in D, such that*

$$|f(z)|\leqq 1 \text{ in } D, \quad |f(z)|\leqq m<1 \text{ on } C.$$

Let $\Delta: 0<\delta\leqq \arg z\leqq \pi-\delta$, $0<|z|\leqq \rho<1$ be a sector, then

$$|f(z)|\leqq m^{\frac{\alpha}{\pi}} \quad \text{in } \Delta_\rho,$$

where $$\frac{\alpha}{\pi}=\frac{2}{\pi}\tan^{-1}\left(\frac{1-\rho}{1+\rho}\tan\frac{\delta}{2}\right).$$

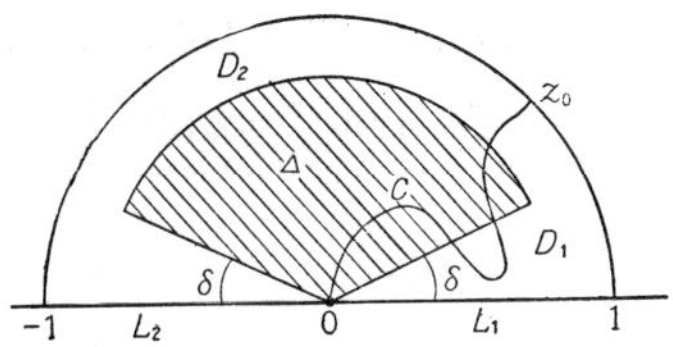

Proof. Let L_1: $0\leqq x\leqq 1$, $y=0$ and L_2: $-1\leqq x\leqq 0$, $y=0$ be two segments on the real axis. Now C decomposes D into two parts D_1, D_2, where D_1 contains L_1 on its boundary and D_2 contains L_2 on its boundary.

Let $\omega_2(z)$ be the harmonic measure of L_2 with respect to D, then by the lemma (p. 303),

$$\omega_2(z)\geqq \frac{\alpha}{\pi} \quad \text{in } \Delta\cdot D_1.$$

Let $\Omega_1(z)$ be the harmonic measure of C with respect to D_1, then $\Omega_1(z)\geqq \omega_2(z)$ in D_1, so that

$$\Omega_1(z)\geqq \frac{\alpha}{\pi} \quad \text{in } \Delta\cdot D_1.$$

Hence

$$|f(z)|\leqq m^{\Omega_1(z)}\leqq m^{\frac{\alpha}{\pi}} \quad \text{in } \Delta\cdot D_1.$$

Similarly considering the harmonic measure of L_1 with respect to D, we have $|f(z)|\leqq m^{\frac{\alpha}{\pi}}$ in $\Delta\cdot D_2$, so that

$$|f(z)|\leqq m^{\frac{\alpha}{\pi}} \quad \text{in } \Delta.$$ ——

From the theorem, we have

Theorem VIII. 10. *Let D: $0<\arg z<\pi$, $|z|<1$ be a half-disc and $f(z)$ be regular and bounded in D. If $\lim_{z\to 0} f(z)=a$, when $z\to 0$ along a Jordan arc C in D, ending at $z=0$, then $\lim_{z\to 0} f(z)=a$ uniformly, when $z\to 0$ from the inside of any angular domain: $0<\delta\leqq\arg z\leqq\pi-\delta$.*

Theorem VIII. 11. *Let E be a closed set, contained in $|z|\leqq 1$, such that any circle $|z|=r$ $(0\leqq r<1)$ intersects E and D be the complement of E with respect to $|z|<1$. Let $L(\theta)$: $z=re^{i\theta}$ $(0\leqq r\leqq 1)$ be a radius of $|z|=1$. Let $\omega(z;E)$ be the harmonic measure of E with respect to D and $\omega(z;L(\theta))$ be that of $L(\theta)$ with respet to its complement Δ_θ to $|z|<1$. Then*

$$\omega(re^{i\theta};E)\geqq\omega(re^{i\theta};L(-\theta))=\frac{2}{\pi}\sin^{-1}\left(\frac{1-r}{1+r}\right)\quad(0\leqq r<1),$$

where the equality holds, only when $E=L(-\theta)$. Nevanlinna [6].

Proof. Let $L=L(0)$: $0\leqq x\leqq 1$, $y=0$ be a segment on the real axis and let

$$u(z)=\int_L \log\left|\frac{1-xz}{z-x}\right|d\mu(x) \tag{1}$$

be its hyperbolic capacity potential, such that $u=1$ on L, $u=0$ on $|z|=1$ and $0\leqq u\leqq 1$ in $|z|\leqq 1$. Hence $u(z)=\omega(z;L)$.

Let $a\in E$ be the point of E, which lies on the circle $|z|=x$, such that $\arg a\geqq 0$ is the smallest. We associate this a to $x=|a|$ and we define a positive mass distribution $d\mu^*(a)$ on E by $d\mu^*(a)=d\mu(x)$ and put

$$v(z)=\int_E \log\left|\frac{1-\bar{a}z}{z-a}\right|d\mu^*(a), \tag{2}$$

then $v=0$ on $|z|=1$ and

$$v(z)\leqq\int_E \log\left|\frac{1-|a||z|}{|z|-|a|}\right|d\mu^*(a)=u(|z|)\leqq 1\quad\text{in } |z|\leqq 1. \tag{3}$$

Since $\omega(z;E)=1$ on E, except a set of capacity zero, we see by (3), $\omega(z;E)-v(z)\geqq 0$ "almost everywhere" on the boundary of D, so that

$$\omega(z;E)\geqq v(z)\quad\text{in } D. \tag{4}$$

Since

$$v(re^{i\theta})\geqq\int_E \log\frac{1+|a|r}{r+|a|}\,d\mu^*(a)=u(-r)=\omega(-r,L)=\omega(re^{i\theta},L(-\theta)),$$

we have by (4)

$$\omega(re^{i\theta};E)\geqq\omega(re^{i\theta};L(-\theta)). \tag{5}$$

We see easily that the equality holds, only when $E=L(-\theta)$.

By $w=\sqrt{z}$, we map Δ_0: $0<\arg z<2\pi$, $|z|<1$ conformally on the half-disc on the w-plane, then we have easily

$$u(-r)=\frac{2}{\pi}\sin^{-1}\left(\frac{1-r}{1+r}\right)^{1)}. \text{ —— } \tag{6}$$

From the theorem, we have

Theorem VIII. 12. *Under the same condition as Theorem* VIII 11, *let* $f(z)$ *be regular and bounded in* D, *such that*

$$|f(z)|\leqq 1 \quad in\ D, \qquad |f(z)|\leqq m<1 \quad on\ E,$$

then

$$|f(z)|\leqq m^{\frac{2}{\pi}\sin^{-1}\left(\frac{1-|z|}{1+|z|}\right)}<m^{\frac{1}{\pi}(1-|z|)} \quad in\ D. \qquad \text{Nevanlinna [6].}$$

Theorem VIII. 5 can be generalized as follows.

Theorem VIII. 13. *Let* $w=f(z)$ *be meromorphic in a sector* $\Delta: 0\leqq\arg z\leqq\theta_0$, $0<|z|\leqq R$ *and* $f(z)\neq\alpha, \beta, \gamma$ *in* Δ.

If $\lim_{z\to 0} f(z)=a$, *when* $z\to 0$ *on the positive real axis, then* $\lim_{r\to 0} f(re^{i\theta})=a$ *uniformly for* $0\leqq\theta\leqq\theta_1$ $(<\theta_0)$.

Proof. By a suitable linear transformation, we may assume that a, α, β, γ are finite. We map the outside of α, β, γ conformally on $|\zeta|<1$ by $\zeta=\Phi(w)$ and put $F(z)=\Phi(f(z))$, then $F(z)$ is regular and $|F(z)|<1$ in Δ.

(i) Suppose that $a\neq\alpha$, $a\neq\beta$, $a\neq\gamma$, then $\Phi(a)=\zeta_0$ lies in $|\zeta|<1$ and $\lim_{z\to 0} F(z)=\zeta_0$, when $z\to 0$ on the positive real axis, so that by Montel's theorem, $\lim_{r\to 0} F(re^{i\theta})=\zeta_0$ uniformly for $0\leqq\theta\leqq\theta_1$ $(<\theta_0)$. Since a neighbourhood of ζ_0 is mapped conformally on a neighbourhood of $w=a$, $\lim_{r\to 0} f(re^{i\theta})=a$ uniformly for $0\leqq\theta\leqq\theta_1$ $(<\theta_0)$.

(ii) Next suppose that a coincides with one of α, β, γ and suppose that $a=\gamma$, then by Lindelöf's artifice, (Lindelöf [1]), we consider $\sqrt{f(z)-a}$, then $\sqrt{f(z)-a}$ is meromorphic in Δ and $\sqrt{f(z)-a}\neq\pm\sqrt{\alpha-\gamma}$, $\pm\sqrt{\beta-\gamma}$, so that the number of exceptional values is 4.

Hence by (i), $\lim_{r\to 0}\sqrt{f(re^{i\theta})-a}=0$, or $\lim_{r\to 0} f(re^{i\theta})=a$ uniformly for $0\leqq\theta\leqq\theta_1$ $(<\theta_0)$. ——

---

1) 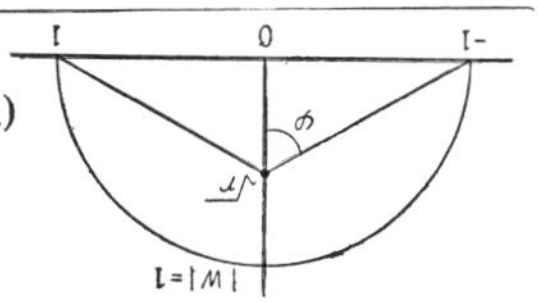

$$u(-r)=\frac{2}{\pi}\left(2\varphi-\frac{\pi}{2}\right) \text{ and } \sin\left(2\varphi-\frac{\pi}{2}\right)=-\cos 2\varphi$$

$$=\frac{\tan^2\varphi-1}{\tan^2\varphi+1}=\frac{1/r-1}{1/r+1}=\frac{1-r}{1+r}.$$

Similarly we can generalize Theorem VIII. 7 and 8 as follows.

Theorem VIII. 14. *Let D be a domain on the z-plane, which is bounded by a Jordan curve C and $z=0$ belong to C. Then by $z=0$, the part of C, which lies in a neighbourhood of $z=0$ is decomposed into two parts C_1, C_2.*

Let $f(z)$ be meromorphic in D and $f(z)\neq\alpha, \beta, \gamma$.

(i) *If $\lim_{z\to 0} f(z)=a$, when $z\to 0$ on C_1 and C_2, then $\lim_{z\to 0} f(z)=a$ uniformly, when $z\to 0$ from the inside of D.*

(ii) *If $\lim_{z\to 0} f(z)=a$, when $z\to 0$ on C_1 and $\lim_{z\to 0} f(z)=b$, when $z\to 0$ on C_2, then $a=b$, so that $\lim_{z\to 0} f(z)=a$ uniformly, when $z\to 0$ from the inside of D.*

Hence

Corollary. *If $f(z)$ is meromorphic in D and $\lim_{z\to 0} f(z)=a$, when $z\to 0$ on C_1 and $\lim_{z\to 0} f(z)=b$, when $z\to 0$ on C_2 and $a\neq b$, then $f(z)$ takes any value infinitely often in D, with two possible exceptions.*

2. Converse of Abel's theorem

1. Let $f(x)=\sum_{n=0}^{\infty} a_n x^n$ be regular in $|x|<1$ and $A=\sum_{n=0}^{\infty} a_n$ be convergent, then by Abel's theorem, $f(x)\to A$, when $x\to 1$ from the inside of any Stloz domain, whose vertex is at $x=1$. Hardy and Littlewood ([1], [2]) proved the following converse.

Theorem VIII. 15. *Let $f(x)=\sum_{n=0}^{\infty} a_n x^n$ be regular in $|x|<1$ and a_n $(n=0, 1, 2)$ be real and $na_n\leqq K$ $(K>0)$ $(n=1, 2, \cdots)$.*

If $\lim_{x\to 1-0} f(x)=A$ exists, when $x\to 1$ on the positive real axis, then $\sum_{n=0}^{\infty} a_n$ is convergent and $\sum_{n=0}^{\infty} a_n=A$.

In the following, we shall give Wielandt's simple proof. First we remark the following. If we consider $F(x)=\dfrac{f(x)-a_0-(A-a_0)x}{K+|A-a_0|}$ instead of $f(x)$, then we may assume that $A=0$, $a_0=0$, $a_n\leqq 1/n$ $(n=1, 2, \cdots)$. Hence we assume that

$$f(x)=\sum_{n=1}^{\infty} a_n x^n,\ a_n\leqq 1/n\ (n=1, 2, \cdots), \qquad \lim_{x\to 1-0} f(x)=0\,. \qquad (1)$$

With Wielandt, we consider a family $\mathfrak{F}$ of real valued functions $\varphi(x)$, defined for $0<x<1$, such that

(i) $\sum_{n=1}^{\infty} a_n \varphi(x^n)$ is convergent for $0<x<1$,

(ii) $\Phi(x)=\sum_{n=1}^{\infty} a_n\varphi(x^n)\to 0$, when $x\to 1-0$.

Then by the hypothesis, $\varphi(x)\equiv x\in\mathfrak{F}$.

It is easily seen that if $\varphi(x)\in\mathfrak{F}$, then $\varphi(x^k)\in\mathfrak{F}$ $(k=1, 2, \cdots)$ and a linear combination of functions $\in\mathfrak{F}$ is also a function $\in\mathfrak{F}$.

Hence, since $x\in\mathfrak{F}$, a polynomial $p(x)$, such that $p(0)=0$, belongs to $\mathfrak{F}$.

2. First we shall prove a lemma.

Lemma. *Let* $\varphi(x)$ $(0<x<1)$ *satisfy* (i). *If for any* $\varepsilon>0$, *there exist two polynomials* $p_1(x)$, $p_2(x)$, *such that*

$$p_i(0)=0, \quad p_i(1)=1 \ (i=1, 2),$$

$$p_1(x)\leqq\varphi(x)\leqq p_2(x)\,, \qquad \frac{p_2(x)-p_1(x)}{x(1-x)}=q(x)>0\,,$$

$$\int_0^1 q(x)dx<\varepsilon\,,$$

then $\varphi(x)$ *satisfies* (ii), *so that* $\varphi(x)\in\mathfrak{F}$.

Proof. If $\Phi(x)$ denotes the function (ii) and $q(x)=\sum_{\rho=0}^{r} b_\rho x^\rho$, then by the hypothesis,

$$\Phi(x)-\sum_{n=1}^{\infty} a_n p_1(x^n)=\sum_{n=1}^{\infty} a_n(\varphi(x^n)-p_1(x^n))\leqq\sum_{n=1}^{\infty}\frac{1}{n}(p_2(x^n)-p_1(x^n))$$

$$=\sum_{n=1}^{\infty}\frac{1}{n}(1-x^n)x^n q(x^n)\leqq(1-x)\sum_{n=1}^{\infty}x^n q(x^n)=(1-x)\sum_{\rho=0}^{r}b_\rho\sum_{n=1}^{\infty}x^{n(\rho+1)}$$

$$=\sum_{\rho=0}^{r}\frac{b_\rho(1-x)x^{\rho+1}}{1-x^{\rho+1}}\to\sum_{\rho=0}^{r}\frac{b_\rho}{1+\rho}=\int_0^1 q(x)\,dx<\varepsilon, \qquad \text{if } x\to 1-0\,.$$

Since $p_1(x)\in\mathfrak{F}$, $\sum_{n=1}^{\infty} a_n p_1(x^n)\to 0$ $(x\to 1-0)$, so that $\Phi(x)<\varepsilon$, if $x>1-\delta(\varepsilon)$.

Similarly considering $p_2(x)-\Phi(x)$, instead of $\Phi(x)-p_1(x)$, we have $\Phi(x)>-\varepsilon$, if $x>1-\delta(\varepsilon)$, so that $\Phi(x)\to 0$, when $x\to 1-0$, hence $\varphi(x)\in\mathfrak{F}$. ——

3. *Proof of the theorem.* (Wielandt [1]).

Let

$$\varphi^*(x)=\begin{cases}0 & \text{for } 0\leqq x<1/2,\\ 1 & \text{for } 1/2\leqq x\leqq 1\,.\end{cases} \tag{1}$$

By Weierstrass' theorem, for any $\varepsilon>0$, there exist two polynomials $q_i(x)$ $(i=1, 2)$, such that

$$q_1(x)\leqq\frac{\varphi^*(x)-x}{x(1-x)}\leqq q_2(x)\,, \qquad \int_0^1(q_2(x)-q_1(x))\,dx<\varepsilon\,. \tag{2}$$

If we put

$$p_i(x)=x+x(1-x)\,q_i(x)\ (i=1, 2)\,, \qquad q(x)=q_2(x)-q_1(x)\,, \tag{3}$$

then $p_i(0)=0$, $p_i(1)=1$,

$$p_1(x) \leqq \varphi^*(x) \leqq p_2(x)\,, \quad \frac{p_2(x)-p_1(x)}{x(1-x)} = q(x)\,, \quad \int_0^1 q(x)\,dx < \varepsilon\,.$$

Evidently $\varphi^*(x)$ satisfies the condition (i), so that by the lemma,

$$\varphi^*(x) \in \mathfrak{F}\,.$$

Now

$$\Phi(x) = \sum_{n=1}^{\infty} a_n \varphi^*(x^n) = \sum_{2x^n \geq 1} a_n = \sum_{n=1}^{N} a_n = s_N, \qquad N = \left[\log 2/\log \frac{1}{x}\right],$$

so that

$$s_N \to 0, \quad \text{as} \quad N \to \infty, \quad \text{or} \quad \sum_{n=1}^{\infty} a_n = 0\,. \text{ ——}$$

4. Next we shall prove a Tauberian theorem for a curvilnear approach.

Theorem VIII. 16. *Let $f(z) = \sum_{n=0}^{\infty} a_n z^n$ be regular in $|z|<1$ and $n|a_n| \leqq K$ $(n=1, 2, \cdots)$. If $f(z) \to A$, when $z \to 1$ along a Jordan arc C in $|z|<1$ ending at $z=1$, then $\sum_{n=0}^{\infty} a_n$ is convergent and $\sum_{n=0}^{\infty} a_n = A$.* Littlewood [1].

Proof. We may assume that C is an analytic Jordan arc and $A=0$, $n|a_n| \leqq 1$ $(n=1, 2, \cdots)$. By $z = e^{-\zeta}$, we map $|z|<1$ on $\Re\zeta > 0$ and put $f(z) = F(\zeta)$, then

$$F(\zeta) = \sum_{n=0}^{\infty} a_n e^{-n\zeta}, \quad \zeta = \sigma + it\,. \tag{1}$$

$F(\zeta)$ is regular in $\Re\zeta>0$ and $F(\zeta) \to 0$, when $\zeta \to 0$ along a Jordan arc Γ, ending at $\zeta=0$, which is the image of C. We may assume that

$$|F(\zeta)| \leqq 1 \text{ on } \Gamma\,.$$

$$|F'(\zeta)| \leqq \sum_{n=1}^{\infty} n|a_n| e^{-n\sigma} \leqq \sum_{n=1}^{\infty} e^{-n\sigma} = \frac{1}{e^{\sigma}-1} < \frac{1}{\sigma}, \quad \sigma > 0\,. \tag{2}$$

Let Δ: $|\arg\zeta| \leqq \alpha < \pi/2$ be an angular domain and $\zeta' = \sigma + it'$, $\zeta'' = \sigma + it''$ be two points of Δ, which lie on the same vertical line, then by (2),

$$|F(\zeta') - F(\zeta'')| \leqq \frac{|t'-t''|}{\sigma} \leqq 2 \tan\alpha\,. \tag{3}$$

We consider two cases: (i) Γ is a Stolz path, (ii) Γ is not a Stolz path.

(i) Γ is a Stolz path.

Then Γ is contained in Δ for a suitable α. Let ζ' be a point of Γ, then by (3), for any $\zeta \in \Delta$, which lies on the same vertical line as ζ', $|F(\zeta)| \leqq |F(\zeta')| + 2\tan\alpha \leqq 1 + 2\tan\alpha$, so that $F(\zeta)$ is bounded in Δ and $F(\zeta) \to 0$, when $\zeta \to 0$ on Γ, so that by Theorem VIII. 10, $F(\zeta) \to 0$, when $\zeta \to 0$ along the positive real axis, hence by Theorem VIII. 15, $\sum_{n=0}^{\infty} a_n = 0$.

(ii) Γ is not a Stolz path.

Then there occurs four cases: $(\text{ii})_1$ Γ intersects any half-line L:

$\arg\zeta=\frac{\pi}{2}-\delta$ $(\delta>0)$ infinitely often, if $0<\delta\leqq\delta_0$, $(\mathrm{ii})_2$ Γ touches the positive imaginary axis, $(\mathrm{ii})_3$ Γ intersects any half-line L': $\arg\zeta=-\frac{\pi}{2}+\delta$ $(\delta>0)$, infinitely often, if $0<\delta\leqq\delta_0$, $(\mathrm{ii})_{44}$ Γ touches the negative imaginary axis.

Since the proof is the same, we consider the case $(\mathrm{ii})_1$, $(\mathrm{ii})_2$. First we consider the case $(\mathrm{ii})_1$, then Γ intersects L at A_ν, B_ν $(\nu=1,2,\cdots)$, such that the part Γ_ν of Γ, which is bounded by A_ν, B_ν lies between L and the positive imaginary axis. Let D_ν be the simply connected domain, which is bounded by Γ_ν and the segment $A_\nu B_\nu$. We shall prove that

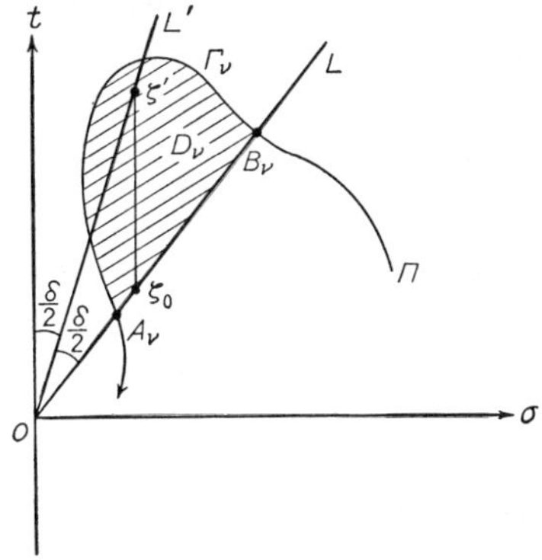

$$|F(\zeta)|\leqq 1+8\cot\delta/2=K \quad \text{in } D_\nu\,. \tag{4}$$

Let $M_\nu=\underset{\zeta\in A_\nu B_\nu}{\mathrm{Max}}|F(\zeta)|$ and

$$M_\nu=|F(\zeta_0)|\,,\quad \zeta_0\in A_\nu B_\nu\,.$$

If $M_\nu\leqq 1$, then since $|F(\zeta)|\leqq 1$ on Γ_ν, by the maximum principle, $|F(\zeta)|\leqq 1$ in D_ν, so that (4) holds.

If $M_\nu>1$, then $\zeta_0\neq A_\nu, B_\nu$. Let L': $\arg\zeta=\frac{\pi}{2}-\frac{\delta}{2}$ be an half-line and ζ' be the point of L', which lies on the same vertical line as ζ_0. If $\zeta'\in D_\nu$, then since $|F(\zeta)|\leqq 1$ on Γ, by Theorem VIII. 3, $|F(\zeta')|\leqq\sqrt{M_\nu}$. On the other hand, by (3), $|F(\zeta')|\geqq|F(\zeta_0)|-2\tan\left(\frac{\pi}{2}-\frac{\delta}{2}\right)=M_\nu-2\cot\delta/2$, so that $M_\nu-2\cot\delta/2\leqq\sqrt{M}$, hence

$$M_\nu\leqq\left(\frac{1+\sqrt{1+8\cot\delta/2}}{2}\right)^2\leqq 1+8\cot\delta/2=K\,.$$

If ζ' lies outside of D_ν, then the segment $\overline{\zeta_0\zeta'}$ intersects Γ at ζ'', so that by (3), $M_\nu=|F(\zeta_0)|\leqq|F(\zeta'')|+2\cot\delta/2\leqq 1+2\cot\delta/2\leqq K$. Hence in any case $M_\nu\leqq K$. Since $|F(\zeta)|\leqq 1$ on Γ_ν, by the maximum principle, we have (4).

By the hypothesis, L': $\arg\zeta=\frac{\pi}{2}-\frac{\delta}{2}$ penetrates into infinitely many D_ν, so that by a suitable change of notation, we assume that L' penetrates into D_ν $(\nu=1,2,\cdots)$.

Now the part of D_ν, which lies between L' and L consists of a finite number of simply connected domains. Let D'_ν be such one, which abutts on $A_\nu B_\nu$ and $\gamma_\nu+A_\nu B_\nu$ be its boundary, where γ_ν consists of a part Γ'_ν of Γ_ν and a part L'_ν of L', which lies in D_ν. Let $|F(\zeta)|\leqq\varepsilon_\nu$ on Γ_ν, then $\varepsilon_\nu\to 0$ as $\nu\to\infty$. Since by (4), $|F(\zeta)|\leqq K$ on $A_\nu B_\nu$, we have by Theorem VIII. 3, $|F(\zeta)|\leqq\sqrt{K\varepsilon_\nu}$ on L'_ν. Hence if $\nu\geqq\nu_0$,

$$|F(\zeta)| \leqq \sqrt{K}\,\varepsilon_\nu \text{ on } \gamma_\nu . \tag{5}$$

We modify Γ into Γ', by replacing Γ_ν by γ_ν ($\nu=1, 2, \cdots$), then Γ' lies between L' and the negative imaginary axis and by (5), $F(\zeta)\to 0$, when $\zeta\to 0$ along Γ'. Hence if Γ' is a Stolz path, then by (i), $\sum_{n=0}^{\infty} a_n=0$. If Γ' is not a Stolz path, then Γ' intersects the real axis infinitely often, then we modify Γ' into Γ'', such that Γ'' lies between L' and the positive real axis and $F(\zeta)\to 0$, when $\zeta\to 0$ along Γ''. Since Γ'' is a Stolz path, we have $\sum_{n=0}^{\infty} a_n=0$.

Next we consider case $(\text{ii})_2$. Then Γ touches the positive imaginary axis. Let L: $\arg\zeta=\frac{\pi}{2}-\delta$ ($\delta>0$) be a half-line and A be one of common points of L with Γ, which lies nearest to the origin O, then the segment OA and Γ bounds a simply connected domain D. We draw a line L', parallel to L, through a point $\zeta=it$ ($t>0$) and let D' be the part of D, which lies between L' and the positive imaginary axis. Then by (4), $|F(\zeta)|\leqq K$ in D', so that making $t\to 0$, we see that $|F(\zeta)|\leqq K$ in D, hence $F(\zeta)$ is bounded in D.

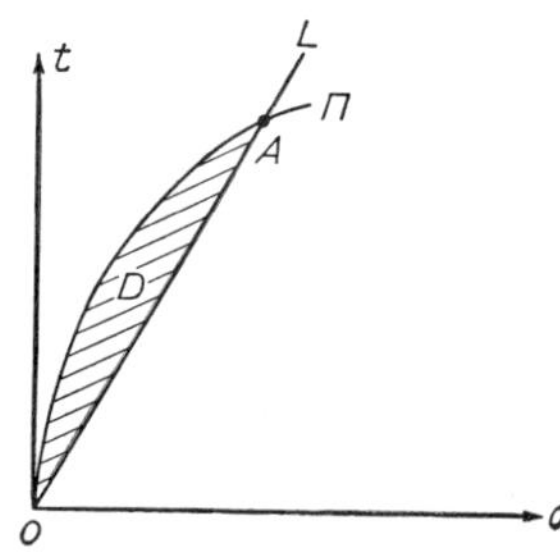

Let L_1: $\arg\zeta=\frac{\pi}{2}-\delta_1$ ($0<\delta_1<\delta/2$) be a half-line and we define the segment OA_1 and a simply connected domain D_1 for L_1 as before.

We reflect D_1 with respect to L_1 and let D_1^* be the thus obtained domain. Then if $\delta_1>0$ is small, D_1^* lies in D. Let ζ be any point of D_1 and ζ^* be its inverse point $\in D_1^*$ with respect to L_1 and put $H(\zeta)=F(\zeta)\overline{F}(\zeta^*)$, then $H(\zeta)$ is regular in $\Delta=D_1+D_1^*$ and is bounded and $H(\zeta)\to 0$, when $\zeta\to 0$ on the boundary of Δ, so that if $\zeta\to 0$ on L_1, $H(\zeta)=|F(\zeta)|^2\to 0$, hence $F(\zeta)\to 0$, if $\zeta\to 0$ on L_1. Since L_1 is a Stolz path, $\sum_{n=0}^{\infty} a_n=0$. Hence the theorem is proved.

5. In Theorem VIII. 16, the order $a_n=O(1/n)$ is the best possible, since the following theorem holds.

Theorem VIII. 17. *Let $\Phi(n)>0$ be any function of n, such that $\lim_{n\to\infty}\Phi(n)=\infty$. Then there exists $f(x)=\sum_{n=0}^{\infty} a_n x^n$, $a_n=O(\Phi(n)/n)$ ($n=1, 2, \cdots$), which is regular in $|x|<1$ and $\lim_{x\to 1-0} f(x)$ exists, but $\sum_{n=0}^{\infty} a_n$ is divergent.*

Littlewood [1].

Proof. We put $\Phi^*(n)=\text{Min}\,(\sqrt{n},\ \Phi(n),\ \Phi(n+1),\ \cdots)$, then

$$\left.\begin{array}{c}\Phi^*(1)\leqq\Phi^*(2)\leqq\cdots\leqq\Phi^*(n)\to\infty\ ,\\ \Phi^*(n)\leqq\sqrt{n}\ .\end{array}\right\} \tag{1}$$

We put $\Delta\chi_n=\chi_n-\chi_{n-1}$ for any sequence $\{\chi_n\}$ and we define the sequence $\{\psi_n\}$ by

$$\Delta\psi_n=\frac{\Phi^*(n)}{n}\,,\qquad \psi_0=0\,. \tag{2}$$

Then we can prove easily that
(a) $\psi_n\to\infty$, (b) $\Delta\psi_n\to 0$, (c) $n\Delta\psi_n$ increases with n and $\to\infty$.

Let $\quad s_n=a_1+\cdots+a_n\,,\qquad \sigma_n=\dfrac{s_1+\cdots+s_n}{n}$, then

$$s_n=n\sigma_n-(n-1)\,\sigma_{n-1}=n\Delta\sigma_n+\sigma_{n-1}\,.$$

We define $\{\sigma_n\}$ by

$$f_n=\frac{\Delta e^{i\psi_n}}{\Delta\psi_n}\,,\qquad \Delta\sigma_n=\frac{f_n}{n}\,,\qquad \sigma_0=0\,, \tag{3}$$

then

$$s_n=f_n+\sigma_{n-1}\,, \tag{4}$$

$$\sigma_n=\sum_{\nu=1}^{n}\frac{f_\nu}{\nu}\,,\qquad e^{i\psi_n}=1+\sum_{\nu=1}^{n}f_\nu\Delta\psi_\nu\,, \tag{5}$$

so that

$$\sum_{n=1}^{\infty}\frac{f_n}{n}=\sum_{n=1}^{\infty}\frac{f_n\Delta\psi_n}{n\Delta\psi_n}=\sum_{n=1}^{\infty}\lambda_n f_n\Delta\psi_n,\qquad \lambda_n=1/n\Delta\psi_n\,.$$

Since by (5), $\sum_{\nu=1}^{n}f_\nu\Delta\psi_\nu=O(1)$ and by (c), $\lambda_1\geqq\lambda_2\geqq\cdots\geqq\lambda_n\to 0$, $\sum_{n=1}^{\infty}\dfrac{f_n}{n}$ is convergent, hence $\lim\limits_{n\to\infty}\sigma_n$ exists, so that $\sum_{n=1}^{\infty}a_n$ is summable $(C,1)$, hence $f(x)=\sum_{n=1}^{\infty}a_nx^n$ is regular in $|x|<1$ and $\lim\limits_{x\to1-0}f(x)$ exists.

Since

$$f_n=\frac{e^{i\psi_n}-e^{i\psi_{n-1}}}{\Delta\psi_n}=\frac{e^{i\psi_n}(1-e^{-i\Delta\psi_n})}{\Delta\psi_n}\sim ie^{i\psi_n}\qquad (n\to\infty),$$

by (a), (b), f_n oscillates finitely, so that by (4), s_n oscillates finitely, hence $\sum_{n=1}^{\infty}a_n$ is divergent, and

$$\begin{aligned}a_n=\Delta s_n=\Delta f_n+\Delta\sigma_{n-1}&=\Delta\left(\frac{e^{i\psi_n}(1-e^{-i\Delta\psi_n})}{\Delta\psi_n}\right)+\frac{f_{n-1}}{n-1}\\ &=\Delta(ie^{i\psi_n}+O(\Delta\psi_n))+O(1/n)=O(\Delta\psi_n)+O(\Delta\psi_{n-1})+O(1/n)\\ =O\left(\frac{\Phi^*(n)}{n}\right)&+O\left(\frac{\Phi^*(n-1)}{n-1}\right)+O(1/n)=O\left(\frac{\Phi^*(n)}{n}\right)=O\left(\frac{\Phi(n)}{n}\right).\end{aligned}$$

3. Function of H^p-class

1. Let $f(z)$ be regular in $|z|<1$ and put

$$I_p(r)=\frac{1}{2\pi}\int_0^{2\pi}|f(re^{i\theta})|^p\,d\theta,\ 0\leqq r<1,\ p>0\,.$$

Theorem VIII. 18. *$I_p(r)$ is an increasing function of r and $\log I_p(r)$ is a convex function of $\log r$.* Hardy [1].

Proof. Since $|f(z)|^p$ is subharmonic, $I_p(r)$ is an increasing function of r. (Theorem II. 10). Since $|z|^a|f(z)|^p\,(a=\text{real})$ is subharmonic in $0<|z|<1$, $r^aI_p(r)$ is a convex function of $\log r$ (Theorem II. 9), so that if $0<r_1<r_2<r_3<1$,

$$r_2^aI_p(r_2)\leqq \operatorname{Max}(r_1^aI_p(r_1),\ r_3^aI_p(r_3))\,. \tag{1}$$

If we define a by the condition $r_1^aI_p(r_1)=r_3^aI_p(r_3)$, then

$$r_2^aI_p(r_2)\leqq r_1^aI_p(r_1) \quad\text{and}\quad a=\frac{\log I_p(r_1)-\log I_p(r_3)}{\log r_3-\log r_1},$$

so that

$$I_p(r_2)\leqq I_p(r_1)^{\frac{\log r_3-\log r_2}{\log r_3-\log r_1}}\cdot I_p(r_3)^{\frac{\log r_2-\log r_1}{\log r_3-\log r_1}} \tag{2}$$

hence $\log I_p(r)$ is a convex function of $\log r$.

We can prove easily that if $f(z)\not\equiv\text{const.}$, then

$$I_p(r_1)<I_p(r_2)\,,\ \text{ if }\ 0\leqq r_1<r_2<1\,. \tag{3}$$

If $I_p(r)=O(1)$ $(0\leqq r<1)$, then we say that $f(z)$ *belongs to H^p-class* and denote $f(z)\in H^p$. Then

$$\lim_{r\to 1}I_p(r)=M_p<\infty,\qquad I_p(r)\leqq M_p\ (0\leqq r<1)\,. \tag{4}$$

Theorem VIII. 19. *Let $f(z)\in H^p$ $(f\not\equiv 0)$, then $f(z)$ is of bounded type, so that for almost all $e^{i\theta}$, $\lim_{z\to e^{i\theta}}f(z)=f(e^{i\theta})$ $(\neq 0,\ \neq\infty)$ exists uniformly, when $z\to e^{i\theta}$ from the inside of any fixed Stolz domain, whose vertex is at $e^{i\theta}$ and $|f(e^{i\theta})|^p$ and $|\log|f(e^{i\theta})||$ are integrable on $|z|=1$.*

Proof. Since $\log^+|f|\leqq|f|^p+\text{const.}$,

$$T(f,r)=\frac{1}{2\pi}\int_0^{2\pi}\log^+|f(re^{i\theta})|d\theta+O(1)\leqq\frac{1}{2\pi}\int_0^{2\pi}|f(re^{i\theta})|^p\,d\theta+O(1)=O(1),$$

so that $f(z)$ is of bounded type, hence the remaining part of the theorem follows from Theorem V. 9. ——

2. Since $T(f,r)=O(1)$, if $z_n\neq 0$ be zero points of $f(z)$ in $|z|<1$, then

$\sum_n (1-|z_n|)<\infty$, so that if $f(z)$ has a zero point of order $\mu\geqq 0$ at $z=0$, then *the Blaschke's product:*

$$B(z)=z^\mu \prod_n \frac{|z_n|}{z_n}\cdot\frac{z_n-z}{1-\bar{z}_n z}=z^\mu \prod_n\left(1-\frac{(1-|z_n|)(z_n+|z_n|z)}{z_n(1-\bar{z}_n z)}\right) \qquad (1)$$

is convergent in $|z|<1$, so that $B(z)$ is regular and $|B(z)|<1$ in $|z|<1$.

Theorem VIII. 20. (F. Riesz' decomposition theorem). *If $f(z)\in H^p$, then*

$$f(z)=B(z)\,F(z)\,,$$

where $B(z)$ is the Blaschke's product, $F(z)\in H^p$ and $F(z)\neq 0$ in $|z|<1$.

F. Riesz [1].

Proof. Let $B_n(z)=z^\mu \prod_{\nu=1}^{n}\frac{|z_\nu|}{z_\nu}\cdot\frac{z_\nu-z}{1-\bar{z}_\nu z}$, then since $|B_n(z)|=1$ on $|z|=1$, $|B_n(z)|\geqq 1-\varepsilon$, if $r_0(\varepsilon)\leqq|z|<1$, so that

$$\frac{1}{2\pi}\int_0^{2\pi}\left|\frac{f(re^{i\theta})}{B_n(re^{i\theta})}\right|^p d\theta\leqq\frac{1}{2\pi(1-\varepsilon)^p}\int_0^{2\pi}|f(re^{i\theta})|^p d\theta\leqq\frac{M_p}{(1-\varepsilon)^p},\quad r_0(\varepsilon)\leqq r<1. \quad (1)$$

Since the left hand side is an increasing function of r, (1) holds for $0\leqq r<1$. Hence if we make $n\to\infty$ and then $\varepsilon\to 0$, we have

$$\frac{1}{2\pi}\int_0^{2\pi}\left|\frac{f(re^{i\theta})}{B(re^{i\theta})}\right|^p d\theta\leqq M_p\,.$$

Hence if we put $f(z)=B(z)F(z)$, then $F(z)\in H^p$ and $F(z)\neq 0$ in $|z|<1$.

Theorem VIII. 21. *If $f(z)\in H^p$, then*

$$f(z)=f_1(z)+f_2(z) \qquad in\ |z|<1,$$

where $f_i(z)\in H^p$, $f_i(z)\neq 0$ in $|z|<1$ $(i=1, 2)$.

Proof. Since $|B(z)|<1$ in $|z|<1$, if we put $f_1(z)=(B(z)-1)F(z)$, $f_2(z)=F(z)$, then $f(z)=f_1(z)+f_2(z)$ and $f_i(z)$ satisfies the required condition.

Theorem VIII. 22. *If $f(z)\in H^p$, then*

$$\lim_{r\to 1}\int_0^{2\pi}|f(re^{i\theta})-f(e^{i\theta})|^p d\theta=0\,,$$

$$M_p=\lim_{r\to 1}\frac{1}{2\pi}\int_0^{2\pi}|f(re^{i\theta})|^p d\theta=\frac{1}{2\pi}\int_0^{2\pi}|f(e^{i\theta})|^p d\theta\,.$$

F. Riesz [1].

Proof. First suppose that $p=2$ and let $f(z)=\sum_{n=0}^{\infty}a_n z^n$, then since

$$\frac{1}{2\pi}\int_0^{2\pi}|f(re^{i\theta})|^2 d\theta=\sum_{n=0}^{\infty}|a_n|^2 r^{2n}\leqq M_2\,,$$

we have $\sum_{n=0}^{\infty}|a_n|^2<\infty$ and from this, we can prove easily that

$$\lim_{r\to 1}\int_0^{2\pi}|f(re^{i\theta})-f(e^{i\theta})|^2 d\theta=0\,. \tag{1}$$

Hence for any measurable set E on $|z|=1$,

$$\lim_{r\to 1}\int_E |f(re^{i\theta})-f(e^{i\theta})|^2 d\theta \leqq \lim_{r\to 1}\int_0^{2\pi}|f(re^{i\theta})-f(e^{i\theta})|^2 d\theta=0\,,$$

so that if $mE<\varepsilon$,

$$\int_E |f(e^{i\theta})|^2 d\theta \leqq \delta(\varepsilon), \qquad \int_E |f(re^{i\theta})|^2 d\theta \leqq \delta(\varepsilon), \qquad r_0(\varepsilon)\leqq r<1\,, \tag{2}$$

where $\lim_{\varepsilon\to 0}\delta(\varepsilon)=0$.

In the general case, let $f(z)=B(z)F(z)$ be the Riesz' decomposition and let $\Phi(z)=[F(z)]^{\frac{p}{2}}$, then $\Phi(z)\in H^2$ and $|f(z)|^p\leqq|F(z)|^p=|\Phi(z)|^2$, hence by (2), if $mE<\varepsilon$,

$$\int_E |f(e^{i\theta})|^p d\theta \leqq \delta(\varepsilon), \qquad \int_E |f(re^{i\theta})|^p d\theta \leqq \delta(\varepsilon), \qquad r_0(\varepsilon)\leqq r<1\,. \tag{3}$$

Since $\lim_{r\to 1} f(re^{i\theta})=f(e^{i\theta})$ almost everywhere, by Egoroff's theorem, there exists a closed set E of measure $>2\pi-\varepsilon$, such that $\lim_{r\to 1} f(re^{i\theta})=f(e^{i\theta})$ uniformly on E, so that

$$\lim_{r\to 1}\int_E |f(re^{i\theta})-f(e^{i\theta})|^p d\theta=0\,. \tag{4}$$

Let E' be the complement of E on $|z|=1$, then $mE'<\varepsilon$ and

$$\begin{aligned}\int_0^{2\pi}|f(re^{i\theta})-f(e^{i\theta})|^p d\theta &= \int_E |f(re^{i\theta})-f(e^{i\theta})|^p d\theta+\int_{E'}|f(re^{i\theta})-f(e^{i\theta})|^p d\theta\\ &\leqq \int_E |f(re^{i\theta})-f(e^{i\theta})|^p d\theta+2^p\int_{E'}(|f(re^{i\theta})|^p+|f(e^{i\theta})|^p)d\theta\\ &\leqq \int_E |f(re^{i\theta})-f(e^{i\theta})|^p d\theta+2^{p+1}\delta(\varepsilon)\,.\end{aligned}$$

Hence by (4)

$$\lim_{r\to 1}\int_0^{2\pi}|f(re^{i\theta})-f(e^{i\theta})|^p d\theta=0 \tag{5}$$

and from this, we have easily

$$\lim_{r\to 1}\int_0^{2\pi}|f(re^{i\theta})|^p d\theta=\int_0^{2\pi}|f(e^{i\theta})|^p d\theta\,. \tag{6}$$

Theorem VIII. 23. *If $f(z)\in H^p$ $(p\geqq 1)$, then $f(z)$ can be expressed by a Poisson integral:*

$$f(z)=\frac{1}{2\pi}\int_0^{2\pi} f(e^{i\theta})\frac{1-|z|^2}{|z-e^{i\theta}|^2}\,d\theta\,, \qquad |z|<1\,.$$

Proof. Since by Theorem VIII. 22, $f(e^{i\theta})\in L^p$ $(p\geqq 1)$, if we put

$$u(z)=\frac{1}{2\pi}\int_0^{2\pi} f(e^{i\theta})\frac{1-|z|^2}{|z-e^{i\theta}|^2}\,d\theta\,, \tag{1}$$

then by Theorem IV. 4,

$$\lim_{r\to 1}\int_0^{2\pi} |u(re^{i\theta})-f(e^{i\theta})|^p d\theta=0\,,$$

and since $\lim_{r\to 1}\int_0^{2\pi} |f(re^{i\theta})-f(e^{i\theta})|^p d\theta=0$, if we put $w(z)=f(z)-u(z)$, then

$$\lim_{r\to 1}\int_0^{2\pi} |w(re^{i\theta})|^p d\theta=0\,. \tag{2}$$

Since $|w(z)|^p$ is subharmonic, $\int_0^{2\pi} |w(re^{i\theta})|^p d\theta$ is an increasing function of r, so that $\int_0^{2\pi} |w(re^{i\theta})|^p\, d\theta\equiv 0$, hence $w\equiv 0$, or $f\equiv u$. ——

Remark. Let $f(z)=u(z)+iv(z)$. If (i) $\int_0^{2\pi}|u(re^{i\theta})|d\theta=O(1)$, and (ii) $\int_0 |v(re^{i\theta})|d\theta=O(1)$, then $\int_0^{2\pi} | f(re^{i\theta})|d\theta=O(1)$, so that $f(z)$ and hence $u(z)$ can be expressed by a Poisson integral, but under the condition (i) only, $u(z)$ can be expressed by a Poisson-Stieltjes integral.

(Theorem IV. 16 and 14).

Theorem VIII. 24. *Let* $0<|a_n|<1$, *such that* $\sum_{n=1}^{\infty}(1-|a_n|)<\infty$ *and*

$$B(z)=\prod_{n=1}^{\infty}\frac{|a_n|}{a_n}\cdot\frac{a_n-z}{1-\bar{a}_n z}$$

be the Blaschke's product formed with $\{a_n\}$, *then* $|B(e^{i\theta})|=1$ *almost everywhere on* $|z|=1$.

Proof. We put $B_n(z)=\prod_{\nu=1}^{n}\frac{|a_\nu|}{a_\nu}\cdot\frac{a_\nu-z}{1-\bar{a}_\nu z}$, $\qquad B(z)=B_n(z)R_n(z)\,.$

Since $|B_n(z)|=1$ on $|z|=1$,

$$\frac{1}{2\pi}\int_0^{2\pi}|R_n(e^{i\theta})|d\theta=\frac{1}{2\pi}\int_0^{2\pi} |B(e^{i\theta})|d\theta\leqq 1\,. \tag{1}$$

$$\prod_{\nu=1}^{\infty}|a_{n+\nu}|=|R_n(0)|\leqq\frac{1}{2\pi}\int_0^{2\pi}|R_n(re^{i\theta})|d\theta\leqq\lim_{r\to 1}\frac{1}{2\pi}\int_0^{2\pi}|R_n(re^{i\theta})|d\theta$$
$$=\frac{1}{2\pi}\int_0^{2\pi}|R_n(e^{i\theta})|d\theta\leqq 1\,.$$

Since $\sum_{n=1}^{\infty}(1-|a_n|)<\infty$. $\prod_{n=1}^{\infty}|a_n|$ is convergent, so that $\lim_{n\to\infty}\prod_{\nu=1}^{\infty}|a_{n+\nu}|=1$, hence

$$\lim_{n\to\infty}\frac{1}{2\pi}\int_0^{2\pi}|R_n(e^{i\theta})|d\theta=1\,, \tag{2}$$

so that by (1),

$$\frac{1}{2\pi}\int_0^{2\pi}|B(e^{i\theta})|d\theta=1\,. \tag{3}$$

Since $|B(e^{i\theta})|\leqq 1$, $|B(e^{i\theta})|=1$ almost everywhere on $|z|=1$. ——

Theorem VIII. 25. *If $f(z)\in H^\alpha$ and $f(e^{i\theta})\in L^\beta$ $(\beta>\alpha)$, then $f(z)\in H^\beta$.*

Smirnoff [1].

Proof. First suppose that $f(z)\neq 0$ in $|z|<1$ and put

$$[f(z)]^{\frac{\alpha}{2}}=\Phi(z)\in H^2,$$

$$\Phi(re^{i\theta})=\frac{1}{2\pi}\int_0^{2\pi}\Phi(e^{i\varphi})\frac{1-r^2}{1-2r\cos(\varphi-\theta)+r^2}d\varphi\,,\qquad 0\leqq r<1\,.$$

Since $\Phi(e^{i\theta})\subset L^{\frac{2\beta}{\alpha}}$, we have by Hölder's inequality,

$$|\Phi(re^{i\theta})|^{\frac{2\beta}{\alpha}}\leqq\frac{1}{2\pi}\int_0^{2\pi}|\Phi(e^{i\varphi})|^{\frac{2\beta}{\alpha}}\frac{1-r^2}{1-2r\cos(\varphi-\theta)+r^2}d\varphi\,,$$

so that

$$\int_0^{2\pi}|\Phi(re^{i\theta})|^{\frac{2\beta}{\alpha}}d\theta\leqq\int_0^{2\pi}|\Phi(o^{i\varphi})|^{\frac{2\beta}{\alpha}}d\varphi\,,$$

hence $\Phi(z)\in H^{\frac{2\beta}{\alpha}}$, so that $f(z)\in H^\beta$.

In the general case, let $f(z)=B(z)F(z)$ be the Riesz' decomposition, where $F(z)\neq 0$, $F(z)\in H^\alpha$. Since $|B(e^{i\theta})|=1$ almost everywhere on $|z|=1$, $F(e^{i\theta})\in L^\beta$, so that $F(z)\in H^\beta$. Since $|B(z)|\leqq 1$, $f(z)\in H^\beta$. ——

4. F. and M. Riesz' theorem on conformal mapping

1. We shall prove the following **F. and M. Riesz' theorem.**

Theorem VIII. 26. *Let D be a simply connected domain on the w-plane, which is bounded by a rectifiable Jordan curve C. If we map D conformally on $|z|<1$ by $w=f(z)$, then $f(e^{i\theta})$ is an absolutely continuous function on $|z|=1$ and a null set on $|z|=1$ is mapped on a null set on C and vice versa, a null set on C is mapped on a null set on $|z|=1$[1]. If L be the length of C, then*

$$L=\int_0^{2\pi}|f'(e^{i\theta})|d\theta\,,$$

where $f'(e^{i\theta})=\lim\limits_{z\to e^{i\theta}}f'(z)$, when $z\to e^{i\theta}$ from the inside of any stolz domain, whose vertex is at $e^{i\theta}$.

F. and M. Riesz [1].

Proof. Let C_r be the image of $|z|=r$ (<1) in D and $L(r)$ be its

1) The measure of a set on C is defined by means of the arc length of C.

length, then

$$L(r)=\int_0^{2\pi}|f'(re^{i\theta})|rd\theta\,. \tag{1}$$

Let $\omega=e^{\frac{2\pi i}{n}}$ and put

$$\Lambda_n(z)=|f(z)-f(\omega z)|+|f(\omega z)-f(\omega^2 z)|+\cdots+|f(\omega^{n-1}z)-f(z)|\,, \tag{2}$$

then $\Lambda_n(z)$ is subharmonic in $|z|<1$ and is continuous in $|z|\leqq 1$, so that it attaines its maximum at a point z_0 on $|z|=1$, so that

$$\Lambda_n(r)\leqq\Lambda_n(z_0)\quad(0\leqq r\leqq 1)\,.$$

Since $\Lambda_n(z_0)$ is the length of a polygon, inscribed in C, $\Lambda_n(z_0)\leqq L$. Hence $\Lambda_n(r)\leqq L$ $(0\leqq r<1)$. Since $\lim_{n\to\infty}\Lambda_n(r)=L(r)$, we have $L(r)\leqq L$ $(0\leqq r<1)$, so that by (1), $f'(z)\in H^1$, hence $\lim_{r\to 1}f'(re^{i\theta})=f'(e^{i\theta})$ exists almost everywhere on $|z|=1$ and by Theorem VIII. 22,

$$L\geqq\lim_{r\to 1}L(r)=\lim_{r\to 1}\int_0^{2\pi}|f'(re^{i\theta})|r\,d\theta=\int_0^{2\pi}|f'(e^{i\theta})|d\theta\,. \tag{3}$$

Since as easily seen, $\underline{\lim}_{r\to 1}L(r)\geqq L$, we have

$$L=\int_0^{2\pi}|f'(e^{i\theta})|d\theta\,. \tag{4}$$

If we put $F(\theta)=f(e^{i\theta})$, then since C is a rectifiable curve, $F(\theta)$ is a function of bounded variation of θ, so that $F'(\theta)=\dfrac{dF(\theta)}{d\theta}$ exists almost everywhere. Since $f(z)$ is continuous in $|z|\leqq 1$, $f(z)$ can be expressed by a Poisson integral with respect to $F(\theta)$, so that by Theorem IV. 5, if $F'(\theta)$ exists at $\theta=\theta_0$, then

$$\frac{\partial f(re^{i\varphi})}{\partial\varphi}=ire^{i\varphi}f'(re^{i\varphi})\to F'(\theta_0)\,, \tag{5}$$

when $z=re^{i\varphi}\to e^{i\theta_0}$ from the inside of any Stolz domain, whose vertex is at $e^{i\theta_0}$, so that $ie^{i\theta}f'(e^{i\theta})=F'(\theta)$ almost everywhere on $|z|=1$, hence by (4),

$$L=\int_0^{2\pi}|F'(\theta)|d\theta\,, \tag{6}$$

so that by Tonelli's theorem (Saks [1]), $F(\theta)=f(e^{i\theta})$ is absolutely continuous on $|z|=1$, hence if $E\in C$ be the image of a set e on $|z|=1$, then

$$mE=\int_e|F'(\theta)|d\theta\,, \tag{7}$$

so that a null set on $|z|=1$ is mapped on a null set on C.

Since by Theorem VIII. 19, $F'(\theta)=ie^{i\theta}f'(e^{i\theta})\neq 0$ almost everywhere, we see from (7), that a null set on C is mapped on a null set on $|z|=1$.——

2. If $f'(e^{i\theta})=\gamma$ $(\neq\infty)$ exists, then $\lim_{z\to e^{i\theta}}\dfrac{f(z)-f(e^{i\theta})}{z-e^{i\theta}}=\gamma$, when $z\to e^{i\theta}$ from

the inside of any Stolz domain, whose vertex is at $e^{i\theta}$, hence we have

Theorem VIII. 27. *Under the same condition as Theorem* VIII. 26, *for almost all* $e^{i\theta}$,

$$\lim_{z\to e^{i\theta}} \frac{f(z)-f(e^{i\theta})}{z-e^{i\theta}} = \lim_{z\to e^{i\theta}} f'(z)=f'(e^{i\theta})=\gamma \ (\neq 0, \ \neq\infty)$$

exists, when $z\to e^{i\theta}$ *from the inside of any Stolz domain, whose vertex is at* $e^{i\theta}$.

$\lim\limits_{z\to e^{i\theta}} \dfrac{f(z)-f(e^{i\theta})}{z-e^{i\theta}}=\gamma$ is called *the angular derivative* of the mapping function $f(z)$ at $e^{i\theta}$.

5. Lusin-Privaloff's theorem

Theorem VIII. 28. *Let $f(z)$ be meromorphic in $|z|<1$. Suppose that there exists a measurable set E of positive measure on $|z|=1$, such that if $e^{i\theta}\in E$, then $\lim\limits_{z\to e^{i\theta}} f(z)=0$ uniformly, when $z\to e^{i\theta}$ from the inside of any fixed Stolz domain, whose vertex is at $e^{i\theta}$, then*

$$f(z)=0.$$

Lusin-Privaloff [1].

Proof. Let $\Delta(e^{i\theta})$ be a Stolz domain, whose vertex is at $e^{i\theta}$ and is bounded by two lines through $e^{i\theta}$, making an angle $\pi/4$ with the inner normale of $|z|=1$ at $e^{i\theta}$. By Egoroff's theorem, we may assume that E is a closed set and $\lim\limits_{z\to e^{i\theta}} f(z)=0$ uniformly on E, when z tends to any point $e^{i\theta}\in E$ from the inside of $\Delta(e^{i\theta})$.

Let $\Delta_\rho(e^{i\theta})$ be the part of $\Delta(e^{i\theta})$, which lies in $0<\rho\leqq|z|<1$ and put $\Delta=\sum\limits_{e^{i\theta}\in E}\Delta_\rho(e^{i\theta})$, added for all $e^{i\theta}\in E$ and we join the disc $|z|\leqq\rho$ to Δ and let D be the resulting domain, then we see easily that D is bounded by a rectifiable Jordan curve C and E belongs to its boundary.

If z tends to any point of E from the inside of D, then $\lim f(z)=0$ uniformly. Hence there is only a finite number of poles of $f(z)$ in D and let $z_1, \cdots, z_k$ be such ones. If we consider $(z-z_1)\cdots(z-z_k)f(z)$, instead of $f(z)$, we may assume that $f(z)$ is regular in D. We map D conformally on $|\zeta|<1$ and put $f(z)=F(\zeta)$, then $F(\zeta)$ is regular and bounded in $|\zeta|<1$. Let E^* be the image of E on $|\zeta|=1$, then by F. and M. Riesz' theorem, $mE^*>0$ and $\lim F(\zeta)=0$, when tends to any point of E^*, hence by Theorem IV. 10, $F(\zeta)\equiv 0$, or $f(z)\equiv 0$.

Lemma. (i) *Let $f_n(x)$ $(n=1, 2, \cdots)$ be a continuous function, defined in $[a, b]$. Then the set of x, such that $\lim\limits_{n\to\infty} f_n(x)$ $(\neq\infty)$ exists, is*

an $F_{\sigma\delta}$-set and the set of x, such that $\lim_{n\to\infty}|f_n(x)|=\infty$ is an $F_{\sigma\delta}$-set.
(ii) *Let $f_r(x)$ be a continuous function of x and r, defined for $a\leqq x\leqq b$, $0\leqq r<1$. Then the set of x, such that $\lim_{r\to 1} f_r(x)$ $(\neq\infty)$ exists, is an $F_{\sigma\delta}$-set and the set of x, such that $\lim_{r\to\infty}|f_r(x)|=\infty$ is an $F_{\sigma\delta}$-set.*

Proof. First we shall prove (i). Let $E_{n,k}$ $(n, k=1, 2, \cdots)$ be the set of x, such that $|f_m(x)-f_n(x)|\leqq 1/k$ $(m=n+1,\ n+2,\ \cdots)$, then since $f_n(x)$ is continuous, $E_{n,k}$ is a closed set, hence $E_k=\sum_{n=1}^{\infty} E_{n,k}$ is an F_σ-set, so that $E=\prod_{k=1}^{\infty} E_k$ is an $F_{\sigma\delta}$-set. If $x_0\in E$, then $x_0\in E_k$ $(k=1, 2, \cdots)$, hence for some n_0, $x_0\in E_{n_0,k}$, so that $|f_m(x_0)-f_{n_0}(x_0)|\leqq 1/k$ $(m\geqq n_0)$. Since k is arbitrary, $\lim_{n\to\infty} f_n(x_0)$ $(\neq\infty)$ exists. Conversely, if $\lim_{n\to\infty} f_n(x_0)$ $(\neq\infty)$ exists, then for any k, $|f_n(x_0)-f_{n_0}(x_0)|\leqq 1/k$ $(n\geqq n_0)$, so that $x_0\in E_{n_0,k}\subset E_k\subset E$, hence E is that set of x, such that $\lim_{n\to\infty} f_n(x)$ $(\neq\infty)$ exists.

Next consider the set of x, such that $\lim_{n\to\infty}|f_n(x)|=\infty$. Let $E_{n,k}$ be the set of x, such that $|f_m(x)|\geqq k$ $(m=n,\ n+1,\ \cdots)$, then $E_{n,k}$ is a closed set, so that $E_k=\sum_{n=1}^{\infty} E_{n,k}$ is an F_σ-set and $E=\prod_{n=1}^{\infty} E_k$ is an $F_{\sigma\delta}$-set. If $x_0\in E$, then $x_0\in E_k$ $(k=1, 2, \cdots)$, so that $x_0\in E_{n,k}$ for some n_0, hence $|f_n(x_0)|\geqq k$ $(n\geqq n_0)$, so that $\lim_{n\to\infty}|f_n(x_0)|=\infty$. If $\lim_{n\to\infty}|f_n(x_0)|=\infty$, then $|f_n(x_0)|\geqq k$ $(n\geqq n_0)$, so that $x_0\in E_{n_0,k}\subset E_k$, hence $x_0\in E$. Hence E is the set of x, such that $\lim_{n\to\infty}|f_n(x)|=\infty$. Hence (i) is proved. To prove (ii), we choose $r_1<r_2<\cdots<r_n\to 1$, $\varepsilon_n\to 0$, such that for $r_{n-1}\leqq r\leqq r_n$, $|f_{r_n}(x)-f_r(x)|<\varepsilon_n$ in $[a, b]$, and apply (i) to $f_{r_n}(x)$ $(n=1, 2, \cdots)$. ——

As an application of Lusin-Privaloff's theorem, we shall prove

Theorem VIII. 29. *Let D be a bounded domain on the w-plane and Γ be its boundary and E be a closed set of capacity zero on Γ, which consists of accessible boundary points. If we map the universal covering surface of D conformally on $|z|<1$, then E is mapped on an $F_{\sigma\delta}$-set E^* of measure zero on $|z|=1$.*

Proof. Let $w=f(z)$ be the mapping function, then since D is a bounded domain, for almost all $e^{i\theta}$, $\lim_{z\to e^{i\theta}} f(z)=f(e^{i\theta})$ exists uniformly, when $z\to e^{i\theta}$ from the inside of any fixed Stolz domain, whose vertex is at $e^{i\theta}$.

Let E^* be the image of E on $|z|=1$.
Since E is a bounded closed set of capacity zero, let

$$u(w)=\int_E \log\frac{1}{|w-a|}\,d\mu(a), \qquad \mu(E)=1\,,$$

be the Evans' function with respect to E, such that $u(w)\to+\infty$, when w

tends to any point of E and let $v(w)$ be its conjugate harmonic function and put $\varphi(w)=u(w)+iv(w)$, $\varphi(f(z))=F(z)$, then $F(z)$ is regular in $|z|<1$ and $F(z)\to\infty$ uniformly, when z tends to any point $e^{i\theta}\in E^*$ from the inside of any Stolz domain, whose vertex is at $e^{i\theta}$. Now $F(re^{i\theta})$ is a continuous function of θ and r for $0\leqq\theta\leqq 2\pi$, $0\leqq r<1$, so that by the lemma, the set A of $e^{i\theta}$, such that $\lim_{r\to 1}|F(re^{i\theta})|=\infty$ is an $F_{\sigma\delta}$-set. We see easily that $A=E^*$, hence E^* is an $F_{\sigma\delta}$-set, so that is measurable. Hence by Lusin-Privaloff's theorem, if $mE^*>0$, then $F(z)\equiv\infty$, which is absurd, hence

$$mE^*=0\,.$$

6. Extension of Lowner's theorem

Theorem VIII. 30. *Let $w=f(z)$ be regular and $|f(z)|<1$ in $|z|<1$, and $f(0)=0$. Then $\lim_{r\to 1}f(re^{i\theta})=f(e^{i\theta})$ exists almost everywhere on $|z|=1$. Let E_0 be the set of $e^{i\theta}$, such that $|f(e^{i\theta})|=1$. Then E_0 is a Borel set. Let $E\subset E_0$ be any Borel subset of E_0 and E^* be its image on $|w|=1$, then E^* is an analytic set, so that is measurable and*

$$mE\leqq mE^*\,.$$

If $0<mE_0<2\pi$, $mE>0$, then $mE<mE^*$. Tsuji [48], Ohtsuka [1].

Löwner [1] proved the case, where E is a circular arc on $|z|=1$, Kawakami [1] the case, where $E=E_0$ and $f(z)$ is schlicht in $|z|<1$, Kametani and Ugaheri [1] proved that for any subset E of E_0, $m_iE\leqq m_eE^*$, where m_i, m_e denotes the inner and the outer measure.

Proof. By the lemma (p. 320), the set A of $e^{i\theta}$, such that $\lim_{r\to 1}f(re^{i\theta})=f(e^{i\theta})$ exists, is a Borel set, so that $f(e^{i\theta})$ is a Borel function, defined on a Borel set A. Hence the set E_0 of $e^{i\theta}$, such that $|f(e^{i\theta})|=1$ is a Borel set. Let E be a Borel subset of E_0 and put for $e^{i\theta}\in E$, $f(e^{i\theta})=e^{i\psi(\theta)}$. Consider on the (θ,ψ)-plane a set M of points $(\theta,\psi(\theta))$, $e^{i\theta}\in E$, then M is a Borel set. Since E^* is the projection of a Borel set M on the ψ-axis, E^* is an analytic set (Sierpinski [1]), so that is measurable (Saks [1]).

We may assume that $mE>0$ and put

$$u(z)=\int_E\frac{1-|z|^2}{|z-e^{i\theta}|^2}d\theta,\qquad u^*(w)=\int_{E^*}\frac{1-|w|^2}{|w-e^{i\psi}|^2}d\psi\,, \tag{1}$$

$$v(z)=u^*(f(z))-u(z)\,. \tag{2}$$

Let O be an open set on $|w|=1$, which contains E^* and we define $u_0^*(w)$ by

$$u_0^*(w)=\int_O\frac{1-|w|^2}{|w-e^{i\psi}|^2}\,d\psi\,, \tag{3}$$

and put

$$v_0(z)=u_0^*(f(z))-u(z)\,. \tag{4}$$

Then $v_0(z)$ is a bounded harmonic function in $|z|<1$. Since $u_0^*=1$ at every point of O, a fortiori, on E^* and $u=1$ almost everywhere on E, we see that $\lim\limits_{r\to 1} v_0(re^{i\theta})=0$ almost everywhere on E and since $\lim\limits_{r\to 1} v_0(re^{i\theta})\geqq 0$ almost everywhere on the complement E' of E, we see that $v_0(z)\geqq 0$ in $|z|<1$. Hence if we make $mO\to mE^*$, then we have $v(z)\geqq 0$ in $|z|<1$, so that

$$v(0)=u^*(0)-u(0)\geqq 0\,, \quad \text{or}$$

$$mE\leqq mE^*\,. \tag{5}$$

Next suppose that $0<mE_0<2\pi$, $mE>0$, then by (5) $mE^*>0$, so that

$$u^*(w)>0 \quad \text{in } |w|<1\,. \tag{6}$$

If $mE=mE^*$, then $v(0)=0$, so that $v(z)\equiv 0$, or $u^*(f(z))\equiv u(z)$.

Since $0<mE_0<2\pi$, there exists a point $e^{i\theta}$ on the complement E_0' of E_0, such that

$$\lim_{r\to 1} u(re^{i\theta})=0\,, \quad \lim_{r\to 1} f(re^{i\theta})=w_0\,, \quad |w_0|<1\,.$$

Hence $\lim\limits_{r\to 1} v(re^{i\theta}) = u^*(w_0)> 0$, which contradicts to $v\equiv 0$, hence

$$mE<mE^*. \text{ ——}$$

7. Function of U-class

1. Let $w=f(z)$ be regular and $|f(z)|<1$ in $|z|<1$, then $\lim\limits_{z\to e^{i\theta}} f(z)=f(e^{i\theta})$ exists almost everywhere on $|z|=1$, when $z\to e^{i\theta}$ from the inside of any Stolz domain, whose vertex is at $e^{i\theta}$. If $|f(e^{i\theta})|=1$ almost everywhere on $|z|=1$, then we say with Seidel [1] that $f(z)$ *belongs to U-class* and denote $f(z)\in U$. If $f(z)$ is regular in $|z|<1$ and $\frac{1}{\rho}(f(z)-a)\in U$, then we denote $f(z)\in U_\rho(a)$.

Theorem VIII. 31. *Let $f(z)\in U$ and F be the Riemann surface, generated by $w=f(z)$ over the w-plane and F_ρ be its connected piece, which lies above a disc $K: |w-a|<\rho$, which is contained in $|w|<1$. Then F_ρ is simply connected and if we map F_ρ conformally on $|\zeta|<1$ by $w=\varphi(\zeta)$, then $\varphi(\zeta)\in U_\rho(a)$.* Noshiro [3].

Proof. Let D be the image of F_ρ in $|z|<1$, then by the maximum principle, we see that D is simply connected, so that F_ρ is simply connected. We may assume that F_ρ is non-compact, so that D has boundary points on $|z|=1$ and let e_0 be the set of such boundary points. We map F_ρ conformally on $|\zeta|<1$ by $w=\varphi(\zeta)$, then D is mapped conformally on $|\zeta|<1$. By Fatou's theorem, $\lim\limits_{r\to 1} \varphi(re^{i\psi})=\varphi(e^{i\psi})$ exists almost everywhere on $|\zeta|=1$. Let e_1 be the set of $e^{i\psi}$, such that $|\varphi(e^{i\psi})-a|<\rho$.

If $\zeta=re^{i\psi}\to e^{i\psi}(r\to 1)$, then $w\to\varphi(e^{i\psi})$ along a curve L. Let C be the image of L in D, then C ends at a point $e^{i\theta}\in e_0$, so that if z tends to $e^{i\theta}$ along C, then $w=f(z)\to\varphi(e^{i\psi})$ along L.

Since $|f(z)|<1$ in $|z|<1$, by Theorem VIII. 10, $\lim_{r\to 1} f(re^{i\theta})=\varphi(e^{i\psi})$.

Since $f(z)\in U$, the set $e_1{}^*$ of such $e^{i\theta}$ is a null set. By a suitable linear transformation, which makes $|z|<1$ invariant, we may assume that D contains $z=0$ and $z=0$ corresponds to $\zeta=0$. Since $e_1{}^*$ is the image of e_1, by Theorem VIII. 30, e_1 is a null set, hence $\varphi(\zeta)\in U(a)$.

Theorem VIII. 32. *Let $w=f(z)\in U$. If $f(z)$ is regular on $|z|=1$, then $f(z)$ is of the form:*

$$f(z)=\varepsilon\prod_{\nu=1}^{n}\frac{z-z_\nu}{1-\bar{z}_\nu z},\quad |\varepsilon|=1,\quad |z_\nu|<1, \tag{A}$$

and $f(z)-a$ ($|a|<1$) has n zero points in $|z|<1$.

If $f(z)$ has a singular point z_0 on $|z|=1$, then $f(z)-a$ ($|a|<1$) has infinitely many zero points in any small neighbourhood $U(z_0)$ of z_0, a fortiori, in $|z|<1$, except a set of a, which is of capacity zero.

Ohtsuka [1][1].

Let F be the Riemann surface, generated by $w=f(z)$ over the w-plane and K: $|w-a|<\rho$ ($|a|<1$) be a disc, contained in $|w|<1$ and F_ρ be a connected piece of F, which lies above K, then either F_ρ covers any point of K exactly a finite number of times, or F_ρ covers any point of K infinitely often, except a set of capacity zero.

Remark. We see easily that the same result holds, if a disc K: $|w-a|<\rho$ is replaced by a simply connected Jordan domain $\varDelta$, contained in $|w|<1$, by mapping $\varDelta$ conformally on a disc.

Proof. First suppose that $f(z)$ is regular on $|z|=1$, then by the principle of reflexion, $f(z)$ can be continued analytically across $|z|=1$ into $|z|>1$ and is a rational function of z and we see easily that $f(z)$ is of the special form (A). Since $|f(z)|=1$ on $|z|=1$, if $|a|<1$, then by Rouché's theorem, $f(z)$ and $f(z)-a$ have the same number of zero points in $|z|<1$, so that $f(z)-a$ has n zero points in $|z|<1$.

Next suppose that $f(z)$ has a singular point z_0 on $|z|=1$ and $U(z_0)$ be the intersection of $|z-z_0|<\rho$ and $|z|<1$. Let E be a set of positive capacity, contained in $|w|<1$ and suppose that for any $a\in E$, $f(z)-a$ has

1) The result is contained implicitly in Ohtsuka's paper. The following direct proof is due to K. Oikawa.

only a finite number of zero points in $U(z_0)$, then we may assume that E is a closed set and by taking a smaller neighbourhood of z_0, we may assume that $f(z)-a$ has no zero points in $U(z_0)$ for any $a\in E$. Let D be the complement of E with respect to $|w|<1$ and let $\chi(w)$ be the solution of the Dirichlet problem for D with the boundary value 0 on $|w|=1$ and 1 on E, then $\chi=0$ at every point on $|w|=1$ and $\chi=1$ "almost everywhere" on E, so that $\chi(w)\not\equiv 0$. If we put

$$u(z)=\chi(f(z)),$$

then $u(z)$ is a bounded harmonic function in $U(z_0)$. Let α be the arc of $|z|=1$, which belongs to the boundary of $U(z_0)$, then since $|f(z)|=1$ almost everywhere on α, $u=0$ almost everywhere on α. Hence if we map $U(z_0)$ conformally on $|t|<1$ and express u by a Poisson integral on $|t|=1$, then we see that $u=0$ at every point of α, so that u can be continued harmonically across α into $U^*(z_0)$, which is the inverse of $U(z_0)$ with respect to $|z|=1$. Let $v(z)$ be the conjugate harmonic function of $u(z)$ in $V(z_0)=U(z_0)+U^*(z_0)$, such that $v(z_0)=0$, then by $\zeta=u(z)+iv(z)$, $V(z_0)$ is mapped conformally on a full neighbourhood $V(\zeta_0)$ of $\zeta_0=0$.

Since $f(z)\in U$, there exist $z_\nu\in U(z_0)$ $(\nu=1, 2, \cdots)$, such that $z_\nu\to z_0$, $(z\ )\to w_0$ $(|w_0|=1)$. Since $\chi=0$ on $|w|=1$, $\chi(w)$ is harmonic at w_0. Let $\psi(w)$ be its conjugate harmonic function in a full neighbourhood $V(w_0)$ of w_0, such that $\psi(w_0)=0$, then by $\zeta=\chi(w)+i\psi(w)$, $V(w_0)$ is mapped conformally on $V(\zeta_0)$, hence a full neighbourhood of z_0 is mapped conformally on a full neighbourhood of w_0, hence $f(z)$ is regular at z_0, which is absurd, hence the theorem.

The part concerning F_ρ can be decluced from the first part and Theorem VIII. 31, by mapping F conformally on $|\zeta|<1$. ——

2. We shall prove a more general theorem as follows.

Theorem VIII. 33. *Let $f(z)\in U$ and Ω be the space of uniformly bounded regular functions $\{g(z)\}$ in $|z|<R$ $(R>1)$, such that $|g(z)|<1$ in $|z|\leqq 1$. If $f(z)$ is regular on $|z|=1$ and of the form(A), then for any $g\in\Omega$, $f(z)-g(z)$ has n zero points in $|z|<1$.*

If $f(z)$ has a singular point z_0 on $|z|=1$, then $f(z)-g(z)$ has infinitely many zero points in any small neighbourhood $U(z_0)$ of z_0, fortiori, in $|z|<1$, except a set of g, which is of capacity zero. Tsuji [64].

Proof. If $f(z)$ is regular on $|z|=1$ and of the form (A), then we can prove by means of Rouché's theorem, that $f(z)-g(z)$ has n zero points in $|z|<1$.

Next suppose that $f(z)$ has a singular point z_0 on $|z|=1$. Let $\mathfrak{M}$ be a set of positive capacity in Ω and suppose that $f(z)-g(z)$ has only a finite number of zero points in $U(z_0)$ for any $g\in\mathfrak{M}$, then we may assume that $\mathfrak{M}$ is a closed set and by taking a smaller neighbourhood of z_0, $f(z)-g(z)$ has no zero points in $U(z_0)$ for any $g\in\mathfrak{M}$.

Since $\mathfrak{M}$ is a closed set, $|g(z)|\leqq\rho<1$ in $|z|\leqq 1$ for any $g\in\mathfrak{M}$.

Let σ be the equilibrium distribution of $\mathfrak{M}$, such that for any ζ and z ($|z|\leqq 1$),

$$\chi_z(\zeta)=\int_{\mathfrak{M}}\log\frac{1}{|\zeta-g(z)|}\,d\sigma(g)\leqq V(\mathfrak{M})<\infty, \qquad \sigma(\mathfrak{M})=1\,. \tag{1}$$

For a fixed $z(|z|\leqq 1)$ let M_z be the bounded closed set of points $\{g(z)\}$ $(g\in\mathfrak{M})$ on the w-plane, contained in $|w|\leqq\rho<1$, then by Theorem III. 74,

$$\gamma(M_z)\geqq\Gamma(\mathfrak{M})>0\,.$$

Let

$$\sup_{\zeta\in M_{z_0}}\chi_{z_0}(\zeta)=V_0\,, \tag{2}$$

then since M_{z_0} is contained in $|w|\leqq\rho<1$, by the maximum principle for logarithmic potential, we have for a small $\varepsilon>0$,

$$V_0-\varepsilon\geqq\operatorname*{Max}_{|\zeta|=1}\chi_{z_0}(\zeta)\,. \tag{3}$$

Let $\zeta_0\in M_{z_0}$, such that $\chi_{z_0}(\zeta_0)>V_0-\varepsilon$, then by (3),

$$\int_{\mathfrak{M}}\log\frac{1}{|\zeta_0-g(z_0)|}\,d\sigma(g)>\operatorname*{Max}_{|\zeta|=1}\chi_{z_0}(\zeta)\,. \tag{4}$$

Since $\gamma(M_{z_0})>0$, by Theorem VIII. 32, we may assume that there exists $z_\nu\in U(z_0)$, such that

$$f(z_\nu)=\zeta_0, \qquad z_\nu\to z_0, \qquad z_\nu\in U(z_0) \quad (\nu=1, 2, \cdots)\,.$$

By Fatou's lemma, we have

$$\begin{aligned}\int_{\mathfrak{M}}\log\frac{1}{|\zeta_0-g(z_0)|}\,d\sigma(g)&\leqq\varliminf_{\nu}\int_{\mathfrak{M}}\log\frac{1}{|\zeta_0-g(z_\nu)|}\,d\sigma(g)\\ &=\varliminf_{\nu}\int_{\mathfrak{M}}\log\frac{1}{|f(z_\nu)-g(z_\nu)|}\,d\sigma(g)\,.\end{aligned} \tag{5}$$

If we put

$$u(z)=\int_{\mathfrak{M}}\log\frac{1}{|f(z)-g(z)|}\,d\sigma(g)\,, \tag{6}$$

then by (1), $u(z)$ is a bounded harmonic function in $U(z_0)$. Let α be the arc of $|z|=1$, which belongs to the boundary of $U(z_0)$. If we map $U(z_0)$ conformally on $|t|<1$ and express u by a Poisson integral on $|t|=1$, then since $|f(z)|=1$ almost everywhere on α and $|g(z)|\leqq\rho<1$, we see that

$$\varlimsup_{\nu}u(z_\nu)\leqq\varlimsup_{e^{i\theta}\to z_0}u(e^{i\theta})\leqq\operatorname*{Max}_{|\zeta|=1}\chi_{z_0}(\zeta)\,,$$

where we consider only such $e^{i\theta}$, that $\lim_{r\to 1} u(re^{i\theta})=u(e^{i\theta})$ exists and $\lim_{r\to 1}|f(re^{i\theta})|=1$, so that by (5),

$$\int_{\mathfrak{M}} \log \frac{1}{|\zeta_0-g(z_0)|}\, d\sigma(g) \leq \operatorname*{Max}_{|\zeta|=1} \chi_{z_0}(\zeta)\ , \tag{7}$$

which contradicts (4), hence the theorem.

Theorem VIII. 34. *Let $w=f(z)\in U$ and $z=z(w)$ be its inverse function, then the projection of direct transcendental singularities of $z(w)$ on the w-plane is of capacity zero.*

Proof. We may assume that $f(z)$ has a singular point on $|z|=1$ and let F be the Riemann surface, generated by $w=f(z)$ over the w-plane. Let K be a disc, contained in $|w|<1$, then the part of F, which lies above K consists of at most a countable number of connected pieces $\{F_\nu\}$ and every F_ν convers any point of K at least once, except a set of capacity zero, so that the set of points of K, which are not covered by some F_ν is of capacity zero. Let K_n $(n=1, 2, \cdots)$ be a disc in $|w|<1$, whose center is a rational point and whose radius is a rational number and let E_n be the set of points of K_n, which are not covered by some connected piece of F, which lies above K_n, then $\gamma(E_n)=0$, so that $E=\sum_{n=1}^{\infty}E_n$ is of capacity zero.

Let (a) be a direct transcendental singularity of $z(w)$, whose projection on the w-plane is a. Then a is not covered by some connected piece of F, which lies above a certain K_n, so that $a\in E_n$, hence $a\in E$, so that the projection of direct transcendental siugularities of $z(w)$ on the w-plane is of capacity zero. ——

Theorem VIII. 35. *Let $f(z)\in U$ and suppose that $f(z)$ has a singular point on $|z|=1$. If $f(z)$ takes a $(|a|<1)$ only a finite number of times in $|z|<1$, then there exists a point $e^{i\theta}$ on $|z|=1$, such that $\lim_{r\to 1} f(re^{i\theta})=a$.* Seidel [1].

Proof. Since $f(z)$ has a singular point on $|z|=1$, the Riemann surface F, generated by $w=f(z)$ over the w-plane consists of infinitely many sheets.

Let K: $|w-a|<\rho$ be a disc, contained in $|w|<1$. Since a is covered only a finite number of times by F, there exists a connected piece F_ρ of F, which lies above K and does not cover a. If we map F_ρ conformally on $|\zeta|<1$ by $w=\varphi(\zeta)$, then $\varphi(\zeta)\in U_\rho(a)$, so that $\frac{1}{|\varphi(\zeta)-a|}=\frac{1}{\rho}$ almost everywhere on $|\zeta|=1$. Hence if $\frac{1}{\varphi(\zeta)-a}$ is bounded in $|\zeta|<1$, then $\frac{1}{|\varphi(\zeta)-a|} \leqq \frac{1}{\rho}$, or $|\varphi(\zeta)-a| \geqq \rho$ in $|\zeta|<1$, which is absurd. Hence $\frac{1}{\varphi(\zeta)-a}$ is unbounded in $|\zeta|<1$, so that $\frac{1}{f(z)-a}$ is unbounded in $|z|<1$. From this, we see easily that there exists a Jordan arc C in $|z|<1$, ending at a certain point $e^{i\theta}$, such that if $z \to e^{i\theta}$ along C, then $f(z) \to a$. Since $|f(z)|<1$ in $|z|<1$, by Theorem VIII. 10, $\lim_{r\to 1} f(re^{i\theta})=a$. ——

3. Next we consider a sub-class of $f(z) \in U$, such that $|f(e^{i\theta})|=1$ on $|z|=1$, except a set of capacity zero and we denote such $f(z)$ by $f(z) \in U^*$.

Theorem VIII. 36. *If $f(z) \in U^*$ and has a singular point on $|z|=1$, then $f(z)$ takes any value a ($|a|<1$) infinitely often, with one possible exception.* Noshiro [3].

Proof. Let $|f(e^{i\theta})|=1$ on $|z|=1$, except a set E of capacity zero and F be the Riemann surface, generated by $w=f(z)$ over the w-plane. Since there is a singlar point of $f(z)$ on $|z|=1$, F consists of infinitely many sheets. Suppose that F covers two points α, β of $|w|<1$ only a finite number of times. We enclose α, β in a domain $\varDelta$, which is bounded by a Jordan curve $\varGamma$ and is contained in $|w|<1$, such that there exists a connected piece F_0 of F, above $\varDelta$, which does not cover α, β. Then F_0 is non-compact and since F_0 covers any point of $\varDelta$ infinitely often, except a set of capacity zero (Theorem VIII. 32, Remark), the area of F_0 is infinite. Let D be the image of F_0 in $|z|<1$, then D is simply connected, so that F_0 is simply connected. Let C be the image of $\varGamma$ in $|z|<1$, then we see easily that C consists of at most a countable number of Jordan arcs $\{\gamma_n\}$, whose end points lie on $|z|=1$, so that the boundary of D consists of $\{\gamma_n\}$ and a certain closed set E_0 on $|z|=1$. We see that E_0 is a sub-set of E, so that E_0 is a closed set of capacity zero. Let $u(z)$ be the Evans' function with respect to E_0 and $\theta(z)$ be its conjugate harmonic function and put $\zeta=e^{u+i\theta}=re^{i\theta}$, $w=\varPhi(\zeta)$. Let C_r be the niveau curve: $r=$const. $(=r)$, which consists of a finite number of Jordan curves $C_r=\sum_{i=1}^{n_r} C_r^i$. Let D_r be the part of D, which lies outside of $\{C_r^i\}$. Since the area of F_0 is infinite, by means of C_r and D_r, we can prove similarly as Theorem VI. 9, that F_0 is regularly exhaustible in Ahlfors' sense and since F_0 is simply connected, F_0 covers any point of $\varDelta$ infinitely often, with one possible exception, which is absurd. Hence the theorem.

Corollary. *Let M be a closed set, contained in $|w|<1$, which contains at least two points. Let D be the domain, bounded by M and $|w|=1$. If we map the universal covering surface of D conformally on $|z|<1$, then M is mapped on a set of positive capacity on $|z|=1$.*

8. Implicit function $y(x)$, defined by an integral relation $G(x, y)=0$

Let $G(x, y)$ be an integral function of two variables x and y and $y(x)$ be the analytic function, defined by $G(x, y)=0$ and F be its Riemann surface, spread over the x-plane. If

$$G(x, y)=A_0(x)y^n+A_1(x)y^{n-1}+\cdots+A_n(x)=0\ ,$$

where $A_i(x)$ are integral functions of x, then $y(x)$ is an algebroid function and F covers any point $x(\neq\infty)$ n-times.

Lemma. *Let (x_0) be an accessible boundary point of F, whose projection on the x-plane is x_0. If x tends to (x_0) on F, then $y(x)\to\infty$.*

Proof. Let F_ρ be a connected piece of F, which lies above a disc: $|x-x_0|<\rho$ and contains (x_0) on its boundary and let D_ρ be its image on the y-plane. We choose $R>0$, such that $G(x_0, y)\neq 0$ on $|y|=R$. If ρ is small, then D_ρ lies outside of $|y|=R$. For, if otherwise, we can choose $\rho_\nu\to 0$, such that the image of $|x-x_0|<\rho_\nu$ on the y-plane contains a point y_ν on $|y|=R$. Let y_0 be one of the cluster points of $\{y_\nu\}$, then $|y_0|=R$ and $G(x_0, y_0)=0$, which is absurd. Hence if ρ is small, D_ρ lies outside of $|y|=R$. Since $R>0$ is arbitrary, $y(x)\to\infty$, if $x\to(x_0)$ on F.

Theorem VIII. 37. *If $y(x)$ is not an algebroid function, then F covers any point x infinitely often, except a set of capacity zero. Let F_ρ be a connected piece of F, which lies above a disc K: $|x-x_0|<\rho$.*
(i) *If we map the universal covering surface of F_ρ conformally on $|\zeta|<1$ by $x=\varphi(\zeta)$, then $\varphi(\zeta)\in U_\rho(x_0)$.*
(ii) *Either F_ρ covers any point of K exactly a finite number of times, or F_ρ covers any point of K infinitely often, except a set of capacity zero.* Tsuji [8], [44].

Since a set of capacity zero does not contain a continuum, the set of x, which is not covered by F does not contain a continuum, which was proved by Julia [2].

Proof. First we shall prove the second part.
(i) Let E be the set of $e^{i\theta}$ on $|\zeta|=1$, such that $|\varphi(e^{i\theta})-x_0|<\rho$, then if ζ tends to $e^{i\theta}$ from the inside of any Stolz domain, whose vertex is at $e^{i\theta}$, $x=\varphi(\zeta)$ tends to an accessible boundary point of F, so that by the lemma, $y(\varphi(\zeta))$ tends to ∞. Since $y(\varphi(\zeta))$ is a meromorphic function of ζ in $|\zeta|<1$ and $y(\varphi(\zeta))\not\equiv\infty$, by Lusin-Privaloff's theorem, E is a null set, so that $\varphi(\zeta)\in U_\rho(x_0)$.
(ii) Suppose that there exists a disc K_0 in K, which is covered exactly $n(1\leqq n<\infty)$-times by F_ρ, then we shall prove that F_ρ covers any point

of K exactly n-times. Let D be the domain $\subset K$, which contains K_0 and is covered exactly n-times by F_ρ. Suppose that D does not coincide with K and let Γ be the part of its boundary, which lies in K and ξ_0 be any point of Γ, then ξ_0 is covered at most n-times by F_ρ. We shall prove that ξ_0 is covered at most $(n-1)$-times by F_ρ. Suppose that ξ_0 is covered n-times by F_ρ, then the part of F_ρ, which lies above a small disc Δ_0 about ξ_0, contains n discs $F_1, \cdots, F_n$, consisting of inner points, where the part of the Riemann surface of $(x-\xi_0)^{\frac{1}{k}}$ is considered as k discs.

If there is no other connected piece of F_ρ above Δ_0, then Δ_0 is covered n-times by F_ρ, so that ξ_0 belongs to D, which is absurd. Hence there exists a connected piece F_0 of F_ρ above Δ_0, other than $F_1, \cdots, F_n$. Since $\varphi(\zeta) \in U_\rho(\xi_0)$, F_0 covers any point of Δ_0 at least once, except a set of capacity zero, but F_0 does not cover $D. \Delta_0$, which is of positive capacity, which is absurd. Hence every point of Γ is covered at most $(n-1)$-times by F_ρ.

Next we shall prove that $\gamma(\Gamma)=0$. Suppose that $\gamma(\Gamma)>0$. Let Γ_k be the sub-set of Γ, every point of which is covered k-times by F_ρ $(0\leqq k\leqq n-1)$. Then for some k, $\gamma(\Gamma_k)>0$. Since F_ρ covers any point of K at least once, except a set of capacity zero, $\gamma(\Gamma_0)=0$, so that $1\leqq k\leqq n-1$. We may assume that Γ_k is a closed set, contained entirely in K, then there exists a point $\xi_0 \in \Gamma_k$, such that $\gamma(\Gamma_k\cdot\Delta_0)>0$ for any small disc Δ_0 about ξ_0. Since $\xi_0\in\Gamma_k$, ξ_0 is covered k-times by F , so that there exist k discs $F_1, \cdots, F_k$ above Δ_0, consisting of inner points of F_ρ.

Since $1\leqq k\leqq n-1$ and $\Delta_0\cdot D\neq 0$, there is a connected piece F_0 above Δ_0, other than $F_1, \cdots, F_k$. F_0 does not cover $\Gamma_k\cdot\Delta_0$, which is of positive capacity, which is absurd, since F_0 covers any point of Δ_0 at least once, except a set of capacity zero. Hence $\gamma(\Gamma)=0$.

Let $y_i(x)\,(i=1, 2, \cdots, n)$ be n branches of $y(x)$ in D and ξ_0 be any point of Γ. Suppose that $y_i(x)\,(i=1, 2, \cdots, k\leqq n)$ are not meromorphic at ξ_0 and consider

$$\prod_{i=1}^{k}\left(\frac{1}{y}-\frac{1}{y_i(x)}\right)=\frac{1}{y^k}+\frac{a_1(x)}{y^{k-1}}+\cdots+a_k(x)=0\ .$$

Since $1/y(x)$ tends to zero, if x tends to any accessible boundary point of F, $a_i(x)$ is one-valued, regular and bounded in a neighbourhood of ξ_0. Since $\gamma(\Gamma)=0$, $a_i(x)$ is regular at ξ_0, so that ξ_0 is covered n-times by F_ρ, which is absurd. Hence D concides with K, so that F_ρ covers any point of K exactly n-times.

Let a be any point of K and be covered $n(a)$-times by F_ρ and put

$$n_0=\sup_{a\in K} n(a)\ . \tag{1}$$

If $n_0<\infty$, then there exists $a_0\in K$, such that $n(a_0)=n_0$, then a small disc K_0 about a_0 is covered exactly n_0-times by F_ρ, so that F_ρ covers any point of K exactly n_0-times. Next suppose that $n_0=\infty$. Let E_n be the set of $a\in K$, such that $n(a)=n(<\infty)$, then we shall prove that $\gamma(E_n)=0$. Suppose that $\gamma(E_n)>0$, then we may assume that E_n is a closed set, contained entirely in K. Then there exists a point $\xi_0\in E_n$, such that $\gamma(E_n\cdot\varDelta_0)>0$ for any small disc $\varDelta_0$ about ξ_0. Since $\xi_0\in E_n$, ξ_0 is covered n-times by F_ρ, so that there is n discs $F_1, \cdots, F_n$ above $\varDelta_0$, consisting of inner points.

If there is no other connected piece of F_ρ above $\varDelta_0$, then $\varDelta_0$ is covered n-times by F_ρ, so that F_ρ covers any point of K exactly n-times, which is absurd, since $n_0=\infty$. Hence there is a connected piece F_0 above $\varDelta_0$ other than $F_1, \cdots, F_n$. F_0 covers any point of $\varDelta_0$ at least once, except a set of capacity zero, but F_0 does not cover $E_n\cdot\varDelta_0$, which is of positive capacity, which is absurd. Hence $\gamma(E_n)=0$, if $n<\infty$, so that F_ρ covers any point of K infinitely often, except a set of capacity zero. Hence the second part is proved.

The first part can be deduced from the second part as follows. Let K_n: $|x|<n$ $(n=1, 2, \cdots)$ be a disc and F_n be a connected piece of F, which lies above K_n, such that $F_1\subset F_2\subset\cdots\subset F_n\to F$. Then by the second part, there are two alternatives: either (i) F_n covers K_n exactly N_n-times $(N_n<\infty)$, or (ii) F_n covers K_n infinitely often, except a set of capacity zero. In case (i), $N_1\leqq N_2\leqq\cdots\leqq N_n$, so that $\lim_{n\to\infty} N_n=N$ exists. If $N<\infty$, then F consists of N sheets and $x=\infty$ is only one singularity of $y(x)$, so that $y(x)$ is an algebroid function of x, which contradicts the hypothesis, hence $N=\infty$ and F covers any point $x(\neq\infty)$ infinitely often. Hence in any case, F covers any point x infinitely often, except a set of capacity zero. ——

Similarly as Theorem VIII. 34, we can prove

Theorem VIII. 38. *The projection of direct transcendental singularities of $y(x)$ on the x-plane is of capacity zero.*

9. Cluster set of a meromorphic function at a non-isolated boundary point

1. Let $\varDelta$ be a bounded domain on the z-plane and Γ be its boundary and $z_0\in\Gamma$ be a non-isolated boundary point. We denote the part of $\varDelta$, which lies in $|z-z_0|<r$ by $\varDelta_r$ and that of Γ in $|z-z_0|\leqq r$ by Γ_r.

Let $w=f(z)$ be one-valued and meromorphic in $\varDelta$ and W_r be the set of values, taken by $w=f(z)$ in $\varDelta_r$ and $\overline{W}_r$ be its closure, then

$$\lim_{r\to 0}\overline{W}_r = H_\Delta(z_0) \tag{1}$$

is called *the cluster of* $f(z)$ *in* Δ *at* z_0.

Let e be a set of capacity zero on Γ, such that $z_0 \in e$ and let

$$V_r(\Gamma-e) = \sum_{\zeta\in\Gamma_r-e_r} H_\Delta(\zeta), \text{ added for all } \zeta\in\Gamma_r-e_r$$

and $\overline{V}_r(\Gamma-e)$ be its closure, then

$$\lim_{r\to 0}\overline{V}_r(\Gamma-e) = H_{\Gamma-e}(z_0) \tag{2}$$

is called *the cluster set of* $f(z)$ *on* $\Gamma-e$ *at* z_0.

Evidently $H_\Delta(z_0)$ and $H_{\Gamma-e}(z_0)$ are closed sets and $H_{\Gamma-e}(z_0)\subset H_\Delta(z_0)$.

Theorem VIII. 39. *Every boundary point of* $H_\Delta(z_0)$ *belongs to* $H_{\Gamma-e}(z_0)$.

Tsuji [7].

The special case, where e consists of only one point z_0, was proved by Iversen [1]. c.f. Kunugui [1].

2. First we shall prove a lemma.

Lemma. *Let* Δ *be a bounded domain on the z-plane and* Γ *be its boundary and* $z_0\in\Gamma$ *be a boundary point, which is a regular point of the Dirichlet problem for* Δ, *and* e *be a set of capacity zero on* Γ *and* $z_0\in e$. *Let* $f(z)$ *be one-valued, regular and bounded in* Δ.

If

$$\overline{\lim_{\zeta\to z_0}}\ \overline{\lim_{z\to\zeta\in\Gamma-e}} |f(z)| \leqq M,$$

then

$$\overline{\lim_{z\to z_0}} |f(z)| \leqq M,$$

when $z\to z_0$ *from the inside of* Δ.

Proof. We may assume that $|f(z)|<1$ in Δ. For any $\varepsilon>0$, we choose $r>0$ so small that $\overline{\lim\limits_{z\to\zeta\in\Gamma_r-e_r}} |f(z)| \leqq M+\varepsilon$. We solve the Dirichlet problem for Δ, with the boundary value $M+\varepsilon$ on Γ_r and 1 on $\Gamma-\Gamma_r$ and let $u(z)$ be its solution, then $u(z)$ takes the given boundary value on Γ, except a set of capacity zero. Hence $\overline{\lim\limits_{z\to\zeta\in\Gamma}} |f(z)| \leqq \lim\limits_{z\to\zeta\in\Gamma} u(z)$, except a set of $\zeta\in\Gamma$ of capacity zero. Since $|f(z)|$ is subharmonic and $u(z)$ is harmonic, by the maximum principle, $|f(z)|\leqq u(z)$ in Δ. Since z_0 is a regular point of the Dirichlet problem, $\lim\limits_{z\to z_0} u(z) = M+\varepsilon$, so that $\overline{\lim\limits_{z\to z_0}} |f(z)| \leqq M+\varepsilon$. Since $\varepsilon>0$ is arbitrary, we have $\overline{\lim\limits_{z\to z_0}} |f(z)| \leqq M$.

3. *Proof of the theorem.*

Suppose that there exists a boundary point w_0 of $H_\Delta(z_0)$, which does

not belong to $H_{\Gamma-e}(z_0)$ and we assume that $w_0=0$.
(i) First suppose that z_0 is a regular point of the Dirichlet problem for Δ. We choose $r>0$ and $\rho>0$ so small that

$$\overline{V}_r(\Gamma-e) \text{ lies outside of } |w|=2\rho\,. \tag{1}$$

Since $w=0$ is a boundary point of $H_\Delta(z_0)$, there exists $w_1(|w_1|<\rho)$, which does not belong to $H_\Delta(z_0)$. Since $H_\Delta(z_0)$ is a closed set, the distance of w_1 to $H_\Delta(z_0)$ is positive, so that $\frac{1}{f(z)-w_1}$ is bounded in a neighbourhood of z_0. By taking r small, we assume that $\frac{1}{f(z)-w_1}$ is bounded in Δ_r. Since $w=0$ belongs to $H_\Delta(z_0)$, there exist $z_\nu\to z_0$ $(z_\nu\in\Delta)$, such that $f(z_\nu)\to 0$, so that

$$\frac{1}{|w_1|}\leqq \overline{\lim_{z\to z_0}}\,\frac{1}{|f(z)-w_1|}\,, \qquad z\in\Delta\,. \tag{2}$$

Hence by the lemma and (1), if $z\to z_0$ in Δ,

$$\frac{1}{|w_1|}\leqq \overline{\lim_{z\to z_0}}\,\frac{1}{|f(z)-w_1|}\leqq \overline{\lim_{\zeta\to z_0}}\;\overline{\lim_{z\to\zeta\in\Gamma-e}}\;\frac{1}{|f(z)-w_1|}$$
$$\leqq \overline{\lim_{\zeta\to z_0}}\;\overline{\lim_{z\to\zeta\in\Gamma-e}}\;\frac{1}{|f(z)|-|w_1|}\leqq\frac{1}{2\rho-|w_1|}\,,$$

so that $|w_1|\geqq\rho$, which is absurd. Hence the theorem is proved, if z_0 is a regular point of the Dirichlet problem for Δ.
(ii) Next suppose that z_0 is an irregular point of the Dirichlet problem for Δ. Then there exists no continuum $\subset\Gamma$, which contains z_0, so that in any small neighbourhood of z_0, there exists a Jordan curve C in Δ, surrounding z_0. We assume that C lies in $|z-z_0|<r$ and there are no zero points of $f(z)$ on C, then by taking ρ small, we assume that

$$\overline{V}_r(\Gamma-e) \text{ lies outside of } |w|=2\rho \quad\text{and}\quad |f(z)|>2\rho \text{ on } C\,. \tag{3}$$

We consider the image of $|w|<\rho$ on the z-plane, which lies in C. Then it consists of at most a countable number of connected domains $\{\Delta_i\}$. We shall prove that there exists a certain Δ_0 among $\{\Delta_i\}$, which contains z_0 on its boundary. For, if otherwise, since $w=0$ is a boundary point of $H_\Delta(z_0)$, there are infinitely many Δ_ν among $\{\Delta_i\}$, such that its boundary γ_ν has common points with Γ and Δ_ν contains a point z_ν, such that $z_\nu\to z_0$, $f(z_\nu)\to 0$. Then Δ_ν converges to z_0. For, if otherwise, γ_ν has a common point ξ_ν with a suitable Jordan curve C' in Δ, surrounding z_0 and lying inside of C. Let ξ_0 be one of the cluster points of $\{\xi_\nu\}$, then ξ_0 belongs to Δ, so that $f(z)$ is regular at ξ_0 and in any small neighbourhood of it, there are infinitely many niveau curves $|f(z)|=\text{const.}=\rho$, which is absurd. Hence Δ_ν converges to z_0.

Let e_ν be the common part of γ_ν and Γ, then by (3), e belongs to

e, so that e_ν is a closed set of capacity zero. If we map the universal covering surface of Δ_ν conformally on $|\zeta|<1$ by $z=\varphi(\zeta)$, then by Theorem VIII. 29, e_ν is mapped on a null set on $|\zeta|=1$, so that $w=f(\varphi(\zeta))=F(\zeta)\in U_\rho(0)$, hence $F(\zeta)$ takes any value $w(|w|<\rho)$ at least once, except a set of capacity zero. Since in any small neighbourhood of z_0, there are infinitely many Δ_ν, clustering to z_0, $f(z)$ takes any value $w(|w|<\rho)$ infinitely often, except a set of capacity zero, so that $|w|\leqq\rho$ belongs to $H_\Delta(z_0)$, which is absurd, since $w=0$ is a boundary point of $H_\Delta(z_0)$. Hence there is a certain Δ_0 among $\{\Delta_i\}$, which contains z_0 on its boundary. Since $w=0$ is a boundary point of $H_\Delta(z_0)$, we can prove similarly that there is only a finite number of such Δ_0, hence one fixed Δ_0 contains infinitely many $z_\nu\to z_0$, such that $f(z_\nu)\to 0$.

If we consider the image of $|w|<\frac{\rho}{n}$ $(n=1, 2, \cdots)$ in Δ_0, we see that there exists a curve L in Δ_0, ending at z_0, such that $f(z)\to 0$, if $z\to z_0$ along L. We take off L from Δ and let $\tilde{\Delta}$ be the remaining domain, then z_0 is a regular point of the Dirichlet problem for $\tilde{\Delta}$. Since $w=0$ is a boundary point of $H_\Delta(z)$, there exists $w_1(|w_1|<\rho)$, which does not belong to $H_\Delta(z_0)$. Since $\lim \frac{1}{|f(z)-w_1|}=\frac{1}{|w_1|}$, if $z\to z_0$ along L and by (3),

$$\overline{\lim_{\zeta\to z_0}}\ \overline{\lim_{z\to\zeta\in\Gamma-e}} \frac{1}{|f(z)-w_1|}\leqq \overline{\lim_{\zeta\to z_0}}\ \overline{\lim_{z\to\zeta\in\Gamma-e}} \frac{1}{|f(z)|-|w_1|}\leqq\frac{1}{2\rho-|w_1|},$$

if we apply the lemma to $\frac{1}{|f(z)-w_1|}$ for $\tilde{\Delta}$, then we have

$$\overline{\lim_{z\to z_0}} \frac{1}{|f(z)-w_1|}\leqq \operatorname{Max}\left(\frac{1}{|w_1|}, \frac{1}{2\rho-|w_1|}\right)=\frac{1}{|w_1|},$$

or

$$\underline{\lim}\,|f(z)-w_1|\geqq|w_1|, \quad \text{if } z\to z_0 \text{ in } \Delta. \tag{4}$$

Hence the disc: $|w-w_1|<|w_1|$ does not belong to $H_\Delta(z_0)$. We choose a point w_2 on the circle $|w|=w_1$, which lies in $|w-w_1|<|w_1|$, then the disc $|w-w_2|<|w_2|$ does not belong to $H_\Delta(z_0)$ and if we proceed similarly, then we see that a pointed disc: $0<|w|<|w_1|$ does not belong to $H_\Delta(z_0)$, so that $w=0$ is an isolated point of $H_\Delta(z_0)$, but this is absurd, since we can prove easily that $H_\Delta(z_0)$ contains a continuum, which connects $w=0$ to $H_{\Gamma-e}(z_0)$. Hence the theorem is proved. ——

Since by the theorem, the boundary of $H_\Delta(z_0)$ belongs to $H_{\Gamma-e}(z_0)$, we have

Theorem VIII. 40. *Let Δ be a bounded domain on the z-plane and Γ be its boundary and z_0 be a non-isolated boundary point and e be a set of capacity zero on Γ, such that $z_0\in e$. Let $f(z)$ be one-valued,* *regular and bounded in Δ, then*

$$\overline{\lim_{z\to z_0}}\,|f(z)|\leqq\overline{\lim_{\zeta\to z_0}}\ \overline{\lim_{z\to\zeta\in\Gamma-e}}\,|f(z)|\,.$$

4. By Theorem VIII. 38, $H_\Delta(z_0)-H_{\Gamma-e}(z_0)$ is an open set, so that if it is not an empty set, it consists of at most a countable number of connected domains (components).

Theorem VIII. 41. *Let Ω be one of components of $H_\Delta(z_0)-H_{\Gamma-e}(z_0)$, then in any small neighbourhood of z_0, $f(z)$ takes any value $\in\Omega$ infinitely often, except a set of capacity zero.* Tsuji [7], [50].

If e consists of only one point z_0, then in any small neighbourhood of z_0, $f(z)$ takes any value $\in\Omega$ infinitely often, with two possible exceptions. Kunugui [2].

This can be deduced easily from the following more general theorem.

Theorem VIII. 42. *Let $w_0\in\Omega$ and K: $|w-w_0|<\rho$ be a disc, contained in Ω. We choose $r>0$ so small that $\overline{V}_r(\Gamma-e)$ lies outside of $|w-w_0|=\rho$ and the circle $|z-z_0|=r$ does not contain a point of e. Then by the image curve of $|z-z_0|=r$ on the w-plane, K is decomposed into a finite number of connected domains $\{D_\nu\}$. Let F be the Riemann surface, generated by $w=f(z)$ over the w-plane and F_ρ be a connected piece of F, which lies above K.*

Then either (i) *F_ρ covers any point of D_ν exactly a finite number n_ν of times, or* (ii) *F_ρ covers any point of K infinitely often, except a set of capacity zero. Especially if e consists of only one point z_0, then F_ρ covers any point of K infinitely often, with two possible exceptions.*

Proof. By the choice of r, the part of circle $|z-z_0|=r$, which lies in a small neighbourhood of Γ is mapped outside of K, so that by the image curve of $|z-z_0|=r$, K is decomposed into a finite number of connected domains D_ν ($\nu=1, 2, \cdots, m$). First we remark the following. Let K' : $|w-a|<\delta$ be a disc, contained in D_ν and F'_ρ be a connected piece of F_ρ, which lies above K'. Then F'_ρ covers any point of K' at least once, except a set of capacity zero. For, let Δ' be the image of F'_ρ on the z-plane, then by the choice of r, the common part e' of the boundary of Δ' and Γ is a closed sub-set of e, so that it is of capacity zero. Hence if we map the universal covering surface of Δ' conformally on $|\zeta|<1$ by $z=\varphi(\zeta)$, then e' is mapped on a null set on $|\zeta|=1$, so that $w=f(\varphi(\zeta))=F(\zeta)\in U_\delta(a)$, hence F_ρ covers any point of K' at least once, except a set of capacity zero.

Suppose that a disc $K'_1\subset D_1$ is covered exactly $n_1(1\leqq n_1<\infty)$-times by F_ρ and let Σ_1 be the domain $\subset D_1$, which contains K'_1 and every point of which is covered exactly n_1-times by F_ρ. If Σ_1 does not coincides with D_1, let Γ_1 be the part of its boundary, contained in D_1, then by the above remark, we can prove similarly as Theorem VIII. 37, that $\gamma(\Gamma_1)=0$ and by means of this, we can prove that Σ_1 coincides with D_1, so that every point of D_1 is covered exactly n_1-times by F_ρ. Let D_2 abutts on D_1 along an image curve λ of $|z-z_0|=r$. Then we see easily that in a small neighbourhood of λ, there exists a disc K_2' in D_2, which is covered exactly n_2-times by F_ρ, where $n_2=n_1+1$, or $n_2=n_1-1$. Hence every point of D_2 is covered exactly n_2-times by F_ρ and we proceed similarly, then we see that if a disc K', contained in one of $\{D_\nu\}$ is covered exactly a finite number of times by F_ρ, then every point of D_ν is covered exactly a finite number n_ν of times by F_ρ.

Next suppose that any disc K', contained in $\{D_\nu\}$ is not covered exactly a finite number of times by F_ρ. Let a be any point of D_ν and be covered $n(a)$-times by F_ρ and put

$$N_\nu=\sup_{a\in D_\nu} n(a)\,,$$

then similarly as Theorem VIII. 37, we can prove that if $N_\nu<\infty$, then F_ρ covers any point of D_ν exactly N_ν-times, which is excluded. Hence $N_\nu=\infty$ and F_ρ covers any point of D_ν infinitely often except a set of capacity zero. Hence F_ρ covers any point of $\sum_{\nu=1}^{m} D_\nu$ infinitely often, except a set of capacity zero. Since F_ρ can be continued to a slightly larger surface, across its boundary curves, we see that F_ρ covers any point of K infinitely often, except a set of capacity zero.

Next we consider the special case, where e consists of only one point z_0. Let Δ_ρ be the image of F_ρ on the z-plane, then by the choice of r, z_0 is the only one common point of the boundary of Δ_ρ with Γ. Since F_ρ covers any point of K infinitely often, except a set of capacity zero, the area of F_ρ is infinite, so that if by $1/z-z_0$, we assume that $z_0=\infty$, then similarly as Theorem VI. 9, we can prove that F_ρ covers any point of K infinitely often, with two possible exceptions. Hence the theorem is proved. ——

Proof of Theorem VIII. 41. Next we shall show that Theorem VIII. 41 can be deduced simply as follows. Suppose that there exists a set E of positive capacity in Ω, every point of which is taken by $w=f(z)$ only a finite number of times in a neighbourhood Δ_r of z_0. Then by taking r small, we may assume that E is a closed set and $f(z)$ does not take any value $a\in E$ in Δ_r.

There exists a point $a\in E$ and a disc K: $|w-a|<\rho$, contained in Ω, such that $\gamma(E.K)>0$. Since $a\in\Omega$, there exists a connected piece

F_ρ of F, which lies above K. Since F_ρ does not cover $E.K$, F_ρ is non-compact, so that by Theorem VIII. 42, F_ρ covers any point of K infinitely often, except a set of capacity zero, which is absurd, since F_ρ does not cover $E.K$ and $\gamma(E.K)>0$. Hence $f(z)$ takes any value $a\in\Omega$ infinitely often, except a set of capacity zero in any small neighbourhood of z_0.

Next suppose that e consists only one point z_0 and $f(z)\neq a, b, c$ in Δ_r, where a, b, c are three points of Ω. We enclose a, b, c in a domain Σ, contained in Ω, which is bounded by a Jordan curve C. Let F_Σ be a connected piece of F, which lies above Σ. Since F_Σ does not cover a, b, c, F_Σ is non-compact. By mapping Σ conformally on a disc, we see by Theorem VIII. 42, that F_Σ covers any point of Σ infinitely often, with two possible exceptions, which is absurd, since F_Σ does not cover a, b, c. Hence $f(z)$ takes any value $\in\Omega$ infinitely often, with two possible exceptions in any small neighbourhood of z_0.

10. Cluster set of a meromorphic function in $|z|<1$

Let $w=f(z)$ be meromorphic in $|z|<1$ and $S(e^{i\theta}, \varphi, \varphi')$ be an angular domain, bounded by two lines L, L' through $e^{i\theta}$, making an angle φ, φ' $(0<\varphi<\varphi'<\pi)$ with the positive tangent of $|z|=1$ at $e^{i\theta}$. Let $H(e^{i\theta}, \varphi, \varphi')$ be the cluster set of $f(z)$ in $S(e^{i\theta}, \varphi, \varphi')$ at $e^{i\theta}$, then $H(e^{i\theta}, \varphi, \varphi')$ is a closed set, so that

$$H^*(e^{i\theta}) = \sum_{0<\varphi<\varphi'<\pi} H(e^{i\theta}, \varphi, \varphi'), \text{ added for all } \varphi, \varphi': 0<\varphi<\varphi'<\pi ,$$

is an F_σ-set, which we call *the angular cluster set of* $f(z)$ *at* $e^{i\theta}$.

If $H^*(e^{i\theta})$ consists of only one point a, then $\lim_{z\to e^{i\theta}} f(z)=a$ uniformly, when $z\to e^{i\theta}$ from the inside of any fixed Stolz domain, whose vertex is at $e^{i\theta}$. We call such $e^{i\theta}$ *a point of the first kind.*

If for any φ, φ', $H(e^{i\theta}, \varphi, \varphi')$ concides with the whole w-sphere, then we call such $e^{i\theta}$ *a point of the second kind.*

Theorem VIII. 43. *Let $w=f(z)$ be meromorphic in $|z|<1$, then the unit circle K: $|z|=1$ is decomposed into three parts:*

$$K=E_1+E_2+N ,$$

where E_1 consists of points of the first kind, E_2 consists of points of the second kind and N is a null set. Plessner [2].

Proof. We decompose the w-sphere Ω into 8 equal (spherical) triangles by great circles of Ω and then we decompose each triangle into 3 equal triangles by great circles of Ω through the middle point and vertices of the triangle and proceed similarly, then after n steps, we decompose Ω into 8.3^n equal triangles Δ_ν^n ($\nu=1, 2, \cdots, 8.3^n$). For a fixed φ, φ' ($0<\varphi<\varphi'<\pi$), let A_ν^n be the set of $e^{i\theta}$, such that $H(e^{i\theta}, \varphi, \varphi')$ lies outside of Δ_ν^n, then we see easily that A_ν^n is a measurable set. We suppose that $mA_\nu^n>0$, then for any $\varepsilon>0$, there exists a closed subset B_ν^n of A_ν^n, such that $mB_\nu^n>mA_\nu^n-\varepsilon>0$.

Let $0<\rho<1$ and $S_\rho(e^{i\theta}, \varphi, \varphi')$ be the part of $S(e^{i\theta}, \varphi, \varphi')$, which lies in $\rho<|z|<1$ and put

$$D_\nu^n=\sum_{e^{i\theta}\in B_\nu^n} S_\rho(e^{i\theta}, \varphi, \varphi'), \text{ added for all } e^{i\theta}\in B_\nu^n \tag{1}$$

and we join the disc $|z|\leqq\rho$ to D_ν^n and let $\tilde{D}_\nu^n$ be the resulting domain, then $\tilde{D}_\nu^n$ is bounded by a rectifiable Jordan curve Γ, on which B_ν^n lies. We map $\tilde{D}_\nu^n$ conformally on $|\zeta|<1$ and put $f(z)=F(\zeta)$ and let M_ν^n be the image of B_ν^n on $|\zeta|=1$, then by F. and M. Riesz' theorem, $mM_\nu^n>0$.

Since the value of $F(\zeta)$ in a neighbourhood of M_ν^n lies outside of Δ_ν^n, by a suitable transformation, we may assume that $F(\zeta)$ is bounded in $|\zeta|<1$, so that $\lim_{\zeta\to e^{i\psi}} F(\zeta)$ exists almost everywhere on M_ν^n, when ζ tends to $e^{i\psi}$ non-tangentially to $|\zeta|=1$.

Now by F. and M. Riesz' theorem, a null set on $|\zeta|=1$ is mapped on a null set on Γ and since Γ is a rectifiable Jordan curve, Γ has a tangent almost everywhere on B_ν^n, which coincides with the tangent of $|z|=1$ and by Lindelöf's theorem (Theorem IX. 5), the angle is preserved by conformal mapping, so that $\lim_{z\to e^{i\theta}} f(z)$ exists almost everywhere on B_ν^n, when $z\to e^{i\theta}$ non-tangentially to $|z|=1$. Since $\varepsilon>0$ is arbitrary, $\lim_{z\to e^{i\theta}} f(z)$ exists almost everywhere on A_ν^n, when $z\to e^{i\theta}$ non-tangentially to $|z|=1$.

We put

$$A^n(\varphi, \varphi')=\sum_{\nu=1}^{8.3^n} A_\nu^n, \quad A(\varphi, \varphi')=\sum_{n=1}^{\infty} A^n(\varphi, \varphi') . \tag{2}$$

Let φ_i ($i=1, 2, \cdots$) be the set of all rational numbers in $(0, \pi)$ and

$$A=\sum_{i,j} A(\varphi_i, \varphi_j) . \tag{3}$$

Then $\lim_{z\to e^{i\theta}} f(z)$ exists almost everywhere on A, when $z\to e^{i\theta}\in A$ non-tangentially to $|z|=1$. Hence if we denote the set of $e^{i\theta}\in A$, such that $\lim_{z\to e^{i\theta}} f(z)$ exists by E_1, then $A=E_1+N$, where N is a null set.

Let E_2 be the complement of A on K: $|z|=1$, then

$$K=E_1+E_2+N . \tag{4}$$

If $e^{i\theta}\in E_2$, then $e^{i\theta}$ belongs to the complement of $A^n(\varphi_i, \varphi_j)$ ($n=1, 2, \cdots$), so that $H(e^{i\theta}, \varphi_i, \varphi_j)$ has common points with every Δ_ν^n ($\nu=1, 2, \cdots, 8.3^n$), so that $H(e^{i\theta}, \varphi_i, \varphi_j)$ coincides with the whole w-sphere. From this, we see easily that for any real φ, φ' $(0<\varphi<\varphi'<\pi)$, $H(e^{i\theta}, \varphi, \varphi')$ coincides with the whole w-sphere, hence E_2 consists of points of the second kind.

Theorem VIII. 44. *Let $w=f(z)$ $(\not\equiv \text{const.})$ be meromorphic in $|z|<1$ and E be a measurable set of positive measure on $|z|=1$ and*

$$E^* = \sum_{e^{i\theta}\in E} H^*(e^{i\theta}),$$

then

$$\gamma(E^*) > 0. \qquad \text{Tsuji [11].}$$

R. Nevanlinna [6] and Frostman [1] proved the case, where $f(z)$ is of bounded type.

Proof. Suppose that $\gamma(E^*)=0$, then E^* does not contain a continuum, so that $\lim f(z)$ exists, when z tends to any point of E non-tangentially to $|z|=1$. By Egoroff's theorem, we may assume that E is a closed set and $\lim_{z\to e^{i\theta}} f(z)$ exists uniformly on E, when $z\to e^{i\theta}\in E$ from the inside of $S(e^{i\theta}, \pi/4, 3\pi/4)$. Hence E^* is a bounded closed set of capacity zero.

As the proof of Theorem VIII. 28, we define a domain D, which is bounded by a rectifiable Jordan curve C, on which E lies. Let A be the set of points $z\in D$, such that $f(z)\in E^*$, which may be empty, then A is an F_σ-set of capacity zero. Let $D=D-A$ and we map the universal covering surface of $\tilde{D}$ conformally on $|\zeta|<1$ and put $f(z)=F(\zeta)$, then by Theorem VIII. 29, A is mapped on a null set on $|\zeta|=1$.

Let $C-E$ be the complement of E on C, then $C-E$ is mapped on an open set on $|\zeta|=1$ and since A is mapped on a null set on $|\zeta|=1$, E is mapped on a measurable set M on $|\zeta|=1$.

Let $\omega(z)$ be the harmonic measure of E with respect to D, then since $mE>0$, $\omega(z)\not\equiv 0$. Hence if we consider $\omega(z)$ as a harmonic function in $|\zeta|<1$, then we see that $mM>0$, hence E is mapped on a measurable set M of positive measure on $|\zeta|=1$. Let $u(w)$ be the Evans' function with respect to E^* and $v(w)$ be its conjugate harmonic function and put $\varphi(w)=u(w)+iv(w)$, $\varphi(F(\zeta))=\Phi(\zeta)$, then $\Phi(\zeta)$ is meromorphic in $|\zeta|<1$ and $\Phi(\zeta)\to\infty$, if ζ tends to any point of M non-tangentially to $|\zeta|=1$, so that by Lusin-Privaloff's theorem, $\Phi(\zeta)\equiv\infty$, or $f(z)\equiv \text{const}\in E^*$, which contradicts the hypothesis. Hence $\gamma(E^*)>0$.

11. Fejér and F. Riesz' theorem and its extension by M. Riesz

Fejér and F. Riesz [1] proved the following theorem:

Theorem VIII. 45. *Let $f(z)$ be regular in $|z|<1$ and belong to*

H^p-class ($p>0$), then

$$\int_{-1}^{1}|f(z)|^p|dz|\leqq\frac{1}{2}\int_{|z|=1}|f(z)|^p|dz|,\quad p>0,$$

where on the left hand side, we integrate on the real axis and on the right hand side, we integrate on the unit circle $|z|=1$.

This is a special case of the following theorem.

Theorem VIII. 46. *Let $f(z)$ be regular in $|z|<1$ and belong to H^p-class ($p>0$) and Γ be a rectifiable curve in $|z|<1$ and*

$$V(z)=\int_{x\in\Gamma}|d_x\arg(z-x)|,\quad |z|=1.$$

Then

$$\int_{\Gamma}|f(z)|^p|dz|\leqq\frac{1}{\pi}\int_{|z|=1}|f(z)|^pV(z)|dz|.$$ M. Riesz [2].

If Γ is the diameter of $|z|=1$, then $V(z)=\pi/2$ on $|z|=1$, so that we have Theorem VIII. 45.

If Γ is a convex curve, then $V(z)\leqq 2\pi$ on $|z|=1$, so that

$$\int_{\Gamma}|f(z)|^p|dz|\leqq 2\int_{|z|=1}|f(z)|^p|dz|.$$

Proof. We may assume that $f(z)$ is regular on $|z|=1$ and first suppose that $p=1$. Let $u(x)$ be harmonic in $|x|\leqq 1$ and put

$$g(x)=\frac{1}{\pi}\int_0^{2\pi}u(e^{i\theta})\frac{\partial\arg(e^{i\theta}-x)}{\partial\theta}d\theta,\quad |x|<1,\tag{1}$$

then $g(x)$ is harmonic in $|x|<1$ and if x tends to $z_0=e^{i\theta_0}$, then

$$g(e^{i\theta_0})=\lim_{x\to e^{i\theta_0}}g(x)=u(e^{i\theta_0})+u(0),\tag{2}$$

which can be proved as follows. Let C_δ be the part of $|z|=1$, which lies in $|z-z_0|<\delta$ and C'_δ be its complement on $|z|=1$, then

$$g(x)=\frac{1}{\pi}\int_{C_\delta}u(e^{i\theta})\frac{\partial\arg(e^{i\theta}-x)}{\partial\theta}d\theta+\frac{1}{\pi}\int_{C'_\delta}u(e^{i\theta})\frac{\partial\arg(e^{i\theta}-x)}{\partial\theta}d\theta=\mathrm{I}+\mathrm{II}.$$

Since

$$\lim_{\delta\to0}\lim_{x\to e^{i\theta_0}}\mathrm{I}=u(e^{i\theta_0}),\qquad \lim_{\delta\to0}\lim_{x\to e^{i\theta_0}}\mathrm{II}=\frac{1}{2\pi}\int_0^{2\pi}u(e^{i\theta})\,d\theta=u(0),$$

we have (2).

Hence $g(x)=u(x)+u(0)$ on $|x|=1$, so that by the maximum principle,

$$g(x)=u(x)+u(0)\quad \text{in } |x|\leqq 1.\tag{3}$$

If we take $u(x)$ as

$$u(x)=\int_0^x f(x)dx\,, \tag{4}$$

where we integrate on the radius of $|x|=1$, then $u(x)$ is harmonic in $|x|\leqq 1$ and $u(0)=0$, so that $u(x)=g(x)$ in $|x|\leqq 1$, hence

$$u(x)=\int_0^x f(x)dx=\frac{1}{\pi}\int_0^{2\pi} u(e^{i\theta})\frac{\partial \arg(e^{i\theta}-x)}{\partial\theta}\,d\theta\,,$$

so that

$$f(x)=\frac{1}{\pi}\int_0^{2\pi} u(e^{i\theta})\frac{\partial}{\partial\theta}\left(\frac{\partial \arg(e^{i\theta}-x)}{\partial x}\right)d\theta$$

$$=\frac{1}{\pi}\left[u(e^{i\theta})\frac{\partial \arg(e^{i\theta}-x)}{\partial x}\right]_0^{2\pi}-\frac{1}{\pi}\int_0^{2\pi}\frac{\partial u(e^{i\theta})}{\partial\theta}\cdot\frac{\partial \arg(e^{i\theta}-x)}{\partial x}\,d\theta\,,$$

hence

$$f(x)=-\frac{1}{\pi}\int_0^{2\pi} f(e^{i\theta})\,ie^{i\theta}\,\frac{\partial \arg(e^{i\theta}-x)}{\partial x}\,d\theta\,, \tag{5}$$

so that

$$\int_\Gamma |f(x)|\,|dx|\leqq\frac{1}{\pi}\int_0^{2\pi}|f(e^{i\theta})|\left(\int_\Gamma\left|\frac{\partial \arg(e^{i\theta}-x)}{\partial x}\right||dx|\right)d\theta$$

$$=\frac{1}{\pi}\int_0^{2\pi}|f(e^{i\theta})|\left(\int_{x\in\Gamma}|d_x \arg(e^{i\theta}-x)|\right)d\theta=\frac{1}{\pi}\int_{|z|=1}|f(z)|\,V(z)|dz|\,. \tag{6}$$

Hence the case $p=1$ is proved. In the general case, if $f(z)\neq 0$ in $|z|<1$, then $(f(z))^p$ is regular in $|z|<1$ and is continuous in $|z|\leqq 1$, so that

$$\int_\Gamma |f(z)|^p|dz|\leqq\frac{1}{\pi}\int_{|z|=1}|f(z)|^p V(z)|dz|,\ p>0\,. \tag{7}$$

If $f(z)$ has zero points $z_1,\cdots,z_n$ in $|z|<1$, let $B(z)=\prod_{\nu=1}^{n}\frac{z-z_\nu}{1-\bar{z}_\nu z}$, then $F(z)=f(z)/B(z)$ is regular in $|z|<1$ and continuous in $|z|\leqq 1$ and $F(z)\neq 0$ in $|z|<1$, so that

$$\int_\Gamma |f(z)|^p|dz|<\int_\Gamma\left|\frac{f(z)}{B(z)}\right|^p|dz|\leqq\frac{1}{\pi}\int_{|z|=1}\left|\frac{f(z)}{B(z)}\right|^p V(z)|dz|$$

$$=\frac{1}{\pi}\int_{|z|=1}|f(z)|^p V(z)|dz|\,.$$

Hence the theorem is proved. ——

If we apply Theorem VIII. 45 to $f'(z)$, $p=1$, then we have

$$\int_{-1}^{1}|f'(z)||dz|\leqq\frac{1}{2}\int_{|z|=1}|f'(z)||dz|\,, \tag{1}$$

hence

Corollary. *Let $w=f(z)$ be regular in $|z|<1$ and continuous in $|z|\leqq 1$ and the length L of the image of $|z|=1$ on the w-plane be finite, then* *the length of the image of a diameter of $|z|=1$ is at most $L/2$.*

Remark. By Theorem VIII. 45, we can prove the following inequalities.

If $a_n \geqq 0$, then

$$\sum_{i,k=0}^{\infty} \frac{a_i a_k}{i+k+1} \leqq \pi \sum_{n=0}^{\infty} a_n^2, \quad \sum_{i,k=1}^{\infty} \frac{a_i a_k}{i+k} \leqq \pi \sum_{n=1}^{\infty} a_n^2 . \quad \text{(Hilbert)}$$

Proof. Let $f(x)=\sum_{n=0}^{\infty} a_n x^n$, then

$$\int_{-1}^{1} f^2(x)dx \geqq \int_0^1 f^2(x)\,dx = \int_0^1 \left(\sum_{i=0}^{\infty} a_i x^i \sum_{k=0}^{\infty} a_k x^k \right) dx = \sum_{i,k=0}^{\infty} \frac{a_i a_k}{i+k+1},$$

and $\int_{|x|=1} |f(x)|^2 |dx| = 2\pi \sum_{n=0}^{\infty} a_n^2$, so that

$$\sum_{i,k=0}^{\infty} \frac{a_i a_k}{i+k+1} \leqq \pi \sum_{n=0}^{\infty} a_n^2 . \tag{1}$$

If we put $b_\nu = a_{\nu+1}$, then

$$\sum_{i,k=1}^{\infty} \frac{b_i b_k}{i+k} \leqq \sum_{i,k=1}^{\infty} \frac{b_i b_k}{(i-1)+(k-1)+1} = \sum_{i,k=0}^{\infty} \frac{a_i a_k}{i+k+1} \leqq \pi \sum_{n=0}^{\infty} a_n^2 = \pi \sum_{n=1}^{\infty} b_n^2 . \tag{2}$$

Theorem VIII. 47. *Let $w=f(z)$ be regular in a sector Δ: $|\theta| \leqq \theta_0$, $0<r\leqq 1$ ($z=re^{i\theta}$) and $f(z) \neq \alpha, \beta, \gamma$ in Δ and put*

$$L(\theta) = \int_0^1 |f'(re^{i\theta})| dr \qquad (-\theta_0 \leqq \theta \leqq \theta_0) .$$

If $L(-\theta_0) \leqq L$, $L(\theta_0) \leqq L$, then

$$L(\theta) \leqq L + \int_{-\theta_0}^{\theta_0} |f'(e^{i\theta})| d\theta \quad (-\theta_0 \leqq \theta \leqq \theta_0) .$$

Proof. Since $L(-\theta_0) \leqq L < \infty$, $L(\theta_0) \leqq L < \infty$, $\lim_{r\to 0} f(re^{-i\theta_0}) = a$, $\lim_{r\to 0} f(re^{i\theta_0}) = b$ exist and since $f(z) \neq \alpha, \beta, \gamma$ in Δ, by Theorem VIII. 14, $a=b$ and $\lim_{r\to 0} f(re^{i\theta}) = a$ uniformly for $|\theta| \leqq \theta_0$.

We map Δ conformally on $|\zeta|<1$, such that three points $z=e^{i\theta_0}$, 0, $e^{-i\theta_0}$ correspond to $\zeta = i, -1, -i$ respectively and put $f(z)=F(\zeta)$.

Let l_θ be the segment: $z=re^{i\theta}$ ($0\leqq r\leqq 1$) and Γ_θ be its image in $|\zeta|<1$, then Γ_θ connects $\zeta=-1$ to $\zeta=e^{i\varphi}$ ($|\varphi| \leqq \pi/2$). Let A be the arc: $\zeta=e^{i\varphi}$ ($\pi/2 \leqq \varphi \leqq \pi$) and B be the arc: $\zeta=e^{i\varphi}$ ($\pi \leqq \varphi \leqq 3\pi/2$) and C be the complement of $A+B$ on $|\zeta|=1$. Then

$$L(\theta) = \int_{l_\theta} |f'(z)||dz| = \int_{\Gamma_\theta} |F'(\zeta)||d\zeta| , \tag{1}$$

$$\left.\begin{aligned} L(\theta_0) &= \int_A |F'(\zeta)||d\zeta| \leqq L, \\ L(-\theta_0) &= \int_B |F'(\zeta)||d\zeta| \leqq L\,. \end{aligned}\right\} \quad (2)$$

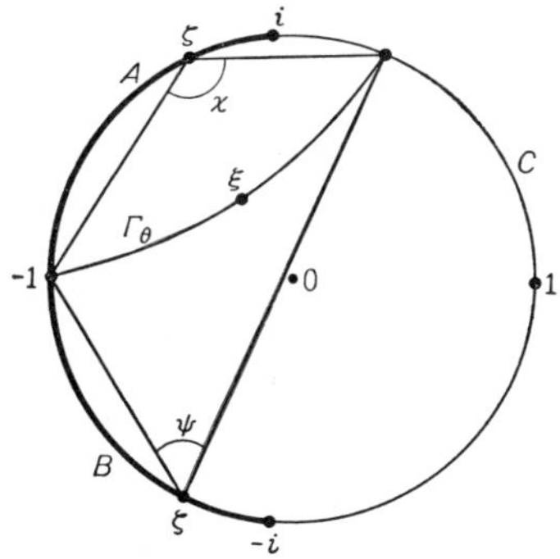

If we put

$$V(\zeta) = \int_{\xi \in \Gamma_\theta} |d_\xi \arg(\xi - \zeta)|, \qquad |\zeta| = 1\,,$$

then

$$\left.\begin{aligned} V(\zeta) &= \text{const.} = \chi \ \text{ on } A\,, \\ V(\zeta) &= \text{const.} = \psi \ \text{ on } B\,, \\ V(\zeta) &\leqq \pi \ \text{ on } C, \end{aligned}\right\} \quad (3)$$

where $\chi + \psi = \pi\,.$

Since $|\zeta| = 1$ is mapped on a rectifiable curve on the w-plane, $F'(\zeta) \in H^1$, so that by Theorem VIII. 46,

$$\begin{aligned} L(\theta) &= \int_{\Gamma_\theta} |F'(\zeta)||d\zeta| \leqq \frac{1}{\pi} \int_{|\zeta|=1} |F'(\zeta)|\, V(\zeta)\, |d\zeta| \\ &= \frac{1}{\pi}\Big(\int_A |F'(\zeta)|\, V(\zeta)|d\zeta| + \int_B |F'(\zeta)|\, V(\zeta)|d\zeta| \\ &\quad + \int_C |F'(\zeta)|\, V(\zeta)\, |d\zeta|\Big) \leqq \frac{1}{\pi}\Big(L\chi + L\psi + \pi \int_C |F'(\zeta)||d\zeta|\Big) \\ &= L + \int_C |F'(\zeta)||d\zeta| = L + \int_{-\theta_0}^{\theta_0} |f'(e^{i\theta})|\, d\theta\,. \end{aligned} \quad (4)$$

Theorem VIII. 48. *Let $w = f(z)$ be meromorphic in a sector Δ: $|\theta| \leqq \theta_0$, $0 < r \leqq 1$ $(z = re^{i\theta})$ and $f(z) \neq \alpha, \beta, \gamma$ in Δ and put*

$$L(\theta) = \int_0^1 \frac{|f'(re^{i\theta})|}{1 + |f(re^{i\theta})|^2}\, dr \qquad (-\theta_0 \leqq \theta \leqq \theta_0)\,.$$

If $L(-\theta_0) < \infty$, $L(\theta_0) < \infty$, then

$$L(\theta) \leqq \text{const.} \qquad (-\theta_0 \leqq \theta \leqq \theta_0)\,.$$

Proof. Since $L(-\theta_0) < \infty$, $L(\theta_0) < \infty$, $\lim_{r \to 0} f(re^{-i\theta_0}) = a$, $\lim_{r \to 0} f(re^{i\theta_0}) = b$ exist and since $f(z) \neq \alpha, \beta, \gamma$ in Δ, $a = b$ and $\lim_{r \to 0} f(re^{i\theta}) = a$ uniformly for $|\theta| \leqq \theta_0$. By a suitable rotation of the w-sphere, we may assume that $a = 0$ and without loss of generality, we may assume that $f(z)$ is regular and bounded in Δ. Hence the theorem is reduced to the preceding theorem.

12. Functions of class (D) and (D*)

1. Let $w = f(z)$ be regular in $|z| < 1$ and

$$D[w] = \iint_{|z|<1} |f'(z)|^2\, r\, dr\, d\theta < \infty\,, \qquad z = re^{i\theta}\,, \quad (1)$$

where $D[w]$is the area of the Riemann surface F, generated by $w=f(z)$ over the w-plane. Then we say that $f(z)$ *belongs to the class* (D) and denote $f(z)\in(D)$.

Let $w=f(z)$ be meromorphic in $|z|<1$ and

$$D^*[w]=\iint\limits_{|z|<1}\left(\frac{|f'(z)|}{1+|f(z)|^2}\right)^2 rdrd\theta<\infty , \tag{2}$$

where $D^*[w]$ is the spherical area of the Riemann surface F, generated by $w=f(z)$ over the w-sphere. Then we say that $f(z)$ *belongs to the class* (D^*) and denote $f(z)\in(D^*)$.

Theorem VIII. 49. *Let $w=f(z)$ be regular in $|z|<1$ and $D[w]<\infty$. Then there exists a set E of capactiy zero on $|z|=1$, such that if $e^{i\theta}$ does not belong to E, then any segment, which connects $e^{i\theta}$ to a point of $|z|<1$ is mapped on a rectifiable curve on the w-plane, so that* $\lim\limits_{z\to e^{i\theta}} f(z)=f(e^{i\theta})$ $(\neq\infty)$ *exists uniformly, when $z\to e^{i\theta}$ from the inside of any fixed Stolz domain, whose vertex is at $e^{i\theta}$.* Tsuji [19].

Beurling [2] proved that the radius of $|z|=1$ through $e^{i\theta}$ is mapped on a rectifiable curve on the w-plane. (*c.f.* Theorem IV. 42).

Theorem VIII. 50. *Let $w=f(z)$ be meromorphic in $|z|<1$ and $D^*[w]<\infty$. Then there exists a set E of capacity zero on $|z|=1$, such that if $e^{i\theta}$ does not belong to E, then any segment, which connects $e^{i\theta}$ to a point of $|z|<1$ is mapped on a rectifiable curve on the w-sphere, so that* $\lim\limits_{z\to e^{i\theta}}f(z)=f(e^{i\theta})$ *exists uniformly, when $z\to e^{i\theta}$ from the inside of any fixed Stolz domain, whose vertex is at $e^{i\theta}$.* Tsuji [19].

Beurling [2] proved that $\lim\limits_{r\to 1} f(re^{i\theta})$ exists, if $e^{i\theta}$ does not belong to E.

Since Theorem VIII. 49 can be proved similarly, we shall prove Theorem VIII. 50.

Proof. Let $e^{i\theta}$ be a point on $|z|=1$ and K: $\left|z-\frac{3}{4}e^{i\theta}\right|=\frac{1}{4}$ be a circle, which touches $|z|=1$ internally at $e^{i\theta}$. Let l_ψ be a segment, which connects a point $\in K$ to $e^{i\theta}$, making an angle $\psi\left(-\frac{\pi}{2}<\psi<\frac{\pi}{2}\right)$ with the inner normal of $|z|=1$ at $e^{i\theta}$ and put

$$L(\psi)=\int_{l_\psi}\frac{|f'(z)|}{1+|f(z)|^2}|dz| , \tag{1}$$

$$\chi(\theta)=\int_{-\frac{\pi}{2}}^{\frac{\pi}{2}} L(\psi)\cos\psi\, d\psi . \tag{2}$$

Let E be the set of $e^{i\theta}$, such that $\chi(\theta)=+\infty$. Then we shall prove that $\gamma(E)=0$. Suppose that $\gamma(E)>0$, then by taking a suitable closed sub-set, we may assume that E is a closed set on $|z|=1$. Let $u(z)$ be the conductor potential of E, such that for any z,

$$u(z)=\int_E \log \frac{1}{|z-e^{i\varphi}|}\, d\mu(\varphi) \leqq V<\infty, \quad \mu(E)=1. \tag{3}$$

Let

$$u(re^{i\theta})=\sum_{n=1}^{\infty} \frac{r^n}{n}(a_n \cos n\theta+b_n \sin n\theta),$$

$$a_n=\int_E \cos n\varphi\, d\mu(\varphi), \quad b_n=\int_E \sin n\varphi\, d\varphi d\mu(\varphi),$$

be the Fourier expansion of u, then since $u\leqq V$,

$$\int_E u(re^{i\theta})d\mu(\theta)=\sum_{n=1}^{\infty} \frac{r^n}{n}(a_n^2+b_n^2) \leqq V,$$

so that

$$\sum_{n=1}^{\infty} \frac{1}{n}(a_n^2+b_n^2) \leqq V,$$

hence

$$\iint_{|z|<1} \left(\frac{\partial u}{\partial r}\right)^2 r dr\, d\theta=\frac{\pi}{2}\sum_{n=1}^{\infty}\frac{1}{n}(a_n^2+b_n^2) \leqq \frac{\pi}{2}V. \tag{4}$$

If we put

$$J=\iint_{|z|<1} \frac{|f'(z)|}{1+|f(z)|^2}\frac{\partial u}{\partial r}\, r dr d\theta, \qquad A=\iint_{|z|<1}\left(\frac{|f'(z)|}{1+|f(z)|^2}\right)^2 r dr d\theta, \tag{5}$$

then

$$J^2 \leqq \iint_{|z|<1}\left(\frac{|f'(z)|}{1+|f(z)|^2}\right)^2 r dr\, d\theta \iint_{|z|<1}\left(\frac{\partial u}{\partial r}\right)^2 r dr d\theta \leqq \frac{\pi}{2}AV. \tag{6}$$

Let $v(z)$ be the conjugate harmonic function of $u(z)$, then

$$r\frac{\partial u}{\partial r}=\frac{\partial v}{\partial \theta}=-\int_E \frac{\partial}{\partial \theta}\arg(re^{i\theta}-e^{i\varphi})\, d\mu(\varphi),$$

so that

$$J=\int_E d\mu(\varphi)\int_0^1 dr\int_0^{2\pi}\frac{|f'(re^{i\theta})|}{1+|f(re^{i\theta})|^2}(-d_\theta \arg(re^{i\theta}-e^{i\varphi})). \tag{7}$$

Hence if we put

$$J(\varphi)=\int_0^1 dr\int_0^{2\pi}\frac{|f'(re^{i\theta})|}{1+|f(re^{i\theta})|^2}(-d_\theta \arg(re^{i\theta}-e^{i\varphi})), \tag{8}$$

then

$$J=\int_E J(\varphi)\,d\mu(\varphi)\,. \tag{9}$$

We shall prove that $J(\varphi)=\infty$, if $e^{i\varphi}\in E$. Suppose that $z=1$ ($\varphi=0$) belongs to E, then $\chi(0)=\infty$. If we put

$$\psi=\pi-\arg(re^{i\theta}-1)\,, \tag{10}$$

then

$$J(0)=\int_0^1 dr\int_{\theta=-\pi}^{\theta=\pi}\frac{|f'(re^{i\theta})|}{1+|f(re^{i\theta})|^2}\,d\psi\,. \tag{11}$$

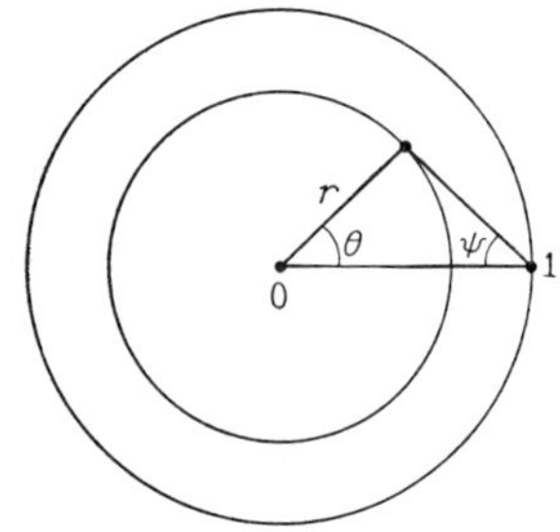

Since $\tan\psi=\frac{r\sin\theta}{1-r\cos\theta}$, we have on the circle $|z|=r$,

$$d\psi=\frac{\cos\theta-r}{1-2r\cos\theta+r^2}\,rd\theta\,. \tag{12}$$

Hence if we put $\cos^{-1}r=\theta_0>0$, then $d\psi\geqq 0$, if $|\theta|\leqq\theta_0$ and $d\psi\leqq 0$, if $\theta_0\leqq|\theta|\leqq\pi$, so that

$$J(0)=\int_0^1 dr\int_{|\theta|\leqq\theta_0}\frac{|f'(re^{i\theta})|}{1+|f(re^{i\theta})|^2}\,d\psi-\int_0^1 dr\int_{\theta_0\leqq|\theta|\leqq\pi}\frac{|f'(re^{i\theta})|}{1+|f(re^{i\theta})|^2}\,|d\psi|\,. \tag{13}$$

We remark that the circular arc: $z=re^{i\theta}$ ($|\theta|\leqq\theta_0$) lies in the circle: $\left|z-\frac{1}{2}\right|=\frac{1}{2}$.

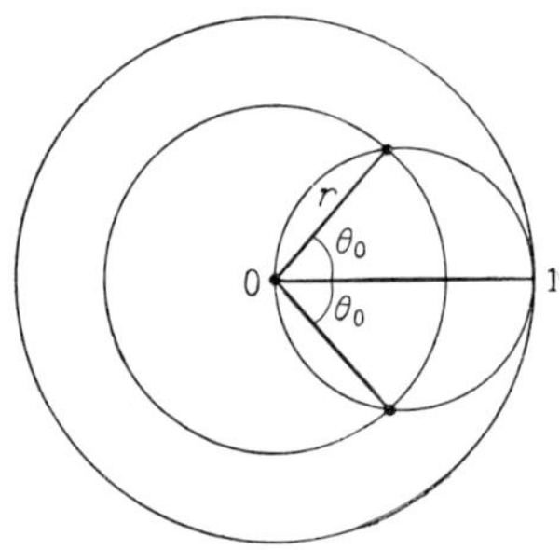

Since for $\theta_0\leqq|\theta|\leqq\pi$, $0\leqq\frac{r-\cos\theta}{1-2r\cos\theta+r^2}\leqq\frac{1}{1+r}<1$, we have $|d\psi|\leqq rd\theta$, so that by Schwarz's inequality,

$$\int_0^1 dr\int_{\theta_0\leqq|\theta|\leqq\pi}\frac{|f'(re^{i\theta})|}{1+|f(re^{i\theta})|^2}\,|d\psi|$$

$$\leqq\iint_{|z|<1}\frac{|f'(z)|}{1+|f(z)|^2}\,rdrd\theta\leqq\sqrt{\pi A}\,,$$

hence

$$J(0)\geqq\int_0^1 dr\int_{|\theta|\leqq\theta_0}\frac{|f'(re^{i\theta})|}{1+|f(re^{i\theta})|^2}\,d\psi-\sqrt{\pi A}$$

$$=\iint_{|z-\frac{1}{2}|\leqq\frac{1}{2}}\frac{|f'(z)|}{1+|f(z)|^2}\,drd\psi-\sqrt{\pi A}\,. \tag{14}$$

Let $t=|z-1|$ and we change the variables in (14) from (r,ψ) to (t,ψ), by $r=\sqrt{1-2t\cos\psi+t^2}$, $\psi=\psi$, then

$$dr d\psi = \frac{\cos\psi - t}{\sqrt{1-2t\cos\psi + t^2}} dt d\psi \qquad (0 \leqq t \leqq \cos\psi).$$

Since $1-2t\cos\psi+t^2<1$ in $\left|z-\frac{1}{2}\right| \leqq \frac{1}{2}$, we have

$$\iint\limits_{|z-\frac{1}{2}|\leqq\frac{1}{2}} \frac{|f'(z)|}{1+|f(z)|^2} dr\, d\psi = \int_{-\frac{\pi}{2}}^{\frac{\pi}{2}} d\psi \int_0^{\cos\psi} \frac{|f'(z)|}{1+|f(z)|^2} \cdot \frac{\cos\psi - t}{\sqrt{1-2t\cos\psi+t^2}} dt$$

$$\geqq \int_{-\frac{\pi}{2}}^{\frac{\pi}{2}} d\psi \int_0^{\cos\psi} \frac{|f'(z)|}{1+|f(z)|^2} (\cos\psi - t)\, dt$$

$$\geqq \frac{1}{2} \int_{-\frac{\pi}{2}}^{\frac{\pi}{2}} \cos\psi d\psi \int_0^{\frac{\cos\psi}{2}} \frac{|f'(z)|}{1+|f(z)|^2} dt$$

$$= \frac{1}{2} \int_{-\frac{\pi}{2}}^{\frac{\pi}{2}} \cos\psi\, d\psi \int_{l_\psi} \frac{|f'(z)|}{1+|f(z)|^2} |dz| = \frac{1}{2} \chi(0) = \infty,$$

so that by (14), $J(0)=\infty$. Similarly $J(\varphi)=\infty$, if $e^{i\varphi}\in E$, hence by (9), $J=\infty$, which contradicts (6), so that $\gamma(E)=0$.

If $e^{i\theta}$ does not belong to E, then $\chi(\theta)<\infty$, so that by (2), $L(\psi)<\infty$ for almost all ψ in $\left(-\frac{\pi}{2}, \frac{\pi}{2}\right)$. Since $A<\infty$, $f(z)$ takes any value only a finite number of times, except a null set, hence by Theorem VIII. 48, $L(\psi)<\infty$ for all ψ in $\left(-\frac{\pi}{2}, \frac{\pi}{2}\right)$. Hence a segment, which connects $e^{i\theta}$ to any point of $|z|<1$ is mapped on a rectifiable curve on the w-sphere. Since $L(\psi)<\infty$, $\lim\limits_{z\to e^{i\theta}} f(z)$ exists, when $z\to e^{i\theta}$ on l_ψ and since $f(z)$ takes any value a finite number of times, except a null set, by Theorem VIII. 14, $\lim\limits_{z\to e^{i\theta}} f(z)=f(e^{i\theta})$ exists uniformly, when $z\to e^{i\theta}$ from the inside of any fixed Stolz domain, whose vertex is at $e^{i\theta}$. Hence Theorem VIII. 50 is proved.

If we replace $\frac{|f'(z)|}{1+|f(z)|^2}$ by $|f'(z)|$ in the proof and use Theorem VIII. 47, then we have Theorem VIII. 49. ——

2. As an application of Theorem VIII. 49, we shall prove the following metrical theorem on conformal mapping.

Theorem VIII. 51. *Let D be a simply connected bounded domain on the w-plane and Γ be its boundary and E be a closed set of capacity zero on Γ, which consists of accessible boundary points. If we map D conformally on $|z|<1$, then E is mapped on a set E^* of capacity zero on $|z|=1$.*

If D is bounded by a Jordan curve Γ and E be a Borel set of capacity zero on Γ, then E is mapped on a Borel set E^* of capacity zero on $|z|=1$. Dufresnoy [3], Tsuji [19].

Proof. Let $u(w)$ be the Evans' function with respect to E:

$$u(w)=\int_E \log\frac{1}{|w-a|}\,d\mu(a), \qquad \mu(E)=1\,,$$

such that $u\to+\infty$, if w tends to any point of E and put

$$t=t(w)=\int_E \log\frac{1}{w-a}\,d\mu(a)=u+iv\,. \tag{1}$$

Since $u\to+\infty$, when w tends to any point of E and D is a bounded domain, there exists a positive constant $c>0$, such that $\Re t+c=u(w)+c\geqq 1$ in D, hence if we put

$$\zeta=(t+c)^{\frac{1}{3}}=\zeta(w)\,, \tag{2}$$

then $\zeta(w)$ is regular in D.

Let A be the area, described by ζ, when w varies in D, then since $d\zeta=\dfrac{1}{3}\dfrac{dt}{(t+c)^{\frac{2}{3}}}$, we have

$$A=\frac{1}{9}\iint_{\Phi}\frac{du\,dv}{((u+c)^2+v^2)^{\frac{2}{3}}}\,, \tag{3}$$

where Φ is the Riemann surface, generated by $t=t(w)$, when w varies in D.

Let C_r: $u(w)=\text{const.}=r$ be the niveau curve of u and C_r^* be its part, contained in D. Since

$$\int_{C_r} dv=\int_{C_r}\frac{\partial u}{\partial\nu}\,ds=2\pi\,,$$

where ν is the inner normal of C_r and ds the arc length and $dv\geqq 0$, we have

$$\int_{C_r^*} dv\leqq 2\pi\,,$$

hence

$$A\leqq\frac{1}{9}\int_{1-c}^{\infty}dr\int_{C_r^*}\frac{dv}{((r+c)^2+v^2)^{\frac{2}{3}}}\leqq\frac{1}{9}\int_{1-c}^{\infty}\frac{dr}{(r+c)^{\frac{4}{3}}}\int_{C_r^*}dv$$
$$\leqq\frac{2\pi}{9}\int_1^{\infty}\frac{d\tau}{\tau^{\frac{4}{3}}}=\frac{2\pi}{3}\,. \tag{4}$$

If we put $\zeta=\zeta(f(z))=F(z)$, then $F(z)$ is regular in $|z|<1$ and the area A, described by ζ, when z varies in $|z|<1$ is finite, so that by

Theorem VIII. 49, the set E^* on $|z|=1$, which corresponds to $\zeta=\infty$ is of capacity zero. Since $\zeta=\infty$ corresponds to E, E is mapped on a set E^* of capacity zero on $|z|=1$. Hence the first part is proved.

Next suppose that D is bounded by a Jordan curve Γ, then by Carathéodory's theorem, the mapping function is continuous on the boundary, so that a closed set on Γ corresponds to a closed set on $|z|=1$. From this, we see that a Borel set on Γ corresponds to a Borel set on $|z|=1$. Let E be a Borel set of capacity zero on Γ and E^* be its image on $|z|=1$ and suppose that $\gamma(E^*)>0$, then E^* contains a closed sub-set E_1^* of positive capacity and let E_1 be its image on Γ, then E_1 is a closed sub-set of E, so that is of capacity zero and corresponds to a closed set E_1^* of positive capacity on $|z|=1$, which contradicts the above result. Hence $\gamma(E^*)=0$.

3. Let $w=f(z)$ be meromorphic in $|z|<1$ and F be the Riemann surface, generated by $w=f(z)$ over the w-sphere K and A be the spherical area of F.

Let a be any point of K and F_ρ be the part of F, which lies above a disc of radius ρ with a as its center and $A(\rho)$ be the spherical area of F_ρ. If

$$\underline{\delta}(a)=\varliminf_{\rho\to 0}\frac{A(\rho)}{\pi\rho^2}<\infty\ , \tag{1}$$

then a is called *an ordinary value of $f(z)$ in Beurling's sense.*

Let $n(a)$ be the number of zero points of $f(z)-a$ in $|z|<1$, then $n(a)\leqq\underline{\delta}(a)$ and if $A<\infty$, then by Lebesgue's theorem, $\underline{\delta}(a)=n(a)<\infty$ almost everywhere on K.

Theorem VIII. 52. *Let $w=f(z)$ be meromorphic in $|z|<1$ and take three values α, β, γ only a finite number of times in $|z|<1$. Let a be an ordinary value of $f(z)$ in Beurling's sense, then the set E of $e^{i\theta}$ such that $\lim_{r\to 1} f(re^{i\theta})=a$ is of capacity zero.* Tsuji [19].

If $f(z)$ is schlicht in $|z|<1$, then every value a is an ordinary value of $f(z)$ in Beurling's sense, so that it corresponds to a set of capacity zero on $|z|=1$, which is a special case of Theorem VIII. 51.

If $A<\infty$, then $f(z)$ takes any value only a finite number of times, except a null set, so that $f(z)$ satisfies the condition of the theorem. Under this condition, the theorem was proved by Beurling [2].

Proof. Without loss of generality, we may assume that $a=0$ and let $w=0$ corresponds to a set E on $|z|=1$. Since $n(0)\leqq\underline{\delta}(0)<\infty$, $f(z)$ has only a finite number of zero points $z_1, \cdots, z_n$ in $|z|<1$.

Let F_ρ be the part of F, which lies above a disc $|w|<\rho$, then if $\rho>0$ is small, the image of F_ρ in $|z|<1$ consists of n compact Jordan domains D_k $(k=1, 2, \cdots n)$, which contains a zero point z_k and at most a countable number of non-compact domains $\varDelta_\rho^\nu$ $(\nu=1, 2, \cdots)$, such that $\varDelta_\rho^\nu$ has boundary points on $|z|=1$. If we put $\varDelta_\rho=\sum_\nu \varDelta_\rho^\nu$, then E lies on the boundary of $\varDelta_\rho$. Let C_r^* be the part of the circle C_r: $|z|=r$, contained in $\varDelta_\rho$.

Since $\underline{\delta}(0)<\infty$, for a suitable $M>0$, there exists an arbitrarily small $\rho>0$, such that

$$\iint_{\Delta_\rho}\left(\frac{|f'(z)|}{1+|f(z)|^2}\right)^2 r dr d\theta \leqq A(\rho) \leqq M\rho^2 , \quad z=re^{i\theta} . \tag{1}$$

Suppose that $\gamma(E)>0$, then we may assume that E is a closed set on $|z|=1$. Let $u(z)$ be its conductor potential and as the proof of Theorem VIII. 50, we put

$$J=\iint_{\Delta_\rho}\frac{|f'(z)|}{1+|f(z)|^2}\cdot\frac{\partial u}{\partial r} r\, dr\, d\theta , \tag{2}$$

then

$$J^2 \leqq \iint_{\Delta_\rho}\left(\frac{|f'(z)|}{1+|f(z)|^2}\right)^2 r\,dr\,d\theta \iint_{\Delta_\rho}\left(\frac{\partial u}{\partial r}\right)^2 r\,dr\,d\theta \leqq M\rho^2 \iint_{\Delta_\rho}\left(\frac{\partial u}{\partial r}\right)^2 r\,dr\,d\theta .$$

Since $\displaystyle\iint_{|z|<1}\left(\frac{\partial u}{\partial r}\right)^2 r\,dr\,d\theta<\infty$, we have $\displaystyle\lim_{\rho\to 0}\iint_{\Delta_\rho}\left(\frac{\partial u}{\partial r}\right)^2 r\,dr\,d\theta=0$, so that if we put

$$|J| \leqq \sqrt{\iint_{\Delta_\rho}\left(\frac{|f'(z)|}{1+|f(z)|^2}\right)^2 r\,dr\,d\theta \iint_{\Delta_\rho}\left(\frac{\partial u}{\partial r}\right)^2 r\,dr\,d\theta}=\varepsilon\rho , \tag{3}$$

then
$$\lim_{\rho\to 0}\varepsilon=0.$$

As before, we have

$$J=\int_E J(\varphi)\, d\mu(\varphi) , \tag{4}$$

where

$$J(\varphi)=\int_0^1 dr \int_{z\in C_r^*}\frac{|f'(z)|}{1+|f(z)|^2}(-d\arg(z-e^{i\varphi})), \quad z=re^{i\theta} . \tag{5}$$

Suppose that $z=1$ $(\varphi=0)$ belongs to E, then $\lim_{r\to 1} f(r)=0$.

Let Ω be a stolz domain, whose vertex is at $z=1$ and is bounded by two lines through $z=1$, making an angle $\pi/4$ with the inner normal of $|z|=1$ at $z=1$.

Since by the hypothesis, $f(z) \neq \alpha, \beta, \gamma$ in a neighbourhood of $z=1$, by Theorem VIII. 13, $\lim_{z\to 1} f(z)=0$ uniformly, when $z\to 1$ from the inside of Ω.

Since $|f(z)|=\rho$ on the boundary of Δ_ρ, the part of Ω, which lies in a small neighbourhood of $z=1$ is contained in Δ_ρ. Hence if $\rho>0$ is sufficiently small, $\Delta_\rho.\Omega$ lies in $\left|z-\frac{1}{2}\right|<\frac{1}{2}$, so that by (1), we have as before,

$$J(0) \geqq \iint\limits_{\Delta_\rho\cdot\Omega} \frac{|f'(z)|}{1+|f(z)|^2}\, dr d\psi - \iint\limits_{\Delta_\rho} \frac{|f'(z)|}{1+|f(z)|^2}\, r dr d\theta$$

$$\geqq \iint\limits_{\Delta_\rho\cdot\Omega} \frac{|f'(z)}{1+|f(z)|^2}\, dr\, d\psi - \sqrt{\iint\limits_{\Delta_\rho} r\, dr\, d\theta \iint\limits_{\Delta_\rho} \left(\frac{|f'(z)|}{1+|f(z)|^2}\right)^2 r\, dr\, d\theta}$$

$$\geqq \iint\limits_{\Delta_\rho\cdot\Omega} \frac{|f'(z)|}{1+|f(z)|^2}\, dr\, d\psi - \varepsilon_1\rho \geqq \frac{1}{2}\int_{-\frac{\pi}{4}}^{\frac{\pi}{4}} \cos\psi\, d\psi \int_{l_\psi} \frac{|f'(z)|}{1+|f(z)|^2}\, |dz| - \varepsilon_1\rho\,,$$

where $\lim_{\rho\to 0} \varepsilon_1=0$ and l_ψ is the part of the line $\psi=$const., contained in $\Delta_\rho.\Omega$. l_ψ contains a segment, which connects $z=1$ to a boundary point of Δ_ρ and is contained entirely in Δ_ρ, so that its image on the w-plane is a curve, which connects $w=0$ to a point of the circle $|w|=\rho$, so that its length is $\geqq \rho$, hence

$$\int_{l_\psi} \frac{|f'(z)|}{1+|f(z)|^2}\, |dz| \geqq \frac{\rho}{\sqrt{1+\rho^2}}\,,$$

so that

$$J(0) \geqq \frac{\rho}{\sqrt{2}(1+\rho^2)} - \varepsilon_1\rho.$$

Similarly for any $e^{i\varphi}\in E$,

$$J(\varphi) \geqq \frac{\rho}{\sqrt{2}(1+\rho^2)} - \varepsilon_1\rho\,, \tag{6}$$

hence from (3), (4), we have

$$\frac{\rho}{\sqrt{2}(1+\rho^2)} - \varepsilon_1\rho \leqq J \leqq \varepsilon\rho, \quad \text{or} \quad \frac{1}{\sqrt{2}} \leqq (\varepsilon+\varepsilon_1)\sqrt{(1+\rho^2)}\,,$$

which is absurd, since $\lim_{\rho\to 0} \varepsilon=0$, $\lim_{\rho\to 0} \varepsilon_1=0$. Hence $\gamma(E)=0$.

CHAPTER IX

CONFORMAL MAPPING

Part I. Simply connected domains

In Theorem I. 17, we haved proved that any simply connected domain on the w-plane, whose boundary Γ is a continuum can be mapped conformally on $|z|<1$. In this chapter, we shall study the boundary behaviour of the mapping function and other problėms concerning conformal mapping.

1. Theorems of Carathéodory and Lindelöf

1. First we shall explain some geometrical notions.

Let D be a bounded simply connected domain on the w-plane and Γ be its boundary, which is a continuum.

(i) *A cross cut* q is a simple Jordan arc in D, whose both end points lie on Γ.

(ii) *A fundamental sequence of cross cuts* is a sequence of cross cuts $\{q_n\}$, which satisfies the following condition: q_m, q_n $(m \neq n)$ are disjoint and q_n converges to a point of Γ and q_{n-1} and q_{n+1} are separated by q_n in D. q_n decomposes D into two parts. Let D_n be one of them, which contains q_{n+1}, then $D_1 \supset D_2 \supset \cdots \supset D_n \supset \cdots$.

(iii) *A boundary element* π is the limiting set of the boundaries of $\{D_n\}$. Let $\{q_n\}$ and $\{q'_m\}$ be two fundamental sequences of cross cuts and D_n, D'_m be the corresponding domains. If D_n contains all q'_m $(m \geqq m_0)$ and D'_m contains all q_n $(n \geqq n_0)$, then $\{q_n\}$ and $\{q'_m\}$ are said *equivalent*. Two equivalent fundamental sequences of cross cuts define the same boundary element. In the proof of Theorem IX. 1, it will be proved that for any point $w_0 \in \Gamma$, there exists a boundary element, which contains w_0.

(iv) *A principal point* (Hauptpunkt) of the boundary element π is such a point $\in \pi$, to which one fundamental sequence of cross cuts c nverges. Other point of π, which is not a principal point, is said *an accessory point* (Nebenpunkt) of π.

(v) A sequence of points $w_\nu \in D$ is said to converge to a boundary element π, if D_n contains all w_ν $(\nu \geqq \nu_0(n))$.

We shall show some examples.

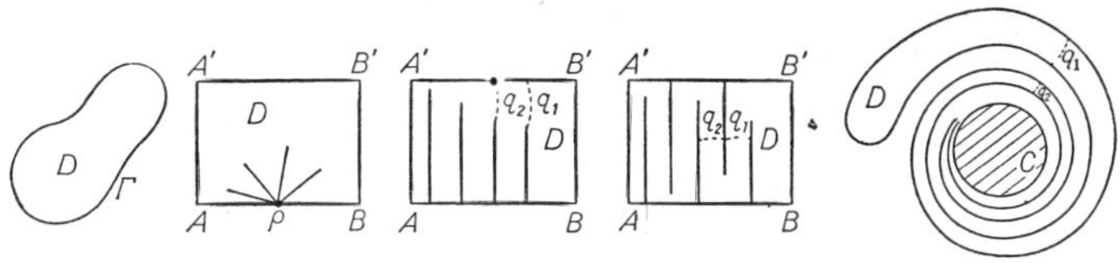

(i) D is bounded by a Jordan curve Γ. (ii) D is obtained from a rectangle $ABA'B'$: $0<x<a$, $0<y<b$ by taking off segments through P. (iii) D is obtained from a rectangle $ABA'B'$ by taking off segments: L_n: $x=\frac{a}{n}$, $0<y<b-\frac{b}{n}$ $(n=2, 3, \cdots)$. (iv) D is obtained from a rectangle $ABA'B'$ by taking off segments: L_{2n}: $x=\frac{a}{2n}$, $0<y\leqq b-\frac{b}{2n}$ and L_{2n+1}: $x=\frac{a}{2n+1}$, $b-\frac{b}{2n+1}\leqq y<b$ $(n=1, 2, \cdots)$. (v) D cluster to a circle C asymptotically from the outside of C.

In (i), every boundary element consists of only one point, in (ii) several boundary elements coincide with P, in (iii) AA' is a boundary element and A' is a principal point and other points of AA' are accesory points, in (iv) AA' is a boundary element and every point of AA' is a principal point, in (v) C is a boundary element and every point of C is a principal point.

2. After this preliminary remarks, we shall prove the following **Carathéodory's theorem.**

Theorem IX. 1. *Let D be a bounded simply connected domain on the w-plane and Γ be its boundary. If we map D conformally on $|z|<1$, then a boundary element π of D and a point P of $|z|=1$ correspond one-to-one, in such a way that to a sequence of points w_ν converging to π, there corresponds a sequence of points z_ν converging to P and vice versa, to a sequence of points z_ν converging to P, there corresponds a sequence of points w_ν converging to π.* Carathéodory [3].

First we shall prove a lemma.

Lemma. *Let a sequence of domains Δ_n be given in $|z|<1$:*

$$\Delta_n:\quad \frac{r_n}{2}\leqq|z-1|\leqq r_n\,,\quad |z|<1\,,\qquad r_{n+1}\leqq r_n/2\quad (n=1, 2, \cdots)\,,$$
$$r_1>r_2>\cdots>r_n\to 0\,.$$

Then we can find a sequence of circular arcs C_{n_k}: $|z-1|=\rho_{n_k}$, $\left(\frac{r_{n_k}}{2}\leqq\rho_{n_k}\leqq r_{n_k}\right)$ in Δ_{n_k} $(k=1, 2, \cdots)$, such that the image γ_{n_k} of C_{n_k} in D form a fundamental sequence of cross cuts and the length of γ_{n_k} tends to zero, as $k\to\infty$.

Proof. Let $w=f(z)$ be the mapping function. Since D is a bounded domain, its area is finite, so that

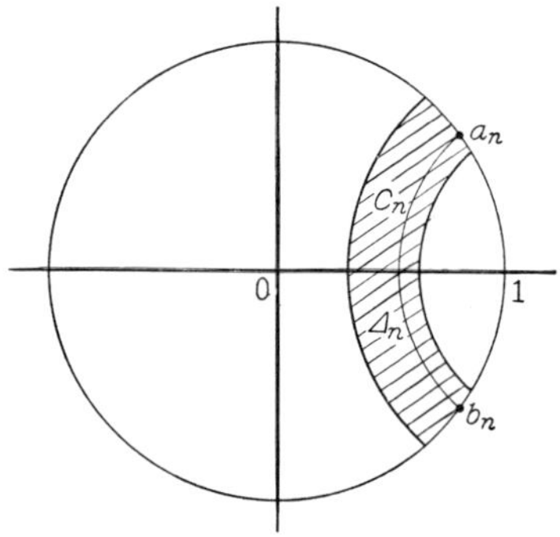

$$\varepsilon(n)=\iint_{\Delta_n}|f'(z)|^2 r dr d\theta \to 0\,,\quad \text{as } n\to\infty\,,$$
$$z=1+re^{i\theta}\,. \quad (1)$$

By Schwarz's inequality, we have

$$\iint_{\Delta_n}|f'(z)|r\,dr\,d\theta \leqq \sqrt{\iint_{\Delta_n}|f'(z)|^2 rdrd\theta \iint_{\Delta_n} rdrd\theta} \leqq r_n\sqrt{\pi\varepsilon(n)}\,. \quad (2)$$

Let C be a circular arc in Δ_n, with its center at $z=1$ and put

$$\eta(n)=\inf_C \int_C |f'(z)|rd\theta\,, \quad (3)$$

then

$$\iint_{\Delta_n}|f'(z)|r\,dr\,d\theta \geqq \eta(n)\int_{\frac{r_n}{2}}^{r_n} dr=\frac{\eta(n)r_n}{2}\,,$$

so that by (2)

$$\eta(n)\leqq 2\sqrt{\pi\varepsilon(n)}\,,$$

hence there exists a circular arc C_n in Δ_n, such that

$$L_n=\int_{C_n}|f'(z)|r\,d\theta \leqq 3\sqrt{\pi\varepsilon(n)}\,, \quad (4)$$

where L_n is the length of the image γ_n of C_n in D.

Since the length L_n of γ_n is finite, γ_n is a Jordan arc, whose both end points α_n, β_n lie on Γ.

Let C be a circular arc in Δ_n $(n\geqq 2)$, with $z=1$ as its center and a, b be its end points on $|z|=1$ and suppose that the image γ of C in D is a Jordan arc, whose end points coincide with one of α_1, β_1. Then by Theorem IV. 9, the measure of such a, b is zero, so that if we take off such $\{C\}$ from Δ_n and let Δ_n' be the remaining set, then

$$\iint_{\Delta_n'}|f'(z)|r\,dr\,d\theta=\iint_{\Delta_n}|f'(z)|r\,dr\,d\theta \qquad (n\geqq 2)\,.$$

Hence if we define $\eta'(n)$ for Δ_n', as before, then $\eta'(n)=\eta(n)$, hence we may assume that the both end points of α_n, β_n of γ_n $(n\geqq 2)$ are different from α_1, β_1, so that γ_n $(n\geqq 2)$ and γ_1 are disjoint.

Next we take off circular arcs $\{C\}$ from Δ_n $(n\geqq 3)$, whose image γ

in D is a Jordan arc, whose end points coincide with one of α_1, β_1, α_2, β_2 and proceed similarly, then we may assume that the both end points α_n, β_n of γ_n are different from α_m, β_m $(m\neq n)$, so that γ_m, γ_n $(m\neq n)$ are disjoint. From (4), $L_n\to 0$ $(n\to\infty)$, so that we can select a partial sequence $\{\gamma_{n_k}\}$ from $\{\gamma_n\}$, which converges to a point of Γ. Hence $\{\gamma_{n_k}\}$ is a fundamental sequence of cross cuts of D.

Proof of the theorem.

Let z_0 be any point of $|z|=1$, then by the lemma, there exists a sequence of circular arcs $\{C_n\}$ about z_0 converging to z_0, which corresponds to a fundamental sequence of cross cuts $\{\gamma_n\}$ of D.

Let π be the boundary element, defined by $\{\gamma_n\}$, then to a sequence of points z_ν converging to z_0, there corresponds a sequence of points w_ν converging to π and vice versa, to a sequence of points w_ν converging to π, there corresponds a sequence of points z_ν converging to z_0.

Hence to any point z_0 of $|z|=1$, there corresponds a boundary element π of D. Next we shall prove that for any point $w_0\in\Gamma$, there exists a boundary element π, which contains w_0. Let $\{w_\nu\}$ be a sequence of points $\in D$, converging to w_0, and let z_ν be the image of w_ν in $|z|<1$. By taking a suitable sub-sequence, we may assume that z_ν converges to a point z_0 on $|z|=1$. Let π be the boundary element of D, which corresponds to z_0, then π contains w_0.

Next we shall prove that to any boundary element π of D, there corresponds a point z_0 on $|z|=1$. Let π be defined by a fundamental sequence of cross cuts $\{\gamma_n\}$ converging to a point $w_0\in\Gamma$ and C_n be the image of γ_n in $|z|<1$. Let α_n, β_n be two end points of γ_n. If w tends to one of α_n, β_n along γ_n, then z tends to a definite point on $|z|=1$. For, if otherwise, there exist two fixed radii of $|z|=1$, which intersect C_n infinitely often, so that by Theorem VIII. 2, $f(z)\equiv$const., which is absurd. Let a_n, b_n be two end points of C_n on $|z|=1$, then $a_n\neq b_n$. For, if $a_n=b_n=z_0$, then C_n is a Jordan curve, so that by Theorem VIII. 7 and 8, $\alpha_n=\beta_n=\omega$ and $\lim f(z)=\omega$ uniformly, when $z\to z_0$ from the inside of C_n, so that $\alpha_m=\beta_m=\omega$ $(m>n)$, which is absurd, since γ_m, γ_n are disjoint. Hence $a_n\neq b_n$. a_n, b_n converge to a point z_0 on $|z|=1$. For, if otherwise, we can prove similarly as above, that $f(z)\equiv$const., which is absurd.

Hence a_n, b_n converge to a point z_0 on $|z|=1$. If we make this z_0 correspond to π, then to any sequence of points converging to z_0, there corresponds a sequence of points converging to π and vice versa. Hence a boundary element of D and a point of $|z|=1$ corresponds one-to-one. We can prove easily that two different boundary elements of D correspond to two different points of $|z|=1$.——

From the theorem, we have

Theorem IX. 2. *Let D be a simply connected domain on the w-plane, which is bounded by a Jordan curve Γ.*

If we map D conformally on $|z|<1$ by $w=f(z)$, then $f(z)$ is continuous in $|z|\leqq 1$ and maps $|z|\leqq 1$ topologically on $\overline{D}$.

Theorem IX. 3. *Under the same condition as Theorem* IX. 1, *let z_0 be a point on $|z|=1$, which corresponds to a boundary element π of D. Let $z_n \to z_0$ be a sequence of points, which lie in a Stolz domain, whose vertex is at z_0, and w_n be the image of z_n in D. Then any cluster point of $\{w_n\}$ is a principal point of π.* Lindelöf [1].

Proof. We assume that $z_0=1$, which corresponds to a boundary element π of D. Let $\{z_n\}$ $(z_n \to 1)$ lie in a Stolz domain APB, which lies symmetric to the real axis and $A'PB'$ be a Stolz domain, which contains APB.

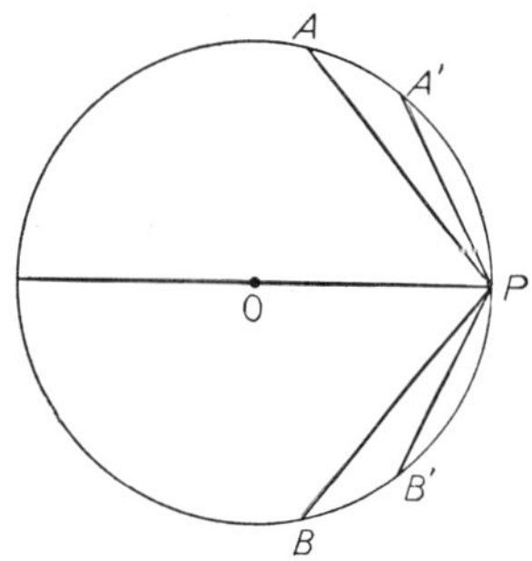

Let $K_n:\ |z-1|=1/2^n$ $(n=1,2,\cdots)$ be a circle about $z=1$, then K_n intersects AP, $A'P$, if $n\geqq n_0$.

Let $\varDelta_n$ be the domain, bounded by AP, BP and K_n, K_{n+1} and $\varDelta_n'$ be that, bounded by $A'P$, $B'P$ and K_{n-1}, K_{n+2}. Let $w_0 \in \Gamma$ be one of cluster points of $\{w_n=f(z_n)\}$, then by taking a partial sequence, we may assume that w_n converges to w_0. We select a partial sequence $\{\varDelta_{n_\nu}\}$ from $\{\varDelta_n\}$, which contains $\{z_n\}$. By changing the notation, we assume that $\varDelta_{n_\nu}$ contains z_{n_ν}. By the lemma (p. 353), we can select a partial sequence from $\{\varDelta_{n_\nu}\}$, which we denote by $\{\varDelta_{n_\nu}\}$ again, such that $\varDelta_{n_\nu}$ intersects a circular arc C_{n_ν} about $z=1$, which corresponds to a fundamental sequence of cross cuts q_{n_ν} of D converging to a principal point α of the boundary element π, defined by $\{q_{n_\nu}\}$. Our theorem is proved, if we can prove that $w_0=\alpha$. By $\zeta-1=2^{n_\nu}(z-1)$, we transform $\varDelta_{n_\nu}$ into $\varDelta_0$, then $\varDelta_{n_\nu}'$ is transformed into $\varDelta_0'$ and put $F_{n_\nu}(\zeta)=f(z)$ in $\varDelta_0'$ and let ζ_{n_ν} be the image of z_{n_ν} in $\varDelta_0$ and γ_{n_ν} be the image (in $\varDelta_0'$) of the part of C_{n_ν}, contained in $\varDelta_{n_\nu}'$. Since $\{F_{n_\nu}(\zeta)\}$ is uniformly bounded in $\varDelta_0'$ and $F_{n_\nu}(\zeta)\to\alpha$ on γ_{n_ν}, by Vitali's theorem, $F_{n_\nu}(\zeta)\to\alpha$ uniformly in $\varDelta_0$. Since $F_{n_\nu}(\zeta_{n_\nu})=f(z_{n_\nu})\to w_0$, we have $w_0=\alpha$.

2. Theorems of Fejér and Lindelöf

Lemma. *Let $f(z)=\sum_{n=0}^{\infty} a_n z^n$ be regular in $|z|<1$ and*

$$\iint_{|z|<1} |f'(z)|^2 r\,dr\,d\theta < \infty , \qquad z = re^{i\theta} .$$

If $\lim_{r\to 1} f(re^{i\theta}) = f(e^{i\theta})$ *exists, then* $\sum_{n=0}^{\infty} a_n e^{in\theta}$ *is convergent and*

$$\sum_{n=0}^{\infty} a_n e^{in\theta} = f(e^{i\theta}) .$$

If $\lim_{r\to 1} f(re^{i\theta}) = f(e^{i\theta})$ *uniformly for* $\theta_1 \leqq \theta \leqq \theta_2$*, then*

$$\sum_{n=0}^{\infty} a_n e^{in\theta} = f(e^{i\theta}) \quad \textit{uniformly for } \theta_1 \leqq \theta \leqq \theta_2.$$

Proof. $M = \frac{1}{\pi} \iint_{|z|<1} |f'(re^{i\theta})|^2 r\,dr\,d\theta = \sum_{n=1}^{\infty} n|a_n|^2 < \infty$.

Let $s_n(e^{i\theta}) = a_0 + a_1 e^{i\theta} + \cdots + a_n e^{in\theta}$, then

$$|f(re^{i\theta}) - s_{n_0}(e^{i\theta})| \leqq |a_1|(1-r) + \cdots + |a_{n_0}|(1-r^{n_0}) + \sum_{n=n_0+1}^{\infty} |a_n| r^n .$$

Since $1 - r^n \leqq n(1-r)$,

$$|f(re^{i\theta}) - s_{n_0}(e^{i\theta})| \leqq (1-r)(|a_1| + \cdots + n_0|a_{n_0}|) + \sum_{n=n_0+1}^{\infty} \sqrt{n}\,|a_n| \frac{r^n}{\sqrt{n}}$$

$$\leqq (1-r)\sqrt{\sum_{\nu=1}^{n_0} \nu \sum_{\nu=1}^{n_0} \nu |a_\nu|^2} + \sqrt{\sum_{n=n_0+1}^{\infty} \frac{r^{2n}}{n} \sum_{n=n_0+1}^{\infty} n|a_n|^2}$$

$$\leqq \sqrt{M}\, n_0(1-r) + \sqrt{\frac{1}{n_0(1-r)} \sum_{n=n_0+1}^{\infty} n|a_n|^2} .$$

We choose r_{n_0}, such that $n_0(1-r_{n_0}) = \sqrt{\sum_{n=n_0+1}^{\infty} n|a_n|^2}$, then $r_{n_0} \to 1$ $(n_0 \to \infty)$,

$$|f(r_{n_0} e^{i\theta}) - s_{n_0}(e^{i\theta})| \leqq \sqrt{M} \sqrt{\sum_{n=n_0+1}^{\infty} n|a_n|^2} + \sqrt[4]{\sum_{n=n_0+1}^{\infty} n|a_n|^2} \to 0 , \quad n_0 \to \infty .$$

Since $\lim_{r\to 1} f(re^{i\theta}) = f(e^{i\theta})$, we have $s_{n_0}(e^{i\theta}) \to f(e^{i\theta})$, so that $\sum_{n=0}^{\infty} a_n e^{in\theta} = f(e^{i\theta})$. The second part can be proved similarly.——

By Theorem VIII. 49 and Theorem IX. 2, we have

Theorem IX. 4. *Let D be a bounded simply connected domain on the w-plane. If we map D conformally on* $|z|<1$ *by* $w = f(z) = \sum_{n=0}^{\infty} a_n z^n$*, then* $\sum_{n=0}^{\infty} a_n e^{in\theta}$ *converges on* $|z|=1$*, except a set of capacity zero.*

If D is bounded by a Jordan curve, then $\sum_{n=0}^{\infty} a_n e^{in\theta}$ *converges uniformly on* $|z|=1$. Fejér [1], [2].

Theorem IX. 5. *Let D be a simply connected domain on the w-plane, which is bounded by a Jordan curve C and we map D* *conformally on* $|z|<1$ *by* $w=f(z)$. *If* C *has a tangent at* $w_0=f(z_0)$, *then* $\lim_{z\to z_0}\arg\left(\frac{w-w_0}{z-z_0}\right)$ *exists, when* $z\to z_0$ *from the inside of* $|z|<1$.

Lindelöf [1].

Hence let γ_1, γ_2 be two Jordan arcs in $|z|<1$, which meet at an angle θ at z_0 and γ_1^*, γ_2^* be its image in D, then γ_1^*, γ_2^* meet at the same angle θ at w_0. Hence

If C *has a tangent at* w_0, *then the angle is preserved at the corresponding points* z_0, w_0.

Proof. Let $u(z)=\arg\left(\frac{w-w_0}{z-z_0}\right)$, then $u(z)$ is a bounded harmonic function in $|z|<1$ and is continuous on $|z|=1$, except at $z_0=e^{i\theta_0}$. Since C has a tangent at w_0, $\lim_{\theta\to\theta_0+0} u(e^{i\theta})=\lim_{\theta\to\theta_0-0} u(e^{i\theta})$ exists, so that if we define $u(e^{i\theta_0})$ by this common limiting value, then $u(e^{i\theta})$ is continuous at e^{θ_0}. Since $u(z)$ is a bounded harmonic function in $|z|<1$, it can be expressed by a Poisson integral and since $u(e^{i\theta})$ is continuous at $e^{i\theta_0}$, by Schwarz's theorem, $\lim_{z\to e^{i\theta_0}} u(z)=u(e^{i\theta_0})$, when $z\to e^{i\theta_0}$ from the inside of $|z|<1$.

Theorem IX. 6. *Under the same condition as Theorem* IX. 5, *suppose that* C *has a tangent at every point, which varies continuously, then* $\arg f'(z)$ *is continuous in* $|z|\leqq 1$.

Proof. Let for $\delta>0$, $u_\delta(z)=\arg\left(\frac{f(ze^{i\delta})-f(z)}{ze^{i\delta}-z}\right)$, then $u_\delta(z)$ is a bounded harmonic function in $|z|<1$, so that it can be expressed by a Poisson integral:

$$u_\delta(z)=\frac{1}{2\pi}\int_0^{2\pi}\arg\left(\frac{f(e^{i(\varphi+\delta)})-f(e^{i\varphi})}{e^{i(\varphi+\delta)}-e^{i\varphi}}\right)\frac{1-|z|^2}{|z-e^{i\varphi}|^2}d\varphi\,. \qquad (1)$$

If $|z|<1$, then $\lim_{\delta\to 0} u_\delta(z)=\arg f'(z)$. Now

$$\lim_{\delta\to 0}\arg\left(\frac{f(e^{i(\varphi+\delta)})-f(e^{i\varphi})}{e^{i(\varphi+\delta)})-f(e^{i\varphi})}\right)=\Theta(\varphi)-\left(\frac{\pi}{2}+\varphi\right),$$

where $\Theta(\varphi)$ is the angle, which makes the positive tangent of C at $w=f(e^{i\varphi})$ with the positive real axis. Hence we define $\arg f'(e^{i\varphi})$ by

$$\arg f'(e^{i\varphi})=\Theta(\varphi)-\left(\frac{\pi}{2}+\varphi\right). \qquad (2)$$

Since $\arg\left(\frac{f(e^{i(\varphi+\delta)})-f(e^{i\varphi})}{e^{i(\varphi+\delta)})-e^{i\varphi}}\right)$ is bounded for $\delta\to 0$, if we make $\delta\to 0$ in (1), then by Lebesgue's theorem, we have

$$\arg f'(z)=\lim_{\delta\to 0} u_\delta(z)=\frac{1}{2\pi}\int_0^{2\pi}\arg f'(e^{i\varphi})\frac{1-|z|^2}{|z-e^{i\varphi}|^2}d\varphi\,,\quad |z|<1. \qquad (3)$$

Since $\arg f'(e^{i\varphi})=\Theta(\varphi)-\left(\frac{\pi}{2}+\varphi\right)$ is a continuous function of φ, by Schwarz's theorem, $\arg f'(z)$ is continuous in $|z|\leqq 1$.

3. Kellogg's theorem

1. First we shall prove some lemmas.

Lemma 1. *Let D be a simply connected domain on the w-plane, which is bounded by a Jordan curve C, which has a tangent at its every point, varying continuously. We map D conformally on $|z|<1$ by $w=f(z)$.*

Let $0<\psi<\pi/2$. Then there exist $\rho=\rho(\psi)>0$, $\delta=\delta(\rho)>0$, $\varepsilon=\varepsilon(\rho)>0$, $\lim_{\rho\to 0}\delta(\rho)=0$, $\lim_{\rho\to 0}\varepsilon(\rho)=0$, which satisfy the following conditions.

(i) *Let w_0 be any point of C and $S_\rho=S_\rho(w_0;\psi)$ be a sector, which is bounded by a circle $|w-w_0|=\rho$ and two lines L_1, L_2 through w_0, making an angle ψ with the inner normal of C at w_0, then S_ρ is contained in D.*

(ii) *Let S_ρ^* be the image of S_ρ in $|z|<1$ and $w_0=f(z_0)$, then S_ρ^* contains a sector $S_\delta(z_0;\psi-\varepsilon)$, which is bounded by a circle $|z-z_0|=\delta$ and two lines l_1, l_2 through z_0, making an angle $\psi-\varepsilon$ with the inner normal of $|z|=1$ at z_0.*

Proof. Since C has a tangent at its every point, which varies continuously, there exists $\rho>0$, such that for any $w_0\in C$, $S_\rho=S_\rho(w_0;\psi)$ is contained in D. Now the boundary Γ of S_ρ consists of three parts: $\Gamma=\gamma+L_1+L_2$, where γ is the circular arc on $|w-w_0|=\rho$. Let $0<\eta<\rho$ and $L_i(\eta)$ $(i=1,2)$ be the part of L_i, which lies in $|w-w_0|\leqq\eta$ and $L_i^*(\eta)$ be its image in $|z|<1$.

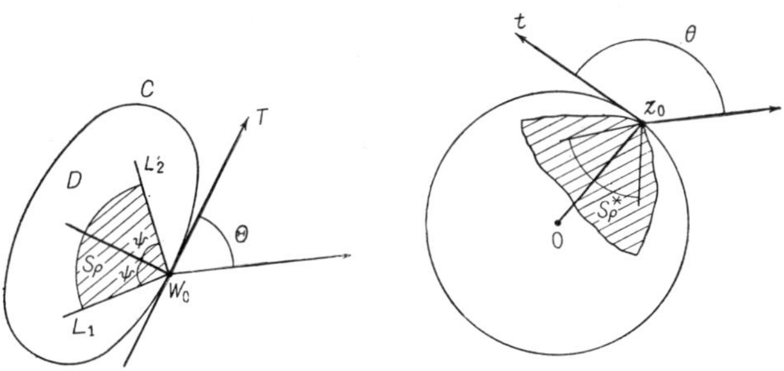

Let T be the positive tangent of C at w_0, which makes an angle Θ with the positive real axis and t be the positive tangent of $|z|=1$ at z_0, which makes an angle θ with the positive real axis.

Then by Theorem IX. 6,

$$\lim_{z\to z_0} \arg f'(z) = \lim_{z\to z_0} (\arg dw - \arg dz) = \arg f'(z_0) = \Theta - \theta . \qquad (1)$$

Hence if $\eta>0$ is small,

$$|\arg dw - \arg dz - (\Theta-\theta)| < \varepsilon \quad \text{on } L_i^*(\eta) \quad (i=1,2) . \qquad (2)$$

Since

$$\arg dw = \Theta + \frac{\pi}{2} + \psi \quad \text{on } L_1(\eta), \qquad \arg dw = \Theta + \frac{\pi}{2} - \psi \quad \text{on } L_2(\eta) ,$$

we have from (2),

$$\left|\arg dz - \left(\theta + \frac{\pi}{2} + \psi\right)\right| < \varepsilon \quad \text{on } L_1^*(\eta) , \quad \left|\arg dz - \left(\theta + \frac{\pi}{2} - \psi\right)\right| < \varepsilon \quad \text{on } L_2^*(\eta) . \qquad (3)$$

From (3) and Rolle's theorem, we see that if $\eta>0$ is small, $L_i^*(\eta)$ $(i=1, 2)$ lies outside of the angular domain, which is bounded by two lines l_1, l_2 through z_0, making an angle $\psi-\varepsilon$ with the inner normal of $|z|=1$ at z_0.

Let $z=\varphi(w)$ be the inverse function of $w=f(z)$, then

$$|z-z_0| \geqq \delta > 0 \qquad \text{on } \gamma ,$$

where $\delta>0$ is independent of w_0, so that a sector $S_\delta(z_0, \psi-\varepsilon)$ is contained in the image S^* of S_ρ.

Lemma 2. *Let $S_{\rho_0}(\theta_0)$: $0<r\leqq\rho_0$, $|\theta|\leqq\theta_0$ be a sector on the $z=re^{i\theta}$-plane and $f(z)$ be regular in $S_{\rho_0}(\theta_0)$.*

(i) *If $f(z)/z$ is bounded in $S_{\rho_0}(\theta_0)$, then $f'(z)$ is bounded in $S_{\rho_0}(\theta_1)$ $(\theta_1<\theta_0)$.*

(ii) *If $\lim_{z\to 0} f(z)/z = \gamma$ uniformly for $|\theta|\leqq\theta_0$, then $\lim_{z\to 0} f'(z) = \gamma$ uniformly for $|\theta|\leqq\theta_1<\theta_0$.*

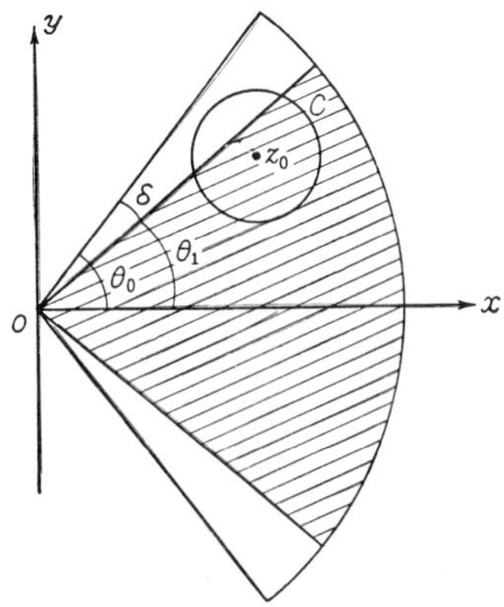

Proof. (i) Let $|f(z)|/|z|\leqq M$ in $S_{\rho_0}(\theta_0)$ and $\delta=\theta_0-\theta_1>0$.

If $z_0=re^{i\theta}\in S_{\rho_0}(\theta_1)$ and r is small, then the circle C: $|z-z_0|=r\sin\delta$ is contained in $S_{\rho_0}(\theta_0)$, hence if $z=z_0+r\sin\delta\cdot e^{i\varphi}$, then

$$|f'(z_0)| \leqq \frac{1}{2\pi}\int_C \frac{|f(z)||dz|}{|z-z_0|^2} \leqq \frac{M}{2\pi}\int_C \frac{|z|d\varphi}{r\sin\delta}$$
$$\leqq \frac{M(r+r\sin\delta)}{r\sin\delta} = \frac{M(1+\sin\delta)}{\sin\delta} . \qquad (1)$$

Hence $f'(z)$ is bounded in $S_{\rho_0}(\theta_1)$.

(ii) Next suppose that $\lim_{z\to 0} f(z)/z=\gamma$ uniformly for $|\theta|\leqq\theta_0$.

If we put $F(z)=\dfrac{f(z)}{z}-\gamma$, then

$$\lim_{z\to 0} F(z)=0 \text{ uniformly for } |\theta|\leqq\theta_0\,, \qquad f'(z)=F(z)+zF'(z)+\gamma\,. \quad (2)$$

For any $\varepsilon>0$, we choose $0<\rho_1\leqq\rho_0$, so small that $|F(re^{i\theta})|<\varepsilon$, if $r\leqq\rho_1$, $|\theta|\leqq\theta_0$, then by (1),

$$|z_0F'(z_0)|\leqq\frac{r}{2\pi}\int_C\frac{|F(z)||dz|}{|z-z_0|^2}<\frac{\varepsilon}{\sin\delta}\,.$$

Hence $\lim_{z\to 0} zF'(z)=0$ uniformly for $|\theta|\leqq\theta_1$, so that by (2)

$$\lim_{z\to 0} f'(z)=\gamma \quad \text{uniformly for } |\theta|\leqq\theta_1.$$

Lemma 3. *$f(z)=z-z^{1+\beta}$ $(\beta>0)$ is regular and schlicht in a half-disc Δ: $x>0$, $|z|\leqq 1/3$, $z=x+iy$.*

Proof. Let $z_1\in\Delta$, $z_2\in\Delta$, $z_1\neq z_2$, then

$$\frac{f(z_2)-f(z_1)}{z_2-z_1}=1-\frac{z_2^{1+\beta}-z_1^{1+\beta}}{z_2-z_1}=1-\frac{1+\beta}{z_2-z_1}\int_{z_1}^{z_2}t^\beta dt\,,$$

where we integrate on the segment $\overline{z_1z_2}$. Since

$$\frac{1+\beta}{|z_2-z_1|}\left|\int_{z_1}^{z_2}t^\beta dt\right|\leqq\frac{1+\beta}{|z_2-z_1|}\int_{z_1}^{z_2}|t|^\beta|dt|\leqq\frac{1+\beta}{|z_2-z_1|}\operatorname{Max}(|z_1|^\beta,|z_2|^\beta)|z_2-z_1|$$

$$\leqq(1+\beta)\left(\frac{1}{3}\right)^\beta<1\,,$$

$f(z_1)\neq f(z_2)$, so that $f(z)$ is schlicht in Δ.

2. Now we shall prove the following **Kellogg's theorem.**

Theorem IX. 7. *Let D be a simply connected domain on the w-plane, which is bounded by a rectifiable Jordan curve C, which is represented by $w=w(s)$ $(0\leqq s\leqq L)$, where L is the length of C and s the arc length of C, measured from a fixed point, such that if s varies from 0 to L, then $w(s)$ makes one turn on C in the positive sense. We suppose that C has a tangent at every point, which varies continuously and $w'(s)$ satisfies the following Hölder's condition:*

$$|w'(s_1)-w'(s_2)|\leqq K|s_1-s_2|^\alpha \quad (0<\alpha<1) \quad K=\text{const.}$$

If we map D conformally on $|z|<1$ by $w=f(z)$, then $f'(z)\neq 0$ in $|z|\leqq 1$ and satisfies the same Hölder's condition:

$$|f'(e^{i\theta_1})-f'(e^{i\theta_2})|\leqq K^*|\theta_1-\theta_2|^\alpha\,, \quad K^*=\text{const.}. \quad \text{Kellogg [1].}$$

Proof. (Warschawski [3]). Let $w_0=w(s_0)$ be a point of C and $w_0=f(z_0)$. Let $w=\xi+i\eta$. We may assume that $w_0=0$ and the inner normal of C at w_0 coincides with the positive η-axis and the positive tangent of C at w_0 coincides with the positive ξ-axis.

Since C has a tangent at every point, which varies continuously, there exists $R>0$, which is independent of w_0, such that any parallel line to the imaginary axis: $\xi=\text{const.}$ ($|\xi|\leqq R$) meets C at one point, so that the part of C, contained in a small neighbourhood of $w_0=0$ can be expressed by

$$\eta=\eta(\xi) \quad (|\xi|\leqq R)\,. \tag{1}$$

Since $w(s_0)=\xi(s_0)+i\eta(s_0)=0$, $\xi'(s_0)=1$, $\eta'(s_0)=0$, we have, if $|h|$ is small, $\xi'(s_0+h)\geqq 1/2$, hence

$$|\xi|=|\xi(s_0+h)|\geqq |h|/2\,.$$

$$|\eta|=|\eta(s_0+h)|=\left|\int_0^h \eta'(s_0+t)dt\right|=\int_0^h |\eta'(s_0+t)-\eta'(s_0)|dt\leqq\int_0^h |w'(s_0+t)-w'(s_0)|dt$$

$$\leqq K\left|\int_0^h t^\alpha dt\right|=\frac{K|h|^{1+\alpha}}{1+\alpha}\leqq\frac{2^{1+\alpha}K|\xi|^{1+\alpha}}{1+\alpha}\,,$$

so that

$$|\eta|\leqq\kappa|\xi|^{1+\alpha}\,,\qquad \kappa=\frac{2^{1+\alpha}K}{1+\alpha}\,. \tag{2}$$

If in $w_1=\zeta_1-\zeta_1^{1+\beta}$ of Lemma 3, we put $\beta=\frac{\alpha}{2}$, $w_1=\frac{w}{i}$, $\zeta_1=\frac{\zeta}{i}$, then

$$w=\xi^*+i\eta^*=\varphi(\zeta)=\zeta-e^{-i\frac{\pi}{2}\beta}\zeta^{1+\beta}\,,\qquad \beta=\alpha/2\,, \tag{3}$$

is regular and schlicht in a half-disc: $\Im\zeta>0$, $|\zeta|\leqq 1/3$. If $\zeta=re^{i\theta}$, then

$$\xi^*=r\cos\theta-r^{1+\beta}\cos\left((1+\beta)\theta-\frac{\pi}{2}\beta\right),$$

$$\eta^*=r\sin\theta-r^{1+\beta}\sin\left((1+\beta)\theta-\frac{\pi}{2}\beta\right). \tag{4}$$

Let Γ: $\left|\zeta-\frac{i}{6}\right|=\frac{1}{6}$ be a circle and Γ^* be its image on the w-plane. If ζ tends to $\zeta=0$ along Γ, then since $\sin\theta=3r$ and $0<\beta<1$,

$$|\xi^*|\sim r,\qquad \eta^*=3r^2+r^{1+\beta}\sin\frac{\pi}{2}\beta\sim r^{1+\beta}\sin\frac{\pi}{2}\beta\,,$$

so that

$$\eta^*\geqq\text{const.}\,|\xi^*|^{1+\beta}\,. \tag{5}$$

Since $\beta<\alpha$, we see by (2), that $\eta^*>\eta$ for the same ξ, so that the part of Γ^*, which lies in a small neighbourhood of $w_0=0$ belongs to D.

Hence if Δ: $|\zeta-ia|\leqq a$ $(a>0)$ be a disc and if $a>0$ is small, then the image Δ^* of Δ on the w-plane is contained in D. It can be proved easily that a can be chosen, independently of w_0.

Let $a^*=\varphi(ia)$ and $G_\Delta(\zeta, ia)$, $G_{\Delta^*}(w, a^*)$ be the Green's function of Δ, Δ^* respectively, then

$$G_{\Delta^*}(w, a^*)=G_\Delta(\zeta, ia)=\log\left|\frac{a}{\zeta-ia}\right|. \tag{6}$$

Let $S_\rho(\psi)$: $|w|\leqq\rho$, $\left|\arg\left(\frac{w}{i}\right)\right|\leqq\psi<\pi/2$ be a sector, which is contained in Δ^* and does not contain a^* and U be its image on the

ζ-plane, then the boundary of U does not touch the circle $|\zeta-ia|=a$ at $\zeta=0$, so that

$$\log\left|\frac{a}{\zeta-ia}\right|\geqq \text{const.}\,|\zeta|\,,\qquad \zeta\in U\,. \tag{7}$$

Since by (3), $|\zeta|\sim|w|$, we have by (6), (7),

$$G_{\Delta^*}(w, a^*)\geqq \text{const.}\,|w|\,,\qquad w\in S_\rho(\psi)\,. \tag{8}$$

Let $G_D(w, a^*)$ be the Green's function of D, then since $\Delta^*\subset D$, $G_D(w, a^*)\geqq G_{\Delta^*}(w, a^*)$, so that

$$G_D(w, a^*)\geqq \text{const.}\,|w|\,,\qquad w\in S_\rho(\psi)\,. \tag{9}$$

Let z^* be the image of a^* in $|z|<1$, such that $a^*=f(z^*)$, then

$$G_D(w, a^*)=\log\left|\frac{1-\bar{z}^*z}{z-z^*}\right|. \tag{10}$$

Now by Lemma 1, the image of $S_\rho(\psi)$ in $|z|<1$ contains a sector

$$S_\delta(z_0;\psi-\varepsilon):\ \ |z-z_0|<\delta\,,\qquad \left|\arg\left(\frac{z_0-z}{z_0}\right)\right|\leqq\psi-\varepsilon\,,\qquad \varepsilon>0\,, \tag{11}$$

where $\delta\to0$, $\varepsilon\to0$ with $\rho\to0$ and δ, ε are independent of w_0.

Since

$$\log\left|\frac{1-\bar{z}^*z}{z-z^*}\right|\leqq \text{const.}\,|z-z_0|\,,\qquad z\in S_\delta(z_0;\psi-\varepsilon)\,,$$

we have by (9), (10),

$$|w|=|f(z)|=|f(z)-f(z_0)|\leqq K_1|z-z_0|\,,\quad z\in S_\delta(z_0;\psi-\varepsilon)\,,\quad K_1=\text{const.} \tag{12}$$

Hence by Lemma 2, if $z_0=e^{i\theta_0}$,

$$|f'(re^{i\theta_0})|\leqq K_2\,,\qquad 1-\delta\leqq r<1\,,$$

where K_2 is a constant, independent of θ_0.

Since θ_0 is arbitrary, by the maximum principle,

$$|f'(z)|\leqq K_2\qquad \text{in}\quad |z|<1\,, \tag{13}$$

so that

$$|f(e^{i\theta_1})-f(e^{i\theta_2})|\leqq K_2|\theta_1-\theta_2|\,. \tag{14}$$

Let $w=w(s)$ correspond to $z=e^{i\theta}$ and put $s=s(\theta)$. Let $\Theta(s)$ be the angle, which makes the positive tangent of C at $w=w(s)$ with the positive real axis and put $\Theta^*(\theta)=\Theta(s(\theta))$. Then by the hypothesis,

$$|\Theta^*(\theta_1)-\Theta^*(\theta_2)|=|\Theta(s_1)-\Theta(s_2)|\leqq \text{const.}\,|s_1-s_2|^\alpha\qquad (0<\alpha<1)\,, \tag{15}$$
$$s_1=s(\theta_1)\,,\quad s_2=s(\theta_2)\,.$$

Evidently, there exists a constant $m>0$, such that for any s_1, s_2,

$$\left|\frac{w(s_1)-w(s_2)}{s_1-s_2}\right|\geqq m>0\,,$$

hence by (14), $|s_1-s_2|\leqq\frac{1}{m}|w(s_1)-w(s_2)|\leqq \text{const.}\,|\theta_1-\theta_2|$, so that by (15),

$$|\Theta^*(\theta_1)-\Theta^*(\theta_2)|\leqq \text{const.}\,|\theta_1-\theta_2|^\alpha\,. \tag{16}$$

By Theorem IX. 6, $\arg f'(z)$ is continuous in $|z|\leqq 1$. If we put

$$v(e^{i\theta})=\arg f'(e^{i\theta})=\Theta-\left(\frac{\pi}{2}+\theta\right),$$

then

$$\Theta^*(\theta)=v(e^{i\theta})+\theta+\frac{\pi}{2}\,.$$

Hence by (16),

$$|v(e^{i\theta_1})-v(e^{i\theta_2})|\leqq \text{const.}\,|\theta_1-\theta_2|^\alpha\,. \tag{17}$$

Since $-\log|f'(e^{i\theta})|$ is the conjugate function of $v(e^{i\theta})$, by Privaloff's theorem,

$$|\log|f'(e^{i\theta_1})|-\log|f'(e^{i\theta_2})||\leqq \text{const.}|\theta_1-\theta_2|^\alpha\,, \tag{18}$$

hence $\log|f'(e^{i\theta})|$ is continuous on $|z|=1$, so that $f'(e^{i\theta})\neq 0$ on $|z|=1$. Since $f'(z)\neq 0$ in $|z|<1$, $f'(z)\neq 0$ in $|z|\leqq 1$.

Since $\log f'(e^{\ \theta})=\log|f(e^{i\theta})|+iv(e^{i\theta})$, we have by (17), (18),

$$|f'(e^{i\theta_1})-f'(e^{i\theta_2})|\leqq \text{const.}\,|\theta_1-\theta_2|^\alpha\,, \tag{19}$$

hence the theorem is proved.

Lemma 4. *Let C_1, C_2 be two analytic Jordan arcs on the z-plane, through $z=0$, such that C_1 touches the positive real axis at $z=0$ and the tangent of C_2 at $z=0$ makes an angle $\alpha\pi$ $(0<\alpha<2)$ with the positive real axis. If by $w=z^{1/\alpha}$, C_1, C_2 become Γ_1, Γ_2 on the w-plane, then Γ_1, Γ_2 touch the real axis at $w=0$ and put $\Gamma=\Gamma_1+\Gamma_2$. Let Γ be represented by $w=w(s)$, where s is the arc length of Γ, measured from $w=0$, then*

$$|w'(s_1)-w'(s_2)|\leqq \text{const.}|s_1-s_2|^\beta\,,\qquad \beta=\text{Min}\,(1,\alpha)\,.$$

Proof. Let C_1 be represented by $z=z(\sigma)=r(\sigma)e^{i\theta(\sigma)}$, where σ is the arc length of C_1, measured from $z=0$.

Since $dw=\dfrac{1}{\alpha}z^{\frac{1}{\alpha}-1}dz$, we have $ds=\dfrac{1}{\alpha}|z|^{\frac{1}{\alpha}-1}d\sigma$, so that

$$w'(s)=e^{i\left(\frac{1}{\alpha}-1\right)\theta(\sigma)}z'(\sigma)\,. \tag{1}$$

Since C_1 is an analytic Jordan arc, $|z'(\sigma_1)-z'(\sigma_2)|\leqq \text{const.}|\sigma_1-\sigma_2|$, so that

$$|w'(s_1)-w'(s_2)|\leqq \text{const.}\,|\sigma_1-\sigma_2|\,, \tag{2}$$

where $w(s_i)$ correspond to $z(\sigma_i)$ $(i=1,2)$.

From $z=w^\alpha$, $d\sigma\sim\alpha s^{\alpha-1}ds$, so that

$$|\sigma_1-\sigma_2|\leqq \text{const.}\left|\int_{s_1}^{s_2}s^{\alpha-1}ds\right|=\text{const.}\,|s_1^\alpha-s_2^\alpha|\,.$$

If $\alpha\geqq 1$, then $|s_1^\alpha-s_2^\alpha|=\alpha|s_1+\theta(s_2-s_1)|^{\alpha-1}|s_1-s_2|\leqq \text{const.}|s_1-s_2|$ $(0<\theta<1)$, and if $0<\alpha<1$, $|s_1^\alpha-s_2^\alpha|\leqq|s_1-s_2|^\alpha$, so that

$$|\sigma_1-\sigma_2|\leqq\text{const.}\,|s_1-s_2|^\beta\,,\qquad \beta=\text{Min}\,(1,\alpha)\,. \tag{3}$$

Hence by (2), (3),

$$|w'(s_1)-w'(s_2)|\leqq\text{const.}\,|s_1-s_2|^\beta\,. \tag{4}$$

Similar relation holds on Γ_2, so that on Γ. ——

Theorem IX. 8. *Let D be a simply connected domain on the w-plane, which is bounded by a Jordan curve C through $w=0$. The part of C, which lies in a small neighbourhood of $w=0$ is divided by $w=0$ into two parts C_1, C_2, which, we assume, are analytic Jordan arcs, making an inner angle $\alpha\pi$ $(0<\alpha<2)$ at $w=0$. We map D conformally on the upper-half $\Im z>0$ of the z-plane by $w=w(z)$, $w(0)=0$.*

Let Δ: $0<|z|\leqq\rho$, $\Im z\geqq 0$ be a half-disc. Then if $\rho>0$ is small,

$$A|z|^\alpha\leqq|w(z)|\leqq B|z|^\alpha\,,\qquad A|z|^{\alpha-1}\leqq\left|\frac{dw}{dz}\right|\leqq B|z|^{\alpha-1}\,,\qquad z\in\Delta\,,$$

where $A>0$, $B>0$ are constants.

Proof. We may assume that C_1 touches the positive real axis at $w=0$ and the tangent of C_2 at $w=0$ makes an angle $\alpha\pi$ with the positive real axis. By $\zeta=w^{1/\alpha}$, we map a part of D, contained in a small neighbourhood of $w=0$, on a domain U on the ζ-plane and let C_1, C_2 become Γ_1, Γ_2 and put $\Gamma=\Gamma_1+\Gamma_2$, then Γ touches the real axis at $\zeta=0$. Let Γ be represented by $\zeta=\zeta(s)$, where s is the arc length of Γ, measured from $\zeta=0$, then by Lemma 4, $|\zeta'(s_1)-\zeta'(s_2)|\leqq\text{const.}\,|s_1-s_2|_\beta$ $(0<\beta<1)$, where we choose $\beta<\text{Min}\,(1,\alpha)$.

Let U^* be the image of U on the z-plane and suppose that $\Delta\subset U^*$. Let $\zeta=\zeta(z)$ be the mapping function, then by Kellogg's theorem, $\zeta'(z)$ is continuous and $\neq 0$ in Δ, so that

$$A_1\leqq\left|\frac{\zeta(z)}{z}\right|\leqq B_1\,,\qquad A_1\leqq|\zeta'(z)|\leqq B_1\,,\qquad z\in\Delta\,, \tag{1}$$

where $A_1>0$, $B_1>0$ are constants. Since

$$\left|\frac{w}{z^\alpha}\right|=\left|\frac{\zeta}{z}\right|^\alpha,\qquad \left|\frac{dw}{dz}\right|=\left|\frac{dw}{d\zeta}\right|\cdot\left|\frac{d\zeta}{dz}\right|=\alpha|\zeta|^{\alpha-1}\left|\frac{d\zeta}{dz}\right|,$$

we have by (1),

$$A|z|^\alpha\leqq|w(z)|\leqq B|z|^\alpha\,,\qquad A|z|^{\alpha-1}\leqq\left|\frac{dw}{dz}\right|\leqq B|z|^{\alpha-1}\,,\qquad z\in\Delta\,, \tag{2}$$

where $A>0$, $B>0$ are constants.

4. Angular derivative

1. Let D be a simply connected domain on the $w=\xi+i\eta$-plane, bounded by a Jordan curve C, which passes through $w=0$ and touches the r al axis at $w=0$ and its inner normal at $w=0$ coincides with the positive imaginary axis. We map D conformally on the upper half $\Im z>0$ of

the $z=x+iy$-plane by $w=w(z)$, $w(0)=0$.

If $\lim_{z\to 0}\frac{w(z)}{z}=\gamma$ exists, then γ is called *the angular derivative of* $w(z)$ *at* $z=0$.

To enunciate the main theorem on the angular derivative, first we shall explain some geometrical notions.

Let C be respresented by a parameter t: $w=w(t)$ $(-1\leqq t\leqq 1)$, $w(0)=0$, such that for a small $\delta>0$, $0<t\leqq\delta$ corresponds to the part of C, which lies on the right of the imaginary axis and $-\delta\leqq t<0$ corresponds to the part of C, which lies on the left of the imaginary axis.

Let C intersect the circle $|w|=r$ and $E(r)$ be the set of t $(0\leqq t\leqq\delta)$, such that $|w(t)|=r$ and put

$$\underline{t}=\underline{t}(r)=\min_{t\in E(r)} t\,,\quad \bar{t}=\bar{t}(r)=\max_{t\in E(r)} t\,,\qquad (1)$$

$$r_1=r_1(r)=\min_{\underline{t}\leqq t\leqq\bar{t}}|w(t)|\,,\quad r_2=r_2(r)=\max_{\underline{t}\leqq t\leqq\bar{t}}|w(t)|\,. \qquad (2)$$

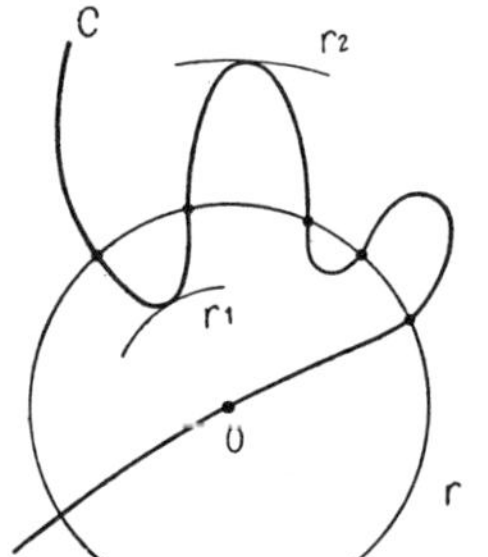

If C satisfies the following condition:

$$\lim_{r\to 0}\frac{r_1(r)}{r}=1\,,\qquad \lim_{r\to 0}\frac{r_2(r)}{r}=1\,,\qquad (3)$$

and the similar relation on the left of the imaginary axis, then we say that C satisfies *the condition* (W) *at* $w=0$, since it is first introduced by Warschawski [2].

2. Now we shall prove the following **Main theorem on the angular derivative.**

Theorem IX. 9. *Let D be a simply connected domain on the w-plane, bounded by a Jordan curve C, which passes through $w=0$ and touches the real axis at $w=0$ and its inner normal at $w=0$ coincides with the positive imaginary axis. We map D conformally on $\Im z>0$ of the $z=x+iy$-plane by $w=w(z)$, $w(0)=0$.*

(i) *If in a neighbourhood of $w=0$, C lies between two curves H and $\overline{H}$, each of which is symmetric to the imaginary axis and whose part on the right of the imaginary axis is as follows:*

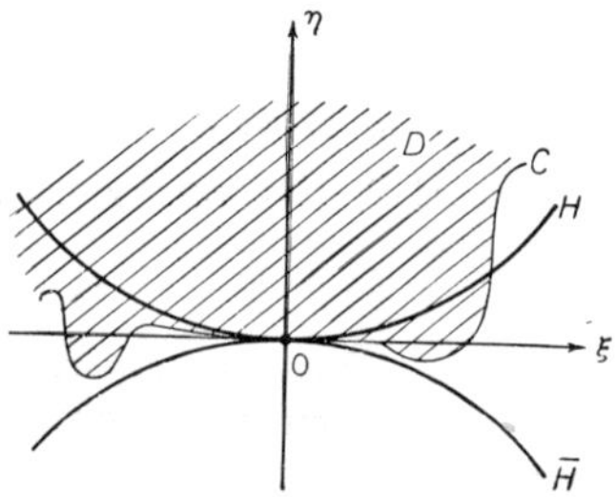

H: $\eta=h(\xi)$, $\overline{H}$: $\eta=-h(\xi)$ $(0<\xi\leqq\delta)$, $h(0)=0$, *where $h(t)>0$ is a continuous increasing function of $t>0$, such that*

$$\int_0^{\delta} \frac{h(t)}{t^2} dt < \infty .$$

Then

$$\lim_{z \to 0} \frac{w(z)}{z} = \lim_{z \to 0} w'(z) = \gamma \qquad (0 < \gamma < \infty) \qquad (*)$$

exists uniformly, when $z \to 0$ *from the inside of any fixed Stolz domain, whose vertex is at* $z=0$.

(ii) *If further,* C *satisfies the condition* (W) *at* $w=0$, *then* $(*)$ *holds uniformly, when* $z \to 0$ *from the inside of* $\Im z \geqq 0$.

Warschawski [2], Tsuji [42].

If C lies between two circles, which touch the real axis at $w=0$, then the condition (i) is satisfied. This special case is proved by Carathéodory [4].

3. Before we begin the proof, we remark the following.

Let C be represented by a parameter t as before, C: $w=w(t)$ $(-1 \leqq t \leqq 1)$. If we make t increase from 0 to 1, then C meets the line $\xi=\delta>0$ at first at w_2, so that the arc $\widehat{0w_2}$ lies between two lines: $\xi=0$ and $\xi=\delta$.

Similarly we define w_1 on the left of the imaginary axis, such that the arc $\widehat{w_1 0}$ lies between two lines: $\xi=-\delta$ and $\xi=0$.

We remark that the existence of the angular derivative is a local property in the following sense. Let D, D_1 be two simply connected domains on the w-plane, each of which is bounded by a Jordan curve, which passes through $w=0$ and touches the real axis at $w=0$. Suppose that D and D_1 have a common boundary Γ in a neighbourhood of $w=0$. We map D conformally on $\Im z>0$ by $z=\varphi(w)$, $\varphi(0)=0$ and D_1 on $\Im z_1>0$ by $z_1=\varphi_1(w)$, $\varphi_1(0)=0$.

If $\lim_{z\to 0} w/z$ exists, then $\lim_{z_1 \to 0} w/z_1$ exists. For, by the hypothesis, if $\rho>0$ is small, then there exists a simply connected domain Ω, which is bounded by a circle $|w|=\rho$ and Γ and is contained in the intersection $D\cdot D_1$ of D and D_1. Let $\varphi(\Omega)$ be the image of Ω on the z-plane and $\varphi_1(\Omega)$ be that on the z_1-plane, then $\varphi(\Omega)$ is mapped conformally on $\varphi_1(\Omega)$. By Schwarz's reflexion principle, the mapping function is regular at $z=0$, so that $\lim_{z\to 0} z_1/z=\alpha$ $(0<\alpha<\infty)$ exists. Since $\frac{w}{z}=\frac{w}{z_1}\cdot\frac{z_1}{z}$, we see that if $\lim_{z\to 0} w/z=\gamma$ exists, then $\lim_{z_1\to 0} w/z_1=\gamma_1$ exists and $\gamma=\alpha\gamma_1$.

By this remark, we may assume that the boundary C of D consists of the following curves:

(i) the arc $\widehat{w_1 0 w_2}$, (ii) a rectilinear segment, which connects w_2 to $w=2\delta$, (iii) a segment on the real axis: $2\delta \leqq \xi \leqq r_0$, (iv) a semi-circle

$|w|=r_0$, $\Im w\geqq 0$, (v) a segment on the real axis: $-r_0\leqq\xi\leqq -2\delta$, (vi) a rectilinear segment, which connects $w=-2\delta$ to w_1.

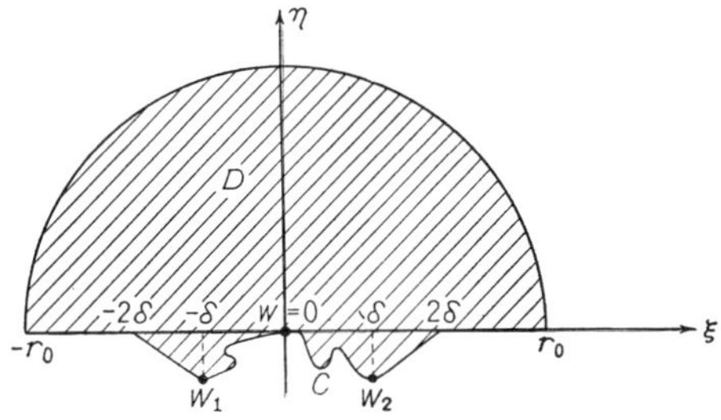

In the following, $S_\rho(\varphi)$ denotes a sector, which is bounded by a circle of radius ρ about the origin O and two lines through O, each of which makes an angle φ $(0<\varphi<\pi/2)$ with the positive imaginary axis.

5. Fundamental lemma

Fundamental lemma. *Under the condition* (i) *of Theorem* IX. 9,

$$A|z|\leqq|w(z)|\leqq B|z|\ ,\qquad z\in S_{\rho_0}(\varphi_0)\ ,$$

where $A>0$, $B>0$ *are constants.*

Proof. We divide the proof in two parts (i), (ii).

(i) $|w(z)|\leqq B|z|$, $z\in S_{\rho_0}(\varphi_0)$, *if the part of* C, *which lies in a small neighbourhood of* $w=0$, *lies below the curve* H.

Proof. Let K: $|w-ia|=a$ $(0<a<\delta)$ be a circle, about $w_0=ia$. We choose $a>0$ so small that $w_0\in D$ and the part of $|w-ia|<a$, which lies above H, belongs to D. Let D_1 be the common part of D and $|w-ia|<a$ and Γ be its boundary, then Γ consists of two parts: $\Gamma=\Gamma_K+\Gamma_H$, where Γ_K is an arc of K and Γ_H is that of H, where if H lies outside of K, then we take K instead of Γ.

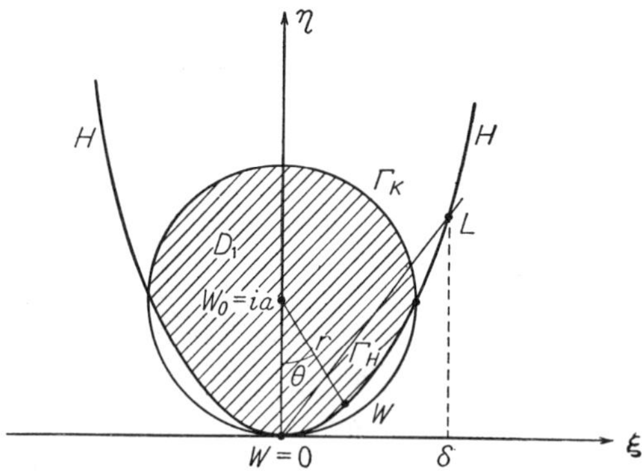

Let L: $\arg w=\varepsilon$ $(\varepsilon=\tan^{-1}h(\delta)/\delta)$ be the line, which connects $w=0$ to $w=\delta+ih(\delta)$.

We take $a>0$ so small that Γ_H lies below L. Let $G_{D_1}(w, ia)$ be the Green's function of D_1, then

$$G_{D_1}(w, ia)=\log\frac{a}{r}-v(w)\ ,\qquad r=|w-ia|\ ,\tag{1}$$

where $v(w)$ is harmonic in D_1, such that $v=\log\dfrac{a}{r}$ on Γ.

Since Γ_H lies below L, if $w\in\Gamma_H$, then $r\geqq a\cos\varepsilon$. Since $\int_0^\delta \frac{h(t)}{t^2}dt<\infty$ and $h(t)$ is an increasing function of $t>0$, $\lim_{t\to 0}\frac{h(t)}{t}=0$, so that we take $\delta>0$ so small that $\cos\varepsilon\geqq 1/2$, then on Γ_H,

$$v(w)=\log\frac{a}{r}=\log\left(1+\frac{a-r}{r}\right)\leqq\frac{a-r}{r}\leqq\frac{a-r}{a\cos\varepsilon}\leqq 2\frac{a-r}{a}\,,\quad w\in\Gamma_H\,. \tag{2}$$

Let $w=\xi+i\eta\in\Gamma_H$ and θ be defined as in the figure, then $\xi=r\sin\theta$, $\eta=a-r\cos\theta$, so that

$$\eta=a-r\cos\theta=h(r\sin\theta)\,,\quad\text{hence}\quad a-r\leqq h(\xi)\,,$$

so that by (2),

$$v(w)\leqq 2\frac{h(\xi)}{a}\,,\qquad w=\xi+i\eta\in\Gamma_H\,. \tag{3}$$

Let

$$u(w)=\frac{1}{\pi}\int_{-a}^{a}h(t)\frac{\eta}{\eta^2+(\xi-t)^2}dt\,,\qquad w=\xi+i\eta\,. \tag{4}$$

Then $u(w)$ is harmonic in $\Im w>0$. The line L intersects K at a point (ξ_0,η_0), where $\xi_0=a\sin 2\varepsilon$, so that if $w=\xi+i\eta\in\Gamma_H$, $\xi>0$, then since $\xi\leqq\xi_0=a\sin 2\varepsilon$, $\eta=h(\xi)$, we have $\xi+\eta\leqq a$, if $\delta>0$ is small, so that

$$u(w)\geqq\frac{1}{\pi}\int_\xi^{\xi+\eta}h(t)\frac{\eta}{\eta^2+(\xi-t)^2}\,dt\geqq\frac{1}{\pi}\int_\xi^{\xi+\eta}h(t)\frac{\eta}{\eta^2+\eta^2}\,dt$$
$$=\frac{1}{2\pi\eta}\int_\xi^{\xi+\eta}h(t)dt\geqq\frac{h(\xi)}{2\pi}\,,$$

hence by (3),

$$u(w)\geqq\frac{a}{4\pi}v(w)\quad\text{on}\quad\Gamma_H\,. \tag{5}$$

A similar relation holds on the left of the imaginary axis.

Since on Γ_K, $v=0$ and $u>0$, we have $u\geqq\frac{av}{4\pi}$ on Γ, so that by the maximum principle,

$$u(w)\geqq\frac{a}{4\pi}v(w)\quad\text{in}\quad D_1\,. \tag{6}$$

Let $w=\rho e^{i\left(\frac{\pi}{2}-\varphi\right)}=\xi+i\eta\in S_{\rho_1}(\varphi_1)$, $\eta=\rho\cos\varphi$, then by (4),

$$u(w)=\frac{\rho\cos\varphi}{\pi}\int_{-a}^{a}\frac{h(t)}{t^2-2t\rho\sin\varphi+\rho^2}dt\,.$$

Since $t^2\cos^2\varphi\leqq t^2-2t\rho\sin\varphi+\rho^2$,

$$u(w)\leqq\frac{\rho}{\pi\cos\varphi}\int_{-a}^{a}\frac{h(t)}{t^2}\,dt=\frac{2\rho}{\pi\cos\varphi}\int_0^a\frac{h(t)}{t^2}\,dt\leqq\frac{2\rho}{\pi\cos\varphi_1}\int_0^a\frac{h(t)}{t^2}\,dt\,,$$

so that by (6),

$$v(w)\leqq\frac{8\rho}{a\cos\varphi_1}\int_0^a\frac{h(t)}{t^2}dt\,,\qquad w\in S_{\rho_1}(\varphi_1)\,. \tag{7}$$

If $w\in S_{\rho_1}(\varphi_1)$ and $|w|=\rho$ is small and $r=|w-ia|$, then

$$\log\frac{a}{r}=\log\left(1+\frac{a-r}{r}\right)\geqq\text{const.}\frac{a-r}{r}\geqq\text{const.}\frac{\rho}{a},$$

so that if $a>0$ is small,

$$G_{D_1}(w,\,ia)=\log\frac{a}{r}-v(w)\geqq\frac{\rho}{a}\left(\text{const.}-\frac{8}{\cos\varphi_1}\int_0^a\frac{h(t)}{t^2}dt\right)\geqq\text{const.}\,\rho\,.$$

Hence

$$G_{D_1}(w,\,ia)\geqq\text{const.}\,|w|\,,\qquad w\in S_{\rho_1}(\varphi_1)\,.$$

Let $G_D(w, ia)$ be the Green's function of D, then since $D_1\subset D$,

$$G_D(w,\,ia)\geqq G_{D_1}(w,\,ia)\geqq\text{const.}\,|w|\,,\qquad w\in S_{\rho_1}(\varphi_1)\,. \tag{8}$$

If by $w=w(z)$, $w_0=ia$ corresponds to z_0 in $\Im z>0$, then

$$G_D(w,\,ia)=\log\left|\frac{z-\bar{z}_0}{z-z_0}\right|\leqq\text{const.}\,|z|\,, \tag{9}$$

so that by (8),

$$|w(z)|\leqq B|z|\,,\qquad z\in S_{\rho_0}(\varphi_0)\,, \tag{10}$$

where $B>0$ is a constant and $\varphi_0=\varphi_1-\varepsilon$ $(\varepsilon>0)$, $\lim\limits_{\rho_1\to0}\varepsilon=0$.

Hence (i) is proved.

(ii) *$A|z|\leqq|w(z)|$, $z\in S_{\rho_0}(\varphi_0)$, if the part of C, which lies in a small neighbourhood of $w=0$, lies above the curve $\bar{H}$.*

Proof. We consider a domain $D_2\supset D$, which is symmetric to the imaginary axis and whose boundary consists of the following curves:

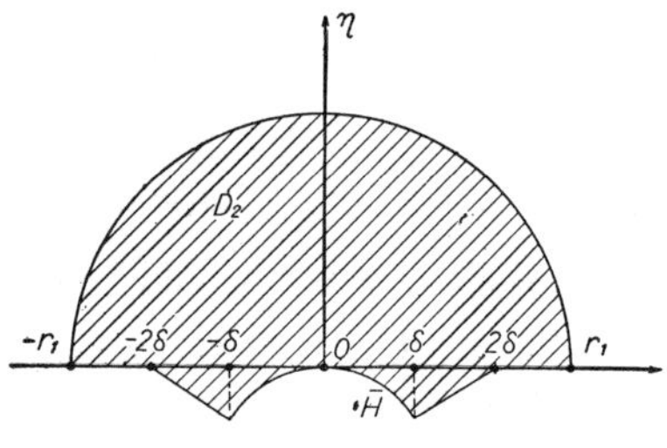

(i) the curve $\bar{H}$: $\eta=-h(\xi)$ $(-\delta\leqq\xi\leqq\delta)$, (ii) a rectilinear segment, which connects $w=\delta-ih(\delta)$ to $w=2\delta$, (iii) a segment on the real axis: $2\delta\leqq\xi\leqq r_1$ $(r_1>r_0)$, where r_0 is defined before. (iv) a semi-circle $|w|=r_1$, $\Im w\geqq0$, (v) a segment on the real axis: $-r_1\leqq\xi\leqq-2\delta$, (vi) a rectilinear segment, which connects $w=-2\delta$ to $w=-\delta-ih(\delta)$.

We map D_2 conformally on $\Im z>0$ by $w=w_2(z)$, $w_2(0)=0$, such that three points $w=0$, $w=r_1e^{i\frac{\pi}{4}}=r_1\dfrac{1+i}{\sqrt{2}}$, $w=ir_1$ correspond to $z=0$, $z=1$, $z=\infty$ and let $w=\delta-ih(\delta)$, $w=2\delta$ correspond to $z=\alpha$, $z=\beta$ $(0<\alpha<\beta<1)$ respectively.

Let $\varDelta$: $|w|\leqq r_1$, $\Im w\geqq0$ be a half-disc, contained in D_2 and $a=r_0/2$. Let $G_\varDelta(w, ia)$ be the Green's function of $\varDelta$ and $G_{D_2}(w, ia)$ be that of D_2, then by (9) of the proof of (i),

$$G_\varDelta(w,\,ia)\leqq G_D\,(w,\,ia)\leqq\text{const.}\,|z|\,.$$

Since $G_\Delta(w, ia) \geqq \text{const.}|w|$, $w \in S_{\rho_1}(\varphi_1)$, we have

$$|w_2(z)| \leqq K|z| , \qquad z \in S_{\rho_0}(\varphi_0) , \tag{1}$$

where $K>0$ is a constant, which is independent of δ, as seen easily. This is important in the sequel.

Let $z_0 = r(0<r \leqq \alpha)$ correspond to $w_0 = w_2(z_0) \in \overline{H}$. Now D_2 is contained in angular domain: $|\arg(w/i)| \leqq \psi_0 = \dfrac{\pi}{2} + \varepsilon$ $(\varepsilon > 0)$, where $\varepsilon \to 0$ with $\delta \to 0$.

We map D_2 conformally on a domain on the ζ-plane, contained in $|\zeta|<1$ by

$$\left.\begin{aligned} v &= i\left(\frac{w}{i}\right)^{\frac{\pi}{2\psi_0}} , \quad v_0 = i\left(\frac{w_0}{i}\right)^{\frac{\pi}{2\psi_0}} , \\ u &= \frac{v}{|v_0|} \quad , \quad u_0 = \frac{v_0}{|v_0|} , \\ \zeta &= \frac{u-i}{u+i} , \end{aligned}\right\} \tag{2}$$

and put $\zeta = \zeta(z)$. Then $u = 1, i, -1$ correspond to $\zeta = -i, 0, i$ respectively, so that the semi-circle $|u|=1$, $\Im u \geqq 0$ is mapped on the diameter Λ of $|\zeta|=1$ through $i, -i$. Since $|u_0|=1$, $\zeta_0 = \zeta(z_0)$ lies on Λ. Let

$$A(t) = \int_0^t \int_0^\pi |\zeta'(z_0 + te^{i\theta})|^2 t\, dt\, d\theta , \quad L(t) = \int_0^\pi |\zeta'(z_0 + te^{i\theta})| t\, d\theta, \quad z_0 = r . \tag{3}$$

Since $A(t)$ is the area of the domain, contained in $|\zeta|<1$, $A(t)<\pi$ and by Schwarz's inequality, we have

$$\int_{\rho^2 r}^{\rho r} \frac{L^2(t)}{t} dt \leqq \int_0^{\rho r} \frac{L^2(t)}{t} dt \leqq \pi A(\rho r) < \pi^2 , \qquad 0<\rho<1 ,$$

so that there exists τ $(\rho^2 r \leqq \tau \leqq \rho r)$, such that $L^2(\tau) \log 1/\rho < \pi^2$, hence if $\rho>0$ is small,

$$L(\tau) < \varepsilon < 1/6 . \tag{4}$$

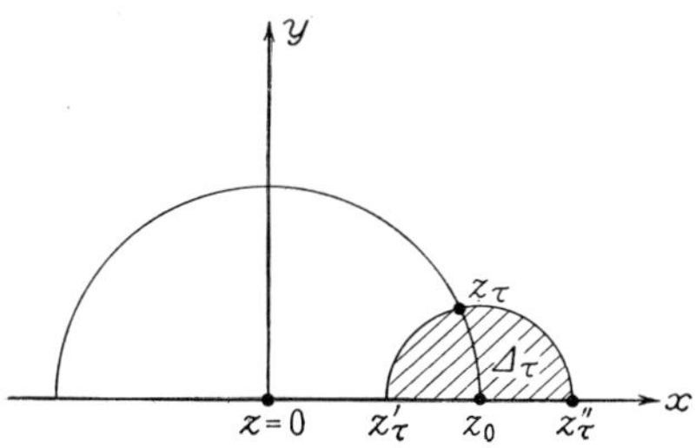

Let z_τ be the common point of two circles $|z|=r$ and $|z-z_0|=\tau$ ad $z'_\tau = z_0 - \tau$, $z''_\tau = z_0 + \tau$, $w_\tau = w_2(z_\tau)$, $w'_\tau = w_2(z'_\tau)$, $w''_\tau = w_2(z''_\tau)$ and $\widehat{w'_\tau w''_\tau}$ be the image of the semi-circle $\widehat{z'_\tau z''_\tau}$ on the w-plane and we define similarly v_τ, v'_τ, v''_τ, $\widehat{v'_\tau v''_\tau}$ etc.

From $|u_0|=1$ and the hypothesis on the curve $\overline{H}$, we see that $\widehat{u'_\tau u''_\tau}$ intersects the unit circle $|u|=1$, so that $\widehat{\zeta'_\tau\zeta''_\tau}$ intersects the imaginary axis and since by (4), its length is $<\varepsilon$, $\widehat{\zeta'_\tau\zeta''_\tau}$ lies outside of the circle $|\zeta-1|=1-\varepsilon$.

Since $u=i\dfrac{1+\zeta}{1-\zeta}$, we have for any $z\in\widehat{z'_\tau z''_\tau}$,

$$\left|\frac{du}{dz}\right|=\frac{2}{|1-\zeta|^2}\left|\frac{d\zeta}{dz}\right|\leqq\frac{2}{(1-\varepsilon)^2}\left|\frac{d\zeta}{dz}\right|,$$

so that by (4),

$$|\widehat{u'_\tau u''_\tau}|=\int_0^\pi|u'(z_0+\tau e^{i\theta})|\tau\,d\theta\leqq\frac{2}{(1-\varepsilon)^2}\int_0^\pi|\zeta'(z_0+\tau e^{i\theta})|\tau\,d\theta\leqq\frac{2\varepsilon}{(1-\varepsilon)^2}<\frac{1}{2}\,.$$

Since $\widehat{u'_\tau u''_\tau}$ intersects the unit circle $|u|=1$ and its length is $<1/2$, the image of the half-disc Δ_τ: $|z-z_0|\leqq\tau$, $\Im z>0$ on the u-plane is contained in a circular ring: $\frac{1}{2}\leqq|u|\leqq\frac{3}{2}$, so that $\frac{1}{2}|v_0|\leqq|v|\leqq\frac{3}{2}|v_0|$, hence

$$A|w_0|\leqq|w|\leqq B|w_0|\,,\qquad z\in\Delta_\tau\,, \tag{5}$$

where $A=(1/2)^{\frac{2\psi_0}{\pi}}$, $B=(3/2)^{\frac{2\psi_0}{\pi}}$. Hence especially,

$$A|w_0|\leqq|w_\tau|\,,\qquad w_\tau=w_2(z_\tau)\,. \tag{6}$$

Since $\arg z_\tau\geqq 2\sin^{-1}\rho^2/2$, we have by (1),

$$|w_\tau|\leqq\text{const.}\,|z_\tau|=\text{const.}\,|z_0|\,,$$

so that by (6)

$$|w_0|=|w_2(z_0)|\leqq K|z_0|\,, \tag{7}$$

where $K>0$ is a constant, independent of δ.

Now for any $\varepsilon>0$, we choose $\delta>0$ in the definition of D_2 so small that

$$\int_0^\delta\frac{dh(\xi)}{\xi}=\frac{h(\delta)}{\delta}+\int_0^\delta\frac{h(\xi)}{\xi^2}\,d\xi<\varepsilon\,. \tag{8}$$

Let $0<x\leqq\alpha$, $w_2(x)=\xi_2(x)+i\eta_2(x)=\xi-ih(\xi)\in\overline{H}$ $(\xi=\xi_2(x),\ h(\xi)=-\eta_2(x))$ and put

$$h(\xi)=h(\xi_2(x))=h^*(x)\,.$$

Since by (7), $0<\xi\leqq Kx$, we have by (8),

$$\int_0^\alpha\frac{dh^*(x)}{x}\leqq K\int_0^\delta\frac{dh(\xi)}{\xi}<K\varepsilon\,. \tag{9}$$

Since

$$\int_0^\alpha\frac{dh^*(x)}{x}=\frac{h^*(\alpha)}{\alpha}+\int_0^\alpha\frac{h^*(x)}{x^2}\,dx\geqq\int_0^\alpha\frac{h^*(x)}{x^2}\,dx\,,$$

we have

$$\int_0^\alpha\frac{|\eta_2(x)|}{x^2}dx=\int_0^\alpha\frac{h^*(x)}{x^2}dx<K\varepsilon\,,$$

and by (8),

$$\int_\alpha^\beta \frac{|\eta_2(x)|}{x^2}dx \leqq h(\delta)\int_\alpha^\beta \frac{dx}{x^2} \leqq \frac{h(\delta)}{\alpha} \leqq K\frac{h(\delta)}{\delta} < K\varepsilon ,$$

so that

$$\int_0^\beta \frac{|\eta_2(x)|}{x^2}dx = \int_0^\alpha \frac{|\eta_2(x)|}{x^2}dx + \int_\alpha^\beta \frac{|\eta_2(x)|}{x^2}dx < 2K\varepsilon . \tag{10}$$

Let $w_2(z)=\xi_2(z)+i\eta_2(z)$, $z=x+iy$, $y>0$, then since D_2 is a bounded domain, $\eta_2(z)$ can be expressed by a Poisson integral, so that

$$\frac{\eta_2(iy)}{y} = \frac{1}{\pi}\int_{-\infty}^\infty \frac{\eta_2(t)}{y^2+t^2}dt = \frac{2}{\pi}\int_0^\infty \frac{\eta_2(t)}{y^2+t^2}dt . \tag{11}$$

Since $\frac{|\eta_2(t)|}{y^2+t^2} \leqq \frac{|\eta_2(t)|}{t^2}$ and by (10), $\frac{|\eta_2(t)|}{t^2}$ is integrable, we have by Lebesgue's theorem,

$$\gamma_2 = \lim_{y\to 0}\frac{\eta_2(iy)}{y} = \frac{2}{\pi}\int_0^\infty \frac{\eta_2(t)}{t^2}dt < \infty . \tag{12}$$

Since $\eta_2(t) \geqq 0$ for $\beta \leqq t \leqq 1$ and $\eta_2(t) \geqq \frac{r_1}{\sqrt{2}}$ for $1 \leqq t < \infty$, we have by (10)

$$\lim_{y\to 0}\frac{\eta_2(iy)}{y} \geqq \frac{2}{\pi}\left(-\int_0^\beta \frac{|\eta_2(t)|}{t^2}dt + \frac{r_1}{\sqrt{2}}\int_1^\infty \frac{dt}{t^2}\right) \geqq \frac{2}{\pi}\left(-2K\varepsilon + \frac{r_1}{\sqrt{2}}\right).$$

Since K is independent of δ, if we choose $\delta>0$ so small that $-2K\varepsilon + \frac{r_1}{\sqrt{2}} > 0$, then

$$\lim_{y\to 0}\frac{\eta_2(iy)}{y} = \gamma_2 > 0 . \tag{13}$$

Since by Theorem IX. 5, $\lim_{y\to 0}\xi_2(iy)/\eta_2(iy)=0$, we have $\lim_{y\to 0}\xi_2(iy)/y=0$, so that

$$\lim_{y\to 0}\frac{w_2(iy)}{iy} = \gamma_2 , \quad 0<\gamma_2<\infty . \tag{14}$$

Since by (1), $w_2(z)/z$ is bounded in $S_{\rho_0}(\varphi_0)$, we have by (14) and Montel's theorem,

$$\lim_{z\to 0}\frac{w_2(z)}{z} = \gamma_2 , \quad 0<\gamma_2<\infty , \quad z \in S_{\rho_0}(\varphi_0) . \tag{15}$$

Hence

$$A_2|z| \leqq |w_2(z)| \leqq B_2|z| , \qquad z \in S_{\rho_0}(\varphi_0) , \tag{16}$$

where $A_2>0$, $B_2>0$ are constants.

Let $G_{D_2}(w, ia)$ $(a>0)$ be the Green's function of D_2. If $w_0=ia$ corresponds to z_0 in $\Im z>0$, then by (16),

$$G_{D_2}(w, ia) = \log\left|\frac{z-\bar{z}_0}{z-z_0}\right| \leqq \text{const.}\,|z| \leqq \text{const.}\,|w| , \qquad w \in S_{\rho_1}(\varphi_1) .$$

Let $G_D(w, ia)$ be the Green's function of D, then since $D \subset D_2$,

$$G_D(w, ia) \leqq G_{D_2}(w, ia) \leqq \text{const.} |w| , \qquad w \in S_{\rho_1}(\varphi_1) . \tag{17}$$

If by $w = w(z)$, $w_0 = ia$ corresponds to z_0^* in $\Im z > 0$, then

$$G_D(w, ia) = \log \left| \frac{z - \bar{z}_0^*}{z - z_0^*} \right| \geqq \text{const.} |z| , \qquad z \in S_{\rho_0}(\varphi_0)$$

so that by (17),

$$A|z| \leqq |w(z)| , \qquad z \in S_{\rho_0}(\varphi_0) , \tag{18}$$

where $A > 0$ is a constant. Hence (ii) is proved.

6. Proof of Theorem IX. 9

(i) *Proof of Part* (i).

Let D_1, D_2 ($D_1 \subset D \subset D_2$), r_0, r_1 ($r_0 < r_1$) be defined as in § 5. We map D conformally on the upper half $\Im z > 0$ of the z-plane by $w = w(z)$, $w(0) = 0$, then by the fundamental lemma,

$$A|z| \leqq |w(z)| \leqq B|z| , \qquad z \in S_{\rho_0}(\varphi_0) . \tag{1}$$

We map D_2 conformally on the upper half $\Im \zeta > 0$ of the ζ-plane by $w = w_2(\zeta)$, $w_2(0) = 0$, then by (15), (16) (p. 373),

$$A_2|\zeta| \leqq |w_2(\zeta)| \leqq B_2|\zeta| , \qquad \zeta \in S_{\rho_0}(\varphi_0) , \tag{2}$$

$$\lim_{\zeta \to 0} \frac{w_2(\zeta)}{\zeta} = \gamma_2 , \quad 0 < \gamma_2 < \infty , \quad \zeta \in S_{\rho_0}(\varphi_0) . \tag{3}$$

Since $D \subset D_2$, D is mapped on a domain D^* in $\Im \zeta > 0$. We map D^* conformally on $\Im z > 0$ by $\zeta = \zeta(z) = \xi(z) + i\eta(z)$, $\zeta(0) = 0$, then since D^* is a bounded domain, $\eta(z)$ can be expressed by a Poisson integral, so that

$$\frac{\eta(iy)}{y} = \frac{1}{\pi} \int_{-\infty}^{\infty} \frac{\eta(t)}{y^2 + t^2} dt , \qquad z = x + iy .$$

Since $\eta(t) \geqq 0$ and $\dfrac{\eta(t)}{y^2 + t^2}$ increases, if y decreases, we have by Lebesgue's theorem,

$$\lim_{y \to 0} \frac{\eta(iy)}{y} = \frac{1}{\pi} \int_{-\infty}^{\infty} \frac{\eta(t)}{t^2} dt = \gamma_1 , \quad 0 < \gamma_1 \leqq \infty , \tag{4}$$

exists. By (1), (2),

$$\left| \frac{\zeta}{z} \right| = \left| \frac{\zeta}{w} \right| \cdot \left| \frac{w}{z} \right| \leqq \frac{B}{A_2} , \qquad z \in S_{\rho_0}(\varphi_0) , \tag{5}$$

so that $\left| \dfrac{\eta(iy)}{y} \right| \leqq \left| \dfrac{\zeta(iy)}{iy} \right| \leqq \dfrac{B}{A_2}$, hence $0 < \gamma_1 < \infty$.

Since by Theorem IX. 5, $\lim_{y \to 0} \xi(iy)/\eta(iy) = 0$, we have $\lim_{y \to 0} \xi(iy)/y = 0$, so that

$$\lim_{y \to 0} \frac{\zeta(iy)}{iy} = \gamma_1 , \quad 0 < \gamma_1 < \infty . \tag{6}$$

Since by (5), $\zeta(z)/z$ is bounded in $S_{\rho_0}(\varphi_0)$, we have by (6) and Montel's theorem,

$$\lim_{z\to 0}\frac{\zeta(z)}{z}=\gamma_1\,, \quad 0<\gamma_1<\infty\,, \quad z\in S_{\rho_0}(\varphi_0)\,. \tag{7}$$

Since $\dfrac{w(z)}{z}=\dfrac{w_2(\zeta)}{\zeta}\cdot\dfrac{\zeta(z)}{z}$, we have by (3), (7), if we put $\gamma=\gamma_1\gamma_2$,

$$\lim_{z\to 0}\frac{w(z)}{z}=\gamma\,, \quad 0<\gamma<\infty\,, \quad z\in S_{\rho_0}(\varphi_0)\,. \tag{8}$$

Since $0<\varphi_0<\pi/2$ is arbitrary, we have by Lemma 2 (p. 360),

$$\lim_{z\to 0} w'(z)=\gamma\,, \qquad z\in S_{\rho_0}(\varphi_0)\,. \tag{9}$$

Hence the first part (i) is proved.

(ii) *Proof of Part* (ii).

Next suppose that C satisfies the condition (W) at $w=0$. Let $z_0=r>0$ and $w_0=w(z_0)$. By (2) (p. 371), we map D conformally on a domain in $|\zeta|<1$. Then by (4) (p. 371), there exists $\tau(\rho^2 r\leqq\tau\leqq\rho r)$, such that

$$|\widehat{\zeta'_\tau\zeta''_\tau}|=\int_0^\pi |\zeta'(z_0+\tau e^{i\theta})|\tau\, d\theta<\varepsilon_1<1/6\,. \tag{1}$$

In §5, we remarked that $\widehat{\zeta'_\tau\zeta''_\tau}$ intersects the imaginary axis, but in the present case, it is not the case in general.

With the same notations as in §5, if the arc $\widehat{\zeta'_\tau\zeta''_\tau}$ has common points with the circle $|\zeta-1|=\dfrac{1}{4}$, then by (1), $\widehat{\zeta'_\tau\zeta''_\tau}$ lies in $|\zeta-1|\leqq\dfrac{1}{2}$, hence on $\widehat{\zeta'_\tau\zeta''_\tau}$, $|u|=\left|\dfrac{1+\zeta}{1-\zeta}\right|\geqq 2$, so that $|u'_\tau|\geqq 2$, hence $|v'_\tau|\geqq 2|v_0|$, so that

$$|w'_\tau|\geqq k|w_0|\,, \qquad k=2^{\frac{2\psi_0}{\pi}}>1\,, \tag{2}$$

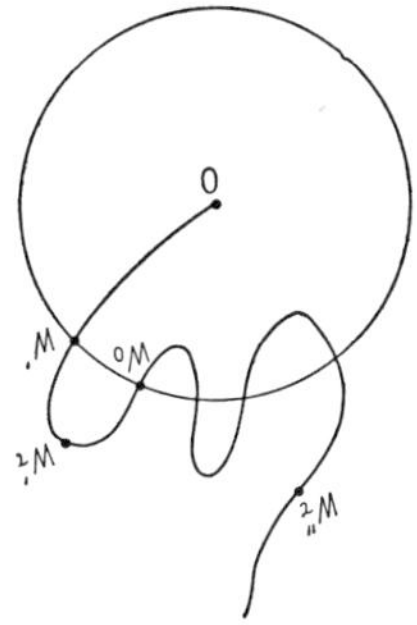

hence the image $\widehat{0w'_\tau}$ of the segment $\overline{0z'_\tau}$ intersects the circle $|w|=|w_0|$ at a point w' ($|w'|=|w_0|$, $w'\neq w_0$). Since by the condition (W), $|w'_\tau|<k|w_0|$, if $|w_0|$ is small, which contradicts (2). Hence $\widehat{\zeta'_\tau\zeta''_\tau}$ lies outside of the circle $|\zeta-1|=\dfrac{1}{4}$, so that for any $z\in\widehat{z'_\tau z''_\tau}$,

$$\left|\frac{du}{dz}\right|=\frac{2}{|1-\zeta|^2}\left|\frac{d\zeta}{dz}\right|\leqq 32\left|\frac{d\zeta}{dz}\right|,$$

hence by (1),

$$|\widehat{u'_\tau u''_\tau}|=\int_0^\pi |u'(z_0+\tau e^{i\theta})|\tau\, d\theta<\varepsilon_2\,, \qquad \varepsilon_2=32\varepsilon_1\,. \tag{3}$$

If $|u''_\tau| \leqq 1$, then the image of the half-line $\overline{z''_\tau \infty}$ intersects the unit circle $|u|=1$ at a point u'' ($|u''|=|u_0|=1$, $u'' \neq u_0$), hence by the condition (*W*), $|u''_\tau| > 1-\varepsilon_2$, if $|w_0|$ is small, so that by (3), $\widehat{u'_\tau u''_\tau}$ lies in a circular ring: $1-2\varepsilon_2 \leqq |u| \leqq 1+\varepsilon_2$.

If $|u''_\tau| > 1$ and $|u'_\tau| \leqq 1$, then $\widehat{u'_\tau u''_\tau}$ intersects the unit circle $|u|=1$, so that $\widehat{u'_\tau u''_\tau}$ lies in a circular ring: $1-\varepsilon_2 \leqq |u| \leqq 1+\varepsilon_2$.
If $|u''_\tau| > 1$ and $|u'_\tau| > 1$, then the image $\widehat{0u'_\tau}$ of the segment $\overline{0z'_\tau}$ intersects the unit circle $|u|=1$ at a point u' ($|u'|=|u_0|=1$, $u' \neq u_0$), so that by the condition (*W*), $|u'_\tau| < 1+\varepsilon_2$, if $|w_0|$ is small, hence $\widehat{u'_\tau u''_\tau}$ lies in a circular ring: $1-\varepsilon_2 \leqq |u| \leqq 1+2\varepsilon_2$.

Hence in any case, $\widehat{u'_\tau u''_\tau}$ lies in a circular ring: $1-2\varepsilon_2 \leqq |u| \leqq 1+2\varepsilon_2$. By this and the condition (*W*), we can prove easily that if z_0 is small, then the image of the half-disc Δ_τ: $|z-z_0| \leqq \tau$, $\Im z \geqq 0$ lies in a circular ring: $1-3\varepsilon_2 \leqq |u| \leqq 1+3\varepsilon_2$, hence for any $z \in \Delta_\tau$,

$$(1-\varepsilon_3)|w_0| \leqq |w(z)| \leqq (1+\varepsilon_3)|w_0| , \tag{4}$$

where $\varepsilon_3 \to 0$ with $\varepsilon_2 \to 0$. Hence especially,

$$(1-\varepsilon_3)|w_0| \leqq |w_\tau| \leqq (1+\varepsilon_3)|w_0| , \qquad w_\tau = w(z_\tau) ,$$

so that by (4),

$$(1-\varepsilon)|w_\tau| \leqq |w(z)| \leqq (1+\varepsilon)|w_\tau| , \qquad z \in \Delta_\tau , \tag{5}$$

where $\varepsilon \to 0$ with $\varepsilon_3 \to 0$.

Since $\arg z_\tau \geqq 2 \sin^{-1} \rho^2/2$, we have by the first part (i),

$$\lim_{z_\tau \to 0} \frac{w_\tau}{z_\tau} = \gamma , \qquad 0 < \gamma < \infty , \tag{6}$$

so that by (5), if $z \in \Delta_\tau$ and $|z|=|z_\tau|$ and $|z|$ is small, then

$$(1-2\varepsilon)\gamma|z| \leqq |w(z)| \leqq (1+2\varepsilon)\gamma|z| . \tag{7}$$

From (7) and the first part (i), we see that if $z \to 0$ from the inside of $\Im z \geqq 0$, then

$$\lim_{z \to 0} \left| \frac{w(z)}{z} \right| = \gamma , \qquad 0 < \gamma < \infty . \tag{8}$$

Since by Theorem IX. 6, $\lim_{z \to 0} \arg \left(\frac{w(z)}{z} \right) = 0$, we have

$$\lim_{z \to 0} \frac{w(z)}{z} = \gamma , \qquad 0 < \gamma < \infty , \tag{9}$$

if $z \to 0$ from the inside of $\Im z \geqq 0$. Hence the theorem is proved.

7. Remark on Theorem IX. 9

1. In Theorem IX. 9, the condition $\int_0^\delta \frac{h(t)}{t^2}dt<\infty$ is essential, since the following theorem holds.

Theorem IX. 10. *Let D be a simply connected domain on the $w=\xi+i\eta$-plane, bounded by a Jordan curve C, which passes through $w=0$ and touches the real axis at $w=0$ and its inner normal at $w=0$ coincides the positive imaginary axis. We assume that in a neighbourhood of $w=0$, C is represented by one of two forms:*

$$\text{(i)}\quad \eta=h(\xi)\,,\quad \textit{or}\quad \text{(ii)}\quad \eta=-h(\xi)\quad (-\delta\leqq\xi\leqq\delta)\,,\ h(0)=0\,,$$

where $h(t)\geqq 0$ is an continuous function of t, which is decreasing for $t\leqq 0$ and increasing for $t\geqq 0$.

If we map D conformally on the upper half $\Im z>0$ of the z-plane by $w=w(z)$, $w(0)=0$, then

$$\lim_{z\to 0}\frac{w(z)}{z}=\gamma$$

exists, if $z\to 0$ from the inside of $\Im z\geqq 0$, where

$$0<\gamma\leqq\infty \ \textit{in case (i)} \qquad \textit{and} \qquad 0\leqq\gamma<\infty \ \textit{in case (ii)}.$$

In each case, $0<\gamma<\infty$, when and only when

$$\int_{-\delta}^{\delta}\frac{h(t)}{t^2}\,dt<\infty\,.$$

Proof. Let $w(z)=\xi(z)+i\eta(z)$, $z=x+iy$, then

$$\frac{\eta(z)}{y}=\frac{1}{\pi}\int_{-\infty}^{\infty}\frac{\eta(t)}{|z-t|^2}\,dt\,. \tag{1}$$

First we consider the case (i), then we may assume that D lies in the upper half-plane, so that $\eta(t)\geqq 0$, hence

$$\frac{1}{\pi}\int_{-\infty}^{\infty}\frac{\eta(t)}{t^2}\,dt=\gamma\,,\qquad 0<\gamma\leqq\infty\,. \tag{2}$$

If $z\in S_{\rho_0}(\varphi_0)$, then $1/|z-t|^2\leqq 1/t^2\cos^2\varphi_0$, so that if $\gamma<\infty$, then by Lebesgue's theorem, we have

$$\lim_{z\to 0}\frac{\eta(z)}{y}=\frac{1}{\pi}\int_{-\infty}^{\infty}\frac{\eta(t)}{t^2}\,dt=\gamma\,,\qquad z\in S_{\rho_0}(\varphi_0)\,. \tag{3}$$

If $\gamma=\infty$, then by Fatou's lemma and (1), we have $\infty=\int_{-\infty}^{\infty}\frac{(\eta)t}{t^2}dt\leqq\varliminf_{z\to 0}\frac{\eta(z)}{y}$, so that

$$\lim_{z\to 0}\frac{\eta(z)}{y}=\infty\,,$$

hence (3) holds in general.

Let $w=\rho e^{i\varphi}$, $z=re^{i\theta}\in S_{\rho_0}(\varphi_0)$, then $\dfrac{w(z)}{z}=\dfrac{\eta(z)(i+\cot\varphi)}{y(i+\cot\theta)}$.

Since by Theorem IX. 5, $\lim\limits_{z\to 0}(\varphi-\theta)=0$, we have by (3),

$$\lim_{z\to 0}\frac{w(z)}{z}=\gamma\,,\quad 0<\gamma\leqq\infty\,,\quad z\in S_{\rho_0}(\varphi_0)\,. \tag{4}$$

By means of (4), we can prove similarly as Theorem IX. 9, that

$$\lim_{z\to 0}\frac{w(z)}{z}=\gamma\,,\quad 0<\gamma\leqq\infty\,,\quad \Im z\geqq 0\,. \tag{5}$$

For $\delta>0$, we choose $\alpha<0$, $\beta>0$, such that $w(\alpha)=-\delta+ih(-\delta)$, $w(\beta)=\delta+ih(\delta)$. For $0<x\leqq\beta$, we put $w(x)=\xi(x)+i\eta(x)=\xi+ih(\xi)$, $h(\xi)=\eta(x)$.

If $0<\gamma<\infty$, then by (2), $\displaystyle\int_0^\beta\frac{\eta(x)}{x^2}dx<\infty$, so that $\displaystyle\int_0^\beta\frac{d\eta(x)}{x}<\infty$.

By (5),

$$A|z|\leqq|w(z)|\leqq B|z|\,,\quad \Im z\geqq 0\,,$$

where $A>0$, $B>0$ are constants, so that $0<x\leqq\dfrac{1}{A}(\xi+h(\xi))\leqq\dfrac{\xi}{A}\left(1+\dfrac{h(\xi)}{\xi}\right)$ $\leqq K\xi$ ($K=$const.), hence $\displaystyle\int_0^\delta\frac{dh(\xi)}{\xi}\leqq K\int_0^\beta\frac{d\eta(x)}{x}<\infty$, so that $\displaystyle\int_0^\delta\frac{h(\xi)}{\xi^2}\,d\xi<\infty$. Similarly, we can prove that $\displaystyle\int_{-\delta}^0\frac{h(\xi)}{\xi^2}d\xi<\infty$. Hence if $0<\gamma<\infty$, then

$$\int_{-\delta}^{\delta}\frac{h(t)}{t^2}dt<\infty\,. \tag{6}$$

Conversely, if (6) holds, then by Theorem IX. 9, $0<\gamma<\infty$, hence the theorem is proved in the case (i).

Next consider the case (ii). Since $\eta(z)/y>0$, if $z\in S_{\rho_0}(\varphi_0)$ and $\eta(t)\leqq 0$ in a neighbourhood of $t=0$, we can prove as in the case (i), that

$$\lim_{z\to 0}\frac{\eta(z)}{y}=\frac{1}{\pi}\int_{-\infty}^{\infty}\frac{\eta(t)}{t^2}dt=\gamma\,,\quad 0\leqq\gamma<\infty\,,\quad z\in S_{\rho_0}(\varphi_0)\,, \tag{7}$$

exists. Since $\eta(t)\leqq 0$ in a neighbourhood of $t=0$, we have from (7),

$$\int_{-\infty}^{\infty}\frac{|\eta(t)|}{t^2}dt<\infty\,. \tag{8}$$

By means of (7), we can prove as in the case (i), that

$$\lim_{z\to 0}\frac{w(z)}{z}=\gamma\,,\quad 0\leqq\gamma<\infty\,,\quad \Im z\geqq 0\,, \tag{9}$$

and $0<\gamma<\infty$, when and only when (6) holds. Hence the case (ii) is proved

2. In Theorem IX. 9, we state a criterion on the existence of the angular derivative by means of Cartesian coordinates. We shall prove a criterion in polar coordinates as follows. If in the following theorem, we put $w_1=\dfrac{i}{w}$, $z_1=\dfrac{i}{z}$, then since $\theta(r)$, $\dfrac{dr}{r}$ are invariant by this transformation, we have a criterion on the existence of the angular derivative

at $z_1=\infty$ by conformal mapping of an infinite domain on the w_1-plane on the right half of the z_1-plane, such that $z_1=\infty$ corresponds to $w_1=\infty$.

Theorem IX. 11. *Let D be a simply connected domain on the $w=\xi+i\eta=re^{i\theta}$-plane, bounded by a Jordan curve C, which passes through $w=0$ and touches the real axis at $w=0$ and its inner normal at $w=0$ coincides with the positive imaginary axis. We assume that in a neighbourhood of $w=0$, C lies between two curves H and $\overline{H}$, each of which lies symmetric to the imaginary axis and whose part on the right of the imaginary axis is as follows.*

$$H:\ \theta=\theta(r)\ , \qquad \overline{H}:\ \theta=-\theta(r) \qquad (0\leqq r\leqq\delta)\ , \quad \theta(0)=0\ ,$$

where $\theta(r)>0$ is a continuous increasing function of $r>0$, such that

$$\int_0^\delta \frac{\theta(r)}{r}dr<\infty\ .$$

If we map D conformally on the upper half $\Im z>0$ of the z-plane by $w=w(z)$, $w(0)=0$, then

$$\lim_{z\to 0}\frac{w(z)}{z}=\gamma \qquad (0<\gamma<\infty)$$

exists uniformly, when $z\to 0$ from the inside of any fixed Stolz domain, whose vertex is at $z=0$. Valiron [4].

Proof. We take $a\ (0<a<\delta)$, so small that $w_0=ia\in D$ and let K: $|w-ia|=a$ be a circle about ia. Let D_1 be the common part of the inside of K and the part of the w-plane, which lies above H and Γ be its boundary.

We may assume that H has points inside of K and let Γ_K, Γ_H be the part of Γ, which belongs to K and H respectively. Let $G_{D_1}(w, ia)$ be the Green's function of D_1, then

$$G_{D_1}(w, ia)=\log\left|\frac{a}{w-ia}\right|-v(w)\ , \tag{1}$$

where $v(w)$ is harmonic in D_1, $v=0$ on Γ_K and by (2) (p. 369),

$$v(w)\leqq\frac{2(a-|w-ia|)}{a} \qquad \text{on} \quad \Gamma_H\ . \tag{2}$$

If $w=re^{i\theta}\in\Gamma_H$, then $\theta=\theta(r)$, so that

$$\frac{a-|w-ia|}{a}=\frac{a-\sqrt{a^2-2ra\sin\theta(r)+r^2}}{a}\leqq\frac{2r\sin\theta(r)}{a}\ ,$$

hence

$$v(w)\leqq\frac{4r\sin\theta(r)}{a} \qquad \text{on} \quad \Gamma_H\ . \tag{3}$$

By taking $a>0$ small, we may assume that the part of Γ_H, which lies on the right of the imaginary axis lies below the line L: $\arg w=\theta(\delta)$.

Since the equation of K is $r=2a\sin\theta$, if $w=r_0e^{i\theta_0}$ be the common point of K and L, then $r_0=2a\sin\theta(\delta)$, so that if $w=re^{i\theta}\in\Gamma_H$, then

$$r\leqq r_0=2a\sin\theta(\delta)\ .$$

We extend the definition of $\theta(r)$ for $-\delta\leqq r\leqq 0$ by putting $\theta(-r)=-\theta(r)$ $(0\leqq r\leqq\delta)$ and put

$$u(w)=\frac{1}{\pi}\int_{-a}^{a} t\sin\theta(t)\frac{\eta}{|w-t|^2}dt\ ,\qquad w=\xi+i\eta\ , \tag{4}$$

then $u(w)$ is harmonic in $\Im w>0$.

Let $w=\xi+i\eta=re^{i\theta}\in\Gamma_H$, then $\theta=\theta(r)$, $\eta=r\sin\theta$, so that

$$u(w)\geqq\frac{r\sin\theta}{\pi}\int_{r}^{r+r\sin\theta}\frac{t\sin\theta(t)}{|w-t|^2}dt\ . \tag{5}$$

If r is small, then for $r\leqq t\leqq r+r\sin\theta$, $|w-t|\leqq r\theta+r\sin\theta$ $\leqq r\frac{\pi}{2}\sin\theta+r\sin\theta\leqq 3r\sin\theta$. Since $t\sin\theta(t)$ is an increasing function of t, we have

$$u(w)\geqq\frac{r\sin\theta}{9\pi r^2\sin^2\theta}\int_{r}^{r+r\sin\theta} t\sin\theta(t)dt\geqq\frac{r\sin\theta(r)}{9\pi}\ .$$

Hence by (3),

$$u(w)\geqq\frac{a}{36\pi}v(w)\qquad\text{on}\quad\Gamma_H\ . \tag{6}$$

Since on Γ_K, $u>0$ and $v=0$, we see that $u\geqq\frac{a}{36\pi}v$ on Γ, so that by the maximum principle, (6) holds in D_1.

If $w=\rho e^{i\left(\frac{\pi}{2}-\varphi\right)}\in S_{\rho_1}(\varphi_1)$, then as p. 369, we have

$$u(w)\leqq\frac{2\rho}{\pi\cos\varphi_1}\int_0^a\frac{\sin\theta(t)}{t}dt\ ,$$

so that

$$v(w)\leqq\frac{72\rho}{a\cos\varphi_1}\int_0^a\frac{\sin\theta(t)}{t}dt\ ,\qquad w\in S_{\rho_1}(\varphi_1)\ . \tag{7}$$

By means of (7), we can complete the proof by modifying suitably the proof of Theorem IX. 9. ——

3. If we assume only the existence of a tangent of C at $w=0$, then we have

Theorem IX. 12. *Let D be a simply connected domain on the $w=\xi+i\eta$-plane, bounded by a Jordan curve C, which passes through $w=0$ and touches the real axis at $w=0$ and its inner normal at $w=0$ coincides with the positive imaginary axis. If we map D conformally on $\Im z>0$ by $w=w(z)$, $w(0)=0$, then for any $\varepsilon>0$, if $z\in S_{\rho_0}(\varphi_0)$,*

$$A|z|^{1+\varepsilon}\leqq|w(z)|\leqq B|z|^{1-\varepsilon}\ ,\qquad A|z|^{\varepsilon}\leqq|w'(z)|\leqq B|z|^{-\varepsilon}\ ,$$

where $A>0$, $B>0$ *are constants.*

Proof. By Theorem IX. 5, the image of $S_0=S_{\rho_0}(\varphi_0)$ on the w-plane is contained in a sector $S_1=S_{\rho_1}(\varphi_1)$ $(\varphi_1=\varphi_0+\delta, \delta>0)$ with $w=0$ as its vertex, which is contained in D, where $\delta\to0$, with $\rho_0\to0$.

By

$$v=i\left(\frac{w}{i}\right)^{\frac{\pi}{2}\varphi_1}, \quad \zeta=i\left(\frac{z}{i}\right)^{\frac{\pi}{2}\varphi_0}, \tag{1}$$

we map S_1 on a half-disc Δ_1: $|v|\leqq\rho_1^{*\frac{\pi}{2}\varphi_1}$, $\Im v>0$ on the v-plane and S_0 on a half-disc Δ_0: $|\zeta|\leqq\rho_0^{*\frac{\pi}{2}\varphi_0}$, $\Im\zeta>0$ on the ζ-plane respectively, then Δ_0 is mapped on a domain $\Omega_1\subset\Delta_1$. Let $v=ia$ $(a>0)$ be contained in Ω_1 and $G_{\Omega_1}(v, ia)$ be the Green's function of Ω_1 and $G_{\Delta_1}(v, ia)$ be that of Δ_1, then

$$G_{\Omega_1}(v, ia)\leqq G_{\Delta_1}(v, ia)\leqq\text{const.}\,|v|\,. \tag{2}$$

Since Ω_1 is mapped on Δ_0, let ζ_0 be the image of $v=ia$ in Δ_0, then

$$G_{\Omega_1}(v, ia)=G_{\Delta_0}(\zeta, \zeta_0)\geqq\text{const.}\,|\zeta|\,,$$

where $G_{\Delta_0}(\zeta, \zeta_0)$ is the Green's function of Δ_0. Hence by (2),

$$|\zeta|\leqq\text{const.}\,|v|\,,$$

or

$$\text{const.}\,|z|^{1+\varepsilon}\leqq|w|\,, \qquad z\in S_{\rho_0}(\varphi_0)\,, \quad \varepsilon=\delta/\varphi_0\,. \tag{3}$$

If we interchange z and w, then we have

$$|w|\leqq\text{const.}\,|z|^{1-\varepsilon}\,,$$

so that

$$A|z|^{1+\varepsilon}\leqq|w(z)|\leqq B|z|^{1-\varepsilon}\,, \qquad z\in S_{\rho_0}(\varphi_0)\,, \tag{4}$$

where $A>0$, $B>0$ are constants.

By means of (4), we can prove similarly as Lemma 2 (p. 360), that

$$A|z|^{\varepsilon}\leqq|w'(z)|\leqq B|z|^{-\varepsilon}\,, \qquad z\in S_{\rho_0}(\varphi_0)\,. \text{ ——} \tag{5}$$

From $A|z|^{\varepsilon}\leqq|w'(z)|\leqq B|z|^{-\varepsilon}$, we have

Corollary. *Under the same condition as Theorem* IX. 12, *a segment through* $z=0$ *is mapped on a rectifiable curve on the* w*-plane and vice versa, a segment through* $w=0$ *is mapped on a rectifiable curve on the* z*-plane.*

8. Variable domains

1. Let D_n $(n=1, 2, \cdots)$ be uniformly bounded simply connected domains on the w-plane, which contain $w=0$. *The kernel* G *of* $\{D_n\}$ is defined as follows. If there exists no disc $|w|<\rho$, which is contained in all D_n, then we put $G=0$. If otherwise, there exists a disc $|w|<\rho$, which is contained in all D_n. Let D be a simply connected domain, which contains $w=0$ and is contained in D_n $(n\geqq n_0)$. Then G is defined

as the sum of all such D. A sequence of simply connected domains D_n is said to *converge to its kernel* G, if any partial sequence from $\{D_n\}$ has the same kernel G.

Theorem IX. 13. *Let $\{D_n\}$ $(n=1, 2, \cdots)$ be a sequence of uniformly bounded simply connected domains on the w-plane, which contain $w=0$ and converges to its kernel $G\neq 0$. If we map D_n conformally on $|z|<1$ by $w=f_n(z)$, $f_n(0)=0$, $f_n'(0)>0$, then $f_n(z)$ converges uniformly in the wider sense in $|z|<1$ and if we put $\lim_n f_n(z)=f(z)$, then $w=f(z)$ maps $|z|<1$ conformally on G.* Carathéodory [2].

Proof. Let $z=\varphi_n(w)$ be the inverse function of $w=f_n(z)$, then $|\varphi_n(w)|<1$ in D_n, so that if a disc $|w|<\rho$ is contained in all D_n, then

$$1/|f_n'(0)|=|\varphi_n'(0)|=\left|\frac{1}{2\pi i}\int_{|w|=\rho}\frac{\varphi_n(w)}{w^2}dw\right|\leqq\frac{1}{\rho},$$

$$|f_n'(0)|\geqq\rho \quad (n=1, 2, \cdots). \tag{1}$$

Since $f_n(z)$ is uniformly bounded in $|z|<1$, we can select a partial sequence $f_{n_k}(z)$ from $f_n(z)$, which converges uniformly in the wider sense in $|z|<1$ and we put

$$f(z)=\lim_k f_{n_k}(z). \tag{2}$$

Next we select a partial sequence from $\varphi_{n_k}(w)$, which converges uniformly in the wider sense in G, which we denote by $\varphi_{n_k}(w)$ again and put

$$\varphi(w)=\lim_k \varphi_{n_k}(w). \tag{3}$$

Since $|f_n'(0)|\geqq\rho$, we have $|f'(0)|\geqq\rho$, so that $f(z)\not\equiv$const., hence $f(z)$ is regular and schlicht in $|z|<1$. Similarly $\varphi(w)$ is regular and schlicht in G.

Now by $w=f(z)$, $|z|<1$ is mapped conformally on a domain G' on the w-plane. We shall prove that $G'=G$.

Let w_0 be any point of G' and $w_0=f(z_0)$ $(|z_0|<1)$. If $|z_0|<r<1$, then by $w=f_{n_k}(z)$, a disc $|z|<r$ is mapped on a domain, which contains w_0, if $k\geqq k_0$, hence $w_0\in D_{n_k}$ $(k\geqq k_0)$, so that $w_0\in G$, hence

$$G'\subset G. \tag{4}$$

Next let w_0 be any point of G and K be a disc: $|w-w_0|<\delta$, contained in G and put $\varphi_{n_k}(K)=\varDelta_{n_k}$, then $\varDelta_{n_k}$ contains $z_0=\varphi(w_0)$, if $k\geqq k_0$, so that $w_{n_k}=f_{n_k}(z_0)\in K$. Since $\delta>0$ is arbitrary, we have $w_0=f(z_0)$. Since w_0 is any point of G, we have

$$G\subset G', \tag{5}$$

so that

$$G'=G. \tag{6}$$

Hence $w=f(z)$ maps $|z|<1$ conformally on G.

Next we shall prove that $f_n(z)$ converges uniformly in the wider sense in $|z|<1$. For, if otherwise, we can select two partial sequences $\{f_{m_k}(z)\}$, $\{f_{n_k}(z)\}$ from $\{f_n(z)\}$, such that

$$f_1(z)=\lim_k f_{m_k}(z)\ ,\quad f_2(z)=\lim_k f_{n_k}(z)\ ,\qquad f_1(z)\not\equiv f_2(z)\ , \tag{7}$$
$$f_i(0)=0\ ,\quad f_i'(0)>0\quad (i=1,2)\ .$$

Since $w=f_i(z)$ $(i=1,2)$ maps $|z|<1$ conformally on G, by the uniqueness of the mapping function, we have $f_1\equiv f_2$, which is absurd. Hence $f_n(z)$ converges uniformly in the wider sense in $|z|<1$ and $w=f(z)$ maps $|z|<1$ conformally on G.

2. Let C and C_n $(n=1,2,\cdots)$ be Jordan curves on the w-plane and T_n be a topological mapping of C on C_n and let $w\in C$ and $w_n\in C_n$ be its corresponding point on C_n by T_n and put

$$\operatorname*{Max}_{w\in C}|w_n-w|=d_n\ .$$

If there exist topological mappings T_n $(n=1,2,\cdots)$, such that $d_n\to 0$, then we say that C_n *converges to* C *in Fréchet's sense.*

Theorem IX. 14. *Let D, D_n $(n=1,2,\cdots)$ be simply connected domains on the w-plane, which contains $w=0$ and is bounded by a Jordan curve C and C_n respectively, such that $C_n\to C$ in Fréchet's sense. If we map D conformally on $|z|<1$ by $w=f(z)$, $f(0)=0$, $f'(0)>0$ and D_n on $|z|<1$ by $w=f_n(z)$, $f_n(0)=0$, $f_n'(0)>0$, then*

$$\lim_{n\to\infty} f_n(z)=f(z)\quad \textit{uniformly in } |z|\leqq 1\ .\qquad \text{Courant [1].}$$

Proof. Since C and C_n are Jordan curves, by Carathéodory's theorem, $f(z)$, $f_n(z)$ are continuous in $|z|\leqq 1$. First we shall prove that $f_n(z)$ are equi-continuous on $|z|=1$. For, if otherwise, we can select a partial sequence $\{f_{n_k}(z)\}$ from $\{f_n(z)\}$, such that there exist z'_{n_k}, z''_{n_k} on $|z|=1$, $z'_n\to\zeta$, $z''_{n_k}\to\zeta$ $(|\zeta|=1)$ and

$$|w'_{n_k}-w''_{n_k}|=|f_{n_k}(z'_n)-f_{n_k}(z''_{n_k})|\geqq\eta>0\quad (k=1,2,\cdots)\ . \tag{1}$$

We may assume that $w'_{n_k}\to w'\in C$, $w''_{n_k}\to w''\in C$, then

$$|w'-w''|\geqq\eta>0\ . \tag{2}$$

We choose a point w_0 in D, which lies in a small neighbourhood of the arc $\widehat{w'w''}$ of C and through w_0, we draw two lines L', L'', such that w' and w'' lie outside of the angular domain Δ, bounded by L' and L''. Now by w', w'', C is divided into two parts: the arc $\widehat{w'w''}$ and its complement. Let Δ_{n_k} be the common part of Δ and D_{n_k}. Then since

$C_n \to C$ in Fréchet's sense, if $k \geqq k_0$, the complement of the arc $\widehat{w'_{n_k} w''_{n_k}}$ on C_{n_k} lies outside of Δ_{n_k}, so that the boundary of Δ_{n_k} consists of sub-arcs of $\widehat{w'_{n_k} w''_{n_k}}$ and segments on L' and L''.

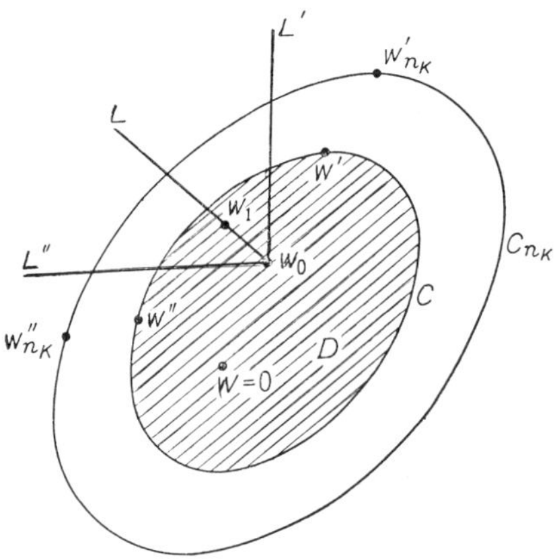

Let $z=\varphi_n(w)$ be the inverse function of $w=f_n(z)$, then $\varphi_{n_k}(w) \to \zeta$ uniformly on $w'_{n_k} w''_{n_k}$, so that if we apply Theorem VIII. 1 to $\varphi_{n_k}(w)-\zeta$ for the domain Δ_{n_k}, then we see that $\varphi_{n_k}(w_1) \to \zeta$ $(k \to \infty)$, where w_1 is any point of D, which lies on the bisector L of L' and L''. Since $\varphi_{n_k}(w)$ is uniformly bounded and w_1 is any point of L, by Vitali's theorem, $\varphi_{n_k}(w) \to \zeta$ uniformly in the wider sense in D, but $\varphi_{n_k}(0)=0$, which is absurd. Hence $f_n(z)$ are equi-continuous on $|z|=1$. Hence by Arzelà's theorem, we can select a uniformly convergent partial sequence $\{f_{n_k}(z)\}$ from $\{f_n(z)\}$ on $|z|=1$ and hence in $|z| \leqq 1$. Let

$$\lim_k f_{n_k}(z)=F(z) \quad \text{uniformly in } |z| \leqq 1 . \tag{3}$$

$F(z)$ is regular and schlicht in $|z|<1$ and is continuous in $|z| \leqq 1$. Since the kernel of $\{D_n\}$ is D, by Theorem IX. 12, $\lim_n f_n(z)=f(z)$ in $|z|<1$, so that $F(z) \equiv f(z)$. From this, we can prove that $f_n(z)$ converges to $f(z)$ uniformly on $|z|=1$. For, if otherwise, then by the above proof, we can select two uniformly convergent partial sequences $\{f_{m_k}(z)\}$, $\{f_{n_k}(z)\}$ from $\{f_n(z)\}$ on $|z|=1$, such that

$$F_1(z)=\lim_k f_{m_k}(z) , \qquad F_2(z)=\lim_k f_{n_k}(z) , \qquad F_1 \not\equiv F_2 \quad \text{on } |z|=1 .$$

But since $F_1(z)=F_2(z)=f(z)$, this is absurd, hence $f_n(z)$ converges uniformly on $|z|=1$, so that in $|z| \leqq 1$ and $\lim_n f_n(z)=f(z)$ uniformly in $|z| \leqq 1$.——

As an application, we shall prove

Theorem IX. 15. *Let D be a simply connected domain on the z-plane, bounded by a Jordan curve C. Let $f(z)$ be regular in D and continuous in its closure $\overline{D}$. Then for any $\varepsilon>0$, there exists a polynomial $P(z)$, such that*

$$|f(z)-P(z)|<\varepsilon \quad \text{in } \overline{D} . \qquad \text{Walsh [1].}$$

Proof. We map the outside of D conformally on $|w|<1$ and let C_r be the image of the circle $|w|=r<1$, then $C_r \to C$ $(r\to 1)$ in Fréchet's sense. Hence there exists a sequence of Jordan curves C_n $(n=1, 2, \cdots)$, which lie outside of C and $C_n \to C$ in Fréchet's sense. Let D_n be the domain, bounded by C_n. We assume that $z=0$ belongs to D. We map D conformally on $|w|<1$ by

$$w=\varphi(z), \quad \varphi(0)=0, \ \varphi'(0)>0 ; \qquad z=\psi(w), \quad \psi(0)=0, \ \psi'(0)>0 , \tag{1}$$

and D_n conformally on $|w|<1$ by

$$w=\varphi_n(z), \quad \varphi_n(0)=0, \ \varphi_n'(0)>0 ; \quad z=\psi_n(w), \quad \psi_n(0)=0, \ \psi_n'(0)>0 . \tag{2}$$

Since C_n lies outside of C, $\varphi_n(\overline{D})$ is a closed domain, contained entirely in $|w|<1$, with its boundary, so that if $z\in\overline{D}$, then

$$z'=\psi(\varphi_n(z))=\psi(w) \qquad (w=\varphi_n(z) , \quad z=\psi_n(w)) \tag{3}$$

is an inner point of D. Since $C_n\to C$ in Fréchet's sense, by Courant's theorem, for any $\delta>0$, if $n\geqq n_0$,

$$|z-z'|=|\psi_n(w)-\psi(w)|<\delta \quad \text{in } |w|\leqq 1 . \tag{4}$$

Since z' is an inner point of D, if we put

$$F(z)=f(z')=f(\psi[\varphi_n(z)]) , \qquad z\in\overline{D} , \tag{5}$$

then $F(z)$ is regular in $\overline{D}$, hence by Runge's theorem, for any $\delta>0$, there exists a polynomial $P(z)$, such that

$$|F(z)-P(z)|<\delta \quad \text{in } \overline{D} . \tag{6}$$

Since $f(z)$ is continuous in $\overline{D}$, we have by (4),

$$|f(z)-f(z')|<\eta(\delta) , \qquad z\in\overline{D} , \tag{7}$$

where $\eta(\delta)\to 0$ with $\delta\to 0$, hence

$$\begin{aligned} |f(z)-P(z)| &\leqq |f(z)-f(z')|+|f(z')-P(z)| \\ &=|f(z)-f(z')|+|F(z)-P(z)|<\eta(\delta)+\delta=\varepsilon , \qquad z\in\overline{D} , \end{aligned} \tag{8}$$

where $\varepsilon\to 0$ with $\delta\to 0$.——

From the theorem, we can prove Cauchy's fundamental theorem in the most general form as follows.

Theorem IX. 16. *Let D be a domain on the z-plane, bounded by a rectifiable Jordan curve C and $f(z)$ be regular in D and continuous in $\overline{D}$, then*

$$\int_C f(z)dz = 0 .$$

Proof. For any $\varepsilon>0$, there exists a polynomial $P(z)$, such that $|f(z)-P(z)|<\varepsilon$ in $\overline{D}$. Since $\int_C P(z)dz=0$, we have

$$\left|\int_C f(z)dz\right|=\left|\int_C (f(z)-P(z))dz\right|\leqq\int_C |f(z)-P(z)|\,|dz|<\varepsilon L\ ,$$

where L is the length of C. Since $\varepsilon>0$ is arbitrary, we have

$$\int_C f(z)dz=0\ .$$

Theorem IX. 17. *Let D, D_n $(n=1, 2, \cdots)$ be a simply connected domain on the w-plane, which contains $w=0$ and is bounded by a rectifiable Jordan curve C and C_n respectively, such that $C_n\to C$ in Fréchet's sence and $L_n\to L$, where L, L_n is the length of C and C_n respectively. If we map D conformally on $|z|<1$ by $w=f(z)$, $f(0)=0$, $f'(0)>0$ and D_n conformally on $|z|<1$ by $w=f_n(z)$, $f_n(0)=0$, $f_n'(0)>0$, then*

$$\lim_{n\to\infty}\int_0^{2\pi}|f_n'(e^{i\theta})-f'(e^{i\theta})|d\theta=0\ . \qquad \text{Warschawski [1].}$$

We remark that by Theorem VIII. 26, $f'(e^{i\theta})$, $f_n'(e^{i\theta})$ exists almost everywhere on $|z|=1$.

Proof. Since $f'(z)\neq 0$, $f_n'(z)\neq 0$ in $|z|<1$, if we put $f'(z)=(\varphi(z))^2$, $f_n'(z)=(\varphi_n(z))^2$, then $\varphi(z)$, $\varphi_n(z)$ are regular in $|z|<1$, so that let

$$\varphi(z)=\sum_{k=0}^{\infty}a_k z^k\ , \qquad \varphi_n(z)=\sum_{k=0}^{\infty}a_k^{(n)} z^k \qquad \text{in}\quad |z|<1\ , \tag{1}$$

then by theorem VIII. 26,

$$L=\int_0^{2\pi}|f'(e^{i\theta})|d\theta=\lim_{r\to 1}\int_0^{2\pi}|f'(re^{i\theta})|d\theta=\lim_{r\to 1}\int_0^{2\pi}|\varphi(re^{i\theta})|^2 d\theta=2\pi\sum_{k=0}^{\infty}|a_k|^2\ ,$$

$$L_n=\int_0^{2\pi}|f_n'(e^{i\theta})|d\theta=2\pi\sum_{k=0}^{\infty}|a_k^{(n)}|^2\ . \tag{2}$$

Since by Courant's theorem, $\lim_n f_n(z)=f(z)$ uniformly in $|z|\leqq 1$,

$$\lim_n a_k^{(n)}=a_k \qquad (k=0, 1, 2, \cdots)\ . \tag{3}$$

For any small $\varepsilon>0$, we choose k_0 so large that

$$L-2\pi\sum_{k=0}^{k_0}|a_k|^2=2\pi\sum_{k=k_0+1}^{\infty}|a_k|^2<\varepsilon\ , \tag{4}$$

then

$$\begin{aligned}2\pi\sum_{k=k_0+1}^{\infty}|a_k^{(n)}|^2&=L_n-2\pi\sum_{k=0}^{k_0}|a_k^{(n)}|^2\leqq|L_n-L|+\left|L-2\pi\sum_{k=0}^{k_0}|a_k|^2\right|\\&\quad+2\pi\sum_{k=0}^{k_0}|(|a_k|^2-|a_k^{(n)}|^2)|\\&\leqq|L_n-L|+\varepsilon+2\pi\sum_{k=0}^{k_0}|a_k^2-a_k^{(n)2}|\ .\end{aligned} \tag{5}$$

By (3), we choose n_0 so large that

$$|L_n-L|<\varepsilon\,,\qquad 2\pi\sum_{k=0}^{k_0}|a_k^2-a_k^{(n)2}|<\varepsilon\,,\qquad \text{if}\quad n\geqq n_0\,,$$

then by (5),

$$\sum_{k=k_0+1}^{\infty}|a_k^{(n)}|^2<3\varepsilon\,,\qquad \text{if}\quad n\geqq n_0\,. \tag{6}$$

Since $\sum_{k=0}^{\infty}|a_k-a_k^{(n)}|^2\leqq\sum_{k=0}^{k_0}|a_k-a_k^{(n)}|^2+2\sum_{k=k_0+1}^{\infty}(|a_k|^2+|a_k^{(n)}|^2)$, we have by (3), (4), (6),

$$\lim_{n\to\infty}\sum_{k=0}^{\infty}|a_k-a_k^{(n)}|^2=0\,. \tag{7}$$

Since

$$\int_0^{2\pi}|\varphi(re^{i\theta})|^2\,d\theta=\int_0^{2\pi}|f'(re^{i\theta})|d\theta\leqq L\,,\quad 0\leqq r<1\,,$$

$\varphi(z)$ belongs to H^2-class, so that

$$\lim_{r\to1}\int_0^{2\pi}|\varphi(re^{i\theta})-\varphi(e^{i\theta})|^2\,d\theta=0\,,$$

hence

$$a_k=\lim_{r\to1}\frac{1}{2\pi}\int_0^{2\pi}\varphi(re^{i\theta})\,e^{-ik\theta}\,d\theta=\frac{1}{2\pi}\int_0^{2\pi}\varphi(e^{i\theta})\,e^{-ik\theta}\,d\theta\,,$$

so that $\varphi(e^{i\theta})\sim\sum_{k=0}^{\infty}a_k\,e^{ik\theta}$. Similarly, $\varphi_n(e^{i\theta})\sim\sum_{k=0}^{\infty}a_k^{(n)}\,e^{ik\theta}$, hence by Parseval's theorem and (7),

$$\int_0^{2\pi}|\varphi(e^{i\theta})-\varphi_n(e^{i\theta})|^2\,d\theta=2\pi\sum_{k=0}^{\infty}|a_k-a_k^{(n)}|^2\to0\,,\qquad n\to\infty\,. \tag{8}$$

By (8) and Schwarz's inequality, we can prove easily that

$$\lim_{n\to\infty}\int_0^{2\pi}|f'(e^{i\theta})-f_n'(e^{i\theta})|d\theta=\lim_{n\to\infty}\int_0^{2\pi}|\varphi^2(e^{i\theta})-\varphi_n^2(e^{i\theta})|d\theta=0\,.$$

9. Löwner's differential equation

1. Let D be a simply connected domain on the w-plane, contained in $|w|<1$, whose boundary consists of the unit circle $|w|=1$ and a Jordan arc C in $|w|<1$, which connects a point w_0 ($|w_0|<1$) to a point w_1 ($|w_1|=1$) and does not pass through $w=0$. Let C be represented by a parameter τ, such that $w=w_\tau=w(\tau)$ $(0\leqq\tau\leqq1)$, $w_0=w(0)$, $w_1=w(1)$.

We denote the part of C, which corresponds to $[\tau_1,\tau_2]$ $(0\leqq\tau_1\leqq\tau_2\leqq1)$ by C_{τ_1,τ_2} and $D_{\tau,1}$ be the simply connected domain, which is bounded by $|w|=1$ and $C_{\tau,1}$, so that $D_{0,1}=D$ and $D_{1,1}$ coincides with $|w|<1$.

We map $D_{\tau,1}$ conformally on $|z|<1$ by

$$w=g_\tau(z), \quad g_\tau(0)=0, \quad g'_\tau(0)>0. \tag{1}$$

Let w_τ correspond to z_τ on $|z|=1$, then $C_{0,\tau}$ is mapped on a Jordan arc $\Gamma_{0,\tau}$ in $|z|<1$, which connects a point z_0 ($|z_0|<1$) to z_τ. Let $\Delta_{0,\tau}$ be the simply connected domain, which is bounded by $|z|=1$ and $\Gamma_{0,\tau}$.

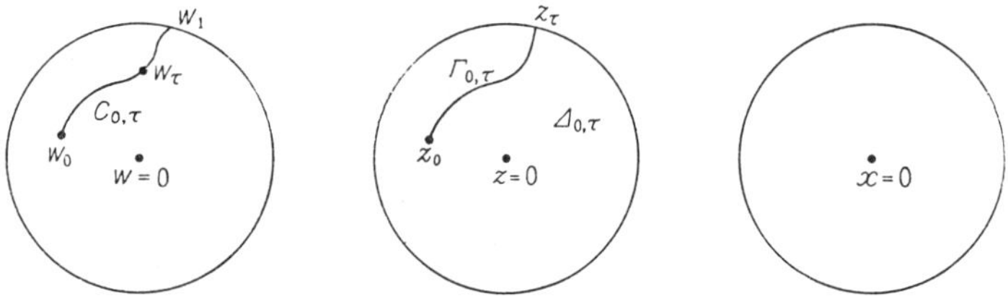

w-plane z-plane x-plane

We map $\Delta_{0,\tau}$ conformally on $|x|<1$ by

$$z=f_\tau(x)=e^{-t}(x+\cdots), \qquad t>0, \tag{2}$$

and we write

$$w=g_\tau(z)=g(z, t), \qquad z=f_\tau(x)=f(x, t), \qquad z_\tau=e^{i\theta(t)}. \tag{3}$$

t is a function of τ, which we denote by $t=t(\tau)$. Evidently $t(0)=0$ and if we put $T=t(1)$, then $w=g_0(z)=g(z, 0)$ maps $|z|<1$ conformally on D and

$$w=g_1(z)=g(z, T)\equiv z. \tag{4}$$

Lemma 1. *$t=t(\tau)$ is a continuous increasing function of τ $(0\leqq\tau\leqq1)$, so that its inverse function $\tau=\tau(t)$ is a continuous increasing function of t $(0\leqq t\leqq T)$.*

Proof. Let $0\leqq\tau_1<\tau_2\leqq1$, $t_1=t(\tau_1)$ $t_2=t(\tau_2)$,

$$f_{\tau_1}(x)=f(x, t_1), \qquad f_{\tau_2}(x)=f(x, t_2). \tag{1}$$

By $w=g_{\tau_2}(z)$, C_{0,τ_1}, C_{τ_1,τ_2} is mapped on a Jordan arc Γ_{0,τ_1}, Γ_{τ_1,τ_2} in $|z|<1$ respectively and let Δ_{τ_1,τ_2} be the simply connected domain, which is bounded by $|z|=1$ and Γ_{τ_1,τ_2}.

We map Δ_{τ_1,τ_2} conformally on $|\zeta|<1$ by

$$z=\varphi(\zeta), \qquad \varphi(0)=0, \qquad \varphi'(0)>0. \tag{2}$$

Then Γ_{0,τ_1} is mapped on a Jordan arc Γ in $|\zeta|<1$. Let Ω be the simply connected domain, which is bounded by $|\zeta|=1$ and Γ.

Next we map Ω conformally on $|x|<1$, then by the definition of $f(x, t)$,

$$w=g_{\tau_1}(\zeta)=g(x, t_1), \qquad \zeta=f(x, t_1). \tag{3}$$

Since by $z=\varphi(\zeta)=\varphi(f(x, t_1))$, $|x|<1$ is mapped on Δ_{0,τ_2},

$$z=\varphi(f(x, t_1))=f(x, t_2)\ , \tag{4}$$

so that if

$$z=\varphi(\zeta)=e^{-\alpha}(\zeta+\cdots)\ , \qquad \alpha>0\ , \tag{5}$$

then $-t_2=-\alpha-t_1$, or

$$\alpha=\Delta t=t_2-t_1=t(\tau_2)-t(\tau_1)>0\ , \tag{6}$$

hence $t(\tau)$ is an increasing function of τ $(0\leqq\tau\leqq 1)$. If $\tau_2-\tau_1\to 0$, then Γ_{τ_1,τ_2} tends to $|z|=1$, so that $\varphi(\zeta)\to\zeta$, hence $\alpha\to 0$, so that $t(\tau)$ is a continuous function of τ $(0\leqq\tau\leqq 1)$.

Lemma 2. *$\theta(t)$ is a continuous function of t $(0\leqq t\leqq T)$.*

Proof. Let in the proof of Lemma 1, w_{τ_1}, w_{τ_2} correspond to z_{τ_1}, z_{τ_2} $(|z_{\tau_1}|<1,\ |z_{\tau_2}|=1)$ respectively, then

$$|z_{\tau_1}-z_{\tau_2}|<\varepsilon\ , \qquad \text{if} \quad |\tau_1-\tau_2|<\delta=\delta(\varepsilon)\ . \tag{1}$$

If $\tau_2-\tau_1\to 0$, then the boundary of Δ_{τ_1,τ_2} tends to the unit circle $|z|=1$ in Fréchet's sense, so that by Courant's theorem, if $|\tau_1-\tau_2|<\delta(\varepsilon)$,

$$|\varphi(\zeta)-\zeta|<\varepsilon\ , \qquad |\zeta|\leqq 1\ . \tag{2}$$

Hence if $z_{\tau_1}=\varphi(\zeta_{\tau_1})$, $(|\zeta_{\tau_1}|=1)$, then $|z_{\tau_1}-\zeta_{\tau_1}|<\varepsilon$, so that by (1),

$$|z_{\tau_2}-\zeta_{\tau_1}|<2\varepsilon\ , \qquad \text{if} \quad |\tau_1-\tau_2|<\delta\ . \tag{3}$$

If $t_1=t(\tau_1)$, $t_2=t(\tau_2)$, then by the definition,

$$z_{\tau_2}=e^{i\theta(t_2)}\ , \qquad \zeta_{\tau_1}=e^{i\theta(t_1)}\ ,$$

so that by (3),

$$|e^{i\theta(t_2)}-e^{i\theta(t_1)}|\to 0\ , \qquad \text{if} \quad |\tau_1-\tau_2|\to 0\ .$$

Since $t=t(\tau)$ is a continuous function of τ,

$$|e^{i\theta(t_2)}-e^{i\theta(t_1)}|\to 0\ , \qquad \text{if} \quad |t_1-t_2|\to 0\ .$$

Hence $\theta(t)$ is a continuous function of t $(0\leqq t\leqq T)$.——

Similarly we can prove

Lemma 3. *For a fixed x $(|x|<1)$, $f(x, t)$ is a continuous function of t $(0\leqq t\leqq T)$.*

2. Now we shall prove that $f(x, t)$ and $g(z, t)$ satisfy the following **Löwner's differential equations.**

Theorem IX 18.

$$\frac{\partial f(x, t)}{\partial t}=-f(x, t)\,\frac{1+\kappa(t)f(x, t)}{1-\kappa(t)f(x, t)}\ ,$$

$$\frac{\partial g(z, t)}{\partial t}=z\,\frac{1+\kappa(t)z}{1-\kappa(t)z}\cdot\frac{\partial g(z, t)}{\partial z}\ ,$$

where $\kappa(t)=e^{-i\theta(t)}$ is a continuous function of t $(0\leqq t\leqq T)$. Löwner [1].

If C is an analytic Jordan arc, such that $w(\tau)$ is a regular function of τ for $0\leqq\tau\leqq 1$ and meets $|w|=1$ orthogonally at w_1, then Komatu [2] proved that $d\theta(t)/dt=\theta'(t)$ exists and is continuous for $0<t<T$ and $\theta'(t)=-3K(t)$, where $K(t)$ is the curvature of $\Gamma_{0,\tau}$ at $z_\tau=e^{i\theta(t)}$. Tsuji [4] proved that $0<t<T$ can be replaced by $\leqq t\leqq T$.

Proof. (Komatu [1]). Let $0\leqq\tau_1<\tau\leqq 1$, $t_1=t(\tau_1)$, $t=t(\tau)$. With the same notations as Lemma 1, with τ instead of τ_2, let

$$z=\varphi(\zeta)=e^{-\alpha}(\zeta+\cdots), \qquad \alpha=\Delta t=t-t_1>0, \tag{1}$$

then

$$\left[\log\frac{\varphi(\zeta)}{\zeta}\right]_{\zeta=0}=-\alpha=-\Delta t.$$

If by $z=\varphi(\zeta)$, $\Gamma_{\tau_1,\tau}$ is mapped on an arc $[e^{i\theta_1}, e^{i\theta}]$ on $|\zeta|=1$, then since $|\varphi|=1$ on its complement on $|\zeta|=1$, we have

$$\log\frac{\varphi(\zeta)}{\zeta}=\frac{1}{2\pi}\int_{\theta_1}^{\theta}\log|\varphi(e^{i\psi})|\frac{e^{i\psi}+\zeta}{e^{i\psi}-\zeta}d\psi. \tag{2}$$

Since $\log|\varphi(e^{i\psi})|\leqq 0$ on $[e^{i\theta_1}, e^{i\theta}]$, we have by the mean value theorem,

$$\log\frac{\varphi(\zeta)}{\zeta}=\frac{A}{2\pi}\int_{\theta_1}^{\theta}\log|\varphi(e^{i\psi})|d\psi=\frac{A}{2\pi}\int_0^{2\pi}\log\left|\frac{\varphi(e^{i\psi})}{e^{i\psi}}\right|d\psi$$

$$=A\log\left|\frac{\varphi(\zeta)}{\zeta}\right|_{\zeta=0}=-A\Delta t, \tag{3}$$

where A is a certain point, contained in the smallest convex set, which contains the curve $\dfrac{e^{i\psi}+\zeta}{e^{i\psi}-\zeta}(\theta_1\leqq\psi\leqq\theta)$.

Since $\zeta=f(x, t_1)$, $z=\varphi(\zeta)=f(x, t)$,

$$\Delta f=f(x, t)-f(x, t_1)=\varphi(\zeta)-\zeta, \qquad \varphi(\zeta)=\zeta+\Delta f,$$

so that if $t_1\to t$, then

$$-A\Delta t=\log\frac{\varphi(\zeta)}{\zeta}=\log\left(1+\frac{\Delta f}{\zeta}\right)\sim\frac{\Delta f}{\zeta}, \tag{4}$$

and

$$A\to\frac{e^{i\theta(t)}+\zeta}{e^{i\theta(t)}-\zeta}, \qquad \zeta\to f(x, t), \tag{5}$$

hence

$$\frac{\partial f(x, t)}{\partial t}=-f(x, t)\cdot\frac{1+\kappa(t)f(x, t)}{1-\kappa(t)f(x, t)}, \qquad \kappa(t)=e^{-i\theta(t)}. \tag{6}$$

Now $w=f(x, T)$ maps $|x|<1$ conformally on D. On the other hand, $w=g(z, t)=g(f(x, t), t)$ $(0\leqq t\leqq T)$ maps $|x|<1$ conformally on D, so that

$$f(x, T)=g(f(x, t), t) \qquad (0\leqq t\leqq T),$$

hence by differentiating the both sides with t, we have

$$0=\frac{\partial g}{\partial z}\cdot\frac{\partial f(x,t)}{\partial t}+\frac{\partial g}{\partial t},$$

hence

$$\frac{\partial g(z,t)}{\partial t}=-\frac{\partial f(x,t)}{\partial t}\cdot\frac{\partial g(z,t)}{\partial z}=f\frac{1+\kappa(t)f}{1-\kappa(t)f}\cdot\frac{\partial g(z,t)}{\partial z}$$

$$=z\frac{1+\kappa(t)z}{1-\kappa(t)z}\cdot\frac{\partial g(z,t)}{\partial z}\,.\text{——} \tag{7}$$

3. Now

$$f(x,T)=e^{-T}(x+\cdots),\qquad f(x,t)=e^{-t}(x+\cdots), \tag{1}$$

$$f(x,T)=g(f(x,t),t)=g(z,t),\qquad z=f(x,t).$$

Hence if

$$g(z,t)=e^{-\alpha}(z+\cdots),$$

then $-T=-\alpha-t$, $\alpha=T-t$, so that

$$g(z,t)=e^{-(T-t)}(z+a_2(t)z^2+a_3(t)z^3+\cdots), \tag{2}$$

$$g(z,0)=e^{-T}(z+a_2z^2+a_3z^3+\cdots),\qquad a_n=a_n(0).$$

Then we shall prove

Lemma. $|a_2|\leqq 2,\quad |a_3|\leqq 3.$

Proof. If we put (2) into (7), then we have

$$a_2'(t)=a_2(t)+2\kappa(t), \tag{3}$$

$$a_3'(t)=2a_3(t)+4\kappa(t)a_2(t)+2\kappa^2(t). \tag{4}$$

Since $g(z,T)\equiv z$, we have $a_n(T)=0$ $(n=2,3,\cdots)$, so that from (3), (4), we have

$$a_2(t)=-2e^t\int_t^T\kappa(t)e^{-t}\,dt,$$

$$a_3(t)=-4e^{2t}\int_t^T\kappa(t)e^{-2t}a_2(t)dt-2e^{2t}\int_t^T\kappa^2(t)e^{-2t}\,dt,$$

hence

$$a_2=a_2(0)=-2\int_0^T\kappa(t)e^{-t}\,dt, \tag{5}$$

$$a_3=a_3(0)=8\int_0^T\kappa(t)e^{-t}\left(\int_t^T k(\tau)e^{-\tau}\,d\tau\right)dt-2\int_0^T\kappa^2(t)e^{-2t}\,dt. \tag{6}$$

Now

$$\int_0^T\kappa(t)e^{-t}\left(\int_t^T\kappa(\tau)e^{-\tau}\,d\tau\right)dt=\int_0^T\kappa(t)e^{-t}\left(\int_0^T\kappa(\tau)e^{-\tau}\,d\tau-\int_0^t\kappa(\tau)e^{-\tau}\,dt\right)dt$$

$$=\left(\int_0^T\kappa(t)e^{-t}\,dt\right)^2-\int_0^T\kappa(t)e^{-t}\left(\int_0^t\kappa(\tau)e^{-\tau}\,d\tau\right)dt.$$

If we put

$$\Phi(t)=\int_0^t \kappa(t)\, e^{-t}\, dt \,,$$

then

$$\int_0^T \kappa(t)\, e^{-t}\left(\int_0^t \kappa(\tau)\, e^{-\tau}\, d\tau\right)dt=\int_0^T \Phi'(t)\,\Phi(t)dt=\frac{1}{2}\Phi^2(t_0)=\frac{1}{2}\left(\int_0^T \kappa(t)\, e^{-t}\, dt\right)^2,$$

so that

$$a_3=4\left(\int_0^T \kappa(t)\, e^{-t}\, dt\right)^2-2\int_0^T \kappa^2(t)\, e^{-2t}\, dt \,. \tag{7}$$

Since $\kappa(t)=e^{-i\theta(t)}$, we have by (5),

$$|a_2|\leqq 2\,, \tag{8}$$

and by (7),

$$\begin{aligned}\Re a_3&=4\left(\int_0^T e^{-t}\cos\theta(t)dt\right)^2-4\left(\int_0^T e^{-t}\sin\theta(t)dt\right)^2-2\int_0^T e^{-2t}\cos 2\theta(t)dt\\ &\leqq 4\left(\int_0^T e^{-t}\cos\theta(t)dt\right)^2-2\int_0^T e^{-2t}\cos 2\theta(t)dt\\ &\leqq 4\left(\int_0^T e^{-t}\cos\theta(t)dt\right)^2-4\int_0^T e^{-2t}\cos^2\theta(t)dt+1\,.\end{aligned}$$

Since

$$\left(\int_0^T e^{-t}\cos\theta(t)dt\right)^2\leqq\int_0^T e^{-t}\cos^2\theta(t)dt\int_0^T e^{-t}\, dt\leqq\int_0^T e^{-t}\cos^2\theta(t)dt\,,$$

we have

$$\Re a_3\leqq 4\int_0^T (e^{-t}-e^{-2t})\cos^2\theta(t)dt+1\leqq 4\int_0^\infty (e^{-t}-e^{-2t})dt+1=3\,.$$

From $\Re a_3\leqq 3$, we have easily

$$|a_3|\leqq 3\,. \tag{9}$$

Theorem IX. 19. *Let* $f(z)=z+a_2 z^2+a_3 z^3+\cdots$ *be regular and schlicht in* $|z|<1$, *then*

$$|a_2|\leqq 2\,,\qquad |a_3|\leqq 3\,.$$ Löwner [1].

Proof. If we consider $f(rz)/r$ $(0<r<1)$ instead of $f(z)$, then we may assume that $f(z)$ is regular in $|z|\leqq 1$. Let $\underset{|z|=1}{\operatorname{Max}}|f(z)|=M$ and put

$$w=\varphi(z)=\frac{f(z)}{M}=\frac{1}{M}(z+a_2 z^2+a_3 z^3+\cdots)\,. \tag{1}$$

By $w=\varphi(z)$, $|z|<1$ is mapped conformally on a domain $\varDelta$ in $|w|<1$. We approximate $\varDelta$ by a sequence of domains $\varDelta_n$ of the form of Theorem IX. 18, then $\varDelta$ is the kernel of $\{\varDelta_n\}$ and $\varDelta_n\to\varDelta$. We map $\varDelta_n$ conformally on $|z|<1$ by

$$w=\varphi_n(z)=e^{-t_n}(z+a_2^{(n)} z^2+a_3^{(n)} z^3+\cdots)\,,\qquad t_n>0\,. \tag{2}$$

then by the lemma, $|a_2^{(n)}|\leqq 2$, $|a_3^{(n)}|\leqq 3$.

Since Δ_n converges to its kernel Δ, by Carathéodory's theorem, $\lim_n \varphi_n(z)=\varphi(z)$, so that $e^{-t_n}\to 1/M$, $a_2^{(n)}\to a_2$, $a_3^{(n)}\to a_3$, hence

$$|a_2|\leqq 2\,, \qquad |a_3|\leqq 3\,. \tag{3}$$

10. Rengel's inequality

1. Let D be a multiply connected domain on the $z=x+iy$-plane, which is obtained from a rectangle R: $0<x<a$, $0<y<b$, by taking off a finite number k of segments L_i parallel to the x-axis:

$$L_i:\ \alpha_i\leqq x\leqq\beta_i\,, \qquad y=\gamma_i \qquad (i=1, 2, \cdots, k)\,,$$

where $0<\alpha_i<\beta_i<a$ and $0<\gamma_i<b$.

Let $w=f(z)$ be regular and schlicht in D and let Δ be the image of D on the $w=u+iv$-plane and J be its area. Suppose that every segment: $0<x<a$, $y=\gamma$ $(0<\gamma<b, \gamma\neq\gamma_i)$, parallel to the x-axis, is mapped on a curve, whose projection on a fixed line g has a length $\geqq\alpha>0$. Then

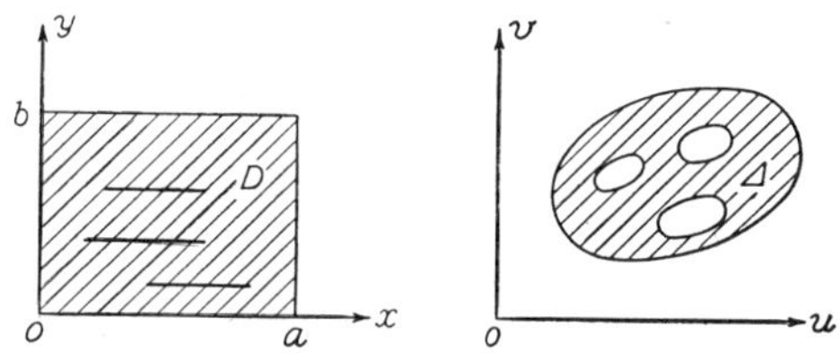

Theorem IX. 20. (Rengel's inequality). $\dfrac{b}{a}\leqq\dfrac{J}{\alpha^2}$, *where the equality holds, only when $f(z)$ is of the form $f(z)=Az+B$ ($A, B=$const.).*

Rengel [1].

Proof. $J=\iint_D |f'(z)|^2\,dx\,dy=\int_a^b dy\int_0^a |f'(z)|^2\,dx\quad (z=x+iy)\,.$

By the hypothesis,

$$\alpha\leqq\int_0^a |f'(z)|\,dx\,, \qquad y\neq\gamma_i\,,$$

so that

$$\alpha^2\leqq\left(\int_0^a |f'(z)|\,dx\right)^2\leqq\int_0^a dx\int_0^a |f'(z)|^2\,dx=a\int_0^a |f'(z)|^2\,dx\,,$$

$$\frac{\alpha^2}{a}\leqq\int_0^a |f'(z)|^2\,dx\,,$$

hence

$$\frac{b}{a}\alpha^2\leqq\int_a^b dy\int_0^a |f'(z)|^2\,dx=J\,,$$

$$\frac{b}{a} \leqq \frac{J}{\alpha^2}. \tag{1}$$

Next we shall study the case, where the equality holds. Suppose that for y_0 $(\neq \gamma_i)$,

$$\int_0^a |f'(x+iy_0)|dx > \alpha + c, \qquad c>0. \tag{2}$$

If $\delta>0$ is small, then for any y in $|y-y_0|<\delta/2$,

$$\int_0^a |f'(x+iy)|dx > \alpha + c,$$

hence from the above proof, we have

$$J > \frac{b-\delta}{a}\alpha^2 + \frac{\delta(\alpha+c)^2}{a} > \frac{b}{a}\alpha^2 + \frac{2\delta\alpha c}{a},$$

so that

$$\frac{b}{a} < \frac{J}{\alpha^2} - \frac{2\delta c}{a\alpha} < \frac{J}{\alpha^2}. \tag{3}$$

Hence if $b/a = J/\alpha^2$, then for any y $(0<y<b,\ y \neq \gamma_i)$,

$$\int_0^a |f'(x+iy)|dx = \alpha. \tag{4}$$

Hence any segment $y=$const. $(\neq \gamma_i)$ is mapped on a curve Γ of length α. Since the projection of Γ on g has a length $\geqq \alpha>0$, Γ is a segment, parallel to g, so that L_i $(i=1,2,\cdots,k)$ is mapped on a segment, parallel to g, hence by the principle of reflexion, $w=f(z)$ is regular on L_i and R is mapped on a rectangle R^* on the w-plane and the image of the segment $x=$const., being the orgothonal trajectory of Γ's, is a segment, orthogonal to g.

Let a^*, b^* be the sides of R^*, then $J=a^*b^*$ and $\alpha=a^*$, so that from $b/a=J/\alpha^2$, we have $b/a=b^*/a^*$, hence R and R^* are similar rectangles.

From this, we can prove that $f(z)=Az+B$ as follows. We decompose R into two rectangles R_1, R_2 by a line parallel to the x-axis and let R_i^* $(i=1,2)$ be the image of R_i on the w-plane, then R^* is a rectangle. Let a_1, b_1 be the sides of R_1 and a_1^*, b_1^* be those of R_1^*, then if we apply (1) to R_1, R_1^*, then we have $b_1/a_1 \leqq b_1^*/a_1^*$ and if we interchange z and w, then we have $b_1^*/a_1^* \leqq b_1/a_1$, so that $b_1/a_1=b_1^*/a_1^*$, hence R_1 and R_1^* are similar rectangles. Similarly we can prove that a rectangle in R, whose sides are parallel to the coordinates axes is mapped on a similar rectangle on the w-plane, hence $f(z)$ is of the form $f(z)=Az+B$.

2. As an application of Rengel's inequality, we shall prove the following theorem, which is an extension of Koebe's 1/4-disc theorem,

which is conjectured by Szegö and proved by Rengel.

Theorem IX. 21. *Let $w=f(z)=z+\cdots$ be regular and schlicht in $|z|<1$ and D be the image of $|z|<1$ on the w-plane and Γ be its boundary.*

We draw n equi-angular half-lines through $w=0$ and let $w_1, \cdots, w_n$ be the points of intersection of these lines with Γ, such that the segment $\overline{0w_i}$ $(i=1, 2, \cdots, n)$, except w_i, belongs to D and put

$$d_n=\operatorname{Max}(|w_1|, \cdots, |w_n|).$$

Then

$$d_n \geqq \sqrt[n]{1/4},$$

where the equality holds, only when $f(z)\equiv e^{i\theta}E_n(e^{-i\theta}z)$, where

$$E_n(z)=\frac{z}{(1+z^n)^{2/n}}. \qquad \text{Rengel [1].}$$

The case $n=1$ is the Koebe's 1/4-disc theorem.

Proof. Since if $-\frac{\pi}{n}\leqq\theta\leqq\frac{\pi}{n}$, $E_n(e^{i\theta})=\frac{\sqrt[n]{1/4}}{(\cos n\theta/2)^{2/n}}$ and $E_n\left(ze^{\frac{2\pi i}{n}}\right)=e^{\frac{2\pi i}{n}}E_n(z)$, by $w=E_n(z)$, $|z|<1$ is mapped on a domain Δ on the w-plane, whose boundary consists of n half-lines

$$L_\nu:\ w=te^{\frac{2\pi\nu i}{n}}\ (\sqrt[n]{1/4}\leqq t\leqq\infty)\ (\nu=0, 1, 2, \cdots, n-1).$$

$z=E_n^{-1}(w)$ is not one-valued and its branch points are $\xi=\sqrt[n]{1/4}\,e^{\frac{2\pi\nu i}{n}}$ $(\nu=0, 1, 2, \cdots, n-1)$. Let $\overset{*}{E}{}_n^{-1}(w)$ be its principal branch, which is defined in Δ and $|\overset{*}{E}{}_n^{-1}(w)|<1$.

Let one of n equi-angular half-lines $\{L_\nu\}$ through $w=0$ coincides with the positive real axis and suppose that

$$d_n\leqq\sqrt[n]{1/4}, \tag{1}$$

then we shall prove that $f(z)=E_n(z)$, so that $d_n=\sqrt[n]{1/4}$. Hence $d_n\geqq\sqrt[n]{1/4}$ in general.

Let $|z|=q\rho$ $(0<\rho<1)$ be the greatest circle, contained in the image of $|z|\leqq\rho$ by $\overset{*}{E}{}_n^{-1}(f(z))$, then since $\overset{*}{E}{}_n^{-1}(f(z))=z+\cdots$, we have

$$\lim_{\rho\to 0} q = 1. \tag{2}$$

By $\zeta=\xi+i\eta=\log z$, we map a circular ring: $\rho<|z|<1$ on a rectangle R: $\log\rho<\xi<0$, $0\leqq\eta\leqq 2\pi$ on the ζ-plane and put

$$g(\zeta)=\log E_n^{-1}(f(e^\zeta)), \quad \zeta\in R. \tag{3}$$

If there exists z_ν in $\rho<|z|<1$, such that $f(z)=\xi_\nu=\sqrt[n]{1/4}\,e^{\frac{2\pi\nu i}{n}}$ and let $\zeta_\nu=\log z_\nu$, then $g(\zeta)$ is not regular at ζ_ν. Through ζ_ν $(\nu=1, 2, \cdots)$, we draw a line, parallel to the η-axis, then R is decomposed into rectangles $\{R_i\}$, in which $g(\zeta)$ is regular. Since $\log \overset{*}{E}{}_n^{-1}(w)$ can be continued analytically across L_ν by the principle of reflexion, we see easily that

$$\iint\limits_{R_i} \left| \frac{d}{d\zeta} \log E_n^{-1}(f(e^\zeta)) \right|^2 r dr d\theta = \iint\limits_{R_i} \left| \frac{d}{d\zeta} \log \overset{*}{E}{}_n^{-1}(f(e^\zeta)) \right|^2 r dr d\theta , \quad \zeta = re^{i\theta} ,$$

hence by the definition of q,

$$J = \iint\limits_{R} |g'(\zeta)|^2 r dr d\theta = \sum_i \iint\limits_{R_i} \left| \frac{d}{d\zeta} \log \overset{*}{E}{}_n^{-1}(f(e^\zeta)) \right|^2 r dr d\theta$$

$$= \iint\limits_{\rho<|z|<1} \left| \frac{d}{dz} \log \overset{*}{E}{}_n^{-1}(f(z)) \right|^2 dx dy \leqq 2\pi \log 1/q\rho , \quad z = x+iy . \tag{4}$$

Since $d_n \leqq \sqrt[n]{1/4}$, any segment l in R_i, parallel to the η-axis, is mapped on a closed curve on the $w=f(e^\zeta)$-plane, which intersects n segments $\overline{0\xi}$ $(\nu=0, 1, 2, \cdots, n-1)$, so that l is mapped on a curve on the $v=\log E_n^{-1}(f(e^\zeta))$-plane, whose projection on the imaginary axis has a length $\geqq 2\pi$. If we apply Rengel's inequality to $v=g(\zeta)$ $(\zeta \in R)$, with $a=2\pi$, $b=\log 1/\rho$, $\alpha=2\pi$, $J \leqq 2\pi \log 1/q\rho$, then we have

$$\frac{\log 1/\rho}{2\pi} \leqq \frac{2\pi \log 1/q\rho}{(2\pi)^2} = \frac{\log 1/q\rho}{2\pi} . \tag{5}$$

If the equality does not hold, then

$$\frac{\log 1/\rho}{2\pi} \leqq \frac{\log 1/q\rho}{2\pi} - \frac{2\delta c}{(2\pi)^2} , \qquad \delta>0 , \quad c>0 ,$$

where $\delta>0$ and $c>0$ are independent of ρ, which is impossible, since $\lim\limits_{\rho\to 0} q=1$. Hence the equality holds, so that $g(\zeta)=\log E_n^{-1}(f(e^\zeta))=A\zeta+B$ and from this, we have easily $f(z)=E_n(z)$, so that $d_n=\sqrt[n]{1/4}$.

If one of n equi-angular half-lines through $w=0$ is a line: $\arg w=\theta$, and $d_n \leqq \sqrt[n]{1/4}$, then $f(z)=e^{i\theta}E_n(e^{-i\theta}z)$ and $d_n=\sqrt[n]{1/4}$, so that

$$d_n \geqq \sqrt[n]{1/4} \quad \text{in general.}$$

Part II. Multiply connected domains

1. Conformal mapping on a parallel slits-plane

Let Δ be a multiply connected domain on the w-plane, which contains $w=\infty$ and every boundary component of Δ, which is a continuum, is a segment, parallel to the real axis, then Δ is called *a parallel slits-plane.*

Lemma 1. *Let D be a simply connected domain on the z-plane,* *which contains $z=\infty$ and whose boundary is a continuum, which is not a segment, parallel to the real axis. If by $w=f(z)$, such that*

$$w=f(z)=z+\frac{a}{z}+\cdots \quad at \quad z=\infty ,$$

D is mapped conformally on a parallel slit-plane Δ on the w-plane, then

$$\Re(a)>0 .$$

Proof. Let the boundary of Δ be a segment: $\alpha\leqq\xi\leqq\beta$, $\eta=\text{const.}$ $(w=\zeta+i\eta)$. Then by $w=Av+B$ with suitable real constants A, B, Δ is mapped on a parallel slit-plane Δ^* on the v-plane, whose boundary is a segment: $-2\leqq\xi^*\leqq2$, $\eta^*=0$ $(v=\xi^*+i\eta^*)$. Next by $v=\zeta+\dfrac{1}{\zeta}$, we map Δ^* on the outside of $|\zeta|=1$, then

$$z=w-\frac{a}{w}+\cdots=Av+B-\frac{a}{Av}+\cdots .$$

Hence if we put $z^*=\dfrac{z-B}{A}$, then

$$z^*=z^*(\zeta)=\zeta+\left(1-\frac{a}{A^2}\right)\frac{1}{\zeta}+\cdots .$$

Since $z^*(\zeta)$ is regular and schlicht for $|\zeta|>1$, by Bieberbach's theorem, $\left|1-\dfrac{a}{A^2}\right|\leqq 1$. Hence $\Re(a)\geqq 0$. If $\Re(a)=0$, then $a=0$, so that $z^*=\zeta+\dfrac{1}{\zeta}=v$, hence $w=z$, which contradicts the hypothesis, hence

$$\Re(a)>0 .$$

Lemma 2. *Let D be a multiply connected domain on the z-plane, which contains $z=\infty$ and is bounded by k analytic Jordan curves C_i $(i=1, 2, \cdots, k)$ and the area, enclosed in $\{C_i\}$ be $\geqq A>0$.*

If by $w=f(z)$, such that

$$w=f(z)=z+\frac{a_1}{z}+\cdots \quad at \quad z=\infty ,$$

D is mapped conformally on a parallel slits-plane Δ on the w-plane, then

$$\Re(a_1)\geqq\frac{A}{2\pi} .$$

Proof. $z=w+\dfrac{b_1}{w}+\dfrac{b_2}{w^2}+\cdots$ at $w=\infty$, where $b_1=-a_1=\alpha+i\beta$. (1)

Hence if we put $z=u+iv$, $w=x+iy$,

$$\varphi_1(w)=\frac{b_2}{w^2}+\cdots=u_1+iv_1 , \tag{2}$$

then

$$z=u+iv=w+\frac{b_1}{w}+\varphi_1(w)=\left(x+\frac{\alpha x+\beta y}{x^2+y^2}+u_1\right)+i\left(y+\frac{\beta x-\alpha y}{x^2+y^2}+v_1\right). \quad (3)$$

Hence if we put

$$U=x+\frac{\alpha x+\beta y}{x^2+y^2}, \qquad V=y+\frac{\beta x-\alpha y}{x^2+y^2}, \quad (4)$$

then

$$u=U+u_1, \qquad v=V+v_1. \quad (5)$$

Let R be a rectangle: $|x|\leqq L$, $|y|\leqq\sqrt{L}$ on the w-plane, which contains all the boundary slits of Δ. Let C be the boundary of R and Γ be its image on the z-plane, then the area J^*, enclosed in Γ, is given by

$$J^*=\int_C u\,dv=\int_C U dV+\int_C u_1 dV+\int_C U dv_1+\int_C u_1 dv_1, \quad (6)$$

where we integrate on C in the positive sense. Since as easily seen

$$\int_C u_1 dV=O(1/L), \qquad \int_C U dv_1=O(1/\sqrt{L}), \quad \int_C u_1 dv_1=O(1/L^2),$$

we have

$$J^*=\int_C U dV+O(1/\sqrt{L}). \quad (7)$$

Now

$$\int_C U dV=\int_C\left(x+\frac{\alpha x+\beta y}{x^2+y^2}\right)\frac{\beta(y^2-x^2)+2\alpha xy}{(x^2+y^2)^2}dx$$
$$+\int_C\left(x+\frac{\alpha x+\beta y}{x^2+y^2}\right)\left(1+\frac{\alpha(y^2-x^2)-2\beta xy}{(x^2+y^2)^2}\right)dy. \quad (8)$$

We remark the following: If $F(x, y)$ is an even function of y, then $\int_C F(x, y)dx=0$ and if $F(x, y)$ is an even function of x, then $\int_C F(x, y)dy=0$, hence

$$\int_C U dV=2\alpha\int_C\frac{x^2y}{(x^2+y^2)^2}dx+2\alpha^2\int_C\frac{x^2y}{(x^2+y^2)^3}dx+\beta^2\int_C\frac{y(y^2-x^2)}{(x^2+y^2)^3}dx$$
$$+\int_C x dy+\alpha\int_C\frac{x(y^2-x^2)}{(x^2+y^2)^2}dy+\alpha\int_C\frac{x}{x^2+y^2}dy+\alpha^2\int_C\frac{x(y^2-x^2)}{(x^2+y^2)^3}dy$$
$$-2\beta^2\int_C\frac{xy^2}{(x^2+y^2)^3}dy. \quad (9)$$

Since

$$\int_C x\,dy = 4L^{\frac{3}{2}}$$

and

$$\int_C \frac{x^2y}{(x^2+y^2)^2}dx=-2\sqrt{L}\int_{-L}^{L}\frac{x^2dx}{(x^2+L)^2}=-2\int_{-\sqrt{L}}^{\sqrt{L}}\frac{t^2dt}{(1+t^2)^2}\to-2\int_{-\infty}^{\infty}\frac{t^2dt}{(1+t^2)^2}$$
$$=-\pi\ , \qquad L\to\infty,$$

and, as easily seen, the other terms of (9) are $O(1/\sqrt{L})$, we have by (7),

$$J^*=4L^{\frac{3}{2}}-\alpha k(L)+O(1/\sqrt{L})\ , \tag{10}$$

where

$$\lim_{L\to\infty} k(L) = 2\pi\ . \tag{11}$$

Let J be the area of the domain, bounded by $C_1, \cdots, C_k, \Gamma$, then by the hypothesis,

$$J \leqq J^* - A\ . \tag{12}$$

For a large L, the projection on the u-axis of the image of the segment: $-L\leqq x\leqq L$, $y=\text{const.}=c$ $(|c|\leqq\sqrt{L})$ has the length $\geqq 2L-K/L$ ($K=\text{const.}$), except k values of c, corresponding to k boundary slits of Δ. Hence if we put

$$a=2L\ , \qquad b=2\sqrt{L}\ , \qquad \alpha=2L-K/L$$

in Rengel's inequality, then we have

$$\frac{1}{\sqrt{L}}\leqq\frac{4L^{\frac{3}{2}}-\alpha k(L)+O(1/\sqrt{L})-A}{(2L-K/L)^2}\ ,$$

or

$$A\leqq-\alpha k(L)+O(1/\sqrt{L})\ .$$

Hence if we make $L\to\infty$, then we have by (11),

$$\Re(a_1)=-\alpha\geqq\frac{A}{2\pi}\ .$$

Lemma 3. *Let D be a multiply connected domain on the z-plane, which contains $z=\infty$.*

Let $\mathfrak{F}$ be the family of $f(z)$, which satisfy the following conditions:

(i) *$f(z)$ is meromorphic and schlicht in D,*

(ii) $f(z)=z+\frac{a_1}{z}+\cdots$ *at* $z=\infty$.

If we put

$$\mu = \sup_{f\in\mathfrak{F}} \Re(a_1)\ ,$$

then there exists $f_0(z)\in\mathfrak{F}$, such that

$$f_0(z)=z+\frac{a_1}{z}+\cdots \quad at\ z=\infty\ , \qquad \Re(a_1^0)=\mu\ .$$

Proof. We choose $r>0$ so large that the boundary of D is contained in $|z|<r$. Let $f(z)\in\mathfrak{F}$ and by $w=f(z)$, the circle $|z|=r$ be mapped on a Jordan curve Γ_r on the w-plane. Inside of Γ_r, we choose a point w_0 and put

$$\chi(\zeta)=\frac{r}{f(z)-w_0}, \qquad \zeta=\frac{r}{z}. \tag{1}$$

Then $\chi(\zeta)$ is regular and schlicht in $|\zeta|<1$ and at $\zeta=0$,

$$\chi(\zeta)=\frac{r}{z+a_1/z+\cdots-w_0}=\zeta+\frac{w_0}{r}\zeta^2+\cdots,$$

so that

$$\left|\frac{w_0}{r}\right|\leqq 2, \qquad |w_0|\leqq 2r. \tag{2}$$

Since by Koebe's theorem, $|\chi(\zeta)|\geqq 1/4$ on $|\zeta|=1$, $\left|\frac{r}{f(z)-w_0}\right|\geqq\frac{1}{4}$ on $|z|=r$, $|f(z)-w_0|\leqq 4r$, $|f(z)|\leqq|w_0|+4r\leqq 6r$, so that

$$|f(z)|\leqq 6r \qquad \text{in} \quad |z|\leqq r. \tag{3}$$

Hence $\mathfrak{F}$ is a normal family.

Now by the definition of μ, there exist $f_n\in\mathfrak{F}$ $(n=1,2,\cdots)$, such that

$$\Re(a_1^{(n)})\to\mu, \tag{4}$$

where

$$f_n(z)=z+\frac{a_1^{(n)}}{z}+\cdots \qquad \text{at} \quad z=\infty.$$

Since $\mathfrak{F}$ is a normal family, by taking a suitable partial sequence from $\{f_n\}$, we may assume that $\lim_n f_n(z)=f_0(z)$ uniformly in the wider sense in D, then $f_0(z)\in\mathfrak{F}$ and let

$$f_0(z)=z+\frac{a_1^0}{z}+\cdots \qquad \text{at} \quad z=\infty,$$

then

$$a_1^{(n)}=\frac{1}{2\pi i}\int_{|z|=r}f_n(z)dz\to\frac{1}{2\pi i}\int_{|z|=r}f_0(z)dz=a_1^0 \quad (n\to\infty),$$

hence

$$\Re(a_1^0)=\mu. \text{ ——}$$

Theorem IX. 22. *Let D be a multiply connected domain on the z-plane, which contains $z=\infty$ and $f_0(z)$ be the function, defined by Lemma 3, then by $w=f_0(z)$,*

(i) *D is mapped conformally on a parallel slits-plane Δ on the w-plane.* R. de Possel [1].

(ii) *The surface measure of the boundary of Δ is zero.* Tsuji [10].

Proof. (i) Let $\varDelta$ be the image of D by $w=f_0(z)$ on the w-plane and suppose that $\varDelta$ is not a parallel slits-plane, then there exists a boundary component γ of $\varDelta$, which is a continuum and is not a segment, parallel to the real axis. Let $\varDelta_0$ be the simply connected domain, which is bounded by γ and we map $\varDelta_0$ conformally on a parallel slit-plane on the ζ-plane by $\zeta=\varphi(w)$, such that

$$\varphi(w)=w+\frac{a}{w}+\cdots \quad \text{at} \quad w=\infty\,, \tag{1}$$

then by Lemma 1, $\Re(a)>0$.

Now $\varphi(f_0(z))\in\mathfrak{F}$ and

$$\varphi(f(z_0))=f_0(z)+\frac{a}{f_0(z)}+\cdots=z+\frac{a_1^0+a}{z}+\cdots \quad \text{at} \quad z=\infty\,,$$

$$\Re(a_1^0+a)=\mu+\Re(a)>\mu\,, \tag{2}$$

which contradicts the definition of μ. Hence $\varDelta$ is a parallel slits-plane.

(ii) Suppose that the surface measure A of the boundary Γ of $\varDelta$ is positive. We exhaust $\varDelta$ by a sequence of domains $\varDelta_n$, whose boundary consists of a finite number of analytic Jordan curves, surrounding Γ.

Then the sum of areas, enclosed in these curves is $\geqq A>0$.

If we map $\varDelta_n$ conformally on a parallel slits-plane on the ζ-plane by $\zeta=g_n(w)$, such that

$$\zeta=g_n(w)=w+\frac{a_1^{(n)}}{w}+\cdots \quad \text{at} \quad w=\infty\,, \tag{1}$$

then by Lemma 2, $\Re(a_1^{(n)})\geqq A/2\pi$.

Since as before, $\{g_n(w)\}$ forms a normal family, we may assume that $\lim_n g_n(w)=g(w)$ uniformly in the wider sense in $\varDelta$ and let

$$g(w)=w+\frac{a_1}{w}+\cdots \quad \text{at} \quad w=\infty\,, \tag{2}$$

then $\Re(a_1)\geqq A/2\pi$.

Since $g(f_0(z))\in\mathfrak{F}$ and

$$g(f_0(z))=z+\frac{a_1^0+a_1}{z}+\cdots \quad \text{at} \quad z=\infty\,,$$

$$\Re(a_1^0+a_1)\geqq\mu+A/2\pi>\mu\,,$$

which contradicts the definition of μ, hence the surface measure of the boundary of $\varDelta$ is zero.

Theorem IX. 23. *If D is of finite connectivity and $w=f(z)$ ($f(z)=z+\frac{a_1}{z}+\cdots$ at $z=\infty$) maps D conformally on a parallel slits-plane on the w-plane, then $f(z)$ is unique.*

Proof. Suppose that there are two such mappings:

$$f_1(z)=z+\frac{a_1}{z}+\cdots, \qquad f_2(z)=z+\frac{b_1}{z}+\cdots \qquad \text{at} \quad z=\infty,$$

then $F(z)=f_1(z)-f_2(z)$ is regular and bounded in D, so that $w=F(z)$ maps D on a bounded domain on the w-plane, whose boundary consists of a finite number of segments, parallel to the real axis, which is impossible. Hence

$$f_1(z)\equiv f_2(z).$$

Remark. If D is of infinite connectivity, then such a mapping is not unique in general, which is proved by Koebe [3]. We shall give a simplified proof, due to Tamura [1] as follows.

Let $f(x)\geqq 0$ be a continuous function, defined in $-1\leqq x\leqq 1$, such that $f(-1)=f(1)=0$, whose graph g consists of a finite number of rectilinear segments. We define a transformation T_n $(n=1,2,\cdots)$ as follows. Let A, B be two consecutive vertices of g and C, D be its projection on the x-axis.

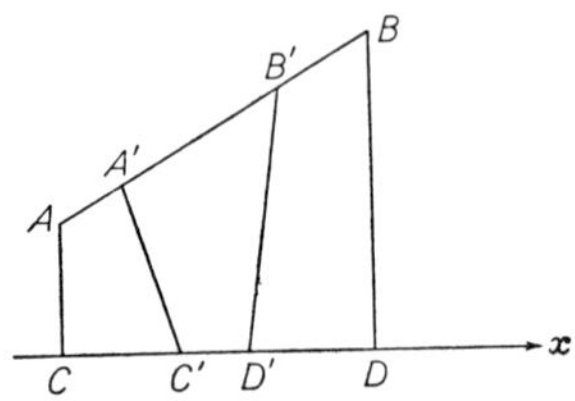

Let $A'B'$ be a sub-segment of AB with the same middle point as AB and $C'D'$ be that of CD, such that

$$A'B'=\frac{AB}{2^n}, \qquad C'D'=\frac{CD}{2^{n+1}}.$$

We replace the segment $AB\in g$ by a broken line $AA'C'D'B'B$ and preform the similar operation on every segment $\in g$ and let g' be the thus obtained graph and $f^*(x)$ be the function of x, whose graph is g' and put

$$f^* = T_n(f). \tag{1}$$

Let

$$f(x)=1-|x|, \quad f_1=T_1(f), \quad f_2=T_2(f_1), \cdots, \quad f_n=T_n(f_{n-1}), \cdots, \tag{2}$$

then since $f_{n+1}(x)\leqq f_n(x)$,

$$\lim_n f_n(x)=\varphi(x), \qquad -1\leqq x\leqq 1, \tag{3}$$

exists, where $\varphi(x)$ is an upper semi-continuous function of x in $[-1,1]$ and $\varphi(-1)=\varphi(1)=0$.

We shall prove that

$$\overline{\lim_{x\to a\pm 0}}\,\varphi(x)=\varphi(a), \qquad -1\leqq a\leqq 1. \tag{4}$$

Let g_n be the graph of $f_n(x)$. If $x=a$ corresponds to a vertex A of g_n for some n, then $f_n(a)=f_{n+1}(a)=\cdots$, so that

$$\varphi(a)=f_n(a). \tag{5}$$

Let AB be a segment $\in g_n$, then on AB, there are vertices $A_{n+\nu}$ of $g_{n+\nu}$ $(\nu=1,2,\cdots)$, clustering to A. Hence if $A_{n+\nu}=(x_{n+\nu}, f_{n+\nu}(x_{n+\nu}))$, then

$$x_{n+\nu}\to a\ , \qquad \varphi(x_{n+\nu})=f_{n+\nu}(x_{n+\nu})\to\varphi(a) \qquad (\nu=1, 2, \cdots)\ ,$$

hence in this case, (4) is proved.

If $x=a$ does not correspond to a vertex of any g_n $(n=1, 2, \cdots)$, then there is a segment $l_n \in g_n$, which intersects the line $x=a$ at a point P_n, such that $P_n\to P=\{a, \varphi(a)\}$. Since the length of l_n tends to zero, as $n\to\infty$, the both end points of l_n, which are vertices of g_n, tend to P. Hence in any case, we have (4).

By the construction, the set E of a, such that $\varphi(a)>0$ is a non-dense perfect set on the x-axis. If $\varphi(a)>0$, let L_a be a segment: $x=a, 0\leqq y\leqq\varphi(a)$. We take off all such L_a from the upper half of the $z=x+iy$-plane, then there remains a simply connected domain D and by (4), every L_a is a boundary element of D in Carathéodory's sense. Hence if we map D conformally on the upper half of the $w=u+iv$-plane, then L_a corresponds to a single point on the u-axis and let M be the image of $\{L_a\}$, then M is a non-dense perfect set on the u-axis. Let $\bar{D}$ be the domain, obtained from D by reflexion with respect to the real axis, then $D+\bar{D}$ is a parallel slits-plane, whose boundary is $\{L_a\}+\{\bar{L}_a\}$ and $D+\bar{D}$ is mapped conformally on the complement of M on the w-plane, which is also a parallel slits-plane, which proves the theorem.

2. Conformal mapping on a circular, or a radial slits-plane

Let $\varDelta$ be a multiply connected domain on the w-plane, which contains $w=0$ and $w=\infty$. If every boundary component of $\varDelta$, which is a continuum, is an arc on a circle $|w|=$const., then $\varDelta$ is called *a circular slits-plane.*

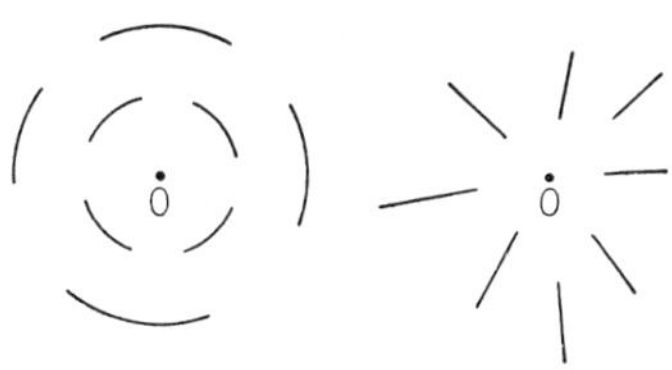

If every boundary component of $\varDelta$, which is a continuum, is a segment on a line $\arg w=$const., then $\varDelta$ is called *a radial slits-plane.*

Lemma 1. *Let D be a simply connected domain on the z-plane, which contains $z=0$ and $z=\infty$. Then by $w=f(z)$, such that*

$$f(0)=0, \qquad w=f(z)=z+a_0+\frac{a_1}{z}+\cdots \qquad \textit{at} \quad z=\infty\ ,$$

(i) *D can be mapped conformally on a circular slit-plane on the w-plane.*

(ii) *D can be mapped conformally on a radial slit-plane on the w-plane.*

Proof. We map D conformally on $|\zeta|>1$ by $\zeta=\varphi(z)$, such that

$$\varphi(z)=\alpha z+a_0+\frac{a_1}{z}+\cdots \qquad \text{at} \quad z=\infty,$$

and $\varphi(0)=-\rho$ $(\rho>1)$. Then

$$w=\frac{\rho\zeta(\zeta+\rho)}{\alpha(1+\rho\zeta)}=\frac{\rho\varphi(z)(\varphi(z)+\rho)}{\alpha(1+\rho\varphi(z))} \tag{1}$$

satisfies the condition (i) and

$$w=\frac{1}{\alpha}\left(\zeta+\frac{1}{\zeta}+\rho+\frac{1}{\rho}\right)=\frac{1}{\alpha}\left(\varphi(z)+\frac{1}{\varphi(z)}+\rho+\frac{1}{\rho}\right) \tag{2}$$

satisfies the condition (ii).

Lemma 2. *Let D be a multiply connected domain on the z-plane, which contains $z=0$ and $z=\infty$ and is bounded by a finite number of continua. Let $w=f(z)$ be meromorphic and schlicht in D and*

$$f(0)=0, \quad w=f(z)=z+a_0+\frac{a_1}{z}+\cdots \qquad at \quad z=\infty .$$

(i) *If $w=f(z)$ maps D conformally on a circular slits-plane on the w-plane, then*

$$|f'(0)|\geqq 1 ,$$

where the equality holds, only when $f(z)\equiv z$.

(ii) *If $w=f(z)$ maps D conformally on a radial slits-plane on the w-plane, then*

$$|f'(0)|\leqq 1 ,$$

where the equality holds, only when $f(z)\equiv z$.

Proof. (i) Let $w=f(z)$ map D conformally on a circular slits-plane Δ on the w-plane and $z=\varphi(w)$ be its inverse function.

Let a circular ring: $0<r<|w|<R$ contain all circular slits of Δ and put

$$m(r)=\operatorname*{Min}_{|w|=r}|\varphi(w)| , \qquad M(R)=\operatorname*{Max}_{|w|=R}|\varphi(w)| . \tag{1}$$

Let $\Delta(r,R)$ be the part of Δ, contained in $r<|w|<R$. By $\omega=\log w$, we map $\Delta(r,R)$ on a domain $\Delta^*(r,R)$ on the ω-plane, which is obtained from the rectangle:

$$\log r<\xi<\log R , \qquad 0<\eta<2\pi , \qquad \omega=\xi+i\eta , \tag{2}$$

by taking off a finite number of segments, parallel to the η-axis. Let $D(r,R)$ be the part of D, which is contained between the images of the circles $|w|=r$ and $|w|=R$.

By $\zeta=\log z$, we map $D(r,R)$ on a domain $D^*(r,R)$ on the ζ-plane, which is contained in a rectangle;

$$\log m(r)<u<\log M(R) , \quad 0<v<2\pi , \quad \zeta=u+iv . \tag{3}$$

We apply Rengel's inequality to $\zeta=\zeta(\omega)$, with

$$a=2\pi\ ,\quad b=\log(R/r)\ ,\quad J\leqq 2\pi\log(M(R)/m(r))\ ,\quad \alpha=2\pi\ ,$$

where we choose the fixed line g as the v-axis, then we have

$$\frac{\log(R/r)}{2\pi}\leqq 2\pi\frac{\log(M(R)/m(r))}{(2\pi)^2}\ ,\qquad \frac{m(r)}{r}\leqq\frac{M(R)}{R}\ . \tag{4}$$

Since $\lim\limits_{r\to 0}\dfrac{m(r)}{r}=|\varphi'(0)|=1/|f'(0)|$, $\lim\limits_{R\to\infty}\dfrac{M(R)}{R}=1$, we have

$$|f'(0)|\geqq 1\ . \tag{5}$$

If $f(z)\not\equiv z$, then from the proof of Rengel's inequality,

$$\log\frac{R}{r}\leqq\log\frac{M(R)}{m(r)}-c\ ,\qquad c>0\ ,$$

where $c>0$ is a constant, independent of r and R. From this, we have $|f'(0)|\geqq e^c>1$, hence if $|f'(0)|=1$, then $f(z)\equiv z$.

(ii) Next let $w=f(z)$ map D conformally on a radial slits-plane Δ on the w-plane and $z=\varphi(w)$ be its inverse function and a circular ring: $0<r<|w|<R$ contain all radial slits of Δ and put

$$m(\rho)=\operatorname*{Min}_{|w|=\rho}|\varphi(w)|\ ,\qquad M(\rho)=\operatorname*{Max}_{|w|=\rho}|\varphi(w)|\ ,\qquad r\leqq\rho\leqq R\ . \tag{1}$$

Similarly as (i), let $\omega=\log w$, $\zeta=\log z$ and apply Rengel's inequality, with

$$a=\log\frac{R}{r}\ ,\qquad b=2\pi\ ,\qquad J\leqq 2\pi\log\frac{M(R)}{m(r)}\ ,\qquad \alpha=\log\frac{m(R)}{M(r)}\ ,$$

then we have

$$\frac{2\pi}{\log(R/r)}\leqq\frac{2\pi\log(M(R)/m(r))}{[\log(m(R)/M(r))]^2}\ ,\quad \left(\log\frac{m(R)}{M(r)}\right)^2\leqq\log\frac{M(R)}{m(r)}\cdot\log\frac{R}{r}\ . \tag{2}$$

If we put

$$\frac{m(R)}{M(r)}=\frac{R(1+\varepsilon)}{r|\varphi'(0)|}\ ,\qquad \frac{M(R)}{m(r)}=\frac{R(1+\eta)}{r|\varphi'(0)|}\ , \tag{3}$$

then $\lim\varepsilon=0$, $\lim\eta=0$, if $r\to 0$, $R\to\infty$.

Hence if we put (3) into (2) and make $r\to 0$, $R\to\infty$, then we have

$$\left|\frac{1}{\varphi'(0)}\right|=|f'(0)|\leqq 1\ . \tag{4}$$

If $f(z)\not\equiv z$, then from the proof of Rengel's inequality,

$$\frac{2\pi}{\log(R/r)}\leqq\frac{2\pi\log(M(R)/m(r))}{[\log(m(R)/M(r))]^2}-\frac{2\delta c}{\log(R/r)\cdot\log(m(R)/M(r))}\ , \tag{5}$$

where $\delta>0$ and $c>0$ are constants, independent of r and R.

Hence if we put (3) into (5) and make $r\to 0$, $R\to\infty$, then we have $|f'(0)|\leqq e^{-\frac{\delta c}{\pi}}<1$, hence if $|f'(0)|=1$, then $f(z)\equiv z$.

Lemma 3. *Let D be a multiply connected domain on the z-plane, which contains $z=0$ and $z=\infty$ and is bounded by k analytic Jordan curves C_i $(i=1,2,\cdots,k)$ and the area, enclosed in $\{C_i\}$ be $\geqq A>0$ and let*

$$K=\underset{z\in C_1+\cdots+C_k}{\mathrm{Max}}|z|\,.$$

Let $f(z)$ be meromorphic and schlicht in D and

$$f(0)=0,\qquad f(z)=z+a_0+\frac{a_1}{z}+\cdots\qquad at\quad z=\infty\,.$$

(i) *If $w=f(z)$ maps D conformally on a circular slits-plane Δ on the w-plane, then*

$$|f'(0)|\geqq e^{\frac{A}{2\pi K^2}}\,.$$

(ii) *If $w=f(z)$ maps D conformally on a radial slits-plane Δ on the w-plane, then*

$$|f'(0)|\leqq e^{-\frac{A}{2\pi K^2}}\,.$$

Proof. (i) Let $w=f(z)$ map D conformally on a circular slits-plane Δ on the w-plane and Γ_i be the image of C_i $(i=1,2,\cdots,k)$ on the $\zeta=\log z$-plane and J^* be the area, enclosed in $\{\Gamma_i\}$, then if we denote the inside of C_i by (C_i) and put $z=\rho e^{i\theta}$, then

$$J^*=\sum_{i=1}^{k}\iint_{(C_i)}\frac{d\rho\, d\theta}{\rho}=\sum_{i=1}^{k}\iint_{(C_i)}\frac{\rho\, d\rho\, d\theta}{\rho^2}\geqq\frac{1}{K^2}\sum_{i=1}^{k}\iint_{(C_i)}\rho\, d\rho\, d\theta\geqq\frac{A}{K^2}\,.$$

Hence with the same notations as the proof of (i) of Lemma 2, we have by Rengel's inequality,

$$\frac{\log(R/r)}{2\pi}\leqq\frac{2\pi\log(M(R)/m(r))-A/K^2}{(2\pi)^2}\,,$$

and from this, we have

$$|f'(0)|\geqq e^{\frac{A}{2\pi K^2}}\,. \tag{1}$$

(ii) Next let $w=f(z)$ map D conformally on a radial slits-plane Δ on the w-plane, then with the same notations as the proof of (ii) of Lemma 2, we have

$$\frac{2\pi}{\log(R/r)}\leqq\frac{2\pi\log(M(R)/m(r))-A/K^2}{[\log(m(R)/M(r))]^2}\,,$$

and from this, we have

$$|f'(0)|\leqq e^{-\frac{A}{2\pi K^2}} \tag{2}$$

Lemma 4. *Let D be a multiply connected domain on the z-plane, which contains $z=0$ and $z=\infty$ and $\mathfrak{F}$ be the family of $f(z)$, which satisfy the following conditions:*

(i) *$f(z)$ is meromorphic and schlicht in D,*

(ii) $f(0)=0, \quad f(z)=z+a_0+\frac{a_1}{z}+\cdots \quad$ *at* $\quad z=\infty$.

If we put

$$\bar{\mu}=\sup_{f\in\mathfrak{F}}|f'(0)|, \qquad \underline{\mu}=\inf_{f\in\mathfrak{F}}|f'(0)|,$$

then there exist $\varphi(z)$, $\psi(z)$ in $\mathfrak{F}$, such that

$$|\varphi'(0)|=\bar{\mu}, \qquad |\psi'(0)|=\underline{\mu}.$$

This can be proved similarly as Lemma 3 (p. 399).

Theorem IX. 24. *Let D be a multiply connected domain on the z-plane, which contains $z=0$ and $z=\infty$ and $\varphi(z)$, $\psi(z)$ be the functions of Lemma 4, then*

(i) *$w=\varphi(z)$ maps D conformally on a circular slits-plane Δ on the w-plane and the surface measure of the boundary of Δ is zero.*

(ii) *$w=\psi(z)$ maps D conformally on a radial slits-plane Δ on the w-plane and the surface measure of the boundary of Δ is zero.*

The first part of (i) and (ii) was proved by Rengel [2] and the second part by Tsuji [10].

Proof. (i) Suppose that $w=\varphi(z)$ maps D conformally on a domain Δ on the w-plane, which is not a circular slits-plane. Let γ be a boundary component of Δ, which is a continuum and is not a circular slit and let Δ_0 be the simply connected domain, bounded by γ. By Lemma 1, we map Δ_0 conformally on a circular slit-plane on the ζ-plane by $\zeta=\zeta(w)$ ($\zeta(0)=0$, $\zeta(w)=w+b_0+\frac{b_1}{w}+\cdots$ at $w=\infty$), then since $\zeta(w)\not\equiv w$, by Lemma 2, $|\zeta'(0)|>1$. $F(z)=\zeta(\varphi(z))$ belongs to $\mathfrak{F}$ and $|F'(0)|=|\zeta'(0)|\cdot|\varphi'(0)|>\bar{\mu}$, which contradicts the definition of $\bar{\mu}$. Hence Δ is a circular slits-plane.

Next we shall prove that the surface measure of the boundary Γ of Δ is zero. Suppose that the surface measure of Γ is $\geqq A>0$. We exhaust Δ by a sequence of domains $\Delta_1\subset\Delta_2\subset\cdots\subset\Delta_n\to\Delta$, such that the boundary of Δ_n consists of a finite number of analytic Jordan curves, surrounding Γ. Then the sum of areas, enclosed in these curves, is $\geqq A>0$.

We map Δ_n conformally on a circular slits-plane on the ζ-plane, by $\zeta=g_n(w)$ ($g_n(0)=0$, $g_n(w)=w+b_0^{(n)}+\frac{b_1^{(n)}}{w}+\cdots$ at $w=\infty$), then by Lemma 3, $|g_n'(0)|\geqq e^{\frac{A}{2\pi K_n^2}}$, where K_n is the maximum distance of the boundary of Δ_n from $w=0$. Since as before, $\{g_n(w)\}$ forms a normal family, we may assume that $\lim_n g_n(w)=g(w)$ uniformly in the wider sense in Δ. Then $|g'(0)|\geqq e^{\frac{A}{2\pi K^2}}$, where K is the maximum distance of the boundary of

Δ from $w=0$.

Since $F(z)=g(\varphi(z))$ belongs to $\mathfrak{F}$ and $|F'(0)|=|g'(0)|\cdot|\varphi'(0)|\geqq e^{\frac{A}{2\pi K^2}}\bar{\mu}>\bar{\mu}$, which is absurd. Hence the surface measure of Γ is zero.

Similarly we can prove the second part (ii).

Theorem IX. 25. *If D is of finite connectivity and $w=f(z)$ ($f(0)=0$, $f(z)=z+a_0+\frac{a_1}{z}+\cdots$ at $z=\infty$) maps D conformally on a circular, or a radial slits-plane on the w-plane, then $f(z)$ is unique.*

Proof. Suppose that there are two such $w=f_i(z)$ $(i=1,2)$, which map D conformally on a circular slits-plane Δ_i on the w-plane. Let $z=\varphi_1(w)$ be the inverse function of $w=f_1(z)$, then $\zeta=f_2(\varphi_1(w))=F(w)$ maps Δ_1 conformally on Δ_2, $F(0)=0$, $F(w)=w+c_0+\frac{1}{w}+\cdots$ at $w=\infty$. Hence by Lemma 2, $\left|\frac{d\zeta}{dw}\right|_{w=0}\geqq 1$. Similarly $\left|\frac{dw}{d\zeta}\right|_{\zeta=0}\geqq 1$, hence $\left|\frac{d\zeta}{dw}\right|_{w=0}=1$, so that $\zeta\equiv w$, or $f_1(z)\equiv f_2(z)$.

3. Conformal mapping on a circular, or a radial slits-disc

1. Let Δ be a multiply connected domain on the w-plane, contained in $|w|<1$, which contains $w=0$ and the unit circle $|w|=1$ belongs to its boundary. If every boundary component of Δ, which is a continuum, is a circular arc on a circle $|w|=\text{const.}<1$, then Δ is called *a circular slits-disc*.

If every boundary component of Δ, which is a continuum, is a segment, which is contained in $0<|w|<1$ and lies on a line $\arg w=\text{const.}$, then Δ is called a *radial slits-disc.*

Lemma 1. *Let D be a doubly connected domain on the z-plane, which contains $z=0$ and is bounded by the unit circle $|z|=1$ and a continuum γ, contained in $0<|z|<1$. Then*

(i) *By $w=f(z)$ ($f(0)=0$, $|f(z)|=1$ on $|z|=1$), we can map D conformally on a circular slit-disc Δ on the w-plane.*

(ii) *By $w=f(z)$ ($f(0)=0$, $|f(z)|=1$ on $|z|=1$), we can map D conformally on a radial slit-disc Δ on the w-plane.*

Proof. (i) Let $\bar{\gamma}$ be the inverse of γ with respect to the unit circle $|z|=1$ and D^* be the doubly connected domain, bounded by $\gamma+\bar{\gamma}$. We map D^* conformally on a circular slits-plane on the w-plane by $w=\varphi(z)$ ($\varphi(0)=0$, $\varphi(z)=z+a_0+\frac{a_1}{z}+\cdots$ at $z=\infty$). Then $F(z)=\bar{\varphi}'(0)/\bar{\varphi}(1/\bar{z})$

is meromorphic and schlicht in D^*, $F(0)=0$, $F(z)=z+b_0+\frac{b_1}{z}+\cdots$ at $z=\infty$, and $\zeta=F(z)$ maps D^* conformally on a circular slits-plane on the ζ-plane.

Let $z=\psi(w)$ be the inverse function of $w=\varphi(z)$ and put $\zeta=F(\psi(w))=G(w)$, then $G(0)=0$, $G(w)=w+c_0+\frac{c_1}{w}+\cdots$ at $w=\infty$. Hence by Lemma 2 (p. 404), $\left|\frac{d\zeta}{dw}\right|_{w=0}\geqq 1$. Similarly $\left|\frac{dw}{d\zeta}\right|_{\zeta=0}\geqq 1$, so that $\left|\frac{d\zeta}{dw}\right|_{w=0}=1$, hence $\zeta\equiv w$, $F(z)\equiv\varphi(z)$, or $\varphi(z)\varphi(1/\bar{z})=\varphi'(0)$. If $|z|=1$, then $z=1/\bar{z}$, hence $|\varphi(z)|^2=\varphi'(0)$ on $|z|=1$, so that the unit circle $|z|=1$ is mapped on a circle $|w|=\sqrt{\varphi'(0)}$, hence $w=\varphi(z)/\sqrt{\varphi'(0)}$ maps D conformally on a circular slit-disc on the w-plane.

(ii) We map D^* on a radial slits-plane on the w-plane by $w=\chi(z)$ ($\chi(0)=0$, $\chi(z)=z+a_0+\frac{a_1}{z}+\cdots$ at $z=\infty$) and we can prove similarly that $w=\chi(z)/\sqrt{\chi'(0)}$ maps D conformally on a radial slit-disc on the w-plane.

Lemma 2. *Let D be a multiply connected domain on the z-plane, contained in $|z|<1$, which contains $z=0$ and is bounded by the unit circle $|z|=1$ and a finite number of continua, contained entirely in $0<|z|<1$.*

(i) *If $w=f(z)$ ($f(0)=0$, $|f(z)|=1$ on $|z|=1$) maps D conformally on a circular slits-disc $\varDelta$ on the w-plane, then*

$$|f'(0)|\geqq 1\,,$$

where the equality holds, only when $f(z)\equiv e^{i\theta}z$.

(ii) *If $w=f(z)$ ($f(0)=0$ $|f(z)|=1$ on $|z|=1$) maps D conformally on a radial slits-disc $\varDelta$ on the w-plane, then*

$$|f'(0)|\leqq 1\,,$$

where the equality holds, only when $f(z)\equiv e^{i\theta}z$.

Proof. (i) Let $w=f(z)$ map D conformally on a circular slits-disc $\varDelta$ on the w-plane. Since $|f(z)|=1$ on $|z|=1$, by the reflexion principle, $f(z)$ is regular on $|z|=1$. Let all circular slits of $\varDelta$ be contained in $0<r<|w|<R<1$.

Let $z=\varphi(w)$ be the inverse function of $w=f(z)$ and

$$m(\rho)=\operatorname*{Min}_{|w|=\rho}|\varphi(w)|\,,\qquad M(\rho)=\operatorname*{Max}_{|w|=\rho}|\varphi(w)|\,,\qquad r\leqq\rho\leqq R\,, \tag{1}$$

then similarly as Lemma 2 (p. 404), we have $\frac{m(r)}{r}\leqq\frac{M(R)}{R}$.

Since $\lim_{r\to 0}\frac{m(r)}{r}=|\varphi'(0)|=1/|f'(0)|$, $\lim_{R\to 1}M(R)=1$, we have

$$|f'(0)| \geqq 1 . \tag{2}$$

If $f(z) \not\equiv e^{i\theta}z$, then we can prove similarly as Lemma 2 (p. 404), that $|f'(0)|>1$, so that if $|f'(0)|=1$, then $f(z)\equiv e^{i\theta}z$.

(ii) Let $w=f(z)$ map D conformally on a radial slits-disc $\varDelta$ on the w-plane and let all radial slits of $\varDelta$ be contained in $0<r<|w|<R<1$. Then similarly as Lemma 2 (p. 404), we have

$$\left(\log \frac{m(R)}{M(r)}\right)^2 \leqq \log \frac{M(R)}{m(r)} \cdot \log \frac{R}{r} .$$

Since $\lim\limits_{r\to 0} \dfrac{m(r)}{r} = \lim\limits_{r\to 0} \dfrac{M(r)}{r} = |\varphi'(0)| = 1/|f'(0)|$, $\lim\limits_{R\to 1} m(R) = \lim\limits_{R\to 1} M(R)=1$, we have

$$|f'(0)| \leqq 1 , \tag{3}$$

and we can prove that if $|f'(0)|=1$, then $f(z)\equiv e^{i\theta}z$.——

Similarly as Lemma 3 (p. 406), we can prove

Lemma 3. *Let D be a multiply connected domain on the z-plane, contained in $|z|<1$, which contains $z=0$ and is bounded by the unit circle $|z|=1$ and a finite number k of analytic Jordan curves C_i $(i=1, 2, \cdots, k)$, contained in $0<|z|<1$ and the area, enclosed in $\{C_i\}$, be $\geqq A>0$.*

(i) *If $w=f(z)$ $(f(0)=0, |f(z)|=1$ on $|z|=1)$ maps D conformally on a circular slits-disc $\varDelta$ on the w-plane, then*

$$|f'(0)| \geqq e^{\frac{A}{2\pi}} .$$

(ii) *If $w=f(z)$ $(f(0)=0, |f(z)|=1$ on $|z|=1)$ maps D conformally on a radial slits-disc $\varDelta$ on the w-plane, then*

$$|f'(0)| \leqq e^{-\frac{A}{2\pi}} .$$

2. Let D be a multiply connected domain on the z-plane, which contains $z=0$ and whose boundary contains a free boundary component γ[1], then we can map the outside of γ conformally on $|w|<1$ by $w=f(z)$, such that $f(0)=0$, $|f(z)|=1$ on γ. Hence the family $\mathfrak{F}$ of $f(z)$, which satisfy the following conditions is not empty.

(i) $f(z)$ is regular, schlicht and $|f(z)|<1$ in D,

(ii) $f(0)=0$,

(iii) $|f(z)|=1$ on γ.

Let

1) i. e., γ is a continuum, such that its suitable neighbourhood U does not contain other boundary components of D. This is always the case, if D is of finite connectivity.

$$\bar{\mu}=\sup_{f\in\mathfrak{F}}|f'(0)|\,,\qquad \underline{\mu}=\inf_{f\in\mathfrak{F}}|f'(0)|\,,\tag{1}$$

then since $\mathfrak{F}$ is a normal family, there exist $\varphi(z)$, $\psi(z)$ in $\mathfrak{F}$, such that

$$|\varphi'(0)|=\bar{\mu}\,,\qquad |\psi'(0)|=\underline{\mu}\,.\tag{2}$$

Theorem IX. 26. *Let D be a multiply connected domain on the z-plane, which contains $z=0$ and whose boundary contains a free boundary component γ and $\varphi(z)$, $\psi(z)$ be defined by* (2), *then*

(i) *$w=\varphi(z)$ maps D conformally on a circular slits-disc Δ on the w-plane, such that γ is mapped on $|w|=1$ and the surface measure of the boundary of Δ is zero.*

(ii) *$w=\psi(z)$ maps D conformally on a radial slits-disc Δ on the w-plane, such that γ is mapped on $|w|=1$ and the surface measure of the boundary of Δ is zero.*

Proof. (i) Let $w=\varphi(z)$ map D conformally on a domain Δ on the w-plane, contained in $|w|<1$. If Δ is not a circular slits-disc, then the boundary Γ of Δ contains a continuum λ, which lies in $0<|w|<1$ and is not a circular slit. Let Δ_0 be the doubly connected domain, bounded by λ and the unit circle $|w|=1$. Then by Lemma 1, we can map Δ_0 conformally on a circular slit-disc on the ζ-plane by $\zeta=\zeta(w)$ ($\zeta(0)=0$, $|\zeta(w)|=1$ on $|w|=1$), then by Lemma 2, $|\zeta'(0)|>1$. Now $F(z)=\zeta(\varphi(z))$ belongs to $\mathfrak{F}$ and $|F'(0)|=|\zeta'(0)|\bar{\mu}>\bar{\mu}$, which is absurd. Hence Δ is a circular slits-disc. Next we shall prove that the surface measure of the boundary Γ of Δ is zero. Suppose that the surface measure of Γ is $\geqq A>0$. We enclose Γ in a finite number of analytic Jordan curves, then the area, enclosed in these curves is $\geqq A>0$. Let Δ_n be the domain, bounded by the unit circle $|w|=1$ and these curves and we map Δ_n conformally on a circular slits-disc on the ζ-plane by $\zeta=g_n(w)$ ($g_n(0)=0$, $|g_n(w)|=1$ on $|w|=1$), then by Lemma 3, $|g_n'(0)|\geqq e^{\frac{A}{2\pi}}$. If we choose Δ_n, such that $\Delta_n\to\Delta$, then since $\{g_n(w)\}$ forms a normal family, we may assume that $\lim_n g_n(w)=g(w)$ uniformly in the wider sense in Δ, then $g(w)$ is regular and schlicht in Δ, $g(0)=0$, $|g'(0)|\geqq e^{\frac{A}{2\pi}}>1$, $|g(w)|=1$ on $|w|=1$. Hence $F(z)=g(\varphi(z))$ belongs to $\mathfrak{F}$ and $|F'(0)|=|g'(0)|\bar{\mu}>\bar{\mu}$, which is absurd. Hence the surface measure of Γ is zero. Hence the part (i) is proved. We can prove the part (ii) similarly.

3. In Theorem IX. 26, we assumed that there is a free boundary component of D. Next we consider the general case, where we do not assume the existence of a free boundary component, but we assume that there exists a boundary component γ of D, which is a continuum

and D contains $z=0$, then we can map the outside of γ conformally on $|w|<1$ by $w=f(z)$ ($f(0)=0$), hence the family $\mathfrak{F}$ of $f(z)$, which satisfy the following conditions contains a function $\neq 0$.

(i) $f(z)$ is regular, schlicht and $|f(z)|<1$ in D,

(ii) $f(0)=0$.

Let

$$\bar{\mu}=\sup_{f\in\mathfrak{F}}|f'(0)|\,, \tag{1}$$

then there exists $\varphi(z)$ in $\mathfrak{F}$, such that

$$|\varphi'(0)|=\bar{\mu}\,. \tag{2}$$

Since $\mathfrak{F}$ contains a function $\neq 0$, $\bar{\mu}>0$, so that $\varphi(z)\neq 0$.

Theorem IX. 27. *$w=\varphi(z)$ maps D conformally on a circular slits-disc Δ on the w-plane and the surface measure of the boundary of Δ is zero.*

Proof. Let $w=\varphi(z)$ map D conformally on a domain Δ on the w-plane, contained in $|w|<1$. First we shall prove that the unit circle $|w|=1$ is a boundary component of Δ. For, if otherwise, there exists a point α ($|\alpha|<1$) on the boundary Γ of Δ, such that there exists no Jordan curve in Δ, surrounding α. If we put

$$w=\varphi(z)\,,\quad w_1=\frac{w-\alpha}{1-\bar{\alpha}w}\,,\qquad w_2=\sqrt{w_1}\,,\qquad w_3=\frac{w_2-\sqrt{-\alpha}}{1-\overline{\sqrt{-\alpha}}\,w_2}=F(z)\,,$$

then $F(z)\in\mathfrak{F}$ and $|F'(0)|=\dfrac{1+\alpha}{2\sqrt{|\alpha|}}\bar{\mu}>\bar{\mu}$, which is absurd. Hence $|w|=1$ is a boundary component of Δ, so that there exists no continuum $\in\Gamma$, which connects a point α ($|\alpha|<1$) to the unit circle $|w|=1$. Hence every boundary component of Γ, which is not the unit circle $|w|=1$, lies entirely in $|w|<1$. From this, we can prove similarly as Theorem IX. 26, that Δ is a circular slits-disc. Next we shall prove that the surface measure of the boundary Γ of Δ is zero.

For, suppose that the surface measure of Γ is $>A>0$. Let Γ_ρ be the part of Γ, which lies in $|w|\leqq\rho<1$, then if $1-\rho$ is small, the surface measure of Γ_ρ is $>A>0$. We enclose Γ_ρ in a finite number of analytic Jordan curves, then the area, enclosed in these curves is $>A>0$. Let Δ_n be the domain, bounded by the unit circle $|w|=1$ and these curves and we choose Δ_n, such that $\Delta_n\to\Delta$ and from this, we can prove similarly as Theorem IX. 26, that the surface measure of Γ is zero. ——

By means of Lemma 2, we can prove easily

Theorem IX. 28. *Let D be a multiply connected domain on the* *z-plane, which contains $z=0$ and is bounded by a finite number of continua γ_i $(i=1, 2, \cdots, k)$.*

If $w=f(z)$ $(f(0)=0,\ f'(0)>0)$ maps D conformally on a circular (or a radial) slits-disc on the w-plane, such that γ_1 is mapped on $|w|=1$, then $f(z)$ is unique.

4. Conformal mapping on a circular, or a radial slits-ring

1. Let $\varDelta$ be a multiply connected domain on the w-plane, contained in $0<\rho<|w|<1$ and the circles $|w|=\rho$ and $|w|=1$ belong to the boundary of $\varDelta$ and every boundary component of $\varDelta$, which is different from $|w|=\rho$ and $|w|=1$, is a circular arc on a circle $|w|=\text{const.}=c$ $(\rho<c<1)$, then $\varDelta$ is called *a circular slits-ring.*

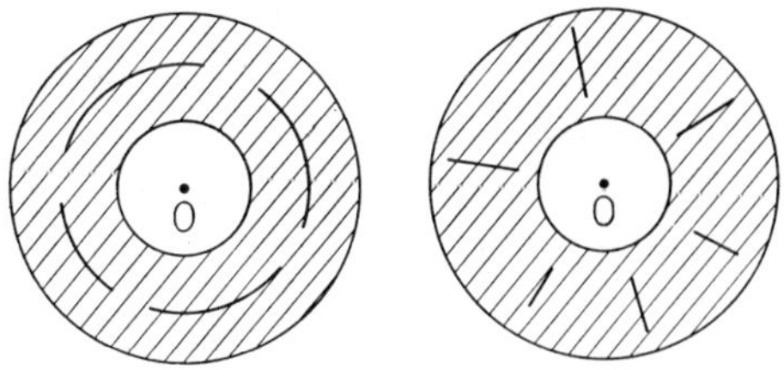

If every boundary component of $\varDelta$, which is different from $|w|=\rho$ and $|w|=1$, is a segment, which is contained in $\rho<|w|<1$ and lies on a line $\arg w=\text{const.}$, then $\varDelta$ is called *a radial slits-ring.*

Lemma 1. *Let D be a triply connected domain on the z-plane, which is bounded by three continua γ_i $(i=1, 2, 3)$. Then*

(i) *D can be mapped conformally on a circular slit-ring on the w-plane, such that γ_1 is mapped on $|w|=1$, γ_2 on $|w|=\rho$ $(0<\rho<1)$ and γ_3 on a circular slit in $\rho<|w|<1$.*

(ii) *D can be mapped conformally on a radial slit-ring on the w-plane, such that γ_1 is mapped on $|w|=1$, γ_2 on $|w|=\rho$ $(0<\rho<1)$ and γ_3 on a radial slit in $\rho<|w|<1$.*

Proof. (i) We may assume that γ_i $(i=1, 2, 3)$ is an analytic Jordan curve[1] and D is an infinite domain, bounded by γ_1, γ_2, γ_3.

Let $u_i(z)$ $(i=1, 2, 3)$ be the harmonic measure of γ_i with respect to D and put $u(z)=k_2 u_2(z)+k_3 u_3(z)$, where k_2 and k_3 are constants and we shall prove that there exist $k_2<k_3<0$, which satisfy the following condition:

1) We map the outside of γ_1 conformally on $|u|<1$ and let γ_2', γ_3' be the image of γ_2, γ_3 respectively. Next we map the outside of γ_2' conformally on $|v|<1$ and let γ_3'' be the image of γ_3'. Next we map the outside of γ_3'' conformally on $|w|<1$, then D is mapped conformally on a domain on the w-plane, which is bounded by three analytic Jordan curves.

$$\left.\begin{aligned}
k_2\int_{\gamma_1}\frac{\partial u_2}{\partial\nu}ds+k_3\int_{\gamma_1}\frac{\partial u_3}{\partial\nu}ds&=-2\pi\ ,\\
k_2\int_{\gamma_2}\frac{\partial u_2}{\partial\nu}ds+k_3\int_{\gamma_2}\frac{\partial u_3}{\partial\nu}ds&=\ \ 2\pi\ ,\\
k_2\int_{\gamma_3}\frac{\partial u_2}{\partial\nu}ds+k_3\int_{\gamma_3}\frac{\partial u_3}{\partial\nu}ds&=\ \ 0\ ,
\end{aligned}\right\}\tag{1}$$

where ν is the inner normal with respect to D.

If we put

$$a_{ij}=\int_{\gamma_i}\frac{\partial u_j}{\partial\nu}ds\qquad(i,j=1,2,3)\,,\tag{2}$$

then (1) becomes

$$a_{12}k_2+a_{13}k_3=-2\pi\ ,\quad a_{22}k_2+a_{23}k_3=2\pi\ ,\quad a_{32}k_2+a_{33}k_3=0\ .\tag{3}$$

Since $u_i(z)$ is harmonic in D,

$$a_{1i}+a_{2i}+a_{3i}=0\qquad(i=1,2,3)\,,\tag{4}$$

and since $u_1(z)+u_2(z)+u_3(z)\equiv1$,

$$a_{i1}+a_{i2}+a_{i3}=0\qquad(i=1,2,3)\,.\tag{5}$$

In virtue of (4), if we add three equations of (3), then we have $0=0$, hence the solution of

$$a_{12}k_2+a_{13}k_3=-2\pi\ ,\qquad a_{22}k_2+a_{23}k_3=2\pi\tag{6}$$

is a solution of (3).

Since $a_{12}>0$, $a_{13}>0$, $a_{22}<0$, $a_{23}>0$, we have $\begin{vmatrix}a_{12}&a_{13}\\a_{22}&a_{23}\end{vmatrix}>0$, so that (6) is solvable, hence there exist k_2, k_3, which satisfy (1). If $k_2\geqq0$, then since $a_{32}>0$, $a_{33}<0$, we have from the last equation of (3), $k_3\geqq0$, so that since $a_{12}>0$, $a_{13}>0$, we have $a_{12}k_2+a_{13}k_3\geqq0$, which contradicts the first equation of (3). Hence $k_2<0$, $k_3<0$.

Since $a_{31}+a_{32}+a_{33}=0$, we have from the last equation of (3), $a_{32}k_2-(a_{31}+a_{32})k_3=0$, or $a_{32}(k_2-k_3)=a_{31}k_3$. Since $a_{32}>0$, $a_{31}>0$, $k_3<0$, we have $k_2<k_3$. Hence

$$k_2<k_3<0\ .\tag{7}$$

Now $u(z)=k_2u_2(z)+k_3u_3(z)$ is harmonic in D, $u=0$ on γ_1, $u=k_2$ on γ_2, $u=k_3$ on γ_3 and by (1),

$$\int_{\gamma_1}\frac{\partial u}{\partial\nu}ds=-2\pi\ ,\qquad\int_{\gamma_2}\frac{\partial u}{\partial\nu}ds=2\pi\ ,\qquad\int_{\gamma_3}\frac{\partial u}{\partial\nu}ds=0\ .\tag{8}$$

Hence if $v(z)$ be the conjugate harmonic function of u, then we see easily that $w=e^{u+iv}$ maps D conformally on a circular slit-ring on the w-plane, such that γ_1 is mapped on $|w|=1$, γ_2 on $|w|=e^{k_2}$ and γ_3 on a circular slit, which lies on $|w|=e^{k_3}$.

(ii) By (i), we may assume that D is a circular slit-ring, which is bounded by $|z|=\rho$, $|z|=1$ $(0<\rho<1)$ and a circular slit L: $z=\rho_0 e^{i\theta}$ $(\rho<\rho_0<1)$ $(-\alpha\leqq\theta\leqq\alpha)$. Let D^+ be the part of D, which lies in the upper half of the z-plane, then D^+ is simply connected. We map D^+ conformally on a rectangle R on the ζ-plane, whose vertices are $\zeta=0$, πi, $-\tau+\pi i$, $-\tau$ $(\tau>0)$, such that $z=1$, -1, $-\rho$, ρ correspond to $\zeta=0$, πi, $-\tau+\pi i$, $-\tau$ respectively[1)], then by $w=e^{\zeta}$, D is mapped conformally on a radial slit-ring on the w-plane. ——

Lemma 2. *Let D be a multiply connected domain on the z-plane, which lies in $|z|<1$ and is bounded by $|z|=\rho$, $|z|=1$ $(0<\rho<1)$ and a finite number of continua, contained in $\rho<|z|<1$.*

(i) *If $w=f(z)$ maps D conformally on a circular slits-ring $\varDelta$ on the w-plane, such that $|z|=\rho$, $|z|=1$ are mapped on $|w|=\rho^*$, $|w|=1$ $(0<\rho^*<1)$ respectively, then*

$$\rho^*\geqq\rho\,,$$

where the equality holds, only when $f(z)\equiv e^{i\theta}z$.

(ii) *If $w=f(z)$ maps D conformally on a radial slits-ring $\varDelta$ on the w-plane, such that $|z|=\rho$, $|z|=1$ are mapped on $|w|=\rho^*$, $|w|=1$ $(0<\rho^*<1)$ respectively, then*

$$\rho^*\leqq\rho\,,$$

where the equality holds, only when $f(z)\equiv e^{i\theta}z$.

Proof. (i) By $\zeta=\log z$, $\omega=\log w$, we map D on D^* and $\varDelta$ on $\varDelta^*$ respectively and apply Rengel's inequality, then we have

$$\frac{\log 1/\rho^*}{2\pi}\leqq\frac{2\pi\log 1/\rho}{(2\pi)^2}\,,\qquad\text{or}\quad \rho\leqq\rho^*\,,$$

and the equality holds, only when $f(z)\equiv e^{i\theta}z$.

Similarly we can prove the part (ii). ——

Lemma 3. *Let D be a multiply connected domain on the z-plane, which lies in $|z|<1$ and is bounded by $|z|=\rho$, $|z|=1$ $(0<\rho<1)$ and a finite number of analytic Jordan curves $\{C_i\}$, contained in $\rho<|z|<1$ and the area, enclosed in $\{C_i\}$ be $\geqq A>0$.*

(i) *If $w=f(z)$ maps D conformally on a circular slits-ring $\varDelta$ on the w-plane, contained in $\rho^*<|w|<1$, such that $|z|=\rho$, $|z|=1$ are mapped on $|w|=\rho^*$, $|w|=1$ respectively, then*

1) We map D^+ conformally on the upper half of the v-plane, such that by taking $0<k<1$ suitably, $z=-1$, $-\rho$, ρ, 1 correspond to $v=-\dfrac{1}{k}$, -1, 1, $\dfrac{1}{k}$ respectively and let $v^*=\displaystyle\int_0^v\frac{dv}{\sqrt{(1-v^2)(1-k^2v^2)}}$, then $\zeta=Av^*+B$ with suitable A, B.

$$\rho^* \geqq \rho e^{\frac{A}{2\pi}} .$$

(ii) *If* $w=f(z)$ *maps* D *conformally on a radial slits-ring* $\varDelta$ *on the* w*-plane, contained in* $\rho^*<|w|<1$, *such that* $|z|=\rho$, $|z|=1$ *are mapped on* $|w|=\rho^*$, $|w|=1$ *respectively, then*

$$\rho^* \leqq \rho e^{-\frac{A}{2\pi}} .$$

Proof. By $\zeta=\log z$, $\omega=\log w$, we map D on D^* and $\varDelta$ on $\varDelta^*$ respectively and apply Rengel's inequality, then in case (i), we have

$$\frac{\log 1/\rho^*}{2\pi} \leqq \frac{2\pi \log 1/\rho - A}{(2\pi)^2}$$

so that

$$\rho^* \geqq \rho e^{\frac{A}{2\pi}} . \tag{1}$$

In case (ii), we have

$$\frac{2\pi}{\log 1/\rho^*} \leqq \frac{2\pi \log 1/\rho - A}{(\log 1/\rho)^2} ,$$

so that since $\rho^* \leqq \rho$,

$$\log 1/\rho \leqq \log 1/\rho^* - \frac{A}{2\pi} \frac{\log 1/\rho^*}{\log 1/\rho} \leqq \log 1/\rho^* - \frac{A}{2\pi} ,$$

$$\rho^* \leqq \rho e^{-\frac{A}{2\pi}} . \tag{2}$$

2. Let D be a multiply connected domain on the z-plane, whose boundary contains two free boundary components γ_1, γ_2, each of which is a continuum and is isolated from other boundary components[1]. By mapping the doubly connected domain, bounded by γ_1, γ_2, conformally on a circular ring, we may assume that γ_1 is a circle $|z|=1$ and γ_2 is a circle $|z|=\rho$ $(0<\rho<1)$ and D is contained in $\rho<|z|<1$ and $|z|=\rho$ and $|z|=1$ are isolated from other boundary components.

Let $\mathfrak{F}$ be the family of $f(z)$, which satisfy the following conditions:

(i) $f(z)$ is regular and schlicht in D,

(ii) $|f(z)|=1$ on $|z|=1$, $|f(z)|=\text{const.}=r(f)$ on $|z|=\rho$, and

$$r(f)<|f(z)|<1 \qquad \text{in } D .$$

Let

$$\bar{r}=\sup_{f\in\mathfrak{F}} r(f) , \qquad \underline{r}=\inf_{f\in\mathfrak{F}} r(f) . \tag{1}$$

Since by the principle of reflexion, $f(z)$ is regular on $|z|=\rho$ and $|z|=1$, there exist $\varphi(z)$, $\psi(z)$ in $\mathfrak{F}$, such that

$$r(\varphi)=\bar{r} , \qquad r(\psi)=\underline{r} . \tag{2}$$

1) This is always the case, if D is of finite connectivity.

Theorem IX. 29. *Let D be a multiply connected domain on the z-plane, which lies in $\rho<|z|<1$ and $|z|=\rho$, $|z|=1$ are isolated from other boundary components and $\varphi(z)$, $\psi(z)$ be defined by* (2). *Then*

(i) *$w=\varphi(z)$ maps D conformally on a circular slits-ring Δ on the w-plane, contained in $\bar{r}<|w|<1$, such that $|z|=\rho$, $|z|=1$ are mapped on $|w|=\bar{r}$, $|w|=1$ respectively and the surface measure of the boundary of Δ is zero.*

(ii) *$w=\psi(z)$ maps D conformally on a radial slits-ring Δ on the w-plane, contained in $\underline{r}<|w|<1$, such that $|z|=\rho$, $|z|=1$ are mapped on $|w|=\underline{r}$, $|w|=1$ respectively and the surface measure of the boundary of Δ is zero.*

Proof. (i) Let $w=\varphi(z)$ map D conformally on a domain Δ on the w-plane, contained in $\bar{r}<|w|<1$. Suppose that Δ is not a circular slits-ring, then there exists a boundary component γ of Δ, which is not a circular slit. Let Δ_0 be the triply connected domain, bounded by $|w|=\bar{r}$, $|w|=1$ and γ. Then by Lemma 1, we can map Δ_0 conformally on a circular slit-ring on the ζ-plane by $\zeta=g(w)$, such that $|w|=\bar{r}$, $|w|=1$ are mapped on $|\zeta|=r^*$, $|\zeta|=1$ respectively and γ is mapped on a circular slit. Then by Lemma 2, $r^*>\bar{r}$. Since $g(\varphi(z))\in\mathfrak{F}$, this is absurd. Hence Δ is a circular slits-ring. By Lemma 3, we can prove as before that the surface measure of the boundary of Δ is zero. Hence the part (i) is proved. The part (ii) can be proved similarly.——

By means of Lemma 2, we can prove easily

Theorem IX. 30. *Let D be a multiply connected domain on the z-plane, which is bounded by a finite number of continua γ_i $(i=1, 2, \cdots, n)$.*

If $w=f(z)$ maps D conformally on a circular (or a radial) slits-ring Δ on the w-plane, which is contained in $\rho<|w|<1$, such that γ_1 is mapped on $|w|=1$ and γ_2 on $|w|=\rho$, then $w=f(z)$ is unique, except a rotation about $w=0$.

5. Conformal mapping on an n-sheeted circular disc

First we shall prove a lemma.

Lemma. *Let a system of n homogeneous linear equations be given:*

$$\left.\begin{array}{l} a_{11}x_1+a_{12}x_2+\cdots+a_{1n}x_n=0\,, \\ \cdots\cdots\cdots\cdots\cdots\cdots\cdots\cdots \\ a_{n1}x_1+a_{n2}x_2+\cdots+a_{nn}x_n=0\,, \end{array}\right\} \qquad \text{(i)}$$

where

$$\left.\begin{array}{l} a_{ii}<0\,, \quad a_{ij}>0 \quad (i\neq j) \quad (i,j=1,2,\cdots,n)\,, \\ a_{11}+a_{21}+\cdots+a_{n1}=0\,, \quad \cdots, \quad a_{1n}+a_{2n}+\cdots+a_{nn}=0\,. \end{array}\right\} \qquad \text{(ii)}$$

Then (i) *has a solution* $x_i^0>0$ $(i=1,2,\cdots,n)$ *and*

$$\Delta_{n-1}=\begin{vmatrix} a_{11} & a_{12} & \cdots & a_{1n-1} \\ \cdots & \cdots & \cdots & \cdots \\ a_{n-1,1} & a_{n-1,2} & \cdots & a_{n-1,n-1} \end{vmatrix} \neq 0 .$$

Proof. By the condition (ii), the determinant of the coefficient matrix of (i) is zero, so that (i) has a solution $(x_1^0,\cdots,x_n^0)\neq(0,\cdots,0)$. We may assume that one of $\{x_i^0\}$ is >0 and suppose that

$$x_1^0>0\,,\quad \cdots,\quad x_k^0>0\,,\quad x_{k+1}^0\leqq 0\,,\quad \cdots,\quad x_n^0\leqq 0 \quad (1\leqq k<n)\,. \tag{1}$$

Consider the last $(n-k)$ equations of (i):

$$\left.\begin{aligned} &a_{k+1,1}x_1^0+\cdots+a_{k+1,k}x_k^0+a_{k+1,k+1}x_{k+1}^0+\cdots+a_{k+1,n}x_n^0=0\,,\\ &\cdots\cdots\cdots\cdots\cdots\cdots\cdots\cdots\cdots\cdots\cdots\\ &a_{n,1}x_1^0+\cdots\cdots+a_{n,k}x_k^0+\quad a_{n,k+1}x_{k+1}^0+\cdots+a_{n,n}\quad x_n^0=0\,. \end{aligned}\right\} \tag{2}$$

If we add these equations, then in virtue of (ii), we have

$$\begin{aligned} &(a_{k+1,1}+\cdots+a_{n,1})x_1^0+\cdots+(a_{k+1,k}+\cdots+a_{n,k})x_k^0\\ &\quad -(a_{1,k+1}+\cdots+a_{k,k+1})x_{k+1}^0-\cdots-(a_{1,n}+\cdots+a_{k,n})x_n^0=0\,, \end{aligned}$$

so that by (1),

$$(a_{k+1,1}+\cdots+a_{n,1})x_1^0+\cdots+(a_{k+1,k}+\cdots+a_{n,k})x_k^0\leqq 0\,,$$

which is absurd. Hence

$$x_1^0>0\,,\ \cdots\cdots\cdots,\ x_n^0>0\,. \tag{3}$$

Suppose that $\Delta_{n-1}=0$, then

$$\left.\begin{aligned} &a_{11}x_1+\cdots\cdots+a_{1,n-1}x_{n-1}=0\,,\\ &\cdots\cdots\cdots\cdots\cdots\cdots\\ &a_{n-1,1}x_1+\cdots+a_{n-1,n-1}x_{n-1}=0 \end{aligned}\right\} \tag{4}$$

has a solution $(x^0,\cdots,x_{n-1}^0)\neq(0,\cdots,0)$. Similarly as above, we can prove that $x_1^0>0,\cdots,x_{n-1}^0>0$, so that (i) has a solution $(x_1^0,\cdots,x_{n-1}^0,0)$, which is absurd. Hence $\Delta_{n-1}\neq 0$.

Theorem IX. 31. *Let D be a multiply connected domain on the z-plane, which is bounded by n continua Γ_i $(i=1,2,\cdots,n)$. Then D can be mapped conformally on an n-sheeted circular disc $|w|<1$ over the w-plane, such that n prescribed points $\zeta_j^0\in\Gamma_j$ $(j=1,2,\cdots,n)$ are mapped on points, whose projection on the w-plane is $w=1$ and under this condition, the mapping function is unique.*

Bieberbach [1], Grunsky [1].

Proof. (Tsuji [57]). As the proof of Lemma 1 (p. 413), we may assume that Γ_i is an analytic Jordan curve and D is an infinite domain, bounded by $\{\Gamma_i\}$. Let $g(z,a)$ be the Green's function of D and put

$$u_i(z)=\int_{\Gamma_i}\frac{\partial g(\zeta, z)}{\partial \nu}|d\zeta| \qquad (i=1, 2, \cdots, n)\ , \tag{1}$$

where ν is the inner normal of Γ_i at $\zeta\in\Gamma_i$ with respect to D, then $\frac{u_i(z)}{2\pi}$ is the harmonic measure of Γ_i with respect to D, such that if we denote any point of Γ_i by ζ_i,

$$u_j(\zeta_j)=2\pi\ , \qquad u_j(\zeta_i)=0 \qquad (i\neq j)\ ,$$

hence

$$\left.\begin{array}{l} u_1(\zeta_1)+\cdots+u_1(\zeta_n)=2\pi\ , \\ \cdots\cdots\cdots\cdots\cdots\cdots\cdots \\ u_n(\zeta_1)+\cdots+u_n(\zeta_n)=2\pi \end{array}\right\} \quad (\zeta_j\in\Gamma_j). \tag{2}$$

Since $u_1(z)+\cdot+u_n(z)\equiv 2\pi$,

$$\left.\begin{array}{l} u_1(z_1)+\cdots+u_n(z_1)=2\pi\ , \\ \cdots\cdots\cdots\cdots\cdots\cdots\cdots \\ u_1(z_n)+\cdots+u_n(z_n)=2\pi\ . \end{array}\right\} \quad (z_j\in D)\ (j=1, 2, \cdots, n). \tag{3}$$

We shall prove that there exists $z_j^0\in D$, which lies in a small neighbourhood of $\zeta_j\in\Gamma_j$, such that

$$\left.\begin{array}{l} u_1(z_1^0)+\cdots+u_1(z_n^0)=2\pi\ , \\ \cdots\cdots\cdots\cdots\cdots\cdots\cdots \\ u_n(z_1^0)+\cdots+u_n(z_n^0)=2\pi\ . \end{array}\right\} \quad (z_j^0\in D)\ . \tag{4}$$

We map the outside of Γ_j conformally on the upper half $y_j>0$ of the $t_j=x_j+iy_j$-plane, such that ζ_j becomes $t_j=0$ and Γ_j becomes the real axis and put for $z_j\in D$, which lies in a small neighbourhood of ζ_j,

$$u_j(z_j)-2\pi=v_j(t_j)=v_j(x_j, y_j)\ , \quad u_i(z_j)=v_i(t_j)=v_i(x_j, y_j) \quad (i\neq j)\ , \tag{5}$$

then $v_i(t_j)$ $(i=1, 2, \cdots, n)$ is harmonic in $|t_j|<\rho$ and $v_i(0, 0)=0$.

Since for $y_j>0$, $v_j(0, y_j)<0$, $v_i(0, y_j)>0$ $(i\neq j)$, if we put

$$\left(\frac{\partial v_i(0, y_j)}{\partial y_j}\right)_{y_j=0}=a_{ij}\ , \tag{6}$$

then by the foot note of p. 15, $a_{ii}<0$, $a_{ij}>0$ $(i\neq j)$ and by (3), we have

$$a_{11}+a_{21}+\cdots+a_{n1}=0\ , \quad \cdots, \quad a_{1n}+a_{2n}+\cdots+a_{nn}=0\ , \tag{7}$$

hence a_{ij} satisfy the condition of the lemma, so that $\Delta_{n-1}\neq 0$.

Now in (4), we assume that z_j^0 is the image of $t_j=(0, y_j)$ $(y_j>0)$, then (4) becomes

$$\left.\begin{array}{l} v_1(0, y_1)+\cdots+v_1(0, y_n)=0\ , \\ \cdots\cdots\cdots\cdots\cdots\cdots\cdots\cdots \\ v_n(0, y_1)+\cdots+v_n(0, y_n)=0\ . \end{array}\right\} \tag{8}$$

(8) is satisfied by $y_1=y_2=\cdots=y_n=0$.

Since

$$\left(\frac{\partial(v_1, \cdots, v_{n-1})}{\partial(y_1, \cdots, y_{n-1})}\right)_{y_1=\cdots=y_{n-1}=0}=\Delta_{n-1}\neq 0\ ,$$

the first $(n-1)$ equations of (8) and in virtue of (3), n equations of (8) can be solved in the form:

$$y_1=\varphi_1(y_n)\,, \qquad \cdots, \qquad y_{n-1}=\varphi_{n-1}(y_n)\,, \tag{9}$$

where $\varphi_j(y_n)$ is analytic and vanishes at $y_n=0$. If we put $\varphi_j'(0)=A_j$, then by (8), $A_1, \cdots, A_{n-1}$, 1 satisfy the equations (i) of the lemma and since $x_n^0=1>0$, we have $A_j>0$ $(j=1, 2, \cdots, n-1)$, hence $\varphi_j(y_n)>0$, $(j=1, 2, \cdots, n-1)$, if $y_n>0$ is small, so that (8) has a solution $y_j>0$ $(j=1, 2, \cdots, n)$, hence (4) has a solution $z_j^0\in D$.

Let $h(z, z_j^0)$ be the conjugate harmonic function of $g(z, z_j^0)$ and put

$$u(z)=\sum_{j=1}^{n} g(z, z_j^0)\,, \qquad v(z)=\sum_{j=1}^{n} h(z, z_j^0)\,, \tag{10}$$

then

$$\int_{\Gamma_i} dv=-\sum_{j=1}^{n}\int_{\Gamma_i}\frac{\partial g(\zeta, z_j^0)}{\partial\nu}|d\zeta|=-(u_i(z_1^0)+\cdots+u_i(z_n^0))=-2\pi \ \ (i=1, 2, \cdots, n),$$

so that $w=e^{-(u+iv)}=f(z)$ is one-valued and regular in D and by the argument principle, we can prove easily that $w=f(z)$ maps D conformally on an n-sheeted circular disc $|w|<1$ over the w-plane.

Let $\zeta_j^0\in\Gamma_j$ $(j=1, 2, \cdots, n)$ be n prescribed points on the boundary of D and let $f(\zeta_j^0)=e^{i\theta_j}$. If we consider $e^{-i\theta_1}f(z)$ instead of $f(z)$, we may assume that $\theta_1=0$, so that $f(\zeta_1^0)=1$. Let $\mathfrak{F}$ be the family of $f(z)$, which maps D conformally on an n-sheeted circular disc $|w|<1$ and $f(\zeta_1^0)=1$ and put

$$\alpha=\inf_{f\in\mathfrak{F}}(|\theta_2|+\cdots+|\theta_2|)\,, \tag{11}$$

and we shall prove that $\alpha=0$.

Since $\mathfrak{F}$ is a normal family, there exists $f_0(z)\in\mathfrak{F}$, such that $|\theta_2^0|+\cdots+|\theta_n^0|=\alpha$. Suppose that $\alpha>0$ and we choose a point ξ on $|w|=1$, which is different from 1, $e^{i\theta_2^0}, \cdots, e^{i\theta_n^0}$.

Let $w_0=f_0(z)$ and we define $w=f(z)$ by

$$\frac{w-1}{w-\xi}=\lambda\frac{w_0-1}{w_0-\xi} \qquad (0<\lambda<1)\,,$$

then $f(z)\in\mathfrak{F}$ and if λ is small, $|\theta_2|+\cdots+|\theta_n|<|\theta_2^0|+\cdots+|\theta_n^0|=\alpha$, which is absurd, hence $\alpha=0$, so that $f_0(\zeta_j^0)=1$ $(j=1, 2, \cdots, n)$. Hence $w=f_0(z)$ maps D conformally on an n-sheeted circular disc $|w|<1$ over the w-plane, such that n prescribed points ζ_j^0 are mapped on points, whose projection on the w-plane is $w=1$. Next we shall prove that such a mapping is unique. Suppose that there are two such mappings $f_1(z)$, $f_2(z)$, then $F(z)=f_2(f_1^{-1}(z))$ maps D conformally on itself, such that

$$F(\zeta_j^0)=\zeta_j^0 \ \ (j=1, 2, \cdots, n)\,.$$

By Theorem IX. 29, we may assume that D is a circular slits-ring, contained in $\rho<|z|<1$ and Γ_1 is $|z|=1$ and Γ_2 is $|z|=\rho$. From $F(\zeta_j^0)=\zeta_j^0$, we see that by $F(z)$, $|z|=1$ is mapped on $|z|=1$ and $|z|=\rho$ on $|z|=\rho$, so that by Theorem IX. 30, $F(z)=e^{i\theta}z$ and from $F(\zeta_j^0)=\zeta_j^0$, we see that $F(z)\equiv z$, or $f_1(z)\equiv f_2(z)$. Hence the mapping is unique.

6. Conformal mapping of an open Riemann surface of planar character on a schlicht domain

1. Let F be a Riemann surface and if F is decomposed into two parts by any Jordan curve on F, then F is called *a Riemann surface of planar character* (schlichtartig). An open Riemann surface F is called *simply connected*, if F is of planar character and is decomposed into two parts F_1, F_2 by any Jordan curve on F, such that at least one of F_1, F_2 is compact.

Theorem IX. 32. (Koebe's Fundamental theorem). *Let F be an open Riemann surface of planar character, spread over the z-plane, then F can be mapped conformally on a schlicht domain on the w-plane.*

Koebe [3].

Proof. We exhaust F by a sequence of compact Riemann surfaces: $F_0\subset F_1\subset F_2\subset\cdots\subset F_n\to F$, where $\overline{F}_n\subset F_{n+1}$ and the boundary Γ_n of F_n consists of a finite number of analytic Jordan curves Γ_n^i $(i=1,2,\cdots,k_n)$. As a sub-surface of F, F_n is of planar character. First we shall prove by Koebe's ingeneous idea, that F_n can be mapped conformally on a schlicht domain on the w-plane. We draw a rectilinear polygon π_n^i in $F-F_n$, which lies in a small neighbourhood of Γ_n^i and bounds with Γ_n^i a doubly connected domain Δ_n^i. We joint $\{\Delta_n^i\}$ to F_n and let $F_n'=F_n+\sum_i \Delta_n^i$. We consider F_n' as a surface in space and we choose a point a_n^i in space, outside of F_n' and connects a_n^i to vertices of π_n^i by rectilinear segments, then we obtain a polyhedron P_n^i, whose vertex is a_n^i and whose base line is π_n^i. We join $\{P_n^i\}$ to F_n' and let $\widetilde{F}_n=F_n'+\sum_i P_n^i$.

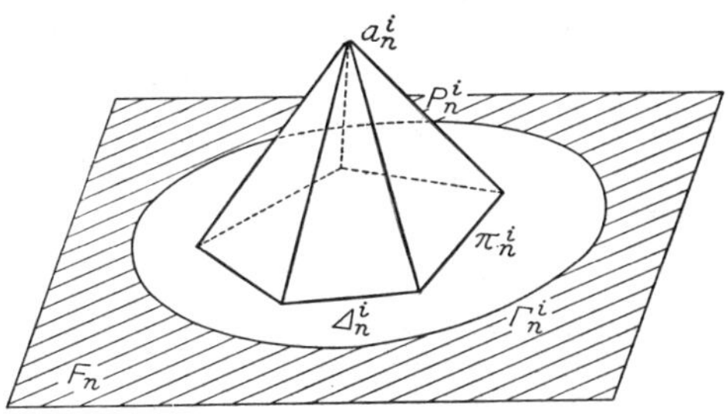

Since F_n' is of planar character, F_n' is topologically equivalent to a domain on a sphere, bounded by a finite number Jordan curves. If we define suitably a parameter neighbourhood of a point of F_n[1)], then $\widetilde{F}_n$ can be considered as a closed Riemann surface of genus

1) If a is a vertex of P_n and U be its neighbourhood, then we cut U along an

$p=0$. Hence $\widetilde{F}_n$ can be mapped conformally on $|w|\leqq\infty$, where the conformity at a point on the sides of P_n^i is suitably defined. Then by this mapping, F_n is mapped conformally on a schlicht domain on the w-plane.

Let $w=f_n(z)$ be the mapping function, which we normalize such that $f_n(z_0)=0$, $f_n'(z_0)=1$, where z_0 is a fixed point of F_0. Let Ω be any compact sub-surface of F, which contains z_0, then by Koebe's distortion theorem, we can prove easily that

$$|f_n(z)|\leqq K=K(\Omega) \qquad \text{in } \Omega\,,$$

where $K(\Omega)$ is a constant, which depends on Ω only.

Hence we can select a partial sequence from $\{f_n(z)\}$, which we denote by $\{f_n(z)\}$ again, such that $\lim\limits_n f_n(z)=f(z)$ uniformly in the wider sense on F, then $f(z)$ is meromorphic and schlicht on F, $f(z_0)=0$, $f'(z_0)=1$ and $w=f(z)$ maps F conformally on a schlicht domain on the w-plane.——

By means of the theorem, we can prove the following **Fundamental theorem in the theory of uniformization.**

Theorem IX. 33. *Let F be a simply connected open Riemann surface, spread over the z-plane, then F can be mapped conformally either* (i) *on* $|w|<\infty$, *or* (ii) *on* $|w|<1$.

Proof. Since F is of planar character, F can be mapped conformally on a schlicht domain Δ on the t-plane. Since F is simply connected, Δ is simply connected, so that its boundary consists of either (i) a single point, or (ii) a continuum. In case (i), Δ can be mapped conformally on $|w|<\infty$ and in case (ii), Δ can be mapped conformally on $|w|<1$.

Corollary. *Let F be a simply connected open Riemann surface, then F can be exhausted by a sequence of simply connected compact Riemann surfaces.*

2. We shall prove a lemma, due to Koebe (Math. Ann. 69 (1910)).

Lemma. *Let E be a bounded closed set on the z-plane and U be its neighbourhood and $f(z)$ be regular and bounded in $U-E$. Suppose that for any $\varepsilon>0$, E can be enclosed into a finite number of analytic Jordan curves C_k $(k=1,2,\cdots,n)$, such that*

$$\text{(i)}\quad L_1^2+\cdots+L_n^2<\varepsilon\,, \qquad \text{(ii)}\quad \omega_1^2+\cdots+\omega_n^2<\varepsilon\,,$$

where L_k is the length of C_k and ω_k is the oscillation of $f(z)$ on C_k, such that $\omega_k=\max\limits_{z_1,z_2\in C_k}|f(z_1)-f(z_2)|$. Then

edge of P_n^i, then U can be developed on a schlicht sector: $|\arg\tau|\leqq\alpha$, $|\tau|<\rho$ on the τ-plane, then by $t=\tau^{\pi/\alpha}$, U can be mapped conformally on $|t|<r$. If a is a point on an edge of P_n, then its neighbourhood U can be developed on a schlicht domain on the t-plane: $|t|<r$.

$f(z)$ is regular on E.

Proof. We may assume that U is bounded by an analytic Jordan curve C and let D be the domain, bounded by $C+\sum_{k=1}^{n} C_k$ and z be any point of D, then

$$f(z)=\frac{1}{2\pi i}\int_C \frac{f(\zeta)}{\zeta-z}d\zeta+\frac{1}{2\pi i}\sum_{k=1}^{n}\int_{C_k}\frac{f(\zeta)}{\zeta-z}d\zeta . \qquad (1)$$

Let ζ_k be a point of C_k, then

$$\int_{C_k}\frac{f(\zeta)}{\zeta-z}d\zeta=f(\zeta_k)\int_{C_k}\frac{d\zeta}{\zeta-z}+\int_{C_k}\frac{f(\zeta)-f(\zeta_k)}{\zeta-z}d\zeta=\int_{C_k}\frac{f(\zeta)-f(\zeta_k)}{\zeta-z}d\zeta ,$$

since z lies outside of C_k, $\int_{C_k}\frac{d\zeta}{\zeta-z}=0$. Hence

$$\left|\sum_{k=1}^{n}\int_{C_k}\frac{f(\zeta)}{\zeta-z}d\zeta\right|\leqq\sum_{k=1}^{n}\int_{C_k}\left|\frac{f(\zeta)-f(\zeta_k)}{\zeta-z}\right||d\zeta|\leqq\sum_{k=1}^{n}\frac{\omega_k L_k}{d}\leqq\frac{1}{d}\sqrt{\sum_{k=1}^{n}\omega_k^2\sum_{k=1}^{n}L_k^2}<\frac{\varepsilon}{d},$$

where d is the shortest distance of z to the boundary of D.

Since $\varepsilon>0$ is arbitrary, the second term of (1) is zero, so that

$$f(z)=\frac{1}{2\pi i}\int_C\frac{f(\zeta)}{\zeta-z}d\zeta , \qquad (2)$$

hence $f(z)$ is regular on E.

3. Let F be a closed Riemann surface of genus $p\geqq 2$, spread over the w-sphere. We cut F by p disjoint ring cuts C_k $(k=1, 2, \cdots, p)$ and let F_0 be the resulting surface. We take infinitely many same samples as F_0 and connect them along the opposite shores of $\{C_k\}$ as in the well known way, then we obtain an open Riemann surface $F^{(\infty)}$, which is of planar character. Hence $F^{(\infty)}$ can be mapped conformally on a schlicht domain Ω on the z-plane. We may assume that Ω contains $z=\infty$.

Let Δ_0 be the image of F_0 and we may assume that Δ_0 contains $z=\infty$. Δ_0 is bounded by $2p$ Jordan curves Γ_i $(i=1, 2, \cdots, 2p)$, where Γ_{2k-1}, Γ_{2k} correspond to the both shores of C_k. We see easily that Ω admitts a group G of conformal transformations $T_\nu(z)$ $(\nu=0, 1, 2, \cdots)$ into itself and Δ_0 is its fundamental domain, such that if we apply all transformations of G to Δ_0, then its equivalents $\{\Delta_\nu\}$ fill up Ω, clustering to the boundary E of Ω.

We may assume that $z=0$ belongs to Δ_0 and let a disc K_0: $|z|\leqq r$ be contained in Δ_0. Let K_ν be the equivalent of K_0 by T_ν, then by Koebe's 1/4-disc theorem, K_ν contains a disc of radius $\frac{r}{4}|T'_\nu(0)|$, so that if we denote the area of K_ν by A_ν, then $\frac{\pi r^2}{16}|T'_\nu(0)|^2\leqq A_\nu$. Since $\{K_\nu\}$ do not overlap, $\sum_\nu A_\nu<\infty$, hence

$$\sum_{\nu}|T'_{\nu}(0)|^2 < \infty . \tag{1}$$

By Koebe's distortion theorem, for any z on the boundary of Δ_0, $|T'(z)|\leqq\text{const.}|T'_{\nu}(0)|$, hence if we denote the equivalent of Γ_i $(i=1,2,\cdots,2p)$ by T_{ν} by $\Gamma_i^{(\nu)}$ and its length by $L_i^{(\nu)}$, then $L_i^{(\nu)}\leqq\text{const.}|T'_{\nu}(0)|$, so that $\sum_{\nu}L_i^{(\nu)2}<\infty$, hence for any $\varepsilon>0$, if ν is large,

$$\sum_{i=1}^{2p}[L_i^{(\nu)}]^2 < \varepsilon . \tag{2}$$

Since $\{\Gamma_i^{(\nu)}\}_{i=1,\cdots,2p}$ enclose E in its inside, the condition (i) of the lemma is satisfied.

Let $T_1\in G$ transform Δ_0 into Δ_1, which abutts on Δ_0 along Γ_1 and we put $f(z)=T_1(z)$. Let $\Gamma_i^{(\nu)}$ be transformed into $\Gamma_i^{(\mu)}$ by T_1, then μ is large, if ν is large, so that $\sum_{i=1}^{2p}[L^{(\mu)}]^2<\varepsilon$. Let $\omega_i^{(\nu)}$ be the oscillation of $f(z)$ on $\Gamma_i^{(\nu)}$, then $\omega_i^{(\nu)}\leqq L_i^{(\mu)}$, hence

$$\sum_{i=1}^{2p}[\omega_i^{(\nu)}]^2 < \varepsilon , \tag{3}$$

so that the condition (ii) of the lemma is satisfied. Hence $T_1(z)$ is regular on E, so that $T_1(z)$ is meromorphic and schlicht on the whole z-plane, hence $T_1(z)$ is a linear function of the form: $T_1(z)=\dfrac{az+b}{cz+d}$.

Similarly we can prove that any transformation $T_{\nu}(z)$ of G is a linear function of z. Hence

Theorem IX. 34. *G is a linear group of Schottky type*[1].

Remark. By Theorem III. 64, the singular set E of G is a bounded closed null set, which is of positive capacity and every point of which is a regular point of the Dirichlet problem.

7. Conformal mapping on a circular domain

A domain on the z-plane is called *a circular domain*, if it is bounded by a finite number of circles.

Theorem IX. 35. *Let D be a multiply connected domain on the z-plane, which is bounded by n continua, then D can be mapped conformally on a circular domain on the w-plane, bounded by n circles.*

Koebe [1].

Proof. We may assume that $n\geqq3$. We map D conformally on a parallel slits-plane F_0 on the t-plane, which is bounded by n slits L_i $(i=1,2,\cdots,n)$, parallel to the real axis. We invert F_0 with respect to L_i $(i=1,2,\cdots,n)$ and let the inverse be superposed to F_0 and joined

1) We shall give another proof in Chapter X. §5. Remark.

along L_i, the opposite banks being brought together. Let F_1 be the surface resulting from the like addition of sheets along all slits $\{L_i\}$ of F_0. F_1 consists of a finite number of sheets and is bounded by a finite number of slits. Let L'_j be one of the slits, lying on a sheet Φ_j of F_1. Let Φ_j be inverted with respect to L'_j and the inverse be joined to F_1 along L'_j. Let F_2 be the surface resulting from the like addition of sheets along all slits, bounding F_1. We continue this process indefinitely, then we obtain a limit surface F. An inversion with respect to any slit carries F into itself. We see easily that F is of planar character, hence F can be mapped conformally on a schlicht domain Ω on the w-plane. We may assume that the boundary E of Ω is a bounded closed set. Let Δ_0 be the image of F_0 on the w-plane, then Δ_0 is bounded by n Jordan curves C_i $(i=1, 2, \cdots, n)$, which is the image of L_i. Now F admits a group G of transformations into itself, alternately inversely and directly conformal, got by carrying F by a sequence of inversions into various sheets of F.

Hence Ω admits a group G of transformations into itself, alternately inversely and directly conformal. Δ_0 is the fundamental domain of G, such that if we apply all transformations of G to Δ_0, then its equivalents fill up Ω. Let $w'=S_i(w)$ be a transformation of G, such that by S_i, Δ_0 is transformed into Δ_i, which is adjacent to Δ_0 along C_i. G is generated by $(S_1, \cdots, S_n)$.

We may assume that $w=0$ belongs to Δ_0 and let a disc K_0: $|w|\leqq r$ be contained in Δ_0 and $\{K_\nu\}$ be its equivalents by G. Then since $\{K_\nu\}$ do not overlap,

$$\sum_\nu A_\nu < \infty , \tag{1}$$

where A_ν is the area of K_ν.

Let $T_\nu=T_\nu(w)$ be any transformation of G, then $T_\nu(w)$, or $\overline{T_\nu}(w)$ is meromorphic and schlicht in Ω and transforms Ω into itself.

Now $f(w) =\overline{S_1}(w)$ is meromorphic and schlicht in Ω and we can prove similarly as Theorem IX. 34, that $f(w)$ is regular on E, so that $f(w)$ is meromorphic and schlicht on the whole w-plane, hence $f(w)$ is a linear function of the form $f(w)=\dfrac{aw+b}{cw+d}$, so that

$$S_1(w) = \frac{\bar{a}\bar{w}+\bar{b}}{\bar{c}\bar{w}+\bar{d}} . \tag{2}$$

We take three points a, b, c on C_1 and through them, we draw a circle Γ, then Γ and C_1 are invariant by S_1, so that C_1 coincides with Γ, hence C_1 is a circle. Similarly C_i $(i=1, 2, \cdots, n)$ is a circle, so that Δ_0 is bounded by n circles. Hence D can be mapped conformally on a circular domain Δ_0.

Theorem IX. 36. *Let D be a circular domain on the z-plane and Δ be that on the w-plane. If $w=f(z)$ maps D conformally on Δ, then $f(z)$ is a linear function of the form:* $f(z)=\dfrac{az+b}{cz+d}$.

Proof. Since the case $n=1,2$ can be proved easily, we assume that $n\geqq 3$. Let D be bounded by n circles C_i $(i=1,2,\cdots,n)$ and Δ be bounded by n circles Γ_i $(i=1,2,\cdots,n)$. We invert D with respect to the bounding circles C_i $(i=1,2,\cdots,n)$ and perform the inversions indefinitely, then we obtain infinitely many circles, clustering to a perfect non-dense set E.

Let Ω be the complement of E with respect to the whole z-plane and we define similarly the perfect non-dense set E' and its complement Ω' on the w-plane, then by the principle of reflexion, $f(z)$ is meromorphic in Ω and maps Ω conformally on Ω'. Similarly as Theorem IX. 34, we can prove that $f(z)$ is regular on E, hence $f(z)$ is a linear function of z.

8. Moduli of plane regions

1. Let D and Δ be two plane regions, each of which is bounded by n continua. If D can be mapped conformally on Δ, then we say that D and Δ belong to the same class. Different classes form a manifold, which is represented by μ real parameters, which are called *moduli of n-ply connected plane regions.* We assume that $n\geqq 2$.

Theorem IX. 37. *If* $n=2$, *then* $\mu=1$,

If $n\geqq 3$, *then* $\mu=3n-6$.

Proof. We may assume that D is a circular domain on the z-plane, which is bounded by n circles C_i $(i=1,2,\cdots,n)$, where C_1 is the unit circle $|z|=1$ and C_2 is a circle $|z|=r_2$ $(0<r_2<1)$ and other C_i are contained in $r_2<|z|<1$. Similarly we assume that Δ is a circular domain on the w-plane, which is bounded by n circles Γ_i $(i=1,2,\cdots,n)$, where Γ_1 is the unit circle $|w|=1$ and Γ_2 is a circle $|w|=\rho_2$ $(0<\rho_2<1)$ and other Γ_i are contained in $\rho_2<|w|<1$.

If $w=f(z)$ maps D conformally on Δ, such that $|z|=1$, $|z|=r_2$ are mapped on $|w|=1$, $|w|=\rho_2$ respectively, then $f(z)$ is a linear function of z and we see easily that $f(0)=0$, $f(\infty)=\infty$. Hence $f(z)=kz$ $(k=\text{const.})$. Since $|f(z)|=1$ on $|z|=1$, $|k|=1$, so that Δ is obtained from D by a rotation about $z=0$, hence $\rho_2=r_2$.

Hence if $n=2$, then r_2 is the essential parameter, so that $\mu=1$. If $n\geqq 3$, let r_i $(i=2,\cdots,n)$ be the radius of C_i and d_i be the distance of the center a_i of C_i from $z=0$, then if $n=3$, r_2, r_3, d_3 are essential parameters, so that if $n=3$, $\mu=3$.

If $n\geqq 4$, let θ_i $(i=4, \cdots, n)$ be the angle, subtained by two vectors $\overrightarrow{0a_3}$, $\overrightarrow{0a_i}$, then r_i $(i=2, \cdots, n)$, d_i $(i=3, \cdots, n)$ and θ_i $(i=4, \cdots, n)$ are essential parameters, so that

$$\mu=(n-1)+(n-2)+(n-3)=3n-6 ,$$

which holds, if $n=3$. Hence if $n\geqq 3$, then $\mu=3n-6$.——

From the proof, we have

Theorem IX. 38. *If $n\geqq 3$, there are only a finite number of conformal mappings of D on itself.*

2. Let D be an n-ply connected domain on the z-plane. If $n\geqq 3$, then D admitts only a finite number $N=N(D)$ of conformal mappings on itself. They form a group G of order N.

We fix $n\geqq 3$ and consider all n-ply connected domains D and put

$$\mathfrak{N}(n) = \operatorname*{Max}_{D} N(D) , \qquad (1)$$

and we shall prove

Theorem IX. 39. $\mathfrak{N}(n)=2n$, *if* $n\neq 4, 6, 8, 12, 20$,

$\mathfrak{N}(4)=12$, $\mathfrak{N}(6)=\mathfrak{N}(8)=24$, $\mathfrak{N}(12)=\mathfrak{N}(20)=60$. Heins [2].

Proof. First we shall prove that

$$\mathfrak{N}(n) \geqq 2n . \qquad (1)$$

Let D_0 be the complement of n points $z_\nu=e^{\frac{2\pi\nu}{n}i}$ $(\nu=0, 1, 2, \cdots, n-1)$ with respect to the z-plane, then D_0 is n-ply connected.

$$S: \quad z' = e^{\frac{2\pi i}{n}} z , \qquad T: \quad z = 1/z \qquad (2)$$

generate a group G_0 of order $2n$ and D_0 is mapped conformally on itself by $2n$ transformations of G_0, so that $\mathfrak{N}(n)\geqq 2n$.

We may assume that D is bounded by n circles, where some of them may reduce to single points. By Theorem IX. 36, G is a linear group.

Since G is a finite linear group, it is isomorphic to a cyclic group, or a group of regular solids (dihedron, tetrahedron, cube, octahedron, dodecahedron, icosahedron)[1]. Hence we assume that G is a cyclic, or a group of regular solids. We shall search such a group G of order $N\geqq 2n$, which makes an n-ply connected domain invariant. If G is a cyclic group, then its order N is at most n, so that the condition $N\geqq 2n$ is not satisfied.

We consider a regular solid, inscribed in a unit sphere K with its center at $z=0$ and we consider the projection $\mathfrak{K}$ of the regular solid from

1) L. Ford: Automorphic functions. New York (1929) p. 133.

$z=0$ on K.

(i) If G is a dihedral group of order $N=2n$, we draw a small spherical disc of equal radius about each vertex of the dihedron and take off these discs from K and let D be the remaining domain, then D is n-ply connected and is invariant by G.

(ii) Next we consider an ordinary regular solid and let $\alpha_2, \alpha_1, \alpha_0$ be the number of its faces, edges and vertices and N be the order of the group of the regular solid, then we have the following table, where $N=2\alpha_1$, since the solid can be brought into coincidence with itself such that a given edge a_0b_0 is made to coincide with any edge a_ib_i or with b_ia_i.

	α_2	α_1	α_0	N
Tetrahedron	4	6	4	12
Cube	6	12	8	24
Octahedron	8	12	6	24
Dodecahedron	12	30	20	60
Icosahedron	20	30	12	60

$(N=2\alpha_1)$

A domain on K, which is invariant by the group of a regular solid $\mathfrak{K}$ can be obtained in three ways as follows.

(a) We draw a small spherical disc of equal radius about the mid-point of each face of $\mathfrak{K}$ and take off these discs from K.

(b) We draw a small spherical disc of equal radius about the mid-point of each edge of $\mathfrak{K}$ and take off these discs from K.

(c) We draw a small spherical disc of equal radius about each vertex of $\mathfrak{K}$ and take off these discs from K.

Hence we have the following table, according as the case (a), (b), (c), where as seen from the table, we consider only a tetrahedron, an octahedron and an icosahedron.

$(\mathrm{ii})_1$ If G is a tetrahedral group,

$$n=4,\ N=12;\ n=6,\ N=12;\ n=4,\ N=12.$$

$(\mathrm{ii})_2$ If G is an octahedral group,

$$n=8,\ N=24;\ n=12,\ N=24;\ n=6,\ N=24.$$

$(\mathrm{ii})_3$ If G is an icosahedral group,

$$n=20,\ N=60;\ n=30,\ N=60;\ n=12,\ N=60.$$

From (i), (ii), we see that

$$\mathfrak{N}(n)=2n\,, \qquad \text{if} \quad n\neq 4, 6, 8, 12, 20\,,$$

$$\mathfrak{N}(4)=12\,, \qquad \mathfrak{N}(6)=\mathfrak{N}(8)=24\,, \qquad \mathfrak{N}(12)=\mathfrak{N}(20)=60\,.$$

CHAPTER X

RIEMANN SURFACE

1. Classification of open Riemann surfaces

Let F be an open Riemann surface, spread over the z-plane. We use the following notations.

$$F \in P_G, \qquad F \in O_G \tag{1}$$

means that on F, a Green's function exists, or not and we say that F is *of positive*, or *of null boundary*, according as $F \in P_G$, or $F \in O_G$.

$$F \in P_{HB}, \qquad F \in O_{HB} \tag{2}$$

means that on F a non-constant one-valued bounded harmonic function exists or not.

$$F \in P_{HD}, \qquad F \in O_{HD} \tag{3}$$

means that on F a non-constant one-valued harmonic function $u(z)$, whose Dirichlet integral $D[u]$ is finite, exists, or not, where

$$D[u] = D_F[u] = \iint_F \left(\left(\frac{\partial u}{\partial x} \right)^2 + \left(\frac{\partial u}{\partial y} \right)^2 \right) dxdy, \qquad z = x + iy \,.$$

$$F \in P_{HBD}, \qquad F \in O_{HBD} \tag{4}$$

means that on F a non-constant one-valued bounded harmonic function, whose Dirichlet integral is finite, exists, or not.

$$F \in P_{AB}, \qquad F \in O_{AB} \tag{5}$$

means that on F a non-constant one-valued bounded regular function exists, or not.

$$F \in P_{AD}, \qquad F \in O_{AD} \tag{6}$$

means that on F a non-constant one-valued regular function $f(z)$, whose Dirichlet integral $D[f]$ is finite, exists, or not, where

$$D[f] = D_F[f] = \iint_F |f'(z)|^2 \, dxdy \,.$$

Similarly we define

$$F \in P_{ABD}, \qquad F \in O_{ABD} \,. \tag{7}$$

Let $f(z)$ be one-valued and meromorphic on F, then

$$D^*[f] = D_F^*[f] = \iint_F \left(\frac{|f'(z)|}{1 + |f(z)|^2} \right)^2 dxdy$$

is called the elliptic Dirichlet integral of f on F.

$$F \in P_{MD^*}, \quad F \in O_{MD^*} \tag{8}$$

means that on F a non-constant one-valued meromorphic function, whose elliptic Dirichlet integral is finite, exists, or not.

Evidently $$O_{MD^*} \subset O_{AD}\ .$$

2. Harmonic measure of the ideal boundary of an open Riemann surface.

1. Let F be an open Riemann surface, spread over the z-plane. We exhaust F by a sequence of compact Riemann surfaces: $F_0 \subset F_1 \subset \cdots \subset F_n \to F$, where $\overline{F}_n \subset F_{n+1}$ and the boundary Γ_n of F_n consists of a finite number of analytic Jordan curves. Let $\omega_n(z)$ be the harmonic measure of Γ_n with respect to $F_n - F_0$, such that $\omega_n(z)$ is harmonic in $F_n - F_0$, $\omega_n = 1$ on Γ_n, $\omega_n = 0$ on Γ_0. Then by the maximum principle, $\omega_{n+1}(z) < \omega_n(z)$ in $F_n - F_0$, so that by Harnack's theorem,

$$\lim_n \omega_n(z) = \omega(z)$$

uniformly in the wider sense in $F - F_0$. $\omega(z)$ is called *the harmonic measure of the ideal boundary of F with respect to $F - F_0$*.

There occurs two cases: either (i) $\omega(z) \not\equiv 0$, or (ii) $\omega(z) \equiv 0$.

In case (i),

$$0 < \omega(z) < 1 \text{ in } F - F_0, \quad \omega = 0 \text{ on } \Gamma_0.$$

Theorem X. 1. $\omega(z) \equiv 0$, *only when $F \in O_G$. If $F \in P_G$, then*

$$\overline{\lim}\, \omega(z) = 1\ ,$$

when z tends to the ideal boundary of F.

Proof. We may assume that $z=0$ belongs to F_0 and let $g_n(z, 0)$ be the Green's function of F_n and put

$$M_n = \operatorname*{Max}_{z \in \Gamma_0} g_n(z, 0)\ . \tag{1}$$

Then by the maximum principle,

$$(1 - \omega_n(z)) M_n \geqq g_n(z, 0) \quad \text{in } F_n - F_0\ ,$$

so that

$$(1 - \operatorname*{Min}_{z \in \Gamma_1} \omega_n(z)) M_n \geqq \operatorname*{Max}_{z \in \Gamma_1} g_n(z, 0) \quad (n \geqq 2)\ . \tag{2}$$

Since $g_n(z, 0) - g_1(z, 0)$ is harmonic in F_1, by the maximum principle,

$$\operatorname*{Max}_{z \in \Gamma_0} (g_n(z, 0) - g_1(z, 0)) \leqq \operatorname*{Max}_{z \in \Gamma_1} (g_n(z, 0) - g_1(z, 0)),$$

so that

$$M_n - \text{const.} \leqq \operatorname*{Max}_{z \in \Gamma_1} g_n(z, 0)\,.$$

Hence by (2),

$$(1 - \operatorname*{Min}_{z \in \Gamma_1} \omega_n(z))\ M_n \geqq M_n - \text{const.},$$

so that

$$0 \leqq \operatorname*{Min}_{z \in \Gamma_1} \omega(z) \leqq \operatorname*{Min}_{z \in \Gamma_1} \omega_n(z) \leqq \text{const.}/M_n\,. \tag{3}$$

If $F \in O_G$, then $\lim_n g_n(z, 0) = \infty$, so that $\lim_n M_n = \infty$, hence $\operatorname*{Min}_{z \in \Gamma_1} \omega(z) = 0$. Since $\operatorname*{Min}_{z \in \Gamma_1} \omega(z) > 0$, if $\omega \not\equiv 0$, we have

$$\omega(z) \equiv 0, \quad \text{if } F \in O_G\,. \tag{4}$$

Next suppose that $F \in P_G$. Then $\lim_n g_n(z, 0) = g(z, 0)$, which is harmonic on F, except at $z=0$, where it has a logarithmic singularity. We shall provo that

$$\underline{\lim}\, g(z, 0) = 0\,, \tag{5}$$

when z tends to the ideal boundary of F. For, if otherwise, $g(z, 0) > \alpha > 0$ on F. We put

$$h(z) = g(z, 0) - \alpha > 0\,. \tag{6}$$

Since $h(z) - g_n(z, 0)$ is harmonic on F_n and $g_n(z, 0) = 0$ on Γ_n, we have by (6) and the maximum principle, $h(z) - g_n(z, 0) > 0$, or $h(z) > g_n(z, 0)$ on F_n, hence $h(z) \geqq g(z, 0)$ on F, or

$$g(z, 0) - \alpha \geqq g(z, 0) \quad \text{on } F\,,$$

which is absurd. Hence $\underline{\lim}\, g(z, 0) = 0$. Let

$$m = \operatorname*{Min}_{z \in \Gamma_0} g(z, 0)\,, \tag{7}$$

then by the maximum principle,

$$0 \leqq 1 - \omega_n(z) \leqq g(z, 0)/m \quad \text{on } F_n - F_0\,,$$

so that

$$0 \leqq 1 - \omega(z) \leqq g(z, 0)/m \quad \text{on } F - F_0\,. \tag{8}$$

Since $\underline{\lim}\, g(z, 0) = 0$, we have

$$\overline{\lim}\, \omega(z) = 1\,, \tag{9}$$

when z tends to the ideal boundary of F. Hence

$$\omega(z) \not\equiv 0, \quad \text{if } F \in P_G\,. \text{ ——} \tag{10}$$

From the proof, we have

Theorem X. 2. *If $F \in P_G$, then*

$$\underline{\lim}\, g(z, 0) = 0\,,$$

when z tends to the ideal boundary of F.

2. Let Φ be a non-compact sub-surface of F and γ be its compact boundary, which consists of at most a countable number of analytic curves, which do not cluster at inner points of F.

Let $F_0 \subset F_1 \subset \cdots \subset F_n \to F$ be an exhaustion of F and Γ_n be the boundary of F_n. Let Φ_n be a connected sub-surface among components of the intersection $\Phi \cdot F_n$, which we choose such that $\Phi_0 \subset \Phi_1 \subset \cdots \subset \Phi_n \to \Phi$.

Let $\Gamma_n^{(\Phi)}$ be the part of Γ_n, which belongs to the boundary of Φ_n and $\omega_n(z, \Phi)$ be the harmonic measure of $\Gamma_n^{(\Phi)}$, with respect to Φ_n, then by the maximum principle, $\omega_{n+1}(z, \Phi) < \omega_n(z, \Phi)$, so that by Harnack's theorem,

$$\lim_n \omega_n(z, \Phi) = \omega(z, \Phi) \tag{1}$$

uniformly in the wider sense in Φ. We call $\omega(z, \Phi)$ *the harmonic measure of the ideal boundary of Φ, with respect to Φ.*

Theorem X. 3. *If $\omega(z, \Phi) \not\equiv 0$, then*

$$\sup_{z \in \Phi} \omega(z, \Phi) = 1.$$

Proof. For, if otherwise, $\omega(z, \Phi) < \alpha < 1$ on Φ, so that by the maximum principle, $\omega_n(z, \Phi) \geqq \frac{1}{\alpha} \omega(z, \Phi)$ on Φ_n, hence $\omega(z, \Phi) \geqq \frac{1}{\alpha} \omega(z, \Phi)$, or $\alpha \geqq 1$, which is absurd, hence the theorem.

Theorem X. 4. *Let $\omega_n(z)$ be the harmonic measure of Γ_n with respect to $F_n - F_0$ and $D[\omega_n]$ be its Dirichlet integral in $F_n - F_0$ and $D[\omega]$ be that of $\omega(z) = \lim_n \omega_n(z)$ in $F - F_0$, then*

$$\lim_n D[\omega_n] = D[\omega] = \int_{\Gamma_0} \frac{\partial \omega}{\partial \nu} ds,$$

where ν is the inner normal of Γ_0 with respect to $F - F_0$. Hence

$$\lim_n D[\omega_n] = 0, \text{ only when } F \in O_G .$$

Proof. Since $\omega_n = 0$ on Γ_0, by the Green's formula, we have

$$D[\omega_n] = -\int_{\Gamma_n} \omega_n \frac{\partial \omega_n}{\partial \nu} ds = -\int_{\Gamma_n} \frac{\partial \omega_n}{\partial \nu} ds = \int_{\Gamma_0} \frac{\partial \omega_n}{\partial \nu} ds . \tag{1}$$

Since Γ_0 is an analytic curve and $\omega_n = 0$ on Γ_0, ω_n can be continued harmonically across Γ_0, so that $\frac{\partial \omega_n}{\partial \nu} \to \frac{\partial \omega}{\partial \nu}$ on Γ_0, hence

$$\lim_n D[\omega_n] = \int_{\Gamma_0} \frac{\partial \omega}{\partial \nu} ds . \tag{2}$$

As easily seen, $D[\omega] \leqq \lim_n D[\omega_n]$, so that

$$D[\omega] \leqq \int_{\Gamma_0} \frac{\partial \omega}{\partial \nu} ds . \tag{3}$$

Now

$$0 \leqq D_{F_n-F_0}[\omega-\omega_n]=D_{F_n-F_0}[\omega]-2D_{F_n-F_0}[\omega, \omega_n]+D_{F_n-F_0}[\omega_n]$$
$$=D_{F_n-F_0}[\omega]+2\int_{\Gamma_n}\frac{\partial\omega}{\partial\nu}ds+\int_{\Gamma_0}\frac{\partial\omega_n}{\partial\nu}ds$$
$$=D_{F_n-F_0}[\omega]-2\int_{\Gamma_0}\frac{\partial\omega}{\partial\nu}ds+\int_{\Gamma_0}\frac{\partial\omega_n}{\partial\nu}ds,$$

so that

$$D_{F_n-F_0}[\omega]\geqq 2\int_{\Gamma_0}\frac{\partial\omega}{\partial\nu}ds-\int_{\Gamma_0}\frac{\partial\omega_n}{\partial\nu}ds,$$

hence if we make $n\to\infty$, then $D[\omega]\geqq\int_{\Gamma_0}\frac{\partial\omega}{\partial\nu}ds$, so that by (3),

$$D[\omega]=\int_{\Gamma_0}\frac{\partial\omega}{\partial\nu}ds .$$

3. Open Riemann surfaces of class O_G and P_G

Theorem X. 5. (Maximum principle). *Let $F\in O_G$ and Φ be a non-compact sub-surface of F and Γ be its relative boundary. Let $u(z)$ be a one-valued bounded harmonic function on Φ.*

If

$$m\leqq u(z)\leqq M \quad \text{on } \Gamma,$$

then

$$m\leqq u(z)\leqq M \quad \text{in } \Phi .$$

Especially, if $u=0$ on Γ, then $u\equiv 0$ in Φ.

Proof. Suppose that there exists a point $z_0\in\Phi$, such that $u(z_0)=M+2\varepsilon$ $(\varepsilon>0)$. Let Ω be the sub-surface of Φ, such that $u>M+\varepsilon$, then $u=M+\varepsilon$ on the boundary of Ω.

Let $F_0\subset F_1\subset\cdots\subset F_n\to F$ be an exhaustion of F, where $F_0\subset F-\Omega$. If $u(z)\leqq K$ $(=\text{const})$ in Φ, then by the maximum principle,

$$u(z)\leqq K\omega_n(z)+(M+\varepsilon) \quad \text{in } \Omega\cdot F_n ,$$

where $\omega_n(z)$ is the harmonic measure of the boundary of F_n with respect to F_n-F_0.

Since $F\in O_G$, $\lim_n \omega_n(z)=0$, so that $u(z)\leqq M+\varepsilon$ in Ω, which is absurd. Hence $u(z)\leqq M$ in Φ. Similarly $u(z)\geqq m$ in Φ.

Theorem X. 6. *If $F\in O_G$, then $F\in O_{HB}$.*

Proof. Suppose that there exists a non-constant one-valued bounded harmonic function $u(z)$ on F, then there exist two points z_1, z_2 on F, such that $u(z_1)>\alpha>u(z_2)$. Let Φ be the sub-surface of F, such that $u(z)>\alpha$ in Φ, then $u(z)=\alpha$ on its relative boundary. Hence by Theorem X. 5, $u\equiv\text{const.}=\alpha$ in Φ, which is absurd. Hence $F\in O_{HB}$.

Theorem X. 7. *The necessary and sufficient condition that $F \in P_G$ is that there exists a non-constant continuous positive superharmonic function on F.* Ahlfors [7].

Corollary. *If* $F \in O_G$, *then* $F \in O_{HP}$[1]. Myrberg [2].

Proof. (i) Suppose that $F \in P_G$ and $F_0 (\subset F)$ be a schlicht circular disc and Γ_0 be its bounding circle. Let $\omega(z)$ be the harmonic measure of the ideal boundary of F with respect to $F-F_0$. Then $\omega=0$ on Γ_0 and $0<\omega<1$ in $F-F_0$.

If we put

$$w(z)=1 \quad \text{in } F_0, \qquad w(z)=1-\omega(z) \quad \text{in } F-F_0, \tag{1}$$

then $w(z)$ is a non-constant continuous positive superharmonic function on F.

(ii) Next suppose that there exists a non-constant continuous positive superharmonic function $w(z)$ on F.

We consider a family $\mathfrak{F}$ of $\psi(z)$, which satisfy the following condition:

(i) $\psi(z)$ is a continuous positive superharmonic function in $F-F_0$.

(ii) $\psi(z) \geqq 1$ on Γ_0.

Evidently $\psi(z) \equiv 1$ belongs to $\mathfrak{F}$, so that $\mathfrak{F} \neq 0$ and if we put

$$u(z)=\inf_{\psi \in \mathfrak{F}} \psi(z), \qquad z \in F-F_0, \tag{2}$$

then $0 \leqq u(z) \leqq 1$ in $F-F_0$ and by Theorem I. 8, $u(z)$ is harmonic in $F-F_0$ and since a barrier exists at every point of Γ_0, $u(z)=1$ on Γ_0 (Theorem I. 10), so that $u \not\equiv 0$.

Let $\operatorname*{Min}_{z \in \Gamma_0} w(z)=m>0$, then $\dfrac{w(z)}{m} \in \mathfrak{F}$, so that $\dfrac{w(z)}{m} \geqq u(z)$ in $F-F_0$.
Hence if $u \equiv 1$, then

$$w(z) \geqq m \quad \text{in } F-F_0. \tag{3}$$

Since $w(z)$ is superharmonic, $w(z) \geqq m$ in F_0, hence,

$$w(z) \geqq m \quad \text{in } F, \tag{4}$$

so that $w(z)$ attains its minimum on Γ_0, hence $w(z) \equiv \text{const.}=m$ on F, which is absurd. Hence $u \not\equiv 1$, so that $0<u(z)<1$ in $F-F_0$ and $u=1$ on Γ_0. Hence by Theorem X. 5, $F \in P_G$.

Theorem X. 8. *If $F \in P_G$ and $\tilde{F}$ be its unramified covering surface, then $\tilde{F} \in P_G$.*

Proof. Let Δ_0 be a schlicht circular disc, contained in F and Γ_0 be its bounding circle. Let $\omega(z)$ be the harmonic measure of the ideal boundary of F with respect to $F-\Delta_0$. Then $\omega \not\equiv 0$ and $\omega=0$ on Γ_0.

1) $F \in O_{HP}$ means that there exists no no-constant one-valued positive harmonic function on F.

Let $\{\Delta_i\}$ be discs on $\widetilde{F}$, which lie above Δ_0. We denote the point of $\widetilde{F}$, which lies above $z\in F$ by $\tilde{z}$, We define $u(\tilde{z})$ in $\Phi=\widetilde{F}-\sum_i \Delta_i$ by $u(\tilde{z})=\omega(z)$. Then $u(z)$ is a bounded harmonic function ($\not\equiv 0$) on Φ and $u=0$ on the curve of $\widetilde{F}$, which lies above Γ_0. Hence by Theorem X. 5,

$$\widetilde{F}\in P_G .$$

Theorem X. 9. **(Maximum principle).** *Let $F\in O_G$ and Φ be a non-compact sub-surface of F, such that its relative boundary γ consists of at most a countable number of compact or non-compact analytic curves. Let $u(z)$ be a one-valued harmonic function on Φ, such that*

$$D_\Phi[u]<\infty .$$

If $m\leqq u(z)\leqq M$ on γ, then $m\leqq u(z)\leqq M$ in Φ.

Proof. First we shall prove that if $u=0$ on γ, then $u\equiv 0$ in Φ. Let $\{F_n\}$ be an exhaution of F with the boundary $\{\Gamma_n\}$, where F_0 is a schlicht circular disc in Φ. Let $\omega_n(z)$ be the harmonic measure of Γ_n with respect to F_n-F_0 and $\overline{\omega}_n(z)$ be its conjugate harmonic function.

We put

$$d_n=\int_{\Gamma_0} d\overline{\omega}_n=\int_{\Gamma_0}\frac{\partial\omega_n}{\partial\nu}ds>0 , \tag{1}$$

where ν is the inner normal of Γ_0 with respect to F_n-F_0. Since $F\in O_G$, $\omega_n>\omega_{n+1}\to 0$, so that $d_1>d_2>\cdots d_n\to 0$, hence if we put $\mu_n=2\pi/d_n$, then

$$\mu_1<\mu_2<\cdots<\mu_n\to\infty . \tag{2}$$

If we put

$$u_n(z)=\mu_n\omega_n(z),\qquad v_n(z)=\mu_n\overline{\omega}_n(z) , \tag{3}$$

then

$$\int_{\Gamma_0} dv_n=2\pi . \tag{4}$$

Let Ω_λ ($0\leqq\lambda\leqq\mu_n$) be the sub-surface of F_n-F_0, such that $0\leqq u_n(z)\leqq\lambda$ and put

$$\Phi_\lambda=\Phi\cdot(\Omega_\lambda+F_0).$$

Let C_λ be the niveau curve: $u_n(z)=\lambda$ and $C_\lambda^{(\Phi)}$ be its part, contained in Φ.

Suppose that $u\not\equiv 0$, then since $u=0$ on γ,

$$D(\lambda)=D_{\Phi_\lambda}[u]=\int_{C_\lambda^{(\Phi)}} u\frac{\partial u}{\partial\nu}ds=\int_{C_\lambda^{(\Phi)}} u\frac{\partial u}{\partial u_n}dv_n>0 , \tag{5}$$

where ν is the outer normal of $C_\lambda^{(\Phi)}$ with respect to Φ_λ.

On the other hand,

$$D(\lambda)=\int_0 d\lambda\int_{C_\lambda^{(\Phi)}}\left(\left(\frac{\partial u}{\partial u_n}\right)^2+\left(\frac{\partial u}{\partial v_n}\right)^2\right)dv_n+\text{const.},$$

so that

$$D'(\lambda)=\int_{C_\lambda^{(\Phi)}}\left(\left(\frac{\partial u}{\partial u_n}\right)^2+\left(\frac{\partial u}{\partial v_n}\right)^2\right)dv_n ,$$

hence by (5),

$$D^2(\lambda)\leqq\int_{C_\lambda^{(\Phi)}}u^2\,dv_n\int_{C_\lambda^{(\Phi)}}\left(\frac{\partial u}{\partial u_n}\right)^2 dv_n\leqq D'(\lambda)\int_{C_\lambda^{(\Phi)}}u^2\,dv_n .$$

Hence if we put

$$m(\lambda)=\int_{C_\lambda^{(\Phi)}}u^2\,dv_n , \tag{6}$$

then

$$D^2(\lambda)\leqq m(\lambda)\,D'(\lambda) . \tag{7}$$

Since $u=0$ on γ,

$$m'(\lambda)=2\int_{C_\lambda^{(\Phi)}}u\frac{\partial u}{\partial u_n}dv_n=2D(\lambda)>0 , \tag{8}$$

so that $m(\lambda)$ is an increasing function of λ.

Since $m''(\lambda)=2D'(\lambda)$ and $m'(\lambda)>0$, we have from (7), (8),

$$\frac{m'(\lambda)}{m(\lambda)}\leqq 2\frac{m''(\lambda)}{m'(\lambda)} . \tag{9}$$

Hence integrating, we have

$$m(\lambda)\leqq k\,[m'(\lambda)]^2 , \qquad k=m(0)/(m'(0))^2=m(0)/4\,(D(0))^2 ,$$

so that from (7), (8), we have

$$d\lambda\leqq 4k dD(\lambda) .$$

Hence integrating on $[0, \mu_n]$, we have

$$\mu_n\leqq 4k\,[D(\mu_n)-D(0)]<4k\,D(\mu_n) . \tag{10}$$

Since by the hypothesis, $D(\mu_n)$ is bounded and $\mu_n\to\infty$, this is absurd. Hence $u\equiv 0$ in Φ.

Now we can prove the general case simply as follows.

Suppose that there exists a point $z_0\in\Phi$, such that $u(z_0)>M+\varepsilon$ $(\varepsilon>0)$ and let Ω be the sub-surface of Φ, such that $u(z)>M+\varepsilon$, then $u=M+\varepsilon$ on the boundary of Ω, so that if we apply the above result to $u-(M+\varepsilon)$, then we have $u\equiv M+\varepsilon$ in Ω, which is absurd. Hence $u(z)\leqq M$ in Ω. Similarly $u(z)\geqq m$ in Φ. ——

From Theorem X. 9, we have easily

Theorem X. 10. *If* $F\in O_G$, *then* $F\in O_{HD}$.

Theorem X. 11. *Let* Φ_i $(i=1, 2)$ *be two disjoint non-compact subsurfaces of* F, *whose boundary* γ_i *consists of at most a countable number of compact, or non-compact analytic curves. Suppose that there exists a non-constant one-valued bounded harmonic function* $u_i(z)$ *in* Φ_i, *such that* $u_i=0$ *on* γ_i, *then*

$$F\in P_{HB}, \quad \textit{so that} \quad F\in P_G .$$

Nevanlinna [5], Bader and Parreau [1].

Proof. (Mori [2]). By dividing $u_i(z)$ by a suitable constant, we may assume that

$$|u_i(z)|<1 \text{ in } \Phi_i, \quad \sup_{z\in\Phi_1} u_1(z)=1, \quad \inf_{z\in\Phi_2} u_2(z)=-1 . \tag{1}$$

Let $\{F_n\}$ be an exhaustion of F with the boundary $\{\Gamma_n\}$ and let $\Gamma_n^{(\Phi_i)}$ be the part of Γ_n, contained in Φ_i. We solve the Dirichlet problem for F_n, with the boundary value u_1 on $\Gamma_n^{(\Phi_1)}$ and u_2 on $\Gamma_n^{(\Phi_2)}$ and 0 on $\Gamma_n-(\Gamma_n^{(\Phi_1)}+\Gamma_n^{(\Phi_2)})$ and let $v_n(z)$ be its solution.

Then $|v_n(z)|\leqq 1$ on F_n. Hence we can select a convergent partial sequence from $\{v_n(z)\}$, which we denote by $\{v_n(z)\}$ again and put

$$\lim_n v_n(z)=v(z) . \tag{2}$$

Then $v(z)$ is harmonic and $|v(z)|\leqq 1$ on F.

Since by the maximum principle,

$$v_n(z)\geqq 2u_1(z)-1 \quad \text{in } F_n\cdot\Phi_1 ,$$

we have

$$v(z)\geqq 2u_1(z)-1 \quad \text{in } F\cdot\Phi_1 . \tag{3}$$

Since $\sup_{z\in\Phi_1} u_1(z)=1$, we have

$$\sup_{z\in\Phi_1} v(z)\geqq 1 ,$$

and similarly we have

$$\inf_{z\in\Phi_2} v(z)\leqq -1 ,$$

so that $v(z)\not\equiv$const. Hence $F\in P_{HB}$.

Theorem X. 12. *Let* $\Delta:|z-a|<\rho$ *be a disc. Suppose that there are no points of* F *above* Δ, *or there exists a connected piece* F_ρ *of* F *above* Δ, *which does not cover a set of positive capacity in* Δ, *then*

$$F\in P_{HB}, \quad \textit{so that} \quad F\in P_G.$$

Hence if $F\in O_G$, *then* F_ρ *covers any point of* Δ *at least once, except a set of capacity zero.* Mori [1].

Proof. If there are no points of F above Δ, then $\Re\left(\frac{1}{z-a}\right)$ is a non-constant one-valued bounded harmonic function on F, so that

$$F \in P_{HB}.$$

Next suppose that there exists a connected piece F_ρ of F above Δ, which does not cover a set E of positive capacity in Δ. We may assume that E is a closed set, contained entirely in Δ. Let Γ be the outer boundary of E, then $\gamma(\Gamma)=\gamma(E)>0$. We choose a point $a_i \in \Gamma$ $(i=1, 2)$ and let Δ_i: $|z-a_i|<\rho_0$ be a disc about a_i, contained in Δ, such that Δ_1, Δ_2 are disjoint each other and the part Γ_i of Γ, contained in Δ_i, is of positive capacity. Let D_i be the complement of Γ_i with respect to Δ_i, which contains a_i on its boundary. Then D_i is bounded by Γ_i and a set C_i of arcs on $|z-a_i|=\rho_0$. We solve the Dirichlet problem for D_i, with the boundary value 0 on C_i and 1 on Γ_i and let $u_i(z)$ be its solution, then since $\gamma(\Gamma_i)>0$, $u_i \not\equiv 0$. Let Φ_i be a connected piece of F_ρ, which lies above D_i, then if we consider $u_i(z)$ as a function on Φ_i, then $u_i(z)$ is a non-constant one-valued bounded harmonic function on Φ_i, such that it vanishes on the boundary curve of Φ_i, which lies above C_i, hence by Theorem X. 11, $\quad F \in P_{HB}$.

Theorem X. 13. *If $F \in O_G$, then F is regularly exhaustible in Ahlfors' sense.* Noshiro [5].

Proof. We consider F as spread over the z-sphere K. Let $\{F_n\}$ be an exhaustion of F with the boundary $\{\Gamma_n\}$. We define $u_n(z)$, $v_n(z)$, μ_n as in the proof of Theorem X. 9 and put

$$\zeta = e^{u_n+iv_n} = re^{i\theta}, \quad r=r(z)=e^{u_n(z)}, \quad \theta=\theta(z)=v_n(z). \tag{1}$$

In the following, we denote $r(z)$, $\theta(z)$ by r, θ respectively.

If we put $r_n=e^{\mu_n}$, then $1<r_1<r_2<\cdots<r_n\to\infty$.

Let $C_n^{(r)}$: $r(z)=\text{const.}=r$ $(1\leqq r\leqq r_n)$ be the niveau curve of $r(z)$, then $C_n^{(r)}$ bounds with Γ_0 a compact sub-surface $F_n^{(r)}$ of F_n-F_0 and

$$\int_{C_n^{(r)}} d\theta = \int_{C_n^{(r)}} dv_n = \int_{\Gamma_0} dv_n = 2\pi. \tag{2}$$

Let $A(r)$ be the area of $F_n^{(r)}$ and $L(r)$ be the length of $C_n^{(r)}$ $(1\leqq r\leqq r_n)$, measured on K, then

$$A(r)=\iint_{F_n^{(r)}} \left(\frac{|z'|}{1+|z|^2}\right)^2 r\,dr\,d\theta, \qquad L(r)=\int_{C_n^{(r)}} |z'|r\,d\theta, \quad z'=\frac{dz}{d\zeta}.$$

Hence by Schwarz's inequality and (2), we have

$$L^2(r)\leqq 2\pi r\frac{dA(r)}{dr}. \tag{3}$$

(i) First suppose that $\lim_n A(r_n)=\infty$.

If $L(r)>(A(r))^{\frac{3}{4}}$ for any r $(\sqrt{r_n}\leqq r\leqq r_n)$, then by (3),

$$\frac{1}{2}\log r_n=\int_{\sqrt{r_n}}^{r_n}\frac{dr}{r}\leqq 2\pi\int_{\sqrt{r_n}}^{r_n}\frac{dA(r)}{(A(r))^{\frac{3}{2}}}\leqq \text{const.}\ (n\to\infty),$$

which is absurd, since $r_n\to\infty$. Hence there exists τ_n $(\sqrt{r_n}\leqq\tau_n\leqq r_n)$, such that

$$L(\tau_n)\leqq(A(\tau_n))^{\frac{3}{4}}. \tag{4}$$

Let D_n be the sub-surface of F_n-F_0, such that $\sqrt{r_n}\leqq r\leqq r_n$ $(r=r(z))$, then since $\log r=u_n(z)=\mu_n\omega_n(z)$, $\log r_n=\mu_n$, we have $\omega_n(z)=\log r/\log r_n$, so that

$$\frac{1}{2}\leqq\omega_n(z)\leqq 1 \quad \text{in } D_n.$$

Since $\omega_n(z)\to 0$ in any compact sub-surface of $F-F_0$, D_n tends to the ideal boundary of F as $n\to\infty$, hence by the hypothesis, $A(\tau_n)\to\infty$, so that by (4), $L(\tau_n)/A(\tau_n)\to 0$, hence in this case, the theorem is proved.

(ii) If $A(r_n)\leqq M$ (=const.) $(n=1, 2, \cdots)$, then

$$\int_{\sqrt{r_n}}^{r_n}\frac{L^2(r)}{r}dr\leqq 2\pi A(r_n)\leqq 2\pi M,$$

so that there exists τ_n $(\sqrt{r_n}\leqq\tau_n\leqq r_n)$, such that $L(\tau_n)\to 0$, hence $L(\tau_n)/A(\tau_n)\to 0$ as $n\to\infty$. ——

Theorem X. 14. *If $F\in O_G$, then F has the Gross-property.*

Yûjôbô [2], Noshiro [5].

Proof. We consider F as spread over the z-plane and $\{F_n\}$ be an exhaustion of F. Let z_0 be a point of F_0, which is not a branch point and $H(z_0)$ be the Mittag-Leffler's principal star-region with respect to z_0 and $H_R(z_0)$ be its part, contained in $|z-z_0|<R$. With the same notations as the proof of Theorem X. 13, let $\overset{*}{F}{}_n^{(r)}$ be the part of $F_n^{(r)}$, contained in $H_R(z_0)$ and $\overset{*}{C}{}_n^{(r)}$ be that of $C_n^{(r)}$ and $A(r)$ be the area of $\overset{*}{F}{}_n^{(r)}$ and $L(r)$ be the length of $\overset{*}{C}{}_n^{(r)}$, measured on the z-plane, then

$$A(r)=\int_1^r dr\int_{\overset{*}{C}{}_n^{(r)}}|z'(\zeta)|^2\,r d\theta,\qquad L(r)=\int_{\overset{*}{C}{}_n^{(r)}}|z'(\zeta)|\,r d\theta,\qquad L^2(r)\leqq 2\pi r\frac{dA(r)}{dr}.$$

Since $A(r)\leqq\pi R^2$, we have

$$\int_{\sqrt{r_n}}^{r_n}\frac{L^2(r)}{r}dr\leqq 2\pi A(r_n)\leqq 2\pi^2R^2.$$

Hence if we put $L_n = \operatorname*{Min}_{\sqrt{r_n} \leq r \leq r_n} L(r)$, then

$$L_n^2 \log r_n \leqq 4\pi^2 R^2 .$$

Since $r_n \to \infty$, we have $L_n \to 0$ and from this, we can prove the theorem, similarly as Theorem V. 36.

Theorem X. 15. *Let $F \in O_G$ be spread over the z-sphere K and F_ρ be a connected piece of F, which lies above a disc Δ: $|z-a_0|<\rho$.*

(i) *If we map the universal covering surface of F_ρ conformally on $|\zeta|<1$ by $z=\varphi(\zeta)$, then* $\varphi(\zeta) \in U_\rho(a_0)$.

(ii) *Let a be any point of Δ and be covered $n(a)$-times by F_ρ and put*

$$n_0 = \sup_{a \in \Delta} n(a) ,$$

then F_ρ covers any point of Δ n_0-times, except a set of capacity zero.

(iii) *Let a be any point of K and be covered $n(a)$-times by F and put $n_0 = \sup_{a \in K} n(a)$, then F covers any point of K n_0-times, except a set of capacity zero.*

Hence if F consists of infinitely many sheets, then F covers any point of K infinitely often, except a set of capacity zero.

Proof. (i) We may assume that F_ρ is non-compact. By Fatou's theorem, $\lim_{\zeta \to e^{i\theta}} \varphi(\zeta)$ exists almost everywhere on $|\zeta|=1$, when $\zeta \to e^{i\theta}$ non-tangentially to $|\zeta|=1$. Let the boundary curves of F_ρ, which lies above the circle C: $|z-a_0|=\rho$ correspond to an open set E on $|\zeta|=1$ and E' be its complement on $|\zeta|=1$. Then almost all points of E' correspond to accessible boundary points of F_ρ. Now $\varphi(\zeta)$ is invariant by a certain Fuchsian group G, which makes $|\zeta|<1$ invariant. Suppose that $mE'>0$ and by means of a Poisson integral, let $v(\zeta)$ be the harmonic function in $|\zeta|<1$, such that $v=1$ on E' almost everywhere and $v=0$ on E, then $v \not\equiv 0$. Since E' is invariant by G, if we put $v(\zeta)=u(z)$, then $u(z)$ is a non-constant one-valued bounded harmonic function on F_ρ, which vanishes on the boundary curves of F_ρ, which lie above C. Hence by Theorem X. 5, $u \equiv 0$, which is absurd. Hence $mE'=0$, so that $\varphi(\zeta) \in U_\rho(a_0)$. The remaining part can be proved similarly as Theorem VIII. 37. ——

From the theorem, we can prove similarly as Theorem VIII. 34, the following

Theorem X. 16. *If $F \in O_G$, then the projection of direct transcendental singularities of F on the z-plane is of capacity zero.*

Next we shall prove **A criterion of the class O_G.**

Let $\{F_n\}$ be an exhaustion of F with the boundary $\{\Gamma_n\}$, which consists of a finite number of analytic Jordan curves: $\Gamma_n=\sum_i \Gamma_n^i$. Suppose that there exist constants $m>0$ and $0<d<1$, which are independent of n and satisfy the following conditions (i), (ii), (iii).

(i) Γ_n^i is covered by a chain of $\nu=\nu(n, i)$ neighbourhoods $U_j=U(z_j)$, $z_j \in \Gamma_n^i$ $(j=1, 2, \cdots, \nu)$, such that U_j, U_{j+1} overlap, where $U_{\nu+1}=U_1$.

If we put $G_n^i=\sum_{j=1}^{\nu} U_j$, then any point of G_n^i belongs to at most m of $\{U_j\}$ and G_n^i and G_n^k $(i \neq k)$ are disjoint. We put $G_n=\sum_i G_n^i$.

(ii) We map U_j conformally on $|t_j|<1$, such that z_j corresponds to $t_j=0$. Then there exists $t_{j,j+1}$ in $|t_j|<d<1$ and $t_{j+1,j}$ in $|t_{j+1}|<d<1$, such that $t_{j,j+1}$ and $t_{j+1,j}$ correspond to the same point of $U_j \cdot U_{j+1}$.

(iii) Let $\lambda_{j,j+1}$ be the rectlinear segment, which connects $t_j=0$ to $t_{j,j+1}$ and $\lambda_{j+1,j}$ be that, which connects $t_{j+1}=0$ to $t_{j+1,j}$ and put $\lambda_j=\lambda_{j,j+1}+\lambda_{j+1,j}$. Let $l_{j,j+1}$ be the image of $\lambda_{j,j+1}$ in U_j and $l_{j+1,j}$ be that of $\lambda_{j+1,j}$ in U_{j+1}, then $l_j=l_{j,j+1}+l_{j+1,j}$ is a curve, which connects z_j to z_{j+1} in U_j+U_{j+1}. If we put $L_n^i=\sum_{j=1}^{\nu} l_j$, then L_n is a Jordan curve in G_n. We put $L_n=\sum L_n$.

If the above condition is satisfied, then we say that $\{G_n\}$ is *a regular covering of* $\{\Gamma_n\}$.

Theorem X. 17. *Suppose that there exists a regular covering $\{G_n\}$ of $\{\Gamma_n\}$, such that $\{G_n\}$ overlap at most N_0-times and let $N(n)$ be the number of the neighbourhoods U_j in G_n.*

If

$$\sum_{n=1}^{\infty} \frac{1}{N(n)}=\infty ,$$

then

$$F \in O_G .$$

Nevanlinna [7].

Proof. For any positive integer k, we choose $n>k$ so large that $G_k \subset F_n$. Let $\omega_n(z)$ be the harmonic measure of Γ_n with respect to F_n-F_0 and $\bar{\omega}_n(z)$ be its conjugate harmonic function and put

$$w(z)=\omega_n(z)+i\bar{\omega}_n(z) . \tag{1}$$

Let $U_j=U(z_j)$ $(j=1, 2, \cdots N(\nu))$ be a neighbourhood, which belongs to G_ν $(1 \leqq \nu \leqq k)$ and L_ν be the system of Jordan curves in G_ν, which are defined above. We map U_j conformally on $|t|<1$, such that z_j corresponds to $t=0$ and put $w(z)=v_j(t)$ $(|t|<1)$ and

$$q_j=\operatorname*{Max}_{|t| \leq d<1} |v_j'(t)| . \tag{2}$$

Then $|v_j'(t_0)|=q_j$ at a certain point t_0 on $|t|=d$.

Since $|v_j'(t)|^2$ is subharmonic and the Dirichlet integral is invariant for conformal mapping,

$$\pi(1-d)^2 q_j^2 \leqq \int_0^{1-d}\int_0^{2\pi} |v_j'(t_0+re^{i\theta})|^2\, r dr d\theta \leqq D_{U_j}[w] .$$

Since at most m of $\{U_j\}$ overlap, we have

$$\pi(1-d)^2 \sum_{j=1}^{N(\nu)} q_j^2 \leqq m\, D_{G_\nu}[\omega_n] . \tag{3}$$

Since Γ_0 and L_ν bound a compact sub-surface of F,

$$D_n = D_{F_n-F}[\omega_n] = \int_{\Gamma_0} \frac{\partial \omega_n}{\partial \nu} ds = \int_{\Gamma_0} d\bar{\omega}_n$$

$$= \int_{L_\nu} d\bar{\omega}_n \leqq \int_{L_\nu} |d\bar{\omega}_n| \leqq \sum_{j=1}^{N(\nu)} \int_{\lambda_j} |v_j'(t)|dt \leqq 2\sum_{j=1}^{N(\nu)} q_j ,$$

so that by Schwarz's inequality and (3),

$$D_n^2 \leqq 4\, N(\nu) \sum_{j=1}^{N(\nu)} q_j^2 \leqq \frac{4m\, N(\nu)\, D_{G_\nu}[\omega_n]}{\pi(1-d)^2} ,$$

hence

$$\sum_{\nu=1}^{k} \frac{1}{N(\nu)} \leqq \frac{4m}{\pi(1-d)^2 D_n^2} \sum_{\nu=1}^{k} D_{G_\nu}[\omega_n] . \tag{4}$$

Since $\{G_\nu\}$ overlap at most N_0-times,

$$\sum_{\nu=1}^{k} D_{G_\nu}[\omega_n] \leqq N_0\, D_n ,$$

so that

$$\sum_{\nu=1}^{k} \frac{1}{N(\nu)} \leqq \frac{4m\, N_0}{\pi(1-d)^2\, D_n} . \tag{5}$$

Since $\sum_\nu 1/N(\nu)=\infty$, we have $D_n \to 0$ $(n\to\infty)$, so that by Theorem X. 4,

$$F \in O_G .$$

4. Theorems of Bader-Parreau and Virtanen

Theorem X. 18. *Let Φ be a non-compact sub-surface of F, whose boundary γ consists of at most a countable number of compact, or non-compact analytic curves. If there exists a non-constant one-valued harmonic function $u(z)$ on Φ, such that $u=0$ on γ and $D_\Phi[u]<\infty$, then there exists a non-constant one-valued bounded harmonic function $v(z)$ on Φ, such that $v=0$ on γ and $D_\Phi[v]<\infty$.*

Proof. Let $\{F_n\}$ be an exhaustion of F with the boundary $\{\Gamma_n\}$, where we assume that $F_0 \subset \Phi$ and put $\Phi_n = \Phi \cdot F_n$.

Let γ_n be the part of γ, contained in F_n and $\Gamma_n^{(\Phi)}$ be that of Γ_n, contained in Φ. We choose a, b in F_0, such that $u(a) \neq u(b)$.

Let $w_n = w_n(z, a)$ be a harmonic function on Φ_n, such that $w_n = 0$ on γ_n and $\dfrac{\partial w_n}{\partial \nu} = \dfrac{\partial g_n(z, a)}{\partial \nu}$ on $\Gamma_n^{(\Phi)}$ (solution of mixed boundary value problem), where $g_n(z, a)$ is the Green's function of Φ_n and ν is the inner normal of $\Gamma_n^{(\Phi)}$ with respect to Φ_n. If we put

$$\chi_n(z, a) = g_n(z, a) - w_n(z, a)\,, \tag{1}$$

then $\chi_n(z, a)$ satisfies the following condition.

(i) $\chi_n(z, a)$ is harmonic in Φ_n, except at a, where $\chi_n(z, a) - \log \dfrac{1}{|z-a|}$ is harmonic.

(ii) $\chi_n = 0$ on γ_n, $\quad \dfrac{\partial \chi_n}{\partial \nu} = 0$ on $\Gamma_n^{(\Phi)}$.

Similarly we define $w_n(z, b)$, $\chi_n(z, b)$.

We put

$$g_n(z; a, b) = g_n(z, a) - g_n(z, b)\,,$$

$$\chi_n(z; a, b) = \chi_n(z, a) - \chi_n(z, b) = g_n(z; a, b) + w_n(z, b) - w_n(z, a)\,, \tag{2}$$

$$v_n(z) = \chi_n(z; a, b) - g_n(z; a, b) = w_n(z, b) - w_n(z, a)\,. \tag{3}$$

Then $v_n(z)$ is harmonic in Φ_n and

$$v_n = 0 \quad \text{on } \gamma_n\,. \tag{4}$$

Since by the hypothesis, $u = 0$ on γ_n and $\dfrac{\partial \chi_n}{\partial \nu} = 0$ on $\Gamma_n^{(\Phi)}$, by the Green's formula, we have

$$D_{\Phi_n}[u, v_n] = -\int_{\Gamma_n^{(\Phi)}} u \frac{\partial v_n}{\partial \nu}\, ds = \int_{\Gamma_n^{(\Phi)}} u \frac{\partial g_n(z; a, b)}{\partial \nu}\, ds$$

$$= \int_{\Gamma_n^{(\Phi)} + \gamma_n} u \frac{\partial g_n(z; a, b)}{\partial \nu}\, ds = 2\pi(u(a) - u(b)) \neq 0\,. \tag{5}$$

Similarly if $m \geqq n$, then

$$D_{\Phi_n}[v_m, v_n] = 2\pi(v_m(a) - v_m(b)) = D_{\Phi_m}[v_m, v_m] = D_{\Phi_m}[v_m]\,, \tag{6}$$

so that

$$0 \leqq D_{\Phi_n}[v_m - v_n] = D_{\Phi_n}[v_m] - 2D_{\Phi_n}[v_m, v_n] + D_{\Phi_n}[v_n]$$

$$\leqq D_{\Phi_m}[v_m] - 2\, D_{\Phi_m}[v_m] + D_{\Phi_n}[v_n] = D_{\Phi_n}[v_n] - D_{\Phi_m}[v_m]\,, \tag{7}$$

$$D_{\Phi_m}[v_m] \leqq D_{\Phi_n}[v_n] \quad (m \geqq n)\,. \tag{8}$$

Hence $\lim_n D_{\Phi_n}[v_n]$ exists, so that by (7),

$$\lim_n D_{\Phi_n}[v_m - v_n] = 0 \quad (m \geqq n)\,. \tag{9}$$

Since $D_{\Phi_n}[v_n] \leqq D_{\Phi_1}[v_1]$ and $v_n=0$ on γ_n, we can select a partial sequence from $\{v_n(z)\}$, which we denote by $\{v_n(z)\}$ again, such that $\lim_n v_n(z) = v(z)$ uniformly in the wider sense in Φ, then $v(z)$ is harmonic on Φ and

$$v=0 \quad \text{on } \gamma . \tag{10}$$

$\lim_n g_n(z, a) = g(z, a)$, where $g(z, a)$ is the Green's function of Φ, which exists, since $g=0$ on γ, hence $\lim_n g_n(z; a, b) = g(z, a) - g(z, b)$, so that

$$\lim_n (g_n(z; a, b) + v_n(z)) = \lim_n \chi_n(z; a, b) = \chi(z; a, b) \tag{11}$$

exists and

$$v(z) = \chi(z; a, b) - g(z; a, b) . \tag{12}$$

From (6), (9), we have

$$D_{\Phi_n}[v, v_n] = \lim_m D_{\Phi_m}[v_m] ,$$

$$0 = \lim_n D_{\Phi_n}[v - v_n] = \lim_n (D_{\Phi_n}[v] - 2D_{\Phi_n}[v, v_n] + D_{\Phi_n}[v_n]) ,$$

so that

$$D_\Phi[v] = \lim_n D_{\Phi_n}[v] = \lim_n D_{\Phi_n}[v_n] , \tag{13}$$

$$|D_{\Phi_n}[u, v - v_n]| \leqq \sqrt{D_{\Phi_n}[u] D_{\Phi_n}[v - v_n]} \leqq \sqrt{D_\Phi[u] D_{\Phi_n}[v - v_n]} \to 0 \ (n \to \infty) ,$$

hence

$$D_\Phi[u, v] = \lim_n D_{\Phi_n}[u, v] = \lim_n D_{\Phi_n}[u, v_n] ,$$

so that by (5),

$$D_\Phi[u, v] = 2\pi(u(a) - u(b)) \neq 0 ,$$

hence $v \not\equiv 0$ and from (13),

$$D_\Phi[v] = \lim_n D_{\Phi_n}[v_n] \leqq D_{\Phi_1}[v_1] < \infty . \tag{14}$$

Since $\frac{\partial \chi_n}{\partial \nu} = 0$ on $\Gamma_n^{(\Phi)}$, χ_n is harmonic on the Schottky prolongation of $\Phi_n - F_0$ with respect to $\Gamma_n^{(\Phi)}$, and since $\chi_n = 0$ on γ_n, $\operatorname{Max} |\chi_n(z; a, b)|$ in $\Phi_n - F_0$ is attained on Γ_0, so that $\operatorname{Max} |\chi(z; a, b)|$ in $\Phi - F_0$ is attained on Γ_0. Since $g(z; a, b)$ is bounded on $\Phi - F_0$, $v(z) = \chi(z; a, b) - g(z; a, b)$ is bounded on $\Phi - F_0$. Since $v(z)$ is harmonic in F_0, $v(z)$ is a bounded harmonic function on Φ, $v=0$ on γ and $D_\Phi[v] < \infty$. —

Theorem X. 19. *Let Φ_i $(i=1, 2)$ be two disjoint non-compact subsurfaces of F, whose boundary γ_i consists of at most a countable number of compact, or non-compact analytic curves. If there exists on Φ_i a non-constant one-valued harmonic function $u_i(z)$, such that $u_i=0$ on γ_i and $D_{\Phi_i}[u_i] < \infty$, then $F \in P_{HBD}$.* Bader and Parreau [1].

Proof. (Mori [2]). By Theorem X. 18, we may assume that $u_i(z)$ is bounded on Φ_i and

$$|u_i(z)|<1 \text{ in } \Phi_i, \quad \sup_{z\in\Phi_1} u_1(z)=1, \quad \inf_{z\in\Phi_2} u_2(z)=-1\,. \tag{1}$$

Let $\{F_n\}$ be an exhaustion of F with the boundary $\{\Gamma_n\}$ and put $F_n^{(i)}=F_n\cdot\Phi_i$ $(i=1,2)$. We define a continuous function $w_n(z)$ on F_n, as follows.

$$w_n(z)=u_1(z) \quad \text{on } F_n^{(1)}, \quad w_n(z)=u_2(z) \quad \text{on } F_n^{(2)}, \tag{2}$$
$$w_n(z)=0 \quad \text{on } F_n-(F_n^{(1)}+F_n^{(2)})\,.$$

Then

$$D_{F_n}[w_n]=D_{F_n^{(1)}}[u_1]+D_{F_n^{(2)}}[u_2]\leqq D_{\Phi_1}[u_1]+D_{\Phi_2}[u_2]=K<\infty\,. \tag{3}$$

Let $v_n(z)$ be the solution of the Dirichlet problem for F_n with the boundary value w_n on Γ_n, then by the Dirichlet principle,

$$D_{F_n}[v_n]\leqq D_{F_n}[w_n]\leqq K\,.$$

Now $v_n(z)$ is the same with $v_n(z)$ in the proof of Theorem X. 11, so that we may assume that $\lim_n v_n(z)=v(z)$ uniformly in the wider sense in F, hence $v(z)$ is harmonic on F and $v(z)\not\equiv$const., $|v(z)|\leqq 1$ on F, $D_F[v]\leqq K$. Hence

$$F\in P_{HBD}\,.$$

Theorem X. 20. *If* $F\in P_{HD}$, *then* $F\in P_{HBD}$. Virtanen [1].

Proof. Let $u(z)$ be a non-constant one-valued harmonic function on F, such that $D_F[u]<\infty$. Let a, b be two points of F, such that $u(a)<\alpha_1<\alpha_2<u(b)$ and Φ_1 be the sub-surface of F, such that $u(z)<\alpha_1$ and Φ_2 be that, such that $u(z)>\alpha_2$, then Φ_i $(i=1,2)$ are two disjoint non-compact sub-surfaces of F and $u=\alpha_i$ on the boundary of Φ_i and $D_{\Phi_i}[u]<\infty$ $(i=1,2)$, so that by Theorem X. 19, $F\in P_{HBD}$.

5. Open Riemann surfaces of class O_{AD}

Theorem X. 21. *Let* $D\in O_{AD}$ *be a schlicht domain on the z-plane. If* $f(z)$ *is one-valued meromorphic and schlicht in* D, *then*

$$f(z)=\frac{az+b}{cz+d}\,.$$

Proof. We may assume that D contains $z=\infty$, so that its boundary E is a bounded closed set. Let $f(z)$ be regular at $z_0\in D$, then

$$\varphi(z)=\frac{1}{f(z)-f(z_0)}-\frac{1}{f'(z_0)(z-z_0)}$$

is regular at z_0, so that $\varphi(z)$ is regular in D. We see easily that the Dirichlet integral of $\varphi(z)$ in D is finite, so that by the hypothesis: $D\in O_{AD}$, $\varphi(z)\equiv$const., hence $f(z)$ is a linear function of z. ——

Now we shall prove **A criterion of the class O_{AD}.**

Theorem X. 22. *Let $\{F_n\}$ be an exhaustion of F with the boundary $\{\Gamma_n\}$. Suppose that there exists a regular covering $\{G_n\}$ of $\{\Gamma_n\}$. Let $G_n=\sum_i G_n^i$ and G_n^i consist of $\nu(n, i)$ neighbourhoods U_j and put*

$$N^*(n)=\operatorname{Max}\nu(n, i) .$$

If

$$\sum_{n=1}^{\infty}\frac{1}{N^*(n)}=\infty,$$

then

$$F\in O_{AD} .$$ Sario [1].

Proof. Suppose that there exists a non-constant one-valued regular function $w=w(z)=u(z)+iv(z)$ on F, whose Dirichlet integral $D[w]=D_F[w]$ is finite. Let $L_n=\sum_i L_n^i$ be the system of Jordan curves in $G_n=\sum_i G_n^i$, defined before, then L_n bounds a compact sub-surface Φ_n of F and

$$D_{\Phi_n}[w]=\sum_i\int_{L_n^i} u\,dv . \tag{1}$$

Since $w(z)$ is one-valued, $\int_{L_n^i} dv=0$, so that if z_n^i is a point of L_n^i and

$$\operatorname*{Max}_{z\in L_n^i}|u(z)-u(z_n^i)|=M_n^i , \tag{2}$$

then

$$\begin{aligned}\Phi_{\Phi_n}[w]&=\sum_i\int_{L_n^i}(u(z)-u(z_n^i))dv\leqq\sum_i M_n^i\int_{L_n^i}|dv|\leqq\sum_i M_n^i\int_{L_n^i}|dw|\\ &\leqq\sum_i\int_{L_n^i}|dw|\int_{L_n^i}|dw|=\sum_i\left(\int_{L_n^i}|dw|\right)^2.\end{aligned} \tag{3}$$

Let $U_j=U(z_j)$ be a neighbourhood, which belongs to G_n^i and we map U_j conformally on $|t|<1$, such that z_j corresponds to $t=0$ and put $w=w(t)$. Let λ_j, l_j be defined as before, then

$$\left(\int_{l_j}|dw|\right)^2=\left(\int_{\lambda_j}|w'(t)||dt|\right)^2\leqq\int_{\lambda_j}|w'(t)|^2|dt| . \tag{4}$$

Since $|w'(t)|^2$ is subharmonic, we have for $|t|\leqq d<1$,

$$|w'(t)|^2\leqq\frac{1}{\pi(1-d)^2}\int_0^{1-d}\int_0^{2\pi}|w'(t+re^{i\theta})|^2\,rdrd\theta\leqq\frac{1}{\pi(1-d)^2}D_{U_j}[w] ,$$

so that by (4),

$$\left(\int_{l_j}|dw|\right)^2 \leqq \text{const. } D_{U_j}[w] ,$$

hence

$$\left(\int_{L_n^i}|dw|\right)^2 = \left(\sum_{j=1}^{\nu(n,i)}\int_{l_j}|dw|\right)^2 \leqq \nu(n,i)\sum_{j=1}^{\nu(n,i)}\left(\int_{l_j}|dw|\right)^2 \leqq$$

$$\text{const. } N^*(n)\sum_{j=1}^{\nu(n,i)} D_{U_j}[w] \leqq \text{const. } N^*(n)\cdot D_{G_n^i}[w] ,$$

so that by (3),

$$0 < a \leqq D_{\Phi_n}[w] \leqq \text{const. } N^*(n)\, D_{G_n}[w] ,$$

hence

$$\sum_n \frac{1}{N^*(n)} \leqq \text{const. } \sum_n D_{G_n}[w] \leqq \text{const. } D[w] < \infty , \tag{5}$$

which is absurd, since $\sum_n 1/N^*(n)=\infty$. Hence $F\in O_{AD}$. ——

Let $\tilde{F}$ be an unramified covering surface of F, such that above any point of F, there are N points of $\tilde{F}$, where $1\leqq N<\infty$. Then we call $\tilde{F}$ *an unramified covering surface of F of finite order N*. If F satisfies the condition of Theorem X. 22, then $\tilde{F}$ satisfies the same condition, hence

Theorem X. 23. *If F satisfy the condition of Theorem* X. 22 *and $\tilde{F}$ be an unramified covering surface of F of finite order, then $\tilde{F}\in O_{AD}$.*

Remark. Let F be a closed Riemann surface of genus $p\geqq 2$, spread over the w-sphere. Let C_k $(k=1, 2, \cdots, p)$ be p disjoint ring cuts of F and we cut F by $\{C_k\}$ and let F_0 be the resulting surface. By means of F_0, we construct a covering surface $F^{(\infty)}$ of F, which is of planar character and we map $F^{(\infty)}$ conformally on a schlicht domain Ω on the z-plane. We see easily that $F^{(\infty)}$ satisfies the condition of Theorem X. 22, hence $F^{(\infty)}\in O_{AD}$, so that $\Omega\in O_{AD}$. Now Ω is invariant by a certain group G of transformations $z'=T_\nu(z)$ $(\nu=0, 1, 2)$, where $T_\nu(z)$ is meromorophic and schlicht in Ω, so that by Theorem X. 21, $T_\nu(z)$ is a linear function of z, hence G is a linear group, which is proved in Theorem IX. 34.

Let E be the boundary of Ω, which is a bounded closed set of positive capacity, so that $\Omega\in P_G$. Hence *there exists an open Riemann surface, which is P_G and O_{AD}.*

6. Open Riemann surfaces of class O_{MD^*} and P_{MD^*}

Theorem X. 24. *Let F be an open Riemann surface, spread over the z-sphere, whose compact boundary Γ consists of a finite number of*

disjoint analytic Jordan curves and the harmonic measure of its ideal boundary be zero.

Let $w(z)$ be a non-constant one-valued meromorphic function on F and Ψ be the Riemann surface, generated by $w=w(z)$ on the w-sphere K.

(i) *If $D_F^*[w]<\infty$, then Ψ consists of a finite number N of sheets and the projection of its ideal boundary on K is a closed set of capacity zero. Hence any ideal boundary point of F is defined by a decreasing sequence of non-compact sub-surfaces of F, whose compact boundary consists of at most N disjoint analytic Jordan curves and* $\lim w(z)$ *exists, if z tends to any ideal boundary point of F.*

(ii) *If $D_F^*[w]=\infty$, then Ψ covers any point of K infinitely often, except a set of capacity zero.* Nagai [1], Tsuji [60].

Hence if $w(z)$ is bounded on F, then $D_F^*[w]<\infty$, so that $\lim w(z)$ *exists, if z tends to any ideal boundary point of F.* Kuramochi [1].

Proof. Let Λ be the image of Γ on Ψ and Λ^* be its projection on K, then by Λ^*, K is decomposed into a finite number of connected domains Ω_ν ($\nu=1, 2, \cdots, m$). Let a be any point of Ω_ν and be covered $n(a)$-times by Ψ and put

$$N_\nu=\sup_{a\in\Omega_\nu} n(a)\,. \tag{1}$$

Since the ideal boundary of F is of harmonic measure zero, we can prove similarly as Theorem VIII. 37 and Theorem X, 15, that Ψ covers any point of Ω_ν N_ν-times, except a set of capacity zero. We put

$$N=\text{Max}\,(N_1, \cdots, N_m)\,. \tag{2}$$

Suppose than $D_F^*[w]<\infty$, then Ψ is of finite spherical area, so that $N<\infty$, hence Ψ consists of at most N sheets and the projection E of its ideal boundary on K is a closed set of capacity zero. Hence E does not contain a continuum, so that if $w_0\in E$ is the projection of an ideal boundary point α of Ψ on K, then there exists a sequence of Jordan curves C_n in $K-E$, surrounding w_0 and tending to w_0 as $n\to\infty$. Let D_n be the inside of C_n, which we may assume that $D_1\supset D_2\supset\cdots\supset D_n\to w_0$ and let Ψ_n be the non-compact sub-surface of Ψ, which lies above D_n, such that $\Psi_1\supset\Psi_2\supset\cdots\supset\Psi_n\to\alpha$. Let γ_n be the compact boundary of Ψ_n, which lies above C_n, then γ_n consists of at most N disjoint analytic Jordan curves. $\{\Psi_n\}$ is a decreasing sequence of non-compact sub-surfaces of Ψ, defining an ideal boundary point α, so that any two ideal boundary points of Ψ are separated each other by at most N disjoint analytic Jordan curves. Hence returning to F, any ideal boundary point of F is defined by a decreasing sequence of non-compact sub-surfaces of F, whose compact boundary consists of at most N disjoint analytic Jordan curves and if z tends to any ideal boundary point A of F, then $w(z)$ tends to the boundary point α of Ψ, which corresponds to A.

If $D_F^*[w]=\infty$, then $N=\infty$ and we see easily that $N_1=\cdots=N_m=\infty$, so that Ψ covers any point of K infinitely often, except a set of capacity zero. ——

From the proof, we have

Theorem X. 25. *Let F satisfy the condition of Theorem* X. 24. *If there exists an ideal boundary point of F, which can not be defined by a decreasing sequence of non-compact sub-surfaces of F, whose compact boundary consists of a bounded number of analytic Jordan curves, then $F\in O_{MD^*}$.*

We can construct such a Riemann surface F easily and since F can be considered as a sub-surface of another Riemann surface of positive, or of null boundary, *there exists an open Riemann surface of positive, or of null boundary, which is O_{MD}^*.*

Let F be an open Riemann surface, obtained from a closed Riemann surface, by taking off a point, then $F\in O_G$, but we if we put $w(z)\equiv z$, then $D_F^*[w]<\infty$, so that $F\in P_{MD}^*$.

Let F be an open Riemann surface, which is a covering surface of planar character of a closed Riemann surface of genus $p\geqq 2$, then $F\in P_G$, (Theorem X. 45), but since F can be mapped conformally on a schlicht domain, $F\in P_{MD}^*$, so that *there exists an open Riemann surface of positive, or of null boundary, which is P_{MD}^*.*

7. Modified Green's function

1. Let F be an open Riemann surface, spread over the z-plane and $\{F_n\}$ be an exhaustion of F with the boundary $\{\Gamma_n\}$. We assume that F_0 is a schlicht disc: $|z|\leqq\rho_0<1$ and Γ_0: $|z|=\rho_0$ be its bounding circle. Let $g_n(z,0)$ be the Green's function of F_n and let

$$g_n(z,0)=\log\frac{1}{|z|}+\omega_n(z)\,, \tag{1}$$

where $\omega_n(z)$ is harmonic at $z=0$. $\omega_n(0)$ is the Robin's constant of F_n at $z=0$. Let

$$M_n=\max_{z\in\Gamma_0} g_n(z,0), \qquad u_n(z)=M_n-g_n(z,0)\,, \tag{2}$$

then $M_n>0$ and by the maximum principle,

$$u_n(z)>0 \quad \text{in } F_n-F_0\ (n\geqq 1)\,. \tag{3}$$

Since $u_1 \geqq 0$ on Γ_0 and $u_n(z_0)=0$ at a certain point $z_0 \in \Gamma_0$, $u_1(z_0)-u_n(z_0) \geqq 0$, and since $u_1(z)-u_n(z)$ $(n \geqq 2)$ is harmonic in F_1, by the maximum principle, $\operatorname{Max}_{z \in \Gamma_1} (u_1(z)-u_n(z)) \geqq 0$, or

$$\operatorname{Min}_{z \in \Gamma_1} u_n(z) \leqq \operatorname{Max}_{z \in \Gamma_1} u_1(z) = M_1 . \tag{4}$$

Hence by Harnack's principle, for any compact sub-surface Ω of $F-F_0$, which has a positive distance from Γ_0, there exists a constant $K=K(\Omega)$, which depends on Ω only, such that if F_n $(n \geqq n_0)$ contains Ω, then

$$|u_n(z)| = |g_n(z, 0)-M_n| \leqq K \quad \text{in } \Omega . \tag{5}$$

Let

$$v_n(z) = g_n(z, 0)-M_n-\log\frac{1}{|z|}, \tag{6}$$

and $|z| \leqq \rho_1$ $(\rho_0 < \rho_1 < 1)$ be contained in F_1, then

$$|v_n(z)| \leqq |g_n(z, 0)-M_n| + \log\frac{1}{|z|} \leqq K+\log\frac{1}{\rho_1} \quad \text{on } |z|=\rho_1 . \tag{7}$$

Since $v_n(z)$ is harmonic in $|z| \leqq \rho_1$, by the maximum principle, (7) holds in $|z| \leqq \rho_1$, so that

$$|v_n(0)| = |\omega_n(0)-M_n| \leqq K+\log\frac{1}{\rho_1},$$

hence by (5),

$$|g_n(z, 0)-\omega_n(0)| \leqq |g_n(z, 0)-M_n| + |M_n-\omega_n(0)| \leqq 2K+\log\frac{1}{\rho_1} \quad \text{in } \Omega . \tag{8}$$

Hence

$$\left| g_n(z, 0)-\omega_n(0)-\log\frac{1}{|z|} \right| \leqq 2K+2\log\frac{1}{\rho_1} \quad \text{on } |z|=\rho_1 .$$

Since the left hand side is harmonic in $|z| \leqq \rho_1$, the same relation holds in $|z| \leqq \rho_1$ and since the left hand side vanishes at $z=0$, we have

$$\left| g_n(z, 0)-\omega_n(0)-\log\frac{1}{|z|} \right| \leqq \text{const.}\ |z| \quad \text{in } |z| \leqq \rho_0 \ (n=1, 2, \cdots).$$

Since $\rho_0 > 0$ is arbitrary, we have

Theorem X. 26. *Let Ω be a compact sub-surface of F, which has a positive distance from $z=0$, then there exists a constant $K=K(\Omega)$, which depends on Ω only, such that if F_n $(n \geqq n_0)$ contains Ω, then*

$$|g_n(z, 0)-\omega_n(0)| \leqq K \quad \textit{in } \Omega,$$

$$\left| g_n(z, 0)-\log\frac{1}{|z|}-\omega_n(0) \right| \leqq \text{const.}\ |z| \quad \textit{in } |z| \leqq \rho_0 \ (n=1, 2, \cdots).$$

Hence we can select a partial sequence from $\{g_n(z, 0)\}$, which we denote by $\{g_n(z, 0)\}$ again, such that

$$\lim_n (g_n(z, 0)-\omega_n(0))=\tilde{g}(z, 0) \tag{1}$$

uniformly in the wider sense in $F-\{0\}$. $\tilde{g}(z, 0)$ is harmonic on F, except at $z=0$, where $\tilde{g}(z, 0)-\log\frac{1}{|z|}$ is harmonic and vanishes. Hence we have

Theorem X. 27. *Let F be an open Riemann surface, spread over the z-plane and $z_0 \in F$. Then there exists $\tilde{g}(z, z_0)$, which is harmonic on F, except at z_0, where $\tilde{g}(z, z_0)-\log\frac{1}{|z-z_0|}$ is harmonic and vanishes.*

We remark that if there exists a Green's function $g(z, z_0)$ of F, then $\tilde{g}(z, z_0)=g(z, z_0)-\omega(z_0)$, where $\omega(z_0)$ is the Robin's constant of F at z_0.

Theorem X. 28. *Let F be a closed Riemann surface and $z=0$, $z=z_0(\neq 0)$ be two points of F. If $\tilde{g}(z, 0)$ is defined for an open Riemann surface $F-\{z_0\}$, then $\tilde{g}(z, 0)$ has a logarithmic singularity $-\log\frac{1}{|z-z_0|}$ at z_0, such that $g(z, 0)+\log\frac{1}{|z-z_0|}$ is harmonic at z_0.*

Hence on a closed Riemann surface F, there exists a harmonic function, which has a logarithmic singularity $\log\frac{1}{|z-z_1|}$ at $z=z_1$ and a logarithmic singularity $-\log\frac{1}{|z-z_2|}$ at $z=z_2$ $(z_1\neq z_2)$.

Proof. Let Δ_n: $|z-z_0|\leqq r_n$ $(r_1>r_2>\cdots>r_n\to 0)$ be a disc about z_0 and $F_n=F-\Delta_n$, then $\{F_n\}$ is an exhaustion of $F-\{z_0\}$ and let $g_n(z, 0)$ be the Green's function of F_n. By taking a suitable partial sequence, we may assume that

$$\lim_n (g_n(z, 0)-\omega_n(0))=\tilde{g}(z, 0)\ . \tag{1}$$

Since $\tilde{g}(z, 0)$ is harmonic in $0<|z-z_0|\leqq r_1$, $\tilde{g}(z, 0)$ can be expressed in the form:

$$\tilde{g}(z, 0)=\log r+a_0+\sum_{k=1}^{\infty}(a_k r^k+a_{-k}r^{-k})\cos k\theta+\sum_{k=1}^{\infty}(b_k r^k+b_{-k}r^{-k})\sin k\theta, \tag{2}$$

$$z-z_0=re^{i\theta},\ 0<r\leqq r_1\ .$$

The coefficient of $\log r$ is 1, since if $\rho>0$ is small, two circles $|z|=\rho$ and $|z-z_0|=r_1$ bound a domain on F, so that

$$\text{the coefficient of }\log r=\frac{1}{2\pi}\int_{|z-z_0|=r_1}\frac{\partial\tilde{g}}{\partial r_1}r_1 d\theta=-\frac{1}{2\pi}\int_{|z|=\rho}\frac{\partial\tilde{g}}{\partial\rho}\rho\, d\varphi=1\ .$$

Since $g_n(z, 0)$ is harmonic in $r_n\leqq|z-z_0|\leqq r_1$,

$$g_n(z, 0)-\omega_n(0)=\log r+a_0^{(n)}+\sum_{k=1}^{\infty}(a_k^{(n)}r^k+a_{-k}^{(n)}r^{-k})\cos k\theta$$

$$+\sum_{k=1}^{\infty}(b_k^{(n)}r^k+b_{-k}^{(n)}r^{-k})\sin k\theta \quad (r_n\leqq r\leqq r_1)\,, \tag{3}$$

where the coefficient of $\log r$ is 1 as (2).

Then by (1),

$$\lim_n a_0^{(n)}=a_0, \qquad \lim_n a_{\pm k}^{(n)}=a_{\pm k}, \qquad \lim_n b_{\pm k}^{(n)}=b_{\pm k}\,. \tag{4}$$

Since $g_n(z, 0)=0$ on $|z-z_0|=r_n$,

$$\log r_n+a_0^{(n)}=-\omega_n(0), \qquad a_k^{(n)}r_n^k+a_{-k}^{(n)}r_n^{-k}=0, \qquad b_k^{(n)}r_n^k+b_{-k}^{(n)}r_n^{-k}=0\,. \tag{5}$$

Since by (4), $a_k^{(n)}$, $b_k^{(n)}$ are bounded for $n\to\infty$ and $r_n\to 0$, we see from (5),

$$a_{-k}=\lim_n a_{-k}^{(n)}=0, \qquad b_{-k}=\lim_n b_{-k}^{(n)}=0\,,$$

so that

$$\tilde{g}(z, 0)=\log r+a_0+\sum_{k=1}^{\infty}(a_k\cos k\theta+b_k\sin k\theta)r^k\,. \tag{6}$$

Hence $\tilde{g}(z, 0)-\log r=\tilde{g}(z, 0)+\log\dfrac{1}{|z-z_0|}$ is harmonic at z_0. ——

Remark. $\lim_n (g_n(z, 0)-\omega_n(0))$ exists, so that we do not need take a partial sequence from $\{g_n(z, 0)-\omega_n(0)\}$, which can be proved as follows. If $\lim_n (g_n(z, 0)-\omega_n(0))$ does not exist, we choose two partial sequences $\{n_\nu\}$, $\{m_\nu\}$ from $\{n\}$, such that

$$\lim (g_{n_\nu}(z, 0)-\omega_{n_\nu}(0))=\tilde{g}_1(z, 0), \qquad \lim_\nu (g_{m_\nu}(z, 0)-\omega_{m_\nu}(0))=\tilde{g}_2(z, 0)$$

exist, such that $\tilde{g}_1\not\equiv\tilde{g}_2$. Since each of $\tilde{g}_1$, $\tilde{g}_2$ has a logarithmic singularity $\log 1/|z|$ at $z=0$ and $-\log 1/|z-z_0|$ at $z=z_0$, $G(z)=\tilde{g}_1(z, 0)-\tilde{g}_2(z, 0)$ is harmonic on F, so that $G\equiv$const., but since $\tilde{g}_1$ and $\tilde{g}_2$ vanish at $z=0$, $G=0$ at $z=0$, so that $G\equiv 0$, hence $\tilde{g}_1\equiv\tilde{g}_2$, which is absurd. Hence

$$\lim_n (g_n(z, 0)-\omega_n(0))=\tilde{g}(z, 0) \text{ exists. —— } \tag{7}$$

Hence from (5),

$$\lim_n\left(\omega_n(0)-\log\frac{1}{r_n}\right)=-a_0 \tag{8}$$

exists.

2. The following lemma plays an important rôle in the sequel.

Fundamental lemma. *Let F be an open Riemann surface, spread over the z-plane and $\{F_n\}$ be its exhaustion with the boundary $\{\Gamma_n\}$, where we assume that F_0 is a schlicht circular disc, which contains $z=0$.*

Let $g_n(z, \zeta)$ be the Green's function of F_n and $\omega_n(\zeta)$ be the Robin's *constant of F_n at ζ.*

(i) *Let a disc Δ_0: $|z-\zeta_0|\leqq\rho_0$ be contained in F_n $(n\geqq n_0)$, then*

$$|\omega_n(\zeta)-\omega_n(\zeta_0)|\leqq K|\zeta-\zeta_0| \quad \text{for any } \zeta\in\Delta_0 \quad (n\geqq n_0),$$

where $K=K(\Delta_0)$ is a constant, which depends on Δ_0 only.

(ii) *Let Ω be a compact sub-surface of F, which has a positive distance from the boundary of Δ_0: $|z-\zeta_0|\leqq\rho_0$. Then*

$$|g_n(z, \zeta)-g_n(z, \zeta_0)|\leqq K_1|\zeta-\zeta_0| \quad \text{for any } z\in\Omega, \quad \zeta\in\Delta_0 \quad (n\geqq n_0),$$

where $K_1=K_1(\Omega, \Delta_0)$ is a constant, which depends on Ω and Δ_0 only.

Proof. Let $\rho_0<\rho_1<\rho_2$ and a disc: $|z-\zeta_0|\leqq\rho_2$ be contained in F and Ω lie outside of $|z-\zeta_0|\leqq\rho_2$. We put

$$\Delta_0: |z-\zeta_0|\leqq\rho_0, \quad \Delta_1: |z-\zeta_0|\leqq\rho_1, \quad \Delta_2: |z-\zeta_0|\leqq\rho_2\,, \tag{1}$$

$$C_0: |z-\zeta_0|=\rho_0, \quad C_1: |z-\zeta_0|=\rho_1, \quad C_2: |z-\zeta_0|=\rho_2\,. \tag{2}$$

Let $\zeta\in\Delta_0$ and $g_{\Delta_2}(z, \zeta)$ be the Green's function of Δ_2, then

$$g_{\Delta_2}(z, \zeta)=\log\left|\frac{\rho_2^2-(\overline{\zeta-\zeta_0})(z-\zeta_0)}{\rho_2(z-\zeta)}\right|. \tag{3}$$

Let Δ_2 be contained in $F_n\,(n\geqq n_0)$ and put for $\zeta\in\Delta_0$ $(\zeta\neq\zeta_0)$,

$$\left.\begin{aligned} M_n(\zeta)&=\operatorname*{Max}_{z\in C_1}\frac{1}{|\zeta-\zeta_0|}(g_n(z, \zeta)-g_n(z, \zeta_0))\,,\\ u_n(z)&=M_n(\zeta)-\frac{1}{|\zeta-\zeta_0|}(g_n(z, \zeta)-g_n(z, \zeta_0))\,,\end{aligned}\right\}\quad(\zeta\in\Delta_0)\,. \tag{4}$$

$$\left.\begin{aligned} M(\zeta)&=\operatorname*{Max}_{z\in C_1}\frac{1}{|\zeta-\zeta_0|}(g_{\Delta_2}(z, \zeta)-g_{\Delta_2}(z, \zeta_0))\,,\\ u(z)&=M(\zeta)-\frac{1}{|\zeta-\zeta_0|}(g_{\Delta_2}(z, \zeta)-g_{\Delta_2}(z, \zeta_0))\,,\end{aligned}\right\}\quad(\zeta\in\Delta_0)\,. \tag{5}$$

We may assume that $M_n(\zeta)\geqq 0$. For if otherwise, we interchange ζ and ζ_0 in $g_n(z, \zeta)-g_n(z, \zeta_0)$, $g_{\Delta_2}(z, \zeta)-g_{\Delta_2}(z, \zeta_0)$.

Since $u_n(z)\geqq 0$ on C_1 and $u_n(z)=M_n(\zeta)\geqq 0$ on Γ_n, we have by the maximum principle,

$$u_n(z)\geqq 0 \quad \text{on } F_n-\Delta_1\,. \tag{6}$$

$u(z)\geqq 0$ on C_1 and at a certain point $z_1\in C_1$, $u_n(z_1)=0$, so that $u(z_1)-u_n(z_1)\geqq 0$, and since $u(z)-u_n(z)$ is harmonic in Δ_2, by the maximum principle, $\operatorname*{Max}_{z\in C_2}(u(z)-u_n(z))\geqq 0$, or

$$\operatorname*{Min}_{z\in C_2}u_n(z)\leqq\operatorname*{Max}_{z\in C_2}u(z)=M(\zeta)\leqq K(\Delta_0)\,. \tag{7}$$

Hence by (6), (7) and Harnack's principle, we have

$$|u_n(z)|=\left|M_n(\zeta)-\frac{1}{|\zeta-\zeta_0|}(g_n(z,\zeta)-g_n(z,\zeta_0))\right|\leqq K(\Omega,\Delta_0),\quad z\in\Omega,\quad \zeta\in\Delta_0\,. \tag{8}$$

Hence for any $z\in C_2$, $\zeta\in\Delta_0$,

$$\left|M_n(\zeta)-\frac{1}{|\zeta-\zeta_0|}\left(g_n(z,\zeta)-\log\frac{1}{|z-\zeta|}-g_n(z,\zeta_0)+\log\frac{1}{|z-\zeta_0|}\right)\right|\leqq K(\Delta_0). \tag{9}$$

Since the left hand side is a harmonic function of z in Δ_2, (9) holds for $z\in\Delta_2$, so that at $z=\zeta_0$, $z=\zeta$, we have

$$\left|M_n(\zeta)-\frac{1}{|\zeta-\zeta_0|}\left(g_n(\zeta_0,\zeta)-\log\frac{1}{|\zeta-\zeta_0|}-\omega_n(\zeta_0)\right)\right|\leqq K(\Delta_0)\,,$$

$$\left|M_n(\zeta)-\frac{1}{|\zeta-\zeta_0|}\left(\omega_n(\zeta)-g_n(\zeta,\zeta_0)+\log\frac{1}{|\zeta-\zeta_0|}\right)\right|\leqq K(\Delta_0)\,.$$

Hence adding and subtracting each other, we have

$$\left|M_n(\zeta)-\frac{\omega_n(\zeta)-\omega_n(\zeta_0)}{2|\zeta-\zeta_0|}\right|\leqq K(\Delta_0)\,, \tag{10}$$

$$\left|\omega_n(\zeta)-\omega_n(\zeta_0)-2\left(g_n(\zeta,\zeta_0)-\log\frac{1}{|\zeta-\zeta_0|}-\omega_n(\zeta_0)\right)\right|\leqq K(\Delta_0)|\zeta-\zeta_0|. \tag{11}$$

Since by Theorem X. 26,

$$\left|g_n(\zeta,\zeta_0)-\log\frac{1}{|\zeta-\zeta_0|}-\omega_n(\zeta_0)\right|\leqq K(\Delta_0)|\zeta-\zeta_0|\,,$$

we have by (11),

$$|\omega_n(\zeta)-\omega_n(\zeta_0)|\leqq K(\Delta_0)|\zeta-\zeta_0|,\qquad \zeta\in\Delta_0\,, \tag{12}$$

so that by (10), $|M_n(\zeta)|\leqq K(\Delta_0)$, hence by (8),

$$|g_n(z,\zeta)-g_n(z,\zeta_0)|\leqq K(\Omega,\Delta_0)|\zeta-\zeta_0|,\qquad z\in\Omega,\ \zeta\in\Delta_0\,. \tag{13}$$

Hence the lemma is proved.

3. Let Ω_0, Ω be two disjoint compact sub-surfaces of F, contained in F_n $(n\geqq n_0)$. Then by the lemma and Borel's covering theorem, for any $\varepsilon>0$, there exist $\delta=\delta(\Omega_0,\Omega,\varepsilon)$ and $\eta=\eta(\Omega_0,\varepsilon)$, which satisfy the following condition:

$$|g_n(z,\zeta_1)-g_n(z,\zeta_2)|<\varepsilon,\quad \text{if } z\in\Omega,\ \zeta_1\in\Omega_0,\ \zeta_2\in\Omega_0,\quad |\zeta_1-\zeta_2|<\delta\,, \tag{1}$$

$$|g_n(z_1,\zeta)-g_n(z_2,\zeta)|<\varepsilon,\quad \text{if } \zeta\in\Omega_0,\ z_1\in\Omega,\ z_2\in\Omega,\quad |z_1-z_2|<\delta\,, \tag{2}$$

$$|\omega_n(\zeta_1)-\omega_n(\zeta_2)|<\varepsilon,\quad \text{if } \zeta_1\in\Omega_0,\ \zeta_2\in\Omega_0,\quad |\zeta_1-\zeta_2|<\eta\,. \tag{3}$$

Hence if we assume that $z=0$ belongs to F, then

$$\tilde{g}_n(z,\zeta)=g_n(z,\zeta)-\omega_n(0) \tag{4}$$

is equi-continuous for $z\in\Omega$, $\zeta\in\Omega_0$. We shall prove that $\tilde{g}_n(z,\zeta)$ $(n=1,2,\cdots)$ is uniformly bounded for $z\in\Omega$, $\zeta\in\Omega_0$.

Let $\zeta_0 \in \Omega_0$ be a fixed point and $\zeta \in \Omega_0$, $z \in \Omega$, then

$$|\tilde{g}_n(z, \zeta)| \leqq |g_n(z, \zeta) - g_n(z, \zeta_0)| + |g_n(z, \zeta_0) - \omega_n(\zeta_0)| + |\omega_n(\zeta_0) - \omega_n(0)|. \quad (5)$$

If $n=1, 2, \cdots$, then by (1), $|g_n(z, \zeta) - g_n(z, \zeta_0)|$ is uniformly bounded for $z \in \Omega$, $\zeta \in \Omega_0$ and by Theorem X. 26, $|g_n(z, \zeta_0) - \omega_n(\zeta_0)|$ is uniformly bounded for $z \in \Omega$ and by (3), $|\omega_n(\zeta) - \omega_n(0)|$ is bounded, hence $\tilde{g}_n(z, \zeta)$ is uniformly bounded for $z \in \Omega$, $\zeta \in \Omega_0$. Hence by Arzelà's theorem, we can select a partial sequence from $\{\tilde{g}_n(z, \zeta)\}$, which we denote by $\{\tilde{g}_n(z, \zeta)\}$ again, such that

$$\lim_n \tilde{g}_n(z, \zeta) = \tilde{g}(z, \zeta) \quad (6)$$

uniformly for $z \in \Omega$, $\zeta \in \Omega_0$. From this, we see easily that we can find a partial sequence from $\{\tilde{g}_n(z, \zeta)\}$, which we denote by $\{\tilde{g}_n(z, \zeta)\}$ again, such that (6) holds uniformly in the wider sense for $z \in F - \{\zeta\}$, for a fixed $\zeta \in F$ and (6) holds uniformly in the wider sense for $\zeta \in F - \{z\}$, for a fixed $z \in F$. Hence $\tilde{g}(z, \zeta)$ is a harmonic function of $z (\neq \zeta)$ for a fixed ζ and is a harmonic function of $\zeta (\neq z)$ for a fixed z.

Since $g_n(z, \zeta) = g_n(\zeta, z)$, we have

$$\tilde{g}(z, \zeta) = \tilde{g}(\zeta, z) . \quad (7)$$

Let Δ_0 be a schlicht disc, contained in F and C_0 be its bounding circle and we put for $z \in \Delta_0$, $\zeta \in \Delta_0$,

$$\chi_n(z, \zeta) = g_n(z, \zeta) - \omega_n(0) - \log \frac{1}{|z - \zeta|} . \quad (8)$$

Then $\chi_n(z, \zeta)$ is a harmonic function with respect to each variable z and ζ in Δ_0 and since for a fixed $\zeta \in \Delta_0$,

$$\lim_n \chi_n(z, \zeta) = \chi(z, \zeta) = \tilde{g}(z, \zeta) - \log \frac{1}{|z - \zeta|} \quad (9)$$

uniformly on $z \in C_0$, (9) holds uniformly for $z \in \Delta_0$, hence for a fixed $\zeta \in \Delta_0$, $\chi(z, \zeta)$ is a harmonic function of $z \in \Delta_0$ and for a fixed $z \in \Delta_0$, $\chi(z, \zeta)$ is a harmonic function of $\zeta \in \Delta_0$.

Hence we have proved the following theorem.

Theorem X. 29. *Let F be an open Riemann surface, spread over the z-plane, then there exists $\tilde{g}(z, \zeta)$ on F, which satisfies the following condition:*

(i) *For a fixed ζ, $\tilde{g}(z, \zeta)$ is a harmonic function of z ($\neq \zeta$) and for a fixed z, $\tilde{g}(z, \zeta)$ is a harmonic function of ζ ($\neq z$).*

(ii) *Let Δ_0 be a schlicht disc, contained in F and put*

$$\tilde{g}(z, \zeta) = \log \frac{1}{|z - \zeta|} + \chi(z, \zeta), \quad z \in \Delta_0, \ \zeta \in \Delta_0 ,$$

then for a fixed $\zeta \in \Delta_0$, $\chi(z, \zeta)$ *is a harmonic function of* $z \in \Delta_0$ *and for a fixed* $z \in \Delta_0$, $\chi(z, \zeta)$ *is a harmonic function of* $\zeta \in \Delta_0$.

(iii) $$\tilde{g}(z, \zeta) = \tilde{g}(\zeta, z) .$$ Tsuji [27].

We call $\tilde{g}(z, \zeta)$ *the modified Green's function of* F *with* ζ *as its pole.*

8. Potential function with two logarithmic singularities

1. We shall prove

Theorem X. 30. *Let* F *be a closed, or an open Riemann surface, spread over the z-plane and* ζ_1, ζ_2 *be two points of* F, *which are not branch points of* F[1]. *Then there exists* $u(z; \zeta_1, \zeta_2)$ *on* F, *which satisfies the following condition:*

(i) $u(z; \zeta_1, \zeta_2)$ *is a harmonic function of* z $(\neq \zeta_1, \zeta_2)$ *and*

$$u(z; \zeta_1, \zeta_2) - \log \frac{1}{|z-\zeta_1|} \text{ is harmonic at } z = \zeta_1 ,$$

$$u(z; \zeta_1, \zeta_2) + \log \frac{1}{|z-\zeta_2|} \text{ is harmonic at } z = \zeta_2 .$$

(ii) *Let* Ω *be a simply connected compact sub-surface of* F, *which contains* ζ_1, ζ_2 *and is bounded by an analytic Jordan curve* Γ, *then the Dirichlet integral of* u *on* $F-\Omega$ *is finite, such that*

$$D_{F-\Omega}[u] \leqq \int_\Gamma u \frac{\partial u}{\partial \nu} ds ,$$

where ν *is the inner normal of* Γ *with respect to* Ω.

(iii) u *is bounded on* $F-\Omega$, *such that if* $|u| \leqq M$ *on* Γ, *then*

$$|u| \leqq M \quad \text{on } F-\Omega .$$

Proof. (i) First we assume that F is an open Riemann surface, then

$$u(z; \zeta_1, \zeta_2) = \tilde{g}(z, \zeta_1) - \tilde{g}(z, \zeta_2) \tag{1}$$

satisfies the condition (i). To prove (ii), let $\{F_n\}$ be an exhaustion of F with the boundary $\{\Gamma_n\}$ and let $\Omega \subset F_n$ $(n \geqq n_0)$. We put

$$u_n = u_n(z; \zeta_1, \zeta_2) = g_n(z, \zeta_1) - g_n(z, \zeta_2) = (g_n(z, \zeta_1) - \omega_n(0)) - (g_n(z, \zeta_2) - \omega_n(0)) ,$$

where g_n is the Green's function of F_n, then by (1)

$$\lim_n u_n(z; \zeta_1, \zeta_2) = u(z; \zeta_1, \zeta_2) .$$

Since $u_n = 0$ on Γ_n,

---

1) For the sake of brevity, we assume that ζ_1, ζ_2 are not branch points of F. If ζ_1 is branch point, then we map a neighbourhood of ζ_1 conformally on $|t| < \rho$ by $t = (z-\zeta_1)^\alpha$ with a suitable $\alpha > 0$ and we consider on the t-plane.

$$D_{F_n-\Omega}[u_n]=\int_{\Gamma} u_n \frac{\partial u_n}{\partial \nu} ds ,$$

so that if $m<n$,

$$D_{F_m-\Omega}[u_n] \leqq D_{F_n-\Omega}[u_n]=\int_{\Gamma} u_n \frac{\partial u_n}{\partial \nu} ds .$$

Hence if we make $n \to \infty$ and then $m \to \infty$, then we have

$$D_{F-\Omega}[u] \leqq \int_{\Gamma} u \frac{\partial u}{\partial \nu} ds . \tag{2}$$

Suppose that $|u| \leqq M$ on Γ, then we shall prove that $|u| \leqq M$ on $F-\Omega$. Since $|u_n|$ is subharmonic on $F_n-\Omega$ and $u_n=0$ on Γ_n,

$$|u_n| \leqq \operatorname*{Max}_{z \in \Gamma} |u_n| \quad \text{on } F_n-\Omega ,$$

so that

$$|u| \leqq \operatorname*{Max}_{z \in \Gamma} |u| \qquad \text{on } F-\Omega . \tag{3}$$

(ii) Next if F is a closed Riemann surface, then we take off a point z_0 from F and construct $u(z;\ \zeta_1,\ \zeta_2)$ for the open Riemann surface $F-\{z_0\}$, then since u is bounded in a neighbourhood of z_0, u is harmonic at z_0, hence $u(z; \zeta_1, \zeta_2)$ satisfies the condition of the theorem. ——

We shall use the following corollary in Chapter XI.

Corollary. *Let Ω, Ω_0 be two compact sub-surfaces of F, such that $\overline{\Omega} \subset \Omega_0$ and ζ_2 be a fixed point of Ω, whose neighbourhood $U(\zeta_2)$ is contained in Ω and ζ_1 be a variable point in $\Omega-U(\zeta_2)$. Then there exists $\rho_0>0$, such that ρ_0-neighbourhood $U(\zeta_1)$ of ζ_1 is contained in Ω_0 and does not contain a branch point of F, except ζ_1. Then*

$$|u(z;\ \zeta_1,\ \zeta_2)| \leqq K \quad \text{on } F-U(\zeta_1)-U(\zeta_2) ,$$

where K is a constant, which is independent of $\zeta_1 \in \Omega-U(\zeta_2)$.

Proof. Let Γ_0 be the boundary of Ω_0, which, we assume, consists of a finite number of analytic Jordan curves and put

$$M(\zeta_1)=\operatorname*{Max}_{z \in \Gamma_0} |u(z;\ \zeta_1,\ \zeta_2)| , \tag{1}$$

then by Theorem X. 30,

$$|u(z;\ \zeta_1,\ \zeta_2)| \leqq M(\zeta_1) \quad \text{for any } z \in F-\Omega_0 . \tag{2}$$

Since $u(z; \zeta_1, \zeta_2)$ is a continuous function of two variables ζ_1 and z for $\zeta_1 \in \Omega-U(\zeta_2)$, $z \in \Gamma_0$, there exists a constant M, such that $M(\zeta_1) \leqq M$ for any $\zeta_1 \in \Omega-U(\zeta_2)$, so that

$$|u(z; \zeta_1, \zeta_2)| \leqq M \quad \text{for any } z \in F-\Omega_0, \quad \zeta_1 \in \Omega-U(\zeta_2) . \tag{3}$$

Next we shall prove that there exists a constant M_1, such that

$$|u(z;\zeta_1,\zeta_2)|\leqq M_1 \text{ for any } z\in\Omega_0-U(\zeta_1)-U(\zeta_2),\ \zeta_1\in\Omega-U(\zeta_2)\,. \quad (4)$$

For, if otherwise, there exist $z^{(n)}\in\Omega_0-U(\zeta_1^{(n)})-U(\zeta_2)$, $\zeta_1^{(n)}\in\Omega-U(\zeta_2)$, $(n=1,2,\cdots)$, such that $\lim_n |u(z^{(n)};\zeta_1^{(n)},\zeta_2)|=\infty$. We may assume that $z^{(n)}\to z_0\in\Omega_0-U(\zeta_1)-U(\zeta_2)$, $\zeta_1^{(n)}\to\zeta_1\in\Omega-U(\zeta_2)$, then $|u(z_0;\zeta_1,\zeta_2)|=\infty$, which is absurd, since $z_0\neq\zeta_1$. Hence there exists a constant M_1, which satisfies (4). Hence if we put $K=\text{Max}\,(M,M_1)$, then K satisfies the required condition.

2. Let ζ_1,ζ_2 be two points of F, which are not branch points of F and we connect them by an analytic Jordan arc C: $z=z(t)$ $(0\leqq t\leqq 1)$, $z(0)=\zeta_1$, $z(1)=\zeta_2$, which does not pass through branch points of F. The positive direction of C is that, for which t increases and the positive normal ν of C is such that if we rotate ν by an angle $+\frac{\pi}{2}$, then it coincides with the positive tangent of C.

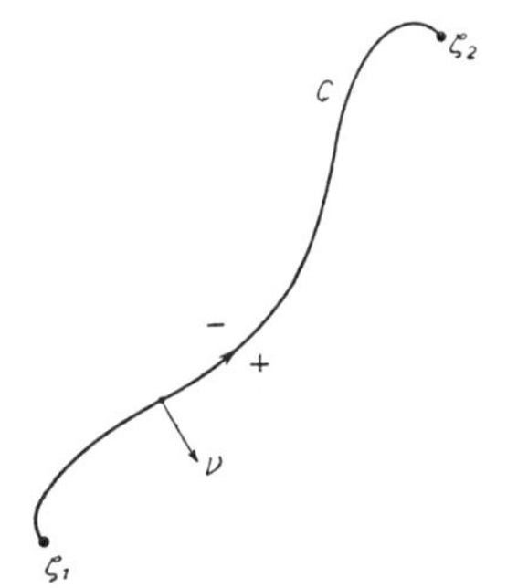

C has two sides $\overset{+}{C}$, $\overset{-}{C}$, where $\overset{+}{C}$ is such that from which the positive normal issues. Let z be any point of C, then we denote it by $\overset{+}{z}$, $\overset{-}{z}$, according as it is considered as belong to $\overset{+}{C}$, or $\overset{-}{C}$ respectively.

Theorem X. 31. *Let F be a closed, or an open Riemann surface, spread over the z-plane and ζ_1,ζ_2 be two points of F, which are not branch points. Then there exists $v(z;\zeta_1,\zeta_2)$ on F, which satisfies the following condition:*

(i) *$v(z;\zeta_1,\zeta_2)$ is a one-valued harmonic function of z on $F-\{C\}$, such that at any point $z\in C$,*

$$v(\overset{+}{z};\zeta_1,\zeta_2)-v(\overset{-}{z};\zeta_1,\zeta_2)=2\pi\,.$$

Hence v can be continued harmonically across C to an infinitely many valued harmonic function on F.

$$v(z;\zeta_1,\zeta_2)-\arg(z-\zeta_1) \text{ is harmonic at } z=\zeta_1\,,$$

$$v(z;\zeta_1,\zeta_2)+\arg(z-\zeta_2) \text{ is harmonic at } z=\zeta_2\,.$$

(ii) *Let Ω be a simply connected compact sub-surface of F, which contains ζ_1,ζ_2 and is bounded by an analytic Jordan curve Γ, then the Dirichlet integral of v on $F-\Omega$ is finite, such that*

$$D_{F-\Omega}[v]\leqq\int_\Gamma v\frac{\partial v}{\partial\nu}ds\,,$$

where ν is the inner normal of Γ with respect to Ω.

(iii) *v is bounded on $F-\Omega$, such that if $|v|\leqq M$ on Γ, then $|v|\leqq M$ on $F-\Omega$.*

Proof. First we assume that F is an open Riemann surface and put

$$v(z;\zeta_1,\zeta_2)=\int_C \frac{\partial\tilde{g}(z,\zeta)}{\partial\nu_\zeta}|d\zeta| =\int_{\zeta_1}^{\zeta_2}\frac{\partial\tilde{g}(z,\zeta)}{\partial\nu_\zeta}|d\zeta|,\qquad z\in F-\{C\}\,, \tag{1}$$

then v is a one-valued harmonic function of z on $F-\{C\}$.

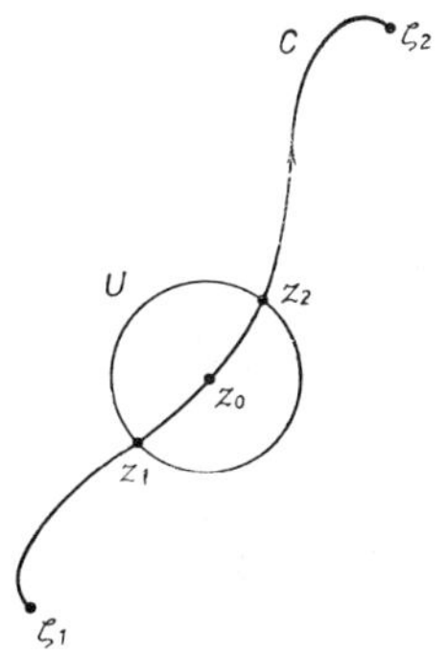

Let $z_0=z(t_0)\,(0<t_0<1)$ be a point of C and U: $|z-z_0|\leqq\rho$ be its schlicht neighbourhood, which does not contain ζ_1, ζ_2.

If $\rho>0$ is small, then the circle $|z-z_0|=\rho$ meets C at two points $z_1=z(t_1)$, $z_2=z(t_2)$ $(t_1<t_0<t_2)$.

Let for $z\in U$, $\zeta\in U$,

$$\tilde{g}(z,\zeta)=\log\frac{1}{|\zeta-z|}+\chi(z,\zeta)\,, \tag{2}$$

then for any $z\in U$, which does not lie on C, we have by the choice of ν,

$$v(z;\zeta_1,\zeta_2)=\int_{\zeta_1}^{z_1}\frac{\partial\tilde{g}(z,\zeta)}{\partial\nu_\zeta}|d\zeta|-\int_{z_1}^{z_2}d\arg(\zeta-z) +\int_{z_1}^{z_2}\frac{\partial\chi(z,\zeta)}{\partial\nu_\zeta}|d\zeta|+\int_{z_2}^{\zeta_2}\frac{\partial\tilde{g}(z,\zeta)}{\partial\nu_\zeta}|d\zeta|\,. \tag{3}$$

Since all terms, except $\int_{z_1}^{z_2}d\arg(\zeta-z)$, are continuous at z_0, we have easily

$$v(\overset{+}{z}_0;\zeta_1,\zeta_2)-v(\overset{-}{z}_0;\zeta_1,\zeta_2)=2\pi\,. \tag{4}$$

By (4), $v(z;\zeta_1,\zeta_2)-\arg(z-\zeta_1)$ is a one-valued bounded harmonic function of z in $0<|z-\zeta_1|<\rho$, so that it is harmonic at $z=\zeta_1$. Similarly $v(z;\zeta_1,\zeta_2)+\arg(z-\zeta_2)$ is harmonic at $z=\zeta_2$. Hence (i) is proved.

To prove (ii), let $\{F_n\}$ be an exhaustion of F with the boundary $\{\Gamma_n\}$ and $g_n(z,\zeta)$ be the Green's function of F_n. We divide C into N arcs of equal length Δs and ξ_k $(k=0,1,2,\cdots,N)$ be the points of division, where $\xi_0=\zeta_1$, $\xi_N=\zeta_2$ and $e^{i\theta_k}$ be the unit vector on the positive normal ν of C at ξ_k and put

$$v_n^N(z;\zeta_1,\zeta_1)=\sum_{k=1}^{N}\frac{g_n(z,\xi_k+\delta e^{i\theta_k})-g_n(z,\xi_k)}{\delta}\Delta s,\qquad \delta>0\,. \tag{5}$$

Then

$$\lim_{n\to\infty}\lim_{N\to\infty}\lim_{\delta\to 0} v_n^N=\int_C \frac{\partial\tilde{g}(z,\zeta)}{\partial\nu}|d\zeta|=v(z;\zeta_1,\zeta_2)\,.$$

Since $v_n^N=0$ on Γ_n,

$$D_{F_m-\Omega}[v_n^N]\leqq D_{F_n-\Omega}[v_n^N]=\int_\Gamma v_n^N\frac{\partial v_n^N}{\partial\nu}ds\qquad(m\leqq n)\,,$$

so that if we make $\delta\to 0$, $N\to\infty$, $n\to\infty$, $m\to\infty$ successively, then we have

$$D_{F-\Omega}[v]\leqq\int_\Gamma v\frac{\partial v}{\partial\nu}ds\,. \tag{6}$$

(iii) can be proved similarly as theorem X. 30.

Hence the theorem is proved, if F is an open Riemann surface. If F is a closed Riemann surface, then we take off a point z_0 from F and we can prove the theorem similarly as Theorem X. 30. ——

From the proof, we have

Theorem X. 32. *Let F be a closed Riemann surface of genus $p\geqq 1$, spread over the z-plane and C be an analytic Jordan curve on F, with the prescribed positive sense, which does not pass through branch points of F and does not decompose F into two parts. Then there exists a one-valued harmonic function $v(z)$ on $F-\{C\}$, such that at any point $z\in C$,*

$$v(\overset{+}{z})-v(\overline{z})=1\,.$$

Hence $v(z)$ can be continued harmonically across C to an infinitely many valued harmonic function on F.

9. Abelian integral of the first kind

Let F be a closed Riemann surface of genus $p\geqq 1$, spread over the z-sphere and C_h, C_h^* $(h=1,2,\cdots,p)$ be p pairs of canonical ring cuts of F, such that C_h, C_h^* meet at a point and C_h, C_k, C_k^* $(h\neq k)$ are disjoint. We define the positive side of C_h, C_h^* as shown in the figure.

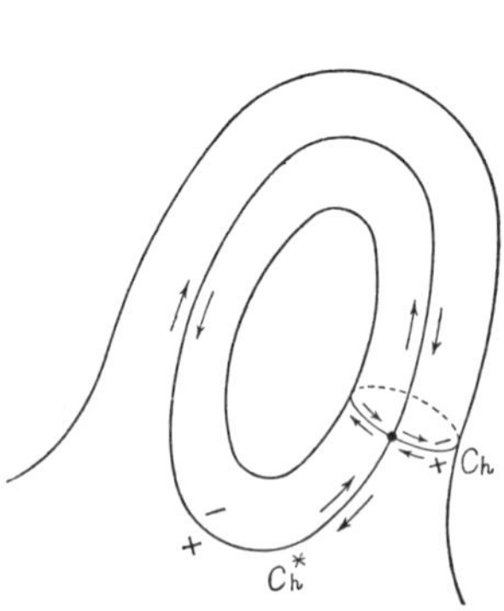

We cut F by $\{C_h\}$, $\{C_h^*\}$ and let F_0 be the resulting surface.

Let $w(z)=u(z)+iv(z)$ be *an Abelian integral of the first kind on F*, such that $w(z)$ is one-valued and regular on F_0 and if $z_h\in C_h$, then

$$\left.\begin{aligned} &w(\overset{+}{z}_h)-w(\bar{z}_h)=\text{const.}=\omega_h=\alpha_h+i\beta_h\,, \\ &u(\overset{+}{z}_h)-u(\bar{z}_h)=\alpha_h, \quad v(\overset{+}{z}_h)-v(\bar{z}_h)=\beta_h\,. \end{aligned}\right\} \ (h=1,2,\cdots,p)\,. \quad (1)$$

If $z_h^* \in C_h^*$, then

$$\left.\begin{aligned} &w(\overset{+}{z}{}_h^*)-w(\bar{z}_h^*)=\text{const.}=\omega_h^*=\alpha_h^*+i\beta_h^*\,, \\ &u(\overset{+}{z}{}_h^*)-u(\bar{z}_h^*)=\alpha_h^*, \quad v(\overset{+}{z}{}_h^*)-v(\bar{z}_h^*)=\beta_h^*\,. \end{aligned}\right\} \ (h=1,2,\cdots,p)\,. \quad (2)$$

ω_h, ω_h^* are called the period of $w(z)$ on C_h, C_h^* respectively.

By Theorem X. 32, there exists a non-constant Abelian integral of the first kind on F.

Theorem X. 33. *Let $w(z)$ be a non-constant Abelian integral of the first kind on F, then*

$$\sum_{h=1}^{p}(\alpha_h\beta_h^*-\alpha_h^*\beta_h)>0\,.$$

Proof. Let Γ be the boundary of F_0, then

$$A=\int_{\Gamma} u\,dv$$

is the area, described by $w=w(z)$ $(z\in F_0)$ on the w-plane, where we integrate on Γ in the positive sense. Hence $A>0$.

We have easily

$$\begin{aligned} A&=\sum_{h=1}^{p}[(u(\overset{+}{z}_h)-u(\bar{z}_h))(v(\overset{+}{z}{}_h^*)-v(\bar{z}_h^*))-(u(\overset{+}{z}{}_h^*)-u(\bar{z}_h^*))(v(\overset{+}{z}_h)-v(\bar{z}_h))] \\ &=\sum_{h=1}^{p}(\alpha_h\beta_h^*-\alpha_h^*\beta_h)>0\,. \end{aligned}$$

Corollary. *Hence if $\omega_h=0$ $(h=1,2,\cdots p)$, or $\Re\omega_h=0$, $\Re\omega_h^*=0$ $(h=1,2,\cdots,p)$, then $w(z)\equiv$const.*

By means of Theorem X. 32, we can construct Abelian integrals of the first kind I_h, I_h^* $(h=1,2,\cdots,p)$, whose real parts of the periods satisfy the following condition.

	α_1	α_2		α_p	α_1^*	α_2^*		α_p^*
I_1	1	0	...	0	0	0	...	0
I_1^*	0	0	...	0	1	0	...	0
I_2	0	1	...	0	0	0	...	0
I_2^*	0	0	...	0	0	1	...	0
...	....				....			
I_p	0	0	...	1	0	0	...	0
I_p^*	0	0	...	0	0	0	...	1

Hence if $\{\alpha_h\}$, $\{\alpha_h^*\}$ be any given real constants, then $w(z)=\sum_{h=1}^{p}(\alpha_h I_h+\alpha_h^* I_h^*)$ is an Abelian integral of the first kind, such that

$$\Re\omega_h=\alpha_h, \quad \Re\omega_h^*=\alpha_h^* \ (h=1, 2, \cdots, p) .$$

Conversely if $w(z)$ is such an Abelian integral, then $v(z) = w(z)-\sum_{h=1}^{p}(\alpha_h I_h+\alpha_h^* I_h^*)$ is an Abelian integral of the first kind, such that the real part of its period on C_h, C_h^* $(h=1, 2, \cdots, p)$ vanishes, so that by Theorem X. 33, Cor., $v(z)\equiv$const., which we take as zero, then

$$w(z)=\sum_{h=1}^{p}(\alpha_h I_h+\alpha_h^* I_h^*) . \tag{1}$$

If $\{\alpha_h\}$, $\{\alpha_h^*\}$ are real numbers and $w(z)=\sum_{h=1}^{p}(\alpha_h I_h+\alpha_h^* I_h^*)\equiv 0$, then $\alpha_h=0$, $\alpha_h^*=0$ $(h=1, 2, \cdots, p)$, since $\alpha_h=\Re\omega_h$, $\alpha_h^*=\Re\omega_h^*$.

Hence $\{I_h, I_h^*\}$ are (in the real sense) linearly independent Abelian integrals of the first kind and since by (1) any Abelian integral of the first kind is a linear combination of $\{I_h\}$, $\{I_h^*\}$ with real coefficients, *there are* $2p$ *(in the real sense) linearly independent Abelian integrals of the first kind on* F.

Let $w_1(z), \cdots, w_{q_1}(z)$ be q_1 (in the complex sense) linearly independent Abelian integrals of the first kind, then $w_1(z), \cdots, w_{q_1}(z), iw_1(z), \cdots, iw_{q_1}(z)$ are (in the real sense) linearly independent Abelian integrals of the first kind, so that $2q_1\leqq 2p$, or $q_1\leqq p$.

Hence there are at most p (in the complex sense) linearly independent Abelian integrals of the first kind. Let q be the maximum number of such Abelian integrals, then $q\leqq p$ and let $w_1(z), \cdots, w_q(z)$ be such q Abelian integrals, then any Abelian integral of the first kind is a linear combination of $w_1(z), \cdots, w_q(z), iw_1(z), \cdots, iw_q(z)$ with real coefficients, so that $2p\leqq 2q$, or $p\leqq q$. Hence $q=p$. Hence we have

Theorem X. 34. *There are* p *(in the complex sense) linearly independent Abelian integrals of the first kind on a closed Riemann surface* F *of genus* $p\geqq 1$.

Theorem X. 35. *Let* $w_1(z), \cdots, w_p(z)$ *be* p *(in the complex sense) linearly independent Abelian integrals of the first kind on* F *and* ω_{kh} *be the period of* $w_k(z)$ *on* C_h $(h=1, 2, \cdots, p)$, *then*

$$\Delta=\begin{vmatrix}\omega_{11}\cdots\omega_{1p}\\ \cdots\\ \omega_{p1}\cdots\omega_{pp}\end{vmatrix}\neq 0 .$$

Proof. If $\Delta=0$, then there exist $(k_1, \cdots, k_p)\neq(0, \cdots, 0)$, such that

$$\left.\begin{array}{l} k_1\omega_{11}+\cdots+k_p\omega_{p1}=0\ , \\ \cdots\cdots \\ k_1\omega_{1p}+\cdots+k_p\omega_{pp}=0\ . \end{array}\right\}$$

Then $w(z)=k_1w_1(z)+\cdots+k_pw_p(z)$ is an Abelian integral of the first kind, whose period on C_h $(h=1, 2, \cdots, p)$ vanishes, so that $k_1w_1(z)+\cdots+k_pw_p(z)\equiv 0$, which is absurd, since $w_1(z), \cdots, w_p(z)$ are linearly independent. Hence $\Delta\neq 0$. ——

Hence for any ω_h $(h=1, 2, \cdots, p)$, there exist $\{k_h\}$, such that

$$\left.\begin{array}{l} k_1\omega_{11}+\cdots+k_p\omega_{p1}=\omega_1\ , \\ \cdots\cdots \\ k_1\omega_{1p}+\cdots+k_p\omega_{pp}=\omega_p\ . \end{array}\right\}$$

Hence $w(z)=\sum_{h=1}^{p} k_hw_h(z)$ has a period ω_h on C_h $(h=1, 2, \cdots, p)$.

Hence

Theorem X. 36. *There exists an Abelian integral of the first kind $w(z)$ on F, whose period on C_h $(h=1, 2, \cdots, p)$, or the real part of the period on C_h, C_h^* $(h=1, 2, \cdots, p)$ is any prescribed value and such $w(z)$ is unique.*

Hence there exist p Abelian integrals J_h $(h=1, 2, \cdots, p)$ of the first kind, whose periods are as the following table.

	C_1	C_2	$\cdots$	C_p	C_1^*	C_2^*	$\cdots$	C_p^*
J_1	1	0	$\cdots$	0	ω_{11}^*	ω_{12}^*	$\cdots$	ω_{1p}^*
J_2	0	1	$\cdots$	0	ω_{21}^*	ω_{22}^*	$\cdots$	ω_{2p}^*
$\cdots$		$\cdots\cdots$				$\cdots\cdots$		
J_p	0	0	$\cdots$	1	ω_{p1}^*	ω_{p2}^*	$\cdots$	ω_{pp}^*

$\{J_h\}$ are called *the normal integrals of the first kind.*

$\{J_h\}$ are (in the complex sense) linearly independent and any Abelian integral of the first kind is a linear combination of $\{J_h\}$ with complex coefficients.

Theorem X. 37. (i) $\omega_{hk}^*=\omega_{kh}^*$ $\quad(\omega_{hk}^*=\alpha_{hk}^*+i\beta_{hk}^*)$.

(ii) $\qquad \sum_{h,k=1}^{p} \beta_{hk}^* x_h x_k$ *is a positive definite quadratic form.*

Proof. Since J_1, J_2 are regular on $F_0=F-\{C_h\}-\{C_h^*\}$, by the principle of residue, we have

$$I=\int_\Gamma J_1 dJ_2=0\ ,$$

where Γ is the boundary of F_0.

As the proof of Theorem X. 33, we have

$$I=\sum_{h=1}^{p}[J_1(\overset{+}{z}_h)-J_1(\overline{z}_h))(J_2(\overset{+}{z}{}_h^*)-J_2(\overline{z}{}_h^*))-(J_1(\overset{+}{z}{}_h^*)-J_1(\overline{z}{}_h^*)(J_2(\overset{+}{z}_h)-J_2(\overline{z}_h))]$$
$$=\omega_{21}{}^*-\omega_{12}{}^*\ ,$$

hence $\omega_{12}=\omega_{21}{}^*$ and similarly $\omega_{hk}^*=\omega_{kh}^*$.

Let x_k be real, such that $\sum_{k=1}^{p}x_k^2>0$, then

$$w(z)=x_1J_1+\cdots+x_pJ_p\not\equiv 0$$

is an Abelian integral of the first kind. Let ω_h, ω_h^* be its period on C_h, C_h^* respectively, then

$$\omega_h=x_h,\qquad \omega_h^*=x_1\omega_{1h}^*+\cdots+x_p\omega_{ph}^*.$$

Hence if we apply Theorem X. 33, then we have

$$\sum_{h=1}^{p}x_h(x_1\beta_{1h}^*+\cdots+x_p\beta_{ph}^*)=\sum_{h,k=1}^{p}\beta_{hk}^*x_hx_k>0\ .$$

10. Potential function with a polar singularity

Theorem X. 38. *Let F be a closed, or an open Riemann surface, spread over the z-plane and ζ be a point of F, which is not a branch point. Then there exist $u^k=u^k(z,\zeta)$ and $v^k=v^k(z,\zeta)$ $(k=1,2,\cdots)$ on F, which satisfy the following condition:*

(i) *u^k and v^k are harmonic on F, except at ζ, where*

$$u^k \text{ has a singularity } \Re\left(\frac{1}{(z-\zeta)^k}\right)=\frac{\cos k\theta}{r^k},$$

$$v^k \text{ has a singularity } \Im\left(\frac{1}{(z-\zeta)^k}\right)=-\frac{\sin k\theta}{r^k},\qquad z-\zeta=re^{i\theta}\ .$$

(ii) *Let Ω be a simply connected compact sub-surface of F, which contains ζ and is bounded by an analytic Jordan curve Γ. Then the Dirichlet integral of u^k and v^k on $F-\Omega$ is finite, such that*

$$D_{F-\Omega}[u^k]\leqq\int_\Gamma u^k\frac{\partial u^k}{\partial\nu}\,ds,\qquad D_{F-\Omega}[v^k]\leqq\int_\Gamma v^k\frac{\partial v^k}{\partial\nu}\,ds\ ,$$

where ν is the inner normal of Γ with respect to Ω.

(iii) *u^k and v^k are bounded on $F-\Omega$, such that if $|u^k|\leqq M$ on Γ, then $|u^k|\leqq M$ on $F-\Omega$ and similarly for v^k.*

Proof. As before, we may assume that F is an open Riemann surface and $\zeta=0$. Let U: $|z|\leqq\rho$ be a schlicht disc, contained in F. Let for $z\in U$, $\zeta\in U$,

$$\tilde{g}(z,\zeta)=\log\frac{1}{|z-\zeta|}+\chi(z,\zeta)\ ,\tag{1}$$

where $\chi(z, \zeta)$ is a harmonic function of each variable z and ζ in U. Let $z=re^{i\theta}$, $\zeta=\xi+i\eta$, then

$$\left[\frac{\partial^k}{\partial\xi^k}\log\frac{1}{|z-\zeta|}\right]_{\zeta=0}=\Re\left[\frac{d^k}{d\zeta^k}\left(\log\frac{1}{z-\zeta}\right)\right]_{\zeta=0}=(k-1)!\frac{\cos k\theta}{r^k},$$

$$\left[\frac{\partial^k}{\partial\xi^{k-1}\partial\eta}\log\frac{1}{|z-\zeta|}\right]_{\zeta=0}=-\Im\left[\frac{d^k}{d\zeta^k}\left(\log\frac{1}{z-\zeta}\right)\right]_{\zeta=0}=(k-1)!\frac{\sin k\theta}{r^k}.$$

Hence if we put[1)]

$$u^k=u^k(z,0)=\frac{1}{(k-1)!}\frac{\partial\tilde{g}^k(z,0)}{\partial\xi^k},\quad v^k=v^k(z,0)=-\frac{1}{(k-1)!}\frac{\partial^k\tilde{g}(z,0)}{\partial\xi^{k-1}\partial\eta},\qquad (2)$$

then by (1), u^k and v^k satisfy the condition (i).

Let $\{F_n\}$ be an exhaustion of F with the boundary $\{\Gamma_n\}$ and $z=0$ be contained in F_0. Let $g_n(z,\zeta)=g_n(z,\xi,\eta)$ $(\zeta=\xi+i\eta)$ be the Green's function of F_n and let for $\delta>0$,

$$\begin{aligned}\Delta^k g_n=\Delta^k g_n(z;0,0)&=g_n(z;k\delta,0)-\tbinom{k}{1}g_n(z;(k-1)\delta,0)\\&\quad+\tbinom{k}{2}g_n(z;(k-2)\delta,0)-\cdots\pm g_n(z;0,0),\end{aligned}\qquad (3)$$

$$\Delta_1^k g_n=\Delta_1^k g_n(z;0,0)=\Delta^{k-1}g_n(z;0,\delta)-\Delta^{k-1}g_n(z;0,0),\qquad (4)$$

and put

$$u_n^k=\frac{1}{(k-1)!}\frac{\Delta^k g_n}{\delta^k},\qquad v_n^k=\frac{-1}{(k-1)!}\frac{\Delta_1^k g_n}{\delta^k},\qquad (5)$$

then

$$\lim_{n\to\infty}\lim_{\delta\to0}u_n^k=\frac{1}{(k-1)!}\frac{\partial^k\tilde{g}(z,0)}{\partial\xi^k}=u^k(z,0),\qquad (6)$$

$$\lim_{n\to\infty}\lim_{\delta\to0}v_n^k=\frac{-1}{(k-1)!}\frac{\partial^k\tilde{g}(z,0)}{\partial\xi^{k-1}\partial\eta}=v^k(z,0).\qquad (6^*)$$

Since $u_n^k=0$ on Γ_n,

$$D_{F_m-\Omega}[u_n^k]\leq D_{F_n-\Omega}[u_n^k]=\int_\Gamma u_n^k\frac{\partial u_n^k}{\partial\nu}ds\qquad(m\leq n),$$

so that if we make $\delta\to0$, $n\to\infty$, $m\to\infty$ successively, then we have

$$D_{F-\Omega}[u^k]\leq\int_\Gamma u^k\frac{\partial u^k}{\partial\nu}ds.\qquad (7)$$

Similarly

$$D_{F-\Omega}[v^k]\leq\int_\Gamma v^k\frac{\partial v^k}{\partial\nu}ds.\qquad (7^*)$$

(iii) can be proved similarly as Theorem X. 30.

---

1) $\dfrac{\partial^k\tilde{g}(z,0)}{\partial\xi^k}=\left[\dfrac{\partial^k\tilde{g}(z,\zeta)}{\partial\xi^k}\right]_{\zeta=0}$, $\dfrac{\partial^k\tilde{g}(z,0)}{\partial\xi^{k-1}\partial\eta}=\left[\dfrac{\partial^k\tilde{g}(z,\zeta)}{\partial\xi^{k-1}\partial\eta}\right]_{\zeta=0}$

Theorem X. 39. *Let F be a closed, or an open Riemann surface, spread over the z-plane and F_0: $|z|\leqq\rho_0$ and F_1: $|z|\leqq\rho_1$ $(\rho_0<\rho_1)$ be contained in F. Let*

$$f(z)=\sum_{k=1}^{\infty}\frac{c_k}{z^k} \qquad (c_k=a_k+ib_k)$$

be regular in $\rho_0\leqq|z|\leqq\rho_1$ and

$$U(z)=\Re(f(z))=\sum_{k=1}^{\infty}\frac{a_k\cos k\theta+b_k\sin k\theta}{r^k}, \qquad z=re^{i\theta},$$

then

(i) *There exists a harmonic function $u(z)$ on $F-F_0$, which can be continued harmonically into F_0, such that*

$$V(z)=u(z)-U(z)$$

in harmonic in F_0. Osgood [1].

(ii) *Let Ω be a simply connected compact sub-surface of F, which contains F_0 and is bounded by an analytic Jordan curve Γ, then the Dirichlet integral of u on $F-\Omega$ is finite, such that*

$$D_{F-\Omega}[u]\leqq\int_{\Gamma}u\frac{\partial u}{\partial\nu}ds,$$

where ν is the inner normal of Γ with respect to Ω.

(iii) *u is bounded on $F-\Omega$, such that if $|u|\leqq M$ on Γ, then*

$$|u|\leqq M \quad on\ F-\Omega.$$

Proof. As before, we may assume that F is an open Riemann surface and $\{F_n\}$ be an exhaustion of F with the boundary $\{\Gamma_n\}$.

(i) Let for $z\in F_1$, $\zeta\in F_1$,

$$\tilde{g}(z,\zeta)=\log\frac{1}{|z-\zeta|}+\chi(z,\zeta), \quad z=re^{i\theta}, \quad \zeta=\xi+i\eta,$$

then

$$\left[\frac{\partial^k}{\partial\xi^k}\log\frac{1}{|z-\zeta|}\right]_{\zeta=0}=(k-1)!\,\frac{\cos k\theta}{r^k},$$

$$\left[\frac{\partial^k}{\partial\xi^{k-1}\partial\eta}\log\frac{1}{|z-\zeta|}\right]_{\zeta=0}=(k-1)!\,\frac{\sin k\theta}{r^k}.$$

Hence if we put

$$u(z)=\sum_{k=1}^{\infty}\frac{1}{(k-1)!}\left(a_k\frac{\partial^k\tilde{g}(z,0)}{\partial\xi^k}+b_k\frac{\partial^k\tilde{g}(z,0)}{\partial\xi^{k-1}\partial\eta}\right), \tag{1}$$

$$V(z)=\sum_{k=1}^{\infty}\frac{1}{(k-1)!}\left(a_k\frac{\partial^k\chi(z,0)}{\partial\xi^k}+b_k\frac{\partial^k\chi(z,0)}{\partial\xi^{k-1}\partial\eta}\right), \tag{2}$$

then

$$u(z)=U(z)+V(z)\,. \qquad (3)$$

Let $|\chi(z,\zeta)|\leqq K(=\text{const.})$ for any $z\in F_1$, $\zeta\in F_1$. For a fixed $z\in F_1$, we consider $\chi(z,\zeta)$ as a harmonic function of ζ and let

$$F(\zeta)=\frac{1}{2\pi}\int_0^{2\pi}\chi(z,\rho_1e^{i\varphi})\frac{\rho_1e^{i\varphi}+\zeta}{\rho_1e^{i\varphi}-\zeta}d\varphi\,,$$

then $\chi(z,\zeta)=\Re(F(\zeta))$, so that

$$\left|\frac{\partial^k\chi(z,0)}{\partial\xi^k}\right|\leqq|F^{(k)}(0)|\leqq\frac{k!\,M}{\rho_1^k}\,,$$
$$\left|\frac{\partial^k\chi(z,0)}{\partial\xi^{k-1}\partial\eta}\right|\leqq|F^{(k)}(0)|\leqq\frac{k!\,M}{\rho_1^k}\qquad(M=\text{const.}). \qquad (4)$$

Since $f(z)=\sum_{k=1}^{\infty}c_k/z^k$ is regular for $\rho_0\leqq|z|\leqq\rho_1$,

$$\sum_{k=1}^{\infty}\frac{k(|a_k|+|b_k|)}{\rho_0^k}<\infty,\qquad\text{hence}\qquad\sum_{k=1}^{\infty}\frac{k(|a_k|+|b_k|)}{\rho_1^k}<\infty\,, \qquad (5)$$

so that in virtue of (4), $V(z)$ converges uniformly in $|z|\leqq\rho_1$, hence

$$u(z)-U(z)=V(z) \qquad (6)$$

is harmonic in $|z|<\rho_1$.

Let Ω_1 be a compact sub-surface of F, which lies outside of $|z|\leqq\rho$ $(\rho_0<\rho<\rho_1)$, then

$$|\tilde{g}(z,\zeta)|\leqq K(\Omega_1)\qquad\text{for }z\in\Omega_1,\ \zeta\in F_0,$$

so that we have as (4)

$$\left|\frac{\partial^k\tilde{g}(z,0)}{\partial\xi^k}\right|\leqq\frac{k!\,M^*}{\rho_0^k}\,,\qquad\left|\frac{\partial^k\tilde{g}(z,0)}{\partial\xi^{k-1}\partial\eta}\right|\leqq\frac{k!\,M^*}{\rho_0^k}\quad(M^*=\text{const.}). \qquad (7)$$

In virtue of (5) and (7), $u(z)$ converges uniformly for $z\in\Omega_1$, hence $u(z)$ is harmonic in Ω_1. Since Ω_1 is arbitrary, $u(z)$ is harmonic on $F-F_0$ by (6).

(ii) Let $\Delta^k g_n$, $\Delta_1^k g_n$ be defined as before and put

$$u_n^N=\sum_{k=1}^{N}\frac{1}{(k-1)!}\left(a_k\frac{\Delta^k g_n}{\delta^k}+b_k\frac{\Delta_1^k g_n}{\delta^k}\right)\qquad(\delta>0)\,, \qquad (8)$$

then since $u_n^N=0$ on Γ_n,

$$D_{F_m-\Omega}[u_n^N]\leqq D_{F_n-\Omega}[u_n^N]=\int_\Gamma u_n^N\frac{\partial u_n^N}{\partial\nu}ds\qquad(m\leqq n)\,,$$

hence if we make $\delta\to0$, $N\to\infty$, $n\to\infty$, $m\to\infty$ successively, then we have

$$D_{F-\Omega}[u]\leqq\int_\Gamma u\frac{\partial u}{\partial\nu}ds\,. \qquad (9)$$

(iii) can be proved similarly as Theorem X. 30.

11. Reciprocal relations

(i) Let $u^k(z,\zeta)$, $v^k(z,\zeta)$ be potential functions, defined in Theorem X. 38, and $\tau_\zeta^k(z)$, $\overset{*}{\tau}{}_\zeta^k(z)$ be analytic functions of z, whose real part is

$(k-1)!\,u^k(z,\zeta)$, $(k-1)!\,v^k(z,\zeta)$ respectively, so that if we denote its imaginary part by $v(z)$, then

$$\tau_\zeta^k(z) = \frac{\partial^k \tilde{g}(z,\zeta)}{\partial \xi^k} + iv(z) = \frac{(k-1)!}{(z-\zeta)^k} + \cdots, \tag{1}$$

$$\overset{*}{\tau}{}_\zeta^k(z) = -\frac{\partial^k \tilde{g}(z,\zeta)}{\partial \xi^{k-1}\partial \eta} + iv(z) = \frac{(k-1)!}{i(z-\zeta)^k} + \cdots. \tag{2}$$

(ii) As before, we may assume that F is an open Riemann surface. Let α be an analytic Jordan curve on F, with the prescribed positive sense. We assume that α does not pass through branch points of F and does not decompose F into two parts.

Let the positive normal ν of α be defined as Theorem X. 31 and put

$$\omega_\alpha(z) = \frac{1}{2\pi}\int_\alpha \frac{\partial \tilde{g}(z,\zeta)}{\partial \nu_\zeta}|d\zeta|, \tag{3}$$

then by Theorem X. 31, $\omega_\alpha(z)$ is one-valued and harmonic on $F-\{\alpha\}$ and at any point $z\in\alpha$,

$$\omega_\alpha(\overset{+}{z}) - \omega_\alpha(\overset{-}{z}) = 1 . \tag{4}$$

Hence $\omega_\alpha(z)$ can be continued harmonically across α to an infinitely many valued harmonic function on F.

Let $w_\alpha(z)$ be the analytic function of z, whose real part is $\omega_\alpha(z)$, so that if we denote its imaginary part by $v(z)$, then

$$w_\alpha(z) = \frac{1}{2\pi}\int_\alpha \frac{\partial \tilde{g}(z,\zeta)}{\partial \nu_\zeta}|d\zeta| + iv(z) . \tag{5}$$

(iii) For a fixed z, let $\tilde{g}_z(\zeta)$ be the analytic function of ζ, whose real part is $\tilde{g}(z,\zeta)$, so that if we denote its imaginary part by $v(\zeta)$, then

$$\tilde{g}_z(\zeta) = \tilde{g}(z,\zeta) + iv(\zeta) . \tag{6}$$

(iv) Let $z_0(\neq\zeta)$ be a fixed point of F and we connect z_0 to a point $z(\neq\zeta)$ by an analytic Jordan arc C, which does not pass through ζ and branch points of F and put

$$\tilde{h}(z,\zeta) = \int_C \frac{\partial \tilde{g}(t,\zeta)}{\partial \nu_t}|dt| = \int_{z_0}^{z} \frac{\partial \tilde{g}(t,\zeta)}{\partial \nu_t}|dt| . \tag{7}$$

For a fixed z, let $\tilde{h}_z(\zeta)$ be the analytic function of ζ, whose real part is $\tilde{h}(z,\zeta)$, so that if we denote its imaginary party by $v(\zeta)$, then

$$\tilde{h}_z(\zeta) = \int_{z_0}^{z} \frac{\partial \tilde{g}(t,\zeta)}{\partial \nu_t}|dt| + iv(\zeta) . \tag{8}$$

Then the following reciprocal relations hold.

Theorem X. 40. (Reciprocal relations).

$$\left.\begin{aligned}\Re\left(\frac{d^l\tau_\zeta^k(z)}{dz^l}\right)=\Re\left(\frac{d^k\tau_z^l(\zeta)}{d\zeta^k}\right), \quad \Im\left(\frac{d^l\tau_\zeta^k(z)}{dz^l}\right)=\Re\left(\frac{d^k\overset{*}{\tau}{}_z^l(\zeta)}{d\zeta^k}\right),\\ \Re\left(\frac{d^l\overset{*}{\tau}{}_\zeta^k(z)}{dz^l}\right)=\Im\left(\frac{d^k\tau_z^l(\zeta)}{d\zeta^k}\right), \quad \Im\left(\frac{d^l\overset{*}{\tau}{}_\zeta^k(z)}{dz^l}\right)=\Im\left(\frac{d^k\overset{*}{\tau}{}_z^l(\zeta)}{d\zeta^k}\right).\end{aligned}\right\} \qquad (\mathrm{I})$$

$$\left.\begin{aligned}\int_\alpha d\tau_z^k(\zeta)=2\pi i\,\Re\left(\frac{d^k w_\alpha(z)}{dz^k}\right),\\ \int_\alpha d\overset{*}{\tau}{}_z^k(\zeta)=2\pi i\,\Im\left(\frac{d^k w_\alpha(z)}{dz^k}\right).\end{aligned}\right\} \qquad (\mathrm{II})$$

$$\Re(\tau_\zeta^k(z))=\Re\left(\frac{d^k\tilde{g}_z(\zeta)}{d\zeta^k}\right), \quad \Re(\overset{*}{\tau}{}_\zeta^k(z))=\Im\left(\frac{d^k\tilde{g}_z(\zeta)}{d\zeta^k}\right). \qquad (\mathrm{III})$$

$$\left.\begin{aligned}\Im(\tau_\zeta^k(z))-\Im(\tau_\zeta^k(z_0))=\int_{z_0}^z d\,\Im(\tau_\zeta^k(t))=\Re\left(\frac{d^k\tilde{h}_z(\zeta)}{d\zeta^k}\right),\\ \Im(\overset{*}{\tau}{}_\zeta^k(z))-\Im(\overset{*}{\tau}{}_z^k(z_0))=\int_{z_0}^z d\,\Im(\overset{*}{\tau}{}_\zeta^k(t))=\Im\left(\frac{d^k\tilde{h}_z(\zeta)}{d\zeta^k}\right).\end{aligned}\right\} \qquad (\mathrm{IV})$$

Proof. If we put $z=x+iy$, $\zeta=\xi+i\eta$, then by (1),

$$\Re\left(\frac{d^l\tau_\zeta^k(z)}{dz^l}\right)=\frac{\partial^{l+k}\tilde{g}(z,\zeta)}{\partial x^l\partial\xi^k}, \qquad \Re\left(\frac{d^k\tau_z^l(\zeta)}{d\zeta^k}\right)=\frac{\partial^{l+k}\tilde{g}(\zeta,z)}{\partial\xi^k\partial x^l}.$$

Since $\tilde{g}(z,\zeta)=\tilde{g}(\zeta,z)$, we have

$$\Re\left(\frac{d^l\tau_\zeta^k(z)}{dz^l}\right)=\Re\left(\frac{d^k\tau_z^l(\zeta)}{d\zeta^k}\right). \qquad (\mathrm{I}_1)$$

If we put

$$\tau_\zeta^k(z)=\frac{\partial^k\tilde{g}(z,\zeta)}{\partial\xi^k}+iv(z)=\frac{\partial^k\tilde{g}(\zeta,z)}{\partial\xi^k}+iv(z)=u+iv,$$

then since $\dfrac{\partial v}{\partial x}=-\dfrac{\partial u}{\partial y}$, we have

$$\Im\left(\frac{d^l\tau_\zeta^k(z)}{dz^l}\right)=\frac{\partial^l v}{\partial x^l}=-\frac{\partial^l u}{\partial x^{l-1}\partial y}=-\frac{\partial^{l+k}\tilde{g}(\zeta,z)}{\partial\xi^k\partial x^{l-1}\partial y}=\Re\left(\frac{d^k\overset{*}{\tau}{}_z^l(\zeta)}{d\zeta^k}\right). \qquad (\mathrm{I}_2)$$

Similarly we can prove the remaining relations of (I).

By (5) and $\dfrac{\partial^k g(\zeta,z)}{\partial x^k}=\Re(\tau_z^k(\zeta))$, we have

$$2\pi\Re\left(\frac{d^k w_\alpha(z)}{dz^k}\right)=\int_\alpha\frac{\partial}{\partial\nu_\zeta}\left(\frac{\partial\tilde{g}(z,\zeta)}{\partial x^k}\right)|d\zeta|=\int_\alpha\frac{\partial}{\partial\nu_\zeta}\left(\frac{\partial^k\tilde{g}(\zeta,z)}{\partial x^k}\right)|d\zeta|$$
$$=\int_\alpha d\Im(\tau_z^k(\zeta)).$$

Since $\Re(\tau_z^k(\zeta))$ is one-valued on F, $\int_\alpha d\Re(\tau_z^k(\zeta)=0$, so that

$$2\pi\Re\left(\frac{d^k w_a(z)}{dz^k}\right)=-i\int_a d\tau_z^k(\zeta),$$

hence

$$\int_a d\tau_z^k(\zeta)=2\pi i\Re\left(\frac{d^k w_a(z)}{dz^k}\right). \tag{II$_1$}$$

Since

$$2\pi w_a(z)=\int_a \frac{\partial\tilde{g}(z,\zeta)}{\partial\nu_\zeta}|d\zeta|+iv(z)=u+iv,$$

$$2\pi\Im\left(\frac{d^k w_a(z)}{dz^k}\right)=\frac{\partial^k v}{\partial x^k}=-\frac{\partial^k u}{\partial x^{k-1}\partial y}=-\int_a \frac{\partial}{\partial\nu_\zeta}\left(\frac{\partial^k\tilde{g}(\zeta,z)}{\partial x^{k-1}\partial y}\right)|d\zeta|$$

$$=\int_a \frac{\partial}{\partial\nu_\zeta}\Re(\overset{*}{\tau}{}_z^k(\zeta))\,|d\zeta|=-i\int_a d\overset{*}{\tau}{}_z^k(\zeta),$$

so that

$$\int_a d\overset{*}{\tau}{}_z^k(\zeta)=2\pi i\Im\left(\frac{d^k w_a(z)}{dz^k}\right). \tag{II$_2$}$$

Since $\tilde{g}_z(\zeta)=\tilde{g}(z,\zeta)+iv(\zeta)=u+iv$,

$$\Re\left(\frac{d^k\tilde{g}_z(\zeta)}{d\zeta^k}\right)=\frac{\partial^k u}{\partial\xi^k}=\frac{\partial^k\tilde{g}(z,\zeta)}{\partial\xi^k}=\Re(\tau_\zeta^k(z)), \tag{III$_1$}$$

$$\Im\left(\frac{d^k\tilde{g}_z(\zeta)}{d\zeta^k}\right)=\frac{\partial^k v}{\partial\xi^k}=-\frac{\partial^k u}{\partial\xi^{k-1}\partial\eta}=\frac{-\partial^k\tilde{g}(z,\zeta)}{\partial\xi^{k-1}\partial\eta}=\Re(\overset{*}{\tau}{}_\zeta^k(z)). \tag{III$_2$}$$

$$\tilde{h}_z(\zeta)=\int_{z_0}^{z}\frac{\partial\tilde{g}(t,\zeta)}{\partial\nu_t}|dt|+iv(\zeta)=u+iv,$$

so that

$$\Re\left(\frac{d^k\tilde{h}_z(\zeta)}{d\zeta^k}\right)=\frac{\partial^k u}{\partial\xi^k}=\int_{z_0}^{z}\frac{\partial}{\partial\nu_t}\left(\frac{\partial^k\tilde{g}(t,\zeta)}{\partial\xi^k}\right)|dt|=\int_{z_0}^{z}d\Im(\tau_\zeta^k(t)). \tag{IV$_1$}$$

Similarly

$$\Im\left(\frac{d^k\tilde{h}_z(\zeta)}{d\zeta^k}\right)=\int_{z_0}^{z}d\Im(\overset{*}{\tau}{}_\zeta^k(t)). \tag{IV$_2$}$$

Remark. If F is a closed Riemann surface, then we take off a point ζ_0 from F and we construct $\tilde{g}(z,\zeta)$ for an open Riemann surface $F-\{\zeta_0\}$. Then $\tau_\zeta^k(z)$, $\overset{*}{\tau}{}_\zeta^k(z)$, $w_a(z)$ are regular at ζ_0 and $g_\zeta(z)$, $\tilde{h}_\zeta(z)$ have logarithmic singularities at ζ and ζ_0 (Theorem X. 28). Hence $w_a(z)$ is an Abelian integral of the first kind, $\tau_\zeta^k(z)$, $\overset{*}{\tau}{}_\zeta^k(z)$ are Abelian integrals of the second kind and $\tilde{g}_\zeta(z)$, $\tilde{h}_\zeta(z)$ are Abelian integrals of the third kind.

12. Riemann-Roch's theorem

Let F be a closed Riemann surface, spread over the z-plane and z_0 be any point of F. If z_0 is not a branch point, then we put

$$t=z-z_0, \text{ if } z_0\neq\infty, \quad t=1/z, \text{ if } z_0=\infty . \tag{1}$$

If z_0 is a branch point of order n, then we put

$$t=(z-z_0)^{\frac{1}{n}}, \text{ if } z_0\neq\infty, \qquad t=1/z^{\frac{1}{n}}, \text{ if } z_0=\infty . \tag{2}$$

Then a neighbourhood of z_0 is mapped conformally on $|t|<\rho$, such that z_0 corresponds to $t=0$. t is called *the local parameter of F at z_0*.

Let $f(z)$ be a one-valued meromorphic function on F and t be the local parameter of F at z_0 and put $f(z)=F(t)$ in $|t|<\rho$.

We say that $f(z)$ is regular and has a zero of order n at z_0, if $F(t)$ is regular and has a zero of order n at $t=0$ and $f(z)$ is meromorphic and has a pole of order n at z_0, if $F(t)$ is meromorphic and has a pole of order n at $t=0$.

Let $f(z)$ be one-valued on F and be meromorphic at every point $z\in F$, then

$$dv=f(z)dz \tag{3}$$

is called *a differential on F*.

If t is the local parameter at z_0 and $z\in U(z_0)$ corresponds to t, then dv can be witten in the form:

$$dv=\varphi(t)dt , \tag{4}$$

where $\varphi(t)$ is meromorphic in $|t|<\rho$.
We say that dv is regular and has a zero of order n at z_0, if $\varphi(t)$ is regular and has a zero of order n at $t=0$ and dv is meromorphic and has a pole of order n at z_0, if $\varphi(t)$ is meromorphic and has a pole of order n at $t=0$.

Let $\{\zeta_\nu\}$, $\{z_\mu\}$ be a finite number of points on F and we define a *divisor* $\mathfrak{d}$ formally as follows.

$$\mathfrak{d} = \prod_{\nu=1}^{r} \mathfrak{q}^{m_\nu}(\zeta_\nu)/\prod_{\mu=1}^{s} \mathfrak{q}^{n_\mu}(z_\mu), \qquad m_\nu>0,\ n_\mu>0 , \tag{5}$$

and $\operatorname{ord}(\mathfrak{d})=\sum_{\nu=1}^{r} m_\nu-\sum_{\mu=1}^{s} n_\mu$ is called *the order of* $\mathfrak{d}$.

Let $f(z)$ be a one-valued meromorphic function on F. If $f(z)$ is regular on F, except at $\{\zeta_\nu\}$ and has a zero of order $\geqq n_\mu$ at z_μ and a pole of order $\leqq m_\nu$ at ζ_ν, formally if $f(z)\mathfrak{d}$ is regular on F, then $f(z)$ is called *a multiple of* $1/\mathfrak{d}$.

Let dv be a differential on F. If dv is regular on F, except at $\{z_\mu\}$ and has a zero of order $\geqq m_\nu$ at ζ_ν and a pole of order $\leqq n_\mu$ at z_μ, formally if $dv/\mathfrak{d}$ is regular on F, then dv is called *a multiple of* $\mathfrak{d}$.

Theorem X. 41. (Riemann-Roch's theorem). *Let F be a closed Riemann surface of genus $p\geqq 1$, spread over the z-plane and*

$$\mathfrak{d}=\prod_{\nu=1}^{r}\mathfrak{q}^{m_\nu}(\zeta_\nu)/\prod_{\mu=1}^{s}\mathfrak{q}^{n_\mu}(z_\mu),\qquad \operatorname{ord}(\mathfrak{d})=\sum_{\nu=1}^{r}m_\nu-\sum_{\mu=1}^{s}n_\mu,$$

be a divisor.

Let $D(\mathfrak{d})$ be the number (in the complex sense) linearly independent differentials $dv=\varphi(t)dt$ on F, which are multiple of $\mathfrak{d}$ and $A(1/\mathfrak{d})$ be the number of (in the complex sense) linearly independent one-valued meromorphic functions $f(z)$ on F, which are multiple of $1/\mathfrak{d}$, then

$$A(1/\mathfrak{d})=D(\mathfrak{d})+(\operatorname{ord}(\mathfrak{d})+1-p)\,.$$

Proof. (Weyl [2]). For the sake of brevity, we assume that ζ_ν, z_μ are different from branch points of F and ∞. If otherwise, we consider on the plane of its local parameter.

Let ζ_0 be a point of F, which is different from a branch point of F and $\{\zeta_\nu\}$, $\{z_\mu\}$ and we take off ζ_0 from F and construct $\tilde{g}(z,\zeta)$ for an open Riemann surface $F-\{\zeta_0\}$. Then by Theorem X. 28, $\tilde{g}(z,\zeta)$ has a singularity $-\log\dfrac{1}{|z-\zeta_0|}$ at $z=\zeta_0$.

Let C_i $(i=1,2,\cdots,p)$ be p disjoint ring cuts of F, which make F into a surface of planar character and C_i^* $(i=1,2,\cdots,p)$ be its conjugate ring cuts, such that C_i, C_i^* meet at a point and C_i, C_k, C_k^* $(i\neq k)$ are disjoint. We write $\alpha_1,\cdots,\alpha_{2p}$ instead of $C_1\cdots,C_p,C_1^*,\cdots,C_p^*$.

We choose $\{\alpha_h\}$, such that on $\{\alpha_h\}$, there are no branch points of F and $\{\zeta\ \}$, $\{z_\mu\}$.

Let F_0 be the surface, obtained from F by cutting along $\alpha_1,\cdots\alpha_{2p}$. Then F_0 is simply connected and the analytic functions, defined in Theorem X. 40 are one-valued on F_0. In the following, $\tau_\zeta^k(z)$ etc. are understood as their one-valued branches on F_0. We write

$$w_{\alpha_h}(z)=w_h(z)\ (h=1,2,\cdots,2p).$$

Let $f(z)$ be a one-valued meromorphic function on F, which is a multiple of $1/\mathfrak{d}$, then $f(z)$ can be expressed in the form:

$$f(z)=\sum_{\nu=1}^{r}\sum_{k=1}^{m_\nu}[\alpha_k^\nu\tau_{\zeta_\nu}^k(z)+\beta_k^\nu\tau_{\zeta_\nu}^{*k}(z)]+(a+ib),\qquad z\in F_0\,, \tag{1}$$

where α_k^ν, β_k^ν, a, b are suitable real constants.

Since $f(z)$ is one-valued on F, $\displaystyle\int_{\alpha_h}df(z)=0$, so that by the reciprocal relations (II),

$$\sum_{\nu=1}^{r}\sum_{k=1}^{m_\nu}\left[\alpha_k^\nu\Re\left(\frac{d^kw_h(\zeta_\nu)}{d\zeta_\nu^k}\right)+\beta_k^\nu\Im\left(\frac{d^kw_h(\zeta_\nu)}{d\zeta_\nu^k}\right)\right]=0\quad(h=1,2,\cdots,2p). \tag{2}$$

Since $f(z)$ is a multiple of $\prod_{\mu=1}^{s} \mathfrak{q}^{n_\mu}(z_\mu)$,

$$f(z_\mu)=0,\ f'(z_\mu)=0,\ \cdots,\ f^{(n_\mu-1)}(z_\mu)=0\,,$$

so that by (1)

$$\left.\begin{aligned}&\sum_{\nu=1}^{r}\sum_{k=1}^{m_\nu}[\alpha_k^\nu\Re(\tau_{\zeta_\nu}^k(z_\mu))+\beta_k^\nu\Re(\tau_{\zeta_\nu}^k(z_\mu)]+a=0,\\&\sum_{\nu=1}^{r}\sum_{k=1}^{m_\nu}[\alpha_k^\nu\Im(\tau_{\zeta_\nu}^k(z_\mu))+\beta_k^\nu\Im(\overset{*k}{\tau}_{\zeta_\nu}(z_\mu)]+b=0.\end{aligned}\right\}\tag{3}$$

$(\mu=1, 2, \cdots, s)$

$$\left.\begin{aligned}&\sum_{\nu=1}^{r}\sum_{k=1}^{m_\nu}\left[\alpha_k^\nu\Re\left(\frac{d^l\tau_{\zeta_\nu}^k(z_\mu)}{dz_\mu^l}\right)+\beta_k^\nu\Re\left(\frac{d^l\overset{*k}{\tau}_{\zeta_\nu}(z_\mu)}{dz_\mu^l}\right)\right]=0\,,\\&\sum_{\nu=1}^{r}\sum_{k=1}^{m_\nu}\left[\alpha_k^\nu\Im\left(\frac{d^l\tau_{\zeta_\nu}^k(z_\mu)}{dz_\mu^l}\right)+\beta_k^\nu\Im\left(\frac{d^l\overset{*k}{\tau}_{\zeta_\nu}(z_\mu)}{dz_\mu^l}\right)\right]=0\,.\end{aligned}\right\}\tag{4}$$

$(l=1, 2, \cdots n_\mu-1)$

If we normalize $\{\tau_{\zeta_\nu}^k(z)\}$, $\{\overset{*k}{\tau}_{\zeta_\nu}(z)\}$, such that at a point $z_0 \neq \{\zeta_\nu\}, \{z_\mu\}, \zeta_0$, $\tau_{\zeta_\nu}^\nu(z_0)=0$, $\overset{*k}{\tau}_{\zeta_\nu}(z_0)=0$, then by the reciprocal relations, (2), (3), (4) can be written in the following form (2), (3′), (4′):

$$\sum_{\nu=1}^{r}\sum_{k=1}^{m_\nu}\left[\alpha_k^\nu\Re\left(\frac{d^k w_h(\zeta_\nu)}{d\zeta_\nu^k}\right)+\beta_k^\nu\Im\left(\frac{d^k w_h(\zeta_\nu)}{d\zeta_\nu^k}\right)\right]=0\,.\tag{2}$$

$(h=1, 2, \cdots, 2p)$

$$\left.\begin{aligned}&\sum_{\nu=1}^{r}\sum_{k=1}^{m_\nu}\left[\alpha_k^\nu\Re\left(\frac{d^k\tilde{g}_{z_\mu}(\zeta_\nu)}{d\zeta_\nu^k}\right)+\beta_k^\nu\Im\left(\frac{d^k\tilde{g}_{z_\mu}(\zeta_\nu)}{d\zeta_\nu^k}\right)\right]+a=0\,,\\&\sum_{\nu=1}^{r}\sum_{k=1}^{m_\nu}\left[\alpha_k^\nu\Re\left(\frac{d^k\tilde{h}_{z_\mu}(\zeta_\nu)}{d\zeta_\nu^k}\right)+\beta_k^\nu\Im\left(\frac{d^k\tilde{h}_{z_\mu}(\zeta_\nu)}{d\zeta_\nu^k}\right)\right]+b=0\,.\end{aligned}\right\}\tag{3′}$$

$(\mu=1, 2, \cdots, s)$

$$\left.\begin{aligned}&\sum_{\nu=1}^{r}\sum_{k=1}^{m_\nu}\left[\alpha_k^\nu\Re\left(\frac{d^k\tau_{z_\mu}^l(\zeta_\nu)}{d\zeta_\nu^k}\right)+\beta_k^\nu\Im\left(\frac{d^k\tau_{z_\mu}^l(\zeta_\nu)}{d\zeta_\nu^k}\right)\right]=0\,,\\&\sum_{\nu=1}^{r}\sum_{k=1}^{m_\nu}\left[\alpha_k^\nu\Re\left(\frac{d^k\overset{*l}{\tau}_{z_\mu}(\zeta_\nu)}{d\zeta_\nu^k}\right)+\beta_k^\nu\Im\left(\frac{d^k\overset{*l}{\tau}_{z_\mu}(\zeta_\nu)}{d\zeta_\nu^k}\right)\right]=0\,.\end{aligned}\right\}\tag{4′}$$

$(\mu=1, 2, \cdots, s,\ l=1, 2, \cdots, n_\mu-1)$

(2), (3′), (4′) is a system of homogeneous linear equations for $2(\sum_{\nu=1}^{r} m_\nu+1)$ unknown quantities α_k^ν, β_k^ν, a, b $(\nu=1, 2, \cdots, r,\ k=1, 2, \cdots, m_\nu)$.

Let R be the rank of the coefficient matrix A of the system, then the system has (in the real sense),

$$N=2(\sum_{\nu=1}^{r} m_\nu+1)-R\tag{5}$$

linearly independent solutions.

N is the number of (in the real sense) linearly independent one-valued meromorphic functions on F, which are multiple of $1/\mathfrak{d}$.

Let A' be the transposed matrix of A, then A' has the rank R. Hence the following system (7), (8), (9) of homogeneous linear equations with the coefficient matrix A' for $2(\sum_{\mu=1}^{s} n_\mu + p)$ unknown quantities $a_h (h=1, 2, \cdots, 2p)$, b_μ, $\overset{*}{b}_\mu$, c_μ^l, $\overset{*}{c}{}_\mu^l$ $(\mu=1, 2, \cdots, s,\ l=1, 2, \cdots, n_\mu-1)$ has (in the real sense)

$$M=2\Big(\sum_{\mu=1}^{s} n_\mu + p\Big)-R \tag{6}$$

linearly independent solutions.

$$\sum_{h=1}^{2p} a_h \Re\left(\frac{d^k w_h(\zeta_\nu)}{d\zeta_\nu^k}\right)+\sum_{\mu=1}^{s}\left[b_\mu \Re\left(\frac{d^k \tilde{g}_{z_\mu}(\zeta_\nu)}{d\zeta_\nu^k}\right)+\overset{*}{b}_\mu \Re\left(\frac{d^k \tilde{h}_{z_\mu}(\zeta_\nu)}{d\zeta_\nu^k}\right)\right]$$
$$+\sum_{\mu=1}^{s}\sum_{l=1}^{n_\mu-1}\left[c_\mu^l \Re\left(\frac{d^k \tau_{z_\mu}^l(\zeta_\nu)}{d\zeta_\nu^k}\right)+\overset{*}{c}{}_\mu^l \Re\left(\frac{d^k \overset{*}{\tau}{}_{z_\mu}^l(\zeta_\nu)}{d\zeta_\nu^k}\right)\right]=0\,. \tag{7}$$

$$\sum_{h=1}^{2p} a_h \Im\left(\frac{d^k w_h(\zeta_\nu)}{d\zeta_\nu^k}\right)+\sum_{\mu=1}^{s}\left[b_\mu \Im\left(\frac{d^k \tilde{g}_{z_\mu}(\zeta_\nu)}{d\zeta_\nu^k}\right)+\overset{*}{b}_\mu \Im\left(\frac{d^k \tilde{h}_{z_\mu}(\zeta_\nu)}{d\zeta_\nu^k}\right)\right]$$
$$+\sum_{\mu=1}^{s}\sum_{l=1}^{n_\mu-1}\left[c_\mu^l \Im\left(\frac{d^k \tau_{z_\mu}^l(\zeta_\nu)}{d\zeta_\nu^k}\right)+\overset{*}{c}{}_\mu^l \Im\left(\frac{d^k \overset{*}{\tau}{}_{z_\mu}^l(\zeta_\nu)}{d\zeta_\nu^k}\right)\right]=0\,. \tag{8}$$

$$(\nu=1, 2, \cdots, r,\ k=1, 2, \cdots, m_\nu)$$

$$\sum_{\mu=1}^{s} b_\mu=0, \qquad \sum_{\mu=1}^{s} \overset{*}{b}_\mu=0\,. \tag{9}$$

From (7), (8), we have

$$\sum_{h=1}^{2p} a_h \frac{d^k w_h(\zeta_\nu)}{d\zeta_\nu^k}+\sum_{\mu=1}^{s}\left[b_\mu \frac{d^k \tilde{g}_{z_\mu}(\zeta_\nu)}{d\zeta_\nu^k}+\overset{*}{b}_\mu \frac{d^k \tilde{h}_{z_\mu}(\zeta_\nu)}{d\zeta_\nu^k}\right]$$
$$+\sum_{\mu=1}^{s}\sum_{l=1}^{n_\mu-1}\left[c_\mu^l \frac{d^k \tau_{z_\mu}^l(\zeta_\nu)}{d\zeta_\nu^k}+\overset{*}{c}{}_\mu^l \frac{d^k \overset{*}{\tau}{}_{z_\mu}^l(\zeta_\nu)}{d\zeta_\nu^k}\right]=0\,. \tag{10}$$
$$(\nu=1, 2, \cdots, r,\ k=1, 2, \cdots, m_\nu).$$

If we put

$$\varphi(\zeta)=\sum_{h=1}^{2p} a_h \frac{dw_h(\zeta)}{d\zeta}+\sum_{\mu=1}^{s}\left[b_\mu \frac{d\tilde{g}_{z_\mu}(\zeta)}{d\zeta}+\overset{*}{b}_\mu \frac{d\tilde{h}_{z_\mu}(\zeta)}{d\zeta}\right]$$
$$+\sum_{\mu=1}^{s}\sum_{l=1}^{n_\mu-1}\left[c_\mu^l \frac{d\tau_{z_\mu}^l(\zeta)}{d\zeta}+\overset{*}{c}{}_\mu^l \frac{d\overset{*}{\tau}{}_{z_\mu}^l(\zeta)}{d\zeta}\right], \tag{11}$$

then $\varphi(\zeta)\,d\zeta$ is a differential and $\varphi(\zeta)$ has a pole of order$\leqq n_\mu$ at z_μ.

At the excepted point $\zeta=\zeta_0$, $\tau_{z_\mu}(\zeta)$, $\overset{*}{\tau}_{z_\mu}(\zeta)$ are regular and $\tilde{g}_{z_\mu}(\zeta)$ has a singularity $-\log\dfrac{1}{\zeta-\zeta_0}$ and we easily that $\tilde{h}_{z_\mu}(\zeta)$ is regular at $\zeta=\zeta_0$ and $\tilde{h}_{z_\mu}(\zeta)$ has a singularity $\dfrac{1}{i}\log(\zeta-z_0)$ at $\zeta=z_0$, but in virtue of (9), the singularity of $\varphi(\zeta)$ vanishes at ζ_0 and z_0, so that $\varphi(\zeta)$ is regular at ζ_0 and z_0. By (10),

$$\varphi(\zeta_\nu)=0,\ \varphi'(\zeta_\nu)=0,\ \cdots,\ \varphi^{(m_\nu-1)}(\zeta_\nu)=0 \quad (\nu=1, 2, \cdots, r),$$

so that $\varphi(\zeta)$ has a zero of order $\geqq m_\nu$ at ζ_ν, hence

$$dv=\varphi(\zeta)d\zeta \tag{12}$$

is a differential, which is a multiple of $\mathfrak{d}$. Conversely if $dv=\varphi(\zeta)d\zeta$ is a a multiple of $1/\prod_{\mu=1}^{s} q^{n_\mu}(z_\mu)$, then $\varphi(\zeta)$ is of the form (11), hence M is the number of (in the real sense) linearly independent differentials, which are multiple of $\mathfrak{d}$. From (5), (6), we have

$$N=M+2(m+1-p), \qquad m=\sum_{\nu=1}^{r} m_\nu-\sum_{\mu=1}^{s} n_\mu=\operatorname{ord}(\mathfrak{d}) . \tag{13}$$

We shall prove that

$$N=2A(1/\mathfrak{d}), \qquad M=2D(\mathfrak{d}) . \tag{14}$$

Let $w_1, \cdots, w_k$ ($k=A(1/\mathfrak{d})$) be (in the complex sense) linearly independent one-valued meromorphic functions on F, which are multiple of $1/\mathfrak{d}$, then $w_1, \cdots, w_k, iw_1, \cdots, iw_k$ are (in the real sense) linearly independent one-valued meromorphic functions on F, which are multiple of $1/\mathfrak{d}$, hence

$$2k \leqq N .$$

On the other hand, any one-valued meromorphic function on F, which is a multiple of $1/\mathfrak{d}$, can be expressed as a linear combination of $w_1, \cdots, w_k, iw_1, \cdots, iw_k$ with real coefficients, so that

$$N \leqq 2k .$$

Hence $N=2k=2A(1/\mathfrak{d})$. Similarly $M=2D(\mathfrak{d})$, so that by (13),

$$A(1/\mathfrak{d})=D(\mathfrak{d})+(m+1-p). \text{ ——} \tag{15}$$

If $\mathfrak{d}=\prod_{\nu=1}^{p+1} \mathfrak{q}(\zeta_\nu)$, then $\operatorname{ord}(\mathfrak{d})=p+1$, so that $A(1/\mathfrak{d})\geqq 2$, hence

Theorem X. 42. *Let ζ_ν ($\nu=1, 2, \cdots, p+1$) be any $p+1$ given points on F, then there exists a non-constant one-valued meromorphic function on F, which is regular, except at $\{\zeta_\nu\}$, where it has at most a simple pole.*

Theorem X. 43. *Let F be a closed Riemann surface of genus $p\geqq 1$ and d be the number of zero points of an Abelian integral of the first kind on F, then*

$$d=2p-2.$$

Proof. Let $\mathfrak{d}$ be the divisor, represented by an Abelian integral of the first kind dv, then the order of $\mathfrak{d}$ is d. We express $\mathfrak{d}$ as a product of two factors $\mathfrak{d}=\mathfrak{d}_1\cdot\mathfrak{d}_2$, then $\operatorname{ord}(\mathfrak{d}_1)+\operatorname{ord}(\mathfrak{d}_2)=\operatorname{ord}(\mathfrak{d})=d$.

By Riemann-Roch's theorem,

$$A(1/\mathfrak{d}_1)=D(\mathfrak{d}_1)+(\mathrm{ord}\,(\mathfrak{d}_1)+1-p)\,. \qquad (1)$$

Let $f(z)$ be a one-valued meromorphic function on F, which is a multiple of $1/\mathfrak{d}_1$, then $f(z)dv$ is a differential, which is a multiple of $\mathfrak{d}_2$ and conversely, if du is a differential, which is a multiple of $\mathfrak{d}_2$, then by the definition of the differential, du/dv is a one-valued meromorphic function on F, which is a multiple of $1/\mathfrak{d}_1$, hence $A(1/\mathfrak{d}_1)=D(\mathfrak{d}_2)$, so that by (1),

$$D(\mathfrak{d}_2)=D(\mathfrak{d}_1)+(\mathrm{ord}\,(\mathfrak{d}_1)+1-p)\,. \qquad (2)$$

Similarly

$$D(\mathfrak{d}_1)=D(\mathfrak{d}_2)+(\mathrm{ord}\,(\mathfrak{d}_2)+1-p)\,. \qquad (3)$$

Hence adding each other, we have

$$d=2p-2\,.$$

13. Royden's theorem

Let F be a closed Riemann surface of genus $p\geqq 1$, spread over the z-sphere and C_i, C_i^* $(i=1, 2, \cdots, p)$ be p pairs of canonical ring cuts of F. Let Φ be an unramified covering surface of F. If we denote F_k a surface, which is obtained from F by cutting along a certain number of C_i, C_i^*, then $\Phi=\sum_{k=0}^{\infty} F_k$. Let Γ_k be the boundary of F_k, then Γ_k consists of a certain number of C_i, C_i^*, which we denote by $\Gamma_k=\sum_{i=1}^{\alpha_k}\sigma_k^i$. Along σ_k, there joins another F_s to F_l.

Theorem X. 44. *Suppose that there exist constants* m_k^i, *corresponding to* σ_k, *which satisfy the following condition:*

(i) *If* $\sigma_k\in\Gamma_k$ *belongs to the boundary* Γ_s *of another* F_s *and* $\sigma_k=\sigma_s^j$, *then*

$$m_s^j=-m_k^i\,.$$

(ii) $\sum_i m_0\neq 0, \quad \sum_i m_k=0 \quad (k=1, 2, \cdots)$.

(iii) $\sum_k M_k^2<\infty, \quad$ where $M_k=\mathrm{Max}_i\,|m_k^i|$.

Then

$$\Phi\in P_G\,.$$

Royden [1].

Proof. (Tsuji) [43]). We may assume that C_i, C_i^* are analytic Jordan curves, which meet at an positive angle at its common point z_i and do not contain ∞. Let $|\sigma_k^i|$ be the length of σ_k.

We define $f_k(z)$ on Γ_k $(k=0, 1, 2, \cdots)$ by

$$f_k(z)=m_k^i/|\sigma_k| \quad \text{on } \sigma_k\ (i=1, 2, \cdots, \alpha_k)\,, \qquad (1)$$

then

$$|f_k(z)| \leqq \text{const.}\ M_k \quad \text{on } \Gamma_k ,$$

and

$$\int_{\Gamma_k} f_k(z)|dz| = \sum_i m_k = 0 \quad (k \geqq 1) . \tag{2}$$

Hence we can solve the Neumann's problem for F_k $(k \geqq 1)$ with the boundary value $f_k(z)$ and let $w_k(z)$ be its solution, such that $w_k(z)$ is harmonic on F_k and

$$\frac{\partial w_k}{\partial \nu} = f_k(z) \quad \text{on } \Gamma_k , \tag{3}$$

where ν is the inner normal of Γ_k with respect to F_k.

Then by Theorem I. 24,

$$D_{F_k}[w_k] \leqq \text{const.}\ M_k^2 \quad (k \geqq 1) . \tag{4}$$

Let Δ_0 be a schlicht circular disc, contained in F_0 and C_0 be its bounding circle. We define $w_0(z)$ in $F_0 - \Delta_0$ as the solution of the mixed boundary value problem, such that $w_0(z)$ is harmonic in $F_0 - \Delta_0$, $w_0 = 0$ on C_0, $\dfrac{\partial w_0}{\partial \nu} = f_0(z)$ on Γ_0, where ν is the inner normal of Γ_0 with respect to F_0.

Now we define $w(z)$ on $\Phi - \Delta_0$ by

$$w(z) = w_0(z) \text{ on } F_0 - \Delta_0, \quad w(z) = w_k(z) \text{ on } F_k \quad (k \geqq 1) , \tag{5}$$

then $w(z)$ is not continuous on σ_k^i, but by the condition (i), its normal derivative is continuous on σ_k^i.

$$\int_{C_0} \frac{\partial w}{\partial \nu} ds = \int_{C_0} \frac{\partial w_0}{\partial \nu} ds = -\int_{\Gamma_0} \frac{\partial w_0}{\partial \nu} ds = -\int_{\Gamma_0} f_0(z)|dz| = -\sum m_0^i \neq 0, \tag{6}$$

where ν is the inner normal with respect to $F_0 - \Delta_0$ and by (4) and the condition (iii),

$$D_{\Phi - \Delta_0}[w] < \infty . \tag{7}$$

Let

$$\Phi_n - \Delta_0 = (F_0 - \Delta_0) + \sum_{k=1}^{n} F_k \tag{8}$$

be a connected sub-surface of $\Phi - \Delta_0$ and $C_0 + \gamma_n$ be its boundary and $u_n(z)$ be the solution of the mixed boundary problem for $\Phi_n - \Delta_0$, with the boundary condition:

$$u_n = 0 \quad \text{on } C_0, \quad \frac{\partial u_n}{\partial \nu} = \frac{\partial w}{\partial \nu} \quad \text{on } \gamma_n , \tag{9}$$

where ν is the inner normal of γ_n with respect to $\Phi_n - \Delta_0$.

Since the normal derivative of $w(z)$ is continuous on σ_k^i, we can apply the Green's formula, so that

$$D_{\Phi_n-\Delta_0}[u_n, w]=-\int_{\gamma_n} u_n \frac{\partial w}{\partial \nu} ds=-\int_{\gamma_n} u_n \frac{\partial u_n}{\partial \nu} ds=D_{\Phi_n-\Delta_0}[u_n, u_n] ,$$

so that

$$D_{\Phi_n-\Delta_0}[u_n, w-u_n]=0 ,$$

hence

$$\infty > D_{\Phi-\Delta_0}[w] \geqq D_{\Phi_n-\Delta_0}[w]=D_{\Phi_n-\Delta_0}[u_n]+D_{\Phi_n-\Delta_0}[w-u_n] \geqq D_{\Phi_n-\Delta_0}[u_n], \quad (10)$$

$$\int_{C_0} \frac{\partial u_n}{\partial \nu} ds=-\int_{\gamma_n} \frac{\partial u_n}{\partial \nu} ds=-\int_{\gamma_n} \frac{\partial w}{\partial \nu} ds=\int_{C_0} \frac{\partial w}{\partial \nu} ds \neq 0 . \quad (11)$$

By (10) and $u_n=0$ on C_0, we can select a partial sequence from $\{u_n(z)\}$, which we denote by $\{u_n(z)\}$ again, such that

$$\lim_n u_n(z)=u(z) \quad (12)$$

uniformly in the wider sense in $\Phi-\Delta_0$, then $u(z)$ is harmonic in $\Phi-\Delta_0$ and

$$u=0 \text{ on } C_0, \quad D_{\Phi-\Delta_0}[u]<\infty, \quad \int_{C_0} \frac{\partial u}{\partial \nu} ds \neq 0 . \quad (13)$$

Hence $u \neq$ const. Since such a harmonic function exists on $\Phi-\Delta_0$,

$$\Phi \in P_G \qquad \text{(Theorem X. 9)} .$$

14. Schottkyan covering surfaces of a closed Riemann surface

Let F be a closed Riemann surface of genus $p \geqq 2$, spread over the z-sphere. We cut F by p disjoint ring cuts C_i $(i=1, 2, \cdots, p)$ and let F_0 be the resulting surface. We take infinitely many same samples as F_0 and connect them along the opposite shores of $\{C_i\}$ as in the well known way, then we obtain a covering surface $\Phi=\Phi^{(p)}$ of F, which is of planar character. Now instead of cutting F by p ring cuts, we cut F by q $(1 \leqq q \leqq p)$ ring cuts C_i $(i=1, 2, \cdots, q)$ and let F_0 be the resulting surface. We take infinitely many same samples as F_0 and connect them along the opposite shores of C_i $(i=1, 2, \cdots, q)$, then we obtain a covering surface $\Phi=\Phi^{(q)}$ of F, which we call *a Schottkyan covering surface of F*.

We can construct Φ as follows. We write

$$C_i=\overset{+}{C}_i, \ C_{-i}=\overset{-}{C}_i \quad (i=1, 2, \cdots, q),$$

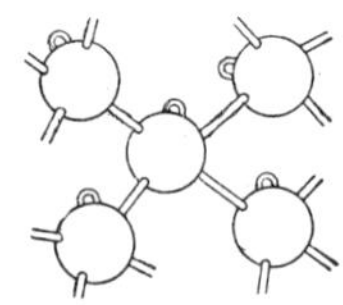

where $\overset{+}{C}_i$, $\overset{-}{C}_i$ are the positive and the negative side of C_i and let $F_{i_1}, F_{i_1 i_2}, \cdots, F_{i_1 i_2 \cdots i_k}, \cdots$ $(i_k=\pm 1, \pm 2, \cdots, \pm q)$ be the same sample as F_0. First we join F_{i_1} to F_0 along C_{i_1}. If C_{i_1} is considered as the boundary of F_{i_1}, then it is C_{-i_1} of F_{i_1}. Next we join $F_{i_1 i_2}$ to F_{i_1} along C_{i_2} $(i_2 \neq -i_1)$ and join $F_{i_1 i_2 i_3}$ to $F_{i_1 i_2}$ along C_{i_3} $(i_3 \neq -i_2)$ and proceed similarly, then

$$\Phi=\Phi^{(q)}=F_0+\sum_{i_1}F_{i_1}+\sum_{i_1,i_2}F_{i_1i_2}+\cdots+\sum_{i_1,\cdots i_k}F_{i_1\cdots i_k}+\cdots, \tag{1}$$

where we sum up for $i_k=\pm1,\ \pm2,\ \cdots\pm q,\quad i_k\neq -i_{k-1}$.

We shall prove

Theorem X. 45. (i) $\Phi^{(1)}\in O_G$. (ii) *If* $q\geqq 2$, $\Phi^{(q)}\in P_{HBD}$, *so that* $\Phi^{(q)}\in P_G$. *Let* $g(z, z_0)$ *be the Green's function of* $\Phi^{(q)}$, *then* $g(z, z_0)$ *tends to zero, if* z *tends to the ideal boundary of* $\Phi^{(q)}$[1].

(iii) *In any case,*

$$\Phi^{(q)}\in O_{AD}\quad (1\leqq q\leqq p)\,. \qquad\text{Tsuji [37].}$$

Theorem III. 63 is a special case of this theorem, where $q=p$.

Proof. Since by Theorem X. 17, $\Phi^{(1)}\in O_G$, we assume that $2\leqq q\leqq p$. We shall prove that $\Phi^{(q)}\in P_{HBD}$, then $\Phi^{(q)}\in P_G$ follows from Theorem X. 6. If we take off F_0 from Φ, then there remains $2q$ non-compact sub-surfaces Φ_i ($i=\pm1, \pm2, \cdots, \pm q$) of Φ, where Φ_i abutts on F_0 along C_i. We consider $\Psi=\Phi_1$ and exhaust it by a sequence of compact sub-surfaces: $\Psi_1\subset\Psi_2\subset\cdots\subset\Psi_n\to\Psi$, where C_1 belongs to the boundary of Ψ_n ($n=1, 2, \cdots$) and let $C_1+\Gamma_n$ be the boundary of Ψ_n. Let $u_n(z)$ be the harmonic measure of Γ_n with respect to Ψ_n, then $u_n(z)$ decreases with n, so that

$$\lim_n u_n(z)=u(z)$$

exists. If $u\equiv 0$, then by means of the niveau curves of u_n, we can prove as Theorem X. 13, that Ψ is regularly exhaustible in Ahlfors' sense, but it is not the case, which will be proved as follows. Suppose that Ψ is regularly exhaustible in Ahlfors' sense, then there exists a sequence of compact sub-surfaces $\Psi^{(n)}$ of Ψ, such that

$$L_n/S_n\to 0\,, \tag{1}$$

where L_n is the length of the boundary Λ_n of $\Psi^{(n)}$ and $S_n=|\Psi^{(n)}|/|F|$, where $|\Psi^{(n)}|$, $|F|$ denotes the area of $\Psi^{(n)}$ and F respectively, the length and the area being measured on the z-sphere. As seen from the proof of Theorem X. 13, Λ_n is the niveau curve of a harmonic measure, so that Λ_n consists of a finite number ν_n of disjoint analytic Jordan curves, which are not homotop null, hence the length of each Jordan curve is $\geqq a>0$, where a is a constant, which depends on F only, so that

$$L_n\geqq a\nu_n\,. \tag{2}$$

1) Hence every ideal boundary point of $\Phi^{(q)}(2\leqq q\leqq p)$ can be considered as a regular point of the Dirichlet problem.

Let ρ_n be the Euler's characteristic of $\Psi^{(n)}$. If $q=p$, then Φ and hence $\Psi^{(n)}$ is of planar character, so that

$$\rho_n \leqq \nu_n \leqq L_n/a . \tag{3}$$

If $q<p$ and C_i $(q+1\leqq i\leqq p)$ be covered N_i-times by $\Psi^{(n)}$, then if we consider a surface, obtained from $\Psi^{(n)}$ by cutting along $\{C_i\}$ $(q+1\leqq i\leqq p)$, we have easily

$$\rho_n \leqq 2(N_{q+1}+\cdots+N_p)+\nu_n \leqq 2(N_{q+1}+\cdots+N_p)+L_n/a .$$

By Ahlfors' covering theorem, $N_i \leqq S_n+hL_n$, where $h>0$ is a constant, which depends on F only, so that

$$\rho_n \leqq 2(p-q)S_n+hL_n . \tag{4}$$

(4) holds for $2\leqq q\leqq p$, in virtue of (3).

Since $\rho_0=2(p-1)$ is the Euler's characteristic of F, by Ahlfors' fundamental theorem, we have $\rho_n \geqq 2(p-1)S_n-hL_n$, so that by (4)

$$2(q-1)\,S_n \leqq hL_n ,$$

which contradicts (1). Hence

$$\lim_n u_n(z)=u(z)\not\equiv 0 . \tag{5}$$

Let

$$d_n=\int_{C_1}\frac{\partial u_n}{\partial \nu}\,ds, \qquad d=\int_{C_1}\frac{\partial u}{\partial \nu}\,ds ,$$

where ν is the inner normal with respect to Ψ_n, then $d_1>d_2>\cdots>d_n\to d$.

$$0<D_\Psi[u]=d<\infty \quad \text{(Theorem X. 4)} , \tag{6}$$

and

$$u=0 \text{ on } C_1, \qquad 0<u<1 \quad \text{in } \Psi=\overset{+}{\Phi}_1 . \tag{7}$$

Similarly we can prove that there exist similar harmonic functions on Φ_i $(i=\pm1, \pm2, \cdots, \pm q)$, so that by Theorem X. 19, $\Phi\in P_{HBD}$.

Next we shall prove that $g(z, z_0)$ tends to zero, if z tends to the ideal boundary of Φ.

Now by the definition,

$$\Phi_j=F_j+\sum_{i_1\neq -j}F_{j\,i_1}+\cdots+\sum_{i_1,\,\cdots,\,i_k}F_{j\,i_1\cdots i_k}+\cdots \quad (i_k\neq -i_{k-1}) . \tag{8}$$
$$(j=\pm1,\ \pm2,\ \cdots,\ \pm q)$$

Let $u_j(z)$ be the harmonic function on Φ_j, defined similarly as (5) and put

$$w_j(z)=1-u_j(z) , \tag{9}$$

then $w_j(z)$ is harmonic and $0<w_j<1$ on Φ_j and $w=1$ on C_j.

To prove that $g(z, z_0)$ tends to zero, if z tends to the ideal boundary of Φ, it suffices to prove that

$$\lim w_j(z)=0 \quad (j=\pm 1, \pm 2, \cdots \pm q)\,, \tag{10}$$

if z tends to the ideal boundary of Φ_j.

Let $C_{j\,i_1}$ be the common boundary of F_j and $F_{j\,i_1}$ and put

$$M_{j\,i_1}=\operatorname*{Max}_{z\in C_{j\,i_1}} w_j(z), \qquad M=\operatorname*{Max}_{j,i_1} M_{j\,i_1}\,, \tag{11}$$

where $\qquad j=\pm 1, \pm 2, \cdots, \pm q,\ i_1=\pm 1, \pm 2, \cdots, \pm q,\ i_1 \neq -j\,.$

Let

$$\Phi_{j\,i_1}=F_{j\,i_1}+\sum_{i_2} F_{j\,i_1,i_2}+\cdots+\sum_{i_2,\cdots,i_k} F_{j\,i_1 i_2\cdots i_k}+\cdots, \tag{12}$$

and we define $w_{j\,i_1}(z)$ on $\Phi_{j\,i_1}$, similarly as $w_j(z)$ on Φ_j, then by the maximum principle, we have easily $w_j(z)/M \leqq w_{j\,i_1}(z)$ on $\Phi_{j\,i_1}$, so that if $C_{j\,i_1 i_2}$ is the common boundary of $F_{j\,i_1}$ and $F_{j\,i_1 i_2}$, then

$$\operatorname*{Max}_{z\in C_{j\,i_1 i_2}} w_j(z)/M \leqq \operatorname*{Max}_{z\in C_{j\,i_1 i_2}} w_{j\,i_1}(z)\,.$$

From the symmetrical structure of Φ and the definition of M, we have

$$\operatorname*{Max}_{z\in C_{j\,i_1 i_2}} w_{j\,i_1}(z) \leqq M\,,$$

so that

$$\operatorname*{Max}_{z\in C_{j\,i_1 i_2}} w_j(z) \leqq M^2\,,$$

hence

$$w_j(z) \leqq M^2 \quad \text{on } \Phi_{j\,i_1 i_2}\,.$$

Similarly we have

$$\operatorname*{Max}_{z\in C_{j\,i_1\cdots i_n}} w_j(z) \leqq M^n\,,$$

so that

$$w_j(z) \leqq M^n \quad \text{on } \Phi_{j\,i_1\cdots i_n}\,. \tag{13}$$

Since $0<M<1$, $\lim w_j(z)=0$, if z tends to the ideal boundary of Φ_j. Hence (10) is proved.

By Theorem X. 22, we see that

$$\Phi^{(q)} \in O_{AD} \quad (1 \leqq q \leqq p)\,. \tag{14}$$

15. Abelian covering surfaces of a closed Riemann surface

1. First we shall prove a lemma on Abelian groups. Let $\pi_1, \cdots, \pi_q$ be q complex numbers $\neq 0$, which we call periods, then $\{\pi_i\}$ generates an Abelian group G, whose element π is represented by a lattice point in the q-dimensional space:

$$\pi=\sum_{i=1}^{q} m_i \pi_i \quad (m_i=0, \pm 1, \pm 2, \cdots)\,. \tag{1}$$

Let $(\pi'_1, \cdots \pi'_s)$ be s periods from $\{\pi_i\}$, then $(\pi'_1, \cdots, \pi'_s)$ is said *linearly independent*, if k_i are integers and $k_1\pi'_1+\cdots+k_s\pi'_s=0$, then $k_1=\cdots=k_s=0$ and *linearly dependent*, if otherwise, so that there exists $(k_1, \cdots, k_s) \neq (0, \cdots, 0)$, such that $k_1\pi'_1+\cdots+k_s\pi'_s=0$.

Let r $(1\leqq r\leqq q)$ be the maximum number of linearly independent periods and let $(\pi_1, \cdots, \pi_r)$ be such ones and G_0 be the group, generated by $(\pi_1, \cdots, \pi_r)$:

$$G_0=\{m_1\pi_1+\cdots+m_r\pi_r\} \quad (m_i=0, \pm 1, \pm 2, \cdots). \tag{2}$$

r is called *the rank of G*.

If π_i is any one from $(\pi_{r+1}, \cdots, \pi_q)$, then since $(\pi_1, \pi_2, \cdots, \pi_r, \pi_i)$ is linearly dependent, there exists the least positive integer $N_i\geqq 1$, such that

$$N_i\pi_i\in G_0, \quad N_i\geqq 1 .$$

We take off all periods from $\{\pi_i\}$, which are $\in G_0$ and let $(\pi_1^{(1)}, \cdots, \pi_{q_1}^{(1)})$ be the remaining ones, then if $\pi_i^{(1)}$ is any one from $(\pi_1^{(1)}, \cdots, \pi_{q_1}^{(1)})$, then

$$N_i^{(1)}\pi_i^{(1)}\in G_0, \quad N_i^{(1)}\geqq 2 . \tag{3}$$

Let $(\pi'_1, \cdots, \pi'_s)$ be s periods from $(\pi_1^{(1)}, \cdots, \pi_{q_1}^{(1)})$, then $(\pi'_1, \cdots, \pi'_s)$ is said "linearly independent", if k_i are integers and $k_1\pi'_1+\cdots+k_s\pi'_s\in G_0$, then $k_i\equiv 0$, mod $N_i^{(1)}$ and "linearly dependent", if otherwise.

Let r_1 $(1\leqq r_1\leqq q_1)$ be the maximum number of "linearly independent" periods among $(\pi_1^{(1)}, \cdots, \pi_{q_1}^{(1)})$ and let $(\pi_1^{(1)}, \cdots, \pi_{r_1}^{(1)})$ be such ones and put

$$G_1=\{m_1^{(1)}\pi_1^{(1)}+\cdots+m_{r_1}^{(1)}\pi_{r_1}^{(1)}\} \ (m_i=0, 1, \cdots, N_i^{(1)}-1), \tag{4}$$

where

$$N_i^{(1)}\pi_i^{(1)}\in G_0, \quad N_i^{(1)}\geqq 2 \quad (i=1, 2, \cdots, r_1).$$

We take off all periods from $\{\pi_i\}$, which are $\in G_0+G_1$ and let $(\pi_1^{(2)}, \cdots, \pi_{q_2}^{(2)})$ be the remaining ones, then if $\pi_i^{(2)}$ is any one from $(\pi_1^{(2)}, \cdots, \pi_{q_2}^{(2)})$, then there exists the least positive integrer $N_i^{(2)}\geqq 2$, such that

$$N_i^{(2)}\pi_i^{(2)}\in G_0+G_1 \quad (i=1, 2, \cdots, q_2). \tag{5}$$

Let $(\pi'_1, \cdots, \pi'_s)$ be s periods from $(\pi_1^{(2)}, \cdots, \pi_{q_2}^{(2)})$, then $(\pi'_1, \cdots, \pi'_s)$ is said "linearly independent", if k_i are integers and $k_1\pi'_1+\cdot+k_s\pi'_s\in G_0+G_1$, then $k_i\equiv 0$, mod $N_i^{(2)}$ and "linearly dependent", if otherwise.

Let $r_2(1\leqq r_2\leqq q_2)$ be the maximum number of "linearly independent" periods among $(\pi_1^{(2)}, \cdots, \pi_{q_2}^{(2)})$ and let $(\pi_1^{(2)}, \cdots, \pi_{r_2}^{(2)})$ be such ones and put

$$G_2=\{m_1^{(2)}\pi_1^{(2)}+\cdots+m_{r_2}^{(2)}\pi_{r_2}^{(2)}\} \ (m_1^{(2)}=0, 1, 2, \cdots, N_i^{(2)}-1), \tag{6}$$

where

$$N_i^{(2)}\pi_i^{(2)}\in G_0+G_1, \quad N_i^{(2)}\geqq 2 \quad (i=1, 2, \cdots, r_2).$$

We take off all periods from $\{\pi_i\}$, which are $\in G_0+G_1+G_2$ and let $(\pi_1^{(3)}, \cdots, \pi_{q_3}^{(3)})$ be the remaining ones and proceed similarly, then after a finits number of steps, we have

$$G=G_0+G_1+\cdots+G_k\ , \tag{7}$$

where

$$\left.\begin{aligned}&G_0=\{m_1\pi_1+\cdots+m_r\pi_r\} \quad (m_i=0,\ \pm 2,\ \cdots)\ ,\\ &\text{and if } 1\leqq j\leqq k,\\ &G_j=\{m_1^{(j)}\pi_1^{(j)}+\cdots m_{r_j}^{(j)}\pi_{r_j}^{(j)}\} \quad (m_i^{(j)}=0,1,2,\cdots,N_i^{(j)}-1),\\ &\qquad N_i^{(j)}\pi_i^{(j)}\in G_0+G_1+\cdots+G_{j-1}, \quad N_i^{(j)}\geqq 2\ .\\ &\qquad\qquad (i=1,2,\cdots,r_j)\end{aligned}\right\} \tag{8}$$

Hence we have

Lemma. *Let G be an Abelian group, generated by $\{\pi_i\}$, then G can be decomposed in the form:*

$$G=G_0+G_1+\cdots+G_k\ ,$$

where G_j is defined by (8).

2. Now we shall define an Abelian covering surface of a closed Riemann surface F of genus $p\geqq 1$, spread over the z-sphere.

Let C_i, C_i^* $(i=1,2,\cdots,p)$ be p pairs of canonical ring cuts of F, such that C_i, C_i^* meet at a point z_i and C_i, C_k, C_k^* $(i\neq k)$ are disjoint.

We denote the both shores of C_i, C_i^* by $\overset{+}{C}_i$, $\overset{-}{C}_i$, $\overset{+}{C}_i{}^*$, $\overset{-}{C}_i{}^*$ respectively.

If z is a point on one of ring cuts, we denote it by $\overset{+}{z}$, $\overset{-}{z}$, according as it is considered to belong to the positive or the negative side.

Let $\varphi(z)$ be a many valued complex valued continuous function, defined on F, where we do not assume that $\varphi(z)$ is analytic, whose many valuedness, as an Abelian integral, is such that

$$\begin{aligned}&\varphi(\overset{+}{z})-\varphi(\overset{-}{z})=\text{const.}=\int_{C_i^*}d\varphi(z)=\omega_i, \quad \text{if } z\in C_i\ ,\\ &\varphi(\overset{+}{z})-\varphi(\overset{-}{z})=\text{const.}=\int_{C_i}d\varphi(z)=\omega_i^*, \quad \text{if } z\in C_i^*\ ,\end{aligned} \tag{1}$$

where we integrate on C_i^* from $\overset{-}{z}_i$ to $\overset{+}{z}_i$ of C_i and similarly we integrate on C_i from $\overset{-}{z}_i$ to $\overset{+}{z}_i$ of C_i^*.

We call $\{\omega_i,\omega_i^*\}$ the periods of $\varphi(z)$. Let $(\pi_1,\cdots,\pi_q)$ be such ones among $\{\omega_i,\omega_i^*\}$, which are not zero, then the general form of the period π of $\varphi(z)$ is of the form:

$$\pi=\sum_{i=1}^{q}m_i\pi_i \quad (m_i=0,\ \pm 1,\ \pm 2,\ \cdots)\ . \tag{2}$$

Then $\{\pi_i\}$ generates an Abelian group G, which, by the lemma, can be decomposed in the form:

$$G=G_0+G_1+\cdots+G_k \tag{3}$$

and let r be the rank of G.

Let Φ be the covering surface of F, on which $\varphi(z)$ is one-valued, then we call Φ *an Abelian covering surface of F*.

By means of (3), Φ can be constructed. Φ is of complicated topological structure as shown in the figure, where we assume that $G=G_0=\{m_1\pi_1+\cdots+m_r\pi_r\}$ $(m_i=0, \pm1, \pm2, \cdots)$. If there are linearly dependent periods among $\{\pi_i\}$, then Φ becomes more complicated.

(i) $r=3$.

(a) $G_0=\{m_1\omega_1+m_2\omega_2+m_3\omega_3\}$. (b) $G_0=\{m_1\omega_1+m_2\omega_1{}^*+m_3\omega_2\}$.

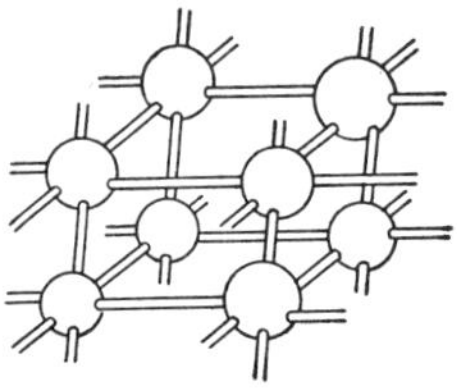

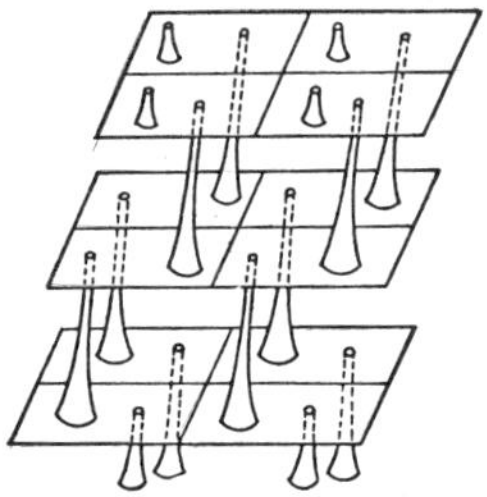

(ii) $r=2$.

(a) $G_0=\{m_1\omega_1+m_2\omega_2\}$. (b) $G_0=\{m_1\omega_1+m_2\omega_1{}^*\}$.

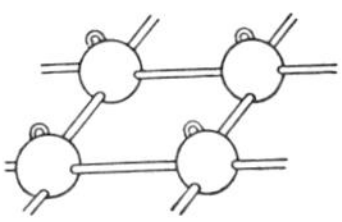

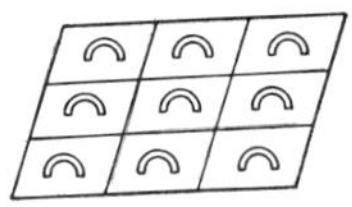

(iii) $r=1$.

$$G_0=\{m_1\omega_1\}.$$

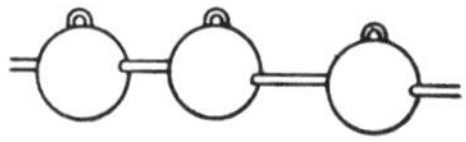

We shall prove

Theorem X. 46. *If* $r=1, 2$, *then* $\Phi\in O_G$.
If $r\geqq3$, *then* $\Phi\in P_G$. Mori [4].

Proof. Let $G=\{\pi_1, \cdots, \pi_q\}$, $G_0=\{m_1\pi_1+\cdots+m_r\pi_r\}$ $(m_i=0, \pm1, \pm2, \cdots)$. If $r+1\leqq j\leqq q$, then there exists a positive integer N_j, such that

$$N_j\pi_j=\sum_{i=1}^{r}k_{ji}\pi_i\,,$$

where k_{ji} are integrers. Let

$$K=\operatorname*{Max}_{j,i}|k_{ji}| \quad (i=1, 2, \cdots, r, \quad j=r+1, \cdots, q)\,, \tag{1}$$

and for $n=1, 2, \cdots$, we consider the totality $\mathfrak{M}_n$ of lattice points π:

$$\pi=\sum_{i=1}^{r} m_i\pi_i+\sum_{j=r+1}^{q} m_j\pi_j\,, \tag{2}$$

where

$$m_i=0, \pm 1, \pm 2, \cdots, \pm nK \quad (i=1, 2, \cdots, r)\,,$$
$$m_j=0, \pm 1, \pm 2, \cdots, \pm(N_j-1) \quad (j=r+1, \cdots, q)\,.$$

We cut F by ring cuts, which correspond to π_i $(i=1, 2, \cdots, q)$ and let F_0 be the resulting surface and $F(m_1, \cdots, m_q)$ be the same sample as F_0, which corresponds to the latice point π and put

$$\Phi_n=\sum_{\pi\in\mathfrak{M}_n} F(m_1, \cdots, m_q)\,, \tag{3}$$

then Φ_n is a connected sub-surface of Φ and $\lim_n \Phi_n=\Phi$.

From the definition of $\mathfrak{M}_n$, if $\pi\in\mathfrak{M}_n$, then $\pi\pm\pi_\sigma\in\mathfrak{M}_{n+1}$ $(\sigma=1, 2, \cdots, q)$, hence if F_i, F_j be any two same samples as F_0 in $\{F(m_1, \cdots, m_q)\}$ and $F_i\in\mathfrak{M}_n$ and F_j abutts on F_i, then $F_j\in\Phi_{n+1}$, so that every boundary curve of Φ_n is contained entirely in Φ_{n+1}, hence if we denote the length of the boundary of Φ_n by L_n, measured on the z-sphere, then

$$L_n\leqq \text{const.}\times\text{number of } F_i \text{ in } \Phi_{n+1}-\Phi_n\leqq$$
$$\text{const.}\,((n+1)^r-n^r)\leqq \text{const.}\, n^{r-1}\,, \tag{4}$$

so that if $r=1, 2$, then $\sum_n 1/L_n=\infty$, hence by Theorem X. 17, $\Phi\in O_G$.

Next suppose that $r\geqq 3$. First consider a special case, where $r=3$ and

$$G\equiv G_0=\{k_1\pi_1+k_2\pi_2+k_3\pi_3\} \quad (k_i=0, \pm 1, \pm 2, \cdots)\,. \tag{5}$$

To a lattice point (k_1, k_2, k_3) in the xyz-space, we associate a unit cube:

$$\Delta(k_1, k_2, k_3):\quad |x-k_1|\leqq 1/2, \quad |y-k_2|\leqq 1/2, \quad |z-k_3|\leqq 1/2\,, \tag{6}$$

and let $Q_i(k_1, k_2, k_3)$ $(i=1, 2, \cdots, 6)$ be the unit square, which belongs to the boundary of $\Delta(k_1, k_2, k_3)$. Let $|m_i(k_1\, k_2, k_3)|$ be the soild angle, subtained by $Q_i(k_1, k_2, k_3)$ at the origin O, where $m_i(0, 0, 0)>0$ $(i=1, 2, \cdots, 6)$ and if $(k_1, k_2, k_3)\neq(0, 0, 0)$, then we define the sign of $m_i(k_1, k_2, k_3)$, such that $m_i(k_1, k_2, k_3)<0$, if the lattice point (k_1, k_2, k_3) is separated from O by the plane, on which $Q_i(k_1, k_2, k_3)$ lies and $m_i(k_1, k_2, k_3)>0$, if otherwise, then $\sum_{i=1}^{6} m_i(k_1, k_2, k_3)=0$. Hence

$$\sum_{i=1}^{6} m_i(0,0,0)>0, \quad \sum_{i=1} m_i(k_1, k_2, k_3)=0, \quad \text{if } (k_1, k_2, k_3)\neq(0,0,0)\,. \tag{7}$$

If (k_1, k_2, k_3) lies on the boundary S_N of a cube Q_N: $|x|\leqq N$, $|y|\leqq N$,

$|z| \leqq N$ ($N=1, 2, \cdots$), then

$$M(k_1, k_2, k_3) = \operatorname*{Max}_{1 \leqq i \leqq 6} |m_i(k_1, k_2, k_3)| \leqq \text{const.}/N^2,$$

so that

$$\sum_{(k_1, k_2, k_3) \in S_N} M^2(k_1, k_2, k_3) \leqq \frac{\text{const.}}{N^4} N^2 = \text{const.}/N^2,$$

hence

$$\sum_{k_1, k_2, k_3} M^2(k_1, k_2, k_3) < \infty, \tag{8}$$

where we sum up for all lattice points (k_1, k_2, k_3).

To (k_1, k_2, k_3), there corresponds $F(k_1, k_2, k_3)$, which is the same sample as F_0. If we make $m_i(k_1, k_2, k_3)$ ($i=1, 2, \cdots, 6$) correspond to $F(k_1, k_2, k_3)$, then $m_i(k_1, k_2, k_3)$ satisfy the condition of Royden's theorem, hence

$$\Phi \in P_G. \tag{9}$$

Next we consider the general case and let Φ^* be the Riemann surface, which corresponds to $G \equiv G_0 = \{m_1\pi_1 + m_2\pi_2 + m_3\pi_3\}$ ($m_i = 0, \pm 1, \pm 2, \cdots$), then by (9), $\Phi^* \in P_G$. Since Φ is an unramified covering surface of Φ^*, by Theorem X. 8, $\Phi \in P_G$. Hence if $r \geqq 3$, then

$$\Phi \in P_G. \tag{10}$$

3. Next we shall study Φ from the stand point of the classification of P_{MD^*} and O_{MD^*} and shall prove

Theorem X. 47.

(i) *If* $r \geqq 3$, *then* $\Phi \in O_{MD^*}$.

(ii) *If* $r=2$, *then there are two cases:*

(a) $G_0 = \{m_1\omega_1 + m_2\omega_2\}$, (b) $G_0 = \{m_1\omega + m_2\omega_1^*\}$.

In case (a), $\Phi \in O_{MD^*}$.

In case (b), $\Phi \in P_{MD^*}$, *when and only when there exists an Abelian integral of the first kind on* F, *whose period* Ω_i, Ω_i^* *on* C_i, C_i^* ($i=1, 2, \cdots, p$) *satisfy the same linear relation as* $\{\omega_i, \omega_i^*\}$.[1)]

If $\Phi \in P_{MD^*}$, *then there exists such an Abelian integral* $\int^z R(z)dz$ *on* F, *that by*

$$\zeta = \zeta(z) = \int^z R(z)dz,$$

Φ *is mapped conformally on an* N ($2 \leqq N < \infty$)*-sheeted Riemann surface* $\sum$, *spread over the* ζ*-sphere, with only one boundary point* ∞, *in*

1) *i.e*, if there exists a linear relation $L(\omega_1, \omega_1^*, \cdots, \omega_p, \omega_p^*)=0$ with integral coefficients, then $L(\Omega_1, \Omega_1^*, \cdots, \Omega_p, \Omega_p^*)=0$.

whose neighbourhood, N sheets of $\sum$ *are connected by infinitely many branch points and any one-valued meromorphic function* $w(z)$ *on* Φ, *such that* $D_{\Phi}^{*}[w]<\infty$, *is a rational function of* ζ *and conversely if* $w(z)$ *is a rational function of* ζ, *then* $w(z)$ *is a one-valued meromorphic function on* Φ, *such that* $D_{\Phi}^{*}[w]<\infty$.

(iii) *If* $r=1$,[1] *then* $\Phi\in P_{MD^*}$, *when and only when there exists an Abelian integral of the first kind on* F, *whose period* Ω_i, Ω_i^* *on* C_i, C_i^* $(i=1, 2, \cdots, p)$ *satisfy the same linear relation as* $\{\omega_i, \omega_i^*\}$, *mod* $2\pi i$[2].

If $\Phi\in P_{MD^*}$, *then there exists such an Abelian integral of the first kind* $\int^z R(z)dz$ *on* F, *that by*

$$\zeta=\zeta(z)=e^{\int^z R(z)dz},$$

Φ *is mapped conformally on an* $N\,(2\leqq N<\infty)$*-sheeted Riemann surface* $\sum$, *spread over the* ζ*-sphere, with two boundary points* 0 *and* ∞, *in each of whose neighbourhood, N sheets of* $\sum$ *are connected by infinitely many branch points and any one-valued meromorphic function* $w(z)$ *on* Φ, *such that* $D_{\Phi}^{*}[w]<\infty$, *is a rational function of* ζ *and conversely if* $w(z)$ *is a rational function of* ζ, *then* $w(z)$ *is a one-valued meromorphic function on* Φ, *such that* $D_{\Phi}^{*}[w]<\infty$. Tsuji [62].

(iv) *In any case* $r\geqq 1$,

$$\Phi\in O_{AD}\,. \qquad \text{Mori [4].}$$

Proof. Suppose that there exists a non-constant one-valued meromorphic function $w(z)$ on Φ, such that $D_{\Phi}^{*}[w]<\infty$ and let Ψ be the Riemann surface, generated by $w=w(z)$ over the w-sphere K.

(i) First suppose that $r\geqq 3$. Since the proof is the same, we assume that $G\equiv G_0=\{m_1\pi_1+\cdots+m_r\pi_r\}$ $(m_i=0, \pm 1, \pm 2, \cdots)$ and first we assume that

$$G_0=\{m_1\omega_1+\cdots+m_r\omega_r\}\ (m_i=0,\ \pm 1,\ \pm 2,\ \cdots)\,. \tag{1}$$

We cut F by $C_1, \cdots, C_r$ and let F_0 be the resulting surface. We take infinitely many same samples as F_0 and connect them along the opposite shores of C_i $(i=1, 2, \cdots, r)$, then we obtain Φ, which can be constructed as follows. Let $F_{m_1\cdots m_r}$ be the same sample as F_0, which corresponds to the lattice point $(m_1, \cdots, m_r)$, then

$$\Phi=\sum_{m_1,\cdots,m_r=-\infty}^{\infty} F_{m_1\cdots m_r}\,. \tag{2}$$

1) The case $r=1$ is a special case of a more general theorem of Ozawa [1].
2) *i.e*, if there exists a linear relation $L(\omega_1, \omega_1^*, \cdots, \omega_p, \omega_p^*)=0$ with integral coefficients, then $L(\Omega_1, \Omega_1^*, \cdots, \Omega_p, \Omega_p^*)\equiv 0$, mod $2\pi i$.

Let

$$\Phi_n = \sum_{m_1, \cdots, m_r = -n}^{n} F_{m_1 \cdots m_r} \tag{3}$$

and Γ_n be the boundary of Φ_n, then Γ_n consists of a certain number k_n of $C_1, \cdots, C_r$, such that $\Gamma_n = \sum_{j=1}^{k_n} \Gamma_n^{(j)}$, where $\Gamma_n^{(j)} = C_i$ $(1 \leqq i \leqq r)$.

Let $\Delta_n^{(j)}$ be a doubly connected domain on Φ, which contains $\Gamma_n^{(j)} = C_i$ and is bounded by two analytic Jordan curves C_i', C_i'', which lie in a small neighbourhood of C_i.

If we choose C_i', C_i'' suitably, then we can map $\Delta_n^{(j)}$ conformally on a circular ring: $\rho_1 < |\zeta| < \rho_2$ $(\rho_1 < 1 < \rho_2)$ on the ζ-plane, such that C_i', C_i, C_i'' corresponds to $|\zeta| = \rho_1$, $|\zeta| = 1$, $|\zeta| = \rho_2$ respectively.

In $\Delta_n^{(j)}$, we write $w(z) = v(\zeta)$ and put

$$A_n^{(j)} = \iint_{\Delta_n^{(j)}} \left(\frac{|w'(z)|}{1+|w(z)|^2} \right)^2 dxdy = \int_{\rho_1}^{\rho_2} \int_0^{2\pi} \left(\frac{|v'(\zeta)|}{1+|v(\zeta)|^2} \right)^2 \rho d\rho d\theta , \tag{4}$$

$$z = x+iy, \ \zeta = \rho e^{\theta} ,$$

$$L(\rho) = \int_0^{2\pi} \frac{|v'(\zeta)|}{1+|v(\zeta)|^2} \rho d\theta \quad (\rho_1 \leqq \rho \leqq \rho_2) . \tag{5}$$

Then by Schwarz's inequality, we have

$$\int_{\rho_1}^{\rho_2} \frac{L^2(\rho)}{\rho} d\rho \leqq 2\pi A_n^{(j)} ,$$

so that there exists ρ' in $[\rho_1, \rho_2]$, such that

$$L^2(\rho') \leqq \text{const.}\ A_n^{(j)} .$$

Let $\gamma_n^{(j)}$ be the image of the circle $|\zeta| = \rho'$ in $\Delta_n^{(j)}$ and we write $L(\rho') = L_n^{(j)}$, then

$$L_n^{(j)} = \int_{\gamma_n^{(j)}} \frac{|w'(z)|}{1+|w(z)|^2} |dz| , \tag{6}$$

$$[L_n^{(j)}]^2 \leqq \text{const.}\ A_n^{(j)} \quad (j = 1, 2, \cdots, k_n) ,$$

so that

$$\sum_{j=1}^{k_n} [L_n^{(j)}]^2 \leqq \text{const.}\ A_n, \qquad A_n = \sum_{j=1}^{k_n} A_n^{(j)} . \tag{7}$$

$\gamma_n = \sum_{j=1}^{k_n} \gamma_n^{(j)}$ bounds a compact sub-surface Φ_n' of Φ. We put $\Delta_n = \sum_{j=1}^{k_n} \Delta_n^{(j)}$ and we assume that $\{\Delta_n\}$ do not overlap, then $\sum_n A_n < D_\Phi^*[w] < \infty$, so that $\lim_n A_n = 0$, hence for any $\varepsilon > 0$, if $n \geqq n_0$,

$$\sum_{j=1}^{k_n} [L_n^{(j)}]^2 < \varepsilon . \tag{8}$$

Let $\Lambda_n^{(j)}$ be the image of $\gamma_n^{(j)}$ on Ψ, then the projection of $\Lambda_n^{(j)}$ on K is contained in a disc $K_n^{(j)}$ of diameter $L_n^{(j)}$, so that by (8),

$$\sum_{j=1}^{k_n} |K_n^{(j)}| \leqq \text{const. } \varepsilon , \tag{9}$$

where $|K_n^{(j)}|$ denotes the spherical area of $K_n^{(j)}$. Since $\varepsilon > 0$ is arbitrary, we see that every sheet of Ψ is covered by Ψ, except a null set. Since $D_\Phi^*[w] < \infty$, Ψ is of finite spherical area, so that Ψ consists of a finite number N of sheets: $\sum_1, \cdots, \sum_N$ and the projection E of the boundary of Ψ on K is a null set.

If n is sufficiently large, then there are two discs $K_n^{(j)}$, $K_n^{(j')}$ $(j \neq j')$, which lie outside each other. For, if otherwise, since $\lim \varepsilon = 0$, E consists of only one point w_0. Let Δ: $|w - w_0| < \rho$ be a disc and Ψ_ρ be a non-compact sub-surface of Ψ, which lies above Δ, then its compact boundary consists of at most N disjoint Jordan curves, which lie above the circle: $|w - w_0| = \rho$. We choose $\rho_1 > \rho_2 > \cdots > \rho_n \to 0$ for ρ and returning to Φ, we see that there exists a decreasing sequence of non-compact sub-surfaces of Φ, defining an ideal boundary point, whose compact boundary consists of at most N disjoint Jordan curves, which is impossible, if $r \geqq 3$, as seen from the topological structure of Φ.

Hence if $r \geqq 3$ and $n \geqq n_0$, there are two discs $K_n^{(j)}$, $K_n^{(j')}$ $(j \neq j')$, which lie outside each other. Then from the topological structure of Φ, there exists a curve on Ψ, whose projection on K connects a point of $K_n^{(j)}$ to a point of $K_n^{(j)}$, which does not belong to $\sum_1, \cdots, \sum_N$, which is absurd. Hence

$$\Phi \in O_{MD^*}, \quad \text{if } G_0 = \{m_1\omega_1 + \cdot + m_r\omega_r\}, \quad r \geqq 3 . \tag{10}$$

Similarly we can prove that

$$\Phi \in O_{MD^*}, \quad \text{if } G_0 = \{m_1\pi_1 + \cdot + m_r\pi_r\}, \quad r \geqq 3 , \tag{11}$$

where π_i is one of ω_i, ω_i^*.

Hence there remains the case, where the both ω_i and ω_i^* are contained in $\pi_1, \cdots, \pi_r$. Since the proof is the same, we assume that

$$G_0 = \{m_1\omega_1 + m_1^*\omega_1^* + m_2\omega_2 + \cdot + m_{r-1}\omega_{r-1}\}, \quad r \geqq 3 . \tag{12}$$

We cut F by $C_1, C_1^*, C_2, \cdots, C_{r-1}$ and let F_0 be the resulting surface and we define Φ_n as before, then its boundary Γ_n consists of two classes of curves. The first class consists of s_n Jordan curves $\gamma_n^{(i)}$ $(i = 1, 2, \cdots, s_n)$, each of which is $(2n-1)\overset{+}{C_1} + (2n-1)\overset{-}{C_1} + (2n-1)\overset{+}{C_1}{}^* + (2n-1)\overset{-}{C_1}{}^*$ and the second class consists of a certain number k_n of $C_2, \cdots, C_{r-1}$, which we denote by $\{\Gamma_n^{(j)}\}$, then

$$\Gamma_n = \sum_{i=1}^{s_n} \gamma_n^{(i)} + \sum_{j=1}^{k_n} \Gamma_n^{(j)} , \tag{13}$$

where

$$\gamma_n^{(i)}=\overset{+}{\gamma}{}_n^{(i)}+\overset{-}{\gamma}{}_n^{(i)}+\overset{+}{\gamma}{}_n^{*(i)}+\overset{-}{\gamma}{}_n^{*(i)},$$

$$\overset{+}{\gamma}{}_n^{(i)}=(2n-1)\overset{+}{C}_1,\quad \overset{-}{\gamma}{}_n^{(i)}=(2n-1)\overset{-}{C}_1,\quad \overset{+}{\gamma}{}_n^{*(i)}=(2n-1)\overset{+}{C}_1{}^*,\quad \overset{-}{\gamma}{}_n^{*(i)}=(2n-1)\overset{-}{C}_1{}^*.$$

Let $\varDelta$ be a doubly connected domain on F, which contains C_1 and we map $\varDelta$ conformally on a circular ring: $\rho_1<|\zeta|<\rho_2$ $(\rho_1<1<\rho_2)$ on the ζ-plane as before.

Next by $\tau=\log\zeta=\xi+i\eta$, we map $\varDelta$ conformally on a rectangle: $\xi_1\leqq\xi\leqq\xi_2$, $0\leqq\eta\leqq 2\pi$ $(\xi_1=\log\rho_1,\ \xi_2=\log\rho_2)$ on the τ-plane. Then a rectangle: $\xi_1\leqq\xi\leqq\xi_2$, $0\leqq\eta\leqq 2\pi(2n-1)$ is mapped conformally on a simply connected domain $\overset{+}{\varDelta}{}_n^{(i)}$, which contains $\overset{+}{\gamma}{}_n^{(i)}$.

In $\overset{+}{\varDelta}{}_n^{(i)}$, we write $w(z)=v(\tau)$ and put

$$\overset{+}{A}{}_n^{(i)}=\iint_{\overset{+}{\Delta}{}_n^{(i)}}\left(\frac{|w'(z)|}{1+|w(z)|^2}\right)^2 dxdy=\int_{\xi_1}^{\xi_2}\int_0^{2\pi(2n-1)}\left(\frac{|v'(\tau)|}{1+|v(\tau)|^2}\right)^2 d\xi\,d\eta\,, \tag{14}$$

$$L(\xi)=\int_0^{2\pi(2n-1)}\frac{|v'(\tau)|}{1+|v(\tau)|^2}d\eta \quad (\xi_1\leqq\xi\leqq\xi_2)\,. \tag{15}$$

Then by Schwarz's inequality, we have

$$\int_{\xi_1}^{\xi_2}L^2(\xi)\,d\xi\leqq 2\pi\,(2n-1)\,\overset{+}{A}{}_n^{(i)}\,,$$

so that there exists ξ' in $[\xi_1, \xi_2]$, such that

$$L^2(\xi')\leqq\text{const.}\ n\,\overset{+}{A}{}_n^{(i)}\,.$$

Let $\overset{+}{\lambda}{}_n^{(i)}$ be the image of the segment: $\xi=\xi'$, $0\leqq\eta\leqq 2\pi\,(2n-1)$ in $\overset{+}{\varDelta}{}_n^{(i)}$ and we write $L(\xi')=\overset{+}{L}{}_n^{(i)}$, then

$$\overset{+}{L}{}_n^{(i)}=\int_{\overset{+}{\lambda}{}_n^{(i)}}\frac{|w'(z)|}{1+|w(z)|^2}|dz|\,, \tag{16}$$

$$[\overset{+}{L}{}_n^{(i)}]^2\leqq\text{const.}\ n.\ \overset{+}{A}{}_n^{(i)}\,. \tag{17}$$

Similarly, considering a neighbourhood of $\overset{-}{\gamma}{}_n^{(i)}$, $\overset{+}{\gamma}{}_n^{*(i)}$, $\overset{-}{\gamma}{}_n^{*(i)}$, we see that there exists a doubly connected domain $\varOmega_n^{(i)}$ on $\varPhi$, which contains $\gamma_n^{(i)}$ and a Jordan curve $\lambda_n^{(i)}$ in $\varOmega_n^{(i)}$, such that if we put

$$A_n^{(i)}=\iint_{\Omega_n^{(i)}}\left(\frac{|w'(z)|}{1+|w(z)|^2}\right)^2 dxdy,\qquad L_n^{(i)}=\int_{\lambda_n^{(i)}}\frac{|w'(z)|}{1+|w(z)|^2}|dz|\,, \tag{18}$$

then

$$[L_n^{(i)}]^2\leqq\text{const.}\ n\,A_n^{(i)}\,,$$

so that

$$\sum_{i=1}^{s_n}[L_n^{(i)}]^2 \leqq \text{const.}\, n A_n, \qquad A_n = \sum_{i=1}^{s_n} A_n^{(i)} . \tag{19}$$

We may assume that $\{\Omega_n^{(i)}\}$ do not overlap, then $\sum_n A_n < D_\Phi^*[w] < \infty$, so that there exists $n_1 < n_2 < \cdots < n_\nu \to \infty$, such that $n_\nu A_{n_\nu} \to 0$, hence $\sum_{i=1}^{s_{n_\nu}} [L_{n_\nu}^{(i)}]^2 \to 0$. This being established, the remaining part of the proof is the same as before and we can prove that $\Phi \in O_{MD}{}^*$. Hence

$$\Phi \in O_{MD}{}^*, \quad \text{if } G \equiv G_0 = \{m_1\pi_1 + \cdot + m_r\pi_r\}, \quad r \geqq 3 . \tag{20}$$

We assumed that $G \equiv G_0$, but the general case $G = G_0 + \cdot + G_k$ can be proved similarly, hence if $r \geqq 3$

$$\Phi \in O_{MD^*}, \quad \text{a fortiori}, \quad \Phi \in O_{AD} . \tag{21}$$

(ii) Next suppose that $r=1$, then we may assume that

$$G_0 = \{m_1\omega_1\} .$$

Then as the case $r \geqq 3$, Ψ consists of a finite number N of sheets and since there are two ideal boundary points of Φ, there are two boundary points (α), (β) of Ψ, whose projection on K is α, β. Hence

$$\Phi \in O_{AD} .$$

By a linear transformation, we may assume that $\alpha \neq \infty$, $\beta \neq \infty$. We cut F by C_1 and let F_0 be the resulting surface. Let F_n be the same sample as F_0, which corresponds to a lattice point $n\omega_1$ ($n=0, \pm 1, \pm 2, \cdots$), then

$$\Phi = \sum_{n=-\infty}^{\infty} F_n{}^{1)} . \tag{1}$$

Let z be any point of F_0 and $\tilde{z}$ be the point of F_1, which lies above z and put

$$\tilde{z} = Tz , \tag{2}$$

then T can be defined on the whole Φ and transforms it into itself and for any $z \in \Phi$,

$$T^\nu z \to (\alpha^*)\ (\nu = 0, 1, 2, \cdots), \qquad T^\nu z \to (\beta^*)\ (\nu = 0, -1, -2, \cdots) ,$$

where (α^*), (β^*) are two ideal boundary points of Φ.

Let $w = w(z)$ be a non-constant one-valued meromorphic function on Φ, such that $D_\Phi^*[w] < \infty$ and put $\tilde{w} = w(\tilde{z}) = w(Tz)\ (z \in \Phi)$ and consider $\tilde{w}$ as a function of w and put

$$\tilde{w} = T(w) , \qquad w \in \Psi , \tag{3}$$

1) We assume that $G \equiv G_0$, but the proof is the same for the general case $G = G_0 + G_1 + \cdot + G_k$.

then $\tilde{w}$ is a point of Ψ and $T(w)$ is meromorphic or has an algebraic singularity on Ψ, except at (α), (β) and $\tilde{w}\to(\alpha)$, (β), if $w\to(\alpha)$, (β). Since Ψ consists of N sheets, above any point $w(\neq a, \beta)$, there are N points of Ψ, so that there are N points z_i $(i=1, 2, \cdots, N)$ of Φ, such that $w(z_i)=w$, hence there are N values $\tilde{w}_i=w(\tilde{z}_i)$ of $\tilde{w}$, so that $T(w)$ is N-valued. If we put

$$\prod_{i=1}^{N}(\tilde{w}-\tilde{w}_i)=\tilde{w}^N+A_1(w)\,\tilde{w}^{N-1}+\cdots+A_N(w)=0\,, \tag{4}$$

then $A_j(w)$ $(j=1, 2, \cdots, N)$ is a one-valued meromorophic function of w outside of a, β. We shall prove that $A_j(w)$ is meromorphic at α, β.

We see easily that in a neighbourhood U of (α), $N_1(\leqq N)$ sheets $\sum_1, \cdots, \sum_{N_1}$ of Ψ are connected by infinitely many branch points and let $\tilde{w}_1, \cdots, \tilde{w}_{N_1}$ be N_1 branches of $\tilde{w}$, such that $w_i\in\sum_i$ $(i=1, 2, \cdots, N_1)$ and put

$$\prod_{i=1}^{N_1}(\tilde{w}-\tilde{w}_i)=\tilde{w}^{N_1}+B_1(w)\tilde{w}^{N_1-1}+\cdots+B_{N_1}(w)=0\,. \tag{5}$$

Then $B_j(w)$ $(j=1, 2, \cdots, N_1)$ is one-valued and regular in a neighbourhood of α and is bounded, since $\tilde{w}_i\to(\alpha)$, if $w\to(\alpha)$, so that $B_j(w)$ is regular at α. Similarly, considering in a neighbourhood of β, we see that $A_j(w)$ $(i=1, 2, \cdots, N)$ of (4) is meromorphic at α, β, so that it is a rational function, hence $\tilde{w}=T(w)$ is an algebraic function of w. We shall prove that $\tilde{w}_1\equiv\cdots\equiv\tilde{w}_{N_1}$, so that the right hand side of (4) is reducible.

If $\sum_1$ and $\sum_2$ have a branch point w_0 in common, then

$$\tilde{w}_1(w)=\tilde{w}_2(w) \qquad \text{at } w=w_0\,.$$

Now among $\sum_2, \cdots, \sum_{N_1}$, there is one, which has infinitely many branch points in common with $\sum_1$ and let $\sum_2$ be such one, then $\tilde{w}_1(w)=\tilde{w}_2(w)$ at infinitely many points. Since $\tilde{w}_1(w)$, $\tilde{w}_2(w)$ are algebraic functions of w, we have

$$\tilde{w}_1(w)\equiv\tilde{w}_2(w)\,.$$

Next among $\sum_3, \cdots, \sum_{N_1}$, there is one, which has infinitely many branch points in common with $\sum_1+\sum_2$ and let $\sum_3$ be such one, then similarly we have

$$\tilde{w}_1(w)\equiv\tilde{w}_2(w)\equiv\tilde{w}_3(w)\,.$$

We proceed similarly and we have

$$\tilde{w}_1(w)\equiv\cdots\equiv\tilde{w}_{N_1}(w)\,.$$

Hence the right hand side of (4) is reducible. Let

$$\tilde{w}^{\mu}+c_1(w)\,\tilde{w}^{\mu-1}+\cdots+c_{\mu}(w)$$

be one of its irreducible factors, then we see easily that

$$\tilde{w}^N+A_1(w)\tilde{w}^{N-1}+\cdot+A_N(w)=[\tilde{w}^{\mu}+c_1(w)\tilde{w}^{\mu-1}+\cdots+c_{\mu}(w)]^{N_1}. \quad N=\mu N_1 \tag{6}$$

By a suitable change of notation, let

$$\prod_{i=1}^{\mu}(\tilde{w}-\tilde{w}_i)=\tilde{w}^{\mu}+c_1(w)\tilde{w}^{\mu-1}+\cdots+c_{\mu}(w)=0\,, \tag{7}$$

then (7) defines the Riemann surface χ of $\tilde{w}=T(w)$, which consists of μ sheets and Ψ is a covering surface of χ of order N_1, such that above any point of χ, there are N_1 points of Ψ. Since T^{ν} ($\nu=1, 2, \cdots$) transforms Φ conformally on itself, we see that χ admitts infinitely many conformal mappings $\{T_{\nu}\}$ on itself. Hence if p is the genus of χ, then by Hurwitz's theorem (Theorem X. 50), $p=0$, or 1. We shall prove that $p=0$. Suppose that $p=1$, then the universal covering surface of χ can be mapped conformally on $|\zeta|<\infty$ and the image of χ consists of infinitely many equivalent parallelograms. By this mapping, T_{ν} becomes a linear transformation L_{ν} of ζ. Since (α), (β) are fixed points of T_{ν}, there are two fixed points of L_{ν} in each parallelogram, so that L_{ν} has infinitely many fixed points, which is absurd.

Hence $p=0$, so that χ can be mapped conformally on $|\zeta|\leqq\infty$, so that $w=w(z)$ is a rational function of ζ and Ψ and hence Φ is mapped conformally on an N_1-sheeted Riemann surface $\sum$, spread over the ζ-sphere, with two boundary points, which we assume that $\zeta=0$, ∞, such that in each of whose neighbourhood, N_1 sheets are connected by infinitely many branch points.

Let $\zeta=\zeta(z)$ be the mapping function and put $\tilde{\zeta}=\zeta(\tilde{z})=\zeta(Tz)=\tilde{\zeta}(z)=\varphi(\zeta)$, then as before, $\varphi(\zeta)$ is an N_1-valued algebraic function of ζ and since N_1 sheets are connected by infinitely many branch points in a neighbourhood of 0 and ∞, we can prove as before, that N_1 branches of $\varphi(\zeta)$ coincide, so that $\tilde{\zeta}=\varphi(\zeta)$ is a rational function of ζ. Similarly ζ is a rational function of $\tilde{\zeta}$, hence $\tilde{\zeta}$ is a linear function of ζ of the form:

$$\tilde{\zeta}=k\zeta \quad (k=\text{const.})\,,$$

so that

$$\frac{\tilde{\zeta}'(z)}{\tilde{\zeta}(z)}=\frac{\zeta'(z)}{\zeta(z)}. \tag{8}$$

Hence $\dfrac{\zeta'(z)}{\zeta(z)}=R(z)$ is a one-valued meromorphic function on F and

$$\zeta=\zeta(z)=e^{\int^{z} R(z)dz}, \tag{9}$$

where $\int^{z} R(z)dz$ is an Abelian integral of the first kind on F, whose period Ω_i, Ω_i^* on C_i, C_i^* $(i=1, 2, \cdots, p)$ satisfy the same linear relation as $\{\omega_i, \omega_i^*\}$, mod $2\pi i$.

Next we shall prove that any non-constant one-valued meromorphic function $w_1(z)$ on Φ, such that $D_\Phi^*[w_1]<\infty$, is a rational function of ζ. As before, $w_1(z)$ is a rational function of ζ_1, where Φ is mapped conformally on an N_1'-sheeted Riemann surface $\sum'$, spread over the ζ_1-sphere, with two boundary points $\zeta_1=0, \infty$, such that in a neighbourhood of $\zeta_1=0, \infty$, N_1' sheets of $\sum'$ are connected by infinitely many branch points. If we put $\zeta_1=\varphi(\zeta)$, then as before, $\varphi(\zeta)$ is an N_1-valued algebraic function of ζ and since N_1 sheets of $\sum$ are connected by infinitely many branch points in a neighbourhood of $\zeta=0, \infty$, $\zeta_1=\varphi(\zeta)$ is a rational function of ζ. Similarly ζ is a rational function of ζ_1, so that ζ_1 is a linear function of ζ of the form $\zeta_1=k\zeta$. Since $w_1(z)$ is a rational function of ζ_1, $w_1(z)$ is a rational function of ζ.

Conversely if $w_1(z)$ is a rational function of ζ, then $w_1(z)$ is a one-valued meromorphic function on Φ and the Riemann surface Ψ_1, generated by $w=w_1(z)$, is a covering surface of finite order of the ζ-sphere, so that $D_\Phi^*[w_1]<\infty$.

Next suppose that there exists an Abelian integral of the first kind $\int^{z} R(z)dz$ on F, whose period Ω_i, Ω_i^* on C_i, C_i^* $(i=1, 2, \cdots, p)$ satisfy the same linear relation as $\{\omega_i, \omega_i^*\}$, mod $2\pi i$ and put

$$w(z)=e^{\int^{z} R(z)dz},$$

then $w(z)$ is a non-constant one-valued regular function on Φ.

Since $G_0=\{m_1\omega_1\}$, there exist positive integers N_j, N_k^*, such that

$$N_j\omega_j\equiv 0, \text{ mod } \omega_1\,(j=2, \cdots, p), \quad N_k^*\omega_k^*=0, \text{ mod } \omega_1\,(k=1, 2, \cdots, p),$$

so that

$$N_j\Omega_j\equiv 0, \text{ mod}\,(\Omega_1, 2\pi i), \quad N_k^*\Omega_k^*\equiv 0, \text{ mod }(\Omega_1, 2\pi i).$$

Hence if $\Omega_j=\sigma_j+i\tau_j$, $\Omega_k^*=\sigma_k^*+i\tau_k^*$, then

$$N_j\sigma_j\equiv 0, \text{ mod } \sigma_1, \quad N_k^*\sigma_k^*\equiv 0, \text{ mod } \sigma_1,$$

so that if $\sigma_1=0$, then $\sigma_j=0$, $\sigma_j^*=0$ $(j=1, 2, \cdots, p)$, which contradicts Theorem X. 33. Cor. Hence $\sigma_1\neq 0$ and we may assume that $\sigma_1>0$, then

$$D_\Phi^*[w]\leqq \text{const.}\sum_{n=-\infty}^{\infty}\frac{e^{2n\sigma_1}}{(1+e^{2n\sigma_1})^2}<\infty.$$

Hence the case $r=1$ is proved.

(iii) Next suppose that $r=2$, then there are two cases:

(a) $G_0=\{m_1\omega_1+m_2\omega_2\}$, (b) $G_0=\{m_1\omega_1+m_2\omega_1^*\}$.

In case (a), $\Phi\in O_{MD^*}$, a fortiori $\Phi\in O_{AD}$, similarly as the case $r\geqq 3$.

In case (b), suppose that there exists a non-constant one-valued meromorphic function $w(z)$ on Φ, such that $D_\Phi^*[w]<\infty$ and let Ψ be the Riemann surface, generated by $w=w(z)$ on the w-sphere. Then as before, Ψ consists of N sheets and since there is only one ideal boundary point of Φ, there is only one boundary point (α) of Ψ, which we may assume that $\alpha=\infty$. Hence $\Phi\in O_{AD}$.

We cut F by C_1 and C_1^*, let F_0 be the resulting surface and F_{mn} be the same sample as F_0, which corresponds to a lattice point $m\omega_1+m\omega_1^*$ $(m, n=0, \pm1, \pm2, \cdots)$, then

$$\Phi=\sum_{m,n=-\infty}^{\infty} F_{mn}\text{[1]}. \tag{1}$$

Let z be any point of $F_0=F_{00}$ and $\tilde{z}$ be the point of F_{10}, which lies above z and put

$$\tilde{z}=T_1z, \tag{2}$$

and similarly, let $\tilde{z}$ be the point of F_{01}, which lies above z and put

$$\tilde{z}=T_2z. \tag{3}$$

Then T_1 and T_2 can be defined on the whole Φ and $\{T_1, T_2\}$ generate an Abelian group $\mathfrak{G}$, which is isomorphic to G and any transformation of $\mathfrak{G}$ transforms Φ into itself.

Let $w=w(z)$, $\tilde{w}=w(\tilde{z})$, $\tilde{z}=T_iz$ $(i=1, 2)$ and consider $\tilde{w}$ as a function of w and put

$$\tilde{w}=T_i(w) \quad (i=1, 2). \tag{4}$$

As before, $T_i(w)$ is an algebraic function of w and its Riemann surface χ consists of μ sheets, where $N=\mu N_1$ and its genus is zero, so that χ can be mapped conformally on $|\zeta|\leqq\infty$ and w is a rational function of ζ. Φ is mapped conformally an N_1-sheeted Riemann surface $\sum$, spread over the ζ-sphere, with one boundary point ∞, in whose neighbourhood N_1 sheets of $\sum$ are connected by infinitely many branch points.

Let $\zeta=\zeta(z)$ be the mapping function and put $\tilde{z}=T_iz$, $\tilde{\zeta}=\zeta(\tilde{z})=\zeta(T_iz)$ $=\tilde{\zeta}(z)=\varphi(\zeta)$, then as before, $\varphi(\zeta)$ is a linear function of ζ of the form:

$$\tilde{\zeta}=\zeta+c_i \quad (i=1, 2), \tag{5}$$

so that $\tilde{\zeta}'(z)=\zeta'(z)$, hence $\zeta'(z)=R(z)$ is a one-valued meromorphic function on F and

1) We assume $G\equiv G_0$, but the proof is the same for the general case $G=G_0+G_1+\cdot+G_k$.

$$\zeta=\zeta(z)=\int^z R(z)dz\ ,$$

where $\int^z R(z)dz$ is an Abelian integral of the first kind on F, whose period Ω_i, Ω_i^* on C_i, C_i^* $(i=1, 2, \cdots, p)$ satisfy the same linear relation as $\{\omega_i, \omega_i^*\}$.

We can prove as before that any one-valued meromorphic function $w_1(z)$ on Φ, such that $D_\Phi^*[w_1]<\infty$ is a rational function of ζ.

Next suppose that there exists an Abelian integral of the first kind $\int^z R(z)dz$ on F, whose period Ω_i, Ω_i^* on C_i, C_i^* $(i=1, 2, \cdots, p)$ satisfy the same linear relation as $\{\omega_i, \omega_i^*\}$ and put

$$w(z)=\int^z R(z)dz\ , \tag{7}$$

then $w(z)$ is a non-constant one-valued regular function on Φ.

Since ω_1, ω_1^* and hence Ω_1, Ω_1^* are linearly independent, we have from the theory of elliptic functions,

$$D_\Phi^*[w]\leqq\text{const.}\sum_{m,n=-\infty}^{\infty}\frac{1}{(1+|m\Omega_1+n\Omega_1^*|^2)^2}<\infty\ .$$

Hence the case $r=2$ is proved. From the proof, we have

$$\Phi\in O_{AD}\ \ (r\geqq 1)\ . \tag{8}$$

16. Conformal mappings of a Riemann surface on itself

Let F be a Riemann surface, spread over the w-sphere K. We denote the spherical distance of two points a, b on K by $[a, b]$. If for any compact sub-surface Ω of F and $\varepsilon>0$, there exists a conformal mapping $\varphi=\varphi_\varepsilon$ of F on itself, such that

$$[\varphi(w),\ w]<\varepsilon\quad\text{for any } w\in\Omega\ , \tag{A}$$

then we say that F *admitts an infinitesimal conformal mapping on itself*.

Theorem X. 48. *If F is not conformally equivalent to one of the following domains: (i) a sphere, (ii) a sphere, excepted one or two points, (iii) a torus*[1]*, (iv) a circular disc, (v) a circular ring: $\rho_0<|t|<1$ $(\rho_0\geqq 0)$, then F admitts no infinitesimal conformal mapping on itself.*

Proof. We assume that F is not conformally equivalent to any of the above five domains.

Let $F^{(\infty)}$ be the universal covering surface of F, then $F^{(\infty)}$ can be

1) *i.e.* a closed Riemann surface of genus $p=1$.

mapped conformally on $|z|<1$. Let $w=w(z)$ be the mapping function, then $w(z)$ is invariant by a certain Fuchsian group G of linear transformations, which make $|z|<1$ invariant and let D_0 be its fundamental domain, which is bounded by orthogonal circles to $|z|=1$ and a certain closed set e_0 on $|z|=1$, which is empty, if F is a closed Riemann surface. The part of the boundary of D_0, which lies in $|z|<1$, consists of equivalent pairs Γ_ν, $\overline{\Gamma}_\nu$ ($\nu=1, 2, \cdots$) of orthogonal arcs to $|z|=1$, such that $\overline{\Gamma}_\nu=S_\nu(\Gamma_\nu)$, where $S_\nu \in G$. G is generated by $\{S_\nu\}$. Suppose that F admitts an infinitesimal conformal mapping φ on itself, which satisfies the condition (A). Then φ becomes a linear transformation T in $|z|<1$, which makes $|z|<1$ invariant.

Let a be any point of Γ_ν ($1\leqq\nu\leqq\nu_0$), then since a and $S_\nu(a)$ correspond to the same point of F, if $\varepsilon>0$ is sufficiently small, then we have $S_\nu^{-1}TS_\nu(a)=T(a)$. Since $S_\nu^{-1}TS_\nu$ and T are linear transformations and a is any point of Γ_ν, we have $S_\nu^{-1}TS_\nu\equiv T$, or

$$TS_\nu=S_\nu T\,. \tag{1}$$

Let α_ν, β_ν be the fixed points of S_ν, which lie on $|z|=1$, which may coincide each other, then by (1),

$$T(\alpha_\nu)=S_\nu(T(\alpha_\nu)), \qquad T(\beta_\nu)=S_\nu(T(\beta_\nu))\,,$$

so that $T(\alpha_\nu)$, $T(\beta_\nu)$ are fixed points of S_ν.

Hence if $\alpha_\nu\neq\beta_\nu$, then

either (i) $T(\alpha_\nu)=\alpha_\nu$, $T(\beta_\nu)=\beta_\nu$, or (ii) $T(\alpha_\nu)=\beta_\nu$, $T(\beta_\nu)=\alpha_\nu$.

If $\varepsilon>0$ is small, the case (ii) does not occur, hence the case (i) occurs, so that S_ν and T have common fixed points, so that if we denote the fixed points of T by α, β, then

$$\alpha_\nu=\alpha, \qquad \beta_\nu=\beta \qquad (\nu=1, 2, \cdots, \nu_0)\,. \tag{2}$$

Since ν_0 is an arbitrary, (2) holds for $\nu=1, 2, \cdots$.

(i) First suppose that $\alpha\neq\beta$. We map $|z|<1$ on the upper half $\Im\zeta>0$ of the ζ-plane, such that α, β correspond to $\zeta=0$, ∞ respectively, then S_ν becomes a linear transformation of the form:

$$S_\nu\colon\ \zeta'=\kappa_\nu\zeta \qquad (\kappa_\nu>0)\,. \tag{3}$$

Since G is properly discontinuous, we see that κ_ν is an integral power of a fixed $k_0>1$, so that by changing the notation, we may assume that

$$S_\nu\colon\ \zeta'=k_0^\nu\zeta \qquad (\nu=0, \pm1, \pm2, \cdots)\,. \tag{4}$$

Hence a half-ring Δ_0: $1\leqq|\zeta|\leqq k_0$, $\Im\zeta>0$ is a fundamental domain of G, which is mapped conformally on $\rho_0<|t|<1$ ($\rho_0=e^{-\frac{2\pi^2}{\log k_0}}$) by $t=e^{\frac{2\pi i\log\zeta}{\log k_0}}$.

Hence F is conformally equivalent to $\rho_0<|t|<1$, which is absurd.

(ii) Next suppose that $\alpha=\beta=\omega$, then we map $|z|<1$ on $\Im\zeta>0$ of the ζ-plane, such that ω corresponds to $\zeta=\infty$, then S_ν becomes

$$S_\nu:\ \zeta'=\zeta+\kappa_\nu \quad (\kappa_\nu=\text{real})\,. \tag{5}$$

Since G is properly discontinuous, κ_ν is an integral multiple of a fixed $k_0>0$, so that by changing the notation, we may assume that

$$S_\nu:\ \zeta'=\zeta+\nu k_0 \quad (\nu=0,\ \pm1,\ \pm2,\ \cdots)\,. \tag{6}$$

Hence Δ_0: $0\leqq\xi<k_0$, $\eta>0$ $(\zeta=\xi+i\eta)$ is a fundamental domain of G, which is mapped conformally on $0<|t|<1$ by $t=e^{\frac{2\pi i\zeta}{k_0}}$. Hence F is conformally equivalent to $0<|t|<1$, which is absurd. Hence the theorem.

Theorem X. 49. *Let F be a closed Riemann surface of genus $p\geqq2$, then F admitts only a finite number of conformal mappings on itself.*

Proof. Let F be spread over the z-sphere and suppose that F admitts infinitely many conformal mappings $z'=f_n(z)$ $(n=1, 2, \cdots)$ on itself. If z varies on K, then $w=f_n(z)$ describes F, so that by Theorem VI. 15, $\{f_n(z)\}$ forms a normal family, hence we can select a partial sequence from $\{f_n(z)\}$, which we denote by $\{f_n(z)\}$ again, such that $f_n(z)$ convegers uniformly on K.

Hence for any $\varepsilon>0$, if $n\geqq n_0$, then $[f_n(z), f_{n+1}(z)]<\varepsilon$ for any z on K, so that if we put $w=f_n(z)$, $z=f_n^{-1}(w)$, $\varphi_n(w)=f_{n+1}(f_n^{-1}(w))$, then φ_n is a conformal mapping of F on itself and $[\varphi_n(w), w]<\varepsilon$ for any $w\in F$, hence F admitts an infinitesimal conformal mapping on itself, which contradicts Theorem X. 48, hence the theorem.

Theorem X. 50. *Let F be a closed Riemann surface of genus $p\geqq2$ and admit N conformal mappings on itself, then*

$$N\leqq84\,(p-1)\,.$$ Hurwitz [1].

Proof. Let F be spread over the z-sphere and N conformal mappings of F on itself be T_i: $z'=T_i(z)$ $(i=0, 1, 2, \cdots, N-1)$, $T_0(z)\equiv z$. We may assume that N is the maximum number of conformal mappings of F on itself, then $\{T_i\}$ form a group G of order N. Hence if we put

$$w=\varphi(z)=T_0(z)+T_1(z)+\cdots+T_{N-1}(z)\,, \tag{1}$$

then $\varphi(z)$ is meromorphic on F and is invariant by G, such that

$$\varphi(T_i(z))=\varphi(z) \quad (i=0, 1, 2, \cdots, N-1)\,. \tag{2}$$

If z varies on F, then $w=\varphi(z)$ describes a closed Riemann surface Φ, spread over the w-sphere. By (2), a point w of Φ corresponds to N points $T_0(z)$, $T_1(z)$, $\cdots$, $T_{N-1}(z)$ of F, so that F can be considered as a covering surface of Φ of order N, such that above any point of Φ, there are N points of F. Suppose that for a point $z_0 \in F$, $T_i(z_0)=z_0$ for some $i \neq 0$ and let $T_0=1, T_1, \cdots, T_{\nu-1}$ be the totality of such T_i, then they form a sub-group of G of order ν. From this, we see that $T_0(z_0)$, $T_1(z_0), \cdots, T_{N-1}(z_0)$ can be grouped into N/ν classes, each class consisting of ν elements, such that if T_i, T_j belong to the same class, then

$$T_i(z_0)=T_j(z_0).$$

Hence if we put $w_0=\varphi(z_0)$, then, considered as a covering surface of Φ, F has N/ν branch points of the same multiplicity $\nu-1$ above w_0, where ν sheets are connected cyclically.

Now $\rho=2(p-1)$ is the Euler's characteristic of F and $\rho_0=2(\pi-1)$ is that of Φ, where π is the genus of Φ. Since F is a covering surface of Φ of order N, by Hurwitz's relation, we have $\rho=N\rho_0+v$, or

$$2(p-1)=2N(\pi-1)+v, \tag{3}$$

where v is the sum of multiplicities of branch points of F relatively to Φ.

Let $w_1, \cdots, w_r$ be the projection of relative branch points of F on Φ, then there are N/ν_i branch points of F of the same multiplicity ν_i-1 above w_i, so that

$$v=\sum_{i=1}^{r}\frac{N}{\nu_i}(\nu_i-1)=\sum_{i=1}^{r}N\left(1-\frac{1}{\nu_i}\right),$$

hence by (3),

$$\frac{2(p-1)}{N}=2(\pi-1)+\sum_{i=1}^{r}\left(1-\frac{1}{\nu_i}\right). \tag{4}$$

Hence if $\pi \geqq 2$, then $\dfrac{2(p-1)}{N} \geqq 2$, or

$$N \leqq p-1. \tag{5}$$

If $\pi=1$, then $\dfrac{2(p-1)}{N} \geqq 1-\dfrac{1}{\nu_1} \geqq 1-\dfrac{1}{2}=\dfrac{1}{2}$, or

$$N \leqq 4(p-1). \tag{6}$$

If $\pi=0$, then

$$\frac{2(p-1)}{N}=-2+\sum_{i=1}^{r}\left(1-\frac{1}{\nu_i}\right)=r-2-\sum_{i=1}^{r}\frac{1}{\nu_i}, \tag{7}$$

hence we have

$$r \geqq 3.$$

If $r \geqq 5$, then since $\nu_i \geqq 2$,

$$\frac{2(p-1)}{N} \geqq r-2-\frac{r}{2}=\frac{r}{2}-2 \geqq \frac{1}{2},$$

so that

$$N \leqq 4(p-1)\,. \tag{8}$$

If $r=4$, then

$$\frac{2(p-1)}{N}=2-\frac{1}{\nu_1}-\frac{1}{\nu_2}-\frac{1}{\nu_3}-\frac{1}{\nu_4}.$$

We may assume that $2\leqq\nu_1\leqq\nu_2\leqq\nu_3\leqq\nu_4$, then the maximum value of $\frac{1}{\nu_1}+\frac{1}{\nu_2}+\frac{1}{\nu_3}+\frac{1}{\nu_4}$ (<2) is $\frac{11}{6}$, which is attained when $\nu_1=2$, $\nu_2=2$, $\nu_3=2$, $\nu_4=3$[1], so that

$$N \leqq 12(p-1)\,. \tag{9}$$

If $r=3$, then

$$\frac{2(p-1)}{N}=1-\frac{1}{\nu_1}-\frac{1}{\nu_2}-\frac{1}{\nu_3}.$$

We may assume that $2\leqq\nu_1\leqq\nu_2\leqq\nu_3$, then the maximum value of $\frac{1}{\nu_1}+\frac{1}{\nu_2}+\frac{1}{\nu_3}(<1)$ is 41/42, which is attained, when $\nu_1=2$, $\nu_2=3$, $\nu_3=7$,[2] so that

$$N \leqq 84\,(p-1)\,. \tag{10}$$

Hence in any case, $N\leqq 84\,(p-1)$. The equality can occur only when $\pi=0$, $\nu_1=2$, $\nu_2=3$, $\nu_3=7$. Whether such a covering surface F of a sphere can be realized or not is not known, except the case $p=3$.

---

1)

ν_1	ν_2	ν_3	ν_4	$\frac{1}{\nu_1}+\frac{1}{\nu_2}+\frac{1}{\nu_3}+\frac{1}{\nu_4}$
$\geqq 3$	$\geqq 3$	$\geqq 3$	$\geqq 3$	$\leqq 4/3$
2	$\geqq 3$	$\geqq 3$	$\geqq 3$	$\leqq 3/2$
2	2	$\geqq 3$	$\geqq 3$	$\leqq 5/3$
2	2	2	$\geqq 3$	$\leqq 11/6$

2)

ν_1	ν_2	ν_3	$\frac{1}{\nu_1}+\frac{1}{\nu_2}+\frac{1}{\nu_3}$
$\geqq 4$	$\geqq 4$	$\geqq 4$	$\leqq 3/4$
3	$\geqq 4$	$\geqq 4$	$\leqq 5/6$
3	3	$\geqq 4$	$\leqq 11/12$
2	$\geqq 5$	$\geqq 5$	$\leqq 19/20$
2	3	$\geqq 7$	$\leqq 41/42$

17. Moduli of closed Riemann surfaces of genus $p\geqq 1$

Let F be a closed Riemann surface of genus $p\geqq 1$, spread over the w-sphere. If another Riemann surface F' can be mapped conformally on F, then F' is called equivalent to F and belongs to the same class as F. Different classes of closed Riemann surfaces of the same genus $p\geqq 1$ form a manifold M, such that a point of M and a class correspond one-to-one. Let μ be the dimension of M in the real sense, such that M depends on μ real parameters, which are called *moduli of closed Riemann surfaces of genus* $p\geqq 1$.

We denote a unit circle by C, a staright line by L and a half-line by L_0 and the product space $\overbrace{C\times\cdots\times C}^{p}$ by C^p etc., then we shall prove

Theorem X. 51. (i) *If* $p=1$, *then* $\mu=2$, $\quad M=C\times L_0$.

(ii) *If* $p\geqq 2$, *then* $\mu=6p-6$. *Let* $\mathfrak{M}=C^p\times L_0^p\times L^{4p-6}$, *then* $\mathfrak{M}$ *is decomposed by singular manifolds into a countable number of connected manifolds* $\{\mathfrak{M}_i\}$, *such that each* $\mathfrak{M}_i$ *can be taken as* M.

The part concerning μ is well known. Riemann [1]. Picard [2].

Proof. (Tsuji [56]). Let $(\gamma_1, \cdots, \gamma_p)$ be p disjoint ring cuts of F, such that if we cut F by $\{\gamma_i\}$, then F becomes a surface F_0 of planar character. We take infinitely many same samples as F_0 and connect them along the opposite shores of $\{\gamma_i\}$, as in the well known way, then we obtain a covering surface $F^{(\infty)}$ of F, which is of planar character, hence $F^{(\infty)}$ can be mapped conformally on a schlicht domain Ω on the z-plane. Let Γ_ν, $\overline{\Gamma}_\nu$ be the image of the both shores of γ_ν on the z-plane, then Γ_ν, $\overline{\Gamma}_\nu$ $(\nu=1, 2, \cdots, p)$ bound a domain Δ_0, where we assume that Δ_0 contains $z=\infty$. Let $w=w(z)$ be the mapping function, then $w(z)$ is invariant by a certain linear group G of Schottky type and Δ_0 is its fundamental domain, such that $\overline{\Gamma}_\nu=S_\nu(\Gamma_\nu)$, where S_ν is a linear transformation of G. Since G is of Schottky type, S_ν is hyperbolic or loxodromic, hence S_ν can be written in the form:

$$S_\nu:\quad \frac{z'-\alpha_\nu}{z'-\beta_\nu}=\rho_\nu e^{i\theta_\nu}\frac{z-\alpha_\nu}{z-\beta_\nu}\quad (\alpha_\nu,\ \beta_\nu\neq\infty), \qquad (1)$$

where we may assume that $\rho_\nu>1$, since if $0<\rho_\nu<1$, then we interchange α_ν, β_ν. α_ν, β_ν are fixed points of S_ν.

Since

$$S_\nu^{-1}:\quad \frac{z'-\beta_\nu}{z'-\alpha_\nu}=\rho_\nu e^{i\theta_\nu}\frac{z-\beta_\nu}{z-\alpha_\nu},$$

if we write S_ν^{-1} in the form:

$$S_\nu^{-1}:\ \frac{z'-\alpha_\nu'}{z'-\beta_\nu'}=\rho_\nu e^{i\theta_\nu}\frac{z-\alpha_\nu'}{z-\beta_\nu'},$$

then $\alpha_\nu'=\beta_\nu$, $\beta_\nu'=\alpha_\nu$, so that

$$\alpha_\nu+\beta_\nu=\alpha_\nu'+\beta_\nu', \quad \alpha_\nu\beta_\nu=\alpha_\nu'\beta_\nu'. \tag{2}$$

Conversely, if there be given p pairs of disjoint Jordan curves Γ^ν, $\overline{\Gamma}_\nu$ ($\nu=1, 2, \cdots, p$), which lie outside each other, bounding an infinite domain Δ_0 and p linear transformations S_ν of the form (1), such that $\overline{\Gamma}_\nu=S_\nu(\Gamma_\nu)$, then if we identify the equivalent points of Γ_ν, $\overline{\Gamma}_\nu$, Δ_0 can be considered as a closed Riemann surface F of genus p, which we denote by $F=(\Gamma_\nu, \overline{\Gamma}_\nu; S_\nu)$. We remark that

$$(\Gamma_\nu, \overline{\Gamma}_\nu; S_\nu)=(\overline{\Gamma}_\nu, \Gamma_\nu; S_\nu^{-1}).$$

Now to represent F, $\{\Gamma_\nu, \overline{\Gamma}_\nu\}$ are not essential. For, let Λ_ν ($\nu=1, 2, \cdots, p$) be a Jordan curve, surrounding α_ν and put $\overline{\Lambda}_\nu=S_\nu(\Lambda_\nu)$. If Λ_ν, $\overline{\Lambda}_\nu$ ($\nu=1, 2, \cdots, p$) lie outside each other and bound an infinite domain, then $(\Gamma_\nu, \overline{\Gamma}_\nu; S_\nu)=(\Lambda_\nu, \overline{\Lambda}_\nu; S_\nu)$.

Let (S_ν) ($\nu=1, 2, \cdots, p$) be a set of p linear transformations of the form (1). If there exists a Jordan curve Γ_ν, surrounding α_ν, such that if we put $\overline{\Gamma}_\nu=S_\nu(\Gamma_\nu)$, then Γ_ν, $\overline{\Gamma}_\nu$ ($\nu=1, 2, \cdots, p$) lie outside each other and bound an infinite domain, then we denote (S_ν) *an ordinary set* and *a singular set*, if otherwise.

Let (S_ν), (S_ν') be two sets of p linear transformations of the form (1). If there exists a linear transformation T of z, such that

$$S_\nu'=TS_\nu T^{-1}, \quad \alpha_\nu'=T(\alpha_\nu), \quad \beta_\nu'=T(\beta_\nu) \quad (\nu=1, 2, \cdots, p), \tag{3}$$

then we say that (S_ν), (S_ν') are *equivalent*.

First suppose that $p\geqq 2$. Then by a linear transformation, we normalize S_1 and S_2 in $(S_1, S_2, \cdots, S_p)$, such that in (1)

$$\alpha_1=0, \quad \beta_1=1, \quad \alpha_2=-1,$$

where β_2, α_ν, β_ν ($\nu=3, \cdots, p$) are any complex numbers.

If (S_ν), (S_ν') are equivalent, then since $\alpha_1=\alpha_1'=0$, $\beta_1=\beta_1'=1$, $\alpha_2=\alpha_2'=-1$, we see by (3), that T has three fixed points, so that $T=1$, $S_\nu=S_\nu'$ ($\nu=1, 2, \cdots, p$), hence if $(S_1, \cdots, S_p)\neq(S_1', \cdots, S_p')$, then (S_ν), (S_ν') are not equivalent. Now we consider all non-equivalent (S_ν).

Let $\mathfrak{M}$ be the manifold of ρ_ν, θ_ν ($\nu=1, 2, \cdots, p$), β_2, $\alpha_\nu+\beta_\nu$, $\alpha_\nu\beta_\nu$ ($\nu=3, \cdots, p$), which correspond to S_ν by (1) and μ be its dimension in the real sense, then

$$\mathfrak{M}=C^p\times L_0^p\times L^{4p-6},\quad \mu=6p-6\,. \tag{4}$$

We denote the point of $\mathfrak{M}$, which corresponds to (S_ν) by the same letter (S_ν). $\mathfrak{M}$ consists of non-equivalent (S_ν) and by (2), (S_ν) and (S_ν^{-1}) are the same point of $\mathfrak{M}$.

Now we except from $\mathfrak{M}$ all singular (S_ν), then there remains connected manifolds $\{\mathfrak{M}_i\}$, consisting of ordinary (S_ν). As is easily seen, $\mathfrak{M}_i$ is an open set, so that its dimension is $6p-6$.

If (S_ν), (S'_ν) are two points of the same $\mathfrak{M}_i$, then they can be connected by a curve in $\mathfrak{M}_i$. Hence if $F=(\Gamma_\nu, \overline{\Gamma}_\nu;\, S_\nu)$, $F'=(\Gamma'_\nu, \overline{\Gamma}'_\nu,\, S'_\nu)$ and $\{\gamma_\nu\}$, $\{\gamma'_\nu\}$ be the ring cuts of F, F', corresponding to $\{\Gamma_\nu\}$, $\{\Gamma'_\nu\}$ respectively and F_0, F'_0 be the surface of planar character, corresponding to $\{\gamma_\nu\}$, $\{\gamma'_\nu\}$, then we see easily that F can be deformed continuously into F', such that $F_0\to F'_0$, $\gamma_\nu\to\gamma'_\nu$ $(\nu=1, 2, \cdots, p)$.

Let $(\gamma_1, \cdots, \gamma_p)$, $(\gamma'_1, \cdots, \gamma'_p)$ be two sets of p disjoint ring cuts of the same Riemann surface F and F_0, F'_0 be the surface of planar character, corresponding to $\{\gamma_\nu\}$, $\{\gamma'_\nu\}$ respectively. By means of F_0, F'_0, we construct $F^{(\infty)}$, $F'^{(\infty)}$ and mapping them conformally on schlicht domains on the z-plane, we represent F in two ways:

$$F=(\Gamma_\nu, \overline{\Gamma}_\nu;\, S_\nu),\qquad F=(\Gamma'_\nu, \overline{\Gamma}'_\nu;\, S'_\nu)\,,$$

and let $(S_\nu)\in\mathfrak{M}_i$, $(S'_\nu)\in\mathfrak{M}_j$. If F can not be deformed continuously, such that $F_0\to F'_0$, $\gamma_\nu\to\gamma'_\nu$ $(\nu=1, 2, \cdots, p)$, then $\mathfrak{M}_i\neq\mathfrak{M}_j$. We call such $\{\gamma_\nu\}$, $\{\gamma'_\nu\}$ non-equivalent sets of p ring cuts of F. Since there are a countable number of non-equivalent (γ_ν) on F, there are a countable number of different $\mathfrak{M}_i$, so that $\mathfrak{M}$ is decomposed into a countable number of connected manifolds $\{\mathfrak{M}_i\}$ by singular manifolds, consisting of singular (S_ν).

Let $\mathfrak{R}$ be the totality of closed Riemann surfaces of genus $p\geqq 2$. We choose one $F\in\mathfrak{R}$ and let $(\gamma_1, \cdots, \gamma_p)$ be p disjoint ring cuts of F and F_0 be the corresponding surface of planar character.

Let F' be any one from $\mathfrak{R}$. Then there exists p ring cuts $(\gamma'_1, \cdots, \gamma'_p)$ of F', such that F can be deformed continuously into F', such that $F_0\to F'_0$. $\gamma_\nu\to\gamma'_\nu$ $(\nu=1, 2, \cdots, p)$. Thus by means of $\{\gamma_\nu\}$ on one fixed $F\in\mathfrak{R}$, we can define $\{\gamma'_\nu\}$ on any $F'\in\mathfrak{R}$. If by means of F_0, F'_0, we represent F and F' as $F=(\Gamma_\nu, \overline{\Gamma}_\nu;\, S_\nu)$, $F'=(\Gamma'_\nu, \overline{\Gamma}'_\nu;\, S'_\nu)$, then if $(S_\nu)\in\mathfrak{M}_1$, (say), then $(S'_\nu)\in\mathfrak{M}_1$. Hence a Riemann surface $\in\mathfrak{R}$ has a representative in $\mathfrak{M}_1$. Next we shall prove that two different points of $\mathfrak{M}_1$ correspond to two Riemann surfaces, which are not equivalent.

Suppose that F, F' belong to the same class, then there exists a conformal mapping φ, such that $F'=\varphi(F)$. Let $(\gamma_1, \cdots, \gamma_p)$ be p disjoint ring cuts of F and F_0 be the corresponding surface of planar character and let $\gamma'_\nu=\varphi(\gamma_\nu)$, $F'_0=\varphi(F_0)$, then $\{\gamma'\}$ are p disjoint ring cuts of F' and F'_0 is the corresponding surface of planar character. By means of F_0 and F'_0, we construct $F^{(\infty)}$, $F'^{(\infty)}$ and we map $F^{(\infty)}$ conformally on a schlicht domain Ω on the z-plane and $F'^{(\infty)}$ on a schlicht domain Ω' on the z'-plane and represent F, F' as

$$F=(\Gamma_\nu, \overline{\Gamma}_\nu; S_\nu), \quad F'=(\Gamma'_\nu, \overline{\Gamma}'_\nu; S'_\nu).$$

Then Ω is mapped conformally on Ω'. Now by Theorem X. 21, $\Omega\in O_{AD}$, $\Omega'\in O_{AD}$, so that the mapping function is a linear function. Hence there exists a linear transformation T, such that

$$S'_\nu=TS_\nu T^{-1}, \quad \alpha'_\nu=T(\alpha_\nu),\ \beta'_\nu=T(\beta_\nu)\ (\nu=1, 2, \cdots, p),$$

hence (S_ν), (S'_ν) are equivalent. Since $\mathfrak{M}$ consists of non-equivalent points, $(S\ )$ and (S'_ν) are the same point of $\mathfrak{M}$. Hence a point of $\mathfrak{M}_1$ and a class correspond one-to-one.

By taking another p ring cuts of F, which are not equivalent to $\{\gamma_\nu\}$, we can represent a class by a point of any $\mathfrak{M}_i$. Hence we can take any $\mathfrak{M}_i$ as M.

Next we consider the case $p=1$. Let $\mathfrak{R}$ be the totality of closed Riemann surfaces of genus $p=1$. By a suitable linear transformation, we normalize S_1, such that $\alpha_1=0$, $\beta_1=1$ and ρ_1, θ_1 are arbitrary real constants. Hence if $\mathfrak{M}$ be the manifold of ρ_1, θ_1 and μ be its dimension, then

$$\mathfrak{M}=C\times L_0, \quad \mu=2. \tag{5}$$

By means of γ_1 on one fixed $F\in\mathfrak{R}$, we can define γ'_1 on any $F'\in\mathfrak{R}$ as before and we can prove similarly that $M=\mathfrak{M}$.

18. Prolongation of a Riemann surface

Let F be a Riemann surface. If F can be mapped conformally one-to-one on a proper part Φ_0 of another Riemann surface Φ, then Φ is called a prolongation of F. If there exists no such a prolongation, then F is called *non-prolongable*, or *maximal*. This concept was first introduced by Radó [1].

If F is a closed Riemann surface, then evidently F is non-prolongable. Radó showed that there exists an open Riemann surface, which is non-prolongable by the following example.

On the z-plane, we choose infinitely many slits: [0, 1], [−2, −1], [2, 3], [−4, −3], $\cdots$ on the real axis. We take two such slits-planes and connect them along the opposite shores of the slits, then we obtain a two-sheeted Riemann surface F, which is of infinite genus. Then

$$F \text{ is non-prolongable}. \qquad (1)$$

This seems evident from the topological consideration, but this is not so as seen as follows. We map the finite z-plane: $|z|<\infty$ topologically on $|\zeta|<1$, such that every point of the unit circle $|\zeta|=1$ corresponds to $z=\infty$. Then F becomes a Riemann surface F^*, spread over the unit disc $|\zeta|<1$. Since the unit circle $|\zeta|=1$ contains an arc, which is free from cluster points of branch points of F^*, F^* can be prolongated topologically to a larger Riemann surface. Hence to prove (1), we have to take account of conformity. We shall prove the following theorem, which contains the Radó's example as a special case.

Theorem X. 52. *Let F be a Riemann surface, spread over the w-sphere, which consists of n-sheets and every boundary point of F is a cluster point of its branch points and the projection of its boundary on the w-sphere is a closed set E of capacity zero, then F is non-prolongable.* Tsuji [9].

Proof. First we shall prove that $F\in O_G$. Suppose that $F\in P_G$. Taking off all points of F, which lie above E, if there exist such ones, we obtain a Riemann surface F^*, which consists of n sheets and is a covering surface of the complement Ω of E with respect to the w-sphere. Since $F\in P_G$, we have $F^*\in P_G$. Let $g(p, p_0)$ be the Green's function of F^*, with a pole p_0, which lies above $w_0\in\Omega$. Let $p_1(w), \cdots, p_n(w)$ be n points of F^*, which lie above $w\in\Omega$. Then $\overset{*}{g}(w, w_0)=\sum_{i=1}^{n} g(p_i(w), p_0)$ is a positive harmonic function in Ω, with a logarithmic singularity $\log\dfrac{1}{|w-w_0|}$ at w_0, hence there exists a Green's function in Ω[1)], so that $\gamma(E)>0$, which is absurd. Hence

$$F\in O_G. \qquad (1)$$

Suppose that F is prolongable and can be mapped conformally on a proper part Φ_0 of another Riemann surface Φ, spread over the z-sphere and let $w=f(z)$ be the mapping function. Then there exists an inner point z_0 of Φ, which is a boundary point of Φ_0. We may assume that z_0 is not a branch point of Φ, since otherwise, we map a neighbourhood of z_0 on a schlicht disc $|t|<\rho$ on the t-plane and consider on the t-plane. We take $\rho>0$ so small that a schlicht disc Δ: $|z-z_0|<\rho$ and its

1) Let $\{F_n\}$ be an exhaustion of Ω, with the boundary $\{\Gamma_n\}$, such that $w_0\in F_0$ and $g_n(w, w_0)$ be the Green's function of F_n, then $\overset{*}{g}-g_n$ is harmonic in F_n and >0 on Γ_n, so that $\overset{*}{g}>g_n$ in F_n. Hence $\lim_n g_n(w, w_0)\leqq\overset{*}{g}(w, w_0)<\infty$, so that the Green's function of Ω exists.

boundary circle C: $|z-z_0|=\rho$ are contained in Φ.

Let $\Phi_0^{(\rho)}$ be the part of Φ_0, contained in Δ and M be the set of boundary points of $\Phi_0^{(\rho)}$, contained in Δ. First suppose that $\gamma(M)=0$, then $D=\Delta-M$ is a connected domain and $f(z)$ is meromorphic in D. If $f(z)$ is meromorphic at a point $z_1 \in M$, then $w_1=f(z_1)$ is a boundary point of F, so that w_1 is a cluster point of branch points of F, which is absurd.

Hence $f(z)$ has an essential singularity at every point $\in M$. Hence $f(z)$ takes any value infinitely often, except a set of capacity zero in any small neighbourhood of M, which is absurd, since F consists of only n sheets. Hence $\gamma(M)>0$, so that by Theorem X. 12, $\Phi_0 \in P_G$, hence $F \in P_G$, which contradicts (1). Hence F is non-prolongable.

Theorem X. 53. *For any Riemann surface F, there exists a maximal prolongation.* Bochner [1].

Bochner used Zermelo's axiom in the proof and Heins [1] proved without it. We shall reproduce Heins' proof in a somewhat simplified form.

Proof. Let F be a Riemann surface, spread over the z-sphere and Φ be its prolongation, such that F is mapped conformally on a proper part Φ_0 of Φ. We assume that Φ is spread over the w-sphere. We enlarge Φ to a Riemann surface $\tilde{\Phi}$ as follows.

Let C be a closed curve on Φ and let w be its initial point and $\tilde{w}$ be its end point, which are the same point as w of Φ. If C is homotop null (in Φ), or homotop (in Φ) to a closed curve $\in \Phi_0$, then we define $\tilde{w}$ and w are the same point of $\tilde{\Phi}$ and if otherwise, $\tilde{w}$ and w are different points of $\tilde{\Phi}$. If we define $\tilde{\Phi}$ in this way, then $\tilde{\Phi}$ is a covering surface of Φ and is a prolongation of F. We call $\tilde{\Phi}$ *simply connected relatively to Φ_0*. A closed curve in $\tilde{\Phi}$ is homotop null (in $\tilde{\Phi}$), or homotop (in $\tilde{\Phi}$) to a closed curve $\in \Phi_0$. In the following, the prolongation Φ of F means always the thus enlarged Riemann surface, so that $\Phi=\tilde{\Phi}$.

We may assume that $z=0$ belongs to F and is different from a branch point of F and $z=0$ is mapped on $w=0$, which is different from a branch point of Φ and by $w=f(z)$ $(f(0)=0)$, F is mapped conformally on a proper part Φ_0 of Φ.

We may assume that F does not admit a sphere, or a torus (a closed Riemann surface of genus $p=1$) as its prolongation, then we can map the universal covering surface $F^{(\infty)}$ of F conformally on $|\mathfrak{z}|<1$, such that $z=0$ is mapped on $\mathfrak{z}=0$. Then we obtain a Fuchsian group $\mathfrak{G}_F$ in $|\mathfrak{z}|<1$.

We map the universal covering surface $\Phi^{(\infty)}$ of Φ conformally on

$|\zeta|<R$ by $\zeta=h(w)$ $(h(0)=0)$, where $R=R_\Phi$ is determined by the condition, that if we put

$$\zeta=h(f(z))=\varphi(z)\,, \tag{1}$$

then

$$\varphi(0)=0, \quad \varphi'(0)=1\,. \tag{2}$$

Then we obtain a Fuchsian group $\mathfrak{G}_\Phi$ in $|\zeta|<R$.

Since $\Phi=\widetilde{\Phi}$ is simply connected relatively to Φ_0, $\mathfrak{G}_F$ is mapped homomorphically on $\mathfrak{G}_\Phi$, such that to an element of $\mathfrak{G}_F$, there corresponds an element of $\mathfrak{G}_\Phi$, but to an element of $\mathfrak{G}_\Phi$, there may correspond several elements of $\mathfrak{G}_F$. By $\zeta=\varphi(z)$, F is mapped conformally on a proper part of the fundamental domain of $\mathfrak{G}_\Phi$.

Let

$$T_\nu:\quad \mathfrak{z}'=e^{i\psi_\nu}\frac{(\mathfrak{z}-\alpha_\nu)}{1-\bar{\alpha}_\nu\mathfrak{z}}, \quad |\alpha_\nu|<1 \quad (\nu=0, 1, 2, \cdots) \tag{3}$$

be an element of $\mathfrak{G}_F$, then by the homomorphism, T_ν corresponds to an element S_ν of $\mathfrak{G}_\Phi$:

$$S_\nu:\quad \zeta'=e^{i\theta_\nu}\frac{R^2(\zeta-a_\nu)}{R^2-\bar{a}_\nu\zeta}, \ |a_\nu|<R \quad (\nu=0, 1, 2\ldots)\,. \tag{4}$$

We consider all prolongations Φ of F and put

$$R^*=\sup_\Phi R_\Phi\,. \tag{5}$$

We shall prove that $R^*<\infty$ and there exists a prolongation Φ^* of F, such that $R_{\Phi^*}=R^*$ and Φ^* is maximal.

By (5), there exist prolongations $\Phi_n\,(n=1, 2, \cdots)$ of F, such that $\lim_n R_n=R^*$, where $R_n=R_{\Phi_n}$ and let $\varphi_n(z)$ $(\varphi_n(0)=0,\ \varphi_n'(0)=1)$ be the corresponding function, defined by (1).

Let a schlicht disc: $|z|<\rho$ be contained in F, then since $\varphi_n(z)$ is schlicht in $|z|<\rho$, by Koebe's distortion theorem, $\varphi_n(z)$, $\varphi_n'(z)$, $1/\varphi_n'(z)$ are uniformly bounded in $|z|\leqq\rho_1<\rho$. Since $\varphi_n(z)$ is locally schlicht on F, we see easily that $\{\varphi_n(z)\}$ are uniformly bounded in any compact sub-surface of F, hence we can select a partial sequence from $\{\varphi_n(z)\}$, which we denote by $\{\varphi_n(z)\}$ again, such that

$$\lim_n \varphi_n(z)=\varphi(z), \quad \varphi(0)=0,\ \varphi'(0)=1\,, \tag{6}$$

uniformly in the wider sense on F, then $\varphi(z)$ is schlicht on F.

Let

$$S_\nu^{(n)}:\ \zeta'=e^{i\theta_\nu^{(n)}}\frac{R_n^2(\zeta-a_\nu^{(n)})}{R_n^2-\bar{a}_\nu^{(n)}\zeta}=e^{\,\theta_\nu^{(n)}}\frac{\zeta-a_\nu^{(n)}}{1-\dfrac{\bar{a}_\nu^{(n)}}{R_n}\cdot\dfrac{\zeta}{R_n}}, \quad |a_\nu^{(n)}|<R_n\,, \tag{7}$$

$(\nu=0, 1, 2, \cdots)$

be the element of $\mathfrak{G}_{\Phi_n}$, which corresponds to T_ν by the homomorphism. By (6), we see that

$$\lim_n a_\nu^{(n)} = a_\nu \quad (\nu=0, 1, 2, \cdots) \tag{8}$$

exists and by taking a suitable sub-sequence, we may assume that

$$\lim_n \theta_\nu^{(n)} = \theta_\nu \quad (\nu=0, 1, 2, \cdots) \tag{9}$$

exists.

If $R^*=\infty$, then $R_n\to\infty$, so that by (7), (8), (9),

$$S_\nu^{(n)}\to S_\nu:\ \zeta'=e^{i\theta_\nu}(\zeta-a_\nu)\ (n\to\infty)\ (\nu=0, 1, 2, \cdots)\ .\ .$$

Since $\mathfrak{G}_F$ has no fixed points, $\theta_\nu=0$, so that

$$S_\nu:\ \zeta'=\zeta-a_\nu\ (\nu=0, 1, 2, \cdots)\ . \tag{10}$$

By Koebe's 1/4-disc theorem, the image of $|z|<\rho$ on the $\zeta=\varphi_n(z)$-plane contains a disc $|\zeta|<\rho/4$, hence the group $\mathfrak{G}$, generated by $\{S_\nu\}$ is properly discontinuous. Hence $\mathfrak{G}$ is either a simply periodic group of translations, or a doubly periodic group of translations. Since F is mapped conformally on a part of the fundamental domain of $\mathfrak{G}$, F admitts a sphere, or a torus as its prolongation, which contradicts the hypothesis. Hence $R^*<\infty$, so that by (7),

$$S_\nu^{(n)}\to S_\nu:\ \zeta'=e^{i\theta_\nu}\frac{R^{*2}(\zeta-a_\nu)}{R^{*2}-\bar{a}_\nu\zeta},\quad |a_\nu|<R^*\ (n\to\infty)\ . \tag{11}$$

$$(\nu=0, 1, 2, \cdots)$$

Since the group $\mathfrak{G}$, generated be $\{S_\nu\}$ is properly discontinuous, let D_0 be its fundamental domain, then F is mapped on a part of D_0. If we identify the equivalent points on the boundary of D_0, then D_0 can be considered as a Riemann surface Φ^*, then $R_{\Phi^*}=R^*$ and Φ^* is a prolongation of F. We shall prove that Φ^* is maximal. Suppose that Φ^* is not maximal and can be mapped conformally on a proper part of another Riemann surface Φ_1, so that Φ_1 is a prolongation of F. As before, we map the universal covering surface $\Phi_1^{(\infty)}$ of Φ_1 conformally on $|\zeta_1|<R_1$. Let $\zeta=\varphi^*(z)\,(\varphi^*(0)=0,\ \varphi^{*\prime}(0)=1)$ and $\zeta_1=\varphi_1(z)\,(\varphi_1(0)=0,\ \varphi_1'(0)=1)$ be the functions (1) corresponding to Φ^* and Φ_1 respectively. Then $|\zeta|<R^*$ is mapped on a proper part of $|\zeta_1|<R_1$, which is the image of Φ^*. Let $\zeta_1=g(\zeta)$ $(g(0)=0)$ be the mapping function, then by Schwarz's lemma,

$$|g'(0)|<R_1/R^*.$$

Since $\varphi_1(z)=g(\varphi^*(z))$, $\varphi_1'(0)=g'(0)\varphi^{*\prime}(0)$, we have $g'(0)=1$, so that

$$R^*<R_1\ ,$$

which contradicts the definition of R^*. Hence Φ^* is maximal.

CHAPTER XI

FUCHSIAN GROUP

1. Non-euclidean metric in $|z|<1$

1. Let S: $w=e^{i\alpha}\dfrac{z-a}{1-\bar{a}z}$ $(|a|<1)$ be a linear transformation, which makes $|z|<1$ invariant, then

$$\frac{|dw|}{1-|w|^2}=\frac{|dz|}{1-|z|^2}. \tag{1}$$

If $w=S(z)$, $e^{i\varphi}=S(e^{i\theta})$, then

$$\frac{1-|w|^2}{|w-e^{i\varphi}|^2}d\varphi=\frac{1-|z|^2}{|z-e^{i\theta}|^2}d\theta . \tag{2}$$

We introduce a non-euclidean metric ds and a non-euclidean surface element $d\sigma$ by

$$ds=\frac{2|dz|}{1-|z|^2}, \qquad d\sigma=\frac{4\,dxdy}{(1-|z|^2)^2}, \qquad z=x+iy , \tag{3}$$

then ds and $d\sigma$ are invariant by S.

Let $L(r)$ be the non-euclidean length of a segment: $z=t$ $(0\leqq t\leqq r<1)$, then

$$L(r)=2\int_0^r\frac{dt}{1-t^2}=\log\frac{1+r}{1-r}, \tag{4}$$

and let $A(r)$ be the non-euclidean area of a disc: $|z|\leqq r<1$, then

$$A(r)=4\int_0^r\int_0^{2\pi}\frac{tdtd\theta}{(1-t^2)^2}=\frac{4\pi r^2}{1-r^2}. \tag{5}$$

Let a, b be two points in $|z|<1$ and $C(a, b)$ be the part of the orthogonal circle to $|z|=1$, through a, b and bounded by a, b, then we can prove easily that the non-euclidean length of $C(a, b)$ is the minimum of non-euclidean length of any curve through a, b.

Theorem XI. 1. *Let Δ be a triangle in $|z|<1$, bounded by three orthogonal circles to $|z|=1$ and A, B, C be the inner angle at its vertex A, B, C. Let $\sigma(\Delta)$ be the non-euclidean area of Δ, then*

$$\sigma(\Delta)=\pi-(A+B+C) .$$

Proof. By a suitable linear transformation of the form S, we may assume that $A=0$ and $\widehat{BC}$ is orthogonal to the real axis, so that $\widehat{AB}$, $\widehat{AC}$ are rectilinear segments. Let $(x-\rho)^2+y^2=a^2$ be the equation of $\widehat{BC}$,

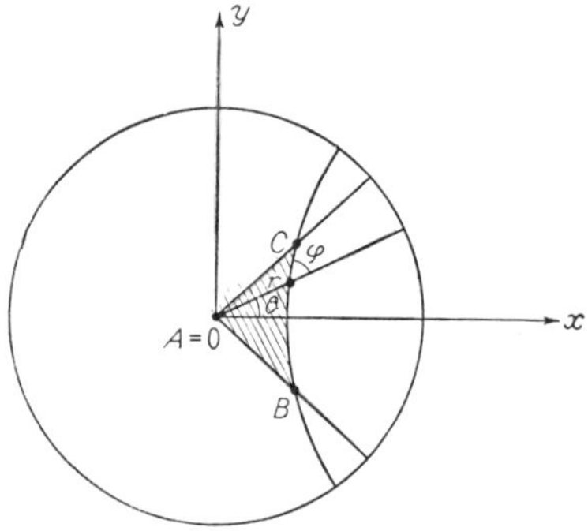

then since $\widehat{BC}$ is orthogonal to $|z|=1$, $\rho^2=a^2+1$, so that the equation of $\widehat{BC}$ in polar coordinates is

$$r^2-2r\rho\cos\theta+1=0 \quad (\rho>1),$$

$$r=\rho\cos\theta-\sqrt{\rho^2\cos^2\theta-1}, \tag{1}$$

hence $1-r^2=2\sqrt{\rho^2\cos^2\theta-1}\,(\rho\cos\theta-\sqrt{\rho^2\cos^2\theta-1})=2r\sqrt{\rho^2\cos^2\theta-1}$, so that

$$\frac{2r^2}{1-r^2}=\frac{1+r^2}{1-r^2}-1$$

$$=\frac{\rho\cos\theta}{\sqrt{\rho^2\cos^2\theta-1}}-1. \tag{2}$$

Let $B=(r_1, \theta_1)$, $C=(r_2, \theta_2)$, then

$$A=\theta_2-\theta_1. \tag{3}$$

By (2),

$$\sigma(\Delta)=4\int_{\theta_1}^{\theta_2}d\theta\int_0^r\frac{t\,dt}{(1-t^2)^2}=2\int_{\theta_1}^{\theta_2}\frac{r^2d\theta}{1-r^2}=\int_{\theta_1}^{\theta_2}\frac{\rho\cos\theta\,d\theta}{\sqrt{\rho^2\cos^2\theta-1}}-(\theta_2-\theta_1)$$

$$=\int_{\theta_1}^{\theta_2}\frac{\rho\cos\theta\,d\theta}{\sqrt{\rho^2\cos^2\theta-1}}-A. \tag{4}$$

Let φ be the angle between the tangent of $\widehat{BC}$ at $P=(r,\theta)$ and $\overrightarrow{AP}$, then

$$\tan\varphi=r\Big/\frac{dr}{d\theta}=\frac{\sqrt{\rho^2\cos^2\theta-1}}{\rho\sin\theta}, \quad \cos\varphi=\frac{\rho\sin\theta}{\sqrt{\rho^2-1}}, \quad \sin\varphi=\frac{\sqrt{\rho^2\cos^2\theta-1}}{\sqrt{\rho^2-1}},$$

so that

$$-d\varphi=\frac{\rho\cos\theta\,d\theta}{\sqrt{\rho^2\cos^2\theta-1}},$$

hence by (4),

$$\sigma(\Delta)=\varphi_1-\varphi_2-A. \tag{5}$$

Since $\varphi_1=\pi-B$, $\varphi_2=C$, we have $\sigma(\Delta)=\pi-(A+B+C)$. ——

2. We introduce another metric $[a, b]$ for two points a, b in $|z|<1$ as follows:

$$[a, b]=\left|\frac{a-b}{1-\bar{a}b}\right|. \tag{1}$$

Then $[a, b]=[b, a]$ and if S is a linear transformation, which makes $|z|<1$ invariant, then

$$[S(a), S(b)]=[a, b]. \tag{2}$$

$[a, b]$ satisfies the following triangle inequality:

$$[a, c] \leqq [a, b]+[b, c]. \tag{3}$$

Proof. By a suitable linear transformation, which makes $|z|<1$ invariant, we may assume that $a=0$, $b>0$, then (3) becomes

$$|c|<b+\left|\frac{b-c}{1-bc}\right|. \tag{3'}$$

Since (3′) is evident, if $|c|\leqq b$, we assume that $|c|>b$.

Let $|c|=\rho$, then on the circle $|z|=\rho$,

$$1-\left|\frac{b-z}{1-bz}\right|^2=\frac{(1-b^2)(1-\rho^2)}{|1-bz|^2},$$

so that $\operatorname*{Max}_{|z|=\mu}\left(1-\left|\frac{b-z}{1-bz}\right|^2\right)$ is attained, when $z=\rho$, hence $\operatorname*{Min}_{|z|=\mu}\left|\frac{b-z}{1-bz}\right|$ is attained, when $z=\rho$ and its value is $\frac{\rho-b}{1-b\rho}$. Hence (3′) holds, if $\rho<b+\frac{\rho-b}{1-b\rho}$, which holds evidently. ——

Let $0<\rho<1$, $0<a<1$, then $C(a, \rho)$: $[z, a]=\left|\frac{z-a}{1-az}\right|=\rho$ is a circle, which meets the real axis at $r=\frac{a-\rho}{1-a\rho}$, $r_1=\frac{a+\rho}{1+a\rho}$ $(r<a<r_1)$.

Since $r_1-a=\frac{\rho(1-a^2)}{1+a\rho}<\frac{\rho(1-a^2)}{1-a\rho}=a-r$, $C(a, \rho)$ contains a disc: $|z-a|\leqq\frac{\rho(1-a^2)}{1+a\rho}$ and is contained in a disc: $|z-a|\leqq\frac{\rho(1-a^2)}{1-a\rho}$. Also $C(a, \rho)$ is contained in a disc: $|z|\leqq r_1=\frac{a+\rho}{1+a\rho}$.

If $a(|a|<1)$ is a complex number, then $C(a, \rho)$: $[z, a]=\rho$ is a circle, such that

(i) $C(a, \rho)$ contains a disc: $|z-a|\leqq\frac{\rho(1-|a|^2)}{1+|a|\rho}$ and is contained in a disc: $|z-a|\leqq\frac{\rho(1-|a|^2)}{1-|a|\rho}$.

(ii) $C(a, \rho)$ is contained in a disc: $|z|\leqq\frac{|a|+\rho}{1+|a|\rho}$, so that if $|a|\leqq r<1$, then $C(a, \rho)$ is contained in a disc: $|z|\leqq r_1=\frac{r+\rho}{1+r\rho}$, where

$$1-r_1=\frac{(1-r)(1-\rho)}{1+r\rho}.$$

From (i), we have easily

Theorem XI. 2. *Let $L(\alpha)$ be a line through $z=1$, making an angle α $(0<\alpha<\pi)$ with the positive tangent of $|z|=1$ at $z=1$ and a be a point on $L(\alpha)$, tending to $z=1$ along $L(\alpha)$. For a fixed ρ $(0<\rho<1)$, let D be the domain, swept by discs $[z, a]\leqq\rho$. Then D contains a part of an angular domain, bounded by two lines L_1, L_2 through $z=1$, making an angle α_1, α_2 $(0<\alpha_1<\alpha<\alpha_2<\pi)$ with the positive tangent of $|z|=1$ at*

$z=1$ respectively.

Similarly D is contained in an angular domain, bounded by two lines L_1', L_2' through $z=1$, making an angle α_1', α_2' $(0<\alpha_1'<\alpha_1<\alpha_2<\alpha_2'<\pi)$ with the positive tangent of $|z|=1$ at $z=1$ respectively.

2. Fundamental domain

1. Let G be a group of linear transformations S_n $(n=0, 1, 2, \cdots)$, $S_0\equiv 1$, which make $|z|<1$ invariant:

$$S_n:\ z'=e^{i\theta_n}\frac{z-a_n}{1-\bar{a}_n z}\quad (|a_n|<1)\quad (n=0, 1, 2, \cdots). \tag{1}$$

z, z' are called *equivalent* by G.

If equivalents of $z\,(|z|<1)$ have no cluster point in $|z|<1$, then G is called *properly discontinuous*. A group of linear transformations, which make $|z|<1$ invariant and is properly discontinuous is called *a Fuchsian group*.

In the following, G denotes always a Fuchsian group.

A domain D_0 in $|z|<1$ is called *a fundamental domain of* G, if any two points of D_0 are not equivalent and any point z $(|z|<1)$ has its equivalent in D_0.

2. We shall prove that there exists a fundamental domain of G. First we shall define a normal domain of G as follows.

Let $z_n=S_n(0)$ $(S_n\in G)$ $(n=0, 1, 2, \cdots)$ be equivalents of $z=0$, where $z_0=0$ and Δ_0 be the set of z, such that

$$[z, 0]<[z, z_n]\quad (n=1, 2, \cdots). \tag{1}$$

Since G is properly discontinuous, if $\rho>0$ is sufficiently small, then a disc $|z|\leqq\rho$ belongs to Δ_0, so that Δ_0 has an inner point.

We call Δ_0 *a normal domain of* G (Bieberbach [2]).

At a boundary point $z(|z|<1)$ of Δ_0, $[z, 0]=[z, z_n]$ for some n, so that

$$|z|=\left|\frac{z-z_n}{1-\bar{z}_n z}\right|,\qquad 1-|z|^2=\frac{(1-|z|^2)(1-|z_n|^2)}{|1-\bar{z}_n z|^2},$$

hence

$$|1-\bar{z}_n z|^2=1-|z_n|^2,\ \text{or}\quad \left|z-\frac{1}{\bar{z}_n}\right|^2=\frac{1}{|z_n|^2}-1. \tag{2}$$

(2) is the equation of an orthogonal circle to $|z|=1$. Hence the boundary of Δ_0 consists of arcs on orthogonal circles to $|z|=1$ and a certain closed set e_0 on $|z|=1$, which may be empty.

Δ_0 has the following properties.

(i) *Any two points of* Δ_0 *are not equivalent.*

Proof. Suppose that two points a, b of Δ_0 are equivalent and $b=S(a)$ $(S\in G)$ and let $z'=S(0)$, $z''=S^{-1}(0)$, then

$$[b, 0]<[b, z']=[S^{-1}(b), S^{-1}(z')]=[a, 0]<[a, z'']=[S(a), S(z'')]=[b, 0]\,,$$

which is absurd.

(ii) *Any point* a $(|a|<1)$ *has its equivalent in* Δ_0, *or on its boundary.*

Proof. We may assume that a is different from equivalents $\{z_n\}$ of $z=0$. Let C: $[z, a]=\rho$ be a circle. If we make ρ increase from 0 to 1, then C meets an equivalent z' of $z=0$ at first, when $\rho=\rho_0$, so that

$$[a, z']\leqq[a, z_n] \quad (n=0, 1, 2, \cdots)\,.$$

Let $z'=S(0)$ $(S\in G)$ and $a'=S^{-1}$ (a), then

$$[a', 0]=[S(a'), S(0)]=[a, z']\leqq[a, z_n]=[S^{-1}(a), S^{-1}(z_n)]=[a', S^{-1}(z_n)]\,.$$

Since the totality of $S^{-1}(z_n)$ $(n=0, 1, 2, \cdots)$ coincides with the totality of z_n $(n=0, 1, 2, \cdots)$, we have

$$[a', 0]\leqq[a', z_n] \quad (n=0, 1, 2, \cdots)\,,$$

hence $a'=S^{-1}(a)$ belongs to Δ_0, or its boundary.

(iii) *Any point* a $(|a|<1)$ *on the boundary* Γ *of* Δ_0 *has its equivalent* a' *on* Γ, *such that* $|a'|=|a|$.

Proof. Since $a\in\Gamma$, $[a, 0]=[a, z']$ for some $z'=S(0)$ $(S\in G)$.

Let $a'=S^{-1}(a)$, $z''=S^{-1}(0)$, then

$$[a, 0]=[S^{-1}(a), S^{-1}(0)]=[a', z''], \quad [a, z']=[S^{-1}(a), S^{-1}(z')]=[a', 0]\,.$$

Hence $[a, 0]=[a, z']=[a', 0]$, so that $|a'|=|a|$ and from $[a', 0]=[a, 0]=[a', z'']$, we see that $a'=S^{-1}(a)$ belongs to Γ.

(iv) Δ_0 *is circular convex, i, e, if* a, b *be two points of* Δ_0 *and* $\widehat{ab}$ *be the arc on the orthogonal circle to* $|z|=1$ *through* a, b *and bounded by* a, b, *then* $\widehat{ab}$ *belongs to* Δ_0.

Proof. If $\widehat{ab}$ does not belong to Δ_0, then it intersects the boundary Γ of Δ_0, which consists of orthogonal arcs to $|z|=1$.

Since two orthogonal circles to $|z|=1$ intersect at most at a point in $|z|<1$, one of a, b does not belong to Δ_0, which is absurd.

3. Now we shall define the fundamental domain D_0 of G as follows.

By (iii), the part of the boundary of Δ_0, which lies in $|z|<1$, consists of equivalent pairs Γ_k, Γ'_k $(k=1, 2, \cdots)$ of arcs on orthogonal circles to

$|z|=1$. We see easily that the fundamental domain D_0 is obtained from Δ_0, by adding Γ_k ($k=1, 2, \cdots$) to Δ_0. Hence the boundary of D_0 consists of equivalent pairs Γ_k, Γ'_k ($k=1, 2, \cdots$) of arcs on orthogonal circles to $|z|=1$ and a certain closed set e_0 on $|z|=1$, which may be empty.

By (iii), equivalent points on Γ_k, Γ'_k are equidistant from $z=0$. Γ_k, Γ'_k are called *the sides of* D_0 and a point on the boundary of D_0, where two sides meet, is called *a vertex of* D_0.

Let D_n ($n=0, 1, 2, \cdots$) be equivalents of D_0 by G, then by (ii), $\{D_n\}$ fill up the inside of $|z|<1$ and by (i), D_i, D_j ($i\neq j$) do not overlap and any two points of D_0 are not equivalent.

3. Type of Fuchsian groups

1. Let G be a Fuchsian group, D_0 be its fundamental domain and let $S_n \in G$ ($n=0, 1, 2, \cdots$), $a_n=S_n(a)$ ($a\in D_0$), $z_n=S_n(0)$ and $D_n=S_n(D_0)$, so that D_n contains a_n and z_n.

Since $|a|=[a, 0]=[a_n, z_n]$, we have

$$1-|a|^2=\frac{(1-|a_n|^2)(1-|z_n|^2)}{|1-\bar{a}_n z_n|^2}\leqq\frac{4(1-|a_n|)(1-|z_n|)}{(1-|z_n|)^2}, \text{ and} \leqq\frac{4(1-|a_n|)(1-|z_n|)}{(1-|a_n|)^2},$$

so that

$$\frac{(1-|a|^2)(1-|z_n|)}{4}\leqq 1-|a_n|\leqq\frac{4(1-|z_n|)}{1-|a|^2}. \tag{1}$$

From this, we have

Theorem XI. 3. *Let* $a\in D_0$, *then*

$$\textit{either } \text{(i)} \quad \sum_{n=0}^{\infty}(1-|a_n|)<\infty, \qquad \textit{or} \quad \text{(ii)} \quad \sum_{n=0}^{\infty}(1-|a_n|)=\infty$$

is independent of $a\in D_0$.
G is called *of convergence*, or *of divergence type*, according as the case (i), or (ii).

Theorem XI. 4. *Let e_0 be the part of the boundary of D_0, which lies on $|z|=1$. If $me_0>0$, then G is of convergence type, such that*

$$\sum_{n=0}^{\infty}(1-|z_n|)<4\pi/me_0 ,$$

where $\{z_n\}$ are equivalents of $z=0$.

Proof. Let $z_n=S_n(0)$, $e_n=S_n(e_0)$ ($S_n\in G$) and put $E=\sum_{n=0}^{\infty} e_n$,

$$u(z)=\int_{e_0}\frac{1-|z|^2}{|z-e^{i\theta}|^2}\,d\theta . \tag{1}$$

Since $\frac{1-|z|^2}{|z-e^{i\theta}|^2}\,d\theta$ is invariant by G,

$$me_0=u(0)=\int_{e_n}\frac{1-|z_n|^2}{|z_n-e^{i\theta}|^2}\,d\theta\leqq\frac{1+|z_n|}{1-|z_n|}\,me_n<\frac{2\,me_n}{1-|z_n|},$$

so that

$$\sum_{n=0}^{\infty}(1-|z_n|)<\frac{2\sum_{n=0}^{\infty}me_n}{me_0}=\frac{2\,mE}{me_0}\leqq\frac{4\pi}{me_0}.$$

Remark. The converse is not true in general. There exists a Fuchsian group G of convergence type, such that $me_0=0$.

For example, let F be a closed Riemann surface of genus $p\geqq 2$ and $F^{(\infty)}$ be its covering surface of planar character. We map $F^{(\infty)}$ conformally on a schlicht domain Ω on the w-plane and let E be its boundary, then E is a bounded closed set of positive capacity. Let $\Omega^{(\infty)}$ be the universal covering surface of Ω and we map $\Omega^{(\infty)}$ conformally on $|z|<1$, then we obtain a Fuchsian group G in $|z|<1$. Since $\gamma(E)>0$, G is of convergence type (Theorem XI. 13) and we can prove easily that $me_0=0$.

Theorem XI. 5. *If G is of divergence type, then there exists no non-constant bounded harmonic function in $|z|<1$, which is invariant by G.* (Theorem X. 6).

Proof. Suppose that there exists such a harmonic function $u(z)$ and let $v(z)$ be its conjugate harmonic function, which we normalize, such that $v(0)=0$. Let $u(0)=k$ and C_0 be the part of the niveau curve: $u=\text{const.}=k$, through $z=0$, which lies in a small neighbourhood of $z=0$ and lies in D_0 and let

$$\int_{C_0}dv=c\,, \tag{1}$$

where we may assume that $c>0$, by taking a suitable positive sense of integration. Let C_n be equivalents of C_0, then $\int_{C_n}dv=c$.

If we put

$$F(z)=e^{\frac{2\pi(u+iv)}{c}},\qquad F(0)=e^{\frac{2\pi k}{c}}, \tag{2}$$

then $F(z)$ is a bounded regular function in $|z|<1$. If z describes C_n, then $w=F(z)$ makes one turn on the circle $|w|=e^{\frac{2\pi k}{c}}=F(0)$, so that there exists a point z_n^* on C_n, such that $F(z_n^*)=F(0)$. Hence by Blaschke's theorem, $\sum_{n=0}^{\infty}(1-|z_n^*|)<\infty$.

Since as easily seen, $1-|z_n|\leqq \text{const.}\,(1-|z_n^*|)$, we have $\sum_{n=0}^{\infty}(1-|z_n|)<\infty$, which is absurd. Hence the theorem.

Corollary. *If G is of divergence type, then there exists no measurable set E on $|z|=1$, which is invariant by G and $0<mE<2\pi$.*

Proof. Suppose that there exists such a set E and put

$$u(z)=\int_E \frac{1-|z|^2}{|z-e^{i\theta}|^2}\,d\theta\,,$$

then $u(z)$ is invariant by G and is a non-constant bounded harmonic function in $|z|<1$, which contradicts the above theorem. ——

By considering $F(z)=e^{-\frac{2\pi}{c}(u+iv)}$, we can prove similarly

Theorem XI. 6. *If G is of divergence type, then there exists no non-constant positive hamonic function in $|z|<1$, which is invariant by G.* (Theorem X. 7. Cor.) Myrberg [2].

2. Let $a\in D_0$ and $n(r, a)$ be the number of its equivalents a_n, contained in $|z|<r<1$, then we see easily that $\sum_{n=0}^{\infty}(1-|a_n|)<\infty$ is equivalent to

$$\int_0^1 n(r, a)\,dr<\infty.$$

Theorem XI. 7.
$$n(r, a)\leqq \frac{\text{const.}}{1-r} \quad (0\leqq r<1)\,,$$
where const. is independent of $a\in D_0$.

Proof. If $|a_n|<r$ and $a_n=S_n(a)$ $(S_n\in G)$, then

$$|a_n|=[a_n, 0]=[S_n(a), 0]=[a, S_n^{-1}(0)]=[S_n^{-1}(0), a]<r\,,$$

so that $n(r, a)$ is the number of equivalents z_n of $z=0$, contained in a disc $[z, a]<r$. Let Δ_0: $|z|<\delta$ be a disc and Δ_n be its equivalent, which contains z_n, where we take $\delta>0$ so small that $\{\Delta_n\}$ do not overlap.

If $[z_n, a]<r$, then Δ_n is contained in a disc $[z, a]<r_1$ $(r<r_1)$, such that $1-r\leqq \text{const.}\,(1-r_1)$.[1] Since $\sigma(\Delta_n)=\sigma(\Delta_0)$ $(n=0, 1, 2, \cdots)$ and the non-euclidean area of $[z, a]<r_1$ is equal to that of $|z|<r_1$, which is $\leqq \frac{\text{const.}}{1-r_1}$, we have $\sigma(\Delta_0)\,n(r, a)\leqq \frac{\text{const.}}{1-r_1}\leqq \frac{\text{const.}}{1-r}$, so that $n(r, a)\leqq \frac{\text{const.}}{1-r}$.

Theorem XI. 8. *If $a\in D_0$, then*

$$\sum_{n=0}^{\infty}(1-|a_n|)^{1+\delta}<\infty, \quad \delta>0\,,$$

1) By a linear transformation $\zeta=\frac{z-a}{1-\bar{a}z}$, the discs $[z, a]<r$ and Δ_n become $|\zeta|<r$, $[\zeta, \zeta_n]<\delta$ $(|\zeta_n|<r)$ respectively and apply (ii) (p. 511) and return to z.

where $\{a_n\}$ are equivalents of a.

Proof.

$$\sum_{|a_n|<r<1}(1-|a_n|)^{1+\delta}=\int_0^r(1-t)^{1+\delta}dn(t,a)\leqq(1-r)^{1+\delta}n(r,a)+$$

$$(1+\delta)\int_0^r(1-t)^{\delta}n(t,a)\,dt\leqq\text{const.}\,(1-r)^{\delta}+\text{const.}\int_0^r\frac{dt}{(1-t)^{1-\delta}}=O(1)\,,$$

hence
$$\sum_{n=0}^{\infty}(1-|a_n|)^{1+\delta}<\infty.$$

Theorem XI. 9. *If G is of convergence type, then for any $a\in D_0$,*

$$\int_0^1 n(r,a)dr\leqq K\,,$$

where K is a constant, independent of $a\in D_0$.

Proof. Since $\sum_{n=0}^{\infty}(1-|a_n|)<\infty$,

$$g(z,a)=\sum_{n=0}^{\infty}\log\left|\frac{1-\bar{a}_nz}{z-a_n}\right| \tag{1}$$

is convergent in $|z|<1$, except at a_n $(n=0,1,2,\cdots)$.

Let $\varDelta_0$: $|z|\leqq\rho<1$ be a disc, contained in D_0, then by (4) of the proof of Theorem XI. 13, for any $a\in D_0-\varDelta_0$,

$$g(0,a)=\sum_{n=0}^{\infty}\log\frac{1}{|a_n|}\leqq K_0=\text{const.}, \tag{2}$$

where K_0 is a constant, independent of $a\in D_0-\varDelta_0$.

Since

$$\sum_{|a_n|<r<1}\log\frac{1}{|a_n|}=\int_0^r\log\frac{1}{t}\,dn(t,a)=\log\frac{1}{r}\cdot n(r,a)+\int_0^r\frac{n(t,a)}{t}dt$$
$$\geqq\int_0^r n(t,a)dt\,,$$

we have

$$\int_0^r n(t,a)dt\leqq K_0,\quad a\in D_0-\varDelta_0\,. \tag{3}$$

If $a\in\varDelta_0$, then

$$\sum_{|a_n|<r<1}(1-|a_n|)=\int_0^r(1-t)dn(t,a)=(1-r)n(r,a)-n(+0,a)+\int_0^r n(t,a)dt$$
$$\geqq\int_0^r n(t,a)dt-1\,. \tag{4}$$

Let D_n be the equivalent of D_0, which contains a_n and z_n, where z_n is an equivalent of $z=0$, then $\left|\frac{a_n-z_n}{1-\bar{a}_nz_n}\right|=[a_n,z_n]=[a,0]\leqq\rho$, so that

$$1-|a_n|\leqq\frac{4(1-|z_n|)}{1-\rho^2},\quad ((1),\text{ p. }514)$$

hence by (4),

$$\int_0^r n(t,a)dt \leqq 1 + \frac{4}{1-\rho^2} \sum_{n=0}^{\infty} (1-|z_n|) = K_1 = \text{const.}, \qquad a \in \varDelta_0. \tag{5}$$

Hence by (3), (5), we have

$$\int_0^r n(t,a)dt \leqq K = \text{const.}, \qquad a \in D_0. \tag{6}$$

Theorem XI. 10. *Suppose that* $\sigma(D_0)<\infty$. *If* $a \in D_0$ *and* $|a|<\rho<1$, *then*

$$\frac{A}{1-r} \leqq n(r,a) \leqq \frac{B}{1-r} \quad (r_0 \leqq r < 1),$$

where $A=A(\rho)$ *is a constant, which depends on* ρ *only and* B *is a numerical constant.*

Hence $\int_0^1 n(r,a)dr=\infty$, so that if $\sigma(D_0)<\infty$, then G is of divergence type

Proof. In virtue of Theorem XI. 7, we have only to prove that $n(r,a) \geqq A/1-r$. First we assume that D_0 has boundary points on $|z|=1$.

We choose r_0, such that $\rho<r_0<1$ and let D_0'' be the part of D_0, which lies outside of $|z|=r_0$, where we take $1-r_0$ so small that $\sigma(D_0'')<\varepsilon$. Let $\{D_n''\}$ be the equivalents of D_0'' and put $\overset{*}{D}''=\sum_{n=0}^{\infty} D_n''$ and let $\overset{*}{D}''(r)$ be the part of $\overset{*}{D}''$, contained in $|z|<r<1$. Then by Theorem XI. 7,

$$\sigma(\overset{*}{D}''(r)) = \int_{D_0''} n(r,a)d\sigma(a) \leqq \frac{\text{const.}\ \sigma(D_0'')}{1-r} \leqq \frac{\text{const.}\ \varepsilon}{1-r}. \tag{1}$$

We put $D_0'=D_0-D_0''$ and let $\{D_n'\}$ be its equivalents and $\overset{*}{D}'=\sum_{n=0}^{\infty} D_n'$. Let $\overset{*}{D}'(r)$ be the part of $\overset{*}{D}'$, contained in $|z|<r<1$ and $\varDelta(r)$: $|z|<r<1$ be a disc, then $\overset{*}{D}'(r)+\overset{*}{D}''(r)=\varDelta(r)$, so that

$$\sigma(\overset{*}{D}'(r)) + \sigma(\overset{*}{D}''(r)) = \sigma(\varDelta(r)) \geqq \frac{\text{const.}}{1-r} \quad ((5),\ \text{p. } 509). \tag{2}$$

If ε is sufficiently small, then by (1),

$$\sigma(\overset{*}{D}'(r)) \geqq \frac{\text{const.}}{1-r}. \tag{3}$$

By (ii) (p. 511), there exists $r_1(>r)$, such that $1-r_1 \geqq \text{const.}\ (1-r)$ and if D_n' intersects the disc $|z|<r$, then D_n' is contained in $|z|<r_1$, so that

$$n(r_1,a)\sigma(D_0') \geqq \sigma(\overset{*}{D}'(r)) \geqq \frac{\text{const.}}{1-r} \geqq \frac{\text{const.}}{1-r_1},$$

hence writing r instead of r_1, we have

$$n(r,a) \geqq \frac{A}{1-r} \qquad (r_0 \leqq r < 1), \tag{4}$$

where A is a constant, which depends on ρ only.

We assumed that D_0 has boundary points on $|z|=1$, but from the proof, we see easily that (4) holds, if D_0 lies entirely in $|z|<1$, with its boundary.

4. Riemann surface F_G

Let G be a Fuchsian group and D_0 be its fundamental domain.

If we identify the equivalent points on the boundary of D_0, then D_0 can be considered as a Riemann surface $F=F_G$, which we call *the Riemann surface, associated to G*. If D_0 lies entirely in $|z|<1$ with its boundary, then F is a closed Riemann surface and if D_0 has only a finite number k of parabolic vertices on $|z|=1$, where two sides of D_0 touch each other and the number of vertices of D_0 in $|z|<1$ is finite, then F is obtained from a closed Riemann surface by taking off k points.

The totality of equivalents $\{D_n\}$ of D_0 filling up $|z|<1$ is considerd as representatives of a simply connected covering surface $F^{(\infty)}$ of F.

Let a $(|a|<1)$ be a vertex on the boundary Γ of D_0 and a_i $(i=0, 1, 2, \cdots, n-1)$ $(a_0=a)$ be the totality of its equivalents on Γ, then $|a_i|=|a|$ and let α_i be the inner angle of D_0 at a_i.

There is a finite number of equivalents D_j $(j=0, 1, 2, \cdots, m-1)$ of D_0, which contains a on its boundary, filling up the full neighbourhood of a.

Let $D_0=S_j(D_j)$ $(S_j \in G)$ and $b_j=S_j(a)$, then $b_j \in \Gamma$ is an equivalent of a. In general, several b_j may coincide with a. Let G_0 be the totality of S_j, such that $S_j(a)=a$ and let N be the number of such S_j. If $b_j=b_k=a_i$ $(0 \leqq i \leqq n-1)$, then $S_j(a)=S_k(a)$, $S_k^{-1}S_j(a)=a$, so that $S_k^{-1}S_j \in G_0$. Hence the number of b_j, which coincides with a_i is N. Since $\{D_j\}$ fill up the full neighbourhood of a, $N(\alpha_0+\alpha_1+\cdot+\alpha_{n-1})=2\pi$, so that

$$\alpha_0+\alpha_1+\cdot+\alpha_{n-1}=2\pi/N . \tag{1}$$

If we consider a as a point of $F=F_G$, then at a, $F^{(\infty)}$ has a branch point of order $(N-1)$ relatively to F. Hence if $F^{(\infty)}$ is unramified relatively to F, i, e, if $F^{(\infty)}$ is the universal covering surface of F, then $N=1$, so that

$$\alpha_0+\alpha_1+\cdot+\alpha_{n-1}=2\pi . \tag{2}$$

Theorem XI. 11. *If the Riemann surface $F=F_G$ is a closed Riemann surface of genus $p \geqq 2$, then*

$$\sigma(D_0) \geqq 4\pi(p-1),$$

where the equality holds, only when $F^{(\infty)}$ is the universal covering surface of F.

Proof. The totality of vertices of D_0 can be grouped into a finite number n of classes of equivalent ones and let $a_1^{(i)}, \cdots, a_{\nu_i}^{(i)}$ $(i=1, 2, \cdots, n)$ be vertices of the i-th class, which are equivalent with each other.

If we put $\nu=\nu_1+\cdots+\nu_n$, then there are ν vertices on the boundary Γ of D_0.

Let $A_i=2\pi/N_i$ be the sum of inner angles of D_0 at $a_1^{(i)}, \cdots, a_{\nu_i}^{(i)}$. We connect $z=0$ to all vertices of D_0 by rectilinear segments, then we obtain a triangulation of F. Let $\alpha_0, \alpha_1 \alpha_2$ be the number of vertices, edges and triangles in the triangulation and ρ be the Euler's characteristic of F, then

$$\rho=-\alpha_0+\alpha_1-\alpha_2=2(p-1)\,. \tag{1}$$

We see easily that $\alpha_0=1+n, \quad \alpha_1=\nu+\nu/2, \quad \alpha_2=\nu$, so that

$$\nu=4p+2n-2\,. \tag{2}$$

By Theorem XI. 1, we have

$$\sigma(D_0)=\nu\pi-(A_1+\cdots+A_n+2\pi)=(4p+2n-2)\pi-2\pi\left(\sum_{i=1}^{n}\frac{1}{N_i}+1\right)$$

$$=4\pi(p-1)+2\pi\sum_{i=1}^{n}\left(1-\frac{1}{N_i}\right). \tag{3}$$

Hence

$$\sigma(D_0)\geqq 4\pi(p-1)\,, \tag{4}$$

where the equality holds, only when $N_i=1$ $(i=1, 2, \cdots, n)$, *i, e,* $F^{(\infty)}$ is the universal covering surface of F.

Theorem XI. 12. *$\sigma(D_0)<\infty$, when and only when $\overline{D}_0$ lies entirely in $|z|<1$, or D_0 has a finite number of sides in $|z|<1$ and a finite number of parabolic vertices on $|z|=1$.*

Hence if $\sigma(D_0)<\infty$, then $F=F_G$ is a closed Riemann surface, or a Riemann surface, obtained from a closed Riemann surface by taking off a finite number of points. Siegel [1], Tsuji [26].

Proof. If D_0 is of the above special type, then by Theorem XI. 1, we see easily that $\sigma(D_0)<\infty$. Hence we assume that D_0 is not of the above special type and we shall prove that $\sigma(D_0)=\infty$.

Then D_0 is one of the following three types (i), (ii), (iii):

(i) The boundary Γ of D_0 contains an arc on $|z|=1$.

(ii) Infinitely many orthogonal arcs $\widehat{a_i b_i}$ to $|z|=1$ $(i=1, 2, \cdots)$ belong to Γ, where a_i and b_i lie on $|z|=1$.

(iii) D_0 has infinitely many vertices a_i $(i=1, 2, \cdots)$ in $|z|<1$.

In case (i), $\sigma(D_0)=\infty$ evidently. In case (ii), we connect $z=0$ to a_i, b_i by rectilinear segments, then we obtain a triangle Δ_i, which is bounded by $\widehat{a_i b_i}$ and $\overline{oa_i}, \overline{ob_i}$. Then by Theorem XI. 1, $\sigma(\Delta_i)=\pi-\theta_i$, where θ_i is the angle of Δ_i at $z=0$.

Hence

$$\sigma(D_0) \geqq n\pi - \sum_{i=1}^{n} \theta_i \geqq n\pi - 2\pi \to \infty, \ n \to \infty ,$$

so that

$$\sigma(D_0) = \infty.$$

Hence we have only to consider the case (iii). We denote the inner angle of D_0 at a_i by α_i, then $0 < \alpha_i \leqq \pi$. If $0 < \alpha_i < \pi$ for only a finite number of i, then $\alpha_i = \pi$ for infinitely many i, so that by case (ii), $\sigma(D_0) = \infty$, hence we assume that $0 < \alpha_i < \pi$ for $i = 1, 2, \cdots$.

Let $0 < r_1 < r_2 < \cdots < r_k \to 1$ and $a_1, \cdots, a_{\sigma_k}$ be the vertices of D_0, contained in $|z| \leqq r_k$, then since equivalent vertices on the boundary of D_0 are equidistant from $z=0$, if a' is an equivalent of a_i $(1 \leqq i \leqq \sigma_k)$ and lies on the boundary of D_0, then a' is contained among $a_1, \cdots, a_{\sigma_k}$.

Let l_i, l' be the sides of D_0, which meet at a_i $(1 \leqq i \leqq \sigma_k)$ and b_i, b_i' be the end vertex of l_i, l_i' respectively, which is different from a_i. We connect $z=0$ to a_i, b_i, b_i' $(i=1, 2, \cdots, \sigma_k)$ by rectilinear segments, then we obtain N_k triangles, which are grouped into ν_k disjoint blocks (shaded parts in the figure), where $a_1, \cdots, a_{\sigma_k}$ are inner vertices of some one of the blocks. Then

$$\sigma_k = N_k - \nu_k . \tag{1}$$

Let $c_1, \cdots, c_{s_k}$ be the end vertices of the blocks and γ_j be the inner angle of the block at c_j $(1 \leqq j \leqq s_k)$. We remark that c_j lies outside of the circle $|z| = r_k$.

Let A_k be the sum of the non-euclidean areas of N_k triangles, then by Theorem X. 1,

$$A_k = \pi N_k - \sum_{i=1}^{\sigma_k} \alpha_i - \sum_{j=1}^{s_k} \gamma_j - \theta_k , \tag{2}$$

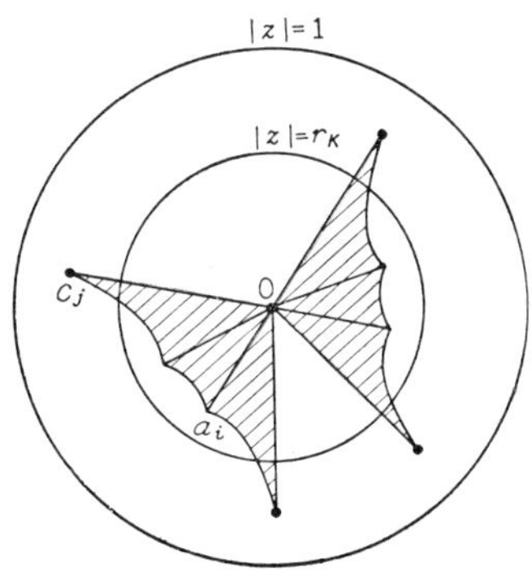

where θ_k is the sum of angles of N_k triangles at $z=0$. Since $0 < \alpha_i < \pi$, the number of equivalents of a_i is at least 3 and since the sum of inner angles at equivalent vertices is $\leqq 2\pi$, we have

$$\sum_{i=1}^{\sigma_k} \alpha_i \leqq \frac{2\pi}{3} \sigma_k = \frac{2\pi}{3} (N_k - \nu_k) . \tag{3}$$

Since c_j lies outside of the circle $|z| = r_k$ and a_i lies in $|z| \leqq r_k$, we have $\gamma_j \leqq \pi/2$, so that

$$\sum_{j=1}^{s_k} \gamma_j \leqq \frac{\pi}{2} s_k = \frac{\pi}{2} \cdot 2\nu_k = \pi \nu_k . \tag{4}$$

Since $\nu_k \leqq N_k/2$, we have

$$\sum_{i=1}^{\sigma_k}\alpha_i+\sum_{j=1}^{s_k}\gamma_j \leqq \frac{2\pi}{3}(N_k-\nu_k)+\pi\nu_k=\frac{2\pi N_k}{3}+\frac{\pi\nu_k}{3}\leqq\frac{2\pi N_k}{3}+\frac{\pi N_k}{6}$$
$$=\frac{5\pi N_k}{6},$$

so that by (2),

$$\sigma(D_0)\geqq A_k \geqq \pi N_k-\frac{5\pi N_k}{6}-\theta_k \geqq \frac{\pi N_k}{6}-2\pi\to\infty,\ k\to\infty,$$

hence
$$\sigma(D_0)=\infty .$$

5. Green's function of F_G

1. We shall prove the following **Myrberg's theorem.**

Theorem XI. 13. *Let $F=F_G$ be the Riemann surface, associated to G. Then $F\in P_G$, only when G is of convergence type and if $F\in P_G$, then its Green's function $g(z, a)$ with $a\in D_0$ as its pole is given by*

$$g(z, a)=\sum_{n=0}^{\infty}\log\left|\frac{1-\bar{a}_n z}{z-a_n}\right|,$$

where $\{a_n\}$ are equivalents of $a\in D_0$. Myrberg [2].

Remark. By Theorem VIII. 24, for almost all $e^{i\theta}$,

$$\lim_{z\to e^{i\theta}} g(z, a)=0 ,$$

when $z\to e^{i\theta}$ from the inside of any Stolz domain, whose vertex is at $e^{i\theta}$.

Proof. (i) Suppose that G is of convergence type, then $\sum_{n=0}^{\infty}(1-|a_n|)<\infty$, hence

$$g^*(z)=\sum_{n=0}^{\infty}\log\left|\frac{1-\bar{a}_n z}{z-a_n}\right| \tag{1}$$

is convergent in $|z|<1$ and $g^*(z)$ is invariant by G, so that $g^*(z)$ is a one-valued harmonic function on $F-\{a\}$. Let $\{F_n\}$ be an exhaustion of F, where F_0 contains a and let $g_n(z, a)$ be the Green's function of F_n.

Since $g^*(z)>0$, by the maximum principle, $g_n(z, a)\leqq g^*(z)$ on F_n, so that

$$g(z, a)=\lim_n g_n(z, a)\leqq g^*(z)<\infty . \tag{2}$$

Hence the Green's function $g(z, a)$ of F exists, so that $F\in P_G$.

Next we shall prove that $g(z, a)=g^*(z)$. We put

$$g_n^*(z)=\sum_{\nu=0}^{n}\log\left|\frac{1-\bar{a}\ z}{z-a_\nu}\right|. \tag{3}$$

Let K_0: $[z, a]\leqq\delta$ be a disc, contained in D_0 and C_0: $[z, a]=\delta$ be its bounding circle and K_n, C_n be its equivalents respectively. We take off $\sum_{\nu=0}^{n} K_\nu$ from $|z|<1$ and let Ω_n be the remaining domain. Then $g_n^*(z)$ is harmonic in Ω_n and $g_n^*(z)=0$ on $|z|=1$, so that by the maximum principle,

$$0<g_n^*(z)\leqq \operatorname*{Max}_{c_0+\cdot+c_n} g_n^*(z)\leqq \operatorname*{Max}_{c_0+\cdot+c_n} g^*(z)=\operatorname*{Max}_{c_0} g^*(z) \quad \text{in } \Omega_n ,$$

since $g^*(z)$ is invariant by G. Hence for $n\to\infty$, we have

$$0<g^*(z)\leqq \operatorname*{Max}_{c_0} g^*(z) \quad \text{in } \Omega=\lim_n \Omega_n . \tag{4}$$

Hence if we put

$$H(z)=g^*(z)-g(z, a) ,$$

then by (2), $0\leqq H(z)<g^*(z)$, so that $H(z)$ is bounded in Ω. Since $H(z)$ is harmonic in $\sum_{\nu=0}^{\infty} K_\nu$, $H(z)$ is harmonic and bounded in $|z|<1$. Hence by Fatou's theorem, for almost all $e^{i\theta}$, $\lim_{r\to 1} H(re^{i\theta})=H(e^{i\theta})$ exists. Since by Theorem VIII, 24, $\lim_{r\to 1} g^*(re^{i\theta})=0$ for almost all $e^{i\theta}$ and $0\leqq H(z)<g^*(z)$, we have $H(e^{i\theta})=0$ for almost all $e^{i\theta}$. Since $H(z)$ is a bounded harmonic function in $|z|<1$, $H(z)\equiv 0$, or $g(z, a)\equiv g^*(z)$. ——

(ii) Next suppose that the Green's function $g(z,a)$ of F exists and let $h(z,a)$ be its conjugate harmonic function and put $f(z)=e^{-(g+ih)}$, then $f(z)$ is regular and bounded in $|z|<1$ and $f(a_n)=0$ $(n=0,1,2\cdots)$, so that by Blaschke's theorem, $\sum_{n=0}^{\infty}(1-|a_n|)<\infty$. Hence G is of convergence type.

2. As an application of the theorem, we shall prove

Theorem XI. 14. *Let F be an open Riemann surface of positive boundary, spread over the z-sphere and $w(z)$ be a one-valued meromorphic function on F, such that $D_F^*[w]<\infty$. If we map the universal covering surface $F^{(\infty)}$ of F conformally on $|\zeta|<1$ and put $w=\overset{*}{w}(\zeta)$, then $\overset{*}{w}(\zeta)$ is a meromorphic function of bounded type in $|\zeta|<1$.* Tsuji [61].

As a special case, if we take $w(z)\equiv z$, then we have

Theorem XI. 15. *Let F be an open Riemann surface of positive boundary, spread over the z-sphere and its spherical area be finite. If we map the universal covering surface $F^{(\infty)}$ of F conformally on $|\zeta|<1$ by $z=\varphi(\zeta)$, then $\varphi(\zeta)$ is a meromorphic function of bounded type in $|\zeta|<1$, so that for almost all $e^{i\theta}$, $\lim_{\zeta\to e^{i\theta}}\varphi(\zeta)=\varphi(e^{i\theta})$ $(\neq\infty)$ exists, when $\zeta\to e^{i\theta}$ non-tangentially to $|\zeta|=1$.*

Proof of Theorem XI. 14. Let $z=\varphi(\zeta)$ be the mapping function, then $\varphi(\zeta)$ is invariant by a certain Fuchsian group G and let D_0 be its fundamental domain, which contains $\zeta=0$. Let a be any point of D_0 and a_n be its equivalents, then since F is of positive boundary, $\sum_{n=0}^{\infty}(1-|a_n|)<\infty$, so that by Theorem X. 9,

$$\int_0^1 n(r, a)\,dr \leqq K=\text{const.} \qquad a\in D_0\,. \tag{1}$$

We define a positive mass distribution $d\mu(\zeta)$ in $|\zeta|<1$ by

$$d\mu(\zeta)=\left(\frac{|\overset{*}{w}'(\zeta)|}{1+|\overset{*}{w}(\zeta)|^2}\right)^2 r\,dr\,d\theta, \qquad \zeta=re^{i\theta}\,,$$

then

$$\int_{D_0} d\mu(\zeta)=\iint_F\left(\frac{|w'(z)|}{1+|w(z)|^2}\right)^2 dx\,dy=D_F^*[w]<\infty, \qquad z=x+iy\,. \tag{2}$$

Since

$$S(r)=\frac{1}{\pi}\iint_{|\zeta|\leqq r}\left(\frac{|\overset{*}{w}'(\zeta)|}{1+|\overset{*}{w}(\zeta)|^2}\right)^2 r dr d\theta=\frac{1}{\pi}\int_{D_0} n(r, a)d\mu(a),$$

we have

$$\int_0^r S(r)dr=\frac{1}{\pi}\int_{D_0} d\mu(a)\int_0^r n(r, a)dr \leqq \frac{K}{\pi}\int_{D_0} d\mu(a)=\frac{K}{\pi}D_F^*[w]\,,$$

so that

$$T(\overset{*}{w}, r)=\int_0^r \frac{S(r)}{r}dr=O(1), \qquad 0\leqq r<1\,. \text{ ——}$$

Similarly we can prove the following theorem, due to Kuramichi [1].

Theorem XI. 16. *Let F be an open Riemann surface, spread over the z-sphere and suppose that there exists a non-constant one-valued harmonic function $u(z)$ on F, whose Dirichlet integral $D_F[u]$ is finite. If we map the universal covering surface $F^{(\infty)}$ of F conformally on $|\zeta|<1$ and put $u(z)=\overset{*}{u}(\zeta)$, then $\overset{*}{u}(\zeta)$ is a harmonic function in $|\zeta|<1$, such that*

$$\int_0^{2\pi}[\overset{*}{u}(re^{i\theta})]^2\,d\theta=O(1), \qquad 0\leqq r<1\,.$$

Proof. Since there exists a non-constant one-valued harmonic function on F, whose Dirichlet integral is finite, F is of positive boundary. Let

$$f(\zeta)=\sum_{n=0}^{\infty}a_n\zeta^n, \qquad \zeta=re^{i\theta}\,,$$

be the regular function in $|\zeta|<1$, such that $\Re(f(\zeta))=\overset{*}{u}(\zeta)$, then

$$\iint_{D_0} |f'(\zeta)|^2 r dr d\theta = D_F[u] .$$

We define a positive mass distribution $d\mu(\zeta)$ in $|\zeta|<1$ by

$$d\mu(\zeta)=|f'(\zeta)|^2 r dr d\theta ,$$

then

$$\int_{D_0} d\mu(\zeta)=\iint_{D_0} |f'(\zeta)|^2 r dr d\theta = D_F[u]<\infty .$$

Since

$$A(r)=\iint\limits_{|\zeta|\leq r} |f'(\zeta)|^2 r dr d\theta = \pi\sum_{n=1}^{\infty} n|a_n|^2 r^{2n} = \int_{D_0} n(r, a) d\mu(a) .$$

we have as before,

$$\int_0^r A(r)dr = \pi\sum_{n=1}^{\infty} \frac{n}{2n+1} |a_n|^2 r^{2n+1} = O(1) .$$

Hence $\sum_{n=0}^{\infty} |a_n|^2<\infty$, so that

$$\int_0^{2\pi} [\overset{*}{u}(re^{i\theta})]^2 d\theta \leqq \int_0^{2\pi} |f(re^{i\theta})|^2 d\theta = 2\pi\sum_{n=0}^{\infty} |a_n|^2 r^{2n} = O(1) . \text{ ——}$$

3. Let F be a compact Riemann surface, spread over the w-sphere, whose boundary Γ consists of a finite number of disjoint Jordan curves $\Gamma=\sum_{i=1}^{k}\Gamma_i$. We map the universal covering surface $F^{(\infty)}$ of F conformally on $|z|<1$ by $w=\varphi(z)$, then $\varphi(z)$ is invariant by a certain Fuchsian group G, whose fundamental domain D_0 is bounded by a finite number of orthogonal arcs to $|z|=1$ and a finite number of arcs on $|z|=1$. We denote the part of the boundary of D_0, which lies on $|z|=1$ by E_0 and let E_n $(n=0, 1, 2, \cdots)$ be its equivalents by G and put $E=\sum_{n=0}^{\infty} E_n$. Then E is the image of Γ on $|z|=1$.

We shall prove

Theorem XI. 17. $mE=2\pi$.

Proof. Let

$$u(z)=\frac{1}{2\pi}\int_E \frac{1-|z|^2}{|z-e^{i\theta}|^2} d\theta ,$$

then $u(z)$ is a bounded harmonic function in $|z|<1$ and is invariant by G and $u=1$ on E. Hence if we put $u(z)=\overset{*}{u}(w)$, then $\overset{*}{u}(w)$ is a one-valued bounded harmonic function on F and $u=1$ on Γ, so that $\overset{*}{u}\equiv 1$, or $u\equiv 1$, hence

$$mE=2\pi u(0)=2\pi . \text{ ——}$$

Let $z\in D_0$ and z_n $(n=0, 1, 2, \cdots)$ be its equivalents, then

$$g(\zeta, z)=\sum_{n=0}^{\infty} \log\left|\frac{1-\bar{z}_n\zeta}{\zeta-z_n}\right| \tag{1}$$

is the Green's function of $F=F_G$ with z as its pole.

If $e^{i\theta}\in E$, then

$$\lim_{\zeta\to e^{i\theta}} g(\zeta, z)=0\,. \tag{2}$$

If $e^{i\theta}\in E$, then $|e^{i\theta}-z_n|\geqq\text{const.}>0$ $(n=0, 1, 2, \cdots)$ and since

$\sum_{n=0}^{\infty}(1-|z_n|)<\infty$, we have

$$\chi(e^{i\theta}, z)=\sum_{n=0}^{\infty}\frac{1-|z_n|^2}{|e^{i\theta}-z_n|^2}<\infty, \quad e^{i\theta}\in E, \quad z\in D_0\,. \tag{3}$$

Theorem XI. 18. (i) *If* $e^{i\theta}\in E$, *then* $\dfrac{\partial g(e^{i\theta}, z)}{\partial\nu}$ *exists and*

$$\frac{\partial g(e^{i\theta}, z)}{\partial\nu}=\sum_{n=0}^{\infty}\frac{1-|z_n|^2}{|e^{i\theta}-z_n|^2},$$

$$\int_{E_0}\frac{\partial g(e^{i\theta}, z)}{\partial\nu}\,d\theta=2\pi, \quad z\in D_0\,,$$

where ν *is the inner normal of* $|z|=1$ *at* $e^{i\theta}$.

If $\zeta=\rho e^{i\varphi}$ *tends to* $e^{i\theta}$, *then*

$$\lim_{\zeta\to e^{i\theta}}\frac{g(\zeta, z)}{1-|\zeta|}=-\lim_{\zeta\to e^{i\theta}}\frac{\partial g(\zeta, z)}{\partial\rho}=\frac{\partial g(e^{i\theta}, z)}{\partial\nu}, \quad z\in D_0\,.$$

(ii) *Let* $u(z)$ *be a bounded harmonic function in* $|z|<1$, *which is invariant by* G *and* $u(e^{i\theta})$ *be its boundary value, existing almost everywhere on* $|z|=1$, *then*

$$u(z)=\frac{1}{2\pi}\int_{E_0}u(e^{i\theta})\frac{\partial g(e^{i\theta}, z)}{\partial\nu}\,d\theta, \quad z\in D_0\,.$$

(iii) *Let* $f(e^{i\theta})$ *be a bounded measurable function on* $|z|=1$, *which is invariant by* G *and put*

$$u(z)=\frac{1}{2\pi}\int_{E_0}f(e^{i\theta})\frac{\partial g(e^{i\theta}, z)}{\partial\nu}\,d\theta, \quad |z|<1\,,$$

then $u(z)$ *is a bounded harmonic function in* $|z|<1$, *which is invariant by* G *and for almost all* $e^{i\theta}$ *on* E_0,

$$\lim_{z\to e^{i\theta}}u(z)=f(e^{i\theta})\,,$$

if $z\to e^{i\theta}$ *from the inside of any Stolz domain, whose vertex is at* $e^{i\theta}$.

For the proof, we use the following lemma.

Lemma. *If* $|a|<1$, $|z|<1$, *then*

$$\frac{(1-|z|^2)(1-|a|^2)}{2|1-\bar{a}z|^2}\leqq\log\left|\frac{1-\bar{a}z}{z-a}\right|\leqq\frac{(1-|z|^2)(1-|a|^2)}{2|z-a|^2}\,.$$

Proof. Since $\log(1+x)\leqq x\ (x\geqq 0)$ and $-\log(1-x)\geqq x\ (0\leqq x\leqq 1)$, we have

$$\log\left|\frac{1-\bar{a}z}{z-a}\right|=\frac{1}{2}\log\left|\frac{1-\bar{a}z}{z-a}\right|^2=\frac{1}{2}\log\left(1+\left(\left|\frac{1-\bar{a}z}{z-a}\right|^2-1\right)\right)$$

$$=\frac{1}{2}\log\left(1+\frac{(1-|z|^2)(1-|a|^2)}{|z-a|^2}\right)\leqq\frac{(1-|z|^2)(1-|a|^2)}{2|z-a|^2}. \tag{1}$$

$$\log\left|\frac{1-\bar{a}z}{z-a}\right|=-\frac{1}{2}\log\left|\frac{z-a}{1-\bar{a}z}\right|^2=-\frac{1}{2}\log\left(1-\left(1-\left|\frac{z-a}{1-\bar{a}z}\right|^2\right)\right)$$

$$=-\frac{1}{2}\log\left(1-\frac{(1-|z|^2)(1-|a|^2)}{|1-\bar{a}z|^2}\right)\geqq\frac{(1-|z|^2)(1-|a|^2)}{2|1-\bar{a}z|^2}. \tag{2}$$

Proof of the theorem.

(i) By the lemma,

$$\frac{1+|\zeta|}{2}\sum_{n=1}^{\infty}\frac{1-|z_n|^2}{|1-\bar{z}_n\zeta|^2}\leqq\frac{g(\zeta,z)}{1-|\zeta|}\leqq\frac{1+|\zeta|}{2}\sum_{n=0}^{\infty}\frac{1-|z_n|^2}{|\zeta-z_n|^2},$$

so that if $\zeta=\rho e^{i\varphi}\to e^{i\theta}\in E$, then

$$\lim_{\zeta\to e^{i\theta}}\frac{g(\zeta,z)}{1-|\zeta|}=\sum_{n=0}^{\infty}\frac{1-|z_n|^2}{|e^{i\theta}-z_n|^2}. \tag{1}$$

Since $g(e^{i\theta},z)=0$, we have by (1),

$$\frac{\partial g(e^{i\theta},z)}{\partial\nu}=\sum_{n=0}^{\infty}\frac{1-|z_n|^2}{|e^{i\theta}-z_n|^2}. \tag{2}$$

Since $\rho\dfrac{\partial}{\partial\rho}\log\left|\dfrac{1-\bar{z}_n\zeta}{\zeta-z_n}\right|=\Re\left(\zeta\dfrac{d}{d\zeta}\log\left(\dfrac{1-\bar{z}_n\zeta}{\zeta-z_n}\right)\right)$, we have

$$-\lim_{\zeta\to e^{i\theta}}\frac{\partial}{\partial\rho}\log\left|\frac{1-\bar{z}_n\zeta}{\zeta-z_n}\right|=\frac{1-|z_n|^2}{|e^{i\theta}-z_n|^2},$$

so that

$$-\lim_{\zeta\to e^{i\theta}}\frac{\partial g(\zeta,z)}{\partial\rho}=\sum_{n=0}^{\infty}\frac{1-|z_n|^2}{|e^{i\theta}-z_n|^2}. \tag{3}$$

If $z\in D_0$, $z_n=S_n(z)$, $E_n=S_n(E_0)$ $(S_n\in G)$, then since $mE=2\pi$,

$$\int_{E_0}\frac{\partial g(e^{i\theta},z)}{\partial\nu}d\theta=\sum_{n=0}^{\infty}\int_{E_0}\frac{1-|z_n|^2}{|e^{i\theta}-z_n|^2}d\theta=\sum_{n=0}^{\infty}\int_{E_n}\frac{1-|z|^2}{|e^{i\theta}-z|^2}d\theta$$

$$=\int_E\frac{1-|z|^2}{|e^{i\theta}-z|^2}d\theta=2\pi. \tag{4}$$

(ii) Let $u(z)$ be a bounded harmonic function in $|z|<1$, which is invariant by G, then $u(z)$ can be expressed by a Poisson integral:

$$u(z)=\frac{1}{2\pi}\int_E u(e^{i\theta})\frac{1-|z|^2}{|e^{i\theta}-z|^2}d\theta=\frac{1}{2\pi}\sum_{n=0}^{\infty}\int_{E_n}u(e^{i\theta})\frac{1-|z|^2}{|e^{i\theta}-z|^2}d\theta$$

$$=\frac{1}{2\pi}\sum_{n=0}^{\infty}\int_{E_0}u(e^{i\theta})\frac{1-|z_n|^2}{|e^{i\theta}-z_n|^2}d\theta=\frac{1}{2\pi}\int_{E_0}u(e^{i\theta})\left(\sum_{n=0}^{\infty}\frac{1-|z_n|^2}{|e^{i\theta}-z_n|^2}\right)d\theta$$

$$=\frac{1}{2\pi}\int_{E_0}u(e^{i\theta})\frac{\partial g(e^{i\theta}, z)}{\partial\nu}d\theta, \qquad z\in D_0\,. \tag{5}$$

(iii) Let $f(e^{i\theta})$ be a bounded measurable function on $|z|=1$, which is invariant by G and put

$$u(z)=\frac{1}{2\pi}\int_{E_0}f(e^{i\theta})\frac{\partial g(e^{i\theta}, z)}{\partial\nu}d\theta, \qquad |z|<1\,, \tag{6}$$

then similarly as (ii), we have

$$u(z)=\frac{1}{2\pi}\int_0^{2\pi}f(e^{i\theta})\frac{1-|z|^2}{|e^{i\theta}-z|^2}d\theta\,,$$

hence by Fatou's theorem, $\lim_{z\to e^{i\theta}}u(z)=f(e^{i\theta})$ almost everywhere on E_0, when $z\to e^{i\theta}$ from the inside of any Stolz domain, whose vertex is at $e^{i\theta}$.

4. As an application of Theorem XI. 18, we shall prove the following

Theorem XI. 19. *Let F be a compact Riemann surface, spread over the z-sphere, whose boundary Γ consists of a finite number of disjoint rectifiable Jordan curves $\Gamma=\sum_{i=1}^{k}\Gamma_i$. Let $g(a, z)$ be the Green's function of F, then*

(i) *For almost all $a\in\Gamma$, $\frac{\partial g(a, z)}{\partial\nu}$ exists and $0<\frac{\partial g(a, z)}{\partial\nu}<\infty$ and*

$$\int_{\Gamma}\frac{\partial g(a, z)}{\partial\nu}|da|=2\pi\,, \qquad z\in F\,,$$

where ν is the inner normal of Γ at a, existing almost everywhere on Γ.

(ii) *Let $u(z)$ be a one-valued bounded harmonic function on F and $u(a)$ ($a\in\Gamma$) be its boundary value, existing almost everywhere on Γ, then $u(z)$ can be expressed by*

$$u(z)=\frac{1}{2\pi}\int_{\Gamma}u(a)\frac{\partial g(a, z)}{\partial\nu}|da|\,.$$

(iii) *Let $f(a)$ ($a\in\Gamma$) be a bounded measurable function, defined on Γ and put*

$$u(z)=\frac{1}{2\pi}\int_{\Gamma}f(a)\frac{\partial g(a, z)}{\partial\nu}|da|\,,$$

then $u(z)$ is a bounded harmonic function on F and for almost all $a\in\Gamma$,

$$\lim_{z\to a}u(z)=f(a)\,,$$

when $z \to a$ from the inside of any Stolz domain, whose vertex is at a.

Proof. (i) We map the universal covering surface $F^{(\infty)}$ of F conformally on $|\zeta|<1$ by $z=\varphi(\zeta)$, then $\varphi(\zeta)$ is invariant by a certain Fuchsian group G. Let D_0 be its fundamental domain and E_0 be the part of the boundary of D_0 on $|\zeta|=1$. Then by M. and F. Riesz' theorem, $\varphi(e^{i\theta})$ is absolutely continuous on E_0 and by Theorem VIII. 27,

$$\lim_{\zeta \to e^{i\theta}} \frac{\varphi(\zeta)-\varphi(e^{i\theta})}{\zeta - e^{i\theta}} = \varphi'(e^{i\theta}), \quad 0<|\varphi'(e^{i\theta})|<\infty, \tag{1}$$

exists, almost everywhere on E_0, when $\zeta \to e^{i\theta} \in E_0$ from the inside of any Stolz domain, whose vertex is at $e^{i\theta}$.

Suppose that $\varphi'(e^{i\theta})$ exists and $0<|\varphi'(e^{i\theta})|<\infty$ at $e^{i\theta}\in E_0$ and let $a=\varphi(e^{i\theta})$.

Since a tangent exists almost everywhere on Γ, let a^* be a point on the inner normal ν of Γ at a and let $a^*=\varphi(\zeta^*)$ and we put $g(a^*, z)$ $=G(\zeta^*, \zeta)$, then

$$G(\zeta^*, \zeta)=\sum_{n=0}^{\infty}\log\left|\frac{1-\bar{\zeta}_n\zeta^*}{\zeta^*-\zeta_n}\right|, \tag{2}$$

where ζ_n are equivalents of ζ by G.

Since if $\overset{*}{a} \to a$ on ν, $\lim\limits_{\zeta^* \to e^{i\theta}} \dfrac{1-|\zeta^*|}{|\zeta^*-e^{i\theta}|}=1$, we have by (1),

$$\begin{aligned}\frac{\partial g(a, z)}{\partial \nu} &= \lim_{a^* \to a}\frac{g(a^*, z)}{|a^*-a|} = \lim_{\zeta^* \to e^{i\theta}}\left(\frac{G(\zeta^*, \zeta)}{1-|\zeta^*|}\cdot\frac{1-|\zeta^*|}{|\zeta^*-e^{i\theta}|}\cdot\frac{|\zeta^*-e^{i\theta}|}{|\varphi(\zeta^*)-\varphi(e^{i\theta})|}\right) \\ &= \lim_{\zeta^* \to e^{i\theta}}\frac{G(\zeta^*, \zeta)}{1-|\zeta^*|}\cdot\frac{1}{|\varphi'(e^{i\theta})|} = \frac{\partial G(e^{i\theta}, \zeta)}{\partial \nu}\cdot\frac{1}{|\varphi'(e^{i\theta})|},\end{aligned} \tag{3}$$

which holds almost everywhere on E_0. Since by M. and F. Riesz' theorem, a null set on E_0 corresponds to a null set on Γ, (3) holds almost everywhere on Γ. Hence $\dfrac{\partial g(a, z)}{\partial \nu}$ exists and $0<\dfrac{\partial g(a, z)}{\partial \nu}<\infty$ almost everywhere on Γ.

Since $\varphi(e^{i\theta})$ is absolutely continuous, $|da|=|\varphi'(e^{i\theta})|d\theta$, so that by (3) $\dfrac{\partial g(a, z)}{\partial \nu}|da| = \dfrac{\partial G(e^{i\theta}, \zeta)}{\partial \nu}d\theta$, hence by Theorem XI. 18,

$$\int_{\Gamma}\frac{\partial g(a, z)}{\partial \nu}|da| = \int_{E_0}\frac{\partial G(e^{i\theta}, \zeta)}{\partial \nu}d\theta = 2\pi. \tag{4}$$

(ii) Let $u(z)$ be a bounded harmonic function on F and put $u(z)=U(\zeta)$, then by Theorem XI. 18,

$$u(z)=U(\zeta)=\frac{1}{2\pi}\int_{E_0}U(e^{i\theta})\frac{\partial G(e^{i\theta}, \zeta)}{\partial \nu}d\theta = \frac{1}{2\pi}\int_{\Gamma}u(a)\frac{\partial g(a, z)}{\partial \nu}|da|. \tag{5}$$

(iii) Let $f(a)$ be a bounded measurable function, defined on Γ and put

$$u(z)=\frac{1}{2\pi}\int_{\Gamma} f(a)\frac{\partial g(a, z)}{\partial \nu}|da|\,. \tag{6}$$

If we put $f(a)=F(e^{i\theta})$, then

$$u(z)=\frac{1}{2\pi}\int_{E_0} F(e^{i\theta})\frac{\partial G(e^{i\theta}, \zeta)}{\partial \nu}d\theta\,. \tag{7}$$

From this, we see easily that $\lim_{z\to a} u(z)=f(a)$ almost everywhere on Γ, when $z\to a\in\Gamma$ from the inside of any Stolz domain, whose vertex is at a.

6. Ideal boundary of F_G

1. Let D_0 be the fundamental domain of a Fuchsian group G and K_0: $|z|\leqq\rho$ be a disc, contained in D_0 and C_0:$|z|=\rho$ be its bounding circle and K_n, C_n be its equivalents respectively. We take off $\sum_{n=0}^{\infty} K_n$ from $|z|<1$ and let Ω be the remaining domain. We map the universal covering surface $\Omega^{(\infty)}$ of Ω conformally on $|\zeta|<1$ and let M be the image of $\sum_{n=0}^{\infty} C_n$ on $|\zeta|=1$, then we shall prove

Lemma. *If G is of convergence type, then $0<mM<2\pi$.*
If G is of divergence type, then $mM=2\pi$.

Proof. Let M' be the complement of M on $|\zeta|=1$ and put

$$u(\zeta)=\frac{1}{2\pi}\int_{M'}\frac{1-|\zeta|^2}{|\zeta-e^{i\theta}|^2}d\theta\,.$$

We consider $u(\zeta)$ as a function of z and put $u(\zeta)=\overset{*}{u}(z)$, then $\overset{*}{u}(z)$ is a one-valued bounded harmonic function in Ω and is invariant by G, so that if we consider $u(z)$ as a function on $F=F_G$, then $u(z)$ is a one-valued bounded harmonic function on $F-K_0$ and vanishes on C_0. Hence if $F\in O_G$, then $\overset{*}{u}\equiv 0$, so that $mM'=0$, $mM=2\pi$.

If $F\in P_G$, let $\omega(z)$ be the harmonic measure of its ideal boundary with respect to $F-K_0$, then $\omega\not\equiv 0$ and vanishes on C_0. Hence if $mM=2\pi$, then $\omega\equiv 0$, which is absurd, hence $0<mM<2\pi$.

2. Let z be any point of $|z|<1$. We denote its equivalent in D_0 by z_0. Let $L(e^{i\theta};\alpha)$ be a segment through $e^{i\theta}$, contained in $|z|<1$, making an angle α $(0<\alpha<\pi)$ with the positive tangent of $|z|=1$ at $e^{i\theta}$. Its equivalent in D_0 consists of at most a countable number of arcs, which we denote by $L_0(e^{i\theta};\alpha)$. Then we shall prove the following

Theorem XI. 20. *There exists a set E of measure 2π on $|z|=1$,* *which satisfies the following conditions:*

(i) *If G is of convergence type, then* $\lim|z_0|=1$, *if* $z\to e^{i\theta}\in E$ *from the inside of any Stolz domain, whose vertex is at* $e^{i\theta}$.

(ii) *If G is of divergence type and* $e^{i\theta}\in E$, *then* $L_0(e^{i\theta};\alpha)$ *is everywhere dense in* D_0 *for any* α $(0<\alpha<\pi)$. Yûjôbô [1], Tsuji [21].

Proof. (i) First Suppose that G is of convergence type and let $g(z,0)$ be the Green's function of $F=F_G$, then

$$g(z,0)=\sum_{n=0}^{\infty}\log\left|\frac{1-\bar{\zeta}_n z}{z-\zeta_n}\right|,$$

where ζ_n are equivalents of $\zeta=0$. Then for almost all $e^{i\theta}$, $\lim_{z\to e^{i\theta}} g(z,0)=0$, when $z\to e^{i\theta}$ from the inside of any Stolz domain, whose vertex is at $e^{i\theta}$. From this, we see that $\lim|z_0|=1$, since otherwise, $\lim_{z\to e^{i\theta}} g(z,0)\neq 0$.

(ii) Next suppose that G is of divergence type.

Let $a\in D_0$ and K_0: $[z,a]<\delta$ be a disc, contained in D_0 and C_0: $[z,a]=\delta$ be its bounding circle and a_n, K_n, C_n be the equivalent of $a=a_0$, K_0, C_0 respectively, such that $a_n\in K_n$.

We take off $\sum_{n=0}^{\infty}K_n$ from $|z|<1$ and let Ω be the remaining domain. We take off $\sum_{\nu=0}^{n}K_\nu$ from $|z|<1$ and let Ω_n be the remaining domain. Let $u_n(z)$ be the harmonic function in Ω_n, such that $u_n=0$ on $\sum_{\nu=0}^{n}C_\nu$ and $u_n=1$ on $|z|=1$, then by the lemma,

$$\lim_n u_n(z)=0 \quad \text{in } \Omega\,. \tag{1}$$

Let K_0': $[z,a]<2\delta$ be a disc, where we take $\delta>0$ so small that K_0' is contained in D_0 and let K_n' be its equivalents.

Let $M_\rho(\alpha)$ $(0<\rho<1)$ be the set of $e^{i\theta}$, such that the part $L_\rho(e^{i\theta};\alpha)$ of $L(e^{i\theta};\alpha)$, contained in $\rho\leqq|z|<1$, does not intersect $\sum_{n=0}^{\infty}K_n'$, then $M_\rho(\alpha)$ is a closed set. We shall prove that

$$mM_\rho(\alpha)=0\,. \tag{2}$$

Suppose that $mM_\rho(\alpha)>0$ and $e^{i\theta}\in M_\rho(\alpha)$ and $\zeta\in L_\rho(e^{i\theta};\alpha)$, then ζ lies outside of $\sum_{n=0}^{\infty}K_n'$, so that $[\zeta,a_n]\geqq 2\delta$ $(n=0,1,2,\cdots)$.

Let ζ' be any point on C_n: $[z,a_n]=\delta$, then

$$[\zeta,\zeta']+[\zeta',a_n]\geqq[\zeta,a_n]\geqq 2\delta\,.$$

Since $[\zeta',a_n]=\delta$, we have $[\zeta',\zeta]=[\zeta,\zeta']\geqq\delta$. Since ζ' is any point of C_n, the disc $[z,\zeta]>\delta$ lies outside of $\sum_{n=0}^{\infty}K_n$. Hence by Theorem XI. 2, there exists a sector $S_\rho(e^{i\theta};\alpha_1,\alpha_2)$, which is bounded by the circle $|z|=\rho$ and two lines L_1, L_2 through $e^{i\theta}$, making an angle α_1, α_2 $(0<\alpha_1<\alpha<\alpha_2<\pi)$ with the positive tangent of $|z|=1$ at $e^{i\theta}$ respectively, such that $S_\rho(e^{i\theta};\alpha_1,\alpha_2)$ lies outside of $\sum_{n=0}^{\infty} K_n$, where α_1, α_2 are constants, which depend on ρ, a, δ only.

Let $\Delta_\rho=\sum S_\rho(e^{i\theta};\alpha_1,\alpha_2)$, added for all $e^{i\theta}\in M_\rho(\alpha)$ and we join Δ_ρ to the disc $|z|\leqq\rho$ and let Δ be the resulting domain, then Δ is bounded by a rectifiable Jordan curve Λ, such that there is only a finite number of C_n in Δ and $M_\rho(\alpha)$ belongs to its boundary. By modifying the part of Δ, which lies in $|z|\leqq\rho$ suitably, we may assume that there are no $C_n's$ in Δ. If we map Δ conformally on $|\zeta|<1$, then by M. and F. Riesz' theorem, $M_\rho(\alpha)$ is mapped on a set of positive measure on $|\zeta|=1$. Hence by means of Poisson integral on $|\zeta|=1$, we see that there exists a bounded harmonic function $v(z)$ in Δ, such that $v=0$ on $\Lambda-M_\rho(\alpha)$ and $v=1$ almost everywhere on $M_\rho(\alpha)$. Now $v\leqq u_n$ on the boundary of Δ, hence by the maximum principle, $v\leqq u_n$ in Δ, so that if we make $n\to\infty$, then by (1), $v\equiv 0$ in Δ, which is absurd. Hence

$$mM_\rho(\alpha)=0\,.$$

Let $0<\rho_1<\rho_2<\cdots<\rho_\nu\to 1$ and put $\sum_\nu M_{\rho_\nu}(\alpha)=M(a,\delta;\alpha)$. Then $M(a;\delta,\alpha)$ is a null set on $|z|=1$. Let $\{a_i\}$ be rational points in D_0 and $\delta_1>\delta_2>\cdots>\delta_j\to 0$ and $\{\alpha_k\}$ be rational numbers in $(0,\pi)$ and put $M=\sum_{i,j,k} M(a_i,\delta_j;\alpha_k)$, then M is a null set on $|z|=1$, so that if we denote its complement by E, then $mE=2\pi$. As is easily seen, if $z\in L(e^{i\theta};\alpha)$, $z'\in L(e^{i\theta};\alpha')$, such that $|z-e^{i\theta}|=|z'-e^{i\theta}|$ and $|\alpha-\alpha'|$ is sufficiently small, then $|z_0-z_0'|$ is small. From this, we see easily that if $e^{i\theta}\in E$, then $L_0(e^{i\theta};\alpha)$ is everywhere dense in D_0 for any α $(0<\alpha<\pi)$. ——

3. Suppose that D_0 has boundary points on $|z|=1$. Let $0<\rho<1$, then the part of D_0, which lies in $\rho<|z|<1$ consists of a finite number of connected domains. We consider only such ones, which have boundary points on $|z|=1$ and let $D_0(\rho)$ be the sum of such connected domains and $D_\nu(\rho)$ be its equivalents and put

$$\Delta(\rho)=\sum_{\nu=0}^{\infty} D_\nu(\rho)\,. \qquad (1)$$

Then $\Delta(\rho)$ consists of a countable number of connected domains $\Delta_n(\rho)$:

$$\Delta(\rho)=\sum_n \Delta_n(\rho)\,. \qquad (2)$$

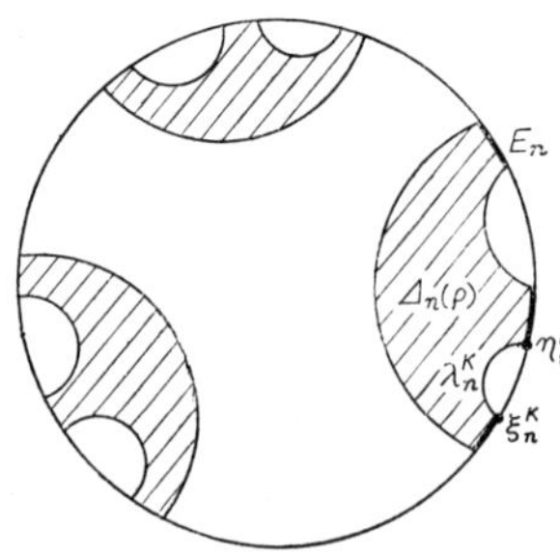

Since the equivalent points on the

boundary of D_0 are equidistant from $z=0$, $\Delta_n(\rho)$ is bounded by Jordan arcs λ_n^k $(k=0, 1, 2, \cdots)$ and a closed set E_n on $|z|=1$.

We put

$$E(\rho)=\sum_n E_n . \qquad (3)$$

λ_n^k ends at two points ξ_n^k, η_n^k on $|z|=1$, which are fixed points of some $S_\nu \in G$. If $\xi_n^k=\eta_n^k$ for some k, then $\Delta_n(\rho)$ is bounded by a single Jordan curve, which touches $|z|=1$ at $\xi_n^k=\eta_n^k$.

Since λ_n^k $(k=0, 1, 2, \cdots)$ can be grouped into a finite number of equivalent classes, we see easily that if $\xi_n^k \neq \eta_n^k$, then λ_n^k is contained between two circular arcs C_n^k, $C_n'^k$ through ξ_n^k, η_n^k, which meet $|z|=1$ at an angle α, α' $(0<\alpha<\alpha'<\pi)$ respectively, as shown in the figure, where α, α' are independent of n and k.

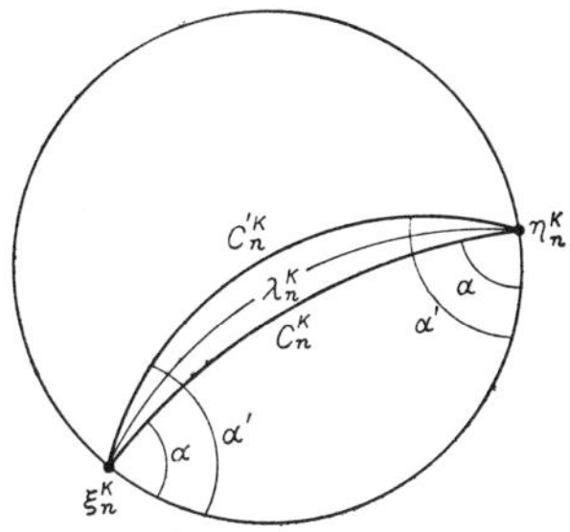

Then we shall prove

Theorem XI. 21. (i) *If G is of convergence type, then $mE(\rho)=2\pi$.*
(ii) *If G is of divergence type, then $mE(\rho)=0$.*

First we shall prove a lemma.

Lemma. *Let K be a circle on the z-plane, which intersects a half-line L: $\arg z=\eta$ $(0<\eta\leqq\pi/2)$ and the real axis at two points A, B, with an angle $\alpha>0$.*

We suppose that A lies to the right of the origin O and B lies to the right of A. Let M be the middle point of AB, then

$$AM \geqq OM \sin\alpha \tan\frac{\eta}{2}.$$

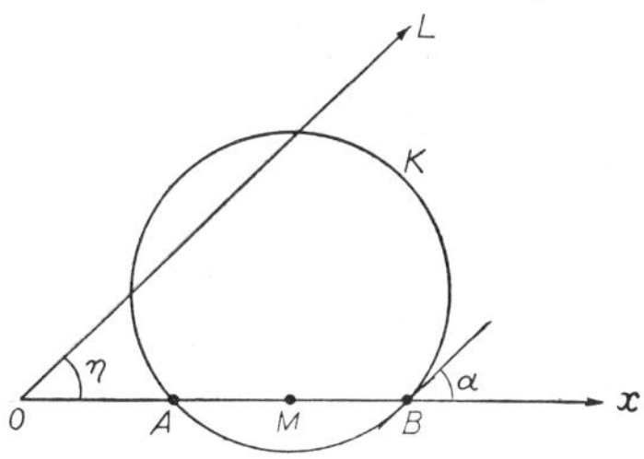

Proof. First denoting by r the radius of K, we have $AM=r\sin\alpha$. Next, while we change the position of K, with fixed r, we see easily that OM attains its maximum, when K is inscribed in the angle between the positive real axis and L and the maximum value is $r\cot\frac{\eta}{2}$. Hence $r\geqq OM\tan\frac{\eta}{2}$, so that

$$AM \geqq OM \sin\alpha \tan\frac{\eta}{2}.$$

Proof of the theorem.

(i) First suppose that G is of convergence type, then D_0 has boundary points on $|z|=1$. Let $g(z,0)$ be the Green's function of F_G, then for almost all $e^{i\theta}$, $\lim_{r\to1} g(re^{i\theta},0)=0$. Let $e^{i\theta}$ be such a point. Since $g(z,0)$ is invariant by G, $g(z,0)\geqq\kappa(\rho)>0$ outside of $\Delta(\rho)$, so that the segment $L(e^{i\theta};\tau,1): z=re^{i\theta}$ $(\tau\leqq r<1)$ belongs to $\Delta(\rho)$, if $1-\tau$ is sufficiently small. Since $\{\Delta_n(\rho)\}$ are disjoint each other, $L(e^{i\theta};\tau,1)$ belongs to a certain $\Delta_n(\rho)$, so that $e^{i\theta}\in E_n$. Since the set of $e^{i\theta}$ is of measure 2π, we have

$$mE(\rho)=2\pi\ .$$

(ii) Next suppose that G is of divergence type and we shall prove that

$$mE_n=0\ (n=0,1,2,\cdots)\ . \tag{1}$$

We may assume that $\xi_n^k\neq\eta_n^k$ for any k, since otherwise, E_n reduces to a single point, as remarked before.

Let $e^{i\theta}\in E_n$. We decompose E_n into three parts: $E_n=E_n^{(1)}+E_n^{(2)}+E_n^{(3)}$, according as (i) $L(e^{i\theta};\tau,1)$ intersects $\sum_k\lambda_n^k$ for any small $1-\tau$, (ii) $L(e^{i\theta};\tau,1)$ belongs to $\Delta_n(\rho)$ for a value τ $(0<\tau<1)$, (iii) $L(e^{i\theta};\tau,1)$ lies outside of $\Delta_n(\rho)$ for a value τ.

If $e^{i\theta}\in E_n^{(1)}$ and $\xi_n^k\neq\eta_n^k$, then we map $|z|<1$ conformally on the upper half of the w-plane, such that $z=0$, $z=e^{i\theta}$ becomes $w=i$, $w=0$ respectively, then $L(e^{i\theta};\tau,1)$ becomes a segment L^* on the imaginary axis.

Let Γ_n^k, $\Gamma_n'^k$ be the image of C_n^k, $C_n'^k$ respectively, then Γ_n^k, $\Gamma_n'^k$ are circles, which meet the real axis at an angle α, α' respectively. Since L^* intersects infinitely many Γ_n^k, $\Gamma_n'^k$, if we apply the lemma to Γ_n^k, $\Gamma_n'^k$ with $\eta=\pi/2$, then we see that the lower density of $E_n^{(1)}$ at $e^{i\theta}$ is $\leqq1-\sigma<1$, where $\sigma=\mathrm{Min}(\sin\alpha,\sin\alpha')$. Since by Lebesgue's theorem, the density is 1 almost everywhere on $E_n^{(1)}$, we have $mE_n^{(1)}=0$.

If $e^{i\theta}\in E_n^{(2)}$, then the equivalents of $L(e^{i\theta},\tau,1)$ in D_0 lies in $\rho<|z|<1$, so that is not everywhere dense in D_0, hence by Theorem XI. 20,

$$mE_n^{(2)}=0\ .$$

If $e^{i\theta}\in E_n^{(3)}$, then $e^{i\theta}$ must coincide with one of ξ_n^k, η_n^k, so that $mE_n^{(3)}=0$. Hence $mE_n=mE_n^{(1)}+mE_n^{(2)}+mE_n^{(3)}=0$, so that

$$mE(\rho)=\sum_n mE_n=0\ .\ \text{——} \tag{2}$$

4. Let $0<\rho_1<\rho_2<\cdots<\rho_\nu<\to1$, then $E(\rho_1)\geqq E(\rho_2)\geqq\cdots\geqq E(\rho_\nu)$, so that if we put

$$E=\lim_\nu E(\rho_\nu)\ , \tag{1}$$

then we have

(i) If G is of convergence type, then $mE=2\pi$.

(ii) If G is of divergence type, then $mE=0$. (2)

Let $F=F_G$ be the Riemann surface, associated to G and F_ν be the image of $\Delta(\rho_\nu)$ on F, then $E(\rho_\nu)$ is the image of the boundary of F_ν on $|z|=1$, so that E is the image of the ideal boundary of F on $|z|=1$.

Hence as a special case, we have the following theorem.

Theorem XI. 22. *Let F be an open Riemann surface, spread over the w-sphere. We map the universal covering surface $F^{(\infty)}$ of F conformally on $|z|<1$ and let E be the image of the ideal boundary of F on $|z|=1$.*

(i) *If $F\in P_G$, then $mE=2\pi$.* (ii) *If $F\in O_G$, then $mE=0$.* Tsuji [21].

7. Hopf's ergodic theorem

1. First we shall prove some lemmas.

Lemma 1. *Let $S: z'=\dfrac{z-a}{1-az}$ $(0<a<1)$ and $J: z=e^{i\theta}$ $(0<\eta\leqq|\theta|\leqq\pi)$ be an arc on $|z|=1$. Then $S(J)$ is contained in an arc*

$$J^*:\ z=e^{i\theta}\ (|\theta-\pi|\leqq 2\pi(1-a)/\sin^2\eta) \quad \text{on } |z|=1.$$

Let e be a measurable set, contained in J, then

$$\frac{(1-a)}{2}\,me\leqq mS(e)\leqq\frac{2(1-a)}{\sin^2\eta}\,me\,,$$

hence

$$\frac{|J^*|}{8\pi}\sin^2\eta\cdot me\leqq mS(e)\leqq\frac{|J^*|}{2\pi}\,me\,,$$

where $|J^*|=\dfrac{4\pi(1-a)}{\sin^2\eta}$ *is the length of J^*.*

Proof. $|dz'|=\dfrac{1-a^2}{|1-az|^2}|dz|$. If $z=e^{i\theta}\in J$, then

$$\frac{1-a}{2}\leqq\frac{1-a^2}{|1-az|^2}\leqq\frac{2(1-a)}{\sin^2\eta}\,,$$

so that from

$$mS(e)=\int_e\frac{1-a^2}{|1-az|^2}\,|dz|\,, \tag{1}$$

we have

$$\frac{1-a}{2}\,me\leqq mS(e)\leqq\frac{2(1-a)}{\sin^2\eta}\,me\,. \tag{2}$$

From this, we see that $S(J)$ is contained in an arc

$$J^*:\ z=e^{i\theta}\ (|\theta-\pi|\leqq 2\pi(1-a)/\sin^2\eta)\,.$$

Lemma 2. *Let D_0 be the fundamental domain of a Fuchsian group G and $\sigma(D_0)<\infty$. Let K_0: $|z|\leqq\delta$ be a disc, contained in D_0 and $\{K_n\}$*

be its equivalents and $K^*=\sum_{n=0}^{\infty} K_n$.

Let $r_\nu=1-\lambda^\nu$ $(0<\lambda<1)$ $(\nu=1, 2, \cdots)$. *If* λ *is sufficiently small, then there exists* ρ_ν $(r_\nu\leqq\rho_\nu\leqq r_{\nu+1})$, *which satisfies the following condition.*

The circle $|z|=\rho_\nu$ *intersects* K^* *in a set of arcs, among them there are arcs* θ_j $(j=1, 2, \cdots, s_\nu)$, *such that*

$$\kappa'(1-\rho_\nu)\geqq|\theta_j|\geqq\kappa\,(1-\rho_\nu), \qquad \sum_{j=1}^{s_\nu}|\theta_j|\geqq\eta>0\,,$$

where $|\theta_j|$ *denotes the angular magnitude of* θ_j *and* $\kappa'>0$, $\kappa>0$, $\eta>0$ *are constants, which are independent of* ν.

Proof. By Theorem XI. 10, $\dfrac{\text{const.}}{1-r}\leqq n(r, a)\leqq\dfrac{\text{const.}}{1-r}$ for any $a\in K_0$. Let $K^*(r_\nu, r_{\nu+1})$ be the part of K^*, contained in $r_\nu\leqq|z|\leqq r_{\nu+1}$, then

$$\sigma(K^*(r_\nu, r_{\nu+1}))=\int_{a\in K_0}(n(r_{\nu+1}, a)-n(r_\nu, a))d\sigma(a)$$

$$\geqq\int_{a\in K_0}\left(\frac{\text{const.}}{1-r_{\nu+1}}-\frac{\text{const.}}{1-r_\nu}\right)d\sigma(a)\geqq\frac{\text{const.}}{\lambda^{\nu+1}}-\frac{\text{const.}}{\lambda^\nu}\geqq\frac{\text{const.}}{\lambda^{\nu+1}}, \quad (1)$$

if λ is sufficiently small.

Let $\theta(r)$ be the angular magnitude of the part of the circle $|z|=r$, contained in K^*, then

$$\sigma(K^*(r_\nu, r_{\nu+1}))=4\iint_{K^*(r_\nu, r_{\nu+1})}\frac{r\,drd\theta}{(1-r^2)^2}=4\int_{r_\nu}^{r_{\nu+1}}\frac{r\theta(r)dr}{(1-r^2)^2}<4\int_{r_\nu}^{r_{\nu+1}}\frac{\theta(r)dr}{(1-r)^2}\,,$$

so that by (1),

$$\int_{r_\nu}^{r_{\nu+1}}\frac{\theta(r)dr}{(1-r)^2}\geqq\frac{\text{const.}}{\lambda^{\nu+1}}\,. \qquad (2)$$

Let the maximum of $\theta(r)$ in $[r_\nu, r_{\nu+1}]$ be attained at $r=\rho_\nu$, then

$$\int_{r_\nu}^{r_{\nu+1}}\frac{\theta(r)dr}{(1-r)^2}\leqq\theta(\rho_\nu)\int_{r_\nu}^{r_{\nu+1}}\frac{dr}{(1-r)^2}<\frac{\theta(\rho_\nu)}{\lambda^{\nu+1}}\,,$$

so that by (2),

$$\theta(\rho_\nu)\geqq2\eta>0, \quad \eta=\text{const.}\ (\nu=1, 2, \cdots)\,. \qquad (3)$$

Let the circle $|z|=\rho_\nu$ intersect K^* in a set of arcs θ_i $(i=1, 2, \cdots, N)$, then

$$\theta(\rho_\nu)=\sum_{=1}^{N}|\theta_i|\geqq2\eta\,. \qquad (4)$$

Since θ_i is contained in an equivalent K_i of K_0, which intersects $|z|=\rho_\nu$ and the radius of K_i is $\leqq$const. $(1-\rho_\nu)$,

$$|\theta_i|\leqq\text{const.}\ (1-\rho_\nu)\,. \qquad (5)$$

For a fixed $\kappa>0$, we decompose $\{\theta_i\}$ into two classes: $\{\theta_i\}=\{\theta_j\}+\{\theta_{j*}\}$, where $|\theta_j|\geqq\kappa(1-\rho_\nu)$ and $|\theta_{j*}|<\kappa(1-\rho_\nu)$, so that

$$\sum_j|\theta_j|+\sum_{j*}|\theta_{j*}|\geqq 2\eta\,. \tag{6}$$

If an equivalent K_s of K_0 intersects the circle $|z|=\rho_\nu$, then K_s is contained in $|z|\leqq\rho_\nu^*$, such that $1-\rho_\nu^*\geqq\text{const.}(1-\rho_\nu)$, so that $N\leqq n(\rho_\nu^*, a)$ $\leqq\dfrac{\text{const.}}{1-\rho_\nu^*}\leqq\dfrac{A}{1-\rho_\nu}$ $(A=\text{const.})$, hence

$$\sum_{j*}|\theta_{j*}|\leqq N\kappa(1-\rho_\nu)\leqq A\kappa.$$

Hence if we choose $\kappa>0$ so small that $A\kappa\leqq\eta$, then by (5), (6), we have

$$\kappa'(1-\rho_\nu)\geqq|\theta_j|\geqq\kappa(1-\rho_\nu), \qquad \sum_j|\theta_j|\geqq\eta>0\,. \text{ ——}$$

2. Hopf's ergodic theorem.

Let $\eta_1=e^{i\theta}$, $\eta_2=e^{i\varphi}$ be two points on $|z|=1$, then the pair (η_1, η_2) can be considered as a point of a torus Θ: $0\leqq\theta\leqq 2\pi$, $0\leqq\varphi\leqq 2\pi$.

For a measurable set E on Θ, we define its measure $\mu(E)$ by

$$\mu(E)=\iint_E d\theta\,d\varphi, \qquad \text{so that} \quad \mu(\Theta)=4\pi^2\,. \tag{1}$$

Let S_ν be any substitution of a Fuchsian group G and

$$T_\nu:\quad \eta_1'=S_\nu(\eta_1), \qquad \eta_2'=S_\nu(\eta_2)\,, \tag{2}$$

then the totality of $\{T_\nu\}$ constitute a group $\mathfrak{G}=G\times G$.

Hopf ([1], [2]) proved the following theorem.

Theorem XI. 23. **(Hopf's ergodic theorem).** *If $\sigma(D_0)<\infty$, then there exists no measurable set E on Θ, which is invariant by $\mathfrak{G}$ and $0<\mu(E)<4\pi^2$.*

Hence if $\mu(E)>0$, then $\mu(E)=4\pi^2$.

Proof. (Tsuji [47]). Suppose that there exists a measurable set E on Θ, which is invariant by $\mathfrak{G}$ and $\mu(E)>0$ and we shall prove that $\mu(E)=4\pi^2$. Let $f(\theta, \varphi)$ be the characteristic function of E and put

$$u(z, w)=\frac{1}{4\pi^2}\iint_\Theta f(\theta, \varphi)\frac{1-|z|^2}{|z-e^{i\theta}|^2}\cdot\frac{1-|w|^2}{|w-e^{i\varphi}|^2}\,d\theta\,d\varphi\,, \tag{1}$$
$$|z|<1,\ |w|<1\,,$$

then $u(z, w)$ is invariant by $\mathfrak{G}$, such that

$$u(S(z),\ S(w))=u(z, w), \qquad S\in G\,. \tag{2}$$

We denote the Stolz domain: $|\arg(1-ze^{-i\theta})|\leqq\pi/4$ by $\Omega(e^{i\theta})$ and $\Omega_\rho(e^{i\theta})$ be its part, contained in $|z-e^{i\theta}|<\rho$, then by Theorem IV. 12, for almost all (θ, φ) on Θ,

$$\lim_{z\to e^{i\theta},\, w\to e^{i\varphi}} u(z, w)=f(\theta, \varphi) \quad \text{uniformly}, \tag{3}$$

if $z\to e^{i\theta}$, $w\to e^{i\varphi}$ from the inside of $\Omega(e^{i\theta})$, $\Omega(e^{i\varphi})$ respectively.

Let $E[\theta]$ be the section of E by the line $\theta=\text{const.}=\theta$ and $E[\varphi]$ be that by the line $\varphi=\text{const.}=\varphi$, then

$$\mu(E)=\int_0^{2\pi} mE[\theta]d\theta>0\ , \tag{4}$$

where $mE[\theta]$ denotes the linear measure.

Since $\sigma(D_0)<\infty$, G is of divergence type (Theorem XI. 10), so that there exists no measurable set e on $|z|=1$, which is invariant by G and $0<me<2\pi$, hence if $me>0$, then $me=2\pi$.

If $mE[\theta]=0$ on a set e of positive measure in $[0, 2\pi]$, then since, by the hypothesis on E, such a set e is invariant by G, $me=2\pi$, so that $mE[\theta]=0$ almost everywhere in $[0, 2\pi]$, hence $\mu(E)=0$, which is absurd. Hence

$$mE[\theta]>0 \text{ for almost all } \theta \text{ in } [0, 2\pi]\ . \tag{5}$$

Hence by Egoroff's theorem, there exists a closed sub-set E_0 of E, which satisfies the following conditions (i), (ii), (iii), where $\delta>0$ and $\eta>0$ are sufficiently small constants.

(i) E_0 lies outside of the strip: $|\theta-\varphi|<\eta$, mod. 2π.

(ii) Let $\overset{*}{E}_0$ be the projection of E_0 on the θ-axis, then

$$m\overset{*}{E}_0>2\pi-\delta \quad \text{and} \quad mE_0[\theta]\geqq\eta>0 \text{ for any } \theta\in\overset{*}{E}_0\ .$$

(iii) $\lim_{z\to e^{i\theta},\, w\to e^{i\varphi}} u(z, w)=1$ uniformly on E_0,

when $z\to e^{i\theta}$, $w\to e^{i\varphi}$ from the inside of $\Omega(e^{i\theta})$, $\Omega(e^{i\varphi})$ respectively.

Hence if $\theta\in\overset{*}{E}_0$, $z\in\Omega_\rho(e^{i\theta})$ $(\rho=\rho(\varepsilon))$, then

$$1-\varepsilon<u(z, e^{i\varphi})<1, \quad \varphi\in E_0[\theta]\ , \tag{6}$$

where $$u(z, e^{i\varphi})=\lim_{w\to e^{i\varphi}} u(z, w).$$

Let K_0: $|z|\leqq\tau$ be a disc, contained in D_0 and $\{K_n\}$ be its equivalents and $K^*=\sum_{n=0}^{\infty} K_n$, then by Lemma 2, there exist $0<\rho_1<\rho_2<\cdots<\rho_\nu\to 1$, such that the circle $|z|=\rho_\nu$ intersects K^* in a set of arcs θ_j $(j=1, 2, \cdots, s_\nu)$, such that

$$\text{const.}(1-\rho_\nu)\geqq|\theta_j|\geqq\text{const.}\ (1-\rho_\nu), \qquad \sum_{j=1}^{s_\nu}|\theta_j|\geqq\text{const.}>0\ (\nu=1, 2, \cdots)\ .$$

Hence if we denote the projection on $|z|=1$ of θ_j from $z=0$ by J_j, then

$$\text{const.}(1-\rho_\nu)\geqq|J_j|\geqq\text{const.}\ (1-\rho_\nu),\quad \sum_{j=1}^{s_\nu}|J_j|\geqq\text{const.}>0\ (\nu=1,2,\cdots). \tag{7}$$

Since $\{J_j\}$ are disjoint each other, if in the condition (ii), $\delta>0$ is sufficiently small, then by the second inequality of (7), we may assume that every J_j contains a point $e^{i\omega_j}(\omega_j\in\overset{*}{E}_0)$.

Let

$$K_j:\ \left|\frac{z-z_j}{1-\bar{z}_jz}\right|\leqq\tau,\quad \arg z_j=\alpha_j\,, \tag{8}$$

be the equivalent of K_0, which contains the arc θ_j, where $z_j\in K_j$ is an equivalent of $z=0$ and let

$$K_0=S_j(K_j),\quad 0=S_j(z_j),\quad S_j\in G\,. \tag{9}$$

Since J_j contains $e^{i\omega_j}$, if the radius τ of K_0 is sufficiently small and $\nu\geqq\nu_0$, then by Theorem XI. 2, z_j is contained in $\Omega_\rho(e^{i\omega_j})$. Hence by (6),

$$1-\varepsilon_\nu<u(z_j,e^{i\psi})<1\qquad\text{for any }\psi\in E_0[\omega_j]\,, \tag{10}$$

where $\varepsilon_\nu\to0$ with $\nu\to\infty$.

Since $u(z, w)$ is invariant by $\mathfrak{G}$, we have

$$1-\varepsilon_\nu<u(0,e^{i\varphi})<1\qquad\text{for any }\varphi\in S_j(E_0[\omega_j])\,.$$

Hence if we put

$$M_\nu=\sum_{j=1}^{s_\nu}S_j\,(E_0[\omega_j])\,, \tag{11}$$

then

$$1-\varepsilon_\nu<u(0,e^{i\varphi})<1\qquad\text{for any }\varphi\in M_\nu\,. \tag{12}$$

We shall prove that

$$mM_\nu\geqq\text{const.}>0\ (\nu=1,2,\cdots)\,. \tag{13}$$

Let $\widetilde{K}_j=S_j(K_0)$, then $\widetilde{K}_j$ is obtained from K_j by a rotation about $z=0$, so that the circle $|z|=\rho_\nu$ intersects $\widetilde{K}_j$ in an arc $\widetilde{\theta}_j$, whose projection on $|z|=1$ from $z=0$ be denoted by $\widetilde{J}_j$, then $|\widetilde{J}_j|=|J_j|$, so that

$$\text{const.}(1-\rho_\nu)\geqq|\widetilde{J}_j|\geqq\text{const.}(1-\rho_\nu),\quad \sum_{j=1}^{s_\nu}|\widetilde{J}_j|\geqq\text{const.}>0\quad(\nu=1,2,\cdots). \tag{14}$$

Since $\omega_j\in\overset{*}{E}_0$, $mE_0[\omega_j]\geqq\eta>0$ by the condition (ii) and by the condition (i), if $\psi\in E_0[\omega_j]$, then $|\psi-\omega_j|\geqq\eta$, mod 2π, so that if $\nu\geqq\nu_0$, then $|\psi-\alpha_j|\geqq\frac{\eta}{2}$, mod 2π in (10), hence by Lemma 1, $S_j(E_0[\omega_j])$ is contained in an arc $\tilde{J}_j^*$ on $|z|=1$, concentric with $\tilde{J}_j$, such that

$$mS_j(E_0[\omega_j]) \geqq \text{const.} |\tilde{J}_j^*| . \tag{15}$$

By Lemma 1, $|\tilde{J}^*| = \text{const.} (1-|z_j|)$ and since

$$\text{const.}(1-\rho_\nu) \leqq 1-|z_j| \leqq \text{const.}(1-\rho_\nu) ,$$

we have

$$\text{const.}(1-\rho_\nu) \leqq |\tilde{J}_j^*| \leqq \text{const.}(1-\rho_\nu) . \tag{16}$$

Hence by (14), (16), $|\tilde{J}_j^*| \geqq \text{const.} |\tilde{J}_j|$, so that by (14), we have

$$\sum_{j=1}^{s_\nu} |\tilde{J}_j^*| \geqq \text{const.} > 0 \quad (\nu=1, 2, \cdots) . \tag{17}$$

Since $|\tilde{J}_j| \geqq \text{const.}(1-\rho_\nu)$ and $\tilde{J}_j, \tilde{J}_{j'}$ $(j \neq j')$ are disjoint, the distance between the centers of $\tilde{J}_j, \tilde{J}_{j'}$ is $\geqq \text{const.}(1-\rho_\nu)$. Since $\tilde{J}_j^*$ is concentric with $\tilde{J}_j$ and $|\tilde{J}_j^*| \leqq \text{const.}(1-\rho_\nu)$, $\{J_j^*\}$ overlap at most N-times, where N is independent of ν.

Since $S_j(E_0[\omega_j]) \tilde{\subset} J_j^*$, $\{S_j(E_0[\omega_j])\}$ overlap at most N-times, so that by (15), (17),

$$mM_\nu \geqq \frac{1}{N} \sum_{j=1}^{s_\nu} mS_j(E_0[\omega_j]) \geqq \text{const.} \sum_{j=1}^{s_\nu} |\tilde{J}_j^*| \geqq \text{const.} > 0 \ (\nu=1, 2, \cdots).$$

Hence if we put

$$M = \overline{\lim_\nu} M_\nu, \tag{18}$$

then $mM > 0$ and by (12),

$$u(0, e^{i\varphi}) = 1, \quad \text{if } \varphi \in M . \tag{19}$$

Now

$$u(0, w) = \frac{1}{2\pi} \int_0^{2\pi} F(\varphi) \frac{1-|w|^2}{|w-e^{i\varphi}|^2} d\varphi , \tag{20}$$

where

$$F(\varphi) = \frac{1}{2\pi} \int_0^{2\pi} f(\theta, \varphi) \, d\theta = \frac{mE[\varphi]}{2\pi}. \tag{21}$$

Hence for almost all φ on M, $1 = u(0, e^{i\varphi}) = F(\varphi) = \dfrac{mE[\varphi]}{2\pi}$, or

$$mE[\varphi] = 2\pi . \tag{22}$$

Since the set e of φ, such that $mE[\varphi] = 2\pi$ is invariant by G and

$mE[\varphi]=2\pi$ on a set M of positive measure, we see that $mE[\varphi]=2\pi$ almost everywhere in $[0,2\pi]$, hence

$$\mu(E)=\int_0^{2\pi} mE[\varphi]d\varphi=4\pi^2 . \text{ ——}$$

3. Next we consider the n-dimensional case. Let $\eta_1=e^{i\theta_1}, \cdots, \eta_n=e^{i\theta_n}$ be n points on $|z|=1$, then the pair $(\eta_1,\cdots,\eta_n)$ can be considered as a point of an n-dimentional torus Θ_n: $0\leqq\theta_1\leqq 2\pi, \cdots, 0\leqq\theta_n\leqq 2\pi$ and the measure $\mu(E)$ of a measurable set E on Θ_n is defined by

$$\mu(E)=\iint_E d\theta_1\cdots d\eta_n , \quad \text{so that} \quad \mu(\Theta_n)=(2\pi)^n . \tag{1}$$

Let S_ν be any substitution of a Fuchsian group G and

$$T_\nu: \quad \eta_1'=S_\nu(\eta_1) , \cdots, \eta_n'=S_\nu(\eta_n) , \quad S_\nu\in G ,$$

then the totality of $\{T_\nu\}$ constitutes a group $\mathfrak{G}_n=\overbrace{G\times\cdots\times G}^{n}$.

Theorem XI. 24. *Let G be a Fuchsian group and D_0 be its fundamental domain, where $\sigma(D_0)$ may be finite or infinite. If $n\geqq 3$, then there exists always a measurable set E on Θ_n, which is invariant by $\mathfrak{G}_n$ and $0<\mu(E)<(2\pi)^n$.* Tsuji [14].

Proof. Since the proof is the same, we assume that $n=3$ and put

$$\theta_1=\theta, \ \theta_2=\varphi, \ \theta_3=\psi .$$

Let

$$Q: \quad \left|\theta-\frac{\pi}{2}\right|\leqq\varepsilon , \quad |\varphi-\pi|\leqq\varepsilon , \quad \left|\psi-\frac{3\pi}{2}\right|\leqq\varepsilon \quad (0<\varepsilon<\pi/16), \tag{1}$$

be a cube on Θ_3 and α, α'; β, β'; γ, γ' be arcs on $|z|=1$, such that

$$\left.\begin{aligned} &\alpha: \left|\arg z-\frac{\pi}{2}\right|\leqq\varepsilon , && \alpha': \left|\arg z-\frac{\pi}{2}\right|\leqq\frac{\pi}{8} ,\\ &\beta: |\arg z-\pi| \leqq\varepsilon , && \beta': |\arg z-\pi| \leqq\frac{\pi}{8} ,\\ &\gamma: \left|\arg z-\frac{3\pi}{2}\right|\leqq\varepsilon , && \gamma': \left|\arg z-\frac{3\pi}{2}\right|\leqq\frac{\pi}{8} , \end{aligned}\right\} \tag{2}$$

so that $\qquad Q=\alpha\times\beta\times\gamma .$

Let K_0: $|z|\leqq\rho$ be a disc, contained in D_0 and K_n: $\left|\frac{z-z_n}{1-\bar{z}_n z}\right|\leqq\rho$, $(z_n=r_n e^{i\theta_n})$ $(n=0,1,2,\cdots)$ be its equivalents and let

$$K_0 = S_n(K_n)\,, \qquad 0 = S_n(z_n)\,, \qquad S_n \in G\,. \tag{3}$$

(i) If $e^{i\theta_n}$ belongs to the complement of $\alpha' + \beta' + \gamma'$, then since $\varepsilon < \pi/16$, we have for any z on α, $|\arg z - \theta_n| \geqq \pi/16$, hence by Lemma 1,

$$mS_n(\alpha) \leqq \text{const.}\ |\alpha|(1-|z_n|) \leqq \text{const.}\ \varepsilon(1-|z_n|)\,.$$

Similarly

$$mS_n(\beta) \leqq \text{const.}\ \varepsilon(1-|z_n|)\,, \qquad mS_n(\gamma) \leqq \text{const.}\ \varepsilon(1-|z_n|)\,,$$

so that

$$\mu(T_n(Q)) \leqq \text{const.}\ \varepsilon^3(1-|z_n|)^3\,. \tag{4}$$

(ii) Next suppose that $e^{i\theta_n}$ belongs to $\alpha' + \beta' + \gamma'$ and let $e^{i\theta_n} \in \alpha'$, then for any z on β, $|\arg z - \theta_n| \geqq \pi/16$, so that

$$mS_n(\beta) \leqq \text{const.}\ \varepsilon(1-|z_n|)\,.$$

Similarly

$$mS_n(\gamma) \leqq \text{const.}\ \varepsilon(1-|z_n|)\,,$$

hence

$$\mu(T_n(Q)) \leqq \text{const.}\ \varepsilon^2(1-|z_n|)^2\,. \tag{5}$$

Hence by (4), (5), in any case,

$$\mu(T_n(Q)) \leqq \text{const.}\ \varepsilon^2(1-|z_n|)^2\,. \tag{6}$$

Since $\sum_n (1-|z_n|)^2 < \infty$ (Theorem XI. 8), if $\varepsilon > 0$ is sufficiently small,

$$\sum_{n=0}^{\infty} \mu(T_n(Q)) \leqq \text{const.}\ \varepsilon^2 \sum_{n=0}^{\infty} (1-|z_n|)^2 \leqq \text{const.}\ \varepsilon^2 < \mu(\Theta_3)\,.$$

Hence if we put $E = \sum_{n=0}^{\infty} T_n(Q)$, then E is invariant by $\mathfrak{G}_3$ and

$$0 < \mu(E) \leqq \sum_{n=0}^{\infty} \mu(T_n(Q)) < \mu(\Theta_3)\,.$$

8. Myrberg's approximation theorem

As an application of Hopf's theorem, we shall prove the following **Myrberg's approximation theorem.**

Theorem XI. 25. *Let G be a Fuchsian group, such that $\sigma(D_0) < \infty$.*

Then there exists a set E of measure 2π on $|z|=1$, which satisfies the following condition. If $e^{i\theta} \in E$, let $\gamma(e^{i\theta})$ be any small circular arc in $|z|=1$, which intersects $|z|=1$ orthogonally at $e^{i\theta}$ and $\{\gamma_n(e^{i\theta})\}$ be its equivalents by G. Let C be any circular arc in $|z|<1$, which intersects $|z|=1$ orthogonally at its two end points on $|z|=1$. Then we can find n_ν, such that

$$\gamma_{n_\nu}(e^{i\theta}) \to C \qquad (\nu \to \infty)\,.$$ Myrberg [1].

Proof. (Tsuji [32]). Let Θ: $0 \leqq \theta \leqq 2\pi$, $0 \leqq \varphi \leqq 2\pi$ be the torus and $d\mu = d\theta\, d\varphi$ and $\mathfrak{G} = G \times G$ be defined as before. Let $(\theta, \varphi) \in \Theta$ and (θ_n, φ_n) $(n = 0, 1, 2, \cdots)$ be its equivalents by $\mathfrak{G}$.

Let M be the set of (θ, φ), such that $\{(\theta_n, \varphi_n)\}$ is not everywhere dense on Θ and we shall prove that

$$\mu(M) = 0\,. \tag{1}$$

First we shall prove that M is measurable. Let r_ν $(\nu = 1, 2, \cdots)$ be the set of all rational numbers in $[0, 2\pi]$ and

$$\Delta_{ijkl}: \quad r_i \leqq \theta \leqq r_j\,, \qquad r_k \leqq \varphi \leqq r_l\,, \tag{2}$$

be a rectangle on Θ and $M_{ijkl}^{(s)}$ be the set of (θ, φ), such that (θ_n, φ_n) $(n = 0, 1, 2, \cdots, s)$ lie outside of Δ_{ijkl}, then $M_{ijkl}^{(s)}$ is an open set, so that $M_{ijkl} = \prod_{s=0}^{\infty} M_{ijkl}^{(s)}$ is a G_δ-set, hence $\widetilde{M} = \sum_{i,j,k,l} M_{ijkl}$ is a measurable set. We see easily that $M = \widetilde{M}$, hence M is a measurable set.

Suppose that $\mu(M) > 0$, then $\mu(M_{ijkl}) > 0$ for some i, j, k, l.

Since as is easily seen, M_{ijkl} is invariant by $\mathfrak{G}$, by Hopf's theorem, $\mu(M_{ijkl}) = 4\pi^2$, but M_{ijkl} lies outside of Δ_{ijkl}, which is absurd. Hence

$$\mu(M) = 0\,.$$

Let M^* be its complement, then $\mu(M^*) = 4\pi^2$ and if $(\theta, \varphi) \in M^*$, then its equivalents (θ_n, φ_n) are everywhere dense on Θ. Since $\mu(M^*) = 4\pi^2$, there exists a set e_0 of measure 2π on the θ-axis, such that if $\theta \in e_0$, then the section $M^*[\theta]$ of M^* by the line $\theta = \text{const.} = \theta$ is of measure 2π.

Let $C(\theta, \varphi)$ be a circular arc in $|z| < 1$, which intersects $|z| = 1$ orthogonally at $e^{i\theta}$, $e^{i\varphi}$ and let $C = C(\theta_0, \varphi_0)$ be the given circular arc. Let $\theta \in e_0$, $\varphi \in M^*[\theta]$, then $(\theta, \varphi) \in M^*$, so that its equivalents (θ_n, φ_n) are everywhere dense on Θ, hence we can find n_ν, such that

$$(\theta_{n_\nu}, \varphi_{n_\nu}) \to (\theta_0, \varphi_0) \qquad (\nu \to \infty)\,. \tag{3}$$

We draw a circle κ of radius r with center at $e^{i\theta}$ and let z be the common point of κ with $C(\theta, \varphi)$. Similarly let κ' be a circle about $e^{i\varphi}$ of radius r and z' be the common point of κ' with $C(\theta, \varphi)$. We choose r, so small that κ, κ' lie outside each other.

Let z_n, z'_n be the equivalents of z, z' by G, then

$$z_{n_\nu} \to e^{i\theta_0}\,, \text{ or } e^{i\varphi_0} \quad \text{and} \quad z'_{n_\nu} \to e^{i\theta_0}\,, \text{ or } e^{i\varphi_0}\,. \tag{4}$$

(i) If $z_{n_\nu} \to e^{i\varphi_0}$, then the equivalents κ_{n_ν} of κ tend to $e^{i\varphi_0}$. Since $e^{i\theta_\nu} \to e^{i\theta_0}$, the equivalents of the outside of κ tend to $e^{i\varphi_0}$, so that $\gamma_{n_\nu}(e^{i\theta})$ tend to $C=C(\theta_0, \varphi_0)$.

(ii) If $z_{n_\nu} \to e^{i\theta_0}$, then $z'_{n_\nu} \to e^{i\theta_0}$. For, if otherwise, $z'_{n_\nu} \to e^{i\varphi_0}$, then the non-euclidean distance of z_{n_ν}, z'_{n_ν} tends to ∞, while it is equal to that of z, z', which is finite. Hence $z'_{n_\nu} \to e^{i\theta_0}$. Then by (i), $\gamma_{n_\nu}(e^{i\varphi})$ tends to $C=(\theta_0, \varphi_0)$.

Suppose that the case (i) occurs for every $\theta \in e_0$ and for at least one $\varphi=\varphi(\theta) \in M^*[\theta]$, then we put

$$E(\theta_0, \varphi_0) = e_0 . \tag{5}$$

If this is not the case, then there exists $\theta' \in e_0$, such that $\kappa_{n_\nu} \to e^{i\theta_0}$, so that $\kappa'_{n_\nu} \to e^{i\theta_0}$ for every $\varphi \in M^*[\theta']$. In this case, we put

$$E(\theta_0, \varphi_0) = M^*[\theta'] . \tag{6}$$

Then in each case, $\quad mE(\theta_0, \varphi_0)=2\pi$.

Let θ_m $(m=1, 2, \cdots)$ be the set of all rational numbers in $[0, 2\pi]$ on the θ-axis and φ_n $(n=1, 2, \cdots)$ be that on the φ-axis and put

$$E = \prod_{m,n=1}^{\infty} E(\theta_m, \varphi_n) , \tag{7}$$

then $mE=2\pi$ and we see easily that E satisfies the condition of the theorem.

9. Analogue of Weyl's theorem on uniform distribution

1. Let G be a Fuchsian group and D_0 be its fundamental domain, such that $\sigma(D_0)<\infty$. Let $z=x+iy \in D_0$. We associate a direction φ at z, which makes an angle φ $(0 \leqq \varphi < 2\pi)$ with the positive real axis, then the line elements (z, φ) $(z \in D_0, 0 \leqq \varphi < 2\pi)$ constitute a space Ω, which is the product space of a unit circle and D_0. We define the volume element $d\mu$ in Ω by

$$d\mu = \frac{4dx\,dy\,d\varphi}{(1-|z|^2)^2} , \qquad z=x+iy , \tag{1}$$

then $\mu(\Omega)=2\pi\sigma(D_0)$. $d\mu$ is invariant by a linear transformation, which makes $|z|<1$ invariant.

Let α be any number, such that $0<\alpha<\pi$. If (z, φ) is a line element, then we see easily that there exists a unique circular arc $C^{(\alpha)}(z, \varphi)$, which satisfies the following condition.

$C^{(\alpha)}(z, \varphi)$ touches the direction φ at z and $\eta_1 = e^{i\theta_1}$, $\eta_2 = e^{i\theta_2}$ be its end points on $|z|=1$, such that if we proceed on $C^{(\alpha)}(z, \varphi)$ in the direction φ, then we meet $|z|=1$ at η_2, where $C^{(\alpha)}(z, \varphi)$ makes an angle α with the positive tangent of $|z|=1$ at η_2.

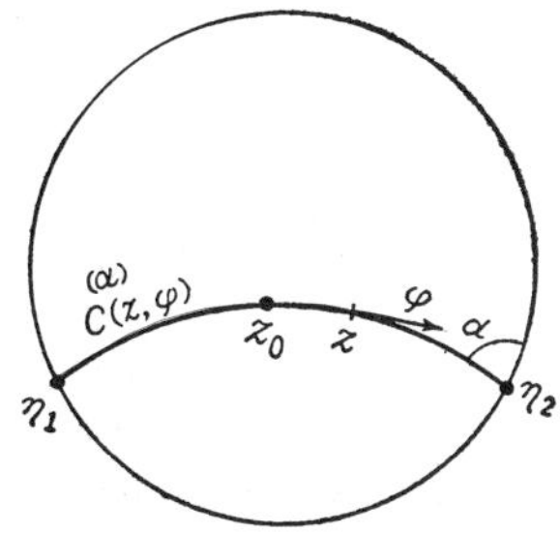

Let z_0 be the middle point of the arc $\widehat{\eta_1\eta_2} = C^{(\alpha)}(z, \varphi)$ and s be the non-euclidean length of the arc $\widehat{z_0 z}$ of $C^{(\alpha)}(z, \varphi)$, where $s>0$, if z lies on $\widehat{z_0\eta_2}$ and $\leqq 0$, if otherwise.

Then we have a one-to-one correspondence between (z, φ) and (η_1, η_2, s), so that we write $(z, \varphi) = (\eta_1, \eta_2, s)$. Then we shall prove

Lemma. $d\mu = \dfrac{4dx\,dy\,d\varphi}{(1-|z|^2)^2} = \dfrac{2\sin\alpha|d\eta_1||d\eta_2|ds}{|\eta_1-\eta_2|^2}$.

Proof. By $z = \dfrac{\zeta - i}{\zeta + i}$, we map $|z|<1$ on the upper half of the $\zeta = u + iv$-plane, then

$$\left.\begin{aligned} ds &= \frac{2|dz|}{1-|z|^2} = \frac{|d\zeta|}{v}\,, \\ d\sigma &= \frac{4dx\,dy}{(1-|z|^2)^2} = \frac{du\,dv}{v^2}\,. \end{aligned}\right\} \quad (1)$$

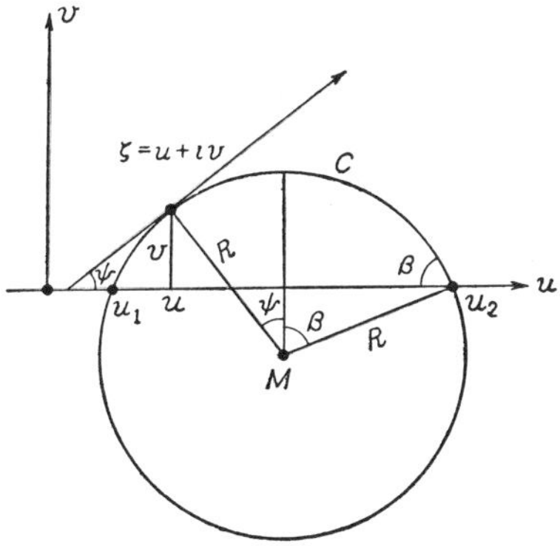

Let $C^{(\alpha)}(z, \varphi)$ become a circle C of radius R and center M and η_1, η_2 become u_1, u_2 $(u_1<u_2)$ on the real axis respectively, then

$$\left.\begin{aligned} R &= \frac{u_2-u_1}{2\sin\beta}\,, \qquad \beta = \pi - \alpha\,, \\ u &= \frac{u_1+u_2}{2} - R\sin\psi = \frac{u_1+u_2}{2} - \frac{u_2-u_1}{2\sin\beta}\sin\psi\,, \\ v &= R\cos\psi - R\cos\beta = \frac{u_2-u_1}{2\sin\beta}(\cos\psi - \cos\beta)\,, \end{aligned}\right\} \quad (2)$$

so that

$$\frac{\partial(u, v)}{\partial(u_1, u_2)} = \frac{1}{2\sin\beta}(\cos\psi - \cos\beta)\,,$$

hence

$$du\,dv=\frac{|\cos\psi-\cos\beta|}{2\sin\beta}du_1\,du_2=\frac{v\,du_1du_2}{u_2-u_1}\,. \tag{3}$$

By (1), $ds=R\dfrac{d\psi}{v}$ and since $d\psi=d\varphi$, we have

$$\frac{4dx\,dy\,d\varphi}{(1-|z|^2)^2}=\frac{du\,dv\,d\psi}{v^2}=\frac{du\,dv\,ds}{vR}=\frac{du_1\,du_2\,ds}{R(u_2-u_1)}=\frac{2\sin\beta\,du_1du_2\,ds}{(u_2-u_1)^2}$$
$$=\frac{2\sin\alpha\,du_1\,du_2\,ds}{(u_2-u_1)^2}\,. \tag{4}$$

Since the anharmonic ratio $[z_1, z_2, z_3, z_4]=\dfrac{z_3-z_1}{z_3-z_2}\Big/\dfrac{z_4-z_1}{z_4-z_2}$ and hence $\dfrac{dz_1\,dz_2}{(z_1-z_2)^2}=-[z_1, z_2, z_1+dz_1, z_2+dz_2]$ is invariant by a linear transformation, we have $\dfrac{|d\eta_1||d\eta_2|}{|\eta_1-\eta_2|^2}=\dfrac{du_1\,du_2}{(u_2-u_1)^2}$, so that by (4),

$$\frac{4dx\,dy\,d\varphi}{(1-|z|^2)^2}=\frac{2\sin\alpha|d\eta_1||d\eta_2|ds}{|\eta_1-\eta_2|^2}\,. \text{——} \tag{5}$$

2. Now for a fixed α $(0<\alpha<\pi)$, we consider a flow $T_t^{(\alpha)}$ $(-\infty<t<\infty)$ in Ω:

$$T_t^{(\alpha)}:\quad P=(\eta_1, \eta_2, s)\to P_t=(\eta_1, \eta_2, s+t)\,. \tag{1}$$

Then by the lemma, $T_t^{(\alpha)}$ is a mass-preserving transformation of Ω into itself. We shall prove

Theorem XI. 26. *If $\sigma(D_0)<\infty$, then the flow $T_t^{(\alpha)}$ is metric transitive*[1].

Proof. Let $(z, \varphi)=(\eta_1, \eta_2, s)$ and we write $C^{(\alpha)}(z, \varphi)=C^{(\alpha)}(\eta_1, \eta_2)$, where $\eta_1=e^{i\theta_1}$, $\eta_2=e^{i\theta_2}$. Then the part of $C^{(\alpha)}(\eta_1, \eta_2)$, which is contained in D_0 corresponds to an interval $s_1(\theta_1, \theta_2)\leqq s\leqq s_2(\theta_1, \theta_2)$ on the s-axis, so that if we denote the projection of Ω on the torus Θ: $0\leqq\theta_1\leqq 2\pi$, $0\leqq\theta_2\leqq 2\pi$ by Ω_Θ, then Ω consists of points $(e^{i\theta_1}, e^{i\theta_2}, s)$, such that

$$\Omega:\quad (\theta_1, \theta_2)\in\Omega_\Theta\,,\qquad s_1(\theta_1, \theta_2)\leqq s\leqq s_2(\theta_1, \theta_2)\,. \tag{1}$$

To prove that $T_t^{(\alpha)}$ is metric transitive, it suffices to prove that if a set M in Ω is invariant by the flow $T_t^{(\alpha)}$ and $\mu(M)>0$, then $\mu(M)=\mu(\Omega)$. If we denote the projection of M on Θ by M_Θ, then since M is invariant by the flow, M consists of points $(e^{i\theta_1}, e^{i\theta_2}, s)$, such that

$$M:\quad (\theta_1, \theta_2)\in M_\Theta\,,\qquad s_1(\theta_1, \theta_2)\leqq s\leqq s_2(\theta_1, \theta_2)\,. \tag{2}$$

Let $\{M_\Theta^{(\nu)}\}$ be equivalents of M_Θ by $\mathfrak{G}=$ [illegible] and $M_\Theta^*=\sum_\nu M_\Theta^{(\nu)}$, then M_Θ^* is invariant by $\mathfrak{G}$. Since $\mu(M)>0$, M_Θ^* is a set of positive measure on Θ, so that by Hopf's theorem, $\Theta-M_\Theta^*$ is a null set, hence $\Omega_\Theta-M_\Theta^*$ is a

1) For the definition of "metric transitive", see Appendix.

null set. Since as is easily seen, $M_\Theta^* = M_\Theta$, $\Omega_\Theta - M_\Theta$ is a null set, hence by (1), (2), $\mu(M) = \mu(\Omega)$. ——

Since $T_t^{(\alpha)}$ is metric transitive, we have by Birkhoff's ergodic theorem (c. f. Appendix), the following theorem.

Theorem XI. 27. *Let $f(P)$ be a bounded measurable function in Ω, then for almost all $P \in \Omega$,*

$$\lim_{L\to\infty} \frac{1}{L}\int_0^L f(P_t)dt = \frac{1}{\mu(\Omega)}\int_\Omega f(P)d\mu(P) .$$

3. Let G be a group of translations on the xy-plane: $x' = x+m$, $y' = y+n$ $(m, n = 0, \pm 1, \pm 2, \cdots)$, then D_0: $0 \leqq x < 1$, $0 \leqq y < 1$ is its fundamental domain. Let E be a set in D_0, which is measurable in Jordan's sence and E_ν $(\nu = 0, 1, 2, \cdots)$ be its equivalents by G and $E^* = \sum_{\nu=0}^{\infty} E_\nu$. Let l: $x = a + \theta_1 r$, $y = b + \theta_2 r$ $(0 \leqq r < \infty)$ be a half-line, where r denotes the length and l_r be its part, which corresponds to $[0, r]$ on the r-axis and $|E_r^*(\theta_1, \theta_2)|$ be the linear measure of the common part of E^* and l_r. Then Weyl [1] proved that if θ_2/θ_1 is irrational, then $\lim_{r\to\infty} |E_r^*(\theta_1, \theta_2)|/r = mE$.

We shall prove an analogue of this theorem for Fuchsian groups as follows.

Theorem XI. 28. *Let G be a Fuchsian group, such that $\sigma(D_0) < \infty$.*

Let M be a set in D_0, which is measurable in Jordan's sense and M_ν be its equivalents and $M^* = \sum_{\nu=0}^{\infty} M_\nu$.

Let $l = l(e^{i\theta}; \alpha)$ $(0 < \alpha < \pi)$ be a line through $e^{i\theta}$, making an angle α with the positive tangent of $|z| = 1$ at $e^{i\theta}$ and l_L be a segment on l of non-euclidean length L, measured from a fixed point of l.

Let $|M_L^*(e^{i\theta}; \alpha)|$ be the non-euclidean linear measure of the common part $M_L^*(e^{i\theta}; \alpha)$ of M^* and l_L.

Then there exists a set E of measure 2π on $|z| = 1$, such that if $e^{i\theta} \in E$, then for any M and α $(0 < \alpha < \pi)$,

$$\lim_{L\to\infty} |M_L^*(e^{i\theta}; \alpha)|/L = \sigma(M)/\sigma(D_0) .$$

Tsuji [47].

We remark that $|M_L^*(e^{i\theta}; \alpha)|$ is the non-euclidean linear measure of the common part of M and the equivalent of l_L in D_0.

Proof. Let Δ: $r_1 \leqq x \leqq r_2$, $r_3 \leqq y \leqq r_4$ be a rectangle, contained in D_0, where r_1, r_2, r_3, r_4 are rational numbers. We call such a rectangle a rational rectangle. Let $\{\Delta_\nu\}$ be its equivalents by G and put $\Delta^* = \sum_\nu \Delta_\nu$. Let Ω be defined as before. We associate directions φ $(0 \leqq \varphi < 2\pi)$ at every point z of Δ, then the line elements $\{z, \varphi\}$ constitute a sub-space

Ω_Δ of Ω, whose volume is $\mu(\Omega_\Delta)=2\pi\sigma(\varDelta)$.

Let $f(P)$ be the characteristic function of Ω_Δ, then by Theorem XI. 27, if $P=(z,\varphi)$ does not belong to a certain null set $N^{(\alpha)}(r_1, r_2, r_3, r_4)$ in Ω, then

$$\lim_{L\to\infty}\frac{1}{L}\int_0^L f(P_t)dt=\mu(\Omega_\Delta)/\mu(\Omega)=\sigma(\varDelta)/\sigma(D_0)\,. \tag{1}$$

Let $N^{(\alpha)}=\sum_{r_1,r_2,r_3,r_4} N^{(\alpha)}(r_1, r_2, r_3, r_4)$, added for all rationals r_1, r_2, r_3, r_4, then $N^{(\alpha)}$ is a null set in Ω, which depends on α. If $P=(z,\varphi)$ does not belong to $N^{(\alpha)}$, then (1) holds for any rational rectangle $\varDelta$ in D_0.

Let $C^{(\alpha)}=C^{(\alpha)}(z,\varphi)=C^{(\alpha)}(\eta_1,\eta_2)$ be the circular arc, defined before, which touches the direction φ at z and meets $|z|=1$ at an angle α $(0<\alpha<\pi)$ at $\eta_2=e^{i\theta_2}$ and $C_L^{(\alpha)}$ be its arc of non-euclidean length L, measured from the middle point z_0 of $\widehat{\eta_1\eta_2}$ and $|\varDelta^*\cdot C_L^{(\alpha)}|$ be the non-euclidean linear measure of the common part $\varDelta^*\cdot C_L^{(\alpha)}$ of $\varDelta^*$ and $C_L^{(\alpha)}$, then

$$\int_0^L f(P_t)dt=|\varDelta^*\cdot C_L^{(\alpha)}|\,, \tag{2}$$

so that if $P=(z,\varphi)$ does not belong to $N^{(\alpha)}$, then for any rational rectangle $\varDelta$ in D_0,

$$\lim_{L\to\infty}|\varDelta^*\cdot C_L^{(\alpha)}|/L=\sigma(\varDelta)/\sigma(D_0)\,. \tag{3}$$

Let $\varDelta$ be any rectangle in D_0, whose sides are parallel to the coordinates axes, then for any $\varepsilon>0$, we can find two rational rectangles $\varDelta_1$, $\varDelta_2$, such that $\varDelta_1\subset\varDelta\subset\varDelta_2$, $\sigma(\varDelta_2-\varDelta_1)<\varepsilon$.

Let l be a line through $\eta_2=e^{i\theta_2}$, which touches $C^{(\alpha)}(\eta_1,\eta_2)$ at η_2.

If $z\in l$, $\zeta\in C^{(\alpha)}$, such that $|z-\eta_2|=|\zeta-\eta_2|$ and $1-|z|<\delta$, then

$$(1-\varepsilon)|dz|\leqq|d\zeta|\leqq(1+\varepsilon)|dz|\,, \tag{4}$$

where $\varepsilon\to0$ with $\delta\to0$.

Since (3) holds for $\varDelta_1$, $\varDelta_2$, if $|\varDelta_L^*(e^{i\theta_2};\alpha)|$ be the non-enclidean linear measure of the common part $\varDelta^*\cdot l_L$ of $\varDelta^*$ and l_L, then we have from (4),

$$(1-\eta)\sigma(\varDelta)/\sigma(D_0)\leqq\overline{\lim_{L\to\infty}}|\varDelta_L^*(e^{i\theta_2};\alpha)|/L\leqq(1+\eta)\sigma(\varDelta)/\sigma(D_0)\,,$$

where $\eta\to0$ with $\varepsilon\to0$. Hence if $P(z,\varphi)$ does not belong to $N^{(\alpha)}$, then for any rectangle $\varDelta$ in D_0, whose sides are parallel to the coordinates axes,

$$\lim_{L\to\infty}|\varDelta_L^*(e^{i\theta_2};\alpha)|/L=\sigma(\varDelta)/\sigma(D_0)\,. \tag{5}$$

Let Ω_Θ be the projection of Ω on the torus Θ: $0\leqq\theta_1\leqq2\pi$, $0\leqq\theta_2\leqq2\pi$ and $\{\Omega_\Theta^{(\nu)}\}$ be its equivalents by $\mathfrak{G}=G\times G$ and $\Omega_\Theta^*=\sum_\nu\Omega_\Theta^{(\nu)}$, then by

Hopf's theorem, Ω_Θ^* is of measure $4\pi^2$. Let $N_\Theta^{(\alpha)}$ be the projection of $N^{(\alpha)}$ on Θ, then since $N^{(\alpha)}$ consists of points $(e^{i\theta_1}, e^{i\theta_2}, s)$, such that $(\theta_1, \theta_2)\in N_\Theta^{(\alpha)}$, $s_1(\theta_1, \theta_2)\leqq s\leqq s_2(\theta_1, \theta_2)$, $N_\Theta^{(\alpha)}$ is a null set on Θ. Let $N_\Theta^{(\alpha)*}$ be the sum of its equivalents by $\mathfrak{G}$, then $N_\Theta^{(\alpha)*}$ is a null set, so that if we put $N_\Theta^*=\sum_\alpha N_\Theta^{(\alpha)*}$, added for all rational numbers α $(0<\alpha<\pi)$, then N_Θ^* is a null set on Θ.

Hence there exists θ_1^0, such that on the line $\theta_1=\text{const.}=\theta_1^0$, there exists a set E of measure 2π, which belongs to $\Omega_\Theta^*-N_\Theta^*$. If $\theta_2\in E$, (θ_1^0, θ_2) does not belong to N_Θ^*, so that if $\theta_2\in E$, then (5) holds for any rectangle Δ in D_0, whose sides are parallel to the coordinates axes and for any rational α $(0<\alpha<\pi)$.

We shall prove that if $\theta_2\in E$, then (5) holds for any irrational α in $(0, \pi)$. Let α be an irrational number in $(0, \pi)$, then for any $\varepsilon>0$, we choose a rational number α', such that $|\alpha-\alpha'|<\varepsilon$ and let l' be a line through $e^{i\theta_2}$, making an angle α' with the positive tangent of $|z|=1$ at $e^{i\theta_2}$.

Let Δ be any rectangle in D_0, whose sides are parallel to the coordinates axes, then we choose two rational rectangles Δ_1, Δ_2, such that $\Delta_1\subset\Delta\subset\Delta_2$, $\sigma(\Delta_2-\Delta_1)<\varepsilon$. Since (5) holds for the pair (Δ_1, α'), (Δ_2, α'), we have easily

$$(1-\eta)\,\sigma(\Delta)/\sigma(D_0)\leqq\varlimsup_{L\to\infty}|\Delta_L^*(e^{i\theta_2};\alpha)|/L\leqq(1+\eta)\,\sigma(\Delta)/\sigma(D_0)\ ,$$

where $\eta\to0$ with $\varepsilon\to0$, so that if $\theta_2\in E$, then

$$\lim_{L\to\infty}|\Delta_L^*(e^{i\theta_2};\alpha)|/L=\sigma(\Delta)/\sigma(D_0)\ , \tag{C}$$

for any rectangle Δ in D_0, whose sides are parallel to the coordinates axes and for any α $(0<\alpha<\pi)$.

Let M be a set in D_0, which is measurable in Jordan's sense, then for any $\varepsilon>0$, we choose two sets M_1, M_2, such that $M_1\subset M\subset M_2$, $\sigma(M_2-M_1)<\varepsilon$, where M_1 and M_2 consist of a finite number of non-overlapping rectangles in D_0, whose sides are parallel to the coordinates axes.

Since (6) holds for M_1 and M_2 and $\varepsilon>0$ is arbitrary, we see that (6) holds for M, such that if $\theta_2\in E$, then for any α $(0<\alpha<\pi)$.

$$\lim_{L\to\infty}|M_L^*(e^{i\theta_2};\alpha)|/L=\sigma(M)/\sigma(D_0)\ . \tag{7}$$

10. Fundamental Theorem

1. Let G be a Fuchsian group and D_0 be its fundamental domain and $n(r, a)$ be the number of equivalents a_n of $a\in D_0$ in $|z|<r<1$ and put

$$N(r, a)=\int_{\frac{1}{2}}^{r} \frac{n(r, a)}{r} dr\,, \qquad \frac{1}{2} \leqq r<1\,. \tag{1}$$

If $\sigma(D_0)<\infty$, then we put

$$T(r)=\int_0^r \frac{S(r)}{r} dr=\frac{1}{\sigma(D_0)}\iint_{D_0} N(r, a)\, d\sigma(a)+O(1)\,, \tag{2}$$

where

$$S(r)=A(r)/\sigma(D_0)\,, \qquad A(r)=\iint_{D_0} n(r, a)\, d\sigma(a)\,.$$

If $\sigma(D_0)=\infty$, let $D_0(\rho)$ be the part of D_0, contained in $|z|<\rho<1$ and put

$$T_\rho(r)=\int_0^r \frac{S_\rho(r)}{r} dr=\frac{1}{\sigma(D_0(\rho))}\iint_{D_0(\rho)} N(r, a)\, d\sigma(a)+O(1)\,, \tag{2*}$$

where

$$S_\rho(r)=A_\rho(r)/\sigma(D_0(\rho))\,, \qquad A_\rho(r)=\iint_{D_0(\rho)} n(r, a)\, d\sigma(a)\,.$$

Then we shall prove the following analogue of Nevanlinna's first fundamental theorem.

Theorem XI. 29. (Fundamental Theorem).

(i) *If* $\sigma(D_0)<\infty$, *then*

$$N(r, a)=T(r)+O(1)\,, \qquad a\in D_0\,,$$

where $O(1)$ *is uniformly bounded for* $a\in D_0$, $1/2\leqq r<1$ *and*

$$T(r)=\frac{2\pi}{\sigma(D_0)} \log \frac{1}{1-r}+O(1)\,.$$

(ii) *If* $\sigma(D_0)=\infty$, *then*

$$N(r, a)=T_\rho(r)+O(1)\,, \qquad a\in D_0(\rho)\,,$$

where $O(1)$ *is uniformly bounded for* $a\in D_0(\rho)$, $1/2\leqq r<1$ *and*

$$T_\rho(r)=o\left(\log \frac{1}{1-r}\right), \qquad r\to 1\,.$$

Tsuji [26], [49].

We call $T(r)$, $T_\rho(r)$ *the characteristic function of* G, which is an increasing convex function of $\log r$.

Proof. By Theorem X. 30. Cor., on any Riemann surface F, there exists a potential function $u(z; a, b)$, which is harmonic, except at a, b, where $u-\log \frac{1}{|z-a|}$ is harmonic at $z=a$ and $u+\log \frac{1}{|z-b|}$ is harmonic at $z=b$. Let $\Omega\subset\Omega_0$ be two compact sub-surfaces of F, such that $\bar{\Omega}\subset\Omega_0$ and b be a fixed point of Ω, whose neighbourhood $U(b)$ is contained in Ω and a be a variable point in $\Omega-U(b)$, hence there exists $\rho_0>0$, such that ρ_0-neighbourhood $U(a)$ of a is contained in Ω_0. Then

$$|u(z; a, b)|\leqq K \qquad \text{on} \quad F-U(a)-U(b) , \tag{1}$$

where K is a constant, which is independent of $a\in\Omega-U(b)$.

Let $F=F_G$ be the Riemann surface, associated to G and $u(z; a, b)$ $(a\in D_0, b\in D_0)$ be that potential function on F, then $u(z; a, b)$ is invariant by G and is harmonic in $|z|<1$, except at equivalents a_ν, b_ν of a, b, where $u-\log\dfrac{1}{|z-a_\nu|}$ is harmonic at $z=a_\nu$ and $u+\log\dfrac{1}{|z-b_\nu|}$ is harmonic at $z=b_\nu$ $(\nu=0, 1, 2, \cdots)$.

Let $b\in D_0(\rho)$ and $U(b)$: $|z-b|<\delta$ be its neighbourhood, contained in $D_0(\rho)$. We consider b as fixed and a be a variable point in $D_0(\rho)-U(b)$. We assume that there are no a_ν, b_ν on the circles $|z|=1/2$ and $|z|=r<1$.

Applying the Green's formula: $\displaystyle\int_C\left(u\frac{\partial v}{\partial\nu}-v\frac{\partial u}{\partial\nu}\right)ds=0$ to $u=u(z; a, b)$, $v=\log r/|z|$ for the domain, bounded by the circles $|z|=1/2$ and $|z|=r$ and small circles about a_ν, b_ν, contained in $1/2<|z|<r$ and then making the radii of these circles tend to zero, we have

$$\frac{1}{2\pi}\int_0^{2\pi}u(re^{i\theta}; a, b)d\theta+\sum_{1/2<|a_\nu|<r}\log r/|a_\nu|-\sum_{1/2<|b_\nu|<r}\log r/|b_\nu|$$
$$+\frac{1}{2\pi}\int_{|z|=1/2}\left(u\frac{\partial v}{\partial\nu}-v\frac{\partial u}{\partial\nu}\right)ds=0 ,$$

where ν is the inner normal with respect to the domain.

In the following, const. denotes a constant, which is independent of $a\in D_0(\rho)-U(b)$, $1/2\leqq r<1$ and $O(1)$ denotes a quantity, such that $|O(1)|\leqq$const., then from (1), we have

$$\frac{1}{2\pi}\int_0^{2\pi}u(re^{i\theta}; a, b)d\theta+N(r, a)=N(r, b)+O(1) . \tag{2}$$

We shall prove that

$$\int_0^{2\pi}|u|d\theta\leqq\text{const.}$$

We assumed that a_ν, b_ν do not lie on $|z|=r$ or on $|z|=1/2$, but we easily that (2) holds, if a_ν, b_ν lie on $|z|=r$, or on $|z|=1/2$.

Let $a\in D_0(\rho)-U(b)$ and $U'(a): |z-a|<\delta/4$ be its neighbourhood, where we take $\delta>0$ so small that the equivalents $\{U'_\nu(a)\}$ of $U'(a)$ do not overlap for any $a\in D_0(\rho)-U(b)$.

Let $U'(b): |z-b|<\delta/4$ and $\{U'_\nu(b)\}$ be its equivalents, then by (1)

$$|u(z; a, b)|\leqq\text{const.} \quad \text{outside of } \sum_\nu U'_\nu(a)+\sum_\nu U'_\nu(b) . \tag{3}$$

The part of the circle $|z|=r$, contained in $\sum_\nu U'_\nu(a)$ consists of a finite number of arcs θ_i $(i=1,2,\cdots,N)$ and that in $\sum_\nu U'_\nu(b)$ consists of a finite number of arcs θ'_j $(j=1,2,\cdots,N')$. Hence by (3), if we integrate $|u|$ on $|z|=r$, then

$$\int_0^{2\pi}|u|d\theta \leqq \sum_{i=1}^{N}\int_{\theta_i}|u|d\theta+\sum_{j=1}^{N'}\int_{\theta'_j}|u|d\theta+\text{const.}. \tag{4}$$

Now

$$\int_{\theta_i}|u|d\theta=\frac{1-r^2}{r}\int_{\theta_i}|u|\frac{|dz|}{1-|z|^2}\leqq 3(1-r)\int_{\theta_i}\frac{|u||dz|}{1-|z|^2} \quad (1/2\leqq r<1)\,. \tag{5}$$

Let θ_i^* be the equivalent of θ_i in D_0, then θ_i^* is a circular arc in $U'(a)$ and since $\dfrac{|u||dz|}{1-|z|^2}$ is invariant by G,

$$\int_{\theta_i}\frac{|u||dz|}{1-|z|^2}=\int_{\theta_i^*}\frac{|u||dz|}{1-|z|^2}\,.$$

Since in $U'(a)$, $|u|\leqq\log\dfrac{1}{|z-a|}+\text{const.}$, we have

$$\int_{\theta_i^*}\frac{|u||dz|}{1-|z|^2}\leqq\text{const.}\,, \tag{6}$$

so that by (5), (6),

$$\sum_{i=1}^{N}\int_{\theta_i}|u|d\theta\leqq\text{const.}\,N(1-r)\,.$$

Similarly

$$\sum_{j=1}^{N'}\int_{\theta'_j}|u|d\theta\leqq\text{const.}\,N'(1-r)\,,$$

so that by (4),

$$\int_0^{2\pi}|u|d\theta\leqq\text{const.}\,N(1-r)+\text{const.}\,N'(1-r)+\text{const.}\,. \tag{7}$$

If $U'_\nu(a)$ intersects the circle $|z|=r$ and $a_\nu\in U'_\nu(a)$, then a_ν lies in $|z|\leqq r'$ $(r<r')$, such that $1-r\leqq\text{const.}\,(1-r')$, hence $N\leqq n(r',a)\leqq \text{const.}/1-r'\leqq\text{const.}/1-r$ (Theorem XI. 7), so that $N(1-r)\leqq\text{const.}$.

Similarly $N'(1-r)\leqq\text{const.}$, hence

$$\int_0^{2\pi}|u|d\theta\leqq\text{const.}\,.$$

Hence by (2),

$$N(r,a)=N(r,b)+O(1)\,,\qquad a\in D_0(\rho)-U(b)\,. \tag{8}$$

If $a\in U(b)$, then we choose $a_0\in D_0(\rho)$, such that $|a_0-a|\geqq 2\delta$ and take a_0 instead of b, then by (8),

$$N(r,a)=N(r,a_0)+O(1)\,,\qquad a\in U(b)\,.$$

Since by (8), $N(r, a_0)=N(r, b)+O(1)$, we have

$$N(r, a)=N(r, b)+O(1)\,, \qquad a\in U(b)\,, \tag{9}$$

hence by (8), (9),

$$N(r, a)=N(r, b)+O(1)\,, \qquad a\in D_0(\rho)\,. \tag{10}$$

Hence we have

$$\iint\limits_{D_0(\rho)} N(r, a)\, d\sigma(a)=\sigma(D_0(\rho))N(r, b)+O(1)\,,$$

so that

$$N(r, b)=\frac{1}{\sigma(D_0(\rho))}\iint\limits_{D_0(\rho)} N(r, a)\, d\sigma(a)+O(1)=T_\rho(r)+O(1)\,. \tag{11}$$

Hence by (10),

$$N(r, a)=T_\rho(r)+O(1)\,, \qquad a\in D_0(\rho)\,. \tag{12}$$

If $\sigma(D_0)<\infty$, then by Theorem XI. 12, the Riemann surface $F=F_G$ is a closed Riemann surface, or is obtained from a closed Riemann surface, by taking off a finite number of points, so that $u(z; a, b)$ is harmonic at each of such points, hence we may take $\rho=1$ in (12), so that

$$N(r, a)=T(r)+O(1)\,, \qquad a\in D_0\,, \tag{13}$$

where

$$T(r)=\frac{1}{\sigma(D_0)}\iint\limits_{D_0} N(r, a)\, d\sigma(a)\,. \tag{14}$$

Since

$$\iint\limits_{D_0} n(r, a)\, d\sigma(a)=4\iint\limits_{|z|\leq r}\frac{r\, dr\, d\theta}{(1-r^2)^2}=\frac{4\pi r^2}{1-r^2}\,,$$

$$T(r)=\frac{1}{\sigma(D_0)}\int_0^r\frac{dr}{r}\iint\limits_{D_0} n(r, a)\, d\sigma(a)=\frac{2\pi}{\sigma(D_0)}\log\frac{1}{1-r}+O(1)\,. \tag{15}$$

Next suppose that $\sigma(D_0)=\infty$, then

$$T_\rho(r)\leqq\frac{1}{\sigma(D_0(\rho))}\int_0^r\frac{dr}{r}\iint\limits_{D_0} n(r, a)\, d\sigma(a)=\frac{2\pi}{\sigma(D_0(\rho))}\log\frac{1}{1-r}+O(1)\,.$$

Hence from $N(r, 0)=T_\rho(r)+O(1)$ and $\lim\limits_{\rho\to 1}\sigma(D_0(\rho))=\infty$, we have $\lim\limits_{r\to 1} N(r, 0)/\log\dfrac{1}{1-r}=0$, so that from $N(r, 0)=T_\rho(r)+O(1)$, we have

$$T_\rho(r)=o\left(\log\frac{1}{1-r}\right)\,, \qquad r\to 1\,. \tag{16}$$

11. Some consequences from the Fundamental Theorem

1. Let E be a measurable set of positive measure in $D_0(\rho)$ and put

$$T(r; E)=\int_0^r \frac{S(r; E)}{r} dr , \quad 0<r<1 , \tag{1}$$

where

$$S(r; E)=A(r; E)/\sigma(E) , \quad A(r; E)=\iint_E n(r, a)\, d\sigma(a) , \tag{2}$$

so that

$$T(r; E)=\frac{1}{\sigma(E)}\iint_E N(r, a)\, d\sigma(a)+O(1). \tag{3}$$

Hence from the fundamental theorem, we have

Theorem XI. 30. $T(r; E)=T_\rho(r)+O(1)$,
where $T_\rho(r)$ *is replaced by* $T(r)$, *if* $\sigma(D_0)<\infty$.

More generally, let $\mu\geqq 0$ be a positive mass distribution in $|z|<1$, which is invariant by G and put

$$T(r; E, \mu)=\int_0^r \frac{S(r; E, \mu)}{r} dr , \tag{1}$$

where

$$S(r; E, \mu)=A(r; E, \mu)/\mu(E) , \quad A(r; E, \mu)=\iint_E n(r, a)\, d\mu(a) , \tag{2}$$

so that

$$T(r; E, \mu)=\frac{1}{\mu(E)}\iint_E N(r, a) d\mu(a)+O(1) . \tag{3}$$

Theorem XI. 31. $T(r; E, \mu)=T_\rho(r)+O(1)$,
where $T_\rho(r)$ *is replaced by* $T(r)$, *if* $\sigma(D_0)<\infty$.

As an application of Theorem XI. 31, we shall prove

Theorem XI. 32. *Let F be a closed Riemann surface of genus $p\geqq 2$, spread over the w-sphere Ω, which consists of N sheets. We map the universal covering surface $F^{(\infty)}$ of F conformally on $|z|<1$ by $w=w(z)$ and let $T(w, r)$ be the Nevanlinna's characteristic function of $w(z)$, then*

$$T(w, r)=\frac{N}{2(p-1)} \log \frac{1}{1-r}+O(1) , \quad 0\leqq r<1 . \qquad \text{Tsuji [26].}$$

Proof. $w(z)$ is invariant by a certain Fuchsian group G, whose fundamental domain D_0 is bounded by a finite number of orthogonal circles to $|z|=1$ and by Theorem XI. 11, $\sigma(D_0)=4\pi(p-1)$.

Let $d\omega(w)$ be the area-element of Ω at w and we define a positive mass distribution $d\mu(z)$ in $|z|<1$ by $d\mu(z)=d\omega(w)$, then $d\mu(z)$ is invariant by G. Let K be a spherical disc on Ω, then since F consists of N sheets, there are N discs on F above K, so that the part of F, which lies above K is mapped on N domains $\Delta_1, \cdots, \Delta_N$ in D_0, such that $\mu(\Delta_i)=\omega(K)$ $(i=1, 2, \cdots, N)$.

Let α be any point of K and $\alpha_1, \cdots, \alpha_N$ be N points of F, which lie above α and let α_i be mapped on a_i in D_0 $(i=1, 2, \cdots, N)$.

Let $\mathfrak{n}(r, \alpha)$ be the number of zero points of $w(z)-\alpha$ in $|z|<r<1$, then

$$\iint_K \mathfrak{n}(r, \alpha)\, d\omega(\alpha)=\sum_{i=1}^{N} A(r; \Delta_i, \mu)\,, \tag{1}$$

where

$$A(r; \Delta_i, \mu)=\iint_{\Delta_i} n(r, a)\, d\mu(a)\,, \tag{2}$$

$n(r, a)$ being the number of equivalents of a in $|z|<r<1$, so that

$$\frac{1}{\omega(K)}\int_0^r \frac{dr}{r}\iint_K \mathfrak{n}(r, \alpha)\, d\omega(\alpha)=\int_0^r\left(\sum_{i=1}^{N}\frac{A(r; \Delta_i, \mu)}{\mu(\Delta_i)}\right)\frac{dr}{r}\,. \tag{3}$$

By Nevanlinna's first fundamental theorem, the left hand side of (3) is $T(w, r)+O(1)$ and the right hand side is, by Theorem XI. 31,

$$NT(r)+O(1)=\frac{2\pi N}{\sigma(D_0)}\log\frac{1}{1-r}+O(1)=\frac{N}{2(p-1)}\log\frac{1}{1-r}+O(1)\,,$$

so that

$$T(w, r)=\frac{N}{2(p-1)}\log\frac{1}{1-r}+O(1)\,. \quad \text{——}$$

2. Theorem XI. 30 concerns the surface measure. Next we shall prove theorems concerning the linear measure.

Theorem XI. 33. *Suppose that $\sigma(D_0)<\infty$ and let E be a measurable set of positive measure in D_0 and $\{E_\nu\}$ be its equivalents and $E^*=\sum_{\nu=0}^{\infty} E_\nu$. Let $L(r, E^*)$ be the non-euclidean linear measure of the intersection of the circle $|z|=r$ and E^* and put*

$$\Lambda(r; E)=\frac{1}{\sigma(E)}\int_0^r \frac{L(r; E^*)}{r}\,dr\,,$$

then

$$\lim_{r\to 1} \Lambda(r; E)/T(r)=1\,.$$

Proof. Let $\varphi(z)$ be the characteristic function of E^*, then

$$L(r;E^*)=2\int_0^{2\pi}\frac{\varphi(re^{i\theta})r\,d\theta}{1-r^2}\,, \tag{1}$$

$$S(r;E)=\frac{A(r;E)}{\sigma(E)}=\frac{4}{\sigma(E)}\int_0^r\int_0^{2\pi}\frac{\varphi(te^{i\theta})t\,dt\,d\theta}{(1-t^2)^2}=\frac{2}{\sigma(E)}\int_0^r\frac{L(t;E^*)dt}{1-t^2}$$
$$=2\int_0^r\frac{t\,d\Lambda(t;E)}{1-t^2}\,. \tag{2}$$

Hence by Theorem XI. 30,

$$T(r)+O(1)=T(r;E)=\int_0^r\frac{S(\tau;E)}{\tau}d\tau=2\int_0^r\frac{d\tau}{\tau}\int_0^\tau\frac{t\,d\Lambda(t;E)}{1-t^2}\,. \tag{3}$$

Now

$$2\int_0^r\frac{d\tau}{\tau}\int_0^\tau\frac{t\,d\Lambda(t;E)}{1-t^2}$$
$$\leqq 2\int_0^r\frac{d\tau}{\tau}\int_0^\tau\frac{\tau\,d\Lambda(t;E)}{1-t^2}=2\int_0^r d\tau\int_0^\tau\frac{d\Lambda(t;E)}{1-t^2}=2\int_0^r\frac{(r-t)\,d\Lambda(t;E)}{1-t^2}$$
$$=-2\int_0^r\Lambda(t;E)\frac{d}{dt}\left(\frac{r-t}{1-t^2}\right)dt=\int_0^r\Lambda(t;E)\frac{d}{dt}\left(\frac{1-r}{1-t}-\frac{1+r}{1+t}\right)dt$$
$$=(1-r)\int_0^r\frac{\Lambda(t;E)}{(1-t)^2}dt+O(1)\int_0^r\Lambda(t;E)dt$$
$$\leqq(1-r)\Lambda(r;E)\int_0^r\frac{dt}{(1-t)^2}+O(1)\int_0^r\Lambda(t;E)dt\leqq\Lambda(r;E)+O(1)\int_0^r\Lambda(t;E)dt\,. \tag{4}$$

Since $L(r;E^*)=O(1/1-r)$, we have $\Lambda(r;E)=O\left(\log\frac{1}{1-r}\right)$, so that

$$\int_0^r\Lambda(t;E)dt=O(1)\,,$$

hence

$$2\int_0^r\frac{d\tau}{\tau}\int_0^\tau\frac{t\,d\Lambda(t;E)}{1-t^2}\leqq\Lambda(r;E)+O(1)\,. \tag{5}$$

We choose $r'<r$, such that $1-r'=(1-r)T(r)$, then since

$$L(r;E^*)=O\left(\frac{1}{1-r}\right),\quad \Lambda(r;E)-\Lambda(r';E)=O\left(\log\frac{1-r'}{1-r}\right)=O(\log T(r))\,,$$

so that

$$2\int_0^r\frac{d\tau}{\tau}\int_0^\tau\frac{t\,d\Lambda(t;E)}{1-t^2}\geqq 2\int_0^r d\tau\int_0^\tau\frac{t\,d\Lambda(t;E)}{1-t^2}=2\int_0^r\frac{t(r-t)\,d\Lambda(t;E)}{1-t^2}$$
$$=-2\int_0^r\Lambda(t;E)\frac{d}{dt}\left(\frac{t(r-t)}{1-t^2}\right)dt=\int_0^r\Lambda(t;E)\left(\frac{1-r}{(1-t)^2}-\frac{1+r}{(1+t)^2}\right)dt$$
$$=(1-r)\int_0^r\frac{\Lambda(t;E)}{(1-t)^2}dt+O(1)\geqq(1-r)\int_{r'}^r\frac{\Lambda(t;E)}{(1-t)^2}dt+O(1)$$
$$\geqq(1-r)\Lambda(r':E)\left(\frac{1}{1-r}-\frac{1}{1-r'}\right)+O(1)$$

$$=(A(r;E)-O(\log T(r))\left(1-\frac{1}{T(r)}\right)+O(1)\,. \tag{6}$$

Hence by (3), (5), (6),

$$(A(r;E)-O(\log T(r))\left(1-\frac{1}{T(r)}\right)+O(1)\leqq T(r)\leqq A(r;E)+O(1)\,, \tag{7}$$

so that

$$\lim_{r\to1}\frac{A(r;E)}{T(r)}=1.$$

3. Let Θ: $\alpha<\arg z<\beta$ ($|\Theta|=\beta-\alpha$) be an angular domain and Θ_r be its part, contained in $|z|<r<1$ and $n(r, a; \Theta)$ be the number of the equivalents a_ν of $a\in D_0$, contained in Θ_r and put

$$N(r, a; \Theta)=\int_{\frac{1}{2}}^{r}\frac{n(r, a; \Theta)}{r}dr\,, \qquad \frac{1}{2}\leqq r<1\,.$$

Theorem XI. 34. *If* $\sigma(D_0)<\infty$, *then*

$$\lim_{r\to1}\frac{N(r, a; \Theta)}{T(r)}=\frac{|\Theta|}{2\pi}\,.$$

Hence $\{\arg a_\nu\}$ are distributed uniformly in Weyl's sense (in somewhat modified form).

Proof. Let Δ be a circular disc, contained in D_0 and $\{\Delta_\nu\}$ be its equivalents and $\Delta^*=\sum_{\nu=0}^{\infty}\Delta_\nu$. Let $l_r(\theta)$: $z=te^{i\theta}$ $(0\leqq t\leqq r<1)$ be a segment and $|\Delta_r^*(\theta)|$ be the non-euclidean linear measure of the intersection of Δ^* and $l_r(\theta)$ and $L(r)$ be the non-euclidean length of $l_r(\theta)$. Then by Theorem XI. 28, for almost all θ,

$$\lim_{r\to1}\frac{|\Delta_r^*(\theta)|}{L(r)}=\frac{\sigma(\Delta)}{\sigma(D_0)}\,. \tag{1}$$

We put

$$L(r;\Delta,\Theta)=\int_\alpha^\beta|\Delta_r^*(\theta)|d\theta\,, \tag{2}$$

then since $|\Delta_r^*(\theta)|/L(r)\leqq1$, we have by Lebesgue's theorem and (1),

$$\lim_{r\to1}\frac{L(r;\Delta,\Theta)}{L(r)}=\int_\alpha^\beta\left(\lim_{r\to1}\frac{|\Delta_r^*(\theta)|}{L(r)}\right)d\theta=\frac{\sigma(\Delta)}{\sigma(D_0)}|\Theta|\,.$$

Since $L(r)\sim\log\dfrac{1}{1-r}\sim\dfrac{\sigma(D_0)}{2\pi}T(r)$,

$$\lim_{r\to1}\frac{L(r;\Delta,\Theta)}{T(r)}=\frac{|\Theta|}{2\pi}\sigma(\Delta)\,. \tag{3}$$

Let $\varphi(z)$ be the characteristic function of Δ^*, then

$$|\Delta_r^*(\theta)|=2\int_0^r \frac{\varphi(te^{i\theta})dt}{1-t^2}\,,\quad L(r;\,\Delta,\,\Theta)=2\int_0^r\int_\alpha^\beta \frac{\varphi(te^{i\theta})dt\,d\theta}{1-t^2}\,. \tag{4}$$

We put

$$A(r;\,\Delta,\,\Theta)=\iint_\Delta n(r,\,a;\,\Theta)\,d\sigma(a)=4\int_0^r\int_\alpha^\beta \frac{\varphi(te^{i\theta})t\,dt\,d\theta}{(1-t^2)^2}=2\int_0^r \frac{t\,dL(t;\,\Delta,\,\Theta)}{1-t^2}\,, \tag{5}$$

$$S(r;\,\Delta,\,\Theta)=A(r;\,\Delta,\,\Theta)/\sigma(\Delta)\,,\qquad T(r;\,\Delta,\,\Theta)=\int_0^r \frac{S(r;\,\Delta,\,\Theta)}{r}dr\,, \tag{6}$$

then

$$T(r;\,\Delta,\,\Theta)=\frac{2}{\sigma(\Delta)}\int_0^r \frac{d\tau}{\tau}\int_0^\tau \frac{t\,dL(t;\,\Delta,\,\Theta)}{1-t^2}\,. \tag{7}$$

From this, we have similarly as Theorem XI. 33,

$$\lim_{r\to 1} L(r;\,\Delta,\,\Theta)/T(r;\,\Delta,\,\Theta)=\sigma(\Delta)\,,$$

so that by (3),

$$\lim_{r\to 1} T(r;\,\Delta,\,\Theta)/T(r)=|\Theta|/2\pi\,. \tag{8}$$

For any $\varepsilon>0$, we choose two angular domains Θ', Θ'', such that $\Theta'\subset\Theta\subset\Theta''$, $|\Theta''|-|\Theta'|<\varepsilon$ and for a given r $(1/2\leqq r<1)$, we choose r_1, r_2, such that $r_1<r<r_2$, $1-r_1=(1+\varepsilon)(1-r)$, $1-r_2=(1-\varepsilon)(1-r)$,

$$r=\frac{r_1+\varepsilon}{1+\varepsilon}=\frac{r_2-\varepsilon}{1-\varepsilon}\,. \tag{9}$$

Let Δ: $\left|\dfrac{z-a}{1-\bar{a}z}\right|\leqq\rho<1$ and $\{a_\nu\}$ be the equivalents of a, such that $a_\nu\in\Delta_\nu$. We choose ρ so small, that if $a_\nu\in\Theta_r$, then $\Delta_\nu\subset\Theta''_{r_2}$ and if Δ_ν has common points with Θ'_{r_1}, then $a_\nu\in\Theta_r$, so that

$$S(r_1;\,\Delta,\,\Theta')\leqq n(r,\,a;\,\Theta)\leqq S(r_2;\,\Delta,\,\Theta'')\,. \tag{10}$$

Since $\dfrac{dr}{r}=\dfrac{dr_1}{r_1+\varepsilon}=\dfrac{dr_2}{r_2-\varepsilon}$ and $\dfrac{1-2\varepsilon}{t}\leqq\dfrac{1}{t+\varepsilon}$, $\dfrac{1}{t-\varepsilon}\leqq\dfrac{1+4\varepsilon}{t}$, if $1/2\leqq t<1$, $0<\varepsilon\leqq 1/4$, we have by (10),

$$(1-2\varepsilon)\int_0^{r_1}\frac{S(r_1;\,\Delta,\,\Theta')}{r_1}dr_1+O(1)\leqq N(r,\,a;\,\Theta)$$

$$\leqq(1+4\varepsilon)\int_0^{r_2}\frac{S(r_2;\,\Delta,\,\Theta'')}{r_2}dr_2+O(1)\,,$$

or

$$(1-2\varepsilon)T(r_1;\,\Delta,\,\Theta')+O(1)\leqq N(r,\,a;\,\Theta)\leqq(1+4\varepsilon)T(r_2;\,\Delta,\,\Theta'')+O(1)\,. \tag{11}$$

By (8),

$$T(r_1;\,\Delta,\,\Theta')\sim\frac{|\Theta'|}{2\pi}T(r_1)=\frac{|\Theta'|}{2\pi}\,\frac{2\pi}{\sigma(D_0)}\log\frac{1}{1-r_1}=\frac{|\Theta'|}{2\pi}\,\frac{2\pi}{\sigma(D_0)}\left(\log\frac{1}{1-r}+O(1)\right)$$

$$=\frac{|\Theta'|}{2\pi}T(r)+O(1)\,, \qquad T(r_1, \varDelta, \Theta')/T(r)\to|\Theta'|/2\pi \quad (r\to 1)\,.$$

Similarly

$$T(r_2, \varDelta, \Theta'')/T(r)\to|\Theta''|/2\pi \quad (r\to 1)\,.$$

Since $\varepsilon>0$ is arbitrary, we have from (11),

$$\lim_{r\to 1} N(r, a; \Theta)/T(r)=|\Theta|/2\pi\,.$$

12. Geometry of numbers for Fuchsian groups

1. The following theorem is an analogue for Fuchsian groups of Minkowski's well-known theorem.

Theorem XI. 35. *Suppose that $\sigma(D_0)<\infty$. Let E be a measurable set in $|z|<1$ and $\{E_\nu\}$ be its equivalents by G. If $\sigma(E)>\sigma(D_0)$, then $\{E_\nu\}$ overlap.*

Proof. Suppose that $\{E_\nu\}$ do not overlap, then

$$A(r; E)\leqq 4\iint_{|z|\leqq r}\frac{r\,dr\,d\theta}{(1-r^2)^2}=\frac{4\pi r^2}{1-r^2}\,,$$

so that

$$\int_0^r\frac{A(r; E)}{r}dr\leqq 2\pi\log\frac{1}{1-r}+O(1)\,.$$

Since by Theorem XI. 30,

$$\int_0^r\frac{A(r; E)}{r}dr=\sigma(E)T(r)+O(1)=\frac{2\pi\sigma(E)}{\sigma(D_0)}\log\frac{1}{1-r}+O(1)\,,$$

we have

$$\frac{2\pi\sigma(E)}{\sigma(D_0)}\log\frac{1}{1-r}\leqq 2\pi\log\frac{1}{1-r}+O(1)\,,$$

so that $\sigma(E)\leqq\sigma(D_0)$, hence if $\sigma(E)>\sigma(D_0)$, then $\{E_\nu\}$ overlap.

2. Let G be a group of translations on the xy-plane: $x'=x+m$, $y'=y+n$ $(m, n=0, \pm 1, \pm 2, \cdots)$, then the unit square $\varDelta_0$: $0\leqq x<1$, $0\leqq y<1$ is its fundamental domain. Let k points be given in $\varDelta_0$. We call the totality of equivalents of these k points by G the lattice points. Then Blichfeldt [1] proved the following theorem.

Theorem XI. 36. *Let E be a bounded measurable set on the xy-plane and mE be its measure. Then we can translate E into E', such that E' contains at least kmE lattice points. Similarly we can translate E into E'', such that E'' contains at most kmE lattice points.*

This theorem can be deduced easily from the following Theorem XI. 37. We translate E by $x'=x+u$, $y'=y+v$ $(u \geqq 0, v \geqq 0)$ and let $E(u, v)$ be the translated set and $\mu(E(u, v))$ be the number of lattice points, contained in $E(u, v)$. Suppose that E is contained in a square: $0 \leqq x \leqq N$, $0 \leqq y \leqq N$, where N is a positive integer. Then

Theorem XI. 37. $\dfrac{1}{N^2}\displaystyle\int_0^N\int_0^N \mu(E(u, v))du\,dv = kmE$.

Proof. We put a mass 1 at every lattice point, then we obtain a positive mass distribution $\mu \geqq 0$ on the xy-plane, so that the number of lattice points, contained in $E(u, v)$ is

$$\mu(E(u, v)) = \iint\limits_{E(u,v)} d\mu(x, y) \,. \tag{1}$$

Let $\varphi(x, y)$ be the characteristic function of E, then $\varphi(x-u, y-v)$ is that of $E(u, v)$, so that

$$\mu(E(u, v)) = \int_0^\infty\int_0^\infty \varphi(x-u, y-v)\, d\mu(x, y) \,,$$

hence

$$\int_0^T\int_0^T \mu(E(u, v))du\,dv =$$

$$\int_0^\infty\int_0^\infty d\mu(x, y)\int_0^T\int_0^T \varphi(x-u, y-v)du\,dv =$$

$$\int_0^\infty\int_0^\infty d\mu(x, y)\int_{x-T}^{x}\int_{y-T}^{y} \varphi(\xi, \eta)d\xi\, d\eta \,. \tag{2}$$

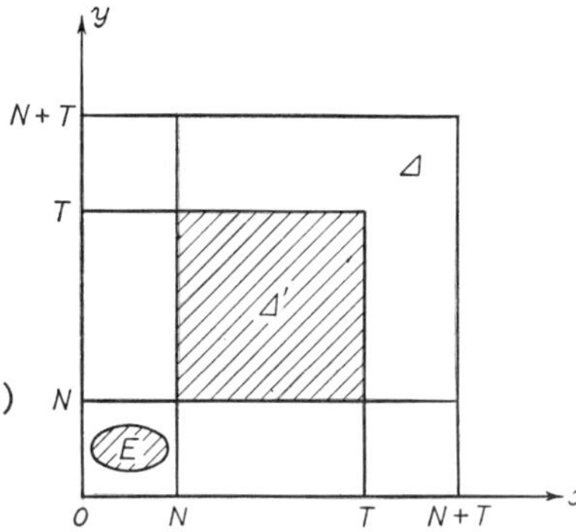

Let Δ: $0 \leqq x \leqq N+T$, $0 \leqq y \leqq N+T$ and Δ': $N \leqq x \leqq T$, $N \leqq y \leqq T$.

If (x, y) lies outside of Δ, then $\displaystyle\int_{x-T}^{x}\int_{y-T}^{y} \varphi(\xi, \eta)d\xi\, d\eta = 0$ and if (x, y) lies in Δ', then $\displaystyle\int_{x-T}^{x}\int_{y-T}^{y} \varphi(\xi, \eta)d\xi\, d\eta = mE$, so that

$$\int_0^T\int_0^T \mu(E(u, v))du\,dv = mE\iint\limits_{\Delta'} d\mu(x, y) + O\Big(\iint\limits_{\Delta-\Delta'} d\mu(x, y)\Big) = kmE\cdot T^2 + O(T) \,.$$

From this, we have

$$\lim_{T\to\infty} \frac{1}{T^2}\int_0^T\int_0^T \mu(E(u, v))du\,dv = kmE \,. \tag{3}$$

We assume that $T=nN$ $(n=1, 2, \cdots)$, then since $d\mu(x, y)$ is invariant, mod 1, we have

$$\frac{1}{(nN)^2}\int_0^{nN}\int_0^{nN} \mu(E(u, v))du\,dv = \frac{1}{N^2}\int_0^N\int_0^N \mu(u, v)du\,dv \,,$$

so that by (3),

$$\frac{1}{N^2}\int_0^N\int_0^N \mu(E(u,v)du\,dv=kmE\,. \text{——} \tag{4}$$

From this, we see that there exists (u',v') $(0\leqq u'\leqq N,\ 0\leqq v'\leqq N)$, such that $\mu(E(u',v'))\geqq kmE$. Similarly there exists (u'',v'') $(0\leqq u''\leqq N,\ 0\leqq v''\leqq N)$, such that $\mu(E(u'',v'')\leqq kmE$. This is Theorem XI. 36.

3. We shall prove an analogue of this Blichfeldt's theorem for Fuchsian groups. Let G be a Fuchsian group, such that $\sigma(D_0)<\infty$. Let k points z_0^i $(i=1,2,\cdots,k)$ be given in D_0 and we call the totality of the equivalents of these k points by G the lattice points.

We call a linear transformation of the form T_a: $z'=\dfrac{z+a}{1+\bar{a}z}$ $(|a|<1)$ a "translation". Let E be a set in $|z|<1$. We "translate" E by T_a and we write $T_a(E)=E(a)$.

Theorem XI. 38. *Let E be a measurable set, contained in $|z|\leqq\rho_0<1$. Then we can "translate" E into E', such that E' contains at least $k\sigma(E)/\sigma(D_0)$ lattice points. Similarly we can "translate" E into E'', such that E'' contains at most $k\sigma(E)/\sigma(D_0)$ lattice points.*

Santaló [1], Tsuji [55].

This theorem can be deduced easily from the following Theorem XI. 39. We put a mass 1 at every lattice point, then we obtain a positive mass distribution $\mu\geqq 0$ in $|z|<1$, then

$$\mu(E(a))=\iint_{z\in E(a)} d\mu(z)$$

is the number of lattice points, contained in $E(a)$.

Theorem XI. 39. *If we put*

$$\omega(r;E)=\iint_{|a|\leqq r}\mu(E(a))\,d\sigma(a)\,,\qquad \Omega(r;E)=\int_0^r\omega(r;E)dr\,,$$

then

$$\lim_{r\to 1}\int_0^r\frac{\Omega(r,E)}{1-r}dr\bigg/\left(\log\frac{1}{1-r}\right)^2=k\pi\frac{\sigma(E)}{\sigma(D_0)}\,.$$

4. For the proof, we use the following lemma.

Lemma. *Let $|\zeta|\leqq\rho_0<1$ and $\zeta_a=T_a(\zeta)=\dfrac{\zeta+a}{1+\bar{a}\zeta}$ $(|a|<1)$. We consider ζ as fixed and a as variable. If $|a|\leqq r_0=r_0(\rho_0)$, then*

$$(1-|\zeta_a|^2)\,d\sigma(\zeta_a)=(1-|a|^2)\,d\sigma(a)+O(1)r\,dr\,d\theta\,,\qquad a=re^{i\theta}\,,$$

where $|O(1)|\leqq K(\rho_0)$, $K(\rho_0)$ being a constant, which depends on ρ_0 only.

Proof. Since $\zeta_a = \dfrac{\zeta+a}{1+\bar{a}\zeta} = \dfrac{\zeta+a}{1+a\bar{\zeta}} \cdot \dfrac{1+a\bar{\zeta}}{1+\bar{a}\zeta}$, if we put

$$\frac{\zeta+a}{1+a\bar{\zeta}} = Re^{i\Phi}, \qquad \frac{1+a\bar{\zeta}}{1+\bar{a}\zeta} = e^{i\varphi}, \qquad a = re^{i\theta}, \qquad \zeta = \rho e^{i\psi}, \tag{1}$$

then

$$\zeta_a = Re^{i\Theta}, \qquad \Theta = \Phi + \varphi, \tag{2}$$

$$R\begin{vmatrix} \dfrac{\partial R}{\partial r} & \dfrac{\partial R}{\partial \theta} \\ \dfrac{\partial \Theta}{\partial r} & \dfrac{\partial \Theta}{\partial \theta} \end{vmatrix} = R\begin{vmatrix} \dfrac{\partial R}{\partial r} & \dfrac{\partial R}{\partial \theta} \\ \dfrac{\partial \Phi}{\partial r} & \dfrac{\partial \Phi}{\partial \theta} \end{vmatrix} + R\begin{vmatrix} \dfrac{\partial R}{\partial r} & \dfrac{\partial R}{\partial \theta} \\ \dfrac{\partial \varphi}{\partial r} & \dfrac{\partial \varphi}{\partial \theta} \end{vmatrix} = I + II. \tag{3}$$

Since $\dfrac{\partial \log R}{\partial \log r} = \dfrac{\partial \Phi}{\partial \theta}$, $\dfrac{\partial \log R}{\partial \theta} = -\dfrac{\partial \Phi}{\partial \log r}$, we have

$$I = \frac{R^2}{r}\left(\left(\frac{\partial \log R}{\partial \log r}\right)^2 + \left(\frac{\partial \Phi}{\partial \log r}\right)^2\right) = \frac{R^2}{r}\left|\frac{d\log\left(\dfrac{\zeta+a}{1+a\bar{\zeta}}\right)}{d\log a}\right|^2 = R^2 r\left|\frac{1-|\zeta|^2}{(\zeta+a)(1+a\bar{\zeta})}\right|^2$$

$$= R^2 r\left|\frac{1-|\zeta|^2}{R(1+a\bar{\zeta})(1+a\bar{\zeta})}\right|^2 = \frac{r(1-\rho^2)^2}{|1+a\bar{\zeta}|^4}. \tag{4}$$

Since $1-R^2 = \dfrac{(1-r^2)(1-\rho^2)}{|1+a\bar{\zeta}|^2}$, we have $R\dfrac{\partial R}{\partial \theta} = O(1-r)$, so that

$$II = R\frac{\partial R}{\partial r} \cdot \frac{\partial \varphi}{\partial \theta} + O(1-r). \tag{5}$$

From $\log(1+re^{i\theta}\bar{\zeta}) - \log(1+re^{-i\theta}\zeta) = i\varphi$, we have easily $\dfrac{\partial \varphi}{\partial \theta} = \dfrac{2\rho r\cos(\theta-\psi)+2\rho^2 r^2}{|1+a\bar{\zeta}|^2}$ and since $R\dfrac{\partial R}{\partial r} = \dfrac{r(1-\rho^2)}{|1+a\bar{\zeta}|^2} + O(1-r)$,

$$II = R\frac{\partial R}{\partial r} \cdot \frac{\partial \varphi}{\partial \theta} + O(1-r) = \frac{r(1-\rho^2)}{|1+a\bar{\zeta}|^4}(2\rho r\cos(\theta-\psi)+2\rho^2 r^2) + O(1-r). \tag{6}$$

Hence

$$R\frac{\partial(R,\Theta)}{\partial(r,\theta)} = I + II = \frac{r(1-\rho^2)^2}{|1+a\bar{\zeta}|^4}\left(1+\frac{2\rho r\cos(\theta-\psi)+2\rho^2 r^2}{1-\rho^2}\right) + O(1-r)$$

$$= \frac{r(1-\rho^2)^2}{|1+a\bar{\zeta}|^4} \cdot \frac{1+2\rho r\cos(\theta-\psi)+\rho^2 r^2}{1-\rho^2} + O(1-r)$$

$$= \frac{r(1-\rho^2)^2}{|1+a\bar{\zeta}|^4} \cdot \frac{|1+a\bar{\zeta}|^2}{1-\rho^2} + O(1-r) = \frac{r(1-\rho^2)^2}{|1+a\bar{\zeta}|^4} \cdot \frac{1-r^2}{1-|\zeta_a|^2} + O(1-r). \tag{7}$$

Hence $\dfrac{\partial(R,\Theta)}{\partial(r,\theta)} > 0$, if $1-r$ is sufficiently small, so that

$$4R\,dR\,d\Theta=\frac{(1-\rho^2)^2}{|1+a\bar{\zeta}|^4}\cdot\frac{1-r^2}{1-|\zeta_a|^2}4r\,dr\,d\theta+O(1-r)r\,dr\,d\theta\,.$$

Since $1-|\zeta_a|^2=O(1-r)$,

$$(1-|\zeta_a|^2)4R\,dR\,d\Theta=\frac{(1-\rho^2)^2}{|1+a\bar{\zeta}|^4}(1-r^2)4r\,dr\,d\theta+O((1-r)^2)r\,dr\,d\theta\,. \quad (8)$$

If we divide the both sides of (8) by $(1-R^2)^2=\dfrac{(1-\rho^2)^2(1-r^2)^2}{|1+a\bar{\zeta}|^4}$, then we have

$$(1-|\zeta_a|^2)\,d\sigma(\zeta_a)=(1-|a|^2)\,d\sigma(a)+O(1)r\,dr\,d\theta\,. \text{ —}$$

5. *Proof of the theorem.*

For a fixed i $(1\leqq i\leqq k)$, we put a mass 1 at every z_ν^i $(\nu=0,1,2,\cdots)$, where z_ν^i is an equivalent of z_0^i, then we obtain a positive mass distribution $\mu^i\geqq 0$ in $|z|<1$, so that for a Borel set E in $|z|<1$,

$$\mu^i(E)=\iint_E d\mu^i(a)$$

is the number of z_ν^i, contained in E, hence if we put $\mu=\sum_{i=1}^k\mu^i$, then

$$\mu(E)=\iint_E d\mu(a)$$

is the number of lattice points, contained in E.

Let $\varDelta$: $\left|\dfrac{z-\zeta}{1-\bar{\zeta}z}\right|\leqq\tau<1$ be a disc, with the non-enclidean center ζ $(|\zeta|\leqq\rho_0<1)$ and $\varDelta_0^i$: $\left|\dfrac{z-z_0^i}{1-\bar{z}_0^i z}\right|\leqq\tau$ be a disc, with the same non-euclidean radius as $\varDelta$ and let $\varDelta_\nu^i$ be its equivalent, which contains z_ν^i.

Let $\psi^i(z)$ be the characteristic function of $\sum_{\nu=0}^\infty\varDelta_\nu^i$. Now $\varDelta(a)=T_a(\varDelta)$ contains z_ν^i, when and only when $\zeta_a\in\varDelta_\nu^i$, so that by the lemma,

$$(1-|\zeta_a|^2)\,\psi^i(\zeta_a)\,d\sigma(\zeta_a)=(1-|a|^2)\,\mu^i(\varDelta(a))\,d\sigma(a)+O(1)rdrd\theta\,, \quad a=re^{i\theta}\,. \quad (1)$$

Since $1-|\zeta_a|^2=\dfrac{(1-|a|^2)(1-|\zeta|^2)}{|1+a\bar{\zeta}|^2}$,

$$\frac{1-\rho}{1+\rho}(1-r^2)\leqq 1-|\zeta_a|^2\leqq\frac{1+\rho}{1-\rho}(1-r^2)\,,\quad |a|=r,\ |\zeta|=\rho,$$

hence if we define r_1, r_2 by

$$1-r_1^2=\frac{1+\rho}{1-\rho}(1-r^2)\,,\qquad 1-r_2^2=\frac{1-\rho}{1+\rho}(1-r^2)\,, \quad (2)$$

then

$$r_1\leqq|\zeta_a|\leqq r_2\qquad(r_1<r<r_2)\,. \quad (3)$$

Hence by (1),

$$\iint\limits_{|\zeta_a|\leq r_1} (1-|\zeta_a|^2)\psi^i(\zeta_a)\,d\sigma(\zeta_a)+O(1)\leq \iint\limits_{|a|\leq r}(1-|a|^2)\mu^i(\Delta(a))\,d\sigma(a)$$

$$\leq \iint\limits_{|\zeta_a|\leq r_2} (1-|\zeta_a|^2)\psi^i(\zeta_a)\,d\sigma(\zeta_a)+O(1)\,. \qquad (4)$$

Let

$$A(r;E)=\iint\limits_{a\in E} n(r,a)\,d\sigma(a)\,,$$

then by Theorem XI. 30,

$$\int_0^r \frac{A(r;E)}{r}dr=\frac{2\pi\sigma(E)}{\sigma(D_0)}\log\frac{1}{1-r}+O(1)\,. \qquad (5)$$

With this notation, we have

$$\iint\limits_{|\zeta_a|\leq r} \psi^i(\zeta_a)\,d\sigma(\zeta_a)=A(r;\Delta_0^i)\,,$$

so that the left hand side of (4) is

$$\int_0^{r_1}(1-t^2)dA(t;\Delta_0^i)=(1-r_1^2)A(r_1;\Delta_0^i)+2\int_0^{r_1}tA(t;\Delta_0^i)dt=2\int_0^{r_1}tA(t;\Delta_0^i)dt+O(1),$$

and if we put

$$\omega^i(r;\Delta)=\iint\limits_{|a|\leq r}\mu^i(\Delta(a))\,d\sigma(a)\,, \qquad (6)$$

then similarly the middle term of (4) is

$$(1-r^2)\omega^i(r;\Delta)+2\int_0^r t\omega^i(t;\Delta)dt=2\int_0^r t\omega^i(t;\Delta)dt+O(1)\,,$$

hence by (4)

$$\int_0^{r_1}tA(t;\Delta_0^i)dt\leq\int_0^r\omega^i(t;\Delta)dt+O(1)\,. \qquad (7)$$

Since $\left(\frac{1}{1-r_1}-\frac{1}{1+r_1}\right)dr_1=\left(\frac{1}{1-r}-\frac{1}{1+r}\right)dr$, we have

$$\int_0^{r_1}\frac{dr_1}{1-r_1}\int_0^{r_1}t^2\frac{A(t;\Delta_0^i)}{t}dt\leq\int_0^r\frac{dr}{1-r}\int_0^r\omega^i(t;\Delta)dt+O\left(\log\frac{1}{1-r}\right), \qquad (8)$$

so that if we put

$$\omega(r;\Delta)=\iint\limits_{|a|\leq r}\mu(\Delta(a))d\sigma(a)=\sum_{i=1}^k\omega^i(r;\Delta)\,, \qquad (9)$$

$$\Omega(r;\varDelta)=\int_0^r \omega(r;\varDelta)dr\ , \tag{10}$$

then from (8), we have if $r_0\leqq r_1<1$,

$$r_0^2\int_{r_0}^{r_1}\frac{dr_1}{1-r_1}\left(\sum_{i=1}^{k}\int_{r_0}^{r_1}\frac{A(t;\varDelta_0)}{t}dt\right)\leqq\int_0^r\frac{\Omega(r;\varDelta)}{1-r}dr+O\left(\log\frac{1}{1-r}\right). \tag{11}$$

Since by (5),

$$\int_{r_0}^{r_1}\frac{A(t;\varDelta_0^i)}{t}dt=\frac{2\pi\sigma(\varDelta_0^i)}{\sigma(D_0)}\log\frac{1}{1-r_1}+O(1)=\frac{2\pi\sigma(\varDelta)}{\sigma(D_0)}\log\frac{1}{1-r_1}+O(1)\ ,$$

the left hand side of (11) is

$$\frac{r_0^2k\pi\sigma(\varDelta)}{\sigma(D_0)}\left(\log\frac{1}{1-r_1}\right)^2+O\left(\log\frac{1}{1-r_1}\right)=\frac{r_0^2k\pi\sigma(\varDelta)}{\sigma(D_0)}\left(\log\frac{1}{1-r}\right)^2+O\left(\log\frac{1}{1-r}\right),$$

so that by (11),

$$\frac{r_0^2k\pi\sigma(\varDelta)}{\sigma(D_0)}\left(\log\frac{1}{1-r}\right)^2+O\left(\log\frac{1}{1-r}\right)\leqq\int_0^r\frac{\Omega(r;\varDelta)}{1-r}dr\ . \tag{12}$$

Similarly from the right half of (4), we have

$$\int_0^r\frac{\Omega(r;\varDelta)}{1-r}dr\leqq\frac{k\pi\sigma(\varDelta)}{\sigma(D_0)}\left(\log\frac{1}{1-r}\right)^2+O\left(\log\frac{1}{1-r}\right), \tag{13}$$

hence

$$\begin{aligned}&\frac{r_0^2k\pi\sigma(\varDelta)}{\sigma(D_0)}\left(\log\frac{1}{1-r}\right)^2+O\left(\log\frac{1}{1-r}\right)\\ &\leqq\int_0^r\frac{\Omega(r;\varDelta)}{1-r}dr\leqq\frac{k\pi\sigma(\varDelta)}{\sigma(D_0)}\left(\log\frac{1}{1-r}\right)^2+O\left(\log\frac{1}{1-r}\right).\end{aligned} \tag{14}$$

Let O be an open set, contained in $|z|\leqq\rho_0<1$, then by Vitali's covering theorem, O is the sum of disjoint discs $\{\varDelta_j\}$ and a null set N, such that $O=\sum_{j=1}^{\infty}\varDelta_j+N$. We put $O^*=\sum_{j=1}^{\infty}\varDelta_j$ and apply (14) to $\varDelta_j$ and sum up, then since $\sigma(O^*)=\sigma(O)$ and $\Omega(r;O^*)=\sum_{j=1}^{\infty}\Omega(r;\varDelta_j)$, we have

$$\begin{aligned}&\frac{r_0^2k\pi\sigma(O)}{\sigma(D_0)}\left(\log\frac{1}{1-r}\right)^2+O\left(\log\frac{1}{1-r}\right)\\ &\leqq\int_0^r\frac{\Omega(r;O^*)}{1-r}dr\leqq\frac{k\pi\sigma(O)}{\sigma(D_0)}\left(\log\frac{1}{1-r}\right)^2+O\left(\log\frac{1}{1-r}\right).\end{aligned} \tag{15}$$

Since $\Omega(r;O^*)\leqq\Omega(r;O)$,

$$\frac{r\ k\pi\sigma(O)}{\sigma(D_0)}\left(\log\frac{1}{1-r}\right)^2+O\left(\log\frac{1}{1-r}\right)\leqq\int_0^r\frac{\Omega(r;O)}{1-r}dr\ . \tag{16}$$

Since $\sigma(N)=0$, for any $\varepsilon>0$, we can cover N by discs $\{\delta_\nu\}$, such that $\sum_{\nu=1}^{\infty}\sigma(\delta_\nu)<\varepsilon$. Since $\Omega(r; O)\leqq\Omega(r; O^*)+\sum_{\nu}\Omega(r; \delta_\nu)$, we have by (14), (15),

$$\int_0^r \frac{\Omega(r; O)}{1-r}dr \leqq \frac{k\pi(\sigma(O)+\varepsilon)}{\sigma(D_0)}\left(\log\frac{1}{1-r}\right)^2 + O\left(\log\frac{1}{1-r}\right).$$

Since $\varepsilon>0$ is arbitrary, we have

$$\int_0^r \frac{\Omega(r; O)}{1-r}dr \leqq \frac{k\pi\sigma(O)}{\sigma(D_0)}\left(\log\frac{1}{1-r}\right)^2 + O\left(\log\frac{1}{1-r}\right). \tag{17}$$

Hence

$$\frac{r_0^2 k\pi\sigma(O)}{\sigma(D_0)}\left(\log\frac{1}{1-r}\right)^2 + O\left(\log\frac{1}{1-r}\right)$$
$$\leqq \int_0^r \frac{\Omega(r; O)}{1-r}dr \leqq \frac{k\pi\sigma(O)}{\sigma(D_0)}\left(\log\frac{1}{1-r}\right)^2 + O\left(\log\frac{1}{1-r}\right). \tag{18}$$

Let F be a closed set, containd in $|z|\leqq\rho_0<1$, then F is the limit of a sequence of open sets O_n, $F=\lim_n O_n$, then $\sigma(O_n)\to\sigma(F)$, $\mu(O_n(a))\to\mu(F(a))$ for any a, so that $\Omega(r; O_n)\to\Omega(r; F)$, hence (18) holds for F, instead of O.

Let E be a Borel set, contained in $|z|\leqq\rho_0<1$, then E is obtained from a closed set after at most a countable number of operations: sum and product, so that (18) holds for a Borel set E, instead of O.

Since $r_0<1$ is arbitrary, we have

$$\lim_{r\to1}\int_0^r \frac{\Omega(r; E)}{1-r}dr \Big/ \left(\log\frac{1}{1-r}\right)^2 = \frac{k\pi\sigma(E)}{\sigma(D_0)}. \tag{19}$$

Next let E be a measurable set, contained in $|z|\leqq\rho_0<1$, then there exists an F_σ-set E_1 and a G_δ-set E_2, such that $E_1\subset E\subset E_2$, $\sigma(E_1)=\sigma(E)=\sigma(E_2)$. Since $\Omega(r; E_1)\leqq\Omega(r; E)\leqq\Omega(r; E_2)$ and (19) holds for E_1 and E_2, we see that (19) holds for E. Hence the theorem is proved.——

6. Now Theorem XI. 38 can be proved simply as follows.

Theorem XI. 39 is equivalent to

$$\lim_{r\to1}\int_0^r \frac{dr}{1-r}\int_0^r dt \iint_{|a|\leqq t}\mu(E(a))d\sigma(a) \Big/ \left(\log\frac{1}{1-r}\right)^2 = \frac{k\pi\sigma(E)}{\sigma(D_0)}. \tag{1}$$

Since $\displaystyle\iint_{|a|\leqq r} d\sigma(a)=\frac{4\pi r^2}{1-r^2}$, we have

$$\lim_{r\to1}\int_0^r \frac{dr}{1-r}\int_0^r dt \iint_{|a|\leqq t} d\sigma(a) \Big/ \left(\log\frac{1}{1-r}\right)^2 = \pi. \tag{2}$$

Hence if we put

$$M=\operatorname*{Max}_{|a|<1}\mu(E(a)), \qquad m=\operatorname*{Min}_{|a|<1}\mu(E(a)), \tag{3}$$

then we have from (1), (2),

$$M \geqq k\sigma(E)/\sigma(D_0)\,, \qquad m \leqq k\sigma(E)/\sigma(D_0)\,, \tag{4}$$

which is Theorem XI. 38, since being a positive integer, M and m are attained at a certain point a in $|z|<1$.

13. Lattice points in a circle

Let G be the group of translations: $x'=x+m$, $y'=y+n$ $(m, n=0, \pm 1, \pm 2, \cdots)$ on the $z=x+iy$-plane, then D_0: $0 \leqq x<1$, $0 \leqq y<1$ is its fundamental domain. Let a be any point of D_0 and $n(r, a)$ be the number of its equivalents a_ν, contained in $|z|<r$ and put

$$n(r, a)=\pi r^2+\Omega(r, a)\,. \tag{1}$$

$n(r)=n(r, 0)$ is the number of lattice points (m, n), contained in $|z|<r$. Then we shall prove

Theorem XI. 40. $$\int_1^r \frac{\Omega(r, a)}{r} dr=O(1)\,,$$

where $O(1)$ is uniformly bounded for any $a \in D_0$, when $r \to \infty$.

Let $\lambda>1$, then for any $r>1$, there exists τ in $[r, \lambda r]$, such that

$$|\Omega(\tau, a)| \leqq \text{const.}. \qquad \text{Tsuji [41].}$$

Proof. If we identify the equivalent points on the sides of D_0, then D_0 can be considered as a closed Riemann surface F of genus $p=1$. Let $a \in D_0$, $b \in D_0$ $(a \neq b)$ and $u(z; a, b)$ be the potential function on F, defined in Theorem X. 30, then $u(z; a, b)$ is invariant by G and

$$u-\log \frac{1}{|z-a_\nu|} \text{ is harmonic at } z=a_\nu,\quad u+\log \frac{1}{|z-b_\nu|} \text{ is harmonic at } z=b_\nu. \qquad (\nu=0, 1, 2, \cdots) \tag{1}$$

Similarly as Theorem XI. 29, we have

$$\frac{1}{2\pi} \int_0^{2\pi} u(re^{i\theta}; a, b) d\theta+N(r, a)=N(r, b)+O(1)\,, \tag{2}$$

where

$$N(r, a)=\int_1^r \frac{n(r, a)}{r} dr\,,$$

and $O(1)$ is uniformly bounded for $r \to \infty$, if $|a-b| \geqq \rho>0$.

We shall prove that

$$\int_0^{2\pi} |u(re^{i\theta}; a, b)| d\theta=O(1)\,.$$

Let $0<\rho<1$ be fixed and $U(a_\nu)$: $|z-a_\nu|<\rho$, $U(b_\nu)$: $|z-b^\nu|<\rho$ $(\nu=0, 1, 2, \cdots)$, then since $|u|\leqq K=\text{const.}$ outside of $\{U(a_\nu)\}$, $\{U(b_\nu)\}$, if we denote the circle $|z|=r$ by C_r, then

$$\int_0^{2\pi}|u|d\theta\leqq\sum_\nu\int_{C_r\cdot U(a_\nu)}|u|d\theta+\sum_\nu\int_{C_r\cdot U(b_\nu)}|u|d\theta+\text{const.}, \tag{3}$$

where we sum up for all a_ν, b_ν, contained in $r_1<|z|<r_2$ $(r_1=r-1, r_2=r+1)$. By (1), we see that

$$\int_{C_r\cdot U(a_\nu)}|u|ds=\int_{C_r\cdot U(a_\nu)}|u|r\,d\theta\leqq\text{const.},$$

so that

$$\int_{C_r\cdot U(a_\nu)}|u|d\theta=O(1/r).$$

Similarly

$$\int_{C_r\cdot U(b_\nu)}|u|d\theta = O(1/r).$$

Since the number of a_ν, b_ν, contained in $r_1<|z|<r_2$ is $O(r)$, we have

$$\int_0^{2\pi}|u|d\theta\leqq\text{const.},$$

hence by (2),

$$N(r, a)=N(r, b)+O(1), \qquad \text{if } |a-b|\geqq\rho>0. \tag{4}$$

Similarly as the proof of Theorem XI. 29, we see that (4) holds for any a, b of D_0.

Let $d\omega=dx\,dy$ be the surface element, then by (4)

$$N(r, a)=\iint_{D_0} N(r, b)d\omega(b)+O(1). \tag{5}$$

Since $\iint_{D_0} n(r, b)d\omega(b)=\pi r^2$,

$$\iint_{D_0} N(r, b)\,d\omega(b)=\int_1^r\frac{dr}{r}\iint_{D_0} n(r, b)\,d\omega(b)=\frac{\pi r^2}{2}+O(1),$$

so that by (5),

$$N(r, a)=\frac{\pi r^2}{2}+O(1). \tag{6}$$

From $n(r, a)=\pi r^2+\Omega(r, a)$, we have

$$N(r, a)=\frac{\pi r^2}{2}+\int_1^r\frac{\Omega(r, a)}{r}dr+O(1),$$

hence by (6),

$$\int_1^r \frac{\Omega(r, a)}{r} dr = O(1)\,, \tag{7}$$

where $O(1)$ is uniformly bounded for any $a \in D_0$, when $r \to \infty$.

Let $\lambda > 1$, then

$$\left| \int_r^{\lambda r} \frac{\Omega(r, a)}{r} dr \right| \leqq \text{const.}. \tag{8}$$

If $\Omega(r, a)$ does not change its sign in $[r, \lambda r]$, then considering $\inf |\Omega(r, a)|$ in $[r, \lambda r]$, we see that there exists τ in $[r, \lambda r]$, such that

$$|\Omega(\tau, a)| \leqq \text{const.}. \tag{9}$$

Next suppose that $\Omega(r, a)$ changes its sign in $[r, \lambda r]$. Now $[r, \lambda r]$ is decomposed into a finite number of disjoint intervals, in each of which $\Omega(r, a)$ is continuous and decreasing and at the common point r' of two adjacent intervals, $\Omega(r'-o, a) < \Omega(r'+o, a)$, where the circle $|z| = r'$ contains a lattice point, so that either in $[r, \lambda r]$, there exists τ, such that $\Omega(\tau, a) = 0$, or in one of intervals $[r, \sqrt{\lambda}\, r]$, $[\sqrt{\lambda}\, r, \lambda r]$, $\Omega(r,a)$ is of constant sign, hence there exists τ, which satisfies (9). Hence in any case, there exists τ in $[r, \lambda r]$, such that

$$|\Omega(\tau, a)| \leqq \text{const.} \quad \text{——}$$

Remark. By the same method, we can prove the following theorem. Let $\sum_{i,k=1}^{n} a_{ik} x_i x_k$ $(n \geqq 2)$ be a positive definite quadratic form, with the determinant $D = |a_{ik}| > 0$, then $\sum_{,k=1}^{n} a_{ik} x_i x_k < r^2$ is the inside of an n-dimensional ellipsoid, whose volume is $V(r) = \pi^{n/2} r^n / \sqrt{D}\, \Gamma(n/2+1)$. Let $n(r)$ be the number of lattice points, contained in this ellipsoid and put $n(r) = V(r) + \Omega(r)$, then

$$\int_1^r \frac{\Omega(r)}{r^{n-1}} dr = O(1)\,.$$

Let $\lambda > 1$, then for any $r > 1$, there exists τ in $[r, \lambda r]$, such that

$$|\Omega(\tau)| \leqq \text{const.}\, \tau^{n-2}\,.$$

Tsuji [41].

APPENDIX

Birkhoff's ergodic theorem

1. Flow

Let Ω be an n-dimensional manifold, such that every point $x \in \Omega$ has a neighbourhood, which is the inside of an n-dimensional sphere. Let dx be the n-dimensional volume element and $d\mu(x)=\varphi(x)dx$, where $\varphi(x) \geqq 0$ is a continuous function of x and we assume that $\mu(\Omega)<\infty$.

Suppose that there exists a family of topological mappings: $x'=T_\lambda x$ $(-\infty<\lambda<\infty)$ of Ω into itself, such that $T_\lambda T_\mu = T_{\lambda+\mu}$, $T_0 \equiv 1$. We call such T_λ *a flow*. T_λ is said *mass-preserving*, if $\mu(T_\lambda E)=\mu(E)$ for any measurable set E in Ω and T_λ is said *metric transitive*, if a measurable set E in Ω is invariant by the flow T_λ, then $\mu(E)=0$, or $\mu(E)=\mu(\Omega)$.

Lemma. *If T_λ is a mass-preserving flow and $f(x)$ is a measurable function of x, then $f(T_\lambda x)$ is a measurable function in the product space $\Omega \times (\lambda)$, where (λ) denotes the line: $-\infty<\lambda<\infty$.*

Proof. First let O be an open set in Ω, then the tube $T_\lambda O$ $(-\infty<\lambda<\infty)$ is an open set in $\Omega \times (\lambda)$, so that it is a measurable set in $\Omega \times (\lambda)$. Hence if A is a G_δ-set in Ω, then the tube $T_\lambda A$ $(-\infty<\lambda<\infty)$ is a measurable set in $\Omega \times (\lambda)$.

Next let E be a measurable set in Ω, then $E=A-N$, where A is a G_δ-set and N is a null set. The tube $T_\lambda E$ $(-\infty<\lambda<\infty)$ is a difference of two tubes $T_\lambda A$ $(-\infty<\lambda<\infty)$ and $T_\lambda N$ $(-\infty<\lambda<\infty)$. We cover N by an open set O, such that $\mu(O)<\varepsilon$. Since T_λ is mass-preserving, if we apply Fubini's theorem and then make $\varepsilon \to 0$, then we see that the tube $T_\lambda N$ $(-\infty<\lambda<\infty)$ is a null set in $\Omega \times (\lambda)$. Since the tube $T_\lambda A$ $(-\infty<\lambda<\infty)$ is a measurable set in $\Omega \times (\lambda)$, the tube $T_\lambda E$ $(-\infty<\lambda<\infty)$ is a measurable set in $\Omega \times (\lambda)$.

Let $f(x)$ be the characteristic function of E, then the set of points of $\Omega \times (\lambda)$, such that $f(T_\lambda x)=1$ coincides with the tube $T_\lambda E$ $(-\infty<\lambda<\infty)$, so that $f(T_\lambda x)$ is a measurable function in $\Omega \times (\lambda)$. From this, we see easily that if $f(x)$ is a measurable function of x, then $f(T_\lambda x)$ is a measurable function in $\Omega \times (\lambda)$.——

Hence for almost all x in Ω, $\int_0^t f(T_\lambda x)d\lambda$ has a meaning.

2. Birkhoff's ergodic theorem

Let T_λ be a mass-preserving and metric transitive flow in Ω and $f(x)$ be a bounded measurable function of x in Ω, then for almost all x,

$$\lim_{t\to\infty}\frac{1}{t}\int_0^t f(T_\lambda x)d\lambda=\frac{1}{\mu(\Omega)}\int_\Omega f(x)d\mu(x)\,. \qquad \text{Birkhoff [1].}$$

Proof. (Khintchine [1]). Let

$$\varphi(x, r)=\int_0^r f(T_\lambda x)d\lambda \qquad (r=1, 2, \cdots)\,, \tag{1}$$

$$\varliminf_{r\to\infty}\varphi(x, r)/r=f_*(x)\,, \qquad \varlimsup_{r\to\infty}\varphi(x, r)/r=f^*(x)\,. \tag{2}$$

We shall prove that $f_*(x)=f^*(x)$ almost everywhere in Ω.

Let $\alpha<\beta$ and S be the set of x, such that

$$S:\ f_*(x)<\alpha\,, \qquad \beta<f^*(x)\,. \tag{3}$$

Then S is invariant by T_r $(r=0, 1, 2, \cdots)$. For, if $x\in S$, then for $k=1, 2, \cdots$,

$$\varphi(T_r x, k)=\int_0^k f(T_{\lambda+r}x)d\lambda=\int_r^{r+k} f(T_\lambda x)d\lambda=\varphi(x, r+k)-\varphi(x, r)\,, \tag{4}$$

$$\frac{\varphi(T_r x, k)}{k}=\frac{\varphi(x, r+k)-\varphi(x, r)}{k}=\frac{\varphi(x, r+k)}{r+k}\left(1+\frac{r}{k}\right)-\frac{\varphi(x, r)}{k}\,.$$

If we make $k\to\infty$, then we see that $T_r x\in S$, so that S is invariant by T_r.

If $x\in S$, then for some $k\geqq 1$,

$$\varphi(x, k)>k\beta\,. \tag{5}$$

Let M_l $(l\geqq 1)$ be the set of x, such that (5) holds at first $k=l$, so that

$$M_l:\ \varphi(x, l)>\beta l\,, \quad \varphi(x, j)\leqq\beta j \quad (j=1, 2, \cdots, l-1)\,. \tag{6}$$

We put

$$S_k=\sum_{l=1}^{k} M_l\,. \tag{7}$$

Evidently

$$S=\sum_{l=1}^{\infty} M_l\,, \qquad M_j\cdot M_{j'}=0 \qquad (j\neq j')\,. \tag{8}$$

The following property of M_l is important in the sequel.

If $x\in M_k$ and $1\leqq r\leqq k-1$, then

$$T_r x\in M_j \qquad \text{for some} \quad j\leqq k-1\,. \tag{9}$$

For, since $x\in M_k$, we have $\varphi(x, k)>\beta k$, $\varphi(x, r)\leqq\beta r$, so that by (4),

$$\frac{\varphi(T_r x, k-r)}{k-r}=\frac{\varphi(x, k)-\varphi(x, r)}{k-r}>\frac{\beta k-\beta r}{k-r}=\beta\,,$$

hence $T_r x\in M_j$ for some $j\leqq k-r\leqq k-1$.

Hence

$$T_r M_k \subset M_1 + \cdots + M_{k-1} \qquad (1 \leqq r \leqq k-1)\,. \tag{10}$$

We put

$$\hat{M}_k = M_k + T_1 M_k + \cdots + T_{k-1} M_k\,, \quad M_{k-1}^* = M_{k-1} - M_{k-1} \cdot \hat{M}_k\,.$$

Since $M_{k-1}^* \subset M_{k-1}$, we have by (10),

$$T_r M_{k-1}^* \subset M_1 + \cdots + M_{k-2} \qquad (1 \leqq r \leqq k-2)\,.$$

Next we put

$$\hat{M}_{k-1} = M_{k-1}^* + T_1 M_{k-1}^* + \cdots + T_{k-2} M_{k-1}^*\,, \quad M_{k-2}^* = M_{k-2} - M_{k-2}(\hat{M}_k + \hat{M}_{k-1})$$

and proceed similarly, then we have the following aggregate of sets $T_r M_l^*$ $(0 \leqq r \leqq l-1)$, $(l = 1, 2, \cdots, k)$, where $M_k^* = M_k$:

$$\left.\begin{array}{llll} M_k^*, & T_1 M_k^*, & \cdots\cdots\cdots\cdots, & T_{k-1} M_k^*, \\ & M_{k-1}^*, & T_1 M_{k-1}^*, \quad \cdots, & T_{k-2} M_{k-1}^*, \\ & \cdots\cdots\cdots\cdots & \cdots\cdots\cdots\cdots & \cdots\cdots\cdots \\ & & M_2^*, & T_1 M_2^*, \\ & & & M_1^*. \end{array}\right\} \tag{11}$$

Then

$$\left.\begin{array}{l} \text{(i)}\ \ \sum_{l=1}^{k} \sum_{r=0}^{l-1} T_r M_l^* = \sum_{l=1}^{k} M_l = S_k\,, \qquad \text{(ii)}\ \ \mu(T_r M_l^*) = \mu(M_l^*)\,, \\ \text{(iii)}\ \ \text{Two sets of (11) are disjoint.} \end{array}\right\} \tag{12}$$

By (10), (i) is evident and since T_λ is mass-preserving, (ii) is evident, so that we shall prove (iii).

Suppose that $T_r M_l^*$, $T_{r'} M_{l'}^*$ $(|r-r'|+|l-l'| \neq 0)$ have a common point x, such that $x \in T_r M_l^*$, $x \in T_{r'} M_{l'}^*$, then

$$T_{-r} x \in M_l^*\,, \qquad T_{-r} x \in T_{r'-r} M_{l'}^*\,. \tag{13}$$

We may assume that $r \leqq r'$. If $r = r'$, then $l \neq l'$ and

$$T_{-r} x \in M_l^*\,, \qquad T_{-r} x \in M_{l'}^*\,,$$

which is absurd, since $M_l^* \subset M_l$, $M_{l'}^* \subset M_{l'}$ and M_l, $M_{l'}$ are disjoint by (8), hence $r < r'$.

If $l' \leqq l$, then $T_{-r} x \in T_{r'-r} M_{l'}^* \subset T_{r'-r} M_{l'}$ and since $r' \leqq l'-1$, we have $1 \leqq r'-r \leqq l'-1$, so that by (9), $T_{-r} x$ belongs to some M_j $(j \leqq l'-1 \leqq l-1)$. Since $T_{-r} x \in M_l^* \subset M_l$, M_j and M_l have a common point $T_{-r} x$, which is absurd by (8).

If $l' > l$, then by the definition of M^*'s, a point $T_{-r} x \subset T_{r'-r} M_{l'}^*$ does not belong to M_l^*.

Hence two sets of (11) are disjoint.

Since T_λ is mass-preserving, we have for any measurable set E in Ω,

$$\int_{T_r E}\varphi(x,1)\,d\mu(x)=\int_E \varphi(T_r x,1)\,d\mu(T_r x)=\int_E \varphi(T_r x,1)\,d\mu(x)$$
$$=\int_E (\varphi(x,r+1)-\varphi(x,r))\,d\mu(x)\quad(\text{by (4)}),$$

so that

$$\int_{S_k}\varphi(x,1)\,d\mu(x)=\sum_{l=1}^{k}\sum_{r=0}^{l-1}\int_{T_r M_l^*}\varphi(x,1)\,d\mu(x)=\sum_{l=1}^{k}\sum_{r=0}^{l-1}\int_{M_l^*}(\varphi(x,r+1)-\varphi(x,r))\,d\mu(x)$$
$$=\sum_{l=1}^{k}\int_{M_l^*}\varphi(x,l)\,d\mu(x)>\beta\sum_{l=1}^{k}l\mu(M_l^*)=\beta\sum_{l=1}^{k}\sum_{r=0}^{l-1}\mu(T_r M_l^*)=\beta\mu(S_k)\,.$$

Hence if we make $k\to\infty$, then we have

$$\int_S \varphi(x,1)\,d\mu(x)\geqq\beta\mu(S)\,. \tag{14}$$

Similarly

$$\int_S \varphi(x,1)\,d\mu(x)\leqq\alpha\mu(S)\,, \tag{15}$$

so that $\beta\mu(S)\leqq\alpha\mu(S)$. Since $\alpha<\beta$, we have $\mu(S)=0$.

Let $\{r_n\}$ be the set of all rational numbers and $S_{m,n}$ be the set of x, such that $f_*(x)<r_m<r_n<f^*(x)$ and put $S^*=\sum_{m,n}S_{m,n}$, then $\mu(S^*)=0$. Since S^* is the set of x, such that $f_*(x)<f^*(x)$, we have $f_*(x)=f^*(x)$ almost everywhere in Ω, hence

$$\lim_{r\to\infty}\frac{\varphi(r,x)}{r}=\chi(x)\qquad(r=1,2,\cdots) \tag{16}$$

exists almost everywhere in Ω.

If t is a continuous variable and

$$\varphi(x,t)=\int_0^t f(T_\lambda x)d\lambda\,, \tag{17}$$

then

$$\frac{\varphi(x,t)}{t}=\frac{[t]}{t}\cdot\frac{1}{[t]}\int_0^{[t]}f(T_\lambda x)d\lambda+\frac{1}{t}\int_{[t]}^{t}f(T_\lambda x)d\lambda\,.$$

Since $f(x)$ is bounded,

$$\lim_{t\to\infty}\frac{\varphi(x,t)}{t}=\chi(x) \tag{18}$$

exists almost everywhere in Ω.

We shall prove that $\chi(x)\equiv$const.. Suppose that $\chi(x)\not\equiv$const., then for some α, the set E of x, such that $\chi(x)\leqq\alpha$ and the set E', such that $\chi(x)>\alpha$ are both of positive measure. Evidently E and E' are invariant by T_λ, which contradicts the hypothesis, that T_λ is metric transitive. Hence

$$\lim_{t\to\infty}\frac{\varphi(x,t)}{t}=\chi(x)\equiv\text{const.}=\kappa\,. \tag{19}$$

Since

$$\int_\Omega \varphi(x,t)\,d\mu(x)=\int_0^t d\lambda\int_\Omega f(T_\lambda x)\,d\mu(x)=t\int_\Omega f(x)\,d\mu(x)\,, \tag{20}$$

and $\varphi(x,t)/t$ is bounded for $t\to\infty$, we have by Lebesgue's theorem,

$$\lim_{t\to\infty}\int_\Omega \frac{\varphi(x,t)}{t}d\mu(x)=\int_\Omega\left(\lim_{t\to\infty}\frac{\varphi(x,t)}{t}\right)d\mu(x)=\kappa\mu(\Omega)\,,$$

hence by (20)

$$\kappa=\frac{1}{\mu(\Omega)}\int_\Omega f(x)\,d\mu(x)\,,$$

so that by (17), (19), for almost all x in Ω

$$\lim_{t\to\infty}\frac{1}{t}\int_0^t f(T_\lambda x)d\lambda=\frac{1}{\mu(\Omega)}\int_\Omega f(x)\,d\mu(x)\,. \tag{21}$$

BIBLIOGRAPHY

AHLFORS, L. V.: [1] Beiträge der meromorphen Funktionen. 7. Congr. Math. Scand. Oslo. (1929).—[2] Untersuchungen zur Theorie der konformen Abbildung und der Theorie der ganzen Funktionen. Acta Soc. Sci. Fenn. (1930).—[3] Über die asymptotischen Werte der meromorphen Funktionen endlicher Ordnung. Acta Acad. Aboensis. Math. et Phys. **6** (1932).—[4] On Phragmén-Lindelöf principle. Trans. Amer. Math. Soc. **41** (1933),—[5] Über eine Methode in der Theorie der meromorphen Funktionen. Soc. Sci. Fenn. Comment. Phys.-Math. 8 (1935). —[6] Zur Theorie der Überlagerungsflächen. Acta Math. **65** (1935).—[7] On the characterization of hyperbolic Riemann surfaces. Ann. Acad. Sci. Fenn. **125** (1952).

AHLFORS, L. V. and HEINS, M.: [1] Questions of regularity connected with the Phragmén-Lindelöf principle. Ann. Math. **50** (1949).

BADER, R. et PARREAU, M.: [1] Domaines non-compacts et classification des surfaces de Riemann. C. R. Paris, **232** (1951).

BEURLING, A.: [1] Études sur un problème de majoration. Thèse. Upsala (1933). —[2] Ensembles exceptionnelles. Acta Math. **72** (1940).

BESICOVITCH, A. S.: [1] On integral functions of order.<1. Math. Ann. **27** (1927).

BIEBERBACH, L.: [1] Über einen Riemannschen Satz aus der Lehre der konformen Abbildung. Sitzungsber. Berliner. Math. Gesellschaft. **24** (1925).—[2] Lehrbuch der Funktionentheorie, II. Leipzig u. Berlin (1931).

BIERNACKI, M.: [1] Sur les directions de Borel des fonctions méromorphes. Acta Math. **56** (1930).

BIRKHOFF, G. D.: [1] Proof of the ergodic theorem. Nat. Acad. USA, **18** (1932).

BLICHFELDT, H. F.: [1] A new principle in the geometry of numbers with some its applications. Trans. Amer. Math. Soc. **15** (1914).

BLOCH, A.: [1] Les théorèmes de M. Valiron sur les fonctions entières et la théorie de l'uniformisation. Ann. Fac. Sci. Univ. Toulouse. **3** (1925).

BOCHNER, S.: [1] Fortsetzung Riemannscher Flächen. Math. Ann. **98** (1927).

BOHR, H. und LANDAU, E.: [1] Über das Verhalten von $\zeta(s)$ und $\zeta_k(s)$ in der Nähe der Geraden $\sigma=1$. Göttinger Nachr. (1910).

BOREL, E.: [1] Sur les zéros des fonctions entières. Acta Math. **20** (1896).

BOULIGAND, C.: [1] Sur le problème de Dirichlet. Ann. Soc. Polonaise de Math. (1926).

BRELOT, M.: [1] Familles de Perron et Problème de Dirichlet. Acta de Szeged **9** (1938).—[2] Études des fonctions souharmoniques au voisinage d'un point. Actualite scientifique et industrielles. **139** (1934).

CARATHÉODORY, C.: Über die Variabilitätsbereich der Fourierschen Konstanten von positiven Funktionen. Rendiconti Palermo. **32** (1911).—[2] Untersuchungen über die konformen Abbildung von festen und veränderlichen Gebieten. Math.

Ann. **72** (1912).—[3] Über die Begrenzung einfach zusammenhängender Gebiete. Math. Ann. **73** (1913).—[4] Über die Winkelderivierten von beschränkten analytischen Funktionen. Sitzungsber. d. Preuss. Akad. Wiss. (1929).

CARLEMAN, T.: [1] Sur une inégalité dans la théorie des fonctions analytiques. C. R. Paris, **196** (1926).—[2] Leçons sur les fonctions quasi-analytiques. Paris (1926).

COLLINGWOOD, E. F.: [1] Sur les valeurs exceptionelles des fonctions entières d'ordre fini. C. R. Paris, **179** (1924).—[2] Exceptinal values of meromorphic functions. Trans. Amer. Math. Soc. **66** (1949).

COURANT, R.: [1] Über eine Eigenschaft der Abbildungsfunktionen bei konformen Abbildung. Göttinger Nachr. (1922).—[2] Dirichlet principle, conformal mapping and minimal surfaces. New York (1950).

DIEUDONNÉ, J.: [1] Recherces sur quelques problèmes relatifs aux fonctions bornées. Ann. Ecole nor. sup. **48** (1931).

DINGHAS, A.: [1] Über das Phragmén-Lindelöfschen Prinzip und den Julia-Carathéodoryschen Satz. Sitzungsber. d. Preuss. Akad. Wiss. (1936).—[2] Eine Bemerkung zur Ahlforsschen Theorie der Überlagerungsflächen. Math Zeits. **44** (1939).—[3] Eine Verallgemeinerung des Picard-Borelschen Satzes. Math. Zeits. **44** (1939).—[4] Über positive harmonischen Funktionen in einem Halbraum. Math. Zeits. **46** (1940).

DUFRESNOY, J.: [1] Sur les domaines couverts par des valeurs d'une fonction méromorphe ou algebroïde. Ann. Ecole nor. sup. **58** (1941).—[2] Sur les fonctions méromorphes et univalentes dans le cercle unité. Bull. Sci. Math. (1945). [3] Remarques complementaires sur deux properietés de la représentation conforme. Bull. Sci. Math. (1945).

ERDÖS, P. and GILLIS, J.: [1] Note on the transfinite diameter. Journ. London Math. Soc. **7** (1937).

EVANS, G. C.: [1] Application of Poincaré's sweeping-out process. Nat. Acad, USA (1933).—[2] Potentials of positive mass. Trans. Amer. Math. Soc. **37** (1935).—[3] Potentials and positively infinite singularities of harmonic functions. Monathefte f. Math. u. Phys. **43** (1936).

FATOU, P.: [1] Séries trigonométriques et séries de Taylor. Acta Math. **30** (1906).

FEJÉR, L.: [1] La convergence sur son cercle de convergence d'une série de puissance effectuant une représentation conforme du cercle sur le plane simple. C. R. Paris, **156** (1913).—[2] Über die Konvergenz der Potenzreihe an der Konvergenzgrenze in Fällen der konformen Abbildung auf die schlichte Ebene. Mathematische Abhandlungen, Herman Amandus Schwarz zu seinem fünfzigjährigen Doktorjubiläum am 6, August 1914 gewidmet von Freunden und Schüllern. Berlin, Springer (1914).—[3] Über trigonometrische Polynome. Journ. f. reine u. angewandte Math. **146** (1916).

FEJÉR, L. und RIESZ, F.: [1] Über eininge funktionentheoretische Ungleichungen. Math. Zeits. **11** (1921).

FEKETE, M.: [1] Über die Verteilung der Wurzelen bei gewisser algebraischen Gleichungen mit ganzzahligen Koeffizienten. Math. Zeits. **17** (1923).

FORD, L.: [1] Automorphic functions. New York (1929).

FROSTMAN, O.: [1] Potentiel d'équilibre et capacité des ensembles. Lund. Univ. Math. Sem. **3** (1935).

GEHRING, F. W.: [1] On the radial order of subharmonic functions. Journ. Math. Soc. Japan, **9** (1957).

GROSS, W.: [1] Über die Singularitäten analytischer Funktionen. Monathefte f. Math. u. Phys. **29** (1918).

GRÖTZSCH, H.: [1] Über einige Extremalprobleme der konformen Abbildung. Jahresber. d. deut. Math. Ver. **44** (1934).

GRUNSKY, H.: [1] Über die konforme Abbildung mehrfach zusammenhängender Bereiche auf mehrblättrige Kreise. Sitzungber. d. preuss. Akad. Wiss. (1937).

HADAMARD, J.: [1] Étude sur les propriétés des fonctions données par leur devéloppement de Taylor. Journ. d. Math. (4) **9** (1893).

HÄLLSTRÖM, G. F.: [1] Über meromorphen Funktionen mit mehrfach zusammenhängender Existenzgebieten. Acta Acad. Aboensis. Math. et Phys. **12** (1940).

HARDY, G. H.: [1] The mean value of the modulus of an analytic function. Proc. London Math. Soc. **14** (1915).

HARDY, G. H. and LITTLEWOOD, J. E.: [1] Tauberian theorem concerning power series and Dirichlet's series, whose coefficients are positive. Proc. London Math. Soc. **13** (1914).—[2] Abel's theorem and its converse (II). Proc. London. Math. Soc. **22** (1923).—[3] Maximal theorem with functiontheoretic applications. Acta Math. **54** (1930).

HAUSDORFF, F.: [1] Über halbstetige Funktionen und deren Verallgemeinerung. Math. Zeits. **5** (1919).

HAYMAN, W. K: [1] Some applications of the transfinite diameter to the theory of functions. J. d'analyse Math. **1** (1951).

HEINS, M.: [1] On the continuation of a Riemann surface. Ann. Math. **43** (1942). —[2] On the number of 1-1 directly conformal maps, which a multiply connected plane region of finite connectivity p>2 admitts onto itself. Bull. Amer. Math. Soc. **52** (1946).—[3] The conformal mapping of simply connected Riemann surfaces. Ann. Math. **50** (1949).—[4] On the Lindelöf principle. Ann. Math. **61** (1955).

HOPF, E.: [1] Fuchsian groups and ergodic theory. Trans. Amer. Math. Soc. **39** (1936).—[2] Ergodentheorie. Berlin (1937).

HURWITZ, A.: [1] Über algebraische Gebilde mit eindeutigen Transformationen in sich. Math. Ann. **41** (1893).

IVERSEN, F.: [1] Sur quelques propriétes des fonctions monogènes au voisinage d'un point singulier. Öfv. af. Finska Vet. Soc. Förh. 58 (1918).

JULIA, G.: [1] Extension d'un lemme de Schwarz. Acta Math. **42** (1920).—[2] Sur le domaine d'existence d'une fonction implicite, définie par une relation entière $G(x, y)=0$. Bull. Soc. Math. (1926).

KAMETANI, S.: [1] The exceptional values of functions with the set of capacity zero of essential singularities. Proc. Imp. Acad. Japan. **17** (1941).—[2]. On Hausdorff's measures and generalized capacities with some of their applications to the theory of functions. Jap. Journ. Math. **19** (1944–48).

KAMETANI, S. and UGAHERI, T.: [1] A remark on Kawakami's extension of Löwner's lemma. Proc. Imp. Acad. Japan, 18 (1942).

KAWAKAMI, Y.: [1] On an extension of Löwner's lemma. Jap. Journ. Math. **17** (1941).

KELLOGG, O, D.: [1] Harmonic functions and Green's integrals. Trans. Amer. Math. Soc. **13** (1912).—[2] Unicité des fonctions harmoniques. C. R. Paris, **187** (1928).

KHINTCHINE, A.: [1] Zur Birkhoffs Lösung des Ergodenproblems. Math. Ann. **107** (1932).

KOEBE, P.: [1] Über die konforme Abbildung mehrfach zusammenhängender Bereiche, inbesondere solche Bereiche, deren Begrenzung von Kreisen gebildet wird. Jahresber. d. deut. Math. Ver. **15** (1906).—[2] Über die konforme Abbildung mehrfachzusammenhängender Bereiche. Jahresber. d. deut. Math. Ver. **19** (1910).—[3] Über die Uniformisierung beliebigen analytischer Kurven, I. Das allgemeine Uniformiserungsprinzip. Crelle **138** (1910).—[4] Zur konforme Abbildung undendlichvielfach zusammenhängenden schlichter Bereiche auf Schlitzbereiche. Göttinger Nachr. (1918).

KOMATU, Y.: [1] Über einen Satz von Herrn Löwner. Proc. Imp. Japan Acad. **16** (1940).—[2] Zur konformen schlichte Schlitzabbildung. Proc. Imp. Acad. Japan, **17** (1941).—[3] The order of the derivative of a meromorphic function. Proc. Japan Acad. **27** (1951).

KUNUGUI, K.: [1] Sur un théorème de MM. Seidel-Beurling. Proc. Imp. Acad, Japan, **15** (1939).—[2] Sur un problème de M. Beurling. Proc. Imp. Acad. Japan, **16** (1940).—[3] Une généralisation des théorèmes de MM. Picard.—Nevanlinna sur les fonctions méromorphes. Proc. Imp. Acad. Japan, **17** (1941).—[4] Sur l'allure d'une fonction analytique uniforme au voisinage d'un point frontière de son domaine de définition. Jap. Journ. Math. 18 (1941—43).—[5] Sur la théorie des fonctions méromorphes et uniforme. Jap. Journ. Math. 18 (1941–43).

KURAMOCHI, Z.: [1] Potential theory and its applications, I. Osaka Math. Journ. **3** (1951).—[2] On the existence of harmonic functions on Riemann surfaces. Osaka Math. Journ. **7** (1955).

KURODA, T.: [1] Some remarks on an open Riemann surface, with null boundary. Tohoku Math. Journ. **3** (1950).

LANDAU, E.: [1] Über die Blochschen Konstante and zwei verwandte Weltkonstanten. Math. Zeits. **30** (1929).

LANDAU, E. and VALIRON, G.: [1] A deduction from Schwarz's lemma. Journ. London Math. Soc. **4** (1929).

LINDEBERG, J. W.: Sur l'existence des fonctions d'une variable complexe et des fonctions harmoniques bornées. Ann. Acad. Sci. Fenn. **11** (1918).

LINDELÖF, E.: [1] Sur un principle générale de l'analyse et ses applications à la théorie de la représentation conforme. Acta Soc. Sci. Fenn. **46** (1915).

LITTLEWOOD, J. E.: [1] The converse of Abel's theorem on power series. Proc. London Math. Soc. **9** (1911).—[2] On a theorem of Fatou. Journ. London Math.

Soc. **2** (1927).—[3] On functions subharmonic in a circle, I. Journ. London Math. Soc. **2** (1927).—[4] On functions subharmonic in a circle, II. Proc. London Math. Soc. **28** (1929).

LOOMIS, L. H.: [1] The converse of Fatou's theorem for positive harmonic functions. Trans. Amer. Math. Soc. **53** (1943).

LÖWNER, K.: [1] Untersuchungen über schlichte konforme Abbildung des Einheitskreises, I. Math. Ann. **89** (1923).

LUSIN, N. et PRIVALOFF, J.: [1] Sur l'unicité et multiplicité des fonctions analytiques. Ann. Ecole nor. sup. **42** (1925).

MARIA, A. J.: [1] The potential of a positive mass and the weight function of Wiener. Proc. Nat. Acad. Sci. USA. **20** (1934).

MINKOWSKI, H.: [1] Diophantische Approximationen. Leipzig u. Berlin (1927).

MONTEL, P.: [1] Sur les familles des fonctions analytiques, qui admettent des valeurs exceptionnelles dans un domaine. Ann. Ecole nor. sup. **23** (1912).

MORI, A.: [1] On Riemann surface on which no bounded harmonic function exists. Journ. Math. Soc. Japan, **3** (1931).—[2] On the existence of harmonic functions on a Riemann surface. Journ. Fac. Sci. Tokyo Univ. Section I. Vol. **6** (1951).—[3] Conformal representation of multiply connected domain on many sheeted disc. Journ. Math. Soc. Japan, **2** (1951).—[4] A note on unramified abelian covering surfaces of a closed Riemann surface. Journ. Math. Soc. Japan, **6** (1954).

MYRBERG, L.: [1] Über die vermischte Randwertaufgabe der harmonischen Funktionen. Ann. Acad. Sci. Fenn. Series A, **103** (1951).

MYRBERG, P. J.: [1] Ein Approximationssatz für Fuchsischen Gruppen. Acta Math. **57** (1931).—[2] Über die Existenz der Greenschen Funktionen auf einer gegebenen Riemannschen Fläche. Acta Math. **61** (1933).—[3] Die Kapazität der singulären Menge der linearen Gruppe. Ann. Acad. Sci. Fenn. Series A **10** (1941).

NAGAI, Y.: [1] On the behaviour of the boundary of Riemann surfaces, II. Proc. Japan Acad. Japan **26** (1950).

NEVANLINNA, R.: [1] Zur Theorie der meromorphen Funktionen. Acta Math. **46** (1925).—[2] Ein Satz über offnen Riemannsche Flächen. Ann. Acad. Sci. Fenn. **54** (1940).—[3] Quadratisch integrierbare Differentiale auf einer Riemannschen Mannigfaltigkeit. Ann. Acad. Sci. Fenn. Series A. **1** (1941).—[4] Über das Anwachsen des Dirichlet-integrals einer analytischen Funktion auf einer offenen Riemannschen Fläche. Ann. Acad. Sci. Fenn. Series A, **45** (1948).—[5] Über die Existenz von beschränkten Potentialfunktionen auf Flächen von unendlichen Geschlecht. Math. Zeits. **52** (1950).—[6] Eindeutige analytische Funktionen. Berlin-Göttingen-Heidelberg. (1953).—[7] Uniformisierung. Berlin (1953).

NOSHIRO, K.: [1] On the theory of the cluster sets of analytic functions. Journ. Fac. Sci. Hokkaido Univ. **6** (1937—38).—[2] On the singularities of analytic function. Jap. Journ. Math. **17** (1940).—[3] Contributions to the theory of the singularities of analytic functions. Jap. Journ. Math. **19** (1948).—[4] Note on the cluster set of analytic function. Journ. Math. Soc. Japan, **1** (1949—50).—[5] Open Riemann surface with null boundary. Nagoya Math. Journ. **3** (1951).

OHTSUKA, M.: [1] On the cluster sets of analytic functions in a Jordan domain. Journ. Math. Soc. Japan, **2** (1950).—[2] Dirichlet problem on Riemann surfaces and conformal mapping. Nagoya Math. Journ. **3** (1951).

OSGOOD, W. F.: [1] Lehrbuch d. Funktionentheorie, II. Leipzig u. Berlin (1932).

OZAWA, M.: [1] On Riemann surface admitting an infinite cyclic transformation group. Kōdai Math. Sem. Rep. (1957).

PENNYCUICK, K.: [1] On a theorem of Besicovitch. Journ. London Math. Soc. **10** (1935).

PERRON, O.: [1] Eine neue Behandlung der ersten Randwertaufgabe für $\Delta u=0$. Math. Zeits. **18** (1923).

PICARD, E.: [1] Démonstration d'un théorème général des fonctions uniforme lies par une relation algébrique. Acta Math. **11** (1887).—[2] Traité d'analyse, II.

PLESSNER, A.: [1] Zur Theorie der konjugierten trigonometrischen Reihen. Mitteilungen d. Math. Sem. Giessen, **10** (1923).—[2] Über das Verhalten analytischer Funktionen am Rande ihres Definitionsbereiches. Journ. f. reine u. angewandte Math. **158** (1927).

PÓLYA, G. und SZEGÖ, G.: [1] Über den transfiniten Durchmesser (Kapazitätskonstante) von ebenen und räumlichen Punktmengen. Journ. f. reine u. angewandte Math. **165** (1931).

POSSEL, R. de.: [1] Zum Parallelschlitztheorem unendlich vielfach zusammenhängender Gebiete. Göttinger Nachr. (1931).

PRIVALOFF, J.: [1] Sur les fonctions conjuguées. Bull. Soc. Math. de France. **44** (1916).

RADÓ, T.: [1] Subharmonic functions. Berlin (1937).—[2] Über eine nicht-fortsetzbare Riemannsche Mannigfaltigkeit. Math. Zeits. **20** (1924).

RADÓ, T. und RIESZ, F.: [1] Über die erste Randwertaufgabe für $\Delta u=0$. Math. Zeits, **22** (1925).

RAUCH, A.: [1] Extensions de théorèmes relatifs aux directions de Borel des fonctions méromorphes. Journ. de Math. **12** (1933).

RENGEL, E.: [1] Über einige Schlitztheoreme der konformen Abbildung. Schr. Math. Sem. Univ. Berlin. **1** Heft 4 (1933).—[2] Existenztheoreme für schlichte Abbildungen mehrfach zusammenhängender Bereiche auf gewisse Normalbereiche. Jahresber. d. deut. Math. Ver. **44** (1934).

RIEMANN, B.: [1] Theorie der Abelschen Functionen. Crelle **54** (1857).

RIESZ, F.: [1] Über die Randwerte einer analytischen Funktion. Math. Zeits. **18** (1923).—[2] Sur les fonctions subharmoniques et leur rapport à la théorie du potentiel, I. Acta Math. **48** (1926). II. Acta Math. **54** (1930).—[3] Sur un théorème de maximum de M. M. Hardy et Littlewood. Journ. London Math. Soc. **7** (1932).

RIESZ, M.: [1] Sur les fonctions conjuguées. Math. Zeits. **27** (1927).—[2] Remarque sur les fonctions holomorphes. Acta Sci. Math. (Szeged) **12** (1950).

RIESZ, F. und M.: [1] Über die Randwerte einer analytischen Funktion. Quatrième congr. d. math. scand. à Stockholm. (1925).

ROYDEN, H. I.: [1] Harmonic functions on open Riemann surfaces. Trans. Amer,

Math. Soc. **75** (1952).

SAKS, S.: [1] Theory of the integral. New York (1937).

SANTALÓ, L. A.: [1] On geometry of numbers. Journ. Math. Soc. Japan, **7** (1955).

SARIO, L.: [1] Über Riemannsche Fläche mit hebbarem Rand. Ann. Acad. Sci. Fenn. **50** (1948).

SCHUR, I.: [1] Über Potenzreihen, die im Innern des Einheitskreises beschränkt sind. Journ. f. reine u. angewandte Math. **147** (1917).

SCHWARZ, H. A.: [1] Zur Integration der partiellen Differentialgleichung $\frac{\partial^2 u}{\partial x^2}+\frac{\partial^2 u}{\partial y^2}=0$. Zusatz. Ges. Abh. **2**.

SEIDEL, W.: [1] On the distribution of values of bounded analytic functions. Trans. Amer. Math. Soc. **36** (1934).

SEIDEL, W. and WALSH, L.: [1] On the derivatives of functions analytic n the unit circle and their radii of univalence and of p-valence. Trans. Amer. Math. Soc. **52** (1942).

SELBERG, H.: [1] Über die ebenen Punktmenge von der Kapazität null. Avh, Norsk Vid. Akad. Oslo, I. No. 10 (1930).—[2] Eine Ungleichung der Potentialtheorie und ihre Anwendung in der Theorie der meromorphen Funktionen. Comment. Math. Helv. **18** (1945-65).—[3] Über eine Ungleichung der Potentialtheorie. Comment. Math. Helv. **18** (1946).—[4] Über einen Satz von Collingwood. Arch. Math. Naturv. BX. LVIII. Nr. 9 (1944).

SHIMIZU, T.: [1] On the theory of meromorphic functions. Jap. Journ. Math. **6** (1929).

SIEGEL, C. L.: [1] Some remarks on discontinuous groups. Ann. Math. **46** (1945).

SIERPINSKI, W.: [1] General topology. Toronto (1952).

SMIRNOFF, V.: [1] Sur les values limites des fonctions analytiques. C. R. Paris. **188** (1929).

STEIN, P.: [1] On a theorem of M. Riesz. Journ. London Math. Soc. 8 (1933).

SZEGÖ, G.: [1] Über die Randwerte einer analytischen Funktion. Math. Ann. **84** (1921).—[2] Bemerkungen zur einer Arbeit von Herrn Fekete "Über die Verteilung der Wurzeln bei gewisser algebraischen Gleichungen mit ganzzahligen Koeffizienten". Math. **21** (1924).

TAMURA, J.: [1] A non-prolongable Riemann surface. Sci. Rep. of the College of general education. Univ. Tokyo. (1956).

TEICHMÜLLER, O.: [1] Untersuchungen über konforme und quasi-konformen Abbildungen. Deut. Math. **3** (1938).

TOEPLITZ, O.: [1] Über die Fouriersche Entwicklung positiver Funktionen. Rendiconti Palermo **32** (1911).

TONELLI, I.: [1] I polinomi d'approximazione di Tchebychev. Annali Math **15** (1908).

TSUJI, M.: [1] On the theorems of Carathéodory and Lindelöf in the theory of conformal representation. Jap. Journ. Math. **7** (1930).—[2] On a positive harmonic function in a half-plane. Jap. Journ. Math. **15** (1939).—[3] A theorem on conformal representation. Jap. Journ. Math. **15** (1941).—[4] On an extension of Löwner's theorem. Proc. Imp. Acad. Japan, **18** (1942).—[5] On an extension of Bloch's theorem. Proc. Imp. Acad. Japan, **18** (1942).—[6] On the Green's function. Jap. Journ. Math. **18** (1942).—[7] On the cluster set of a meromorphic function. Proc. Imp. Acad. Japan, **19** (1943).—[8] On the domain of existence of an implicit function defined by an integral relation $G(x, y)=0$. Proc. Imp. Acad. Japan, **19** (1943).—[9] On non-prolongable Riemann surfaces. Proc. Imp. Acad. Japan (1943).—[10] Theory of conformal mapping of a multiply connected domain. Jap. Journ. Math. 18 (1943).—[11] Theory of meromorphic functions in a neighbourhood of a closed set of capacity zero. Jap. Journ. Math. **19** (1944).—[12] On the boundary value of a bounded analytic function of several complex variables. Proc. Imp. Acad. Japan, **20** (1944).—[13] On a regular function, whose real part is positive in a unit circle. Proc. Imp. Acad. Japan, **21** (1945).—[14] On Hopf's ergodic theorem. Jap. Journ. Math. **14** (1945).—[15] On Löwner's differential equation in the theory of univalent functions. Jap. Journ. Math. **19** (1947).—[16] Some metrical theorems on Fuchsian groups. Jap. Journ. Math. **19** (1947).—[17] On Blichfeldt's theorem in the geometry of numbers. Jap. Journ. Math. **19** (1948).—[18] Wiman's theorem on integral functions of order$<1/2$. Proc. Imp. Acad. Japan, **26** (1950).—[19] Beurling's theorem on exceptional sets. Tohoku Math. Journ. 2 (1950).—[20] Borel's directions of meromorphic functions of finite order, I. Tohoku Math. Journ. **2** (1950), II. Kōdai Math. Sem. Rep. (1950), III. Kōdai Math. Sem. Rep. (1950).—[21] Some metrical theorems on Fuchsian groups. Kōdai Math. Sem. Rep. (1950).—[22] A theorem on the majoration of harmonic measure and its applications. Tohoku Math. Journ. **3** (1951).—[23] On a regular function, which is of constant absolute value on the boundary of an infinite domain. Tohoku Math. Journ. **3** (1951).—[24] On the order of the derivative of a meromorphic function. Tohoku Math. Journ. **3** (1951).—[25] Some theorems on open Riemann surfaces. Nagoya Math. Journ. **3** (1951).—[26] Theory of Fuchsian groups. Jap. Journ Math. **20** (1951).—[27] Existence of a potential function with a prescribed singularity on any Riemann surface. Tohoku Math. Journ. **4** (1952).—[28] On F. Riesz' fundamental theorem on subharmonic functions. Tohoku Math. Journ. **4** (1952).—[29] Maximal continuation of a Riemann surface. Kōdai Math. Sem. Rep. (1952).—[30] Fundamental theorems in potential theory. Journ. Math. Soc. Japan, **4** (1952).—[31] Theorems in the geometry of numbers for Fuchsian groups. Journ. Math. Soc. Japan, **4** (1952).—[32] Myrberg's approximation theorem on Fuchsian groups. Journ. Math. Soc. Japan, **4** (1952).—[33] Extension of Bloch's theorem and its applications to normal family. Tohoku Math. Journ. **4** (1952).—[34] On a direct transcendental singularity of an inverse function of a meromorphic function. Journ. Math. Soc. Japan, **5** (1953).—[35] On the converse of Abel's theorem. Journ. Math. Soc. Japan, **5** (1953).—[36] Theory of meromorphic functions on an open Riemann surface with null boundary. Nagoya Math. Journ. **6** (1953).—[37] On convering surfaces of a closed Riemann surface of genus $p\geqq 2$. Tohoku Math. Journ. **5** (1953).—[38] On criteria of the regularity of Dirichlet problem. Journ. Math. Soc. Japan. **5** (1853).—[39] A remark on Rengel's theorem concerning Szegö's conjecture. Kōdai Math. Sem. Rep. (1953).—[40] On the capacity of general Cantor sets. Journ. Math. Soc. Japan, 5 (1953).—[41] On lattice points in an n-dimensional ellipsoid. Journ. Math. Soc. Japan, 5 (1953).—[42] The boundary distortion on conformal mapping. Journ. Math. Soc. Japan, **6** (1954).—[43] On Royden's theorem on a covering surface of a closed Riemann surface. Journ. Math. Soc. Japan, **6** (1954).—[44] Functions of U-class and its applications. Journ. Math. Soc. Japan, **7** (1955).—[45] On the modulus of a ring domain. Comment. Math. Univ. Sancti Pauli (Tokyo) 4 (1955).—[46] Borel's directions of a meromorphic function in a unit circle. Journ. Math. Soc. Japan, **7** (1955).—[47] Hopf's ergodic theorem on Fuchsian groups. Journ. Math. Soc. Japan, **7** (1955).—[48] Remark on my former paper "On an extension of Löwner's theorem." Comment. Math. Univ. Sancti Pauli (Tokyo) **5** (1955).—[49] A remark on my former paper "Theory of Fuchsian groups". Journ. Math. Soc. Japan, **7** (1955).—[50] On the cluster set of a meromorphic function. Comment. Math. Univ. Sancti Pauli (Tokyo) **4** (1955).—[51] On a positive harmonic function in a half-plane. Journ. Math. Soc. Japan. **7** (1955).—[52] A simple proof of Dirichlet principle. Journ. Math. Soc. Japan. **7** (1955).—[53] Canonical product for a meromorphic function in a unit circle. Journ. Math. Soc. Japan, 8 (1956).—[54] Littlewood's theorem on subharmonic functions in a unit circle. Comment. Math. Univ. Sancti Pauli (Tokyo) 5 (1956).—[55] Analogue of Blichfeldt s theorem for Fuchsian groups. Comment. Math. Univ. Sancti Pauli (Tokyo) 5 (1956).—[56] On the moduli of closed Riemann surfaces. Comment. Math. Univ. Sancti Pauli (Tokyo) **5** (1956).—[57] A simple proof of a theorem of Bieberbach-Grunsky. Comment. Math. Univ. Sancti Pauli (Tokyo) **6** (1956).—[58] A simple proof of a theorem of Erdös and Gillis on the transfinite diameter. Comment. Math. Univ. Sancti Pauli (Tokyo) **5** (1956).—[59] Solution of Neumann's problem without use of integral equation. Comment. Math. Univ. Sancti Pauli (Tokyo) **5** (1956).—[60] On a non-negative subharmonic function in a half-plane. Kōdai Math. Sem. Rep. (1956).—[61] On a Riemann surface, with a finite spherical area. Comment. Math. Univ. Sancti Pauli (Tokyo) **6** (1957).—[62] On Abelian and Schottkyan covering surfaces of a closed Riemann surface. Comment. Math. Univ. Sancti Pauli (Tokyo) **6** (1957).—[63] On Hardy-Littlewood's maximal theorem. Comment. Math. Univ. Sancti Pauli (Tokyo) **6** (1957).—[64] On the capacity of a set in the space of regular functions and its applications. Comment. Math. Univ. Sancti Pauli (Tokyo) **7** (1958).

VALIRON, G.: [1] Sur la distribution des valeurs des fonctions méromorphes. Acta Math. 47 (1926).—[2] Recherches sur le théorème de M. Borel dans la théorie des fonctions méromorphes d'ordre fini. Acta Math. **52** (1928).—[3] Sur les directions de Borel des fonctions méromorphes d'ordre fini. Journ. de Math. Seéie 10 (1931).—[4] Sur la derivée angulaire dans la représentation conforme. Bull. Sci. Math. (1932.)—[5] Points de Picard et points de Borel des fonctions méromorphes dans un cercle. Bull. sci. Math. (1932).

VALLÉE-POUSSIN, C de la.: [1] Points irreguliers. Determination de masses par potentiels. Bull. de la classe d. Sci. Acad. Belgique. **24** (1938).

VIRTANEN, K. J.: [1] Über die Existenz von beschränkten harmonischen Funktionen auf offenen Riemannschen Flächen. Ann. Acad. Sci. Fenn. Serie A I. **75** (1950).

WALSH, J. L.: [1] Über die Entwicklung einer analytischen Funktion nach Polynomen. Math. Ann. **96** (1927).

WARSCHAWSKI, S.: [1] Über einige Konvergenzsätze aus der Theorie der konformen Abbildung. Göttinger Nachr. (1930).—[2] Über die Randverhalten der Ableitung der Abbildungsfunktionen bei konformer Abbildung. Math. Zeits. **35** (1932).—[3] Über einen Satz von O. D. Kellogg. Göttinger Nachr. (1932).

WEYL, H.: [1] Über die Gleichverteilung von Zahlen mod. 1. Math. Ann. **77** (1916). —[2] Die Idee der Riemannschen Flächen. Stuttgart (1955).

WHITTAKER, J. M.: [1] The order of the derivative of a meromorphic function. Journ. London Math. Soc. **11** (1936).

WIELANDT, H.: [1] Zur Umkehrung des Abelschen Steitigkeitssatzes. Math. Zeits. **56** (1952).

WIENER, N.: [1] Certain notions in potential theory. Journ. Math. Massachusetts Inst. Tech. (1924).—[2] The Dirichlet problem. Journ. Math. Massachusetts Inst. Tech. (1924).

WIMAN, A.: [1] Sur une extension d'un théorème de M. Hadamard. Arkiv f. Math. Ast. Phys. **2** (1905).

WOLF, J.: [1] Sur la représentation conforme des bandes. Compositio Math. **1** (1935).

YÛJÔBÔ, Z.: [1] A theorem on Fuchsian groups. Mathematica Japonicae 1 (1949). —[2] On Riemann surfaces, no Green's function of which exists. Mathematica Japonicae, **2** (1952).—[3] An application of Ahlfors' theory of covering surfaces. Journ. Math. Soc. Japan, **4** (1952).

ZYGMUND, A.: [1] On a theorem of Littlewood. Summa Brasiliensis Mathematicae, **2** (1949).

LIST OF AUTHORS

INDEX

H

L

M

O

P

R

S

T

W